CONSTRUCTION

**PRINCIPLES, MATERIALS,
AND METHODS**

CONSTRUCTION
PRINCIPLES, MATERIALS, AND METHODS

SIXTH EDITION

HAROLD B. OLIN, AIA

JOHN L. SCHMIDT, AIA

WALTER H. LEWIS, AIA

revised by

H. LESLIE SIMMONS,
AIA, CSI

VAN NOSTRAND REINHOLD

I⟨T⟩P
A Division of
International Thomson Publishing Inc.

New York • Albany • Bonn • Boston
Detroit • London • Madrid • Melbourne
Mexico City • Paris • San Francisco
Singapore • Tokyo • Toronto

Previously published jointly by the Institute of Financial Education, 111 East Wacker Drive, Chicago, Illinois 60601, and by the Interstate Printers and Publishers, Inc. © 1983, 1980, 1975, 1972, 1969, 1966, 1965, 1963, 1962 by the United States League of Savings Institutions.

Copyright © 1995 by Van Nostrand Reinhold

This book is printed on acid-free recycled paper.

I(T)P ™ A division of International Thomson
    Publishing Inc.
    The ITP logo is a trademark under license.

Printed in the United States of America

For more information, contact:

Van Nostrand Reinhold
115 Fifth Avenue
New York, New York 10003

International Thomson Publishing
Berkshire House
168-173 High Holborn
London WC1V 7AA
England

Thomas Nelson Australia
102 Dodds Street
South Melbourne, Victoria 3205
Australia

Nelson Canada
1120 Birchmount Road
Scarborough, Ontario
M1K 5G4, Canada

International Thomson Publishing GmbH
Königswinterer Strasse 418
53227 Bonn
Germany

International Thomson Publishing Asia
221 Henderson Road
#05 10 Henderson Building
Singapore 0315

International Thomson Publishing Japan
Hirakawacho Kyowa Building, 3F
2-2-1 Hirakawa-cho, Chiyoda-ku
Tokyo 102
Japan

International Thomson Editores
Campos Eliseos 385, Piso 7
Col. Polanco
11560 Mexico D.F. Mexico

2 3 4 5 6 7 8 9 10 BANHR 01 00 99 98 97 96 95

Library of Congress Cataloging in Publication Data

Olin, Harold Bennett.
        Construction : principles, materials & methods / Harold B. Olin,
    John L. Schmidt, Walter H. Lewis. — 6th ed. / revised by H. Leslie Simmons
            p.   cm.
        Includes bibliographical references (p.   ) and index.
        ISBN 0-442-00605-5
        1. Building.   I. Schmidt, John L.   II. Lewis, Walter H., 1928–
        III. Simmons, H. Leslie.   IV. Title.
    TH145.043   1994
    690--dc30   94-16486
                                                                 CIP

CONTENTS

PREFACE

Previous editions of this book were directed toward the home construction portion of the construction industry. Those editions covered wood construction, masonry, interior and exterior finishes, heating, ventilating, air conditioning, plumbing, electrical, and many other subjects as they related to small residential construction. The sixth edition includes most of these subjects but supplements them with information about materials and methods currently used in the construction of other types of buildings. The discussions about mechanical and electrical systems are still limited to requirements for residential construction. The other subjects have been expanded to include most kinds of commercial, larger residential, and institutional buildings. As a result, the book now introduces the materials and systems used in most types of building construction.

The organization of the first nine chapters of the sixth edition in a format similar to that used in *Masterformat* ties the material to the current system used for organizing specifications. The chapter titles are similar to those of the divisions of *Masterformat*, and the material discussed therein is similarly related.

An attempt has been made to cover every principle, material, and method used to design and construct most buildings. The information presented includes the background and the history of materials and systems. Construction systems covered include precast and cast-in-place concrete, steel, wood, and masonry. In each case, materials and their manufacture are discussed first, followed by the methods of construction used to erect those materials.

The text also covers doors and windows, glazed curtain walls, glazing, and finishes, including plaster, gypsum board, tile, terrazzo, acoustical ceilings and other acoustical treatments, wood flooring, resilient flooring, carpet, and painting and finishing.

General coverage has been given to industry standards, codes, land survey and descriptions, properties of materials, barrier-free design, metrication, siteworks (including excavation, grading, shoring and sheeting, and other earthwork), and sound control in buildings.

The discussion of mechanical and electrical systems is related mostly to residential construction. Private sewage disposal systems also are covered.

This book contains an extensive glossary of terms. It also contains, in Appendix C, a list of the names, addresses, and telephone numbers of the organizations, associations, and agencies that contributed either directly or indirectly to the book.

It takes the help of many people and organizations to produce a book such as this, including those who prepared previous editions. First and foremost are the original authors, Harold B. Olin, John L. Schmidt, and Walter H. Lewis, without whom this book would not exist. The preface for the fifth edition, written by the three authors, acknowledges the contributions of several persons, including senior research associate Christina Farnsworth, for her contributions in revising old chapters and in authoring the chapters on barrier-free design, home security and safety, air plenum systems, and metrication; Richard Laya, who prepared the chapters on heat control, passive solar heating, and pole foundations; and production specialist Betsy Pavichevich, who was responsible for the typing, layout, and preparation of graphics.

Even though the material these people

prepared has, in some cases, been extensively revised in the sixth edition to bring it up to date with current materials, methods, and standards, their contributions in producing the original data are to be commended.

I would like also to gratefully acknowledge the contributions of the many manufacturers, trade and professional associations, standards-setting bodies, government agencies, periodicals, book publishers, and individuals who provided valuable research material and illustrations, as well as text reviews and comments. Most of their names are listed in Appendix B, References, which is subdivided by book chapter and contains the names of the publications and publishers from which the material in the related chapter was derived. Appendix B is a source of additional reading and study for anyone who wants or needs to delve further into a particular subject.

My work was aided immensely by Brenda Bertozzi, my primary research associate, who located sources and obtained permission to use their materials, and by Irene Demchyshyn, who helped me gather the research data necessary to update and expand this book.

I am indebted also to the professionals, both practicing architects and teachers, who lent their time, energy, and expertise to review the manuscript so that it would be accurate and more useful as a text for students of architecture and allied fields. These professionals include Harold B. Olin, AIA, the original primary editor of previous editions; Daniel W. Halpin, Ph.D., Professor and the head of the Division of Construction Engineering and Management, Purdue University; Larry Grosse, Ph.D., Associate Professor, College of Architecture, Texas A & M University; and Terry L. Patterson, AIA, Professor, College of Architecture, University of Oklahoma, whose extensive, well-informed, and detailed comments contributed greatly to making this book better than it otherwise would have been. Professor Patterson is also the author of the *Student Workbook* and the *Instructor's Manual*, both available from the publisher for use with this book.

The *Student Workbook* is designed to expand and intensify the educational experience provided by this text. The *Instructor's Manual* is designed to support the activities described in the *Student Workbook* and to provide examinations for use by the instructor. Issues for class discussion also are suggested. Together, the two companion manuals are designed to embellish a student's textbook learning experience.

H. Leslie Simmons, AIA, CSI

INTRODUCTION

The sixth edition of *Construction: Principles, Materials, and Methods* is the result of more than 32 years of research and editorial effort costing several million dollars. Much of this work was carried out in an ongoing membership education program initiated in 1964 by the United States League of Savings Institutions. The sixth edition is the result of that research and a major program carried out during the three years prior to its publication by its new publisher, Van Nostrand Reinhold. This education program consisted of completely revising the previous edition to reflect the many changes that have occurred in recent years and to bring that information up to date with latest industry standards. In addition, the sixth edition introduces construction materials and methods not in general use when the previous edition was prepared. And finally, materials and construction methods have been added that relate to commercial construction, including high-rise buildings. Previous editions primarily covered residential construction. The sixth edition addresses materials and methods used in both.

The sixth edition contains more than 1100 pages divided into 12 chapters. The organization of these chapters is similar to that used in *Masterformat*, which makes the sixth edition easier to relate to the specifications and data-filing formats currently used in the construction industry. An *Instructor's Manual* and a *Student Workbook* are available for use with the sixth edition. Accordingly, the "Work File" sections of previous editions have been deleted.

Through the years, *Construction* has been widely adopted as a course text in nearly 200 colleges and universities that offer architectural and building technology curricula. The text also is becoming a standard reference source in professional offices.

DISCLAIMER

The information in this book was derived from data published by trade associations, standards-setting organizations, manufacturers, and government organizations and statements made to the author by their representatives; from model codes; and from related books and periodicals. The author and the publisher have exercised their best judgment in selecting the material to be presented, have reported the recommendations of the sources consulted in good faith, and have made every reasonable effort to make the information presented accurate and authoritative. But neither the author nor the publisher warrants the accuracy or the completeness of the data, nor does either assume liability for the data's fitness for a particular purpose. Users of the data contained in this book are expected to apply their own professional knowledge and experience or consult with someone who has such knowledge and experience. Users also are expected to consult the original sources of the data, obtain additional and updated information as needed, and seek expert advice when appropriate.

A few manufacturers and products are mentioned in this book. Such mention is intended merely to indicate the existence of such products and manufacturers, not to imply endorsement by the author or the publisher of the mentioned manufacturer or product. Endorsement is not implied of other products made by a mentioned manufacturer or of any statement made by a mentioned manufacturer or associated in any way with any product, its accompanying literature, or in advertising copy.

Similarly, literature and standards produced by various manufacturers, associations, and other organizations are mentioned. Such mention does not imply that a mentioned item is the only available data or standard, the best of its kind, or the legal requirement or accepted standard in a particular locality. The authors and the publisher expect the readers to seek out other manufacturers, appropriate associations, or other organizations to ascertain whether they have similar literature or will make similar data available and to determine whether the information presented conforms with the codes and other legal requirements in a particular locale.

CONSTRUCTION
PRINCIPLES, MATERIALS, AND METHODS

1

GENERAL REQUIREMENTS

INTRODUCTION

GENERAL DESIGN REQUIREMENTS

INDUSTRY STANDARDS

CODES

LAND SURVEY AND DESCRIPTION

PROPERTIES OF MATERIALS

BARRIER-FREE DESIGN

METRICATION

Photo: Honvest Corporation, Honolulu, Hawaii. (Architect: Leo A. Daly and Associates; photo courtesy of Bethlehem Steel Corporation.)

INTRODUCTION

A *project manual* is a single volume produced in the office of the designer of a building construction project. It contains the written portion of that project's contract documents and usually includes bidding requirements, forms, conditions of the contract, information about alternates and allowances, and specifications. The number of possible requirements in project manuals is large, but not all requirements are needed for every project. For example, one project may be for concrete construction and have no requirement for wood trusses. Another project may be framed with wood and contain no structural steel.

Project manuals are complex and often need a numbering system to allow users to find individual requirements easily. Without a standard to follow, design firms tend to invent their own numbering systems and methods of organizing these materials. This makes it difficult for contractors to find particular requirements, thus increasing the possibility of errors in estimating, which affect project cost, and errors in the field, which affect the completed building. For instance, in one project the requirements for structural steel may be in the third section of the manual, while in another project manual similar requirements may be in the twelfth section. To unify the location of the parts of project manuals, the Construction Specifications Institute (CSI) and Construction Specifications Canada (CSC) jointly produce a publication titled *Masterformat*. CSI was organized in 1948 by a group of construction specifiers to improve specifications practices. Its current membership includes primarily specifiers, product manufacturers' representatives, and government agency representatives interested in specifications. CSC is a similar organization operating in Canada.

Masterformat has become the industry standard for naming and numbering the material in project manuals. It has been adopted by the American Institute of Architects (AIA) in its publications, and by most federal government agencies for their construction projects. It is the format generally used in manufacturers' product literature. It is the basis of organization for *Sweet's Catalog Files* (McGraw-Hill's yearly collection of manufacturers' catalogs), which are a major source of product literature in most architects' offices. It is also the basis for data filing in many design offices. More important for our purposes here, *Masterformat* is used on the vast majority of private sector and government building construction projects in the United States.

Because of its almost universal use in the U.S. construction industry, students and professionals need to be familiar with *Masterformat*. *Masterformat* divides the work associated with construction projects into 16 *divisions*. The first nine chapters of this book are named and numbered the same as *Masterformat*'s first nine divisions. The material in Chapters 2 through 9 is roughly equivalent to the material in the corresponding divisions of *Masterformat*. Furthermore, in most cases, the organization of subjects in these first nine chapters mimics the organization of *sections* in *Masterformat*. For example, in this book mortar and grout are addressed first in Chapter 4 because that's the location for them recommended by *Masterformat*. The exact order of sections in *Masterformat* tends to change slightly from edition to edition, however, so they may differ slightly from the order used here. Chapters 1, 10, 11, and 12 are exceptions to this order. *Masterformat* Division 1 covers bidding requirements, contract forms, and conditions of the contract for construction, none of which is within the scope of this book. Therefore, Chapter 1 is devoted to a discussion of general requirements related to building design. Chapters 10, 11, and 12 contain material similar to that found in *Masterformat*, but its organization in *Masterformat* is inappropriate for the presentation of those subjects in this particular book.

1.1 GENERAL DESIGN REQUIREMENTS

The general objectives of the building industry should be to build structures of good quality at as low a cost as is practicable, while making the individual building and the total built environment more beautiful in the process. Buildings should be efficient in their use and operation. They should also serve their intended purpose and permit the activities that take place in them to proceed with appropriate dispatch and ease. In addition, commercial buildings should be planned and built so that they are capable of producing a profit.

Buildings take many forms and shapes, depending on their use, the materials used to build them, the needs and desires of the owners, the building construction budgets, the buildings' potential operating costs, and the designers' preferences. Buildings other than single-family residences and townhouses are so varied in size and shape as to make simplification of their types difficult (Fig. 1.1-1). However, some basic types and construction methods can be identified (Fig. 1.1-2).

The simplest building is a one-story, single-span, slab-on-grade structure with a flat roof (Fig. 1.1-2a). Similar buildings with basements are also commonly built (Fig. 1.1-2b). Single-story structures with more than one structural span (Fig. 1.1-2c), in which one or more intermedi-

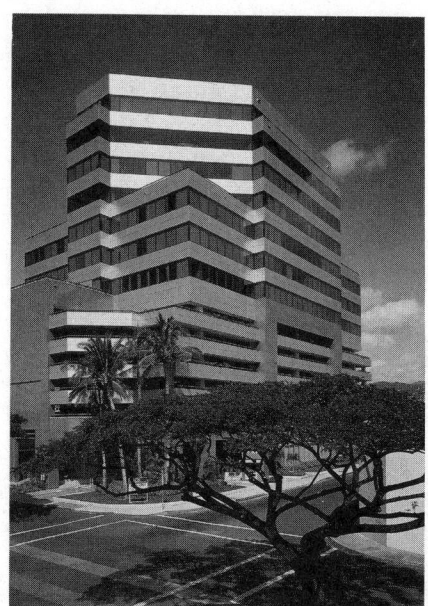

FIGURE 1.1-1 *Modern buildings are seldom rectangles.* (*Honvest Corporation, Honolulu, HI. Architect Leo A. Daly and Associates. Photo courtesy Bethlehem Steel Corporation.*)

ate rows of walls or columns support the roof structure, enclose more space per unit of exterior wall cladding than do smaller buildings.

Single-story buildings may also have full or partial basements. The structural systems in these types of buildings may be concrete, masonry, steel-framed, or wood-framed bearing walls with steel, concrete, or wood roof-framing systems; steel, concrete, or wood interior and exterior columns with steel, concrete, or wood roof framing; or a combination of these systems. Foundations are usually poured concrete, but rarely treated wood foundations may be used (see Section 6.4). The roof of a single-story building may either be flat or have one of a wide variety of shapes (1.1-2d, 1.1-2e, and 1.1-2f). Roof decks may be wood, steel, or concrete. Basements may have either poured concrete or reinforced masonry walls, depending on the level of the earth against the wall and the height and the *hydrostatic head* of adjacent underground water. Floors above basements may be steel framed with concrete or wood floors; concrete framed with concrete floors; steel framed with concrete floors; wood framed with wood floors; or a combination of these systems.

The same principles apply to multistory structures (Fig. 1.1-2g). The construction materials and structural systems in multistory structures and the height of such buildings are usually dictated by economic factors, such as land cost, but may be affected by codes and laws that restrict building heights, land area coverage, or the materials that may be used. Fire codes, for example, may restrict the types of construction systems and the materials that may be left exposed. Many fire codes do not permit wood construction or the exposure of wood finishes on the exterior of buildings in certain locations.

Multistory buildings require less roof surface than do single-story buildings with the same floor area. This results in a savings in the cost of roofing materials. In addition, multistory buildings require less land per unit of usable space. They are also generally more energy efficient, because of their higher ratio of interior space to building shell area, than single-story buildings. Except in rare instances, these advantages increase with the number of stories. The lower costs are somewhat offset by the increased maintenance cost of the exterior surfaces of multistory buildings, the relatively high costs of materials that can be used there, and the increased cost of construction associated with moving materials to high levels and working with them far above ground level.

Low-rise multistory buildings may be of steel or concrete construction, or a combination of both. Some even have masonry bearing walls. Steel columns and concrete floors are common. Foundations are usually poured concrete spread footings, although poor soil conditions sometimes dictate the use of *piles* or *caissons*.

High-rise buildings are usually framed in steel, with thin concrete floor slabs, because concrete structures of great height have heavier and larger framing members than steel structures, which reduces the amount of usable space and increases the cost of construction. Foundations are either poured concrete footings or pads, piles, or caissons, depending on the soil conditions and the size and load imposed on the soil by the building.

Sometimes the desire to create a statement for ego or advertising purposes affects the size, height, and appearance of a building. For example, a corporation may want its headquarters building to be a symbol, or it may just want to own the tallest, largest, or most spectacular building in town.

Multistory buildings need elevators or escalators to make their use practicable. In most types of uses, federal accessibility laws and regulations make elevators or wheelchair lifts a legal requirement in every building that is not inherently accessible to the handicapped (see Section 1.6), which of course includes every multistory building. The additional cost of such vertical transportation must be considered when deciding whether to build a multistory building.

The basic building types used in single-family and townhouse construction are easier to define. Figure 1.1-3 shows some common types. Most of these types are also used for buildings other than single-family homes or townhouses, so they should not be thought of only for these restricted applications. The most prevalent type is the one-story building (Fig. 1.1-3a), because it provides the most size and shape variations. One-story buildings may or may not contain a basement, and their roofs may be sloped (as shown) or flat.

One-and-one-half-story buildings, with or without basements (Fig. 1.1-3b), are sometimes used for housing. They offer more living space than single-story buildings with minimum additional cost. The second-floor space varies with the building size and roof slope. Light, ventilation, and view can be provided by dormers. One-and-one-half story buildings are seldom built for other types of uses because their inherently small, second-floor rooms, with their sloped ceilings, are adequate for sleeping rooms but often do not make satisfactory work spaces.

Two-story (Fig. 1.1-3c) or taller buildings, with or without basements, provide the maximum usable area at relatively low cost. Two- and three-story single-family houses and townhouses are common. These types of buildings can reduce construction cost, depending on the value of the land. When they must be accessible to the handicapped (see Section 1.6), buildings of more than one story require elevators.

Bilevel buildings (Fig. 1.1-3d) are well suited for single-family houses, townhouses, or small commercial buildings in hillside locations. They provide habitable space at both grade levels when connected with full flights of stairs. In certain types of uses, accessibility restrictions may

FIGURE 1.1-2 *Basic building types for other than single-family residential buildings: (a) one-story; (b) one-story with basement; (c) one-story with multiple framing bays; (d, e, and f) typical roof shapes; and (g) multistory.* (Drawings by HLS.)

require that elevators be included. This type of configuration can also be used for two different occupancies, such as an apartment on one level and a small store on the other. In that case, both levels can easily be made independently accessible to the handicapped. Roofs may be either sloped (as shown) or flat.

Split-level (Fig. 1.1-3e) and bilevel/split-entry (Fig. 1.1-3f) buildings are used mostly for single-family houses and townhouses. They are infrequently used for other purposes because of the difficulty of making them accessible by the handicapped. Split-level houses offer distinct separation of functions, either on three

levels or on four, including a basement. These buildings are best suited for sloping lots. They offer numerous design possibilities but can have awkward proportions when not carefully designed.

Bilevel/split-entry buildings also are best suited to sloping lots. They are characterized by a split foyer between two full living levels. This configuration can provide either a sunken two-story (or more) house without a basement or a raised one-story (or more) house with a finished basement.

The selection of construction methods and systems that produce the basic building types shown in Figures 1.1-2 and

1.1-3 is usually governed by three criteria: functional requirements, cost, and desired appearance. These basic criteria may require a consideration of a number of factors, including climate; site topography; initial costs; maintenance costs; building codes, zoning ordinances, and other laws; availability of materials and labor; builder resourcefulness and size; owner taste; and local custom.

The selection of methods and systems is further complicated by the thousands of materials, products, and construction system choices available, many of which are interdependent. In fact, the interrelationship of these building elements may create

FIGURE 1.1-3 *Basic single-family residential building types: (a) single-story; (b) one-and-one-half-story; (c) two-story or higher; (d) bilevel; (e) split-level; and (f) bilevel/split-level.* (Drawings by HLS.)

a situation in which the construction method or system is the major influence on a building's design. For example, the selection of a dome as the means to roof a coliseum may dictate that the shape of the building be circular or near-circular. Conversely, design requirements may dictate the framing system. A dome, for example, may be a poor choice for roofing a theater, because of the inherent acoustical difficulties of a dome and because a circular building may not be preferable for the kind of theater that is desired. The cost and availability of very large laminated-wood (*gluelam*) structural elements, compared to smaller gluelam units, may influence the width or even the shape of a church or the way in which the gluelam units are fitted together to make a roof structure. The permissible span of the available wood decking may further influence the spacing of the gluelam units or the design of the roof structure and the placement of purlins.

While there have been experiments with a few revolutionary construction systems since World War II, most new homes and many small commercial and institutional buildings in the United States are still built using conventional light-wood-platform framing (see Chapter 6), often with wood-truss-framed roofs. In many areas, the use of preassembled components, such as those discussed in Chapter 6, is common. In the future, advanced industrialization

techniques using new materials and methods may offer new construction forms far different from those typical today.

Other basic systems in use today include wood-post-and-beam framing and wood-pole construction (see Chapter 6); masonry bearing-wall construction (see Chapter 4), sometimes with concrete floors (see Chapter 3), often supported on metal framing or bar joists; concrete-framed construction; and steel-framed construction (see Chapter 5).

In small construction, conventional wood framing still offers many advantages. As a complete construction system, it is one of the most economical ways to build. The ease of working and fastening wood together with simple tools provides flexibility that permits job changes without extensive reengineering. Wood framing is still the basis for most building codes and labor practices and will probably remain so for some time. Conventional framing is adaptable to site fabrication by the smallest builder handling each member piece by piece, as well as to off-site fabrication of individual pieces into larger preassembled components that require additional manpower or machinery for erection.

Further industrialization, using more and larger prefinished and prefabricated components, appears essential to help offset the rising costs of land, labor, and materials. Off-site fabrication permits

maximum utilization of labor and materials under factory-controlled conditions with little loss in on-site time due to bad weather. Efficiency may be increased with the use of power tools and machinery; volume purchasing of materials and stockpiling of finished parts are possible; greater convenience for workers and better protection for finished materials are provided; and site erection of components usually can be accomplished more economically and in less time by semiskilled or even unskilled labor.

To save costs, mechanical components for small buildings have been developed that combine the furnace, air conditioner, water heater, and electric power panel into one package. Larger mechanical components include completely furnished kitchens and bathrooms. The concept of prefinishing complete rooms has been extended to prefabricating up to half a small building, such as a house, so that upon setting and joining two halves, an entire building is completed. Future developments may include assembling an entire building and completely finishing it prior to site placement.

Some future building construction methods will be highly sophisticated and closely integrated systems. For instance, integrated floor and ceiling systems available for use in commercial construction include structure, lighting, acoustical control, heating, cooling, and air distribution in a single system.

Components should be capable of satisfying varying design requirements, permit simple modifications in the field in case of errors, and be sized for ease of shipping, storage, and assembly. As component size increases, design and construction problems increase. Design flexibility is lessened as the size of the component increases. The dimensions of large units are restricted to what can be transported physically and legally over the highways, and larger components usually require more manpower and larger erection equipment at the site.

Accordingly the design, engineering, and selection of preassembled components require judgments between size and flexibility. The most useful systems will combine the advantages of fully standardized, factory-built modular units, which capitalize on the inherent savings resulting from repetitive production, and those that offer the design advantages of custom fabrication in the field.

Unfortunately there are also certain disadvantages associated with prefabrication that have so far limited its use. For example, to be profitable, large components require a large market willing to accept a standarized design, which has not been forthcoming. In comparison, there is a huge market for prefabricated roof trusses, because they can be adapted to many building sizes, shapes, and designs,

making them relatively inexpensive and readily available.

There are also potential disputes among construction trade unions and between trade unions and manufacturers about the right to do certain work. Union-member plumbers, for example, are not likely to be pleased when the plumbing piping and fixtures in a prefabricated building are installed by nonunion factory workers.

Other problems with prefabrication include consumer and builder resistance to prefabricated structures due to the preconceived notion that prefabricated buildings will be shabbily constructed and look like house trailers. In fact, the construction may actually be superior to stick-built units, but the appearance possibilities are somewhat limited, and design variety is difficult to achieve.

A final deterrent to prefabrication is the lack of conformity in building codes. These differences can require slight, but costly, modifications in prefabricated units to comply with the codes in different jurisdictions.

Construction practices vary widely across the United States. In the main, conventional methods and systems are not carefully engineered but are based on a combination of long-established custom, rules of thumb, and arbitrary building code requirements. These practices have resulted in buildings that have usually

provided satisfactory performance over the years. However, when these practices are overly conservative, as is often the case, they foster excessive waste and therefore higher cost. New performance standards for methods and systems are constantly being established. Design criteria based on laboratory and field testing can materially reduce overdesign, waste, and cost. Members of the building construction industry should continue to encourage the development of criteria based on performance rather than the requirement that a specific product or system be used. They should also take steps to further the design of methods and systems based on these performance criteria. In addition, the industry needs to encourage quicker acceptance of innovations as they occur.

This book discusses building materials and design, touching on most basic building components and many conventional construction materials and methods. Some new and unique materials, methods, and systems are covered because they offer strong potential for future acceptance, but because of space limitations, many methods and systems cannot be described. Their omission in no way indicates or infers a lack of suitability.

It is hoped that this book will help to improve the reader's understanding of the principles and criteria on which construction methods and systems are based.

1.2 INDUSTRY STANDARDS

The building industry is made up of so many diverse interest groups that it has been impossible to develop a single comprehensive set of criteria or standards acceptable to all concerned. Groups that are more intimately involved in a product or construction system are generally best qualified to establish standards for that product or system. The result is a myriad of organizations that establish standards. This section discusses the types of organizations and agencies that set construction standards in the United States.

1.2.1

SPECIFICATIONS, STANDARDS, AND CODES

Before beginning a discussion of construction industry standards, it is necessary to clarify some terms. There is much confusion in the industry about the use of the terms *specifications*, *standards*, and *codes*. Unfortunately, the three are often erroneously used interchangeably.

1.2.1.1

Specifications

In the sense that they are detailed, precise descriptions of products or practices, some construction industry standards could be called *specifications*. In fact, they sometimes are called that by their producers. Some of the manufacturers' data that designers, builders, and owners rely on are also specifications in the dictionary sense, and such literature is sometimes called specifications by the manufacturers.

But using the term *specifications* in both instances can lead to confusion. Therefore, this book refers to construction industry standards as *standards* and to manufacturers' data as *product descriptions* or *product literature*. Unless specifically modified in the text, the term *specification* is used in its narrow construction industry meaning, that is, a portion of the written construction documents prepared in the office of a building's designer. Specifications should include clear and accurate descriptions of the technical requirements related to materials, products, and services. They should also state a project's requirements for quality and the prescribed use of materials and methods to produce a desired product, system, application, or finish.

Guide specifications are commercially available today from associations, such as the AIA and the CSI, and from several private organizations in either paper or electronic form. Guide specifications are also produced by federal agencies, such as the General Services Administration (GSA) and the various military services. These specifications are intended for use only on government projects, but they may contain data that can be valuable to a specifier on private sector projects.

1.2.1.2

Standards

When selecting materials or determining the suitability of materials and methods, specifiers and builders refer to a variety of *industry standards*, sometimes called *reference standards*.

■ TYPES OF STANDARDS

Some standards result from the efforts of manufacturers, professionals, and tradespeople to simplify and increase the efficiency of their work or to ensure a minimum level of quality. Other standards are the work of governmental agencies and other groups interested in establishing minimum levels of safety and performance. Therefore, standards take a variety of forms depending on their source and purpose.

MATERIAL STANDARDS Standards that define the properties of a material are called *material standards*. Examples include standards for extruded aluminum bars, rods, shapes, and tubes and for a particu-

lar type of steel item. They usually specify the constituents of a material, its physical properties, and its performance under stress and varying climatic conditions.

PRODUCT STANDARDS The requirements of a specific product, such as aluminum windows, are defined by *product standards*. These standards often define terms, classify constituent materials, and state acceptable thicknesses, lengths, and widths. They may also spell out the acceptable methods of joining separate materials, of fabricating various parts of a product, or of assembling systems.

DESIGN STANDARDS Design standards define the requirements for sound design using a particular material, product, or system. They include publications by such organizations as the American Concrete Institute (ACI), the Architectural Woodwork Institute (AWI), and the U.S. Department of Housing and Urban Development (HUD).

WORKMANSHIP STANDARDS Workmanship standards are standards for installing materials, products, and systems. The American Society for Testing and Materials (ASTM) produces many of these standards.

TEST METHOD STANDARDS Test method standards spell out acceptable criteria for testing materials and systems. Again, ASTM is a prime producer of this type of standard.

■ OBJECTIVES OF STANDARDS

Construction standards have two basic objectives: (1) to establish levels of quality that a user, specifier, approver, or buyer of a material, product, or system can recognize; and (2) to standardize or simplify such variables as dimensions, varieties, and other characteristics of specific products so as to minimize variations in their manufacture and use.

■ USES OF STANDARDS

Construction standards are used by manufacturers, specifiers, consumers, and communities. Standards may be used and referred to either separately or within collections, such as in the HUD *Minimum Property Standards for Housing*. Stan-

dards may also be incorporated into municipal or state building codes by inclusion or reference. Such larger works may have broader objectives than to act only as construction standards. Codes, for example, are concerned basically with minimum acceptable standards of public health, safety, and welfare; the HUD *Minimum Property Standards for Housing* establishes minimum requirements of design and construction for the insurance of mortgage loans. Standards are often incorporated by reference into construction document specifications to help establish requirements for materials, equipment, finishes, and workmanship for a particular construction project.

1.2.1.3

Codes

Some so-called codes are actually standards because they are voluntary and have no status in law. Only when standards are adopted by legislative bodies and incorporated into law, do they become *codes* in the true sense of the word. In one sense, the model codes discussed in Section 1.3.3 are actually standards that have been adopted in many, but not all, jurisdictions as codes. Other examples of documents called codes that actually are not include various American National Standards Association (ANSI) and National Fire Protection Association (NFPA) codes.

Standards that are called codes by their producers often fall in the design standards category (see Section 1.2.1.2) but may encompass one or more of the other types.

Because some documents called *codes* are universally recognized as such, we will accede to that convention in this book. Therefore, publications such as ANSI A17.1, "Safety Code for Elevators and Escalators," and NFPA 101, "Life Safety Code" and "National Electrical Code©" are called codes, even though they truly are not.

Codes are discussed more fully in Section 1.3.

1.2.2

PRIVATE INDUSTRY STANDARDS

Standards are established by two kinds of private industry organizations: trade associations and standards-setting and testing agencies.

1.2.2.1

Trade Associations

Manufacturers, tradespeople, and suppliers, working through trade associations, prepare most private industry standards. Some trade associations are called societies or institutes. Appendix C, Data Sources, contains the names and addresses of some of the industry groups that are active in developing technical information and promulgating construction standards.

A *trade association* is an organization of individual manufacturers or businesses engaged in the production, supply, or installation of materials or services of a similar nature. A basic function of a trade association is to promote the interests of its membership. Those interests generally are best served by the proper use of the group's materials, products, or services. The proper use of building materials and methods is guided largely by the development of suitable levels of quality for their manufacture, use, and installation. Some of the most important activities of trade associations are directed toward research into the use and improvement of materials and methods and toward the formulation of performance standards. In many instances, trade associations also sponsor programs of certification, in which labels, seals, or other identifiable marks are placed on materials or products manufactured to particular standards.

■ STANDARDS-SETTING TRADE ASSOCIATIONS

Many trade associations engage in developing standards for materials and products their members manufacture and services they perform. The American Plywood Association (APA) is an example of a trade association active in developing standards. This association is supported by a large membership of manufacturers producing softwood plywood. The production of various types and grades of plywood requires the selection and classification of wood veneers according to strength and appearance. Producing plywood from these veneers demands careful manufacturing control of such factors as moisture content

and adhesive type. The completed product, properly assembled, bonded, and finished, must conform to a body of material standards, manufacturing procedures, and performance testing standards.

Plywood manufacturers, through their own research and combined efforts within the APA, financially and technically support the research and development of criteria that the association formulates into industry standards. The members of the association then agree to produce plywood that conforms with these standards.

■ TRADE ASSOCIATION STANDARDS AS A BASE FOR OTHER STANDARDS

The standards developed by trade associations often are used as a base from which other groups, such as ANSI, may develop standards. For example, industry standards may be promulgated as *ANSI standards*, which are often incorporated into the HUD *Minimum Property Standards for Housing* or building codes.

■ CERTIFICATION

Many trade associations and other industry groups provide assurance that materials and manufactured products have met established standards. Certification of quality may take the form of *grade marks*, *labels*, or *seals*. For example, the APA maintains a continuing program of product testing during and after manufacture with grade markings applied directly to the plywood. A *grade mark* is a visible statement that the appropriate product standard has been met.

The reliability of certifications issued by manufacturers or their associations varies. Some are excellent, while others are worthless. The most reliable certification is one issued by an independent testing agency.

1.2.2.2

Standards-Setting and Testing Agencies

In addition to trade associations, organizations have been established whose primary purpose is the setting of standards or the testing of materials and products to ensure that they comply with established standards. When a material or product is com-

pliance tested by its manufacturer, its trade association, or an independent testing agency, a *testing standard* produced by one of the agencies discussed in this subsection is often used as the standard for the testing procedure.

■ AMERICAN SOCIETY FOR TESTING AND MATERIALS (ASTM)

The American Society for Testing and Materials (ASTM) is an international, privately financed, nonprofit, technical, scientific, and educational society. The objectives of the society are "the promotion of knowledge of the materials of engineering, and the standardization of specifications and methods of testing."

The ASTM membership consists of about 32,200 active members, including individuals—engineers, scientists, and educators—and organizations—companies, government agencies, and universities. About 135 technical committees formulate and recommend ASTM standards covering many types of materials; a number of administrative committees deal with publications, research, testing, consumer standards, and other activities. The society is supported mainly by membership dues, with some income from the sale of publications.

ASTM started in 1898 as the American Section of the International Society for Testing Materials. It incorporated in 1902 and became the American Society for Testing Materials. In 1961, the name was changed to the American Society for Testing *and* Materials to emphasize its interest in basic information about materials.

Two general categories of information and publications are available from ASTM: (1) *ASTM standards*, which include definitions of terms, materials standards, workmanship standards, and methods of test standards used throughout the industry (Fig. 1.2-1); and (2) data dealing with research and testing of materials, including *Materials Research and Standards*, published monthly; *Journal of Materials*, published quarterly; and special technical publications that cover symposia and collections of data.

ASTM standards are designated by the initials ASTM followed by a code number and the year of last revision. For example, ASTM C 55-85 refers to ASTM Document C 55, "Standard Specification

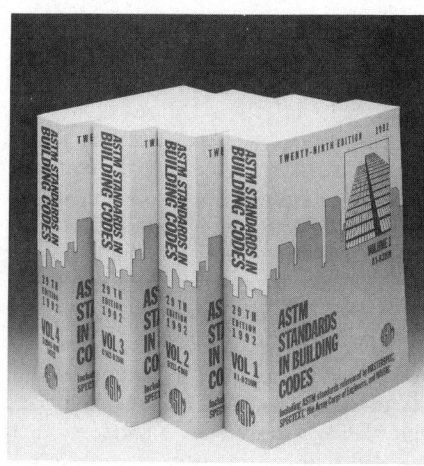

FIGURE 1.2-1 ASTM Standards in Building Codes. *(Copyright ASTM, reprinted with permission.)*

for Concrete Building Brick," last revised in 1985.

Publications are sold directly by ASTM (see Appendix C, Data Sources, for the address and telephone number) and by bookstores that handle technical publications. Lists of publications and their prices may be obtained from ASTM.

■ AMERICAN NATIONAL STANDARDS INSTITUTE (ANSI)

The American National Standards Institute (ANSI) is the name adopted by the United States of America Standards Institute (USASI) in October 1969. USASI was created in 1966 by the complete reorganization of the earlier standards organization, the American Standards Association (ASA). This restructuring provided for a Member Body Council (the former Standards Council of ASA), a Company Member Council, and a Consumer Council.

The Member Body Council, which comprises national trade, professional, and scientific associations, is responsible for establishing and maintaining procedures for the approval of standards developed by other associations, agencies, or groups as American National Standards. In this way, a standard developed by a trade association, such as the American Architectural Manufacturers Association (AAMA) can become an American National Standard. Unlike ASTM, ANSI does not formulate its own standards or provide testing services.

The Company Member Council consists of representatives of industrial firms, labor, and government. The Company Member Council and the Consumer Council provide membership on the Board of Directors and work closely with the Member Body Council in recommending areas of standardization deemed essential and in reviewing standards.

ANSI is privately financed, its financial support being derived from voluntary membership dues and from the sale of the published American National Standards. These national standards, which are arrived at by common consent, are available for voluntary use and often are incorporated in regulations and codes.

ASA, the forerunner of USASI and ANSI, was founded during World War I to avoid duplication and waste in war production. In 1918, five leading American engineering societies, including the American Society of Mechanical Engineers, the American Society of Civil Engineers, and the American Society for Testing Materials, and three departments of the federal government, Commerce, War, and Navy, formed the American Engineering Standards Committee to coordinate the development of national standards. This committee was reorganized in 1928 into the American Standards Association, which was later incorporated into USASI and subsequently renamed ANSI.

More than 3000 American National Standards have been developed and approved under the procedures of the Institute. These standards apply in the fields of engineering, industry, safety, and consumer goods. An American National Standard is designated by a code number and the date of last revision, for example, ANSI Y32.9-1972, "Graphic Symbols for Electrical Wiring and Layout Diagrams Used in Architecture and Building Construction."

An ANSI catalog, price list, literature about ANSI, and standards can be obtained from ANSI at the address listed in Appendix C, Data Sources.

■ UNDERWRITERS LABORATORIES, INC. (UL)

Underwriters Laboratories, Inc. (UL) is chartered as a nonprofit organization to establish, maintain, and operate laboratories for the examination and testing of devices, systems, and materials. The

stated objectives of UL are (1) to determine the relation of various devices, systems, and materials to life and property and (2) to ascertain, define, and publish standards, classifications, and specifications for materials, devices, products, equipment, constructions, methods, and systems affecting such hazards.

UL contracts with manufacturers and other parties to examine, classify, test, and inspect buildings, materials, devices, and methods with reference to life, fire, and casualty hazards. Findings are circulated by various publications to insurance organizations, other interested parties, and the public. UL also contracts to provide product certification by attaching certificates or labels to examined, tested, and inspected materials, devices, and products.

UL, formed in 1894, was originally subsidized by stock insurance companies. Before the turn of the century, as new electrical devices and products came rapidly into the market, it became necessary to test and inspect them to ensure public safety. The National Board of Fire Underwriters (now the American Insurance Association) organized and sponsored UL to meet this demand.

UL became self-supporting in about 1916. To sustain its testing program, UL contracts with product submitters for testing, reporting, and listing of devices, systems, and materials on a time and material basis. The cost of the inspection service is provided for either by an annual fee or by service charges for labels, depending on the type of service. Materials and products carrying UL labels and certificates must meet published standards of performance and manufacture and are subjected to UL inspection during manufacture.

Although it is primarily interested in public safety, UL's policy is to list and label only products that perform their intended function. A product that does not perform with reasonable efficiency, even though it may be perfectly safe, does not qualify for a UL label.

UL standards are designated by the initials UL, followed by a code number (Fig. 1.2-2). No date of last revision is indicated by the number. For example, UL 55A, "Materials for Built-up Roof Coverings," was issued in 1983 and last revised in 1989.

UL's large staff includes engineers and inspectors, four testing stations, and

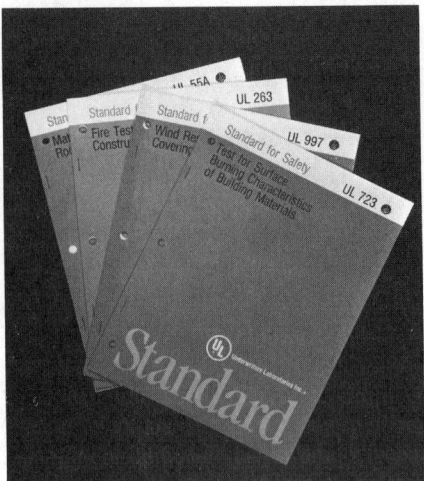

FIGURE 1.2-2 *UL Standards.* *(Courtesy Underwriters Laboratories Inc. [UL].)*

FIGURE 1.2-3 *UL Product Directory.* *(Courtesy Underwriters Laboratories Inc. [UL].)*

representatives in more than 200 cities throughout the United States.

UL's "Standards for Safety" catalog of standards and the standards themselves can be obtained from the Northbrook, Illinois, office listed in Appendix C, Data Sources. Single copies of standards can be obtained by calling one of UL's offices in Northbrook, Illinois, (708) 272-8800; Melville, New York, (516) 272-6200; Santa Clara, California, (408) 985-2400; or Research Triangle Park, North Carolina (919) 549-1400. Lists of submitters whose products are listed and labeled are published annually (Fig. 1.2-3). These lists are available from the office in Northbrook, Illinois.

■ NAHB RESEARCH FOUNDATION, INC.

The NAHB Research Foundation is a wholly owned subsidiary of the National Association of Home Builders. The objectives of the foundation are as follows:

1. To conduct and disseminate the results of research and development with respect to homes, apartments, and light commercial structures for the purposes of lowering the cost and improving the quality of buildings constructed by the U.S. homebuilding industry
2. To conduct, for itself or by contract for others, tests and investigations into and on materials, products, systems, and other matters related to the design, construction, or occupancy of homes and other buildings

3. To the extent permitted by law, to do what is necessary and proper to encourage lower construction costs and improved quality in the design and construction of residential and related structures
4. To the extent permitted by law, to provide a system for labeling material and other products and to grant a seal or certificate of quality or similar device

The NAHB Research Foundation was created in 1965 as an expansion of the NAHB Research Institute, which was founded in 1952. The Foundation continues research activities initiated by the Institute. These activities have included the design and construction of a number of "research houses" that incorporated many of the newest methods and materials available. The Research Institute also developed the successful TAMAP system, which marked the first time that industrial engineering techniques were used to improve productivity in the design and construction of new homes.

As it became more evident that the work of the Research Institute was having far-reaching effects throughout the homebuilding industry, a number of manufacturers and related industry associations asked the Institute to perform research, development, and testing services on a sponsored basis. Since its creation the Foundation has carried out product, standards, and systems research, development, and evaluation studies for many different clients. These clients include many leading companies

in the building product and chemical industries, as well as trade associations.

The Research Foundation's activities are directed by a Board of Directors, with representatives from NAHB and the fields of engineering, science, education, labor, finance, building codes, government, architecture, and public interest. The organization is divided into four major operating portions: laboratory, industrial engineering, research houses, and special studies. The Foundation's professional staff consists of civil and architectural engineers, architects, wood technologists, physicists, industrial engineers, and related building industry specialists.

The NAHB Research Foundation is supported by its clients, which include the National Association of Home Builders, a number of building industry manufacturers, and other associations and organizations. Its facilities are available to individual manufacturers, builders, trade associations, and combinations of these on a group basis. The Foundation's laboratory is equipped with a broad range of facilities, including equipment for ASTM standard tests, vibration studies, acoustical measuring, and temperature-humidity control testing.

Results of the tests and studies are either published by the Foundation through NAHB or remain the sole property of the client. Published reports of research and development of standards are available from the NAHB Research Foundation at the address listed in Appendix C, Data Sources.

■ NATIONAL FIRE PROTECTION ASSOCIATION (NFPA)

The National Fire Protection Association (NFPA) was organized in 1896 to promote the science and advance the methods of fire protection. NFPA is a nonprofit educational organization that publishes and distributes various publications on fire safety, including model codes, materials standards, and recommended practices. These technical materials, aimed at minimizing losses of life and property by fire, are prepared by NFPA Technical Committees and are adopted at the annual meeting of the Association. All are published as *National Fire Codes*, a 12-volume compilation of NFPA's official technical material. (*The National Electrical Code* is Volume 3.)

THE AMERICAN SOCIETY OF HEATING, REFRIGERATING AND AIR CONDITIONING ENGINEERS (ASHRAE)

Since 1894, ASHRAE and its predecessor societies have pursued their objective of advancing the arts and sciences of heating, ventilating, and air conditioning buildings. ASHRAE conducts an extensive research program, publishes meeting transactions, and establishes standards.

ASHRAE standards are established to assist industry and the public by offering a uniform method of testing for rating purposes, by suggesting safe practices in designing and installing heating, ventilating, and air conditioning equipment and systems, and by providing other information that may serve to guide the industry. The creation of ASHRAE standards is determined by the need for them. Conformance is voluntary.

ASHRAE standards are updated on a five-year cycle; each title is preceded by a hyphenated number. The digits before the hyphen are the standard's numerical designation; the digits after the hyphen are the year of approval, revision, or update. For example, ASHRAE Standard 90.1-1989, "Energy Efficient Design of New Buildings Except Low-Rise Residential Buildings," describes a standard of designation 90.1, approved in 1989.

ASHRAE has developed standards not only in the traditional areas of heating, ventilating, and air conditioning equipment, but on such diverse subjects as fire safety in buildings, energy conservation, solar energy, pollution control, and ozone depletion.

ASHRAE's catalog of publications, research reports, standards, transactions, and the *ASHRAE Handbook, Fundamentals* are available from ASHRAE at the address listed in Appendix C, Data Sources.

1.2.3

FEDERAL GOVERNMENT STANDARDS

Many federal agencies either develop standards themselves or commission their development by other federal agencies or private sector organizations.

1.2.3.1

Department of Commerce

Manufacturers seek to encourage product acceptance and improve their own effi-

ciency by establishing basic levels of quality for materials and products and by coordinating dimensions, terminology, and other variables such as type and style.

Manufacturers cannot legally agree to establish unreasonable standards that might rule out the success of individual competitors, nor can they engage in price-fixing agreements. Therefore, in recognition of the desirability of certain types of industry-supported standardization, the U.S. Department of Commerce provides for the development of Product Standards.

■ PRODUCT STANDARDS (PS)

Product Standards (PS) are developed by manufacturers, distributors, and users in cooperation with the Office of Engineering Standards Services of the National Institute of Standards and Technology. The purpose of a Product Standard may be either (1) to establish standards of practice for sizes, dimensions, varieties, or other characteristics of a specific product or (2) to establish quality criteria, including standard methods of testing, rating, certifying, and labeling the manufactured products.

The adoption and use of a Product Standard is voluntary. However, when reference to a Product Standard is made in contracts, labels, invoices, or advertising literature, the provisions of the standard are enforceable through usual legal channels as a part of the sales contract.

A Product Standard usually originates with the manufacturing segment of the industry. The sponsors may be manufacturers, distributors, or users of the specific product. One of these three elements of industry (the proponent) submits to the Office of Engineering Standards Services the necessary data to be used as the basis for developing a Product Standard. The Office by means of assembled conferences, letter referenda, or both assists the proponent in arriving at a tentative standard of practice and thereafter refers it to the other elements of the same industry for approval or for constructive criticism to help make necessary adjustments. The regular procedure of the Office ensures continuous servicing of each Product Standard through review and revision whenever, in the opinion of the industry, changing conditions warrant such action.

A Product Standard is designated by the letters PS, followed by an identification number and the last two digits of the year

of issuance or last revision. For example, PS 1-83 is the Product Standard for "Construction and Industrial Plywood." It was originally issued in 1966 as the first Product Standard and revised in 1983. Product Standards will gradually replace the older Commercial Standards (CS) and Simplified Practice Recommendations (SPR), previously published by the Department of Commerce.

A list of the Product Standards, Commercial Standards, and Simplified Practice Recommendations currently available, a price list, and ordering instructions can be obtained from the Office of Engineering Standards Services, National Institute of Standards and Technology at the address listed in Appendix C, Data Sources. Also available from that Office are copies of "Procedures for the Development of Voluntary Product Standards," which explain the process through which such voluntary standards are developed.

1.2.3.2

General Services Administration (GSA)

The General Services Administration (GSA) of the U.S. government develops *Federal Standardization Documents*, including *Federal Specifications*, *Interim Federal Specifications*, and *Federal Standards*, through the cooperation of federal agencies and industry groups. The purpose of these documents is to standardize the variations and quality of materials and products being purchased by governmental agencies. Approximately 5600 Federal and Interim Federal Specifications have been developed by the GSA.

The "Index of Federal Specifications, Standards, and Commercial Item Descriptions" lists documents alphabetically by title and numerically and can be purchased from the Superintendent of Documents, U.S. Government Printing Office, at the address listed in Appendix C, Data Sources.

■ FEDERAL SPECIFICATIONS

In the pure sense of the dictionary definition, Federal Specifications can be called specifications. According to the definitions we are using in this book, however, and according to the standard practices of the construction industry, even on projects for the government, they are actually used as standards. They are not permitted, for example, to be part of a *project manual* for a building construction project, except by reference. It is not possible to enter Federal Specifications intact into a project manual, and editing them for that purpose is neither permissible nor desirable. However, since GSA calls them specifications and this terminology is generally accepted in the building industry, we will accede to this convention.

A new Federal Specification is developed when a government procurement need arises, when a present specification becomes obsolete, or when revision is required for other reasons.

GSA may assign the development of a particular Federal Specification to a federal agency that has specialized technical competence and has available the necessary facilities. However, nationally recognized industry, technical society, and trade association standards, such as those by ASTM, are used and adopted in Federal Specifications to the maximum extent practicable.

Federal Specifications are designated by letter and number codes. For example, FF-D-396H is the Federal Specification for soap dispensers. The term Federal Specification is often abbreviated as Fed. Spec. or simply FS.

■ INTERIM FEDERAL SPECIFICATIONS

Federal Specifications are often issued in interim or temporary form by individual federal agencies with optional use by all agencies. These Interim Federal Specifications are coordinated with federal agencies and industry groups as much as possible. If justified by usage in more than one agency, an Interim Federal Specification may be changed to a Federal Specification. Before this is done, however, the proposed specification is further coordinated with industry and with all federal agencies having an interest.

An Interim Federal Specification is designated by a double zero in front of the serial number. For example, TT-E-002124A is the code for the Interim Federal Specification for "Enamel (Vinyl-Alkyd, Exterior)."

1.2.3.3

Military Agencies

The federal government is one of the world's largest buyers of equipment, mate-

rials, and supplies, with annual purchases in the billions of dollars. Various departments within the Department of Defense have developed specifications covering materials, products, and services used predominately by military activities.

Military Specifications may be used by any interested civilian organization or specifier. Military Specifications are indexed by the title and the code-letter prefix MIL.

1.2.3.4

Department of Housing and Urban Development (HUD)

The National Housing Act, enacted by Congress in 1934 and amended from time to time, created the Federal Housing Administration (FHA) to stimulate home construction by insuring mortgage loans. The functions of this agency were transferred by Congress in 1965 to the newly created Department of Housing and Urban Development (HUD), and FHA became part of this larger cabinet-level department.

The overall purpose of HUD is to assist in the sound development of the nation's communities and metropolitan areas. Encouragement of housing production through mortgage insurance and various subsidies has been one of HUD's chief objectives. Improvement in housing quality and land planning standards has been another HUD objective mandated by Congress.

■ FHA/HUD HOUSING PROGRAMS

FHA/HUD makes no loans, nor does it plan or build housing. It functions mainly as an insuring agency for mortgage loans made by private lenders, such as savings associations and commercial banks. For instance, through the Section 203(b) program, FHA/HUD encourages lenders to make loans with low down payments and long maturities on one- to four-family dwellings. The borrower pays an annual insurance premium of a small percentage of the average principle outstanding over the premium year. The Secretary of HUD sets the interest rate ceiling on FHA/HUD loans at a level required to meet market conditions. Another frequently used section, 221(d)(4), provides for mortgage insurance of new or rehabilitated low- or middle-income rental housing.

The traditional role of the FHA was transformed when the agency became the administrator of interest-rate subsidy and rent-supplement programs authorized by Congress since 1965. The Section 235 program combines insurance with interest-assistance payments for owner-occupied homes. In addition to insuring the loan, FHA/HUD pays part of the interest the borrower owes the mortgage lender. Section 236 offers insurance and interest assistance for rental projects. Section 237 provides insurance on loans to borrowers with poor credit histories. Section 238 authorizes insurance for mortgage loans in high-risk situations, such as transitional urban areas, not covered by other programs.

■ HUD MINIMUM PROPERTY STANDARDS

Because not all housing programs authorized by Congress involve mortgage insurance, they are not all administered by FHA/HUD. For instance, Section 8 of the Housing Act of 1974 authorizes rental subsidies for leased low-income housing. The housing may be existing or new and may be financed either conventionally or with FHA/HUD mortgage insurance.

Before 1973, FHA-insured private housing had to conform to the FHA *Minimum Property Standards* (MPS), while subsidized public housing was regulated by a different set of standards. With the adoption in that year of the MPS, uniform standards became applicable to all HUD housing programs.

The MPS were intended to provide a sound technical basis for the planning and design of housing under the numerous HUD programs. The standards described those characteristics in a property that would provide initial and continuing utility, durability, desirability, economy of maintenance, and a safe and healthful environment.

Environmental quality was considered throughout the MPS. As a general policy, property development was required to be consistent with the national program for conservation of energy and other natural resources. Care had to be exercised to avoid air, water, land, and noise pollution and other environmental hazards.

The MPS consisted of three volumes of mandatory standards: (1) MPS for One- and Two-Family Dwellings, HUD 4990.1; (2) MPS for Multifamily Housing, HUD 4910.1; and (3) MPS for Care-Type

Housing, HUD 4920.1. Variations and exceptions for seasonal homes intended for other than year-round occupancy were listed in HUD 4900.1. Exceptions for elderly housing were listed in HUD 4900.1 and 4910.1. A fourth volume, the MPS Manual of Acceptable Practices, HUD 4930.1, contained advisory and illustrative material for the three volumes of mandatory standards.

Today, these documents have been withdrawn and replaced by a single document, HUD 4910.1, *Minimum Property Standards for Housing*. This document is intended to supplement the requirements of the applicable local and state building codes and the Council of American Building Officials (CABO) publication *One- and Two-Family Dwelling Code* (see Section 1.3.3.8).

■ MATERIALS RELEASES

The Architectural Standards Division of HUD issues *Materials Releases* for specific proprietary products for which no recognized standards exist. Materials Releases relate only to the technical elements of a product. Each release describes a product and its use and is issued to HUD field offices for guidance in determining the acceptance of the product.

The absence of a release for a particular product does not preclude its use. Materials Releases are not intended to indicate approval, endorsement, or acceptance by HUD. Manufacturers of materials and products for which Materials Releases have been issued are not authorized to use them in any manner for sales promotion. Copies of Materials Releases are on file in HUD field offices but are not available for general distribution.

HUD has additional provisions for the review of special materials, products, and construction methods that it may be asked to insure. Design, materials, equipment, and construction methods other than those described in *Minimum Property Standards for Housing* are considered for use, provided complete substantiating data satisfactory to HUD are submitted. Local HUD field offices are authorized to accept variations from the standards for specific cases, subject to conditions outlined in the standards.

Variations on an area or regional basis and variations involving a substantial number of properties on a repetitive basis are authorized only after consideration of recommendations by the HUD field office and approval by the Architectural Standards Division. Under certain conditions some variations are established and published as Local Acceptable Standards (LAS) for a specific area.

1.2.4

EUROPEAN STANDARDS

In January of 1993, 12 European countries joined economically into a Single European Market (SEM). In addition, the European Community (EC) is moving steadily toward the opening of borders and the establishment of free trade of ideas and goods among the member nations. These major changes in Europe will greatly affect trade and other relationships between the United States and the EC. It also will affect the U.S. building industry in many ways.

One effect that will be greatly felt is the change in European standards. As a part of the establishment of SEM, the EC member states have deemed it necessary to unify their existing 12 separate groups of national standards into common European standards. The European Committee for Standarization (CEN) and the Committee for European Electrotechnical Standarization (CENELEC) have been charged with publishing standards for the EC member states.

The International Organization for Standardization (IOS) is a nongovernmental organization made up of representatives from the standards institutions from 91 countries. The United States is a member, represented by ANSI, but it is not very influential. IOS approves and publishes standards produced by its members. Most of the standards published by IOS are European ones.

The new EC standards will be based either on reconciling the differences between existing standards in the 12 EC countries or by adopting IOS standards. The United States could exert greater influence in this area by increasing its involvement in IOS, but there seems little movement in that direction. Although development is in progress, single European standards for construction products do not yet exist. Until they do, EC member countries will continue to enforce either their own standards or interim ones.

United States firms doing work in Europe and U.S. products sold in Europe today have to comply with the standards of the country in which the work is being done. Eventually they will all have to comply with the unified EC standards. In the interim, there will be a morass of conflicting and possibly overlapping standards in the various countries. Some standards will be the old national ones, some will be the new EC standards, and others will be IOS standards. Probably the best first step of anyone contemplating working in the EC would be to contact the U.S. Department of Commerce Office of European Community Affairs for advice.

What eventual effect the new EC standards will have on U.S. standards is unknown at this point. There will probably be some changes in our standards relatively soon. Down the road there may be extensive changes as the United States tries to compete economically with a unified Europe.

1.3 CODES

The planning, construction, location, and use of buildings are regulated by a variety of laws enacted by local, state, and federal governments. These statutes and ordinances include building, plumbing, electrical, and mechanical codes that are intended to protect the health, safety, and general welfare of the public.

Such laws are not new. The ancient code of the Babylonian emperor, Hammurabi, dating back to about 1800 B.C., is often cited as the first recorded building code.

It provided severe penalties for construction practices that violated the health or safety of citizens. For example, if a building collapsed killing the occupants, its architects and builders were put to death. This ancient code, based on the idea that the strong should not injure the weak, sets the principle for today's construction regulations: the public has a right to be protected from the harmful acts of others.

Modern building regulation evolved over time, starting in the early 19th century with the adoption of fire laws in some large cities. These laws prohibited the construction of wooden buildings in congested areas. About the same time, cities also began to adopt health regulations to improve the living conditions of the poor that are the forerunners of today's minimum housing codes. Some courts, however, resisted the enactment of such laws as infringements on personal property rights.

As the validity of fire and health laws was slowly established, courts began to accept government control of all aspects of building construction involving the health and safety of the public. Today, most states and cities and a great many counties and towns regulate the planning, construction, and installation of building systems through a variety of laws and ordinances.

The right to regulate building construction constitutionally rests in the states, but before 1960 few state governments exercised that right. Usually the states delegated to local governments the power to regulate buildings to protect the public health, safety, and welfare. Today, many states take an active role in building regulation. Over half the states have now adopted some form of statewide building regulations concerned with the construction of industrialized buildings, mobile homes, and conventional construction.

Federal legislation in the areas of occupational health and safety, environmental protection, pollution controls, and consumer protection, coupled with increased state legislative activity, offers evidence that the building regulatory process will become ever more complex.

Regulatory codes incorporate many recognized construction industry standards. This section discusses some of the various types and purposes of industry standards and the organizations active in establishing them.

This section also describes the code regulatory system, including the efforts of the major building officials' organizations and the numerous federal, state, and local groups that are working to improve the code regulatory system.

Building codes establish minimum requirements to protect the public health, safety, and welfare. They do not necessarily contain criteria that ensure efficient, convenient, or adequately equipped buildings. A buiding code establishes requirements for the construction and occupancy of buildings. It contains standards of performance and requirements for materials, methods, and systems. Codes also cover structural strength, fire resistance, adequate light and ventilation, egress, occupant safety, and other considerations determined by the design, construction, alteration, and demolition of buildings.

1.3.1

ADOPTION

A collection of building requirements becomes a code when it is adopted by a municipality as a public ordinance or law. Local communities may write their own codes or legally adopt other codes, such as state building codes or one of the model codes described in this section.

In fact, the construction industry is enmeshed in an extraordinary network of building codes that attempts to ensure that buildings and their environs will be safe. They generally accomplish that purpose, but codes and code administration are criticized widely as being restrictive to building progress by retarding the acceptance of new and improved uses of materials and methods, thereby unnecessarily increasing construction costs. In some instances, the adoption of improved codes has stimulated better building practices. But the existing complex and chaotic building code situation is recognized as one of the problems facing the building industry today because of the use of descriptive type codes, the lack of code uniformity, the multiplicity of codes, the slow response of codes to change, and inadequate performance standards in the codes.

1.3.2

DESCRIPTIVE AND PERFORMANCE CODES

Building codes are either descriptive or performance oriented. A *descriptive* (sometimes called *specification* or *prescription*) type building code establishes construction requirements by reference to a particular material or method. For example, an unrevised and out-of-date code might have its requirements for wood framing based on the use of balloon framing (see Section 6.5), which is seldom used today. Such a code may not mention balloon framing per se but may retain some aspects of that type of framing system. It might, for instance, require that exterior walls be built of 2×4 wall studs spaced 16 in. on centers and covered by 1-in.-thick board sheathing applied diagonally. A builder seeking to space studs at 24 in. on centers, which might be structurally sound and safe, or to use another type of sheathing, even one that is better than the diagonal sheathing, would be in violation of the specifically stated code requirements.

A *performance* type code does not limit the selection of methods and systems to a single type but establishes requirements for the performance of building elements. Such a code states design and engineering criteria without reference to specific materials or methods of construction. For example, an outside wall might be required to support loads (wind, dead, live, etc.) with specific defined values and to meet or exceed stated insulating and permeability requirements. Any system performing as required by the code would be acceptable, regardless of the materials or methods used.

True performance codes are idealistically excellent but impractical in use. To enforce them, local code administrators would have to be extraordinarily competent and equipped to interpret performance criteria and evaluate proposed methods, uses, and systems. Such people and equipment are rare.

A workable solution lies somewhere between descriptive and performance codes. Both types of codes should adequately provide for the accpetance of alternative methods and systems.

Local Administration

Local administration and enforcement of codes are usually done by a building inspector or engineer who has the authority to approve materials and methods that may not be directly referenced in the code. Qualified people are necessary to administer a building code properly. No matter how good a code may be, it must be enforced by someone who is experienced, informed, and objective. Most builders complain about part-time inspectors who do not understand construction and may be arbitrary and inconsistent. They are seldom upset by a careful and competent inspector who is consistent even though tough. A competent code enforcer knows when the letter of the code should prevail and when subjective interpretation should be made.

Local code administrators, faced continually with difficult decisions, may well argue that their job is solely to check compliance. An often drawn, and reasonably correct, analogy is that a local police officer is not expected to make judgments about whether a law is right or just but is charged solely with determining compliance. Perhaps it is best that technically qualified people, not local code administrators, must determine performance criteria and judge whether methods and systems perform suitably. The model code groups (see Section 1.3.3) are qualified to offer this type of service.

Model code groups, which are supported by building officials themselves, have performed a great service to the building industry. But model codes have not fully solved the problems of code uniformity. This is partly because some communities have not adopted one of the model codes, and those that do use them have adopted them with modifications. The National Association of Home Builders (NAHB) has said that many of the changes related to housing made to model codes to adapt them to local conditions come from local codes prepared 20 or more years ago. Sometimes progressive time- and money-saving requirements of a model code are revised so that antiquated provisions apply instead, simply to agree with local custom.

The modification of model codes by local communities is often influenced by self-interest groups. A major step toward uniformity would occur if local communities would not change the model codes when those codes are enacted as local statutes.

Lack of Uniformity

Because establishing and enforcing building codes are local functions, the designer or builder who works in more than one community is often faced with a frustrating variety of requirements. This lack of uniformity in building codes complicates the jobs of all concerned.

In developing products, a manufacturer must consider the problem of winning acceptance by thousands of local building code administrators instead of concerning itself only with performance and public acceptance. Much effort has been made to unify the codes used by communities, and considerable improvement has been gained through the local adoption of model codes.

Now and again one hears of an attempt to develop a single national building code, but these efforts have not been successful. There is, however, a single model code for one- and two-family dwellings (see Section 1.3.3.8).

Multiplicity of Codes

If designers and builders were dealing with a single code, their problems would be greatly simplified. However, in addition to local building codes that tell them how a structure must be assembled, they also have to satisfy a number of additional codes covering a variety of subjects such as plumbing, electrical wiring, traffic, utilities, health and sanitation, land planning, building occupancy, access by the handicapped, and zoning.

The lack of code uniformity and the multiplicity of additional regulations frequently so complicate matters that buildings are designed and built using very conservative and often expensive criteria and methods. Designers and builders alike are often unable to improve their techniques because of the restrictive nature of one or more of the applicable codes.

Slow Response to Change

Codes often are criticized for failing to recognize new materials and methods. Judging new products on the basis of performance criteria must be in the hands of technically qualified and equipped organizations.

This function is one in which model code groups can exercise great leadership and provide a major stimulus to progress. The groups have responded reasonably to progressive change. However, by their very size, they are subject to many pressures that deter progress and sometimes have been slow in acting on proposed changes. For example, it took almost four years for all the major code groups to accept the NAHB's proposal to eliminate floor bridging, a proposition well documented through extensive research and testing.

Inadequate Performance Standards

The building industry is made up of so many diverse industries and serves such a wide proliferation of interest groups that no single group of performance criteria, testing procedures, or standards exists. In selecting materials or determining the suitability of materials and methods, specifiers and builders refer to ASTM standards, Federal Specifications, American Standards, Commercial Standards, HUD's *Minimum Property Standards for Housing*, a variety of trade association certification programs, grade marks and trademarks, UL labels, and many other construction standards (see Section 1.2).

Although there are independent testing agencies, such as UL, and standards-setting bodies, such as ASTM, and some manufacturers have substantial commitments in research and testing facilities, no single agency concerned with the development of performance criteria and testing methods is currently recognized by all groups interested in building construction.

As long as codes continue to restrict designer's prerogatives, complicate builders' methods and operations, and deter manufacturers from developing new materials and products, unnecessary costs will be added to buildings.

MODEL CODES

To help fill the need for workable codes, three major organizations of industry and professional groups and several states have developed building codes that may be adopted into law for use in local communities. These codes are commonly referred to as model codes. The major model code groups are the Building Officials and Code Administrators International (BOCA), the International Conference of Building Officials (ICBO), and the Southern Building Code Congress International (SBCCI). The major national model codes are the BOCA *National Building Code*, the ICBO *Uniform Building Code*, and the SBCCI *Standard Building Code*.

Model codes have been widely accepted by local communities and now are used by more than 80% of those communities with a population of more than 10,000. The organizations that prepare model codes also provide the continuous updating necessary to include recommendations based on industry research and the development of new materials and methods.

Several model codes that deal with specialized subjects not fully covered in the major model codes are also available. These codes include, but are not limited to, the *National Standard Plumbing Code*, the *Standard Plumbing Code*, the *Uniform Plumbing Code*, the *National Mechanical Code*, the *Standard Gas Code*, the *Standard Mechanical Code*, the *Uniform Mechanical Code*, and the *National Electrical Code*. Many local communities have adopted these codes.

The nationally recognized model codes and their sponsoring organizations are shown in Figure 1.3-1.

1.3.3.1

National Building Code

The first model building code was introduced in 1905 by the National Board of Fire Underwriters, which was later called the American Insurance Association (AIA). This organization was concerned primarily with public protection against fire hazards, and the original code was designed to guide communities in setting up fire safety standards. With subsequent additions, its *National Building Code* laid

FIGURE 1.3-1
Model Building Codes and Their Sponsoring Organizations

Building Codes

National Building Code	Building Officials and Code Administrators International, Inc.
Standard Building Code	Southern Building Code Congress International
Uniform Building Code	International Conference of Building Officials
One and Two Family Dwelling Code	Council of American Building Officials

Electrical Codes

National Electrical Code	National Fire Protection Association
One and Two Family Electrical Code	National Fire Protection Association

Elevator Codes

Safety Code for Elevators and Escalators	American National Standards Institute
Safety Code for Manlifts	American National Standards Institute

Fire Prevention Codes

National Fire Prevention Code	Building Officials and Code Administrators International, Inc.
Fire Uniform Code	International Conference of Building Officials

Housing Codes

Property Maintenance Code	Building Officials and Code Administrators International, Inc.
Standard Housing Code	Southern Building Code Congress International
Uniform Housing Code	International Conference of Building Officials

Plumbing Codes

National Plumbing Code	Building Officials and Code Administrators International, Inc.
National Standard Plumbing Code	National Association of Plumbing-Heating-Cooling Contractors
Standard Plumbing Code	Southern Building Code Congress International
Uniform Plumbing Code	International Association of Plumbing and Mechanical Officials

Mechanical Codes

National Mechanical Code	Building Officials and Code Administrators International, Inc.
Uniform Mechanical Code	International Conference of Building Officials
Standard Gas Code	Southern Building Code Congress International
Standard Mechanical Code	Southern Building Code Congress International

Miscellaneous Codes

Boiler and Unfired Pressure Vessel Code	American Society of Mechanical Engineers
Flammable and Combustible Liquids Code	National Fire Protection Association
Safety Code for Mechanical Refrigeration	American Society of Heating, Refrigerating, & Air Conditioning Engineers, Inc.
Life Safety Code	National Fire Protection Association

the groundwork for the development of building codes throughout the country. AIA produced its last edition of the *National Building Code* in 1976.

In 1950, BOCA published the *Basic Building Code*. When AIA ceased publishing the *National Building Code*, BOCA picked it up and has published it since. BOCA also publishes the *National Fire Prevention Code*, the *National Property Maintenance Code*, the *National Plumbing Code*, and the *National Mechanical Code* (Fig. 1.3-2).

FIGURE 1.3-2 *BOCA codes.* *(BOCA National Building Code/1990, copyright 1989, Building Officials and Code Administrators International, Inc. Published by arrangements with author. All rights reserved.)*

1.3.3.2

Standard Building Code

SBCCI drafted the *Southern Standard Building Code* in 1945. Now renamed the *Standard Building Code* (Fig. 1.3-3), this model code is designed to recognize the special problems, such as high winds, in the southern region of the country and is used extensively there. SBCCI also publishes the *Standard Plumbing Code*, the *Standard Gas Code*, the *Standard Mechanical Code*, and the *Standard Housing Code*.

1.3.3.3

Uniform Building Code

In 1927, ICBO (initially the Pacific Coast Building Officials Conference) published its *Uniform Building Code*. Since that time, the code has gained wide acceptance, particularly in the West (Fig. 1.3–4). ICBO also publishes the *Uniform Building Code Standards*, the *Uniform*

FIGURE 1.3-3 *SBCCI codes.* *(Photograph of the Standard Codes™. Printed with permission of the copyright holder, Southern Building Code Congress International.)*

Housing Code, the *Uniform Code for the Abatement of Dangerous Buildings*, and the *Uniform Sign Code*. With the Western Fire Chiefs Association, it copublishes the *Uniform Fire Code* and the *Uniform Fire Code Standards*. ICBO also jointly publishes the *Uniform Mechanical Code* with the International Association of Plumbing and Mechanical Officials (see Section 1.3.3.5).

(see Section 1.3.3.5)

1.3.3.4

National Standard Plumbing Code

The *National Standard Plumbing Code* is sponsored by the National Association of the Plumbing-Heating-Cooling Contractors (NAPHCC). The purpose of the *National Standard Plumbing Code* is to protect the public health. The code, therefore, covers the proper design, installation, and maintenance of plumbing systems based on principles of sanitation and safety, and not necessarily on efficiency, convenience, or adequacy for good service or future expansion of system use. Standards for materials and fixtures are based largely on industry standards such as Commercial Standards, American Standards, and ASTM standards.

1.3.3.5

Uniform Plumbing and Uniform Mechanical Codes

In 1929, the Western Plumbing Officials Association (now the International Association of Plumbing and Mechanical Officials (IAPMO) established the first *Uniform Plumbing Code* and continues to publish it today. IAPMO also publishes the *Uniform Mechanical Code* jointly with ICBO.

FIGURE 1.3-4 *ICBO codes.* *(Photo courtesy of ICBO.)*

These codes have been developed in the interest of public health and safety and cover the design, installation, and maintenance of plumbing, heating, and air conditioning systems. Materials and equipment standards are based largely on other industry standards. These codes have gained widespread use in the western states, as well as in other communities across the country.

1.3.3.6

National Electrical Code

The *National Electrical Code* is produced by the National Fire Protection Association (NFPA). Its purpose is to ensure the safeguarding of persons and of buildings and their contents from hazards arising from the use of electricity for light, heat, power, and other purposes. The code contains basic minimum provisions for safety that, with proper maintenance, will result in installations free from hazard, but not necessarily efficient, convenient, or adequate for good service or future expansion of electrical use.

The *National Electrical Code* makes reference to many industry standards, such as ANSI/ASME A17.1, "Safety Code for Elevators and Escalators," and to UL labels. It is revised periodically and is adopted by reference into many building codes and federal and state laws and regulations.

1.3.3.7

Life Safety Code

The *Life Safety Code* began in 1912 with a pamphlet entitled "Exit Drills in Factories, Schools, Department Stores and Theaters" and has evolved over the years to cover most building types. It is published by NFPA to establish minimum requirements for a reasonable degree of safety from fire in structures. It addresses requirements for egress; design and construction to prevent undue danger from fire, smoke, and gases; fire detection and alarm systems; and automatic sprinklers and other fire extinguishing systems. It includes requirements for building design and construction and for heating, ventilating, air conditioning, and electrical systems.

1.3.3.8

One and Two Family Dwelling Code

In 1971, the AIA, BOCA, ICBO, and SBCCI adopted a consensus code entitled the *One and Two Family Dwelling Code*. This model code is now published by the Council of American Building Officials (CABO), which is a national forum of the three major model code organizations, BOCA, ICBO, and SBCCI (see Section 1.3.5.2). It includes a compilation of data from the following codes:

BOCA: *National Building Code, National Plumbing Code, National Mechanical Code*

SBCCI: *Standard Building Code, Standard Plumbing Code, Standard Gas Code, Standard Mechanical Code*

ICBO: *Uniform Building Code*

ICBO and IAPMO: *Uniform Mechanical Code*

IAPMO: *Uniform Plumbing Code*

NFPA: *National Electrical Code*

The CABO *One and Two Family Dwelling Code* contains requirements for building planning and construction, including requirements for heating, cooling, and plumbing. HUD now requires compliance with the CABO *One and Two Family Dwelling Code* in addition to its own *Minimum Property Standards for Housing* and local and state building codes for consideration for HUD/FHA loan guarantees.

Publication of this single national model code for one- and two-family dwellings constitutes a significant step toward uniform minimum regulations in the housing industry, but it still has not been recognized and adopted by every state and local government.

1.3.4

ENERGY CODES

In the United States, energy consumed in buildings amounts to about one-third of all energy consumption. Building energy consumption has been increasing rapidly for many years. The U.S. Department of Commerce estimated recently that energy consumption will have increased by 25% between 1984 and 2000. That increase would amount to an annual growth rate of 1.14% for housing and 1.66% for commercial use. Slowing the rate of increase through increased efficiency in building systems design can significantly reduce overall energy demand.

The *National Energy Plan* calls for substantial decreases in energy consumption through conservation. A major element of an energy conservation program is enforcement by state and local governments of thermal efficiency standards and insulation requirements in new and renovated buildings.

The National Conference of States on Building Codes and Standards (NCSBCS) in 1973 recognized the need for a model national, uniform energy conservation code. NCSBCS requested the National Bureau of Standards (NBS), now the National Institute of Standards and Technology (NIST), to develop standards that could be used by all states for energy-efficient design in new buildings. NBS completed these standards in 1974, but they were considered too complex for most state and local code enforcement officials to administer. As a result, NCSBCS requested the American Society of Heating, Refrigerating and Air Conditioning Engineers (ASHRAE) to translate the NBS standards into enforceable language.

In 1975, ASHRAE published its Standard 90-75, "Energy Conservation in New Building Design." This standard was the first nationally recognized standard for energy-efficient design applicable to all building types. It was, however, written for use by design engineers. State and local practitioners still needed a document that could be administered through the traditional code enforcement system. In 1976, NCSBCS translated ASHRAE Standard 90-75 into the Model Code for Energy Conservation in New Building Construction. Under a grant from the Energy Research and Development Administration (now the Department of Energy), NCSBCS worked with building officials from all parts of the country who administer model codes so that the Model Code for Energy Conservation would be applicable to all geographic areas in the country and could be incorporated easily into existing codes. The NCSBCS Model Code for Energy Conservation in New Building Construction was published in January 1977.

The *Energy Policy and Conservation Act* (EPCA) became law in 1975. The law provided for substantial grants to those states that developed and implemented statewide energy conservation plans aimed at reducing statewide energy consumption by 5% by 1980. State energy conservation plans had to include at least five elements: (1) mandatory lighting efficiency standards for public buildings; (2) programs to promote the availability and use of car pools and van pools; (3) mandatory standards and policies related to procurement practices of state and local governments; (4) a traffic law that permits a right turn on red after stopping; and (5) mandatory thermal energy standards and insulation requirements for new and renovated buildings.

In 1976, Congress passed the *Energy Conservation and Production Act* (ECPA). Title III of ECPA contained the following stipulations: (1) HUD was directed to develop performance-oriented thermal energy efficiency standards; (2) HUD is to monitor the progress of state and local governments in implementing the standards; (3) if a substantial amount of new housing is not constructed in conformance with the standards, HUD can recommend to Congress that the standards be made mandatory. If Congress concurs that the standard should be mandatory, the following provisions will take effect:

1. All construction assisted by federal money will have to be consistent with the standards. This will include federal grants, loans, and loan guarantees and will apply to all Savings and Loan Associations that are federally insured.
2. Local governments will not be required to adopt the standards into their building code. HUD and the states will be responsible for ensuring that all proposed construction is in conformance with the standards. However, local governments that elect to adopt the standards will be reimbursed for the cost of the certification.
3. If a local government incorporates the federal standards into its codes, no further approvals will be required. Money will be available to communities to assist them in incorporating and enforcing the federal standards.

Current requirements for insulation are addressed in Section 7.3; there are also currently accepted standards for energy conservation and thermal design criteria.

1.3.5

CODE ADVANCEMENT

The growing complexity of building design and construction and the subsequent increase of regulatory controls have induced several organizations to improve the effectiveness of the regulatory process with better code documents, administration, and enforcement.

1.3.5.1

National Institute of Building Sciences

The Housing and Community Development Act of 1974 authorized the creation of the National Institute of Building Sciences (NIBS). NIBS was initiated by the federal government with the assistance of the National Academy of Science and the National Academy of Engineering Research Council to help improve the way building construction is regulated.

NIBS gives the United States a national center for (1) assembly, storage, and dissemination of technical data and related information on construction; (2) development and promulgation of nationally recognized performance criteria, standards, test methods, and other evaluating techniques, and (3) evaluation and prequalification of existing and new building techniques.

1.3.5.2

Council of American Building Officials

In 1972, the governing bodies of three national building officials organizations—BOCA, ICBO, and SBCCI—established the Council of American Building Officials (CABO).

CABO's policies are determined by a board of trustees composed of representatives of the governing bodies of the member organizations. Between annual meetings of the board of trustees, the Council is managed by a nine-member executive committee. The executive directors of the three organizations serve as exofficio members.

Examples of CABO's work include

(1) advancing the *One and Two Family Dwelling Code* as a recognized national standard through the procedures of ANSI; (2) sponsoring the development of model performance standards that will complement the requirements of model codes; (3) sponsoring a national research activity that provides a single approval agency for manufacturers of building components, systems, and materials; and (4) developing and adopting a model mechanical, plumbing, and fire prevention code as the standard for all three organizations.

1.3.5.3

Associated Major City Building Officials

In 1974, building code officials from the nation's 30 largest cities created an organization called Associated Major City Building Officials (AMCBO). Its purpose is to exchange ideas, experience, and information on the growing complexities of code administration and enforcement in densely populated urban areas.

AMCBO is a membership organization and the technical arm of the National League of Cities, the International City Management Association, and the National Association of Counties.

1.3.5.4

National Conference of States on Building Codes and Standards

The National Conference of States on Building Codes and Standards (NCSBCS) was formed in 1967. The organization is composed of state delegates designated by their respective governors to represent the 50 states in discussions and programs pertaining to building regulation; state and local building officials; various representatives of the building industry's design, manufacturing, and construction sectors; federal government officials; and consumers. It has affiliations with governmental agencies, such as the National Institute of Standards and Technology and the Consumer Product Safety Commission. Its members participate in the activities of many organizations, including CABO, BOCA, ICBO, SBCCI, NFPA, ASHRAE, HUD, NIBS, the American Institute of Architects (AIA), ANSI, Council of State Community Affairs Agencies (COSCAA), and ASTM.

NCSBCS was established to (1) provide a forum for states to discuss and develop solutions to code regulatory problems; (2) promote the adoption and administration of uniform building codes and standards that regulate construction in and between states; (3) create an effective voice for state input on the committees of nationally recognized standards-writing organizations; (4) support comprehensive training and educational programs for code enforcement personnel; and (5) foster cooperation and encourage innovation between code administration officials and the design, manufacturing, and consumer interests affected by the code regulatory system.

1.4 LAND SURVEYS AND DESCRIPTIONS*

Before a legal description of a piece of property can be prepared, someone must first survey, that is, measure and mark, the land. The legal description is then made based on the survey.

Land descriptions and surveys have been used for centuries. It is not possible to assign the birth of the science of surveying to a particular year or even country, but the earliest known records refer to the making of skillful measurements and calculations with respect to land. The Chinese at an early date and the ancient Egyptians practiced the art of surveying. In Egypt, it appears that it was necessary every spring to reestablish corners and boundary lines obliterated by floods of the Nile River.

Today, the transfer of land ownership and the mortgaging of property require legal land descriptions, by which the location and the boundaries of land parcels can be determined. Descriptions must be referenced to established systems of field marks and measures. Therefore, at least a general knowledge of survey systems is necessary to understand the description of land in writing.

This section describes the several general systems of survey and types of legal land descriptions used in the United States today.

1.4.1

SURVEY SYSTEMS

A survey is the measure and marking of land, accompanied by maps and field

*Portions of the text in Section 1.4 have been reprinted with the permission of the Chicago Title and Insurance Company.

notes that describe the measures and marks actually made in the field. The lengths and directions of boundary lines are established between reference points on the ground. The reference points may be natural marks, such as rivers, or they may be stone, concrete, or other artificial markers located and set in the field by a surveyor. Natural points, noted as permanent, and permanent markers set in the field by surveyors are called *monuments*.

Today, survey measure of lengths is commonly made in feet and decimal parts of feet. However, terminology based on measures no longer used is encountered often in surveys and *legal descriptions*. For example, early surveys were based on lengths such as *chains* and *links*. Surveyors actually made field measurements with a metal chain made of 100 links. The chain was 66 ft. long. The combined length of 80 chains was equal to one statute mile, or 5280 ft. Some of the various measures used and their equivalents are shown in Figure 1.4-1.

Two general systems of survey are used in the United States. The *metes and bounds system*, which is based on tracing the boundary lines surrounding an area, evolved from colonial days. The *rectangular system* of the Federal Government Survey is used in most states. It is based on a modified grid of north-south and east-west lines.

The survey systems discussed in this section may appear to be exact methods of describing land, but in fact they are not. Often unavoidable inaccuracies in surveying may present difficulty and disagreement in the legal description of land. Some of the uncertainties in land description based on problems inherent in surveying are discussed in Section 1.4.2.4.

1.4.1.1

Metes and Bounds System

Survey by *metes and bounds* (Figs. 1.4-2 and 1.4-3) consists of beginning at a known point, *running out* the boundaries of the area by *courses* (directions) and *distances* (lengths), and fixing natural or artificial *monuments* at the corners. The *place of beginning* (P.O.B.) must be a known point that can be readily identified. The P.O.B. must be established and witnessed so that it can be again located with certainty if the monument that identifies it is destroyed or removed. The survey description must "close," that is, the courses and the distances of the description traced in order return to the starting point. An example of a modern legal description based on a metes and bounds survey is shown in Figure 1.4-3.

FIGURE 1.4-2 *Early metes and bounds surveys were marked by natural and artificial markers, as shown at the left. Modern surveys are established more carefully with accurate distances and angles. In the example at the right, the starting point of the survey, or place of beginning (P.O.B.), has been referenced to the rectangular system.* (Reprinted with permission of Chicago Title Insurance Company.)

The metes and bounds system is one of the oldest known manners of surveying and describing land. The system evolved from the early practice of "settlement first, survey afterward." Land being bought, claimed, or inherited would be identified by known natural points, such as trees and rivers, and artificial points, such as roads, structures, or other man-made marks.

For example, in 1784, when Simon VanderCook sold his Massachusetts land, the survey on which a legal description was based started at a "white pine stump," proceeded north to a "small white oak tree," and continued with a series of measures by "chains and links" between markers such as trees, stakes, and piles of stones, finally returning to the "middle of the stump where it first begun" (see Fig. 1.4-2).

Early metes and bounds surveys and descriptions often were based on monuments that lacked permanency. Surveying of large and irregular tracts of land without regard to any system or uniformity and the failure of surveyors to make their survey notes a matter of public record gave rise to frequent boundary line disputes and litigation.

Metes and bounds surveys no longer have the uncertainty as to P.O.B. that existed in colonial days. Present-day surveys and descriptions may refer to government survey section lines and corners (see Section 1.4.1.2) as monuments, in addition to permanent artificial marks placed by surveyors.

Metes and bounds surveys and descriptions are still the accepted method in many states and may be used anywhere to survey sites that are irregular in size and shape. A discussion of legal description by the metes and bounds method is found in Section 1.4.2.1.

1.4.1.2

Rectangular System

After the Revolutionary War, the federal government found itself with vast tracts of undeveloped and uninhabited land, with few natural characteristics suitable for use as monuments in metes and bounds descriptions. It was necessary to devise a new standard system of describing land so that parcels could be located readily and permanently for land office sales. A committee headed by Thomas Jefferson evolved a plan, which the Continental Congress adopted in 1785, for dividing the land into a series of rectangles. This plan, designated the *rectangular system* (also called the *government system*) of survey, is in use today in the majority of the 50 states (Fig. 1.4-4).

■ MERIDIANS AND BASE LINES

The world map in Figure 1.4-5 shows a few of the imaginary north-south lines called *meridians of longitude*, which extend from the North Pole to the South Pole. The meridian that passes through the Royal Observatory at Greenwich, England, is designated as 0 (zero) degrees longitude and called the *Greenwich*, or *prime, meridian*. Locations on the globe east or west of Greenwich are measured in degrees from the prime meridian.

The reference lines running around the globe due east and west are called *parallels of latitude*. The *equator*, which circles the center of the globe, is designated as 0 (zero) degrees latitude. Locations on the globe that fall north or south of the equator are measured in *degrees* from the equator.

The Greenwich meridian and the equator are the lines of reference used in navigation. The position of a ship or a plane relative to these lines of reference can be determined by taking an instrument reading of the position of celestial bodies. The position, or location, is stated in degrees, minutes, and seconds (see Appendix D, Glossary) north or south of the equatorial line (*latitude*); and in degrees, minutes, and seconds east or west of the Greenwich meridian (*longitiude*). For example, the O'Hare Airport tower west of Chicago is

Commencing at the North West corner of Section 12, thence South along the section line 21 feet; thence East 10 feet for a place of beginning; thence continuing East 34 feet; thence South 62 degrees, 30 minutes, East 32 feet; thence Southeasterly along a line forming an angle of 8 degrees, 04 minutes, to the right with a prolongation of the last described course 29 feet; thence South 13 degrees, 0 minutes, to the left with a prolongation of the last described line a distance of 49 feet; thence East to a line parallel with the West line of said section and 180 feet distant therefrom; thence South on the last described line a distance of 65 feet; thence due West a distance of 82 feet to a point; thence North 1 degree West 39 feet; thence North 58 degrees West a distance of 49 feet; thence Northwesterly along a line forming an angle of 163 degrees as measured from right to left with the last described line a distance of 49 feet; thence North to the place of beginning.

FIGURE 1.4-3 *A metes and bounds description traces the perimeter of a land parcel. Starting from a reference point, in this case one tied to a rectangular survey, the direction and the distances of segments of the boundary are described in order.* (Reprinted with permission of Chicago Title Insurance Company.)

located at 87 degrees, 57 minutes, and 28 seconds west longitude and 41 degrees, 59 minutes, and 10 seconds north latitude.

The rectangular system of survey is based on a series of *prime meridians* and *base lines* located by surveyors using methods similar to those used in navigation. The meridians run north and south, the base lines run east and west. Their positions are established by longitude and latitude, as shown in Figure 1.4-4.

PRINCIPAL MERIDIANS In establishing the rectangular system of survey, surveyors first selected a substantial landmark from which a start could be made in surveying an area of land. Usually a place was selected that could be identified readily, such as the mouth of a river, and whenever possible a monument was placed. From that point the surveyors ran a line due north and south and designated it as the principal meridian for that particular state or area. The location of the *principal meridian* (its longitude) was fixed by a reading measured in degrees, minutes, and seconds west of the Greenwich meridian.

As territories were opened and surveyed by the government, principal meridians were established for each area opened. Some of the principal meridians were referred to by number, such as the First, Second, and Third principal meridian. The Third principal meridian, for example, is the line located 89 degrees, 10 minutes, and 15 seconds west of Greenwich. It extends from the mouth of the Ohio River to the northern boundary of Illinois.

Other principal meridians were named for the states, such as the Michigan meridian, which covers the survey of that state. The 35 principal meridians and the 32 base lines established in the United States are shown in Figure 1.4-4.

BASE LINES After the principal meridian for a particular area was fixed, a point on the meridian was selected from which a line was run at right angles, due east and west. This line was designated as the *base line* for the area, and its location (its latitude) was established in degrees, minutes, and seconds north of the equator (see Fig. 1.4-4).

In certain instances the base line for a new territory was established by extending the base line of an adjoining territory, previously surveyed. For example, the base line for both the Second and Third principal meridians is a parallel of latitude 38 degrees, 28 minutes, and 20 seconds north of the equatorial line (see Fig. 1.4-4).

CORRECTION LINES All meridians meet at the North and South Poles. *They are not parallel lines.* This fact is not observable from a point on the earth's surface without

Rectangular System is used in states shown unshaded, Hawaii and Alaska. Each principal meridian and its base line is shown on the map with location by longitude and latitude in table below; shaded areas use metes and bounds surveys.

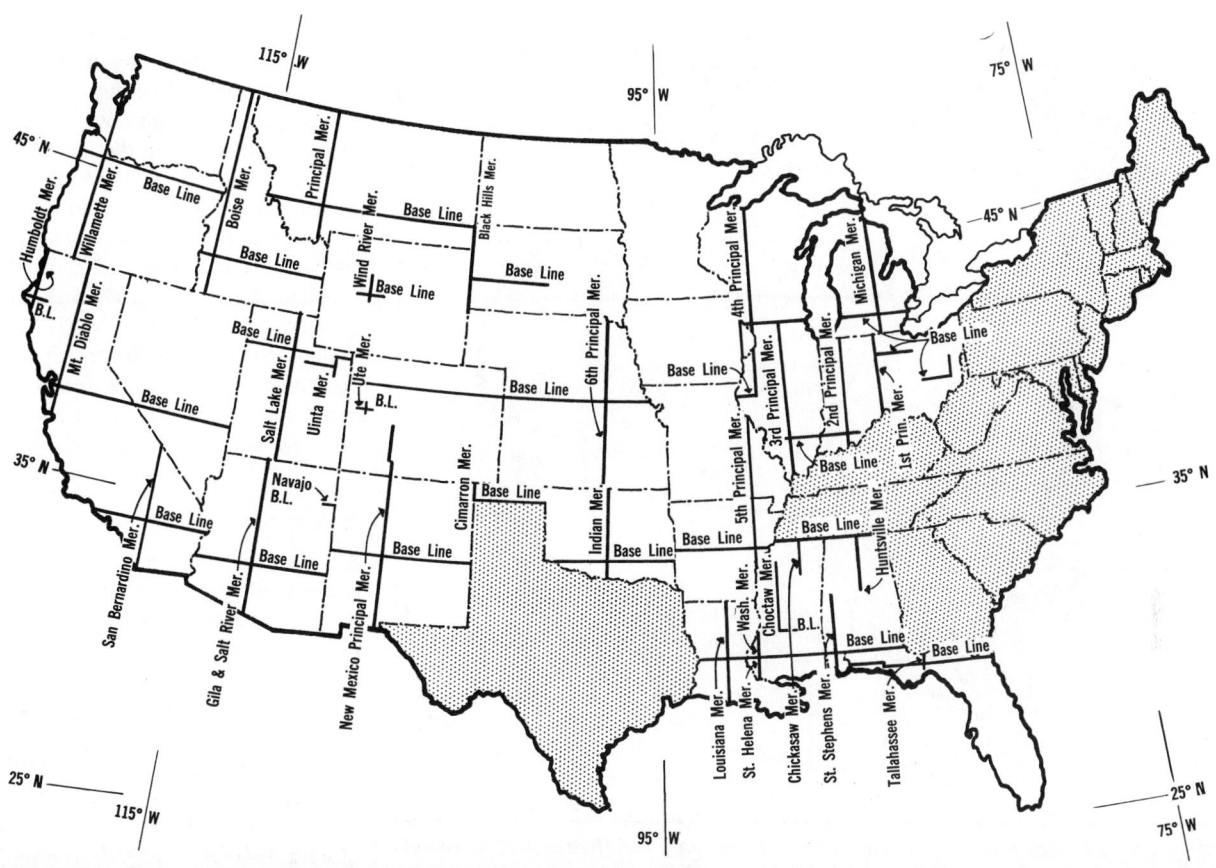

Meridians.	Governing surveys (wholly or in part) in States of—	Longitude of principal meridians west from Greenwich. ° ' "	Latitude of base lines north from Equator. ° ' "	Meridians.	Governing surveys (wholly or in part) in States of—	Longitude of principal meridians west from Greenwich. ° ' "	Latitude of base lines north from Equator. ° ' "
Black Hills	South Dakota	104 03 00	44 00 00	Navajo	Ariz. & N. Mex.	108 32 45	35 45 00
Boise	Idaho	116 24 15	43 22 31	New Mex. Principal	New Mexico	106 52 41	34 15 25
Chickasaw	Mississippi	89 15 00	34 59 00	New Mex. Principal	Colorado	106 53 36	
Choctaw	Mississippi	90 14 45	31 54 40	Principal	Montana	111 38 50	45 46 48
Cimarron	Oklahoma	103 00 00	36 30 00	Salt Lake	Utah	111 54 00	40 46 04
Copper River	Alaska	145 18 42	61 49 11	San Bernardino	California	116 56 15	34 07 10
Fairbanks	Alaska	147 38 33	64 51 49	Second Principal	Ill. & Ind.	86 28 00	38 28 20
Fifth Principal	Ark., Iowa, Minn., Mo., N. Dak., & S. Dak.	91 03 42	34 44 00	Seward	Alaska	149 21 53	60 07 26
				Sixth Principal	Colo., Kans., Nebr., S. Dak. & Wyo.	97 23 00	40 00 00
First Principal	Ohio & Indiana	84 48 50	41 00 00				
Fourth Principal	Illinois[1]	90 28 45	40 00 30	St. Helena	Louisiana	91 09 15	31 00 00
Fourth Principal	Minn. & Wisc.	90 28 45	42 30 00	St. Stephens	Ala. & Miss.	88 02 00	31 00 00
Gila and Salt River	Arizona	112 18 24	33 22 33	Tallahassee	Florida	84 16 42	30 28 00
Humboldt	California	124 07 11	40 25 04	Third Principal	Illinois	89 10 15	38 28 20
Huntsville	Ala. & Miss.	86 34 45	35 00 00	Uinta	Utah	109 57 30	40 26 20
Indian	Oklahoma	97 14 30	34 30 00	Ute	Colorado	108 33 20	39 06 40
Louisiana	Louisiana[2]	92 24 15	31 00 00	Washington	Mississippi	91 09 15	31 00 00
Michigan	Mich. & Ohio	84 22 24	42 26 30	Willamette	Ore. & Wash.	122 44 20	45 31 00
Mount Diablo	Calif. & Nev.	121 54 48	37 51 30	Wind River	Wyo.	108 48 40	43 01 20

1. Numbers are carried to fractional township 29 north in Ill., and are repeated in Wisc., beginning with south boundary of the State; range numbers are in regular order. 2. Latitude doubtful; is to be verified.

Note: East boundary of Ohio, known as "Ellicott's Line," longitude 80°32'20", was the first reference meridian, with township numbers counting from Ohio River, and range numbers in regular order. Township and range numbers within U. S. military land in Ohio are counted from south and east boundaries of tract.

FIGURE 1.4-4 *Meridians and base lines of the U.S. rectangular surveys.* *(Reprinted with permission of Chicago Title Insurance Company.)*

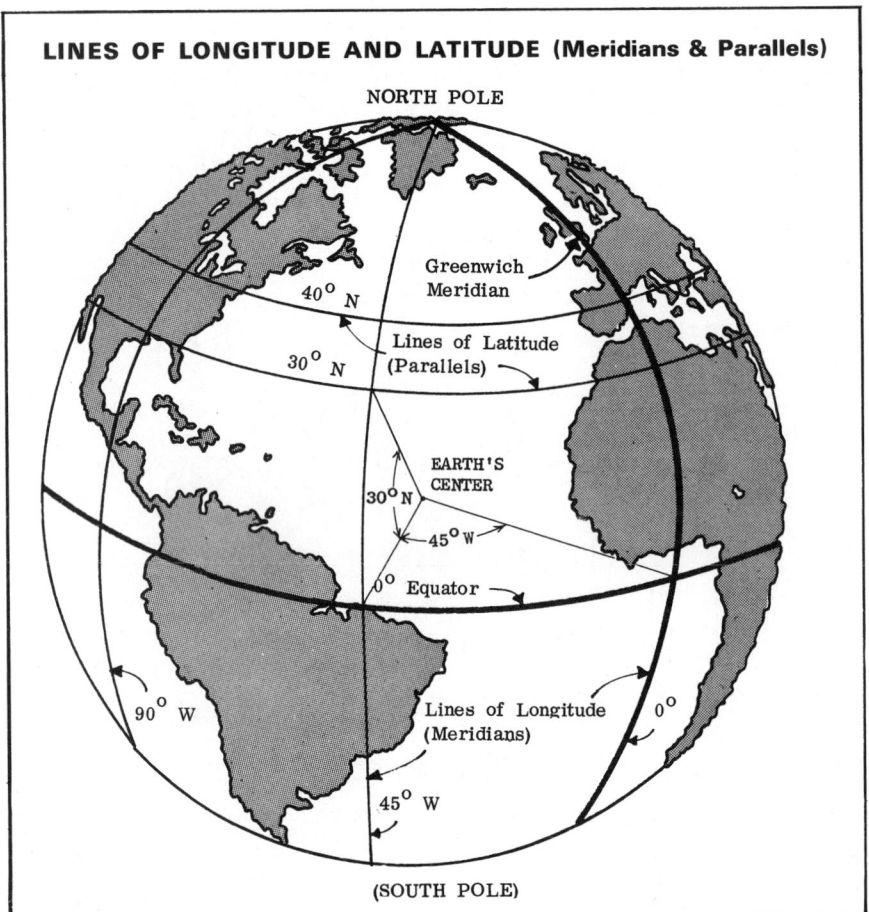

LINES OF LONGITUDE AND LATITUDE (Meridians & Parallels)

NORTH POLE

Greenwich Meridian

40° N

Lines of Latitude (Parallels)

30° N

EARTH'S CENTER

30° N

45° W

0° Equator

90° W

Lines of Longitude (Meridians)

0°

45° W

(SOUTH POLE)

FIGURE 1.4-5 *The rectangular system of survey is based on a modified grid of meridians and parallels measured from the Greenwich meridian and the equator.* (*Reprinted with permission of Chicago Title Insurance Company.*)

the aid of surveying instruments, but an accurate measure of a midwestern area six miles square would show its north line to be about 50 ft. shorter than its south line. To compensate for the convergence of the meridians, it was necessary to establish additional reference lines to measure equal east-west distances. East and west lines, parallel to the base line, were located at intervals of 24 miles north and south of the base line. These lines were designated as *correction lines*.

GUIDE MERIDIANS After the east-west correction lines were set, *guide meridians* running due north and south at 24-mile intervals on each side of the principal meridians were established. Guide meridians extend from the base line to the first correction line and then from correction line to correction line. The guide meridians and the correction lines form squares approximately 24 miles on each side, as shown in Figure 1.4-6.

■ **TOWNSHIP AND RANGE LINES**

The 24-mile squares were then divided into 16 smaller tracts by *township lines* and *range lines*. Township lines were run east and west (parallel to the base line) at 6-mile intervals. Range lines were run north and south at 6-mile intervals. This cross-hatching resulted in a grid of squares, called *townships*, approximately 6 miles on each side (Fig. 1.4-6).

Two *reference numbers* were assigned to each square: a *township number* and a *range number*. Rows of squares (*tiers*) were numbered consecutively to the north or south of the base line. Thus, each square in the first row north of the base line is called Township 1 North (T.1N.); each square in the second row, Township 2 North (T.2N.), and so on. Similarly, each square in the first row south of the base line is called Township 1 South; each square in the second row, Township 2 South, and so on. In the same manner, rows of squares (*ranges*) were numbered consecutively to the east and west of the principal meridian. Thus, each square in the first row east of the principal meridian is called Range 1 East (R.1E.), those in the second row, Range 2 East (R.2E.), and so on. On the west of the principal meridian, those squares in the first row were numbered Range 1 West, those in the second row, Range 2 West, and so on.

■ **SECTIONS**

Townships were the smallest divisions of land provided for in the act of 1785, which created the rectangular system of survey. In early surveys, the outside boundaries of the townships were surveyed, and monuments were placed at every mile on the township lines.

It soon became apparent that a township 6 miles square was too large an area in which to describe and locate a given tract of land. In 1796, Congress passed an act directing that townships already surveyed be subdivided into 36 sections, each to be approximately 1 mile square containing "as nearly as may be" 640 acres. The corner of each section was to be monumented, and the sections numbered consecutively from 1 to 36, beginning with number 1 in the northeast corner of the township, ending in the southeast corner with number 36, as shown in Figure 1.4-6. This manner of numbering sections in a township has continued to this date. A given tract of land can be located within a particular square mile by giving the section number, the township number north or south of the base line, and the range number east or west of the controlling principal meridian.

FRACTIONAL SECTIONS Due to the convergence of the meridians and to other causes, each township does not form a perfect square. A township cannot be surveyed to form 36 identical sections in shape and area. Therefore, rules were

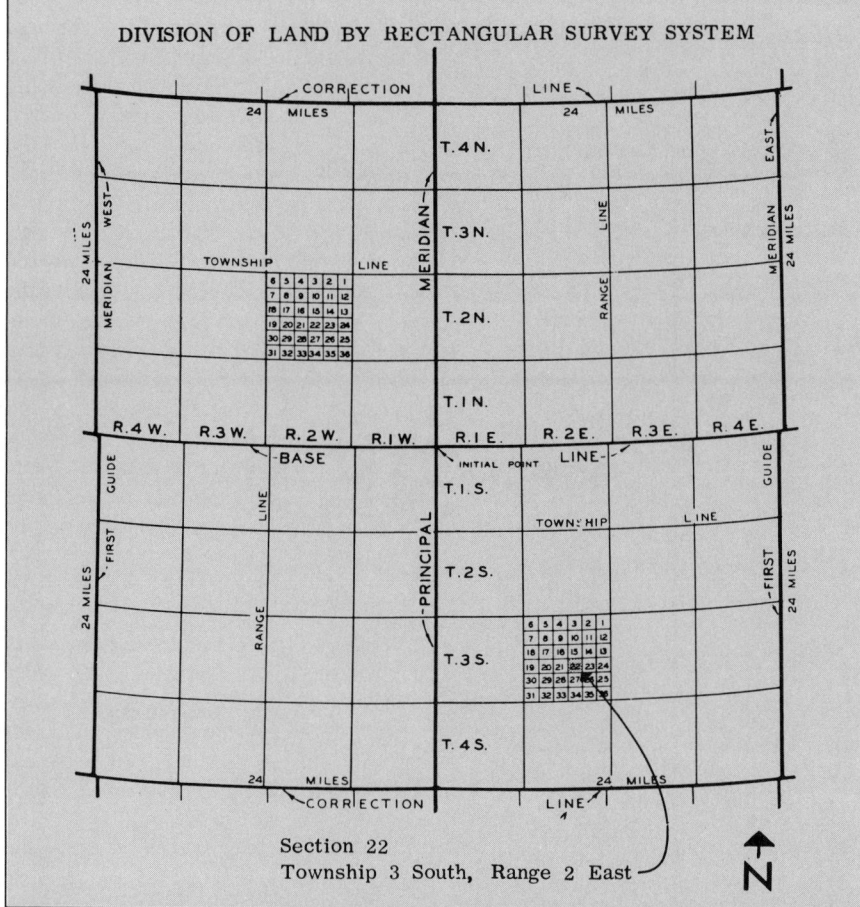

DIVISION OF LAND BY RECTANGULAR SURVEY SYSTEM

Section 22
Township 3 South, Range 2 East

FIGURE 1.4-6 *Division of land by the rectangular survey system.* *(Reprinted with permission of Chicago Title Insurance Company.)*

enacted directing that any excess or deficiency in south-to-north measurements should be added to or deducted from the north portion of the sections in the north row in the township (namely, sections 1 through 6). Any excess or shortage in the east to west measurements should be added to or deducted from the west portion of the west row in the township (namely, sections 6, 7, 18, 19, 30, and 31). Accordingly, such adjusted sections bordering on the north and west sides of a township are classified as *fractional sections*.

Fractional sections may also result from geographical factors, such as land area bordering on lakes. They may also occur where the surveys of separately surveyed areas do not tie together perfectly, and sections result that are less than one mile square.

Government lots were created from fractional sections. Regulations directed that fractional sections be divided into equal fractional parts so far as possible. Any remaining portion was to be divided into parts called government lots and numbered consecutively. The intent of the regulations was to divide fractional sections into as many regular parts as possible according to their relative position in the section, with any remaining portions divided into lots of approximately the same size.

■ **DIVISION OF SECTIONS**

Congress later provided for the division of sections into still smaller units (Fig. 1.4-7). In 1800, regulations directed the subdivision of sections into east and west halves of approximately 320 acres by running lines north and south through the center of the sections. In 1805, regulations provided for *quarter sections* by running east and west lines through the center of

sections with the corners of all quarter sections to be monumented.

This act further provided that all corners marked in the public surveys would be accepted as the proper corners of the sections or quarter sections. That is, monumented corners set by the government surveyors would stand as true corners whether or not the monument was actually placed as described in the field notes (see Section 1.4.2.4).

Congress later provided for the further division of quarter sections in *half-quarters* and finally *quarter-quarters*. These subdivisions were made by running east-west lines and north-south lines through the quarter sections. A quarter-quarter section of 40 acres is the smallest statutory division of regular sections in the rectangular survey system (Fig. 1.4-7).

1.4.2

LEGAL DESCRIPTIONS

The objective of a *legal description* is to identify, locate, and define a specific piece of land and distinguish it from all other tracts of land. Legal descriptions in the United States may be based on (1) the metes and bounds survey system, (2) the rectangular (government) survey system, and (3) reference to recorded maps and plats. It is not uncommon to find a combination of these methods used in one description. For example, a lot in a subdivision may be described by reference to a recorded plat of a land, which in turn is described by a metes and bounds survey within the framework of the rectangular system.

1.4.2.1

Metes and Bounds Description

A metes and bounds description consists of a series of statements describing each portion of the boundary around the parcel of land being described. Starting at a *place of beginning* (P.O.B.), the direction and distances of the property lines are *called* (stated in the description) until the perimeter has been traced around the entire property, returning to the starting point. Each leg of the perimeter is described first by direction, then by length (see Fig. 1.4-3).

FIGURE 1.4-7 *A section in the rectangular system of survey, showing the various subdivisions, acreages, and distances.* (*Reprinted with permission of Chicago Title Insurance Company.*)

■ DIRECTION

When directions (*courses*) agree with the major compass directions, they are so named. Courses other than due north, south, east, or west (*cardinal points*) are called *angular courses*. The direction of an angular course is stated in degrees, minutes, and seconds as an angular deviation east or west of due north or south (Fig. 1.4-8). (An exception to this convention is explained below under "Prolongation.")

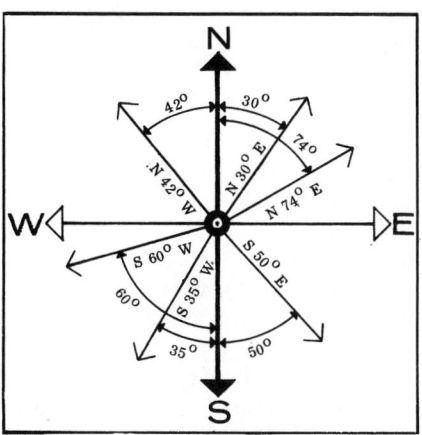

FIGURE 1.4-8 *Boundary line directions are designated by their eastward and westward deviation from north and south.* (*Reprinted with permission of Chicago Title Insurance Company.*)

The direction is always written with the general statement north or south, followed by the angle of eastward or westward deviation from north or south. For example, "north 30 degrees east" defines an angular course direction 30 degrees eastward of due north. A call "south 42 degrees 15 minutes west" defines a direction 42 degrees and 15 minutes westward of south.

PROLONGATION Sometimes course direction is described by stating the amount of deviation from the preceding course line. An angle is measured from an extension, a *prolongation*, of the preceding line and might be stated as "thence northerly 30 degrees to the left of an extension of the last described line." The third and fourth legs of the description shown in Figure 1.4-3 are examples of prolongation.

■ CURVED LINES

A curved line is sometimes referred to in a description, for example, "thence south 80 degrees east 100 feet to a point of tangency; thence along a curve convex northeasterly and having a radius of 30 feet; a distance of 40 feet" (Fig. 1.4-9). The point of tangency, or the point of contact between the last described straight line and the curve, is the starting place of the curve. *Convex* refers to the outside of a curve; *concave*, if used, refers to the inside of a curve. "Convex northeasterly" means that the outside of the curve lies to the northeast.

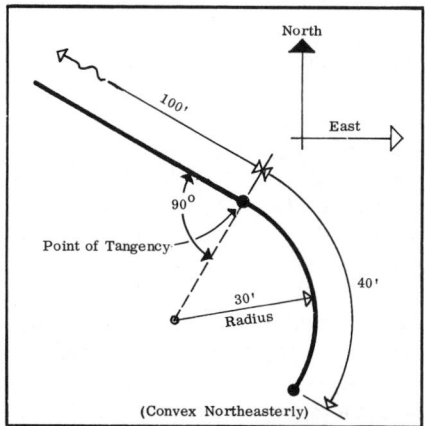

FIGURE 1.4-9 *Curved lines are described by giving the radius and the length of the arc and the direction of the curve.* (*Reprinted with permission of Chicago Title Insurance Company.*)

■ DISTANCE

Distance is usually stated in feet and decimal parts of feet, sometimes in feet and inches or fractions of feet, or (rarely) in chains, rods, and links (see Fig. 1.4-1). Measure is made from point to point along straight lines and along the curves of curved lines.

1.4.2.2

Rectangular System Descriptions

As a result of land being surveyed by the rectangular system, large unmarked areas of land are reduced to series of small squares; basic lines of survey are established by astronomical measurements; permanent monuments have been located

on the ground by the surveyors to establish quarter-quarter section corners; and careful survey notes of the descriptions and locations of all monuments have been made and filed with the government.

A tract of land in the government survey can be described legally by reference to its appropriate principal meridian and base line, township, and range numbers, section number, and, if necessary, its location within the section.

Accordingly, such a description for a parcel of land in Cook County, Illinois, might read, "the southwest quarter of the northeast quarter of Section 6, Township 39 North, Range 13 East of the Third Principal Meridian" (Fig. 1.4-10).

Government lots (see Section 1.4.1.2) are described by identifying the lot number and the quarter section in which the lot is located, for example, "Government Lot 1 in the northwest quarter of the fractional Section 19, Township 38 North, Range 11 East of the Third Principal Meridian."

The rectangular system often is used to describe the boundaries of a subdivision in which individual lots may then be described by reference to a recorded plat or map. The plat is prepared by a surveyor and recorded in the appropriate county records.

1.4.2.3

Reference to Recorded Plat

Curved street patterns often produce irregular shaped blocks and lots. An entire subdivision may be described within the rectangular system or by a metes and bounds description, but it may be impractical to use these systems to describe individual blocks or lots. Therefore, many states provide for the legal description of property by the reference to plats properly recorded, generally in the office of the appropriate county.

A surveyor surveys the tract and prepares a *plat*. The tract or subdivision may be divided into blocks and then into lots. Lots and blocks may be given consecutive numbers or letters enabling identification of a particular area of land within the subdivision by a lot and block number. For further identification it is common practice to give the subdivision a name. An example of a recorded *subdivision plat* is shown in Figure 1.4-11.

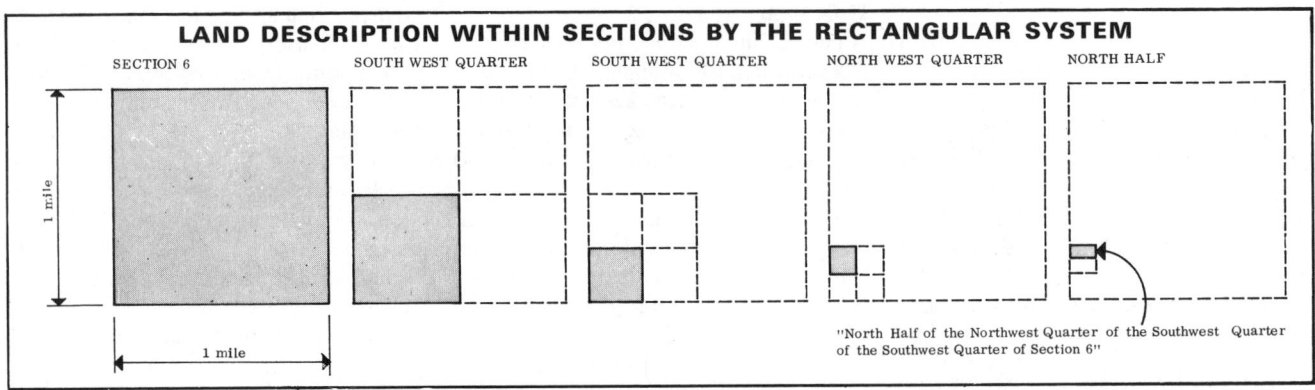

LAND DESCRIPTION WITHIN SECTIONS BY THE RECTANGULAR SYSTEM

SECTION 6 — SOUTH WEST QUARTER — SOUTH WEST QUARTER — NORTH WEST QUARTER — NORTH HALF

"North Half of the Northwest Quarter of the Southwest Quarter of the Southwest Quarter of Section 6"

FIGURE 1.4-10 *To understand a description such as "the north half of the northwest quarter of the southwest quarter of the southwest quarter of Section 6," it is sometimes easier to locate the tract of land within the section by reading the description in reverse, that is, "Section 6, southwest quarter, southwest quarter, northwest quarter, north half."* (Reprinted with permission of Chicago Title Insurance Company.)

The surveyor's plat is captioned with the legal description of the land. It shows all boundary lines, all necessary monuments and dividing lines for blocks, lots, and streets, and the numbering and dimensions of each lot. *Easements* and *restrictions* may also be indicated on a plat. The plat is then certified by the surveyor and signed by the owners. Upon approval by the proper authorities (such as municipal or county zoning boards and planning and building commissions), it is recorded, generally in the appropriate county recorder's office.

When recorded, the lots in a subdivision can be easily identified legally by reference to lot and block numbers, the subdivision name, and the section, township, and range in which the entire subdivision boundaries are described within a rectangular survey or a metes and bounds description.

1.4.2.4

Uncertainties

Making a survey is not an exercise of pure science by a technician in the manner in which a mathematician would solve a problem, with an exact answer to be arrived at inevitably by every surveyor. Mistakes, errors, and judgments may result in conflicts and uncertainties in the legal description of land. Whether a recorded subdivision plat, a rectangular survey, or a metes and bounds description is involved, there may be differing opinions as to the precise location of the boundary lines of a tract of land. The exact location of a point on the land can be a matter of opinion and subject to interpretation.

◾ LOCATION OF MARKERS

When the original government surveys, upon which resurveys should be based, were made, section corners were marked. Through the years, however, monuments may have been repositioned or uprooted during earth-moving projects, such as road construction, or in farming. Relocating a missing corner may require measuring from monuments in existing surveys (which also may contain uncertainties) or from other references.

The disappearance of monuments constantly causes uncertainties in surveying. Metes and bounds surveys monumented to a tree or a post that no longer exists may require the surveyor to dig carefully in the probable vicinity of the lost monument in an attempt to find the decayed remains of the tree or post. Locating a point originally marked by the corner of a barn that no longer exists may require asking older inhabitants of an area for information concerning the original location of the barn.

Roads described in an old survey may have been abandoned and offer little or no trace of their former location. Even though title companies may have unrecorded evidence of such roads, surveying judgment is required to fix a line to substitute for a lost road.

Many other uncertainties are caused by conjecture as to the location of lakes and rivers used as monuments that have receded, dried up, or changed course and location over the years.

◾ CONVERGENCE

Sixteen townships, each roughly six miles square, were laid out in original government surveys in areas approximately 24 miles square (see Section 1.4.1.2). This was a practical but not perfect effort to adjust for the convergence of meridians of longitude as they travel north. Sections were marked by lines spaced at one-mile intervals. These lines were run parallel to the range line closest to the principal meridian. This causes a divergence in direction that becomes progressively more pronounced in each successive tier of sections. Therefore, an ambiguity may result when a cardinal point is used as a course in a legal description. For instance, the call "thence north 150 feet to a point," unless monumented, usually is interpreted as indicating a course that runs parallel with the east line of the section. If the section line deviates from true north, an uncertainty exists. Other clues in the description or perhaps a reading of the description in reverse order may result in the course being more firmly defined.

◾ INACCURACIES

The measurement of distances called for in some legal descriptions is virtually impossible. Written in a description, the distance "126.01 feet" gives an assuring illusion of precision. However, if two surveying crews measured the same distance, they rarely would observe the same measurement. Minute differences in the tautness of the measuring tape and changes in the length of the tape due to the effect of temperature fluctuations are

FIGURE 1.4-11 *A recorded plat showing an entire subdivision, including lot dimensions, easements, and other information. This example includes the surveyor's certification and plan commission and county approvals. The note at the top center ties the subdivision to the rectangular system; as noted at the upper right, this plat has been filed with the Lake County Recorder, plat book 34, page 147. A lot in the subdivision can be described by number within the Nixon-Blietz Deerfield Development, Unit # 1, as recorded.*
(Reprinted with permission of Chicago Title Insurance Company.)

chain made of 100 metal links. Weather changes resulted in expansion and contraction of these chains, and wear and use caused them to stretch. In some instances, surveyors would count the revolutions of the wheel of the vehicle in which they rode to measure long, straight stretches. Distances also were "stepped off" by the measured stride. Even though these measurements were far from accurate, markers were placed accordingly. Such monuments are considered legal points of reference, regardless of whether they were accurately placed.

In addition to errors in distance, the perimeter lines of a section sometimes deviate from true east-west or north-south courses. In locating the south boundary of a tract described as the "north 10 acres of the north west quarter" of a section in which the north line deviates from a true east and west course, some judgment is required. The south line might be considered to be parallel with the north line of the section, or it might run due east and west. Again, the surveyor must search for physical evidence, such as a stake set by an earlier surveyor or a fence indicating a line of occupation, to decide which line was intended.

Courses may show an appreciable error in direction. For example, a metes and bounds course shown as "south 7 degrees west" in an earlier survey might be observed to be actually "south 6 degrees west," indicating that an error could have been made by an earlier surveyor or that faulty equipment was used.

Mistakes, such as careless observations and inaccurate measurements, are random and unpredictable. And though the best surveying instruments are subject to a certain degree of error, these errors are predictable and can be allowed for in observing measurements. Again, however, this calls for some human judgment.

1.4.2.5

Recorded Plats

When an individual lot is described by reference to a recorded subdivision plat, the plat together with all information and data shown thereon are part of the lot description. It is possible that the lot size may not be exactly as shown on the plat.

unavoidable. Mathematically defined, a point does not have length, breadth, or depth. The ".01" is equivalent to a little less than 1/8 in. It would be amazing if two surveyors would locate this invisible point at exactly the same place. Yet a surveyor must express an opinion as to its location, and the opinions of individual surveyors will differ unless the point is monumented in the first place.

In surveying a tract, it is necessary to locate the boundaries of abutting tracts as shown in earlier surveys and to reconcile with them the perimeter lines of the tract being surveyed. The surveyor must check back through the records to find the legal descriptions of these abutting tracts.

Section corners and other marks in the rectangular system sometimes were placed incorrectly. Early government surveyors sometimes used crude tools and methods. Their measuring instrument was a 66-ft.

The incorporation by reference is of the entire plat and all data appearing thereon. This includes not only dimensions and expressions of quantity and area, but also the notations of the courses and distances of the lines of the subdivision and of the identity and locations of the monuments marking the boundaries.

When this information is incorporated into a legal description, it is, in effect, a metes and bounds description and is subject to rules of interpretation, should any conflict or discrepancy be contained in the data of the plat. For example, monuments prevail over courses and distances, courses prevail over distances, and expressions of quantity or area are generally the least reliable.

It is possible that the overall recorded measurements of a block may not equal the recorded widths of the lots in the block. If such an error is found, the widths of some of or all the lots must be adjusted accord-ingly. If the entire row of lots is vacant and unmonumented, the overage or deficiency could be distributed among all the lots. Lots marked by a subdivider's stakes and sold with reference to them probably would escape adjustment. If improvements have been erected close to the boundaries of a lot as platted, without a surveyor checking the location, disputes may arise.

Errors and mistakes, fairly common when land values were low, have led to boundary disputes when land values greatly increased.

Plats are subject to errors and mistakes, which may result in uncertainties and conflicts. A plat is drawn by a surveyor from field notes, and the field notes are secondary evidence of the surveyor's acts. The primary evidence consists of monuments placed in the ground to mark lines and corners.

The systems of survey on which legal descriptions are based are as precise as is possible. There is always the possibility of uncertainties in separate surveys, and mistakes may occur in the writing and rewriting of legal descriptions. Descriptions may identify land parcels adequately, but surveys are necessary to actually locate boundary lines.

The laws governing questions of legal description and survey are complex and include federal, state, and local regulations. Court decisions and opinions pertaining to land disputes may be based on whichever regulation is deemed appropriate, or on principles dating back to the real estate laws of England. This section is to acquaint the reader with the basic systems of survey and forms of legal descriptions. The complexities of resolving uncertainties in either land survey or legal land description are a matter of real estate law and interpretation and are outside the scope of this book.

1.5 PROPERTIES OF MATERIALS

In this century there has been a huge increase in the number of construction products and even in the number of totally new materials used for construction purposes. The entire array of plastic materials we use today is one example. More than 300 new products continue to appear annually. Some of these are made by merely varying the ingredients or the formulation of existing materials, such as sealants, or by adding new shapes and finishes to a line of masonry units, cladding panels, and other products. A few are entirely new products, often made by combining existing materials. Coated-glass and glass products made by mixing other substances into the molten glass are examples of new products with characteristics that did not exist in older products. In addition, many new uses for existing materials are developed annually. The introduction of glass-fiber insulation in poured-in-place concrete is an example. To analyze and use these innovations effectively, designers and builders need a basic understanding of the general properties of materials.

A material's suitability for a particular use depends on its ability to perform satisfactorily under the conditions to be encountered. When a material's properties are known, its performance in a particular situation can be predicted. Four general groups of properties are of interest to those in the construction industry: mechanical, thermal, electrical, and chemical properties. Each property depends on the composition of the material in question.

The millions of materials in our universe are all made from just three basic building blocks: *atoms*, *ions*, and *molecules*. These building blocks are bonded together by one of several basic chemical bonds. Materials can be classified according to these basic structural elements and bonds into three major categories: (1) *ceramics and glasses*, which are hard, brittle, and poor conductors of heat and electricity; (2) *metals*, which are more ductile than ceramics and good conductors of heat and electricity; and (3) *molecular materials*, which have low melting temperatures and fair strength and which are poor conductors of heat and electricity.

Since all construction materials fall into one of these three categories, a greater understanding of such materials can be obtained by recognizing in which category a particular material belongs and remembering the properties of that category. For instance, glass wool and foamed polystyrene might have identical insulating capabilities, but the glass wool falls into the ceramic category and the foam into the molecular category. From that knowledge a designer can surmise, without seeing manufacturers' specifications for the materials, that the foam will have a lower melting temperature than the glass wool, because the melting temperatures of molecular materials are lower than that of ceramic materials.

This section is intended to expand and sharpen the reader's instinct for categorizing materials. First, the basic structure of matter is discussed, then the four sets of properties that determine a material's performance are considered in some detail. For more information about the properties of a particular material, refer to the chapter in which that specific material is discussed.

STRUCTURE OF MATTER

Matter is formed by the chemical bonding of atoms, ions, and molecules. The particular combination of components and bonding methods determines final material properties.

1.5.1.1

Building Blocks of Matter

There are 102 chemical substances that cannot be subdivided into other substances. These basic materials, such as carbon, iron, hydrogen, and oxygen, are called *chemical elements*. Just two chemical elements, oxygen and silicon, make up more than three-quarters of all the matter on earth; just eight chemical elements constitute more than 97% of all earthly substances (Fig. 1.5-1).

A given quantity of any element can be divided into smaller and smaller quantities until a minute particle is reached. This entity, the smallest particle that retains the properties of the original material, is called an *atom*.

Ordinarily, atoms are neutral, that is they have no electrical charge. When atoms acquire a positive or negative charge by losing or attracting negatively charged particles, they are called *ions*. Atoms may combine into simple units called *molecules*.

■ ATOMS

The simplest representation of an atom looks like a miniature solar system (Fig. 1.5-2). The atom has at its core a *nucleus*, which consists of positively charged *protons* and electrically neutral *neutrons*. Orbiting about the nucleus are negatively charged *electrons*.

Because protons and neutrons are much denser than electrons, almost all the weight of an atom resides in the nucleus. The sum of the weights of the protons and the neutrons in one of its individual atoms is called the *atomic weight* of an element. In a normal electrically neutral atom, the positive charge of the protons in the nucleus is balanced by an equal negative charge from the electrons. The number of electrons in each of its atoms, which is equal to the number of protons in a single atom, is the *atomic number* of an element.

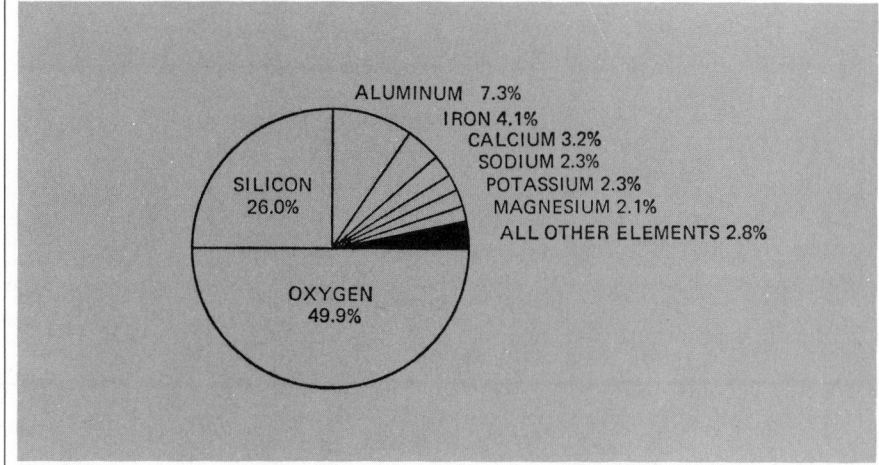

FIGURE 1.5-1 *A handful of elements make up most earthly matter.*

FIGURE 1.5-2 *Electrons move in orbits around a nucleus. Helium's two electrons move in the same orbit, but fluorine's nine electrons move in different orbits. Groups of electrons with identical or neighboring orbits make up a shell; hydrogen and helium have one shell, fluorine two.*

Groups of electrons moving in neighboring orbits are said to belong to the same *shell* (see Fig. 1.5-2). The shell closest to the nucleus can accommodate only two electrons; the second, a maximum of eight; the third, 18; and the fourth, 32. The distribution of electrons around the nucleus gives elements some of their inherent chemical properties. Figure 1.5-3 shows the chemical symbols and common names of all 102 elements, as well as their atomic numbers (number of electrons) and shell distribution.

The elements also can be classified according to ascending atomic numbers and number of electron shells into an arrangement called the *periodic table* (Fig. 1.5-4). The *period number* represents the number of electron shells in each element; thus, all elements with one electron shell (H and He) are listed horizontally opposite Period 1; those with two shells, opposite Period 2, and so on up to 7. For each element, the chemical symbol is shown, with its atomic number above the symbol and its atomic weight below.

Each vertical column of the table is labeled with a roman numeral, and elements within a column possess similar properties. For example, copper (Cu), silver (Ag), and gold (Au) in column IB are dense, soft, and fairly chemically inert metals. The elements on the right of the table exhibit nonmetallic behavior, while those on the left exhibit metallic behavior to various degrees.

■ IONS

Under certain conditions electrons may leave an atom, resulting in a net positive charge. The atom then becomes a *positive ion*. This tendency is especially prevalent

FIGURE 1.5-3
The Elements: Their Atomic Numbers, Chemical Symbols, Common Names, and Shell Distributions

Atomic No.	Element Symbol	Element Name	1	2	3	4	5
1	H	Hydrogen	1				
2	He	Helium	2				
3	Li	Lithium	2	1			
4	Be	Beryllium	2	2			
5	B	Boron	2	3			
6	C	Carbon	2	4			
7	N	Nitrogen	2	5			
8	O	Oxygen	2	6			
9	F	Fluorine	2	7			
10	Ne	Neon	2	8			
11	Na	Sodium	2	8	1		
12	Mg	Magnesium	2	8	2		
13	Al	Aluminum	2	8	3		
14	Si	Silicon	2	8	4		
15	P	Phosphorus	2	8	5		
16	S	Sulfur	2	8	6		
17	Cl	Chlorine	2	8	7		
18	A	Argon	2	8	8		
19	K	Potassium	2	8	8	1	
20	Ca	Calcium	2	8	8	2	
21	Sc	Scandium	2	8	9	2	
22	Ti	Titanium	2	8	10	2	
23	V	Vanadium	2	8	11	2	
24	Cr	Chromium	2	8	13	1	
25	Mn	Manganese	2	8	13	2	
26	Fe	Iron	2	8	14	2	
27	Co	Cobalt	2	8	15	2	
28	Ni	Nickel	2	8	16	2	
29	Cu	Copper	2	8	18	1	
30	Zn	Zinc	2	8	18	2	
31	Ga	Gallium	2	8	18	3	
32	Ge	Germanium	2	8	18	4	
33	As	Arsenic	2	8	18	5	
34	Se	Selenium	2	8	18	6	
35	Br	Bromine	2	8	18	7	
36	Kr	Krypton	2	8	18	8	
37	Rb	Rubidium	2	8	18	8	1
38	Sr	Strontium	2	8	18	8	2
39	Y	Yttrium	2	8	18	9	2
40	Zr	Zirconium	2	8	18	10	2
41	nb	Niobium	2	8	18	12	1
42	Mo	Molybdenum	2	8	18	13	1
43	Tc	Technetium	2	8	18	14	1
44	Ru	Ruthenium	2	8	18	15	1
45	Rh	Rhodium	2	8	18	16	1
46	Pd	Palladium	2	8	18	16	1
47	Ag	Silver	2	8	18	18	1
48	Cd	Cadmium	2	8	18	18	2
49	In	Indium	2	8	18	18	3
50	Sn	Tin	2	8	18	18	4
51	Sb	Antimony	2	8	18	18	5

Atomic No.	Element Symbol	Element Name	1	2	3	4	5	6	7
52	Te	Tellurium	2	8	18	18	6		
53	I	Iodine	2	8	18	18	7		
54	Xe	Xenon	2	8	18	18	8		
55	Cs	Cesium	2	8	18	18	8	1	
56	Ba	Barium	2	8	18	18	8	2	
57	La	Lanthanum	2	8	18	18	9	2	
58	Ce	Cerium	2	8	18	20	8	2	
59	Pr	Pra'mium	2	8	18	21	8	2	
60	Nd	Neodymium	2	8	18	21	8	2	
61	Pm	Promethium	2	8	18	23	8	2	
62	Sm	Samarium	2	8	18	24	8	2	
63	Eu	Europium	2	8	18	25	8	2	
64	Gd	Gadolinium	2	8	18	25	9	2	
65	Tb	Terbium	2	8	18	27	8	2	
66	Dy	Dysprosium	2	8	18	28	8	2	
67	Ho	Holmium	2	8	18	29	8	2	
68	Er	Erbium	2	8	18	30	8	2	
69	Tm	Thulium	2	8	18	31	8	2	
70	Yb	Ytterbium	2	8	18	32	8	2	
71	Lu	Lutetium	2	8	18	32	9	2	
72	Hf	Hafnium	2	8	18	32	10	2	
73	Ta	Tantalum	2	8	18	32	11	2	
74	W	Tungsten	2	8	18	32	12	2	
75	Re	Rhenium	2	8	18	32	13	2	
76	Os	Osmium	2	8	18	32	14	2	
77	Ir	Iridium	2	8	18	32	17	0	
78	Pt	Platinum	2	8	18	32	17	1	
79	Au	Gold	2	8	18	32	18	1	
80	Hg	Mercury	2	8	18	32	18	2	
81	Tl	Thallium	2	8	18	32	18	3	
82	Pb	Lead	2	8	18	32	18	4	
83	Bi	Bismuth	2	8	18	32	18	5	
84	Po	Polonium	2	8	18	32	18	6	
85	At	Astatine	2	8	18	32	18	7	
86	Rn	Radon	2	8	18	32	18	8	
87	Fr	Francium	2	8	18	32	18	8	1
88	Ra	Radium	2	8	18	32	18	8	2
89	Ac	Actinium	2	8	18	32	18	9	2
90	Th	Thorium	2	8	18	32	18	10	2
91	Pa	Pr'tinium	2	8	18	32	20	9	2
92	U	Uranium	2	8	18	32	21	9	2
93	Np	Neptunium	2	8	18	32	22	9	2
94	Pu	Plutonium	2	8	18	32	23	9	2
95	Am	Americium	2	8	18	32	24	9	2
96	Cm	Curium	2	8	18	32	25	9	2
97	Bk	Berkelium	2	8	18	32	26	9	2
98	Cf	Californium	2	8	18	32	27	9	2
99	Es	Einsteinium	2	8	18	32	28	9	2
100	Fm	Fermium	2	8	18	32	29	9	2
101	Md	Mendelvium	2	8	18	32	30	9	2
102	No	Nobelium	2	8	18	32	31	9	2

FIGURE 1.5-4
Periodic Table of the Elements

Period	Group IA	Group IIA											Group IIIA	Group IVA	Group VA	Group VIA	Group VIIA	Group O
							Transition Elements											
1	1 **H** 1.00797					Metals											1 **H** 1.0079	2 **He** 4.0026
												Non-Metals						
2	3 **Li** 6.939	4 **Be** 9.012											5 **B** 10.811	6 **C** 12.011	7 **N** 14.007	8 **O** 15.9994	9 **F** 18.998	10 **Ne** 20.183
3	11 **Na** 22.990	12 **Mg** 24.312	Group IIIB	Group IVB	Group VB	Group VIB	Group VIIB		Group VIII		Group IB	Group IIB	13 **Al** 26.98	14 **Si** 28.086	15 **P** 30.97	16 **S** 32.064	17 **Cl** 35.453	18 **Ar** 39.95
4	19 **K** 39.102	20 **Ca** 40.08	21 **Sc** 44.96	22 **Ti** 47.90	23 **V** 50.94	24 **Cr** 52.00	25 **Mn** 54.94	26 **Fe** 55.85	27 **Co** 58.93	28 **Ni** 58.71	29 **Cu** 63.54	30 **Zn** 65.37	31 **Ga** 69.72	32 **Ge** 72.59	33 **As** 74.92	34 **Se** 78.96	35 **Br** 79.91	36 **Kr** 83.80
5	37 **Rb** 85.47	38 **Sr** 87.62	39 **Y** 88.91	40 **Zr** 91.22	41 **Nb** 92.91	42 **Mo** 95.94	43 **Tc** 99	44 **Ru** 101.07	45 **Rh** 102.91	46 **Pd** 106.4	47 **Ag** 107.87	48 **Cd** 112.40	49 **In** 114.82	50 **Sn** 118.69	51 **Sb** 121.75	52 **Te** 127.60	53 **I** 126.90	54 **Xe** 131.30
6	55 **Cs** 132.90	56 **Ba** 137.34	57–71 **La** series*	72 **Hf** 178.49	73 **Ta** 180.95	74 **W** 183.85	75 **Re** 186.2	76 **Os** 190.2	77 **Ir** 192.2	78 **Pt** 195.1	79 **Au** 196.97	80 **Hg** 200.59	81 **Tl** 204.37	82 **Pb** 207.19	83 **Bi** 208.98	84 **Po** 210	85 **At** 210	86 **Rn** 222
7	87 **Fr** 223	88 **Ra** 226	89– **Ac** series†															

*Lanthanide Series
†Actinide Series

58 **Ce** 140.12	59 **Pr** 140.91	60 **Nd** 144.24	61 **Pm** 147	62 **Sm** 150.35	63 **Eu** 151.96	64 **Gd** 157.25	65 **Tb** 158.92	66 **Dy** 162.50	67 **Ho** 164.93	68 **Er** 167.26	69 **Tm** 168.93	70 **Yb** 173.04	71 **Lu** 174.97
90 **Th** 232.04	91 **Pa** 231	92 **U** 238.03	93 **Np** 237	94 **Pu** 239	95 **Am** 241	96 **Cm** 242	97 **Bk** 249	98 **Cf** 252	99 **Es** 254	100 **Fm** 253	101 **Md**	102 **No**	103 **Lw**

in elements in groups IA and IIA, which have just a few electrons in their outermost shells, since the attraction of the positive nucleus for these remote electrons is limited. Metallic elements exhibit this tendency for positive ionization most markedly.

Certain other elements show a strong tendency to acquire excess electrons in their outer orbits and to become *negative ions*. These elements, which usually have five or more electrons in their outermost shells to begin with, are almost invariably nonmetallic and appear at the right of the periodic table (groups VIA and VIIA).

■ MOLECULES

A material may consist of billions of one or more kinds of atoms (*atomic substances*) or of billions of several kinds of ions (*ionic substances*). Still other substances are formed by the joining together of units, each unit consisting of a number of atoms. These units are called *molecules* and can only be broken into their constituent atoms with difficulty. A molecule of methane, for example, is composed of one carbon atom and four hydrogen atoms (Fig. 1.5-5).

Substances composed of molecules are called *molecular materials*. Water, for instance, is a molecular material. Each

water molecule contains two hydrogen atoms and one oxygen atom, and has a triangular shape. When large numbers of molecules are bonded together, the water exists as ice. As the tightness of bonding of the triangular molecule is decreased, water assumes the liquid form. When bonding is very loose, water exists as water vapor. In all three forms of the material the triangular units remain intact, and the individual hydrogen and oxygen atoms are not free to leave their respective molecules.

Molecules can vary greatly in size, as shown in Figure 1.5-6. A molecule of water contains only three atoms, whereas

FIGURE 1.5-5 *Models of a molecule of methane, CH_4: (a) a two-dimensional representation; (b) a three-dimensional model.*

FIGURE 1.5-6 *Three molecules of varying sizes: (a) water; (b) ethane; and (c) butane.*

a molecule of ethane contains eight and butane 14. Some molecules contain thousands of individual atoms; molecules of this size are called *polymers*, which form the basis of materials such as plastics (see Section 6.9).

1.5.1.2

Bonding of Matter

One of the major reasons that atoms, ions, and molecules bond together to form substances is the strong tendency in nature for outermost electron shells to contain as many pairs of electrons as possible. When a shell contains the maximum number of electron pairs, it is said to be *filled*. A filled shell results in a low energy condition on the atomic scale, meaning that the matter is relatively inert.

Because systems in nature tend to remain in the state of lowest energy, materials with filled outer shells are stable and resist change. Elements that occur naturally with filled outer shells are inert gases, such as helium (He), neon (Ne), and argon (Ar), found in group O at the far right of the periodic table. These elements refuse to combine with other elements, including each other.

Whenever the outer shell of an atom or ion of a noninert gas element is filled, the atom or ion is said to have achieved the *inert gas configuration* (IGC), in which it is content to remain because of its stability in that situation. Nature moves atoms and ions so strongly toward the IGC that these particles can be thought of as employing many tricks, including capturing electrons, surrendering electrons, and sharing electrons with their neighbors in an effort to achieve chemical stability.

In exchanging or sharing electrons, atoms and ions interact with their neighbors, are attracted to them, and bond into matter. There are four types of bonding: *ionic, metallic, covalent,* and *secondary.*

◼ IONIC BONDING

When atoms with almost empty outer electron shells exist in the vicinity of atoms with almost filled shells, the natural tendency of each set of atoms to achieve the IGC can be fulfilled. The atoms with few outer electrons surrender their "free" electrons and become positive ions, whereas atoms with many outer electrons accept these electrons to fill their outer orbits and become negative ions. The situation now exists in which positive ions and negative ions are in close proximity. The oppositely charged particles are attracted and bonded into matter by *ionic bonding.* Metals tend to have almost empty outer orbits, and nonmetals tend to have them almost filled; thus ionic substances usually consist of elements from the left and right hand groups of the periodic table, bonded together.

Probably the most familiar ionically bonded substance is common table salt, NaCl. Atoms of sodium (Na) have just a single outer electron and therefore have a strong tendency to form positive Na^+ ions. Atoms of chlorine (Cl) have seven outer electrons and upon receiving one more electron fill their outer orbits and achieve the IGC by becoming Cl^- ions. The two sets of ions are then bonded into NaCl (Fig. 1.5-7).

Ceramic materials, being combinations of metallic atoms and nonmetals, are bonded primarily through the ionic mechanism. Examples of this category of materials are brick, tile, portland cement, con-

crete, and natural stone. Their properties follow from the quality of the ionic bond: they have high melting temperatures and are chemically inert because both ions have achieved the IGC and are tied by a strong ionic bond. Melting involves separating the bonds, while chemical attack requires separation and recombination of ions. Both processes are difficult when the original substance is stable to begin with.

Ceramic materials tend to be brittle, because the regular and extremely rigid arrangement of positive and negative ions resists shape changes. They tend to shatter, rather than be forced from one shape into another. Brittleness generally is associated with good strength in compression but low strength in tension. These materials also are poor conductors of electricity and heat, because the electrons are tightly bound into the stable ions and are not free to conduct electrical or thermal energy in response to an applied voltage or temperature.

The typical properties of ceramic materials follow logically from the nature of their constituent ions and of the ionic bond. Hundreds of new ceramic materials with enhanced properties may be developed in the future. It is difficult to predict all their properties, but in general each of them will have a high melting temperature, a high compressive strength, chemical inertness, and good thermal and electrical resistances.

◼ METALLIC BONDING

Metallic atoms tend toward the IGC by attempting to surrender their few outer electrons, thus becoming positive ions. When large numbers of atoms are brought together, they each contribute these electrons to a mobile "sea" of electrons that circulates near the original parent atoms.

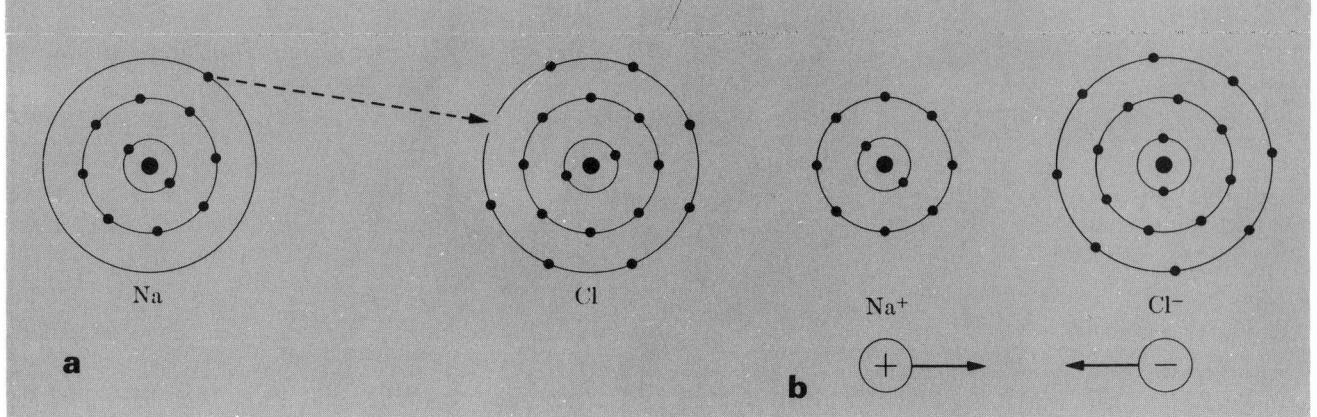

FIGURE 1.5-7 *Ionic bonding in NaCl: (a) Na gives up a single electron, becoming a positive ion, while Cl accepts a single electron, becoming a negative ion; (b) positive and negative ions are then attracted to each other and bond together to become NaCl.*

The metal ions are now in a stable situation since they no longer possess their few outer electrons. They exist in a three-dimensional shape, as a metal crystal, with an electron "sea" dispersed throughout the solid (Fig. 1.5-8).

This peculiar cooperative surrendering of electrons by metal ions results in a *metallic bond*. The characteristic properties of metals follow from the nature of this bond. They are strong and have fairly high melting temperatures because the metallic bond is strong. Unlike ceramics, metals are good conductors of heat and electricity because the electrons in the sea are extremely mobile and free to transport thermal and electrical energy.

Metals may or may not be chemically inert. Their ability to transport electrical current makes them susceptible to chemical degradation in the form of corrosion. The corrosion tendency sometimes outweighs the intrinsic strength of the metallic bond.

■ COVALENT BONDING

Many elements found in groups IVA and VA of the periodic table (see Fig. 1.5-4), such as carbon (C) and nitrogen (N), lack the strong tendency to form either positive ions or negative ions. These elements, with moderate numbers of electrons in their outer shells, reach the IGC by sharing electrons with similar elements. The process of mutual sharing of outer (*valence*) electrons by a cluster of atoms to create a stable entity is known as *covalent bonding*.

Covalent bonding is the most common mechanism whereby small numbers of atoms are bound into molecules. These molecules are then joined together by

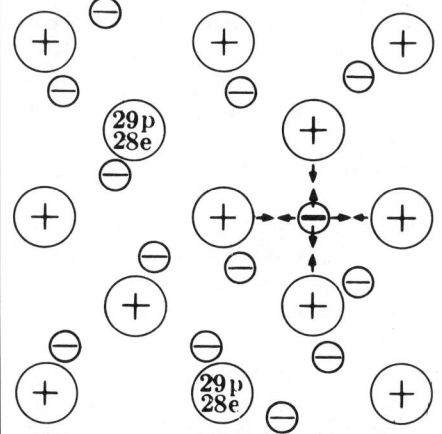

FIGURE 1.5-8 *A metallic bond results from the attraction of the positive cores of the atom (copper) to the negative electrons in the surrounding "sea."*

weak secondary bonds, to produce molecular materials. The covalent bond is very strong, as is evidenced by the fact that it is very difficult to break up a molecule into its constituent atoms.

A few substances are produced entirely by covalent bonding. A diamond is a familiar example. These substances are extremely hard and have very high melting temperatures, due to the strength of the covalent bonds. However, they seldom occur in nature, as suggested by the fact that diamonds are rarely found and then only in small amounts.

■ SECONDARY BONDING

To become molecular materials, atoms are bound into molecules by covalent bonding, and the molecules then are joined by

means of weak *secondary bonds*. These bonds arise because positive nuclei or negative electrons in one molecule feel the weak attraction of their opposites in neighboring molecules. The weak attractions bind adjacent molecules into the substance.

Molecular materials, such as wood, plastics, and bituminous products, derive their properties from the behavior of the secondary bond. They have low strength levels and low melting temperatures compared to metals and ceramics, because the weak secondary bond can be easily overcome by the application of force or heat. They are poor conductors of heat and electricity, because individual electrons remain tightly constrained within covalently bonded molecules.

Surprisingly, the weak secondary bond is not broken by many of the strong chemical compounds that attack metals and ceramics. Thus, molecular materials are chemically inert in a large number of environments. They are attacked by molecular solvents, such as acetone, but their resistance to attack by most salts, acids, and industrial atmospheres makes them complementary to metals and ceramics.

1.5.2

PROPERTIES OF MATTER

Anyone who has been alone in a building on a quiet evening knows that materials are "alive." The creaks and groans are in fact material responses to external stimuli such as heat, wind, and gravity loads. These stimuli fall into four categories, which define the four major properties exhibited by materials: *mechanical*, *thermal*, *electrical*, and *chemical*.

Mechanical Properties

Mechanical properties describe the response of a material to *static* (continuous) or *dynamic* (intermittent) loads.

■ STRESS AND STRAIN

When a force (load) of *x* pounds is applied to an object, the object *deflects* (sags) or *deforms* (changes shape) *y* inches. However, knowing the load in pounds or the deflection in inches does not provide enough information to predict the likely response.

For instance, a load of 100 lb. may induce a negligible reaction when distributed uniformly on the surface of a brick. The same load, concentrated at the point of a chisel, may shatter the brick. A deformation of 6 in. may be negligible when averaged over the height of an entire building, but the same deformation may have catastrophic consequences if localized. It is therefore necessary to consider forces in relation to the area over which they act and to view deformations with respect to the size of the region being deformed.

Stress (σ) is equal to the applied load (L) divided by the area (A) on which the load is acting:

$$\sigma = \frac{L}{A}$$

A 100-lb. load acting downward on the face of a typical 4-by-8-in. brick exerts a stress of only 3 pounds per square inch (psi), which is well below the brick's inherent compressive strength. The same load concentrated at the point of a chisel 0.5 in. wide and sharpened to 0.02 in. thick exerts a stress of 10,000 psi, well beyond the strength of the typical brick. In engineering texts, stress is usually designated by the Greek symbol σ when the applied load is at right angles to the area in question, as in *tension* (stretching) or *compression* (squeezing).

Stress takes into account the fact that an applied load may or may not be shared by a substantial area of a body. Strain (ϵ) correspondingly accounts for the fact that deformation may be distributed over a large area or localized in a small region. If the original dimension of a body along a certain line is designated *d*, and the deformation (change in that dimension) is designated Δd, then,

$$\delta = \frac{\Delta d}{d}$$

Strain is stated generally in percentage of dimension change; thus, sometimes,

$$\epsilon = \frac{\Delta d}{d}(100)$$

ELASTIC DEFORMATION An applied load, no matter how small, always generates both a stress and a strain in a solid object. A feather floating downward and landing on a steel beam causes the beam to deflect minutely. The resulting deflection cannot be seen, and the resulting strain cannot be measured even with today's sophisticated instruments, but the effect can be theoretically calculated and is really there. When the feather is removed, the strain disappears, and the beam returns to its original dimensions.

In this case, deflection is present only so long as the load is present, and its effect can be reversed by simply removing the load. This reversible type of strain is called *elastic deformation*. Many forces acting on structures cause only elastic deformation. For instance, the swaying of a tall building under the action of wind is an example of elastic deformation. In general, elastic deformation is harmless, since the structure is restored to its original configuration when the force is removed.

Most building criteria contain factors of safety that ensure that a structure will not be stressed beyond its elastic range. The most common measure of a material's *stiffness*, or ability to resist elastic deformation, is its modulus of elasticity, *E*.

PLASTIC DEFORMATION Permanent or irreversible deformation is called *plastic deformation*. This type of deformation is essential when an aluminum ingot is formed into structural shapes. It would be inconvenient, to say the least, if the shapes sprang back elastically into the shape of the original ingot. In final structures, load-bearing areas are made large enough so that applied loads will not generate large enough stresses to cause permanent plastic deformation.

■ THE STRESS-STRAIN TEST

Many of a material's mechanical properties can be accurately determined from the material's performance in a *stress-strain tensile test*. In this test a sample of material is stretched to fracture (breaking), and a continuous record is made of both the applied stress and the resulting strain. The universal acceptance of certain parameters in a stress-strain test as illustrating mechanical behavior requires that such testing be conducted according to rigid standards. Description of test procedures for materials, ranging from structural steel to plastic electrical insulation, are described in various ASTM standards for stress and strain testing.

A typical ductile metal, such as pure aluminum, has a stress-strain behavior as plotted in Figure 1.5-9. The behavior in region OA in the figure is elastic and linear. If any stress between O and A is applied and then removed, the strain in the body returns to 0%. The straight line shows that for each unit of increase in stress, the strain increases by a constant amount, in other words, strain is proportional to stress.

The proportionality constant between stress (σ) and strain (ϵ) is the modulus of elasticity (E) and is represented by the slope of line OA. The steeper the stress-strain curve, the stiffer the material. It follows then that for stresses within the elastic range

$$E = \frac{\sigma}{\epsilon}$$

The modulus E is an important property because it enables calculation of the elastic strain accompanying a particular stress or loading condition, and it enables building structural designers to calculate deflection. Figure 1.5-10 compares the *E*-values of construction materials belonging to the metals, ceramics, and molecular categories. Metals are generally quite stiff, ceramics slightly less stiff, and plastic still less stiff.

When the stress rises above A, into the nonlinear portion of the curve, permanent deformation will appear in the body. If the stress is then reduced to 0, the elastic strain (ϵ_e) in the body will disappear, but some plastic strain (ϵ_p) will remain. The stress value on the curve marking the transition from elastic to plastic behavior is called the *yield point*, or *yield strength* (σ_y).

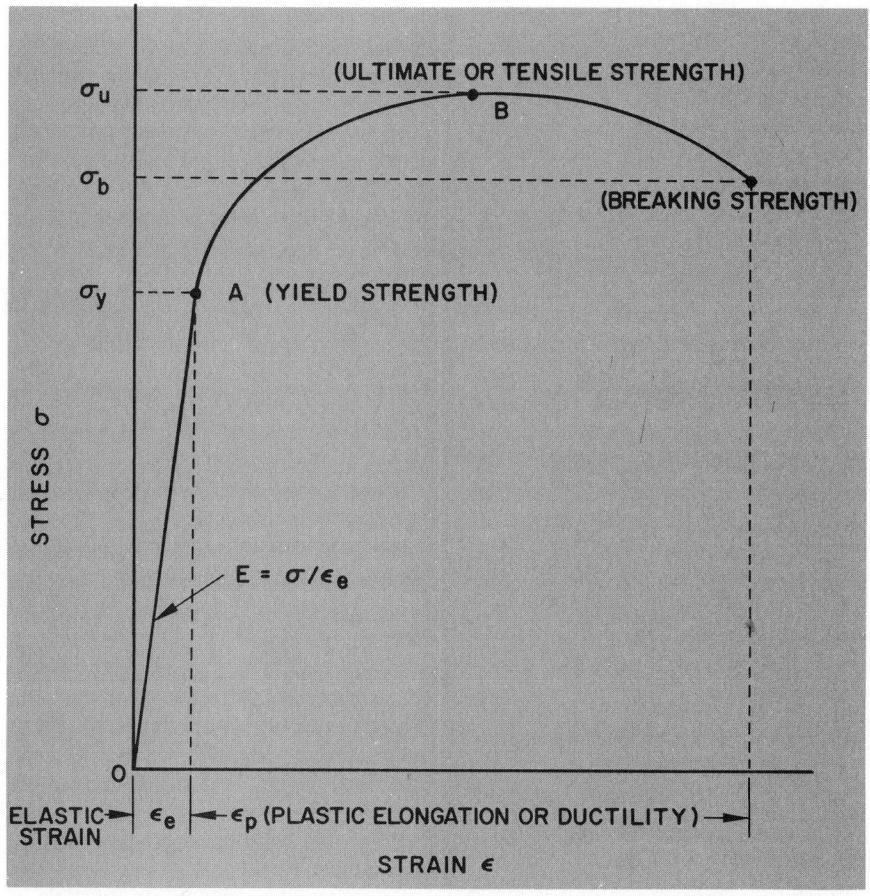

FIGURE 1.5-9 *Typical stress-strain behavior of a ductile metal.* *(Courtesy Addison-Wesley Publishing Company.)*

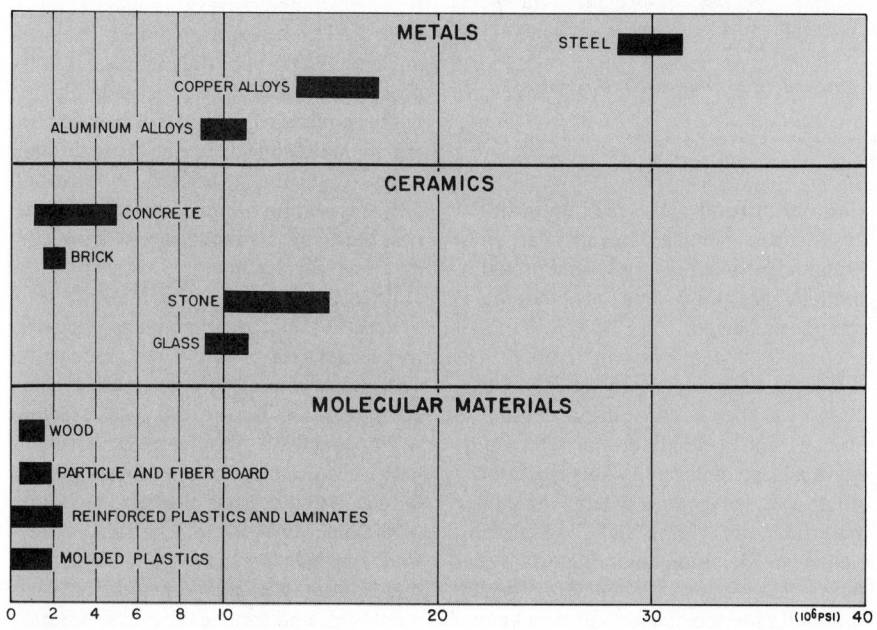

FIGURE 1.5-10 *Comparative stiffnesses of construction materials.* *(Courtesy Addison-Wesley Publishing Company.)*

The stress-strain curves for most materials do not show clearly defined yield points. For these materials, a theoretical yield strength is calculated from the stress-strain curve. It is useful to determine the yield strength of a ductile material, because it is the basis for a common structural design method. The precise definition of σ_y and detailed methods for determining it are given in ASTM standards.

If stresses greater than σ_y are applied in the test, the material behaves as shown by the AB portion of the curve, with the entire region AB representing plastic behavior. At the stress marked by point B, the material rapidly begins "necking down," like a piece of chewing gum stretched to the breaking point. The quantity σ_u is the *ultimate* or *tensile strength*.

Beyond point B, deformation continues even with application of increasingly smaller loads for a ductile material and thus leads to the eventual fracture of the material at stress σ_b, which is the *fracture*, or *breaking strength*, of the material. The total plastic stretching of the material to failure is given by ϵ_p, the *elongation*, which is a measure of the material's ductility.

Figure 1.5-11 shows the stress-strain behavior of four typical construction materials: regular structural steel, annealed copper, portland cement concrete, and polyethylene plastic.

Steel has a much higher E than copper and, therefore, deforms much less in the elastic range. The yield strength of steel is also much higher than that of copper, which indicates that steel can withstand greater stresses before it undergoes permanent deformation. On the other hand, steel stretches only about one-half as much as does copper, and its greater strength is obtained at the price of higher *brittleness* (less ductility). The behavior of steel and copper is typical of the metallic category of materials and is compared with that of plastics in Figure 1.5-12.

Because of the strength of a metallic bond compared with that of a secondary bond (see Section 1.5.1.2), metals resist permanent deformation more effectively than do plastics. Note in Figure 1.5-12 the range of yield strengths in the metals due to appropriate processing such as annealing, heat-treating, and work-hardening.

Ceramics were omitted from Figure 1.5-12 because they do not deform plastically before fracture and are, therefore,

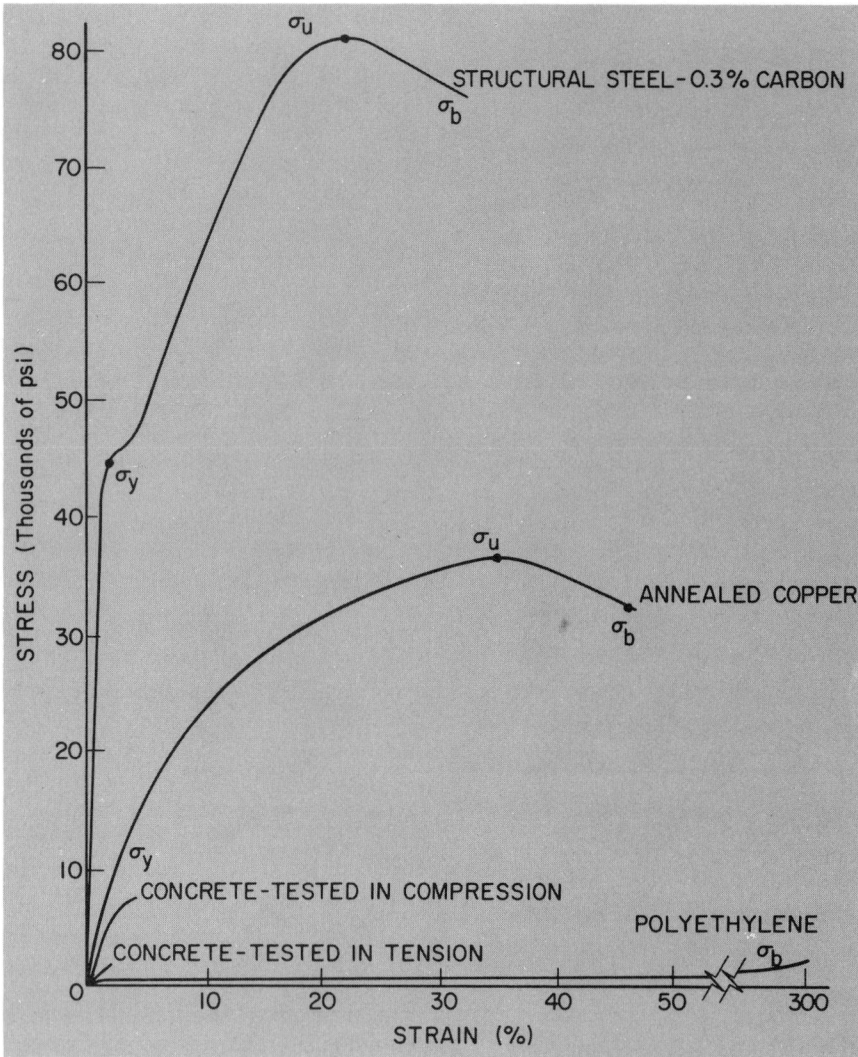

FIGURE 1.5-11 *Stress-strain behavior of four construction materials.* *(Courtesy Addison-Wesley Publishing Company.)*

Under the most severe conditions, metals and plastics can be stressed to ultimate strength, σ_u, before catastrophic fracture occurs, whereas ceramics can be stressed to breaking strength, σ_b, which for them is identical to yield strength, σ_y, and ultimate strength, σ_u. Figure 1.5-13 compares the extreme limits of strength for construction materials in each of the three categories.

Figures 1.5-11, 1.5-12, and 1.5-13 substantiate the generalizations about material properties derived from their structure. Metals are strong, elastically rigid, and fairly ductile. Ceramics are rigid, strong in compression, and brittle, especially in tension. Molecular materials have adequate mechanical properties for many construction applications but are limited by other properties.

■ MISCELLANEOUS MECHANICAL PROPERTIES

Tabulated stress-strain data provide a quick means for gauging a material's mechanical performance. There are however, a number of special situations, in which materials are subjected to mechanical loading, where knowledge of a number of special properties is necessary to predict performance.

Hardness is a measure of a material's ability to resist indentation or penetration. It is determined by various tests, described in ASTM standards, in which indenters are forced into materials under carefully prescribed conditions. In general, the harder a material, the greater its wear and abrasion resistance. Materials with large values of E and σ_y tend to be hard and wear resistant, so in the absence of hardness data, stress-strain data give a measure of wear resistance.

Fatigue resistance is a measure of a material's ability to withstand cyclic (repeated) stresses. When repeatedly stressed, even at stresses below yield strength, many materials will fracture without warning. Some airplane accidents have been caused by failures due to fatigue. The lifetimes of pumps and other mechanical devices in a building's electrical and mechanical systems depend on the fatigue behavior of the materials employed, and the use of adequate shock-absorbent mountings deserves careful consideration.

not generally characterized by a specific yield strength. Ceramic materials are typified by the curves for concrete shown in Figure 1.5-11. They have moderately high E values, represented by slopes greater than those for plastic but not as great as those for steel. They respond rigidly to elastic stresses, showing small deformations for even fairly large stresses. At the same time they are very brittle, providing little plastic deformation before fracture. The brittleness of ceramic materials is accentuated when they are stressed in tension, which accounts for the common use of concrete, brick, tile, and stone to resist compressive loads, but not tensile loads. The absence of an elastic-

plastic transition and a maximum in the stress-strain behavior mean that yield strength, ultimate strength, and breaking strength are the same for individual ceramic materials.

The polyethylene curve in Figure 1.5-11 is typical of materials in the molecular category. The linear elastic region is limited, and a small elastic stress produces a large deflection. This behavior is often a disadvantage when molecular materials are used as load-bearing elements. The molecular materials that show the greatest tendency to stretch elastically are the class of rubbers known as elastomers, which are used in rubber bands and sealants.

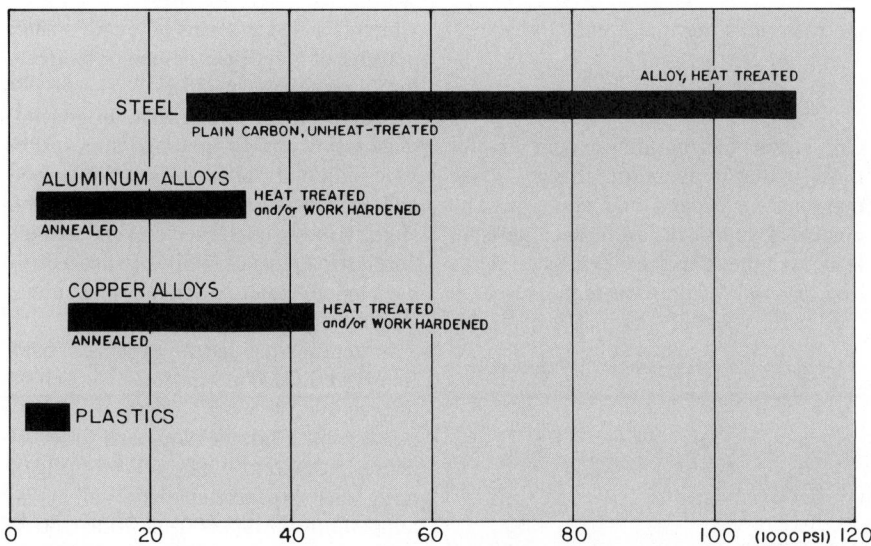

FIGURE 1.5-12 *Yield strengths of metals and plastics.* *(Courtesy Addison-Wesley Publishing Company.)*

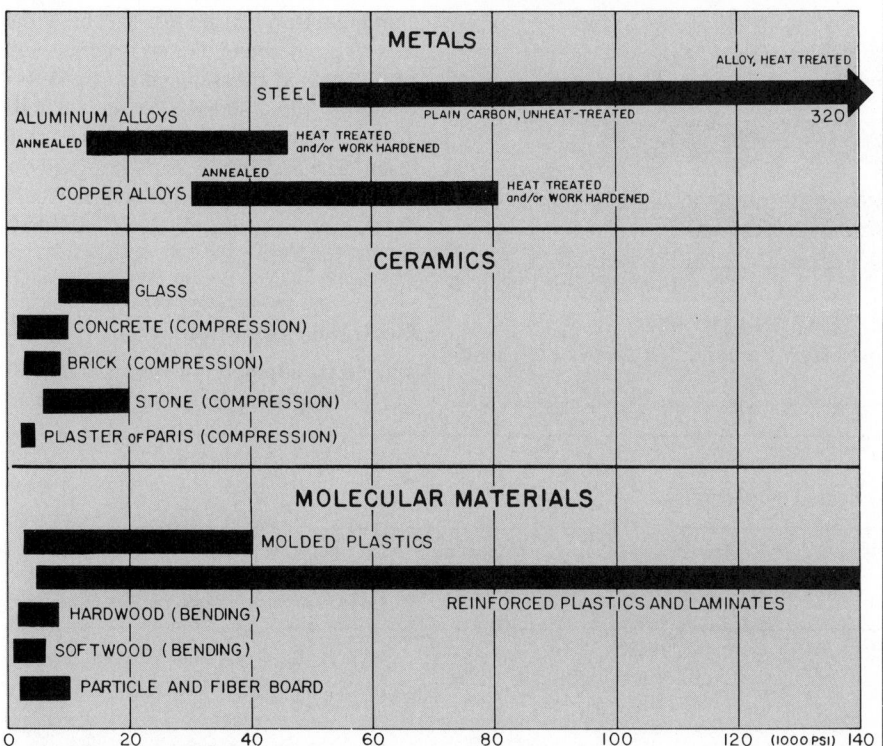

FIGURE 1.5-13 *Ultimate strengths of representative materials.* *(Courtesy Addison-Wesley Publishing Company.)*

Damping capacity is a measure of a material's ability to dissipate or deaden mechanical vibration. Since sound is mechanical vibration, a material's ability to absorb sound is directly related to its damping capacity.

Impact strength, or *toughness*, marks a material's capacity to absorb impact without fracturing. It is defined as the total energy, from elastic deformation to fracture, that a material can absorb before breaking under impact. A material's strength and ductility affect this property. Ceramics are strong but do not deform significantly under static or dynamic load, so they lack toughness and shatter under impact. Plastics can stretch up to several hundred percent under load, but their strength is relatively low, so they absorb little energy under impact. Metals, with their good strength and ductility, are the toughest of the common construction materials.

1.5.2.2

Thermal Properties

When subjected to temperature changes, a material may change its state (solidify, melt, or vaporize), expand or contract, and conduct or reflect heat.

■ MELTING TEMPERATURE

Certain mechanical properties, such as tensile strength, give an indirect measure of other properties, such as hardness. A material's melting temperature can be similarly employed. Materials with high melting points usually retain their mechanical properties over a greater temperature range. In addition, they tend to be stronger and more chemically inert than materials with lower melting points.

Materials with strong ionic bonding and high melting temperatures, such as ceramics, tend to perform best at high temperatures. Metals perform moderately well, and molecular materials, with their weak secondary bonding, perform less well. Figure 1.5-14 compares the temperature limits for continuous service of common construction materials. A more quantitative summary of the thermal properties of typical construction materials is given in Figure 1.5-15.

■ THERMAL CONDUCTIVITY

A building designer is concerned with a solid material's response to heat. Of special importance is the material's thermal conductivity, which is its ability to transfer heat from a region of high temperature to a region of lower temperature (see Fig. 1.5-15). Because the free electrons in metals transport heat effectively, metals typically exhibit the highest thermal conductivities; ceramics have much lower conductivities, and molecular materials the lowest.

A material's *thermal conductivity* is the number of British thermal units (Btu's) conducted in an hour through 1 sq. ft. of 1-in.-thick material for each degree Fahrenheit of temperature differential. Conductivity is measured by placing a 1-sq.-ft. by 1-in.-thick slab of the material to be tested in a laboratory oven with the materials "hot" face 1°F hotter than its "cold" face. The heat energy in Btu's conducted through the material in 1 hour is the material's thermal conductivity.

■ **THERMAL EXPANSION**

One reason that buildings seem alive at times is that temperature changes cause materials to expand and contract. The crackle of ductwork as a furnace starts up is an example of thermal expansion in action. *Thermal expansion* is measured by the coefficient of thermal expansion, which is stated in terms of length change per unit of length per degree of temperature change (in./in./°F).

Figure 1.5-15 shows that, in general, materials with low melting temperatures have high thermal expansion coefficients and vice versa. Ceramics tend to have slightly lower coefficients of thermal expansion than metals, while plastics usually have much higher coefficients than either ceramics or metals.

Structural steel has an expansion coefficient of 0.0000065 in./in./°F, which indicates that each 1-in. length of steel expands 6.5 millionths of an inch for every degree temperature rise. A steel bridge 1 mile long expands about 12 in. as its temperature is raised from 70 to 100°F.

Thermal expansion is a critical property in applications where several materials are joined. For example, aluminum has twice the thermal expansion of steel. Thus, if aluminum trim is fixed to a steel base, the uneven expansion of the two materials will cause internal stresses and elastic deformation in the assembly. Noises, as well as unattractive buckling effects, may result. The need for properly designed expansion joints in such cases is obvious.

1.5.2.3

Electrical Properties

The electrical property of primary interest to a building designer is electrical conduc-

MAXIMUM SERVICE TEMPERATURES

FIGURE 1.5-14 *Service temperatures of construction materials.* *(Courtesy Addison-Wesley Publishing Company.)*

FIGURE 1.5-15
Thermal Properties of Typical Construction Materials

Category	Material	Melting Temperature (°F)	Thermal Conductivity (Btu/sq. ft./°F/in./hr.)	Thermal Expansion (10⁻⁶ in./in./°F)
Metals	Steels (carbon or low alloy)	2700–2800	300–500	6.5
	Stainless steels	2700–2800	100–150	7
	Aluminum alloys	1100–1200	500–1500	9
	Copper alloys	1800–2000	500–2500	12.5
Ceramics	Brick	2000–5000	3–6	5
	Concrete	3000	5–10	7
	Plate glass	1500	5–6	1
Molecular materials	Wood	400 (chars)	1–2	10–30
	ABS plastic	200–300	1–2	50
	Acrylics (PMMA)	200–300	1	30–50
	Vinyls (PV)	200–300	0.5–1	30–100
	Polyethylene (PE)	200	2–5	60
	Epoxies (EP)	200–500	1–2	30
	Silicones (SI)	400–600	1	1–100

tivity, a characteristic closely related to a material's thermal conductivity. The free electrons in metals allow heat and electrical energy to be conducted easily, so conductivity is high. Ceramics have lower electrical conductivities, and molecular materials have the lowest. This property accounts for the extensive use of molecular materials, such as plastics, in electrical insulation products.

Electrical conductivity is usually given in *mho* (ohm spelled backward) units per foot of conductor length. It is the reciprocal of electrical resistivity, which is expressed in *ohms per square foot*. For example, metals have *low* electrical *resistivity* and *high* electrical *conductivity*. Figure 1.5-16 summarizes the resistivities of typical construction materials and illustrates the excellent insulating capabilities of ceramics and molecular materials.

1.5.2.4

Chemical Properties

Air and airborne moisture contain small amounts of active chemical compounds, which, under certain conditions, react with building materials and degrade their properties.

Ceramics resist chemical attack in most normal environments, and plastics resist attack from all but a few organic solvents. The situation is more complicated, however, in the case of metals. Metals degrade or corrode through the transport of minute amounts of electricity from regions called *anodes* to other regions called *cathodes*. A cathode accepts electrons and remains intact, but an anode is degraded by the chemical reaction.

The essential elements in a corroding system are the anode, the cathode, and the current-carrying medium (*electrolyte*) that connects them. An electrolyte is usually a solution of water and a gas, such as carbon dioxide or sulfur dioxide. Preventing corrosion, therefore, is simply a matter of removing one of the three elements from the system. For example, most paints are molecular materials, which are poor conductors of electricity. Painting metallic surfaces not only shelters them directly from corrosive environments but also provides a nonconductive barrier to electrical current flow.

Corrosion can occur when an entire material or given region in a material is sub-

FIGURE 1.5-16
Electrical Resistivity of Typical Construction Materials

Category	Material	Electrical Resistivity (ohm/ft.)
Metals	Steel (carbon or low alloy)	.000006
	Stainless steel	.00003
	Aluminum alloys	.000001
	Copper alloys	.0000005
Ceramics	Brick	4,000,000
	Concrete	4,000,000
	Plate Glass	10,000,000,000,000
Molecular materials	ABS plastic	1,000,000,000,000,000
	Acrylics (PMMA)	100,000,000,000,000
	Vinyls (PV)	100,000,000,000
	Polyethylene (PE)	1,000,000,000,000
	Epoxies (EP)	100,000,000,000,000
	Silicones (SI)	1,000,000,000,000

jected to a situation conducive to anodic behavior. Such situations develop in the presence of certain impurities in the metal (acting as cathodes) and when metals with different *galvanic potentials* are placed in close proximity. Galvanic tables that rank materials according to their tendency to become anodic can be used to predict potentially corrosive situations (Fig. 1.5-17).

Materials at the top of the list in Figure 1.5-17 have a strong tendency to become anodic and those toward the bottom cathodic. If a material from near the top of the list, such as carbon steel, is placed in contact with a material lower on the list, such as brass, corrosion is likely to result. The steel, acting as an anode, and the brass, acting as the cathode, will cause an electric current to flow. Corrosion of the steel is inevitable.

Several points regarding galvanic reactions and corrosion problems should be remembered:

1. Intimate contact between materials widely separated in the galvanic table should be avoided.
2. Zinc can be used as a protective coating (galvanizing) on steel. Because it is higher than steel in the galvanic series, the zinc serves as a "sacrificial anode" and is intentionally permitted to corrode, thus saving the steel. As long as the anode is present, the cathode (the steel) will not corrode.
3. Tin is cathodic with respect to steel, so as a coating over steel it simply acts as a physical barrier to the corrosive environment. Should the tin coating be

FIGURE 1.5-17
Galvanic Series of Common Metals and Alloys

Electrolytic Tendency	Metal or Alloy
Anodic (read up)	Magnesium alloys
	Zinc
	Aluminum alloys
	Carbon steel
	Stainless steel (active)
	Lead
	Tin
	Brass
	Copper
Cathodic (read down)	Bronze
	Stainless steel (passive)
	Gold

scratched or removed in spots, the exposed steel will corrode quickly. Care must be taken, for example, with *terneplate* (a tin and lead coating over steel), as with painted surfaces, to maintain a continuous film of the protective coating.

Surprisingly, aluminum and stainless steel are inherently "actively" anodic and susceptible to corrosion. The corrosion resistance normally associated with these metals develops when they are exposed to oxygen. The oxygen forms a thin, impervious, transparent oxide film, which protects the underlying metal from further contact with atmospheric oxygen. The anodizing of aluminum alloys is simply an electrochemical process that produces a thicker oxide coating than forms naturally.

1.6 BARRIER-FREE DESIGN

According to ANSI A117.1-1986, "Providing Accessibility and Usability for Physically Handicapped People," a *physically handicapped person* is "an individual who has a physical impairment, including impaired sensory, manual, or speaking abilities, that results in a functional limitation in gaining access to and using a building or facility."

Historically, most of the buildings, facilities, transportation systems, and other structures and spaces we work, play, and live in (the built environment) were designed for use by average, able-bodied, young adults who walk without aid or assistance (ambulatory). People who were older, younger, smaller, taller, or who had physical handicaps were at a disadvantage in most places. Many buildings were inaccessible to them and some were even dangerous. To a certain extent, this situation continues today. Fortunately, however, this picture is rapidly changing.

Physical conditions that make a building or facility unsafe or confusing or that prevent physically handicapped people from using them are called *architectural barriers*. *Barrier-free*, or *accessible*, design eliminates or avoids creating such barriers. Barrier-free design is frequently thought of as a way to accommodate a few special people called the *handicapped*. The wheelchair symbol, used to designate public parking spaces, toilets, telephones, and water fountains, tends to foster the belief that barrier-free design offers accessibility mainly to those in wheelchairs. Actually, barrier-free design benefits everyone, because it makes facilities safer and more convenient to use.

In fact, those who benefit from barrier-free design include not only the physically handicapped, as defined earlier, but also children, who are physically and mentally immature; pregnant women, who often have reduced agility, stamina, and balance; people who care for children and must carry them, hold their hands, or maneuver baby carriages; older people, who may suffer progressive degeneration of physical, perceptual, and mental faculties; those disabled by size-related

disorders such as giantism, dwarfism, or obesity; able-bodied people who suffer temporary illness or injury; and even able-bodied people who are carrying large packages.

Some people are born with a limited or unusable physical, mental, or sensory function (*disability*), but even a normally able-bodied person can become either temporarily or permanently disabled due to an illness or injury. *Disabilities* include visual impairment, hearing impairment, mental and perceptual impairment, confinement to a wheelchair, and a coordination disability.

Nationwide, about 22% of those between the ages of 15 and 64 and about 59% of those 65 and older are disabled to some degree. That adds up to about 20% of the total population over 15 years old, or more than 37 million disabled people. Some sources put the number of disabled at more than 43 million. It is likely that the percentage of disabled people will rise in the future, because today people survive illnesses and injuries that were once fatal. In fact, the average life span has increased dramatically during this century, from about 47 years in 1900 to more than 75 years today. In addition, the U.S. population as a whole is getting older. The number of people more than 65 years old is projected to have grown from a little more than 10% of the total population in 1986 to more than 20% by 2040.

This section divides barrier-free design into four broad categories: (1) safety and (2) general accessibility criteria applicable to all buildings; (3) recommendations applicable to buildings for people with specific disabilities; and (4) requirements for making buildings adaptable for future use by the handicapped.

1.6.1

THE LAW AND APPLICABLE STANDARDS

In the late 1950s the *President's Committee on Employment of the Handicapped*

joined with consumer groups, such as the National Easter Seal Society, and standards makers to develop standards for making buildings accessible to handicapped persons. The result was ANSI A117.1, which was first published in 1961 by the American Standards Association, now the American National Standards Institute (ANSI). ANSI A117.1 quickly became, and remains today, a major standard for design for the handicapped. It is a voluntary standard, of course, having force only when adopted by a governing body.

In 1968 Congress passed the Architectural Barriers Act, which required that access for the handicapped be achieved in accordance with standards to be established by the General Services Administration (GSA), the Department of Housing and Urban Development (HUD), and the Department of Defense (DOD). The rules Congress developed established ANSI A117.1 as the generic standard of access for buildings owned or leased by the federal government.

Enforcement, however, was spotty at best. As a result Congress passed the Rehabilitation Act of 1973, which established the *Architectural and Transportation Barriers Compliance Board* (ATBCB) to enforce the Architectural Barriers Act. The ATBCB issued a document called *ATBCB Minimum Guidelines and Requirements*.

The various states followed the federal government in passing laws requiring that buildings be made accessible to the handicapped. Unfortunately, many of them did not require compliance with ANSI A117.1 or the *ATBCB Minimum Guidelines*, instead developing their own standards. In an attempt to gain wider acceptance, ANSI expanded and reissued A117.1, but it still failed to gain universal acceptance. To make matters more complicated, different federal agencies required compliance with different editions of ANSI A117.1.

In 1984, in the midst of all this confusion, GSA, HUD, DOD, and the U.S. Postal Service jointly issued the *Uniform*

Federal Accessibility Standard (UFAS). The purpose of UFAS was to cause all federal agencies to follow the same technical requirements for complying with the Architectural Barriers Act. It was based on the *ATBCB Minimum Guidelines* but incorporated the basic requirements of ANSI A117.1 with some changes and additions.

As a result, various agencies of the federal government were requiring compliance with four different standards, two versions of ANSI A117.1, UFAS, and the *ATBCB Minimum Guidelines*. In some states, those four standards were being used in addition to the state's own standards. Some states actually require compliance with more than one standard. Even at the federal level, compliance with more than one standard was required by some agencies.

Gradually the two major standards have become the latest edition of ANSI A117.1, which at the time of this writing was the 1986 edition, for private sector work and the UFAS for federal government work, although the others mentioned are still required by some state and local authorities. More than half of the states have adopted some version of ANSI A117.1 or incorporated it into their own standards. The Building Officials and Code Administrators International (BOCA) *National Building Code*, the Southern Building Code Congress International (SBCCI) *Standard Building Code*, the International Conference of Building Officials (ICBO) *Uniform Building Code*, and the National Fire Protection Association (NFPA) NFPA 101, *Life Safety Code*, all incorporate or reference some edition of ANSI A117.1.

The *Federal Fair Housing Act of 1988, Title VIII of the 1968 Civil Rights Act* prohibits housing discrimination based on a physical handicap in housing projects with four or more dwelling units. Basically that act says that everything in such dwelling units must be adaptable for accessibility by the handicapped, and all public areas serving such dwelling units must be accessible. Compliance must be in accord with the *Final Fair Housing Accessibility Guidelines* published in the Federal Register on March 6, 1991. Copies are available from HUD.

The *Americans With Disabilities Act* of 1990 prohibits discrimination on the basis of disability in places of employment, public service (transportation facilities), public accommodations (restaurants, hotels, theaters, doctors' offices, retail stores, museums, libraries, parks, private schools, day care centers, and others), or telecommunications services. The standards developed by the ATBCB to implement this law differ slightly from and expand on ANSI A117.1 and the UFAS, but the ATBCB may ultimately incorporate one or the other as the basis of the new standard.

The recommendations in this section come from many sources, but most of them comply with the requirements of ANSI A117.1-1986, "Providing Accessibility and Usability for Physically Handicapped People," the April 1988 edition of *Uniform Federal Accessibility Standards*, the 1984 edition of HUD *Minimum Property Standards for Housing*, and the 1987 edition of the HUD document *Adaptable Housing*. Keep in mind that some differences exist among these standards and that state and local requirements may contain some other differences. In addition, these standards and the *Fair Housing Accessibility Guidelines* issued in conjunction with the *Federal Fair Housing Act of 1988* are subject to modification at any time. Therefore, in dealings with an actual project, whether for a federal or state agency or for a private-sector owner, it is necessary to determine the specific requirements of the project and local, state, and federal laws and ordinances that dictate the requirements for accessibility.

Rules for providing accessibility have developed slowly and will no doubt continue to expand and change as more is understood about the needs of the handicapped. Until a single standard emerges, it is a mistake to follow any one standard alone or any edition of a standard before verifying the actual requirements. It is necessary also to remember that both the *Federal Fair Housing Act of 1988* and the *Americans With Disabilities Act of 1990* are civil rights laws rather than code compliance laws. Accordingly, it is best to comply with their spirit and not try to circumvent their intent. In the event of a compliance dispute, a court is likely to lean in favor of the intent of the law rather than to the wording of the standards with which it requires compliance.

1.6.2

SAFETY

Many accidents in buildings can be traced to obvious causes, such as slippery floors, the lack of grab bars, or inappropriate or faulty stair railings. Accident prevention requires elimination of these obvious causes as well as other measures. Surfaces, such as those of approach walkways, stoops, entryways, and corridor floors, which are likely to get wet due to the weather, should have slip-resistant surfaces. In addition, floors in toilet rooms, bathrooms, shower rooms, locker rooms, kitchens, and other rooms that are likely to get wet should have slip-resistant surfaces. Throw rugs and area rugs should not be used in public spaces; where they are appropriate, they should have a nonslip backing.

1.6.2.1

Doors and Sidelights

Doors and sidelights are a major source of accidents. *Safety glazing*, consisting of tempered glass, laminated glass, or plastic, should be used in all glazed or glazing-insert doors and sidelights. Most building codes and regulations require such glazing.

Rounded door and jamb edges or resilient door edges minimize injury to fingers inadvertently stuck between them.

1.6.2.2

Stairs

A great many accidents occur on stairs. The following measures help reduce these accidents: (1) the number of risers in a series should be at least three, because people are careless on fewer; (2) treads and risers should be uniform; (3) treads should be no less than 11 in. (280 mm) wide (TW), measured from riser to riser, and should have a slip-resistant surface; and (4) risers should be not more than 7 in. (180 mm) high and should be closed.

■ NOSINGS

Nosings (the edges of the stair treads that project over the risers) should project not more than 1-1/2 in. (38 mm), and their undersides should be rounded. The radius at the leading edge of the tread should not be greater than 1/2 in. (13 mm).

■ HANDRAILS

Handrails should run continuously along both sides of a stairway, extend parallel to the floor at least 12 in. (305 mm) beyond the top and bottom of the staircase, and be free of protrusions that might snag clothing. Rails should be securely mounted at a height of 30 to 34 in. (815 to 865 mm) from the floor. If children regularly use the stairs, an extra railing should be mounted at a height of 24 in. (610 mm) from the floor. The space between balusters, if present, should be not more than 5 in., since wider spacings may allow a child's head to become trapped.

■ LANDINGS

Because long flights of stairs can be tiring, a landing should be provided for flights with more than 16 risers.

1.6.2.3

Toilet Rooms, Bathrooms, Shower Rooms, Locker Rooms

Most accidents in toilet rooms, bathrooms, shower rooms, and locker rooms involve falls caused by slipping and burns caused by scalding.

■ WATER

A person can receive a third-degree burn in just two seconds in water 150 °F, six seconds in water 140 °F, and 30 seconds in water 130 °F. Many people have limited sensitivity to heat and may not be able to adequately gauge the temperature of water in a lavatory, sink, tub, or shower. Others may not be able to react quickly enough to a sudden surge of hot water.

To prevent scalding, the temperature of hot water should be regulated by (1) setting the temperature of the heated water to below 115 °F (40.6 ° Celsius), (2) providing a temperature-regulating device on tub fillers and shower heads, or (3) providing a temperature-regulating device on the hot water supply lines. Individual fixture-mounted temperature controls or a temperature-regulating device on supply lines are preferred because other uses, such as dishwashing and clothes washing require temperatures from 120° to 140 °F. Fittings are available that can be set to maintain a specific water temperature at either the fixture or the supply pipe.

In addition, exposed hot water supply and drain pipes should be wrapped with insulation to prevent contact burns.

■ SHOWERS AND TUBS

Nonslip strips should be provided on the bottom surface of tubs and showers. Soap dishes, tub fillers, and controls should be recessed so that users cannot fall on them.

■ GRAB BARS

Firmly mounted grab bars should be placed in and near tubs and showers and adjacent to water closets to help wheelchair users transfer from the chair to the fixture and to help those with other *disabilities* and the elderly get up and down and maintain their balance.

The width of a grab bar gripping surface should be 1-1/4 to 1-1/2 in. (32 to 38 mm). The shape of the bar should allow a natural grip. There should be no protrusions or rough surfaces to catch clothing or cause injury.

Bars should be spaced 1-1/2 in. (38 mm) from the wall, to allow proper grasping but prevent the arm from slipping behind the bar.

1.6.2.4

Kitchens

The number of kitchen accidents, such as burns and falls, can be reduced by proper planning.

■ APPLIANCES

More accidents are related to cooktops and ranges than any other kitchen appliance. To prevent such injury, cooktop and range controls should be in the front so a cook does not have to reach over hot burners to adjust them. When dwelling residents include small children, controls should be out of sight on a horizontal surface and either have a protective lid or require an extra movement, such as pushing, to activate a burner. Electric cooktops should be equipped with a light to indicate that burners are on.

■ CABINETS

Reaching for cabinets above refrigerators requires standing on a stool, which presents a falling hazard. Reaching for cabinets over cooktops or stoves may cause burns. Therefore, placing cabinets over refrigerators and stoves is not desirable.

1.6.2.5

Protruding Objects

Objects, such as telephones, located between 27 and 80 in. (685 and 2030 mm) above a finished floor should not project more than 4 in. (100 mm) into a room or circulation path (see Section 1.6.3.1). Objects mounted at or below 27 in. (685 mm) may project any amount because they are detectable by a moving cane. Freestanding objects mounted between 27 and 80 in. (685 mm and 2030 mm) on posts or pylons may overhang a maximum of 12 in. (305 mm). No projection should reduce the clear width of an accessible route or maneuvering space.

When objects project into circulation spaces, a change in texture or a contrast in color from the floor surface can serve as a warning. Such warning devices should, however, be used with discretion, because overuse diminishes their effectiveness.

1.6.2.6

Lighting

Sufficient glare-free lighting should be available everywhere to meet the varying intensity needs of different tasks and to eliminate dark shadows that might conceal hazards. Changes in light intensity should be gradual to give the eyes time to adjust.

The paths people follow to get from one place to another in the built environment are called *circulation paths*. Good, glare-free lighting in circulation paths, especially those in potentially hazardous areas such as stairways, is particularly important in the creation of a safe environment.

GENERAL ACCESSIBILITY

An *accessible* site, building, facility, or portion thereof is one that complies with current standards and that can be approached, entered, and used by physically handicapped people.

Accessible Route

A *circulation path* is an exterior or interior way of passage from one place to another for pedestrians, including but not limited to walks, hallways, courtyards, stairways, and stair landings.

An *accessible route* is a continuous unobstructed circulation path that connects every accessible element and space in a facility that can be negotiated by a person with a severe disability using a wheelchair and that is also safe for and usable by people with other disabilities.

Site Planning

Barrier-free design should include the planning of accessible site facilities. Individuals with disabilities should have as much freedom of movement outside as inside.

Barrier-free buildings demand barrier-free routes of access, including (1) convenient parking, (2) accessible walks and curb ramps, (3) suitable paving, and (4) logical organization.

Direct routes should be provided into the building from public transportation, parking lots, and passenger loading zones.

■ PARKING

Parking spaces and passenger loading zones for people with disabilities should be located as close as possible to an accessible entrance and should be identified by the *international symbol of accessibility* (Fig. 1.6-1). If an accessible entrance cannot be seen from the parking area, signs should direct users to it.

Accessible parking spaces should be at least 96 in. (2440 mm) wide and have an adjacent access aisle at least 60 in.

FIGURE 1.6-1 *International symbol of accessibility.*

(1525 mm) wide. Two adjacent spaces may share one access aisle.

Passenger loading zones should have an *access aisle* at least 48 in. (1220 mm) wide and 20 ft. (6100 mm) long adjacent and parallel to the vehicle pull-up space.

Curbs between access aisles and walks should have *accessible curb cuts* and *curb ramps*, as described in "Curb Ramps."

■ WALKS

Walks and other access paths should be at least 36 in. (915 mm) wide for single passage and 60 in. (1525 mm) wide for passing. Paths that are less than 60 in. (1525 mm) wide should have periodically spaced areas where a pedestrian or wheelchair user can move over to allow passing. The clear *passage space* should not be reduced by protruding objects.

Walks should be smooth and nonslip, with gaps of not more than 1/4 in. (6 mm) and level changes not more than 1/2 in.

(13 mm) (see Section 1.6.3.4). Textural changes should warn users of level changes, projecting objects, and other hazards.

■ CURB RAMPS

Curbs define and separate walks from vehicular traffic areas. A *curb ramp* is a short ramp cutting through a curb or built into it to remove barriers for wheelchair travel. There should be a curb ramp wherever an accessible route crosses a curb and the path changes level more than 1/2 in. (13 mm).

If anyone must cross perpendicular to a *curb ramp*, the side should be flared at a maximum slope of 1:10 (see Fig. 1.6-2a) or be protected by planting strips or some other nonwalking surface (see Fig. 1.6-2b).

FIGURE 1.6-2 *Curb ramps may either (a) start at the curb or walkway and project into the roadway or (b and c) be curb cuts.*

FIGURE 1.6-3 *(a) Planting strips help organize and separate street furniture from the pedestrian path. (b) Rest areas may be integrated with planting strips to provide convenient stopping places outside the circulation path.*

Curb ramps should be at least 36 in. (915 mm) wide and should not project into vehicular traffic (Fig. 1.6-2c), unless there is a *safety zone* on the roadway. Their slope should be 1:12 or less.

Curb ramps should be of a different texture from the surrounding passage to warn visually handicapped users of hazards. This *textural warning* area should extend for the entire width and depth of the curb ramp.

■ PLANTING STRIPS

Figure 1.6-3 shows typical *planting strips*. These strips are a good way to keep protruding objects and street furniture from intruding on a clear accessible path (see Fig. 1.6-3a). Planting strips can be combined with benches to provide convenient resting places for those with limited stamina (see Fig. 1.6-3b). Hazards should be marked with *detectable warnings*, as recommended in Section 1.6.4.1.

■ LIGHTING

Proper illumination is just as important outdoors as it is indoors. Weatherproof fixtures should provide a minimum of 5 foot-candles of light on the travel surfaces at entrances, ramps, and steps. Weatherproof surface or post lights or floodlights should provide 1 foot-candle and at least 1/2 foot-candle of light along paths of travel and in parking areas.

■ ORGANIZATION AND ARCHITECTURAL HARMONY

Frequently used public spaces inside and outside a building should be easy to find and use. Elements that make buildings accessible, such as ramps, curb ramps, and signs, should blend harmoniously with other building elements, especially in rehabilitated older buildings. Careful design and site development in new construction can often reduce the need to add special elements that single out the handicapped. For example, a single-level building with appropriately sized doors and grade-level entrances would require no ramps, elevators, platform lifts, or special entrances.

Landscaping also contributes to a site (Fig. 1.6-4). Care should be taken to choose plants that do not obstruct the view or leave excessive debris in walkways. Use of poisonous and thorny plants should be avoided.

1.6.3.3

Entries and Passageways

Entries should be at grade level or made accessible by means of elevators or lifts (see Section 14.1). Doorways should have a minimum clear opening of 32 in. (815 mm) with the door open 90 degrees, measured from the face of the door to the opposite stop.

Hallways and corridors should be at least 36 in. (915 mm) wide. When a door opens into a corridor, the corridor should be 54 in. (1370 mm) wide. Corridors where handicapped people must pass each other or where wheelchairs must turn around should be 60 in. (1525 mm) wide.

1.6.3.4

Ramps

Passageways and walks are said to *blend to a common level* when their various portions meet in such a way that there is no abrupt rise or drop in the surface. Level changes up to 1/4 in. (6 mm) high need no edge treatment; those between 1/4 in. (6 mm) and 1/2 in. (13 mm) should be beveled with a slope no greater than 1:2. Level changes greater than 1/2 in. (13 mm) require a ramp to be accessible by wheelchair users (Fig. 1.6-5) and to maintain more easy passage for those who must use crutches, canes, or walkers. Ramps are the primary means by which wheelchair users enter buildings that do not have level entryways.

A *ramp* is a walking surface in an accessible space that has a running slope greater than 1:20. A *running slope* is the slope of a pedestrian way that is parallel to the direction of travel. Conversely, a *cross slope* is the slope of a pedestrian way that is perpendicular to the direction of travel.

Ramp slopes should be as gradual as possible. The maximum slope should be

FIGURE 1.6-4 *An attractively landscaped, fully accessible space provides a pleasant meeting area for everyone.*

FIGURE 1.6-5 *Ramps can be made to blend with other building elements.*

1:12 unless the ramp is part of a remodeled existing facility that has space limitations. More permissive ramp dimensions are sometimes permitted for existing construction, where space limitations prevent the use of the preferred lower slope of 1:12. The maximum rise for a ramp is 30 in. (760 mm). Ramps should have a minimum clear width of 36 in. (915 mm).

Level *landings* at least as wide as the ramp itself and at least 60 in. (1525 mm) long should be provided at the top and bottom of each ramp. Landings are also necessary at intermediate levels where the *rise* of a ramp exceeds 30 in. (760 mm). If a ramp changes direction, the landing should be a minimum of 60 in. by 60 in. (1525 mm by 1525 mm). The width of the platform at a *switch-back ramp* should be at least as wide as the width of the two ramp portions plus the width of the space between them (Fig. 1.6-6).

The surfaces of ramps and landings should be *slip resistant* and should not collect water. Their edges should have curbs, walls, railings, or other protection to prevent people from falling.

Handrails should be provided on both sides of ramps that rise more than 6 in. (150 mm) or that are longer than 72 in. (1830 mm). Handrails should (1) extend at least 12 in. (305 mm) beyond the top and the bottom of a ramp, (2) be parallel with the ramp, (3) have a clear space of 1-1/2 in. (38 mm) between the handrail and any adjacent wall, and (4) be 30 to 34 in. (760 to 865 mm) above the ramp surfaces. Inside handrails on switch-back or *dogleg ramps* should be continuous.

Because many handicapped people find it easier to negotiate stairs than ramps, both should be provided in buildings with an above-ground-level entry.

1.6.3.5

Stairs

Stairs should comply with the requirements stated in Section 1.6.2. In addition, there should be *detectable warnings* (see "Texture" in Section 1.6.4.1) at least 36 in. (915 mm) wide at the top of every stair run. Stairs should have *closed risers* of uniform height, and treads should be slip resistant. Also, nosings should not be square or abrupt.

Steps and landings should have sharp color contrasts to help people with limited light-and-dark sensitivity or depth perception. It is also helpful if the edge of a stair contrasts with the rest of the tread.

1.6.3.6

Signage

Signage is defined as verbal, symbolic, and pictorial information presented in a graphic, two-dimensional format.

Informational and warning signs should be easy to find and to read, that is, (1) they should be positioned as close as possible and perpendicular to the path of travel; (2) the information on them should contrast with the background; and (3) they should be made of glare-free materials. A sign's legibility depends on (1) its location and viewing distance, (2) the design and proportion of the characters on the sign, and (3) color contrast.

Signage should also have a consistent format or use *international symbols*.

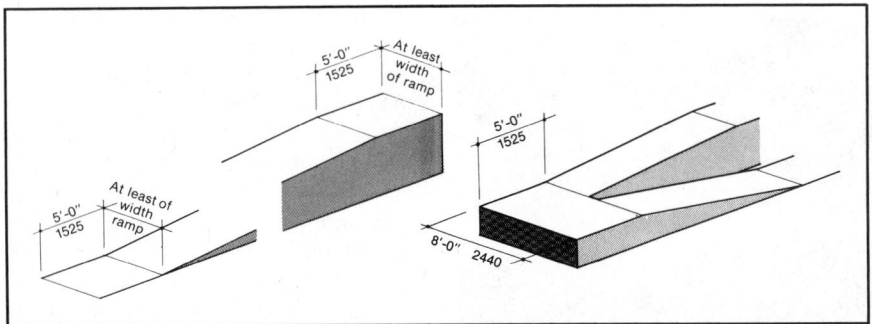

FIGURE 1.6-6 *Layout and sizes of flat platforms at ramps. (Railings are not shown for clarity.)*

■ LOCATION AND VIEWING DISTANCE

Most able-bodied people can see within an angle of 30 degrees to either side of the centerline of their faces without moving their heads. People with disabilities often have limited head movement or reduced peripheral vision. Therefore, signs should be positioned as close as possible and perpendicular to the path of travel. Signs close enough to touch should be positioned for easy hand detection by a standing person, 54 to 60 in. from the floor.

The distance from which a sign will be read determines the size of the letters, numbers, and symbols used on it.

■ DESIGN AND PROPORTION OF LETTERS, NUMBERS AND SYMBOLS

Letters and numbers should be in sans serif (block) type and proportioned as recommended in Figure 1.6-7.

Characters on signs that must be read from a distance should be large enough to be read easily from the maximum viewing distance expected. Characters on signs close enough to touch should be raised at least 1/32 in. (0.8 mm) from their background. They should also be at least 5/8 in. (16 mm) tall but no taller than 2 in. (50 mm) and have a stroke width of at least 1/4 in. (6 mm) to be easily detectable by one finger (Fig. 1.6-8).

Braille is a special raised touch alphabet for the blind that uses a cell of six dots. Unfortunately, only about 10% of those with visual impairments can read braille. Raised Arabic symbols can be read by almost anyone, regardless of visual handicap. To aid those who can read it, braille characters in the standard dot size and spacing may be added to the left of standard characters.

Borders, which might confuse a reader with a visual impairment, are not recommended on signs.

■ COLOR CONTRAST

The background and characters on a sign should contrast. Dark characters may be used on a light background; but light characters on a dark background are more easily readable. Surfaces should be glare free.

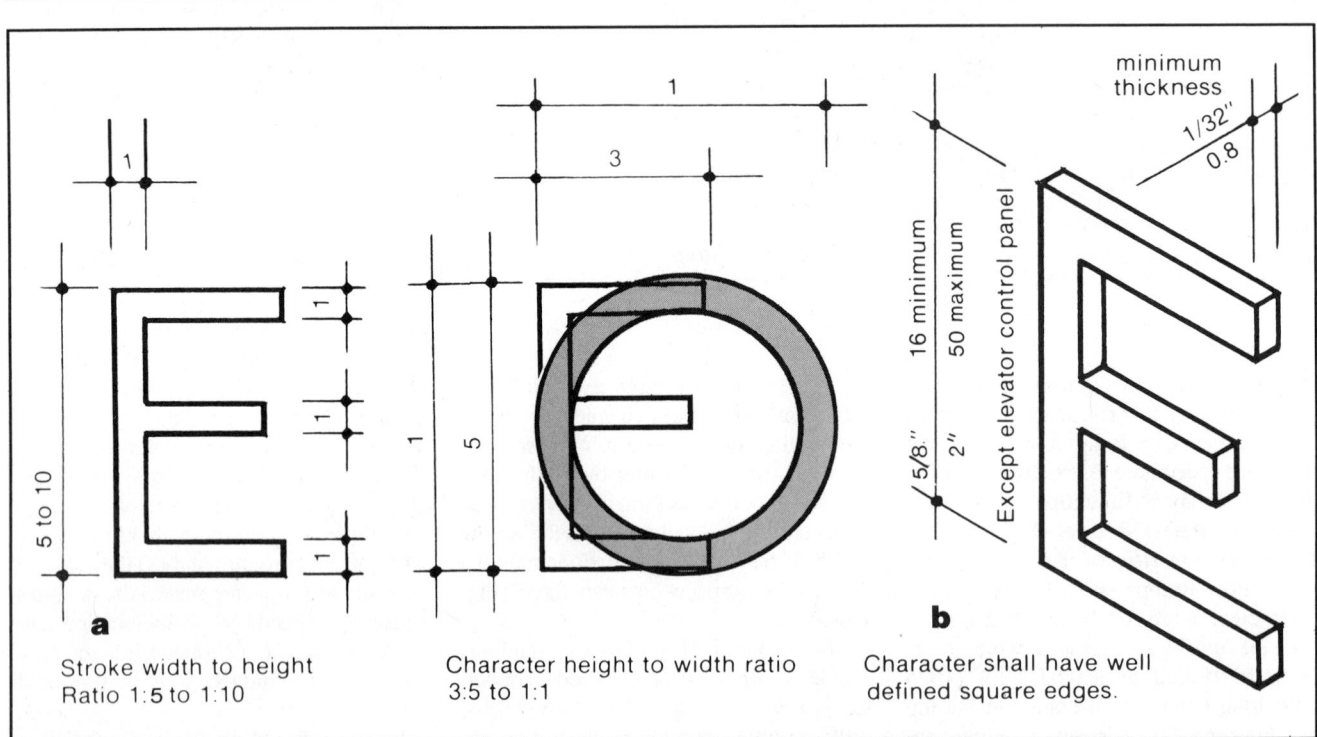

FIGURE 1.6-7 *(a) Proportions of letters expressed as a ratio; (b) dimensions of letters for signs close enough to be touched.*

FIGURE 1.6-8 *A sign with raised lettering should be readable both those with normal sight and those with visual impairments.*

1.6.3.7

Surfaces

Floor surfaces should be nonslip. When practicable, they should be carpeted to add the advantage of noise reduction, which helps those who must use hearing aids.

Carpet should be (1) securely attached, (2) have either a firm backing or no backing, and (3) have pile height not in excess of 1/2 in. (13 mm) so it does not hinder wheelchair travel. Shag carpets should be avoided. Carpets should be of *antistatic* materials or treated to be static-free, because static electricity interferes with hearing aid operation. Humidifiers may be used to help control static electricity. Exposed carpet edges should have carpet trim along the entire carpet length.

1.6.3.8

Electrical

Electrical outlets should be centered at least 15 in. (380 mm) and preferably 24 in. (610 mm) above the floor.

The best height for *wall switches* is centered 48 in. (1220 mm) above the floor, but they should never be placed higher than 54 in. (1370 mm) above the floor. Where persons in wheelchairs will approach from a forward position, as is the case when switches are located over counters with knee spaces, switches should be centered between 36 and 40 in. (915 and 1016 mm) above the floor. In no case should switches and outlets over counters be placed more than 48 in. (1220 mm) above the floor.

Lighting should be as described in Sections 1.6.2 and 1.6.3.2.

1.6.4

SPECIFIC HANDICAPS

The kinds of specific handicaps discussed in this section include (1) *visual impairments*, (2) *hearing impairments*, (3) *mental* and *perceptual impairments*, (4) *disabilities* that require the use of a wheelchair, and (5) *coordination disabilities*.

Designs that remove *barriers* for people with one kind of *disability* sometimes create *barriers* for those with a different set of *disabilities*. For example, *curb ramps* (see Fig. 1.6-2) installed to permit easy street crossings for wheelchair users can cause a person with a *visual impairment* to wander unknowingly into automobile traffic. Both environmental needs can be met by providing a textured pavement that will act as a warning to those with visual impairments but will not impede wheelchair travel.

This section includes recommendations, in addition to those in Sections 1.6.2 and 1.6.3, applicable to those with specific handicaps. The recommendations there apply to accessible spaces for those with specific handicaps, except when more demanding provisions are necessary to accommodate specific disabilities.

1.6.4.1

Visual Impairment

Only about 10% of those with visual disabilities are totally blind. Legal blindness is usually defined as vision at or below 20/200. Some of those with visual disabilities have impaired peripheral visions; others cannot distinguish light and dark; some cannot perceive the full spectrum of colors.

Many people who have a visual impairment rely on memory instead of their eyes. They use long canes, guide dogs, or other devices such as an *infrared detector* as aids in traveling to specific des-

tinations (they often have explored these destinations with the aid of someone with sight). Dwelling units may need little adaptation because those with visual impairments become familiar with their surroundings.

Persons with visual impairments learn to rely to a certain extent on *tactile response* (touch). A *tactile object* is one that can be perceived using the sense of touch.

■ SIGNAGE

The proper type of signage is a very important communication means for visually handicapped people and should follow the rules in Section 1.6.3.6.

■ TEXTURE

A *cue* is a device that alerts a handicapped person to an upcoming condition. Cues may be *audible*, *visual*, or *textural*. A cue that consists of a standarized surface texture applied to or built into a walking surface or other element to warn visually impaired people of a hazard in the path of travel is called a *detectable warning*.

Detectable warnings help those with visual impairments to avoid hazards such as automobile traffic, level changes, and physical obstacles. Detectable warnings on walking surfaces include (1) exposed aggregate concrete, (2) rubber or plastic cushioned surfaces, (3) raised strips, and (4) grooves. Textures used for warning should contrast with surrounding surfaces but be uniform in design within any building or site. Grooves should be used indoors only. Handles on doors leading to hazardous areas (such as spaces housing mechanical equipment) should be knurled or roughened as a warning.

■ OBSTACLES

Benches and other obstacles in the direct path of travel should be surrounded by detectable warnings to aid those with visual impairments, unless the obstacles are encircled by other indicators such as curbs (Fig. 1.6–9).

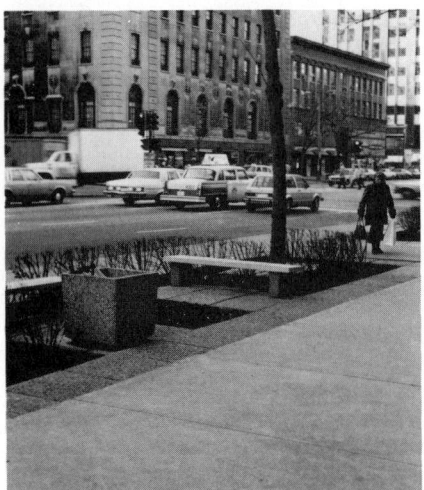

FIGURE 1.6-9 *The textured surface around the planter and the bench serves as a warning to those with visual impairments.*

detectable warnings to aid those with visual impairments, unless the obstacles are encircled by other indicators such as curbs (Fig. 1.6–9).

■ PROJECTIONS

Many people with severe visual impairments use white canes to aid their mobility. However, canes are useful only if hazards are detectable. Refer to Section 1.6.2.5 for requirements related to protruding objects.

■ SOUND

An *audible cue* is a sound or verbal alert and is another means of helping those with visual impairments. *Audible alarms* should be incorporated into emergency warning systems. Alarm signals should be at least 15 decibels (dB) louder than the room's normal sound level, but should not exceed 120 dB.

1.6.4.2

Hearing Impairment

More than 13,000,000 people in the United States have partial or total hearing loss. They need some modifications in their built environment. Just as sound compensates those with visual impairments for the loss of sight, *visual cues*

help those with hearing impairments make up for their loss.

■ WARNING AND INDICATION LIGHTS

Emergency signals, such as fire and burglar alarms, should have *visual* as well as *audible warning systems*. These should be located so that the signal or its reflection is clearly visible and should flash with a frequency of not more than 5 cycles per second (Hz).

Elevator directional lights should be at least 2-1/2 in. (63 mm) in the smallest dimension and should be mounted at least 72 in. (1830 mm) above the floor.

■ SOUND

Control of background noises and amplification of sound can help those with partial hearing loss to communicate.

For hearing aids to function properly, reduction of background noise, including reverberation, and control of sound frequencies, is important. Use of ultra-high-frequency sound security systems and low-cycle electric transformers should be avoided.

Telephones for those with hearing impairments should be fitted with adjustable volume controls and other devices. Such telephones in public areas should be labeled as shown in Figure 1.6-10.

FIGURE 1.6-10 *Signs identifying a special phone for those with hearing impairments.*

1.6.4.3

Mental and Perceptual Impairment

Clear *signage* and good organization are especially important for those who are

mentally immature or who have *perceptual* or *cognitive disabilities*.

If people must frequently ask for the location of a commonly used facility, such as a washroom, that environment may have a barrier. Many individuals with disabilities are self-conscious and will not ask for help; if they feel uncomfortable in a facility, they often will not use it. Public facilities should be easy to locate and to utilize.

■ SIGNAGE

It is especially important to those with mental impairments that signs be easy to locate and to read. The basic requirements for signs are discussed in Section 1.6.3.6.

Many individuals with mental impairments who are unable to read learn to recognize information by word length. For example, the word *women* is a longer than the word *men*. Use of the words *ladies* and *gentlemen* might confuse those familiar with the three- and five-letter words.

■ ORGANIZATION

The location of specific rooms within an area should be clearly identified. If corridors in different areas are similar in appearance, efforts should be made to create *landmarks* using color or other decoration. Stairways and other circulation areas should be clearly identified or in be highly conspicuous locations.

1.6.4.4

Wheelchair Users

The abilities of wheelchair users differ widely. Disability may affect only the legs or the entire body. Pressure sores, muscle atrophy, and blood pooling are frequent problems. Paralysis reduces the muscle tone of the affected area. Barrier-free environments encourage mobility, and the exercise gained helps prevent additional medical problems.

■ WHEELCHAIR ACCESS

Most average adult-sized wheelchairs are of a standard size; some special wheelchairs may be larger. Wheelchairs need an accessible route from parking spaces or bus stops to the interior of a building and within the building to the final destination of the wheelchair user.

SITE ACCESS Refer to Sections 1.6.3.2 and 1.6.3.4 for requirements.

ELEVATORS AND LIFTS Unless every space is accessible by level movement, ramps, or another vertical access means, *elevators*, or *platform lifts* should be provided for wheelchair users. In some applications, *chair-type stair lifts* can be provided to assist persons who cannot readily climb stairs, when the stairs cannot be eliminated, or when *elevators* or *platform lifts* cannot be provided.

ENTRIES AND PASSAGEWAYS Refer to Sections 1.6.2 and 1.6.3.3 for requirements for doors, hardware, and thresholds.

Figure 1.6-11 shows the recommended minimum approach spaces necessary for convenient maneuvering of wheelchairs at doorways.

Wheelchair users frequently use the wheelchair itself to push a door open, scuffing or scratching the door. Therefore, the bottom 10 in. (255 mm) of hinged accessible doors should have a smooth, hard-surface *kickplate* (Fig. 1.6-12).

■ ROOM SIZE AND ARRANGEMENT

Accessible spaces for wheelchair users should be on one level and have adequate circulation and maneuvering space. Windows and doors should be located so that furniture arrangements do not obstruct wheelchair circulation.

The minimum clear floor area that will accommodate a wheelchair is 30 in. by 48 in. (760 mm by 1220 mm). The minimum diameter needed for a wheelchair user to execute a 180-degree turn is 60 in. (1525 mm).

Figure 1.6-13 shows examples of accessible bedroom plans. In an accessible dwelling at least two bedrooms should be on the main level.

■ TOILET ROOMS, SHOWER ROOMS, AND BATHROOMS

Wheelchair users should be able to enter and use toilet, shower, and bathroom

FIGURE 1.6-11 *Minimum clear floor space at doors: (a) front approach, (b) hinge side approach, and (c) latch side approach.*

FIGURE 1.6-12 *An abrasion-resistant kickplate of stainless steel or hard plastic should protect the lower 10 in. of doors intended for wheelchair users.*

facilities without having to squeeze past obstructing fixtures or doorways.

CLEAR FLOOR SPACE Accessible toilet rooms, shower rooms, and bathrooms should have the recommended clearances around plumbing fixtures and controls. Ideally, an accessible facility should have a clear floor area 60 in. (1525 mm) in diameter, so a wheelchair user can make a 180-degree turn. The clear floor spaces of the fixtures, controls, and turning space may overlap.

DOORS Doors should have a minimum clear opening of 32 in. (815 mm). Doors should not swing into the clear floor area of a fixture or obstruct entrance to the room. If a door must swing out, it should not obstruct passage through a corridor or any other circulation space.

WATER CLOSETS Water closets in an accessible public toilet room should be 17 to 19 in. (430 to 484 mm) high measured to the top of the toilet seat. In a dwelling, the seat height should be at least 15 in. (380 mm). The bowl should be centered at least 18 in. (455 mm) from each side wall.

LAVATORIES Lavatories should be mounted so as to provide at least 29 in. (735 mm) of knee space to the underside of the apron. Exposed water pipes that intrude into the knee space should be padded and insulated.

MIRRORS Mirrors should be mounted with the bottom edge not higher than 40 in. (1015 mm) from the floor. If there is a medicine cabinet, it should be mounted with a usable shelf not higher than 44 in. (1120 mm) above the floor.

BATHTUBS AND SHOWERS Depending on the location of the bathroom door, a bathtub should have a clear floor space 30 to 48 in. (760 to 1220 mm) wide and 60 to 72 in. (1525 to 1905 mm) long. Showers should have a clear floor space at least 36 in. (915 mm) wide and 48 in. (1220 mm)

Single bed   Double bed

FIGURE 1.6-13 *Minimum dimensions for accessible bedrooms.*

a

b

FIGURE 1.6-14 *Minimum clearances between opposing cabinets in (a) parallel and (b) U-shaped kitchens.*

long for *transfer stalls* and 60 in. (1525 mm) long for *roll-in stalls.*

Seats should be provided in both transfer shower stalls and tubs. In shower stalls the seat should extend the full depth of the stall on the wall opposite the controls.

A shower spray unit that can be used as either a fixed shower head or a hand-held shower should be provided. The hose should be at least 60 in. (1525 mm) long.

Curbs in shower stalls 36 in. (915 mm) square should be not higher than 4 in. (100 mm). Stalls, such as those 30 in. by 60 in. (760 mm by 1525 mm), that allow a wheelchair to be rolled in; these should not have curbs.

Bathtub and shower enclosures should not obstruct the use of controls or the transfer from wheelchairs onto seats.

FAUCETS   The controls on bathtubs, showers, and lavatories should be within comfortable reach of a wheelchair user. Controls such as handle and drain mechanisms should be operable with one hand and not require tight grasping or twisting.

■ **STORAGE**

A clear floor space at least 30 in. by 48 in. (760 mm by 1220 mm) should be provided near storage areas. Closet rods should be mounted not higher than 54 in. (1370 mm) above the floor.

■ **KITCHENS**

Suitable kitchen appliances, proper

countertop heights, and adequate circulation space are the primary prerequisites of a wheelchair user's kitchen.

CLEAR FLOOR SPACE   The minimum clear floor space between opposing base cabinets, countertops, appliances, and walls should be 40 in. (1015 mm), unless the kitchen is U-shaped. U-shaped kitchens should have 60 in. (1525 mm) of clear floor space between opposing cabinets (Fig. 1.6-14). Clear floor space of 30 in. by 48 in. (760 mm by 1220 mm) should be provided at appliances such as cooktops, ovens, refrigerators, and dishwashers.

CONTROLS   Knobs, faucets, and appliance controls must be within comfortable reach and operable with one hand.

COUNTERS   One 30-in. (760 mm) section of counter, intended for food preparation, should be adjustable to a range of heights, including at least these three positions: 28, 32, and 36 in. (710, 815, and 915 mm).

There should be no cabinets under the adjustable counter, which should have a clear knee space 30 in. (760 mm) wide by 19 in. (485 mm) deep over a finished floor space. Counter thickness and supports should be 2 in. (50 mm) maximum over the necessary clear space. The underside of the counter should not have any sharp protrustions or rough surfaces.

APPLIANCES   Many wheelchair users have problems using appliances designed for standing rather than seated users.

Cooktops should have controls in front; wheelchair users risk fire and burns when they have to reach across hot burners. People with limited arm coordination must drag rather than lift pots and pans. A *cooktop*, if separate from the *oven*, should be adjustable with the same knee space and height dimensions as sinks and work counters. The knee space should be finished like the rest of the floor. To protect a user from injury, the underside of a *cooktop* should be smooth, insulated, and free of sharp projections.

Ovens should have *side-hinged* doors and be *self-cleaning.* Ovens without a *self-cleaning* feature and the *latch side* of those that are *side-hinged* should be located next to a lowered counter area with *accessible* knee space. *Side-hinged ovens* should have a *heat-proof pull-out shelf* below the *oven*, running the full width of the *oven*; the depth of the shelf should not be less than 10 in. (255 mm).

An ideal cooking arrangement consists of (1) a lowered, front-controlled *cooktop*, (2) a lowered *sink*, (3) a lowered *work counter* at least 30 in. (760 mm) wide, and (4) an *accessible oven.*

Refrigerators with *freezers* (side-by-side and some over-under styles) are the most accessible type for wheelchair users. *Over-under refrigerators* should have at least 50% of the *freezer* and most of the *refrigerator* space and the *controls* at a height lower than 54 in. (1370 mm). *Freezers* that have any of their *storage volume* above the 54-in. (1370 mm) reach limit should be *self-defrosting.*

Dishwashers should open in the front, and the racks should be within easy reach. Pull-out racks are the easiest to use.

Washers and *dryers* should be the *front-loading* type and hinged on the side rather than the bottom.

CABINETS Both base and wall cabinets should be reachable from a wheelchair. At least one continuous shelf in the cabinets above a work counter should not be higher than 48 in. (1220 mm). Upper cabinet doors should have pulls located as close to the bottom as possible; base cabinets should have pulls located as close to the top as possible. Because of the necessity for knee space, cabinets should not be built under counters. Instead, pantries should be used for storage. They should have storage below the 54-in. (1370 mm) height and a clear space of 30 in. by 48 in. (760 mm by 1220 mm).

Pull-out shelves in base cabinets and lazy susan corner units are highly recommended.

SINKS Like work counters and cooktops, sinks should be adjustable to 28, 32, and 36 in. (710, 815, and 915 mm). The minimum width of a counter surrounding a sink should be 30 in. (760 mm).

Plumbing should be roughed in for *sinks* mounted at the 28-in. (710 mm) height. *Counter* thickness and supports should be 2 in. (50 mm) maximum over the necessary clear space. A *sink bowl* should be not more than 6-1/2 in. (165 mm) deep; in a *double-bowl sink* only one of the *bowls* needs to be so shallow.

Faucets should be at the side of the *sink* and operated by a *single lever control*.

■ ELECTRICAL

The location of *electrical outlets* and *switches* should be as described in Section 1.6.3.8. *Lighting* requirements are in Section 1.6.2.6.

1.6.4.5

Coordination Disabilities

Coordination disabilities include balance, leg, upper-limb, and flexibility impairments. Many people have problems with balance, stamina, and coordination that do not confine them to a wheelchair but do limit their use of the environment. They may have difficulty walking and may need aids such as walkers or crutches, which need approximately the same amount of space as wheelchairs. Others may have difficulty bending, sitting, kneeling, or rising; still others may find it difficult to push open doors or turn knobs and faucets.

The needs of those with coordination disabilities and the needs of the elderly frequently overlap. However, younger people with disabilities usually do not want to be associated or housed solely with the elderly.

■ ENTRY AND CIRCULATION

People who use walking aids need the same amount of *circulation space* as those who use wheelchairs. Many people with coordination disabilities prefer stairs to ramps because ramps are more tiring to use. However, ramps may be beneficial to crutch and walker users who cannot negotiate stairs as easily. Ramps should be constructed according to the criteria established in Section 1.6.3.4. Wherever possible, both stairs and ramps should be provided.

■ SURFACES

Unevenness, raised joints, and debris can be hazardous to those with walking disabilities. Surfaces should be kept free of debris; joints and level changes should be less than 1/4 in. (6 mm).

1.6.2.6

The Elderly

Accessibility recommendations for the elderly overlap those for wheelchair users and persons with coordination disabilities. Sensory losses accelerate at various ages. For example, hearing starts to diminish at age 40; vision at 50; taste at 55 to 60; and smell after 70. The actual ages and degrees of loss vary considerably.

The elderly have a higher rate of visual impairment than do those in other age groups. Age impedes the eyes' ability to change focus, adjust to sudden changes in light intensity, and function at low levels of light intensity. For example, an 80-year-old might need three times as much light for comfortable reading as does a 16-year-old. As it ages, the lens of the eye becomes increasingly rigid, yellow, and opaque. This process affects the ability to discern color, especially blues and greens, and to distinguish between colors such as pink and lavender. With aging, depth perception decreases, and glare becomes a greater problem.

Aging also increases hearing loss, reducing the ability to hear sounds in general and high-frequency sounds in particular.

The cushioning of the foot degenerates with advancing age. This loss can cause pain and contributes to the shuffling gait characteristic of many elderly people (Fig. 1.6-15).

The mobility of the elderly varies from complete ambulatory freedom to wheelchair confinement. Entries and other access facilities should be barrier-free for elderly persons, including wheelchair users. Only those criteria that add to or differ from those established for wheelchair users will be mentioned in this section.

■ SIGNAGE

Many elderly people suffer visual impairment, and others may be easily confused. Signage and lighting should conform to the recommendations outlined in Sections 1.6.3.6 and 1.6.4.1.

■ ROOM SIZE AND ARRANGEMENT

Recommended room sizes for the elderly are greater than the minimums specified for wheelchair users. This is partly because the elderly sometimes have coordination difficulties that make it difficult for them to maneuver a wheelchair within the minimum spaces recommended for other wheelchair users.

The criteria for corridor widths and floor surfaces should follow those indicated in Sections 1.6.3.3 and 1.6.3.7, respectively.

Pressure point — Pressure point

a

Pressure point — Pressure point

b

FIGURE 1.6-15 *The foot of a younger person (a) contains more cushioning material at pressure points than that of an older person (b).*

■ COLOR AND TONE

Changes in the eye lens caused by aging often affect color perception. It may be difficult for some elderly people to distinguish between the light tints of pastel colors, such as blue, lavender, and pink, and sometimes even between dark colors, such as navy, brown, and charcoal. Therefore, the selection of colors for elderly persons should emphasize objects against their background. Examples include light entry–dark door jamb; and light floor–dark furniture.

■ ELECTRICAL

Electrical outlet and switch locations and lighting requirements should be as described in Section 1.6.3.8.

■ BATHROOMS

Since the elderly may have to use a wheelchair at times, bathroom requirements should be the same as for wheelchair users.

■ KITCHENS

The requirements for kitchens used by the elderly vary. Some elderly persons are reasonably able-bodied; others use wheelchairs. Kitchens for the elderly generally should follow the recommendations for adaptable kitchens (see Section 1.6.5), thus permitting ready alterations to suit the specific needs of the occupants.

Kitchen storage should be accessible without endangering the user's safety. Because not all elderly persons are wheelchair users, the use of wall cabinets is acceptable. However, these are not recommended above a refrigerator or a cooktop. Pantries are an excellent means of providing safe and accessible storage. Base cabinets should have pull-out shelves and corner lazy susans to make items in them more accessible.

1.6.5

ADAPTABILITY

Most people with disabilities resent being segregated into special housing built specifically for the handicapped. They would prefer *mainstreaming*, a process that (1) equips people who have disabilities with the personal devices and adaptive skills needed to function effectively in the built environment, and (2) removes physical barriers that prevent handicapped people from functioning like able-bodied individuals.

Current laws and regulations require that most types of buildings be made fully accessible to handicapped people. Other laws and regulations have been proposed that would mandate making new housing also fully accessible. When such laws and regulations take effect, the following discussion may become moot. Until then, however, one way to prevent the segregation of people who become disabled into housing prepared specifically for handicapped people is to build adaptable buildings. An adaptable building is one that can be modified to satisfy the needs of an occupant's physical limitations with ease and without excessive costs. The elements that may be modified or added in an adaptable house include but are not limited to kitchen counters, sinks, and grab bars.

Adaptability for wheelchair users requires the most modification and therefore should receive the greatest attention in the design process.

The recommendations in Sections 1.6.2 and 1.6.3 apply fully to adaptable buildings. In general, adaptability requires that the following design requirements be followed.

■ ENTRIES

When practicable, entrances should be at grade level, so that ramps or stairs will not be needed. When this is not practicable, an adaptable dwelling can be on a level accessible by a ramp or an elevator or platform lift.

■ DOORS

Passage doors should have a minimum clear opening of 32 in. (815 mm) with the door open 90 degrees, measured from the face of the door to the opposite stop.

HARDWARE Door knobs, handles, pulls, latches, and locks should have a shape that is easy to hold with one hand and that does not require tight grasping, pinching, or twisting. Extensions are available that adapt a regular door knob for people with manual disabilities such as arthritis.

THRESHOLDS Thresholds on exterior sliding glass doors should not be higher than 3/4 in. (18 mm). A recessed track is preferred where suitable, such as for interior sliding doors. Thresholds on hinged passage doors should not exceed 1/2 in. (13 mm) in height and should be beveled. Exterior doors should have weather stripping applied to the bottom edge of the door, because weather stripping applied to thresholds can present a tripping hazard (Fig. 1.6-16).

■ ROOM SIZE AND ARRANGEMENT

A single-level dwelling is the preferred type for adaptability. Rooms should be large enough to permit convenient furniture arrangement and wheelchair maneuvering. Most living rooms, dining rooms, and bedrooms already meet the requirements for adaptability. Bathrooms and kitchens require the most modification.

■ CORRIDORS

Hallways and corridors should be a minimum of 36 in. (915 mm) wide. If a door

½″ 13mm Max.

½″ 13mm Max.

FIGURE 1.6-16 *Thresholds.*

opens into a corridor, the passage width should be 54 in. (1370 mm); corridors where people must pass each other should be 60 in. (1525 mm) wide.

■ BATHROOMS

An adaptable bathroom has the appearance of a conventional bathroom but with the layout and clearances suitable for later adaptation to the needs of those with disabilities.

CLEAR FLOOR SPACE   A typical bathroom should have a minimum width of 60 in. (1525 mm). Clear floor spaces around fixtures should be as shown in Figure 1.6-17.

DOORS   Bathroom doors should have a minimum clear opening of 32 in. (815

mm); doors should not swing into the clear floor space around fixtures or obstruct entrance into the bathroom. Bathroom doors can be rehung to open out to gain the necessary clear door space, but they should not obstruct corridors or other circulation space outside the bathroom.

LAVATORIES   Lavatories should be mounted to provide at least a 29-in. (735 mm) clearance from the floor to the underside of the apron. The clear floor space should be at least 30 in. (760 mm) wide by 48 inches (1220 mm) deep; the lavatory basin should take no more than 19 in. (485 mm) of that depth, and pipes should not protrude into the knee space.

WATER CLOSETS   Toilet seats should be at least 15 in. (380 mm) above the floor. Clear floor space and structural wall reinforcement for grab bars should be provided.

MIRRORS   The bottom edge of the mirror should be mounted not higher than 40 in. (1015 mm) above the floor. If there is a medicine cabinet, it should be mounted with at least one usable shelf not higher than 44 in. (1120 mm) above the floor.

BATHTUBS AND SHOWERS   Depending on the location of the bathroom door, bathtubs and showers should have the adequate clear floor space to accommodate wheelchairs.

Structural wall support and floor space should be provided so that grab bars and

a seat can be readily installed for a resident with a handicap. A flexible-hose shower unit that can be used as either a fixed or a hand-held head should be provided. Shower stalls 36 in. by 36 in. (915 mm by 915 mm) should have curbs not higher than 4 in. (100 mm). Stalls 30 in. by 60 in. (760 mm by 1525 mm) should not have curbs, so a wheelchair may be rolled in; the stall floor should be pitched slightly toward the drain to help avoid drainage problems.

GRAB BARS   Structural reinforcement or other provisions should be provided near bathroom fixtures in the locations where grab bars would be installed.

■ KITCHENS

Adjustable-height counters, sinks, and cooktop and adequate *circulation space* are the key elements of an adaptable kitchen that allow a wheelchair user to maneuver freely within the space.

CLEAR FLOOR SPACE   The minimum clearance between opposing base cabinets, countertops, appliances, and walls should be 60 in. (1525 mm) in U-shaped kitchens (see Fig. 1.6-14) and 40 in. (1015 mm) in other kitchens (see Fig. 1.6-14). Clear floor space of at least 30 in. by 48 in. (760 mm by 1220 mm) should occur at major appliances.

COUNTERS   One section of counter 30 in. (760 mm) wide should be capable of being repositioned to heights of 28, 32, and 36 in. (710, 815, and 915 mm), to provide a comfortable area for food preparation. Cabinets may be mounted under the adjustable counter, but they should be readily removable to provide the necessary knee space for a wheelchair user.

APPLIANCES   Appliances should fulfill the following requirements for adaptability.

Stoves should have controls in the front. Cooktops, if separate from the oven, should be adjustable to the same dimensions as the work counter. Ovens should have side-hinged doors with a pull-out shelf underneath and should be self-cleaning. Another desirable feature is to position the oven adjacent to an adjustable counter.

FIGURE 1.6-17 *Recommended minimum dimensions for accessible bathrooms.*

Refrigerators with freezers should have at least 50% of the freezer and most of the refrigerator space below 54 in. (1370 mm) in height.

Dishwashers, washers, and dryers should be front loading.

CABINETS Wall cabinets should have pulls located as close to the bottom as possible; base cabinets should have pulls located as close to the top as possible.

Cabinet pulls should be smooth and rounded, without any sharp edges or pointed projections. Because storage under adjustable spaces may have to be removed, pantry storage is desirable. Other helpful devices are lazy susans for corner cabinets, pull-out shelves in base

cabinets, and cutting boards with mixing bowl cutouts.

SINKS Like work counters and cooktops, sinks should be adjustable to several heights: 28, 32, and 36 in. (710, 815, and 915 mm). Plumbing should be roughed in for the 28-in. (710 mm) height. A sink bowl should be not more than 6-1/2 in. (165 mm) deep; in a double-bowl sink only one bowl needs to meet this requirement. Cabinets may be mounted under the sink but should be removable, to provide the necessary knee space for a wheelchair user.

FLOORS Flooring should be nonslip even when wet.

■ ELECTRICAL

Lighting should be as described in Section 1.6.2.6. Electrical outlet and switch locations should be as indicated in Section 1.6.3.8.

■ COST

When practicable, new dwellings should be made adaptable, because the cost of making adaptable dwellings suit the needs of wheelchair users is small compared to the cost of retrofitting existing conventional houses. In addition, making houses adaptable does not in any way harm their use by the able-bodied.

1.7 METRICATION

Metrication is a term that was coined in Britain to describe *metric conversion*, which is the process of changing from the *customary* to the *metric* system of measurement, including the planning and coordination necessary for the change.

The *customary* system of measurement referred to here is the collection of English and other nonmetric units currently used in the United States. The *metric* system is a system originally developed in France that has been adopted by more than 90% of the nations of the world. It is based on the *metre* and six other standardized *base units*. *Metric units* already in general use include, but are not limited to, the following: *second, ampere, candela, watt, volt, ohm, farad, coulomb,* and *lumen*.

Much of the material in this section is based on the American Society for Testing and Materials (ASTM) Standard E 621-84, "Standard Practice for the Use of

Metric (SI) Units in Building Design and Construction," which was originally based on National Bureau of Standards (NBS) (now the National Institute of Standards and Technology) publication 938, "Recommended Practice for the Use of Metric (SI) Units in Building Design and Construction." NBS 938, which was authored by world renowned metrication expert Hans J. Milton while he was on loan to the NBS by the Australian government, has been withdrawn in favor of ASTM E 621. ASTM E 621 is an ASTM Committee E-6 supplement to ASTM E 380-89a, "Standard for Metric Practices," on which other portions of this section are based.

This section describes the history of measurement and the metric system, the properties of the *International System of Units* (SI) version of the metric system, and its applications and proper usage. Appendix A, Tables, includes detailed tables

that show lists of units in the SI system; rules and recommendations for the presentation of SI units and symbols; presentation of numerical values in conjunction with SI; application for SI quantities and associated names and symbols; fraction, decimal, and metric equivalents tables; and conversion tables for selected customary and SI measurement units.

If the events in the rest of the world can be used as a guide, conversion to the metric system of measure in this country will probably change the entire measurement base of the construction industry, including guidelines, codes, standards, drawings, specifications, and associated documents. The most extensive use of metric measure in the building industry will be for linear measurement, area, volume, and mass.

Metrication disturbs some people. Manufacturers worry about the cost of retooling and introducing new product sizes. People in general worry about learning many new names and numbers. As its critics point out, metrication will force a complete revision in the accepted units of measurement and calculation, but at the same time it will offer a logical opportunity to revise standard product sizes, performance criteria, and methods of testing. That in turn will reduce inventory, rationalize criteria, and simplify methods of testing and reporting test results.

Experience in other countries has shown that metrication takes less time and costs less than estimated. The cost of metrication occurs once, but the benefits last forever. Approximately one-third of the contracts awarded to the top U.S. construction firms have been for work outside the United States. American know-how is one of the exports not usually figured into foreign trade discussions. The change to the metric system of measure will help the United States maintain a role of technical leadership and keep it from lagging behind the metric world.

1.7.1

THE HISTORY OF MEASUREMENT

Measurement is a comparison between an unknown quantity and a known quantity. It is also a ratio of how many times a known thing can be divided into something unknown. Measurement tells how large, how many, how fast, how much of almost anything imaginable.

A unified system of measurement is a means of communication. Throughout history, various number systems and representational symbols have been used. Though remnants exist of systems based on the numbers 5, 12, and 20, only systems based on the numbers 2, 10, and 60 are still in wide use. The metric system uses the base-10 (*decimal*) system of numbers.

It took many centuries to develop accurate units of measurement. The first were based on parts of the body. People used the human hand or the human foot (Fig. 1.7-1); they also used the length of a furrow made by a plow and the distance a man could walk in a day. Since it is harder and takes longer to walk uphill than downhill, the Chinese even made an uphill mile shorter.

All these forms of measurement were inexact. The size of a human hand or foot varied among individuals. Different people plowed different furrows and walked different distances in a day. There was a need for a unified standard system of weights and measures—one that everybody could agree on.

The Romans established a unified system of weights and measures. They made a bronze rod of specific length and called it a *pes*, Latin for foot; they made a bronze weight called a *libra*, Latin for pound. Both the *pes* and the *libra* were kept under guard in a temple and used as standards of measurement throughout the Roman Empire. The U.S. customary abbreviation for pound, lb., is derived from the Latin *libra*.

After the fall of the Roman Empire, units of measurement once again became localized, and confusion existed when different regions tried to communicate numerically.

1.7.1.1

Noses, Thumbs, and Barleycorns

Because local measurements were often based on the king's body measurements, they differed from country to country and depended on the size of the current king (Fig. 1.7-2).

The U.S. customary system uses the *yard* as a standard unit of measurement. Legend says that King Henry I of England established by royal decree the length of the *yard* as 36 in., which was the distance from Henry's nose to his thumb (Fig. 1.7-3).

Another arbitrary and unrelated unit of the customary system is the system of 13s, which is used to measure shoes. It is based on the size of a barleycorn, which is about one-third of an inch. Shoe sizes vary about one-third of an inch from size to size. A size 7 shoe, for example, is 1 in. smaller than a size 10.

1.7.1.2

Number Systems in Current Use

There are several numbering systems in current use around the world.

■ SEXAGESIMAL SYSTEM

The *sexagesimal* system is based on the number 60. First used by the Babylonians, it survives in the customary system in the subdivision of time and angular measurement. For instance, an hour has 60 minutes, and a circle has 360 degrees. Both minutes and degrees can be subdivided into 60 seconds.

■ THE BINARY SYSTEM

The *binary* system is based only on the numerals 0 and 1; it is used in electronic computers, because it can represent yes/no or true/false decision logic.

| THE PALM | THE HAND | THE SPAN | THE FOOT |
| 3" | 4" | 9" | 12" |

FIGURE 1.7-1 *Measurements based on parts of the human body.*

FIGURE 1.7-2
Cultural Variability of the Foot as a Unit of Linear Measure

Country	Local Name	Metric Equivalent (mm)
Mexico	Pie	279
Netherlands	Voet	283
Poland	Stopa	288
Sweden	Fot	296
Switzerland	Fuss	300
Great Britain	Foot	305
Germany	Rheinisher Fuss	313
Denmark	Fod	314
Austria	Wiener Fuss	316
Norway	Fod	318
France	Pied	325
Italy	Piede Librando	513

FIGURE 1.7-3 *King Henry I inventing the yard.*

FIGURE 1.7-4
Preferred and Nonpreferred Multiples, Submultiples, and Prefix Names for Some Powers of 10

Multiplication Factor		Prefix Name	Symbol	Pronunciation
Preferred				
10^{12} or	1 000 000 000 000	tera	T	as in *terrace*
10^{9} or	1 000 000 000	giga	G	*jig'a*
10^{6} or	1 000 000	mega	M	as in *megaphone*
10^{3} or	1 000	kilo	k	as in *kilowatt*
10^{-3} or	0.001	milli	m	as in *military*
10^{-6} or	0.000 001	micro	μ	as in *microphone*
10^{-9} or	0.000 000 001	nano	n	*nan'oh*
10^{-12} or	0.000 000 000 001	pico	p	*peek'oh*
Nonpreferred				
10^{2} or	100	hecto	h	*heck'toe*
10^{1} or	10	deka	da	*deck'a*
10^{-1} or	0.1	deci	d	as in *decimal*
10^{-2} or	0.01	centi	c	as in *sentiment*

THE DECIMAL SYSTEMS

The *decimal* system, which is based on the number 10, is the system on which the metric system is founded. It was probably derived from the early use of human fingers to count. The Greeks and Romans first used a decimally based word structure for numbers, then adopted letter numerals for their representation. Roman numerals, which were used exclusively in Europe until about A.D. 1100, were gradually replaced by Arab numerals of Hindu-Arabic origin, which were much simpler to use.

The *Roman numeral* system consists of *quinary* (base-5) symbols: I equals 1; V equals 5; X equals 10; L equals 50; C equals 100; D equals 500; M equals 1000. In-between numbers are formed by addition to or subtraction from the seven basic symbols. When a smaller number precedes a larger, it is subtracted; when the smaller follows the larger number, it is added. Thus, 1978 is written MCMLXXVIII: C before M means 900; all the other numbers retain their absolute values and are added.

The *Hindu-Arabic* system we use today expresses all numbers by combinations of 10 symbols, each having an *absolute* value and a *position* value (Fig. 1.7-4).

Decimal position notation relates each number's position value as a power of 10. For example, 10^1 equals 10, 10^2 equals 100, 10^3 equals 1000. Numbers to the right of the decimal are expressed in negative powers of 10. For example, 10^{-1} equals 0.1, 10^{-2} equals 0.01, and 10^{-3} equals 0.001. Each multidigit number is written as a linear combination of powers of 10. For example,

$$456 = 4 \times 10^2 + 5 \times 10^1 + 6$$
$$37.5 = 3 \times 10^1 + 7 + 5 \times 10^{-1}$$

1.7.2

THE METRIC SYSTEM

The scientific evolution of the 18th and 19th centuries, the industrial revolution of the 19th century, and the desire for more international trade accelerated the movement toward a universal system of weights and measures. In 1795, under pressure from leading scientists of the era, the French National Assembly adopted the metric system of measure.

The French metric system was based on a standard unit of length called a *metre* (not meter). The word *metre* was derived from the Greek word *metron*, meaning measure. The French metre was expressed in geodetic terms, that is, related to the size and shape of the earth. A metre was the 10-millionth part of an imaginary line running through Paris from the equator to the North Pole. Eventually that distance was marked on a platinum-iridium bar and became the international standard (prototype). It was kept at the International Bureau of Weights and Measures in France.

This system, as does its modern equivalent, established reference values, called *units*, for each measurable attribute of a physical phenomenon, which is called a *quantity*. A unit, then, was the reference value of a given quantity. For example, the metre was the unit for the given length (quantity) established by the prototype.

Additional units in the system included those for the quantities of *area*, *volume*, *capacity*, and *mass*. These quantities were related decimally to the metre through the properties of water. For example, 1 *litre* represented 1/1000 of 1 cubic metre. When filled with water at its maximum density, a *litre* had a mass (weight) of 1000 *grams*, or 1 *kilogram*.

The system also contained a set of decimal prefixes that could be attached to reference units to alter their magnitude. For example, *milli* meant 1/1000, and *kilo* meant 1000 times.

In 1881, the International Electrical Congress adopted a unit of time to produce the *centimetre-gram-second* (cgs) system. About 1900, metric measurements began to be based on the *metre-kilogram-second* (MKS) system. In 1935, the *ampere* became the fourth *base unit*, resulting in the *metre-kilogram-second-ampere* (MKSA) system. The *General Conference on Weights and Measures* (CGPM), an international treaty organization, added the *kelvin* as a unit of temperature and the *candela* as the unit of *luminous intensity*.

All three metric systems were in use in different countries, making international trade and conversation difficult, even though many nations were using a form of the metric system of measurement.

1.7.3

THE INTERNATIONAL SYSTEM OF UNITS (SI)

In 1960, to help further unify the system of measurement internationally, the CGPM formally introduced a new metric system called the *International System of Units*, the abbreviation of which is SI in all languages. The letters SI represent the French *Système International d'Unités*. SI was derived from earlier decimal metric systems but supersedes them.

In SI, in response to demands by scientists and technologists for a means of precisely defining the metre in a way that could be accurately reproduced without having to refer to the prototype, the CGPM introduced a new definition based on the electromagnetic emission of a certain *krypton* atom.

1.7.3.1

Adoption of SI

SI has been adopted by every country that has changed to metric measurement since 1960. Countries that were using the cgs, MKS, or MKSA system before 1960 are changing those units that have been superseded by the SI system. The decimal metric system is now used in more than 90% of the world's nations. Paradoxically, the United States, the first nation to introduce a system of decimal currency in 1785 and an original party of the 1875 *Treaty and International Metric Convention* that established the CGPM, is one of the last nations to adopt SI. It shares this dubious distinction with only a few small countries.

The CGPM meets at least every six years. It controls the *International Bureau of Standards*, which preserves the metric standards and compares national standards with them. The CGPM also conducts research to establish new standards. The U.S. *National Institute for Standards and Technology* (NIST) represents the United States in these activities.

The metric system was legalized by Congress in 1866, but it was not made mandatory. Renewed interest at about the time SI appeared in the 1960s led to publication of the *ASTM Metric Practice Guide*, which in 1976 became ASTM E-380, "Standards for Metric Practice." Meanwhile, the government, responding to Public Law 90-472, signed in 1968, conducted a survey to determine the impact, desirability, feasibility, implications, and difficulties of converting the United States to the metric system of measurement.

At about the same time, the American National Standards Institute (ANSI) established a Metric Advisory Committee, and in 1973, in response to the new-found interest in conversion, ANSI established an independent group called the American National Metric Council (ANMC) to provide assistance in the conversion to metric measurement through coordination, planning, and information services to its membership and all segments of society in the United States. Three years later, ANMC separated itself from ANSI, becoming a totally independent, private, and self-supporting national organization.

In 1975, in response to the completed study required by Public Law 90-472, Congress passed Public Law 94-168, the Metric Conversion Act. It stated that voluntary metrication was to be national policy and established the U.S. Metric Board to plan, publicize, and implement a voluntary change to the metric system based on SI. The results of this watered-down law were predictably less than were needed to make conversion a reality. A timetable for the changeover to SI units was established by the construction industry. The target date for metric conversion, called M-Day, was January 1, 1985, which came and went with little accomplished.

Then, with enactment of the Omnibus Trade and Competitiveness Act of 1988, metrication was again off and running. This law stated that the metric system is the "preferred system of weights and measures for United States trade and commerce" and required that, when it is practicable and economically feasible, federal agencies use the metric system when making grants and procurements and in other business-related activities by the end of fiscal year 1992. Implementation is the responsibility of the Department of Commerce. Citing the Omnibus Trade and Competitiveness Act of 1988, the Department of Commerce announced on January 2, 1991, that metrication is no longer voluntary for federal agencies.

Accordingly, by the middle of 1991 the various federal agencies began to make plans to convert. It remains to be seen whether states and local governments will follow, but their response to other federal government initiatives indicates that they will. The private sector will not be far behind.

1.7.3.2

The Structure of SI

A lifetime of using the customary system gives people a feel for the length of an inch or a foot. New associations or recognition points need to be developed to be comfortable using SI (Fig. 1.7-5). This conversion in thinking happens more quickly than most people expect. It can be equated with the relative ease of learning a foreign language when one is living in the nation of its origin as opposed to learning the same language in a classroom far removed from its general use. In addition, many SI and related units are already in our vocabulary, for example, *volt, ampere, watt, ohm, second,* and *lumen.*

FIGURE 1.7-5
Some Metric Recognition Points

Quantity	Unit	Description
Linear measurement	1 mm	Approximate diameter of a paper clip wire
	25 mm	Vertical dimension of an ordinary U.S. postage stamp; nearest equivalent to 1 in.
	100 mm	International cigarette length; basic metric building module; nearest metric preferred dimension to 4 in. (1.6% less)
	600 mm	Height of three courses of concrete blocks including mortar joints
	2 m	Approximate height of a standard door opening (6 ft. 8 in.)
Area measurement	500 mm^2	Face area of an ordinary U.S. postage stamp [20 mm × 25 mm]
	1 m^2	1 square metre: approximate area of a shower base
Volume measurement	1 L	1 litre: 5.7% more than a U.S. quart; new soft drink bottle size
Mass [weight]	1 g	Approximate weight of a paper clip or a dollar bill; artificial sweetener package size
	5 g	mass [weight] of a nickel [5¢ coin]
	1 kg	1 kilogram: the base unit of mass in SI; mass of water in a cube with 100 mm sides (approximately 2.2 lb.)
Temperature	0 °C	Freezing point of water (32 °F)
	20 °C	Comfortable thermostat setting in winter (68 °F)
	37 °C	Normal body temperature (98.6 °F)
	100 °C	Boiling point of water (212 °F)

The application of SI units and new *preferred* numeric values (see Section 1.7.6) will simplify measurements in construction and reduce errors in calculations.

Regardless of the surrounding language or script, SI symbols have an agreed-upon form and the same meaning worldwide. There is only one recognized SI unit for each physical quantity. All units are derived from seven *base* and two *supplementary* units (Fig. 1.7-6). The entire system consists of *base units*, *supplementary units*, and *derived units*.

SI is a universal, rational, coherent, and preferred measurement system that uses the base-10 (decimal) system of numbers. Relations between the units in the system contain as a numerical factor only the number 1, or "unity," because all derived units have a unity relationship to the constituent base and supplementary units. For example, 1 litre is equal to 1/1000 of 1 cubic metre (0.001 m^3) and (when filled with water) has a mass of 1 *kilogram*.

■ BASE UNITS

SI has fewer units than the customary system. One unit of length, the metre (m), and its decimally related multiples and submultiples, such as kilometre (km) and millimetre (mm), will replace a variety of customary units. The replaced units include the *mile, furlong, chain, link, rod, fathom, yard, foot, hand, inch,* and *mil*. In addition, the change to the *metre* will correct the difference between the *standard foot* and the *survey foot* (which is two parts per million longer) by replacing both.

SI originally addressed the base units for only six physical quantities: *length, mass* (which is closely related to weight), *time, temperature, electrical current,* and *luminous intensity*. In 1971, the growth of thermal dynamics, nuclear physics, and electronic science caused the addition of a seventh quantity, *amount of substance*. Figure 1.7-7 compares the names for the base units for these seven quantities in SI and in the customary system.

The base units shown in Figure 1.7-8 are the building blocks of SI. Except for the kilogram, they are reproducible under controlled conditions anywhere (Fig. 1.7-9). The *kilogram* is a platinum-iridium *prototype* kept by the International Bureau of Weights and Measures in Sevres, France.

■ SUPPLEMENTARY UNITS

In addition to the seven base units, SI has two supplementary units, the *radian* (plain angle) and the *steradian* (solid angle).

■ DERIVED UNITS

All other quantities are derived from the seven base and two supplementary units. *Derived units* are formed by combining base and supplementary units and other derived units according to algebraic relations linking corresponding quantities (Fig. 1.7-10). For example, the metre is a base unit, the radian is a supplementary unit, and the pascal is a derived unit.

Standard prefixes allow a change in magnitude from the subatomic scale (10^{-18}) to the astronomic scale (10^{18}) (see Fig. 1.7-4a). The prefixes in Figure 1.7-4b may also be used, but are not preferable. Use of multiples of SI units that cannot be handled by using the SI prefixes is discouraged.

Certain other units may also be used with SI.

■ NON-SI UNITS FOR USE WITH SI

To preserve the advantages of SI as a *coherent* system, the use of units from other systems should be minimized. There are, however, acceptable but noncoherent traditional units retained for use with SI because of practical significance in general application. Non-SI units accepted for general use with SI include units for volume (*litre*), mass (*metric ton*), time (*minute, hour, day, year*), plane angle (*degree of arc*), and velocity (*kilometer per hour*). Three other non-Si units are accepted for limited use. These include units for area (*hectare*), energy (*kilowatt-hour*), and speed of rotation (*revolutions per minute*). These quantities and their associated units, along with their symbols and their relationships to SI units, are given in Figure A-2 in Appendix A.

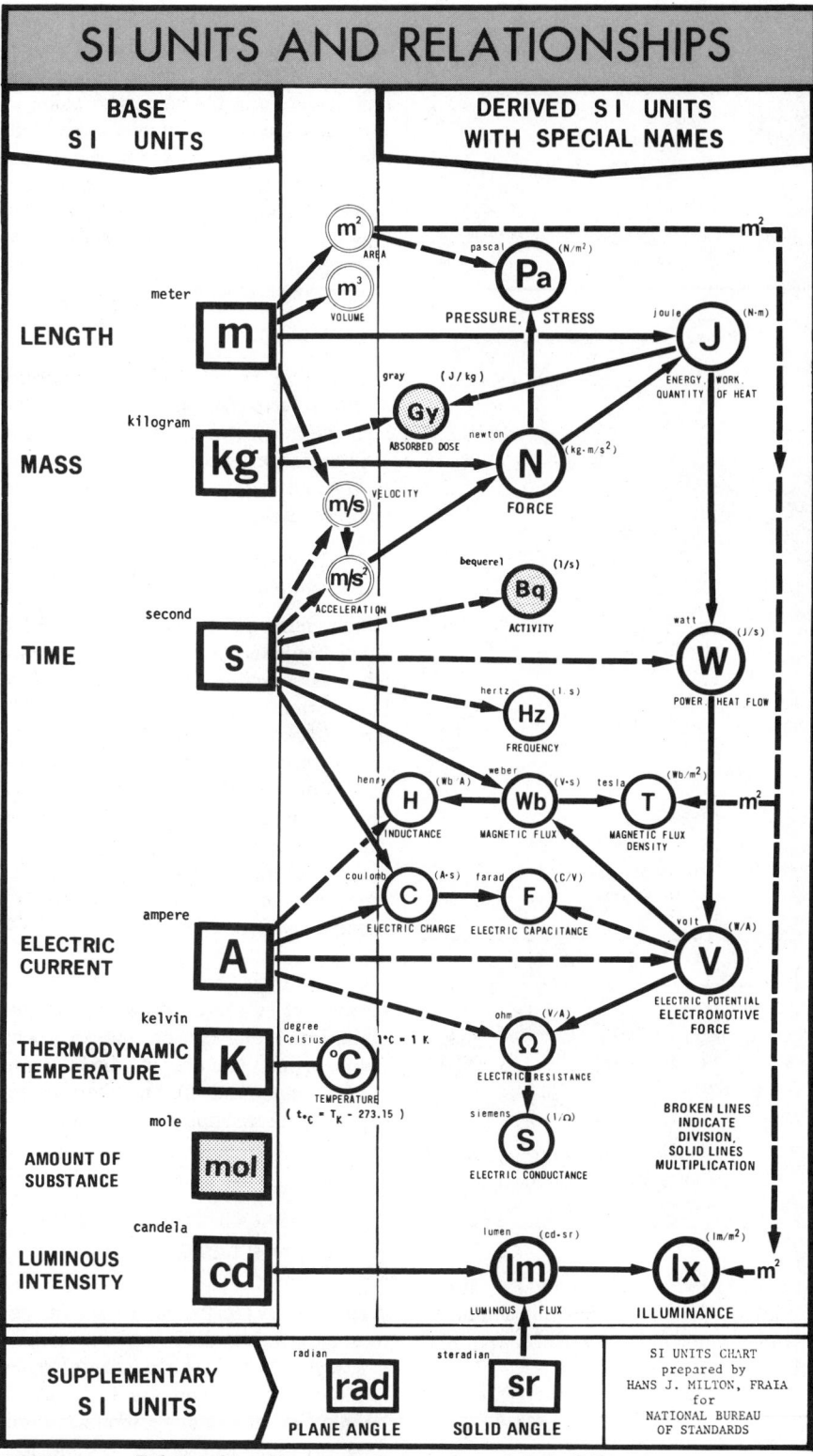

SI UNITS AND RELATIONSHIPS

BASE SI UNITS

DERIVED SI UNITS WITH SPECIAL NAMES

LENGTH — meter — **m**

MASS — kilogram — **kg**

TIME — second — **s**

ELECTRIC CURRENT — ampere — **A**

THERMODYNAMIC TEMPERATURE — kelvin — **K**

AMOUNT OF SUBSTANCE — mole — **mol**

LUMINOUS INTENSITY — candela — **cd**

m² AREA
m³ VOLUME
pascal (N/m²) **Pa** PRESSURE, STRESS
joule (N·m) **J** ENERGY, WORK, QUANTITY OF HEAT
gray (J/kg) **Gy** ABSORBED DOSE
newton (kg·m/s²) **N** FORCE
m/s VELOCITY
m/s² ACCELERATION
bequerel (1/s) **Bq** ACTIVITY
watt (J/s) **W** POWER, HEAT FLOW
hertz (1·s) **Hz** FREQUENCY
henry (Wb·A) **H** INDUCTANCE
weber (V·s) **Wb** MAGNETIC FLUX
tesla (Wb/m²) **T** MAGNETIC FLUX DENSITY
coulomb (A·s) **C** ELECTRIC CHARGE
farad (C/V) **F** ELECTRIC CAPACITANCE
volt (W/A) **V** ELECTRIC POTENTIAL ELECTROMOTIVE FORCE
degree Celsius 1°C = 1 K **°C** TEMPERATURE (t°C = T_K - 273.15)
ohm (V/A) **Ω** ELECTRIC RESISTANCE
siemens (1/Ω) **S** ELECTRIC CONDUCTANCE
lumen (cd·sr) **lm** LUMINOUS FLUX
(lm/m²) **lx** ILLUMINANCE

BROKEN LINES INDICATE DIVISION, SOLID LINES MULTIPLICATION

SUPPLEMENTARY SI UNITS

radian **rad** PLANE ANGLE

steradian **sr** SOLID ANGLE

SI UNITS CHART prepared by HANS J. MILTON, FRAIA for NATIONAL BUREAU OF STANDARDS

FIGURE 1.7-6 *Relationship of base, supplementary, and derived SI units.* (*Copyright ASTM, reprinted with permission.*)

FIGURE 1.7-7
Customary and SI Base Units

Quantity	Customary	SI
Length	foot	metre (m)
Mass (weight)	pound	kilogram (kg)
Time	second	second (s)
Temperature	degree Fahrenheit	kelvin (K)
Electric current	ampere	ampere (A)
Luminous intensity	candela	candela (cd)
Amount of a substance		mole (mol)

FIGURE 1.7-8
Base Units in SI

Quantity	Unit	Symbol
Length	metre	m
Mass	kilogram	kg
Time	second	s
Electric current	ampere	A
Thermodynamic temperature	kelvin	K
Amount of substance	mole	mol
Luminous intensity	candela	cd

Copyright ASTM, reprinted with permission.

■ METRIC UNITS TO BE ABANDONED

Many metric units in use before SI was adopted have been abandoned. SI supersedes all previous metric systems.

Centigram-gram-second (cgs) units such as the *abampere* and the *statvolt* have been superseded by SI and are to be avoided.

Decimal multiples of SI units that cannot be handled by using the SI prefixes should not be used. For example, the *angstrom* (0.1 mm) should not be used.

■ SPECIAL NAMES

Except for *litre*, *metric ton*, and *hectare*, special names or multiples and

FIGURE 1.7-9
SI Technical Definitions

Base Units

METRE

The length equal to 1 650 763.73 wavelengths in vacuum of the radiation corresponding to the transition between the levels $2p_{10}$ and $5d_5$ of the krypton-86 atom.

KILOGRAM

The unit of mass, it is equal to the mass of the international prototype of the kilogram. The kilogram is the only base unit whose name for historic reasons contains a prefix.

SECOND

The measure of time with a duration of 9 192 631 770 periods of the radiation corresponding to the transition between the two hyperfine levels of the ground state of the cesium-133 atom.

AMPERE

The measure of current flow equal to that flow that, if maintained in two straight parallel conductors, would produce between those conductors a force equal to 2×10^7 newtons per metre of length. The conductors are assumed to be of infinite length, of negligible circular cross section, and placed 1 metre apart in vacuum.

KELVIN

The unit of thermodynamic temperature equal to the fraction 1/273.16 of the thermodynamic temperature of the triple point of water.

The temperature interval of the degree Celsius and the kelvin are identical and the scales using these two units are related by a diference of exactly 273.15 kelvins.

MOLE

The amount of substance of a system that contains as many elementary entities as there are atoms in 0.012 kilogram of carbon-12. When the mole is used, the elementary entities must be specified and may be atoms, molecules, ions, electrons, other particles, or specified groups of such particles.

CANDELA

The luminous intensity in the perpendicular direction of a surface of 1/600 000 square metre of blackbody at the temperature of freezing platinum under a pressure of 101 325 newtons per square metre.

Supplementary Units

RADIAN

The plane angle between two radii of a circle that cut off on the circumference an arc equal in length to the radius. The radian is equal to $360/2\pi$ and is an angle of approximately $57° 17' 44.6''$.

STERADIAN

The solid angle that, having its vertex in the center of a sphere, cuts off an area of the surface of the sphere equal to that of a square with sides of length equal to the radius of the sphere. The total solid angle of a sphere is 4π steradians.

---

submultiples of SI units should be avoided, Figure 1.7-11 shows examples of units to be avoided, together with their SI equivalents.

1.7.3.3

Recommendations for the Use of SI

SI symbols are internationally agreed upon and many rules exist for their proper use. Refer to Figures A-3 and A-4 in Appendix A for details. SI units should be written following the style conventions shown there.

■ WRITING UNIT SYMBOLS

In general, unit symbols should (1) be printed in upright type regardless of the type style of the surrounding text; (2) be unaltered in the plural; (3) not be followed by a period except when used at the end of a sentence; (4) be written in lowercase unless the unit is derived from a proper name; and (5) have a space between the unit symbol and the numerical value.

Because a comma is used in some countries as a decimal marker, it should not be used to separate digits into groups of three. Instead, digits should be divided into groups of three by using spaces at least equal to the width of the letter i to separate the groups. In four-digit numbers a space is not necessary. Examples are 0.133 47, 2.141 596, 73 722, and 7372.

The word *billion* means a *thousand million* (prefix *giga*) in the United States, but a *million million* (prefix *tera*) in many other countries. Therefore, terms such as *billion* and *trillion* should not be used in technical writing.

■ PREFERRED RANGE OF VALUES

Whole numbers between 1 and 1000 should be used with the proper prefix wherever possible. For example, 725 m is preferred over 0.725 km or 725 000 mm. However, the prefix that adequately covers the whole quantity range should be used. For example, if 725 m is part of a group of numbers shown in kilometres, it should be expressed as 0.725 km.

Exponential notation, instead of prefixes, is preferred in calculations. For example, 900 mm^2 equals 0.9×10^{-3} m^2.

One measurement unit should be used throughout a set of drawings so that numerical values can be represented by numbers only. In a drawing dimensioned in millimetres, five-digit numbers are acceptable, for example, 32 845 mm.

1.7.4

SI UNITS FOR DESIGN AND CONSTRUCTION

There are two principal factors in any measurement statement: a *number*, which is always written first, and a *measurement unit*. Correct selection of units for use in building construction is essential to minimize errors and optimize coordination. Figure 1.7-12 lists SI units needed for building construction in terms of activity.

FIGURE 1.7-10
Derived SI Units with Special Names

Physical Quantity	Unit	Symbol	Definition
Absorbed dose	gray	Gy	The energy imparted by ionizing radiation to a mass of matter corresponding to one joule per kilogram.
Activity	becquerel	Bq	The activity of a radionuclide having one spontaneous nuclear transition per second.
Electric capacitance	farad	F	The capacitance of a capacitor between the plates of which there appears a difference of potential of one volt when it is charged by a means of electricity equal to one coulomb.
Electric conductance	siemens	S	The electric conductance of a conductor in which a current of one ampere is produced by an electric potential difference of one volt.
Electric inductance	henry	H	The inductance of a closed circuit in which an electromotive force of one volt is produced when the electric current in the circuit varies uniformly at a rate of one ampere per second.
Electric potential difference, electromotive force	volt	V	The unit of electric potential difference and electromotive force that is the difference of electric potential between two points of a conductor carrying a constant current of one ampere, when the power dissipated between these points is equal to one watt.
Electric resistance	ohm	Ω	The electric resistance between two points of a conductor when a constant difference of potential of one volt, applied between these two points, produces in this conductor a current of one ampere, this conductor not being the source of any electromotive force.
Energy	joule	J	The work is done when the point of application of a force of one newton is displaced a distance of one metre in the direction of the force.
Force	newton	N	The force that, when applied to a body having a mass of one kilogram, gives it an acceleration of one metre per second squared.
Frequency	hertz	Hz	The frequency of a periodic phenomenon of which the period is one second.
Illuminance	lux	lx	The illuminance produced by a luminous flux of one lumen uniformly distributed over a surface of one square metre.
Luminous flux	lumen	lm	The luminous flux emitted in a solid angle of one steradian by a point source having a uniform intensity of one candela.
Magnetic flux	weber	Wb	The magnetic flux that, linking a circuit of one turn, produces in it an electromotive force of one volt as it is reduced to zero at a uniform rate in one second.
Magnetic flux density	tesla	T	The magnetic flux density given by a magnetic flux of one weber per square metre.
Power	watt	W	The power that gives rise to the production of energy at the rate of one joule per second.
Pressure or stress	pascal	Pa	The pressure or stress of one newton per square metre.
Quantity of electricity	coulomb	C	The quantity of electricity transported in one second by a current of one ampere.

Figures A-5 through A-10 in Appendix A contain lists of recommended units for use in building design and construction in terms of quantity and application.

1.7.4.1

Linear Measurement

The preferred SI units for linear measurement in building design and construction are the millimetre (mm) and the metre (m). These have an inherently greater precision than units in the customary system, which uses decimalized fractions in site work but feet, inches, and fractions in construction work. Using the millimetre as the principal unit of construction measurement makes fractions unnecessary. One millimetre provides an accuracy of 1/25 in.; five digits can express any length up to 328 ft. The use of whole numbers greatly speeds calculations and helps to avoid errors; calculations can be done twice as fast with only one-fifth the rate of error (Fig. 1.7-13). In the customary system fractions have to be reconciled to a common denominator and then added, inches have to be added and then converted to feet and inches, and finally, the feet must be added.

FIGURE 1.7-11
Special Names to Be Avoided

Unit	Designation and Equivalent
fermi	$1 \text{ fm} = 10^{-15} \text{ m}$
micron	$1 \ \mu\text{m} = 10^{-6} \text{ m}$
millimicron	$1 \text{ nm} = 10^{-9} \text{ m}$
are	$1 \text{ dam}^2 = 100 \text{ m}^2$
gamma*	1 nT
γ (mass)	$1 \ \mu\text{g}$
λ (volume)	$1 \ \mu\text{l} = 1 \text{ mm}^3$
mho	1 S
candle	1 cd
candlepower	1 cd

Copyright ASTM, reprinted with permission.
*Magnetic flux density

Sizes and dimensions should not be listed in centimetres, because (1) the order of magnitude between millimetres and centimetres is only 10, and using both units causes confusion and (2) using millimetres permits using whole numbers rather than fractions within appropriate tolerances for all building and product dimensions.

Unit symbols may be deleted on a drawing if a note such as "Dimensions are shown in millimetres (or metres), unless otherwise noted" appears on the drawing.

Whole numbers always indicate millimetres. A five-digit number can indicate a length up to 328 ft.; 327 ft. 10-11/16 in. equals 99 941 mm. A length up to 32 ft. 9 in. can be shown by a four-digit number; a length of 3 ft. 3-5/16 in. by a three-digit number; 1/25 in. is 1 millimetre.

Numbers carried to three decimal places always indicate metres, for example, 3.600, 0.300, and 0.025.

Using millimetres and metres saves time and space on drawing, typing, and computer applications. It also improves clarity in drawings that have many dimensions.

The drawing scales used by architects and engineers have traditionally differed. Architects have generally used scales such

FIGURE 1.7-12
SI Units Used in Building Construction

Activity	Quantity	Unit	Symbol
Land surveying	Linear measure	kilometre, metre	km, m
	Area	square kilometre	km²
		hectare (10 000 m²)	ha
		square metre	m²
Excavating	Linear measure	metre, millimetre	m, mm
	Volume	cubic metre	m³
Concreting	Linear measure	metre, millimetre	m, mm
	Area	square metre	m²
	Volume	cubic metre	m³
	Temperature	degree Celsius	°C
	Water capacity	litre	L
Constitutents Reinforcement	Mass (weight)	kilogram, gram	kg, g
	Cross section	square millimetre	mm²
Trucking	Distance	kilometre	km
	Mass (weight)	metric ton (1000 kg)	t
Paving and plastering	Linear measure	metre, millimetre	m, mm
	Area	square metre	m²
Bricklaying	Linear measure	metre, millimetre	m, mm
	Area	square metre	m²
	Mortar-volume	cubic metre	m³
Carpentry/joinery	Linear measure	metre, millimetre	m, mm
Steelworking	Linear measure	metre, millimetre	m, mm
	Mass (weight)	metric ton (1000 kg)	t
		kilogram, gram	kg, g
Roofing	Linear measure	metre, millimetre	m, mm
	Area	square metre	m²
	Slope	millimetre/metre	mm/m
Painting	Linear measure	metre, millimetre	m, mm
	Area	square metre	m²
	Capacity	litre, millilitre	L, mL
Glazing	Linear measure	metre, millimetre	m, mm
	Area	square metre	m²
Plumbing	Linear measure	metre, millimetre	m, mm
	Mass (weight)	kilogram, gram	kg, g
	Capacity	litre	L
	Pressure	kilopascal	kPa
Drainage	Linear measure	metre, millimetre	m, mm
	Area	hectare (10 000 m²)	ha
		square metre	m²
	Volume	cubic metre	m³
	Slope	millimetre/metre	mm/m
Electrical services	Linear measure	metre, millimetre	m, mm
	Frequency	hertz	Hz
	Power	watt, kilowatt	W, kW
	Energy	megajoule (1 kWh = 3.6 MJ)	MJ
	Electric current	ampere	A
	Electric potential	volt, kilovolt	V, kV
	Resistance	ohm	Ω
Mechanical services	Linear measure	metre, millimetre	m, mm
	Volume	cubic metre	m³
	Capacity	litre	L
	Airflow	metre/second	m/s
	Volume flow	cubic metre/second	m³/s
		litre/second	L/s
	Temperature	degree Celsius	°C
	Force	newton, kilonewton	N, kN
	Pressure	kilopascal	kPa
	Energy, work	kilojoule, megajoule	kJ, MJ

FIGURE 1.7-13

Comparison of Calculations in Customary Units and in SI Units

DIVISION OF A LINEAR DIMENSION INTO EQUAL PARTS

A detailed drawing shows a height of 8' 10" between two floors connected by a stair with 14 risers, but not the height of each riser. What is the target height of risers so that none of them varies by more than 1/16"?	A detail drawing shows a height of 2690 mm between two floors connected by a stair with 14 risers, but not the height of each riser. What is the target height of risers so that none of them varies by more than 1 mm?

8' 10" = 106" 106 ÷ 14 = 7.57" 7.57" ≅ 7-9/16" (Check) 7-9/16" × 14 = 105-7/8" (−1/8") Target height of riser: 7-9/16" with 2 risers at 7-5/8" (+ 1/16" each)	2690 ÷ 14 = 192.1 mm (Check) 192 × 14 = 2688 mm (+2 mm) Target height of riser: 192 mm with 2 risers at 193 mm each

Comment: Division of measurement in customary units requires decimalization and change to and from decimal fractions.

AREA AND VOLUME CALCULATIONS TO DETERMINE ORDER QUANTITY

A flat concrete slab for an industrial structure has the following dimensions:	A flat concrete slab for an industrial structure has the following dimensions:
Length: 25' 0"; width: 26' 6"; thickness: 5"	Length: 7.5 m; width: 8.4 m; thickness: 125 mm (0.125 m).
Determine the order quantity of concrete to the nearest half of a cubic yard.	Determine the order quantity of concrete to the nearest half of a cubic meter.

$$\frac{25 \times 26.5 \times 5}{12} = 276 \text{ cu. ft.}$$

276 ÷ 27 = 10.22 cu. yd.
Order: 10-1/2 cu. yd.

7.5 × 8.4 × 0.125 = 7.85 m³
Order: 8 m³

Comment: Customary units need to be decimalized, and factors are required with inches and conversion of cubic feet to cubic yards. Metric volume is derived directly by multiplication of dimensions.

as 1/4 in. equals 1 ft. which results in nondecimal-scale ratios; even when engineers used seemingly metric scales, such as 1 in. equals 100 ft., the ratio is nondecimal (Fig. 1.7-14). Metrication will eliminate these differences.

The consistent use of fewer scales has definite benefits: (1) drawings prepared on transparent media drawn to the same scale may be easily superimposed, making drawing faster and easier and reducing the possibility of errors and omissions; (2) drawings lend themselves more to computer-aided reduction and enlargement techniques; and (3) consistently sized drawings are easier to understand and speed construction.

1.7.4.2

Area

In general, the preferred unit for area is the derived SI unit *square metre* (m²).

Large areas can be expressed in square *kilometres* (km²) or *hectares* (ha). *Hectare* is an alternative name for *square hectometre*, which is equal to 10 000 m². However, the hectare (ha) should be used only for land and water surface area measurements. Small areas should be expressed in *square millimetres* (mm²). The *square centimetre* (cm²) should be avoided to minimize confusion. Sometimes the area of buildings or products is indicated by linear dimensions (e.g., 40 mm by 90 mm); generally the width is indicated first and the height second.

1.7.4.3

Volume and Fluid Capacity

The derived SI unit *cubic metre* (m³) is the preferred measurement unit for volume in construction and for large storage-tank capacities.

The non-SI units *litre* (L) and *millilitre* (mL) are the preferred measurement units of *fluid capacity* (liquid volume) and gas volume. A *litre* is equal to 1/1000 of a cubic metre or 1 cubic *decimetre* (dm³) and can be expressed as 1 L = 10^{-3} m³, 1 L = 1 dm³, or 1 m³ = 1000 L.

A cubic metre contains 1 billion (10^9) cubic millimetres. Cubic centimetres and decimetres should be converted to the preferred cubic metres or millimetres whenever possible. Figure 1.7-15 lists the SI units for volume and fluid capacity.

1.7.4.4

Geometric Cross-Sectional Properties

The expression of *geometric cross-sectional* properties of structural sections involves raising the unit of length to the third, fourth, or sixth power with *exponential notation* (see Fig. 1.7-4), as follows:

Modulus of section (S) is written in units of mm³ or m³. For example, 1 mm³ = 0.000 000 001 = 10^{-9} m³.

Second moment of area (I) and torsional constant (*J*) are written in units of mm⁴ or m⁴. For example, 1 mm⁴ = 10^{-12} m⁴.

Warping constant (C) is written in units of mm⁶ or m⁶. For example, 1 mm⁶ = 10^{-18} m⁶.

Figure 1.7-16 shows the cross-sectional properties of a wide flange beam 460 mm deep and weighing 82 kg/m per unit of length.

1.7.4.5

Plane Angle

The preferred unit for expressing *plain angles* is the SI supplementary unit *radian* (rad). The radian should be used in engineering calculations when possible, because it is a *coherent unit*. However, use of the customary *degree of arc* and its decimal submultiples (as in 26.250) is permissible when using the *radian* is not convenient, especially on maps and engineering drawings and in surveying and construction operations. Use of *minutes*

FIGURE 1.7-14
Metric and Nonmetric Drawing Scales

Traditional Scales (Expressed as Ratios)		Metric Scales	
		Preferred	Other
Full size	(1:1)	1:1	
Half full size	(1:2)		1:2
4" = 1' 0"	(1:3)		
3" = 1' 0"	(1:4)		
		1:5	
2" = 1' 0"	(1:6)		
1-1/2" = 1' 0"	(1:8)		
		1:10	
1" = 1' 0"	(1:12)		
3/4" = 1' 0"	(1:16)		
		1:20	
1/2" = 1' 0"	(1:24)		
			1:25*
3/8" = 1' 0"	(1:32)		
1/4" = 1' 0"	(1:48)		
		1:50	
1" = 5' 0"	(1:60)		
3/16" = 1' 0"	(1:64)		
1/8" = 1' 0"	(1:96)		
		1:100	
1" = 10' 0"	(1:120)		
3/32" = 1' 0"	(1:128)		
1/16" = 1' 0"	(1:196)		
		1:200	

*Limited use

FIGURE 1.7-15
Units for Volume and Fluid Capacity

All Volumes	Fluid Volume Only	Limited Application	Relationships
m^3			$1\ m^3 = 1000\ L = 1000\ dm^3$
	L	dm^3	$1\ L = 1\ dm^3 = 10^{-3}\ m^3 = 1000\ mL$ $10^6\ mm^3$
	mL	cm^3	$1\ mL = 1\ cm^3 = 10^{-6}\ m^3$ $10^3\ mm^3$
mm^3			$1\ mm^3 = 10^{-9}\ m^3$

Copyright ASTM, reprinted with permission.

and *seconds* is discouraged, with the recognition that they will probably continue in use for some time.

1.7.4.6

Time Intervals

The *second* (s) is the SI base unit for time measurement. The *day* (d), *hour* (h), and *minute* (min) are non-SI alternatives. The minute should be avoided as much as possible, however, to minimize the variety of units in which time is a dimension.

Flow rates should be expressed in *cubic metres per second, litres per second,* or *cubic metres per hour*:

1 m^3/s = 1000 L/s (do not use 60 m^3/min)

1 L/s = 3.6 m^3/h (do not use 60 L/min)

1 m^3/h = 1000 L/h (do not use 1667 L/min).

The *month*, which varies in length, should not be used, unless a specific calendar month is indicated.

The *calendar year* (a, for *annum*) represents 365 days, equal to 31 536 000 *seconds*.

1.7.4.7

Temperature

The *kelvin* (K) is the SI base unit of temperature. The derived SI unit *degree Celsius* (°C) is acceptable and will replace the customary unit *degree Fahrenheit* (°F) in everyday use (Fig. 1.7-17). The earlier name for Celsius, *centigrade*, is no longer used.

The temperature interval of 1 degree Celsius exactly equals 1 kelvin. A temperature expressed in degrees Celsius is equal to the temperature in kelvins less 273.15: 0 K = −273.15°C; 0°C = 273.15 K.

The zero point of the kelvin scale is *absolute zero*, which is the point at which all molecular movement stops. Therefore, there are no negative (minus) temperatures in the kelvin scale. The lack of negative numbers makes the kelvin scale useful for scientific applications, but the large numeric values make it cumbersome for everyday use. For example, room

FIGURE 1.7-16
Properties of a 460-mm-Deep Wide Flange Beam Weighing 82 kg/m

Property	Designation and Units
Plastic modulus	$Z_x = 1.835 \times 10^6$ mm^3 (1.835×10^{-3} m^3)
Second moment of area	$I_{x\text{-}x} = 0.371 \times 10^9$ mm^4 (0.371×10^3 m^4)
Torsional constant	$J = 0.691 \times 10^6$ mm^4 (0.691×10^{-3} m^4)
Warping constant	$C_w = 0.924 \times 10^{12}$ mm^6 (0.924×10^{-6} m^6)

FIGURE 1.7-17 *Temperature equivalents.*

temperature may be expressed as 68 *degrees Fahrenheit*, 20 *degrees Celsius*, or 293.15 *kelvin*; body temperature as 98.6 *degrees Fahrenheit*, 37 *degrees Celsius*, or 310.15 *kelvin*; boiling water as 212 *degrees Fahrenheit*, 100 *degrees Celsius*, or 373.15 *kelvin*.

1.7.4.8

Mass, Weight, and Force

SI differs from both the customary system and other forms of metric measurement in the use of explicit and distinctly separate units for *mass* and *force*. Considerable confusion exists in the use of the term *weight* to mean either *force* or *mass*.

■ MASS

In everyday use, *weight* generally means *mass*. When one speaks of a person's weight, the quantity referred to is really mass.

Mass is the amount of matter there is in a body. An iron ball and a baseball of exactly the same size take up the same amount of space but there is more matter (greater density or mass) per unit volume in the iron ball; therefore, the iron ball has more mass. *Density* is the ratio of mass to volume.

The *kilogram* (kg) is the SI *base unit* for mass, the amount of matter in an object that is constant and independent of gravitational attraction. Multiples of the kilogram are formed by attaching an SI prefix to *gram* (*milligram*, *megagram*). A *megagram* (Mg), which is equal to 1 000 000 grams, should be used to measure large masses formerly expressed in tons.

The name *ton* has been given to several large mass units that are widely used in commerce and technology: the *long ton* of 2240 lb., the *short ton* of 2000 lb., and the *metric ton* of 1000 kg, which is also called a *tonne*. None of these terms is applicable to SI. The term *metric ton* may be used but should be restricted to commercial usage and never used with *prefixes*. Use of the term *tonne* is discouraged.

■ FORCE

In technical papers, *weight* has often been used to mean *force of gravity*, which is a particular force related to *gravitational acceleration*. The force of gravity varies from place to place on the earth and on other planets. A body of the same mass will weigh more or have more force on the earth than on the moon (Fig. 1.7-18).

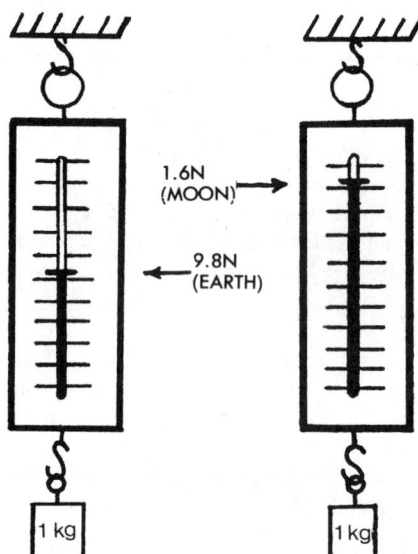

FIGURE 1.7-18 *Although the mass is the same, a kg of matter will weigh more than six times as much on the Earth as it does on the moon.*

To alleviate confusion, the term *weight* should be avoided in technical practice. *Force of gravity* should denote *force*. The *newton* (N) is the derived SI unit of force. It equals *mass* times *acceleration*: $F = kg \times m/s^2$.

Previous metric systems have used the term *kilogram-force of newton*, often dropping the suffix *force*, which caused confusion. The use of *kilogram-force* should be avoided. The *newton* also supersedes the term *kilopond* (kp). The *newton* can also be used to designate *pressure*, *stress*, *energy*, *work*, *quantities of heat* and *power*, and many *electrical units*.

1.7.4.9

Pressure, Stress, and Modulus of Elasticity

The derived unit *pascal* (Pa), which is the equivalent of *newtons per square metre*, is the unit for pressure, stress, and elastic modulus.

Pressure is a force applied uniformly over a surface and measured as a force per unit of area.

Stress is the effect of a force on a body, such as squeezing (compression), pulling (tension), and bending (compression plus tension). Stress is generally accompanied by *strain*, a physical change in the body's shape or dimension.

The pascal and its prefixed multiples—*megapascal* (MPa), *kilopascal* (kPa), and *gigapascal* (GPa)—replace a large number of customary units such as *pounds per square inch* (lb./in.^{2} or psi). It also supersedes a few non-SI metric units such as the *bar* (equal to 100 kPa) and the *millibar* (equal to 100 Pa). Old *metric gravitational units* for pressure and stress such as *kilogram-force per square centimetre* (kgf/cm^2) should not be used; neither should units such as *bar* and *torr*. Use of the *millibar* will continue to permit meteorologists to communicate within their profession, but even they should use *kilopascal* (kPa) to present data to the public.

Though it may sometimes be useful to report test results in N/mm^2 (identical to MN/m^2), calculations and results should be shown in MPa or kPa.

1.7.4.10

Energy, Work, and Quantity of Heat

The *joule* (J) with its multiples is the derived SI unit of *energy*, *work*, and *quantity of heat*. It is equal to a *newton metre* (N · m) and a *watt second* (W · s). The joule supersedes many traditional units such as, *Btu*, *therm*, *calorie*, *kilocalorie*, and *foot pound-force*.

The *kilowatthour* (kWh) has been the customary unit of *energy consumption*; the preferred unit is the *megajoule* (MJ). Since it would be needlessly costly to recalibrate electric meters, the kWh will continue to be used in electrical applications, but it should not be introduced into new areas.

The *joule* should never be used for *torque*; for this property the designated unit is *newton metre* (N · m).

For dimensional consistency in rotational dynamics, torque should be expressed as *newton metre per radian* (N · m/rad) and *moment of inertia* as *kilogram square metre per radian squared* (kg · m^2/rad^2).

1.7.4.11

Power and Heat Flow Rate

The *watt* (W), the derived SI unit for *power* and *heat flow rate*, is already in worldwide use as a unit for *electrical power*. It will replace some traditional units of power and *heat flow rate*: (1) *horsepower* (electric, boiler); (2) *foot pound-force per hour* (or *minute* or *second*); (3) *Btu per hour*; (4) *calorie* (or *kilocalorie*) *per minute* (or *second*); and (5) the *ton of refrigeration*.

1.7.4.12

Electrical Engineering

The only changes in electrical engineering will be the renaming of the unit of conductance to the *siemens* (S) instead of *mho* and the use of *hertz* (Hz) for frequency instead of *cycles per second* (cps).

1.7.4.13

Lighting Units

The *candela* (cd) is the SI *base unit* for *luminous intensity*; it is already in common use and replaces *candle* and *candlepower*.

Luminance will be expressed in *candela per square metre* (cd/m^2), which replaces *candela per square foot*, *foot lambert*, and *lambert*.

The derived SI unit, *lumen* (lm) is also in common use already. *Illuminance* is expressed in *lux* (lx), which is equal to *lumens per square metre* (lm/m^2), and replaces *lumens per square foot* and *footcandle*.

1.7.4.14

Velocity

There is no base, supplementary, or derived SI unit expressing *velocity*. The accepted non-SI unit is *kilometre per hour*.

1.7.4.15

Speed of Rotation

There is no SI unit for *speed of rotation*. The accepted non-SI unit is *revolutions per minute*.

1.7.4.16

Dimensionless Quantities

Ratios such as *relative humidity* (RH),

specific gravity (SG), and *decibel* (dB) remain unchanged in the conversion to SI.

1.7.5

CONVERSION OF NUMERICAL VALUES

Metrication is the process of change to a new set of reference units. It involves a change in numerical values by the ratio between customary and SI reference units. This ratio is known as a *conversion factor*.

Types of conversion are (1) *exact*, (2) *soft*, (3) *hard*, and (4) *rationalization*.

1.7.5.1

Exact Conversion

An *exact*, also called direct, *conversion* is a change from a customary value to its precise SI equivalent, generally expressed to a number of decimal places. For example, 24 in. is exactly 609.6 mm (Fig. 1.7-19).

Exact conversion is often needed in the early transitional period of a change to SI units. It should be thought of as only the first step to the other types of conversion. Exact conversions will rarely, if ever, be needed in design and construction, but they are an essential step before rounding or *rationalization* of traditional values can begin. Figure A-11 in Appendix A shows exact conversion of fractions and decimals to three places. Figure A-12 shows conversion tables for the most common SI units used in building design and construction.

Conversion of quantities should be handled carefully so that accuracy is neither sacrificed nor exaggerated. *Conversion factors* and quantities should not be rounded prior to multiplication or division.

Calculated values should be rounded to the minimum number of digits that will maintain the appropriate accuracy. For example, if a metric tape measure shows values to the nearest millimetre, showing lineal values more precisely would be unnecessary if the final dimensions are to be applied using a tape measure.

For calculations using numbers of differing precision, the final answer should be no more precise than the least precise number (Fig. 1.7-20).

FIGURE 1.7-19
Exact, Soft, and Hard Conversions of Customary Values

Customary Value	Exact Conversion	Soft Conversion	Hard Conversion	% Change for Hard Conversion
24 inches (2 ft.)	609.6 mm	610 mm	600 mm	−1.6
1 square (100 sq. ft.)	9.2903 m²	9.3 mm²	10 m²	+7.6
4 cu. yd.	3.058 m³	3.1 m³	3 m³	−1.9
1 quart	0.946 L	0.95 L	1 L	+5.7

FIGURE 1.7-20
Calculating Exact Conversion

Addition or Subtraction

Original numbers:	153 787 400
	647 200 000
	21 000 000
	821 987 400

Rounded to the least precise and added:

	154 000 000
	647 000 000
	21 000 000
	822 000 000

Multiplication or Division

$175.5 \times 3.475 = 609.8625$

Round the answer to 609.9 because the least precise number, 175.5, has the decimal carried to only one place.

1.7.5.2

Soft Conversion

Soft conversion is a change in the way a product's size is described without significant change in its actual size. In general, a soft conversion is the rounding of an exact conversion to a more workable numerical value. For example, a piece of 24-in.-wide gypsum lath would become 609.6 mm wide if exact conversion were desired. A soft conversion would simply round off the number to 610 mm. Such a practice would give the appearance of metric conversion but would not result in the most ideal product width, 600 mm, which would be attained through hard conversion (see Fig. 1.7-19).

Since soft conversion is only a metric veneer, a change made only on paper, soft converted products are in production now. Some are already labeled with both customary and metric measures. While soft conversions may be of some value in specifications, standards, codes, and other technical data, it is inappropriate in product sizing because it perpetuates customary practices and diminishes the advantages of SI.

1.7.5.3

Hard Conversion

In a *hard conversion* there is a definite change in physical size or magnitude from the customary values to preferred SI values (see Fig. 1.7-19). Such a conversion often leads to incompatibility between products made using the SI system and those earlier ones produced in customary measure. For example, in a hard conversion, the 24-in.-wide gypsum lath mentioned in Section 1.7.5.2 might actually be changed to 610 mm wide, instead of just being called that. This would be within normal tolerances and create little interface problems. More likely, however, the width would be changed to an even 600 mm wide, which is 1.6% smaller. Additional examples include a 1-quart container would be replaced by a 1-litre container, which is a 5.7% increase in volume; or a 12 ft. length would become a preferred 3600 mm, which is a 1.6% decrease in length.

Though the changes are more difficult to make, hard conversion is better than the other ways to convert because it creates a more simple SI system. Figure 1.7-19 shows the differences in the conversion of customary values using exact, soft, and hard conversions.

Unfortunately, hard conversion is unrealistic and unacceptable in some cases, such as changing the standard railroad gauge. The U.S. railroad gauge currently has a nonpreferred dimension of 4 ft. 8-1/2 in. The soft conversion is 1435 mm. Since railroad tracks must conform to the large number of existing railroad cars, the soft conversion to 1435 mm is acceptable and would cost little when compared with a hard change to 1450 mm or 1500 mm.

In general, hard conversion should be used when possible, because it results in preferred values, which are easier to work with. However, the decision to make a hard conversion must always be weighed against the cost of changing the many products or components that are intended to work together.

1.7.5.4

Rationalization

Rationalization, which often occurs during the conversion from customary to SI units of measure, is the development of standards for greater product compatibility and cost effectiveness. It results from selection of the most rational, preferred, and economical alternatives, after research into the technical implications and an appraisal of the economic impact. Rationalization usually results in changes in product sizes and often in a reduction in the number of a particular product's available sizes. For instance, during metrication, the number of reinforcing-bar sizes has been reduced from 11 to 10 in Australia, from 11 to 9 in Britain, and from 11 to 8 in Canada.

Rationalization can also be applied to product standards, combining them and reducing their number.

The objectives of rationalization are to (1) reduce the variety of products by eliminating unnecessary or uneconomical sizes; (2) gain optimal steps in a new product range; (3) harmonize various technical data, standards, and codes by using preferred values; and (4) simplify practices and procedures through less complicated numerical description.

1.7.6

PREFERRED DIMENSIONS AND COORDINATION

The internationally recommended basic *building module* is 100 mm. It is small enough to provide flexibility in design, large enough to promote simplification of the number of component sizes, and acceptable worldwide. It is slightly smaller than the most common module of the customary system, 4 in. For example, the width of a wood stud is nominally 4 in.; the width of a brick and the height of a block are nominally 8 in.; the spacing of studs and the nominal length of a block are 16 in.; the width of gypsum lath and insulating sheathing is 24 in.

Preferred SI sizes are those that are preferred over others for building components and assemblies. These are usually selected multiples of the basic 100-mm building module. During the change to SI, it is important to select preferred dimensions for building products to ensure that they will work together when combined into walls, floors, and other building assemblies.

Dimensional coordination is the systematic application of preferred and related dimensions in the design of buildings and the manufacture and positioning of building components, assemblies, and elements.

Modular coordination is dimensional coordination using multiples of the international 100-mm building module.

1.7.6.2

Advantages of Preferred Dimensions

Selecting preferred building dimensions, such as floor-to-floor heights, provides an incentive to produce components of a standardized height for use in high-rise buildings, such as precast concrete stairs, ducts, shaft walls, and cladding panels.

Preferred building dimensions will also (1) simplify distributors' inventory by reducing product-size variety, (2) make it easier to combine interrelated components into a building, and (3) reduce the time

and waste normally associated with assembling uncoordinated products.

In addition to reducing construction cost and time, the application of uniform preferred dimensions will simplify communications and give additional impetus to the use of computers for design and detailing.

1.7.6.2

The 100-mm Module

The 100-mm *building module* has many compelling advantages:

1. It is 1.6% smaller than the common 4-in. module of the customary system, which means that existing equipment can often be used to manufacture new metric products and that new products will often fit into existing openings.

2. The use of the *millimetre* as the working unit in building design and construction makes multimodular dimensions visible immediately. For example, 6400 mm is exactly 64 basic 100-mm modules; it is also 16 modules of 400 mm each, as opposed to sixteen 16-in. stud spaces. In customary units, the use of feet and inches hides the multiplier in modular dimensions. For example, it is not obvious that 21 ft. 4 in. is 64 whole 4-in. modules or that 26 ft. 8 in. is a whole multiple of 16-in. stud spacings.

3. The modular square, 100 mm × 100 mm, has an area of 0.01 m². This relationship is useful in cost comparisons because the unit cost per modular square, in cents, has the same number as the unit cost per square metre in dollars. For example, carpet costing 30 cents a modular square would cost $30.00 a square metre.

4. The modular cube, 100 mm × 100 mm × 100 mm, represents a volume of exactly one litre, or 0.001 m³. One litre of water has a specific gravity of 1 and a mass of exactly 1 kilogram, a

relationship that is useful in many calculations. For example, a modular cube of concrete, with a specific gravity of 2.4, has a mass of 2.4 kilograms. The calculation of the mass of a 100-mm-wide concrete wall, 2800 mm high and 7200 mm long, is $28 \times 72 \times 2.4 = 4\ 834.4$ kg. The mass per unit length is 28×10 (1 metre) $\times 2.4 = 672$ kg/m.

5. One dimension, the *millimetre*, expresses all building dimensions, including tolerances and limitations, which will enable much greater accuracy in construction.

1.7.7

INDUSTRY STANDARDS

Figure 1.7-21 shows some of the standards and recommendations for implementing SI published by the National Institute of Standards and Technology (NIST) and the American Society for Testing and Materials (ASTM).

FIGURE 1.7-21
Industry Standards for Metric Practice

Number	Title
NIST Special Publication 504	"Metric Dimensional Coordination—The Issues and Precedent"
NIST Special Publication 530	"Metrication in Building Design, Production and Construction"
NIST Technical Note 990	"The Selection of Preferred Metric Values for Design and Construction"
ASTM E 380-89a	"Practice for Use of the International System of Units"
ASTM E 621-84	"Standard Practice for the Use of Metric (SI) Units in Building Design and Construction"
ASTM E 713-80	"Guide for the Selection of Scales for Metric Building Drawings"

2

SITEWORK

INTRODUCTION

SOILS

SITE PREPARATION

EARTHWORK

**SURFACE- AND GROUND-
WATER PROBLEMS**

**LAWNS AND
LANDSCAPING**

*Photo: Manual and hand-driven hand augers for
preliminary field investigation of soils. (Courtesy of ELE
International, Inc., Soiltest Products Division.)*

INTRODUCTION

This chapter discusses some aspects of sitework, which is as important to a building's success as any of its other features. Sitework includes work to be done on the grounds, accessways, parking, lawns, plantings, and other site features. Sitework has important aesthetic and site-use implications and greatly affects access to a building. It also includes building location, orientation, and relationship to the finish grade. Besides their obvious aesthetic implications, a building's location and orientation and the lawns and plantings around it can alter the climate immediately at the building and somewhat control the effects of sun and wind on the building. These aspects can increase the comfort of the building's occupants and reduce not only the size of the heating and air conditioning systems but their cost of operation as well.

Sitework also includes subsurface investigation; site clearing; dewatering; excavation and backfilling for site improvements and buildings; ground- and surface-water control, including dewatering during construction and prevention of water intrusion into the completed building; and grading. These components can have important effects on construction cost, and how they are done can change the final look of a building site. They can also be crucial to the structural stability of a completed building. Improper excavation and backfilling can be dangerous during the work and damaging to the completed structure later if supporting materials are improperly disturbed or fill is not compacted as

it should be. Proper grading ensures that water drains away from a building rather than toward it, thus preventing potentially disastrous leaks and flooding.

Sitework also includes water distribution (see Section 11.1.2), sewage and drainage (see Sections 11.1.3), and onsite electrical transmission lines (see Section 12.2).

Site characteristics that may affect the design and construction of slabs on grade, building foundations, and site improvements include the type and the properties of the underlying soil, moisture conditions, thermal conditions, and geographic factors that may require construction precautions to control termites or unusual conditions.

2.1 SOILS

Some buildings rest on a continuous mass of solid crystalline *bedrock*, or a continuous sedimentary rock, but some type of soil underlies most building sites and usually supports the loads imposed by most low-rise and many high-rise building foundations. Since the structural integrity of a building depends on the material that supports its foundations, a basic understanding of the nature of soils and how they perform under loads is important.

2.1.1

CLASSIFICATION

Proper classification is a critical factor in identifying problem soils. Several soil classification systems are available, but the one most commonly used for soils associated with buildings is that developed by the American Society for Testing and Materials (ASTM). The applicable ASTM stan-

dards are D 2487, "Classification of Soils for Engineering Purposes," and D 2488, "Practice for Description and Identification of Soils (Visual-Manual Method)."

In the ASTM system, soils are classed broadly as being either *granular* or *cohesive* (Fig. 2.1-1). Granular soils, such as boulders, gravel, and sand, consist of relatively large particles visible to the eye. In cohesive soils the particles are relatively small, and in some soil types, such

FIGURE 2.1-1 *Sand grains are large and angular, and the volume of the voids is relatively small. Clay grains are minute and scaly, and the volume of the voids is quite large.*

as fine-grained clays, the particles cannot be seen even through a low-powered microscope. Silts are soils made up of particles finer than sand but coarser than clay. Figure 2.1-2 lists the different soil types and the size limits of their particles. Actual soil samples usually consist of mixtures of many soils of different particle sizes. They are described as combinations of the predominate soils, for example, silty sand, clayey sand, and gravely sand.

The soil classification chart in Figure 2.1-3 identifies soils and divides them into coarse-grained and fine-grained soils according to the ASTM system. Each type of soil is given a descriptive name and a letter symbol indicating its principal characteristics. As indicated by the typical names in Figure 2.1-3, field classification of soils includes a variety of different materials. When soil is a combination of two types, it is described as a combination of both. For example, clayey sand describes a soil that is predominantly sand

FIGURE 2.1-2
Soil Types

Soil Types	Approximate Size Limits of Soil Particles
Boulders	Larger than 3 in. diameter
Gravel	Smaller than 3 in. diameter but larger than #4 sieve
Sand	Smaller than #4 sieve[a] but larger than #200 sieve[b]
Silts	Smaller than 0.02 mm diameter but larger than 0.002 mm diameter
Clay	Smaller than 0.002 mm diameter

[a] Approximately 1/4 in. in diameter.

[b] Particles less than #200 sieve are not visible to the naked eye.

but that contains an appreciable amount of clay. A sandy clay has the properties of clay but contains an appreciable amount of sand.

2.1.2

PROPERTIES

The stability and the structural properties of soil in relation to foundations are determined in large measure by the *shearing resistance* of the soil. Shearing resistance in most subgrade materials is the combined effect of internal friction and cohesion.

2.1.2.1

Internal Friction

The *internal friction* of a soil can be described as the resistance of the soil grains to their tendency to slide over one another.

2.1.2.2

Cohesion

The binding force that holds soil grains together is called *cohesion*. It varies with the moisture content of the soil. Cohesion may be very high in clays, but sand and silt have little, if any, cohesion.

2.1.2.3

Other Properties

In addition to internal friction and cohesion, other important characteristics of soil influence building foundation and slab-on-grade construction. These characteristics include soil classification and composition, grain size, moisture content, density, capillarity, drainage characteristics, stratification, consistency, compressibility, and climatic variations.

■ DENSITY AND CONSISTENCY

The most significant property of cohesionless soils is *relative density*, which is described in terms such as very loose, loose, medium, dense, and very dense. Most coarse-grained soils are cohesionless.

The most significant property of cohesive soils is *consistency*, described in terms such as very soft, soft, medium, stiff, very stiff, and hard. Most fine-grained soils are cohesive. Purposely densifying or increasing the unit weight of a soil mass by compaction with rollers, tampers, or vibrators can improve its structural properties (see Section 2.3.7).

■ PLASTICITY

A soil within some range of water content that can be rolled into thin threads is called *plastic*. All clay minerals are plastic at some range of water content. Most very-fine-grained soils contain clay minerals; therefore, practically all of them are plastic. The degree of plasticity may be defined by the terms *fat* and *lean*. Lean clays are only slightly plastic because they usually contain a large proportion of silt or sand. The *plasticity index* (PI) is a value that expresses the numerical difference between the *liquid limit* and the *plastic limit* of the soil as determined by test, such as ASTM D 4318, "Test Method for Liquid Limit, Plastic Limit, and Plasticity Index of Soils." The plasticity index is equal to the liquid limit less the plastic limit ($PI = LL - PL$). The liquid limit (*LL*) is the amount of moisture present when the soil changes from a plastic to a liquid state. The plastic limit (*PL*) is the amount of moisture present when the soil changes from semisolid to plastic.

A clay soil with a plasticity index greater than 15 may expand when its moisture content increases, for example, as a result of a rising water table. If the soil beneath

a building is not consistent, as is commonly the case, expansion of the part that is of such clay while adjacent, more stable materials remain stable could result in uneven heaving beneath a building. This situation will cause cracks in the foundation and the supported structure, which can be severe enough to result in serious structural damage. Structural damage can also occur when the moisture content of such clay beneath exterior foundations is increased by groundwater seeping into the clay, while the same clay beneath interior foundations remains stable.

2.1.3

BEARING CAPACITY

Since the loads transmitted to soils by most small buildings are not large, the sizing of their foundations usually is based on local rules of thumb rather than on soil tests and engineering design. Building codes often list maximum bearing capacities for soils of several general classes that may be used without testing (Fig. 2.1-4). These bearing capacities normally are conservative and rarely take into account compressibility of soils. In addition, the general descriptions of the soils are subject to different interpretations, and the use of generalized tables may lead to erroneous conclusions regarding the true load-carrying capability of the soil at a specific site. As a result, using the loads the code permits often results in overdesigned foundations.

Where numerous individual house foundations are being planned, as in a subdivision, and in the case of commercial and institutional buildings and apartments, where taller buildings and heavier live loading are present, soil samples and tests to determine foundation type and size always should be made (see Section 2.1.7). Even in the case of an individual house, where foundations often are overdesigned, the resulting savings in smaller foundation costs may offset the cost of soil sampling and testing.

2.1.4

SETTLEMENT

When the soil being loaded is dense and granular, such as coarse sand or gravel,

FIGURE 2.1-3
ASTM Soil Classification System

	Major Divisions	Group Symbols	Typical Names	Presumptive Bearing Capacity, tons/sq. ft.[a]	Modulus of Subgrade reaction, k, lb./sq. in./in.
Coarse-grained soils more than 50% retained on #200 sieve[b]	Gravels 50% or more of coarse fraction retained on #4 seive				
	Clean gravels	GW	Well-graded gravels and gravel-sand mixtures, little or no fines	5	300 or more
		GP	Poorly graded gravels and gravel-sand mixtures, little or no fines	5	300 or more
	Gravels with fines	GM	Silty gravels, gravel-sand-silt mixtures	2.5	200 to 300 or more
		GC	Clayey gravels, grave-sand-clay mixtures	2	200 to 300
	Sands more than 50% of coarse fraction passes #4 sieve				
	Clean sands	SW	Well-graded sands and gravelly sands, little or no fines	3.75	200 to 300
		SP	Poorly graded sands and gravelly sands, little or no fines	3	200 to 300
	Sands with fines	SM	Silty sands, sand-silt mixtures	2	200 to 300
		SC	Clayey sands, sand-clay mixtures	2	200 to 300
Fine-grained soils 50% or more passes #200 sieve[b]	Silts and clays, liquid limit 50% or less	ML	Inorganic silts, very fine sands, rock flour, silty or clayey fine sands	1	100 to 200
		CL	Inorganic clays of low to medium plasticity, gravelly clays, sandy clays, silty clays, lean clays	1	100 to 200
		OL	Organic silts and organic silty clays of low plasticity		100 to 200
	Silts and clays, liquid limit greater than 50%	MH	Inorganic silts, micaceous or diatomaceous fine sands or silts, clastic silts	1	100 to 200
		CH	Inorganic clays of high plasticity, fat clays	1	50 to 100
		OH	Organic clays of medium to high plasticity		50 to 100
	Highly organic soils	PT	Peat, muck, and other highly organic soils		

Courtesy Portland Cement Association.

[a]National Building Code, 1976 Edition, American Insurance Association.

[b]Based on the material passing the 3-in. (75-mm) sieve.

FIGURE 2.1-4
Presumptive Bearing Values of Foundation Materials

Class of Material	Bearing Pressure, lb./sq. ft.*
1. Crystalline bedrock	12,000
2. Sedimentary rock	6,000
3. Sandy gravel or gravel	5,000
4. Sand, silty sand, clayey sand, silty gravel and clayey gravel	3,000
5. Clay, sandy clay, silty clay, and clayey silt	2,000

*1 psf = 47.88 P.

little settlement is likely and occurs quickly after the load is applied (Fig. 2.1-5). When the soil is a cohesive clay, settlement is likely to be greater, but it occurs slowly, taking perhaps several years.

The two principal causes of settlement are (1) reduction of the volume of soil voids (spaces between soil particles containing air or water) and (2) lateral displacement of the supporting soil.

2.1.4.1

Reduction of Voids

The reduction in the volume of the voids within soil that is loaded is related to the

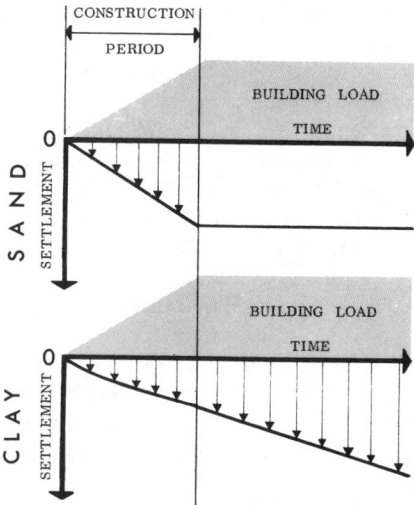

FIGURE 2.1-5 *Building settlement on sand tends to be quick and slight; on clay it tends to be greater and may occur over a long period of time.*

soil's compressibility. Since sand contains large angular grains and a relatively small percentage of void spaces (see Fig. 2.1-1), foundation pressures can cause little decrease in the volume of the voids. Clay, however, has a scale-like grain structure and a relatively large percentage of voids. Therefore, decreases in the volume of the voids in clays subjected to foundation loads can be quite large.

When soil voids are filled with water, this water must be pressed out if the volume of the voids is to be decreased. Sand is permeable and offers little resistance to the passage of water, so total void reduction and settlement occur almost simultaneously with the application of a load. Impermeable clays, on the other hand, require a long time for compression (*consolidation*) to take place, since water cannot pass out of them easily.

Air contained in the voids of sand offers no resistance to compression because it is expelled as soon as the load is applied. A certain amount of compression may occur instantaneously in clays if air is present in the voids, since air itself is compressible.

2.1.4.2

Lateral Displacement

Lateral displacement in soil is the actual horizontal movement of soil under heavy foundation loads. Settlement due to lateral displacement is not common and generally offers no problem with normal residential foundation loads, unless the lateral support to the soil under the foundation areas is reduced or disturbed by adjoining excavation.

2.1.5

NONPROBLEM SOILS

Nonproblem soils are defined as those that are dense, coarse-grained, and properly drained (Fig. 2.1-6). These soils cause few problems for slab-on-grade construction or for light-building foundations. They possess excellent bearing capacity because they are not subject to significant volume change. Gravels, for instance, usually occur in a dense state and are unlikely to cause any settlement problems. Usually, the only consideration is that the natural confinement of the soil not be disturbed by trenching or grading operations, which would reduce the necessary lateral support.

The load-bearing capabilities of both cohesive and cohesionless soils can be improved by compaction.

When sand or silt is encountered, it should be determined whether it is dense, medium dense, or loose. Loose sand or silt frequently can be compacted economically to a dense state before slab construction.

Loess soils, which are unstratified loamy materials in the ML or MH group (see Fig. 2.1-3), usually should be treated in the same manner as other soils in the same groups. The loads in light-building construction generally are not sufficient to cause problems with these soils unless the slab or the foundation supports masonry or other heavy, concentrated loads. Heavily loaded slabs and foundations may settle unduly when loess soils become saturated; therefore, proper drainage is extremely important with these soils.

2.1.6

PROBLEM SOILS

Problem soils possess less than suitable bearing characteristics for the slab or foundation to be supported. Special construction precautions are necessary with such soils as highly compressible or highly expansive clays; highly plastic soils, sands, and silts that are very loose; and loess soils that are heavily loaded or not well drained. If a supported slab or foundation is not sufficiently stiff or does not possess the necessary strength, detrimental effects to the slab, foundation, or supported superstructure may result. These effects are due to settlement or differential movement caused by lack of density of, or a change in the moisture content of, the supporting soil.

FIGURE 2.1-6
Description of Nonproblem Soils

Soil Classification	Soil Description	Density or Consistency*
GW, GP	Gravels or sand-gravel mixtures, little or no fines	All densities
GM, GC	Gravels or gravel-sand mixtures, with clay or silt	Medium-dense to dense
SW, SP	Sands with little or no fines	Medium-dense to dense
SM, SC	Silty or clayey sands	Medium-dense to dense
ML	Silty or clayey fine sands	Medium-dense to dense
ML, MH	Inorganic silts	Medium-dense to dense
CL	Inorganic and silty clays of low to medium plasticity	Stiff to hard

*See Figure 2.1-3 for definition of terms.

SUBSURFACE INVESTIGATION

Soil classification and descriptions defining the density, consistency, plasticity, bearing capacity, and other properties of soil depend on the skill of the individual making the observation and the extent of the controls under which the observations are made. As a result, classification by field observation is subject to uncertainties and inaccuracies. The extensiveness of soil investigation will vary with the type of building, local knowledge of the subsurface conditions, and the conclusions drawn from preliminary investigations.

Preliminary Investigation for Small Buildings

For single-family homes and other small buildings, where the applicable building code permits, a preliminary investigation should be made to determine general information about the nature and the origin of existing soils. This investigation should determine the need for and the necessary extent of further investigation. In some cases, it can eliminate the expense of extensive soil boring and testing. Such a preliminary examination usually consists of at least one test boring at the site to determine the acceptability of the soil for ground-supported slabs and foundations. These test borings can be made with simple tools to determine soil types and the extent of each to the depth usually necessary to predict the performance of small-building foundations (Fig. 2.1-7).

Obtaining the professional advice of a soils engineer experienced in a given locality may preclude the need for even a single test boring at a single-family home site. This is especially true if the engineer's advice is coupled with examination of other structures in the immediate area for evidence of apparent settlement or expansion of the soil, provided that similar conditions of soil, topography, and proposed construction prevail.

No additional soil investigation is required for single-family home sites if the preliminary site investigation indicates that the site has nonproblem soils consisting of firm, nonexpansive material, that

FIGURE 2.1-7 *Manual and power-driven hand augers for preliminary field investigation of soils.* *(Courtesy ELE International, Inc., Soiltest Products Division.)*

no surface- or groundwater problems exist, and that fill thickness greater than 3 ft. will not be required.

When preliminary investigation reveals problem soils at a single-family dwelling site, the procedures, test borings, and soil investigation procedures discussed in Section 2.1.7.2 for commercial, institutional, and other building types should be followed.

Subsurface Investigation for Medium-Sized and Large Buildings

An engineering study should be conducted by a qualified soils engineer for every building other than a single-family dwelling. This study should be used to determine necessary design and construction procedures.

Depending on the extent of the project, the subsurface investigation may be made by digging test pits, driving sounding rods, or making test borings. Though limited in

depth, test pits are effective investigative means when the footings will not go deeper than about 8 ft. Load tests can be made at their bottoms, soil samples can be taken in them, and the water table can be measured easily after water fills the pit to the maximum level of the table.

Bearing capacities can be measured by the standard penetration test of driving rods with a standard hammer and counting the number of blows needed to drive the rod a given distance.

The most versatile method of obtaining soil test samples and determining bearing capacities is by using soil borings. Borings can be made much deeper into the soil than can either test pits or sounding rods.

■ DENSITY AND CONSISTENCY TESTS

The relative density of coarse-grained soils can be estimated using a standard penetration test with a split-barrel sampler and power-driven equipment (Fig. 2.1-8). This test is not always necessary for single sites for one- or two-family residences, but it should be required for other types of buildings. The relative density may be estimated from the number of blows that cause the sampler to penetrate 1 ft. and is classed from very loose to very dense (Fig. 2.1-9).

FIGURE 2.1-8 *Test borings to determine soil properties are made with power-driven equipment and (a) a split-barrel sampler for coarse-grained soil, or (b) a thin-walled sampler for fine-grained soils.* *(Courtesy ELE International, Inc., Soiltest Products Division.)*

FIGURE 2.1-9
Relative Density of Coarse-Grained Soils

Soil Description	Rule-of-Thumb Field Guide[a]	ASTM Method No. of Blows[b]	Relative Density (%)
Very loose		0–4	0–20
Loose	Easily penetrated with 1/2-in. reinforcing rod pushed by hand	4–10	20–40
Medium	Easily penetrated with 1/2-in. reinforcing rod driven with 5-lb. hammer	10–30	40–60
Dense	Penetrated 1 in. with 1/2-in. reinforcing rod driven with 5-lb. hammer	30–50	60–80
Very dense	Penetrated only a few inches with 1/2-in. reinforcing rod driven with 5-lb. hammer	Over 50	80–100

[a]Field guide intended as an example of one of many field procedures currently in use for indicating density and is not necessarily a preferred method.

[b]Number of blows required to penetrate to a depth of 1 ft., as measured in accordance with ASTM D 1586, "Tentative Method for Penetration Test and Split-Barrel Sampling of Soils."

The consistency of clay (cohesive) soils should be determined by test from undisturbed samples using a thin-walled sampler (see Fig. 2.1-8), or the assumed characteristics may not be valid.

A direct numerical value of consistency determined by laboratory test is the load per unit of area that causes failure of an unconfined soil sample in a simple compression test. The value obtained is called the *unconfined compressive strength* (q_u) of the soil. Where CL, CH, OL, or OH soils (see Fig. 2.1-3) are encountered, this laboratory test should be performed to determine the unconfined compressive strength to select the type of slab used. The consistency of cohesive soils is given in Figure 2.1-10.

■ TESTS FOR OTHER PROPERTIES
Soil samples should be tested in a soils laboratory to establish their other properties. These properties include the soil's particle sizes, plasticity, cohesiveness, shear strength, water content, shrinkage coefficient, and compressive strength and the amount that the soil can be expected to consolidate (compress) under load.

2.1.8

EFFECTS OF WATER

Moisture due to either surface water from precipitation or groundwater, including the effects of seepage fields, may (1) result in volume change or reduction of the bearing capacity of a soil; (2) impose additional requirements for site preparation with regard to grading and drainage; and (3) necessitate special precautions, such as a base course or a vapor retarder (see Sections 2.3.5 and 2.4.3.2).

The amount and the frequency of precipitation have a primary influence on problem soils, because they may cause changes in the moisture content of soil supporting a slab or a foundation. Exact values for the amount of precipitation, its seasonal variation, the amount and duration of each occurrence, and its effect on supporting soil under a slab or a foundation are difficult to measure and almost impossible to predict. The most important considerations are whether precipitation or the absence of it will change the moisture content of the soil intended for support of slabs or foundations during and after construction and whether the change will have a detrimental effect on the building.

When variations in moisture content occur on a type of soil that expands as it absorbs moisture and shrinks as it loses moisture, the slab or the foundation it supports is subjected in turn to cycles of uplift as the soil expands and settling as the soil shrinks. When a period of high moisture content is followed by drought, soil moisture evaporates more rapidly around the perimeter of a building, and soil moisture under the center of the building, which may be due to capillary action and migration, becomes trapped and sealed from direct exposure. Moisture can be retained under the center of a building long after extended periods of drought.

Trees and shrubbery immediately adjacent to a building's perimeter also can

FIGURE 2.1-10
Consistency of Undisturbed Cohesive Soils

Soil Description	Rule-of-Thumb Field Guide[a]	Unconfined Comprehensive Strength (q_u)[b]
Very soft	Core (height = 2 × diameter) sags under own weight	0.25
Soft	Can be pinched in two between thumb and forefinger	0.25–0.50
Medium	Can be imprinted easily with fingers	0.50–1.00
Stiff	Can be imprinted with considerable pressure from fingers	1.00–2.00
Very stiff	Barely can be imprinted by pressure from fingers	2.00–4.00
Hard	Cannot be imprinted by fingers	4.00

[a]Field guide is only an indication of soil consistency. Values of q_u are given as basic values of consistency by which field classification can be verified and should not be used for design purposes without laboratory verification.

[b]q_u is unconfined compressive strength in tons/sq. ft. (not the ultimate bearing capacity of the soil for design).

affect the soil moisture content by providing a shield from natural precipitation and by extracting soil moisture during the growing season. Also, watering of gardens and lawns adjacent to a building can increase the underlying soil's moisture content.

When soil changes from a low to a high moisture content, a similar but opposite phenomenon develops. If prolonged periods of alternating drought and moisture occur, considerable differential moisture can develop in different areas under a building. If the soil is a type that undergoes substantial volume changes with changes in moisture, one of two conditions may result. If the building's slabs and foundations are relatively flexible, as in the case of unreinforced or only lightly reinforced slab Types I and II (see Section 3.3), they will follow the uneven contour of the supporting soil caused by uneven volume change, resulting in distortions and possible damage to both the slabs and the foundations and the supported superstructure (Fig. 2.1-11a).

On the other hand, a slab or a foundation that is sufficiently rigid, as is the case with a properly structurally reinforced and stiffened Type III slab or concrete foundation, will resist deformations even though the contour of the supporting soil is uneven. Higher soil pressures develop under such a slab or foundation over the plateaus

FIGURE 2.1-11 *Slab distortion and soil movements (highly exaggerated): (a) a flexible slab will distort with volume changes in the supporting soil; (b) a rigid slab is designed to span uneven contours caused by volume changes in the supporting soil.*

with reduced soil pressures at the valleys. The slab or foundation will be subjected to bending between these uneven contours and the soil at the plateaus may deform due to higher bearing pressures, distributing loads to adjacent areas (Fig. 2.1-11b). Type III slabs are designed to resist bending stresses in order to span between plateaus without damage to the slab, as are properly reinforced concrete footings.

2.1.9

EFFECTS OF TEMPERATURE

The combined effects of temperature and moisture influence the depth of perimeter grade beams and foundations, as well as the selection of slab types, because some

soils are subject to volume changes called *frost heaves*, which result from the formation of ice layers when water in soil freezes. Unless a Type III slab is used or nonheaving soils (GW, GP, SW, SP; see Fig. 2.1-3) are present and the water level is below the frost line, foundations always should be carried below the frost line (Fig. 2.1-12).

The temperature of a building slab can affect the comfort of occupants, especially during cold periods. Insulation is often introduced to help keep slabs warmer. Some buildings are partially heated by radiant heat generated by pipes carrying warm water or ducts carrying heated air built integrally with a slab on grade. This configuration warms the slab as well as the air above it, which can produce a more comfortable environment for occupants than if the air alone is heated.

2.2 SITE PREPARATION

Site preparation is that part of the work associated with new building construction that includes removal of existing site features such as buildings, site improvements, vegetation, and top soil that would interfere with the new construction. Movable structures, such as house trailers, simply may be towed away. Houses and other light buildings are sometimes removed from the site intact.

2.2.1

DEMOLITION

A major aspect of some projects is the

demolition and removal of existing buildings or sections of existing buildings that are not to be a part of the new work. When structures are to be removed as a part of the general construction contract, the term *demolition* is often used to refer only to those parts of such structures that are above ground. In that case, removal of below-grade portions of structures is usually thought of as part of the excavation work (see Section 2.3.2). When demolition is done under a separate contract, however, removal of the walls and slabs of basements and of building foundations is usually a part of the separate contract.

Demolition may be either complete or

selective. In *complete demolition*, an existing structure is demolished completely, and the debris is removed from the site. *Selective demolition* consists of removal of a portion of an existing structure, usually to make way for remodeling or an addition.

Demolition is potentially hazardous to the workers involved, the public, adjacent property, and the environment. Therefore, extraordinary measures should be taken to protect all of them from those dangers. Most heavily populated jurisdictions have code and other legal requirements for public and property safety during demolition. There are also usually restrictions on

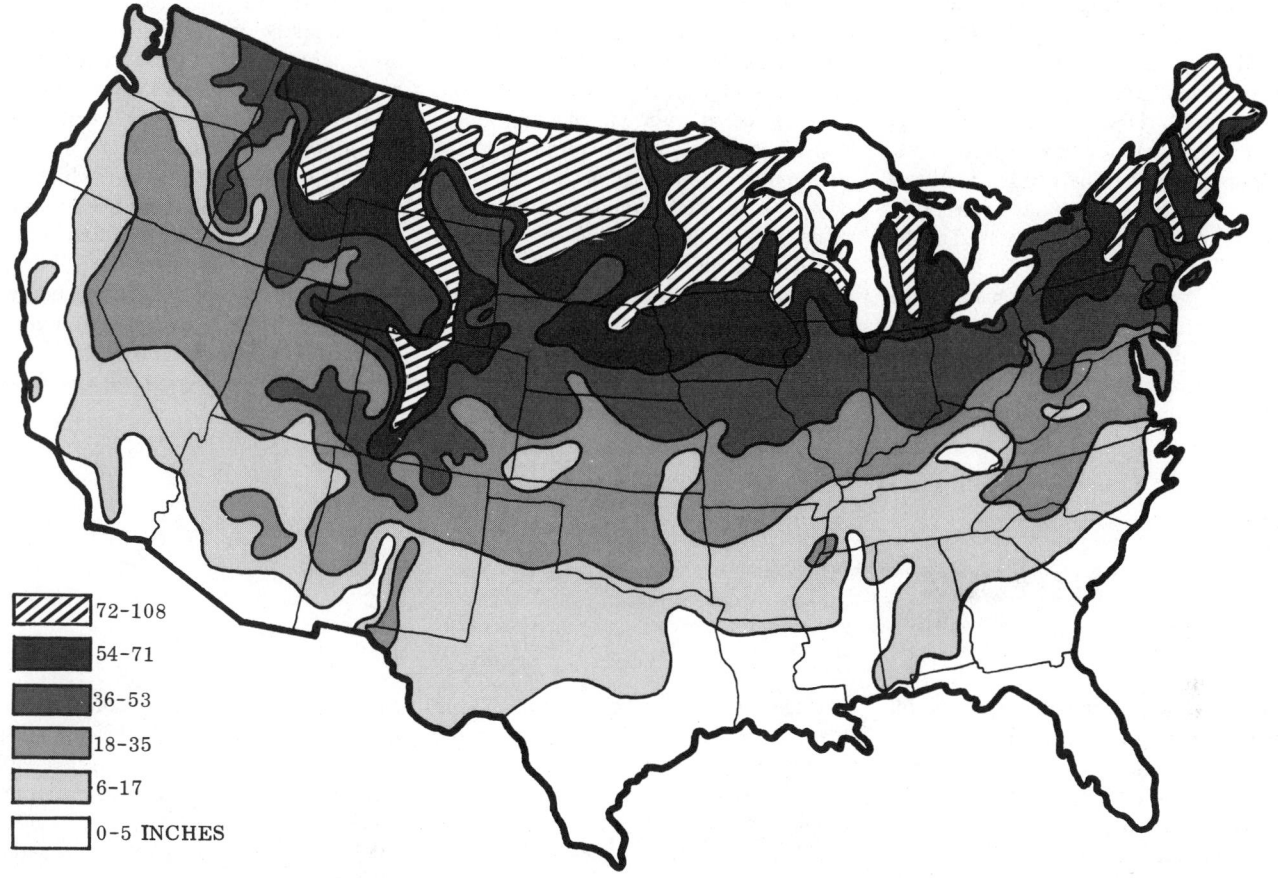

▨	72-108
■	54-71
■	36-53
■	18-35
■	6-17
□	0-5 INCHES

FIGURE 2.1-12 *Maximum depth of frost penetration in inches.* (*Courtesy National Weather Service.*)

practices that affect the health and welfare of the workers and the public, as well as damage to the environment. An example is the handling of hazardous materials like asbestos, toxic chemicals, explosive materials, flammable materials, and radioactive waste.

There are also often extensive requirements relative to the bracing and shoring of buildings during demolition. These restrictions are usually more stringent when a building has been damaged in a fire, a storm, or an earthquake. Restrictions are likely to be extensive regarding the use of explosives.

Even when safety and environmental restrictions are not mandated by code or by laws, precautions should be observed. Sprinkling with water, temporary enclosures, and other appropriate means should be used to limit the spread of dust and dirt in the air. Pedestrian and vehicular traffic should be routed around the area of danger. Rodents and insects should be controlled before and during demolition using legal

and environmentally sound means so they are not driven into adjacent structures.

In selective demolition there usually is a need to protect portions of the building that will not be demolished. Cutting and patching of existing materials often are required. Protective enclosures should be erected between areas to be demolished and areas to be left alone. Onlookers will often occupy spaces adjacent to the area in which selective demolition is to take place. They should be protected from danger and from exposure to fumes, dust, dirt, loud noises, and toxic materials by whatever means are appropriate. These means include temporary dustproof walls, sound barriers, barricades, and covered passageways.

Portions of a building that are to remain should similarly be protected. In addition they should be protected from the weather by temporary walls and roofs or by weatherproof tarpaulins or plastic sheeting.

When portions of a building undergoing selective demolition remain occupied,

utility service interruptions should be kept to a minimum and should be done so that the occupants are not harmed, placed in danger, or caused significant discomfort. For example, fire sprinklers should not be interrupted during times when the building is occupied. Exits should be maintained at the number and size required by code. Electric service, water, sewer, and other services should not be interrupted for extended periods while the building is occupied.

2.2.2

SITE CLEARING

Site clearing consists of the removal of site improvements and vegetation that will not be a part of the new work. This includes removing trees and other vegetation, clearing and grubbing, topsoil stripping, and removing above- and below-grade site improvements.

Before site clearing is started, adjacent public and private property should be protected from harm by fences and barricades. Before trees and other vegetation are removed, those to remain should be identified, marked, and also protected by temporary fences or barricades. Similarly, water, electric, sewer, and other utility lines should be marked and staked so that they will not be damaged inadvertently.

Clearing and grubbing consists of removing trees, shrubs, and other vegetation, including their roots and stumps. Depressions caused by such removal should be filled with satisfactory backfill material, as discussed in Section 2.3.4.

Topsoil should be stripped from areas where new construction is to take place. Topsoil is often stockpiled on the site and later reused in lawn and planting areas. Topsoil should not be removed from within the *dripline* of a tree (the area around a tree that extends out the length of the longest branch), because doing so will damage the root structure of the tree.

Roadways, walkways, site walls, and other structures that will not be used in the new work should be demolished, and the debris removed from the site.

One means of disposing of materials removed from a building site is dumping them in public or private *landfills* or *spoil areas*. Landfills often accept organic materials and construction debris, whereas spoil areas usually accept only earth materials, such as excavated soil and unusable topsoil. Burning on the site, which was once the most prevalent means of disposing of combustible waste, often is prohibited in heavily populated jurisdictions. Even when it is legal, some owners will not permit burning because of environmental and safety concerns.

2.3 EARTHWORK

Earthwork includes the excavation and filling necessary to make way for a building, backfilling of the remaining excavation after the building has been built, and the grading needed to finish the site preparatory to installing paving, walkways, lawns, and landscaping. Related sitework components include underpinning of existing structures to protect them from damage caused by adjacent excavation, slope protection and erosion control, and chemical treatment of the soil to protect the building against subterranean termites.

2.3.1

UNDERPINNING

Underpinning is the time-consuming and expensive process of supporting the foundations of an existing building by imposition of new footings, pilings, or caissons beneath the existing footings. Underpinning is often needed when the excavation for a new building or an addition to an existing building will undercut the foundations of an existing adjacent building (Fig. 2.3-1). Underpinning may also be needed when an existing footing is too small to support a vertical addition to a building, an existing undersized footing has resulted in undue settlement and damage, or additional support is needed.

Underpinning can be accomplished in many different ways, depending on the nature of the problem. Only two of the many possibilities are covered here: (1) placing a new footing and foundation wall below the existing ones and (2) piles.

When new building foundations are lower than those of an existing building, the existing footings can be supported by placing new footings lower than the existing ones (Fig. 2.3-2). One way to accomplish this is to place a series of *needle beams* through the existing building walls to support them until the new foundation wall and footing have been built (Fig. 2.3-3).

Another way to place a lower foundation wall and footing is to dig intermittent pits at the existing foundation down to the level of the new footing (Fig. 2.3-4). These pits must be small enough to permit the existing footing and wall to span the pit with support by temporary jacks and shoring. The footing between the pits remains supported by the original soil. After the underpinning foundation wall and footing have been placed (see Fig. 2.3-2), another section of pit is dug, until the entire wall has been underpinned.

Underpinning also can be accomplished using piles (see Section 2.3.3). In general, the method is to place new piles adjacent to the existing footings. A section of the foundation wall is then removed at the top of each set of piles, and a concrete cap is poured in place so that it rests on the piles and penetrates the wall (Fig. 2.3-5). A proprietary system is available that substitutes a foundation support and a jack assembly for the cap. This assembly rides down on a steel pipe pile and is rotated toward the existing footing until it extends beneath the footing. It is then raised hydraulically until the loads transmitted by the footing are picked up by the support assembly and transmitted to the pile. This type of assembly is particularly useful when an existing footing has settled.

2.3.2

EXCAVATING

Even for small single-family residences and other small buildings with floor slabs on grade, excavation is necessary for trenches to contain footings and utilities. For most buildings, much more excavation than this minimum is required.

Building excavation usually is defined as the removal of materials down to the level required by the contract documents and disposal of the materials excavated.

In some small applications, excavation is done using hand tools, but most excavation requires power equipment (Fig. 2.3-6). This equipment ranges from the

FIGURE 2.3-1 *The existing footing here most likely will slide into the adjacent new excavation unless it is underpinned.* (Drawing by HLS.)

bulldozers, bucket loaders, trench excavators, and backhoes common in earth excavation to the rippers, power shovels, pneumatic hammers, and drop balls used to excavate in rock.

Before excavation is permitted to begin, underground utilities that are to remain should be located and staked to prevent damage to them.

The materials that underlie a site are often referred to as *unclassified*, meaning that it is unknown whether they consist of all soil or whether rock is present. This common practice is unfair to the contractors, who, when they present their bids, must gamble as to the nature of the material that will have to be excavated. Excavation of rock is far more expensive than excavation of earth. It takes more time, requires different equipment, and often necessitates blasting, which is restricted

FIGURE 2.3-2 *The lower-footing method of underpinning.* (Drawing by HLS.)

in many jurisdictions to certain hours and days. In addition, the size of the explosive is sometimes restricted, requiring the use of smaller and more charges. A far fairer method would be to identify the materials to be excavated as either earth or rock.

The definitions of the terms *earth* and *rock* vary with the source. One way to define rock is to say that it includes all material that cannot be removed using a specific piece of earth-moving equipment. This method is suggested in AIA's *Masterspec* Section 02200, "Earthwork." Earth, then, is any material that is not rock.

Excavation generally includes removal of above-ground site improvements such as pavements, walks, and site walls, as well as underground items like old foundations, piping that will not be used in the new work, wells, and cisterns.

Materials that are suitable for fill and backfill (see Section 2.3.4) should be stockpiled for reuse. Excess and unsuitable materials should be disposed of.

Excavations are classified as *open excavations*, *trenches*, or *pits*. Open excavations are those that cover the large portion of a building excavation. They are usually excavated with power equipment to the level of the lowest floor slab.

Trenches and pits are dug into an open excavation or into soil that has not been

otherwise excavated. Trenches for footings and utilities usually are dug by machine and finished by hand. Pits for many isolated footings, sumps, and mechanical devices are dug with hand tools. Larger pits for elevators and large mechanical equipment are usually excavated by machine and finished by hand.

The bottoms of excavations should be protected against erosion, softening due to water saturation, and freezing. Disturbed material at the bottom of trenches and pits should be removed. Disturbed material beneath slabs that is suitable for the slabs should be compacted.

2.3.3

EXCAVATION SUPPORT SYSTEMS

Except for shallow trenches and pits, the side walls of all excavations, including open excavations, trenches, and pits, need to be supported until the permanent construction is in place and able to support them. Support can be obtained by simply sloping the sides of an excavation when the site is large enough to permit it. The slope should be at such an angle that the soil will not slide down it. The angle must be shallow for soils with a large percentage of sand and gravel, but it may be steeper for soil that contains clay. The maximum slope should be determined by actually testing the soil to determine the angle at which it will not slide. Applicable codes and other legal restrictions may dictate the permissible slopes. As excavations become deeper, it is sometimes necessary to step the sides back (Fig. 2.3-7). Excavated materials stockpiled for use in backfill should be stored away from the edges of the excavation to reduce the height of the material adjacent to the excavation.

When the space to store excavated soil is small or excavations are too close to property lines or other structures to properly slope the sides, other types of support are necessary. Architects are seldom called on to design excavation support systems; this problem usually is left to the contractor. The type of support to be used depends on the conditions at the site and the preferences of the contractor. An architect may be required by law to review a contractor's plans for excavation support systems.

FIGURE 2.3-3 *The exterior ends of needle beams may be supported at the existing grade level or on footings at the bottom of the new excavation. In either case, the soil against the existing wall must not be removed until the beams are in place.* (Drawing by HLS.)

FIGURE 2.3-4 *Pits dug to permit underpinning must themselves be braced and shored to prevent collapse and injury to workers.* (Drawing by HLS.)

On shallow excavations (12 to 15 ft.) bracing may be done with *sheet piles*. Most sheet piles are wood or metal (Fig. 2.3-8), but precast concrete is sometimes used. Sheet-pile sections are interlocked and driven into the ground before excavation begins.

Deep excavations usually are braced with *soldier piles and lagging* (Fig. 2.3-9). Soldier piles are H-shaped steel beams, usually wide-flange sections. Lagging consists of heavy wood timbers. Soldier piles are driven into the ground before excavation starts. Lagging is placed from the bottom as the excavation is made. Lagging may be placed behind the flanges of the soldier piles, as shown in Figure 2.3-9a, or placed on the face of the soldier piles and held in place by battens, which are bolted through the joints in the lagging to the flanges of the soldier piles (Fig. 2.3-9b).

On very shallow excavations, sheet piles driven to sufficient depth to act as cantilevers are sometimes strong enough to support the loads encountered without additional support. When excavations are deeper, however, both sheet piles and soldier piles and lagging need additional support to withstand the loads imposed by the soil and the groundwater behind them. This support is provided by *shoring and bracing*. On very deep excavations, several tiers of bracing may be needed.

Sheet piles and soldier piles and lagging may be supported using *rakers*, *cross-bracing*, or *tiebacks*. All three systems are designed to support horizontal continuous *wales* that extend along the face of the sheet piles or soldier piles. Wales are usually wide-flange steel beams, but other steel shapes may be used.

The two bracing systems shown in Figure 2.3-10 have a common drawback: they both obstruct activities in the excavation. For example, rakers can be placed only after the excavation has been extended to the level of the concrete footer or heel. But the sheet piles or soldier piles and lagging can be exposed only after the bracing is in place. This means that excavation at the location of the footer or heel must be made to the level of the piles and the rakers be in place before the earth can be removed from the face of the sheet piles

FIGURE 2.3-5 *Vertical piles used as underpinning.* (Drawing by HLS.)

Labels in figure: Existing wall, Pile cap, Existing footing, Pile

FIGURE 2.3-7 *Stepping the sides of an excavation when there is room may preclude the need for shoring.* (Drawing by HLS.)

a **b**

FIGURE 2.3-8 *(a) Timber sheet piling and (b) one type of steel sheet piling. Other configurations and other joint types are also available.* (Drawings by HLS.)

or soldier piles and lagging. The heavy dashed line labeled "Limit of excavation" in Figure 2.3-10 indicates the level of the excavation until the rakers have been placed. If more than one line of rakers is needed, the earth beneath them must be removed in stages, extending just below each level. The proper angle of slope must be maintained so that the remaining soil will stay in place and support the load imposed by the soil and the groundwater on the opposite side of the sheet piling or the soldier piles and lagging. In addition, excavation of the soil remaining against the sheet piles or the soldier piles and lagging is difficult, because the rakers obstruct movement of excavation equipment.

Cross-bracing also obstructs the movement of equipment, but it does allow excavation to continue uninterrupted from one

FIGURE 2.3-6 *Many types and sizes of earth-moving equipment are used in building excavation.* (Source unknown.)

FIGURE 2.3-9 *(a) Soldier piles and (b) lagging.* *(Drawings by HLS.)*

are frequently used. When the soil tends to be somewhat unstable, or a type of soil, such as sand or gravel, is present that would collapse into the hole, a steel pipe may be inserted to keep the hole open until the grout has been placed.

Cross-bracing and rakers are removed as soon as permanent construction makes them unnecessary. Tiebacks are left in place, but are not relied on for support after permanent construction is in place. Soldier piles and lagging often are removed as soon as permanent construction negates their need. However, when their removal would disturb new or existing construction, soldier piles and lagging are left in place. Sometimes they are covered with a drainage board or mat and used as the outer form for concrete walls (see Section 2.4.3.3). Sheet piling may be either cut off below ground level and left in place or removed completely, although wood sheet piling is usually removed.

2.3.4

FILL AND BACKFILL

Compacted fills are used for building sites in undulating or hilly terrain to create reasonably level ground for slab construction. *Engineered fill*, which is fill whose material selection, installation, and compacting are done according to engineering principles, is also sometimes used below foundations in small buildings when the existing soil is unsuitable. In any case, the bad soil is removed and replaced with engineered fill. Soils used for fill should contain no vegetation or foreign material that would cause uneven settlement. Beneath slabs, a base course should be provided when necessary.

Topsoil and other unsuitable materials should be removed from areas beneath foundations and slabs. There should be no soil or other materials present that would cause a slab or footing to settle unevenly or excessively.

Fills should be used beneath slabs to establish the desired finished slab elevation and compacted to the maximum practical density. The slab bed for Types I and II ground-supported slabs (see Section 3.3) should not expand or contract due to moisture. Therefore, organic or active clays and silts of the OL, CH, CL, OH,

wale level to the next lower one without slopes. In wide excavations, cross-bracing requires additional support, which is provided by driving steel columns at intermediate points across the span and installing additional wales on the columns. As the excavation is lowered, additional levels of braces can be added as needed.

The third method of bracing uses earth or rock anchors called tiebacks. This method does not obstruct activities in the excavation. Tiebacks are constructed by drilling a hole at each line of wales through the sheet piling or lagging into stable rock or soil. Tiebacks are not appropriate in unstable or very loose soils. An anchor rod or cables are inserted into the drilled hole and the space around the rod or cables is filled with grout under pressure. When the grout has set, the rod or cables are *post-tensioned* (stretched) using a hydraulic jack and anchored securely to a continuous wale so that the tension is maintained. Figure 2.3-11 shows a wale made from steel channels, but other shapes, such as wide-flange steel beams,

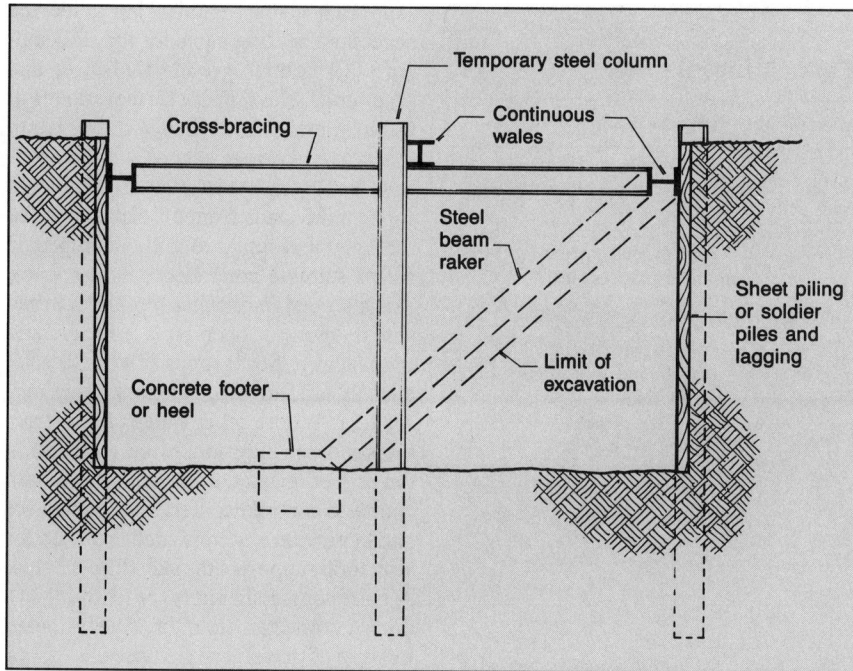

FIGURE 2.3-10 *Two foundation-bracing systems. The solid lines show cross-bracing. The dashed lines represent a raker system. The two systems generally are not used together.* (Drawing by HLS.)

FIGURE 2.3-11 *A tieback anchoring system for excavation supports.* (Drawing by HLS.)

and PT groups, as classified in ASTM Standard D 2487, "Classification of Soils for Engineering Purposes," are not suitable for fills under these types of slabs. Soil groups GW, GP, SW, and SP (Fig. 2.3-12) may be used as fill material beneath slabs without a base course. These latter soil groups are also those used in engineered fill and in most fill in other locations. Groups GM and SM are also suitable under some conditions.

Before fill material is placed, clearing and grubbing should have been completed where the fill will be placed. Sloped surfaces should be plowed or otherwise broken up so that fill material will mix with existing material and not slip from it.

Suitable excavated soils may be used to backfill excavations, including soils classified in ASTM Standard D 2487 as GW, GP, GM, SM, SW, or SP (see Fig. 2.3-12), but some jurisdictions may limit the acceptable types. In addition, some other qualifications may apply in some projects. For example, a soil that is not in the preceding list may be acceptable if its liquid limit and plastic limit are in a range that is acceptable to the project's soils engineer and its use does not violate the building code or other restrictions. When insufficient quantities of suitable excavated materials exist, acceptable soil should be brought in from off site (*borrow material*) to complete the required backfilling and grading.

Filling and backfilling materials should be placed in layers, or *lifts*, of about 8-in. loose thickness when compaction will be done by power equipment and about 4-in. loose thickness when hand tampers will be used. When the fill or backfill material has a lower or higher than acceptable moisture content for proper compaction, it should moistened or aerated until it reaches the optimum moisture content. The required moisture content varies with the soil material and should be established by testing.

Backfill should not be placed until underground utilities to be covered are in place and have been tested, concrete formwork has been removed, shoring and bracing have been removed, and sheet piling to remain has been cut off below grade level. In addition, backfilling against walls should not be done until building slabs are in place and the walls have been braced by either temporary or permanent construction.

2.3.5

BASE COURSE FOR SLABS ON GRADE

When slab-on-grade construction is intended, some preparation must be made to the subgrade, depending on the type of soil and other conditions encountered at the site. At the least, the area beneath the slab must be cleared and grubbed, and unsuitable soils, such as topsoil, must be removed. The slab bed for ground-

FIGURE 2.3-12
Requirements for Slab Beds According to Soil Classification

Major Division		Group Symbols	Typical Names
Coarse-grained soils (More than half of material is larger than the smallest particle visible to the naked eye)	Gravels (more than half of coarse fraction is larger than 1/4 in.) Clean gravels (little or no fines)	GW	Well-graded gravels, gravel-sand mixtures, little or no fines
		GP	Poorly graded gravels or gravel-sand mixtures, little or no fines
	Gravels with fines (appreciable amount of fines)	GM	Silty gravels, gravel-sand-silt mixtures
		GC	Clayey gravels, gravel-sand-clay mixtures
	Sands (more than half of coarse fraction is smaller than 1/4 in.) Clean sands (little or no fines)	SW	Well-graded sands, gravelly sands, little or no fines
		SP	Poorly graded sands or gravelly sands, little or no fines
	Sands with fines (appreciable amount of fines)	SM	Silty sands, sand-silt mixtures
		SC	Clayey sands, sand-clay mixtures
Fine-grained soils (More than half of material is smaller than the smallest particle visible to the naked eye)	Silts and clays (liquid limit is less than 50)	ML	Inorganic and very fine sands, rock flour, silty or clayey fine sands or clayey silts with slight plasticity
		CL	Inorganic clays of low to medium plasticity, gravely clays, sandy clays, silty clays, lean clays
		OL	Organic silts and organic silty soils, elastic silts
	Silts and clays (liquid limit is greater than 50)	MH	Inorganic silts, micaceous or diatomaceous fine sandy or silty soils, elastic silts
		CH	Inorganic clays of high plasticity, fat clay
		OH	Organic clays of medium high plasticity, organic silts
	Highly organic soils	PT	Peat and other highly organic soils

□ Generally may be used for fill without a base course

▨ Fill that requires a base course

▨ Not permitted for foundation fill

Note: Based on ASTM D 2487, "Classification of Soils for Engineering Purposes."

---

supported slabs should (1) provide the necessary bearing capacity for slab support; (2) control ground moisture; and (3) establish the proper slab elevation. Fill material and undisturbed soil that is not in a suitably dense state should be properly compacted. Materials that will not support the loads from the slab should be removed and replaced with a compacted fill of suitable materials.

A slab bed may consist of undisturbed soil if of the proper type, density, and consistency. Soil groups GW, GP, SW, and SP are in this category (see Fig. 2.3-12). Where other types of soil form the subbase, they should be removed to the proper level and replaced with a base course to control capillary rise of moisture and to provide a uniform and level bed for structural support of the slab (Fig. 2.3-13). A base course should be at least a 4-in.-thick compacted layer of a well-graded mixture of gravel, crushed stone, sand, or crushed slag.

2.3.6

GRADING

Grading is that part of earthwork that establishes the rough and final grades on a building site. It is accomplished with a combination of power equipment and hand tools. Grading is necessary where lawns, plantings, pavement, walks, and building slabs will be placed.

The site around a building should be graded to form protective slopes that will drain surface water away from the foundation so water will not collect under the slab. Surface runoff should be directed to streets or drainage structures. Finish grade should slope downward in all directions away from a building at a minimum fall of 6 in. in each 10 ft., which is a slope of about 5%. A percent of slope is the unit of rise (or fall) in each 100 units of run (e.g., 1 ft. in 100 ft.). When the area adjacent to a building is paved, a 1% slope, which is about 1/8 in. in each foot, may be sufficient to direct water away. However, when a paved area has less than a 5% slope, positive drainage, such as by yard drains or catch basins, should be introduced at sufficient intervals to prevent ponding due to imperfections in paving

Concrete slab
Undisturbed soil

Concrete slab
Compacted fill
Undisturbed soil

Concrete slab
Base course
Compacted fill
Undisturbed soil

Vapor retarder
Concrete slab
Base course
Compacted fill
Undisturbed soil

FIGURE 2.3-13 *Slab beds for ground-supported slabs. A base course or a vapor retarder may be needed to control moisture.*

slope and build-up of water during heavy rainfall. Hillside sites should be graded to divert surface water around and away from structures.

Areas beneath lawns and walks should be graded to within 1/10 ft. above or below the design subgrade elevations, taking into account the thickness of sod and walkway surface. Areas beneath pavements and building slabs should be graded to within 1/2 in. above or below the design subgrade elevations.

2.3.7

COMPACTING

Compacting applies energy to soil to consolidate it by compressing air voids to increase the soil's dry density. Proper compacting (1) minimizes settling, (2) increases the soil's load-bearing characteristics, (3) increases soil stability, and (4) reduces water penetration. Proper compacting will

usually prevent slab and pavement cracking caused by differential settlement in the soil. It will also reduce settling of backfill placed against foundation and retaining walls and in other locations to prevent low spots where water would pond.

Compaction requirements generally are specified as a percentage of the maximum dry density obtainable in the test procedures in ASTM Standard D 1557, "Test Methods for Moisture-Density Relations of Soils and Soil-Aggregate Mixtures Using 10-lb. (4.54-kg) Rammer and 18-in. (457-mm) Drop," or ASTM Standard 698, "Test Methods for Moisture-Density Relations of Soils and Soil-Aggregate Mixtures Using 5.5-lb. (2.49-kg) Rammer and 12-in. (304.8-mm) Drop." The relaive density can be specified for fills constructed with free-draining sand and gravel soils (GW, GP, SW, SP) in accordance with ASTM Standard D 4253, "Test Methods for Maximum Index Density of Soils Using a Vibratory Table," and ASTM Standard D 4254, "Test Methods for Minimum Index Density of Soils and Calculation of Relative Density." Varying degrees of compaction for the many different types of soils that may be encountered cannot be stated without engineering analysis. Generally, however, clean sand fill (SW, SP) should be compacted to 95% of the optimum dry density and all other fine-grained, nonexpansive fills to 90% of the optimal dry density in accordance with ASTM D 1557. Compacting requirements for expansive soils should be determined from engineering analysis and may require somewhat less than 90% of optimum dry density.

Using suitable hand or power equipment (Fig. 2.3-14), compacting of fill material should be performed in 4- to 8-in. layers (lifts) of soil. Regardless of the type of soil being compacted, the thinner the lift, the better the compaction.

Structural slabs (Type IV) that receive their support at the edges and through the center by intermediate foundations require only sufficient compacting beneath the slab portion to provide temporary support of the dead load imposed by the slab ntil the slab develops the necessary strength to span between its foundations.

Backfill adjacent to foundation walls should be uniformly compacted to at least 90% of optimum dry density unless other requirements dictate a greater degree of

FIGURE 2.3-14 *(a) Soil compacting at most building sites is done using power equipment. (b) Hand tamping is occasionally used in locations inaccessible to power-driven equipment. Hand tamping also may be used exclusively in small residential buildings and other small buildings when the soil is an easily compactible type. (Source unknown.)*

compacting. For example, backfill beneath walks, steps, and pavements should be compacted to at least 95% of optimum dry density.

2.3.8

SLOPE PROTECTION AND EROSION CONTROL

Exposed earth on a construction site should be protected from erosion; indeed, most jurisdictions have codes, ordinances, or rules that dictate such controls. These controls may be as simple as the staking in place of hay bales to prevent eroded materials from fouling waterways, or they may be more elaborate systems with filter fabrics or riprap. In addition, providing diversion ditches (Fig. 2.3-15) and terracing of slopes (Fig. 2.3-16) may be needed to reduce the speed of water running down a slope and direct it to drainage channels protected with filter fabrics or riprap.

Filter fabrics are open woven mats of nylon or other synthetic fibers. They are staked down over a slope to permit water to flow without eroding the underlying soil (Fig. 2.3-17). Some of these products can

FIGURE 2.3-15 *A diversion ditch at the top of a slope will slow down the flow of ground water and can be used to direct the flow to a paved outfall or into an interceptor drain that will conduct it to the lower level through pipes. (Drawing by HLS.)*

FIGURE 2.3-16 *Terraces cut into a slope will reduce the velocity of ground water and reduce erosion. (Drawing by HLS.)*

FIGURE 2.3-17 *Filter fabric installed to protect a slope. (Drawing by HLS.)*

FIGURE 2.3-18 *Riprap can be either temporary or permanent. It can be used to cover a slope, as shown here, or to line drainage channels or ditches. It is also used to protect roadway embankments and stream beds. (Drawing by HLS.)*

be left in place and later covered with topsoil to receive lawns or other vegetation.

Riprap is a layer of stone or broken concrete designed to form a pathway for the flow of water and to prevent erosion of underlying soil (Fig. 2.3-18). Riprap consists of pieces large enough to resist their movement down the slope by the force of water. Riprap is used sometimes as a drainage channel and sometimes as protection for an entire slope.

2.3.9

TERMITE CONTROL

Two types of termites cause damage to wood in the United States: subterranean and nonsubterranean. As Figure 2.3-19

shows, the hazard is greater in areas having mild temperatures.

Subterranean termites account for about 95% of all termite damage in the United States. These termites (Fig. 2.3-20) live in nests and develop their colonies underground, but build tubes of earth to reach wood and other cellulose material such as fiberboard, fabrics, and paper. Termites must have a constant source of moisture or they die. They become most numerous in moist, warm soil containing an abundant supply of food.

Subterranean termites are not carried into buildings in lumber but establish themselves by entering from ground nests after the building has been constructed, using three basic routes: (1) attacking wood in direct contact with the ground (porch stairs, trellises, and siding installed too close to grade) (Fig. 2.3-21); (2) entering through cracks and voids as small as 1/32 in. in slabs, piers, or foundation walls; and (3) building mud tubes over materials they cannot go through (Figs. 2.3-22 and 2.3-23). During certain seasons, winged forms swarm (fly) from the nest and, if successful, develop new colonies.

The control of subterranean termites may consist of one or more of the following protective methods: (1) properly installed metal shields that isolate termites from wood or force them to build tubes around the shield; (2) chemically treated soil, which forms a barrier through which

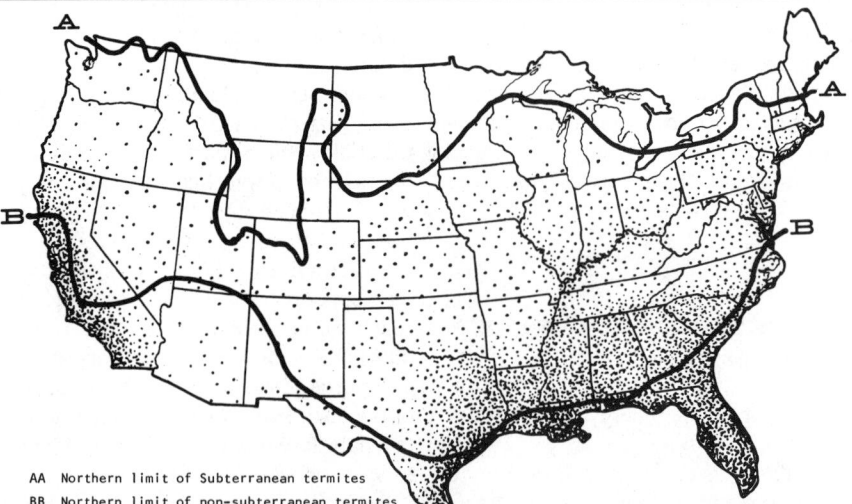

AA Northern limit of Subterranean termites
BB Northern limit of non-subterranean termites

FIGURE 2.3-19 *Relative hazard of termite infestation. Subterranean termites have been found as far north as line A-A. The northern limit of nonsubterranean termites is indicated by line B-B.*

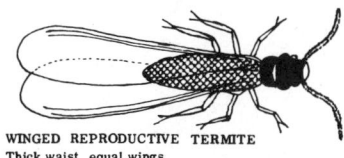

WINGED REPRODUCTIVE TERMITE
Thick waist equal wings

ANT

Thin waist
unequal wings

FIGURE 2.3-20 *Ants are often mistaken for termites. Subterranean termites can be identified by their thick waists and wings of equal length. Ants are thin-waisted and have unequal-length wings.*

FIGURE 2.3-21 *Subterranean termite infestation of foundation plates. Improper clearance between wood siding (removed for photo) and the ground permitted termite entry. (Source unknown.)*

FIGURE 2.3-22 *Termites build mud tubes to permit access from the ground to the wood. The tube shown here is on a concrete foundation pier. Note that the tube has been built around the metal shield at the top of the pier. (Source unknown.)*

FIGURE 2.3-23 *The tubes shown here have been built from the wood above down to the ground. Tubes can be built up foundation walls or posts many feet from the ground to the wood. (Source unknown.)*

termites cannot pass to reach the construction; (3) chemically treated, pressure-impregnated wood, which is not subject to termite attack (see Section 6.1); (4) poured concrete foundations with no cracks greater than 1/64 in.; (5) poured reinforced concrete caps, at least 4 in. thick, on unit masonry foundations, with no cracks greater than 1/64 in. occur.

Regardless of the protective method used, its design and construction should permit visual inspection wherever possible.

Nonsubterranean termites are found only south of line B-B shown in Figure 2.3-19. Their total damage is much less than that of subterranean termites, but they may present a problem in the areas where they do occur. Also called *drywood termites*, because of their ability to live in damp or dry wood without moisture or contact with the ground, their presence is often recognized by slightly compressed pellets of partially digested wood pushed through surface openings. Where this type of termite is found, lumber should be inspected to see that it is not infested before arrival at the site. When construction is underway during the swarming season, lumber should be inspected for infestation.

Slab-on-grade construction requires special considerations in areas where sub-terranean termite hazard is a significant problem (see Figure 2.3-19). Slabs are expected to crack in varying degrees, and termites can penetrate through cracks as small as 1/32 in., through joints between the slab and the foundation wall, through control joints, and through openings made for plumbing and conduit.

Some degree of protection of wood structural elements is provided by positive site and foundation drainage and separation of wood elements from the ground. Foundations and piers of hollow masonry units should have solid masonry caps in which all joints or voids are filled with mortar. Unless the foundation fill is of limited capillarity, at least a 4-in. base course of limited capillarity and a suitable vapor retarder should be provided under the slab.

Where local experience indicates that additional protection against termites is necessary, termite barriers should be provided. Additional protection can be afforded by using pressure-treated wood, which will resist decay as well. See Sections 6.1.5 and 6.5.2.

Soil poisoning is the most effective termite barrier. The area under and around a building should be treated, as well as all possible points of entry. The termiticides registered with the U.S. Environmental Protection Agency (EPA) and currently most commonly used are chlorpyrifos, cypermethrin, fenvalerate, isofenphos, and permethrin. The recommended concentrations of these chemicals in treatment solutions range from 0.25% to 1%. The actual concentration and application of the chemicals and the precautions involved in their use should be in accordance with the manufacturer's recommendations, U.S. Department of Agriculture Home and Garden Bulletin 64, "Subterranean Termites, Their Prevention and Control in Buildings," and applicable building codes.

The effective performance of termite shields depends on proper construction and workmanship and long-term integrity of the slab without cracking, which is unpredictable. Therefore, the effectiveness of termite shields is questionable.

2.4 SURFACE- AND GROUNDWATER PROBLEMS

Provisions must be made to solve problems associated with groundwater and surface water and to prevent them from entering a building through the walls or floors. If water penetrates the walls of a basement, the habitable space there may become undesirable. Water in a crawl space or other enclosed space may cause conditions conducive to decay, musty odors, mold growth, and corrosion of some metals. Water below an improperly constructed slab on grade can migrate through it to cause delamination of floors and can support condensation and frost accumulation at the perimeter of associated walls. In some soils, groundwater can rise high enough to fill heating ducts embedded in or buried beneath a concrete slab.

In addition to the measures discussed here, foundation walls enclosing basement spaces usually require additional protection by applications of *parging* and *dampproofing* on the exterior surface below grade (see Sections 4.7.3 and 7.2). When water under pressure (*head*) occurs above a footing, *waterproofing* may be necessary (see Section 7.2).

2.4.1

SURFACE WATER

Surface water is water that runs along the surface of the ground as a result of rain, downspout discharge, melting snow, or another source. Most surface-water problems come from improper grading of the site adjacent to a building.

2.4.2

GROUNDWATER

Groundwater is water that is in the ground most of the time, although its level (the *water table*) may fluctuate during dry and rainy seasons. Most groundwater problems result from locating foundations too

close to the subsurface water table. Problems then occur when (1) the level of the water in the ground temporarily rises due to a seasonal increase in precipitation, or (2) the water in the ground rises due to the capillarity of the soil. When the water table is so high that a foundation would be submerged constantly, either the site should be abandoned or more extensive waterproofing methods must be used to make the foundation watertight (see Section 7.2).

2.4.2.1

Water-Table Rise

Problems caused by seasonal fluctuations in precipitation that cause the water table to rise temporarily above the level of the foundation may be solved by subsurface drainage. The most effective means of resisting water infiltration through foundations is to collect the water and discharge it away from the site before it reaches the foundation.

2.4.2.2

Capillary Rise

Water will rise in certain soils in opposition to gravity due to capillary action. This *capillarity* occurs when the molecular attraction of water molecules to adjacent soil molecules is greater than the molecular attraction of water molecules to each other. This causes water in a small space (capillary tube), such as the voids between soil grains, to spread out upward on the sides of the space (soil grain surfaces), forming a concave water surface (Fig. 2.4-1). The water's surface tension, which always tries to reduce the surface area, pulls the top of the water upward following the edges. The capacity of the water to resist forces tending to pull it apart holds the tube of water together as it creeps upward.

In general, capillarity increases with a decrease in the size of the space between the grains in a soil. In fact, moisture can

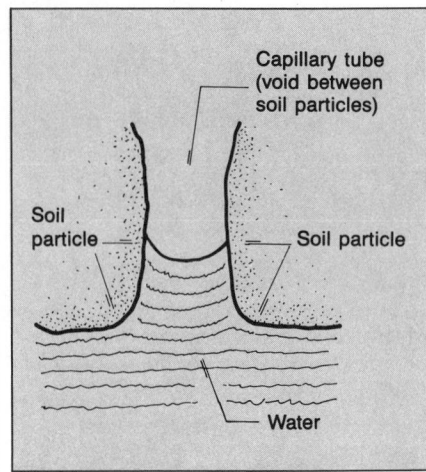

FIGURE 2.4-1 *Water rises in the voids between soil grains in some soils where the grains are tightly packed.* (Drawing by HLS.)

move upward through the voids within some clays as much as 11 ft. above the water table (Fig. 2.4-2). However, because of the large spaces between soil particles, capillarity does not occur in coarse granular soil types, such as gravel. Therefore, one method of control is to replace soil having high capillarity with granular materials.

Water vapor will also rise through the voids between soil particles, whether or not granular soil materials are used. This rise of water vapor can be interrupted and prevented from entering a crawl space or a building by covering the ground surface with a vapor retarder (Fig. 2.4-3). Section 7.1.3.2 discusses vapor retarders in crawl spaces.

2.4.3

CONTROL OF WATER PROBLEMS

Surface- and groundwater can cause problems both during construction and in a completed building. Means of controlling surface water in the completed project by grading are discussed in Section 2.3.6.

CAPILLARY RISE	SOIL TYPE	SATURATION ZONE
Greater Than 8'	CLAY	Greater Than 5'
Greater Than 8'	SILT	Greater Than 5'
3' to 8'	FINE SAND	1' to 5'
1' to 3'	COARSE SAND	0 to 1'
0	GRAVEL	0

FIGURE 2.4-2 *Water rises in soils by capillary action. Clays and silts may become fully saturated to almost 6 ft. above a water table, and some water may rise more than 11 ft. Note that even coarse sand may permit a rise of up to 3 ft. Coarse gravel prevents the rise of water by capillarity.*

FIGURE 2.4-3 *General moisture control: a sloped finished grade provides run-off of surface water from the foundation; a vapor retarder resists the rise of water vapor due to capillarity; and the foundation drains provide subsurface drainage.* (Drawing by HLS.)

This section covers removal of ground water both during construction and from the completed building.

2.4.3.1

Dewatering

Dewatering is the prevention of water from entering an excavation when possible and the removal of water that does find its way in. Groundwater can be prevented from entering an excavation by means of berms and drainage channels or ditches.

Water that enters an excavation as rain, snow, or other precipitation can often be removed by means of portable pumps. Such water should be conducted to pump locations by depressions or ditches. Foundation trenches should not be used as drainage ditches, however, because the water will erode or otherwise damage the bearing surface at the bottom of the trench.

When the water table is higher than the bottom of excavations, it is necessary to continuously lower the level of the water while the work is going on. The most common way of doing this is to sink a series of pipes with screened ends (*well points*) around the perimeter of the excavation to a level below the foundations. These well points may each be fitted with a pump, but more often they are interconnected by *suction lines* (pipes to one or more larger pumps that empty them all simultaneously). In very deep excavations two rows of well points may be needed. Well points and pumps lower the water table and keep it lower until pumping stops. The *discharge* from such wells should be conducted to storage ponds or positive outfalls that lead to a stream or storm sewer, depending on which is permitted or required in the jurisdiction of the project.

2.4.3.2

Slabs on Grade

Settlement or differential movement of slabs and foundations caused by moisture fluctuations in problem soils is discussed in Section 2.1.8. Additional provisions for moisture control are necessary for long-term performance of concrete slab-on-grade construction. Slabs should be protected from water and water vapor, to prevent damage to finish flooring materials and pipes or heating ducts embedded in the slab and to guard against reduced effectiveness of thermal insulation. Problems may be caused by either surface or groundwater.

The methods and the extent of protection depend on (1) the slab elevation with respect to the elevation and the slope of the finished grade; (2) the location of the groundwater table or other man-made water sources, such as seepage fields; (3) the drainage properties of the soil or fill beneath the slab; and (4) the type of finish flooring used.

Many of the moisture problems associated with slab-on-grade construction can be minimized or eliminated by proper preliminary grading, correct selection of fill or base course materials, and installation of a vapor retarder. Slab base courses are discussed in Section 2.3.5; water control by grading in Section 2.3.6.

■ SLAB ELEVATION

The tops of slabs should be located at least 8 in. above the exterior finished grade level. The bottoms of heating ducts placed in or under a slab should be at least 2 in. above the adjacent exterior finished grade, unless the ducts are concrete, plastic, mineral-fiber-cement, or ceramic.

■ VAPOR RETARDER

Where no drainage or soil problems exist and in regions where irrigation and heavy sprinkling are not done, a vapor retarder may not be necessary. However, because it is impossible to correct moisture problems after construction and because a vapor retarder usually can be installed at nominal cost during construction, a vapor retarder is recommended for all slab-on-grade construction when impermeable floor coverings will be used and when floor coverings, furnishings, or equipment must be protected from damage by moisture conditions (see Fig. 2.3-13). A vapor retarder membrane should have a permeableness of less than 0.30 perm and should resist deterioration as well as puncture from heavy traffic during construction. A vapor retarder should be installed with a 6-in. lap and be fitted tightly around utility and other service openings.

Vapor retarder materials suitable for slab-on-grade construction include single-layer membranes, such as polyethylene sheet, and multiple-layer membranes, such as glass fiber–reinforced waterproof paper with polyethylene film extrusion coated on both sides. Polyethylene sheet used over sand or firmly compacted soil under unreinforced slabs should be at least 4 mils in nominal thickness; when used over gravel or crushed stone or under any structurally reinforced slab, it should be at least 6 mils in nominal thickness.

Except when a high-strength concrete with a lower than normal slump is used, a vapor retarder should not be placed directly beneath a slab. Instead a 2- to 3-in. layer of sand should be placed over the vapor retarder. This method permits water that bleeds from a high-slump or low-strength concrete to pass into the sand rather than migrate upward through the slab where it will cause the slab to crack more than usual and weaken the surface, resulting in dusting and spalling.

2.4.3.3

Subdrainage Systems

Subdrainage systems include *foundation drains* and *underslab drains*. Some codes require that all buildings having basements or crawl spaces below grade have foundation drains. This is a good idea in any case, whether or not required by code, to prevent surface- or groundwater from finding its way into crawl spaces or into basements through the walls. Foundation drains frequently are used in conjunction with waterproofing (see Section 7.2). In addition, a grid of drains should be installed to prevent the build-up of hydrostatic head beneath large buildings and buildings where the water table is above the slab on grade or above the grade inside a crawl space.

Many drain pipe materials are available, including nonperforated concrete drain tile, clay drain tile, cast iron soil pipe, and polyvinyl chloride pipe; porous concrete pipe; and perforated concrete, clay, polyvinyl chloride, and bituminous fiber pipe. Foundation and underslab drains are usually made from 4- or 6-in. pipe.

Nonperforated pipes are laid with open joints. These joints are then covered with a layer of screening to prevent fine particles from the fill above the pipe from entering the pipe through the joints. This screening material is usually heavy mesh burlap, coal tar–saturated felt, copper mesh screen, or synthetic filter fabric.

Perforated pipes are laid with the perforations facing down and the joints closed in accordance with the pipe manufacturer's instructions.

Drainage fill material is gravel or crushed stone evenly graded so that all of it passes a 1-1/2-in. sieve and not more than 5% passes a #50 sieve.

Foundation drains are placed in a bed of compacted drainage fill material around the exterior perimeter of a building's foundation (see Fig 2.4-3). Drain pipe should be placed far enough below the floor slab to properly drain and with its lowest *invert* (bottom of pipe) at least 4 in. above the bottom of the footing to prevent possible washout of the soil beneath the footing. Footings should be lowered if necessary to meet these parameters.

Underslab drains are placed in trenches filled with drainage fill material (Fig. 2.4-4). Pipes usually are placed in parallel rows across the building. Depending on the distances involved, interceptor pipes may be placed perpendicular to these rows to reduce the depth of the piping required to establish a positive drainage slope.

In traditional installations, foundation drain pipes are set in a compacted 4-in.-thick bed of drainage fill material laid with the bottom on the undisturbed subgrade at the bottom of the footing. The pipes are then covered with drainage fill material that extends from 4 to 6 in. on each side of the pipe up to a level about 12 in. below the surface of the ground (see Fig. 2.4-3). The drainage fill should be placed in 3-in.-thick layers, and each layer should be compacted before the next layer is placed. The drainage fill material is then covered with a layer of 15-lb. asphalt- or tar-saturated felt or filter fabric, to prevent fine materials from infiltrating the drainage fill. An impervious fill material (earth) is then applied, sloping away from the building, to bring the grade to that required.

Another method is being used today in some installations to overcome the difficulty in a traditional installation of keeping the drainage fill material and adjacent backfill materials separated. The preceding discussion about the traditional method

FIGURE 2.4-4 *Underslab drain.* *(Drawing by HLS.)*

Finish grade

Insulation and finish

Floor finish and base

Concrete slab

Drainage board

Sand

Vapor retarder

Slab bed

Filter fabric

Drainage fill

Foundation drain

Underslab drain beyond

FIGURE 2.4-5 *Foundation drain with in-plane drain, showing the connections of under-slab drains where they occur.* *(Drawing by HLS.)*

applies to this method as well, except that the drainage fill material is stopped 4 to 6 in. above the drain pipe and an *in-plane drain* is placed against the founda-

tion wall from the bottom of the drainage fill to a level about 12 in. below grade (Fig. 2.4-5).

In-plane drains (also called *drainage boards*) are thin boards (usually about 1 in.) consisting of two parts. The drain portion is made of a plastic material configured so that it forms multiple open pathways, which permit water to flow freely through them. The side facing the earth is a layer of synthetic filter fabric that permits water to enter the drain but excludes fine soil. These boards can be installed with adhesives or mechanical fasteners as directed by the manufacturer.

Whether the traditional or in-plane drainage system is used, the drainage system should be tested to be sure that it properly drains before it is covered by drainage fill materials. The method of test is usually dictated by code or local practice.

Collected water may drain by gravity to a natural outfall at a lower elevation, or it may be discharged by positive machine pumping from one or more collecting sumps located within the foundation area. Some codes dictate the minimum size of pits and pumps, but they may be insufficient for a particular project. Therefore, it is important that both pit size and pump capacity be based on engineering design.

2.5 LAWNS AND LANDSCAPING

Lawns and landscaping serve multiple purposes around a building. Most people think first of their aesthetic appeal, but lawns and landscaping also serve as buffer zones between adjacent buildings and between buildings and roadways or other property. When properly placed, plants can help keep a building cooler in summer and warmer in winter, thereby reducing dependence on mechanical air conditioning and heating equipment.

2.5.1

PLANNING REQUIREMENTS

Plantings can prevent reflection and radiation of solar energy from paved areas and adjacent buildings (Fig. 2.5-1). They can shield exterior walls from the sun, thus reducing heat transmission through them. Earth berms can also be used to shield walls and outdoor living spaces.

Coniferous (evergreen) trees can be located to block prevailing winter winds. *Deciduous* (leaf-dropping) trees on south and west exposures can shade walls and windows in summer but allow desirable winter sun penetration (Fig. 2.5-2). Ponds and fountains fed by collected rainwater act as natural coolants to reduce air temperature around building exteriors (Fig. 2.5-3) and should not be blocked by vegetation.

Plants must be selected and placed carefully for several reasons. For example, placing deciduous trees on northern expo-

FIGURE 2.5-1 *Shading paved areas prevents heat build-up and channels breezes.*

sures can leave a building unprotected from cold winter winds. In addition, plants that hide a dwelling can also hide an intruder. Trees should be far enough away from a dwelling so that a burglar cannot climb them and enter the second story. Bushes should be low enough so

WINTER: LEAFLESS TREES LET SUN THROUGH. **SUMMER: LEAVES SHADE HOUSE.** **WINTER: EVERGREENS PROTECT HOUSE.** **SUMMER: LEAVES DIRECT BREEZE INTO HOUSE.**

FIGURE 2.5-2 *Proper trees in the right places will reduce energy needs.*

FIGURE 2.5-3 *Air flowing over cool water acts as natural air conditioning. Plants should not block this flow.*

they do not obscure windows or doors. Hedges used as fences should be no more than 40 in. high. Landscaping should not obscure a house from the street. Small plants and floral borders can serve to define the borders between private and public spaces.

Refer to Section 11.3.2 for additional site-planning recommendations related to thermal considerations.

2.5.2

TREES AND SHRUBS

Tree and shrub quality should be based on the American Association of Nursery-

men publication "American Standard for Nursery Stock," which has been accepted by ANSI as Standard Z-60.1. This publication contains a standardized system of sizing and describing plants. It covers shade and flowering trees, deciduous shrubs, coniferous evergreens, broadleaf evergreens, rose shrubs, young plants, fruit trees, bulbs, corms, and tubers. It describes and limits the dimensions of container plants, balled and burlapped plants, bareroot plants, balled and potted plants, and processed balled plants. It also lists the trees and shrubs that fit into each of its categories.

The U.S. Department of Agriculture has identified 11 zones of plant hardiness in the United States. Figure 2.5-4 is an adaptation of the USDA Plant Hardiness Zone Map in the USDA Agricultural Research Service Miscellaneous Publication Number 1475. USDA has also identified many plants as they relate to the plant hardiness zone map. Figure 2.5-5 lists 10 of the 11 zones and examples of plants that will succeed in each. (Zone 11 includes areas where the average annual temper-

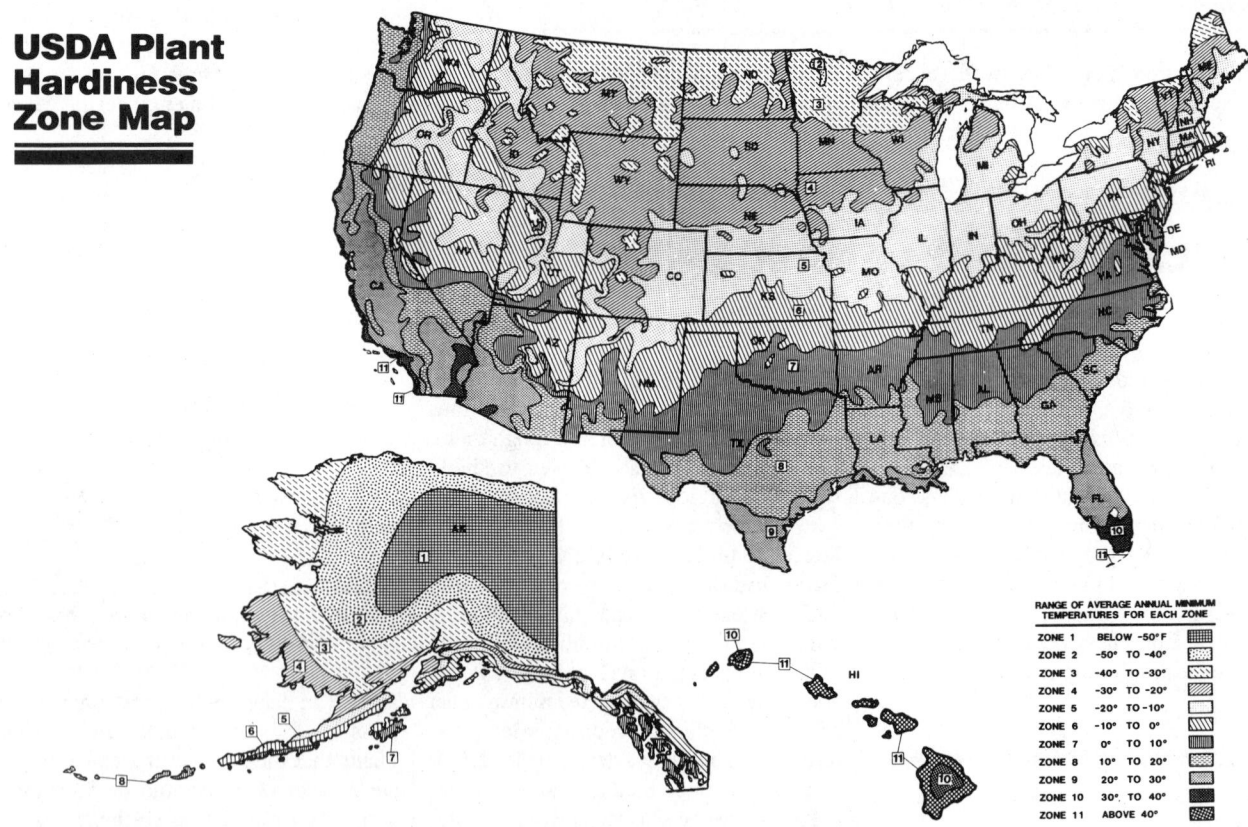

USDA Plant Hardiness Zone Map

RANGE OF AVERAGE ANNUAL MINIMUM TEMPERATURES FOR EACH ZONE	
ZONE 1	BELOW −50° F
ZONE 2	−50° TO −40°
ZONE 3	−40° TO −30°
ZONE 4	−30° TO −20°
ZONE 5	−20° TO −10°
ZONE 6	−10° TO 0°
ZONE 7	0° TO 10°
ZONE 8	10° TO 20°
ZONE 9	20° TO 30°
ZONE 10	30° TO 40°
ZONE 11	ABOVE 40°

FIGURE 2.5-4 *USDA plant hardiness zone map.* (From U.S. Department of Agriculture, Agricultural Research Service Miscellaneous Publication 1475, "USDA Plant Hardiness Zone Map.")

FIGURE 2.5-5
Indicator Plant Examples

Zone	Botanical Name	Common Name
1 °F Below −50 °C Below −45.6	Betula glandulosa Empetrum nigrum Populus tremuloides Potentilla pensylvanica Rhododendron lapponicum Salix reticulata	Dwarf birch Crowberry Quaking aspen Pennsylvania cinquefoil Lapland rhododendron Netleaf willow
2 °F −50 to −40 °C −45.6 to −40	Betula papyrifera Cornus canadensis Elaeagnus commutata Larix laricina Potentilla fruticosa Vibrunum trilobum	Paper birch Bunchberry dogwood Silverberry Eastern larch Bush cinquefoil American cranberry bush
3 °F −40 to −30 °C −40 to −34.5	Berberis thunbergii Elaeagnus angustifolia Junipercus communis Lonicera tatarica Malus baccata Thuja occidentalis	Japanese bayberry Russian olive Common juniper Tatarian honeysuckle Siberian crabapple American arborvitae
4 °F −30 to −20 °C −34.5 to −28.9	Acer saccharum Hydrangea paniculata Juniperus chinensis Ligustrum amurense Parthenocissus quinquefolia Spiraea x vanhouttei	Sugar maple Panicle hydrangea Chinese juniper Amur River privet Virginia creeper Vanhoutte spirea
5 °F −20 to −10 °C −28.9 to −23.3	Cornus florida Deutzia gracilis Ligustrum vulgare Parthenocissus tricuspidata Rosa multiflora Taxus cuspidata	Flowering dogwood Slender deutzia Common privet Boston ivy Japanese rose Japanese yew
6 °F −10 to 0 °C −23.3 to −17.8	Acer palmatum Buxus sempervirens Euonymus fortunei Hedera helix Ilex opaca Ligustrum ovalifolium	Japanese maple Common box Winter creeper English ivy American holly California privet
7 °F 0 to 10 °C −17.8 to −12.3	Acer macrophyllum Rhododendron Kurume hybrids Cedrus atlantica Cotoneaster microphylla Ilex aquifolium Taxus baccata	Bigleaf maple Kurume azalea Atlas cedar Small-leaf cotoneaster English holly English yew
8 °F 10 to 20 °C −12.3 to −6.6	Arbutus unedo Choisya ternata Olearia haastii Pittosporum tobira Prunus laurocerasus Viburnum tinus	Strawberry tree Mexican orange New Zealand daisy-bush Japanese pittosporum Cherry-laurel Laurestinus
9 °F 20 to 30 °C −6.6 to −1.1	Asparagus setaceus Eucalyptus globulus Syzygium paniculatum Fuchsia hybrids Grevillea robusta Schinus molle	Asparagus fern Tasmanian blue gum Australian bush cherry Fuschia Silk-oak California pepper tree
10 °F 30 to 40 °C −1.1 to 4.4	Bougainvillea spectabilus Cassia fistula Eucalyptus citriodora Ficus elastica Ensete ventricosum Roystonea regia	Bougainvillea Golden shower Lemon eucalyptus Rubber plant Ensete Royal palm

From U.S. Department of Agriculture, Agricultural Research Service Miscellaneous Publication 1475, "USDA Plant Hardiness Zone Map."

Note: Representative persistent plants listed under the coldest zones in which they normally succeed. Such plants may serve as useful indicators of the cultural possibilities of each zone.

ature is above 40 °F and therefore essentially frost free.) Figure 2.5-6 is a list of additional plants and the zones in which they will survive. The zones listed for each plant are optimum; listed plants may survive in warmer or colder climates, but their performance probably will not be as satisfactory.

Other factors that affect plant growth and hardiness include the amount of rainfall and the soil and its amendments. Plants can be grown in areas that have less than optimum rainfall for them, as long as they are watered artificially.

Before selecting plants for use in a particular area, it is advisable to verify their use with a local horticulturist or nurseryman or with a state or local agricultural agent. In some states agricultural extension services are good sources for such information.

2.5.3

LAWNS

There are some 600 grass genera, of which about 40 are suitable for lawns. However, not all of these can be used in every climatic zone. For example, bluegrass (*Poa pratensis*) grows very well in many areas but often is not successful in subtropical zones. St. Augustine grass (*Stenotaphrum secundatum*) needs a warm, moist environment and will not fare well in cold, dry zones.

Some grasses are more suitable for use in shaded areas than others, but not one is really suitable for use in deep shade. Those that sometimes work in light shade include Bahia (*Paspalum notatum*), bent (*Agrostis*), carpet (*Axonopus offinis*), centipede (*Eremochloa ophiuroides*), fescue (*Festuca rubra*), rye (*Lolium perenne*), St. Augustine, and zoysia (*Zoisia matrella*). Those suitable for deeper shade include fescue, St. Augustine, and zoysia. Successful growth of a particular grass in shade depends on other factors such as the amount of moisture present or delivered artificially, the drainage of the area, the fertility of the soil, and the type of vegetation providing the shade. All shading vegetation will draw some moisture from adjacent grass, but trees and shrubs with shallow roots may present more of a problem than deep-rooted ones. Local sources

FIGURE 2.5-6

Cold Hardiness Ratings for Some Additional Wood Plants

Botanical and Common Name	Zone	Botanical and Common Name	Zone
Abeliophyllum distichum (white forsythia)	5b	*Jacaranda acutifolia* (green ebony)	10
Acer platanoides (Norway maple)	4	*Juglans regia* (English or Persian walnut)	6b
Aesculus x *carnea* (red horsechestnut)	4	*Juniperus horizontalis* (creeping juniper)	3
Araucaria araucana (monkeypuzzle)	7b		
Arctostaphylos uva-ursi (bearberry)	2b	*Koelreuteria paniculata* (goldenrain-tree)	6
Aristolochia durior (Dutchman's-pipe)	4b		
Aucuba japonica (Japanese aucuba)	7b	*Laburnum* x *watereri* (Waterer laburnum)	5b
		Lagerstroemia indica (crapemyrtle)	7
Bauhinia variegata (purple orchid tree)	9b		
Berberis darwinii (Darwin barberry)	8	*Mahonia aquifolium* (Oregon hollygrape)	5b
Betula pendula (European white birch)	3	*Malus* x *arnoldiana* (Arnold crabapple)	4
Bouvardia 'Coral' (Coral bouvardia)	9	*Melia azedarach* (chinaberry)	7b
Butia capitata (Pindo palm)	8b	*Metasequoia glyptostroboides* (dawn redwood)	5b
		Myrtus communis (true myrtle)	8b
Camellia reticulata (*reticulata camellia*)	9		
Camellia sasanqua (*sasanqua camellia*)	7b	*Nandina domestica* (heavenly bamboo)	7
Carya illinoinensis 'Major' (pecan)	5 (grows) 6 (fruits)	*Nerium oleander* (oleander)	8b
Casuarina equisetifolia (Australian pine)	9b	*Olea europaea* (common olive)	9
Ceanothus impressus (Santa Barbara ceanothus)	8	*Osmanthus heterophyllus* (holly osmanthus)	7
Cedrus dodara (deodar cedar)	7b		
Cercis chinensis (Chinese redbug)	6b	*Picea abies* (Norway spruce)	3
Chamaecyparis lawsoniana (Lawson cypress)	6b	*Pieris japonica* (Japanese andromeda)	6
Chamaecyparis pisifera (Sawara cypress)	5	*Pinus mugo* var. *mughus* (Mugo pine)	3
Cinnamomum camphora (camphor tree)	9	*Pinus radiata* (Monterey pine)	7b
Cistus laurifolius (laurel rockrose)	7	*Pinus strobus* (eastern white pine)	3b
Cistus x *purpureus* (purple rockrose)	8	*Prunus yedoensis* (Yoshino cherry)	6
Cornus alba (Tatarian dogwood)	3		
Cornus kousa (Japanese dogwood)	5b	*Rhaphiolepis indica* 'Rosea' (Indian hawthorn)	8
Cunninghamia lanceolata (cunninghamia)	7	*Rhododendron* 'America' (hybrid rhododendron)	5
Cytisus x *praecox* (Warminster broom)	6	*Rhododendron* 'Loderi King George' (hybrid rhododendron)	8
		Rhododendron mollis hybrids (mollis azalea)	5
Elaeagnus multiflora (cherry elaeagnus)	5	*Rhododendron prinophyllum* (*roseum*) (roseshell azalea)	4
Elaeagnus pungens (thorny elaeagnus)	7	*Rhododendron* 'Purple Splendor' (hybrid rhododendron)	7
Eriobotrya japonica (loquat)	8	*Rhododendron* southern Indian hybrids (Indian azalea)	8b
Euonymus alatus (winged euonymus)	3b	*Rosa rugosa* (rugosa rose)	3
Euphorbia pulcherrima (poinsettia)	10		
		Schinus terebinthifolius (Brazilian pepper-tree)	9b
x *Fatshedera lizei* (botanical-wonder)	8	*Sequoia sempervirens* (redwood)	8
Forsythia ovata (early forsythia)	4b	*Sequoiadendron giganteum* (giant sequoia)	7
Forsythia suspensa (weeping forsythia)	5b	*Stewartia pseudocamellia* (Japanese stewartia)	6
Fremontodendron mexicanum (flannel bush)	9	*Syringa vulgaris* (common lilac)	3b
Ginkgo biloba (ginkgo, maidenhair-tree)	5	*Ulmus americana* (American elm)	2
Hibiscus rosa-sinensis (Chinese hibiscus)	9b	*Viburnum burkwoodii* (Burkwood viburnum)	5b
Hibiscus syriacus (shrub althea)	5b		
Hypericum 'Hidcote' (Hidcote St. Johnswort)	6	*Zelkova serrata* (Japanese zelkova)	5b
Iberis sempervirens (evergreen candytuft)	5		
Ilex crenata 'Convexa' (convexleaf Japanese holly)	6b		

From U.S. Department of Agriculture, Agricultural Research Service Miscellaneous Publication 1475, "USDA Plant Hardiness Zone Map."

should be contacted for advice before selecting a grass for use in shaded areas.

Some species, such as bluegrass and fescue, propagate successfully from seed. Many southern species and some northern creeping species, however, should be established from living material such as sod, plugs, sprigs, or stolens. These species include zoysia, St. Augustine, centipede, and many northern bent grasses. Nonseeding methods should be used regardless of the grass selected on slopes of 1:6 and steeper to prevent loss of seed through erosion and when an "instant lawn" is desired.

The name of the proper grass to use in a particular location should be obtained from a local horticulturist or nurseryman or from a state or local agricultural agent. In general, the following grasses work well in cool, humid zones: bluegrass, both red and alta fescue, and rye. Grasses for dry areas include beach (*Ammophila*), buffalo (*Buchloë*), blue gamma (*Bouteloua*), and wheat (*Agropyron*). In some cases, Bermuda (*Cynodon dactylon*), centipede, fescue, and zoysia grasses may also work in dry zones. Bent grass and bluegrass may be used in dry zones when irrigation or sprinklers are used. Bahia, Bermuda, carpet, centipede, St. Augustine, and zoysia work well in warm, moist zones.

2.5.4

PLANTING SOIL

Planting soil for trees, shrubs, and lawns should be topsoil that is fertile, friable, natural loam. It should not contain subsoil, clay lumps, brush, weeds, litter, stumps, roots, rocks, or any other material that would harm the plants or hinder their growth.

In addition, planting soil for trees, shrubs, and lawns must contain the nutrients each particular plant needs. In general, plants need the proper amounts of some 10 to 14 nutrients. Trees and shrubs need a pH value between 6 and 7. The numbers in the pH scale, which runs from 0 to 14, represent the relative level of alkalinity or acidity. Seven is neutral; above is alkaline, and below is acidic. The farther away from 7, the more alkaline or acidic. The required pH value for grasses varies with the species. The actual nutrients needed and their percentage of the soil material vary with the plant species. In addition, some trees and shrubs require a pH lower or higher than 6 to 7 level. It is necessary to determine the exact requirements for each particular plant from a local horticulturist or nurseryman or a state or local agricultural agent.

Soil samples taken from the site should be tested in a qualified laboratory to determine its pH level and the amounts of necessary nutrients present. Some state agricultural extension services will perform such tests. Private laboratories also are available in some places. The levels of the proper nutrients can be adjusted as necessary with such soil amendments as humus, mulch, sawdust, manure, and fertilizer. The pH of an acidic soil can be lowered by the addition of lime in the form of natural dolemitic limestone. Alkaline soils can be adjusted by adding aluminum sulfate. Heavy or cohesive soils may be loosened or lightened by the addition of perlite, vermiculite, or in some cases, sand. Amended soils should be retested to confirm that they have the correct amounts of the proper nutrients and the correct pH value for the plant material to be used.

3

CONCRETE

Photo: Associates Center, Chicago, Illinois. (Architects: A. Epstein and Sons; photo courtesy of Portland Cement Association.)

INTRODUCTION

Concrete is used in almost every type and size of architectural and engineering structure. Its major uses are for foundation walls and footings, building superstructures, floor slabs, stairs, sidewalks, paving, curbs and gutters, streets, highways, bridges, and site structures such as storm water inlets and piping, septic tanks, and drainage swales. Evidence is that the use of concrete is increasing. This increase is due partly to the inherent fire resistance of concrete, which eliminates the expensive fireproofing required on steel structures. Ease of erection has increased the use of precast concrete units in all types of buildings. Concrete is an old material, but the technology associated with it continues to evolve. Recent developments in fiber-reinforced concrete are being specified increasingly. Depending on the fibers used, these materials toughen slabs and help to control cracking, thus reducing the need for control joints in slabs. Recently developed super plasticizers make concrete more fluid, which makes delivering it to its final location by pumping more practicable.

The term *concrete of a suitable quality* and other phrases to that effect refer to concrete that will perform satisfactory service in the use for which it is intended. To do so, it must possess four essential functional properties: (1) the strength to carry superimposed loads; (2) sufficient watertightness to prevent water penetration; (3) the durability to resist wear and weather; and (4) workability, to ensure proper handling, placing, finishing, and curing. These properties can be achieved only with good materials and careful workmanship.

Required surface treatments may range from smooth-troweled surfaces for finished floors to highly decorative geometric patterns for terraces, walls, and other surfaces. The plastic quality of concrete offers almost unlimited design possibilities of form, pattern, and texture.

The in-place cost of concrete is from three to six times the cost of its ingredients. The in-place cost as well as the long-term cost of maintenance depend largely on proper care in handling, placing, finishing, and curing. Skillful selection and combining of ingredients in the mix are necessary to achieve the most economical concrete. These same factors also influence the quality of the concrete. No concrete mix is ever better than the workmanship it receives in the field.

The essence of producing concrete of suitable quality can be summarized as follows: Sound materials must be carefully selected, proportioned, and combined into the correct mixture with a consistency that can be worked into place to provide the required surface finish, followed by adequate curing and protection.

Unless quality control measures are observed, undesirable properties in the hardened concrete may occur, especially on the wearing surface. These undesirable properties may lead to a soft or *dusting* surface, permeable concrete, cracking, or poor durability.

3.1 HISTORY

The Romans were using a form of concrete made with lime, broken stones, and sand to construct temples and other buildings as early as the third century B.C. At first, they left the surface of concrete rough and finished it with a form of stucco. Soon they began to embed small stones in the surface to give their concrete a finish. Eventually they incorporated broken terra-cotta roof tiles, embedding these at the surface with their smooth side outward. This innovation eventually led to the manufacture of clay bricks.

Roman architecture was altered forever by the accidental discovery near Mt. Vesuvius of a native natural cement called *puzzolana*. This was a silica and alumina-bearing material that, when combined with crushed limestone, produced a slow-drying cement that was much stronger, harder, and more adhesive than the cements the Romans had been making using lime. In addition, it was *hydraulic*, meaning that it would set both in the air and under water. The name, now spelled *pozzolana*, is used today to describe a variety of siliceous or siliceous and aluminous materials used as a substitute for a part of the portland cement in some modern portland cement concrete.

Better concrete soon led the Romans to use more of this material and to build larger structures, especially arches, vaults, and domes. Concrete then, as now, had no permanent form of its own before it set and needed some sort of form or mold to hold it in the desired shape. The Romans used wooden forms often held by temporary support (*centering*). To conserve materials and labor, they invented reusable wood forms and centering that could be used repeatedly. They also used brick or stone to form the outer layers of a structure and filled the voids between the outer layers with concrete. They built impressive domed and vaulted structures using unreinforced concrete to house temples, baths, and other large spaces, some with spans as large as 150 feet.

With the collapse of the Roman Empire, concrete technology largely fell into disuse and remained virtually unknown until the rebirth of learning during the Renaissance. In the 15th century, the 10-book series *De architectura*, written by Roman architect and engineer Marcus Vitruvius Pollio in the first century B.C., became a popular topic of study.

It was not until near the end of the 18th century, however, that research related to concrete technology resumed and not until 1824 that the essential ingredient in modern concrete was discovered. In that year, a brick mason named Joseph Aspdin, living in Leeds, England, patented a new type of cement. He made it by treating, heating (*calcining*), and cooling raw materials into small rock-like shapes called *clinker*, which consisted essentially of calcium silicates, and crushing them into a fine powder. Aspdin called his discovery *portland cement* because of its resemblance to a gray rock found on the Isle of Portland, in the English Channel. Aspdin's cement was an overnight success. However, even though ample raw materials were present in the United States, manufacture of portland cement did not begin in this country until 1872.

Concrete has many uses on its own, but modern concrete buildings would not be possible using concrete alone. This is because, while concrete has high strength in compression, it has little tensile strength. Even though other forces are also involved, the primary forces acting on a beam or an unsupported slab are those that tend to push it together (compression) at the top of the beam or slab and those that pull it apart (tension) at the bottom. Therefore, a concrete slab or beam of any appreciable length will pull apart at the bottom and collapse, especially if a weight (load) is applied to it (Fig. 3.1-1a). This problem can be solved by placing in the concrete unit a material, such as iron or steel, that is strong in tension. Such a *reinforced concrete* unit is capable of safely spanning large distances (Fig. 3.1-1b).

In the 1850s, the race was on to develop a practical method of embedding iron and steel in concrete shapes. Early experimentation was not only with beams but also with boats and other devices. In 1857, a

FIGURE 3.1-1 *Dashed lines show beam shape (exaggerated) after load is applied. (a) Unreinforced beam bottom will separate, leading to collapse. (b) Reinforced beam also deflects, but magnitude is less. Concrete cracks at bottom, but does not collapse.* (Drawing by HLS.)

French gardener named F. Joseph Monier patented a system for producing reinforced concrete. He used it to produce flower pots, water tanks, and garden bridges. His techniques were quickly adapted to building construction. In 1857, the first application of reinforced concrete in the United States occurred in Port Chester, New York. By 1900, the use of reinforced concrete in buildings was widespread.

The strength of a reinforced concrete unit can be increased by reducing tensile stresses in the concrete. This is done by stretching the reinforcing steel, which in turn places the concrete in compression before it is loaded. The scientific basis for modern *prestressing* of concrete was developed in the late 1920s by a French engineer named Freyssinet for use in long-span bridge construction. Prestressing is used extensively in precast concrete units.

The production of modern prestressed precast concrete requires the use of a high-strength concrete and special curing methods that were developed in this century. Advances in concrete technology continue at a rapid pace, with the relatively recent development of plastic and metal reusable forms, some of which impart textured finishes to concrete; *flying forms*, which permit rapid form erection and removal; advanced chemical formulas for portland cement that produce different cement formulations (high early strength, air-entrained, etc.) for different uses; admixtures to control the setting time, ease of delivery, and other characteristics of concrete mixes; high-strength steel strands, which make prestressing possible; steel and plastic fiber reinforcement, which make slab surfaces stronger and reduce the need for joints; machine delivery devices, such as transit mixing trucks, batch-mixing plants, and concrete pumps; and machine placing and finishing devices, such as internal and form vibrators, motor-driven screeds, and mechanical riding trowels.

In addition to more conventional structures built with columns, beams, and slabs, concrete is effective in erecting thin-shell structures, some with very large unsupported spans. Because concrete is placed as a fluid, many configurations of such structures are possible. Figure 3.1-2 depicts some of the more common shapes.

Concrete has been used by such renowned architects as Frank Lloyd Wright, LeCorbusier, Pier Luigi Nervi, Felix Candela, Eero Saarinen, Paul Rudolph, and I. M. Pei. The sketches in Figures 3.1-3 through 3.1-6 depict some of their modern concrete masterpieces. Contemporary architects continue to use concrete today, in both high-rise and smaller buildings (Figs. 3.1-7 and 3.1-8).

3.2 CONCRETE MATERIALS AND ACCESSORIES

Concrete is made by combining four materials into a mixture of basically two parts: aggregate and portland cement paste. The paste, composed of portland cement and water, binds the fine aggregate (sand) and coarse aggregate (gravel or crushed stone) into a rock-like mass when the paste hardens. Concrete hardening (*setting*) results from a chemical reaction (*hydration*)

a The **concrete arch** is curved in one direction only and is simple and pleasing in appearance. The arch thickness is variable, being thinner at the center and gradually increasing in thickness toward the supports.

CROWN

BEAM or WALL

b The **barrel-shell** roof is also curved in one direction but is more complicated in construction and spans between rigid frame supports.

c The **dome roof** is a slab of double curvature.

d **Groined vaults** are curved cross vaults often used in large structures such as convention halls or airport buildings.

FIGURE 3.1-2 *(a) Concrete arch; (b) barrel shell; (c) dome; (d) groined vaults.* (*Courtesy Concrete Reinforcing Steel Institute.*)

FIGURE 3.1-3 *The Guggenheim Museum, New York City.* (*Architect: Frank Lloyd Wright; drawing by HLS.*)

FIGURE 3.1-4 *The chapel of Notre Dame du Haut at Ronchamp, France.* (*Architect: LeCorbusier; drawing by HLS.*)

FIGURE 3.1-5 *TWA Terminal, John F. Kennedy Airport, New York.* (*Architect: Eero Saarinen; drawing by HLS.*)

between portland cement and water. The portland cement and fine and coarse aggregates used should conform to the requirements of the standards of the American Society for Testing and Materials (ASTM), which are listed in Figure 3.2-1 and summarized in this section.

Concrete is often reinforced with steel mesh or rods to increase its tensile strength.

3.2.1

PORTLAND CEMENT

Portland cement is a finely pulverized material that consists principally of compounds of lime, silica, alumina, and iron. It is manufactured from selected materials under closely controlled processes (Fig. 3.2-2). The process starts by mixing limestone or marl with such other ingredients as clay, shale, or blast furnace slag in proper proportions and then burning the mixture in a rotary kiln at a temperature of approximately 2700 °F to fuse it into *clinker* (rock-like shapes of varying size, averaging 3/4 in. in diameter). The clinker is cooled and then pulverized with a small amount of gypsum, which is added to slow down setting time. The finished product, portland cement, is so fine that nearly all of it will pass through a sieve with 40,000 openings to the square inch.

When portland cement is mixed with water, a paste is formed that first *sets* (becomes stiff) and then hardens into a solid mass. The setting and hardening is called *hydration* and is brought about by chemical reactions between the portland cement and water. The reaction is not simply one of evaporation or drying.

FIGURE 3.1-6 *Art and Architecture Building, Yale University, New Haven, Connecticut.* *(Architect: Paul Rudolph; drawing by HLS.)*

FIGURE 3.1-8 *Commissioners of Public Works Administrative Offices, Charleston, South Carolina.* *(Architect: Lucus Stubbs Pascullis Powell and Penny, Ltd.; photo courtesy Portland Cement Association.)*

FIGURE 3.1-7 *One Magnificent Mile, Chicago, Illinois.* *(Architect: Skidmore Owings and Merrill; photo courtesy Portland Cement Association.)*

FIGURE 3.2-1
Concrete Materials

Material	Standard
Portland cement	
Types I, II, III, IV, V	ASTM C 150
Types IA, IIA, IIIA	
(air entraining)	ASTM C 150
Pozzolan and blast	
furnace	ASTM C 595
Ready-mixed concrete	ASTM C 94
Aggregates	
Normal weight	ASTM C 33
Lightweight	ASTM C 330
Admixtures	
Air-entraining agents	ASTM C 260
Accelerators:	
calcium chloride	ASTM D 98

Courtesy Portland Cement Association.

3.2.1.1

Types of Portland Cement

Portland cements are made to meet different physical and chemical requirements for specific purposes. ASTM Standard C 150 describes eight types of portland cement: Type I, Normal; Type IA, Normal, Air-entraining; Type II, Moderate; Type IIA, Moderate, Air-entraining; Type III, High Early Strength; Type IIIA, High Early Strength, Air-entraining; Type IV, Low Heat of Hydration; Type V, Sulfate Resisting. These types are summarized in Figure 3.2-3 and discussed in more detail in the following paragraphs.

■ TYPE I, NORMAL

Type I is a general-purpose portland cement suitable for most uses when the special properties of other types are not required. Type I portland cement should not be used where the conrete will be in contact with high-sulfate soils or subject to excessive temperatures during hydration.

■ TYPE II, MODERATE

Type II portland cement is used where precaution against moderate sulfate attack is important, as in drainage structures

FIGURE 3.2-2 *Steps in the manufacture of portland cement, showing both dry and wet processes.* *(Courtesy Portland Cement Association.)*

FIGURE 3.2-3
Portland Cement Types and Uses

ASTM C 150	Use	ASTM C 175
Type I	In general construction when special properties of other types are not required	Type IA*
Type II	In general construction where moderate heat of hydration is required	Type IIA*
Type III	When high early strength is required	Type IIIA*
Type IV	When low heat of hydration is required	
Type V	When high sulfate resistance is required	

Courtesy Portland Cement Association.
*Air-entraining portland cement.

where sulfate concentrations in the groundwater are higher than normal. Type II portland cement generates less heat of hydration and cures at a slower rate than Type I. This moderate heat of hydration reduces temperature rise, which is especially important when concrete is placed in warm weather in structures of con-siderable mass, such as in large piers or heavy retaining walls.

■ TYPE III, HIGH EARLY STRENGTH

Type III portland cement is used when high strengths are desired at a very early time, usually in a week or less. It is used when (1) early form removal is desirable; (2) the concrete must be put into service quickly; (3) the weather is cold, to reduce the period required for protection against low temperatures to control curing; and (4) high early strengths can be secured more satisfactorily or more economically than by using richer mixes of Type I portland cement.

■ TYPE IV, LOW-HEAT

Type IV is a special portland cement for use where the amount and rate of heat generated during hydration must be kept to a minimum. The development of strength is also at a slower rate. Type IV is intended for use in large masses of concrete, such as in large gravity dams, where the temperature rise resulting from the heat generated during hardening is a critical factor.

■ TYPE V, SULFATE RESISTING

Type V is a special portland cement intended for use only in construction exposed to severe sulfate action, such as in some western states that have soils and waters with high alkali content. It has a slower rate of strength gain than normal portland cement.

■ AIR-ENTRAINING PORTLAND CEMENTS

Types IA, IIA, and IIIA, which correspond in composition to Types I, II, and III, respectively, were developed to produce concrete that is resistant to severe frost action and to scaling caused by applications of chemicals for snow and ice removal. In these portland cements, small quantities of air-entraining additions have been incorporated by grinding them in along with clinker during manufacture. These additions should conform with the requirements of ASTM Standard C 226. Concrete made with these portland cements contains tiny, well-distributed, and completely separated air bubbles. Air-entraining portland cements cost no more than the corresponding non-air-entraining types in ASTM C 150.

3.2.1.2

White Portland Cement

White portland cement is manufactured to conform to ASTM Standard C 150, although it is not specifically mentioned there. It is used primarily in precast wall and facing panels, terrazzo, stucco, cement paint, tile grout, and decorative concrete. It is recommended wherever white or colored concrete or mortar is desired. It is made of selected raw materials that contain only negligible amounts of iron and manganese oxide, the substances that give concrete its gray color.

3.2.1.3

Portland Blast-Furnace Slag Cement

In portland blast-furnace slag cement, granulated and selected blast-furnace slag, which is obtained by rapidly chilling or quenching molten slag in water, steam, or air, is added to and ground along with portland cement. This type of portland cement should meet the requirements of ASTM Standard C 595, Type IS and Type IS-A, Air-entraining. This ASTM standard requires the slag constituent to be from 25% to 65% of the total weight of the portland cement.

Portland blast-furnace cement is usually available only in areas where it can be obtained economically, such as near blast furnace sites.

3.2.1.4

Portland-Pozzolan Cement

In portland-pozzolan cement, pozzolan, which consists of siliceous (or siliceous and aluminous) material, is blended with ground portland cement. Typical pozzolan materials are fly ash, volcanic ash, and calcined shale and clay. This type of portland cement should meet the requirements of ASTM Standard C 595, Type IP and Type IP-A, Air-entraining. This ASTM standard requires the pozzolan constituent to be from 15% to 40% of the total weight of the portland cement.

Concrete with pozzolan has a higher cured strength than does concrete that contains only portland cement, aggregate, and water. Pozzolan also decreases the heat of hydration, which is important in mass concrete structures. Pozzolans should not be used where high early strength is a requirement, because they tend to slow down the curing process.

3.2.2

WATER

Mixing water should be clean and free from oil, alkali, and acid. In general, water that is fit to drink is suitable for use in concrete. Water with excessive quantities of sulfates, however, should be avoided even when it is potable.

3.2.3

AGGREGATES

Aggregates constitute 60% to 75% of the volume of concrete (Fig. 3.2-4). Therefore, both the cost and the quality of a concrete are affected by the kind of aggregates used in it. Aggregates should be obtained from reliable dealers whose materials are of a quality known to make suitable concrete. They should comply with the requirements of ASTM Standard C 33, which limits the allowable amounts of damaging substances and specifies requirements for grading, strength, and soundness of normal-weight aggregate.

The most commonly used aggregates—sand, crushed stone, and air-cooled blast-furnace slag—produce normal-weight

FIGURE 3.2-4 *Range in proportions of materials used in concrete, by absolute volume. Bars 1 and 3 represent rich mixes with small aggregate. Bars 2 and 4 represent lean mixes with large aggregate.* (Courtesy Portland Cement Association.)

LOW DENSITY CONCRETE | MODERATE STRENGTH CONCRETE | STRUCTURAL CONCRETE

Expanded slag, shale, clay and slate

Pumice and Scoria

Perlite, Vermiculite and Foamed Plastics

Cellular

400 600 800 1000 1200 1400 1600 1800

20   40   60   80   100   120

28-day-air-dry unit weight, pounds per cubic foot (above: kilograms per cubic meter)

FIGURE 3.2-5 *Concrete made with lightweight aggregate.* *(Courtesy Portland Cement Association.)*

concrete, which weighs between 135 and 160 lb./cu. ft. Aggregates of expanded shale, clay, slate, and slag are used to produce structural lightweight concrete, which weighs from 85 to 115 lb./cu. ft. Other lightweight materials, such as pumice, scoria, perlite, vermiculite, and diatomite, are used to produce insulating concretes, which range in weight from 15 to 90 lb./cu. ft. (Fig. 3.2-5).

Both fine and coarse aggregates should be uniformly graded from their finest particles up to their largest. They should also be clean and free of loam, clay, and vegetable matter, because these foreign particles prevent portland cement paste from properly binding aggregate particles together. Concrete that contains these objectionable materials will be porous, have low resistance to weathering, and have low strength. It could also develop surface defects such as popouts.

In addition, the most economical concrete results when both fine and coarse aggregates are uniformly graded, thus requiring the least amount of portland cement paste to surround the aggregate and fill all the spaces between the particles (Fig. 3.2-6). Less paste means a reduction in shrinkage, because most shrinkage occurs in the paste.

3.2.3.1

Fine Aggregate

Fine aggregate consists of sand or another suitable fine material. A good concrete sand will contain particles that vary uniformly in size from very fine up to 1/4 in. in diameter (Fig. 3.2-7). In well-graded

FIGURE 3.2-6 *Cross section of concrete showing cement paste completely surrounding each aggregate particle and filling the spaces between particles. The quality of its paste largely controls the quality of a concrete.* *(Courtesy Portland Cement Association.)*

FIGURE 3.2-7 *Well-graded sand before and after separation into its various sizes. The particles vary from dust to about 1/4 in. in diameter.* *(Courtesy Portland Cement Association.)*

sand the finer particles help fill the spaces between the larger particles. Sufficient fine particles are necessary for good workability and smooth surfaces, but an excess of "fines" requires an increase in the amount of portland cement–water paste to achieve a desired strength, thus increasing cost. Failure to increase the portland cement content when there is an excess of fines in the aggregate will lower the strength of a concrete. A poor concrete sand is shown in Figure 3.2-8.

Figure 3.2-9 shows a typical grading of fine aggregates for light-concrete slabs on grade.

FIGURE 3.2-8 *A poor concrete sand before and after being separated into four sizes. This sand lacks particles larger than 1/16 in. More cement is needed with fine sand.* *(Courtesy Portland Cement Association.)*

3.2.3.2

Coarse Aggregate

Coarse aggregate consists of gravel, crushed stone, or another suitable material larger than 1/4 in. in diameter (Fig. 3.2-10). Coarse aggregate that is sound, hard, and durable is best suited for making concrete. Material that is soft or flaky or that wears away rapidly is unsatisfactory.

In most concrete, normal-weight coarse-aggregate sizes are limited to those between 3/8 in. and 1-1/2 in. Thin toppings and slabs may be limited to 3/8-in. aggregate. Coarse aggregate as large as 6 in. in

FIGURE 3.2-9

Gradings of Fine Aggregate for Concrete Slabs

| Sieve Size | Percent of Aggregate Passing Sieve | |
	Normal Weight	Lightweight
3/8 in. (10 mm)	100	100
No. 4	95–100	85–100
No. 8	80–90	—
No. 16	50–75	40–80
No. 30	30–50	30–65
No. 50	10–20	10–35
No. 100	2–5	0–5

Courtesy Portland Cement Association.

FIGURE 3.2-10 *(a) A well-graded coarse aggregate for concrete. Notice how the small pieces fit among the larger ones. (b) The same material separated, from left to right, into 1/4-in. to 3/8-in., 3/4-in., and 1-1/2-in. sizes. (Courtesy Portland Cement Association.)*

diameter is often found in massive concrete structures, such as dams. Light-

weight aggregate is usually limited to a maximum of 1 in. In walls, the largest piece of any type of coarse aggregate should never be more than one-fifth the overall wall thickness. The maximum size in slabs should be about one-third the thickness of the slab. The largest piece of aggregate should never be larger than three-quarters the width of the narrowest space through which the concrete will be required to pass during placing. This space is usually that between the reinforcing bars or between the bars and the forms.

Coarse aggregate is well-graded when particles range uniformly from 3/8 in. up to the largest size that may be used on the kind of work to be done. Use of the maximum allowable particle size usually results in less drying shrinkage and more economical concrete.

The size of coarse aggregate used in concrete has a bearing on economy. More water is usually required for small aggregates than for large. The water required for a slump of 3 to 4 in. is shown in Figure 3.2-11 for a wide range of coarse aggregate sizes. For a given water-cement ratio (see Section 3.5.1), the amount of portland cement required decreases as the maximum size of coarse aggregate increases. However, the increased cost of obtaining and handling aggregates larger than about 2 in. may offset the savings in portland cement.

3.2.4

ADMIXTURES

Admixtures are materials other than portland cement, aggregates, and water that are added to concrete either immediately before or during its mixing to alter the properties of the concrete in a variety of ways. For example, admixtures can be used to improve workability, to reduce the separation of coarse and fine aggregates due to settling out of the heavier coarse aggregate, to entrain air, or to accelerate or retard setting and hardening.

Concrete should be workable, finishable, strong, durable, watertight, and wear resistant. Before deciding to use admixtures in a concrete mix, one should be aware that (1) a change in the type or the amount of portland cement used or a modification of aggregate gradation or mix proportions may offer a better, surer, and more economical way to achieve the desired objectives than will using an admixture; (2) many admixtures affect more than one property of concrete, sometimes hurting desirable properties; (3) the effect of some admixtures is changed by such things as wetness and richness of the mix, aggregate gradation, and the character and length of mixing; and (4) some admixtures will not react the same with all portland cements, even of the same type.

Accordingly, admixtures should be used

FIGURE 3.2-11 *The effect of aggregate size on required water content. (Courtesy Portland Cement Association.)*

with caution, since specific effects that will result from their use seldom can be predicted accurately. In addition, two or more admixtures should not be used in the same concrete mix, unless their combined effect is well known and will not produce an undesirable condition.

ASTM Standard C 494 classifies some chemical admixtures by function, as follows:

Type A: Water-reducing
Type B: Retarding
Type C: Accelerating
Type D: Water-reducing and retarding
Type E: Water-reducing and accelerating
Type F: High-range water-reducing
Type G: High-range water-reducing and retarding

Of these, the most frequently used are air-entraining agents, water-reducing agents, retarders, and accelerators.

3.2.4.1

Water-Reducing Admixtures

Water-reducing admixtures, such as hydroxylated carboxylic acid, should be used in almost all concrete. They permit the use of a lower water content and improve workability. They also increase the efficiency of the portland cement in a mix, which lowers the cost of the concrete relative to its performance.

High-range water-reducing admixtures (*super plasticizers*) are used mostly in concrete that is to be pumped. They produce a mix that flows easily with no increase in its water content.

3.2.4.2

Retarders

Admixtures that have a retarding effect on the set of portland cement are used in concrete to overcome the accelerating effect that temperature has on setting during hot weather and in large masses of concrete and to delay the early stiffening of concrete placed under difficult conditions. Retarder solutions are sometimes applied directly to the surface of concrete to retard the set of a surface layer of mortar so it can be readily removed by brushing, thus exposing the aggregate and producing textured surface effects.

Because most retarders also act as water reducers, they are frequently called *water-reducing retarders*. Retarders may also entrain some air in concrete. Many chemicals are mentioned in current literature as having a retarding influence on the normal setting time of portland cement. Some of these chemicals have been found variable in action, retarding the set of certain portland cements and accelerating the set of others. Unless experience with a retarder has determined the extent of its effect on the setting time and other properties of a concrete, its use as an admixture should not be attempted without technical advice or advance experimentation with the portland cement and other concreting materials involved to determine its effects on the properties of the concrete.

3.2.4.3

Accelerators

Accelerators increase the rate of early strength development in concrete to (1) reduce the waiting time before finishing operations can be started; (2) permit earlier removal of forms and screeds; (3) reduce the required period for curing in certain types of work; (4) advance the time when a structure can be placed in service; (5) partially compensate for the slow gain in strength of a concrete even with proper protection during cold weather; and (6) reduce the period of protection required for initial and final set in making emergency repairs and other work.

Under most conditions, commonly used accelerators cause an increase in the drying shrinkage of concrete. In many cases, it must be decided whether to use an admixture, increase the portland cement content, use a portland cement with high early strength, provide greater protection or longer curing period, or use a combination of these.

Calcium chloride is used to accelerate the time of set and to increase the rate of strength gain. It should meet the requirements of ASTM Standard D 98. The amount used should not exceed 2% by weight of the portland cement used. Greater amounts may cause rapid stiffen-ing, increase drying shrinkage, and corrode reinforcement. Calcium chloride should always be added in solution as part of the mixing water to ensure uniform distribution. If it is added in dry form, all the dry particles may not dissolve completely during mixing. Undissolved lumps can cause popouts or dark spots in hardened concrete. Calcium chloride should never be used as an antifreeze. To lower the freezing point of concrete appreciably would require using so much calcium chloride that the concrete would be ruined. Instead, protective measures should be taken to prevent the concrete from freezing (see Section 3.8.3.1).

Using calcium chloride or admixtures that contain soluble chlorides is not recommended under the following conditions: (1) in prestressed concrete, because of possible corrosion hazards; (2) in concrete containing embedded aluminum, such as conduit, because serious corrosion can result, especially if the aluminum is in contact with embedded steel and the concrete is in a humid environment; (3) in concrete exposed to soils or water containing sulfates; (4) in floor slabs intended to receive dryshake metallic finishes; or (5) in hot weather generally.

3.2.4.4

Air-Entraining Agents

Air-entrained concrete can be made either with admixtures added to the mix or with air-entraining portland cement as listed in ASTM Standard C 150, which already contains the admixtures. Air-entraining admixtures produce many microscopic stable air bubbles in concrete. These materials should conform with the requirements of ASTM Standard C 260, which includes specifications and methods of testing. These admixtures will, at little or no additional cost, improve the workability and durability of concrete and produce a hardened concrete that is resistant to severe frost action and the effects of applications of salt for snow and ice removal. Common air-entrainment admixtures include wood resin, sulfonated hydrocarbons, and fatty resinous acids. Because careful control is essential in producing air-entrained concrete, its use should be restricted to plant or transit-mixed batches.

Properly proportioned air-entrained concrete contains less water per cubic yard than non-air-entrained concrete of the same slump (see Section 3.5.3.1 for a discussion of slump tests) and has better workability. This results in a more solid, weather-resistant, blemish-free surface. Air-entrained concrete can be handled and placed with less segregation of materials and less tendency to bleed (see Section 3.7.1.1). These properties indirectly aid in promoting durability by increasing the uniformity of the concrete.

When using air-entraining portland cement or admixtures containing air-entraining agents, care must be taken that the quantity of water in the concrete mix is adjusted to maintain the desired slump. Such mix designs and adjustments should be made only by a qualified technician.

Because of its proven record, both in the laboratory and in the field, and because it increases workability and durability, all concrete would benefit from entrained air regardless of exposure conditions. Generally, however, air entrainment is specified only for concrete that is exposed to cycles of freezing and thawing because, except in very lean and harsh mixtures, air entrainment reduces both compressive and flexural strengths by from 2% to 6% for each 1% of air added. However, when freezing and thawing will be encountered, and when watertightness is necessary, air entrainment is essential. Whether by use of air-entraining portland cement or admixture, a proper percentage of entrained air in a concrete mix should be obtained, depending on the maximum size of aggregate, as shown in Figure 3.2-12.

3.2.5

ACCESSORIES

Many types of accessories are used in concrete construction, including forming accessories (see Section 3.3), reinforcement accessories (see Section 3.4), anchors and fasteners, sleeves, joint inserts, reglets, and waterstops.

3.2.5.1

Anchors and Fasteners

Anchors and fasteners are used to anchor other materials, such as wood, stone, and masonry, to concrete. They range from

---

FIGURE 3.2-12
Recommended Air-Entrainment Amounts for Various Maximum Coarse Aggregate Sizes

Nominal Maximum Size Coarse Aggregate	Total Air Content Percentage by Volume of Concrete
3/8 in.	6 to 10
1/2 in.	5 to 9
3/4 in.	4 to 8
1 in.	3-1/2 to 6-1/2
1-1/2 in.	3 to 6

Courtesy Portland Cement Association.

simple threaded anchor bolts to elaborate adjustable inserts. Figures 3.2-13 through 3.2-18 show a few of the many types of anchors and fasteners available for use in concrete.

Figure 3.2-13 shows a simple anchor bolt. The bent portion turns out into the concrete to hold the bolt in place. Anchor bolts are used for fastening precast concrete, metal, and wood construction to concrete. Examples are precast concrete columns, steel column plates, and wood plates. The diameter and the length of such bolts vary depending on the loads that will be applied and the construction materials they will anchor.

Figure 3.2-14 shows two of the many configurations of concrete inserts. Figure

FIGURE 3.2-13 *Threaded anchor bolt with nut and washer. (Drawing by HLS.)*

FIGURE 3.2-14 *Concrete inserts are installed in formwork so they finish flush with the concrete: (a) shelf-angle insert and anchor; (b) two types of light-duty internally threaded inserts. (Drawing by HLS.)*

3.2-14a is a *shelf angle insert* designed to receive a special bolt; Figure 3.2-14b is an internally *threaded insert*, which receives a standard bolt. In each case, the insert is fastened to the inside face of the form before the concrete is placed. The shelf angle insert shown is just one of many available configurations, all of which work in essentially the same way. An anchor bolt with a special head designed to fit the insert is placed into the cross slot on the insert and dropped into the vertical slot. The device to be supported is slipped over the bolt, and the washer and nut are installed and tightened, pulling the anchor head tightly against the inside of the insert. With a threaded insert, a bolt is passed through the item to be supported into the threaded insert. Then a washer and nut are installed, and the nut is tightened.

Figure 3.2-15 shows a dovetail slot and two of the many types of anchors that can be used in such a slot. Dovetail slots are used to anchor masonry and stone to concrete. The slot is formed from galvanized steel, copper, or stainless steel. The an-

FIGURE 3.2-15 *Dovetail anchor slots are designed to receive anchors for stone and masonry.* *(Drawing by HLS.)*

FIGURE 3.2-17 *Powder-actuated fasteners: (a) drive pin and (b) stud.* *(Drawing by HLS.)*

chors may be galvanized steel or stainless steel. Dovetail slots are available with a felt or plastic-foam filler, to prevent intrusion of concrete and other materials, or with no filler. Dovetail slots are nailed to the inside of the form before the concrete is placed. The wedge-shaped ends of the anchors are slipped into the slot vertically and twisted to lock them in place.

Figure 3.2-16 shows two types of expansion anchors. Figure 3.2-16a is representative of the many types of heavy-duty expansion anchors available. They all work in essentially the same way. A hole

is drilled in the concrete after it has set, and the anchor is placed into the hole. The bolt furnished with the anchor is passed through the item to be supported and into the anchor shell. As the bolt is tightened, the expanding portion of the anchor is forced apart, wedging the anchor securely into the drilled hole.

Figure 3.2-16b shows one of the many types of light-duty expansion anchors. The one shown has a *drive pin*; others have bolts or screws. Light-duty expansion anchors are placed in drilled holes in set concrete. A pin, screw, or bolt is passed through the item to be supported into the anchor. Drive pins are driven with a hammer. Screws are tightened with a screwdriver; bolts with a wrench. As the pin, screw, or bolt is forced into the anchor, the anchor expands, locking itself into the drilled hole.

Figure 3.2-17 shows two types of powder-actuated fasteners. These fasteners are driven into concrete by an exploding charge of gunpowder in a gun specially designed for the purpose. Figure 3.2-17a is a drive pin that is driven through the item to be supported into the concrete. Figure 3.2-17b is a stud, which has threads so that a supported item can be held in place with washers and nuts.

Figure 3.2-18 shows several configurations of anchor plates and angle clips typically used in concrete construction. These devices are fastened to the forms before the concrete is placed, with the anchors shown toward the concrete. Bolts may be welded to plate and angle clips to support or anchor other construction, or other steel shapes may be welded to them.

Other anchor types not shown include hammer and power-driven nails and screws.

FIGURE 3.2-16 *Expansion anchors: (a) heavy-duty and (b) light-duty.* *(Drawing by HLS.)*

FIGURE 3.2-18 *Various types of anchor plates and angles.* *(Drawing by HLS.)*

3.2.5.2

Sleeves

Sleeves of metal, cardboard, or plastic are used to produce holes in concrete for the insertion of pipes, railings, and other items that pass through or are set into concrete. Openings for banks of several pipes, concrete. Openings for banks of several pipes, conduits, and the like often are formed with a wood box set into the form. Openings for single pipes usually are formed with galvanized steel pipes or cardboard or plastic tubes. When railing standards or other items that must be wedged in tightly for support are placed in sleeves, paper and concrete materials are removed; metal sleeves are left in place (Fig. 3.2-19).

3.2.5.3

Joint Inserts

Joint inserts are used to form control joints in slabs. Figure 3.2-20 shows a typical tongue-and-groove control-joint insert for use in slabs on grade. These inserts are made of either plastic or galvanized steel. Similar inserts are used in supported slabs.

FIGURE 3.2-19 *A sleeve for use in concrete.* *(Drawing by HLS.)*

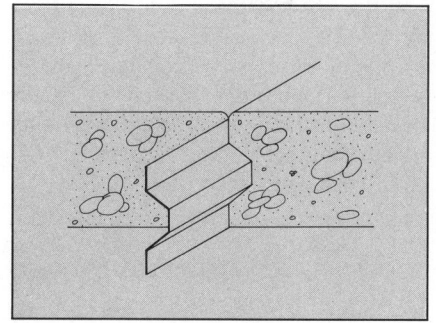

FIGURE 3.2-20 *Control-joint insert.* *(Drawing by HLS.)*

FIGURE 3.2-21 *Keyway insert.* *(Drawing by HLS).*

Similar inserts are used to form keyways in concrete to permit bonding to adjacent concrete or masonry (Fig. 3.2-21). Keyways often are placed in the tops of footings to key in the foundation walls, whether the wall is concrete or masonry (Fig. 3.2-22).

Plastic strips are used as inserts to create control joints and these may be just flat strips or flanged units that remain in place. Some of these strips have removable tops to permit installation of sealant on top of the insert (Fig. 3.2-23).

Plastic caps are sometimes used over fibrous joint fillers to permit easier application of sealant in the joint (Fig. 3.2-24).

3.2.5.4

Reglets

Reglets are devices that form slots in concrete for the insertion of flashings. Figure

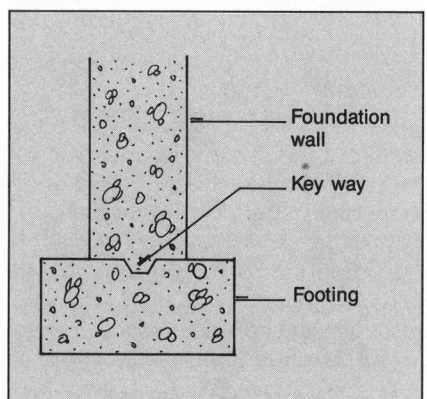

FIGURE 3.2-22 *Keyway used with a concrete wall.* *(Drawing by HLS.)*

FIGURE 3.2-23 *Control joint insert with removable plastic cap.* *(Drawing by HLS.)*

FIGURE 3.2-24 *Removable cap for joint filler.* *(Drawing by HLS.)*

FIGURE 3.2-25 *A reglet.* *(Drawing by HLS.)*

3.2-25 shows one reglet configuration. Others are available, but the general characteristics of all reglets are similar. They consist essentially of a slot into which flashing is inserted. Reglets are available in galvanized steel, copper, stainless steel, and plastic.

3.2.5.5

Waterstops

Waterstops are rubber or vinyl inserts designed to be placed in concrete joints to prevent water from penetrating the joints. There are many available configurations; a few of the more commonly used ones are shown in Figure 3.2-26.

FIGURE 3.2-26 *Waterstops.* *(Drawing by HLS.)*

3.3 FORMWORK

Because concrete is unable to maintain a particular shape before it sets, it must be placed in a form or a mold. In building construction, this form or mold is called *formwork*. Architects usually select the form liners and inserts often used to impart special finishes or patterns to concrete. Architects and structural engineers, working together, also select special forms for ribbed slabs. But the design of the structural portions of formwork and the necessary external supports should be, and usually is, the responsibility of the building contractor. Codes require certain minimum provisions for formwork, and a particular circumstance may cause a structural engineer to place minimum limits on form design. The minimum location and number of slab form supports and shoring, for example, may be dictated. Unfortunately, in some jurisdictions the architect is legally prevented from disclaiming responsibility for form design.

The discussion here is general in nature. Additional requirements for formwork are mentioned in the sections of this chapter where a particular type of concrete unit is discussed. For example, there are discussions about forming slabs, including lift-slab forming, in Section 3.11.4 and some additional requirements for precast concrete formwork in Section 3.13.

3.3.1

MATERIALS

Forms are made from wood, metal, and plastic. They can be divided into two classes: temporary forms and permanent forms.

3.3.1.1

Temporary Forms

Most formwork is temporary, for two reasons: (1) permanent forms would detract from the desired finish, and (2) it is more economical to reuse formwork.

Most temporary concrete forms are made from wood. The parts of the form in contact with the concrete are often high-grade lumber or plastic-overlaid plywood. Wood structural forms are also used to support metal and plastic pans and domes and metal- and plastic-form textured liners. Temporary forms for pan slabs and ribbed slabs (Section 3.11.4.1) are either steel or plastic.

Round column forms made of metal, molded glass-fiber-reinforced plastic, and multilayered paper or fiber are available. These forms often are plastic-coated on the interior to improve the smoothness of the cast column and to eliminate the lines imparted by the spiral winding of the form.

An almost unlimited number of textures and patterns can be imparted to concrete using metal or plastic *form liners*. Most of these textures and patterns are proprietary items, but custom designs are possible.

3.3.1.2

Permanent Forms

Permanent corrugated-steel forms are used to make supported concrete slabs. A supported slab is one that rests on beams or columns rather than on earth (slab on grade). Permanent steel forms are made from corrugated sheet steel.

3.3.2

DESIGN AND CONSTRUCTION

The potential failure of formwork represents a great hazard to people working around and on it and to the property adjacent to or supported by it. Therefore, formwork must be designed and built with the same skills and practices that would be applied to a permanent structure. It must be strong enough to withstand the great pressures imposed by heavy, wet concrete and the additional stresses that result from placing and consolidating the concrete. In addition, formwork must be rigid enough to maintain the desired shape under these stresses. It also must remain tight, so that cement paste and water are not lost from the concrete during hydration—such loss would affect the quality of the cured concrete.

Forms must be accessible so that concrete can be placed easily, moved in the form, and consolidated. They also must be designed so they can be easily removed without damaging the concrete or the form. For example, there must be no undercuts, intrusions, or sharp edges left in the concrete, because these either will be broken off or will tear the forms apart as the forms are removed. Corners should be rounded off or beveled by inserting wood, rubber, or plastic chamfering strips in the forms.

The surface of forms must impart the desired degree of smoothness and texture to the concrete. Before concrete is placed in a form, the surfaces of the form that will be in contact with the concrete should be coated with a form-release agent to prevent the concrete from sticking to the form. This agent may be wax, oil, or a plastic compound formulated for this purpose. Where the concrete will be exposed to view, the parting agent should be nonstaining.

The following discussions assume that forms will be custom-built on the site. However, prefabricated proprietary wood, steel, and aluminum systems for forming footings, columns, and walls are available (Fig. 3.3-1). These systems are similar in concept to the custom-built systems described here, although they may use different materials for the various components. For example, a column form may use angle-shaped steel clamps for yokes instead of wood members or steel bands to hold the form together; footing forms may use corrugated-steel sides rather than boards or plywood; wall forms may be framed with aluminum or steel studs instead of wood ones; and wales and braces may be metal as well, or they may not be necessary at

FIGURE 3.3-1 *Prefabricated formwork.* (*Courtesy the Burke Company.*)

FIGURE 3.3-2 *Prefabricated formwork, some with bracing, some without.* (*Courtesy the Burke Company.*)

all (Fig. 3.3-2). When these types of systems are used, it is necessary to follow the manufacturers' instructions with regard to erection, safety, and removal.

3.3.2.1

Forming Footings

Footings should be formed, unless they are in very stable soil and have been cut cleanly and accurately to the shape of the footing, which is a rare condition. Continuous footing may be formed with plywood, as shown in Figure 3.3-3, or with wood boards. Small footing forms can be braced with stakes and spreaders, as shown. On small footings, steel rods are sometimes used instead of stakes.

Large footings are framed in a similar manner, except that multiple boards or plywood are used for the side forms and are braced by vertical boards (*cleats*) and continuous horizontal framing (*wales*) (Fig. 3.3-4). Tie wires are passed through the form from side to side and fastened to the cleats. Stakes may be used to prevent the formwork from spreading. Steel straps or lumber spreaders are sometimes necessary across the top of larger forms to prevent the side walls from spreading apart when the concrete is placed.

3.3.2.2

Forming Walls

Custom forms for walls are built at the site with plywood supported by studs, tied together by continuous wales, and braced diagonally to the ground (Fig. 3.3-5).

To erect a wall form, the inner liner and its framing are installed first and braced back to the ground. The reinforcing steel is placed next. Then the other liner and its framing are placed and then braced (Fig. 3.3-6). As the formwork is erected, the wales are tied together using form ties to keep the form from spreading. There are two types of ties: *snap ties* and *screw ties* (Fig. 3.3-7).

Snap ties are inserted through holes in the forms as the forms are erected. A washer on the tie rod at the interior side of each form surface maintains the width of the wall and prevents concrete from leaking out through the hole. The ends are then clamped at the wales with slotted tapered wedges. The rod is deformed just inside each washer, so that when the forms have been removed the rod can be twisted until it breaks at the weakened spot. This process leaves a ragged hole in the concrete, which can be filled with mortar.

The cones on *screw ties* have interior threads. These ties are inserted into the forms with the cones against the inner lining of the form. Threaded rods are then passed through holes in the form and screwed into the cones at each wall face. The cones prevent concrete from leaking out through the holes. The threaded rods are then locked into place at the wales by slotted tapered wedges or plates. When the forms are removed, the threaded rods are unscrewed from the cones and the cones unscrewed from the tie, leaving a cone-

FIGURE 3.3-3 *A typical footing form.* (Drawing by HLS.)

FIGURE 3.3-4 *One of several methods commonly used to form large column footings.* (Drawing by HLS.)

shaped hole in the concrete. The cone leaves a clean hole in the concrete, which can be left open or filled with mortar or a sealant.

3.3.2.3

Forming Columns

Columns require a liner, bracing, and yokes, straps, or clamps to hold the form together (Fig. 3.3-8). The straps shown may be replaced by wood yokes held to-

gether with threaded rods and bolts or by a proprietary system that uses two overlapping angle brackets (Fig. 3.3-9). Larger columns may require additional rows of bracing inside the strapping or brackets.

3.3.2.4

Forming Beams and Girders

Beams and girders are formed with wood or plywood. One way to form light beams is with braces, kickers, and spreaders, as

shown in Figure 3.3-10. The bottoms of these forms are supported on the shoring system. A deep beam or girder form is built with ties and diagonal braces much like a wall form. The bottom is supported on continuous wood cleats turned with the long edges horizontal, as shown, or vertical depending on the size of the beam or girder. Forms for large beams and girders may require additional bracing.

3.3.2.5

Forming Floors

Both slabs on grade and supported floors require forming.

■ SLABS ON GRADE

The edges of and openings through slabs on grade require forming. Such forming usually involves using wood or metal forms staked to the ground, similar to the wall footing form shown in Figure 3.3-3.

■ SUPPORTED FLOORS

In general, beams and girders are formed first, then the slab is formed. Flat slabs are formed with plywood supported on joists. If pans or domes are used, they are supported on a wood frame of joists and plywood called *centering* (Fig. 3.3-11). Alternatively, the plywood beneath pans or domes is omitted, and wide boards are placed across the joists at the location of each rib in the floor. The forms for beams and floors are supported by a system of temporary wood or metal beams or joists. These beams or joists are in turn supported by adjustable-length steel columns called *shores*, wood posts, or a series of adjustable ladder-like steel assemblages called *staging*. The carrier shown in Figure 3.3-11 is a part of this form-support system. The entire support system for forms is called *shoring*.

3.3.2.6

Flying Forms

Flying forms are made by building a large section of form as described for centering and supporting the entire section on deep steel trusses. This assembly can then be lifted by crane and positioned in the next

FIGURE 3.3-5 *Typical wall forms.* (Courtesy Portland Cement Association.)

FIGURE 3.3-6 *Detail of one type of wall form.* (Drawing by HLS.)

location without dismantling it, thus saving cost and speeding up construction.

Slip Forming

Slip forming is a method of continuously moving a form for vertical structures, such as elevator or stair shafts, upward on jacks as new concrete is placed on top of the old.

FORM REMOVAL

Leaving forms in place as long as possible helps in the curing process and protects freshly placed concrete from damage (see Section 3.8.2.1). However, it is sometimes desirable to remove forms from beams and slabs as soon as it can be done safely. For example, patching and repairing of formed surfaces should be done as early as possible, which requires removal of the forms. It is also often necessary to remove forms quickly to permit their reuse.

In any case, forms should not be removed until the concrete has attained sufficient strength to ensure structural stability and to carry safely both the dead load and construction loads that may be imposed on it. Concrete should be hard enough so its surfaces will not be injured in any way when reasonable care is used in removing forms.

When their forms are removed, slabs, beams, and girders, while strong enough to support themselves, often will not be sufficiently strong for several more weeks to support construction loads that will be imposed on them. This is especially true when additional concrete will be placed on higher floors, whose forms will be at least partially supported by the concrete from which the forms have just been removed. So that work may continue uninterrupted, wood posts or adjustable metal columns called *reshores* are placed beneath slabs and beams from which the forms have been recently removed. These reshores are left in place until the beams and slabs are able to support the imposed loads.

FIGURE 3.3-7 *Wall form ties: (a) a snap tie and (b) a screw tie.* (Drawing by HLS.)

FIGURE 3.3-8 *A column form with steel straps.* (Drawing by HLS.)

FIGURE 3.3-9 *A column form with overlapping angle brackets.* (Drawing by HLS.)

FIGURE 3.3-10 *A typical beam form. Heavier beams may need wall ties or diagonal braces back to the header. Additional framing members will also be necessary to support floor slab forms. Floor-slab form framing will usually eliminate the need for spreaders.* (Drawing by HLS.)

FIGURE 3.3-11 *Typical centering for a supported floor system.* (Drawing by HLS.)

3.4 REINFORCEMENT

Until the idea of *reinforced concrete* was developed, concrete use was limited to cases where it was required to resist forces that tended to compress it (compressive loads) (Fig. 3.4-1). Concrete is strong in resisting *compressive* (pushing) loads, but its ability to resist *tensile* (pulling) forces is, for all practical purposes, nonexistent. All slabs, beams, girders, and even some columns and walls are required to resist not only compressive loads (*compression*), but also tensile loads (*tension*) caused by bending (see Fig. 3.1-1), drying shrinkage, temperature changes, and vertical and horizontal shear. The net result of vertical and horizontal shear forces is called *diagonal tension*. Tensile forces that result from bending will cause an unreinforced concrete beam or slab to crack along the bottom and collapse (see Fig. 3.1-1a). Diagonal tension will cause a crack at the support. This type of cracking will act at an approximate 45-degree angle (Fig. 3.4-2). Diagonal tension at a column or other support is called *punching shear*.

The requirements in this section are general in nature, applying in principle to all reinforced concrete. Specific require-

FIGURE 3.4-2 *Diagonal tension (shear) failure.* (Drawing by HLS.)

ments for reinforcing steel in different types of slabs, beams, girders, walls, and columns are addressed in Sections 3.10 and 3.11.

3.4.1

PURPOSE OF CONCRETE REINFORCEMENT

The concept of reinforced concrete is a simple one. Concrete is strong in resisting compressive forces, while steel bars effectively resist tensile forces and shear. When both are placed in the same structural member (slab, wall, beam, girder, column), the steel resists the tensile, bending, and shear forces, which would tear the concrete apart, and the concrete resists applied compression forces, which would bend the steel.

The theory is simple; the practice, however, is not. The location and the amount of steel in a reinforced concrete structural component are critical and vary with the type of component, the loads that will be applied, whether the unit is prestressed and whether the unit is cast in place (Section 3.11) or precast (Section 3.13). For example, the reinforcement required in a

simple (single-span) beam varies considerably from the reinforcement in a beam that is continuous across several spans.

Other complications also occur. For instance, because steel is also strong in compression, it is sometimes used to resist a portion of the applied compression loads. This often occurs in columns and in beams and girders when limitation of their depth is important.

In addition, steel is used in concrete even when it is not required to resist tensile and shear forces. For example, slabs on grade are not structural elements. They simply pass loads through to the underlying soil. Structural cracks in slabs on grade are due to a failure in the base and not in the concrete or its lack of reinforcement. However, slabs on grade do tend to crack due to drying shrinkage and when they are subjected to widely varying temperatures. Therefore, a layer of light welded wire fabric is often added to them. This is not structural reinforcement and should not be counted on to help a slab span depressions in the slab base. Neither will it prevent cracking. All it does is hold cracks together so they are less unsightly. Some practitioners believe that welded wire fabric is not necessary in slabs on grade and routinely omit it. The Portland

FIGURE 3.4-1 *Concrete is best at supporting compression loads.* (Drawing by HLS.)

Cement Association's "Concrete Floors on Ground" suggests that welded wire fabric may not be necessary when the slab is uniformly supported and joints are located close together. The spacing of joints that would permit deletion of fabric is not specified.

Reinforcing bars are needed in other types of slabs to resist cracks caused by drying shrinkage and temperature changes. This steel is called *temperature steel*. When structural reinforcement occurs, it acts to resist these stresses, and temperature steel is not required. When structural reinforcement is placed in one direction only, temperature steel is required in the perpendicular direction (see Section 3.11.4).

The marriage of concrete and steel is a natural one, made possible by the chemical compatibility of the two materials. In fact, the alkalies in concrete actually protect steel from corroding. Concrete bonds very strongly to steel. This bond is not broken by thermal movement, because the coefficients of expansion of steel and concrete are close to the same.

3.4.2

MATERIALS

Steel reinforcement can be classified as either *structural reinforcement* or *temperature reinforcement*. Structural reinforcement resists primary tensile, shear, and compression forces. It consists mainly of steel bars, but welded wire fabric is used in structural slabs and walls and in some precast concrete elements.

Temperature reinforcement is added to hold together cracks that occur from drying shrinkage and from expansion and contraction due to temperature variations. It consists of light welded wire fabric in slabs on grade and of small bars in supported slabs.

3.4.2.1

Reinforcing Bars

Reinforcing bars are either plain or *deformed* (Fig. 3.4-3). Deformations improve the bond between steel and concrete. Reinforcing bars are available in various sizes (Fig. 3.4-4) and several grades and should comply with the ASTM standards listed in Figure 3.4.5. Most concrete reinforcement for building construction is made from Grade 40 or Grade 60, deformed new billet or ingot steel bars, conforming with ASTM Standard A 615.

Reinforcing bars typically have numbers and letters stamped on them to identify the producing mill, bar size, type of steel, and grade (Fig. 3.4-6).

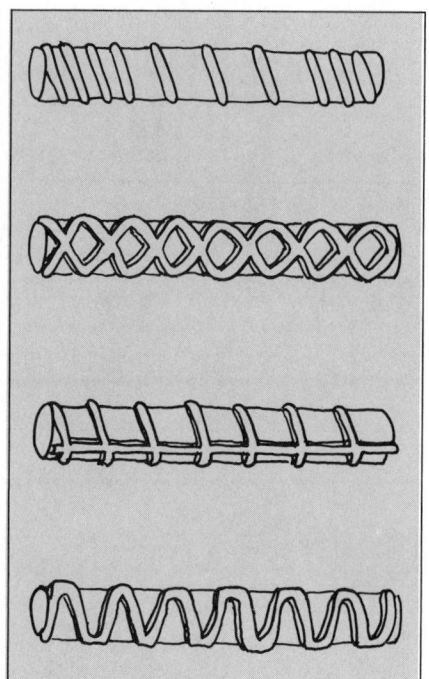

FIGURE 3.4-3 *Typical deformed steel bars. (Drawing by HLS.)*

3.4.2.2

Welded Wire Fabric

Welded wire fabric is made by welding reinforcing bars into a square- or rectangular-shaped mesh. Both smooth and de-

---

FIGURE 3.4-4

Deformed Bar Designation Numbers, Nominal Weights, Nominal Dimensions, and Deformation Requirements

Bar Designation No.[b]	Nominal Weight, lb./ft.	Nominal Dimensions[a]			Deformation Requirements, in.		
		Diameter, in.	Cross-Sectional Area, sq. in.	Perimeter, in.	Maximum Average Spacing	Minimum Average Height	Maximum Gap (Chord of 12-1/2% of Nominal Perimeter)
3	0.376	0.375	0.11	1.178	0.262	0.015	0.143
4	0.668	0.500	0.20	1.571	0.350	0.020	0.191
5	1.043	0.625	0.31	1.963	0.437	0.028	0.239
6	1.502	0.750	0.44	2.356	0.525	0.038	0.286
7	2.044	0.875	0.60	2.749	0.612	0.044	0.334
8	2.670	1.000	0.79	3.142	0.700	0.050	0.383
9	3.400	1.128	1.00	3.544	0.790	0.056	0.431
10	4.303	1.270	1.27	3.990	0.889	0.064	0.487
11	5.313	1.410	1.56	4.430	0.987	0.071	0.540
14	7.65	1.693	2.25	5.32	1.185	0.085	0.648
18	13.60	2.257	4.00	7.09	1.58	0.102	0.864

Copyright ASTM, reprinted with permission.

[a]The nominal dimensions of a deformed bar are equivalent to those of a plain round bar having the same weight per foot as the deformed bar.

[b]Bar numbers are based on the number of eighths of an inch included in the nominal diameter of the bars.

FIGURE 3.4-5
ASTM Standards Covering Various Bars by Grade

Material	ASTM Standard	Grade	Yield Strength (psi)	Bar Sizes Available
Billet or ingot steel	A 615	40	40,000	3–6
	A 615	60	60,000	3–11, 14 and 18
	A 615	75	75,000	11, 14, 18
Rail steel	A 616	50	50,000	3–11
	A 616	60	60,000	3–11
Axle steel	A 617	40	40,000	3–11
	A 617	60	60,000	3–11

forcement Institute as common stock styles. Figure 3.4-8 shows the nominal wire diameters and weights of steel associated with size numbers. The normal welded wire fabric used in light slabs on grade is WWF 6×6−W1.4×W1.4.

Lighter gauges of welded wire fabric are available in rolls, heavier gauges in sheets (see Fig. 3.4-7).

3.4.2.3

Accessories

Accessories are used with structural steel to support it in the forms and to make splices in it. The devices shown in Figure 3.4-9 are used to support reinforcing steel in forms. Welded wire fabric in slabs on grade should be supported on similar devices having flat plates at their legs to prevent the legs from sinking into the slab base (*sand plates*). The common practice of supporting reinforcement in slabs on grade with pieces of concrete or concrete bricks is discouraged because concrete often does not adhere to such materials. This produces cracks, which permits water to reach and rust the reinforcement,

GRADE 60 AND A 706

GRADE 75

GRADE 40 AND 50

FIGURE 3.4-6 *Standard markings on steel bars.* (*Courtesy Concrete Reinforcing Steel Institute.*)

make it. An example of the designation system is WWF 6×12−W16×W8, which denotes welded wire fabric (WWF) with a 6-in. spacing between longitudinal wires and a 12-in. spacing between transverse wires, with smooth W16 longitudinal wires and smooth W8 transverse wires. The W represents smooth wire; the letter D would represent deformed wire. The number following the W (or D) is the cross-sectional area of the wire in hundredths of an inch. Figure 3.4-7 shows WWF styles designated by the Wire Rein-

formed bars are used. Welded wire fabric is designated by the size and the spacing of the reinforcing bars (*wires*) used to

FIGURE 3.4-7
Common Stock Styles of Welded Wire Fabric

Style Designation		Steel Area, sq. in./ft.		
New Designation (by W-number)	Old Designation (by Steel Wire Gauge)	Longitudinal	Transverse	Approximate Weight, lb./100 sq. ft.
Rolls				
6×6−W1.4×W1.4	6×6−10×10	.03	.03	21
6×6−W2×W2	6×6−8×8[a]	.04	.04	29
6×6−W2.9×W2.9	6×6−6×6	.06	.06	42
6×6−W4×W4	6×6−4×4	.08	.08	58
4×4−W1.4×W1.4	4×4−10×10	.04	.04	31
4×4−W2×W2	4×4−8×8[a]	.06	.06	43
4×4−W2.9×W2.9	4×4−6×6	.09	.09	62
4×4−W4×W4	4×4−4×4	.12	.12	86
Sheets				
6×6−W2.9×W2.9	6×6−6×6	.06	.06	42
6×6−W4×W4	6×6−4×4	.08	.08	58
6×6−W5.5×W5.5	6×6−2×2[b]	.11	.11	80
4×4−W4×W4	4×4−4×4	.12	.12	86

Courtesy Concrete Reinforcing Steel Institute.
[a]Exact W-number size for 8-gauge is W2.1.
[b]Exact W-number size for 2-gauge is W5.4.

FIGURE 3.4-8
Sectional Area and Weight of Welded Wire Fabric

Wire Size Number[a]		Nominal Diameter,[b] in.	Nominal Weight, lb./lin. ft.	Area of Width for Various Spacings, sq. in./ft. Center-to-Center Spacing						
Plain	Deformed			2 in.	3 in.	4 in.	6 in.	8 in.	10 in.	12 in.
W20	D20	0.505	.680	1.20	.80	.60	.40	.30	.24	.20
W18	D18	0.479	.612	1.08	.72	.54	.36	.27	.216	.18
W16	D16	0.451	.544	.96	.64	.48	.32	.24	.192	.16
W14	D14	0.422	.476	.84	.56	.42	.28	.21	.168	.14
W12	D12	0.391	.408	.72	.48	.36	.24	.18	.144	.12
W11	D11	0.374	.374	.66	.44	.33	.22	.165	.132	.11
W10.5		0.366	.357	.63	.42	.315	.21	.157	.126	.105
W10	D10	0.357	.340	.60	.40	.30	.20	.15	.12	.10
W9.5		0.348	.323	.57	.38	.285	.19	.142	.114	.095
W9	D9	0.338	.306	.54	.36	.27	.18	.135	.108	.09
W8.5		0.329	.289	.51	.34	.255	.17	.127	.102	.085
W8	D8	0.319	.272	.48	.32	.24	.16	.12	.096	.08
W7.5		0.309	.255	.45	.30	.225	.15	.112	.09	.075
W7	D7	0.299	.238	.42	.28	.21	.14	.105	.084	.07
W6.5		0.288	.221	.39	.26	.195	.13	.097	.078	.065
W6	D6	0.276	.204	.36	.24	.18	.12	.09	.072	.06
W5.5		0.265	.187	.33	.22	.165	.11	.082	.066	.055
W5	D5	0.252	.170	.30	.20	.15	.10	.075	.06	.05
W4.5		0.239	.153	.27	.18	.135	.09	.067	.054	.045
W4	D4	0.226	.136	.24	.16	.12	.08	.06	.048	.04
W3.5		0.211	.119	.21	.14	.105	.07	.052	.042	.035
W3		0.195	.102	.18	.12	.09	.06	.045	.036	.03
W2.9		0.192	.099	.174	.116	.087	.058	.043	.035	.029
W2.5		0.178	.085	.15	.10	.075	.05	.037	.03	.025
W2.1		0.162	.070	.126	.084	.063	.042	.031	.025	.021
W2		0.160	.068	.12	.08	.06	.04	.03	.024	.02
W1.5		0.138	.051	.09	.06	.045	.03	.022	.018	.015
W1.4		0.134	.048	.084	.056	.042	.028	.021	.017	.014

Courtesy Reinforcing Steel Institute.

Note: The above listing of smooth and deformed wire sizes represents wires normally selected to manufacture welded wire fabric styles to specific areas of reinforcement. Wire sizes other than those listed here, including larger sizes, may be available if the quantity required is sufficient to justify manufacture.

[a]The number following the prefix W or the prefix D identifies the cross-sectional area of the wire in hundredths of a square inch.

[b]The nominal diameter of a deformed wire is equivalent to the diameter of a smooth wire having the same weight per foot as the deformed wire.

destroying its effectiveness. The plastic caps shown on the feet of the support devices in Figure 3.4-9 are used when the underside of the slab will be exposed and in other locations to prevent the supports from rusting and transferring this rust to the reinforcement.

In beams, girders, walls, and slabs, when bar space is not at a premium, splices usually are made by overlapping the bars a prescribed number of bar diameters and tying them together with wire to keep them in place until the concrete has set. Sometimes, in columns when the space is limited, bars are butted end to end and connected with special devices made for the purpose or welded together.

3.4.3

DESIGN AND PLACEMENT

Reinforcing steel should be designed, fab-ricated, and installed in accordance with the Concrete Reinforcing Steel Institute publication "Manual of Standard Practice." Other applicable documents are the American Welding Society publications AWS D1.4, "Structural Welding Code-Reinforcing Steel" and "Connections in Reinforced Concrete Construction."

In most cases, reinforcing steel is fabricated in a mill or a shop according to shop drawings prepared by the fabricator from the structural engineer's drawings and

SYMBOL	BAR SUPPORT ILLUSTRATION	BAR SUPPORT ILLUSTRATION PLASTIC CAPPED OR DIPPED	TYPE OF SUPPORT	TYPICAL SIZES
SB			Slab Bolster	¾, 1, 1½, and 2 inch heights in 5 ft. and 10 ft. lengths
SBU*			Slab Bolster Upper	Same as SB
BB			Beam Bolster	1, 1½, 2, over 2" to 5" heights in increments of ¼" in lengths of 5 ft.
BBU*			Beam Bolster Upper	Same as BB
BC			Individual Bar Chair	¾, 1, 1½, and 1¾" heights
JC			Joist Chair	4, 5, and 6 inch widths and ¾, 1 and 1½ inch heights
HC			Individual High Chair	2 to 15 inch heights in increments of ¼ inch
HCM*			High Chair for Metal Deck	2 to 15 inch heights in increments of ¼ in.
CHC			Continuous High Chair	Same as HC in 5 foot and 10 foot lengths
CHCU*			Continuous High Chair Upper	Same as CHC
CHCM*			Continuous High Chair for Metal Deck	Up to 5 inch heights in increments of ¼ in.
JCU**			Joist Chair Upper	14" Span Heights – 1" thru +3½" vary in ¼" increments
CS			Continuous Support	1½" to 12" in increments of ¼" in lengths of 6'-8"

*Usually available in Class 3 only, except on special order.
**Usually available in Class 3 only, with upturned or end bearing legs.

FIGURE 3.4-9 *Typical types and sizes of wire bar supports.* *(Courtesy Concrete Reinforcing Steel Institute.)*

approved by the structural engineer and the architect. The fabricated steel is bundled, tagged as to its location in the building, and delivered to the construction site. Details of bends and hooks are standard (Figs. 3.4-10 and 3.4-11). These bundles are delivered to their place in the building by hand or crane, where the reinforcement is wired to hold its place and alignment until the concrete has set. In some cases, sections of reinforcement, such as those for a column, are preassembled at the mill or the shop and delivered to the site in one piece. Such assemblies may also be preassembled at the site on the ground and lifted into place by crane.

Steel should be accurately positioned, secured to prevent its displacement during the placing of concrete, and cleaned of rust and foreign matter before concrete is placed against it. Steel should be protected from possible harm by fire and corrosion by a covering at least the thickness of concrete shown in Figure 3.4-12.

Welded wire fabric should be installed in the largest practicable length. Adjoining pieces should be lapped at least one full mesh and tied together with wire to prevent movement during concrete placement.

3.5 MIXTURES AND MIXING

The mixture of its ingredients determines whether the consistency, workability, and finishability of plastic concrete will be in a desirable range. It also determines whether the essential properties of hardened concrete (strength, durability, watertightness, and wear resistance) are obtained and whether the concrete will be economical.

In properly made concrete, every particle of aggregate, no matter how large or how small, is surrounded completely by portland cement and water paste, and the spaces between aggregate particles are filled completely with paste (see Fig. 3.2-6). Aggregates are considered inert materials, while paste is the cementing medium that binds the aggregate particles into a solid mass. Therefore, the quality

NOTES:
1. All dimensions are out-to-out of bar except "A" and "G" on standard 180° and 135° hooks.
2. "J" dimension on 180° hooks to be shown only where necessary to restrict hook size, otherwise ACI standard hooks are to be used.
3. Where "J" is not shown, "J" will be kept equal to or less than "H" on Types 3, 5, and 22. Where "J" can exceed "H", it should be shown.
4. "H" dimension stirrups to be shown where necessary to fit within concrete.
5. Where bars are to be bent more accurately than standard fabricating tolerances, bending dimensions which require closer fabrication should have limits indicated.
6. Figures in circles show types.
7. For recommended diameter "D", of bends, hooks, etc., see tables, page 6-4.
8. Type S1-S6, S11, T1-T3, T5-T9 apply to bar sizes #3 through #8.
9. Unless otherwise noted, diameter D is the same for all bends and hooks on a bar (except for bend types 11 and 13).

Where slope differs from 45° dimensions "H" and "K" must be shown.

ENLARGED VIEW SHOWING
BAR BENDING DETAILS

FIGURE 3.4-10 *Typical bar bends.* (*Courtesy Concrete Reinforcing Steel Institute.*)

of a concrete greatly depends on the quality of its paste. It is important that the paste itself have the required strength, durability, and watertightness.

3.5.1

WATER-CEMENT RATIO

A concrete mixture's water-cement ratio is the ratio by weight of water to portland cement in the mixture. This means pound of water per pound of cement (or ounce per ounce). A water-cement ratio of 0.31 could be thought of as 0.31 lb. of water for each pound of cement. The units (pounds, ounces) are not used, because they cancel out mathematically (pound/pound = 1). Formerly, the proportion of water to portland cement was stated in gallons of water per 94-lb. sack of cement. The cementing or binding properties of a cement paste are due to a chemical reaction between the cement and the water (hydration) that requires time and favorable conditions of temperature and moisture. The reaction takes place rapidly at first and then more slowly for a long time under favorable conditions. Only a relatively small amount of water, about 3-1/2 gal. per 94 lb. of portland cement, is actually required to complete the chemical reaction. This amount is equivalent to a water-cement ratio of 0.31. But a water-cement ratio ranging from 0.35 to 0.80 is usually maintained for the sake of workability and placeability. Another advantage of using more water is that more aggregate can be used, with resulting economy. All the water that is not required for hydration eventually evaporates, leaving the small holes that account for the porosity of concrete.

For successful results, a proper proportion of water to portland cement is essential. If there is a high water-cement ratio, meaning that too much water has been added, the portland cement paste becomes thin or diluted and will be weak and porous when it hardens. Too little water may result in a mix that cannot be properly placed and finished. Portland cement paste made with the correct amount of water has strong binding qualities and is watertight and durable when cured. If its portland cement paste and aggregates are strong and durable, a concrete will be strong and durable. The strength, durability, freeze-thaw resistance, watertightness, and wear resistance of a concrete are largely controlled by a sufficiently low ratio of water to portland cement. Additional control can be gained by adding an adequate amount of entrained air.

The water-cement ratio selected should be the lowest value required to meet design considerations, such as durability, strength, and impermeability. For instance, concrete that will be exposed to a combination of wet-dry and freeze-thaw cycling and de-icing chemicals, requires the following conditions for durability: (1) a low water-cement ratio, (2) air entrainment, (3) suitable materials, (4) adequate curing, and (5) good construction practices.

A guide for selecting the water-cement ratios for various exposure conditions is given in Figure 3.5-1.

3.5.1.1

Strength

There is a direct link between the strength of concrete and its water-cement ratio. In fact, for given materials and handling conditions, strength is determined primarily by the water-cement ratio, as long as the mixture is plastic and workable. The age-strength relationships for Types I and III portland cements are shown in Figure 3.5-2.

For a given water-cement ratio in a concrete mixture, the strength at a certain age is practically fixed, if the mixture is plastic and workable, the aggregates are strong, clean, and sound, and the concrete is properly cured. More water will result in less strength, and less water will result in greater strength.

All specific sizes recommended by CRSI below meet minimum requirements of ACI 318

RECOMMENDED END HOOKS
All Grades

D=Finished bend diameter

Bar Size	D	180° HOOKS		90° HOOKS
		A or G	J	A or G
# 3	2¼	5	3	6
# 4	3	6	4	8
# 5	3¾	7	5	10
# 6	4½	8	6	1-0
# 7	5¼	10	7	1-2
# 8	6	11	8	1-4
# 9	9½	1-3	11¾	1-7
#10	10¾	1-5	1-1¼	1-10
#11	12	1-7	1-2¾	2-0
#14	18¼	2-3	1-9¾	2-7
#18	24	3-0	2-4½	3-5

180°

90°

STIRRUP AND TIE HOOKS

90°   135°

STIRRUPS (TIES SIMILAR)

135° SEISMIC STIRRUP/TIE HOOKS

135°   135°

STIRRUP AND TIE HOOK DIMENSIONS
Grades 40-50-60

Bar Size	D (in.)	90° Hook	135° Hook	
		Hook A or G	Hook A or G	H Approx.
#3	1½	4	4	2½
#4	2	4½	4½	3
#5	2½	6	5½	3¾
#6	4½	1-0	8	4½
#7	5¼	1-2	9	5¼
#8	6	1-4	10½	6

135° SEISMIC STIRRUP/TIE HOOK DIMENSIONS
Grades 40-50-60

Bar Size	D (in.)	135° Hook	
		Hook A or G	H Approx.
#3	1½	4¼	2½
#4	2	4½	3
#5	2½	5½	3¾
#6	4½	8	4½
#7	5¼	9	5¼
#8	6	10½	6

NOTES:
1. 180° hook J dimension (sizes #10, #11, #14 and #18), and A or G dimension (#14 and #18) have been revised to reflect test research using ASTM/ACI bend test criteria as a minimum.
2. Tables for Stirrup and Tie Hook dimensions have been expanded to include sizes #6, #7, and #8 to reflect current design practices.

FIGURE 3.4-11 *Standard hooks. (Courtesy Concrete Reinforcing Steel Institute.)*

FIGURE 3.4-12
Recommended Minimum Concrete Cover for Reinforcing Steel

Location and Use	Minimum Cover
Against ground without forms	3 in.
Exposed to weather or ground but placed in forms	
Greater than 5/8-in.-diameter bars	2 in.
Less than 5/8-in.-diameter bars	1-1/2 in.
Slabs and walls (no exposure)	3/4 in.
Beams, girders, columns (no exposure)	1-1/2 in.

Courtesy Portland Cement Association.

After concrete has been placed, its strength continues to increase if moisture and optimum temperatures are present to promote hydration and as long as the evaporation of its water (*drying*) is prevented. When concrete is permitted to dry, the water that would cause hydration is gone, and the chemical reactions between the portland cement and the water cease. This results in lower strengths than would occur if the concrete were kept moist. Therefore, it is desirable to keep concrete constantly moist as long as possible after placing (Fig. 3.5-3).

The temperature at which concrete is made and cured also affects the rate of the chemical reactions between the portland cement and the water. Seventy-three degrees Fahrenheit is considered the normal curing temperature. Most specifiers require that the range be between 50° and 85°F. Some will permit temperatures as high as 90°F. At higher temperatures, concrete increases in strength faster than normal during the first few days of curing but much slower than normal during later periods (Fig. 3.5-4). The final strength of concrete cured at temperatures above 90°F is also lower than that for concrete cured at between 50° and 90°F. Concrete made and cured at 40°F will gain strength slowly during curing and perhaps never reach the strength of concrete cured at between 50° and 90°F. As the temperature in concrete rises, the slump decreases. The tendency is to add water to the mix to maintain a desirable slump, but doing so will decrease the strength of the concrete and adversely affect its other properties.

3.5.1.2

Durability

There is a direct link between the durability of a concrete and the relative quantities of water and portland cement in its mixture. The most destructive natural weathering force is a freeze-thaw cycle while concrete is wet or moist, because of the expansion of water as it changes to ice. A portland cement paste made with a low water-cement ratio will be much more resistant to damage by freezing and thaw-

FIGURE 3.5-1
Recommended Concrete Mixtures for Various Exposure Conditions

Concrete Construction Element	Exposure Condition[a]	Minimum Compressive Strength at 28 days,[b] Plant or Transit, psi	Practical Water-Cement Ratio by Weight		Normal Maximum Coarse Aggregate Size,[e] in.	Minimum Cement[f] Content, lb./cu. yd.	Air-Entrainment by Volume, %	Slump, in.
			Non-Air-Entrained Concrete	Air-Entrained Concrete				
Foundations, basement walls, and slabs not exposed to weather	Severe Moderate Negligible	3000[c]	See footnote c	0.55	1	564 520	See footnote f	5 in. maximum for hand methods of strike-off and consolidation; 3 in. maximum for mechanical strike-off and consolidation
Foundations, basement walls, exterior walls, and other concrete work exposed to weather	Severe	3500[c]	See footnote c	0.45	1	564	6–8	
	Moderate	3000	0.58	0.50		520	5–7	
	Negligible	2500	0.67	0.55			See footnote f	
Concrete exposed to sulfate attack	Severe	5000[c]	See footnote d	See footnote d	1	564	6–8	
	Moderate					520	5–7	
	Negligible	5000					See footnote f	
Driveways, garage floors, walks, porches, patios, and stairs exposed to weather	Severe	4000[c]	See footnote c	0.45	1	564	6–8	
	Moderate	3000[c]		0.50		520	5–7	
	Negligible	2500	0.67	0.55			See footnote f	

Courtesy Portland Cement Association.

[a]See Figure 3.5-8 for severity of exposure, regional weathering areas.

[b]With most materials, the water-cement ratios shown will provide average strengths greater than required.

[c]Use air-entrained concrete only.

[d]Proportions should be established by the trial batch method.

[e]Maximum size of coarse aggregate should not be larger than one-fifth the narrowest dimension between forms, nor larger than three-quarters the minimum clear spacing between reinforcing bars.

[f]2% to 3% air entrainment, to improve cohesiveness and reduce bleeding, is recommended but not essential.

ing than one with a higher water-cement ratio. This property is demonstrated by the concrete cubes in Figure 3.5-5, which have been subjected to 70 cycles of freezing and thawing while saturated with water. Those in the upper portion of the figure, which were made with a water-cement ratio of 0.80, show much more disintegration than those in the lower portion, which were made with a water-cement ratio of 0.64. Furthermore, introducing entrained air into concrete will significantly increase its resistance to freezing and thawing damage and the application of de-icer salts.

3.5.1.3

Watertightness

Impervious concrete requires a portland cement paste formulated to produce watertightness, or imperviousness. As with strength and durability, there is a direct link between the watertightness of a concrete and the relative quantities of water and portland cement in its mixture.

Tests show that the permeability or watertightness of a portland cement concrete depends on the amount of mixing water used in its water paste and the extent to which the chemical reactions between the portland cement and the water have progressed. The results of subjecting discs made with portland cement mortar, fine aggregate, and water to 20 psi water pressure are shown in Figure 3.5-6. In these tests, mortar cured moist for seven days had no leakage when made with a water-cement ratio of 0.50, but there was considerable leakage with mortars made with the higher water-cement ratios. Also, in each case, leakage became less as the length of the curing period was increased. Mortar discs with a water-cement ratio of 0.80 leaked, even when moist-cured for a month.

Air entrainment improves watertightness by allowing reduction of the water-cement ratio. To be watertight, concrete must be free from cracks and honeycombing.

As a result of testing and experience in the field, definite recommendations can be made regarding the maximum amount of mixing water that should be used for various construction applications (see Fig. 3.5-1).

3.5.2

SELECTING A CONCRETE MIX

The best proportions for a concrete mixture are those that result in the most economical combination of portland cement, available aggregate, and other materials that will produce concrete of the required durability, abrasion resistance, and strength that can be placed and finished

FIGURE 3.5-2 *The relationship between age and compressive strength for Types I and III portland cement concretes. A majority of the results of tests for compressive strength made by several laboratories, using a variety of materials complying with ASTM standards, fall within the shaded areas.* (Drawing by HLS.)

properly. For slab construction, the properties of the plastic concrete, which affect placing and finishing, are equally as important as the properties of the hardened concrete, such as abrasion resistance, durability, and strength. The plastic properties will greatly affect the quality of the top 1/16 to 1/8 in. of the slab surface. However, because these properties are not easily measured, there is a tendency to emphasize the importance of more readily determined properties, such as compressive strength.

Recommendations for slump, strength, maximum size of coarse aggregate, and minimum portland cement content required for each size aggregate are given in Figure 3.5-7. The portland cement content of a mix should be properly proportioned to produce concrete that can be easily placed and finished and that, in addition, will possess adequate strength and durability. Mixtures with low portland cement content result in harsh concretes and produce slabs that tend to bleed excess water to the surface, are difficult to finish, and have poor surface characteristics. Mixtures with excessive portland cement content can result in increased drying shrinkage.

Except when concrete mixtures are established by tests before construction work begins, the concrete mixtures shown

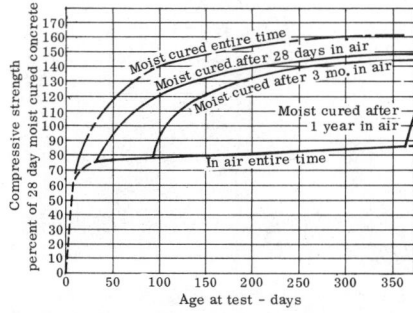

FIGURE 3.5-3 *The strength of a concrete continues to increase as long as moisture is present to promote hydration of the cement. Notice that a resumption of moist curing after a period of drying also increases the strength. However, the test specimens used in this study were small compared with most concrete members. Because resaturation is difficult in the field, moist curing should be continuous. (Courtesy Portland Cement Association.)*

FIGURE 3.5-4 *The effect of temperature on the strength of a concrete. Other characteristics also are affected. (Courtesy Portland Cement Association.)*

in Figure 3.5-1 should be used. It is first necessary to determine the exposure condition of the concrete construction from the regional weathering areas shown in Figure 3.5-8. Areas shown as severe exposure are subject to many freeze-thaw cycles per year and de-icer chemicals; moderate-exposure areas are subject to few freeze-thaw cycles and no de-icer chemicals; negligible-exposure areas have no freeze-thaw cycles, and de-icer chemicals are not used.

When the exposure condition of a concrete construction element is known, it is possible to use Figure 3.5-1 to determine the recommended minimum compressive strength. It is also possible to determine the maximum permissible water-cement

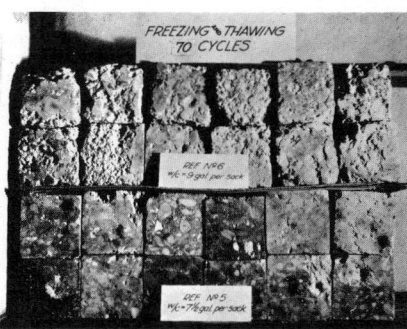

FIGURE 3.5-5 *The durability of a concrete exposed to freezing and thawing is affected by the quality of its paste. The specimens at the bottom of this photograph were made with a lower water-cement ratio (0.64) than those in the upper part (0.80) and were, therefore, more resistant to damage. The same aggregate was used in all the specimens. (Courtesy Portland Cement Association.)*

FIGURE 3.5-6 *The effect of water-cement ratio on watertightness. Leakage is reduced as the water-cement ratio is decreased and the curing period is increased. Specimens were 1-in. × 6-in. (25-mm × 150-mm) mortar discs. Pressure was 20 psi (0.14 MPa). Values are 48-hour averages. (Courtesy Portland Cement Association.)*

ratio necessary to obtain the specified minimum compressive strength. The minimum portland cement content is included, in addition to the lowest maximum water-cement ratio. Minimum portland cement requirements ensure satisfactory finishability, improved water resistance, and suitable appearance of vertical surfaces. The percentages of air entrainment for various maximum aggregate sizes are also given; slumps should be in the ranges shown.

An example showing how to use Figure 3.5-1 follows. The proposed project is in the Chicago area, which, according to the map in Figure 3.5-8, is in a severe-exposure region. According to Figure 3.5-1, the minimum compressive strength required for basement walls that will be exposed to weather is 3500 psi. Reading across the table shows a practical water-cement ratio of 0.45, and 6% to 8% air-entrained concrete. The maximum portland cement content likely to be required to achieve that strength at the recommended air-entrainment level is 564 lb./cu. yd.

Even though 2500 psi concrete might satisfactorily support typical vertical and lateral loading in a residential basement wall, this lower-strength concrete would be less watertight than the concrete recommended here. A higher-strength concrete provides a margin of safety against the possibility of foundation cracking and leaking.

3.5.2.1

Allowance for Water in the Aggregate

In selecting a proper water-cement ratio, it is important to note that most fine aggregates contain some water. Therefore, allowance must be made for this fact in determining the amount of water to be added to the mix. A simple test for determining whether sand is *damp*, *wet* or *very wet* is to press some together in your hand. If the sand falls apart after your hand is opened, it is damp (Fig. 3.5-9a); if it forms a ball that holds its shape, it is wet (Fig. 3.5-9b); if the sand sparkles and leaves water on your hand, it is very wet (Fig. 3.5-9c).

The most practical procedure is to make trial mixes. On large jobs where many materials or unusual materials may be under consideration or where strength is particularly important, the trials may be full-sized batches checked in the field. The first trial mix may be selected on the basis of experience or established relationships, such as those given in Figure 3.5-1. This table has been developed from experience and data from several sources and indicates the suggested proportions of all the various ingredients that should result in good workability.

FIGURE 3.5-7
Recommended Mixtures for Light Slabs on Grade[a]

| Maximum Size of Coarse Aggregate[b] | Minimum Cement Content, lb./cu.yd. | Maximum Water/Cement by Weight | | Slump | Probable Compressive Strength[c] |
		Non-Air-Entrained Concrete	Air-Entrained Concrete		
1-1/2 in.	470				
1 in.	520				
3/4 in.	540	0.51	0.40	2–4 in.	3500 psi
1/2 in.	590				
3/8 in.	610				

[a]For both air-entrained and non-air-entrained concretes.

[b]Normal-weight aggregate; different mixes may be required for lightweight aggregate concretes.

[c]Strength refers to approximate compressive strength of cylinders made and tested according to applicable ASTM standards at 28 days that have been continuously moist cured.

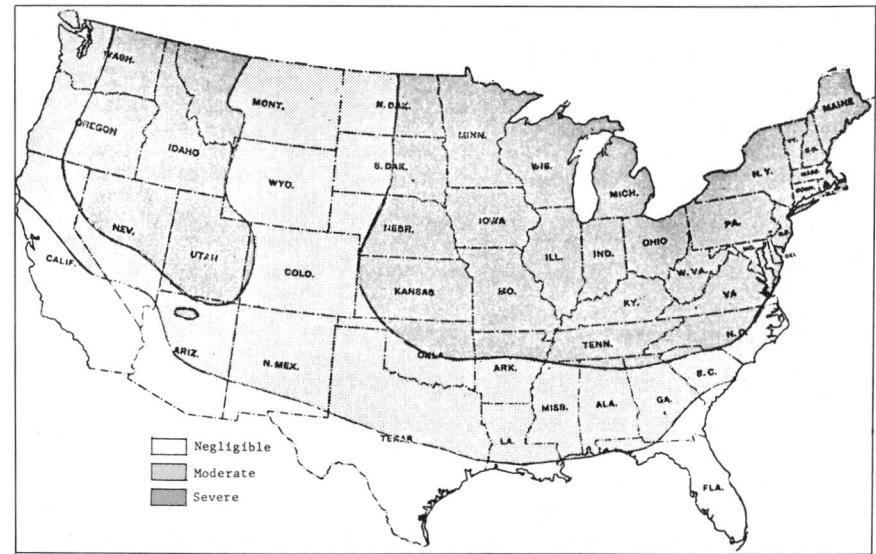

FIGURE 3.5-8 *Regional weathering areas.**

*Areas are based upon ASTM C 62 weathering index map adjusted to conform with an experience survey by HUD and the Portland Cement Association. Local conditions may be more or less severe than indicated by regional classification.

FIGURE 3.5-9 *(a) Damp sand; (b) wet sand; (c) very wet sand.* *(Courtesy Portland Cement Association.)*

3.5.3

CONSISTENCY OF THE MIX

With a given amount of portland cement paste, more aggregate is used in stiff mixes than in more fluid mixes; consequently, stiff mixes are more economical in cost of materials. Stiff mixes require more labor in placing, however, and when a mixture is too stiff the additional cost of placing may offset savings in materials. A mix should always be of a consistency and workability that will permit the concrete to be worked into the angles and corners of forms and around reinforcement without permitting the materials to segregate or excess free water to collect on the surface. Thin members and heavily reinforced members require more fluid mixes than large members or flat surfaces that contain little reinforcing.

A *plastic concrete* is one that is readily molded and yet will change its form only slowly if the mold is immediately removed. Mixes of plastic consistency are required for most concrete work. Concrete of such consistency does not crumble but flows sluggishly without segregation. Thus, neither very stiff, crumbly mixes nor very fluid, watery mixes are of plastic consistency (Fig. 3.5-10).

The ease or difficulty of placing concrete in a particular location is referred to as *workability.* A stiff but plastic mix with large aggregate would be workable in a large open form but not in a thin wall with closely spaced reinforcement.

Excessive slump and consequent bleeding and segregation are primary causes of poor concrete performance. For slabs, the slump may vary from 2 to 4 in. (see Fig.

FIGURE 3.5-10 *(a) A concrete mixture with insufficient cement-sand mortar to fill the spaces between the coarse aggregate. This mixture will be difficult to handle and place and will result in honeycombed surfaces and porous concrete. (b) A concrete mixture with the correct amount of cement-sand mortar. With light troweling, the spaces between aggregate are filled with mortar. Note the appearance on the edge of the pile. This is a good workable mixture that will give the maximum yield of concrete. (c) A concrete mixture with too much cement-sand mortar. This mixture is plastic and workable and will produce smooth surfaces, but the yield of concrete will be low and uneconomical, and the concrete will probably be porous. (Courtesy Portland Cement Association.)*

3.5-7), which will produce slabs that can be easily placed and properly consolidated yet have a consistency not conducive to bleeding or segregation during placing and finishing. Concrete for slabs where the appearance is important and concrete for ramps and sloping surfaces should have a slump that does not exceed 3 in. Different batches of concrete placed in the same slab, ramp, or sloping surface should have approximately the same slump. When lightweight concrete is placed at slumps in excess of 3 in., the coarser lightweight aggregate may rise to the surface, and excessive bleeding may also occur.

Reinforced foundation systems may have slumps as low as 1 in., but should not exceed 3 in. Concrete that will be struck off and consolidated by hand may have slumps of up to 5 in. The slump of concrete that will be struck off and consolidated with mechanical devices should be limited to 3 in. Concrete that contains a high-range water-reducing admixture may have slumps as large as 8 in. after the admixture has been added. But the slump before the admixture is added must be between 2 and 3 in. Slumps for other concrete should be not less than 1 in. or more than 4 in.

3.5.3.1

Slump Test

A slump test conforming with the requirements of ASTM Standard C 143 may be used as a rough measure of the consistency

of concrete. This test is not a measure of workability, and it should not be used to compare mixes of entirely different proportions or that contain different kinds of aggregate. Any change in slump on the job indicates that changes have been made in grading or proportioning the aggregate or in water content. The mix should be corrected immediately to get the proper consistency by adjusting amounts and proportions of sand and coarse aggregate, care being taken not to change the total specified water-cement ratio.

In a slump test, a test specimen is made in a mold called a *slump cone*, which is made of 16-gauge galvanized metal to the shape shown in Figure 3.5-11a. The base and the top are open. The mold is provided with foot pieces and handles, as shown.

For a slump test, a concrete sample is taken just before its batch is placed in the forms. A slump cone is placed on a flat surface, such as a smooth plank or a slab of concrete, held firmly in place by the tester standing on the foot pieces, and filled with concrete to about one-third its volume. Then the concrete is puddled with 25 strokes of a 5/8-in.-diameter rod about 24 in. long and bullet pointed at the lower end (Fig. 3.5-11b). The filling is completed in two more layers, each about one-third the volume of the cone. Each new layer is rodded 25 times with the rod penetrating into the underlying layer on each stroke. After the top layer has been rodded, it is struck off with a trowel so that the cone is filled completely. The

cone is removed, by gently raising it vertically, immediately after the top layer has been struck off.

The slump of the concrete is measured, as shown in Figure 3.5-11c, immediately after the cone is removed. If the top of the slump pile is, for example, 5 in. below the top of the cone, the slump for that concrete is 5 in.

3.5.3.2

Compression Test

Compression tests in accordance with ASTM Standard C 31 are made to determine if a concrete has the specified compressive strength. In these tests, specimens of the concrete are tested at seven days and at 28 days after they are taken to determine the rate of strength gain of the concrete.

Standard procedures provide for curing the specimens either in the laboratory or in the field. Laboratory curing gives an indication of the potential quality of the concrete when properly cured under ideal conditions. Field-cured specimens may give a more accurate interpretation of actual strength in the structure or slab under the less than ideal curing conditions that always exist at a job site, but they offer no explanation as to whether lack of strength is due to error in proportioning, poor materials, or unfavorable curing conditions. Sometimes, both methods are used, especially when the weather is unfavorable, in order to interpret the tests

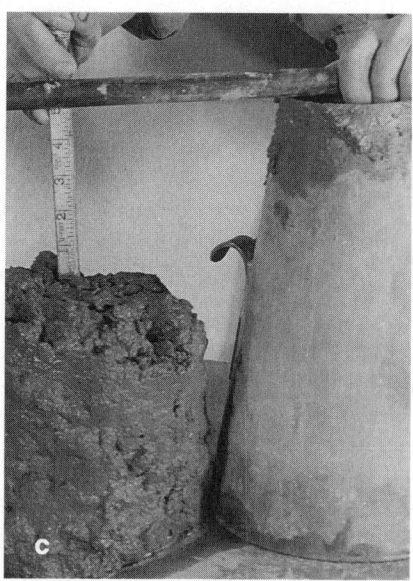

FIGURE 3.5-11 *(a) Slump cone (mold) for slump test.* (Courtesy ELE International, Inc., Soil Test Products Division.) *(b) Rodding the concrete in a slump cone to ensure complete filling.* (Copyright ASTM, reprinted with permission.) *(c) Measuring slump by the depression below a rod laid across the cone. This amount of slump indicates a medium-wet mixture.* (Copyright ASTM, reprinted with permission.)

properly. When differences occur, the laboratory test results prevail.

To make a compressive strength test, a sample of the concrete to be tested should be taken at three or more regular intervals throughout the discharge of an entire batch, but not at the beginning or the end of discharge. The batch of concrete sampled should be noted as to its location in the work, the air temperature, and any unusual conditions at the time. The usual number of test specimens in a sample is four: one is tested at seven days, two at 28 days, and the fourth held in reserve in case additional testing is necessary. If both laboratory and field tests are required, duplicate samples are taken.

A compressive test specimen is placed in a cylindrical mold that is watertight to prevent loss of water. Standard cylindrical molds, like those shown in Figure 3.5-12, are available 3 in. in diameter by 6 in. long, 4 in. in diameter by 8 in. long, and 6 in. in diameter by 12 in. long. The 6-in. by 12-in. cylinders are generally used when the coarse aggregate does not exceed 2 in. in nominal size. A mold is filled in three equal layers. Each layer is puddled with 25 strokes of a 5/8-in.-diameter steel rod about 24 in. long with a bullet-pointed tip. Reinforcing bars and other tools should not be used as puddle rods. After the top layer has been rodded, the surface of the concrete is struck off

FIGURE 3.5-12 *Standard plastic test cylinders. Waxed cardboard cylinders with metal bottoms also are available.* (Courtesy ELE International, Inc., Soil Test Products Division).

with a trowel and covered to prevent evaporation.

The cured specimens are crushed in a laboratory testing machine to determine their strength (Fig. 3.5-13). If the strength of the specimens is too low, test cores should be cut from the concrete structural unit that contains the suspect concrete. These cores should then be tested for compression. If this test indicates unsatisfactory strength, the concrete should be removed and new concrete of adequate strength placed.

MATERIALS MEASUREMENT

If uniform batches of concrete of proper proportions and consistency are to be produced, it is essential that the ingredients be measured accurately for each batch. The effects of the water-cement ratio on the qualities of concrete make it necessary to measure the water as well as the other ingredients. A troublesome factor is the effect of the varying amounts of moisture nearly always present in aggregate, particularly in natural sand. The amount of free moisture introduced into the mixture with the aggregates must be determined and allowance made for it if accurate control is to be obtained.

Measuring Portland Cement

On small projects where sacked portland cement is used, the batches of concrete should be of such size that only full sacks are used. If partial sacks of portland cement are used, they should be weighed for each batch. It is not satisfactory to divide sacks of portland cement on the basis of volume.

Bulk (unsacked) portland cement, which is used today for most concrete, should be measured by weight for each batch.

FIGURE 3.5-13 *A concrete compression tester.* *(Courtesy ELE International, Inc., Soil Test Products Division.)*

3.5.4.2

Measuring Water

Dependable and accurate means for measuring mixing water are essential. Portable mixers generally are equipped with water tanks and measuring devices that are fairly accurate when properly operated. The measuring device most generally used operates on the principle of the siphon. The tank is filled, and the desired amount of water is siphoned off. When the water level reaches the point at which the bottom of the siphon is set, the seal is broken and the water is shut off automatically.

3.5.4.3

Measuring Aggregates

Measurement of aggregates by volume cannot be depended on except under the most careful supervision. A small amount of moisture in fine aggregate, which is nearly always present, causes the aggregate to bulk up (increase in volume), as shown in Figure 3.5-14. The amount of bulking varies with the amount of moisture present and the grading; fine sands bulk more than coarse sands for a given

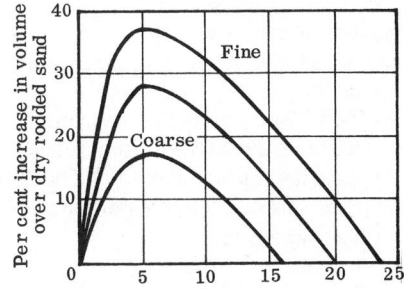

FIGURE 3.5-14 *Bulking in fine aggregate caused by surface moisture.* *(Courtesy Portland Cement Association.)*

amount of moisture, because a given quantity of smaller particles presents more surface to which water can adhere than does the same amount of aggregate with larger particles. Note in Figure 3.5-14 that a fine sand with a 5% moisture content increases in volume almost 40% over its dry volume. Sand often contains almost enough moisture to produce the maximum bulking. Small changes in moisture can cause large changes in the amount of bulking, until all particles have been surrounded with water. After that, the addition of water causes a lesser degree of increased bulking. Therefore, the relationship be-

tween moisture and bulk is not linear. For these reasons, aggregates should always be measured by weight.

To reduce segregation of aggregates and to produce uniformity from batch to batch, the coarse aggregate should be weighed in two or more sizes, especially if the maximum size exceeds 1 in. Generally, the ratio of the maximum-sized particle to the minimum size for coarse aggregate separations should not exceed 2:1 for materials larger than 1 in., or 3:1 for finer materials. Therefore, 1-1/2-in. aggregate would be separated into two batches, one with 1/4- to 3/4-in.-sized aggregate and the other with 3/4- to 1-1/2-in. sizes.

3.5.5

MIXING

The ingredients should be in the proper condition when they are charged into a mixer. The aggregates should have the correct moisture content, and the portland cement should be free of lumps. If lumps are present that cannot be pulverized between the thumb and a finger, the portland cement should not be used.

Concrete should be mixed thoroughly until its appearance, including its color, is uniform and its ingredients uniformly distributed. The time required for thorough mixing depends on several factors. Specifications usually require a minimum of 1 minute mixing for mixtures up to 1 cu. yd. capacity with an increase of 15 seconds mixing for each 1/2 cu. yd. or fraction thereof of additional capacity. In general, longer mixing gives more uniform results. The mixing period should be measured from the time all solid materials are in the mixer drum, provided that all the water has been added before a quarter of the mixing time has elapsed. Many mixers have timing devices, some of which can be set for a given mixing time and locked so that the batch cannot be discharged until the designated time has elapsed.

Mixers should not be loaded above their rated capacity and should be operated at approximately the speeds for which they are designed. If the blades of the mixer become worn or coated with hardened concrete, the mixing action will be less efficient. Badly worn blades should be replaced, and hardened concrete should be removed before each run of concrete.

Under usual operating conditions, up to about 10% of the mixing water should be placed in the drum before the aggregates and the portland cement are added. Water should then be added uniformly with the dry materials, leaving about 10% to be added after all other materials are in the drum. When heated water is used during cold weather, this order of charging may require some modification to prevent flash setting of the portland cement. In that case, addition of the portland cement should be delayed until most of the aggregate and water have intermixed in the drum.

The concrete should be discharged completely from a mixer before the mixer is recharged.

3.5.5.1

Ready-Mixed Concrete

Most concrete used in building construction originates in a central ready-mix plant (Fig. 3.5-15). Ready-mixed concrete should comply with ASTM Standard C 94, which requires, among other things, that the hauling be done in agitator trucks or transit mixer trucks operated at agitator speed.

In some ready-mixed operations, the materials are dry-batched at the central plant and then mixed en route to the job in truck mixers. In others, the concrete is mixed in a stationary mixer at the central plant only enough to intermingle the ingredients, generally about a half-minute. The mixing is completed in a truck mixer en route to the job.

FIGURE 3.5-15 *Loading trucks at a ready-mix plant.* (Courtesy Portland Cement Association [PCA slide VIII-2].)

Truck mixers consist essentially of a mixer with a separate water tank and a water-measuring device mounted on a truck chassis. They usually are made with capacities of from 1 to 5 cu. yd. Agitator trucks are similar but do not have separate water tanks.

ASTM C 94 requires that concrete be delivered and discharged from a truck mixer or agitator truck within 1-1/2 hours after water was added to the portland cement and aggregate.

3.5.5.2

Remixing Concrete

The initial set of concrete does not usually take place until 2 or 3 hours after the portland cement is mixed with water, but fresh concrete that is left standing tends to dry out and stiffen somewhat before the portland cement sets. Such concrete may be remixed and used within 1-1/2 hours of the time water was added to the mix, if when remixed it becomes sufficiently plastic to be compacted in the forms. However, adding water to make the mixture more workable (*retempering*) should never be permitted, because it lowers the quality just as would adding a larger amount of water in the original mixing.

3.6 HANDLING, TRANSPORTING, PLACING, AND CONSOLIDATING

Each step in handling, transporting, placing, and consolidating concrete should be controlled carefully to maintain uniformity within each batch and from batch to batch so that the complete structure has uniform quality throughout. It is essential to avoid separation of coarse aggregate from mortar and of water from the other ingredients. Equipment for conveying concrete from mixer to place of final deposit should do so without loss or separation of the materials, to provide a practically continuous flow of concrete at the point of delivery. Segregation at the point of discharge from the mixer can be avoided by a down pipe at the end of the chute so the concrete will drop vertically into the center of the receiving bucket, hopper, or form.

Concrete is handled and transported by many methods, such as chutes, push buggies, buckets handled by cranes, and pumping through a pipeline. The methods of handling and transporting concrete and the equipment used should not place restrictions on the consistency of the concrete, which should be governed by the placing conditions. If placing conditions permit the use of a stiff mix, the delivery equipment should be designed and arranged to facilitate handling and transporting such a mix.

3.6.1

PREPARATION

Before concrete is placed, the surfaces to receive the concrete should be properly prepared, forms should be erected, and reinforcement should be installed as required for the structure.

3.6.1.1

Requirements Common to All Concrete

Concrete should not be placed on a frozen subgrade. Snow, ice, and debris must be removed from within forms and excavations and from slab beds before concrete is placed.

When concrete is to be placed on rock, loose material should be removed and cut faces nearly vertical or horizontal rather than sloping.

Chlorides should not be used in concrete in which dissimilar metals are embedded, such as reinforcing steel and heating ducts.

Reinforcing steel should be clean and free of loose rust and mill scale at the time concrete is placed. Coatings of hardened mortar should be removed from the steel.

3.6.1.2

Slabs on Grade

Subgrades should be trimmed to specified elevations, uniformly compacted, and moist when the concrete is placed (Fig. 3.6-1). Except when a vapor retarder has been placed immediately below a slab, a moist subgrade is important to prevent too rapid extraction of water from the concrete when it is placed, especially in hot weather.

FIGURE 3.6-1 *Wetting the subgrade before concrete is placed.* (*Courtesy Portland Cement Association.*)

When a vapor retarder, waterproof membrane, or separator is not used, the bed to receive concrete should be moistened to minimize extraction of water from the concrete. However, at the time of placing, there should be no mud, soft spots, or free water standing where concrete will be placed.

The Portland Cement Association (PCA) recommends that a vapor retarder be placed under every concrete slab on grade where an impervious floor finish will be installed and in every other location where the passage of water vapor through a slab on grade is undesirable. However, a vapor barrier placed in contact with the concrete will cause excess moisture in the concrete to bleed to the surface rather than downward, increasing the number of capillaries in the concrete and contributing to weakening the surface, more cracking, and higher permeability of the slab. Therefore, the PCA recommends that a vapor retarder not be placed directly beneath a slab, but rather covered by a 3-in.-thick layer of granular, self-draining material, such as sand. Refer to Section 2.3 for a detailed discussion.

When a vapor retarder is not used beneath a slab on grade, and a base course or fill is provided, a separator of building paper or other material that will withstand handling and construction traffic is sometimes installed. When such separators are used, they are usually installed between the base course or fill and the slab to prevent the fines or paste of plastic concrete from seeping down into the base course or fill. However, if such a separator is complete and impervious, it should not be placed directly in contact with the slab. If it is, it could act in the same way a vapor retarder often acts and contribute

to warping (*curling*) of the slab and exacerbate drying shrinkage cracking (see Section 2.3).

Steel reinforcing, ductwork, heating pipes, and other items embedded in a slab should be set and supported at the proper elevation before concrete is placed.

3.6.1.3

Formed Concrete

Forms should be tight, adequately braced, and constructed of material that will impart the desired texture to the finished concrete. Before concrete is placed in them, forms should be cleaned of any dirt, nails, sawdust, wood chips, and other foreign materials. To facilitate form removal, forms should be treated with a release agent, such as oil or lacquer. For architectural concrete, the agent should be a non-staining lacquer or emulsified stearate. Wood forms should be moistened before concrete is placed, or they will absorb water from the concrete and swell.

3.6.2

PLACING

Concrete should be placed as nearly as practicable in its final position, to avoid segregation due to rehandling or flowing. It should not be placed in large quantities at a given point and allowed to run or be worked over a long distance in the form. This practice results in segregation because the mortar tends to flow out ahead of the coarser material. It also results in sloping work planes and potential weak bonding between successive layers of concrete. In general, concrete should be placed in horizontal layers of uniform thickness, with each layer thoroughly compacted before the next is placed.

Layers should be 6 to 24 in. thick; thinner layers may be preferred if the forms are narrow and heavily reinforced. Each layer must be placed before the previous one stiffens.

3.6.2.1

Placing Systems

On small projects, concrete is delivered to the site by truck or mixed at the site and delivered to its final position either directly from a truck's chute or by buggy. On larger projects, several additional delivery methods are available. Concrete may be delivered from a truck's chute into a large bucket, which is then lifted by crane and deposited in its proper location (Fig. 3.6-2). A third method is to dump the truck's load into a concrete pump, which delivers the mix to its final location through a hose. Pumps are capable of delivering concrete over long distances and up many floors (Fig. 3.6-3).

Regardless of the method used to get concrete to its final location, it should not be dropped freely more than 4 ft. If it is, segregation will result. Neither should concrete be permitted to strike against the forms as it is being placed. Instead, drop chutes should be used to prevent spattering of mortar on reinforcement and forms (Fig. 3.6-4). If placement can be completed before the mortar dries, drop chutes may not be needed.

CORRECT

Separation is avoided by discharging concrete into hopper feeding into drop chute. This arrangement also keeps forms and steel clean until concrete covers them.

INCORRECT

Permitting concrete from chute or buggy to strike against form and richochet on bars and form faces causes separation and honeycomb at the bottom.

FIGURE 3.6-4 *Placing concrete in the top of a narrow form.* *(Courtesy Portland Cement Association.)*

FIGURE 3.6-2 *Placing concrete by crane-delivered bucket.* *(Courtesy Portland Cement Association [PCA slide 43733].)*

3.6.2.2

Order of Placing

To prevent water from collecting at the end of forms, in corners of forms, or along form faces, the order of placing should be as follows: on walls, start at the ends and work toward the center; on slabs, start around the perimeter.

On flat surfaces, such as in slab construction, the placing of concrete should be started at the far end of the work so that each batch will be dumped against previously placed concrete, not away from it (Fig. 3.6-5). Concrete should not be dumped in separate piles and then leveled and worked together. If stone pockets

FIGURE 3.6-3 *Placing concrete by pump hose.* *(Courtesy Portland Cement Association [PCA slide VIII-36].)*

a

CORRECT

Concrete should be dumped into face of previously placed concrete.

INCORRECT

Dumping concrete away from previously placed concrete causes separation.

b

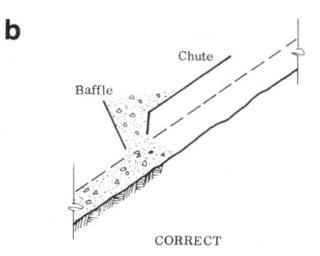

CORRECT

A baffle and drop at end of chute will avoid separation and concrete remains on slope.

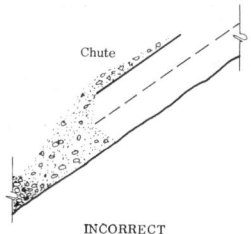

INCORRECT

Discharging concrete from free end chute onto a slope causes separation of rock which goes to bottom of slope. Velocity tends to carry concrete down the slope.

FIGURE 3.6-5 *(a) Placing slab concrete; (b) placing concrete on a sloping surface.* (Courtesy Portland Cement Association.)

occur, some of the excess large particles should be removed and distributed to areas where there is more mortar present to surround them.

In walls, the first batches should be placed at either end of the section; the placing should then progress toward the center. This same procedure should be used for each layer.

In large flat open areas, the first batches should be placed around the perimeter. In all cases, the procedure should prevent water from collecting at the ends and the corners of forms and along form faces.

3.6.2.3

Placing on Hardened Concrete

To secure good bond and a watertight joint, when fresh concrete is placed on hardened concrete, the hardened concrete should be fairly level, reasonably rough, clean, and moist, and some aggregate particles should be exposed. *Laitance* (soft layers of mortar) should be removed from the top surface of hardened concrete. An appreciable thickness of laitance indicates that segregation and bleeding have occurred or that too much water was used in the concrete. Laitance can be prevented by using a stiffer mix or more fine material, particularly in the upper part of the lift.

On floor slabs to be built in two courses, the top of the lower course may be broomed just before it sets with a stiff fiber or steel broom. The surface should be level, scored, and free of laitance. It must then be protected and thoroughly cleaned just before the second course is placed.

When new concrete is to be placed on old concrete, the old concrete must be roughened and cleaned thoroughly. In most cases, it is necessary to remove the entire surface to expose a new surface satisfactory for bonding.

Hardened concrete should be moistened thoroughly before new concrete is placed on it. Where concrete has dried out, it is necessary to saturate it for several days. There should be no pools of water, however, when the new concrete is placed.

Where concrete is to be placed on hardened concrete or on brick, a layer of mortar on the hard surface is necessary to provide a cushion against which the new concrete can be placed; this layer prevents stone pockets and ensures a tight joint. The mortar should be made with the same content as the concrete and a slump of about 6 in. It should be placed to a thickness of 1/2 to 1 in. and be well worked into the irregularities of the hard surface. In two-course floor construction, a paste of portland cement and water the consistency of thick paint should be brushed onto the hard surface just before the second course is placed.

3.6.3

CONSOLIDATING

Concrete should be compacted by a method appropriate to the material and its location. The method used should eliminate stone pockets and large air bubbles, consolidate each layer with that previously placed, completely embed reinforcing and fixtures, and bring just enough fine material to the faces and top surfaces to produce the desired finish.

3.6.3.1

Spading and Puddling

Medium- to high-slump (3 to 8 in.) concrete should be compacted and worked into place by spading or puddling. Spades or sticks long enough to reach the bottom of the forms and thin enough to pass between the reinforcing steel and the forms should be used.

3.6.3.2

Vibrating

Low- to medium-slump (1 to 6 in.) concrete should be compacted with mechanical vibrators either within the concrete (internal vibrators) or on top of the concrete or on the forms (external vibrators). Relatively thin slabs that are not heavily reinforced may be vibrated by mechanical beam or truss screeds (see Section 3.7.1.2). Thick slabs, slabs with structural reinforcement, and thicker concrete members, like footings, walls, columns, beams, and girders, should be vibrated with internal vibrators (Fig. 3.6-6).

When concrete is vibrated, the internal friction between the coarse aggregate particles is temporarily eliminated and the

FIGURE 3.6-6 *Using an internal vibrator.* *(Courtesy Portland Cement Association [PCA slide IX-15].)*

concrete behaves like a liquid; it settles in the forms under the action of gravity and the entrapped air bubbles rise more easily to the surface. Friction is reestablished as soon as vibration stops. Therefore, mixes of lower water content or leaner mixes for a given water content can be used. If less water is used per unit volume of portland cement, the concrete will be of better quality; if less portland cement per unit volume of water is used, the concrete will be more economical. Vibrators should not be used to push or move concrete laterally over long distances within a form, a practice that too often is allowed. The concrete should be deposited as near its final position as possible. It should be distributed in layers and then vibrated (Fig. 3.6-7). Some hand spading or puddling may be necessary along with the vibration to secure smooth surfaces and reduce pitting on formed surfaces.

Precautions should be taken not to over-vibrate to the point that segregation results, especially when the concrete has a higher slump than is necessary for vibration. Precaution and judgment must be used to be sure that complete consolidation is secured without segregation and that no areas are missed. Sufficient vibration usually is indicated by a line of mortar along the forms and by the submerging of the coarse aggregate particles in the mortar.

a

Correct: Start placing at bottom of slope so that compaction is increased by weight of newly added concrete. Vibration consolidates the concrete.

Incorrect: When placing is begun at top of slope, the upper concrete tends to pull apart, especially when vibrated below because this starts flow and removes support from concrete above.

b

Correct: Vertical penetration of vibrator a few inches into previous lift (which should not yet be rigid) at systematic regular intervals will give adequate consolidation.

Incorrect: Haphazard random penetration of the vibrator at all angles and spacings without sufficient depth will not assure intimate combination of the two layers.

FIGURE 3.6-7 *(a) Consolidation of concrete on slopes; (b) consolidation of concrete in multiple lifts. (Courtesy Portland Cement Association.)*

3.7 FINISHING

Concrete may be finished in several ways, depending on its location, the concrete material used, and the aesthetic effect desired. In this section, general finishing procedures are summarized first, followed by techniques to produce special surface finishes.

3.7.1

FINISHING STANDARD-WEIGHT CONCRETE SLABS

The finishing of standard-weight concrete slabs proceeds through several steps in a defined order. These steps are bleeding, screeding, leveling, edging, jointing, floating, troweling, and surface texturing.

3.7.1.1

Bleeding

Generally, the dry materials used in making quality concrete are heavier than water. Thus, shortly after placement, they have a tendency to settle to the bottom and displace the mixing water to the surface, which is called *bleeding*. Bleeding occurs more readily when the concrete is not air entrained.

If any operation is performed on the surface while bleed water is present, serious scaling, dusting, or crazing of the hardened concrete can result. This point cannot be overemphasized and is the basic rule for successful finishing of concrete surfaces. Section 3.7.6 contains a detailed discussion of scaling, crazing, and dusting.

It is of the utmost importance that placing, screeding, consolidating, and leveling of the placed concrete be performed before any bleeding takes place. Therefore, concrete should not be placed faster than it can be spread, struck off, and leveled or allowed to remain in wheelbarrows, buggies, or buckets any longer than is absolutely necessary. It should be dumped and spread as soon as possible and immediately struck off to proper grade and leveled.

3.7.1.2
Screeding

The surface of newly placed concrete is struck off (*screed*) by moving a straight-edge back and forth with a saw-like motion across the tops of the forms and screeds. If mechanical vibrating equipment is used in the striking-off process, the need for leveling may be eliminated. Whether boards, vibratory screeds, or roller screeds are used, a small amount of concrete always should be kept ahead of the straightedge to fill in low spots and maintain a plane surface (Fig. 3.7-1).

FIGURE 3.7-1 *Striking off (screeding).* (*Courtesy Portland Cement Association.*)

Of all the placing and finishing operations, screeding the surface to a predetermined grade has the greatest effect on surface tolerances. Concrete should be struck off immediately after it is placed, and screed stakes used to establish surface elevation should be removed as the work progresses so that workers do not have to walk back into areas that have already been struck off.

3.7.1.3
Leveling

Leveling is the bringing of a concrete surface to true grade with enough mortar to produce the desired finish. After a concrete slab has been screeded, it should be immediately smoothed with a *darby* to level raised spots and fill depressions left after screeding (Fig. 3.7-2). Long-handled floats, called *bull floats*, of either wood or metal, are sometimes used instead of darbies to smooth and level concrete surfaces (Fig. 3.7-3). Because it is hard to produce surfaces in plane near the edges of a slab using bull floats, leveling with a darby is sometimes needed after the bull floating has been done.

FIGURE 3.7-2 *Leveling using a darby.* (*Courtesy Portland Cement Association.*)

FIGURE 3.7-3 *Leveling using a bull float.* (*Courtesy Portland Cement Association.*)

Leveling is sometimes called darbying or bull floating, but this terminology requires that both terms be used or that the architect dictate the method to be used, which is inappropriate. In addition, the term *bull floating* confuses this activity with the later finishing step *floating*, which is an entirely different operation. The PCA in its literature and the AIA in its master guide specifications series, *Masterspec*, both use the broader term *leveling*.

The purpose of leveling is to eliminate the ridges and voids left by screeding. In addition, it should slightly embed the coarse aggregate, thus preparing the surface for the subsequent finishing opera-

tions of edging, jointing, floating, and troweling.

A slight stiffening of the concrete is necessary after leveling before further finishing operations are started. No subsequent operations should be performed until the concrete will sustain foot pressure with only about 1/4-in. indentation.

3.7.1.4
Edging

When all bleed water and water sheen have left the surface and the concrete has started to stiffen, other finishing operations, such as edging, may be started (Fig. 3.7-4). Edging rounds off the formed edge of a slab to prevent chipping or damage. An edger should be run back and forth until a finished edge is produced. All coarse aggregate particles should be covered, and the edger should not leave too deep an impression in the top of the slab. If it does, the indentation may be difficult to remove with subsequent finishing operations.

FIGURE 3.7-4 *Edging.* (*Courtesy Portland Cement Association.*)

Usually, edging is not required for most interior slabs on grade; it is more commonly performed on sidewalks, driveways, and steps. An edger should not be used when a slab is to be finished with resilient flooring requiring a smooth, level subfloor. Edges at construction joints in the slab may be ground lightly with a silicon carbide stone to remove irregularities after the forms are stripped and before the adjacent slab is placed.

3.7.1.5
Jointing

Except when joints will later be sawed, immediately following or during edging, premolded inserts are placed in concrete slabs to control cracking in the concrete due to shrinkage. Refer to Section 3.12 for a detailed discussion about joints.

3.7.1.6

Floating

After edging and hand-jointing operations, a slab should be floated. Many variables, such as concrete temperature, air temperature, relative humidity, and wind, affect the process, making it difficult to set a definite time to begin floating. This knowledge comes only through job experience. In general, floating may be done when the water sheen has disappeared and the concrete will support the weight of the finisher.

The purpose of floating is fourfold: (1) to embed large aggregate just beneath the surface; (2) to remove slight imperfections, humps, and voids to produce a level or plane surface; (3) to consolidate mortar at the surface in preparation for other finishing operations; and (4) to open the surface to permit excess moisture to escape.

Aluminum or magnesium floats should be used, especially on air-entrained concrete. Use of a metal float greatly reduces the amount of work required by the finisher because the float slides more readily over the concrete surface, has a good floating action, and forms a smoother surface texture than a wood float. A wood float tends to stick to and "tear" the concrete surface.

The marks left by edgers and jointers should be removed by floating (Figs. 3.7-5 and 3.7-6), unless such marks are desired for decorative purposes. In that case, the edger or jointer should be rerun after the floating operation.

3.7.1.7

Troweling

Troweling is done on slabs that are to be left exposed or to receive thin finishes, such as resilient flooring, carpet, tile, or paint. When troweling is required, the surface should be steel-troweled immediately after floating (Figs. 3.7-7 and 3.7-8). It is customary for a cement mason using hand tools to float and then steel-trowel an area before moving the knee boards. If neces-

FIGURE 3.7-5 *Hand floating.* (Courtesy Portland Cement Association.)

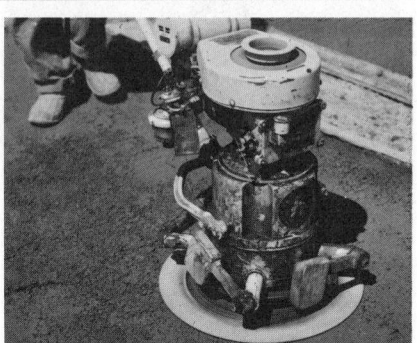

FIGURE 3.7-6 *Power floating.* (Courtesy Portland Cement Association.)

FIGURE 3.7-7 *Hand troweling.* (Courtesy Portland Cement Association.)

sary, tooled joints and edges should be rerun before and after troweling to maintain true lines, proper depths, and uniformity and to remove kinks.

The purpose of troweling is to produce a smooth, hard surface. For the first troweling, whether by power or by hand, the trowel blade must be kept as flat against the surface as possible. If tilted or pitched at too great an angle, an objectionable "washboard" or "chatter" surface will result. For the first troweling, a new trowel is not recommended. An older trowel that has been broken in can be

FIGURE 3.7-8 *Power troweling. Also available are self-propelled power trowels on which the operator rides in a seat.* (Courtesy Portland Cement Association.)

worked quite flat without the edge digging into the concrete. The smoothness of a surface can be improved by timely additional trowelings. Intervals between successive trowelings will permit the concrete to increase its set. As the surface stiffens, each successive troweling should be made by a smaller-sized trowel tipped at a progressively higher angle so that sufficient pressure can be applied for proper finishing.

For exposed slabs, additional troweling increases the compaction of fines at the surface, giving greater density and better wear resistance. A second troweling is recommended even if the slab is to be finished with resilient flooring, because it results in closer surface tolerances and a better surface to receive the flooring.

3.7.1.8

Broom Finish

A steel-troweled concrete surface is very smooth. Unfortunately, such surfaces become quite slippery when wet. They can be slightly roughened to produce a nonslip surface by brushing or brooming them. A brushed surface is made by drawing a broom over the surface after steel troweling. Notice in Figure 3.7-9 that the worker is drawing the broom right over the edge joint, but the joint is not being marred. This indicates that the concrete surface has

FIGURE 3.7-9 *Broom finishing. (Courtesy Portland Cement Association.)*

been steel-troweled properly and is hard enough. Other types of surface finishes, such as exposed-aggregate finishes, are discussed in Section 3.7.4.

3.7.2

FINISHING AIR-ENTRAINED CONCRETE SLABS

The microscopic air bubbles in air-entrained concrete tend to hold the ingredients, including water, in suspension. This type of concrete requires less mixing water and still has good workability with the same slump. Since there is less water and it is held in suspension, little or no bleeding occurs. There is, therefore, no need to wait for the evaporation of free water from the surface, and floating and troweling can and should be started as soon as the slab can support the finisher and equipment. Many horizontal surface defects and failures are caused by performing finishing operations while bleed water or excess surface moisture is present. Therefore, better results are generally accomplished with air-entrained concrete.

As with regular concrete, if floating is done by hand, using an aluminum or a magnesium float is essential. A wood float drags and greatly increases the amount of work necessary to accomplish the same result. If floating is done by power, the only major difference between the finishing procedures for air-entrained concrete and those for concrete that is not air-entrained is that floating may be started sooner on air-entrained concrete.

3.7.3

FINISHING LIGHTWEIGHT STRUCTURAL CONCRETE SLABS

Finishing operations for slabs made with lightweight structural concrete containing coarse aggregates of expanded clay, shale, or slag vary somewhat from those used on slabs made with normal-weight concrete. When the surface of the concrete is worked, there is a tendency for the coarse lightweight aggregate rather than the mortar to rise to the surface. Lightweight concrete can be easily finished if the following precautions are observed: (1) the mix should be properly proportioned and not be over- or undersanded in an attempt to meet unit weight requirements, and (2) finishing should not be started too early, and the concrete should not be overworked or overvibrated. A well-proportioned mix can generally be placed, struck off, leveled, and floated with less effort than is necessary for normal-weight concrete. Excessive leveling and floating are principal causes of finishing problems, because the heavier mortar is driven down and the coarse aggregate brought to the surface.

3.7.4

SPECIAL SLAB FINISHES

Due to the plastic quality of concrete, many surface finishes can be applied. The surface may be scored or tooled with a jointer in decorative and geometric patterns. Some of the more common special finishes are discussed here.

3.7.4.1

Exposed Aggregate

An exposed-aggregate surface often is chosen for any area where a special textural effect is desired (Fig. 3.7-10). A surface that is ground and polished is suitable, especially for such places as entrances, interior terraces, and recreation rooms.

Selection of aggregates is so important that test panels should be made before the

FIGURE 3.7-10 *Exposed-aggregate surface. (Courtesy Portland Cement Association.)*

job is started. Colorful gravel aggregate that is quite uniform in gradation and in sizes ranging from 1/2 to 3/4 in. is recommended. Flat, sliver-shaped particles or aggregate less than 1/2 in. in diameter should be avoided, because they become dislodged during exposing operations. Exposing the aggregate used in ordinary concrete generally is unsatisfactory since a high percentage of coarse aggregates is not necessarily revealed.

A 5-1/2- to 6-sack concrete with a maximum slump of 3 in. should be used. Immediately after the slab has been screeded and leveled, the selected aggregate should be scattered by hand and evenly distributed so the entire surface is completely covered (Fig. 3.7-11a). The initial embedding of the aggregate usually is done by patting with a darby or the flat side of a 2×4 (Fig. 3.7-11b). After the aggregate is quite thoroughly embedded and as soon as the concrete will support the weight of a mason on kneeboards, the surface should be hand-floated with a float or a darby (Fig. 3.7-11c). This operation should be performed so thoroughly that all aggregate is embedded entirely just beneath the surface. The grout should completely surround and slightly cover all

FIGURE 3.7-11 *Producing an exposed-aggregate surface: (a) decorative aggregate is distributed to cover the entire surface; (b) initial embedding of aggregate; (c) floating the surface to completely cover aggregate with grout; (d) exposing aggregate is done by simultaneously brushing and hosing with water.* (Courtesy Portland Cement Association.)

aggregate, leaving no holes or openings in the surface.

Shortly following this floating, a reliable retarder may be sprayed or brushed over the surface according to the manufacturer's recommendations. On small jobs a retarder may not be necessary. Retarders generally are used on large jobs for better control of exposing operations. Where a retarder has been used, exposing the aggregate is usually done some hours later by brushing and hosing with water. However, the manufacturer's recommendations should be followed closely.

Whether or not a retarder has been used, the proper time for exposing aggregate is critical. It should be done as soon as the grout covering the aggregate can be removed by simultaneously brushing and hosing (Fig. 3.7-11d), without overexposing or dislodging the aggregate. If during exposing it is necessary for masons to move about on the surface, kneeboards should be used gently. If possible, however, movement on the surface should be avoided because of the risk of breaking aggregate bond.

For interior areas and where a smooth surface is desired, no retarder is used, and exposure of the aggregate is accomplished entirely by grinding. This procedure may be followed by polishing, which will produce a surface similar to terrazzo.

Because the aggregate completely covers the surface, tooled joints in this type of work are impractical. Decorative or control joints are accomplished best by sawing. Control joints should be cut from 4 to 12 hours after the slab is placed and should be at least one-fifth the depth of the slab. A small-radius edger should be used before and immediately after the aggregate has been embedded to provide a more decorative edge to the slab. Another method of providing control joints is to install permanent strips of redwood before placing concrete.

Another method of placement specifies a top course that contains the special aggregate and that is usually 1 in. thick or thicker.

As in all concrete work, exposed aggregate slabs should be cured thoroughly.

3.7.4.2

Dry-Shake Color Surface

Dry-shake colored-concrete surfaces are used for interior areas, terraces, decorative walks, driveways, and other areas where a colored surface is desired. These surfaces are made by applying a dry-shake material that may be purchased ready for use from various reliable manufacturers. Its basic ingredients are mineral oxide pigment, white portland cement, and specially graded silica sand or fine aggregate. Other pigments are inappropriate and should not be used. Selecting the ingredients, proportioning, and mixing of a dry-shake material on the job are not recommended, because controlling the proportions is difficult, and the results cannot be predicted with accuracy. Field-mixed colors are likely to produce a mottled, uneven coloring.

After concrete has been struck off and leveled, and free water and excess moisture have evaporated from its surface, the surface should be floated either by power or by hand. If floated by hand, a magnesium or aluminum float should be used. Preliminary floating should be done before the dry-shake material is applied, to bring up enough moisture for combining with this dry material. Floating also removes ridges and depressions that might cause variations in color intensity. Immediately following floating, the dry-shake material is shaken evenly by hand over the surface. Care must be taken to follow the dry-shake material's manufacturer's instructions. If the wrong amount of the dry-shake material is applied in one spot, non-uniformity in color and possibly surface peeling will result.

Unless the manufacturer recommends otherwise, the first application of a colored dry-shake should use about two-thirds the total amount needed. Amounts are usually specified in pounds per square foot of surface. In a few minutes, this dry material will absorb some moisture from the plastic concrete and should then be thoroughly floated into the surface, preferably with a power float. Immediately following, the balance of the dry-shake material should be distributed evenly over the surface, thoroughly floated again, and made a part of the surface, taking care that a uniform color is obtained.

Tooled edges and joints should be run both before and after the dry-shake application.

Shortly after final floating, the surface should be power-troweled. If done by hand, troweling should immediately follow final floating. After the first troweling, whether by power or by hand, there should be time to allow the concrete to increase its set. The length of elapsed time should depend on the temperature and the humidity present. Concrete may be troweled a second time to improve the texture and also produce a denser, harder surface.

For exterior surfaces, two trowelings are usually enough. Then a fine, soft-bristled pushbroom may be drawn over the surface to produce a roughened texture for better traction under foot. For interior surfaces, a third hard troweling might be appropriate. This final troweling should be done by hand to eliminate washboard or trowel marks, and to produce a smooth, dense, hard-wearing surface.

Dry-shake colored concrete, as with other types of freshly placed concrete, must be cured thoroughly. After thorough curing and surface drying, interior surfaces may be given at least two coats of special concrete floor wax, which is available from various reliable manufacturers. It contains the same mineral oxide pigment used in the dry-shake. Care should be taken to avoid staining, such as by dirt or foot traffic, during the curing or drying period and before waxing.

3.7.4.3

Swirl Design

A swirl design, nonskid surface texture can be produced on a slab with a magnesium or aluminum float or a steel finishing trowel. When a float is used, the finish is called a swirl float finish; when a trowel is used, it is a swirl trowel finish.

After a concrete surface has been struck off, leveled, floated, and steel-troweled, it is ready to be given either a swirl float or a swirl trowel finish. A float should be worked flat on the surface in a semicircular or fan-like motion. Pressure applied on the float with such a motion will give a rough-textured swirl design, as shown in Figure 3.7-12a.

With the same motion and using a steel trowel held flat, a cement mason can obtain a finer textured swirl design on a concrete surface (Fig. 3.7-12b). Moist curing of the slab is the final operation.

3.7.4.4

Keystone Finish

A keystone finish is a special finish with a travertine-like texture. It can be used on

FIGURE 3.7-12 *Producing a swirl design on a concrete surface: (a) using an aluminum float; (b) using a steel finishing trowel.* (Courtesy Portland Cement Association.)

terraces, garden walks, driveways, decks around swimming pools, and in any other locations where an unusually decorative flat concrete surface is desired. It should not be used where it will be subjected to freezing weather.

After a concrete slab has been screed, leveled, and edged in the usual manner, it is broomed with a stiff-bristled broom to ensure that the finish (mortar) coat will bond.

A mortar coat is made by mixing one sack of white portland cement and 2 cu. ft. of sand with about 1/2 lb. of a color pigment. Yellow is usually selected to tint the mortar coat, but any mineral oxide color may be used. Care must be taken to keep the proportions exactly the same for all batches. Enough water is added to make a mixture with the consistency of thick paint.

Mortar-coat material is placed in pails and thrown vigorously on the slab with a stiff bristle brush called a *dash brush* to produce an uneven surface with ridges and

depressions. The ridges should be about 1/4 to 1/2 in. high.

When the surface is hard enough to permit the use of kneeboards, it is troweled with a steel trowel to flatten the ridges, leaving it smooth in the high spots and rough- or coarse-grained in the low ones. Many interesting textures can be produced, depending on the amount of troweling (Figs. 3.7-13a and 3.7-13b). A textured slab can then be scored into random geometrical designs before it cures.

FIGURE 3.7-13 *(a) Keystone or travertine finish; (b) random-scored keystone finish.* (Courtesy Portland Cement Association.)

3.7.5

FINISHING FORMED SURFACES

The way in which formed surfaces are finished depends on their exposure and the aesthetic effect the designer is trying to achieve. Several classifications are used in the industry to describe the various finishes.

3.7.5.1

Rough Form Finish

A *rough form finish* is the finish left by the formwork in locations that will be concealed from view in the completed building. The surface is improved only by the repair of tie holes and defective areas, removal of fins left by concrete leaking out through joints in the forms, and chipping off or grinding down of projections larger than 1/4 in. in height.

Smooth Form Finish

A *smooth form finish* is used in two instances. The first is when form liners are used to impart a pattern or a texture to the concrete. The second is for concrete that will receive paint, waterproofing, damp-proofing, veneer plaster, or another thin coating or covering material. After the forms are removed, fins and projections are removed completely, and imperfections are repaired. The surface is made smooth or of the texture imparted by the forms to adjacent areas, as applicable. Tie holes may be patched or left alone, depending on the designer's intent. Sometimes tie holes are used as a decorative element. *Screw ties* with cones are usually used when tie holes will not be filled (see Section 3.3.2.2).

Smooth Rubbed Finish

A *smooth rubbed finish* is applied when a smooth form or form liner is used. This finish is applied by moistening the new concrete within one day after the forms have been removed and rubbing it with an abrasive, such as a carborundum brick. Cement grout is not used during this finishing process.

Grout-Cleaned Finish

A *grout-cleaned finish* is also used on concrete that has been placed against a smooth form or form liner. It consists of applying to the surface a thick, paint-consistency grout of 1 part portland cement and 1-1/2 parts fine sand, mixed with equal parts of an acrylic or styrene butadiene bonding material and water. A blend of standard (gray) and white portland cement is used to achieve the color and tone desired by the designer. After the grout has been applied, the finish is completed by rubbing it with burlap.

SURFACE DEFECTS

There are many causes for concrete surface defects. Some of the major defects, their causes, and the construction techniques that should be used to prevent them are described here. Many defects result from improper curing or errors in mixing or finishing.

Scaling

Scaling is the breaking away of the hardened concrete surface of a slab to a depth of about 1/6 to 3/16 in. It usually occurs at an early age of the slab.

Scaling of a slab might occur if it is subjected to cycles of freezing and thawing soon after the slab has been placed. A favorable temperature must be maintained long enough to prevent injury (see Section 3.8.3.1). Cycles of freezing and thawing and applications of de-icing salts on non-air-entrained concrete can also cause scaling. This is why air-entrained concrete is recommended for all severe exposure conditions.

Performing a finishing operation while free excess water or bleed water is on the surface causes segregation of the surface fines (sand and portland cement) and also brings a thin layer of neat portland cement, clay, and silt to the surface, leaving another layer of nearly clean, washed sand that is not bonded to the concrete under it. To prevent scaling from this cause, water should be allowed to evaporate from the surface or be forced to evaporate by fans or blower-type heaters, or it should be removed by dragging a rubber garden hose over the surface before finishing operations begin.

Crazing

Crazing is the occurrence of numerous fine cracks in the surface of a newly hardened slab due to surface shrinkage. These cracks form an overall pattern similar to that of a crushed eggshell.

Crazing can be the result of rapid surface drying, usually caused by high air temperatures, hot sun, drying winds, or a combination of these. It can be lessened by curing with water, because water will maintain or lower the concrete surface temperature (see Section 3.8.3.3).

Premature floating and troweling when there is an excess amount of moisture at the surface or while the concrete is still too plastic brings an excess amount of fines and moisture to the surface. A rapid loss of this moisture will cause shrinkage at the surface, which may result in crazing. Floating and troweling should be delayed until excess moisture has evaporated from the surface, and the concrete has started its initial set. Excess moisture can be avoided by reducing the slump and using air-entrained concrete.

Overuse of a vibrating screed, darby, or bull float may also contribute by working an excess of mortar to the surface, which tends to cause additional surface shrinkage.

Dusting

Dusting is the appearance of a powdery material on the surface of a newly hardened concrete slab.

An excess of harmful fines (clay or silt) in a concrete mix with the sand and portland cement at the surface can result in dusting. This condition emphasizes the need to use clean and well-graded coarse and fine aggregates.

Premature troweling and floating mix excess surface water with surface fines, weakening the portland cement paste. Troweling should be delayed until all free water and excess moisture have disappeared and concrete has started its initial set.

When carbon dioxide, as might be emitted from open portable fossil-fuel heaters and gasoline engines, comes into contact with the surface of plastic concrete, a reaction takes place that impairs proper hydration. Such fumes should be vented to the outside and sufficient fresh-air ventilation provided.

Condensation sometimes occurs on a concrete surface before floating and troweling have been completed, usually in the spring and fall, when materials have become cold due to low night temperatures. If possible, this condition should be anticipated, and the concrete should be heated or at least hot water should be used for mixing. If this is impossible, blower-type heaters should be used to lower the humidity directly over the slab, and fans should be used to increase air circulation. If heaters and fans are not available, windows and doors should be opened. When condensation is present, floating and initial troweling should be held to a minimum, and a concrete surface should not be given a second troweling.

Neat portland cement and mixtures of portland cement and fine sand should never be used as a dry-shake. Because condensation may occur for several hours while concrete is beginning to harden, emergency measures may have to be taken to finish a slab. A well-mixed dry mixture of 1 part portland cement and 1 part well-graded concrete sand may be evenly and lightly distributed over a concrete surface, if it is followed at once by troweling. There should be no second troweling, because additional condensation may take place after the first troweling.

Winter-protection heaters may lower the relative humidity around concrete excessively and inhibit proper hydration of the portland cement. Water jackets should be placed on heaters to increase the relative humidity by evaporation, and moist curing methods should be employed. Heaters should be moved periodically so that no area will be subjected to an extreme or harmful amount of heat.

Proper curing for the correct length of time is essential. Concrete that is not cured properly will often be weak, and its surface will be easily worn by foot traffic.

3.8 CURING AND PROTECTION

The curing of concrete is an important construction operation that is often neglected. Even when concrete has been properly mixed, carefully placed, and correctly finished, a poor job will result if proper curing techniques have not been followed.

3.8.1

PURPOSE OF CURING

The strength and the watertightness of concrete improve with age as long as conditions are favorable for continued hydration of its portland cement. Other qualities, such as weathering and resistance to freezing and thawing, are affected similarly. Improvement is rapid early and continues more slowly for an indefinite period as long as moisture and favorable temperatures are present.

Fresh concrete contains more than enough water for hydration, but under most job conditions much of this water is lost by evaporation unless certain precautions are taken. Hydration proceeds at a much slower rate as the temperature drops until it stops altogether at about 14 °F. However, at about 40 °F, hydration becomes too slow for concrete work to proceed without special provisions to heat the ingredients so that the concrete will be at least 50 °F when it is placed. Therefore, concrete should be protected so that moisture is not lost during the early stages of hardening, and it should be kept at a temperature that will promote hydration and also protect against injury from subsequent construction activities.

3.8.2

CURING METHODS

Concrete can be kept moist in a number of ways, including leaving its forms in place, sprinkling, ponding, and using moisture-retention covers or a seal coat that is applied as liquid and then hardens to form a thin membrane.

Curing methods that use water, such as sprinkling, ponding, and using wet coverings, are the most effective and should be used whenever practicable. Although not usually as effective as curing with wet coverings, moisture retention membranes are widely used because of their convenience. In any case, the most important consideration is that the surface of the concrete be kept uniformly wet or moist. Partial drying of any part of the surface of a slab can result in crazing and cracking. Curing compounds will affect the bonding properties of adhesives and should not be used on slabs that are to receive resilient flooring or other adhesive-applied finishes unless they are known to be compatible with the adhesive. Curing compounds also are not recommended for slabs that will be given a further surface treatment.

3.8.2.1

Leaving Forms in Place

Forms left in place are of great assistance in retaining moisture. In hot, dry weather wood forms will dry out and should be kept moist by sprinkling. In all cases, exposed surfaces must be protected from moisture loss.

3.8.2.2

Sprinkling

When concrete is kept moist by sprinkling (Fig. 3.8-1), care should be taken to prevent drying of the surfaces between applications of water. Alternating cycles of wetting concrete are conducive to crazing or cracking of the surface. A fine spray of water applied continuously provides a

FIGURE 3.8-1 *Sprinkling.* (*Courtesy Portland Cement Association.*)

more constant supply of moisture and is better than copious applications of water with periods of drying between.

3.8.2.3

Ponding

Ponding is sometimes used on flat surfaces, such as pavements, sidewalks, and floors. A small dam of earth or other water-retaining material is placed around the perimeter of a surface, and the enclosed area is kept flooded with water. Ponding gives a more constant condition than does sprinkling.

3.8.2.4

Curing Covers

Moisture-retaining covers, such as wet sand, burlap, or polyethylene film, are sometimes used. Care should be taken to cover the entire concrete surface, including exposed sides of members and the sides of pavements and sidewalks where the forms have been removed. Covers should be kept constantly moist enough to provide a film of moisture on the concrete surface.

■ WATERTIGHT COVERS

Watertight paper is used on floors and other horizontal areas. It should be nonstaining and strong enough to withstand the wind and abrasive action of workers walking over it. Seams should be overlapped several inches and covered with glued tape. Polyethylene films often are used as watertight curing covers.

■ SEALING COMPOUNDS

Sealing curing compounds are available in black, red, colorless, and white pigmented coatings. Some can be applied in one coat, but two coats will give better results. Their application should be made immediately after the concrete has been finished. If there is a delay, the concrete should be kept moist until application of the compound. In extremely hot weather it is advisable to cover a slab with water for 12 hours before using curing compounds. On formed surfaces, such as beams and columns, the forms should be removed,

the concrete sprayed lightly with water, and a sealing compound applied. Where the membrane must be protected against traffic, it should be covered with at least 1 in. of sand, earth or some other cover. Such a protective cover should be placed not sooner than 24 hours after the sealing compound.

3.8.3

CURING TEMPERATURES

Temperature affects the rate of chemical reactions between portland cement and water. Consequently, temperature affects the rate at which concrete hardens, as well as its strength and other properties.

The temperatures and time periods in the following discussions of cold-, mild-, and hot-weather construction are approximations to introduce the subject. The actual required temperatures and times vary with concrete thickness and air temperature. Refer to American Concrete Institute (ACI) publication ACI 305R, "Hot Weather Concreting," and ACI 306, "Standard Specification for Cold Weather Concreting," for more specific requirements.

3.8.3.1

Cold-Weather Construction

In cold weather it is often necessary to heat a concrete's ingredients and cover fresh concrete, provide a heated enclosure, or both. Hydration of portland cement causes some heat to be generated and, if this heat is retained, it raises the temperature of a concrete. In cold weather, concrete at the time of placing should be between 50° and 70°F. In no case should its ingredients be heated to the point where the temperature of fresh concrete is above 70°F or built-in thermal stresses will result.

When the temperature of the surrounding air is below 40°F, the water and the aggregate used in a concrete should be heated so that the temperature of the mixed concrete at the time of placing is between 50° and 70°F. The temperature of the air adjacent to the placed concrete should then be kept at between 50° and 70°F during the curing period. This may require covering the concrete with an insulating material like straw, surrounding the con-

crete with a heated enclosure, or both until sufficient time has passed for the concrete to remain undamaged by the prevailing conditions. This amount of time varies with the thickness of the concrete, the air temperature, and other conditions, such as high winds, that may be present. The normal range is above 70°F for three days or above 50°F for five days when normal portland cement (Type I) is used; and above 70°F for two days or 50°F for three days when high-early-strength portland cement (Type III) is used. This range may vary if high winds occur and will have to be increased if the concrete has not properly cured within the specified times.

Fresh concrete should not be placed on a frozen subgrade. When such a subgrade thaws, there is likely to be uneven settlement and cracking of the concrete member it supports. The insides of forms and reinforcing steel and embedded fixtures should be free of ice at the time concrete is placed. A thin layer of warm concrete should not be placed on cold, hardened concrete, because the thick upper layer will shrink as it cools, and the lower layer will expand as it warms. Bond failure will result, which should be guarded against, especially in slabs.

No dependence should be placed on salt or other chemicals to prevent freezing.

3.8.3.2

Mild-Weather Construction

In relatively mild weather, that is, when the temperature is generally above 40° to 45°F with only short periods below this range, heating only the mixing water will usually provide the desired temperatures in the concrete. The mixing water should be heated or other steps taken to raise the temperature of the concrete to between 50° and 70°F at the time of placing.

In-place concrete should be kept at a favorable temperature long enough to avoid injury by exposure to the atmospheric temperature. In general, the air surrounding placed concrete made with normal portland cement (Type I) should be kept either at 70°F or above for the first three days or above 50°F for the first five days. When high-early-strength concrete (Type III) is used, air temperatures should be kept at 70°F for two days or 50°F for three days.

In mild weather, a covering of tarpaulins may be enough protection. A layer of straw covered with tarpaulins will protect against more severe conditions, but an enclosure of tarpaulins or other watertight material with artificial heat may be necessary to provide proper protection in many cases. To prevent rapid drying of concrete, especially near heating elements, it may be necessary to elevate these elements and protect concrete near them with sand kept wet constantly.

3.8.3.3

Hot-Weather Construction

In extremely hot weather, extra care is required to avoid high temperatures in newly placed concrete and to prevent its rapid drying. Keeping aggregate stockpiles moistened with cool water will help to lower the temperature. In very hot weather,

mixing water should be chilled by refrigeration or by using ice as part of the mixing water. The ice should be melted by the time the concrete leaves the mixer.

A subgrade on which concrete will be placed should be saturated some time in advance and then sprinkled just before the placing operation, so it will not absorb water from the mix. Wood forms should be wetted thoroughly if they have not been treated otherwise. There should be no delay in placing concrete, and it should be struck off and leveled at once. Temperature covers, such as burlap kept constantly wet, should be placed over fresh concrete immediately after it has been leveled. During final finishing, only a small section immediately ahead of the finishers should be uncovered. The finished surface should then be re-covered and the cover kept wet. Delays in finishing air-entrained concrete in hot weather usually lead to the

formation of a surface that is difficult to finish without leaving ridges.

Fresh concrete should be shaded as soon as possible after it has been finished, and moist curing should be started as quickly as can be done without marring the surface. Concrete surfaces should be kept constantly wet during the curing period.

3.8.4

LENGTH OF CURING

Sufficient curing time should always be allowed for concrete to develop adequate strength before it is loaded. Curing should be started as soon as it is possible to apply a curing medium without damaging the surface and should continue as long as is practicable. Some curing times and surface temperatures for normal mild-weather conditions are summarized in Figure 3.8-2. During hot weather, however, curing should continue for at least seven days, regardless of the type of concrete or the temperature. In mild weather, the curing time can sometimes be reduced to two days when high-early-strength portland cement (Type III) is used. Even this time may have to be increased, though, in very hot weather.

In any case, concrete temperature should not be allowed to fall below 50°F at any time during curing. Rapid cooling may induce thermal cracking, especially during the first 12 to 24 hours.

FIGURE 3.8-2
Length of Curing

Type of Portland Cement	Curing Time, days	Concrete Surface Temperature, °F
Type I	5	70 or higher
	7	50–70
Type II	3	50
Other than Type I or Type II	7	70
	14	50–70

3.9 CONCRETE FOUNDATION SYSTEMS

A building's foundations form its *substructure* and support its *superstructure*, transmitting loads to the earth. The foundations also resist *lateral* (horizontal) loads from the ground and the superstructure and provide anchorage for the superstructure against uplift (Fig. 3.9-1).

A foundation system is composed of several load-carrying elements that may include various types of footings, caissons, piles, grade beams, walls, piers, columns, pilasters (Fig. 3.9-2), and the turned-down portion of a combination footing-and-

floor system. Foundation systems are constructed of concrete, masonry, steel, wood, and combinations of these. Some of the more common variations are shown in Figure 3.9-3.

Foundations take three basic forms: basements, crawl spaces, and slabs on grade. Figure 3.9-3a,b shows a basement and a crawl space with spread footings, concrete or masonry walls, and steel-joist floor framing. The floor might also be framed in wood, cast-in-place concrete, or precast concrete. Figure 3.9-3c,d shows

a similar basement or crawl space supported on piles and caissons (the floor framing has been omitted for clarity). Basements and crawl spaces provide underfloor space for easier installation of mechanical, electrical, and plumbing systems.

Slabs on grade may turn down at the edges to form combination slab-and-foundation systems, as shown in Figure 3.9-3e, or rest on foundation walls and footings, as in Figure 3.9-3f. Slabs may also be supported on footings or grade beams resting on piles (Fig. 3.9-3g).

VERTICAL LOADS   LATERAL LOADS   UPLIFT AND RACKING LOADS

FIGURE 3.9-1 *Foundation walls must (1) resist vertical loads from the weight of the building and its occupants; (2) resist lateral loads from the soil, water, and other superimposed loads, such as earth-moving equipment and earthquakes; and (3) anchor the building to its foundation.*

Floating foundations (see Section 3.9.3.2) are also combination slab-and-foundation systems.

This section discusses concrete foundation systems. Chapter 2 covers specific requirements for soil materials and earthwork needed to provide supports for foundations, control of surface and groundwater, requirements for thermal controls, and precautions recommended to prevent damage by termites. Masonry foundation walls, columns, piers, and pilasters are addressed in Chapter 4; wood foundations, additional precautions necessary to prevent termite damage, and foundation anchorage are in Chapter 6. Waterproofing, dampproofing, and the insulation of foundation walls are in Chapter 7.

Spread foundations, such as footings and mat and raft foundations, resting on the soil receive loads from foundation walls, piers, pilasters, and columns and must be able to distribute those loads directly to the soil without exceeding its allowable bearing capacity. Piles and caissons receive loads from foundation walls, isolated columns, or grade beams and transfer them to the soil or into bedrock.

Foundations form the substructure for buildings whether or not there is a basement. Basements are habitable spaces that lie mostly below the adjacent finished grade. Buildings without basements include those with a crawl space or a slab

on grade. Some such buildings have limited basement space, but many do not.

Many of the recommendations in this section are based on the American National Standards Institute (ANSI) publication A56.1, "Building Code Requirements for Excavation and Foundations," and the Underground Space Center's *Building Foundation Design Handbook.*

3.9.1

DESIGN

A foundation system should be economical and carry all expected loads with a reasonable factor of safety. The design of foundation elements is based on assumptions as to the kinds of loads the foundation must carry.

Loads associated with foundation design are *vertical* or *lateral* (horizontal); *static* (nonmoving) or *dynamic* (moving); and *concentrated* at a point or *uniform* along a surface. These loads may be further divided into dead loads, live loads, wind loads, and other loads.

3.9.1.1

Dead Loads

Dead loads act constantly. They include the weight of the superstructure and of the foundation itself. Dead loads are calculated by summing up the total weight of the materials used in the construction. Dead loads are usually assumed to be uniformly distributed over the length and the thickness of a foundation wall.

3.9.1.2

Live Loads

Live loads are variable, periodic, and sometimes moving. They include, but are not limited to, the weight of snow or water on the roof, furniture, equipment, and people. The type and the magnitude of live loads acting on a structure vary according to locality, occupancy, and other factors. Most building codes specify minimum live loads that may be used in sizing foundation walls, footings, and other elements of the substructure.

FIGURE 3.9-3 *Basic foundation forms: (a) basement and (b) crawl space, both with spread footings, concrete or masonry walls, and steel-joist floor framing; (c) basement and (d) crawl space, both supported on piles and caissons; (e) slab on grade that turns down at the edges to form a combination slab-and-foundation system; (f) slab on grade that rests on foundation walls and footings; and (g) slab supported on footings. (Drawings by HLS.)*

3.9.1.3

Wind Loads

Wind loads result from the intermittent effect of wind on the superstructure. They may act downward, laterally at any angle, or upward.

3.9.1.4

Other Loads

Some loads that affect foundations are difficult to categorize as purely live, dead, or wind loads. Sometimes they incorporate all three. They are usually calculated and accounted for in the design separately. Among these are lateral pressures exerted on foundation walls by the adjacent soil and water in that soil; multidirectional loads imparted by seismic disturbances (earthquakes); thrust loads transmitted into the foundations from domes, rigid frames, vaults, and arches; uplift caused by expansion due to freezing and subsequent *frost heave*; and uplift resulting from the tendency of a building with its foundations below the water table to float (*buoyancy*).

3.9.1.5

Settlement

A building's substructure must distribute building loads so that settlement will be either negligible or uniform under all parts of the building. *Differential* (uneven) settlement under different parts of a foundation may cause substantial problems. Uneven settlement can occur when a foundation is out of plumb or not level and can cause such damage as cracks in foundation walls and in finished walls and ceilings, sloping floors, binding doors and windows, and open joints in woodwork, as well as functional annoyances, such as binding doors and windows. When differential settlement is extreme, partial failure of the structural integrity of the building can result.

In time, all foundations built on soil will settle to some extent. The extent of settling depends on the compressibility of the underlying soil and the loads placed on it. One objective in the design and construction of foundations to minimize settlement, especially differential settlement, so that neither the structure itself nor its use will be impaired.

In most buildings, differential settlement should be controlled by proportioning the size of footings so they will transmit an equal load per unit of area to the soil. The only exception is in the foundations for single-family residences and some other small single-story buildings where the loads are so light that footings are often sized by rule of thumb. For small buildings, however, when the soil is poor or if unusual load variations exist, footing sizes should be proportioned for equal settlements; even in good soils, it may be necessary to verify that applied loads are balanced where concentrated loads occur, as might happen at a chimney or a column. Refer to Section 2.1.4 for a more detailed discussion of the causes and effects of settlement in different types of soil.

The bearing capacity and compressibility of soils can be calculated by site and laboratory tests on actual soil samples (see Section 2.1.7). With a knowledge of bearing capacity and the magnitude of superimposed loads, designers can use engineering analysis to size foundations and predict settlement. Although engineering analysis is used to size foundations for commercial, institutional, industrial, and multifamily housing structures, this procedure is seldom followed in designing foundations for single-family dwellings or small buildings of other types. Those footings are usually sized using rules of thumb that are based on established practices and the performance of other foundations in the locality.

3.9.2

FOUNDATION TYPES

Foundations are of two general types: (1) *spread foundations*, which distribute a building's loads directly to a sufficient area of soil to obtain adequate bearing capacity; and (2) *pile* and *caisson* foundations, which transmit a building's loads through soils that have inadequate bearing capacity for spread footings to deeper layers of soil or rock that have adequate bearing value.

Most foundations in single-family dwellings and other one- and two-story buildings are spread foundations. Many buildings of three or more floors also are supported on spread foundations. Piles and caissons are used when the soil's bearing capacity is insufficient to support the loads imposed on the soil by a building.

3.9.2.1

Spread Foundations

Spread foundations support transmitting elements, such as walls, pilasters, columns, piers, or grade beams, on an abruptly enlarged base called a *footing*, a *mat foundation*, or a *raft foundation*. This base in turn spreads the load directly to the supporting soil.

▪ FOOTINGS

Except on bedrock or in the extremely rare instances when the bearing capacity of the soil is inadequate to support a foundation wall or pier alone, a footing is required to spread the loads on the supporting soil. Foundation walls built of masonry units bonded with mortar should have footings, regardless of the soil's bearing capacity. Refer to Section 2.1.3 for a discussion of the bearing capacities of various soil types.

Footings are classed according to the way in which they are loaded, shaped, and constructed. Several types are shown in Figure 3.9-4. Unreinforced, or plain, footings carry light loads. Reinforced footings contain embedded steel to strengthen them in tension and in shear. Any footing type can be either plain or reinforced.

FOOTING MATERIAL Footings have been made from many materials, including wood, stone, brick, and concrete masonry units. Most footings today, however, are monolithic cast-in-place concrete. Concrete for unreinforced footings in foundations should have a minimum compressive strength of 2000 psi at 28 days. Higher strengths may be required when determined by engineering analysis. Sufficient time should be allowed for the strength of the concrete to develop before subjecting a footing to loads. In cold weather, concrete should be placed as soon after excavation as possible and must be protected from freezing until it cures (see Section 3.8.3).

When a large excavation has been carried deeper than desired, the depth should be refilled with concrete rather than with earth fill material. If it is necessary to excavate beneath a footing for a sewer or water line or other small opening, the excavation should be backfilled with compacted stone or gravel.

CONTINUOUS FOOTINGS Continuous footings support foundation walls of either concrete or masonry (Fig. 3.9-4a). They also may be used to support a row of several metal or wood columns or masonry piers to minimize differential settlement. When a single footing supports more than one column or masonry pier, it is called a combined footing (Fig. 3.9-4d).

STEPPED FOOTINGS Stepped footings change levels in stages to accommodate a sloping grade (Fig. 3.9-4b). Typically, the number of steps necessary to traverse a slope is minimized by making them large both vertically and horizontally. A few large steps are generally less expensive to form than many small ones.

The vertical part of the step in a stepped footing should be of the same thickness and width as the horizontal footing and should be placed at the same time. If more than one step is necessary, the horizontal distance between steps should not be less than 2 ft., and the vertical step should not exceed three-quarters the horizontal distance between steps (see Fig. 3.9-4b).

ISOLATED FOOTINGS Isolated footings are independent footings that receive the loads of free-standing columns or piers (Fig. 3.9-4c). Masonry piers are bonded to a footing with mortar. Steel and precast concrete columns are usually anchored to a footing with embedded bolts. Wood columns frequently are held in position on a footing by protruding steel pins embedded in the concrete.

FOOTING SIZE Footings should be sized for existing soil conditions and the magnitude of dead and live loads. The usual rule-of-thumb size for normal residential footings resting on soil of average bearing capacity (2000 psf) is that the width of the footing should equal twice the thickness of the wall, and the depth of the footing should equal the wall thickness. In other words, given a wall thickness t, the footing

FIGURE 3.9-4 *Various types of footings spread building loads over soil area. All the types shown can be reinforced.* *(Drawings by HLS.)*

width will be 2*t* and its depth will be *t*. In the absence of specific building code requirements or engineering analysis, the footing sizes given in Figure 3.9-5 have been used successfully. However, their applicability to a particular project should be verified in every instance. In no case should the sizes be smaller than those shown in Figure 3.9-5. These sizes are based on a conventionally loaded wall footing and on soil of average bearing value (approximately 2000 lb./sq. ft. or greater). Footings should not be less than 6 in. thick and, unless transversely reinforced, their thickness should not be less than 1-1/2 times the footing projection.

Footings for piers, columns, and masonry chimneys should be at least 8 in. thick and, unless reinforced, not less than 1-1/2 times the footing projection. In addition, the minimum footing for interior pier or column footings should be 24 in. × 24 in.

FOOTING LOCATION In most cases, foundations should be extended below fill material so that footings rest on undisturbed earth. Footings for small buildings are occasionally supported on *engineered fill* (see Section 2.3.4) when other types of foundations are inappropriate or would be more expensive than placing spread footings on engineered fill. In those cases, the footings must be designed by engineering analysis for the conditions. In addition, more than usual care must be taken in compacting and testing the compacted engineered fill to ensure that the necessary compaction has been achieved.

Footings should not be placed on frozen ground. As frozen soil thaws, excess water can transform the soil into liquid mud that is unusable to support loads. Footings should be located far enough below ground level to protect them from possible frost heaving caused by the freezing of soil moisture (see Section 2.1.9).

In areas where soils expand due to moisture changes, footings should be deep enough to be unaffected by seasonal moisture changes or specifically engineered for these conditions.

REINFORCED FOOTINGS In reinforced footings, steel bars are placed transversely (perpendicular to the wall) when loads must be spread over wider areas. When its projection exceeds two-thirds its thickness, a footing should be reinforced transversely with steel as determined by engineering analysis.

Footings are reinforced longitudinally when they must bridge weak spots, such as excavations for service or sewer connections, and when the soil has low bearing capacity or varies in compressibility.

■ MAT AND RAFT FOUNDATIONS

Mat and raft foundations (Fig. 3.9-6) are

Based on average soil bearing value = 2000 P. S. F.

NUMBER OF FLOORS		FRAME CONSTRUCTION		MASONRY OR VENEER CONSTRUCTION	
		t$^{\prime\prime}$	p$^{\prime\prime}$	t$^{\prime\prime}$	p$^{\prime\prime}$
1	BASEMENT	6″	2′	6″	3″
1	NO BASEMENT	6″	3″	6″	4″
2	BASEMENT	6″	4″	8″	5″
2	NO BASEMENT	6″	3″	6′	4″

FIGURE 3.9-5 *Dimensions for continuous wall footings.*

FOUNDATION WALL

FROST LINE

p

t

P = FOOTING PROJECTION
t = FOOTING THICKNESS

FIGURE 3.9-6 *Mat and raft foundations.* *(Drawing by HLS.)*

used over soils of low bearing capacity when other foundations would be inadequate. They are made of concrete and heavily reinforced with steel, so that the entire foundation will act as a unit.

A *mat foundation* is a thickened slab that supports loads and transmits them as a single structural unit over the entire slab and soil surface area. Mat foundations are used when the foundations for a building become so large, usually due to poor soil, that it is more economical to join them into a single unit than to cast them each separately.

A mat foundation becomes a *raft* or *floating foundation* when it is combined with basement walls and placed beneath a building in such a way that the weight of the soil removed to make the basement equals the weight of the building's superstructure and substructure combined. In such a structure, settlement is minimal even on low-strength soils, because the building load on the soil is no more than the weight of the excavated soil.

3.9.2.2

Piles and Caissons

Piles and caissons are used as foundation support for buildings when the soil is not capable of supporting the loads that will be imposed by a building on spread foundations.

■ **DRIVEN PILES**

Driven piles are column-like units that transmit loads through poor soil to rock or lower levels of soil that have adequate bearing capacity (Fig. 3.9-7). In some regions, short piles are called *piers*. Piles serve the same purpose as footings in that they transmit loads to subbase strata capable of carrying the load. Piles usually are placed in clusters of two or more spaced from 30 to 48 in. on centers. This arrangement permits them to act together and produces a higher load-carrying capacity than can be achieved with isolated piles. The load capacities mentioned in this section are based on the piles being distributed in clusters. These load capacities are also the maximum permitted. Optimum load capacities are somewhat lower in each instance. Piles receive building loads from isolated columns and from grade beams by means of reinforced concrete pile caps (Fig. 3.9-8). Pile caps are sometimes simply a widened section of a grade beam (Fig. 3.9-9).

Pile foundations are either (1) *point bearing* types, which transmit loads to lower, stronger soil or rock through their points, or (2) *friction* types, which develop the necessary bearing capacity through surface friction between the pile and the ground (see Fig. 3.9-7).

Piles are driven with heavy *hammers* in large machines called *pile drivers*. *Drop hammers* are simply raised and dropped on a pile by force of gravity. *Differential-acting steam hammers* are rammed into

FIGURE 3.9-7 *Point bearing piles reach to strong soils or rock. Friction piles depend on friction between the piles and the soil for their support.*

FIGURE 3.9-8 *A cluster of two piles with an isolated pile cap. Similar caps are also used when the supported element is a grade beam instead of a column. Caps may be square, rectangular, or multisided, depending on the number of piles and the size of the cap.* *(Drawing by HLS.)*

the top of a pile by steam pressure or compressed air. Modern vibratory hammers and diesel-driven hammers are also used.

Friction piles are driven to a predetermined depth or resistance based on soil boring analysis and field tests. To verify the design analysis, test piles are usually driven and loaded before the remainder of the piles are driven. Point bearing piles are driven until additional blows of the hammer produce very little movement in the pile (*refusal*).

Piles may be made of wood, concrete, steel, or a combination (*composite*) of these. Timber piles have been used for at least 2000 years, probably longer. Some below-water timber piles beneath bridges in Europe are known to have remained in continuous service for more than 1000 years. These piles are, however, suitable only for relatively light loads (40 tons maximum) and must be treated with pre-

FIGURE 3.9-9 *Pile caps formed integrally with a grade beam are sometimes used when the load-distributing element is a bearing wall.* (*Drawing by HLS.*)

servatives when they will extend above the water table. They are also limited in length to the effective height of the tree from which they are cut (45 to 65 ft.), because they cannot be spliced.

Steel piles may be H-shaped sections or pipes (Fig. 3.9-10). H-shaped piles are heavy wide-flange sections varying in size from 8 to 14 in. in both depth and flange width. They can carry loads of from 50 to 200 tons each and may be as much as 150 ft. long. To produce such lengths, sections are welded together as they are being driven.

Steel-pipe piles are later filled with concrete. They are available in several types and shapes (Fig. 3.9-11). These piles may be heavy-walled types, which can be driven directly, or thinner-walled types, which require a tight-fitting, heavy inner lining (*mandrel*) that is withdrawn before the concrete is placed. They may be smooth-walled or corrugated, round or fluted. Sizes range from 8 to 24 in. in diameter, and maximum load capacity

ranges from 75 to 200 tons, depending on the type. Pipes with closed ends may be driven to as much as 120 ft. Other pipe piles are limited to about 80 ft.

Precast concrete piles are either solid concrete or open cylinders that are later filled solid with concrete. They may be square, round, or octagonal in cross section. All precast concrete piles are reinforced, and most are also prestressed. Solid precast concrete piles with simple reinforcement can be up to 80 ft. long. This increases to 150 ft. for cylinder piles and up to 200 ft. when they are prestressed. Maximum capacity ranges from 100 tons for simple reinforced piles to as much as 500 tons for prestressed cylinder piles. Sizes vary from 10 to 54 in. across.

Composite piles are constructed in several configurations. Some have timber or concrete-filled steel-pipe lower sections and concrete-filled shell with mandrel upper sections. Others have H-shaped steel lower sections and precast concrete upper sections. Composite piles sometimes permit less expensive applications. For example, the combination of timber and concrete-filled steel shell permits the use of relatively inexpensive timber piles at much lower depths than is ordinarily possible (up to 150 ft.). A disadvantage of using composite piles is that their load-carrying potential is limited to that of the lower of the two elements. For example, a composite pile of timber and concrete-filled steel shell is limited to 40 tons, instead of the 75 to 80 tons the concrete-filled steel piles could carry.

■ CAISSONS

Caissons are concrete columns placed in auger-drilled or excavated holes. They serve the same purpose as piles. While some caissons are only 6 to 12 in. in diameter, most are much larger and capable of carrying heavy loads. They can reach as much as 6 ft. in diameter and carry 3500 tons or more. Caissons extend through unsatisfactory soil to a firm soil-bearing stratum or to bedrock. As a caisson hole is drilled or dug, a steel casing is lowered into it to keep the hole from caving in. This casing is raised and removed as the concrete is placed. Small, relatively short, and lightly loaded caissons, which are sometimes called piers, are frequently poured into cardboard or composition

FIGURE 3.9-10 *Steel-pile shapes.* (*Drawing by HLS.*)

FIGURE 3.9-11 *Round- and fluted-pile shapes. Fluted steel piles are usually tapered.* (*Drawing by HLS.*)

casings that are not removed as the concrete is placed.

Caissons come in four types: (1) *rock caissons* simply rest directly on solid rock; (2) *high-capacity rock caissons*, sometimes called *socket caissons*, rest in a socket cut into rock, thus supporting the load both by resting on the rock and by friction against the sides of the pocket; (3) *hardpan* or *clay caissons* rest on a soil-bearing stratum; and (4) *friction caissons*, sometimes called *cast-in-place piles*, are supported by the friction acting on the sides of the caisson, much as friction piles are supported.

Hardpan or clay caissons have either straight shafts or an enlarged base called a *bell* (Fig. 3.9-12). Bells are produced by hand excavation or with a special device, a *belling bucket*, on the auger.

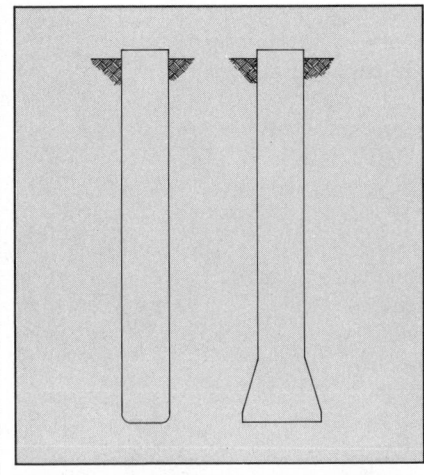

FIGURE 3.9-12 *Typical caisson configurations.* (Drawing by HLS.)

3.10 CONCRETE SLABS ON GRADE

The suitability of a particular slab-on-grade construction is generally influenced by the geographic location of the building, the building site, and the construction of the superstructure. Specific influences on the type and design of a slab include site preparation, topography, soil properties, climate, the type and support of the superstructure, and comfort factors such as thermal and moisture control. Producing satisfactory slabs requires properly designing the slabs, correctly selecting the concrete materials and mixtures, and carefully placing, finishing, and curing the concrete.

In this section recommendations are given for selecting and constructing concrete slabs on grade. If care is exercised to control concrete quality and the necessary precautions dictated by the characteristics of the site, superstructure, and geographic location are taken, slabs on grade are suitable for floors and combination foundation-and-floor systems in all parts of the United States. The provisions stated earlier in this chapter are applicable to concrete slabs on grade.

The recommendations given here are based on the Portland Cement Association's "Concrete Floors on Ground"; the American Concrete Institute Standard 302, "Guide for Concrete Floor and Slab Construction"; the Building Research Advisory Board, National Academy of Sciences's *Criteria for Selection and Design for Residential Slabs-on-Ground*; and the Underground Space Center's *Building Foundation Design Handbook*.

3.10.1

DESIGN

Slabs on grade may be either independent floors or combination foundation-and-floor systems. In independent floors, the slab is separate from the building's load-carrying elements, such as walls, columns, and footings, which transmit their loads directly to the supporting soil (Fig. 3.10-1a). In combination foundation-and-floor systems, the slab is an integral part of the foundation and carries superstructure loads and transmits them to the soil (Fig. 3.10-1b). Concrete foundation elements, such as walls and footings, that are common to many types of foundation

FIGURE 3.10-1 *(a) Ground-supported floor slab built independently of foundation walls; (b) ground-supported slab built integrally with foundation-grade beam.* (Drawing by HLS.)

systems are discussed in Section 3.9. Masonry walls, columns, piers, and pilasters are covered in Chapter 4.

Slab-on-grade construction requirements are determined largely by site characteristics, such as drainage and soil type, and the methods used to build the slab. Slabs may be ground supported or supported structurally. They may be unreinforced (plain), contain welded wire fabric reinforcement, or be structurally reinforced using welded wire fabric and steel reinforcing bars. Four types of slabs are described here.

Slab construction depends on the slab type; site preparation, including preparation of the slab bed, grading, and backfilling; other precautions to control either ground or surface moisture; and thermal control when required. The performance of slabs depends on the quality of concrete materials and workmanship.

3.10.2

SLAB TYPES

Concrete slabs for floors and combination foundation-and-floor systems can be classed according to construction as ground-supported slabs that rest directly on a slab bed consisting of undisturbed soil, compacted fill, or a base course or as structurally supported slabs that are supported independently of the ground and rest on foundations consisting of walls, piers, grade beams, or piles.

The appropriateness of a concrete slab floor or combination foundation-and-floor system should be based on soil properties, climate, functional and aesthetic requirements, and economy.

3.10.2.1

Ground-Supported Slabs

Ground-supported slabs can be designated as Type I, II, or III.

■ TYPE I SLABS

A Type I slab is mostly unreinforced and is separated from the elements that support the superstructure (Fig. 3.10-2) so that it carries no superstructure loads. It has a nominal thickness of at least 4 in. and is cast directly on a dense or com-

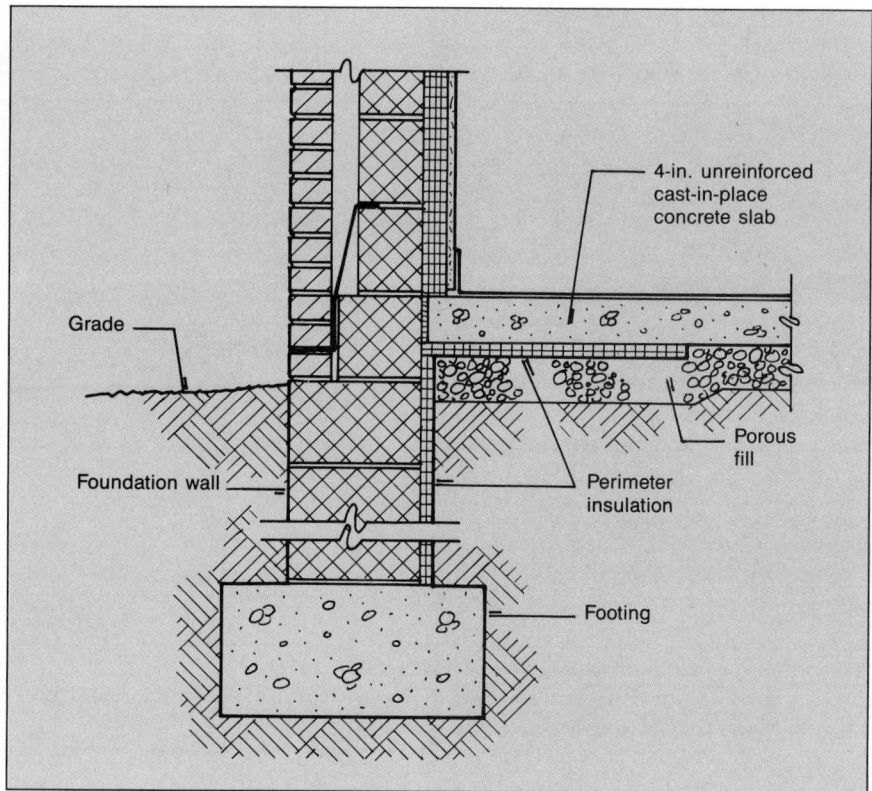

FIGURE 3.10-2 *Type I slab.* *(Drawing by HLS.)*

pacted slab bed. It is unreinforced but may contain welded wire fabric or bars in localized areas. Type I slabs should not be placed on a slab bed that can develop significant changes in volume over time. Differential movement in the slab bed may produce enough tensile stress in the concrete to crack the slab. Random cracking due to drying shrinkage is controlled by isolation joints, by construction joints, and by limiting the area between control joints, which are placed to induce cracking at preselected locations. Superstructure loads should be supported directly on foundations independent of the slab.

The controlling factors in the suitable performance of a Type I slab are selection of the correct slab bed, concrete quality, location, control joint spacing, and other details.

Because a Type I slab is unreinforced, except in specific localized areas, the slab possesses only compressive strength and cannot tolerate appreciable amounts of tension or warping. Such a slab may crack from shrinkage during drying, but when used under the conditions recommended below, cracks that do occur should not

open excessively or prove detrimental to the serviceability of the slab.

Type I slabs should be placed only on well-drained and properly graded coarse-grained soils (Fig. 3.10-3). Such soils are not affected significantly by climate or moisture content changes and should develop no appreciable volume change that would be detrimental to the slab. The soil must be capable of supporting a Type I slab and be uniformly and adequately compacted to ensure that warping and tensile stresses that could contribute to cracking are not induced in the slab.

Type I slabs can be applied successfully to gravely soils of any density (Types GW, GP), sandy soils with or without silts and clays (Types GM, GC, SW, SP, SM, SC), and silts (Types ML, MH), provided they are dense or medium dense (see Section 2.1.1 for a discussion about soil types). Type I slabs can be applied also to loose sandy and silty soils if the soil can be compacted to a dense or medium-dense state to its entire depth before slab construction (see Fig. 3.10-3).

Type I slabs should be rectangular or square in shape (Fig. 3.10-4a). To control

FIGURE 3.10-3
Recommended Slab Bed for Slabs on Grade

Slab Type	Soil Type	Density	Plasticity[a]	Consistency, q_u/w[b]	Climatic Rating
TYPE I	GW, GP	All	—	—	All
	GM, GC, SW, SP, SM, SC, ML, MH	Dense or medium dense[c]			
TYPE II	GW, GP, GM, GC, SW, SP, SM, SC, ML, MH	All	—	—	All
	CL, CH,	—	PI < 15	$q_u/w \geq 7.5$	All
	OL, OH	—	PI > 15	$q_u/w \geq 7.5$	$C_w \geq 45$
TYPE III	CL, CH,	—	PI > 15	$q_u/w \geq 7.5$	$C_w < 45$
	OL, OH	—	—	$7.5 > q_u/w > 1.5$	All
TYPE IV	CL, CH, OL, OH	—	—	$q_u/w < 1.5$	All
	PT	—	—	All	

[a]PC = Plasticity Index, determined in accordance with ASTM D 4318.

[b]q_u/w = unconfined compressive strength/ton/sq. ft. of undisturbed sample determined in accordance with ASTM D 2166.

[c]May be loose if properly compacted to its entire depth.

FIGURE 3.10-4 *Maximum dimensions for Type I and II slabs.*

shrinkage cracking during drying and subsequent thermal volume changes, no dimension of a slab perimeter should exceed 32 ft. However, if a slab is to be left exposed, control joints spaced 15 to 20 ft. apart are recommended (Fig. 3.10-4b).

Irregularities in shape, such as a non-rectangular T- or L-shape, need control joints to divide the slab into squares or rectangles (see Fig. 3.10-4b). If control joints are not provided, diagonal cracks, which may be objectionable, are likely to occur where the slab changes in shape.

Irregularities in thickness that reduce a slab to less than 4 in. in nominal thickness should not be permitted. Stresses caused by drying shrinkage or thermal change would tend to cause cracking at these locations. When the upper surface of a slab must be lowered, such as to permit proper installation of ceramic tile or another floor finish, the uniform slab thickness should be maintained by lowering the underside of the slab beginning at least 24 in. away from the slab depression (Fig. 3.10-5a). In addition, when the vertical displacement is greater than 1-1/2 in., WWF 6×6—W2.9×W2.9 welded wire fabric should be placed in the slab extending 25 in. on either side of the point of displacement (Fig. 3.10-5b).

Heating coils or pipes should not be embedded in Type I slabs, because thermal stresses will be induced that the slab is unable to resist without cracking. Heating ducts may be embedded in Type I slabs if WWF 6×6—W2.9×W2.9 welded wire fabric is placed to extend 19 in. on either side of the centerline of the duct or the slab edge, whichever is closer.

Superstructure loads should be supported directly on foundations independent of the slab, but Type I slabs can accommodate partition loads up to 500 lb./lin. ft. Because a Type I slab is unreinforced, it cannot carry greater loads without the possibility of cracking. Where this loading must be exceeded, reinforcement or special construction should be provided to minimize cracking that would result. Methods of supporting partitions and concentrated loads are illustrated in Figure 3.10-6.

Openings in Type I slabs should be kept to a minimum because they can introduce stresses that are not uniform. When an opening 12 in. or wider must be provided for a column or other penetration, a slab should be reinforced with 25-in.-wide WWF 6×6—W1.4×W1.4 welded wire fabric around the opening (Fig. 3.10-7).

TYPE II SLABS

A Type II slab also is separated from the elements that support the superstructure

FIGURE 3.10-5 *(a) Method of maintaining minimum slab thickness; (b) location of welded wire fabric reinforcement.* *(Drawing by HLS.)*

(Fig. 3.10-8) but contains welded wire fabric or reinforcing bars in localized areas. The controlling factors in the suitable performance of a Type II slab include correct preparation of the slab bed, concrete quality, and correct size and placement of the welded wire fabric reinforcement.

Type II slabs are cast on a dense or compacted bed and contain welded wire fabric to hold drying-shrinkage and thermal-movement cracks together once they form. This reinforcement does not prevent cracking or add to the load-carrying capacity of a slab. Grading and loading are similar to those for Type I slabs. As with Type I slabs, the design of a Type II slab is not influenced by the type of superstructure that is supported directly on independent foundations.

Because they contain welded wire fabric, Type II slabs can be cast onto some soils that are unsuitable for Type I slabs. However, Type II slabs cannot be expected to resist differential movement of the soil, because the amount of reinforcement provided is based on the premise that the entire slab will remain uniformly supported by the underlying soil.

Type II slabs can be constructed on all soils suitable for Type I slabs under all climatic conditions, and these soils can be of any density, including loose (see Fig. 3.10-3). In addition, Type II slabs can be applied to certain fine-grained soils (CL, OL, CH, OH) (see Section 2.1.1) under certain climatic conditions if the unconfined compressive strength and plasticity index fall within the limits shown in Figure 3.10-3.

Type II slabs must be at least 4 in. in uniform thickness, and no dimension of the slab perimeter should exceed 75 ft. (see Fig. 3.10-4).

The same recommendations apply as for Type I slabs with respect to irregularities in thickness (see Fig. 3.10-5) and shape (see Fig. 3.10-4), with the added requirement that reinforcement be continuous across control joints or that keyed joints be provided to eliminate the possibility of vertical displacement between adjacent sections.

FIGURE 3.10-6 *(a) In Type I and II slabs a heavy partition or concentrated load should be carried on a separate footing. (b, c) Alternative methods that may be used when the slab is designed using engineering methods and structural reinforcement.* *(Drawing by HLS.)*

FIGURE 3.10-7 *Location of reinforcing around a hole in a Type I or II slab.*

Welded wire fabric reinforcement in Type II slabs controls only the *size* of cracks. Cracks that are expected to occur will be held tightly closed by the tensile reinforcement and should not be objectionable. To control crack size, the slab should be provided with the minimum welded wire fabric reinforcement recommended in Figure 3.10-9. Reinforcement should be placed in the center of the slab, midway between the top and the bottom and should be supported while the concrete is being placed, preferably on metal chairs designed for the purpose. Edges should be lapped at least one mesh space. The common practice of laying the mesh on the bed and pulling it up using hooks (hooking) after the concrete as been

FIGURE 3.10-9
Minimum Fabric Reinforcement for Type II Slabs

Maximum Slab Dimension, ft.	Wire Spacing, in.	Wire Gauge, number
Up to 45	6×6	W10×W10
45–60	6×6	W8×W8
60–75	6×6	W6×W6

FIGURE 3.10-8 *Type II slab.* *(Drawing by HLS.)*

placed is not recommended because accurate and level placement of the mesh within the slab using this method is almost impossible. The practice of laying the mesh on top of a poured slab and embedding it by having workers walk on it is also discouraged for the same reasons.

Because the reinforcement in Type II slabs will hold together cracks that form due to temperature changes, heating pipes or coils and heating ducts can be embedded in them. However, the construction recommendations given in Section 3.12 should be followed.

Type II slabs are expected to deflect with slight soil movement; therefore, superstructure loads should be supported independently on foundations. However, like Type I slabs, Type II slabs can accommodate partition loads of up to 500 lb./lin. ft. and equivalent concentrated loads. An additional layer of reinforcement can be provided to extend at least 25 in. on either side of the partition or concentrated load to distribute added stresses (see Fig. 3.10-6). Heavy concentrated loads, such as masonry walls, should be supported on

independent or structurally reinforced footings (see Fig. 3.10-6a,c).

The recommendations for Type I slabs relative to openings apply also for Type II slabs, except that the additional layer of reinforcing placed around the opening should be 6×6-W2.9×W2.9 welded wire fabric. The additional reinforcing around the opening will prevent the concentration of stresses that are likely to crack the slab at these points (see Fig. 3.10-7).

■ **TYPE III SLABS**

Type III slabs are combination foundation-and-floor systems (Fig. 3.10-10) that are structurally reinforced to make them usable over problem soils that undergo substantial volume changes with time and climate. The use of foundations with spread footings is not advisable on such soils. Type III slabs receive superstructure loads and transmit them to the foundation soil under the entire slab area. Figure 3.10-11 shows an example of a Type III slab.

Depending on its sensitivity to differential settlement, the superstructure imposes

FIGURE 3.10-10 *Type III slab.* *(Drawing by HLS.)*

Cast-in-place concrete slab
Structural reinforcement
Grade
Grade beam
Porous fill
Perimeter insulation

limits on the maximum tolerable deflection in a Type III slab. Because the superstructure is supported on the slab itself, walls and other load-bearing superstructure elements will deform when the slab deforms. Different materials tolerate deformations of differing intensities before developing undesirable defects, such as cracking and distortions. For example, wood frame construction generally can sustain more deformation than masonry construction before warping and cracking create mechanical and esthetic problems. Comparatively brittle cementitious materials, such as concrete, stucco, plaster, and mortar, can tolerate only small deformations before cracking develops.

The objective of a Type III slab is to cause the foundation-and-floor system and the superstructure to act as one monolithic unit that restricts differential movements in both slab and superstructure. To ensure that the slab and the superstructure actually act in this manner, the slab must be designed with the necessary rigidity and strength. Therefore, in addition to concrete quality control, Type III slabs should be designed by a professional engineer in accordance with good engineering practice to provide the proper dimensions for stiffness and the correct amount and size of steel reinforcement for strength. *Criteria for Selection and Design for Residential Slabs-on-Ground* by the Building Research Advisory Board, National Academy of Sciences, is applicable.

Because a Type III slab is structurally reinforced, it can be used over certain problem soils, such as highly compressible, highly expansive, and highly plastic soils. The unconfined compressive strength, plasticity index, and climatic ratings for soils on which Type III slabs should be used are given in Figure 3.10-3.

Highly compressible soils include those that are high in organic content (PT), such as peat, which are so compressible that they are not suitable for the support of ground-supported slabs.

Highly compressible clays are soils with sufficient moisture content to have a soft or very soft consistency or high organic content. Excessive settlement may result even under relatively light residential loading or from the weight of fill material.

Very loose sands and silts can cause serious problems beneath a slab. Fine sands and silts are particularly troublesome when very loose. Vibration of very loose sands and silts may cause a decrease in the volume of soil voids with resultant settlement, depending on the amplitude and the frequency of the vibration. When it is not practical to compact very loose sand and silt to a suitably dense state for ground-supported slabs, a Type IV slab should be used.

Highly expansive and plastic soils are problem soils because they are greatly affected by climatic conditions. Any soil with a plasticity index greater than 15 may cause problems by expanding under exposure to severe climatic conditions. However, under favorable or intermediate conditions this material may be considered for Type II slabs.

A soil with a plasticity index of 30 or higher will react vigorously to changes in moisture content and will present the most difficult design problems because of the great volume change that can be expected. Loads associated with small buildings are not usually sufficient to control such expansion by loading. In any case, when such soils are encountered, professional advice should be sought.

Loose soils can become problem soils when they are heavily loaded and become saturated. When slab construction is desirable over these soils and the slab must be heavily loaded but problems of surface– or groundwater exist, a Type III slab should be selected.

3.10.2.2

Structurally Supported Slabs

Type IV slabs are structurally reinforced and can be used over very poor soils, such as highly expansive clay, which are extremely sensitive to moisture, and soils of negligible bearing capacity or high organic material content, and where Type I, II, and III slabs would not be suitable. Type IV slabs require professional design based on engineering analysis.

Like Type III slabs, Type IV slabs receive and transmit superstructure loads to the soil, but unlike Type III slabs, they do not depend on soil for support. The slab bed functions as a form and is used

NOTES:

1. Panel size may be increased by a maximum of 15'-0" both ways and with same reinforcement by increasing slab thickness to 5", or by using a 4" slab with #3 Ø bars 10" o.c. both ways.

2. All beams shall extend minimum of 12" into undisturbed soil.

3. If garage is to be built in lieu of carport, sec. D, the perimeter beams will be of the same size and same reinforcement as main house beams.

4. Bar laps or splice shall be a min. of 24 bar diameters.

5. This design is for 1-story construction only.

6. Foundation concrete shall be minimum 2500 psi and shall be cured by an acceptable and positive method.

7. Slab reinforcement shall be accurately supported in upper one-third of slab.

8. Bottom bars in perimeter beams shall extend around corners a min. of 24 bar diameters.

9. Porch slabs beams and reinforcement shall be same as carports.

FIGURE 3.10-11 *An example of Type III slab construction. This design should not be used until reviewed by a professional engineer.*

only for temporary support of the slab until it develops sufficient strength to be supported on beams, piers, or intermediate foundations walls. These foundation elements in turn are carried by piles or footings, which distribute the loads to stable ground below the level of the slab (Fig. 3.10-12).

Type IV slabs can be used over very poor soils (see Fig. 3.10-3) that cannot be treated economically or practically to receive ground-supported slabs. These soils might include expansive clays, highly organic soils, and very loose sands and silts. Foundation elements such as grade beams should be designed so that swelling of the soils beneath the slab will not lift them off the foundations or pull the foundations out of the ground.

One method used in some areas to prevent the swelling of soils from lifting grade beams is to cast the beam between its supports (piles, caissons, grade beams,

etc.) on rectangular cardboard tube forms. These forms are folded from a flat sheet with a diagonal across the center and are about 6 in. thick. When the earth swells, it crushes the cardboard instead of lifting the beam.

Due to local practice, variations of Type IV slabs are sometimes used over unstable soils that may not necessarily be considered extreme problem soils (Fig. 3.10-13). Foundation piles, piers, or footings should be designed to penetrate to a depth that provides adequate support below unstable soil.

3.10.3

MOISTURE CONTROL

Refer to Section 2.3 for a discussion of the control of surface– and groundwater, including provisions for slab beds and vapor

retarders directly under slabs on grade.

3.10.4

THERMAL CONTROL

For the most comfortable conditions and to conserve energy, slab perimeters should be insulated in all except the warmest climate regions. Whenever walls are insulated, slabs should be insulated. Slab insulation is discussed in Section 7.3.

Slabs are sometimes used to deliver heat throughout buildings.

3.10.5

SLAB FLATNESS AND LEVELNESS

Flatness is a measure of the degree to which the surface of a slab deviates from

FIGURE 3.10-12 *Type IV slab.* (Drawing by HLS.)

FIGURE 3.10-13 *Construction details for a variation of a Type IV slab.*

a plane. Slab *levelness* is a measure of the degree to which a slab deviates from horizontal. A slab can be flat (having no depressions or rises in its surface) and still not be level. For example, a slab that is flat within the allowable tolerances may have one corner that is higher than the others, or one side that is lower than the opposite side. Conversely, a level slab may have depressions, rises, or pockets in its surface.

Until recently, slab surface tolerances were specified as a dimension between 1/8 in. and 1/2 in. in every 10 ft., depending on the activities that would take place where the slab was located and the finishes that would be applied. Historically, these deviations were measured using a 10-ft.-long straight board, with its long edge vertical, placed across the slab in each primary direction and diagonally. The slab was to be within the specified tolerance dimension of the board along its entire length. This method was inaccurate, especially in measuring levelness.

A more accurate method has now been developed and is fast coming into general use. It is called the *F-curve* or *F-number* system. *F*-numbers are designated F_F for flatness and F_L for levelness and have no units. The method for determining a slab's *F*-numbers is defined in ASTM Standard E 1155, "Standard Test Method for Determining Floor Flatness and Levelness Using the *F*-number System." Basically, the system depends on marking multiple lines on a slab and measuring the deviation at 12-in. intervals along those lines. *F*-numbers are obtained mathematically from these measurements using formulas contained in ASTM Standard E 1155.

Several devices are used to determine the deviations. One is a 10- or 12-ft.-long highway straightedge laid along the lines. The deviations are measured down from the straightedge. This method produces an accurate reading when done properly, but it is costly in terms of the time it takes. Devices such as laser levels, optical levels, floor profilometers, and inclinometers are being used more and more today because they give quick and accurate readings.

The degrees of flatness and levelness increase as the *F*-numbers increase. Slabs requiring a closer tolerance must have higher F-numbers. The required F-numbers for a particular slab depend on the activities that will take place on it and the finish that will be applied. In general, the required F_F is 15, and the required F_L is 13 for a slab with a scratch finish, as is permitted on slabs that are to receive concrete floor toppings, mortar setting beds for tile or terrazzo, and other bonded cementitious finishes. Slabs with a float finish that are to receive membrane waterproofing, elastic roofing, or sand-bed terrazzo should have an F_F of 18 and an F_L of 15. Trowel-finished slabs that are to receive resilient flooring, carpet, paint, and other thin-film finishes should have an F_F of 20 and an F_L of 17.

The F_F for slabs in buildings ranges from 15 to 100; the F_L from 13 to 50. The higher numbers are required when extraordinary degrees of flatness and levelness are needed, such as in skating rinks.

3.11 CAST-IN-PLACE STRUCTURAL CONCRETE

The cast-in-place structural concrete discussed in this section includes walls, columns, piers, pilasters, beams, girders, and slabs. The discussion here related to the use and placement of reinforcing steel should be considered introductory only. The actual size and placement of bars, ties, stirrups, and accessories should be determined from structural analysis by a structural engineer. In addition, forming, installing reinforcement, and placing concrete are the responsibility of the contractor. The methods described here may not be followed on a particular project and in fact may be inappropriate under certain circumstances. For example, the procedures described here do not apply to mass concrete structures.

3.11.1

WALLS

Foundation walls, which rest on footings, must be able to support vertical loads from the superstructure, resist lateral (horizontal) loads from the ground and the superstructure, and provide anchorage for the superstructure against uplift. Wall construction should be delayed until the footing is sufficiently cured to support the loads. This time will vary depending on the air temperature and the magnitude of the loads to be supported.

In addition to their structural functions, foundation walls must be durable, resist penetration of free water and water vapor, provide a barrier against fire, and control air and heat flow. A foundation wall that encloses a habitable space must present a suitable appearance or be able to accept finishes and must accommodate windows, doors, and other openings.

Reinforced concrete and concrete masonry are used extensively for foundation walls. Foundation walls for most commercial, institutional, and industrial buildings are designed based on engineering principles, which in turn are based on theoretical analysis of the forces and stresses involved. Foundations for single- and two-family homes and other small buildings are more commonly designed using minimum requirements based on past experience and observation of foundations for such buildings or on similar requirements listed in the applicable building code. Such requirements generally are empirical, particularly in designing basement walls to resist lateral earth pressures. The usual minimum thickness for concrete foundation walls for small buildings is 10 in. for concrete walls and 12 in. for concrete masonry walls. These thicknesses should not be used, however, until they have been verified as acceptable under local conditions. The type of soil and the height of the water table affect the lateral loads acting on foundation walls and must be taken into account.

Figure 3.11-1 shows a typical concrete foundation wall. The key shown on the top of the footing is sometimes omitted. The procedure for building such a wall is as follows: The footing is formed. Reinforcement, including dowels, is installed, and the concrete is placed and screed. When the concrete in the footing has cured sufficiently (about one day), one side of the wall forms is erected and coated with a form-release agent. Vertical and horizontal reinforcing steel are placed in the wall and wired to the dowels and to each other. Figure 3.11-1 shows two layers of steel, but walls carrying lighter loads often require only one layer. The reinforcing ties shown are necessary only for walls that bear heavy vertical loads.

After the steel has been placed and inspected, wall ties are inserted through holes in the side of the form that is in place. The remaining side of the formwork is coated with a form-release agent and slipped into place over the wall ties. When screw-type wall ties are used, the threaded sections are screwed into the cones. The wales and braces are then installed and the wall ties are anchored to the wales. The formwork is inspected to ensure that it has been installed properly and that it is in the proper position, tight, and securely braced.

Foreign material is removed from the inside of the forms, the level to which the concrete will be placed is marked on the forms using surveying instruments, and concrete of the proper consistency is placed. Placing of the concrete is done by dumping it from a bucket or by hose from a concrete pump. In a low wall, the concrete may be placed directly from a transit-mix truck's chute. As the concrete is placed, it is consolidated using internal vibrators. When it reaches the line on the forms, the concrete is smoothed with hand floats. The top is then covered with canvas

Projecting rods to tie to
slab or wall above

Ties used in
walls carrying
heavy loads

Reinforcing rods

Dowels

Key way

Footing reinforcing rods

FIGURE 3.11-1 *Typical concrete foundation wall.* *(Drawing by HLS.)*

or sheet plastic to prevent drying, and the concrete is permitted to cure.

Several days later, the forms are removed, the exposed wall ties are twisted off (unscrewed in the case of screw ties), and finishing is begun.

Most cast-in-place concrete walls higher in a building are built in essentially the same way as foundation walls. Their vertical steel is tied to the steel projecting from below rather than to dowels. Exposed-to-view joints between levels of concrete in walls should be treated in some manner, such as with a grooved joint or other positive articulation. This is needed to prevent an unsightly condition that might occur because of slight differences in the concrete or by the difficulty of obtaining clean horizontal joints at such locations.

To save time and money by eliminating most of the usual formwork, concrete wall sections are sometimes cast on a poured-in-place concrete slab and tilted or lifted into place (Fig. 3.11-2). This procedure

FIGURE 3.11-2 *A tilt-up panel being lifted into place.* *(Courtesy the Burke Company.)*

is called *tilt-up construction*. Some additional reinforcement is necessary in tilt-up panels to permit attachment of lines from a crane and to resist the stresses generated by lifting the panels into place. The greatest use of tilt-up construction is

on single-story industrial buildings, but the method is also used in other types of buildings.

Joining the sections of tilt-up panels is accomplished in several ways. The space between panels may be filled with a poured-in-place concrete pilaster or column. Panels may be joined using welded-together steel angles cast into the tilt-up panels. Tilt-up panels may be supported by separate concrete columns, concrete columns cast as a part of the tilt-up panel, or structural steel columns. Panels may themselves be bearing walls, supporting a roof structure without columns.

3.11.2

COLUMNS, PIERS, AND PILASTERS

Columns, piers, and pilasters act in compression to transmit loads to footings. Columns and piers are freestanding; pilasters are built integrally with the wall and also may function as stiffeners to provide necessary lateral support to a wall. Masonry piers and pilasters are discussed in Section 4.6.7.

Column, pier, and pilaster construction should be delayed until the footing is sufficiently cured to support the loads. Curing time will vary depending on the air temperature and the magnitude of the loads to be supported.

Figure 3.11-3 shows the basic types of reinforcement in a concrete column. Piers and pilasters are similar. Several parts of this drawing have been stylized for reasons of clarity. For example, single vertical bars are shown. When more steel is required, bundles containing several individual bars are often used in each location. Only four vertical bar locations are shown, but additional bar locations are often necessary. Eight bars are common, with one falling halfway between the corner bars. Sometimes many more locations are required. In Figure 3.11-3 vertical steel is shown in a rectangular pattern, but round patterns are also used, and either may be used in either square or round columns. The reinforcement shown is that for a tied column. When the steel is in a circular pattern, circular ties are often used, but more expensive spiral ties are sometimes used (Fig. 3.11-4) when the structural engineer deems their use justified. Spiral ties are

FIGURE 3.11-3 *Typical tied column reinforcement.* (Drawing by HLS.)

FIGURE 3.11-4 *Typical spiral column reinforcement.* (Drawing by HLS.)

3.11.3

BEAMS AND GIRDERS

There are several types of beams and girders. *Grade beams* are located approximately at grade level. In contrast to a continuous footing, which is supported uniformly along its length, a grade beam receives building loads and transfers them to spread, pile, or caisson foundations. Grade beams may take many forms. The examples in Figure 3.11-5 are (a) a reinforced concrete beam placed integrally with the concrete slab and supported on piles and (b) a precast reinforced wall carried on footings and supported laterally by concrete or masonry piers.

Superstructure beams are located above grade level, where they serve to support floors and roofs. Beams are of three general types: simple beams, continuous beams, and beams that are part of a slab. Beams usually carry only uniform loads that act over the entire length of the beam.

Girders also may be simple or continuous. They also often carry a uniform load, but in addition they carry beams. Large transfer girders sometimes also carry columns.

shipped to the job site coiled like a spring. They are stretched and wired in place at the job.

The process for building a concrete column, pier, or pilaster is similar to that described for a wall. There are, however, the following exceptions:

Dowels are set to line up with the vertical steel.

Reinforcement ties serve as horizontal reinforcement. The spacing of reinforcement ties is not necessarily the same throughout the height of a column, being doubled where splices occur and increased in number where loads so dictate. Their spacing and size are selected by the structural engineer.

The reinforcement is often wired together in the horizontal position into a *column cage*, which is then hoisted into location by a crane.

The reinforcement is placed and wired to the dowels before any formwork is installed. When space is limited within a column, bars may be spliced in line rather than lapped.

Form ties usually are not needed.

a GRADE BEAM SUPPORTED ON PILES   b GRADE BEAM SUPPORTED ON FOOTINGS

FIGURE 3.11-5 *Grade beams: (a) reinforced concrete beam and (b) precast reinforced wall.*

3.11.3.1

Simple Beams

Simple beams (Fig. 3.11-6) span only one bay from support to support. They are not often found in large concrete-framed buildings, but an understanding of them will help in understanding the more complex beams and girders usually found in actual construction. As discussed earlier (see Fig. 3.1-1), when a uniform load, including its own weight, is applied to a simple beam, compression stresses are imposed in the top portion of the beam and tensile stresses occur in the lower portion. Reinforcing steel is added near the bottom of the beam to absorb the tensile stresses (see Fig. 3.1-2). This reinforcement is supported at the proper height in the form by chairs or bolsters (see Fig. 3.4-10).

Tensile stresses in a simple beam are concentrated at the bottom of the beam in the center of its span and are strongest at the center. Near the ends of the beam, the stresses move diagonally upward, where they tend to cause the beam to crack diagonally. While weaker than the stresses at

FIGURE 3.11-6 *Simple beam.* (*Drawing by HLS.*)

the center of the beam, these diagonal stresses are still sufficient to cause damage. *Stirrups* are added and wire tied to the bottom reinforcing bars near the supports to absorb this diagonal stress (Fig. 3.11-7). No stirrups are needed near the center of a simple beam. Stirrups may be U-shaped (*U-stirrups*), as shown in the main portion of Figure 3.11-7; closed (*closed stirrup-ties*), as shown in Figure 3.11-7a; or capped (two-piece ties), as shown in Figure 3.11-7b. Light reinforcing rods are placed near the top of a beam with stirrups to hold the stirrups in place until the concrete has set. When heavy loads are present and the depth of a beam is limited by factors other than structural considerations, compression steel may be added to the top of a beam. Then this compression steel is used to hold the stirrups in place.

Reinforcing steel dissipates the forces it absorbs into the concrete by means of its bond with the concrete, until the stresses have been neutralized. However, when the reinforcement reaches the end of a beam, it typically has some remaining stress, but there is no more concrete. This remaining stress is dissipated by bending each bar into a U shape called a *hook*.

Beams are formed, reinforcement installed, and concrete placed in the beams in a sequence similar to that used in walls. The major differences are that the steel reinforcement in beams is supported on chairs or bolsters resting on the bottom of the formwork, and wall ties usually are not required.

3.11.3.2

Continuous Beams

A continuous beam spans more then one span (Fig. 3.11-8). The dashed lines in Figure 3.11-8 show a greatly exaggerated view of the bending that tends to take place in such a beam. Reinforcing steel is added to resist this bending. Continuous beams are formed and steel and concrete are placed in them as described for simple beams in Section 3.11.3.1. Some ways in which continuous beams are different from simple beams follow:

The reinforcement in a simple beam tends to be the same throughout. In a continuous beam, more cross-sectional area of reinforcement is required at the bottom of the beam near the center of each bay than is required to pass through the columns. This usually means fewer bottom bars at the columns.

Top steel is required in a continuous beam at every column to resist the tendency of the beam to bend upward at those locations.

Stirrups are required at each end of a continuous beam and, in addition, in the beam on both sides of each intermediate column.

Right-angle bends, with the steel turning downward into the columns at the ends of a continuous beam, may be used in lieu of hooks, when there is sufficient room for them.

3.11.3.3

Beams That Are Part of Floor Slabs

Beams are often part of a concrete slab (Fig. 3.11-9). They do not act like simple beams or even continuous beams but rather have some unique characteristics. For example, they act as a part of the slab, rather than as separate elements, and they are affected by forces acting in the slab. Therefore, they are discussed in Section 3.11.4.

3.11.3.4

Girders

Girders are essentially large beams, and the principles discussed in Sections 3.11.3.1 and 3.11.3.2 apply to them as well. Their

FIGURE 3.11-7 *Reinforcement and forming for a typical beam.* (*Drawing by HLS.*)

FIGURE 3.11-8 *A continuous beam.* (*Drawing by HLS.*)

FIGURE 3.11-9 *Beam and floor slab working together.* (*Drawing by HLS.*)

steel reinforcement must be specifically designed by a structural engineer, but they usually have both top and bottom steel reinforcement and heavier bars than do beams. Additional steel is required where concentrated loads are carried.

3.11.3.5

Post-Tensioning

Tensile forces near the bottoms of loaded beams tends to pull the concrete apart, making it crack (see Fig. 3.1-1). Even when reinforcing steel is added, these cracks often are large enough to be seen.

Such cracks can be eliminated altogether by placing the concrete in compression before an external load is applied. If enough compression is applied and held there, when a load is applied, the amount of compression will be lessened but will never disappear completely. Compressing the concrete in a beam will also permit it to carry the same loads with less concrete, making it lighter and less expensive.

Compression loads can be added to a beam by stretching the steel and locking it to the concrete at the ends. As the steel tries to return to its normal length, the resulting stress is transferred to the concrete. There are two procedures used to

produce this result. *Prestressing* is done before concrete is placed around the steel. This process is best done under controlled conditions in a shop or a factory, because it is very expensive to do anywhere else. Prestressing is used almost exclusively in precast concrete work and is described in Section 3.13.3.

A process called *post-tensioning* is used to add permanent tension to the reinforcing steel in cast-in-place concrete and subsequently to place the concrete under compression. Essentially it is done using *tendons*, which are bundles of high-strength, cold-drawn steel wire strands or steel bars. The tendons are coated with oil or placed in a steel tube to prevent bonding between the concrete and the tendons. The concrete is placed and permitted to cure. Generally, the tendons are fastened to one end of the beam and stretched from the other end using hydraulic jacks. When the desired amount of tension has been achieved, the jack end of the tendon is secured to the concrete, usually with a bearing plate of some sort. When the strands are very long, they must be stretched from both ends simultaneously to ensure uniform tensioning.

Tendons can be placed level, like conventional reinforcing bars. Post-tensioned beams are more efficient, however, if the tendons are placed in a shape that approximates the lines of tensile force in the beam. This is near the bottom of the beam in the center and sloping upward toward the top of the beam at the ends in a V-shape.

Post-tensioned beams tend to shorten due to elastic compression, shrinkage, and creep. These movements must be accommodated in the building design and construction. Adjacent elements that would be affected by these changes should be built after the post-tensioning has been completed, or they should be isolated from the post-tensioned unit. In addition, the amount of tension to be applied can sometimes be adjusted slightly upward to at least partially compensate for these movements.

3.11.4

STRUCTURAL SLABS

In this book the term *structural slabs* refers to slabs that are not cast against the earth (slabs on grade). Structural slabs fall into one of two categories: one-way slabs and two-way slabs.

3.11.4.1

One-Way Slabs

One-way slabs span in one direction from support to support. There are two types of one-way slabs: *solid* and *ribbed* or *joist* slabs (Figs. 3.11-10 and 3-11-11).

■ SOLID SLABS

The solid slab shown in Figure 3.11-10 is cut to show its shape. Such slabs usually have beams, girders, or bearing walls on each of their sides (Fig 3.11-12).

The first step in constructing a one-way solid slab is to form the supporting walls and columns, place the concrete in them, and permit it to cure sufficiently to support the loads that will be imposed. The second step is to build the forms for the beams and girders, then the forms for the slabs. The forms should be supported on temporary supports and shores of sufficient strength and stability to resist the large loads that will be imposed by the forms and the concrete. When all the forms have been built, the side of the forms that will be in contact with concrete should be coated with a form-release agent, and the steel should be placed in them, following the shop drawings.

Structural reinforcement in a one-way solid slab is similar to that described in Section 3.11.3.2 for continuous beams, generally with bottom bars and top bars over the beams and girders. The structural reinforcement in one-way slabs is placed parallel to the span (see Fig. 3.11-10). Temperature steel, which usually consists of Number 3 bars, is placed perpendicular to the structural bars. Temperature steel is not needed parallel to the span, because the structural bars there will serve to resist shrinkage and temperature cracks in that direction.

When the steel is in place and has been inspected, concrete may be placed into

FIGURE 3.11-10 *A one-way solid slab. (Drawing by HLS.)*

FIGURE 3.11-11 *A one-way ribbed slab. (Drawing by HLS.)*

the forms and consolidated. The slab then should be cured and finished, as described in Section 3.7.1. When the slab, beams, and girders are sufficiently strong to support themselves, the forms should be removed and reshores placed (Fig. 3.11-13). *Reshoring* consists of vertical metal or wood props designed to help the new concrete support construction loads for several additional weeks until it has cured sufficiently to stand alone. After reshoring has been done, formed surfaces should be finished. If there is another floor above the one just completed, formwork should be moved up to it and the process repeated.

FIGURE 3.11-12 *Solid one-way slab framing.* *(Drawing by HLS.)*

FIGURE 3.11-13 *Formwork is under construction on the top floor. On the next-to-the-top floor, the formwork is still in place. The formwork has been removed from the lower floor, and reshoring has been placed.* *(Courtesy Portland Cement Association [PCA slide 50333].)*

■ RIBBED SLABS

One-way solid slabs become uneconomical when slab spans are long, because the slab must become thick and heavy to support the loads imposed and the steel must be placed lower in the slab. Eventually, the weight of the slab itself becomes a problem. The solution is to eliminate part of the concrete. This is possible, because much of the concrete near the bottom of a deep slab normally is not needed to resist loads. Most of the stress there is tensile, which is resisted by the steel. The solution then is to remove part of this nonworking concrete; one way is by forming the slab using metal or plastic *pans* (Fig. 3.11-14), which produces the effect of a thick slab but uses less concrete.

Standard pans come in two widths: 20 in. and 30 in. They range in depth from 8 in. to 20 in. Pans are tapered to permit their easy removal after the concrete has set. Tapered ends are used to increase the amount of concrete at the beams.

The ribs formed between pans are called *joists*. Joist width can be varied by widening the spacing of the pans. Ribs are supported on beams. When these beams are kept at the same depth as the pans, they are called *joist bands*. Joists are often interrupted at midspan by a beam known as a *distribution rib*, which helps to distribute concentrated loads to several joists.

Ribbed one-way slabs are constructed by first building the supporting elements, such as columns and walls, and permitting them to cure sufficiently to support the loads that will be imposed. Then the supporting structure (centering) is built for the pans, as described in Section 3.3.2.5. The centering decking forms the bottom of the joists. The pans should be placed on this centering and coated with a form-release agent. The steel should then be placed.

After the steel has been installed and inspected, the concrete should be poured and the slab surface cured and finished, as described in Section 3.7.1. When the ribbed one-way slab has set sufficiently to support its weight, the centering should be removed and the pans dropped from the concrete and moved to the next forming area. The formed surfaces should then be finished.

FIGURE 3.11-14 *Typical one-way ribbed-slab pan.* (Drawing by HLS.)

3.11.4.2

Two-Way Slabs

Two-way slabs span in two perpendicular directions at the same time. They generally are selected when the columns in a building can be located to create square or almost square bays. In that case, they are usually less expensive than comparable one-way slabs. It is possible to construct a two-way slab that is supported on all four sides by beams, similar to the one-way solid slabs described in 3.11.4.1, but they are seldom used. Two-way slabs without beams are easier to form and thinner when the beam depth is taken into account.

Essentially two types of two-way slabs are in general use: *flat-plate* and *waffle* slabs (Figs. 3.11-15 and 3.11-16). Two-way flat-plate slabs are used for buildings with relatively short spans. When spans become longer, the same thing happens with two-way slabs that happens with one-way slabs: the slabs get thicker until they become uneconomical. The solution is similar. The two-way equivalent of a pan slab is a *waffle slab*, which has ribs in both directions (see Fig. 3.11-16).

3.11.4.3

Flat-Plate Slabs

Flat-plate slabs that carry light loads, such as those in office buildings, apartment buildings, hotels, and dormitories, usually have no thickened areas (see Fig. 3.11-15). Flat-plate slabs in more heavily loaded storage and industrial buildings often have dropped panels at the columns to resist the heavy shear loads associated with heavier loading (Fig. 3.11-17).

The first step in constructing a two-way flat-plate slab is to form the supporting walls and columns, place the concrete in them, and permit it to cure sufficiently to support the loads that will be imposed. The second step is to build the forms for the dropped panels, when they occur, and then the forms for the slabs. The forms should be supported on temporary supports and shores of sufficient strength and stability to resist the large loads that will be imposed by the forms and the concrete. When all the forms have been built, the side of the forms that will be in contact with concrete should be coated with a form-release agent, and the steel should be placed in them, following the shop drawings.

Structural reinforcement in each of the two perpendicular directions in a two-way flat slab varies depending on its location relative to the columns. The slab is divided into zones: the zone between the columns is called the *middle strips*; the other zone is the *column strips*, which lie along the column rows in both directions (Fig. 3.11-18). The least amount of reinforcement is placed in the middle strips, where the loads are lightest. Additional steel and closer spacing of bars occur in the column strips to accommodate the higher bending stresses there. Steel is located near the top of the beam in the column strips, where upward bending stresses are present. Temperature steel is not needed in two-way slabs, because there is already steel in both directions. There is usually no additional steel in column caps when they are used. Rather, the additional concrete is relied on to resist the additional shear stress.

When the steel is in place and has been inspected, concrete may be placed into the forms and consolidated. The slab then should be cured and finished, as described in Section 3.7.1. When the slab is sufficiently strong to support itself, the forms should be removed and reshores placed. The formed surfaces should then be finished. If there is another floor above the one just completed, formwork should be moved up to it and the process repeated.

A variation in two-way flat-slab construction that eliminates most of the formwork is *lift-slab* construction. In this process the slabs for the floors of a building

FIGURE 3.11-15 *Two-way flat-plate slab.* (Drawing by HLS.)

FIGURE 3.11-16 *Waffle slab.* *(Courtesy Portland Cement Association [PCA slide 34775].)*

FIGURE 3.11-17 *Two-way flat-plate slab with dropped panels.* *(Drawing by HLS.)*

above the ground floor are cast on the ground floor in a stack and then lifted into place by a system of hydraulic jacks. The slabs have cast-in sleeves that ride up the columns. When the lowest slab of a stack has reached its proper level, the sleeves are welded to the columns and the jacks are used to pull the remaining slabs up the column to the next level. When the last slab has been welded into place, the jacks are removed.

3.11.4.4

Waffle Slabs

Two-way waffle slabs are formed using metal or plastic pans called *domes* (Fig. 3.11-19). The ribs formed between domes are called *joists*. There are two basic sizes of standard domes. The smaller of the two forms joists 5 in. wide at the bottom on 24-in. centers, and from 8 to 14 in. deep. The second standard dome size produces joists 6 in. wide at the bottom on 36-in. centers and from 8 to 20 in. deep. Larger sizes are also available. Domes are tapered to permit their easy removal after the concrete has set. Some domes have compressed air fittings to permit easier removal.

Waffle slabs are most efficient when allowed to cantilever at the perimeter. When cantilever is not practicable or desirable, a perimeter beam (*slab band*) must be placed there (see Fig. 3.11-20). In addition, *column heads* are created around the columns to accommodate the shear stresses there (see Fig. 3.11-20). Slab bands and column heads are created by omitting domes.

Waffle slabs are constructed by first building the supporting elements, such as columns and walls, and permitting them to cure sufficiently to support the loads that will be imposed. Then the supporting structure (*centering*) is built for the domes, as described in Section 3.3.2.5. The domes should be placed on this centering and coated with a form-release agent. The steel should then be placed.

After the steel has been installed and inspected, the concrete should be poured and the slab surface cured and finished, as described in Section 3.7.1. When the waffle slab has set sufficiently to support its weight, the centering should be removed and the domes dropped from the concrete and moved to the next forming area. The formed surfaces should then be finished.

3.11.4.5

Slabs on Permanent Forms

In steel structures, concrete slabs are often placed on a permanent corrugated steel sheet decking (Fig. 3.11-21). The steel decking acts as a form until the concrete sets. Such slabs are usually reinforced with welded wire fabric. This produces a finished lightweight deck. Such slabs are finished and cured as described in Section 3.7.1.

Heavier-gauge steel decking is also used with a concrete fill to produce *composite decking* (see Chapter 5).

3.11.4.6

Toppings

Concrete toppings are thin slabs of concrete

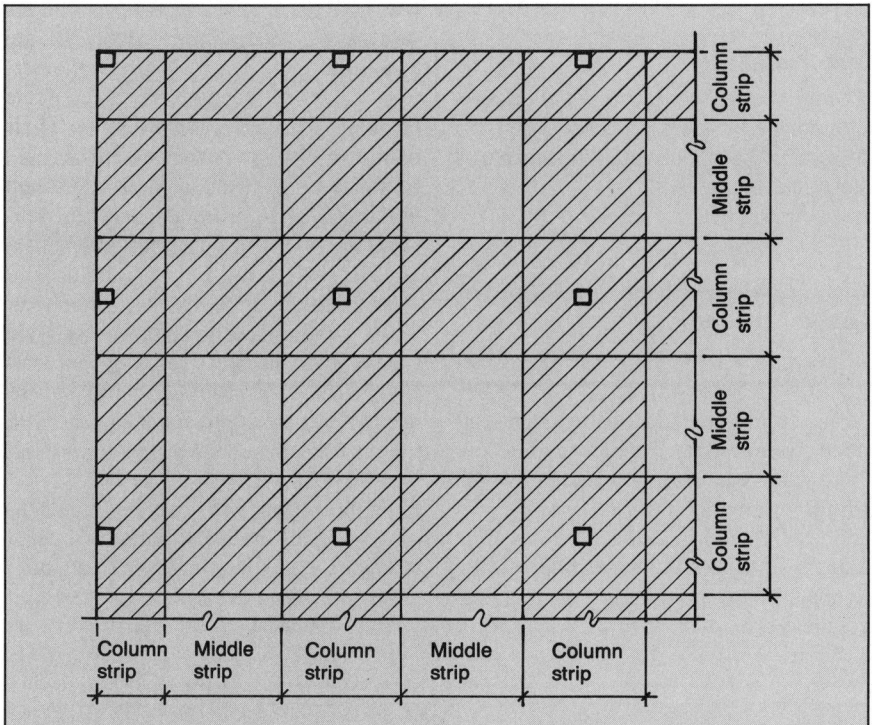

FIGURE 3.11-18 *Plan showing locations of middle strips and column strips in a two-way flat slab.* (Drawing by HLS.)

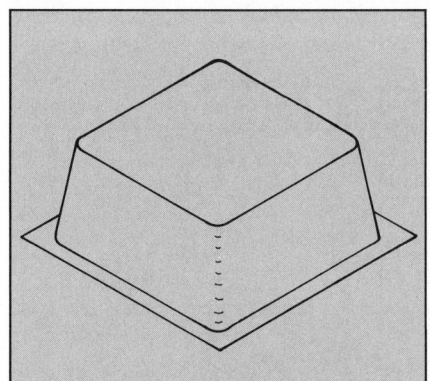

FIGURE 3.11-19 *A waffle-slab dome.* (Drawing by HLS.)

placed on top of other concrete, waterproofing, or insulation. They are of two basic types: *bonded* and *unbonded*.

Most toppings are composed of portland cement, coarse aggregate, and fine aggregate mixed with water. Sometimes a plasticizing agent is used instead of the water or in addition to it. The aggregates may be either normal-weight or heavy-duty types. The latter are especially hard aggregates, such as granite, quartz, or corundum, which are used when abrasive resistance is needed. Aggregates are limited to a 1/2-in.-diameter maximum size, which is generally smaller than those used in normal concrete. However, aggregates in 3-in.-thick toppings may be somewhat larger. Aggregates may be of selected colors if desired. Sometimes iron particles are added to the mix to improve surface wear and to increase the compressive strength of the concrete.

Toppings may be machine- or ready-mixed. They are placed, cured, and finished as described for slabs in Sections 3.6 and 3.7. Toppings that contain special selected aggregate are sometimes given a ground surface finish with grinding machines to expose the aggregate.

■ **BONDED TOPPINGS**

A *bonded topping* is a layer of concrete placed over an underlying concrete slab or precast-concrete structural element (see Section 3.13). Bonded toppings may be placed at the same time as the structural slab or later.

Bonded toppings are used for several purposes. Over monolithic concrete slabs, they can create a special finish, such as an abrasive one, where it would be impracticable, too expensive, or otherwise undesirable to include the aggregate necessary to produce this finish in the structural

FIGURE 3.11-20 *Pan slab in plan showing slab bands and column heads.* (Drawing by HLS.)

FIGURE 3.11-21 *A concrete slab on a permanent steel form.* *(Drawing by HLS.)*

concrete. Unreinforced bonded toppings are also used over monolithic concrete in heavy industrial buildings, where the movement of equipment might damage the finished floor. The toppings are placed after the equipment has been placed. Bonded toppings over monolithic concrete slabs are usually 1 in. to 1-1/2 in. thick and contain neither reinforcing steel nor welded wire fabric.

Bonded toppings are also used to make a level and smooth surface over structural precast-concrete elements and to make a composite structural unit with them (see Section 3.13.1). Such toppings often contain a layer of welded wire fabric or steel reinforcing bars to help tie together a series of precast elements. Toppings over precast-concrete elements are usually 2 in. to 4 in. thick.

■ UNBONDED TOPPINGS

Unbonded toppings are thin concrete slabs used to produce wearing surfaces over sheet membrane waterproofing or insulation (see Chapter 7). They are usually 3 in. thick but are sometimes thicker. They always contain a layer of welded wire fabric.

3.11.4.7

Slab Flatness

The principles concerning flatness requirements for slabs on grade discussed in Section 3.10.5 also apply to structural slabs and toppings.

3.11.4.8

Post-Tensioning

A process similar to that described in Section 3.11.3.5 for post-tensioning beams is sometimes used to post-tension flat-plate concrete slabs. Post-tensioning concrete slabs reduces their thickness, which results in smaller floor-to-floor heights and less cladding of the building. The amount of steel required can be as little as one-third that needed for a conventional concrete slab.

3.11.5

CONCRETE STAIRS

Concrete stairs are of two types. Each run of the first type consists of an inclined reinforced slab with integral risers and horizontal landings at the top and the bottom (Fig. 3.11-22). The bottoms, sides, and risers are formed. The risers usually slope outward at the top. All concrete in such stairs is placed simultaneously. This type of stairs usually occurs in concrete-framed buildings. Similar stairs are also used at entrances and site locations with the underneath supported by earth rather than by formwork.

The second type of concrete stairs consists of steel pans supported on steel stringers. These pans are filled with concrete. This type of stairs is most often used in steel-framed buildings.

FIGURE 3.11-22 *Typical stair reinforcement.* *(Courtesy Concrete Reinforcing Steel Institute.)*

3.12 JOINTS

It is impossible to entirely prevent cracks in concrete, but good jointing practice will help reduce their number and encourage them to occur in more acceptable locations. Joints in concrete work permit movement and volume changes; the handling, placing, and finishing of conveniently sized areas; and the separation of independent elements. The types of joints required in concrete include control joints, construction joints, isolation and separation joints, and building expansion joints.

The amount of thermal movement in concrete can be calculated by using its approximate coefficient of expansion (and contraction) of 0.0000055 in./in./°F. For example, a 100-ft. length of unrestrained concrete would expand 0.66 in. in length for each 100°F rise in temperature. If the temperature were to drop 100°F, the same amount of contraction would occur. Concrete would be exposed to a 100°F rise in temperature if it were placed in the spring when the temperature was 50°F and then in the summer the temperature rose to 150°F.

Concrete shortens about 0.72 in./100 ft. while drying from its saturated condition at placing to an average hardened state, with a moisture content equilibrium with air at 50% relative humidity. This shrinkage slightly exceeds the expansion caused by such an extreme increase in temperature as 100°F. Furthermore, in most outdoor applications concrete reaches its maximum moisture content (and moisture expansion) during the season of low temperature (and thermal contraction). Thus, the volume changes due to moisture and temperature variations frequently tend to offset each other. There is, therefore, no need for expansion joints in most concrete associated with buildings. The common reference to *expansion* is misleading because it implies that an increase in size after placement must be allowed for. There may be rare instances, in very large structures and in highways or other large paved surfaces, where extensive surfaces of concrete are subject to large temperature variations and sufficient expansion may occur to justify expansion joints. In

most instances, however, concrete is at its greatest mass when it is placed.

Nevertheless, some practitioners in warm regions insist that expansion joints are needed and include them as a safety factor, especially in concrete paving. Where they are used, expansion joints are similar to the isolation joints discussed in Section 3.12.3.

3.12.1

CONTROL JOINTS

As concrete sets or hardens, its excess mixing water is lost through evaporation and through hydration of the portland cement. This initial loss of water in the early life of a concrete section results in shrinkage that is of greater magnitude than any subsequent increase in the size of the hardened mass due to temperature or moisture content rise.

Shrinkage is inevitable as water eventually dries from freshly placed concrete, and consequently large areas can develop jagged, irregularly spaced shrinkage lines and cracks. If shrinkage is anticipated and joints installed to limit areas and control where cracking occurs, concrete work can be made more attractive, serviceable, and relatively free from unsightly and expected random cracking. Such joints are called *control joints*. These joints provide a break or a reduction in slab thickness and thus create a weakened section that encourages movement and cracking to occur at that location. When concrete shrinks, the cracks in these joints can open slightly, reducing the number of irregular and unsightly random cracks.

The maximum spacing between control joints depends on the concrete's thickness and shrinkage potential, the curing environment, and the absence or presence of distributed reinforcement.

Control joints in sidewalks should be spaced at intervals generally equal to the width of the slab (Fig. 3.12-1) but should not be more than 6 ft. apart. Joints in driveways should be spaced generally

FIGURE 3.12-1 *Jointing using a jointing tool. (Courtesy Portland Cement Association.)*

equal to the width of the slab, but never more than 20 ft. apart.

In large slab areas that do not contain structural reinforcement (Types I and II, as described in Section 3.10.2.1), control joints should be provided so that the slab is divided into approximately square panels, with a maximum spacing of 20 ft. The actual spacings depend on the slab's thickness (Fig. 3.12-2c). A rule of thumb for plain concrete slabs is that joint spacing in feet should not exceed two slab thicknesses in inches for unreinforced concrete made with coarse 3/4-in. maximum aggregate. Thus, a 4-in.-thick plain slab would require control joints at intervals not to exceed 8 ft. In residential construction, such spacing usually permits the joints to be placed under partitions. Additional control joints should be provided at the juncture of slab intersections (see Fig. 3.10-4).

Control joints are unnecessary in a slab that is to receive a floor finish, such as resilient flooring or carpeting, because the flooring will cover unsightly cracking resulting from shrinkage. Also, control joints are usually unnecessary in structurally reinforced slabs (Types III and IV, as described in Sections 3.10.2.1 and 3.10.2.2, respectively).

As shown in Figure 3.12-2a,b, control joints can be made by sawing a groove in the hardened but not yet fully cured con-

FIGURE 3.12-2 *(a) Several types of control joints for slabs on grade. (b) Keyed joint details for slabs on grade. Using too large a key may cause a crack parallel to the joint; if the key is too small, it may shear off, resulting in loss of transfer capacity. A horizontal saw cut permits the wood key strip to swell slightly without damaging the concrete and allows for easy removal of the strip. The joint should be filled with joint filler. (c) Isolation joint for interior column. Column is boxed out so the corners meet the control joints. This eliminates cracks radiating from the column corners. Circular fiberboard forms can be used instead of square wood forms. (d) Components of a slab on grade.* (All courtesy Portland Cement Association.)

crete, by using a power saw, or by installing a keyed joint. Sawed joints are provided by cutting approximately 1/8-in.-wide grooves in the concrete to a depth equal to one-fourth the total slab thickness, but not less than 3/4 in., and at least equal to the maximum size of the aggregate. Tooled control joints should be of a similar minimum depth. If a slab is to be grooved only for decorative purposes, joints may be shallower. Tensile stresses generated will be relieved at a control joint, thus reducing the likelihood of cracks occurring where they might be objectionable.

It is good practice to use a straight board as a guide when making a groove in a concrete slab (see Fig. 3.12-1), because tooled joints should be straight and perpendicular to the edge of the slab. The same care must be taken in running joints as in edging, for a tooled joint can add to or detract from the appearance of the finished slab.

On large concrete flat surfaces it may be more convenient to cut joints with an electric or gasoline-driven power saw fitted with an abrasive or diamond blade. Joints can be cut as soon as the concrete surface is firm enough not to be torn or damaged by the blade (within 4 to 12 hours) and before random shrinkage cracks can form in the slab.

3.12.2

CONSTRUCTION JOINTS

In large construction work, construction joints are necessary because it is not possible to handle, place, and finish a large area in one operation. When construction joints are installed, concrete can be conveniently and practically placed in several operations with no loss in the appearance or performance of the complete job. Although they may double as control or isolation joints, construction joints must allow no vertical movement in the completed floor and are thus often keyed joints, as shown in Figure 3.12-2b.

Plastic or rubber waterstops are inserted in construction joints below grade and in other locations to prevent water penetration through a concrete structural element (Fig. 3.12-3).

3.12.3

ISOLATION AND SEPARATION JOINTS

Isolation and separation joints often are necessary to separate concrete sections and to prevent the bonding of one concrete section with another or to separate a concrete section from another material or structural part so that one can move independently of the other. *Isolation joints* are sometimes called *expansion joints*.

Isolation and separation joints are usually formed by installing 1/8-in.-thick (or slightly thicker) asphalt-impregnated fiber sheets in floors at columns, footings, the junctures between floors and walls, and anywhere else that adjacent surfaces are required to move independently of each other. Figure 3.12-2c shows how isolation joints are formed around a column. Similar consideration around the perimeter of a slab should be given as shown in Figure 3.12-2d. Figure 3.12-4 shows an isolation joint at a column.

3.12.4

BUILDING EXPANSION JOINTS

While expansion of the concrete itself is seldom a problem, expansion of an entire structure often requires the insertion of *building expansion joints*, which pass through both structural elements and finishes. Concrete that occurs at one of these types of joints must be interrupted. Various types of joints have been developed to cover such joints.

In slabs and in walls, where water is likely to pass through a building expansion joint, waterstops are placed in the concrete (Fig. 3.12-5).

FIGURE 3.12-3 *Waterstop in a control joint.* (Drawing by HLS.)

FIGURE 3.12-4 *A construction joint used as an isolation joint.* (Drawing by HLS.)

3.13 PRECAST CONCRETE

The cost of forming and placing concrete in the field and the need to cope with unfavorable weather conditions while doing so has led to *plant-cast* precast-concrete elements that are produced in a factory environment. Units are cast at ground level in the plant, shipped to the construction site, and set in place using cranes in much the same way as structural steel is erected (Fig. 3.13-1).

Precasting concrete elements in a plant reduces their cost and makes it easier to control their quality. For example, concrete in beds in a plant can be easily consolidated by mechanical vibration, which results in fewer voids and better surfaces than can be achieved in concrete vibrated in the field. The ease of prestressing structural units in an industrial plant allows units to be made smaller. It is also possible in a plant to cure units using steam. Steam-cured precast-concrete units made using high-early-strength portland cement can be removed from the casting bed

FIGURE 3.12-5 *Waterstop in a building expansion joint.* (Drawing by HLS.)

FIGURE 3.13-1 *A precast-concrete beam being erected by crane.* (Photo courtesy Precast/Prestressed Concrete Institute.)

within 24 hours after the concrete has been placed.

Plant-cast precast-concrete units, which are heavy and bulky, must be shipped from plant to job site by truck. Legal limitations on the size of trucks and the use of roads dictate the size of precast units that can be shipped. These limits influence the types of units that can be effectively precast in a plant. For these and other reasons, plant-cast precast-concrete units are usually modular in nature and rela-

tively small. For example, it is impracticable to precast several bays of a two-way structural slab in a plant. Entire walls and even rooms have been produced, but most plant-cast precast-concrete is made in relatively small modular units in a few standard shapes.

Concrete units are sometimes precast at a job site. An example is the *tilt-up construction* wall units discussed in Section 3.11.1. Job-cast units are not subject to the size limitations imposed on plant-cast units. Therefore, it is sometimes advantageous to precast large elements, such as

transfer beams, on the ground and hoist them into position. Such site-cast precast units are subject to the same disadvantages as cast-in-place concrete relative to curing time and adverse weather conditions.

Once plant-cast precast-concrete units arrive at a job site, they can be erected quickly in most kinds of weather (Fig. 3.13-2). There is no need to wait for them to cure or to provide the moisture conditions required to cure cast-in-place concrete.

Precast concrete units can be divided into two basic categories: *structural* precast-concrete units and *architectural* precast-concrete units.

3.13.1

STRUCTURAL PRECAST CONCRETE

Structural precast-concrete units are so-called because they form a part of the structural system of the building and carry loads other than their own weight.

The Prestressed Concrete Institute (PCI) publication MNL-116, "Manual for Quality Control for Plants and Production of Precast Prestressed Concrete Products," establishes standard practices for the manufacture of structural precast-concrete elements. PCI also administers a plant certification program.

FIGURE 3.13-2 *A precast-concrete structure being erected in the snow.* (Photo courtesy Precast/Prestressed Concrete Institute.)

Types of Units

Structural precast-concrete elements include walls, girders and beams, columns, and floor and roof systems.

■ WALLS

Structural precast walls are used occasionally, but most precast wall panels are architectural concrete. Therefore, this category is discussed in Section 3.13.2.

■ BEAMS, GIRDERS, AND COLUMNS

The usual shapes for precast beams are rectangular, L-shaped, and inverted tee (Fig. 3.13-3). The projections serve as support ledges for floor and roof systems.

Girders are usually L-shaped or inverted T-shaped, but heavier girders may be the American Association of State Highway and Transportation Officials (AASHTO) shape shown in Figure 3.13-4. These girders were originally used on bridges, but they work as well in buildings.

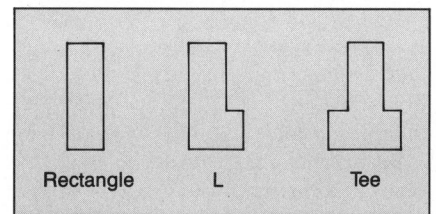

FIGURE 3.13-3 *Typical precast beam shapes.* *(Drawing by HLS.)*

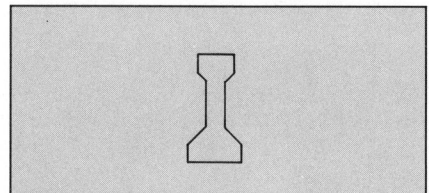

FIGURE 3.13-4 *AASHTO bridge girder shape.* *(Drawing by HLS.)*

Precast columns are usually either square or rectangular. While most structural precast concrete is post-tensioned, columns often have conventional reinforcement. Columns may be either single-story units or be cast in multistory sections with corbels to support beams and girders (Figs. 3.13-5 and 3.13-6).

FIGURE 3.13-5 *Typical precast concrete haunched column.* *(Drawing by HLS.)*

■ FLOOR AND ROOF ELEMENTS

Floor and roof elements rest on bearing walls, beams, or girders and are either left exposed to become the floor or roof slab or covered with a concrete topping (see Section 3.11.4.6).

Floor and roof elements fall into one of two general categories: flat slab or tee. Each type may be made from either normal-weight or lightweight concrete. A balance must be drawn in this selection between the decreased weight of lightweight concrete and its greater cost.

FLAT-SLAB PRECAST ELEMENTS Flat-slab precast elements may be either solid or hollow-core. *Solid flat-slab units* are simple flat concrete slabs (Fig. 3.13-7). They are used where spans can be held short (about 20 ft. maximum) and maximum headroom is needed.

Hollow-core flat-slab units are cast with round voids running their entire length (Fig. 3.13-8). They are appropriate for intermediate spans (about 50 ft. maximum) where solid flat units would be too thick and heavy. The principle is the same as that discussed in Section 3.11.4 as justification for selecting pan slabs instead of flat-plate slabs. The cores eliminate the nonworking concrete, lightening the slab section. The steel is contained in the top and the bottom of the cored slab.

Both solid and hollow-core slabs offer the advantage of being relatively thin, which saves on floor-to-floor height. They can also be painted or given a thin textured coating and left exposed as a finished ceiling.

T-SHAPED PRECAST ELEMENTS When spans are longer than can be accommodated by cored slabs, the cores are elimi-

FIGURE 3.13-6 *Concrete-framed building under construction, showing two-story columns and several types of beam haunches.* *(Photo courtesy Precast/Prestressed Concrete Institute.)*

FIGURE 3.13-7 *Solid flat precast-concrete slab.* (Drawing by HLS.)

FIGURE 3.13-8 *Cored flat precast-concrete slab.* (Drawing by HLS.)

nated, the slab thinned, and stems added to give stiffness to the system. The result is T-shaped floor and roof units. When the units have a single stem, they are called *single tees*; two-stemmed elements are called *double tees* (Fig. 3.13-9).

Double tees can be designed to span as much as 100 ft. They can also be used on shorter spans where solid slabs and cored slabs will also work. Where headroom is

FIGURE 3.13-9 *Stemmed precast-concrete elements: (a) double tee; (b) single tee.* (Drawing by HLS.)

not a problem and their appearance is desirable, double tees might be appropriate on short spans.

Single tees have the disadvantage of having to be braced during construction because they are subject to tipping until they are finally fixed in place. They are, therefore, seldom used except on long spans. Their normal spans of between 60 and 150 ft. require depths from 36 to 60 in., which is appropriate in few buildings.

Tees usually have steel weld plates cast in to their edges so that adjacent units can be welded together to develop a diaphragm action in the entire system. This process helps to reduce differential deflection in adjacent units.

TOPPING   Precast-concrete floor and roof elements are usually finished rough and covered with a 2- to 4-in.-thick topping, as described in Section 3.11.4.7.

3.13.1.2

Manufacture

Most structural precast-concrete elements are manufactured in a plant on casting beds that range from 200 to 400 ft. in length. However, some are cast into molds. The process is essentially as follows:

Prestressing steel is placed in a casting bed and tensioned using hydraulic jacks. Bulkheads are placed in the bed to separate individual units. Bulkheads are sometimes omitted for solid and cored slabs, which are sawed apart later. Core forms and opening forms are placed in the beds. Concrete

is placed in the beds, consolidated by vibration, and leveled. The top surface is finished as appropriate with floats or trowels. Steam or radiant heat is then used to cure the unit. In some plants, some units, especially flat-slab units, are wet-cured for seven days, in lieu of steam curing.

Test cylinders are taken at the same time the concrete is placed and cured. When the units are steam-cured, about 24 hours after the concrete is placed, the cylinders are tested to verify the strength of the concrete. When it is found to be satisfactory, the prestressing strands are cut, and the units are separated and removed from the casting bed.

3.13.1.3

Erection

Precast structural concrete elements are handled and erected in much the same way as structural steel. Units are hoisted into place by crane (Fig. 3.13-10). Some units are held in place by gravity, others by the welding together of inserts in the elements.

■ **CONNECTIONS**

Precast concrete columns usually have a cast-in steel plate at the bottom. This plate is fastened to anchor bolts set into a concrete footing (Fig. 3.13-11). Column sections are joined using similar plates and bolts (Fig. 3.13-12).

Beams and girders are anchored to columns in a variety of ways. A typical system is the use of corbels to support beams and girders (see Fig. 3.13-5).

FIGURE 3.13-10 *Installing a precast hollow-core flat-slab section.* (Photo courtesy Precast/Prestressed Concrete Institute.)

FIGURE 3.13-11 *A typical precast-concrete column base at footing. Pockets are filled with nonshrink grout after anchor nuts are set.* (Drawing by HLS.)

FIGURE 3.13-12 *A typical bolted column-to-column connection. Pockets and spaces around shims are packed solid with nonshrink grout after anchor nuts are set.* (Drawing by HLS.)

FIGURE 3.13-13 *Tees resting on a beam and on masonry.* (Photo courtesy Precast/Prestressed Concrete Institute).

Tees usually rest on L- or T-shaped beams or girders (Fig. 3.13-13). Tees are often, but not always, given a concrete topping (see Section 3.11.4.6).

Flat slabs, both solid and hollow-core, usually rest on beams or girders (Fig. 3.13-14. Such slabs are usually given a concrete topping (see Section 3.11.4.6).

Bearing pads are usually inserted between concrete elements and other concrete elements supported by them so that the elements may move to accommodate thermal changes and other structure movement. Bearing pads are usually synthetic rubber for beams, girders, and tees and plastic for slabs.

Precast structural elements must be held together to resist seismic and wind loads. This is sometimes accomplished by installing reinforcing rods in a topping or by welding together plates cast into the elements. However, the stems of tees are never fastened solidly to beams or columns because doing so would prevent them from expanding and contracting freely.

3.13.2

ARCHITECTURAL PRECAST CONCRETE

Architectural precast-concrete units include panels and other members used as a part of a building's exterior cladding. They usually are not load bearing.

The Prestressed Concrete Institute (PCI) publication MNL-117, "Manual for Quality Control for Plants and Production of Architectural Precast Concrete Products," establishes standard practices for the manufacture of architectural precast concrete elements. PCI also administers a plant certification program.

Most architectural precast-concrete panels are made from standard-weight concrete and reinforced conventionally. However, glass-fiber-reinforced concrete panels are being used today that require no conventional steel reinforcement. These panels are lighter than conventional panels and often have steel studs cast into them, thus providing a cavity for insulation.

Architectural precast-concrete panels are available in many finishes, ranging from flat, as-cast finishes to ornate articulation made with form liners. Exposed aggregate, bush-hammered, and abrasive-blasted finishes are possible, as are panels

FIGURE 3.13‑14 *Hollow-core flat slab resting on a beam.* *(Photo courtesy Precast/Prestressed Concrete Institute.)*

3.13-15, 3.13-16, and 3.13-17). Joints are usually treated using sealants.

3.13.3

PRESTRESSING

Most structural precast-concrete units and some architectural precast-concrete units are prestressed. *Prestressing* is done in precast concrete for the same reasons post-tensioning is done in cast-in-place beams and slabs (see Section 3.11.3.5). The process is also similar to the post-tensioning described in Section 3.11.3.5, except that the tendons are stretched between *abutments* before the concrete is placed around them. When the concrete has cured, the tension is released from the steel, which recoils, transferring the stresses in it to the concrete in the form of compression. This load transfer is done through the bond between the steel and the concrete. Bearing plates are not required.

Usually, precast units are cast end to end over continuous tendons. When the concrete has cured, the tendons are cut at both ends of each concrete element. This reduces the need for a separate set of costly abutments for each precast element.

faced with granite, marble, brick, and other materials. Some units have insulation cast into them.

Many panel configurations are possible, from flat panels to those with deep shadow lines and recesses. Architectural precast-concrete column covers are also available.

Architectural precast panels are usually installed by crane and fastened in place using cast-in angles and plates (Figs.

3.14 STRUCTURAL INSULATING ROOF DECKS

Two kinds of structural insulating roof decks are addressed here: poured-in-place gypsum concrete decks and prefabricated board decks. The material, gypsum, is described in Section 9.1.1.

3.14.1

POURED-IN-PLACE GYPSUM CONCRETE ROOF DECKS

A poured-in-place gypsum concrete roof deck is a reinforced gypsum concrete slab placed on permanent formboards that may or may not be supported by subpurlins.

3.14.1.1

Materials

■ GYPSUM CONCRETE

Gypsum concrete mix is a factory-controlled mixture of gypsum and wood chips or mineral aggregate. It requires only the addition of water at the job site. A gypsum concrete slab, which weighs from 35 to 55 lb./cu. ft., depending on the type and the amount of aggregate, is poured to an average thickness of 2 to 3-1/2 in. Gypsum concrete slabs are reinforced with galvanized-steel welded wire mesh or woven wire mesh to provide at least a 0.26-sq.-in. cross-sectional area of steel in each foot of slab width. Figure 3.14-1 shows two examples of gypsum concrete roof construction.

■ FORMBOARDS

Formboards are nonstructural components of a gypsum concrete roof deck and usually are not considered in the load-bearing capacity of the slab. They serve as a form and remain in place to provide a functional underside of the deck. Various formboards are available to provide fire resistance, light reflection, insulation, sound absorption, and economy.

Cast-in-place
concrete frame

Angle tieback

Angle
tieback clip

Precast
concrete
panel

Angle seat clip

Angle seat

Cast-in
anchor

FIGURE 3.13-15 *Architectural precast cladding panel supported by a cast-in-place concrete structural frame, showing a typical method of supporting and anchoring panels. Finishes, flashing, caulking, roofing, and thermal and safing insulation have been omitted for clarity. Seat and tieback clips usually have slotted bolt holes to permit adjustment and are shimmed as necessary and bolted to panel and to seat and tieback angles. Plates are sometimes used in place of seat and tieback angles.* (Drawing by HLS.)

Gypsum formboards, also called *roof boards*, are surfaced on the exposed face and longitudinal edges with glass-fiber sheets or with calendered manila paper specially treated to resist fungus growth. The back face is surfaced with gray liner paper as is used on regular gypsum wallboard (see Section 9.5), or with glass-fiber sheets. Gypsum formboards are usually 1/2 or 5/8 in. thick, 32 in. wide, and either 8 or 10 ft. long, although other sizes may be available.

Wood-, glass-, and mineral-fiber boards and various types of precast planks are also used as forms for gypsum concrete. Their widths generally coincide with the standard 24-, 32-, and 48-in. nominal subpurlin spacing.

■ SUBPURLINS

Steel subpurlins (rails, bulb tees, or other tee sections) are welded transversely to the primary framing to support formboards and roof slab (Fig. 3.14-2). They span between main purlins, beams, or joists. Subpurlins also anchor the deck against uplift forces, restrict deck movement due to temperature changes, and provide lateral bracing for the main roof purlins. Subpurlins, which vary in size, weight, and shape, are selected on the basis of the span between the main purlins or joists supporting them and the required safe load-bearing capacity of the deck.

When subpurlins are not used, the formboards rest directly on purlins, beams, or joists.

3.14.1.2

Use and Construction

Gypsum concrete decks are used when noncombustible roof construction is required and other types are less practicable or more costly. They can be installed on virtually any roof shape, size, or configuration, even on those that are warped, sawtoothed, curved, or sloped. The primary framing can be wood, steel, or concrete. The light weight of gypsum concrete, compared to portland cement concrete, often allows the use of a lighter structural frame. Gypsum concrete sets rapidly, and the roof slab may be used as a working surface 30 minutes after pouring.

Decks of gypsum concrete poured over gypsum- or mineral-fiber formboards are classified as noncombustible. When combustible wood-fiber formboards are used, there generally is an increase in insurance premiums. Gypsum concrete decks, 2 and 2-1/2 in. thick on gypsum formboards without a suspended ceiling below, qualify for 1- and 2-hour fire resistance ratings, respectively. The materials used in gypsum concrete should conform with the requirements of ASTM Standard C 317; installation should conform with ASTM Standard C 956.

3.14.2

PREFABRICATED STRUCTURAL INSULATING ROOF DECKS

Prefabricated structural insulating roof deck boards are designed to provide a structural deck as well as thermal insulation and interior ceiling finish. They are used primarily on flat or sloping roofs with exposed beams. The underside of the

FIGURE 3.13-16 *A light precast panel being installed.* *(Photo courtesy Precast/Prestressed Concrete Institute.)*

a

b

FIGURE 3.14-1 *Gypsum concrete roof construction: (a) with subpurlins; (b) without subpurlins.* *(Courtesy Gypsum Association.)*

decking is factory-finished in various designs or textures, some with acoustical value.

Two types of structural insulating roof decks are discussed here: lightweight aggregate boards and wood-fiber boards (Fig. 3.14-3). These boards are available in a variety of thicknesses and sizes, in either butt-joint or tongue-and-groove edge profiles. Because the underside of the deck is also the interior finish, care must be taken not to damage the surface during erection. In areas with high interior humidities, such as bathrooms and laundries, adequate roof insulation and ventilation should be provided to prevent condensation.

FIGURE 3.13-17 *A heavyweight cored precast wall panel being installed. These panels are designed to carry the weight of the panels above.* *(Photo courtesy Precast/Prestressed Concrete Institute.)*

FIGURE 3.14-2 *Subpurlins and reinforcing: (a) bulb tee and welded wire; (b) trussed tee and woven wire.* *(Courtesy Gypsum Association.)*

FIGURE 3.14-3 *Insulating deck: (a) lightweight perlite aggregate; (b) wood fiber.*

4

MASONRY

Photo: Aligning the faces of a course of blocks and leveling the units. Courtesy of Portland Cement Association.)

185

INTRODUCTION

Stone, clay masonry units, and concrete masonry units are all used extensively in construction today. Brick, which is one of the oldest manufactured building materials, is an example of clay masonry. Concrete masonry units include concrete block, a relatively new material. To take advantage of their compressive strength, durability, and fireproof character, they are used primarily in wall assemblies, often appearing in conjunction. In addition, all three are also used as decorative screening. Stone and clay masonry units can be suitable for floor surfacing and exterior paving. Lightweight aggregates have made lighter concrete masonry units possible and permitted the fabrication of larger masonry panels than earlier technology permitted. The development of high-bond-strength mortars has made prefabricated brick masonry panels practicable. Continuing improvement of sealants, adhesives, and masonry units will ensure continued use of these time-proven masonry materials.

To serve their intended function, masonry walls must perform as they were designed to do. Exterior walls, for example, must possess at least three essential functional properties. They must be strong enough to carry the loads that will be applied. They must be watertight enough to prevent ground- and storm-water penetration. And they must be durable enough to withstand severe weather and other atmospheric conditions. Some masonry walls must also resist heat, sound, or fire transmission.

Aesthetic considerations often influence the selection of masonry units that will be exposed to view. The color of both the units and the mortar, the texture of the units themselves, glazes applied to the units, the texture created by the method of laying the units, and, finally, the pattern produced by the type of bond and the size of the units are all factors involved in the final choice of materials.

With a good design in place, the procedures for producing satisfactory masonry assemblies can be summarized as follows: Materials must be sound, and they must be properly handled, stored, protected, and prepared. The selected mortar type must be the one that is best suited for the use and exposure of the assembly. Mortar constituents must be proportioned and mixed correctly. Units must be laid with full head and bed joints, and mortar joints must be tooled properly. Masonry must be protected from damage by the weather during construction. Properly placed and installed flashing and weep holes for drainage will prevent water from penetrating to the inside of the assembly after it has been built.

This chapter describes materials and procedures that will help to produce masonry of suitable quality. Appendix A contains a figure that summarizes the fire rating, sound transmission class, heat transmission characteristics, wall weight, and estimated comparative cost figures for a selection of masonry walls.

4.1 HISTORY

Masonry structures today are made from brick and block, which together are called masonry units, or from stone. Masonry structures predate written history. The earliest structures were huts made from unshaped native field stones piled upon one another without mortar or other materials in the joints. Where stone was not available, sod or dried mud served the same purpose. Later, clay and silt were mixed with water and formed by hand into bricks. The spaces between these bricks were sometimes packed with mud to keep out the wind and rain and to make it easier to build level walls with irregular bricks. Later still, it was discovered that clay bricks placed in or adjacent to a fire became harder and more weather resistant. The Romans used this knowledge to build kilns to produce burned clay roofing tiles, and eventually burned clay bricks.

The Mesopotamians built stone and sun-dried brick buildings about 4000 B.C. A thousand years later, the Egyptians began building temples and pyramids of cut stone. By laboriously cutting the stone using bronze tools, they made all pieces fit closely together.

Early stone buildings were limited because openings and column spacings could be only as large as could be spanned with a single stone. Roofs were made of wood. The Babylonians built small arches over windows and other small openings.

When it became possible to make stone-working tools from iron, the art of stone building developed to a high order. The Greeks refined the process to produce fine details in stone. The Romans were able to build for the first time buildings with large open spaces. They were the first to build arches large enough to sustain bridges and large buildings. They used arches and vaults to build temples, forum buildings, theaters, amphitheaters, baths, aqueducts, and homes.

The building of brick and stone structures without joint materials changed when the Etruscans developed a lime mortar that could be used to fill gaps between masonry units. Later, the Romans discovered how to make a hydraulic (literally, "that which will set under water") cement by burning and grinding a type of volcanic rock. This led to stronger and more watertight stone and brick structures. This same discovery led to the expanded use of concrete.

The sun-baked bricks used extensively by ancient peoples began to disappear after the Romans invented kilns. Clay could then be baked hard enough so that its resistance to the elements expanded dramatically.

The fall of the Roman Empire was followed by the emergence of the semi-oriental Byzantine Empire, centered in Constantinople (Istanbul). The Byzantine Empire flourished for a thousand years until it was overrun in 1453 by the Ottoman Turks. Byzantine architecture combined Roman arch forms with other shapes and added detail and color. The Romans had been limited to circular supporting shapes (for example, the Pantheon), but the development of the pendative permitted the construction of a stone dome over a square space. A pendative is a spherical triangle whose lower point rests on a pier and whose curved upper edge in combination with three other pendatives forms a circle that can support a dome. Using this device, Byzantine architecture placed large stone domes over square buildings and so constructed buildings able to support several domes. The Byzantines also used broad arches and massive buttresses to add aisles and galleries to their buildings.

Beginning in the eleventh century, a 200-year-long revival called the Romanesque period saw the beginnings of forms of stone and architecture that would be refined in later periods. These included ribbed vaults, membered piers, and pier buttresses.

The Gothic period, which followed, saw the development of flying buttresses that could withstand the outward thrust from large vaulted and domed structures. These produced even larger and more complex structures and, when combined with vaults and piers, produced what has been called a "cage of stone." Walls could be replaced with windows, leading to the development of ornate stained glass windows especially in the cathedrals of western Europe.

Beginning in the fourteenth century, the Renaissance (Age of Discovery) stirred an interest in the neoclassical. This was a period of great strides in the arts, humanities, and science.

In the late eighteenth and early nineteenth centuries, the "age of revivals," architects actually measured Greek and Roman ruins and reproduced them as new buildings. Gothic buildings and other former styles were experimented with, also. During this period, not much was done to further the science of building.

In the late eighteenth century the industrial revolution ushered in the modern era. Machines began to replace much of the handwork necessary to quarry and cut stone and to mold and fire bricks. Stone units became more uniform in shape and size. Bricks became more consistent in color, strength, and size.

Blocks made from cement and various aggregates appeared after the development of portland cement and portland cement concrete in the late eighteenth and early nineteenth centuries. They were cheaper and lighter than cut stone and larger than bricks, which reduced the time needed to lay them. They were more versatile than concrete in that they could be easily placed in small quantities and did not require formwork.

Until the development of the theory of elasticity in the nineteenth century, unit masonry and stone construction were based solely on experience. After that, masonry structures could be built using rational design based on calculated stresses.

Until the late nineteenth century, masonry and stone were the materials of choice throughout the world for constructing buildings, bridges, and viaducts. The advent of cast iron and steel structural elements quickly replaced masonry and stone as the major structural support elements in large buildings. Reinforced concrete had the same effect in smaller buildings and in bridges and other arched structures. The building of large unit masonry and stone arches reached its climax about the beginning of the twentieth century; then they were replaced by reinforced concrete. Concrete was less labor intensive because it was easier to form and place, and it could be made to span distances three to four times that possible in stone.

Although they have been largely replaced by steel and concrete as primary load carrying elements in larger buildings, unit masonry and stone remain in great use today for cladding, partitions, and flooring. They are especially valuable where fire and weather resistance is required. Their varied appearance is used to create a variety of aesthetic effects. They are adaptable to current standards and methods for insulating buildings to make them more comfortable and to increase their ability to conserve energy. Modern masonry cavity walls resist water penetration and provide the space to hold insulation. Masonry partitions are fire resistant and help to restrict sound penetration. New mortars and construction adhesives reduce the labor needed to place masonry walls. Some walls are built from prefabricated masonry panels and held together with adhesives.

Further advances are certain. In spite of innovations in metal and concrete construction techniques and materials, masonry and stone construction are likely to be with us for a very long time to come.

4.2 MORTAR AND GROUT

Mortar is a combination of one or more cementitious materials (portland cement, lime, or masonry cement); a clean, well-graded aggregate, such as sand; and enough water to give the mixture a plastic, workable quality. The term *grout* refers to several different types of construction products, depending on the material and its use. Here we are concerned with two types of grout only. One type is used to fill voids in both reinforced and unreinforced masonry. This type is made from the same ingredients as is mortar, but is produced in a pourable consistency. The various ingredients must be proportioned properly to give the mortar or grout a good balance of essential properties. The second type of grout is a tile setting grout used in filling the joints in unit masonry or stone paving and flooring. In this section, the term *grout* refers to the material used to fill voids unless specifically identified as the type used in paving and flooring joints.

Mortar is placed in the joints between individual masonry or stone units in a wall or other building element to cushion the units and to provide a level setting bed for them. It also seals the spaces between the units, compensates for size variations in the units, and provides an aesthetic quality by creating shadow lines and color effects. Grout is poured into voids between masonry unit wythes and into the cores in hollow masonry units to produce solid filled and reinforced masonry walls. Mortar and grout also grip metal ties and reinforcements, helping them to act integrally with the other parts of a wall. Tile setting grout is used to fill the joints between brick pavers.

4.2.1

MORTAR AND GROUT PROPERTIES

Because mortar and grout are placed while plastic and then harden, they must have two sets of properties: (1) those present when they are in their plastic state, and (2) those that result after they have

hardened. Proper plastic properties and hardened properties are both necessary for a mortar or grout to be suitable for use in building construction. Both sets of properties affect a finished wall's strength, durability, and watertightness.

The basic constituents of most modern masonry mortars and grouts are portland cement, lime, sand, and water. Masonry cement also contains other ingredients (see Section 4.2.2.4). Sometimes additives are used in portland cement–lime mortars and grouts, even though some additives are harmful to the hardened mortar or grout. Since both the plastic and hardened properties of a mortar or grout depend largely on its ingredients, these should conform to the appropriate American Society for Testing and Materials (ASTM) standards and be properly proportioned. Because a mortar's or grout's properties also depend on the proportion of its ingredients, different mortar types (see Section 4.2.4.1) and grout mixes also have different properties.

4.2.1.1

Plastic Properties

The properties of mortars and grouts while they are still plastic include *workability, water retention, initial flow,* and *flow after suction.* Mortar and grouts must have suitable plastic properties so that in their hardened states they can perform adequately.

■ WORKABILITY

A workable mortar is uniform, cohesive, and of a consistency that makes it usable by a mason. A mortar is workable when particles in the mix do not segregate and when it spreads easily, holds the weight of the units, makes alignment easy, clings to the vertical surfaces of masonry units, and easily extrudes from the mortar joints without dropping or smearing.

Water retention, flow, resistance to segregation, and other factors affect a mortar's workability. These in turn are affected

by the properties of the mortar ingredients. This complex relationship makes quantitative estimates of workability difficult. There is, in fact, no standard laboratory test for measuring it.

Grout is designed to pour and therefore must flow freely without being too soupy. It also must be uniform.

■ WATER RETENTION

Water retention in a mortar or grout prevents rapid loss of water and a resultant loss of plasticity when the mortar or grout contacts a masonry unit with a high absorption rate. A high degree of water retention also prevents a mortar from bleeding when it comes into contact with a masonry unit that has a low absorption rate. *Bleeding* is a process in which water leaves the mortar and is deposited in a thin layer between the masonry unit and the mortar. When this happens, the unit is said to *float.* This floating materially reduces bond.

■ INITIAL FLOW AND FLOW AFTER SUCTION

The water retention of a mortar is the ratio of a plastic characteristic called *flow* immediately after *mixing* to the *flow* of the same mortar after *suction.* The flow of a mortar immediately after mixing (initial flow) is measured by a simple laboratory test (Fig. 4.2-1). A truncated cone of mortar 4 in. in diameter is placed on a testing apparatus called a *flow table.* The flow table is then mechanically raised and dropped 25 times in 15 seconds through a height of 1/2 in. This movement causes the mortar to flow outward, increasing the diameter of the cone. The flow is the ratio of the final diameter to the original diameter of the cone, expressed as a percentage. For example, if the diameter of the mortar mass is 8 in. after the test, the flow is 100%. ASTM Standard C 270 requires an initial flow of 110% ±5% for specimens tested for water retention or compressive strength.

A *flow-after-suction* test is then used to predict the theoretical flow of a mortar

FIGURE 4.2-1 *The laboratory test for measuring flow.* *(Courtesy Portland Cement Association.)*

after it has lost water to an absorbent masonry unit. In this test, a sample of the same mortar previously tested for initial flow is placed over an absorbent filter in a perforated metal dish mounted over a vacuum device. Operation of the vacuum device removes some of the water from the mortar. This mortar is then removed from the suction device and tested for flow in the same way as the first test. The laboratory value of water retention is the ratio of flow after suction to initial flow, expressed as a percentage. The formula is $(A/B) \times 100$, where A is the flow after suction and B is the flow immediately after mixing. ASTM Standard C 270 requires a water retention of not less than 75% percent.

4.2.1.2

Hardened Properties

Among the properties of hardened mortar are *bond strength, durability, compressive strength, volume change, watertightness, rate of hardening,* and *color.* The most important characteristics of a hardened grout are bond strength and compressive strength.

■ BOND STRENGTH

Bond strength is perhaps the most important property of hardened mortar and an important characteristic of hardened grout, especially of grout used in conjunction with the type of terra cotta called ceramic veneer (see Section 4.7.7). Some of the variables that can affect bond strength are addressed in this section.

A mortar's or grout's other properties often determine its hardened bond strength. These properties include the type and amount of cementitious materials and the amount of water in the mix (see Sections 4.2.4.3 and 4.2.5), the water retention characteristics of the mortar or grout (see Section 4.2.1.1), the entrained air content of the mortar or grout, and the mortar's or grout's compressive strength.

The amount of water in a mortar mix and its water retention affect its flow (see Section 4.2.1.1). The more water, the greater the flow. A mortar's tensile bond strength is a mechanical function rather than a chemical one. The mortar flows into pores in the masonry unit and interlocks with them. Therefore, bond strength always increases as the mortar's flow increases because wetter mortar flows more readily into these pores. Adding lime to portland cement mortar increases its workability and therefore facilitates the interlocking (bonding) process. Too little water can also affect a bond (see Section 4.2.6).

When the air content of a mortar or grout exceeds 12% there is an accompanying decrease in bond.

The type of masonry unit in contact with a mortar or grout can affect its bond strength. Masonry unit characteristics that can affect bonding include surface texture, the suction of clay masonry units (see Section 4.4.2.2), and the moisture content of concrete masonry units (see Section 4.4.2.3).

Workmanship can also affect the bond strength of a mortar. For example, movement of a masonry unit after it has been laid can cause loss of bond, as can water on the surface of a clay masonry unit, or a delay in laying masonry after the mortar has been placed.

The air temperature and relative humidity during a mortar's curing period can also affect its bond strength.

■ DURABILITY

The *durability* of mortar is measured principally by its ability to resist water penetration. Water entering the mortar through a masonry joint into the adjacent masonry can cause damage to the mortar and the adjacent masonry in several ways. Repeated cycles of freezing and thawing under natural weather conditions will cause the surface of the mortar or masonry unit to pop off (spall). Water entering a mortar will dissolve salts in the mortar. These salts will then leach out into or onto the surface of adjacent masonry. Upon drying, these salts will crystallize. Salts that crystallize on the exterior face of a wall (efflorescence) are unsightly but do little harm. However, when the crystallization occurs within a wall (subflorescence or cryptoflorescence) considerable damage can result. Subsequent wetting followed by drying cause larger and larger crystalline deposits to form, until the deposits grow large enough to cause the surface of the mortar or masonry units to spall.

An increase in entrained air content may increase the durability of mortar, but will substantially decrease its bond strength and adversely affect other desirable properties. Because portland cement–lime mortar is sufficiently durable without air entrainment, admixtures should not be used to increase its entrained air content. Masonry cement mortars tend to be less durable, so air entrainment admixtures are used in them.

■ COMPRESSIVE STRENGTH

The *compressive strength* of a mortar or grout depends largely on the amount of portland cement in its mix, rising when the cement content is increased. Conversely, a larger flow (see Section 4.2.1.1) brought about by a rise in the water content of a mortar will decrease its compressive strength.

Since there are few reports of structural distress or failure as a result of low compressive strength problems, it is not necessary to use greater-than-moderate-strength mortars in normal construction. Bond strength, workability, and water retention are more important and are generally given a higher priority when mortar is being selected. Conversely, using mortar with too high a strength can cause significant problems, especially in repointing older masonry. A high-strength mortar used in an existing joint may be stronger and less flexible than the existing mortar or even the masonry units. Normal thermal movements can cause the stronger mortar to tear away from the weaker mortar, causing cracks that may leak. These movements may even cause the masonry units to chip or crack. Therefore, mortar strength should never be higher than is necessary for the conditions, and in existing walls new mortar should match the existing mortar in composition and materials.

■ VOLUME CHANGE

Some people believe that volume change, or mortar shrinkage, can be extensive and can cause leaky walls. Actually, the maximum possible shrinkage in a properly tooled mortar joint containing sound mortar that has a good balance of all desirable properties is so small that any resultant crack would not be seen with the naked eye.

■ WATERTIGHTNESS

When a wall leaks significantly, and there are no major holes in it, fine cracks between the mortar and the masonry or stone units, especially in the vertical joints, are usually the culprit. The watertightness of the masonry units themselves, or of mortars commonly used today, is seldom a factor in wall leaks.

■ RATE OR HARDENING

The rate of hardening of mortar or grout is the speed at which it develops a resistance to indentation and crushing. Too rapid hardening may interfere with a mason's use of a mortar or pouring of a grout. Hardening too slowly may impede the progress of the work or may subject a mortar or grout to early damage from frost action during winter. A well-defined, consistent rate of hardening allows a mason to tool joints at the same degree of hardness and thus obtain uniform joint color.

■ COLOR

Color is an inherent characteristic of hardened mortar. The materials needed to produce colored mortar are discussed in Section 4.2.3.4. Regardless of what the color is, color uniformity affects the overall appearance of masonry. There are several factors other than the materials used that can affect the uniformity of color and shade of a hardened mortar. They include atmospheric conditions, the moisture content of the masonry units, the admixtures used, the control exercised over the mixture, and the timing of the mason's tooling of the joints.

Controlling the mix from batch to batch and day to day is necessary if a mortar is to have the same color throughout. Careful measurement of materials and thorough mixing are important in providing uniformity.

The timing of joint tooling influences the shade of a hardened mortar. If the mason tools the joint when the mortar is relatively hard, a darker color results than if he or she tools the joint when the mortar is relatively soft. To ensure a uniform mortar color in the finished wall, the mason should tool all the joints at like degrees of mortar hardness (Fig. 4.2-2).

FIGURE 4.2-2 *Joints should be tooled when the mortar is at a like degree of hardness to prevent variations in color.* (Courtesy Portland Cement Association.)

4.2.2

KINDS OF MORTARS AND GROUTS

Four kinds of mortars and two kinds of grout are described in this section. The mortars are sand-lime, portland cement, portland cement–lime, and masonry cement. All of the discussions about mortar elsewhere in this chapter refer either to portland cement–lime or masonry cement mortars, and they are the only ones covered by ASTM Standard C 270. Lime mortars and portland cement mortars are seldom used today except in small, special cases, such as in patching a like mortar.

The two kinds of grout discussed here are portland cement grout and portland cement–lime grout. Lime mortars are not suitable for grout. Masonry cement is usually not permitted in grout.

4.2.2.1

Sand-Lime Mortars

Until relatively recently, all mortar was a mixture of lime, sand, and water. The lime was the cementing material. Sand-lime mortars harden at a slow, variable rate, develop low compressive strength, and have poor durability when subjected to freeze-thaw cycles. Therefore, the joints in many old masonry installation that were built using sand-lime mortar have deteriorated so much that there is almost no binder left. The sand can be raked from the joint with a finger.

However, sand-lime mortars have high workability and high water retention and they are more flexible than are portland cement mortars. Because lime hardens only when in contact with air, complete hardening of a lime mortar occurs very slowly, over a long period of time. This is a disadvantage in terms of project progress. The advantage of this slow hardening is that even when cracks occur they often "heal" spontaneously during the hardening process, which tends to help keep water penetration to a minimum.

Sand-lime mortar is seldom used today except in renovation work where an existing masonry was laid using a similar material.

4.2.2.2

Portland Cement Mortars and Grouts

Portland cement mortars and grouts (portland cement, sand, and water) harden quickly at a consistent rate and develop high compressive strengths. They also have good durability when subjected to freeze-thaw cycles, but their workability and water retention are low.

Portland cement grouts are used in the installation of the kind of architectural terra cotta called ceramic veneer (see Section 4.7.7). They may also be used in other types of masonry but are not permitted in engineered unit masonry walls.

(see Section 4.7.7)

4.2.2.3

Portland Cement–Lime Mortars and Grouts

To combine the advantages of portland cement and lime and to compensate for their disadvantages, and because their mixture produces the most predictable of mortars and grouts, portland cement–lime mortars and grouts (portland cement, lime, sand, and water) are widely used. In portland cement-lime mortar and grout, the portland cement contributes durability, high early strength, a consistent rate of hardening, and high compressive strength. Lime adds workability, water retention, and plasticity. Both contribute bond strength. Sand acts as a filler, making the mix economical, and adds strength. Water is the mixing vehicle. It creates plastic workability and initiates the cementing action.

4.2.2.4

Masonry Cement Mortars

Masonry cement mortars (masonry cement, sand, and water) are made using proprietary mortar mixes called masonry cement. Masonry cement manufacturers combine in one bag such ingredients as portland cement, portland blast-furnace slag cement, portland-pozzolan cement, natural cement, and slag cement. Masonry cement may also contain hydrated lime, limestone, chalk, calcareous shell, talc, or slag, and an air-entraining agent or other admixtures.

The principal advantages of using masonry cement mortar are that it is readily available, convenient to mix, and produces a mortar with generally good workability. The first of those advantages is being reduced today, however, by the availability (at least in some areas) of ready-mixed portland cement–lime mortars. These resemble ready-mixed concrete in that the ingredients are batched and mixed at a central location, and delivered to a project site ready to use.

The primary disadvantage of using masonry cement is that the standard for it, ASTM Standard C 91, does not limit the ingredients or their proportions, or control the types or quantities of admixtures a producer can include in it. Therefore, the performance of masonry cement is relatively unpredictable when compared with portland cement–lime mortars made in compliance with ASTM Standard C 270, which does limit the ingredients and admixtures in mortar. Although it is true that laboratory test results furnished by individual manufacturers for specific masonry cements can serve as a guide to performance, masonry cements are not recommended for use in reinforced masonry or in masonry that may have to withstand lateral forces or heavy loads.

4.2.3

MORTAR AND GROUT MATERIALS

ASTM Standard C 270, "Standard Specification for Mortar for Unit Masonry," which is the currently recognized standard of mortar for unit masonry work and stone, lists the required standards controlling the ingredients for mortar. ASTM Standard C 476, "Standard Specification for Grout for Reinforced Masonry," similarly lists the controlling standards for grout ingredients for use in engineered masonry. The listed standards are based on laboratory tests, field use, and many years of experience.

To comply with ASTM Standards C 270 and C 476, mortar and grout materials must conform to the individual ASTM standards given there. Many of the numbers of those standards are repeated in this section. The following list is a summary of those standards:

Portland cement (Types I or II. Type II may also be used for cold-weather construction. Air-entraining portland cement is also covered in ASTM C 150, but should not be used in mortars.)	ASTM C 150

Masonry cement	ASTM C 91
Quicklime	ASTM C 5
Hydrated lime (Type S)	ASTM C 207
Aggregates (sand) for use in mortar	ASTM C 144
Aggregates for use in grout	ASTM C 404

Some mortars and grouts, such as those used in terra cotta, may not be required by a code or other governmental regulation, or recommended by an industry standard, to comply with ASTM standards. Nevertheless, it is prudent to require that their ingredients comply with the individual ASTM standards listed in ASTM Standards C 270 and C 476 for mortar and grout, respectively.

4.2.3.1

Cementitious Materials

Cementitious materials used in mortar and grout include portland cement, masonry cement, quicklime, and hydrated lime.

■ PORTLAND CEMENT

Portland cement should comply with ASTM Standard C 150. There are three types recommended for use in mortar and grout. Type I is used when no special properties are required; Type II when moderate sulfate resistance or moderate heat of hydration is desired; and Type III when high early strength is needed, as might be the case in cold weather.

ASTM Standard C 150 also covers several air-entraining types of cement, but these should not be used in masonry mortar or grout because the amount of air entrained is often difficult to predict. Excess entrained air in a mortar reduces the bond between the mortar and masonry units laid with it.

■ MASONRY CEMENT

Masonry cements should conform to the requirements of ASTM Standard C 91. Masonry cements should not be used in grout. Where their use is permitted in mortar, more satisfactory results may be obtained by requiring the use of a product that has been specifically manufactured for the use intended. As an alternative, masonry cements may be specified by listing the acceptable properties and limiting the ingredients. Admixtures should also be limited to those that are permitted in mortars by ASTM Standard C 270.

Masonry cements are available with color pigments added at the factory. The advantage of using them when colored mortar is desired is that it is much easier to obtain uniformity of color when the pigments are added to the mix in the plant, rather than at the job site. Masonry cement is often selected for architectural projects for this very reason. Care must be taken, however, to ensure that the colored masonry cement selected will produce a mortar with the other properties required.

ASTM Standard C 91 permits the air content of masonry cement to go as high as 22%. When masonry cement is allowed, its air content is limited to 12%, since amounts above the lower level result in loss of bond strength.

▪ QUICKLIME

Quicklime (which is essentially calcium oxide, CaO) should conform to ASTM Standard C 5. It must be carefully mixed with water (slaked) and stored for as long as 2 weeks before use, and have the consistency of putty. Though used extensively in the past, quicklime is used in that form today only in preservation work to match an existing mortar.

▪ HYDRATED LIME

Hydrated lime must conform with ASTM Standard C 207. Hydrated lime is quicklime that has been slaked before packaging, converting the calcium oxide into calcium hydroxide, $Ca(OH)_2$. Hydrated lime, which can be mixed and used without delay, is much more convenient than quicklime. Hydrated lime is available in both Types S and N, but only Type S is recommended for masonry mortar and grout, because ASTM Standard C 207 does not control the amount of unhydrated oxides in Type N or its plasticity. Air-entraining hydrated lime is also available, but is not suitable for masonry mortars or grout.

Aggregates

The only aggregates suitable for use in masonry mortars are materials that comply with ASTM Standard C 144. The standard for aggregates for use in grout is ASTM Standard C 404. Either natural or manufactured aggregates may be used, but both must be clean, sound, and well graded. Manufactured aggregate may be ground stone, such as marble, granite, or other sound stone of the desired color. The aggregate used has an important effect on the workability and durability of a mortar or grout.

▪ AGGREGATE GRADATION FOR MASONRY- AND STONE-SETTING MORTAR

For the best workability and most economy, aggregate for general use should be graded uniformly to include all particle sizes from very fine up to 1/4 in. Aggregates with less than 5% very fine particles produce harsh, hard-to-work mortars that need additional cement or lime to become workable. Aggregates deficient in large particles generally result in weaker mortars. Nevertheless, aggregates for joints narrower than 1/4 in. wide should have 95% of its particles able to pass a No. 16 sieve and 100% able to pass a No. 8 sieve.

Many commercially available sands do not comply with the gradation limits set in ASTM Standard C 144 and should not be used in masonry mortar. Fortunately, the gradation of many commercial sands can be altered easily and inexpensively by adding fine or coarse sands as appropriate.

▪ AGGREGATE GRADATION FOR STONE JOINT-POINTING MORTAR

Aggregates for stone joint-pointing mortar should be very fine with no granules retained after passing through a No. 16 sieve.

▪ AGGREGATE GRADATION FOR GROUT

Grout for use in engineered unit masonry is divided by ASTM Standard C 476 into two categories, fine and coarse. The aggregate controlled by ASTM Standard

C 404 is similarly graded, and there are several subgrades within each category. The gradation of the aggregate for a particular grout is determined by the size of the grouted space and the height of the pour.

The aggregate recommended for use in grout associated with the kind of terra cotta called ceramic veneer (see Section 4.7.7) is pea gravel. This pea gravel is usually defined as aggregate that is uniformly graded from 1/4 in. size down to that which can pass through a No. 8 sieve. It should comply with the other requirements of ASTM Standard C 404.

Water

The water used in mortar and both kinds of grout should be clean and free from deleterious amounts of acids, alkalies, and organic materials. Water that is fit to drink is generally suitable for use in mortar or grout unless the water contains minerals that would be harmful to the mortar itself.

Materials for Colored Mortar

Mortar may be colored either by controlling the color of the ingredients or by adding pigments to the mix.

▪ COLORED AGGREGATE MORTAR

The preferred method of obtaining colored mortar is by using colored aggregate and cement. This method will usually give permanent color and will not weaken the mortar. Either white or gray portland cement may be used. The aggregates may be white or brown sand, or ground granite, marble, limestone, or other stone. For example, white sand, ground limestone, or ground marble may be used with white portland cement and lime to produce white joints. The disadvantage of this method, of course, is that the number of possible mortar colors is limited.

▪ COLORED PIGMENTED MORTAR

The second method available for producing colored mortar is to add pigments to the mix. Masonry cement is available with the pigments already added. However, for colored portland cement–lime mortars pigments must be added in the field.

Mortar pigments must be fine enough to disperse throughout the mix, be capable of imparting the desired color when used in permissible quantities, and must not react with other ingredients to the detriment of the mortar. It is usually possible to meet these requirements using metallic oxides. For example, iron, manganese, and chromium oxides have been used successfully. Carbon black and ultramarine blue have also been used successfully. Organic pigments should be avoided, as should colors containing Prussian blue, cadmium lithopone, and zinc and lead chromates.

The minimum quantity of pigment that will produce the desired results should be used, because any excess may seriously impair strength and durability. The American Cement Institute's (ACI) 530.1-88/ASCE 6-88, "Specifications for Masonry Structures," limits the amount of mineral oxide pigments to 10% by weight of the cement in portland cement–lime mortar and 5% of the weight of masonry cement in masonry cement mortar. It limits carbon black to not more than 2% of the weight of the cement in portland cement mortar and 1% of the weight of the masonry cement in masonry cement mortar. It permits no other pigments. Therefore, if black mortar is desired, the best pigment to use is black iron oxide.

For best results pigments should be premixed with portland cement in large, controlled quantities. Premixing large quantities will ensure a more uniform color than can be obtained by mixing smaller batches at the project site.

4.2.3.5

Admixtures

Some admixtures are permissible in mortar and grout. For example, color pigments are an admixture. Other available admixtures, some of which are permitted in some mortars and grouts, include air-entraining agents, water-repellent agents, accelerators, retarders, and antifreeze compounds. ASTM Standard C 270 mentions the use of calcium chloride as an accelerator, but ACI 530.1/ASCE 6-88 specifically prohibits the use of admixtures that contain chlorides in mortar and grout to be used in engineered masonry. Nonchloride accelerating admixtures are available, but if these are necessary, specific products should be selected that are known to give satisfactory results without damage to the properties of the mortar or grout.

ASTM Standard C 476 prohibits using any kind of antifreeze compound in grout used in engineered masonry and permits other admixtures only with specific approval from the purchaser of the grout. All admixtures should be laboratory tested to ensure that they will not be harmful to a mortar or grout.

Except for latex additives and colored pigments, additives should not be used in mortar that will come into contact with stone. Some additives will discolor or otherwise harm stone.

4.2.4

MORTAR TYPES, USES, AND PROPORTIONS

Mortar types are defined in ASTM Standard C 270, "Standard Specifications for Mortar for Unit Masonry."

4.2.4.1

Mortar Types

The four mortar types described in ASTM Standard C 270 (M, S, N, and O) are those recommended for use with unit masonry and stone. Of those, only Types M, S, and N are permitted in engineered unit masonry. The four types are as follows:

■ TYPE M

Type M mortar is a high-strength mortar that has somewhat greater durability than other mortar types. It is specifically recommended for use in reinforced masonry and where high compressive strength is necessary. A reinforced masonry wall that is below grade, such as a foundation or retaining wall, is a typical example. It is also used extensively for walks, sewers, and manholes.

Type M mortar is used to set stone in paving installations.

■ TYPE S

Type S mortar is a medium-high-strength mortar recommended for use where bond and lateral strength are more important than high compressive strength. Tensile bond strength between brick and Type S mortar approaches the maximum obtainable with portland cement–lime mortars. Type S mortar is recommended for use in reinforced masonry and in unreinforced masonry where maximum flexural strength is required.

Type S mortar is used in setting beds for granite, marble, and slate in installations other than paving.

■ TYPE N

Type N mortar is a medium-high-strength mortar recommended for general use in exposed masonry above grade where high compressive and lateral strengths are not required. It is specifically recommended for chimneys, parapet walls, and exterior walls subject to severe exposure.

Type N mortar should not be used to set limestone in paving installations.

■ TYPE O

Type O mortar is a low-strength mortar suitable only for general interior use in non-load-bearing masonry and for tuck pointing. Type O mortar should never be used where it will be subjected to freezing conditions.

4.2.4.2

Mortar Uses

No single mortar type is best for all purposes. The basic rule to follow in selecting a mortar type is never to select one that is stronger in compression than it needs to be to satisfy the structural requirements of the wall.

Selection of the proper mortar type for each unit masonry use should be based on the type of masonry and its location in the building (Fig. 4.2-3). No mortar type rates highest in all desirable properties. Adjustments in the mix to improve one property often are made at the expense of others. For this reason, the properties of each mortar type should be evaluated and

FIGURE 4.2-3
Recommended Mortar Types for Various Construction Applications

Construction Application	Recommended Minimum ASTM Mortar Types	Order of Relative Importance of Principal Properties		
		Plasticity[a]	Compressive Strength	Weather Resistance
Foundations, basements, walls, isolated piers[b]	M, S[c]	3	2	1
Exterior walls	S, N	2	3	1
Solid masonry unit veneer over wood frame	N	2	3	1
Interior walls, load bearing	S, N	1	2	3
Interior partitions, non–load bearing	N, O	1	—	—
Reinforced masonry (columns, pilasters, walls, beams)	M, S[c]	3	1	2

[a] Adequate workability and a minimum water retention (flow after suction of 70%) assumed for all mortars.

[b] Also any masonry wall subject to unusual lateral loads for earthquakes, hurricanes, etc.

[c] Only portland cement-lime Type S and M mortars.

Courtesy Portland Cement Association.

the type chosen that will best satisfy the end-use requirements. Figure 4.2-3 shows the recommended mortar types for six common construction applications. Additional information about selection of mortar types can be found in the Brick Institute of America's (BIA) Technical Notes, specifically Technical Notes Nos. 8, 8A, and 8B.

In addition to the general uses suggested in Section 4.2.4.1, some specific cases should be mentioned. The following paragraphs suggest the mortar types usually selected for the specific conditions indicated. Mortar types for brick paving and flooring and for stone are addressed in Sections 4.2.8, 4.2.9, and 4.2.10.

■ CAVITY WALLS

In cavity walls where wind velocities will exceed 80 miles per hour, Type S mortars should be used. In locations where winds of lower velocity are expected, either Type S or Type N may be used.

■ FACING TILE

For facing tile, M, S, N, or O type is usually acceptable. ASTM Standard C 144 allows the use of sand particle sizes as large as 1/4 in. However, facing tile is often laid with 1/4-in.-wide joints. In that case, the aggregate used must pass through a No. 16 sieve, which allows aggregates to pass that are slightly less than 1/16 in. in size.

■ DIRT-RESISTING MORTAR

For dirt resistance, and when medium resistance to staining is needed, aluminum tristerate, calcium stearate, or ammonium stearate is sometimes added to a masonry mortar or to the mortar in the form of terra cotta called ceramic veneer (see Section 4.7.7). When this is done, the amount should be limited to that amount dictated by code, if applicable, or the amount recommended by the masonry unit manufacturer. In the absence of a code restriction or manufacturer's recommendation, the amount should be limited to not more

than 3% of the weight of the portland cement in the mortar.

Where maximum dirt resistance is desired, as for facing tile, the mortar should be raked out to a depth of about 3/8 in. and the joints pointed with a mortar consisting of 1 part portland cement, 1/8 part lime, and 2 parts graded fine (80 mesh) sand, by volume, to which has been added aluminum tristerate, calcium stearate, or ammonium stearate in amounts equal to about 2% of the weight of the portland cement in the mortar.

Admixtures should not be used in mortar for engineered masonry unless specifically allowed by ACI 530-88/ASCE 5-88, "Building Code Requirements for Masonry Structures," and ACI 530.1/ASCE 6-88, "Specifications for Masonry Structures," and by the applicable building code.

■ TUCK-POINTING MORTAR

Tuck pointing is the refilling of an existing joint from which some of the mortar has been removed. Tuck-pointing mortar must be prehydrated. To prehydrate a mortar, the ingredients are thoroughly mixed without water, then mixed again, adding only enough water to produce a damp, unworkable compound that will retain its form when pressed into a ball. After 1 to 2 hours, enough water is then added to make the mortar workable; the resulting mortar should be used immediately.

For best results, tuck-pointing mortar should have the same ingredients and mixture as the original mortar. When the original mortar type is unknown, prehydrated Type N mortar is often used.

■ TWO DIFFERENT MORTARS IN THE SAME JOINT

Sometimes it is helpful to place two different mortars in the same joint. Both these mortars should be the same type, however, and masonry cement and portland cement–lime mortars should not be used together. Two different mortars might be installed, for example, where white, black, or colored mortar is desired. One option, of course, is to use the same mortar throughout. However, white, black, and colored mortars are usually more expensive than natural colored mortars, and their properties are not as consistent or as predictable as are those of mortars that do not contain pigments. An alternative is to lay the unit masonry in a normal, natural colored mortar. After the units have been laid, rake the joints back to a depth of about 1/2 in., and then point them with the desired white, black, or colored mortar.

4.2.4.3

Mortar Proportions

The ingredient proportions required by ASTM Standard C 270 for the mortar types listed in Section 4.2.4.1 are shown in Figure 4.2-4.

The mortar used in the type of terra cotta called adhesion-type ceramic veneer is usually required to have a mixture consisting of 1 part portland cement, 1/2 part lime putty or hydrated lime, and 4 parts sand, by volume.

4.2.5

GROUT TYPES, USES, AND PROPORTIONS

In ASTM Standard C 476, "Standard Specification for Grout for Reinforced Masonry," grout is classified as either fine or coarse. The only difference between the two is the gradation of the aggregate. They both contain 1 part portland cement, 0 to 1/10 part hydrated lime or lime putty, and damp, loose fine aggregate in an amount between 2-1/4 and 3 times the sum of volumes of the cementitious materials. In addition, coarse grout contains coarse aggregate in an amount equal to between 1 and 2 times the sum of the volumes of the cementitious materials. Both grout types are used to fill reinforced masonry walls, tie beams, lintels, and other reinforced masonry items.

Grout for use in anchored-type ceramic veneer is usually required to have 1 part portland cement, 1 part sand, and 5 parts pea gravel.

4.2.6

MORTAR AND GROUT MIXING

Thorough mixing is important to the development of desirable properties in a

FIGURE 4.2-4
Mortar Proportions by Volume[a]

Type[b]	Portland Cement	Hydrated Lime or Lime Putty	Masonry Cement	Maximum Damp Loose Aggregate[c]	Minimum Compressive Strength, 2-in. Cubes at 28 Days, psi
M	1 or	1/4	—	3	2500
	1	—	1	6	2500
S	1 or	1/2	—	4-1/2	1800
	1/2	—	1	4-1/2	1800
N	1 or	1	—	6	750
	—	—	1	3	750
O	1	2	—	9	350

[a] The weight of 1 cu. ft. of materials used is considered to be: Portland cement, 94 lb. (1 bag); hydrated lime, 40 lb.; lime putty, 80 lb.; dry sand, 80 lb.; masonry cement, weight printed on bag.

[b] Shaded figures show proportions for portland cement–lime mortar. Mortars made with masonry cement are unshaded.

[c] The damp loose aggregate should be not less than 2-1/4 times, nor more than 3 times, the sum of the cementitious material used.

mortar or grout. Because some properties begin to deteriorate soon after a mortar or grout is mixed, no more should be mixed than can be placed before hydration (curing) begins.

Water is essential to the development of bond strength in mortar (see Section 4.2.1.2). If a mortar contains too little water, the bond will be weak and spotty. Therefore, a mortar should be mixed with the maximum amount of water possible, as long as it does not become so soupy that it will not stay where placed, extrude from the joint under the weight of the masonry units, or otherwise become unusable by a mason.

Grout should be mixed with sufficient water to produce a slump of between 8 in. and 11 in. Retempering should not be necessary. Grout should be used within 1-1/2 hours after water was first added to the mix.

ASTM Standards C 270 and C 476 give the proportions of the ingredients in mortars and grouts in terms of parts by volume. Ingredients may be measured in the field either by volume or by weight. In either case the proportions of the ingredients must be accurately controlled if a predictable mortar or grout is to be

produced. Both ASTM Standards C 270 and C 476 contain estimated weights per cubic foot of mortar ingredients.

Whether volume or weight measure is employed, containers of known measure must be used. Cement is usually added directly from the bag as full or half bags, depending on the amount of mortar or grout needed. The materials must also be in the condition required by ASTM Standards C 270 and C 476. For example, moisture in sand causes it to bulk or fluff up, which can make a cubic foot of wet sand contain a different amount of sand than assumed by the design tables in ASTM Standards C 270 and C 476. ASTM Standards C 270 and C 476 therefore both require that sand be in a damp loose condition.

Oversanding mortar occurs because the sand is usually shoveled rather than carefully added using a measured container. Oversanded mortar is harsh and unworkable, provides a weak bond with masonry units, and tends to erode. The BIA suggests several accurate methods for measuring sand. Refer to BIA Technical Note 8B, "Mortar for Brick Masonry, Selection and Controls," for specifics.

Machine Mixing

Except on very small projects, mortar should always be mixed in a mechanical batch mixer designed for the purpose. Grout should always be mixed in a mechanical batch mixer regardless of the size of the project. The size of the mixer needed will depend on the amount of mortar to be mixed.

In machine mixing, mortar and grout ingredients should be added in small enough batches to mix properly. Water should be added slowly until the proper consistency has been reached. After all the ingredients have been added, the mortar or grout should be mixed a minimum of 3 minutes and a maximum of 5 minutes. The mixer drum should be completely empty before it is recharged for the next batch.

4.2.6.2

Hand Mixing

If mortar must be mixed by hand, the sand should first be spread in the mortar box, the cement and other cementitious materials on top of the sand, and then be mixed well with a hoe from both ends of the box. About three-quarters of the required water should then be added and mixed until the cementitious ingredients and aggregates are uniformly damp. Water should be added in small amounts and mixing continued until the mortar reaches a suitable workability. The batch should be allowed to stand for about 5 minutes and remixed thoroughly with the hoe without additional water.

4.2.6.3

Cold-Weather Mixing

The temperature of a mortar or grout when applied should be between 70° and 120°F when the outside temperature is below 40°F. Higher mortar temperatures may result in fast hardening, which can make it impossible to lay masonry units with that mortar. Even when the units can be laid, the mortar may have reduced compressive and bond strengths.

Heating the mixing water is one of the easiest methods of raising the temperature of a mortar or grout. Mixing water should

not be heated above 160°F, however, because a flash set can occur when very hot water comes in contact with cement.

In freezing weather, the moisture that is almost always present in masonry sand will freeze. Therefore, sand must be heated, slowly and carefully to prevent scorching, to remove this ice.

The use of calcium chloride and other antifreeze admixtures to lower the freezing point of mortar or grout is not permitted by most present-day codes and standards. The BIA also discourages their use. The quantity of such materials necessary to lower the freezing point of mortar appreciably is so large that mortar strength and other desirable properties would be seriously impaired. Such materials not only hasten the corrosion of metal anchors and reinforcement, they weaken mortar, contribute to efflorescence, and may even cause the masonry to spall when the chemicals dissolve in water and then recrystallize.

For additional information concerning cold-weather masonry construction, refer to BIA Technical Notes Nos. 1 and 1A.

4.2.7

RETEMPERING

When mortar is not used immediately after it is mixed, as is often the case, some of its water may evaporate. Adding water, which is called retempering, is sometimes prohibited because it is thought to have detrimental effects. Although it is true that retempering slightly reduces a mortar's compressive strength, failing to retemper mortar that needs it may result in a materially lowered bond strength. Therefore, retempering should be permitted to replace water lost by evaporation, providing that the mortar has not begun to set (Fig. 4.2-5). Regardless of whether it has been retempered, mortar should be placed within 2-1/2 hours after its initial mixing, and no mortar should be used after it has begun to set.

A change in the time between the spreading of a mortar and the placing of masonry units into it will affect the flow of the mortar, especially if the masonry units have a high suction. The highest bond strength is achieved when this time interval is minimal. Varying this interval to any large extent will result in a wall with varying tensile bond strengths.

FIGURE 4.2-5 *Retempering of mortar. (Courtesy Portland Cement Association.)*

4.2.8

BRICK PAVING AND FLOORING MORTAR AND GROUT

Thick-set portland cement mortar beds for brick paving and flooring should be a mixture of portland cement, hydrated lime, and fine aggregate. This material is mixed with water, but a latex additive often replaces some or all of the water to provide better adhesion between the brick and the setting bed and to produce a more flexible setting bed than can be accomplished using water alone. Setting bed mortar should comply with ASTM Standard C 270, either Type S or Type N; Type S is preferable in most instances for interior use, and Type M for exterior use.

Thin-set brick flooring is set in a bed of either dry-set portland cement mortar conforming with American National Standards Institute ANSI A118.1 or latex portland cement mortar conforming with ANSI A118.4.

Grout for brick pavers should be a mixture of portland cement and aggregate mixed with water or a latex additive; a dry-set grout complying with ANSI A118.6; or a portland cement grout complying with ANSI A118.6.

4.2.9

STONE FLOORING MORTAR AND GROUT

Stone flooring is installed using either thin-set or thick-set beds in accordance with ANSI standards. Thin stone tiles are, however, usually set using thin-set methods.

Setting beds for stone flooring are usually portland cement mortar that complies with ANSI A108.1. Grout may be either sand-portland cement, commercial portland cement, dry-set, or latex portland cement type.

4.2.10

STONE PAVING

Stone paving bed mortar should be either Type M portland cement–lime mortar conforming to ASTM Standard C 270 or a mixture of 1 part portland cement and 3 parts sand mixed with water. Part or all of the water is sometimes replaced with a latex admixture to make the mortar bond more readily to the stone, to make the setting bed somewhat more flexible, and to reduce water penetration through the mortar. Adding a latex admixture increases the setting time of a bed and may not be advisable in all cases, especially if the additive also contains a retarder.

Pointing mortar for stone paving should be a mixture of 1 part portland cement and 2 parts aggregate. The aggregate may be natural or ground white sand, ground marble, granite, or other stone to produce the desired color.

4.3 REINFORCEMENT, TIES, ANCHORS, AND FLASHING

Depending on materials, configuration, and location, masonry and stone walls require various types of reinforcement, ties, anchors, and flashing.

4.3.1

JOINT REINFORCEMENT, TIES, AND ANCHORS FOR UNIT MASONRY

Many types of joint reinforcement, ties, and anchors are used in unit masonry construction. Some of them are shown in Figure 4.3-1. The use of these items is discussed and illustrated in Sections 4.7.2.2 and 4.7.2.3.

Metal joint reinforcement, ties, and anchoring devices in masonry or stone walls that are exposed to salt-laden water will tend to corrode. Four inches or more of masonry cover may offer some protection, but copper-coated steel, hot-dip galvanized (zinc coated) steel, or stainless steel should be employed. Even though stainless steel is more expensive than the other two, it offers the highest degree of safety and is essential in stone construction and preferable in unit masonry construction. The uses, types, and relative ASTM standards for masonry joint reinforcement, anchors, and ties are summarized in Figure 4.3-2.

Requirements for some types of joint reinforcement, ties, and anchors in unit masonry walls are as follows:

Joint reinforcement. Joint reinforcement for use in horizontal courses of interior masonry and other masonry protected from the weather should be fabricated from zinc-coated cold-drawn steel wire (ASTM Standard A 82) or Type 304 stainless steel wire (ASTM Standard A 580). Joint reinforcement for exterior walls should be fabricated from Type 304 stainless steel (ASTM Standard A 580). Joint reinforcement should consist of two smooth or deformed longitudinal wires 9 gauge or larger, and the weld connected with 12-gauge wires. The distance between weld contacts of the cross-wires should not exceed 6 in. for smooth longitudinal wire or 16 in. for deformed longitudinal wire. The width of the joint reinforcement should be such that it can be embedded in a mortar joint with its wires not closer than 5/8 in. to the masonry face surface. The configuration of joint reinforcement may be ladder design, truss design, or tab design (see Fig. 4.3-1).

Wire mesh ties. Wire mesh ties or hardware cloth are used to anchor non-load-bearing partitions to intersecting masonry. They should be made from at least 15-gauge zinc- or copper-coated steel wire, have a 1/2-in. mesh, and be at least 12 in. long. Their width should be 1 in. less than the wall thickness.

Cavity wall ties. Ties for use in exterior cavity walls should be 3/16-in.-diameter, type 304 stainless steel wire (ASTM Standard A 580) formed either in a rectangular shape not less than 4 in. wide (for use with hollow units having vertical cells) or in a Z shape with 2-in. legs (for use with solid units or hollow units with horizontal cells).

Ties should be long enough to permit their ends to be embedded in the outer face shell mortar beds of hollow units or the center of mortar beds of solid units.

Joint reinforcement comprising 9-gauge or larger cross-wires may be used instead of metal ties, provided the cross-wires are spaced to render strength and stiffness per square foot of wall area equivalent to that of 3/16-in.-diameter Z ties used at the maximum spacing.

Faced wall ties. Ties used in faced walls should meet the same requirements given above for cavity wall ties.

Veneered wall ties, wood studs. The ties usually used for brick veneer over wood studs are 22-gauge, 6-in.-long, 7/8-in.-wide corrugated galvanized steel. These are usually bent up and nailed through the sheathing into the underlying studs.

Veneered wall ties, metal studs. Ties for masonry veneer over metal studs should be triangular, flexible, stainless steel wire ties similar to those used to anchor masonry to steel structural members. The fixed portion of the ties should be screwed through the sheathing to the studs. The flexible triangular portion of the tie should then be threaded through the support and built into the masonry as it is erected.

Dovetail anchors. Dovetail anchors for use with embedded slots or inserts in concrete

FIGURE 4.3-1 *Joint reinforcement, ties, and anchoring devices.*

FIGURE 4.3-2
Joint Reinforcement, Ties, and Anchoring Device Materials and Uses

Use	Material	ASTM Standard
Interior wall joint reinforcement	Copper-coated wire	B 227
	Zinc-coated wire	A 116
Exterior wall joint reinforcement	Stainless steel wire	A 580, Type 304
Reinforced masonry reinforcement	Billet steel bars	A 615
	Epoxy-coated billet steel bars	A615 and A 775
	Deformed wire	A 496
	Welded wire fabric	
	Plain	A 185
	Deformed	A 497
Interior wall anchors and ties	Galvanized carbon steel wire	A 82
	Galvanized steel sheet	A 526, G60
Exterior wall anchors and ties	Stainless steel wire	A 580, Type 304
	Galvanized steel sheet	A 366 and A 153
	Stainless steel sheet	A 167, Type 304
Rigid anchors	Heavy-thickness galvanized steel sheet	A 635 and A 525
	Plates and bars	
	Steel	A 36
	Painted	2 coats epoxy paint
	Galvanized	A 153
	Stainless steel	A 666, Type 304

should be 16-gauge, zinc-coated sheet steel, 7/8 in. wide, and long enough to permit the ends of the tie to be embedded in the outer face-shell mortar beds of hollow units or the center of mortar beds in solid units.

Rigid steel anchors. Rigid steel anchors for anchorage of load-bearing walls and fire walls should be 1-1/4 in. × 1/4 in. steel with the ends turned up not less than 2 in. They should be not less than 16 in. long for 8-in.-thick walls and not less than 24 in. long for 12-in.-thick walls.

4.3.2

STRUCTURAL REINFORCEMENT FOR UNIT MASONRY WALLS

Structural reinforcement in unit masonry walls consists mainly of steel bars and welded wire mesh.

Anchors should be Stainless Steel

strap anchors
(Recommended
1/8" & 3/16" x 1" or 1 1/4" wide)

rod anchors
(Recommended 3/8" Ø min.)

dovetail anchors
(Recommended
1/8" & 3/16" x 1" or 1 1/4" wide)

miscellaneous anchors

Epoxy Anchors

expansion anchors
follow manufacturer's
recommendations on use.

special anchors
Bracket mounted to
masonry or steel

FIGURE 4.3-3 *Stone anchors.* (*Courtesy Indiana Limestone Institute.*)

FIGURE 4.3-4 *Through-wall flashing in a concrete masonry wall and a rope-wick weep hole.*

4.3.3

REINFORCEMENT, TIES, ANCHORS, AND SUPPORT SYSTEMS FOR STONE

Various types of reinforcement, ties, anchors, and grid support systems are used in stone construction depending on the type of stone and its use. All such devices that come into contact with stone should be fabricated from American Iron and Steel Institute (AISI) Type 304 stainless steel (ASTM Standard A 666). Devices and support components that are not in direct contact with stone may be fabricated from hot-dip galvanized steel (ASTM Standard A 36 for materials, ASTM Standard A 123 for galvanizing). Some of the various types of devices used are shown in Figure 4.3-3.

4.3.4

FLASHING

Flashing is installed at vulnerable locations in masonry to help prevent water from entering and to collect water that has already entered and to divert it through weep holes to the exterior (Fig. 4.3-4).

Flashings are formed from sheet metal, bituminous fabric membranes, plastic sheet materials, or a combination of these materials. Flashing materials are discussed in Section 7.9.1.

Because replacement costs can exceed the initial installed costs, it is prudent to select a durable flashing material in the first place. Recommendations for locating and installing flashing are discussed in Section 4.7.2.8.

4.4 CLAY MASONRY UNITS

Clay masonry units are burned clay units used in buildings. They include brick, hollow clay tile, and architectural terra cotta. They do not include ceramic tile, flue linings, drain tile, or other ceramic products. Ceramic tile and related materials are discussed in Section 9.6.

Solid clay masonry units, such as brick, and hollow clay masonry units, such as tile, are composed of burned clay, shale, fire clay, or a mixture of these, formed into the desired shape while plastic and then burned in a kiln.

Terra cotta is Italian for "cooked earth." It is composed of an aged clay to which admixtures and sand or pulverized previously fired clay, called grog, are added. It is extruded or formed into the desired shape by hand packing and then fired at a high temperature to produce a material that is harder and more compact than most brick.

Both the raw materials used in making clay masonry and the way in which it is manufactured have a marked influence on the properties of the units in a finished wall. Therefore, a general understanding of both materials and methods is essential to the selection and use of brick and tile.

4.4.1

RAW MATERIALS

To satisfy production requirements, the clays used in making clay masonry units must have enough plasticity so that they can be shaped or molded when mixed with water. They must also have sufficient tensile strength to maintain their shape after they have been formed. And their particles must fuse when they are subjected to high temperatures in a kiln. The types of clays used to manufacture clay masonry units take three principal forms having similar chemical compositions but different physical characteristics.

1. *Surface clays*. These are found near the surface of the earth and are of a sedimentary character.

2. *Shales*. These are clays that have been subjected to high pressures until they have hardened almost to the form of slate.

3. *Fire clays*. These are found deeper in the ground than the other types of clays used. They have refractory qualities, which makes them resistant to high temperatures and useful in such applications as fire brick. They usually have fewer impurities and more uniform chemical and physical properties than surface clays or shales.

All are compounds of silica and alumina with varying amounts of metallic oxides and other impurities. The metallic oxides act as fluxes, which promote fusion at lower temperatures than would be possible without them. They also influence the color of finished units.

There are always slight differences in the properties of any two finished clay masonry units, regardless of how hard their manufacturer tried to minimize variations. Some of the methods used to attempt standardization of the properties of units include mixing clays from different locations in a pit, mixing clays from different pits, and making compensations during the manufacturing processes. None of these is completely successful.

MANUFACTURE OF CLAY MASONRY UNITS

In general, the manufacturing process (Fig. 4.4-1) consists of mining (also called winning), preparing, and storing the raw materials; and forming, drying, glazing, burning, cooling, drawing, and storing the finished units.

Mining (Winning), Preparation, and Storage

In a process called *winning*, raw clay is taken from either open pits or underground mines. Raw clay that does not contain large lumps or stones is immediately passed through an inclined vibrating screen to control its particle size, and then stored.

The clay is crushed to break up lumps, and stones are removed. Then the clay is pulverized and mixed by large grinding wheels, each weighing 4 to 8 tons, that revolve in a circular pan. This pulverized clay is then screened and stored for later use.

The clay from a single pit is often placed in several storage areas to permit later blending. Blending produces more uniformity in raw clay, helps to control the color of finished units, and permits some control over the suitability of the raw material for manufacturing a given type of unit.

Forming

The first step in the forming process, tempering, produces a homogeneous, plastic mass ready for molding. Tempering is usually achieved by adding water to the clay in a mixing chamber called a pug mill, which contains one or two revolving shafts with blades (Fig. 4.4-2a). Three methods are used to form clay masonry units: the stiff-mud, the soft-mud, and the dry-press processes.

■ STIFF-MUD PROCESS

The stiff-mud process is the most commonly used method for forming brick and clay tile. All structural clay tiles and many

FIGURE 4.4-1 *Manufacturing flow chart.* (*Courtesy Brick Institute of America.*)

FIGURE 4.4-2 *(a) Mixing powdered clay with water. (b) Clay emerging from a de-airing machine. (c) Mixture being extruded through a die that molds it into a long, flat column. (Courtesy Brick Institute of America.)*

bricks are made this way, and a similar method is used for extruded terra cotta. In the stiff-mud process, clay is mixed with sufficient water (12% to 15% by weight) to produce the desired degree of

plasticity. After thorough mixing, called pugging, the tempered clay goes through a de-airing machine in which a partial vacuum is maintained. De-airing removes air, thus eliminating holes and bubbles. This improves the clay's plasticity and workability and results in greater strength in finished units (Fig. 4.4-2b).

After de-airing, the clay is forced (extruded) through a die, which produces a clay column of the desired shape (Fig. 4.4.2c). Here, various textures may be applied. The column then passes through an automatic cutter (Fig. 4.4-3). Die sizes and the spacing of cutter wires must be calculated to compensate for normal shrinkage during drying and burning.

After they are cut from the column, individual units are inspected on a continuously moving belt. Acceptable units are removed from the belt as it passes and placed on dryer cars (Fig. 4.4-4). Imperfect units are returned to the pug mill.

■ SOFT-MUD PROCESS

The soft-mud process is used only for producing brick. It is particularly suitable for clays that contain too much natural water (20% to 30% by weight) to be satisfactorily extruded in the stiff-mud process. In the soft-mud process, raw clay is mixed with water and the mixture is then formed

FIGURE 4.4-3 *Forming and cutting a clay column in the stiff-mud process. (Courtesy Brick Institute of America.)*

into units in molds. To prevent this clay from sticking, the molds are lubricated with sand and water, or water only. When sand is used, the brick are called *sand-struck* (Fig. 4.4-5h); if water is used they are called *waterstruck* (see Fig. 4.4-5g).

A method similar to the soft-mud process for forming brick has been used to make molded terra cotta shapes, except that a high water content is not a prerequisite.

FIGURE 4.4-4 *Bricks on a belt. Core holes of various sizes are cut through the body of bricks as they are extruded from the die to make them lighter and enable them to burn more uniformly. (Courtesy Brick Institute of America.)*

a. Smooth

b. Matt-vertical markings

c. Matt-horizontal markings

d. Rugs

e. Barks

f. Sandmold

g. Waterstruck

h. Sandstruck

FIGURE 4.4-5 *Typical brick textures. These might not be available in all locations.*
(Courtesy Brick Institute of America.)

4.4.2.4

Ceramic Glazing

Ceramic glazing is the process used to produce the finish on glazed facing tile, glazed facing brick, and glazed terra cotta. It is a specialized, controlled procedure in which glazes are sprayed onto masonry units and then fired. Spray glazes are composed of mineral ingredients that fuse into a glasslike coating at a given temperature. They are available in almost any color. There are two basic variations of the process: high-fired glazing and low-fired glazing.

■ HIGH-FIRED GLAZING

In the high-fired process, the glazes are sprayed onto the masonry units sometimes before and sometimes after drying. The units are then kiln-fired at normal firing temperatures.

■ LOW-FIRED GLAZING

Some glazed colors cannot be produced at high temperatures, so low-fired glazing is used. In this process, the glazes are applied after the units have been burned to maturity and cooled. Then the units with the glazes applied are fired again at a lower temperature.

4.4.2.5

Burning and Cooling

Burning is the process of baking or firing dried clay masonry units in kilns to produce finished units. Cooling is the process of returning the masonry units to atmospheric temperature. The two chief types of kilns used are *periodic kilns* and *tunnel kilns* (Fig. 4.4-6a and b). Both types can be fueled by natural gas, oil, or coal.

It takes 2 to 5 days to burn brick and clay tile units, depending on the kiln used and other variables. The burning time for terra cotta may be a little shorter.

During burning and cooling the temperatures in a kiln must be raised and lowered at controlled rates. To permit careful control of temperatures and their rates of change, kilns are equipped with devices to permit the constant monitoring and regulating of their interior temperatures.

■ DRY-PRESS PROCESS

The dry-press process is the method used when clay has a very low plasticity. This clay is mixed with a minimum of water (up to 10% by weight) and then formed in steel molds under pressure of between 500 and 1500 pounds per square inch (psi).

4.4.2.3

Drying

When wet clay masonry units come from molding or cutting machines, they contain from 7% to 30% moisture, depending on the forming method. Before brick or clay tile units are burned, most of this water is evaporated in dryer kilns at temperatures ranging from 100 to 300°F. Drying times vary from 24 to 48 hours depending on the type of clay used in the units. During this time, heat and humidity must be regulated to avoid excessive cracking.

Terra-cotta units must be thoroughly dried before they are fired. Unless these units are bone dry at the time of firing, they will crack or explode.

FIGURE 4.4-6 *Until recent years bricks were fired in (a) periodic (beehive) kilns. Today, most bricks are fired in (b) tunnel kilns.* *(Courtesy Brick Institute of America.)*

times subjected to a reducing atmosphere (one that contains insufficient oxygen for complete combustion) near the end of the burning process. The colors and shades produced by this so-called flashing will vary with the type of clay used in the units.

Cooling begins after the temperature in a kiln has reached the maximum needed to properly burn the units in it. The rate of cooling must be controlled, because it has a direct effect on the color of the units, and because excessively rapid cooling will cause the masonry units to crack or check. In tunnel kilns, cooling seldom requires more than 48 hours, but in periodic kilns, proper cooling will usually require at least 48 hours and sometimes as much as 72 hours.

4.4.2.6

Drawing

Drawing is the process of unloading a kiln after cooling (Fig. 4.4-7). During drawing, units are sorted and graded. They are then taken to a storage yard or loaded directly onto rail cars or trucks for delivery.

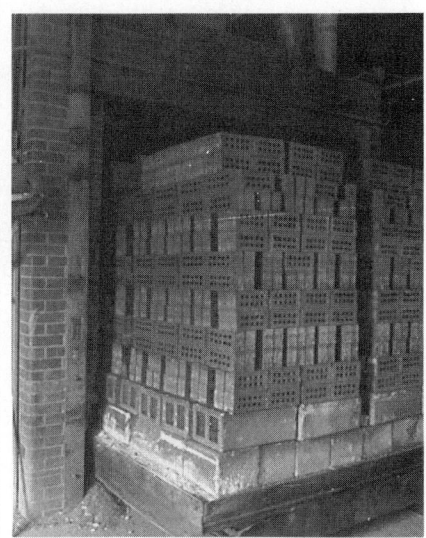

FIGURE 4.4-7 *Unloading a kiln.* *(Courtesy Brick Institute of America.)*

4.4.3

PROPERTIES OF CLAY MASONRY UNITS

The properties of finished units depend on the characteristics of the raw materials and the effects of the manufacturing process.

The burning process begins when dried masonry units are loaded into kilns. In periodic kilns (see Fig. 4.4-6a) the units are placed in the kiln in a pattern that permits free circulation of hot gases. In tunnel kilns (see Fig. 4.4-6b) the units are placed on special cars that pass through various temperature zones as they traverse the tunnel.

Burning may be divided into six stages: the evaporation of free water (water-smoking); dehydration; oxidation; vitrifi-

cation; flashing; and cooling. The first four are associated with rising temperatures in the kiln. The actual temperatures used in the burning process will vary somewhat depending on the type of clay. In general, for brick and clay tile units water-smoking takes place at about 400 °F, dehydration at between 300° and 1800 °F, oxidation at between 1000° and 1800 °F, and vitrification at between 1600° and 2400 °F.

In order to produce different colors and shades of colors, masonry units are some-

Compressive Strength

The clay, the method of manufacturing, and the degree of burning affect compressive strength. With some exceptions, the plastic clays used in the stiff-mud process have higher compressive strengths when burned than do the clays used in the soft-mud or dry-press process. For a given clay and method of manufacture, higher compressive strengths are associated with higher burning temperatures. The compressive strength of brick varies from 1500 to more than 20,000 psi, resulting mainly from the wide variation in the properties of the clays used.

4.4.3.2

Absorption

The degree of water absorption also depends on the clay, manufacturing method, and temperature of burning used. Plastic clays and higher burning temperatures generally produce units having lower absorption. Generally, the stiff-mud process produces units with lower absorption than either the soft-mud or dry-press process. A brick's initial rate of absorption, which is called suction, results from pores or small openings in the burned clay that act as capillaries to draw or suck water into the unit. Suction has little bearing on the transmission of free water through a brick, but it has an important effect on the bond between brick and mortar. Maximum bond strength and minimum water penetration are obtained when the suction of the brick at the time of laying does not exceed 20 grams (0.7 oz.) of water per minute. Bricks having suction of more than 20 grams per minute should be wetted before they are laid to reduce the suction. Section 4.7.1.2 contains more details about acceptable degrees of absorption and describes a field test that will help determine suction.

4.4.3.3

Durability

The durability of clay masonry units is affected by pore size within the units and by the amount and completeness of fusion during burning. In general, higher burning temperatures produce harder clay masonry units.

A major weathering action that affects burned clay masonry units is alternate freezing and thawing in the presence of moisture. The ability of clay masonry units to withstand freeze-thaw cycles without disintegration is important in every cold climate, but is especially so where the annual average precipitation exceeds 20 in. of water and alternate freezing and thawing is common. Units produced from the same raw materials and by the same method of manufacture may have different resistances to damage by freeze-thaw cycles depending on their compressive strengths and absorption. Units with high compressive strength and units with low absorption are usually the more resistant to damage during freeze-thaw cycles.

A second factor that affects the durability of clay masonry units is their soluble salts content. Such salts will dissolve when water is present and crystallize either on the surface of the masonry units or within them. Surface deposits (efflorescence), although unsightly, are not damaging. Internal crystallization (cryptoflorescence), however, will build up until the surface of the masonry units spalls. For this reason, brick that will be exposed to water in any form should be tested and rated *not effloresced* in accordance with ASTM Standard C 67.

4.4.3.4

Natural Color

Burned clays produce a wide range of colors, from the pure tones of pearl grays and creams through buff, and in a descending scale of reds to purple, maroon, and gunmetal black. The clay's chemical composition, the burning temperatures, and the method of burning all control color. Terracotta clays produce units that vary widely in color from red to brown to white.

Clays used in masonry units contain varying degrees of many metallic oxides. Of these, iron oxides probably have the greatest effect on color. Clay containing iron in almost any form will burn red when exposed to an oxidizing fire because the iron combines with the oxygen to form ferrous oxide, which is the same as red rust. When clay containing iron is burned in a reducing atmosphere (flashed), it will take on a purple cast.

Lighter colors are produced when a clay is underburned. This is the method used to produce salmon-colored brick from a clay that would ordinarily burn red. Underburned clay masonry is softer, more absorptive, and has a lower compressive strength than units that are burned completely.

Overburning clay produces clinker brick, which is dark red to black when the clays used are those that produce red units when burned normally, and dark speckled brown when the clay used would produce buff units when burned normally. Clinker bricks are ordinarily used for aesthetic reasons and for their charm.

With few exceptions, the natural color of clay masonry units comes from mixtures of shades rather than pure colors. Therefore, the accepted practice in specifying color is to require that the color of the delivered units match an approved sample panel made up of several units representative of the color range expected.

By using applied coatings, such as glazes or nonlustrous finishes, and by introducing chemicals into the kiln that vaporize and combine with the clay, color effects can be produced in almost the complete range of the spectrum.

4.4.3.5

Texture

Textures result from the dies or molds used in forming masonry units. The pressure exerted by a steel die normally produces smooth units. Many textures may be applied, however, by mounting at the outlet end of a die an attachment that cuts, scratches, rolls, brushes, or otherwise roughens the surface of the uncut clay column as it leaves the die. The possible textures range from fine through medium to coarse. Some typical textures are smooth, matt with vertical or horizontal markings, rugs, barks, sandmold, waterstruck, and sandstruck (see Fig. 4.4-5).

4.4.3.6

Size Variation

The clays used to manufacture brick and clay tile shrink 4.5% to 15% during drying and burning. Terra-cotta clay shrinks 6% to 11%. In general, this shrinkage can be compensated for by increasing die size and adjusting the spacing of cutter wires so that the units will approximate the desired final size after they are burned. Molds are simply made larger to compensate for shrinkage.

A more serious problem exists when units to be used together do not shrink the same amount. Units shrink at differing rates for several reasons. For example, units made with clays having different chemical compositions may not all be the same size. In spite of the manufacturer's attempts to limit it, this kind of variation will always occur at least to a certain extent. It is simply not possible to have every unit made from exactly the same clay. Large differences in size can be prevented, however, by careful monitoring to ensure that all units to be used together are made from very similar clays.

The amount of shrinkage during burning increases as temperatures rise to produce darker shades of color. As a result, when a wide range of color tone is desired, some variation between the size of the dark and light units is inevitable.

Although manufacturers attempt to control shrinkage, variations in raw materials and kiln temperatures make absolute uniformity impossible. Therefore, specifications should spell out permissible variations in size.

4.4.4

CLASSIFICATION OF CLAY MASONRY UNITS

Clay masonry units are classified (Fig. 4.4-8) as solid masonry units, hollow masonry units, and architectural terra cotta. The many individual products are qualified further for size, grade, type, color, and texture.

Clay masonry units should conform to the current standards of the ASTM as follows:

Building (common) brick	ASTM C 62
Facing brick, unglazed	ASTM C 216
Facing brick, ceramic glazed	ASTM C 126
Hollow brick	ASTM C 652

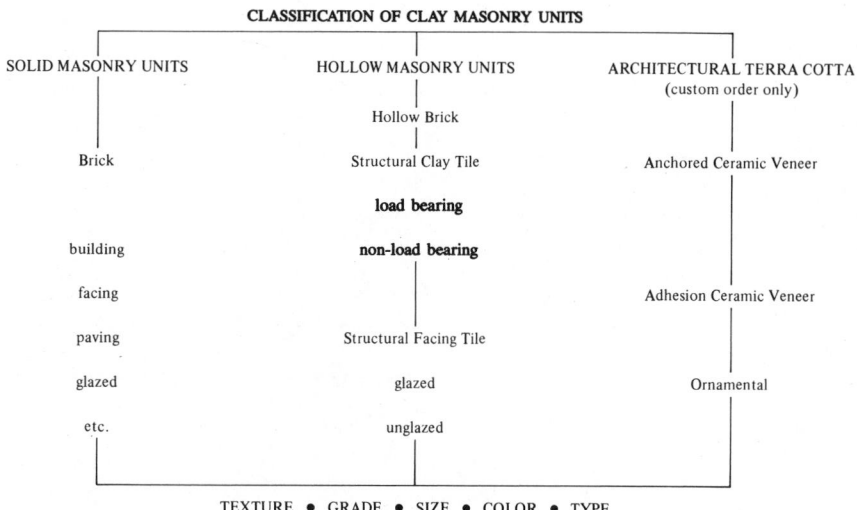

FIGURE 4.4-8 *Classification of clay masonry units.*

Structural clay tile, load bearing	ASTM C 34
Structural clay tile, non–load bearing	ASTM C 56
Structural clay facing tile, unglazed facing tile	ASTM C 212
Structural clay facing tile, ceramic glazed	ASTM C 126
Architectural terra cotta	None
Pedestrian and light-traffic paving brick	ASTM C 902

ASTM standards do not fix the required size, color, or texture of masonry units for a specific use, but rather cover a range of sizes and a wide variety of colors and textures. When construction documents require that masonry units be of a certain type and grade as defined in an ASTM standard, it is usually necessary to add specific requirements related to the units' size, color, and texture, and to the quality of workmanship required and the specific finished appearance desired. This can be done by requesting that these variables match approved samples.

4.4.4.1

Solid Masonry Units

Usually the term *brick* refers to a solid masonry unit. Cored brick, including those with large cores and those with as many as 21 small "pencil" cores, are considered solid masonry units unless the core area exceeds 25% of the total cross-sectional area of the unit. Cores are introduced in brick to promote even drying during burning and to reduce their weight. Brick has been produced in many shapes and sizes, including those listed in Figures 4.4-9 and 4.4-10. Most brick, however, is produced in modular sizes having nominal dimensions based on a 4-in. module.

The term *modular coordination* refers to a system of components sized to fit together easily with a minimum of cutting and fitting in the field. Coordinated assemblies often finish with an overall dimension in even units without fractions. Designing masonry units with modular coordination in mind tends to increase productivity during construction and can result in significant cost savings. Masonry units so designed are referred to as modular units. Common designations and sizes of brick are shown in Figure 4.4-11. Not many manufacturers produce all the sizes shown, so it is important to determine during the design phase which sizes are available in the locality of the project.

The nominal dimensions of a modular brick are equal to the manufactured (actual) dimensions plus the thickness of the mortar joint for which the unit is designed. In general, joint thicknesses in brick are either 3/8 in. or 1/2 in.

The manufactured height for modular brick designed to be laid with three courses in every 8 in. of height is about 2-1/4 in. This is to ensure that the vertical

FIGURE 4.4-9
Standard Nomenclature for Brick Sizes[a]

Unit Designation	Nominal Dimensions, in.			Joint Thickness,[b]	Specified Dimensions,[c] in.			Vertical Coursing
	w	h	l		w	h	l	
Modular Brick Sizes								
Modular	4	2-2/3	8	3/8 1/2	3-5/8 3-1/2	2-1/4 2-1/4	7-5/8 7-1/2	3C = 8 in.
Engineer modular	4	3-1/5	8	3/8 1/2	3-5/8 3-1/2	2-3/4 2-3/16	7-5/8 7-1/2	5C = 16 in.
Closure modular	4	4	8	3/8 1/2	3-5/8 3-1/2	3-5/8 3-1/2	7-5/8 7-1/2	1C = 4 in.
Roman	4	2	12	3/8 1/2	3-5/8 3-1/2	1-5/8 2-1/4	11-5/8 7-1/2	2C = 4 in.
Norman	4	2-2/3	12	3/8 1/2	3-5/8 3-1/2	2-1/4 2-1/4	11-5/8 11-1/2	3C = 8 in.
Engineer Norman	4	3-1/5	12	3/8 1/2	3-5/8 3-1/2	2-3/4 2-13/16	11-5/8 11-1/2	5C = 16 in.
Utility	4	4	12	3/8 1/2	3-5/8 3-1/2	3-5/8 3-1/2	11-5/8 11-1/2	1C = 4 in.
Nonmodular Brick Sizes								
Standard				3/8 1/2	3-5/8 3-1/2	2-1/4 2-1/4	8 8	3C = 8 in.
Engineer standard				3/8 1/2	3-5/8 3-1/2	2-3/4 2-13/16	8 8	5C = 16 in.
Closure standard				3/8 1/2	3-5/8 3-1/2	3-5/8 3-1/2	8 8	1C = 4 in.
King				3/8	3 3	2-3/4 2-5/8	9-5/8 9-5/8	5C = 16 in.
Queen				3/8	3	2-3/4	8	5C = 16 in.

[a] 1 in. = 25.4 mm; 1 ft = 0.3 mm

[b] Common joint sizes used with length and width dimensions. Joint thicknesses of bed joints vary based on vertical coursing and specified unit height.

[c] Specified dimensions may vary within this range from manufacturer to manufacturer.

(Courtesy Brick Institute of America.)

coursing in an addition built with modular brick can be made to match the coursing in an existing building constructed with either modular or nonmodular brick. Those manufacturers who produce modular brick have agreed to keep the manufactured face height 2-1/4 in., even when the length and thickness of their units are modular. They consider as minimal the differences in mortar bed thickness necessary to maintain the modular coursing of three courses in 8 in.

The brick unit dimensions in all the figures in this chapter are shown with the thickness first, followed by the face dimensions (height and length), a practice that has become the industry standard. For example, if the nominal dimensions of a Roman brick are 4 in. × 2 in. × 12 in., and the units are designed to be laid with 3/8 in. mortar joints their actual size will be 3-5/8 in. × 1-5/8 in. × 11-5/8 in., with allowance, of course, for permissible size variations in manufacture.

4.4.4.2

Hollow Masonry Units

Hollow masonry units include hollow brick and hollow clay tile. The many types

FIGURE 4.4-10
Other Brick Sizes[a]

				Modular Brick Sizes				
Nominal Dimensions, in.			Joint Thickness,[b] in.	Specified Dimensions,[c] in.				Vertical Coursing
w	h	l		w	h	l		
4	6	8	3/8	3-5/8	5-5/8	7-5/8		2C = 12 in.
			1/2	3-1/2	5-1/2	7-1/2		
4	8	8	3/8	3-5/8	7-5/8	7-5/8		1C = 8 in.
			1/2	3-1/2	7-1/2	7-1/2		
6	3-1/5	12	3/8	5-5/8	2-3/4	11-5/8		5C = 16 in.
			1/2	5-1/2	2-13/16	11-1/2		
6	4	12	3/8	5-5/8	3-5/8	11-5/8		1C = 4 in.
			1/2	5-1/2	3-1/2	11-1/2		
8	4	12	3/8	7-5/8	3-5/8	11-5/8		1C = 4 in.
			1/2	7-1/2	3-1/2	11-1/2		
8	4	16	3/8	7-5/8	3-5/8	15-5/8		1C = 4 in.
			1/2	7-1/2	3-1/2	15-1/2		
			Nonmodular Brick Sizes					
			3/8	3	2-3/4	8-5/8		5C = 16 in.
				3	2-5/8	8-5/8		

[a] 1 in. = 25.4 mm; 1 ft. = 0.3 mm.

[b] Common joint sizes used with length and width dimensions. Joint thicknesses of bed joints vary based on vertical coursing and specified unit height.

[c] Specified dimensions may vary within this range from manufacturer to manufacturer.

Courtesy Brick Institute of America.

of hollow clay tile are categorized as either structural clay tile or structural facing tile. In all hollow masonry units the cores, cells, and hollow spaces exceed 25% of the gross cross-sectional area of the unit. Tile designed to be laid with its cells in a horizontal plane is called side construction or horizontal cell tile; tile designed to be laid with its cells placed vertically is called end construction or vertical cell tile. The size, number, shape, and thickness of cells will vary greatly depending on the manufacturer.

Hollow brick is defined by ASTM Standard C 652 as a hollow clay masonry unit whose net cross-sectional area in every plane parallel to the bearing surface is not less than 60% of the gross cross-sectional area measured in the same plane. Hollow brick units are cored between 25%, which is the upper limit for solid brick units, and 40%. At the higher limit, they are only 60% solid.

Generally, hollow brick can be used where solid brick is used; it is the same size and looks the same on the outside as solid brick. However, solid brick design standards are not applicable to hollow brick construction. Instead, hollow brick structures should be individually designed according to accepted engineering practice.

Technological improvements in masonry construction and lower production, transportation, and installation costs have given hollow brick advantages over solid brick in both speed of erection and economy of construction. Unfortunately, hollow brick is not readily available in some parts of the country.

Engineered brick masonry design and construction have created a trend toward larger brick units. Typical hollow brick units are outlined in Figure 4.4-12. Their greater core area makes them lighter and handling larger sizes than solid brick easier. Larger sizes permit faster, more

economical construction. For example, three units each 8 in. × 4 in. × 12 in. produce a square foot of 8-in.-thick wall, while 13 standard-size brick units would be required.

Structural clay tile is available in both load-bearing and non-load-bearing units. A few of the available shapes and sizes are included in Figure 4.4-13.

Structural facing tile is available either glazed or unglazed. Glazed tile is made from a high-grade, light-burning fire clay to which a ceramic glaze has been applied. During burning, the glaze fuses into a glasslike coating on the unit. Unglazed tile is produced from light- or dark-burning clays and shales and may have either a smooth or rough textural finish. The nominal modular sizes of structural facing tile are summarized in Figure 4.4-14. Figure 4.4-13 includes some of the common and special shapes currently produced.

FIGURE 4.4-11 *Designations for brick: (a) and (b) modular brick sizes (dimensions are nominal); (c) nonmodular brick sizes (dimensions shown are actual size).* *(Courtesy Brick Institute of America.)*

that are produced by casting. Earlier terra-cotta tile often had an egg crate structure on the back side to lighten the units. Some modern tiles have vertical voids similar to those in concrete block, whereas some are flat and ribbed on the back like ceramic tile. Tiles may be surfacing units with repetitive patterns, or they may be decorative elements with letters or numbers engraved in them. They may also be sculpted.

4.4.5

STANDARDS CONTROLLING CLAY MASONRY UNITS

Standards developed by the ASTM for the many types and grades of brick and clay tile are widely used to control the quality of clay masonry units. Every architect's specification and every masonry contractor's order for brick, hollow brick, structural clay tile, and structural facing tile should include a requirement that the units delivered comply with the applicable ASTM standard. The standard should be specifically named.

ASTM standards classify clay masonry units in types, grades, and classes based on the weathering index (Fig. 4.4-15), the appearance of the completed wall, as well as other factors. When ASTM standards are used to control masonry quality it is essential that the grade, type, and class of the units be spelled out.

4.4.5.1

Weathering Index

The weathering index is a measure of the potential effect of weathering on clay masonry units. The various regions shown in Figure 4.4-15 were established by multiplying the average annual number of freezing cycle days by the annual winter rainfall.

A freezing cycle day is a day when the air temperature passes above or below 32°F. The average annual number of freezing cycle days equals the difference between the mean numbers of days when the minimum temperature is 32°F or below and the mean number of days when the maximum temperature is 32°F or below. Winter rainfall is the sum, in

4.4.4.3

Architectural Terra Cotta

Terra cotta has been used to make sewer pipes, fireproofing for metal structural elements, floor vaulting, and flue liners, but it is chiefly used today as coping and decorative cladding. It is made in practically every color possible and in a variety of shapes and forms. The surface may be flat, grooved, or raised, and either smooth or textured. It may be glazed or natural.

One form of terra cotta used for exterior decoration is called ceramic veneer, and is made in a process similar to that for extruded brick. There are two types. The first, called adhesion-type, is designed to be installed using mortar as one would install ceramic tile. The maximum thickness of this type is 1-1/8 in. The maximum face size is 540 sq. in., which usually results in pieces that are about 18 in. × 30 in. or 20 in. × 27 in.

The second type of ceramic veneer is called anchored-type ceramic veneer. It is designed to be held in place by anchors tied to reinforcing rods installed in a thick bed of grout. This type of ceramic veneer can be made in larger panels than the adhesion type. The panels have ribs or scoring on their backs, making their total thickness 2 to 2-1/2 in.

The second form of terra cotta in general use today is tile. The term *tile* is used here to designate those terra-cotta shapes

FIGURE 4.4-12 *Typical hollow brick units.* *(Courtesy Brick Institute of America.)*

inches, of the mean monthly precipitation occurring between the first and last killing frosts in the fall and spring.

A severe weathering region has an index greater than 500; a moderate region, 50 to 500; and a negligible region, less than 50. Figure 4.4-15 shows regions where clay masonry units will be subjected to severe, moderate, and negligible weathering.

Masonry in contact with the earth is considered to be in a severe weathering condition, regardless of the climate.

4.4.5.2

Summary of Physical Characteristics and Uses

A summary of the physical requirements and normal uses for various types of clay masonry units follows. The accompanying figures show more detail and include guides to selecting the proper types, classes, and grades of units based on their use.

Building (common) brick. ASTM Standard C 62 covers three grades, SW (severe weathering), MW (moderate weathering), and NW (negligible weathering), based on resistance to weathering (see Fig. 4.4-15). The physical requirements for these grades are given in Figure 4.4-16, the uses in Figure 4.4-17.

Facing brick. ASTM Standard C 216 covers two grades, SW and MW, based on resistance to weathering, and three types, FBX, FBS, and FBA, based on the appearance of the finished wall. FBX units are allowed small dimension tolerances, small distortion tolerances, little chippage, and small color range differences from unit to unit. In fact, they require such a high precision and uniformity that they are seldom specified. FBS units require less precision and uniformity, and FBA units require even less than FBS units. The most common are FBS units. The physical requirements for facing brick are the same as for building brick grades SW and MW (see Fig. 4.4-16). The uses of facing brick based on both weathering and appearance are shown in Figure 4.4-18.

Hollow brick. ASTM Standard C 652 covers two grades, SW and MW, based on resistance to weathering, and four types, HBS, HBX, HBA, and HBB, based on the appearance of the finished wall. The physical requirements for hollow brick are the same as for building brick grades SW and MW (see Fig. 4.4-16). The uses of hollow brick are shown in Figure 4.4-19.

Structural clay tile, load bearing. ASTM Standard C 34 covers two grades, LBX and LB, based on compressive strength and resistance to weathering. The physical requirements for load-bearing structural clay tile are given in Figure 4.4-20, uses in Figure 4.4-21.

Structural clay tile, non–load bearing. ASTM Standard C 57 covers one grade only, NB. The single physical requirement is a 28% limitation on maximum water absorption. Non-load-bearing structural clay tile is used in non-load-bearing walls and partitions and as fireproofing and furring. These units have a core area greater than 25% of their total cross-sectional area.

Structural clay tile, unglazed facing tile. ASTM Standard C 212 covers two classes, standard and special duty, based on the thickness of the face shells, and two types, FTX and FTS, based on the appearance of the finished wall. The physical requirements for unglazed structural clay facing tile are given in Figure 4.4-22, the uses in Figure 4.4-23.

Facing brick and structural clay facing tile, ceramic glazed. ASTM Standard C 126 covers ceramic glazed facing brick and other solid units, as well as facing tile having a finish consisting of a ceramic color glaze or a clear glaze. It includes two grades, S (select), for use with comparatively narrow mortar joints, and SS (select sized or ground edge), for use where variation of face dimension must be very small, and two types, I (single-faced units), for use where only one finished face will be exposed, and II (two-faced units), for use where two opposite finished faces will be exposed (Fig. 4.4–24). The physical requirements include minimum compressive strength; tests of finish (imperviousness, chemical resistance, and crazing); limitations on distortion; and limitations on variation in dimensions.

Architectural terra cotta. There is no ASTM standard for terra cotta. Terra-cotta units of every type are custom products. Many stock dies are available, but in general terra-cotta sizes and shapes are produced to meet specific project requirements.

4" WALL THICKNESS

12 X 12   8 X 8 or 12   10 2/3 X 12   5 1/3 X 12   5 1/3 X 12   5 1/3 X 12

6" WALL THICKNESS

12 X 12   12 X 12   8 X 12   8 X 12

8" WALL THICKNESS

12 X 12   12 X 12   8 X 12   5 1/3 X 12   8 X 8   5 1/3 X 12

8 X 12   5 1/3 X 12   10 2/3 X 12   6 2/3 X 12   8 X 12 or 16   8 X 12

10" WALL THICKNESS

5 1/3 or 8 X 12 or 16   12 X 12   12 X 12

12" WALL THICKNESS

12 X 12   12 X 12   8 X 12   8 X 12   8 X 12

FIGURE 4.4-13 *(a) Structural clay tile. (b) Structural facing tile.*

FIGURE 4.4-14
Nominal Modular Sizes of Structural Facing Tile*

Thickness	Face Dimension in Wall	
	Height	Length
2, 4, 6, and 8 in.	4 in.	8 and 12 in.
2, 4, 6, and 8 in.	5-1/3 in.	8 and 12 in.
2, 4, 6, and 8 in.	6 in.	12 in.
2, 4, 6, and 8 in.	8 in.	12 and 16 in.

*Nominal sizes include the thickness of the mortar joint for all dimensions.

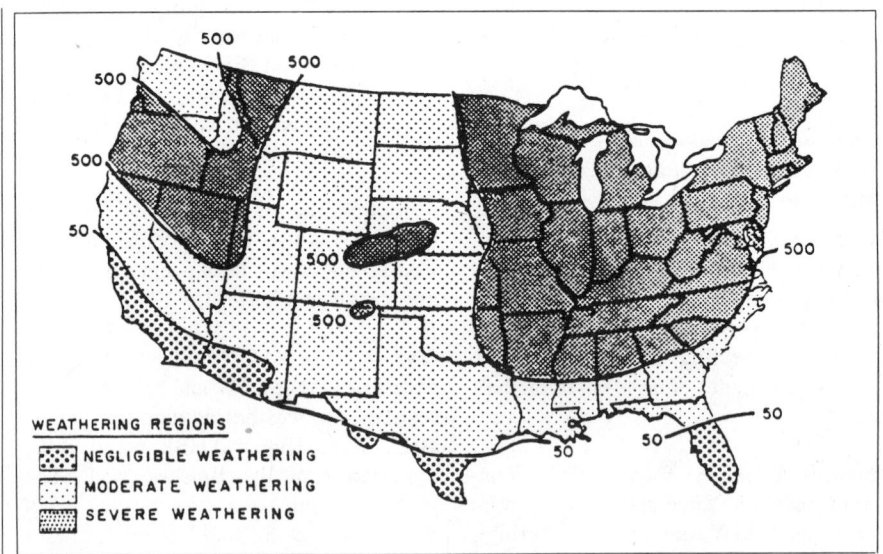

WEATHERING REGIONS

▦ NEGLIGIBLE WEATHERING
▨ MODERATE WEATHERING
▩ SEVERE WEATHERING

FIGURE 4.4-15 *U.S. weathering indexes. (Copyright ASTM, reprinted with permission.)*

FIGURE 4.4-16
Physical Requirements for Solid Clay Masonry Units by Grade

Designation	Minimum Compressive Strength (brick flatwise), Gross Area, psi		Maximum Water Absorption by 5-Hour Boiling		Maximum Saturation Coefficient*	
	Average of 5 Bricks	Individual	Average of 5 Bricks	Individual	Average of 5 Bricks	Individual
Grade SW	3000	2500	17.0%	20.0%	0.78	0.80
Grade MW	2500	2200	22.0%	25.0%	0.88	0.90
Grade NW	1500	1250	No limit	No limit	No limit	No limit

*The saturation coefficient is the ratio of absorption by 24-hour submersion in cold water to that after 5-hour submersion in boiling water.

Copyright ASTM, reprinted with permission.

FIGURE 4.4-17
Uses of Building (Common) Brick[a] (ASTM C 62) by Grade

Grade[b] (based on weathering index)	Uses
SW (severe weathering)	Foundations or other structures below grade or in contact with earth, especially when subjected to freezing temperatures
MW (moderate weathering)	Exterior walls and other exposed vertical masonry surfaces above grade
NW (negligible weathering)	All interior masonry or as backup

[a]Units with core areas less than 25% of unit's gross cross-sectional area.
[b]See Fig. 4.4-15.

FIGURE 4.4-18
Uses of Facing Brick[a] (ASTM C 216) by Grade and Type

Grade[b] (based on weathering index)	Uses
SW (severe weathering)	Masonry in contact with earth, especially when subject to freezing temperatures
MW (moderate weathering)	Exterior walls and other exposed masonry above grade

Type (based on appearance of finished wall)	Uses
FBX	Masonry where high degree of mechanical perfection, minimum variation in color range, and minimum variation in size are required
FBS	Where wider color ranges and greater size variations are permissible than specified for FBX
FBA	Where architectural effects are desired resulting from nonuniformity in size, color, and texture of units

[a]Units with core areas less than 25% of unit's total cross-sectional area.
[b]See Fig. 4.4-15.

FIGURE 4.4-19
Uses of Hollow Brick[a] (ASTM C 652) by Grade and Type

Grade[b] (based on weathering index)	Uses
SW (severe weathering)	Where a high and uniform degree of resistance to frost action and disintegration by weathering is desired and where brick may be frozen when permeated with water
MW (moderate weathering)	Where a moderate and somewhat nonuniform degree of resistance to frost action is permissible and where brick is unlikely to be permeated with water when exposed to below-freezing temperatures

Type (based on appearance of finished wall)	Uses
HBS	Exposed exterior and interior walls where wider color ranges and greater variations in size are permitted than specified for HBX
HBX	Exposed exterior and interior walls where narrow color range and minimum permissible variation in size are required
HBA	Where architectural effects are desired resulting from nonuniformity in size, color, and texture of units
HBB	Where color and texture are not a consideration and greater size variation is permitted than specified for HBX

[a]Units with core area greater than 25% of unit's total cross-sectional area.
[b]See Fig. 4.4-15.

FIGURE 4.4-20
Physical Requirements for Load-Bearing Structural Clay Tile

Grade	Maximum Water Absorption[a] by 1-Hour Boiling, %		Minimum Compressive Strength (based on gross area),[b] psi			
			End Construction Tile		Side Construction Tile	
	Average of 5 Bricks	Individual	Average of 5 Bricks	Individual	Average of 5 Bricks	Individual
LBX	16	19	1400	1000	700	500
LB	25	28	1000	700	700	500

[a]The range in percentage absorption for tile delivered to any one job should be not more than 12.
[b]Gross area of a unit should be determined by multiplying the horizontal face dimension of the unit, as placed in the wall, by its thickness.
Copyright ASTM, reprinted with permission.

FIGURE 4.4-21
Uses of Load-Bearing Structural Clay Tile[a] (ASTM C 34) by Grade

Grade[b] (based on weathering index)	Uses
LBX	Masonry exposed to weathering and direct application of stucco
LB	Masonry not exposed to frost action or earth; may be used in exposed masonry if protected with a facing of 3 in. or more of other masonry

[a]Units with core area greater than 25% of unit's total cross-sectional area.
[b]See Fig. 4.4-17.

FIGURE 4.4-22
Physical Requirements for Unglazed Structural Clay Facing Tile

| | Maximum Water Absorption | | | |
| | By 24-Hour Submersion in Cold Water, % | | By 1-Hour Boiling, % | |
Type	Average	Individual	Average	Individual
FTX	7	9	9	11
FTS	13	16	16	19

| | Compressive Strength Based on Gross Area | | | |
| | End Construction Tile | | Side Construction Tile | |
Class	Minimum Average of 5 Tests, psi	Individual Minimum psi	Minimum Average of 5 Tests, psi	Individual Minimum psi
Standard	1400	1000	700	500
Special duty	2500	2000	1200	1000

Copyright ASTM, reprinted with permission.

FIGURE 4.4-23
Uses of Unglazed Structural Clay Facing Tile* (ASTM C 212) Based on Type

Type	Uses
FTX	Exposed masonry: smooth face, low absorption, and resistance to staining. Has high degree of mechanical perfection, minimal variation in color and face dimensions
FTS	May be smooth or rough textured, with moderate absorption and variation in dimensions, medium color range, and minor surface finish defects

*Units with core area less than 25% of unit's total cross-sectional area.

FIGURE 4.4-24
Uses of Ceramic-Glazed Facing Tile and Brick* (ASTM C 126) Based on Grade and Type

Grade	Uses
S (select)	Masonry with narrow mortar joints (1/4 in.)
SS (select sized or ground edge)	Masonry where face dimension variation must be very small

Type	Uses
I (single-face units)	Masonry where only one finished face will be exposed
II (two-faced units)	Masonry where two opposite finished faces will be exposed

*Units with core area greater than 25% of unit's total cross-sectional area.

Pedestrian and light-traffic paving brick. ASTM Standard C 902 covers paving units designed to support pedestrian and light vehicular traffic on patios, walkways, floors, plazas, and driveways. It lists three classes, SX, MX, and NX, based on the severity of weather exposure, and three types, I, II, and III, based on the severity of traffic end exposure. The physical requirements for the classes are given in Figure 4.4-25, the uses in Figure 4.4-26. Paving brick is also classified in ASTM Standard 902 by use as either Application PS, Application PX, or Application PA. PS is a general-use classification for brick used with mortar- and grout-filled joints, or in patterns (such as running bond) laid without mortar where a close tolerance in the units is necessary. PX is a classification for units to be laid without mortar in patterns that require a high degree of dimensional accuracy. PA is a classification for units with special appearance characteristics where dimensional tolerances are not restricted.

FIGURE 4.4-25
Physical Requirements of Pedestrian and Light-Traffic Paving Brick[a] (ASTM C 902)

Designation Class	Minimum Compressive Strength (brick flatwise), Gross Area, psi		Maximum Cold Water Absorption		Maximum Saturation Coefficient[b]	
	Average of 5 Bricks	Individual	Average of 5 Bricks	Individual	Average of 5 Bricks	Individual
SX	8000	7000	8%	11%	0.78	0.80
MX	3000	2500	14%	17%	No limit	No limit
NX	3000	2500	No limit	No limit	No limit	No limit

[a]Minimum modulus of rupture values should be considered by the purchaser for uses of brick where support or loading may be severe.

[b]The saturation coefficient is the ratio of absorption by 24-hour submersion in room temperature water to that after 5-hour submersion in boiling water.

Copyright ASTM, reprinted with permission.

FIGURE 4.4-26
Uses of Pedestrian and Light-Traffic Paving Brick (ASTM C 902) Based on Class and Type

Class (based on weathering index)	Uses
SX	Where brick may be frozen while saturated with water
MX	Where resistance to freezing is not a factor
NX	Interior use when an effective sealer, wax, or other suitable surface coating will be applied

Type (based on traffic)	Uses
I	Where brick will be exposed to extensive abrasion, such as in driveways and entranceways to public or commercial buildings
II	Where brick will be exposed to intermediate traffic, such as floors in restaurants, stores, and exterior walkways
III	Where brick will be exposed to low traffic, such as floors or patios in single-family houses

4.5 CONCRETE MASONRY UNITS

Concrete masonry units are molded concrete units that are used in building construction as an integral part of the structure, as facing for or filler panels between structural elements, and as partitions. Concrete masonry units include concrete brick, hollow concrete block, slump block, split-face block, and other special units. They do not include concrete splash blocks, manhole blocks, paving units, sills, or similar concrete products (see Chapter 3), or cast stone (see Section 4.8).

Solid and hollow masonry units are made from a relatively dry mixture of portland cement (and sometimes other cementitious materials), aggregates, and water. Admixtures are often also used. After the ingredients have been thoroughly mixed they are molded into the desired shapes by compaction and vibration, both by machine. They are then cured under controlled degrees of moisture and temperature. After a period of aging, during which the required strength, moisture content, and other desired properties are attained, the units are ready for use. Because both the raw materials and manufacturing methods influence the strength, among other properties, of concrete masonry units in a finished wall, a general understanding of each will help in the selection and use of such units.

RAW MATERIALS

The raw materials used to manufacture concrete masonry units include cementitious materials, water, aggregates, and various admixtures.

Cementitious Materials

The primary cementitious material used in concrete masonry units is portland cement. Most of the portland cement used is normal portland cement without an air-entraining admixture. High-early-strength portland cement and both normal and high-early-strength portland cements containing an air-entraining admixture are also used, although not to the same extent as normal portland cement. Units formed using high-early-strength cement have greater strength during the early stages of setting and curing than do their counterparts made with normal portland cement. They are also less likely to break during handling and delivery.

Portland blast-furnace slag cements, fly ash, silica flour, and other pozzolanic or hydraulic materials may also be used as part of the cementing medium. Pozzolanic materials, such as fly ash, have little or no cementitious value in themselves, but combine at ordinary temperatures with lime liberated from portland cement during hydration to form cementing compounds. This reaction is slow, which contributes to strength development after several months. Materials such as silica flour react similarly at the elevated temperatures used in high-pressure (autoclave) curing. Both fly ash and silica flour are used as a portion of the cementing material with portland cement. The correct proportioning of these materials depends on the curing temperatures and type of aggregate used.

Although its use is not common, natural cement can be used satisfactorily to replace up to 20% of the portland cement in concrete masonry units.

The types of cementitious materials used in concrete masonry units and the applicable ASTM standards covering them are shown in Figure 4.5-1.

FIGURE 4.5-1
Standards Covering the Raw Materials Used in Concrete Masonry Units

Cementitious Materials	
Portland cement	ASTM C 150
Portland blast-furnace cement	ASTM C 595
Fly ash	ASTM C 618
Hydrated lime	ASTM C 207 — Type S
Siliceous and pozzolanic	No ASTM standards. Siliceous and pozzolanic materials should be shown by test or experience not to be detrimental to the durability of the concrete.
Aggregates	
Normal weight	ASTM C 33
Lightweight	ASTM C 331
Admixtures	
Air-entraining agents	ASTM C 260
Accelerators and calcium chloride	ASTM D 98
Coloring pigments, integral water repellents, etc.	No ASTM standards. These admixtures should be previously established as suitable for use in concrete or be shown by test or experience not to be detrimental to the durability.

Aggregates

Aggregates constitute about 90% of the weight of a concrete masonry unit, and therefore have an important effect on the desirable properties as well as the cost of the finished units. In general, local availability is the primary determinant in the use of any one type of aggregate.

Desirable aggregate properties include: (1) the toughness, hardness, and strength necessary to resist impact, abrasion, and loading; (2) the durability needed to resist freezing and thawing and the expansion and contraction resulting from moisture and temperature changes; (3) a uniform gradation of fine and coarse aggregate sizes to produce an economical, workable mixture and uniform appearance—the maximum size of aggregate should not exceed one-third of the smallest section of a block's shell; and (4) cleanliness and lack of foreign particles that could impair strength or cause surface imperfections.

The applicable ASTM standards for aggregates are shown in Figure 4.5-1.

Aggregates are classified according to their weight either as dense or lightweight. Dense aggregates, which are also called normal weight aggregates, include sand, gravel, crushed limestone, and air-cooled slag. Lightweight aggregates include expanded shale, clay, slag, coal cinders, pumice, and scoria. The effect of aggregate type on weight, strength, and other characteristics of concrete block is shown in Figure 4.5-2.

Admixtures

Numerous admixtures are marketed for use in concrete mixes, but only a few have been found beneficial or desirable in concrete masonry units. The admixtures used include air-entraining agents, accelerators, coloring pigments, integral water repellents, and a few others. The applicable

FIGURE 4.5-2
Properties of Concrete Masonry Units Based on the Aggregates Used

Aggregate (graded; 3/8 in. to 0)		Weight, lb./cu. ft. of Concrete	Weight, 8-in. × 8-in. × 16-in. Unit	Compressive Strength (gross area), psi*	Water Absorption, lb./cu. ft. of Concrete	Thermal Expansion Coefficient (per °F) × 10⁻⁶
Type	Density (air-dry) lb./cu. ft.					
Normal weight						
Sand and gravel	130–145	135	40	1200–1800	7–10	5.0
Limestone	120–140	135	40	1100–1800	8–12	5.0
Air-cooled slag	100–125	120	35	1100–1500	9–13	4.6
Lightweight						
Expanded shale	75–90	85	25	1000–1500	12–15	4.5
Expanded slag	80–105	95	28	700–1200	12–16	4.0
Cinders	80–105	95	28	700–1000	12–18	4.5
Pumice	60–85	75	22	700–900	13–19	4.0
Scoria	75–100	95	28	700–1200	12–16	4.0

*Multiply these values by 1.80 to obtain approximate corresponding values of strength of the concrete (strength of unit on net area).

ASTM standards are shown in Figure 4.5-1. Admixtures should be investigated individually to determine whether they are appropriate and to ensure that they will perform satisfactorily under the individual manufacturing plant conditions.

Air-entraining agents increase the plasticity and workability of a concrete mixture and distribute minute air bubbles uniformly throughout the hardened concrete, which increases its ability to withstand frost action. The addition of an air-entraining admixture to the mix for concrete masonry units usually permits greater compaction of the mix and produces a denser unit. In addition, the appearance of the units is improved, a closer reproduction of the contours of the mold is obtained, and the likelihood of breakage of freshly molded units is reduced.

Water-repellent agents, such as metallic stearates, are used, but to a limited extent. They are somewhat effective in reducing the rate of absorption and capillarity of the units.

Accelerators, the most common type being calcium chloride, are used by some producers to speed up the hardening of units during cold weather operations.

Workability agents, many of which are of the air-entraining type, are sometimes used to correct the harshness of concrete mixes made with poorly graded aggregates.

4.5.2

MANUFACTURE

Although block manufacturing plants vary in size, layout, equipment, type of automation, and manufacturing details, the basic operations are essentially the same. They include: (1) receiving and storing raw materials, (2) batching and mixing, (3) molding units, (4) curing, (5) cubing and storing cured units, and (6) delivering finished units (Fig. 4.5-3).

4.5.2.1

Receiving Raw Materials

Materials are delivered in bulk to a plant by truck or rail and elevated into steel storage bins above the batching and mixing area (Fig. 4.5-4). At some plants, cement is delivered in special tank trucks or railroad cars and forced by compressed air through pipes directly into the overhead cement storage bin. Different aggregates are handled and stored separately to permit proper blending and proportioning.

4.5.2.2

Batching and Mixing

Materials in the concrete mix are proportioned by weight with a weigh batcher. The weigh batcher is placed under the storage bins and the correct amount of each material for a batch is discharged into the batcher. The weigh batcher is then positioned over the mixer and the materials are discharged into the mixer, where they are mixed with a controlled quantity of water for a period of 6 to 8 minutes (Fig. 4.5-5). An electronic water meter senses the water contained in the aggregate and adjusts the amount of mixing water. Admixtures in liquid form may be added as part of the mixing water.

4.5.2.3

Molding Units

The mixed batch is then discharged into a hopper above the molding machine. Controlled quantities of the mix are fed into the machine for molding into desired shapes and sizes. Concrete is consolidated in the mold by a combination of vibration and pressure (Fig. 4.5-6). Typically, three 8 in. × 8 in. × 16 in. concrete blocks or their equivalent in volume of concrete

FIGURE 4.5-3 *Flowchart for manufacturing concrete masonry units.*

FIGURE 4.5-4 *Placing materials into an overhead storage bin above a batching and mixing machine.*

FIGURE 4.5-5 *A weigh batcher positioned over a mixer.*

are molded per machine cycle at a machine speed of about 5 cycles per minute.

The freshly molded, or green, concrete units emerge from the machines on steel

FIGURE 4.5-6 *A molding machine.*

pallets and are placed on a steel curing rack. This operation, called off bearing, may be done manually (Fig. 4.5-7) or by using an automated rack loader. At this stage, the units are sufficiently dry to hold their shape outside the mold, but are fragile and can be broken in handling.

FIGURE 4.5-7 *Freshly molded concrete units being placed on a steel curing rack.*

4.5.2.4

Curing

After a curing rack has been filled with pallets of green units it is transported either to a steam curing kiln or to an autoclave for accelerated hardening. The principal curing methods used are low-pressure steam curing in a kiln, which accounts for more than 80% of the concrete masonry units produced in the United States, and high-pressure steam curing in an autoclave (Fig. 4.5-8). Occasionally, where the climate is favorable, units are moist cured at normal temperatures of 70° to 100°F.

FIGURE 4.5-8 *An autoclave.*

■ LOW-PRESSURE STEAM CURING

Steam curing at atmospheric pressure accelerates the hardening process using steam. The main benefit of steam curing is the rapid strength gain of units, which permits their placement in inventory within hours after they are molded. Two to 4 days after molding, the compressive strength of low-pressure steam-cured block will be 90% or more of the final ultimate strength. Moist-cured block cured at normal temperatures, on the other hand, develops strength slowly and has less than 40% of its ultimate strength 4 days after molding. Low-pressure steam curing produces units of lighter color than is usually obtained with moist curing.

■ HIGH-PRESSURE STEAM CURING

Applying saturated steam under a pressure between 125 and 150 psi is also sometimes used to accelerate the hardening process of concrete masonry units. Units cured in this way are called autoclaved units. This curing process is not used as frequently today as it was in the past. In fact, autoclaved units are not made in some parts of the country.

The principal benefits of high-pressure steam curing are high early strength and greater stabilization against volume changes caused by varying moisture conditions. Within 24 hours after molding, usable units can be produced that have a permanent compressive strength equal to that of moist-cured units only after they have been cured for a full 28 days.

4.5.2.5

Cubing and Storage

When curing has been completed the curing racks are removed from the steam kiln or autoclave and are conveyed to the cubing station, where the units are assembled into cubes, usually consisting of six layers of 15 to 18 blocks, 8 in. × 8 in. × 16 in., or an equivalent number of other sized units (Fig. 4.5-9). Cubes are stacked so that there are two low openings able to accommodate the forks of a forklift. They are then secured together, either by hand or machine, with steel bands and transported to the storage area where they are placed in inventory in stacks that are three or four cubes high. Depending on the curing methods used, the units remain in inventory anywhere from a few days to several weeks before they are released for use.

FIGURE 4.5-9 *Cured concrete masonry units being assembled into a cube.*

4.5.2.6

Delivery

When units are ready to be released, they are loaded onto flat-bed trucks by the cube using a forklift or other loading device (Fig. 4.5-10a). At a construction project's site, cubes of units are picked up one or two at a time by a mechanical unloader and set at convenient locations for use (Fig. 4.5-10b).

FIGURE 4.5-10 *(a) Units being loaded on a flat-bed truck at the manufacturing plant. (b) A truck equipped with a mechanical device for unloading units at a construction site.*

4.5.3

PHYSICAL PROPERTIES

As with concrete, the physical properties of concrete masonry units are determined largely by the properties of the hardened cement paste and the aggregate. In general, much of the technical knowledge relating to concrete is applicable to concrete masonry units. However, differences do occur as a result of variations in mix composition and consistency, the consolidation method used, textural requirements, method of curing, and other factors.

Concrete masonry units usually have a substantially lower cement factor than does structural concrete. Between 2.75 and 4 sacks of cement are used per cubic yard in masonry units compared with between 4 and 7 bags of concrete. The ratio of water to cement is much lower too, running between 2 and 4 gallons per sack of cement in concrete masonry and between 5 and 7 gallons per sack in concrete. The aggregate in concrete masonry is graded finer than that used in concrete; the largest size seldom exceeds 3/8 in. Concrete masonry units are more often made with porous or lightweight materials than concrete is. Concrete masonry units contain a relatively large volume of interparticle void spaces (spaces between aggregate particles that are not filled with cement paste). They are also cured at higher temperatures than is cast-in-place structural concrete.

Typical ranges in compressive strength, water absorption, density, and other physical properties of commercial-grade hollow concrete masonry units are shown in Figure 4.5-2.

4.5.3.1

Compressive Strength

The strength of concrete masonry units is difficult to predict from mix data because a simple ratio of water to cement is not valid for harsh dry mixes, and because each type of aggregate exhibits different characteristics during mixing. Unlike cast-in-place concrete, in concrete masonry units the highest strengths are obtained from the wettest mixes. However, the maximum amount of mixing water is not usually added, in order to cut down on breakage of freshly molded units during the handling period before curing. Dry mixes do not compact as well as wet ones and therefore have lower strengths. Through experimentation, each manufacturing plant develops mixes and techniques to produce units of optimal strength compatible with other desirable properties.

The compressive strengths shown in Figure 4.5-2 are based on the gross bearing area of the concrete masonry unit, including core spaces. The compressive strength of the unit based on net area, excluding core space, is about 1.8 times the values shown. Solid face and veneer units (such as concrete brick and split block) when made with sand and gravel or crushed stone aggregate, will generally exceed 3000 psi in compression.

Major influences on compressive strength are: (1) the type and gradation of the aggregate, (2) the type and amount of cementitious material, (3) the degree of compaction attained in molding the units, and (4) the moisture content and temperature of the units at the time of testing. As shown in Figure 4.5-2, units made with lightweight aggregates have lower compressive strengths than units made with sand and gravel, limestone, or air-cooled slag.

4.5.3.2

Tensile Strength, Flexural Strength, and Modulus of Elasticity

Tensile strength, flexural strength, and modulus of elasticity all vary in proportion to compressive strength. Tensile strength will range from 7% to 10% of the compressive strength, flexural strength from 15% to 20% of compressive strength, and the modulus of elasticity from 300 to 1200 times the compressive strength.

4.5.3.3

Water Absorption

Absorption tests are used to determine the density of concrete. Absorption is a measure of the number of pounds of water in each cubic foot of concrete. In hollow concrete masonry units this measure varies over a wide range, from as little as 4 or 5 pounds per cubic foot for the heaviest sand and gravel units to as much as 20 pounds per cubic foot for the very porous, lightweight aggregate types. Solid units made with sand and gravel will normally have less than 7 pounds per cubic foot absorption. The absorption of the aggregate (which may vary from 1% to 5% of the dry weight of aggregate for dense aggregates to as much as 30% or more for the lightest aggregate) has a strong influence on the absorption of the units.

The porosity and pore structure of concrete influence other properties, such as permeability, thermal conductivity, and sound absorption, but the influence is not always predictable, especially when comparing units made with different types of aggregates. High water absorption is not a desirable or purposely incorporated property, but it is accepted as a natural consequence when properties such as lightness in weight, higher sound absorption, or thermal insulation are desired.

A high initial rate of absorption (suction rate) results when concrete contains a large portion of relatively large, interconnected pores and voids. When a high suction rate is combined with high absorption, units will have high permeability to water, air, and sound, and may also have less resistance to frost action. On the other hand, unconnected, air-filled pores that are present in lightweight aggregate and in air-entrained cement paste impart the advantages of porosity yet minimize permeability to water, air, and sound.

An abnormally high suction rate adversely affects the structural bond of mortar to units, but this can be minimized by using mortars with high water retention properties.

Concrete masonry that is intended for use in exterior walls and that will not be painted should have low absorption, and mortar joints should be tooled for weathertightness.

4.5.3.4

Volume Changes

Concrete masonry units undergo small dimensional changes resulting from changes in temperature, changes in moisture content, and a chemical reaction called carbonation. Of these factors only moisture changes can be controlled conveniently and appreciably by preshrinking the units and adequately drying them before they are used.

Carbonation causes irreversible shrinkage in masonry units. This is a result of a chemical reaction within the concrete occurring when it absorbs carbon dioxide from the air. Though little test data exist, it appears that the magnitude of this change under certain conditions and over extended periods may approximate that of moisture content changes.

Temperature changes cause concrete masonry units to expand when they are heated and to contract when cooled. These volume changes are fully reversible. Units return to their original shape when the temperature returns to its original level. The coefficient of thermal expansion for concrete masonry units depends largely on the coefficient of the aggregate, since it comprises 80% or more of the concrete volume. Coefficients of thermal expansion for concrete masonry units are shown in Figure 4.5-2.

Moisture changes cause concrete masonry units to expand when wetted and contract when dried. Moisture volume changes that occur when units are newly made may not be fully reversible because of concrete's tendency to assume a permanent contraction during the first few cycles of wetting and drying. Later moisture volume changes are usually fully reversible, however.

An important contributor to the development of cracks in concrete masonry walls is the volume change in the units from original shrinkage during drying. If the units are laid up in a wall before they have become dimensionally stable and have been allowed to shrink, tensile stresses will develop where the wall is restrained from shrinking, and cracking can be expected. Drying shrinkage can be greatly reduced with proper curing and drying so that when units are placed in a wall their moisture content is in equilibrium with the surrounding air, that is, dried down to the average air-dry condition to which the finished wall will be exposed. For a given aggregate type, autoclave curing will sometimes reduce drying by as much as 50%, compared with that of units cured by low-pressure steam or cured at normal temperatures.

The amount of residual shrinkage, or movement, a concrete unit will undergo because of moisture change after placement is the product of its shrinkage potential and its moisture loss after placement. In other words, residual shrinkage = shrinkage potential × moisture loss. Shrinkage potential depends on manufacturing technique, raw materials used, age of unit, among other factors. Moisture loss depends on the moisture content of the units when laid in the wall and the relative humidity and temperature of the drying environment. Although measurable in the laboratory, the small magnitudes make it impractical to measure shrinkage on a routine basis as a means of demonstrating compliance with specifications. Instead, shrinkage is controlled by limiting the moisture content of the units at the time of delivery.

Figure 4.5-11 shows the correlation between moisture content and drying environment, expressed in terms of relative humidity. Masonry units complying with the limits for moisture content shown will have a tolerable residual shrinkage, and cracking of the wall will be minimized. The linear shrinkage classifications shown depend on aggregate type and curing method. Figure 4.5-12 shows the average values of linear shrinkage. The term *normal weight* refers to units made using sand, gravel, limestone, and other dense aggregates. *Lightweight* refers to all common lightweight aggregate types except pumice. Aggregate classifications are not necessarily exclusive and may overlap since many lightweight blocks contain minor amounts of normal-weight aggregate.

4.5.3.5

Surface Texture

Texture may be greatly varied (Fig. 4.5-13) to satisfy aesthetic requirements or to suit a desired physical requirement, such as a coarse texture to provide mechanical bond for direct application of plaster. Normal-weight and lightweight aggregate units can be given many different surface textures by controlling (1) the gradation of the aggregates, (2) the amount of water used, or (3) the degree of compaction at the time of molding. In addition, the surface of a unit may be ground to produce the exact texture desired (Fig. 4.5-14), and the face mold can also be changed to produce

FIGURE 4.5-11
Maximum Moisture Content of Type I Units

| Linear Shrinkage Classification,[a] % | Maximum Moisture Content, % of Total Absorption (average 5 units) | | |
| | Average Annual Relative Humidity,[b] % | | |
	Over 75	75–50	Under 50
Up to 0.03	45	40	35
0.03 to 0.045	40	35	30
0.045 to 0.065	35	30	25

[a]As determined by ASTM C 426 test for drying shrinkage of concrete within previous 12 months.
[b]As reported by the U.S. Weather Service nearest source of manufacture.

FIGURE 4.5-12
Linear Shrinkage of Concrete Masonry Units

Aggregate Type	Steam-Curing Method	Average Linear Shrinkage,* %
Normal weight	High pressure	0.019
	Low pressure	0.027
Lightweight	High pressure	0.023
	Low pressure	0.042
Pumice	High pressure	0.039
	Low pressure	0.063

*As determined by test, ASTM C 426.

FIGURE 4.5-13 *The surface texture of a concrete masonry unit can affect its insulating value and sound absorption ability. As shown here, surface textures may range from fine to coarse.*

FIGURE 4.5-14 *Ground-face units laid in stack bond.*

numerous contours in the face shell of the unit (Fig. 4.5-15).

Units with open surface textures offer some value in absorbing sound. Exposed concrete masonry walls built with ordinary commercial units will abosrb between 18% and 68% of the incident sound. The textures shown in Figure 4.5-13 illustrate this range. Surface finishes, such as paint, tend to close the surface pores, and thereby reduce the absorption significantly.

FIGURE 4.5-15 *The decorative concrete blocks shown are called* **Shadowal.** *They may be laid in other patterns to produce a variety of effects.*

4.5.3.6

Color

The usual aggregates and portland cements used in commercial production produce units with a range of color from off-white to gunmetal gray, and from light tan to shades of brown. A wider color range, including blues, browns, buffs, greens, reds, and slate grays, can be obtained using pure mineral oxide pigments, singly or in combination, mixed integrally with the concrete before it is molded.

4.5.4

CLASSIFICATION OF CONCRETE MASONRY UNITS

Concrete masonry units can be classified as concrete brick, consisting of solid units; concrete block, consisting of both solid and hollow units; and special units (Fig. 4.5-16).

4.5.4.1

Concrete Brick

This category includes concrete building brick and slump brick.

■ BUILDING BRICK

Building bricks are either completely solid or have a shallow depression called a frog (Fig. 4.5-17). The frog's purpose is to reduce weight and provide for better mechanical bond when these units are laid in mortar. Concrete building brick are modular units sized to finish 4 in. wide

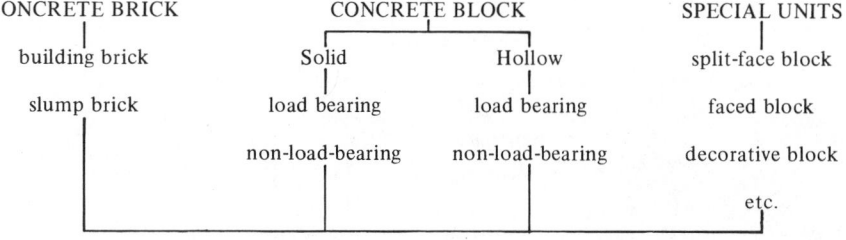

TEXTURE ● GRADE ● SIZE ● COLOR ● TYPE ● WEIGHT

FIGURE 4.5-16 *Classifications of concrete masonry units.*

and 8 in. long when laid with 3/8-in.-thick mortar joints. The thickness varies, but in most units is such that by slightly varying the bed joints a mason can lay the units with three brick and three mortar joints in every 8 in. of wall height. Some manufacturers produce oversized jumbo (see Fig. 4.5-17) and double brick units.

■ SLUMP BRICK

The concrete mixture used in making slump brick has a consistency that causes the units to sag or slump when they are removed from the molds, resulting in irregularly faced units that vary in height, surface texture, and appearance (Fig. 4.5-18). The slump may occur from the effects of gravity alone or may be enhanced by a mechanical device that presses down on the units after they have been removed from the molds. Slump brick units are used when their special and unusual appearance is desired.

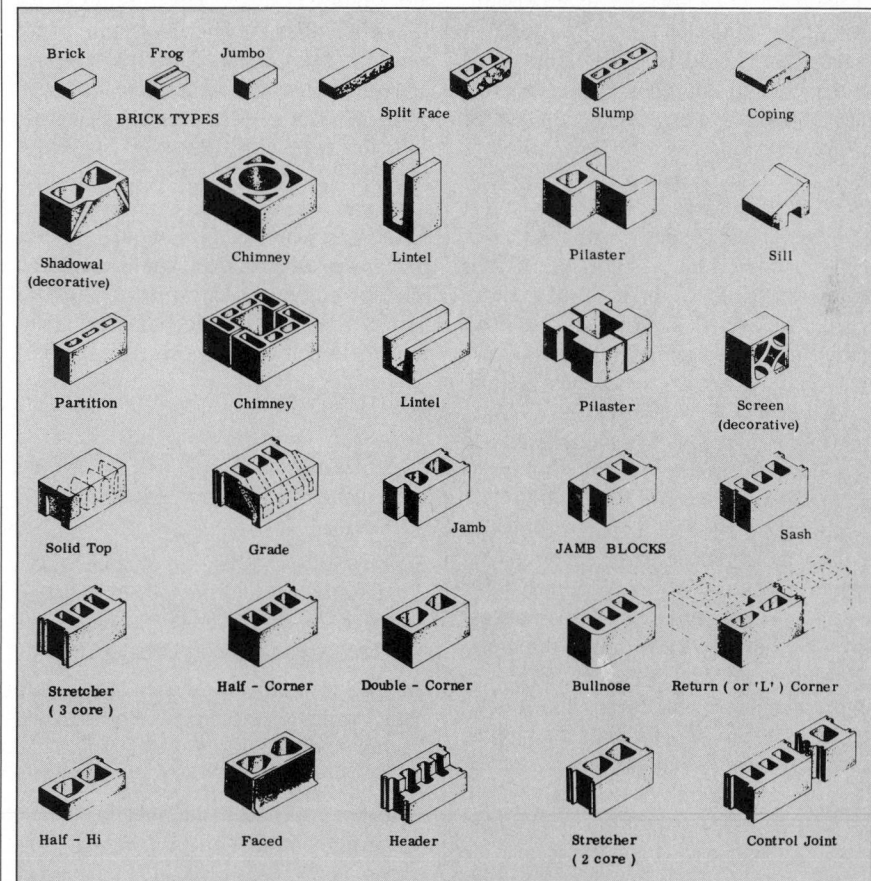

FIGURE 4.5-17 *Typical shapes of concrete masonry units.*

FIGURE 4.5-18 *Slump blocks and slump bricks can produce bold, irregular wall textures.*

FIGURE 4.5-19 *A typical 8-in. × 8-in. × 16-in. concrete corner block.*

4.5.4.2

Concrete Block

Concrete blocks are either solid load-bearing units, hollow load-bearing units, or hollow non-load-bearing units. Regardless of the grade or type of the unit or whether it is load bearing or non–load bearing, a solid unit has a core area less than 25% of the total cross-sectional area of the unit, whereas a hollow unit has a core area greater than 25% of the total cross-sectional area of the unit.

Some typical shapes of concrete block are shown in Figure 4.5-17, but it is seldom possible to obtain all of them in a given locality. Most producers also make half-length and half-height units for use with the full-sized units they manufacture. Concrete blocks are made in modular sizes for use with 3/8-in.-wide head and bed joints. Therefore, an 8 in. × 8 in. × 16 in. block is actually only 7-5/8 in. × 7-5/8 in. × 15-5/8 in. The typical dimensions and component parts of a three-core corner unit are shown in Figure 4.5-19. Face shells and webs increase in thickness from the bottom of the unit to the top to permit easy removal of the core form during manufacture and to provide greater bedding area for mortar in the bed joints. The end flanges of most units have mortar grooves that are 3/4 in. wide × 3/16 in. deep to increase transverse strength in a wall laid up with them and to provide a vertical mortar key, which helps create a weathertight joint.

Figure 4.5-19 shows a three-core unit, but units are also made with two cores. These offer some advantages over the three-core design. The principal advantage is the resulting graduated increase of the face shell thickness at the center web, which increases the strength of the unit in tension and reduces the tendency of units to crack from drying shrinkage and temperature changes. This is particularly true when units are laid in running bond, since the head joints of the course above and below a block occur over the thickened center web rather than over an open core, as is the case with three-core blocks. A two-core block has three webs instead of four, which reduces heat conduction from 3% to 4%. They are also about 4 pounds lighter in weight than the same-sized three-cored block. And finally, they have more vertical space to accommodate reinforcement and conduit or other utilities.

The uses of some of the available types of concrete block follow:

Corner blocks have one flush end for use in pilasters and piers and on exposed corners.

Bullnose blocks have one or more small radius-rounded corners. They are used instead of square-edged corner units where a rounded corner is desirable.

Jamb or sash blocks are used to facilitate the installation of windows, louvers, and other opening fillers.

Solid top blocks have solid tops for use as a bearing surface in the finishing course of a foundation wall.

Header blocks have a recess to receive the header unit in a masonry bonded wall.

Lintel blocks are U- or W-shaped. They are used to construct horizontal lintels or beams.

Partition blocks are usually 4 or 6 in. thick. They are used to build non-load-bearing partition and walls.

Return (L) corner blocks are used in 6-in.-, 10-in.-, and 12-in.-thick walls at corners to maintain a consistent horizontal coursing with the appearance of full-length units and half-length units without cutting of the corner units.

Control joint blocks of various types are used to construct vertical shear-type control joints.

4.5.4.3

Special Units

Some special units (see Fig. 4.5-17) are stock units, whereas others are custom made. Some of the types available follow.

Split-face blocks are solid or hollow units that are produced in parallel pairs and fractured (split) lengthwise by machine after they have hardened to produce two units, each with a rough surface texture on one face (Fig. 4.5-20). Color variations are obtained by introducing mineral colors and by using aggregates of different gradation and color in the concrete mixture. The units are laid in a wall with the fractured faces exposed. Units are available in random sizes, which are often larger than brick sizes, so that broken joint patterns can be used.

FIGURE 4.5-20 *Split-face blocks have a rough and natural stonelike texture.*

Slump blocks are similar to slump brick but larger (see Fig. 4.5-18).

Faced blocks are units with ceramic glazed, plastic, polished, or ground faces. These facings are applied in separate operations on cured concrete masonry units. One type of glazed unit uses a glazing material that is a thermo-setting, resinous binder combined with glass silica and color pigments or colored granules. This mixture, in a semi-liquid state, is cast onto each block in individual molds. The units are then heat treated to set the glaze and bind it integrally into the pores of the concrete unit.

Decorative blocks are manufactured in many different types and configurations. Some have beveled face-shell recesses. Others have surfaces with more than one plane (see Fig. 4.5-15). And still others have clear rectangular or curved holes through them. Different effects can be created by using the same block design in different pattern bonds. Standard units can also be used to produce decorative effects. A few decorative and other special units, custom units, and standard units used in decorative applications are shown in Figure 4.5-21.

Chimney blocks are designed for use with a flue lining in chimney construction.

Pilaster blocks are used in plain or reinforced pilasters and columns.

4.5.5

STANDARDS CONTROLLING CONCRETE MASONRY UNITS

The four types of concrete masonry units that are controlled by ASTM standards are listed with the applicable standard in Figure 4.5-22.

4.5.5.1

Grades

Concrete masonry units intended for load-bearing construction are classified in ASTM standards as Grade N and Grade S. Grade N units are suitable for general use. They may be used in exterior walls, whether below or above grade, that may be exposed to moisture or weather, and for interior walls and backup.

Grade S units are suitable for use only where they will not be exposed to the weather or other moisture. They may not be used below grade. They may be used in exterior walls only when there is a weather-protective coating applied over them. They may be unprotected in walls that are not exposed to the weather or other moisture.

4.5.5.2

Types

Concrete building brick and both load-bearing and non–load-bearing concrete block can be manufactured with a controlled moisture content. *Moisture-controlled* units are designated by the ASTM as Type I units. They are available in both Grades N and S. The two designations are written together (N-I or S-I). The moisture content of Type I units is regulated during manufacture to conform with the limits established by the applicable ASTM standard. Type II units as designated in the ASTM standards are not restricted to a definite moisture content.

The maximum moisture contents for Type I units, based on various average annual relative humidities and the linear shrinkage classification of the masonry units, are shown in Figure 4.5-11. The selection should be based on the local

FIGURE 4.5-21 *Both standard and special concrete masonry units can be used for decorative purposes. In this figure, (a) is an interior wall made using double-jamb block; (b) is a screen wall made using sash blocks; (c) and (d) are screen walls made using decorative blocks; (e) is an exterior wall made using standard pilaster block; and (f) is an exterior screen made using standard solid concrete masonry units.*

FIGURE 4.5-22
Strength and Water-Absorption Requirements for Concrete Masonry Units

Unit Type and Standard	Grade*	Minimum Compressive Strength on Average Gross Area, psi		Maximum Water Absorption for Units of Different Densities, lb./cu. ft.			
				Lightweight Concrete		Medium-Weight Concrete	Normal-Weight Concrete
		Average of 3 Units	Individual Unit	Less than 85 lb./cu. ft.	Less than 105 lb./cu. ft.	105 to 125 lb./cu. ft.	More than 125 lb./cu. ft.
Hollow load-bearing units, ASTM C 90	N-I, N-II	1000	800	—	18	15	13
	S-I, S-II	700	600	20	—	—	—
Solid load-bearing units, ASTM C 145	N-I, N-II	1800	1500	—	18	15	13
	S-I, S-II	1200	1000	20	—	—	—
Concrete building brick, ASTM C 55	N-I, N-II	3500	3000	15	15	13	10
	S-I, S-II	2500	2000	18	18	15	13
Hollow non-load-bearing block, ASTM C 129	—	350	300	—	—	—	—

*Grades S-I and S-II are limited to walls not exposed to the weather and to exterior above-grade walls protected with weather-resistant coatings. Grades N-I and N-II may be used below grade as well as above; protective coatings are recommended below grade and where required above grade.

Copyright ASTM, reprinted with permission.

humidity conditions as reported by the U.S. Weather Service nearest to the site of manufacture, and the shrinkage classification of the particular masonry units used.

Units with limited moisture content should be used, in most cases, to minimize in-place shrinkage of masonry units and the wall cracks that result from it. As shown in Figure 4.5-11, when the construction is located in an area of low relative humidity and a low equilibrium moisture condition is expected, the maximum moisture content at the time of use should be lower than when the construction is located in a less arid region. A lower maximum moisture content is also required when the shrinkage potential of the masonry unit is greater. The shrinkage classification for different units can be determined from Figure 4.5-12.

In addition to regulating moisture content, ASTM standards also establish requirements for the strength and water absorption of masonry units (see Fig. 4.5-22). The effect of moisture absorption and of other properties on masonry construction is described in Section 4.5.3.

4.6 UNIT MASONRY DESIGN

Masonry walls provide excellent structural performance and offer long-term, low-maintenance finishes. They are a basic element of buildings of all kinds, from enormous skyscrapers to single-family dwellings. They are used as part of the structural bearing system in small and large structures and as a part of the enclosing and finishing system of many large and small steel-, concrete-, and wood-framed buildings. To be successful, the design, selection, and construction of a masonry wall must take into consideration the required level of functional performance, the conditions of loading, the allowable costs, the exposure, the type of occupancy, and the desired finished appearance. These considerations are often influenced by the systems and finishes used in other basic elements of the construction.

The following are the usual functional requirements of a masonry wall:

1. It must have adequate strength to support the applied *dead load*, which is the weight of the wall and of the other building elements if it carries any, such as floors and roofs; to resist forces applied directly to the wall, such as those that result from wind and earthquakes; and to support *live loads* transmitted into the wall by building occupants, furniture, and equipment.

2. It must be sufficiently weathertight to control the flow of heat, moisture, air, and water vapor.

3. It must produce satisfactory levels of sound reduction and visual privacy.

4. It must have an appropriate fire resistance.

5. It must be able to accommodate heating, air conditioning, electrical, and plumbing equipment.

6. It must be suitable for the application of various finish materials.

7. It must be economical.

8. Finally, it must have provision for the installation of doors, windows, louvers, and other opening closers.

When requirements other than these exist, the materials, design, and construction methods described in this chapter may be inadequate.

Wall systems must safely support superimposed vertical loads and resist horizontal and racking loads resulting from lateral forces, such as wind and earthquake, and, where they are load bearing, transmit all loads to the foundation. Masonry walls on small structures are rarely specifically engineered. Ordinarily, building codes and rules of thumb based on established practices determine the materials and methods used in their construction.

The costs of a masonry wall include in-place costs, maintenance costs, and building operating costs associated with the masonry. In-place costs are the sum of the costs of materials, labor, and overhead required to erect a wall. The selection of a particular wall type will affect other costs, such as that for installing insulation, vapor retarders, and siding and other finishes, and for providing openings for windows and doors. The necessary time and costs to enclose a building and provide its heating, air conditioning, electrical, and plumbing systems may also be influenced by the wall system selected. The costs of maintaining the masonry and the costs of maintaining the heating, ventilating, and air conditioning systems that are attributable to a wall system (including heat loss or gain through the construction) have an important influence on the ultimate cost of a wall and should not be overlooked if long-term economy is to be achieved.

The aesthetics of a building are influenced by the location, size, and type of doors, windows, and other openings. Also, the wall construction should be compatible with the roof and floor systems. Usually masonry walls are left exposed, but they may be finished on the exterior with stucco or other coatings. Interior surfaces can be finished with applications such as plaster or gypsum board, or can be left unfinished.

Many of the industry's currently recognized minimum requirements for masonry wall design are based on experience and observation of the performance of masonry assemblies over many years. It naturally follows that building codes and published standards often require minimum wall thicknesses and maximum lateral support spacings, instead of demanding that walls be designed using theoretical analyses of the forces and stresses involved.

Most of the recommendations for the design of masonry walls in this section are based on the provisions of the *Building Code Requirements for Masonry Structures*, ACI 530-88/ASCE 5-88, and *Specifications for Masonry Structures*, ACI 530.1-88/ASCE 6-88, which have been developed jointly by the American Concrete Institute (ACI) and the American Society of Civil Engineers (ASCE). These documents have been adopted by reference in the Standard Building Code published by the Southern Building Code Congress International and in the BOCA/National Building Code published by the Building Officials and Code Administrators International, Inc. These new standards consolidate and supersede several masonry codes that were the standards used for masonry for many years, but which are no longer valid.*

Section 4.2 of this book explains how to choose the correct mortar type for the use and exposure of a wall. Sections 4.4 and 4.5 explain how to select the right masonry units. Section 4.7 illustrates the techniques of proper masonry installation. This section (4.6) describes the various types of masonry wall assemblies and their properties when built, and relates many details essential to satisfactory design. Taken together, the sections of this chapter explain how to produce masonry construction of suitable quality.

The need for insulating masonry walls and the materials and application methods generally used for that insulation are discussed in Section 7.3.

4.6.1

WALL TYPES

Masonry walls may be classified as solid walls, cavity walls, veneered walls, and reinforced walls (Fig. 4.6-1). Each type may be constructed of either solid or hollow masonry units. Solid and cavity walls are built to be either load bearing or non–load bearing; veneered walls are non–load bearing; reinforced walls are load bearing. Figure 4.6-2 shows one load-bearing condition; Figure 4.6-3 shows another. No wall type is suitable for all exposures or loading conditions, nor is any one type appropriate for all localities or types of occupancy. The final selection of a wall type should be made only after weighing the various requirements of function, economy, and appearance.

4.6.1.1

Solid Walls

A solid wall is built up of units laid close together with all joints between units filled solidly with mortar. Solid walls may be built of either solid or hollow masonry units (separately or in combination) in any required thickness, in either load-bearing or non-load-bearing construction. The structural bond of solid walls is provided by metal ties, masonry headers, or joint reinforcement as shown in Figure 4.6-1. Walls that utilize masonry headers or have units overlapping in alternate courses (see Fig. 4.6-1g) are called masonry bonded. These are used in areas of slight weather exposure, for economical foundation walls, and for exterior load-bearing walls. There has been some recent indication that masonry headers may break because of wall movement and may not, therefore, provide as good a bond as has previously been supposed.

---

*The superseded standards are: American Standard Building Code Requirements for Masonry, ANSI A41.1-1953 (reaffirmed 1970), by the American National Standards Institute (ANSI); American Standard Building Code Requirements for Reinforced Masonry, ANSI A41.2-1960/NBS G74 (reaffirmed 1970); Building Code Requirements for Engineered Brick Masonry, BIA-69, by the Brick Institute of America (BIA); Specification For the Design and Construction of Load-bearing Concrete Masonry Construction, NCMA TR-75B, by the National Concrete Masonry Association (NCMA); and Building Code Requirements for Concrete Masonry Structures, ACI 531-79 (revised 1983), by the American Concrete Institute (ACI).

A - WOOD FRAME

VENEERED WALLS

B - HOLLOW UNITS   **C - SOLID UNITS**

REINFORCED WALLS

D - CAVITY   **E - CAVITY**   **F - CAVITY**

CAVITY WALLS

G - MASONRY BONDED   **H - MASONRY HEADER**   **I - METAL TIES**   **J - COMPSITE (FACED)**

SOLID WALLS

FIGURE 4.6-1 *Masonry wall types.*

Solid walls may be further classified depending on the type of units used in their construction. They may consist of (1) solid units of brick or concrete brick and block; (2) hollow units of concrete block, hollow brick, or structural clay tile; and (3) composite (faced) walls consisting of facing and backup units of different materials bonded so that both facing and backup are load bearing.

Typical details of 8-in. solid brick walls and composite walls are shown in Figures 4.6-1h, i, and j. Solid and composite walls bonded with masonry headers are less resistant to rain penetration than metal-tied walls, and should be used only in areas where they will not be exposed to wind and rain (Fig. 4.6-4) (see also Section 4.6.3). However, the addition of furring to the inner wall reduces moisture penetration and so permits use of masonry headers in areas of severe exposure. In areas subjected to hurricanes or wind-driven rains, the exterior surface of concrete block walls is usually stuccoed to resist rain penetration.

Nationally recognized building codes permit the use of exterior load-bearing 6-in.-thick masonry walls for one-story, single-family dwellings where the wall height does not exceed 9 ft. to the eaves or 15 ft. to the peak of the gable. Figure 4.6-5 is a typical wall section using SCR brick. SCR brick is a solid unit designed to form a nominal 6-in.-thick wall laid up in running bond with full 1/2-in.-thick bed and head joints. The length and width of the SCR brick make it possible to turn corners and maintain the half-bond. An air space, formed by 2 × 2 wood furring strips, is flashed at the base of the wall to provide a barrier to moisture penetration. The furred space also permits installation of electrical wiring and outlets, and provides room for wall insulation if it is required. Similar construction using solid or hollow concrete masonry units may also be permitted.

Where solid masonry walls are used, insulation and mechanical equipment are often installed within furred spaces on the interior side of the wall. On below-grade basement and foundation walls, insulation is often placed on the exterior side of the wall.

FIGURE 4.6-2 *Many types of clay and concrete unit masonry load-bearing walls are possible. This drawing shows concrete masonry walls that have been reinforced to resist lateral forces.*

FIGURE 4.6-3 *This building has load-bearing masonry walls. The floor framing is supported directly on the masonry.*

A cavity offers two advantages in areas of severe exposure. The continuous air space provides insulation value to the wall and permits insulation to be installed within the wall to further reduce heat transfer (see Fig. 4.6-7; also Section 7.3) and it acts as a barrier to moisture, eliminating the need for furring since significant rain penetration to the interior is practically impossible if proper flashing and weep holes are installed. A cavity also may be used to conceal mechanical and electrical utilities (Fig. 4.6-8).

For a cavity wall to be effective, certain precautions must be observed during its construction. The cavity must be kept free of mortar droppings that could form a bridge allowing moisture to penetrate to the interior face. Two methods for doing this are shown in Figure 4.6-9. Flashings and weep holes must be provided to drain the cavity of water that passes through the exterior wythe. Metal cavity wall ties, designed to resist both tension and compression, should be used to bond the exterior and interior wythes and should be solidly embedded in mortar so that the two wythes will act together to resist wind and other lateral forces. These ties may be separate items or may be part of continuous joint reinforcement.

A vapor retarder is unnecessary in a cavity wall when the cavity is insulated with fill materials that will not retain excessive moisture, such as water-repellent vermiculite or silicon-treated perlite, or with rigid board materials, like foam glass and foamed plastics, that are at least 1 in. less in thickness than the cavity and are installed next to the inner wythe.

In cavity walls, insulation and mechanical equipment are often installed between thicknesses of masonry or within furred spaces on the interior side of the wall.

4.6.1.2

Cavity Walls

A cavity wall consists of two wythes of either solid or hollow masonry, each of which is at least 4 in. thick, separated by a continuous air space not less than 2 in. nor more than 4-1/2 in. wide, and bonded together with metal ties or joint reinforcement (see Figs. 4.6-1d, e, and f, and 4.6-7). The exterior wythe may be brick, hollow brick, or concrete masonry units. The interior wythe may be brick, hollow brick, structural clay tile, or solid or hollow concrete masonry units. The exterior wythe is usually a nominal 4-in. thickness. The interior wythe may be 4 in., 6 in., or 8 in., depending on the height or length of the wall and the loads to be carried. The nominal overall thickness of the cavity wall will be 10 in., 12 in., or 14 in. when the air space is maintained at a nominal 2 in. The cavity may be left as an air space or insulated in a variety of ways.

4.6.1.3

Veneered Walls

Installing masonry units as a facing material (veneer), without using their load-bearing properties, is common in residential construction. Figure 4.6-10 shows typical brick veneer over wood studs and metal studs. The facing of a veneered wall is attached but not bonded to the studs and does not act structurally with the load-bearing portion of the wall.

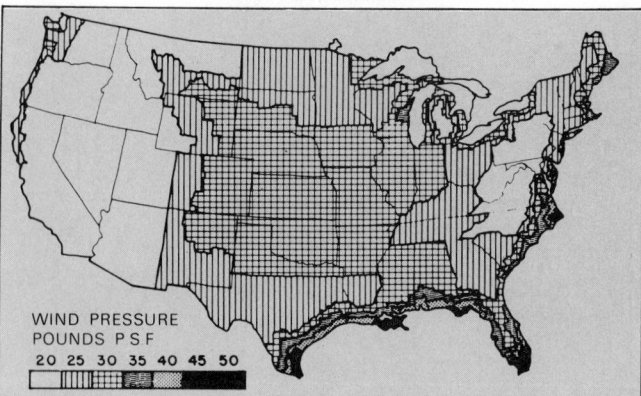

FIGURE 4.6-4 *The weather exposure of masonry units and assemblies is based on precipitation and wind pressure. These maps show the annual precipitation rates and wind pressures that may be expected in the United States.* (Courtesy Brick Institute of America.)

FIGURE 4.6-5 *Section through a nominal 6-in.-thick load-bearing wall.* (Drawing by HLS.)

FIGURE 4.6-6 *Section through a nominal 12-in.-thick load-bearing cavity wall.* (Drawing by HLS.)

FIGURE 4.6-7 *Cavity wall with rigid insulation in the cavity.* (Courtesy DOW USA.)

FIGURE 4.6-8 *The cavity in this wall is being used to conceal plumbing pipes and electrical wiring runs.*

In masonry veneer construction, wood or metal framing carries the vertical loads while the masonry performs mostly as an exterior finish surface and provides a weather barrier. The space between the studs typically is used to contain insulation and to conceal heating, air conditioning, electrical, and plumbing lines.

Key factors in constructing brick veneer walls include correct selection of masonry materials, adequate anchorage to the backup wall, and proper installation of flashing and weep holes.

Since the veneer is isolated from the backup wall by an air space and exposed to the exterior, severe weathering (SW) grade facing brick should be used. Since a masonry veneer carries only its own weight, Type N mortar should be used in it, consistent with the principle that

FIGURE 4.6-9 *This figure demonstrates two methods of keeping mortar out of a cavity. The two left-hand drawings show the beveling of a mortar bed to prevent mortar from being squeezed into the cavity when the next unit is placed on it. The right-hand drawing shows how a wood strip can be placed on each tier of metal ties or continuous joint reinforcement to catch mortar droppings. The wood strip is raised as the work progresses. (Courtesy Brick Institute of America.)*

mortar be of the lowest strength that is compatible with structural requirements.

Refer to Section 4.7.2.2 for a description of the bonding together of brick veneer walls and a short discussion about the controversy surrounding the use of brick veneer with metal studs.

Flashing and weep holes to provide resistance to rain penetration should be located at the bottom of the wall and at all openings.

4.6.1.4

Reinforced Masonry Walls

Following the principles used in reinforced concrete, steel reinforcement is sometimes placed in a masonry wall and embedded in grout to give the wall a greatly increased resistance to forces that produce tensile, shear, and compressive stresses. The chief uses of reinforced masonry are as walls, columns, lintels, and bond beams (Fig. 4.6-11).

There are two ways to reinforce masonry walls. In the first method, which is sometimes called reinforced hollow masonry, the reinforcing and grout are placed in the cores of concrete masonry units (see Fig 4.6-1b) or hollow brick. In this method, concrete masonry units are laid with full face, shell bed joints. The head joints are filled with mortar to a distance in from the face at least equal to

the face shell thickness. Cross webs that are adjacent to cores that will be filled with grout should be filled solidly with mortar. Hollow brick should be laid with full head joints, but the cores must be left open in the bed joints. Both concrete masonry units and hollow brick must be laid and lapped with the cores lined up vertically to permit the installation of reinforcing steel and grout.

In the second method, which is called reinforced solid masonry or reinforced grouted masonry, the reinforcement and grout are placed in the space between two wythes of masonry (see Fig. 4.6-1c). In this method the two masonry wythes are laid with completely filled bed and head joints. Cores and joints adjacent to spaces to be filled with grout should also be filled with mortar.

In both methods, care must be taken to prevent mortar droppings from entering the spaces to be filled with grout and to remove mortar fins from those spaces. Sometimes, cleanouts are placed in walls to be grouted so that foreign materials can be removed.

Reinforcing bars are placed vertically in walls and tied to horizontal bars. Continuous joint reinforcement is usually also placed in reinforced masonry walls.

The kind of grout used varies with the size of the spaces to be filled. When the least clear dimension of the space is 2-1/2 in. or less, fine grout (see Section 4.2.5)

should be used. Coarse grout should be used when the spaces to be filled are larger. The height of pour of fine grout should be limited to 12 in., whereas coarse grout can be placed in layers as much as 4 ft. thick. Fine grout must be agitated or puddled to ensure that it fills the grout space completely. Coarse grout should be vibrated.

Refer to the BIA's *Technical Notes on Brick Construction*, Nos. 17 through 17L and 41, for additional information about reinforced masonry. Reinforced masonry design should be executed in accordance with the requirements of the ACI and ASCE *Building Code Requirements for Masonry Structures* (ACI 530-88/ASCE 5-88) and *Specifications for Masonry Structures* (ACI 530.1-88/ASCE 6-88).

Insulation and mechanical equipment are usually installed in furred spaces on the interior side of reinforced masonry walls. Insulation is often placed on the exterior side of below-grade basement and foundation walls.

4.6.2

PATTERN BONDS

The pattern formed by the masonry units and mortar joints is called pattern bond. Pattern bond may result either from the type of structural bond used or may be purely decorative. Structural bond is the method by which individual masonry units are interlocked together, using masonry headers or metal ties, to cause the entire wall assembly to act as a single structural unit.

4.6.2.1

Brick Pattern Bonds

Traditional brick pattern bonds still commonly used today are briefly described below and are illustrated in Figure 4.6-12.

■ RUNNING BOND

Running bond is the simplest of the basic pattern bonds in that there are no headers at all. Since there are no masonry headers, metal ties are usually used for structural bond. Running bond is used largely in cavity and veneer walls of solid units and

FIGURE 4.6-10 *Sections through brick veneer walls. Part (a) is through a wood stud wall. Part (b) is through a metal stud wall. Insulation has been omitted from the drawings for clarity.*

often in concrete masonry units and clay tile walls where masonry bonding may be accomplished by wider stretcher units (see Fig. 4.6-1g).

COMMON (AMERICAN) BOND

Common bond is similar to running bond except that there is a course of full-length headers at every fifth, sixth, or seventh course, depending on the structural bonding requirements. In some historic buildings, header courses are placed at even wider spacings. It is important in laying out every bond pattern that the corners be started correctly. For common bond, a three-quarter brick starts each header course at the corner.

FLEMISH BOND

Flemish bond has alternate stretchers and headers in each course. Where the headers are not used for structural bonding, half-brick, called blind headers, may be used. There are two methods of starting the corners in a Flemish bond pattern. In the *Dutch* corner, a three-quarter brick starts each course. In the *English* corner 2-in. or quarter-brick closures are used.

ENGLISH BOND

In the English bond pattern, alternate courses are composed of headers and stretchers. The joints between stretchers in all courses line up vertically. Blind headers are used, except in structural bonding courses.

ENGLISH CROSS OR DUTCH BOND

The English cross or Dutch bond pattern is a variation of the English bond. Vertical joints between the stretchers in alternate courses do not line up vertically, but center on the stretchers in the courses above and below.

STACK BOND

In a stack bond pattern all the vertical joints are aligned. Since units do not overlap, this pattern is usually bonded to the backing with rigid steel ties. In large wall areas and in load-bearing construction, such walls should be reinforced using continuous joint reinforcement or steel rods placed in the horizontal mortar joints. If the vertical mortar joints are to align, the masonry units must be dimensionally accurate, or prematched units must be carefully selected for size.

CONTEMPORARY BONDS

Many traditional pattern bonds have been modified by projecting and recessing units or by leaving out units to form perforated walls and screens. A few contemporary brick bonds are shown in Figure 4.6-13. The effect of a particular pattern bond can be altered by varying the pattern through variations in the color and texture of individual units; changing the joint types used; or recessing or projecting individual units, courses, or wythes.

The basket weave and diagonal bond patterns shown for concrete masonry units in Figure 4.6-14 are also used in brick paving.

4.6.2.2

Concrete Masonry Unit Pattern Bonds

Many bond patterns currently used for concrete masonry units are shown in Figure 4.6-14. The more commonly used ones are the running bond patterns in

FIGURE 4.6-11 *Typical methods of reinforcing concrete unit masonry walls.*

A - Continuous reinforced concrete bond beam. Lap bars at corners
B - Reinforced concrete studs tied to footing
C - Reinforcement in horizontal mortar joints
D - Reinforced concrete footing

which the units overlap the units in the next course by half a unit and the vertical joints in every second row are in alignment. Although some of the patterns shown are decorative, they all have been tested to be suitable for load-bearing construction.

4.6.3

MOISTURE CONTROL

Refer to Section 7.1 for a general discussion about moisture control in buildings and some specific requirements for moisture control in foundations and basements with masonry walls. The principal sources of moisture in masonry are rain or melting snow penetration, capillary action from contact with free water or moisture in the ground, and condensation of water vapor within the masonry. Other possible sources are faulty drainage and leaking plumbing systems.

Preventing this moisture from entering a masonry wall is important to prevent the formation of efflorescence and cryptoflorescence and to prevent freeze-thaw cycle damage (see Sections 4.2.1.2 and 4.4.5.1). Excluding moisture from walls is also necessary to prevent corrosion of reinforcement and ties, and resultant damage to the wall and other building components.

Methods used to control rain and melting snow penetration of masonry walls include providing adequate flashing; properly tooling the mortar joints; parging and dampproofing the walls; painting the walls; filling the joints between the masonry and doors, windows, and other openings and penetrations with sealants; providing a sufficient slope or wash to readily drain horizontal surfaces, such as those at sills and copings; providing an overhang on sills, copings, and the like and including drips on them to keep rain and melted snow away from the walls; and installing and maintaining adequate gutters and downspouts.

Methods used to control ground water penetration into masonry foundation, basement, and retaining walls include parging and dampproofing (see Sections

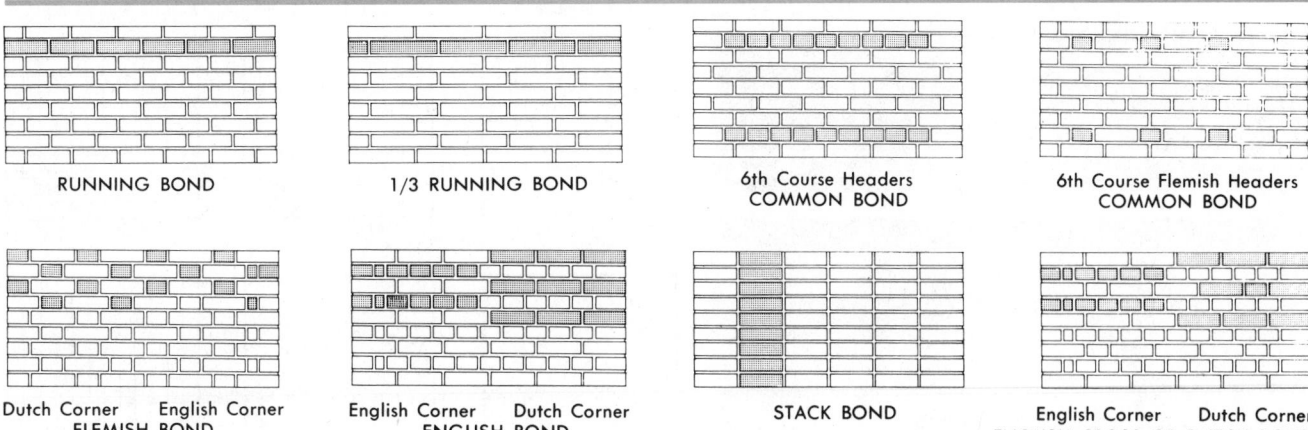

RUNNING BOND   1/3 RUNNING BOND   6th Course Headers COMMON BOND   6th Course Flemish Headers COMMON BOND

Dutch Corner   English Corner FLEMISH BOND   English Corner   Dutch Corner ENGLISH BOND   STACK BOND   English Corner   Dutch Corner ENGLISH CROSS OR DUTCH BOND

FIGURE 4.6-12 *Traditional brick pattern bonds.* (Courtesy Brick Institute of America.)

FIGURE 4.6-13 *Some contemporary brick pattern bonds.* (*Courtesy Brick Institute of America.*)

4.7.3 and 7.2), waterproofing (see Section 7.2), and sloping the exterior grade to drain water away from the walls.

Water vapor that may condense within uninsulated masonry can often be controlled by installing a vapor retarder on the warm side of the wall (see Section 7.1). A water-emulsion asphalt paint applied to the wall's surface is one method used to provide such a retarder. Masonry insulated with proper types of insulation may not require an additional vapor retarder.

4.6.4

CRACK CONTROL

Differential movements within masonry walls and conditions that prohibit necessary movement can produce stresses that may cause serious cracking unless precautions are taken to minimize them. Conditions that can generate stress concentrations large enough to cause cracking of masonry walls include the following: (1) expansion or contraction of the masonry as a result of temperature changes; (2) expansion or contraction as a result of changes in moisture content of the units, especially of concrete masonry units; (3) contraction in concrete masonry units as a result of chemical reactions called carbonation (see Section 4.5.3.4); (4) structural movements caused by unequal settlement of the foundation or at the point of concentration of applied loads; (5) concentration of stress that develops where doors, windows, and other openings occur; and (6) members built into the wall that locally restrain or prevent the wall from moving, as at the junction of floors, roofs, columns, and intersecting walls.

The potential for cracking in masonry walls is exacerbated by the following facts: (1) Materials, especially dissimilar materials as often occur in composite walls, may expand or contract at different rates. Metal lintels, windows, and door frames have substantially greater coefficients of expansion than does masonry, which causes relatively large differential movements. (2) Exterior walls that have a cold exterior surface and a warm interior surface have a tendency to warp. (3) The degree of restraint to movement imposed by roofs, floors, and intersecting walls, and the effect of openings are difficult to estimate.

RUNNING BOND (8" units)

RUNNING BOND (4" units)

BASKET WEAVE BOND B

HORIZONTAL STACK BOND

VERTICAL STACK BOND

COURSED ASHLAR

DIAGONAL BASKET WEAVE

DIAGONAL BOND

BASKET WEAVE BOND A

FIGURE 4.6-14 *Concrete masonry unit pattern bonds.*

Expansion and control joints are built into masonry walls to reduce cracking caused by movement. The location of expansion and control joints cannot be established with mathematical accuracy because the many variables involved may act singly or in combination in any given case. Therefore, these joints are usually located using experience in and observation of existing buildings. The recommendations that follow have evolved in just that way.

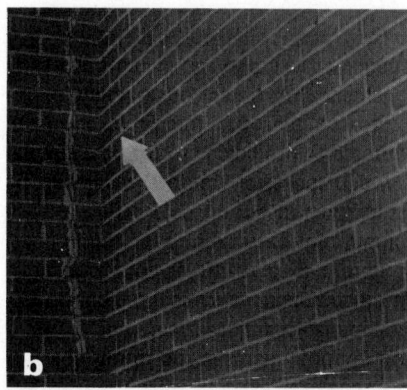

FIGURE 4.6-15 *Cracks in masonry caused by movement. The crack in (a) is in a foundation wall. The one in (b) is in a building wall. (Courtesy Brick Institute of America.)*

4.6.4.1

Expansion Joints for Clay Masonry

An unrestrained clay masonry wall 100 ft. long will tend to expand 3/8 in. as a result of a temperature increase of 100 °F. If this expansion is cumulative it could cause the wall to extend beyond its foundation (Fig. 4.6-15a). When the wall contracts, its tensile strength is not sufficient to overcome frictional resistance, and one or more cracks will appear. Cumulative expansion causes failure in walls at offsets where the masonry in the offset at right angles to a long wall is placed in bending and shear by the expansion of the long wall (Fig. 4.6-15b).

Expansion joints are installed to provide a complete separation through the structure. Figure 4.6-16 shows several of the many methods used to build them. Expansion joints are expensive, may be difficult to maintain, and should be avoided when it is possible to do so without having the walls crack.

In general, expansion joints should be located at offsets when the wall expanding into the offset is 50 ft. or more in length and at junctions of walls in L-, T-, and U-shaped buildings. Recommended locations are shown in Figure 4.6-16. The table in Figure 4.6-17 gives recommended spacings.

FIGURE 4.6-16 *Expansion joint design and location in masonry walls. (Courtesy Brick Institute of America.)*

4.6.4.2

Control Joints for Concrete Masonry

Most initial movement in a concrete masonry wall results from drying shrinkage. As a new masonry wall tries to contract, tensile forces act on it that must be resisted by the tensile strength of the wall. The problem is made worse when a foundation, roof, or other restraining element offers resistance to the normal shrinkage of the masonry. A crack will occur when the tensile forces exceed the tensile strength of either the masonry units or the mortar.

Cracking because of shrinkage can be controlled by one or more methods. (1) Moisture-controlled Type I concrete masonry units manufactured according to ASTM standards can be used to reduce the shrinkage potential of the wall. (2) Horizontal steel reinforcement can be installed to increase the wall's tensile resistance to cracking and to minimize the width of cracks. (3) Control joints located to accommodate wall movements can be built into the wall. Sometimes, more than one method must be used in the same wall.

Control joints are continuous vertical joints built into concrete masonry walls as they are laid up to minimize shrinkage cracks and to control cracks at locations where points of weakness or concentration of stresses are expected. Properly built control joints should prevent slight wall movements from cracking the wall, be themselves sealed against the weather, and stabilize the wall laterally across the joint by means of a shear key.

■ LOCATION AND SPACING

The layout of control joints will depend on the following variables: the length and height of the wall, the type of wall, the moisture content of the units at time of laying, the spacing of joint reinforcement, the use of bond beams, architectural details, and individual experience within the locality where the building is built. In the absence of specific local data or practical experience, the locations and spacings of control joints shown in Figures 4.6-18 and 4.6-19 have been used successfully. Other spacings and locations may be necessary, however, in a particular project, especially if unusual circumstances are present.

Control joints should be placed in both load-bearing and non-load-bearing walls at changes in wall height or thickness, at the junction of walls with columns and pilasters, where openings occur, at wall intersections, and at return angles in L-, T-, and U-shaped buildings.

In composite walls, where concrete masonry is used as a backup for other materials, control joints should be extended through the facing when the wall is laid with masonry bonding. Control joints do not need to extend through the facing material when it is bonded to the backup with metal ties.

When stucco or plaster is applied directly to masonry units, control joints should extend through the stucco or plaster.

Because the volume change due to temperature and moisture changes of below-grade concrete masonry units is usually quite low, and because it is difficult to construct a watertight joint in masonry below grade, control joints are not recommended

FIGURE 4.6-17
Recommended Spacing of Expansion Joints

Outside Temperature Ranges[a]	Maximum Length of Wall[b]			
	Unheated or Insulated		Heated, Not Insulated	
	Solid, ft.	Openings, ft.[c]	Solid, ft.	Openings, ft.[c]
100 and over	200	100	250	125
Less than 100	250	125	300	150

Courtesy Brick Institute of America.

[a] The range from the lowest average temperature to the highest in degrees F.

[b] For insulated cavity or veneer walls, distances are approximately one-half the lengths shown.

[c] Openings 20% or more of wall area.

FIGURE 4.6-18 *Typical control joint locations in concrete masonry unit walls and partitions.*

FIGURE 4.6-19
Maximum Spacings of Control Joints in Unreinforced Masonry[a]

Panel Dimensions[b]	Vertical Spacing of Joint Reinforcements[c]			
	None	24 in.	16 in.	8 in.
Ratio of panel, length to height (L/H)[d]	2	2-1/2	3	4
Panel length (L) not to exceed	40 ft.	45 ft.	50 ft.	60 ft.

[a] Moisture controlled, Type I concrete masonry units assumed. If nonmoisture control units, reduce joint spacing by 50%; if wall is solid grouted, spacing should be 33% less.

[b] A panel is a wall element in one plane lying between (a) corners or wall ends, (b) control joints, or (c) a control joint and end wall.

[c] Continuous metal ties with a minimum of two No. 9 gauge longitudinal wires.

[d] Maximum panel length (L) is applicable regardless of height (H).

for basement or foundation walls. The movements at a control joint in above-grade masonry may cause cracks to develop in the foundation wall below. To minimize this possibility, a continuous bond beam can be provided at the top of the foundation walls. Then the control joints in the above-grade wall can be stopped at the bond beam with no damage to the wall below it.

■ CONSTRUCTING A CONTROL JOINT

Control joints should be laid up in mortar just like any other vertical joint. If the control joint is to be exposed to the weather or to view, a recess should be provided to receive a joint sealant by raking the mortar out to a depth of about 3/4 in. after it has become quite stiff (Fig. 4.6-20). Sometimes a pressure gun and a thin, flat caulking trowel are used to force a knife-grade caulking compound into the joint (Fig. 4.6-21). In other cases, the methods

discussed in Section 7.10 are used to apply an elastomeric sealant. To keep them as inconspicuous as possible, control joints should be built plumb and of the same thickness as the other mortar joints.

One type of control joint can be built with regular stretcher block. Building paper or roofing felt is inserted in the end core of full- and half-length block and extended the full height of the control joint. The core is then filled with mortar for lateral support (Fig. 4.6-22). The paper, or felt, cut to convenient lengths and wide enough to extend across the joint, prevents the mortar from bonding on one side of the joint, thus permitting the control joint to function (Fig. 4.6-23). Control joint blocks with tongue-and-groove shaped ends, which are available in many areas, provide lateral support to the wall sections on each side of the control joint (Fig. 4.6-24). These control joint blocks are made in both full- and half-length units (Fig. 4.6-25).

FIGURE 4.6-24 *Special control joint units provide lateral stability by means of tongue-and-groove-shaped ends.* (*Courtesy Portland Cement Association.*)

FIGURE 4.6-25 *Laying a half-length control joint unit.* (*Courtesy Portland Cement Association.*)

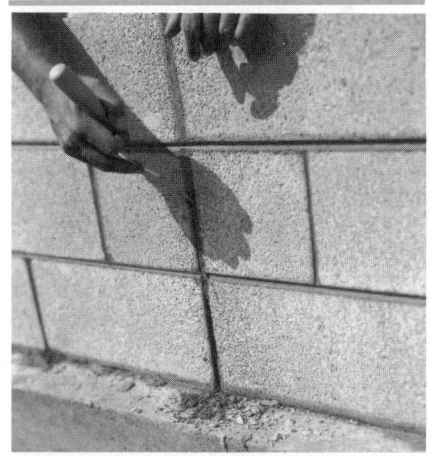

FIGURE 4.6-20 *The mortar in this control joint has been raked out to a depth of about 3/4 in. preparatory to applying a sealant.* (*Courtesy Portland Cement Association.*)

FIGURE 4.6-22 *This control joint is being built by inserting building paper in the end core of the block. Roofing felt may also have been used.* (*Courtesy Portland Cement Association.*)

FIGURE 4.6-21 *This control joint is being sealed with a knife-grade caulking compound.* (*Courtesy Portland Cement Association.*)

FIGURE 4.6-23 *A control joint with building paper or roofing felt and a mortar-filled core.* (*Courtesy Portland Cement Association.*)

Because control joints can be unsightly and also may require maintenance, they should be avoided when possible. Methods used to increase the tensile strength of a wall and to improve the dimensional stability of the masonry units used in it will reduce the number of control joints required, as will designing to eliminate the conditions that dictate using them.

Other methods are also available to increase the crack resistance of walls. For example, two-core blocks with thickened fillets and three-core blocks with thickened center cores have higher tensile strengths than standard units. Their use can increase the crack resistance of a wall about 50% and 25%, respectively, over standard three-core blocks. Another example is the inclusion of continuous joint reinforcement in a wall. As Figure 4.6-19 shows, when joint reinforcement is introduced, the spacing of control joints may be increased.

STRUCTURAL DESIGN OF MASONRY WALLS

Masonry walls are used both as load-bearing walls to support floor or roof structures and as non-load-bearing walls and partitions. The same basic principles apply to the design of masonry walls regardless of their use.

This section covers many strength properties and stability measures for masonry walls. Included are compressive and transverse strengths, lateral support, wall thicknesses, bonding and anchoring, and the support of walls over openings.

Masonry should always be supported, either on other masonry, on concrete, or on steel. Wood girders and other types of wood framing should never be used to support masonry.

4.6.5.1

Compressive Strength

The strength at the point of failure, which is called the ultimate compressive strength, of masonry walls is closely related to the compressive strength of the masonry units themselves and of the mortar. The quality of workmanship, thickness of mortar joints, regularity of the units' bearing surfaces, and workability of the mortar are also important.

High compressive strength is developed by installing high-strength units so that every joint is completely filled with mortar. The strengths used for designing purposes (which are called allowable compressive stresses) and recommended for various wall types, unit strengths, and mortar types are given in Figure 4.6-26.

The principal loads that produce compressive stresses in a wall are (1) the weight of the wall itself and the weights of associated floors and roofs that are transferred into the wall, which are collectively called the dead load; and (2) the weights of occupants, furniture, and equipment, which are collectively called the live load. Both dead and live loads can be determined with reasonable accuracy, and once they have been determined, the resultant compressive stresses can be computed. Such stresses should not exceed the allowable stresses given in Figure 4.6-26

FIGURE 4.6-26
Allowable Compressive Stresses in Unit Masonry

Construction; Compressive Strength of Unit, Gross Area, psi	Allowable Compressive Stresses,[a] Gross Cross-Sectional Area, psi	
	Type M or S Mortar	Type N Mortar
Solid masonry of brick and other solid units of clay or shale; sand-lime or concrete bricks		
8000 or greater	350	300
4500	225	200
2500	160	140
1500	115	100
Grouted masonry, of clay or shale; sand-lime or concrete		
4500 or greater	225	200
2500	160	140
1500	115	100
Solid masonry of solid concrete masonry units		
3000 or greater	225	200
2000	160	140
1200	115	100
Masonry of hollow load-bearing units		
2000 or greater	140	120
1500	115	100
1000	75	70
700	60	55
Hollow walls (noncomposite masonry bonded[b]) Solid units		
2500 or greater	160	140
1500	115	100
Hollow units	75	70
Stone ashlar masonry		
Granite	720	640
Limestone or marble	450	400
Sandstone or cast stone	360	320
Rubble stone masonry		
Coarse, rough, or random	120	100

[a] Linear interpolation for determining allowable stresses for masonry units having compressive strengths that are intermediate between those given in the table is permitted.

[b] Where floor and roof loads are carried upon one wythe, the gross cross-sectional area is that of the wythe under load; if both wythes are loaded, the gross cross-sectional area is that of the wall minus the area of the cavity between the wythes. Walls bonded with metal ties shall be considered as noncomposite walls unless collar joints are filled with mortar or grout.

Courtesy American Concrete Institute.

for the type of wall, kinds of units, and type of mortar being used. When composite walls or other structural masonry elements are composed of different kinds or types of units or mortars, the maximum compressive stress should not exceed the allowable stress for the weakest of the combinations of either the masonry unit(s) or the mortar type(s) of which the wall is composed.

Fortunately, typical construction today does not require most masonry walls to carry large vertical loads. For example, the superimposed load on the foundation wall of a typical one-story residence rarely exceeds 15 psi. About 75% of the bricks produced today have compressive strengths of more than 4500 psi; concrete block, more than 1000 psi. Typical mortars used today have compressive strengths in the range of 750 to 2500 psi. Therefore, it is evident that most masonry walls for one-story buildings have far more compressive strength than they need.

Tests have established the relationship between the compressive strength of walls and the masonry units in them. The compressive strength of walls constructed of hollow concrete masonry units is approximately 42% of the compressive strength of the units when they are laid with face-shell mortar bedding, and 53% when they are laid with full mortar bedding (Fig. 4.6-27). So a wall built with full mortar bedding and concrete masonry units with compressive strengths of 1000 psi would have a compressive wall strength of 0.53 × 1000, or 530 psi. Using an allowable compressive stress of 70 psi (from Fig. 4.6-26), a wall having a strength of 530 psi will have a factor of safety of 530/70, or about 7-1/2. A factor of safety of 4 is ample for masonry walls.

4.6.5.2

Transverse Strength

The transverse strength of a wall is a measure of its lateral stability, which is its

FIGURE 4.6-27 *The top drawing shows face-shell mortar bedding; the bottom, full mortar bedding.* (Courtesy American Concrete Institute.)

resistance to such lateral forces as earth pressures, as occur in foundation, basement, and retaining walls; wind pressures; and earthquake forces. Transverse loads induce tensile stresses in masonry that must be resisted. The lateral stability of masonry walls depends on the tensile strength of the bond between mortar and masonry units. Actions that increase tensile bond strength also increase transverse strength. These include using Type S mortar mixed with the maximum amount of water consistent with workability, maintaining the suction of clay masonry units below 20 grams/min./30 sq. in. (a simple field test to determine the suction of brick is described in Section 4.7.1.2), and making sure that both head and bed joints are completely full of mortar.

The lateral stability of a masonry wall is also increased by vertical loads, which produce compressive stresses that offset tensile stresses caused by transverse loads. The addition of reinforcing steel also greatly increases the lateral strength of a masonry wall.

4.6.5.3

Effect of Flashing on Wall Strength

The effect of flashing on wall strength depends on where mortar is placed and on the bond of the mortar to both the flashing and the masonry. Through-wall flashing, such as base and head flashing, does not affect a wall's compressive strength, but will reduce bending and shearing strengths. If flashing is placed directly on masonry with no mortar beneath it, the flexural strength at that point is 0. Fortunately, through-wall flashing usually occurs where bending resistance is least significant.

Test data suggest that when mortar is placed immediately above and below copper flashing, flexural strength is 30% to 70% of a comparable unflashed wall. There is no sufficient data that permit a similar generalization for shearing strength under similar conditions. Assuming that there is no mortar under a base flashing, a wall's shear resistance, its ability to withstand lateral forces, depends on friction. The wall still retains some shearing resistance since the coefficient of friction between the flashing and the masonry is in the order of 0.25 to 0.50.

4.6.5.4

Lateral Support of Walls

Masonry walls require lateral support both while they are under construction and after they have been completed.

■ LATERAL SUPPORT DURING CONSTRUCTION

Masonry walls are seldom designed to be freestanding, without support from columns, piers, or cross walls. Wind pressures can create four times as much bending stress in a freestanding wall as they do in a wall that is supported by other structural parts of the building. This stress often occurs at the bottom of the wall where flashing and lack of bond in fresh mortar decrease the wall's strength to resist tensile wind forces (see Section 4.6.5.3).

Bracing should be installed to resist wind pressure. Figure 4.6-28 shows commonly accepted limits on wall heights based on various peak wind velocities. Where bracing is used, the heights shown are the safe heights above the bracing. For example, to withstand wind gusts of 50 mph, the freestanding, unbraced height of a 10-in.-thick wall weighing 67 psi or less should not exceed 7-1/2 ft. The curves in

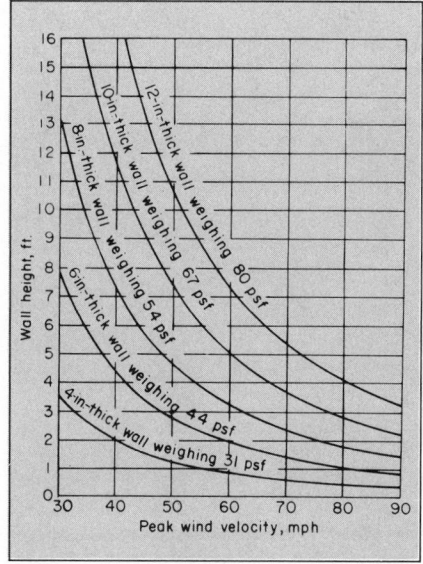

FIGURE 4.6-28 *Maximum unsupported height of nonreinforced, ungrouted concrete masonry walls during construction.*

the figure assume that the mortar has no tensile strength and the walls are free-standing, unreinforced, ungrouted concrete masonry. In cavity walls, the thickness is assumed to be two-thirds of the sum of the thickness of the two wythes. Table values may not be adequate under some circumstances.

Temporary bracing should be left in place until permanent bracing, such as that afforded by piers, cross walls, floors, and roof structures, have been placed and permanently brace the wall.

■ LATERAL SUPPORT OF COMPLETED WALLS

Masonry walls must be permanently, laterally supported or braced at certain intervals either by vertical or horizontal supports. When the limiting distance is length, vertical supports may take the form of columns, piers, pilasters, buttresses, or cross walls. When the limiting distance is height, horizontal supports may be floors, beams, or roofs. The spacing of these supports is primarily dependent on the thickness of the wall. Possible spacing of supports and other lateral support requirements are summarized in Figure 4.6-29. Individual codes often require different spacings than those shown and should be consulted before limits are imposed. In walls of different classes of units or mortars, the ratio of wall height or length to wall thickness should not exceed that allowed for the weakest of the combinations of units and mortars in the member. Sufficient bonding and anchorage must be provided to transfer the loads from the wall to its supports.

If vertical supports are provided, there is no limit on the height of walls between floors or between floor and roof. When horizontal supports are provided, there is no limit on the length of a wall, and lateral supports in the form of columns, pilasters, or cross walls are not required.

Neither the ratio of the unsupported wall height to wall thickness nor the ratio of unsupported wall length to wall thickness should be permitted to exceed the values given in Figure 4.6-29 for various wall types.

4.6.5.5

Wall Thicknesses, Heights, and Lengths

Of course, it is possible to design every wall by a theoretical analysis of the forces

FIGURE 4.6-29 *Lateral support requirements for completed masonry walls.*

involved, and such an analysis is imperative when unusual conditions occur or when a code requires that it be done. However, theoretical analysis is seldom needed or required by code for houses or other such small buildings. Therefore, the design of masonry walls in most small projects is based on empirical requirements, as are the suggestions in this section.

In general, walls should not vary in thickness between their lateral supports. An exception is made, however, in residential buildings not over two stories in height, where vertical recesses may be built into 8-in.-thick walls if the recess is not more than 4 in. deep and does not occupy more than 4 sq. ft. of the wall's area.

■ FOUNDATION WALLS

Unreinforced concrete masonry is used extensively for foundation walls, especially in single-family dwellings and other single-story buildings. The combination of experience and observation has resulted in the development of minimal requirements to govern their design. The requirements are empirical, particularly in designing basement walls to resist lateral earth pressures. Therefore, foundation design for small buildings is rarely based on theoretical analyses of the forces and stresses involved. In the absence of specific building code requirements or individual engineering analysis, the wall thicknesses shown in Figure 4.6-30 have been used when the height of unbalanced fill against the wall did not exceed the limits given in the figure and no other unusual circumstances existed. However, the figure does not take into account the presence of hydrostatic pressure against the wall as a result of ground water. In addition, some practitioners believe that masonry foundation walls should not be less than 12 in. thick, regardless of the height of the earth against them.

■ OTHER WALLS

This section includes empirical design data for walls *other* than single-family, residential foundation walls. The data here should be verified with applicable code requirements and local experience in every case.

FOUNDATION WALL CONSTRUCTION	MAXIMUM HEIGHT OF UNBALANCED FILL H (FEET)	MINIMUM WALL THICKNESS T (INCHES)	
		FRAME CONSTRUCTION	MASONRY OR VENEER
HOLLOW MASONRY	3'	8"	8"
	5'	8"	8"
	7'	12"	10"
SOLID MASONRY	3'	6"	8"
	5'	8"	8"
	7'	10"	8"

FIGURE 4.6-30 *Foundation wall thicknesses based on wall types and the height of unbalanced fill.*

LOAD-BEARING WALLS Load-bearing walls in residential buildings three stories or less in height may be 8 in. thick when not more than 35 ft. high, if the roof is designed to impart no horizontal thrust.

Except where earthquake design is required, exterior walls in one-story residential buildings may be 6 in. thick when they are not more than 9 ft. high, the height to the peak of the gable does not exceed 15 ft., and girders and concentrated loads are supported on not less than 8 in. × 12 in. integral piers or pilasters.

Interior bearing walls in dwellings not more than 1-1/2 stories or 20 ft. high may be 6 in. thick.

Load-bearing walls more than 35 ft. high should be at least 12 in. thick for the top 35 ft. of their height, and should increase 4 in. in thickness for each successive 35 ft.

Cavity walls built of solid masonry units and more than 25 ft. high above the support of the wall may be 10 in. thick. The facing and backing of cavity walls should each be at least 4 in. thick and the cavity should not be less than 2 in. nor more than 4-1/2 in. wide.

NON-LOAD-BEARING WALLS Non-load-bearing walls on the exterior may be 4 in. less in thickness than is required for bearing walls, but not less than 8 in. in thickness except where 6-in.-thick walls are permitted in one-story residences.

4.6.6

LINTELS AND ARCHES

The design of a structural member to support masonry over an opening should not be taken lightly. It is true that so-called rule-of-thumb methods often result in overdesign. It is also true that arbitrary selection of steel shapes or reinforcing steel can result in inadequate sizing, causing unsightly structural cracks that lead to leaks. Underdesigned supports may also fail completely, with disastrous results. Therefore, both masonry lintels and arches should be designed in accordance with engineering practice.

The deadweight of a masonry wall that must be supported over an opening can safely be assumed as the weight of a triangular section whose height is one-half the clear span of the opening (Fig. 4.6-31). Arching action of the masonry above the top of the opening may be counted on to support the additional weight. A concentrated load over an opening may be distributed along a wall length

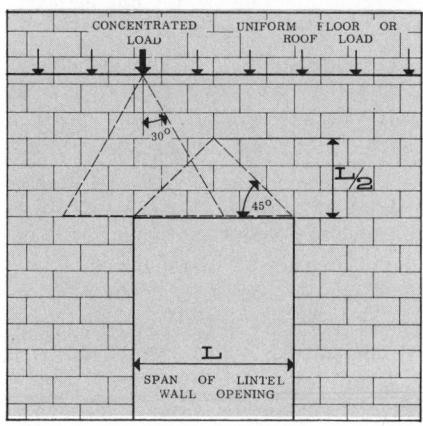

FIGURE 4.6-31 *Assumptions for determining the loads imposed on lintels in masonry walls.*

equal to the base of a triangle whose sides make an angle of 30° with the vertical from the point of application of the load. The horizontal thrust resulting from an arching action must be resisted by sufficient mass in the adjoining wall.

Historically, various forms of arches have been used to span openings. Figure 4.6-32 shows some typical brick arch types. Arches built in contact with adjacent arches form a barrel vault. Domes may be thought of as arches rotated about their vertical centerlines.

Brick arch construction requires extensive labor, especially when the bricks are ground into wedge shapes with an abrasive stone to produce the curve. Most arches today are made using tapered joints between rectangular bricks to save the labor cost of producing wedge-shaped bricks. These are called rough arches. Most arches are built in solid walls because they are difficult to construct and even more difficult to flash properly in cavity walls and veneer walls. The arch must extend across the cavity to present a uniform appearance across the bottom of the arch. This not only blocks the cavity, it also ties the inner and outer wythes together, producing difficulties based on the differential movement between the wythes.

In North America, where labor costs are high, arches have for the most part been replaced by lintels. The most common lintel for supporting masonry over openings in concrete unit masonry is a reinforced concrete masonry lintel. Such lintels may be rectangular, concrete units made in either the shop or the field from U-shaped lintel block, reinforcing steel, and grout (Fig. 4.6-33).

The most common method of supporting masonry over exterior openings is by using steel angles or structural shapes (see Fig. 4.6-33). Using galvanized steel lintels will reduce the possibility of rusting, but will slightly increase the initial costs of the lintels. Stainless steel lintels would eliminate maintenance and the possibility of failure as a result of corrosion altogether, and, even though the initial costs are higher, may well be worth it.

When the steel reinforcement can be completely protected from the elements, reinforced masonry lintels have certain advantages over steel ones. Their maintenance costs are lower, because the steel

FIGURE 4.6-32 *Typical brick arches.* *(Courtesy Brick Institute of America.)*

FIGURE 4.6-33 *Typical lintels for clay and concrete masonry.*

does not require periodic painting. Even the initial costs are lower because less steel is required. However, if the wall around the lintel permits water to penetrate to the reinforcement, it will rust.

Lintels should bear on the wall not less than 4 in. at each end and should be of sufficient stiffness to carry the super-imposed load without deflection of more than 1/360 of the clear span.

4.6.7

PIERS AND PILASTERS

Piers are freestanding columns; pilasters are columns that are bonded to and are built as integral parts of a wall. Both carry concentrated vertical loads. In addition, pilasters often provide necessary lateral support to walls.

Pier size is usually governed by code and is often a ratio of the height of the pier to its least cross-sectional dimension (h/t). The minimum pier section for solid units is usually 8 in. by 12 in. and for hollow units is usually 8 in. by 16 in.

The unsupported height of a solid masonry pier should not exceed 10 times its least cross-sectional dimension. The same is true of piers constructed with hollow masonry units, provided that the cells are filled with Type M or Type S mortar (see Section 4.2.4.1). Piers constructed with unfilled hollow masonry units may have unsupported heights not exceeding 4 times their least dimension.

4.6.8

FIRE RESISTANCE

The fire resistance of walls, floors, and partitions is usually based on the ASTM Standard E 119, "Method for Fire Tests of Building Construction and Materials." The standard fire test does not measure the fire resistance in terms of actual performance in a real fire. Rather, it gives a comparison of the measure of performance between assemblies tested under similar conditions.

The test consists of exposing one side of a wall to a fire of controlled intensity measured by the standard time-to-temperature curve (Fig. 4.6-34). Immediately

FIGURE 4.6-34 *Standard time-to-temperature curve (ASTM E 119 Fire Test).* (*Copyright ASTM, reprinted with permission.*)

after firing, the hot face of the wall is subjected to a blast of water from a fire hose. The wall must withstand the fire and the hosing without permitting flame or gases to pass through it, and the heat transmitted through the wall must be limited to the extent that there is less than a 250 °F rise in temperature. A load-bearing wall must also be able to resist its design stresses without failing. Within 72 hours after the tests have been completed, bearing walls must safely sustain loads twice those applied during the test. Such tests provide the basis for estimating the performance of selected construction types.

Although some walls have ultimate fire-resistance test periods of more than 10 hours, they are often rated lower than their ultimate fire-resistance test periods because most building codes require ratings in multiples of 1 hour with a 4 hour maximum. Figure 4.6-35 gives ultimate fire-resistance periods for solid brick load-bearing walls. Figure 4.6-36 lists the equivalent thickness of concrete masonry units required for different aggregate types to give the fire ratings found in model building codes (Fig. 4.6-37).

4.6.9

FIREPLACES AND CHIMNEYS

The design and construction of an efficient, functional fireplace and chimney require adherence to some basic rules concerning dimensions and the placement of the various component parts. The objectives of a correctly designed fireplace are to: (1) ensure proper combustion of the

fuel; (2) deliver smoke and other products of combustion up the chimney; (3) radiate the maximum amount of heat into the room; and (4) afford simplicity and fire safety in construction.

The first two of those objectives are closely related and depend mainly on the shape and dimensions of the combustion chamber, the proper location of the fireplace throat and the smoke shelf, and the ratio of the flue area to the area of the fireplace opening. The third objective depends on the dimensions of the combustion chamber, whereas the fourth depends on the size and shape of the masonry units and their ability to withstand high temperatures without warping, cracking, or deteriorating. Typical details of fireplace construction are shown in Figure 4.6-38.

4.6.9.1

Separation from Other Construction

Fireplace walls should be separated from combustible construction as follows:

1. *Framing members*: 2-in. air space. This air space should be fire-stopped at floor level with an extension of ceiling finish, strips of fire-resistant board, or other noncombustible material.

2. *Subfloor and flooring*: 3/4-in. air space. Trim in contact with back of fireplace should be of noncombustible material or should be fire-retardant wood.

3. *Wall sheathing*: 2-in. air space.

4. *Furring strips*: 1-in. air space, except that furring may be applied directly to masonry at sides or corners of fireplace.

Combustible materials should not be placed within 8 in. of the top or side edges of a fireplace opening.

When the floor construction or finish is combustible, an incombustible hearth should be installed. This hearth should extend at least 16 in. in front and 8 in. beyond each side of the fireplace opening when the opening is less than 6 sq. ft. in area, and 20 in. in front and 12 in. to the sides when the opening is larger.

Wood framing and furring members should never be placed closer than 2 in.

FIGURE 4.6-35
Ultimate Fire-Resistance Periods for Load-Bearing Clay and Shale Brick Walls

Nominal Wall Thickness, in.	Wall Type	Ultimate Fire Resistance Period, hours				
		Incombustible Members Framed into Wall or No Framed-in Members			Combustible Members Framed into Wall	
		No Plaster	Plaster[a] on One Side	Plaster[a] on Two Sides	No Plaster	Plaster on Exposed Side[a]
4	Solid	1-1/4	1-3/4	2-1/2	—	—
8	Solid	5	6	7	2	2-1/2
12	Solid[b]	10	10	12	8	9
12	Solid[c]	12	13	15	—	—
10	Cavity	5	6	7	2	2-1/2

[a] To achieve these ratings, each plastered wall face must have at least 1/2 in., 1:3 gypsum-sand plaster.
[b] Based on load failure.
[c] Based on temperature rise (for non-load-bearing walls).
Courtesy Brick Institute of America.

FIGURE 4.6-36
Estimated Fire-Resistance Ratings for Concrete Masonry Unit Walls of Equivalent Thickness[a]

Concrete Masonry Units	Members Framed into Wall or Partition						
	None or Noncombustible, in.[b]						
	4 hours	3 hours	2 hours	1-1/2 hours	1 hour	3/4 hour	1/2 hour
Expanded slag or pumice aggregates	4.7	4.0	3.2	2.7	2.1	1.9	1.5
Expanded clay or shale aggregates	5.7	4.8	3.8	3.3	2.6	2.2	1.8
Limestone, cinders, or unexpanded slag aggregates	5.9	5.0	4.0	3.4	2.7	2.3	1.9
Calcareous gravel aggregates	6.2	5.3	4.2	3.6	2.8	2.4	2.0
Siliceous gravel aggregates	6.7	5.7	4.5	3.8	3.0	2.6	2.1

[a] Equivalent thickness is the solid thickness that would be obtained if the same amount of concrete contained in a hollow unit were recast without core holes.
[b] Where combustible members are framed into the wall, the wall must be of such thickness or be so constructed that the thickness of solid material between the end of each member and the opposite face of the wall, or between members set in from opposite sides, will not be less than 93% of the thickness shown in this table.

from the walls of a chimney. The space between chimneys and floor members should be filled with an incombustible material providing a firestop to prevent the transfer of fire from below.

4.6.9.2

Size and Shape

The size of a fireplace depends not only on aesthetic considerations, but also on the size of the room to be heated. For a room with 300 square feet of floor area, a fireplace with an opening 30 in. to 36 in. wide is sufficient. Approximate dimensions for sizing a fireplace are given in Figure 4.6-39.

■ COMBUSTION CHAMBER

The shape of the combustion chamber influences the draft and the amount of heat that will be radiated into the room. A slope at the back throws the flame forward and leads the gases, with increasing velocity, through the throat. Sloping the back and sides also helps radiate the maximum amount of heat into the room. The dimensions, wall angles, and slopes shown in Figures 4.6-38 and 4.6-39 may be varied slightly to correspond with brick coursing, but no major changes should be made in the dimensions shown.

The combustion chamber, unless it is a preformed metal one, should be lined with firebrick at least 2 in. thick laid in fireclay mortar. The back and end walls, including the thickness of the fire brick, should be at least 8 in. thick to support the weight of the chimney above.

■ THROAT

Because of its effect on draft, a fireplace's throat must be carefully designed. It should be not less than 8 in. above the highest point of the fireplace opening. The sloping back should extend to the same height and support the back of the damper.

A metal damper, preferably of a design in which the valve plate opens upward and toward the back, should be placed in the throat. It should extend the full width of the fireplace opening. When open, the damper's valve plate must form a barrier against downdrafts, and deflect them upward with the ascending column of

FIGURE 4.6-37
Fire Resistance Ratings of Structural Elements in Hours from the 1990 BOCA National Building Code

	Type of Construction Section 401.0									
	Noncombustible					Noncombustible/Combustible			Combustible	
	Type 1 Section 402.0		Type 2 Section 403.0			Type 3 Section 404.0		Type 4 Section 405.0	Type 5 Section 406.0	
	Protected		Protected		Unprotected	Protected	Unprotected	Heavy timber[c]	Protected	Unprotected
Structure element[a]	1A	1B	2A	2B	2C	3A	3B	4	5A	5B
1 Exterior walls Load bearing	4	3	2	1	0	2	2	2	1	0
	← Not less than the rating based on fire separation distance (see Section 905.2)									
Non-load bearing	← Not less than the rating based on fire separation distance (see Section 905.2)									
2 Fire walls and party walls (Section 907.0)	4	3	2	2	2	2	2	2	2	2
	← Not less than the rating required by Table 907.1									
3 Fire separation assemblies (Section 909.0) Fire enclosure of exits (Sections 817.11, 909.0)[b]	2	2	2	2	2	2	2	2	2	2
Shafts (other than exits) and elevator hoistways (Sections 909.0, 915.0)[b]	2	2	2	2	2	2	2	2	1	1
Mixed-use separation (Section 313.0)	← Fire-resistance rating corresponding to the rating required by Table 313.1.2 →									
Other separation assemblies[i]	1	1	1	1	1	1	1	1	1	1
	← Note d →									
4 Fire partitions (Section 910.0) Exit access corridors[f,g]	1	1	1	1	1	1	1	1	1	1
	← Note d →									
Tenant spaces separations[f]	1	1	1	1	0	1	0	1	1	0
	← Note d →									
5 Dwelling unit separations (Sections 910.0, 913.0)[f,j]	1	1	1	1	1	1	1	1	1	1
	← Note d →									
6 Smoke barriers (Section 911.0)[g]	1	1	1	1	1	1	1	1	1	1
7 Other nonbearing partitions	0	0	0	0	0	0	0	0	0	0
	← Note d →									
8 Interior bearing walls, bearing partitions, columns, girders, trusses (other than roof trusses) and framing (Section 912.0) Supporting more than one floor	4	3	2	1	0	1	0	See Sec. 405.0	1	0
Supporting one floor only or a roof only	3	2	1-1/2	1	0	1	0	See Sec. 405.0	1	0
9 Structural members supporting wall (Section 912.0)[g]	3	2	1-1/2	1	0	1	0	1	1	0
	← Not less than fire-resistance rating of wall supported →									
10 Floor construction including beams (Section 913.0)[h]	3	2	1-1/2	1	0	1	0	See Sec. 405.0[c]	1	0
11 Roof construction, including beams, trusses and framing, arches and roof deck (Section 914.0)[e,i] 15 ft. or less in height to lowest member	2	1-1/2	1	1	0	1	0	See Sec. 405.0[c]	1	0
			← Note d →							
More than 15 ft. but less than 20 ft. in height to lowest member	1	1	1	0	0	0	0	See Sec. 405.0	1	0
	← Note d →									
20 ft. or more in height to lowest member	0	0	0	0	0	0	0	See Sec. 405.0	0	0
	← Note d →									

Note: Other codes may vary.

[a] For fire-resistance rating requirements for structural members and assemblies that support other fire-resistance-rated members or assemblies, see Section 912.1.

[b] For reductions in the required fire-resistance rating of exit and shaft enclosures, see Sections 817.11 and 915.3.

[c] For substitution of other structural materials for timber in Type 4 construction, see Section 1703.1.1.

[d] Fire-retardant-treated wood permitted; see Sections 904.3 and 1702.4.

[e] For permitted uses of heavy timber in roof construction in buildings of Types 1 and 2 construction, see Section 914.4.

[f] For reductions in required fire-resistance ratings of exit access corridors, tenant separations, and dwelling unit separations, see Section 810.4 and 810.4.1.

[g] For exceptions to the required fire-resistance rating of construction supporting exit access corridor walls, tenant separation walls in covered mall buildings, and smoke barriers, see Sections 910.4 and 911.2.

[h] For buildings having habitable or occupiable stories or basements below grade, see Section 807.3.1.

[i] Not less than the rating required by code.

[j] For Use Group R-3; see Section 309.4.

BOCA National Building Code/1990, copyright 1989, Building Officials and Code Administrators International, Inc. Published by arrangements with the author, all rights reserved.

smoke. When the fireplace is not in use, the damper should be kept closed to prevent heat loss from the room and to keep dirt out of the flue.

SMOKE SHELF AND CHAMBER

The position of the smoke, or downdraft, shelf is established by the location of the throat. The smoke shelf should be located directly under the bottom of the flue and extend horizontally the full width of the throat. The space above the shelf is called the smoke chamber. The back wall of the chimney should be built straight, and the other three sides should be sloped uniformly toward the center to meet the bottom of the flue lining.

Metal linings are available for smoke chambers to give the chamber its proper form. They provide smooth surfaces and simplify masonry installation.

FLUE

It is desirable to obtain relatively high velocities of flue gases and smoke through the throat and flue. The velocity is affected by both the area of the flue and the height of the chimney. The proper sizes of fireplace flues are given in Figure 4.6-39.

Each fireplace should have a separate flue with no other openings or connections. A flue lining should be supported on at least three sides by a ledge of protecting brick that finishes flush with the inside of the lining.

HEARTH

A widely used practice in hearth construction is to form a cantilevered and reinforced concrete underhearth, as shown in Figure 4.6-38. An alternative (more widely used in the past than today) is to construct an arch of brick from the chimney base to the floor construction. The arch is then filled to a level surface to receive the finished hearth (see Fig. 4.6-38).

EXTERIOR AIR SUPPLY

Some codes now require that each fireplace be provided with an exterior air intake to ensure proper combustion. This air intake should permit air to reach the fireplace either directly from the exterior

FIGURE 4.6-38 *Typical fireplace and chimney construction details. Hearths are often cantilevered concrete slabs rather than the arch-supported or floor-framing-supported structures shown. Modern codes and regulations often require fresh air intakes near a fireplace to provide combustion air.*

or from an enclosed space open to the exterior, such as a crawl space or attic. The air intake should open into the side wall of the fireplace chamber, or within 24 in. of the fireplace opening on or near the floor.

4.6.9.3

Fireplace Types

There are several types of masonry fireplaces being used today, but the basic principles involved in their design and construction are the same. The single-faced fireplace is the most common and the oldest variety, and therefore most of the standard design information is based on this type.

The multifaced fireplace, if used properly, may be effective and attractive. It presents certain problems concerning draft and opening size that must usually be solved on an individual basis. Certain standard sizes that have been found to work satisfactorily under most conditions are shown in Figure 4.6-39.

Most fireplaces are not energy efficient, but can be made more so by the following practices:

1. Provide a source of outside air for fireplace combustion so that heated air is not drawn out of the house.

2. Adjust the damper to suit the type of fire. For example, a slow-burning log may need only a 1 in. or 2 in. damper opening to provide adequate combustion without smoking.

3. Provide a glass screen with an adjustable damper, so that the damper in the throat of the fireplace can be left open while the entrance of combustion air is controlled by the damper in the glass screen. When the fireplace is not in use, the glass screen can be closed, preventing heated indoor air from being wasted.

4.6.9.4

Chimneys

Because of the large mass and weight of a chimney, it is quite possible that its foundation will have to support a greater load per square foot than the surrounding foundations. Therefore, special attention must be given to a chimney's footing and the rest of the foundation's design so the unit load (psi) on the soil under all portions

TYPE	OPENING HEIGHT h"	HEARTH SIZE w" x d"	FLUE SIZE (standard modular)
Single Face	29"	30" x 16'	12" x 12"
	29"	36" x 16"	12" x 12"
	29"	40" x 16"	12" x 16"
	32"	48" x 18"	16" x 16"
Two Face (adjacent)	26"	32" x 16"	12" x 16"
	29"	40" x 16"	16" x 16"
	29"	48" x 20"	16" x 16"
Two Face (opposite)	29"	32" x 28"	16" x 16"
	29"	36" x 28"	16" x 20"
	29"	40" x 28"	16" x 20"
Three Face (2 long, 1 short)	27"	36" x 32"	20" x 20"
	27"	36" x 36"	20" x 20"
	27"	44" x 40"	20" x 20"

FIGURE 4.6-39 *Fireplace types and dimensions.*

of the building's foundation will be nearly the same. If not, there may be uneven settling of the foundation, which could result in masonry cracking or other damage to the building.

The thickness of the masonry in the base of a chimney that contains only flues may be 4 in. thick, but if a fireplace is included it must be at least 8 in. thick. The ash pit in the base should be fitted with a tight cast-iron door and a frame of the same material anchored securely at approximately 16 in. above the floor. If there is no basement, the door should be located one or two courses above the grade line. A flue lining is not required 8 in. or more below the point where the lowest smoke pipe thimble or flue ring enters the flue.

Each flue must be built as a separate unit entirely free from other flues or openings. Joints in flue walls should be completely filled with mortar. In masonry chimneys with walls less than 8 in. thick, liners should be separated from the chimney wall and the space between the liner and the masonry should be left unfilled;

only enough mortar should be used to make a good joint and hold the liners in position. Exposed joints inside a flue should be struck smooth. The interior surface should not be plastered.

Although unlined fireplace chimneys with masonry walls at least 8 in. thick may perform adequately, the use of a fireclay flue liner is recommended. A flue lining provides a smooth interior surface that permits more efficient operation of the flue.

The top of chimneys without flue linings may have to be rebuilt every few years because of the disintegrating effect of smoke and gases on the mortar. When rebuilding is necessary, a chimney must be taken down to a point where the mortar joints are solid and a new top built.

Each section of flue lining should be set in place before the brick has been laid to the top of the lining section below, and then the brick should be built up around the just-placed lining to a level below the next section. Offsets and bends should be made by equally mitering both ends of abutting sections of lining. This prevents

FIGURE 4.6-40 *Smoke pipe connections to chimney.*

reduction of the flue area, so that the same effective area is maintained for the entire height of a flue.

Smoke pipe connections should enter the side of a flue at a thimble or flue ring, which should be built as the work progresses. A metal smoke pipe should not extend beyond the inside face of the flue (Fig. 4.6-40). The top of a smoke pipe should be not less than 18 in. below the ceiling and no wood or combustible materials should be placed within 6 in. of a thimble.

A flue lining should project at least 4 in. above the top course, or capping, of the chimney and should be surrounded with portland cement mortar at least 2 in. thick, finished with a straight or concave slope to direct the air currents upward at the top of the flue. This wash also serves to drain water from the top of the chimney and should be constructed with a drip to keep the walls dry and clean.

Unless the applicable code has more stringent requirements, chimneys should extend at least 2 ft. above every part of the roof that lies within 10 ft. of the chimney and at least 3 ft. above the roof at the point of penetration.

After a chimney has been completed and the masonry cured, a test of each flue should be conducted. The test is made by building a smudge fire at the top of the flue and, when smoke is pouring freely from the top, covering the top tightly. Any escaping smoke indicates openings that must be repaired. Since such repair is usually difficult and expensive, it is far more satisfactory and economical to see that construction is properly executed as the work progresses.

Smoke passages and chimney flues should be kept clean. The most efficient method of cleaning a chimney flue is with a weighted brush or bundle of rags lowered and raised from the top. An accumulation of soot and resins may cause a chimney fire whose sparks might ignite the roof. A chimney fire may also damage the chimney, which may, in turn, permit the passage of more fire and cause significant damage to the building.

4.7 UNIT MASONRY ERECTION

Unit masonry and mortar preparation, unit laying, mortar bedding and jointing, joint finishing, and flashing installation must all be done correctly if a satisfactory appearance is to be obtained and retained and if a completed masonry assembly is to be watertight, strong, and durable. They are discussed in this section. Sections 4.1 and 4.2 cover the selection of the correct materials. Section 4.3 addresses the proper design of masonry assemblies.

4.7.1

PREPARATION OF MATERIALS

Protection during shipping, handling, and storage of masonry and associated materials and accessories is essential to prevent damage to them. Proper preparation is necessary before masonry and related materials and accessories are placed in a masonry assembly.

4.7.1.1

Storage and Protection of Materials

All materials that will be used in masonry assemblies should be stored under cover, off the ground, in a dry place, where they are well protected from water and freezing. They should also be kept free of soil, ice, frost, and other contaminants. If they become contaminated during storage or handling, the contaminants must be removed before the materials are used.

It is common to place clay and concrete masonry units and cementitious mortar materials outside on skids or platforms and to cover them during inclement weather with tarpaulins, clear plastic film, or other suitable material. It is better to store these materials indoors if at all possible.

Structural clay tile units, either glazed or unglazed, and terra cotta units should be covered at all times. It is best to store these units indoors.

Sand and other aggregates are often not protected as carefully as they should be, which results in their becoming soaked with water or contaminated with soil or other foreign materials. Foreign materials, excess water, and ice must be removed before aggregates are used. If different aggregates become mixed together during storage, it is necessary to grade them again, which is difficult, or to discard them, which is expensive.

Metal ties, anchors, and reinforcements must be kept dry and protected from contact with soil and away from conditions that would harm them. Before they are placed, these items must be free from loose rust and other coatings that would destroy or reduce a bond.

4.7.1.2

Clay Masonry

The suction, or initial rate of absorption, of a brick is the amount of water in grams absorbed by 30 sq. in. of surface, when a brick is placed in water to a depth of 1/8 in. for one minute as tested in accordance with ASTM Standard C 67 (Fig. 4.7-1). In almost every case, mortar bonds best to brick whose suction is 5 to 20 grams of water.

FIGURE 4.7-1 *Apparatus for testing the suction of a brick.* (*Courtesy Brick Institute of America.*)

Bricks having a suction in excess of 20 grams of water per minute should be wetted sufficiently so that the suction when the brick is laid does not exceed that amount. If high-suction bricks are not wetted, they will absorb water from the mortar before a bond has developed. Merely sprinkling brick to wet it will not produce the desired results. Rather, a hosed stream of water should be played on each brick pile until water runs from all sides. Wetting should be done at least several hours before the units are laid; a full day beforehand is most preferable. The surfaces of wetted brick must be allowed to dry before the brick is laid. Water on the surface of a brick will cause it to float on the mortar bed, preventing it from bonding with the mortar.

Conversely, if bricks with a low suction are wetted or if other bricks are wetted too much they may become saturated. Saturated brick will not absorb enough moisture from the mortar and will float, which will usually not permit the formation of a sufficient bond between the brick and

the mortar. When brick suction exceeds 60 grams of water at the time of laying, bond may be extremely poor, regardless of the quality of workmanship or mortar type used.

An effective on-the-job method of determining whether a brick should be wetted before it is laid is to sprinkle drops of water on its flat side. If these drops are absorbed completely in less than 1 minute, the brick should be wetted. A more accurate method is to draw a circle on the flat side of a brick using a wax pencil and a quarter as a guide. With a medicine dropper apply 20 drops of water to the surface of the brick inside the circle. If the water is completely absorbed in less than 1-1/2 minutes, the brick should be wetted before it is laid.

4.7.1.3

Concrete Unit Masonry

The ACI and ASCE *Specifications for Masonry Structures* (ACI 530.1-88/ASCE 6-88) specifically prohibits the wetting of concrete masonry units. Therefore, care must be taken to keep these units dry on the job. They should be stockpiled on planks or other supports, free from contact with the ground, and covered for protection against getting wet (Fig. 4.7-2). Concrete units should never be wetted before or during laying in the wall.

FIGURE 4.7-2 *One means of protecting concrete masonry units.*

The suction test described in Section 4.7.1.2 is not applicable to concrete masonry units. A concrete masonry unit with a moisture content of less than 20% of the total amount it can absorb can draw water out of mortar on hot, dry days, but concrete masonry unit suction is usually

insignificant because the units generally contain sufficient moisture to retard suction effects.

4.7.1.4

Mortar

The strength and resistance of masonry walls to water penetration are both dependent on the completeness and strength of the bond between the mortar and the masonry units. Mortar characteristics that affect bond strength are discussed in Section 4.2.1. Mortar preparation includes mixing the mortar and maintaining the proper water ratio until the mortar is applied. Mixing is discussed in Section 4.2.6, retempering to control water content in Section 4.2.7. Mortar application techniques that affect bond strength and water penetration resistance are discussed in Section 4.7.2.5.

4.7.2

INSTALLATION OF SOLID AND HOLLOW MASONRY UNITS

The installation of masonry units requires laying the units, bonding them together, and anchoring them to the structure. It also includes installing continuous joint reinforcement, building in bearings for supported structural members, mortar bedding and jointing, backparging, joint finishing, and installing flashing and weep holes. All of these are addressed in this section.

Some specific terms used to describe the positions of masonry units in a wall are shown in Figure 4.7-3.

4.7.2.1

Laying Masonry Units

Many general techniques are applicable to the laying of both solid and hollow masonry units, including clay brick, hollow clay brick, structural clay tile, concrete brick, and concrete block. Many of these general procedures are discussed in this section. Most of the illustrations used here show concrete block, but the same principles apply to other masonry units.

Masonry should be laid *plumb* (exactly vertical) and *true* (accurately placed) to lines. Each unit should be adjusted to its final position while the mortar is still soft and plastic. A mason should avoid

FIGURE 4.7-3 *The names of the various positions in which masonry units may be placed in walls, and the names given to various joint locations.*

FIGURE 4.7-4 *Moving masonry units after they have been laid can result in cracks. Even hammering a brick back in line, as shown in (a), can break the bond. Cracks caused by realigning units often occur at corners, as shown in (b).* (*Courtesy Brick Institute of America.*)

overplumbing, tapping or pounding on, or otherwise attempting to realign units after they have been set into position (Fig. 4.7-4a and b). Such movements will often break the bond between the unit and the mortar. When the bond has been broken in this manner it usually remains broken, regardless of a mason's attempts to reestablish it. When a unit must be adjusted after the mortar has started to harden, the units and the mortar should be removed and the units should be laid again using new mortar.

The highest bond strength is achieved when the time interval between placing a mortar bed and laying a masonry unit into it is kept to a minimum. In addition, a change in the time interval between spreading mortar and placing masonry units into that mortar will affect the flow of the mortar (see Section 4.2.1.1), especially if the masonry units have a high suction rate (see Section 4.7.1.2). Varying this interval to any large extent will result in a wall with varying tensile bond strengths.

Concrete masonry units and some other hollow masonry units have thicker face

shells on one side than on the other and should be laid with the thicker face shell up to provide a larger mortar bedding area.

Anchors, accessories, flashings, and other items required to be built into masonry should be built in as the work progresses.

When fresh masonry joins masonry that is partially or totally set, the exposed surface of the set masonry should be cleaned and, in the case of clay masonry, wetted to obtain the best possible bond with the new work.

A story or course pole (a board marked with the courses) provides an accurate method of finding the top of the masonry for each course (Fig. 4.7-5). The same story pole should be used throughout the work.

■ STARTING THE WORK

After locating the corners, a mason often spaces out the units dry (without mortar) to determine the extent to which they must be cut or have their joint sizes varied to accomplish accurate horizontal coursing (Fig. 4.7-6).

A mortar bed is then spread and the units embedded in it (see Section 4.7.2.5). The corner unit is laid first and carefully placed in its correct position.

After several units have been laid, a mason's level is used as a straightedge to verify that they are in correct alignment at the proper grade and are plumb. Corrections are made by tapping the units with a trowel handle (Fig. 4.7-7). The first course of masonry should be laid carefully to ensure that it is properly

FIGURE 4.7-5 *Using a story pole.* (*Courtesy Portland Cement Association.*)

FIGURE 4.7-6 *Spacing units at a corner without mortar before laying them.* (*Courtesy Portland Cement Association.*)

aligned, leveled, and plumbed. This will assist the mason in laying succeeding courses and in building a straight, true wall.

■ LAYING UP THE CORNERS

After the first course has been laid, the corners of the wall are built before the rest of the wall is laid. The corners are started by laying several courses higher than the center of the wall. Each course is stepped back by one half unit. As each course is laid, it is checked with a level for alignment and plumb (Fig. 4.7-8), and for being level (Fig. 4.7-9). At the same time, each course is checked carefully with a level or straightedge to make certain that the faces of the units are all in the same plane so that the wall will be true and straight (Fig. 4.7-10).

FIGURE 4.7-7 *Aligning the faces of a course of blocks and leveling the units.* (Courtesy Portland Cement Association.)

FIGURE 4.7-8 *Aligning units at a corner and checking for plumb.* (Courtesy Portland Cement Association.)

FIGURE 4.7-9 *Checking to see if a unit is level.* (Courtesy Portland Cement Association.)

FIGURE 4.7-10 *Checking to see if units are in line.* (Courtesy Portland Cement Association.)

■ LAYING BETWEEN THE CORNERS

Before filling in the wall between the corners, a mason stretches a line from corner to corner for each course. He or she then lays the top outside edge of the units to this line. In laying concrete masonry units to a mason's line, the way a mason handles and grips the units is important. By tipping the units slightly toward him or her, a mason can see the upper edge of the course below, thus enabling the placement of the block's lower edge directly over the course below (Fig. 4.7-11). By rolling the block slightly to a vertical position and shoving it against the adjacent block, the mason can lay it to the line with minimal adjustment.

As stated earlier, adjustments to final position must be made while the mortar is soft and plastic. By tapping lightly with a trowel handle a mason can level and adjust each unit to the mason's line. The use of a level between corners is necessary only to check the alignment of each unit with the face of the wall.

4.7.2.2

Bonding and Anchoring

The units in a single-wythe masonry wall are bonded together by the wall's pattern bond.

Generally, two methods are used to bond together structurally the several wythes of wall assemblies that have more than one wythe of masonry. One method, which here is called *masonry bonded*, is accomplished by overlapping and interlocking the masonry units. The second method, which is the preferred one in most cases, bonds the masonry together using metal ties embedded in the mortar

FIGURE 4.7-11 *A mason positions units between the corners by laying them to a line.* (Courtesy Portland Cement Association.)

joints. This text calls this method *metal tied*.

Masonry must also be anchored to other structural elements when it relies on them for support. Most such anchoring is done using metal anchors.

■ MASONRY BONDED SINGLE-WYTHE WALLS

The units in a single-wythe brick wall are bonded together by building the wall in the running bond pattern shown in Figure 4.6-12. Other bond patterns can be used in single-wythe walls, of course, but the headers must be cut to a length that matches the runner's thickness. This method is unsatisfactory when both sides of the wall are exposed, because the finish of the cut units cannot be made to match the face texture of the brick. When a stack bond pattern is used, it is usually necessary to provide continuous joint reinforcement (see Section 4.7.2.3).

The units in a single-wythe concrete unit masonry wall are bonded together by building the walls in one of the patterns shown in Figure 4.6-14. Stack bond patterns require the use of continuous joint reinforcement (see Section 4.7.2.3).

Corners and intersections in a single-wythe brick or concrete unit masonry non-bearing wall or partition can be bonded by laying at least 50% of the units in an overlapping pattern. Each overlapping unit should bear at least 3 in. on the unit below.

Masonry bonded single-wythe brick and concrete unit masonry walls that are load bearing may be bonded at intersections by overlapping 50% of the units, with alternate units bearing at least 3 in. on the unit below. This method is not recommended by some experts, however, except in earthquake zones. In other areas, they would prefer that such walls be masonry bonded only at corners, and be built up separately at other intersections and be tied together there with metal connectors as discussed later in this section.

■ MASONRY BONDED MULTIPLE-WYTHE WALLS

The bonding of the units in a masonry bonded multiple-wythe wall is based on variations of the two traditional methods of bonding: (1) English bond, consisting of alternating courses of headers and stretchers; and (2) Flemish bond, consisting of alternating headers and stretchers in every course (see Fig. 4.6-12). In this regard, common bond is but a variation of English bond. English and common bonds are seldom used today in the United States in contemporary construction. Flemish bond sometimes appears in contemporary churches. Of course, any of the three may be needed to match existing construction.

Overlapping stretchers, laid parallel with the length of the wall, develop longitudinal bonding strength; headers laid across the width of the wall bond with the wall transversely. Some types of masonry bonded walls are shown in Figure 4.6-1, using brick (4.6-1h), overlapping stretchers (4.6-1g), brick and block or tile backup (4.6-1j).

Masonry bonded brick walls should be built so that not less than 4% of the wall's surface is composed of headers. The headers should extend not less than 4 in. into the backing, and should not be more than 24 in. apart either vertically or horizontally. This means that there must be at least one header course in every seven brick courses. In multiple-wythe tile or block walls, structural bonding is achieved by overlapping stretcher units rather than by using header units (see Fig. 4.6-1g). The stretcher units may be bonded at vertical intervals not exceeding 34 in. when the bonding course can overlap one-half of the unit below and the lap is at least 4 in. (Fig. 4.7-12a). When these conditions cannot be met, the bond may be achieved by overlapping headers that are at least 50% greater in thickness than the units below, and placing the bond courses at vertical intervals of 17 in. (Fig. 4.7-12b).

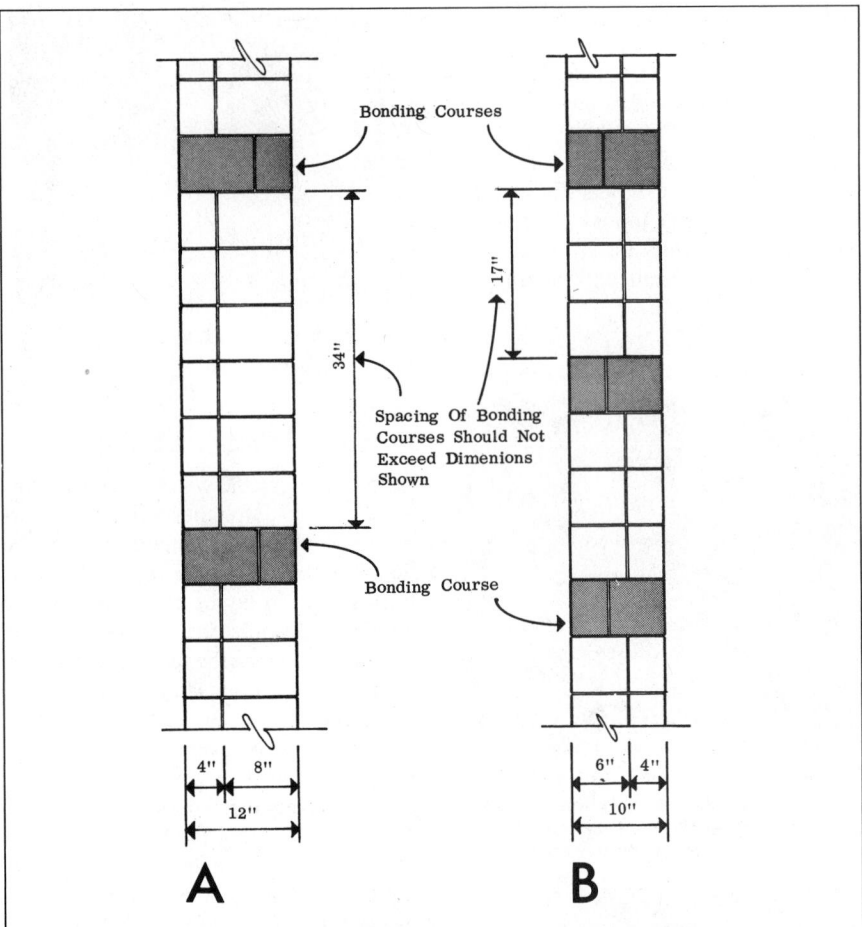

FIGURE 4.7-12 *The overlapping stretcher method of bonding.*

Unfortunately, water can penetrate a masonry bonded multiple-wythe wall fairly easily, for several reasons. First, it is sometimes difficult for a mason to obtain full mortar joints between brick headers. The resultant space permits water to find its way to the inside face of the wall along the headers, which are continuous through the full wall thickness.

Another reason that water can often penetrate a masonry bonded wall is that it is difficult to bond brick and larger, heavier backup units without impairing the bond between the headers and the mortar. If the backup unit is set on the header before the mortar has had time to stiffen, the header brick will settle unevenly and a crack may develop (Fig. 4.7-13).

Perhaps the most important reason that water may find its way through a masonry bonded wall is that there are differences in thermal, moisture, and other movements between different facing and backup materials. The differences are greatest between brick and concrete masonry units where opposing movement of the concrete masonry unit backup, which shrinks as it dries, and the brick, which progressively expands over time, creates an eccentric load on the brick headers that tends to rupture the bond between the headers and the mortar at the external face. The flexibility inherent in metal ties negates this disadvantage.

■ METAL-TIED, SINGLE-WYTHE WALLS

The units in the field of metal-tied, single-wythe walls are not bonded using metal ties. Metal ties are used, however, where such walls abut other walls.

When a single-wythe, load-bearing brick or concrete masonry unit wall is not masonry bonded to the walls that it abuts, except at the corners, it should terminate

FIGURE 4.7-13 *(a) The effect of a heavy backup unit on a masonry header. (b) A mason lays a header course in an 8-in. solid brick and block wall. In this case, the header courses occur in every seventh brick course.* (Photo courtesy Brick Institute of America.)

FIGURE 4.7-17 *This hardware cloth strip was left protruding from the completed wall at the right and is now being bedded into a joint in an intersecting wall.* (Courtesy Portland Cement Association.)

at the faces of these other walls in control joints. For lateral support, bearing walls should be tied to walls they intersect with rigid steel tie bars 1/4 inch thick, 1-1/4 in. wide, and 28 in. long, with a 2-in. right angle bend on each end (Fig. 4.7-14). Tie bars should be placed not more than 4 ft. apart vertically. The bends at the ends of the tie bars should be embedded in cores filled with mortar (Fig. 4.7-15).

Single-wythe non-load-bearing masonry unit walls or partitions that are 6 in. or less in thickness may be tied to walls they intersect using strips of metal lath or 1/4-in. mesh, galvanized hardware cloth placed across the joint between the two walls (Fig. 4.7-16). These metal strips should be placed in alternate courses (not

FIGURE 4.7-15 *The core into which the upturned end of the metal tie bar shown in Figure 4.7-14 projects is here being filled with mortar.* (Courtesy Portland Cement Association.)

FIGURE 4.7-16 *A strip of galvanized hardware cloth being used to tie a non-load-bearing wall to an intersecting wall.* (Courtesy Portland Cement Association.)

more than 16 in. apart vertically) in the wall. When one wall is constructed first, the metal strips should be built into this

FIGURE 4.7-14 *A metal bar anchor being installed to tie a bearing wall to an intersecting wall. Metal lath was used to support the mortar filling the core where the turned-down end of the bar is being inserted.* (Courtesy Portland Cement Association.)

wall and later embedded into the mortar joint in the second wall as it is built (Fig. 4.7-17). Corrugated metal ties may be used in the same way. Where the two walls meet, a control joint should be provided if the joint is exposed to view.

■ METAL-TIED, MULTIPLE-WYTHE WALLS

Both multiple-wythe solid and cavity walls can be structurally bonded using metal ties (see Fig. 4.6-1). For exterior walls, especially where weather resistance is important or the physical characteristics of the facing and backing materials are different, bonding with metal ties is better than the masonry bonding method. This is because metal-tied walls have greater resistance to moisture penetration and the ties allow slight differential movement between the facing and backing units, which may help to relieve stress and reduce cracking.

The usual requirements for metal-tie bonding are for 3/16-in.-diameter steel ties, or metal ties of equivalent stiffness and strength, to be spaced so that at least one tie occurs in every 4-1/2 sq. ft. of wall area. Except where different spacings are required by code or local practice, and unless another spacing is recommended later in this section for a specific wall type or condition, in solid, composite, and faced walls, the maximum vertical distance between ties should not exceed 18 in. and the maximum horizontal distance should not exceed 36 in. (Fig. 4.7-18). Ties in alternate courses should be staggered (see Fig. 4.7-18). Ties should be embedded in the horizontal joints of the facing and

FIGURE 4.7-18 *Wall tie locations in multiple-wythe walls.*

backing. Additional bonding ties should be provided at openings. These should be spaced not more than 3 ft. apart around the perimeter, within 12 in. of the openings, and on both sides of expansion and control joints.

General and specific recommendations for some types of walls and conditions are summarized below. Notice that some conditions require masonry bonding or a combination of masonry bonding and metal ties.

REINFORCED MASONRY WALLS The units in a single-wythe hollow reinforced masonry wall are bonded together by their pattern bond (see Figs. 4.6-12 and 4.6-14) and by the reinforcing steel and the grout in their cores (see Section 4.6.1.4).

The units in the separate wythes of reinforced solid masonry walls are bonded together by their pattern bond (see Figs. 4.6-12 and 4.6-14). The wythes are structurally bonded together by the reinforcing steel and grout in the cavity (collar joint) between the wythes of the wall (see Section 4.6.1.4).

FACED WALLS The facing should be bonded to the backing with metal ties.

CAVITY WALLS Cavity walls should be bonded with metal ties. Their spacings should not exceed 24 in. vertically or 36 in. horizontally.

BRICK VENEER WALLS In wood frame construction (see Fig. 4.6-10a) brick veneer is often anchored to the backup wall by nailing corrosion-resistant corrugated metal ties through the sheathing to the studs. This type of construction has been used for many years, but recently some experts have begun to question its longevity. Failure may result from rusting through leaks or from condensation from high humidity. Professional guidance should be sought before electing to use this method of construction and local practice should ultimately be followed.

There is a great deal of controversy about whether brick veneer should be used over metal studs (see Fig. 4.6-10b). There are more questions about using this combination on tall structures than about its use on one- and two-story residences and other small buildings. Its supporters claim that brick veneer over metal studs will give satisfactory long-term results if the proper materials are selected and are applied correctly. Detractors claim that brick veneer applied over metal studs will eventually crack, regardless of the materials used or the configuration of the wall. This will allow water to enter the wall and corrode the metal supports, resulting in veneer failure.

The key elements in the controversy are the required stiffness of the studs, the type of anchors to use, and whether any available anchoring system will perform satisfactorily in the long run. The most authoritative current advice about the methods needed to construct such a brick veneer wall are offered by the BIA. However,

before deciding to use brick veneer over metal studs it would be wise to contact industry sources, such as the BIA, the Portland Cement Association, the Metal Lath/Steel Framing Association, and the manufacturers of the products that will be used, and obtain their latest recommendations concerning this controversial use of brick. It is also necessary to determine if this type of construction is permitted by the applicable local codes and the owner. There has been some movement by some federal agencies and some states to ban its use altogether.

■ **ANCHORAGE**

Masonry assemblies must be securely anchored to elements that are used for vertical and horizontal support. These include steel and concrete columns and beams; concrete walls; concrete and masonry bond beams; and floors, other walls, and roofs. Flexible metal ties that permit at least some differential movement usually supply the anchorage.

Exterior walls facing or abutting concrete members should be anchored by dovetailed anchors inserted in slots set into the concrete (see Fig. 4.3-1). These anchors should be spaced not more than 18 in. vertically and 24 in. horizontally.

4.7.2.3

Continuous Joint Reinforcement

Continuous joint reinforcement is often used today to control cracks in masonry walls (see Figs. 4.3-1, 4.6-1, and 4.6-7). When joint reinforcement is used, it often replaces metal wall ties to do the same job. In wythes laid in a stacked bond pattern, continuous joint reinforcement is necessary to bond the units together. Continuous joint reinforcement is also sometimes used in lieu of horizontal reinforcement in reinforced masonry walls.

Continuous joint reinforcement should be installed in every masonry wall and partition that is more than 20 ft. long, and is usually installed regardless of length. Using continuous joint reinforcement not only reduces the amount of cracking associated with thermal movements, it also permits a greater spacing of control joints (see Fig. 4.6-20).

Joint reinforcement should be embedded in horizontal mortar joints so that it is at least 5/8 in. from the masonry surface and should be continuous, except through control and expansion joints. Lengths of joint reinforcement should be lapped at least 6 in. and the lap should contain at least one cross-wire of each length.

Continuous joint reinforcement should be installed, in most cases, at not more than 16 in. on center vertically. In reinforced walls, continuous reinforcement is sometimes more closely spaced. In parapet walls and foundation walls, reinforcement is often spaced every 8 in. Local codes sometimes govern this placement. To compensate for loss of reinforcement at wall openings, joint reinforcement should be installed in the first and second bed joints immediately above and below openings and should extend at least 24 in. beyond the ends of sills and lintels.

4.7.2.4

Masonry Support for Structural Members

The bearing course of masonry under structural members such as floor joists, concrete slabs, beams, and lintels should be at least 8-in.-thick solid masonry or hollow masonry, with the cells filled solidly with mortar or fine grout.

4.7.2.5

Mortar Bedding and Jointing

Laboratory tests and field observations of masonry indicate that good masonry construction demands that the joints be completely filled with mortar as the units are laid. Partially filling joints results in leaky walls, reduces the strength of masonry from 50% to 60% percent, and contributes to disintegration and cracking when water penetrates the wall and expands as it freezes.

■ BRICK

Clay and concrete brick and other solid masonry units should generally be embedded and jointed in similar ways. Of primary importance is that every head, bed, and collar joint in them be filled solidly with mortar.

Mortar joints in contemporary brick walls usually vary from 1/4 to 1/2 in. in thickness, depending on the size of the brick and the bonding. Thicker joints are sometimes used, however.

BED JOINTS The mortar in bed joints in brick should be spread to a uniform thickness and furrowed only slightly (Fig. 4.7-19). As the brick is laid excess mortar will fill the furrow and ensure full bed joints. If the bed is too thin or the furrow is too deep there will not be enough mortar to fill the furrow completely, and the resulting void will enable water to enter an opening in the mortar or between the mortar and the brick and penetrate the wall (Fig. 4.7-20).

FIGURE 4.7-19 *In brick, the mortar furrow in the bed joint should be shallow.* (*Courtesy Brick Institute of America.*)

FIGURE 4.7-20 *The furrow in the bed joint shown here is too deep.* (*Courtesy Brick Institute of America.*)

To ensure a good bond, mortar should not be spread too far ahead of the actual laying of block or brick, or it will stiffen and lose its plasticity (Fig. 4.7-21). This is especially true in hot weather when absorbent bricks are being laid. One way to determine if a bond is developing is to remove a freshly laid brick from the wall (Fig. 4.7-22). If the mortar is soft and plastic, it will stick to the brick placed on

FIGURE 4.7-21 *Mortar should be spread only a few units ahead of the masonry.* (*Courtesy Brick Institute of America.*)

FIGURE 4.7-22 *This mason has removed a newly laid brick to determine if adequate bond is being achieved.* (*Courtesy Brick Institute of America.*)

top of the bed as well as to the brick on which the bed is spread, and adequate bond will be attained.

HEAD JOINTS Head joints are more vulnerable to water penetration than bed joints because the weight of the wall above tends to close cracks in bed joints. Therefore, every head joint in both the facing brick and backup units should be completely filled with mortar. One method is to spread the mortar on the end of a brick before it is laid (Fig. 4.7-23). When the brick is shoved into place, mortar should squeeze out at the sides and top, indicating that the head joint is completely filled (Fig. 4.7-24). An alternative method is to throw plenty of mortar on the end of the brick already in place (Fig. 4.7-25), giving the desired result (see Fig. 4.7-24) when the next brick is shoved into place. A third method is to place a full trowel of mortar on the wall (Fig. 4.7-26) and shove the next brick into this deep mortar bed, which will squeeze mortar out at the sides and top of the joint (Fig. 4.7-27).

Dabs of mortar spotted on both corners of the brick (Fig. 4.7-28) will not completely fill the head joints (Fig. 4.7-29).

FIGURE 4.7-23 *Loading a brick before laying it.* (Courtesy Brick Institute of America.)

FIGURE 4.7-24 *Mortar oozing out indicates that the joint has been filled completely.* (Courtesy Brick Institute of America.)

FIGURE 4.7-25 *Buttering the end of a brick that has already been laid.* (Courtesy Brick Institute of America.)

FIGURE 4.7-26 *Throwing a full trowel of mortar on an already installed brick to receive the next unit.* (Courtesy Brick Institute of America.)

FIGURE 4.7-27 *A brick being shoved into the mortar that was thrown on in Figure 4.7-26.* (Courtesy Brick Institute of America.)

FIGURE 4.7-28 *Spotting the corners of a brick. This will produce unsatisfactory head joints and must not be permitted.* (Courtesy Brick Institute of America.)

FIGURE 4.7-29 *The result of spotting the corners as shown in Figure 4.4-28. Note that the head joint is not filled completely.* (Courtesy Brick Institute of America.)

FIGURE 4.7-30 *Attempting to fill a head joint by slushing.* (Courtesy Brick Institute of America.)

Attempting to fill the joints from above after the brick is placed (Fig. 4.7-30), which is called slushing, cannot be relied upon to fill voids left in head joints (Fig. 4.7-31).

Cross joints in header courses must also be completely filled with mortar. This is accomplished by spreading mortar over the entire side of the header brick before it is placed, and then shoving it into place to fill all voids (Fig. 4.7-32).

CLOSERS   The last brick in a course is called a closer. Before laying a closer brick, plenty of mortar should be placed

FIGURE 4.7-31 *A freshly laid brick has been removed after slushing (see Figure 4.7-30). Note that the head joint was not adequately filled.* (Courtesy Brick Institute of America.)

FIGURE 4.7-32 *Filling a cross joint in a header course with mortar. The side of a header brick is buttered (top) and shoved into place (bottom).* (Courtesy Brick Institute of America.)

both on the brick already in place (Fig. 4.7-33) and on both sides or ends as applicable for the closer brick (Fig. 4.7-34). The closer should then be laid without disturbing the brick already in place (Fig. 4.7-35).

■ CONCRETE MASONRY UNITS

Mortar joints for concrete masonry units should be 3/8 in. thick.

FIGURE 4.7-33 *Placing mortar on a brick that is already in place. (Courtesy Brick Institute of America.)*

FIGURE 4.7-34 *A closer unit being buttered. (Courtesy Brick Institute of America.)*

FIGURE 4.7-35 *The unit in Figure 4.7-34 is here being placed in the wall. (Courtesy Brick Institute of America.)*

For the first course of both solid and hollow units, a full, thick mortar bed should be spread on the foundation or other support under the entire unit, including the cells, and furrowed with a trowel to ensure that there will be enough mortar along the bottom edge of the face shells and webs of the block to completely fill the joint (Fig. 4.7-36).

FIGURE 4.7-36 *A furrowed bed joint and three blocks with their face shells buttered prior to laying. Note the full mortar bedding. (Courtesy Portland Cement Association.)*

In columns, piers, pilasters, and where adjacent cells will be filled with grout, the face shells and webs should be set in a full and complete bed of mortar. This is called full mortar bedding (Fig. 4.7-37).

FIGURE 4.7-37 *Full mortar bedding. (Courtesy Portland Cement Association.)*

Solid units should be laid in full mortar beds regardless of their use or location.

In all other concrete unit masonry work, mortar is usually applied only to the horizontal face shells of the units. This is called face-shell mortar bedding (Fig. 4.7-38).

HEAD JOINTS In vertical joints in concrete block, only the ends of the face

FIGURE 4.7-38 *Face-shell mortar bedding. (Courtesy Portland Cement Association.)*

shells are embedded in mortar. This embedding should, however, be complete and solid for the entire depth of the face shell.

Mortar for vertical joints in the block can be applied to the vertical face shells of either the block to be placed or the block previously laid. Some masons "butter" both edges to ensure filled joints. By placing three or four blocks on end, a mason can butter the vertical face shells of all of them in one operation (see Fig. 4.7-36). Each block is then brought to its final position and pushed downward into the mortar bed and against the previously laid blocks, thus producing full vertical mortar joints (Fig. 4.7-39).

FIGURE 4.7-39 *Positioning a block before setting it in place. Note the full vertical joints between the blocks that are already in place. (Courtesy Portland Cement Association.)*

CLOSERS The last unit in a course is called a closer. Before laying closer units, plenty of mortar should be placed both on the unit already in place and on both ends of the closer unit. The closer should then be laid without disturbing the unit already in place (Fig. 4.7-40).

FIGURE 4.7-40 *Setting a closer block.*
(Courtesy Portland Cement Association.)

FIGURE 4.7-41 *Backparging a collar joint.*
(Courtesy Portland Cement Association.)

■ STRUCTURAL CLAY TILE AND OTHER HOLLOW MASONRY UNITS

In general, the requirements discussed above for hollow concrete masonry units apply as well to hollow brick, structural clay tile, and other hollow masonry units. This section includes several exceptions.

Vertical-cell structural clay tile units should be laid with divided (not full) head joints and, when exposed to the weather, care should be taken to prevent continuous mortar joints through the wall.

Horizontal-cell structural clay tile units that are more than 4 in. thick should be laid with divided head joints. Head joints of horizontal-cell units should be placed on both sides of the tile and sufficient mortar used so that excess mortar will be squeezed out of the joints as the units are placed in position.

Backparging

The vertical, longitudinal joints between the wythes of multi-wythe masonry walls, which are called collar joints, should be completely filled by parging either the face of the backing or the back of the facing. This provides some additional assurance against leakage (Fig. 4.7-41). Mortar extruding into a collar joint should be cut flush before it hardens. Parging placed over hardened mortar will not bond to it and may result in a leaky wall (Fig. 4.7-42). Filling collar joints from above, which is called slushing, will not fill the joint completely and should not be permitted (Fig. 4.7-43). For best results, parging should be done as the walls are

FIGURE 4.7-42 *Mortar joints that have been cut flush before they hardened.*
(Courtesy Portland Cement Association.)

FIGURE 4.7-43 *A freshly laid wythe of masonry has been removed to show the spotty and incomplete backparging that resulted from slushing a collar joint.*
(Courtesy Portland Cement Association.)

laid up. Careful workmanship is necessary because the pressure required to spread a parge coat of mortar on newly laid units can break the bond between the units and their mortar joints, producing cracks that may contribute to future leakage.

Joint Finishing

When masonry walls leak, the water often enters through improperly sealed cracks and voids around openings, through flashings between the wall and a roof, or through cracks where the wall intersects with other materials or construction. Even when water actually penetrates a wall, the leak is almost never through the masonry units themselves. Most of the time, water enters a masonry wall through cracks and voids between the masonry units and the mortar in the joints. Many such leaks can be prevented by properly finishing the joints. Mortar joint finishes fall loosely into three types: troweled joints, tooled joints, and sealant-filled joints.

Regardless of the type of joint finishing used, care should be taken to prevent smearing mortar or sealants onto the finished surface of masonry units that will be exposed to view. Hardened, embedded mortar often cannot be removed. Mortar droppings that adhere to a wall should be allowed to dry before their removal is attempted. Removal should be done with a trowel so that the mortar will not become embedded into the face of the unit.

■ TROWELED JOINTS

In a troweled joint, excess mortar is simply cut off (struck) with a trowel and no further finishing is used. Joints should be struck flush on surfaces that will be parged, plastered, stuccoed, or covered with other masonry.

The best troweled joint is a weathered joint (Fig. 4.7-44), which will shed water. Unfortunately, producing this type of joint does not force enough mortar against the brick above to form a watertight joint. Therefore, it should not be used where exposed to the weather.

A flush joint (see Fig. 4.7-44) produces an uncompacted joint with a hairline crack where the mortar is pulled away from the masonry by the cutting action of the trowel (Fig. 4.7-45).

A struck joint (see Fig. 4.7-44) and its kin, a raked joint, which is created by removing the surface of the mortar while it is still soft, both leave a shelf on which water can collect. Raked joints are popular with some architects in spite of their drawbacks, however, because they produce crisp, dark joint lines.

FIGURE 4.7-44 *Types of mortar joints.*

TOOLED | TROWELLED
1 Concave | 2 Vee | 3 Weathered | 4 Flush | 5 Struck | 6 Raked

FIGURE 4.7-45 *A mortar joint being cut flush.* (*Courtesy Brick Institute of America.*)

■ TOOLED JOINTS

Because tooled joints provide excellent protection against water penetration they should be used in every wall that is exposed to the weather, except where sealant-filled joints are required. Joints in below-grade walls that will not be parged should also be tooled. For example, slab-on-grade and crawl space foundations need not be parged when moisture penetration is not a problem, and if they are not, then those joints should be tooled. Most exposed interior joints are also tooled to give them a finished appearance, even though water penetration is not usually a problem on the interior.

The most common tooled joints used today are either concave or V-shaped (see Fig. 4.7-44). Historically, many other tooled joint shapes have been used, however, and may be encountered in renovation work. To make a tooled joint, a special tool, called a jointer, is used to compress and shape the mortar. Tooling compacts the mortar and forces it tightly against the masonry on each side of the joint (Fig. 4.7-46). Tooling should be done as soon as a section of the wall has been laid and the mortar has become "thumbprint" hard. Every joint should be tooled at the same degree of mortar hardness so that the joints will all be the same color and uniform in appearance with clean, sharp lines. The jointer should be slightly larger than the joint itself so that complete contact is made along the edges of the masonry unit (thus compressing and sealing the surface of the mortar), and upturned on the ends to prevent gouging the mortar (Fig. 4.7-47a, b).

■ SEALANT-FILLED JOINTS

The joints between masonry and the perimeter of door and window frames,

FIGURE 4.7-46 *A mortar joint before and after tooling.* (*Courtesy Brick Institute of America.*)

louvers, and other penetrations, and joints where masonry abuts other materials should be sealant-filled joints. Expansion and control joints are also usually sealant filled. Where joints to be sealant filled contain mortar or grout, they should be cleaned out to a uniform depth of at least 3/4 in., and then filled solidly with an elastic sealant compound forced into place with a pressure gun. Sealant-filled joints should not be less than 1/4 in. wide. Sealant materials and methods of installation are discussed and illustrated in Section 7.10.

4.7.2.8

Flashing

The general principles, materials, and methods used to install flashing are discussed in detail in Section 7.9. Wall flashing in masonry is included here, however, because, unlike many other flashing installations, it is usually installed by the mason as the masonry is erected, and its installation is generally thought of in the industry as part of the masonry work. Because most of the recommendations about flashing in Section 7.9 apply as well to the flashing discussed here, those requirements have not been detailed here.

Except in areas of very slight exposure to precipitation (see Fig. 4.6-4) where local experience recommends the omission of it, flashing should be installed at the following vulnerable locations: (1) at the base of a wall; (2) under the sills of wall openings; (3) over the heads of wall openings; and (4) where masonry penetrates a roof surface (Fig. 4.7-48).

■ PURPOSE OF FLASHING

Masonry wall flashing (Fig. 4.7-49) is installed to collect the moisture that may penetrate the wall and to divert it through weep holes to the exterior. The amount of exposure of masonry to rain, which varies greatly in the United States, is a basic consideration when designing masonry walls to resist moisture penetration. Exposure may be defined as *severe*, *moderate*, or *slight*, roughly in terms of the resultant wind pressures and precipitation maps shown in Figure 4.6-4. *Severe* exposure means an annual precipitation of 30 in. or more, with a wind pressure of 30 lb./sq.

FIGURE 4.7-47 *Here the joints are being tooled (a) in a concrete masonry unit wall with a concave jointer* (courtesy Portland Cement Association) *and (b) in a brick wall with a V-jointer* (courtesy Brick Institute of America).

ft. or higher. *Moderate* exposure means an annual precipitation of 30 in. or more and a wind pressure of 20 to 25 lb./sq. ft. *Slight* exposure means an annual precipitation less than 30 in. and a wind pressure of 20 to 25 lb./sq. ft., or an annual precipitation less than 20 in.

The following recommendations apply to areas of severe exposure, such as the Gulf Coast, Atlantic Seaboard, and Great Lakes region, and to areas of moderate exposures, which are typical in the Midwest. Where exposure is slight, completely concealed flashing, such as lintel or sill flashing, is often eliminated, and in some areas even exposed flashing is reduced to a minimum.

■ **FLASHING LOCATIONS**

The vulnerable locations in masonry where wall flashing is necessary are described below (see Fig. 4.7-48). Flashing should be provided under horizontal masonry surfaces such as sills and copings, at intersections of masonry walls with horizontal surfaces such as the roof and a chimney, over the heads of openings such as doors and windows, and, frequently, at floor lines, depending on the type of construction.

Base flashing and damp checks are used at the base of a wall. Base flashing diverts to the exterior any moisture that enters the wall above the location of the flashing. A damp check stops the upward capillary travel of ground moisture. If installed properly, metal shields used for termite protection of wood joists (see Sections 2.2 and 6.5.2.2) may double for these purposes.

Sill flashing should be placed under and behind every sill. The ends of sill flashing

should extend beyond the sides of an opening and turn up at least 1 in. into the wall. Sills should slope and project from the wall to drain water away from the building and to prevent staining. When the undersides of sills do not slope, a drip notch should be provided, or else the flashing may be extended and bent down to form a drip.

Head flashing should be placed over all openings except those completely protected by overhanging projections. At steel lintels the flashing should be placed under and behind the facing material, with its outer edge bent down over the lintel to form a drip.

Roof flashing at chimneys consists of base flashing turned up against the masonry (see Fig. 4.7-48 and Sections 7.6, 7.7, and 7.9) with counterflashing overlapping the base flashing and extended and embedded securely into the mortar joints.

■ **FLASHING INSTALLATION**

When flashing is to be laid on or against masonry, the surface of the masonry must be smooth and free from projections that would puncture the flashing material or otherwise destroy its effectiveness. Through-wall flashing should be placed on top of a thin mortar bed and then another thin mortar bed should be placed on top of the flashing to receive and bond with the next masonry course. Metal flashing should be deformed with ridges or depressions to help lock the mortar in place.

Flashing seams must be overlapped and thoroughly bonded or sealed to prevent water penetration. Although most sheet metal flashing materials can be soldered, lockslip joints (interlocking joints that can

move) are required at intervals to permit thermal expansion and contraction. Many plastics can be permanently and effectively joined by heat or an appropriate adhesive. The elastic pliability of plastic flashings eliminates the need for expansion and contraction seams.

4.7.2.9

Weep Holes

The water collected by flashing must be drained to the outside. Concealed flashings in tooled mortar joints are not self-draining and so require weep holes. Actually, such flashing acts as a trap to collect water and, by concentrating it in one place, can do more harm than good. Therefore, except at windowsill flashing where little water collects, weep holes should be provided in the head joint immediately above all flashing.

Weep holes may be formed in several ways, and although no one way provides better weep holes than the others, some have a few disadvantages. Weep holes may be formed by simply omitting the mortar in the head joints at the desired intervals, or by inserting oiled rods or pins and then removing them when the mortar is ready for tooling. Weep holes may also be made by inserting plastic or metal tubes, or metal or plastic vents, in the head joints and leaving them in place. Another way is to insert short lengths of sash cord and leave them in place to disintegrate over time. In any of these methods, weep holes should be placed at 24-in. intervals horizontally. Unfortunately, these methods all produce open weep holes, which provide access for insects and, especially when placed over lintel flashing, may contribute to wall staining. Care must be taken to prevent open weep holes from being clogged with mortar droppings, dirt, or other foreign materials.

Closed weep holes can be created by inserting wicks in the head joints (see Fig. 4.3-4). These wicks are made of 1/4-in. glass-fiber rope or a similar inorganic material. Another way is to insert 1/2-in. glass-fiber insulation in open head joints. Since they do not drain as well as open joints, closed weep holes should be spaced every 16 in. Care must be taken to ensure that mortar droppings do not block the holes behind the wick or insulation,

CHIMNEY FLASHING

FLUE LINING

COUNTERFLASHING TURN INTO JOINTS 1" AND LAP OVER BASE FLASHING

CHIMNEYS WIDER THAN 30" REQUIRE A SADDLE (or cricket) SEE DETAIL 'A' IF CHIMNEY IS LESS THAN 30" WIDE

METAL SADDLE

BASE FLASHING TURN UP UNDER COUNTERFLASHING 3" MINIMUM

3" MIN. FOR SLOPES UNDER 5 in 12

4" MIN. FOR SLOPES 5 in 12 AND OVER

FACE OF CHIMNEY (high side) level

EXTEND FLASHING TO A POINT LEVEL WITH Ⓐ (not less than 1 1/2 times shingle exposure)

DETAIL 'A'

SOLID MASONRY WALL

CAULKING

FLASHING

STEEL LINTEL

WEEP HOLE

CAVITY WALL

FLASHING

STEEL LINTEL

CAULKING

MASONRY VENEER

FLASHING

CAULKING

HEAD FLASHING

CAULKING

STONE SILL

FLASHING

BRICK SILL

MASONRY VENEER

WOOD FRAME WALL

CAULKING

STONE SILL

FLASHING

CAVITY WALL

SILL FLASHING

MASONRY VENEER

CONTINUOUS BASE FLASHING

WEEP HOLE

EXTEND BASE FLASHING 6" BEHIND SHEATHING OR SHEATHING PAPER

WOOD FLOOR SYSTEM

BASE FLASHING MAY BE EXTENDED TO FORM A TERMITE SHIELD

CONCRETE SLAB

FOUNDATION WALL

BASE FLASHING

FIGURE 4.7-48 *Details showing flashing installations in masonry assemblies.*

FIGURE 4.7-49 *Through-wall flashing.*
(*Courtesy Portland Cement Association.*)

and that the materials in the holes do not themselves become clogged with mortar or dirt.

4.7.3

PARGING AND DAMPPROOFING

When the groundwater table and subsoil conditions do not cause water to build up hydrostatic pressure against masonry foundation and basement walls below grade, but moisture is present in the soil, then penetration of that moisture can usually be controlled by parging and dampproofing the wall. Dampproofing is a coating applied to resist moisture penetration as a result of capillary action (see Section 7.2). When hydrostatic pressures may develop, concrete foundation walls are normally used rather than masonry, and water penetration is controlled by waterproofing. Waterproofing is a continuous, watertight membrane capable of resisting water under pressure (see Section 7.2).

Parging consists of Type M mortar or portland cement plaster applied to the exterior face of a below-grade wall and troweled to a smooth, dense surface (Fig. 4.7-50a). Portland cement mortar applied above grade is called stucco (see Section 9.4). Parging should be thickened and turned out onto the footing to form a cove (see Fig. 4.7-50) and should extend upward from the footing to a line about 6 in. above finished grade. Its top should be beveled to form a slope (wash).

Parging should be applied in two coats. The first coat should be roughened when partially set, hardened for 24 hours, then

FIGURE 4.7-50 *(a) Applying parging; (b) a wall section showing parging and dampproofing. Note that the parging turns out onto the footing to form a cove.* (Courtesy Portland Cement Association.)

moistened before the second coat is applied. Parging should be moist cured for at least 48 hours and dry before dampproofing is applied over it. In single-family residential buildings, parging is sometimes applied in only one 1/4-in.-thick coat and is often applied in two 1/4-in.-thick coats to a completed thickness of 1/2 in. In larger projects, two 3/8-in.-thick coats are usually applied for a total thickness of 3/4 in. The latter method produces a better job and should be used even on small projects. A single 1/4-in.-thick coat is much too thin to form a credible parging. Even a 1/2-in. coat is too thin; the additional cost of the extra 1/4 in. of parging is small when compared with the benefits. One problem with thin coats of parging is that walls are seldom straight and never smooth. The parging tends to average the projections and depressions. Over the projections in a wall, a thin coat of parging tends to get even thinner. Very thin coats may flake off the wall when they are subjected to wall movement. When pieces of parging fall off, the covering dampproofing goes with them.

Bituminous dampproofing products (see Section 7.2) may be applied either over parging or directly onto the masonry to increase resistance to moisture penetration. Parging is sometimes omitted when the wall to be dampproofed is smooth and the soil is well drained and dry.

Dampproofing of interior masonry walls is also discussed in Section 7.2.

4.7.4

BRICK PAVING AND FLOORING

Brick *paving* is installed exteriorly, where it is generally exposed to weather. Brick *flooring* is installed interiorly, where it is not exposed to weather. However, even on the interior a brick traffic surface is exposed to severe wear and must therefore be highly durable. Because a unit's resistance to freeze-thaw cycles is a good measure of its durability, units with this characteristic are often chosen even for indoor use and for use outdoors where severe weather is not a problem.

Brick paving and flooring are laid in any one of many possible patterns. Figure 4.7-51 shows some of the more popular ones.

Brick paving is sometimes laid in a mortar setting bed with grouted joints, but is more often laid in a leveling bed without mortar. This leveling bed may be sand, gravel, crushed stone, or a mixture of sand and cement. Brick flooring is also sometimes laid in a leveling bed, but more often is set in mortar, because of the difficulties of sealing unfilled joints to prevent liquids from entering them when the floor is cleaned. Brick paving is also sometimes set in bituminous setting beds because these are less vulnerable to damage from freezing.

The mortar and grout used in paving are usually in the form of a thick-set applica-

tion of portland cement mortar mixed with a latex additive. Flooring may be laid in a thin-set mortar bed. A thin-set bed is lighter and requires less floor-to-floor height because it is thinner than a thick-set method. A thin-set method, however, requires that the substrate be sufficiently level and properly sloped to permit the thinner application. In addition, curing compounds cannot be used on a slab that will receive flooring set in a thin-set mortar bed. Thick-set beds may be bonded to a slab on grade, but should be placed over a *cleavage membrane* on slabs subject to bending or deflection, so that these movements do not disrupt the flooring or paving.

Brick pavers may be set in either thin-set or thick-set beds with open joints or with hand-tight closed joints. Open joints are filled with latex-modified cement grout. Hand-tight closed joints are filled by sweeping a dry mixture of portland cement and sand into them and fogging the surface with water.

4.7.5

PROTECTION DURING CONSTRUCTION

Masonry under construction must be protected from rain and, in cold weather, from snow and freezing. Methods of providing this protection are discussed in this section. Lateral support of walls during construction is covered in Section 4.6.5.4.

4.7.5.1

Protection from Rain and Snow

Unprotected, partially finished masonry that becomes saturated with water during a rain- or snowstorm may take months to dry out, and during this drying period efflorescence may appear on the surface (see Section 4.2.1.2).

At the end of each day or shutdown period when there is a threat of precipitation, partially completed masonry should be protected by covering it with securely weighted-down canvas or other strong waterproof membrane. This covering should overhang at least 2 ft. on each side of a wall and be securely anchored (Fig. 4.7-52).

FIGURE 4.7-51 *Brick flooring patterns. Patterns may be laid parallel with the borders or on a diagonal.* *(Courtesy Brick Institute of America.)*

FIGURE 4.7-52 *One way to protect the exposed top of a masonry wall under construction.* *(Courtesy Portland Cement Association.)*

4.7.5.2

Cold-Weather Construction

As the air temperature falls, mason productivity and masonry strength are affected.

Hydration, here the chemical reaction between portland cement and water, slows significantly at temperatures below 40 °F. Below 32 °F hydration proceeds only when sufficient water is available. Hydration ceases altogether at about 14 °F. Therefore, masonry should not be laid when the air temperature is 40 °F or lower, unless the proper precautions are taken.

The procedures for mixing mortar in cold weather are addressed in Section 4.2.6.3. Figure 4.7-53 summarizes the other precautions required in cold weather. The basic principle is that at air temperatures below 40 °F, masonry materials must be heated to achieve satisfactory masonry assemblies. Masonry construction may continue even when temperatures drop below freezing, but only when the proper measures (such as heating them) are taken to make the materials act as they would

at temperatures above 40 °F.

It is also very important that mortar and masonry units not be laid on frozen materials and that the masonry units be at a temperature of at least 20 °F and contain no visible frost or ice at the time of laying. Stored masonry units must be protected and warmed if necessary. If wetting the brick is necessary (see Section 4.7.1.2) in cold weather, then the water used must be heated.

Masonry must be protected from precipitation and winds in excess of 15 miles per hour and maintained at above-freezing temperatures during construction and for at least 24 hours after construction. Even if air temperature is between 32° and 40 °F, completed masonry must still be kept covered and protected from rain and snow for at least 24 hours after construction.

FIGURE 4.7-53
Recommended Procedures for Masonry Construction in Cold Weather[a]

Work Day Temperature	Construction Requirement[b]	Recommended Protection[b]
All temperatures	Normal masonry procedures.	Masonry materials should be received, stored, and protected in ways that prevent water damage. Cover tops of newly laid masonry with plastic or canvas to prevent water damage.
40° to 32°F	Heat mixing water to produce mortar temperatures between 40° and 120°F. Never place masonry on frozen, snowy, or ice-covered surfaces.	Cover walls and materials with plastic or canvas to prevent wetting and freezing.
32° to 25°F	Heat mixing water and sand to produce mortar temperatures between 40°F and 120°F.	Provide windbreaks for wind velocities over 15 mph. Cover walls to prevent freezing. Maintain masonry above 32°F for 24 hours using auxiliary heat for insulating blankets.
25° to 20°	Mortar on boards should be maintained above 40°F. Masonry units below 20°F must be heated above 20°F.	
20° and below	Heat mixing water and sand to produce mortar temperatures between 40° and 120°F.	Provide enclosures and supply heat to maintain masonry enclosure above 32°F for 24 hours.

[a] For unenclosed construction sites. These recommendations do not apply where the complete work area, including materials storage space, is enclosed and heated.

[b] Each requirement is in addition to the one above it.

FIGURE 4.7-54 *Nail and line-pin holes should be filled as soon as possible.* (Courtesy Brick Institute of America.)

4.7.6

POINTING AND CLEANING MASONRY

Exposed masonry should be protected against staining by covering it, and excess mortar should be wiped off the surface as the work progresses. When the masonry work has been otherwise completed, the joints should be pointed and the entire masonry should be carefully cleaned. Large particles of excess mortar should be removed using a putty knife or chisel and the masonry should be cleaned of dirt, dust, efflorescence, stains, and other foreign materials.

4.7.6.1

Pointing

Holes in exposed masonry, except for weep holes, should be *pointed* (filled), and defective joints should be cut out and re-pointed with fresh mortar.

Holes left by nails or line pins should be pointed while the adjacent mortar is green enough to allow an adequate bond with the patch. Otherwise, water may enter through them into the wall (Fig. 4.7-54).

4.7.6.2

Cleaning Clay Masonry

Unglazed brick and tile should be cleaned to remove general dirt, grime, and stains. Cleaning should not be attempted until at least 48 hours after completion.

GENERAL CLEANING

Unglazed brick and tile on which no efflorescence or other stain appears should first be cleaned with stiff brushes and clean water. Soil not removed by that cleaning should be scrubbed off using stiff brushes, clean water, and soap powder, or another mild solution recommended by the manufacturer of the masonry unit.

Only after such mild cleaning has been unsuccessful should stronger measures be used. With the concurrence of the masonry unit manufacturer, stubborn dirt may be removed with a mild solution of acid similar to that suggested under "Mortar Stains," later in this section. The surface should be thoroughly wetted with clean water and then scrubbed with the acid, followed immediately by a thorough rinsing with clean water.

The exposed surfaces of glazed brick, glazed structural clay tile, and terra cotta should be washed with detergent powder and warm water, applied with a scrubbing brush, and then rinsed thoroughly with clean water. Metal cleaning tools and brushes, acid solutions, and abrasive materials should not be used.

Regardless of the cleaning solution, the cleaning should begin at the top and proceed downward. The wall should be wetted with water before a cleaning solution is applied, and adjacent construction should be protected. Stiff fiber brushes should be used. Only a small area should be cleaned at one time. The cleaning should then be followed immediately by a thorough rinsing with clean water.

EFFLORESCENCE AND OTHER STAINS

Large amounts of water-soluble salts in mortar and masonry units can cause efflorescence or lime run to appear on the face of the masonry if excessive water penetrates the masonry. The most common form of efflorescence is a white powdery deposit of alkali salts and carbonates (Fig. 4.7-55), but all white deposits on masonry are not efflorescence, and

FIGURE 4.7-55 *A bad case of efflorescence.* *(Courtesy Brick Institute of America.)*

salts of other colors sometimes appear as well. For example, a white stain may be lime run, as described below, or a silicate deposit called scumming or white scum. Other stains on brick masonry may be so-called green stain caused by vanadium salts or brown, tan, or gray manganese dioxide salts. These water-soluble salts are originally present in either the mortar or the masonry units. They are carried to the surface by the movement of any water that has entered the wall. The dissolved salts are left on the face of the masonry when the water evaporates.

White efflorescence will often be washed away through normal weathering, especially if it is the result of water in a wall that dried after completion. When needed it can usually be removed by dry brushing alone. When that fails, a cleaning with water and stiff brushes will usually do the job. In severe cases, a solution of 1 part muriatic acid (a solution of hydrochloric acid and water) to 12 parts water may be used.

White scum is very difficult and sometimes impossible to remove. Green stain is also difficult to remove. Magnesium stains, although not difficult to remove, do require special techniques. All such stains should be removed in the manner and using the materials recommended by the manufacturer of the masonry units that are affected. Different methods may be needed for light-colored masonry than those used for darker masonry units. In the absence of specific recommendations regarding removal of stains from brick masonry, the BIA's Technical Notes 20 and 23 offer excellent guidance.

The condition called lime run, which is actually a deposit of carbonate, presents itself as a vertical white crusty area that has dried below a hole in a mortar joint. It is often mistaken for efflorescence. Despite its name, lime run is not associated with the lime in a mortar, but is thought to be a deposit of one or more calcium compounds that have been dissolved in a water stream that moved through the joint to react with carbon dioxide in the air. Lime run can be removed using a weak solution of hydrochloric acid. The condition will return, however, unless the water source is removed.

■ MORTAR STAINS

When possible, mortar stains should be removed from brick by scrubbing with detergent and warm water. More harsh cleaning methods should not be used on buff, gray, or other light-colored bricks without the concurrence of their manufacturer. Unless specifically not recommended by the manufacturer, when milder methods have failed, mortar may be removed from darker colored bricks using a mild acid solution.

When use of acid is unavoidable, muriatic acid is often used. Muriatic acid should be further diluted until it is no stronger than 1 part muriatic acid to 9 parts water. Even this dilute a solution should not be used until the mortar in the wall to be cleaned is well hardened and at least 7 days old. Acid solutions should always be tested in an inconspicuous location on a wall.

Lumps of mortar should be removed from terra cotta using sharpened wood paddles. Acid should never be used on terra cotta.

4.7.6.3

Cleaning Concrete Unit Masonry

Because concrete masonry should not be cleaned with an acid wash, particular care should be taken to prevent smearing mortar into the surface of an exposed unit. Mortar droppings should be allowed to dry before removal is attempted. Trying to remove wet mortar will probably result in the mortar being smeared into the surface of the unit. Hardened embedded mortar smears can never be removed and they will detract from the otherwise neat appearance of a finished wall. Conversely, most dry, hard mortar can be removed with a trowel. A final brushing will then remove practically all the remaining mortar (Fig. 4.7-56).

FIGURE 4.7-56 *Final brushing.* *(Courtesy Portland Cement Association.)*

4.7.7

INSTALLATION OF ARCHITECTURAL TERRA COTTA

Historically, architectural terra cotta was tied to a building's structure using metal anchors. Then it was backfilled to make it more fireproof and to protect the concealed anchors.

Present-day terra cotta is applied in two ways. The first method is used to install adhesion-type ceramic veneer and terra cotta tile (see Section 4.4.4.3). It is similar to the method used to install ceramic tile. In this method the wall is dampened and the terra-cotta units are soaked in water for 1 hour. Then a brush coat of portland cement and water is applied to the back of the units. Mortar (see Section 4.2.4) is then applied both to the units and the wall and the units are pressed into the mortar. The units should be set plumb and level with even 1/4-in.-wide mortar joints. The units are supported by wood wedges placed in the joints until the mortar hardens. The joints are then tooled or raked back 1/2 in. deep, pointed, and then tooled.

The second method of installing terra cotta is used to install anchored-type ceramic veneer. In this method the ceramic veneer may be mounted with or without vertical pencil rods, depending on the design and size of the units and the design of the supporting elements. Pencil rods are 1/4-in.-diameter steel reinforcing rods. When they are used, they are mounted vertically and held in place by passing them through the loops of dowel anchors that are attached to the supporting construction. The ceramic veneer units are set in place with wood wedges in the joints to support their weight. Wire anchors are installed in slots in the units and tied to the pencil rods. When rods are not used, the anchors are tied to other anchors set into or attached to the supporting construction. After one horizontal course has been laid, the space between the units and the supporting construction is filled with the grout described in Section 4.2.5. The installation is completed by removing the wood wedges, pointing the joints, and cleaning the units.

4.8 STONE

Stone played an important part in the lives of people long before recorded history. Undoubtedly, early humans used naturally occurring stones as both tools and weapons. About half a million years ago, our predecessors began to flake chips off flint and other stone to make more useful implements and weapons. Early dwellings were, in many instances, natural stone caves. With the development of farming came more permanent settlements and the associated need for permanent dwellings. When loose stones were available, they were piled up to form crude stone huts.

Shaped stone was not used extensively until iron stoneworking tools were developed. Before the Etruscans introduced lime mortar, cut stone buildings were made by accurately cutting and fitting stones together with no joint filler, which was time consuming and tedious work. The use of mortar joints reduced the need for extreme accuracy in stone cutting and speeded up the construction process. When the Romans developed hydraulic mortar, much stronger stone structures could be made and much longer spans built.

Stone was gradually replaced, to a large extent in smaller and residential buildings, by lighter and easier to handle brick and, later, concrete masonry units. However, stone remained in extensive use in decorative and ceremonial buildings, such as cathedrals and large public buildings.

In the eighteenth and nineteenth centuries, lighter steel and concrete frameworks virtually replaced mass stone as the structural support element in buildings. Stone structures cannot compete with these newer materials either in terms of economy or speed of construction. The large structures possible in those other materials simply cannot be built in stone.

However, stone still remains in great use as an important modern building material, but it is used mostly as cladding, trim, and paving. Though it is still sometimes used in mass, most stone today is hung on concrete or steel structural framing. Massive stone structures and stone bridges and aqueducts are things of the past.

4.8.1

MATERIALS

The types of natural stone used today are usually limestone, marble, granite, and slate.

4.8.1.1

Limestone

Limestone is sedimentary rock that is composed either of calcium carbonate (calcite) or dolomite, which is a mixture of calcium and magnesium, or of a mixture of calcite and dolomite. It is classified as either oolitic or dolomitic. Oolitic limestone is quarried mostly in Indiana and dolomitic mostly in Minnesota, but both can be found in other locations. Dolomitic limestone is more crystalline and stronger than oolitic limestone, and some will take a polish, whereas oolitic limestone will not. Finishes for oolitic limestone include smooth finishes, and textured and coarse finishes such as plucked, split-face, tooled, chat sawed, and rock-face. Finishes for dolomitic limestone include honed, sand blasted, and polished smooth. Sometimes different finishes are used in combination.

Limestone occurs in colors ranging from a light creamy buff to a brownish buff, and from silver gray to bluish gray. Stone primarily of one of these colors may contain various spots and streaks of the others. Variegated limestone contains both buff and gray colors, either in the same stone or in adjacent stones. Limestone ranges in texture from very fine grained to coarse grained with pit holes, reedy formations, honeycomb, iron spots, and travertine-like formations.

Limestone is not suitable for use in acidic environments or where chemicals will be used to melt ice and snow, nor should it be cleaned with harsh chemical cleaners. Acids and de-icing and cleaning chemicals will damage the stone.

4.8.1.2

Marble

Marble is metamorphic stone composed principally of calcite and dolomite (limestone) that has been recrystallized by heat and pressure. Marble is classified into four categories: I, calcite; II, dolomite; III, serpentine; and IV, travertine. Marble comes in so many colors and with such a variety of veining that it is necessary to specify it by its source or to require that the material supplied match a sample furnished by the architect.

A requirement of marble is that it be capable of taking a polish, and marble used on interiors is often so finished. However, highly polished finishes are inappropriate on the exterior where weathering will dull them or on traffic surfaces.

Typical finishes used on exterior surfaces are honed, sand blasted, abrasive, axed, hammered, rock faced, rough sawn, and tooled.

Granite

Granite is granular igneous rock consisting mostly of quartz or feldspar. It is hard, strong, durable, and almost impervious to water. True granite ranges in color from pink to dark gray. Other igneous stones that do not strictly meet the chemical and mineralogical requirements for granite are nevertheless sold as granite. These come in black, brown, green, and buff. Because there are so many possible colors and grain characteristics in granite, the best way to specify it is either by its specific source or as a match for samples furnished by the architect.

Common granite finishes include mirrorlike polished, honed, thermal, split-face, and jet honed. It can be used under severe weathering conditions and in contact with the ground.

Slate

Slate is a microcrystalline metamorphic rock formed originally from clay. It consists of thin plates that can be easily split into sheets. It is used primarily for paving and roofing. Slate comes in red, purple, green, blue, and black. Normal finishes are natural cleft, which is the untreated face left by splitting the slate; sand-rubbed, in which the texture left by splitting is completely removed to a smooth surface; and honed.

MANUFACTURE

Stone for building is obtained by *quarrying* and prepared for use by *milling*. Quarrying is simply the removal of the stone from the ground. Stone that will be used for rubble (see Section 4.8.3.1) may be blasted out of the ground or broken off using picks and crowbars. Most stone, however, is cut from the quarry using wire saws that are drawn through the rock to separate it into large blocks.

In the milling process, quarry blocks are sawn into smaller blocks or slabs in a shop using a different type of wire saw, according to shop drawings that are based on the architect's drawings and then reviewed and accepted by that architect. The wires that actually cut the stone are continuously lubricated by a slurry of water and silicone carbide or other abrasive.

Cut stones are finished by grinding, polishing, or other treatment necessary to produce the desired finish. A diamond saw often does the final cutting. Cylindrical shapes are turned on a lathe. Articulation is sculpted by machine or by hand depending on the final configuration. Each stone is marked with a unique code so that its correct location in the building can be ascertained when the stone is delivered to the construction site.

Most stone blocks and slabs are too heavy to be moved and placed by hand. Cranes and other mechanical lifting devices are used extensively in the quarry, at the mill, and at the building site. Individual stones must have the proper holes and sinkages to permit proper lifting of them without damage (Fig. 4.8-1) and for placing them without interference or damage to mortar or other bed joints.

Natural stone contains moisture (quarry sap) when quarried. When the moisture dries and the stone stabilizes, it is said to be *seasoned*. Stone is often cut and used before seasoning is complete, but until seasoned, which can take as long as a year, stone will not reach its final color and should not be coated with water repellants of any kind.

DESIGN AND CONSTRUCTION

Stone may be laid in mortar beds and supported by the stone below, or be supported by metal. It may also be adhered to backing panels of concrete or metal. It may be used to build solid walls, arches, and vaults, as a facing over masonry, or as cladding. As cladding, it may be supported by the structure or mounted on a metal framework.

Traditional Stone Setting

Stone structures may be built in the traditional method using mortar beds either as a complete wall or as a facing for masonry. Finished joints may be either mortar or sealant filled. Traditional mortar bed setting methods are used today primarily to build stone retaining walls, planters, and the like; for installing stone copings and trim in masonry walls; and in restoration work on existing stone structures.

Stone laid using traditional mortar bed methods are placed either as *rubble* or as *ashlar*. Rubble consists of natural stone shapes that may be laid either in random or in coursed patterns (Fig. 4.8-2).

Ashlar is cut stone with squared corners. It may be laid in either random or coursed patterns (Fig. 4.8-3). Large blocks with square corners are called *dimension stone*.

The methods used are first to place a setting bed, then set the stone in this bed. The joints are raked back. When the bed has set, the joints are filled either with mortar and tooled or with a sealant.

Mechanical Fastening

Stone cladding may be set on a field-erected grid system of stone anchors, fasteners, and metal struts and braces (Figs. 4.8-4, 4.8-5, and 4.8-6). Very thin stone cladding may also be installed as panels in aluminum frames. Thin stone cladding panels may be applied as a facing on concrete panels. They may also be preassembled into panels on a prefabricated metal support system that is then lifted into place and mechanically attached to metal frames or to the structure. Panels may have insulation sprayed on their interior surfaces. Joints in mechanically fastened stone cladding are usually sealant filled.

Stone cladding may be almost any thickness, but the trend today is to use stone that is 2 in. thick or less. The performance requirements for stone cladding are determined in much the same way as are those for aluminum curtain walls or concrete cladding. Thin stone is not likely to perform as well as thicker stone. It is more

The supplier of Cut Indiana Limestone will provide holes and sinkages for both anchoring and lifting. However, it is the responsibility of the general contractor or erector to correlate the types of equipment he plans to use with the matching types of lifting holes to be provided by the supplier. Where no arrangements are made, the supplier will provide those sinkages, if any, which suit his own handling requirements. Do not use lewis devices in stones under 3½" thick.

NOTE: ILI has included this page for illustrative purposes only. This information is not intended as a guide to lifting procedures. Commercial chain, clamps, pins and other devices are available. Their manufacturers will provide instructions for their safe and proper use.

Lewis Pins

Slings

Damage

Incorrect

Too-short slings cause damage at stone edges. Loads lifted by such slings tend to be unstable.

Correct

Box Lewis

Lifting Pulley allows self-leveling of stone.

Sometimes two clamps can handle heavy stones better.

Stone Clamp

On textured finishes, a ribbed pressure pad evenly distributes the clamp pressure to the smooth surface.

Pressure Pad

Button into Stone Back

Tightening Screw

Chain retains ribbed pressure pad from falling when loosened.

Note: Pins, box lewises and other devices attached to or near the tops of stones should not be used to raise stone panels to vertical position from the horizontal. They may, however, be used in conjunction with slings or other devices which support the main weight of panels while turning them to the position for installation. **In contemporary stone design and usage, lewis devices are unreliable and dangerous.** ILI recommends against their use by persons unfamiliar with safe practices.

FIGURE 4.8-1 *Some methods of lifting large stones. Other devices are also used..* (Indiana Limestone Handbook, *19th ed., p. 37, reprinted with permission of Indiana Limestone Institute.*)

susceptible to water penetration; to damage because of chemical weathering because of its thinness; to damage resulting from relatively light-impact forces, such as those from wind-driven rain or sleet; and to bowing from internal stresses that would be less likely to cause damage in thicker stone. Therefore, greater factors of safety must be used when selecting and installing thin stone veneer. However, the savings realized from the reduction in cost of the material itself, the ease of erection using lighter stone, and the reduction of the load on the structure itself often make using thin stone cladding seem worthwhile.

4.8.3.3

Paving and Flooring

Stone is used as both exterior paving and as flooring. Granite is particularly suited for exterior paving. The abrasive resistance of stone used as paving or flooring is important, as is the slip resistance of its finish.

The usual method of placing stone pav-ing is to set the stone in a thick portland cement mortar bed having a latex additive. Joints are pointed with grout or filled with sealant.

Stone flooring may also be set in a thick-setbed, but thin stone tile flooring is often set in a thin-set method similar to that used to install ceramic tile (see Section 9.6).

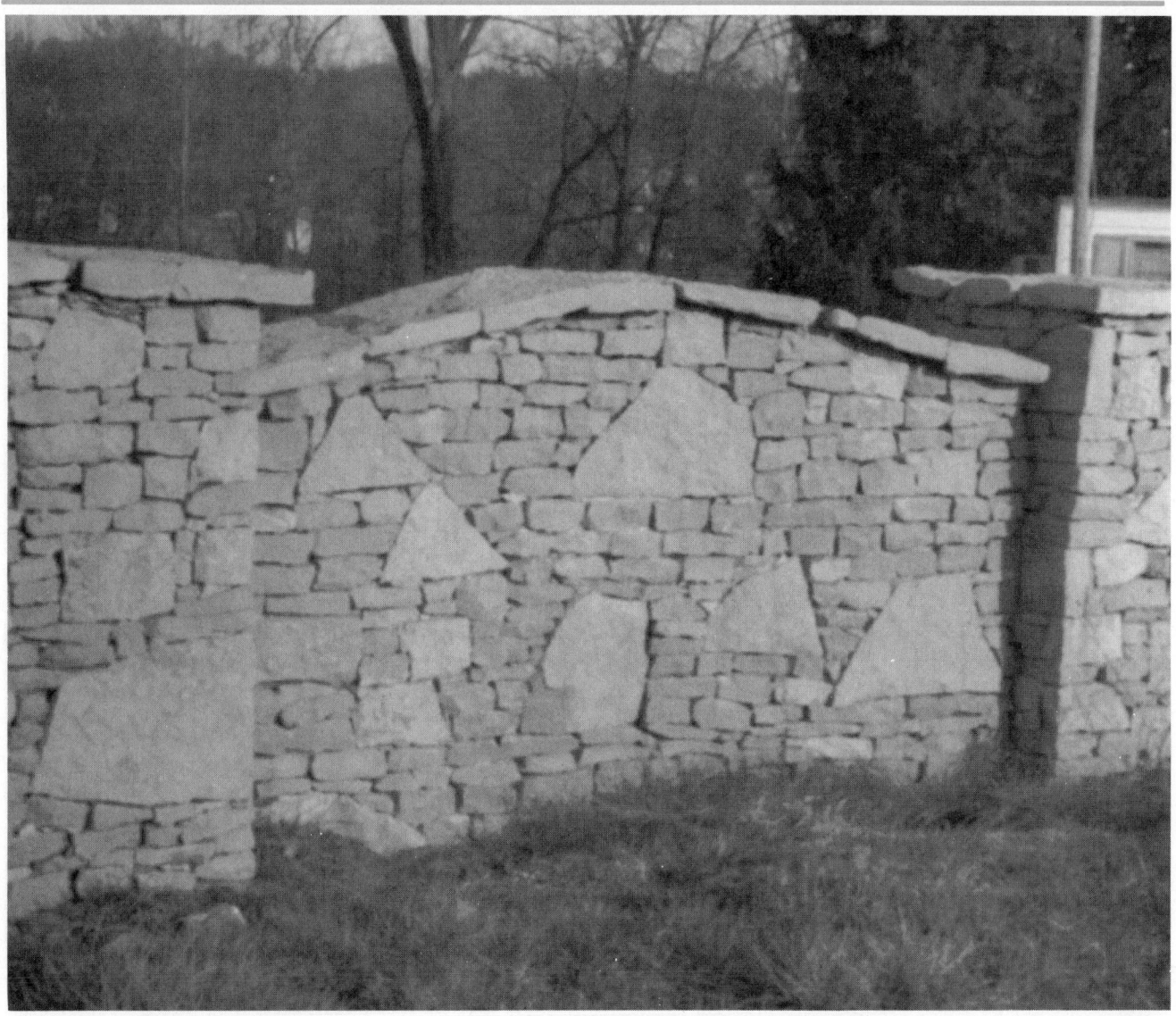

FIGURE 4.8-2 *A limestone rubble wall.* (*Photo courtesy Indiana Limestone Institute.*)

elevations

1

2

1   **random ashlar**   2

3

4

3   **coursed ashlar**   4

Random ashlar and coursed ashlar are both made from strips of limestone with lengths cut as desired at the job site. The only real difference between them is that the mason lays to a line with one (coursed) and does not with the other (random).

Standard course heights are 2¼", 5", 7¾" (and sometimes 10½" & 13¼") based upon ½" beds and joints.

The most common and economical finish for this stone is split face. However, other standard industry finishes (chat or sand sawn, and shot sawn) are also available.

Either pattern may be varied further by the use of more than one thickness as shown in Section at right, thereby creating a three dimensional effect (staccato pattern).

Splitface strip ashlar may have either sawed or split backs. Sawed backs allow a close tolerance in setting. Split backs allow the mason to select a face for greater boldnesss.

sections

1-1   2-2

3-3   4-4

FIGURE 4.8-3 *Ashler stone patterns.* *(Courtesy Indiana Limestone Institute.)*

FIGURE 4.8-4 *Multistory stone cladding anchored to a steel frame without masonry backup. Three different connection systems are shown.* (Drawings courtesy Indiana Limestone Institute.)

Adapter Channel Welded to Steel

Built into Masonry

1" Minimum 2" Preferred

Grout Anchor in C.M.U.

S.S. Split Anchor

"a"   "d"   "e"

Shim as Req'd.

Expansion Anchor

S.S. Split Anchor

Bond Beam

Plate with Bar

Shims

Adjustable

"b"   "f"

"c"

Adjustable

alt. "c"

Gusseted Plate with Dowels

alt. "c"

"a" "d" "e" "g" "b" "h" "c" "f" "g" "h"

alt. "g"

Anchors shown are at Horiz. Joints only.

Note: It is permissible to pierce blocks to insert anchors. Block must be dammed and well-filled with grout.

FIGURE 4.8-5 *Some of the many possible ways to anchor multistory stone cladding with masonry backup to a steel structural system. The number of types of anchors in a single building should be the fewest practicable. Panels should be supported at least at each floor level by a positive structural element (angle, plate).* (Drawings courtesy Indiana Limestone Institute.)

Labels within figure: "a", "d", "j", "f", "b", "j", "b", "g", "h", "c", "e", Dovetail Insert, Fill Block Around Anchor Plate with Bar, Adjustable Insert, Askew Bolt & Nut, Weld Plate, "j", "h", "g", alt. "f" alt., alt. "c", Support Angle with Dowel, Support Angle with Bar

FIGURE 4.8-6 *Anchoring multistory stone cladding with masonry backup to a concrete frame is similar to anchoring it to a steel frame (see Figure 4.8-5) and the same principles apply.* (*Drawings courtesy Indiana Limestone Institute.*)

5

METALS

INTRODUCTION

IRON AND STEEL

ALUMINUM

OTHER METALS

Photo: An ironworker high on a steel-frame building.
(Courtesy of Bethlehem Steel Corporation.)

INTRODUCTION

Metals have been important to humans from our beginnings. The materials for very early stone structures were cut and shaped laboriously, using tools made of hard stone and flint. The first known metals were those found in nature, such as gold and silver. Later, other metals, probably discovered accidentally, were obtained by melting ores. The first relatively hard metals produced were mostly copper alloys, such as brass and bronze. These metals were vastly superior to flint and stone for some purposes. They are still used today in building construction. Copper is used in roofing, flashing, piping, wiring, and in the manufacture of equipment. Copper alloys are used to manufacture doors and their frames, windows, door and window hardware, cladding panels, plumbing fixtures, and for many decorative purposes.

The knowledge that metals could be obtained by heating ores naturally led to a search for other metals. Iron was discovered fairly early in history because the process of making it was relatively easy. Iron and its modern derivative, steel, are used extensively in buildings today.

Aluminum, which was not produced until the nineteenth century, is another common metal in most of today's buildings. It is used to make cladding panels, storefronts, curtain walls, doors and their frames, windows, siding, roofing, flashing, wiring, piping, railings, hardware, and many other products.

Constructing a modern building containing no metal products is impossible.

Products made from iron, steel, aluminum, copper, bronze, lead, alloys of these metals, and other metals are contained in a vast array of building components, from those forming the underlying structure to those in the shell. Modern mechanical and electrical systems cannot function without metals. Masonry structures contain metal reinforcements, anchors, and fasteners. Concrete structures are reinforced with steel. Wood framing is held together with metal nails, lag bolts, and screws. Finishes are hung on metal furring and framing. This chapter is an overview of some of the metal products used in buildings today. Covering them all in detail would take an entire book larger than this one.

5.1 IRON AND STEEL

The chief ingredient of the metals known as iron and steel is the chemical element iron (*ferrum*), from which iron-based alloys get the generic name *ferrous metals*. Although about 5% of the earth's crust is composed of iron, it is rarely found in its metallic form in nature. It is almost always combined with oxygen (iron oxide) and sulfur silicon and is mixed with other minerals. It is found in rock, gravel, sand, clay, and mud. In this form it is mined as iron ore, which is the chief source of iron. Iron- and steelmaking involve extracting iron from ore, combining it with carbon and other elements, and then forming it into shapes.

Pure iron is inherently soft, ductile, easily shaped, and relatively weak. However, *wrought iron*, which is nearly pure, is strong enough for many construction purposes. In fact, it has been used for many bridges and buildings, and as the structural frame for the Eiffel Tower. The many irons and steels used in modern construction and manufacture get their strength and hardness mainly from the element carbon. Most steels have the best combination of properties when carbon is present in amounts less than 1.2%. Ferrous metals containing substantially larger amounts of carbon become so hard and brittle that they cannot be readily shaped by either hot or cold working methods. These metals are formed by being cast into molds, and are referred to as *cast iron*.

5.1.1

HISTORY

Iron was discovered at different times in different parts of the world. Iron implements that are more than 4000 years old have been found in Egypt; the earliest discovered in China are more than 2700 years old. By the time of Rameses II, who reigned from 1292 to 1225 B.C., iron was being widely used in Egypt. In the following thousand years, the use of iron spread throughout most of the known world. Iron was not used extensively in western Europe, however, until around 100 B.C.

Fairly high-quality steel was made in ancient Syria and in Toledo, in what is now Spain. The methods used were lost, however, during the Dark Ages. As a result, metals we would now classify as steel were not manufactured again until the 1730s.

The known steelmaking processes between the beginning of the Dark Ages and the 1730s could not control the impurities and carbon content sufficiently to produce metal with the properties of modern steel. The metal produced, called *blister steel*, was nothing more than carburized wrought iron. Benjamin Huntsman changed that by inventing the crucible process. The crucible process permits the melting of blister steel into a homogeneous mass, which makes it steel in the modern sense.

Even Huntsman's crucible process, which produced very high-quality steel, was quite labor intensive and yielded only small quantities of steel with each charging. In addition, modern rolling, drawing, and heat-treating processes that improve the mechanical properties of steel were unknown. Therefore, the earliest ferrous metals used in construction were essentially wrought iron and cast iron, shaped by hand working or by casting hot metal into molds.

The first successful North American iron works was built in Saugus, Massachusetts, in the late 1640s. Its earliest products were cast-iron cooking pots and wrought-iron nails. Nails were so expensive that old buildings were often burned so that the nails could be recovered.

The first high-tonnage steelmaking process was discovered almost simultaneously by an American, William Kelly, and an Englishman, Sir Henry Bessemer. The Civil War interrupted Kelly's work and

slowed the development of commercial steelmaking here until after 1864. This gave Bessemer time to develop the process and make it practical. Therefore, the process and the furnace used in it were given his name. In the Bessemer (pneumatic) process, air was forced through molten iron contained in a pear-shaped vessel. As the air rose through the iron it burned out impurities. The addition of manganese caused chemical changes, which converted the iron to steel.

After 1864, the Bessemer process and the discovery of abundant iron ore deposits in the Lake Superior region made the United States a world leader in steel production. The economical production of iron and steel is what allows advances in construction, industry, commerce, transportation, and communications.

The term *steel* covers a variety of ferrous metals of different chemical compositions and properties. These properties are affected by the alloying elements in-

cluded, the production processes used, and the methods of manufacture. This section discusses iron- and steelmaking and the resulting properties.

5.1.2

RAW MATERIALS

In this section the term *raw material* is used to designate all of the materials employed in making iron and steel up to the point of producing metal ingots used for product manufacture.

The general steps in the mining and processing of raw materials, smelting them into pig iron, and refining that into steel ingots are illustrated in Figure 5.1-1. Iron- and steelmaking start with the separation of the metal iron from its oxide, which is the way iron appears in iron ores. This is done by heating iron ore while it is in contact with coke, a form of carbon. The car-

FIGURE 5.1-1 *Flow chart of operations in the production of steel ingot.* (*Courtesy American Iron and Steel Institute.*)

bon then combines with both the oxygen present and the iron in the ore, forming a volatile gas and a carbon-rich crude iron called *pig iron*. The term *pig iron* comes from the shape of the trenches into which the molten iron from early furnaces ran. The iron was poured into a main trench, then ran into smaller connected trenches where it hardened into ingots. These smaller trenches, lined up against the main trench, looked to some like piglets lined up against their mother for suckling.

The primary raw materials used to make pig iron are: (1) iron ore, which furnishes the iron; (2) fuel (coke), which furnishes the carbon for combustion and for alloying with the iron; (3) air, which supplies the oxygen for combustion and oxidation (burning out) of impurities (unwanted elements) in the raw materials; (4) flux, which provides a molten slag to separate the ash of combustion and the impurities from the molten iron; and (5) refractories, which form the linings of furnaces, flues, and vessels used to contain the molten metal.

In steelmaking, pig iron and metal scrap replace iron ore as the primary source of iron, and electricity may replace fossil fuel as a source of heat. In some processes, gaseous oxygen supplements or completely replaces ordinary combustion air. The emphasis in steelmaking is on removing some or most of the carbon and other impurities from pig iron that are present in the raw material.

5.1.2.1

Mining and Processing

The primary raw materials—iron ores, fuels, fluxes, and refractories—are often mined in areas that are far from iron- and steelmaking plants. To reduce shipping costs, raw materials are sometimes processed near the mines to increase their yield and reduce their bulk. These are then shipped by rail and water (Fig. 5.1-2) to steel mills or plants for direct use or further processing. Mills and plants are typically located along navigable water routes, which permits the low-cost transportation of bulk materials and provides an abundant supply of the fresh water that is essential in steelmaking.

■ IRON ORES

Iron ore is mined either by underground or open pit method (Fig. 5.1-3). The iron

FIGURE 5.1-2 *Ore and limestone are often delivered to a steelmill (background) by boat.* (Courtesy Republic Steel Corporation.)

FIGURE 5.1-3 *Typical terraced effect of open pit ore mining.* (Courtesy American Iron and Steel Institute.)

ore of the Lake Superior region dominates steel production; other deposits are mined in Alabama, New Jersey, New York, California, and a few other states. Iron ore is also imported, with Canada providing more than half of all imports. The most common iron minerals are hematite (FE_2O_3), siderite $(FeCO_3)$, and pyrite (FeS), which have iron contents ranging as high as 70%. Limonite may be considered as magnetite plus water; taconite and jasper are names for a wide variety of iron-bearing rocks with iron contents ranging between 22% and 30%.

The largest tonnage of ore mined in the Lake Superior region is taconite. This ore lies very close to the surface but is so hard that superheated flames must be used to drill holes for explosive charges. After it has been blasted out of the ground, taconite is processed by *beneficiation*, a treatment that increases its iron content per ton of raw material. Ores with iron contents near 50% are usually shipped as is. Ores like taconite, with a 23% to 25% iron content, can be improved to nearly 66% iron content by beneficiation. Richer ores and beneficiated taconite can be further improved by direct reduction to an iron content of more than 90%.

BENEFICIATION The beneficiation of ores generally consists of (1) grinding and

concentrating them to increase the iron content, and (2) *agglomeration*, which increases the ore's particles to a size suitable for blast furnace use. Two forms of agglomeration are *pelletizing*, used for jasper and taconite, and *sintering*, used to recycle iron ore dust.

Pelletizing by both magnetic separation and concentration is usually done near mine sites to avoid transporting large tonnages of non-iron-bearing minerals.

In one pelletizing process, magnetic taconite is crushed into a fine powder the consistency of flour. The powder is passed through a magnetic separator, which removes the iron particles. The iron powder then goes to a balling drum, where it is mixed with coal dust and a binder material to form pellets with 60% iron content. The pellets, each about 1 in. in diameter, are baked to a hard finish in a kiln or furnace.

In another pelletizing process, high-grade nonmagnetic ores are crushed and then concentrated in flotation cells. Flotation cells contain liquids in which the heavier iron-bearing particles sink to the bottom of the cell while the lighter particles remain suspended and are skimmed away. After filtering, the iron-bearing particles are formed into pellets and baked hard.

Sintering is a process by which iron ore dust, reclaimed from blast furnace gases and other manufacturing processes, is fused with coke and fluxes into a clinker-like product containing 52% to 65% iron.

DIRECT REDUCTION Direct reduction includes various processes in which ore is concentrated to a high iron content that is comparable to that of molten iron from a blast furnace. Some ores are made into pellets, others into briquettes. Direct-reduced ores with over 90% iron contents are able to bypass the blast furnaces in which iron for steelmaking has been traditionally produced for centuries. The pellets and briquettes produced by direct reduction are so rich that they can be charged directly into electric steelmaking and basic oxygen furnaces.

■ FUELS

Of the three main fuels used in the iron and steel industry—coal, oil, and natural gas—coal supplies more than 65% of the industry's heat and energy requirements.

FIGURE 5.1-4 *Hot coke is forced from one of a battery of slot-shaped ovens into a quenching car for cooling with water.* (Courtesy Bethlehem Steel Corporation.)

The chief fuel used in a blast furnace is coke, which is produced in coking ovens from selected types of bituminous coal (Fig. 5.1-4). Most coal of coking quality is mined in West Virginia, Pennsylvania, Kentucky, and Alabama. In addition to furnishing heat and carbon, coke also acts as a reducing element to separate iron from its oxide in the ores to make pig iron.

■ FLUXES

Fluxes are minerals that have an affinity for the impurities in iron ore or pig iron. They combine with these impurities and separate them from the molten metal in a furnace by forming a liquid slag. Some fluxes are used mainly to separate the impurities, others to make the slag more fluid, thus floating it to the surface of the metal sooner. Depending on the affinity of a flux to combine chemically with certain impurities, it is classed either as *basic* or as *acid*. Limestone and dolomite are examples of basic fluxes; sand, gravel and quartz rock are acid fluxes. Fluorspar is a neutral flux used mainly to make the slag more fluid.

■ REFRACTORIES

Refractories are nonmetallic materials with superior heat and abrasion resistance used as linings of steelmaking furnaces, flues, and vessels. Like fluxes, refractories may be *acid* or *basic*, depending on their chemical reaction with impurities. Theoretically, acid and basic refractories must be used with like fluxes to minimize chemical reaction between them, but in practice this rule is not always followed.

A common acid refractory material is *ganister*, a form of quartzite rock obtained from quarries and mines in Pennsylvania,

Wisconsin, Alabama, Utah, and California. The most common basic refractory material is *magnesia*, which is obtained from the mineral magnesite and from seawater (Fig. 5.1-5).

FIGURE 5.1-5 *Magnesia is extracted from seawater or brines for use as a basic refractory in steel furnaces.* (Courtesy American Iron and Steel Institute.)

■ ALLOY ORES

The chemistry of steel requires many elements for alloying purposes; manganese, chromium, nickel, silicon, tungsten, molybdenum, vanadium, and boron are discussed below. Although the United States is a major steel producer, it depends on imports to furnish many of these elements. Most alloys are imported as ores and then processed here into useful forms called ferroalloys.

MANGANESE   All steel contains manganese, both as a scavenger to remove oxygen from molten metal and as an alloying element to improve hardness and wear resistance. An average of 12 pounds of manganese goes into each ton of steel.

CHROMIUM   Added in amounts of 1% to 5%, chromium provides stainless steel with resistance to heat, rust, and corrosion. To other alloy steels and cast iron, it provides hardness and wear resistance. Chromium is also important as a coating metal.

NICKEL   Cryogenic steels (those produced at low temperatures) use nickel to improve hardness. Steels containing nickel are especially suitable for case hardening, the process of surface hardening through the absorption of carbon at high temperatures.

SILICON   In steelmaking, silicon serves as a deoxidizer and as an alloying element. When used in large percentages, silicon enables cast irons to withstand highly corrosive acids.

TUNGSTEN   Tungsten provides heat resistance and is used in cutting tools of high-speed steel and to make tungsten carbide, an abrasion-resistant material almost as hard as diamonds. It is also used as a welded-on deposit on machine parts exposed to extreme wear.

MOLYBDENUM   One of the more versatile alloying elements, molybdenum increases the hardness and corrosion resistance of steels.

VANADIUM   Vanadium is often used in conjunction with other alloying elements: with chromium in making steel springs, with manganese in making special plates and other structural forms, and with molybdenum in making certain high-temperature steels. Small amounts (often measured in hundredths of a percent) provide the desired effect in construction steels.

BORON   Although known widely as a cleansing agent, boron is used as a hardening agent in steel. Minute quantities are added to obtain the desired effect, but these must be protected by other deoxidizers or must be added to a completely deoxidized molten metal bath.

■ AIR AND OXYGEN

Air and oxygen can be considered as much a raw material as the specially prepared iron ore, coke, and lime that the steel industry consumes. Air must be forced into furnaces, heated or cooled, and cleaned to control air pollution. Oxygen is a product of those plants that also take nitrogen and argon from ordinary air. It takes about 4 tons of air to make 1 ton of iron.

■ WATER

The steel industry uses 35,000 gallons of water for each ton of raw steel produced. Often water must be cleaned both before and after use. Its primary function in steelmaking is as a coolant in closed systems.

Iron Making

Smelting is the metallurgical operation by which metal ore is heated to separate the metal in it from impurities with which the metal may be chemically combined or physically mixed. The separation of iron from the oxygen with which it is chemically combined is called *reduction*. Pig iron is made in a blast furnace by smelting, which extracts the iron by reducing its oxide and by physically separating it from other impurities.

■ BLAST FURNACE

A *blast furnace* is a cylindrical structure, approximately 120 ft. tall, lined with refractory brick and encased in a steel shell (Fig. 5.1-6). A *charge*, consisting of ore, coke, and limestone, is loaded at the top and a blast of hot air is introduced near the bottom. A charge is usually made up of about 1-1/2 tons of ore, 3/4 ton of coke, and 1/4 ton of limestone for each ton of iron to be produced. Oxygen in the air combines with carbon in the hot coke, causing combustion and producing heat. The charge is smelted by the heat and, as the gases of combustion pass through the hot ore, another chemical reaction takes place that frees the iron from its oxide. The hot combustion gases are cleaned in dust catchers and recycled through brick checkerworks (stoves) to heat the incoming air blast (Fig. 5.1-7). A blast furnace is operated 24 hours a day.

The impurities in the ore and the flux combine with oxygen to form a floating slag that is 4 to 5 ft. deep and that is tapped from the furnace at about 2-hour intervals. The molten crude iron, which contains about 93% iron and 7% extraneous elements, is tapped at a lower level at about 3-hour intervals. This is then conveyed in ladle cars to *mixers* (holding furnaces) where the liquid iron is mixed to equalize its chemical composition and kept hot pending further refinement.

Most of this molten iron is sent to open hearth or basic oxygen steelmaking facilities. Some goes to a casting machine where it solidifies into molds, known as *pigs* (ingots).

Most iron from pig casting (pig iron) is made to individual specifications, which vary according to end use. Pig iron that

FIGURE 5.1-6 *Diagrammatic illustration of a blast furnace and its auxiliaries.* (Courtesy American Iron and Steel Institute.)

Labels in figure:
Iron ore, coke and limestone are the principal materials used in making iron.

IRON ORE STOCKPILE

STOCK HOUSE
receives raw materials which are carried up the inclined ramp (skip hoist) and charged into the blast furnace.

STOVES
pre-heat air for the blast furnace.

MOLTEN SLAG for disposal.

BLAST FURNACE

DUST CATCHER

MOLTEN IRON

LADLE CAR

FIGURE 5.1-7 *Overall view of a blast furnace and its auxiliaries. A battery of cylindrical checkerworks appears at center left.* (Courtesy American Iron and Steel Institute.)

is intended to be sold as an end product is called *merchant* pig iron.

End products made from pig iron vary from soil pipe to fancy dutch ovens. A producer of merchant pig iron may have several hundred piles of different grades of iron and may blend a dozen or more grades of iron ore to arrive at a desired chemical composition. While 30,000 cu. ft. of solids weighing just under 1000 tons are being processed in a furnace, the chemical composition of the mix is controlled so that the balance of carbon, silicon, sulfur, phosphorus, and manganese is maintained to within fractions of a percent of the desired amounts.

■ FOUNDRY

A small percentage of pig iron is remelted in *foundries* and formed into special shapes by casting it in sand or loam molds. This process is called *iron founding*. Since no major change in chemical composition takes place when pig iron is cast, it is most often melted together with scrap metal and alloying elements before it is cast in order to impart desirable properties. The carbon content of cast iron therefore is somewhat lower than that of pig iron, ranging between 1.5% and 4%.

The types of cast iron produced are *gray, white, malleable, chilled, alloyed,* and *nodular.* Only the first three are important in construction and only these will be discussed here.

GRAY AND WHITE CAST IRON When molten cast iron solidifies, carbon in the iron may remain chemically combined in the form of iron carbide, or may separate out as graphite. *White cast iron* contains carbon mainly in the chemically combined form (iron carbide), which makes this metal hard and nonductile. In *gray cast iron*, the carbon is chiefly in the form of graphite flakes, which impart its characteristic gray fracture and promote machinability and resistance to wear.

MALLEABLE CAST IRON   White cast iron may be made somewhat softer and more ductile, though not strictly malleable, by *annealing* (controlled heating and cooling). The carbon in the iron carbide is converted to a more dispersed graphite form, which provides greater ductility than does the flaky graphite of gray iron, or the carbide of white iron.

■ DIRECT REDUCTION

There are direct-reduction processes reported to be capable of bypassing the blast furnace. These include processes using solid reductants, processes using gaseous reductants, and direct-steel processes. However, none of these processes is really ready to replace efficient blast furnaces.

5.1.2.3

Steelmaking

Steelmaking requires lowering the carbon content of pig iron, controlling impurities in the pig iron and in other raw materials, and adding alloying elements to obtain desired properties.

Carbon is beneficial when contained within certain ranges, which vary with the type of steel. Carbon in quantities up to about 1.2% of the total material increases hardness and strength (at a lesser rate between 0.5% and 1.2%). Because there are accompanying increases in brittleness and loss of ductility as the carbon content increases, lower contents are preferable in most steels.

Carbon, phosphorus, sulfur, manganese, and silicon are classed as impurities when present in uncontrolled or excessive amounts. Pig iron usually contains these elements as impurities originating from the raw materials of the blast furnace. Up to certain limits or within specified ranges, the effect of these elements on steel may be beneficial (see Section 5.1.2), and steelmaking processes attempt to regulate their contents within these limits or ranges. Phosphorus and sulfur impart some desirable properties, but must be limited to less than 0.05% in most steels, because of their adverse effects on malleability (ability to be shaped). Silicon and manganese are strengthening elements desired in ordinary steels in the range of 0.01% to 0.35% and 0.20% to 2.00%, respectively, and may be

added when raw materials contain insufficient quantities. All of these elements are present in controlled amounts in most steels. When they occur naturally, they are referred to as residual elements; when they are deliberately added to achieve certain properties, these and other elements are called alloying elements.

Steelmaking processes generally start with hot pig iron delivered in ladles by overhead cranes to the steel furnace, where scrap metal and fluxes are added. After the melting operation, the hot steel is tapped from the furnace into other ladles and moved to awaiting molds for casting into ingots or to a strand-casting machine. The chief steelmaking processes are *open hearth*, *electric*, and *pneumatic* (Fig. 5.1-8).

■ OPEN HEARTH PROCESS

The majority of open hearth furnaces are of the basic type, capable of controlling the phosphorus present in pig iron. Both ordinary carbon-grade and high-grade alloy steels are usually produced by this process.

The name *open hearth* is applied because the saucer-shaped hearth, or floor of the furnace, is exposed to the sweep of the flames. The furnace itself is a masonry

FIGURE 5.1-8 *Hot pig iron may be converted to steel in one of the steelmaking furnaces shown or in induction furnaces or basic oxygen furnaces.* (*Courtesy American Iron and Steel Institute.*)

structure lined with refractory materials arranged on two levels; the upper level is for charging the raw materials, the lower for tapping the molten steel and slag by gravity. Furnaces may hold from 50 to 600 tons and produce from 25 to 50 tons of steel per hour. The charge generally consists of molten pig iron, scrap metal, and fluxes (Fig. 5.1-9).

FIGURE 5.1-9 *An open hearth furnace. Hot pig iron is being poured from a ladle. A charging machine (foreground) adds scrap metal and limestone flux. Open hearth furnaces are seldom used today. (Courtesy American Iron and Steel Institute.)*

The fuel (gas, oil, tar, or a combination of these mixed with hot air) is blown through ports at the ends of the furnace to produce combustion and heat over the hearth. As in a blast furnace, the hot gases of combustion are recycled through checkerworks, which preheat the incoming air. In addition, gaseous oxygen is usually introduced to speed the melting operation.

Refinement of the metal results when the impurities and carbon in the pig iron and the oxygen and fluxes in the mix react to each other chemically to form a slag that floats to the surface. The molten steel flows through a hole in the bottom of the hearth into ladles. When desired, the carbon content may be increased by adding pig iron in the furnace or coke or coal in the ladle. Oxidizable ferroalloys, iron-based compounds rich in desired alloying elements, may also be added in the ladle.

ELECTRIC PROCESSES

Most high-grade steels, such as alloy, stainless, and heat-resisting steel, are pro-

duced in electric furnaces capable of developing the high temperatures and reducing conditions necessary for melting these metals. Here electrical resistance and arc radiation, rather than combustion, are used to produce heat, and so air or oxygen is not necessary for combustion. High-purity oxygen is introduced at appropriate times to oxidize impurities, which results in better control of the oxidation process and a smaller loss of expensive alloying elements.

The charge usually consists of scrap metal, mill scale, fluxes, and often alloying elements. Most electric furnaces produce low-phosphorus steel. Many grades of basic steel can be produced at a rate of 30 to 40 tons per hour in an electric arc furnace, which may hold up to 300 tons. Special grades involving valuable alloying elements are generally produced in smaller quantities, often in an electric induction furnace.

ELECTRIC ARC FURNACE The body of an electric arc furnace is a circular steel shell resembling a huge kettle, lined with refractory materials and mounted so that it may be tilted to pour off the molten steel. Three or more cylindrical electrodes project into the furnace from above. When high-voltage current is passed through the electrodes, an electric arc is produced between them and the charge, generating heat. Many of the impurities are oxidized during the melting operation and float to

the surface as slag, allowing the steel to be poured off from under it (Fig. 5.1-10).

INDUCTION FURNACE An induction furnace consists of a magnesia pot, insulated by other refractory materials and surrounded by windings of copper tubing (Fig. 5.1-11). When high-voltage alternating electric current is passed through the copper tubing around the pot, an induction current is generated in the metal charge of the furnace. The resistance of the metal to the induced current creates the heat for the melting operation.

PNEUMATIC PROCESS

Steelmaking operations that use high-speed oxygen or air to oxidize impurities in the charge without fuel are called *pneumatic processes*. Their chief advantages are shortened heat time and high output. With furnace capacities of 300 tons and tap-to-tap heat cycles of 45 minutes, it is possible to produce steel at a rate 6 to 10 times that of open hearth and electric arc furnaces. The pneumatic process, which takes place in a *basic oxygen furnace*, is the predominant method of producing steel in the United States.

In a basic oxygen furnace, which is a steel, pear-shaped vessel lined with refractory materials (Fig. 5.1-12), a jet of high-purity oxygen is directed onto molten metal. The charge, consisting of molten pig iron, some scrap, and fluxes, is added

FIGURE 5.1-10 *An electric arc furnace. (Courtesy Lukens Inc.)*

FIGURE 5.1-11 *A 1500-ton channel induction furnace.* (Courtesy Ajax Magnethermic Corporation.)

FIGURE 5.1-12 *A basic oxygen furnace. A ladle pours hot pig iron into the mouth of a tilted furnace.* (Courtesy American Iron and Steel Institute.)

from the top. Larger amounts of hot pig iron and lesser amounts of scrap metal are used in the oxygen furnace than are used in other steel furnaces.

No fuel is required in this process because the heat of the large mass of molten pig iron is sufficient to start a chemical reaction between the oxygen, the carbon, and the other impurities. Oxidation, in turn, produces sufficient heat to continue the process without the introduction of fuel.

If the as-tapped metal has a lower carbon or manganese content than is required, *ferromanganese* (a manganese-rich ferro-

alloy containing carbon) may be added in the ladle. The carbon gives added strength and hardness, whereas the manganese, in addition to strength, ensures proper ductility at elevated temperatures by combining with excess sulfur.

■ **STEEL CASTING**

No matter which steelmaking process is used, steel must be formed into a specific shape before further processing. To do this, it is either poured into *ingot molds* or put through a *strand-casting machine*.

INGOTS Raw steel from a furnace is typically *teemed* (poured) from a ladle into cast-iron ingot molds, which may be more than 8 ft. high and 3 ft. in diameter. These molds are coated on the inside with a releasing agent that prevents splashed metal from marring the ingot's surface quality. Often stripped ingots are moved to furnaces called soaking pits while the outside is solid but the interior is still molten.

STRAND CASTING Strand casting produces a continuous ribbon of steel rather than separate molds. It can result in semi-finished solid products, thus bypassing many of the steps necessary to produce ingots. At some point the continuous ribbon of steel is cut into desired lengths (Fig. 5.1-13).

5.1.3

PROPERTIES

The properties of various irons and steels are closely related to their chemical compositions and grain structure, that is, the size and arrangement of the crystalline microscopic particles making up the metal. In the initial crude refining processes, the most important factors influencing grain structure are the carbon and its chemical reaction with iron as it cools. Further refining processes, alloying elements, heat treatments, and hot and cold working also affect grain structure. In ordinary steels, carbon and the other residual elements are important. In higher grades of steel, alloying elements contribute special properties to the metal.

The essential physical property that distinguishes steel from iron is its malleabil-

ity as it comes from the furnace and is initially cast. Other unique properties of steel are its ability to resist high stresses by deforming without breaking (toughness), and its ability to be hardened by hot and cold working. The important physical and mechanical properties of the most common iron and steel types used in construction are shown in Figure 5.1-14. The effects of the several residual elements originating in the raw materials and of the added alloying elements are outlined in Figure 5.1-15.

5.1.3.1

Cast Iron

Only slight variations in properties or in the grain structure of the metal occur in cast iron as a result of the reaction between the carbon in the cast iron and the iron. Most ordinary white and gray cast irons, therefore, like pig iron, are hard and brittle, and have relatively high compressive strength and low tensile strength. However, with the proper combination of selected pig irons, scrap metals, and alloying elements, cast iron can be made with increased hardness, toughness, and corrosion and wear resistance. Casting also permits the formation of complex and massive shapes not readily produced by machining or rolling.

Drainage, vent, and waste pipe, ornamental railings, lamp posts, handrail brackets, stair treads, manhole covers, and concrete inserts for anchors are familiar cast-iron products. In sanitary ware involving enameling and baking at elevated temperatures, cast iron's resistance to warping and cracking at high heat is an important advantage.

Malleable cast iron has improved breakage resistance as a result of increased toughness and ductility. It is also easier to weld than is gray or white cast iron. It is used extensively for small articles of door hardware, concrete inserts, brackets, flanges, and anchors. Special *alloy cast irons*, containing between 1% and 5% alloying elements, and *high-strength cast irons*, which contain molybdenum, chromium, and nickel, are also produced for special uses. *Nodular cast iron*, which is produced in this country by adding magnesium, has greater strength, ductility, toughness, machinability, and corrosion resistance than the other cast irons.

1. **Molten steel pours from a ladle into a reservoir called a tundish.**

2. **The metal flows out the bottom of the tundish at a carefully regulated rate into the mold, which is moving up and down to prevent the hot metal from sticking. The interior of the mold is hollow—just the size, in width and thickness, of the slab to be formed. Lining the walls are pipes through which water flows, chilling the metal. A thin shell of steel begins to solidify around the molten metal.**

3. **The gradually solidifying slab moves down through the secondary cooling zone. A series of rollers support the slab and gradually turn it into a horizontal position. Sprays of water under high pressure cool and harden the metal still further.**

4. **The ribbon of steel moves on to a level table.**

5. **A flame-cutting torch slices down through the metal. When the slab is cut off, it is carried on rollers to a cooling bed. The entire trip from the ladle has taken less than one-half hour.**

Hot steel is transported rapidly by ladle from electric and open hearth furnaces to the casting unit and is fed into the tundish.

Hot Metal Ladle

The refractory-lined tundish controls the flow and distribution of metal into the molds.

Operator's Console

In the water-cooled mold the steel begins to solidify. A solid shell is formed.

Gantry Service crane

Roller Aprons and secondary cooling

Traveling Slab Cut-off Torch

Solidifying steel enters the secondary cooling zone. Cooling is accomplished by direct water spray. Roller aprons are arranged to guide and support the strands and simultaneously take up the ferrostatic pressure exerted by the liquid metal core upon the strand shell.

Slabs are cut into predetermined lengths and removed by roller tables.

Roller Leveler

Here rolls withdraw and level the strands.

Slab Run Out Table

FIGURE 5.1-13 *Strand, or continuous, casting.* *(Courtesy American Iron and Steel Institute.)*

5.1.3.2

Carbon Steels

Carbon steels are those in which the residual elements, such as carbon, manganese, phosphorus, sulfur, and silicon, are controlled, but in which no alloying elements are added to achieve special properties. They contain up to 1.2% carbon, and other elements are controlled within specified limits or ranges. More than 90% of the steel manufactured into finished products is carbon steel.

■ CLASSIFICATION AND USES

In the past, steels have been classified according to carbon content as follows:

Soft steel. 0.2% maximum
Mild steel. 0.15% to 0.25%
Medium steel. 0.25% to 0.45%
Hard steel. 0.45% to 0.85%
Spring steel. 0.85% to 1.15%

These broad classifications have become less useful as the need for precisely compounded steel types has increased. Today, the American Iron and Steel Institute (AISI) and the American Society for Testing and Materials (ASTM) designations are more commonly used in manufacturing and construction.

AISI DESIGNATIONS   The AISI designation system assigns code numbers that indicate the chemical composition and the steelmaking process of various types of steel. For carbon steels, each four-numeral code number specifies the permissible maximums or ranges for residual elements, including carbon, manganese, phosphorus, sulfur, and sometimes silicon. Some types of alloy steels are designated by five numerals.

The first two digits in the AISI system indicate the type of steel (carbon or alloy) (Fig. 5.1-16). The last two digits of the four-numeral series and the last three digits of the five-numeral series indicate

the approximate mean of the carbon range. For example, in Series designation 1035, 35 represents a carbon range of 0.32% to 0.38%. It is necessary, however, to deviate from this system and to interpolate numbers in the case of some carbon ranges, and for variations in manganese, sulfur, or other elements with the same carbon range.

ASTM DESIGNATIONS   Structural steels are commonly specified according to ASTM standards. The carbon steels typically used in building construction are listed in Figure 5.1-17. The steel most used for construction purposes is that complying with ASTM Standard A 36.

ASTM standards indicate certain partial chemical compositions and specified minimums for strength and ductility. However, they regulate only those chemical elements that directly affect the fabrication and erection of steel, making it possible to produce proprietary steels of different chemical compositions con-

FIGURE 5.1-14
Selected Physical and Mechanical Properties of Representative Iron and Steel Types

	Metal	Specific Gravity	Coefficient of Thermal Expansion (in./in./°F)[a]	Modulus of Elasticity (psi)	Yield Point (ksi)[b,c,e,f]	Ultimate Strength (ksi)[c,d,e,f]	Ductility (Elongation in 2 in.)[d,e]
Irons	Gray cast iron	7.2	5.9×10^{-6}	15×10^6	22	30	(not ductile)
	Malleable cast iron	7.4	6.6×10^{-6}	25×10^6	30	54	18%
Carbon Steels	ASTM A 373	7.85	6.5×10^{-6}	29×10^6	32	58 to 75	24%
	A 283	7.85	6.5×10^{-6}	29×10^6	24 to 33	45 to 60	23% to 30%
	Merchant quality A 663 bars	7.85	6.5×10^{-6}	29×10^6	22.5 to 40	45 to 90	33% to 17%
	A 675 Bars	7.85	6.5×10^{-6}	29×10^6	22.5 to 55	45 to 90	33% to 14%
	ASTM A 36	7.85	6.5×10^{-6}	29×10^6	36	58 to 80	23%
High-Strength Steels	A 572 Columbium-vanadium	7.85	6.5×10^{-6}	29×10^6	42 to 65	60 to 80	17% to 24%
	A 588	7.85	6.5×10^{-6}	29×10^6	42 to 50	63 to 70	21%
	ASTM A 441	7.85	6.5×10^{-6}	29×10^6	40 to 50	60 to 70	21%
	ASTM A 242	7.85	6.5×10^{-6}	29×10^6	42 to 50	63 to 70	21%
Stainless Steels	AISI 410 (Martensitic)	7.75	5.5×10^{-6}	29×10^6	45	70	25%
	AISI 430 (Ferritic)	7.75	5.8×10^{-6}	29×10^6	50	75	25%
	AISI 302 (Austenitic)	8.0	9.6×10^{-6}	28×10^6	40	90	50%

Courtesy American Iron and Steel Institute.

[a]Within temperature range of 32° to 212°F for stainless steels.

[b]Yield strength or yield point is a measure of usable strength (i.e., maximum stress that produces deformation without increase in load).

[c]These mechanical properties of ASTM steels are specified minimums that vary according to shape and thickness of the material.

[d]Ultimate strength is a measure of total reserve strength (i.e., maximum stress that produces rupture in a test specimen).

[e]Values given for stainless steels are for annealed sheet and strip only. Higher-strength values are common for products and alloys hardened by cold working or heat treating.

[f]1 kip/sq. in. (ksi) equal 1000 lb./sq. in.

forming with the required ASTM performance standards.

UNIFIED NUMBERING SYSTEM Alloy and carbon steels have been assigned a designation in the Unified Numbering System (UNS) for Metals and Alloys. The UNS was established in 1975 by the ASTM and the Society of Automotive Engineers (SAE). These designations appear in ASTM Standard E 527 and SAE Standard J1086.

A UNS number consists of a single-letter prefix followed by five digits. The prefix E designates steels made by the basic electric furnace process with special practices; the absence of a prefix indicates open hearth steels. The prefix M designates a series of merchant-quality steels; G indicates standard alloy or carbon steels; and H indicates standard hardenability steels. The first four digits correspond to AISI, ASTM, or SAE steel designations and the last digit usually relates to an additional chemical requirement, such as lead or boron. The use of 6 as the last digit designates steels that are made by the basic electric furnace process

with special practices. An example of the UNS is "Series Designation E52100," which is the same steel as the 1.45% chromium steel (52) shown in Figure 5.1-16, made by the basic electric furnace process (E), which has a carbon content between 0.98% and 1.10% (100).

COMMON USES Carbon steels are used for most products in the building industry, from structural shapes, concrete reinforcing bars, sheets, plates, and pipes to the smallest items of builders' hardware. Carbon steels generally possess adequate

FIGURE 5.1-15
Effects of Alloying Elements

Element	Common Content	Effects
Carbon	Up to 0.90%	Increases hardness, tensile strength, and responsiveness to heat treatment with corresponding increases in strength and hardness
	Over 0.90%	Increases hardness and brittleness; over 1.2% causes loss of malleability
Manganese	0.50% to 2.0%	Imparts strength and responsiveness to heat treatment; promotes hardness, uniformity of internal grain structure
Silicon	Up to 2.50%	Same general effects as manganese
Sulfur	Up to 0.050%	Maintained below this content to retain malleability at high temperatures, which is reduced with increased content
	0.05% to 3.0%	Improves machinability
Phosphorus	Up to 0.05%	Increases strength and corrosion resistance, but is maintained below this content to retain malleability and weldability at room temperature
Aluminum	Variable	Promotes small grain size and uniformity of internal grain structure in the as-cast metal or during heat treatment
Copper	Up to 0.25%	Increases strength and corrosion resistance
Lead	0.15% to 0.35%	Improves machinability without detrimental effect on mechanical properties
Chromium	0.50% to 1.50%	In alloy steels, increases responsiveness to heat treatment and hardenability
	4.0% to 12%	In heat-resisting steels, causes retention of mechanical properties at high temperatures
	Over 12%	Increases corrosion resistance and hardness
Nickel	1.0% to 4.0%	In alloy steels, increases strength, toughness, and impact resistance
	Up to 27.0%	In stainless steels, improves performance at elevated temperatures and prevents work hardening
Molybdenum	0.10% to 0.40%	In alloy steels, increases toughness and hardenability
	Up to 4.0%	In stainless steels, increases corrosion resistance and strength retention at high temperatures
Tungsten	17% to 20%	In tool steels, promotes hardness at high cutting temperatures; in stainless steels, smaller amounts ensure strength retention at high temperatures
Vanadium	0.15% to 0.20%	Promotes small grain size and uniformity of internal grain structure in the as-cast metal or during heat treatment; improves resistance to thermal fatigue shock
Tellurium	Up to .05%	Improves machinability when added to leaded steels
Titanium	Variable	Prevents loss of effective chromium through carbide precipitation in 18/8 stainless steels
Cobalt	17.0% to 36.0%	Increases magnetic properties of alloy steels; in smaller amounts, promotes strength at high temperatures in heat-resisting steels

Courtesy American Iron and Steel Institute.

strength, hardness, stiffness, malleability, and weldability, but poor resistance to corrosion.

Occasionally, copper and other elements may be added to improve atmospheric corrosion resistance and, by varying the chemical compositions, heat treatments, and subsequent hot and cold working operations, other desirable properties can be enhanced.

5.1.3.3

Alloy Steels

Alloy steel is so classified when the content of alloying elements exceeds certain limits or when a definite range of alloying elements is specified. However, the term *alloy steel* does not cover stainless steel, tool steel, specialty steel, and some other classifications, even though those steels contain specified amounts of alloying elements.

FIGURE 5.1-16
AISI Designations

Series Designation*	Type and Approximate Percentages of Identifying Elements
Alloy Steels	
13xx	Manganese 1.75
40xx	Molybdenum 0.20 or 0.25; or molybdenum 0.25 and sulfur 0.042
41xx	Chromium 0.50, 0,80, or 0.95; molybdenum 0.12, 0.20, or 0.30
43xx	Nickel 1.83, chromium 0.50 or 0.80, molybdenum 0.05
46xx	Nickel 0.85 or 1.83, molybdenum 0.20 or 0.25
47xx	Nickel 1.05, chromium 0.45, molybdenum 0.20 or 0.35
48xx	Nickel 3.50, molybdenum 0.25
51xx	Chromium 0.80, 0.88, 0.93, 0.95, or 1.00
51xxx	Chromium 1.03
52xxx	Chromium 1.45
61xx	Chromium 0.60 or 0.95, vanadium 0.13 or min. 0.15
86xx	Nickel 0.55, chromium 0.50, molybdenum 0.20
87xx	Nickel 0.55, chromium 0.50, molybdenum 0.25
88xx	Nickell, 0.55, chromium 0.50, molybdenum 0.35
92xx	Silicon 2.00; or silicon 1.40 and chromium 0.70
50Bxx	Chromium 0.28 or 0.50
51Bxx	Chromium 0.80
81Bxx	Nickel 0.30, chromium 0.45, molybdenum 0.12
94Bxx	Nickel 0.45, chromium 0.40, molybdenum 0.12
Carbon Steels	
10xx	Nonresulfurized, manganese 1.00% maximum
11xx	Resulfurized
12xx	Rephosphorized and resulfurized
15xx	Nonresulfurized, manganese maximum over 1.00%

Courtesy American Iron and Steel Institute.
*B denotes boron steel.

■ CLASSIFICATION AND USES

Alloy steels are generally grouped by the element or combination of elements used to obtain desired properties. Manganese-molybdenum steels, for example, are known for hardenability and resistance to fatigue; silicon-manganese steels for high relative strength and shock resistance; nickel steels for toughness, impact, and corrosion resistance.

AISI DESIGNATIONS The AISI has developed a numerical designation system for identifying alloy steels. These steels are designated by a four-digit system, from 1200 to 9800, according to chemical composition (see Fig. 5.1-16). There are specific allowable ranges for carbon, manganese, silicon, nickel, chromium, and molybdenum; limits for sulfur and phosphorus are also usually specified. The prefix *E* is used to designate electric furnace steel; the absence of a prefix indicates open hearth steels. The suffix *H* is sometimes used to identify steels of controlled hardenability; *B* or *BV* between the second and third digits indicates boron or boron-vanadium alloys, respectively.

These steels are grouped in series according to the most important alloying element(s), identified by the first two digits; the last two digits indicate the approximate carbon content. For example, AISI type 4023 steel indicates a molybdenum alloy steel (40) with an average carbon content of 0.23% (23), made in an open hearth furnace (no *E* prefix).

COMMON USES Alloy steels have properties essentially similar to or exceeding those of the carbon steels. In general, particular properties are improved by certain elements. For example, a high silicon content gives steel excellent magnetic permeability and low core loss, making it useful in motors, generators, and transformers; cobalt improves magnetic properties for such uses as permanent magnets in electrical apparatus and sealing gaskets in modern refrigerators; nickel imparts great toughness, which is useful in rock-drilling and air-hammer equipment.

Because of their higher cost, alloy steels are not often used in construction materials. However, a particular group known as *heat-treated constructional alloy steels* have been rolled in a limited number of structural shapes and used successfully in large structures where unusually large loads or temperatures are encountered. These steels are capable of developing very high strengths (ultimate strength up to 135,000 psi, yield point up to 100,000 psi) and show promise of increased structural use.

FIGURE 5.1-17
ASTM Standards for Carbon Steels Used in Building Construction

Title	Standard
Specification for General Requirements for Rolled Steel Plates, Shapes, Sheet Piling, and Bars for Structural Use	A 6
Specification for Structural Steel	A 36
Specification for Pipe, Steel, Black and Hot-dipped, Zinc-coated Welded and Seamless	A 53
Specification for Steel Wire for Concrete Reinforcement	A 82
Specification for Steel Welded Wire, Fabric, Plain, for Concrete Reinforcement	A 185
Specification for Low and Intermediate Tensile Strength Carbon Steel Plates	A 283
Specification for Welded Deformed Steel Wire Fabric for Concrete Reinforcement	A 497
Specification for Cold-formed Welded and Seamless Carbon Steel Structural Tubing in Rounds and Shapes	A 500
Specification for Hot-formed Welded and Seamless Carbon Steel Structural Tubing	A 501
Specification for Deformed and Plain Billet-steel Bars for Concrete Reinforcement	A 615
Specification for Deformed and Plain Billet-steel Bars for Concrete Reinforcement (Metric)	A 616
Specification for Axle-steel Deformed and Plain Bars for Concrete Reinforcement	A 617
Specification for Steel Bars, Carbon, Merchant Quality, Mechanical Properties	A 663
Specification for Steel Bars, Carbon, Hot-wrought, Special Quality, Mechanical Properties	A 675
Specification for Zinc-coated (Galvanized) Bars for Concrete Reinforcement	A 767

5.1.3.4

High-Strength, Low-Alloy Steels

High-strength, low-alloy steels are a group with chemical compositions specially developed to impart better mechanical properties and greater resistance to atmospheric corrosion than are obtainable from conventional carbon structural steels.

■ **CLASSIFICATION AND USES**

The AISI classifies high-strength, low-alloy steels according to chemical composition. ASTM publishes recommended practices and standards for them.

STRUCTURAL STEELS   The following list shows some ASTM standards for high-strength, low-alloy structural steels:.

Specification for high-strength, low-alloy structural steel	ASTM A 242
Specification for high-strength, low-alloy structural maganese vanadium steel	ASTM A 572
Specification for high-strength low-alloy structural steel with 50 ksi (345 MPa) minimum yield point to 4 in. thick	ASTM A 588

SHEET STEEL   The most widely produced and ordered high-strength, low-alloy sheet steel products are:

1. *Conventional high-strength, low-alloy steel*, which should conform with the requirements of ASTM A 607, "Specification for Steel, Sheet and Strip, High-strength, Low-alloy Columbium or Vanadium, or Both, Hot-rolled and Cold-rolled." This steel is produced to various strength levels where atmospheric corrosion resistance and maximum formability are not a requirement.

2. *Improved atmospheric corrosion-resistant high-strength, low-alloy sheet steel*, which should conform to the requirements of ASTM Standard A 606, "Specification for Steel Sheet and Strip, Hot-rolled and Cold-rolled, High-strength, Low-alloy, with Improved Atmospheric Corrosion Resistance." This steel is produced to 45 or 50 kips per square inch (ksi) minimum yield points (1 kip equals 1000 lb.). The types of steel with enhanced corrosion resistance include Type 2 with a corrosion resistance twice that of plain carbon steel, and Type 4 with a corrosion resistance at least four times greater.

3. *Improved formability high-strength, low-alloy steel*, which should conform with the requirements of ASTM A 715, "Specification for Steel Sheet and Strip, High-strength, Hot-rolled, and Steel Sheet, Cold-rolled, High-strength, Low-alloy, with Improved Formability." This type of steel is produced with special forming properties to various levels of strength where maximum formability is required. This type is generally furnished for making parts too difficult for the fabricating properties of the other two types.

In general, the minimum yield points for these steels vary within a range of 45 to 80 ksi and their minimum tensile strength varies from 55 to 90 ksi.

STEEL BARS   The most frequently specified high-strength, low-alloy steel bars have a minimum yield point ranging from 42 to 80 ksi and a minimum tensile strength that varies between 60 and 95 ksi.

■ **COMMON USES**

High-strength, low-alloy steels are commonly used where high strength in relation to weight is important. The steels listed in the previous paragraphs are often used in bridge and high-rise building construction. These and other high-strength steels are also used extensively in various transportation vehicles, heavy construction equipment, industrial containers, and wherever resistance to corrosion and impact damage is important.

Some proprietary low-alloy steels can be made with increased corrosion resistance, when intended for exposed architectural use. These steels, sometimes called *weathering steels*, form natural rust-colored, self-healing oxide coatings that inhibit further corrosion when left unpainted. This rusting may not stop, however, if certain pollutants are present in the atmosphere. They are used increasingly in commercial and industrial construction, but are not used extensively in residential work.

Unfortunately, the rust-colored oxide coating that forms to protect weathering steels will wash off if water runs across it. Special detailing is necessary to prevent this runoff from staining adjacent materials.

5.1.3.5

Stainless and Heat-Resisting Steels

The outstanding characteristics of the two groups of steels referred to as *stainless* and *heat resisting* are implied in their names. Stainless steels possess excellent corrosion resistance at varying temperature ranges. Heat-resisting steels retain their essential physical and mechanical properties at elevated temperatures. Stainless and heat-resisting steels account for about 1% of total steel production.

Chromium is the alloying element mainly responsible for corrosion-resisting and heat-resisting properties, although other elements, such as nickel, manganese, and molybdenum, also contribute to these properties. The stainless qualities of steel are derived from a self-healing chromium oxide that forms a transparent skin and prevents further oxidation. Steels with less than 4% chromium are regarded as alloy steels. Heat-resisting steels normally contain from 4% to 12% chromium, and stainless steels more than 12%. The chief physical and mechanical properties of representative stainless steels are shown in Figure 5.1-18.

Stainless and heat-resisting steels are classified according to chromium content and internal grain structure into three major groups: *martensitic*, *ferritic*, and *austenitic*, and a minor group under the martensitic group, *precipitation hardening* (see Fig. 5.1-18).

MARTENSITIC STEELS   Heat-resisting and stainless steels with a chromium content between 4% and 18% are known as *martensitic*. These may be hardened by heat treatment and are therefore sometimes referred to as hardenable chromium alloys. They also have adequate cold-forming characteristics and are satisfactory for hot working or forging, but must be slowly cooled or annealed after forging to prevent cracking. They are easily welded but require subsequent annealing or tempering to restore ductility.

FERRITIC STEELS   Stainless steels that contain between 12% and 27% chromium are called *ferritic*. These steels do not respond readily to heat treatment and are sometimes referred to as nonhardenable chromium alloys. However, their hardness may be increased somewhat by cold working. Ductility may be improved by hot or cold working followed by annealing. These steels are not welded as easily as are the martensitic steels, but have greater corrosion resistance. Fair ductility after annealing allows product fabrication by a variety of processes such as cold forming, spinning, and light drawing. Ferritic steels possess low strength at elevated temperatures and a low coefficient of thermal expansion.

AUSTENITIC STEELS   The chromium content of austenitic steels ranges between 16% and 26%. They also contain between 3.5% and 22% nickel. These steels are characterized by a greater ability to harden as a result of cold working than do the ferritic types. Austenitic steels include both nickel-chromium and nickel-chromium-manganese groups of alloys. Both groups provide superior resistance to corrosion and are suited to many fabrication techniques. They have high ductility, which is important for deep drawing and forging operations, are easily welded, and have good oxidation resistance at elevated temperatures. A relatively high coefficient of

thermal expansion is characteristic of these steels and must be considered in the design equipment used in conjunction with high temperatures. They are also more susceptible to corrosive attack by deoxidizing sulfur gases than other stainless steels.

PRECIPITATION HARDENING STEELS   Precipitation hardening steels are iron-chromium-nickel alloys with additional elements and are hardenable by solution treating and aging. These steels combine high strength, hardness, and corrosion resistance.

AISI DESIGNATIONS   Heat-resisting and stainless steels are grouped by the AISI into five series that bear number designations from 200 to 500 as shown in Figure 5.1-18.

UNIFIED NUMBERING SYSTEM (UNS)   As mentioned earlier the ASTM and the SAE developed the six-digit UNS as a method of designating steel types. Stainless steels are identified in the UNS with an *S* followed by five digits (see Fig. 5.1-18).

COMMON USES   There are many applications for stainless steels and a wide variety of alloys from which to choose.

Martensitic chromium steels are suitable for applications requiring high strength, hardness, and resistance to abrasion, such as steam and gas turbine parts, bearings, and cutlery. Type 410 (the basic type) is one of the most widely used alloys in this category.

Ferritic chromium steels are used mostly for automotive trim, applications involving nitric acid, high-temperature service requiring resistance to scaling, and uses that call for low thermal expansion. Types 405 and 430 steels are the most economical and widely used alloys in this category. Type 430 is sometimes used in building construction for gutters and downspouts, architectural trim, sills, and column covers. It is also used in kitchen applications such as range hoods, range tops, and appliance fronts.

Austenitic nickel-chromium steels are the most numerous and popular category of stainless steels. They are sometimes referred to as the 18/8 alloys because several of the most common types (302, 303, and 304) contain approximately 18% chromium and 8% nickel. Among these,

FIGURE 5.1-18
AISI Designations and Characteristics of Stainless and Heat-Resisting Steels

Grain Structure	Chief Alloying Elements	AISI Series	Representative Type	UNS Designation	Characteristics	Steel Type
Martensitic (ferromagnetic and hardenable by heat treatment)	Chromium 4% to 12%	500	502	S50200	Retain their mechanical properties at high temperatures	Heat-resisting steels
	Chromium 12% to 18%	400	410	S41000	Moderate corrosion resistance, high strength, and hardness	Stainless steels
Ferritic (ferromagnetic and nonhardenable)	Chromium 12% to 27%	400	430	S43000	Very good corrosion resistance, particularly at high temperatures	Stainless steels
Austenitic (nonmagnetic and hardenable by cold working)	Chromium 16% to 26% Nickel 6% to 22%	300	302	S30200	Excellent corrosion resistance, high strength, and ductility; suitable for many fabrication techniques	Stainless steels
	Chromium 16% to 19% Nickel 3.5% to 6% Manganese 5.5% to 10%	200	202	S20200	Excellent corrosion resistance at high temperatures, high strength, and toughness	Stainless steels
Precipitation hardening	Chromium 12% to 18% Nickel 3% to 8.5%	—	—	S15500	Combines high strength and hardness with excellent corrosion resistance	Stainless steels

Courtesy American Iron and Steel Institute.

Types 302 (the basic 18/8 grade) and 304 are used most often for such building applications as fascias, curtain walls, storefronts, doors and windows, column covers, and railings. Their relative chemical inertness and finish-retention qualities also make them useful in food preparation and surgical equipment, hospital clean rooms, laboratories, and for such familiar kitchen applications as sinks, countertops, and appliances.

In addition to a variety of automotive and transportation equipment uses, Type 301 steel is sometimes used for structural members because it work hardens rapidly to a high strength and toughness. Because it has good cold-forming properties it is used for making roof drainage products.

Type 316 is the most resistant to attack by salt spray and corrosive industrial fumes, and is therefore used in seacoast and industrial areas where protection from such exposures is essential.

Austenitic nickel-chromium manganese steels Types 201 and 202 are similar to their nickel-chromium counterparts, Types 301 and 302, and may be used in-terchangeably with them. In addition to automotive and food-storage equipment, they are used for flatware, cooking utensils, kitchen sinks, and countertops.

5.1.4

MANUFACTURE

Steel products are manufactured by casting in foundries, and by forging, extruding, rolling, and cold drawing in mills. The products of the foundry are referred to as cast products; all others are wrought products. Wrought products account for more than 90% of the total steel production, and by far the largest part of that is produced by hot rolling and cold finishing.

Steelmill products are regarded as semi-finished when in the form of blooms, billets, and slabs; and as finished when in a form suitable for further fabrication into components or for use directly in construction. Figure 5.1-19 shows the variety of finished mill products produced at steelmills.

5.1.4.1

Cast-Steel Products

Castings are generally made by pouring molten steel into sand molds. Before it is cast, pig iron may be further refined and melted with scrap metal and ferroalloys at the foundry to obtain desired casting and mechanical properties.

Casting is used in preference to other forming operations when desired shapes are of a size or complexity not readily obtained by rolling or machining. Complex and massive shapes possessing great strength and impact resistance at high and low temperatures usually are cast (Fig. 5.1-20). Other cast-steel products used in construction are concrete inserts, anchors, and items of hardware. In a typical year, iron and steel castings contribute many millions of tons to a broad range of uses. Of this, most are in gray and malleable cast iron, whereas some are in steel. Of the cast-steel products, approximately two-thirds are carbon steels; the rest are alloy and special steels.

The transportation industry relies on

FIGURE 5.1-19 *Operations in the manufacturing sequence for steel products. Blooms, billets, slabs, bars, plates, and pipes are shown as stacks of individual units.* (Courtesy American Iron and Steel Institute.)

FIGURE 5.1-21 *Continuity of flow lines in a forged metal structure results in greater strength and toughness.* (Courtesy American Iron and Steel Institute.)

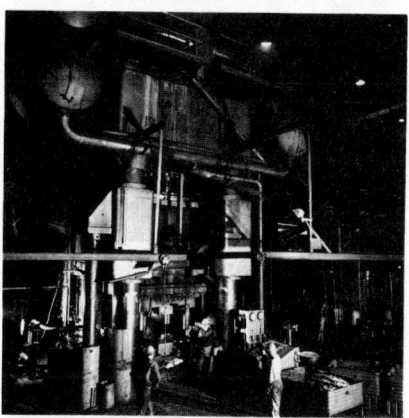

FIGURE 5.1-22 *A hydraulic press, which may exert a force of up to 18,000 tons, is particularly suitable for forging high-strength steels.* (Courtesy Kaiser Aluminum & Chemical Corporation.)

FIGURE 5.1-20 *Steel castings may range in weight from a few ounces to several hundred tons, as in this bridge saddle.* (Courtesy Bethlehem Steel Corporation.)

castings for many engine and auto-body parts, as does the excavating and mining equipment industry. Castings are also used extensively in many parts of steelmaking furnaces and rolling mills.

5.1.4.2

Wrought-Steel Products

A few steel products that are complex and have an irregular cross section are made by *forging*. Linear products of constant cross section, such as bars, tubes, and rods, and complex shapes may be made by *extruding*. Less complex and larger linear forms and sheets, strips, rods, bars, and most structural shapes used in construction are produced by hot rolling and cold finishing or by hot rolling without cold finishing.

FORGING

Forging is a method of forming hot metal into desired shapes by pressing it between heat-resistant dies. When metal is forged, toughness, strength, and ductility increase significantly along the lines of flow (Fig. 5.1-21), which is an important advantage of this forming technique. The metal to be forged is preheated to between 2200° and 2400°F in special furnaces, and in some operations the dies are heated as well. The two major types of forging are *open die* and *closed die*.

OPEN-DIE FORGING   In open-die forging, a large press squeezes (rather than strikes) steel between two flat surfaces. Temperature must be carefully controlled; this requires frequent reheating of the steel between shaping actions. Some hydraulic open-die forging presses may accept ingots weighing several hundred tons (Fig. 5.1-22).

CLOSED-DIE FORGING   In closed-die forging, a hammer pounds a section of steel between carved dies until it reaches the desired shape. Closed-die forging most often uses steam hammers. Drop hammers develop forging force by the fall of a heavy weight or ram. In double-action hammers the speed of the falling weight is usually given additional impetus.

EXTRUDING

Semifinished shapes may be converted into lengths of uniform cross section by *extruding*. In this process an advancing ram (Fig. 5.1-23) forces preheated, plastic metal through a tough, heat-resistant die of the desired profile (Fig. 5.1-24). Presses vary in size from a small 700-ton, 2-1/2-in. maximum circle die press designed for fast extrusion speeds of up to 25 ft./sec., to 12,000-ton installations capable of extruding sections up to 21 in. in diameter at slower speeds. Stainless steel is generally extruded in shapes up to 6-1/2 in. in diameter.

Extruding can produce more complex sections with better surface characteristics than rolling can. It can be used to shape

FIGURE 5.1-23 *Hot steel is forced through a lubricated flow extrusion press.* (Courtesy American Iron and Steel Institute.)

FIGURE 5.1-24 *Bars and tubing can be extruded into a variety of open and hollow shapes.* (*Courtesy American Iron and Steel Institute.*)

certain alloys without undesirable residual effects (such as excessive hardness) and is more economical than other forming methods for small quantities. However, most steel extrusions are limited to 35 ft. in length and 21 in. in diameter.

■ HOT ROLLING AND COLD FINISHING

By far the largest proportion of wrought-steel products are manufactured by *hot rolling* and *cold finishing*.

Hot rolling is used to make semifinished shapes as well as some finished products. Hot rolling may be done as part of the continuous casting process on a strand caster or by using hot ingots made by traditional steelmaking processes.

In a hot-rolling process, hot steel passes through a system of rolls, which gradually imparts rectangular bloom, billet, or slab shapes. These are cooled, cleaned of surface irregularities, inspected, and reheated for further rolling. Semifinished shapes may be hot rolled directly into finished hot-rolled products such as structural shapes, plates, sheets, strip, and bars; or hot rolling may be used as an intermediate step before cold finishing, as in the manufacture of bar, sheet, strip, and wire. The hot rolling of finished products is sometimes referred to as *hot finishing*.

Hot-rolled products intended for cold finishing are *descaled* (cleaned of surface oxide scale) usually by *pickling*, which involves passing the steel through sulfuric or hydrochloric acid, followed by rinsing with hot and cold water, steam drying, and, usually, oiling.

Cold finishing consists of cold rolling, cold reduction, and cold drawing of pre-viously hot-rolled, descaled shapes. Cold rolling involves passing metal at room temperature through sets of rolls to impart the desired shape, finish, and mechanical properties. *Cold reduction* is a form of cold rolling that drastically reduces the thickness of a flat product (sheet, strip, blackplate, etc.) and improves the strength, surface finish, and flatness. Cold reduction often hardens the metal excessively, so that it must be annealed to soften it and further temper (skin) rolled to obtain the proper strength and stiffness and the desired surface texture.

Cold drawing is used to make smaller or more complex bar shapes from hot-rolled annealed bars by pulling the bars through a hard, abrasion-resistant die. Wire is similarly cold drawn from wire rods. This operation results in improved machinability and strength, smoother finish, and greater dimensional accuracy.

SLABS, BLOOMS, AND BILLETS Slabs, blooms, and billets are semifinished shapes from which other products are made. *Slabs* are flat rectangular shapes with a width more than twice the thickness, which is usually less than 10 in. *Blooms* are rectangular shapes generally larger than 36 sq. in. *Billets* are less than 36 sq. in., more nearly square in cross section, and longer.

The type of semifinished shape that is rolled depends on its intended final use. Slabs are commonly used in the rolling of flat products such as plates, sheets, strip, and skelp; blooms in the rolling of rails, tube rounds, and structural shapes; and billets for making bars, tube rounds, and wire rods (Fig. 5.1-25). Skelp, tube rounds, and wire rods are also semifinished shapes. Skelp is used to make welded pipe, tube rounds for piercing into seamless pipe, and wire rods for drawing into wire.

SHEET AND STRIP Much of the steel shipped annually is the product of continuous-sheet mills. The definition of these and other flat-rolled products is complex, and varies depending on whether they are stainless or carbon steel. Generally, the dimensions of sheet and strip lie between

FIGURE 5.1-25 *Intermediate mill processes used to convert semifinished steel shapes into finished mill products.* (*Courtesy American Iron and Steel Institute.*)

those of bars and plates (Fig. 5.1-26). For carbon steels, the method of rolling (whether hot or cold) is also a factor.

Hot-rolled sheet and strip are manufactured from semifinished slabs in sheet and strip mills. Preheated slabs are squeezed and shaped by progressive hot rolling, until the final desired thickness and width are obtained. Sheets and strip emerge from the roller tables at speeds up to 3500 ft./min. and are wound into coils for shipment, or for transfer to other departments for further working.

Modern cold-reducing mills may receive hot-rolled steel about as thick as a

FIGURE 5.1-26
General Classification of Flat-Rolled Steel Products by Size[a]

Carbon Steel, Hot Rolled						
	Width					
Thickness	To 3-1/2 in. incl.	Over 3-1/2 in. to 6 in. incl.	Over 6 in. to 8 in. incl.	Over 8 in. to 12 in. incl.	Over 12 in. to 48 in. incl.	Over 48 in.
0.2300 in. & thicker	Bar	Bar	Bar	Plate	Plate	Plate
0.2299 in. to 0.2031 in.	Bar	Bar	Strip	Strip	Sheet	Plate
0.2030 in. to 0.1800 in.	Strip	Strip	Strip	Strip	Sheet	Plate
0.1799 in. to 0.0449 in.	Strip	Strip	Strip	Strip	Sheet	Sheet
0.0448 in. to 0.0344 in.	Strip	Strip				
0.0343 in. to 0.0255 in.	Strip					

Carbon Steel, Cold Rolled						
	Width					
Thickness	To 15/32 in. incl.		Over 15/32 in. to 12 in. incl.	Over 12 in. to 23-15/16 in. incl.		Over 23-15/16 in.
0.2500 in. & thicker	Bar		Bar	Strip[c]	Sheet[e]	Sheet
0.2499 in. to 0.0142 in.	Flat Wire[b]	Strip	Strip	Sheet[d]	Srip[c]	Sheet[e]	Sheet
0.0141 in. & thinner	Flat Wire[b]	Strip	Strip	Strip	Blackplate[e]	Blackplate

Stainless Steel, Hot and Cold Rolled				
	Width			
Thickness	1/4 in. to 3/8 in.	3/8 in. to 10 in.	10 in. to 24 in.	24 in. & Over
3/16 in. & over	Bar[f]	Bar	Plate	Plate
1/8 in. to 3/16 in.	Flat Wire[g]	Bar[g]	Strip	Sheet[h]
Up to 1/8 in.	Flat Wire	Strip	Strip	Sheet[h]

Courtesy American Iron and Steel Institute.

[a]Dimensional ranges for flat-rolled alloy steels are customarily slightly different from those for carbon or stainless steel.

[b]When the material has rolled or prepared edges.

[c]When special temper, edge, finish, or single-strand rolling is specified.

[d]Cut lengths or sheet coils slit from wider coils (with resulting No. 3 edges), in thicknesses 0.0142 in. to 0.0821 in. inclusive, and widths 2 in. to 12 in., carbon 0.20% maximum.

[e]When no special temper, edge, or finish (other than Dull or Luster) is specified.

[f]Only hot rolled.

[g]May also be rolled as strip.

[h]In polished finishes Nos. 3, 4, 6, 7 and 8.

half dollar and three-quarters of a mile long; just 2 minutes later, it will be rolled to the thickness of two playing cards and more than 2 miles long.

Cold-reduced sheet and strip are made by cold reducing previously hot-rolled and pickled flat products. Much sheet and strip are cold rolled to meet consumer requirements for thickness, surface finish, or mechanical properties (see Fig. 5.1-26). Cold-reduced products may also be annealed and temper rolled or zinc or tin coated.

Blackplate is the classification for cold-reduced products thinner than 0.0141 in. and wider than 12 in. (see Fig. 5.1-26).

PLATES Flat-rolled finished products thicker than sheet and strip, produced from slabs or slab ingots by hot rolling, are called plates (see Figs. 5.1-26 and 5.1-27). The two principal classes of plates are *sheared* and *universal*, depending on the type of mill on which they are rolled. Sheared-plate mills have only horizontal rolls and produce plates with uneven edges and ends, which must later be flame cut or sheared straight to the desired size. Universal-plate mills have vertical rolls that are capable of producing smooth, straight edges so that only the ends must be trimmed. Lighter plates may also be rolled on some sheet and strip mills from slabs produced in blooming or slabbing mills.

STRUCTURAL SHAPES Structural members, such as W-shaped (wide flange) sections, angles, channels, tees, zees, and piling, are hot rolled from blooms or billets. As many as 40 passes through grooved rolls may be required to produce a desired shape (Fig. 5.1-28). The first sets of rolls drastically reduce the cross section and impart the approximate shape; the finishing rolls gradually form the final shape and, for deep sections such as channels and angles, bend the legs into position. Wide-flange shapes are rolled on universal mills that have vertical rolls capable of shaping the vertical surfaces. Different weights of a shape are produced by varying the spacing of the rollers, wider spacings producing heavier sections.

Structural shapes are identified by standard designations (Fig. 5.1-29). Wide-flange shapes are available in a great many sizes, a few of which are shown in Figure 5.1-30. Those used for beams and girders tend to be relatively tall with narrow flanges; those for columns and piles tend to be more square in shape.

Angles are available with legs of either equal or unequal length (Fig. 5.1-31). They may be used as structural lintels over masonry openings, supports for precast concrete and stone cladding elements, floor opening framing, concrete slab opening edging, structural cross bracing, brackets to connect structural beams to girders and columns, components of trusses, and supports for finish furring.

Channels have many of the same applications. They are also used as light beams, wall framing, and purlins. Tees are also used as light beams and in such applications as purlins in roof framing and as support for cementitious and plank decking.

Each structural shape is marked at the mill with its shape designation (*W, S, L, C, WT*), mill identification, and a code that identifies the batch of steel from which it was produced. This marking provides a convenient means of tracing a shape back to its source in the event of its failure. A certificate showing the chemical analysis of the batch of steel accompanies steel shapes from the mill to the fabricator

ANGLE Slab	First Pass									Final Pass	

| W-SHAPE Beam Blank | First Pass | | | | | | | | | | | Final Pass | |

| SHEET PILING Beam Blank | First Pass | | | | | | | | | Final Pass | |

FIGURE 5.1-28 *Successive steps in rolling structural shapes. Hot blanks are roughly shaped by one set of rolls, then other rolls impart the final shape.* (Courtesy American Iron and Steel Institute.)

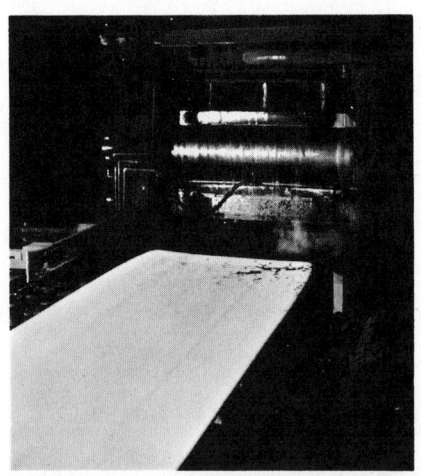

FIGURE 5.1-27 *A hot plate is rolled back and forth in a reversing mill until the desired thickness is achieved.* (Courtesy American Iron and Steel Institute.)

FIGURE 5.1-29
Some Common Structural Steel Shape Designations

Shape	Designation	Example (as used on drawings)*	
Wide flange	W	W24 × 62	24 is depth, 62 is weight
Angle	L	L5 × 3 × 1/4	5 and 3 are leg lengths 1/4 is leg thickness
Channel	C	C10 × 15.3	10 is depth, 15.3 is weight
Structural tee	WT	WT12 × 52	12 is depth, 52 is weight

*Dimensions are in inches; weights are in pounds per linear foot.

FIGURE 5.1-30
Some Wide-Flange (W) Shapes*

Designation	Area A (in²)	Depth d (in)	Depth d (in)	Web Thickness t_w (in)	t_w (in)	$\frac{t_w}{2}$ (in)	Flange Width b_f (in)	b_f (in)	Flange Thickness t_f (in)	t_f (in)	T (in)	k (in)	k_1 (in)	Nom. Wt. per Ft. (lb)	$\frac{b_f}{2t_f}$	F_y' (ksi)	$\frac{d}{t_w}$	F_y''' (ksi)	r_T (in)	$\frac{d}{A_f}$	X-X I (in⁴)	X-X S (in³)	X-X r (in)	Y-Y I (in⁴)	Y-Y S (in³)	Y-Y r (in)	J (in⁴)	Z_x (in³)	Z_y (in³)
W18×119	35.1	18.97	19	0.655	5/8	5/16	11.265	11-1/4	1.060	1-1/16	15-1/2	1-3/4	15/16	119	5.3	—	29.0	—	3.02	1.59	2190	231	7.90	253	44.9	2.69	10.6	261	69.1
×106	31.1	18.73	18-3/4	0.590	9/16	5/16	11.200	11-1/4	0.940	15/16	15-1/2	1-5/8	15/16	106	6.0	—	31.7	—	3.00	1.78	1910	204	7.84	220	39.4	2.66	7.48	230	60.5
× 97	28.5	18.59	18-5/8	0.535	9/16	5/16	11.145	11-1/8	0.870	7/8	15-1/2	1-9/16	7/8	97	6.4	—	34.7	54.7	2.99	1.92	1750	188	7.82	201	36.1	2.65	5.86	211	55.3
× 86	25.3	18.39	18-3/8	0.480	1/2	1/4	11.090	11-1/8	0.770	3/4	15-1/2	1-7/16	7/8	86	7.2	—	38.3	45.0	2.97	2.15	1530	166	7.77	175	31.6	2.63	4.10	186	48.4
× 76	22.3	18.21	18-1/4	0.425	7/16	1/4	11.035	11	0.680	11/16	15-1/2	1-3/8	13/16	76	8.1	64.2	42.8	36.0	2.95	2.43	1330	146	7.73	152	27.6	2.61	2.83	163	42.2
W18×71	20.8	18.47	18-1/2	0.495	1/2	1/4	7.635	7-5/8	0.810	13/16	15-1/2	1-1/2	7/8	71	4.7	—	37.3	47.4	1.98	2.99	1170	127	7.50	60.3	15.8	1.70	3.48	145	24.7
× 65	19.1	18.35	18-3/8	0.450	7/16	1/4	7.590	7-5/8	0.750	3/4	15-1/2	1-7/16	7/8	65	5.1	—	40.8	39.7	1.97	3.22	1070	117	7.49	54.8	14.4	1.69	2.73	133	22.5
× 60	17.6	18.24	18-1/4	0.415	7/16	1/4	7.555	7-1/2	0.695	11/16	15-1/2	1-3/8	13/16	60	5.4	—	44.0	34.2	1.96	3.47	984	108	7.47	50.1	13.3	1.69	2.17	123	20.6
× 55	16.2	18.11	18-1/8	0.390	3/8	3/16	7.530	7-1/2	0.630	5/8	15-1/2	1-5/16	13/16	55	6.0	—	46.4	30.6	1.95	3.82	890	98.3	7.41	44.9	11.9	1.67	1.66	112	18.5
× 50	14.7	17.99	18	0.355	3/8	3/16	7.495	7-1/2	0.570	9/16	15-1/2	1-1/4	13/16	50	6.6	—	50.7	25.7	1.94	4.21	800	88.9	7.38	40.1	10.7	1.65	1.24	101	16.6
W18×46	13.5	18.06	18	0.360	3/8	3/16	6.060	6	0.605	5/8	15-1/2	1-1/4	13/16	46	5.0	—	50.2	26.2	1.54	4.93	712	78.8	7.25	22.5	7.43	1.29	1.22	90.7	11.7
× 40	11.8	17.90	17-7/8	0.315	5/16	3/16	6.015	6	0.525	1/2	15-1/2	1-3/16	13/16	40	5.7	—	56.8	20.5	1.52	5.67	612	68.4	7.21	19.1	6.35	1.27	0.81	78.4	9.95
× 35	10.3	17.70	17-3/4	0.300	5/16	3/16	6.000	6	0.425	7/16	15-1/2	1-1/8	3/4	35	7.1	—	59.0	19.0	1.49	6.94	510	57.6	7.04	15.3	5.12	1.22	0.51	66.5	8.06
W16×100	29.4	16.97	17	0.585	9/16	5/16	10.425	10-3/8	0.985	1	13-5/8	1-11/16	15/16	100	5.3	—	29.0	—	2.81	1.65	1490	175	7.10	186	35.7	2.51	7.73	198	54.9
× 89	26.2	16.75	16-3/4	0.525	1/2	1/4	10.365	10-3/8	0.875	7/8	13-5/8	1-9/16	7/8	89	5.9	—	31.9	64.9	2.79	1.85	1300	155	7.05	163	31.4	2.49	5.45	175	48.1
× 77	22.6	16.52	16-1/2	0.455	7/16	1/4	10.295	10-1/4	0.760	3/4	13-5/8	1-7/16	7/8	77	6.8	—	36.3	50.1	2.77	2.11	1110	134	7.00	138	26.9	2.47	3.57	150	41.1
× 67	19.7	16.33	16-3/8	0.395	3/8	3/16	10.235	10-1/4	0.665	11/16	13-5/8	1-3/8	13/16	67	7.7	—	41.3	38.6	2.75	2.40	954	117	6.96	119	23.2	2.46	2.39	130	35.5
W16×57	16.8	16.43	16-3/8	0.430	7/16	1/4	7.120	7-1/8	0.715	11/16	13-5/8	1-3/8	7/8	57	5.0	—	38.2	45.2	1.86	3.23	758	92.2	6.72	43.1	12.1	1.60	2.22	105	18.9
× 50	14.7	16.26	16-1/4	0.380	3/8	3/16	7.070	7-1/8	0.630	5/8	13-5/8	1-5/16	13/16	50	5.6	—	42.8	36.1	1.84	3.65	659	81.0	6.68	37.2	10.5	1.59	1.52	92.0	16.3
× 45	13.3	16.13	16-1/8	0.345	3/8	3/16	7.035	7	0.565	9/16	13-5/8	1-1/4	13/16	45	6.2	—	46.8	30.2	1.83	4.06	586	72.7	6.65	32.8	9.34	1.57	1.11	82.3	14.5
× 40	11.8	16.01	16	0.305	5/16	3/16	6.995	7	0.505	1/2	13-5/8	1-3/16	13/16	40	6.9	—	52.5	24.0	1.82	4.53	518	64.7	6.63	28.9	8.25	1.57	0.79	72.9	12.7
× 36	10.6	15.86	15-7/8	0.295	5/16	3/16	6.985	7	0.430	7/16	13-5/8	1-1/8	3/4	36	8.1	64.0	53.8	22.9	1.79	5.28	448	56.5	6.51	24.5	7.00	1.52	0.54	64.0	10.8
W16×31	9.12	15.88	15-7/8	0.275	1/4	1/8	5.525	5-1/2	0.440	7/16	13-5/8	1-1/8	3/4	31	6.3	—	57.7	19.8	1.39	6.53	375	47.2	6.41	12.4	4.49	1.17	0.46	54.0	7.03
× 26	7.68	15.69	15-3/4	0.250	1/4	1/8	5.500	5-1/2	0.345	3/8	13-5/8	1-1/16	3/4	26	8.0	—	62.8	16.8	1.36	8.27	301	38.4	6.26	9.59	3.49	1.12	0.26	44.2	5.48

Courtesy American Institute of Steel Construction, Inc.

*The shapes listed here are only a few of the many available. This table shows every detail needed by a designer.

FIGURE 5.1-31
Some of the Available Structural Steel Angles*

Size and Thickness			k	Weight per Foot	Area	Axis X-X				Axis Y-Y				Axis Z-Z	
						I	S	r	y	I	S	r	x	r	Tan α
in.			in.	lb.	in.²	in.⁴	in.³	in.	in.	in.⁴	in.³	in.	in.	in.	
L9×4	×	5/8	1-1/8	26.3	7.73	64.9	11.5	2.90	3.36	8.32	2.65	1.04	0.858	.847	0.216
		9/16	1-1/16	23.8	7.00	59.1	10.4	2.91	3.33	7.63	2.41	1.04	0.834	.850	0.218
		1/2	1	21.3	6.25	53.2	9.34	2.92	3.31	6.92	2.17	1.05	0.810	.854	0.220
L8×8	×1-1/8		1-3/4	56.9	16.7	98.0	17.5	2.42	2.41	98.0	17.5	2.42	2.41	1.56	1.000
	1		1-5/8	51.0	15.0	89.0	15.8	2.44	2.37	89.0	15.8	2.44	2.37	1.56	1.000
		7/8	1-1/2	45.0	13.2	79.6	14.0	2.45	2.32	79.6	14.0	2.45	2.32	1.57	1.000
		3/4	1-3/8	38.9	11.4	69.7	12.2	2.47	2.28	69.7	12.2	2.47	2.28	1.58	1.000
		5/8	1-1/4	32.7	9.61	59.4	10.3	2.49	2.23	59.4	10.3	2.49	2.23	1.58	1.000
		9/16	1-3/16	29.6	8.68	54.1	9.34	2.50	2.21	54.1	9.34	2.50	2.21	1.59	1.000
		1/2	1-1/8	26.4	7.75	48.6	8.36	2.50	2.19	48.6	8.36	2.50	2.19	1.59	1.000
L8×6	×1		1-1/2	44.2	13.0	80.8	15.1	2.49	2.65	38.8	8.92	1.73	1.65	1.28	0.543
		7/8	1-3/8	39.1	11.5	72.3	13.4	2.51	2.61	34.9	7.94	1.74	1.61	1.28	0.547
		3/4	1-1/4	33.8	9.94	63.4	11.7	2.53	2.56	30.7	6.92	1.76	1.56	1.29	0.551
		5/8	1-1/8	28.5	8.36	54.1	9.87	2.54	2.52	26.3	5.88	1.77	1.52	1.29	0.554
		9/16	1-1/16	25.7	7.56	49.3	8.95	2.55	2.50	24.0	5.34	1.78	1.50	1.30	0.556
		1/2	1	23.0	6.75	44.3	8.02	2.56	2.47	21.7	4.79	1.79	1.47	1.30	0.558
		7/16	15/16	20.2	5.93	39.2	7.07	2.57	2.45	19.3	4.23	1.80	1.45	1.31	0.560
L8×4	×1		1-1/2	37.4	11.0	69.6	14.1	2.52	3.05	11.6	3.94	1.03	1.05	0.846	0.247
		3/4	1-1/4	28.7	8.44	54.9	10.9	2.55	2.95	9.36	3.07	1.05	0.953	0.852	0.258
		9/16	1-1/16	21.9	6.43	42.8	8.35	2.58	2.88	7.43	2.38	1.07	0.882	0.861	0.265
		1/2	1	19.6	5.75	38.5	7.49	2.59	2.86	6.74	2.15	1.08	0.859	0.865	0.267
L7×4	×	3/4	1-1/4	26.2	7.69	37.8	8.42	2.22	2.51	9.05	3.03	1.09	1.01	0.860	0.324
		5/8	1-1/8	22.1	6.48	32.4	7.14	2.24	2.46	7.84	2.58	1.10	0.963	0.865	0.329
		1/2	1	17.9	5.25	26.7	5.81	2.25	2.42	6.53	2.12	1.11	0.917	0.872	0.335
		3/8	7/8	13.6	3.98	20.6	4.44	2.27	2.37	5.10	1.63	1.13	0.870	0.880	0.340
L6×6	×1		1-1/2	37.4	11.0	35.5	8.57	1.80	1.86	35.5	8.57	1.80	1.86	1.17	1.000
		7/8	1-3/8	33.1	9.73	31.9	7.63	1.81	1.82	31.9	7.63	1.81	1.82	1.17	1.000
		3/4	1-1/4	28.7	8.44	28.2	6.66	1.83	1.78	28.2	6.66	1.83	1.78	1.17	1.000
		5/8	1-1/8	24.2	7.11	24.2	5.66	1.84	1.73	24.2	5.66	1.84	1.73	1.18	1.000
		9/16	1-1/16	21.9	6.43	22.1	5.14	1.85	1.71	22.1	5.14	1.85	1.71	1.18	1.000
		1/2	1	19.6	5.75	19.9	4.61	1.86	1.68	19.9	4.61	1.86	1.68	1.18	1.000
		7/16	15/16	17.2	5.06	17.7	4.08	1.87	1.66	17.7	4.08	1.87	1.66	1.19	1.000
		3/8	7/8	14.9	4.36	15.4	3.53	1.88	1.64	15.4	3.53	1.88	1.64	1.19	1.000
		5/16	13/16	12.4	3.65	13.0	2.97	1.89	1.62	13.0	2.97	1.89	1.62	1.20	1.000
L6×4	×	7/8	1-3/8	27.2	7.98	27.7	7.15	1.86	2.12	9.75	3.39	1.11	1.12	0.875	0.421
		3/4	1-1/4	23.6	6.94	24.5	6.25	1.88	2.08	8.68	2.97	1.12	1.08	0.860	0.428
		5/8	1-1/8	20.0	5.86	21.1	5.31	1.90	2.03	7.52	2.54	1.13	1.03	0.864	0.435
		9/16	1-1/16	18.1	5.31	19.3	4.83	1.90	2.01	6.91	2.31	1.14	1.01	0.866	0.438
		1/2	1	16.2	4.75	17.4	4.33	1.91	1.99	6.27	2.08	1.15	0.987	0.870	0.440
		7/16	15/16	14.3	4.18	15.5	3.83	1.92	1.96	5.60	1.85	1.16	0.964	0.873	0.443
		3/8	7/8	12.3	3.61	13.5	3.32	1.93	1.94	4.90	1.60	1.17	0.941	0.877	0.446
		5/16	13/16	10.3	3.03	11.4	2.79	1.94	1.92	4.18	1.35	1.17	0.918	0.882	0.448
L6×3-1/2 ×		1/2	1	15.3	4.50	16.6	4.24	1.92	2.08	4.25	1.59	0.972	0.833	0.759	0.344
		3/8	7/8	11.7	3.42	12.9	3.24	1.94	2.04	3.34	1.23	0.988	0.787	0.767	0.350
		5/16	13/16	9.8	2.87	10.9	2.73	1.95	2.01	2.85	1.04	0.996	0.763	0.772	0.352
L5×5	×	7/8	1-3/8	27.2	7.98	17.8	5.17	1.49	1.57	17.8	5.17	1.49	1.57	0.973	1.000
		3/4	1-1/4	23.6	6.94	15.7	4.53	1.51	1.52	15.7	4.53	1.51	1.52	0.975	1.000
		5/8	1-1/8	20.0	5.86	13.6	3.86	1.52	1.48	13.6	3.86	1.52	1.48	0.978	1.000
		1/2	1	16.2	4.75	11.3	3.16	1.54	1.43	11.3	3.16	1.54	1.43	0.983	1.000
		7/16	15/16	14.3	4.18	10.0	2.79	1.55	1.41	10.0	2.79	1.55	1.41	0.986	1.000
		3/8	7/8	12.3	3.61	8.74	2.42	1.56	1.39	8.74	2.42	1.56	1.39	0.990	1.000
		5/16	13/16	10.3	3.03	7.42	2.04	1.57	1.37	7.42	2.04	1.57	1.37	0.994	1.000

Courtesy American Institute of Steel Construction, Inc.

*Angles in shaded rows may not be readily available. Availability is subject to rolling accumulation and geographical location, and should be checked with material suppliers.

to display compliance with the required specifications.

BARS Bars may be round, square, hexagonal, or multifaceted long shapes, generally thicker than wire. They are produced by hot rolling and cold drawing. In hot rolling, between 10 and 15 roll passes, taking less than 2 minutes, are required to change a square hot billet to a bar. Bars more than 3/4 in. in diameter are usually sheared into lengths from 16 to 20 ft. long. Smaller bars are typically wound in coils for ease of handling and shipping. Carbon steel bars may be galvanized or treated with other corrosion-resistant coatings. Flat rectangular shapes (flats), which are narrower than sheet and strip, are also known as bars (see Fig. 5.1-26) and may be either hot or cold rolled.

Hot-rolled bars may be cold drawn to further reduce their cross section and produce complex shapes (Fig. 5.1-32) of closer dimensional tolerances and smoother finishes. Improved mechanical properties, such as greater tensile strength and hardness, also result from cold drawing. The degree of reduction depends on the type of metal used and the desired mechanical properties of the end product.

WIRE Cold-drawn products of round, square, or multifaceted cross section, which are smaller than bars, are called wire. Round wire is drawn from 0.005 in. up to 1 in. in diameter. Flat wire is a cold-rolled product, rectangular in shape, and generally narrower than bars (see Fig. 5.1-26).

Semifinished wire rods of low carbon steel may be drawn directly into wire by *angle draft drawing*. For finer wire sizes or harder steels, the rods must be softened by annealing before the continuous *multiple draft drawing* (Fig. 5.1-33) through a series of dies. Wire may also be annealed after drawing to make extremely soft *annealed wire*.

Cold-drawn wire of carbon steel can be produced with very high tensile strengths ranging to more than 500,000 psi. Corrosion resistance can be achieved by galvanizing or coating with other corrosion-resistant metals.

TUBULAR PRODUCTS Tubular products are long, hollow metal products of round, oval, square, rectangular, or multifaceted

FIGURE 5.1-32 *Irregular as well as round and rectangular bars can be easily cold drawn.* (Courtesy American Iron and Steel Institute.)

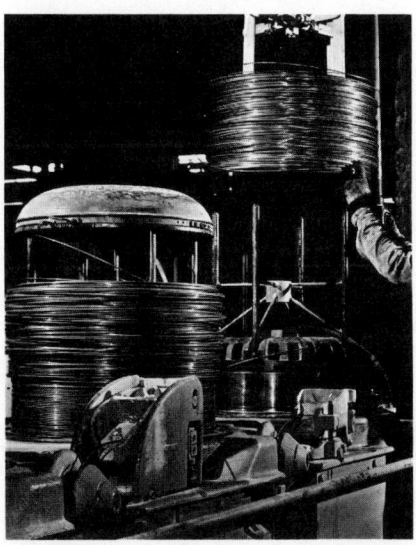

FIGURE 5.1-33 *Fine wire can be made by drawing rods through successively smaller die openings.* (Courtesy American Iron and Steel Institute.)

cross section. Tubular products can be either pipe or tubing, the terms being applied somewhat interchangeably. Round pipe in a wide range of sizes and square or rectangular structural tubing are widely employed in building construction. Tubular products that are mechanically worked and produced on *welded* and *seamless* mills are referred to as *wrought pipe*, to distinguish them from cast pipe. Wrought tubular products may be marketed, after being milled as either hot or cold finished, as *black pipe* or as *galvanized pipe* when hot dipped in zinc.

Welded pipe is made chiefly by the butt weld and the electric weld processes. In

the *butt weld* process, a strip of preheated, semifinished skelp is continuously formed into a tubular shape between concave rolls, and the edges are joined by mechanical pressure during rolling (Fig. 5.1-34a). Pipe sizes between 1/8 in. and 4 in. are commonly produced by this process.

The majority of steel pipe in this country is produced by *electric weld* processes. Preheating of the metal is not required, and localized heat necessary for joining the edges is generated during the welding process itself. Resistance welding is used to make *electric-resistance welded* (ERW) tubular products up to 20 in. in diameter. In this process, a strip of cold metal is formed into a tubular shape by rolling, and the edges are joined by heat and pressure (Fig. 5.1-34b). Welding heat is generated in the seam by the resistance of the metal to an electric current introduced by a wheel-shaped roller that acts as the electrode.

For larger diameter and thick-walled pipe, one of several electric welding processes may be used. In most, steel strip or plates are die formed cold by several successive forming operations, consisting of crimping (bending the edges), U forming, and O forming. The last forming operations result in a nearly round shape with edges touching and ready for welding. Submerged arc welding and metal inert-gas-shielded arc (MIG) welding are often used for this purpose. The final shape and diameter are obtained by expanding the welded shell with hydraulic power against a retaining jacket.

Seamless pipe and tubing are made by hot piercing and hot extrusion. Piercing is accomplished by continuously feeding a hot billet between pairs of tapered rolls that are skewed to apply uneven pressure, which causes a pilot hole to form in the center of the billet as it rolls between them. A piercing point enlarges the pilot hole and a sizing mandrel shapes the size of the final opening and the thickness of the wall (Fig. 5.1-35a). Most seamless products are made by hot working, although some seamless tubing is made by cold drawing a prepierced billet through a die over an internal mandrel (Fig. 5.1-35b). Stainless and alloy steel tubing up to 11 in. in diameter are both often made in this manner. Hot-piercing methods can produce seamless pipe to about 26 in. in diameter.

FROM SKELP TO BUTT WELD PIPE

(continuous rolled)

TRAVELLING SAW
CUTS PIPE
TO LENGTH

FROM SKELP TO ELECTRIC WELD PIPE

DC
WATER
CUTTING TOOLS
PRESSURE ROLLS

b

FIGURE 5.1-34 *(a) The butt-weld process can produce pipes up to 4 in. in diameter; (b) the electric-weld process can produce pipes up to 8 in. in diameter. (Courtesy American Iron and Steel Institute.)*

FROM BILLET TO SEAMLESS PIPE

a

CENTER PUNCHED
END VIEW
SIZING MANDREL

FROM PIPE TO COLD DRAWN TUBING

MANDREL
DIE
CHAIN PULLS GRIP DRAWING PIPE THROUGH DIE
LUBRICANT
MANDREL
DIE

b

FIGURE 5.1-35 *(a) Most seamless carbon steel pipe is made by hot piercing and extruding. (b) Stainless and alloy steel tubing is cold drawn. (Courtesy American Iron and Steel Institute.)*

Hot extrusion produces a tubular shape by forcing hot predrilled billets through a die over an internal mandrel. This method is used primarily to make small sizes of stainless or high-alloy pipe for cold reduction. It is also the most economical method of producing complex tubular shapes in small quantities.

5.1.4.3

Fabrication

Most steels can be fabricated by hot forming, cold forming, and machining. Some forming methods may be the same as those described above; however, fabrication as used here implies reworking the products of the mills and foundries in a fabricating plant. Cold-forming techniques are by far the more common ones used in the manufacture of products for building construction.

■ HOT FORMING

Hot-forming operations are used on metals too thick or too hard to be formed cold. These operations include: (1) forging (described above); (2) high-temperature forming (electroforming), which uses heat generated by the workpiece's resistance to an electric current passed through it by the forming roll; and (3) high-energy rate forming, a method of hot (although sometimes cold) shaping materials by the sudden impact of a small explosive charge, a shock wave generated by an electric spark in a fluid, or sudden release of a compressed gas through a system of valves.

■ COLD FORMING

Cold-forming operations are performed on most flat-rolled and tubular products by shaping, bending, or drawing the metal beyond its yield point so that a permanent set is achieved. The most common cold-

forming methods are *roller, stretch, shear,* and *brake forming,* and *deep drawing.*

ROLLER FORMING   A variety of shapes (Fig. 5.1-36) are formed by passing sheet and strip through a series of roller dies that gradually impart the desired shape (Fig. 5.1-37). Many types of corrugated and ribbed decking and wall panels are made in this manner.

FIGURE 5.1-36 *Sheet and strip may be roller formed into a variety of shapes. (Courtesy American Iron and Steel Institute.)*

FIGURE 5.1-37 *Wall and roofing panels typically are roller formed from galvanized or stainless steel sheets. (Courtesy American Iron and Steel Institute.)*

STRETCH FORMING   In stretch forming, light-gauge flat products are molded over form blocks into shallow curved shapes by pressure. The steps in stretch forming a particular shape are illustrated in Figure 5.1-38.

SHEAR FORMING   In shear forming, flat round blanks are made into curved surfaces, such as cylinders, domes, and

FIGURE 5.1-38 *In stretch forming, an oversized piece of sheet metal is gripped (a) and bent over a form block (b) to produce the desired contour (c), and then is trimmed to the final shape (d).* (Courtesy American Iron and Steel Institute.)

cones, by turning the workpiece on a mandrel and shaping it over a block with external pressure. *Spinning* is a type of shear forming in which the deformation and reduction in original blank thickness are not quite as severe (Fig. 5.1-39). Tank

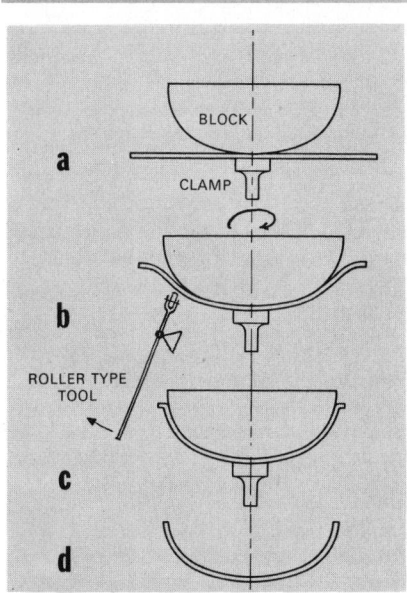

FIGURE 5.1-39 *In spinning, a round blank is clamped on a form block (a) and spun while pressure is applied (b), gradually imparting the shape of the form block (c), then is trimmed to the desired profile (d).* (Courtesy American Iron and Steel Institute.)

heads, television tube cones, and stainless steel, railing base flanges and caps are examples.

BRAKE FORMING A common method of fabricating sheet metal components for construction is brake forming. In this method, sheet steel is shaped on a press brake by squeezing it between a punch and a die block in one or several consecutive operations (Fig. 5.1-40). Heating and air conditioning ducts, steel roofing products, and many of the sections used in architectural stainless steel storefront construction are fabricated by this method.

FIGURE 5.1-40 *Both carbon steel and stainless steel can be bent sharply in a press brake to form various products.* (Courtesy American Iron and Steel Institute.)

DEEP DRAWING Flat, thin-gauge blanks are often formed in a draw press between an operating punch and a stationary die. Many household utensils, such as pots and pans, and commercial vessels and containers are made this way. Stainless steel can be drawn to various degrees, but the austenitic group lends itself most readily to deep-drawing operations.

■ MACHINING

Most steels can be milled, sawed, drilled, punched, sheared, tapped, and reamed with appropriate equipment. Tool steels containing tungsten, molybdenum, and chromium that have been subjected to tempering and other heat treatments are used for this purpose. Sulfur and lead improve the machining properties of carbon steels and are added to steels requiring extensive machining. Among the stainless steels, the ferritic alloys are the easiest to machine.

Protective Coatings and Mechanical Finishes

Ordinary cast iron, most carbon steels, and some alloy steels oxidize under ordinary atmospheric conditions. Elevated temperatures and moisture accelerate this oxidation, which is known as atmospheric corrosion. The resulting surface scale (rust) deteriorates the metal surface and, if not prevented, progresses until the metal reverts to the oxide state found in nature. The chief function of protective coatings used on iron and carbon steel is to inhibit corrosion.

Not all ferrous metals require equal protection against progressive corrosion. Copper-bearing steels, some cast irons, and some high-strength, low-alloy steels acquire excellent corrosion resistance from a tight surface oxide that inhibits further corrosion. Sometimes the appearance of the natural patina is objectionable and finish coatings are applied on these steels for decorative purposes. Decorative and protective coatings are used less often on stainless steels because of their superior corrosion resistance and attractive natural finishes.

Finishes for iron and steel may be classified as: (1) *mechanical*, including the as-rolled and as-drawn natural mill finish, and those imparted by further grinding, polishing, or patterning; (2) *chemical*, generally consisting of cleaning or preparatory operations for further finishing; and (3) *organic and inorganic coatings*, such as metallic, vitreous, laminated, and painted finishes. Organic coatings commonly sprayed on or brushed on in the field are addressed in Chapter 9.

■ MECHANICAL FINISHES

The *hot-rolled mill finish* (black, as-rolled finish) on carbon steel is characterized by a mill scale and rust powder, which is sometimes acceptable as a base for organic coatings applied by brush. For spray painting, this rust powder must be removed by sandblasting or wire brushing.

Cold-finished surfaces are usually smoother and contain little or no surface corrosion, but must be degreased of oil coatings and sometimes roughened to provide a good base for organic coatings. Steel products that will not be coated may

FIGURE 5.1-41 *Steel sheets can be stiffened and decorated with patterned finishes.* *(Photographs furnished by Rigidized Metals Corp.)*

be sent to a *temper mill*, which rolls the steel to a desired flatness and surface quality. Cold-reduced sheets and strips have a natural mill finish acceptable for many uses, and as a base for subsequent galvanizing, tinning, or terne coating. *Black pipe* has a natural dark finish often left uncoated. Stainless steel sheet is produced in a variety of patterned finishes (Fig. 5.1-41) imparted by roll forming for greater stiffness and decorative effect, and to reduce differential light reflection that emphasizes the surface waviness of thin, flat products.

STAINLESS STEEL SHEET AND STRIP   Because stainless steels are inherently corrosion resistant and can be made strong and rigid in thin sheets, these products are the ones most often not protected by coatings.

They are most often either left with their natural cold-rolled finish, such as Nos. 1, 2D, and 2B sheet finishes, or given a polished finish, such as Nos. 3, 4, 6, 7, or 8 (Fig. 5.1-42). An exception is terne-coated stainless steel, which is used in roofing.

STAINLESS BARS, PIPES, AND TUBING
Stainless bars, pipes, and tubing are manufactured in several neutral and polished finishes similar to those of sheets and strips described in the preceding paragraph. In industrial applications, they may be used in one of the natural finishes; for architectural work where uniformity of appearance is important, polished finishes are generally preferred.

■ **CHEMICAL FINISHES**
Chemical finishes are used chiefly as cleaning and conditioning steps for other coating applications. Pickling removes oxide scale; other operations remove lubricants in product manufacture. *Conversion coatings* are sometimes used as decorative finishes on carbon and stainless steels.

FIGURE 5.1-42
Common Stainless Sheet Steel Mechanical Finishes

	Designation	Description	Uses
Natural Finishes (unpolished)	No. 1 Finish	Rough dull surface resulting from hot rolling, annealing, and descaling	Generally available on heavy gauge sheets; not used in architectural applications
	No. 2D Finish (No. 1 Strip Finish)	Smooth dull surface resulting from cold rolling, annealing, and descaling	Some architectural applications requiring low luster, as in roofing and drainage products; suitable for cold forming and further polishing
	No. 2B Finish (No. 2 Strip Finish)	Bright, smooth cold-rolled surface produced either by highly polished rolls or bright annealing	General-purpose natural finish for transportation equipment; limited use in curtain walls and storefronts; used for further polishing
Polished Finishes	No. 3 Finish	A surface produced by grinding and polishing with slightly finer abrasives than for finish No. 4	General-purpose natural finish for transportation equipment; limited use in curtain walls and storefronts; used for further polishing
	No. 4 Finish	Bright, polished surface produced by grinding and polishing either at the mill or fabricator plant	Most common for architectural uses; directional grit line permits blending of fabricated or field joints with mill-polished surface
	No. 6 Finish	Dull matte finish produced by brushing No. 4 finish	Architectural applications where softer, less reflective finish is desired
	No. 7 Finish	A highly reflective finish produced by further polishing and buffing	Chiefly in architectural applications
	No. 8 Finish	Mirrorlike finish produced by polishing with buffing rouges until surface is free of grit lines	Infrequent in architectural applications

Courtesy American Iron and Steel Institute.

CARBON STEEL CONVERSION COATINGS
Conversion coatings are mostly chemical treatments (as with an acid phosphate solution) that alter the chemical nature of the surface film to improve its bonding properties with paint and other applied coatings. In one decorative finish, textured steel is zinc plated and dyed using a chromate conversion coating, followed by buffing. Buffing causes the bright zinc to stand out in relief, highlighted by the color coating in the recesses.

STAINLESS CONVERSION COATINGS Surface blackening, a common conversion coating, on stainless steel produces a decorative finish ranging from a bluish hue to dark brown or dead black. It is produced by encouraging the growth of a thin oxide, either by controlled heat treatment or by dipping the steel in a hot chemical oxidizing bath.

■ **ORGANIC AND INORGANIC COATINGS**

Carbon steels are most often finished for protective and decorative purposes with organic or inorganic coatings. These may be (1) metallic, applied by hot dipping, electroplating, or other methods; (2) vitreous, fused-on glassy materials, such as porcelain enamel; or (3) laminated, involving the application of adhesive inert plastic films.

METALLIC COATINGS Metallic coatings provide protection against corrosive elements, either by: (1) acting as a so-called sacrificial metal (one that is purposely permitted to corrode), as is the case for zinc coatings; or (2) being relatively inert and corrosion resistant (thus protecting the underlying base metal), as in nickel or chromium coating. The metals most often used for protecting carbon steels are: zinc, tin, terne metal, aluminum, cadmium, chromium, and nickel. These coating metals are most commonly applied by hot dipping, electroplating, metallizing, or cladding. Galvanizing (coating with zinc) is usually done by hot dipping or electroplating. Aluminum coatings are generally applied by hot dipping. Tin plating is coating with tin, either by hot dipping or electroplating. Bonderizing on galvanized sheets and galvannealing are light metallic coatings with a suitable surface for paint coatings.

Hot dipping is a process in which steel is immersed in a molten bath of the coating metal (Fig. 5.1-43). The most prevalent method is continuous hot-dip galvanizing, which accounts for more than 10 times as much galvanized steel as does the electrolytic process. The hot-dip process, which results in heavier coatings, is the chief method of galvanizing tubular and flat-rolled products. Hot-dipped galvanized sheets and strips can be made with zinc coatings of designation G90 (0.90 oz./sq. ft.), which are used frequently in architectural applications. Heavier coatings of approximately 2 oz./sq. ft. are made for severe corrosive applications. These coating weights are the combined weights on both surfaces.

Hot-dipped *aluminized* sheets are usually coated with an aluminum-silicon (5 % to 11 % percent) coating known as type 1.

Aluminum-silicon coatings provide excellent resistance to atmospheric and high-temperature (up to 1250 °F) corrosion, and are extensively used in the automotive industry.

Electroplating employs an electric current and an electrolytic solution to deposit metallic coatings on steel or iron (Fig. 5.1-44). Tin, zinc, and cadmium are commonly applied by this method, as are nickel and chromium either directly or over copper. Electroplating is the chief method used in making modern tinplate, and the second most important process for making galvanized products. Lighter coatings of 0.10 oz./sq. ft. (for a minimum of corrosion resistance) and greater variations in coating thicknesses are possible by electroplating. The lighter coatings are normally recommended on surfaces to be painted or to be used in mildly corrosive exposures.

FIGURE 5.1-43 *A continuous hot-dip galvanizing unit.* (*Courtesy American Iron and Steel Institute.*)

FIGURE 5.1-44 *Steps in the production of tinplate and galvanized products by the electrolytic process. The looping pits shown are necessary to ensure that the proper tension is applied to the metal throughout the process.* (*Courtesy American Iron and Steel Institute.*)

Metallizing consists of spraying molten metal against the surface to be coated. It is used extensively for applying zinc and aluminum coatings. The coating adheres to the base metal by a combination of mechanical interlocking and metallurgical bonding. Metallizing is the chief method of metallic coating practical for field use.

Cladding produces bimetallic "sandwich" products consisting usually of a carbon or low-alloy steel core, covered with a thin layer of coating metal. Stainless steel and aluminum coatings are applied by hot rolling under controlled, nonoxidizing conditions. Copper may also be used for cladding steel plates by hot dipping or electroplating.

Zinc-based paint pretreatments are common on steel to improve the bond of subsequent paint finishes.

Bonderizing is the most common process for producing a suitable paint base on fabricated products. Zinc-coated products are dipped in a hot phosphate solution, resulting in a thin (0.02 to 0.03 oz./sq. ft.) crystalline surface film of zinc phosphate.

Although there are several types of vitreous coatings used on metal, including glass-on-steel linings applied to hot-water tanks and process piping, *porcelain enamel* is the type most common in architectural applications. Common uses are porcelain enamel on cast iron or steel in bath tubs, sinks, and other sanitary ware (Fig. 5.1-45). Porcelain-enameled building panels are used in commercial and institutional buildings.

Porcelain enamels have the same general inert abrasion- and corrosion-resistant properties as glass. They can be produced in a great variety of lightfast colors and textures. Porcelain enamel is applied in one or two coats to sheets that have been prepared by etching. In the case of the two-coat system, the first enamel coat contains adherence promoters, such as cobalt or nickel oxides, and is applied by spraying, dipping, or flow coating, followed by firing at high temperatures. The second (and sometimes third) coat contains the coloring elements, and is applied and fired in a similar manner. Coating thickness ranges between 4 and 20 mils, although durability is not necessarily a function of thickness; specifically, where flexure is a possibility, thinner or fewer coatings may be preferred.

FIGURE 5.1-45 *A tub shell is heated in an oven before application of glass frit, which melts on the hot surface. Further baking produces the enamel coating.* (Courtesy Rheem Manufacturing Co.)

Laminated coatings are thin (2 to 12 mils thick), tough plastic films applied with thermosetting (heat-hardening) adhesives to steel products. The most common films used are polyvinyl chloride (PVC) and polyvinyl fluoride (PVF), both of which possess good abrasion resistance and sufficient tenacity, toughness, and flexibility to withstand severe forming operations. They have excellent resistance to weathering and to attack by a variety of chemical agents, and good color retention. Although color selections are more limited here than they are in porcelain enamels, the potential color variations are substantial.

5.1.4.5

Joints and Connections

Steel and iron shapes can be joined with (1) *mechanical fastenings*, such as bolts, screws, and rivets; (2) *mechanically formed joints*, using the workpiece itself to form the connection; (3) a variety of *welding* techniques; and (4) *adhesive bonding* methods. The joints and connections used in the fabrication of iron and steel products are generally similar to those of other metals.

■ MECHANICAL FASTENINGS

Mechanical fastenings are available in a variety of permanent and semipermanent types, and are used to connect metal to metal as well as to other building materials. The permanent types, such as welds and rivets, involve permanent deformation so that destruction or cutting the fastener or the component must take place for dis-

assembly. The familiar and often used semipermanent fastenings (screws, bolts, and nails) allow disassembly without permanent damage either to the component or to the fastener.

Bolting and welding are the most significant joining methods for steel construction. Riveting is almost never employed for that purpose today and is used only occasionally in shop-fabricated steel structural components. One of the disadvantages of riveting is the high noise level of riveting machines.

Riveting involves the joining of components by means of a rivet with a prefabricated head on one side, and deformation of the shank to form a head on the other side. Structural rivets are generally deformed by a pneumatic or hydraulic riveter while red hot. As the rivet cools it shortens in length, drawing the components together. The resulting compressive and frictional forces between the contact surfaces are important in resisting tensile and shear stresses developed in structural joints.

Bolting is a semipermanent mechanical method of fastening with nuts, bolts, and washers. Bolting is easier to accomplish in the field than is welding, especially in locations where access is difficult and when adverse weather conditions exist. Bolted connections work best when it is necessary for them to resist only shearing forces. When moment forces also occur, bolted joints often become complex and bulky when compared with welded joints. Bolts are sometimes used to hold joints together during welding.

Specified tensile values must be developed in bolts by tightening them with calibrated torque wrenches, by turning the nuts a specified number of turns from a snug-tight position, or using a special power tool that locks onto the tip of the bolt and breaks it off when the nut is sufficiently tight, thus preventing overtightening. Proper tightening is important to develop frictional resistance in a joint to counteract shearing forces. Keeping the threaded part of a bolt out of the structural parts being joined is also essential in developing the full strength of the connection.

Nuts and bolts are made in a range of sizes up to 4 in. in diameter and with several types of heads (square, round, hexagonal, countersunk, etc.). Since most bolts have square or hexagonal heads and most nuts are square or hexagonal as well, this allows the use of wrenches. Although bolts are available in both carbon and alloy steels, using high-strength bolts for most major structural applications is becoming more prevalent (Fig. 5.1-46). Ordinary carbon steel bolts are used in securing many components in small building construction, such as sill plates to foundations and of beams to columns and to each other.

STUD WELDING   In stud welding, which is a semipermanent fastening method, one end of a stud is arc welded by a gun-shaped unit to a steel component, and a nut is applied to the stud's threaded end (Fig. 5.1-47). Studs are used most often when nonstructural materials must be attached to steel, as in the application of architectural finishes to a structural frame-

FIGURE 5.1-46 *Air-impact wrenches speed high-strength bolting of steel connections in modern buildings. (Courtesy American Iron and Steel Institute.)*

FIGURE 5.1-47 *Special stud welding equipment permits fast and efficient joining of many materials to steel. (Courtesy TRW Nelson Co.)*

work. Stud welding is also used in composite structural design as a means of transferring stresses between structural steel and reinforced concrete to produce a combined resistance to stresses. Studs are also available unthreaded or internally threaded, in a variety of shapes such as bent and rectangular eyebolts and j-bolts.

POWDER-ACTUATED FASTENINGS   Powder-actuated fastenings are permanent or semipermanent fastenings which employ a gun that is powered by a small explosive charge to drive a pin into hard and tough materials. In permanent fastenings, a specially hardened drive pin is used to connect components in a fashion similar to nailing. Often, however, a pointed stud, threaded on one end, is driven into one of the components and assembly is completed with a nut. The availability of inexpensive guns and a variety of stud sizes and shapes make this a useful fastening procedure in all types of construction (Fig. 5.1-48).

FIGURE 5.1-48 *Powder-actuated fasteners are often used to secure steel members to concrete and masonry surfaces.*

THREADED FASTENERS AND NAILS   The most familiar semipermanent fastening devices are screws, bolts, and threaded and grooved nails. These are manufactured from carbon, alloy, and stainless steels in many shapes and sizes. Their uses include ordinary household applications, assemblies for product fabrication, and joining structural trim components in construction.

EXPANSION BOLTS   Expansion bolts are semipermanent fasteners generally used to join materials to concrete and masonry

surfaces. They are also referred to as cinch anchors, expansion anchors, and many other proprietary names. These devices consist of two or more units that develop holding power in concrete or masonry through wedge action and friction with the walls of a pilot hole by the element of expansion. This action is produced by striking or turning the threaded machine bolt insert. Expansion bolts act in a manner similar to screws used with lead, fiber, or plastic shields, but usually develop much greater withdrawal strengths (Fig. 5.1-49).

FIGURE 5.1-49 *Self-drilling expansion anchors speed the joining of steel to concrete. (Courtesy ITW Ramset/Red Head.)*

ADHESIVE-HELD FASTENERS   A new type of fastener is designed to hold fasteners in place by the action of an adhesive. A device consisting of two glass containers, each holding one component of a two-part adhesive, is inserted into a hole drilled in concrete. The desired fastener is driven into the hole, breaking both parts of the glass container and permitting the components of the adhesive to mix and begin curing. When the adhesive has fully cured, the fastener is held securely in place.

MECHANICALLY FORMED JOINTS   The joints in thin sheets are often made without the use of fasteners. Instead, a part of the material itself is used to make the connection. Two common methods are staking and lock-seam joining.

Staking is a method of joining flat materials by fold-over tabs inserted through matching slots in the elements to be joined. It is often used in product fabrication.

Lock-seam joining is a mechanical method of joining sheet metals in both shop and field by bending and interlocking the elements (Fig. 5.1-50). This is a common method of fabricating air ducts for heating and air conditioning, as well as roofing and roof drainage components.

FIGURE 5.1-50 *Typical methods of lock-seam joining of sheet metal components in product manufacturer.* (*Courtesy American Iron and Steel Institute.*)

■ WELDING, BRAZING, AND SOLDERING

Welding, brazing, and soldering are methods of joining two or more metal components. Welded joints are made by melting the edges of the components to be joined and adding filler metal if required. Brazing and soldering generally employ nonferrous filler metals that have melting points below those of the base metals and do not require melting the base metal to effect a bond. When the filler metal has a melting point below 800 °F the joining method is considered soldering; if above, it is brazing. Brazed connections can be as strong as welded joints, depending on service requirements. The brazing filler metal flows into a joint of any configuration by capillary attraction. Soldering and brazing are most often used in product manufacture; welding is used both in manufacturing and in structural applications.

Steel-product manufacturers employ many different welding, brazing, and soldering processes. The following discussion, however, is limited to common techniques used in building construction.

Structural joints in building frames may be either welded or bolted. The decision whether to use welding or bolting to form a particular joint rests with the structural designer. Welding is probably best for most shop joints because control of the surrounding environment is easier than it is in the field and welding is usually more economical in the shop than is bolting. Computer-driven robots can sometimes be used for shop welding, for example, reducing labor costs. When joints must resist both shear and moment forces, welding usually produces cleaner and less complicated joints.

The thickness and number of welds in a particular joint is decided by the structural designer based on the forces at the joint.

WELDING TECHNIQUES The majority of welding related to building construction can be classified into two main processes: arc and resistance, both of which are electric welding processes. *Arc welding* depends on the heat of an electric arc established between an electrode and the components to be welded. It is most commonly used in the field for connecting steel structural components. Steel decking is also welded in place using arc welding.

Welding is an exacting process requiring care and skill. Welders should therefore be well trained and periodically tested to ensure their competence. In addition, each weld should be inspected to ensure that it was properly made. These inspections may be done visually for minor conditions, but in structural joints more sophisticated inspection methods are usually employed. These include radiographic, ultrasonic, magnetic-particle, and dye-penetrant testing, each of which will detect hidden flaws and voids within welds. The decision about which type of testing is appropriate rests with the structural designer.

Resistance welding is generally a shop process in which the components are butted together and heat is generated in the seam by the resistance of the metal to the passage of an electric current. An electrode in contact with the component introduces the current. Welded wire fabric is usually produced using resistance welding.

Oxygen and nitrogen in the air have a detrimental effect on hot metal in a weld area. Most arc processes *shield* the weld area from atmospheric contamination with inert gases or granular fusible materials. *Shielded metal arc welding* (also known as *manual metal arc* or *coated stick elec-*

trode welding) is the most common arc process (Fig. 5.1-51a). This process uses a consumable metal stick electrode that is gradually melted by the arc and deposits filler metal in the weld. A chemical coating on the electrode releases an inert gas that forms a shielding envelope around the weld area. This process depends on manual replacement of the stick electrode as it is consumed.

In *semiautomatic* processes using self-advancing wire electrodes, the wire comes

FIGURE 5.1-51 *(a) Shielded metal arc welding; (b) metal inert-gas-shielded welding (MIG); (c) submerged arc welding.* (*Courtesy American Iron and Steel Institute.*)

off a spool at a predetermined rate and passes through a handheld welding gun. The two chief processes of this type are *metal inert-gas-shielded arc* (MIG), or *gas-shielded metal arc*, welding (Fig. 5.1-51b) and *submerged arc* welding (Fig. 5.1-51c). In MIG, an inert gas or mixture of gases (argon, helium, a combination of both, or carbon dioxide) from an external supply is directed through a gun held on the weld area to shield it from atmospheric contaminants. In submerged arc welding, protection is achieved by burying the arc and immediate weld area under a blanket of fusible granular material. Arc heat melts the material to form a layer of protective slag over the molten metal. The slag and loose material are removed after welding has been completed. In MIG, filler metal is provided by a consumable electrode; in submerged arc welding either a consumable electrode or a separate filler rod may be used.

The most common welded joint types are shown in Figure 5.1-52. Exposed welds are often ground smooth in welds that will permit it. The welding processes used with various types of steel are outlined in Figure 5.1-53. Symbols are used on drawings to denote the types of welds required (Fig. 5.1-54).

Spot welding is a type of resistance welding in which the components are joined by a series of spot (unjoined) welds, rather than a continuous line of weldment, as in other techniques. It is produced by the resistance heating of a small area between two cylindrical electrodes under pressure (Fig. 5.1-55).

Seam welding is actually a series of overlapping spot welds produced under pressure by a pair of disc electrodes, as in welded pipe manufacture.

Projection welding is also similar to spot welding but makes use of preapplied projective embossments on one of the components to localize the heat and effect a weld in the desired spot.

FIGURE 5.1-52 *Welded joints common in structural assemblies.* *(Courtesy American Iron and Steel Institute.)*

FIGURE 5.1-53
Welding Processes Used to Join Various Types of Steel

Type of Steel	Metal-Shielded Arc	Submerged Arc	Metal-Inert Gas-Shielded (MIG) Arc	Spot, Seam, and Projection
Low Carbon				
Sheets	Common	Common	Common	Common
Plates and bars	Common	Common	Common	Occasional
Medium Carbon*				
Sheets	Common	Common	Common	Common
Plates and bars	Common	Common	Common	Rare
High Carbon*				
Sheets	Common	Occasional	Occasional	Common
Plates and bars	Common	Occasional	Occasional	Not used
Low Alloy*				
Sheets	Common	Common	Common	Occasional
Plates and bars	Common	Common	Common	Rare
Stainless*				
Sheets	Common	Occasional	Common	Common
Plates and bars	Common	Common	Common	Rare

Courtesy American Iron and Steel Institute.

*Preheating and/or post-heating (stress relieving) structures may be required in the weld zones, to prevent embrittlement and excessive residual stresses on certain carbon and low alloy steels, and to stabilize in order to prevent corrosion on stainless steels.

EFFECTS OF WELDING ON STEEL Since heat is an essential element of all welding techniques and metals respond to heat exposure in varying degrees, some steels are more weldable than others. The degree of weldability is related to the sensitivity of the metal to heat, as reflected in the soundness of the weld and the degree of undesirable side effects produced by the opera- tion. The possible detrimental side effects are: (1) the creation of cracks in the base and weld metals; (2) changes in the physical properties of the base metal, such as decreased strength, ductility, and corrosion resistance; and (3) oxidation of the finished surface.

The most important element affecting the weldability of steel is its carbon con- tent. Low-carbon steels (less than 0.15%) are the most weldable; mild steels (0.15% to 0.30% carbon) are next in weldability and can be welded by most processes up to a section thickness of 1/2 in. Thicker sections require greater input of heat to effect a weld, and this may have detrimen- tal results unless proper procedures and care are used. All ASTM structural steels

WELDED JOINTS: Standard symbols

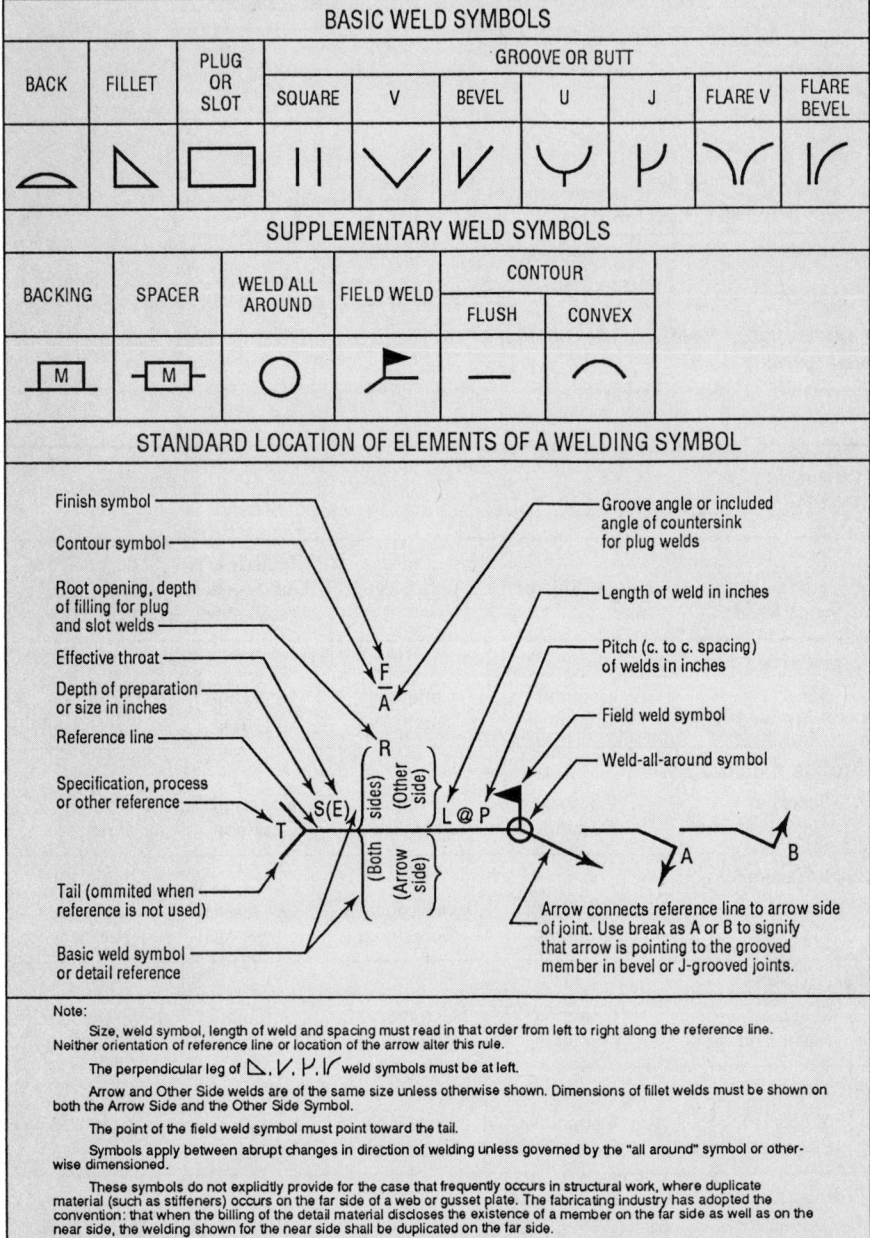

BASIC WELD SYMBOLS									
BACK	FILLET	PLUG OR SLOT	GROOVE OR BUTT						
			SQUARE	V	BEVEL	U	J	FLARE V	FLARE BEVEL

SUPPLEMENTARY WELD SYMBOLS					
BACKING	SPACER	WELD ALL AROUND	FIELD WELD	CONTOUR	
				FLUSH	CONVEX

STANDARD LOCATION OF ELEMENTS OF A WELDING SYMBOL

Finish symbol

Contour symbol

Root opening, depth of filling for plug and slot welds

Effective throat

Depth of preparation or size in inches

Reference line

Specification, process or other reference

Tail (ommited when reference is not used)

Basic weld symbol or detail reference

Groove angle or included angle of countersink for plug welds

Length of weld in inches

Pitch (c. to c. spacing) of welds in inches

Field weld symbol

Weld-all-around symbol

Arrow connects reference line to arrow side of joint. Use break as A or B to signify that arrow is pointing to the grooved member in bevel or J-grooved joints.

Note:
Size, weld symbol, length of weld and spacing must read in that order from left to right along the reference line. Neither orientation of reference line or location of the arrow alter this rule.

The perpendicular leg of △, V, ⊢, ⊩ weld symbols must be at left.

Arrow and Other Side welds are of the same size unless otherwise shown. Dimensions of fillet welds must be shown on both the Arrow Side and the Other Side Symbol.

The point of the field weld symbol must point toward the tail.

Symbols apply between abrupt changes in direction of welding unless governed by the "all around" symbol or otherwise dimensioned.

These symbols do not explicitly provide for the case that frequently occurs in structural work, where duplicate material (such as stiffeners) occurs on the far side of a web or gusset plate. The fabricating industry has adopted the convention: that when the billing of the detail material discloses the existence of a member on the far side as well as on the near side, the welding shown for the near side shall be duplicated on the far side.

FIGURE 5.1-54 *Standard welding symbols.* *(Courtesy American Institute of Steel Construction, Inc.)*

FIGURE 5.1-55 *Multiple-head spot welder joins a metal skin to the continuous metal reinforcement in a hollow metal door.*

mized by limiting the heat input, promoting heat dissipation, and selecting low-heat resistance welding methods. Annealing the finished assembly and then following it with fast cooling also restores corrosion resistance.

■ ADHESIVE BONDING

Many decorative and protective laminated coatings are applied to metal base materials with special adhesives. Adhesive bonding is also used to secure thin carbon, galvanized, and stainless steel sheets and foils to nonferrous materials to provide the required rigidity, flatness, strength, or insulating qualities.

Carbon steel surfacing adds abrasion and indentation resistance to softer core materials. Steel-clad honeycomb doors (Fig. 5.1-56) and steel bonded to plywood panels for column and radiator covers are common examples; galvanized or vitreous enamel-coated steel sheets, bonded with rigid urethane foam to form

FIGURE 5.1-56 *Stainless steel skins are adhesively bonded to honeycomb cores to produce metal doors.* *(Courtesy American Iron and Steel Institute.)*

have a carbon content of less than 0.35% and, therefore, can be welded.

Although all stainless steels can be welded by some method, austenitic steels are the most weldable. Almost any commercial welding method can be used to join these steels without causing problems in soundness and porosity, or loss in toughness or ductility. However, the high rate of thermal expansion in austenitic steels must be considered in order to control distortion. Another hazard is reducing corrosion resistance in the weld area, because some of the chromium becomes ineffective through a reaction with carbon. This often occurs when stainless steels are exposed to high welding temperatures over extended periods. This effect can be minisandwich panels, are often used in building construction.

Stainless steel sheets are also bonded to rigid or insulating materials in the manufacture of many products with corrosion-resistant, attractive finishes. Very thin stainless steel foil is laminated to kraft paper for use in packaging where additional strength and impermeability to vapor are important considerations.

The chief advantages of adhesive bonding are uniform distribution of stresses, absence of surface fasteners, and economy of mass production. Ease of joining small parts not accessible for welding, and low weight in relation to strength make this a valuable structural joining method in the aircraft industry. However, high initial cost of specialized equipment, low peel and creep (flow under stress) strengths, sensitivity to certain temperatures (characteristic of low-cost adhesives) at present limit their use in building construction mainly to the manufacture of panelized products.

The most common adhesives used in the lamination of steel to other materials are epoxies, phenolic resins, rubber polymers modified by resins, polyvinyl acetates, and chemically cured neoprenes. The most promising adhesives for laminating steel are the polyurethanes, which, with epoxies and phenolics, withstand considerable loads at a greater temperature range and come closest to the concept of *structural* adhesives. These adhesives often possess high peel strength, chemical inertness, resistance to wetting action, and shear and tensile strengths greater than the material bonded. Under certain conditions they may be used for joining structural members in building construction.

Adhesives are available in a variety of forms, such as pressure-sensitive tapes, films, liquids, pastes, and solids. They can be applied manually with ordinary building tools or with specialized mass-production equipment, which sprays, rolls, or flows the adhesives onto component surfaces. Some acquire permanent mechanical properties when heated (thermosetting adhesives); others set at moderate temperatures but can be softened repeatedly at high temperatures (thermoplastics). Other adhesives depend on the chemical reaction between two agents to acquire their permanent characteristics.

Although many adhesives can perform well under a given set of service conditions, there is no universal adhesive and each must be selected to meet the anticipated operating requirements.

5.1.5

STEEL STRUCTURES AND CONSTRUCTION SYSTEMS

Conventional steel-framed structures are built from hot-rolled beams, columns, and girders, combined with bar joists and steel decking. Sometimes these components are combined into complete preengineered buildings that are shipped to the site broken down into units and fastened together in the field to form a complete building frame and shell. Small structures and parts of larger structures carrying light loads are sometimes erected using lightweight cold-formed framing.

Long spans are achieved in steel using several different types of systems. Conventional framing can be stretched using plate girders and castellated beams. Roofs may be framed with long-span trusses. Special areas, such as arenas, stadium roofs, and airline and train terminals, may spanned with domes, space frames (Fig. 5.1-57), or steel cable systems.

5.1.5.1

Design

Steel building frames may be designed by either architects or structural engineers, but architects seldom design frames for large buildings or space frames, steel-cable roof systems, or trusses.

■ **THE PROCESS**

The designer produces a set of working drawings (Fig. 5.1-58) and specifications detailing the type of steel, basic requirements for fasteners, the location and sizes of the basic elements of the frame (columns, girders, beams, etc.), the basic dimensions of the frame (column centers, floor heights, etc.), and unusual erection details.

Using the designer's drawings and specifications as a guide, a fabricator, employed by a building contractor or by a structural steel subcontractor, produces a set of shop drawings showing the same basic information as the designer's drawings, but adding information about joints,

FIGURE 5.1-57 *Large-span atrium spaceframe supporting a tension fabric cladding.* (Hyatt Regency Hotel, Burlingame, California; architect, Hornberger Worstell and Associates; photograph courtesy MERO Structures, Inc.)

FIGURE 5.1-58 *Part of a designer's steel-framing plan. Squares with Xs are elevator shafts. Column sizes and types are usually shown in tables.* (Drawing by HLS.)

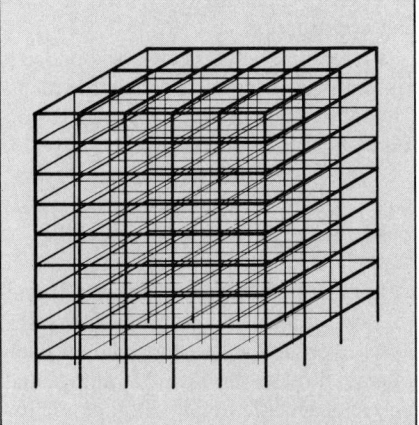

FIGURE 5.1-59 *Typical steel-framing grid.* (Drawing by HLS.)

member lengths, and all other details necessary to fabricate and erect the frame. The shop drawings also show details of methods and bolts necessary to anchor the frame to the foundations.

These shop drawings are then sent by way of the contractor to the designer for review. If revisions are necessary, the designer marks the drawings and returns them to the contractor for revision and resubmittal. This process continues until the designer finds the shop drawings acceptable. The designer then places a stamp on the drawings stating that the drawings are acceptable and returns them to the contractor for return to the fabricator.

■ CONVENTIONAL STEEL FRAMES

The structural frame for a conventional steel-framed building is composed of columns, girders, and beams fastened together to form a three-dimensional grid (Fig. 5.1-59).

Under normal conditions, the joints in conventional structural steel building grids are subjected to both shear and moment forces. Unless some means is employed to resist these forces, the frame will be inherently unstable and subject to wracking and ultimate collapse. The American Institute of Steel Construction (AISC) lists three methods for solving these inherent difficulties as types of steel-framed construction. In Type 1, connections must be made sufficiently rigid to resist all forces that will occur without significant change in the angle between framing members (beams, girders, columns). In Type 2, diagonal bracing or shear panels are introduced to resist moment forces. Thus, most joints need resist only shear forces. Type 3 is a semirigid system in which the joints are sufficiently rigid to resist the applicable moment forces, but not as rigid as those required in Type 1.

Joints that are required to resist only shear forces are usually only bolted (Fig. 5.1-60). Joints required to resist both moment and shear forces can be fastened using bolts alone, but such joints are both complicated and expensive. Therefore, joints resisting both shear and moment forces are either welded or, more fre-

quently, both bolted and welded (Fig. 5.1-61). Such joints are more expensive to produce than those designed to resist only shear forces. For this reason, an attempt is usually made, especially on large buildings, to reduce the number of moment joints in a steel frame.

Such a reduction in the number of moment joints in a building is often accomplished by making part of the frame rigid and tying the remainder of the frame to the rigid portions by means of rigid floor systems, such as steel decking with a concrete topping. These floor systems must be designed structurally as diaphragms, meaning that they act as a rigid unit to distribute the horizontal forces acting on the building to the rigid portion of the frame, which is, in turn, capable of resisting them and transmitting them harmlessly to the ground.

The rigid portion of a building frame may be either in the perimeter or in the core. The core is that portion of a building that houses elevator and stair towers and service facilities, such as toilets. The amount of the frame that must be made rigid and its location are functions of the design and are the responsibility of the structural designer.

The three basic ways of creating the rigid core or perimeter of a steel building frame are diagonal bracing, shear panels, and moment connections. Most of the remainder of the frame can usually be fastened with shear connections only, although there are some exceptions. For example, cantilever beams usually require moment connections.

SUGGESTED DETAILS
Beam Framing

SHEAR CONNECTIONS

2 angles

Tee

Shop bolt or weld

Tee

2 angles

Note: Check web shear and moment
in coped beam.

SHEAR SPLICES

4 framing angles

1 or 2 plates

2 plates

Note: Of the above types, 4 framing angles is most flexible.

BOLTED MOMENT SPLICES

Shim as required

Splice plate

4 angles

Mom. Mom.

M

Backing
bar

M

Finger

Strip

TYPICAL SHIMS

FIGURE 5.1-60 *Details of shear connections.* *(Courtesy American Institute of Steel Construction, Inc.)*

Diagonal bracing consists simply of steel members jointing opposite corners of a structural bay (Fig. 5.1-62) and thus forming rigid triangles within the bay. Because the triangular structure itself cannot distort, moment joints are not necessary to attach the bracing to the frame.

Shear panels often consist of rigid concrete walls, but are sometimes constructed of braced steel framing. In either case, they must resist the forces imposed on them without distorting.

Like diagonal bracing or shear panels, moment connections can be used to form a rigid portion of the frame. Although structural design and safety must be considered, the final decision is often based on which method is the least expensive to stabilize a steel-framed building.

■ **STEEL JOISTS AND JOIST GIRDERS**

Standard open-web steel joists (Fig. 5.1-63) are used for floor and roof framing mostly in masonry or steel-framed buildings (Fig. 5.1-64). Joist girders are open-web trusses designed for longer spans and as primary framing members in lieu of beams. Both are usually built from high-strength steel to standard sizes. Selection is made by a structural designer from tables produced by the Steel Joists Institute (SJI).

Joists are classified as K-Series, LH-Series, or DLH-Series. K-Series joists range in depth from 8 to 30 in. in 2-in. increments. They are designed to span up to 60 ft. Except where their ends abut over steel beams, where bolts and welding can be used to provide additional support,

K-Series joists must be supported at least 2-1/2 in. on steel-bearing plates.

LH-Series (longspan) joists are available in depths from 18 to 48 in. The deeper ones can span up to 96 ft. LH-Series joists must be supported by at least 4 in. of steel beam or 6 in. of steel-bearing plate when resting on concrete or masonry.

DLH-Series (deep longspan) joists range in depth from 56 to 72 in. Some can span as much as 144 ft. Bearing requirements for DLH-Joists are the same as those for LH-Series joists.

Joists are usually spaced about 2 ft. apart, but spacings as wide as 10 ft. sometimes occur when the floor or roof decking can accommodate such width. Joists usually require bridging to prevent them from twisting or overturning. This can be accomplished by horizontal angles or rods running across the bottom chords or by diagonal steel bracing from joist to joist. Bridging is usually welded in place.

■ **ROOF AND FLOOR DECKING**

Many types of products are used as roof decks on steel-framed buildings. Examples include the poured-in-place gypsum concrete decks and the lightweight aggregate board and wood fiber board decks described in Section 3.14. A major type of roof decking used on steel-framed buildings is corrugated steel decking, either with or without a concrete topping (Fig. 5.1-65).

Concrete slabs are often used as floor decking in steel-framed buildings. Sometimes such slabs are formed using temporary formwork as described in Chapter 3. More frequently, floor slabs are placed over corrugated steel decking.

Many different configurations of corrugated steel decking are available, depending on the use and span. The *Design Manual for Composite Decks, Form Decks, and Roof Decks* by the Steel Deck Institute (SDI) spells out the minimum requirements for each type.

Steel decking is usually fastened in place by welding through it to the supporting steel. In thin metal decking, and where greater strength is required, welding washers are used. Occasionally, decking is fastened in place using mechanical fasteners. When a deck must act as a diaphragm, its panels are sometimes fastened together at the laps between panels with

FIGURE 5.1-61 *Details of moment connections.* *(Courtesy American Institute of Steel Construction, Inc.)*

Text in figure:

SUGGESTED DETAILS
Beam Framing
MOMENT CONNECTIONS

Shim as required
Provide driving clearance
Finish to bear
Tee or angles
Stiffeners (if req'd)
Shim as required
Shear bolts
Backing bar
Shim as required
Tee

Cope for fire protection (if req'd)
Sec. A
Shim as required
Stiffeners (if req'd)
Shear bolts
Accessibility may limit weld length
Shim as required
Backing bar

Slotted holes as permitted
Backing bar
Slotted holes as permitted
Backing bar
Backing bar
1/16
Oversize holes as permitted
Erection seat

For additional information on moment connections of beams to column webs, refer to ASCE *Journal of the Structural Division*, Vol. 105, No. ST5, May 1980, pg. 1005.

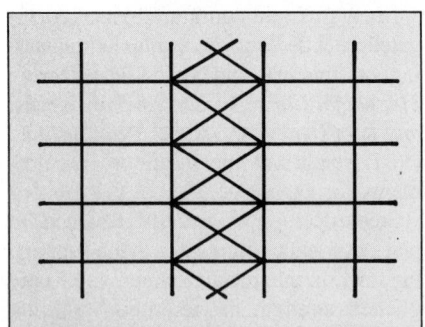

FIGURE 5.1-62 *Diagonal bracing.* *(Drawing by HLS.)*

screws or welds. Whether deck laps need fastening and the method of attachment to be used are the responsibility of the structural designer. The important consideration is to ensure that the deck and its fastenings will adequately resist any forces that might act on them.

ROOF DECKING Roof decking is made from 22-, 20-, 18-, or 16-gauge steel in depths from 1-1/2 in. to 7-1/2 in. It can span from 4 ft. 6 in. for the lightest decking to as much as 30 ft. for heavy, deep, cellular decking. Rib openings in 1-1/2 in. roof deck are 1 in. wide (*narrow rib*), 1-3/4 in. wide (*intermediate rib*), or 2-1/2 in. wide (*wide rib*). Corrugated sheet steel roof decks are either *open flute* (Fig.

5.1-66a) or cellular (Fig. 5.1-66b), meaning that they have a sheet steel plate affixed to the bottom of the decking. *Acoustical roof decking* is also available. Some of it is of the open-flute type and some is of the cellular type. The open-flute type has openings in the vertical webs. Openings in the cellular type are in the bottom plate. The concealed spaces in acoustical roof decking are filled with sound-absorptive matter, usually glass-fiber insulation.

FLOOR DECKING Corrugated steel floor decking is available in three different types: *form deck*, *composite deck*, and *composite cellular deck*.

Form deck is used as a permanent form for a concrete slab (Fig. 5.1-67a). It varies in thickness from 1/2 in. to 2 in. Spans range from 1 ft. to 2 ft. for light 1/2-in. form decking to as much as 12 ft. for heavy 2-in.-thick decking with 6-in. rib spacings. Form deck adds no strength to the slab. It simply reduces the amount of work involved in forming and eliminates the need to strip the forms from a slab.

Composite decking is designed not only to support a concrete slab while it is being placed, but also to bond with it, providing tensile reinforcement for the composite structure (Fig. 5.1-67b). The resulting longer slab spans and greater strength produce a more economical structure than does noncomposite construction. Sometimes the composite nature of the system is further enhanced by welding *shear connectors* to the tops of supporting beams, thus bonding the concrete slab to the steel structural members as well as to the decking (Figs. 5.1-68 and 5.1-69).

Cellular composite decking works in essentially the same way as does noncellular composite decking (see Fig. 5.1-67c). In addition, cellular decking can be designed to house electrical wiring or to provide acoustical treatment in the same way as can cellular roof decking.

■ STEEL TRUSSES

Steel trusses are manufactured from steel structural shapes to produce units that are lighter in weight than comparable solid steel construction, such as welded steel plate girders. Most trusses form roof structures rather than floors.

FIGURE 5.1-63 *Bar joist.* *(Drawing by HLS.)*

web joists or joints and girders, and sheet steel roofing and siding.

5.1.5.2

Fabrication

After receipt of shop drawings bearing the designer's stamp of acceptance, a fabricator produces full-size templates of wood or cardboard to assist in the positioning of bolt holes and cutouts, and begins the fabrication process.

Fabrication consists of measuring and cutting beams, columns, girders, base plates, and other components to the proper length and configuration; punching or boring holes for bolts; welding on stiffeners, connection devices such as angles, and smaller base plates; beveling joints for welding; flattening and squaring bearing surfaces for a tight fit; cambering (curving slightly upward) beams and girders where the design requires it; and assembling large components, such as plate girders, trusses, and built-up columns.

5.1.5.3

Erection

The various components of most steel-framed structures are fabricated in the shop, delivered to the construction site as individual components (beams, girders,

are also available. These provide a non-flammable and dimensionally stable structural system for light loads. Although compatible with conventional stud or bearing wall construction, lightweight steel framing also introduces the possibility of longer spans and large-scale prefabrication (Fig. 5.1-70).

■ PREENGINEERED BUILDING

Preengineered buildings are used primarily as industrial facilities, but may also be used as commercial buildings. They typically combine rolled steel sections, open-

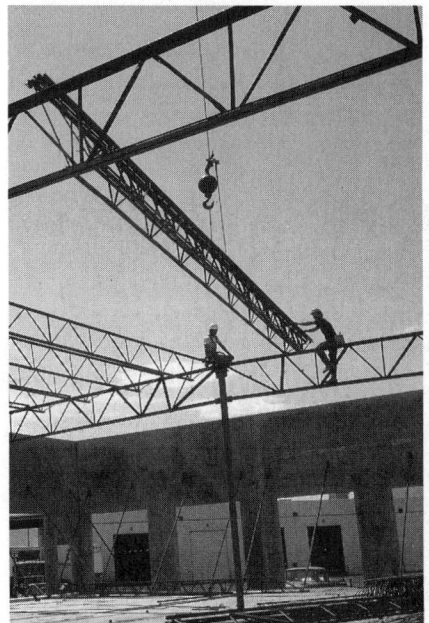

FIGURE 5.1-64 *Bar joist being placed on joist girders.* *(Courtesy of the Vulcraft Division of Nucor Corporation.)*

The structures known as *space frames* are actually three-dimensional trusses, designed to span in two directions simultaneously.

■ LIGHTWEIGHT STEEL FRAMING

Section 9.5.3 addresses steel studs used primarily for non-load-bearing partitions. Heavier *cold-formed* steel studs, joists, and accessories, capable of carrying loads other than themselves and their finishes,

FIGURE 5.1-65 *Steel deck being placed on steel joists and joist girders.* *(Courtesy of the Vulcraft Division of Nucor Corporation.)*

FIGURE 5.1-66 *Roof decks: (a) open flute type; (b) cellular type.* (Drawing by HLS.)

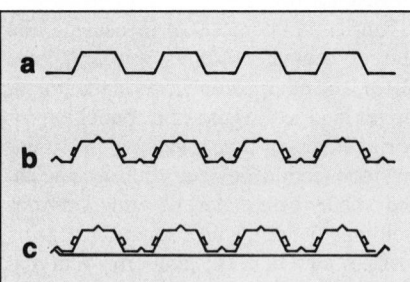

FIGURE 5.1-67 *Floor decks: (a) Noncomposite form deck. (b) Composite deck. The indentations and projections act to produce a bond with the concrete slab as it is placed. The configuration of these indentations and projections varies from manufacturer to manufacturer. Some composite decking has steel rods welded across the top of the flutes. (c) Composite cellular deck. Many configurations are available.* (Drawing by HLS.)

FIGURE 5.1-68 *Composite construction shear studs.* (Drawing by HLS.)

joists, decking, etc.), and fastened together in the field (Fig. 5.1-71).

The procedures and equipment used in the erection of steel-framed buildings are the responsibility of the *erector.* The actual work is done by *ironworkers* (some-

FIGURE 5.1-69 *Concrete being placed on a composite steel deck.* (Courtesy of the Vulcraft Division of Nucor Corporation.)

FIGURE 5.1-70 *Prefabricated steel stud wall assembly can often be lifted in place by one or two workers.* (Courtesy American Iron and Steel Institute.)

FIGURE 5.1-71 *A complicated steel frame under construction.* (Courtesy Bethlehem Steel Corporation.)

times called *steelworkers*) (Fig. 5.1-72). The designer should require that the layout and positioning of the steel, positioning of anchor bolts, and establishing and checking of bearing levels and other elevations be done by a licensed surveyor. The erection should be accomplished in accordance with the requirements and within the tolerances established by the AISC, unless the structural designer specifies or approves deviations from the AISC requirements.

Structural steel is placed in position in the field using cranes or other lifting devices. On fairly low buildings ground-mounted cranes are used to erect the entire structure (Fig. 5.1-73). On taller structures, either guy derricks (Fig. 5.1-74) or climbing cranes (Fig. 5.1-75), both of which are mounted on the building, are employed.

The general process usually followed in erecting a steel structure begins with the general contractor placing anchor bolts and leveling plates, and grout beneath the leveling plates. Leveling plates are often omitted, however, especially for larger columns.

Each of the lowest level of columns is then transported to its proper position and lowered by crane so that the anchor bolts pass through the corresponding holes in the base plate. On multistoried buildings, the lowest (first-*tier*) columns are usually two stories high.

After the first-tier columns have been bolted in place and are plumb, the beams and girders for the first two floors are placed and fastened in place (Fig. 5.1-76).

The next tier of columns, which is also usually two stories high, is erected and the beams and girders for the next two floors are installed (Fig. 5.1-77). This process continues until the entire structure has been completed (Fig. 5.1-78).

5.1.5.4

Fireproofing

The Underwriters Laboratories, Inc., (UL) publication *Fire Resistance Directory* contains a listing and description of various floor and ceiling, and roof and ceiling assemblies, and structural member fire protection systems and their fire ratings.

The UL ratings for assemblies limit the type, size, and spacing of the floor framing (including bar joists); the type and size

FIGURE 5.1-72 *(a) Ironworkers wrestling a steel column into place; (b) ironworkers welding steel members together.* *(Courtesy Bethlehem Steel Corporation.)*

FIGURE 5.1-73 *Steel being erected using a ground-mounted crane.* *(First Indiana Plaza, Indianapolis, Indiana; Architect, CSO Architects, Inc.; design architect, 3DI International; photograph courtesy of the Vulcraft Division of Nucor Corporation.)*

(material thickness, depth, and configuration) of floor and roof decking; and the methods of fastening framing and decking in place.

The UL ratings for columns, beams, and girders spell out the materials and thicknesses of the fireproofing that may be used to accomplish the various hourly ratings. Among the many materials that may be used to fireproof steel are concrete, masonry, gypsum board, plaster, mineral fiber, cementitious materials, and intumescent mastics.

Before the advent of modern lightweight fireproofing materials, structural steel requiring a fire-resistance rating was encased in poured concrete or masonry (Fig.

FIGURE 5.1-74 *Guy derricks are lifted onto the structure by a ground-mounted crane when the building has reached the practical limit of the ground-mounted crane. A guy derrick rests on the structure and is stabilized by guys fastened to the structure. Components of guy derricks are moved upward to a new level as the frame progresses. To move a guy derrick upward, the boom is disconnected from the mast, temporarily supported on the structure, and guyed to stabilize it. The mast is then lifted by the boom onto a higher level, where it is supported on the structure and stabilized. The crane is then reassembled.* (Courtesy Bethlehem Steel Corporation.)

5.1-79). These methods were effective but considerably increased the weight of the building and therefore increased the required size of framing members, which also increased the cost of the construction.

Attempts to decrease the weight of the fireproofing led first to the use of plaster as an encasing material (Fig. 5.1-80). The weight was further reduced by a change from sand aggregate to lightweight aggregates such as vermiculite.

Plaster-encased fireproofing is used only occasionally today, having been largely replaced by gypsum board enclosures (Fig. 5.1-81). Plaster and gypsum board encasements have the advantage of being able to double as a finish. When the structure to be fireproofed is concealed from view, other types of fireproofing are more often used today. These include mineral fiber and gypsum slabs, which are held against the steel by welded-on attachments (Fig. 5.1-82), and sprayed-on contour materials.

Sprayed-on contour fireproofing products (Fig. 5.1-83) are applied following the characteristic shape of the structural element being protected. They are also used to protect steel decking and sometimes bar joists and trusses. These uses should be approached with caution, however, because thermal and other movement of the substrate, as might happen in a steel floor or roof deck, can damage sprayed-on fireproofing. Both cementitious and sprayed-fiber products are available. Cementitious materials include cement-mineral aggregate cementitious fireproofing, mineral-fiber cementitious fireproofing, foamed magnesium oxychloride fireproofing, and magnesium cement fireproofing. Sprayed-fiber products are made from cement and mineral fiber.

Intumescent coatings include both paints and mastics. The paints are intended for locations where the steel is exposed to view. They work by expanding when heated to form a thick carbonaceous foam that thermally insulates the steel and resists further fire damage. These coatings are available in colors and some of them can serve as a base coat for other paints.

Other methods are also used to protect steel from fire. One is to hang a fire resis-

FIGURE 5.1-75 *A climbing crane rises through an opening in the structure, such as an elevator shaft. Additional sections are added at the top of the crane as the structure grows.* (Courtesy Bethlehem Steel Corporation.)

FIGURE 5.1-76 *Installing beams.* (Courtesy Bethlehem Steel Corporation.)

tant ceiling, such as plaster, beneath the structure. Exterior steel members may be partly protected from contact with a fire by steel *flame shields* (Fig. 5.1-84).

5.1.6

OTHER BUILDING PRODUCTS

Steel is used for may other building products. Those listed in the paragraphs below are among the most important of these.

5.1.6.1

Sitework

Products in this category include iron castings for fences, barricades, drains, manholes, guards, corrugated pipe for drainage, as well as many others.

5.1.6.2

Roofing, Siding, and Cladding

Standing-seam steel roofing has a long history of use in several styles. These products offer a broad choice of finishes and coatings for durability, texture, and color.

Siding, in traditional clapboard designs as well as in vertical and other patterns, is an increasingly popular steel product.

Wall panel systems (cladding), often in combination with lightweight steel framing, are used in many types of buildings today.

5.1.6.3

Doors and Windows

Many types and styles of exterior and interior doors and various steel windows are available. Steel doors provide fire safety, security, durability, and dimensional stability. Modern exterior swinging steel doors contain an insulating core surrounded by steel skins and sides. Designs range from smooth flush faces to variations that use glazing, moldings, textured surfaces, and vinyl laminations.

FIGURE 5.1-77 *An ironworker guiding a beam into place.* (*Courtesy Bethlehem Steel Corporation.*)

FIGURE 5.1-78 *Erecting a portion of a sloping steel structure.* (*Courtesy Bethlehem Steel Corporation.*)

Steel-framed windows offer a complete range of designs for residential, commercial, and institutional uses.

Refer to Chapter 8 for additional information about steel doors and windows.

5.1.6.4

Interior Finishes

Steel interior finish products include non-load-bearing studs and acoustical ceilings.

Non-load-bearing studs provide non-combustible, dimensionally stable framing for gypsum wallboard, plaster, and other interior surfaces. These products are discussed in Section 9.5.

Steel acoustical ceilings are found in many types of buildings, sometimes as an integral part of a steel deck floor and ceiling system.

5.1.6.5

Other Artchitectural Uses

Many other steel products are used in buildings, among them the following:

Prefabricated stairs, such as steel pan stairs

Spiral stairs

Prefabricated fireplaces and flues

Laboratory, kitchen, and other cabinets

Unit kitchens

Rainware (flashing, gutters, and downspouts)

Roof accessories (skylights, stacks, ventilators, roof hatches, etc.)

5.1.6.6

Mechanical Systems

Iron and steel are used extensively in plumbing fixtures and piping; for hot water and steam piping; for radiators, valves, and accessories; as well as in heating and cooling ductwork. Plumbing fixtures include both cast-iron and sheet steel products, such as tubs, shower and tub enclosures, toilets, bidets, and sinks. These are often finished in porcelain enamel. Stainless steel is widely used for kitchen, laboratory, and bar sinks. Sheet

FIGURE 5.1-79 *Concrete and masonry fireproofing.* (Drawing by HLS.)

FIGURE 5.1-81 *Gypsum board fireproofing.* (Drawing by HLS.)

FIGURE 5.1-80 *Plaster fireproofing.* (Drawing by HLS.)

FIGURE 5.1-82 *Fireproofing slab enclosure.* (Drawing by HLS.)

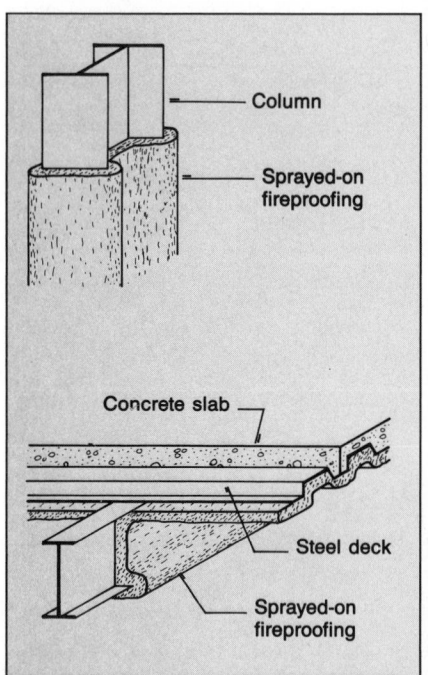

FIGURE 5.1-83 *Sprayed-on fireproofing.* *(Drawing by HLS.)*

FIGURE 5.1-84 *Flame shields.* *(Drawing by HLS.)*

steel ductwork, galvanized for corrosion resistance, is part of most heating and cooling systems that employ moving air. Such systems generally have fans, vents, louvers, and other sheet steel accessories.

5.2 ALUMINUM

Aluminum was first detected in 1807 as an ingredient in common clay, but it was not until 1825 that the first metallic aluminum was produced, and not until 1845 was aluminum powder successfully transformed into solid particles. In 1852, because of costly production methods, pure aluminum sold for $545 per pound and was used only for the finest of jewelry and tableware.

In 1886, it was discovered that metallic aluminum could be produced by dissolving alumina (aluminum oxide) in molten cryolite and then passing an electric current through the solution. From that discovery arose the modern aluminum industry.

Today, building and other construction make up a large market for aluminum in the United States. Products that contain large amounts include windows, doors, storefronts, building cladding systems, screens, siding, and mobile homes.

A unique combination of properties makes aluminum one of the most versatile engineering and construction materials. It is lightweight, yet some of its alloys have strengths greater than that of structural steel. It has high resistance to corrosion and will not rust red to stain adjacent surfaces or discolor products.

It has no toxic action, and possesses high electrical and thermal conductivities and high reflectivity of both heat and light. Aluminum can be cast or worked into a large variety of shapes, and readily accepts many surface finishes. These characteristics give aluminum its extreme versatility. In the majority of applications, two or more of these characteristics are prominent.

One of aluminum's most important attributes is that it functions as designed and maintains its finish over extended periods in a variety of buildings and construction. For example, an aluminum cap installed on the top of the Washington Monument in 1884 remains in good condition today. A great many large and small buildings have been clad with aluminum compo-

nents, and millions of homes and other small buildings have been sided with aluminum during the past 60 years.

As with other materials, however, aluminum's properties must be understood to ensure its proper use in service. This section will discuss the production of aluminum and aluminum alloys, their various properties, and methods by which aluminum products are manufactured. Individual aluminum products are discussed in Chapters 7, 8, and 9.

5.2.1

RAW MATERIAL

Aluminum forms 8% of the earth's crust, making it the most plentiful metal, but it is never found free in nature. Unlike gold and silver, aluminum is always combined chemically with other elements. Though nearly all common rocks and clays contain some aluminum, it has not been historically practical to extract aluminum unless the raw ore has an aluminum oxide content of at least 45%. This percentage may change from time to time, however, as economic conditions vary.

Such ores are called *bauxites*, after the town of Les Baux in France, where one of the first ore deposits was found. Most of the world's bauxite deposits exist outside of the United States, making it necessary for the United States to import most of its supply. Ninety-five percent of the bauxite mined in the United States comes from Arkansas; the remainder from small, lower-grade deposits in Georgia and Alabama.

Work is now under way in the United States to develop methods of using domestic ores, such as kaolin and alunite, as a replacement for imported bauxite. The U.S. Bureau of Mines, which is coordinating the project, believes the technology can be developed to extract aluminum commercially from these domestic ores.

5.2.1.1

Mining

Strip or open pit mining methods are generally employed to obtain the aluminum ore, which is then crushed, washed, and screened. The ore is ground and then dried in kilns (large revolving steel drums) at temperatures up to 250°F. The general steps in the mining and refining of aluminum are shown in Figure 5.2-1.

5.2.1.2

Refining

Commercial bauxite is refined by the Bayer process, which separates aluminum oxide (alumina) from unwanted minerals.

The dried ground bauxite is mixed with a solution of caustic soda (sodium hydroxide), which dissolves the alumina to form sodium aluminate. The silica in the bauxite reacts and precipitates out of solution. Iron oxide and other impurities are not affected chemically and, being solids, settle out.

This "green liquor" is now highly supersaturated sodium aluminate. Previously prepared hydrated alumina crystals are added to the solution. These form larger crystals, which gradually settle out. After being washed to remove remaining traces of impurities, the concentrated aluminum hydrate crystals are roasted at temperatures of more than 2000°F. The water is thus driven off and the resulting alumina remains as a fine white powder, similar to sugar in appearance and consistency.

5.2.1.3

Reduction

Reduction is the electrolytic separation of aluminum from its oxide, alumina. The general reduction procedure is described in Figure 5.2-1.

This process employs a carbon-lined vessel (pot) containing molten cryolite (sodium aluminum fluoride) in which alumina is dissolved. Metallic aluminum is separated from the alumina by passing an electric current through this solution at a temperature of about 1775°F.

Molten aluminum collects at the bottom of the pot and is tapped or siphoned off into large ladles and then poured into molds to form ingots, or transferred to holding furnaces for alloying. Aluminum, as it comes from the pots, is about 99.5% pure. By additional refining, super purity aluminum, which is about 99.99% pure, can be produced.

5.2.1.4

Recycling

Recycled aluminum, metal that has already had a useful life in a product and is used over again, is playing an increasingly important role (see Fig. 5.2-1). Recycling is important because it saves 95% of the energy required to make aluminum from ore. Ingots that come from recycling are called *secondary ingots*.

Old aluminum windows, doors, and siding are among the many products that are being brought to scrap dealers for recycling.

5.2.1.5

Aluminum Alloy

In high-purity form, aluminum has a tensile strength of about 7000 psi and is relatively soft and ductile. Most applications, however, require greater strength and other properties, achieved by the addition of other elements to produce various aluminum alloys.

Aluminum alloys can be divided into two general categories: those whose strength may be improved by heat treatment (*heat-treatable* alloys), and those whose strength may be improved by mechanical methods of cold working (*non-heat-treatable* alloys).

■ HEAT-TREATABLE ALLOYS

The initial strength of heat-treatable alloys is improved by the addition of alloying elements such as copper, magnesium, zinc, and silicon. Since these elements (singly or in various combinations) show increasingly solid solubility in aluminum with increasing temperature, it is possible to subject them to thermal treatments that will impart further strengthening.

The first step, called *solution heat treatment*, is a process that uses elevated temperatures. This is followed by rapid quenching, usually in water, which momentarily "freezes" the structure and for a short time renders the alloy very workable. It is at this stage that some fabricators retain this more workable structure by storing the alloys at below-freezing temperatures until they are ready to be formed. At room or elevated temperatures the alloys are not stable after quenching, and after a period of several days at room temperature, the alloy is considerably stronger. This is called *natural aging* or *room temperature precipitation*. Some alloys, particularly those containing magnesium and silicon or magnesium and zinc, continue to age and harden for long periods of time at room temperature.

By heating the alloy at a controlled temperature, even further strengthening is possible and properties are stabilized. This process is called *artificial aging* or *precipitation hardening*. Heat-treated alloys carry the temper designation T (Fig. 5.2-2).

The highest strengths are obtained by the proper combination of solution heat treatment, quenching, cold working, and artificial aging.

■ NON-HEAT-TREATABLE ALLOYS

Non-heat-treatable (*common*) aluminum alloys have alloying elements that do not

HOW ALUMINUM IS MADE

ALLOYING & CASTING

EXTRUSION BILLETS

ALLOY INGOTS

ROLLING INGOTS

CASTING MOLDS

HOLDING AND ALLOYING FURNACE

ALUMINUM PRODUCTS

CANOES
AIRPLANES
WHEELS
CANS
SKYSCRAPERS
FOOD TRAYS
SPACE SHUTTLE
TENNIS RACKETS
WINDOW SIDING

RECYCLING

ALLOYING FURNACE

EXTRUSION BILLETS

CASTING MOLDS

USED ALUMINUM

RECYCLED ALUMINUM PRODUCTS

SMELTING

SIPHON

CRUCIBLE

ALUMINA

TO REDUCTION

CARBON ANODE
CRYOLITE BATH
MOLTEN ALUMINUM
CARBON LINING

STEEL SHELL

BUS BAR

HALL-HEROULT REDUCTION PROCESS

$$2Al_2O_3 + 3C \xrightarrow[Na_3AlF_6]{E} 4Al + 3CO_2$$

MINING & REFINING

STORAGE
SPRAY WATER
MINING BAUXITE
CRUSHER
CLAY, SILICA, WASTE
WASTE TREATMENT
DRYING KILN
CRUSHED BAUXITE
COVERED HOPPER CARS
SODA ASH
CRUSHED LIME
CRUSHED BAUXITE
MIXER
DIGESTER
PRESSURE REDUCER
SETTLING TANK
FILTER
COOLING TOWER
WASTE TREATMENT
THICKENER
ROTARY CALCINATING KILN
PRECIPITATOR
COVERED HOPPER CARS
FILTER

the Aluminum Association ®
818 Connecticut Avenue, N.W., Washington, D.C. 20006

FIGURE 5.2-1 *The mining, refining, and reduction of aluminum.* (Courtesy Aluminum Association.)

FIGURE 5.2-2
Aluminum Association's Temper Designations

Temper Designations

-F	As fabricated
-O	Annealed
-H[a]	Strain hardened (wrought only)
	-H1 Plus one or more digits[b]: strain hardened only
	-H2 Plus one or more digits[b]: strain hardened and partially annealed
	-H3 Plus one or more digits[b]: strain hardened, then stabilized
	-H4 Plus one or more digits*: strain hardened and lacquered or painted
-W	Solution heat treated—unstable temper
-T[c]	Thermally treated to produce stable tempers other than -F, -O, or -H
	-T1 Cooled from an elevated temperature shaping process and naturally aged to a substantially stable condition
	-T2 Cooled from an elevated temperature shaping process, cold worked, and naturally aged to a substantially stable condition
	-T3 Solution heat treated, cold worked, and naturally aged to a substantially stable condition
	-T4 Solution heat treated and naturally aged to a substantially stable condition
	-T5 Cooled from an elevated temperature shaping process, then artificially aged
	-T6 Solution heat treated, then artificially aged
	-T7 Solution heat treated and stabilized
	-T8 Solution heat treated, cold worked, then artificially aged
	-T9 Solution heat treated, artificially aged, then cold worked
	-T10 Cooled from an elevated temperature shaping process, cold worked, then artificially aged

[a]Temper letter H is used for non-heat-treatable alloys.

[b]Second digit indicates final degree of strain hardening, i.e., 2 is 1/4 hard, 6 is 3/4 hard, 8 is full hard.

[c]Temper letter T is used for heat-treatable alloys.

show an increase in strength with heat treatment. The initial strength of non-heat-treatable alloys depends upon the hardening effects of elements such as manganese, silicon, iron, and magnesium, singly or in various combinations. These alloys can be strengthened by *strain hardening, cold rolling,* and other mechanical working. Strain-hardened alloys are designated by the *H* series of temper designations (see Fig. 5.2-2).

■ ANNEALING

Annealing is a thermal treatment that is applicable to both heat-treatable and non-heat-treatable alloys and to cast and wrought products. To anneal a product, it is necessary first to elevate the temperature of the metal to between 600° and 800°F and then to cool it slowly. The purpose of this treatment is to relieve internal stresses and to return the alloy to its softest and most ductile condition. Annealing is usually performed to condition the metal for severe forming operations. Subsequent working processes increase the mechanical properties of non-heat-treatable alloys. The mechanical properties of heat-treatable alloys may be increased by appropriate thermal treatments, after forming operations have been completed.

■ CLADDING

To increase the corrosion-resistant properties of some products or to provide for special surface appearance or preparation for additional surface treatment, an aluminum alloy or other metal coating may be metallurgically bonded to an aluminum product. Most cladding is performed on sheet and plate products, but some tube, rod, and wire products are also clad.

Aluminum clad with high-purity aluminum or a non-heat-treatable alloy to improve its surface corrosion resistance is designated by its four-digit alloy number (see below) preceded by the word *Alclad*. For example, Alclad 3004 is a common sheet and plate material used in building applications.

■ ALLOY DESIGNATION SYSTEMS

Aluminum products fall into two general groups: *wrought* products and *cast* products.

Alloy and temper are designated by numerical systems for both cast and wrought products. Figure 5.2-3 shows some typical uses for various alloys.

TEMPER DESIGNATIONS A notation or temper designation is used following the alloy. The letter *H* is used for non-heat-treatable alloys and the letter *T* for heat-treatable alloys. Some temper designations apply only to wrought products, others to cast products, but most apply to both. Additional digits after the *T* or *H* indicate variations in the basic treatment (see Fig. 5.2-2).

WROUGHT ALLOYS Almost all aluminum products used in construction are made from wrought alloys, which are designated by a four-numeral index system (Fig. 5.2-4). The first digit of the number indicates an alloy group or series.

The 1000 Series designates aluminum of 99% or higher purity. These alloys have excellent corrosion resistance, high thermal and electrical conductivity, low mechanical properties, and excellent workability. Moderate increase in strength may be obtained by strain hardening. Alloys used in electrical and chemical fields and aluminum foils for insulation and vapor retarders are in this group (1235) and alloy 1100 is frequently used for sheet metal work.

The 2000 Series alloys contain copper as the principal alloying element. These alloys require heat treatment to attain optimal properties, and in the heat-treated condition their mechanical properties are similar to and may exceed those of mild steel.

Sometimes artificial aging is employed to further increase strength. The alloys in the 2000 Series do not have as good cor-

FIGURE 5.2-3
Examples of Aluminum Alloys Used for Some Building Applications

Application	Alloy Number	Product
Acoustic ceiling	3003-H14	Sheet
Air duct	3003-H14, 1100-H16, Duct stock	Sheet
Awning	3003-H14	Sheet
Builder's hardware	5050-0 B443.0, 356.0 6063-T5	Sheet Casting Extrusion
Curtain wall	3003-H14, Alclad 4043, Alclad 1235 B443.0 6063-T5, 4043-F	Sheet Casting Extrusion
Deck (corrugated)	3003-H14	Sheet
Door frame	6063-T5, 6063-T6	Extrusion
Fascia plate	6063-T42 3003-H14	Extrusion Sheet
Flashing	3003-0	Sheet
Flue lining	1100-H16	Sheet
Garage door	3003-H14	Sheet
Grating	6061-T6	Bar
Gravel stop	6063-T5, 6063-T42	Extrusion
Grille	B443.0 3003-H14 6063-T5	Sand casting Sheet Extrusion
Gutter	3003-H14	Sheet
Hood	3003-H14	Sheet
Insulation	1235	Foil
Kick plate	3003-H18	Sheet
Letters	B443.0, 514.0 3003-H14	Sand casting Sheet
Louver	3003-H14 6063-T5	Sheet Extrusion
Mullion	3003-H14 6063-T5, 6063-T6	Sheet Extrusion
Nails	6061-T913	Wire
Railing	6063-T5, 606 1-T83	Extrusion
Relief panel (sculptural)	B443.0	Casting
Roofing	Special roofing alloys	Sheet
Screen	Alclad 5056-H392	Wire
Screws	2024-T4, 2017-T4	Screw machine stock
Shade screening	5052-H38	Sheet
Shingle	3003-H14	Sheet
Siding	3003-H14	Sheet
Structural shape	2014-T6, 6061-T6 6061-T6, 2014-T6	Extrusion Rolled shape
Termite shield	3003-H14	Sheet
Terrazzo strip	3003-H14	Sheet
Threshold	6063-T5	Extrusion
Vapor barrier	1235-0	Foil
Venetian blind	5052-H18	Sheet
Weatherstrip	5052-H38	Sheet
Window	6063-T5, 6063-T6, 6061-T6	Extrusion

rosion resistance as most other aluminum alloys and in sheet form are usually clad with a high-purity alloy or a magnesium-silicon alloy of the 6000 Series. Alloy 2024 is used for fasteners. Structural shapes for use in construction are sometimes made from 2014.

The 3000 Series alloys with manganese as the major alloying element are generally non-heat treatable. Because only a limited percentage of manganese (up to about 1.5%) can be effectively added to aluminum, it is used as a major element in only a few alloys.

Alloy 3003 is a general-purpose alloy used for moderate-strength applications requiring good workability. In sheet form, it is used for air ducts, acoustical ceiling pans, awnings, corrugated decks, garage doors, flashings, gutters, siding, and shingles, among other uses.

The 4000 Series alloys employ silicon as the major alloying element, which substantially lowers the melting point of the alloy without producing brittleness. Most alloys in this series are non-heat treatable. Those containing appreciable amounts of silicon become dark gray when anodic oxide finishes are applied and hence are used for architectural applications.

The 5000 Series alloys use magnesium as the major alloying element. Magnesium is one of the most effective and widely used alloying elements, and when used as the major alloying element (or with manganese), the result is a moderate to high-strength, non-heat-treatable alloy. Magnesium is more effective than manganese as a hardener and can be added in considerably greater quantities. These alloys have especially good resistance to corrosion in contact with sea air and seawater. Shade screening, venetian blinds, and weather stripping are often made from alloy 5052.

The 6000 Series alloys are composed of heat-treatable alloys containing silicon and magnesium. Though not as strong as most alloys of the 2000 or 7000 Series, the 6000 Series alloys possess good formability and corrosion resistance and are of medium strength. They may be formed in the solution of heat-treated temper (T4 temper) and then reach higher strengths by artificial aging (T6 temper). This series is the one most commonly used in products for residential construction. For example, 6061 is used in structural shapes, nails,

FIGURE 5.2-4
The Aluminum Association's Four-Digit Numbering System for Designating Wrought Aluminum Alloys

First Digit	Second Digit	Third and Fourth Digits
Major Alloying Elements	**Modification of Alloy**	**Arbitrary Numbers**
2 xxx — Copper	x0xx — Original alloy	Used to identify alloys within the series
3 xxx — Manganese	x1xx	
4 xxx — Silicon	x2xx	
5 xxx — Magnesium	x3xx	
6 xxx — Magnesium and silicon	x4xx	
7 xxx — Zinc	x5xx ⟩ Alloy modifications	
8 xxx — Other elements	x6xx	
9 xxx — Unused series	x7xx	
	x8xx	
	x9xx	

and bars; 6063 in window and door frames, hardware, louvers, tubing, and curtain walls.

The 7000 Series has zinc as the major alloying element, which, when combined with a smaller percentage of magnesium, results in heat-treatable alloys of very high strength. Usually other elements, such as copper and chromium, are also added in small quantities. Among the highest-strength alloys available is 7075. It is often used in aircraft structures and other highly stressed parts.

Experimental alloys are also designated with the groups described above, but carry the prefix *X*, which is dropped when the alloy becomes standard.

Cast alloy designations use a four-digit system similar to that for wrought alloys, shown in Figure 5.2-4.

5.2.2

PROPERTIES

Aluminum has a specific gravity of about 2.70; a density of approximately 0.170 lb./cu. ft.; and weighs about 0.1 lb./cu. in., as compared with 0.28 for iron and 0.32 for copper.

Pure aluminum melts at about 1200 °F. Melting points for various alloys range from 900° to 1250 °F.

Aluminum and aluminum alloys have relatively high coefficients of thermal expansion. For structural calculation, the coefficient of expansion is assumed to be 13×10^{-6}. This means, for example, that an unrestrained piece of aluminum 18 ft. long will expand or contract 0.259 in. (about 1/4 in.) in a 100 °F temperature change.

The two common metals used as electrical conductors are copper and aluminum. The electrical conductivity of EC grade (electrical-conductor) aluminum is about 62% that of the International Annealed Copper Standard. However, with a specific gravity of less than one-third that of copper, 1 pound of aluminum will conduct twice as much electricity as 1 pound of copper. Unfortunately, aluminum is stiffer than copper; therefore, connections are more difficult to make, resulting in inadequate and unsafe connections.

The high thermal conductivity of aluminum makes it well fitted for use wherever a transfer of thermal energy from one medium to another is desired. However, high thermal conductivity is generally not required in most building material applications, and, in fact, is often a problem in components such as window frames and in storefronts and curtain wall sections where *thermal breaks* (other nonconducting materials) are often introduced to reduce the flow of heat through the aluminum.

Aluminum is an excellent reflector of radiant energy through the entire range of wavelengths, from ultraviolet through the visible spectrum to infrared (heat) waves. Aluminum's light reflectivity of more than 80% accounts for its wide use in lighting fixtures, and its high heat reflectivity and low emissivity have led to its use as reflective insulation and improved performance in applications such as paint, roofing, and ductwork.

Aluminum's nonmagnetic properties make it useful for electrical shielding purposes in electrical equipment and its nonsparking characteristics make it an excellent material for use around flammable or explosive substances.

The ease with which aluminum may be fabricated is one of its most important properties. The metal can be cast, rolled to many thicknesses (down to a foil that is thinner than paper), extruded, stamped, drawn, spun, roll formed, hammered, or forged; it can be turned, milled, bored, or machined at maximum speeds.

Almost any method of joining is applicable to aluminum and a wide variety of mechanical aluminum fasteners simplifies the assembly of products.

5.2.2.1

Strength

Although pure aluminum has a tensile strength of only about 7000 psi, by working the metal, as by cold rolling, its strength can be about doubled. Much larger increases in strength can be obtained by alloying aluminum with small percentages of one or more other metals such as manganese, silicon, copper, magnesium, or zinc. Like pure aluminum, the alloys are also made stronger by cold working, and heat-treatable alloys are further strengthened and hardened by heat treatments, so that tensile strengths approaching 100,000 psi are possible (Fig. 5.2-5). However, such high-strength aluminums are seldom used in building construction products.

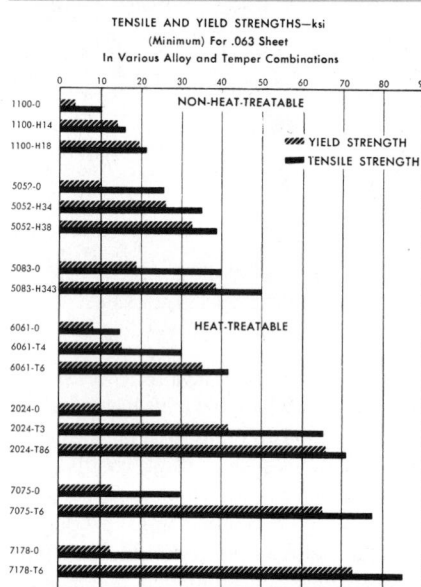

FIGURE 5.2-5 *Mechanical properties of representative aluminum alloys.*

Strength decreases at elevated temperatures, although some alloys do retain good strength at temperatures from 400° to 500°F. At subzero temperatures strength increases without loss of ductility so that aluminum is a particularly useful metal for low-temperature applications.

5.2.2.2

Corrosion Resistance

Aluminum has a high resistance because of the very thin, inert surface film of aluminum oxide that forms rapidly and naturally when exposed to the air. After it reaches a thickness of about a ten-millionth of an inch, this film effectively halts further atmospheric oxidation of the metal and thus protects the surface. If this natural oxide film is broken, as by a scratch, a new protective film forms immediately. The oxide film increases in thickness with temperature and remains protective of the underlying aluminum, even at the melting point.

It may be necessary to protect aluminum in certain severely corrosive environments by the addition of protective coatings such as organic paints, by cladding, or by increasing the thickness and effectiveness of the oxide film by anodizing (see Section 5.2.3.4).

■ WEATHERING

The extent of corrosion from weathering depends on the type and extent of the contamination in the atmosphere. More corrosive elements are present in industrial and coastal areas than in rural areas.

In tests and observations (in part under the direction of the National Institute of Standards and Technology) made over a 10-year period in different atmospheric conditions on alloys 1100, 3003, and 3004, the depth of penetration of atmospheric corrosion was less than 1 mil (1/1000 in.) in rural areas, 4 mils in industrial areas, and up to 6 mils in coastal locations.

■ GALVANIC CORROSION

When two dissimilar metals are connected by a solution that conducts electricity (an electrolyte), *galvanic corrosion* may occur. As in a battery, an electric current is created, and one metal corrodes away while the other is being plated. Galvanic attack can occur when moisture condenses from the air and contaminating elements act together as an electrolyte. The threat of such corrosion is greatest in industrial areas because the atmospheric moisture is more contaminated, and in coastal areas because of the salt and moisture in the air.

The severity of galvanic attack varies, depending on how far apart the two metals are in the galvanic table (Fig. 5.2-6). For example, coupling mild steel to aluminum would produce less corrosion than would coupling copper to aluminum under identical conditions of exposure. The metal that has a higher voltage potential is sacrificial and corrodes. Thus, zinc provides cathodic protection to aluminum and aluminum alloys. That is, when zinc and aluminum are in the presence of an electrolyte, the zinc usually corrodes. Thus, zinc-plated steel parts are commonly used with aluminum to prevent the harmful effect of galvanic action on the aluminum.

Cadmium-plated steel in contact with aluminum alloys is also satisfactory. Since there is no appreciable difference in the

FIGURE 5.2-6
Electrolytic Solution Potentials*

Metal or Alloy	Potential in Volts
Magnesium	−1.73
Zinc	−1.00
520.0-T4 alloy (cast)	−0.92
7072 (wrought)	−0.96
514.0 alloy (cast)	−0.87
Pure aluminum and 5052-0, 1100-0, 1100-H18, 3003-H18, 6061-T6 alloys (wrought) and 43 alloy (cast)	−0.83
Cadmium	−0.82
356.0-T4 alloy (cast)	−0.81
2017-T4, 2024-T4 alloys (wrought)	−0.68
Mild steel	−0.58
Lead	−0.55
Tin	−0.49
Brass (60–40)	−0.28
Copper	−0.20
Stainless steel (18–8)	−0.15

*Measured in a 5.85% sodium chloride solution containing 0.3% hydrogen peroxide. The values vary somewhat depending on the particular lot of material and on the surface preparation employed. Coupling metals with dissimilar electrolytic potentials in an electrolyte results in galvanic corrosion of the metal with the highest negative potential. Degree of attack generally increases as potential difference increases.

electrolytic potential of these two metals, practically no galvanic action occurs.

Aluminum is compatible with some stainless steels, chromium, zinc, and small areas of white bronze. Where permanent contact with other metals cannot be avoided, the risk of galvanic corrosion can be greatly reduced by painting the other metal and the aluminum at the contact area with zinc chromate, followed by one coat of a nonleaded paint, such as aluminum paint, or a coat of a heavy-bodied bituminous paint. When the dissimilar metal cannot be painted, the aluminum may be given the insulating treatment. As an alternative, a strip of plastic or a similar insulator may be placed between the two metals.

In severely corrosive atmospheres and in areas of high moisture the edges of a dissimilar metal joint may be sealed with compatible building mastic or caulking.

A dangerous source of electrolytic attack may be water drainage over metals such as copper. Therefore, aluminum gutters on buildings with copper flashings must be protected by painting either the copper flashing or both the copper and the aluminum.

DIRECT CHEMICAL ATTACK

Some chemicals will dissolve metals on contact. For example, solutions of strong alkalies, sulfuric acid, hydrochloric acid, carbonates, and fluorides will attack aluminum because they can dissolve or penetrate its protective oxide coating. Some chemical attack may also occur when aluminum is placed in contact with wet alkaline materials such as mortar, concrete, and plaster. Therefore, surface protection of aluminum products is often necessary during construction.

Where aluminum is buried in concrete, it is usually not necessary to paint the parts. However, if chlorides are present from additives or other sources in steel-reinforced concrete, the aluminum parts should be coated with bituminous paint or another suitable coating.

Aluminum in contact with concrete, masonry, or other absorbent materials, even under intermittently wet conditions, should be protected with a coating of bituminous paint, zinc chromate primer, or a separating layer of plastic or other gasketing material. Creosote and tar coatings should not be used, however, because of their acidic contents. Where wet concrete or another alkaline material may splash against aluminum during construction, the aluminum surfaces should be protected. A coating of clear acrylic-type lacquer or a suitable strippable coating may be used when maintenance of appearance is important.

Although aluminum is inert when buried in many types of soil, some soils and the presence of stray electrical currents in the ground may cause corrosion. Therefore, buried aluminum pipe may have to be coated or otherwise protected.

5.2.3

MANUFACTURE

Wrought mill products make up the greater part of aluminum products used in the building industry. They include plate, sheet, foil, bar, rod wire, extruded sections, structural shapes, tubes, pipe, and forging stock. The mechanical properties of such materials are developed by alloying, hot working, cold working, heat treatment, or combinations of these processes.

Cast products include sand castings, permanent mold castings, and die castings. These are produced in a foundry by pouring molten metal into molds of a desired shape. The mechanical properties of some castings may be improved by heat treatment.

The general categories of wrought and cast products are outlined in Figure 5.2-7.

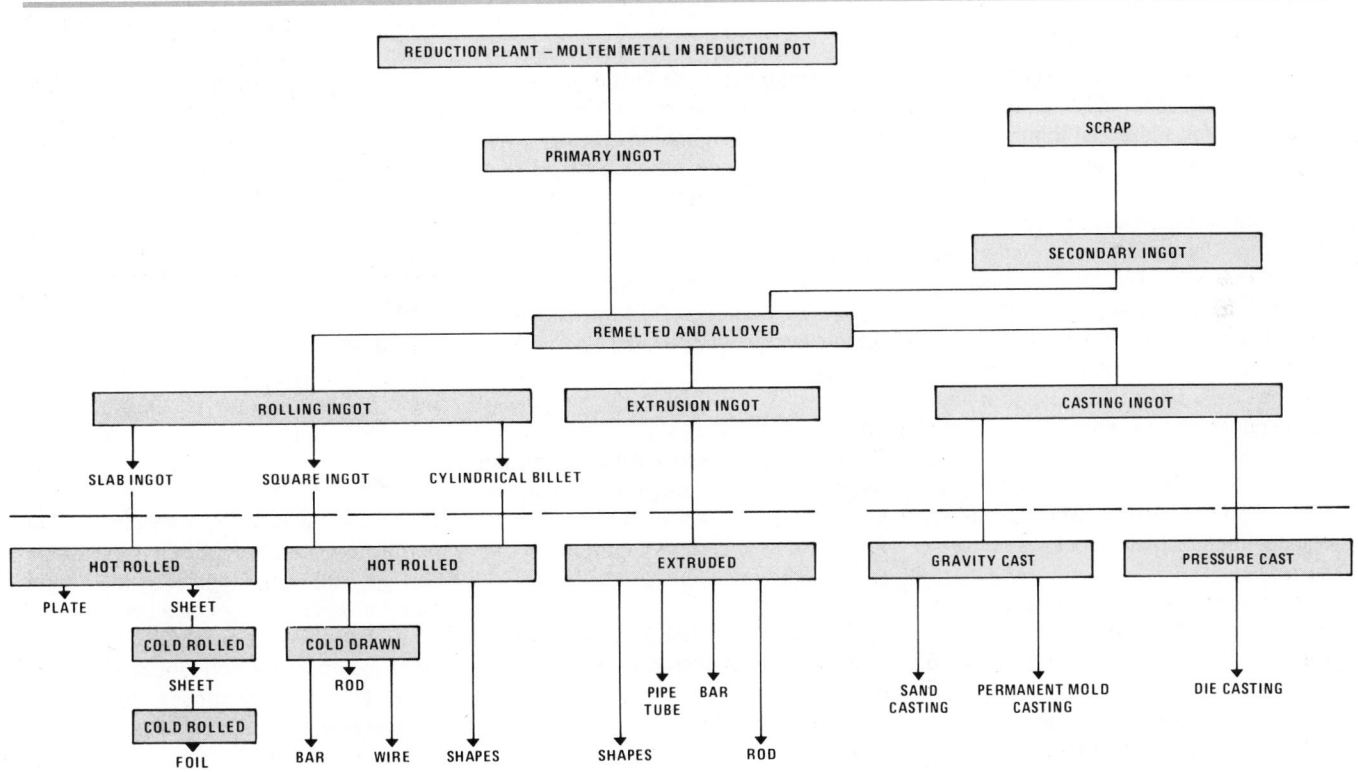

MILL OPERATIONS AND MILL PRODUCTS (WROUGHT PRODUCTS)   FOUNDRY OPERATIONS AND CAST PRODUCTS

FIGURE 5.2-7 *Classification of aluminum-mill operations used to manufacture wrought products and foundry operations used to make cast products.*

Wrought Products

Wrought mill products are those shaped by hot and cold rolling, cold drawing, extruding, and forging.

■ HOT AND COLD ROLLING

Hot rolling is performed in mills, in rolling machines, through which an ingot, heated to soften it, is passed until reduced to the desired shape. Hot rolling can be continued until a thickness of about 0.125 in. is obtained. After that, further thinning is usually accomplished by cold rolling. Cold-rolled material has a better finish and, in the case of non-heat-treatable alloys, increased strength and hardness.

PLATE   Plate has a minimum thickness of 0.250 in. To make it, a slab of aluminum is hot rolled in a reversing mill (acting like a rolling pin) until its thickness has been reduced to from 1 to 3 in. Further rolling operations are performed in a hot mill.

SHEET   Sheet is material with a thickness between 0.006 in. and 0.249 in. It is rolled in a hot mill to about 0.125 in. thick, then cold rolled for additional thinning and to give it more strength and a better finish.

Coiled sheet is rolled to its finished gauge and width in a continuous strip up to several hundred feet in length and wound into coils. *Flat sheet* is sheared to proper lengths from coiled sheet. It may be stretcher leveled for improved flatness.

Brazing sheet is a special type of sheet that has a cladding that melts at a lower temperature than the core. Assemblies of parts made from this sheet can be brazed together by subjecting them to a temperature of about 1000 °F, which melts the cladding but not the core alloy, joining the surfaces.

Aluminum foil is less than 0.006 in. thick. It is made by rolling sheet in cold-rolling mills, and is usually supplied in coil form. Its most common applications in the building industry are as thermal insulation and vapor retardant.

STRUCTURAL SHAPES   Though most structural aluminum shapes are extruded, some are hot rolled by a series of specially shaped rolls. Complicated structural shapes require many roller passes to prevent the undue straining that is possible when too much force is applied in a single pass. Because they require considerable metal and a high degree of machining and finishing to reach the desired contour, rolls are expensive. Therefore, lower tooling costs often make extruded structural aluminum shapes less expensive than rolled ones. Finishing operations include roller or stretch straightening and, when required, the rolled shapes may be drawn through dies to improve surface finish and to obtain closer dimensional tolerances.

Standard rolled structural shapes, usually supplied in 2014 and 6061 alloys, include angles, channels, and beams.

ROLLED ROD AND BAR   Hot rolling is used to produce rod and bar shapes. Cold-finished rod and bar are produced by hot rolling to a size larger than specified and then by cold drawing through a die for a better finish and closer dimensional tolerances. Bar is commercially available as square (round or square edged), square-edged rectangular, and hexagonal.

FORGING STOCK ROD   Rod, bar, and other sections suitable for further shaping by impact or pressure in a die are called forging stock. Forging rod is nearly always rolled, but forging bar may be rolled or extruded, and forging shapes are usually extruded.

REDRAW ROD   Redraw rod is typically 3/8 in. in diameter. This rod is suitable for further drawing to produce wire.

■ COLD DRAWING

Drawn products are formed by pulling (drawing) metal through a die. The purpose of cold drawing is to obtain close dimensional tolerances and a better surface finish.

WIRE PRODUCTS   Wire, which must be kept within close tolerances, is cold drawn by pulling redraw rod (3/8-in. diameter) through a series of progressively smaller dies.

TUBE DRAWING   Hollow tube sections are drawn in a process similar to drawing bar and rod, except that the tube is threaded over a mandrel (shaper) positioned in the center of the die. When the tube is drawn, close tolerances both inside and out are maintained.

■ EXTRUDING

Extruding is a process in which a heated aluminum ingot is pushed by hydraulic pressure through a die opening. The extruded aluminum assumes the same size and cross section as the die opening. Extruded products include shapes of varying cross section, rod, bar, and seamless tube. Aluminum window sash is an example of an extruded shape. Expansion joint covers are others.

Shapes that cannot be extruded as a single section because of dimensional limitations or complexity of cross section can be achieved by fitting (mating) more than one section together.

The extruding process provides great flexibility in design and shape. The economical manufacture of practically any shape is possible as a result of the relatively low cost of dies. Close tolerances in manufacture can be obtained, and small parts of desired shapes can be sliced from a longer extrusion, often minimizing or eliminating machining. Extrusions are straighter, dimensionally more accurate, and have a better surface finish than rolled shapes.

Cast Products

Cast products are manufactured in a foundry. They include sand casting, permanent mold casting, and die casting.

■ SAND CASTING

Molds made of sand are lowest in cost and permit maximum flexibility in design changes, although labor costs are higher than for other castings. Sand castings require heavier section thicknesses than other methods, and tolerances are not as close. They are used for spandrel panels, decorative panels, lighting standard bases, and general fittings.

■ PERMANENT MOLD CASTING

Permanent mold casting employs metal molds with a metal core. It is used where a large number of castings from the same mold are required. If a sand core is used, the casting is called a semipermanent mold casting. Molten metal is poured, not injected, into the mold.

DIE CASTING

Die castings are permanent mold castings in which the metal is forced into the mold under pressure. Die castings have smoother surfaces than permanent mold castings, and permit greater repetitive use of the molds. Die casting is generally used for the mass production of small items such as handles, latches, and other door and window fittings.

The main advantages of permanent mold and die castings over sand castings are higher strength, better surface appearance, and closer dimensional tolerances.

5.2.3.3

Fabrication

Fabrication includes forming and machining operations performed on aluminum products and castings after they leave the mill or foundry.

FORMING

Aluminum, one of the most workable of common metals, can be fabricated into a variety of shapes by bending, stretching, rolling, drawing, and forging.

Formability varies greatly, depending on the aluminum alloy and temper.

BENDING   Temper and thickness greatly affect the minimum bend radius that can be used in forming. Certain aluminum alloys can be cold formed to a sharp 90-degree bend; others must be formed on stipulated radii.

STRETCH FORMING   Sheets are stretched to flatten them or to form them into special shapes. The most suitable alloys for stretch forming are those with high levels of elongation (the ability to stretch without fracturing). The advantage of stretch forming is that there is hardly any tendency of the material to assume its original shape after forming stresses are removed.

ROLL FORMING   Products are roll formed by passing strips through a series of roller

dies (Fig. 5.2-8). Each successive pair of rolls shapes the metal slightly—only a small deformation is made by any single pair of rolls. No straightening is required since roll-formed products ordinarily are straight and free from distortion.

FORGING   High-strength aluminum parts are often produced by forging (Fig. 5.2-9). Most forgings are made by hammering or pressing heated aluminum between closed steel dies. Press forging forces the metal under mechanical or hydraulic pressure into a die cavity of the desired shape. Hammer forging produces the same results by repeated blows from a drop hammer. Aluminum forgings may be heat treated after fabrication.

FIGURE 5.2-8 *A series of cold rolls form corrugated decking from embossed sheet.* (*Courtesy Kaiser Aluminum and Chemical Corporation.*)

FIGURE 5.2-9 *A large 8000-ton hydraulic press produces forgings for use as architectural panels.* (*Courtesy Kaiser Aluminum and Chemical Corporation.*)

Since forging refines and provides desirable directional characteristics to the grain structure of the metal, improving strength and fatigue resistance, it is used to make various types of door hardware.

MACHINING

The machining characteristics of aluminum and its alloys, especially of wrought products, are excellent. Sawing, drilling, tapping, and threading methods vary little from those employed for other metals.

5.2.3.4

Finishes

The surfaces of many aluminum products require no further finishing. The natural oxide film makes protective coatings unnecessary under ordinary conditions. However, finishes are frequently applied in order to (1) change or improve the appearance for decorative purposes; (2) improve its abrasion or corrosion resistance; (3) provide a base for subsequent surface treatment such as painting.

Natural finishes and other common surface treatments are briefly described in this section and summarized in Figure 5.2-10.

UNCONTROLLED NATURAL FINISHES

Hot-rolled products show a certain amount of discoloration or darkening, whereas cold-finished products have a whiter, brighter surface. Extruded products have an intermediate appearance nearer that of cold-finished items and show traces of longitudinal striations (*die lines*) caused by the extrusion process. Castings vary little in their surface appearance, except that die castings and permanent mold castings are smoother than sand castings.

CONTROLLED NATURAL FINISHES

Natural finishes of cold-rolled sheet may be controlled by the degree of polishing given to the rollers of the rolling mill. *Mill finish* has an uncontrolled, slightly variable finish that may not be entirely free from stains or oil. *One-side-bright mill finish* has a moderate degree of brightness on one side; the reverse side is uncontrolled and may have a dull, nonuniform

FIGURE 5.2-10 *Principal aluminum-finishing processes.*

appearance. *Standard bright finish* has a relatively bright, uniform appearance on both sides, but is somewhat less lustrous than the standard one-side-bright sheet. *Standard one-side-bright sheet* has a uniformly bright and lustrous surface on one side; the reverse side is uncontrolled and may have a dull nonuniform appearance.

Embossed sheet is obtained by passing a mill finish sheet through a pair of rolls with a matched design or between a design roll and a plain roll. Embossed surfaces increase the stiffness and tend to hide wear and scratches. Patterns include wood grain, leather grain, fluted, stucco, diamond, notched, ribbed, bark, and pebble designs (Fig. 5.2-11).

■ MECHANICAL FINISHES

Aluminum surfaces may be mechanically treated by *grinding* with a dry abrasive wheel, primarily to remove surface roughness from castings; *buffing* with a fine abrasive, resulting in a highly lustrous surface; *polishing* with a finer abrasive than for buffing, and with a very light pressure, resulting in a mirrorlike surface; *scratch brushing* with power-driven rotating wire brushes to give coarse or smooth-lined surface textures, depending on the size of the wire bristles used; *shot blasting* and *sand blasting* with various combinations of pressure, steel shot, and sand to produce matte and coarse surfaces. Rough

FIGURE 5.2-11 *Aluminum sheet and plate can be embossed to produce a variety of surface textures.*

textures may be given a lacquer coat or may be anodized to minimize dust collection.

Satin finishing is accomplished with fine wire bristles, very fine abrasives, or a muslin or felt wheel to produce a soft, smooth sheet.

Spin finishing is achieved by pressing an oily abrasive cloth against the aluminum piece while it is revolved rapidly. Subsequent applications of fine stainless steel wool and fine mesh emery give a

bright, silvery finish with a pattern of concentric circles.

Highlighting provides a two-tone effect for relief surfaces. The elevated areas are masked while the recessed areas are sandblasted. Then the masking is reversed and the elevated areas are highlighted by painting or polishing.

The Aluminum Association's designations for various mechanical finishes are shown in Figure 5.2-12.

■ CHEMICAL FINISHES

Many methods of treating aluminum surfaces employ the reaction of various chemicals on the metal. Some of the most common are discussed here. The Aluminum Association's designations for various chemical finishes are shown in Figure 5.2-13.

ETCH FINISH   A decorative, silvery white finish, also known as frosted finish, may be produced by etching the metal with caustic soda (sodium hydroxide) or another chemical. Several methods permit partial or *design* etching to produce varying patterns and colored backgrounds.

CONVERSION COATINGS   Although chemical conversion coatings are generally used on aluminum to prepare the surface for painting, some types may also be used as a final finish.

FIGURE 5.2-12
Aluminum Association's Designations for Various Mechanical Finishes*

As Fabricated	Buffed	Directional Textured	Nondirectional Textured
M10—Unspecified	M20—Unspecified	M30—Unspecified	M40—Unspecified
M11—Specular as fabricated	**M21—Smooth specular**	M31—Fine satin	M41—Extra fine matte
M12—Nonspecular as fabricated	M22—Specular	**M32—Medium satin**	**M42—Fine matte**
M1x—Other	M2x—Other	M33—Coarse satin	**M43—Medium matte**
		M34—Hand rubbed	**M44—Coarse matte**
		M35—Brushed	M45—Fine shot blast
		M3x—Other	**M46—Medium shot blast**
			M47—Coarse shot blast
			M4x—Other

Reprinted from Metal Finishes Manual for Architectural and Metal Products *with permisson of the National Association of Architectural Metal Manufacturers (NAAMM), Chicago.*

*Finishes printed in **boldface type** are those most frequently used for architectural work.

FIGURE 5.2-13
Aluminum Association's Designations for Various Chemical Finishes*

Nonetched Cleaned	Etched	Brightened	Conversion Coatings
C10—Unspecified	C20—Unspecified	C30—Unspecified	C40—Unspecified
C11—Degreased	C21—Fine matte	C31—Highly specular	C41—Acid chromate-fluoride
C12—Chemically cleaned	**C22—Medium matte**	C32—Diffuse bright	C42—Acid chromate-fluoride-phosphate
C1x—Other	C23—Coarse matte	C3x—Other	C43—Alkaline chromate
	C2x—Other		C44—Nonchromate
			C45—Nonrinsed chromate
			C4x—Other

Reprinted from Metal Finishes Manual for Architectural and Metal Products *with permisson of the National Association of Architectural Metal Manufacturers (NAAMM), Chicago.*

*Finishes printed in **boldface type** are those most frequently used for architectural work.

Since the natural oxide film on aluminum surfaces may not provide a suitable bond for paints and other organic coatings and laminates, its chemical nature is often altered to improve adhesion. *Conversion films*, which are also called *conversion coatings*, are generally applied by use of phosphate or chromate solutions, many of which are proprietary in nature. One of the simplest methods of surface preparation for painting is to etch the surface with a phosphoric acid solution, thus providing a good mechanical bond.

CHEMICAL OXIDIZING Chemical oxide films have improved resistance to corrosion. As a base for painting, they provide additional protection in the event that the finish is scratched or otherwise broken. They are more economical and easier to apply than anodic processes (described below) that also produce oxide films, but are thinner and softer than anodic finishes. Chemical oxide coatings may be dyed, but

anodic films are preferred if a superior color quality is desired.

ZINCATING In the zincating process, aluminum is immersed in a zincate bath providing a very thin, firmly adherent coating of zinc. This base is often used for aluminum that is to be electroplated.

CHEMICAL BRIGHTENING A mirror-bright surface finish can be provided on aluminum by immersing the parts in a hot chemical solution, followed by rinsing. Depending on the alloy and its mechanical preparation, reflectivity values up to 85% may be obtained.

■ ANODIZED FINISHES
Anodized electrolytic oxide coatings are readily applied to aluminum alloys and are widely used. Thinner anodic flash films may be used as a prepaint treatment.

The distinction between a coating and a film is thickness. A film is less than

0.1 mil; any thicker covering is referred to as a coating.

Ordinarily, clear anodic coatings on architectural products are preceded by either a chemical etch or some form of mechanical finish, or both. Etching is the most economical and common prefinish, imparting a frosted appearance that helps to conceal variations in the mill finish and die lines on extrusions. Mechanical finishes also leave a pleasing texture, which is retained after anodizing.

If neither etching nor mechanical finishing is specified, the aluminum still must be chemically cleaned before it is anodized to ensure satisfactory results. Contaminants deposited on the metal surfaces during fabrication processes may affect the anodic coating and should be completely removed before anodizing.

Anodizing consists of immersing a product in an acid solution (*electrolyte*) and passing an electric current, usually of the direct type, between the metal and the solution, with the aluminum serving as the anode. This results in the controlled formation of a relatively thin but durable oxide or coating on the surface of the metal.

In general, such coatings are many times thicker than the natural oxide film, and may be transparent, translucent, or opaque, depending on the alloys and the electrolyte used. Anodic coatings do not change the surface texture of the metal, but they greatly increase resistance to corrosion and provide increased resistance to abrasion.

Several anodizing processes are commonly used. The principal differences between these processes are the type of solution used for the electrolyte, the voltage and current required, and the temperature of the bath.

The *sulfuric acid process*, because of its relatively low cost, is most widely used for architectural anodizing. It produces comparatively thick, transparent, and absorptive coatings, which are suitable for receiving coloring dyes and pigments (see below). Thin films may be produced by this process, which are excellent pretreatments for paint, enamel, and lacquer coatings.

COLOR ANODIZING There are three methods of color anodizing in common use: sulfuric acid anodizing and dyeing; integral-color anodizing; and sulfuric acid anodizing and electrolytically depositing color (Fig. 5.2-14).

In *sulfuric acid anodizing and conventional dyeing*, the work is rinsed after it leaves the anodizing bath, and is dipped in another bath containing either an organic dye or an inorganic pigment. To set the dye, the work is dipped in a nickel-acetate solution and is sealed (usually in boiling deionized water) to produce a clear film that provides minimal interferences with, and optimal lightfastness of, the color,

A wide range of colors is possible with this process, but not all of them are satisfactory for outdoor service. As indicated in Figure 5.2-14, the color is concentrated in the top third of the anodic film; therefore, it is necessary to adjust dyeing conditions (time, temperature, dye concentration, and pH) to get similar colors with different film thicknesses.

In *integral-color anodizing*, the coloring material is spread uniformly through the anodic film. Integral-color anodic films

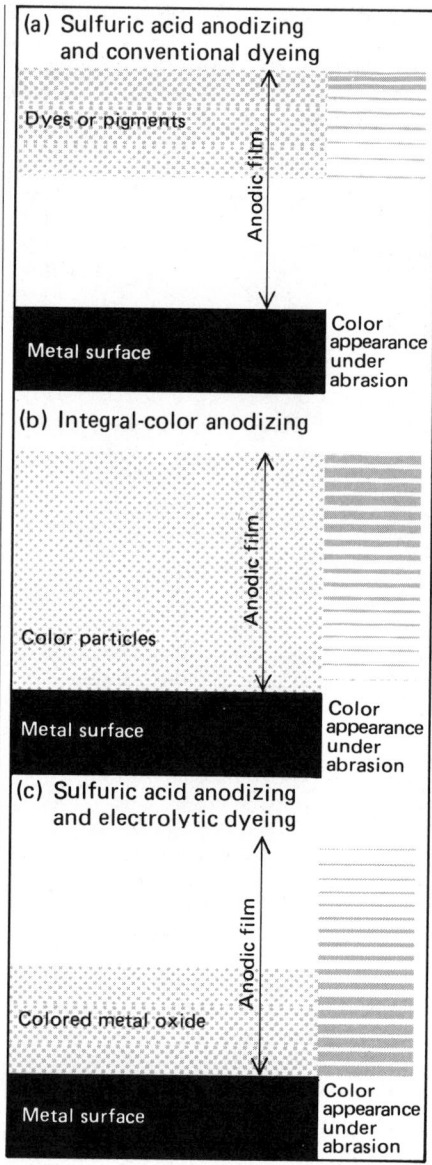

FIGURE 5.2-14 *Distribution of coloring elements in the anodic film of the three main color anodizing processes.*

may be produced by one of two different methods, or by a combination of both. The first method makes use of an alloy with constituents that color the anodic film as it grows. The second method makes use of an electrolyte that yields a colored precipitate during anodizing, again coloring the anodic film as it grows.

A combination of these two methods increases the range of possible colors. The colors produced by integral-color anodizing are very lightfast but, like sulfuric acid anodizing and electrolytically deposited

color, the depth of color varies with the anodic film thickness. In addition, the kind of alloy is a major variable; some colors are restricted to specific alloys.

Sulfuric acid anodizing and electrolytically depositing color concentrate the color at the base of the anodic film (see Fig. 5.2-14). Consequently, the color receives a maximum of protection from sunlight and weathering.

In most applications, the pores in the anodic film are sealed off to increase resistance to staining and corrosion and to improve color durability. Because most applications require a clean film, immersion of the work in boiling deionized water with a pH of 5.5 to 6.5 is the usual sealing method. When maximum corrosion resistance is required, potassium dichromate is added to the water.

The thicknesses of anodic films and coatings range from less than 0.1 mil, in the case of primer films for receiving organic finishes, to 5.0 mils for some of the engineering hardcoats.

For architectural applications the thickness and weight of coatings are of prime importance; no coating thinner than 0.4 mil should be used even for interior purposes. Coatings for exterior parts that will be subject to contaminated atmospheres, or will not receive regular maintenance care, should have at least a 0.7-mil thickness and a minimal coating weight of 27 mg/sq. in.

CLASSIFICATION OF ANODIC COATINGS The Aluminum Association identifies the various anodic finishes under four general types: General Anodic Coatings, Protective and Decorative Coatings, Architectural Class I Coatings, and Architectural Class II Coatings. The first two classifications apply to general industrial work. Therefore, only the last two are discussed here. The Aluminum Association's designations for anodic coatings are shown in Figure 5.2-15.

Architectural Class I Coatings are more than 0.7 mil thick, weigh more than 27 mg/sq. in., and include the hardest and most durable anodic coatings available. They are suitable for exterior items that receive no regular maintenance care, and for interior architectural items subject to normal wear.

Architectural Class II Coatings range in thickness from 0.4 to 0.7 mil, with corre-

FIGURE 5.2-15
Aluminum Association's Designations for Anodic Finishes[a]

General	Protective and Decorative	Architectural Class II	Architectural Class I
A10—Unspecified anodic coating	A21[b]—Clear (natural)	**A31—Clear (natural)**	**A41—Clear (natural)**
A11—Preparation for other applied coatings	A22[b]—Integral color	**A32—Integral color**	**A42—Integral color**
A12—Chromic acid anodic coating	A23[b]—Impregnated color	**A33—Impregnated color**	**A43—Impregnated color**
A13—Hard coating	A24—Electrolytically deposited color	**A34—Electrolytically deposited color**	**A44—Electrolytically deposited color**
A1x—Other	A2x—Other	A3x—Other	A3x—Other

Reprinted from Metal Finishes Manual for Architectural and Metal Products *with permisson of the National Association of Architectural Metal Manufacturers (NAAMM), Chicago.*

[a]Finishes printed in **boldface type** are those most frequently used for architectural work.

[b]Third digit (1, 2, or 3) added to designate min. thickness in 1/10 mils.

sponding weights of from 17 to 27 mg/sq. in. These are appropriate for interior items not subject to excessive wear or abrasion, and for exterior items such as storefronts and entrances that are regularly cleaned and maintained.

PAINTING

Field-applied paints are discussed in Section 9.12. The discussion here relates to coatings applied in the shop or factory.

A substantial volume of aluminum is painted for both decorative and protective purposes. The most important applications are siding, cladding panels, and doors.

The trend in painting aluminum is toward continuous processing that integrates pretreatment, coating, and curing into a single operation. *Roller coating* enables the painting of sheet at line speeds of up to 600 ft./min.

First the bare coiled sheet is passed through a series of cleaning and surface-preparation baths to provide a surface to which paint will strongly adhere; then it is roller coated with an organic paint; finally, it is placed in an oven where the paint is baked hard. The resulting finish has an excellent combination of hardness, ductility, adhesion, and gloss, and can undergo further fabrication without harm.

Castings and extrusions are commonly sprayed. Electrostatic spraying, in which electrically charged coating particles are attracted to the surface by maintaining it at the opposite charge, provides a more uniform finish, with less waste, than conventional spraying.

VITREOUS COATINGS

Porcelain enamel is a vitreous inorganic coating bonded to metal by fusion at high temperatures. A porcelain enamel coating significantly improves the existing properties of aluminum, increasing durability and corrosion resistance. Practically any texture and lightfast color are available.

ELECTROPLATING

Chromium and copper plating are the most common and easiest to apply electroplated finishes for aluminum. Chromium is readily plated on aluminum either after zincating, or after copper, brass, or nickel plating. Lustrous chrome plate is easy to obtain by buffing. Chromium plates are generally applied for decorative effect or resistance to abrasion.

Aluminum may be copper plated after a zincate treatment or by other methods, including anodizing in phosphoric acid. Copper may also be electroplated after first cleaning and zincating the aluminum. Copper plates on aluminum are used where soldering operations are required, for certain electrical requirements, for a base for chromium or silver plating, and for decorative items.

Silver plating decreases electrical resistance and is used for some electrical connections. Tin and brass plates are especially good as a base for soldering. Zinc and cadmium plates provide corrosion protection. Chemically treated nickel plates yield a black decorative finish.

ELECTROPOLISHING

High-purity aluminum may be treated by electropolishing to obtain a light reflectivity of about 85%. The metal surface is anodically smoothed to a highly polished surface suitable for use in aluminum light-fixture reflectors.

Electropolished reflecting surfaces are usually given a final anodizing treatment for surface protection. Final reflectivity depends on the purity of the aluminum, the thickness of the oxide film, and the particular procedure employed.

LAMINATED COATINGS

For special use and design applications, many other materials may be laminated to aluminum surfaces. For example, polyvinyl chloride and vinyl films are used to provide nonmetallic color and long-lasting finish to exterior applications such as aluminum siding. The film is adhesive bonded to the aluminum surfaces that have been suitably prepared, usually by the application of a chemical conversion coating.

TEMPORARY PROTECTIVE COATINGS

Waxes and lacquers applied to aluminum surfaces provide some weather protection and make cleaning easier. Aluminum so treated can usually be kept clean by wiping it with a damp cloth or rinsing it with clear water.

LACQUER Clear lacquer, applied by a fabricator or contractor, provides the most common type of temporary surface protection for aluminum. Methacrylate lacquers, which are commonly used, are resistant to alkalies such as lime mortar, plaster, and concrete, and provide some protection against abrasion. Under normal conditions, a methacrylate coating lasts for several years and weathers by chalking rather than by unsightly peeling. Methacrylate lacquer should not be used on color-anodized aluminum, however, because lacquer never fully adheres to an anodized coating. Where adhesion is incomplete, the lacquer will appear white. On clear anodizing, this effect is largely unnoticeable. On a color-anodized surface, however, the white spots produce a blotchy, unappealing effect.

WAX Liquid waxes or wax base cleaners offer some protection against atmospheric weathering and retard the accumulation of grime on aluminum surfaces. However, after repeated waxings, surfaces may tend to yellow. If necessary, the wax may be removed with a solvent or mild abrasive cleaner before rewaxing.

STRIPPABLE COATINGS Strippable coatings are rather expensive coatings that are usually applied by the fabricator to protect surfaces from the normal hazards encountered during shipping and construction. They are ideal for finer finishes that may be scratched or damaged more easily. Strippable coatings should be left in place until final cleanup. Aluminum with such coatings should not be stored in places exposed to sunlight, since the coatings may tend to become tacky and difficult to remove. If they cannot be easily removed, a solvent should be used.

TAPES For protection against abrasion or staining on smaller areas, various types of tape may be used. They may be applied either by the fabricator or by the contractor.

■ MAINTENANCE OF SURFACES

Cleaners that may be used on aluminum vary in strength and abrasiveness. The type of cleaner used should be determined by the aluminum finish, the degree of soiling, and the size of the area to be cleaned.

In general, aluminum should be washed frequently with mild cleaners rather than occasionally with abrasive, heavy-duty cleaners. Products should never be mixed. Abrasives may alter the appearance of the aluminum finish. To avoid streaking, cleaning work should be done in the shade or on cloudy days. Applying cleaners on sun-warmed metal or on cold days should be avoided. All cleaners should be spot tested before general use. The following cleaners are listed in increasing order of harshness.

Mild cleaners should be used on aluminum that is lightly soiled or subject to frequent cleaning. These include most soaps and detergents that can be applied without gloves. Even mild cleaners should not be permitted to remain on the metal for long. They should be sponged on and removed with a clean water rinse.

Solvent and emulsion cleaners are effective on oil and grease stains and some types of soil. Like mild cleaners, solvents have little effect on weathered aluminum and heavy grime encrustation. Some solvents, however, can soften or dissolve certain paints and lacquers. Cleaners should be applied with one clean cloth and removed with another. Remaining residue should be removed with mild soap and water.

Abrasive cleaners may contain water, oil, wax, soap, silicones, acids, or alkalies, in addition to the abrasive itself. They are packaged under such names as polish, cleaner, waxpolish, metal brightener, and scouring powder. In general abrasive cleaners are sprinkled on a clean, damp pad and lightly rubbed onto the metal. A separate cloth is used to remove the cleaner. In some cleaners a wax or silicone remains, which produces a shine and protects the metal to a certain extent.

Steel wool (size 00 or finer) or abrasive nylon pads may be used alone or with mild or abrasive cleaners. Steel wool particles must be removed since these may rust and stain.

Etching cleaners are strong chemicals that may change the appearance of the finish and permanently mar the aluminum surface unless thoroughly rinsed off. The cleaner concentration, application time, and safety precautions suggested by the manufacturer should be followed carefully.

Cleaning should always start at the top and work down. The solution should be evenly rubbed on one small area at a time with a sponge or brush. The effectiveness of etching cleaners can be increased by using fine steel wool or nylon pads or by adding a fine abrasive. Care should be taken to prevent drips and spills and to wipe up immediately any that occur. A cleaned area, especially one with any cracks, grooves, and holes, should be thoroughly rinsed and dried as soon as the work is complete.

Special cleaners include steam cleaners and power-driven wire brushes. Steam, used alone or with a cleaning compound, is useful for removing grease or oil-based deposits and for cleaning large areas. The area to be cleaned should be spot tested.

Stainless steel or German silver wire brushes are used in cleaning large, heavily weathered areas. Since power-driven wire brushes remove metal, they must be used carefully.

5.2.3.5

Joints and Connections

Practically every method of joining and connecting is used in assembling aluminum products, including mechanical fastenings; various methods of mechanically forming connections; welding, brazing, and soldering; and bonding with adhesives.

■ MECHANICAL FASTENINGS

A great variety of permanent and semipermanent fasteners are used to join aluminum to itself and to other materials. Permanent fastening requires the destruction of the fastener or the product itself for disassembly. Semipermanent fastening includes methods such as screwing and bolting that permit nondestructive disassembly.

RIVETING Riveting is a method of permanent mechanical fastening in which a headed pin (rivet) inserted through the pieces being joined is mechanically deformed so that a head is created on both sides of the work, tightly gripping the pieces together. Riveting is probably the most widely used aluminum fastening method (particularly in the aircraft industry) and makes use of a great variety of rivet shapes, sizes, and assembling techniques.

Aluminum rivets are typically used in joining aluminum for their neat appearance and uniform finish, avoidance of galvanic corrosion, low cost, and light weight.

Aluminum rivets are available in all standard designs and sizes with the usual round, brazier, flat, and countersunk heads.

Rivets can be made in practically any alloy. The following alloys meet most needs: 1100, 2024, 2117, 5056, and 6053. Alloys 1100 (commercially pure aluminum) and 5056 are non-heat treatable. The others are heat treatable. In general, rivets to be used with parts of a given alloy should match that alloy in mechanical and physical properties.

Although standard rivets meet requirements, many special types are available, including tubular and other designs for blind riveting (where the work is accessible from one side only).

STITCHING Joining various metals to other metals and other materials by wire sewing is called *stitching*. Aluminum's relatively low density and softness make it very suitable for stitching.

Stitching requires no backing material, prepunching, or drilling. The work need not be clamped nor the metal cleaned. Inspection is simple since it can be determined visually whether a stitch has penetrated the work and is properly clinched. Stitching is used in assembling aluminum garage doors, on sheet metal jobs such as ductwork and blower installations, and in joining aluminum to rubber, fabrics, and other materials.

THREADED FASTENERS Threaded fasteners are semipermanent fasteners that are widely used in assembling aluminum products. They include a multitude of screw and bolt types. Whatever type is used, however, it should be aluminum or another metal compatible with aluminum, such as stainless steel.

Screws for use with aluminum are usually made from alloy 2024-T4 and are manufactured in practically any head type, size, and standard thread (Fig. 5.2-16).

Bolts, nuts, and washers are usually made from alloy 2024-T4, 6061-T6, or 7075-T73. They are manufactured in practically any head style (typically hexagonal), size, and standard machine thread.

Because of their freedom from corrosion and staining, aluminum nails are popular fasteners for roofing and siding made of aluminum or of nonmetallic materials. These nails are manufactured from alloys 6061 and 5056 that, in this form, have tensile strengths in excess of 60,000 psi.

FIGURE 5.2-16 *Extruded aluminum window and door frame sections are often connected using screw fasteners. A common method is to thread the screws into continuous slots.*

SPECIAL FASTENERS Many other special fasteners and rivets are available for blind fastening and other connections. Blind fasteners include pop rivets, explosive rivets, and others where only one side of the work is accessible. Other special fasteners such as speed nuts and quick release fasteners are also commonly used.

■ MECHANICALLY FORMED JOINTS

Connections obtained by forming or cutting the parts to be joined in such a manner that they interlock are called mechanically formed joints. Those used for aluminum work do not basically differ from such joints in other metals. Where high strength is of primary importance, other joining methods will usually be preferred. But the

FIGURE 5.2-17 *Mechanically formed slip joints are used to assemble aluminum gutters and downspouts.*

building industry offers many applications, such as roof work, ducts, gutters, downspouts, and railings where mechanically formed joints are quite satisfactory (Fig. 5.2-17).

Other mechanically formed joints are made by *staking*. A staked joint is commonly used in assembling extruded aluminum members. Tabs or prongs formed at the end of one member are inserted through mating holes in the member to be joined and are twisted, bent over, or spun down (Fig. 5.2-18).

■ WELDING, BRAZING, AND SOLDERING

Welding, brazing, and soldering are methods of forming a metallurgical bond between two or more members of an assembly. The selection of a specific process for a particular job depends on many variables including conditions encountered in service, shape and thickness of parts, volume of production, and the alloys employed.

The choice of filler metal alloy for welding depends on joint design, service conditions, surface finish, welding process, the alloy or alloy combinations to be joined, and the desired appearance. Therefore, filler alloys vary considerably and must be selected for each particular circumstance.

There are many welding processes suitable for use with aluminum alloys. Some of the more important processes are discussed here.

GAS WELDING The earliest welding process used on aluminum was gas welding. In gas welding, oxygen, acetylene, oxyhydrogen, or another gas flame is the heat

FIGURE 5.2-18 *Staked joints are often used in assembling aluminum frame members of windows and doors.*

source. The gas flame melts the parent metal and a flux-coated filler rod is used to complete the weld.

The disadvantages of gas welding include the flux residue, which is a potential source of corrosion or weld defects in multipass welds, and which must be removed; distortion, which is greater than in arc welding; a wide heat-affected zone; and slow welding speeds. These conditions generally result in higher costs, when compared with gas-shielded arc methods.

Aluminum from about 1/32 in. to 1 in. thick may be gas welded. Heavier material is seldom gas welded because heat dissipation is so rapid that it is difficult to apply sufficient heat to melt the metal with a torch.

Gas welding finds its greatest use today where few or occasional welds are required. It is also the obvious choice in the field, when electric power lines or portable generators for arc welding are not accessible.

Most forms of arc welding depend on the production of heat from high-voltage electric flow (arc) from an electrode to the parent metal.

Of all the fusion welding processes, the inert-gas-shielded arc processes are by far the most important and most widely used. In the tungsten inert-gas (TIG) arc process, an arc is drawn between the work and a nonconsumable tungsten electrode, the arc being shielded by a flow of inert argon or helium. Filler metal is added manually. In metal inert-gas (MIG) arc welding, the arc in inert gas is drawn between an automatically fed filler wire and the work. Fluxes, used to dissolve the aluminum oxide surface film on the metal in other welding processes, are not required for these because of the action of the arc and the inert gas shield.

These processes are applicable to all welding positions. Thinner gauges are commonly welded by TIG. MIG is particularly appropriate for automatic welding; TIG is also appropriate there when filler metal is not required. Since welding speeds are high, the effect of heat input on strength is minimized.

Resistance welding uses an electric current conducted through the parts to be joined. Heat is generated at the juncture of the two parts (a result of the resistance to the passage of heavy current), and when combined with applied pressure, fusion is produced. Resistance welding has practically no effect on the alloy temper since the weld is made so quickly that the adjoining area does not heat appreciably.

Flash welding is a type of resistance welding used to butt weld shapes such as tubes, extruded shapes, rods, bars, and wire. Typical is the mitering of aluminum window sections. In general, resistance welding is used where technically practical and where mass production is involved.

Spot and *seam welding* are resistance welding methods especially important in fabricating high-strength, heat-treated sheet alloys that can be so joined with practically no loss in strength. Spot welding is widely used to replace riveting, joining sheets at intervals as required. *Seam welding* is merely spot welding with the spots spaced so closely that they overlap, producing an airtight joint where desired.

When the welding process uses a filler metal that is nonferrous, or an alloy with a melting point higher than 800 °F, but lower than that of the materials joined, it is called brazing.

In brazing the molten metal is called *brazing filler metal*; in soldering it is called *solder*.

Brazing is employed to join millions of aluminum parts each year, ranging in thickness from foil to heavy plate and castings. It is used for joints that because of their complexity of contour, limited access, or great ratio of length to part size, are unsuitable for welding.

Filler alloys for brazing aluminum include such aluminum-silicon alloys as 4047 and 4145. They have liquidus temperatures much closer to the solidus temperature of the base metal than filler alloys for brazing most other metals. Consequently, temperature control is critical, and only those aluminum alloys with relatively high melting points are brazable. These include alloys 1050, 1100, 3003, 6063, and the self-aging foundry alloy Alcan 432.

Although alloy 6061 can be brazed, the melting range of this alloy is very close to that of the alloy 4047, the commonest brazing filler alloy. Alloys with high magnesium content are not considered able to be brazed because of poor wetting by the filler alloys. Flux is essential to remove the oxide film.

Aluminum may be brazed by several methods, differing mainly in the way that heat is applied to the work. These include torch brazing, furnace brazing, and salt bath brazing. *Torch brazing* is employed where the joints are accessible both for heating with the torch flame and for feeding filler metal into the joint.

In *furnace brazing* the entire work assembly is held in a furnace for 1 to 5 minutes. Because of the capillary action of a brazing alloy, the furnace brazing process is particularly useful for inaccessible joints.

In the *salt bath process* (also known as dip brazing) the work is submerged in molten salt, which acts as the heating medium and as the flux. Time in the salt bath varies from 30 to 180 seconds.

Whatever the process, it is essential that surfaces be thoroughly cleaned before they are brazed. For 1100 and 3003 alloys, degreasing is usually sufficient; for other alloys, an etching treatment is recommended.

Under some conditions, brazed joints develop strengths comparable to welded joints, make more economical use of filler metal, can be made faster, and have a neater appearance (requiring little finishing work, if any). Brazing makes possible the joining of extremely thin sections. In general, the corrosion resistance of brazed joints is comparable to that of welded joints.

SOLDERING   Filler metals (*solders*) with melting points under 800°F are used in soldering.

Difficulties in soldering aluminum are caused by the great affinity of aluminum for oxygen and the lack of bond between the solder and the natural oxide film present on aluminum. Ordinarily a suitable flux is applied to dissolve the oxide and prevent it from reforming, or otherwise react, so that the solder may bond directly to the underlying aluminum. Another method to get below the oxide film is to abrade it mechanically down through an overlying layer of molten solder, using a soldering iron, a wire brush, or another tool.

When soldering aluminum to other metals, the aluminum surface should be tinned (precoated with solder) and then joined to the other metal with solder and flux. If any difficulty is encountered, the other metal surface should also be tinned first.

The essentials of soldering aluminum are to remove the surface oxide and to use a filler metal that will not encourage galvanic corrosion. Therefore, appropriate fluxes, solders with higher melting points, pure zinc, and zinc-aluminum are used where corrosion is a factor. Where corrosion resistance is less important, zinc-tin and lead-tin alloy solders with lower melting points are used.

In corrosive atmospheres, lacquer coatings may be used to improve the corrosion resistance of soldered aluminum.

■ ADHESIVE BONDING

Adhesive bonding utilizes adhesives for bonding aluminum to aluminum or other materials. Design possibilities are limited only by extremes of stress, temperature, corrosive environment, and cost. One special consideration in designing joints for the use of adhesives is that peeling stresses should be kept as low as possible. The pretreatments of aluminum used in adhesive bonding are concerned mainly with cleaning the surface. Where the adhesive bond must develop the greatest strengths, a chromic-sulfuric acid dip is often used as a chemical cleaning method.

Adhesives themselves may be applied by any of several techniques. Roller coating is particularly effective for the high-speed application of a thin coat of adhesive, as in the production of laminates. Currently, various building materials are adhesively laminated with aluminum foil to increase their thermal insulation properties.

For more limited areas, adhesives may be applied by brush, blade, or spray. Certain adhesives are adapted to mass production by manufacturing them in film form. The film is cut into sheets, ribbons, or special shapes that follow the contours of specific joint areas and are assembled with the metal parts before curing.

Curing processes for adhesives vary considerably. The length of curing time ranges from 3 seconds to over 76 hours, depending on the type of adhesive used and whether heat and pressure are applied to the assembled joint. Either heat, pressure, or both are used to obtain the highest bond strengths.

There is no universal adhesive. The nature of the material joined and the ultimate conditions of service determine the type of adhesive to be used. Therefore, there are many proprietary adhesives developed for varying bonding processes. Natural and synthetic rubbers, polyvinyl acetates, phenolic blends, and epoxies are among the commonly used adhesive bases.

Bonding aluminum is becoming more common in building products manufacture where plastics, cork, rubber, wood veneer, and other materials are used with aluminum. The processes offer great opportunities for aluminum fabrication.

5.3 OTHER METALS

Other metals besides iron and aluminum are used in buildings. The chief among these are copper and its alloys, brass and bronze. Other metals include zinc, lead, chromium, nickel, tin, and gold.

5.3.1

COPPER AND ITS ALLOYS

Copper is probably the world's second most widely used metal, following the various alloys of iron, including steel. Copper occurs in nature in its metallic state, which probably explains its use as early as 5000 B.C. Today, it is mostly obtained by refining the ores cuprite, which is copper oxide; malachite, which is basic copper carbonate; or chalcocite or chalcopyrite, both of which are copper sulfide. The process is a complicated one, requiring roasting, smelting, and converting the ore, and then refining the resulting metal by electrolysis.

Newly refined copper is about 99.95% pure. It has a reddish yellow color, and is soft, malleable, ductile, relatively strong, and a very good conductor of electricity. Copper does not corrode in dry air, but will oxidize when heated. In normal atmospheric conditions, it oxidizes slowly, forming a thin surface coating of verdigris, which is similar to the ore malachite.

Copper has many uses, including making copper sulfate for printing, electroplating, and purifying public water supplies. It is also used in tanning leather, in manufacturing pigments, insecticides, and dyes for cotton, and even as an astringent. Copper is also used extensively in minting coins and making cooking utensils, art objects, wire, and various sheet metal items, such as gutters, downspouts, and roofing. One of the primary uses for copper is to make various alloys, including the two most interesting to those in the building industry, brass and bronze.

5.3.1.1

The Copper Alloys

There are more than 1100 recognized alloys of copper. Of these, more than half are alloys of copper and zinc or copper and tin. The first has traditionally been known as *brass*; the latter as *bronze*. Today, most copper alloys used for other than utilitarian purposes in building and related construction, regardless of whether containing zinc or tin, are called bronze.

Bronze was almost certainly the first industrial alloy. In some places in the world, copper and tin ore are found naturally intermixed, which probably explains the appearance of bronze as early as 3500 B.C. Brass followed much later, but even it has been in use since about 50 B.C.

■ ALLOY DESIGNATIONS

The Copper Development Association (CDA) has a numbering system for copper alloys. CDA is also the administrator of the Unified Numbering System (UNS), which was developed jointly by the ASTM and the SAE. The CDA system consists of three digits. The UNS adds a letter *C* prefix to the three digits and a 00 suffix. The alloy 280 (*Muntz metal*) in the CDA system is C28000 in the UNS.

■ COMMON ALLOYS

The CDA places the most common copper alloys used in the construction industry into three groups. The A group contains copper, including alloy 110, which is 99.9% copper, and alloy 122, which is 0.02% phosphorous.

The B group contains the common brasses and extruded *architectural bronze*. It includes alloy 220 (*commercial brass*), which is 10% zinc; alloy 230 (*red brass*),

which is 15% zinc; alloy 260 (*cartridge brass*), which is 30% zinc; alloy 280 (*Muntz metal*), which is 40% zinc; and alloy 385 (*architectural bronze*), which is 40% zinc and 3% lead.

The C group contains the so-called *white bronzes*. It includes alloy 651 (*low silicon bronze*), which has between 1.5% and 3% silicon; alloy 655 (*high silicon bronze*), which has 3% silicon; alloy 745 (*nickel silver*), which has 25% zinc and 10% nickel; and alloy 796 (*leaded nickel silver*), which has 42% zinc, 10% nickel, 2% manganese, and 1% lead.

Casting alloys are special alloys made specifically for casting. They fall in the 800 and 900 Series of CDA alloy designations.

Other alloys are also used, of course, such as when color matching is required. The CDA booklet *Copper Brass Bronze Design Handbook, Architectural Applications* includes a chart showing which alloys match others in color. When a nonstructural metal item contains several different shapes or types of components, such as sheets, pipes, extrusions, fasteners, castings, and so on, the different parts are most likely made from different alloys, even though the colors match.

5.3.1.2

Copper Alloy Products Production

Many copper alloy products are used in producing nonstructural metal items for use in buildings and related construction. Figure 5.3-1 is a partial list of such products, showing one possible

series of color-matched alloys that might be used.

■ FORMING METHODS

The production of final copper alloy items for use in buildings starts with the forming or fabricating of copper alloy materials into useful products, such as those listed in Figure 5.3-1. There are four basic methods generally used to form these products: casting, rolling, extruding, and drawing.

CASTING Intricate shapes are often made using the casting method. Castings may later be drilled, milled, or otherwise machined to produce necessary openings and close fits with adjacent materials.

ROLLING Many copper alloy items used in buildings and related construction are at some point in their development run through a hot-rolling mill. Many of them are then passed through a cold-rolling process to make their surfaces brighter and smoother. Items produced by the rolling process include plates, sheets, shapes, rods, bars, and forging stock.

EXTRUDING Many copper alloy products used in buildings and related construction are produced by extrusion. Extruded items include structural shapes, rods, bars, seamless tubes and pipes, and the kinds of special shapes often used in nonstructural metal items.

Extrusions are superior in virtually every way to rolled products. They are

---

FIGURE 5.3-1
Copper Alloy Products That Are Compatible in Color[a]

Product	ASTM Standard	CDA Alloy Number and Name
Extrusions	B 455	385, architectural bronze
Sheets, plates, and bars	B 370	280, Muntz metal
Round tubing	B 135	230, red brass
Castings	B 271	857
Sand castings	B 584	857
Wire and rods	—[b]	280, Muntz metal

[a]The alloys listed here represent only a single group of materials that generally match in color and are by no means the only alloys used for the products listed. Refer to the CDA publication *Copper Brass Bronze Design Handbook, Architectural Applications* for other alloys used for the purposes listed.

[b]No current universally recognized standard for alloy 280 wire rods.

straighter, more accurate in dimension, and smoother. They are, however, limited to a diagonal cross section of 6 in. and a thickness of about 1/8 in.

DRAWING Tubes and wire can be made by drawing a copper alloy through a die, rather than pushing it, as in the extrusion process.

■ FABRICATION

Some rolled, extruded, and drawn products are used exactly as they come from the mill to fabricate items used in buildings. Others require only additional finishing. Many other components, however, are made by fabricating these products into usable shapes. The methods employed include additional roll forming, forging, bending, break forming, explosive forming, and spinning. Not all methods are applicable to all alloys. Refer to the CDA publication *Copper Brass Bronze Design Handbook, Architectural Applications* for an indication of the alloys that can be formed by each method.

ROLL FORMING The rolled products mentioned earlier are passed through an additional roll-forming process to make the final product. Roll forming at this stage consists of passing sheets through a series of shaped rollers to produce the desired end shapes.

FORGING Forging is done using one of four methods. The first two methods require closed steel dies. In the first method, *hammering*, the copper alloy is hammered into the die cavity by repeated blows of a *drop hammer.*

The second method, *pressing*, involves forcing the metal into the die cavity under hydraulic pressure.

In the third method, *hydroforming*, a punch or formed steel die is forced under hydraulic pressure against the metal, which is backed up by rubber.

In the fourth method, *stamping*, the metal is bent, shaped, cut, indented, embossed, coined, or formed by means of a press or hammer that is faced with a shaped die.

BENDING Many copper alloy sheet metal and plate components used in buildings, such as door frames and the like, and many roofing-related copper items are shaped by bending.

BRAKE FORMING Brake forming is probably the most widely used method for shaping copper alloy sheets used in buildings. Many copper alloy components used in nonstructural metal items, shapes such as door frames and the like, and many roofing-related copper alloy items are shaped by brake forming.

In essence, brake forming consists of squeezing sheet metal between a punch and a die block. Several consecutive operations are usually required to make a final shape.

EXPLOSIVE FORMING High-energy-rate forming forces metal against a single die using a shock wave generated by the detonation of small explosive charges.

SPINNING A type of shear forming, *spinning*, is executed by clamping the sheet metal to a form block and spinning the block and the metal against a roller, which gradually forces the metal to fit the shape of the form block. Either power or hand tools can be used in spin forming..

5.3.1.3

Finishes

In its publication *Metal Finishes Manual for Architectural and Metal Products*, the National Association of Architectural Metal Manufacturers (NAAMM) specifies the requirements for three types of basic finishes for copper alloys. These are *mechanical finishes, chemical finishes*, and *coatings* (Fig. 5.3-2).

■ MECHANICAL FINISHES

Mechanical finishes include those left by the manufacturing process and those created by grinding, polishing, sandblasting, and rolling. The NAAMM designates the types of mechanical finishes for use on copper alloys as: *as fabricated, buffed, directional textured, nondirectional textured*, and *patterned*. Within these types are many finishes (Fig. 5.3-3). Each of these finishes is given a designation that consists of the letter *M* (mechanical) followed by a two-digit number. In the first four types, the first digit corresponds to the four types of finish. The second digit designates the specific finish within that type. Thus, M21 is a smooth specular buffed finish, and M31 is a fine satin, directional textured finish.

The fifth type, *patterned*, is designated M4x with the x to be specified. The x usually denotes a proprietary finish.

AS-FABRICATED FINISH An *as-fabricated finish* is simply the natural finish imparted by the casting, rolling, or extruding process. It is the finish that the copper alloy has when it leaves the initial manufacturing stage. It is not the final finish after the material has been fabricated into a metal item. The term *as-fabricated finish* corresponds to the term *mill finish* used for iron and steel.

Most copper alloys with an as-fabricated finish will display some imperfections, which must be taken into account when deciding whether to use them without further finishing. Castings, for example, have a rough, matte finish. Sand castings are rougher than die castings. Hot-rolled copper alloy items will have a dull surface, may be darker than cast, cold-rolled, or extruded copper alloy items, and will usually have some random discolorations as well. Extrusions are usually the same color as cold-rolled material, but are often left with parallel striations (*die lines*). Most imperfections will tend to worsen when the metal is further formed and assembled into metal items.

Cold-rolled copper alloy items usually come closer to the final desired appearance than those produced by the other methods. The process permits some control by varying the amount of polishing used on the rollers. The smoother the rollers are, the closer the cold-rolled surface will be to a desirable finish. Even cold-rolled copper alloy items, however, may have stains, or a coat of oil, or both.

The NAAMM divides as-fabricated finishes on copper alloys into four classes (see Fig. 5.3-3). An *unspecified* as-fabricated finish is the natural finish imparted by casting, hot rolling, cold rolling with unpolished rollers, or extruding.

A *specular* as-fabricated finish is a mirrorlike finish that is produced by cold rolling with polished rollers. The specular finish may be applied to one or both sides. Castings, forgings, and extrusions cannot be given a specular finish.

FIGURE 5.3-2
Summary of Standard Designations for Copper Alloy Finishes[a]

Mechanical Finishes (M)

As Fabricated	Buffed	Directional Textured	Nondirectional Textured
M10—Unspecified	M20—Unspecified	M30—Unspecified	M40—Unspecified
M11—Specular as fabricated	M21—Smooth specular	M31—Fine satin	M41—(Unassigned)
M12—Matte finish as fabricated	M22—Specular	**M32—Medium satin**	**M42—Fine matte**
M1x—Other (to be specified)	M2x—Other (to be specified)	**M33—Coarse satin**	M43—Medium matte
		M34—Hand rubbed	M44—Coarse matte
		M35—Brushed	M45—Fine shot blast
		M36—Uniform	M46—Medium shot blast
		M3x—Other (to be specified)	M47—Coarse shot blast
			M4x—Other (to be specified)

Chemical Finishes (C)

Nonetched Cleaned	Conversion Coatings	
C10—Unspecified	**C50—Ammonium chloride**	(patina)
C11—Degreased	**C51—Cuprous chloride—hydrochloric acid**	(patina)
C12—Cleaned	**C52—Ammonium sulfate**	(patina)
C1x—Other (to be specified)	C53—Carbonate	(patina)
	C54—Oxide	(statuary)
	C55—Sulfide	(statuary)
	C56—Selenide	(statuary)
	C5x—Other (to be specified)	

Coatings, Clear Organic (O)

Air Dry (General Architectural Work)	Thermoset (Hardware)	Chemical Cure
060—Unspecified	070—Unspecified	080—Unspecified
06x—Other (to be specified)	07x—Other (to be specified)	08x—Other (to be specified)

Coatings, Laminated (L)

	Thermoset (Hardware)	
	L90—Unspecified	
	L91—Clear polyvinyl fluoride	
	L9x—Other (to be specified)	

Coatings, Vitreous and Metallic

Since the use of these finishes in architectural work is rather infrequent, it is recommended that they be specified in full, rather than being identified by number.

Coatings, Oils and Waxes

These applied coatings are primarily used for maintenance purposes on site. Because of the broad range of materials in common use, it is recommended that, where desired, such coatings be specified in full.

Reprinted from Metal Finishes Manual for Architectural and Metal Products *with permisson of the National Association of Architectural Metal Manufacturers (NAAMM), Chicago.*

[a]Those printed in **boldface type** are most frequently used for architectural work.

FIGURE 5.3-3
Standard Designations for Mechanical Finishes

Type of Finish	Designation	Description	Examples of Method of Finishing
As fabricated	M10	Unspecified	Optional with finisher
	M11	Specular as fabricated	Cold rolling with polished steel rolls
	M12	Matte finish as fabricated	Cold rolling followed by annealing; hot rolling, extruding, casting
	M1x	Other	To be specified
Buffer	M20	Unspecified	Optional with finisher
	M21	Smooth specular	Cutting with aluminum oxide or silicon carbide compounds, starting with relatively coarse grits and finishing with 320 grit, using a peripheral wheel speed of 6000 fpm; followed by buffing with aluminum oxide buffing compounds using a peripheral wheel speed of 7000 fpm
	M22	Specular	Cutting with compounds as for the M21 finish, followed by a final light buffing
	M2x	Other	To be specified
Directional textured	M30	Unspecified	Optional with finisher
	M31	Fine satin	Wheel or belt polishing with aluminum oxide or silicon carbide abrasives of 180–240 grit, using a peripheral speed of 6000 fpm
	M32	Medium satin	Wheel or belt polishing with aluminum oxide or silicon carbide abrasives of 120–180 grit, using a peripheral wheel speed of 6000 fpm
	M33	Coarse satin	Wheel or belt polishing with aluminum oxide or silicon carbide abrasives of 80–120 grit, using a peripheral wheel speed of 6000 fpm
	M34	Hand rubbed	Hand rubbing with #0 stainless steel wool and solvent, #0 pumice and solvent, nonwoven abrasive mesh pad or Turkish oil and emery
	M35	Brushed	Brushing with rotary stainless steel, brass or nickel silver wire wheel; coarseness of finish controlled by diameter and speed of wheel and pressure exerted
			Wheel polishing with aluminum oxide or silicon carbide abrasive compounds of 50–150 grit on packed or loose muslin buffs at peripheral speeds of from 3500–5600 fpm
	M36	Uniform	Wheel or belt polishing in a single pass with aluminum oxide or silicon carbide abrasives of 60–80 grit, using a peripheral speed of 6000 fpm
	M3x	Other	To be specified
Nondirectional textured	M40	Unspecified	Optional with finisher
	M41	(number unassigned)	
	M42	Fine matte	Air blast with #100–#200 mesh silica sand or aluminum oxide; air pressure 30–90 lb.; gun 12 in. from work at an angle of 60–90 degrees
	M43	Medium matte	Air blast with 40#–80 mesh silica sand or aluminum oxide; air pressure 30–90 lb.; gun 12 in. from work at an angle of 60–90 degrees
	M44	Coarse matte	Air blast with #20 mesh silica sand or aluminum oxide; air pressure 30–90 lb.; gun 12 in. from work at an angle or 60–90 degrees
	M45	Fine shot blast	Air blast with S-70 metal shot
	M46	Medium shot blast	Air blast with S-230 metal shot
	M47	Coarse shot blast	Air blast with S-550 metal shot
	M4x	Other	To be specified

Reprinted from Metal Finishes Manual for Architectural and Metal Products *with permisson of the National Association of Architectural Metal Manufacturers (NAAMM), Chicago.*

A *matte* as-fabricated finish is a dull finish produced by annealing copper alloy items that have been produced by hot rolling, cold rolling with unpolished rolls, casting, or extruding.

The *other* category is reserved for special finishes that do not fit into one of the above three categories.

BUFFED FINISHES Copper alloy finishes created by a process of grinding, polishing, and buffing are called buffed finishes. There are four standard buffed finishes for copper alloys (see Fig. 5.3-3). An *unspecified* buffed finish is optional with the copper alloy finisher.

A *smooth specular* buffed finish, which is the brightest and most lustrous mechanical finish that can be given to copper alloy items, is produced by successively grinding, polishing, and buffing.

A *specular* buffed finish is produced by the same methods as a smooth specular buffed finish, but is not as smooth.

The *other* category is reserved for

special finishes that do not fit into one of the above three categories.

DIRECTIONAL TEXTURED FINISHES The standard designation system recognizes eight finish types in the directional textured finish category (see Fig. 5.3-3). An *unspecified* directional textured finish is optional with the copper alloy finisher.

The three *satin* finishes represent different degrees of fineness. They are all produced by wheel or belt polishing. *Spin finishing*, in which an abrasive cloth is held against the metal and rotated, followed by stainless steel wool or emery polishing, produces a bright finish with a fine pattern of concentric circles.

A *hand-rubbed* finish is an expensive finish type that is produced by rubbing the metal with a fine brass brush or a non-woven abrasive mesh pad, abrasive cloths, or stainless steel wool. It is seldom used for general finishing, but is often used for final touch-up on the other satin finishes.

A *brushed* finish is a directional finish characterized by fine parallel scratches that are produced by stainless steel, nickel steel, or brass wire brushes, sander heads, impregnated plastic discs, or other means.

A *uniform* finish is produced by a single pass of a No. 80 grit belt.

The *other* category is reserved for special finishes that do not fit into one of the other seven categories.

NONDIRECTIONAL TEXTURED FINISHES Finishes that are designated as nondirectional textured finishes are produced by abrasive blasting (see Fig. 5.3-3). Most of these finishes are *matte*, with varying degrees of texture roughness. They are not widely used on metal items for building construction. Their most extensive use is on castings. They should not be used on thin metal (1/4 in. or less) because they tend to distort the metal too much.

An *unspecified* nondirectional textured finish is an optional choice for the finisher.

The categories that have *matte* in their names are produced by blasting with silica sand or aluminum oxide. Those with *shot blast* in their names are produced by steel shot blasting.

The *other* category is reserved for special finishes that do not fit into one of the other eight categories.

Combinations are sometimes used on surfaces that have relief; the incised areas are blasted with abrasives while the raised areas are masked. Then the raised areas are polished.

PATTERNED FINISHES Thin copper alloy sheets can be given a patterned finish by rolling as-fabricated sheet between rollers that have been shaped to the desired design. There is no specific standard category for patterned finishes.

■ CHEMICAL FINISHES

Chemicals are used on copper alloys for two primary purposes: to clean the copper alloys of foreign matter, and to change the metal's color and provide a final finish. There are some other chemical treatments used on copper alloys, but they are not extensively used on building construction items. The two chemical finishes we are interested in are called *nonetch cleaning* and *conversion coatings*.

Chemical finishes for use on copper alloys are designated in a way similar to that for mechanical finishes, except that the letter used is a *C* (Fig 5.3-4). The first digit corresponds to the types, the second digit to the specific finish. Thus C11 is a degreased cleaning, and C51 is a cuprous chloride-hydrochloride acid patina.

Conversion coatings are used to change the color of a copper alloy and to provide a final finish for the metal. As a copper alloy ages, it oxidizes, changing its color and producing a coating that not only changes the appearance of the material, but protects it from further oxidization. Conversion coatings are not always completely successful in duplicating, by accelerated chemical means, the natural appearance and protective coating that result from aging. The coatings produced are oxides or sulfides of the metal.

There are two basic types of conversion coatings commonly used: those that produce a *patina* (verd antique) finish and those that produce an oxidized finish known as *statuary bronze*. Figure 5.3-4 lists the seven types of materials used to produce the more common conversion coatings. Patinas, which are produced using acid chlorides, are more difficult to control than statuary bronze finishes. Patinas often have variations in color,

especially over large surface areas, and sometimes fail to adhere to the metal. They are also likely to stain adjacent materials.

Statuary finishes are somewhat more stable than patinas, and the color is easier to control. They come in three tones: light, medium, and dark. A range of color should be expected, however, even within a single tone group.

■ COATINGS

As with chemical and mechanical finishes, coatings are similarly denoted. The letter *O* is used for clear organic coatings and *L* is used for laminated coatings (see Fig. 5.3-2).

■ THE COMPLETE SYSTEM

All the appropriate designations are used together when describing a particular finish. For example, the finish M31-M34-07*x* has a fine satin (M31), hand-rubbed (M34) directional textured mechanical finish, and a clear thermoset organic coating (07*x*). To make the designation complete, the organic coating, represented by the *x*, must be completely described, including its characteristics and manufacturer.

When an *x* appears in a designation, it is necessary to follow that designation with a description, usually of a proprietary finish.

5.3.2

ZINC

Zinc is a bluish-white metal that is plastic and highly corrosion resistant. It is not decorative, however, and is usually given an applied finish, usually paint and other organic coatings.

Nonstructural metal items whose strength is not paramount are sometimes made from zinc. Most zinc is used, however, as an alloying metal or as a coating for other metals. Refer to Section 5.1.4.4 for a discussion about zinc coatings, such as galvanizing on iron and steel.

Another major use of zinc is as an alloying metal. It is alloyed with copper to make brass and with aluminum to make high-strength, heat-treatable alloys. It is

FIGURE 5.3-4
Standard Designations for Chemical Finishes

Type of Finish	Designation	Description	Examples of Method of Finishing
Nonetched cleaned	C10	Unspecified	Optional with finisher.
	C11	Degreased	Treatment with organic solvent.
	C12	Chemically cleaned	Use of inhibited chemical cleaner.
	C1x	Other	To be specified.
Conversion coatings	C50	Ammonium chloride (*patina*)	Saturated solution of commercial sal ammoniac, spray or brush applied. Repeated applications may be required.
	C51	Cuprous chloride Hydrochloric acid (*patina*)	In 500 ml of warm water, dissolve 164 g cuprous chloride crystals, 117 ml hydrochloric acid, 69 ml glacial acetic acid, 80 g ammonium chloride, 11 g arsenic trioxide. Dilute to 1 liter. Apply by spray, brush, or stippling. Repeated applications may be required. Avoid use of aluminum containers.
	C52	Ammonium sulfate (*patina*)	Dissolve in 1 liter of warm water, 111 g ammonium sulfate, 3.5 g copper sulfate, 1.6 ml concentrated ammonia. Spray apply. Six–eight applications may be required under high humidity conditions.
	C53	Cargonate (*patina*)	Various formulations utilizing copper carbonate as the major constituent.
	C54	Oxide (*statuary*)	Principal fomulations utilize aqueous solutions of copper sulfates and copper nitrates at temperatures of from 85°C to boiling; or permanganate solutions at temperatures of from 80°C to boiling, using immersion periods of from 30 sec. to 5 min.
	C55	Sulfide (*statuary*)	2%–10% aqueous solutions of ammonium sulfide, potassium sulfide, or sodium sulfide. Solutions swabbed or brushed on. Repeated application increases depth of color.
	C56	Selenide (*statuary*)	Principally proprietary formulations. Because the solutions are toxic, user preparation should be avoided. Follow manufacturer's directions for use without deviation.
	C5x	Other	To be specified.

Reprinted from Metal Finishes Manual for Architectural and Metal Products *with permission of the National Association of Architectural Metal Manufacturers (NAAMM), Chicago.*

also added to magnesium-aluminum alloys to increase their ductility.

Zinc is also used as a pigment in some paints, particularly in primer paints for galvanized surfaces and in paints for rusted steel surfaces.

5.3.3

LEAD

The heavy metal lead is soft and weak because of its ductility. Its properties, however, especially its ductility, resistance to corrosion, and its density, which make it able to stop X ray penetration, cause it to be used extensively in the construction industry.

Lead can be easily rolled into sheets and extruded to make pipes. It is used extensively in the chemical industry, because of its resistance to corrosion by many chemicals, including sulfuric acid. Its chief uses in buildings are as radiation protection, sound barriers and baffles, and as a coating for other metals to provide corrosion protection. Sheet lead is also used in roofing and associated rainwear, but lead-coated copper is a more common roofing material.

Some paints contain lead, especially primer coatings on steel. This use may diminish in the future because of the concern about lead poisoning, especially of children, but lead paints will be found on existing metals for many years, even if their use is eventually banned altogether.

5.3.4

CHROMIUM

Chromium, which is very hard and highly corrosion resistant, is used mostly as a coating on steel or other metals, and as an alloying metal for steel. All stainless steel, for example, contains chromium.

Chromium is so hard that it resists polishing. Therefore, to produce those familiar, highly polished chrome-plated surfaces, such as those on plumbing fixtures, it must be deposited over a highly polished steel surface, usually with an intermediate layer of nickel.

5.3.5

NICKEL

The primary use of nickel in buildings is as an alloying metal. *Monel metal*, for example, is an alloy of nickel with 31% copper. *Nickel silver* is an alloy of nickel and brass. Perhaps half the nickel produced in this country ends up in steel alloys.

At one time, nickel was a popular finish coating for steel. Today it is more likely to be used as an undercoating over steel for the more popular chromium.

5.3.6

TIN

Probably the best-known use of tin in the building industry today is as an alloying metal with copper to make bronze. Tin is also used as a coating for steel and iron. In earlier days, much roofing was made from tinplate. Alloys of tin and other metals, such as zinc, are also used for coatings on steel. Solder contains tin.

Tin is used less frequently than zinc as a protective coating for steel, because tin is much rarer than zinc, especially in North America, and because tin does not protect steel from corrosion as effectively as zinc does.

5.3.7

GOLD

Gold is seldom used in the building industry, except to gild decorative items. It is worth mentioning here, however, because there are many older buildings on which gold leaf was used extensively. Gilding with gold is not quite as expensive as it might appear, because of the metal's great malleability. Gold leaf is only 1/250,000 in. thick. At that thickness, an ounce of gold will cover 120 sq. ft. of surface.

5.3.8

METAL-COATED METAL

Some metals are plated or coated with another metal. The base metal may be either ferrous or nonferrous, but the coating or plating metal is usually nonferrous. In most cases, the two metals are bonded together, but remain separate metals, and do not form an alloy. In an *alloy*, the constituent materials are melted together to form a single material. For example, the various types of steel are all alloys of iron and other materials. Bronze is an alloy of copper.

Finishes are applied to metals for two purposes: to protect the metal and to decorate it. Some metallic finishes, such as galvanizing and aluminum plating on steel, are primarily intended to protect the metal, and do not serve to decorate it,

although they may be left as the final finish in inconspicuous locations. When decoration is desired, those kinds of metallic finishes are usually given a further finish coat of another material. Other metallic finishes, such as chromium plating, both decorate and protect the underlying metal.

Metal coatings for steel are discussed in Section 5.1.3.4.

Metal coatings on nonferrous metals are used mostly to give the underlying metal a desired finish. Some, however, also protect the underlying metal. Chromium is much harder and more scratch resistant, for example, than either aluminum or copper, which it is often used to coat. Some metallic coatings protect other materials from the coated metals. Lead-coated copper, for example, prevents water runoff over the copper from staining, corroding, or otherwise damaging adjacent materials.

5.3.8.1

Methods

There are six basic methods of coating metal with other metals: hot dipping, electroplating, spraying (metallizing), cladding, alloying (cementation), and fusion welding.

In the *hot-dip* process, the underlying metal is coated with a second metal by being immersed in a molten bath of it. Zinc, aluminum, tin, and terne are applied to steel, for example, by the hot-dip process.

Zinc, cadmium, aluminum, and nickel are often deposited on steel, aluminum, and copper alloy by the *electroplating* process. Chromium and copper are similarly deposited over nickel that has been deposited on steel or aluminum. Chromium is also deposited over nickel that has been deposited over copper over steel.

Most metal coatings may be applied to steel and nonferrous metals by spraying, but probably the most frequent use of that method is to apply zinc and aluminum to steel. This is the only method, other than touch-up galvanizing repair by brush, that may be effectively used in field applications.

Steel is covered with copper, stainless steel, and aluminum by the *cladding* process. This process, which may include dipping, rolling, electrolytic deposit, or electrowelding, differs depending on the

metal that forms the cladding, but in all cases the cladding metal and the steel remain as separate materials.

Cementation is the formation of an alloy between a coating metal and the coated metal.

Fusion welding is, as it sounds, the fusing of the cladding metal to the base metal by welding.

The most common methods used in the building industry to coat steel with other metals are the hot-dip and the electroplating processes. *Galvanizing* (zinc coating) and *aluminizing* (aluminum coating) are the most common coatings used on steel.

The most common method used to coat nonferrous metals with metal for use in the building industry is electroplating, although some dipping is also used.

5.3.8.2

Types

There are many types of metal-coated metal used in building construction. In addition to those discussed in Sections 5.1 and 5.2 and earlier in this section, other common types include those covered in the following paragraphs.

■ **CADMIUM-PLATED STEEL**

Steel fasteners and hardware items are sometimes electroplated with cadmium to provide an electrolytic separation between the steel and other materials to prevent galvanic corrosion. Since cadmium is toxic, cadmium-plated steel items should not be used where they will come in contact with people, especially children, or food. Cadmium-plated steel should comply with the requirements of ASTM Standard A 165.

■ **CHROMIUM-PLATED METAL**

Chromium is used to plate steel, aluminum, and copper. Also, iron, carbon steel, and stainless steel are all chromium plated to produce products used in buildings. Applicable standards include ASTM Standards B 177, B 254, B 320, and B 650.

Chromium is plated onto aluminum after zincating (see "Zinc-Plated Aluminum," below) or plating the aluminum with copper, brass, or nickel.

Copper can be best plated with chromium after a coating of nickel has been applied. Chromium-plated copper has many uses in the construction industry, especially on hardware, toilet accessories, and fasteners.

NICKEL-PLATED METAL

Nickel plating is seldom used alone in the building industry. It is sometimes used on steel as an intermediate coat beneath chromium or copper plating. Nickel plating on copper is used primarily as a base for chromium plating.

BRASS-PLATED STEEL

Brass-plated steel is sometimes used as door hardware and similar items.

COPPER-COATED STEEL

Copper-coated steel sheets are sometimes used in flashing and other sheet metal applications.

COPPER-PLATED ALUMINUM

Copper plating may be used as a base for chromium plating or as a finish.

TERNEPLATE

Terne is an alloy of lead and tin, usually in a ratio of 4 parts lead to 1 part tin. Terneplate, which is the terne used in older buildings, is sheet iron or steel coated with a layer of terne. Today's terne-coated steel (TCS) is copper-bearing steel or stainless steel coated with terne. Terneplate and TCS are used mostly in roofing, but may occasionally be found in other nonstructural metal items.

LEAD-COATED COPPER

Sheets of lead-coated copper are used extensively in roofing and flashing applications, and in other nonstructural metal items, especially when water running across bare copper would be harmful to adjacent materials. Lead-coated copper is produced by electroplating or hot dipping. It may be on both sides of the sheet or on one side only. The standard for lead-coated copper is ASTM Standard B 101, which lists two types and two classes. Type I denotes application by dipping. A Type II coating is electrodeposited. Class A Standard requires a lead coating weight of 6 to 7-1/2 lb./100 sq. ft. of surface. Class B Heavy requires 10 to 15 lb./100 sq. ft. The weights given above are the total weight of lead on both sides of the sheet. Thus, the total weight of lead on a Class A Standard sheet with lead on both sides, which is the way B 101 lists it, is from 12 to 15 lb./100 sq. ft. of sheet.

ZINC-PLATED ALUMINUM

In a process called zincating, aluminum plate is immersed in a bath of zinc to coat it with a thin film of the zinc. Zincating is generally used to prepare aluminum for electroplating. The standard for zincating on aluminum is ASTM Standard B 253.

GOLD PLATING (GILDING)

Gilding with gold is an ancient art at least 4000 years old, and is still practiced today. Gold plating is the application of gold leaf to wood, plaster, metal, glass, or other material to simulate the look of solid gold. It does not include the application of gold paint. There are some imitation golds (called Dutch leaf) that are used to simulate real gold leaf, but that material tarnishes (oxidizes), even when protected by varnish. Pure gold needs no protection, since gold, for all practical purposes, does not oxidize. Even extremely thin high-carat gold leaf, when properly maintained and protected, will withstand most climates and pollution levels for between 30 and 90 years.

The purity of gold is measured by its weight in carats. Pure gold weighs 24 carats. Golds of lesser carat weight are alloys that include silver, copper, or both. Gold leaf can weigh as much as 23-1/2 carats, but most is lighter, because pure gold does not wear as well as its lower-carat alloys.

Gold leaf is available today in four forms. *Loose leaf* comes in loose sheets for use on interior surfaces only. *Patent leaf* comes in books of sheets adhered to backing paper. It is the form used for most architectural gilding, especially on exteriors. *Surface leaf* is used where close inspection is not critical, since it is made from scrap gold and has flaws and discolorations. Finally, *glass leaf* is used on signs and glass. Among these forms there are many qualities and grades.

There are essentially three methods of gilding in general use today, although there are others. The three are *oil* (mordant) *gilding*, *water gilding*, and *glass gilding*.

Different gilders have their own ways of using each method. In essence, however, both the *oil* and *water* methods are begun by applying several coats of a material known as *gesso*, which is a mixture of calcium carbonate (*gilders' whiting*) and glue. In oil gilding, the gesso is then coated with shellac or marine varnish. The gold leaf is next laid into a coat of oil sizing, which acts as an adhesive. In water gilding, the gold leaf is laid over the gesso in a wet mixture of alcohol and water called *gilders' liquor*. If the gold leaf is to be burnished, a layer of a special clay, called *burnishing clay*, is laid just ahead of the gold. In glass gilding, an *oil size* is sometimes used, but usually the gold is held in place by a *size* that is a mixture of gelatin and water.

6

WOOD AND
PLASTICS

*Photo: Treated poles combined with timbers to form a
structural frame and to enhance a design. (Courtesy of
American Wood Preservers Institute.)*

INTRODUCTION

Wood is a renewable resource. Wise forest management ensures a permanent supply of timber to meet foreseeable future needs. Foresters can grow trees faster than nature can without their assistance. As a result of reforestation and other conservation practices, the yearly growth of timber nationwide exceeds that harvested or lost to fire or disease.

Aside from its broad availability, wood possesses many properties that make it suitable for hundreds of end uses. It has a high degree of strength relative to its weight in compression, tension, bending, and resistance to impact. It can be worked easily to desired shapes with simple tools, is highly durable when simple precautions are observed, and has excellent insulating qualities. Wood is highly prized for its decorative character, which is derived from an infinite variety of grain, figure, texture, hue, and other natural markings. These properties are important in lumber, plywood, and laminated timber, as well as in other wood products, such as flooring, windows, doors, and millwork.

Lumber is the product of a sawmill and planing mill. It varies widely in quality and therefore must be divided into commercial lumber grades. Each grade has a narrow range of quality characteristics to provide a basis for selecting lumber to suit a particular use.

Plywood is a form of wood used both structurally and decoratively. Structural benefits result from its thin, crossbanded veneer sheets that are glued together into panels. This construction tends to balance the positive and negative aspects of wood's strength properties, improves its dimensional stability, and increases its resistance to both checking and splitting. Decoratively, face veneers may be arranged to form panels with a large variety of visual effects.

Laminated timber makes sizes and shapes possible that are difficult and sometimes impossible to obtain in solid timber. Arches, beams, trusses, and other laminated members are custom fabricated from pieces of stress-graded lumber bonded together with structural glues stronger than the wood itself.

6.1 PROPERTIES OF WOOD

The wide range of properties in wood makes it suitable for use in many structural, nonstructural, and decorative products. The general properties of wood that set it apart from other materials, such as concrete, metals, and plastics, should be understood so that wood and wood products are used properly to perform satisfactorily.

6.1.1

IMPORTANT TREE SPECIES

Figure 6.1-1 shows important commercial species of timber and the areas in which the trees grow. The division of species into two classes, *softwood* and *hardwood*, is a botanical difference and is not always descriptive of the softness or hardness of the wood. Because of fundamental differences in microstructure and properties, softwoods and hardwoods differ in use, size standards, and method of grading.

6.1.1.1

Softwoods

About 75% of the lumber produced in this country comes from softwood trees, the rest comes from hardwoods. This lumber is used as structural and framing lumber, sheathing, roofing, subflooring, exterior siding, flooring, trim, and interior paneling. Softwoods are called *conifers* or *coniferous* because most species bear cones. With few exceptions, softwoods have scalelike or needlelike leaves and are evergreen (Fig. 6.1-2).

6.1.1.2

Hardwoods

Lumber produced from hardwood trees is used as flooring, interior paneling, and for building cabinets and furniture. Hardwood trees, which are broad leaved, are called *deciduous* because they shed their leaves at the end of each growing season, except in warm climates (see Fig. 6.1-2).

6.1.2

WOOD GROWTH

A tree grows by the production of new cells in the *cambium* layer, a 3-to-5-cell-thick zone of active cell growth just beneath the inner bark (Fig. 6.1-3). As the cells mature, the inner layer of the cambium becomes sapwood and the outer layer becomes inner bark. The inner bark changes to outer bark as the cells become inactive. A wood cell does not develop further after it is formed and matures. All tree growth, consisting of increases in diameter and height, results entirely from new cell formation.

Each year completely new layers of wood cells envelop the preceding year's growth beneath the bark over the entire tree. Each year's growth can be seen as an annual growth ring consisting of a light-colored layer and a darker layer. In many species growth is more active during the spring, and cells formed then

Summer Wood   One Annual
                        Growth Ring
Wood Rays   Spring Wood
                        Summer
                        Wood

FIGURE 6.1-4 *The cell structure of wood. This minute cube represents a block of wood about 1/32 in. on a side. Its top surface is parallel with the end surface of a log.* (Reprinted from Forest Products Laboratory, U.S. Forest Service, USDA publication. Not subject to copyright in the United States.)

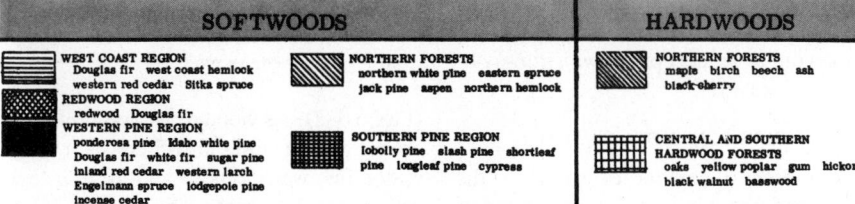

SOFTWOODS		HARDWOODS
WEST COAST REGION Douglas fir west coast hemlock western red cedar Sitka spruce REDWOOD REGION redwood Douglas fir WESTERN PINE REGION ponderosa pine Idaho white pine Douglas fir white fir sugar pine inland red cedar western larch Engelmann spruce lodgepole pine incense cedar	NORTHERN FORESTS northern white pine eastern spruce jack pine aspen northern hemlock SOUTHERN PINE REGION loblolly pine slash pine shortleaf pine longleaf pine cypress	NORTHERN FORESTS maple birch beech ash black-cherry CENTRAL AND SOUTHERN HARDWOOD FORESTS oaks yellow poplar gum hickory black walnut basswood

FIGURE 6.1-1 *Areas of timber growth in the United States.*

a   **b**

FIGURE 6.1-2 *(a) Typical growth from a coniferous (softwood) tree; (b) representative leaf from deciduous (hardwood) tree.*

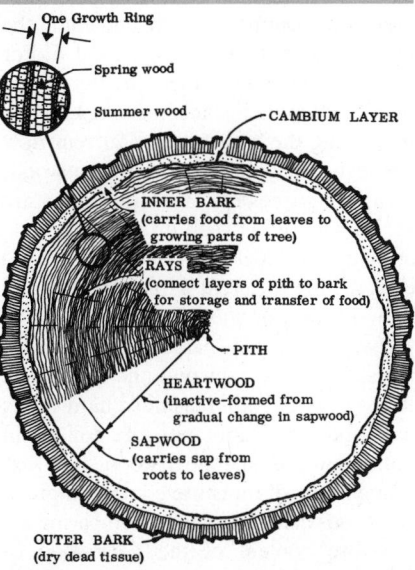

FIGURE 6.1-3 *Cross section of a living tree, showing active and inactive cell layers.*

(*springwood*) tend to be large and thin walled, compared with cells formed later in the year when growth is slower (*summerwood*). The smaller, thick-walled cells of summerwood are denser and darker than those of springwood (Figs. 6.1-3 and 6.1-4). Because of its density, summerwood is both stronger and harder than springwood. In species such as Douglas fir and southern yellow pine, the amount of summerwood can be used to estimate the wood's strength and density.

The wood between the *pith* at the center of a tree and the *cambium* is divided into light-colored *sapwood*, whose cells are active in storing food and carrying sap from the roots up to the leaves, and generally darker *heartwood*, whose cells are active in the growth process but provide structural support for the three. Generally speaking, sapwood and heartwood are of comparable strength. The gradual change from sapwood to heartwood involves the depositing of chemicals in the cellular structure of the heartwood, sometimes filling the cell cavities. In some species these chemicals are toxic, providing heartwood with a higher degree of resistance to wood-destroying fungi and insects.

Most wood cells run parallel to the height of the tree and are called *fibers* because of their needlelike shape. However, running transversely from the pith to the bark are cells called rays that convey food across the grain (Fig. 6.1-4). It is these rays that produce the decorative grains of lumber and plywood cut from species such as oak, where the rays are conspicuous.

6.1.2.1

Growth Characteristics

Certain growth characteristics affect the grade and use of wood. These include *knots*, *shakes*, and *pitch pockets* (Fig. 6.1-5).

■ KNOTS

Knots are common growth characteristics that are formed when branches of a tree, which originate in the center at the pith, become enclosed within the wood during subsequent growth. If a branch dies and later falls off, the dead stubs are incorporated into the wood, become overgrown

FIGURE 6.1-5 *Growth characteristics of wood.*

with new wood, and appear as knots when the wood is cut into lumber. Knots are detrimental to the structural properties of wood.

■ SHAKES

A shake is a grain separation occurring between annual growth rings, running along the grain parallel to the height of the tree (see Fig. 6.1-5). Shakes seldom develop in lumber unless they are present in the tree before it is felled.

■ PITCH POCKETS

A pitch pocket is a small, well-defined grain separation that may contain solid or liquid resin (see Fig. 6.1-5). Pitch pockets may be found in Douglas fir, western larch, pine, and spruce.

6.1.3

PHYSICAL PROPERTIES

The properties of wood are determined by its physical and chemical composition. Most characteristics of wood are related to its cellular structure.

6.1.3.1

Chemical Composition

Wood cells, or fibers, are primarily *cellulose* cemented together with *lignin*. The wood structure is about 70% cellulose, between 12% and 28% lignin, and up to 1% percent ash-forming materials. These constituents give wood its hygroscopic properties, its susceptibility to decay, and its strength. The bond between individual fibers is so strong that when tested under tension they commonly tear apart rather than separate. The remainder of wood, although not part of its structure, consists of extractives that give different species distinctive characteristics such as color, odor, and natural resistance to decay.

It is possible to dissolve the lignin in wood chips using chemicals, thus freeing the cellulose fibers. By further processing, these fibers can be turned into a *pulp* from which paper and paperboard products are made. It is also possible to chemically convert cellulose so that it may be used to make textiles (such as rayon), plastics, and other products that depend on cellulose derivatives.

6.1.3.2

Hygroscopic Properties

Wood is *hygroscopic*, meaning that it expands when it absorbs moisture and shrinks when it dries or loses moisture. This property affects the end use of wood. Although the wet (green) condition is normal for wood throughout its life as a tree, most products made of wood require that it be used when dry; therefore, *seasoning* by drying to an acceptable *moisture content* is necessary.

■ MOISTURE CONTENT

The moisture content of wood is the weight of water it contains, expressed as a percentage of the weight of the wood when oven dry. The oven-drying method, used to determine moisture content in the laboratory, is shown in Figure 6.1-6. For field use, a moisture meter is more practical; it gives an instantaneous reading by measuring the resistance to current flow between two pins driven into the wood.

In living trees the amount of moisture varies widely between different species, among individual trees of the same species, among different parts of a tree, and between sapwood and heartwood. Many softwoods have a large proportion of moisture in the sapwood and far less in the heartwood, whereas most hardwoods have about the same moisture content in both sapwood and heartwood. The extreme limits of moisture content in green softwoods can be shown by comparing the moisture content of the heartwood of Douglas fir and southern pine, which may be as low as 30%, with the moisture content of the sapwood of cedars and redwoods, which may be as high as 200%.

■ FIBER SATURATION POINT

Moisture in green wood is present in two forms: in the cell cavities as free water and within the cell fibers as absorbed water

STEP ONE
Sample (1" cube) is cut and immediately weighed.
W_{wet} = weight wet

STEP TWO
Sample is dried @ 212° in oven until weight is constant.

STEP THREE
Oven-dry sample is weighed.
W_{dry} = weight dry

STEP FOUR
The difference between W_w and W_d equals the weight of water in the wood; if divided by W_d and multiplied by 100, the moisture content is given as a percent of the oven-dry wood.

$$\frac{W_{wet} - W_{dry}}{W_{dry}} = \frac{Wt\ of\ Water}{Wt\ of\ Wood\ (oven-dry)} \times 100 = M.C.\%$$

FIGURE 6.1-6 *Oven-drying method of measuring the moisture content of wood.*

(Fig. 6.1-7). As wood dries, its cell fibers do not give off their absorbed water until all the free water is gone and the adjacent cell cavities are empty. The point at which the fibers are still fully saturated, but the cell cavities are empty, is called the *fiber saturation point*. In most species this occurs at about 30% moisture content. The significance of this condition is that it represents the point at which shrinkage begins. Even lumber cut with a green moisture content as high as 200% can dry to the fiber saturation point with no shrinkage of the wood. Only when the cell fibers begin to give off their absorbed water and start to constrict does the wood shrink.

Therefore, all of the shrinkage that wood can experience takes place between its fiber saturation point and a theoretical moisture content of 0% (oven-dry condition). Within this range, shrinkage is proportional to moisture loss. For every 1% loss or gain in moisture content below 30%, the wood shrinks or swells, respectively, about one-thirtieth of the total expansion or contraction. For example, at 15% moisture content wood will have experienced half of its total possible shrinkage. However, wood in service almost never reaches a 0% moisture content because of the influence of water vapor in the surrounding atmosphere. Therefore, the total possible shrinkage is far less

FIGURE 6.1-7 *Moisture content and wood-cell shrinkage.*

FIGURE 6.1-8 *Equilibrium moisture content for various temperatures and relative humidities.*

important than the probable shrinkage under ordinary conditions.

■ EQUILIBRIUM MOISTURE CONTENT

After a tree is felled and cut into lumber, its moisture content begins to drop as moisture in the wood is lost to the surrounding air. Wood will then continue to give off or take on moisture until the moisture within the wood has reached a point of equilibrium with the moisture in the air. The moisture content of the wood at this point is called the *equilibrium moisture content*. The relation of atmospheric temperature and humidity to the equilibrium moisture content of wood is given in Figure 6.1-8, which shows, for example, that wood kept constantly in air at 70 °F with a 60% relative humidity even-

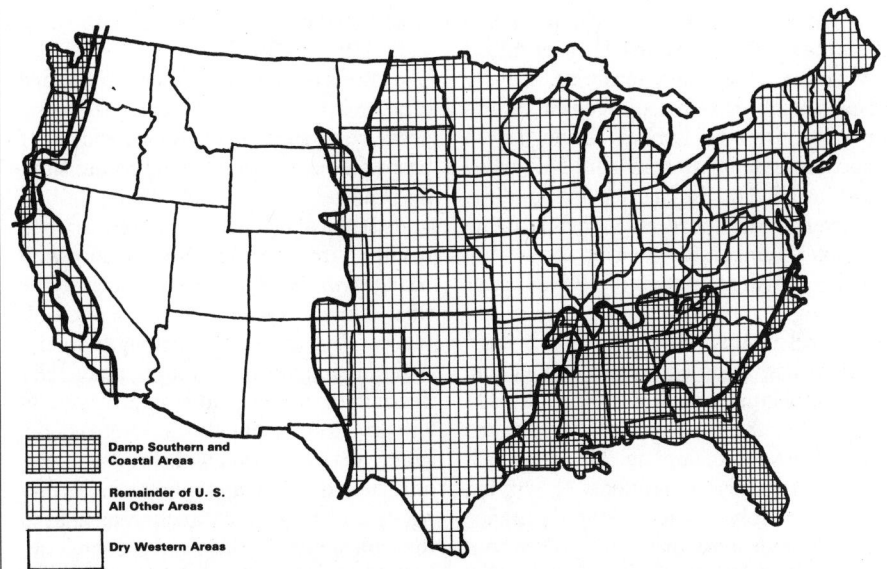

FIGURE 6.1-9 *Atmospheric humidity regions of the United States.*

tually will balance at a moisture content of about 11%. Since temperature and humidity are not constant, wood is subject to variations in moisture content as it seeks an equilibrium with the surrounding air. These variations tend to be gradual and seasonal. In addition, the equilibrium moisture content of wood is usually less on the interior of heated buildings than in unheated buildings or in an exterior exposure.

Knowing the equilibrium moisture content in a particular location is significant because it permits one to predict the moisture content a wood will attain in service there. To ensure that wood will experience only minor dimensional changes, it should be fabricated and installed at a moisture content as close as possible to the equilibrium moisture content it will attain in use (Fig. 6.1-9). This is especially critical when dealing with strip flooring, wood doors, and items of millwork, where a proper fit between adjoining wood elements is important. Since the equilibrium moisture content of interior wood items in most of the country varies between 6% and 12%, these products are usually fabricated with a moisture content in this range. However, since wood may lose or absorb moisture during shipment, it may end up at the installation site with a moisture content substantially higher or lower than the local equilibrium moisture content. Therefore, it is advisable to condition such products by storing them at the site for several days before installation.

Exterior trim, siding, and board sheathing are subject to greater variations in atmospheric humidity and temperature, and therefore also in equilibrium moisture content. In addition, a given wood's equilibrium moisture content is usually higher on the exterior than it is indoors, and may differ from season to season within the same area, as well as from region to region (see Fig. 6.1-9). However, the conditions of installation and service are not as demanding for exterior elements as they are for interior ones, so these are normally fabricated with a moisture content between 12% and 15%, which is considered adequate for exterior use in all locations and for all purposes.

Framing lumber is surfaced with a moisture content above 19% (grade marked S-GRN, meaning surfaced green), at maximum 19% (grade marked S-DRY), or at maximum 15% (grade marked MC 15). These grade marks should be those of an agency approved by the Board of Review of the American Lumber Standards Committee.

To increase the chances that no individual piece exceeds the maximum moisture content allowable by a particular grade stamp, lumber is seasoned to an average moisture content several percentage point slower than the maximum. Therefore, S-DRY lumber is dried to an average moisture content of 15%, and MC 15 lumber is dried to an average moisture content of 12%. Since the equilibrium moisture content of wood in most of the country is less than 15%, lumber that has been milled while green may, after air drying, actually have a much lower moisture content. In arid regions, framing lumber should have a maximum moisture content of 15% when it is installed on the interior of a building. In every other location, including damp regions, lumber for interior use should have a moisture content of 19% or lower (see Fig. 6.1-9).

In arid regions, either S-GRN or S-DRY lumber may achieve the desired maximum 15% moisture content if it is permitted to air dry at the yard or site, because the lumber loses moisture quickly under such dry atmospheric conditions. Specifying MC 15 merely provides visual assurance from a grade mark that this moisture content has been achieved at the mill. In other parts of the country, where air drying is not so rapid, it is best to specify dry lumber (19% maximum moisture content) for all 2-in.-thick or thicker framing lumber.

Large quantities of lumber less than 2 in. thick are used in residential construction. Boards and dimension lumber 2 in. thick or thinner should be purchased either as S-DRY or MC 15, and should be so identified by the appropriate grade mark.

In most woods, boards and 2-in.-thick dimension lumber are graded and sold either dry or green. Southern pine lumber that has been graded according to Southern Pine Inspection Bureau (SPIB) rules is usually surfaced at 15% maximum moisture content and carries the KD or KD 15 grade mark. Lumber 3 in. and 4 in. thick is not readily available in dry grades in many locations, because most mills fabricate these sizes from green wood. However, such lumber, even when bearing an S-GRN grade mark, often has an acceptable moisture content because it has been air dried at the yard. Larger pieces, 3 in. and 4 in. thick, should have a 19% maximum moisture content and, in the absence of a grade mark, should be checked to ensure that the moisture content is indeed at that level.

Moisture contents should be checked in accordance with the procedures outlined in the American Society for Testing and Materials (ASTM) D 4442, "Test Methods for Direct Moisture Content Measurement of Wood and Wood Base Materials," and ASTM D 4444, "Test Method for Use and Calibration of Hand-held Moisture Meters."

◼ EFFECTS OF MOISTURE CONTENT ON PROPERTIES

The effects of moisture content on size and dimensional stability, strength, stiffness, decay resistance, paintability, and other properties such as glue bonding are summarized below.

As stated above, size variations in wood resulting from shrinkage occur only with changes in moisture content below the fiber saturation point. Shrinkage or swelling parallel to the grain (longitudinally with the height of the tree) is practically negligible, and has little significance in construction applications. Across the grain, however, wood will shrink appreciably in both width and thickness. Shrinkage is greatest in the direction tangent to the annual growth rings and is about one-half to two-thirds as much across these rings (radial). Figure 6.1-10 shows the combined effects of tangential and radial shrinkage for various shapes as a result of drying from the green condition.

In general, hardwoods shrink more than softwoods and heavier species shrink more than lighter ones. Members with a large cross section do not shrink as much proportionately because drying is not simultaneous in the inner and outer parts of the piece. Since the inner part dries more slowly than the outer portion, the wood near the surface is prevented from shrinking normally. The outer layers become set and tend to keep the inner portions from shrinking as they dry out. A softwood structural lumber member that is 6 in. × 6 in. or larger will shrink about 1/32 in. per inch of face width as it dries from its green state and equalizes to its surroundings.

Softwood structural lumber 2 in. and 3 in. thick is usually partially or fully seasoned when marketed. The average amount of shrinkage in service for material of these sizes usually does not exceed 1/32 in. per inch of face width (Fig. 6.1-11).

The strength of wood increases as its moisture content decreases. This increase in strength is caused by (1) the strengthening and stiffening of the cell fibers as they dry out, and (2) the increase in the compactness of the wood in a given volume. Wood dried from green to a 5% moisture content may add from 2.5% to 20% to its density (weight of wood per unit of volume) and, in small pieces, end-

FIGURE 6.1-10 *Characteristic shrinkage and distortion of members as affected by the direction of the annual rings.*

crushing strength and bending strength may easily be doubled or, in some woods, even tripled. The increase in strength in small, clear specimens is much greater than in large timbers because the increase is offset by the influence of defects that develop in large members during seasoning.

All strength properties are not equally affected by decreases in moisture content. For example, crushing and bending strengths increase greatly, but some, such as stiffness, increase only moderately. Still others, like shock resistance, may show a very slight decrease. This is because shock resistance and toughness depend on pliability as well as strength, and,

FIGURE 6.1-11 *Approximate shrinkage during seasoning of a 2 × 10 from green to a theoretical 0% percent moisture content.*

although drier wood is stronger, it does not bend as far before failure as does green wood. The increase in various structural properties as a result of a change in moisture content from 30% to 19% is shown in Figure 6.1-12.

9 % Greater Stiffness

10 % Greater Load Without Splitting

22 % Greater Strength

32 % Greater Load Without Crushing

FIGURE 6.1-12 *Approximate increase in wood structural properties with a decrease in moisture content from 30% to 19%.*

The decay resistance of wood is ensured at a moisture content below 20%. In fact, the most effective and practical method of preventing decay is to control the moisture content of wood. Optimal conditions for decay occur when moisture content is about 25%. Wood that is installed and maintained below a moisture content of 20% will not decay. Most wood in use under unprotected conditions is below this level. If the moisture content will exceed 20% because of conditions of use or exposure, the wood should be treated with chemicals that are toxic to decay (see Section 6.1.5), or the naturally durable heartwood of certain species may be used (see Section 6.1.3.6).

Paintability varies slightly with moisture content. The caveat that wet wood should not be painted refers to free water present on the surface and not to the moisture content of the wood itself. Wood painted at 16% to 20% moisture content holds paint slightly longer than wood painted at 10% moisture content.

Although the time required for paint to dry (harden) depends greatly on the paint, intensity of sunlight, temperature, and relative humidity, drying is retarded when the moisture content of a wood substrate is excessive. Therefore, wood to be painted should have a moisture content of not more than 20%.

Many wood products depend extensively on gluing. The moisture content of the wood at the time of gluing affects both the strength of the bond and, later, the checking and warping of glued members. In general, glue bond improves at reduced moisture contents. Wood glues adhere well to wood with moisture contents up to 15% and even higher, but when large changes in moisture content occur after gluing, the stresses of shrinking or swelling may weaken both the wood and the glued joints. The moisture content of wood products at the time of gluing should be increased by the moisture in the glue, and the total should approximate the average moisture content that the glued member will attain in service.

Lumber used for interior millwork can be satisfactorily glued at a 5% to 6% moisture content, exterior millwork at 10% to 12% moisture content. Veneers used to make plywood must be dried to 2% to 6% moisture content at the time of gluing.

6.1.3.3

Specific Gravity

Specific gravity is a ratio comparison of the weight of a material to water. It is obtained by dividing the weight of the material by the weight of an equal volume of water. Water has a specific gravity of 1. Wood fiber, with a specific gravity of about 1.5, is heavier than water. However, dry wood of most species floats in water because a proportion of the volume is occupied by air-filled cell cavities. The specific gravity of most species when oven dry ranges from about 0.36 to 0.70 (Fig. 6.1-13). Variation among species in the size of cells and in the thickness of cell walls affects the amount of solid wood substance present and, therefore, the specific gravity. Specific gravity is an approximate measure of solid wood substance and a general index of strength properties, although specific gravity values may be affected somewhat by gums, resins, and extractives, which contribute little to strength. The relationship of specific gravity to wood strength is evident in the practice of assigning higher basic stress values to lumber such as Douglas fir and southern yellow pine, which are designated as dense.

FIGURE 6.1-13
Average Specific Gravities of Some Wood Species

Species	Specific Gravity*
Aspen	0.39
Balsam fir	0.36
Beech–birch–hickory	0.71
Coast Sitka spruce	0.39
Cottonwood	0.41
Douglas fir–larch	0.50
Douglas fir–larch (north)	0.49
Douglas fir (south)	0.46
Eastern hemlock	0.41
Eastern hemlock–tamarack	0.41
Eastern hemlock–tamarack (north)	0.47
Eastern softwoods	0.36
Eastern spruce	0.41
Eastern white pine	0.36
Hem–fir	0.43
Hem–fir (north)	0.46
Mixed maple	0.55
Mixed oak	0.68
Mixed southern pine	0.51
Mountain hemlock	0.47
Northern pine	0.42
Northern red oak	0.68
Northern species	0.35
Northern white cedar	0.31
Ponderosa pine	0.43
Red maple	0.58
Red oak	0.67
Red pine	0.44
Redwood, close grained	0.44
Redwood, open grained	0.37
Sitka spruce	0.43
Southern pine	0.55
Spruce–pine–fir	0.42
Spruce–pine–fir (south)	0.36
Western cedars	0.36
Western cedars (north)	0.35
Western hemlock	0.47
Western hemlock (north)	0.46
Western white pine	0.40
Western woods	0.36
White oak	0.73
Yellow poplar	0.43

*The specific gravities are based on weight and volume when oven dry.

■ POROSITY OF WOOD

Void space in wood occurs as (1) coarse, microscopic capillaries within the basic cell system, and (2) submicroscopic capillaries within the cell walls, which are at a maximum when the wood is saturated with moisture, for example, when green or freshly cut. A single cubic inch of green wood has about 15 sq. ft. of internal cell wall surface area and about 22,000 sq. ft. (about half the surface area of a football field) of internal submicroscopic capillary surface area in the cell walls (see Fig. 6.1-4).

This enormous internal surface area gives wood many of its unique properties: (1) ease of impregnation with chemicals toxic to decay and insects, and with moisture repellents for slowing down moisture exchange to minimize shrinkage; (2) excellent insulating properties; (3) shrinking and swelling in response to variations in moisture content; and (4) ease of adherence of paint films and adhesives. In fact, some synthetic resins adhere with such tenacity that they resist many exposure conditions.

6.1.3.4

Structural Properties

Unlike crystalline materials, such as steel and concrete, wood is fibrous. The length of these fibers varies from about 1/25 in. in hardwoods to between 1/8 in. and 1/3 in. in softwoods. The strength of wood, however, does not depend on the length of the fibers but on the thickness of their walls and on the direction of the fibers relative to applied loads. The fibers in wood are oriented so that their length is essentially parallel to the vertical axis of the tree, and the strength of wood parallel to these fibers (parallel to the grain) is quite different from its strength perpendicular to the fibers (perpendicular to the grain).

External forces, called *loads* (such as wind, gravity, and those caused by such natural phenomena as earthquakes) produce internal resisting forces in structural members known as *stresses*. *Tensile stresses* result from stretching forces; *compressive stresses* from squeezing forces. When a member is loaded so that the force is acting more or less perpendicular to its length, as is the case in a beam, the resulting stress is called *bending*. In bending, a member develops compressive stresses in the upper part and tensile stresses in the lower part of its cross section.

The strength values of wood are determined by testing small clear specimens in bending, compression, shear, etc., and assigning basic stresses (basic strength values) to different species. Since small clear pieces are seldom used in construction, lower working stresses (also called allowable unit stresses) must be derived to take into account the effect of knots, slope-of-grain, checks, and other characteristics that reduce the strength of a member to a value less than that for clear wood.

Accordingly, the *working stress* for a grade of lumber is determined by the effect of the maximum size and most unfavorable position of any strength-reducing characteristics permitted in the grade. Figures A-14 and A-15 include selected tables from the American Forest and Paper Association (AFPA) publication *NDS Supplement, Design Values for Wood Construction,* showing working stresses for bending (F_b); tension parallel to grain (F_t); shear parallel to grain (F_v); compression both perpendicular to the grain ($F_{c\perp}$) and parallel to the grain (F_c); and modulus of elasticity (E) for various wood species.

The allowable unit stresses shown in Figures A-14 and A-15, and adjusted using size, repetitive member, flat use, wet service, and shear stress factors (see Appendix A, "Wood Design Tables") are based on normal duration of loading. The AFPA publication *National Design Specification for Wood Construction* (NDS) defines this condition as fully stressing a member to its allowable design value by the application of the full design load for a cumulative duration of approximately 10 years, or applying 90% of the full design load continuously throughout the remainder of the life of the structure. Values for normal loading generally are used for designing members to carry dead loads (such as floor or roof construction) and live loads (occupants, equipment, and furnishings). For loads of brief duration, such as those produced by snow, earthquake, or wind, higher working stresses are used than are required for normal loads (Fig. 6.1-14).

■ BENDING

Wood develops a high degree of fiber strength in bending. When a wooden beam, loaded as in Figure 6.1-15a, deflects, bending stresses are produced in the fibers of the member. If the load is great enough to produce stresses that exceed the fiber strength of the wood, the beam will fail by breaking (Fig. 6.1-15b). The stresses produced in the fibers are greater in magnitude the farther they are from the central axis. If a fiber is located twice as far from the central axis as another fiber, it will have twice the stress. Therefore, in a beam, bending stresses are

FIGURE 6.1-14 *Load duration factors. The graph plots duration of load on the horizontal axis and load duration factors on the vertical axis. The tabular allowable unit stresses in the* **NDS** *Supplement (see Fig. A-14) are multiplied by the load duration factor determined using this graph. The duration factors plotted are for load duration other than the* **normal duration** *of 10 years, which has a load duration factor of 1. (Courtesy American Forest and Paper Association.)*

FIGURE 6.1-15 *Stresses produced in a beam subjected to bending.*

at a maximum in the outermost fibers (called the *extreme fiber*) at the top and bottom of the beam. In beams subjected to bending, the fiber stresses in bending should be held below the working stress values (F_b) for the grade of wood used (see Fig. A-14).

■ HORIZONAL SHEAR

A beam also tends to fail by dropping down between supports (see Fig. 6.1-15c) as a result of *vertical shear*, the tendency for one part of the beam to move vertically with respect to an adjacent part. Wood beams are more likely to fail because of the tendency of the top fibers in the beam to move horizontally with respect to the bottom fibers. This produces stresses in fibers called *horizontal shear* (see Fig. 6.1-15d). Horizontal shearing stresses are largest at each end of a beam along the central (neutral) axis. In beams the fiber stress in horizontal shear should be held below the working stress values (F_v) for the grade of wood used (see Fig. A-14).

■ MODULUS OF ELASTICITY

The stiffness of a material is measured by its modulus of elasticity (E) (see Fig. A-14 and A-15). This measure of wood's stiffness determines a beam's resistance to deflection. The modulus of elasticity is also used in computing the load on long columns, because such columns depend on stiffness to resist buckling.

Mechanical grading techniques based on the modulus of elasticity permit greater utilization of wood's strength properties (see Section 6.2.4.3).

■ TENSION

Theoretically, tensile strength parallel to the grain is the strongest property of wood, but in practice this is greatly affected by knots, splits, and checks. Since the allowable unit stress in *tension parallel to grain* (F_t) is a value applied to the entire cross section (less holes) of a piece of wood, it must reflect an average internal resistance to loading that is much more affected by wood characteristics than a value applied to a narrow band of the cross section, such as the extreme fiber in bending stress (F_b). For most species, the allowable F_t values are approximately two-thirds of F_b values. Wood is weak in *tension perpendicular to the grain*, which permits the fibers to be pulled apart under heavy loading. Therefore, wood should not be loaded to create this type of stress. Allowable values for tension perpendicular to the grain usually are not tabulated.

■ COMPRESSION

The allowable unit stress in compression parallel to the grain is 2 to 5 times greater than it is perpendicular to the grain (Fig. 6.1-16). There are only minor differences in the strength properties in the two directions perpendicular to the grain (radially and tangentially). The values for compressive strength perpendicular to the grain, ($F_{c\perp}$) and parallel to the grain (F_c) are shown in Figures A-14 and A-15.

6.1.3.5

Gluing Properties

The gluing properties of the woods most widely used in glued products are shown in Figure 6.1-17. These classifications are based on the average quality of typical joints of wood when glued with animal, casein, starch, urea resin, and resorcinol resin glues. A joint is considered to be glued satisfactorily when its strength is approximately equal to the strength of the wood.

Whether it will be easy or difficult to obtain a satisfactory joint depends upon the density of the wood used, its structure, the presence of extractives or infiltrated materials in the wood, and the kind of glue. In general, hardwoods are more

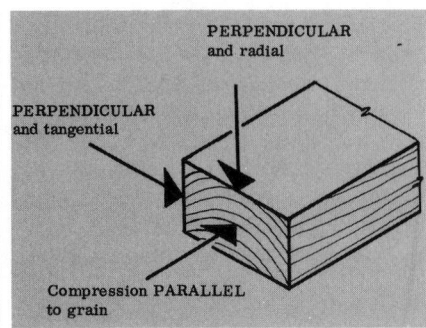

FIGURE 6.1-16 *Compression parallel and perpendicular to the grain, both radially and tangentially.*

FIGURE 6.1-17 **Gluing Properties of Some Timber Species**		
Glueability*	**Hardwoods**	**Softwoods**
Excellent	Aspen Chestnut Cottonwood Black willow Yellow poplar	Bald cypress Western red cedar White fir Western larch Redwood Sitka spruce

Good	Alder
	Basswood
	Butternut
	Rock elm
	Hackberry
		Eastern
		  red cedar
		Douglas fir
		West Coast
		  hemlock
		Northern
		  white pine
		Southern
		  yellow pine
		Ponderosa
		  pine

Satisfactory	White ash Black cherry Soft maple Oak Pecan Sycamore Tupelo Black walnut	Alaska cedar
Poor	Beech Birch Hickory Hard maple	

*Ratings based on the ability of the wood to be glued under a wide range of conditions and with a variety of glues.

difficult to glue than softwoods, and heartwood is more difficult than sapwood. Several species vary considerably in their gluing characteristics with different glues (see Fig. 6.1-17).

Nonstructural Properties

Wood's many unique qualities result in its selection as a material for many nonstructural uses. These qualities include decorative characteristics, weathering qualities, thermal expansion and conductivity, and natural decay resistance.

■ DECORATIVE CHARACTERISTICS

Factors influencing the decorative value of wood include color, figure patterns, texture, and luster; ability to take filler, stains, bleaches, and other finishes; and the method of cutting or sawing. The terms *figure, grain,* and *texture* are often incorrectly used interchangeably.

FIGURE   Figure is the pattern or design created more by the abnormal than by the normal growth of a tree (wavy or swirly) and is incorrectly used synonymously with grain.

GRAIN   The term *grain* is often used in referring to the annual rings (fine grain and coarse grain), but it is also used to indicate the direction of the fibers (straight grain, vertical grain). Finishers refer to the relative size of the pores (open grained, close grained), which determines the need for a filler.

TEXTURE   The term *texture*, although often used synonymously for *grain*, usually refers more to the finer structure and the visual dimension of depth in the wood than to the annual growth rings.

In plain-sawn boards and rotary-cut veneer, the annual rings often form ellipses and parabolas, making striking figures. In quarter-sawn boards and sliced veneers, the growth rings form stripes, and, in species such as oak, the rays add conspicuous figure. Examples of plain- and quarter-sawn boards are shown in Figure 6.2-14. In open-grained hardwoods (such as oak, mahogany, and walnut) and sliced veneers, the appearance can be varied greatly by using fillers of different colors; in softwoods the annual growth layers can be accentuated by applying stain.

■ WEATHERING

Unless they have a protective coating, exposed boards may be subject to the following from weathering (1) change in color, ranging from light gray with a silvery sheen (in cedar and cypress) to dark gray with no sheen (in Douglas fir and redwood); (2) roughening and checking of the surface (except cedar, cypress, and redwood, on which weather checks are inconspicuous); and (3) warping tendencies, from cedar, cypress, and redwood, in which this tendency is slight, to a distinct tendency in most of the other softwoods and a pronounced tendency in most hardwoods. Warping can be restricted to some extent by keeping the width of an exposed, unpainted board to not more than 8 times its thickness.

■ THERMAL EXPANSION

Like most materials, wood expands when heated and contracts when cooled. Except in special applications, such as buildings kept dry but subject to wide temperature ranges, the thermal expansion and contraction of wood is ignored, because it is much smaller than the swelling and shrinkage that result from moisture content variations that occur under normal conditions of exposure.

■ THERMAL CONDUCTIVITY

The thermal conductivity of wood is influenced by several of its properties, the most important of which are specific gravity and moisture content. Lighter-weight woods are better insulators, and insulation value increases as the moisture content decreases. Because of its good insulating characteristics, wood fiber is a principal constituent in insulating fiberboards.

■ NATURAL DECAY RESISTANCE

Naturally occurring substances that are toxic to decay and termite attack are deposited in the cell structure of certain species and make their heartwood resistant to decay and termite attack. Decay-resistant species are bald cypress (tidewater red), cedars, redwood, black locust, and black walnut. Termite-resistant species are redwood, bald cypress (tidewater red), and eastern red cedar.

Naturally durable heartwood of the above species may be used in all conditions of exposure except for members embedded in the ground to support permanent structures. For maximum durability, 100% heartwood of resistant species selected during grading should be used. For such species as redwood and cedar, several grading authorities, including the Redwood Inspection Service, West Coast Lumber Inspection Bureau, and the Western Wood Products Association, have foundation grades selected from 100% heartwood with a high toxic extractive content. Lumber with natural decay and termite resistance should be marked with the grade mark of an inspection agency authorized by the American Lumber Standards Committee. Where wood is embedded in the ground for the support of permanent structures, it should be pressure treated as discussed in Section 6.1.5.

FUNGI AND INSECT HAZARD

When lumber is used under continuously dry conditions and is adequately protected from attack by insects and fungi, preservative treatment is unnecessary. Preventive measures against these hazards during the manufacturing and merchandising stages are well established and widely practiced by reputable lumber producers. Serious decay or insect problems in buildings result most often from poor construction techniques.

Decay

Fungi (microscopic plants) cause decay, molds, and stains. Fungus growth can develop in wood only under the following conditions: (1) there must be an adequate food supply of organic material, which for some forms of fungi is the wood itself; (2) temperatures must range from 41° to 104°F; (3) there must be sufficient oxygen, which is always present unless the wood is completely below the groundwater line or completely submerged in water; and (4) the wood's moisture content must be in excess of 20%.

■ DRY ROT

The term *dry rot* is an often-used misnomer, because the fungi that cause decay must have access to water in order to function. Decay will not occur in wood that is maintained below a 20% moisture content, which is typical of air-dried wood. Therefore, even wood that is dry when decay is detected must have been wet earlier. Most decay in wood occurs when its moisture content is above the fiber saturation point (30% moisture content). But wood that is continuously water soaked will not decay either, because insufficient oxygen will be present.

In the later stages of decay a wood-destroying fungus seriously reduces the strength of wood. Therefore, most grading rules limit the amount of decay that may be present in structural lumber.

■ WHITE POCKET

Also known as white speck, *white pocket* is caused in living softwood trees by a relatively harmless fungus that ceases to develop when a log is cut into lumber and put into normal use. This fungus is as harmless as other natural characteristics normally found in lumber, such as knots, shake, or wane. It will not spread from one piece to another or even within an individual piece. Lumber with white pocket is serviceable, and certain amounts of this fungus are accepted by grading authorities in all but the higher grades of lumber (Fig. 6.1-18). However, white pocket does reduce the strength of lumber, and, more importantly, nail-holding values are greatly reduced in lumber having white pocket.

6.1.4.2

Molds and Stains

Molds and some stains are caused by fungi that do not destroy the infected wood. Stock with such stains and molds is practically unimpaired except where appearance is important.

Powdery surface molds vary from white or light colors to black. These discolorations are largely superficial and are often easily brushed off or removed by surfacing the wood.

Stains appear as specks, spots, streaks, or patches that vary most commonly from bluish to bluish black, or brown. These discolorations usually cannot be removed by brushing or surfacing because they extend into the sapwood.

FIGURE 6.1-18 *White pocket (white speck) is permitted in certain lumber grades and is serviceable and acceptable for many uses.* (Courtesy West Coast Lumber Inspection Bureau.)

6.1.4.3

Insect Damage

Holes caused by insects may occur in standing trees and in either unseasoned or seasoned lumber. These are generally classified as pinholes or grub holes. Such damage is taken into account when lumber is graded, but the strength of a piece of wood damaged by insects after it has been installed cannot be estimated visually. Therefore, when strength is an important factor, every installed piece found containing insect holes should be removed and replaced by a new member.

Many precautions taken to prevent decay are also effective against various insects that attack and destroy wood.

Subterranean termites account for about 95% of all termite damage in the United States and a significant portion of the total insect damage. Nonsubterranean termites (called dry-wood termites) occur only in the southern part of the country and the lower East and West Coasts, but they are a hazard where they do exist. Refer to Section 2.2 for a discussion about the location and control of termites; to Section 6.1.5 for information about the preservative treatment of wood to prevent termite damage; to Section 6.5.2 for ways to separate wood from the ground and the installation of termite barriers; and to Sections 6.4 and 6.7.4 for some specific requirements for antitermite treatment of wood foundations and wood pole construction.

6.1.5

PRESERVATIVE TREATMENTS

Preservative treatments protect wood against decay and insect attack. Pressure treatments, in which preservative chemicals are applied under pressure to obtain maximum penetration, afford the greatest protection for lumber and plywood against decay or termite attack. Nonpressure treatments should be restricted for use only for the protection of exterior millwork not in contact with the ground.

Wood preservatives may be divided into three major classes: (1) water-borne preservatives, (2) oil-borne preservatives, and (3) creosote and solutions containing creosote.

In all areas of the country, lumber and wood products used in contact with the ground or embedded in the ground should be pressure treated.

In areas of severe termite infestation (see Section 2.2) or with a high degree of decay likelihood, such as in damp coastal areas (see Fig. 6.1-9), lumber and wood products should be pressure treated in the following locations:

1. Wood members in contact with masonry or concrete

2. Wood members at grade, below grade, or less than 8 in. above grade

3. In a crawl space, floor joists less than 18 in. above interior grade and wood girders less than 12 in. above interior grade

4. Wood members used in conjunction with roofing or flashing

5. Wood siding closer than 6 in. to exterior finish grade.

Other wood members and products may require preservative treatment where dictated by local experience. Soil poisoning (see Section 2.2), termite shields (see Section 6.5), monolithic concrete slab construction, and other construction procedures can minimize likelihood of termite damage.

Except for structural members embedded in the ground and used as part of the foundation of permanent buildings, and except in other locations where the applicable code requires the use of pressure-treated wood, heartwood grades of naturally resistant species can be used in place of pressure-treated lumber. Some acceptable grades and species of naturally resistant dimension lumber are: California redwood—Select Heart, Construction Heart, or Foundation 2 Grade; tidewater red cypress—Heart Common; and eastern red cedar—Foundation Grade.

Exterior millwork items such as doors, window units, and casing trim, which are subjected to intermittent wetting and drying, should be preservative treated by nonpressure methods (see Section 6.1.5.4), unless more effective treatment for protection against termites is dictated by local conditions.

Job-cut ends of preservative-treated lumber should be brush coated with not less than a 3% solution of an approved preservative, usually the same one used in the factory process.

6.1.5.1

Water-Borne Preservatives

Water-borne preservatives are most often used to pressure treat lumber and plywood for construction because they leave the wood clean, odorless, and easy to paint, and because they present less of an environmental hazard than do other preservatives. Two types have proven to be very durable and will protect wood to be used in contact with the ground, supporting permanent building structures. These are ammoniacal copper arsenate (ACA) and chromated copper arsenate (CCA). The latter is available in three different formulations, but each is considered to be equivalent in performance. Water-borne preservatives are relatively odorless and are paintable, provided that the wood has been dried to the moisture content required for untreated wood. These preservatives will not cause discoloration in interior finishes unless a wet-process finish, such as plaster, is applied to treated wood without a waterproof backing paper.

6.1.5.2

Oil-Borne Preservatives

Lumber treated with oil-borne preservatives can be used in any installation that does not place it in contact with saltwater, except where its use is not permitted by code, law, or other restriction. Less lumber treated with oil-borne preservatives is used each year, and its use may eventually be banned altogether because of environmental concerns. Such lumber is often prohibited for use where it will come into contact with people, and especially children.

The oil-borne preservative Penta (pentachlorophenol), which had been for years the most widely used oil-borne preservative, is highly toxic to both fungi and insects, but is insoluble in water and so is permanent. It is this permanency that has caused such great concern about its use. Because it is almost impossible to break down and because of its high toxicity, Penta has been designated a toxic environmental hazard and is no longer used in the building construction industry.

6.1.5.3

Creosote

Creosote is a widely used preservative for some applications, but it is also starting to be prohibited in some jurisdictions where it is likely to come into contact with people. Creosote and solutions containing it are used where protection against wood-destroying organisms is of prime importance, where painting is not required, and where its odor is not objectionable. Wood may be protected in severe exposures with high retentions and in less severe exposures with lower retentions.

Creosote–coal tar mixtures are the most widely accepted preservatives for marine or saltwater installations and are ideally suited for treating pilings used for shore dwellings.

Creosote–petroleum mixtures are used where economy is of first importance, but they should not be used for marine installations.

6.1.5.4

Treatment Methods

There are two methods of preservative treatment: (1) pressure processes and (2) nonpressure processes.

■ PRESSURE PROCESSES

There are two pressure preserving processes in general use: the *full-cell process*

and the *empty-cell process*. Both are dependable means of ensuring uniform penetration and distribution of preservative. Timber to be treated by either of these pressure processes is loaded on tram cars and rolled into a long steel cylinder called a *retort*, which is then sealed.

The *full-cell process* is used for treating wood with creosote, creosote mixtures, water-borne preservatives, and fire-retarding chemicals (see Section 6.1.6). It provides a high degree of chemical retention, which is required for wood that must be installed in a marine environment, for example.

In the full-cell process much of the air is removed from the cylinder and consequently from the wood. Then the cylinder is filled with preservative under various combinations of temperature and pressure to force the preservative deep into the wood. In this process, preservative is not only absorbed by the cell walls, but the cell cavities are also are filled. After the required penetration and retention are achieved, the preservative charge is removed. This process leaves the maximum amount of preservative in the wood but, when used with creosote or its solutions, will not leave as clean a surface as do the other processes. In addition, especially when water-borne preservatives are used, the wood is left wet and must be air or kiln dried before shipment.

The *empty-cell process* is primarily used for treatment with oil-borne preservatives. In this process the wood is subjected to high air pressures before the preservative is introduced. When this pressure is released, the expanding entrapped air expels the excess preservative from the wood's cell cavities, thus reducing the net retention of preservative. This method provides treatment of the cell walls but leaves the cell cavities empty. Thus it yields a drier product, gives deeper and more uniform penetration, and permits maximum penetration for a given weight.

■ NONPRESSURE PROCESSES

Wood treated with nonpressure processes should not be used in contact with the ground or under severe conditions of decay or termite attack, because the preservative is not sufficiently impregnated in the wood to afford adequate protection. Nonpressure treatment can be effective for exterior millwork, such as precut door, window, and trim components, where the wood is generally dry but may be exposed to moisture or wetting intermittently.

Nonpressure-treated lumber should be seasoned to a maximum moisture content of 19%, or lower if the intended use dictates. Surfaces that are cut or milled after treatment should be treated again with a brushed-on application of preservative before final assembly or installation. *Vacuum* and *brief dip* nonpressure treatments usually provide clean, paintable surfaces.

In the *vacuum process*, a vacuum or partial vacuum is introduced to exhaust air from the cells and pores of the wood. Atmospheric pressure is then reintroduced to force the preservative into the wood.

The *brief dip* or 3-minute immersion process consists of completely submerging the wood in an open tank of the preservative solution for a period of at least 3 minutes.

6.1.5.5

Standards and Certification of Quality

Standards have been established for the wood preserving industry and certification seals are available for treated products.

■ **PRESSURE-TREATED WOOD**

The American Wood Preservers Association (AWPA) is a technical and research organization that develops and promulgates requirements for preservative formulations, methods of treatment, and minimum chemical retentions, which are the recognized industry standards for the pressure-preservative treatment of wood. The most commonly used AWPA standards are shown in Figure 6.1-19.

A related organization, the American Wood Preservers Institute (AWPI) performs the promotional and educational functions for the industry.

■ **NONPRESSURE-TREATED WOOD**

The National Wood Window and Door Association (NWWDA) administers a

FIGURE 6.1-19
Frequently Used Preservative Treatment Standards*

AWPA Standards	
C1	All timber products, pressure treatment
C2	Lumber, timbers, bridge ties, and mine ties, pressure treatment
C9	Plywood, pressure treatment
C15	Wood for commercial-residential construction, pressure treatment
C22	Lumber and plywood for permanent wood foundations, pressure process
C23	Pole building construction, pressure treatment

*The listed standards and procedures are the more frequently used ones. Standards and procedures related to Penta (pentachlorphenol) have been omitted from the list. Refer to AWPA literature for the applicable standards and procedures if Penta preservatives are to be used. Note that these are not permitted in some jurisdictions.

FIGURE 6.1-20 *National Wood Window and Door Association seal. (Courtesy National Wood Window and Door Association.)*

testing and plant inspection program, resulting in certification and issuance of its seal (Fig. 6.1-20). The NWWDA seal ensures that a product conforms to NWWDA IS-4, "Water-Repellent Preservative Nonpressure Treatment for Millwork." NWWDA issues a list of products that comply with IS-4. IS-4 is based on the performance of the oil-borne preservative, Penta, which was for years the most widely used product for nonpressure preservative treatment of wood, but is no longer used for this purpose.

6.1.6

■ **FIRE-RETARDANT TREATMENT**

The natural fire resistance of wood can be greatly enhanced by its impregnation with inorganic salts that react chemically at temperatures below the ignition point of wood. This reaction causes the combustible vapors normally generated in the wood to break down into nonflammable water and carbon dioxide.

Fire-retardant-treated wood has been marketed since World War II. Today's treatment methods and chemicals provide a uniformity that was earlier unattainable. Products tested and approved by the Underwriters Laboratories, Inc. (UL) bear its label and are recognized by model building-code authorities and federal agencies as suitable for structural use in fire-resistive and noncombustible construction.

After treatment, plywood and lumber, 2 in. nominal or less in thickness, should be dried to a moisture content of 19% or less. Designers may want to require a lower moisture content for fire-retardant-treated wood to be used for millwork, cabinets, office paneling, and other special uses. Fire-retardant-treated wood is generally not suitable for transparent finishes. Most of it is painted when exposed to view.

Except where exterior grade treatment complying with rain-testing requirements has been employed, fire-retardant-treated wood should not be used where it will be exposed directly to the weather. The presence of fire-retardant chemicals in treated wood may prohibit the use of some wood preservatives, such as CCA.

6.2 LUMBER

Lumber is the designation given to the many products produced in a sawmill and planing mill. Lumber is manufactured by sawing, edging, and trimming logs into square or rectangular pieces of wood. Veneers, barrel staves, and hammer handles, however, are not lumber. Yard lumber includes (1) *boards* used for sheathing, siding, flooring, paneling, trim, and patterned millwork; (2) *dimension* stock used for light framing, joists, planks, and roof and floor decking; and (3) *timbers* used for posts, beams, and stringers.

6.2.1

MANUFACTURING PROCESS

The manufacture of lumber begins in the forest. After trees are felled, the logs are transported to a lumber mill for processing. Following manufacture by a sawmill, a wholesaler locates lumber and resells it to wood-using plants and to retail lumberyards that distribute about 75% of all lumber to the ultimate user. Many retail yards and some wholesalers maintain a planing mill and other equipment to satisfy individual user requirements.

6.2.1.1

Logging Practice

Modern logging methods make it possible to transport most of the trunk of a tree from a forest to a mill. Portable power saws are used for felling trees and cutting them into logs. Depending on the terrain and other variables, logs are removed from the forests by trucks, tractors, trains, or by fast, powerful cable logging systems. The logs are grouped at a landing where mobile or other types of loaders place them onto trucks for haulage to the mill (Figs. 6.2-1 and 6.2-2).

6.2.1.2

Sawmill Practice

Operations vary from mill to mill, depending on the rate of output, the type of

FIGURE 6.2-1 *Logs are loaded aboard trailer trucks for transport to a sawmill. (Courtesy Weyerhaeuser.)*

FIGURE 6.2-2 *At a sawmill, a log unloader can remove an entire load from a truck in one bite. (Courtesy Weyerhaeuser.)*

machinery, and the end product desired, but in the main, the pattern of operations is as follows:

Automatic machinery handles much of the sawmill process (Fig. 6.2-3), but plants still depend heavily on the judgment of skilled workers at each step during manufacture.

FIGURE 6.2-3 *A sawmill operator. (Courtesy Weyerhaeuser.)*

FIGURE 6.2-4 *Operator making sure boards do not pile up. He is also tossing culls. (Courtesy Weyerhaeuser.)*

In some mills, logs are cut to a shorter length. In some cases, low-quality parts are removed. In other cases, logs are cut to fit the mill's machinery or to produce lumber of certain lengths. Usually, logs are sent first to a debarker, where their bark is removed by peeling, scraping, or blasting with high-pressure water jets. This reveals the shape of each log and its knots and irregularities. The log is then placed on a carriage and fed into a *head saw*, which may be either a circular saw or a band saw. The main saw mechanism, called a *head rig*, is the point where a log is broken down into rough cants, timbers, planks, and boards. Some mills employ a gang saw as the head saw with as many as 40 straight saws. Others use a gang saw in combination with a band or circular head saw to break large cants rapidly into boards and dimension lumber.

The next step is the *edger*, a series of small saws that rip the lumber into the desired width and remove the rounded edge. The edger also rips wide pieces into narrow widths. Edged lumber moves onto the trimmer, where the saws trim off the rough ends and defects to produce the most desirable lengths.

The final step in the sawmill is called a *green chain*, where the edged and trimmed pieces are sorted visually according to grade, species, and size (Figs. 6.2-4 and 6.2-5).

From the green chain, lumber may be sent to a seasoning yard or dry kiln (Fig. 6.2-6), or to remanufacturing plants for resawing, dressing, ripping, planing, or other treatment.

FIGURE 6.2-5 *Lumber grading. (Courtesy Weyerhaeuser.)*

6.2.1.3

Seasoning Practice

Lumber seasoning is the process of reducing the moisture content of the wood. Kiln drying and air drying are the methods used. In either case, the advantages to be gained over unseasoned, or green, lumber include increased strength; reduction in shrinkage, checking, and warping in service; less susceptibility to fungus attack in storage; and improvement in the capacity to receive pressure-preservative treatment. Kiln drying can further reduce the moisture content to almost any desired value; can result in even greater reduction in weight; hastens the drying time over that required in air seasoning; and kills stain, decay fungi, and insects in the wood. In the case of hardwoods, air drying in outside yards is often combined with kiln drying. Following seasoning, lumber is sent to a planing mill for sizing and surfacing, after which it is graded, tallied, and prepared for shipment.

Under the American Softwood Lumber Standard (Product Standard PS 20-70), mills can fabricate dimension lumber in either the green or seasoned state, provided that green lumber is surfaced to slightly larger sizes to compensate for eventual shrinkage (Fig. 6.2-7). PS 20-70 also requires that lumber up to, but not including, 5 in. thick be grade marked S-DRY if it was surfaced at a moisture content of 19% or less; if it was surfaced at a higher moisture content it must be marked S-GRN. Although not required by the Standard, lumber milled at a 15% moisture content is identified in most species as MC 15; in grading southern

FIGURE 6.2-6 *Stacks of lumber being loaded into a restraint dryer.* *(Courtesy Weyerhaeuser.)*

pine, the term KD 15 (kiln dried) is used in place of MC 15.

KILN-DRIED LUMBER

In a dry kiln, heat, humidity, and air circulation are controlled to reduce the moisture content to any practical percentage. With few exceptions, the appearance grades of lumber intended for exterior and interior finish are kiln dried, and some mills also kiln dry dimension and other yard grades. Specifying kiln-dried lumber does not necessarily ensure that it is dry, since drying may be terminated in the kiln at any time. The moisture content desired should always accompany the words *kiln dried* unless the moisture content established by the grading rules is suitable for the use intended. This is usually 6% to 12% in the finish grades of softwoods and hardwoods, and 15% to 19% in the common yard grades of softwoods. Kiln drying takes a matter of days.

AIR-DRIED LUMBER

In the air-dried method, lumber dries simply by evaporation when exposed to the atmosphere, either outdoors or in unheated storage over a period of several months. Ordinarily, dimension and lower grades of softwood lumber used for framing are sent to a yard for air drying but sometimes are shipped unseasoned (green). Structural timbers are not usually held long enough to be seasoned, although some drying may take place between their sawing and the time of shipment.

The moisture content of thoroughly air-dried lumber averages between 12% and 15% in the United States, but varies widely from as low as 6% in summer in the arid Southwest to 24% in winter in the Pacific Northwest. The various grading rules governing softwood lumber establish the maximum moisture content for each species of seasoned lumber, whether it is air or kiln dried.

Hardwoods are commonly classified according to grade and size at the time of sawing, with all stock then sent to the air-drying yard. After air drying, it may be kiln dried at the mill or shipped to a remanufacturing plant where it is kiln dried before being made into flooring, interior finish, cabinetwork, or other finished products.

UNSEASONED LUMBER

Lumber more than 2 in. thick is rarely seasoned at the mill. Lumber 3 in. and 4 in. thick requires longer periods of air drying to achieve a 19% moisture content. This thicker lumber is usually fabricated green and is not often available with an S-DRY grademark in a retail lumberyard.

However, such lumber frequently achieves an acceptable moisture content of 19% or less after air drying at the lumberyard. In the absence of a grade mark, the wood can simply be measured. If a representative sample is close to the thickness and width requirements of PS 20-70 for dry lumber of that size (see Fig. 6.2-7), the lumber is likely to have an acceptable moisture content. An electric moisture meter, of course, provides a more reliable estimate of moisture content.

6.2.1.4

The Planing Mill

As previously stated, lumber can be surfaced (planed) either unseasoned or seasoned. Except for some framing lumber, most lumber is surfaced after seasoning because the planing removes some of the imperfections caused by shrinkage during seasoning.

Surfacing is done at a planing mill, where the lumber is planed and worked to patterns. Equipment varies from a few machines for surfacing and resawing in some mills to complete fabricating operations in others. Dried, rough-sawn products are converted into boards and dimension, flooring, ceiling, partition, siding, casing, and other finished stock.

After going through a planing mill, unseasoned lumber is returned to be seasoned; seasoned lumber may be removed to storage sheds to await shipment, but most of it is usually moved directly to cars and loaded for immediate shipment to a wholesale or retail yard or to a woodworking or industrial plant (Fig. 6.2-8).

6.2.1.5

Reforestation

The lumber industry has had to adapt production to its source of supply—trees. Raw material requirements must be geared to the volume of timber that the forests can grow. This adaptation has been accomplished by greater wood research and utilization, coupled with the application of forest conservation principles such as tree farming.

FIGURE 6.2-7
American Standard Dressed Sizes of Green and Dry Boards, Dimension Lumber, and Timbers

		Thicknesses					Face Widths			
			Minimum Dressed					Minimum Dressed		
	Nominal	Dry		Green		Nominal	Dry		Green	
Item	in.	in.	mm	in.	mm	in.	in.	mm	in.	mm
						2	1-1/2	38	1-9/16	40
						3	2-1/2	63	2-9/16	65
						4	3-1/2	89	3-9/16	90
						5	4-1/2	114	4-5/8	117
	1	3/4	19	25/32	20	6	5-1/2	140	5-5/8	143
						7	6-1/2	165	6-5/8	168
Boards*	1-1/4	1	25	1-1/32	26	8	7-1/4	184	7-1/2	190
						9	8-1/4	210	8-1/2	216
	1-1/2	1-1/4	32	1-9/32	33	10	9-1/4	235	9-1/2	241
						11	10-1/4	260	10-1/2	267
						12	11-1/4	286	11-1/2	292
						14	13-1/4	337	13-1/2	343
						16	15-1/4	387	15-1/2	394
						2	1-1/2	38	1-9/16	40
						3	2-1/2	63	2-9/16	65
						4	3-1/2	89	3-9/16	90
	2	1-1/2	38	1-9/16	40	5	4-1/2	114	4-5/8	117
Dimension	2-1/2	2	51	2-1/16	52	6	5-1/2	140	5-5/8	143
	3	2-1/2	63	2-9/16	65	8	7-1/4	184	7-1/2	190
	3-1/2	3	76	3-1/16	78	10	9-1/4	235	9-1/2	241
						12	11-1/4	286	11-1/2	292
						14	13-1/4	337	13-1/2	343
						16	15-1/4	387	15-1/2	394
						2	1-1/2	38	1-9/16	40
						3	2-1/2	63	2-9/16	65
						4	3-1/2	89	3-9/16	90
						5	4-1/2	114	4-5/8	117
Dimension	4	3-1/2	89	3-9/16	90	6	5-1/2	140	5-5/8	143
	4-1/2	4	102	4-1/16	103	8	7-1/4	184	7-1/2	190
						10	9-1/4	235	9-1/2	241
						12	11-1/4	286	11-1/2	292
						14	13-1/4	337	13-1/2	343
						16	15-1/4	387	15-1/2	394
Timbers	5 & thicker	1/2 off	13 off	1/2 off	13 off	5 & wider	1/2 off	13 off	1/2 off	13 off

Source: Table 3, Proposed Voluntary Product Standard PS 20-94, "American Softwood Lumber Standard," NIST, p. 12. Not subject to copyright in the United States.

*Boards less than the minimum thickness for nominal 1 in. but 5/8 in. (16 mm) or greater thickness dry [11/16 in. (17 mm) green] shall be regarded as American Standard Lumber, but such boards shall be marked to show the size and condition of seasoning at the time of dressing. They shall also be distinguished from nominal 1-in. boards on invoices and certificates.

FIGURE 6.2-8 *Lumber waiting to be loaded aboard ship.* (*Courtesy Weyerhaeuser.*)

There are four approaches generally used for timber harvesting and management: *clear-cutting, selective cutting, shelter-wood cutting,* and *seed-tree cutting.* Clear-cutting means harvesting all the trees in a given area of forest at one time. The lumber industry prefers the clear-cutting practice for many species because it is more efficient and results in quicker regeneration. Conservationists, on the other hand, charge that this method is unsightly, causes erosion, and destroys wildlife habitats.

Clear-cutting is advisable only where there is a great deal of rainfall. Without it, seedlings will burn up under the rays of the hot sun. Many West Coast forests receive the amount of rainfall needed for clear-cutting. Clear-cutting is also done regardless of rainfall amounts when a disease is discovered in a forest or when one is moving to the forest from adjacent forests. The wood is then said to have been saved from the disease, which would have destroyed the living trees. Clear-cutting is also done when part of a forest has become very ragged from selective cutting. As soon as a stand of trees is clear-cut, forestry crews start planting new trees. Unfortunately, this replanting must be done by hand, which is expensive.

Entire forests are not usually clear-cut all at once. Lumber companies instead lease a limited number of acres of a government-owned forest to cut, or they cut a limited number of acres of their own land at a time. Some people call this *patch clear-cutting* because it some-times resembles a checkerboard or patch quilt pattern.

Selective cutting implies taking only the older, slower growing, and defective trees, leaving behind enough trees for natural re-seeding, wildlife shelter, and erosion control. However, regardless of the method of cutting and reseeding, seedlings require 60 years or more before they produce harvestable trees. Unfortunately, selective cutting is often performed too frequently and sometimes too extensively for forest self-perpetuation. Therefore, selective-cut forests tend eventually to run out of harvestable wood and the trees that would soon produce it.

Shelter-wood cutting and *seed-tree cutting* are processes in which some trees are left standing while others are harvested. The difference between the two is that in shelter-wood cutting, 10 to 20 trees are left per acre, whereas in seed-tree cutting only about four or five trees are left per acre. Shelter-wood cutting is employed when there is insufficient rainfall for clear-cutting, when there are trees that cannot tolerate continuous direct sun, and when the subject tree species has a shallow root system (for example, spruce) and therefore cannot withstand significant wind. The trees left standing reseed the cut area, eliminating the need for reseeding by hand. The 10 to 20 trees left per acre in shelter-wood cutting also provide some shade and wind protection for the seedlings. The fewer trees left by seed-tree cutting do not afford this same protection in any significant way.

MARKETING AND MEASURE

PS 20-70 states that lumber should be "tallied board measure," which is defined as using the board foot as the unit of measure, and most lumber is marketed and sold that way. A board foot is the equivalent of a piece one inch thick and one foot square. Some products such as moldings, which have been machined further, are sold by the lineal foot (running foot). Wood shingles are packaged in bundles and sold according to the surface area they are intended to cover. Posts, poles, and pilings are sold by the piece or the lineal foot.

Sizes

Lumber is bought and sold by its nominal dimensions. The nominal size of lumber is its rough unfinished size, sold commercially as 1 × 6, 2 × 4, 4 × 8, etc. (Fig. 6.2-9). The dressed size is smaller than the nominal dimensions as a result of seasoning and surfacing. Thus, a 2 × 4 is actually 1-1/2 in. by 3-1/2 in. Members up to 6 in. wide lose 1/2 in. Wider members lose 3/4 in. Wood in buildings constructed before 1973 will be larger. A 2 × 4, for example, will be 1-5/8 in. by 3-5/8 in. Even-older buildings may have full-sized members: a 2 × 4 may actually be 2 in. × 4 in. In worked lumber

Nominal Size of lumber is its unfinished size, known commercially, as 1x6, 2x4, etc. Seasoning and planing bring it to its Dressed Size. The Face Size is that portion of piece exposed to view when in place.

FIGURE 6.2-9 *Nominal, dressed, and face sizes of worked lumber.*

DRESSED AND MATCHED (D & M)
Used for sub-flooring, flooring,
sheathing or where tight joints are
required between boards

SHIPLAP
Used for sub-flooring, wall and roof
sheathing or low-cost siding

Vertical siding   Drop siding   Mouldings

PATTERNED

FIGURE 6.2-10 *Typical cross sections of worked lumber.*

(Fig. 6.2-10) the face is that portion of the piece that will be exposed to view after it has been installed.

Under the provisions of PS 20-70, the dressed size of dimension lumber less than 5 in. thick is related to the moisture content of the wood at the time of manufacture. Therefore, lumber fabricated at a moisture content of more than 19% is called green and must be dressed to slightly larger dimensions (see Fig. 6.2-9) than lumber fabricated at a moisture content of 19% or less, which is called dry. The objective of this requirement is to ensure that both green and dry lumber achieve approximately the same actual dimensions (and strength) after air drying in service.

Softwood lumber is usually sold in multiples of 2-foot lengths, ranging from 6 to 24 ft., but many mills and lumberyards will precut members to exact dimensions to meet the demands of preassembled component construction and other special customer requirements. Hardwood lumber is usually sold by odd- and even-foot lengths or in random lengths for remanufacture into other products.

6.2.2.2

Board Measure

As stated earlier *board measure* is used to indicate quantities of lumber. A board foot is the amount of lumber in a piece with nominal dimensions of 1 in. thick, 1 ft. wide, and 1 ft. long. Therefore, a 1-in.-thick board that is 1 ft. wide and 10 ft. long has 10 board feet of lumber in it. The actual dimensions of this board are

about 3/4 in. × 11-1/4 in. × 10 ft. To calculate the board feet in a piece of lumber, multiply the thickness in inches by the width in feet by the length in feet, using all nominal dimensions. Since the width of a board is usually expressed in inches, it is necessary to divide by 12 to convert inches to feet (see example, Fig. 6.2-11). If the nominal thickness is less than 1 in., the number of board feet equals the width in feet multiplied by the length in feet.

ONE BOARD FOOT

$T = 1''$

$L = 1'-0''$ $W = 12''$

2×4

12'

$T \times W \times L$

$1'' \times \dfrac{12''}{(12)} \times 1' = 1 \text{ Board Foot}$

$2'' \times \dfrac{4''}{(12)} \times 12' = 8 \text{ Board Feet}$

FIGURE 6.2-11 *Board foot measurement.*

6.2.3

LUMBER QUALITY

As cut from a log, lumber varies widely in quality, because of either natural growth characteristics or the manufacturing process. The product of a log is divided into

commercial lumber grades, each having a relatively narrow range of quality. The visual grading of lumber should be conducted in accordance with ASTM D 245, "Method Establishing Structural Grades and Related Properties for Visually Graded Lumber," and provisions of PS 20-70.

In terms of quality, or number of undesirable growth characteristics, lumber can be classed as common or select. Common grades are suitable for general construction purposes. Select grades, as the name implies, provide a surface suitable for painting or natural finishes. The bulk of yard lumber is graded as common lumber, with only boards (lumber up to 1-1/2 in. thick, and at least 2 in. wide) available in select grades. Lumber over 1-1/2 in. thick may also be graded as select but it is used for fabrication into millwork and generally is not stocked as yard lumber. The word *select* in *Select Structural* and *Select Decking* does not identify a select grade. Timbers and dimension lumber grades are classed only as common grades.

Quality-reducing characteristics are faults that detract from the appearance, strength, or general utility of a piece. The definition of such faults is often difficult in grading because a slight or small defect in *common* lumber is often a medium or large defect in *select* (finish) lumber. In *select* lumber the defects are light, small, or serious, depending on the size of the piece or as they come into combination with other defects. Most of the natural and manufacturing characteristics having an effect on the quality of lumber as it is graded are briefly described below.

6.2.3.1

Natural Characteristics

Some characteristics that affect the appearance, grading, or use of wood include knots, shakes, pitch and pitch pockets, decay, molds and stains, and insect damage. Reference should be made to the grading rules covering each species for the specific limitations of these characteristics in a given grade.

■ KNOTS

Contrary to common notion, lumber containing knots is not seriously defective for ordinary use. Knots may be undesirable

ROUND KNOT
(sawn perpendicular
to the branch)

OVAL KNOT
(sawn diagonally
to the branch)

SPIKE KNOT
(sawn parallel
to the branch)

Pin Knot
Less than 1/2 in.

Small Knot
Less than 3/4 in.

Medium Knot
Less than 1 1/2 in.

Large Knot
More than 1 1/2 in.

KNOTS may be termed <u>Sound</u> if free from decay, or <u>Unsound</u>
if decay has occured; <u>Tight</u> if it will firmly retain its place,
or <u>Loose</u> if it cannot be relied upon to remain in place.

Knots are measured by averaging
the maximum and minimum diameters

FIGURE 6.2-12 *Knots are limbs embedded in a tree that have been cut through during lumber manufacture.*

because of their appearance in select grades, but they are characteristic of all common grades (Fig. 6.2-12). Knots are restricted by size and location in grading because they do depreciate the strength value of wood used for structural purposes. Strength reduction is a result of the combined effects of the local cross grain, produced by knots interrupting the direction of the grain, plus checking that may develop in and around knots during drying. Strength reduction is not caused by an inherent inferiority of the material comprising the knots. Therefore, knots in lumber are more accurately defined as strength-reducing characteristics than as defects.

SHAKES

The effect of shakes (Fig. 6.2-13) on wood is largely in appearance, since they only slightly affect the compressive and tensile strength of members. Shakes do reduce the shear resistance of wood used for beams, however, so limitations are imposed during grading.

PITCH POCKETS

Whether pitch pockets affect strength depends on their number, size, and location in the piece. Their effect on strength has often been overestimated. Certain restrictions are imposed in the lumber grading rules.

DECAY

The presence of decomposed or destroyed wood as a result of decay is restricted by

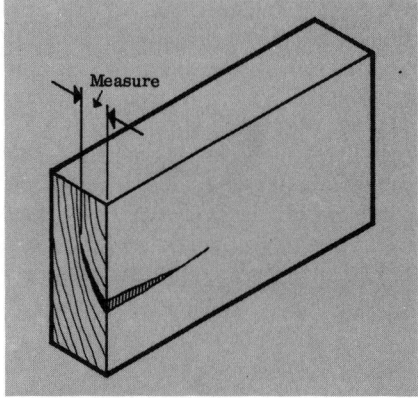

FIGURE 6.2-13 *Shake is a lengthwise separation between, not across, the annual rings.*

grading rules in structural grades of lumber to specific proportions of the cross section.

MOLDS AND STAINS

Molds and stains that cannot be removed by surfacing are quality-reducing factors only for appearance grades.

INSECT DAMAGE

The effect on strength of pinholes, grub holes, and other insect damage is taken into account during grading. The strength of such lumber is acceptable when properly graded. Such insect damage is of primary concern in the appearance grades.

Manufacturing Characteristics

Some characteristics that affect the appearance, grading, or use of lumber are the result of the manufacturing process. Manufacturing faults may be the result of sawing practices or finishing operations at either the sawmill or the planing mill, or they may be caused during seasoning.

EFFECTS OF SAWING

Theoretically, lumber is cut from a log in one of two distinct ways (Fig. 6.2-14): (1) tangential to the annual growth rings, producing lumber called *plain sawn* in hardwoods and *flat grained* or *slash grained* in softwoods; or (2) on a radial to the rings (parallel to the rays), producing lumber called *quarter sawn* in hardwood and *edge grained* or *vertical grained* in softwoods. Not all lumber can be cut to fit the above definitions exactly. In commercial practice, a piece sawn so that the growth rings when viewed from the end of the piece form an angle of 45 degrees or more with the wide face is classified as quarter-sawn, and when the rings form an angle less than 45 degrees it is classified as plain sawn.

For many uses, including structural applications, plain-sawn or quarter-sawn lumber is equally satisfactory, but each has its unique advantages. When compared with *quarter-sawn* lumber, *plain-sawn* lumber has the following characteristics:

It is is less expensive to produce because less labor and waste are involved.
Some figure patterns involving growth rings and other figure sources are more conspicuously exhibited.
Shrinking and swelling are less in thickness.
Knots are round or oval in shape, usually have a lesser weakening effect, and affect surface appearance less than the spike knots more common in quarter-sawn lumber.
Shakes and pitch pockets extend through fewer pieces.

Conversely, *quarter-sawn* lumber has the following characteristics:

Some figure patterns peculiar to the radial surface are more conspicuously displayed in it.

Lumber cut tangent to the annual rings is:
Plain sawn for hardwoods
Flat sawn for softwoods

Lumber cut radially to the annual rings and parallel to the ray is:
Quarter-sawn for hardwoods
Edge grained or vertical grained for softwoods

FIGURE 6.2-14 *The two basic ways to cut lumber from a log.* (*Drawing by HLS.*)

It shrinks and swells less in width.

It tends to wear more evenly because the radial surface is more uniform.

In some species, paint adheres to the radial surface better.

It works smoother for items run to patterns or profiles.

It twists and cups less and has less tendency to develop surface checks and splits in seasoning and in use.

A *cross grain* results when the fibers in a piece of wood are not parallel to the edges. One type of cross grain, called *diagonal grain*, occurs when a log is sawn at an angle not parallel to the bark. If the center of the log is lined up with the center of the head rig carriage, the first few cuts do not cut the full length of the log because most logs are tapered, and cross grain may result. Cross grain may cause warpage, end shrinkage, rough faces, torn or lifted grain, and weak lumber.

Like knots, cross grain has a depreciating influence on the strength of lumber because of the deviation in grain alignment. But unlike knots, which are of a highly localized nature, cross grain may involve an entire piece. The various grading rules control cross grain in

structural grades by a measurement of the slope-of-grain. Figure 6.2-15 shows how this is measured.

x = the horizontal ratio

one

Average line of the direction of fibers

FIGURE 6.2-15 *The slope of the grain affects strength.*

■ **EFFECTS OF SURFACING**

Occasionally, imperfections in lumber are created during surfacing or finishing. Depending on their size and location in the piece, the imperfections or blemishes resulting from manufacture that are defined below are limited by the grading rules so that they will have no more effect on the utility of the piece than will

other characteristics permitted in the same grade.

Torn grain is a roughened area caused by a machine's tearing out bits of the wood during dressing.

Skip is an area on a piece that is not surfaced smoothly as it passes through a planer.

Mismatched lumber results when adjoining pieces in worked or patterned lumber do not fit tightly or will not align in the same plane.

Wane is the presence of bark or the absence of wood on the edge or corner of a piece. It is restricted in grading rules because of limitations on bearing area, nailing edge, and appearance rather than because it has any direct effect on strength (Fig. 6.2-16).

Machine burn is the darkening of the wood as a result of overheating by machine knives or rolls when pieces are stopped in a machine.

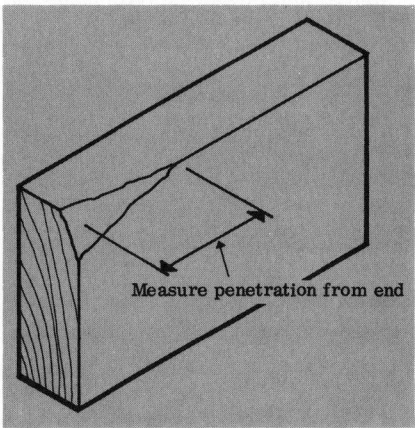

Measure penetration from end

FIGURE 6.2-16 *Wane.*

■ **EFFECTS OF SEASONING**

The lowering in quality of lumber resulting from seasoning, which is called *de-grade*, is usually a result of unequal shrinkage, which includes surface and end checking, loosening of knots, and various types of warping. Seasoning de-grade can be largely eliminated by proper practice in either kiln drying or air drying, not only at the point of manufacture but also in retail yard storage and at the construction site. Lumber dried too rapidly will check and split. If it is dried too slowly, and the temperature is favorable for fungal growth, stain or decay may develop.

The several grading rules specify the amount of seasoning de-grade permitted in the various grades, some of which are described and illustrated below.

Warp is a deviation from a true or plane surface and includes bow, crook, twist, and cup or combinations of each (Fig. 6.2-17). Warp is measured from the point of greatest deviation from a straight line.

Checking is the separation of the wood normally across the growth rings, occurring as a surface check if present on only one surface or as a through check if it extends entirely through the piece. Checking is measured by depth and length (Fig. 6.2-18). The effect of checks is mostly one of appearance rather than strength. This is particularly true in the select grades of softwoods where large grain separations are objectionable. In structural lumber, checks exert little, if any, effect on tension and compression members. They may reduce shear resistance, and limitations in the grading rules are based on this reduction.

Splits are grain separations that extend lengthwise through the piece from one surface to the other. They are measured as the penetration of the split from the end of the piece and parallel to the edge of the piece (Fig. 6.2-19).

6.2.4

SOFTWOOD LUMBER

In the early days, sawmilling was entirely a local industry without need for definite standards of size or quality; manufacturers produced lumber to local customers' requirements. As the lumber industry and transportation systems expanded, however, markets were no longer so localized. Lumber of various timbers originating in widely separated areas produced by different mills could not be sold and used successfully without common standards of quality.

Around 1900, several lumber manufacturers' associations were formed, largely to develop grading rules for the various timbers manufactured by the producers in their region. Although these grading rules had much in common, they differed in detail. Industry standardization conferences, begun in 1919, were held and by 1924 had established countrywide soft-

FIGURE 6.2-17 *Warp.*

FIGURE 6.2-18 *Check.*

FIGURE 6.2-19 *Split.*

wood lumber standards. The consensus document, known as the American Lumber Standard, was revised several times and stayed in force for many years as "Simplified Practice Recommendation SPR 16-53." This standard underwent a major revision in 1965, and was adopted and published in 1970 as the National Bureau of Standards voluntary Product Standard PS 20-70, American Softwood Lumber Standard. Under this standard, for the first time dressed lumber sizes were related to moisture content, in order to ensure reliability of actual in-service dimensions and strengths. PS 20-70 was amended again in 1986 and stands today as the industry standard for softwood lumber.

A major advance of PS 20-70 was the establishment of an American Lumber Standards Committee and, under it, a National Grading Rules Committee. The

function of the latter was to establish and to maintain the National Grading Rule for *dimension* lumber. Its members include representatives from all the major grading associations, model code groups, designers, consumers, and many other organizations representative of the industry.

Dimension lumber, as defined in PS 20-70 is "lumber from 2 in. to, but not including 5 in. in nominal thickness, and 2 or more inches in nominal width. Dimension may be classified as framing, joists, planks, rafters, studs, small timbers, etc."

Under the National Grading Rule, all grading authorities apply the same grade names and descriptions to dimension lumber of similar quality, regardless of species of timber. However, the Rule does not restrict grading authorities from including additional grades of special products in their own grading rules. For example, the Southern Pine Inspection Bureau grades and its members market many grades not covered by the National Grading Rule. In addition, the National Grading Rule does not cover decking lumber, so grade names for it are not the same in all associations. Nevertheless, PS 20-70 and the National Grading Rule considerably simplify grade recognition and the assignment of working stresses for softwood lumber.

6.2.4.1

Major Regions of Production

The majority of softwood used in construction is produced in three major regions. The timber species, approximate yearly volume of production, and the association that establishes and supervises the lumber grading practices for the different species are shown for each region in Figure 6.2-20.

■ WESTERN WOODS REGION

The western woods region is the largest of the producing areas, comprising the thirteen states of Oregon, Washington, California, Idaho, Arizona, New Mexico, Texas, Colorado, Montana, South Dakota, Nevada, Utah, and Wyoming. This region produces lumber cut from ponderosa pine, Douglas fir, Engelmann spruce, Idaho white pine, incense cedar, lodgepole pine, sugar pine, Sitka spruce, western hemlock, western larch, western red cedar, and white fir.

■ REDWOOD REGION

The redwood region is the smallest in area, encompassing but 12 northern California counties. About half of the total production there consists of lumber cut from redwood; the remainder comes from other softwood species.

■ SOUTHERN PINE REGION

The southern pine region covers the entire South, from Texas to Virginia, with some production in Maryland and Missouri. This region produces lumber cut from shortleaf, loblolly, longleaf, slash, Virginia pine, and pond pine.

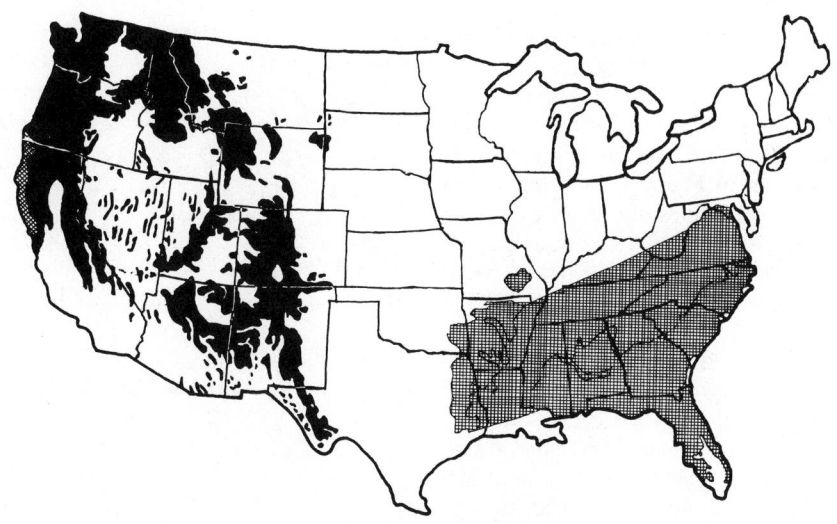

FIGURE 6.2-20
Major Softwood-Producing Regions of the United States.

Key	Region	Species	Approximate Yearly Production	Lumber Association	Grading Authority
■	Western Wood Region	Douglas fir Ponderosa pine Western red cedar Incense cedar Western hemlock White fir Engelmann spruce Sitka spruce Western larch Lodgepole pine Idaho white pine Sugar pine	16.9 billion board feet	Western Wood Products Association	Western Wood Products Association West Coast Lumber Inspection Bureau
▨	Redwood	Redwood Douglas fir	2.7 billion board feet	California Redwood Association	Redwood Inspection Service
▦	Southern Pine Region	Longleaf pine Shortleaf pine	6.7 billion board feet	Southern Forest Products Association	Southern Pine Inspection Bureau

6.2.4.2

Standards of Quality

PS 20-70 does not contain detailed requirements for selecting or specifying lumber. It does, however, establish the basic principles, provisions, and procedures for developing commercial grading rules for various species.

The commercial grading rules for each region not only conform to PS 20-70 and the National Grading Rule, but also reflect the properties of individual species and recognize the specific qualities of lumber desired by users. The grading rules are detailed requirements used to facilitate the manufacture, specifying, sale, and use of lumber. Their basic aim is to separate into marketable and usable classifications the wide range of lumber qualities obtained from a log. Grading is based mainly on quality-reducing characteristics that affect the appearance, strength, durability, or utility of wood.

Commercial grading rules are formulated and maintained by regional manufacturers' associations covering the species and individual products manufactured in a given producing region. These agencies publish grading rules for the species produced in their regions and maintain staffs of qualified lumber inspectors who supervise grading at the mills.

Mills that manufacture lumber according to PS 20-70 usually stamp each piece of dimension lumber with an appropriate grade mark. Examples of official grade marks of the four chief grading authorities are shown in Figure 6.2-21. American grading authorities are approved by the Board of Review of the American Lumber Standards Committee operating under procedures established by PS 20-70. The Board of Review reviews the adequacy, competency, and reliability of the U.S. agencies that it has approved. The Board of Review has approved agencies in Canada on the same basis. Therefore, Canadian lumber that has been graded and stamped by an authorized agency is compatible with PS 20-70 and U.S. standard engineering practice.

6.2.4.3

Classifications

Softwood lumber is classified by use, by manufacture, by size, and by grade. A knowledge of these classifications is important in selecting and using lumber.

■ USE CLASSIFICIATIONS

The use classifications of softwood lumber include yard lumber, structural lumber, and factory and shop lumber.

Yard lumber consists of grades, sizes, and patterns generally intended for ordinary construction and general building purposes.

Structural lumber is lumber that is 2 or more in. in nominal thickness, and is intended for use where working stresses are required. Structural lumber is sometimes called *stress grade* lumber. Working stresses (allowable unit stresses) for several important commercial species are given in Figure A-14. Complete tabulations for commercially available species and grades are published by the National Forest Products Association as *NDS Supplement* to its *National Design Specification for Wood Construction* (NDS). NDS is the industry-recognized standard of proper engineering practice for structural wood design.

Factory and shop lumber is intended primarily for further manufacture into windows, doors, millwork, and similar items. Therefore, the grade of a piece is

---

FIGURE 6.2-21
Typical Grade Marks for Dimension Lumber

West Coast Lumber Inspection Bureau
Construction grade light framing
Douglas fir maximum 19% m.c.
Mill No. 10

Western Wood Products Association
Select grade decking
Incense cedar maximum 15% m.c.
Mill No. 12

Southern Pine Inspection Bureau
No. 1 grade light framing
Kiln dried less than 19% m.c.
Mill No. 7

Redwood Inspection Service
Stud grade light framing
Redwood over 19% m.c.
Mill No. 50

(Courtesy West Coast Lumber Inspection Bureau, Western Wood Products Association, Southern Pine Inspection Bureau, and Redwood Inspection Service, a Division of the California Redwood Association.)

---

based on the amount of usable clear wood that can be cut from it.

■ MANUFACTURING CLASSIFICATIONS

The extent to which lumber is processed is covered in its classification, which includes *rough* lumber, *dressed* (surfaced) lumber, and *worked* lumber.

Rough lumber is lumber that has not been dressed (surfaced) but is sawed, edged, and trimmed at least to the extent of showing saw marks in the wood on four longitudinal surfaces of each piece for its overall length.

Dressed (surfaced) lumber is dressed by a surfacing machine to attain a smooth surface and uniform size. It is designated as follows: when surfaced on one side (S1S), two sides (S2S), one edge (S1E), two edges (S2E), or a combination of sides and edges (S1S, S1S2E, S2S1E, or S4S).

Worked lumber, in addition to being dressed, is matched, shiplapped, or patterned. *Matched lumber* has been worked with a tongue on one edge of each piece and a groove on the opposite edge to provide a close tongue-and-groove joint by fitting two pieces together. For end matching, the tongue and groove are worked in the ends (see Fig. 6.2-10).

Shiplapped lumber has been rabbeted on both edges of each piece to provide a close-lapped joint by fitting two pieces together (see Fig. 6.2-10).

Patterned lumber is shaped to a pattern or to a molded form in addition to being dressed, matched, or shiplapped and includes siding and other millwork.

■ SIZE CLASSIFICATION

Softwood lumber is subdivided into three major categories by size: nominal size, rough size, and dressed size.

NOMINAL SIZE   The nominal size classification is further subdivided into three categories: boards, dimension lumber, and timbers. Sizes here are given in nominal dimensions.

Boards are pieces less than 2 in. thick and 2 or more in. wide. They include sheathing, subflooring, roofing, finish, trim, siding, and paneling. Boards less than 6 in. wide may be classified as strips.

Dimension lumber, as defined earlier, is used principally for the structural elements of wood-framed buildings. It makes up most common lumber.

Timbers are lumber pieces 5 in. or more in the least dimension. These may be further classified as beams, stringers, posts, caps, sills, girders, purlins, and other members. Lumber thicker than 5 in. in either face is seldom used in ordinary construction and, when stocked by a retail yard, is frequently found in a green, rough (undressed) condition. When using heavy timbers, shrinkage of the assembly should be expected, and the design of the structure should be planned to minimize the effects of this shrinkage. Laminated timber, made by gluing up smaller pieces that can readily be kiln dried, should be used when seasoned timbers are mandatory.

Most dimension and timber lumber is stress graded, with working stresses assigned to each grade. Such lumber, intended for load-bearing conditions, is also called structural.

■ GRADE CLASSIFICATIONS

Another method of classifying softwood lumber is to grade it as *Select, Finish,* or *Common. Select* and *Finish* grades have good appearance and are used where a natural or high-quality paint finish is required. They are used for the highest quality of interior and exterior finish, trim, moldings, paneling, flooring, and siding. In most species the two highest grades, A and B, are combined as *B and Better* (B&B or B&Btr). Grades C&Btr, C Select, and D Select are more economical and serve many finishing purposes where cost savings are more important than perfect appearance. *Finish* grades include superior, Prime, and E (Fig. 6.2-22). *Common* grades are suitable for general construction and utility purposes.

The various grades of softwood framing lumber are grouped according to size and intended use into several categories, as follows:

Light framing is lumber 2 to 4 in. thick and 2 to 4 in. wide, and consists of

FIGURE 6.2-22 *Grade stamps for finish and select grades.* (*Courtesy Western Wood Products Association.*)

several categories of products. Ordinary-light-framing grades (Construction, Standard, and Utility) are intended for uses not requiring high strength, such as general framing, plates, and blocking.

Structural light-framing grades (Select Structural, No. 1, No. 2, and No. 3) are designed to fit engineering applications where higher bending strength ratios are needed, such as in trusses and in box beams.

Structural joists and planks are lumber of rectangular cross section 2 to 4 in. thick and 5 in. wide or wider, graded with respect to strength in bending when loaded either on the narrow face of a joist or on the wide face of a plank. The grades are the same as in structural light framing: Select Structural, No. 1, No. 2, and No. 3.

Studs are suitable for use in both non-load-bearing and load-bearing walls. Studs (2 to 4 in, thick, 2 to 4 in. wide) are made in lengths 10 ft. and shorter primarily for use in walls. Strength in bending is comparable to No. 3 structural light-framing lumber.

Decking is lumber 2 to 4 in. thick, 4 or more in. wide, and is intended for exposed view as planking in ceilings and roofs. The moisture content usually is limited to 15% and the lumber is worked to a tongue-and-groove pattern. It may be ordered with V-shaped or rounded edges, and with striated or grooved patterns on the face. Typical grades are Select and Commercial.

Timbers are classified as either stress rated or non-stress rated. The grades of non-stress-rated timbers are Square Edge

and Sound (SE&S), No. 1, No. 2, and No. 3. Stress-rated grades are Select Structural SR, No. 1SR, and No. 2SR. Timbers that are nearly square (that is, those in which the width does not exceed the thickness by more than 2 in.) are generally intended for use as posts and have slightly lower allowable bending strengths than rectangular timbers, which are intended for use mainly as beams.

Most lumber is graded visually by trained inspectors. In stress-grade lumber, limitations are imposed on slope-of-grain and the size, number, and placement of knots. Some structural grades in some species (for example, Douglas fir and southern pine) have additional grade requirements based on density and growth-ring count. Although most dimension lumber is graded visually, some lumber 2 in. or less in thickness is now being mechanically stress rated at the sawmill, allowing manufacturers to standardize the load-bearing capacity of each piece. This more refined method of selection permits assignment of higher stresses for much of the lumber in a grade that had been subjected to visual inspection.

In mechanical rating, lumber is fed through a machine that subjects each piece to bending and measures the modulus of elasticity, *E*. The machine then electronically computes the stress grade, taking into account the interaction of knots, slope-of-grain, growth rate, density, and moisture content. As the piece leaves the machine it is automatically stamped MACHINE RATED, with an assigned fiber stress in bending (F_b on the grade mark) and corresponding modulus of elasticity (*E*) rating (Fig. 6.2-23).

FIGURE 6.2-23 *Machine stress-rated (MSR) lumber is automatically stamped with an identifying grade mark.* (*Courtesy Western Wood Products Association.*)

The close relationship between E and F has led to the establishment of categories of stress grades for lumber 2 in. or less in thickness (Fig. 6.2-24) by the Southern Forest Products Association (SFPA) (see also Fig. A-15 in Appendix A). Machine stress-rated lumber is also visually graded to limit certain characteristics (checks, splits, wane, warp) that might adversely affect horizontal shear strength.

6.2.4.4

Size and Grade Standardization

Historically, lumber grade names and grade requirements have been developed independently on a regional basis and, even within regions, have varied among some species. Methods of establishing stress values for stress-grade lumber also have differed among lumber-producing regions, and changes in structural grading are continuing as the newer methods of machine rating are refined.

Under provisions of PS 20-70, a National Grading Rule Committee (NGRC) was established for the purpose of developing uniform countrywide grade requirements for dimension lumber. The functions of the NGRC, according to PS 20-70, were to "establish, maintain, and make fully and fairly available grade strength ratios, nomenclature, and descriptions of grades for dimension lumber conforming to American Lumber Standards." The requirements developed by the NGRC are known as the National Grading Rule for dimension lumber and constitute a part of all association grading rules.

Individual association grading rules are much more extensive and detailed than the National Grading Rule, which primarily covers nomenclature and bending strength ratios for light-framing and joist and plank grades. Bending strength ratios are standard relationships between the basic stresses of an ideal clear-wood piece of lumber free of strength-reducing characteristics (knots, sloping grain, splits) and the stresses developed in a particular grade of lumber in that species.

Since lumber shrinks in direct proportion to loss of moisture in drying from its green condition to the moisture content it ultimately attains in service, it is possible to establish, with reasonable tolerances, separate sizes for green and dry-surfaced lumber so that both types achieve essentially the same size in service and provide equal strength and stiffness for the same spans and loading conditions.

The supplement to the *National Design Specifications for Wood Construction*, which lists working stresses for all commercial species and grades, takes into account this slight oversizing of green lumber and allows the same values for green as for dry-surfaced lumber, provided both are used at a moisture content of 19% or less. Higher values are permitted when lumber is surfaced at 15% and used at the same moisture content; lower values are assigned to lumber that will be used under exposed conditions where the moisture content is likely to exceed 19% for extended periods.

FIGURE 6.2-24
SPIB Machine Stress-Rated (MSR) Lumber, 2 in. Thick, Recommended Design Values (psi)

Extreme Fiber Stress in Bending, F_b[a]	f-E Classification	Modulus of Elasticity, E	Tension Parallel to Grain F_t	Compression Parallel to Grain, $F_{c\perp}$
1200	1200f-1.2E	1.2	600	
1350	1350f-1.3E	1.3	750	1600
1400	1400f-1.2E	1.2	800	1600
1500	1500f-1.3E	1.3	900	1650
	1500f-1.4E	1.4		
1600	1600f-1.4E	1.4	950	1675
1650	1650f-1.4E	1.4	1020	1700
	1650f-1.5E	1.5		
1800[b]	1800f-1.5E	1.5	1300	1750
1800	1800f-1.6E	1.6	1175	1750
1950	1950f-1.5E	1.5	1375	1800
	1950f-1.7E	1.7		
2000	2000f-1.6E	1.6	1300	1825
2100	2100f-1.8E	1.8	1575	1875
2250	2250f-1.6E	1.6	1750	1925
	2250f-1.7E	1.7		
	2250f-1.9E	1.9		
2400	2400f-1.7E	1.7	1925	1975
	2400f-1.8E	1.8		
	2400f-2.0E	2.0		
2550	2550f-2.1E	2.1	2050	2025
2700	2700f-2.2E	2.2	2150	2100
2850	2850f-2.3E	2.3	2300	2150
3000	3000f-2.4E	2.4	2400	2200
3150	3150f-2.5E	2.5	2500	2250
3300	3300f-2.6E	2.6	2650	2325

Horizontal shear F_v — 90 psi
Compression perpendicular to grain $F_{c\perp}$ — 565 psi

Courtesy Southern Pine Inspection Bureau.

[a] The tabulated Extreme Fiber in Bending values, F_b, are applicable to lumber loaded on edge. When loaded flatwise, these values may be increased by multiplying by the following factors:

Nominal Width (in.)	4	6	8	10	12	14
Factor	1.1	1.15	1.15	1.2	1.2	1.2

[b] Grade requires F_t certification and quality control.

6.2.5

HARDWOOD

Hardwood accounts for about 25% of the total board foot production of lumber. It is supplied by many species and has uses ranging from small specialty items to the heaviest structural work. The predominant hardwood species are oaks and walnut, which constitute more than 50% of total hardwood lumber production.

6.2.5.1

Areas of Growth

Most of the commercial hardwood species in the United States grow east of the Great

Plains (see Fig. 6.1-1). The hardwoods of the west, which grow principally in California, Oregon, and Washington, amount to about 2% of the total hardwood produced.

6.2.5.2

Standards of Quality

The National Hardwood Lumber Association (NHLA) publishes size and grading rules for sawn hardwood lumber. These rules, however, are used within the lumber industry to describe hardwood lumber used for remanufacture into other products and are not normally used to specify hardwood used in building construction. In the construction industry, hardwood lumber is made to special order for the specific requirements of the purchaser, and the grade descriptions and characteristics are those that are agreed on by both buyer and seller.

The grading rules set down by the Architectural Woodwork Institute in their *Architectural Woodwork Quality Standards, Guide Specifications, and Quality Certifi-*

cation Program can be used to control hardwood lumber for cabinets, paneling, shelving, miscellaneous ornamental items, stairs and handrails, millwork, doors, windows, and casework.

6.2.5.3

Hardwood Dimension

A large proportion of the hardwood lumber produced at sawmills is used in paneling, furniture, flooring, and other finished or semifinished products. *Dimension stock*, or *hardwood dimension*, produced by hardwood sawmills is shipped to furniture, tool, piano, cabinet, and similar factories.

Hardwood *dimension* lumber is normally kiln dried and processed to a point where maximum utility is delivered to the user. It is manufactured from rough boards and flitches to the specific requirements of a particular plant or industry, in specified thicknesses, widths, and lengths. It may be solid or glued-up and is classified as rough dimension, surfaced dimen-

sion, semifabricated dimension, or completely fabricated dimension.

Rough hardwood dimension consists of planks sawn and ripped to certain sizes.

Surfaced and *semifabricated* are rough dimension carried one or more steps further, which may include one or more of several operations such as surfacing, molding, tenoning, drum sanding, equalizing, trimming, or mitering. However, such operations do not make the product a completely fabricated one ready for assembly.

Completely fabricated is lumber ready for assembly into whatever type of product is desired. Hardwood parts are frequently prefabricated at the sawmill site in adjoining hardwood dimension plants. Also, many hardwood flooring plants are operated in conjunction with hardwood sawmills.

The manufacture of hardwood lumber and its by-products is one of the most complex branches of the lumber industry because of the variety of uses for hardwood and the many species involved.

6.3 PLYWOOD AND OTHER PANELS

Plywood is an engineered panel composed of several wood veneer sheets generally placed at right angles to one another (or with veneer sheets in combination with a lumber, mineral, or oriented strand board core) and bonded together under high pressure using either a waterproof or water-resistant adhesive.

This section describes the manufacture of plywood and the types, grades, and sizes in which it is produced. The general characteristics of plywood and its properties are defined, and the principles governing its application are outlined along with a description of its most common uses in residential construction.

Both decorative and construction grades of plywood are manufactured. Construction plywood is used generally wherever strength is an important factor, but occasionally for paneling, too; decorative plywood (chiefly hardwood) is used for wall paneling, cabinetwork, and furniture.

6.3.1

CONSTRUCTION PLYWOOD

Construction plywood is manufactured primarily of softwoods, but hardwood species are also employed. However, a plywood panel need not be composed entirely of the same species. Strength and stiffness requirements for inner plies are less demanding and these frequently are of a different species than the outer plies. Structural strength or stiffness is not significantly changed since the outer faces of a plywood panel provide most of its strength and stiffness lengthwise, which is the normal direction to apply a plywood panel when it is used structurally.

6.3.1.1

Manufacture

The manufacturing process begins by debarking a tree, cutting it into peeler

logs, and cutting these into shorter log-shaped *blocks*. These blocks are placed in a giant lathe and rotated against a long knife that peels the wood into long, continuous veneer sheets, mainly 1/10 to 3/16 in. thick (Fig. 6.3-1). Veneer cut in this manner is called *rotary cut*. Veneer primarily intended for decorative plywood

FIGURE 6.3-1 *At a lathe, a sharp steel blade peels a block into thin layers.* (*Courtesy American Plywood Association.*)

is sometimes *sliced* rather than peeled from the block (Fig. 6.3-2), but an insignificant amount of softwood plywood is manufactured of sliced veneer.

FIGURE 6.3-2 *Flitch is held in clamps and moved against a knife.*

The veneer is next conveyed to clippers, cut into desired widths, then run through long drying ovens that reduce its moisture content to between 2% and 6% to provide the best possible glue bond and panel stability (Fig. 6.3-3).

FIGURE 6.3-3 *A dryer oven reduces the moisture content of veneer. (Courtesy American Plywood Association.)*

The 4 ft. × 8 ft. veneers are then graded and placed on a long 4-ft.-wide conveyer where they pass under a device that drops glue in lines or sprays glue downward onto the veneers as they move along. A parallel conveyer above this lower system delivers other veneers onto the top of the glued veneers. The interior veneers, which have their grain running perpendicular to the face veneers, are not 4 ft. × 8 ft. as are the face veneers. Because they are made from poorer-quality veneers and because of their grain orientation, they are 4 ft. measured parallel to their grain and some dimension less than 4 ft. in the other direction, which is the direction of their movement. Consequently it takes several of these smaller veneers side by side to match the length of the 8-ft. veneers moving down the conveyor. Workers standing beside the conveyor quickly position the inner veneers dropping on the line from the upper conveyor into a 4 ft. × 8 ft. format. Once these inner veneers have been positioned, they pass under a glue dropping machine. The last veneer to drop from the upper conveyor onto the stack is another 4 ft. × 8 ft. face veneer.

These laid-up panels are stacked in contact with each other and inserted into a hydraulic press, where pressure of about 150 psi is exerted to compress the veneer and glue together. The panels are then separated and each is placed onto a separate shelf of a final pressing machine (Fig. 6.3-4). The panels are not in contact with each other. The machine applies heat at about 300°F to each panel and presses up from the bottom closing the shelves against each other. This lasts from 10 to 15 seconds. The bottom of the machine then lowers and tilts the shelves. The panels slide out one at a time onto an elevated moving table that collects the panels. After removal from the press, these panels are trimmed to exact size, run through large drum sanders if they are of a sanded grade, and are finally checked for grade and appearance, emerging as finished plywood ready for use.

FIGURE 6.3-4 *Laid-up panels are placed in a press that presses veneers into plywood panels. (Courtesy American Plywood Association.)*

Standards of Quality

Construction plywood, except for a few specialty panels, should conform with the requirements of the latest version of "U.S. Product Standard PS 1 for Construction and Industrial Plywood with Typical APA Trademarks." PS 1 contains the recognized grading rule for construction plywood and the accepted bond, panel construction, dimensional tolerance, and workmanship requirements for maintaining a high level of quality. It also includes procedures for testing, grading, labeling, and identifying plywood. Conformance with the requirements of PS 1 is indicated by the grade trademark or other suitable designation of a qualified testing and certifying agency performing the functions outlined in the Product Standard.

About 85% of all construction plywood is produced by mills whose owners belong to the American Plywood Association (APA). This organization maintains a quality control program to ensure that plywood carrying an official grade trademark conforms with PS 1. Typical APA grade trademarks are shown in Figure 6.3-5.

Figure 6.3-6, which is excerpted from a more extensive APA table, can be used to help make a selection of plywood for general use. The grade of plywood needed can be determined by examining the data in the "Description and Use" column. It is then necessary to select a thickness. When engineering design is needed, such as for stress skin panels or unusual loading conditions, a designer should refer to the more extensive design data produced by the APA.

The Timber Engineering Company (TECO) and the Pittsburgh Testing Laboratories are two other qualified independent testing agencies for plywood.

Plywood is graded and marked according to species, exposure durability, veneer grade, panel grade, and application.

Wood Species

PS 1 classifies about 70 species of wood for use in construction and industrial plywood. These are divided into five groups according to their strength, stiffness, and other properties (Fig. 6.3-7) with Group 1 being the strongest and stiffest. The group number shown in the grade stamp (see Fig. 6.3-5) refers to the species used in the face and back veneer plies.

Exposure Durability

Two basic types of construction plywood are produced to comply with the requirements of PS 1. These are *exterior* type and *interior* type, which refer to the exposure durability of the panel. Within each type there are several classifications based on the types of veneers and adhesives used.

■ EXERIOR TYPE (WATERPROOF)

The veneers used in exterior panels must be C grade or better. Exterior plywood panels are designed to retain their glue bond under one of three levels of exposure. The three exterior exposure durability classifications are: Exterior, Exposure 1, and Exposure 2. *Exterior* panels will maintain their integrity when repeatedly wetted and dried or otherwise subjected to weathering in a permanent exterior exposure.

Exposure 1 panels are designed for applications where they may be exposed for a long time during construction, but finally protected. They have the same adhesives as are used in exterior type panels, but are not designed for permanent exposure.

Exposure 2 panels are intended for use where exposure time will be limited before the panel is protected.

■ INTERIOR TYPE (MOISTURE RESISTANT)

Interior type plywood is made with D grade veneer or better. There are three levels of adhesive durability within the interior type classification. (1) Plywood bonded with interior glue is called moisture resistant, and is intended for interior applications where it may be temporarily exposed to the elements. (2) Plywood bonded with intermediate glue has a lower moisture resistance than that bonded with exterior glue but greater than interior glue.

Intermediate glues have high levels of bacteria, mold, and moisture resistance. It is suitable for use in protected construction where moderate delays in providing protection may be expected or conditions of high humidity and water leakage may exist. (3) Plywood bonded with exterior glue is intended for protected construction where maximum performance is required for protection against moisture exposure from long construction delays or other conditions of similar severity.

Grades

The term *grade* refers both to veneer grade and panel grade. The veneer grade is a definition of the quality of the face veneer. The panel grade is identified in terms of the veneers used, by a name identifying the panel's intended use, or both.

■ VENEER GRADES

Each type of plywood panel is manufactured with several grades of face veneers. A veneer's grade designation is based on the species group and the quality of the veneer used on the face and back. Panels are built up of veneers glued together at right angles to each other, using one or more of five veneer grades ranging from N (highest) through A, B, C and D (lowest). Veneer may be repaired to raise its quality to a higher grade. Repairs consisting of patches or plugs, usually oval or circular, along with long, narrow shims, fill voids in the veneer sheet or replace characteristics such as knots and pitch pockets. There are also free-form synthetic (plastic) patches. These are put into a panel after the veneers have been bonded together. A flaw is routed out, the synthetic patch is squirted into the void using a handheld device resembling a glue gun. The synthetic material is then cured and sanded. In the grade stamp designation for speciality panels in Figure 6.3-5, the 303-6-S/W designation means the following: 303 is the number code for a particular line of siding. The S stands for synthetic patches. The W stands for wood patches. The 6 is the number of patches permitted in the face. In this case, six synthetic or wood patches are permissible.

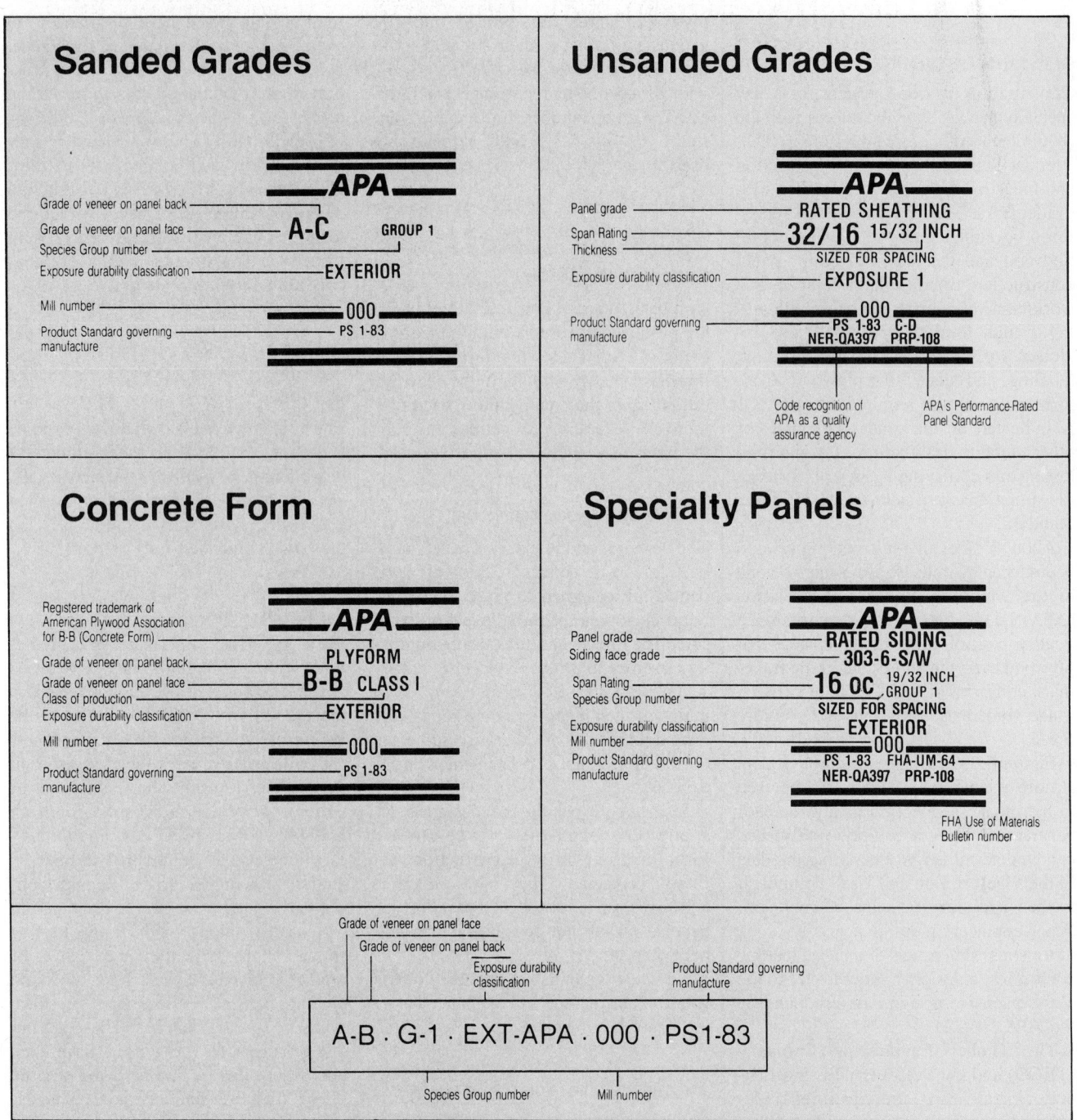

FIGURE 6.3-5 *Typical grade trademarks of the American Plywood Association.* *(Courtesy American Plywood Association.)*

Veneer grades are illustrated in Figure 6.3-8 and described below.

N Grade: A specialized veneer available on special order, used for natural finish. It is select, made of all heartwood or all sapwood, and free of open defects with only a few small well-matched repairs (not illustrated). N grade veneer is of such high quality and so expensive that it is rarely specified.

A Grade: A smooth paint grade permitting a certain number of neatly made repairs, plugs, or patches with certain restrictions. This is the grade usually specified when high-quality work is desired.

B Grade: A solid surface veneer, except for specified minor characteristics. It may have a considerable number of neatly made repairs and is suitable for concrete form work where a smooth surface is desired.

FIGURE 6.3-6
Guide to Use of Plywood Panels

Interior or Protected Applications

Plywood Grade	Description and Use	Typical Trademarks	Veneer Grade			Common Thicknesses
			Face	Back	Inner	
APA-Rated Sheathing Exp 1 or 2[a]	Unsanded sheathing grade for wall, roof, subflooring, and industrial applications such as pallets and for engineering design, with proper stresses. Manufactured with intermediate and exterior glue.[b] For permanent exposure to weather or moisture only. Exterior type plywood is suitable.	APA RATED SHEATHING 32/16 15/32 INCH SIZED FOR SPACING EXPOSURE 1 000 PS 1-83 C-D NER-QA397 PRP-108	C	D	D	5/16, 3/8, 15/32, 1/2, 19/32, 5/8, 23/32, 3/4
APA Structural I Rated Sheathing Exp 1[a] or APA Structural II[c] Rated Sheathing Exp 1	Plywood grades to use where strength properties are of maximum importance, such as plywood-lumber components. Made with exterior glue only. STRUCTURAL I is made from all Group 1 woods. STRUCTURAL II allows Group 3 woods.	APA RATED SHEATHING STRUCTURAL I 24/0 3/8 INCH SIZED FOR SPACING EXPOSURE 1 000 PS 1-83 C-D NER-QA397 PRP-108	C	D	D	5/16, 3/8, 15/32, 1/2, 19/32, 5/8, 23/32, 3/4
APA Rated Sturd-I-Floor Exp 1 or 2[a]	For combination subfloor-underlayment. Provides smooth surface for application of carpet. Possesses high concentrated and impact load resistance during construction and occupancy. Manufactured with intermediate and exterior glue. Touch-sanded.[d] Available with tongue and groove.[e]	APA RATED STURD-I-FLOOR 20 OC 19/32 INCH SIZED FOR SPACING T&G NET WIDTH 47-1/2 EXPOSURE 1 000 PS 1-83 UNDERLAYMENT NER-QA397 PRP-108	C plugged	D	C & D	19/32, 5/8, 23/32, 3/4, 1-1/8 (2-4-1)
APA Underlayment Exp 1, 2, or INT	For underlayment under carpet. Available with exterior glue. Touch-sanded. Available with tongue and groove.[e]	APA UNDERLAYMENT GROUP 1 EXPOSURE 1 000 PS 1-83	C plugged	D	C & D	1/2, 19/32, 5/8, 23/32, 3/4
APA C-D Plugged Exp 1, 2, or INT	For built-ins, wall and ceiling tile backing, NOT for underlayment. Available with exterior glue. Touch-sanded.[e]	APA C-D PLUGGED GROUP 2 EXPOSURE 1 000 PS 1-83	C plugged	D	D	1/2, 19/32, 5/8, 23/32, 3/4
APA Appearance Grades Exp 1, 2, or INT	Generally applied where a high quality surface is required. Includes APA N-N, N-A, N-B, N-D, A-A, A-B, A-D, B-B, and B-D INT grades.[e]	APA A-D GROUP 1 EXPOSURE 1 000 PS 1-83	B or better	D or better	C & D	1/4, 11/32, 3/8, 15/32, 1/2, 19/32, 5/8, 23/32, 3/4

FIGURE 6.3-6 — Continued

Exterior Applications

Plywood Grade	Description and Use	Typical Trademarks	Veneer Grade			Common Thicknesses
			Face	Back	Inner	
APA-rated Sheathing Ext[a]	Unsanded sheathing grade with waterproof glue bond for wall, roof, subfloor, and industrial applications such as pallet bins.	APA RATED SHEATHING 48/24 23/32 INCH SIZED FOR SPACING EXTERIOR 000 PS 1-83 C-C NER-QA397 PRP-108	C	C	C	5/16, 3/8, 15/32, 1/2, 19/32, 5/8, 23/32, 3/4
APA Structural I Rated Sheathing Ext[a] or APA Structural II[c] Rated Sheathing Ext	"Structural" is a modifier for this unsanded sheathing grade. For engineered applications in construction and industry where full exterior-type panels are required. STRUCTURAL I is made from Group 1 woods only.	APA RATED SHEATHING STRUCTURAL I 24/0 3/8 INCH SIZED FOR SPACING EXTERIOR 000 PS 1-83 C-C NER-QA397 PRP-108	C	C	C	5/16, 3/8, 15/32, 1/2, 19/32, 5/8, 23/32, 3/4
APA Rated Sturd-I-Floor Ext[a]	For combination subfloor-underlayment where severe moisture conditions may be present, as in balcony decks. Possesses high concentrated and impact load resistance during construction and occupancy. Touch-sanded.[d] Available with tongue and groove.[e]	APA RATED STURD-I-FLOOR 20 oc 19/32 INCH SIZED FOR SPACING EXTERIOR 000 PS 1-83 C-C PLUGGED NER-QA397 PRP-108	C plugged	C	C	19/32, 5/8, 23/32, 3/4
APA Underlayment Ext and APA C-C Plugged Ext	Underlayment for floor where severe moisture conditions may exist. Also for controlled atmosphere rooms and many industrial applications. Touch-sanded. Available with tongue and groove.[e]	APA C-C PLUGGED GROUP 2 EXTERIOR 000 PS 1-83	C plugged	C	C	1/2, 19/32, 5/8, 23/32, 3/4
APA B-B Plyform Class I or II[c]	Concrete-form grade with high reuse factor. Sanded both sides, mill-oiled unless otherwise specified. Available in HDO. For refined design information on this special-use panel see *APA Design/Construction Guide: Concrete Forming,* Form No. V345. Design using values from this specification will result in a conservative design.[e]	APA PLYFORM B-B CLASS I EXTERIOR 000 PS 1-83	B	B	C	19/32, 5/8, 23/32, 3/4
APA Marine Ext	Superior Exterior-type plywood made only with Douglas fir or western larch. Special solid-core construction. Available with MDO or HDO face. Ideal for boat hull construction.	MARINE · A-A · EXT APA · 000 · PS1-83	A or B	A or B	B	1/4, 3/8, 1/2, 5/8, 3/4
APA Appearance Grades Ext	Generally applied where a high quality surface is required. Includes APA A-A, A-B, A-C, B-B, B-C, HDO, and MDO Ext.[e]	APA A-C GROUP 1 EXTERIOR 000 PS 1-83	B or better	C or better	C	1/4, 11/32, 3/8, 15/32, 1/2, 19/32, 5/8, 23/32, 3/4

Courtesy American Plywood Association.

[a] Properties and stresses apply only to APA-rated Sturd-I-Floor and APA-rated Sheathing manufactured entirely with veneers.

[b] When exterior glue is specified, i.e., Exposure 1, stress level 2 (S-2) should be used.

[c] Check local suppliers for availability before specifying Structural II and Plyform Class II grades, as they are rarely manufactured.

[d] APA-rated Sturd-I-Floor 2-4-1 may be produced unsanded.

[e] May be available as Structural I.

FIGURE 6.3-7
Classification of Species

Group 1	Group 2		Group 3	Group 4	Group 5
Apitong[a,b]	Cedar, Port Orford	Mengkulang[a]	Alder, red	Aspen	Basswood
Beech, American	Cypress	Meranti, red[a,d]	Birch, paper	Bigtooth	Poplar, balsam
Birch	Douglas fir 2[c]	Mersawa[a]	Cedar, Alaska	Quaking	
Sweet	Fir	Pine	Fir, subalpine	Cativo	
Yellow	Balsam	Pond	Hemlock, eastern	Cedar	
Douglas fir 1[c]	California red	Red	Maple, bigleaf	Incense	
Kapur[a]	Grand	Virginia	Pine	Western red	
Keruing[a,b]	Noble	Western white	Jack	Cottonwood	
Larch, western	Pacific silver	Spruce	Lodgepole	Eastern	
Maple, sugar	White	Black	Ponderosa	Black (western poplar)	
Pine	Hemlock, western	Red	Spruce	Pine	
Caribbean	Lauan	Sitka	Redwood	Eastern white	
Ocote	Almon	Sweetgum	Spruce	Sugar	
Pine, southern	Bagtikan	Tamarack	Engelmann		
Loblolly	Mayapis	Yellow poplar	White		
Longleaf	Red lauan				
Shortleaf	Tangile				
Slash	White lauan				
Tanoak	Maple, black				

[a] Each of these names represents a trade group of woods consisting of several closely related species.

[b] Species from the genus *Dipterocarpus* are marked collectively: Apitong if originating in the Philippines; Keruing if originating in Malaysia or Indonesia.

[c] Douglas fir from trees grown in the states of Washington, Oregon, California, Idaho, Montana, Wyoming, and the Canadian Provinces of Alberta and British Columbia shall be classed as Douglas fir No. 1. Douglas fir from trees grown in the states of Nevada, Utah, Colorado, Arizona, and New Mexico shall be classed as Douglas fir No. 2.

[d] Red Meranti shall be limited to species having a specific gravity of 0.41 or more based on green volume and oven-dry weight.

C Plugged Grade: A special, improved C Grade intended for underlayment for floors and combination subfloor underlayment (not illustrated).

C Grade: The lowest quality of veneer permitted in exterior type panels. It is commonly used as the face of sheathing grade panels. Permitted defects include: tight knots up to 1-1/2 in.; 1-in. knotholes; occasional 1-1/2-in. knotholes under certain conditions; splits up to 1/2 in. tapering to a point; and other minor defects.

D Grade: The lowest quality of veneer permitted in interior type panel. This grade permits: 2-1/2-in. knotholes; occasional 3-in. knotholes under certain conditions; pitch pockets 2 in. to 2-1/2 in. wide and tapering splits up to 1 in.

▪ PANEL GRADES

The grade of a panel is indicated by two letters, such as A-D, describing a panel with an A grade front face veneer and a D grade back face veneer. The inner plies for an interior type panel are D or better, and those for an exterior type panel are C or better. Unless a decorative or other surface is specified, panels in most grades are run through large drum sanders where they are sanded smooth on both sides for appearance and for balanced construction.

6.3.1.6

Performance-Rated Panels

Performance-rated panels are also sometimes called *structural* or *structural-use panels*. These are panels designed for use where design properties, such as tension, compression, shear, cross-panel flexural properties, and nail bearing may be significant. Included in this group are panels designed for use as floor, wall, or roof sheathing; for subflooring, underlayment, or both; for siding; for construction of stress skin panels, box beams, and other structural elements; and for other structural uses. This category includes most of the plywood used in building construction. It also includes oriented strand board (see Section 6.3.3) and Com-ply, which is a product made with alternating plies of wood veneer and reconstituted wood.

Performance-rated panels possess stiffness, strength, nail-holding ability, durability, dimensional stability, lateral bracing ability, and resistance to puncture and impact. Some of them must also provide a suitable base for a covering material, such as shingles, built-up roofing, siding, or flooring.

In small buildings with conventional framing, sheathing, and subflooring, no special engineering design is required if recognized standards and construction details are followed. Correct use of plywood involves selecting the appropriate panel type, grade, and span rating for the spans and loads to be carried, in addition to using proper nailing.

Performance-rated panels are identified according to their application, the quality of the face and back veneer, or both (see Fig. 6.3-6). When necessary, they are also identified according to their structural grade (Structural I or Structural II). They are classified in one of the three exterior exposure durability classifications discussed earlier (Exterior, Exposure 1, or Exposure 2).

"A" Grade

"B" Grade

"C" Grade

"D" Grade

FIGURE 6.3-8 *Plywood veneer grades.*
(Courtesy American Plywood Association.)

Performance-rated panels for use as sheathing, subflooring, underlayment, or siding are assigned a *span rating* to set the support spacing required for each. This rating is included as part of the grade trademark (see Fig. 6.3-5) as a single number or as two numbers separated by a slash. When a single number appears, that is the required maximum spacing of the supports in inches when used as a flooring material. When two numbers separated by a slash appear, the two numbers denote different spacings, depending on the use of the panel. For example, a panel may be used as either wall or roof sheathing, but the support spacing is not the same in the two different uses. The first number denotes the maximum spacing in inches when the panel is used as roof decking. The second number is the maximum spacing in inches when the panel is used for flooring. Wall sheathing has its own single-number system that denotes the maximum span in inches for the supporting studs when sheathing spans in its weak direction.

Some performance-rated panels are left unsanded for greater stiffness and strength as well as economy. However, some panels, such as underlayment, are either sanded or touch sanded to bring adjacent panels to a uniform thickness, and some, such as siding, are sanded for appearance.

6.3.1.7

Specialty Plywoods

Special manufacturing processes and treatments produce several variations from standard panels, each designed to meet specific use requirements.

■ OVERLAID PLYWOOD

Overlaid plywood panels are high-grade panels of *exterior* plywood to which cellulose-fiber sheets containing prescribed amounts of resin have been bonded to one or both faces. Most overlaid plywood has an overlay on both sides. The resin is usually phenolic, but other resins may also be used if they meet the requirements stated in PS 1. The overlay sheets are bonded to the veneers with adhesives under pressure. Overlaid plywood is classified as *high density* (HDO), *medium density* (MDO), or *special overlay* panels.

HIGH-DENSITY OVERLAY   The face veneer quality of HDO plywood may be either A-A or B-B. The inner plies are C-plugged veneer. Concrete form overlay HDO has B-B face and back veneers. The overlay sheets on HDO are impregnated with at least 45% resin solids based on the volatile-free weight of fiber and resin exclusive of glueline. The total thickness of a resin-impregnated overlay sheet is 0.012 in. before pressing and the total weight is at least 60 lb./1000 sq. ft.

HDO has a hard, smooth, chemically resistant, durable surface that requires no further finishing by paint or varnish. It is available in several basic colors, but also provides an excellent paint base and can be painted if color variation is preferred. HDO is especially suitable and widely used in concrete form work for architectural finishes, highway signs, and in other severe exposures.

MEDIUM-DENSITY OVERLAY   The face veneer quality of MDO plywood is B-B. The inner plies are C veneer. The overlay sheets on MDO must contain at least 17% resin solids for a beater-loaded sheet and 22% resin solids for an impregnated sheet, both based on the volatile-free weight of fiber and resin exclusive of glueline. The total thickness of a resin-impregnated overlay sheet is 0.012 in. before pressing and the total weight is at least 58 lb./1000 sq. ft.

MDO is intended to receive high-quality paint finishes and provides an opaque, smooth, nonglossy surface that generally blanks out the grain, although some evidence of the underlying grain is permitted. Several mills manufacture special MDO precut siding, frequently mill-primed, in widths up to 16 in.

SPECIAL OVERLAYS   Overlaid plywood surfacing panels that do not fit the definitions of either HDO or MDO are called special overlays. These must meet the glue bond requirements for overlaid plywood and the panels must be of the exterior type.

■ TEXTURED SIDING PANELS

Textured plywood siding panels are made by several mills under their own trade names. These panels may have a grooved or other patterned surface (Fig. 6.3-9). They are used for exterior siding, gable ends, and fencing, and as forms for special effects in concrete.

■ PANELING

Most paneling plywoods are manufactured from hardwood species (see Section

FIGURE 6.3-9 *Three available textures of plywood that serve as combination siding and sheathing.* (*Courtesy American Plywood Association.*)

6.3.2.1). Some softwood species, such as California redwood, knotty pine, and western red cedar, are used for interior paneling. Other manufacturers assemble softwood veneers into paneling plywood with surface textures such as embossing, striations, or relief grain. Some manufacturers produce softwood panels having printed overlays simulating hardwood grains.

■ **TONGUE-AND-GROOVE PANELS**

Plywood in 1/2-, 5/8-, and 3/4-in. thicknesses, with tongue-and-groove joints, is produced by a number of mills in underlayment grades and occasionally other grades. Tongue-and-groove panels are used to eliminate the need for blocking at joints. One performance-rated panel is available in thicknesses from 19/32 in. up to 1-3/32 in. for use as a combination subfloor and underlayment, for use over joist spacings varying from 16 in. to as wide as 48 in., based on the panels being continuous over at least two spans. These types of panels are tongued and grooved (Fig. 6.3-10) to eliminate the need to block the joints.

FIGURE 6.3-10 *Tongue-and-groove panels.* (*Photograph Courtesy American Plywood Association.*)

■ **CONCRETE FORM PANELS**

The face veneers for plywood panels designed for concrete formwork are not less than B grade. The inner plies are not less than C grade. These panels are usually sanded on both sides and oiled at the mill. They are classified as either Class I or Class II according to the combination of wood species groups used in faces, crossbands, and center veneer plies. Class I panels require the use of stronger and stiffer species groups in faces and crossbands than do Class II panels. Concrete form panels are designed for reuse. HDO panels are also sometimes used as concrete forms.

■ **MARINE PLYWOOD**

Marine plywood panels are exterior grade plywood of either graded A-A, A-B, B-B, HDO, or MDO, but with certain additional requirements. For example, the only wood species permitted are Douglas fir 1 and western larch; the number of repairs in the face veneers is more limited than usual; inner plies must be B grade or better; wood repairs must be glued with an exterior adhesive; and there are additional restrictions as well.

6.3.1.8

Sizes

Plywood is most readily available in panels 4 ft. wide by 8 ft. long. Some mills also produce 5-ft.-wide panels and lengths up to 12 ft. but these are not always stocked. Beyond these limits widths up to 10 ft. and any reasonable length (which has been more than 50 ft.) can be made by scarf jointing or finger jointing panels together. Scarf- or finger-jointed construction consists of gluing panels either end-to-end or side-to-side continuously along a tapered or fingered cut to form a single panel (Fig. 6.3-11).

Plywood panels are permitted by PS 1 to be up to 1/16 in. smaller but may be no larger in horizontal dimension than the

FIGURE 6.3-11 *Panels can be scarf jointed (top) or finger jointed (bottom.)*

specified size. Nominal available thicknesses and thickness tolerances for plywood panels are shown in Figure 6.3-12.

Plywood panels always have an odd number of *layers* for balanced construction, the minimum number being three. A layer may consist of a single thickness of veneer, or it may be a lamination of two veneers with their grains running in the same direction. Each single thickness of veneer is called a ply. The minimum number of plies for various thicknesses of plywood is given in Figure 6.3-13. A diagram of how the layers are typically arranged is given in Figure 6.3-14.

6.3.1.9

Physical and Mechanical Properties

Aside from the characteristics of the veneer used, the most important properties of plywood derive from cross-laminated construction. Plywood utilizes the strength and dimensional stability of wood along the grain by placing alternate layers at right angles. This cross lamination provides resistance to splitting and warping, improved dimensional stability, and other properties as discussed below.

■ DURABILITY

In order to comply with the PS 1 requirements for exterior type plywood, the glueline used to bond the veneers must be as durable as the wood itself. Theoretically, exterior panels may be exposed outside without paint or other treatment and without affecting the durability of the glue bond or the life of the panels. In practice, exposing the edges of any plywood to weather is not a good idea. The glueline at the edge of a plywood panel is the most vulnerable part of the panel to damage by water and weather. Usually, therefore, plywood is painted or stained to protect it against weathering and checking. Preservative treatments are available for plywood that is to be used in areas of high decay or termite hazard.

■ BENDING STRENGTH AND STIFFNESS

The cross lamination of layers in plywood panels capitalizes on the more desirable physical and mechanical properties of

FIGURE 6.3-12
Plywood Panel Thicknesses and Tolerances According to U.S. Product Standard PS 1

Plywood Type	Normally Available Nominal Thicknesses, in.*	Tolerances (plus or minus)
Unsanded panels	5/16, 3/8, 15/32, 1/2, 19/32, 5/8, 23/32, 3/4	1/32 in.
	7/8, 1, 1-1/8	5% of specified thickness
Sanded panels	1/4, 9/32, 11/32, 3/8, 15/32, 1/2, 19/32, 5/8, 23/32, 3/4	1/64 in.
	7/8, 1, 1-1/8	3% of specified thickness
Touch-sanded panels	1/2, 19/32, 5/8, 23/32, 3/4	1/32 in.
	1-1/8	5% of specified thickness
Marine plywood	1/4, 3/8, 1/2, 5/8, 3/4	1/64 in.
Overlaid plywood	11/32, 3/8, 15/32, 1/2, 19/32, 5/8, 23/32, 3/4	1/32 in.

*All listed thicknesses may not be available in every panel grade or from every mill.

wood. The strength of wood parallel to the grain is many times greater than that perpendicular to the grain. Cross-laminated construction provides longitudinal grain crosswise and improves significantly on solid wood's crosswise strength and stiffness. However, plywood panels are both stiffer and stronger along the face grain than across the face grain and should be applied with the face grain perpendicular to its supports for maximum resistance to bending.

■ RACKING RESISTANCE

Plywood functions in its own plane as a structural bracing material when used on a wall, roof, or floor. Although 5/16 in. is the minimum thickness for sheathing, even 1/4-in.-thick plywood on a standard wood stud-framed wall provides twice the stiffness and more than twice the strength furnished by 1-in.-thick diagonal boards, which has been long recognized as adequate. The racking strength of plywood and other sheathing materials is further discussed later in this chapter.

Two properties give plywood superior bracing qualities: (1) high shear strength in all directions for loads applied perpen-

dicular to panel faces, and (2) relatively high nail-bearing strength, allowing nails to be driven as close as 1/4 in. from the edge of a panel. Splitting is virtually impossible lengthwise or crosswise since no cleavage plane exists.

■ IMPACT AND CONCENTRATED LOAD RESISTANCE

Stresses resulting from impact or concentrated loading are distributed over a wide area by the cross veneers in a plywood panel. This is an important property in sheathing, subflooring, siding, and wall paneling.

■ DIMENSIONAL STABILITY

Like solid wood, plywood is subjected to dimensional changes resulting from variations in moisture content, temperature, and other causes.

■ HYGROSCOPIC EXPANSION AND CONTRACTION

Expansion and contraction in solid wood is generally insignificant parallel to the grain, even with large moisture content

FIGURE 6.3-13
Panel Constructions

Panel Grades	Finished Panel Nominal Thickness Range, in.	Minimum Number of Plies	Minimum Number of Layers
Exterior			
Marine	Through 3/8	3	3
Special exterior	Over 3/8, through 3/4	5	5
B-B concrete form	Over 3/4	7	7
High-density overlay			
High-density concrete form overlay			
Interior			
N-N, N-A, N-B, N-D, A-A, A-B, A-D, B-B, B-D			
Structural I (C-D, C-D plugged, and underlayment)	Through 3/8	3	3
Structural II (C-D, C-D plugged, and underlayment)	Over 3/8, through 1/2	4	3
	Over 1/2, through 7/8	5	5
	Over 7/8	6	5
Exterior			
A-A, A-B, A-C, B-B, B-C			
Structural I and Structural II (C-C and C-C plugged)			
Medium density and special overlays			
Interior (including grades with exterior glue)			
Underlayment	Through 1/2	3	3
	Over 1/2, through 3/4	4	3
Exterior	Over 3/4	5	5
C-C plugged			
Interior (including grades with exterior glue)			
C-D	Through 5/8	3	3
C-D plugged	Over 6/8, through 3/4	4	3
	Over 3/4	5	5
Exterior			
C-C			

FIGURE 6.3-14 *Each layer is perpendicular to adjacent layers for strength and stiffness.*

changes. Perpendicular to the grain, dimensional changes can be considerable (see Section 6.1.3.2). In plywood, any tendency for veneers to expand or contract crosswise is greatly minimized by one or more adjacent veneers running length-wise. The total expansion both across the width and along the length of a 4-ft. by 8-ft. panel in service under normal variations in moisture content (8% to 14%) should average between 0.03 in. (1/32 in.) and 0.05 in. (1/20 in.).

■ THERMAL EXPANSION

Expansion and contraction resulting from thermal changes in solid wood is about 10 times as much across the grain as it is parallel to the grain, where changes are quite small. In plywood thermal expansion along the face is somewhat greater than in plain wood because of the influence of the crossbands. In an 8-ft.-long Douglas fir panel with five equal plies the expansion caused by a 120°F rise in temperature should average approximately 0.021 in.

(1/50 in.). This is about 40% more than for plain Douglas fir but still insignificant for most uses. Across the grain, expansion resulting from a 120°F rise in temperature is approximately 0.029 in., or only about one-sixth that for plain wood.

■ WARP

Bow, cup, or twist may occur in a plywood panel if it is subjected to unequal absorption of moisture along its edges, different exposure to moisture on opposite faces, or unbalanced painting or coating of its faces. However, plywood that is flat at the time of manufacture will remain so unless conditions of storage or use subject it to uneven changes in moisture content.

■ CHECKING

Checking may appear on any wood surface, plywood included, in the form of hairline cracks or even slightly open splits. Under severe moisture or dry conditions, checks may, over time, become open cracks that penetrate virtually the full thickness of the face. With moisture pickup, the exposed veneer surface expands while the undersurface glued to the crossband is held in a relatively fixed position. The expanding outer fibers push against one another and acquire a *compression set*. As the fibers dry out and shrink they pull away from each other, creating checks of different intensity.

Remedies to control checking within acceptable limits include protection from dampness or moisture during transit and storage, protection during work at the job site, and the use of protective coatings, such as conventional paint, and water-repellent dips to minimize moisture absorption and loss.

■ WORKABILITY

Plywood is easily worked, cut to various sizes and shapes, and may be bent to fairly sharp radii (Fig. 6.3-15).

■ FIRE AND DECAY RESISTANCE

A high degree of resistance to either fire or decay can be imparted to plywood through a pressure-impregnation treatment similar to that used for lumber (see Section 6.1.6). Only plywood bonded with exterior glue is treated, because the adhesive used in it is not affected by pressure impregnation of fire-retardant chemicals. Fire-retardant treatments have little effect on the rate of heat transfer through a panel,

FIGURE 6.3-15
Minimum Plywood Bending Radii

| Panel Thickness, in. | Bending Radii for Panel Bend in Direction, ft. | |
	Across Grain	Parallel to Grain
1/4	2	5
5/16	2	6
11/32 and 3/8	3	8
15/32 and 1/2	6	12
19/32 and 5/8	8	16
23/32 and 3/4	12	20

Courtesy American Plywood Association.

but virtually eliminate fuel contribution to a fire and prevent the wood from burning by causing it to char.

Plywood used indoors may be protected by the application to its surface of fire-retardant paints. When exposed to high temperatures, these paints blister or bubble, forming air pockets that insulate the plywood, retard combustion, and greatly reduce flame spread along the surface.

6.3.2

DECORATIVE PLYWOOD

The division of plywood products into two categories, *construction and industrial* and *decorative*, is based on the two governing standards: the previously discussed "U.S. Product Standard PS 1 for Construction and Industrial Plywood" and the Hardwood Plywood and Veneer Association's (HPVA) interim standard HP-1 1993, "Interim Voluntary Standard for Hardwood and Decorative Plywood." HP-1 1993 supersedes the previous standard, the American National Standards Institute's (ANSI) Standard HP-1983, "American National Standard for Hardwood and Decorative Plywood."

Construction and industrial plywood covered by PS 1 is made mostly of softwood species, although a few hardwood species are permitted. Decorative plywood covered by HP-1 1993 consists entirely of hardwood-faced panels. The interior layers of these panels are often of softwood or particleboard, however, to help lower their cost and to conserve the more rare hardwood species.

Decorative plywood panels are most often used as wall paneling, in cabinetwork and furniture making, as thin face skins in door manufacture, and as face veneer for block flooring.

The species of wood used for the face veneers of panels identifies the name of hardwood plywood panels. For example, birch plywood has a face veneer and, probably, back veneer of birch.

6.3.2.1

Manufacture

Differences between the manufacture and appearance of hardwood and softwood plywood, as well as differences among hardwood plywoods, may result from: (1) the tree species; (2) the part of the tree used; (3) the cutting methods used to produce the veneer; (4) the methods used to match the veneers; and (5) the type of core construction employed.

■ TREE SPECIES

More than 99,000 species of hardwoods have been classified. About 240 of these are available in commercial quantities in the United States, and about 50 are widely used for hardwood plywood. These include lauan, walnut, birch, gum, oak, maple, ash, cherry, elm, beech, African mahogany, Honduras mahogany, butternut, pecan, cottonwood, sycamore, hackberry, teak, and rosewood. Popularity of any one waxes and wanes with the public's changing preferences. Some typical imported and domestic hardwood veneer species and their characteristics are listed in Figure 6.3-16.

■ PARTS OF THE TREE

The portion of the tree used greatly affects hardwood plywood figure patterns. Most plywood comes from relatively straight portions of the tree, but some of the more interesting and valuable patterns are cut from a crotch, a burl, or a stump. Veneers cut from a crotch have a plumelike design. A burl produces veneers with a pattern of swirls. A stump yields rippling patterns with rich contrasts (Fig. 6.3-17).

FIGURE 6.3-16
Visual Characteristics of Representative Hardwood Species

Commercial Name	Color	Type of Figure
Ash, American	White to light brown	Medium open grain
Basswood	White to pale brown	
Beech	White to reddish brown	
Birch	White to light reddish brown	Curly grained, figured flat cut, plain rotary
Butternut	Pale brown	Leafy grain
Cherry, American	Light to dark reddish brown	Plain to rich mottle
Cottonwood	White to light grayish brown	
Elm	Light brownish red	Strong
Gum, black and tupelo	White, grayish white Greenish to grayish black	
Gum, red	Pink to reddish brown	Medium to highly figured
Hackberry	Pale yellow, yellowish gray to light brown	
Lauan	Pale grayish, yellowish brown to reddish brown	
Mahogany, African	Pink to reddish brown	Plain stripe to highly figured
Mahogany, tropical American	Pink to gold brown	Straight to rich mottle
Mahogany, crotch and swirls	Pink to reddish brown	Moon and feather crotch, plain and figured swirl
Maple	White to tan	Plain, curly burls
Oak, red	Pink tan to ochre	Plain to flake
Oak, white	Gray tan to ochre	Plain to flake
Rosewood, Brazilian	Pink brown and violet	Wide range figure
Rosewood, East Indian	Purple to straw	Striped and figured
Teak	Light tan, dark brown	Plain, ripple, mottle, stripe
Walnut, American	Soft gray brown	Typical figure or stripe

Used with permission of the Hardwood Plywood & Veneer Association.

FIGURE 6.3-17 *Portions of a tree from which various types of figure patterns are obtained. Some species produce several figure types.* (Used with permission of the Hardwood Plywood & Veneer Association and of the Fine Hardwood Veneer Association.)

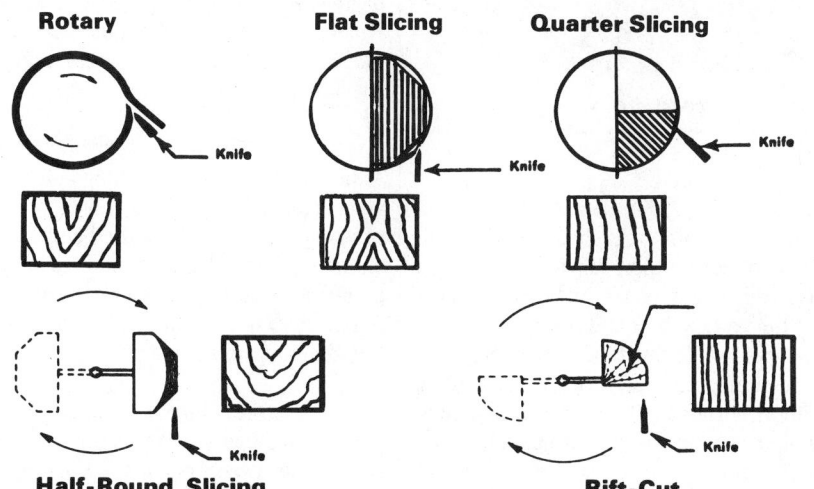

FIGURE 6.3-18 *Visual characteristics of veneers depend on the method of manufacture.*

■ VENEER TYPES

The manner in which veneers are cut is important in producing various visual effects. Two logs of the same species that are cut differently will have entirely different visual characteristics even though their veneer colors are similar. Five principal methods of cutting veneers are used (Fig. 6.3-18).

ROTARY CUT To produce a rotary cut, a log is mounted centrally in a lathe and turned against a blade, producing exceptionally wide veneer compared with that cut by other methods (see Fig. 6.3-2). Since this cut follows the log's annular growth rings, the resulting pattern has a bold, variegated ripple. This is the type of cut used for most softwood veneers.

FLAT SLICING Flat slicing results in a variegated, wavy figure pattern. To produce it a half log, or flitch, is mounted with the heart side flat against the guide plate of a slicer, and the slicing is done parallel to a line through the center of the log (see Fig. 6.3-2).

QUARTER SLICING Quarter-sliced veneer has a series of stripes, straight in some woods, varied in others. To produce this type of cut a quarter log (flitch) is mounted on a guide plate so that the growth rings of the log strike the knife at approximately right angles.

HALF-ROUND SLICING Half-round sliced veneer shows modified characteristics of both rotary and plain-sliced veneers. It is a variation of rotary cutting in which segments of a log (flitches) are mounted off center in a lathe. This results in a cut slightly across the annual growth rings.

RIFT-CUT VENEER Various species of oak as well as some other species are sometimes cut using a rift cut. In these woods the ray cells that radiate from the center of the log are conspicuous, and a rift or comb grain effect can be obtained by cutting perpendicularly to these rays either on a lathe or on a slicer.

■ VENEER MATCHING

The methods used to match face veneers also produce different visual characteristics. Basic and special matching methods are shown in Figure 6.3-19.

BOOK MATCHING Any type of veneers can be book matched. In book matching every other sheet is turned over like the leaves of a book. Thus, the back of one veneer meets the front of the adjacent veneer, producing a matching joint design. Book matching is also called edge matching.

Basic Matching Effects

Book Match

Random Match

Slip Match

Vertical Butt and Horizontal Bookleaf Match

Special Matching Effects

Diamond

Reverse Diamond

Four-way Center and Butt

"V"

Herringbone

FIGURE 6.3-19 *Basic and special matching effects for wood veneers.*

SLIP MATCHING Quarter-sliced veneers are often slip matched, but this method may also be used for other types of veneer cuts. In this type of match, adjacent sheets are joined side by side, without turning, to repeat the flitch figure. Slip-matched veneers are common in *premium*, *good*, and *speciality* hardwood plywood grades, and are often used in doors and matched ungrooved wall panels.

RANDOM MATCHING The joining of veneers with the intention of creating a casual unmatched effect is called random matching. In this type of matching, veneers from several logs may be used in the manufacture of a single panel. Random matching is generally used in prefinished V-grooved wall panels.

VERTICAL BUTT AND HORIZONTAL BOOK-LEAF MATCHING Flitches whose length does not permit their fabrication into the desired height of panel can be matched using the vertical butt and horizontal bookleaf method. In this type of matching, the veneer may be matched horizontally as well as vertically. This type of matching is often used in custom wall paneling and in furniture.

FOUR-WAY CENTER AND BUTT MATCHING Several *special* matching effects are used in furniture manufacture and in custom wall paneling. One of these, four-way center and butt matching, is ordinarily applied to crotch or stump veneers to reveal the figure. Occasionally, flat-sliced veneers are matched in this manner when

panel length requirements exceed the length of available veneers.

■ **CORE CONSTRUCTION**

The innermost ply in a hardwood plywood panel is called the *core*. The plies between the core and the face plies are called *crossbands*. There are four basic types of cores used in hardwood plywood: *veneer*, *lumber*, *particleboard*, and *mineral*. When panels are exposed on one surface only, back and face plies are of the same species and thickness to balance the face veneers.

VENEER CORE The veneer core method is the same as that for softwood plywood. In hardwood plywood, three, five, or more plies are laid with the grain direction of adjacent plies at right angles to each other. The total number of plies is always odd so that the panels are balanced.

LUMBER CORE Some panels have a core of sawed lumber to which crossband plies and face veneers are glued (Fig. 6.3-20). Lumber core construction is often used in cabinetwork and furniture construction in which exposed edges or edge treatment such as doweled, splined, or dovetailed joints are desired, or where butt hinges are to be used. A lumber core is made of narrow wood strips edge glued together. Panels with hardwood bands that match the face applied on the edges may be specially ordered.

PARTICLEBOARD CORE Plywood for use as table, desk, and cabinet tops often has a particleboard core. This is an aggregate of wood particles (also called chipboard, chipcore, etc.) fused under heat and pressure. The face veneers are usually glued directly to this core, although crossbanding is sometimes used.

MINERAL CORE Fire-resistant panels are made with mineral cores. In this construction, the face veneers are bonded to a core of hard, noncombustible material.

■ **SIZES**

Hardwood plywood is most commonly sold in 4-ft. by 8-ft. panels although panels are custom-made up to 40 ft. long and up to 7 ft. wide. Stock panels may also be purchased in lengths of 10 and 12 ft.

Face veneer (of fine hardwood)

Crossband (usually poplar or gum)

Lumber core (gumwood, poplar, basswood, maple, etc.)

Crossband (poplar etc.)

Back veneer (fine hardwood if to be exposed)

FIGURE 6.3-20 *Typical 5-ply lumber core panel. Each ply (indicated by arrows) runs at right angles to adjacent plies.* *(Used with permission of the Hardwood Plywood & Veneer Association and of the Fine Hardwood Veneer Association.)*

Hardwood plywood is manufactured with an uneven number of plies in thicknesses ranging from 1/8 in. up to as much as 3 in. in custom panels. Common thicknesses include 1/4 in., 3/8 in., 1/2 in., and 3/4 in. in veneer core and 3/4 in. in lumber and particleboard core. Figure 6.3-21 shows common thicknesses for varying numbers of plies.

FIGURE 6.3-21
Standard Thicknesses of Veneer-Core Hardwood Plywood*

Plies	Thickness, in.
3	1/8, 3/16, 1/4
5	5/16, 3/8, 1/2
5 and 7	5/8
5, 7, and 9	3/4

Used with permission of the Hardwood Plywood & Veneer Association.

*Hardwood plywood is square within 1/16 in. measured on short dimension. Particleboard and MDF-core hardwood plywood panels are typically 3- and sometimes 5-ply in thicknesses up to 3/4 in. and thicker.

■ PATTERN AND FIGURE CHARACTERISTICS OF VENEER

Figure refers to the highlights of so-called crossfire running at right angles to grain direction. The grain character and direc-
tion are described by using the word *pattern*. Those in the industry may describe figure by saying it "has a great deal of crossfire," or "has a straight or broken stripe," or "is highly figured." Figure 6.3-22 shows some common veneer figures and the terms often used to describe them.

6.3.2.2

General Classifications

Hardwood plywood falls generally into two categories: (1) *specialty* plywoods, which are not typically or necessarily manufactured to meet industry standards, and (2) *standard* plywoods, which are produced to meet ANSI HT-1983 (Fig. 6.3-23).

■ SPECIALTY PLYWOOD

There are no industry-recognized standards covering *specialty plywoods*. Species, type, match, core construction, and other requirements are as agreed on by the seller and the buyer. Generally, panels containing the least amount of blemishes are sold at a higher price than those with large knots and filled knotholes, splits, and other surface defects. Often, the intention is that color and grain vary.

Each manufacturer develops its own grade terminology. However, such termin-
ology is rarely by a coded number or letter but by an arbitrary descriptive system. For example, *aristocratic* and *homestead* are names selected to appeal to different tastes rather than to establish degrees of quality. The aristocratic panels might contain few blemishes and very few filled places, whereas the homestead panels could contain more knots, large filled knotholes, and more character. There may be little, if any, price differential between the so-called aristocratic and the homestead panels.

V-GROOVED WALL PANELS V-grooved panels, either natural or prefinished, account for much of the hardwood plywood used in residential construction. There are two types of V-grooved panels: random matched, and gang grooved.

In random-matched panels the individual veneer sheets, of which there may be four to nine in a 4-ft. by 8-ft.-panel, are selected to simulate lumber planking. The V-grooving at the joints of the veneer sheets produces a mismatched panel with veneer sheets dissimilar in grain and color.

Gang-grooved panels are produced by passing one- or multipiece faces under a machine with grooving knives set at spaced intervals. When multipiece face veneers are used, no attempt is made to groove on the joints of the veneer. In gang-grooved panels, the spacing of grooves is identical from one panel to the next.

| CROSS-FIRE | SWIRL | MOTTLE | FIDDLE-BACK |
| RIBBON STRIPE | FLAKED | BLISTERED | PENCIL STRIPE |

FIGURE 6.3-22 *Commonly used figure descriptions.* (*Used with permission of the Hardwood Plywood & Veneer Association and of the Fine Hardwood Veneer Association.*)

FIGURE 6.3-23 *Classification of hardwood plywood.*

ARCHITECTURAL CUSTOM PANELS Panels made to special order are called architectural custom panels. The method is to select the desired hardwood face veneer from flitch samples before the plywood is manufactured. The selected flitch is then fabricated at a plywood plant into panels to meet the custom specification. The only limitations of architectural custom panels are those imposed by the patterns that are available in the length and width of the selected veneer.

MISCELLANEOUS USES Increasing amounts of hardwood plywood are produced for many uses too numerous to discuss here. Among many others are wall panels of varying widths less than 4 ft., and flooring products such as laminated block.

■ **STANDARD PLYWOOD**

Plywood that conforms with the require-

ments of HPVA HP-1 1993 is called hardwood plywood. It is manufactured primarily for use as stock panels, door skins, and furniture. HPVA maintains a quality control program that ensures that plywood carrying its official grade trademark conforms with HPVA HP-1 1993 (Fig. 6.3-24).

HARDWOOD PLYWOOD & VENEER ASSOCIATION		
FORMALDEHYDE EMISSION 0.2 PPM CONFORMS TO HUD REQUIREMENTS	SIMULATED DECORATIVE FINISH ON PLYWOOD	FLAME SPREAD 200 OR LESS ASTM E84
	hpva	
LAY UP 16 3.6 MM THICK HP-SG-86	MILL SPECIALTY GRADE	GLUE BOND TYPE II HPVA HP-1 1993

FIGURE 6.3-24 *Grade stamp of the Hardwood Plywood & Veneer Association.* (*Used with permission of the Hardwood Plywood & Association.*)

Stock panels, as the name implies, are fabricated and stocked by a plywood manufacturer and sold through local distribution warehouses to many consumers. The character and quality of face veneers are dependent on market conditions, and buyer selection is made from the available stock at the time of purchase. Panels are typically 4 ft. by 8 ft. of varying thicknesses, such as 1/4, 1/2, 3/4 in., and may be put to a variety of different uses.

Door-skin plywood comprises a large percentage of hardwood veneer and is used as door skins to face solid and hollow-core doors (see Section 8.3).

The furniture industry and other wood-product manufacturers make extensive use of hardwood plywood. Outside the construction industry, hardwood plywood finds almost endless uses in automobile, ship, boat, aircraft and railroad car construction, and in sporting goods, musical instruments, caskets, and many other items.

■ **STANDARD TYPES**

The word *type* is used here to indicate a panel's durability, that is, the ability of the plies to stay together under different exposure conditions. Three basic types of hardwood plywood are produced to conform with the requirements of HPVA HP-1 1993.

TYPE I (EXTERIOR) Type I hardwood plywood is made with waterproof adhesive and is used where it may come in contact with water or must withstand full weather exposure. These adhesives are usually resins of one of the following types: phenol, resorcinol, phenolresorcinol, melamine, or melamine urea. Veneer core construction is preferred for Type I plywood.

TECHNICAL TYPE The technical type is made with the same adhesive requirements as Type I but varies in thickness and arrangement of plies to provide approximately equal tensile and compressive strength in length and width.

TYPE II (INTERIOR) Type II plywood is made with water-resistant adhesive (urea resin) and is used where it would not be subjected to contact with water, although it should retain practically all of its strength if occasionally subjected to thorough wetting and drying. The majority of hardwood plywood is Type II.

■ **STANDARD GRADES**

A plywood panel's *grade* designates the quality of the face, back, and inner plies. Hardwood veneer grades used under HPVA HP-1 1993 are designated as face grades, speciality grades, back grades, and inner ply grades.

FACE GRADES Face veneers may be made from more than one piece. With most species, multipiece faces must be book matched or slip matched. The quality of these veneers is high: only a few small burls, occasional pin knots, slight color streaks, and inconspicuous small patches are permitted. No other defects, such as splits, shakes, worm holes, or decay, are allowed. Face grades include AA, A, B, C, D, and E, in descending order of quality.

SPECIALTY GRADE Speciality grade (SP) veneers includes face veneers that are unlike those in the face grades. Examples given in HPVA HP-1 1993 include wormy chestnut, birdseye maple, and English brown oak, all of which have unusual decorative features.

BACK GRADES Back-grade veneers are labeled 1, 2, 3, and 4, with 1 being the most restrictive. These grades are used on surfaces that will not be exposed in the finished work.

INNER PLY GRADES There are four grades used to identify the inner plies of hardwood plywood panels, J, K, L, and M. J is the most restrictive. M is reserved for plies not directly adjacent to facing plies.

6.3.3

NONVENEER PANELS

Particleboard panels are manufactured from small fragments of wood bound together with an adhesive. These types of products have been variously called *flakeboard*, *waferboard*, and *particleboard*. Today, most of them are called *oriented strand board* (OSB). Most structural nonveneer wood panels produced today are of the OSB type, in which panels are made using strands or wafers of wood mostly oriented in the long direction of the panel. Several individual panels are laid with the direction of their strands or wafers perpendicular to each adjacent sheet and bonded together to form a final panel. This cross lamination makes such panels strong and stable for the same reasons that plywood has those characteristics.

In general, particleboard should comply with the requirements of ANSI A208.1, "Mat-formed Wood Particleboard," but some such products are actually extruded. Particleboard may be either fire-retardant treated or left untreated.

In some panel products particleboard panels are used in combination with wood panels to produce a board. The particleboard sheets are used as crossbands with a wood core and wood face veneers. The panels are bonded in the same manner as is standard plywood. Some such products are made in a one-step pressing operation in which the voids in the veneers are filled by the wood particles in the particleboard layers.

6.3.4

FIBERBOARD

Wood- or cane-fiber panels are used primarily for sheathing wood frame walls. They are of two types: regular density and intermediate density. Panels are coated or impregnated with asphalt, or given some other treatment to provide water resistance. Occasional wetting and drying during construction will not damage this material.

Refer to Section 6.6.5 for additional data about fiberboard panels.

6.4 TREATED WOOD FOUNDATIONS

The permanent wood foundation system described in this section was developed cooperatively by the National Forest Products Association (NFPA), the American Wood Preservers Institute (AWI), and the Economics and Marketing Division of the U.S. Forest Service.

Most local and all major model building code authorities recognize, and the U.S. Department of Housing and Urban Development/Federal Housing Administration (HUD/FHA) and the Farmers Home Administration accept, the permanent wood foundation system on the basis of the NFPA's Technical Note No. 7, "The Permanent Wood Foundation System: Basic Requirements." The text of this section is based on the January 1987 editions of Technical Note No. 7 and the NFPA publication *Permanent Wood Foundation System, Design, Fabrication, and Installation Manual.*

Specific design and construction recommendations for a permanent wood foundation system are based on experience with houses utilizing preservative-treated wood foundations built in the United States and Canada. These studies were conducted by the National Association of Home Builders (NAHB) Research Foundation.

The system provides a practical foundation for light-wood-framed buildings that has several advantages over more conventional foundation systems. These advantages include reduction in construction delays because of inclement weather, savings in cost and in erection time, and creation of more comfortable below-grade living area. The system is approved as a permanent foundation in homes insured by HUD/FHA, the Veterans Administration (VA), and the Farmers Home Administration. It is accepted under the National Building Code, the Uniform Building Code, and the Standard Building Code.

Unfortunately, the system also has a few disadvantages. It requires good quality control to prevent seemingly minor errors in judgment becoming serious structural problems later. For example, a worker might substitute a piece of untreated lumber where treated lumber is required. Careless backfilling can damage the protective film layer and expose the wood to below-grade earth.

6.4.1

GENERAL DESCRIPTION

A permanent wood foundation system is fabricated of pressure-treated lumber and pressure-treated plywood. The structural elements of the substructure are wood. Foundation wall sections of nominal 2-in. lumber and plywood cladding may be factory fabricated or constructed at the job site.

Either conventional concrete footings or wood footing plates may be used. Wood footing plates are nominal 2-in. or thicker pressure-treated wood planks resting on a 4-in. or thicker bed of fine gravel or crushed stone.

The walls are of wood stud construction with a plywood facing on the earth side. Where a basement occurs, a finish on the interior of the wall is optional. When habitable space is enclosed, the wall is insulated.

The system can be used for both basement and crawl space construction. The below-grade portions of exterior basement walls are covered with a polyethylene film that is bonded to the plywood at the top. Joints are lapped and sealed with adhesive. Concrete basement floors are poured over a gravel base.

Typical details of a wood foundation system used for basement and crawl space construction are shown in Figures 6.4-1 and 6.4-2. There are several variations permitted, however, so details different from those shown may be stipulated.

6.4.2

SITE PREPARATION

For basement construction, a house site should be excavated and leveled to the required elevation. Utility lines located below the basement floor should then be installed and provision should be made for foundation drainage in accordance with local requirements.

When habitable space is planned below grade, unless the soil is very porous, a sump is required to ensure adequate drainage (Fig. 6.4-3). A sump can be drained by gravity or by pump to a storm sewer or drainage swale.

After grading and installation of plumbing lines and drainage system, the basement site should be covered with a layer of gravel. For crawl-space construction, trenches should be dug to the required depth and gravel placed as a base for the footing plates.

6.4.3

FOOTINGS

The width of footing plates is determined by loads on the foundation wall and the bearing capacities of the gravel base and soil. When brick veneer is used, larger footing plates are required to provide added width for support of the veneer. Brick height should not exceed 8 ft. To reduce the amount of brick required, such veneer is typically supported below grade on a wood stud knee wall that rests on and is nailed to the treated wood footing.

Where basement construction is involved, footings under exterior foundations should be placed on a strip of polyethylene film laid directly on the gravel base.

A load-bearing wall may be used as center support under the first floor (Fig. 6.4-4). If a girder support is used instead, the footings under the foundation ending walls and under the posts should be designed for concentrated loads.

Footing plates should always be located below the frost line.

6.4.4

FOUNDATION WALLS

Detailed structural requirements, including member spacings and sizes, for

Flashing

Plywood may overlap field-applied top plate for shear transfer

Field-applied 2 × top plate

2 × top plate

Caulk

Finish grade slope 1/2 in./ft. minimum 6 ft. from wall

1 × or plywood strip protecting top of polyethylene film

2 × stud wall insulated as appropriate with warm-side vapor retarder

Asphalt or polyethylene film strip on crushed stone or gravel backfill

Optional interior finish

Plywood

Polyethylene film

3-in. or 4-in. concrete slab with optional 1 × screed board & vapor retarder

Bottom plate

Footing plate

4-in. gravel or crushed stone fill

Below frost line

8 in. minimum

3/4 d

d

2d

☐ Pressure-treated wood

FIGURE 6.4-1 *Basement wall.* *(Courtesy American Forest and Paper Association.)*

Floor joist

Field-applied 2 × top plate

2 × top plate

Finish grade slope 1/2in./ft. minimum 6 ft. from wall

Plywood

Polyethylene film

Stud wall

Bottom plate

Footing plate

Gravel or crushed stone footing

Below frost line

18 in. minimum

8 in. minimum

3/4 d

d

2d

▬ Pressure-treated wood

FIGURE 6.4-2 *Crawl-space wall.* *(Courtesy American Forest and Paper Association.)*

the foundation wall of typical one- and two-story wood frame houses are shown in the NFPA publication *Permanent Wood Foundation System, Design, Fabrication, and Installation Manual.*

6.4.4.1

Design Method

Lumber framing members in exterior walls enclosing a basement are designed to resist the lateral pressure of fill and the vertical forces resulting from live and dead loads on the structure. The plywood sheathing is designed to resist the maximum inward soil pressures at the bottom of the wall.

Foundation walls for crawl space construction are designed to resist only vertical loads when the difference between the outside grade and ground level in the crawl space is 12 in. or less. Framing is covered with 1/2-in. plywood to keep out vermin and provide additional stability.

6.4.4.2

Construction

Foundation walls are built over a previously laid wood footing. Common practice is to prefabricate a foundation wall into modular sections. Components are fastened together using corrosion-resistant nails. Treated foundation wall studs are end nailed to treated top and bottom plates. Treated plywood panels, 1/2 in. thick or thicker, are attached to studs. Usually, panels are applied with their face grain vertical and parallel to the studs. Vertical joints between panels within wall sections should be located over studs.

Foundation wall sections should be nailed to the wood footing and to adjacent sections. Wall panel joints should not occur over joints in the footing. An upper top plate should be face-nailed to the lower top plate of the foundation wall to tie the wall sections together.

An alternative method of attaching the upper top plate is to fabricate the foundation wall sections with the edge of the plywood wall panels extending 3/4 in. above the top of the treated top plate. The field-applied top plate is then face nailed to the top plate of the wall section and is secured to the overlapping edge of the plywood.

Note: 1. Use of sump pump is optional.
2. 24-in. diameter or 20-in.-square terra cotta tile or treated wood box may be substituted for precast concrete tile.

3 in. or 4-in. concrete slab

Precast concrete or steel manhole cover or pressure-treated plywood (1-1/2 in. laminated of two 3/4-in. pieces)

Polyethylene film

24-in.-diameter precast concrete tile 30 in. high

Minimum 2-in. gravel fill on all sides

Floor drain cover

To storm sewer or daylight

90-degree ell — 4-in.-diameter pipe (trap required when drained to sewer)

FIGURE 6.4-3 *Sump for poorly drained soils.* (Courtesy American Forest and Paper Association.)

Field-applied top plate 2 × top plate

Stud wall with cladding on at least one side

Polyethylene film or other vapor retarder

Concrete slab

Double 2 × bottom plates

Footing plate

d

2d

Pressure-treated wood

FIGURE 6.4-4 *Interior load-bearing wall.* (Courtesy American Forest and Paper Association.)

Joints in the field-applied top plate should be staggered with respect to joints in the treated wood plate of the wall panels. At corners and wall intersections, the field-applied upper plate of one wall should be extended across the treated plate of the intersecting wall to tie the building together. The first floor is then installed following standard platform frame construction practices.

Plywood panel joints in the foundation wall should provide the minimum space recommended by the manufacturer of the sealant compound selected. In both basement and crawl-space construction, the exterior side of foundation walls should be covered below grade with a 6-mil polyethylene sheet to act as a vapor retarder. This film should be bonded to the plywood, and should have its joints lapped and sealed (see Figs. 6.4-1 and 6.4-2).

The top edge of the polyethylene sheeting should be adhered to the plywood wall with adhesive. Film areas near grade level can be protected from mechanical damage and exposure using treated plywood, brick, stucco, or other covering appropriate to the architectural design. Where plywood is used, the top edge of a 12-in.-wide panel of treated material is attached to the wall several inches above the finish grade level. The top inside edge of the panel should be caulked its full length before the panel is fastened to the wall.

Where appropriate, suitable insulation should be installed within the wall stud space. When habitable space occurs, an interior vapor retarder and wall finish are usually applied to the stud framing.

6.4.5

FLOORS

For basement construction, a 6-mil polyethylene film is applied over the gravel bed, and a concrete slab at least 3 in. thick is placed over the film. Placement of the film and the slab can be delayed until after the building is under roof. Backfill should not be placed against the foundation walls, however, until after the concrete slab is in place and set and the top of the wall is adequately braced. Where the height of backfill exceeds 4 ft., gravel should be used for the lower portion.

For crawl-space construction, a 6-mil polyethylene film ground cover should be used. Film edges should be lapped and sealed.

The first-floor framing on a wood foundation is installed according to conventional platform framing methods. Floor joists should be suspended at least 18 in. above the ground.

6.4.6

CENTER SUPPORT

Framing members in load-bearing walls acting as center support for a first-floor joist system are similar to exterior foundation walls (see Fig. 6.4-4), except that only vertical loads need be considered during design.

6.4.6.1

Load-Bearing Walls

In basement construction, load-bearing interior walls should be made with pressure-treated bottom plates and, in areas of low degrees of termite hazard, with untreated studs and top plates. A double bottom plate is used to provide clearance between the top of the concrete slab floor and the untreated vertical framing in the wall (see Fig. 6.4-4). In areas of a high degree of termite hazard, pressure-treated studs should be used. The interior wall framing can be covered on both sides with gypsum wallboard or other interior finish, as desired.

Interior load-bearing walls in crawl space construction should be made of treated framing and covered on one side with 1/2-in.-thick treated plywood.

6.4.6.2

Post-and-Girder Support

Where a post-and-girder center support system is used, the posts should be supported on blocks of sufficient size to ensure that the vertical concentrated loads will not exceed the bearing capacity of the gravel base and soil. Foundation walls supporting a center girder should also be designed to carry and distribute the concentrated load from the girder to a wood footing and gravel base, so that the allowable gravel- and soil-bearing capacities are not exceeded. Design of a suitable post-and-girder support system requires engineering calculations and should not be attempted without benefit of professional advice.

6.4.7

LUMBER AND PLYWOOD PRESERVATIVE TREATMENT

Lumber and plywood used in exterior foundation walls below an upper top plate and wood footings or bottom plates and placed in contact with concrete or the ground should be pressure treated. Treatment should be in accordance with AWPA Standard C22 "Lumber and Plywood for Permanent Wood Foundations, Pressure Process."

Treated lumber and plywood should bear the appropriate American Wood Preservers Bureau (AWPB) Quality Mark or that of an acceptable independent inspection agency that maintains continuing control, testing, and inspection over the quality of the product. Treated lumber should be dried to a moisture content not exceeding 19% after treatment. To the extent practicable, any fabricating, cutting, boring, and trimming of lumber and plywood should be done before treatment. Lumber or plywood cut after treatment should have any edges exposed brush coated with a 3% solution of the same preservative used in the original treatment.

6.5 GENERAL FRAMING REQUIREMENTS

A variety of methods and systems are used in the construction of wood-framed buildings across the country. Local climatic conditions, economic considerations, and aesthetic preferences may influence the selection of the system to be used. This section includes some of the requirements common to the various systems. Sections 6.6 and 6.7 address the more specific aspects of various framing systems.

6.5.1

ANCHORAGE REQUIREMENTS

The superstructure of a wood-framed building should be anchored to the foundation to provide resistance to lateral forces, primarily from wind. Wind pressures produce compressive, tensile (uplift), shearing, and racking loads on a foundation (Fig. 6.5-1).

6.5.1.1

Overturning, Translation, and Rotation

Overturning forces may produce critical stresses in foundations that support light superstructures subjected to high wind pressures. Connections anchoring a building's superstructure to its foundation, and the foundation itself, must have adequate strength in tension to resist these forces.

FIGURE 6.5-1 *Wind forces tend to lift or overturn light superstructures, move them laterally, or rotate them on their foundations.*

FIGURE 6.5-2 *Minimum separation of wood from the ground.*

When the ratio of exposed building surface to the width of the building is large, high wind forces tend to cause translation (lateral movement) of the superstructure. When the wind forces are not symmetrical, building rotation can occur. Forces tending to cause translation and rotation are transferred to a foundation because the first floor acts as a diaphragm. Connections anchoring the superstructure must develop enough shearing strength to transmit these forces to the foundation. When a first floor does not have enough rigidity to provide diaphragm action, high shear and bending stresses may be produced in the foundation walls. In this case, consideration should be given to the use of shear walls, pilasters, or bond beams to stiffen the walls.

6.5.1.2

Anchorage Recommendations

Superstructures should be anchored to both masonry and concrete foundations as applicable. These anchoring requirements are covered in the model codes and in the Council of American Building Officials' (CABO) *One and Two family Dwelling Code.* In general the following requirements apply, but each should be verified with the requirements in the applicable code.

The sill plates of a wood-joist floor system should be anchored to masonry walls with 1/2-in. bolts, extending at least 15 in. into the filled cores of masonry units. Anchor bolts should be spaced not more than 6 ft. apart, with one bolt not more than 12 in. from each end of the sill plate.

For concrete walls, sill plates should be anchored with 1/2-in. bolts embedded

7 in., with the same maximum spacing as for masonry walls. Hardened steel studs driven by powder-actuated tools also may be used, except when earthquake design is required. Their maximum spacing should not exceed that required by code.

6.5.2

Decay and Termite Control

The design and construction of foundations that support buildings using wood structural floor, wall, and roof systems must include precautions for the protection of the wood from the effects of decay and attack by subterranean termites (see Sections 2.2, 2.3, 6.1.5, and 7.1).

Suitable protection of wood structural elements is provided by (1) positive site and building drainage, (2) treatment of the soil with pesticides, (3) separation of wood elements from the ground, (4) preservative treatment of the wood, and (5) ventilation and condensation control in enclosed spaces.

6.5.2.1

Separation of Wood from the Ground

When possible, wood should be separated from direct contact with soil. Figure 6.5-2 shows recommended distances for the separation of wood from the ground. These distances will (1) maintain wood at a safe moisture content for protection from decay, (2) provide a termite barrier, and (3) facilitate periodic visual inspection. When it is impracticable to comply with the clearances recommended,

all heartwood of a naturally durable species or pressure-treated lumber should be used. However, where termites are known to exist, access space should always be provided for periodic visual inspection. Where mechanical equipment is located in a crawl space, greater clearances may be required.

6.5.2.2

Termite Barriers

A *termite barrier* is any material that cannot be penetrated by termites and that drives the insects into the open where they can be detected and eliminated. Where local experience indicates that additional protection against termites is required, one or more of the following barriers is recommended, depending on the degree of hazard: (1) chemical soil treatment (see Section 2.2), (2) preservative-treated lumber for all floor framing up to and including the subfloor (see Section 6.1.5), or (3) properly installed termite shields.

Additional precautions include using poured concrete foundations in which no cracks greater than 1/64 in. are present, or poured reinforced concrete caps at least 4 in. thick on unit masonry foundations. Such caps should have no cracks greater than 1/64 in.

6.5.3

Nailing Practices

Of primary consideration in the construction of a light-wood-framed building is the method used to fasten the various wooden members together. These connections are most commonly made with nails, but in certain cases metal straps, lag screws, bolts, staples, and adhesives may be used.

Proper fastening of frame members and covering materials provides the rigidity and strength to resist windstorms, earthquakes, and other natural disasters. Adequate nailing is also important for the suitable performance of wood parts. In general, nails make stronger joints when driven into the side grain (perpendicular to wood fibers) than into the end grain of wood. For instance, the allowable lateral (shear) loads for nails driven into the side grain are 50% higher than those allowed when the nails are driven into the end grain. Whenever possible, connections should be designed in such a way that nails are loaded in shear, exerting a sideways pull on the nail, rather than in direct withdrawal, pulling straight out. Connections should never be designed in a way that would load an end-nailed joint in direct withdrawal (Fig. 6.5-3).

The various codes summarize in tables the required nailing practices for the framing of a conventional wood-frame building. Penny designations and actual sizes of common wire nails are shown in Figure 6.5-4.

FIGURE 6.5-3 *Common nailing methods: (a) slant nailing at a 30-degree angle to the face or edge of the attached piece (toe nailing); (b) nailing into the end grain of the supporting piece (end nailing); (c) nailing into side grain through the wide dimension of the attached piece (face nailing); and (d) toe nailing through the edge of a plank to conceal the nail head (blind nailing).*

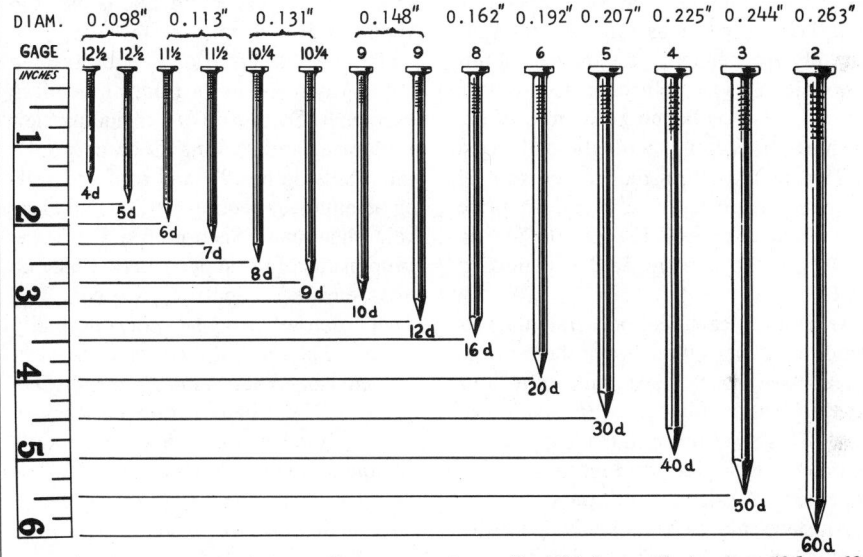

FIGURE 6.5-4 *Standard sizes of common wire nails. Thicker nails in sizes 10d to 60d are known as spikes.*

6.6 CONVENTIONAL FRAMING AND FURRING

Modern, light wood framing is an outgrowth of the *balloon frame*, which originated around the middle of the nineteenth century in Chicago. It was a revolutionary development that drastically changed the direction of the housing industry, and contributed substantially to the growth of the West. Historians have claimed that, if not for the balloon frame, Chicago and San Francisco could not have grown from villages to great cities in a single year. Houses and other wood-frame structures were built with a rapidity possible only with this revolutionary type of construction.

Before the balloon frame evolved, frame buildings were built of heavy timbers, mortised and tennoned together and pinned with hardwood dowels. The principle of balloon framing involves the repetitive use of slender structural members, called *studs* for walls or *joists* for roofs and floors, all tied together with simple nailed connections. It changed house construction from a highly skilled craft to a relatively simple, easily learned mechanical procedure. For that reason, perhaps, it was regarded with suspicion and acquired its derisive name on the assumption that a house built by such methods would blow away like a balloon the first time a strong wind hit it.

The development of the balloon frame is closely associated with the level of industrialization reached in the United States in the middle of the nineteenth century. The evolution of sawmill technology and the ability to mass produce cut nails cheaply are credited as the impetus of this revolutionary idea. When the manufacture of cut nails was begun at the turn of the eighteenth century, wrought nails cost $.25 a pound. With the introduction of more sophisticated machinery, the price was reduced to $.08 in 1820, to $.05 in 1833, and to a mere $.03 a pound by 1842.

Over the years, balloon framing was modified to permit the use of shorter (and cheaper) wall studs spanning from floor to floor rather than from foundation to roof. This new approach made it possible to erect a floor platform first, then to nail together a full-length wall on a horizontal surface, more conveniently and faster than in an upright position. This framing system evolved as wood framing moved west with the early settlers and came to be known as *Western* or *platform framing*. Most new houses today are platform framed.

The search for economy in labor and materials is a continuing process. As the cost of labor became a more significant factor in total housing cost, plank-and-beam construction (see Section 6.7.2) and other floor and roof assemblies evolved, utilizing greater spacing of fewer and larger structural members. Such systems were more compatible with post-and-beam wall framing (see Section 6.7.1) and, as with the skeleton structure of steel and concrete buildings, enabled designers

to treat the space between posts as a curtain wall, intended mainly to keep out the weather rather than to resist loads. The use of prefabricated trusses (see Section 6.7.3) for roof support permitted fewer, and sometimes no, interior walls.

In spite of innovations in light wood framing, the essential construction principles inherent in the balloon frame persist. The nail still remains an essential connective element in light wood-framing systems and the exterior covering between framing members, the sheathing, is relied on to provide structural continuity to an assembly of light members.

Much of the material in this and related sections on wood framing is based on data in the U.S. Department of Agriculture, Agriculture Handbook No. 73, *Wood-Frame House Construction*; the publications of the NFPA; the publications of the APA; and the other publications listed in Appendix B and noted in the text.

The superstructure of a wood-framed building is made up of floor and ceiling assemblies (Section 6.6.1), including their subflooring and decking (Section 6.6.4), walls (Section 6.6.2), and roof and ceiling assemblies (Section 6.6.3), including their sheathing (Section 6.6.5). Some components of the superstructure may be preassembled (Section 6.6.6). The superstructure may be conventionally framed (Section 6.6) or may be constructed using other framing systems (Section 6.7). The superstructure may be supported directly on foundation walls (Chapters 3 and 4; Section 6.4), or may rest on posts, beams, and girders (Section 6.6.1.2) that in turn are supported by the foundations.

6.6.1

FLOORS AND FLOOR AND CEILING ASSEMBLIES

The floor framing in a conventional wood-framed building consists of posts, beams (or girders), sill plates, and joists. Floor sheathing (subflooring) ties the various framing elements together and stabilizes the joists to prevent twisting and buckling. When these elements are assembled properly on a foundation, they form a level, anchored platform for the erection of partitions, additional floors, and the roof of

a building. Posts and center beams, which support the inside ends of the joists, sometimes are replaced with a wood frame or masonry wall when the basement area is divided into rooms. Wood-framed buildings may also be constructed over a crawl space with floor framing similar to that used for a fill basement. In a slab-on-ground house, wood floor framing would be limited to the upper floors.

In multilevel buildings, the combination of each intermediate floor and the ceiling system below it is usually recognized as a single basic element of the superstructure. A combined floor and ceiling system is called a *floor and ceiling assembly*. The design, selection, and construction of a floor and ceiling assembly is based on (1) the required level of functional performance, (2) cost, and (3) the desired finished appearance. These considerations are often influenced by the systems and finishes used in other basic elements of the construction.

The functional performance requirements of a floor and ceiling assembly usually include (1) adequate strength to support its actual weight and other building elements that it supports (dead load) and occupants and furnishings (live load); (2) provision for lateral support of walls; (3) satisfactory resistance to the transmission of airborne and structure-borne sound; (4) suitable fire resistance; (5) suitability for the application of finish materials; (6) adaptability to economical methods of assembly and erection; (7) space to accommodate heating, air conditioning, electrical, and plumbing equipment; and (8) control of heat loss and the flow of water vapor.

Floor and ceiling assemblies must support superimposed loads safely and without excessive deflection or vibration. Designs for dead and live loads are usually established by building codes that require structural members to be sized so that an allowable fiber stress in bending is not exceeded when the assembly is fully loaded. Codes do not always limit deflection, because it seldom has a bearing on structural failure or safety. Deflection limitations should be established in floor and ceiling assembly design, however, to ensure the integrity of applied finish materials and to limit vibration to acceptable levels of human perception. Deflection, rather than bending strength, generally is the controlling factor in the structural design of wood floor and ceiling assemblies.

Depending on the design of the foundation system, a first-floor system may be required to provide lateral support to foundation walls; and floor and ceiling assemblies also may be required to provide lateral support for other elements of the superstructure against loads caused by wind or earthquakes.

The ultimate costs of a floor and ceiling assembly include in-place costs, maintenance costs, and operating costs. In-place costs are the sum of the costs of materials, labor, and overhead required to assemble and install a floor and ceiling assembly. The selection of the system may affect other costs, such as the cost of applying floor and ceiling finishes, constructing interior partitions, and providing space to accommodate heating, air conditioning, electrical, and plumbing equipment.

Maintenance and operating costs attributable to a floor and ceiling assembly, such as heat loss or gain, have an important influence on the ultimate cost of the assembly and should not be overlooked if long-term economy is to be achieved.

The makeup of a floor and ceiling assembly is influenced by the desired finish applications. The finish materials selected (1) may influence the spacing and span of structural members when several options are available; (2) may determine the need for floor underlayment or wall or ceiling furring when it is impracticable to vary the spacing of structural components, as is often the case, or when such variation would violate code requirements; and (3) may even dictate methods of construction.

Prefinished floor or ceiling surfaces may not require the application of an additional finish after the system has been placed. However, the floor should provide a working platform for finishing operations during the construction period. Therefore, a floor and ceiling assembly is often selected that has a subfloor or sheathing application that is later provided with finished surfaces.

Wood-framed floor and ceiling assemblies include wood floor joists (in this section), plank-and-beam construction (Section 6.7.2), and panelized construction (Section 6.6.6). Concrete slab-on-grade floor construction is discussed in Section 3.3.

6.6.1.1

Lumber Materials

Framing lumber should be of the proper grade for the intended use, graded by a recognized grading authority. The moisture content of framing lumber is important in the performance of finishes. It should be as near as possible to the equilibrium moisture content (see Section 6.2) for the particular area and should not exceed 19% at the time the finish material is installed. However, a moisture content not in excess of 15% is better in all regions of the country, and essential in arid regions, especially when a finish, such as gypsum board, will be applied directly to the framing. Dimension lumber generally can be obtained at a moisture content of 19% or less when S-DRY is specified, or at 15% or less when MC 15 is specified. In arid parts of the country, lumber of higher moisture contents can be specified if the lumber is allowed to approach its equilibrium moisture content before applying interior finishes and trim, such as baseboard, shoe mold, and door casings.

In light wood-framed buildings, wood studs are typical framing members for walls and partitions; wood joists and trusses are the most common ceiling- and roof-framing members. These framing members generally are spaced 16 in. or 24 in. on centers, which permits direct application of most finishes, such as gypsum board and plaster. However, furring (Section 6.6.7) may be required when member spacing exceeds the maximum recommended for the particular finish material or when special performance, such as increased sound isolation, is desired.

The National Grading Rule for dimension lumber, established in accordance with Product Standard PS 20-70 (see Section 6.2), provides a series of uniform grade designations applicable to all species. Joists and planks (2 to 4 in. thick and 5 in. wide or wider) are graded as Select Structural, No. 1, No. 2, and No. 3 grades. All grades are suitable for floor joists if they have adequate strength characteristics for the intended spacing, span, and loading conditions (see Section 6.6.1.3). Light framing (2 to 4 in. thick and 2 to 4 in. wide) is sorted into up to seven grades. These grades, in order of decreasing strength, are: Select Structural,

No. 1, No. 2, Construction, No. 3, Standard, and Utility. An eighth, Stud grade, is similar to No. 3 grade, but is cut to 10 ft. or shorter lengths, is straighter, and has a better nailing surface. The lower light-framing grades should be used for sills and plates, rather than for studs and posts.

6.6.1.2

Posts, Beams, and Girders

Wood or steel posts are generally used in a basement to support wood girders or steel beams. Masonry piers can also be used in basements but most commonly are employed in houses with crawl spaces.

Round steel posts, known as pipe columns or lally columns (steel pipes filled with concrete), can be used to support both wood girders and steel beams. Columns should be provided with a steel bearing plate at each end. Secure anchoring of a girder or beam to a post or column is essential (Fig. 6.6-1).

FIGURE 6.6-1 (a) Steel top plate provides a means of anchoring a beam; (b) base plate distributes the load and anchors the steel post to the footing.

Wood posts for free-standing use in a basement should be of solid or built-up lumber and not less than the equivalent cross-sectional area of a solid 6 × 6. When combined with a frame wall, posts may be 4 × 6s to conform to the depth of the studs. Other sizes may be required, depending on the loads. Spaced columns may also be used. Refer to *Wood Structural Design Data* by the NFPA for design requirements for wood columns. Wood posts should be squared at both ends and securely fastened to the girder (Fig. 6.6-2a). The bottom of a post should rest on and be pinned to a concrete pedestal 2 to 3 in. above the finish floor (Fig. 6.6-2b). It is good practice to treat the bottom of a post with a decay preservative, or use a vapor-resistant covering, such as heavy roll roofing, over the concrete pedestal. A sheet metal covering is preferred in areas subject to termite infestation.

FIGURE 6.6-3 *At least one member of a built-up wood girder should be continuous across a supporting post. In termite-infested areas, a beam pocket should be lined with metal to form a termite shield.*

FIGURE 6.6-2 *(a) A wood post should be connected to a girder with metal angles of framing anchors; (b) the base of a wood post should be pinned and protected against decay and termites.*

Both wood girders and steel beams are used in present-day house construction. Standard steel S-shaped and W-shaped (wide flange) beams are the most commonly used steel beam shapes. Wood girders may be either solid or built up (Fig. 6.6-3). A built-up beam is preferred because it can be made from drier dimension lumber and is more dimensionally stable. Solid lumber or commercially available glue-laminated girders may be desirable when a beam will be exposed in a finished room.

The use of the terms *beam*, in connection with steel members, and *girder*, with wood members, is based on common usage, rather than on a technical difference in meaning. In engineering terminology a beam is any member with loads applied perpendicular to its direction of span, and a girder is a principal beam, frequently supporting other beams. In this broader sense, joists, girders, headers, and trimmers are types of beams. Wood joists are seldom called beams, but both built-up and solid wood members carrying joists frequently are referred to as beams or girders.

A built-up girder (see Fig. 6.6-3) is usually made up of two or more pieces of 2-in. nominal dimension lumber nailed together, with the ends of the pieces joined over a supporting post. Building codes dictate the required type, location, size, and number of fasteners necessary in built-up girders. For example, a code might require that a two-piece girder be face nailed from one side with 10d nails, two nails at the end of each piece and others driven near top and bottom in two rows 16 in. on centers, with the nails staggered. A three-piece girder might be required to be face nailed from each side with 20d nails, two nails near each end of each piece and others in two rows driven 32 in. on centers staggered. An example of member sizes and number of members in built-up girders for various spans is shown in Figure A-16. Be sure to read the text in Appendix A related to this figure. Solid beams with an equivalent cross section are usually permitted in lieu of built-up beams.

The ends of wood girders are generally required to bear at least 4 in. on foundation walls and pilasters. When the lumber is not preservative treated, a 1/2-in. airspace is often required at each end and at each side of wood girders framing into a masonry or concrete wall (see Fig. 6.6-3). In termite-infested areas, these pockets should be lined with sheet metal, which should extend on the inside about 3-1/2 in. to form a termite shield. A girder should be accurately aligned (true) and level. Wood shims should not be used to level girders or beams. Where adjustment is required, slate, metal, or another durable noncompressible material should be used to support the girder ends. Shim space under a beam or girder should be grouted solid.

Girder-Joist Connection

The simplest method of floor framing is one in which joists bear directly on a wood girder or steel beam. Then, the top of the beam coincides with the top of the anchored sill plate. This method is used when adequate headroom exists and a dropped girder is not objectionable. It is simple to frame and has the advantage of providing ample joist cavities for ductwork and electrical and mechanical equipment.

■ WOOD GIRDERS OR BEAMS

For more uniform shrinkage and to provide greater headroom, joists can be framed into a wood girder with joist hangers or ledger strips (Fig. 6.6-4). A ledger strip should be at least 2 × 2 in size and should be face nailed to the girder with three 16d nails at each point where a joist frames into the girder.

Depending on the sizes of the joists and girders, joists may be supported on a ledger strip in several ways. In the rare case where individual boards rather than plywood are used for subflooring, continuity across the supporting beam or girder must be provided by elements other than the subfloor. A continuous horizontal tie between exterior walls can be obtained by nailing notched joists together (Fig. 6.6-4a). Instead of notching the joists, a connecting scab at each pair of joists can provide this continuity and also serve as a nailing area for the subfloor (Fig. 6.6-4b). A steel strap can be used to tie joists together when the tops of a beam and the joists are level (Fig. 6.6-4c). When notched joists or scabs ride over the top of a girder, it is important that a small space be allowed above the beam to provide for shrinkage of the joists. Sometimes a ledger strip is positioned higher on a girder and the joists are notched so that the joist shoulders bear on the ledger strip and the joist bottoms are level with the girder. In such situations the notch depth should not exceed one-quarter of the depth of the joist.

When a space is required for heating ducts in a partition supported on a girder, a spaced wood girder sometimes is necessary (Fig. 6.6-5). Solid blocking is used at intervals between the two members. A

FIGURE 6.6-4 *Typical joist-girder connection showing how joists can be tied together when wood board subfloor is used.*

FIGURE 6.6-5 *Spaced girder detail showing connections to steel post and floor joists.*

single post support for a spaced girder usually requires a bolster (wide capital) of sufficient width to support the spaced girders.

■ STEEL BEAMS

Joists may be supported over or framed into a steel beam (Fig. 6.6-6). In general, connections between joists and a beam should provide (1) lateral stability for the beam; (2) resistance against twisting and overturning of the joists; (3) structural continuity and diaphragm action to resist lateral pressures against the foundation walls; and (4) anchorage for the entire floor assembly. This is accomplished generally by bolting a 2 × 4 or 2 × 6 plate to the flange of the beam (upper or lower) on which the joists are supported (Fig. 6.6-6a). Each joist should be toe nailed to the wood plate and should be bolted at 24 in. on centers, either through the beam's web or its flange.

When it is not feasible to use a nailing plate, special clips can be nailed to the sides of the joists to develop the kind of connection described above. In Figure 6.6-6b, the nailing plate has been omitted, and a metal strap tie provides continuity at each joist and anchors the floor assembly, while 2 × 4 blocking that is fitted between the joists keeps them from twisting or overturning. Since strap ties prevent the joists only from pulling apart, the joists should be installed so that the ends are in contact with the beam web to provide compressive resistance as well.

6.6.1.4

Sill Plate Construction

The sill assembly over a foundation wall should conform to either platform- or balloon-framing methods. In either case, a sill plate should be securely anchored (see Section 6.5.1.2) and should be separated from the finish grade by at least 8 in. In the Gulf states and Florida, the sill plate, and all wood members within 18 in. of grade, should be preservative treated against decay and termites. In areas of high humidity but where termite infestation is not a hazard, inherently decay-resistant grades of redwood, cypress, or red cedar can be used.

FIGURE 6.6-6 *Wood joists can be supported (a) on wood ledger or (b) directly on lower steel beam flange.*

Both balloon- and platform-frame systems (Figs. 6.6-7 and 6.6-8) are readily adaptable to multistory construction. Because of its adaptability to the use of tilt-up and component construction methods, and because it provides a stable working platform during construction, platform framing provides the greatest flexibility for the wide variety of possible design requirements in modern light wood-framed buildings. Therefore, balloon framing is seldom used today. It is discussed here because it will still be encountered in renovation projects on older buildings.

■ PLATFORM FRAMING

A *box sill assembly* is commonly used in platform framing (Fig. 6.6-9). It consists generally of a 2 × (nominal 2 in. thick of some width) *band joist* and a 2 × 4 or 2 × 6 *sill plate*. The plate should be wide enough to provide a full base for the band joist, plus at least a 1-1/2-in.-wide bearing ledge for the floor joists. When the

band joist is perpendicular to the floor joists it is also called a *header*; when parallel, a *stringer*. To differentiate between the two conditions, this terminology has been used in this text.

Each floor joist should be toe nailed to the sill plate with three 8d or two 10d nails; a header should be end nailed to each joist with two 16d nails. If suitable rigid wall sheathing is not nailed to the sill plate, the header should be toe nailed to the sill plate with 8d nails 16 in. on centers. A stringer similarly should be toe nailed to the sill plate with 8d nails at 16 in. on centers. A sill plate should be anchored to the foundation wall over a resilient sill sealer or bed of grout. A termite shield should be used in areas subject to termite infestation.

■ BALLOON FRAMING

In *balloon framing* both the joists and the wall studs rest on a 2 × or thicker *sill plate* (Fig. 6.6-10). The joists should be

FIGURE 6.6-7 *Traditional balloon framing, characterized by individual joists, rafters, and building-height studs spaced 12 to 24 in. apart; second-floor joists carried on ribbon let into studs; wood boards for sheathing, subflooring, floor bridging, and corner bracing (in more modern versions, sheet materials would usually be substituted for these).*

FIGURE 6.6-8 *Customary platform framing: the most common system, characterized by floor joists and story-height studs spaced 16 or 24 in. on centers; roof trusses spaced up to 4 ft. on centers; panel sheathing, such as plywood, gypsum, or fiberboard; and elimination of unnecessary floor bridging and corner bracing. The floor construction forms a work platform, permitting efficient tilt-up methods. This system adapts easily to panelized construction.*

nailed to the sill plate as in platform construction, but, in addition, should be face nailed to each stud with two 10d nails. Also, each stud should be toe nailed to the sill plate with two 10d nails. A board subfloor can be laid diagonally or at right angles to the joists. A fire-stop is required between the studs at the floor line. When a diagonal board subfloor is used, a nailing member, such as a 2 × 4, is normally required between the joists and the studs at the wall lines. If this member is wide enough to extend fully into the stud cavity, it can serve also as a fire-stop.

FIGURE 6.6-9 *Platform-framed box sill assembly. The studs rest on a sole plate. The subflooring may be boards as shown, but is more likely to be a sheet material.*

FIGURE 6.6-10 *Balloon-framed fire-stopped sill assembly. The studs rest on a sill.*

joists of nominal 2 × lumber, spaced 12, 16, or 24 in. on centers. Contemporary floor joists, however, are frequently composite units made up of plywood webs and lumber top and bottom chords (Fig. 6.6-11a). Subflooring of wood boards or plywood is applied to form a structural floor and working platform (Figs. 6.6-11b and 6.6-12). Generally, joists span from the exterior walls to center load-bearing walls or beams. Interior load-bearing walls may restrict the flexibility of room planning, however. For relatively narrow buildings (approximately 16 ft. or less), joists may span from exterior wall to exterior wall without a center support.

FIGURE 6.6-11 *(a) Truss type wood joists; (b) a plywood subfloor over wood joists.* (*Courtesy American Plywood Association.*)

Joist floor systems provide space for heating, air conditioning, electrical, and plumbing equipment and devices, and are easily insulated to minimize heat loss and improve sound transmission.

Some types of floor finishes (such as hardwood flooring) may be installed directly over wood or plywood subflooring; other finishes (such as resilient flooring) may require an underlayment. A variety of ceiling finishes may be applied, usually directly to the underside of the joists.

6.6.1.5

Floor Joists

A wood joist floor is part of the floor and ceiling assembly portion of a building's superstructure, which also includes the walls and a roof and ceiling assembly. A joist floor is essentially the same whether the walls are wood framed, solid masonry, or concrete. It also makes no difference whether wood-framed walls have brick veneer or some other exterior finish.

Both balloon- and platform-frame constructions typically include wood floor

FIGURE 6.6-12 *Wood joist floor/ceiling construction. Subflooring may be boards as shown, but is more likely to be a sheet material.*

MATERIALS

Floor joist lumber is selected primarily to meet strength and stiffness requirements, which depend on the dead and live loads to be carried. For average spans encountered in residential construction, stiffness requirements are more critical than strength requirements. In other words, typical joist sizes are selected on the basis of their stiffness (resistance to deflection) rather than strength (ability to carry loads without failure). Stiffness requirements place an arbitrary maximum deflection under loads: 1/360 of the joist span, in inches. For instance, the maximum allowable deflection for a 15-ft. joist span would be 180 in. (15 × 12) divided by 360 (180/360), or 1/2 in.

Other desirable qualities for joist lumber are good nail-holding ability and freedom from warp. Wood floor joists are usually 2-in.-thick nominal lumber, 6 to 12 in. deep. The size depends on loading, length of span, joist spacing, and the species and grade of lumber used. Suitable floor joists can be selected by referring to tables for allowable unit stress and maximum allowable spans in *Span Tables for Joists and Rafters* by the NFPA (see Fig. A-17 and the accompanying text). Joist grades should be selected for the specific design conditions. Where several combinations of joist size and spacing satisfy the given requirements, a deeper joist at greater spacing generally is preferred, provided, of course, the subfloor and the ceiling finish are selected to span the greater spacing. Using deeper joists usually requires less lumber, less labor, and results in a stiffer floor and ceiling assembly.

JOIST INSTALLATION

Floor and ceiling assemblies intended for the direct application of gypsum board or any other board finish should have all joists evenly spaced, with bottom faces aligned in a level plane.

After the sill plates have been anchored to the foundation walls, joists are laid out over the sill plates and center supports. The floor joists are spaced uniformly at 12, 16, or 24 in. on centers as determined by the design requirements. Extra joists

are added where required to support bearing partitions (Fig. 6.6-13) or to frame floor openings (Fig. 6.6-14). Joists are braced temporarily, with wood strips nailed across the top, until the subflooring is applied.

It is important when designing a wood floor system to equalize shrinkage of the framing at the exterior walls and at the center support. By using approximately the same total depth of cross-grained wood at the center girder as at the exterior wall, differential shrinkage can be minimized. This reduces plaster cracks, prevents sticking doors, and avoids other inconveniences caused by uneven shrinkage. For instance, if there is a total of 12 in. of cross-grained wood at the foundation wall (including joists and sill plate), this should be balanced with an equal amount of cross-grained wood at the center girder. To avoid uneven shrinkage, a steel beam center support generally is preferred. If a wood girder is used, MC 15 lumber should be specified, or the joists should be supported on ledger strips, rather than over the top of the girder.

Excessively bowed or crooked joists should not be used, but joists with a slight crown may be used if they are installed with the crown up. After the subfloor and normal floor loads are applied, the joists will tend to straighten out. The largest edge knots also should be placed on top, since the upper side of the joist is in compression and the knots will have a lesser effect on strength and deflection.

In platform construction, the header joist should be fastened by nailing into the end of each joist with two 16d nails. In addition, the header joists and the stringer joists (parallel to the exterior walls) should be toenailed to the sill with 8d nails spaced 16 in. on centers, unless rigid wall sheathing covers the sill plate and is nailed to it with 8d nails 8 in. on centers.

Each floor joist should be toe nailed to the sill and center beam or girder with two 10d or three 8d nails; joists should also be nailed to each other with at least three 16d nails when they lap over the center beam. The lap should be not less than 4 in. and the overhang of either joist beyond the edge of the girder or beam should not exceed 12 in. The limit on overhangs is imposed in order to reduce the possibility of lifting the subfloor resting on the overhang when a heavy, concentrated load

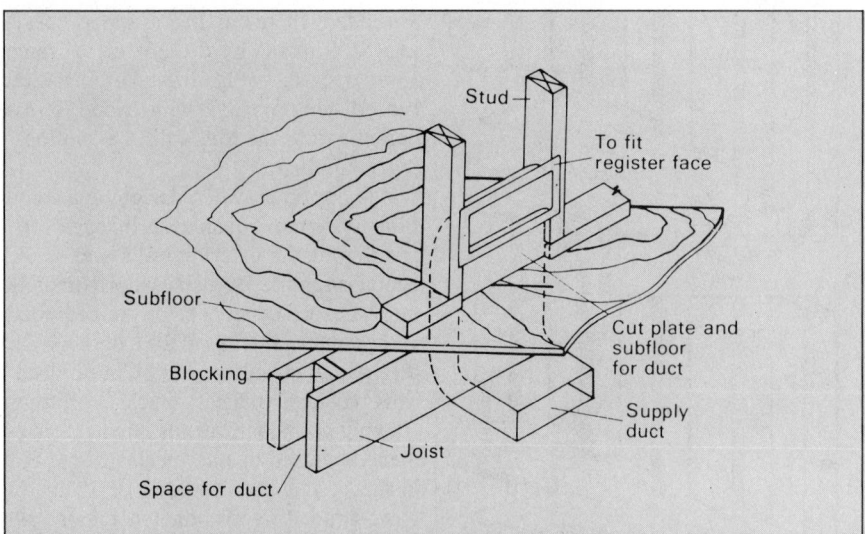

FIGURE 6.6-13 *Spaced joist construction under a parallel bearing partition.*

is placed on the joist near its midspan. If a nominal 2-in. scab is used across butt-ended joists, it should be nailed to each joist with at least three 16d nails at each side of the joint.

For greater lumber economy an in-line joist system sometimes is used in floor and roof framing when center supports are present. This system normally allows the use of joists one size smaller than in conventional floor framing. Briefly, the system consists of joints of uneven length with the long, overhanging joist cantilevered over the center support and connected to the supported joist with a metal connector or plywood splice plate (Fig. 6.6-15). In typical single-family construction, the overhang varies between about 1 ft. 10 in. and 2 ft. 10 in. For Douglas fir and southern pine No. 1 lumber, total spans (between foundation walls) of up to 22 ft. are possible with 2 × 6s; spans up to 30 ft. with 2 × 8s; and spans up to 34 ft. with 2 × 10 joists. Other innovative floor systems that depend on the special attachment of a plywood subfloor are discussed in Section 6.6.4.

Floor joists under parallel bearing partitions should be doubled; if access into the partition is required for air ducts, the double joists should be spaced and solid blocking should be used between them as in spaced wood-girder construction (see Figs. 6.6-5 and 6.6-13).

6.6.1.6

Bridging

Cross-bridging between joists, which has been used in house construction for many years and which is still required by some building codes, has been demonstrated by the National Association of Home Builders Research Foundation and several other independent laboratories to be virtually useless under most circumstances. In fact, even when bridging is tight fitting and well installed there is no significant improvement in a floor's ability to transfer vertical loads after the subfloor and finish flooring have been installed. Therefore, in most wood-framed buildings, bridging between floor joists is not necessary. Properly nailed subflooring and the nailing of the ends of joists to band joists or headers is sufficient to keep the joists aligned and to prevent them from twisting.

However, where the depth-to-thickness ratio of joists exceeds 6, or when joist twisting has been experienced, bridging is often added. This bridging may consist of either metal struts or of 1 × 3 or larger wood members installed at midspan in intervals of not more than 8 ft. (Fig. 6.6-16).

In addition, either cross bridging or solid blocking using 2 × 4 or larger lumber installed vertically between joists may improve the diaphragm action of a

floor, and may increase its resistance to lateral (horizontal) loads imparted to it by foundation walls. Therefore, bridging may be required in situations where engineering design so dictates.

Solid bridging between joists should be used to provide a more rigid base for non-bearing partitions that happen to fall between joists (see Fig. 6.6-16). It also provides a fire-stop against lateral flame spread. Solid blocking should be well fitted and securely nailed to the joists. Load-bearing partitions running parallel to the joist direction should be supported by doubled joists, either by placing the partition over a normally spaced joist that has been doubled, or by providing an additional pair of joists under the partition.

6.6.1.7

Floor Openings

Large openings are often necessary in floors for ducts, chimneys, and stairs (see Fig. 6.6-14). Where joists have been cut away, the remaining *tail joists* are framed into single or double cross joists, called *headers*. The headers in turn are framed into abutting full-length joists, known as *trimmers*.

A single header can be used if the dimension of the opening along the header is less than 4 ft.; a single trimmer can be used to support a single header when the opening is near the end of the joist span and less than 4 ft. long. Tail joists shorter than 6 ft. should be connected to headers with three 16d end nails and two 10d toe nails. Longer tail joists should be connected with framing anchors or supported on a ledger strip as in joist-girder connections. Headers should be connected to trimmers with three 16d end nails and two 10d toe nails.

When framing openings larger than 4 ft. by 4 ft., both the trimmers and headers should be doubled. Connections between headers and trimmers should be made with framing anchors, joist hangers, or ledger strips. Tail joists more than 6 ft. long should be similarly attached to headers; shorter tail joists can be attached as recommended above for smaller openings. The nailing sequence for double headers is shown in Figure 6.6-17; the opening is first framed out using single headers and

A, *Opening perpendicular to floor joists*

Labels (top figure A):
- Double trimmer
- Joist
- Stair
- Joist hanger
- Beam or wall
- 10'-0" Maximum header length (when designed as beam)
- Double header
- Double trimmer
- Joist hanger
- Tail joist
- Post support for header and trimmer

Labels (bottom figure B):
- Load-bearing wall or beam
- Stairway opening
- Width
- Header (double if tail joists exceed 4')
- Length
- Tail joist (if over 6', attach with joist hangers)
- Trimmer joist (double if tail joists exceed 4')
- Joist
- Wall or foundation

B, *Opening parallel to floor joists*

FIGURE 6.6-14 *Framing for a large floor opening (a) with the long dimension of the opening parallel with the joists, and (b) with the long dimension of the opening perpendicular to the joists.*

trimmers; then the second joists are applied and face nailed to the first.

The largest floor openings are usually for stairs; for straight runs and typical floor heights, they are usually about 10 ft. long and 3 to 4 ft. wide. The long dimension of stairway openings may be either parallel or at right angles to the joists. However, it is much easier to frame a stairway opening when its length is parallel to the joists (see Fig. 6.6-14a).

When the length of the stair opening is perpendicular to the length of the joists, a long doubled header is required (see Fig. 6.6-14b). A header under these conditions without a supporting wall beneath it is usually limited to a 10-ft. length and should be designed as a beam. A load-bearing wall under all or part of this opening substantially simplifies the framing, as the joists then bear on the top plate of a wall rather than on a header.

6.6.1.8

Floor Projections

The framing for floor projections beyond the supporting wall below, such as for a bay window or balcony, generally consists of extended floor joists, if they run perpendicular to the supporting wall (Fig. 6.6-18a). This extension normally should not exceed 24 in. unless it is specifically structurally designed for a greater projection. The joists forming each side of such a bay should be doubled. Nailing procedures should conform to those used in framing stair openings. The subflooring should be extended flush with the outer framing member.

In framing the roof of such a projection, rafters often are supported on a header over the bay area, in the plane of the supporting wall below. The header thus supports the roof load and transmits it to the foundation wall below, so that the wall of the bay has less load to support.

Projections parallel to the direction of the floor joists should generally be limited to small areas and extensions of not more than 24 in. (Fig. 6.6-18b). In this construction, the stringers should frame into doubled joists. Joist hangers or a ledger strip can be used to frame joists into the double header, as in a joist-girder connection.

6.6.1.9

Framing for Mechanical and Electrical Equipment

It is frequently necessary to notch, drill, or cut framing members to accommodate heating, plumbing, and electrical equipment. A little preconstruction planning will generally minimize the amount of cutting required later. If mechanical and electrical distribution plans are worked out in advance, many time- and cost-saving adjustments in the framing can be made.

Floor joists should be notched or drilled in a way that will minimize the effect on their structural integrity. The top or bottom of a joist should be notched only in the end one-third of the span and to a depth of not more than one-sixth of the joist depth. Holes may be bored in joists, provided the hole is not more than 2 in. in diameter and is not less than 2-1/2 in. from either edge (Fig. 6.6-19).

Where notching or drilling to a greater extent than recommended above is necessary, the weakened joist should be reinforced with scabs on both sides or by adding another joist. If reinforcement is not

FIGURE 6.6-15 *In-line joist system framed with pairs of overhanging joists and supported joists.*

FIGURE 6.6-16 *Typical platform-framed floor construction, showing alternative subfloor types (plywood and wood boards). The cross bridging shown is not necessary in most residential floors.*

possible because of conflict with the mechanical equipment, the joist should be cut away and the area treated as a floor opening, with tail beams supported on headers.

When a plumbing wall adjacent to a bathroom also serves as a load-bearing partition, the supporting double joists should be spaced and blocked (Fig. 6.6-20) as recommended for a spaced girder. The joist nearest the front apron of a tub should also be doubled to prevent excessive deflection when the tub is full of water.

6.6.2

WALL ASSEMBLIES

Wall assemblies are a basic element of a framed building's superstructure. The design, selection, and construction of a wall assembly considers (1) the required level of functional performance, (2) costs, and (3) the desired finished appearance. These considerations often are influenced by the finishes used and other basic

elements of the building.

The functional performance requirements of a wall assembly usually include (1) adequate structural strength to support any dead and live loads; (2) weathertightness to control the flow of heat, moisture, air, and water vapor; (3) satisfactory levels of sound insulation and visual privacy; (4) suitable fire resistance; (5) the ability to accommodate any heating, air conditioning, electrical, and plumbing equipment that might require thicker than normal walls; (6) suitability for the application of various finish materials; (7) adaptability to economical methods of assembly and erection; and (8) provision for the installation of doors, windows, and other openings. Where there are requirements other than these, the conventional constructions described in this section may not be adequate. Structural walls must safely support all superimposed vertical loads and resist horizontal and racking loads resulting from lateral forces, such as wind and earthquake, and transmit all loads to the foundation. Rarely are walls specifically engineered. Ordinarily, building codes or rules of thumb based on established practices determine materials and methods of construction.

The ultimate costs of a wall assembly include in-place costs, maintenance costs, and operating costs. In-place costs are the sum of the costs of materials, labor, and overhead required to assemble and install the system. The selection of a wall assembly may affect other costs, such as installing insulation and a vapor barrier, applying siding or other finishes, and providing for window and door openings. The necessary time and costs to enclose the building and the costs of installing heating, air conditioning, electrical, and plumbing equipment also may be influenced by the type of assembly. Maintenance costs and operating costs attributable to a wall assembly (such as heat loss or gain through the construction) have an important influence on the ultimate cost of the wall and should not be overlooked if long-term economy is to be achieved.

Occasionally some of the structural elements in a wood-framed building are left exposed, but usually wood-framed wall assemblies are covered with applied materials to provide the desired finished

FIGURE 6.6-17 *Nailing sequence for floor openings: (1) nailing trimmer to first header; (2) nailing header to tail beams; (3) nailing header together; (4) nailing trimmer to second header; and (5) nailing trimmers together.*

FIGURE 6.6-18 *Framing for wall projections (a) when joists are perpendicular to wall; (b) when joists are parallel with wall.*

FIGURE 6.6-19 *Recommended limits for boring of holes in load-bearing floor joists.*

FIGURE 6.6-20 *Recommended framing to minimize floor deflection under tub.*

appearance and performance. The selection of finish materials (1) may influence the spacing and unsupported span of the primary structural members; (2) may determine the need for wall sheathing and other finish bases; (3) may dictate the kind, type, and physical properties of the subsurface material, and (4) may dictate methods of construction.

The appearance of a building is influenced greatly by the location, size, and type of doors, windows, and other openings that form a part of a wall assembly. In addition, a wall assembly should be compatible with the roof and floor assemblies used in the construction.

Typical wall assemblies include conventional stud wall construction (this section), panelized construction (Section 6.6.6), masonry construction (Chapter 4), and post-and-beam construction (Section 6.7.1).

6.6.2.1

Materials

Wall framing lumber should be of the proper grade for the intended use, graded by a recognized authority. Its moisture is important in the performance of applied finishes, and should be as near as possible to the equilibrium moisture content for the particular area. In no case should it be more than 15% when finishes are applied.

6.6.2.2

Stud Walls and Partitions

In light wood-framed construction, wood studs are the typical framing members for walls and partitions; wood joists and trusses are the most common ceiling and roof framing members. These framing members generally are spaced 16 or 24 in. on centers, depending on the unsupported height of the member and the number of floor and roof structures supported. Member sizes and spacings are often dictated by the building code. The usual spacings permit direct application of most finishes, such as gypsum board and plaster. However, furring may be required when member spacing exceeds the maximum recommended for the particular finish material or when special performance, such as increased sound isolation, is desired.

The following discussion of conventional stud wall construction refers to stick-built erection methods, where carpenters on the job handle individual studs rather than prefabricated panels.

Both balloon-frame and platform-frame constructions (see Section 6.6.1.2) typically use wood stud wall assemblies (Fig. 6.6-21). Walls may be sheathed with wood boards, plywood, fiber, or gypsum boards, which provide rigidity, form a weather barrier, and may be necessary as a nailing base to receive exterior finishes (see Section 6.6.5). Some types of exterior and interior finishes can be applied directly to studs; other finishes must be applied over sheathing or other bases.

In load-bearing partitions, studs should be 2 × 4s or 2 × 6s, usually spaced at 16 or 24 in. on centers, depending on the loads to be encountered. In non-load-bearing or staggered-stud partitions, 2 × 3 or 2 × 4 studs may be used, spaced 16 in. on centers when openings are present or cabinets and other elements are supported on them, and 24 in. on centers when no openings are included and no additional loads are imposed on them. Backup members should be provided at interior corners for support and as a nailing base for gypsum board and other finish materials. Warped or crooked studs should not be used.

Because of the existing body of building codes and other regulations and practices, stud spacing is rarely specifically engineered. The conventional 16- and 24-in. spacing of studs has evolved from years of practical experience and is based more on accommodating finish materials applied to the framework than on actual engineering for imposed loads. For example, in one-story, light wood-framed construction, 2 × 4 studs could be placed at 6-ft. intervals and adequately support the generally imposed roof loads. But few exterior or interior finish materials can span such intervals economically. The 16- and 24-in. spacings permit the use of standard widths and lengths of panel sheathing and siding, and therefore remains the convention in use. Stud spacings of 12 in., though rare in low-rise buildings, are sometimes used when walls are tall or lateral loads are to be superimposed. The use of 12-in. spacings is a structural decision based on engineering. There is currently a trend toward using 2 × 6 studs at 24 in. on

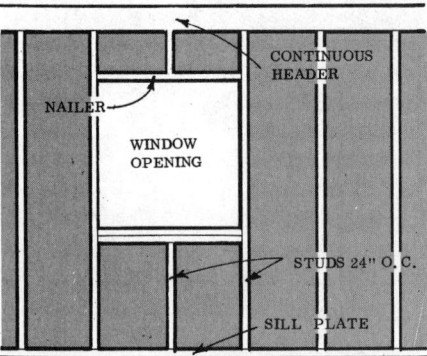

FIGURE 6.6-21 *Typical wood stud construction, showing framing members. Structural wall sheathing or diagonal bracing, which is required to resist wind and other stresses, is not shown here.*

centers for exterior walls to take advantage of the increased depth to install more thermal insulation.

Suitable performance of most types of siding depends in part on the wall structure that supports the siding. Wall studs should be adequately sized and spaced to carry superimposed loads. Conservatively, structural load requirements normally are met using (1) 2 × 4 studs at a maximum spacing of 24 in. on centers when

a wall supports only a ceiling and roof, (2) 2 × 4 studs at a maximum spacing of 16 in. on centers when the wall supports a floor in addition to the roof, and (3) 2 × 6 studs at a maximum spacing of 16 in. on centers when the wall supports two floors in addition to the roof. Codes permit the use of smaller members and wider spacings under certain conditions.

Often the spacing of exterior wall studs is dependent on the type and thickness of sheathing rather than on structural load requirements. Stud spacings for sheathing support are discussed in Section 6.6.5.

In stud systems, the roof and other vertically imposed loads are carried by the studs. Sheathing and applied finishes resist lateral forces but are not assumed to share in carrying vertical loads.

Wood stud assemblies lend themselves to on-site or off-site fabrication. Pre-framed panel wall assemblies are, in fact, wood stud assemblies fabricated in sections away from the job site. Stud wall assemblies provide flexibility in design and erection and require only the use of simple hand tools and fastenings. Changes can be made easily during construction, particularly with on-site fabrication. Space between studs provides for the installation of insulation and heating, air conditioning, electrical, and plumbing equipment. Studs may be erected individually in the traditional manner, but most stud walls today are framed on the subfloor, with sheathing and siding applied and windows and doors installed. The assembly then is tilted into place.

Wood stud walls may be framed, erected, and sheathed quickly and the building can be enclosed quickly. Interior finishing operations and mechanical installations then may take place under cover, giving maximum flexibility to job scheduling.

Wood stud wall systems are economical and are widely used. They provide excellent structural performance, make use of relatively inexpensive, basic materials, permit the application of many finishes, and are adaptable to practically any method of on- or off-site fabrication (see Section 6.6.6).

In masonry veneer construction (see Chapter 4), wood framing assumes all vertical loads, and the masonry performs essentially as an exterior finish surface whose basic function is to provide a weather barrier to the assembly. The space between the studs typically is used for the installation of insulation and heating, air conditioning, electrical, and plumbing lines.

Post-and-beam walls are discussed in Section 6.7.1.

6.6.3

ROOF AND CEILING ASSEMBLIES

In many cases, especially in wood-framed buildings, the ceiling construction is an integral part of the roof construction. When it is, the entire system is called a *roof and ceiling assembly*. Such assemblies are a basic element of a building's superstructure.

A light wood-framed building's roof and ceiling assemblies include joist-and-rafter construction (in this section), plank-and-beam construction (see Section 6.7.2), and panelized construction (see Section 6.6.6).

The design, selection, and construction of a roof and ceiling assembly considers (1) the required level of functional performance, (2) costs, and (3) the desired finished appearance. These considerations often are influenced by the systems and finishes used in other basic elements of the construction.

The functional performance requirements of a roof and ceiling assembly usually include (1) adequate structural strength to support dead loads (the roof and ceiling assembly) and live loads (such as from snow, ice, and wind); (2) the control of the flow of heat, water, air, and water vapor; (3) satisfactory levels of sound isolation; (4) suitable fire resistance; (5) suitability for the application of finish materials; (6) adaptability to economical methods of assembly and erection; and (7) accommodation of heating, air conditioning, electrical, and plumbing equipment.

The ultimate costs of a roof and ceiling system include in-place, maintenance, and operating costs. In-place costs are the sum of the costs of materials, labor, and overhead required to assemble and install a roof and ceiling system. The type of assembly selected may affect other costs such as those for installing insulation, applying roofing finishes, and providing interior wall support, and also may affect

the time needed to enclose the building. Maintenance costs and operating costs attributable to a roof and ceiling assembly (such as heat loss or gain) have an important influence on the ultimate cost of the building and should not be overlooked if long-term economy is to be achieved.

Visually and structurally, the form of a roof is determined by the roof and ceiling assembly selected, but generally the assembly is covered with applied materials that provide the desired finished appearance and performance. The finish materials selected may (1) affect the spacing, span, and slope of structural members; (2) determine the need for roof sheathing or underlayment; and (3) even dictate methods of construction.

A roof and ceiling assembly should be compatible with the wall assembly and floor and ceiling assemblies used in the construction.

In light wood-framed building construction, wood joist-and-rafter construction (this section) and trusses (Section 6.7.3) are the most common ceiling and roof framing members. These framing members generally are spaced 16 or 24 in. on centers, depending on the loads to be carried, which permits direct application of most finishes, such as gypsum board and plaster. However, furring may be required when member spacing exceeds the maximum recommended for the particular finish material or when special performance, such as increased sound isolation, is desired.

Other light wood-framed roof and ceiling assemblies include plank-and-beam construction (Section 6.7.2) and panelized construction (Section 6.6.6).

Ceilings intended for the direct application of gypsum board or any other board finish should have all joists evenly spaced with bottom faces aligned in a level plane. Excessively bowed or crooked joists should not be used, but joists with slight crown may be used if they are installed with the crown up. Slightly crooked or bowed joists can be straightened and leveled by nailing stringers approximately at midspan (Fig. 6.6-22; see also Section 6.7.3).

6.6.3.1

Materials

Framing lumber should be of the proper grade for the intended use, graded by a recognized grading authority. The moisture content of framing lumber is important in the performance of finishes. It should be as near as possible to the equilibrium moisture content for the particular area and should not exceed 15% at the time the finish material is installed.

6.6.3.2

Joist-and-Rafter Roofs

Joist-and-rafter assemblies consist of nominal 2-in. lumber ceiling joists and roof rafters usually spaced 16 or 24 in. on centers. Typical construction for joist-and-rafter roofs is shown in Figure 6.6-23. Joists span roughly one-half the width of the building from the exterior walls to a center load-bearing wall or beam. The rafters are supported on exterior walls and bear against each other or a ridge board

STRONG-BACK

FIGURE 6.6-22 *Joist spacing can be regulated with 2 × 4 stringers at midpoints, and can be leveled with 2 × 6 stringers set on edge. (Courtesy Gypsum Association.)*

at the high point of the roof. When loaded, the ridge tends to lower and the rafters exert a thrusting force that tries to spread the outside walls. This thrust is resisted by the ceiling joists acting as ties from wall to wall. Suitable ceiling joists and rafters can be selected from the tables in *Span Tables for Joists and Rafters*, published by the NFPA (see Figs. A-18, A-19, and A-20).

Joist-and-rafter construction generally requires longer erection time than trussed construction and may expose materials longer to possible weather hazards.

Interior load-bearing walls somewhat restrict the flexibility of room planning. However, for relatively short spans of 16 ft. or less, load-bearing interior walls or beams generally are not required. If a building's plan includes several wings of different spans, joist-and-rafter framing may be more economical than trussed construction.

Joist-and-rafter assemblies provide overhead space for the installation of insulation and mechanical equipment; in addition, with adequate rafter slopes, economical attic storage or expansion space can be allowed.

Interior and exterior finishes may be applied as for truss roofs.

FIGURE 6.6-23 *Wood joist-and-rafter roof and ceiling assembly.*

FIGURE 6.6-24 *Wood joist roof and ceiling assembly.*

6.6.3.3

Joist Roofs

In joist roof and ceiling assemblies, roof joists alone provide the function of both ceiling joists and roof rafters (Fig. 6.6-24). These joist members generally are of nominal 2-in. dimension lumber spaced 12, 16, or 24 in. on centers, depending on the spans and the loads to be carried. Joists may have a clear span between exterior walls or they may be centrally supported on load-bearing partitions or beams. When the roof is sloped rather than flat, the construction also may be called a *ridge-supported rafter assembly*, which often is used to achieve an interior sloped ceiling effect. Roof sheathing materials may be attached directly to the tops of the joists, and ceiling finishes applied directly to the underside. Joists and rafters may be selected from the tables in *Span Tables for Joists and Rafters* by the AFPA.

Ductwork can be accommodated in the joist space if it is run parallel to the joists.

When it is necessary to run ductwork perpendicular to the joists, it is common to use dropped-ceiling areas over hallways. Wiring can be concealed in the joist space and the system is relatively easy to insulate. Interior and exterior finishes are applied in the same manner as for trussed and joist-and-rafter roofs.

6.6.4

SUBFLOORING AND PLYWOOD UNDERLAYMENT

Subflooring is required beneath most kinds of finish flooring. An additional underlayment is also needed for some types of flooring.

6.6.4.1

Subflooring

Subflooring serves as a working platform and as a base for floor finishes. The two main types of subfloor typically used as a base for flooring in light wood-framed construction are wood and concrete. Concrete floors are discussed in Chapter 3. Additional requirements for both wood and concrete subfloors are discussed in the various sections where floor finishes are covered, such as Sections 9.6, 9.7, 9.9, 9.10, and 9.11.

Tile, terrazzo, resilient flooring, and carpet may be installed over a double-layer base composed of a separate subfloor and an underlayment, or may be a single-layer construction consisting of a combination subfloor-underlayment material specifically designed for the purpose. Wood flooring is usually installed over a single-layer subfloor.

Regardless of whether it is covered with underlayment, wood subflooring should be firmly fastened and structurally adequate to carry the intended loads without excessive deflection. The maximum deflection of such subflooring under full dead and live loads after the finished flooring has been installed should not exceed 1/360 of any span.

Carpet, Tile, or
Resilient Flooring

Plywood Subfloor
Perpenticular to Joists

Lumber Blocking at
Edge Joints Except
for T&G Plywood

SINGLE-LAYER WOOD
FLOOR CONSTRUCTION

Carpet, Tile, or
Resilient Flooring

Panel Underlayment
(Plywood, Hardboard
or Particle board)

Lumber or Plywood
Subfloor

Wood Joists

DOUBLE-LAYER WOOD
FLOOR CONSTRUCTION

FIGURE 6.6-25 *Typical subflooring construction for the application of carpet, tile, or resilient flooring.*

Figure 6.6-25 shows wood joist floor construction suitable for the application of resilient flooring, carpet, and ceramic tile. The only acceptable wood material base for terrazzo is plywood. Wood subflooring for wood floors may be boards or plywood.

Suspended wood floors above grade usually are sufficiently removed from excessive external moisture sources, but floors over crawl spaces at or near grade may require the crawl space to be ventilated and the ground inside the crawl space to be covered with a suitable vapor retarder (see Section 2.3).

The discussion in this section describes conventional board subflooring, conventional nailed plywood subflooring, glued plywood subflooring, and combination subflooring-underlayment.

■ BOARD SUBFLOORING

Board subflooring is permissible without underlayment beneath structural flooring such as wood-strip flooring; beneath tile set in a mortar bed; and beneath terrazzo with an underbed. However, underlayment is recommended over board subflooring

beneath thin-set tile, thin-set terrazzo, resilient flooring, and carpet.

Conventional board subflooring consists of nominal 1-in.-thick boards 4 to 8 in. wide (see Fig. 6.6-16). These boards may be applied either diagonally or at right angles to the joists. When the boards are placed at right angles to the joists, wood strip or plank flooring must be laid at right angles to the boards. Diagonal board installation is preferred for wood finish flooring because it permits the finish flooring to be laid either parallel or at right angles to the joists. Boards may have square, shiplap, or tongue-and-groove edges. Board ends should always rest over the joists (see Fig. 6.6-16) unless the board ends are tongued and grooved. Subfloor boards less than 8 in. wide should be nailed to each joist with two 8d nails; those 8 in. wide or wider with three 8d nails. Where desired, 7d annularly threaded nails can be substituted for the 8d smooth-shank nails.

In balloon framing, blocking should be installed between the ends of joists at the wall for nailing the ends of diagonal subfloor boards (see Fig. 6.6-10).

The moisture content of lumber boards used for subflooring at the time of installation should be not more than 15%.

■ PLYWOOD SUBFLOORING

Plywood is a suitable subflooring material for tile, terrazzo, wood flooring, resilient flooring, and carpet. Additional details are covered in Sections 9.6, 9.7, 9.9, 9.10, and 9.11. A single-layer subfloor without underlayment may be used beneath structural flooring such as wood-strip flooring; beneath tile with a thick-set mortar bed; beneath terrazzo with an underbed; and beneath carpet with a pad. Plywood subflooring requires an underlayment for receipt of thin-set tile, thin-set terrazzo, and resilient flooring.

A conventional plywood subfloor should be APA-rated sheathing either Exposure 1 or 2; Exterior; Structural I, Exposure 1; or Structural I, Exterior; depending on its exposure to moisture. The more moisture-resistant grades should be selected when a probability of prolonged exposure to moisture exists, such as in bathrooms, in floors adjacent to plumbing fixtures, or when the subfloor may be exposed to the weather for long periods during construction. The thickness varies from 1/2 to 3/4 in., depending on the span rating, type of finish floor, and spacing of the supporting joists. Plywood subfloor panels to be covered with underlayment may be left unsanded.

Plywood subfloor panels should be installed with their surface grain or long dimension perpendicular to the joists and their end joints centered over the joists (see Fig. 6.6-25). Panels should be staggered so that the ends of adjacent panels do not fall on the same support. The longitudinal edges should be supported on 2 × 3 or heavier wood blocking. The blocking may be eliminated, however, when tongue-and-groove plywood is used (Fig. 6.6-26). Supports should be spaced and panels should be nailed as shown in Figure 6.6-27.

Subfloor panels should be spaced approximately 1/8 in. apart at end and edge joints, and placed about 1/8 in. from adjacent vertical surfaces, such as walls and pipes, to allow for expansion and contraction from changes in humidity. Nailing should start in the middle of the panel and proceed outward to the edges, with nail heads set flush or slightly countersunk. Perimeter nailing should be located not more than 3/8 in. from the edges. Panel joints should be sanded smooth and flush

FIGURE 6.6-26 *Tongue-and-groove APA-rated Sturd-I-Floor with the proper span rating can be installed on spacings of up to 48 in., and tongue-and-groove APA-rated sheathing with the proper span rating can be installed on spans of up to 24 in., both without edge blocking, but should be protected during construction.* (Courtesy American Plywood Association.)

before the finish flooring is installed. It is important to protect single-layer floors from damage during construction. However, the difficulty of furnishing adequate protection during construction may warrant the use of unsanded or touch-sanded plywood and then power sanding the entire floor area just before the flooring is installed.

6.6.4.2

Plywood Underlayment

An underlayment is required over both wood board and plywood subflooring that are beneath thin-set tile, thin-set terrazzo, and resilient flooring. Using an underlayment is often desirable even when not absolutely required for the type of finish

flooring, because the subfloor, which provides a convenient working platform during most of the construction period, can be easily damaged, and installation of an underlayment can be scheduled more closely to the installation of the flooring.

Underlayment beneath resilient flooring is often either Underlayment or C-C Plugged panel grades marked *sanded face*. Other grades may also be used, however, when acceptable to the flooring manufacturer. Thin underlayment for other purposes may be APA Underlayment Interior, Exposure 1; C-C Plugged Exterior; or APA C-C Plugged Exterior. Underlayment thicker than 19/32 in. may be APA-rated Sturd-I-Floor Exposure 1 or 2, or APA-rated Sturd-I-Floor, Exterior. The more moisture-resistant grades should be used in conditions of greater moisture, such as encountered in bathrooms and laundry areas, and beneath tile and terrazzo.

Underlayment panels should be installed with the finished face up, and when over plywood subflooring, with the face grain or long edge across the supports. When laid over lumber-board subflooring, the face grain of plywood underlayment should be laid perpendicular to the boards (see Fig. 6.6-25). The ends and edges should be staggered and should not fall within 2 in. of the joints in the subflooring. However, it is not necessary to stagger the joints in underlayment beneath carpet with a pad. In general, joints between panels should be left open 1/32 in. to permit some movement. However, wider spacings between panels may be

---

FIGURE 6.6-27
APA Panel Subflooring (APA-Rated Sheathing)

Panel Span Rating	Minimum Panel Thickness, in.	Maximum Span, in.	Nail Size[a] and Type	Maximum Nail Spacing, in.	
				Supported Panel Edges	Intermediate Supports
24/16	7/16	16	6d common	6	12
32/16	15/32	16[b]	8d common[c]	6	12
40/20	19/32	20[b,d]	8d common	6	12
48/24	23/32	24	8d common	6	12
60/32	7/8	32	8d common	6	12

Courtesy American Plywood Association.

[a] Other code-approved fasteners may be used.

[b] Span may be 24 in. if 3/4-in. wood strip flooring is installed at right angles to joists.

[c] 6d common nail permitted if panel is 1/2 in. or thinner.

[d] Span may be 24 in. if a minimum 1-1/2 in. of lightweight concrete is applied over panels.

required by some flooring manufacturers for certain applications, and their recommendations should be followed. Joints between underlayment and other material, such as adjacent vertical surfaces and pipes, should be left open 1/8 in. beneath resilient flooring and carpet and 1/4 in. beneath tile and terrazzo, unless the manufacturers of these products recommend different spacings. Surfaces of adjacent panels should be level within 1/32 in.

Underlayment should be of sufficient strength to withstand the loads to be encountered. In general, panels beneath tile should not be less than 5/8 in., except that 1/2 in. is sometimes acceptable in small residential applications, beneath terrazzo not less than 3/8 in.

Panels 1/2 in. thick or thinner should be nailed with 3d ring-shank nails at 6 in. on centers along the edges and at 8 in. on centers each way throughout the remainder of each panel. Thicker panels should be nailed using 4d nails at 6 in. on center along the edges and at 12 in. on center throughout the remainder of each panel. Fastener length should be the same as the total thickness of the subflooring and the underlayment.

6.6.4.3

Combination Subfloor and Underlayment

APA-rated Sturd-I-Floor is a proprietary *combination subfloor and underlayment* plywood panel. It is designed specifically for use as a single-layer subfloor beneath carpet with a pad, but is also usable under other flooring types, and may be used even when a separate underlayment is required. It is available in thicknesses from 19/32 in. to 1-1/8 in. and with span ratings from 16 to 48. It is available in exposure durability classifications Exterior, Exposure 1, and Exposure 2.

The support system for APA-rated Sturd-I-Floor panels may be a conventional floor framing system with the framing members (joists or floor trusses) at 16, 20, or 24 in. on centers (Fig. 6.6-28); a joist system with the joists spaced 32 in. on centers (Fig. 6.6-29); or a girder system with the girders at 48 in. on centers (Fig. 6.6-30).

Sturd-I-Floor panels may be manufactured either as conventional plywood, as a composite, or as a nonveneer panel.

FIGURE 6.6-28 *APA-rated Sturd-I-Floor, 16, 20, and 24 in. on centers.* (*Courtesy American Plywood Association.*)

FIGURE 6.6-29 *APA-rated Sturd-I-Floor 32 and 48 in. on centers (over supports 32 in. on centers).* (*Courtesy American Plywood Association.*)

Panels are available both square edged and tongued and grooved. The face ply can be either sanded or touch sanded; the latter is used when field sanding is intended at the end of construction, just before the floor finish is installed.

The installation of APA-rated Sturd-I-Floor panels is similar to that for other plywood subfloors: face plies perpendicular to supports; end joints staggered; panels continuous over at least two supports; panel ends resting on the supports;

1/8" spacing is recommended at all edge and end joints unless otherwise indicated by panel manufacturer

APA RATED STURD-I-FLOOR 48 oc

Stagger end joints

Girder

Strength axis

2x band joist

2x sill

T&G joint (or block edges)

12" min.

4x4 post

Note: Provide adequate ventilation and use ground cover vapor barrier in crawl space. Panels must be dry before applying finish floor.

FIGURE 6.6-30 *APA-rated Sturd-I-Floor 48 in. on centers (over supports 48 in. on centers).* *(Courtesy American Plywood Association.)*

end and edge joints spaced 1/8 in.; solid 2-in. nominal blocking beneath unsupported panel edges of square-edged panels (blocking is not required at tongue-and-groove joints).

A glued installation is recommended for Sturd-I-Floor panels (see Section 6.6.4.4), but this material can also be installed with nailing only. Nail spacings may be required by local codes. Unless the code requires a different spacing, the spacings should be in accordance with Figure 6.6-31.

6.6.4.4

Glued Floor System

In the APA glued floor system the flooring and the joists are glued together to become so strong and stiff that they act as a series of integral T-beams. In addition, the gluing virtually eliminates the squeaks that can result when lumber shrinks (Fig. 6.6-32). Under some circumstances, using the APA glued floor systems will permit the use of smaller joists or longer spans for the same-sized joists than will a

conventional nailed system. The glued system also has better creep resistance. The APA glued subfloor system can be installed quickly using readily available materials and techniques.

The subflooring in the APA glued floor system should be either APA-rated Sheathing with a plywood overlay or structural flooring, such as a wood-strip floor, or a one-piece combination subfloor and underlayment of APA-rated Sturd-I-Floor panels. The subflooring will vary in thickness from 1/2 to 1-1/8 in. depending on the joist spacing and the finish flooring material to be supported. Tongue-and-groove material is recommended. If square-edged material is used, it is necessary to provide solid blocking under the panel edges between joists. The plywood is glued to this blocking. The basic joist framing is the same as for a conventionally nailed subfloor.

When the finish floor is composed of a resilient nontextile material a separate underlayment is required, even over Sturd-I-Floor panels. This underlayment may be either veneer-faced Sturd-I-Floor panels with a sanded face or another underlayment (see Section 6.6.4.3).

The plywood subflooring panels in a glued system should be placed with the face grain or long dimension perpendicular to the joists, with end joints staggered and spaced 1/8 in. at both end and edge. Before each panel is set in place, a bead of glue is applied to the supports with a caulking gun; for extra stiffness, the edge groove also should be glued. The subflooring panel then is secured with ring- or screw-shank nails. These nails should be 6d for 3/4-in. and thinner panels and 8d for thicker panels. Nail spacing varies from 6 to 12 in. along panel ends and edges and in the field of the panels at each support, depending on the joist spacing and the panel thickness (see Fig. 6.6-31).

More detailed information on the glued floor system, including joist span tables, nailing sizes and spacings, application sequence, and recommended adhesives, is contained in the booklet *APA Residential and Commercial Design/Construction Guide.*

6.6.5

WALL AND ROOF SHEATHING

Most exterior wall finishes and roofing materials that are generally applied over wood-framed construction require some sort of sheathing and some require sheathing paper. The wall sheathing may be lumber, plywood, oriented strand board, or a nonwood material such as fiberboard or gypsum board. Roof sheathing is usually lumber, plywood, or oriented strand board.

Wood shakes and shingles can be attached to wood nailing strips fastened to the wall studs through a sheathing material. These nailing strips are discussed in Sections 7.6 and 7.8).

6.6.5.1

Wall Sheathing and Air Infiltration Barriers

Wall sheathing is applied to framing members for one or more of the following reasons:

1. When properly fastened, it is believed to contribute to the racking (being forced out of shape or plumb) resistance of a

FIGURE 6.6-31
APA-Rated Sturd-I-Floor[a]

Span Rating (Maximum Joist Spacing), in.	Minimum Panel Thickness,[b] in.	Fastening: Glue-Nailed[c]			Fastening: Nailed Only		
		Nail Size and Type	Maximum Spacing, in.		Nail Size and Type	Maximum Spacing, in.	
			Supported Panel Edges	Intermediate Supports		Supported Panel Edges	Intermediate Supports
16	19/32	6d ring- or screw-shank[d]	12	12	6d ring or screw-shank	6	12
20	19/32	6d ring- or screw-shank[d]	12	12	6d ring- or screw-shank	6	12
24	23/32	6d ring-or screw-shank[d]	12	12	6d ring- or screw-shank	6	12
	7/8	8d ring- or screw-shank[d]	6	12	8d ring- or screw-shank	6	12
32	7/8	8d ring- or screw-shank[d]	6	12	8d ring- or screw-shank	6	12
48	1-3/32	8d ring- or screw-shank[e]	6	[f]	8d ring- or screw-shank[e]	6	[f]

Courtesy American Plywood Association.

[a] Special conditions may impose heavy traffic and concentrated loads that require construction in excess of the minimums shown.

[b] Panels in a given thickness may be manufactured in more than one Span Rating. Panels with a Span Rating greater than the actual joist spacing may be substituted for panels of the same thickness with a Span Rating matching the actual joist spacing. For example, 19/32-in.-thick Sturd-I-Floor 20 on centers (oc) may be substituted for 19/32-in. thick Sturd-I-Floor 16 oc over joists 16 in. oc.

[c] Use only adhesives conforming to APA Specification AFG-01, applied in accordance with the manufacturer's recommendations. If OSB panels with sealed surfaces and edges are to be used, use only solvent-based glues; check with panel manufacturer.

[d] 8d common nails may be substituted if ring- or screw-shank nails are not available.

[e] 10d common nails maybe substituted with 1-1/8-in. panels if supports are well seasoned.

[f] Space nails maximum 6 in. for 48-in. spans and 12 in. for 32-in. spans.

structural frame. However, tests by the Division of Building Research, National Research Council of Canada indicate that interior wall finishes, like plaster and gypsum board, provide about 4 times as much resistance to racking as does lumber sheathing applied horizontally with corner bracing. Therefore, the National Building Code of Canada does not require sheathing except when the exterior finish materials require it as a nailing base.

2. It contributes insulation, which reduces heat transmission and resists infiltration of wind and moisture.

3. It functions in many cases as a nailing base for exterior finish materials.

When wall sheathing is used as a nailing base, it should be smooth and securely attached to its supports. Shingles may be applied over nailable solid sheathing consisting of wood boards, plywood, or nailable fiberboard. They may be applied also to wood nailing strips over almost any type of sheathing. Wood shakes and shingles can also be applied over spaced sheathing. Plywood siding can be installed either directly over the framing or over plywood sheathing. The types of sheathing required and additional details concerning it may be found in the sections for each exterior finishing material (see Sections 7.4, 7.6, and 7.8).

■ LUMBER WALL SHEATHING

Lumber sheathing may be solid, spaced, or nailing strips, depending on the type of finish siding to be applied.

Boards used as lumber sheathing beneath wall shingles should be seasoned to a moisture content of from 12% to 18% (see Section 6.2). They should be free from loose knots and large knotholes.

Warped boards that would prevent the siding from lying flat should not be used.

Lumber sheathing boards up to 8 in. wide should be face nailed with two 2-1/8-in. × 0.109-in. (7d) annularly threaded nails or two 2-1/2-in. × 0.131-in. (8d) common wire nails at each framing member. Boards 10 and 12 in. wide require three nails at each support. The maximum spacing of supports for nominal 1-in. lumber sheathing is 24 in. on centers.

SOLID SHEATHING Boards for *solid lumber sheathing* should be of nominal 1-in. thickness. Boards used for solid sheathing beneath mineral fiber cement shingles should be not more than 8 in. in nominal width and should have tongue-and-groove or shiplapped edges. Boards used for sheathing beneath wood shakes and shingles should be preferably not more than 10 in. in nominal width and never more than 12 in. in nominal width.

FIGURE 6.6-32 *(a) APA glued floor system; (b) applying adhesive.* *(Courtesy American Plywood Association.)*

These may be square edged, shiplapped, or tongued and grooved.

Should lumber sheathing boards be placed horizontally, it would be necessary to provide structural corner bracing. Therefore, to eliminate the need for this bracing, solid sheathing boards are usually placed diagonally, except where the finish requires horizontal placement (see Chapter 7).

SPACED SHEATHING   For the application of wood shingles and shakes, wood spaced sheathing should be at least 1 × 3, 1 × 4, or 1 × 6 boards, as required for the finish material (see Chapter 7).

■ **PLYWOOD WALL SHEATHING**

Plywood sheathing should be APA-rated sheathing of Exposure 1 or 2; Exterior; Structural I, Exposure 1; or Structural I, Exterior. It should have the appropriate *span rating* according to the stud spacing. The wall's span rating may be either 16 or 24 in. for a 16-in. stud spacing, but must be 24 in. for a 24-in. stud spacing

(Fig. 6.6-33). These materials may include particleboard.

Strictly speaking, *nailable plywood sheathing* is sheathing to which a finish material, such as siding, will be directly nailed, but sheathing on brick-veneered and stuccoed walls should be installed in the same manner. Such sheathing may be

placed with its long edge either horizontal or vertical, with two exceptions: (1) three-ply panels at studs spaced 24 in. on centers must be placed with the long edge across the studs, and (2) panels 3/8 in. thick or less at studs spaced 16 in. on centers must be placed with the long edge across the studs. Some codes require the

FIGURE 6.6-33
APA Panel Wall Sheathing Material

Panel Span Rating	Maximum Stud Spacing, in.	Nail Size[a,b]	Nail Spacing, in.	
			Supported Panel Edges	Intermediate Supports
12/0, 16/0, 20/0, or Wall-16 oc	16	6d for panels 1/2 in. thick or less; 8d for thicker panels	6	12
24/0, 24/16, 32/16, or Wall-24 oc	24			

Courtesy American Plywood Association.
[a] Common, smooth, annular spiral thread, or galvanized box.
[b] Other code-approved fasteners may be used.

FIGURE 6.6-34 *APA panel wall sheathing installation.* (*Courtesy American Plywood Association.*)

1/8" spacing is recommended at all edge and end joints unless otherwise indicated by panel manufacturer

"Block" horizontal joints in panels used for bracing

Filter strip if required

APA RATED SHEATHING applied with long dimension parallel to studs

APA RATED SHEATHING applied with long dimension across studs

Siding

6" minimum clearance, siding to grade

---

sheathing should be installed and attached to its supports according to its manufacturer's recommendations. The suggestions in this section for installing and fastening such products do not supersede the recommendations of the manufacturer.

Fiberboard, gypsum board, and insulating sheathing are not adequate nailing bases. When these sheathing materials are used, siding should be applied by nailing through the sheathing and into framing members or furring strips.

For the most part, these materials also do not provide adequate frame bracing. When they are used, another form of bracing must be provided. One way to brace such a frame is by using 15/32- or 1/2-in. APA-rated sheathing plywood at the corners (see Fig. 6.6-35).

FIBERBOARD SHEATHING Fiberboard sheathing is a homogeneous cellulosic material with square edges. It is available in 1/2- and 25/32-in. thicknesses and in 4-ft. widths and in 8- and 9-ft. lengths. This material should comply with the requirements of American National Standards Institute/American Hardboard Association (ANSI/AHA) Standard A 194.1, "Cellulosic Fiber Board, Type IV." Two densities are available: Class 1, Regular Density, and Class 2, Intermediate Density. Class 3, High Density, or nail-base fiberboard, is no longer made. Class 1

---

application of nailers (back blocking) at joints in plywood sheathing that do not fall oversupports (Fig. 6-6.34). Because greater-than-normal stiffness is required in sheathing eneath stucco, panels there are often applied with the long side across the studs, and blocking is usually installed at all panel edges, regardless of code requirement.

Plywood wall sheathing 1/2 in. thick or less should be applied with 6d common, smooth-shank, annular, spiral-thread, or galvanized box nails. Thicker panels should be installed with similar 8d nails. Nails should be spaced 6 in. on centers along supported panel edges and 12 in. on centers at intermediate supports.

■ PLYWOOD CORNER BRACING

When nonwood sheathing is used that does not provide corner bracing, such as fiberboard and gypsum board, the corners of a building must be braced. This bracing may be plywood sheathing (Fig. 6.6-35). Such sheathing should be one of the materials mentioned above for plywood wall sheathing. The thickness should match the adjacent sheathing material.

■ NONLUMBER SHEATHING

Nonlumber sheathing materials in com-

mon use include fiberboard, gypsum board, and insulating board sheathing.

When fiberboard, gypsum board, insulating board, or any sheathing materials other than wood or plywood are recommended by the manufacturer for use as wall sheathing under a type of siding, the

Supports 16" or 24" o.c.

1/2" gypsum or fiberboard

15/32" or 1/2" APA RATED SHEATHING nailed to studs with 6d common nails spaced 6" o.c. along panel edges and 12" o.c. at intermediate supports*

*When corner panels are plywood, 1-1/2" roofing nails spaced 4" o.c. along panel edges and 8" o.c. at intermediate supports may be used.

FIGURE 6.6-35 *APA panel corner bracing.* (*Courtesy American Plywood Association.*)

requires the addition of frame corner bracing. Class 2 may or may not require corner bracing, depending on the building's structural requirements.

Studs for both thicknesses and both classes of fiberboard sheathing should be placed 16 in. on centers. Nailers must be provided and panels must be nailed along all edges.

Unless otherwise recommended by its manufacturer, fiberboard sheathing should be applied with its long dimension parallel to the framing members (vertically) with sufficient bearing for satisfactory nailing along all edges. About 1/8 in. should be left open between the edges and ends of adjoining boards.

Fiberboard sheathing should be installed according to its manufacturer's instructions. Usually these boards are nailed or stapled at 3 in. on centers along the edges and at 6 in. on centers at intermediate supports. Fasteners for 1/2-in. sheathing should be 11-gauge galvanized roofing nails, 6d common galvanized nails, or 16-gauge galvanized staples. Fasteners for 25/32-in. material should be correspondingly larger.

GYPSUM BOARD SHEATHING Gypsum sheathing may be used beneath wood, vinyl, composition, metal, and plywood siding, and beneath masonry veneer, stucco, and exterior insulation and finish systems. It is available in 1/2-in. regular and 5/8-in. Type X (fire-resistant) boards. The 1/2-in. boards come in 2-ft. by 8-ft. and 4-ft. by 8-ft. sizes. The 5/8-in.-thick boards are 4 ft. by 8 ft. The 4-ft. by 8-ft. panels have square edges; the 2-ft.-wide ones have tongue-and-groove edges. All sizes may have either a plain or a water-resistant core. Their faces, backs, and edges are covered with either paper or a glass-fiber mat. Gypsum sheathing boards should comply with ASTM Standard C 79.

The maximum framing spacing for gypsum sheathing is 24 in. on centers. Sheathing of the 2-ft. by 4-ft. size should be installed with the long edge across the framing (horizontal) and with the tongue edge up. The ends should occur over framing members. The 4 × 8 sheathing boards should be installed with the long edges vertical, and with each edge occurring over a framing member. A continuous nailer should be installed when the ends

of vertically applied boards do not fall over a sill or plate.

Gypsum sheathing may be fastened in place using nails for the 1/2- and the 5/8-in. thicknesses. Staples may also be used for the 1/2-in. thickness. Nails should be 11 gauge with a 7/16-in. head, and 1-1/2 in. long for 1/2-in.-thick boards and 1-3/4-in. long for 5/8-in.-thick boards. Staples should be 15 or 16 gauge, 7/16 or 1/2 in. wide and 1-1/2 in. long. Fasteners should be placed not closer than 3/8 in. from the edge or end of a board. Boards may also be applied using screws, and often are, especially when the wall framing consists of metal studs. The size and spacing of drywall screws should be in accordance with the board manufacturer's recommendations.

Fasteners should be located in 4-ft. by 8-ft. panels at not more than 4 in. on centers along the edges and ends, and not less than 8 in. on centers along each intermediate support. Fasteners in 2-ft. by 4-ft. tongue-and-groove boards should be located where recommended by the manufacturer of the board, but not less than four fasteners should be used in each 2-ft. width at each framing member when diagonal bracing is used and not less than seven fasteners in each 2-ft. width when bracing is not used.

Codes vary on the use of diagonal bracing with gypsum board sheathing. Some do not require bracing under some circumstances. Some permit the gypsum sheathing to be counted as bracing under some conditions. Others require bracing, but do not permit gypsum sheathing to serve in that capacity. In these cases, plywood panels or wood or metal corner bracing may be used.

INSULATED SHEATHING Several types of insulated sheathing are available. Common insulating materials in them include polystyrene and isocyanurate. Some are plain boards without coverings. Others are covered with reflective foil or other materials that increase their thermal effect and, in some cases, they themselves provide air and moisture barriers, or can be made to function as such by taping the joints, thus eliminating the need for a separate layer of material to provide these functions. It is necessary, however, to take special precautions, such as installing vapor-relief

strips, to ensure that moisture does not become trapped in the stud cavities behind insulated sheathing.

One aspect of insulated sheathing products that bears some mention is their combustibility. For the most part, these boards cannot be left exposed and their use is limited by codes to wall assemblies that protect them. They are often used, for example, in exterior insulation and finish systems (see Section 7.4). Another limitation of these products is that they are not usable or do not perform well under some kinds of finishing materials. The manufacturer of the siding or other finishing material to be used should be consulted and its advice followed regarding the specific use of insulated sheathing products.

The installation and fastening of insulated sheathing boards vary from product to product and they should be installed according to their manufacturers' instructions.

■ AIR INFILTRATION BARRIERS

Air infiltration barriers reduce the number of air changes in a building and help create a more energy-efficient interior environment. They also protect sheathing against deterioration and resist the further penetration of moisture that finds its way through the siding or other finish material. Air infiltration barriers in common use over wood-framed walls include the traditional asphalt-saturated organic felt (building paper) and more modern products such as polyethylene sheets and woven polyolefin sheet.

These materials may be applied directly to studs to resist the infiltration of air and moisture when sheathing is not used or when spaced sheathing is used (see Section 7.6). They may also be applied to the studs behind water-resistant, non-nailable sheathing to make locating the studs easier for nailing finishing materials. They may also be installed over the sheathing. Although not strictly required with plywood sheathing or other water-resistant sheathing materials, polyethylene and polyolefin materials may be installed as an air infiltration barrier only. Local codes may require a vapor retarder beneath masonry veneer, and may or may not permit the more modern plastic materials to be used in this application. Building paper is required beneath stucco. An air infiltration barrier is not appropriate in exterior insulation and finish systems.

Sheathing paper should be a material such as No. 15 or No. 30 asphalt-saturated felt, which has low vapor resistance (see Chapter 7). Materials such as coated felts or laminated waterproof papers that have high vapor resistance and act as vapor retarders should not be used. Such materials may permit moisture condensation between the sheathing paper and the surface of the wall sheathing. Coal tar-saturated felts should not be used because they may stain siding.

When sheathing paper is used, it should be applied over the entire wall as soon as possible with at least a 4-in. top lap at horizontal joints and a 4-in. side lap at end joints. It should be lapped 6 in. from both sides around corners. Only sufficient fasteners need be used to hold the sheathing paper securely in place until the finish materials are applied. Wall finish materials should not be applied over wet sheathing paper.

Installing, fastening, and sealing air infiltration barriers should be done in accordance with the recommendations of their manufacturers. Fastening is done using staples, nails, or adhesives. Overlaps should be sealed. These products degrade in ultraviolet light and should not, therefore, be left exposed for more than one month.

6.6.5.2

Roof Sheathing

Suitable performance of roofing is dependent on the entire roof structure. Roof joists, rafters, and other supports should be adequately sized and spaced to carry the necessary superimposed loads over the required spans (see Sections 6.6.3, 6.7.2, and 6.7.3).

Many types of roofing require a bituminous sheet underlayment for their proper performance. These underlayments are discussed in the section where each roofing type is covered (Sections 7.6, 7.7, and 7.9).

Roof sheathing should be smooth, securely attached to its supports, and should provide an adequate base to receive roofing nails and fasteners. The types of roofing discussed in this book (Chapter 7) are applied over solid sheathing composed of wood boards or plywood. Wood shingles and hand-split wood shakes are also applied over spaced sheathing.

■ LUMBER SHEATHING

Solid sheathing of boards should be seasoned to a maximum moisture content of 15%. Boards should be tongued and grooved, in a nominal 1-in. thickness, and usually not more than 6 in. in nominal width, to minimize shrinkage. Boards should be tightly matched and securely face and edge nailed with two 2-1/8- by 0.109-in. (7d) annularly threaded or 2-1/2- by 0.131-in. (8d) common wire nails at each framing member. Maximum spacing of supports for nominal 1-in. lumber boards is 24 in. on centers.

Spaced sheathing of 1 × 3, 1 × 4, or 1 × 6 wood boards may be used in blizzard-free areas (generally where the outside design temperature is warmer than 0°F) to receive the application of wood shingles or wood shakes (see Chapter 7).

The center-to-center spacing of these boards should be equal to the exposure of the shingle or shake, but not more than 10 in., and the openings between boards should not be wider than the individual boards. Seasoning and face nailing spaced boards should be the same as described for solid wood board sheathing.

■ PLYWOOD SHEATHING

Plywood sheathing panels should be APA-rated sheathing of one of the following grades: Exposure 1 or 2; Exterior; Structural I, Exposure 1; or Structural I, Exterior. Panels exposed to the exterior should be Exterior grade. Plywood sheathing should have a span rating, panel thickness, support spacing, and edge support appropriate to the load supported (Fig. 6.6-36). Panel edges may be square or tongued and grooved.

Plywood sheathing should ordinarily be applied with its long sides perpendicular to the framing and with a 1/8-in. gap between panels to permit thermal movement. Square-edged boards should be held level and supported with back blocking or panel clips. The number and spacing of panel clips depends on the support spacing, but one between each pair of supports that are 24 in. apart and two in 48-in.-wide spaces is normal.

Plywood sheathing up to 1/2-in. thick should be fastened in place with 6d nails; from 1/2-in. to 1 in. thick, with 8d nails. These nails may be either common smooth-shank or deformed-shank nails. Panels 1-1/8 in. thick should be nailed with 8d ring- or screw-shank nails or 10d common smooth-shank nails. Nails should be

FIGURE 6.6-36
Recommended Uniform Floor Live Loads for APA-Rated Sturd-I-Floor and APA-Rated Sheathing with Long Dimension Perpendicular to Supports

Sturd-I-Floor Span Rating	Sheathing Span Rating	Maximum Span, in.	Allowable Live Loads, psf*						
			Joist Spacing, in.						
			12	16	20	24	32	30	48
16 oc	24/16, 32/16	16		100					
20 oc	40/20	20		150	100				
24 oc	48/24	24				100			
32 oc	60/32	32					100		
48 oc		48						100	55

Courtesy American Plywood Association.

*10 psf dead load assumed. Live-load deflection limit is ℓ/360. Note: Shaded joist spacings meet Code Plus recommendations.

spaced at 6 in. on centers around panel edges and 12 in. on centers at intermediate supports. However, in panels over support spacings of 48 in. or more, all nails should be spaced at 6 in. on centers.

■ NONWOOD SHEATHING

When nonwood sheathing materials are recommended by their manufacturers for use as roof sheathing, the sheathing manufacturer should specify the type of fastening required to attach sheathing to supports. Such materials should not be used without the concurrence of the roofing manufacturer.

6.6.6

PREASSEMBLED COMPONENTS

Preassembled components are parts of a building's framing, floor and ceiling assemblies, wall assemblies, or roof and ceiling assemblies that are manufactured in a location other than that in which they will finally be used. They may range from small wall panels to complete structures.

Elements used in preassembled components include units made from plywood and lumber, including stress skin panels, sandwich panels, preframed panels, rigid frames, box beams, and various curved units. Plywood is also used to construct roof trusses and truss joists, which could be considered as preassembled components, but are discussed in Section 6.7.3.

Panelized assemblies for floors, walls, and roofs consist of either normal framing systems that are built as a unit and then erected, or of one or more of the components listed in the preceding paragraph, or both.

6.6.6.1

Elements Used in Preassembled Components

The development of off-site, shop-fabricated building components has led naturally to elements made of dimension lumber and plywood. These integral units may consist of plywood panels either nailed to precut lumber or fastened with glue, using either pressure- or nail-glued techniques under controlled shop condi-

tions. Full-pressure gluing, which requires rather expensive equipment and careful control, produces excellent results (Fig. 6.6-37). Nail gluing, where nails furnish the pressure required for an adequate glue bond, also requires careful and precise techniques but less equipment. Nail gluing leads to satisfactory, predictable results in the laboratory but has not always been successful in the field. This is the result of poor gluing techniques, unfavorable temperatures, use of lumber not properly dried or surfaced, insufficient nailing for proper pressure, and dried-out glue lines. However, with proper supervision and control of these variables, field gluing can be successful.

FIGURE 6.6-37 *Inserting insulation and foil vapor retarder in a stress skin panel.* *(Courtesy American Plywood Association.)*

Aside from the advantages that preassembled components supply in erection time and in-place costs, plywood components derive added benefits from the plywood itself. Plywood's tensile and compressive strength, panel-shear strength, split resistance, and dimensional stability enable it to function structurally as part of the component instead of serving just as a base for covering materials.

Practically any flooring or ceiling finish that can be applied to a conventional assembly can be applied to a panelized floor and ceiling, wall, or roof and ceiling assembly. Finishes can be field applied or may be an integral, prefinished part of the panelized assembly (Fig. 6.6-38).

■ STRESS SKIN AND SANDWICH PANELS

Stress skin panels are panels with stressed covers that consist of plywood sheets glued to the top and, in most cases, the

bottom faces of longitudinal framing members so that the assembly will act integrally in resisting bending stresses (Fig. 6.6-39). Structurally, the plywood faces carry the compressive and tensile stresses induced by bending, whereas the framing members principally carry shear stresses. The stressing of the skin allows a reduction in the size of the framing members. For example, wall and partition studs, which normally are 2 × 4s, can be reduced to 1 × 3s or 2 × 3s with stress skin design. The plywood must be bonded firmly with glue to the frame in order to transfer applied shearing stresses (Fig 6.6-40). Stress skin panels may be used as walls, floors, and roofs (Fig. 6.6-41; see also Section 6.6.6.2).

Sandwich panels follow the general principles of stress skin panel design, but the faces are glued to and separated by weaker, lightweight core material such as resin-impregnated paper honeycomb or a plastic foam core designed for this use (Fig. 6.6-42). Sandwich panels possess many of the advantages of stress skin panels but are lighter in weight, needing no inner panel framing except, perhaps, around the perimeter. The cores provide a uniform, stabilizing support for the strong, relatively thin faces. The cores usually have little strength in flexure but have adequate compressive strength for normal floor panels. Some cores, particularly plastic foams, also provide excellent insulation.

Though panelized construction provides many of the advantages of component fabrication, the incorporation of heating, air conditioning, electrical, and plumbing equipment requires careful design of the system. Designs have been made to incorporate mechanical and electrical equipment into stress skin and sandwich panels. These require special handling and erection procedures in the field, and may not be acceptable to local code authorities. Therefore, they have had limited success to date.

In addition, many local building codes specifically require 2 × 4 studs spaced at 16 or 24 in. on centers, which would negate the advantage of using stress skin or sandwich panels unless an exception could be obtained. Some codes, however, permit the acceptance of any system that performs as well as the system specified in the code. In this case, a properly engineered and constructed panelized system might be accepted.

FIGURE 6.6-38 *(a) Sandwich panel with foamed plastic core; (b) an assembly panel made of a foamed plastic core sandwich panel; (c) an entire house structure made of sandwich panels.* *(Courtesy American Plywood Association.)*

■ PREFRAMED PANELS

Preframed panels consist of fabricated, precut lumber and plywood that are nailed or mechanically fastened together. They may be used for walls, floors, or roofs (Fig. 6.6-43).

■ RIGID FRAMES

Rigid frames are specialized construction elements consisting of arch-shaped frames, usually comprising four pieces of straight lumber (two for posts and two for rafters). Plywood gussets, rigidly nailed or glued, connect adjacent members, transmit stresses across the joints, and consolidate the frames into integral units.

■ BOX BEAMS

Box beams are hollow structural units consisting of two or more vertical plywood webs that are attached, usually by gluing or nailing, to lumber flanges. Typical box beam cross sections are shown in Figure 6.6-44. As in a steel beam, the flanges carry the bending forces, while the plywood webs transmit the shear. In this manner each of the materials is employed more effectively. At intervals along the beam, stiffeners may be inserted between flanges and attached to the webs to distribute concentrated loads and resist web buckling.

Box beams have been used successfully for many purposes, including rather spectacular 48-in.-deep girders for factory roof systems and other members spanning up to 120 ft., as well as in many conventional applications (Fig. 6.6-45).

Although either nail gluing or nailing alone may be satisfactory, pressure gluing with accurate controls is the preferred method to give maximum stiffness and strength to box beams.

■ CURVED AND MISCELLANEOUS ROOF UNITS

Curved and other roof units permit construction of unique design shapes. Curved panels consist of built-up panels with plywood faces glued top and bottom to various core materials. These usually form an arc of a circle as shown in Figure 6.6-46. Other design shapes, some of which are shown in Figure 6.6-47, also can be produced.

FIGURE 6.6-39 *Typical two-faced stress skin panel.* *(Courtesy American Plywood Association.)*

Labels in figure:
- Butt joint between plywood panels
- Plywood top skin
- Plywood splice plate
- Vent holes
- Lumber header may be continuous or as shown on opposite end
- Glue joint
- Ventilation openings
- Blanket insulation as required
- Vent holes
- Lumber headers
- Lumber blocking (not req. if pre-spliced skins are used)
- Plywood lower skin
- Lumber stringers
- Scarf joint in lower skin is preferred method (alternate: spliced butt joint)

Panelized Assemblies

Floor and ceiling assemblies, wall assemblies, and roof and ceiling assemblies can be manufactured and brought to the erection site as panelized units.

■ PANELIZED FLOOR AND CEILING ASSEMBLIES

Panelized floor and ceiling assemblies take many forms. One common assembly uses 1-1/8-in.-thick APA-rated Sturd-I-Floor combination subfloor and underlayment plywood panels spanning between beams spaced 48 in. on centers (see Fig. 6.6-30). Other systems consist of simple preframed panels in which the members are precut and to which subflooring is applied to form a preassembled component (Figs. 6.6-48 and 6.6-49). Frame members of nominal 2-in. *dimension* lumber function as floor joists. Panels may be fabricated on or off site in practically any size, with insulation and vapor retarder installed.

Panelized floor and ceiling assemblies may also utilize stress skin panels or sandwich panels (see Section 6.6.6.1). Panelized floor assemblies can be used in a variety of ways: panels may span between transverse or longitudinal beams or be centrally supported by bearing walls or beams, as with conventional floor and ceiling assemblies.

■ PANELIZED WALL ASSEMBLIES

Panelized wall assemblies may consist of simple preframed panels using precut wood studs to which sheathing has been applied (Figs. 6.6-50 and 6.6-51). Fabrication may include the installation of insulation, air infiltration barriers, interior and exterior finishes, windows, and doors (Figs. 6.6-52 and 6.6-53); and panels may be fabricated off site in any size that is conveniently handled.

Panelized wall assemblies may also include stress skin and sandwich panels (see Section 6.6.6.1).

■ PANELIZED ROOF AND CEILING ASSEMBLIES

Panelized roof and ceiling assemblies may consist of simple preframed, precut, and sheathed panels (Fig. 6.6-54). Nominal 2-in. *dimension* frame members of the panels function as rafters or ceiling joists. Panels may be fabricated on or off site in any size that can be conveniently handled.

Panelized roof and ceiling assemblies may also include stress skin or sandwich panel construction (see Section 6.6.6.1). Panelized roof and ceiling assemblies may be used in a variety of ways to form sloped or flat roofs and ceilings (Fig. 6.6-55). Panels may span between transverse or longitudinal beams or can be centrally supported by bearing walls or beams, as in joist roofs.

Membrane or shingle roofing finishes may be applied to panelized systems. Interior finishes may be field applied or may be an integral prefinished part of the panel.

FURRING

Wood furring strips most commonly are used with wood-framed and masonry or concrete construction (Fig. 6.6-56) where incombustibility, decay resistance, or termites are not a consideration. When the framing is of irregular spacing or the spacing is too great for the intended wall finish material, cross furring should be applied perpendicular to framing members (Fig. 6.6-57). Recommendations for wood cross furring are given in Figure 6.6-58.

Cross furring of wood smaller than 2 × 2s is not sufficiently stiff to permit the nailing of gypsum wall board or gypsum plaster lath without excessive hammer rebound, which may result in loose board attachment. Thinner cross furring is acceptable for screw-attached board because little or no impact is involved when screws are driven. Thinner furring also may be acceptable if the board is applied with stud adhesive and nailing is limited to locations where the furring is fully supported by underlying framing.

When the primary purpose of furring is to increase the wall or ceiling assembly thickness in order to conceal piping or to accept additional thermal insulation, 1 × 2 or larger parallel furring can be applied directly over the primary framing, nailed 16 in. on centers with nails providing penetration recommended for cross furring (see Fig. 6.6-58).

Furring strips over masonry or concrete can be secured with cut nails, helically threaded concrete nails, or powder-actuated fasteners of appropriate size. Maximum fastener spacing should be 16 in. on centers for 1 × 2 strips and 24 in. on centers for 2 × 2 or larger strips.

FIGURE 6.6-40 *(a) Stress skin panels with plywood surfaces glued to a 2-in. lumber frame. After the glue has been placed and the plywood positioned, the entire panel is treated with heat and pressure in a press. (b) Stress skin panel being lowered into place by crane.* (Courtesy American Plywood Association.)

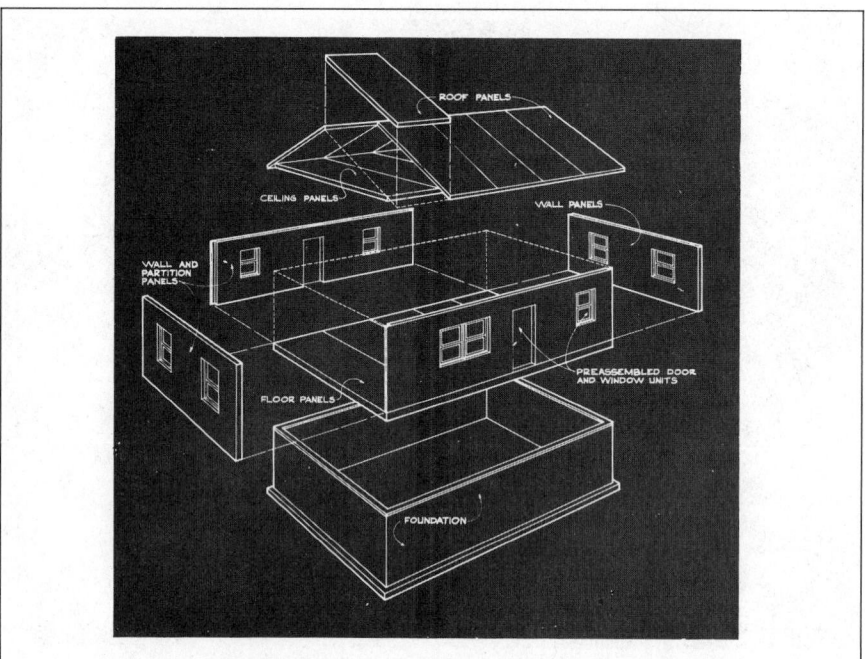

FIGURE 6.6-41 *Typical uses for stress skin panels.*

FIGURE 6.6-42 *Sandwich panel with (a) honeycomb core; (b) foam core. (Courtesy American Plywood Association.)*

FIGURE 6.6-43 *Preframed panels being placed (a) by crane and (b) by hand. (Courtesy American Plywood Association.)*

FIGURE 6.6-44 *Typical plywood box beams.* (*Courtesy American Plywood Association.*)

FIGURE 6.6-45 *A plywood box beam.* (*Courtesy American Plywood Association.*)

Typical Panel Using
Curved Plywood Ribs

Preferred method: Scarf joint in plywood skin

Alternate: Spliced butt joint where permitted by design.

Glue joint

Set out edge member to form tongue of tongue & groove joint

Depth

Headers

Lumber ribs may be used if desired

Width

Lower plywood skin

Plywood upper skin

Insulation (optional)

Set in edge member to form groove

Length

Plywood laminated ribs (laminated & curved prior to panel assembly)

Typical Panel Using
Solid Plywood Core

Scarf joint in plywood

Glue joint

Plywood core set out to form tongue of tongue & groove joint

Width

Lower plywood skin

Plywood core

Plywood upper skin

Plywood core set in to form groove

Tight butt joint in plywood core unless scarf joint required

Length

Typical Panel Using
Lightweight Sandwich Core

Scarf joint in plywood

Lumber edge member set out to form tongue

Depth

Resin-impregnated paper honeycomb core

Lumber headers

Width

Scarf joint in plywood

Upper plywood skin

Edge members & headers may be plywood laminated sections

Length

Lumber edge member set in to form groove of tongue & groove joint

Glue

Lower plywood skin

FIGURE 6.6-46 *Curved stress skin panels.* (*Courtesy American Plywood Association.*)

FIGURE 6.6-47 *Plywood stress skin construction permits use of many geometric shapes.* (Courtesy American Plywood Association.)

FIGURE 6.6-48 *This preframed floor panel consists of plywood sheathing attached to 2 × 4 stiffeners that span 4 ft. from beam to beam.* (Courtesy American Plywood Association.)

FIGURE 6.6-49 *Preframed floor panel with top and bottom plywood surfaces applied to 2 × 8 frame members.* (Courtesy American Plywood Association.)

FIGURE 6.6-50 *Panels may be built off site in sizes than can be conveniently handled by two workers.* (Courtesy American Plywood Association.)

FIGURE 6.6-51 *Preframed panels 4 ft. wide being erected. Studs are spaced 24 in. on centers. The space at the top of the panels is to receive a continuous header.* (Courtesy American Plywood Association.)

FIGURE 6.6-52 *Large wall panel with sheathing applied and windows installed can be fabricated off site.* (Courtesy American Plywood Association.)

FIGURE 6.6-53 *Off-site fabrication may also include (a) panels with exterior finish and windows installed; (b) panels with siding, windows, and insulation installed.* (Courtesy American Plywood Association.)

FIGURE 6.6-54 *This preframed roof panel has been assembled on the ground, partially sheathed, and lifted into position.* (Courtesy American Plywood Association.)

FIGURE 6.6-55 *Vaulted stress skin panels can create unusual roof and ceiling forms.* *(Courtesy American Plywood Association.)*

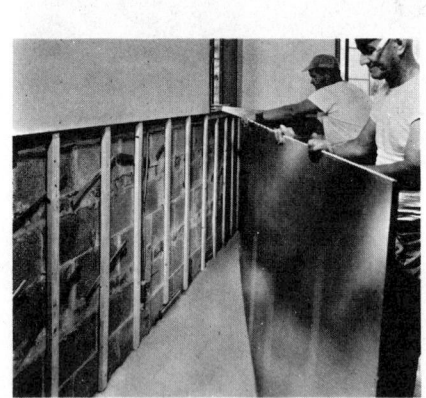

FIGURE 6.6-56 *Wood furring on masonry is shimmed to provide a plumbed, true, and firm base for gypsum wallboard attachment.* *(Courtesy Gypsum Association.)*

FURRING STRIPS ATTACHED TO STUDS

FIGURE 6.6-57 *Wood cross furring supporting gypsum board.* *(Courtesy Gypsum Association.)*

FIGURE 6.6-58
Minimum Requirements for Wood Cross Furring*

| Location | Support Spacing | | | Nails | |
	in.	mm	Furring	Spacing	Penetration
Ceilings	16 o.c.	406 o.c.	2 in. × 2 in.	2 at each support	7/8 in. (22 mm) (threaded)
	24 o.c.	610 o.c.	2 in. × 2 in.		1-1/4 in. (32 mm) (smooth shank)
Walls	16 o.c.	406 o.c.	1 in. × 2 in.	1 at each support	3/4 in. (19 mm) (threaded or smooth shank)
	24 o.c.	610 o.c.	1 in. × 3 in.		

Courtesy Gypsum Association.

*Wallboard screw attached; if nailed, 2-in. × 2-in. furring should be used in all cases.

6.7 OTHER FRAMING SYSTEMS

Other wood-framing systems include post-and-beam, plank-and-beam, trusses, and wood-pole construction.

6.7.1

POST AND BEAM

Post-and-beam systems (Fig. 6.7-1) consist of posts that support a series of beams; the supported beams, which in turn carry the imposed loads of floors, roofs, and ceilings; and non-load-bearing wall sections placed between the posts (Fig. 6.7-2).

In a post-and-beam system the posts typically are 4 × 4 lumber members or are built up of several 2 × 4s and are generally spaced 4 to 8 ft. apart. Wall areas between posts are framed, sheathed, and finished similar to other wood wall systems. They serve primarily as weather barriers, provide resistance to racking, and give lateral support to the posts. Resistance to lateral forces generally is achieved with sheathing materials or diagonal bracing. Since the wall areas between posts do not carry vertical loads, framing members may be placed horizontally to receive vertical exterior siding; and areas between posts can accommodate large windows or doors easily. Glass panels often are actually set (stopped) into the posts. No headers are required, but the placement of windows and doors is determined by the spacing and location of the posts.

Post-and-beam systems are often integrated with plank-and-beam roof and ceiling assemblies. Planks and beams generally are left exposed to form a finished ceiling, and posts may be partially exposed on the interior and exterior surfaces as part of the finished wall. In such designs, the expression of the basic structural system largely determines the interior and exterior appearance and character.

6.7.2

PLANK AND BEAM

Plank-and-beam construction is similar to wood joist construction except that the

FIGURE 6.7-1 *Post-and-beam framing.*

FIGURE 6.7-2 *Wall construction for use with post-and-beam framing.*

subfloor is 2 in. or more in nominal thickness and structural members are larger and spaced farther apart. Underlayment is recommended over plank subfloors.

When heavier structural members such as beams are used for framing, the support spacing can be substantially increased and heavier 2-in.-thick boards generally are utilized. When 2-in.-thick lumber is laid flat and is stressed in bending across its 2-in. dimension, it is referred to as a *plank*, and hence this type of subfloor also is called planking. The planking generally is tongued and grooved to facilitate the transfer of concentrated loads from one plank to another. In lumber grading, such tongue-and-groove material is called decking.

It should be noted that the meanings of the terms *beam*, *plank*, and *decking* as used in lumber grading do not completely agree with general construction terminology. In light wood-framed construction, 4-in.-thick framing members generally are referred to as *beams*; yet in lumber grading these are classed as joists and planks, and the term *beams and stringers* is restricted to rectangular members 5 in. thick or thicker. In lumber grading, *decking* means lumber 2 to 4 in. thick, 4 in. wide or wider, with a maximum moisture content of 15% or 19% and with tongue-and-groove edges. However, in common usage, both installations with this special product and ordinary surfaced four-sided lumber joists and planks are frequently referred to as *planking*, or as *decking*, when used in a roof assembly, which is called a *roof deck*. For instance, a traditional plank-and-beam framing system utilizes planking, which is really decking, and beams, which are really joists, according to lumber grading terminology.

6.7.2.1

Plank-and-Beam Floors

Plank-and-beam floor assemblies are characterized by the use of nominal 2 × 6 or 2 × 8 tongue-and-groove or splined *planks* spanning between beams generally spaced 4 to 8 ft. apart (Fig. 6.7-3). The planks serve as subflooring, as a working platform, and to transmit floor loads to fewer but larger members than occur in conventional wood joist floor systems.

Floor finishes such as wood-strip flooring may be applied directly over such planks, but sheet and tile flooring require an underlayment. When used in multistory construction, beams and the underside of planks can be left exposed to provide the finished ceiling. However, without the application of an additional ceiling finish, a plank-and-beam floor and ceiling assembly may offer less resistance to sound transmission than other systems, and does not provide space for the installation of heating, air conditioning, electrical, or plumbing equipment. Therefore, plank-and-beam systems generally are used as floor systems and are suited particularly to one-story construction. When there are no deep girders beneath them, they permit a lower first-floor-to-grade elevation and result in a lower building height. Design of interior wall placement should consider the location of beams.

Plank-and-beam floor and ceiling assemblies often are integrated with plank-and-beam roof and ceiling assemblies and post-and-beam wall systems. In such

FIGURE 6.7-3 *Plank-and-beam construction.*

design, the expression of the basic structural system largely determines the interior appearance and character.

Conventional joist framing requires a support spacing at a maximum of 24 in. on centers. Using APA-rated Sturd-I-Floor combination subfloor and underlayment can push this spacing up to 48 in. on centers. But plank-and-beam framing permits support spacings of 6 ft. or more. Besides the economy in materials and labor, tongue-and-groove or splined plank subfloors provide an acceptable finish surface on the underside of the floor, which can be left exposed. The disadvantage of such a system, however, is that it does not provide the usual cavities for mechanical and electrical distribution equipment, and, in the absence of a separate ceiling finish, relatively poor sound isolation, particularly from impact noise. For these reasons, plank floors are most suitable over unfinished spaces or where impact sound isolation from above is not an important consideration. In light wood-framed construction, plank-and-beam construction generally is more adaptable to roof assemblies.

A plank subfloor generally requires underlayment for resilient flooring and adhesively applied ceramic tile or thin-set terrazzo. However it can be used as a base for carpeting with suitable cushioning (padding), if the subfloor surface is reasonably level and smooth. Heavy cushioning and deep pile carpeting, with or without underlayment, also considerably improves impact sound isolation and is recommended where sound transmission is a factor.

Plank subfloors can be built with tongue-and-groove (T & G) or surfaced-four-sides (S4S) lumber. T & G lumber is graded in most species as Select Decking or Commercial Decking, both of which are suitable for subfloors. Decking is manufactured 2 to 4 in. in nominal thickness and in widths ranging from 4 to 12 in. The maximum width for subfloors should not exceed 8 in. at a 19% moisture content, but wider boards can be used when their moisture content is limited to a maximum of 15%. Thinner (2-in.) planks generally are made with a single tongue and groove, whereas 3- and 4-in.-thick planks have two. Planks with grooves on both edges, for use with splines, also are available. One face usually is of better quality and is intended for exposed view, such as in a paneled ceiling. This face also may be worked to a special pattern with V joints, rabbets, or rounded edges.

S4S lumber suitable for planking is the same as for joists and is graded as Select Structural No. 1, No. 2, or No. 3.

A plank-and-beam system is essentially a skeleton framework. Planks generally are intended to support moderate, uniformly distributed loads. These are carried to the beams, which in turn transmit their loads to posts or walls supported on foundations. Foundations for plank-and-beam framing are similar to those used for other wood-framing systems (Fig. 6.7-4). With beams (and sometimes posts) spaced 6 ft. or more on centers, this system is well adapted to pier and crawl-space foundations.

Posts should be of adequate size to carry the load and large enough to provide full bearing for the ends of beams. In general, posts should be at least 4 × 4, but 4 × 6 posts are preferred, particularly where beams join or where spaced beams are used. Where the ends of beams abut over a post, the 6-in. dimension should be parallel to the beams (Fig. 6.7-5). Suitable

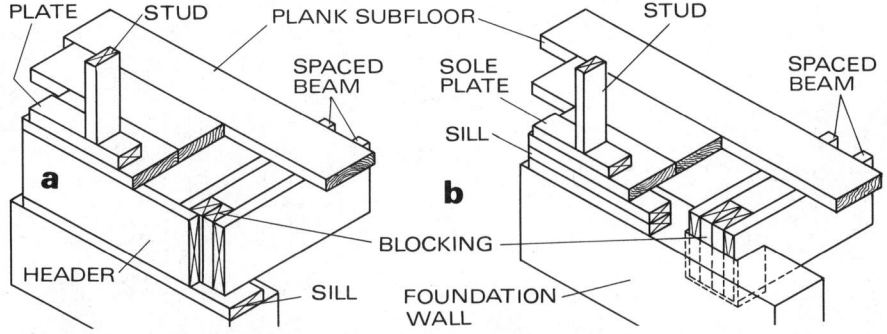

FIGURE 6.7-4 *Spaced beam (or girder) for plank floors can be (a) supported on a foundation sill, or (b) framed into a wall pocket.* (*Courtesy American Forest and Paper Association.*)

FIGURE 6.7-5 *Intermediate support for plank floors can be provided by a wood post under (a) a solid beam or (b) a spaced beam.* (*Courtesy American Forest and Paper Association.*)

FIGURE 6.7-6 *A plank floor at an upper floor can be supported by (a) a built-up beam or (b) a spaced beam, framed into an exterior wall over a post or top plate.* (*Courtesy American Forest and Paper Association.*)

PLANK CONTINUOUS ACROSS ONE SPAN

PLANK CONTINUOUS ACROSS TWO SPANS

FIGURE 6.7-7 *The resultant deflection under similar loads is less when individual planks are long enough to reach across two spans than it is in single-span floors.* (*Courtesy American Forest and Paper Association.*)

beam sizes vary with grade, species, span, loading, and spacing. Beams may be solid lumber, glue-laminated members, or built up of several joists securely nailed to each other or to spacer blocks between them. The minimum required fiber stresses in bending and modulus of elasticity are given in the AFPA publication *Plank and Beam Framing for Residential Buildings* (see Fig. A-21 and the accompanying text). Beams should be securely fastened to posts with framing anchors or clip angles. Details of upper floor beams, framing into an exterior frame wall, are illustrated in Figure 6.7-6.

In laying out plank subfloor, greater advantage can be taken of the strength and stiffness of the material by making the planks continuous over more than one span. For example, using the same span and uniform load in each case, a plank that is continuous over two spans is nearly 2-1/2 times as stiff as a plank that extends over a single span (Fig. 6.7-7). Required values for fiber stresses in bending and modulus of elasticity for different loading conditions of 2-in. planks are given in Figure A-22. An explanation of the use of this table appears in the text accompanying the figure. Regardless of their width, planks should be blind nailed to each support with one 10d nail; in addition, planks 4 or 6 in. wide should be face nailed also with one 16d nail at each support; 8-in. planks should be face nailed with two 16d nails; 10- and 12-in. planks with three 16d nails.

A wood-strip floor should be laid at right angles to a plank subfloor, using the same procedure followed in conventional construction. Where the underside of the plank is to serve as a ceiling, care is needed to make sure that flooring nails do not penetrate through the plank.

In a plank-and-beam system, interior partitions usually are non–load bearing. Where bearing partitions occur, they should be placed over beams and the beams enlarged to carry the added load. If this is not possible, supplementary beams must be placed in the floor framing. Non-load-bearing partitions, which are parallel to the planks, should be supported on a sole plate consisting of double 2 × 4s set on edge and spiked together (Fig. 6.7-8b). Where door openings occur, double 2 × 4s may be placed under the plank floor and framed into the beams with framing anchors (Fig. 6.7-8a). Non-load-bearing partitions, located at right angles to the planks, do not require supplementary framing, since the partition load is distributed across several planks.

6.7.2.2

Plank and Beam Roofs

Plank-and-beam roofs typically are used on one-story buildings and for low-slope and flat roofs. These systems are characterized by the use of 2 × 6 and 2 × 8 tongue-and-groove or splined planks spanning between either longitudinal or transverse beams (Fig. 6.7-9). Beam spacing and the thickness of planking are determined by imposed loads, spans, and strength of the materials. Nominal 2-in. planks generally may be used on spans up to about 8 ft. The AFPA publication *Plank and Beam Framing for Residential Buildings* includes tables for the minimum required stresses in bending and modulus of elasticity for beams and planks used in plank-and-beam roofs. (See Fig. A-21 and A-22.)

Longitudinal beams span parallel to the long dimension of a building and are supported on end walls or columns. In buildings up to about 36 ft. wide, beams typically are placed at quarter points. This divides the width of the building into two equal spans for planking on each side of the center ridge beam. Planks may be supported along the exterior sides by walls or beams. Interior planning flexibility is slightly more restricted than when transverse beams are used because of the greater number of beam supports required.

Transverse beams, generally spaced 6 to 8 ft. apart, span perpendicular to the long dimension of a building. They may be supported by load-bearing walls or beams along the ridge or center of the building.

The system may be used with wood stud exterior walls and is often used with post-and-beam exterior walls.

A plank ceiling and roof deck does not provide overhead space for the convenient concealment of ductwork and wiring. If the system requires insulation, it is necessary to use rigid insulation applied to the top of the deck, which in turn may limit the selection of roofing finishes.

FIGURE 6.7-8 *Interior non-load-bearing partitions running parallel to a plank floor can be supported on (a) a suitable underfloor beam or (b) a double vertical sole plate.* (Courtesy American Forest and Paper Association.)

LONGITUDINAL BEAMS

TRANSVERSE BEAMS

FIGURE 6.7-9 *Wood plank-and-beam roof and ceiling assemblies.*

Plank-and-beam roof assemblies, in contrast to trussed and joist-and-rafter roof and ceiling assemblies, require different design and construction methods and more precision in erection, since the finished appearance generally is an integral part of the system. Planks, beams, and columns generally are left exposed as the finished interior surface.

6.7.3

ROOF TRUSSES

Trusses for light wood-framed buildings are composed of many wood-framing members (chords) arranged in a framework of triangles to form a supporting structural element for both a roof and a

FIGURE 6.7-10 *A wood trussed roof and ceiling assembly.*

ceiling (Fig. 6.7-10). Individual members, generally 2 × 4s or 2 × 6s, are usually connected by split ring connectors or metal gang-nail plates (Fig. 6.7-11). Older trusses were often connected by wood or plywood gusset plates.

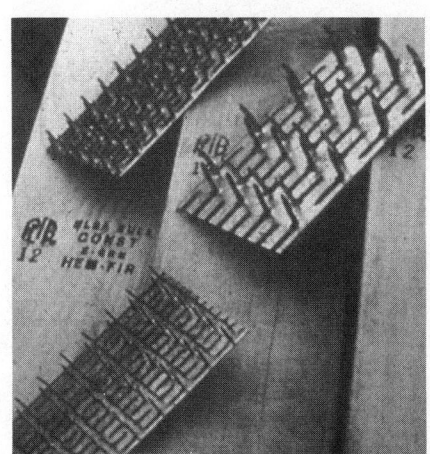

FIGURE 6.7-11 **The types of metal plates used to connect wood truss components.** *(Courtesy Wood Truss Council of America.)*

Roof trusses usually are constructed with metal plates. Such connections permit members to be assembled without having to be overlapped at the joints (Fig. 6.7-12).

A truss acts as a unit to support roof and ceiling loads and to span the entire width of the building between exterior walls. This clear span allows interior walls to be non–load bearing and permits almost complete flexibility in placing walls (Fig. 6.7-13).

FIGURE 6.7-12 **Spliced joints may have metal plates on one or both sides.** *(Courtesy Wood Truss Council of America.)*

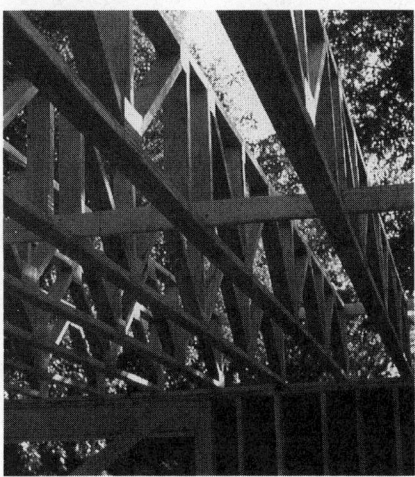

FIGURE 6.7-13 **Trusses provide clear spans and eliminate the need for interior load-bearing walls.** *(Courtesy Wood Truss Council of America.)*

In a light truss, individual diagonal members primarily resist compression and tension forces. The top and bottom chords also resist bending because roof and ceiling loads are applied along their entire length. However, since the chords span only between panel points, their spans are much shorter than those of normal rafters and ceiling joists. Therefore, the loads they carry are lighter and the bending stresses they must resist are smaller than those of conventional framing members. As a result, truss top and bottom chords are much smaller than conventional joists and rafters with the same span. Trusses require stress-rated lumber and must be built from engineered designs. They are commonly spaced from 24 to 48 in. on centers; and like other preassembled components, they can be erected easily, permitting the building to be enclosed quickly (Fig. 6.7-14).

FIGURE 6.7-14 **Individual trusses are positioned easily by hand. Trusses shown will produce a flat roof and ceiling.** *(Courtesy Wood Truss Council of America.)*

Trussed roof systems may be designed for flat (Fig. 6.7-14) or sloped roofs (Fig. 6.7-15) to receive membrane or shingle roofing finishes. Plywood or board sheathing typically is fastened to a truss to receive roofing. Ceilings can be flat, sloped (see Fig. 6.7-14), or even curved (Fig. 6.7-16).

These systems provide overhead space for the installation of insulation and heating, air conditioning, electrical, and plumbing equipment.

Almost any interior finish may be applied to the lower chord member at the ceiling line of a truss. Plaster bases and plaster board finishes commonly are applied directly to trusses. When trusses are spaced more than 24 in. on centers and when other than board finishes are used, it may be necessary to apply nailing strips or furring strips across the trusses to receive the interior finish.

Trusses are economical and are widely used, even for large structures (Fig. 6.7-17). They provide excellent structural performance, make efficient use of materials, are easy to assemble and to erect, and provide maximum interior flexibility.

Irregularities in spacing and leveling often are magnified in wood trussed construction because relatively small members are used for bottom chords and larger spans are involved than in joist construction. Cross furring generally is recommended to remedy these problems. However, when gypsum board is nailed directly to the bottom chords, the chords can be aligned and leveled by using stringers approximately at one-third points continuously across the entire ceiling. Built-in camber to compensate for future deflection of bottom chords also is recommended (Fig. 6.7-18).

With trussed roof construction, exterior walls and the ceiling are sometimes finished before interior partitions are erected and finished. When this method is used, the roofing and other construction elements that would increase roof loads should be installed before interior partitions are erected. If substantial roof loads are introduced after partitions are installed, ceiling distortion may result near interior partitions as trusses deflect (Fig. 6.7-19).

6.7.4

WOOD-POLE CONSTRUCTION

Pole construction has been used for millennia. From prehistoric lake dwellers to twentieth-century farmers, people have been embedding poles in the ground and building simple elevated structures to protect themselves from unfriendly humans, ravaging animals, fierce winds, and engulfing floods.

Over time people have learned to recognize naturally decay-resistant species such as redwood, cypress, and cedar. However, it was not until after World War II, when pressure-treated poles became widely

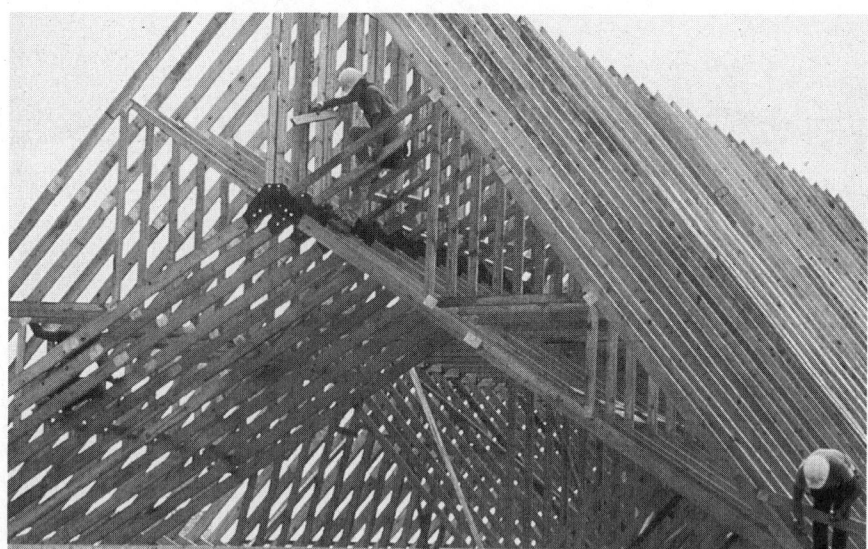

FIGURE 6.7-15 *Trusses with sloped roof and ceiling. Ceilings could also be flat.* (*Courtesy Wood Truss Council of America.*)

FIGURE 6.7-16 *Sloped roof trusses with segmented bottom chords to permit installation of a curved ceiling.* (*Courtesy Wood Truss Council of America.*)

available, that pole construction of farm buildings and beach cottages became commonplace.

The use of chemically treated poles gained impetus from the favorable experience of West Coast builders in the 1950s. They erected houses on steep slopes, on periodically flooded lowlands, and in areas subject to earthquakes, and yet they were able to build them at

reasonable costs and with a high degree of permanence.

The following technical information is based on a study by the American Wood Preservers Institute (AWPI) done for the Department of Housing and Urban Development (HUD). Some recommendations and illustrations were also obtained from publications of the Forest Products Laboratory of the U.S. Department of

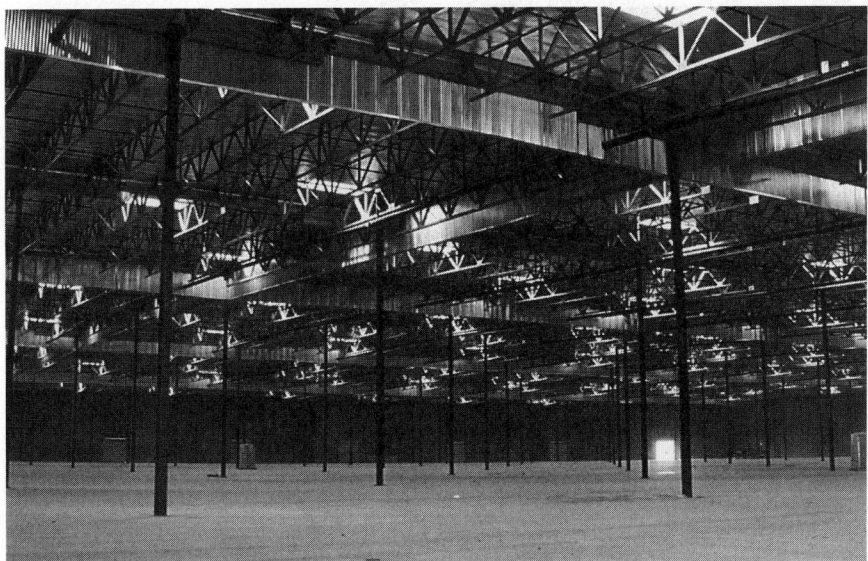

FIGURE 6.7-17 *Large open spaces can be created economically using wood trusses.* (Courtesy Wood Truss Council of America.)

FIGURE 6.7-18 *When cross furring is not used, 1 × 6 or 2 × 4 stringers at one-third points may be installed to regulate truss spacing. Additional 2 × 6 stringers may also be used to level the bottom chords when necessary.*

FIGURE 6.7-19 *In trussed construction, ceiling distortion may result if substantial loads are introduced after interior partitions have been erected.* (Courtesy Gypsum Association.)

Agriculture and from *Pole Building Design* by Donald Patterson for the AWPI.

6.7.4.1

General Description

A pole building is a structure whose contact with the ground is through a series of decay- and termite-resistant round tree trunks (poles) or squared wood members (timbers). Poles are cheaper and longer lasting, are more frequently used, and, therefore, give their name to the system (Fig. 6.7-20).

Although cedar, redwood, and cypress have inherent decay resistance, their availability as long thick poles is limited and their decay resistance varies with local growing conditions. Chemical treatment of Douglas fir, southern pine, ponderosa pine, and red pine in accordance with American Wood Preservers Association's (AWPA) standards ensures long-term resistance to infestation and in-ground durability.

In typical pole construction, poles ranging in diameter from 6 to 12 in. are spaced 6 to 12 ft. apart to support floor and roof areas up to 144 sq. ft. (Fig. 6.7-21). The poles typically are dropped into predug holes, made either with a power auger or by hand digging. Occasionally, poles are "jetted" with a stream of water or driven with a pile driver. Proper embedment of a pole in the ground and backfilling of the hole are critical to the structural soundness of a pole building. A concrete pad under the pole, or a concrete necklace around the pole, may be required to increase the contact area with the soil. The pad and necklace help to distribute the load over a larger area.

Poles may extend substantially above ground to form the load-bearing frame of a superstructure or may terminate at the first-floor level, thus permitting conventional stud-and-joist construction above the pole-framed platform (Fig. 6.7-22).

Embedded poles can support the superstructure either by spreading the loads with a concrete footing or by bearing on rock or very stiff soils. The following recommendations are made only for situations where a concrete footing pad, necklace, or backfill will be used to spread the building loads.

FIGURE 6.7-20 *Treated poles are frequently combined with timbers to form a structural frame and to enhance a design.* (*Courtesy American Wood Preservers Institute.*)

FIGURE 6.7-21 *A plan for a small house with all poles inside the house. Short poles could be eliminated and poles placed outside the walls if floor girders were used.* (*Courtesy American Wood Preservers Institute.*)

The services of a registered professional engineer should be retained whenever a wood pole structure is contemplated.

6.7.4.2

Advantages and Disadvantages

The advantages of pole construction are most pronounced where conventional methods are either uneconomical or do not meet the desired objectives of minimum disturbance to the terrain, freedom from flooding, or resistance to strong winds and earthquakes.

Therefore, on steep slopes, pole buildings are generally more economical because they use relatively light wood members. Because pole construction does not require heavy equipment for excavation, it retains existing drainage patterns and preserves ground cover (Fig. 6.7-23).

On level ground subject to periodic flooding (as in coastal areas from storms or in river plains from spring floods), pole buildings permit raising the lower floor safely above flood levels. When there is adequate embedment and proper connections, flood waters can flow under the building with minimum damage to the structure (Fig. 6.7-24). However, pole buildings in flood-prone areas should be carefully engineered for specific local conditions and should not be based on the generalized guidelines of this section.

In hurricane-plagued areas, whether on flat or sloping ground, pole construction permits secure anchorage of a structure from roof to footings. Proper connections and adequate embedment are especially important in areas of high wind, which are often subject to flooding.

In most parts of the country, it is necessary to insulate the first floor and to provide a finish to protect the insulation.

In heavily wooded areas, any vegetation and combustible materials should be removed from the area that will be beneath the building when it is raised above grade only enough to create a crawl space. A skirt may be desirable to keep out animals and prevent accumulation of debris under the building.

The underfloor cavity should remain accessible to permit periodic inspections of the poles and the underside of the floor.

6.7.4.3

Site Preparation

The use of pole construction provides a great opportunity to preserve the natural features of a site. Buildings can be fitted to the terrain while observing any setbacks and sideyards as dictated by codes and good design principles.

FIGURE 6.7-22 *In a pole-frame building (a) the poles continue up past the main floor. In a pole-platform building (b) the poles stop at the main floor. Lag bolts and joists are used for relatively light loads. Timbers and through-bolted shear connectors usually are required for efficient pole loading.*

FIGURE 6.7-24 *Pole substructure elevates house above the natural terrain, permitting unobstructed runoff of storm and flood waters.* (Courtesy American Wood Preservers Institute.)

FIGURE 6.7-23 *A pole house "floats" above the terrain.* (Courtesy American Wood Preservers Institute.)

Drainage can be crucial on hillside sites. Pavements, roofs, road cuts, and utility trenches may concentrate surface waters into erosive streams. Heavy machinery should be used as little as possible and necessary roads should be terraced along contour lines. It is generally preferable to dig holes by hand and to remove as little natural vegetation as possible. Hand digging theoretically is possible for holes up to 6 ft. deep, but may be uneconomical except for shallow poles, as described in Section 6.7.4.6.

On relatively flat sites, machine digging is more practical, but truck-mounted augers generally are limited to 18-in.-hole diameters. Although these hole sizes are adequate for close pole spacing in good soils, larger diameters are required for most soil conditions.

The minimum number of poles should be used that is consistent with other design and structural requirements. The most efficient use of poles usually occurs when the vertical load per pole is about 8000 lb. Heavier loads require special engineering; one-half the indicated loading is generally inefficient.

Dug holes should be at least 8 in. larger than the diameter of the pole, to permit accurate aligning and plumbing (Fig. 6.7-25). In most cases the hole is wider than this minimum in order to develop a bearing area of sufficient size for the loads.

6.7.4.4

Pole Selection

The main considerations in pole selection are (1) type of treatment and (2) size and length.

Poles should be pressure treated in accordance with AWPA Standard C23 (see Section 6.1.5). This standard specifies how much preservative should be retained

FIGURE 6.7-25 *Poles are aligned and plumbed, then braced temporarily before holes are backfilled. Heavier horizontal members are likely to be used in typical house construction.*

in each pole and how deep it should penetrate into the wood. Treated poles are frequently branded to provide users with certain basic information. Information includes the manufacturer's name and location, and the year of treatment, pole size, length, wood species, and type of preservative.

Preservative treatments are discussed in Section 6.1.5.

6.7.4.5

Constructions Methods

There are two principal methods of pole construction—pole frame and pole platform. The size of the footing and depth of embedment are partly dependent on the method of construction.

In *pole-frame* buildings, the main vertical members extend from below the ground surface to the roof. The poles not only support the vertical weight of the building, but give it considerable lateral resistance to the forces of wind, flood, and earthquake.

In *pole-platform* buildings, treated poles form the substructure, permitting conven-

tional construction methods to be used above. Pole-platform structures generally require greater pole embedment than pole-frame types.

■ GENERAL RECOMMENDATIONS

Following a few general rules will avoid problems in matching conventional framing methods with pole construction. Poles should not be placed within stud walls, because it is costly to apply finishes around poles that vary in diameter and shape. Beams, girders, and posts should be spaced so that standard, readily available sizes of lumber and panel materials can be used with a minimum of waste.

On level ground and on steep slopes with good soils (6000 lb./sq. ft. and better), noncementitious, granular backfill can be used, provided a concrete pad or necklace is also used (Fig. 6.7-26). Common backfill materials include sand, gravel, and crushed rock. Clean sand often is the least expensive of these and can achieve 100% compaction when properly tamped. Excavated soil from the site also can be used, provided it is free of organic matter and is compacted by tamping.

Under these conditions, backfilling with concrete or with soil-cement (excavated soil mixed with portland cement in a 5 to 1 ratio) will generally reduce the required hole depth by 1-1/2 to 2 ft. (see Fig. 6.7-26).

On steep slopes with average or below-average soils (3000 lb./sq. ft. and lower) only concrete or soil-cement backfills are suitable. When concrete or soil-cement is used, a separate pad or necklace is not necessary.

■ DESIGN ASSUMPTIONS

Pole embedment depends on a variety of factors. To reduce the variables to a smaller number that can be tabulated readily, certain design assumptions are made. These are included in the following list. Note that other design assumptions would produce different results from those outlined in the following paragraphs.

Soil bearing values fall in three general categories, as shown in Figure 6.7-27.

Building height of pole-frame structure is 9 ft. from floor to top plate; 4 ft. from top plate to ridge (Fig. 6.7-28).

Embedment Conditions:

Condition A: each pole either rests on a concrete pad or has a concrete necklace and is backfilled with granular material (see Fig. 6.7-28).
Condition B: Each pole is completely backfilled to grade either with concrete or soil-cement (see Fig. 6.7-28).

■ POLE-FRAME BUILDINGS

Embedment depths vary according to the slope of the site, as well as the soil, pole spacing, and unsupported pole height.

FLAT SLOPE On sites with a slope of less than 1 in 1, poles should be embedded and supported as recommended in Figure 6.7-29.

STEEP SLOPE On sites with a slope of 1 in 10 and up to 1 in 1, poles should be embedded and supported as recommended in Figures 6.7-30 and 6.7-31.

It should be noted that embedment recommendations differ for uphill and downhill poles. Uphill poles, which have a short unsupported height, provide the

FIGURE 6.7-26 *Recommended pole footings: (a) concrete pad; (b) concrete necklace; (c) concrete or soil-cement backfill. Dimensions A, B, and D can be obtained from Figures 6.7-29 through 6.7-33. Pad thickness C should be determined by engineering calculations.*

FIGURE 6.7-27
Soil Classification for Pole Construction

Rating	General Description	Soil Groups	Maximum Bearing Value, psf	Lateral pressure per ft. of embedment, psf
Good	Compact, well-graded sand and gravel; hard clay or graded fine and coarse sand	GW, GP, SW, and SP	6,000	400
Average	Loose gravel, medium clay, or more compact compositions	GM, GC, SM, SC, and SL	3,000	200
Below average	Sand or clay containing large amounts of silt	CL, OL, MH, CH, and OH	150	100

Courtesy American Wood Preservers Institute.

FIGURE 6.7-28 *Building height design assumptions and embedment conditions: (a) concrete pad or necklace and granular backfill; (b) concrete or soil-cement backfill. (Courtesy American Wood Preservers Institute.)*

necessary rigidity for the structure. In order to transfer this rigidity to the longer downhill poles, the floor should be designed as a diaphragm. Properly nailed conventional joists and 1/2-in. or thicker plywood floors provide such diaphragm action.

■ **POLE-PLATFORM BUILDINGS**

Embedment depths of poles vary according to the slope of the site, as well as the other factors tabulated for pole-frame buildings.

FLAT SLOPES On sites with a slope of less than 1 in 10, poles should be embedded and supported as recommended in Figure 6.7-32.

STEEP SLOPES On sites with a slope of 1 in 10 and up to 1 in 1, uphill poles should be embedded and supported as recommended in Figure 6.7-33. Recommendations for downhill poles are given in Figure 6.7-30. The floor should be designed as a diaphragm to transfer the rigidity of the uphill poles to the rest of the structure.

■ **CONNECTIONS**

It is generally better to frame poles with pairs of beams by lag bolting, through bolting, or nailing them to the supporting pole (Fig. 6.7-34). To make a stronger connection, poles can be faced or notched (Fig. 6.7-35). Use of a spike grid (Fig. 6.7-36) is preferred to squaring or notching

FIGURE 6.7-29
Embedment Depths for Poles in Pole-Frame Buildings on Sites with Slopes Less Than 1:10

H, ft.	Pole Spacing, ft.	Good Soil Embedment Depth, ft. A	B	D, in.	Tip Size, in.	Average Soil Embedment Depth, ft. A	B	D, in.	Tip Size, in.	Below-Average Soil Embedment Depth, ft. A	B	D, in.	Tip Size, in.
1-1/2 to 3	8	5.0	4.0	18	6	6.5	5.0	24	6	*	6.0	36	6
	10	5.5	4.0	21	7	7.0	5.0	30	7	*	6.5	42	7
	12	6.0	4.5	24	7	7.5	5.5	36	7	*	7.0	48	7
3 to 8	8	6.0	4.0	18	7	7.5	5.5	24	7	*	7.0	36	7
	10	6.0	4.5	21	8	8.0	6.0	30	8	*	7.5	42	8
	12	6.5	5.0	24	8	*	6.0	36	8	*	8.0	48	8

Courtesy American Wood Preservers Institute.

*Embedment depth is greater than 8 ft., considered excessively expensive.

Note: Where a concrete floor slab is used at grade, embedment depths may be reduced to 70% of those shown.

FIGURE 6.7-30
Embedment Depths for Longer, Downhill Poles of Pole-Frame Buildings on Sites with Slopes up to 1:1

Soil Strength	Embedment Depth, ft. Slope of Grade Up to 1:3	Up to 1:2	Up to 1:1
Below average	4.5	6.0	*
Average	4.0	5.0	7.0
Good	4.0	4.0	6.0

Courtesy American Wood Preservers Institute.

*Embedment depth is greater than 8 ft., considered excessively expensive.

because it minimizes the damage to the preservative-treated outside layer and increases the load-bearing value of the connection.

Accepted design practices are outlined in the AWPI's *Pole Building Design* and in *Timber Construction Manual* by the American Institute of Timber Construction (AITC). Strength values used in designing bolted and nailed connections should conform to the *National Design Specifications for Stress-Grade Lumber and Its Fastenings* by the National Forest Products Association (NFPA).

6.7.4.6

Special Construction on Rocky Slopes

The following discussion of pole bracing as an alternative to embedment should not be attempted without competent engineering advice.

On rocky slopes, where deep embedment cannot be developed without expensive machine digging or rock drilling, it is possible to use shallow, hand-dug holes, provided that the pole structure is adequately braced and anchored.

Alternatives to deep embedment include (1) pole seating and cross bracing and (2) shear walls and floor diaphragms.

■ POLE SEATING AND CROSS BRACING

If the understructure is adequately braced, the poles need to be embedded just far enough to prevent slipping downhill. The embedment should be at least 4 ft. and the poles should be placed in specially dug seats in the rock. Since a rock surface is likely to be irregular, tamped earth or

FIGURE 6.7-31
Embedment Depths for Shorter, Uphill Poles in Pole-Frame Buildings on Sites with Slopes up to 1:1

H, ft.	Pole Spacing, ft.	Good Soil Embedment Depth, ft. A	B	D, in.	Tip Size, in.	Average Soil Embedment Depth, ft. A	B	D, in.	Tip Size, in.	Below-Average Soil Embedment Depth, ft. A	B	D, in.	Tip Size, in.
1-1/2 to 3	6	7.0	5.0	18	8	a	6.5	18	8	a	a	a	a
	8	7.5	5.5	18	9	a	7.0	24	9	a	a	a	a
	10	a	6.0	21	9	a	8.0	30	9	a	a	a	a
	12	a	6.5	24	10[b]	a	a	a	a	a	a	a	a
3 to 8	6	7.5	5.5	18	8	a	7.0	18	8	a	a	a	a
	8	8.0	6.0	18	9	a	8.0	24	9	a	a	a	a
	10	a	7.0	21	10[b]	a	a	a	a	a	a	a	a
	12	a	7.0	24	11[b]	a	a	a	a	a	a	a	a

Courtesy American Wood Preservers Institute.

[a] Embedment depth is greater than 8 ft., considered excessively expensive.

[b] These tip diameters maybe decreased 1 in., provided embedment is increased by 1/2 ft.

FIGURE 6.7-32
Embedment Depths for Poles in Platform Buildings on Sites with Slopes Less Than 1:10

H, ft.	Pole Spacing, ft.	Good Soil Embedment Depth, ft. A	B	D, in.	Tip Size, in.	Average Soil Embedment Depth, ft. A	B	D, in.	Tip Size, in.	Below-Average Soil Embedment Depth, ft. A	B	D, in.	Tip Size, in.
1-1/2 to 3	8	4.0	4.0	18	5	5.5	4.0	24	5	7.0	5.0	36	5
	10	4.5	4.0	21	5	6.0	4.0	30	5	8.0	5.5	42	5
	12	5.0	4.0	24	5	6.5	4.5	36	5	*	5.5	48	5
3 to 8	8	5.0	4.0	18	6	6.5	4.5	24	6	*	6.0	36	6
	10	5.5	4.0	21	7	7.0	5.0	30	7	*	6.5	42	7
	12	6.0	4.5	24	7	7.5	5.5	36	7	*	7.0	48	7

Courtesy American Wood Preservers Institute.

*Embedment depth is greater than 8 ft., considered excessively expensive.

FIGURE 6.7-33
Embedment Depths for Shorter, Uphill Poles in Platform Buildings on Sites with Slopes up to 1:1

H, ft.	Pole Spacing, ft.	Good Soil Embedment Depth, ft. A	B	D, in.	Tip Size, in.	Average Soil Embedment Depth, ft. A	B	D, in.	Tip Size, in.	Below-Average Soil Embedment Depth, ft. A	B	D, in.	Tip Size, in.
1-1/2 to 3	6	5.5	4.0	18	6	7.5	5.0	18	6	a	7.0	24	6
	8	6.5	4.5	18	7	a	6.0	24	7	a	8.0	36	7
	10	7.0	5.0	21	7	a	6.5	30	7	a	a	a	a
	12	7.5	5.5	24	8	a	7.0	36	8	a	a	a	a
3 to 8	6	7.0	5.0	18	8	a	6.5	18	8	a	8.0	24	8
	8	7.5	5.5	18	9	a	7.0	24	9	a	a	a	a
	10	a	6.0	21	10[b]	a	7.5	30	10[b]	a	a	a	a
	12	a	6.5	24	10[b]	a	a	a	a	a	a	a	a

Courtesy American Wood Preservers Institute.

[a]Embedment depth is greater than 8 ft., considered excessively expensive.

[b]These tip diameters maybe decreased 1 in., provided embedment is increased by 1/2 ft.

FIGURE 6.7-34 *Double headers at exterior end walls frame ridge-high pole and are bolted to it. (Courtesy American Wood Preservers Institute.)*

FIGURE 6.7-35 *Poles are faced or notched to provide better contact with beams and make suitable connections with shear connectors such as the spike grid shown in Figure 6.7-36. (Courtesy American Wood Preservers Institute.)*

FIGURE 6.7-36 *A single curved spike-grid with a 3/4-in. bolt has a carrying capacity of 3800 lb.; with a 1-in. bolt, 4100 lbs. (Courtesy American Wood Preservers Institute.)*

wetted sand should be used to make a flat bearing surface under each seated pole. If a pole seat cannot be dug into the rock, a concrete pad may be used over the rock, and the pole anchored to the pad with lag bolts or standard post anchors. The pad should be 8 in. thick and of the same diameter as the hole, which may be 8 in. larger than the pole.

In one cross-bracing method, 5/8- or 3/4-in. steel rods are fitted through drilled holes, flooded with preservative, then fastened with nuts, using beveled washers to avoid crushing the wood. Steel-rod cross bracing usually is not needed between every pole; several diagonal rods often are adequate.

A pole structure also can be knee braced with 2 × 6 wood members, nailed or bolted to form a diagonal connection from the poles to the floor beams. The bracing

can be left exposed or covered with skirt boards around the perimeter of the house.

■ SHEAR WALLS AND FLOOR DIAPHRAGMS

A masonry or concrete fireplace block, which is supported on concrete footings at least 4 ft. below grade, or a masonry or concrete wall can act as a shear wall to resist horizontal forces. Similarly, a masonry or concrete utility room below the first floor can act to brace the structure and to anchor it.

Key walls also can be used to resist the horizontal forces that work to push a structure down a slope (Fig. 6.7-37). For both shear walls and key walls, floors should be designed as diaphragms to transfer the lateral rigidity of the walls to the poles. Such floors can be built of conventional wood joists and 1/2-in. or thicker plywood, with nailing as recommended by the APA.

FIGURE 6.7-37 *Key wall bracing.* *(Courtesy American Wood Preservers Institute.)*

6.8 WOOD CABINETS

Wood cabinets have been used in commercial, institutional, and residential buildings for centuries. The origins of residential bath and kitchen cabinets, for example, can be traced back to the sixteenth century, when *ambries* (cupboards) were used to keep rats and other vermin away from food. More extensive built-in cabinets emerged in the seventeenth century, when the space between the chimney and the side walls was filled with open or closed shelves, giving the effect of built-in kitchen cabinets.

Wood is still used extensively in the manufacture of cabinets. However, the growing cost of wood has caused the substitution of plywood and particleboard in most wood cabinetry and the increased use of metal cabinets, especially in commercial and institutional buildings.

Primarily because of the high cost of manufacturing them, custom cabinets are seldom used in residential buildings today, but still appear in commercial and institutional buildings. Many wood cabinets today are mass-produced in factories, some of which can manufacture thousands of units a day using sophisticated milling

machines, presses, sanders, laminating devices, conveyors, and finishing equipment.

This section broadly outlines some of the kinds of cabinets currently available, including the applicable quality standards and certification programs, and typical cabinet and countertop construction, assembly, and installation. The descriptions here represent basic cabinet fabrication and types in common use. These are not the only details used today, however, and the reporting of them here is not meant to imply that they are the only acceptable ones or that other methods are inferior. Any construction that results in cabinets that comply with the indicated quality standards will be satisfactory.

6.8.1

QUALITY STANDARDS

Wood cabinets in commercial and institutional buildings are often required to comply with the requirements of *Architectural Woodwork Quality Standards, Guide*

Specifications and Quality Certification Program, published by the Architectural Woodwork Institute (AWI), especially in Section 400, "Architectural Cabinets," and its subsections. This standard is also sometimes applied to cabinets in more expensive residential buildings. It is intended to apply to custom-fabricated cabinetwork. It lists three grade options, *Premium, Custom,* and *Economy,* and requires the designer to select the desired grade based on appearance and service requirements. The higher grade is usually the stronger and more durable, but is also usually more expensive.

In parts of the western United States, compliance with the standard *Manual of Millwork,* published by the Woodwork Institute of California (WIC), is sometimes required in lieu of compliance with the AWI standard mentioned in the previous paragraph. The WIC standard has the same grade listings as the AWI standards, but adds a fourth grade, *Laboratory.*

The AWI and WIC standards differ in some aspects and use different names for the same cabinet components. It behooves a designer to obtain one of these two standards that will be applied and ascertain its requirements on each aspect of a cabinet.

Sometimes standard manufactured cabinets are selected by specifying one or more acceptable manufacturers and product lines. This type of selection is often used for cabinets for laboratories, hospitals, and schools. Sometimes cabinet manufacturers are required to modify their stock units to comply with either the AWI Standard Section 1600, "Modular Casework," or the equivalent WIC standards. Even more stringent requirements, such as thicker materials, are sometimes necessary for situations requiring special heavy-duty items.

The current standard usually applied to residential kitchen and bath vanity cabinets is *Recommended Performance and Construction Standards for Kitchen and Vanity Cabinets*, published by the Kitchen Cabinet Manufacturers Association (KCMA). This standard has been adopted by the ANSI as ANSI/KCMA A161.1. It applies to manufactured, factory-finished bath and kitchen cabinets whether made of metal or made of wood and other nonmetallic materials, such as plastic laminate, plastics, hardboard, and particleboard. It emphasizes performance rather than construction, allowing considerable freedom in the selection of materials and methods of fabrication so long as the final product meets stringent testing requirements for endurance and quality.

To ensure a high quality installation, bath and kitchen cabinets in houses should also comply with the requirements of HUD 4910.1, *Minimum Property Standards for Housing*, whether or not the housing will be financed with an FHA guaranteed loan. HUD 4910.1 requires compliance with ANSI/KCMA A161.1.

ANSI/KCMA A161.1 is a performance standard that does not prescribe specific requirements for cabinet design or materials. It does, however, require that all lumber and plywood parts be kiln dried to a moisture content of 10% or less at the time of fabrication; that metal cabinets be rust resistant; and that all materials be suitable for use in kitchen and bath environments. It also requires that cabinets be enclosed with sides, backs, and bottoms, with a few exceptions. Backs are not required on vanity units, or on kitchen-sink fronts, sink bases, oven cabinets, or

refrigerator cabinets. Bottoms are not required on vanity bowl fronts or drawer fronts; or on kitchen-drawer units, oven cabinets, refrigerator cabinets, or sink fronts. In addition, countertops must enclose the tops of base cabinets; wall cabinets should have individual tops.

Swinging doors should have a catch or other device to hold them closed. When installed, adjacent cabinets and doors should be in proper alignment with each other. Doors should be of balanced (warp-resistant) construction and should operate freely. Miscellaneous hardware such as drawer slides, shelf standards, brackets, and rotating shelf hardware must support the design loads and operational functions described in ANSI/KCMA A161.1. Hardware should comply with ANSI and the Builders Hardware Manufacturers Association (BHMA) Standard ANSI/BHMA A156.9, "Cabinet Hardware."

Any manufacturer of prefinished, factory-engineered kitchen cabinets may apply for a license to participate in the KCMA Certification Program. KCMA membership is not required. To become a participant in the KCMA Certification Program, a manufacturer must enter into an agreement with the KCMA, spelling out the conditions under which it may affix a KCMA Certification Seal to its cabinets. The KCMA Certification Seal (Fig. 6.8-1) offers assurance that cabinets bearing it have been manufactured in conformance with KCMA standards. However, some manufacturers of high-quality cabinets, even ones who are KCMA members, do not participate, because a KCMA Certification Seal represents a

minimum standard and they do not want to have their products identified with it. To qualify for a Certification Seal, cabinets must meet performance tests that simulate many years of normal use. These include tests for structural integrity, tests for door operation, drawer tests, and tests for exterior cabinet finish.

6.8.2

CABINETS

Figure 6.8-2 shows the normal work heights and mounting heights and clearances for cabinets. These may vary, especially in commercial and institutional uses such as in laboratories.

Figure 6.8-3 shows a few of the many available types and sizes of stock cabinets intended for residential kitchen use. Manufacturers of cabinets for laboratories, hospitals, and schools make cabinets specifically designed for these applications, including laboratory fume hoods, specimen cases, work benches with racks, balance tables, chemistry and physics tables, speciality storage cabinets, reagent racks, glass-front cabinets, and many more. They also provide standard base and wall cabinets not unlike those shown in Figure 6.8-3 in general appearance, though usually quite different in construction. For example, they are likely to have heavier and thicker sides, backs, fronts, and tops, and stronger types of joints.

6.8.2.1

Types and Styles

Cabinets may be classed generally according to function as *base cabinets*, *wall cabinets*, and *miscellaneous cabinets*. The third category includes items such as laboratory cabinets, school science tables, medicine cabinets, broom cabinets, oven cabinets, and other special pieces.

Style refers to both *design style* and cabinet *face style*. Cabinets are manufactured in many decorative design styles, some with glass doors, others with raised panels, applied panel molds, and other articulation.

In standard manufactured kitchen and vanity cabinets there are three basic face styles available—*reveal overlay, flush*

FIGURE 6.8-1 *The KCMA Certification Seal is placed on the inside of a cabinet door or drawer.* (Courtesy Kitchen Cabinet Manufacturers Association.)

FIGURE 6.8-2 *Typical work heights and cabinet clearances.*

COUNTER HEIGHT FOR STOOLS | COUNTER HEIGHT FOR CHAIRS | HIGHEST SHELF ADVISABLE | ABOVE SINK | ABOVE RANGE | ABOVE CABINETS

WALL CABINETS

BLANK CORNER CABINET | CORNER CABINET | WALL COMBINATION, CABINET

9"- 12"- 15"
18"- 21"- 24"
DEPTHS:12"-13" VARIES ACCORDING
TO MANUFACTURER.

24"- 27"- 30"
33"- 36"- 39"
42"- 45"- 48"

24"- 27"- 30"
33"- 36"- 39"
40"- 42"- 45"
48"

24"- 30"- 36"
42"- 48"- 60"
66"- 72"- 84"
MULTIPLE DOORS ON CABINETS
42" AND OVER.

21"- 24"- 27"
REQUIRED ALONG
EACH WALL.

54"- 60"- 66"- 72"- 84"- 90"- 96"

BASE CABINETS

SINGLE DOOR DRAWER | DOUBLE DOOR DRAWER | SINK OR RANGE BASE | BLANK CORNER CABINET | | SINK OR RANGE BASE COMBINATION

9" (NO DRAWER)
12'- 15"- 18"
21"- 24"
DEPTHS: 23"- 23-1/2"- 24"- 24-1/2" VARIES ACCORDING
TO MANUFACTURER.

27"- 30"- 33"- 36"
39"- 42"- 45"- 48"

24"- 30"- 33"- 36"
39"- 42"- 45"- 48"

24"- 36"- 39"
42"- 48"- 60"
66"- 72"- 84"
MULTIPLE DOORS &
DRAWERS ON 60"
AND OVER.

33"- 36"- 42"
REQUIRED ALONG
EACH WALL.

54"- 60"- 66"- 72"- 84"- 90"- 96"

MISCELLANEOUS CABINETS

ALL DRAWER CABINET | SINK OR RANGE FRONT | STACK-ON RANGE BASE | UTILITY OR BROOM CABINET | OVEN CABINET | REFRIGERATOR CABINET

12"- 15"- 18"
21"- 24"
BASE UNITS ALSO AVAILABLE WITH THREE OR FIVE DRAWERS.

24"- 30"- 36"
42"- 48"

24"- 30"- 39"- 40"
VANITIES SIMILAR

18" SINGLE DOORS
24"- 30"- 36"
AVAILABLE IN, 12"- 13"
18" & STANDARD BASE
DEPTHS.

24'- 27"- 30"- 33"
* 66" HIGH (NO TOP DOORS)
CUT OUT SIZE VARIES FOR
DIFFERENT OVEN MODELS.

36"- 39"- 42"
OPENING SIZE VARIES FOR
DIFFERENT REFRIGERATOR
MODELS.

VANITY CABINETS

24"- 30"- 39"- 40" | 54"- 60"- 66"- 72"- 84"- 90"- 96"

FIGURE 6.8-3 *Summary of kitchen and bath cabinet types and sizes.*

overlay, and *flush* (Fig. 6.8-4). Other styles may be used in custom cabinets.

FIGURE 6.8-4 *Cabinet face styles. (Drawing by HLS.)*

6.8.2.2

Materials and Finishes

Cabinets may be made either from wood and plywood with a wood finish, from various materials with a plastic laminate finish, or from metal. Hardboard and particleboard are also frequently used for some components in some types of cabinets. Common species of hardwood used in cabinet manufacture are ash, oak, birch, maple, and walnut; common softwood species are ponderosa (western)

pine, Douglas fir, and western hemlock. Other wood species are also used, however.

The AWI standard lists eight transparent finish systems and five opaque finish systems for the wood portions of custom cabinets. These same finishes can be made to apply to standard fabricated units, but all manufacturers will not be able to supply them. ANSI/KCMA A161.1 lists no specific finishes. Rather, it requires that the finish meet certain performance requirements. This leaves the finish up to the manufacturer. The designer must then select from the manufacturer's standard finishes. These include various wood grains and stains and several opaque colors. The ones available at a particular time vary with the popularity of the particular color or other finish.

6.8.2.3

Construction

There are two basic wood cabinet construction types—*wood finished* and *plastic laminate finished*. Both kinds may have either *flush-overlay*, *reveal-overlay*, or *flush* face styles. Most cabinets used today are one of these wood cabinet types. Therefore, this discussion will primarily address these types. Performance standards for metal cabinets are basically the same as those discussed here; the construction details, of course, differ somewhat. Unfortunately, at this time there is no nationally recognized organization that represents metal cabinet manufacturers and no coordinated consensus for construction criteria. Therefore, it is best when specifying or purchasing metal cabinets to rely on the reputation of the manufacturer.

The most common type of wood cabinet construction is *face-frame* construction. In it, each cabinet has a face frame such as that shown in Figure 6.8-5. The other type is called *frameless*, because the face frame is omitted and the cabinet ends are finished on their exposed edges. In frameless cabinets, components, other than the face frame itself, are similar to those in cabinets that have face frames.

Some manufacturers use panel components similar to those generally used in plastic-laminate-finished cabinets in concealed or semiexposed (visible only with doors or drawers open) locations in their

wood-finished cabinets. Typical locations are end panels, back panels, and bottoms. Sometimes plastic-laminate components are used for shelves in wood-finished cabinets.

The AWI standard requires that some components be a *panel* product. This is defined as plywood, particleboard, fiberboard, or hardboard faced with a hardwood plywood veneer.

■ WOOD-FINISHED CABINETS

The main components of *wood-finished* cabinets are *face frames*; *back, top, and bottom rails*; *backs*; *tops and bottoms*; *end panels*; *doors*; *drawers*; *shelves*; and *hardware* (see Fig. 6.8-5).

FACE FRAMES Face frames are made of 1/2- to 3/4-in.-thick hardwood. *Rails*, *stiles*, and *mullions* are *doweled* (or *mortised and tenoned*), as well as glued and stapled or nailed for rigidity (Fig. 6.8-6).

BACK, TOP, AND BOTTOM RAILS Rails are made from 3-in.-wide, 3/4-in.-thick solid wood or panel product. *Side frames* are often tongued and grooved into the face frames. *Rails* are rabbeted to receive top and bottom panels. *Back rails* are screwed and glued to secure them in place.

BACKS Backs are generally 1/8-in. thick hardboard or 3/8-in.-thick hardwood plywood. They are fastened to end panels and top and bottom rails. Backs for cabinets complying with the AWI standard are 1/4-in.-thick panel product.

TOPS AND BOTTOMS The bottoms of base cabinets and the tops of wall and special cabinets are usually 3/8-in.-thick hardwood plywood, 1/2-in.-thick particleboard, or 1/4-in.-thick hardboard in residential cabinets. Cabinets complying with the AWI standard are 3/4-in.-thick lumber or panel product. They should be *dadoed* (let) into the sides and interlocked into the hanging rails of wall cabinets.

END PANELS End panels typically consist of 1/2-in.-thick hardwood plywood, glued and mechanically fastened to side frames. End panels for cabinets complying with the AWI standard are 3/4-in.-thick panel product.

FIGURE 6.8-5 *Typical base cabinet construction, showing component parts. Cabinets may have two end panels as shown, no end panels for use within a series of cabinets, or one end panel to be placed at the end of a series.*

FIGURE 6.8-6 *Rails and stiles of frames can be joined by (a) mortise and tenon or (b) dowel joint.*

DOORS Door construction varies with the face style of the cabinet. Flush-overlay doors are usually 1/2-in. thick plywood with hardwood face veneers.

Doors for other face styles have solid-wood stiles and rails. These may be either 5/8 in. thick with 3/4-in.-thick solid wood center panels, or 3/4 in. thick with applied 1/4-in.-thick center panels. In some cabinets, the doors are hollow; in others they have particleboard, wood, or high-density fiber cores, with hardwood or plastic veneer. Sometimes the core is plastic with hardwood veneer overlays.

Other doors have particleboard or plywood cores with plastic faces, and in still others the entire door is plastic.

The required thickness for doors for cabinets complying with the AWI standard vary with the composition of the core of the door and its grade, width, and height. This thickness ranges from 3/4 in. to 1-1/4 in. for particleboard core doors, and 3/4 in. thick for veneer core doors. Face veneers are hardwood plywood.

DRAWERS Drawers frequently have hardwood lumber sides and backs, and ply-wood bottoms, but the sides may also be plywood or particleboard. Bottoms are often particleboard.

Drawer construction also varies depending on cabinet quality. In high-quality cabinets, the sides are connected to the front and back with multiple *dovetail* joints; in others, with *lock-shouldered* or *square-shouldered* joints (Fig. 6.8-7). Drawer bottoms are *dadoed* into the sides, front, and back for more rigid construction. Some drawer units are molded plastic (except for the front), with rounded inside corners for ease of cleaning.

FIGURE 6.8-7 *Drawer construction: (a) multiple dovetail; (b) lock-shouldered; and (c) square shouldered joints.*

Drawer fronts usually match the *door fronts* and are of the same or similar construction. They should be positively fastened to a wood *subfront* using screws. The *subfront* should be *rabbeted* into the sides and fastened to them with mechanical fasteners and glue. There is as wide a variation in drawer-front styling as there is in door design.

SHELVES Most shelves are lumber, plywood, or particleboard, 1/2 to 3/4 in. thick with rounded front edges; plywood or particleboard shelves may be banded with hardwood front edges. The AWI standard requires shelves to be a minimum of 3/4 in. thick for short spans (32 to 36 in.) and 1 in. to 1-1/16 in. for larger spans (42 to 48 in.).

■ PLASTIC-LAMINATE-CLAD CABINETS

Some *plastic-laminate-clad* cabinets are constructed in the same way as the *wood-finished* cabinets described above, except that the finish veneers are made from any suitable grade of sound hardwood or, in some cases, from softwood and covered with plastic laminate. The types and grades of plastic laminate used are usually similar to those discussed in this subsection.

A second type of plastic-laminate-clad cabinet is made from panels prefinished with plastic laminate and assembled into a finished cabinet. The main components of this type of plastic-laminate-clad cabinet are similar to those of wood-finished cabinets, but because of the similar construction of some components it is convenient to group them somewhat differently here.

Frameless types are more likely to be used in plastic-laminate-clad cabinets. In *reveal-overlay* plastic-laminate-clad cabinets, *face frames are usually not required.*

The *concealed* (never visible) surfaces of plastic-laminate-clad cabinet components are plywood or particleboard. The *exposed* and *semiexposed* faces of components are finished in *high-pressure decorative laminate* (HPDL) or *thermoset decorative panels.* The HPDL faces should comply with the requirements of the National Electrical Manufacturers Association (NEMA) Standard LD-3, "High Pressure Decorative Laminates."

NEMA LD-3 designates laminate types as *general purpose* (GP), *postforming* (PF), and *cabinet liner* (CL). In specifying HPDL, these designations are followed by a number representing the thickness of the material. For example, GP 50 is a general-purpose HPDL that is 0.050 in. thick.

The HPDLs usually found in cabinet body construction are GP 50, which is used for horizontal applications; GP 28, which is used for vertical applications where exposed to wear; PF 42, which is used for curved surfaces; and CL 20, which is used on the insides of cabinets where subjected to little or no wear. Edges in semiexposed locations are often finished with *high-impact plastic edging.* Exposed edges are usually finished with the same plastic laminate as is used on the adjacent surface.

Thermoset, or *low-pressure,* decorative laminate panels are used in semiexposed locations where they are not subjected to wear. They should comply with the American Laminators Association (ALA) Standard 1985, "Performance Standard for Thermoset Decorative Panels," which is roughly equivalent to GP 20 as defined in NEMA LD-3.

FACE FRAMES Face frames are made of 1-in. by 5/8-in. solid wood, with mortise-and-tenon joints.

SIDES, DIVIDERS, TOPS, BOTTOMS, SHELVES, AND STRETCHERS Stretchers are needed at the top of base cabinets to support the countertop. Sides, dividers, tops, bottoms, shelves, and stretchers are usually 1/2-in.-thick particleboard covered on both sides with GP 28 HPDL. The core for these components for cabinets that comply with the AWI standard are usually 3/4 in. thick with shelves being 1 in. thick for spans more than 36 in.

BACKS Backs are generally 1/8-in.-thick hardboard with a thermoset decorative laminate on the surface that will be inside the cabinet. Backs are fastened to end panels and top and bottom rails. The core of backs for cabinets complying with the AWI standard are 1/4-in.-thick panel product.

DOORS Doors are usually 5/8-in.-thick particleboard with a GP 28 plastic laminate finish on both sides. The doors for

cabinets complying with the AWI standard are 3/4 to 1-1/4 in. thick, depending on the core type and the height and width of the door. Core types permissible are particleboard and plywood veneer.

DRAWERS Drawers usually have 3/8-in.-thick particleboard sides and backs. Bottoms are often 1/4-in.-thick particleboard. Fronts, backs, and bottoms should be *rabbeted* into the sides and fastened with screws and glue. Some drawer units are molded plastic (except for the front), with rounded inside corners for ease of cleaning.

Drawer fronts usually match the door fronts and are of the same construction. They should be positively fastened to a 3/8-in.-thick particleboard subfront using glue and fasteners. The subfront should be rabbeted into the sides and fastened to them with mechanical fasteners and glue.

Drawers for cabinets that comply with the AWI standard are constructed with 1/2-in.-thick lumber, plywood, or particleboard; bottoms from 1/4-in.-thick panel product.

■ HARDWARE

Cabinet hardware includes *pulls, catches, drawer-suspension devices,* and *adjustable shelf supports.* Many types of each are available. In every case, they should comply with ANSI/BHMA A156.9.

Drawer-suspension devices should provide for a minimum capacity of 50 lb., and should have nylon rollers.

Shelves may be either fixed or adjustable, with metal hardware of various types to permit the adjustment.

6.8.2.4

Manufacture

There are two basic kinds of cabinet manufacturing operations: *integrated plants* and *assembly plants.* In an *integrated wood-finished-cabinet plant,* wood is processed from raw material (logs) to finished product (cabinets) with minor purchases of plywood, particleboard, plastic components, and hardware. Such facilities have vast milling operations for the many wood-processing steps needed to produce the final product.

In an *integrated plastic-laminate-clad cabinet plant*, plastic laminates are manufactured and processed on site. Wood may be either milled on site or purchased already milled. Plastic components, plywood, particleboard, and hardware are usually purchased.

In an *assembly plant*, raw wood materials have already been either completely or partially milled, and plastic laminate and other materials are purchased from their manufacturing source. Here, production is mainly the assembly of components produced elsewhere.

6.8.3

COUNTERTOPS

Countertops may be one of many materials, including, but not necessarily limited to, plastic laminate, filled polymer, cultured marble, ceramic tile, wood, natural stone, such as granite and marble, stainless steel, and the special tops used in laboratories and the like, such as soapstone, synthetic stone, and epoxy resin.

6.8.3.1

Plastic Laminate

The most usual type of countertop in residential and general commercial and institutional buildings is plastic laminate. The materials for this type of top are NEMA LD-3 HPDL 50.

Some of the plants that make cabinets also make plastic-laminate countertops. Regardless of where it is made, in a plastic-laminate countertop, the backsplash, top, and front piece are assembled first onto a wood or particleboard blank. Then adhesive is applied to the wood or particleboard surface and a plastic-laminate sheet is laid over the top and pressed into intimate contact with the surface. If the countertop blank has curved surfaces such as a rounded front edge and coved backsplash, the plastic sheet is first heated, then *postformed* (molded) to the blank surface by a special press. Openings for sinks are cut and holes for pipes are drilled at the factory or shop.

6.8.3.2

Filled Polymer

Recently *filled polymer*, or *acrylic-modified polyester*, tops have gained favor. One reason is the nature of the material. It is a rigid-cast acrylic plastic that is homogeneous, solid, and nonporous. It can be cut and shaped with ordinary woodworking tools, and sanded to remove burns and scratches. Probably the best-known example of this material is *Corian* by the DuPont Company, but similar materials are made by other manufacturers.

6.8.3.3

Other

Other materials are also used for bath and kitchen countertops. Tops made from synthetic materials called *cultured marble*, *synthetic stone*, and various other names are manufactured by proprietary methods and shipped to a project site ready for final cutting and assembly. These materials may or may not have real marble dust in their mix and may be either homogeneous or just have a decorative surface applied. Real stone, usually marble or granite, countertops are cut to fit in the shop and shipped to the project site ready for installation. In both real stone and synthetic tops, cutouts for plumbing fixtures and pipes are usually made at the factory or shop.

Ceramic tile countertops are usually built at the construction site from standard tile components. Tops supporting the ceramic tile are shop cut and field fitted.

Wood tops are usually manufactured to fit, but some cutting may be permissible in the field.

Special tops for use in laboratories are usually manufactured or cut to fit at the factory.

6.8.4

INSTALLATION

Cabinets and countertops should be installed in accordance with the requirements of the standards referenced above.

Cabinets, countertops, and devices built into them should be installed level, plumb, and true, in accordance with the manufacturers' instructions.

6.8.4.1

Cabinets

Cabinets should be attached with screws (not nails) to studs or other framing members. Wall cabinets should also be attached with screws to supporting framing rather than merely to the interior finish.

Wall and floor irregularities can prevent proper attachment. To ensure a plumb and level installation, high spots should be removed, and low spots should be shimmed. Wall base and chair rail should be removed behind cabinets to ensure a flush fit.

Wall cabinets are usually installed first, beginning with a corner unit. Screws should be inserted through *hanging rails* built into the backs of cabinets at both top and bottom, but not tightened immediately. After the cabinets have been leveled and plumbed, and doors perfectly aligned, the screws can be tightened.

Installation of base cabinets is also usually started in a corner. After the base corner unit has been installed, other units are added using C clamps to connect adjacent cabinets for proper alignment. T nuts should be used to fasten cabinets together. Each cabinet should be checked front to back and across the front edge with a level as it is attached to the wall. The front frame should be plumb and the top should be level. Base cabinets should be attached with screws into wall studs, preferably through the back frame. After base cabinets have been secured in place, the countertop should be installed and the toe strip trimmed out with a tile, resilient, or wood base.

6.8.4.2

Countertops

Plastic-laminate countertops should be installed using screws through *corner blocks* in the base unit into the top. Joints in the countertop should be *splined* and glued, then pulled up tight with mechanical fasteners.

Filled polymer tops should be similarly fastened in place. Seams should be finished with the manufacturer's joint adhesive.

Other tops should be installed and finished in accordance with their manufacturers' instructions. Refer to Section 9.6 for information about installing ceramic tile countertops.

6.9 PLASTIC FABRICATIONS

Plastics are relative newcomers to the construction field compared with traditional materials such as wood, glass, and concrete. The first commercial plastic in the United States, cellulose nitrate, was developed just a little over a century ago. It was created by John Wesley Hyatt to help overcome a shortage of ivory, from which billiard balls were made.

Progress in plastics technology was slow at first. The next commercial plastic to be developed was phenol-formaldehyde in 1909. Gradually, a few others were introduced, such as casein (1919), alkyds (1926), and polyvinyl chloride (1927). It was not until the 1930s, however, that the modern commercial plastics industry was really started; plastics were not widely used in building construction until after World War II.

Despite their relative newness, plastics have become important materials in the construction field. As a group, they offer a wide range of properties possessed by few other building materials: they are moisture and corrosion resistant, lightweight, tough, and easily molded into complex shapes. Furthermore, the great variety of plastic materials available, each of which offers slightly different advantages, makes plastics well suited for a multitude of purposes.

Sheet and tile flooring is perhaps the most common use for plastics in construction. Vinyl flooring is noted for excellent abrasion and stain resistance and good resilience. Plastic siding is becoming increasingly popular because of its good durability and freedom from maintenance; plastic-covered wood windows and doors offer similar advantages. Plastic laminates are standard materials for counter- and tabletops; they are produced in many decorative designs and bright colors, and have good heat and chemical resistance. Plastics also are used for break-resistant glazing, skylights, and roof domes. Plastic films make excellent vapor retarders; plastic foams can be used for a variety of insulation purposes.

Plastics do have limitations, however, and they can create serious problems if used incorrectly. For satisfactory service, a suitable plastic for the intended use should be chosen. The plastic product should be correctly designed and properly integrated with other products.

This section discusses the composition and generic properties of plastics, and describes the chief applications of the most important types. For more detailed information on molecular structure and how it affects the specific properties of materials, see Section 1.5.

6.9.1

COMPOSITION

Twenty to 30 families of materials are classed as plastics. Within each of these families, a multitude of variations and combinations is possible. The complex details of the chemistry involved in the production of these compounds and the terminology used to classify them are beyond the scope of this book. A brief explanation of their composition, molecular structure, and basic type is included, however, to help the reader gain a basic understanding of what plastics are, how they behave, and the ways in which they can be used in construction.

Despite widely varying properties, all plastics have several characteristics in common, which warrants classifying them as a single materials group.

1. At one stage in their formation they exhibit plastic behavior; that is, they will soften and can be formed into desired solid shapes, with or without the use of heat and pressure.

2. They are made up of molecules built around a carbon atom and, therefore, are classed as organic materials. There are some exceptions to this, however. Some plastics are built around a silicon atom rather than a carbon atom, yet still are considered plastics.

3. They are *synthetic* materials; that is, they are man-made and do not occur in nature.

4. They are high polymers, which are giant molecules made of numerous, relatively simple repeating units linked together into large aggregations.

Not all materials that possess the above-listed characteristics are necessarily plastic. For example, synthetic rubbers have all of these characteristics, but are classed as rubbers.

6.9.1.1

Molecular Structure

Some plastics are soft and flexible; others are hard and brittle. Some will become soft and may melt or even evaporate when exposed to intense heat; others are virtually unaffected by extremes of temperature. An examination of the molecular structure of high polymers can help to explain many of these differences in the properties and behaviors of plastics.

■ HIGH POLYMERS

Most plastics are based on carbon chemistry. This means that they are dependent on the peculiarities and properties of the carbon atom (Fig. 6.9-1). Carbon atoms form bonds with other atoms through covalent bonding (see Section 1.5). They have a valence of 4, meaning that there are 4 points to which other atoms can attach themselves to form covalent bonds.

Atoms of other elements have different valences. For example, hydrogen atoms

FIGURE 6.9-1 *Two molecules based on the carbon atom.*

have a valence of 1, and oxygen atoms have a valence of 2. Therefore, four hydrogen atoms can attach themselves to a single carbon atom (one at each valence point) to produce CH_4, the gas known as methane. Two oxygen atoms can attach themselves to a single carbon atom (each oxygen atom taking up two valence points) to produce CO_2, carbon dioxide (see Fig. 6.9-1).

Two adjacent carbon atoms can attach to each other. If this attachment involves only a single valence point, thus creating a single bond between them, and the other valence points are taken up by other atoms, the molecule is said to be *saturated*. In other words, a saturated molecule contains only single bonds between the carbon atoms and has no free (unattached) valence point.

Two carbon atoms may also attach to each other at two valence points, creating a double bond; or at three valence points, creating a triple bond (Fig. 6.9-2). In these cases, the molecules are said to be *unsaturated*. Under certain conditions the double or triple bonds of an unsaturated molecule can be opened up (activated), creating free valence points. Other atoms, or groups of atoms, can then attach themselves to these free valence points. For example, if the double bonds in many ethylene molecules are activated, they can join together to form a chain (Fig. 6.9-3), which is then known as polyethylene. Such chain formation is called *polymerization* and is essential in the production of plastics. Because only unsaturated molecules will react together to form chains, these are the only kinds of molecules that are useful in producing plastics.

A single molecule such as ethylene is known as a *monomer*. Several of these

FIGURE 6.9-3 *Several ethylene molecules can be linked to form polyethylene, a polymer.*

molecules linked together become a *polymer*. When the number of units in a chain is very large (between several hundred and several thousand) it is called a *high polymer*. The number of monomeric units in different chains can vary widely. When chain growth stops, free valence points are left at each end of the chain. These may be taken up by single atoms, such as hydrogen, or by groups of atoms with a single net valence. In some cases, the two ends of a chain loop around and connect, forming a ring.

Plastics are composed of these giant chainlike molecules, and are therefore classed as high polymers. This is evidenced by the fact that many of the chemical names for plastics start with the prefix *poly*, which means *many*. Examples are polyethylene, polystyrene, and polyvinyl chloride.

■ BASIC CATEGORIES

Plastics can be divided into two basic categories based on their molecular structure: (1) *thermoplastic* materials, which can be repeatedly softened and hardened by heating and cooling; and (2) *thermosetting* materials, which are set into permanent shape during forming, and cannot be softened again by reheating.

THERMOPLASTICS Thermoplastic materials are made up of long chainlike molecules such as the polyethylene molecule described above. These molecules are characterized by being attached to each other. The chemical attraction along the chain is very strong, but the attraction between adjacent chains is relatively weak. Furthermore, these chains are generally saturated, having no double or triple bonds that can be activated. This means

that no free valence points can be created to cause a strong attraction to adjacent chains.

At any temperature above absolute zero, molecules are in constant random motion. As a result of the unconnected molecular structure of thermoplastics, the molecules can slide past each other and change shape, as in a liquid. At room temperature, this random motion is slight and the plastic holds its shape. However, as the temperature rises, molecular motion increases and the attractive forces between molecules weaken, causing the material to expand and become soft and flexible.

When a plastic reaches a high enough temperature, it will flow and can then be molded into new shapes. The temperature at which this occurs is called the *softening point*, which will vary according to the type of plastic involved. When the material is cooled again, it assumes its former stiffness and holds its new shape at normal-use temperatures. With thermoplastic materials, the process of heating to soften and cooling to harden can be repeated again and again without affecting the properties of the plastic.

The configuration of the molecules in thermoplastics also affects the final properties. Some molecules are arranged in neat, continuous chains (linear molecules); others are highly branched. Highly branched chains do not pack together well, whereas linear molecules can be stacked together closely. As a result, plastics composed of branched molecules tend to be softer, less strong, and lower in density than those composed of linear molecules.

The molecular configuration in a given plastic can be controlled to a certain extent by the method used in its formation. For example, when ethylene is polymerized

ETHYLENE   ACETYLENE

FIGURE 6.9-2 *Molecules can have double bonds (ethylene) or triple bonds (acetylene) between the carbon atoms. Such molecules are called unsaturated.*

by high temperature and pressure, the molecules are often highly branched. However, ethylene can also be polymerized through the use of catalysts, which are substances added to the basic chemicals to speed up the chemical reaction. When polyethylene is made by this process, the molecules are generally linear. This means that variations in strength and stiffness can be produced in different batches of the same type of plastic simply by changing the method of polymerization.

There are hundreds of different thermoplastics. A brief description of the most important types is given in Section 6.9.3. However, a few words about the way these different types are developed may be helpful here.

Starting with the ethylene molecule, which is the basic building block of many thermoplastics, substitutions can be made for one or more of the hydrogen atoms involved. Single atoms, such as chlorine or fluorine, or whole groups of atoms, such as the styrene monomer, can be substituted for the hydrogens (Fig. 6.9-4). For example, if one hydrogen atom is replaced by a chlorine atom, a plastic known as polyvinyl chloride results. If one hydrogen atom is replaced by a phenyl group, polystyrene results. If one hydrogen atom is replaced by a methyl group (CH_3) and a second hydrogen atom is replaced by a methoxycarbonyl group ($COOCH_3$), polymethyl methacrylate (acrylic) results.

Hundreds of such variations are possible and each variation produces a plastic with slightly different properties. A large group of plastics called *vinyls* is based on the ethylene molecule. Other types are based on different molecules. For example, *cellulosics* are based on modifications of the basic celluloid molecule.

Still other variations are possible by combining different monomeric units into the same molecular chain. These substances then are called *copolymers*. Therefore, instead of having a single repeating monomer along a polymer chain, two different monomers may alternate along the same chain. Numerous variations in the properties of plastics are made possible through such copolymerization. For example, vinyl chloride may be copolymerized with vinyl acetate, producing a substance with some of the properties of each of the original polymers (Fig. 6.9-5). Acrylonitrile-butadiene-styrene (ABS) is

FIGURE 6.9-4 *Starting with an ethylene molecule, substitutions can be made for a hydrogen atom, producing (a) polyvinyl chloride or (b) polystyrene.*

another copolymer that has proven to be very useful because of its unique combination of properties.

THERMOSETS Thermosets differ from thermoplastics in that once they are set (cured), they cannot be softened again. This is because during forming, they go through an irreversible chemical process that develops strong chemical bonds between adjacent molecules. Therefore, the molecules cannot slide past each other as they can in thermoplastic material, but instead form a strongly interconnected molecular structure.

An example of this process is the production of thermosetting plastics from unsaturated polyesters. Being unsaturated, the chainlike polyester molecules contain at least occasional double carbon bonds. These double bonds can be activated, creating free valence points. If other activated molecules are present in the mixture, strong cross links can be developed between them. For example, activated styrene molecules will react with activated polyester molecules, forming a continuous, cross-linked network. The degree of cross linking that occurs determines the final properties of the material; that is, whether it will be resilient and rubbery, hard and glassy, or somewhere in between.

FIGURE 6.9-5 *Two different monomers may alternate along a polymer chain, producing a copolymer.*

Another type of thermosetting reaction, called *condensation polymerization*, produces a by-product such as steam. An example of this is the reaction between phenol and formaldehyde, producing the thermoset known as phenol-formaldehyde. During this reaction, the phenol molecules each give up one or more hydrogen atoms and the formaldehyde molecules give up an oxygen atom. This leaves open valence points on both molecules, allowing them to join together to form a strongly interlinked structure. The leftover hydrogen and oxygen atoms join together to form water, which is given off as steam.

Thermosetting reactions are irreversible. Therefore, these plastics go through a pliable stage only once, and after they have set they remain rigid and hard unless actual chemical decomposition (breaking down) occurs because of extreme heat, pressure, or environmental effects.

6.9.1.2

Modifiers

Many plastics in their pure form are difficult to work with, expensive to produce, or lack some property needed for a particular application. These difficulties can sometimes be overcome by the addition of one or more of several types of modifiers.

These modifiers are added to the basic plastic ingredient (the resin) before fabrication, in a process known as *compounding*. It is important that the different ingredients are thoroughly compounded into a homogeneous mass, because the uniformity of the mix contributes greatly to the final properties of the plastic.

The type and amount of modifiers used with different plastics vary greatly. Some mixes consist of 90% to 95% basic resin, with only small amounts of modifiers. Other mixes may contain only 20% to 30% resin, the remainder being composed of various plasticizers, fillers, stabilizers, and other additives.

■ PLASTICIZERS

At ordinary-use temperatures, many thermoplastics are inherently hard and rigid. In order to produce sheets, tubes, films, and other flexible plastic products, plasticizers must be added to these materials (Fig. 6.9-6).

FIGURE 6.9-6 *The addition of plasticizers imparts flexibility to vinyl sheet flooring, making it easier to handle and install.*

Plasticizers are liquid or solid materials that are blended with plastic resin to impart increased flexibility, and produce an impact on resistance, resiliency, and moldability. Basically, a plasticizer produces the same effect in a thermoplastic as does high temperature; that is, it lessens the forces of attraction between the long chainlike molecules, allowing them to slide more freely past each other. Unfortunately, this greater flexibility also often reduces the strength of the plastic and lowers its heat resistance and dimensional stability.

To obtain the optimum effects from a plasticizer, it is important that it be compatible with the plastic resin being used. This means that the plasticizer should mix well with the resin and should not separate out during the life of the object. A poor plasticizer may migrate to the surface of the object, creating a greasy film and causing the original plastic to become hard and brittle again. Furthermore, a plasticizer should not deteriorate with age or cause the plastic itself to deteriorate.

■ FILLERS

Fillers are used most commonly with thermosets, although there are some filled thermoplastics as well. Fillers are used to improve a particular property of a plastic, such as moldability, heat resistance, or toughness. However, some fillers are used primarily to increase the bulk and therefore decrease the cost of a plastic. These latter ones are called *extenders*.

A wide variety of material may be used as fillers, depending on the properties desired in the finished product. The amounts used generally vary between 10% and 50% of the total plastic mix. Fillers are used to increase the following properties:

- *Mobility and bulk.* Wood flour, made of finely ground hardwood or nut shells, is commonly added to thermosets to make a plastic that is more easily molded and less expensive to produce.
- *Hardness.* Mineral powders and metallic oxides may be added to a plastic to increase its hardness. However, if too much of these is added, the plastic may become brittle.
- *Toughness.* Fibrous fillers are often added to plastics to decrease their brittleness and increase their impact resistance. Natural fibers such as cotton, sisal, or hemp may be used, as may be synthetic fibers such as rayon, nylon, and polyester.
- *Heat resistance.* Certain kinds of mineral fiber fillers, and other inorganic fillers such as silica, clay, and finely ground limestone, are used to increase the heat resistance of some plastics.
- *Electrical resistance.* Finely ground mica or quartz may be added to increase the electrical resistance of a plastic.

■ REINFORCING AGENTS

Reinforced plastics are a large and important group in the construction industry. Their strengths and stiffnesses have been significantly increased through the addition of high-strength fillers that act as reinforcing agents. Generally, thermosetting resins such as polyester and epoxy are used in reinforced plastics. Fibrous fillers such as sisal and mineral fibers can be used as reinforcing agents, but the strength of glass fiber is so superior that it is almost always the material selected for this purpose.

Sheet glass is quite brittle, but when drawn into fibers, its tensile strength rises enormously. In fact, pound for pound, glass fiber is the strongest commercially available construction material. Therefore, adding glass fibers to a plastic greatly increases its strength and stiffness. Glass fiber can be used either as chopped strands, yarns, pressed matting, or woven fabrics, depending on the design and properties desired for the finished product.

■ STABILIZERS

The effects of weathering on construction materials is an important consideration. Many plastics degrade when exposed to the ultraviolet rays of the sun and to oxygen and other gases in the air. As a result, small amounts of *stabilizers* are added to these plastics to act as antioxidants and ultraviolet light absorbers.

Several different substances may be used as stabilizers. Carbon black is a commonly used one. For example, carbon black changes polyethylene from a quickly degrading material to one that stands up well under extended outdoor exposure.

■ COLORANTS

The addition of colorants to plastics is a complex subject. Some plastics, such as acrylic and polystyrene, are inherently clear and can be produced in a wide range of colors. Others, such as the phenolics, are dark and opaque and therefore have a more limited range of possible colors.

Both pigments and dyes can be used to color plastics. Pigments produce opaque colors; dyes, transparent ones. The pigment or dye chosen must be compatible with the other ingredients in the plastic mix. It should not bleed out of the plastic to discolor or stain adjacent materials. Furthermore, it should have good light and heat stability so that it will not fade or change in appearance with age.

6.9.2

PROPERTIES

The range of properties of various plastics is quite broad. They can be flexible or stiff, tough or brittle, transparent or opaque. However, plastics also have certain limitations, which must be considered when choosing them for applications in buildings. These limits can be seen most clearly when the general properties of plastics are compared with such other construction materials as steel, concrete, wood, and glass.

6.9.2.1

Strength and Stiffness

Strength can be measured in terms of tension, compression, and bending; tensile strength is the most important. The tensile strength of most plastics is near that of wood, which is much lower than that of steel. However, some plastic laminates and reinforced plastics approach steel in strength. The highest levels of tensile strength are found in glass-fiber-reinforced plastics.

Stiffness is more important than strength in most plastics applications. Stiffness is measured by the modulus of elasticity. The higher the stiffness, the greater will be the modulus of elasticity. Generally, plastics are less stiff than steel, glass, and other structural materials and, therefore, they deflect more than these other materials under a given applied load. Thermosets are somewhat stiffer than thermoplastics. Reinforced plastics have the highest modulus of elasticity (about the same as wood and concrete) but even they are only about 10% as stiff as steel.

Plastics weigh considerably less than other structural materials. Unmodified plastic resins range from a specific gravity of about 0.9 for polypropylene to about 1.5 for polyvinyl chloride. Both filled and reinforced plastics are more dense. In contrast, steel has a specific gravity of about 7.5. Therefore, when strength comparisons are made between plastics and metals on an equal-weight basis, plastics appear to have the advantage.

For this reason, plastics are employed in applications where strength combined with lightness is of prime importance, such as in spacecraft. In other construction, they can provide very strong, lightweight components (such as in sandwich panels), which can appreciably decrease the overall weight of a structure (Fig. 6.9-7).

6.9.2.2

Toughness and Hardness

Toughness, which is strength combined with resilience, is an outstanding characteristic of many plastics. This property is difficult to measure precisely, but it is generally expressed in terms of a material's *impact strength*, or ability to resist impact, such as that from a falling ball or a striking pendulum.

Plastics differ greatly in impact strength, and different formulations of the same plastic also show considerable varia-

FIGURE 6.9-7 *Low-density polyurethane foam is injected between two strong, rigid facings to produce a lightweight sandwich panel. (Courtesy DOW USA.)*

tion. For example, regular polystyrene is quite brittle, whereas copolymers of styrene, such as acrylonitrile-butadiene-styrene, are noted for toughness.

Compared with other materials, such as glass, concrete, and wood, plastics are quite resistant to impact. For example, a 1/8-in.-thick sheet of acrylic can withstand 25 times as much impact as can a 1/8-in.-thick sheet of double-strength window glass. This is illustrated by the widespread use of plastics for break-resistant glazing and lighting fixtures.

Another property of importance is hardness (abrasion resistance). In this respect, plastics do not perform as well as glass or steel. Plastic glazing scratches much more readily than glass, for example. Conversely, many plastics are more resistant to abrasion than is wood.

6.9.2.3

Creep

Creep (cold flow) is a permanent change in the dimensions of a material as a result of prolonged stress. All materials are subject to creep, but in steel the rate of deformation at room temperature is so slight that it is negligible.

Because of their molecular structure, many plastics exhibit creep at room temperature, and the amount of deformation that results can be large. This is particularly true of thermoplastics, whose unconnected molecules begin to slide past each other and assume new forms when an external force is applied. If the force is applied only for a short time, the plastic may recover, but if the force is long term, deformation can be permanent and may eventually result in failure of the material.

Thermosets, laminates, and reinforced plastics have greater resistance to creep, but even they cannot resist prolonged high stresses the way that steel and concrete can. Nevertheless, plastics can be used in self-supporting and even some load-bearing applications, if the proper formulation is chosen.

6.9.2.4

Thermal Properties

The important thermal properties for plastics are expansion and contraction, conductivity, and maximum service temperatures.

■ EXPANSION AND CONTRACTION

All building materials expand and contract with changing temperatures. Thermal expansion is much higher in plastics than in most building materials. Thermosets exhibit a lower rate of expansion than thermoplastics, whereas reinforced plastics and laminates generally show the lowest rates among plastics.

Changes in dimension caused by expansion and contraction can cause buckling or cracking of plastic components used in assemblies. These difficulties can be avoided through proper design of the component and through the use of expansion joints to allow for movement. Particular care must be taken when plastics are joined to metals because of the wide difference in rates of expansion of the two materials. Generally, thermosets such as phenolics or reinforced plastics work best in such applications.

■ CONDUCTIVITY

Plastics are generally good heat insulators. They have a much lower thermal conductivity than metals. Pure, unmodified plastics also compare favorably with concrete, glass, and brick. Wood, however, has a lower conductivity than do most plastics.

The exception is foam plastics, which outperform even wood in this respect, and are among the best heat insulators available. Polyurethane foam can have a con-

ductivity as low as 0.15 (see Fig. 6.9-7), which is lower even than mineral wool with its conductivity of 0.27. Values for other foams vary according to density, type of foaming agent used, and whether the foam is of the open-cell (interconnecting) or closed-cell variety.

Because of these low conductivity values, plastic foams make excellent insulation materials. However, some problems have been encountered with their use in construction, because of their flammability, the toxic fumes they can produce when burned, and the tendency of some of them to reduce in their insulating value over time. Usually, these problems can be overcome with proper material selection and good design. Codes require that some plastic insulations not be left exposed, but rather be covered with more fire-resistant material.

■ SERVICE TEMPERATURES

The maximum continuous service temperature of a material is the highest prolonged temperature at which it can be used without softening. Service temperatures for plastics are low compared with other building materials. Many thermoplastics begin to soften at temperatures near the boiling point of water. Thermosets have better heat resistance and some can be used at temperatures up to the wood char point (380° to 400°F). Fluorocarbons have the highest continuous service temperature of unmodified plastics and can be used in non-load-bearing applications at 500°F.

Because additives can substantially improve the heat resistance of plastics, many filled and reinforced plastics can withstand higher temperatures than the pure, unmodified resins.

Service temperatures are important in determining the suitability of a plastic for a particular application. Most plastics can withstand the temperatures encountered in ordinary building structures; for more demanding applications, such as piping for hot-water supply lines, special heat-resistant formulations may be required.

6.9.2.5

Corrosion Resistance

Plastics have excellent corrosion resistance. They resist attack by most chemicals, with the exception of certain organic solvents. As with other properties, performance varies according to composition. Fluorocarbons are extremely inert and can be attacked only by the most powerful chemical reagents. Other plastics are attacked selectively by certain groups of solvents. Therefore, care must be taken to choose a suitable plastic for the corrosive conditions involved in the application.

6.9.2.6

Fire Resistance

Flammability of materials is an important consideration in construction. Being organic in nature, plastics will burn. Some plastics, such as cellulose nitrate, are extremely flammable and are generally not permitted as construction materials by most building codes. Others are slow burning, and some will stop burning after the source of flame has been removed. Flammability cam be somewhat controlled through the use of additives. For example, compounds of chlorine, fluorine, bromine, and phosphorus are commonly added in reinforced plastics to retard flame spread.

The production of smoke and toxic fumes is another element of fire hazard. The amount of smoke and type of fumes produced depend on the composition of the plastic; those plastics rendered flame retardant through the use of additives generally produce the heaviest smoke. Furthermore, if a plastic contains compounds of chlorine, fluorine, nitrogen, or sulfur, they will also be present in the gases given off during combustion. These gases, in combination with carbon monoxide and carbon dioxide, which are by-products of all organic combustion, may prove highly toxic.

In short, plastics are no more flammable than wood, fabrics, or other organic materials. However, because plastics often substitute for traditional nonflammable materials like glass and metals, they should be used with discretion and only after careful consideration of their potential fire hazard.

To deal with these and other problems associated with the use of plastics in construction, most of the major building codes now have chapters devoted specifically to plastics applications. Tests developed or accepted by the ASTM and published as ASTM standards can be used to rate plastics according to their burning rates, fire endurance, and flame spread. Building codes generally require that plastics suitable for installation in buildings be able to pass the tests in one or more of these ASTM standards. In addition, codes often restrict the use of plastics in certain areas by special requirements, such as requiring that foamed plastics be covered with noncombustible materials, requiring that plastic glazing be spaced with intermittent noncombustible surfaces, and limiting the amount of plastic materials that may be used as interior finish.

6.9.3

MAJOR TYPES AND USES

The multitude of commercially available plastics can be broken down into several major groups. This section includes a brief description of the chief characteristics and applications of plastics commonly used in construction. A listing of plastics and their abbreviations compiled by ASTM is given in Figure 6.9-8.

6.9.3.1

Thermoplastics

Thermoplastics are by far the largest group of plastics. Although thermoplastics can be softened by heating, they are dimensionally stable at normal-use temperatures.

■ ACRYLONITRILE-BUTADIENE-STYRENE (ABS)

ABS copolymers are noted for their unique combination of properties. They are an outgrowth of the high-impact polystyrenes, which are a combination of styrene and butadiene-rubber compounds. Polystyrene plastics alone are very brittle, but their toughness can be greatly improved by adding a butadiene-rubber compound to them. The further addition of acrylonitrile to the mixture improves other properties, such as chemical resistance, rigidity, and tensile strength.

ABS plastics are widely used in the construction industry for piping and pipe fittings (Fig. 6.9-9). They are used for

FIGURE 6.9-8
ASTM Abbreviations for Plastics

Term	Abbreviation	Thermoplastic	Thermosetting
Acrylonitrile-butadiene-styrene	ABS	x	
Acrylic			
Poly (methyl methacrylate)	PMMA	x	
Cellulosics			
Cellulose acetate	CA	x	
Cellulose acetate-butyrate	CAB	x	
Cellulose acetate-propionate	CAP	x	
Cellulose nitrate	CN	x	
Ethyl cellulose	EC	x	
Epoxy, epoxite	EP		x
Fiberglass-reinforced plastics	FRP		x
Fluorocarbons			
Polytetrafluoroethylene	PTFE	x	
Fluorinated ethylene propylene	FEP	x	
Melamine-formaldehyde	MF		x
Nylon			
Polyamide	PA	x	
Phenolic			
Phenol-formaldehyde	PF		x
Polycarbonate	PC	x	
Polyester	—		x
Polyethylene	PE	x	
Polypropylene	PP	x	
Polyurethane			
Urethane plastics	UP		x
Styrene			
Polystyrene	PS	x	
Styrene-acrylonitrile	SAN	x	
Styrene-butadiene plastics	SBP	x	
Silicone plastics	SI		x
Urea-formaldehyde	UF		x
Vinyls			
Poly (vinyl acetate)	PVAc	x	
Poly (vinyl alcohol)	PVAL	x	
Poly (vinyl butyral)	PVB	x	
Poly (vinyl chloride)	PVC	x	
Poly (vinyl fluoride)	PVF	x	

Courtesy ASTM, reprinted with permission.

water and gas supply lines, and in drain, waste, and vent systems. They are tough and will withstand rough usage and fairly high temperatures for extended periods.

ABS plastics are also used to make hardware items such as doorknobs, handles, and latches, and to manufacture handrails, furniture, and appliance and machine parts.

■ ACRYLICS

Polymethylmethacrylate (PMMA) is the best known of the acrylic plastics. These transparent plastics are noted for their superior optical properties and can be readily used for glazing. Unlike glass, they also have the ability to *pipe* light.

This means that a beam of light can be bent around a corner because of the internal reflections of the light by the outer layer of the plastic.

Acrylics have a good resistance to breakage and outstanding weathering properties. Their transparency and physical stability are not adversely affected by long outdoor exposure. However, they cannot withstand high temperatures and begin to soften at 200 °F. In addition, their low abrasion resistance makes them more susceptible to scratching than glass.

Acrylics are used for skylights and roof domes and for glazing in public buildings where the likelihood of breakage is high. They also are used in lighting fixtures, as translucent ceiling panels,

FIGURE 6.9-9 *Plastic piping made of ABS is light, easy to handle, and resists corrosion. (Courtesy Plastic Pipe and Fittings Association.)*

lamp enclosures, and light diffusers. Their ability to pipe light is utilized in surgical instruments and in lighted outdoor signs. Other applications include clear or translucent corrugated sheets for roofing; films or sheets bonded to wood or metal for exterior finishes; and various molded pieces of hardware.

In addition, acrylics can be dispersed as finely divided particles in a liquid or mastic medium, producing a *latex*. In this form they are used widely for making paints (see Section 9.12) and building sealants.

■ CELLULOSCIS

Cellulosics are based on modifications of the cellulose molecule. Cellulose acetate (CA) and cellulose-acetate-butyrate (CAB) are the most common cellulosics.

Cellulosics are exceedingly tough and will withstand a great deal of rough usage. They can be produced as clear or translucent, although their optical properties are not as good as those of the acrylics. CA is not suitable for extended outdoor exposure, but CAB can be formulated to be quite resistant to weathering.

Cellulose acetate is known primarily for its use in photographic film and recording tape. Cellulose-acetate-butyrate is used for piping and pipe fittings (primarily for gas and chemicals), as well as for outdoor

lighting fixtures. Because of their toughness, both types of cellulosics are also used for a variety of hardware items, for handrails, and for tool handles. Cellulosics also find use as adhesives and coating compounds.

■ FLUOROCARBONS

Polytetrafluoroethylene (PTFE), which is commonly known under the trade name *Teflon*, is the primary fluorocarbon used in the construction industry. It is remarkably inert to chemical attack. In fact, strong chemicals such as boiling aqua regia, chlorine, and bromine do not affect it at all. Furthermore, it remains serviceable over an extensive range of temperatures (from $-450°$ to $+500°F$), which is unmatched by most other plastics. It also has one of the lowest coefficients of friction of any known solid, which means that it has superior antistick properties. However, it is quite difficult to form and is consequently expensive to produce.

PTFE is used in particularly demanding applications such as piping for extremely corrosive chemicals at high temperatures; for low-friction slider pads to permit movement in steam lines; on the surface of bearing pads between beams and columns of steel structures to permit movement; and as nonstick linings for cooking utensils.

■ NYLONS

Nylon, originally a trade name, is now the common name for a group of plastics known as polyamides (PA). These plastics have good all-around toughness, high strength, and good chemical resistance. They are also resistant to abrasion, but they are not recommended for continuous outdoor exposure.

Nylons are perhaps best known for their use in synthetic fibers and filaments. They produce high-strength fabrics that are durable and tear resistant. These fabrics are used for sails, parachutes, and many articles of clothing. In the construction industry, nylon fabrics, which may be coated or reinforced as necessary, have been used in air-supported structures similar to that shown in Figure 6.9-10. Nylon fibers are also used to make carpets.

FIGURE 6.9-10 *Air-supported structures are lightweight and portable, making them ideal for such uses as temporary construction enclosures. (Courtesy Thermo-Flex, Inc.)*

Because of its toughness, nylon is also used to produce molded parts such as locks, latches, rollers for sliding windows and drawers, gears, and cams.

■ POLYCARBONATES

Polycarbonates (PC) are transparent plastics with exceptionally high impact strength, good heat resistance, and good dimensional stability. They are used in break-resistant safety glazing, lighting fixtures, lighted signs, and various pieces of molded hardware.

■ POLYETHYLENE

Polyethylene (PE) is a well-known plastic made by polymerization of the ethylene molecule. It is strong, light, and flexible and it retains its flexibility even at low temperatures. It is noted for its good water resistance and low moisture-vapor transmission. It is produced in several different formulations, including linear and branched molecular configurations, and these may vary considerably in strength, density, and stiffness.

Polyethylene is used widely for vapor retarders in floors and walls (Fig. 6.9-11), for dampproofing of basement walls, including wood foundation walls, and for sealant backer rods. It is also used for some types of piping (cold water, gas, and chemicals), but it is not suitable for use under severe conditions involving high temperatures or extremely corrosive chemicals. Another common application of polyethylene is for wire and cable insulation.

FIGURE 6.9-11 *A polyethylene vapor retarder on the interior of a wall assembly reduces vapor condensation inside the wall. (Photograph furnished by Owens-Corning Fiberglas Corporation.)*

■ POLYPROPYLENE

Polypropylene (PP) is similar to polyethylene in many respects—both are classed as polyolefines, for example—but it is much stiffer and more heat resistant than polyethylene. This makes it suitable for piping and pipe fittings that can be used under more severe conditions, such as for hot-water supply lines and drainage systems. Although polypropylene lacks the flexibility of polyethylene and is also more expensive to produce, it is used for some films and sheeting, primarily in the packing industry.

Another common use for polypropylene is in the production of high-strength fibers for carpeting.

■ POLYSTYRENE

Polystyrene (PS), made from the styrene monomer, is a transparent, water-resistant, and dimensionally stable plastic. It is not affected by extended low temperatures, but it begins to soften at temperatures near that of boiling water (212 °F). The main drawbacks of polystyrene are that it is quite brittle and not very weather resistant, becoming even more brittle and turning yellow with outdoor exposure.

To improve its properties, polystyrene is often copolymerized with other substances. One such copolymer is high-impact polystyrene, produced by adding small amounts of butadiene to the styrene. This results in a much tougher plastic with greater impact resistance and greater flexibility. Another well-known copolymer of styrene is ABS, discussed earlier.

Polystyrene is used for lighting fixtures and lighted signs and also for various molded pieces of hardware. Expanded (foamed) polystyrene, of which Styrofoam is a well-known example, is used widely in the construction industry for duct and pipe insulation; insulation for freezers and walk-in refrigerators; and in insulation for walls, floors, and ceilings of residences and commercial buildings (Fig. 6.9-12). Polystyrene foam is also used as the core material in the manufacture of doors and sandwich panels.

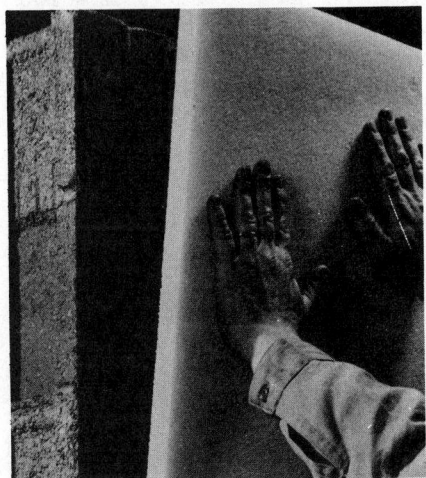

FIGURE 6.9-12 *Polystyrene foam insulation is available in blocks or boards, which can be bonded directly to masonry walls. (Courtesy DOW USA.)*

■ VINYLS

Vinyls are a large group of plastics based on modifications of the ethylene molecule. Included in this group are such plastics as polyvinyl acetate, polyvinyl alcohol, polyvinyl butyral, polyvinyl chloride, polyvinylidene chloride, and many others. These plastics exhibit a wide range of properties. They generally are characterized by good strength and toughness, fair chemical resistance, and a slow rate of water absorption. They perform well at normal-use temperatures as well as at relatively low (freezing) temperatures, but cannot withstand prolonged exposure to high temperatures. In fact, some vinyls begin to soften at 130 °F. Most types are recommended for indoor use, but some are suitable for outdoor exposure.

The most widely used vinyl in the construction industry is polyvinyl chloride (PVC). This is a very versatile plastic, with high impact resistance, good dimensional stability, abrasion resistance, and good aging characteristics. It is used to produce a large variety of products such as sheet and tile flooring, gutters and downspouts, moldings, siding, window frames, piping, and drainage systems. In sheet form it is used for facings in sandwich construction, and is bonded to wood or metal for exterior building finishes, doors, and window frames.

PVC is also available as a foam (either rigid or flexible) and is used for the core material in sandwich construction and for cushions in seating. PVC is copolymerized with many other plastics to produce a variety of adhesives, binders for terrazzo, and other floor toppings.

Some other vinyls encountered in construction are polyvinyl butyral (PVB), which is used for the interlayer in safety glass and also as a coating for upholstery fabrics; polyvinyl acetate (PVAc), which is used in adhesives, mortars, and paints; and vinyl-coated polyester, which is used in demountable air-supported structures.

6.9.3.2

Thermosets

Thermoset plastics are permanently set during forming and cannot be softened again by reheating. In general, they have better heat resistance and greater stiffness than do thermoplastics, but are more difficult to form and more brittle. They have a more limited range of applications than do thermoplastics.

■ EPOXIES

Epoxies (EP) were introduced in the 1950s and were quickly adopted for a wide variety of uses because of their excellent adhesive strength, combined with good chemical and moisture resistance. They will adhere to and can be used to bond metals, glass, masonry, other plastics, and most other building materials.

Epoxies are used primarily in coating compounds and adhesives. They are often produced as two-component systems, consisting of a resin and a hardener, which must be mixed together just before application. Because they bond well to metals, they are used as protective coatings for appliances, automobiles, and piping. As adhesives, they are used in the construction of sandwich panels; for bonding facings to walls, doors, and other assemblies; and for applying tile finishes. They make excellent high-strength mortars for concrete masonry units, and for patching cracked monolithic concrete. Compounds are available for patching deteriorated wood members in existing construction. Epoxy compounds also can be mixed with mineral aggregate or plastic chips to produce terrazzo.

Most glass-fiber-reinforced plastic products are made with polyesters, but sometimes epoxies are used instead. Cured, pure, unmodified epoxy is hard and brittle, but when reinforced with glass fibers, it becomes a strong, tough material that can be used in structural and semi-structural applications.

■ MELAMINES

Melamines (MFs) are produced by the reaction between melamine and formaldehyde. They are classified as amino plastics, as are ureas. Melamines are hard and scratch resistant, and can be used at temperatures approaching the boiling point of water. They also stand up well to chemical attack.

The most important application for melamine plastics is in high-pressure laminates, which are used widely for countertops and cabinet finishes. They are also used as adhesives (notably for plywood) and as a protective treatment for fabrics and paper.

As molding compounds, unmodified melamines are somewhat expensive and difficult to work with. Therefore, they are usually mixed with mineral or wood flour fillers. In this form, they are used for molded hardware and for electrical fittings.

■ PHENOLICS

Phenolics (PFs) are produced by the reaction of phenol and formaldehyde. Like many of the thermosets, pure phenolics are quite brittle and hard, and are generally mixed with fillers to improve their strength and impact resistance. Many different formulations are available, with varying properties. Phenolics have ex-

cellent heat resistance, and glass-fiber-reinforced types will perform well at temperatures up to 400 °F. They also have excellent thermal and electrical insulating properties.

Phenolics are among the more widely used thermosetting plastics. As molding compounds, they are used for electrical parts (sockets, switch boxes, circuit breakers), for a great variety of hardware items, and for appliance cabinets and parts.

Phenolic resins are important in the production of high-pressure laminates. They are used to impregnate the sheets of kraft paper that form the base of most laminates. These resins are also used to stiffen paper honeycomb, which is then bonded to a variety of facings to produce lightweight, strong sandwich panels for walls, doors, and room dividers. Adhesives and protective coatings are also made from phenolic resins.

Another major use for phenolics is as foamed insulation. In this form they are used as the core for sandwich panels, and for insulation around piping and ducts.

■ POLYESTERS

Polyester is the generic name for a large group of plastics that are used for a wide variety of purposes. Some examples of different types of polyesters are oil-modified polyesters (sometimes referred to as alkyds), used in paints and protective coatings; saturated polyesters, used as plasticizers in other plastic compositions; and saturated polyesters of high molecular weight, used to form fibers and film. The most important polyesters for building purposes are the unsaturated polyesters, which form the basic constituent of reinforced plastics.

Unsaturated polyesters can be cross-linked with a compatible monomer (such as styrene) to form a hard, strong thermosetting material. They have become the most widely used resin for the production of reinforced plastics because of an important combination of properties: they are easy to work with and combine readily with a variety of fibers and reinforcing agents; they cure rapidly at ordinary room temperatures without giving off volatile by-products; and they have excellent dimensional stability, combined with high strength and toughness. Further-

more, they can be specially formulated to provide good weathering properties and good heat and flame resistance.

For maximum strength and stiffness, glass fiber is used as the reinforcing agent in polyester compounds. This produces an exceptionally strong, lightweight material that can be used in many structural building applications. Some of the common uses for glass-fiber-reinforced polyesters are prefabricated sections for roof structures; translucent sheets for roofing and interior partitions; window frames and sash; facings for sandwich panels; and many large, molded articles such as bathtubs, shower stalls, sinks, and cabinets. Products such as bathtubs and countertops may be made by molding reinforced polyester within an acrylic shell to produce a combination of rigidity, strength, and attractive appearance.

■ POLYURETHANES

Polyurethanes (UPs), including polyisocyanurate, are a class of plastics known primarily for their use as low-density foams, which can be produced with widely varying properties. These range from soft, flexible open-cell types to tough, rigid closed-cell types. Urethane foams have excellent heat and chemical resistance and good tensile strength. They can be formulated to be fire resistant.

Rigid urethane foams make excellent insulation material. They have the lowest thermal conductivity of any building material, including mineral wool. They are used widely in construction for wall, floor, and ceiling insulation, for insulation around pipes and ducts, and as the core material in sandwich panels. Flexible urethane foams are used for cushions, upholstery, and padding.

Urethane resins also are used in the production of elastomers (synthetic rubbers), adhesives, sealing and glazing compounds, and no-seam flooring.

■ SILICONES

Silicone plastics (SIs) are based on the silicon atom rather than the carbon atom. They are therefore an exception to the general rule that plastics are organic compounds.

Silicones are very stable compounds, characterized by excellent corrosion resistance and good electrical insulating properties. They can be used over a wide range of temperatures (from −80° to +500°F. They also stand up extremely well to most weathering conditions.

Silicones, noted for being extremely water repellent, are used as water-resistant coatings on concrete masonry and brick (Fig. 6.9-13), although they are somewhat controversial in this use. They also make excellent sealants, and are used in that capacity in metal curtain walls, and as joint sealers around windows and doors, and in other joints in exterior walls and in wet locations in kitchens, bathrooms, showers, and toilets. Small amounts of silicone are frequently added to reinforced plastics to improve the adhesion of the polyester resin to the glass fibers, and to increase the strength and moisture resistance of the product.

FIGURE 6.9-13 *Silicone water repellent being applied to a masonry wall.* (Courtesy SIKA Corporation.)

■ UREAS

The most common *urea* plastic, urea-formaldehyde (UF), is produced by the reaction between urea and formaldehyde. Ureas are chemically related to melamines and are classed with them as amino plastics.

Ureas have lower heat and chemical resistance than melamines, but they are more scratch resistant and have good electrical resistance. They are used for molded hardware items and electrical fittings. Urea resins provide adhesives for use in sandwich panels, and they are also available as insulating foams.

7

THERMAL AND MOISTURE PROTECTION

MOISTURE CONTROL

WATERPROOFING AND DAMPPROOFING

INSULATION

EXTERIOR INSULATION AND FINISH SYSTEMS

BITUMINOUS ROOFING PRODUCTS

STEEP ROOFING AND RELATED WALL FINISHES

LOW-SLOPE ROOFING

ALUMINUM AND VINYL SIDING

FLASHING AND SHEET METAL

JOINT SEALING

Photo: A sheet copper roof. (Courtesy of Copper Development Association.)

7.1 MOISTURE CONTROL

Moisture in its visible liquid state—water—is relatively simple to understand and to cope with in construction. Roofs are built to shed it, earth is graded to divert it, pipes are used to conduct it, and flashing is used to divert it back to the exterior when it does penetrate the outer layer of a building. The technology necessary to prevent water penetration into buildings is available; unfortunately, many designers do not give adequate consideration to preventing this penetration. No competent architect will say that leaks are acceptable, but many will say privately that they do not permit potential maintenance problems to interfere with their design goals. The result is that more than half the lawsuits involving architects are related to water leaks.

In addition to its liquid form, moisture also is present as an invisible gas—water vapor—in the air all around us and in cavities of construction assemblies, such as stud and joist spaces, cores and furred spaces of masonry, and chases containing plumbing or heating, ventilating, and air conditioning ducts. As absorbed water, moisture is present in most construction materials and affects their properties, especially dimensional stability. Moisture may be present also in its solid form—ice—within the range of temperatures encountered in most parts of the United States.

Moisture is not damaging to construction in its vapor form. However, it is very changeable and becomes dangerous as it condenses (liquefies) or freezes (solidifies). Over time, hidden condensation can cause the decaying of organic materials, corroding of metal products, blistering of paint coatings, and, if it freezes, cracking or spalling of concrete and masonry.

Condensation and freezing may occur as a result of either a temperature drop or the migration of water vapor to areas of lower temperature. Therefore, moisture problems occur (1) in cold climates in heated occupied buildings, which usually generate substantial amounts of water vapor and (2) in warm humid climates in cooled buildings.

Buildings built 50 or more years ago were not as well built or sealed against the weather as they are today. This permitted considerable air leakage, dissipating a lot of moisture harmlessly through the structure. Considerations of economy and comfort require that modern buildings be built tighter and be kept warmer in winter and cooler in summer.

In addition, just as dehumidification is an important aspect of summer comfort conditioning, so is deliberate humidification an integral part of typical winter heating systems. All these factors create conditions conducive to moisture problems in modern buildings, unless suitable provisions to cope with them are included in the original design and construction.

The properties and behavior of moisture in its various forms are important considerations in many aspects of building construction, ranging from soil mechanics to roofing. This section addresses the principles of water vapor movement, problems associated with its condensation to liquid water, and methods of controlling condensation in buildings.

7.1.1

WATER VAPOR PROPERTIES

Moisture can exist in several states: as a liquid (water), as a solid (ice), and as a gas (water vapor). *Condensation* is the change of water vapor to liquid water. The change of water from a liquid to a gas is called *evaporation*. The transition from ice directly into water vapor is called *sublimation*.

The change of state from water or ice to vapor requires considerable heat input, called *latent heat*, measured in British thermal units, (Btu's). Therefore, when water evaporates from a surface, that surface is cooled. Perspiring skin, for example, is an effective cooling system for a human body. Like any warm body, the skin also gives off heat to the surrounding atmosphere. The heat absorbed by a substance without changing its state is known

as *sensible heat* and also is measured in Btu's.

When water turns to vapor, it mixes with the air and occupies all the available space. In many ways water vapor acts independently of the air because its general properties do not depend on the presence of air. However, when the air is moved by wind or a fan or is heated or cooled, associated water vapor is similarly affected. It is therefore convenient to consider moist air as a mixture of dry air and water vapor.

7.1.1.1

Relative Humidity

The measure of the amount of water vapor in air is known as *relative humidity* (RH), which is stated as a percentage. This percentage is a ratio of the actual amount of water vapor in the air, compared to the maximum amount the same air can contain at a specific temperature and atmospheric pressure. For instance, air with a 50% relative humidity contains half the amount of vapor it could hold at that air temperature and barometric pressure. At 100% relative humidity air contains the maximum amount of water vapor possible. Such air is said to be *saturated*.

An observable law of physics is that the warmer a mass of air, the more water vapor it can hold. Therefore, if the temperature of saturated air is lowered, some of the vapor must be given up by condensation. Up to the point of condensation the relative humidity increases as the temperature of the air is lowered, while maintaining the same amount of moisture.

As shown in Figure 7.1-1, air at 70°F can hold twice as much moisture as the same volume of air at 50°F, when the air is completely saturated. Therefore, at 70°F, the air is said to be half-saturated or to have a relative humidity of 50%. If the temperature were lowered below the saturation point, say to 40°F, the relative humidity would remain at 100%, but about 18 grains of moisture per pound of dry air would condense into visible water

FIGURE 7.1-1
Temperature and Humidity Relationships[a]

Temperature	Actual Moisture Content at Given Temperature, grains/lb. dry air	Maximum Moisture Control at Dew Point, grains/lb. dry air	Relative Humidity, %RH	Condensed Moisture,[b] grains/lb. dry air
70 °F	54	108	54/108 = 50	None
60 °F	54	77	54/77 = 70	None
50 °F	54	54	54/54 = 100	None
40 °F	36	36	36/36 = 100	54 − 36 = 18
30 °F	24	24	24/24 = 100	54 − 24 = 30

Reprinted with permission of the American Society of Heating, Refrigerating and Air-Conditioning Engineers.

[a] At 29.92 in. Hg pressure.

[b] As air at 50% RH and 70°F drops in temperature (read down in first column).

on cool surfaces. Further cooling to 30°F would result in the removal of a total of 30 grains per pound of air (see Fig. 7.1-1).

■ HUMAN COMFORT

Air movement, relative humidity, and temperature all affect the rate at which perspiration evaporates from the skin and at which a human body loses heat. Excessively low air humidity encourages more evaporation and increases the sensation of coolness, which is undesirable in the winter. On the other hand, too high a humidity reduces skin evaporation, retards body cooling, and makes people feel warmer, which makes them uncomfortable when the air temperature is near the upper limit of a comfortable condition. This can happen in any season but is more likely to occur in the summer. Either too little or too much humidity can be uncomfortable.

High humidities at higher temperatures can contribute to fatigue and reduced working efficiency. Air that is too dry can irritate nasal and sinus passages, contribute to chapped and dry skin, and encourage the generation of annoying and, in locations like computer rooms, damaging static electricity.

The specific "comfort" level of humidity or temperature is largely an individual subjective judgment that depends on many factors, such as prevailing climate, seasonal variations, and the individual's age, health, living habits, clothing, and activities.

As mentioned, cold air cannot hold nearly as much water vapor as warm air. Therefore, when cold outside air, which has a low moisture content (even though its relative humidity may be high), is brought inside and heated, a low relative humidity results. For instance, as shown in Figure 7.1-1, saturated air at 30°F can hold 24 grains of moisture per pound of air, and half-saturated air at 30°F has a moisture content of 12 grains per pound of air. If such air were to be heated to 70°F, it would be capable of holding 108 grains, and its relative humidity would be 100% times 12/108, or 11.1% relative humidity. The resulting humidity level is too low for comfort. Relative humidity of 25% to 40% is usually considered conducive to good health, but such humidity levels are difficult to maintain in cold weather.

■ EFFECTIVE TEMPERATURE

The term *effective temperature* (ET) defines an arbitrary index that combines into a single value the effects of temperature, humidity, and air movement on the sensation of warmth or cold felt by a human body. It was developed by the American Society of Heating, Refrigerating and Air-Conditioning Engineers (ASHRAE) in cooperation with the U.S. Public Health Service and based on extensive polls of trained individuals subjected to various temperature and humidity conditions.

Effective temperature is the temperature of saturated air (100% relative humidity) that gives the same physical sensation of warmth as various other combinations of air temperature and relative humidity, with relatively still rates of air movement ranging from 15 to 25 ft. per minute (fpm).

WINTER COMFORT It has been found that 98% of the people tested were comfortable in the winter season at an effective temperature of 68, which is 100% relative humidity and 68°F. Combinations of humidity and temperature that give this same feeling of warmth are tabulated in Figure 7.1-2. This table shows that in winter, as the relative humidity is reduced, higher temperatures are necessary to maintain the same sensation of warmth; conversely a higher relative humidity produces warmth with lower temperatures. However, the thermal and vapor permeance characteristics of exterior construction impose practical limits on interior humidity, which vary with the severity of winters.

In areas of low winter design temperatures (0°F or lower), few buildings can tolerate 40% relative humidity without developing condensation and moisture problems. At this interior humidity and exterior temperature combination, even sealed insulating glass will develop condensation (Fig. 7.1-3). Single glazing and highly conductive surfaces such as metal wall panels and metal door and window frames without thermal break features will show condensation even at 15% relative humidity. To minimize condensation and yet provide for human comfort, about

FIGURE 7.1-2

Temperature-Humidity Equivalents of Optimum Summer and Winter Effective Temperatures[a]

Winter Optimum, ET 68°F[b]		Summer Optimum, ET 71°F[b]	
Relative Humidity	Dry Bulb Temperature	Relative Humidity	Dry Bulb Temperature
100%	68°F	100%	71°F
80%	69°F	80%	73.1°F
70%	70°F	70%	74.3°F
60%	71.1°F	60%	75.6°F
50%	72.2°F	50%	76.9°F
40%	73.5°F	40%	78.3°F
30%	74.8°F	30%	79.7°F
20%	76.1°F	20%	81.2°F
10%	77.5°F	10%	83.0°F

Reprinted with permission of the American Society of Heating, Refrigerating and Air-Conditioning Engineers.

[a]Air movement of 15 to 25 fpm.

[b]Considered comfortable by 98% of people tested.

25% relative humidity should be maintained when the outside temperature is 0°F or lower. The humidity can be maintained at 30% when the outside temperature is 10°F, at 35% when 20°F, and at 40% when 30°F outside (see Fig. 7.1-3). Relative humidities above 40% generally are not recommended during the winter heating season.

Although a relative humidity of 25% or higher is considered desirable during the winter season for health reasons, the effect of relative humidity on occupant comfort has been overestimated in the past.

Fig. 7.1-4 shows both the results of older tests that measured the effect of temperature and humidity on human comfort immediately upon entering a room (curved lines top and bottom) and recent tests on occupants after a three-hour stay (solid vertical lines). Unlike the optimum summer and winter ET lines derived from earlier studies (see Fig. 7.1-2), the newer data suggest that (1) a dry-bulb temperature of 77.5°F is most comfortable and (2) humidity has no perceptible effect on the sensation of warmth within a range of 25% to 60% relative humidity. (See Section 11.3.1.3 for an explanation of dry-bulb temperature and wet-bulb temperature.)

SUMMER COMFORT   In summer, 98% of those tested in the older test mentioned above were comfortable at an ET of 71 (100% percent relative humidity and 71°F). Other combinations of relative humidity and temperature that give the same physical sensation in summer are tabulated in Figure 7.1-2.

The dashed lines in the ASHRAE comfort chart (see Fig. 7.1-4) represent effective temperatures. Each line represents graphically equivalent combinations of temperature and humidity. By following the 68 and 71 ET lines to their intersections with the solid diagonal relative humidity lines, the corresponding dry-bulb temperatures (horizontal coordinate) can be selected. The temperature-humidity equivalents shown in Figure 7.1-2 thus can be derived from this chart.

Note that the optimum summer ET 71 line strikes near the top of the summer curve representing the distribution of responses by the subjects tested. From this curve it can be established that in the summer nearly all subjects were comfortable at an effective temperature of 71, while half were comfortable within an effective temperature range of 68 to 75. Similarly, by reading the winter comfort curve, it can be seen that nearly all subjects were comfortable at an effective temperature of 68, while about 86% were comfortable at effective temperatures between 65 and 70.

The ET index is useful only in establishing human response to temperature-humidity conditions immediately upon entering a room. Where subjects were tested after a three-hour stay, it was found that humidity played a much lesser role in determining comfort. These findings are represented by solid lines 3, 4, 5, and 6 in Figure 7.1-4. For both summer and winter, it was found that a dry-bulb temperature of 77.5°F is equally comfortable for relative humidities from 25% to 60%. Only above 60% relative humidity (curved portion of line 4) must the temperature be reduced to maintain the same degree of comfort. The lines represent averaged responses for both summer and winter conditions; actually there is a spread of about 1.2°F between winter and summer responses. The maximum comfort line, therefore, would be at about 77°F in winter and 78°F in summer.

7.1.1.2

Vapor Movement

Water vapor becomes a hazard when it condenses (liquefies) inside the structural cavity of a floor, wall, or roof assembly. Here it can cause reduced effectiveness of insulating materials, decay of organic materials, corrosion of metals, and deterioration of protective coatings. Wall, floor, and ceiling assemblies should be designed and constructed to minimize vapor migration into structural cavities.

Vapor moves largely independently of the movement of air, from areas of higher vapor concentration and higher temperatures to areas of lower concentration and lower temperatures. Water vapor establishes a pressure proportional to the amount of water vapor present in the air mix. Air with more vapor has a higher vapor pressure. Vapor moves through air by *diffusion* from regions of high vapor pressure to regions of lower pressure, without relying on air circulation to carry it.

Vapor pressure is commonly expressed in inches of mercury. At 70°F and 100% relative humidity, vapor pressure is 0.739 in. of mercury (Hg), while at 0°F and

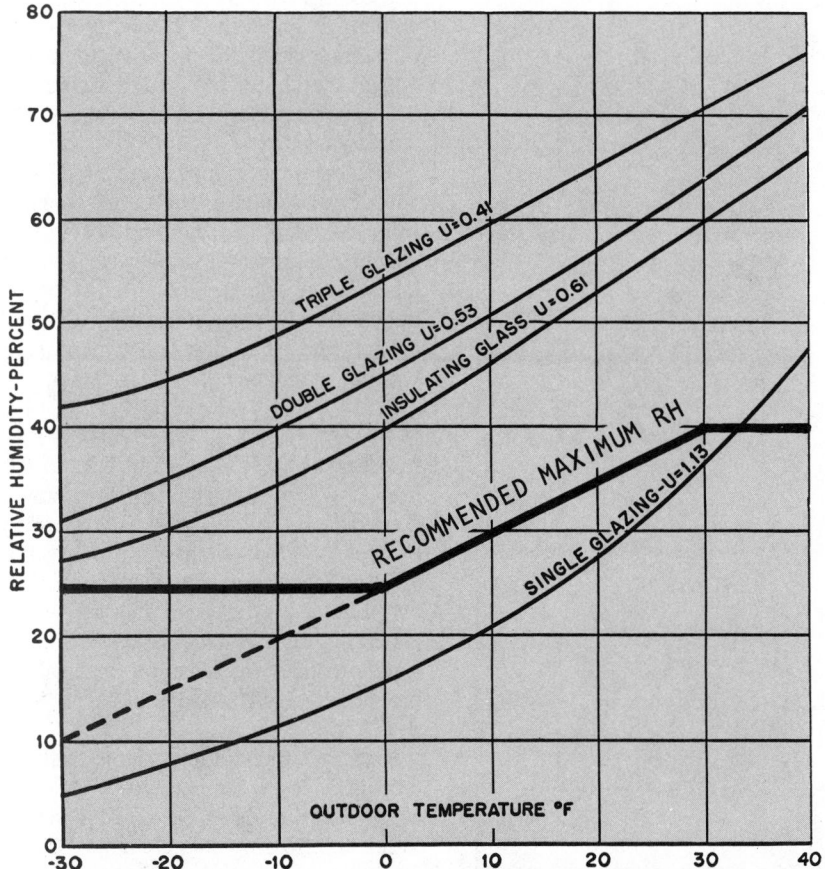

Below 0°F, RH values maintained at 25% to provide acceptable degree of occupant comfort (solid line); however, to minimize condensation on single glazing, use RH values represented by dashed line.

FIGURE 7.1-3 *Recommended maximum interior relative humidity to prevent condensation damage.* (*Reprinted with permission of the American Society of Heating, Refrigerating and Air-Conditioning Engineers.*)

100% relative humidity, vapor pressure is only 0.0377 in. Hg, or about 1/20 of the saturated condition.

With a winter condition of 0°F and 75% relative humidity, an outside vapor pressure of 0.027 in. Hg would exist, while inside a building heated to 70°F and with 35% relative humidity, vapor pressure would equal 0.259 in. Hg. The vapor pressure inside would be nearly 10 times the pressure outside. Like other gases, water vapor moves from an area of high pressure to an area of low pressure until equilibrium is established. During cold weather, the difference in pressure between inside and outside causes vapor to move out through every available crack and directly through permeable materials.

■ **PSYCHROMETRICS**

The behavior of moist air (mixture of dry air and water vapor), involving saturation of air and condensation of water vapor at varying temperatures, generally is referred to as *psychrometrics*. Changes in the properties of moist air with heating and cooling can be followed readily with the aid of a psychrometric chart (Fig. 7.1-5). The curved saturation line (100% relative humidity) represents the maximum concentrations of water vapor (pounds of water per pound of dry air) that can exist as vapor, without condensing, at various temperatures.

Figure 7.1-6 is a simplified psychrometric chart for a specific set of conditions. A possible winter condition inside buildings, 70°F and 40% relative humidity, is

represented by point A. This is a condition of partial saturation (less than 100% relative humidity). As air is cooled from 70°F (point A) to 44.6°F (point B on the saturation curve), the relative humidity keeps increasing until it becomes 100%.

■ **DEW POINT TEMPERATURE**

The temperature at which saturation occurs (point B in Fig. 7.1-6) is called the *dew point temperature* (*frost point*, if below 32°F). With further cooling to 35°F the original amount of vapor (44 grains per pound of dry air) is reduced through condensation to the condition represented by point C (30 grains per pound of dry air). The progression from point A to B to C illustrates what an air-vapor mixture experiences when it comes in contact with a cool window surface. Cooling from point B to point C results in visible condensation on the glass surface. If point C is below 32°F, the condensation is in the form of frost.

Once the temperature drops below the dew point (or frost point), the vapor pressure at a window surface is also reduced. This reduction in vapor pressure causes diffusion currents within the room, moving water vapor continuously to the window surface to be condensed as long as the concentration of vapor is maintained in the room.

The move from point D to point E (see Fig. 7.1-6) illustrates a common winter phenomenon. Outside air at 20°F and 100% relative humidity is heated to 70°F, with a resulting decrease in relative humidity to 14%. This example explains why the relative humidity in a building is greatly reduced in extremely cold weather, when cold outside air enters the building and is heated to room temperature.

High dew point temperatures are common in the summertime. The psychrometric chart shows that for a dry-bulb temperature of 80°F, the dew point temperature is 60°F if the relative humidity is 50% (see Fig. 7.1-5). However, if the relative humidity is raised to 90%, the dew point rises to 78°F. As the humidity rises, the dew point gets closer to the dry-bulb temperature but cannot exceed it. If, however, the temperature is raised to 90°F and the relative humidity is maintained at

FIGURE 7.1-4 *ASHRAE comfort chart showing ETs and subjective responses to temperature and humidity combinations.* *(Reprinted with permission of the American Society of Heating, Refrigerating and Air-Conditioning Engineers.)*

90% (by adding moisture to the air), the dew point rises to 87 °F.

7.1.1.3

Adsorption and Absorption

The surfaces of most materials have an affinity for water, and a film of water molecules often is held to the surface by molecular forces of attraction. This phenomenon is known as *adsorption*. Surface films at low humidities may be only one molecule thick; at moderate humidities, multimolecular films may be established. The film thickness, and therefore the amount of water held in equilibrium with the surrounding atmosphere, is roughly proportional to the relative humidity. At humidities close to 100%, the films become so relatively thick that small surface pores become completely filled and larger

capillaries may be partially filled. At saturation conditions, all voids in the material are completely filled.

Fibrous materials such as wood, plywood, and fiberboards present very large effective surface areas to water molecules, so the amount of water held on the surface, and subsequently absorbed into these materials through surface pores, is relatively large even at moderate humidities. These materials are said to be *hygroscopic*. Materials that present relatively small surfaces, such as most metals, are not penetrated by water molecules and take on only small amounts of water, as surface film.

■ DIMENSIONAL INSTABILITY

Changes of moisture content can cause significant dimensional changes in many construction materials. The dimensional changes occurring in wood are an obvious example. With a change from an air-

dry condition of 12% or 15% moisture content to an oven-dry condition (zero moisture content), the dimensional changes (shrinkage) that occur are approximately 0.1% longitudinally, 2% radially, and 4% tangentially.

The continuous movement of dimensionally unstable materials must be carefully considered in building design and detailing. For example, millwork elements lap one another to permit movement without developing unsightly cracks; air gaps around wood strip floors allow expansion without buckling in humid weather.

■ DETERIORATION OF MATERIALS

Moisture can cause or contribute to the breakdown of materials in three ways: (1) chemical changes, such as the rusting of steel; (2) physical changes, such as the spalling of masonry by frost action; and (3) biological processes, such as the decaying of wood. Measures to control liquid water include waterproofing and dampproofing (Section 7.2), and flashings (Section 7.9).

Because air saturation is more likely to take place at low temperatures, when the risk of freezing also is enhanced, condensation and freezing both are major design considerations. Also, condensation may occur over a period of time in concealed locations in wall, floor and ceiling, and roof assemblies and not be apparent until a conspicuous failure occurs.

■ CONDENSATION TOLERANCE OF MATERIALS

Some materials are able to tolerate condensation with little damage. Each material should be assessed individually. The metal panels in refrigerated storage boxes, for example, almost always contain a coating of frost with no resulting damage. Some materials can store condensation with little or no effect. Concrete masonry units, for example, can easily have their level of saturation raised from 40% to 50% by absorbing condensation without notice. Metal panels absorb little moisture but can rapidly release collected moisture or frost through evaporation when weather conditions change.

■ EFFECT ON HEAT FLOW

Moisture can have a marked effect on the transmission of heat through building materials. When present in liquid form,

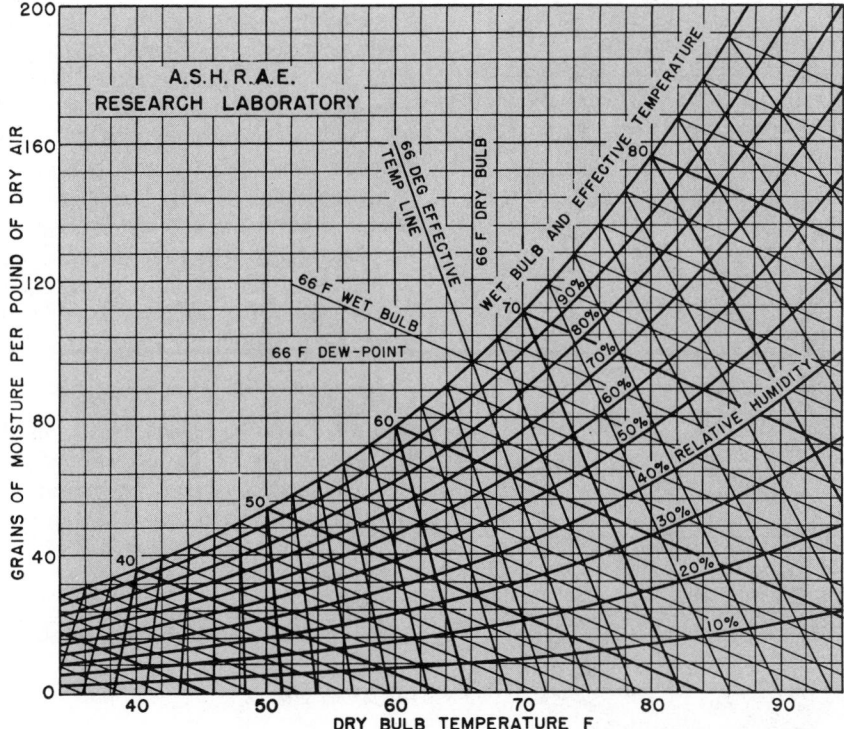

FIGURE 7.1-5 *Psychrometric chart.* (*Reprinted with permission of the American Society of Heating, Refrigerating and Air-Conditioning Engineers.*)

FIGURE 7.1-6 *Use of a psychrometric chart illustrated.* (*Reprinted with permission of the American Society of Heating, Refrigerating and Air-Conditioning Engineers.*)

moisture increases the conductivity of materials by increasing the available paths of heat flow.

Hygroscopic insulations such as vegetable fiber products, which readily absorb and hold moisture, do not dry out as quickly as nonabsorbent fiber insulations and lose their thermal effectiveness for longer periods. Nonabsorptive products, however, have limited moisture storage

capacity and begin "dripping" sooner, possibly damaging interior finishes.

7.1.1.4

Water Vapor Transmission

The water vapor permeance of component materials should be considered in the design of exterior walls, floors, and roofs of habitable spaces. The amount and the rate of water vapor transmitted through building materials depends on the following: (1) the *gradient* (differential) in vapor pressure from one side of the material to the other, (2) the area of the material, and (3) its *permeance* (ability to permit vapor passage).

■ PERMEANCE AND PERMEABILITY

Water vapor *permeance* is a measure of the flow of water vapor through a material of specific thickness or through an assembly of several materials. The unit of permeance, a *perm*, states the amount of vapor flow in grains per hour per square foot of surface per inch Hg (mercury) vapor pressure gradient.

Water vapor *permeability* is the permeance of a 1-in. thickness of a homogeneous substance. Permeability is designated in perm-inches and states the amount of vapor flow through 1 in. of material, in grains per hour per square foot of surface per inch Hg vapor pressure gradient.

Most building materials are permeable to a degree. Some, such as metals and glass, are almost completely impermeable, permitting almost no passage of water vapor. Such materials have a perm rating (permeance) approaching zero. Materials that either are impermeable or have a permeance of 1 perm or less were formerly referred to as *vapor barriers*. However, in recognition that even the best of these materials permit some transmission of vapor, ASHRAE has recently started calling these materials *vapor retarders*. This text uses the latter, more accurate term throughout. In addition, modern construction methods demand better vapor retarders than those approaching 1 perm. Therefore, vapor retarders frequently are required to have a rating of 0.5 perm or less. Permeances for representative construction materials are given in Figure 7.1-7.

Many of the permeance numbers in Figure 7.1-7 were obtained by different test methods, which can produce somewhat varying ratings. Therefore, it is not valid to compare the relative performance of those materials with small variations in their permeances. For instance, it is not meaningful to compare hot melt asphalt, 2 oz./sq. ft. (perm, 0.5) with asphalt-saturated paper (perm, 0.2–0.3). Either of these two or any one of the vapor retarder materials (under 1 perm) can be more readily compared to products with permeances substantially higher than 1 perm.

The overall permeance (M_t) of an assembly of materials can be calculated from the permeances of the individual components. The calculation is similar to that used in determining the overall coefficient of heat transmittance (U) from individual material conductances (C):

$$M_t = \cfrac{1}{\cfrac{1}{M_1} + \cfrac{1}{M_2} + \cfrac{1}{M_3} + \ldots + \cfrac{1}{M_n}}$$

where M_t = overall, total permeance, and M_1, M_2, M_3, through M_n = the individual permeances of materials in an assembly.

FIGURE 7.1-7
Water Vapor Permeance of Building Materials

Material	Permeance (Perm)
Materials used in construction	
1 in. concrete (1:2:4 mix)	3.2
Brick masonry (4-in. thick)	0.8
Concrete block (8-in. cored, limestone aggregate)	2.4
Plaster on metal lath (3/4 in.)	15.0
Plaster on plain gypsum lath (with studs)	20.0
Gypsum wallboard (3/8-in. plain)	50.0
Gypsum sheathing (1/2-in. asphalt impregnated)	10.0
1-in. structural insulating board (sheathing quality)	20–50
Structural insulating board (interior, uncoated, 1/2 in.)	50–90
Hardboard (1/8-in. standard)	11.0
Hardboard (1/8-in. tempered)	5.0
Built-up roofing (hot-mopped)	0.0
1-in. wood sugar pine	0.4–5.4
Plywood (douglas-fir, interior, blue, 1/4-in. thick)	1.9
Thermal insulations, 1-in. thick	
Air (still)	120.0
Cellular glass	0.0
Corkboard	2.1–9.5
Mineral wool (unprotected)	116.0
Expanded polyurethane (R-11 blown)	0.4–1.6
Expanded polystyrene, extruded	1.2
Expanded polystyrene, bead	2.0–5.8
Unicellular synthetic flexible rubber foam	0.02–0.15
Plastic and metal foils and films	
Aluminum foil (1 mil)	0.0
Aluminum foil (0.35 mil)	0.05
Polyethylene (2 mil)	0.16
Polyethylene (4 mil)	0.08
Polyethylene (6 mil)	0.06
Building papers, felts, roofing papers*	
Duplex sheet, asphalt laminated, aluminum foil one side (43)	0.002
Saturated and coated roll roofing (326)	0.05
Kraft paper and asphalt laminated, reinforced 30-120-30 (34)	0.3
Blanket thermal insulation backup paper, asphalt coated (31)	0.4
Asphalt saturated and coated vapor-barrier paper (43)	0.2–0.3
Asphalt saturated but not coated sheathing paper (22)	3.3
15-lb. asphalt felt (70)	1.0
15-lb. tar felt (70)	4.0
Single-kraft, double infused (16)	31.0
Protective Coatings	
Paints, 2 coats	
Aluminum varnish on wood	0.3–0.5
Enamel on smooth plaster	0.5–1.5
Various primers plus 1 coat flat oil paint on plaster	1.6–3.0
Water emulsion on interior insulating board	30.0–85.0
Styrene-butadiene latex coating (2 oz./sq. ft.)	11.0
Polyvinyl acetate latex coating (4 oz./sq. ft.)	5.5
Asphalt cut-back mastic (1/16 in. dry)	0.14
Hot melt asphalt (2 oz./sq. ft.)	0.5

Reprinted with permission of the American Society of Heating, Refrigerating and Air-Conditioning Engineers.

*Numbers in parentheses are weights in lb. per 500 sq. ft.

■ WATER VAPOR TRANSMISSION

Sometimes it is more significant to report the vapor resistance of materials at a specific pressure difference rather than at 1 in. Hg, as in permeance and permeability. Water vapor transmission (WVT) states the weight of vapor (grains) transmitted per unit of time (typically 1 hour) per unit of area (1 sq. ft.); values may be high or low depending on the vapor pressure gradient chosen for the test. Water vapor transmission data can be converted to perms by the following formula:

$$Permeance = WVT \ rating/, \ \Delta p,$$

where *WVT rating* equals the weight of vapor transmitted in grains per hour per square foot and Δp equals vapor pressure gradient in the test in inches of mercury.

7.1.2

CONDENSATION: CAUSES AND PREVENTION

Condensation occurs when vapor-laden air at a high enough humidity contacts surfaces that are sufficiently cool. Condensation problems may occur in cooler climates in occupied, heated buildings and in humid warm climates in air conditioned buildings. Production of condensation-causing moisture in buildings depends on the type of activities there. Commercial kitchens and shower rooms, for example, can generate enormous quantities of vapor. Four people in a normal home environment generate between 22.5 and 50 lb. of vapor in 24 hours. In winter, condensation may collect in the form of water or frost on glass and other cold surfaces, such as metal wall panels, metal window frames, and metal doors and frames.

Although condensation forming on a surface may enter surface pores as fast as it forms and thus become invisible, any condensation on a visible surface is referred to as surface condensation to distinguish it from concealed condensation, which occurs inside a material or assembly.

7.1.2.1

Interior Ventilation

In many areas of the country, summer cooling does not always create serious concealed condensation problems. In these areas, cooled interior air seldom produces conditions in a wall that are conducive to condensation. However, during periods of high humidity, condensation may form on cool basement walls and floors. In humid climates, condensation may occur in concealed spaces (see Section 7.1.3).

In areas where high levels of water vapor are generated indoors, the likelihood of both surface and concealed condensation can be greatly minimized by introducing dry night air or by exhausting excessive water vapor directly at its source, such as in a shower room, kitchen, laundry, or residential bathroom, using exhaust fans. In many buildings of all types and in most homes, a sufficient amount of fresh air is available through infiltration to make up the air lost by exhaust. Where this is not the case, it will be necessary to introduce a fresh air supply to balance this loss.

During cold weather, interior relative humidity often becomes too low for human comfort, and mechanical humidification of occupied spaces is required. Even then, ventilation in areas where high humidity is present, such as those mentioned earlier, is desirable to reduce excessively high local humidities, which could easily result in surface condensation.

The effectiveness of ventilation in reducing humidity is shown in Figure 7.1-8. For simplicity, the figure uses a 2000-sq.-ft. residence as its example. It shows the small amount of water vapor escaping into well-insulated walls and ceilings of a building provided with vapor retarders (the floor is not considered). With the indoor temperature at 70 °F and a relative humidity of 40%, ventilation at the rate of 2000 cu. ft./hr. removes about 21 lb. of moisture per day. Assuming a production of 22.5 lb./day, less than 1.5 lb. escapes into the walls and ceilings of the structure. As mentioned earlier, 22.5 lb. in 24 hours is about average for a family of four at home.

In a residence, ventilation at the rate of 2000 cu. ft./hr. is near the minimum for odor control during cooking. Therefore, mechanical ventilation may not be required just to get rid of excess moisture generated in a home. Often in older buildings of all types, the action of exhaust fans is supplemented by substantial *exfiltration*

FIGURE 7.1-8 *Water vapor balance maintained by ventilation.* (*Reprinted with permission of the American Society of Heating, Refrigerating and Air-Conditioning Engineers.*)

(air leakage) through walls, windows, and doors to keep moisture in check. Tight newer buildings often require more extensive ventilation systems, particularly in periods of cold weather, when the relative humidity must be limited to prevent condensation (see Fig. 7.1-3).

7.1.2.2

Surface Condensation

Condensation occurs when a surface is colder than the dew point of the nearby air. The temperature of interior surfaces such as walls, ceilings, and floors depends on (1) the air temperatures both inside and outside the building and (2) the heat transmittance of the wall, ceiling, or floor assembly. Ratings of overall heat transmittance for an assembly can be derived by formula from individual transmittances of the materials in the assembly.

Surface condensation occurs when a surface loses heat fast enough to reach the dew point (condensation temperature). Adequate insulation reduces heat loss and increases surface temperature, thereby reducing condensation hazard. Therefore, when enough insulation is provided in exterior walls, ceiling and floor assemblies, and roof and ceiling assemblies to keep

surfaces that are in contact with moisture at a temperature below the dew point, there will be no surface condensation. Where installing sufficient insulation to accomplish this condition is not feasible, warm air currents can be induced to increase surface temperature and dissipate moisture. Keeping the interior relative humidity within the range recommended in Figure 7.1-3 also will minimize surface condensation.

Figure 7.1-9 shows the relative humidities at which condensation will appear on interior surfaces with various U-values in a room with an air temperature of 70 °F. Since condensation can appear at any sufficiently cold spot, the values given in Figure 7.1-9 should be used with caution. U-values constitute an average for large surface areas, within which there may be spots where the heat transmittance is higher, such as near structural supports or metal outlet boxes. Also, the inside surface temperature of a stud wall will often be lower at the base due to air stratification inside the stud cavity. As a result, the maximum limit of relative humidity for a nonhomogeneous wall is lower than might be inferred from its average U-value.

Glass areas normally are the most likely surfaces in heated buildings on which con-

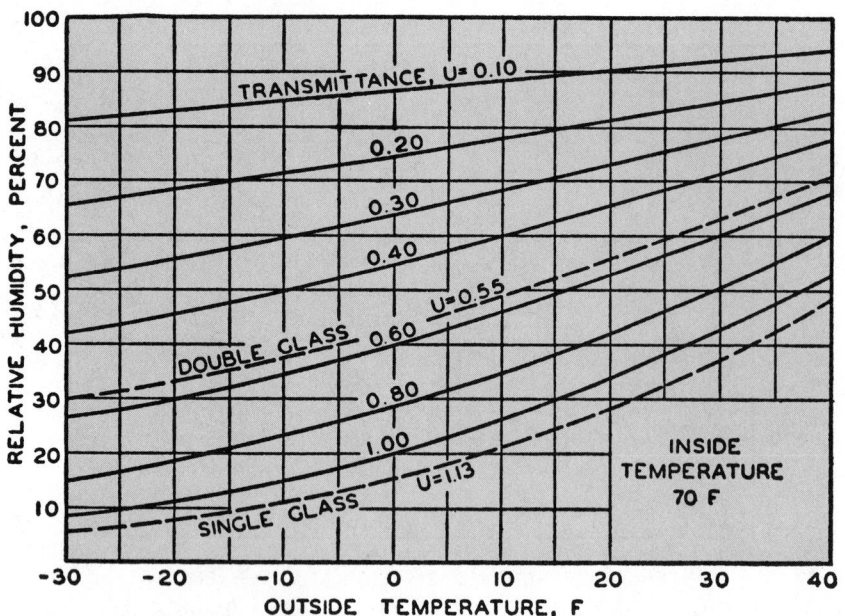

FIGURE 7.1-9 *Relative humidity and condensation.* *(Reprinted with permission of the American Society of Heating, Refrigerating and Air-Conditioning Engineers.)*

densation might appear; however, an uninsulated wall surface shielded from radiation and convection by furniture may also be quite cold. Over time, surfaces with temperatures near the dew point temperature of the adjacent air may experience swelling, mold, staining, or decay.

Condensation on interior room surfaces can be controlled both by suitable construction and by operating precautions such as (1) reducing the interior dew point temperature or (2) raising the temperatures of interior surfaces above the dew point.

The interior air dew point temperature can be lowered by removing moisture from the air, either through ventilation or dehumidification. Surface temperatures in winter can be maintained high enough by incorporating adequate thermal insulation, using double glazing, circulating warm air over the surfaces, or heating the surface directly.

■ **BELOW-GRADE CONSTRUCTION**

Visible surface condensation may occur in summer on concrete basement walls and floors and exposed cold-water pipes. Because of their large mass and the insulating effect of the adjacent soil, basement floors and walls tend to maintain a rela-

tively cool and constant temperature. If the dew point temperature within the space rises above the temperature of walls, floors, and pipes, condensation occurs. Finish surfaces in below-ground spaces can be seriously damaged by condensation.

Condensation on interior surfaces can be controlled by (1) reducing the air intake from the outside at times of high outdoor dew point temperatures, (2) warming the surfaces, or (3) most successfully, by deliberate dehumidification. For instance, pipes to water closets can be warmed by using lukewarm water for flushing; in climates having low temperatures at night, it may be feasible to ventilate only at night when the air contains less moisture and thus reduce the moisture absorption of hygroscopic materials.

■ **SLAB-ON-GRADE CONSTRUCTION**

Like basement walls, slabs on grade tend to maintain a relatively constant temperature. Surface condensation can develop when such a slab is sufficiently cold in relation to a warm interior atmosphere that has a high dew point. In climates where high dew point temperatures frequently occur in summer, condensation may damage finish flooring laid on a slab on grade. Thick rugs or carpets aggravate the prob-

lem because they insulate the slab from interior warmer air and retain the slab at a low temperature. Because woven, knitted, and tufted floor coverings are often vapor permeable, moisture may condense under the covering and promote decay, mold, and a musty odor.

Methods of controlling surface condensation are similar to those for above-grade construction: deliberate dehumidification and increasing surface temperature. A slab on grade can be kept warm by insulating the slab from cooler ground temperatures with rigid insulation at the perimeter and with a granular, well-drained base course under the entire slab.

7.1.2.3

Concealed Condensation

In the winter in most parts of the country, the continual production of water vapor inside an occupied building normally raises the vapor pressure above that outside the building. This pressure difference provides the force that causes vapor diffusion into exterior walls and ceilings. The vapor pressure rise in the building is directly proportionate to the amount of vapor produced and inversely proportionate to its ability to escape. In other words, the pressure rises as more vapor is produced, but the pressure is relieved if vapor is permitted to diffuse freely to areas of lower concentration.

Exterior walls, floors over unheated spaces, and roof and ceiling assemblies should be protected with suitable vapor retarders and ventilation provisions. In most areas, because the sources of moisture typically are within a building, a vapor retarder should be placed near the interior finish to prevent moisture migration into concealed structural spaces. However, a suspended floor above a heated crawl space need not be insulated—the ground constitutes the chief source of moisture (Fig. 7.1-10); therefore, the vapor retarder should be used as a ground cover.

As water vapor moves through a portion of a structure, such as a wall, floor, or roof and ceiling assembly, it may reach a point in the assembly at which it is cooled to the point of condensation. In older frame buildings with little or no insulation, so much heat was lost through a building's

BARE EARTH, HIGH M.C.		16.25 GAL.
STANDING WATER		14.8 GAL.
BARE EARTH, AVERAGE M.C.		10.3 GAL.
VAPOR BAR OVER EARTH		0.24 GAL.

FIGURE 7.1-10 *Daily evaporation from a crawl-space floor. (U.S. Forest Products Laboratory. Not subject to copyright in the United States.)*

FIGURE 7.1-11 *When there is no insulation in a wall cavity (a), so much heat is lost that the temperature in the cavity often remains above the dew point. When insulation is added (b), the dew point will usually fall within the cavity.*

shell that the entire structural cavity often was warm enough to be above the dew point (Fig. 7.1-11a). The dew point temperature was reached somewhere in the sheathing or siding, but condensation was rarely a problem because moisture was carried off by air leakage through joints in the sheathing and siding, which were usually of the board type.

Although the theoretical dew point line in an insulated frame wall generally lies well within the cavity or even within the insulation (Fig. 7.1-11b), the vapor does not condense at this plane but moves on through the vapor-permeable fibrous insulation until it reaches the sheathing. Modern sheathing and siding products frequently are made in broad panels rather than narrow boards, thus reducing the opportunity of moisture diffusion by air leakage through joints. In addition, many siding and sheathing materials such as plywood and hardboard are inherently vapor resistant, so that only small amounts of vapor can move directly through the

material itself. All these factors, coupled with the generally higher interior humidity levels maintained by modern heating systems, account for the greater incidence of condensation problems in modern construction if adequate vapor retarders are not provided.

■ WALL CONSTRUCTION

When water vapor is allowed to enter a wall assembly and condensation occurs within its outer cold elements, frost or water may develop, depending on the outside temperature (Fig. 7.1-12). In weather that is continuously cold for a long period, the frost may build back into the wall cavity or fibrous insulation forming a mass of ice. If the outside temperature rises frequently, the frost or ice may melt and be absorbed by hygroscopic materials such as wood, or it may run down nonabsorbing or already-saturated surfaces (Fig. 7.1-13).

In masonry walls, water seepage to the outside may occur harmlessly when the

weather is above freezing, but water seepage into the building must be prevented. The inclusion of weep holes and base flashings encourages runoff to the outside and bars seepage to the interior (see Section 4.3).

An accumulation of moisture within insulation lowers the insulation's thermal efficiency, causing greater heat loss and other condensation problems. Trapped moisture in siding and sheathing is one of the principal causes of paint failure. In seeking to escape from behind vapor-resistant paint films, moisture can cause paint blistering and peeling, especially when warmed by solar radiation (Fig. 7.1-14). In extreme cases, moisture trapped in wall cavities will eventually cause decay (Fig. 7.1-15).

The installation of an adequate vapor retarder is the most important moisture control device in exterior wall construction.

■ ROOF AND CEILING ASSEMBLIES

The principles of vapor movement and control within a roof and ceiling assembly are much the same as for walls. Water vapor moves from the side with the higher vapor pressure toward the side with the lower vapor pressure. When the side with the lower pressure is also colder, the water vapor may condense in concealed spaces of the roof and ceiling assembly. Some roofing materials, such as wood shingles, can "breathe" and are affected little by trapped vapor; others, such as built-up roofs, are likely to be ruptured by excessive moisture. When a built-up roof blisters and cracks, permitting water to enter the roofing, continual and rapid deterioration is imminent.

In building spaces with high relative humidities, such as laundries, saunas, and pool enclosures, special consideration must be given to properly placed vapor retarders and sufficient insulation.

In unventilated roof and ceiling assemblies, condensation may collect in the form of ice, frost, or water on the framing members, roof deck, and projecting fasteners. Condensation problems seldom exist in roof and ceiling assemblies with insulation above the ceiling, adequate ventilation, and a properly installed vapor retarder on the warm side of the insulation. But roof constructions with rigid insulation above a wood or metal deck may require a vapor retarder or details that

FIGURE 7.1-12 *Sheathing reversed to show accumulation of frost on inside surface. This cavity was insulated but did not have a vapor retarder.* (U.S. Forest Products Laboratory. Not subject to copyright in the United States.)

FIGURE 7.1-14 *Blistering and deterioration of paint surfaces caused by vapor pressure behind a paint film.* (U.S. Forest Products Laboratory. Not subject to copyright in the United States.)

FIGURE 7.1-13 *Condensation formed frost and ice behind this siding during cold weather, then weeped through laps in the siding as the weather moderated, staining the siding.* (U.S. Forest Products Laboratory. Not subject to copyright in the United States.)

FIGURE 7.1-15 *Decay of studs and sheathing due to frequent condensation in the wall cavity.* (U.S. Forest Products Laboratory. Not subject to copyright in the United States.)

FIGURE 7.1-16 *Repeated wetting caused this plaster ceiling to crack and produced a visible stain.* (U.S. Forest Products Laboratory. Not subject to copyright in the United States.)

permit venting of the insulation. On sunny days, even at low temperatures, frost or ice may melt and seep to the ceiling below, causing staining and possible damage to ceilings (Fig. 7.1-16). In extreme cases rapid deterioration and decay of wood framing and rusting of steel roof structure members also are possible.

Adequate vapor retarders on the warm side of the insulation and the ventilation of roof and ceiling assemblies above the insulation are essential (see "Roof and Ceiling Assemblies" in Section 7.1.3).

■ CRAWL SPACE CONSTRUCTION

The presence of water in a crawl space may produce considerable amounts of water vapor. This accumulation of water vapor may increase the possibility of decay and termite attack on wood members and rusting of steel members and result in unpleasant odors in the building. Therefore, it is important to ventilate enclosed foundation spaces adequately to provide for the dissipation of water vapor. The amount and the methods of ventilation depend on the type of space being ventilated, the degree of water vapor present, and the type of vapor retarder installed.

Water vapor rising from the ground below a crawl space is usually a greater problem than vapor coming from inhabited spaces above. Moisture in the soil can rise as much as 11 ft. above the water table. Depending on the level of the groundwater table, the character of the soil, and drainage around the foundation, large amounts of water may be introduced into the crawl space by evaporation from the ground (Fig. 7.1-17).

Excessive vapor can cause condensation on structural members that have been cooled by heat loss to the outside. Water vapor from a crawl space may also enter walls, move upward within them, and even reach roof construction when the wall construction permits. These problems can be prevented by simply reducing evaporation from the ground with a vapor retarder and providing adequate ventilation. A vapor retarder properly placed over the ground in a crawl space can be used to prevent capillary moisture and vapor rise.

FIGURE 7.1-17 *Beads of condensation formed on joists in a crawl space without foundation vents or ground cover.* (U.S. Forest Products Laboratory. Not subject to copyright in the United States.)

7.1.3

DESIGN AND CONSTRUCTION RECOMMENDATIONS

Moisture condensation in walls, floors, and roof and ceiling assemblies can be controlled by (1) providing a vapor retarder on the interior side of the assembly to limit vapor entrance into the construction and (2) ventilating structural spaces such as attic areas and crawl spaces to dissipate vapor to the outside. This double line of defense is desirable in all areas where severe or even moderate condensation hazard exists. However, in wall assemblies, reliance is placed on a retarder, and air movement through the cavity is discouraged to minimize heat transfer. The continental United States can be divided into three condensation zones, based mainly on the isotherms of the −20°F, 0°F, and 20°F winter design temperatures (Fig. 7.1-18).

Humid climates present slightly differ-

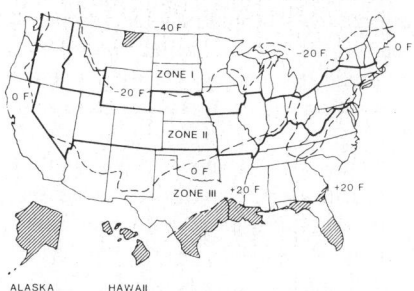

Zones include areas with design temperatures about as follows: Zone I, −20°F and lower; Zone II, 0 to −20°F; and Zone III, above 0°F. Cross-hatched areas are outside Zones I and III.

FIGURE 7.1-18 *Condensation zones in the United States.* (Reprinted with permission of the American Society of Heating, Refrigerating and Air-Conditioning Engineers.)

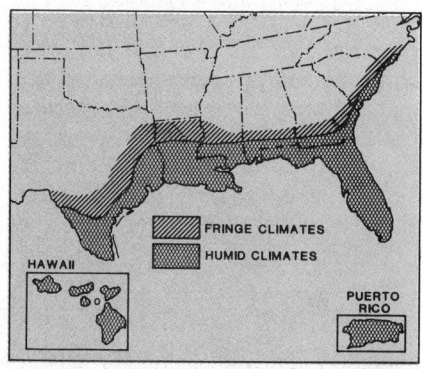

FIGURE 7.1-19 *Humid climates in the continental United States.* (Reprinted with permission of the American Society of Heating, Refrigerating and Air-Conditioning Engineers.)

ent condensation problems, because in air conditioned buildings located in such climates the flow of water vapor is from outdoors to indoors. This means that vapor retarders should be located on the exterior side of insulation rather than on the inside. Figure 7.1-19 is a map produced by ASHRAE that shows the location of humid climates in the United States.

When construction limitations make it impractical to provide a vapor retarder, as would be the case, for example, in a masonry wall with interior insulation and

no applied finish, reliance must be placed on ventilation. Conversely, when adequate ventilation is hard to achieve, vapor-resistant interior construction is essential.

In this discussion, the term *ventilation* is used in its broad, all-inclusive sense as any attempt to dissipate moisture by using air as the vehicle, as well as in a more specific sense. In the narrower meaning, ventilation describes typical methods of dissipating moisture from structural spaces by positive air movement through continuous air passages linked to the general atmosphere. In this use, the term *ventilation* is often accompanied by a quantitative recommendation, such as a ratio of vent area to total area. This narrower meaning can be contrasted with another type of ventilation, *venting*, which describes the inclusion of air leaks or gaps in an otherwise vapor-resistant material, finish, or membrane, such as vinyl siding, as a means of vapor diffusion. In venting, airspaces are not necessarily interconnected, and air movement is nonexistent or sluggish; ventilation implies much more positive air movement and more rapid moisture dissipation.

7.1.3.1

Vapor Retarder Selection

Although a vapor retarder by definition is any material with a permeance of 1 perm or less, not all materials are equally suitable for use in a particular assembly. Some materials are inherently better vapor retarders than others; however, the effectiveness of a retarder depends on the method of installation, as well as its physical properties. For instance, 1-mil-thick aluminum foil has a permeance of 0.00; the permeance of typical 0.35-mil paper-backed product is likely to be about 0.05 when installed in large sheets across the face of studs. The practice of stapling aluminum-faced batts to the sides of wood studs may increase the installed permeance about twentyfold, to as high as 1 perm.

Selection of a suitable retarder depends on a variety of factors such as prevailing interior and exterior humidities, length and severity of the winter season, and materials and methods of construction. In general, the severity of the winter season, as expressed by the winter design temperature, is a prime factor in determining the degree of vapor protection required (see Fig. 7.1-18).

Figure 7.1-20 summarizes recommendations for vapor retarder and ventilation for roof, wall, and floor assemblies in the three zones. Structural space ventilation is recommended for attic spaces and unheated or cooled crawl spaces in all zones. Vapor retarders are recommended for most assemblies in Zones I and II.

In Zone III, the warm southern part of the United States, average winter temperatures are high enough so that large temperature and vapor pressure differentials between the inside and the outside do not develop. Therefore, vapor retarders are often omitted from walls and roof and ceiling assemblies in Zone III; however, a retarder should be installed beneath slabs on grade and in crawl spaces as a general precaution against moisture, as in Zones I and II.

Figure 7.1-20 shows three levels of retarder efficiency: (I), permeance of 0.1 perm or less, usually aluminum foils or plastic films; (II), 0.5 perm or less, asphalt-coated felts or asphalt-laminated papers; (III), 1.0 perm or less, same as (II), but with less asphalt per square foot.

Depending on the type, vapor retarders may be incorporated with or installed separately from other materials in the construction. Typical vapor retarders in frame construction include the following: (1) a separate plastic or metallic membrane, job-installed over framing members and under the interior finish material; (2) a metallic foil, factory-applied on the back of an interior finish material such as gypsum wallboard; (3) aluminum-faced paper or asphalt-laminated paper, factory-applied over batt or blanket insulation; or (4) insulations made of inherently vapor-resistant materials, such as layers of reflective metallic foil or closed-cell plastic foam.

■ RETARDER-FACED INSULATION

Retarder-faced insulation is discussed in Section 7.3.2. Where the vapor retarder material on such insulation is integral with the batts or blankets and the flanges are nailed to stud sides, as recommended by most manufacturers, it is almost impossible to avoid air and vapor leaks. Therefore, this installation method should be restricted to areas of moderate condensation hazard, as in Zone II. In Zone I, the vapor retarder flanges should be secured

FIGURE 7.1-20
Vapor Retarders (V.R.) and Structural Space Ventilation in Different Condensation Zones

Condensation Zone	Atticless Joist Roof/Ceiling		Attic-Type Joist Roof/Ceiling		Crawl Space		Floor Over Open Unheated Area, V.R., Perms	Exterior Walls, V.R., Perms	Slab on Ground V.R., Perms
	Ceiling V.R., Perms	Structural Ventilation[a]	Ceiling V.R., Perms	Structural Ventilation[a]	Ground V.R., Perms	Structural Ventilation[a,b]			
Zone I	0.1	1/300	0.5	1/300	1.0	1/1500	1.0	1.0[c]	3.0
Zone II	0.5	1/300[d]	1.0	1/300[d]	1.0	1/1500	1.0	1.0[c]	3.0
Zone III	1.0	1/300[d]	None	1/300[d]	1.0[e]	1/1500[e]	None	None	3.0

Reprinted with permission of the American Society of Heating, Refrigerating and Air-Conditioning Engineers.

[a]Ratio of net free ventilating area (NFA) to total floor or ceiling area.

[b]Ventilation to atmosphere not required when crawl space is heated or cooled and foundation wall is insulated.

[c]Venting of walls may be desirable where highly impermeable exterior finish is used.

[d]Ratio of 1/150 preferred to relieve high summer solar heat gain.

[e]When crawl space is not heated or cooled, vapor retarder may be omitted provided vent area is increased to 1/150.

across the stud faces, or a separate continuous vapor retarder should be used.

FOIL-BACKED BOARDS

Foil-backed gypsum boards are discussed further in Section 9.5. Foil-backed gypsum lath and gypsum wallboard have an aluminum foil retarder laminated to the back of each board (Fig. 7.1-21). However, breaks exist at board joints, which typically are 16 in. apart vertically and 48 in. apart horizontally for lath and 48 in. apart horizontally and 96 in. apart vertically for wallboard. These joints permit some vapor leakage (especially where the joint is not backed up by a structural member), and the efficiency of the retarder is reduced.

Foil-backed gypsum lath boards generally are installed horizontally with the long edges perpendicular to the studs. This method permits a great deal of vapor leakage and is not recommended as the sole means of vapor control in either Zone I or II.

Using foil-backed wallboard results in effective vapor isolation if the boards are installed vertically, with all joints over studs. It is suitable for vapor control in exterior walls in all zones.

SEPARATE SHEETS

The most efficient vapor retarder installation is one in which the vapor retarder is applied separately from the insulation or the interior finish. It is installed in the largest possible sheets and with joints backed up by framing members. Insulation can be provided as pressure-fit blankets or rigid boards. When combined with a suitable insulation, a separate-sheet vapor retarder is particularly suitable for severe condensation conditions, as in Zone I.

Polyethylene film and kraft paper–supported aluminum foil are typical products available in roll widths from 3 to 20 ft. Best results are obtained when joints are made over a structural member. This *vaporproofing* method is highly recommended for Zone I.

INSULATING BOARDS

A number of synthetic insulating boards, such as foamed glass, polystyrene, and polyurethane, are being marketed; many

FIGURE 7.1-21 *Vapor control can be provided by foil-backed gypsum board.* (Courtesy USG Corporation.)

of these materials have a closed-cell construction and are inherently good vapor retarders. However, their permeability varies with their manufacturing process and should be checked in each case to determine that the required permeance can be achieved in the thickness contemplated. Also, boards should be butted tightly together and joints should be sealed with a vapor-resistant mastic.

PAINTS AND OTHER COATINGS

Although many types of paints have proved to be good vapor retarders under laboratory conditions, the uncertainties of field application make reliance on paint coatings questionable. In new construction, the more reliable factory-produced vapor retarders are recommended. Paints and other surface coatings should not be used as the sole means of vapor control in Zones I and II.

Paint coatings can sometimes be used correctively in older structures and as an adjunct to other vapor control methods. Several manufacturers market products for specific use as vapor retarders. Alkyd, aluminum, and bituminous coatings generally have low permeability (high resistance) to vapor. Because vapor properties of specific products depend on formulation, the manufacturer should be consulted. Most coatings are more vapor resistant when their finish appearance is shiny or glossy.

7.1.3.2

Vapor Retarder and Ventilation Design

Ventilating openings in exposed locations often are provided with louvers to keep out wind and rain; screening usually is included to bar insects and vermin. Ventilating requirements generally are given as net free area, which is the space actually available to the moving air, after deducting obstructions such as the screening and

FIGURE 7.1-22
Required Gross Area of Vent Openings

Vent Covering	RGA[a] NFA[b]
Hardware cloth, 1/4-in. mesh	1
Screening, 1/8-in. mesh	1-1/4
Insect screen, 1/16-in. mesh	2
Louvers and hardware cloth, 1/4-mesh	2
Louvers and screening, 1/8-in. mesh	2-1/4
Louvers and insect screen, 1/16-in. mesh	3

Reprinted with permission of the American Society of Heating, Refrigerating and Air-Conditioning Engineers.

[a]Required gross area.

[b]Net free area (which can be determined from Fig. 7.1-20).

louvers. Figure 7.1-22 shows how to calculate the required gross area of various types of ventilating openings.

■ WALLS

Ventilating exterior wall cavities may provide adequate vapor control in a moderate climate. However, wall openings designed to move substantial amounts of air and vapor may waste considerable heat in colder climates. Therefore, in walls, as in most other assemblies, vapor retarders are the primary means of vapor control; wall ventilation (more typically venting), when provided, is an adjunct to vapor control with suitable retarders.

Venting of an exterior surface should be considered when the siding is a vapor-resistant material in sheet or panel form. Recognizing the high vapor resistance of plastic and metal sidings, manufacturers of these products usually include vent holes. No special venting efforts are required with wood and plywood horizontal board sidings; these are self-venting at the overlap. However, plywood and hardboard panel sidings require greater attention to effective interior vapor retarders and may require special venting details.

In older buildings that have no vapor retarders, painted siding may resist vapor flow sufficiently to trap moisture in stud spaces. In Zones I and II, the resulting

FIGURE 7.1-23 *Prefabricated circular vent. (Courtesy Easy-up Home Ventilation from The Solar Group.)*

vapor pressure under the paint film may cause blistering and peeling. In these cases, ventilating stud spaces with individual vents at top and bottom (Fig. 7.1-23) will help dry out the stud cavities. However, if vents become clogged with cobwebs and dust, it will be necessary to provide an interior vapor retarder.

It is essential to form a continuous retarder envelope on the inside of all walls that contain any material that would be damaged by moisture or its freezing. This includes most forms of thermal insulation, wood and metal studs, and most sheathing. This vapor retarder material should have a permeance of 1.0 perm or less. Relatively permeable materials should be used on the outside of a wall. Sheathing paper, for example, should be a "breathing" type, having a permeance of 6 perms or more.

Special emphasis is placed on sheathing paper because the 15-lb. asphalt-saturated felt typically used for this purpose also is used for built-up roofing and often is too vapor resistant for use in a wall. Figure 7.1-7 shows 15.-lb. asphalt felt and tar felt as having a permeance of 1.0 perm and 4.0 perms, respectively. These are average figures. Permeance of a particular product can be higher or lower, depending on the degree of saturation with bitumens.

An often used rule of thumb to ensure safe outward diffusion of vapor is that the combination of exterior materials should have a total installed permeance of at least five times the total permeance of the interior vapor retarder plus the interior finish materials. Therefore, if the interior retarder-and-finish permeance is 1 perm, the total exterior permeance should be 5 perms or more. Conversely, when it is

known that vapor-resistant finishes such as plywood or hardboard panels will be used on the exterior, a more effective retarder and installation method should be used.

In air conditioned buildings located in the humid areas shown in Figure 7.1.19, water vapor diffuses from the exterior to the interior. Therefore, in these areas the exterior portions of a wall should have a higher vapor resistance than the interior, and when vapor retarders are used they should be on the outside. Water vapor is then directed toward the interior rather than the exterior, as is normal in other climate zones, where it can be removed by the air conditioning system. The air conditioning system must be designed with sufficient dehumidification potential to accomplish this removal. Normal systems employed in other areas of the country may not be appropriate. Reheating of circulating air may be necessary. Ventilating air must be dehumidified as well to prevent introducing high outside humidity levels into a building.

■ CRAWL SPACES

When a crawl space underlies a heated or air conditioned area, it is generally more economical to limit heat loss (or gain) by insulating the foundation wall rather than the floor above. Under these conditions, there is no need to ventilate the crawl space to the outside, but the crawl space should be heated or cooled. Introducing nominal amounts of air from the supply side (or conducting air to the return side) of an air-handling unit will usually ensure adequate ventilation of a crawl space.

In closed crawl spaces not equipped with means of ventilation, moisture rising by capillary action from the ground can cause condensation problems. To stop the moisture at its source, buildings with either heated or cooled crawl spaces in all zones should be provided with a vapor retarder laid directly over the soil as a ground cover.

A typical vapor retarder for this purpose should be a material not subject to decay or insect attack. Typical materials are 4- or 6-mil polyethylene film, a coated 30-lb./sq. ft. or heavier roofing felt, or roll roofing. This retarder should be installed in the largest possible widths, with edges lapped 6 in. and taped or sealed (unless ballast is used) and the sides turned up approximately 6 in. against the foundation walls. A ballast of pea gravel or sand over the retarder is desirable to keep it in place and to prevent mechanical damage (see Fig. 2.3-1).

Ventilation to the atmosphere is recommended in all zones when the foundation wall is not insulated, heated, or cooled. At least four vents should be provided, with one not more than 3 ft. from each corner. They should have a total net free area not less than 1/1500 of the crawl space area. Vents should be placed as high as possible. Vents are not required if the crawl space is completely open on one side to a basement. If the crawl space area is greater than the basement, however, it should be separately ventilated by supplying or returning air.

In older buildings or where the crawl space is not cooled or heated, the ground cover can be omitted, provided the net free area of vents is increased 10 times, to 1/150 of the crawl space area. In the absence of a ground cover, more positive ventilation is required, and at least four vents should be provided, with at least one in each wall.

In Zone I, crawl spaces are usually heated and ventilation is limited. When a crawl space will be ventilated in the winter, it is necessary to insulate the floor above it to reduce heat loss and improve occupant comfort. During periods of cold weather, the air in the crawl space would have a low temperature and a low moisture content, so there would be vapor pressure outward from the building. To prevent condensation on the framing, a vapor retarder should be installed on the upper, warm side of the insulation. Normal retarder-faced blankets have nailing flanges on the same side as the retarder. If the insulation is installed from below the joists, special reverse flange blankets or retarder-wrapped blankets should be used (see Section 7.3). Wide polyethylene sheets laid over the subfloor or joists are most effective, but retarder-faced or retarder-wrapped blankets are acceptable.

ROOF AND CEILING ASSEMBLIES

For purposes of this discussion, roof and ceiling assemblies can be divided into three categories: *insulated framed systems*, *compact conventional systems*, and *protected membrane systems*. Unfortunately the industry is not in complete agreement about the use of ventilation and vapor retarders in each of these systems.

INSULATED FRAMED SYSTEMS In insulated framed systems, an air space separates the ceiling and insulation from the roof deck. Such assemblies may have the roof deck in contact with the framing (Fig. 7.1-24) or may have an attic space above the ceiling framing. Of the three roof and ceiling assembly categories, this one is the most likely to experience condensation problems. Not only are such roofs more susceptible to condensation, many of them are framed in wood, so the result is more damaging.

FIGURE 7.1-24 *Insulated framed roofing system.* *(Courtesy H. Leslie Simmons.)*

In most areas of the United States, the ceiling in an insulated framed system should have a continuous vapor retarder membrane on the warm side of the insulation. In the humid areas shown in Figure 7.1-19, however, a vapor retarder should be installed only if local experience so dictates. In these areas, condensation may form on the outside of a vapor retarder whose temperature is below the dew point of the outside air.

Ventilation of an insulated framed system is particularly important because the area above the ceiling is a natural depository of moisture from inside the building. Convection currents can carry moisture-laden air up wall cavities and into the roof space through openings inadvertently left at the top of partitions or even through the voids in concrete masonry units. Light fixtures and other equipment installed in ceilings can also create air and vapor leaks.

Ideally, all vapor leaks should be eliminated; in practice this is not possible, and a fair amount of moisture will penetrate past the ceiling. Fortunately, suitable ventilation details can dissipate this moisture safely to the atmosphere. In fact, ventilation alone can take care of all such moisture, except in the coldest part of Zone I, with winter design temperature of −10°F or lower.

In such systems having attic spaces, wind direction and velocity, as well as the vent type, greatly affect the distribution and volume of air movement in the attic spaces. Dead spots of minimum air flow create pockets of potential condensation. Therefore, ventilation based only on the ratio of vent opening to ceiling area may be inadequate. Vents should be selected and located for maximum air flow and uniform air distribution; generally, at least one-half the required vent area should be located at the ridge or upper half of the roof, the balance at the eaves.

The amount of ventilation required is 1/300 the area of the ceiling below when gable end ventilators can be used on both ends of the area. When gable end vents cannot be used, a vent area equal to 1/600 the attic area should be uniformly distributed at the eaves, and an additional 1/600 the area of the attic should be provided at the ridge. All attic spaces should be interconnected.

In older buildings that do not have vapor retarders, vent openings should have a net free area not less than 1/150 the ceiling area. This larger volume of ventilation may also be desirable in moderate and southern zones, regardless of the presence of retarders. The effectiveness of the ventilation system in removing both heat and moisture depends on the volume of air moved and the extent to which the air flow reaches into all parts of the void space.

Roof louvers (Fig. 7.1-25a) when used alone contribute little to effective air flow through an attic space. However, when placed in the upper half of the attic space and combined with soffit vents (Fig. 7.1-25b) in equal amounts, the air distribution is improved and air flow is almost doubled. The combined system also becomes more responsive to wind, increasing air volume in direct proportion to wind velocity.

FIGURE 7.1-25 *Typical roof ventilators: (a) roof louvers, (b) soffit vent, (c) gable end louver, and (d) continuous ridge vent.* (*a and c courtesy Easy-up Home Ventilation from The Solar Group; b and d courtesy Alcoa Building Products.*)

Continuous soffit vents (see Fig. 7.1-25b) produce substantial air flow regardless of wind direction, and the flow is directly proportionate to wind velocity. However, most of the air flow remains near the attic floor, indicating a need for vent openings in the upper roof or ridge for good circulation.

Gable end louvers (Fig. 7.1-25c) are most effective for wind perpendicular or diagonal to the louver face. However, air flow is one-half to one-third this amount when the wind direction is parallel to the

louver face, perpendicular to the ridge. Although the addition of soffit vents does not improve air flow for the ideal condition, under adverse wind conditions, the addition of continuous soffit vents increases air flow by 50%.

Continuous ridge vents (Fig. 7.1-25d) are unique in that they provide air flow due to temperature differences with roof slopes as low as 4/12. They provide about the same air flow at zero wind velocity as at 10 mph; however, air flow volume is several times larger with wind paral-

lel to the ridge as it is with the wind perpendicular. The combination of continuous ridge vents with continuous soffit vents produces the greatest volume and most uniform air flow under most wind conditions.

Ventilation of insulated framed systems with no attic space (see Fig. 7.1-24) should include outlets from every joist space, with either individual vents (see Fig. 7.1-23) or with continuous soffit ventilation on the sides perpendicular to the joists. Soffit ventilation can be provided with either prefabricated vents (see Fig. 7.1-25b) or job-built vent details (Fig. 7.1-26). A continuous 3/4-in.-wide screened opening is usually adequate for buildings of average width, but the net free area should be verified as discussed earlier.

COMPACT CONVENTIONAL SYSTEMS
Compact conventional systems may have either metal decks, as shown in Figure 7.1-27, or wood plank decks.

Because built-up roofing and single-ply roofing membranes, which are typically used with compact conventional roofing systems, are nearly perfect vapor retarders (0.00 perm), it is impossible for vapor to be diffused through them. A vapor retarder on the underside of insulation beneath such roofing actually can cause more problems than it solves, by trapping moisture within the insulation. If a vapor retarder is used in such systems, it should be located so that the calculated dew point will fall on the low-vapor-pressure side of the retarder. At any rate, the circumstances under which a retarder should be used are controversial.

FIGURE 7.1-26 *Continuous soffit vents can be used to ventilate both attic-type and atticless insulated framed roofs.*

FIGURE 7.1-27 *Compact conventional roofing system.* (Drawing by HLS).

Until recently, it was believed that a vapor retarder generally is not needed in a conventional compact system when the interior winter relative humidity averages less than 40%. A retarder may be needed when the interior relative humidity exceeds 40% and the average January temperature is 35°F or lower. Recently developed vapor drive maps are a better way to determine the need for a vapor retarder, however.

The 35°F January isotherm coincides approximately with the 0°F isotherm shown in Figure 7.1-18; humidities over 40% are encountered in public shower rooms, kitchens, laundries, and pool areas. In single-family structures, only pool areas merit special consideration; other high-moisture areas, such as baths and kitchens, usually are equipped with exhaust fans to draw off the moisture.

Unless a vapor retarder is used, it is not necessary to dissipate moisture with structural ventilation or venting. The insulation generally is absorptive enough to hold the vapor for short periods of time without condensation. The roof deck generally is permeable enough to permit dissipation of vapor back into the interior, when the vapor pressure is reversed.

However, if a vapor retarder is provided in cold climates and with high interior relative humidities, the possibility of trapping moisture between the retarder and the roofing is greatly increased. The moisture may be present during construction, or it may enter the insulation later through leaks from above or below. In any case, moisture cannot readily escape through the roofing or the vapor retarder by diffusion, and peripheral air leaks (venting) must be provided to permit dissipation of moisture.

Edge venting is a method of connecting roof insulation with the exterior atmosphere through built-in vent openings at the perimeter of the roof (Fig. 7.1-28a). Edge venting is adequate for roofs not exceeding 40 ft. in width; for wider roofs, stack venting as well as eave venting is required.

Stack venting consists of simple sheet metal stacks about 3 in. in diameter, reaching through the roofing membrane into the insulation (Fig. 7.1-28b). One stack should be provided for each 900 sq. ft. of roof area.

Unfortunately, vents can produce more problems than they solve. They themselves can admit water vapor and even free water from drifting snow into the roofing system. In addition, water leaks can occur where the vents are flashed into the roof. Vents may also act as chimneys carrying moisture-laden air into the roof system from faults in the vapor retarder.

PROTECTED MEMBRANE SYSTEMS A protected membrane roofing system is a compact system in which the roofing membrane is below at least a portion of the insulation (Fig. 7.1-29). The roofing lies on the warm side and acts itself as a vapor retarder. In most areas of the country, there is no condensation associated with this type of assembly when it has been properly designed and installed. However, in the humid areas shown in Figure 7.1-19, where the indoor vapor pressure may be lower than that outdoors, condensation may occur in such roofing assemblies over air conditioned buildings. Calculations are required to determine their suitability.

■ **SLABS ON GRADE**

Whether at grade or below grade (basements), slabs on grade are in direct contact with the soil, which often contains large amounts of vapor or free water. Although 4 in. of concrete has good resistance to vapor (perm rating of less than 1.0), the vapor pressure from the moist slab underside to the interior may be sufficiently high and prolonged enough to

FIGURE 7.1-28 *Insulated wood plank deck showing (a) edge venting and (b) stack venting. Insulated steel deck systems are similar.* (U.S. Forest Products Laboratory. Not subject to copyright in the United States.)

FIGURE 7.1-29 *Protected membrane roofing system.* *(Courtesy H. Leslie Simmons.)*

tarders: (1) the retarder should be as near as possible to the warm (heated in winter) side of the insulation or warm face of the assembly; (2) the retarder should be as continuous as possible, with a minimum of joints, breaks, openings, and air leaks; and (3) the retarder should be installed by a method suitable for the specific condensation hazard.

Since a vapor retarder is frequently an integral part of the insulation, effective vapor control generally depends on the proper selection and attachment of the insulation. Insulation and retarder should be fitted neatly behind piping and equipment. The flanges of retarder-faced insulating blankets should be snugly stapled about 6 in. on centers to framing members, to avoid air leaks.

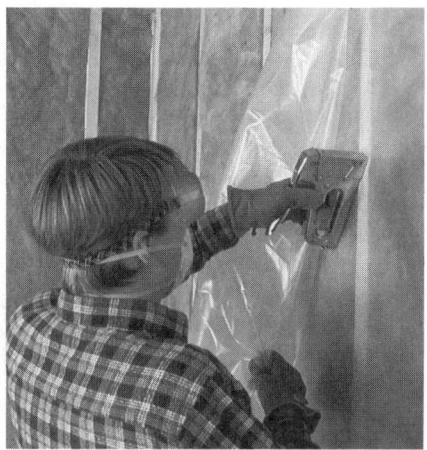

FIGURE 7.1-30 *Polyethylene vapor retarder being installed across an entire wall surface.* *(Photograph furnished by Owens-Corning Fiberglas Corporation.)*

cause large amounts of vapor to be diffused through the slab. Additional moisture, in vapor or liquid form, can seep through cracks and structural joints in the slab. Therefore, a separate vapor retarder should be placed under every building slab on grade. It may seem that a vapor retarder is unnecessary when a slab on grade is underlain by well-drained granular fill and is located where irrigation and heavy sprinkling are not practiced. Nevertheless, because a retarder can be installed at nominal cost during construction and because it is costly to correct moisture problems later, a retarder should always be placed under slabs of habitable rooms. The vapor retarder should have a permeance of less than 0.30 perm, be installed with 6-in. laps, and be fitted carefully around utility and other service openings.

Vapor retarder materials suitable for slab-on-ground construction include single-layer membranes, such as polyethylene film, and multiple-layer membranes, such as reinforced waterproof paper with a polyethylene film extrusion coated onto both sides. Polyethylene film used over gravel or crushed stone or under a reinforced slab should be at least 6 mil in nominal thickness; thinner, 4-mil film can be used over sand or firmly compacted soil.

7.1.3.3

Vapor Retarder Installation

Three fundamental principles should be observed in the application of vapor re-

■ WALLS

Gaps in insulation near the bottom or the top of a wall cavity encourage convection currents, which carry heat and vapor from the inside to the outside face of the cavity. Therefore, insulation should be pushed tightly against the back of sheathing and should fill the cavity from top to bottom.

Batts and blankets are made to fit a standard 14-1/2-in.-wide cavity resulting from 16-in.-on-centers wood stud or joist spacings. Nonstandard spaces less than 14-1/2 in. wide should be fitted with blankets cut 1 in. oversize to form a stapling flange on the cut side. Nonstandard spaces wider than 14-1/2 in. can best be protected by a separate retarder (Fig. 7.1-30) or by foil-faced gypsum board.

If the standard spacing of wood members in a structure is 12 or 24 in., pressure-fit insulation with a separate retarder should be used. Cracks around window and door frames should be stuffed with loose insulation and covered with a separate retarder.

Metal studs do not lend themselves to the use of retarder-faced blankets, due to the difficulty of maintaining a continuous retarder. When metal studs are used, retarders should be formed either with foil-faced gypsum board or with separate sheet retarders.

■ FLOORS

The perimeter of the floor assembly just behind the band joist in wood framing or

at the ends of steel joists is frequently overlooked when insulation and a vapor retarder are installed. This is particularly true when the floor is between two habitable spaces and no insulation is used in the joist cavities.

When a floor is over an open-air area or an unheated crawl space, insulation is likely to be required in the floor. Here too, care should be taken to provide thermal and vapor control near the perimeter. The vapor retarder should be on the upper (heated in winter) side of the insulation. When the blankets are installed from above, they can be readily stapled to the joists. When installed from below, either reverse flange insulation should be used, or the blankets should be placed with both retarder and flanges facing up, using wire supports below.

Retarder-faced blankets usually are not the best way to provide a vapor retarder in a metal-framed (joist) floor system. A continuous sheet retarder is likely to be more successful. When retarder-faced blankets are used in metal-framed floors, they can be supported with wire supports or mesh, but providing a continuous retarder using this type of installation is difficult.

■ ROOF AND CEILING ASSEMBLIES

Blankets in wood ceiling joists should be extended part way over the top plate but not enough to block eave ventilation.

When blankets are installed from below with recessed stapled flanges, the gap at the end should be stuffed with loose insulation. Insulation and vapor retarder should extend over dropped soffits in bathrooms and kitchens, to provide thermal and vapor control. Tightly wedged glass fiber or rock wool insulation above the soffit also isolates the attic area from the soffit space and prevents the spread of fire. Holes for mechanical equipment through top plates should be filled with loose insulation to minimize air and vapor leaks into the roof space.

A separate vapor retarder should be provided when poured insulation is used between ceiling joists.

7.2 WATERPROOFING AND DAMPPROOFING

Below-grade walls of buildings, tunnels, and pits and horizontal decks below paving or earth are subject to groundwater penetration. Waterproofing and dampproofing are used to prevent this penetration. *Waterproofing* is a membrane, coating, or sealer of one of several materials used in concealed locations to prevent water from entering such structures. Waterproofing will work even when the adjacent water is under hydrostatic head. Vertical waterproofing is usually used in conjunction with a foundation drainage system (see Section 2.3). Horizontal waterproofing systems also require a positive means of draining away water that penetrates the wearing surface.

Dampproofing is usually a bituminous coating applied to prevent materials from absorbing adjacent moisture or to prevent absorbed moisture from penetrating farther into an invaded system. Bituminous dampproofing is not appropriate when the adjacent water is under hydrostatic pressure. Bituminous dampproofing may be used on either the interior or exterior, but in either case it is usually concealed. Interior uses include interior surfaces of exterior above-grade walls; interior surfaces of the exterior wythe of exterior cavity walls; and for coating structural steel subject to high-moisture conditions.

7.2.1

WATERPROOFING

There are three basic types of waterproofing materials: membrane, clay, and cementitious.

7.2.1.1

Membrane Waterproofing

On vertical surfaces, a membrane waterproofing system usually consists of a membrane, a protection course, a drainage medium, and a layer of filter fabric (Fig. 7.2-1). Parging may also be required over a masonry substrate.

Several arrangements are possible on horizontal surfaces, but they all have similar characteristics to the system shown in Figure 7.2-2: a membrane, a protection course, a drainage medium (or space), a layer of filter fabric (when a drainage membrane is used), and a wearing surface.

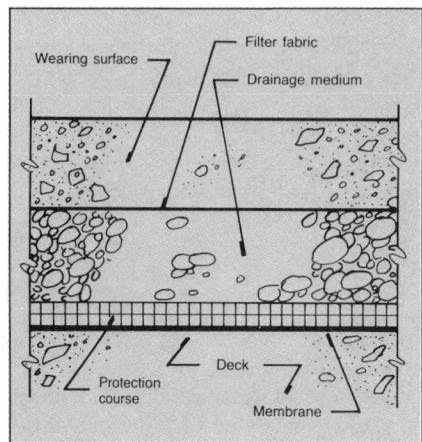

FIGURE 7.2-2 *Horizontal membrane waterproofing system.* (*Courtesy H. Leslie Simmons.*)

face. Some systems may require a layer of insulation, which sometimes also serves as the protection course. Some systems require a working slab beneath the wearing surface.

Membranes may be fluid-applied, sheet, or bituminous, which is a combination product. Fluid-applied membranes are usually thicker than sheet membranes and more foolproof when properly installed. In addition, it is easier to locate leaks in them. The materials used in fluid-applied membranes tend to be less expensive than sheet membranes, but they cost more to install because installation is more difficult. A major advantage of fluid-applied membranes is that they conform easily to irregular surfaces.

Connections between two portions of fluid-applied membrane are often easily made using the same material as in the membrane. Separate sheets of the same or

FIGURE 7.2-1 *Vertical membrane waterproofing system.* (*Courtesy H. Leslie Simmons.*)

compatible material may be bonded at intersections in sheet membranes. Flashing of the edges of fluid-applied and sheet membranes is essentially the same process. Sheet materials are sometimes finished in a reglet, but usually the edges of both should finish beneath a compatible flashing material. Metal flashing is not appropriate when the two materials must be bonded, because of the differential rates of thermal expansion between the waterproofing materials and the metal. Metal can be used, however, as a cap flashing where appropriate.

A major disadvantage of fluid-applied membranes is that they require considerable skill to obtain a uniform coating thickness. Also, these materials are incompatible with some substrate materials.

Most bituminous membranes are built-up courses of felts and bitumen. Some bituminous membranes, however, are totally fluid-applied systems consisting of bitumens reinforced with nylon or glass fiber filaments.

■ FLUID-APPLIED MEMBRANES

Most concealed fluid-applied horizontal waterproofing membranes and many concealed vertical waterproofing membranes used today are either polyurethane or hot rubberized asphalt. Fluid-applied silicone and neoprene systems are also available. These systems have wearing and weathering abilities beyond those of polyurethane or rubberized asphalt systems, but they are more expensive and therefore are most often used where they will be exposed, such as on roofs and decks. They may, however, also be used as concealed waterproofing where conditions warrant their use.

POLYURETHANE Fluid-applied polyurethane waterproofing is available either as one- or two-part materials. One-part materials are convenient to use and offer more quality control than two-part ones. The proportions of two-part materials can be adjusted by skilled applicators to suit the peculiarities of a particular project, and they tend to cure faster. The in-place performance of one- and two-part systems are comparable.

HOT RUBBERIZED ASPHALT Rubberized asphalt waterproofing is not as elastic as polyurethane and requires a thicker application. This is desirable over some substrates but not over others. The manufacturer's advice should be sought before selecting this system.

■ SHEET MEMBRANES

Sheet membrane materials include butyl synthetic rubber sheet, ethylene propylene diene monomers (EPDM) sheet, premolded bituminous sheet, rubberized asphalt sheet, and self-adhesive butyl sheet.

Butyl synthetic rubber sheet is uniform flexible sheet material about 60 mils thick. *EPDM* sheet is a flexible sheet material that is available in several thicknesses, the minimum being 1/16 in. *Bituminous* sheet waterproofing is a proprietary seven-ply material with a water vapor transmission rating of zero. *Rubberized asphalt* sheet waterproofing is a self-adhering membrane bonded to a polyurethane sheet, with a total thickness of about 56 mils. *Self-adhering butyl* waterproofing sheet consists of a butyl rubber core bonded to polyethylene plastic to form a sheet of about 60 mils thick. It requires an adhesive primer.

■ BITUMINOUS MEMBRANES

Bituminous waterproofing is usually applied on vertical below-grade surfaces, but in some instances it may be appropriate for horizontal surfaces as well. It is usually either hot coal-tar pitch or hot asphalt. Its number of plies, overall thickness, and content will vary, depending on hydrostatic head (Fig. 7.2-3). Because it has a proven record of successful use and is highly resistant to deterioration, coal-tar pitch should be used when the melting point can be kept low and there is no incompatibility between that material and adjacent materials. Hot asphalt is necessary when a higher melting point is needed.

Cold asphalt materials are less costly for small applications, and they reduce the danger of air pollution present with hot-applied materials, eliminate the danger of injury due to fire, and produce less objectionable odor. They are not, however, as reliable or as foolproof as the hot-applied materials, and they are suitable only when the hydrostatic head is of short duration and never exceeds 3 ft. Cold asphalt waterproofing is usually composed of two plies of glass fiber mat and three coats of emulsion, each applied at a rate of 3 gal./100 sq. ft.

Materials used in bituminous waterproofing include the following:

Coal tar–based priming oil conforming with ASTM Standard D 43

Coal-tar bitumen conforming with ASTM Standard D 450

Number 15 coal-tar felt conforming with ASTM Standard D 227

Cut-back type asphalt primer conforming with ASTM Standard D 41

Waterproofing asphalt conforming with ASTM Standard D 449 (generally Type I for vertical surfaces, Type II for unexposed horizontal surfaces, and Type C for exposed surfaces)

Number 15 asphalt-saturated organic felt conforming with ASTM Standard D 226

Bituminous-saturated organic fabric conforming with ASTM Standard D 173

FIGURE 7.2-3
Composition of Reinforced Hot Membrane*

Hydrostatic Head, ft.	Number of Fabric Plies	Number of Felt Plies	Bitumen, lb.	Number of Moppings
3	1	1	75	3
10	1	2	100	4
25	2	2	125	5
50	2	3	150	6
100	3	3	175	7

*Possible to use one ply fewer when a coated base sheet is used as an underlayment.

Glass fiber fabric conforming with ASTM Standard D 1668

Glass fiber mat conforming with ASTM Standard D 2178

Base sheet conforming with ASTM Standard D 2626

Five-pound rosin-sized building paper as a slip sheet

Either 1/8-in.-thick semirigid asphalt-saturated and coated laminations or felt; 1/2-in.-thick fiberboard protection board

■ INSTALLATION

Substrates should be installed and prepared, and fluid-applied, sheet, and bituminous membrane waterproofing materials should be applied in accordance with their manufacturers' recommendations.

Decks to receive waterproofing membranes should be sloped at least 1/8 in./ft., preferably 1/4 in./ft. or more. This slope should be present even after the substrate has deflected. The best surface to receive horizontal waterproofing is a structural concrete slab. Toppings are more likely to move and crack. Precast concrete decks should be covered with a reinforced topping 1-1/2 in. thick or thicker.

Concrete substrates should be cured the number of days required by the waterproofing manufacturer, usually at least 14 days. Substrates should be prepared by chipping or grinding off projections that would interfere with or puncture the waterproofing. Substrates should be cleaned of curing compounds, dirt, dust, and other materials that would interfere with placement or bond of the membrane.

A primer should be applied to substrates when so recommended by the waterproofing material manufacturer. Primed areas should be covered with waterproofing within 24 hours.

Waterproofing materials should be thoroughly bonded to vertical substrates and, where recommended by the manufacturer, to horizontal substrates as well. Seams and edges should be sealed and flashed to prevent water penetration. Expansion joints should be provided at each expansion joint in the substrate, where the substrate material changes, and in other locations recommended by the manufacturer.

Except over slabs on grade, horizontal waterproofing should be tested for leaks by covering it with 1 in. of water for 24 hours and observing the drop in water level, if any. Repairs should be made before a protection course is installed.

■ PROTECTION COURSE

Most applications of membrane waterproofing should be covered with a layer of a protecting material to prevent damage to the waterproofing during subsequent construction operations, including backfilling. This *protection course* is usually a layer of insulating material or other board materials. Each waterproofing material requires a different type of protection course. The waterproofing manufacturer should be consulted to determine the correct type.

7.2.1.2

Clay Waterproofing

Clay waterproofing is a layer of an expanding material called bentonite clay. When wetted, bentonite expands to between 10 and 15 times its dry volume. It is available loose for mixing with water and spraying in place. It is also available in dry sheets between cardboard, adhered to a drainage board, and adhered to a plastic sheet.

Bentonite is applied below grade to the exterior surfaces of building walls and tunnels. It should not be installed beneath a wearing surface or under earth or any other material that is less than 18 in. thick. A masonry substrate requires a layer of parging before bentonite is installed over it. Cracks wider than 1/8 in. must be filled in all substrates because bentonite cannot bridge wider gaps.

Bentonite is especially useful in tight locations where access is not available to install membrane waterproofing. This is often the case where existing construction occurs.

Bentonite sheets are usually nailed in place or adhered with bitumen. But installation methods vary, depending on hydrostatic pressures, substrate materials, edge conditions, and other factors. Therefore, the manufacturer should be consulted in each instance.

7.2.1.3

Cementitious Waterproofing

Some forms of cementitious materials are primarily decorative, even though they may have some waterproofing characteristics. One type that is specifically used as waterproofing is composed of portland cement, aggregate, and an acrylic or plastic admix. It may also contain pulverized iron fillers, in which case it is known as *iron oxide waterproofing*. An iron-oxidizing agent in the mix causes the iron to rust quickly and expand to fill the pores in the plaster. This type of waterproofing is used on the exterior or interior of below-grade walls, pits, sumps, and the like.

A second form of cementitious waterproofing is *hydraulic cement*. It is a compound of cement and rapid-setting nonshrinking hydraulic material. It is used to seal holes, cracks, and open joints and will set even in free-flowing water.

Nondecorative cementitious waterproofing tends to be brittle and unable to span even hairline cracks; therefore, its substrate must be smooth and crack-free.

7.2.2

DAMPPROOFING

Bituminous dampproofing is available in cold and hot-applied forms.

7.2.2.1

Hot-Applied Bituminous Dampproofing

Hot-applied bituminous dampproofing consists of (1) an asphalt primer conforming with ASTM Standard D 41 and dampproofing asphalt conforming with ASTM Standard 449, Type I or (2) a coal tar-based primer conforming with ASTM Standard D 43 and coal-tar bitumen conforming with ASTM Standard D 450, Type II or Type III.

Hot-applied bituminous dampproofing is used on the exterior of buildings, usually underground, and especially on large projects when the use of a hot system with solvents is permitted. It is seldom used on the interior, because of the problem of transporting and using hot products indoors. It cannot be used when the dampproofing must be able to transmit water vapor (to breath). Coal-tar systems are almost never used on the interior of buildings because they have a distinctively unpleasant odor that never completely disappears.

Hot-applied dampproofing may be applied directly on smooth concrete walls or on a coat of parging over masonry.

7.2.2.2

Cold-Applied Dampproofing

Cold-applied dampproofing may be either a coal tar– or an asphalt-based system. It may be used on the interior or the exterior and is especially useful on small projects where the use of hot products would be difficult or too costly. As mentioned earlier, however, coal-tar products are seldom used on the interior of buildings.

Asphalt-based cold-applied bituminous dampproofing systems consist of cut-back asphalt or asphalt emulsion materials. Both types are available as liquid, semifibrated, or heavy fibrated types. Fibrated types are reinforced with mineral fibers. Most of them also contain various fillers.

Dampproofing that must breath should be an asphalt emulsion type.

Coal tar–based cold-applied bituminous dampproofing systems consist of a coal-tar bitumen compound including ASTM Standard D 450, Type II coal-tar pitch and solvent compound, an inorganic fiber reinforcement, and various fillers.

Cold-applied dampproofing should be applied on exterior wall surfaces over smooth concrete or masonry that has been given a parging coat. The use of coal-tar products or fibrated asphalt products may permit application over a rougher concrete surface, but it is better to parge rough concrete surfaces as well.

Cold-applied dampproofing is also sometimes applied over the interior face of exterior above-grade walls or on the interior of the exterior wythe of masonry cavity walls. However, the Brick Institute of America says that mortar coatings are more effective in such locations and cau-

tions against the use of dampproofing there. The problem is that dampproofing applied on an interior surface allows water to enter brick, but prohibits its escape, thus contributing to efflorescence and spalling of the masonry when the water in it freezes.

The National Concrete Masonry Association recommends that an impermeable barrier, such as a cut-back asphalt dampproofing, be used on the interior face of concrete masonry unit walls. Under these circumstances, the exterior finish must be completely impermeable. Even oil- or alkyd-based paints may not be used there.

Liquid cold-applied dampproofing materials may be applied with brush or spray, semifibrated materials by brush or spray, and heavy fibrated ones by trowel. In every case, they should be concealed from view.

7.3 INSULATION

In some climates, using structural or finish materials with high thermal capacities is sufficient to adequately control heat gain or loss through a building's envelope and thereby maintain a healthy and comfortable indoor environment. In most cases, though, a building's structural and finish materials permit excessive heat flow, and insulating products should be added in the roof, walls, and floors to make them more energy efficient.

7.3.1

MATERIALS

The most common insulation products can be classified as (1) rigid board, (2) mineral fiber, (3) loose fill, and (4) foamed-in-place. Other materials that provide an insulating function are also sometimes used, although they are not all intended specifically to provide that function. These include insulated sheathing, fiberboard, reflective insulation, and structural insulating roof deck.

The trend today is toward manufacturers listing their thermal insulation products according to their ability to resist heat flow, which is called *thermal resistance* or *R-value*. Thermal resistance is the reciprocal of *thermal conductivity* (k-value). Some manufacturers still list their products according to their k-values, and some even list U-factors. R-values are the preferable listing, however, because they can be added directly to calculate the overall thermal resistance of an assembly. The overall R-value is the reciprocal of the assembly's *thermal coefficient* (U-factor), which is finally used to calculate the heat loss through the assembly.

The R-values for some insulation products are given in Figure 7.3-1. Note that in most cases the values listed are averages taken from several competitive products. Values for specific products may vary somewhat from those shown.

Acceptable R-values have increased significantly since the large jump in energy costs that occurred in the 1970s. Figure 7.3-2 shows some typical values. The

actual values used for a particular project should be calculated, taking into account such economic factors as costs of the insulation, the increased construction cost that may be necessary to provide more insulation, the cost of heating equipment, and the cost of building operation. In addition, some codes dictate minimum required R-values.

7.3.1.1

Rigid Insulation

Rigid insulations are usable in all parts of a building. They are made in board form for use in walls, foundations, and roofs and in special shapes for use in the cores of concrete masonry units. The majority of rigid insulations are polystyrene, polyurethane, polyisocyanurate, perlite, phenolic, cellular glass block, organic fiber, or glass fiber. Various composites made from more than one of the listed materials and occasionally other materials are also available.

FIGURE 7.3-1
Resistance (R) Values of Some Insulating Products[a]

Product	R-Value per 1-in. Thickness
Loose Fill	
Mineral fiber	2.20–3.00
Cellulose	2.80–3.70
Perlite	2.70
Vermiculite	2.13
Batts and blankets	
Mineral fiber batts	3.00–3.70
Glass fiber batts	3.00–4.00
Rigid	
Cellular glass	2.78
Expanded polystyrene, extruded	5.00
Expanded polystyrene, molded	3.60–4.40
Expanded polyurethane	6.25
Polyisocyanurate	5.60–7.70
Phenolic	8.30
Organic fiberboard	2.50–2.80
Mineral fiberboard	3.45
Glass fiberboard	3.80–4.80
Reflective	
Aluminum foil[b]	3.48
Foamed-in-place	
Polyurethane	6.25
Polyisocyanurate	7.00–8.00
Phenolic based	4.50

[a]Values are for 75 °F mean temperature.

[b]Thickness of foil not a factor; 3/4-in. air space on room side.

POLYSTYRENE

Polystyrene is a preformed nonstructural plastic cellular material. It is made in board form for use over structural roof decks, beneath promenade walking surfaces, at foundation perimeters, and in masonry cavities (Fig. 7.3-3). These boards are also used as wall sheathing when wind bracing is provided by other means.

Polystyrene board products are made in two forms: extruded and molded. *Extruded* polystyrene is a closed-cell material made in five densities, ranging from 1.35 to 3 lb./cu. ft. *Molded* polystyrene is an open-cell material made in four densities, which range from 0.9 to 1.8 lb./cu. ft. The extruded material then is more dense and stronger. Its closed-cell structure also

means that it is more resistant to moisture absorption and less permeable to water vapor than the molded form. The molded type is commonly called *beadboard* because it is made up of compressed pellets or beads.

Polystyrene insulations have similar properties to the other plastic foam insulations but lower R-values. They are sensitive to daylight and will deteriorate after prolonged exposure. Like other cellular plastics, they will give off carbon monoxide gas while burning. They do not, however, smolder or add a significant amount of fuel to a fire.

Polystyrene is also produced in tapered form. When used under a waterproof membrane, the tapered form permits positive drainage of a roof or promenade.

POLYURETHANE

Urethane foam is a cellular plastic product noted for its excellent resistance to heat transmission and deterioration from chemicals. When formed into rigid boards, its low-density closed-cell structure produces a lightweight and thermally efficient product. Urethane foam has a resistance per inch of about 6.25. Polyurethane boards are made with asphalt-impregnated felt bonded to both sides for use in roofing systems.

Urethane will burn and must be protected by some type of finish material if used for interior applications or on roofs. It may also expand after prolonged exposure to moisture.

POLYISOCYANURATE

Polyisocyanurate foam (a form of urethane) is manufactured in rigid boards faced with reflective aluminum foil, foil-kraft material, or glass fiber felt. It is used for wall sheathing and in roofing systems. It has a high R-value per inch (7.2). Like other urethanes, it must be protected from flame in interior applications.

PERLITE

Expanded perlite, mineral binders, and waterproofing agents are combined to make boards with weight characteristics similar to the plastic insulations. Unlike the cellular plastics, however, perlite boards are inorganic and will not burn or

smolder. However, the R-value per inch is only about 2.78, which is considerably lower than that of the plastics. Perlite should be kept dry because of its tendency to absorb water.

PHENOLIC

Phenolic insulation is a closed-cell foam core material with a density of about 2.5 lb./cu. ft. It is made in board form with a foil-kraft-liner or a glass fiber scrim. It can be used in walls and in roofing systems. It has a very high R-value (8.3) and will tend to hold that value better than the other foam plastic insulations, but it must be handled carefully because it tends to break easily. It is also reported to corrode metal.

CELLULAR GLASS BLOCK

Cellular glass is also used to produce a noncombustible and lightweight board insulation. Its closed-cell structure makes it impervious to water vapor and moisture absorption. Its R-value per inch is about 2.5, which is comparable to expanded perlite insulation.

This product is used on roof decks, exterior decks, and promenades and as perimeter insulation for slabs and foundations. It is available in thicknesses ranging from 1-1/2 to 4 in. and can be produced in tapered form on special order. When used under a waterproof membrane, the tapered form permits positive drainage of a roof or promenade. Cellular glass must be handled carefully because it breaks easily, even though it has a high compressive strength. Because it will deteriorate severely if exposed to freezing and thawing, it should be protected by a coating of bituminous compound. In severe conditions it should be coated on all faces and set with adhesive-filled joints. The manufacturer's advice should be sought and followed on this issue.

ORGANIC FIBER

Organic fiber rigid insulation is composed of wood fibers or other organic fibers and water-resistant binders. They are asphalt impregnated and chemically treated to prevent their deterioration and treated on the top surface to reduce asphalt absorption from roofing. They are used primarily

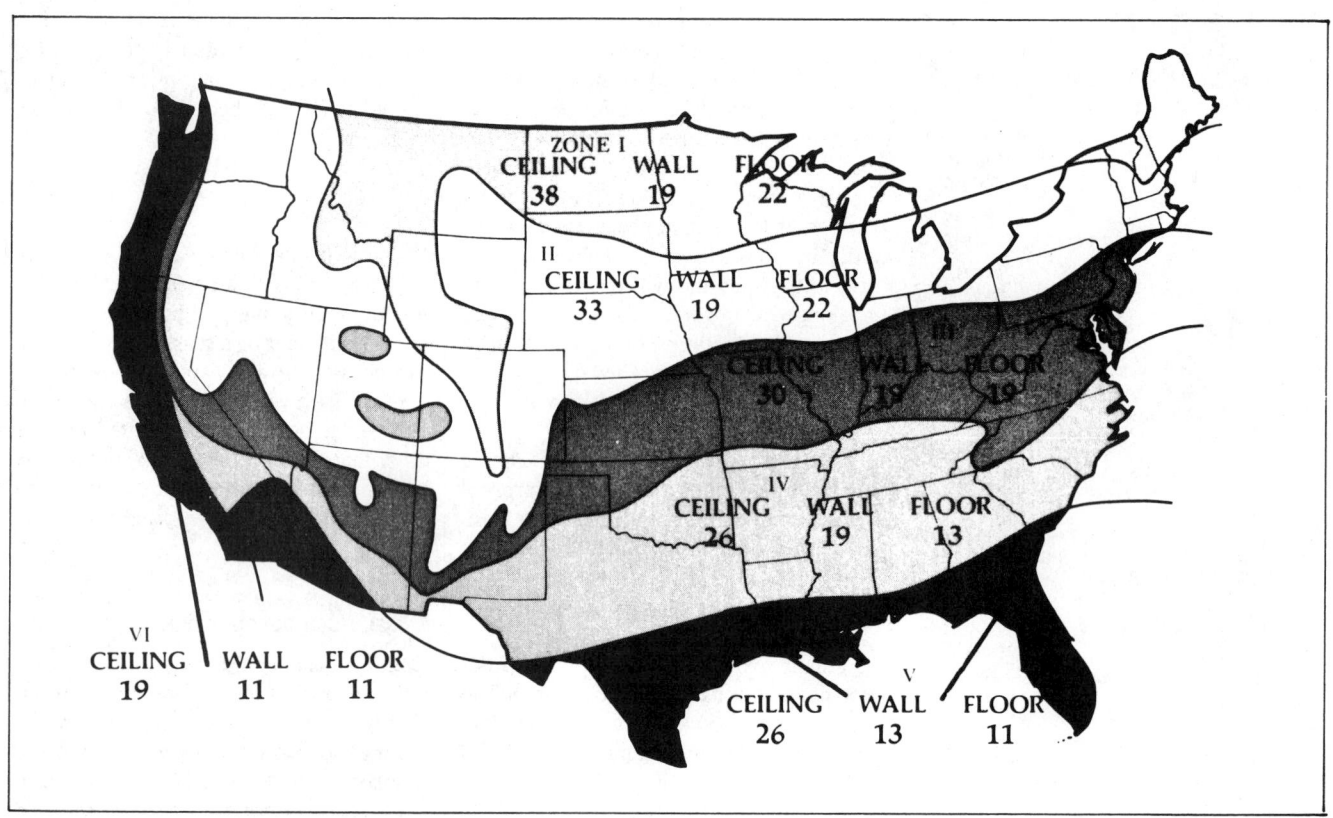

ZONE I
CEILING WALL FLOOR
38 19 22

II
CEILING WALL FLOOR
33 19 22

III
CEILING WALL FLOOR
30 19 19

IV
CEILING WALL FLOOR
26 19 13

V
CEILING WALL FLOOR
26 13 11

VI
CEILING WALL FLOOR
19 11 11

FIGURE 7.3-2 *Target insulation R-values.*

FIGURE 7.3-3 *Insulation is inserted into cavity as wall is built.* *(Courtesy Dow USA.)*

in roofing systems, and are available tapered to provide a positive roof slope. They are available in both regular and high-density versions.

■ GLASS FIBER

Rigid glass fiber boards are made from inorganic glass fibers bonded together by a resinous binder. They are used primarily in roofing systems. For this use, they are given an integrally bonded asphalt-impregnated felt or kraft paper top covering. They are available both flat and tapered and in varying thicknesses from 1/2 to 3 in.

Semirigid glass fiber boards and glass fiber batts and blankets are covered in Section 7.3.1.2.

■ COMPOSITES

Composites of urethane or polyisocyanurate and perlite, waferboard, or fiberboard take advantage of the high R-values of the plastic foams and the stability and higher

heat resistance of the other materials. Composites are also available with gypsum board and glass fiber boards.

7.3.1.2

Mineral Fiber Insulation

Mineral fiber insulation includes products made from fibers manufactured from rock, slag, or glass. Mineral fiber insulations produced in this country are made from either slag or glass. Products produced from slag are sometimes called rock wool or mineral wool but are more properly termed "semirefractory fiber insulations" because of their natural resistance to high temperatures. Because of the inorganic nature of both glass fiber and semirefractory fiber insulations, they are naturally resistant to fire, moisture, and vermin.

Mineral fiber insulation is made in rigid and semirigid boards and in blankets or batts. This section addresses semirigid boards, blankets, and batts. Rigid mineral fiber insulation used mostly for roofing is covered in Section 7.3.1.1.

▪ SEMIRIGID BOARDS

Semirigid mineral fiber insulation boards are available 2 ft. by 4 ft. and 1 to 3 in. thick. They can be cut to size in the field. They are used in curtain walls, between exterior wall furring members, as horizontal support panels for other insulation, and in other locations where a semirigid product is needed.

▪ BATTS AND BLANKETS

Batts and blankets are similar in appearance and composition but differ in length. Batts are 48 in. long; blankets are longer and are packaged in large rolls (Fig. 7.3-4). Batts and blankets are used in areas where it is not practical to install loose fill or where the attached foil or kraft paper facing is desired as a vapor retarder. Typical uses include as thermal insulation in walls and partitions, below floors in crawl spaces, above ceilings in attics and the like, and as sound insulation.

Batts and blankets are produced in several types: (1) wrapped with kraft paper on the edges and a vapor retarder on one or both sides; (2) faced with a vapor retarder on one side only; and (3) friction-fit without any covering, in which the interlaced fibers have sufficient resilience to remain upright in a cavity. Insulation with kraft paper or foil paper facing should be protected from flame.

Batts and blankets are manufactured to fit between wood studs placed at standard spacings of 16 and 24 in. on centers. Faced and wrapped products are provided with a stapling flange for secure installation. Product thickness ranges from 3-1/3 to 12 in. Thickness, however, is not a reliable measure of insulating effectiveness and should be accompanied by an R-number stamped on the wrapper or facing.

7.3.1.3

Loose-Fill Insulation

Fibers, granules, or chips are the usual components of loose-fill insulations. They are either poured or blown into attics or wall cavities. Most loose-fill insulation in use today in attics and walls is manufac-

FIGURE 7.3-4 *Batt insulation (a) is produced in panels. Blanket insulation (b) is made in rolls that can be cut to size.*

tured from mineral wool, glass wool, or cellulosic fibers. Loose perlite and vermiculite are used to fill the voids in concrete masonry units and the spaces between wythes in masonry walls.

Loose-fill insulation that is other than free flowing should not be used in walls or other closed cavities, since the large

fibers may catch on nails or other obstructions and leave uninsulated pockets.

▪ POURING AND BLOWING WOOL

The types of mineral matter included under this heading are slag and glass wool. Both are moisture and vermin resistant. They are less often used in walls than semirigid boards, batts, or blankets because of a tendency for the large clumps of fiber to catch on nails or other obstructions, leaving unfilled voids. The most frequent application is between ceiling joists in attics, where the wool can be poured from bags or blown by machine (Fig. 7.3-5).

▪ VERMICULITE AND PERLITE

Vermiculite and perlite are used in areas where they can be readily poured or blown, such as in masonry cavities (Fig. 7.3-6). Vermiculite is manufactured by expanding aluminum magnesium silicate, a form of mica. Perlite is an expanded volcanic glass. Both are noncombustible,

FIGURE 7.3-5 *Large areas can be insulated quickly with a pressurized hose.*

FIGURE 7.3-6 *Vermiculite and perlite can be poured directly from the sack into a masonry cavity.*

are vermin and moisture resistant, and do not compress in the cavity.

■ CELLULOSIC FIBER

Cellulosic fiber insulations are composed of recycled newsprint, wood chips, and other organic fibers. They can be installed by blowing or pouring into cavities and can be combined with adhesives for sprayed application.

The fibers are chemically treated to resist fire, minimize smoke contribution, and resist vermin. However, certain clearances adjacent to heat-producing fixtures must be maintained. The fibers have a neutral pH to prevent reaction with metal and will not decay after extended exposure to moderate dampness. Cellulosic insulation should be UL listed and labeled to ensure that the material is fire resistant.

7.3.1.4

Foamed-in-Place Insulation

In general, foamed-in-place insulations are polyurethane-, polyisocyanurate-, or phenolic-based materials. Each is created by a chemical reaction that expands a mixture of components by as much as 30 times. When cured, which takes from 24 to 72 hours, they solidify into a cellular plastic.

■ URETHANE FOAM

This foam has a high resistance per inch (R-6.25 or higher), is resistant to moisture and most chemicals, and has fire-retardant capabilities if chemically treated. It has

a wide variety of applications, including floor and wall cavities, roofs, and structural sandwich panels. The R-value per inch may vary with field conditions at the time of placement.

Urethane foam is particularly effective where a high R-value must be achieved in limited space or on irregularly shaped construction. Care must be taken not to overfill a cavity, because the foam expands to its ultimate volume after injection.

This type of insulating foam should be used in open cavities or where it is possible to accurately compute the amount of foam needed. Even when fire-retardant treated, it will burn and should be protected from flame by a fire-resistant finish.

■ POLYISOCYANURATE FOAM

Polyisocyanurate foam can be used in the same kinds of locations as polyurethane foams. It has a higher R-value than polyurethane (7 to 8) and a comparable density.

■ PHENOLIC-BASED FOAM

Phenolic-based foams are relatively new products that may be used in most of the same kinds of locations as polyurethane foams. They have a lower R-value (about 4.5) and a lower density than polyurethane.

7.3.1.5

Other Insulation Materials

Some other materials used as insulation are fiberboard, reflective insulation, insulated sheathing (see Section 6.6.5.1), and structural insulating roof deck (see Section 3.1.4).

■ FIBERBOARD

This general-purpose, low-cost insulation board sometimes is used as an interior wall finish in cottages and temporary buildings. However, because of their softness, fiberboard wall panels are seldom used in permanent buildings.

Fiberboard for wall use is available with flame-resistant, predecorated factory finish. It is manufactured 1/2 and 3/4 in. thick in tongue-and-grooved plank or panel form. The 1/2-in. thickness can be

used when the supporting members are spaced 16 in. on centers; 3/4-in. thickness is recommended when 24-in. spacing is used.

Fiberboard ceiling tile is typically stapled to furring strips or applied with mastic to a planed smooth surface. It can also be installed with wood or metal hangers to form a suspended ceiling.

■ REFLECTIVE INSULATION

Reflective insulation is composed of one or more layers of metallic foil. Reflective insulation is different from other insulations in that its insulating effectiveness depends on how shiny the reflective surface remains over time, whether a 3/4-in. airspace is provided on the warm side, and whether it is intended to reduce heat flow in or out of the structure. Therefore, reflective insulation should face an airspace of at least 3/4 in. and should remain free of dust or other materials that could reduce its reflective qualities.

7.3.1.6

Standards of Quality

Insulating products are manufactured to various standards, including those established by the American Society for Testing and Materials (ASTM) and those set by the General Services Administration and known as Federal Specifications. Figure 7.3-7 summarizes current standards of quality.

7.3.2

DESIGN AND INSTALLATION

The flow of heat through ceilings, walls, and floors can be controlled by one or more of the following techniques: (1) by constructing the assembly with materials that are inherently good insulators; (2) by using concrete, stone, solid masonry, or other dense materials that have the capacity to store heat and slow down heat flow; or (3) by adding thermal insulation to the assembly. The first and second methods are discussed in Section 11.3; the third method in this section. Whichever method is employed, it must be correctly installed to achieve the full benefit.

FIGURE 7.3-7
Summary of Insulating Products

Material	Standard	Description
Polystyrene	ASTM C 578	"Preformed Cellular Polystyrene"
Polyurethane	ASTM C 591	"Unfaced Preformed Rigid Cellular Polyurethane"
Polyisocyanurate	Fed. Spec. HH-I-1972/1	"Insulation Board, Polyurethane or Polyisocyanurate"
Phenolic	ASTM C 1126	"Faced or Unfaced Rigid Cellular Phenolic Thermal Insulation"

Perlite
  Loose Fill
  Board
	ASTM C 549
	ASTM C 728
		"Perlite Loose Fill Insulation"
		"Perlite Thermal Insulation Board"

Vermiculite	ASTM C 516	"Vermiculite Loose Fill Insulation"
Cellular glass	ASTM C 552 ASTM E 136	"Cellular Glass Block and Pipe Thermal Insulation" "Behavior of Materials in a Vertical Tube Furnace"
Organic fiber	ASTM C 208 ASTM C 209	"Cellulosic Fiber Insulation Board" "Test for Cellulosic Fiber Insulating Board"
Mineral fiber	ASTM C 553 ASTM C 612 ASTM C 665 ASTM C 687 ASTM C 726 ASTM C 764	"Mineral Fiber Blanket and Felt Insulation" "Mineral Fiber Block and Board Thermal Insulation" "Mineral Fiber Blanket Thermal Insulation" "Determination of Thermal Performance of Loose-Fill Building Insulation" "Mineral Fiber and Mineral Fiber, Rigid Cellular Polyurethane Composite Roof Insulation Boards" "Mineral Fiber Loose Fill Thermal Insulation"

7.3.2.1

General Recommendations

For conventionally heated and cooled buildings, the objective of heat control is to block the flow of heat through the ceilings, walls, and floors that surround habitable areas.

The building envelope should be designed to minimize heat flow through floors, walls, and ceilings. In most circumstances, except in the areas of high humidity discussed in Section 7.1.3.2, to most effectively block the flow of heat, the insulation should be placed near the inside of the building envelope. Insulated sheathing, of course, is an exception.

In addition, in some passive solar-heated buildings, the insulation is placed on the outside face of the wall, rather than the inside, as in conventional systems. The basic objective of such solar systems is to absorb solar heat in masonry or concrete walls during sunshine hours and to release (radiate) this heat slowly into the interior during sunless periods. To prevent rapid cooling of the warm masonry during sunless periods, insulation is placed on its outside face.

In exterior insulation and finish systems (see Section 7.4), the insulation is also placed on the exterior of the building.

Some insulation materials are manufactured from flammable materials or materials that produce toxic combustion discharges when they are heated or burned. These materials should be protected by fire-resistant finishes, such as gypsum wallboard. In fact, most building codes require this protection.

Polyurethane, polyisocyanurate, and polystyrene should not be exposed to daylight, to prevent ultraviolet deterioration. Molded polystyrene and expanded perlite boards also must be protected from contact with water, to prevent absorption. As mentioned earlier, cellular-glass boards must be protected from freezing.

7.3.2.2

Slabs on Grade

Heat loss of floor slabs built on ground is more nearly proportional to the perimeter than to the area of the slab. Heat loss to the ground below is minor. Therefore, it is important that, in areas where heat-ing systems are required, insulation be installed around the entire slab perimeter, to prohibit excessive heat loss directly to areas of low outside air or ground temperature. Uninsulated perimeter surfaces of slabs may be uncomfortably cold, produce condensation, and result in uneconomical heating efficiency.

This *perimeter insulation* should be (1) virtually nonabsorptive of moisture (noncapillary), (2) not harmed permanently by wetting or contact with wet concrete, (3) sufficiently rigid to resist damage and compression during placing of concrete, and (4) resistant to decay and insect damage.

Perimeter insulation should be at least 1 in. thick, preferably 2 in. Even greater thicknesses may be common in very cold climates. It should be provided around the entire perimeter of the slab and extend either vertically 24 in. below grade level (or below the frost line if it is deeper) or horizontally for a distance of 24 in. under the slab. Alternative methods of perimeter slab insulation are shown in Figure 7.3-8. Insulation may also be placed on the exterior of the foundation wall (Fig. 7.3-9).

FIGURE 7.3-8 *Interior locations for perimeter insulation at slabs.*

FIGURE 7.3-9 *Exterior location for perimeter insulation.*

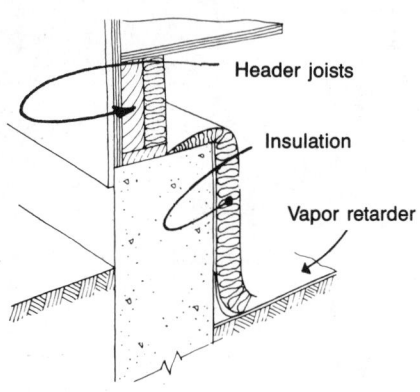

FIGURE 7.3-10 *Insulation location on the perimeter walls of a heated crawl space.*

FIGURE 7.3-11 *Insulation location over an unheated crawl space. Insulation may be stapled in place if installed from above or if reverse flange blankets are used. It may be held up by wires, as shown here, when installed from below. Wires can also be used when the framing is metal.*

FIGURE 7.3-12 *Sill sealer insulation is placed on top of foundation wall. Insulation should be fitted snugly against band joist.*

FIGURE 7.3-13 *Small cantilevered areas should not be overlooked but should be carefully insulated comparable with the remaining building.*

Glass fiber insulation board is sometimes used for perimeter insulation, but most perimeter insulation today is one of the plastic foam boards, with polystyrene being the most used.

7.3.2.3

Crawl Spaces and Floors

Floors over crawl spaces can be protected from heat loss by installing insulating material against the foundation walls when the crawl space is heated (Fig. 7.3-10). If the space is unheated, the insulation should be placed between the floor joists (Fig. 7.3-11). In either case, the vapor retarder should face the living space.

In basements and heated crawl spaces, sill sealer insulation should be used to prevent air infiltration between sill and foundation or between header joists and foundation (Fig. 7.3-12). In addition, the insulating sheathing should extend to the base of the plate.

The space behind the header joist should be covered with short pieces of flexible insulation. If the wall is insulated,

as in Figure 7.3-10, the end of the blanket can be extended upward and pushed against the header. Cantilevered floors should be insulated as shown in Figure 7.3-13. In every case, the outside edge of a floor cavity should be insulated with the

same insulation value and vapor retarder as the adjacent walls (Fig 7.3-14).

7.3.2.4

Walls

Heat transmission through wall assemblies can be reduced by selecting materials that

Thermal insulation

Vapor retarder

FIGURE 7.3-14 *Detail showing insulation and vapor retarder at the outside edge of a floor cavity.* (U.S. Forest Products Laboratory. Not subject to copyright in the United States.)

FIGURE 7.3-15
Estimated U-Values for 8-in. Hollow Concrete Masonry Walls

Wall Details	U-Value Based on Density of Concrete Used in Block				
	60 pcf	80 pcf	100 pcf	120 pcf	140 pcf
No insulation	0.32	0.34	0.38	0.43	0.55
No insulation, 1/2-in. gypsum board on furring strips	0.21	0.23	0.25	0.27	0.31
No insulation, 1/2-in. foil-backed gypsum board on furring strips	0.15	0.15	0.16	0.17	0.19
Loose-fill insulation in cores	0.12	0.14	0.18	0.21	0.35
Loose fill in cores, 1/2-in. gypsum board on furring	0.10	0.12	0.14	0.17	0.24
Loose fill in cores, 1/2-in. foil-backed gypsum board, furring	0.08	0.10	0.11	0.12	0.16
1-in. rigid glass fiber, 1/2-in. gypsum board	0.14	0.14	0.15	0.15	0.17
1-in. polystyrene, 1/2-in. gypsum board	0.12	0.12	0.12	0.13	0.14
1-in. polyurethane, 1/2-in. gypsum board	0.10	0.10	0.11	0.11	0.12
Loose fill in cores plus 1-in. rigid glass, 1/2-in. gypsum board	0.08	0.09	0.10	0.11	0.14
Loose fill in cores plus 1-in. polystyrene, 1/2-in. gypsum board	0.07	0.08	0.09	0.10	0.12
Loose fill in cores plus 1-in. polyurethane, 1/2-in. gypsum board	0.07	0.07	0.08	0.09	0.11
R-7 blanket insulation, 1/2-in. gypsum board, furring	0.09	0.10	0.10	0.10	0.11

are inherently good insulators and by adding insulation. The amount of thermal insulation installed in a wall depends on considerations of human comfort, the effect of total heat loss and heat gain, and, therefore, the cost of heating and cooling the building. Estimated savings in both heating and air conditioning costs may be used as a guide to determine the amount of insulation that can be installed economically. In a few mild climates, insulation may be necessary only to raise or lower the inside surface temperature of a wall to increase human comfort. Cold wall surfaces absorb radiant heat from the body and may produce a feeling of chill even though the inside air temperature is 70 °F or above. In most climates, insulation is added to every exterior wall.

Hourly heat loss or gain through an exterior wall is measured by its U-value. U-value is the amount of heat in British thermal units (Btu) transmitted in one hour per square foot of wall area, for a difference in temperature of 1°F between the inside and the outside air. Figure 7.3-15 gives U-values of 8-in. masonry walls with various combinations of finish and insulation. R-value is the reciprocal of the U-value (R = I/U). R-values are useful for estimating the effect of the individual components of a building section on the total heat flow, because R-values can be directly added. Figure 7.3-16 lists R-values for single-wythe concrete masonry walls.

Figure 7.3-17 gives R- and U-values for some typical concrete masonry walls.

The insulation values currently recommended in most locations are often difficult to accomplish using a single layer of insulation. For example, the cavity in a 2×4 wood stud wall will permit only 3-1/2 in. of blanket insulation with an R-value ranging from about 11 to about 17, depending on the insulation material and its density. A similar situation exists when exterior walls are framed with metal studs. In most areas of the country, total recommended wall R-values are 19 or higher. This leaves a gap between the R-value of the insulation alone and the required value for the entire wall. If insulation in the higher R-value range is selected, the other portions of the wall, such as interior and exterior finishes and sheathing, may produce sufficient total wall R-value. When such is not the case, one way to increase the insulation thickness in wood- or metal-stud-framed structures is to use 6-in. wall studs or staggered studs, but even this is not always enough. Insulations with higher R-values and other construction methods must be used when high R-values are needed. Batt and blanket insulation most likely will not be the proper selection in such cases.

■ RIGID INSULATIONS

When high R-values are required, polyurethane foam insulations may appear to be the proper choice, because of their relatively high R-values. However, all plastic

FIGURE 7.3-16
R-Values of Single-Wythe Concrete Masonry Walls[a]

Nominal Wall Thickness, in.	Insulation in Cells[b]	R-Value Based on Concrete Unit Weight				
		60 pcf	80 pcf	100 pcf	120 pcf	140 pcf
4	Filled	3.36	2.79	2.33	1.92	1.14
	Empty	2.07	1.68	1.40	1.17	0.77
6	Filled	5.59	4.59	3.72	2.95	1.59
	Empty	2.25	1.83	1.53	1.29	0.86
8	Filled	7.46	6.06	4.85	3.79	1.98
	Empty	2.30	2.12	1.75	1.46	0.98
10	Filled	9.35	7.45	5.92	4.59	2.35
	Empty	3.00	2.40	1.97	1.63	1.08
12	Filled	10.98	8.70	6.80	5.18	2.59
	Empty	3.29	2.62	2.14	1.81	1.16

[a]R-values do not include the sums of the effect of air film or surface conductance on the inside of the walls ($1/f_i = 0.68$) and on the outside ($1/f_o = 0.17$).

[b]Loose-fill insulation such as perlite, vermiculite, or others of similar density.

foam insulations are good vapor retarders and will prevent moisture that finds its way into a wall from escaping unless certain precautions are taken. In most climates, a continuous polyethylene vapor barrier on the room side of the assembly and venting of the wall cavity are highly recommended. Specific venting methods vary with the manufacturers, but their recommendations should be followed. Refer to Section 7.1.3.2 for a discussion about the placement of vapor retarders and venting in warm humid climates.

BATT AND BLANKET INSULATIONS

When doing so will not interfere with the installation of finish materials, insulation flanges in wood stud framing may be attached to the stud face to make a continuous vapor retarder. For stud spaces narrower than 14-1/2 in., standard barrier-faced blankets may be cut down to form a stapling flange. These flanges should then be stapled to the inside faces of the studs. However, facing the studs with an insulation flange will interfere with the installation of board products, such as gypsum board, and this type of application should not be used when such products are to be installed. Batts and blankets in stud spaces should touch the sheathing or siding and fit tightly against the top and bottom plates.

When a stud face must be kept clear, friction-fit insulation or other insulation without a vapor retarder is used, or the stud spacing is less than 16 in., the inside face of the wall should be completely covered with a vapor retarder such as polyethylene film (see Fig. 7.1-30) or aluminum foil. Necessary openings should be cut out later. The retarder material should be installed in the largest possible sheets to minimize vapor leakage. Retarder joints should be made over framing members.

Cracks around window and door openings should be packed with loose insulation and covered with polyethylene to reduce air infiltration (Fig. 7.3-18). Insulation should also be packed around pipes, ducts, and electrical boxes to prevent cold spots in the wall (Fig. 7.3-19).

Gaps in the insulation at the top or bottom of the cavity may allow convection currents that carry heat and vapor from the inside to the outside of the cavity.

INSULATING MASONRY WALLS

Masonry walls can be insulated by installing semirigid, batt, or blanket insulation between wood or metal furring strips (Fig. 7.3-20) or by applying mineral fiber or plastic foam board insulation directly to the wall (Fig. 7.3-21). Board insulation that is not inherently fire resistant should be protected with gypsum board.

Gypsum board may be adhesively attached to a mineral fiber semirigid board, but it should not be glued directly to plastic foam materials. If it is, heat will pass through the gypsum board in a fire and melt the plastic, permitting the gypsum board to fall away, leaving the foam board exposed. Instead, such gypsum board should be attached to furring.

Closed cell plastic boards have a high degree of resistance to moisture and vermin and can also be installed on the outer faces of masonry and frame walls. However, polystyrene and polyisocyanurate board insulation must be painted or otherwise protected from exposure to daylight. If used in an exposed area, all plastic boards should also have some type of protection to prevent dents and chips.

In general, the methods used to add insulation to masonry walls include (1) batt or blanket insulation placed in a furring space on the interior face of a wall (Fig. 7.3-22; also see Fig. 7.3-20); (2) rigid or semirigid insulation placed in or applied to the face of a wall (Figs. 7.3-23, 7.3-24, and 7.3-25; also see Fig. 7.3-21); and (3) fill insulation poured or blown into the cavity or core space of masonry (Fig. 7.3-26; also see Fig. 7.3-6).

Loose-fill insulation used in cavity walls and the cores of hollow masonry should possess these properties: (1) insulation should not transmit moisture from exterior to interior wythe; (2) insulating efficiency should not be impaired by retained quantities of moisture; (3) loose fills should support their own weight without settlement; (4) insulation should be inorganic or have comparable resistance to decay, fire, and vermin; and (5) granular fills should be pourable in lifts of at least 4 ft.

Types of loose fill meeting the above criteria are (1) water-repellent, expanded vermiculite and (2) silicone-treated, expanded perlite.

Rigid boards having the required properties include (1) expanded or molded polystyrene, (2) polyurethane, (3) polyiso-

FIGURE 7.3-17
R- and U-Values of Some Typical Concrete Masonry Walls

Components	Heat-Resistance Values, R							
	A	B	C	D1	D2	E	F1	F2
Surface film (outside)	0.17	0.17	0.17	0.17	0.17	0.17	0.17	0.17
8-in. hollow concrete masonry at 100-pcf density (cores open)	1.75			1.75	1.75			
Cores filled with bulk insulation			4.85				4.85	4.85
Concrete brick at 140-pcf density		0.44				0.44		
2-in. air cavity		0.97						
4-in. hollow concrete masonry at 100-pcf density		1.40				1.40		
2-in. bulk insulation in cavity						4.00		
Batt insulation between furring strips: 1 in.				3.70			3.70	
2 or 2-1/4 in.					7.00			7.00
1/2-in. gypsum board interior finish				0.45	0.45		0.45	0.45
5/8-in. plaster, lighweight aggregate		0.39				0.39		
Surface film (inside)	0.68	0.68	0.68	0.68	0.68	0.68	0.68	0.68
TOTAL resistance, R	2.60	4.05	5.70	6.75	10.05	7.08	9.85	13.15
U-value, 1/R	0.384	0.247	0.175	0.148	0.100	0.141	0.102	0.076

A — 8-in. hollow concrete block

B — Concrete brick, 4-in. hollow concrete block, 5/8-in. plaster

C — 8-in. concrete block, insulation in hollow cells

D1 — 8-in. concrete block, gyp. board with 1-in. insulation in furring space

D2 — 2-in. insulation used

E — 2-in. bulk insulation in cavity

F1 — 8-in. concrete block, gyp. & 1-in. insulation and filled cores

F2 — 2-in. insulation used

cyanurate, (4) cellular glass, (5) preformed glass fiber, and (6) perlite board.

■ MASONRY WALL CAPACITY INSULATION

In warm climates and in the summer months in cooler climates, the heat gain characteristics of walls are important. Heat gain in buildings with masonry walls depends on the *capacity insulation*, which is the heat storage capacity of the walls. The greater the capacity insulation, the lower the heat gain and consequently the cooler the inside temperatures. In general, heavier walls have a greater capacity insulation. They absorb and retain heat during the day and dissipate it to the outside during the cooler hours of the night. Of course, the wall must be thick enough so that its entire thickness does not reach a temperature greater than the interior space, in which case the wall will radiate heat into the cooler space. The more difference there is between the daytime and the nighttime outdoor temperatures, the more efficiently a wall will radiate the heat built up in it in the daytime.

7.3.2.5

Ceilings and Roofs

■ CEILINGS

Except in the humid areas discussed in Section 7.1.3.2, flexible insulation with vapor retarders should be installed in ceilings so that the retarder faces the living space. Loose fill or batt or blanket insula- tion without a vapor retarder should have separate polyethylene film placed below the insulation facing the living space.

For maximum effectiveness against vapor transmission the kraft vapor retarder facing on the insulation should overlap the faces of the framing members. If such a procedure is not appropriate because it interferes with tight installation of finish materials such as gypsum board, a separate vapor retarder should be installed.

The insulation should extend entirely across the top plate to prevent heat loss at the joint. If necessary, the junction between the insulation and the plate should be stuffed with loose insulation. When eave vents are used, the insulation should not block the air movement between the vent and the space above the insulation.

FIGURE 7.3-18 *Detail showing stuffing of cracks around a window with insulation and covering them with a vapor retarder.*

INSIDE
Vapor retarder

FIGURE 7.3-19 *Insulation should be packed behind ducts, pipes, and electrical boxes, filling the spaces between them and the exterior finish or sheathing.*

FIGURE 7.3-20 *Batts, blankets, or semi-rigid insulation can be applied between furring strips behind a wall finish. This illustration shows blankets or batts with a nailing flange attached to the face of the furring.*

Eave baffles or special cardboard ducts should be installed to prevent loose fill insulation from covering the vents and ensure air movement from the eaves to the general attic. Cellulosic insulation also should be kept from contact with potential sources of heat, such as electrical fixtures,

FIGURE 7.3-21 *(a) Insulation board is bonded to the interior face of masonry using adhesives; (b) gypsum board is bonded to glass fiber insulation using adhesives.* (Courtesy Dow USA.)

FIGURE 7.3-22 *Friction-fit glass fiber blanket insulation being installed between wood furring at a masonry wall.* (Photograph furnished by Owens-Corning Fiberglas Corporation.)

by providing at least 3 in. of clearance along each face of the fixture.

■ ROOFS

Heat transmission through a roof can be reduced by (1) use of optimum amounts

FIGURE 7.3-23 *Rigid foam insulation board installed in a cavity wall.* (Courtesy Brick Institute of America.)

FIGURE 7.3-24 *Glass fiber insulation being installed in a cavity wall. Brick ties hold insulation against inner masonry wythe, leaving an airspace between the outer wythe and the insulation.* (Courtesy Brick Institute of America.)

FIGURE 7.3-25 *(a) Insulation inserted in the face shell of specially formed blocks; (b) specially formed insulation inserted in cores of concrete masonry units.* (Courtesy Concrete Block Insulating Systems, Inc.)

of insulation; (2) use of shiny or light-colored roof materials to reflect heat;

FIGURE 7.3-26 *Granular insulation fills a brick cavity wall. It can also be used to fill the cores in concrete masonry units.* (Courtesy Brick Institute of America.)

(3) ventilation of attic spaces; (4) increasing the thermal mass to achieve a time lag in the heat flow; (5) use of louvered rather than solid overhangs to free heated air trapped beneath eaves; and (6) use of steeply pitched roofs on the windward side of the building to deflect cold winds and reduce the amount of roof exposure.

All the listed rigid insulations are used over structural roof decks as a base for roofing as well as to provide thermal resistance.

7.3.3

REINSULATION OF EXISTING BUILDINGS

Although the concepts of heat control discussed here apply to existing construction, it is not always possible to insulate an older structure in the same way and to the same level as a new building. Often it is not economical to add insulation to less accessible parts of a completed building, such as walls and floors.

In general, attics with no insulation should be insulated to the same levels as new construction. If the attic already has 6 in. of batt insulation or 8 in. of loose-fill insulation, the benefits of additional insulation should be examined on a case-by-case basis. Other parts of the building envelope, such as floors and walls, should be reinsulated to those levels when it is economically practical.

Whenever insulation is added to an existing structure, proper vapor control is essential. This involves installation of a vapor retarder facing the living space, provision for adequate cavity ventilation, or both. See Section 7.1 for a detailed description of proper techniques.

7.3.3.1

Foundations

Existing foundations of concrete or masonry can be insulated from either inside or outside the wall. Outside application is most frequently used when the interior already has a wall finish that would be covered by the new insulation.

■ INSIDE THE WALLS

Mineral wool batts or blankets or rigid plastic foam boards are the typical materials used inside walls.

Batts or blankets are installed by furring out the wall with wood strips and stapling the insulation to the strips (Fig. 7.3-27). The added wall thickness may not be acceptable for inhabited above-ground rooms, but it can easily be accommodated in basements that are not used as primary living space.

Mineral wool has the advantage of being fireproof if a paper backing is not attached. If the batts are paper-faced, they should be protected from flame by some type of wall covering (Fig. 7.3-28).

FIGURE 7.3-27 *If 2×4 studs are used, a wall can be insulated to R-11; 2×3 studs will accommodate only R-7 insulation.*

FIGURE 7.3-28 *A wall covering should be applied over paper-faced insulation.*

Rigid plastic boards can be installed on the interior wall between traditional 1×2 and 2×2 wood furring strips with the wall finish attached to the strips. There are also various proprietary systems that use mastics, adhesives, or metal channels in place of furring. Adhesives or mastics are thermally more efficient than the furring system and can be installed at a lower cost by contractors familiar with their use. However, these systems can be used only if the wall is straight and true, which basement walls often are not. In addition, mastics and adhesives can be used only on clean, dry, unpainted surfaces.

A major advantage of using plastic boards is that they often require less space to achieve a particular R-value. Extruded polystyrene has an aged R-value per inch of 5 or more. Mineral wool and glass fiber boards have an R-value per inch of about 4, batts of about 3. Thus, a required R-value of 19 would demand 5 in. of mineral wool or glass fiber boards, 6-1/2 in. of mineral wool or glass fiber batts, but only 4 in. of polystyrene. Some densities of glass fiber boards have R-values per inch of 4.4, which is close to, but still not as large as, that of extruded polystyrene. The real difference comes when the foam is polyurethane, polyisocyanurate, or closed-cell phenolic. With their R-values of 7 to 8, only 2-1/2 to 3 in. are required to reach the 19 or more in our example. A major disadvantage of foamed plastic board is that it must be covered with gypsum board or another fire-resistant material to be safe from fire.

■ OUTSIDE THE WALLS

Only the plastic foam boards, such as urethane, polyisocyanurate, and extruded polystyrene, are suitable for installation outside walls because of their resistance to moisture. The mastic or adhesive normally used to secure the board to the masonry requires a sound, unpainted masonry, stucco, or parged surface for a good bond. To be most effective, the insulation should extend below the frost line but not less than 2 ft. below ground.

Plastic foam boards are soft materials that can be dented and chipped if not protected by some harder material (Fig. 7.3-29). Several commercial systems available involve bonding a layer of glass fiber reinforcement to the insulation and

FIGURE 7.3-29 *Plastic foam board should be protected at the top edge with metal flashing and on its face with a hard water-proof covering.*

FIGURE 7.3-30 *Strips of insulation should be stuffed behind header joists.*

FIGURE 7.3-31 *Installing insulation in an existing wall is a three-step process: (a) open the wall cavity, (b) add insulation, and (c) close the wall cavity.*

applying a stuccolike finish. Polystyrene and polyisocyanurate are also sensitive to daylight and must be protected to prevent deterioration.

Outside-the-wall reinsulation is often less economical than that inside the wall, because of higher installation and material costs, but it has the advantage that installation can proceed without disrupting interior spaces or covering up interior finishes.

7.3.3.2

Crawl Spaces

Accessible crawl spaces can be insulated in the same way as for new construction. Special attention should be given to the area between the floor joists and behind the header joist—an often ignored or underinsulated spot (Fig. 7.3-30). If the crawl space is not accessible, it can be insulated on the outside of the wall, as described for basements.

7.3.3.3

Walls

Three basic types of walls—conventional stud, brick veneer, and masonry cavity—are most economically insulated at the time of construction. They can be effectively insulated later only if there was no insulation installed initially and if there are no major mechanical or electrical obstructions in the wall.

■ STUD WALLS

The thermal efficiency of stud walls can be improved by adding insulation to the stud cavity or to the exterior wall face before adding new siding.

Stud cavities are typically reinsulated with blown mineral wool, cellulose, or foamed-in-place plastic foam.

Before the insulation is added, a wall should be checked for existing insulation by removing the face plate of a wall receptacle. If no insulation is present, the cavity should be probed with a plumb-bob or other weight to detect obstructions that might cause blockages. The insulation should be blown or squirted into the wall through holes drilled between the studs (Fig. 7.3-31). Exterior installation is usually preferable, because most often siding can more readily be removed and replaced than can parts of the interior finish.

Exterior re-siding can provide an opportunity for adding plastic foam sheathing board under the new finish. The simple procedure requires only that the insulation be attached to the existing siding before new siding is applied (Fig. 7.3-32). Special metal channels accommodate the extra wall thickness around door and window frames. Although this procedure is not as thermally beneficial as fully insulating the stud cavity, the addition of 1-in. rigid polystyrene board, for example, can add 3.57 to the total R-value of the wall.

■ BRICK VENEERS

Insulation can be added to brick veneer walls by blowing or foaming. It is often difficult, however, to gain access to a stud cavity unless the fascia board or a few bricks near the eaves can be removed to expose an opening into the cavity.

■ MASONRY CAVITY WALLS

Masonry cavity walls are rarely reinsulated because of the difficulty of gaining access to the cavity. However, it may be possible to remove several bricks or a fascia board and squirt the insulation into the cavity. In addition to plastic foams and blown mineral wool, vermiculite and perlite can be used for such applications.

7.3.3.4

Attics

Attics are often the most economical area to reinsulate because normally they are accessible and can account for a substantial part of the building's heat loss or gain. Where the insulation is placed usually

FIGURE 7.3-32 *Foam insulation board completely covers the existing walls, thus reducing air infiltration as well as thermal transmission.*

FIGURE 7.3-33 *Cavities should not be over-filled to the extent that the insulation is compacted when the floor boards are reinstalled.*

FIGURE 7.3-34 *Insulation over an existing floor should be fitted snugly together to prevent thermal leaks.*

FIGURE 7.3-35 *A layer of insulation should be provided between heated or air conditioned spaces and unheated or unconditioned spaces.*

depends on whether the attic has been left unfinished or whether finish materials have been added to floors, walls, and ceilings.

■ UNFINISHED ATTICS

When the attic is unfloored, loose-fill insulation can be blown into the joist cavities, or mineral wool batts can be placed between the joists in the same way that new attics are insulated. If the attic already has some insulation, additional loose fill can be blown or poured over the old, or another layer of unfaced batts can be added. If the old insulation was wrapped or if the new insulation is faced, the kraft paper or aluminum foil facing on the top face should be punctured at 12-in. intervals. This prevents moisture condensation within the insulation. Condensed moisture could reduce insulating effectiveness and cause decay of structural members.

If the attic is floored, it is often possible to remove several boards and blow insulation under the remaining floor before replacing the boards (Fig. 7.3-33). If the space is not used for storage, unfaced batts or blankets can be laid directly on the floor (Fig. 7.3-34). If there is a possibility that the space will be finished and heated

in the future, it may be preferable to attach the insulation to the sloping rafters.

■ FINISHED ATTICS

A finished attic usually has three parts that may need insulation: upper ceiling, sloping ceiling, and outer attic (Fig. 7.3-35).

Upper ceilings can be insulated like unfloored attics, with either blown loose-fill insulation or flexible batts or blankets if the space is accessible.

Proprietary systems are also available that provide insulation and ceiling finish

in one integral panel. In such systems, mineral wool batts or blankets are attached to the back side of typical fiberboard ceiling panels and fastened to the ceiling with metal grids. They typically require 3 in. of ceiling clearance and provide an R-value of about 12.

Sloping ceilings can be insulated if either the outer attic or the upper attic is accessible, to permit stuffing the rafter cavities with loose fill or batts or blankets. It may also be possible to install the thermal ceiling panels described above. Such panels can be applied to the face of existing sloping or cathedral ceilings if the panel thickness does not reduce room dimensions below acceptable levels.

Outer attics can be insulated either at the knee wall and outer attic floor or in the sloping rafters. It is generally preferable to insulate the knee wall and outer attic floor when there are eave vents (or eave vents can be installed) and if the outer attic is sealed from the living space and is not to be used for storage.

7.4 EXTERIOR INSULATION AND FINISH SYSTEMS

Exterior insulation and finish systems (EIFS) are often called synthetic stucco, because they look like stucco. Most of them consist of a layer of a reinforced finish (lamina) over a layer of insulation, which is fastened to a layer of sheathing applied to furring or framing. Other configurations are possible, however, and many have been used since EIFS were introduced into this country from Europe in the 1960s. EIFS are available either as field-applied systems or prefabricated into panels.

There is no standard for EIFS published by a nationally recognized standards-setting body such as ASTM or ANSI. The only overall standards for them are those published by the Exterior Insulation Manufacturers Association (EIMA). These standards are "EIMA Guideline Specification for Exterior Insulation and Finish System Class PB" and "EIMA Guideline Specification for Exterior Insulation and Finish System Class PM." There is also an "EIMA Test Method And Standard 101.86, Standard Test Method for Resistance of Exterior Insulation Finish Systems to the Effects of Rapid Deformation (Impact)."

FIGURE 7.4-1 *Typical Class PB EIFS.* *(Courtesy H. Leslie Simmons.)*

7.4.1

SYSTEM CLASSES

EIFS are classed by EIMA as either Class PB or Class PM. Both classes consist of a textured protective finish, a reinforced base coat, a layer of insulation, a layer of sheathing, and necessary accessories. The configuration and materials, however, are slightly different (Figs. 7.4-1 and 7.4-2).

7.4.1.1

Finish Coats

Both classes use either a polymer-based or polymer-modified finish coat. There are many colors and texture combinations available. Class PB finishes are, however, softer and more flexible than PM finish coatings. For this reason and because of the difference in their insulation material, PB systems are sometimes called *softcoat* systems and Class PM systems are called *hardcoat* systems. There is no mention in the EIMA documents about the composition of the finish coat. Some coating mixes contain random fibers; others do not. Some contain sand.

Some finish coat materials may reemulsify if allowed to become saturated or stand in free water. Therefore, horizontal surfaces should be sloped to provide adequate drainage.

7.4.1.2

Base Coats

In PB systems the base coat varies from 1/16 in. to 1/4 in., depending on the man-

FIGURE 7.4-2 *Typical Class PM EIFS.* *(Courtesy H. Leslie Simmons.)*

ufacturer. It is typically reinforced with a glass fiber mesh embedded at the time of installation.

In Class PM systems base coats vary from 1/4 in. to 3/8 in. depending on the number of layers of reinforcement. A typical method of reinforcing a base coat for Class PM systems is to fasten a layer of wire lath or glass fiber mesh to the insulation. The base coat is then applied over the reinforcement. In some, but not all, systems, another layer of reinforcement (usually glass fiber mesh) is embedded in the base coat as it is applied. Different weights of glass fiber mesh are selected according to the abuse expected for different segments of an installation.

Additional reinforcement should be provided at the corners of openings, such as windows and doors, in both classes.

There is no mention in the EIMA documents about the composition of a base coat.

7.4.1.3

Insulation

The type of insulation varies with the manufacturer. Most Class PB systems use molded rigid cellular polystyrene. Most Class PM systems contain extruded rigid cellular polystyrene, which is more expensive, less vapor permeable, and more dense than the molded product. However, some Class PM systems have molded polystyrene.

Traditionally, the insulation in Class PB systems was adhesively attached to the sheathing, and that in Class PM systems was mechanically fastened. Today, however, some Class PB systems are mechanically attached, and some are attached both mechanically and with adhesives. Conversely, the insulation in Class PM systems is always mechanically fastened.

7.4.1.4

Sheathing

Failures and fear of failures have created some controversy about the type of sheathing that should be used in EIFS. The traditional sheathing for both Class PB and Class PM systems has been gypsum, because it is relatively inexpensive, fire resistant, light, and thin. Unfortunately, failures have occurred in adhesively applied EIFS that relate to the adhesive interface with the sheathing. EIFS manufacturers claim that these failures are small in number and are due to failures in the adhesives themselves.

Gypsum sheathing represents a small portion of the cost of an EIFS wall system, but some gypsum products manufacturers and the Gypsum Association, apparently fearing that such systems will fail because the paper facing delaminates from the core of their sheathing when used in these systems, have taken the position that they will not be responsible for the use of their products in adhesively attached EIFS systems. One sheathing manufacturer has specifically stated that its gypsum sheathing is not endorsed for use in EIFS other than with insulation that is mechanically fastened through the sheathing into the support system. The Gypsum Association's GA 252 states that gypsum sheathing can be used with EIFS having either adhesive or mechanical fastening of the insulation, but the "performance of these systems and recommendation of the proper method of attachment are the sole responsibility of the EIFS manufacturer."

Other types of sheathing can also be used in EIFS, but they are more expensive than gypsum sheathing and are not likely to replace the gypsum product in spite of the current controversy.

The sheathing used should be selected, installed, and protected as recommended by its manufacturer and the manufacturer of the EIFS. Conflicts between the two should be resolved before a system is installed.

7.4.2

CHARACTERISTICS OF EIFS

The two basic EIFS (PB and PM) vary in their appearance and performance characteristics. Failures may occur for several reasons related to failure to recognize the characteristics of the system being used and to accommodate them in the design and installation of the system.

7.4.2.1

Vapor Permeability and Condensation

Both classes of EIFS have protective coatings that are vapor permeable. This permeability is reduced considerably, however, in systems with a second layer of reinforcement added to increase impact resistance. EIFS systems should always have a complete vapor retarder on the interior. Vapor that enters the system in spite of the retarder can cause the coating to separate from the insulation, the insulation to separate from the sheathing, or the sheathing to deteriorate. This condensation can also cause loss of thermal efficiency if it saturates the insulation either in the EIFS or in the wall behind it.

Both types of polystyrene insulation act as a vapor retarder. However, extruded polystyrene is a much better vapor retarder than the molded product. In addition, the extruded material has a much lower rate of absorption, which may help reduce loss of insulation effectiveness in the event that condensation does occur.

To help prevent condensation from forming, the dew point should be calculated to fall within the cellular plastic board insulation in an EIFS.

7.4.2.2

Cracking

The finish coats on Class PM systems are comparatively brittle and the insulation usually employed (extruded polystyrene) more rigid than the materials used in a typical Class PB system. Therefore, Class PM systems must be divided into panels by control joints, much as stucco does. Class PB systems do not require such division and need joints only where expansion or control joints are necessary in the substrates.

7.4.3

SELECTION OF A SYSTEM

The building, including its construction, use, environment, and construction budget, should be analyzed to determine the most effective EIFS to use. EIFS are proprietary products, and each is different from the next, even when in the same class. It is necessary either to name a specific product and design around it or to specify the required characteristics. The physical properties used include a system's accelerated weathering characteristics, resistance to wind-driven rain, salt-spray resistance, absorption-freeze resistance, mildew resistance, abrasion resistance, water vapor transmission, impact resistance, and negative wind load performance.

7.5 BITUMINOUS ROOFING PRODUCTS

Four basic materials are used in the manufacture of bituminous roofing products. These are (1) a reinforcing base that is either organic dry felt or glass fiber mat; (2) bitumen (asphalt or coal tar); (3) mineral stabilizers; and (4) either fine or coarse surfacings.

In general, the manufacture of organic-felt bituminous roofing products includes processing cellulose fibers into dry felt, saturating and coating the felt with asphalt or coal tar, and then, depending on the type of finished product, surfacing the coated felt with selected mineral aggregates. When glass fiber mat is used as a reinforcing base instead of organic felt, the saturating process is bypassed. The mat goes directly from a dry looper into a coater.

7.5.1

RAW MATERIALS

The raw materials and some of the processing steps required to manufacture finished products are diagrammed in Figure 7.5-1.

7.5.1.1

Dry Organic Felt

Dry organic felt, used as a base for making reinforcing felts for built-up roofing and waterproofing, underlayment, smooth and mineral-surfaced roll roofing, and asphalt shingles, is made from combinations of cellulose fibers, such as those derived from rags, paper, and wood, on a machine similar to the type used for making paper (Fig. 7.5-2). The fibers are prepared by various pulping methods, then blended and proportioned to produce felt with the necessary weight, tensile strength, absorptive capacity, and flexibility required to make a suitable roofing product. The felt must be able to absorb from one and a half to two times its weight in asphalt saturants and be strong and flexible enough to withstand strains placed on it during the manufacturing process.

FIGURE 7.5-1 *Raw materials and processing steps used to make asphalt roofing products.* *(Courtesy Asphalt Roofing Manufacturers Association.)*

As the felt comes off the end of the machine in a continuous sheet, it is cut into specified widths and wound into individual rolls from 4 to 6 ft. in diameter, weighing up to a ton or more.

7.5.1.2

Glass Fiber Mat

Dry glass fiber mats are used as a base for reinforcing felts for built-up roofing and waterproofing, coated smooth rolls and mineral-surfaced roll roofing, and asphalt shingles. To make glass fiber mats, sand, soda ash, and limestone are first combined to make chopped strand glass fibers. These fibers then are mixed with a binding agent and cured to produce a mat with the thickness, tensile strength, tear strength, weight, and flexibility required to produce roofing and waterproofing products.

7.5.1.3

Bitumens

The bitumens used in bituminous roofing products are asphalt and coal tar. They are used to saturate and coat dry organic felts and coat glass fiber mats in the manufacture of roofing felt and to weld or fuse roofing or waterproofing felts together to

FIGURE 7.5-2 *Flow diagram of a typical felt mill.* *(Courtesy Asphalt Roofing Manufacturers Association.)*

form a built-up roof or waterproofing membrane. Both asphalt and coal tar bitumens are thermoplastic, which means they become more fluid when heated and return to their former solid state when they are cooled.

A bitumen's equiviscous temperature (EVT) is the optimum temperature at which it should be used. Each asphalt and coal tar bitumen has its own EVT. The lowest type number has the lowest EVT.

ASPHALT

Asphalt used to make roofing and waterproofing products, known as asphalt flux, is a petroleum product obtained from the fractional distillation of crude oil. Asphalt flux is processed to produce roofing grades of asphalt called saturants or, when combined with mineral stabilizers, coating asphalts. Both forms combine with dry felt to make asphalt roofing or waterproofing membranes.

The four types of asphalt used in built-up roofing are classified in ASTM D 312 as Type I, *dead level asphalt*; Type II, *flat asphalt*; Type III, *steep asphalt*; and Type IV, *special steep asphalt*. Type I is limited to slopes of 1/2 in./ft. and lower; Type II may be used on slopes up to 1-1/2 in./ft.; Type III up to 3 in./ft.; and Type IV up to 6 in./ft. However, steep asphalts are not as durable or as easy to apply as the lower-slope types and should be used only when the slope is higher.

The preservative and waterproofing characteristics of asphalt reside largely in certain oily constituents. In the making of roofing products, the body of the highly absorbent felt sheet first is impregnated (saturated) to the greatest possible extent with saturants that are oil-rich asphalts. The saturant then is sealed in with an application of a harder, more viscous coating

asphalt, which in turn can be further protected by a covering of opaque mineral granules. A primary difference between saturants and coating asphalts is the temperature at which they soften. The softening point of saturants varies from 100 to 160°F; that of coatings is as high as 260°F. Saturant asphalt is not used with glass fiber mat. The coating mixture is used to completely surround the glass fibers and to provide a layer of coating on both sides of the mat.

Asphalt flux is also the base material used to make asphalt plastic cement, quick-setting roof adhesives, asphalt primers, other roof coatings, and adhesives.

COAL TAR

Coal tar used to make roofing products is a distillation of bituminous coal. It is considered basically superior to asphalt for use in low-slope roofing. It is most effective at a slope of 1/4 in./ft. and can be used on slopes up to 1/2 in./ft., but no higher.

The two types of coal tar recommended for use in built-up roofing are classified in ASTM D 450 as Type I, *coal-tar pitch*, and Type III, *coal-tar bitumen*. Type III has a higher softening point than Type I.

7.5.1.4

Mineral Stabilizers

Finely ground minerals called stabilizers, such as silica, slate dust, talc, micaceous materials, dolomite, and trap rock, when combined with coating bitumens control hardness, elasticity, adhesion, and weathering. Coating bitumens that contain stabilizers resist weather better, are more shatterproof and shockproof in cold weather, and significantly increase the life of roofing products.

7.5.1.5

Surfacings

The surfaces of bituminous roofing products may be either smooth or surfaced with coarse mineral granules.

FINELY GROUND MINERALS

Powders, usually talc and mica, are dusted on the surface of smooth roll roofings and on the backs of mineral-surfaced roll roofings and shingles to prevent layers in packages from sticking together. They are not a permanent part of the finished product and gradually disappear from exposed surfaces after application.

COARSE MINERAL GRANULES

Coarse granules are usually colored-slate or rock granules (either natural or colored by a ceramic process) that are used on shingles and on certain roll products. These mineral-surfaced products have increased weather and fire resistance and provide a wide range of colors and color blends. To protect the underlying bitumen from the impact of light rays, the granules should be opaque, dense, and well graded for maximum coverage.

7.5.2

MANUFACTURE OF ROOFING

Raw materials are processed into finished roofing products in a continuous sequence (Fig. 7.5-3). The principal steps are described below.

7.5.2.1

Dry Looper

A roll of dry felt or glass fiber mat installed on the felt reel is unwound onto a dry looper. This looper provides a reservoir

FIGURE 7.5-3 *Asphalt roofing products are produced in a continuous process, as shown in this diagram.* (*Courtesy Asphalt Roofing Manufacturers Association.*)

of felt material that can be drawn on by the machine as circumstances demand, eliminating stoppages when a new roll must be put on the felt reel or when imperfections in the felt must be cut out.

7.5.2.2

Saturation of Felt

The felt is then subjected to a hot saturating process, which eliminates moisture and fills the felt fibers and intervening spaces as completely as possible with saturant. Glass fiber mat does not undergo this saturating process; instead, coating bitumen is used to fill the spaces between the glass fibers.

7.5.2.3

Wet Looper

Following the saturating process, an excess saturant usually remains on the surface of the sheet. Therefore, the sheet is held for a time on a wet looper so the natural shrinkage of the bitumen upon cooling will draw the excess into the felt, resulting in a high degree of saturation. At this point, uncoated products, such as saturated underlayment felt, move directly to the cooling looper and onto the winding mandrel for final packaging.

7.5.2.4

Coater

Material to be coated continues to the coater, where coating bitumen is applied

to both the top and the bottom surfaces. Coating thickness is regulated by rollers, which can be brought together to reduce the amount of coating or separated to increase its thickness.

The final weight of the product is controlled by the roofing machine operator. Often machines are equipped with automatic scales that weigh the sheets during manufacture to warn the operator when the material is running over or under weight specifications. When smooth roll roofing is being made, talc or mica is applied to both sides of the sheet by spreading and pressing it through a pressure roll.

7.5.2.5

Mineral Surfacing

When mineral-surfaced products, such as shingles, are made, granules of specified color or color combinations are applied from a hopper and spread thickly on the hot coating bitumen, and talc or mica is applied to the back (smooth) side. The sheet is then run through a series of press and cooling rollers, subjecting the sheet to a controlled pressure that embeds the granules in the coating to the desired depth.

7.5.2.6

Texture

At this stage some products are textured by the pressure of an embossing roller, which forms a pattern in the surface of the sheet.

7.5.2.7

Cooling Looper

The sheet then travels into a cooling (or finish) looper, where the sheet is cooled so it can be cut to length or into shingles and packed.

7.5.2.8

Shingle Cutter

When shingles are being made, the material is fed from the finish looper into a shingle-cutting machine. The sheet is cut from the back (smooth) side as it passes between a cutting cylinder and an anvil roll. Shingles are then separated into stacks of the proper number for packaging. The stacks are moved to packaging equipment, where the bundles are prepared for warehousing and shipment.

7.5.2.9

Roll Roofing Winder

When roll roofing is made, a sheet is drawn directly from the finish looper to a winding mandrel. Here it is wound and the material measured to length as it turns. After sufficient material accumulates, it is cut off, removed from the mandrel, and wrapped for warehousing and shipment.

7.5.2.10

Care of Materials

To ensure suitable performance, the following simple rules should be observed in storing finished products in the warehouse and on the job:

1. Roll goods should be stored on end in an upright position; if stored in tiers, boards should be placed between the tiers to prevent damage to the ends of rolls;
2. To eliminate the possibility of sticking or discoloration, shingle bundles should not be stored to a height of more than 4 ft.;
3. Warehouse stock, particularly white and pastel shades of roofing, should be rotated by moving the material held longest out of storage first;
4. Bituminous roofing and waterproofing materials stored outside should be covered for weather protection. Although these products provide a waterproof membrane after application, they can be harmed if stored outside without protection, particularly if they are applied wet.

7.5.3

CLASSIFICATION OF PRODUCTS

Bituminous roofing products include (1) bitumens (see Section 7.5.1.3), (2) asphalt coatings and cements, (3) sheet products made on an organic felt base, and (4) sheet products made on a glass fiber mat base.

7.5.3.1

Coatings and Cements

Coatings and cements are bituminous materials that are combined with special ingredients and either (1) mixed with suitable solvents of the cut-back type or (2) emulsified in water. These materials are processed to various consistencies depending on their use. The materials are flammable and should never be warmed over an open fire or placed in direct contact with a hot surface. If necessary, they may be softened by placing the container, unopened, in hot water or by storage in a warm place.

Cut-back asphalt coatings and cements should be applied to dry, clean surfaces and troweled or brushed vigorously to force the material into cracks and openings and to eliminate air bubbles. Emulsions may be applied to damp or wet surfaces but should never be applied in an exposed location if rain is anticipated within 24 hours.

These materials may be classified into several groups, including plastic roofing cements, flashing cements, lap cements, quick-setting adhesives, roof coatings, asphalt water emulsions, primers, and roofing tape. These may be used in accordance with the information provided in Figure 7.5-4.

■ PLASTIC ROOFING CEMENT (ROOFING MASTIC) AND FLASHING CEMENT

These materials are processed with mineral fibers and cut-back asphalt or coal tar, which are especially plastic and elastic, combined with nonoxidizing minerals, oils, and solvents to produce black plastic repair and flashing cements. They will not flow at summer temperatures or become brittle at low temperatures. They are sufficiently elastic after setting to compensate for normal expansion and contraction of the roof deck or movement between the deck and other elements of the structure. These cements have excellent resistance to acids and alkalies, will not absorb water, and provide good adhesion to all surfaces. They are similar but not interchangeable. Flashing cement is the stiffer of the two because it has more and longer fibers. It should be used for vertical applications and on cants. Plastic roofing cement is suitable for horizontal applications in flashing assemblies, patching, and roof repairing.

FIGURE 7.5-4
Uses and Properties of Asphalt Coatings and Cement Products

Product	Uses	Properties
Plastic roofing cement	Flashing assemblies; patching; roof repairing; caulking	Remains elastic after setting; will not flow at summer temperatures or become brittle at low temperatures; excellent resistance to acids and alkalies; will not absorb water; good adhesion to all surfaces
Lap cement	Watertight bond between lapping elements of roll roofing	Not as thick and heavy as plastic cements
Quick-setting adhesive	Cementing down the free tabs of strip shingles and laps of roll roofing	Available in brush, trowel, or gun consistency; same consistency as plastic asphalt cement, but more adhesive; sets quickly
Roof coating	Resurfacing old roll roofing and built-up asphalt roofs	Thin enough to be applied with spray or brush
Asphalt water emulsion	Roof coating	Must be kept from freezing and applied when no rain is forecast for 24 hours
Primer	Preparing masonry surfaces to receive other bituminous products; ensures good bond between the masonry and built-up roofing, plastic cement, or roof coatings	Very fluid; applied with brush or spray

LAP CEMENTS

These cements are of various consistencies but are not as thick and heavy as plastic cements. They should be used as recommended by the manufacturer. Lap cements are intended to make a watertight bond between lapping elements of roll roofing. When lap cement is applied, it should be spread over the entire lapped area. Nails used to secure the roofing should pass through the cement so that the shank of the nail will be sealed where it penetrates the deck material.

QUICK-SETTING ASPHALT ADHESIVE CEMENT

This material, available in brush, trowel, and gun consistencies, is used for cementing down the free tabs of strip shingles and for laps of roll roofing.

Quick-setting asphalt adhesive cement is about the same consistency as plastic asphalt cement, but it is more adhesive. It contains a solvent that evaporates rapidly on exposure, causing the cement to set quickly.

ROOF COATINGS

Roof coatings are of various types and consistencies but are usually thin enough to be applied with a spray or a brush. Coatings are used to resurface existing roll roofing and built-up asphalt roofs.

ASPHALT WATER EMULSIONS

Asphalt water emulsions are a special type of roof coating made with asphalt, sometimes combined with other ingredients, and emulsified with water. Emulsions must be stored where they will not freeze in cold weather and should be used at times when they will not be rained on for a period of at least 24 hours after being applied.

PRIMERS

Primers are very fluid bituminous materials that are applied with a brush or spray. They must be thin enough to penetrate rapidly into the surface pores of masonry or concrete without leaving a continuous surface film. Their purpose is to prepare masonry or concrete surfaces to receive other bituminous products and ensure a good bond between the masonry or concrete and built-up roofing, plastic cement, or bituminous coatings.

ROOFING TAPE

Some manufacturers make a roofing tape of cotton fabric or other porous material saturated with asphalt or coal tar. It is available in varying widths from 4 to 36 in. and is provided in rolls up to 50 yd. long. It is used with bituminous coatings and plastic flashing cements and in patching seams, breaks, and holes in bituminous and metal roofing.

7.5.3.2

Products Made on Organic-Felt or Glass-Fiber-Felt Base

Bituminous roofing and siding products made on an organic-felt or a glass-fiber-felt base may be classified broadly into two groups: (1) shingles and (2) rolls.

Sizes, exposures, and other information covering products typical of these two classifications are given in Figure 7.5-5. Products within each group may differ as to weight, dimension, and design. For example, the weight of a three-tab square butt strip shingle ranges from 200 to 300 lb. per square. (See Sections 7.6 and 7.7 for methods of installing these products.)

Bituminous roll products are usually referred to by a number, (No. 15, No. 30, No. 40, etc.), which represents the approximate weight in pounds as shipped of the amount of that material needed to cover a square (100 sq. ft.) with one ply of the material.

BUILT-UP ROOFING FELTS

Built-up roofing felts are broadly called reinforcing felts. They consist of base sheets, ply sheets, and cap and flashing sheets. They may be saturated, saturated and coated, impregnated, or prepared in some special way for a specific purpose.

Dry felts impregnated with asphalt or coal-tar saturants, but otherwise untreated, are called *saturated felts*. They are also used as underlayment for shingles and as sheathing paper. Saturated felts are made in several weights; the most common are No. 15 and No. 30, weighing about 15 and 30 lb. per square respectively.

Although glass fiber felts, which are impregnated with bitumen but otherwise untreated, may be classified under this broad category, they are not actually saturated, since glass fibers do not absorb bitumen. In reality the mat is lightly coated so that all fibers are thoroughly surrounded by bitumen.

Saturated and coated sheets are saturated felts that are then surfaced on one or both sides with a fine mineral material to prevent adhesion before application. These are generally used as base sheets in built-up roofing systems.

Impregnated sheets have less coating material than coated sheets, which makes them lighter and able to ventilate vapors during their application.

Prepared roofing materials are saturated and coated or impregnated with fine decorative mineral granules. Some are combination products made of various layers of glass fiber felt, organic fiber felt, or both.

BASE SHEETS Base sheets are used in some, but not all, built-up roofing systems. Usual materials are No. 30 asphalt-saturated organic felt, No. 40 asphalt-saturated and coated organic felt, asphalt glass fiber mat roll roofing, and venting asphalt-saturated and coated inorganic sheet.

PLY SHEETS Ply sheets are usually No. 15 asphalt-saturated organic felt, No. 15 coal-tar-saturated organic felts, asphalt-impregnated glass fiber felts Type III, or asphalt-impregnated glass fiber felts Type IV. Type IV sheets are heavier and have a breaking strength roughly twice that of Type III sheets.

CAP AND FLASHING SHEETS Cap and flashing sheets are usually wide-selvage asphalt roll roofing cap sheet surfaced with granules, No. 90 mineral-surfaced organic roll roofing, No. 50 or No. 60 smooth-surfaced asphalt roll roofing, inorganic cap sheet with mineral granules, or reinforced bituminous flashing material.

ROLL ROOFING MATERIALS

Roll roofing and siding are made by adding a coating of a more viscous, weather-resistant asphalt to a felt that has first been impregnated with a saturant asphalt or to a glass fiber mat. Some roll roofing and

FIGURE 7.5-5
Summary of Typical Asphalt Roofing Products

(a) Shingles

Product	Configuration	Per Square			Size			ASTM Fire and Wind Ratings
		Approximate Shipping Weight, lb.	Number of Shingles	Number of Bundles	Width, in.	Length, in.	Exposure, in.	
Self-sealing random-tab strip shingle Multi-thickness	Various edge, surface-texture, and application treatments	240–360	64–90	3, 4, or 5	11-1/2–14	36–40	4–6	A or C; many wind resistant
Self-sealing random-tab strip shingle Single-thickness	Various edge, surface-texture, and application treatments	240–300	65–80	3 or 4	12–13-1/4	36–40	4–5-5/8	A or C; many wind resistant
Self-sealing square-tab strip shingle Three-tab	Three-tab or four-tab	200–300	65–80	3 or 4	12–13-1/4	36–40	5–5-5/8	A or C; all wind resistant
Self-sealing square-tab strip shingle No-cutout	Various edge and surface-texture treatments	200–300	65–81	3 or 4	12–13-1/4	36–40	5–5-5/8	A or C; all wind resistant
Individual interlocking shingle Basic design	Several design variations	180–250	72–120	3 or 4	18–22-1/4	20–22-1/2	—	A or C; many wind resistant

FIGURE 7.5-5 — Continued

	(b) Asphalt Rolls							

| | Approximate Shipping Weight | | Number of Squares per Package | Length, ft. | Width, in. | Selvage, in. | Exposure, in. | ASTM Fire and Wind Ratings |
Product	Per Roll, lb.	Per Square, lb.						
Mineral surface roll	75–90	75–90	1	36–38	36	0–4	32–34	C
Mineral surface roll (double coverage)	55–70	110–140	1/2	36	36	19	17	C
Smooth surface roll	50–86	40–65	1–2	36–72	36	N/A	34	None
Saturated felt underlayment (nonperforated)	35–60	11–30	2–4	72–144	36	N/A	17–34	*

Courtesy Asphalt Roofing Manufacturers Association.

*May be a component in a complete fire-rated system. Check with manufacturer for details.

all siding are surfaced with mineral granules providing a wide color range, while some styles are furnished in split rolls designed to give an edge pattern when applied to the roof.

■ SHINGLES

Shingles are surfaced with mineral granules. There are many weights and patterns of both individual and strip shingles designed for both new construction and re-roofing of existing roofs. The most prevalent are three-tab strip shingles.

7.5.4

STANDARDS OF QUALITY

Asphalt shingles and sheet roofing should conform to Underwriters Laboratories, Inc. (UL) standards. The UL specifications constitute the recognized standard for the asphalt roofing industry and include provisions for (1) fire resistance (Fig. 7.5-6), (2) wind resistance, (3) tests of the saturation of felt to determine quantity of saturant and efficiency of saturation, (4) thickness and distribution of coating asphalts, (5) adhesion and distribution of granules, (6) weight, count, size, coloration, and other characteristics of finished products before and after packaging, and (7) instructions for application.

The UL maintains a check on labeled products by making periodic factory inspections. UL inspectors use the manufacturers' control laboratories and visit, inspect, and analyze product samples. When necessary, they recommend to the manufacturer measures that are required to maintain the standards of the product to remain eligible to use the UL label.

FIGURE 7.5-6

Classifications Established by Underwriters Laboratories, Inc. Standard UL 790, "Test Methods for Fire Resistance of Roof Covering Materials"

Class A	Includes roof coverings that are effective against severe fire exposures. Under such exposures, roof coverings of this class are not readily flammable and do not carry or communicate fire; afford a fairly high degree of fire protection to the roof deck; do not slip from position; possess no flying brand hazard; and do not require frequent repairs to maintain their fire-resisting properties.
Class B	Includes roof coverings that are effective against moderate fire exposures. Under such exposures, roof coverings of this class are not readily flammable and do not readily carry or communicate fire; afford a moderate degree of fire protection to the roof deck; do not slip from position; possess no flying brand hazard. May require infrequent repairs to maintain their fire-resisting properties.
Class C	Includes roof coverings that are effective against light fire exposure. Under such exposures, roof coverings of this class are not readily flammable and do not readily carry or communicate fire; afford at least a slight degree of fire protection to the roof deck; do not slip from position; possess no flying brand hazard. May require occasional repairs or renewals to maintain their fire-resisting properties.

Courtesy of Asphalt Roofing Manufacturers Association.

 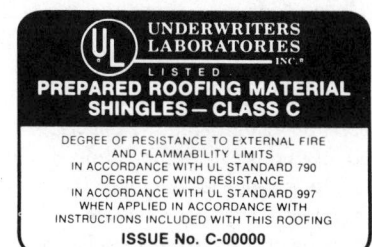

FIGURE 7.5-7 *Products conforming to UL Standard 55B carry labels similar to these.* *(Courtesy Underwriters Laboratories Inc.)*

Products that conform to UL standards are identified by UL labels similar to those illustrated in Figure 7.5-7. Asphalt shingles should conform to no less than UL Standard UL55B Class C for asphalt organic-felt sheet roofing and shingles. Wind-resistant shingles are recommended in areas subject to hurricane-force winds, which the U.S. Weather Service defines as those in excess of 75 mph. These shingles are provided with factory-applied adhesive or integral locking tabs, and the bundles carry the UL Class C label, which includes the words "Wind Resistant."

7.6 STEEP ROOFING AND RELATED WALL FINISHES

Steep roofs are those whose slope exceeds 1-1/2 in./ft. Roofs with slopes less than 1-1/2 in./ft. are called *low-sloped* roofs (see Section 7.7). Roofing for steep roofs comes in many forms, including ancient, traditional shapes and newer configurations, and in materials ranging from the stone and clay materials that have been used for centuries to modern day equivalents made from concrete, mineral fiber–cement, or metal.

Stone and clay tile and slate shingles were early forms of hard roofing materials. Concrete and mineral fiber–cement tiles were originally produced and are still used primarily to mimic clay tile, slate roofing, or wood shingles or shakes. The newer materials share several general advantages. They are often lighter, more readily available, and less susceptible to damage than the earlier forms. In many cases, the newer materials cost less than the roofing they mimic. Wood shingles are an exception, usually costing less than clay, concrete, or mineral fiber–cement substitutes, but wood shingles do not have the fire-resistance characteristics shared by the others.

The newer forms of hard roofing materials share the disadvantage of being machine made. The hand-made forms have somewhat irregular surfaces, which appear more rustic and have greater charm than roofs made with their modern, mass-produced counterparts. Another disadvantage is that, in most cases, the modern materials will not stand up to close visual comparison with the products they are imitating and are immediately apparent when installed along side the real thing.

When used alone, though, some of the modern materials are often mistaken for the materials they mimic when viewed from the ground by people who are not experts in recognizing the materials imitated. By contrast, composition shingles that are supposed to look like wood shingles, do not fool anybody.

Perhaps the earliest form of heavy roofing, stone tile is generally used today only when repairing historic buildings. Stone tile was never used much in the United States and is not found often enough to warrant detailed discussion here.

This section discusses the materials and methods needed to provide steep roof finishes using composition shingles, wood shakes and shingles, and the hard materials: slate, clay and concrete tile, and mineral fiber–cement shingles. It also covers wall finishes using mineral fiber–cement shingles and wood shingles and shakes.

The primary function of steep roofing and of wall finishes of the same materials is to shed water. By contrast, low-sloped roofs are covered with built-up or single-ply roofing that provides a waterproof membrane (see Section 7.7). Mineral-surfaced roll roofing, which can be applied on roofs with slopes of 1 in./ft. or more, is also discussed in Section 7.7. Metal roofing is covered in Section 7.9. Other roofing materials and exterior wall sidings are described in other sections of Chapters 7 and 9.

Due to the large amount of exposed surface, shingled and tiled roofs and walls contribute greatly to the appearance of a building, both in its shape and form, and by adding color, texture, and pattern.

In addition to cost, other criteria to be considered in selecting steep roofing, related wall materials, and the method of their application include the slope of the roof, the expected service life of the finish, wind resistance, fire resistance, and local climate. Depending on the color and kind of material selected, the total heat loss or gain of a roof or wall may also be affected.

7.6.1

GENERAL RECOMMENDATIONS

Individual shingle and tile units are applied to steep roofs and exterior walls in an overlapping fashion to shed water and resist weather penetration. They can be applied effectively on most walls and on most roofs that have sufficient slope to ensure good drainage. Wall shingles also are used on the interior for aesthetic purposes. In either case, they may contribute color and texture to a roof or wall surface, and when properly applied and maintained, they provide a suitable finish offering long service.

Depending on the type of material and the method of application selected, aesthetic effects range from wall surfaces having relatively smooth, uniform coursing to those with random coursing, which give a heavily textured appearance. Color can be achieved by a variety of factory- or field-applied coatings or by leaving the material in its natural color.

In addition to the shingle or tile units themselves, many accessory materials are required to prepare a roof or wall surface for their application and to apply them. Depending on the type of shingles or tile and their methods of application, accessory materials may include sheathing, underlayment, sheathing paper, backer board, undercoursing, channel and corner moldings, flashing, drip edges, roofing cement, caulking, and nails and other fasteners. Depending on the kind of shingle or tile, other accessories, such as starter shingles and hip and ridge units, may also

be required. While most materials and typical methods of applying shingles and tile on roofs and walls are discussed in the following text, the instructions and recommendations of the shingle or tile manufacturer should always be consulted to ensure the most suitable performance.

In this section, the roof construction is assumed to be correctly and adequately ventilated, a practice that is essential if condensation of water vapor is to be prevented. Condensation, the actual cause of many unjust complaints of roof leakage, can be eliminated as a hazard by providing sufficient ventilation in attic areas or enclosed airspaces beneath the roof. In addition, exterior walls are assumed to have been insulated correctly and adequately where climatic conditions require it. It is further assumed that precautions have been taken to prevent condensation of water vapor within walls. The hazard of condensation within walls normally can be eliminated by a suitable vapor retarder on the warm side of the insulation. Refer to Section 7.1 for a discussion of vapor retarders and ventilation.

7.6.1.1

Nomenclature

Some definitions of roof, roofing, and shingle and tile siding terminology are given here.

■ ROOF TERMINOLOGY

Figure 7.6-1 shows several common types of sloped roofs. These simple shapes are often complicated by intersecting walls or roofs and projections through the roof such as dormers, chimneys, plumbing vents, and roof ventilators.

The composite drawing in Figure 7.6-2 names the elements of a sloped roof.

SLOPE AND PITCH   The terms *slope* and *pitch* are often incorrectly used synonymously when referring to the incline of a sloped roof. Both are defined below and in Figure 7.6-3, which also compares some common roof slopes to corresponding roof pitches. *Roof slope* is the term used throughout this text.

Slope indicates the incline of a roof as a ratio of vertical rise to horizontal run.

FIGURE 7.6-1 *Types of sloped roofs.*

GABLE

HIP

GAMBREL

SHED

EAVE DETAIL

UNDERLAYMENT

SHEATHING

DRIP EDGE

RAFTER

FACIA

SOFFIT

SADDLE DETAIL

METAL SADDLE

COUNTER FLASHING

RAKE

RIDGE

STACK

VALLEY

VALLEY

RIDGE

RAKE

SOFFIT

DORMER

GABLE

SADDLE

EAVE

HIP

EAVE

FIGURE 7.6-2 *Composite diagram of sloped roof types showing terminology.*

It is expressed sometimes as a fraction but typically as *x* in 12. For example, a roof that rises at the rate of 4 in. for each foot (12 in.) of run is designated as having a 4 in 12 slope. The triangular symbol above the roof in Figure 7.6-3 conveys this information.

Pitch indicates the incline of a roof as a ratio of the vertical rise to twice the horizontal run. It is expressed as a fraction. For example, if the rise of a roof is 4 in. and the run is 12 in., the roof is designated as having a pitch of 1/6.

Assume:
Rise = 4 ft.; Run = 12 ft.

Slope: 4/12, or 4 in 12

Pitch: $\dfrac{4}{2 \times 12} = \dfrac{4}{24} = \dfrac{1}{6}$

$Slope = \dfrac{Rise}{Run}$

$Pitch = \dfrac{Rise}{2 \times Run}$

Slope	Pitch
2 in 12	1/12
3 in 12	1/8
4 in 12	1/6
5 in 12	5/24
6 in 12	1/4
7 in 12	7/24
8 in 12	1/3
10 in 12	5/12
12 in 12	1/2

FIGURE 7.6-3 *Slope and pitch.*

■ SHINGLE AND TILE ROOFING AND SIDING TERMINOLOGY

The following terms are peculiar to shingle roofing and siding. Other terms are defined when they first appear in the text or in the Glossary.

Backer board. An undercoursing material, typically fiberboard, used beneath siding materials such as mineral fiber–cement shingles (Fig 7.6-4a). It adds insulation value, increases resistance to impact, and provides a heavier shadowline at the butt than shingle siding applied directly without the use of a backing material.

Coverage. The amount of weather protection provided by the overlapping of shingles or tile. Depending on the kind of shingle and the method of application, shingles may furnish one (single coverage), two (double coverage), or three (triple coverage) thicknesses of material over the surface of a roof.

E = Exposure
TL = Toplap
HL = Headlap
SL = Sidelap
W = Width for Strip Shingles
    or Length for Individual Shingles

FIGURE 7.6-4 *Double-coursed walls with shingles applied over an undercourse: (a) mineral fiber–cement shingles; (b) wood shingles or shakes.*

Shingles providing single coverage are suitable for reroofing over existing roofs, in effect providing a new surface for an old roof that is still serviceable except for isolated trouble spots. Shingles providing double and triple coverage are used for new construction, both having increased weather resistance and a longer service life.

Double coursing. A method of applying shingles over an undercourse of lower-grade shingles or other suitable material (Fig. 7.6-4b). Nails are exposed as a result of face (butt) nailing, and greater shingle exposures are possible than with single coursing. Double coursing usually results in greater wall coverage at lower costs.

Endlap. The shortest distance in inches that adjacent shingles, sheets, or tiles horizontally overlap each other (Figs. 7.6-5 and 7.6-6).

Exposure. The shortest distance in inches between exposed edges of overlapping shingles or tiles (see Figs. 7.6-5 and 7.6-6).

Headlap. The shortest distance in inches from the lower edges of overlapping shingles to the upper edge of the unit in the second course below (see Figs. 7.6-5 and 7.6-6).

Shingles. Relatively small individual roofing or siding units designed to overlap other similar units to provide weather pro-

tection. Typically, shingles are applied to a nailing base, such as sheathing or horizontal nailing strips, which supports the shingles between structural framing members.

Shingle butt. The lower, exposed edge of a shingle.

Sidelap. See *endlap*.

Single coursing. A method of applying shingles without the use of an undercourse (Fig. 7.6-7). In the case of wood shingles, this method results in concealed nails and a smaller exposure than is found in double-coursed walls.

Square. Roofing is estimated and sold by the square, although packages of one-third square are usually available so that individuals or contractors do not have to buy a full square of roofing when the amount needed is not an exact multiple of one full square. A square of roofing is the amount required to cover 100 sq. ft. of roof surface.

Toplap. The shortest distance in inches from the lower edge of an overlapping shingle or sheet to the upper edge of the lapped unit in the first course below (that is, the width of the shingle minus the exposure) (see Figs. 7.6-5 and 7.6-6).

7.6.1.2

Roof Slope Limitations

Free and effective drainage of water from

AMERICAN METHOD

DUTCH LAP METHOD

FIGURE 7.6-5 *Shingle roofing terminology.*

E - Exposure

TL - Toplap

HL - Headlap

W - Width for Strip Shingles or Length for Individal Shingles

FIGURE 7.6-6 *Shingle siding terminology.*

sloped roof surfaces is essential for long shingle life. As the roof slope decreases,

FIGURE 7.6-7 *Single-coursed walls have shingles applied without an underlayment: (a) mineral fiber–cement shingles; (b) wood shingles and shakes.*

the run-off of water is slower and the susceptibility to leakage from wind-blown rain or snow being driven up underneath the roofing increases. Accordingly, the roof slope governs or may impose limitations on the following: (1) roofing material selection, (2) exposure, (3) underlayment requirements, (4) eave flashing requirements, and (5) methods used in roofing materials application.

In this text, when a reduction in the slope of the roof requires precautions other than normal practice, the roof is referred to as being *of lower slope than normal*; otherwise, the term *normal slope* is used.

The minimum slope recommended for each kind of shingle and tile is shown in Figure 7.6-8. Slope limitations for underlayment and eave flashing requirements vary with each kind of shingle and tile and are discussed later in this section. Application methods are discussed in the following subsections for each kind of shingle.

7.6.1.3

Wall and Roof Framing and Sheathing

The performance of steep roofing and siding of the same materials is dependent on the supporting structure. Roof joists, rafters, studs, headers, and other supports should be adequately sized and spaced to carry the necessary superimposed loads over the required spans.

Sheathing should be smooth, securely attached to supports, and provide an ade-

FIGURE 7.6-8
Summary of Recommended Minimum Roof Slope

Type of Roofing	Minimum Slope[a]
Mineral fiber–cement shingles	3 in 12[b]
Asphalt shingles	2 in 12[c]
Wood shakes	3 in 12[d]
Wood shingles	3 in 12[e]
Slate shingles	4 in 12[b]
Clay tile	
Flat	5 in 12[b]
Other types	3 in 12[b]
Concrete tile	
Flat	5 in 12[b]
Other shapes	4 in 12[b]
Mud set	4 in 12

[a]Where eave flashing is required, it should be cemented.

[b]Requires double underlayment.

[c]Strip shingles only; requires double underlayment and wind-resistant shingles or cemented tabs.

[d]Requires solid sheathing, underlayment in addition to interlayment and reduced weather exposure.

[e]Requires reduced weather exposure.

quate base to receive nails and other fasteners. All types of shingles are applied over solid sheathing composed of wood boards or plywood. Wood shingles and shakes are also applied over spaced sheathing.

Roof and wall framing and sheathing are discussed in Chapter 6.

7.6.1.4

Roofing Cements

Roofing cements include plastic asphalt cements, lap cements, quick-setting asphalt adhesives, roof coatings, and primers. They are used for installing eave flashing, flashing assemblies, cementing tabs of asphalt shingles and laps in sheet materials, and roof repairing. They are discussed in Section 7.5.

The type and the quality of materials and the methods of application of steep roof and related wall finishes should be as recommended by the manufacturer of the finish materials.

7.6.1.5

Roofing Underlayment

Roof underlayment performs several functions: (1) it protects the sheathing from moisture absorption until the roofing is applied; (2) it provides important additional weather protection by preventing the entrance of wind-driven rain below the roofing onto the sheathing or into the structure; (3) in the case of asphalt shingles, it prevents direct contact between shingles and resinous areas in wood sheathing, which, because of chemical incompatibility, may damage the shingles.

Underlayment should be a material such as asphalt-saturated felt, which has low vapor resistance. Materials such as coated felts or laminated waterproof papers, which act as vapor retarders, should not be used. Such materials may permit moisture or frost to accumulate between the underlayment and the surface of the sheathing.

Underlayment requirements for different kinds of steep roofing for various roof slopes are summarized in Figure 7.6-9.

As soon as roof sheathing has been installed, underlayment should be applied over the entire roof. It should lap at least 2 in. at horizontal joints and 4 in. at end joints (Fig. 7.6-10) and should be lapped 6 in. from both sides over hips and ridges. Only sufficient fasteners need be used to hold the underlayment securely in place until roofing is applied. Roofing should not be applied over wet underlayment.

Sheathing paper for walls is discussed in Section 6.6.5.1.

FIGURE 7.6-9
Summary of Underlayment for Shingle Roofs

Material	Underlayment Requirements				
	15 lb.	30 lb.	1 Layer	2 Layers	Remarks
Roll roofing	X		X	X	a, b, c
Composition shingles					
4 in./ft. and steeper	X		X		b
3 in./ft. to 4 in./ft.	X			X	c
2-1/2 in./ft. to 3 in./ft.	X				d
Slate					
4 in./ft. and steeper					
1/4 in. thick and thinner		X	X		b
More than 1/4 in. thick		X		X	c, e
Less than 4 in./ft.					i
Graduated, all slopes		X		X	c
Wood shingles		X	X		f
Wood shakes		X	X		f, g
Clay tile		X	X		h
Concrete tile		X	X		h
Mineral fiber–cement tile					
4 in./ft. and steeper		X	X		f, g
3 in./ft. to 4 in./ft.		X	X		g
Class A roofs					j
Metal tile		X	X		b

Courtesy H. Leslie Simmons.

Remarks:

[a] Use one layer on slopes 4 in./ft. and more. Use two layers on flatter slopes.

[b] Where one layer of underlayment is required, install two layers of 15-lb. felt or one 50-lb. felt from eave line to 24 in. inside wall line as an ice shield when mean January temperature is 30°F or lower. At slate roofing and metal tile, use two layers of 30-lb. felt. Set ice-shield felt in hot asphalt or mastic. As an alternative, use adhered bituminous membrane manufactured for the purpose as an ice shield.

[c] Where two underlayment layers are required, apply underlayment from eave to 24 in. inside wall using hot asphalt or mastic when mean January temperature is 30°F or less.

[d] Entire underlayment applied using hot asphalt or mastic.

[e] Alternative: One layer of 55- or 65-lb. roll roofing over entire roof, with portion from eave to 24 in. inside wall applied using hot asphalt or mastic when mean January temperature is 30°F or less.

[f] Underlayment scheduled is optional except in snow areas (see text).

[g] Felt eave strips and felt interlayers between shakes are essential (see text).

[h] Underlayment scheduled is often used for roof slopes of 4 in./ft. and higher, but there is disagreement in the industry about underlayment requirements for clay and concrete tile (see text). In applicable areas, the requirements in notes b and c apply.

[i] Slate on slopes less than 4 in./ft. should be installed over membrane roofing or waterproofing (see text).

[j] See text.

FIGURE 7.6-10 *Underlayment requirements for steep roofing.*

tunities for leakage. Flashing must be installed at these locations and be made watertight and water-shedding to weatherseal the wall.

■ FLASHING MATERIALS

Corrosion-resistant metal flashing should be at least 26-gauge galvanized steel, 0.019-in. aluminum, or 16-oz. copper. Refer to Section 7.9 for a discussion of these and other flashing materials.

Additional requirements and typical methods of flashing are described in the following subsections for each kind of shingle.

■ ROOF FLASHINGS

VALLEY FLASHING The joint formed by two sloping roofs meeting at an angle is called a *valley*. Because drainage concentrates in a valley, causing runoff toward and along the joint, valleys are especially vulnerable to leakage. An unobstructed drainage way must be provided with enough capacity to carry away the water rapidly.

CHIMNEY FLASHING A chimney is usually built on a foundation separate from the foundation that supports the structure and is normally subject to some differential settling. To permit this movement between a chimney and associated roof structure, without damage to the waterseal, it is necessary to secure base flashings to the roof deck and secure counter flashing, which is bent down over the base flashing, to the masonry.

A saddle (see Fig. 7.6-2) should be built between a roof and the back face of a chimney that falls below the ridge. The purpose of this saddle is to prevent the accumulation of snow and ice behind the

7.6.1.6

Flashing and Caulking

Roofs often are intersected by other roofs, walls, and projections, creating opportunities for leakage. Special construction, called flashing, must be installed at these locations to weatherseal the roof. Care must be used to make flashing watertight and water-shedding.

Walls may contain window or door openings, intersect with sloped roof surfaces, or terminate in exposed locations against other materials, creating opporchimney and to deflect water down around the chimney.

EAVE FLASHING Eave flashing is recommended in areas where the outside design temperature is 30°F or colder (Fig. 7.6-11) or wherever there is a possibility of ice forming along the eaves. This protection is important to avoid leaks and moisture penetration caused by water backing up under shingles behind dams of ice and snow that collect at the eaves.

Different methods of flashing are used to prevent leakage from this cause, depending on the slope of the roof, the severity of icing conditions, and the kind of roofing.

For normal slopes, eave flashing is typically formed by a course of smooth or mineral-surfaced coated roll roofing applied over the underlayment. It extends up the roof to cover a line at least 12 in. inside the interior wall line of the building (Fig. 7.6-12a).

On slopes that are lower than normal and in areas subject to severe icing, a double layer of underlayment should be applied over the entire roof. The two layers should be cemented together with asphalt cement up to the roof to a line that is at least 24 in. inside the interior wall line of the building (Figs. 7.6-12b and 7.6-13).

Specific recommendations for eave flashing are described in the following subsections for each kind of shingle.

DRIP EDGE A continuous corrosion-resistant, nonstaining material such as 26-gauge galvanized steel, formed to provide a drip and nailed along the eaves and rakes is recommended for most shingle roofs. Drip edges should be designed and installed to protect the edges of the deck and prevent leaks at this point by allowing water to drip free of underlying eave and cornice construction. Typical shapes of preformed drip edges are shown in Figure 7.6-14.

A drip edge is applied to the sheathing and *under* the underlayment at the eaves, but *over* the underlayment up the rake (Fig. 7.6-15).

■ **WALL FLASHING**

FLASHING AT OPENINGS Heads, sills, and jambs of openings should be flashed.

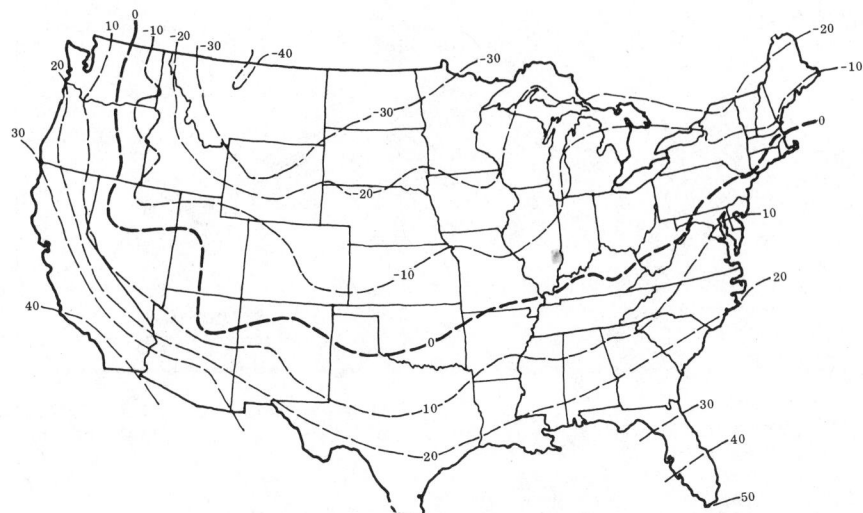

FIGURE 7.6-11 *Outside design temperatures.* (Reprint courtesy National Weather Service.)

FIGURE 7.6-12 *Eave flashing for normal slopes (a) extends up the roof at least 12 in. inside the exterior wall; for lower than normal slope (b) an additional course of underlayment should be cemented down and extend up the roof at least 24 in. inside the interior face of the wall.*

However, in areas not subject to wind-driven rain, head flashing may be omitted when the vertical height between the top of the finish trim of openings and the bottom of eave soffits is equal to or less than one-fourth of the eave overhang.

At heads and jambs of openings in wood framed walls, metal flashing should be installed extending from at least 2 in. above trim and turned down over the outside edge of drip caps forming a drip. Where sheathing paper is omitted, flashing should extend up behind the sheathing.

In unsheathed walls, jambs should be flashed with a 6-in.-wide strip of metal, 3-oz. copper-coated paper, or 6-mil polyethylene film.

FLASHING AT WALL AND ROOF INTERSECTIONS Sheet metal flashing should be installed at horizontal intersections and at vertical intersections when the exterior finish material does not provide a self-flashing joint.

■ **CAULKING**

A nonshrinking caulking compound, in either white or matching color, should be

FIGURE 7.6-13 *Application of double-layer felt underlayment and eave flashing.* *(Reprint courtesy National Roofing Contractors Association.)*

FIGURE 7.6-16 *Types of smooth and threaded shank nails used for application of shingle roofing other than asphalt shingles (see Fig. 7.6-18).*

FIGURE 7.6-14 *Typical drip-edge shapes.*

A RAKE

B EAVE

FIGURE 7.6-15 *Application of drip edge at (a) rake and (b) eave.*

used to weatherseal joints where siding abuts wooden trim, masonry, or other projections.

7.6.1.7

Nails and Fasteners

No single step in applying steep roofing shingles or related siding materials is more important than proper nailing. Suitability depends on (1) selecting the correct nail for the kind of finish material and type of sheathing (Fig. 7.6-16); (2) using the correct number of nails; (3) locating the nails correctly; and (4) choosing a nail metal compatible with the metal used for flashings.

Specific recommendations for the type, size, number, and spacing of roofing nails are given in the following subsections for each roofing and related wall finish material. Some general recommendations are given below.

■ LUMBER BOARDS, PLANK DECKING, AND PLYWOOD SHEATHING

Roofing nails should be long enough to penetrate through the shingle and through the lumber boards or plywood sheathing. They should penetrate at least 1 in. into plank roof decking. Nails for applying shingles over plywood sheathing should have threaded shanks.

■ NONWOOD SHEATHING

When gypsum, concrete, fiberboard, or other nonwood sheathing materials are used, special fasteners and special details for fastening are often necessary to provide adequate anchorage. Recommendations of the manufacturer of the nonwood material should be followed to ensure suitable performance of the installation.

7.6.2

ASPHALT SHINGLES

The following text describes materials and methods needed to provide a roof covering of asphalt shingles over both new and existing construction.

Reference should be made to Section 7.5 for a description of asphalt products and to Section 7.6.1 for definitions of terms, limitations imposed by the slope of the roof, sheathing guidelines (see also Chapter 6), and general guidelines for typical accessory materials, such as underlayment, flashing, roofing cements, eave flashings, drip edges, and nails, for preparation of the roof or wall for application of shingles. Unless specific exceptions are made in the following text, the provisions in Section 7.6.1 apply to asphalt shingles and the method of their application.

7.6.2.1

Materials

■ SHINGLES

Asphalt roof shingles are manufactured in three basic kinds of units: (1) strip shingles of the *square-butt* or *hexagonal* type,

(2) individual shingles of the *interlocking type* or *staple-down* type; and (3) *giant individual shingles* for application by either the *American* or *Dutch lap* methods. *Square-butt strip shingles* are by far the most used today, and the majority of these are of the three-tab type. The other types may be found in older buildings.

For use in high-wind areas, wind-resistant strip shingles have tabs with factory-applied adhesive or integral locking tabs. Lightweight individual shingles applied by the Dutch lap method are intended primarily for reroofing existing roofs.

Each kind of shingle is applied in a distinct way, giving a different appearance to a roof. Starter units, hip and ridge shingles, and valley roll are also manufactured to match asphalt shingles.

Shingles should conform to the UL Standard UL55B Class C, and each bundle of shingles should be identified with the UL Class C label. In areas subject to winds of hurricane force, *wind-resistant* shingles are recommended. The United States Weather Service defines hurricane-force winds as those of 75 mph or more. Such shingles are provided with factory-applied adhesive or integral locking tabs, and the bundles carry the UL Class C label and include the words "Wind Resistant."

The types of asphalt shingles are summarized in Figure 7.5-5.

NAILS AND FASTENERS

Nails (Fig. 7.6-17) for applying asphalt roofing should be of corrosion-resistant metal, such as hot-dipped galvanized steel or aluminum, with sharp points and large flat heads, 3/8 to 7/16 in. in diameter. Shanks should not exceed 0.135 in. or be less than 0.105 in. in overall outside diameter (10- to 12-gauge wire). Shanks may be smooth or threaded; however, threaded nails are preferred for increased holding power in lumber board sheathing and definitely should be used in plywood sheathing. Aluminum nails should have screw threads with approximately a 12-1/2-degree thread angle. Galvanized steel nails, if threaded, should have annular threads. Nail lengths typically required are given in Figure 7.6-18.

The number, spacing, and pattern of nails are described in Section 7.6.2.3.

FIGURE 7.6-18
Asphalt Shingle Nail Lengths

Application	1-in. Sheathing	3/8-in. Plywood
Strip or individual shingle (new construction)	1-1/4 in.	7/8 in.
Over asphalt roofing (reroofing)	1-1/2 in.	1 in.
Over wood shingles (reroofing)	1-3/4 in.	—

7.6.2.2

Preparation for Shingles

Limitations imposed by the slope of the roof are shown in Figure 7.6-3. Asphalt shingles may safely be applied using normal application methods on roofs with slopes of 4 in 12 or greater. Square-butt strip shingles may be used on slopes as low as 2 in 12 when applied using the lower than normal slope methods described in Section 7.6.2.3.

ROOF SHEATHING

Sheathing types, thicknesses, spans, and nailing requirements are discussed in Chapter 6. If lumber boards are used, they should not be over 6 in. in nominal width. Wider boards may swell or shrink in width enough to buckle or distort the shingles. Badly warped boards and those containing excessively resinous areas or loose knots should be rejected. If any of these defects appear after the sheathing has been applied, they should be covered with 26-gauge galvanized steel patches before the underlayment is placed.

UNDERLAYMENT

Underlayment should be No. 15 asphalt-saturated felt, applied over the roof as described in Sections 7.6.1.5. No. 30 asphalt felt should not be used as general underlayment beneath asphalt shingles, because it is too stiff and may cause wrinkling of shingles applied over it. However, on lower than normal slope applications (2 in 12 to 4 in 12) two layers of No. 15 felt should be installed over the entire roof.

EAVE FLASHING

In areas where the outside design temperature is 30 °F or colder (see Fig. 7.6-11) or where there is a possibility of ice forming along the eave and causing a backup of water, eave flashing is recommended.

NORMAL SLOPE (4 IN 12 OR GREATER) A course of 90-lb. mineral-surfaced roll roofing or not less than 55-lb. smooth roll roofing should be installed at the eaves over the underlayment. It should overhang the underlayment and drip edge from 1/4 to 3/8 in. and extend up the roof to cover a line at least 12 in. inside the interior wall line of the building (see Fig. 7.6-12a). When the eave overhang requires the flashing to be wider than 36 in., the necessary horizontal lap joint should be cemented and located outside the exterior wall line of the building.

LOWER THAN NORMAL SLOPE (AS LOW AS 2 IN 12) Square-butt strip shingles may be applied on slopes as low as 2 in 12. For slopes less than 4 in 12 down to 2 in 12 or in areas subject to severe icing, a double layer of underlayment should be used over the entire roof. Eave flashing may be formed by cementing the two layers of underlayment together with asphalt cement up the roof to a line that is at least 24 in. inside the interior wall line of the building (see Figs. 7.6-12b and 7.6-13).

This eave flashing should be formed by applying a continuous layer of plastic asphalt cement at the rate of 2 gal./100 sq. ft., to the surface of the underlayment starter course before the first full course

FIGURE 7.6-17 *Asphalt shingle nails: (a) smooth, (b) annular threaded, and (c) screw threaded.*

of underlayment is applied. Cement should also be applied to the 19-in. underlying portion of each succeeding course that lies within the eave flashing area before the next course is placed. It is important to apply this cement uniformly with a comb trowel, so that underlayment does not touch underlayment at any point when the application has been completed. The overlying sheet should be pressed firmly into the entire cemented area (see Fig. 7.6-13).

■ DRIP EDGES

A corrosion-resistant, nonstaining drip edge should be installed along the eaves and rakes (see Fig. 7.6-15).

■ FLASHING

See Section 7.6.1.6.

VALLEY FLASHING Either the open or closed method may be used to construct valley flashing. A valley underlayment strip of No. 15 asphalt-saturated felt, 36 in. wide, is applied first as shown in Figure 7.6-19, centered in the valley and secured with enough nails to hold it in place. The horizontal courses of underlayment are cut to overlap this valley strip a minimum of 6 in. Where eave flashing is required, it is applied over the valley underlayment strip.

Open valleys may be flashed with the metals described in Section 7.6.1.6 or with 90-lb. mineral-surfaced asphalt roll roofing in a color to match or contrast with the roof shingles. The latter method is shown in Figure 7.6-20. An 18-in.-wide strip of mineral-surfaced roll roofing should be placed over the valley underlayment, centered in the valley with the surfaced side down, and the lower edge cut to conform to and be flush with the eave flashing. When splicing is necessary, the ends of the upper segments should overlap the lower segments 12 in. and be secured with asphalt plastic cement. Only enough nails to hold the strip smoothly in place should be used. The nails should be placed in rows 1 in. in from each edge.

Another strip, 36 in. wide, should be placed over the first strip, centered in the valley with the surfaced side up, secured with nails, and lapped if necessary, just as was the underlying 18-in. strip.

FIGURE 7.6-19 *Flashing underlayment in a valley.* *(Reprint courtesy National Roofing Contractors Association.)*

A chalk line should be snapped on each side of the valley for its full length before shingles are applied. These chalk lines should start 6 in. apart at the ridge and spread wider apart at the rate of 1/8 in./ft. down to the eave. They should serve as a guide in trimming the last shingle units to fit the valley, ensuring a clean, sharp edge. The upper corner of each end shingle should be clipped as shown in Figure 7.6-20 to direct water into the valley and prevent water penetration between courses. Each shingle should be cemented to the valley lining with asphalt cement to ensure a tight seal. No exposed nails should appear along the valley flashing.

Closed (woven) valleys can be used only with strip shingles. Individual shingles cannot be used because nails might be required at or near the center of the valley lining. This method has the advantage of doubling the coverage of the shingles throughout the length of the valley, increasing weather resistance at this vulnerable point.

A valley lining made from a 36-in.-wide strip of 55-lb. or heavier roll roofing should be placed over the valley underlayment and centered in the valley (Fig. 7.6-21).

Valley shingles should be laid over the lining by either (1) applying them on both roof surfaces at the same time, weaving each course in turn over the valley, or (2) covering each surface first to the point approximately 36 in. from the center of the valley and weaving the valley shingles in place later. In either case, the first course at the valley should be laid along the eaves of one surface over the valley lining and extended along the adjoining roof surface for a distance of at least 12 in. The first course of the adjoining roof surface should then be carried over the valley

Underlayment

12" Lap

36" Strip Face up

Spot of Plastic Asphalt Cement

18" Strip — Face Down

Plastic Asphalt Cement

12" Lap

Chalk Lines

Corner of Shingle Clipped

All weights and dimensions are APPROXIMATE.

FIGURE 7.6-20 *Open-valley flashing using mineral-surfaced roll roofing.* *(Reprint courtesy National Roofing Contractors Association.)*

on top of the previously applied shingle. Succeeding courses should then be laid alternately, weaving the valley shingles over each other.

The shingles should be pressed tightly into the valley and nailed normally (see Section 7.6.2.3), except that no nail should be located closer than 6 in. from the valley center line, and two nails should be used at the end of each terminal strip.

VERTICAL WALL FLASHING   When the rake of a roof abuts a vertical wall, the most satisfactory method of protecting the joint is to use stepped metal base flashing applied over the end of each course of shingles (Fig. 7.6-22).

This metal flashing is rectangular in shape, approximately 6 in. long and 2 in. wider than the exposed face of the shingles. With strip shingles laid 5 in. to the weather, the flashing would be 6 in. by 7 in. The flashing should be bent with one end extending 2 in. onto the roof underlayment, the remainder extending up the wall surface. Each flashing piece should be placed just up-roof from the exposed edge of the shingle that overlaps it and secured to the wall sheathing with one nail in the top corner. When the shingles are laid 5 in. to the weather, each 7-in. width of flashing will lap the next flashing piece 2 in. The wall siding should be finished over the flashing to serve as counter (cap) flashing but be held far enough away from the shingles so that the ends of the siding may be painted for weather protection.

CHIMNEY FLASHING   Before flashings are placed, shingles should be applied over the underlayment up to the front face of a chimney and a coat of asphalt primer applied to the masonry to seal the surface at all points where plastic cement will be applied.

Base flashing may be 90-lb. mineral-surfaced roll roofing. The chimney front (Fig. 7.6-23a) should be applied first by securing the lower section over the shingles in a bed of asphalt plastic cement. The vertical upper section should be secured against the masonry with plastic cement and nails driven into the mortar joints; the triangular ends should be bent around the corners of the chimney and cemented into place.

The base flashing for chimney sides should be applied as shown in Figure 7.6-23b. It should be secured to the deck with plastic cement and to the brick with plastic cement and nails; the triangular ends of the upper section should be turned around the chimney corners and cemented in place over the front base flashing.

The base flashing covering a saddle should be secured in plastic cement over the underlayment. The corners should be installed as shown in Figure 7.6-24a. The flashing over the cricket should be installed as shown in Figure 7.6-24b, with the standing portion extending up the masonry 6 to 12 in. and capped as shown in Figure 7.6-24c. The capping should be bedded tightly in plastic cement, centered over the saddle flashing, and extend up the roof. This provides added protection at the point where the ridge of the saddle joins the roof.

Counter flashing of metal should be applied over base flashing to complete the flashing installation (Figs. 7.6-25 and 7.6-26). One method of securing counter flashing to masonry is to rake out the mortar joint to a depth of 1-1/2 in. and insert the bent back edge of the flashing in the raked joint between bricks. The joint should be then refilled with portland cement mortar or plastic cement, and the flashing bent down over the base flashing.

The counter flashing on the chimney front should be one continuous piece. On the sides and rear, the sections should be cut to conform to the locations of brick joints and the pitch of the roof, lapping each other at least 3 in.

Each Strip to
Extend at Least
12" Beyond Center
of Valley

36" Roll Roofing
Material — No. 15
or Heavier

Nails 6" Min. from
Valley Center

Extra Nail in
End of Strip

*All weights and dimensions are APPROXIMATE.

FIGURE 7.6-21 *Closed-valley flashing using woven strip shingles.* (Reprint courtesy National Roofing Contractors Association.)

Application of Shingles

Shingles should be laid so they overlap each other to shed water. The extent of overlapping, which depends on their shape and the method of application, furnishes single, double, and sometimes triple coverage over different areas of a roof. Square butt strip shingles, some interlocking shingles, and giant individual shingles applied by the American method, providing either double or triple coverage, are suitable for new construction, but as stated earlier, square-butt strip shingles are the only ones in general use today.

Strip and individual hexagonal shingles and giant individual shingles, applied by the Dutch lap method, provide only single coverage and are used only to reroof existing construction. Refer to manufacturer's literature for details of their application.

Regardless of the method of application, on new construction shingles should be applied only to a tight roof deck providing a suitable nailing base, with underlayment, drip edge, and flashings in place. Chimneys should be completed and counter flashing installed. Stacks and other equipment requiring openings through the roof should be in place with provision for counter flashing where necessary; gutters should be hung.

■ STRIP SHINGLES (NORMAL SLOPE)

On small roofs, strip shingles may be laid from either rake. On roofs 30 ft. and longer they should be started at the center and worked both ways from a vertical line to ensure more accurate vertical alignment and provide for meeting and matching above a projection, such as a dormer or chimney.

When a roof surface is broken, shingles should be started from a rake and laid toward a dormer or into a valley. Where unbroken, they should be started at the rake that is most visible. If both rakes are equally visible, shingles should be started in the center and worked both ways.

Minor variations in the dimensions of asphalt shingles are unavoidable. Variations seldom exceed ±1/4 in. in either length or width in a 12-in. by 36-in. strip shingle and usually will be no more than ±1/8 in. However, to control the placement of shingles so that cutouts will be

STACK FLASHING Pipes projecting through a pitched roof require special flashing methods. Either corrosion-resistant metal sleeves that slip over stacks and have adjustable flanges to fit any roof slope (Fig. 7.6-27) or asphalt products may be used successfully.

Roofing should be applied up to the point where the stack projects and be fitted around it (Fig. 7.6-28). When metal flashing is used, the sleeve should be fitted down over the stack, adjusted to the slope of the roof, and the top of the sleeve turned into the stack opening (Fig. 7.6-29). When asphalt products are used, an oval opening should be cut in a rectangular

flange of 55-lb. or heavier roll roofing. This flange should be large enough to extend 4 in. below, 8 in. above, and 6 in. on each side of the stack when installed. It should be slipped over the pipe and applied flat on the roof. A collar of plastic cement should be formed around the pipe, extending up the pipe and out over the flange about 2 in. (Fig. 7.6-30). The cement should be vigorously and thoroughly troweled in place to give proper adherence and eliminate air pockets. The shingles should be continued up the roof, fitted around the stack, and bedded in plastic cement where they overlay the flange at the side and back of the stack (Fig. 7.6-31).

FIGURE 7.6-22 *Vertical wall flashing using individual metal flashing units.* *(Reprint courtesy National Roofing Contractors Association.)*

Labels in figure:
No. 15 Asphalt-Saturated Felt Turned Up on Vertical Walls 3" to 4"
4" Up Vert. Wall
Nail
Flashing 2" Side Lap
Finish Siding Serves as Counter Flashing
7"
2" Out Over Roof
Flashing Placed Just Back from Exposed Edge of Shingle
*All weights and dimensions are APPROXIMATE.

Labels in figure:
Coat of asphalt primer
Asphalt plastic cement
Underlayment
Base flashing applied over shingles and set in asphalt plastic cement
Coat of asphalt primer
Asphalt plastic cement
Step flashing
Nail flashing to deck
Embed end shingle in cement
a
b

FIGURE 7.6-23 *Application of chimney base flashing: (a) at front of chimney; (b) at side of chimney.* *(Courtesy Asphalt Roofing Manufacturers Association.)*

accurately aligned horizontally, vertically, and diagonally, horizontal and vertical chalk lines should be used to control alignment.

STARTER COURSE The purpose of a starter course is to back up the first regular course of shingles and fill in the spaces between tabs. The starter course may be either a 9-in.-wide (or wider) *starter strip*

of mineral-surfaced roll roofing of a color to match the shingles or a row of inverted *starter course* shingles or strips made from the top lap portion of the shingles with the tabs cut off.

A starter strip should be applied over an eaves flashing strip, even with its lower edge along the eave. The starter strip should be fastened with roofing nails 3 or 4 in. above the eave edge and spaced so

that the nail heads will not be exposed at the cutouts between tabs on the first course (Fig. 7.6-32).

Starter course shingles should be applied with tabs facing up the roof over the eave flashing strip. Nails should be spaced so they will not be exposed at the cutouts of the first course (Fig. 7.6-33).

FIRST AND SUCCEEDING COURSES (SQUARE-BUTT SHINGLES) The first course should be started with a full shingle, while succeeding courses should be started with full or cut strips depending on the pattern desired. Major variations for square-butt strip shingles are discussed in the following paragraphs.

Cutouts break joints on halves when the second course is started with a full shingle cut 6 in. short (half a tab), the third with a full shingle with the entire first tab removed, the fourth with 1-1/2 tabs cut from the shingle, and so on, causing the cutouts to be centered on the tabs of the course below (see Fig. 7.6-32). This pattern can also be made with less cutting by starting the third course with a full shingle.

Cutouts break joints on thirds when the second course is started with a full shingle cut 4 in. short (a third of a tab), the third course with a full shingle cut 8 in. short (two-thirds of a tab), and the fourth with a full shingle (see Fig. 7.6-33). The shingles should be placed so that the lower edges of the butts are aligned with the top of the cutouts of the underlying course.

Random spacing can be achieved by removing different amounts from the starting tab of succeeding courses in accordance with the following general principles: (1) the width of any starting tab should be at least 3 in.; (2) cutout center-lines of any course should be located at least 3 in. laterally from the cutout center-lines in both the course above and the course below; and (3) starting tab widths should be varied sufficiently so the eye will not follow a cutout alignment. Figure 7.6-34 shows the length of the starting tab at the rake of each succeeding course that will produce a suitable random spacing when the first course is started with a full-length strip.

VERTICAL APPLICATION Research findings by the NAHB Research Foundation have shown that strip shingles can be applied faster by laying one 36-in. strip of

FIGURE 7.6-24 *Application of saddle (cricket) flashing: (a) at corner; (b) over cricket; (c) over ridge of cricket.* *(Courtesy Asphalt Roofing Manufacturers Association.)*

FIGURE 7.6-26 *Metal counter (cap) flashing must turn down over base flashing to complete the flashing assembly.* *(Courtesy Asphalt Roofing Manufacturers Association.)*

FIGURE 7.6-25 *Application of chimney cap flashing: (a) at front and sides; (b) at sides and rear.* *(Courtesy Asphalt Roofing Manufacturers Association.)*

FIGURE 7.6-27 *Adjustable metal flashing sleeves fit over stacks.* *(Courtesy Asphalt Roofing Manufacturers Association.)*

FIGURE 7.6-28 *Shingles are laid to fit around stack.* *(Courtesy Asphalt Roofing Manufacturers Association.)*

shingles in vertical courses from the eave to the ridge, rather than laying courses horizontally from rake to rake. This method results in an equally weathertight roof at labor cost reductions as high as 25%. However, this method should be used only for solid colored shingles, because color blends laid in this manner may develop a striped effect on the roof. Most manufacturers, in fact, recommend against application of any shingles in a vertical manner. Such instructions are printed on some manufacturers' shingle packages.

NAILING AND FASTENING STRIP SHINGLES
The number of nails per strip and the placing of the nails are vital for suitable roof application. To prevent buckling, each

shingle should be in perfect alignment before nails are driven. Nailing should be started at the end nearest the shingle last applied and proceed to the opposite end, with nails driven straight to avoid cutting the fabric of a shingle with the edge of a nail head. Nail heads should be driven flush, not sunk below the surface of the shingles. When a nail fails to penetrate the sheathing, an additional nail should be driven in a new location.

Three-tab square-butt shingles require four nails for each three-tab square-butt strip. When such shingles are applied with a 5-in. exposure, the nails should be placed 5/8 in. above the top of the cutouts and located horizontally with one nail 1 in. back from each end and one nail on the

centerline of each cutout (see Figs. 7.6-32 and 7.6-33).

Windy locations require additional protection. In areas where wind velocities are 75 mph or greater, shingles manufactured with factory-applied adhesives or integral locking tabs that conform to the UL Stan-

FIGURE 7.6-29 *Shingles are laid, and metal sleeve is turned into top of stack. (Courtesy Asphalt Roofing Manufacturers Association.)*

FIGURE 7.6-30 *Roll roofing flange and plastic asphalt cement collar form flashing. (Courtesy Asphalt Roofing Manufacturers Association.)*

FIGURE 7.6-31 *Shingles applied over flange complete flashing assembly. (Courtesy Asphalt Roofing Manufacturers Association.)*

dard for Class C wind-resistant shingles should be used.

If UL Class C wind-resistant shingles are not used, an effective way to obtain protection against high winds is to cement the free tabs as shown in Figure 7.6-35. A spot of quick-setting cement, about 1 sq. in. in area for each tab, should be pressed on the underlying shingle with a putty knife or caulking gun. The free tab should then be pressed against the cement. In applying the cement, shingles should not be bent back farther than necessary.

Three-tab square-butt strip shingles should receive six nails instead of four in high-wind areas. One nail should be placed at each end, as described previously, with two nails 5/8 in. above and

FIGURE 7.6-32 *Application of square-butt strip shingles; cutouts break joints on halves. (Reprint courtesy National Roofing Contractors Association.)*

FIGURE 7.6-33 *Application of square-butt strip shingles; cutouts break joints on thirds. (Reprint courtesy National Roofing Contractors Association.)*

positioned just to either side of the centerline of each cutout.

HIPS AND RIDGES Hips and ridges may be finished by using hip and ridge shingles furnished by the manufacturer or by cutting pieces at least 9 in. by 12 in. either from 12-in. by 36-in. square-butt shingle strips or from mineral-surfaced roll roofing of a color to match the shingles. They should be applied by bending each shingle lengthwise down the center with equal ex-

posure on each side of the hip or ridge. In cold weather the shingles should be warmed before bending. They should be applied beginning at the bottom of a hip or at one end of a ridge with a 5-in. exposure. Each shingle should be secured with one nail at each side, 5-1/2 in. back from the exposed end and 1 in. up from the edge.

Metal ridge roll subject to corrosion may discolor the roof and should never be used with asphalt roofing products.

FIGURE 7.6-34 *Random spacing of square-butt strip shingles.* (*Courtesy Asphalt Roofing Manufacturers Association.*)

FIGURE 7.6-35 *Free tabs may be cemented down for wind protection.* (*Courtesy Asphalt Roofing Manufacturers Association.*)

■ STRIP SHINGLES (LOWER THAN NORMAL SLOPE)

Square-tab strip shingles may be used on roof slopes less than 4 in 12 but not less than 2 in 12. Application methods for a lower than normal slope require (1) double underlayment, (2) cemented eave flashing strip, (3) shingles provided with factory-applied adhesives conforming to the UL Standard for Class C wind-resistant shingles, or each free tab of square-butt strip shingles should be cemented as shown in Figure 7.6-35.

Application of strip shingles over double underlayment and cemented eave flashing is shown in Figure 7.6-36. Any shingle arrangement described under normal-slope application methods may be used.

7.6.2.4

Reroofing Existing Construction

When a roof is to be reroofed, existing wood shingles, asphalt shingles, or roll roofing need not be removed before applying a new asphalt roof, provided (1) the strength of the existing roof framing is adequate to support the new dead and live loads and (2) the existing roof sheathing is sound and will provide adequate anchorage for nails used to apply new roofing.

If either the framing or the sheathing is inadequate, the old roofing, regardless of type, must be removed to correct the defects.

■ EXISTING ROOFING TO BE REMOVED

When old roofing must be removed, a roof should be prepared for new roofing by (1) repairing and reinforcing existing roof framing where required to provide adequate strength; (2) removing existing sheathing that is damaged and installing new sheathing in its place; (3) filling spaces between boards with securely nailed wood strips of the same thickness or removing existing sheathing altogether and resheathing; (4) removing protruding nails; and (5) covering large cracks, slivers, knotholes, loose pitchy knots, and excessively resinous areas with sheet metal securely nailed to the sheathing.

■ EXISTING ROOFING TO REMAIN

When inspection discloses that existing roofing need not be removed, the surface of the existing roofing should be prepared to receive new roofing, as described below for various types of existing roofing.

EXISTING WOOD SHINGLES   Preparation includes removing loose or protruding nails and nailing the shingles in a new location; nailing down loose shingles; splitting badly curled or warped shingles and nailing down the segments; and replacing missing shingles with new ones. When shingles and trim at eaves and rakes are badly weathered or the location is subject to high winds, the shingles there should be cut back and 1×4 or 1×6 boards nailed with their outside edges projecting beyond the edges of the deck the same distance as did the wood shingles (Fig. 7.6-37). Beveled wood "feathering strips" should be used along the butts of each course of old shingles to provide a

FIGURE 7.6-36 *Low-slope application of strip shingles over double underlayment and cemented eave flashing.* (*Reprint courtesy National Roofing Contractors Association.*)

FIGURE 7.6-37 *Preparation of existing roof to receive new shingles.* (*Reprint courtesy National Roofing Contractors Association.*)

A RAKE

B EAVE

FIGURE 7.6-38 *Application of drip edge (a) at rake and (b) at eave.* (*Courtesy Asphalt Roofing Manufacturers Association.*)

FIGURE 7.6-39 *Wall flashing (reroofing).* (*Courtesy Asphalt Roofing Manufacturers Association.*)

relatively smooth surface to receive new asphalt roofing.

EXISTING ASPHALT SHINGLES Preparation includes nailing loose and protruding nails and removing and replacing badly weathered edging strips.

EXISTING ROLL ROOFING Preparation includes splitting buckles, nailing segments down smoothly, and removing loose and protruding nails.

EXISTING BUILT-UP ROOFS If the slope of the roof is not less than 2 in 12 and if the roof sheathing is sound and can be expected to provide good nail-holding power, new shingles can be applied. Old slag, gravel, and other coarse surfacing materials should first be removed, leaving the surface of the underlying felts smooth and clean. New asphalt shingles should be applied directly over the felts as described in "Strip Shingles (Lower Than Normal Slope)" in Section 7.6.2.3.

■ UNDERLAYMENT (REROOFING)

When existing roofing must be removed before new material is applied, underlayment is required over the entire roof as it would be prescribed for new construction.

When new shingles can be applied directly over old roofing, no underlayment is required since the old roofing adequately serves that purpose.

■ DRIP EDGE AND EAVE FLASHING STRIP (REROOFING)

When installing new asphalt shingles over an existing wood-shingled roof, drip edges and eave flashing strip should be installed as for new construction (Fig. 7.6-38).

■ FLASHING (REROOFING)

Existing metal flashing, unless perforated or deteriorated, may be left in place and reused.

FLASHING AGAINST A VERTICAL WALL When new asphalt shingles are being applied over an old roof, the joint between a vertical wall and the rake of an abutting roof may be treated as shown in Figure 7.6-39.

A 6- or 8-in. strip of smooth roll roofing should be applied over shingles abutting a wall surface, using a row of nails placed about 4 in. on centers along each edge. As the work proceeds, the strip should be covered with asphalt plastic cement and the end of each course of new shingles firmly secured by bedding it in the cement. Improved appearance and a tight joint can be achieved by using a caulking gun to apply a final bead of cement between the ends of the shingles and the siding.

CHIMNEY FLASHING Figure 7.6-40 shows a method of chimney flashing using existing flashing.

■ APPLICATION OF SHINGLES (REROOFING)

After the existing roof has been prepared, the methods of applying the various types of shingles are similar to those described above for new construction.

7.6.3

SLATE SHINGLES

Slate shingles are not widely used today, especially on commercial projects, but so much slate roofing exists that some discussion is appropriate.

FIGURE 7.6-40 *Chimney flashing (reroofing). Old metal counter flashing may be reused if still serviceable. It should extend down over new base flashing made by applying plastic asphalt cement over end of shingles and against masonry and covering with a strip of mineral-surfaced roofing.* (Courtesy Asphalt Roofing Manufacturers Association.)

7.6.3.1

Materials

■ SLATE

Slate is hard, practically nonabsorbent rock, characterized by its natural proclivity to split along a single plane. This single direction cleavage characteristic permits slate to be split into thin sheets. Roofing slate is usually split in the direction of its natural grain.

Roofing slate is classified as *commercial standard*, *textural*, or *graduated*. Commercial standard roofing slate is smooth and about 3/16 in. thick. Textural roofing slate is 3/8 in. thick or thinner. Its color is more variable than that of commercial standard slate, and its surface texture is rougher. Graduated roofing slate varies in thickness from 3/16 in. to 1-1/2 in. or thicker. It also varies more in color and texture than either commercial standard or textural slate. Some slate used in roofing (*ribbon stock*) has strips of other rock embedded in it. Slate without ribbons is called *clear slate*.

Individual slate roofing pieces (*slates*) vary in size from 10 in. by 6 in. to 24 in. by 16 in. Most slates are square, but corners are sometimes clipped to produce special effects. Most slate is split so that the grain runs with the length of the slate.

Slate is available in several colors and shades of colors, including those classified in the industry as black, blue black, gray, mottled gray, purple, variegated purple (mottled purple and green), green, and red. Other colors (*specials*) may also be used. Existing slate will almost certainly have changed color over time and may be difficult to match with new material. New slate will, eventually, also change color, but there may be a time when new slates are apparent. Slate is classified as *permanent* (or *unfading*) or *weathering*, denoting the degree of color fading expected.

Before material is selected to match existing slate, it is important to determine the source of the existing slate and try to find the same or similar slate from the same source. Slate is currently mined in Virginia, New York, Maine, Vermont, and Pennsylvania, but existing slate may have come from another location. Slate from different states, even slate from different quarries in the same state or from the same quarry may not match. Fortunately, some variation is normal, especially in some colors, so that an acceptable match is often possible.

Every slate should have at least two nail holes machine-punched at the quarry. Slates 3/4 in. or more in thickness and 20 in. or more in length should have four nail holes. Hand punching should be done only where absolutely necessary, such as for fitting hips.

■ FASTENERS

Zinc nails and copper, brass, yellow metal, and other nails with copper content have all been used to install slate with success. For repairing slate work, however, the best choice is large-head, copper slaters' nails. Under no circumstances should nails intended for any other purpose, such as roofer's nails or galvanized steel nails, be used with slate roofing.

Nails should be at least 1 in. longer than the slate thickness but never less than long enough to penetrate the roof sheathing. Nails should not, however, project on the other side of the sheathing, unless they will be concealed in the finished work.

Nail gauge depends on the weight of the slates and their locations on the roof. For example, 3-penny nails can be used for commercial standard slates up to 18 in. long. Longer commercial standard slates require 4-penny nails. Hips and ridges of roofs covered with commercial standard slate should be installed using 6-penny nails. Heavier slates require heavier nails.

7.6.3.2

Preparation for Slate Roofing

■ SHEATHING

In the past, slate was often installed using wood pegs hooked over wood lath. More recently, some slate was hung from wood lath or battens or directly from steel structural elements by heavy wire hooked to the wood or structural elements. Sometimes slate was held to steel sections using long, bent copper nails. Today, most slate roofing is installed over continuous nailable substrates, often wood, plywood, or a wood fiber product, sometimes over a nailable portland cement–based material, and fastened in place with nails.

■ UNDERLAYMENT

Surfaces to receive slate roofing 1/4 in. or less in thickness where the roof slope is 4 in./ft. or greater should be covered with an underlayment consisting of at least one layer of 30-lb. asphalt-impregnated, unperforated roofing felt.

Surfaces to receive slate roofing with a thickness greater than 1/4 in., where the roof slope is 4 in. or more per foot should receive an underlayment consisting of two layers of 30-lb. asphalt-impregnated, unperforated roofing felt or a single layer of 55- or 65-lb. asphalt-impregnated, unperforated roofing felt.

Surfaces to receive slate roofing where the roof slope is less than 4 in./ft. cannot rightly be called steep roofs. In such conditions, slate should be installed as tile (without laps) as a ballast over membrane roofing or as a promenade surface over membrane waterproofing.

Under the conditions outlined earlier in this section, slate roofs should have ice shields. The ice shield should consist of at least two layers of 30-lb. felt.

7.6.3.3

Slate Installation

Slates installed with nails should each receive not less than two nails. Four nails should be used in larger slates. Nails should be driven so that the head just touches the slate. The slate should not be clenched by the nail but should hang on the nail.

FIGURE 7.6-41 *Slate jointing.* (*Courtesy H. Leslie Simmons.*)

Slates should be set so that joints fall as closely as possible to the midpoint of the length of the slate immediately above or below, but never less than 3 in. from the end of those adjacent slates (see "Joint break" in Fig. 7.6-41). Joints should never be open clear through to the underlayment below.

Slate weather exposure depends on the size of the slate and the headlap (see Fig. 7.6-41). Exposure is the shingle length minus the headlap divided by 2. The exposure for a 24-in. shingle with a 3-in. headlap is 10-1/2 in.

The amount of headlap varies with roof slope and region. On slopes from zero to 4 in./ft., slate should be laid as tile, with no lap. On slopes from 4 to 8 in./ft., headlap is often the "standard" 3 in., but roofers in some parts of the country consider 4 in. necessary. On slopes from 8 to 20 in./ft., the standard 3-in. headlap is generally accepted as sufficient. In some southern areas, a headlap of 2 in. is considered adequate, but over most of the country 2-in. headlap is restricted to mansards and other roofs with slopes more than 20 in./ft.

Slate is sold in squares. A square of slate is the amount needed to cover a square of roof (100 sq. ft. of roof surface) when the headlap is 3 in. The actual headlap to be used must be considered when calculating the amount of slate necessary to cover a given area of roof. More headlap will result in less coverage for a square of slate.

Random-appearing slate roofs using standard and textural slate are achieved by varying slate width while keeping length equal or by varying both widths and lengths. Equal-width slates are seldom used.

Variation in graduated slate roofing is achieved by varying lengths, widths, and thicknesses. Larger and thicker slates are used in lower courses and graduate to thinner and smaller slates at the ridge. All courses contain various slate thicknesses.

Repaired hips and ridges should match the existing ones. The National Roofing Contractors Association's 1986 *The NRCA Steep Roofing Manual* and the National Slate Association's *Slate Roofs* each contain detailed descriptions and drawings showing how several common types of hips and saddles should be formed.

Ridges may be a *saddle* type or a *comb* type. A saddle ridge is one in which the top slates abut evenly. In a regular saddle ridge, the top slates overlap the adjacent top slate (Fig. 7.6-42a). Two nails, concealed by the next higher slate, are used in each top slate. In a strip saddle ridge, the top slates do not overlap; they abut, and each is fastened down with four nails. Figure 7.6-42b is a section through either a regular or strip saddle ridge.

Comb ridges are formed similarly to strip saddle ridges, except that the top of the ridge slates do not abut but overlap by between 1/16 in. and 1/8 in. (Fig. 7.6-43).

FIGURE 7.6-42 *(a) Regular saddle ridge; (b) section through a saddle ridge.* (*Courtesy H. Leslie Simmons.*)

FIGURE 7.6-43 *Detail at a comb ridge.* (*Courtesy H. Leslie Simmons.*)

The projection may occur all on one side, or the projecting slates may alternate from one side to the other in a configuration called a *coxcomb ridge.*

There are several ways to form slate roofing hips. Four common methods are known as *saddle, mitered, fantail,* and *Boston* hips. *Saddle hips* and *strip saddle hips* are similar to saddle and strip saddle ridges, except that each slate is fastened in place using four nails.

In *mitered hips,* the slates forming the hip are in the same plane as the regular roofing slates and are cut to the slope of the hip (Fig. 7.6-44).

Fantail hips are a variation of mitered hips in which the bottom edge of the hip shingle is cut at an angle to form a fantail (Fig. 7.6-45).

Boston hips are similar in appearance to saddle hips, but they are constructed by the weaving of three shapes of cut shingles together (Fig. 7.6-46).

Flashing beneath hips of the kinds mentioned here is usually not necessary. Some existing conditions, however, may have been flashed with metal strips woven in with each course.

Valleys may be *open, closed,* or *round.* Open valleys are protected with metal flashing. They should start with the slate 4 in. apart at the top and widen uniformly

FIGURE 7.6-44 *Mitered hip.* (*Courtesy H. Leslie Simmons.*)

FIGURE 7.6-45 *Fantail hip.* *(Courtesy H. Leslie Simmons.)*

FIGURE 7.6-46 *Boston hip.* *(Courtesy H. Leslie Simmons.)*

from top to bottom at a rate of 1 in. in every 8 ft.

In closed valleys, the slate is cut to abut in the valley, and metal flashing is woven in as the slate is installed.

Flashing is not required in properly installed round valleys. Round valleys require considerable underlying support and special slate sizes. Considerable expertise is needed to properly install round valleys. Round-valley slates are sometimes bedded in elastic cement. Round valleys built by inexperienced or insufficiently skilled personnel should be flashed. Round valleys susceptible to ice damage should always be flashed. Round-valley flashing may be metal or bituminous materials.

In all types of ridge and hip conditions, the top regular slate courses should have their edges set in elastic cement to resist wind uplift. The joint between ridge slates and the overlap in comb ridges should be sealed with elastic cement. Some slate roofing is installed with elastic cement in other locations, such as along flashing lines and in the field of the roofing. Some slate roofing is installed with no elastic cement whatever and experiences no leaks as a result. The absence of elastic cement

does not in itself mean that there is a problem or that the installation in inferior.

In every case, however, exposed nails should be coated with elastic cement. Where slates are nailed near flashing, the nails should not penetrate the flashing.

7.6.4

TILE ROOFING

Unlike some other kinds of roofing, industry standards for tile roofing are scarce. Sheathing, underlayments, fasteners, and tile should be installed in accordance with the tile producers' recommendations, applicable building code requirements, and the highest standards of the trade. The requirements in this chapter are a conglomerate of those generally recommended by the producers. Recommendations of industry-recognized sources, such as the National Roofing Contractor's Association, have also been taken into account, where they were found. Individual producers and their associations' recommendations and building codes may differ, however.

7.6.4.1

Materials

■ CLAY TILE

Clay tile is dense, hard, and nonabsorbent material produced by baking molded clay units in a kiln. It may be either glazed or unglazed.

Clay tile was used before the Golden Age of Greece. Since at least the early part of the twelfth century, some clay tile shapes have remained virtually the same. Today, clay tile for roofing comes in four general styles: "S" (Spanish), half-round (Mission), pan and cover (Roman), and flat tile. Figure 7.6-47 shows some general clay tile shapes, including three in the flat category. Available shapes may vary slightly from those shown, especially in the flat and pan tile types. It is also possible that an existing roof will have a shape not shown.

Flat shingles may have special butt shapes. These are sometimes called *Persian* tile. Special butt shapes include those

FIGURE 7.6-47 *Clay tile shapes.* *(Courtesy H. Leslie Simmons.)*

with rounded corners, round ends, half-hexagonal ends, half-octagonal ends, and single-pointed ends.

Some flat interlocking tile comes in half-, quarter-, and one-and-one-half-wide units to permit finishing at rakes.

Special shapes are usually used for finishing hips and ridges, and for trim and closures at ridges, hips, rakes, eaves, and valleys. Special shapes include regular tiles with the ends closed and specially shaped pieces to fit within or between tiles. Shapes vary to work with each tile type.

Round building elements, such as towers and turrets, may be roofed using *graduated* tiles. Graduated tiles are units made to match the regular types but are specially shaped to fit a rounded surface. Tile manufacturers furnish such tile routinely, but matching an existing graduated tile may be a problem. If the particular existing shape is no longer available, it may be necessary to have a match manufactured.

Some clay tile is finished to mimic wood shingles or slate. Broken or cut tiles will show clay's characteristic red color, however, and are easily identified as clay.

CONCRETE TILE

Concrete tile is made from a mixture of portland cement, aggregate, and water and cured under pressure at a controlled temperature and humidity. Some tile is cast, other tile units are extruded. Concrete mixes vary from manufacturer to manufacturer. Some manufacturers use sand as the aggregate. Others use lightweight materials, such as perlite. Some concrete tile is finished on the exposed surface while still wet, using a pigmented cement slurry. Some is colored throughout by pigment added to the concrete mix. Some is the natural concrete color with no added coloration. Concrete tile may be glazed or unglazed.

Some concrete tile is made to mimic wood shingles, clay tile, or slate in shape, color, and texture. Except under close scrutiny, the match with clay tile is fairly successful when the tile is new, differing only in individual tile shape. The attempt to match other materials, however, often fails. For example, some concrete tile is too large to successfully ape wood shingles. Broken edges identify concrete tile made to look like slate or wood. Some concrete tile matches the color of new wood shingles effectively but, of course, does not age as wood does. In addition, the fungus that grows on wet concrete does not look like the fungus that grows on aging wood. Concrete tile does, however, enjoy the major advantage over wood of being fire resistant. And some concrete tile is colored during manufacture to match closely the appearance of mossy, aged wood. A great advantage of using concrete tile instead of old wood is that the concrete tile will have the same appearance 50 to 75 years later, while wood that far gone will soon need replacing. Such colored concrete tile can be made to look more natural by using different colors placed randomly.

All concrete tile absorbs some moisture. Tile made with some lightweight aggregates absorbs more moisture than tile made with sand aggregate. The moisture absorption does not seem to harm the tile or its reinforcement, but it is sometimes unsightly and can contribute to rusting metal or rotting wood in the underlying structure. Some concrete tile producers include water-absorption inhibitors in their mixtures to prevent such damage from occurring.

Most concrete tile used for roofing is either the standard concrete type barrel tile shape (Fig. 7.6-48) or a flat interlocking shape similar to the unit called "Flat with Interlocking Feature" in Figure 7.6-47. Shapes used on an existing building may vary slightly from those shown. A common variation, for example, is in the width of the valley in the barrel tile shape. Concrete tile on an existing roof may also be similar to one of the other styles shown in Figure 7.6-47 for clay tiles or another shape not shown here.

FIGURE 7.6-48 *Concrete barrel tile type.*
(Courtesy H. Leslie Simmons.)

Special shapes are usually used for finishing hips and ridges and for trim and closures at ridges, hips, rakes, eaves, and valleys. Special shapes include regular tiles with the ends closed and specially shaped pieces to fit within or between tiles. Shapes vary to work with each tile type.

Round building elements, such as towers and turrets, may be roofed using graduated tiles as described for clay tile.

FASTENERS

Tile may be nailed in position or hung from wire hangers, depending on the tile type, substrates, and code requirements.

Fasteners should be noncorrosive, such as stainless steel, aluminum, or a copper-bearing metal, such as hard copper, silicon copper, or yellow metal. Nails should be at least 11-gauge, large-headed (minimum 5/16-in. diameter), and long enough to penetrate completely through plywood sheathing, not less than 3/4 in. into other wood nailers, and the distance into other nailable materials recommended by the tile manufacturer. Nails should not, however, penetrate completely through board plank decks. Nails for use in plywood should be ring-shank nails. Nails for use in boards should be smooth shank. Nails for use in gypsum plank or nailable concrete should be stainless steel or silicon bronze screw shank nails long enough to penetrate one-half and three-quarters their length, respectively, into deck. They should never penetrate fully through the deck. For very hard decks, it may be necessary to use smooth-shank nails. Some so-called nailable decks may not have sufficient nail holding power to support concrete tile.

Wire hangers are sometimes used to hang tile from wood battens but are more often used on concrete and steel substrates. Hangers are also used, even where the substrate is wood, in locations where driven nails would penetrate flashings and in locations where driving nails is difficult. Wire hangers and associated ties are available in stainless steel, brass, copper, and galvanized steel. One system employs a 1-1/2-in. by 1/2-in. wire strip hanger and 14-gauge tie wire, but a variety of hanger and tie types are available that may be acceptable to a particular tile manufacturer.

Hurricane clips, also called *storm clips*, are manufactured from brass or galvanized steel to shapes necessary for use with each tile type. Clips should not be visible in the completed roof. Hurricane clips are fastened positively to the roof deck and clip over tile edges to hold tile in place during high winds.

MORTAR AND CEMENT

Most mortar used in clay tile roofing is composed of a mixture of one part portland cement, four parts sand, and coloring to produce a mortar that matches the color of the tile.

Concrete tile manufacturers do not agree on the mix for mortar for use with concrete tile, or even on the materials to be used. It may be difficult, even impossible, to determine the actual mortar mix that was used in an existing application. Mortar color usually matches tile color.

In "mud-set" tile applications, concrete tile manufacturers usually recommend using Type M mortar as defined in ASTM Standard C 270.

Plastic cement should be a heavy-bodied asphalt roofing cement. Some tile manufacturers prefer that a silicone sealant be used in lieu of plastic cement. Plastic cement and sealant should be clear or colored to match the tile.

7.6.4.2

Preparation for Tile Roofing

■ SUBSTRATES, BATTENS, AND SUPPORT STRIPS

Clay and concrete tile may be installed over wood, metal, or concrete decks. Fastener requirements differ depending on the tile configuration and the deck type.

Wood sheathing for clay or concrete tile should be either exterior-grade plywood or boards. Plywood thickness should be selected to comply with structural requirements and fastener requirements but should not be less than 1/2 in. Plywood joints should be left open 1/16 in. to permit expansion.

Board sheathing should be at least 1×6s, spanning not more than 24 in. Supporting structure spans wider than 24 in. require heavier boards. Spaced sheathing is not generally recommended for clay tile, but it may have been used on some existing roofs. Some sources say that spaced sheathing is not recommended in concrete tile installations, but some manufacturers recommend installation over spaced board sheathing, and some concrete tile roofs have been installed in that manner. Concrete tile applied over spaced sheathing should be installed in accordance with the tile manufacturer's recommendations.

Some manufacturers and many tile workers would prefer that concrete tile be installed over wood battens on every roof, regardless of slope. A major reason for that recommendation is that using battens makes replacing damaged tile much easier than if battens are not used. Another reason is the general feeling that nails do not hold in plywood well enough to support heavy concrete tile over an extended period. Many low-slope concrete tile roofs

have been installed without battens, of course, but battens should be used on slopes of 7 in./ft. and steeper, regardless of tile type or manufacturer's recommendations. Battens are usually 1×2s, fastened to the substrate at spacings recommended by the tile manufacturer. Spacing depends on tile type and size.

Clay tiles are seldom installed over cast-in-place concrete decks, because few cast-in-place concrete decks slope enough to receive clay tile roofs. Where clay tile is applied over cast-in-place concrete, it is usually fastened in place using wood battens in a configuration known as the *counter batten* system. In a counter batten system, wood battens, usually 1×2s, are nailed to wood nailers cast into the concrete deck from eave to ridge at about 20 in. on centers. Underlayment is installed over the deck and cast-in nailers before battens are applied. Batten spacing depends on tile size.

Concrete tile is seldom installed over cast-in-place concrete decks using the nail-on method, because few cast-in-place concrete decks slope as much as 3 in./ft., which is the minimum at which concrete tile can be nailed to a roof. When concrete tile is applied over cast-in-place concrete by nailing, it is usually fastened to wood battens in the same counter batten configuration described in the preceding paragraph.

Clay and concrete tile installed over precast concrete decks (and sometimes over cast-in-place concrete) is usually fastened to a counter batten system, in which the wood members running up the slope are through-bolted to the concrete. Sometimes the batten running up the slope is omitted, and the horizontal battens are themselves fastened directly to the deck by through-bolting.

A system of nailers and wood battens may also be used over metal decks. Often, however, fire ratings or code requirements will dictate that a fire-retardant sheathing be applied over the metal decking. Nailable fire-retardant sheathing, when the sheathing manufacturer so recommends and the tile manufacturer concurs, can be used the same as a plywood deck.

Battens can also be fastened over underlayment to nailers set flush with nonnailable insulation over any deck. Insulating products that are also nailable can be used like plywood when their manufacturer so

recommends and the tile manufacturer concurs.

Manufacturers of some clay tile types recommend that the tile be applied over battens under all or sometimes just certain selected conditions. Such recommendations should be followed.

A means should be provided for water that finds its way through the tile to drain out of the roofing system. One way to do this is to separate the end joints between battens about 1/2 in., with end joints about 40 in. on centers. The problem with that method is that since joints must fall over the nailers below, the battens must be in many relatively short pieces. A better way is to raise the battens at least 1/4 in. above the underlayment using cut shingles or wood lath strips laid beneath the battens along the line of the nailers and fastened through the underlayment to the nailers.

Tile may also be attached directly to concrete or steel decks using a system of wire hangers.

"S" and half-round clay tiles require nailers at ridges, hips, rakes, and projections. Flat clay and concrete tiles require nailers at ridges and hips. Nailer size varies with tile type and rise, which may be different for different manufacturers and is definitely different for different tile types. Nailers are either 1-in. or 2-in. nominal lumber, depending on tile configuration and tile manufacturers' recommendations. Hip and ridge nailers are usually set in cement mortar.

Eave strips are required, except for some types of flat clay tile and for concrete tile where starter course elevation is achieved using a specially made tile strip and other tile where special tile accessories are used to elevate the first course. Eave strips are usually 1×2 wood members. Another way to raise the starter course is to elevate the fascia board. In any case, a tapered cant just above the nailer strip or tile starter course or some other device is necessary to prevent water from ponding at the eave line.

■ UNDERLAYMENT

Surfaces to receive clay tile should be completely covered with an underlayment of at least one layer of 30-lb. asphalt-impregnated roofing felt. Additional layers of underlayment are required in some circumstances, but industry sources do not agree on which circumstances dictate additional layers. For example, some sources say that roof slopes of 3 in./ft. or steeper can receive any type clay tile when completely covered with an underlayment consisting of two layers of 30-lb. asphalt-impregnated felt set in hot asphalt or mastic. These same sources imply that flat clay roofing tile needs a double layer of 30-lb. felt underlayment on all slopes less than 5 in./ft., but that can be installed on a single layer of felt for steeper slopes.

Other sources, including many tile manufacturers, do not agree on these requirements. Some manufacturers agree that any clay tile shape can be used on any slope exceeding 3 in./ft., when there is a double layer of 30-lb. felt as underlayment. Some manufacturers, however, recommend a double layer of 30-lb. felt beneath flat tile on all slopes. Some see no need for double layers of felt beneath clay tile except for flat clay tile, regardless of slope. Some say that one layer of 40-lb. felt can be substituted for the two 30-lb. layers. Some say that underlayment on steep slopes should consist of a single 40-lb. felt layer, and that underlayment on low slopes should consist of 60-lb. felt, or two layers of 30-lb. felt.

The level of disagreement related to underlayment for concrete tile is similar. Some sources say that, to be waterproof, concrete tile should be installed on slopes of 4 in./ft. or steeper, that flat shingle tile requires a slope of 5 in./ft. or more, and that any slope below 4 in./ft. should have an underlayment consisting of two layers of roofing felt set in hot asphalt or mastic.

Other sources, however, do not agree with these requirements, nor do they agree with each other. Some say that concrete tile can be nail-applied on slopes of 3 in./ft. or steeper. Some manufacturers recommend that a single-ply 40-lb. asphalt-impregnated, nonperforated roofing felt underlayment be applied as a minimum. Some recommend that surfaces with slopes between 3 in./ft. and 4 in./ft. that are to receive nailed-on concrete tile be completely covered with an underlayment consisting of two layers of 15-lb. felt with hot asphalt between the plies, or a water-proofing membrane, or a self-adhering bituminous membrane. Some say that slopes of 4 in./ft. or steeper should receive a single layer of 30-lb. asphalt-impregnated roofing felt. Consequently, an existing clay or concrete tile roof may have been installed using any one of those underlayment systems or some other system entirely.

There is general agreement that either clay or concrete tile on slopes below 3 in./ft. must be installed only over built-up roofing or proper membrane waterproofing. Some manufacturers say that concrete tile on slopes less than 4 in./ft. should be set in mortar beds over membrane roofing or membrane waterproofing. This method is called *mud set*. Some manufacturers recommend that mud-set tile applications be made over mineral-surfaced roll roofing weighing not less than 83 lb. per square. Others have different requirements.

All sources agree that valleys should have an additional 36-in.-wide layer of 30-lb. roofing felt beneath the flashing, and that felts should be carried 6 in. up vertical surfaces.

Under the conditions outlined earlier, clay and concrete tile roofs should have ice shields. The ice shield should consist of at least two layers of 30-lb. felt.

7.6.4.3

Installation

The following paragraphs assume that clay and concrete tile, other than in mud-set applications of concrete tile, will be installed using nails. Similar requirements also apply to tile installed using wire hangers. Wire hanger installation should be done in strict accordance with code requirements and the recommendations of the manufacturers of the tile and the wire hanging devices.

In some areas of the country, on roof slopes below 4 in./ft., concrete tile is sometimes installed without nails in mortar beds laid directly over roll roofing. These mud set applications should conform to the tile manufacturer's recommendations.

Clay tile should be installed with one, two, or three nails. Nail-applied concrete tile should be hooked over battens (where battens are used) and nailed in place with one, two, or three nails. In each case, the number of nails depends on the type of tile and the tile manufacturer's recommendations. Most field tile will require only one nail. Hip and ridge covers require one or more nails. Tile overlapping flashing should be supported with wire hangers and set in plastic cement. Nail heads should just clear the tile and never be driven down to the tile. Nails must be driven into nailable material, never into joints in the substrates. Some codes require that every concrete tile be nailed. Other codes permit nailing alternate tile rows or even every third tile row. Sometimes requirements vary, depending on roof slope.

The first course of tile should be elevated to proper height by special tile units manufactured for the purpose, by wood strips, or by a raised fascia. Provisions should be made to permit water to drain from the roof beneath the tile. The first tile course should extend beyond the edge of the sheathing or fascia board by 1-1/2 to 3 in., according to the tile manufacturer's recommendations.

Tile is usually installed with butts parallel to eaves and with tile perpendicular to eaves.

Tile should be cut to follow the line of hips and valleys, or special tile may be used. Valleys may be open, closed, or rounded. Most valleys in tile roofs, however, are open. Open ends of tile at valleys should be sealed with special units or mortar, but tile at valleys should not be sealed in a way that would prevent water that has penetrated through the tile to the underlayment surface from draining into the valley. Fan-shaped valley tile should not be sealed at the laps. Battens along valleys should be broken to permit trapped water to escape onto the valley flashing. Tile should lap valley flashing by at least 4 in.

Open valleys should be left open 6 in. at the top and should taper wider at the rate of 1 in. for every 8 ft. of length.

Roofing tile should be cut and fitted close to and sealed against hip and ridge boards with plastic cement or silicone sealant. Similarly, beads of cement should be placed between tiles at hips and ridges. Hip and ridge covers should be nailed in place and lapped 3 in. Laps should be sealed with plastic cement. Open space within hip and ridge covers should not, however, be sealed against air circulation. The space between roofing tiles at ridges should be sealed using special units made for the purpose or by mortar. Additional nailers should be applied if special ridge filler tile units are used. Ends of hip and ridge covers should be closed with special units designed for the purpose or with cement mortar. Where cement mortar is used, care should be exercised to prevent water intrusion beneath the tile, where it can lead to efflorescence on the face of the tile.

On flat-tile roofs, one-and-one-half-width tiles rather than half-tile units are usually used at gables. The smaller tiles are too light and tend to crack or blow off in windstorms.

Most tile is set with uniform exposure, but some is set with varying exposure. During tile installation, spacing of tiles may need adjustment to achieve uniform exposure. Exposure is determined by tile size and headlap (the distance a tile laps over the preceding tile course). Recommended headlap may vary from manufacturer to manufacturer, but most recommend that headlap not be less than 3 in.

Hurricane clips are required on the nose end of each eave-course tile in areas classified by code as wind hazard areas and, whether so classified or not, where winds frequently exceed 70 mph. On steep roofs, hurricane clips should also be used periodically throughout the roof, or the butt of each tile may be embedded in a bead of plastic cement or silicon sealant. Some codes may require additional provisions.

Tile on steep or vertical roofs should be protected against lifting by wind, even when not in wind hazard areas, by means of wind lock clips.

Other precautions are necessary in high-wind areas. For example, tile headlap should be increased to 4 in., and a bead of mastic should be applied over nail heads at gable, rake, and ridge tiles.

Open-end tile should be sealed at the eaves with special units made for the purpose or sealed with mortar. Either method must permit trapped water to drain out.

7.6.5

MINERAL FIBER–CEMENT SHINGLES

Because of possible health hazards, roofing and siding shingles that were earlier made with asbestos fibers are now manufactured using nonasbestos mineral fibers. The term used here for such shingles is *mineral fiber–cement shingles*. Similar materials found in older buildings, however, may contain asbestos. Readers are, therefore, strongly advised to consult current OSHA, EPA, state, and local regula-

tions and suggestions with regard to safety precautions and to obtain professional assistance when dealing with such existing materials.

This section describes materials and methods needed to provide a suitable roof covering or wall finish of mineral fiber–cement shingles over both new and existing construction.

Reference should be made to Section 7.6.1 for definitions of terms, limitations imposed by the slope of the roof, sheathing guidelines (see also Chapter 6), and general guidelines for typical accessory materials, such as underlayment, flashing, roofing cements, eave flashings, drip edge, nails, and caulking for preparation of a roof or wall for application of shingles. Unless specific exceptions are made in the following text, the provisions of Section 7.6.1 apply to mineral fiber–cement shingles and the methods used to apply them.

7.6.5.1

Materials

Mineral fiber–cement shingles can be discolored by moisture and dampness while still in packages and until applied. Therefore, they should be kept clean and dry and completely protected from the weather until they have been applied. Shingles should not be applied over wet sheathing, underlayment, or sheathing paper.

ROOF SHINGLES

Mineral fiber–cement roof shingles are manufactured in four basic types: *Individual units* and *multiple units*, for application by the *American method*, and *Dutch lap* units and *ranch style* units, for application by the *Dutch* (or *Scotch*) *method*. Each type of shingle should be applied in a distinct way giving a different appearance. Starter units, hip and ridge shingles, and ridge roll of mineral fiber–cement are also manufactured.

Roof shingles are made with prepunched holes correctly located to receive concealed roofing nails. Additional holes are provided in Dutch lap and ranch style units to receive storm anchors.

Mineral fiber–cement roof shingles should conform to the requirements of ASTM C 222. Data covering the four types are summarized in Figure 7.6-49.

WALL SHINGLES

Mineral fiber–cement wall shingles are usually supplied in rectangular units having straight or wavy butt edges. These units range in size from 8 to 16 in. in width and from 24 to 48 in. in length and are designed to be applied with a minimum nominal toplap of 1 in. Depending on their size, these units range up to approximately 3/16 in. in thickness (Fig. 7.6-50).

Wall shingles are manufactured with prepunched nail holes correctly sized and located to receive exposed face nails. The holes are aligned the proper distance above the lower edge of the units to establish the recommended amount of toplap (Fig. 7.6-51) and to minimize damage to shingles during their installation. For standard application, face nails driven through these holes and resting on the top edge of the next lower course serve as guides for the correct amount of toplap, exposure, and unit alignment.

Mineral fiber–cement wall shingles should conform to the requirements of ASTM C 223. Data covering types of such units are summarized in Figure 7.6-50.

NAILS AND FASTENERS

ROOFING NAILS Roofing nails for applying mineral fiber–cement shingles should be corrosion resistant, such as hot-dip galvanized steel, aluminum, or stainless steel, and should be needle or diamond pointed, with large flat heads approximately 3/8 in. in diameter. Shanks should not exceed 0.140 in. in overall outside diameter, so as not to exceed the diameter of prepunched holes in the shingles. Shanks may be smooth or threaded; threaded nails are preferred, to increase holding power. Aluminum nails should have screw threads with approximately a 12-1/2-degree thread angle; galvanized steel nails, if threaded, should have annular threads.

Nails for applying shingles over solid lumber sheathing should be approximately 1-1/4 in. by 0.120 in., and have threaded shanks.

Nails for application of shingles over plywood sheathing should have threaded shanks and be approximately 1-1/4 in. by 0.120 in., or long enough to penetrate through the sheathing.

For information about nails for plank decking and nonwood sheathing, refer to Section 7.6.1.7.

FIGURE 7.6-49
Mineral Fiber–Cement Roof Shingles*

Type	Width, in.	Height, in.	Thickness, in.	Exposure, in.	Weight per Square, lb.
Individual unit	8	16	5/32	7	350
			1/4	7	585
	9	18	1/4	8	570
Multiple unit	30	14	5/32	6	440
Dutch lap	16	16	5/32	12×13	265
Ranch style	24	12	5/32	20×9	255

*Products of individual manufacturers may vary slightly from the dimensions listed.

FIGURE 7.6-50
Mineral Fiber–Cement Wall Shingles

Sizes,[a] in.	Exposure, in.	Pieces per Bundle	Pieces per Square[b]
12×24	11	18	54
9×32	8	19	57
9-5/8×32	8-5/8	17-17-18	52
11-3/4×32	10-3/4	14	42
14-5/8×32	13-5/8	11	33
14-3/4×32	13-3/4	11	33
8-3/4×48	7-3/4	13	39
12×48	11	9	27

[a]Three nail holes per shingle.

[b]Weight per square ranges from 153 to 206 lb.

FIGURE 7.6-51 *Prepunched nail holes in mineral fiber–cement siding shingles.*

In addition to nails, *storm anchors* should be used to apply ranch style and Dutch lap units. Storm anchors should be copper, aluminum, or lead, approximately 3/4 in. long with a flat base and have sufficient ductility to permit repeated clinching and bending upward without cracking

FIGURE 7.6-52 *Storm anchors.*

the shank (Fig. 7.6-52).

The number, spacing, and pattern of nails and fasteners are described in Section 7.6.5.3.

WALL NAILS Face nails should be long enough to enter into and hold securely to the nailing base. Face nails are usually furnished by the siding manufacturer to ensure the use of the proper kind and type. Nails and fasteners should be permanently corrosion resistant and stain resistant. Only those furnished with the siding or recommended by the siding manufacturer should be used.

Figure 7.6-53 is a summary of face nail types recommended for use in applying mineral fiber–cement wall shingles. The number, spacing, and pattern of nails are described in Section 7.6.5.6.

7.6.5.2

Preparation for Shingle Roofing

■ **SLOPE LIMITATIONS**

Limitations imposed by roof slopes are shown in Figure 7.6-8. Mineral fiber–cement shingles are applied by normal slope methods on roofs with slopes of 5 in 12 or greater. Lower than normal-slope methods may be used on slopes as low as 3 in 12.

Mineral fiber–cement shingles are not recommended for use on roof slopes less than 3 in 12.

■ **ROOF SHEATHING**

Sheathing types, thicknesses, spans, and nailing requirements are discussed in Section 7.6.1.3 and in Chapter 6.

■ **UNDERLAYMENT**

Underlayment material should be unperforated No. 15 asphalt-saturated inorganic roofing felt or No. 30 asphalt-saturated organic felt. Coal tar–saturated felt, which may stain the shingles, should not be used. Underlayment should be applied over the entire roof as described in Section 7.6.1.5 and in Section 7.6.5.3.

■ **CANT STRIPS**

When the American method is used with individual-unit and multiple-unit shingles, a 1/4-in. by 2-in. wood strip should be applied along the eaves to pitch (cant) the starter strip at the proper angle (Figs. 7.6-54 and 7.6-55).

FIGURE 7.6-53
Face Nails for Use with Mineral Fiber–Cement Wall Shingles

	New Construction	
Sheathing Type	Nailed Directly to Sheathing or Wood Nailing Strips	Nailed Through Fiberboard Shingle Undercoursing
Board lumber	1-1/8-in. screw thread 1-7/16-in. screw thread 1-1/2-in. annular thread	2-in. annular thread
Plywood	1-1/8-in. screw thread	1-3/4-in. screw thread
Wood nailing strip	1-1/8-in. screw thread	1-3/4-in. screw thread
Nailbase fiberboard	1-3/8-in. special annular thread	2-in. special annular thread

	Residing (over existing wood siding or shingles)
Nailing Base	Nail Types
Sound board lumber	1-3/4-in. or longer annular thread, screw thread, or helical thread
Wood nailing strip	1-1/8-in. screw thread

FIGURE 7.6-54 *Application of individual shingles with straight butt lines.*

■ DRIP EDGE

A corrosion-resistant nonstaining drip edge should be installed along the eaves and rakes. When cant strips occur, the drip edge should be applied over the cant strips.

■ EAVE FLASHING

In areas where the outside design temperature is 30°F or colder (see Fig. 7.6-11) or where the possibility of ice forming along the eaves could cause a backup of water, eave flashing should be installed as described below.

NORMAL SLOPE (5 IN 12 OR GREATER) Eave flashing should be formed by cementing an additional strip of underlayment over the first underlayment course. A continuous layer of plastic asphalt cement should then be applied, at the rate of 2 gal. per square, from the eave to cover a line at least 12 in. inside the interior wall line of the building (see Fig. 7.6-13). This layer should be covered with the additional ply of underlayment.

LOWER THAN NORMAL SLOPES (3 IN 12 TO 5 IN 12) A cemented eave flashing for lower than normal slopes should be formed just as described above for normal slopes. However, this flashing should extend from the eave to cover a line at least 24 in. inside the interior wall line (see Fig. 7.6-13).

■ FLASHING

The following requirements are in addition to those in Section 7.6.1.6.

VALLEY FLASHING The open or closed method may be used to construct valley flashing, as shown in Figure 7.6-56.

CHIMNEY AND SIDE WALL FLASHING When shingles butt against vertical surfaces, such as side walls, dormers, and chimneys, individual flashing units of corrosion-resistant metal should be inserted between each course of shingles. These metal flashing units should be folded from a width of 19-in. material and have a standing leg 6 in. high and a flat leg 4 in. wide on the shingle. The length of this flashing should be 3 in. greater than the exposure of the shingles. Side wall ex-

FIGURE 7.6-55 *Application of multiple-unit shingles.*

terior finish materials should cover the standing leg. At masonry chimneys, a stepped counter flashing of corrosion-resistant metal should cover the standing leg of the flashing units and be trimmed just above the surface of the shingles.

7.6.5.3

Application of Shingle Roofing (New Construction)

The *American method* should be used to apply individual and multiple shingles without a sidelap. The *Dutch method* with sidelap and toplap should be used to apply Dutch lap and ranch style shingles. In both methods the underlayment felt should be rolled back and starter shingles applied directly on top of the metal drip edges pre-viously nailed. Starter shingles should overhang the rakes and the eaves about 3/4 in. When starters match the color of the field shingles and are used without a gutter, they are usually applied with the colored side down for better appearance from below.

■ **AMERICAN METHOD (NORMAL SLOPE)**

INDIVIDUAL SHINGLES (STRAIGHT BUTT LINE) After replacing the underlayment over the starter shingles, the first shingle course should be applied starting with a half-shingle overhanging the rake about 3/4 in. and in line at the butt with the starter shingles. Each shingle should be applied with nails through two of the prepunched holes. The second course should be started with a full shingle, and

the third course, like the first course, should start with a half-shingle. Chalk lines should be marked on the underlay-ment for each shingle course for accurate alignment (see Fig. 7.6-54).

INDIVIDUAL SHINGLES (STAGGERED BUTT LINE) Individual shingles have two sets of prepunched holes to permit a staggered application. The first course should be applied as described above for straight butt line. In starting the second course, the ex-posure of every second shingle should be reduced by dropping it down and nailing through the upper set of nail holes. Stag-gering can be varied, depending on the ap-pearance desired, but is limited to the dis-tances between the upper and lower holes. The third course should be started with a half-shingle, and alternating shingles should be staggered as in the previous course.

Individual shingles with either straight or staggered butt line should have a mini-mum headlap and sidelap of 2 in. (Fig. 7.6-57; see also Fig. 7.6-54).

MULTIPLE-UNIT SHINGLES After replac-ing the underlayment over the starter shin-gles, the first shingle course should be ap-plied starting with a half-shingle over-hanging the rake about 3/4 in. and in line at the butt with the starter shingles. The second course should be started using a full shingle, and the third course, like the first course, should be started with a half-

FIGURE 7.6-56 *(a) Open-valley flashing; (b) closed-valley flashing.*

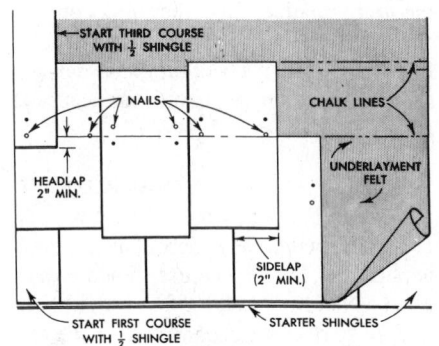

FIGURE 7.6-57 *Staggered butt line can be achieved with individual shingles by varying the exposure.*

shingle. Chalk lines are not necessary if the visual alignment design of the shingles is properly utilized (see Fig. 7.6-55). Overdriving of nails causes the butts of the shingles to cock up and should be avoided to prevent damage to the shingles.

Because the design of multiple-unit shingles may vary among manufacturers, application instructions of the manufacturer for nailing, positioning, and alignment should be followed.

■ DUTCH METHOD (NORMAL SLOPE)

The Dutch method is used to apply Dutch lap shingles (Fig. 7.6-58a) and ranch style shingles (Fig. 7.6-58b). The following application methods refer to shingles having a 4-in. sidelap.

Starter shingles should be applied as shown in Figure 7.6-59a and 7.6-59b. After the felt has been replaced over the starter shingles, a 4-in.-wide piece cut from a full shingle should be applied to form a rake starter. As shown in Figure 7.6-59c, storm anchor A should be inserted from below and the rake starter secured in place using two nails. A full-sized shingle should be applied over the rake starter by first inserting storm anchor B and positioning the shingles so that storm anchor A appears through the free corner (Fig. 7.6-59d). Storm anchor A should be clinched and the shingle secured with two nails (Fig. 7.6-59e). Each succeeding shingle should be applied in the same manner, inserting first a storm anchor in one corner and then engaging the other corner over the previous storm anchor. The second shingle course should be started with a shingle 4 in. less than full width and punched with a new hole as required to hold the shingle along the rake and completed using full shingles. The third course should be started with a shingle 8 in. less than full width; the fourth course with a shingle 12 in. less than full width, and so on up the rake. Automatic alignment of the shingles is usually provided by the location of the prepunched holes.

■ HIPS AND RIDGES

Before hip and ridge shingles (Fig. 7.6-60) are applied, wood nailing strips equal in thickness to the roof shingle assembly should be applied at the hips and ridges to provide a flat surface. Hips or ridges should then be covered with two thicknesses of underlayment (Fig. 7.6-61), or a double thickness of 8-in.-wide felt can be adhered with plastic cement over the nailing strips. Application of hip and ridge shingles should be started with a half-length shingle, covered with a full-length shingle. Thereafter, full-length shingles should be applied with a maximum of 8-in. exposure, depending on the design of the shingle used, by nailing each shingle through two nail holes. Alternating shingles should be lapped on either side of hip or ridge. Finally, the ridge joint should be pointed with plastic asphalt roofing cement.

■ RIDGE ROLL

A tapered half-round ridge roll is used in some localities (Fig. 7.6-62). Before a ridge roll is applied, a wood nailing strip should be installed along the apex of hips and ridges. The nailing strip should then be covered with a double thickness of underlayment. Whenever possible, the application of a ridge roll should start at the end of the ridge and be laid in a direction opposite to that of the prevailing winds; at hips, a ridge roll should be started at the eaves.

The first section of a ridge roll should be placed with the large end at the starting point overhanging the roof shingles 1/4 in. Each end should be fastened to the nailing strip by nailing at the head end through a corrosion-resistant clip designed for this purpose. The next section of ridge should overlay the starter at least 2 in. These clips should be bent back to fasten the exposed end of the next ridge roll section. This method should be continued until the entire hip or ridge has been covered. Ridge pieces should be embedded in asphalt plastic cement, and exposed nails should be pointed with plastic cement. The openings at end units may be closed with colored cement mortar or with a closure consisting of a half-round piece cut from a shingle that is nailed or screwed into the end of the nailing strip.

FIGURE 7.6-58 *(a) Dutch lap shingles and (b) ranch style shingles. Holes marked A and B are for nails; holes D and E are for storm anchors.*

FIGURE 7.6-59 *Application of Dutch lap and ranch style shingles. (a, b) Apply starter shingles. (c) Insert storm anchor A from below and secure the rake starter in place using nails. (d) Insert storm anchor B and position shingles so that storm anchor A appears through free corner. (e) Clinch storm anchor A and secure shingle with nails.*

FIGURE 7.6-60 *Hip and ridge shingles.*

■ **LOWER THAN NORMAL SLOPE METHOD (3 IN 12 TO 5 IN 12)**

Methods for applying shingles on roof slopes less than 5 in 12 but not less than 3 in 12 are the same as described above for normal slopes 5 in 12 and over, with certain exceptions with regard to underlayment.

For slopes 3 in 12 up to 4 in 12, two layers of underlayment should be applied over the entire roof. For slopes 4 in 12 up to 5 in 12, one layer of underlayment is sufficient, except that two layers are required in areas where the outside winter design temperature is 30°F or lower.

When eave flashing is required on lower than normal slope roofs, the additional layer should be cemented at the eaves as described in Section 7.6.1.6.

7.6.5.4

Reroofing Existing Roofs

When mineral fiber–cement shingles are applied to an existing roof, the surface must be made as smooth as possible. Mineral fiber–cement shingles may be applied directly over old asphalt shingles, provided the sheathing and roof framing members are sound and strong enough to support the increased loads.

Loose or curled wood shingles should be split and nailed down and missing shingles replaced. Beveled wood strips should be installed below the butts of each course of shingles (Fig. 7.6-63). When existing wood shingles fail to retain nails with at least the force of 40 lb. per fastener, the shingles should be removed and the existing sheathing prepared as it would be in new construction. If old wood shingles are applied over spaced sheathing, new sheathing should be fitted between the strips to fill the spaces and provide a solid deck. In that case, it is often simpler to remove the old sheathing strips and construct a new deck.

Nails for applying shingles to existing roofs should be long enough to penetrate through the existing sheathing. Threaded shank roofing nails provide improved holding power.

■ **REPLACING DAMAGED SHINGLES**

Damaged shingles may be removed and replaced as shown in Figure 7.6-64 and as described below.

First, the damaged shingle should be shattered and removed. The nails may then be withdrawn or cut off by hooking a slater's ripper over the nails and striking the offset handle of the ripper with a hammer (Fig. 7.6-64a). A new shingle should then be positioned.

With the American method of application, a hole should be drilled through the joint between the overlying shingles and the head of the new shingle. One nail should be driven through that hole, and a 3-in.-wide corrosion-resistant metal flashing strip should be placed over the nail head and under the overlying shingle (Fig. 7.6-64b).

For the Dutch method of application, the new shingles should be anchored in place with two spots of roofing cement placed on the head of the underlying shingle and by engaging the storm anchors. Nailing is not required (Fig. 7.6-64c and 7.6-64d).

FIGURE 7.6-61 *Application method for finishing hips and ridges.*

Labels in figure 7.6-61:
- UNDERLAYMENT FELT
- DECK
- ROOF SHINGLES BUTT AGAINST WOOD STRIPS
- HIP & RIDGE SHINGLES OVERHANG WOOD STRIPS
- ROOF SHINGLES
- PLASTIC ASPHALT CEMENT
- UNDERLAYMENT FELT 12" WIDE STRIP
- WOOD STRIP
- 8" WIDE STRIP OF UNDERLAYMENT FELT
- NAILS
- 5" EXPOSURE

FIGURE 7.6-62 *Ridge roll may be used on ridge as an alternative to ridge shingles.*

Labels in figure 7.6-62:
- OPEN RIDGE ROLL
- RIDGE CLIP
- NAIL
- NAILING STRIP
- FELT
- UNDERLAYMENT

FIGURE 7.6-63 *Preparation of existing roof for application of new shingles.*

Labels in figure 7.6-63:
- SPLIT AND NAIL DOWN WARPED SHINGLES
- BEVELED STRIPS (NEW)
- 1/4" x 2" CANT STRIP
- CUT BACK SHINGLES AT EAVES AND RAKES. REPLACE WITH NEW WOOD

7.6.5.5

Preparation for Wall Shingles

■ WALL SHEATHING

Siding should be applied to surfaces that are smooth and dry and that will provide adequate support and an adequate nailing base. Mineral fiber–cement shingle siding can be applied over lumber, plywood, and other suitable types of rigid sheathing. Sheathing material thicknesses, spans, and nailing requirements are discussed in Section 7.6.1.3 and in Chapter 6.

■ SHEATHING PAPER

Sheathing paper for use over lumber sheathing should be water repellent yet vapor porous (permeable to water vapor), such as No. 15 asphalt-saturated felt. Coal tar–saturated felt, which may stain the shingles, should not be used. Vapor retarder membranes should not be used for sheathing paper between siding and sheathing.

Sheathing paper for walls is discussed in Section 6.6.5.1.

■ JOINT FLASHING STRIPS

Asphalt (not coal tar) -saturated and -coated, water-repellent joint flashing strips are supplied by manufacturers with all types of mineral fiber–cement shingle siding and should be applied behind every vertical joint between wall shingles (Fig. 7.6-65).

■ CANT STRIPS

A 1/4-in. by 1-1/2-in. wood strip should be nailed level along the bottom edge of the sheathing so that it overhangs the top of the foundation wall a sufficient amount to seal the joint between the top of the foundation, the wood sill, and the bottom of the sheathing (see Fig. 7.6-65). The bottom edge of the shingles should align with the bottom edge of the cant strip. This strip gives the necessary cant (pitch) to the first course of shingles and provides solid bearing to prevent breakage.

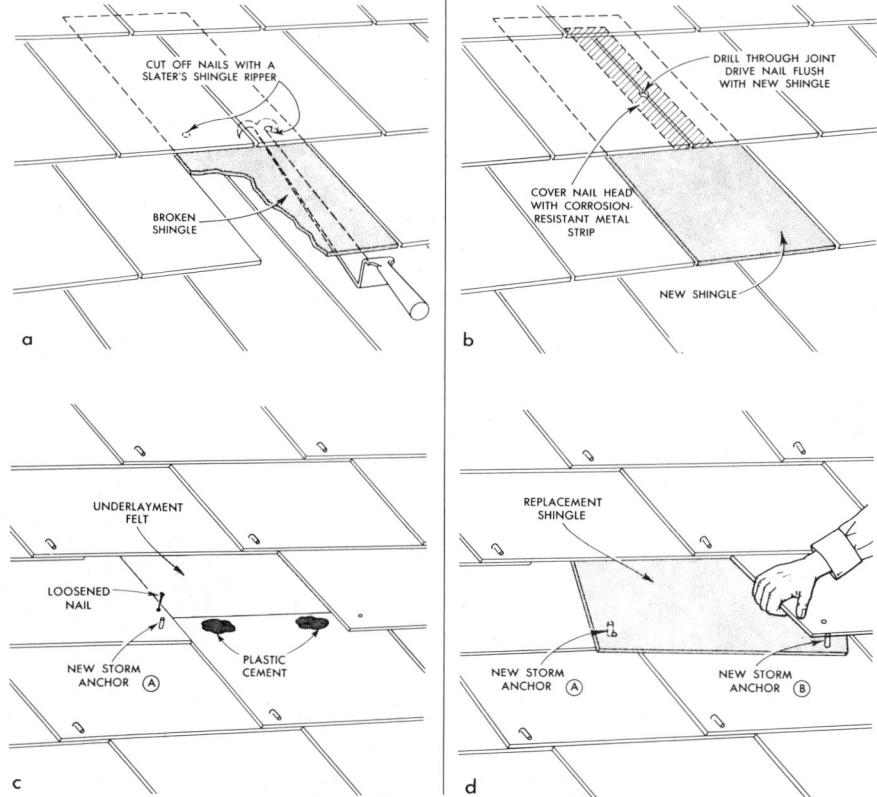

FIGURE 7.6-64 *To replace damaged shingle. (a) Remove damaged shingle and nails.* *(b) Nail new shingle through side joint and cover nail with corrosion-resistant metal strip inserted into lap. (c) In Dutch method, remove damaged shingle and shear off nails and storm anchor by engaging them with the hook end of a slater's ripper and striking offset on handle with hammer. Loosen nail in exposed shingle, remove sheared-off storm anchor, and insert new storm anchor. Drive loosened nail home and apply plastic cement. (d) Place storm anchor B in new shingle. Raise overlapping shingle to allow storm anchor to align with holes. Embed shingle in cement and bend storm anchor down.*

■ **FLASHING**

Flashing should be done in accordance with Section 7.6.1.6.

7.6.5.6

Application of Wall Shingles (New Construction)

Depending on the type of sheathing, mineral fiber–cement shingles may be applied by several methods: (1) *direct application* to nailable sheathing, (2) the *shingle backer method*, (3) the *channel method* of application over nailable sheathing or directly over framing members without sheathing, and (4) the *wood nailing strip method* for use over any type of sheathing

FIGURE 7.6-65 *Direct application of mineral fiber–cement shingles showing application of joint flashing strips.*

or directly over framing members without sheathing.

The methods described are based on applying 12-in. by 24-in. wall shingles from left to right on new construction. The manufacturers' instructions should be followed when applying other sizes of shingles.

Careful layout should establish the location of the first horizontal course and ensure that all courses will meet and match at corners and are level all around the building. The total number of horizontal courses to complete one wall should be laid out, and adjustments should be made so that the top courses at the eaves will not be too narrow. This may require uniform adjustments in reducing the exposure of each shingle course. A level chalk line should be marked all around the building to fix the location of the top edge of the first course of shingles. Additional horizontal chalk lines should mark the top of each succeeding course. The lines should be spaced to provide the desired exposure (see Fig. 7.6-65).

■ **DIRECT APPLICATION METHOD**

The direct application method is used for sheathing types that provide a nailing base, such as wood boards or plywood.

FIRST AND ODD-NUMBERED COURSES A full-sized shingle should be started at the left corner of the wall. This unit should be carefully applied plumb, level, and in alignment with a chalk line, because it guides the laying of all other units.

The correct face nails for the type of sheathing and application method being used are supplied by the manufacturer with the shingles, as described in Figure 7.6-53. Nails should be driven snug but not too tight.

Before the last nail is driven at the right end of a unit, a felt joint flashing strip should be inserted in place and secured with the last nail. Joint flashing strips should always be used, centered behind the butted joint between wall shingles and with the lower end overlapping the cant strip or head of the next lower course.

Application of full-sized shingles should be continued in the first course, with the top edges aligned to the chalk line. The last unit in a course or at an opening should not be less than 6 in. wide; if smaller spaces will remain to be filled, a few inches should be cut from the units applied earlier in the course. Necessary holes for face nails in less than full-sized pieces of siding should be punched. Ends of units should be butted tightly end to end with no space between. Wedge-shaped spaces between ends of units indicate incorrect application.

SECOND AND SUCCEEDING EVEN-NUMBERED COURSES In succeeding courses of 12-in. by 24-in. shingles, vertical joints should break on *halves*; therefore, the second course should be started with a half-unit. When 32-in.-long units are used, vertical joints should also break on halves. When 48-in.-long units are used, they should break on *thirds* and the second course should be started with a two-thirds unit. Some bundles of shingles contain the necessary shorter pieces for starting courses.

A partial unit should be started at the left corner with its head edge aligned to the chalk line and its lower (butt) edge overlapping the head of the next lower course the correct distance, to provide the necessary toplap between courses. A nail should be inserted in a face nail hole and the siding positioned so that the shank of the nail is resting on the top edge of the next lower course, thus establishing the proper amount of toplap. The unit should then be nailed in place with a felt joint flashing strip installed as described above. The course should be continued with full-sized units.

The face nails described in Figure 7.6-53 should always be used. The nails should penetrate completely through the sheathing. A standard threaded face nail should be used when the siding is applied directly to nominal 1-in. wood board sheathing or 5/16-in. and thicker plywood without wood nailing strips or shingle backer.

■ SHINGLE BACKER METHOD

The shingle backer method uses rigid insulating fiberboard as a backer for the shingles and may be used over wood boards or plywood, since they provide a nailing base for both the shingle backer and the shingles (Figs. 7.6-66, 7.6-67, and 7.6-68). Special methods may be used to apply shingles directly over framing members.

A rigid, water-resistant, insulating fiberboard shingle backer should be at

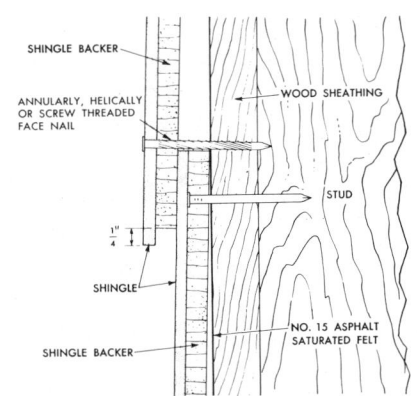

FIGURE 7.6-66 *Application of mineral fiber–cement wall shingles using the shingle backer method.*

FIGURE 7.6-67 *Application of mineral fiber–cement wall shingles using rabbeted nailing strips.*

FIGURE 7.6-68 *Application of mineral fiber–cement wall shingles using a grooved shingle backer.*

least 5/16 in. thick and should underlay the full width and length of each course of shingles completely. The shingle backer should be approximately 1/4 in. narrower than the shingles. The shingles should be applied with their head edges flush with the head edges of the shingle backer and their butt edges extended down and beyond the butt edges of the underlying shingle backer to provide a drip edge. A shingle backer should be at least 4 ft. long and applied with staggered joints in succeeding courses. Vertical joints between shingles should not occur over the ends of the shingle backer. The shingle backer should be applied with its butt edges overlapping the head of the lower course of shingles about 1/4 in. less than the headlap used between courses of shingles.

The shingle backer and the shingles should be secured to the wall with face nails driven through face nail holes in the shingles and into the underlying sheathing. The shingle backer should be held in position before shingles are applied by nailing each panel about 3 in. from the top edge with two or three 1-1/4 in. galvanized roofing nails. The type and the length of the face nails should be as described in Figure 7.6-53.

Succeeding courses of shingles should be applied in a similar manner. Felt flashing joint strips are not required if the shingle backer board is water repellent and stain resistant.

Direct attachment to framing members is possible using a rabbeted nailing strip with a special shingle backer that also serves as sheathing (see Fig. 7.6-67). This method provides firm backing and a heavy shadowline. A similar method utilizing a grooved shingle backer is shown in Figure 7.6-68.

■ CHANNEL METHOD

The channel method uses one of several styles of metal channel molding designed for securing mineral fiber siding to nailable sheathing or directly to framing members. Metal channel should be corrosion-resistant and stain-resistant metal of sufficient thickness to provide adequate support and rigidity. Moldings usually are prepainted to match shingle color and also provide either a standard or heavy butt shadow (Fig. 7.6-69).

A 1-3/16-in. channel permits standard prepunched shingles to be used when units are reversed head to butt (Fig. 7.6-70). Channels should be secured at 16-in. intervals either to the framing member or to nailable sheathing with 10- (0.135 in.) to 12- (0.106 in.) gauge, hot-dip galvanized steel, large-head roofing nails long enough to penetrate and hold securely in the nailing base. Nails should be driven at an angle and set well to provide sufficient pressure on the nailing flange to ensure a tight, vibration-free assembly (Fig. 7.6-71).

Channel molding should have a vertical nailing flange at least 3/8 in. high that can bear against the face of the sheathing and be strong and rigid enough to secure the siding to the wall sheathing.

Direct attachment to framing members requires shingles wide enough to span from center to center of studs. Face nail holes should be punched within 3/8 to 1/2 in. from the vertical edge of the shingles to ensure nailing into the stud when two shingles are butted together. Intermediate face nail holes should be provided for nailing each stud, and face nails should be long enough to penetrate at least 1 in. into the stud.

When this method of nailing is used, studs must be located accurately on 16-in. centers wherever possible. Warped studs should not be used.

When it is not possible to nail directly into a stud, either a *twist nail* or a *self-clinching* nail should be used to attach the corner of the siding to nonlumber sheathing. A twist nail is a slender copper or

FIGURE 7.6-69 *Typical profiles of metal channel moldings for installing mineral fiber–cement shingles by the channel method.*

FIGURE 7.6-70 *Application of mineral fiber–cement shingles using the channel method. See also Figure 7.6-71.*

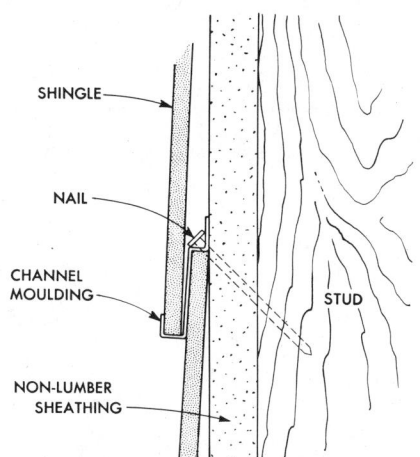

FIGURE 7.6-71 *Channels are nailed directly to framing members when sheathing does not provide a nailing base.*

aluminum nail with a flat head and a medium needle point, designed for twist clinching, that is, manual twisting of the exposed shank. A self-clinching nail is designed so that its shank clinches automatically when the nail has been fully driven.

The inside of exterior walls should be left exposed for final inspection to ensure that shingle nails have been properly driven into the studs and that twist nails and self-clinching nails are properly clinched.

■ WOOD NAILING STRIP METHOD

The wood nailing strip method may be used over lumber boards, plywood, and any type or thickness of nonlumber sheathing that does not have adequate nail-holding power, because the strips rather than the sheathing serve as the nail base. No special nails, fasteners, or channels are required, and the standard face nails shipped with the siding can be used.

In addition to providing maximum security for the siding and economy in application, this method produces a heavy shadowline along the butt edge of each course of shingles.

Wood nailing strips for the application of mineral fiber–cement shingles should be at least nominal 3/8-in. by 3-1/2-in. boards that are straight, sound, and free from knots or knotholes that might cause splitting or hinder nailing. Strips should be not less than 4 ft. long; preferably they should be longer, so that they can span continuously over several framing members. Seasoning conditions and face nailing of spaced boards should be the same as described for solid sheathing.

Wood nailing strips should be placed to overlay the head of the lower course of shingles in an amount equal to the desired toplap less 1/4 in. (Fig. 7.6-72). This firmly clamps the head of the course of shingles to the wall. At each framing member, the wood strip should be nailed through the sheathing and to each underlying stud with two 2-in. by 0.113-in. (6d) or longer annular threaded nails, penetrating the stud at least 1 in. The wood strips should end over framing members; joints in the strips should be staggered in succeeding courses. Vertical joints between shingle units should not occur over the end joints of wood strips.

The next course of shingles should be applied with its butt edge placed over the wood strip, with the lower edge of the shingles extending down over the lower edge of the strip not more than 1/4 in., to provide a drip. The siding should then be nailed through the face nail holes to the

FIGURE 7.6-72 *Application details of mineral fiber–cement shingles using the wood nailing strip method.*

FIGURE 7.6-75 *Head flashing over an opening in a mineral fiber–cement shingle wall.*

wood nailing strips with the threaded face nails supplied with the shingles.

■ CORNER FINISHING

When wall surfaces join at inside corners, the courses of shingles are usually butted, although metal molding or wood corner strips can be used. Outside corners can be finished by any one of the following three methods.

METAL CORNER FINISH Several styles of individual stain-resistant metal corners and finish moldings are available (Fig. 7.6-73). They should be applied as recommended by the manufacturer.

WOOD CORNER BOARD FINISH Corners finished with wood strips can be made of previously primed or painted nominal 1-in. lumber applied along one corner. A 1×4 should be applied along one corner so that its side edge projects a distance beyond the corner equal to the thickness of the board applied on the adjoining wall. A 1×3 should then be applied with its edge butting against the back of the 1×4 first applied. Therefore, the wood corner strips will appear the same width on each wall (Fig. 7.6-74). When wood nailing strips or shingle backer, or both, are used, wood corner boards thicker than nominal lumber or blocking may be required.

LAPPED OR WOVEN CORNER FINISH This method is appropriate for white or gray shingles only. For lapped adjoining courses at outside corners, a whole shingle should be placed temporarily against each

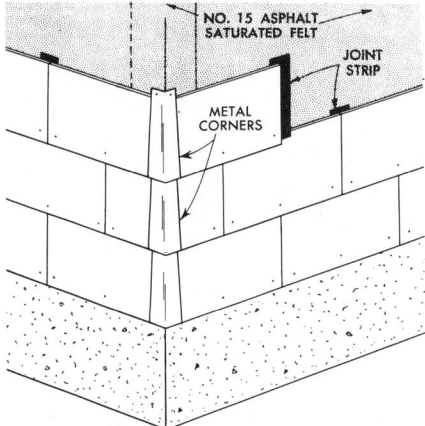

FIGURE 7.6-73 *Metal corner for mineral fiber–cement shingles.*

FIGURE 7.6-74 *Wood board corner for mineral fiber–cement shingles.*

adjoining wall, so that shingles meet and join at the corner. The adjoining vertical edges of the shingles are scribed and cut at the proper angle to form a lapped or woven corner joint, and the edges are dressed with a file. Corner joints should be lapped to the left and right in alter-

nate courses. Courses on adjoining walls should meet and match at corners.

■ FLASHING AT OPENINGS AND CORNERS

Metal flashing should be applied over exposed door and window openings (Fig. 7.6-75). On old structures, the flashing must be renewed if not in good condition. Drip edges should be provided at heads of openings, under sills, and at projections to keep water away from the finished wall surface below.

Outside and inside corners should be flashed with a 12-in.-wide strip of underlay material centered in or around the corner when corner boards are used (see Fig. 7.6-74).

In other corner treatments, the underlayment should be carried around the corner of each side wall so that a double thickness of felt results over the corner (see Fig. 7.6-73).

7.6.5.7

Reshingling Existing Walls

An old wall that is to be covered with new mineral fiber–cement shingles should be repaired so that it has a smooth, substantially true surface with adequate nail-holding capacity and the ability to support the new shingles. When the existing exterior finish consists of relatively thin bevel siding applied directly to widely spaced framing members and does not provide an adequate nailing base for new shingles, beveled wood strips should be placed at nailing locations along the wall. The face nails described in Figure 7.6-53 should be used for application of new shingles.

■ RESHINGLING OVER NONLUMBER SHEATHING

When mineral fiber–cement shingles are to be applied to a structure that originally was sheathed with a nonnailable sheathing, and the existing siding to be covered will not provide an adequate nailing base, the only recommended attachment method is the wood nailing strip method discussed in Section 7.6.5.6. Using any other method will require that both the existing siding and the existing sheathing be removed down to the existing framing members and replaced with new sheathing to receive new shingles.

■ RESHINGLING OVER STUCCO

When new shingles are to be applied over old stucco, the best practice is to remove the old stucco, nails, lath, and protruding elements and apply the new shingles directly over the existing sheathing, if any exists. If the stucco is in reasonably good condition or is removed to bare studs, new shingles can be applied by the wood nail strip or metal channel method described in Section 7.6.5.6, modifying the steps as necessary to meet special conditions.

7.6.6

WOOD SHINGLES AND SHAKES

The following text describes materials and methods recommended to provide a suitable roof covering and wall finish of wood shingles and shakes over both new and existing construction.

Reference should be made to Section 7.6.1 for definitions of terms, limitations imposed by the slope of the roof, sheathing guidelines (see also Chapter 6), and general guidelines for typical accessory materials, such as underlayment, flashing, roofing cements, eave flashings, drip edge, nails, and caulking for preparation of a roof or wall for application of shingles. Unless specific exceptions are made in the following text, the provisions of Section 7.6.1 apply to wood shingles and shakes and methods of their application.

7.6.6.1

Materials

Most wood shingles and shakes used for roofs and side walls are produced from western red cedar trees, giant, slow-growing coniferous trees of the Pacific Northwest. Western red cedar has extremely fine and even grain, exceptional strength in proportion to weight, low expansion and contraction with changes in moisture content, and high impermeability to water. Its outstanding characteristics are great durability and resistance to decay.

Shingles are sawn, have smooth faces and backs, and are produced in four grades and a variety of types. Shakes are split, have rough split faces and either sawn or split backs, and are produced in one grade and in three types.

Both shingles and shakes are made from selected cedar logs cut to proper lengths— 16, 18, and 24 in. for shingles and 18, 24, and 32 in. for shakes. These lengths are again cut or split into the proper size blocks for the shingle-cutting machine or a convenient size for splitting of shakes. In reducing the large log sections into blocks, each is quartered, split, and requartered to produce blocks having a true edge-grained face (Fig. 7.6-76).

Each bundle of wood shingles and shakes should carry the official grade-marked label of the Cedar Shake and Shingle Bureau.

■ CEDAR SHINGLES

Roof shingles are cut from cedar blocks on an upright shingle machine (Fig. 7.6-77). This cutting machine carries a block past a thin, razor-sharp circular saw, which slices a shingle smoothly at each stroke. Shingles cut from one block are jointed or edged and graded by a sawyer. This sawyer can adjust the block in the machine from time to time so that the highest possible grade of shingles will be produced.

A sawyer picks up each shingle as it falls from the saw and places its butt firmly against a guide. Using a circular saw set at right angles to the butt, the sawyer clips one edge of the shingle off smoothly. Flipping the shingle over, the sawyer repeats the process, making another smooth edge parallel to the one first cut. If defects are present, the sawyer trims the shingle in such a way that a narrower shingle or two shingles without defects can be obtained. Depending on the grade, the sawyer then drops the shingle into a chute that leads to the packing bins.

Shingles are graded again as they are packed into bundles in a packing frame. Bundles of shingles are held together by two bandsticks connected by metal straps. Packaged in this manner, bundled shingles are seasoned and remain perfectly flat in the process.

GRADES, TYPES, AND SIZES Shingles are marketed according to grade, thickness, and type. Nominal shingle lengths are 16, 18, and 24 in., within 1/4 in. In addition, a tolerance of 1 in. over or under the specified length is permitted in 10% of the

FIGURE 7.6-76 *Logs are cut to length (a), then quartered to produce edge-grained blocks for sawing or splitting into shingles or shakes (b).* *(Courtesy Cedar Shake and Shingle Bureau.)*

FIGURE 7.6-77 *A sawyer edges shingles after they have been cut from cedar blocks by circular saw at sawyer's left.* (Courtesy Cedar Shake and Shingle Bureau.)

shingles. All grades are produced in these lengths, except that undercoursing grade is not produced in the 24-in. length. A 16-in. shingle is known as XXXXX, or FiveXs; an 18-in. shingle as Perfection; the 24-in. as Royal.

Most cedar shingles conform to one of three grades: No. 1 (Blue Label), No. 2 (Red Label), and No. 3 (Black Label). Other available grades include the following: No. 4, undercoursing; No. 1 and No. 2, rebutted-rejointed; No. 1, machine-grooved; No. 1 and No. 2, dimension; No. 1 and No. 2, preformed; and factory-built hip and ridge units.

Except for dimension shingles, all shingles are produced in random widths from 3 in. up to a maximum of 14 in. Shingle thickness varies depending on length. A 16-in.-long shingle must be thick enough so that the combined butt thicknesses of five shingles will measure at least 2 in. Thus, a shingle thickness is designated as 5 in 2 in. or 5/2 in. An 18-in. shingle has a thickness so that the combined butt thicknesses of five shingles will measure at least 2-1/4 in. (5/(2-1/4) in.); and 24-in. shingles have a thickness so that four shingles will measure at least 2 in. (4/2 in.). These dimensions are for unseasoned shingles measured green. If the shingles are measured after seasoning has occurred, a 3% allowance is made for shrinkage.

Shingles with extreme cross grain are not permitted in any grade.

Grade No. 1 (Blue Label) is the best, or premium, grade of shingle for roofs and for single-coursed and the surface course of double-coursed walls. Intended primarily for roofing, they are 100% edge grain, consist entirely of heartwood, and contain no knots.

On 8 in 12 sloped roofs or steeper, Grade No. 1 shingles are estimated to have a service life of more than 35 years under normal climatic conditions; on roof slopes of 6 in 12, a service life of at least 25 years can be anticipated; and on 4 in 12 slopes, 20 years. On walls they will have an indefinite life, usually greater than the useful life of the structure.

Grade No. 2 (Red Label) is an excellent grade for single-coursed and the surface course of double-coursed walls. It is a good grade for roofs on secondary buildings and may be used on major building roofs when exposures are reduced about 20%. Shingles of this grade must be clear, or free from blemishes, for 10, 11, and 15 in. for 16-, 18-, and 24-in. shingles, respectively. A maximum width of only 1 in. of sapwood is permissible in the first 10 in. Mixed edge and flat grains are allowed.

Grade No. 3 (Black Label) is a utility grade to be used where economy is the prime consideration. It may be used as the undercourse in wall double-coursing and for roofs and walls of secondary buildings. These shingles may contain flat grain and sapwood, but they must be 6 in. clear from the butt for 16- and 18-in. shingles, and 10 in. clear for 24-in. shingles.

Undercoursing shingles are a lower grade than No. 3 and available in 16- and 18-in. lengths. They are produced expressly for use in the inner and completely concealed course of double-coursed walls (Fig. 7.6-78a). They should not be used for any other purpose. Some mills mix Grade No. 3 and undercoursing shingles and call them *special undercoursing shingles.*

Rebutted and rejointed shingles are machine trimmed for exacting parallel

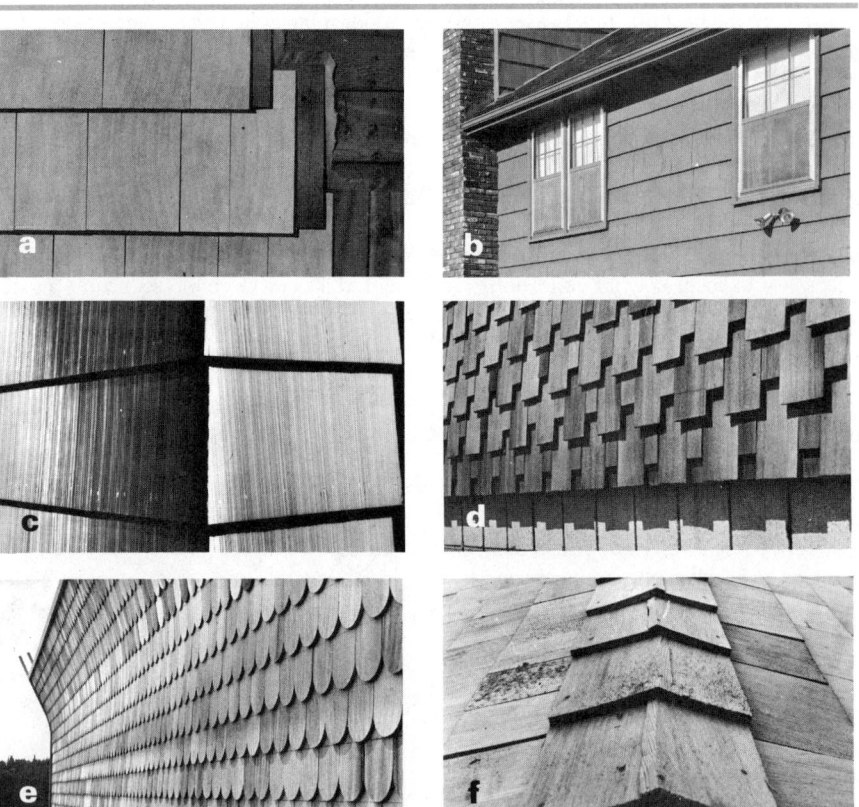

FIGURE 7.6-78 *(a) A double-coursed wall showing complete concealed undercover shingles; (b) a wall made using rebutted and rejointed shingles; (c) machine-grooved shingles; (d) dimension shingles; (e) fancy-butt shingles (many other shapes are available); (f) preassembled hip and ridge units.* (Courtesy Cedar Shake and Shingle Bureau.)

edges, with butts cut to exact right angles. They are also available with smooth-sanded faces (Fig. 7.6-78b). Produced in No. 1 and No. 2 Grades in 16- and 18-in. lengths, they can be used on either single- or double-coursed walls, where tight joints between shingles are desired.

Machine-grooved shingles (also known as machine-grooved shakes) are similar to rebutted and rejointed shingles, except that the face is machine striated to give a corrugated appearance (Fig. 7.6-78c). They are produced in No. 1 Grade only, in 16- and 18-in. lengths, and are natural, prime-coated, or finished coated. These units can be used for the surface course in double-coursed wall applications over low-grade undercoursing. They should not be used for roofing.

Dimension shingles are cut to specific uniform widths where special effects are desired (Fig. 7.6-78d). They are available in 5- or 6-in. widths in 16-in. lengths in No. 1 and No. 2 Grades; in 5- or 6-in. widths in 18-in. lengths in No. 1 Grade; and 6-in. width in 24-in. lengths in No. 1 Grade. *Fancy-butt* shingles (Fig. 7.6-78e) are produced from dimension shingles. The butt designs are produced by trimming the shingles on a band saw.

Hip and ridge units are factory cut, mitered, and assembled to fit hips and ridges of various roof slopes. These units, which reduce costs by eliminating on-the-job trimming and fitting (Fig. 7.6-78f), are produced in either No. 1 or No. 2 Grades in 16- and 18-in. lengths for a 6 in 12 roof slope and are adjustable to fit any slope between 4 in 12 and 8 in 12.

A summary of cedar shingle types, grades, and coverage is given in Figure 7.6-79 and Figure 7.6-80. The labels shown in Figure 7.6-79 are assurance that the shingles meet the quality standards of the Cedar Shake and Shingle Bureau. The maximum exposure for shingles is shown in Figure 7.6-81.

■ CEDAR SHAKES

Wood shakes are manufactured in three types of units: (1) hand-split-and-resawn, (2) taper-split, and (3) straight-split. Shakes are produced in three lengths: 18, 24, and 32 in. Each is 100% clear wood and 100% heartwood. Both taper-split and straight-split shakes, produced largely by

hand, are of 100% edge grain. Hand-split-and-resawn shakes have split faces and sawn backs and may include up to 10% flat grain. Factory-assembled hip and ridge units are also available. Each bundle of shakes should carry the official grade-marked label of the Cedar Shake and Shingle Bureau.

Today, most shakes are split by machine, although some are still split by hand with a hardwood mallet and steel froe (Fig. 7.6-82a). As a result of being split, no two shakes are exactly alike.

Taper-split and straight-split shakes (Figs. 7.6-83a and 7.6-83b) are made by machine- or hand-riving individual shakes from the block. To obtain the butt-to-tip taper for taper-split shakes, blocks are turned end-for-end after each split. Straight-split shakes are split from the same end of the block and therefore are of uniform thickness.

Hand-split-and-resawn shakes (see Fig. 7.6-83c) are produced by splitting blocks into boards of the desired thickness. The boards are then passed diagonally through a thin bandsaw to form two shakes, each with a split face and a sawn back having thin tips and thick butts.

Packers assemble finished shakes into bundles in a standard-sized frame, compressing the bundles slightly and binding them with wooden bandticks and steel strapping (Fig. 7.6-84).

Each type of shake may be used for roof shingles. Hand-split-and-resawn shakes have the heaviest butt lines and give the most rugged textured appearance to a roof, while taper-split and straight-split shakes give a more uniform texture. All three types may be used for walls, but taper-split and straight-split types are usually preferred because of their more uniform texture.

On 8 in 12 sloped roofs or steeper, shakes are estimated to have a service life of more than 60 years. On lesser slopes, to a minimum of 4 in 12, a life of about 40 years may be expected.

GRADES, TYPES, AND SIZES There is only one grade, No. 1, for cedar shakes. Shakes of this grade are 100% clear wood, graded from the split face in the case of hand-split-and-resawn shakes, and from the best face in the case of taper-split and straight-split shakes. All are 100% heartwood,

free of bark and sapwood. Taper-split and straight-split shakes are 100% edge grain; hand-split-and-resawn shakes may have up to 10% flat grain. Nominal shake lengths are 18, 24, and 32 in., with a maximum tolerance of 1/2 in.

Shake thickness is determined by measuring the area within 1/2 in. from each edge. If corrugations or valleys exceed 1/2 in. in depth, a minus tolerance of 1/8 in. is permitted in the minimum specified thickness.

Shakes of random width, none narrower than 4 in., are packed in straight courses in 18- and 20-in.-wide frames. Curvatures in the sawn face of hand-split-and-resawn shakes should not exceed 1 in. from the level plane in the length of the shake. Excessive grain sweeps on the split face are not permitted.

Starter-finish shakes, produced in a 15-in. length, are used as a beginning or an underlay course at the eave and as the final course at the ridge (Fig. 7.6-85).

Factory cut, mitered, and assembled, hip and ridge units are available to fit hips and ridges and eliminate on-the-job trimming and fitting (Fig. 7.6-86). They are produced in 18- and 24-in. lengths for a 6 in 12 roof slope and are adjustable to fit any slope between 4 in 12 and 8 in 12.

A summary of types, grades, and coverage of red cedar shakes is given in Figures 7.6-87 and 7.6-88. The label shown in Figure 7.6-87 is assurance that the shakes meet the quality standards of the Cedar Shake and Shingle Bureau. The maximum exposure for shakes is shown in Figure 7.6-89.

■ STANDARDS OF QUALITY

The manufacture of cedar shingles and shakes should conform to the specifications of the Cedar Shake and Shingle Bureau. These specifications constitute the recognized standards of the industry.

In addition to formulating specifications, a primary function of the Cedar Shake and Shingle Bureau is to establish and enforce rigid quality controls covering the manufacture and grading of red cedar shingles and shakes. To ensure compliance with these standards, products made by members of the Cedar Shake and Shingle Bureau are inspected at frequent unannounced intervals by Bureau inspectors.

Grade	Size	Bundles or Cartons* Per Square		Description	Labels
		No.	Weight		
No. 1 BLUE LABEL	24″ (Royals) 18″ (Perfections) 16″ (XXXXX)	4 bdls. 4 bdls. 4 bdls.	192 lbs. 158 lbs. 144 lbs.	The premium grade of shingles for roofs and sidewalls. These shingles are 100% heartwood, 100% clear and 100% edge-grain.	CERTIGRADE Red Cedar SHINGLES BLUE 1 LABEL
No. 2 RED LABEL	24″ (Royals) 18″ (Perfections) 16″ (XXXXX)	4 bdls. 4 bdls. 4 bdls.	192 lbs. 158 lbs. 144 lbs.	A good grade for most applications. Not less than 10″ clear on 16″ shingles, 11″ clear on 18″ shingles and 16″ clear on 24″ shingles. Flat grain and limited sapwood are permitted.	CERTIGRADE Red Cedar SHINGLES RED LABEL
No. 3 BLACK LABEL	24″ (Royals) 18″ (Perfections) 16″ (XXXXX)	4 bdls. 4 bdls. 4 bdls.	192 lbs. 158 lbs. 144 lbs.	A utility grade for economy applications and secondary buildings. Guaranteed 6″ clear on 16″ and 18″ shingles, 10″ clear on 24″ shingles.	CERTIGRADE Red Cedar SHINGLES BLACK 3 LABEL
No. 4 UNDER-COURSING	18″ (Perfections) 16″ (XXXXX)	2 bdls. 2 bdls.	60 lbs. 60 lbs.	A low grade for undercoursing on double-coursed sidewall applications.	RED CEDAR SHINGLE BUREAU UNDERCOURSING Red Cedar SHINGLES
No. 1 or No. 2 REBUTTED-REJOINTED	18″ (Perfections) 16″ (XXXXX)	1 carton 1 carton	60 lbs. 60 lbs.	Same specifications as No. 1 and No. 2 Grades above but machine trimmed for exactly parallel edges with butts sawn at precise right angles. Used for side-wall application where tightly fitting joints between shingles are desired. Also available with smooth sanded face.	No. 1 & No. 2 labels (above)
No. 1 MACHINE GROOVED	18″ (Perfections) 16″ (XXXXX)	1 carton 1 carton	60 lbs. 60 lbs.	Same specifications as No. 1 and No. 2 Grades above; these shingles are used at maximum weather exposures and are always applied as the outer course of double-coursed sidewalls.	CERTIGROOVE CEDAR SHAKES NUMBER 1 GRADE RED CEDAR SHINGLE BUREAU
No. 1 or No. 2 DIMENSION	24″ (Royals) 18″ (Perfections) 16″ (XXXXX)	4 bdls. 4 bdls. 4 bdls.	192 lbs. 158 lbs. 144 lbs.	Same specifications as No. 1 and No. 2 Grades above, except they are cut to specific uniform widths and may have butts trimmed to special shapes.	No. 1 & No. 2 labels (above)
No. 1 or No. 2 HIP AND RIDGE	18″ (Perfections) 16″ (XXXXX)	—		Same specifications as No. 1 and No. 2 Grades above; factory-cut, mitered and assembled units produced for a 6 in 12 slope and adjust to fit slopes between 4 in 12 and 8 in 12.	No. 1 & No. 2 labels (above)

* Nearly all manufacturers pack 4 bundles to cover 100 sq. ft. when used at maximum exposures for roof construction; Undercoursing, rebutted-and-rejointed and machine grooved shingles typically are packed one carton to cover 100 sq. ft. when used at maximum exposure for double-coursed side-walls.

FIGURE 7.6-79 *Types of red cedar shingles. (Courtesy Cedar Shake and Shingle Bureau.)*

Mills that conform to these standards are privileged to use on their products the *Certigrade* shingle or *Certi-Split* shake labels of the Cedar Shake and Shingle Bureau.

■ NAILS

Nails for applying shakes and shingles should be corrosion resistant, such as hot-dipped galvanized steel or bright or colored aluminum-alloy. Common wire nails or even electrogalvanized nails may shorten the life of a shake or shingle roof or wall and should not be used. Nails may have smooth or threaded shanks and medium or blunt diamond points. Threaded nails provide increased holding power, although smooth-shank nails are used in the major-ity of applications over wood sheathing and give suitable service. Nails for apply-ing wood shingles or shakes to plywood sheathing should be threaded. If threaded, galvanized steel nails should have annular threads; aluminum nails should have screw threads with approximately a 12-1/2-degree thread angle.

Box nails, having smaller diameters and longer lengths than common nails, are also suitable. For new roofs, nail lengths typically required are 6d (2 in.) for shakes with 3/4-in. to 1-1/4-in. butt thickness, and 5d (1-3/4 in.) for shakes with 1/2- to 3/4-in. butt thickness. Nails 2-penny (2d) sizes larger should be used to apply hip and ridge units. Nails should be long enough

FIGURE 7.6-80
Coverage of Wood Shingles at Varying Exposures

Length and Thickness[a]	Approximate Coverage of 4 Bundles or 1 Carton, sq. ft.[b]																		
	Weather Exposures, in.																		
	3-1/2	4	5	5-1/2	6	7	7-1/2	8	8-1/2	9	10	11	11-1/2	12	13	14	15	15-1/2	16
Random width and dimension																			
16 in. × 5/2 in.	70	80	100[c]	110	120	140	150[d]	160	170	180	200	220	230	240[e]	—	—	—	—	—
18 in. × 5/(2-1/4) in.	—	72-1/2	90-1/2	100[c]	109	127	136	145-1/2	154-1/2[d]	163-1/2	181-1/2	200	209	218	236	254-1/2[e]	—	—	—
24 in. × 4/2 in.	—	—	—	—	80	93	100[c]	106-1/2	113	120	133	146-1/2	153[d]	160	173	186-1/2	200	206-1/2	213[e]
Rebutted and rejoined																			
16 in.	—	—	—	—	50	59	63[d]	67	72	76	84	93	—	100[e]	—	—	—	—	—
18 in.	—	—	—	—	43	50	54	57	61[d]	64	72	79	—	86	93	100[e]	—	—	—
Machine-grooved																			
16 in.	(Normally applied at maximum exposure)												100[e]						
18 in.	(Normally applied at maximum exposure)															100[e]			

Courtesy Cedar Shake and Shingle Bureau.

[a] Sum of the thickness; e.g., 5/2 in. means 5 butts equals 2 in.

[b] Nearly all manufacturers pack 4 bundles to cover 100 sq. ft. when used at maximum exposures for roof construction; rebutted and rejointed and machine-grooved shingles typically are packed one carton to cover 100 sq. ft. when used at maximum exposure for double-coursed sidewalls.

[c] Maximum exposure recommended for roofs.

[d] Maximum exposure recommended for single-coursed sidewalls.

[e] Maximum exposure recommended for double-coursed sidewalls.

FIGURE 7.6-81
Maximum Exposure for Wood Shingles

	Maximum Exposure, in.		
Roof Slope	16-in. Shingle	18-in. Shingle	24-in. Shingle
5 in 12 or steeper	5	5-1/2	7-1/2
4 in 12	4-1/2	5	6-3/4
3 in 12	3-3/4	4-1/4	5-3/4

Courtesy Cedar Shake and Shingle Bureau.

FIGURE 7.6-82 *Splitting shakes (a) by hand, (b) by machine. (Courtesy Cedar Shake and Shingle Bureau.)*

Taper-split

Straight-split

Hand-split-and-resawn

FIGURE 7.6-83 *The three types of shakes: (a) taper-split, (b) straight-split, (c) hand-split-and-resawn. (Courtesy Cedar Shake and Shingle Bureau.)*

FIGURE 7.6-84 *Grade labels beneath the bandsticks of shingles and shakes ensure that they meet or exceed Cedar Shake and Shingle Bureau quality standards. (Courtesy Cedar Shake and Shingle Bureau.)*

FIGURE 7.6-85 *Starter-finish shakes are more economical than regular shakes for use in the initial undercourse at the eave and in the final course at the ridge. (Courtesy Cedar Shake and Shingle Bureau.)*

to penetrate completely through the sheathing. For reroofing, nails should be long enough to penetrate through the existing sheathing.

FIGURE 7.6-86 *Hip and ridge shake units. (Courtesy Cedar Shake and Shingle Bureau.)*

Nail dimensions typically required for roof installations are given in Figure 7.6-90. Nails 2-penny (2d) sizes larger should be used to apply hip and ridge units. For overroofing, nails should be

Grade	Length and Thickness	Bundles Per Square*	Weight (lbs. per square)	Description	Label
No. 1 HANDSPLIT & RESAWN	18" x 1/2" to 3/4" 18" x 3/4" to 1-1/4" 24" x 3/8" 24" x 1/2" to 3/4" 24" x 3/4" to 1-1/4" 32" x 3/4" to 1-1/4"	4 5 4 4 5 6	220 250 260 280 350 450	These shakes have split faces and sawn backs. Cedar blanks or boards are split from logs and then run diagonally through a bandsaw to produce two tapered shakes from each.	
No. 1 TAPERSPLIT	24" x 1/2" to 5/8"	4	260	Produced largely by hand, using a sharp-bladed steel froe and a wooden mallet. The natural shingle-like taper is achieved by reversing the block, end-for-end, with each split.	CERTI-SPLIT Handsplit Red Cedar Shakes NUMBER 1 GRADE
No. 1 STRAIGHT-SPLIT (BARN)	18" x 3/8" (True-Edge) 18" x 3/8" 24" x 3/8"	4 5 5	200 200 260	Produced in the same manner as taper-split shakes except that by splitting from the same end of the block, the shapes acquire the same thickness throughout.	
No. 1 STARTER-FINISH	15" x 1/2" to 1-1/4"	—	—	These shakes are used as the starting or underlay course at the eaves and as the final course at the ridge.	
No. 1 HIP & RIDGE	18" x 1/2" to 1-1/4" 24" x 1/2" to 1-1/4"	—	—	Factory cut, mitered and assembled units are produced for a 6 in 12 slope; they adjust to fit slopes between 4 in 12 and 8 in 12.	

* Generally, the number of bundles required to cover 100 sq. ft. when used for roof construction at the maximum recommended weather exposure.

FIGURE 7.6-87 *Types of cedar shakes.* *(Courtesy Cedar Shake and Shingle Bureau.)*

long enough to penetrate through the existing sheathing; the 5d size (1-3/4 in.) is normally adequate.

Nail lengths typically required for wall installations are given in Figure 7.6-91 and are shown in Figure 7.6-18.

The number, spacing, and pattern of nails are described in Sections 7.6.6.3 and 7.6.6.6.

7.6.6.2

Preparation for Roof Shingles and Shakes

■ SLOPE LIMITATIONS AND EXPOSURE

Roof slope limitations are discussed in Section 7.6.1.2. The exposure of wood shingles and shakes on a roof depends on the slope of the roof.

SHINGLES Standard exposures of 5, 5-1/2, and 7-1/2 in. for shingle lengths of 16, 18, and 24 in., respectively, may be used on slopes 5 in 12 or greater. Four bundles of shingles will cover 100 sq. ft. (one square)

of roof area when the shingles are applied at standard exposures. Using reduced exposures, wood shingles may be used on slopes as low as 3 in 12 (see Fig. 7.6-81). They are not recommended on slopes less than 3 in 12.

On 4 in 12 slopes, exposures should be reduced to 4-1/2, 5, and 6-3/4 in., for 16-, 18-, and 24-in. shingles, respectively. On 3 in 12 slopes they should be reduced further to 3-3/4, 4-1/4, and 5-3/4 in. for 16-, 18-, and 24-in. shingles, respectively, which will ensure four layers of shingles over the roof area.

Shingle exposure should not be larger than the length of the shingle minus 1 in., divided by 3. This will provide at least triple coverage (three layers of wood at every point) to ensure complete freedom from leakage in heavy wind-driven rain or snowstorms. Approximate coverage of four bundles of shingles at varying exposures is shown in Figure 7.6-80.

SHAKES The maximum recommended exposure for double-coverage roofs is 13 in. for 32-in. shakes, 10 in. for 24-in.

shakes, and 7-1/2 in. for 18-in. shakes. A triple-coverage roof is achieved by reducing these exposures to 10 in. for 32-in. shakes, 7-1/2 in. for 24-in. shakes, and 5-1/2 in. for 18-in. shakes. Approximate roof coverage for shakes using different exposures is shown in Figure 7.6-88. Maximum exposures to provide double and triple coverage are summarized in Figure 7.6-89.

Shakes are recommended only on slopes 4 in 12 or steeper. However, suitable installations have been achieved on slopes as low as 3 in 12 by taking the following extra precautions: (1) reducing exposures to provide triple coverage and (2) using solid sheathing with an underlayment of No. 30 asphalt-saturated felt applied over the entire roof with No. 30 asphalt-saturated felt interlayment between each course (see Section 7.6.6.3).

■ ROOF SHEATHING

Sheathing types, thicknesses, spans, and nailing are discussed in Section 7.6.1.3 and in Chapter 6.

FIGURE 7.6-88
Coverage of Shakes at Varying Weather Exposures

Length and Thickness	Number of Bundles	Approximate Coverage, sq. ft. Weather Exposures, in.											
		5-1/2	6-1/2	7	7-1/2	8	8-1/2	10	11-1/2	13	14	15	16
Hand-split-and-resawn													
18 in. × 1/2 in. to 3/4 in.	4	55[a]	65	70	75[b]	80	85[c]	—	—	—	—	—	—
18 in. × 3/4 in. to 1-1/4 in.	5	55[a]	65	70	75[b]	80	85[c]	—	—	—	—	—	—
24 in. × 3/8 in.	4	—	65	70	75	80	85	100	115[c]	—	—	—	—
24 in. × 1/2 in. to 3/4 in.	4	—	65	70	75[a]	80	85	100[b]	115[c]	—	—	—	—
24 in. × 3/4 in. to 1-1/4 in.	5	—	65	70	75[a]	80	85	100[b]	115[c]	—	—	—	—
32 in. × 3/4 in. to 1-1/4 in.	6	—	—	—	—	—	—	100[a]	115	130[b]	140	150[c]	—
Taper-split													
24 in. × 1/2 in. to 5/8 in.	4	—	65	70	75[a]	80	85	100[b]	115[c]	—	—	—	—
Straight-split													
18 in. × 3/8 in. (true-edge)	4	—	—	—	—	—	—	—	—	—	100	106	112[d]
18 in × 3/8 in.	5	65[a]	75	80	90	95	100[c]	—	—	—	—	—	—
24 in. × 3/8 in.	5	—	65	70	75[c]	80	85	100	115[c]	—	—	—	—
15-in. Starter-finish course	Use supplementary with shakes applied not over 10-in. exposure												

Courtesy Cedar Shake and Shingle Bureau.

[a]Recommended maximum weather exposure for three-ply roof construction.

[b]Recommended maximum weather exposure for two-ply roof construction.

[c]Recommended maximum weather exposure for single-coursed sidewalls.

[d]Recommended maximum weather exposure for double-coursed sidewalls.

FIGURE 7.6-89
Maximum Exposures of Wood Shakes

Type	Length, in.	Maximum Exposure	
		Double Coverage, in.	Triple Coverage, in.
Hand-split-and-resawn shakes	18	7-1/2	5-1/2
	24	10	7-1/2
	32	13	10
Taper-split shakes	24	10	7-1/2
Straight-split shakes	18	—	5-1/2
	24	—	7-1/2

Courtesy Cedar Shake and Shingle Bureau.

Note: In areas not subject to wind-driven snow, sheathing such as nominal 1×4 or 1×6 boards may be spaced equal to the exposure of the shake, but not more than 10 in.

SHINGLES Either spaced or solid sheathing may be used with wood shingles. When solid sheathing is preferred, either nominal 1-in. boards or plywood may be used. When spaced sheathing, such as nominal 1×3 or 1×4 boards, is used, it should be applied so that the space between boards is not greater than the width of the boards themselves.

SHAKES Either spaced or solid sheathing may be used with wood shakes, depending on the climatic conditions of the region. In areas not subject to wind-driven snow, spaced sheathing, such as nominal 1×4 or 1×6 boards, is suitable (Fig. 7.6-92). The center-to-center spacing of boards should be equal to the exposure of the shake but not more than 10 in.

In areas subject to wind-driven snow (generally where the outside design tem-

FIGURE 7.6-90
Nail Sizes for Application of Wood Shingles

	Nail Size			
Size	Length, in.	Gauge	Head, in.	Shingle Length, in.
3d[a]	1-1/4	14-1/2	7/32	16 and 18
4d[a]	1-1/2	14	7/32	24
5d[b]	1-3/4	14	7/32	16 and 18
6d[b]	2	13	7/32	24

Courtesy Cedar Shake and Shingle Bureau.
[a]3d and 4d nails are used for new construction.
[b]5d and 6d nails are used for overroofing.

FIGURE 7.6-91
Minimum Nail Lengths for Wood Shingles and Shakes[a]

	Nail Lengths[b]	
Type	Double-Coursed Sidewalls	Single-Coursed Sidewalls
Shingles	1-3/4 in. (5d)	1-1/4 in. (3d)
Shakes[c]	2-1/2 in. (8d)	1-1/2 in. (4d)

[a]For application over plywood sheathing, nails should be threaded.
[b]Nails should be long enough to penetrate through sheathing.
[c]Because thickness and exposure of shakes vary, larger nails may be necessary to obtain suitable penetration in the sheathing.

FIGURE 7.6-92 *(a) Spaced sheathing. (b) Maximum recommended exposure produces a two-ply roof; reducing exposure produces a three-ply roof.* *(Courtesy Cedar Shake and Shingle Bureau.)*

perature is 30°F or lower, solid sheathing of nominal 1-in. boards or plywood is recommended.

■ UNDERLAYMENT AND INTERLAYMENT

SHINGLES Underlayment typically is not required between shingles and spaced or solid sheathing, but it may be desirable for the protection of sheathing and to ensure against air infiltration.

When underlayment is desired, No. 15 asphalt-saturated felt may be used.

SHAKES Interlayment strips and the underlayment course used at eaves should be No. 30 asphalt-saturated felt. The installation of interlayment between each shake course, essential to provide a baffle against infiltration of wind-driven rain or snow, is described in Section 7.6.6.3.

■ EAVE FLASHING

In areas where the outside design temperature is 30°F or colder (see Fig. 7.6-11) or where there is a possibility of ice forming along the eaves and causing a backup of water, eave flashing is recommended. Sheathing should be applied solidly above the eave line to cover a line at least 24 in. inside the interior wall line of the building.

SHINGLES For 4 in 12 and steeper slopes, eave flashing should be formed by applying a double layer of No. 15 asphalt-saturated felt to cover the solid sheathing. When the eave overhang requires the flashing wider than 36 in., the horizontal joint between felts should be cemented and located outside the exterior wall line.

For slopes 3 in 12 up to 4 in 12 or in areas subject to severe icing, eave flashing may be formed as described above for 4 in 12 slopes, except that the double layer of No. 15 asphalt-saturated underlayment should be cemented. The eave flashing should be formed by applying a continuous layer of plastic asphalt cement, at the rate of 2 gal./100 sq. ft. to the surface of the underlayment starter course before the second layer of underlayment is applied. Cement should also be applied to the 19-in. underlying portion of each succeeding course that lies within the eave flashing area before placing the next course. It is important to apply the cement uniformly with a comb trowel, so that at no point does underlayment touch underlayment when the application has been completed. The overlying sheet should be pressed firmly into the entire cemented area (see Fig. 7.6-13).

SHAKES On slopes 4 in 12 or steeper, eave flashing should be formed by applying an additional course of No. 30 asphalt-saturated felt over the underlayment starting course at the eaves. The eave flashing should extend up the roof to cover a line at least 24 in. inside the interior wall line of the building. When the eave overhang requires the flashing to be wider than 36 in., the necessary horizontal joint should be cemented and located outside the exterior wall line of the building (see Fig. 7.6-13).

For slopes 3 in 12 up to 4 in 12 or in areas subject to severe icing, eave flashing may be formed as described above for 4 in 12 slopes, except that the double layer of No. 30 asphalt-saturated underlayment should be cemented. The eave flashing should be formed by applying a continuous layer of plastic asphalt cement at the rate of 2 gal./100 sq. ft. to the surface of the underlayment starter course before the second layer of underlayment is applied. Cement should also be applied to the 19-in. underlying portion of each succeeding course that lies within the eave flashing area before the next course is placed. It is important to apply the cement uniformly with a comb trowel, so that at no point does underlayment touch underlayment when the application has been completed. The overlying sheet should be pressed firmly into the entire cemented area.

■ DRIP EDGE

Wood shingles and shakes should extend out over the eave and rake from 1 in. to 1-1/2 in. to form a drip. Accordingly, a corrosion-resistant drip edge is not necessary at these points, since the shingle starter course and rake shingles serve the same purpose.

■ FLASHING

Refer to Section 7.6.1.6 for general comments about flashing.

Unless special precautions are taken (Fig. 7.6-93), copper flashing materials are not recommended for use with red cedar shingles or shakes. Premature deterioration of the copper may occur when the metal and wood are in intimate contact in the presence of moisture.

FIGURE 7.6-93 *Surface contact of copper with red cedar must be minimized. (Courtesy Cedar Shake and Shingle Bureau.)*

VALLEY FLASHING FOR WOOD SHINGLES
Only the open method should be used to construct valley flashing for wood shingles. Butting shingles together in the valley at the centerline to form a closed valley is not recommended.

On slopes up to 12 in 12, metal valley sheets should be wide enough to extend at least 10 in. on each side of the valley centerline (Fig. 7.6-94). On roofs of steeper slope, narrower sheets may be used extending on each side of the centerline of the valley for a distance of at least 7 in.

To avoid restricting the delivery of a valley and to prevent splashing of water

FIGURE 7.6-94 *Method for constructing open-valley flashing. (Courtesy Cedar Shake and Shingle Bureau.)*

from a steep slope on the shingles of a lower slope, the edges of the shingles on the lower slope should be lined up at least 1 in. farther back from the centerline of the valley than those on the steeper slope. As an alternative, a vertical ridge, or waterstop, can be made by crimping the metal up in the center of the valley. It is desirable to edge-crimp valley sheets by turning the edges up and back approximately 1/2 in. toward the valley centerline. This provides an additional waterstop, directing water down the valley. The open portion of the valley should be at least 4 in. wide, but valleys may taper from a width of 2 in. where they start and increase at the rate of 1/2 in. per 8 ft. of length, to become wider as they descend.

In areas where the outside design temperature is 30 °F or colder, underlayment should be installed under metal valley sheets.

VALLEY FLASHING FOR WOOD SHAKES
The open method, using roofing felt and sheet metal, or the closed method, using hand-fitted shakes, may be used to construct valley flashing for wood shakes. For longest service life, the open method is highly recommended.

Open valleys are first covered with a valley underlayment strip of No. 30 asphalt-saturated felt, at least 20 in. wide,

FIGURE 7.6-95 *Open-valley flashing for shakes. (Courtesy Cedar Shake and Shingle Bureau.)*

centered in the valley and secured with enough nails to hold it in place (Fig. 7.6-95). Metal flashing strips, 20 in. wide, are then nailed over the valley underlayment. If galvanized steel flashing is used, it preferably should be 18 gauge (but not less than 26 gauge), center-crimped, and painted on both surfaces. It is desirable to edge-crimp valley sheets by turning the edges up and back approximately

1/2 in. toward the valley centerline. This provides an additional waterstop, directing water down the valley. Shakes laid to finish at the valley should be trimmed parallel with the valley to form a 6-in.-wide gutter.

Closed valleys should first be covered with a 1×6 wood strip, nailed flat into the saddle and covered with roofing felt as discussed above for open valleys. Shakes in each course should be edge-trimmed to fit into the valley, then laid across the valley with an undercourse of metal flashing having a 2-in. headlap and extending 10 in. under the shakes on each side of the saddle.

■ CHIMNEY AND WALL FLASHING

Individual flashing units of corrosion-resistant metal should be inserted between each course of shingles or shakes that butt against vertical surfaces, such as walls, dormers, or chimneys. One leg of this flashing should extend at least 6 in. under the shingles or shakes with the upward leg covered by the vertical wall finish or by metal counter flashing, as shown in Figure 7.6-96.

On new construction, chimney flashing should be built in by the mason (Fig. 7.6-97). If masonry work has been completed, flashing should be inserted between bricks to a depth of 3/4 in. by removing mortar and filling over the flashing with bituminous mastic.

Galvanized-steel flashing should be painted on both surfaces before it is placed. Flashing strips, which must be bent to sharp angles, should be painted after bending; if crimped to form a waterstop in the centers of valleys, the crimped surface should be given a double coat of paint.

7.6.6.3

Application of Roof Shingles (New Construction)

The first course of shingles at the eaves should be doubled or tripled and should project 1 to 1-1/2 in. beyond the eaves to provide a drip. The second layer of shingles in the first course should be nailed over the first layer to provide a minimum sidelap of at least 1-1/2 in. between joints (Fig. 7.6-98). If possible, joints should be "broken" by a greater margin. A triple

FIGURE 7.6-96 *Flashing at a wall.* (*Courtesy Cedar Shake and Shingle Bureau.*)

FIGURE 7.6-97 *Flashing at a chimney.* (*Courtesy Cedar Shake and Shingle Bureau.*)

FIGURE 7.6-98 *Application of wood shingles over spaced or solid sheathing.* (*Courtesy Cedar Shake and Shingle Bureau.*)

layer of shingles in the first course provides additional insurance against leaks at the cornice. No. 3 Grade shingles frequently are used for the starter course.

Shingles should be spaced at least 1/4 in. apart to provide for expansion. Joints between shingles should be separated not less than 1-1/2 in. from joints in adjacent courses above and below, and joints in alternate courses should not be in direct alignment. When shingles are laid with the recommended exposure, triple coverage of the roof results (see Fig. 7.6-98).

When the roof terminates in a valley, the shingles for the valley should be carefully cut to the proper miter at the exposed butts and should be nailed in place first so that the direction of shingle application is away from the valley. This permits valley shingles to be carefully selected and ensures that shingle joints will not break over the valley flashing.

The application of wood shingles can be speeded by the use of a special shingler's or lather's hatchet (Fig. 7.6-99). The hatchet is used for nailing, carries an exposure gauge, and has a sharpened heel that expedites trimming.

■ NAILING

To ensure that shingles will lie flat and for maximum service, only two nails should be used to secure each shingle on roofs. Nails should be placed not more than 3/4 in. from the side edge of shingles, at a distance of not more than 1 in. above the exposure line. Nails should be driven flush but not so that the nail head crushes the wood.

■ HIPS AND RIDGES

Hips and ridges should be of the modified "Boston" type with protected nailing. Either site-applied or preformed factory-constructed hip and ridge units may be used. Nails should be increased at least two sizes larger than the nails used to apply the shingles.

Application on hips and ridges is shown in Figure 7.6-100. Hips and ridges should begin with a double starter course.

■ GABLE RAKES

Shingles should project 1 to 1-1/2 in. over the rake. End shingles can be canted

FIGURE 7.6-99 *A shingler's hatchet.* (Courtesy Cedar Shake and Shingle Bureau.)

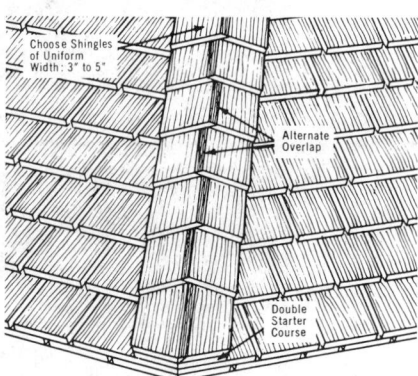

FIGURE 7.6-100 *Application on hips and ridges.* (Courtesy Cedar Shake and Shingle Bureau.)

to eliminate drips, as shown in Figure 7.6-101.

■ ALTERNATIVE APPLICATION METHODS

Several alternative methods of applying shingles, which give different appearances, are shown in Figure 7.6-102.

■ REROOFING EXISTING CONSTRUCTION

It is not necessary to remove existing wood or asphalt shingles to install new wood shingles. Existing slate, tile, asbestos-cement, or mineral fiber–cement shingles, which do not provide an adequate nailing base, must be removed.

However, the first course of existing shingles at the eaves should be removed just below the butts of the shingles in the second course. Old shingles should also be cut back and removed along the gable rakes (Fig. 7.6-103). The removed shingles along the eaves and gable rake edges should be replaced with boards having a thickness approximately equal to the thickness of the old roof section. These boards provide a strong base at the perimeter of

FIGURE 7.6-101 *Beveled strip not only eliminates drips but also provides decorative termination at gables.* (Courtesy Cedar Shake and Shingle Bureau.)

the new roof, concealing the old roof from view. Ridge shingles should be replaced with strips of beveled siding, laid with the thin edges down the roof.

If necessary, new flashing should be installed over existing flashing.

New shingles should be applied by the same methods used in new construction. If the existing roof sheathing is spaced, it is not necessary for all new nails to strike the sheathing.

7.6.6.4

Application of Roof Shakes (New Construction)

First, a 36-in.-wide starter strip of No. 30 asphalt-saturated felt underlayment should be laid along the eave line over the sheathing. The starter course of shakes at the eave line should be doubled, using an undercourse of 24-, 18-, or 15-in. shakes, the latter being made expressly for this purpose. Striking visual effects can be achieved by using a double undercourse at the eave line. After each course of shakes has been applied, an 18-in.-wide strip of No. 30 asphalt-saturated felt interlayment should be applied over the top portion of the shakes extending onto the sheathing (Fig. 7.6-104). The bottom edge of the interlayment should be positioned at a distance above the butt equal to twice the exposure. For example, if 24-in. shakes are being laid with a 10-in. exposure, the bottom edge of the felt should be positioned at a distance 20 in. above the shake butts; the strip will then cover the top 4 in. of the shakes and extend 14 in. onto the sheathing.

a

Thatch. Shingles are positioned above and below a hypothetical course line, with deviation from the line not to exceed 1".

b

Serrated. Courses are doubled every third, fourth, fifth or sixth course. Doubled courses can be laid butt edge flush or with slight overhang.

c

Dutch Weave. Shingles are doubled or superimposed at random throughout the roof area.

d

Pyramid. Two extra shingles, narrow shingle over a wide one, are superimposed at random.

FIGURE 7.6-102 *Alternative shingle applications.* *(Courtesy Cedar Shake and Shingle Bureau.)*

Individual shakes should be spaced approximately 1/4 to 3/8 in. apart to allow for possible expansion due to moisture absorption. The joints between shakes should be offset at least 1-1/2 in. in adjacent courses. The joints in alternate courses also should be kept out of direct alignment when a three-ply roof is being built.

When straight-split shakes, which are of equal thickness throughout, are applied, the froe-end of the shake (the smoother end, from which it has been split) should be laid undermost.

The application of shakes can be speeded by the use of a special shingler's or lather's hatchet (see Fig. 7.6-99). This hatchet is used for nailing, carries an exposure gauge, and has a sharpened heel that expedites trimming.

■ **NAILING**

Only two nails should be used to apply each shake on roofs, regardless of the width of the shake. Nails should be placed about 1 in. from each edge and from 1 to 2 in. above the butt line of the succeeding course. Nails should be driven until the heads meet the shake surface but no farther (Fig. 7.6-105).

■ **HIPS AND RIDGES**

The final course at the ridge line, as well as shakes that terminate at hips, should be secured with additional nails and should be composed of smoother-textured shakes. A strip of No. 30 asphalt-saturated felt, at least 12 in. wide, should be applied over the crown of hips and ridges, with an equal exposure of 6 in. on each side. Prefabricated hip and ridge units can be used, or the hips and ridges may be cut and applied on the site.

In site construction of hips, shakes about 6 in. wide should be sorted out. Two wooden straightedges should be tacked on the roof, 6 in. from the centerline of the hip, one on each side. The starting course of shakes should be doubled. The first shake on the hip should be nailed in place, with one edge resting against the guide strip. The edge of the shake projecting over the center of the hip should be cut back on a bevel. The shake on the opposite side should then be applied and the projecting edge cut back to fit. Shakes in the following courses should be applied alternately in reverse order (Fig. 7.6-106). Ridges should be constructed in a similar manner. Exposure of hip and ridge shakes is normally the same as the shakes on the roof.

Ridge shakes should be laid along an unbroken ridge that terminates in a gable at each end. They should be started at each gable end and terminated in the middle of the ridge. At that point, a small saddle should be face-nailed to splice the two lines. The first course of shakes should always be doubled at each end of a ridge.

■ **APPLICATION VARIATIONS**

GRADUATED EXPOSURES A variation in roof appearance may be achieved with hand-split shakes by reducing the exposure of each course from eaves to ridge. This requires shakes of several lengths.

A roof with such a graduated exposure is shown in Figure 7.6-107. It can be built by starting at the eaves with 32-in. shakes, laid 13 in. to the weather. One-third of the roof area should be covered with shakes of this length, reducing the exposure gradually to about 10 in. The next third of the area should be covered with 24-in. shakes, laid with a 10-in. exposure and diminishing to about 8-1/2 in. The roof should be completed with 18-in. shakes, starting with 8-in. exposure and diminishing to 5 in.

STAGGERED LINES Irregular and random roof patterns can be achieved by laying shakes with butts placed slightly below or

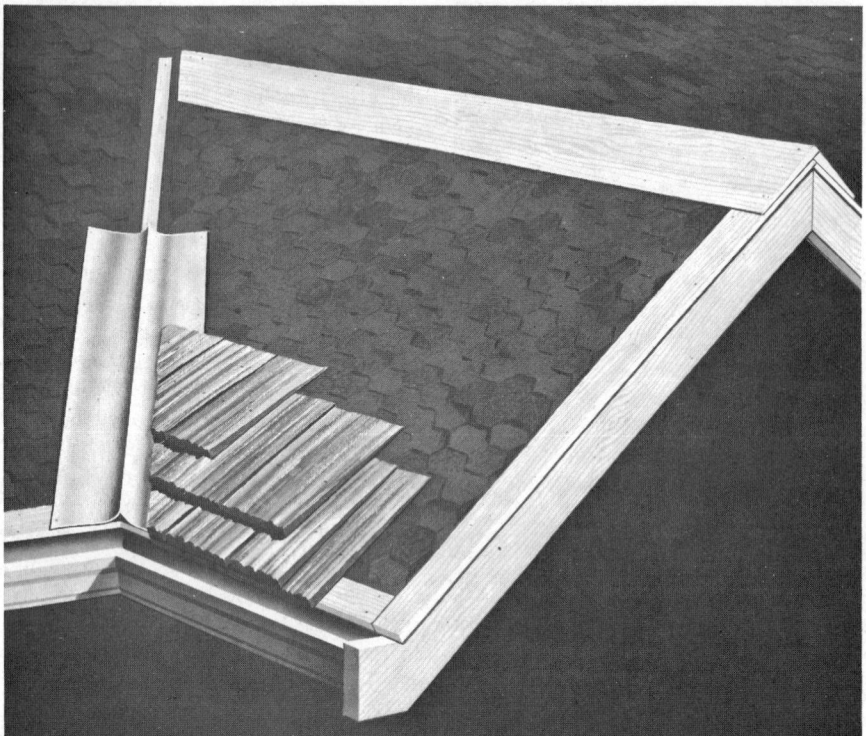

FIGURE 7.6-103 *Preparation of existing roof to receive new wood shingles or shakes.* (*Courtesy Cedar Shake and Shingle Bureau.*)

FIGURE 7.6-104 *Application of shakes over spaced sheathing.* (*Courtesy Cedar Shake and Shingle Bureau.*)

above the horizontal lines governing each course (Fig. 7.6-108). If an extremely irregular pattern is desired, longer shakes may be interspersed over the roof with their butts several inches lower than the course lines.

Mixtures of fairly smooth and very rough shakes will produce a more rugged appearance than a roof composed entirely of rough shakes.

TILTED STRIP AT GABLES Dripping water may be eliminated by inserting a single strip of beveled siding the full length of each gable rake with the thick edge flush with the sheathing edge (see Fig. 7.6-101). The resulting inward pitch of the roof surface diverts water away from the gable edge.

■ REROOFING EXISTING CONSTRUCTION

Structures that require new roof coverings, either for appearance or serviceability, may be covered with any of the three types of hand-split cedar shakes. Except over slate, tile, asbestos-cement, or mineral fiber–cement shingles, shakes may be applied directly over the existing roof.

When the existing roof is to be overroofed with shakes, a 6-in.-wide strip of the old roofing should be removed along the eaves and gable rakes and then filled in with boards having a thickness approximately equal to the thickness of the old roof section (see Fig. 7.6-103). These boards provide a strong base at the perimeter of the new roof, concealing the old roof from view. The old ridge covering should be removed and replaced with beveled siding, overlapping the butt edges at the peak. If necessary, new flashing should be installed over existing flashing. Nails for applying a new roof should be long enough to penetrate the underlying existing sheathing.

7.6.6.5

Preparation for Wall Shingles and Shakes

■ EXPOSURE

The exposure of shakes and shingles depends on their length and whether the walls are to be single- or double-coursed. For single coursing, the weather exposure

FIGURE 7.6-105 *Nails should be correctly placed, and heads should not break the shake surface.* (Courtesy Cedar Shake and Shingle Bureau.)

FIGURE 7.6-106 *Hips and ridges. Starter course should be doubled, as shown.* (Courtesy Cedar Shake and Shingle Bureau.)

FIGURE 7.6-107 *A roof with graduated exposure is created by using all three shake sizes.* (Courtesy Cedar Shake and Shingle Bureau.)

FIGURE 7.6-108 *Staggered butt lines produce varied appearance.* (Courtesy Cedar Shake and Shingle Bureau.)

of the shakes and shingles should never be greater than half the length of the shingle or shake minus 1/2 in., so that two layers of wood occur over the entire surface of the wall.

SHINGLES Maximum recommended weather exposures for single-coursed shingle wall construction are 7-1/2 in. for 16-in. shingles, 8-1/2 in. for 18-in. shingles, and 11-1/2 in. for 24-in. shingles. When double-coursed, 16-in. shingles can be laid at weather exposures up to 12 in.; 18-in. shingles up to 14 in.; and 24-in. shingles up to 16 in. Approximate wall coverages for shingles using different exposures are shown in Figure 7.6-80. Maximum exposures to produce single and double coursing are shown in Figure 7.6-109.

SHAKES Maximum recommended weather exposures for single-coursed shake wall construction are 8-1/2 in. for 18-in.

shakes, 11-1/2 in. for 24-in. shakes, and 15 in. for 32-in. shakes. When double-coursed, 18-in. shakes can be laid at weather exposures up to 14 in., and 24-in. shakes up to 20 in. When straight-split shakes are double-coursed, maximum weather exposure is 16 in. for 18-in. shakes and 22 in. for 24-in. shakes. Approximate wall coverages for shakes using different exposures are shown in Figure 7.6-88. Maximum exposures to produce single and double coursing are shown in Figure 7.6-109.

▪ WALL SHEATHING

Sheathing types, thicknesses, spans, and nailing recommendations are discussed in Section 7.6.1.3 and in Chapter 6.

▪ SHEATHING PAPER

To provide moisture protection when lumber boards are used for sheathing, No. 15 or No. 30 asphalt-saturated felt should be applied over the entire wall as described in Section 7.7.1.3 and in Chapter 6. When spaced (open) sheathing is used, sheathing paper should be applied first to the framing members (Fig. 7.6-110).

7.6.6.6

Applications of Walls Shingles and Shakes

The two basic methods used to apply shakes and shingles are (1) single coursing and (2) double coursing. When single-coursed (Fig. 7.6-111), units can be applied in the same manner as wood roof shakes and shingles except that greater weather exposures can be used. When double-coursed (Fig. 7.6-112), units can be applied at much greater weather exposures over a course of either No. 3-Grade or undercoursing-grade shingles.

Using either method, at least two layers of shingles (double coverage) occur at every point on the wall surface. Except where specifically noted, the application methods described and illustrated below are applicable to both shakes and shingles.

▪ LAYING OUT

The exposure and the number of courses should be determined by measuring the

FIGURE 7.6-109
Maximum Exposure for Wood Shakes and Shingles

Type	Length, in.	Single Coursing, in.	Double Coursing, in.
Hand-split-and-resawn shakes	18	8-1/2	14
	24	11-1/2	20
	32	15	—
Taper-split shakes	24	11-1/2	20
Straight-split shakes	18	8-1/2	16
	24	11-1/2	22
Random-width and dimension shingles	16	7-1/2	12
	18	8-1/2	14
	24	11-1/2	16
Rebutted and rejointed shingles	16	7-1/2	12
	18	8-1/2	14
Machine-grooved shingles	16	—	12
	18	—	14

Courtesy Cedar Shake and Shingle Bureau.

FIGURE 7.6-111 *Single coursing of wood shingle or shake wall. (Courtesy Cedar Shake and Shingle Bureau.)*

FIGURE 7.6-110 *Spaced sheathing being applied over sheathing paper. (Courtesy Cedar Shake and Shingle Bureau.)*

height of the dominant wall from a point 1 in. below the top of the foundation to the top of the windows, underside of soffits, or other vertical reference.

The height is then divided into equal parts as near as possible to the maximum recommended weather exposure of the shingle or shake. This measurement and the number of courses can be transferred to a story pole or be established with

chalk lines all around the building (Fig. 7.6-113a, b). The exposure of courses can vary slightly to line up with the bottoms and tops of openings and at the eave. Story-pole marks or chalk lines should be made after sheathing paper has been applied.

■ SINGLE COURSING

The first course at the foundation should be doubled. Each shake or shingle should be secured with the correct nails (see Fig. 7.6-91), driven about 3/4 in. from each edge and not to exceed 1 in. above the butt line of the succeeding course. Joints between shingles in one course should be separated at least 1-1/2 in. from joints in adjacent courses, and joints in alternate courses should not be in direct alignment.

When fiberboard or gypsum sheathing occurs, shingles and shakes should be applied over 1×3 nailing strips (see Fig. 7.6-110) spaced according to shingle exposure, or special fasteners designed for this purpose should be used.

■ DOUBLE COURSING

The bottom course should be laid with two undercoursings and one outer course. This triple coursing eliminates the need for a costly drip cap and makes the bottom course as thick as the succeeding courses (Fig. 7.6-114).

A straightedge, such as a piece of wood siding or 1×4 lumber, should be tacked tightly along a chalk line snapped between the bottom story-pole marks. Undercourse shingles, spaced about 1/8 in. apart, should be laid on the upper lip of the straightedge and fastened with a single staple near the top (Fig. 7.6-115).

The outer course should then be applied, laid tight without spacing. As each double course is laid, a chalked string should be stretched from opposite corner marks to indicate the level of the next course and to relocate the position of the straightedge.

Each shingle or shake on walls should be nailed with a minimum of two nails (see Fig. 7.6-91), applied 2 in. above the butt line and 3/4 in. from the edge of each

FIGURE 7.6-112 *Double coursing of wood shingle or shake wall.* (Courtesy Cedar Shake and Shingle Bureau.)

Labels in figure:
- Unsaturated Sheathing Paper between Shingles & Sheathing
- Outer course ½" lower than Undercourse.
- No. 3 or Undercoursing Grade Shingles.
- Apply nails in straight line 2" above Shingle Butts.
- Tight Joints

FIGURE 7.6-113 *Courses are laid out around the building using (a) a story pole or (b) a chalk line.* (Courtesy Cedar Shake and Shingle Bureau.)

FIGURE 7.6-114 *Two layers of undercoursing under the first course.* (Courtesy Cedar Shake and Shingle Bureau.)

FIGURE 7.6-115 *Straightedge forms guide for nailing undercoursing and exposure courses.* (Courtesy Cedar Shake and Shingle Bureau.)

shingle or shake. On shingles wider than 8 in., two additional nails should be driven near the center and about 1 in. apart.

■ CORNERS

Outside corners may be lapped alternately. One outer course should protrude slightly past the corner, butting the shake on the other side of it. The protruding corner should then be trimmed with a knife and plane and touched up with stain. Figure 7.6-116 shows several different corner treatments. Metal corner molding designed for this purpose may also be used.

Inside corners should be fitted with a lumber strip about 1-1/2 in. square to receive the undercourse and the outer shakes or shingles with a minimum of fitting. The strip should be stained after application.

7.6.6.7

Re-Siding Existing Walls

Shingles or shakes can be applied directly to existing siding or to new sheathing; building paper should be applied over old siding, and other steps are the same as described above (Fig. 7.6-117). Before application of courses around windows and doors, molding strips to which shakes and shingles should be joined should be nailed on the face of old casings, flush with the outer edges. When re-siding over stucco, nailing strips first should be nailed over the wall to provide a good base for nailing double coursing (Fig. 7.6-118). These horizontal strips should be spaced with centers 2 in. above the butt lines of each course to receive nails.

The lath strip application shown in Figure 7.6-119 should be used to apply shingles and shakes over insulating sheathing. Lath nailed at the studs provides a nailing base for double-coursing nails. As an alternative for attaching to nonwood sheathing, 3/8-in. by 1-1/2-in. treated or decay-resistant lath may be used. Lath should be spaced according to shingle exposure, with the butts of the undercoursing shingles resting on the lath.

Lath should be nailed to the studs through the sheathing with 2-1/8-in. by 0.109-in. (7d) annular threaded or 2-1/2-in. by 0.131-in. (8d) common wire nails. The outer course of shakes or shingles should be nailed to the lath with small-headed corrosion-resistant annular threaded nails penetrating the lath. The butts of the outer course should project about 1/2 in. below the lath (see Fig. 7.6-119).

Joints should not be closer than 1-1/2 in. to joints of the undercoursing in the same course. Joints in the following courses should be staggered at least 1-1/2 in.

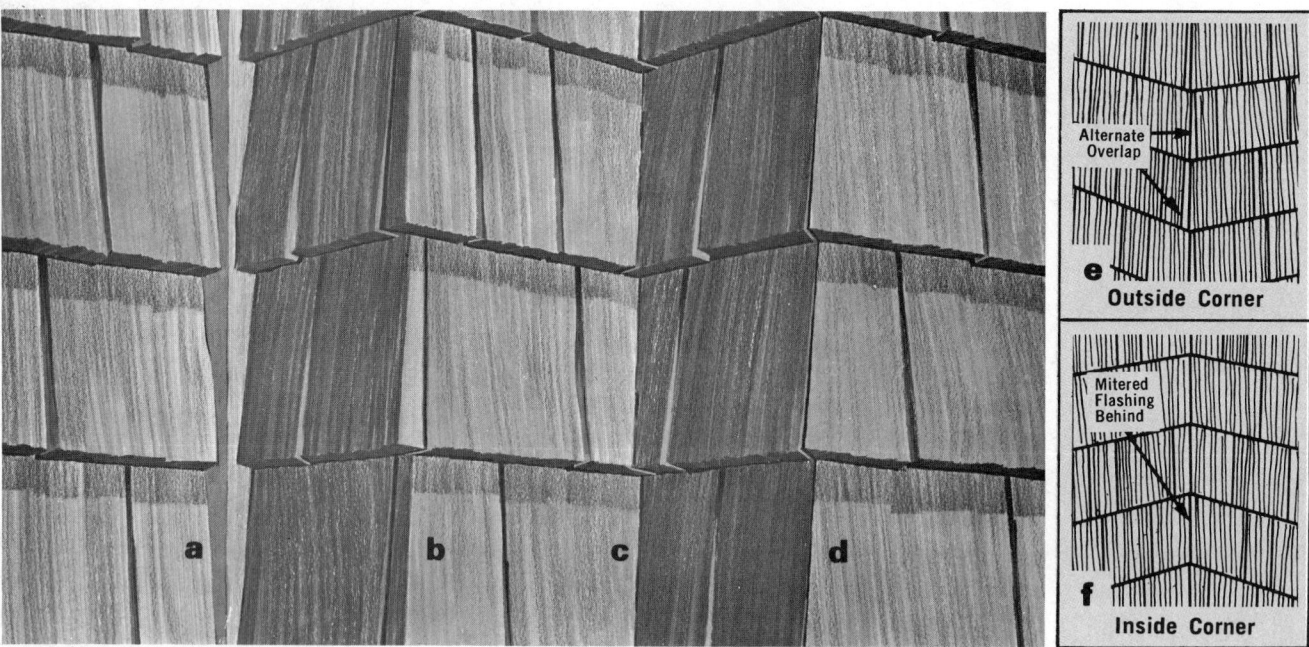

FIGURE 7.6-116 *Corner treatments for shakes: (a) jointed inside corner; (b) woven outside corner; (c) mitered inside corner; (d) mitered outside corner. Corner treatments for shingles: (e) overlapping outside corner; (f) mitered inside corner using metal flashing in the joint.* *(Courtesy Cedar Shake and Shingle Bureau.)*

FIGURE 7.6-117 *Reshingling is similar to new wall construction.* *(Courtesy Cedar Shake and Shingle Bureau.)*

FIGURE 7.6-118 *Horizontal nailing strips must be used over nonnailable sheathing or stucco.* *(Courtesy Cedar Shake and Shingle Bureau.)*

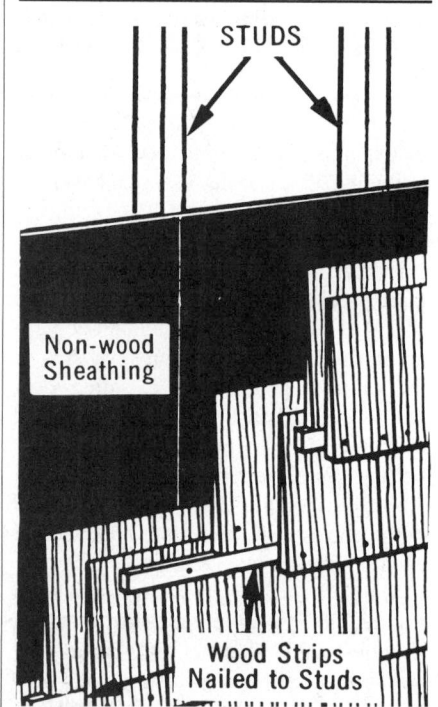

FIGURE 7.6-119 *Lath strips are used with nonnailable sheathing for double-coursed walls.* *(Courtesy Cedar Shake and Shingle Bureau.)*

7.7 LOW-SLOPE ROOFING

This section discusses two types of roofing used on low-slope roofs: built-up bituminous roofing and single-ply sheet roofing. *Low-slope roofs* are those whose slope is less than 1-1/2 in./ft., but some of the roofing systems addressed in this section can actually be used on higher slopes. This section also covers mineral-surfaced roll roofing, which can be used on slopes of 1 in./ft. or more. Liquid membrane systems are not covered here.

The function of both built-up roofing and single-ply roofing is to form a waterproof barrier. Roll roofing is intended to shed water in a manner analogous to shingles, but it uses sheet materials instead of small individual units and incorporates bituminous materials between the sheets. Refer to Section 7.5 for a discussion of bituminous roofing products, including built-up bituminous and roll roofing membranes; the various types of surfacings used on them; and associated asphaltic materials used with them. Shingles are discussed in Section 7.6.

Low-slope roofing is a system of compatible components selected to work together to form a waterproof membrane. Each component has a unique function. Possible components include a vapor retarder, underlayment, insulation, waterproof membrane, and ballast.

7.7.1

ROOFING MATERIALS

7.7.1.1

Built-Up Bituminous Membrane and Roll Roofing

The bituminous felts used in roll and built-up roofing are as described in Section 7.5.

■ ROLL ROOFING

Roll roofing may be either mineral surfaced or smooth. Organic felts are not appropriate, because roll roofing is designed to be water-shedding rather than waterproof, and organic felts tend to absorb water.

One entire side of standard mineral-surfaced roll roofing is coated with granules. Most of it weighs between 75 and 100 lb. per square (100 sq. ft.). *Double-coverage roll roofing* is coated with granules for only the first 17 in. of a 36-in.-wide sheet. The remainder is smooth. This type weighs between 55 and 75 lb. per square.

■ BITUMENS

Bitumens used in roll and built-up roofing systems include asphalt and coal tar. Refer to Section 7.5 for a discussion of these materials.

7.7.1.2

Single-Ply Membranes

Single-ply sheet roofing technology was first developed in Europe, and many of the currently available systems first appeared there. This section covers single-ply membranes that fall into one of the following categories: vulcanized elastomers, nonvulcanized elastomers, thermoplastics, and polymer-modified bitumen (Fig. 7.7-1). The materials covered here are probably the most popular in use today, but they are not the only ones. Single-ply roofing technology has developed rapidly and continues to develop. Some materials that were available just a few years ago have disappeared from the market, and new ones appear from time to time. In addition, some of the materials discussed here may also disappear in time and new systems will most surely be developed.

The thicknesses given here are standard today, but the trend is toward heavier membranes.

■ VULCANIZED ELASTOMERS

Factory-cured (vulcanized) elastomers include EPDM and vulcanized neoprene. EPDM is made from ethylene (*E*), propylene (*P*), a diene monomer (*DM*), and plasticizers, accelerators, and fillers. Neoprene (polychloroprene) is a synthetic

FIGURE 7.7-1
Summary of Single-Ply Roofing Types

Type	Materials	Thickness, mils
Vulcanized (cured) elastomers	EPDM	45–60
	Neoprene[a]	60
Nonvulcanized (uncured) elastomers	CSPE[b]	60
	CPE	40–48
	PIB	100
	Neoprene[a]	60
Thermoplastics	PVC	40–60
Polymer-modified bitumen	APP	40–60
	SBS	40–60
	SBR	40–60

[a]Neoprene is available cured and uncured.

[b]CSPE (Hypalon) is also available in liquid form for use as a coating.

rubber. Because vulcanized elastomers cannot be molded, reshaped, or welded by heat, they can only be bonded to themselves with adhesives.

Most EPDM is 45 mils thick, but 60-mil material is also available. Neoprene is usually 60 mils thick.

Some formulations of neoprene will accept a chlorosulfonated polyethylene (CSPE) color coating.

■ NONVULCANIZED ELASTOMERS

Nonvulcanized (uncured) elastomer sheet materials include those made from chlorosulfonated polyethylene (CSPE), chlorinated polyethylene (CPE), polyisobutylene (PIB), and nonvulcanized neoprene. Uncured elastomers can be molded, shaped, and welded by heat. Therefore, seams in them can be heat- or chemically welded. Most of these materials cure in place after they have been installed and are then said to be *thermoset*. Thermoset materials can be installed by using adhesives in the seams.

CSPE (DUPONT HYPALON) CSPE is a synthetic rubber produced in many colors. Sheets are made of CSPE laminated to a mineral fiber backing or a polyester scrim. Sheets are usually 60 mils thick. It is also available in liquid form for use as a coating. It is weather, chemical, and pollutant resistant, which makes it usable as an exposed finish.

CPE CPE is an unreinforced or polyester-reinforced sheet material, usually 40 to 48 mils in thickness. It can be pigmented with many colors. The raw material used to make CPE is manufactured by the Dow Chemical Company. This material is compatible with bitumen and can, therefore, be installed directly over an existing asphalt or coal tar pitch roof. It is also resistant to weathering, chemicals, and pollutants and is suitable for exposed locations.

PIB PIB is formulated from isobutylene and other polymers, carbon black, and other additives. A sheet consists of a 60-mil PIB membrane laminated to a 40-mil polyester backing. Sealing edges are left unbacked. PIB can be used in exposed locations and in contact with asphalt, but it must be protected from petroleum distillates, organic oils and fats, and tar.

■ THERMOPLASTICS

Thermoplastic materials can be softened and reformed throughout their life. Polyvinyl chloride (PVC) when combined with plasticizers forms a thermoplastic roofing membrane that is usually between 40 and 60 mils thick. It is available unreinforced and reinforced with polyester or glass fiber. It can be used as exposed roofing but is not compatible with bitumens. PVC comes without a finish for field finishing with coatings or with a factory-applied acrylic coating for weather exposure.

■ POLYMER-MODIFIED BITUMEN

Polymer-modified bitumen sheets are made from asphalt that has been modified. Typical modifiers include atactic polypropylene (APP), styrene-butadiene-styrene (SBS), and styrene-butadiene-rubber (SBR). They are reinforced with glass fiber mats or scrim, polyester mats or scrim, polyethylene sheets, or copper or aluminum foils. Most are between 40 and 60 mils thick.

Some polymer-modified bitumen sheets are surfaced at the factory with mineral granules to protect them from the elements and to improve their appearance. APP-modified bitumen sheets *must* have a coating of granules. Some sheets are not factory-coated but are suitable for field application of granules.

APP-modified bitumen sheets are not compatible with asphalt. SBS and SBR sheets are compatible with asphalt and can be used in built-up roofing applications. All three can be made suitable for torch application. Some ABS and SBR sheets are self-adhering. In general, APP sheets work better in warm climates. The other two are better in cold climates.

7.7.1.3

Insulation

Roofing insulation materials are discussed in Section 7.3; criteria for their selection in Section 7.7.6.2.

7.7.2

SUBSTRATES

Proper functioning of low-slope roofing depends not only on the roofing material and its application, but on the roof structure as well. This structure must be sound, smooth, have sufficient slope to prevent ponding of water on the roof, and be designed to support the loads that will be imposed. There must be enough proper-sized roof drains to remove water from the roof.

There are essentially three types of substrates used beneath built-up bituminous and single-ply roofs: nailable, insulated, and concrete.

7.7.2.1

Nailable Substrates

Nailable decks include wood planks, plywood, cement–wood fiber planks, lightweight insulating concrete, poured-in-place gypsum, and precast gypsum planks.

7.7.2.2

Insulated Substrates

Insulated decks are covered with one or more layers of insulating material. The underlying deck may be nailable, metal, or concrete. Insulation may be boards, sprayed-in-place plastic foams, or a thermosetting insulating fill. Steel decks are covered with a layer of insulation (usually boards), even when the insulation is not necessary for thermal purposes, because low-slope roofing materials are not designed to span the flutes in steel decking.

7.7.2.3

Concrete Substrates

Concrete decks may be either poured-in-place or precast.

7.7.3

ROOF CONFIGURATIONS

The three typical roof configurations are called *insulated framed*, *compact conventional*, and *protected membrane*. Refer to Section 7.1.3.2 for a discussion of these three configurations.

Most low-slope roofing systems are insulated. In a framed system, the insulation is separated from the roofing by the roof deck or sheathing. When a ventilated airspace is provided between the insulation and the deck or sheathing, such a system is sometimes called a *ventilated* system (see Fig. 7.1.23).

In compact systems (see Fig. 7.1-27) the insulation, membrane, and vapor retarder, if there is one (see Section 7.1), are all in contact. The conventional order of components in a compact system, starting from the exterior, are ballast, membrane, insulation, vapor retarder (if there is one), insulation, and roof deck. In some systems, the ballast is replaced by another surfacing material (see "Surfacing" in Section 7.7.6.4).

In a protected membrane (PMR) system, the order is ballast, pervious fabric, insulation, membrane, underlayment, and deck (see Fig. 7.1-29). A vapor retarder is not needed because the membrane serves that purpose. Ballast must be used to hold the insulation in place.

A system in which there is insulation both above and below the roof deck is called a *combination framed and compact* system.

7.7.4

FLASHING

Flashing is necessary in roofing systems to seal the gaps between the roofing and adjoining materials, such as walls, pipes, and other penetrations. Flashing may be one of several products, including metal, fabric, and plastics. Regardless of the material selected, it must be compatible with the roofing chemically and in its degree of expansion and contraction due to temperature changes. Flashing materials are discussed in Section 7.9.

7.7.5

WARRANTIES AND MAINTENANCE AGREEMENTS

Two types of warranties are issued covering low-slope roofing: a manufacturer's warranty, or guarantee, and a roofing contractor's warranty. There is considerable controversy about the value of manufacturers' warranties. Some experts believe that some warranties issued by roofing manufacturers limit the rights an owner would otherwise have under current law, rather than expand the owner's protection.

One current method of attempting to ensure quality in a low-slope roofing system is to require the roofing contractor to issue a warranty and to sign a maintenance agreement. The effectiveness of such warranties and maintenance agreements is controversial, however, because they are only as good as the organizations that sign them, and because some standard warranty agreements are biased toward the roofing contractor, offering it more protection than is granted to the building owner. A normal period of such warranties is 2 years, because about 65% of roofing problems occur during that period.

The purpose of a maintenance agreement is to ensure that the roof is properly maintained and repaired if a problem appears. Such agreements usually cover about a 5-year period and are usually renewable.

To be more sure of proper protection, a building owner should seek legal council before agreeing to any roofing warranty or maintenance agreement.

7.7.6

SELECTING SYSTEMS

Important factors in selecting a low-slope roofing system include the configuration of the system and the components used in it. Two overriding selection criteria are (1) compliance with the applicable codes and (2) compliance with the applicable insurance requirements.

The basic decision about whether to use a built-up roofing system or a single-ply one should be influenced by the advantages of each in the particular circumstances. For example, single-ply roofing can be installed in a much wider range of temperatures than can a built-up bituminous system. The availability of competent installers is another factor. The success of a single-ply roofing system in preventing leaks and other problems depends to a much larger extent on the ability of the installer than is true for conventional built-up roofing. In fact, the sealing of single-ply roofing seams must be perfect. Even a slight error will usually result in a leak. In addition, roof decks that will receive single-ply membranes should be sloped to provide positive drainage, because they are susceptible to damage from standing water and ice. Some single-ply systems can be installed directly over existing roofing.

Figure 7.7-2 is a summary of criteria for selecting low-slope roofing types.

7.7.6.1

Configuration

Which components a roofing system should contain depends on the building's design, construction, and location. Other factors influencing this decision include energy considerations, budget considerations, aesthetics, and codes. For example, the applicable code may dictate which type of insulation can be used in contact with a steel or wood deck.

Whether a vapor retarder is required should depend on the local weather conditions and the roof configuration selected. Refer to Section 7.1 for a discussion about vapor retarders in roofing assemblies.

7.7.6.2

Insulation

The type of roof insulation selected depends on the requirements of the applicable codes, the thermal characteristics desired, the selected configuration of the roofing system, economical considerations, fire resistance, traffic-indentation resistance, weight, and available insulation panel dimensions. Different insulation

FIGURE 7.7-2
Selection Criteria for Low-Slope Roofing Types

Selection Item	Criteria
Overriding criteria	Code compliance; insurance requirements
Built-up vs. single-ply	Climate; availability of competent installers; presence of existing roofing; condition of the substrates
Configuration	Building's design, construction, and location; energy use; construction budget; aesthetics; code compliance; insurance requirements
Vapor retarder	Climate; roofing configuration; insulation material; roofing membrane material
Insulation type	Code compliance; climate; desired thermal characteristics; economy; roofing system configuration; fire resistance; traffic indentation resistance; presence of a vapor retarder; use of hot-applied or solvent-applied roofing membrane; roofing materials used

materials may be selected depending on whether the material should be vapor permeable or relatively impermeable. This may depend on whether the selected system contains a vapor retarder. If a vapor retarder is included, the insulation should not be impermeable, because there would then be two vapor retarders, which could trap moisture between them and not permit its escape.

Another criterion affecting insulation selection is that the application of hot-applied built-up roofing directly over polyisocyanurate, polyurethane, and phenolic foam insulations sometimes leads to delamination and blistering problems. Therefore, the NRCA (National Roofing Contractors Association) recommends that a layer of wood fiber, perlite, or glass fiber board insulation be placed over these plastic foam insulations when a hot-applied built-up roofing system will be applied.

Insulation beneath loose-laid single-ply roofing can be installed without adhesives, and the roofing membrane is not attached using adhesives. Therefore, compatibility problems are reduced, and the selection of usable insulation types is much wider.

Because some roofing materials are not compatible with certain insulation products, the roofing membrane manufacturer's recommendations must be followed in the selection of roof insulation that will be in contact with roofing membranes. This is especially true for roofing membranes that are installed using the adhered method discussed in Sections 7.7.6.5 and 7.7.7.5.

7.7.6.3

Roll Roofing

Either of the materials described in Sections 7.5 and 7.7.1.1 may be used. Ballast is seldom applied over roll roofing.

7.7.6.4

Built-Up Membrane Roofing

The NRCA has developed a four-letter system of generic designations to describe low-slope roofing systems, depending on their make-up. The four letters stand for the deck type, the bitumen used, the felt used, and the surface type (Fig. 7.7-3). For example, a NAAG system has a nailable deck, asphalt bitumen, glass fiber felt, and aggregate surfacing.

■ FELT SELECTION

Most felts used in built-up roofing today are glass fiber base felts, because they are stronger and more resistant to slippage than the organic felts that were the favorite in the past.

The polymer-modified bitumen sheets discussed in Section 7.7.1.2 may also be used in built-up roofing.

■ BITUMEN SELECTION

Coal tar is considered superior to asphalt bitumen, but it is limited to use on slopes of 1/2 in./ft. or less in the coldest areas of the United States and to even lower slopes in more southern climates. In the hottest areas, the slope should be less than 1/4 in./ft.

Asphalt bitumen can be used over a wide range of slopes and climatic conditions. Refer to Section 7.5 for a discussion about the various types of asphalt bitumen and associated roof slope limitations. In general, the lowest numbered type that will work should be selected for a given slope.

Soft asphalt and coal tar bitumens are incompatible and should not be used in the same roof system. Some roof systems, however, have successfully combined hard asphalt in the felts with coal tar bitumen between the felts, but such systems must be applied more carefully than fully compatible systems.

■ BASE SHEET

Some systems include base sheets and some do not. The function of base sheets is supposed to be to protect the roofing system against moisture in the substrates and from condensation on the bottom of the membrane. Their effectiveness in this regard is not universally accepted, however. To follow local practice is probably the best advice when deciding whether to include a base sheet in a roofing system. When the base sheet is omitted, an additional ply sheet should be added.

■ PLY COUNT

A built-up roofing system's ply count is the number of felts in the system. Coats of bitumen between plies and topping flood coats and ballast above the top ply do not count as plies. Normal built-up roof systems for commercial projects have three or four plies.

■ SURFACING

In most areas of the country, built-up roofing is usually made up of smooth-surfaced felts covered with stone or unit-paver type ballast. Stone ballast, however, is limited to slopes of 3 in./ft. or less. In some areas of the country, mineral-surfaced felt is often used as the topping sheet over built-up roofing, and no ballast is applied.

When ballast cannot be used because of the slope of a roof, the top ply is often coated with either bitumen or another coating material, such as a fibered or nonfibered aluminum coating.

FIGURE 7.7-3
NRCA Four-Letter Designations for Built-Up Roofing Systems

First Letter (Deck Type)
N = nailable
I = insulated
C = concrete

Second Letter (Bitumen Type)
A = asphalt
C = coal tar

Third Letter (Surfacing)
A = aggregate surface
E = emulsion coating
M = mineral surface

Fourth Letter (Ply Felt Type)
O = organic
G = glass fiber

Example

NAAO = nailable deck type (N), asphalt bitumen (A), aggregate surfaced (A), organic felt (O)

Single-Ply Membrane

■ MATERIAL SELECTION

The selection of which material to use depends on the selection of the desired configuration and is influenced by the factors mentioned in Section 7.7.6.1.

■ INSTALLATION METHOD

The methods available for installing single-ply roofing include *loose laid and ballasted, mechanically fastened,* and *adhered.* The correct method to use mostly depends on the selected material. Each manufacturer issues specific instructions concerning application of their product, which must be followed if a warranty is to be issued.

Regardless of the system selected, plies should be laid shingle fashion starting at low points in the deck so that water drains across the roof without interruption.

LOOSE LAID AND BALLASTED EPDM, neoprene, hypalon, PIB, PVC, and some modified bitumen sheets may be laid using the loose-laid and ballasted system. In this system, the roofing is independent of the roof deck, which allows the structure to move without affecting the roofing. This same feature, however, means that such roofing will blow off quite easily due to wind uplift. This must be prevented by a layer of ballast. Most installations include washed river gravel as the ballast, but concrete pads are also sometimes used.

MECHANICALLY FASTENED When reducing the dead load on a roof is desirable or the roof is more steeply sloped than can be accommodated by ballast, single-ply roofing can be installed using the mechanically fastened system. Because the fasteners penetrate into the decking and, therefore, through an installed vapor retarder (when a vapor retarder is required), mechanically fastened systems should not be used.

ADHERED Adhering a single-ply membrane to the roof deck or to insulation permits applications over complex shapes, curves, and steep slopes. Membranes may be either fully or partially adhered.

Flashing

Flashing associated with low-slope roofing systems should be a part of the roofing. The materials selected should be compatible with the roofing both chemically and in thermal expansion characteristics. This is especially true when there is a large differential in the coefficient of expansion of the roofing material and the flashing. Therefore, metal flashing is not appropriate as a base flashing with most roofing materials. Conditions where metal flashings, such as gravel stops, are built into roofing membranes are particularly susceptible to failure. Refer to Section 7.9 for a detailed discussion about flashings.

APPLICATION OF ROOFING SYSTEMS

In general, the application of the various low-slope roofing systems should follow the specifications of the manufacturer of the system and the recommendations of the NRCA. Installation of single-ply roofing should follow the recommendations of the Single-Ply Roofing Institute (SPRI), and metal flashing and accessories should be fabricated and installed in accordance with the recommendations of the Sheet Metal and Air Conditioning Contractors National Association, Inc. (SMACNA). Conflicting recommendations should be reconciled before installation of a roofing system.

Bitumen Temperature

Temperature affects the rate of flow of a bitumen. Bitumens that have been heated at too high a temperature flow too readily, which affects the amount of bitumen in a roofing system. They may also separate and leave voids between sheets of membrane, which are potential leak points. Conversely, not raising the temperature of a bitumen high enough can also cause problems. Bitumen that is not heated enough will not flow as it should and may not fuse with and thereby adhere properly to the membranes.

The NRCA recommends that each bitumen package be labeled with its *equiviscous temperature* (EVT) and its *flash point* (FP). A bitumen's EVT is its optimum temperature. A bitumen should never be heated above its flash point, and it should be applied at its EVT. In addition, bituminous materials should not be kept at high temperatures or allowed to stand in luggers for prolonged periods.

Application of Insulation

Most roof insulation should be applied in at least two layers with the joints staggered in the two layers, to reduce thermal bridges through the joints, reduce ridging that sometimes occurs in single-layer applications, and make mechanical attachment easier (Fig. 7.7-4).

Beneath built-up and adhered single-ply systems, insulation should be mechanically fastened to nailable decks, except that it may be laid in strips of hot bitumen when a vapor retarder has been installed over the deck.

Over a steel deck, single-layer insulation and the first layer of two-layer board insulation beneath built-up or adhered single-ply roofing should be mechanically fastened in place (Figs. 7.7-5 and 7.7-6). When fasteners are not used, the boards should be bedded in strips of hot bitumen laid along the deck flutes (Fig. 7.7-7). Boards should be placed with the long dimension parallel to the deck flutes and perpendicular to the roof slope (Fig. 7.7-8).

Board insulation beneath built-up or adhered single-ply roofing should be applied over nonnailable decks in hot bitumen or adhesive.

Board insulation may be loose laid beneath loose-laid single-ply systems. The second layer of board insulation should be laid in a full mopping of bitumen (see Fig. 7.7-6).

Application of Roll Roofing

■ SHEATHING

Most roll roofing is applied over plywood or wood board sheathing (see Chapter 6),

NOTE: Steel deck "ribs" are often referred to as "flutes."

FIGURE 7.7-4 *Typical insulation over steel deck. Vapor retarder is not always present.* (*Reprint courtesy National Roofing Contractors Association.*)

FIGURE 7.7-5 *Single-layer insulation over steel deck.* (*Reprint courtesy National Roofing Contractors Association.*)

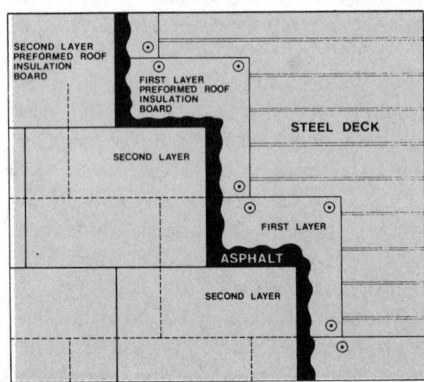

FIGURE 7.7-6 *Two-layer insulation over steel deck.* (*Reprint courtesy National Roofing Contractors Association.*)

but it can also be applied over most nailable decks.

■ UNDERLAYMENT

Roll roofing requires the application of a continuous underlayment. This is usually

FIGURE 7.7-7 *Placing strips of bitumen on flutes of steel deck.* (*Reprint courtesy National Roofing Contractors Association.*)

No. 15 asphalt-impregnated unperforated roofing felt applied with the long edge parallel with the eaves. On slopes of 4 in./ ft. and steeper, one layer is sufficient. Two layers should be used on lower slopes.

Layers of underlayment should be laid shingle fashion so they shed water, and they should be nailed with sufficient nails to hold them in place. Ends should be lapped at least 6 in. Edge joints should lap at least 2 in. The top layer in double-layer applications should lap 19 in. over the lower layer. Underlayment should lap over hips and ridges so that at least two layers cover them.

When the mean January temperature is 30°F or lower, an ice shield should be applied. When one layer of underlayment is used, this ice shield should consist of an additional layer of No. 15 felt laid and set in mastic from the eave to a line 24 in.

FIGURE 7.7-8 *Laying insulation boards over steel deck in bitumen strips.* (*Reprint courtesy National Roofing Contractors Association.*)

inside the inner face of the exterior wall. When two layers of underlayment are used, the ice shield should be formed by applying the underlayment layers in hot asphalt from the eave to a line 24 in. inside the inner face of the exterior wall.

■ DRIP EDGE

A metal drip edge should be placed along rakes and eaves (Fig. 7.7-9). The underlayment should be placed beneath the drip edge at the rakes and above it at the eaves.

■ FLASHING

Flashings at penetrations are usually metal. Base flashings at walls, chimneys, curbs, and the like are usually bituminous, but may also be metal. Valley flashings may be either bituminous or metal. Metal flashings are discussed in Section 7.9.

■ FASTENERS

Roll roofing is usually applied using 11- or 12-gauge annular or screw-threaded, galvanized steel or aluminum roofing nails, with 3/8-in.-diameter heads and lengths between 7/8 in. and 1-1/4 in. Nails should penetrate at least 3/4 in. into the underlying nailable deck or sheathing.

FIGURE 7.7-9 *Metal drip edge at eaves and rake beneath roll roofing.* (*Reprint courtesy National Roofing Contractors Association.*)

FIGURE 7.7-10 *Standard roll roofing installed by the concealed nailing method, with sheets laid parallel with eaves.* (*Reprint courtesy National Roofing Contractors Association.*)

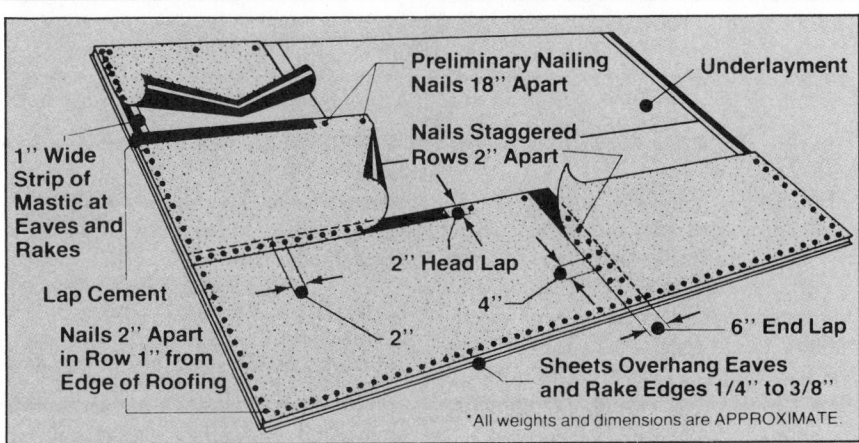

FIGURE 7.7-11 *Standard roll roofing installed by the concealed nailing method, with sheets laid perpendicular to eaves.* (*Reprint courtesy National Roofing Contractors Association.*)

■ **APPLICATION**

Roll roofing sheets may be placed either parallel or perpendicular to the eaves. When perpendicular, the sheets should be tilted toward the rake at a rate of 1/8 in./ft. of rake.

STANDARD MINERAL-SURFACED ROLL ROOFING Sheets may be applied with either concealed or exposed nails. The concealed nailing method (Fig. 7.7-10) should be used when (1) the sheets are applied parallel with the eaves and (2) the slope is 3 in./ft. or more, even when the sheets are laid perpendicular to the eaves (Fig. 7.7-11).

The exposed nailing system is limited to (1) slopes exceeding 2 in./ft. when the roofing is applied parallel with the eaves, and (2) slopes 4 in./ft. or more, when the roofing is applied perpendicular to the eaves (Fig. 7.7-12). Endlaps are usually 6 in.; edgelaps 2 in.

In the concealed nailing system, a 12-in.-wide starter strip is applied completely around the perimeter of the roof and subsequently covered with roofing. Roofing cement is applied over the entire area of each lap of roofing material, whether over the starter strip or over other roofing material.

In the exposed nailing system, edges at rakes and eaves and laps of roofing material should be sealed with lap cement before nailing.

Ridges and hips should be covered with 12-in.-wide strips of roofing. In concealed nailing systems, these strips are nailed, then cemented, and the next section lapped over the succeeding one to conceal the nails. In exposed nailing systems, the sheets are laid with a 2-in.-wide band of roofing cement along each edge and nailed in place. In either case, laps should be fully bedded in roofing cement.

DOUBLE-COVERAGE MINERAL-SURFACED ROLL ROOFING Sheets may be applied with the long dimension either parallel or perpendicular to the eave (Figs. 7.7-13 and 7.7-14). Nailing should be done in the selvage area. Laps should be fully bedded in roofing cement. Sheets should be laid so that only the areas with a granular finish is exposed.

FIGURE 7.7-12 *Standard roll roofing installed by the exposed nailing method.* (*Reprint courtesy National Roofing Contractors Association.*)

FIGURE 7.7-13 *Double-coverage roll roofing laid parallel with eaves.* (*Reprint courtesy National Roofing Contractors Association.*)

7.7.7.4

Application of Built-Up Roofing

The application of built-up roofing should be scheduled and carried out so that neither insulation nor roofing is left directly exposed to precipitation or condensation at any time. Built-up roofing plies should be laid shingle fashion with laps sufficient to provide the selected number of plies throughout (Fig. 7.7-15). The shingles should be placed in the proper direction to shed water. On slopes in excess of 2 in./ft., the sheets should run parallel with the slope.

Built-up roofing plies should be nailed to nailable substrates as directed by the system manufacturer and the standards noted at the beginning of Section 7.7.7 and to built-in nailers where appropriate. Nailers should occur at locations of flashing and in other locations as needed. Nails should be concealed in the finished roofing.

Forty-five-degree cant strips and tapered edge strips should be installed where the

roofing abuts vertical surfaces and at gravel stops.

The base sheet should be applied first in systems where one will be used. It should be mopped and fastened in place.

The selected number of interply sheets should be applied in full beds of hot bitumen having the weight of bitumen per square specified for the selected system. A glaze-coat top surface of bitumen should then be applied over the top roofing ply. Bitumen should be heated to a temperature that will permit its application at its EVT, but never to its flash point.

Plies should be applied over cants and tapered edge strips and extend at least 2 in. above the top of the cant or tapered edge (Fig. 7.7-16). A folded-back envelope should be built in at the edges of built-up roofing that does not terminate at a cant or tapered edge strip to prevent bitumen from flowing over the edge of the roofing onto adjacent surfaces (Fig. 7.7-17). This envelope should be turned back over the plies and bedded in hot bitumen.

Expansion joints (Fig. 7.7-18), area dividers (Fig. 7.7-19), accessories, and flashings (Fig. 7.7-20) should be built into the roofing according to the standards listed at the beginning of Section 7.7.7. Counter flashings are discussed in Section 7.9.

As soon as possible after the roofing has been installed, the top should be flood-coated with hot bitumen, and the selected aggregate ballast should be laid in it (Figs. 7.7-21 and 7.7-22). Where aggregate ballast will not be used, the selected aggregate-surfaced asphalt felt, cap sheet, emulsion top coating, or aluminum reflective coating should be promptly applied.

7.7.7.5

Application of Single-Ply Membranes

In both loose-laid and ballasted and mechanically fastened systems, the membrane should be laid over a separator, such as rosin-sized building paper. If the manufacturer requires another type of separator, its recommendations should be followed.

■ LOOSE-LAID AND BALLASTED SYSTEMS

Loose-laid roofing is placed over the substrates with only minimal fastening around the edges and at penetrations (Fig. 7.7-23). Adjoining sheets should be lapped and

FIGURE 7.7-14 *Double-coverage roll roofing laid perpendicular to eaves.* (*Reprint courtesy National Roofing Contractors Association.*)

bonded together using a sealant recommended by the roofing manufacturer.

Aggregate ballast should be applied uniformly at a rate of 10 to 12 lb./sq. ft. Precast concrete paver ballast should be applied in accordance with the roofing manufacturer's recommendations. Care should be taken to ensure that the membrane is not damaged during installation of ballast.

■ MECHANICALLY FASTENED SYSTEMS

Single-ply membranes may be applied with either penetrating or nonpenetrating fasteners. Penetrating fasteners pass through the membrane (Fig. 7.7-24). Nonpenetrating fasteners are anchored to the structural deck, and the membrane is fastened to them using clamps or snap-on caps (Fig. 7.7-25).

The spacing, location, and type of fasteners should be in accordance with the

Diagram A

Diagram B
Alternate Procedure

FIGURE 7.7-15 *Built-up membrane roofing (a) without a base sheet and (b) with a base sheet.* (*Reprint courtesy National Roofing Contractors Association.*)

FIGURE 7.7-16 *Base flashing for wall-supported deck. When the deck is not supported by the wall, built-up roofing should be applied against an expansion joint to prevent wall movement from damaging the roofing membrane.* (*Reprint courtesy National Roofing Contractors Association.*)

FIGURE 7.7-17 *Built-up roof edge at a heavy metal gravel stop.* *(Reprint courtesy National Roofing Contractors Association.)*

roofing manufacturer's recommendations (Figs. 7.7-26 and 7.7-27).

■ ADHERED SYSTEMS

Adhered membrane may be either fully adhered or partially adhered.

In a *fully adhered* system, the membrane is completely attached to the substrates by hot- or cold-applied bitumen, cold-applied adhesives, solvents by heating (torching) the back of the membrane, or pressing a self-adhering membrane in place (Figs. 7.7-28 and 7.7-29).

In a *partially adhered* system, the roofing membrane is laid into beads or strips of bitumen, adhesive, or solvent and rolled, or is adhered by similar materials placed on the top plates of the fasteners that hold down the insulation.

■ JOINTS

Seams should be made by lapping adjacent sheets and sealing the joint. Different joining methods are required for each type of roofing. The manufacturer's recommendations should be followed. Before laps can be sealed, separation coatings, such as talc or mica must be removed from the area to be joined. This can be done using a solvent.

Joints in nonvulcanized elastomers can be sealed using either heat- or chemical-welding.

Vulcanized and nonvulcanized materials that have been cured must be sealed using adhesives (Fig. 7.7-30). The type of adhesive should be the one recommended by the roofing manufacturer. Some sheets require contact adhesives, some require adhesives made from the same material as the membrane, and others will work with generic adhesives. Some sheets are naturally self-adhering. Some are made self-adhering by the addition of an adhe-

sive coating or self-adhering strips bonded to the sheets.

Joints in thermoplastic materials can be either heat- or solvent-welded (Fig. 7.7-31).

Joints in self-adhering modified bitumen sheets should be made by pressing the sheets together and rolling. Joints in non-self-adhering APP sheets should be torch-welded (Fig. 7.7-32). Joints in non-self-adhering SBS and SBR sheets can be either torch-welded or sealed with hot asphalt.

■ FLASHING

Base and penetration flashings are usually the same material as the sheet membrane or a nonvulcanized compatible elastomeric sheet material (Fig. 7.7-33). Metal base flashings are inappropriate, and joints between elastomeric base flashings and metal cap (counter) flashings should be avoided when possible. Gravel stops may be metal, made from compatible plastic materials, or coated with the same elastomer used in the roofing membrane to permit proper bonding with the roofing.

Joints between roofing membrane and elastomeric flashings should be sealed with the same material used in the seams in the roofing membrane (Fig. 7.7-34). Expansion joint covers should be made with compatible materials (Fig. 7.7-35).

When the roofing manufacturer so requires, nailers should be provided, and elastomeric flashings should be securely fastened in place.

Metal cap flashings are addressed in Section 7.9.

■ COATINGS AND SURFACINGS

Liquid hypalon (CSPE) can be used to coat EPDM and some formulations of neoprene to provide color. Silica sand coatings are also sometimes used on EPDM.

PVC and some modified bituminous membranes are sometimes coated with acrylic emulsions to improve their appearance.

Some modified bitumen membranes that do not have a factory-applied finish require a field-applied protective layer of mineral granules or aggregate to make them fire resistant. Aluminum coatings are also sometimes used over modified bitumen membrane roofing.

FIGURE 7.7-18 *Expansion joint with high curb in built-up roofing. Expansion joints should be carried through the roofing whenever they occur in the structure. Expansion joints may also be covered with preformed elastomeric or combination metal and elastomeric expansion caps. Expansion joints bonded into built-up roofing without high curbs are not recommended.* (Reprint courtesy National Roofing Contractors Association.)

FIGURE 7.7-19 *Area dividers should be located in built-up roofing between expansion joints at 150- to 200-ft. intervals to provide expansion relief in the membrane. They should not restrict water flow to drains.* (Reprint courtesy National Roofing Contractors Association.)

FIGURE 7.7-20 *Installation of bituminous base flashing at a roof penetration.* (Reprint courtesy National Roofing Contractors Association.)

FIGURE 7.7-21 *Spreading aggregate ballast into hot bitumen adjacent to roof penetrations.* (Reprint courtesy National Roofing Contractors Association.)

FIGURE 7.7-22 *Placing aggregate ballast over built-up roofing.* (Reprint courtesy National Roofing Contractors Association.)

FIGURE 7.7-23 *The home court of the NBA's Minnesota Timberwolves is protected by a ballasted, 45-mil EPDM membrane system.* (Courtesy Firestone Building Products Company.)

FIGURE 7.7-24 *Penetrating-type fastener for single-ply roofing. Several different types are available, but their principle is similar to the device shown.* (Drawing by HLS.)

FIGURE 7.7-25 *Nonpenetrating fasteners do not pass through the membrane. These are proprietary systems.* (Drawing by HLS.)

FIGURE 7.7-28 *Fully adhered single-ply roofing systems are particularly useful on complex roof shapes, such as domes, barrels, and parabolas, and of course on flat roofs, like the one shown here.* (Courtesy Goodyear Tire and Rubber Company.)

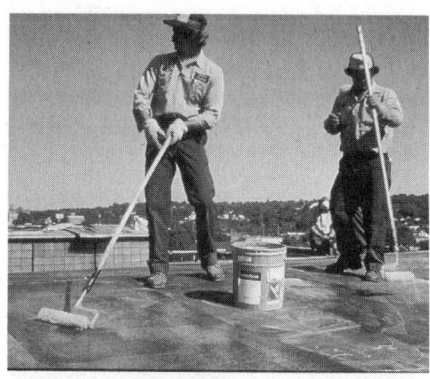

FIGURE 7.7-29 *Installing a fully adhered single-ply roofing system.* (Courtesy Goodyear Tire and Rubber Company.)

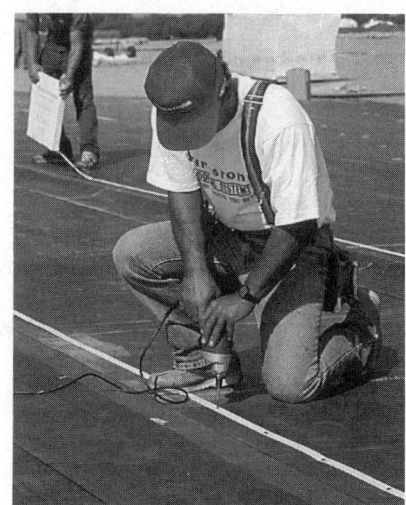

FIGURE 7.7-26 *A polymer batten bar is used to mechanically attach sheets of single-ply EPDM roofing membrane to the substrate atop JVC Company of America's distribution center in Aurora, Illinois.* (Courtesy Firestone Building Products Company.)

FIGURE 7.7-27 *Fasteners are installed along the leading edge of this white thermoplastic alloy roofing membrane to securely attach it to a 2-3/4-in.-thick layer of polyisocyanurate insulation.* (Courtesy Firestone Building Products Company.)

FIGURE 7.7-30 *Applicator ensures that adjoining EPDM sheets are lapped a minimum of 3 in., cleaned, and spliced with splice adhesive to form a continuous, watertight membrane atop the Target Center in downtown Minneapolis.* (Courtesy Firestone Building Products Company.)

FIGURE 7.7-32 *Torch-down application of APP-modified bitumen systems.* (Courtesy Firestone Building Products Company.)

FIGURE 7.7-31 *This white thermoplastic alloy roofing system uses heat-weldable seams to form a continuous watertight membrane.* (Courtesy Firestone Building Products Company.)

SMOOTH CONCRETE—
EXPOSED SURFACES MUST BE WATERPROOFED

CAULK WITH
ELASTOMERIC SEALANT

ANGLE CLAMPING BAR
WITH SLOTTED
ANCHOR HOLES

FASTENERS IN
EXPANSION SHIELDS

COMPRESSIBLE
ELASTOMERIC TAPE TO
SPAN IRREGULARITIES

CURED OR UNCURED FLASHING
MEMBRANE ADHERED TO WALL AND
MEMBRANE WITH APPROPRIATE
ADHESIVE

EPDM MEMBRANE

PRIME CONCRETE, IF REQUIRED

SEAL EDGE OF FLASHING WITH
LAP EDGE SEALANT

FASTENING STRIP APPLIED TO
DECK OR WALL—FASTENED
12" O.C.

FIGURE 7.7-33 *Wall flashing for a single-ply membrane.* (Reprint courtesy National Roofing Contractors Association.)

SEALANT

3" MINIMUM

JOINT COVER 4" TO 6" WIDE—
SET IN COMPATIBLE MASTIC

METAL GRAVEL STOP

SEAL EDGE OF FLASHING MEMBRANE WITH
LAP EDGE SEALANT

CURED OR UNCURED FLASHING MEMBRANE
ADHERED TO METAL AND
MEMBRANE WITH APPROPRIATE
ADHESIVE

1½"

FASTENERS APPROX. 4" O.C.—
DO NOT STAGGER

EPDM MEMBRANE

12" TO 18" TAPERED
EDGE STRIP

CONTINUOUS CLEAT

NOTES:

ATTACH NAILER TO MASONRY WALL. REFER TO FACTORY MUTUAL DATA SHEET 1-49.

THIS DETAIL SHOULD BE USED ONLY WHERE DECK IS SUPPORTED BY THE OUTSIDE
WALL.

THIS DETAIL SHOULD BE USED WITH LIGHT-GAUGE METALS, SUCH AS 16-OZ. COPPER,
24-GAUGE GALVANIZED METAL OR 0.040" ALUMINUM. A TAPERED EDGE STRIP IS USED
TO RAISE THE GRAVEL STOP. FREQUENT NAILING IS NECESSARY TO CONTROL THER-
MAL MOVEMENT.

WOOD BLOCKING MAY BE SLOTTED FOR VENTING WHERE REQUIRED.

FIGURE 7.7-34 *A light-metal gravel stop at the edge of a single-ply membrane. Some roofing manufacturers recommend that a heavy metal gravel stop coated with an elastomeric material compatible with the roofing be used in such a condition. When this is not acceptable for aesthetic or other reasons, the detail shown can be used.* (Reprint courtesy National Roofing Contractors Association.)

FLEXIBLE VAPOR RETARDER TO SERVE AS INSULATION RETAINER—ATTACHED TO TOP OF CURB

CHAMFER EACH SIDE OF WOOD CURB TO DRAIN

COMPRESSIBLE INSULATION

WOOD NAILER EACH SIDE SECURED TO DECK WITH APPROPRIATE FASTENERS APPROX. 24″ O.C.

SEALANT

3″ MINIMUM

CURED OR UNCURED FLASHING MEMBRANE ADHERED TO BELLOWS AND MEMBRANE WITH APPROPRIATE ADHESIVE

SEAL EDGE OF FLASHING WITH LAP EDGE SEALANT

EPDM MEMBRANE

FASTENERS APPROX. 4″ O.C.—DO NOT STAGGER

FIGURE 7.7-35 *An expansion joint in an elastomeric single-ply roofing membrane. Some manufacturers recommend other types of expansion joints, but some type of expansion provision should occur in roofing over every expansion joint in the structure. However, separate expansion joints may not be needed in loose laid systems if a metal plate is placed over the joint beneath the insulation and fastened to the deck at one side of the joint only. The roofing manufacturer's advice should be followed.* (Reprint courtesy National Roofing Contractors Association.)

7.8 ALUMINUM AND VINYL SIDING

Aluminum siding was first introduced to the American public following World War II. Vinyl siding came along a little later. Early aluminum siding was made in the natural aluminum finish, but soon afterward an enamel finish was applied, making it available in a choice of colors.

The prominent features of both aluminum and vinyl siding include minimal maintenance, good durability, light weight, and high resistance to corrosion. For multiple design possibilities, these sidings are available in a variety of colors, textures, sizes, and shapes. They are used in both new and existing construction (Figs. 7.8-1 and 7.8-2).

7.8.1

ALUMINUM SIDING

The recommendations in this subsection are based largely on the American Architectural Manufacturers Association (AAMA) publication 1402, "Standard Specifications for Aluminum Siding, Soffit, and Fascia," and AAMA publication RA-15, "Aluminum Siding Application Manual." These publications constitute the recognized standard of quality for this product.

7.8.1.1

Material

Aluminum sheet is fabricated into horizontal and vertical panels by cold rolling. The typical alloy used in aluminum siding is known as 3003H14, a magnesium-containing alloy distinguished by moderate

FIGURE 7.8-1 *Vinyl siding on a new building.* *(Courtesy Wolverine Technologies.)*

FIGURE 7.8-2 *Vinyl siding on an existing building.* *(Courtesy Alcoa Building Products.)*

strength and good workability. See Section 5.2 for additional information about aluminum fabrication techniques. The finishes used on aluminum siding are generally organic enamels, such as vinyls, acrylics, or alkyds. They are baked on at the factory after proper pretreatment of the surface.

■ SIDING TYPES

Horizontal aluminum siding is available in several typical widths. The most commonly used siding type is a horizontal panel with 8-in. exposure (face). Other types include double 4-in. and double 5-in. exposure effects, which are created by grooves bisecting a single 8- or 10-in. horizontal panel.

Aluminum siding is plain or insulated. Some insulated board has a polystyrene or fiber backer board installed at the factory. Others are supplied with separate backer boards for field installation. Thicknesses vary from 3/8 to 1/2 in. Insulated siding is available in 8-in., double 4, and double 5 styles.

Vertical siding is available in both 12- and 16-in. exposures. Both exposures lend a board-and-batten effect, while the 16-in. exposure is often bisected by a V-groove for additional interest.

Like other metals, aluminum is an effective vapor barrier. Therefore, aluminum siding should be equipped with vents or weep holes to permit the escape of water vapor and prevent the accumulation of moisture in the wall.

■ ACCESSORIES

Accessories for use with aluminum siding include starter strips, J-channels, window head flashing, and inside corner posts, outside corner posts, and individual corner caps. Other accessories include aluminum nails, caulking, touch-up paint, aluminum breather (perforated) foil, and adhesive. Adhesives used to bond backer boards to siding should demonstrate sufficient moisture resistance, strength, and flexibility to maintain a permanent bond under service conditions. The AAMA specifications outline specific performance tests for the adhesives as well as for the finish coating. The AAMA specifications also require compliance with other industry standards for accessory products employed in the manufacture or installation of siding.

7.8.1.2

Installation

In refinishing projects the quality of the finished job depends heavily on the initial step of surface preparation. Loose boards should be nailed down, and irregular surfaces should be leveled. Unbacked siding cannot bridge low spots as readily as the backed type, so furring of low spots may be required to provide adequate support. In some cases, the whole wall must be furred. Good initial preparation permits siding joints to close up and avoids the wavy look of poor siding installations.

In new construction, backed siding can be applied directly to stud framing over sheathing paper. Unbacked siding should be installed only over a solid sheathing surface. When wood board sheathing is used, sheathing paper should be provided; when panel sheathing materials are applied in a way that minimizes air leakage, the sheathing paper can be omitted.

Aluminum siding is often applied over an insulated sheathing to increase a wall's R-value, especially in existing construction. This insulated sheathing may be a faced panel with a core of cellular kraft paper or a foamed plastic such as polystyrene or polyisocyanurate. Care must be taken to ensure that the use of an impervious sheathing will not trap moisture in the wall cavity. If it will, a pervious board should be used (see Section 7.1).

Aluminum foil laminated to kraft paper is sometimes used as sheathing paper under aluminum siding to increase a wall's R-value. It may be installed over an insulated sheathing or directly on an existing wall or on furring where there is no insulated sheathing. This foil should be applied with its shiny side facing an adjacent airspace. Aluminum foil is an excellent vapor retarder and will trap moisture unless it is deliberately perforated to allow vapor to escape from the wall cavity. Foil products intended for use on an existing wall should be factory perforated and have a perm rating of at least 10. An efficient vapor retarder, with a permeance of 0.5 perm or less, should be used on the room side of the wall assembly. Some manufacturers also recommend the installation of vent cups at the top and bottom of each stud space (see Section 7.1). In new construction, the foil should be perforated or unperforated according to the wall design and the location of a vapor retarder.

Careful application of accessories such as starter strips and corner trim is essential to a level and square installation. In most cases, starter strips are installed first to ensure proper alignment and fastening of siding.

Continuous inside and outside corner posts are used to receive siding at corners, insets, projections, and wall junctions. Since aluminum expands and contracts with temperature variations, a space should be allowed where siding ends abut a post or any rigidly supported surface. Outside corners can also be completed with the use of outside corner caps installed after siding is in place.

In the installation of horizontal siding, extra care should also be used in the application of the first course, making certain that it is level and securely locked into the starter strip. Each length of siding should be lapped over the preceding to create the effect of a single panel extended across the building. Factory-finished ends should be exposed and cut ends lapped under the edge of adjoining panels.

To fit siding around areas such as windows, eaves, and doors, panels can be cut to desired size. Joints should be planned so that, wherever possible, no piece is less than 20 in. long.

Proper nailing is essential to successful installation. Panels should be secured with aluminum nails driven through the slotted holes in the siding at intervals of about 16 in. along its length. Siding should be suspended on the nails, not nailed rigidly to the wall, because locked siding will not be able to expand or contract with temperature changes.

Matching gutters, downspouts, fascia boards, and soffit panels are also available. Because of the natural corrosion resistance of aluminum and the durability of the baked-on enamel finish, these products are particularly well adapted to their severe exposures.

To prevent the entry of water and to improve the appearance of the job, caulking should be used at junctions where siding abuts wood, brick, stone, or other metals. Caulking compounds should be durable synthetic materials, compatible with the aluminum and paint finish.

7.8.1.3

Grounding

Although the possibility is remote, some building codes require that aluminum siding be grounded as a precaution against electrical shock from faulty wiring. This can be done by connecting a No. 8 copper wire to a cold-water pipe or a steel rod embedded in the earth. Grounding allows stray electrical fault currents to be dissipated safely into the ground and eliminates shock hazard to persons touching the siding.

7.8.1.4

Maintenance

Baked-on siding finishes are generally resistant to the effects of intense sunlight, snow, sleet, hail, ice, and salt spray. In addition to the natural cleaning by normal rainfall, siding surfaces should be kept clean by periodic rinsings with a garden hose. Persistent stains resulting from industrial fallout, tree sap, or insecticides can be removed with a nonabrasive household cleaner or special siding cleaner.

Panels damaged by storm, fire, and other hazards can be replaced individually, without disturbing adjacent panels. Aluminum siding may also be repainted, if desired or when necessary due to damage. A special paint intended for aluminum surfaces is recommended, although ordinary house paint can also be used.

7.8.2

VINYL SIDING

The recommendations in this subsection are based largely on the Vinyl Siding Institute (VSI) publication "Rigid Vinyl Siding Application Instructions." This publication constitutes the recognized standard for installation of this product. In addition, vinyl siding should comply with ASTM Standard D 3679, "Specification for Rigid Poly (Vinyl Chloride) (PVC) Siding."

7.8.2.1

Material

Vinyl siding and accessories are made from a polyvinyl chloride (PVC) compound and additives, such as lubricants, stabilizers, impact modifiers, ultraviolet inhibitors, and colors. PVC is a thermoplastic material, meaning that it can be reformed upon heating. It is flexible and has a high-impact resistance. The finishes on vinyl siding are integral.

Siding finishes range from almost smooth, through matte, to deeply embossed wood grain surfaces that simulate wood clapboard. Many colors are available.

■ SIDING TYPES

Horizontal and vertical vinyl siding is available in several widths and configurations. Figure 7.8-3 shows some of the more popular sizes and shapes. Special shapes and even vinyl lattice work are also available (Fig. 7.8-4; see also Fig. 7.8-1).

Drop-in backer board is supplied with some vinyl siding, but the VSI does not recommend its use, preferring instead a continuous sheathing backup. The manufacturer should be consulted for advice about drop-in backer board use.

■ ACCESSORIES

Accessories for use with vinyl siding include starter strips, J-channels, window head flashing, inside corner posts, outside corner posts, and others. Their sizes and shapes vary with the manufacturer. Fasteners for installing vinyl siding should be aluminum, galvanized steel, or other corrosion-resistant nails or staples.

Nails should be long enough to penetrate at least 3/4 in. into a solid nailable backing surface. Nail heads should be 5/16 in. in diameter and 1-1/2 in. long for general use. Nails should be longer for residing.

Power-driven staples may be used to install vinyl siding. A special fitting must be used on power equipment, however. The manufacturer's instructions should be followed.

7.8.2.2

Installation

In refinishing projects the quality of the finished job depends heavily on the initial step of surface preparation. Loose boards should be nailed down, and irregular surfaces should be leveled. Loose caulking should be removed and joints recaulked to make them watertight.

In new construction, vinyl siding should be applied over solid sheathing. Installing a new layer of sheathing over an existing wall surface will also produce the best vinyl siding job. However, furring or strapping can be used over an existing wall to even out the face. This furring or strapping should be shimmed where necessary to produce an even face.

Vinyl siding can be applied over insulated sheathing to increase a wall's R-value. The advice of the insulation and siding manufacturers should be sought and followed in making this installation.

Careful application of accessories, such as starter strips and corner trim, is essen-

FIGURE 7.8-3 *Vinyl siding: (1) 4-in. double; (2) 6-, 8-, 9-, or 10-in. single; (3) 5-in. double; (4) 3-in. triple; (5) Dutch lap; (6) vertical is available in 5-in. double, 3-in. triple, and 4-in. quad.* (Courtesy Vinyl Siding Institute.)

pands and contracts with temperature variations, a space should be allowed where the siding ends abut a post or any rigidly supported surface.

In the installation of horizontal siding, extra care should also be used in the application of the first course, making certain that it is level and securely locked into the starter strip. Each length of siding should be lapped over the preceding one to create the effect of a single panel extended across the building. Factory-finished ends should be exposed and cut ends lapped under the edge of adjoining panels (Fig. 7.8-6).

To fit siding around areas such as windows, eaves, and doors, panels can be cut to desired size with saws or snips. Joints should be planned so that, wherever possible, no piece is less than 20 in. long.

Proper nailing is essential to successful installation. Panels should be secured with nails or staples driven through the slotted holes in the siding at intervals of about 16 in. along its length. Siding should be suspended on the nails, not nailed rigidly to the wall, because locked siding will not be able to expand or contract with temperature changes (Figs. 7.8-7 and 7.8-8).

Matching gutters, downspouts, fascia boards, and soffit panels are also available. Because of the natural corrosion resistance of vinyl and the durability of vinyl siding, these products are particularly well adapted to their severe exposures.

Siding overlaps should not be caulked, and siding panels should not be caulked where they meet receivers, corners, or J-trim; rather they should be permitted to

tial to a level and square installation. In most cases, starter strips are installed first to ensure proper alignment and fastening of siding.

Continuous inside and outside corner posts are used to receive siding at corners, insets, projections, and wall junctions (Fig. 7.8-5a, b, and c). Since vinyl ex-

FIGURE 7.8-4 *Shingle shapes over the entranceway are only one of several shapes available in vinyl siding material.* (Courtesy Alcoa Building Products.)

FIGURE 7.8-5 *(a) Inside corner posts and (b) and (c) outside corner posts.* (Courtesy Vinyl Siding Institute.)

FIGURE 7.8-6 *Lapped end joints. (Courtesy Vinyl Siding Institute.)*

FIGURE 7.8-7 *Nails should be driven into the center of the slot. It should be possible to move the siding 1/2 in. each way after fastening to allow for normal expansion and contraction. (Courtesy Alcoa Building Products.)*

FIGURE 7.8-8 *Staples should be driven so that the panels remain loose and movable. (Courtesy Alcoa Building Products.)*

float free to account for expansion and contraction.

7.8.2.3

Maintenance

Vinyl siding is generally resistant to the effects of intense sunlight, snow, sleet, hail, ice, and salt spray. In addition to the natural cleaning by normal rainfall, siding surfaces should be kept clean by periodic washing with a long-handled car-washing brush mounted on the end of a water hose. Hard-to-remove soil may be removed with a solution of 1/3 cup detergent and 2/3 cup household cleaner in a gallon of water. Mildew may be removed using that same solution but substituting a quart of liquid laundry bleach for one quart of the water.

More aggressive means and chemicals such as solvents, cleaning fluids, oxalic acid, and lacquer thinner can be used to remove more stubborn stains, but the advice of the siding manufacturer or the VSI should be sought first.

7.9 FLASHING AND SHEET METAL

Flashing is installed at vulnerable locations to help prevent water from entering a building and to collect water that has entered the exterior shell and divert it back to the exterior. Flashings are formed from sheet metal, bituminous fabric membranes, plastic sheet materials, or a combination of these materials.

Sheet metal roofs are used on all types of buildings. They are made from a variety of sheet metals (Fig. 7.9-1).

7.9.1

MATERIALS

Because replacement costs can exceed the initial installed costs, it is prudent to select a durable flashing or metal roofing material for the original installation.

Flashing materials must possess at least four important properties: (1) imperviousness to moisture penetration; (2) resistance to corrosion caused by exposure to the atmosphere or to caustic materials in adjacent materials, such as the alkalies present in mortars; (3) toughness to resist puncture, abrasion, and other damage during installation; and (4) formability so that the flashing can be worked to the correct shape and retain that shape after installation.

Metal roofing must (1) shed water and prevent its penetration into the building; (2) resist corrosion caused by exposure to the atmosphere and pollutants; (3) resist puncture, abrasion, and other damage during installation and use; and (4) be made from a formable metal that can be worked to the correct shape and retain that shape after installation.

7.9.1.1

Sheet Metals

The metal materials generally used in flashing for walls and roofs include copper, lead-coated copper, lead, galvanized steel, aluminum, and stainless steel.

The metal materials used in metal roofing include lead, copper, lead-coated copper, terne-coated stainless steel, stainless steel, galvanized steel, zinc, and aluminum. Lead and copper have been used for roofing for centuries and are still used today. Steel and aluminum sheets are also available with flexible and colorful factory-applied coatings that can be shaped on a brake without damaging the coatings.

■ **ALUMINUM**

Aluminum is a light, bluish-silver-white metal that is ductile and malleable. It is

FIGURE 7.9-1 *A sheet copper roof.* *(Courtesy Copper Development Association.)*

resistant to oxidization, but flashings made of it are subject to attack by the caustic alkalies present in fresh, unhardened mortar. Although dry, seasoned mortar will not affect aluminum, corrosion can occur if the adjacent mortar becomes wet. Therefore, aluminum should not be used as a masonry or stone wall flashing. If it must be so used, it should be coated to completely protect it from contact with both fresh and cured mortar.

Aluminum sheet flashing exposed to weather should be at least 0.019 in. thick and should have a tensile strength of 16,000 lb./sq. in. Aluminum sheet that is completely concealed should be not less than 0.015 in. thick and should have a tensile strength of 14,000 lb./sq. in.

Aluminum is seldom used to make field-formed roofing. Preformed aluminum roofing, however, is quite common. The usual thickness of such aluminum roofing is about 0.032 in. Roof accessories such as copings, gravel stops, gutters, and downspouts are usually thicker, varying with the manufacturer.

■ COPPER, PLAIN AND LEAD-COATED

Copper is a reddish ductile and malleable metal. The three types of sheet copper available are cold-rolled copper, soft copper, and lead-coated copper. Cold-rolled copper sheet is less malleable than soft copper, but much stronger. Accordingly, it is used much more often. The copper in lead-coated copper is similar to cold-rolled copper.

Though more costly than most other flashing materials, copper is durable, available in special preformed shapes, and an excellent moisture barrier. Because it is not materially affected by the caustic alkalies present in masonry and stone mortars, copper can safely be embedded in fresh mortar and not deteriorate in continuously saturated, hardened mortar unless excessive chlorides are present. With copper flashing, the use of chloride-based additives, such as calcium chloride, in the mortar should be prohibited.

Copper sheet used for flashing should be cold-rolled copper conforming with ASTM Standard B 370 or lead-coated copper conforming with ASTM Standard B 101, Type I or Type II, Class A. Type I has the lead coating applied by hot dipping; Type II by electrodepositing. Class A denotes a lead coating on both sides of the sheet with the total weight of the lead on both sides at between 12 and 15 oz. of lead on each 100 sq. ft. of sheet.

Copper thicknesses for various uses should be in accordance with the recommendations of the Copper Development Association publication "Copper, Brass, Bronze Design Handbook, Sheet Copper Applications." In general, copper flashing that will be completely concealed should weigh not less than 10 oz./sq. ft.; when exposed to the weather not less than 16 oz./sq. ft. Base, valley, gable end, and some other flashings and gravel stops, gutters, and downspouts should weigh between 20 and 24 oz./sq. ft. depending on their size and use. Most batten-seam, flat-seam, and standing-seam roofings should weigh 20 oz./sq. ft. Some types of roofing may weigh less (12 to 16 oz.) when the pans are 16-3/4 in. wide or smaller.

One disadvantage of using copper flashing and roofing is that when it is exposed to weather the normal oxidization on copper will wash across adjacent surfaces. This may stain or discolor those surfaces, a condition that is particularly objectionable when copper flashing is installed in light-colored masonry or stone or when the wash-off from copper will pass over other metals, such as aluminum, or flow down walls. Where such copper staining would be objectionable, lead-coated copper can be used.

■ GALVANIZED STEEL

Galvanized steel is steel that has been coated with a layer of zinc, either electrolytically or by the hot-dip process. It will corrode in fresh mortar and may corrode even when placed in contact with cured mortar. Therefore, it should not be used in masonry or stone where such use is avoidable; when it is so used, it should be protected with a bituminous coating. It will also corrode eventually in the air unless protected with a coating such as a bituminous compound or paint. Unfortunately, because of its low cost, galvanized steel is widely used for both roofing and flashing, especially in small commercial and residential buildings. While it can perform adequately in such a use when it is well protected, exposed galvanized steel requires periodic maintenance and should be accessible for inspection and painting.

Galvanized steel for wall flashings should conform with the requirements of ASTM A 525 and should be 28 gauge where completely concealed and 26 gauge where exposed.

Galvanized steel roofing should be 26 gauge for panels up to 24 in. wide and 24 gauge for wider panels.

■ LEAD

Lead is a bluish, soft, heavy, malleable, ductile, plastic (but not elastic) metal. Some lead roofs have been in constant use for hundreds of years. Lead will corrode in fresh mortar and may corrode when placed in contact with cured mortar. Therefore, it should not be used in masonry or stone where such use is avoidable; when so used, it should be protected with a bituminous coating. For this reason, and because elastomeric flashings are less expensive and easier to work with, lead is seldom used today for wall flashing.

Most sources recommend that sheet lead roofing and associated flashings and roof specialties be formed of not lighter than 3-lb. hard lead. Small pieces may be formed from 2-1/2-lb. hard lead. Batten-seam roofing may also be formed of 2-1/2-lb. lead when the battens are kept not more than 18 in. apart.

■ STAINLESS STEEL

Stainless steel is an alloy of steel that contains more than 12% chromium and sometimes also nickel, magnesium, or molybdenum. Most stainless steel will not rust and corrodes only under very severe conditions. It is available in either hard or soft temper.

Stainless steel flashing materials are available in several gauges and finishes. They are durable, highly resistant to corrosion, excellent moisture barriers, and surprisingly workable. There are a number of types available, the varieties depending basically on the proportions of chromium, nickel, and steel. These varieties have differing degrees of resistance to corrosion and are priced accordingly. Stainless steel for masonry wall flashing should

be ASI Type 304 with a No. 2B finish; it should be specified by number and not by the generic term alone.

Stainless steel flashing that will be exposed to the weather should be 28 gauge; 30 gauge when completely concealed. Stainless steel roofing should be 28 gauge when the roofing panels do not exceed 18 in. and 24 gauge for wider panels. Associated fascias should be formed from 22-gauge material, but base, cap, and counter flashings may be 28 gauge.

In addition to its other favorable properties, stainless steel will not stain adjacent areas, resists rough handling, and can be soldered, welded, or brazed.

■ TERNEPLATE

Terne is an alloy of lead and tin, usually in a ratio of four parts lead to one part tin. Terne-coated steel or stainless steel (terneplate), which is the terne used in roofing, is made by coating steel or stainless steel sheet with a layer of terne metal.

Terneplate used for roofing should be 0.015 in. thick for panels up to 24 in. wide and 0.178 in. thick for wider panels.

■ ZINC

Zinc is a bluish-white crystalline metal. It is brittle at ordinary temperatures but ductile when heated. The zinc sheets used in roofing are an alloy of zinc, copper, and titanium.

Preformed zinc roofing panels are usually 0.027 in. thick but are sometimes available up to 0.020 in. thick.

■ SHEET METAL FINISHES

Sheet metal materials associated with prefinished metal roofing are usually finished to match the roofing. Those not associated with prefinished roofing are usually painted when exposed to view or when a protective coating is required. Copings, gravel stops, gutters, fascias, and downspouts are often factory finished. Those used with single-ply roofing are often coated with the same material used in the roofing.

Copper, lead, aluminum, stainless steel, and galvanized steel do not require painting other than for aesthetic purposes. However, if left unpainted, galvanized steel will eventually rust. Terneplate made

with stainless steel does not need painting. That made with normal steel does, because the terne coating contains pinholes that will permit the underlying steel to rust.

Painted sheet metals are usually painted with an oil- or alkyd-based paint. Some are factory primed; others must be primed in the field. Factory-finished sheet metals may be finished with porcelain enamel, fluoropolymer, acrylic enamel, siliconized polyester, or the same materials used in single-ply roofing membranes. Aluminum may be clear or color anodized.

Flexible Sheet Flashings

Flexible sheet flashings include elastic sheet materials made from polyvinyl chloride (PVC) or modified polymers and membranes composed of glass fiber or organic felt fabrics saturated with bitumen. These flashings are used as damp checks and as low-cost substitutes for metal flashing at the base of walls, heads of openings, window sills, floor lines, and in other locations where wall flashings are needed. Elastic sheet materials are also used in conjunction with single-ply roofing and as base flashings and penetration flashings in some types of built-up bituminous roofing (Fig. 7.9-2). Elastomeric sheets, usually plain or reinforced chlorinated polyethylene, are used as roof expansion joint covers. Often this joint covering is bonded to a layer of foam insulation.

Some elastic flashings are tough, resilient, and highly resistant to corrosion, acid, and salt attack. This is particularly true of flashing materials, such as EPDM

FIGURE 7.9-2 *Elastic flashing installs easily around a pipe vent.*

synthetic rubber, recommended by the manufacturers of single-ply roofing membranes for use with their products. Elastomeric products are also able to withstand greater movement without failure in walls and roofs than metal flashings and can be more easily formed into loops or bellows to accommodate movement. Since the chemical compositions of plastics varies so widely, they cannot all be used in every location. Some plastic flashings, for example, cannot withstand the corrosive effects of masonry mortars, and others, such as PVC ones, cannot be exposed to sunlight.

Properly installed, both bituminous and plastic flashings can be effective. However, flashing must permanently exclude the passage of water. Unfortunately, bituminous and plastic fabric flashings are often inadvertently cut during construction by trowels or abraded by masonry units so that they do not provide a continuous membrane. In addition, thin fabrics may sag and not retain the desired shape as a result of changes in temperature or poor workmanship. This is a particular problem when such flashings are concealed, because the changes will not be apparent until the wall begins to leak or the water held in pockets in the deformed flashing freezes and damages the wall.

Since recognized standards have not yet been established for either bituminous or plastic wall flashing materials, it is necessary to rely on performance and test records of the material and the reputation of its manufacturer to ensure suitable quality and performance.

7.9.1.3

Combination Sheet Flashing

Combination flashings consist of materials combined to utilize the best properties of each. Figure 7.9-3 shows two combination flashings. Other examples are a copper sheet bonded with asphalt between two layers of glass fiber cloth and a rubberized asphalt compound sheet bonded to a polyethylene film. There are many others. The combination of materials can provide a lower-cost flashing by reducing the thickness of metal required, or it may permit the use of corrodible metals that might otherwise prove unsatisfactory unless coated.

Some combination flashing materials are made by coating sheet metal with

FIGURE 7.9-3 *Two types of combination wall flashing: (a) a sheet of copper bonded to asphalt-saturated kraft paper (aluminum may be used in lieu of copper); (b) a five-layer assembly composed of a copper or aluminum sheet sandwiched between layers of asphalt-saturated cotton fabric, which is then faced with outer layers of ductile mastic.*

another material. Coating materials commonly used include bitumen, kraft paper, bituminous saturated cotton fabrics, glass fiber fabrics, and combinations of these materials. The sheet metals commonly used for combination flashing include copper in weights of 3, 4, or 5 oz./sq. ft. and aluminum 0.004 or 0.005 in. thick.

Composites consisting of a reinforced insulated chlorinated polyethylene joint cover bonded on each edge to a galvanized steel, stainless steel, or copper flange are used as roof expansion joints.

While no recognized standards exist for combination flashing, the weights of the copper and the thicknesses of the aluminum sheet metal given above when coated on both sides with any of the coatings listed above and when installed with unbroken skins provide suitable concealed flashings.

7.9.1.4

Fasteners

Aluminum sheets should be fastened in place with screws. Other metals should be fastened in place with either screws or twisted or threaded nails. Screws are often used in roof specialties; nails in flashings.

Fasteners should extend not less than 1-1/2 in. into solid, pressure-preservative-treated wood.

Cleats in lead roofing should be installed with hard copper wire nails. Most other fasteners in sheet metal flashings, roof accessories, and roofing should be of the same material as the metal being fastened. When that is impracticable, the fasteners should be noncorrosive and compatible with the metal being fastened so their contact will not produce galvanic corrosion.

Lead sheets, regardless of their size, and other metal sheets more than 18 in. wide, should not be nailed. Rather they should be held in place with cleats of the same metal and thickness as the sheets, spaced about 12 in. on centers (Fig. 7.9-4). However, cleats should be continuous in the following instances: (1) on slopes less than 3 in./ft.; (2) where the sheets are subject to high winds; and (3) in other locations where the metal manufacturer so recommends. Cleats should not be less than 2 in. wide and 3 in. long. They should be nailed in place with at least two nails and lock-seamed into the sheet being held.

Narrow sheets should be nailed on only one edge. In general, nails should be staggered at about 3 in. apart and 1/2 in. from the edge of the sheet.

The flanges of metal flashings should be nailed to wood members (nailers) at a maximum spacing of 6 inches.

7.9.2

FLASHING LOCATIONS

Refer to Section 4.7.2.8 for additional discussion about the locations where wall flashing is required in masonry walls. Refer to Section 7.6 for similar information about flashing in wall and roof shingles and to Section 7.7 for information about flashing associated with built-up and single-ply roofing.

7.9.2.1

Walls

Except in areas of very slight exposure to precipitation where local experience supports omitting it, flashing should be installed at the following locations: (1) at the

RECTANGULAR PATTERN
SHEETS MAY BE NOTCHED AND FOLDED IN SHOP
DIMENSIONS: 15"X 20", 16"X18", 18"X24" OR 20"X28"

STAINLESS STEEL PAN

STAINLESS STEEL CLEAT
NAILED TO DECK WITH
NO. 12 BY 1" FLAT HEAD
ANNULAR THREAD
STAINLESS STEEL NAILS

STAINLESS STEEL
CLEAT

ROOF DECK

METHOD OF CLEATING

STEP 1

STEP 2

FIGURE 7.9-4 *A typical cleat.* *(Courtesy American Iron and Steel Institute.)*

bottom of a wall; (2) under the sills of wall openings; (3) over the heads of wall openings; (4) frequently at floor lines, depending on the type of construction; (5) where a wall penetrates a roof surface; (6) beneath copings in masonry and stone walls; and (7) at expansion joints (Fig. 7.9-5).

■ BASE OF WALLS

Flashing and damp checks are used at the base of walls. Flashing diverts to the exterior moisture that enters the wall above the location of the flashing. A damp check stops the upward capillary travel of ground moisture. If installed properly, metal shields used for termite protection of wood joists (see Sections 2.3.9 and 6.5.2) may double for these purposes.

■ WALL-OPENING SILLS

Sill flashing should be placed under and behind every sill. The ends of sill flashing should extend beyond the sides of an opening and turn up at least 1 in. into the wall. Sills should slope and project from the wall to drain water away from the building and to prevent staining. When the undersides of sills do not slope, a drip notch should be provided, or the flashing may be extended and bent down to form a drip.

■ WALL-OPENING HEADS

Head flashing should be placed over all openings except those completely protected by overhanging projections. At steel and wood lintels the flashing should be placed under and behind the facing material with its outer edge bent down over the lintel to form a drip.

■ FLOOR LINES

Flashing should be used at floor lines where water entering a wall might be able to find its way into the building through the floor edge construction. It should extend from the exterior face of the wall below the floor construction, up the exterior face of the floor construction, pass beneath the wall above, and turn up. This flashing may be any of the materials described in Section 7.9.1.

■ WALLS AT ROOFS

At chimneys and other walls passing through a roof, the base flashing described in Section 7.9.2.2 should be capped with metal counter (cap) flashing that turns down over the base flashing and is anchored securely in the mortar joints.

■ COPINGS

Flashing should be carried beneath copings in masonry and stone walls and in other walls where the construction will not prevent water penetration. Coping flashing may consist of sheets of any of the materials listed in Section 7.9.1.

■ EXPANSION JOINTS

Flashings are required at expansion joints in walls where water might enter the building. These flashings may be preformed of metal or plastic, or they may be formed in the field, usually from metal fashioned into a bellows.

Roofs

Flashings should be installed (1) at the eaves and rakes of pitched roofs; (2) in

valleys in pitched roofs; (3) at points where walls, pipes, curbs, and other elements penetrate a roof; and (4) at roof drains.

■ EAVE, RAKE, AND VALLEY FLASHINGS

Metal eave and rake flashings and both metal and bituminous valley flashings are described and illustrated in Section 7.6 for each type of pitched roofing covered there.

■ ROOF PENETRATION FLASHINGS

Roof penetrations include major penetrations, such as walls and chimneys, and minor penetrations, such as pipes, curbs, guy wires, and roof drains.

Essentially, flashings at penetrations consist of a base flashing turned up against the penetration and a counter, or cap, flashing turned down over the base flashing. Several variations of such flashings are used, depending on the roofing and the element to be flashed.

BASE AND CAP FLASHINGS Separate base and cap flashings are used with walls, chimneys, and curbs. Metal base flashings of the same material and thickness as the roofing are used with metal roofs. Both metal base flashings and bituminous base flashings are used with shingles (see Section 7.6). Sheet metal base flashings are not usually used with built-up or single-ply roofing, because metals have coefficients of expansion that are not compatible with those of most roofing membranes. Base flashings for built-up roofing are usually bituminous or an elastomeric material compatible with the roofing. Base flashings for single-ply roofing are usually the same materials as the roofing membranes or a compatible elastomeric material. Refer to Section 7.7 for additional discussion about base flashings for use with built-up and single-ply roofing.

Metal base flashings in steep roofing may be either continuous or pieced. Pieced base flashing is woven into the roofing as it is placed, with the pieces overlapped by at least 3 in. in the direction of water flow. Sloped-roof base flashing should turn out onto the roofing and up the wall not less than 4 in. when there is a cap flashing, 6 in. when a wall finish forms the cap flashing, and 8 in. where there is no cap flashing. When the slope is away from the

FIGURE 7.9-5 *Wall flashing locations: (a) in masonry construction; (b) in wood stud construction.* (*Drawing by HLS.*)

The labels within the image (a):

Stone coping
COPING
Flashing
Cap flashing
Roofing and base flashing
ROOF
Flashing
Flashing
FLOOR
Flashing
Flashing
Ceiling
OPENING HEAD
Flashing
OPENING SILL
Flashing
Brick veneer
BOTTOM OF WALL
Concrete slab
Flashing
Foundation wall
Insulation
a

The labels within the image (b):

Metal reinforcement must extend down over flashing
Counter flashing
Flashing
ROOF
Flashing over drip cap
HEAD
SILL Flashing under sill
Where wood contacts stucco, cover with waterproof paper
Metal lath must extend down over flashing
Flashing
WATERTABLE
Grade
b

wall, the base flashing should extend up the wall at least 5 in.

Every base flashing requires some kind of counter (cap) flashing. A wall finish may act as a cap flashing, but most cap flashings are metal. The metal material used in cap flashings varies depending on the base flashing material and the type of penetration being flashed. Figure 7.9-6 shows a typical cap flashing, but many other configurations are possible. A cap flashing may extend beneath a window sill, for example. Cap flashings at sloped roofs should be in sections, with no section more than 8 in. high. Cap flashings should overlap base flashings by at least 4 in.

FIGURE 7.9-6 *A typical cap flashing in a reglet.* *(Courtesy H. Leslie Simmons.)*

FLASHINGS FOR SMALL PENETRATIONS Small penetrations such as pipes, conduits, ducts, and columns are sometimes flashed by building a curb around them and providing base and cap flashings. More often, though, such elements are flashed with sleeves, formed metal housings, or collars.

Flashings for some penetrations consist of a factory-formed collar of an elastomeric material shaped to fit the penetrating element. This is especially true in single-ply roofs, where such a collar can be sealed to the roofing membrane with the same materials used to seal seams in the roofing membrane.

Pipes are flashed with metal in several ways. One common way consists of a lead-sheet flange that extends out onto the roofing about 4 in. to be flashed in with roofing membranes. A sleeve is soldered or welded to this flange, extends up the pipe for its full height, and is then turned down into the pipe at least 2 in.

Penetrations such as shaped steel columns, conduit, and ducts are sometimes flashed using a two-piece metal housing. The lower portion of the housing is a sleeve that extends 8 in. up the penetrating element. It may be either fitted to the penetration's shape or be square or rectangular. This sleeve has a 4-in.-wide flange that is flashed into the roofing with a roofing membrane. The upper portion of the housing is a hood attached firmly to the penetrating element and extending down over the sleeve. The joint between the hood and the penetrating element should be filled with an elastomeric sealant.

Difficult-to-flash penetrating items, such as guy wires and sloping-roof top-mounted sign supports, are sometimes flashed with a pitch pan, which is a two-piece soldered metal pan set over the roofing membrane and flashed into the roofing by the roofer. This pan is then filled with a layer of mortar or a mixture of roofing cement and sand and topped off with a layer of a polyurethane sealant, coal tar pitch, or another self-leveling topping. All materials used must be compatible with the adjacent roofing. The use of pitch pans is discouraged by most authoritative sources, but their use persists, nevertheless, because there is sometimes no other way to provide flashing for a penetrating element.

ROOF DRAIN FLASHING Roof drains in low-slope roofs are usually flashed by extending the roofing membrane into the flange ring of the drain. Then a 30-in. square of 4-lb. lead or 16-oz. copper is placed on the roofing membrane, extending into the drain's flashing ring and bedded in roofing cement. This metal pan should be stripped into the roofing by the roofer. In some single-ply roofing systems, the roofing membrane is extended into the drain's flashing flange, and the metal sheet is omitted.

Sometimes, the roofing insulation is tapered down to the drain. When the flashing material can be exposed to sunlight, a gravel stop is sometimes installed to surround the drain.

7.9.2.3

Roofing Accessories

Roofing accessories include ridge vents, curbs, gravity ventilators, skylights, heat and smoke vents, and roof hatches.

Roof accessories often come with built-in, 4-in.-wide roof flanges and counter flashings. Base flashing should be extended up the accessory's curb and beneath the built-in cap flashing. Roofing accessory flanges for use with metal roofing should conform with the roofing's profile.

7.9.2.4

Roofing Specialties

Roof specialties include copings, expansion joint covers, gravel stops, fascias, scuppers, gutters, and downspouts.

■ COPINGS

Most metal coping covers are either formed sheet metal or extruded aluminum. They may be factory finished or primed for field painting. Most are installed with open joints backed by flashing or with joint covers. Sections are usually 8 or 10 ft. long. Most slope down toward the roof side so that water falling on them will not run down the face of the building.

Installation methods vary with the coping, but some characteristics apply to most. Some come complete with their own compatible support system. Others are supported on pressure-preservative-treated wood nailers. Most are installed over a layer of building paper. Outer edges are snapped down in place over continuous cleats. Inner edges are screwed to wood framing or snapped over metal cleats.

■ EXPANSION JOINT COVERS

Roof expansion joints may be either preformed metal, field-formed metal, preformed elastomeric (see Fig. 7.7-35), or a combination preformed unit of metal and an elastomeric material.

Except in some single-ply roofing systems, roof expansion joints should be installed over pressure-preservative-treated wood curbs. These wood curbs should be flashed into the roofing with base flashing. The expansion joint cover acts as a cap flashing.

Elastic expansion joints are sometimes installed flush with the roofing (see Fig. 7.7-35), but this is discouraged by most roofing experts. This type of installation should be used only in single-ply roofing where specifically recommended by and warranted by the roofing membrane manufacturer and where an independent professionally competent roofing expert concurs.

■ GRAVEL STOPS AND FASCIAS

Gravel stops and fascias are usually prefabricated from either formed sheet metal or from extrusions. Section length is usually 8 ft. Roof flanges should extend 4 in. out onto the roof, be securely nailed in place, and be stripped into the roofing by the roofer. Light fascias and gravel stops should have hook strips at the bottom to hold them in place. Heavy extruded items may not require a hook strip.

■ SCUPPERS

Scuppers are outlets through a parapet or gravel stop that permit water to flow from a roof without traveling through a pipe inside the building. Water flowing through a scupper may fall to the ground with no connection to a drain system or fall directly into a conductor head, which is a short section of gutter with the ends sealed. Conductor heads are usually connected to a downspout.

■ GUTTERS

Gutters collect water from a roof and distribute it to a downspout. They are built in many shapes of formed sheet metal. Some are built-in, but most are hung from a fascia. Most hung gutters are made from aluminum, copper, galvanized steel, or stainless steel in rectangular, half round, or ogee shapes. Built-in gutters are usually made from one of the metals used for flashing and sheet metal roofing. Built-in gutters are seldom made from aluminum, however, because it cannot be soldered, and making joints watertight is difficult.

Hung gutters are supported by straps, brackets, or spikes and ferrules of the same metal as the gutter. The size, thickness, and spacing of these hangers should be based on the size and the material of the gutter supported. Spacings are usually 30 in. maximum on centers.

Gutters should be set with a slight slope toward the downspouts to permit them to drain. Screens should be used over gutters to prevent them becoming clogged with leaves.

Expansion should be accounted for in gutter design, especially in built-in gutters. In built-in gutters, the method depends on the design of the gutter. At the least, watertight expansion joints must be installed. Expansion in hung gutters can be accommodated by lapping sections of the gutter with a water barrier between the sections. Expansion can also be accomplished by building separate sections of gutter with ends and closing the gap between them with a cover plate.

■ DOWNSPOUTS

Downspouts that drain built-in gutters are plumbing items. Downspouts for hung gutters are usually built of the same material as the associated gutters. Their size depends on their spacing and the area of roof drained.

Downspouts should be supported at intervals depending on their size, but never more than 5 ft. apart.

Downspouts may terminate with elbow-type fittings draining out onto splash blocks or pans. An alternative used on many commercial buildings is to terminate downspouts into boots connected to the site storm sewer.

7.9.3

SHEET METAL INSTALLATION

Additional information about installing flashing in masonry walls is discussed in Section 4.4.2.8; in wall and roof shingles in Section 7.6; in built-up and single-ply roofing in Section 7.7.

7.9.3.1

Preparation

Surfaces to receive flashing or roofing should be smooth and free from projections that would puncture the metal material or destroy its effectiveness.

Through-wall flashing in masonry and stone should be placed on a thin mortar bed, and another thin mortar bed should

be placed on top of the flashing to receive and bond the next masonry or stone course. Metal flashing should be deformed to help lock the mortar in place.

Metal flashings over a wood or cementitious substrate should be installed over a red-rosin-paper slip sheet and an underlayment of polyethylene sheet. The flanges of metal flashing should be bedded in roofing cement.

An underlayment of not less than No. 15 asphalt-impregnated unperforated roofing felt should be installed beneath metal roofing. Edges should be lapped 2 in.; ends 6 in. Felt should be nailed in place with large-headed roofing nails. In addition to the underlayment, an ice shield should also be provided where the January mean temperature is 30°F or less. This ice shield should consist of two layers of at least No. 15 asphalt-impregnated roofing felt, installed with hot asphalt, laid to cover an area from the eave to a line 24 in. inside the interior face of the exterior wall. As an alternative, a single-layer membrane produced specifically for this purpose may be used as an ice shield.

The underlayment beneath metal roofing should be covered by a rosin-sized building-paper slip sheet to prevent the roofing from sticking to the felts. This paper is sometimes omitted from beneath roofing installed with flat seams, since some designers do not demand that such roofing be able to move.

7.9.3.2

Installation

Sheet metal flashings and roofing should be installed in accordance with the SMACNA *Architectural Sheet Metal Manual.* Elastic flashings should be installed as recommended by the manufacturer.

■ JOINTS AND EDGES

Seams in metal flashing and roofing must be overlapped and thoroughly bonded or sealed to prevent water penetration. There are many variations of such joints. Although most sheet metal flashing and roofing materials can be soldered, lockslip joints are required at intervals to permit thermal expansion and contraction. Joints in lead sheets are sometimes welded. Most joints in sheet metals are soldered with a tin and lead solder. The exact composition of the solder depends on the metal being soldered. Aluminum sheets are usually lapped, lock seamed, or butted and back flashed. Aluminum gutter and downspouts are often lapped and riveted. Joints in aluminum should be filled with sealant.

Many plastics can be permanently and effectively joined by heat or an appropriate adhesive. The elastic pliability of plastic flashings eliminates the need for expansion and contraction seams.

Metal flashings to be sealed with sealants should be overlapped at least 3 in. Others should be connected with seams to form a continuous watertight system.

The edges of sheet metal that are not attached or locked to adjacent metal should be hemmed for strength.

■ EXPANSION AND CONTRACTION

There is usually a large differential in the expansion coefficient of most sheet metals and that of their supports. Therefore, it is necessary to place expansion joints in sheet metal. The normal intervals are 32 ft. for aluminum and 40 ft. for most other metals. Additional joints are also required when the distance from the last joint to the end of the sheet metal is more than half the normal interval.

Extruded aluminum gravel stops, copings, and fascias should have expansion joints about 12 ft. apart.

The spacing of expansion joints in metal roofing depends on the type of roofing and the seaming method. In general, when the ends of roofing are fastened, joints should be at about 15 ft. on centers; where ends are free to move, joints may be at 30-ft. intervals. Locations where roofing abuts other materials or penetrating walls should have provisions for expansion.

■ FLASHING RECEIVERS

Most metal counter flashings should be set into mortar joints as the masonry is built or later inserted into flashing receivers. Reglets should be used at concrete walls. Flashing receivers are made from the same metal as the cap flashing. Reglets are made from various metals and plastics. The material used should be compatible with the flashing.

Flashing should be snapped or inserted into metal flashing receivers and held in place by the receiver's snaplock feature or by fasteners.

Depending on the design of a reglet, flashing may be snapped into it or inserted and wedged in place with lead wedges or lead wool (see Fig. 7.9-6). After the flashing has been inserted into a reglet, the reglet should be filled with an elastomeric sealant. Lead flashing is an exception. It should be locked to a separate lead strip, which can then be caulked directly into a reglet.

Cap flashings are sometimes inserted into a *raggle* (cut slot) left by the mason or cut into a finished masonry joint and held in place by lead wedges. The raggle should then be filled with sealant.

7.9.4

SHEET METAL ROOFING

The requirements discussed in the previous parts of this section apply to metal roofing except where they are specifically modified in the following discussion.

Metal roofing can be either structural or architectural.

7.9.4.1

Structural Sheet Metal Roofing

Structural metal roofing is a structural roof that also forms a weather barrier. It is designed to span between structural roof supports. No roof deck is required. It can be used on slopes as low as 1/4 in./ft. Most such roofs consist of steel or aluminum panels joined by raised seams, which provide stiffness. Clips are used to attach such roofing to structural supports, usually purlins. Each manufactured product has its own specific installation methods. The manufacturer's instructions should be followed.

7.9.4.2

Architectural Sheet Metal Roofing

Architectural metal roofing is solely a weather barrier. It has no structural component and must be installed over a supporting, nailable roof deck. It is designed to shed water rather than act as a total bar-rier to it and is, therefore, usually installed on slopes of 3 in./ft. or steeper. Lower slopes require special precautions to prevent water intrusion. Among these are welding every joint in lead roofing, filling every joint in aluminum roofing with a nonhardening sealant, and soldering every joint in other types of metal.

Architectural sheet metal roofing may be either preformed or formed in place. Installation methods are basically the same for both.

■ FASTENERS AND CLEATS

Sheet metal roofing should be fastened in place using cleats generally of the same material as the roofing sheets. Lead roofing, however, should be installed using 16-oz. hard-copper cleats, except that exposed cleats in lead roofing may be of 3-lb. lead. Cleats should be fastened to the roof deck and locked into the roofing sheets at the joints. Fasteners should be concealed from view and protected from the weather by the roofing sheets.

Cleats should be placed about 8 in. apart generally. On very steep slopes, cleats should be continuous.

■ INSTALLATION OF FACTORY-FORMED METAL ROOFING

Factory-formed roofing should be installed in accordance with its manufacturer's instructions. Methods may differ from manufacturer to manufacturer, depending on the design and the materials used in a particular roofing system. The general procedures, including the use of cleats, are similar to those used to install formed-in-place metal roofing. Different profiles are made by different manufacturers (Fig. 7.9-7). Many profiles are available. The building in Figure 7.9-8 was roofed with factory-formed standing-seam metal roofing.

■ INSTALLATION OF FORMED-IN-PLACE METAL ROOFING

Seaming methods most used are standing seams, flat seams, and batten seams. All three may be used when the roof slope exceeds 3 in./ft. Most seams today are made using mechanical seaming machines.

Standing seams (Fig. 7.9-9) should be at least 1 in. high.

FIGURE 7.9-7 *A portion of a roof using one manufacturer's standing-seam roof profile. These seams are finished with a mechanical seamer.* (Courtesy Metal Sales Manufacturing Corp.)

FIGURE 7.9-8 *A standing-seam metal roof.* (Courtesy Metal Sales Manufacturing Corporation.)

FIGURE 7.9-9 *Standing seam.* (Courtesy H. Leslie Simmons.)

FIGURE 7.9-10 *Flat-lock seam.* (Courtesy H. Leslie Simmons.)

Flat seams (Fig. 7.9-10) are made by interlocking adjoining sheets. They are used in both directions in flat-seam roofing and as cross seams in standing-seam and batten-seam roofing. Flat-seam roofing may be laid with the seams parallel and perpendicular to the eaves or in a diamond-shaped pattern with the seams at a 45-degree angle with the eaves.

In *batten-seam* roofing (Fig. 7.9-11) the battens may run up and down the roof from eaves to ridge or form ornamental patterns.

■ FLASHING

Except for counter (cap) flashings, flashings associated with field-formed metal roofing are often locked into the roofing using a lock seam joint.

Flashing for preformed roofing should be made specifically to fit the configuration of the roofing or should be flexible

FIGURE 7.9-11 *Batten seam.* (Courtesy H. Leslie Simmons.)

enough to fit the particular roofing's shape. Flashings should be manufactured or supplied by the manufacturer of the preformed roofing.

■ ROOFING SPECIALTIES

Specialties such as ventilators, fans, roof hatches, and skylights for use with formed-in-place roofing usually have the portions that contact the roofing made from the same material as the roofing.

Specialties for use with preformed roofing should be specifically designed to fit the roofing profile. They should be made from the same materials as the roofing and be finished to match it. Using roofing specialties with standard curbs not specifically designed for use with the particular preformed roofing product will usually lead to leaks.

7.10 JOINT SEALING

Sealants are materials that are placed in joints in buildings and pavings where a weathertight seal is needed and in other locations to close building and paving joints for aesthetic reasons or as a base for painting. Before the advent of curtain wall construction in the 1950s, most buildings were finished in stone, brick, or terra cotta. These materials generate relatively small movements across joints in them. In addition, most openings in walls were small. Most windows and doors were wood framed. Oil-based caulking compounds were the only joint sealants around, and they were adequate in most cases.

Curtain walls, however, were made of metals (mostly aluminum) and glass, with their high coefficients of thermal expansion. Movement across joints in and around them was larger than oil-based caulking compounds could accommodate. Those compounds would harden, crack, and fall out of the moving joints, leaving openings through which wind and water could easily pass. Another solution had to be found. The answer was synthetic rubber (elastomeric) sealants.

The first elastomeric sealants were based on polysulfide polymers. These were quickly followed by elastomeric sealants based on acrylics, silicone, and polyurethane.

Each sealant type has its own characteristics, and the preparation necessary for the application of each of them is different. A designer must understand these differences and select the proper sealant for the job at hand. Three other steps are necessary to ensure proper functioning of a sealant: (1) proper joint design, (2) proper preparation of the joint surfaces, and (3) proper installation and finishing of the sealant. This section discusses these four aspects of joint sealing.

Sealants are here classified as either *building sealants*, which includes those designed for use on horizontal and vertical surfaces in or on buildings, or *paving joint sealants*, which are designed for use in joints in concrete paved roads, curbing, and walkways and between asphalt and concrete pavements (see Section 7.10.3). Glazing sealants are addressed in Section 8.7.

7.10.1

BUILDING SEALANTS

The building sealants addressed here are (1) elastomeric (rubberlike) sealants based on polysulfides, silicone, or polyurethane; (2) solvent-release-curing sealants; (3) latex emulsion sealants; (4) tape sealants; (5) acoustical sealants; (6) preformed foam sealants; and (7) oil-based caulking compounds.

Until recently, federal specifications have been the standards for most sealants. Today, these have been largely replaced by ASTM standards, but manufacturers still list the old federal specifications as the standards by which their products are manufactured. Fortunately most manufacturers also list the current ASTM standards. The following descriptions list the applicable standards, which are summarized in Figure 7.10-1.

7.10.1.1

Elastomeric Sealants

Elastomeric sealants are resin-based, rubberlike compounds, intended primarily for use in expansion joints and other moving joints, such as those in curtain walls and between stone panels or precast concrete panels. Three polymers are currently used to make most of them: polysulfide, silicone, and polyurethane. A few other polymers are also used, but they appear mostly in proprietary sealants and are not widely used.

■ STANDARDS

Until recently, the standards for elastomeric sealants were the following federal specifications:

TT-S-00227E(3), Sealing Compound, Elastomeric Type, Multicomponent (for Caulking, Sealing, and Glazing in Buildings and Other Structures)

TT-S-00230C(2), Sealing Compound, Elastomeric Type, Single Component (for Caulking, Sealing, and Glazing in Buildings and Other Structures)

TT-S-001534A, Sealing Compound, Silicone Rubber Base (for Caulking, Sealing, and Glazing in Buildings and Other Structures)

ASTM Specification C 920, "Specification for Elastomeric Joint Sealants," has replaced the three listed federal specifications. ASTM C 920 classifies elastomeric sealants by type, grade, class, and use.

TYPE Elastomeric sealants are available as single-component (one-part) and multi-component types. ASTM C 920 classifies one-part sealants as Type S; multicomponent as Type M. Type S sealants are more expensive than Type M ones, because they are prepackaged in small cartridges or other units for direct field application. Type M sealants can be packaged in bulk form. Type M sealants must be mixed in the field, which requires more skill than is necessary for Type S sealants, but Type M sealants cure faster than Type S ones and often are made with better components, which results in a superior finished sealant. In addition, the color of three-or-more-part Type M sealants can be more closely controlled because their coloring is a separate component that is added in the field.

GRADE The flow characteristics of a sealant are defined by Grades P and NS. Grade P sealants are also called *self-leveling* because they seek a level and remain level and smooth when applied at temperatures as low as 49 °F. They are used on flat horizontal surfaces.

Grade NS sealants are used on vertical surfaces and in sloping horizontal surfaces where a Grade P sealant would flow down the slope. They are called *non-sag* because they will stay in place even when applied in temperatures as high as 122 °F.

CLASS There are two classes in ASTM C 920: Class 12-1/2 and Class 25. The

FIGURE 7.10-1
Standards Governing Sealants

Sealant	Standard
Elastomeric	
Polysulfide	ASTM C 920
Silicone	ASTM C 920
Urethane	ASTM C 920
Solvent-release-curing	
Acrylic	FS TT-S-00230; AAMA 800 Series
Butyl	ASTM C 1805; FS TT-S-001657
Small joint	AAMA 800 Series
Latex-emulsion	ASTM C 834
Tape	AAMA 800 Series
Acoustical	None
Oil-based caulking compound	FS TT-C-00598
Cold-applied paving joint	
Elastomeric	ASTM C 920
Jet-fuel resistant coal-tar modified polysulfide	FS SS-S-200
Jet-fuel resistant urethane	FS SS-S-200
Hot-applied paving joint	
Concrete to concrete	ASTM D 3406
Concrete to asphalt	ASTM D 3405
Jet-fuel resistant	
Concrete to concrete	ASTM D 3569
Concrete to tar-concrete	ASTM D 3581

numbers relate to the amount of joint movement a sealant can withstand without cohesive failure. The numbers are the percentage of the joint at the time of sealant application and are plus or minus measurements. A Class 25 sealant, then, would be able to withstand a 25% expansion and a 25% contraction without failure, in other words, a total of 50% of the original joint width.

Some elastomeric sealants withstand greater joint movement than 25% in laboratory tests, and their manufacturers claim greater capability because of it. However, most authoritative sources believe that joints should be designed and spaced so that the movement across them does not exceed 25% in each direction, because actual sealant performance can be affected by several imponderables. These factors include the character of joint preparation and sealant application, varying joint widths in the field due to allowable construction tolerances and other factors, and the effect of the temperature at the time of application on later joint movement. A sealant may perform exactly within the rated range when the application temperature is about 70°F, but not perform the same if the application temperature is far below or above that midrange temperature.

USE There are two use classifications in ASTM C 920. The first is related to exposure and includes Use T, which covers sealants intended for use on traffic surfaces, and Use NT, which is designed for nontraffic areas.

The second use classification has to do with the ability of a sealant to adhere to various substrates. The classifications are Use M (mortar), Use G (glass), Use A (aluminum), and Use O (everything else). Since the designations M, G, and A refer to specific laboratory samples of those materials, it is advisable to select materials that fall into Use O. Even then, it is necessary to obtain verification that a selected sealant is satisfactory for use with the specific substrates to be found on a given project. This should be evidenced by actual tests of the selected sealant on the joint substrates that will be encountered.

■ POLYSULFIDE SEALANTS

Polysulfide sealants were the first elastomeric sealants and have a proven record of good performance. Unfortunately, they tend to be more expensive than sealants made with silicone or urethane polymers, and some manufacturers have made them with less polymer to remain competitive. In addition, silicone and urethane sealants are available that have recovery and resistance to ultraviolet light and ozone properties that are equal or superior to those of polysulfide products. As a result the percentage of the market is lower today for polysulfides then it once was. Polysulfide sealants can still be made that will give the kind of excellent performance found in earlier polysulfide products, so they should not be disregarded. However, when selecting a polysulfide sealant today it is prudent to compare its polymer content and other properties with those of earlier polysulfide products that have a proven record of dependability.

■ SILICONE SEALANTS

Silicone sealants perform well in terms of weather resistance, color fading, and durability. They also display low shrinkage and no increase in hardness over time, regardless of the temperatures encountered. Some silicone sealants are marketed for use in traffic-bearing joints, but most manufacturers limit the use of their products to nontraffic joints.

Silicone sealants are available in either high-, medium-, or low-modulus formulations. Unfortunately, these terms are useful only in relative terms, since there are no universally recognized standards defining them. In general terms, high-modulus silicone sealants are to be used where joint movement is 25% (each way) or less. They are intended for use in stopless glazing, curtain wall sealing, and structural glazing (see Section 8.7).

Medium-modulus silicone sealants can be used in the same applications as high-modulus silicone sealants, but they have a higher recovery capability and greater tear resistance. Some of them have demonstrated movement resistance capabilities far above ASTM C 920's Class 25 category. Some can withstand movement as much as 50% each way.

Some low modulus silicones appear to be able to withstand a joint movement equal to 100% each way without failure.

Both acid-curing and neutral-curing formulations are available. A major difference is that the acid-curing formulations have shown a tendency not to adhere well to certain substrates, such as marble, cement-based materials, copper, galvanized steel, and materials that corrode under the influence of the acids in the sealant.

■ URETHANE SEALANTS

Urethane sealants rate highly in terms of weather resistance and durability. Their high performance level will persist for a considerable time, but they do tend to harden with age, which eventually leads to a diminishing of their ability to withstand movement in the joints without failure. However, urethanes are available that show a resistance to movement without failure of up to 50% each way. Formulations are available for use in vertical joints and others for use in horizontal ones. Some urethane sealants may be used underwater.

7.10.1.2

Solvent-Release-Curing Sealants

Solvent-release-curing sealants include acrylic, butyl, and pigmented small-joint sealants. These sealants are used primarily in joints that are fastened together (non-working joints), such as those in building cladding.

■ ACRYLIC

Solvent-release-curing acrylic sealants can be used in exterior joints, but are limited to use where the joints will not move more than 12-1/2% either way, and their performance is inferior to the elastomeric sealants discussed in Section 7.10.1.1. However, they are able to adhere to a great many surfaces, mostly without application of a primer and with little joint cleaning necessary. There is no current ASTM standard for solvent-release-curing acrylic sealants. The applicable standards are Federal Specification TT-S-00230 and

American Architectural Manufacturers Association (AAMA) 800 Series, "Voluntary Specifications and Test Methods for Sealants."

■ BUTYL

Solvent-release-curing butyl sealants are somewhat more limited than solvent-release-curing acrylic sealants in that they are limited to a joint movement of 7-1/2% each way. Butyl sealants are covered by ASTM Standard C 1085, "Specification for Butyl Rubber-based Solvent Release Sealants," and Federal Specification TT-S-001657, "Sealing Compound, Butyl Rubber Base, Solvent Release Type (for Buildings and Other Structures)."

■ SMALL-JOINT SEALANTS

Solvent-release-curing small-joint sealants are more limited still. They cannot be used in joints wider than 3/16 in. Movement must be almost nil. Some of these come in spray cans with small nozzles. Others can be applied in very small joints with needles.

7.10.1.3

Latex-Emulsion Sealants

Latex-emulsion sealants are usually used to seal small interior joints that have little or no movement. They should comply with ASTM C 834, "Specification for Latex Sealing Compounds." These sealants are used to fill joints and crevices around doors and windows in wood, concrete, brick, and other masonry surfaces.

7.10.1.4

Tape Sealants

Tape sealants consist of a bead of butyl- or butyl-polyisobutylene-based sealant with a release paper on one side. Tapes may be reinforced with string or unreinforced. They are used for a variety of purposes, the majority being associated with glazing and concealed lap joints. Tape sealants should conform with the applicable portions of AAMA 800 Series, "Voluntary Specifications and Test Methods for Sealants."

7.10.1.5

Acoustical Sealants

Acoustical sealants are used mostly in concealed locations to seal off sound-transmission paths around partitions, electrical outlets, and other openings. They should be nondrying synthetic rubber formulations. No ASTM standard is available for acoustical sealant materials. Acoustical sealant application should be in accordance with ASTM C 919, "Practice for Use of Sealants in Acoustical Applications."

7.10.1.6

Preformed Foam Sealants

Preformed polyurethane foam sealants are installed in joints in a compressed state and expand to fill the joint while remaining compressed to a specified degree. There are no current standards covering preformed foam sealants. The manufacturer's instructions should be followed.

7.10.1.7

Oil-Based Caulking Compound

Oil-based caulking compounds have an oil or oleoresinous base and are used in relatively fixed joints subject to very limited movement. The standard for them is Federal Specification TT-C-00598C, "Caulking Compound, Oil and Resin Base Type (for Building Construction)."

Oil-based caulking compounds can be used to fill joints and crevices around doors and windows in wood, concrete, brick, and other masonry surfaces that are subject to little or no movement. They are supplied in a gun grade suitable for use with a caulking gun and a knife grade, which can be spread with a putty knife. Oil-based caulking compounds tend to dry on the surface but remain soft and tacky inside. The exposed surface of the caulking bead should be painted each time the surrounding area is painted, to help extend its life.

Other types of sealants are preferred in most instances to oil-based caulking compounds, because of their better adhesion and greater elasticity, even at low temperatures. Although most are more expensive than oil-based caulking compounds, their greater reliability and durability often are worth the difference in cost.

7.10.2

PAVING JOINT SEALANTS

Because paving joints are exposed to sun, weather, ice-melting chemicals, and vehicular traffic, they are subject to more severe conditions of expansion, contraction, and wear than most building joints. For this reason, most sealants that will work quite well in building joints will not perform adequately in paving joints. Special sealants have been developed to handle such conditions.

Basically, paving joint sealants fall into two categories: cold-applied sealants and hot-applied sealants. The selection of one over the other often hinges on the amount of work to be done. Hot-applied sealants require the use of kettles and special installation equipment and therefore are usually restricted to those projects where there are a large number of joints to be sealed, such as state or federal highway paving.

Several of the currently available products for use in paving joints are listed in the following two subsections. These sealants represent the types of products available, but they are not the only ones. The list will surely grow longer in time and the products available will become better. There is a trend, for example, toward products with higher resistance to joint movement and abrasion and the ability to withstand both higher and lower temperatures without loss of properties.

7.10.2.1

Cold-Applied Sealants

Cold-applied paving joint sealants include pourable elastomeric sealants complying with ASTM C 920, which may be a jet fuel–resistant sealant for concrete made with urethane, coal tar–modified polymer, or bitumen-modified urethane.

Jet fuel–resistant, coal tar–modified polysulfide and jet fuel–resistant urethane sealants for concrete joints are also available. These sealants should comply with Federal Specification SS-S-200E(1), "Sealants, Joint, Two-component, Jet-blast, Cold-applied, for Portland Cement Concrete Pavement."

Cold-applied silicone paving joint sealants are also available, one for concrete-only joints and another for joints between concrete and asphalt. Both should comply with ASTM C 920.

7.10.2.2

Hot-Applied Sealants

Hot-applied elastomeric paving joint sealants are available for concrete-to-concrete and concrete-to-asphalt joints. Jet fuel–resistant formulations are also available for concrete to concrete and concrete to tar-concrete. Sealants for concrete-to-concrete joints should comply with ASTM D 3406, "Specification for Joint Sealant, Hot-applied, Elastomeric-type, for Portland Cement Concrete Pavements"; concrete-to-asphalt joint sealants with ASTM D 3405, "Specification for Joint Sealant, Hot-poured, for Concrete and Asphalt Pavements"; jet fuel–resistant concrete-to-concrete joint sealants with to ASTM D 3569, "Specification for Joint Sealant, Hot-applied, Elastomeric, Jet-fuel–resistant Type for Portland Cement Concrete Pavements"; and jet fuel–resistant concrete-to-tar-concrete sealants with ASTM D 3581, "Specification for Joint Sealant, Hot-poured, Jet-fuel–resistant Type for Portland Cement Concrete and Tar-concrete Pavements."

7.10.3

ACCESSORY MATERIALS

Accessories for sealed joints include joint backings and bond breakers for building sealants and joint fillers and bond breakers for paving joints.

Joint backings and joint fillers serve as a bond breaker, limit the depth of the sealant to that desired, and support the sealant during installation and tooling. In traffic joints, they also tend to support the sealant under applied loads.

7.10.3.1

Backings and Joint Fillers

Joint backings, also called *backer rods*, include plastic foams and elastomeric tubing. Foams may be of either open-cell poly-

urethane, closed-cell polyethylene, or another plastic material that is acceptable by the sealant manufacturer. Tubing may be neoprene, butyl, EPDM, or silicone complying with ASTM D 1056, "Specification for Flexible Cellular Material—Sponge or Expanded Rubber." Closed-cell foams and elastomeric tubing tend to support a sealant better during installation and tooling. Open-cell foams are better able to accommodate variations in joint width and may help sealants applied over them cure faster. Elastomeric tubing can provide a temporary weather seal until the sealant has been applied and act as a secondary seal after the sealant has been applied.

Joint fillers for use in joints in traffic surfaces include expanding cork, cork sponge rubber, bituminous-saturated fiber board, and proprietary products that are composites of cork or fiber board encased in organic or glass fiber felt.

Joint backing and joint filler materials must be compatible with the sealant and the substrate and should be acceptable to the sealant manufacturer. Backer rods and fillers for use with hot-applied sealants should be those specifically selected by the manufacturer of the sealant.

7.10.3.2

Bond Breakers

Bond breakers are polyethylene or other plastic tapes designed for installation in joints that have no backer rod or filler to prevent the sealant from bonding to the substrate at the back of the joint. The actual material used should be acceptable to the sealant manufacturer.

7.10.4

JOINT DESIGN

The proper design of joints is an essential element in ensuring that sealants will perform satisfactorily.

7.10.4.1

Joint Types

Two types of joints occur in buildings and pavements: *construction joints* and *separation joints*.

■ CONSTRUCTION JOINTS

Construction joints that require sealants are planned joints that divide a building into buildable segments and prevent damage due to building movement after erection. They include expansion joints and control joints. Another type of construction joint is used to interrupt pours of concrete and break slabs into panels for ease of construction (see Section 3.12.2), but these are seldom left open and do not usually require a sealant. When they do require a sealant, the principles discussed here apply.

Expansion joints are relatively large joints intended to permit a structure to expand and contract without damage to the structure. These joints are sometimes called isolation joints. They usually pass completely through the structure and its finishes.

Control joints are relatively small joints used to divide thin materials such as pavements, floors, walls, and finish materials such as ceramic tile into panels of a size that will not be damaged by thermal or other movement.

■ SEPARATION JOINTS

Separation joints are joints between materials, such as between windows or doors and adjacent masonry or between two portions of the same material, such as in sheet metals.

7.10.4.2

Joint Movement

Joints can be classified as dynamic or static. *Dynamic joints*, which are also called *movement joints* or *working joints*, are joints that move with changes in temperature or other movements that occur in a building. This includes all construction joints and some, but not all, separation joints. Dynamic joints in exterior walls are sealed to prevent the passage of air, water, and sound through them into the building. Dynamic joints in both interior and exterior horizontal and vertical surfaces are sealed to prevent the passage of light, air, water, sound, and vermin through them and to make them more aesthetically pleasing.

Static joints have little or no movement. They are also called *nonmovement joints* or *nonworking joints*. These joints include those between interior door frames and adjacent masonry, plaster, or gypsum board. Static joints are usually sealed to prevent passage of light, air, or sound through them and to provide a base for painting.

7.10.4.3

Joint Design Procedure

Poor joint design is a major cause of the failure of joint sealants. The first step in designing dynamic joints is to estimate the amount of movement that will occur across them. To do this, it is necessary to determine the expected temperature extremes and the coefficient of expansion of the materials associated with the joint. Using these, it is possible to calculate the amount of theoretical movement. Actual movement, however, is very difficult to determine accurately, because of the many imponderables such as the exact thermal movement that will occur when more than one material is involved and construction tolerances that result in variable joint widths. Therefore, it is necessary to determine an approximate range of anticipated movement and select a joint width that will accommodate that movement. The temperature at the time a sealant will be applied in a joint must be taken into account in joint width selection to ensure that the joint will be of the necessary size at that time. It is usual to assume a temperature range of between 40° and 100 °F as the temperature at the time the sealant will be installed and select the joint width so that it will work over that entire range.

Two basic joint types are used in buildings: *butt joints* and *shear joints* (Fig. 7.10-2). Of the two, shear joints offer the advantage of protecting the sealant and of placing less strain on the sealant. They are, however, difficult to design, and their appearance might not always be desirable. In addition, it is difficult to prepare shear joints to receive sealant and to install the sealant. Shear joints are sometimes used in such applications as sheet metal joints, certain joints in curtain walls, and in articulated stone walls.

For both dynamic and static joints, the minimum joint width at any time during or after sealant application must be wider than the selected sealant's minimum performance width. For most sealants, this is 1/4 in., but each sealant's minimum perfor-

FIGURE 7.10-2 *Joint types: (a) butt type and (b) shear type.*

mance width should be determined from that sealant's manufacturer and not assumed. Small-joint sealants, for example, cannot even be used in joints wider than 3/16 in. Their minimum joint widths are sometimes very small but still significant in the design and selection of a sealant.

Joints that are wider than the limitation of a sealant should be avoided. It is better to reduce the width of joints by using more joints, but if this is impracticable or undesirable, the joints should be protected by metal plates or preformed elastomeric and metal expansion joints. This is applicable both for joints in traffic surfaces and for vertical joints.

The depth of the space allotted for a sealant must be such that the sealant has the proper shape. If a sealant is too thin, it might tear when it expands to fill a joint. If a sealant is too thick, it might fail to compress properly and may protrude from the joint, or even be squeezed out of it. Under normal conditions, the thickness should be one-half the width but not less than 1/4 in., nor more than 1/2 in. This rule of thumb limits the desirable width of joints to be closed with a sealant to 1 in.

7.10.5

SEALANT SELECTION

A sealant should be selected that will provide the necessary aesthetic and functional performance. Some sealants are available in many colors, others only in white or black, and others are clear. The type and size of joints is a prime consideration when selecting a sealant. The sealant must be capable of performing within the confines of the joint in which it is installed. For example, high-modulus silicone sealants should not be used in joints subject to movement of more than $\pm 25\%$. Narrow joints require sealants specifically formulated for them.

Sealants must also be selected to suit the location of the joint to be sealed. Traffic joints demand self-leveling sealants; vertical joints, non-sag ones. Traffic joints also require sealants with a higher hardness number. Traffic sealants should have a hardness of at least 25 but not more than 50 when tested in accordance with ASTM Standard C 920.

Special conditions of use must also be considered in the selection of a sealant. Some such special conditions include the presence of high temperatures, the need for a sealant to be submersed, or the presence of vehicular traffic.

Selecting elastomeric sealants using only the ASTM C 920 classification system is possible, but often impracticable. A perfectly good sealant product may have slightly different characteristics than those required by the ASTM C 920 system. In addition, the ASTM C 920 system is not a completely adequate gauge of sealant performance. Products that exceed its limits should not be rejected on that basis alone. A product that has the desired features for a particular project may be selected even if it does not comply with another aspect of the ASTM C 920 system. The usual practice is to select specific proprietary sealant products. When competition is required, several comparable products may be specified.

7.10.6

INSTALLATION

The sealant manufacturer's instructions should be followed in the preparation and installation of sealants.

7.10.6.1

Joint Preparation

Cleaning the surfaces to receive sealants is essential. *There must be no contaminants of any kind in joints to receive sealants.* This includes, but is not limited to, incompatible paint and other coatings, rust and other corrosion, mill scale, water, frost, grease, oil, laitance, and form-release agents. Joints to receive sealant also must be dry and free from dust and grit, which should be removed using dry compressed air or brushes. The sealant manufacturer's instructions should be followed in determining if a particular substrate is compatible with its sealant and in cleaning and treating such substrates as rubber, plastics, wood, bituminous materials, water-repellent coatings, ceramic materials, glass, paints, and other decorative coatings. Chemical cleaners and solvents must not leave a residue. A corrosion-resistant coating may be used on corrodible joint surfaces, but it must be compatible with the sealant, and it must be dry when the sealant is applied.

Joints should have no protrusions into them, such as aggregate projecting from concrete, and should be of the same width throughout. Projections should be ground down. Recesses should be filled solidly and permanently.

Joints that are too narrow should be cleaned out or cut wider with power tools to produce the proper width. Under no circumstances should sealant be placed in a joint that is narrower than the minimum performance width of the sealant.

Deep joints should be filled with a joint filler, backer rod, or both to create a sealant pocket of the correct depth (Fig. 7.10-3).

When a backer rod is not installed, a bond breaker tape should be placed over the joint filler or in the back of the joint, as applicable, to prevent the sealant from adhering there (Fig. 7.10-4). If a sealant adheres to the back of a joint, it will not be able to contract properly when it stretches with an expanding joint and will tear apart internally or separate from the sides of the joint.

When the sealant manufacturer recommends doing so, joints to receive a sealant should be primed with a primer recommended by the sealant manufacturer.

Before sealants are applied, adjacent materials should be protected with tape or

FIGURE 7.10-3 *A backer rod should be used to control the joint depth. The depth of the sealant should not be more than one-half the width of the joint.*

FIGURE 7.10-4 *Bond breaker tape is necessary when the joint depth is shallow.*

other means to ensure that sealant does not contaminate them and that the lines of an unconfined sealant run straight and true.

7.10.6.2

Sealant Application

In general, sealants should be applied in accordance with the recommendations of the manufacturers in such a manner that they completely fill the joint and completely *wet* the sides of the joint. Hot-pour sealants should be melted within the manufacturers' recommended temperature limits.

Sealants should be confined within a joint when possible. If they finish below the face of the adjoining materials (Fig. 7.10-5), they will be less subject to damage from contact and deterioration by being somewhat shielded from the sun and weathering.

When a sealant must be applied as a triangular fillet seal (Fig. 7.10-6), it should be applied so that the fillet is not less than 3/8 in. across the face. The face of a fillet seal should be convex to ensure an adequate volume of sealant.

Mixing procedures for multicomponent sealants should be those recommended by the sealant manufacturers. Care should be taken to prevent raising the temperature of cold-applied sealants during mixing and to prevent air from becoming entrapped in the sealant.

Application devices should be those recommended by the sealant manufacturer. Non-sag sealants should be extruded into a joint so that they are forced against the sides. Caulking guns are used to do this with most sealants. Figure 7.10-7 shows a hand-operated caulking gun used with one-part sealants. Similar hand-operated guns (Fig. 7.10-8) and pneumatic guns (Fig. 7.10-9) are used to apply

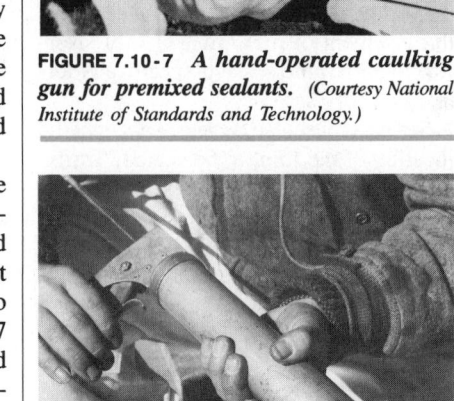

FIGURE 7.10-7 *A hand-operated caulking gun for premixed sealants. (Courtesy National Institute of Standards and Technology.)*

FIGURE 7.10-8 *A hand-operated caulking gun for bulk sealants. (Courtesy Sealant Waterproofing and Restoration Institute.)*

FIGURE 7.10-5 *Sealant depressed below the surface.*

FIGURE 7.10-6 *A triangular fillet seal. If the joint is more than 3/16 in. wide, a bond breaker tape should be used to prevent sealant from adhering to the sides of the joint.*

FIGURE 7.10-9 *A pneumatic caulking gun for bulk sealants. (Courtesy Sealant Waterproofing and Restoration Institute.)*

multicomponent bulk sealants.

Hot-pour sealants are placed through a pouring nozzle of a size that just fits in the joint. Cold-pour sealants can be poured directly from the mixing vessel or placed using a special pouring device.

Tape sealants are placed in open joints by first removing the paper slip sheet and laying the tape into the joint. The second layer of substrate is then installed and pressed into the tape sealant.

Preformed foam sealants are forced into a joint and allowed to expand to fill it. Some preformed strips are heated after they have been inserted in a joint to cause them to achieve continuous contact with the joint sides.

Care must be taken with both tape sealants and preformed tape sealants to ensure that they make positive contact where sections join.

7.10.6.3

Tooling

Non-sag sealants should be tooled before skinning (curing of the top layers of sealant) occurs. Tooling should be done with the devices recommended by the sealant manufacturer. Tooling should compact the sealant and force it into contact with the sides of the joint. The surface of tooled joints should be left slightly concave.

7.10.6.4

Protection

Sealants must be protected from contact with people and traffic until they completely cure.

8

DOORS AND WINDOWS

Photo: Peachtree Hotel, Atlanta, Georgia. *(Architects: John Portman and Associates; photograph courtesy Bruce Wall Systems Corporation.)*

INTRODUCTION

This chapter discusses both interior and exterior windows and doors of steel, wood, aluminum, stainless steel, and copper-alloy construction. It also covers aluminum entrances, store fronts, curtain walls, storm doors, and windows.

Exterior doors and aluminum entrances, of course, provide access openings into a building. They minimize heat gain and loss through these access openings by keeping the weather outside, and they provide privacy for the building occupants and security for persons and property. They also help control the passage of sound through access openings.

Interior doors control passage between interior spaces. They also provide privacy and security, reduce sound transmission between spaces, and screen storage and utility areas from view.

Windows, store fronts, and curtain walls keep out the weather while admitting light and providing view. Windows also facilitate ventilation.

8.1 GENERAL RECOMMENDATIONS

The information in this section applies to most of the doors, windows, entrances, store fronts, and glazed curtain walls addressed in the other sections of this chapter, regardless of their type or the material of their construction. The other parts of this chapter contain more specific information.

8.1.1

HEAT LOSS AND GAIN

In a typical building, windows and doors lose and gain more heat than any other part of the structure. In some climates, this heat loss or gain can account for as much as 70% of the total heating load and 50% of the cooling load, according to a study by the Department of Housing and Urban Development.

Most heat is lost through *infiltration* (air leakage) through cracks around the frame and *conduction* from warm air coming in contact with the cold window and door surfaces. Most heat is gained through *radiation*, which is transmission of heat by invisible electromagnetic waves through the glass. To conserve energy, create a more comfortable indoor environment, and reduce building operation costs, window and door openings should be constructed so that unwanted heat loss and heat gain are minimized.

This section contains some strategies that make heat-gain and heat-loss reduc-

tions possible. Many have been suggested by the National Institute of Standards and Technology publication *Window Design Strategies to Conserve Energy*.

Three ways to reduce heat loss through windows is with tightly sealed openings, multiple glazing (storm sash or insulating glass), and insulating shutters or tight-fitting roll shades. Besides the other measures discussed here, heat-loss reduction through doors is often accomplished by means of vestibules or storm doors.

In most parts of the country, these methods can reduce heating costs, minimize condensation on interior surfaces, and improve interior comfort by reducing drafts and radiant body heat loss. In Zones IV and V of Figure 7.3-2, multiple glazing may not be justified, so its use should be decided on on a case-by-case basis, taking into account fuel and material costs.

8.1.1.1

Tightly Sealed Windows and Doors

Doors and windows should be tightly sealed in all climates, regardless of whether the need for cooling or heating is predominant.

To achieve a tight seal around windows and doors, a sealant should be placed at the joint between their frames and the adjoining walls. In addition, they should be weatherstripped where practicable, and

doors should be closely fitted at the bottom and sealing devices installed there.

■ SEALING JOINTS

Joint sealants should be used wherever a window or door frame meets a wall or other construction (see Section 7.10). Sealants should be applied to a clean, dry surface on a warm day. Sealants for this purpose are usually latex-emulsion sealants (see Section 7.10.1), but silicone and other sealants may also be used. Sealants have a limited life and should be renewed periodically as necessary.

■ WEATHERSTRIPPING

Weatherstripping is applied to window and door frames to prevent air leakage. It typically has a resilient part that prevents air penetration when it is pressed between the door or window and its sash.

The most common types of weatherstripping are (1) foam rubber stripping, (2) felt stripping, (3) rolled vinyl, (4) casement stripping, (5) spring metal, and (6) interlocking metal.

Foam rubber weatherstripping has either an adhesive or a wood backing. It is easy to install but lacks durability. The adhesive type can be used wherever there is no friction, such as at the top and the sides of frames and at the top and the bottom of double hung windows (Fig. 8.1-1a).

FIGURE 8.1-1 *Common types of weather-stripping: (a) adhesive-backed foam rubber; (b) wood-backed foam rubber; (c) rolled vinyl; (d) V-strip.*

Wood-backed foam rubber is more conspicuous than the adhesive type and can be used effectively only on doors (Fig. 8.1-1b).

Felt stripping has the same advantages and disadvantages as foam rubber, but it also collects dirt more readily. Metal-backed felt, which provides more durability, is also available.

Rolled vinyl is available with or without a metal backing. Some types are manufactured with tubular gaskets that can be either hollow or filled with foam (Fig. 8.1-1c). Rolled vinyl is very durable and can be installed on the exterior of double hung, sliding, and casement windows, as well as on doors.

Casement window stripping can be one of several kinds depending on whether the window is wood or metal. Wood casements can be sealed with felt, adhesive-backed foam, or spring metal. For metal casements, a special vinyl gasket with a deep groove can be used to slip onto the edges of the frame. A vinyl-to-metal adhesive must be used to bond the gasket to the frame so the window closes flat against the vinyl strips. Most such weatherstripping is factory applied.

Spring metal stripping is available in two types: flat stripping, which is most commonly used for double hung windows, and V-strip, which is used for doors. Flat stripping is inserted into the narrow side channels of a window sash, on the bottom rails of the upper and lower sash, and on the top rail of the upper sash.

A V-strip is a double-over strip of springy metal that fits between the door edge and the jamb (Fig. 8.1-1d). When the door is open, the metal springs apart;

when the door is closed, it is forced together to fit tightly between door and jamb. Both types are durable and inconspicuous after installation, and each provides an excellent seal. They will last indefinitely but are more costly to install than other types of weatherstripping.

Interlocking metal stripping is produced as interlocking channels and interlocking J-strips. Both types are difficult to install, because alignments must be exact for a secure interlock. They provide excellent, durable seals. The channels are subject to damage, however, because they are exposed. The J-strips are not visible when installed; therefore, they are not exposed and not subject to damage.

■ **DOOR BOTTOMS**

Air leakage through cracks at door bottoms can be stopped with (1) sweeps attached to the door bottoms, (2) door shoes, or (3) weatherproof thresholds.

Sweeps are easy to install because they do not require removal of the door. The most common type consists of a felt or rubber flap mounted in a metal channel (Fig. 8.1-2a). They seal most effectively on flat thresholds. They drag on carpets, however, and will eventually wear down from friction. To avoid carpet drag, automatic door sweeps can be used. These are spring-loaded and flip up when the door is opened, down when it is closed.

An interlocking threshold consists of a metal sill that is shaped to interlock tightly with a metal channel attached to the door bottom (Fig. 8.1-2b). It is durable and provides an excellent weather seal, but it is difficult to install because proper alignment is critical.

Weatherproof thresholds can be either the vinyl-bulb or interlocking type. These thresholds are difficult to install because a good fit requires the door bottom to be trimmed at or manufactured with a 1/8-in. bevel. A vinyl-bulb threshold contains a flexible vinyl strip that is compressed when the door is closed (Fig. 8.1-2c). The threshold itself is metal and very durable; the vinyl bulb will wear out, but it can be replaced.

Door shoes are metal frames with attached vinyl bulbs that press against the threshold to form a tight seal (Fig. 8.1-2d). They are very durable because the bulbs can be replaced when worn. Door shoes

FIGURE 8.1-2 *Door bottoms: (a) sweeps; (b) interlocking threshold; (c) vinyl-bulb threshold; (d) door shoe.*

are best suited for flat, unworn thresholds. They are difficult to install, however, because the door must be removed and trimmed to fit.

8.1.1.2

Multiple Glazing

Because of its thinness, a single sheet of glass is a good conductor of heat. One square foot of 1/4-in. clear glass can conduct 6 to 10 times more Btu's per hour than a square foot of a typical frame wall. The rate of heat transfer through the glass is so rapid that added thickness is of almost no value. However, if layers of glass are separated by an air space, the path of conduction is interrupted, and the rate of heat flow is reduced. Heat can then pass through the air space primarily by radiation and convection and only minimally by conduction.

Storm windows can be effective in reducing conduction and infiltration with an air separation between panes of glass of as much as 6 in. However, an air space is most effective in improving thermal performance when it is between 3/16 in. and 5/8 in. wide. Within this range, the heat flow is reduced markedly as the width is increased from 3/16 in. towards 5/8 in. Increasing the air space beyond 5/8 in. does not reduce heat flow to the same extent, because a wider space allows the air in it to circulate more freely and develop convection currents. These currents transport heat from the warmer glass to the colder glass. This increased convection heat loss can more than offset the slight reduction in conductive heat loss through the air. Conversely, a space of less than 3/16 in. is of little value because heat is readily conducted by the air across such a short distance.

A thermal air space can be created with storm sash and with insulating glass.

■ STORM SASH

Storm sash are commonly called *storm windows*. They control heat flow by making windows more airtight to reduce infiltration and by providing an insulating layer of air between the window sash and the exterior to reduce conductive transfer.

Storm windows are recommended in Zones I, II, and III of the map in Figure 7.3-2. In these areas, storm windows can significantly reduce heating costs and minimize condensation on interior windows.

Storm windows can be of the conventional, combination, or fixed-in-place interior types. They can be manufactured with frames of wood, metal, or rigid plastic. Storm doors are most often made of aluminum, although wood is still sometimes used.

Storm windows and doors are discussed in Section 8.6.

■ SPECIAL GLASS TYPES

Sealed insulating glass units, like storm sash, reduce conductive heat flow. Insulating glass eliminates condensation and has the convenience of being part of the prime window or door. It does not, however, have the added protection against infiltration that a separate storm panel provides.

When interior or exterior shading is not practical, it is possible to use special types of glass to help keep the sun out. The most common are heat-absorbing glass and reflecting glass.

Insulating and other special glasses are discussed in Section 8.7.

■ THERMAL BREAK

A metal frame that holds insulating glass should have a thermal break to prevent a thermal "short circuit," that is, heat flow being resisted by the insulating glass but flowing freely through the aluminum frame. This is important because metal conducts heat many times faster than wood. The thermal break is usually polyurethane, vinyl, or another low-conductive material that provides a separation between the outside and the inside frames (Fig. 8.1-3).

FIGURE 8.1-3 *The shaded thermal break reduces conductive heat transmission through the metal frame.*

An aluminum frame with a good thermal break has an R-value of 1.66, which is approximately equal to that of double-pane insulating glass. A comparable frame with no thermal break would have an R-value of 1.18.

In some areas it may be economically practical to use three layers of glass separated by air spaces. Three layers of sealed glass with two 1/4 in. air spaces, for example, has an R-value of 2.13, compared to 1.72 for typical insulating glass. This technique of triple glazing can also be accomplished by installing storm sash over double-glazed windows.

8.1.1.3

Exterior Door Protection

Entry doors are another source of great heat loss. The two most effective techniques to reduce such heat loss are storm doors and insulated doors.

■ STORM DOORS

Like storm windows, storm doors reduce heat loss by reducing both conduction and infiltration. For example, a typical 1-3/4 in. solid wood exterior door has an R-value of about 2. When a wood storm door is added, the R-value is increased to approximately 3.70 (Fig. 8.1-4).

FIGURE 8.1-4 Door R-Values	
Door Type	**R-Value**
1-3/4 in. exterior door	2.04
1-3/4 in. exterior door and aluminum storm door (50% glass)	3.03
1-3/4 in. exterior door and wood storm door (50% glass)	3.70
Metal-clad insulated door	8.00–15.00

■ INSULATED DOORS

Insulated doors have an inner core of plastic foam insulation, have high R-values (R-8 to R-15), and are even more effective than the combination of prime and storm doors in reducing heat flow. Insulated doors are available in both metal and wood. The wood-framed ones often have metal skins. Insulated doors are often installed with magnetic weatherstrips that are relatively airtight and minimize heat loss by infiltration (Fig. 8.1-5).

FIGURE 8.1-5 *Plastic foam insulation fills the inner core of insulated doors.*

8.1.1.4

Window Treatment

For maximum reduction of heat flow through windows, more thermal protection than that provided by insulating glass or storm panels is sometimes desirable. In that case, windows can be protected from heat loss by window treatment. *Window treatment* includes all types of interior and exterior shading devices and interior blinds, shades, draperies, and shutters. This section includes a discussion of only that window treatment intended to control the loss or gain of heat through a building's windows. The types of shutters, blinds, and draperies used primarily for decoration are beyond the scope of this book, even though some of them may also incorporate some features of the devices covered here. A lined drapery, for instance, may well reduce heat transfer.

The first measure usually proposed to reduce heat gain and loss through glass is special glazing techniques, such as tightly sealed openings, multiple glazing, insulating glass, and storm windows. Such measures produce much more effective windows with less heat loss and gain than conventional single-glazed sash and are worthwhile for that reason. Unfortunately, even double-glazed windows (including glass and frame) lose heat 6 to 10 times faster than do typical, insulated wood-framed walls. Therefore, for maximum reduction of heat flow through windows, additional thermal protection is often desirable. Heat gain can be reduced by the installation of exterior shading devices. Summer heat gain and winter heat loss by interior spaces can be reduced by interior blinds, shades, thermal shutters, and draperies. Integral blinds that are encased between the glazing panes of a window are also helpful.

Light colors should be chosen for most exterior and interior shading devices, because part of their job is to reflect sunlight. An exception is a roof overhang, which works more effectively when its underside is dark because it will reflect less sunlight through the glass.

■ EXTERIOR SHADING DEVICES

Summertime sun-control devices are more effective when installed on the exterior of a building where their shade can prevent sunlight from entering it. This is critical because only a portion of the solar radiation that enters a building can be reflected back to the exterior. The remainder is absorbed and then radiated and conducted to the interior (see Section 10.2).

The most effective shading devices are *exterior shutters, exterior blinds and shades, awnings,* and *architectural projections.* Interior shading devices, though less effective, are still worthwhile.

SHUTTERS   A shutter that effectively blocks out all direct sunlight can reduce solar heat gain through a window by 80%. The overall shading performance, however, depends on how well heat absorbed into the shutter itself is dissipated to the outside air. Shutters that absorb large amounts of heat radiate a significant percentage of that heat into the building. *Light-colored shutters, which reflect most of the sunlight before it can be absorbed, are superior to dark shutters.*

The most desirable shutters are those that have operable-tilt slats that can be adjusted to block the sun and also let air circulate through them. Nonoperable louvered shutters will also permit natural air flow but cannot be adjusted to efficiently track the sun's angle. For shutters to prevent heat gain most effectively, air should be allowed to circulate between the window and the shutter. Such air circulation dissipates heat absorbed by the shutters. Most shutters are manufactured of wood or aluminum. Vinyl is most frequently used to make decorative shutters. Some shutters are not suitable for use as operable shutters because they are molded with a finished surface on one side only. Others have simulated louvers with no open spaces between slats. Several common shutter types are shown in Figure 8.1-6.

Because they reduce air flow, tight-fitting shutters provide more resistance to heat loss than do loose-fitting ones. However, even when the fit is not tight or the louvers cannot be completely closed, shutters will still reduce heat loss by reducing infiltration and preventing the scouring action of the wind from disturbing the insulating film of air at the outer surface of the glass.

ROLL BLINDS AND SHADES   Roll blinds and shades range from simple bamboo or canvas shades to motorized roll blinds that can be operated from the interior. Like shutters, such devices are highly effective because they stop the sunlight from entering a building. Because they provide an opaque covering of the window itself, they also effectively block low-angle sun from

FIGURE 8.1-6 *Common shutter types: (a) rolling shutters move in horizontal tracks; (b) Bahama shutters cover entire window opening; (c) Sarasota shutters extend over the top half of the window opening and are typically combined with side-hinged shutters.*

west elevations. *A light color is most effective because it reflects the heat-producing sunlight.*

Motorized roll blinds are usually made of aluminum, vinyl, or wood and therefore offer the added benefit of year-round use and durability (Fig. 8.1-7). They can also reduce heat loss in winter by as much as 50% by acting like a tight-fitting storm window. Independent shade control of each window and ease of operation are other advantages. Their prime disadvantage is that exterior view and natural ventilation are reduced. Blinds are available, however, that can be projected away from

FIGURE 8.1-7 *(a) Soffits above the windows house the motorized roller that stores the blinds. (b) Even larger windows can be protected, such as on this screened porch.* *(Courtesy Levolor.)*

FIGURE 8.1-8
Heat Gain Through Single-Glazed Windows with Awnings

Window Orientation	Awning Type	Heat Gain/ 100 sq. ft. Glass Surface, Btu/day	Heat Excluded Total, Btu/day	% Reduction
South	No awning	62,200	—	—
	White canvas	22,500	39,700	64
	Dark green canvas	27,700	34,500	55
	Dark green plastic	35,600	26,600	43
West	No awning	84,200	—	—
	White canvas	19,500	64,700	77
	Dark green canvas	23,900	60,300	72
	Dark green plastic	34,800	49,400	59

FIGURE 8.1-9
Roof Overhang for Complete Summer Window Shading*

Latitude • North, Degrees	Overhang Width South Exposure	Southeast/Southwest Exposure
28	1 ft. 5 in.	3 ft. 10 in.
32	1 ft. 11 in.	4 ft. 2 in.
36	2 ft. 5 in.	4 ft. 8 in.
40	3 ft. 0 in.	5 ft. 4 in.
44	3 ft. 6 in.	6 ft. 0 in.

*Assumes window sill is 2 ft. from floor.

the window like an awning, to provide view and ventilation.

AWNINGS How well an awning shades a window depends on how opaque the material is to both direct sunlight and diffuse light from the sky. The more opaque a material, the more complete is the shade provided.

The exposed surfaces of an awning should be a light color to minimize the amount of sunlight absorbed. When sunlight is absorbed by an awning, heat is produced and the awning's temperature is raised. The result is heat transfer through the shaded window.

Clean, white canvas and slatted, light aluminum awnings, for example, reflect between 80% and 91% of the sunlight that strikes their surfaces. A dirty awning is less reflective and therefore absorbs more sunlight. By comparison, a dark green canvas awning reflects only 21% and a dark green plastic awning reflects 27% of the sunlight.

A typical light-colored awning can reduce heat gain by 64% on south-facing windows and 77% percent on west-facing windows. Figure 8.1-8 lists the net effec-

tiveness of several types of awnings.

Heat from sunlight absorbed by a dark-colored fabric will build up under the awning and be transferred to the window unless air is permitted to circulate under the awning. Fabric awnings are commonly installed with a continuous gap between the top of the awning and the wall to prevent hot air from being trapped under the awning. Slatted aluminum awnings provide air circulation between the horizontal slats.

Awnings should provide adequate coverage of a window's area for its specific orientation. South-facing windows require only a minimal horizontal projection to be completely shaded throughout the day. In addition, the sides of awnings on south-facing windows should be closed to prevent sunlight from angling in behind the awning. Awnings on east- and west-facing windows should extend down a substantial

distance to block out early morning and late afternoon sun, as applicable.

ARCHITECTURAL PROJECTIONS Windows can be shaded by horizontal or vertical planes that extend beyond the face of the glass to intercept the summer sun. If properly designed, these projections will not interfere with the penetration of the beneficial winter sun or block the view.

A roof edge that overhangs the wall beneath it is one of the most common ways to shield the windows in small buildings, especially houses, from sunlight. Figure 8.1-9 lists the horizontal projection necessary to provide complete summer window shading with a roof overhang.

A deeply recessed window or one with closely spaced horizontal or vertical fins can also be effective. In general, east- and west-facing windows are more effectively shaded by vertical projecting planes.

South-facing windows are best shaded by horizontal planes.

The underside of an overhang should be dark because it will transmit less ground-reflected sunlight through the glass. The net result of such shading is that heat gain through a completely protected window can be reduced by 80%.

■ INTERIOR HEAT GAIN/LOSS DEVICES

Although they are not as effective as exterior shading devices, interior heat gain/loss devices have the advantages of greater accessibility and ease of operation. Interior devices used to reduce heat gain in the summer and heat loss in the winter include blinds, shades, thermal shutters, and thermal draperies.

BLINDS The most common of such devices are venetian blinds (Fig. 8.1-10). They are slatted horizontally and can be tilted to reflect a maximum amount of sunlight out the window or to direct daylight toward the ceiling for glare-free uniform daylighting. Blinds, however, provide only minimal reduction in heat loss in winter because they do not effectively entrap a layer of air.

Vertical blinds offer similar protection against heat gain. They can be significantly more efficient in reducing heat loss in winter, however, because they more effectively trap an insulating layer of air

FIGURE 8.1-10 *Venetian blinds can be made to cover large glazed areas such as sliding glass doors.* *(Courtesy Levolor.)*

between themselves and the glass (Fig. 8.1-11).

Some proprietary systems include venetian blinds encased between layers of glass to reduce heat gain and heat loss (Fig. 8.1-12). When fully closed this system can provide an R-value of 2.50, compared to an R-value of 2.04 for a similar window with no blinds.

ROLL SHADES The most effective type of interior protection against window heat gain is an opaque or translucent *roll*

FIGURE 8.1-11 *Vertical blinds extend to the floor, thus reducing heat-robbing convection currents.* *(Courtesy Levolor.)*

FIGURE 8.1-12 *With blinds fully closed, solar heat gain is reduced by about 82%.* *(Courtesy Pella Corporation.)*

shade. A roll shade can be easily installed with simple brackets on the window frame and, when not in use, rolled up for unobtrusive storage.

A completely opaque shade with a white outer face can reflect 80% of the sunlight back through the glass.

A roll shade can have a dark color on one side, which effectively absorbs sunlight, and a white surface on the reverse side, which effectively reflects it. By simply reversing the shade, from dark side facing out in winter, to reflective side out in summer, the shade can perform both as a solar collector and a shading device.

THERMAL SHUTTERS Covering windows with a rigid insulating material such as polyurethane or polystyrene is one of the most effective ways to control heat loss. Figure 8.1-13 lists several common insulating materials and the resulting R-values when they are combined with 1/4-in.-thick plate glass. Thermal shutters are usually installed on the interior, but they can be placed on the exterior if the insulation is protected from the weather.

A common way to achieve protection from the elements is by facing the insulation with aluminum, galvanized steel, or exterior-grade plywood. Foam plastic must also be protected from fire. Plywood or metal cladding can serve this function.

Thermal shutters can be hinged, tracked, or mounted with clips. Removable shutters can be taken off and stored during sunny periods or warm weather. Hinged or tracked shutters can be folded or pushed back out of the way when they are not needed.

If the insulating material is placed in direct contact with the glass, the heat loss is reduced only in proportion to the R-value of that material.

If a thermal shutter fits so that a gap exists between it and the sash, conductive heat loss is more effectively reduced because of the insulating dead air space. Infiltration can also be reduced if cracks around the frame are covered. For such a reduction, however, a tight fit is crucial. When a shutter fits loosely, air can circulate between the glass and the shutter. Such air movement can carry heat away when the air passes over the glass and result in a dramatic reduction in overall insulating effectiveness.

FIGURE 8.1-13
Thermal Shutter Materials

Material	R-Value/ 1 in.	Window R-Value	
		With 1 in. Insulation	With 2 in. Insulation
Expanded polystyrene (extruded)	4.00	5.00	9.00
Expanded polystyrene (molded)	3.57	4.54	8.11
Polyurethane	6.25	7.14	13.39

THERMAL DRAPES   Drapes offer a more traditional way of controlling window heat flow. Because the surface temperature of even an ordinary drape is usually much higher than that of the adjacent window glass, their existence reduces the radiant heat loss from occupants and therefore improves thermal comfort.

For maximum benefit from window coverings, the room air must be prevented from passing between the glass and the covering material. When air is allowed to pass over the glass, the improvement from a loose-fitting drape is nominal, amounting to only an increase from the R-0.85 of an undraped window to R-0.94. If the heated air from a wall or floor register passes between the covering and the glass, heat loss is actually more rapid than if no covering were provided.

By contrast, when warm room air is prevented from circulating behind the drapes, the R-value of the window and drape can increase to 1.13. To achieve this, drapes must be tightly sealed at the bottom and sides. Thermal drapes can be fastened to a window frame with hooks, snaps, or velcro fasteners. If long drapes are used, they should touch the floor.

Thermally efficient drapes are usually made of tightly woven fabric. The drape can be further improved by adding a liner of reflective material, foam backing, or additional layers of fabric.

8.1.2

BURGLARY PROTECTION

In most localities it is necessary to make a special effort to reduce the vulnerability of buildings to unauthorized entry by thieves, vandals, and pranksters, or burglary.

Burglars do not like challenges. They prefer to work quickly, quietly, and in the dark. If more than four minutes are required to gain entry, they will often give up the attempt. No lock will stop a determined professional burglar, but good burglary protection hardware on doors and windows and design of doors and windows to resist intrusion can slow a burglar down. Care should be taken, however, to ensure that devices intended to deter intrusion do not endanger occupants by preventing their exit in an emergency.

To be effective, the devices discussed in this section need to be combined with proper site planning, warning devices, and adequate lighting. Site planning measures to deter unwanted intruders and devices intended to make noise when triggered so that a burglar will not be able to work quietly and to warn occupants and neighbors of an attempted intrusion are discussed in Section 12.6. Outdoor lighting designed to deny a burglar a dark place to work is discussed in Section 12.4.

8.1.2.1

Unauthorized Entry

Before it is possible to design effective measures to prevent intrusion, it is necessary to know how burglars gain entry. Even locked doors and windows offer only a slight challenge to experienced burglars.

Simple lightweight tools can be used to pry open doors equipped with automatic latching devices of the type commonly used in many cylinder locks (Fig. 8.1-14). Deadbolts are somewhat harder to pry open and require more force and heavier tools.

A credit card or a similar thin piece of plastic can be slipped between a door and

FIGURE 8.1-14 *Some door locks are easy to pry open.*

its frame to push back the latch and open the door.

A pair of pliers can be used to grip a lock cylinder and pull it out of the lock, making the mechanism accessible. Burglars sometimes use a screwdriver to twist and wrench the core of the locking pins away from the cylinder.

A crowbar may be used to jimmy doors or windows. By inserting the crowbar between the sash and frame, a burglar can pop the bolt out of place. Even if the lock holds, the sash may break.

8.1.2.2

Doors and Frames

It is unfortunate that adequate locks are frequently installed on inadequate doors. A lock can be only as good as the door on which it is installed. In general, exterior entry doors should be carefully selected and properly installed. Selection of door and frame types and materials and the type of locks and other security devices used depends on the project budget and the level of security required. In some cases the company insuring a building's contents may dictate the level of security required.

■ HINGED DOORS

Flush wood solid-core or metal doors, 1-3/4 to 2 in. thick, are the least vulnerable to attack. Wood hollow-core doors and stile-and-rail doors with wood or glass panels are more vulnerable. A burglar can easily break through a hollow door or panels and reach the lock from inside. Refer to Section 8.3 for a discussion about wood doors.

Exterior metal doors consist of a core and a metal frame surrounded by steel skins. Most security experts prefer the use of metal doors over any type of wood door. Doors should fit in their frames snugly, with only enough clearance to ensure easy operation.

■ SLIDING DOORS

Security problems with siding glass doors include (1) inadequate locks, (2) flimsy frames, (3) door panels that can be lifted out of their frames, and (4) vulnerability of the glass.

A series of screws along the upper track of a sliding glass door can prevent the door from being lifted out of its track (Fig. 8.1-15). A piece of wood or metal slightly shorter than the door width can be laid in the lower track (Fig. 8.1-16) to supplement the lock and prevent the door from being forced open. To keep the wood or metal bar from being lost, it can be hinged at the frame and leaned against the side jamb when not in use. Commercial devices are available to serve this purpose.

■ VIEWING DEVICES

One-way viewing devices (peepholes) in doors allow occupants inside a building to see outside without having to open the door.

■ DOOR FRAMES

Even a strong door with a good lock and proper hinges can be opened easily if the door's frame is inadequate. If the rough opening is oversized and the door frame is inadequately shimmed, a burglar can often pry the frame from the lock to release the door. A space as small as 1/16 in. may be large enough to insert a crowbar, screwdriver, or hacksaw. The space between the door and its frame should be 1/64 in. or smaller. The same tolerance should be observed at the side jambs and at the top of the door, because an automobile jack can be used to spring the frame at a point just above the lock.

FIGURE 8.1-15 *Screws at periodic intervals along the upper track will prevent a would-be intruder from lifting a sliding glass door off its floor track.*

FIGURE 8.1-16 *A metal or wood bar in the track prevents a sliding glass door from being forced open from outside. The bar can be hinged to prevent its loss.*

8.1.2.3

Locks

Local fire codes should be checked before a new lock is installed or an old lock upgraded. Fire codes sometimes restrict the use of locks that might slow down escape in the case of fire. Lock types include key-in-knob, mortise, interconnected, and auxiliary locks.

■ KEY-IN-KNOB LOCK

One of the most common and least secure locks is the key-in-knob type. Such locks are relatively inexpensive but have several disadvantages: (1) the locking mechanism is in the exterior knob and can be easily stolen; (2) the latch is often short and tapered and can easily be pried out of position with a credit card or other similar object.

A trigger bolt attached to the latch bolt can make it impervious to credit card attack. Some models feature longer latches to make prying less effective, but none deals with the problem of the primary locking mechanism in the exterior knob.

Existing key-in-knob locks do not provide adequate burglar protection and should be replaced or supplemented with one of the auxiliary locks discussed next.

■ MORTISE LOCK

A mortise lock offers more security than a key-in-knob lock because the locking mechanism is inside a metal enclosure mounted within the door. It not only has a latch, like the key-in-knob type, but also a deadbolt, which, when in place, cannot be violated by a credit card. Push buttons located under the latch allow the exterior knob to lock in a stationary position or unlock and turn freely.

The disadvantages of mortise locks are (1) a user may fail to use the deadbolt;

(2) the large cavity within a door needed to house this lock may impair the structural integrity of the door; (3) the keyed cylinder is often poorly anchored and vulnerable to forceful removal; (4) a mortise lock may be defeated by loosening the set screw that holds the cylinder in place.

Some locks have faceplates that protect the cylinder. Because of the large cavity occupied by the lock case, the mortise lock is difficult to replace with any other type.

There are also locks that combine elements of both the key-in-knob and the mortise lock types.

■ INTERCONNECTED LOCK

An interconnected lock has a key-in-knob cylinder, which makes it look like a mortise lock from the outside. However, it lacks the push buttons under the latch that lock and unlock the knob and differs in its method of installation. An interconnected lock requires the boring of holes. It is a good replacement for a key-in-knob lock because the old lock hole can be used and the other necessary holes added.

As with a mortise lock, the user must remember to engage the bolt. Also if the cylinder is unprotected, it is vulnerable to pulling. Since an interconnected lock removes less of the door structure than a mortise lock, the door may stand up to more abuse before breaking.

■ AUXILIARY LOCK

A second lock is often the easiest and cheapest method of adding security to a door. A cylinder deadbolt fits within a door. As with other internal locks, this may weaken the door. Another disadvantage is that two locks may keep out an intruder, but they may also slow escape in the case of emergency.

A vertical deadbolt lock is also called a *rim lock* because it is surface-mounted on the inside rim of the door. This is an obvious, but effective, type of lock. If installed with long enough screws, it resists kicking and picking. Some locks have special shutters that freeze the bolt in place if someone tries to pry it open.

■ PADLOCK

A good padlock consists of (1) a hardened steel shackle, (2) a brass or laminated steel lockcase, and (3) a five-pin cylinder. A padlock can add protection to any door, but it is most often used on garage or shed doors. Only the side of the door to which the padlock is attached is accessible.

■ REINFORCEMENT

Many adequate locks come with inadequate strikes or screws, or they are installed on inadequate doors.

The *strike* is the jamb fastening that receives the bolt in the locked position. Consumer Union (CU) testers found that several kicks could tear away a conventional strike along with its screws and part of the jamb, leaving the lock intact but the door open. A high-security strike consists of a small metal box that fits into the frame's bolt hole and fastens securely to the building framing (Fig. 8.1-17).

The screws that attach the various kinds of locks to the door and the strikes to the jamb often are too short. In general, the longer the screw, the more securely the lock or strike will be fastened. The screws for the average rim lock should be 2 in. long. The screws on a strike should be at least 2 in. long, preferably 3 in. long, so they will secure the strike to the framing studs.

On soft-wood doors, a wraparound channel made of metal fits on the door and reinforces the area around the lock, preventing the wood from splitting (Fig. 8.1-18).

■ SPECIAL LOCKS

Doors such as sliding glass doors and garage doors need special locks.

FIGURE 8.1-17 *This type of strike fastens securely to the studs and helps prevent a locked door from being kicked open.*

FIGURE 8.1-18 *A wraparound channel helps prevent a door from splitting and breaking in a kicking assault.*

Sliding glass doors can be protected in many ways. A piece of wood or pivoted metal bar will prevent the door from opening (see Fig. 8.1-16). A keyed deadbolt at the door base can also secure the door (Fig. 8.1-19). The disadvantage of these methods is that they can be activated only from inside the building. They may also slow escape in case of fire.

FIGURE 8.1-19 *A keyed deadbolt prevents a door from sliding open.*

Some manufacturers make special locks that replace and upgrade the normal locks provided on siding glass doors. One manufacturer makes a door with a double deadbolt operated by both a handle and a key. Unlike many other siding glass doors protected by security methods, this door can be kept secure yet opened from the outside with a key.

Garage doors can be fitted with a padlock in a hole drilled through the end of a garage door bolt (Fig. 8.1-20). The padlock will secure the door even if the door's regular lock mechanism is destroyed.

8.1.2.4

Windows

The construction of windows generally makes them more vulnerable than doors

FIGURE 8.1-20 *A supplementary padlock can be installed on a garage door.*

to unauthorized entry. Nine out of ten times a potential intruder will find an open window.

Even when windows are closed and locked, the glass can be broken if an intruder does not mind making noise. When the panes are too small to crawl through, an intruder will be forced to spend more time and make more noise in the process of enlarging the opening. Windows can also be forced open with a small crowbar.

In lieu of regular glass one of the following security glasses may be used: (1) laminated glass, which consists of a vinyl panel sandwiched between two glass panels, (2) wired glass, or (3) tempered glass. Acrylic or polycarbonate glazing can also be used. Unfortunately, plastics are subject to surface scratches, which can detract from their appearance.

Exterior ornamental grillwork and interior window gates can be used to protect windows. Gates or grillwork should be installed with nonretractable screws. Grillwork should never be installed on windows needed as fire exits. Interior window grates should have an integral locking-knob system rather than a padlock, so they do not become obstacles in case of fire.

■ LOCK TYPES

Most window locks are vulnerable to prying and therefore are equipped with various locking devices (Fig. 8.1-21). Window locks include friction type, pin type, wedge type, stop type, key-operated cam latches, and eyebolt.

A friction-type device has a pin and a receptacle into which the pin is screwed. Friction is supposed to keep the window

FIGURE 8.1-21 *Window locks include (a) friction type, (b) pin type, (c) wedge type, (d) stop type, and (e) key-operated cam latches.*

in place, but it requires very little effort to nudge the pin from the receptacle.

A pin-type lock is a keyed device. Its bolt slips into a hole drilled into the upper window frame. Though stronger than the friction type, this lock is vulnerable to prying. Because of the keyed lock, it may represent a safety hazard in the case of fire.

A wedge-type ventilating lock is self-defeating. It works to hold the window shut by spreading the upper and lower sash apart. A potential burglar can easily slip a knife blade between sashes and cut off the wedge.

A stop-type ventilating lock uses a knob or other protrusion to limit the distance that a window can be opened. Though it does not spread the window sashes apart, it too is vulnerable to prying.

A key-operated cam latch replaces an ordinary double-hung window cam latch. Like a pin-type lock, it is vulnerable to prying and may represent a safety hazard in the case of fire.

An eyebolt, installed according to the recommendations made by Consumer's Union, is the most effective device to prevent the forcing of a window.

Holes should be drilled on sides of the interior of the window. They should go all of the way through the inside sash and three-quarters of the way through the outside sash at a slight downward angle, as shown in Figure 8.1-22. Two sets of holes can be drilled in the outside sash so that the window can be partially opened for ventilation. Eyebolts, 5/16 in. in diameter, inserted into the holes act as locking pins.

FIGURE 8.1-22 *A simple pry-proof window lock: a 5/16-in. eyebolt is inserted through holes drilled through the inside and outside sashes.*

8.1.3

SOUND CONTROL

Doors and windows are acoustically weak. Maximum acoustical performance requires that the edges and the frames of doors and windows be sealed airtight. Packing and caulking around window and door frames are necessary to reduce sound transmission through cracks.

8.1.3.1

Doors

Solid-core or mineral-filled doors are generally better sound isolators than hollow-core doors (Fig. 8.1-23). However, the amount of air space around the edges of the door is usually the controlling acoustical factor. Whenever possible, doors should be gasketed if sound privacy is required (see Fig. 8.1-23). The relative performances of solid- and hollow-core doors, gasketed and ungasketed, are given in Figure 8.1-24.

Soft weatherstripping provides a good seal at the top and the sides of a door, but automatic threshold closers are needed to seal the bottom. While it may be costly to use gasketed doors, provision of a cushion such as sponge rubber between the door jamb will reduce the impact energy of a door slam and make a door closure quieter. Unfortunately, properly sealed doors, because they are tight fitting, are harder to close and more prone to warping.

FIGURE 8.1-23 *Solid-core doors should be used in sound-conditioned walls, with the door sealed at the jambs, head, and sill.*

Undercut doors to allow for rugs or return air movement in forced-air ventilation systems essentially destroy the acoustical effectiveness of the doors (see Section 10.4).

Sliding doors generally provide little sound privacy, because a relatively tight air seal cannot be maintained.

Heavy-duty doors made of metal or metal clad over very heavy core materials, such as fire-rated metal doors, will not necessarily provide good acoustical performance in spite of their mass, even when two doors are installed, unless they are sealed with gaskets and door sills (Fig. 8.1-25a). Where such doors are installed in locations where loud noises occur, such as in mechanical equipment rooms, a second door with better acoustical performance can be installed (Fig. 8.1-25b).

8.1.3.2

Windows

About 80% of the acoustical problems occurring with windows result from the quality of airtightness of construction and about 20% from the selection of the glass itself. Generally, sturdy, well-weather-stripped, and tightly closed windows are better sound insulators than loosely fitting ones; fixed sash performs better than operating sash. The relative acoustical performances of various types of windows are given in Figure 8.1-26.

Operating single-glazed wood windows in grooved wooden tracks with weatherstripping, which are pressure sealed when closed, perform best. Wood windows in metallic or plastic slides are fair against air infiltration but generally poor against resonant frequencies (see Chapter 10). Metal windows are available in many grades and quality levels. The better grades will perform acceptably if they are tight fitting, properly sealed, correctly installed, and sufficiently rigid to reduce air infiltration and resonant vibration.

In operating double-glazed (storm windows), the greater the separation between two panes of glass, the more efficient the sound isolation. To be significant, however, the separation must be relatively large. For example, a separation of from 2 to 4 in. increases the sound transmission class (STC) rating from 33 to only 34. The minimum useful distance is 4 in.; 8-in. spacing is effective but not always practical, and 12 in. is required to isolate severe noise. If only a 2-in. space is possible, it is considered more economical to double the weight of the single glass. Storm windows attached to casements are credited for added weight of glass only.

Each glass pane should be mounted in a heavy frame and securely fastened in place. Mounting glass in a frame with flexible or preextruded sealant improves the damping characteristics. Each glass should be of a different thickness or weight per square foot, so that one can offset the deficiency of the other and they will not vibrate together or resonate.

Fixed single-glazed plate glass in fixed windows, compared to operating windows, is itself a fairly good sound barrier material, but the window is still the weakest part of the wall construction. The greater acceptance of year-round air conditioning allows glass to be sealed in less-expensive frames and reduces air infiltration.

The thickness of glass should be selected to meet the project's requirements of decibel loss. Two layers of insulating glass with a 1/2-in. air space should be considered for fuel savings and prevention of condensation. Acoustically, two layers

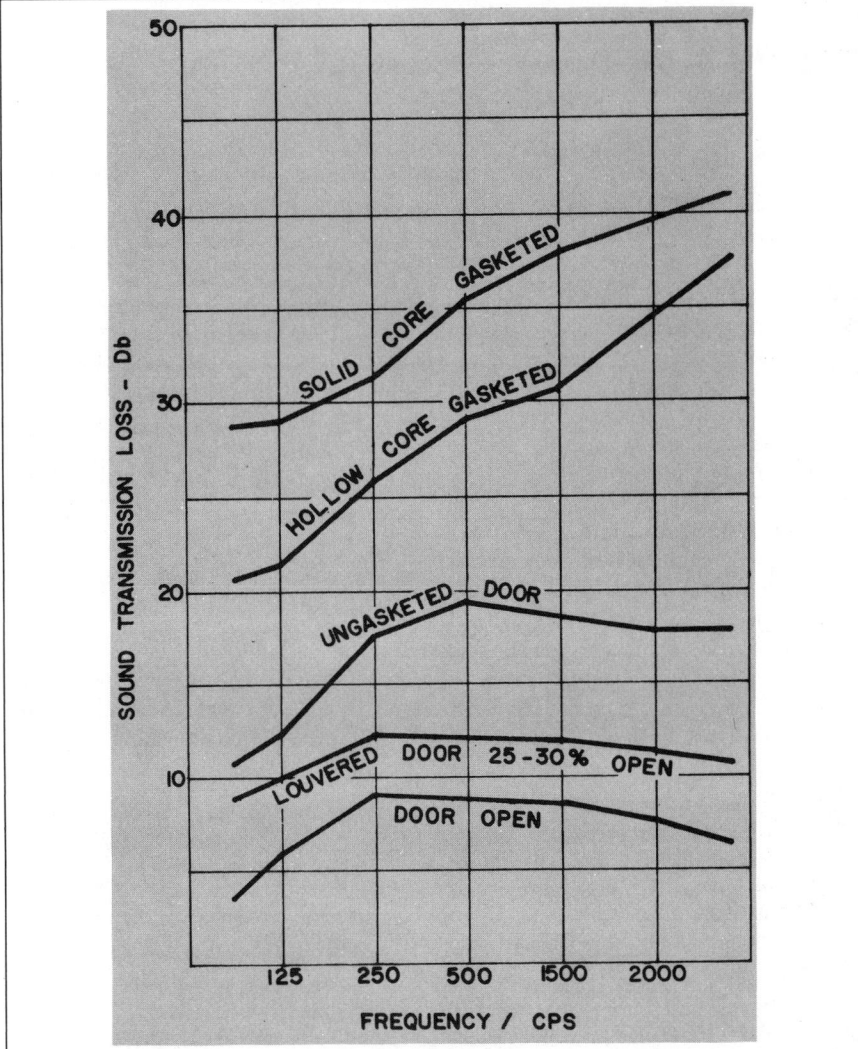

FIGURE 8.1-24 *Relative acoustical performances of hollow- and solid-core wood doors.*

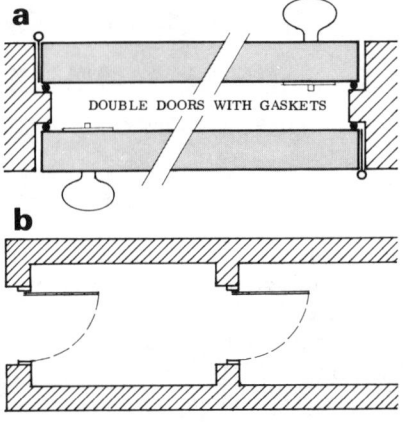

FIGURE 8.1-25 *Acoustical traps can be used to reduce sound transmission from high-noise sources, such as service rooms and mechanical equipment rooms.*

FIGURE 8.1-26 *Relative acoustical performances of various types of windows.*

of closely spaced 1/4-in. glass are approximately equal to one layer of 1/2-in. glass.

Windows should be installed with a flexible mounting, with the mounting installed in a firm frame and airtight to prevent flanking sound from bypassing the glass.

Fixed double-glazed sash should be used for the most demanding noise problems. This consists of two panes of glass separated by at least 8 in. of enclosed air space. The addition of a sound absorber, such as a 1-in.-thick resilient material around the perimeter between the two glass panes, will generally improve performance by absorbing some of the sound energy in the air space.

8.2 METAL DOORS, FRAMES, AND STOREFRONTS

Interior and exterior doors, frames, and associated sidelights are made from formed carbon steel, stainless steel, and copper alloys and from extruded aluminum. Most commercial and institutional entrances and store fronts are made from extruded aluminum, but the other metals mentioned also occur. Most metal doors are of the swinging type, but metal revolving doors are also used. Aluminum sliding glass doors are covered in Section 8.4; aluminum storm doors in Section 8.6. Glass for doors, entrances, and storefronts is addressed in Section 8.7; glazed curtain walls in Section 8.8.

8.2.1

GENERAL REQUIREMENTS

There may be some confusion about the use of the terms *doors and frames* and *entrances*. Doors and frames are usually designed as separate units, which may or may not be combined with storefront components to form entranceways.

Entrances are packaged systems, consisting of doors, frames, sidelights, and usually hardware. To simplify this distinction, this text will refer to such entrances as *packaged entrances*.

8.2.1.1

Standards

There are two types of door and frame manufacturers. One makes so-called *standard* doors and frames; the other makes *custom* units. Standard units are mass produced and sold to retail outlets such as lumber yards and home centers. Standard units are also made for a particular project where they have been specified. Custom units are manufactured specially for a particular project to suit an owner's or an architect's design. Both types should meet the same standards of quality. Carbon steel, stainless steel, and aluminum doors may be standard or custom. Copper-alloy doors and frames are custom.

Carbon steel doors and frames should comply with the American National Standards Institute/Steel Door Institute's publication ANSI/SDI-100, "Recommended Specifications for Standard Steel Doors and Frames." Custom doors and frames should comply with this same standard. In addition, a good guide for custom steel door and frame work is *Hollow Metal Manual*, a collection of National Association of Architectural Metal Manufacturers (NAAMM) publications related to steel doors and frames. It includes the publications of NAAMM's Hollow Metal Manufacturers Association division and the NAAMM *Metal Finishes Manual* series.

Stainless steel, aluminum, and copper-alloy doors should conform to the applicable standards of the appropriate organization, which include the Aluminum Association (AA), the American Architectural Manufacturers Association (AAMA), the American Society for Testing and Materials (ASTM), the National Association of Architectural Metal Manufacturers (NAAMM), the American National Standards Institute (ANSI), the Copper Development Association (CDA), and the American Iron and Steel Institute (AISI).

In most cases, formed stainless steel and copper-alloy doors and frames should also comply with the same standards as carbon steel doors and frames. Extruded aluminum components should comply with the AAMA publication *Aluminum Store Front and Entrance Manual*.

Fire doors and frames should have been tested and labeled by Underwriters Laboratories, Factory Mutual, or another independent testing laboratory and should conform with the requirements of the National Fire Protection Association (NFPA) 80, "Fire Doors and Windows," and ASTM Standard E 152, "Methods of Fire Tests of Door Assemblies."

Sound-rated assemblies should conform with ASTM Standard E 90, "Method for Laboratory Measurement of Airborne-sound Transmission Loss of Building Partitions."

Insulated doors should conform with ASTM Standard C 236, "Test Method for Steady-state Thermal Performance of Building Assemblies by Means of a Guarded Hot Box."

Door hardware should conform to ANSI standards A156.1 through A156.21.

8.2.1.2

Materials

Refer to Chapter 5 for a general discussion about the metals addressed in this section.

Carbon steel for *interior* door facings and frames should be either hot-rolled commercial-quality steel sheet and strip conforming with ASTM Standards A 568, "Specification for General Requirements for Steel, Sheet, Carbon, and High-strength, Low Alloy, Hot-rolled, and Cold-rolled," and A 569, "Specification for Steel, Carbon (0.15 Maximum Percent) Hot-rolled, and Strip, Commercial Quality." It should have a *stretcher-level* standard of flatness.

Carbon steel for *exterior* door facings and frames should be hot-dip, galvanized, commercial-quality, carbon steel sheets complying with ASTM Standards A 525, "Specification for General Requirements for Steel Sheet, Zinc Coated (Galvanized) by the Hot-dip Process," and A 526, "Specification for Steel Sheet, Zinc Coated (Galvanized) by the Hot-dip Process, Commercial Quality." The zinc coating should be not less than class G60. Sheets should be *mill phosphatized* and have a stretcher-level standard of flatness.

A variety of aluminum alloys and tempers are used to make extruded door and frame components. Alloys should conform with ASTM Standard B 221, "Specification for Aluminum-alloy Extruded Bars, Rods, Wire, Shapes, and Tubes." Common alloys used are 6061, 6063, and 6463. The temper should be suited to the finish to be applied. A common temper is T5.

Sheet aluminum is often used for door faces. Sheet aluminum and bent plate components should comply with ASTM Standard B 209, "Standard Specification for Aluminum and Aluminum-alloy Sheet and Plate."

Stainless steel is used to make formed doors and frames, sidelights, store fronts, and revolving doors. Door facings are usually flat sheets, but some door facings are patterned sheets.

Copper-alloy materials are used to make formed doors and frames for both exterior and interior locations. Most door facings on such doors are also copper-alloy sheets. Revolving doors are also made from copper-alloy materials.

Concealed components, such as subframes, stiffeners, edge channels, reinforcements, and the like should be fabricated from the same materials as the exposed portions of the door or frame.

8.2.1.3

Finishes

Carbon steel doors are either given a factory finish, such as a baked enamel or fluorocarbon finish, or painted in the field (see Section 9.11).

The mill finish is usually left on concealed surfaces and sometimes even on the exposed surfaces of aluminum doors. However, most exposed portions of aluminum entrance doors are anodized (see Section 5.2).

Most exposed stainless steel is polished, but organic coatings are also applied to some assemblies. Most exposed copper alloys are given a patina or architectural-bronze finish. Exposed fasteners are usually finished to match the adjacent metal.

8.2.1.4

Supports, Anchors, Fasteners, and Flashing

Supports, anchors, fasteners, and flashing materials should be compatible with the material being anchored or fastened and with the substrate, so galvanic and chemical reactions do not occur between different materials. Carbon steel should be anchored and fastened with galvanized steel, stainless steel, or cadmium-plated steel devices; aluminum and stainless steel with

nonmagnetic stainless steel, cadmium-plated steel, or galvanized steel; copper alloy with copper alloy.

Inserts for use in concrete may be cast iron, malleable iron, or hot-dip galvanized steel.

8.2.1.5

Weatherstripping

The function of weatherstripping is to control air infiltration, prevent rain penetration, minimize dust infiltration, and withstand atmospheric conditions. Weatherstripping must be durable under extended use, resist corrosion, and be readily replaceable.

8.2.2

CARBON STEEL DOORS

Carbon steel doors and frames are often called *hollow metal*. The association representing many members of this industry is called the Hollow Metal Manufacturers Association (HMMA). It is a division of the National Association of Architectural Metal Manufacturers (NAAMM). Its manual of practice is called *Hollow Metal Manual*.

8.2.2.1

Construction

Carbon steel doors consist of a core, faces, tops, and bottoms. There are two types of core construction. In one, cores are hollow

and braced with steel stiffeners. Most are filled with the manufacturer's standard fireproof, rotproof, sound-deadening material. Insulated doors are filled with glass fiber or plastic foam insulation.

In the second form of core construction, a door's face sheets are laminated to a core of impregnated kraft paper honeycomb, plastic foam insulation, or mineral fiber blocks (Fig. 8.2-1).

The tops and the bottoms of standard doors are finished with a recessed welded-in closure. Exterior doors have tops and bottoms formed flush as an integral part of the door construction or are finished with a welded-in inverted channel. Bottoms may also be fitted with a recess for weatherstripping (Fig. 8.2-2).

Doors should be sized so that the *maximum* clearances between the door and the frame, between doors in a pair, and between door and sill are in accordance with the recommendations of the Steel Door Institute's publication ANSI/SDI-100. These clearances, which are subject to a tolerance of ±1/32 in. are shown in Figure 8.2-3. As discussed in Section 8.1.3.1, security requirements may dictate less space between door and frame than that shown in Figure 8.2-3.

8.2.2.2

Door Types, Styles, and Designs

Most of the steel door types, styles, and designs available today fall into the classification system described in ANSI/SDI

FIGURE 8.2-1 *Steel door construction.* (*Reprinted with permission of NAAMM, the National Association of Architectural Metal Manufacturers.*)

STANDARD   FLUSH (closing channel)   AUTOMATIC WEATHERSTRIP   STANDARD Inverted channel   FILLER CHANNEL

sealed if desired

TOP EDGE DETAILS

Other designs available as required

BOTTOM EDGE DETAILS

FIGURE 8.2-2 *Bottom and top details of a typical steel door.* *(Reprinted with permission of NAAMM, the National Association of Architectural Metal Manufacturers.)*

FIGURE 8.2-3
Door Clearances

Joint	Clearance, in.
Door head to frame	1/8
Door jambs to frame	1/8
Between meeting edges of pairs (unrated doors)	1/8–1/4
Between meeting edges of pairs (fire rated doors)	1/8
Door bottom	3/4
Face of door to stop	1/16

100 and SDI 108, "Recommended Selection and Usage Guide for Standard Steel Doors." SDI classifies doors in three grades: Grade I, standard; Grade II, heavy duty; and Grade III, extra heavy-duty. In Grades I and II there are two models: Model 1, full-flush design; and Model 2, seamless design. In Type III, there are five models: Model 1 and 1A are full-flush design; Models 2 and 2A are seamless design; and Model 3 is stile-and-rail-flush panel.

Figure 8.2-4, which comes from SDI 108, shows the uses of the three grades recommended by SDI. It also shows eight basic door designs, their recommended uses, and the abbreviations and acronyms used to identify them. Steel doors that imitate the look of wood stile-and-rail panel doors are also available.

8.2.3

CARBON STEEL FRAMES

Carbon steel frames are brake-formed of steel sheets into one of two basic shapes (Fig. 8.2-5). However, many variations of these basic shapes are made, including cased opening frames that have no stops. Some special frames have tapered or molded shapes. Regardless of their shape or size, frames should be rigid, neat in appearance, and free from defects, warp, and buckle. When possible, they should be fitted and assembled in the manufacturer's plant.

8.2.3.1

General Requirements

The metal thickness of frames should be dependent on the door grade and model. In Grade I the thickness should be either 18 or 16 gauge. In Grade II the thickness should be 16 gauge. In Models 1, 2, and 3 in Grade III the thickness should be either 16 or 14 gauge. In Grade III, Models 1A and 2A, the frame gauge should be either 14 or 12.

Frames that are not weatherstripped or for double-acting, smoke, or fire doors should be drilled and have three rubber or plastic *silencers* (small resilient bumpers) installed on the strike jambs. Frames for double doors should have silencers on the head.

Some codes and regulations require that interior doors in hospitals, nursing homes,

and some other medical buildings have *terminated stops* (also called *cutoff* or *sanitary* stops) (Fig. 8.2-6). These stops must be omitted, of course, from fire, smoke, lead-lined, and sound-rated doors. The bottom of such terminated stops are closed with a filler plate.

Plaster guards or mortar boxes are required on frames where plaster or mortar might intrude into the operating hardware. These are sheet metal boxes welded to the frame.

Glazing beads used to hold glass panels in steel frames should be removable. They may be either mitered or butted in the corners but should be factory fitted. Removable beads should occur on the security side of openings so they cannot be removed for access.

Frames wider than 49 in. should be reinforced with a 14-gauge steel channel. Lead-lined frames and frames for power-operated doors usually also require reinforcement to support the loads.

8.2.3.2

Frame Types

Frames are available in several types. There are two basic frame types, welded and knockdown, and several special frame types, including drywall slip-on, store front, and lead-lined frames.

In *welded* frames, corners and intersections are fully welded or attached together using concealed welded splice plates. Corners should be mitered.

Knockdown frames are made with mechanical joints. Most, but not all, have mitered corners. They are similar in design and appearance to welded frames, except that they are delivered to a project site disassembled for field assembly.

Drywall slip-on frames are designed for installation in drywall partitions after the partitions are in place. Several designs are available.

Store front-type frames are essentially welded frames. Frame sections that are too large to be shipped in one piece are shipped in several pieces and assembled in the field.

Lead-lined frames are fabricated with lead linings or are of a modified design that will receive lead sheeting in the field. They are used in x-ray rooms and wherever else radiation is a hazard.

FIGURE 8.2-4
Basic Door Grades and Applications

Building Types	Grade I Standard Duty, 1-3/4 in. or 1-3/8 in.	Grade II Heavy Duty, 1-3/4 in. only	Grade III Extra Heavy Duty, 1-3/4 in. only	Door Thickness 1-3/4 in.	Door Thickness 1-3/4 in. or 1-3/8 in.	F	G	V	FG	NL	L
Apartment											
Main entrance			•	•					•	•	•
Unit entrance	•	•		•	•						
Bedroom	•		•		•						
Bathroom	•		•		•						
Closet	•		•		•						•
Stairwell		•	•	•					•		
Dormitory											
Main entrance			•	•					•	•	•
Unit entrance	•	•		•	•						
Bedroom	•			•	•						
Bathroom	•			•	•						
Closet			•	•							•
Stairwell		•	•	•					•		
Hotel, Motel											
Unit entrance	•	•		•	•						
Bathroom	•		•		•						
Closet	•		•		•						•
Stairwell		•	•	•					•		
Storage and utility	•	•		•	•						•
Hospital, Nursing Home											
Main entrance			•	•					•	•	•
Patient room		•		•	•						
Stairwell		•	•	•					•		
Operating and exam room		•	•	•	•						
Bathroom	•			•	•						
Closet	•		•		•						•
Recreation		•		•	•				•		
Kitchen		•	•	•					•		
Industrial											
Entrance and exit			•	•					•	•	•
Office	•	•		•	•				•		
Production			•	•					•		
Toilet		•	•	•	•						•
Tool		•	•	•	•						
Trucking			•	•					•		
Monorail			•	•	•				•		
Office											
Entrance			•	•					•	•	•
Individual office	•			•	•	•			•		
Closet	•		•		•						•
Toilet		•	•	•	•						
Stairwell		•	•	•					•		
Equipment		•	•	•	•						
Boiler		•	•	•	•						•
School											
Entrance and exit			•	•					•	•	•
Classroom		•		•					•	•	
Toilet		•	•	•	•						
Gymnasium		•	•	•					•		
Cafeteria		•	•	•					•		
Stairwell		•	•	•					•		
Closet	•			•	•						•

Basic Door Designs

- Flush (F)
- Half glass (G)
- Narrow lite (NL)
- Vision lite (V)
- Bottom louver (L)
- Dutch door (D)
- Full glass (FG)
- Full louver (FL)

FIGURE 8.2-5 *Profiles of basic steel door frames and their parts.* *(Reprinted with permission of NAAMM, the National Association of Architectural Metal Manufacturers.)*

FIGURE 8.2-6 *Terminated stops.* *(Reprinted with permission of NAAMM, the National Association of Architectural Metal Manufacturers.)*

8.2.4

EXTRUDED ALUMINUM DOORS, FRAMES, AND STORE FRONTS

Most main entrance doors and their frames and packaged entrances and store fronts in commercial and institutional buildings are made from extruded aluminum components.

8.2.4.1

Design Requirements

Extruded aluminum doors, frames, and store fronts should be tested at the manufacturer's factory or at the facility of an independent testing agency to ascertain their resistance to deflection due to wind loads, air infiltration, and water penetration. These characteristics are determined by the effect on a door and frame, packaged entrance, or store front system when tested at specific levels of wind pressure. Since these elements are located on the ground floor, they are not subjected to the same high levels of wind pressure as may be windows and glazed curtain walls. The wind pressures used in the tests are, therefore, smaller.

Selection of a door, frame, packaged entrance, or store front begins with the type desired (stick framing or package, type of door, frame profile, etc.). Then the required design pressure is determined.

Design pressure is determined by the fastest wind speed likely to be encountered at a building's location. A map in ANSI A58.1, "Minimum Design Loads for Buildings and Other Structures," can be used to estimate this maximum wind speed. After the expected wind velocity has been determined, the resulting wind load to be used can be found in load tables in ANSI publications.

The assembly selected should have been tested using the expected wind loading and the procedures listed in ASTM Standard E 330, "Test Method for Structural Performance of Exterior Windows, Curtain Walls, and Doors by Uniform Static Air Pressure Difference." Permanent deflection normal to the plane of the unit should be no more than 0.2% of the clear span, and there should be no glass breakage or permanent damage to fasteners, anchors, or hardware. Deflection during the test should not exceed 1/175 of any span. This limit may be decreased to 1/360 of any span if a nonflexible finish, such as plaster, is to be used.

The water resistance of an assembly should be determined by testing it in accordance with ASTM Standard E 331, "Test Method for Water Penetration of Exterior Windows, Curtain Walls, and Doors by Uniform Static Air Pressure Difference," and ASTM Standard E 547, "Test Method for Water Penetration of Exterior Windows, Curtain Walls, and Doors by Cyclic Static Air Pressure Differential." The test pressure is usually 6.24 lb./sq. ft. For a unit to pass this test, no water may pass through the unit during the test.

Air infiltration should be tested in accordance with ASTM E 283, "Test Method for Rate of Air Leakage Through Exterior Windows, Curtain Walls, and Doors." The usual pressure is 1.57 lb./sq. ft.; the permitted infiltration is 0.06 cu. ft./min./sq. ft. of fixed area.

A *thermal break* separates the metal parts on the warm side of a frame from the metal on the cold side using an insulating material such as plastic (Fig. 8.2-7). This keeps heat from being conducted through the frame or sash. The insulating separator is sometimes called a *thermal barrier*.

Condensation resistance should be tested for frames having a thermal break feature. Condensation resistance should be determined by procedures outlined in AAMA 1503.1, "Voluntary Test Methods for Thermal Transmittance and Condensation Resistance of Windows, Doors, and Glazed Wall Sections." An assembly's condensation resistance gives an indication of that assembly's ability to resist condensation. The higher the condensation resistance factor, the less likely condensation is to occur.

As discussed in Section 8.1, heat will flow quite readily through doors, packaged

Thermal break
is formed by a
glazing gasket

Thermal break is
formed by an
imposed barrier

FIGURE 8.2-7 *Two typical frames with a thermal break feature. Many other configurations are also available.* *(Courtesy H. Leslie Simmons.)*

a   b

FIGURE 8.2-8 *(a) A stile-and-rail door; (b) a flush panel door.* *(Drawings by HLS.)*

entrances, and store fronts unless some means are taken to reduce this flow. The methods for determining conductive heat transmission through these assemblies is a part of AAMA 1503.1, "Voluntary Test Method for Thermal Transmittance and Condensation Resistance of Windows, Doors, and Glazed Sections."

8.2.4.2

Manufacture

The manufacture of extruded aluminum doors and packaged entrances consists of fabricating aluminum extrusions of the correct sizes and shapes and assembling them with glass, glazing materials, hardware, weatherstripping, screws, and other fasteners into finished, preassembled components.

8.2.4.3

Doors

Each manufacturer has its own standards for doors, even within the restraints of the standards listed above. However, there are only two basic types of swinging doors: (1) flush-panel doors and (2) stile-and-rail doors.

■ STILE-AND-RAIL DOORS

Aluminum stile-and-rail doors consist of a tubular extruded aluminum frame infilled usually with glass (Fig. 8.2-8a). Stiles may be thin (less than 1-3/4 in. wide), narrow (usually 2 in. wide), medium (4 in. wide), or wide (5 or more in. wide). The thickness of all is usually 1-3/4 in.

Thermal break doors are available, but most aluminum stile-and-rail doors are

standard. The usual glazing is 1/4-in.-thick glass, but most thermal break doors are glazed with 1-in.-thick insulating glass.

■ FLUSH-PANEL DOORS

Aluminum flush-panel doors consist of an extruded channel or angle frame and bracing members and sheet-aluminum facing panels (Fig. 8.2-8b). Most such doors are filled with a core made of a kraft paper honeycomb, rigid plastic foam, or mineral insulation. Faces may be either flat or embossed. Flush-panel doors may have glazed openings or louvers.

8.2.4.4

Frames and Store Fronts

Most packaged entrances and store fronts in commercial and institutional buildings are made from extruded aluminum sections.

Extruded aluminum frames come complete with accessories such as glazing gaskets, glazing stops, clips, trim, fins, flashings, anchors, and fasteners. Incidental components, such as closers, channels, flats, trim, sleeves, and the like, are usually delivered with the frames.

Sections should be accurately cut and fitted to make hairline joints. Where required, joints should be designed and erected to be watertight. Joints between two pieces of metal should be filled with a nonhardening sealant. Welds should be made on the inside so they will not be visible in the finished installation.

Frames are usually glazed with 1/4-in. glass, except where insulating glass is desired. Glazing stops and gaskets should be of the correct size to receive the glass to be used.

Frames are usually either tubes or channels (Fig 8.2-9). They are typically 1-3/4 or 2 in. thick and vary in depth from 4 to 7 in. Other sizes are available.

Frames may be either standard (see Fig. 8.2-9) or thermal break (see Fig. 8.2-7). They may be either flush glazed or with applied stops (see Fig. 8.2-9).

8.2.4.5

Interior Extruded Frames

Extruded aluminum frames are also used for interior doors. They are available in

Recessed glazing   Applied glazing stops

FIGURE 8.2-9 *Typical extruded aluminum entrance and store front frames. Many other types and shapes are available.* (*Courtesy H. Leslie Simmons.*)

a variety of sizes and shapes to suit varying conditions. The requirements for interior frames are similar to those for exterior frames, except that watertightness is not required.

8.2.5

FORMED STAINLESS STEEL AND COPPER-ALLOY DOORS, FRAMES, AND STORE FRONTS

Most stainless steel and copper-alloy doors and frames are custom-made formed shapes. Therefore, most doors and frame designs available for carbon steel doors are also available in stainless steel or copper alloy.

8.2.6

FINISH HARDWARE

Finish hardware for formed carbon steel, stainless steel, and copper-alloy doors and interior extruded aluminum doors is usually supplied by a hardware supplier. Extruded aluminum exterior doors and packaged entrances usually come equipped with all operating hardware except cylinder locks.

Hardware may be standard, handicapped access, automatic, or security type. Standard hardware includes all the devices necessary to operate, lock, and weatherstrip a door, including thresholds.

Handicapped-access hardware includes counterbalanced closers that permit easy operation of a door and specially designed door knobs with extended or oversized grips to permit easy operation by a handicapped person.

Automatic operators open doors when an electromechanical, hydraulic, or pneumatic operator is triggered by an actuating device. These devices include floor treadles, motion sensors, sonic sensors, push-buttons, photoelectric cells, and remote switches. Automatic hardware also includes smoke or fire detectors that automatically close doors when smoke or fire is present.

Security hardware consists of high-security cylinders and special control devices such as electromechanical, electromagnetic, or electronic locks and three-way locks in which a single key operates bolts in the head, jamb, and sill of a door. Security locks may operate by keys, push buttons, or card readers.

8.2.7

INSTALLATION

The finest door, frame, packaged entrance, or store front design and the highest quality of manufacture will not compensate for poor installation. Weathertightness, although simple to achieve, can be ensured only if the installation is done by experienced workers following the guidelines and instructions supplied by the manufacturer and the appropriate industry standards. A rough opening should be prepared to receive each unit in such a manner that it can be installed and will finish out square, plumb, level, straight, and true. The design of most assemblies provides for minor adjustments at the job site, but no unit will operate or weather properly if it is twisted and misaligned during installation.

A properly selected and manufactured door and frame or packaged entrance will be weather resistant as a unit. Properly assembled store front sections will also be weather resistant. Other things must be done, however, to finish with a completely weathertight installation. For example, if the construction is insulated, the space between the rough opening and the unit should be filled with insulation. In addition, the joints between units and adjacent materials must be closed with a sealant, trim members, or both to prevent the passage of air, dust, and water around the frame.

Door and packaged entrance frames and store front sections are seldom designed to support loads other than themselves. Therefore, unless the frame is specially reinforced to support other loads, a separate lintel or other support is usually required.

Door frames, packaged entrances, and store front frames should be supported around the entire perimeter and anchored securely to the supporting construction in a firm and rigid position. They should also be flashed at their heads and sills to provide a path back to the exterior for water that finds its way into the wall.

Some metals, such as aluminum and stainless steel, are compatible and can be used in contact with each other. Other metals, however, react with one another chemically or electrolytically and should be insulated from direct contact by waterproof, nonconductive materials, such as neoprene, waxed paper, or coated felt. Such dissimilar metals that are located where water passing over them may contact another surface should be painted to prevent staining. Refer to Section 5.2 for a discussion about the effect of dissimilar metal materials.

Concealed aluminum in contact with concrete, masonry, or an absorbent material, such as wood, paper, or insulation, should be permanently protected by coating either the aluminum or the adjacent material with a bituminous or aluminum paint or by coating the aluminum with a zinc chromate primer, to minimize the chance of chemical corrosion of the aluminum by acids, alkalies, and salts leached out of the adjacent materials. Creosote and tar coatings, which might damage the aluminum, should not be used.

Sealants should be used in locations where frames, packaged entrances, and storefront sections adjoin other materials to make these joints airtight and watertight.

8.2.7.1

Formed Frames and Doors

Formed door frames should be installed in accordance with SDI 105, "Recommended Erection Instructions for Steel Frames." Fire-rated frames should be installed in accordance with NFPA Standard No. 80, "Fire Doors and Windows."

When practicable, frames should be placed before the surrounding construction is built. Each frame should be anchored with jamb anchors near the top and the bottom and not more than 24 in. apart in between (Fig. 8.2-10). Jambs should be anchored at the floor with adjustable floor anchors (Fig. 8.2-11).

Anchors in masonry should be long enough to extend at least 8 in. into the masonry. Anchors for stud walls should be clips designed to attach to the types of studs used. Several different methods are used to anchor frames in previously placed masonry or concrete (Fig. 8.2-12).

Carbon steel and stainless steel frames in masonry should be filled solid with mortar. Copper-alloy frames, however, should not be filled with mortar, because the mortar will corrode the metal.

8.2.7.2

Extruded Aluminum Frames, Doors, and Store Fronts

Interior extruded aluminum frames should be installed in accordance with the requirements discussed above.

Exterior extruded aluminum doors, packaged entrances, and store front sections should be installed in accordance

FIGURE 8.2-10 *Common types of jamb anchors for conditions where the frames are installed before the wall.* *(Reprinted with permission of NAAMM, the National Association of Architectural Metal Manufacturers.)*

FIGURE 8.2-11 *Adjustable floor anchor.* *(Reprinted with permission of NAAMM, the National Association of Architectural Metal Manufacturers.)*

with AAMA's *Aluminum Store Front and Entrance Manual* and the manufacturer's recommendations.

8.2.7.3

Glazing

Glass selection depends on wind load requirements, thermal requirements, aesthetics, and safety. The type and thickness are often established by the applicable building code. For example, some codes require that glass in entrance doors and the lower portions of sidelights must meet the requirements for safety type and be in compliance with ANSI Z-97.1, "Safety Glazing Materials Used in Buildings," or another similar standard. Refer to Section 8.7 for a discussion of glass and glazing.

Where design calculations indicate the need to reduce the heat transmittance of store fronts, significant reduction can be accomplished by incorporating double or triple glazing. Where frame design permits, thermal breaks in the frames will also help. At any rate, these assemblies

FIGURE 8.2-12 *Anchorage of frames in previously placed openings.* *(Reprinted with permission of NAAMM, the National Association of Architectural Metal Manufacturers.)*

They are made from stainless steel, aluminum, and copper alloys. Most of them are between 6 and 8 ft. in diameter, but some are larger. Both three- and four-wing designs are available. Many options are available. Custom design is frequent.

Revolving doors operate either fully manually or with a power assist. All have speed controls for the safety of users and protection of the door itself. Their optimal speed is 10 revolutions per minute, but slower operation is often desirable in such applications as hospitals, airports, and train stations.

8.2.9

ADJUSTMENT, CLEANING, AND PROTECTION

Because doors and frames, packaged entrances, and store fronts are prefinished building components, their installation should be withheld as long as the construction schedule will permit.

Once installed, doors, frames, packaged entrances, and store fronts should be protected to prevent damage to their finish, appearance, and ability to operate properly. Units should be protected from splashing of plaster, mortar, and cleaning solutions.

when located where the winter design temperature is 10 °F or less should incorporate such devices regardless of calculations.

Glazing materials vary depending on the component being glazed. Formed door and frame units are usually glazed using stops and channel glazing compound or tape sealant or gaskets. Most extruded aluminum store front sections and doors are glazed using compressible gaskets and tape.

When windows or sliding glass doors are glazed at the job site, the manufacturer's recommendations should be followed to ensure a satisfactory result.

8.2.8

REVOLVING DOORS

Revolving doors are proprietary packaged units that come complete with hardware.

8.3 WOOD DOORS

Because of its ready availability, easy workability, and insulating properties, wood was an early material used in the manufacture of doors. Before the advent of plywood, doors were assembled from a number of solid-wood members. The methods used to conceal joints resulted in the familiar panel treatment of surfaces.

Later designs substituted thinner wood or plywood panels in a supporting framework to make interior panel doors and glass to make sash doors.

Mass production of plywood introduced the widespread use of flush doors. Laminating plywood faces to a particleboard, wood, hollow, or mineral core provides a dimensionally stable, economical door. Today, flush doors account for most wood doors used in all types of construction.

The following discussion outlines the variety of types and uses, methods of manufacture, standards, certification of quality, and installation of wood doors.

TYPES AND USES

Wood doors may be classified according to method of construction into (1) flush doors, which generally consist of plywood, hardboard, or plastic-laminate face panels bonded to solid or hollow cores; (2) stile-and-rail doors, composed of vertical and horizontal members enclosing wood, glass, or louver inserts, creating panel, sash, storm, screen, or louver designs; and (3) accordion folding doors, assembled from narrow slats, which give a long, drape-like appearance.

Figure 8.3-1 summarizes common door types, typical sizes, and applicable industry standards that regulate door construction and manufacture. Standards for various door types are based on construction grades, face material selection, core construction, vertical and horizontal edge, and core interface. Wood accordion folding doors are made of a variety of woods and are covered with vinyl, plastic laminate, wood veneer, fabric, and other materials. Plastic accordion folding doors may have a metal or a wood frame. Accordion folding doors are not currently covered by industry standards.

Flush Doors

In addition to providing the desired surface appearance, face panels of flush doors act as stressed skins to impart strength, rigidity, and stiffness to a door. Face panels may be hardwood veneer, plywood, high-pressure plastic laminate, or hardboard. Wood face panels are usually of two or three plies with a total thickness of at least 1/12 in. Door construction is described as 3-, 5-, or 7-ply by counting the total number of plies in both face panels and counting the core as 1 ply.

Flush doors are available with glazed lights and louver inserts. The U.S. Consumer Product Safety Commission (CPSC) "Safety Standard for Architectural Glazing Materials" requires that all glazing materials in doors be safety glazing material. In addition, industry standards limit the size of the cutout to a maximum of 40% of the door area (Fig. 8.3-2).

■ SOLID-CORE DOORS

Solid-core construction increases the dimensional stability of a door, provides thermal insulation, and may add sound isolation and fire resistance. Solid-core doors are commonly used as entrance and other exterior doors, but they may also be used where sound or fire ratings are required. Doors for which specific fire ratings are required must carry a fire door label from an approved testing agency, such as Underwriters Laboratory (UL). Fire ratings for wood doors range from 20 minutes to 1-1/2 hours, depending on materials and constructions. Many regular solid-core doors are capable of a 20-minute fire rating. Applicable codes should be consulted for specific requirements and applications.

Solid-core doors provide sufficient sound isolation for most uses. Where sound isolation is more critical, properly gasketed flush doors with a sound transmission class (STC) rating of up to 42 may be required.

Solid cores may be of wood, particleboard, mineral composition, or a combination of these materials. Wood cores are made of strips or blocks of wood forming a glued-block core, framed-block core, or stile-and-rail core (Fig. 8.3-3).

GLUED-BLOCK CORE Most solid-core doors have a glued-block core construction. They consist of wood blocks bonded together with the end joints staggered. The stiles and rails of these doors are bonded to the core, and the entire assembly is sanded to a smooth, uniform thickness. The face panels are bonded to the assembly and must be a minimum of 1/12 in. thick.

FRAMED-BLOCK CORE A framed-block core door is a stile-and-rail frame surrounding an interior core of wood blocks, which may or may not be bonded to each other, but which are bonded to the face panels. Face panels must be a minimum of 1/12 in. thick.

STILE-AND-RAIL CORE Stile-and-rail core doors consist of wood blocks bonded into stile-and-rail panel units. These units are assembled to make up the core of the door and are bonded to the face panels.

Again, face panels must be a minimum of 1/12 in. thick.

PARTICLEBOARD CORE Particleboard core doors are constructed of one or more pieces of particleboard, each the full thickness of the core, which may or may not be edge- or end-glued together. The stiles and rails of the door may or may not be bonded to the core, depending on the type of construction. Open spaces between core pieces or between the core and the frame are controlled by industry standards. Face panels must be a minimum of 1/12 in. thick.

MINERAL-COMPOSITION CORE Doors with a mineral-composition core normally are not available in other than fire-rated doors. These doors are constructed similar to particleboard core doors, except the core comprises inorganic, noncombustible materials formed into a rigid slab. Special blocking and framing normally are required in these doors to accommodate hardware, glazed openings, and other cutouts.

WOOD-BLOCK, LINED CORE Wood-block, lined core doors are a combination of the constructions described above. This type of door consists of a central wood block core with a 3/16 in. or thinner liner of particleboard or other material bonded to each face to make up the full core thickness. Face panels are then bonded to this assembly and to the stile-and-rail frame. These doors may provide added sound isolation, fire resistance, or other performance features, depending on the properties of the liner material.

■ HOLLOW-CORE DOORS

Flush doors with hollow cores are used for interior locations in residential and some light-commercial buildings. They are occasionally used as exterior doors in residential buildings and can serve satisfactorily in that use if bonded with waterproof adhesives and protected from the weather. They generally do not provide as much heat or sound insulation as solid core-doors, but can be used where these factors are not critical.

Face panels like those used on solid-core doors are available for use on hollow-

FIGURE 8.3-1
Summary of Wood Doors

Type and Industry Standard	Typical Sizes	Description
Flush NWWDA I.S.1-A, NWWDA I.S.1-R, AWI, WIC	*Hollow:* 1-3/8 in. and 1-3/4 in. thick 6 ft. 8 in. and 7 ft. 0 in. high 1 ft. 6 in.–3 ft. 0 in. wide *Solid:* 1-3/8 in., 1-3/4 in., and 2-1/4 in. thick 6 ft. 8 in. and 7 ft. 0 in. high 2 ft. 4 in.–3 ft. 6 in. wide	Constructed from *hollow* or *solid* wood or composition cores supporting face panels of wood veneers, hardboard, or plastic laminates. Presents flush smooth appearance unless interrupted by glass lights or louver openings.
Panel NWWDA I.S.6, AWI, WIC	1-3/8 in. and 1-3/4 in. thick 6 ft. 8 in. and 7 ft. 0 in. high 1 ft. 6 in.–3 ft. 4 in. wide	Assembled from stiles and rails (vertical and horizontal components), which frame and support one or more panels. Has characteristic paneled appearance resulting from raised and recessed areas, patterned molds, etc.
Sash NWWDA I.S.6, AWI, WIC	1-3/8 in. and 1-3/4 in. thick 6 ft. 8 in. and 7 ft. 0 in. high 2 ft. 0 in.–3 ft. 6 in. wide	Similar in construction and appearance to panel door, except that one or more panels are replaced with glass. Completely glazed doors without panels are *casement* (french) doors.
Storm and Screen NWWDA I.S.6 AWI, WIC	1-1/8 in. thick 6 ft. 7 in., 6 ft. 9 in. and 7 ft. 1 in. high 2 ft. 6 in.–3 ft. 0 in. wide	Ligher, thinner stile-and-rail construction supporting screening (*screen doors*), glass panels (*storm doors*), or interchangeable screen and storm panel inserts (*combination doors*).
Louver NWWDA I.S.6 AWI, WIC	1-1/8 in. and 1-3/8 in. thick 6 ft. 6 in., 6 ft. 8 in. and 7 ft. 0 in. high 1 ft. 3 in.–3 ft. 0 in. wide	Composed of stile-and-rail frame with integral louver construction, mortised into stiles or vertical dividing bars.
Accordion Folding (No industry standard)	6 ft. 8 in.–10 ft. 0 in. high Practically unlimited length	Assembled from narrow wood strips or single wood slats, 3-1/2 in. to 5 in. wide, with fabric, plastic, or metal hinges, resembling long drapeline doors.

FIGURE 8.3-2 *Flush doors with glass and louver inserts.* *(Courtesy National Wood Window and Door Association [NWWDA].)*

core doors. Wood face panels must be at least two plies and must have a total thickness of at least 1/10 in. The basic hollow-core constructions are mesh (cellular) core and ladder core (Fig. 8.3-4).

MESH (CELLULAR) CORE Mesh core doors consist of wood or wood-derivative material joined, interlocked, or woven to form a mesh or grid throughout the core area. Honeycomb cores of expanded paper or other material are the most common types of hollow cores. Face panels are bonded to the core and to the stiles, rails, and lock blocks of the door.

LADDER CORE Ladder core doors are made of wood or wood-derivative strips placed horizontally or vertically throughout the core area. Face panels are bonded to these strips and to the stiles, rails, and lock blocks of the door.

Lock blocks are provided in hollow-core doors for lock mortising and backset up to 2-5/8 in. Doors with special blocking may be obtained for other hardware mounting.

8.3.1.2

Stile-and-Rail Doors

Stile-and-rail doors derive their strength mainly from a supporting framework of vertical members (*stiles*) and horizontal members (*rails*) interlocked and glued together. Stiles and rails may be solid or veneered and are usually of softwoods.

The chief categories of stile-and-rail doors are panel, sash, louver, storm, screen, and combination.

■ PANEL AND SASH DOORS

Panel doors consist of stile-and-rail frames enclosing flat plywood or raised wood

FIGURE 8.3-3 *Typical solid-core constructions for flush doors: (a) glued-block core; (b) framed-block glued core; (c) framed-block nonglued core; (d) stile-and-rail core; (e) mat-formed composition-board core; (f) wood-block, lined core.* *(Courtesy National Wood Window and Door Association [NWWDA].)*

panel fillers. They are available in a variety of designs (Fig. 8.3-5) and may be used for interior or exterior application.

Sash doors are similar to panel doors in appearance and construction but have one or more glass lights replacing the wood panels. They are used often for main entrance and exterior service doors (Fig. 8.3-6). Casement (French) doors are fully glazed sash doors and are referred to as *rim* doors when they have a single large glass panel and as *divided light* doors when the glass is divided into smaller lights (Fig. 8.3-7).

Stiles and rails are assembled with glued dowel or blind mortise-and-tenon construction (Fig. 8.3-8a and 8.3-8b). Together with mullions, muntins, and bars, they frame the glass lights or wood panels (Fig. 8.3-8c). The contoured profiles, grooves, or rabbets (sticking) milled on the edges of these members hold the glass or wood panels (Fig. 8.3-8d).

■ LOUVERED DOORS

Wood louvered doors are made of horizontal slats in stile-and-rail frames. The

FIGURE 8.3-4 *Typical hollow-core construction for seven-ply flush doors: (a) mesh or cellular core; (b) ladder core.* *(Courtesy National Wood Window and Door Association [NWWDA].)*

FIGURE 8.3-5 *Panel door designs.* *(Courtesy National Wood Window and Door Association [NWWDA].)*

FIGURE 8.3-6 *Sash door designs.* *(Courtesy National Wood Window and Door Association [NWWDA].)*

slats usually are blind-mortised. They are available in several common designs and two basic slat styles, chevron and standard (Fig. 8.3-9).

Wood louvered doors are used in interior locations primarily as visual screens and where sound privacy is not required. Appropriate uses include closet doors, space dividers, and closures around residential laundry areas, where ventilation and dissipation of moisture are desirable.

■ STORM, SCREEN, AND COMBINATION DOORS

Storm, screen, and combination storm and

Rim   Divided Light
FIGURE 8.3-7 *Casement (French) doors.* *(Courtesy National Wood Window and Door Association [NWWDA].)*

screen doors are lightly constructed glazed doors used with exterior doors to improve weather resistance. Screen doors are of similar construction with screening, which provides ventilation while excluding insects. Combination doors combine the functions of both storm and screen doors with interchangeable screen and glass panels (Fig. 8.3-10).

Some self-storing combination doors are equipped with a screen and a pair of glass panels that operate in a vertical track similar to a double hung window. Either ventilation or weather protection is possible without removing the panel inserts, but the ventilating area available is limited to one half that of a full screen door. Glass jalousie inserts in combination doors have similar self-storing advantages with practically no loss of ventilating area. However, even when closed, jalousies permit a high rate of air infiltration and leak water in high-wind storm conditions.

8.3.1.3

Accordion Folding Doors

Wood accordion folding doors are made in several styles. One is made from either wood slats or wood strips connected with cord or fabric tapes, forming flexible drapelike doors (Fig. 8.3-11). These doors are generally designed to fold to a width less than the thickness of a normal interior partition.

Solid-wood slats usually are connected to each other with continuous metal, vinyl, or nylon fabric hinges. Woven doors are commonly of the basketweave type in which vertical strips of 3/8 to 1 in. wide are interwoven with nylon-reinforced vinyl tape. Corded doors are made with similar wood slats connected with cotton cord. Accordion doors usually are suspended from ceiling-mounted tracks and operate on nylon rollers or glides.

These doors are primarily visual screens for locations such as closets, space dividers, residential laundry enclosures, and where air circulation is desirable. They are not suitable where sound isolation or privacy are prime requirements. The folding operation also makes them useful where door swings are objectionable.

Other accordion folding doors are made with a framework or core of plastic, metal, or wood and a covering of vinyl, fabric, plastic laminate, or another plastic material.

FIGURE 8.3-8 *Typical construction of stile-and-rail doors: (a) glued dowel; (b) blind mortise-and-tenon; (c) shown together with mullions, muntins, and bars; (d) edges milled wtih contoured profiles, grooves, or rabbets.*

FIGURE 8.3-9 *Typical louver doors: (a) door designs; (b) slat styles.*

Screen   **Combination**

FIGURE 8.3-10 *Storm and screen doors give added weather protection and ventilation.* (Courtesy National Wood Window and Door Association [NWWDA].)

Solid   **Woven**   **Corded**

FIGURE 8.3-11 *Typical accordion folding doors.*

The folding doors discussed here should not be confused with *folding partitions*, which are similar to folding doors, but tend to be much sturdier. Folding doors, for example are usually limited to heights of about 14 ft. Folding partitions can be much taller. Some folding partitions are designed to provide both privacy and sound transmission control. These are usually constructed of wood or metal frames covered with plywood or hardboard and a wood, fabric, or plastic-laminate panel. They often have a plastic-foam or other solid core. Folding partitions are used where visual and sound privacy are needed, such as to separate meeting rooms or classrooms.

DOOR OPERATION

The operation of wood doors can be classified as swinging, sliding, or folding (Fig. 8.3-12). The use and the location of a door determine the most suitable manner of operation. Fully-glazed wood sliding glass doors are addressed in Section 8.5.

8.3.2.1

Swinging Doors

Swinging doors typically operate on hinges secured to the side jambs, although they may also operate on pivots supported by the door frame head and the floor. The majority of wood swinging doors are hinged (Fig. 8.3-12a). Either pivot hardware or double-acting hinges are used for double-acting swinging doors (doors that swing in both directions).

Hinged operation provides the greatest degree of security, weather resistance, heat insulation, and sound isolation. Where these features are particularly important, doors may be weatherstripped or gasketed for increased effectiveness.

Swinging door operation is most convenient for frequently used entry and passage doors but may be used in any location unless door swing into usable space is objectionable.

Door swing is either right hand, left hand, right hand reverse, or left hand reverse. Assume that the viewer is standing outside a door (on the exterior of an entrance door, in the corridor at a room door, or on the side opposite the hinges on a connecting door). A door that swings away from the viewer with the hinges on the right side is a right-hand door; the same door swinging toward the viewer is a right-hand-reverse door. When the hinges are on the left and the door swings away from the viewer, it is a left-hand door; if it swings toward the viewer, it is a left-hand-reverse door.

8.3.2.2

Sliding Doors

Sliding doors are suspended on overhead tracks and operate on rollers attached to the door. These rollers may be nylon or

FIGURE 8.3-12 *Door operation: (a) swinging; (b, c, d) sliding; and (e, f) folding. Surface sliding doors (c) are used mostly in industrial and agricultural (usually metal prefab) buildings. They are seldom found in modern commercial, institutional, or residential buildings.*

another plastic or metal with plastic bearings. Floor guides usually are provided to prevent sliding doors from swinging laterally. Rarely, sliding doors are mounted on rollers operating on floor tracks.

Sliding wood doors are used on closets and as room dividers in all kinds of buildings and in storage areas in residential construction, where they are suitable for visual screens. Sliding doors used on closets and other storage areas are usually of the bypass type (Fig. 8.3-12b), but pocket doors are also sometimes used in these locations. Closet sliding doors are often louvered but may also be flush or panel doors.

Surface sliding and pocket sliding doors are more likely to be found as passage doors (Fig. 8.3-12c and d). Biparting doors (two doors sliding in opposite directions) are used for openings 3 ft. or wider and may be either surface or pocket type. For extremely large openings, multitrack installations permit stacking of several sliding doors at one or both ends. Surface and pocket sliding doors are usually either flush or panel doors but may also be louvered.

Sliding operation ordinarily does not give a high degree of privacy, sound isolation, or weather resistance, unless custom hardware or preengineered door units are used. The chief advantage of sliding operation is that it eliminates door swings, which might interfere with the use of interior space.

8.3.2.3

Folding Doors

Folding doors are hung on overhead tracks with rollers similar to those used on sliding doors. Some types also require a floor track. Folding operation, like sliding, is generally appropriate for locations requiring primarily visual screening.

Flush, louver, and panel doors are often used for closets in single or paired bifold arrangements (Fig. 8.3-12e) or as space dividers and movable partitions with special hardware for multifolding action and stacking (Fig. 8.3-12f). Accordion folding doors are used as closet doors or as space dividers and may be pocketed or end-stacked within the opening.

8.3.3

MANUFACTURE

The following discussion of door manufacture presumes conformance with industry standards.

8.3.3.1

Materials

Some common species of hardwoods for door manufacture are oak, ash, elm, birch, butternut, cherry, maple, mahogany, walnut, basswood, and lauan. Softwood species are ponderosa (western) pine, Douglas fir, sitka spruce, and western hemlock.

The majority of flush doors are made with hardwood veneered faces. Most stile-and-rail doors are manufactured from ponderosa pine, but some also are produced from other softwoods. Hardwood veneered stile-and-rail doors and softwood veneered flush doors are available, though not common. Accordion folding doors are made from both hardwoods and softwoods.

The lumber used in door manufacture is kiln dried to a moisture content of 12% or less before being assembled into a door.

The solid wood core for a flush door should contain a single species of wood that has a specific gravity of 0.42 or less. Composition core materials range in density up to about 37 lb./cu. ft. at the same moisture content.

Doors for exterior use should be made with waterproof (Type I) adhesives, such as thermosetting melamine urea or phenolic resins. Interior doors are made with water-resistant (Type II) adhesives, such as casein (sometimes fortified with urea resins and formaldehyde), capable of resisting a limited number of wetting cycles.

8.3.3.2

Fabrication and Assembly

Fabrication varies somewhat depending on the type of door.

■ FLUSH DOORS

Stiles, end rails, lock rails, and lock blocks for hollow-core flush doors are cut from kiln-dried, surfaced lumber. Stiles and rails are joined either with corrugated metal fasteners or by machined dovetail joints. The stile-and-rail frame is laid up in adhesive with one face panel. Cores, lock blocks, and sometimes lock rails are then set and glued in place. Cores may be ladder, mesh, or implanted blanks (Figs. 8.3-13 and 8.3-14). Assembly is completed by bonding the second skin to the frame and the core.

Hollow-core flush doors usually are bonded with water-resistant adhesives that do not require heat to set. These doors are cold pressed (Fig. 8.3-15) and allowed to

FIGURE 8.3-13 *Implanted-blanks core for hollow-core door is made by feeding wood spirals into an assembled frame.*

FIGURE 8.3-14 *Cellular-core construction is made by gluing paper honeycomb into an assembled frame.*

cure overnight, prior to trimming and further working.

Most solid-core doors are made with *thermosetting* adhesives. Continuous block cores are assembled from wood blocks, glued together, and cured. After additional air drying for 24 hours, face panels are adhesive-bonded to the core and the entire assembly is *hot pressed* to set the glueline (Fig. 8.3-16).

Further operations on both solid- and hollow-core doors may include routing for louver and glass openings and installing muntins and bars prior to glazing (Figs. 8.3-17 and 8.3-18).

Glass and louver openings in flush doors should not be more than 40% of the

FIGURE 8.3-15 *Hollow-core flush doors are cold pressed to ensure proper bonding of face panels to core.*

door area, and openings should not be closer than 5 in. from the door edge. The height of openings in a hollow-core door should not exceed half the door, and face skins should be supported with blocking, glued in place, to frame the opening.

■ STILE-AND-RAIL DOORS

After surfacing kiln-dried lumber to exact thickness, boards are cut to proper widths and lengths. Pieces for stiles are bored for dowels on a boring machine (Fig. 8.3-19); a molding machine shapes the sticking on stile and rail edges (Fig. 8.3-20); and a double-end tenoner cuts rails to exact lengths, coping both ends to fit the stacking on the stiles. Another machine bores holes in rails, inserts glue, and drives hardwood dowels into the holes (Fig. 8.3-21).

For panel doors, wood or plywood panels are cut to size and may be shaped to form raised panel designs (Fig. 8.3-22). Stiles and rails are assembled with water-resistant adhesives around the panels and clamped in a jig. When the glue has set, the doors are sanded on both sides to a smooth finish (Fig. 8.3-23).

Stile-and-rail frames and sometimes panels, muntins, and bars for sash doors

FIGURE 8.3-16 *Hot pressing sets the glue line of solid-core doors bonded with thermosetting adhesives.*

FIGURE 8.3-17 *Flush doors are routed for glass lights and louvers.*

are fabricated and assembled in a similar manner. Glass is cut to size and bedded in glazing compound, and stops are accurately mitered and nailed in place (Fig. 8.3-24).

Prefabrication

Doors may be further worked and finished in the plant in varying degrees, to reduce job-site preparation and installation. They may be seal-coated to prevent moisture absorption and soiling or be completely surface-finished. Standard doors and pre-finished custom doors are usually prefit to exact opening dimensions so that the finished edges will not be removed by job-fitting. Doors for prefabricated frames sometimes are machined for locks and hinges (Fig. 8.3-25), and bifold doors may have all operating hardware installed.

Plastic laminate–surfaced doors are usually prefabricated and prefit.

Prehung door units are available assembled complete with frames, trim, and hardware. The stop may be part of the frame or left loose to be applied during installation. Frames are braced at jamb bottoms and carefully packed to prevent racking and distortion in shipment.

Finishes

Door finishes may be opaque or transparent. *Opaque* finishes are essentially paint. There are several types of *transparent* finish systems used on doors, including lacquers, conversion varnishes, catalyzed vinyls, polyesters, polyurethanes, and oils.

Factory finishes are graded as standard, select, or super.

■ STANDARD FINISHES

Standard finishes are simply the manufacturer's standard. They usually include lacquers, conversion varnishes, and catalyzed vinyls.

■ SELECT FINISHES

Select finishes are similar to standard finishes but permit the designer to select colors and sheens. Select finishes are usually more expensive than standard finishes.

■ SUPER FINISHES

Super finishes are generally superior to standard and select finishes. They include polyester, polyurethane, and oil finishes. A designer may select color and sheen. Super finishes are usually more expensive then either standard or select finishes.

STANDARDS OF QUALITY

The National Wood Window and Door Association (NWWDA), The Architectural Woodwork Institute (AWI), and the Woodwork Institute of California (WIC) have developed standards applicable to wood doors. The applicable standards are as follows:

Flush doors	NWWDA, I.S.1-A, "Architectural Wood Flush Doors"
	NWWDA, I.S.1-R, "Residential Flush Doors"
	AWI, *Architectural Woodwork Quality Standards, Guide Specifications, and Quality Certification Program*, Section 1300, "Architectural Flush Doors"
	WIC, *Manual of Millwork*

FIGURE 8.3-18 *Muntins and bars are fitted and installed before glazing.*

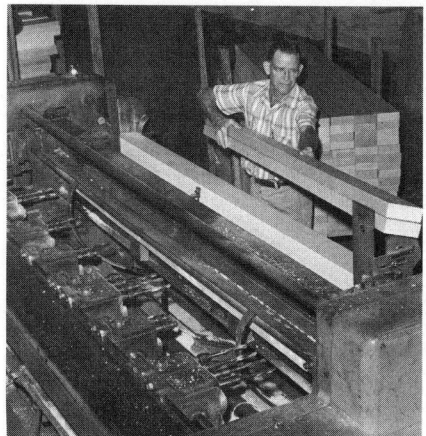

FIGURE 8.3-19 *For stile-and-rail doors, holes for dowels are bored in stiles automatically fed from a hopper.*

Stile-and-rail doors	NWWDA, I.S.6, "Wood Stile and Rail Doors"
	AWI, *Architectural Woodwork Quality Standards, Guide Specifications, and Quality Certification Program*, Section 1400, "Stile and Rail Doors"
	WIC, *Manual of Millwork*

These publications are the recognized standards of quality for the wood door industry. They are similar but vary slightly in detail. The AWI and WIC standards act as supplements to the NWWDA standards. Before making choices related to door selection and design, the designer should examine the requirements of the selected standard and not rely solely on citing the name of the standard preferred. It may be necessary to modify the standards to suit a particular project. For example, one standard may require a particular face veneer thickness for one facing grade, while another standard may require a different minimum thickness or simply be silent on the issue.

In addition to establishing minimum requirements for material, design, construction, and waterproof performance under test conditions, these standards also provide a basis for quality grading of wood doors.

8.3.4.1

Grades

Doors are graded primarily according to their exposed face veneers. However, the standards also include in the grades requirements for vertical edge bands, pair matching, transom matching, beads for light and louver openings, and quality of workmanship. The grades limit the materials that may be used in each grade based on the type of finish to be used (opaque, transparent, or laminate). Flush doors are also limited by core construction.

■ WOOD FLUSH DOORS

Solid- and hollow-core flush doors are available in premium, custom, and economy grades, depending on the quality of face panels. Flush doors are not recommended for exterior use. Interior doors may be made with either *waterproof* (Type I)

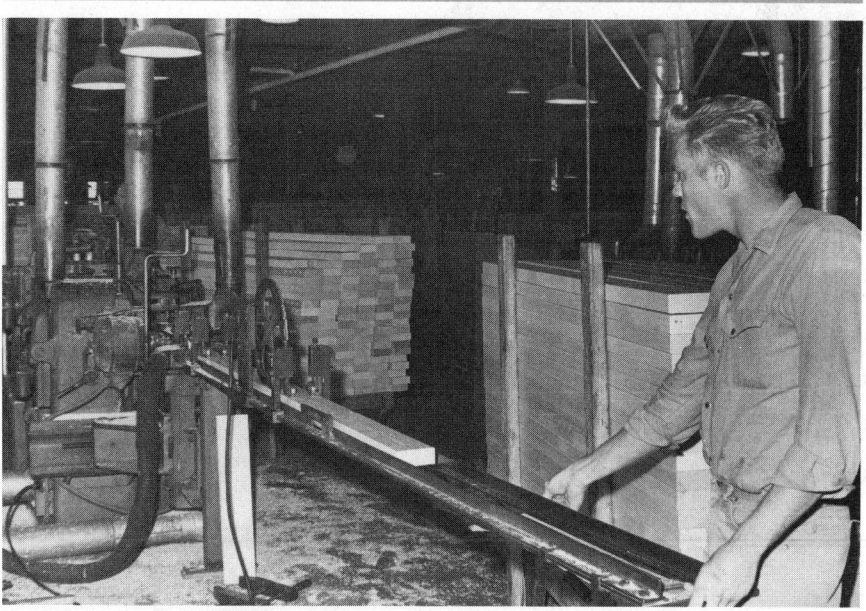

FIGURE 8.3-20 *An endless feeder chain carries rails or stiles through a molding machine to form sticking on edges.*

FIGURE 8.3-21 *A single machine bores holes, inserts glue, and drives dowels.*

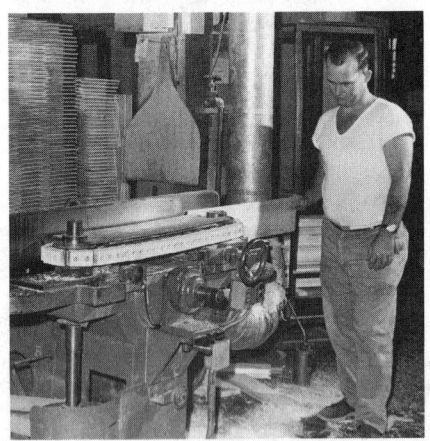

FIGURE 8.3-22 *After being cut to size, panels are shaped to make raised designs.*

or *water-resistant* (Type II) adhesives, depending on their location. Doors in wet or humid areas should be made using Type I adhesives.

PREMIUM GRADE Premium-grade face panels are available for transparent, opaque, and laminate finishes. This is the highest grade available from commercial stock.

For opaque finishes, premium-grade doors have *medium-density overlay* plywood faces.

For transparent finishes, the face veneers are tight and smoothly cut and free from knots, worm holes, splits, shakes, and torn grain. When the face consists of more than one piece, the pieces are of approximately equal widths and are matched at the joints for color and grain. Several different matching patterns are used. There are many species used for transparent finishes (see Section 8.3.3.1).

Laminate faces and edges are nominal 0.050 in.-thick high-pressure decorative laminate. The laminate may be solid or have a grain or texture.

CUSTOM GRADE Flush doors with custom-grade face veneers are also intended for either transparent or opaque finishes. They are intended for use in high-quality projects.

For an opaque finish the face veneer may be any close-grained hardwood. Grain pattern may show through and split veneer joints and other wood characteristics may show through the finish unless special precautions such as filling are taken to prevent this.

For transparent finishes, the face panels for custom-grade flush doors are made of veneers essentially similar to those used in premium-grade doors, but they need not be matched at joints for color and grain if sharp contrasts between adjacent pieces are avoided.

Laminate faces are nominal 0.050-in.-thick high-pressure decorative laminate. The laminate may be solid or have a grain or texture. Edges are close-grained hardwood.

ECONOMY GRADE The face veneers in economy-grade flush doors are intended for painting and may contain a variety of color, grain, and surface defects that will not be visible after two coats of paint have been applied. Their selection is generally left up to the mill producing the doors. Some manufacturers make economy-grade doors with hardboard face veneers.

FIGURE 8.3-23 *Stile-and-rail doors are sanded to a smooth finish.*

FIGURE 8.3-24 *After glazing, accurately mitered stops are nailed in place.*

FIGURE 8.3-25 *Doors are mortised for locksets on an automatic router.*

There is no economy grade for laminate-finish doors.

■ STILE-AND-RAIL DOORS

Grades for stile-and-rail doors are *premium*, sometimes called *select*, and *standard*, also called *custom*. Both grades typically are made with water-resistant adhesives and are often treated with water repellents.

PREMIUM Premium-grade stile-and-rail doors are made from stock practically clear of defects. They are intended to receive a transparent finish.

STANDARD Standard-grade stile-and-rail doors are made from lumber stock that may contain stains and various other minor defects. They are intended to receive a paint finish.

8.3.4.2

Certification of Quality

Doors that conform to the industry standards may be grademarked or labeled by the manufacturer, when requested by the buyers. *Grademarks* usually are stamped or branded on the edge of a door; *labels* are written statements attached to the door and containing similar information to grademarks.

NWWDA administers an independent certification program for hardwood-veneered doors based on continuous, impartial inspection and testing of doors. A NWWDA seal ensures that the materials, construction, and manufacture of doors conform to applicable industry standards. A red plastic plug in the top edge of a door indicates a door bonded with waterproof exterior Type I adhesives.

Exterior doors should be treated with a water-repellent preservative for increased dimensional stability and resistance to decay, fungus, and insect attack. The NWWDA administers a certification program for preservative-treated millwork products that includes softwood doors. The NWWDA seal is assurance of adequate preservative treatment in accordance with NWWDA I.S.4, "Industry Standard for Water-repellent Preservative Treatment for Millwork."

8.3.5

INSTALLATION

The satisfactory performance of doors depends on proper handling, storage, hanging, and finishing.

8.3.5.1

Handling and Storage

Unfinished doors should be handled with clean gloves to prevent soiling. They should be stored flat (not on edge) on a level surface in a clean, dry, well-ventilated space. They should not be stored in excessively hot, dry, or humid areas and should be conditioned for several days to the average prevailing local humidity before hanging or finishing. If doors appear warped, they should be stacked flat under uniformly distributed weights to restore flatness.

8.3.5.2

Hanging

Finished door frames should be square and plumb, and doors should be fitted in the openings with a total clearance of approximately 3/16 in. in both horizontal and vertical dimensions to prevent binding under humid conditions. Doors may be trimmed for fitting, but should not be cut to smaller nominal sizes. If a substantial amount of the perimeter is removed, the construction balance of the door could be destroyed, and warping could result.

■ HARDWARE

Exterior doors and interior solid-core, mineral-core, and other types 7 ft. or more in height should be hung with a minimum of three hinges per door, aligned so the door will not twist out of shape. Interior hollow-core doors less than 7 ft. in height may be hung with two hinges.

■ FLASHING

When exterior doors swing out with no overhead protection from the elements, the top edge of the door should be protected with flashing.

8.3.5.3

Finishing

Unfinished wood doors should receive a seal coat of the finish to be used or a suitable sealer as soon as they are delivered to the job site, to prevent undue moisture absorption and soiling. Finishing should be delayed until doors have been trimmed and fitted in the openings. Door surfaces, including both faces and top, bottom, and side edges, should be sealed against moisture with two coats of paint, varnish, or sealer before final hanging.

Door surfaces should be clean and dry before a finish is applied. Soil marks and other surface defects should be removed by light sanding. Doors should not be finished in excessively humid weather. The proposed finish should be tested on a sample of the same species of wood to determine its appearance and suitability.

8.4 ALUMINUM WINDOWS AND SLIDING GLASS DOORS

The use of aluminum for window construction has increased significantly since aluminum windows were introduced for industrial application in the United States about 50 years ago. In contemporary buildings, glass areas have increased—the size of aluminum windows allows for entire walls of glass and, in the case of aluminum sliding glass doors, movable walls of glass.

This section covers prime and architectural aluminum windows and aluminum sliding glass doors. Aluminum entrances and store fronts are addressed in Section 8.2; aluminum storm windows and doors in Section 8.6; glazed aluminum curtain walls in Section 8.8; and wood windows and sliding glass doors in Section 8.5.

8.4.1

MANUFACTURE

The manufacture of aluminum windows and sliding glass doors consists of fabricating wrought aluminum alloys into the correct size and shape, using various techniques of die cutting, punching, notching, piercing, and sawing. The aluminum members thus fabricated are assembled with glass, glazing materials, hardware, weatherstripping, screws, and other fasteners into finished preassembled components.

The workability of aluminum coupled with the ease and simplicity with which aluminum windows and sliding glass doors can be manufactured have contributed immeasurably to their widespread availability and use. Modern machinery and production techniques produce a finished product whose parts can be fabricated to extremely close tolerances, resulting in weathertight assemblies having precision operation at relatively low cost.

8.4.1.1

Materials

Pure aluminum has few commercial applications, but when it is combined with other metals—manganese, silicon, iron, magnesium—aluminum alloys are formed, making it useful in a variety of applications. Refer to Section 5.2 for additional information about aluminum.

The principal materials used in aluminum windows and sliding glass doors are wrought (worked into shape by mechanical means) alloys. The alloys used for windows and sliding glass doors must possess strength, be resistant to corrosion, be readily worked and machined for efficient fabrication and simple assembly, and provide surfaces suitable to receive protective and decorative finishes. Although other alloys are used, the wrought aluminum alloy known as 6063 is a common one for making aluminum extrusions. It contains silicon and magnesium and possesses good formability, corrosion resistance, and medium strength. To make aluminum products, the aluminum alloy is formed by the extrusion process and artificially aged by heat treatment. After heat treatment, usually in temper designation T5 (or T6), this alloy, known commercially as 6063-T5 (or 6063-T6), is commonly used for the main frame, sash, and ventilator members in both windows and sliding glass doors. Extruded shapes permit infinite possibilities of cross-sectional configuration to satisfy the requirements of glazing, weatherstripping, condensation control, and assembly.

Window and door locks, hinges, handles, and corners are often made from castings (formed by pouring or forcing molten metal into a mold) of aluminum alloy.

Other aluminum wrought alloys made by roll forming (passing strips of flat aluminum sheet through a series of forming rolls that bend it into a tube or other shape) also have a variety of uses in aluminum window and sliding glass doors. These include such shapes as screen frames and glazing strips. Aluminum sheet is also cut and formed by stamping in a punch press, where accessories such as corner locks and screen clips can be made.

8.4.1.2

Finishes

The silvery sheen of the natural, mill-finish surface of aluminum is a common finish used on low-cost aluminum windows and doors. This natural oxide coating, which forms on aluminum provides an adequate degree of protection against corrosion in normal atmospheres and can be readily restored by rubbing with stainless steel wool if damaged or discolored. Normal oxidation over time will convert the silvery sheet to a soft gray. Other finishes often used may be either protective or decorative.

■ PROTECTIVE FINISHES

Protective finishes are necessary when abnormal atmosphere is present, such as found in heavy industrial and seacoast areas. Protective finishes are also necessary over certain types of decorative finishes, particularly where the surface is etched or polished.

Anodizing, an electrolytic process that provides a heavy oxide film on the surface, is a permanent protective finish against moderate abrasion and atmospheric attack. Clear anodizing, which usually has a frosty appearance when an etching material is used during the cleaning cycle, is the most economical finish when corrosion protection is the primary concern.

Certain permanent color tones on aluminum are possible using the anodizing process. These include several shades of gray, bronze, gold, and black. They are produced by varying the alloy constituents, the electric current, the electrolyte, and the time cycle. Practically any color is possible by impregnating the porous surface of an anodic film with dye prior to sealing. But since dye colors are not always permanent, they are not recommended for exterior use. (See Section 5.2 for a discussion about anodizing and other finishing processes.)

Organic coatings, such as paint, that are often applied to aluminum are generally for decorative purposes, but they may also impart a certain degree of protection (see Section 9.11).

■ DECORATIVE FINISHES

Decorative finishes are colored or opaque coatings (such as a backed synthetic resin enamel) similar to those used on aluminum siding. These coatings can be expected to have a life of up to 20 years when the coating is applied and baked on at the factory under controlled conditions. Organic coatings should meet AAMA 603.8, "Voluntary Performance Requirements and Test Procedures for Pigmented Organic Coatings on Extruded Aluminum."

8.4.1.3

Fabrication

Fabrication of aluminum alloy sections into component parts of a window or sliding glass door is accomplished by sawing, drilling, punching, and other traditional methods of metal working.

One of the machines used most is the small-punch press. Aluminum's machinability permits the use of a variety of intricate punches and dies in the press to cut, form, shape, notch, and pierce the metal. Thus, components can be joined accurately and easily to make weathertight assemblies. A multiple-punch press can cut a long length extrusion to the proper length and punch the required notches, holes, and slots, producing a complete framing member in a single operation.

8.4.1.4

Assembly

Materials other than aluminum, such as glass, plastics, and steel, are required to complete the assembly of windows and sliding glass doors.

■ GLASS AND GLAZING MATERIALS

Many aluminum windows and most sliding glass doors have the glass installed at the factory, which eliminates glazing on the job site and allows controlled methods of glazing not feasible with on-site glazing.

Glass selection by size and grade for windows depends primarily on wind load requirements and is often established by the applicable building code. In the absence of code requirements, recommendations of ANSI/AAMA 101, "Voluntary Specifications for Aluminum Prime Windows and Sliding Glass Doors," can be used. For sliding glass doors, in addition to wind load requirements, the glass must meet requirements for safety type and be in compliance with ANSI Z-97.1, "Safety Glazing Materials Used in Buildings," or other similar standards as regulated by codes or laws.

Other functional or aesthetic requirements often dictate the selection of different types and higher grades of glass. These include double-glazing to minimize heat loss and heat gain; tinted glass for privacy and decorative purposes; tempered glass for additional strength; and tempered, wire, or laminated glass for additional safety.

Glazing materials for both aluminum windows and sliding glass doors include gaskets made from neoprene or polyvinyl chloride and mastic-type compounds such as butyl- or rubber-base materials that are compatible with aluminum and do not require painting.

■ WEATHERSTRIPPING

The function of weatherstripping is to control air infiltration, prevent rain penetration, minimize dust infiltration, and withstand atmospheric conditions.

Weatherstripping must be durable under extended usage, resist corrosion, and be readily replaceable. Many aluminum windows and sliding glass doors are equipped with fabric-pile weatherstripping (similar to that used in automobiles) or with flexible vinyl plastic neoprene gaskets or fins. In some cases these materials are combined. Extremely thin stainless steel, which compresses to accomplish a seal, is sometimes used.

■ FASTENERS AND HARDWARE

A variety of fasteners and hardware are required to hold window and sliding glass door parts together and to make them operable. Although aluminum is used extensively for this purpose, other materials may also be desired or required. Other metals used with aluminum must be either compatible with aluminum or insulated from direct metal-to-metal contact with the aluminum. Otherwise, electrolytic corrosion (galvanic action) may result. Therefore, when mild steel fasteners are used, they must always be plated with zinc, cadmium, or another suitable material. Stainless steel need not be plated and can be used in exposed locations if it is of the nonmagnetic type and has a chromium content of not less than 16%. Nylon, polypropylene, and other plastic materials are used for slides, bushings, bearings, and similar contact surfaces.

■ FRAME ASSEMBLY

The main frame, sash, and ventilators of aluminum windows and sliding glass doors are assembled by screwing or welding.

Screwed assemblies are shipped from the factory KD (knocked down) and later assembled by a distributor, dealer, or contractor. A slot to receive the screw threads is often extruded as an integral part of the aluminum member and holes are prepunched to ensure perfect alignment of the corner joints (see Fig. 5.2-24). Lower corners of the main frame are usually sealed with a mastic to make them watertight. Units that are completely assembled at the factory are either staked or welded together.

Welded assemblies utilize several welding methods. The flash (butt) welding method produces a continuous weld across the entire cross-section of a shape and is made by forcing two members together at their intersection under pressure and applying a voltage that melts the parent metal. Flash welding uses no filler material or flux and forms a rigid, strong, neat, watertight joint.

The inert-gas welding method uses filler material and is similar to conventional gas welding used on other metals. It is seldom continuous and is generally used to join members that are of dissimilar shape. While joints made by inert-gas welding can be made reasonably watertight, the use of a mastic sealing material at the lower corners may be required to prevent water leakage.

■ ACCESSORIES

Window and door accessories are assembled and attached during all stages of fabrication and assembly. For example, weatherstripping frequently is pulled into retaining grooves in long lengths of extruded frame members before cutting and punching take place. Operating hardware such as sash balances, pivot arms, guides, and guide followers are usually assembled after the main frames are together. Accessories such as locking hardware and handles are attached by riveting, bolting, screwing, staking, or interlocking with frame members. Manual rotating operators on casements, awning, and jalousie windows are usually the last accessories to be assembled, sometimes being left until after installation of the unit at the job site.

■ GLAZING

In one method of factory glazing, a glazing compound is applied to the glazing leg of a window as a back-bedding material; the glass is then dropped in and held in place by aluminum or rigid plastic face stops. The stops snap into place to serve as trim molding and to hold tension on the assembly until the glazing compound has set. Another method widely used in the factory to glaze units is to apply a channel-shaped gasket over the edge of the glass and then force the channel shape into the window or door frame around it. Flexible vinyl plastic is most often used for this application.

When windows or sliding glass doors are glazed at the job site, the manufacturer's recommendations should be followed to ensure a satisfactory result.

8.4.2

STANDARDS OF QUALITY

Two types of aluminum windows are covered in AAMA documents: *prime windows* and *architectural windows*. Sliding glass doors are covered in the same documents as prime windows.

8.4.2.1

Prime Windows and Sliding Glass Doors

Prime windows constitute most of the windows used in most low-rise buildings. Their design and construction and that of aluminum sliding glass doors should conform to ANSI/AAMA 101, "Voluntary Specifications for Aluminum Prime Windows and Sliding Glass Doors," developed by the Architectural Aluminum Manufacturers Association (AAMA) and adopted by the American National Standards Institute (ANSI). This document is the recognized standard for the aluminum window and sliding glass door industry and includes minimum provisions for frame strength and thickness, corrosion resistance, air infiltration, water resistance, wind load capability, and other requirements necessary to ensure adequate performance.

■ TYPES

Various types of aluminum windows and doors covered by ANSI/AAMA 101 are illustrated and described in Fig. 8.4-1.

■ DESIGNATION SYSTEM

Prime aluminum windows and sliding glass doors are given designations that describe the type of window or door, its grade, its design pressure, and other information. For example, DH-DW-R-15 conveys the following information: the first letters describe the window or door type (DH means double-hung window, C means casement window, SGD means sliding glass door, etc.); the DW signifies a dual window; the R means residential grade (C means commercial grade, HC means heavy-commercial grade); and the number designates the design pressure (DP).

■ PERFORMANCE REQUIREMENTS

AAMA-certified windows and sliding glass doors are required to have certain performance characteristics. These characteristics are determined by the effect on the window or sliding glass door when tested at specific levels of design pressure (DP), structural test pressure (STP), water resistance test pressure (WTP), and air infiltration test pressure and rate of infiltration.

Condensation resistance and heat transmission are two other factors that should be taken into account in the selection of

FIGURE 8.4-1
Summary of Aluminum Windows and Sliding Glass Doors

Type	Designation and Application	Description
Double Hung	DH-R15 Residential DH-DW-R15 Dual Window Residential DH-C20 Commercial DH-DW-C20 Dual Window Commercial DH-HC40 Heavy Commercial DH-DW-HC40 Dual Window Heavy Commercial	*Double-hung windows* have two operable sash. The sash move vertically within the main frame with the assistance of mechanical balancing mechanisms that minimize the effort required to raise the sash. 50% of the window area is available for ventilation.
Casement	C-R15 Residential C-C20 Commercial C-HC40 Heavy Commercial	*Casement windows* contain side hinged ventilators that operate individually and swing outward. They frequently have nonoperating fixed lites (shown) which the ventilators close to. Casements may be open 100% for ventilation.
Projected	P-R15 Residential P-C20 Commercial P-HC40 Heavy Commercial	*Projected windows* consist of horizontally mounted ventilators that may project out or in. They differ from awning and hopper windows in that the hinged side of the ventilator moves in a track up or down when the ventilator is operated. Used commonly where large glass areas are desired and where it is desired to deflect incoming air in an upward direction.
Awning	A-R15 Residential A-C20 Commercial	*Awning windows* consist of a number of top hinged ventilators operated by a single control device that swings the lower edges outward giving an "awning" effect when open. Operation is through a roto operator which also serves as the lock. Ventilating area is 100% of the window area.
Horizontal Sliding	HS-R15 Residential HS-DW-R15 Dual Window Residential HS-C20 Commercial HS-DW-C20 Dual Window Commercial HS-HC40 Heavy Commercial HS-DW-HC40 Dual Window Heavy Commercial	*Horizontal sliding windows* have one or more operable sash arranged to move horizontally within a main frame. Ventilating area is 50% of the window area. This type of window is sometimes combined with fixed lites and constructed in large sizes reaching the proportion of a "window wall."

FIGURE 8.4-1
Summary of Aluminum Windows and Sliding Glass Doors *(Continued)*

Type	Designation and Application	Description
Jalousie	J-R15 Residential	*Jalousie windows* (louver windows) consist of a series of overlapping horizontal glass louvers that pivot together in a common frame. Ventilating area is 100% of the window area. Jalousies are not as resistant to air and dust infiltration as other types, but are used in moderate climates and as porch enclosures in all climates. They should be used only where large heat losses or heat gains can be tolerated. Insulating effectiveness may be improved by inside storm panels, designed to be interchangeable with screens.
Jal-Awning	JA-R15 Residential	*Jal-awning windows* resemble an awning window in appearance but use different operating and locking mechanisms. These mechanisms are usually separate and require individual operation. Ventilating area is 100% of the window area.
Vertical Sliding	VS-R15 Residential VS-DW-R15 Dual Window Residential	*Vertical sliding windows* resemble double hung windows in appearance but use no sash balancing devices. They are operated manually by lifting and are held in various open positions by mechanical catches that engage in the jamb or hold by friction. Ventilating area is 50% of the window area.
Vertical Pivot	VP-C20 Commercial VP-HC40 Heavy Commercial	*Vertically-pivoted windows* consist of a ventilator sash mounted on pivots located in the center of the top and bottom main frame members. This allows the ventilator to be reversed by rotating and permits the glass to be cleaned from the interior. This window is not used for normal ventilation and is operated solely for cleaning purposes or emergency ventilation. It is used generally in completely air conditioned high rise buildings.
Top-Hinged	TH-C20 Commercial TH-HC40 Heavy Commercial	*Top-hinged, in-swinging windows* consist of a ventilator hinged to the main frame at the top that swings into the room. This permits the glass to be cleaned from the interior. This window is not used for normal ventilation and is operated solely for cleaning purposes and emergency ventilation. Its use is generally in completely air conditioned high-rise buildings.

FIGURE 8.4-1
Summary of Aluminum Windows and Sliding Glass Doors *(Continued)*

Type	Designation and Application	Description
Sliding-Glass Door 	SGD-R15 Residential SGD-C20 Commercial SGD-HC40 Heavy Commercial	*Sliding glass doors* consist of two or more framed glass panels containted in a main frame designed so that one or more of the panels move horizontally in the main frame. The most common arrangement is of 2 panels. "Single-Slide" has one operable panel, a "Double-Slide" has two. The ventilation area of a single and double slide door is 50% of the main frame area. Doors are also made in 3 and 4 panel arrangements: 3 panel arrangement can provide 1/3 or 2/3 of its area for ventilation, and 4 panel can provide 1/2 or in some cases 3/4 of its area for ventilation.
Fixed 	F-R15 Residential F-DW-R15 Dual Window Residential F-C20 Commercial F-DW-C20 Dual Window Commercial F-HC40 Heavy Commercial F-DW-H40 Dual Heavy Commercial	*Fixed windows* may utilize any of the other types. They consist of a frame with no operating sash. Fixed windows may have mullions or muntins.

windows, depending on a building's location, size, and design.

Selection of a window or sliding glass door begins by determining the type desired (awning, double-hung, projected, sliding glass door, etc.). Then the required *design pressure* (DP) is calculated. *Residential* windows and sliding glass doors are rated at a minimum DP of 15 pounds per square foot (psf); *commercial* windows and sliding glass doors at a minimum of 20 psf; and *heavy-commercial* units at a minimum of 40 psf. This rating is called a *performance class*. Windows in high-rise structures may require much higher performance classes than these minimums, varying up to as high as 100 psf.

Design pressure (DP) is determined by using the fastest wind speed likely to be encountered at a building's location. A map in ANSI A58.1, "Minimum Design Loads for Buildings and Other Structures," which is also reproduced in ANSI/AAMA 101, can be used to estimate this maximum wind speed. After the expected wind velocity has been determined, the resulting VP, DP, STP, and WTP are determined from the load tables in ANSI/AAMA 101. These load tables are based on wind velocities from 70 mph to 110 mph and the height of a window above the ground, from zero (at grade) to 500 ft. The *velocity pressure* (VP) taken from these load tables varies with building height above grade and the wind speed. For example, a 70-mph wind will produce a VP of 10 at grade and 17.3 at 100 ft. The VP is determined by the formula

$$VP = 0.00256\, K_Z (IV)^2$$

where

V = the wind velocity in miles per hour taken from the map in ANSI A58.1.

I = the importance factor as described in ANSI A58.1.

K_Z = the velocity pressure exposure coefficient as described in ANSI A58.1. This value depends on the height above ground and the exposure category.

The DP is defined as 1.25VP, and the *uniform load structural test pressure* (STP) for aluminum windows and sliding glass doors is 1.25DP.

The *water resistance* of AAMA-rated windows and sliding glass doors is determined by testing them in accordance with ASTM Standard E 331, "Test Method for Water Penetration of Exterior Windows, Curtain Walls, and Doors by Uniform Static Air Pressure Difference," and ASTM Standard E 547, "Test Method for Water Penetration of Exterior Windows, Curtain Walls, and Doors by Cyclic Static

Air Pressure Differential." The *water resistance test pressure* (WTP) is established at 15% of DP but never more than 2.86 psf. This is equal to a wind velocity of 33.4 mph. At this test pressure, water is applied to the test sample at a rate of 5 gal./hr., which is equivalent to 8 in. of rain per hour.

The *air infiltration* of AAMA-certified windows and sliding glass doors is tested in accordance with ASTM E 283, "Test Method for Rate of Air Leakage Through Exterior Windows, Curtain Walls, and Doors." The usual pressure is 1.57 psf for residential and commercial sliding glass doors and windows of all types; 1.57 psf for double-hung and sliding heavy-commercial windows; and 6.24 psf for other heavy-commercial units. The normal maximum rate of air infiltration is 0.37 cubic feet per minute (cfm) per foot of operating ventilator or sash crack for most windows. The rate is 1.50 cfm for jalousie windows, 0.15 cfm for fixed windows, and 0.37 cfm for sliding glass doors, all based on the total area of the unit in square feet.

The *condensation-resistance factor* (CRF) is determined by procedures outlined in AAMA 1503.1, "Voluntary Test Methods for Thermal Transmittance and Condensation Resistance of Windows, Doors, and Glazed Wall Sections." It gives an indication of a window's ability to resist condensation. The higher the CRF, the less likely condensation is to occur.

As discussed in Section 8.1, heat will flow readily through windows and sliding glass doors unless some means are taken to reduce this flow. The methods for determining conductive heat transmission through windows and doors is a part of AAMA 1503.1, "Voluntary Test Method for Thermal Transmittance and Condensation Resistance of Windows, Doors, and Glazed Wall Sections."

Where design calculations indicate the need to reduce the heat transmittance of aluminum windows and sliding glass doors, significant reduction can be accomplished by incorporating double- or triple-glazing and by thermalizing the frames and sash. At any rate, windows and sliding glass doors in areas where the winter design temperature is 10 °F or less should incorporate such devices regardless of calculations. In fact, double- or triple-glazing is used in most new windows in-stalled today, especially in commercial applications.

Thermalizing involves separating the aluminum parts on the warm inside portion of a window from those on the cold outside portion through the use of an insulating material such as plastic. This keeps heat from being conducted through the frame or sash. The insulating separator is called a *thermal break* or *thermal barrier*. Thermal-break windows not only minimize heat conduction through the glass area but through the frame and sash as well.

8.4.2.2

Architectural Windows

Architectural windows are covered in a separate AAMA document, GS-001, "Voluntary Guide Specifications for Aluminum Windows." Unfortunately GS-001 is not yet a standard and cannot be referenced as such. It includes high-quality windows that were previously known as *monumental windows*.

Since GS-001 covers essentially the same provisions as does ANSI/AAMA 101, much of the discussion in Section 8.4.2.1 also applies to architectural windows. There are some differences, however, in types, grades, and performance criteria.

■ TYPES

Architectural window types are similar to prime window types, with some exceptions. For example, there are no awning, jal-awning, jalousie, or greenhouse windows in GS-001, as there are in ANSI/AAMA 101. Neither are sliding glass doors included. However, GS-001 includes several windows that are not covered in ANSI/AAMA 101. These are horizontally pivoted windows, which are similar to vertically pivoted windows turned on their side, and side-hinged inswinging windows, which are similar to casement windows that swing inward. Some of the features of windows in GS-001 with the same names as those covered by ANSI/AAMA 101 are slightly different.

■ GRADES

There are no grades for architectural windows in GS-001. Instead they are divided into three categories: standard, modified, and custom. Standard architectural windows are made entirely from the manufacturer's stock components. In modified architectural windows the manufacturer's stock components may be altered slightly. Custom architectural windows are entirely custom-designed units.

■ PERFORMANCE REQUIREMENTS

The major difference between the requirement of ANSI/AAMA 101 and AAMA GS-001 is in performance requirements. For example, the minimum test unit size required by GS-001 is much larger than most of the sizes required by ANSI/AAMA 101. In addition, the static air pressure required in the air infiltration test is 6.24 psf for all units, which is larger than that required for all except a few heavy-commercial units in ANSI/AAMA 101. The permitted infiltration rates are also much smaller.

The air pressure difference for the water resistance test in GS-001 is 8.00 psf, which is much higher that that required by ANSI/AAMA 101.

Most of the requirements in GS-001 are higher than those in ANSI/AAMA, but not all. For example, the minimum structural test pressure (STP) in GS-001 is 30 psf; in ANSI/AAMA 101, 15 psf. However, this test pressure for heavy-commercial windows in ANSI/AAMA 101 is 40 psf, which is higher than the minimum required by GS-001.

8.4.2.3

Certification of Quality

In addition to developing standard specifications, AAMA sponsors a program of independent testing, evaluating, inspecting, and labeling aluminum products to certify compliance with ANSI/AAMA standards. The certification program is accredited by ANSI. Products that conform to the requirements of the applicable specifications are identified by the AAMA label.

The manufacturer's code number, appropriate standard, and maximum size tested are indicated on the label. The label must be visible on the metal after installation.

The AAMA certification program is operated in accordance with American National Standard Practice for Certification Procedures ANSI Z34.1, "Practice for Certification Procedures," which provides for certification to the public on an impartial, independent, and continuing basis. All certifications guarantee compliance with the ANSI specifications.

8.4.3

INSTALLATION

The finest window or door design and the highest quality of manufacture will not compensate for poor installation. Weathertight installation, although simple, can be ensured only if an aluminum window or sliding glass door is installed by an experienced worker following the guidelines and instructions supplied by the manufacturer. A rough opening should be prepared to receive the unit in such a manner that it can be installed, and will finish out square, plumb, level, straight, and true. The design of most windows and doors provides for minor adjustments at the job site, but no unit will operate or weather properly if it is twisted and misaligned during installation.

A properly selected and manufactured window or sliding glass door will be weather resistant as a unit, but if the building is insulated, the space between the rough opening and the windows or sliding glass doors should be filled with insulation. In addition, the spaces between the unit and adjacent materials must be closed with a sealant, trim members, or both to prevent the passage of air, dust, and water around the frame.

Windows and sliding glass doors should be supported (nailed, screwed, or otherwise fastened) around the entire frame and anchored securely to the supporting construction in a firm and rigid position. They should also be flashed at their heads and sills to provide a path back to the exterior for water that finds its way into the wall.

Anchoring materials and flashing should be of aluminum or material compatible with aluminum, such as stainless steel, cadmium-coated steel, or galvanized steel. Dissimilar materials, such as copper or bronze, should be insulated from direct contact with aluminum by waterproof, nonconductive materials, such as neoprene, waxed papers, or coated felts. Dis-

similar metals located where water passing over them may contact a window or door should be painted to prevent staining of the aluminum. Refer to Section 5.2 for a discussion about the effect dissimilar materials can have on aluminum.

Concealed aluminum in contact with concrete, masonry, or absorbent material, such as wood, paper, or insulation should be permanently protected by coating either the aluminum or the adjacent material with a bituminous or aluminum paint or by coating the aluminum with a zinc chromate primer, to minimize the chance of chemical corrosion of the aluminum by acids, alkalies, and salts leached out of the adjacent materials. Creosote and tar coatings, which might damage the aluminum, should not be used.

Because aluminum windows and sliding glass doors are prefinished building components, they should be handled and treated with care to prevent damage. Once installed, they should be protected to prevent damage to their finish, appearance, and ability to operate properly. Units should be protected from splashing of plaster, mortar, and cleaning solutions.

8.5 WOOD WINDOWS AND SLIDING GLASS DOORS

Wood was one of the earliest materials used in the manufacture of windows, and over the years it has proven its suitability for this purpose. Wood's insulating properties, ready availability, ability to take either natural or painted finishes, and ease of fabrication and repair with simple tools all contribute to its widespread use in window manufacture.

Recent technological innovations have further improved wood as a material for window and door construction. Kiln drying of lumber reduces shrinkage, resulting in better operation and weathertightness. Water-repellent preservative chemical treatments reduce swelling and warping, improve paint retention, and increase wood's resistance to decay and insect

attack under all climatic conditions. Windows that cover entire walls consist of combinations of fixed glass for light and view and operating sash for light and ventilation. In some window walls, sliding glass doors also provide access to the outdoors, in addition to light and ventilation.

For maximum comfort at minimum heating and air conditioning costs, windows and doors must be weathertight, insulated, and condensation free. Also, they must be designed and built to permit ready adaptation to varying installation conditions.

This section outlines the manufacture, nomenclature, standards of quality, and installation of stock wood windows and sliding glass doors. Stock windows and

doors are manufactured at mills in a variety of common types and sizes.

8.5.1

MANUFACTURE

Most stock wood windows are manufactured in accordance with standards developed by the National Wood Window and Door Association (NWWDA) and adopted by the American National Standards Institute. The publication containing these standards is ANSI/NWWDA Industry Standard I.S. 2, "Wood Windows." Sliding door manufacture employs materials and methods generally similar to window manufacture. The manufacture of sliding doors is regulated by NWWDA Industry Standard I.S. 3, "Wood Sliding Glass Doors." The discussion in this section presumes a level of quality conforming with these standards.

8.5.1.1

Materials

Most stock windows and sliding glass doors are manufactured from ponderosa (western) pine. This species has excellent workability, gluability, nail-holding capacity, and a uniform light color suitable for natural or painted finishes. Other species commonly used are southern yellow pine, Idaho white pine, sugar pine, and Douglas fir. Lumber must be sound, free from defects such as loose knots and excessive checking, and kiln dried to a moisture content between 6% and 12%.

8.5.1.2

Fabrication and Assembly

Kiln-dried lumber is delivered to a mill-work shop for fabrication into window and sliding glass door components. There it is cut to specified lengths and widths on ripsaws and crosscut saws, leaving cut stock substantially clear and without defects. Cut lengths are then sorted and marked to indicate the window or door component for which they are intended. Like pieces are conveyed to molders or lineal milling machines, which shape the pieces to the desired profile. The next step is cross-milling, where necessary grooves and channels (dadoes, mortises, and tenons) (Fig. 8.5-1) are cut across each piece to facilitate weathertight joining of units. Milling of components is completed by routing machines, which cut notches to receive recessed hardware.

When shaping has been completed, milled components and assembled sash are preservative-treated with a solution of toxic, water-repellent agents. Nonpressure treatments, such as the vacuum and immersion processes, are commonly used. Besides imparting water-repellent properties and increasing resistance to decay fungus and insect attack, preservative treatment gives the wood increased dimensional stability, minimizing inter-

FIGURE 8.5-1 *(a) Sash-component terminology; (b) common millwork methods of joining sash corners.*

mittent swelling and shrinking due to varying atmospheric conditions or contact with water. Preservative treatment should be in accordance with NWWDA I.S. 4, "Industry Standard for Water Repellent Preservative Treatment for Millwork." Storm sash and exterior components may also be paint-primed or factory-finished for additional protection.

Following preservative treatment, weatherstripping is pressure-fit into grooves or surface-tacked, and some hardware is installed. Sash and frames are then assembled from components on automatic nailing machines. A continuous sander gives the assembled sash three separate sandings—coarse, medium, and fine—in one trip through the machine.

The final steps consist of glazing, fitting sash to the frames, and assembly with all operating hardware. Completed window units are usually provided with angle and spacer braces to prevent racking and distortion in shipment.

■ WEATHERSTRIPPING

The function of weatherstripping is to provide a seal against infiltration (leakage of air) and penetration of dust and wind-

blown rain through a window or sliding glass door. Weatherstripping should be incorporated on all ventilating sash to reduce infiltration substantially below the maximum established by NWWDA standards. The maximum is 0.34 cubic feet per minute (cfm) for Grade 20 (see Section 8.5.3.1) windows and doors; 0.25 for Grade 40; and 0.10 for Grade 60. The measurement is per linear foot of sash crack length for operating sash and per square foot of surface for fixed sash. The test pressure is a uniform 1.57 psf.

The most common weatherstripping used in wood windows is the *spring-tension type* of bronze, aluminum, rigid vinyl, stainless, or galvanized steel (Fig. 8.5-2a). Extruded aluminum, cold-formed galvanized steel, or rigid plastic is used as sash guides and integral weatherstripping in many double-hung windows. These may be either fixed to the window frame or cushioned with plastic foam or springs to provide constant pressure against the sash stiles. In the fixed type, *woven felt* in the stiles sometimes engages a projecting fin in the sash guide for improved weathertightness (Fig. 8.5-2b). The cushioned *compression sash guide*, in addition to weather protection, often provides for easy removal of the sash (Fig. 8.5-2c). Metal or plastic sash guides eliminate the need for finishing sash runs and the possibility of becoming clogged with paint or varnish.

Compressible bulb-type weatherstripping of vinyl or neoprene is found most often in storm sash, screens, and sliding glass doors. It is employed by some manufacturers in main sash (Fig. 8.5-2d). In addition to flexible vinyl, woven fabric weatherstripping also is common in sliding glass doors.

■ HARDWARE

Hardware provides effective closure, operates a sash, and holds it stationary at the desired degree of openness. Hardware types range from simple hinges and locks on basement windows to complex arrangements on windows that require interior operation without removing screens. Underscreen rotary gear, push bar, and lever-type operators are commonly used for awning windows, in combination with sliding friction hinges, which provide projected action (Fig. 8.5-3a). Operators are

FIGURE 8.5-2 *Common types of weatherstripping: (a) spring-tension; (b) woven felt; (c) compression sash guide; (d) compressible bulb.*

usually capable of weathertight closure, but on larger sash, additional awning locks may be provided. Outswinging casement windows often utilize similar rotary gear operators, with either sliding hinges or extension hinges (Fig. 8.5-3b); one or more casement fasteners usually are necessary to draw up relatively long sash stiles against the frame.

Double-hung and horizontal sliding windows generally employ cam-action sash locks, which draw the sash together at the meeting stiles (or rails) while pushing the sash outward against the frames to compress the weatherstripping. Sliding windows often are equipped with a spring-cushioned metal or plastic track at the head jamb, similar to the sash guide of the double-hung window; fixed metal or plastic tracks are commonly used at the sill (Fig. 8.5-3c, d).

Double-hung windows usually are equipped with balancing hardware that assists in raising the sash and holds it in position. Spiral spring, spring, and coiled tape are the most common types of balancing devices (Fig. 8.5-4). Smaller and lighter sash may not require mechanical assistance for raising, because the friction fit provided by a compression sash guide may be sufficient to hold sash in place (see Fig. 8.5-3c).

Exposed hardware such as sash locks, sash lifts, and various operators are available in corrosion-resistant standard (zinc, bronze, aluminum, chrome) and proprietary finishes.

■ GLASS AND GLAZING

Single-strength (3/32 in.) glass is limited to sizes less than 76 *united inches* (width plus height); double-strength (1/8 in.) glass to sizes less than 100 united inches. Additional information on grades and sizes of glass is given in Section 8.7. Most stock windows and sliding glass doors are available with either insulating glass or single glazing; tinted and patterned glass are sometimes available by special order. The Consumer Product Safety Commission (CPSC) Safety Standard for Architectural Glazing Materials requires safety glazing material in sliding doors.

Much wood window sash is face-glazed by bedding (back putting the glass in a rabbet, securing it with glazing points, and sealing it with a bevel face-putty bead (Fig. 8.5-5a). Sometimes pointing is eliminated by using adhesive bedding materials, which cushion the glass, seal it, and secure it in place.

Wood stop glazing replaces both pointing and face puttying (but not bedding) and provides a glass seal of greater strength, durability, and better appearance (Fig. 8.5-5b). Many wood sliding glass doors and large fixed windows are glazed in this manner.

Groove glazing, employed by some manufacturers, has the same advantages as wood stop glazing, but it eliminates the need to handle a separate stop (Fig. 8.5-5c). In this method, grooves in the sash are filled with glazing compound, and sash members are assembled around the glass, clamped in a press, and nailed or pinned together. To facilitate reglazing, one of the sash members can be removed by drawing out the pins that join it at the corners. The sash is reassembled after the new glass has been placed in the groove.

■ SCREENING

Aluminum, galvanized steel, bronze, or vinyl-coated glass-fiber insect screening may be used. Mesh size is limited to 18 by 14 or 18 by 16 (number of openings per inch).

Screening is secured to wood sash by rolling it into a groove and splining, by tacking, or by stapling. A sash is usually trimmed out with a screen mold to conceal tacks, staples, and splines. In metal-rimmed screens, the screening is rolled into a groove in the rim and held in place by a vinyl or metal spline.

8.5.2

TYPES AND NOMENCLATURE

Wood windows may be classified according to design and manner of operation into *fixed*, *double- (or single-) hung*, *awning*, *hopper*, *casement*, *horizontal sliding*, and *basement* types.

Sliding glass doors are essentially large horizontal sliding windows. However, their increased size, weight, and use result in greater demands on frame and sash construction and on sliding hardware. Sliding panels usually are roller supported and operate on metal tracks built into the threshold. Metal parts are covered with wood on the interior to prevent condensation and present a uniform appearance. Door panels are sometimes reinforced with steel to provide necessary strength while retaining a slim profile.

A summary of the most common window and sliding glass door types is given in Figure 8.5-6.

8.5.2.1

Window and Door Components

A window unit consists of a frame, one or more sash, and hardware and weatherstripping necessary to make a complete

a AWNING

crank operator   push bar

HOPPER   sash lock   snap lock

b CASEMENT

crank operator

extension hinge   sash lock

C VERTICAL SLIDING

Sash lock

Sash lift

compression sash guides

d HORIZONTAL SLIDING

Compression sash guide

sill track

FIGURE 8.5-3 *Commonly used hardware for various window types. Note in (a) and (b) that sliding friction hinges provide projected action; for example, when awning sash is opened, the top rail moves downward.*

a SPIRAL SPRING   **b** SPRING

c COILED TAPE   **d** COUNTERWEIGHT

FIGURE 8.5-4 *Balancing devices for double-hung windows.*

a FACE

b WOOD STOP

c GROOVE

FIGURE 8.5-5 *With face glazing (a), glazing compound holds glass in wood seat; with wood stop glazing (b), glass is bedded in compound and secured by wood stop; with groove glazing (c), glass is bedded in compound and set in continuous groove.*

operating unit. Storm sash and screens may be included as integrated elements of the main frame or contained in a separate combination subframe (Fig. 8.5-7).

■ SASH COMPONENTS

A wood window sash is made up of horizontal rails, vertical stiles, and sometimes muntins and bars (see Fig. 8.5-1). A wood sliding glass door is composed of a main frame containing one or more operating or fixed door panels. Like window sash, door panels are composed of stiles and rails and may hold either single glass or insulating glass. Most doors provide self-storing sliding screens in a separate track of the door frame.

■ FRAME COMPONENTS

Figures 8.5-1 and 8.5-8 show the parts of a double-hung window sash. The components of other window types vary some-what from those shown, of course, but the other basic elements—*sill*, *side jambs*, *head jamb*, *stops*, and *exterior casing*—are present in most window and door designs. *Blind stops* are included in some window frames between jambs and outside casings and may vary in thickness for proper window positioning in wood or metal stud walls with 1/2- or 3/4-in. sheathing. *Jamb extenders* adapt windows to varying interior finish thicknesses. They are not usually a part of the window frame assembly but may be ordered loose as optional items. *Interior trim* is not usually supplied with stock windows. Typically it is selected separately and in some installations may be replaced by a gypsum board or plaster reveal.

CLEANING AND MAINTENANCE   To make exterior surfaces accessible for cleaning, double-hung and horizontal sliding win-dows usually have special hardware that permits easy removal of sash (see Figs. 8.5-3c, d). Awning window sash may be removable, but more often awning windows have sliding friction hinges, which create an access space of several inches between sash and frame when the top rail is moved down. Casement windows similarly provide cleaning accessibility by sliding or extension hinges. The in-swinging operation of hopper sash makes both surfaces automatically accessible from the inside. Most sash not ordinarily considered removable may be removed with a few simple tools.

■ STORM SASH AND SCREENS

For greater heat control, any window sash may be glazed with insulating glass (Fig. 8.5-9a) or provided with *storm sash*. Clip-on metal-rimmed *storm panels* commonly are used with hinged sash such as casements, awnings, and hoppers (Fig. 8.5-9b). Sliding and double-hung sash may be equipped with either clip-on storm panels or an external *combination frame*, which holds both storm sash and screens (Fig. 8.5-9c).

Combination frames have the advantage of being self-storing, thus eliminating the need to remove external screens for winter storage and providing a year-round choice of storm panel or screen. However, combination frames provide for screening only half the window area, so only one sash may be opened at a time. When it is desirable to open both top and bottom sash of a double-hung window (or right and left sash of a horizontal sliding window), full screens may be used to cover the entire window.

Outswinging hinged sash (casements and awnings) typically have interior removable screens; hopper sash have exterior screens. Basement windows may have either interior or exterior screens, depending on whether sash swing in or out. Many storm panels and screens are framed with aluminum or stainless steel rims of thinner cross-section, which provide greater flexibility in positioning within the limited thickness of a typical frame wall.

8.5.3

SELECTION

The selection of windows and sliding glass doors depends on the use and the location

TYPE AND INDUSTRY STANDARD	DESCRIPTION
FIXED ANSI/NWWDA I.S.2	*Fixed windows* usually consist of a frame and glazed stationary sash. They are often flanked with double hung and casement windows, or stacked with awning and hopper units, to make up windows of custom designs. To keep the sight lines (width or height of view area) consistent, fixed sash members are made to same appropriate cross sectional dimensions as adjacent operating sash.
DOUBLE HUNG ANSI/NWWDA I.S.2 **SINGLE HUNG** ANSI/NWWDA I.S.2	*Double hung windows* have two operating sash; single hung have only the lower sash operative. The sash move vertically within the window frame and are maintained in the desired position by friction fit against the frame or with balancing devices. Balancing devices also assist in raising the sash. 50% of the window area is available for ventilation.
CASEMENT ANSI/NWWDA I.S.2	*Casement windows* have side-hinged sash, generally mounted to swing outward. They may contain one or two operating sash and sometimes a fixed light between the pair of sash. When fixed lights are used, a pair of casements may close on a mullion, or against themselves, providing an unobstructed view when open. Operating sash may be opened 100% for ventilation.
HORIZONTAL SLIDING ANSI/NWWDA I.S.2	*Horizontal sliding windows* have two or more sash of which at least one moves horizontally within the window frame. In three-sash design, the middle sash is usually fixed; in two-sash units, one or both sash may be ventilating. This type of window sometimes is increased in size to door proportions. Ventilating area is 50% of the window area in most designs.
AWNING ANSI/NWWDA I.S.2	*Awning windows* have one or more top-hinged, outswinging sash. Single awning sash often is combined with fixed and other types of sash into larger window units. Several sash may be stacked vertically and may close on themselves or on meeting rails which separate individual sash. When awning windows have sliding friction hinges, which move the top rail down as the sash swings out, they are said to have *projected action*. Ventilating area is considered to be 100% of operating sash area.
HOPPER ANSI/NWWDA I.S.2	*Hopper windows* have one or more bottom-hinged, inswinging sash. Hopper sash is similar in design and operation to the awning type, and may actually be an inverted awning sash with minor hardware and weatherstripping modifications. For this reason, windows with hopper sash sometimes are referred to as awning windows. Operating sash provides 100% ventilating area.
BASEMENT ANSI/NWWDA I.S.2	*Basement windows* generally are single sash units of simplified design intended for less demanding installations, particularly in masonry and concrete foundations. The sash may be of awning type, hopper or top-hinged inswinging (shown) type. 100% of the window is available for ventilation.

FIGURE 8.5-6 *Summary of wood windows and sliding glass doors.*

TYPE AND INDUSTRY STANDARD		DESCRIPTION
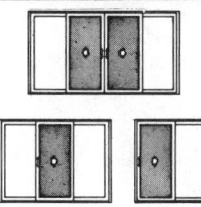	**SLIDING GLASS DOORS** NWWDA I.S.3	*Sliding glass doors* contain at least one fixed and one operating panel in a main frame. Door types are commonly designated by the number and arrangement of panels (as viewed from outside) using the letter "X" for operating panel, and "O" for fixed panel. The two panel door is most common, but three and four panel designs are also available; two and four panel doors (OX & OXXO) have 50% ventilating area, the three panel unit (OXO) has 33% ventilating area.

FIGURE 8.5-6 *Summary of wood windows and sliding glass doors. (Continued.)*

FIGURE 8.5-7 *The principal parts of window units, as shown here for a double-hung window.*

FIGURE 8.5-8 *Terminology for the typical double-hung window. Jamb extenders are also called* jamb liners.

of the units. The first step is to decide on the types of unit desired (see Section 8.5.2). Then the unit's grade and other required performance criteria should be selected.

8.5.3.1

Performance Criteria

As mentioned previously, ANSI/NWWDA I.S.2 and NWWDA Industry Standard I.S.3 constitute the accepted industry standards for wood windows and sliding glass doors. They establish certain minimum performance criteria and rate the products while broadly regulating material, design, and manufacture. For example, both standards recognize three grades of performance: Grade 20, Grade 40, and Grade 60. The numbers represent the minimum structural performance test pressure in pounds per square foot that a unit must be able to sustain without damage. Under ordinary circumstances, Grade 20 units are satisfactory for residential use; Grade 40 for light-commercial use; and Grade 60 for heavy-commercial construction. However, in high-wind zones (Fig. 8.5-10), higher grades should be selected regardless of the use. The grades also dictate the acceptable operating force, air infiltration rates, and water penetration test pressures. The acceptable levels are different in each grade, being more stringent as the grades get higher.

Other items covered by these standards are moisture content of the lumber, preservative treatment, screening, and glass requirements. Appropriate industry standards for wood windows and siding glass doors are shown in Figure 8.5-6.

8.5.3.2

Certification of Quality

The industry standards mentioned do not provide for inspection, testing, or certification of products. These functions are performed by certification programs, sponsored by the National Wood Window and Door Association (NWWDA) or by independent testing laboratories. Issuance of a label ensures that the design, manufacture, and tested performance of the product meet or exceed the appropriate standard. This type of certification program is based on periodic inspection of the manufacturing facilities, quality control procedures, and laboratory testing.

NWWDA offers three certification programs. Wood windows are certified for conformance to NWWDA I.S.2; wood sliding glass doors are certified for conformance with NWWDA I.S.3; and water-repellent preservative treatment is certified for conformance to NWWDA I.S.4. Each program consists of initial qualification testing, in-plant inspection, and periodic follow-up testing and inspection to verify that products are manufactured in conformance to the applicable standards. Conformance to the standards results in certification and issuance of seal. This seal is rubber stamped on various parts of the sash and frame or imprinted on labels affixed to the product. The seal indicates the fabricator of the product and the industry standard under which the product was produced and is evidence of compliance with the standard.

8.5.4

INSTALLATION

Even the best manufactured windows and sliding glass doors will not perform satisfactorily if they are not installed properly. Although these units are commonly squared and braced before shipping, they may be distorted in transit and should always be checked for squareness and equal spacing of jambs before they are installed. Braces should be renailed if necessary and left in place until the unit has been installed and is ready for trimming. The frame should be plumbed and fastened securely in the rough opening. Adequate

FIGURE 8.5-9 *Methods of double glazing: (a) insulating glass shown here groove glazed; (b) clip-on storm panel; (c) combination storm and screen sash.*

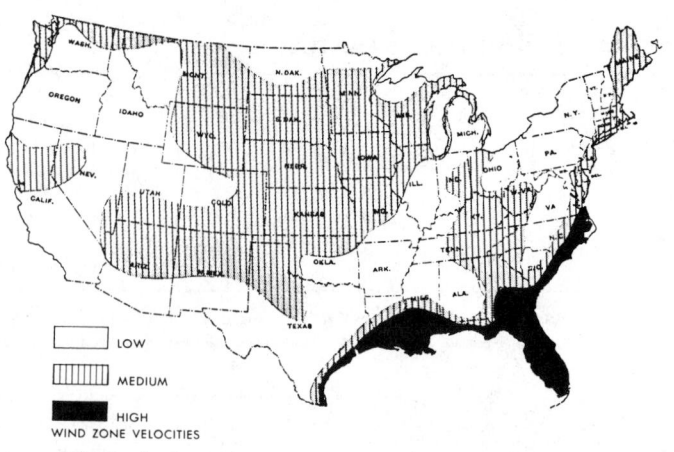

LOW

MEDIUM

HIGH

WIND ZONE VELOCITIES

FIGURE 8.5-10 *Wind velocity zones.*

shimming should be provided so as not to distort the frame during nailing. The spaces between finished frame and rough frame should be completely filled with insulation. Drip caps and flashing over the cap or exterior casing are strongly recommended to minimize the possibility of leakage. The entire unit should be well caulked at the juncture of the exterior casing and the wall finish, after both elements have been prime-coated but before finish painting.

When windows and doors are shipped with loose interior stops, it is essential that these be installed carefully for a snug, weathertight fit against the sash. Weatherstripping and hardware should be adjusted for proper operation and closure, according to the manufacturer's instructions. Interior trim should be applied to seal the wall cavity between the rough frame and the finish frame.

Wood windows and sliding glass doors should be handled carefully to prevent damage and be protected with a prime coat as soon after arrival at the building site as possible. They should be stored out of the weather in a well-ventilated space. During finishing, it is important to keep paint and varnish off weatherstripping and interior finish hardware.

Exterior hinges and storm sash hardware may be painted for improved corrosion resistance, but the metal or plastic sash guides of double-hung and siding windows should not be painted. Oil-based glazing compound should be painted, with the paint lapping slightly over the glass on both face-putty and back-putty runs.

Other sealants may or may not be painted, depending on the sealant type and use.

In addition to installation instructions, most manufacturers of stock windows provide simple operating and maintenance recommendations. For continued trouble-free and effective operation, these should be followed.

8.6 STORM WINDOWS AND DOORS

Storm windows and doors are glazed panels or inserts that are sometimes added to windows and doors as an extra line of defense against the elements. Their major use is in residential and small commercial buildings.

Earlier storm panels were relatively inexpensive sash, usually made of wood or metal frames with single glazing. They were hung in place with special clips attached to each sash or to the frame. (Fig. 8.6-1). They were completely indepen-

dent of the screens, so they typically were installed in the fall and taken down in spring. If the building was air conditioned and not properly oriented for natural ventilation, they could be left up year-round to reduce cooling loads. This type of window can still be found on older residential buildings but is not often used in new construction.

Contemporary storm windows are usually combination types that are permanently installed and include not only storm sash but screens as well (Fig. 8.6-2). Frames are usually aluminum and have an anodized or baked enamel finish. The window and screen sash should move smoothly in the frame tracks and seal tightly when closed. They should have tight joints and suitable weatherstripping to ensure a tight seal.

Fixed-in-place interior storm windows are inexpensive temporary installations used almost exclusively in existing residential buildings. This interior storm window usually consists of a frameless sheet

FIGURE 8.6-3 *These temporary storm panels attach directly to the window frame.*

of rigid plastic held in place with tape or snap-on moldings (Fig. 8.6-3). It is also possible to use polyethylene film taped to the frame. Temporary plastic windows provide an amount of insulating value that is comparable to ordinary storm windows because it is the air films along the glazing surfaces and the air space between that reduce the heat loss, not the material thickness.

8.6.1

GENERAL REQUIREMENTS

Properly installed storm doors and windows provide an effective barrier to the transfer of heat. They are highly recommended for use in the northern part of the country, above the 4500-winter-degree-day line, and are desirable down to the 2500-degree-day line (Fig. 8.6-4). In these areas, they can significantly reduce heating costs and minimize condensation on interior windows. By preventing cold air from leaking around prime windows and doors, they also reduce drafts and improve interior comfort. In both northern and southern areas, storm windows can be retained over glass areas during the summer to help reduce heat gain through the glass and frames.

Although the term *storm doors and windows* generally refers to units intended for weather protection, it is often used collec-

FIGURE 8.6-1 *(a) Clip-on metal-rimmed storm panels are typically used with awning or casement windows; (b) storm panels for double hung windows are attached to the frame.*

FIGURE 8.6-2 *These windows are also called* triple-track *because of the three panels that move in separate tracks.*

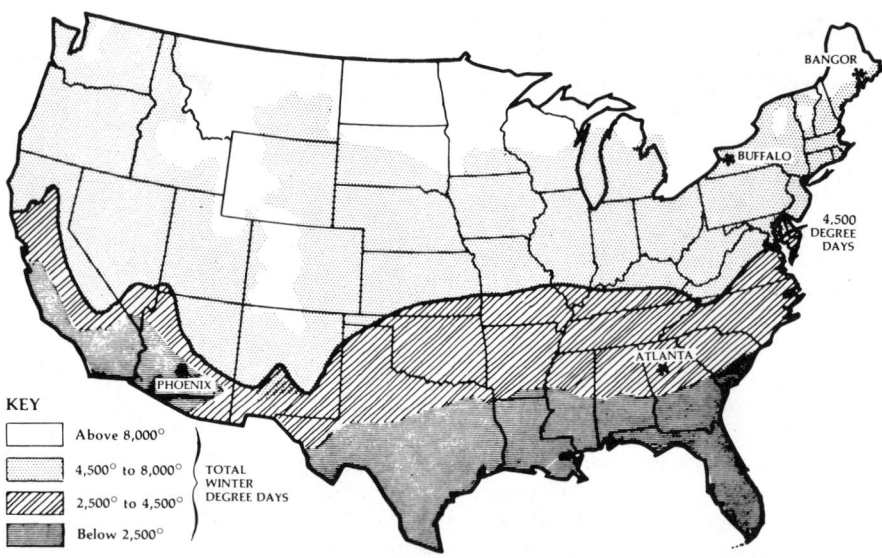

FIGURE 8.6-4 *Zones of heating demand (winter degree days).*

tively to include screen units as well. Accordingly, the following discussion covers both combination units, containing interchangeable glass and screen inserts, as well as single-purpose units, with fixed glass or screening. The recommendations for storm doors and windows are based largely on standards developed by the American Architectural Manufacturers Association (AAMA), while those for screen doors and screen inserts follow standards established by the Screen Manufacturers Association (SMA).

Most storm doors and windows sold today are made of aluminum. Aluminum's popularity is based on its high resistance to corrosion and decay and its ease of maintenance. Durable anodized and paint finishes minimize the need for periodic refinishing and offer a selection of colors to complement any design.

8.6.2

MANUFACTURE

The manufacture of aluminum storm doors and windows is similar to that of aluminum prime windows and sliding glass doors (see Section 8.4). Besides the basic aluminum sash and frame parts, many other accessory products are used in the assembly of window and door units. These include glass, glazing materials, weatherstripping, screening, screen frames, and fasteners.

Storm windows and doors are made in a variety of finishes and colors. A natural mill finish is adequate for many moderate-cost installations, where appearance is not a prime consideration. For more demanding projects, where both corrosion resistance and appearance are important, either an anodized or paint finish is appropriate. Anodized finishes can be of natural aluminum color or deeper shades of bronze or brown-black. Paint finishes are available in a variety of colors.

8.6.3

SELECTION

There are many types of storm doors and windows available today. Both windows and doors can be of the *self-storing* type, or they may depend on a glazed *storm panel* that is applied and removed seasonally as needed.

Self-storing storm windows generally have both glass and screen inserts sliding vertically or horizontally in a frame. In self-storing *combination* (storm and screen) doors, glass and screen inserts slide vertically.

8.6.3.1

Storm Windows and Panels

The most widely used storm windows are of the triple-track and two-track type.

When practicable, storm windows should be installed on the outside of the prime sash. Where the prime sash swings out, as in casement and awning windows, storms may have to be on the room side.

To prevent rapid heat transfer and condensation on the interior window frame, an insulating thermal break should be provided between the prime window and the storm window frame. A thermal break is recommended regardless of whether the storm window is installed on the room side or outside. Some windows are provided with a plastic or wood liner that serves as the thermal break. An air gap or a small wood molding can be used instead.

Triple-track windows are intended for use mainly with vertically sliding double-hung and horizontally sliding triple-light windows. A typical unit consists of a frame, with each insert riding in its own track. This self-storing feature eliminates the need for switching storms and screens every spring and fall. Instead, selection of glass or screen can be made easily from inside the house.

Two-track windows are generally used with vertically sliding single-hung windows and with two-light horizontal sliders. These units also have three inserts, which are stored in the frame. In vertical windows, the bottom light is usually screened; in horizontal sliders, the screen is placed right or left, depending on the prime vent sash location.

Single-track windows have two glass inserts and one screen insert, but only two inserts can be housed in the track at any one time. This makes changing from weather protection to ventilating somewhat less convenient, because one insert must always be stored elsewhere.

Clip-on storm panels are available for outswinging prime windows such as casements, awnings, and jalousies. These consist of single glass panels that are attached either to the individual sash or across the entire prime window, including both fixed and vent sash. When the panel is clipped to the sash only, heat transfer through the glass is reduced, but heat lost through the frame and by air leakage is not affected.

The larger panels covering the entire window control heat loss more effectively, but they cannot be left in place for the summer. These panels must be stored elsewhere and interchanged seasonally with screens. Storm panels should be separated from the prime window frame by an insulating gasket or weatherstripping and should be held tightly in place by pivoting clips or other fasteners.

8.6.3.2

Window Construction

Extrusions used for the main frame and insert frames should be of wrought aluminum alloy with a minimum tensile strength of 22,000 pounds per square inch (psi) and yield strength of 16,000 psi. Reinforcing members that are not in plain view can be of cadmium, zinc-plated steel, or other noncorrosive materials compatible with aluminum.

Storm windows should be provided with weatherstripping to minimize air infiltration and reduce rattling. Glass inserts generally consist of a light-gauge aluminum frame into which the glass is set with bedding compound, plastic channel gaskets, or glazing tape. These materials should cover the glass edge continuously so that there will be no glass-to-metal contact.

Glass used for storm inserts in windows should not be less than B quality and of adequate thickness. Glass inserts in vertically sliding windows should have self-activating latches to hold the insert level in different ventilating positions. Insert frames should be designed to permit reglazing without special tools.

Screen inserts consist of insect screening stretched on an extruded or roll-formed frame. Screening can be of aluminum, vinyl-coated glass fiber, or other synthetic fibers, in 18 by 14 or finer mesh.

8.6.3.3

Storm and Screen Doors

Storm doors are made with either (1) fixed glass lights, exclusively for weather protection or (2) both screen and glass inserts, to permit ventilation with insect control as well. In parts of the country with moderate year-round temperatures, screen doors (without glass inserts) are frequently used. Screen door construction is similar to that of storm doors. Both must be constructed rigidly to resist sagging, bowing, and rattling from repeated impact.

Many storm doors are of the combination type, providing both weather protection and insect control. Such doors are in fact stile-and-rail assemblies, framing a storm window with glass and screen inserts, which can be interchanged seasonally. Doors with openings limited to the upper half of the door generally accommodate only one insert at a time, requiring that the other be stored elsewhere.

Self-storing storm doors contain the equivalent of a two-track window, accommodating two inserts in one track and another in the adjacent track. The glass and screen inserts slide up and down, just as they do in a vertical storm window.

Both prime doors and storm doors can be of the jalousie type. To provide ventilation and storm protection, interchangeable screen and glass panels can be clipped to the inside of the door. Clip-on glass storm panels not only provide added insulation, but also reduce air leakage between the louvers, a common problem with this type of door.

8.6.3.4

Door Construction

Storm and screen doors receive heavy usage in most buildings. They are opened and closed many times a day and are subject to wind buffeting and other stresses. It is therefore important that they be durably constructed and properly installed. Rigid corner connections are particularly important, to resist racking forces.

Standard door frames are 1 or 1-1/4 in. thick, with top rail and stiles 2 to 2-1/2 in. wide. The bottom rail (kickplate) varies in width depending on the size of the opening. To avoid the hazards of broken glass, storm doors should be safety glazed with transparent plastic or with tempered, laminated, or wired glass. In some locations laws stipulate this requirement.

Doors currently are manufactured in hundreds of types and styles, utilizing many different glass sizes. To minimize waste, storm windows and doors should be required to accommodate standard glass sizes adopted by the industry. Limiting the number of glass sizes not only reduces production costs but also facilitates replacement, particularly of tempered glass.

Construction of glass and screen inserts for doors is similar to that of windows, described above.

8.6.4

STANDARDS OF QUALITY

Aluminum storm doors and windows should conform to the following standards developed by the Architectural Aluminum Manufacturers Association (AAMA) and adopted as American National Standards by the American National Standards Institute (ANSI): (1) ANSI/AAMA 1002.10, "Voluntary Specifications for Aluminum Insulating Storm Products for Windows and Sliding Glass Doors," and (2) ANSI/AAMA 1102.7, "Voluntary Specifications for Aluminum Storm Doors." These standards constitute the recognized industry standards and establish minimum performance requirements for air infiltration, water leakage, wind load, and structural strength.

Screen inserts and doors should conform to the following requirements of the Screen Manufacturers Association, adopted as American National Standards by ANSI: (1) ANSI/SMA 1004, "Specifications for Aluminum Tubular Frame Screens for Windows," and (2) ANSI/SMA 2005, "Specifications for Aluminum Sliding Screen Doors."

The AAMA also sponsors an independent certification program of testing, random sampling, and product evaluation to ensure that aluminum storm doors and windows meet ANSI standards. Products that meet all requirements of the standards are identified with an AAMA certification label. The AAMA certification program is administrated in accordance with ANSI Z34.1, "American National Standard Practice for Certification Procedures," which provides for certification to the public on an impartial, independent, and continuing basis.

8.6.5

INSTALLATION AND MAINTENANCE

Above the 4500-degree-day line (see Fig. 8.6-4), storm doors or an entrance vestibule should be provided as part of the initial construction. Storm windows can be omitted in original construction if the prime windows are equipped with insulating glass and a thermal break in the sash and frames. Existing buildings in these northern areas that are not so equipped should be provided with storm doors and storm windows or panels. The economics of adding storm windows in the temperate parts of the country should be determined on a case-by-case basis, depending on construction and energy costs.

Most storm windows are sold as complete units, with all accessories and hardware included with the unit. This ensures smooth, convenient operation and adequate security. Installation consists simply of attaching the window frame in a previ-ously prepared opening.

Storm doors are marketed both prehung on an aluminum frame, with hardware attached, and unassembled, with frame parts and hardware supplied loose. The unassembled type provides somewhat more flexibility in adapting to special situations but requires field assembly and attachment of suitable hardware.

A spring-loaded safety chain or other device should be provided to limit door swing and cushion the impact of a sudden opening, as by a gust of wind. A door closer is recommended to close the door when not in use. A durable latch to maintain the door in a closed position and to permit locking the door from the inside is also desirable.

Relatively little maintenance is required with aluminum storm doors and windows. Guide tracks and sash should be cleaned periodically to keep them operating smoothly. Door closers may need periodic oiling. Factory-applied paint finishes generally last many years. However, if repainting is desired, special paints are available.

8.7 GLAZING

This section discusses glass and plastic glazing materials and their application (glazing). The term *glazing* refers to the installation of vision panels of glass or plastic in a door, formed frame, window, packaged entrance, store front, or glazed curtain wall. Spandrel glass in curtain walls is also glazing. The other types of opaque panels in a curtain wall are also sometimes said to be *glazed* into the system, but they are not generally regarded as glazing materials.

8.7.1

GLASS

Glass results from the fusion of certain common minerals at high temperatures and subsequent controlled cooling. This process causes solidification without crystallization. The chief ingredients of glass are silica, sodium oxide, and calcium oxide (sand, soda, and lime), all of which are abundant minerals throughout the United States.

Glasslike mineral formations resulting from the accidental fusion of silica sand and metallic oxides by the heat of a volcano or lightning (obsidian and lightning stone) can be readily found in nature and may have served as the impetus for the early invention of glass. Glass has been made artificially for more than 4500 years.

The first window glass we know of was made by casting and grinding during Roman times. In 11th-century Germany, glass was hand blown in spheres, elongated into cylinders by a swinging process, then cut apart and flattened to make a product called *broad glass*. *Crown glass* was produced in France during the 14th century by a blowing and twirling process. Thomas Jefferson's Monticello contains Bohemian broad glass, and there is crown glass in Williamsburg. Plate glass was produced in the 17th century mainly to produce the smooth surfaces needed to make mirrors. It was cast and then ground and polished. In the 19th century a product called *cylinder glass* was made by a blowing and swinging process. It could achieve a maximum size of 10 by 49 in. It was this size restriction that governed the building module (48 in.) of the Crystal Palace built in London in 1851, which contained 900,000 sq. ft. of cylinder glass. Later innovations led to the drawing of molten glass cylinders vertically to a length of up to 50 ft.

The real revolution in glassmaking dates from early in the 20th century, when the Fourcault and Colburn processes for mass producing large sheets of flat glass were developed.

Although they were all known earlier, glass ranks with concrete, steel, aluminum, and plaster as a true material of the 20th century. Glass has exerted a tremendous influence on our technology, our modern way of life, and architectural expression. From the divided-light window of colonial days to modern floor-to-ceiling glazing, glass technology has progressed with building design needs and has in turn influenced them.

Just as colonial builders limited the size of windows because of high glass cost and shelter needs, today's designers continually increase glass areas as the cost of glass declines and our ability to control the interior environment improves (Fig. 8.7-1). Largely glass or all-glass buildings are commonplace and create psychological impacts on their inhabitants. The ever-changing panorama of the natural outdoor environment is psychologically satisfying to most people. Possible adverse effects from overexposure to the elements can be minimized by proper orientation and judicious selection of special glass types, such as glare-reducing, heat-absorbing, reflective-coated, or insulating glass.

8.7.1.1

Properties of Glass

Glass is an unusual material in terms of its internal structure. It is mechanically rigid and thus behaves like a solid. However, most mineral solids have a crystalline structure, with the atoms arranged in a definite geometric pattern. In glass, the atoms are arranged in a random or disordered fashion, which is characteristic of a liquid. But the atoms, though arranged at random, are "frozen" in position by the rapid cooling process used to make glass. Therefore, glass combines some of the

FIGURE 8.7-1 *University Tower in Durham, North Carolina, is an example of the kind of glass use that influences architecture today.* *(Architect Pickle/Thomas; photograph courtesy Guardian Industries.)*

aspects of a solid and some of the aspects of a liquid. Because of this, it is sometimes called a *supercooled* liquid.

The physical properties of glass can be varied over an extensive range by modifying the composition of the glass, the production techniques used, or both. In selecting an individual glass for a particular product, a combination of properties must be taken into consideration, because one property usually cannot be changed without causing a change in other properties. The following is a brief discussion of some of the properties of glass that are important in architectural applications.

■ MECHANICAL PROPERTIES

Mechanical properties deal with the action of forces on a material and the effects that those forces produce within the material.

One of the important mechanical properties of glass is that it is *fatigue resistant*. This means that if pressure is applied to

glass, causing it to bend or stretch, the glass will return exactly to its original shape when the bending or stretching force is removed. Of course, if increasing force is applied, the glass will eventually break, but at any point short of breakage, it will not be permanently deformed.

The mechanical strength of glass is determined by its ability to withstand forces that cause breakage. Glass generally breaks from stretching or bending (tensile forces); thus, strength is the chief factor used in determining its mechanical strength.

Unlike most metals, glass does not have a clearly defined tensile strength. The *theoretical* tensile strength of glass has been estimated to be as high as 4,000,000 psi. The *actual* tensile strength of annealed glass ranges between 3000 and 6000 psi.

The main reason for the wide variation between theoretical and actual strength in glass is that actual strength is highly dependent on the surface condition of the glass. No matter how carefully it is handled, glass will acquire some small nicks and scratches in the course of manufacture and later in use. These surface defects then create weak planes in the material; when a force is applied to glass, stresses tend to concentrate at the weak planes rather than distribute themselves uniformly, as is assumed in theoretical strength calculations. Because of variations in the surface condition of different lights of glass, a group of seemingly identical glass specimens will exhibit a wide range of breaking strengths when tested. Thus, the mechanical strength of glass cannot be stated as a clearly defined figure; it must always be calculated on a probability/stress curve.

The mechanical strength of glass can be improved by both heat and chemical treatments. These processes are discussed in Section 8.7.1.2.

■ OPTICAL PROPERTIES

One of the obvious and important properties of glass is its transparency. Despite appearances, however, no glass is completely transparent. When light falls on a piece of glass, some of that light is reflected and some is absorbed. The amount of light that does pass through the glass is called *transmittance*. Commonly used clear commercial glass has a transmittance of about 85 to 90% of visible light.

Glass may be treated in many different ways to reflect or absorb varying degrees of light. Glass that has been tinted absorbs a higher percentage of light than clear glass and, depending on the color and the thickness of the glass, may transmit anywhere from 21 to 75% of visible light. Glass that has been treated with a thin reflective coating reflects a high proportion of light and may transmit from 8 to 50% of visible light, depending on the nature of the coating. The metallic coating on the back of mirrors produces the maximum amount of reflectance possible, allowing no visible transmittance.

Another important optical property of glass is its *fidelity*, or lack of distortion. When the two surfaces of a sheet of glass are perfectly parallel, the image that is seen through, or is reflected from, the glass will be free of waves or distortion (Fig. 8.7-2). However, small variations in the thickness of a sheet of glass can cause considerable distortion in the image seen. For example, in a large area of glass, a variation in thickness of only a few hundred-thousandths of an inch can be quite apparent because the variation acts as a lens to magnify or diminish portions of the reflection.

FIGURE 8.7-2 *The uniform thickness and flatness characteristics of float glass ensure distortion-free views.*

■ THERMAL PROPERTIES

The amount of heat that passes through the glass areas of a building is an important consideration in determining heating and air conditioning loads. Heat can be transferred by conduction, convection, and radiation. Glass has a lower thermal conductivity than most metals, but a great deal of solar energy can be transferred directly through glass by radiation, and air currents that come into contact with glass surfaces can transfer heat by convection. Therefore, overall heat loss in winter and heat gain in summer is generally greater through the glass areas of a building than through ordinary masonry or metal- or wood-framed walls.

Heat gain (or loss) through windows can be controlled by insulation, reflection, absorption, and shading. Special glass products such as insulating glass, reflective-coated glass, and tinted glass make use of these principles and can substantially reduce heat gain. In addition, shading devices such as blinds, draperies, and awnings can be used with glass to further reduce heat gain (see Section 12.2).

The thermal expansion of glass is another important consideration in the design of buildings. This factor is particularly important when tinted or reflective-coated glasses are used, because these glasses absorb a higher percentage of solar energy than ordinary clear glass and therefore will expand and contract more.

When different portions of a single pane of glass heat up at different rates, they expand at different rates, causing thermal stresses to develop within the pane. What often happens under ordinary glazing conditions is that the center portion of the pane heats up faster than the edges (which are shaded by the edge framing), causing the center portion to expand more than the cooler edge regions. This produces stress differentials within the glass, which tend to concentrate at weak points in the glass and may eventually cause breakage. Most thermal failures can be traced to minor edge damage or imperfections, since that is where weak points in a sheet of glass are most likely to occur. The best way to ensure against such thermal breakage is to be sure that glass edges are strong and clean-cut.

The thermal expansion of glass also is important when two glasses or a glass and a metal are to be sealed together, as in insulating glass and some types of coated glass. The rates of expansion of the two materials must be closely matched to avoid damaging differential stresses in the finished product.

■ CHEMICAL PROPERTIES

Glass is much more resistant to corrosion than most metals and other common materials. It is nonporous and nonabsorptive and thus is impervious to many common elements that affect other materials. This is the reason that glass windows, after many years of exposure to the elements, remain clear and apparently unchanged.

Nevertheless, glass will corrode and even dissolve under someconditions. For example, hydrofluoric acid will attack soda-lime glass, causing extreme corrosion; similarly, hot concentrated alkali solutions and superheated water will cause soda-lime glass to dissolve. However, these chemicals are usually encountered only in laboratory situations; under ordinary environmental conditions, glass is an extremely inert and durable substance.

■ ELECTRICAL PROPERTIES

Because it offers high resistance to the passage of electricity, glass is a good electrical insulator. For this reason, glass is used in making many electrical products.

8.7.1.2

Manufacture of Glass

The progress in glass manufacture and the development of glass technology has been outstanding since the turn of the century. Today glass is manufactured in a wide variety of compositions and can be fabricated by several different techniques into a tremendous array of products.

■ RAW MATERIALS

Glass is manufactured in several generic types, which vary according to chemical composition (Fig. 8.7-3). The most common and least expensive type is *soda-lime glass*, made from the oxides of silicon, sodium, and calcium. This type of glass is used for virtually all glazing purposes in commercial and residential buildings and automobiles.

Soda-lime glass is composed of approximately 72% silica, 15% sodium, 9% calcium, and 4% various minor ingredients. Other glass types, which are derived from this basic glass, have slightly different ingredients. These types will be mentioned

only briefly here, because they are of minor importance to the construction industry.

SILICA   Silica, the principal ingredient of glass, is obtained from sand. It is one of the most common elements of the earth's crust. Unlike the sand found at the seashore, sand for making glass must be more than 99% silica, perfectly white and not too fine. Sand of this quality usually is obtained from sandstone deposits, which are abundant in the United States, particularly in Pennsylvania, West Virginia, Illinois, and Missouri.

Silica is the glass former in the glass-making process. It can be used alone to produce glass, but it requires a very high melting temperature (3100°F) and is very difficult to form. It is usually combined with other ingredients to make a more workable glass.

SODIUM   Sodium for glassmaking is usually obtained from soda ash made from salt, but sodium nitrate and sodium sulfate may also be used. Sodium is added in glassmaking as a flux, to lower the melting point and to make a more workable mix. When sodium is added to silica, it forms a mixture that will melt at 1460°F, a considerably lower temperature than for silica alone.

Silica and sodium alone form a glass that is not very durable. In fact, it slowly dissolves when placed in contact with water. It also has a low resistance to chemical attack. Therefore, an ingredient that will act as a stabilizer must be added to every glass batch.

CALCIUM   The stabilizer usually added to glass is calcium. It is obtained from either limestone or dolomite. When added to silica and sodium, calcium produces a glass that is both durable and easily worked.

ADDITIVES   Special elements may be added to or substituted for the basic ingredients in glassmaking to produce glass with specific properties or characteristics.

Potash, a form of potassium, was used instead of soda ash in the early days of glass manufacture in the United States. At that time, wood was the principal fuel for glassmaking and potash was a byproduct

FIGURE 8.7-3
Representative Glass Types, Their Composition, and Their Uses

| Type of Glass | Composition[a] | | | | | | | | Applications or Uses |
| | Sand | Soda Ash or Salt Cake | Lime-stone | Dolomite | Potash | Boric Acid | Litharge | Feld-spar | |
	SiO_2	Na_2O	CaO	MgO	K_2O	B_2O_3	PbO	Al_2O_3	
100% silica	100%[b]	—	—	—	—	—	—	—	Glass laboratory ware
Soda-lime silica[c]	68–74	11–16	8–12	1–4	—	—	—	0–4	Float glass
Borosilicate	65–80	2–4	3	—	0–4	13–28	—	1–3	Heat-resistant glasses
Lead-alkali-silicate	35–63	0–10	0–1	—	10–13	0–1	15–58	0–1	Radiation-shielding products
Optical (flint)[d]	35	—	—	—	7	—	58	—	Opthalmic lenses
Aluminosilicate	55–57	0–1	5	8–12	4	4–8	—	20–23	High-thermal-impact glasses

[a]In addition to raw materials shown, small amounts of other elements may be added to help oxidizing, coloring, reducing, etc. Cullet (crushed glass) is also used to help melting.

[b]All figures are percentages of the ultimate oxides in glass compositions.

[c]Most important type based on volume and value.

[d]Optical glass often contains oxides of zinc, titanium, zirconium, and boron in addition to oxides shown.

of wood ash. Potassium is used infrequently today because it is much more costly than sodium. Glass made with potassium is finer, harder, and more brilliant than that made with sodium.

In soda-lime glass, lead is sometimes used to replace the calcium and sometimes to replace part of the silica. It produces a soft, easy-to-melt glass and when used in quantity, imparts a canary yellow color to the glass. Lead glass is relatively expensive, but it has excellent electrical insulating and radiation-shielding properties and is more brilliant in appearance than soda-lime glass. It is used primarily in making electrical products, objects of art, and fine tableware.

Boron and aluminum are used to produce types of glass that are both heat and shock resistant. Borosilicate glass and aluminosilicate glass, as they are called, are used primarily for laboratory ware and oven ware.

Several additives are used specifically to produce color in glass (Fig. 8.7-4). For example, nickel oxide produces a tint ranging from yellow to purple depending on the glass base; cobalt oxide gives an intense blue; and cobalt in combination with nickel produces a gray tone. Red glass is made with gold, copper, or selenium oxides. Arsenic is sometimes added

to glass in small quantities to counteract the greenish color caused by iron impurities in the glass. Some of these colored and tinted glasses are used widely in architectural applications.

■ MIXING AND MELTING

The first major step in glassmaking is mixing and melting the raw materials. They are granulated to nearly uniform particle size, then carefully weighed and mixed in the proper proportions. Some *cullet*, or scrap glass, from a previous melt of the same kind of glass is added to the mixture to assist in the melting process and to conserve materials. The mixed batch is then loaded into a melting unit. The type of unit used depends on the type and quantity of glass needed.

Some glasses that are required in small quantities are melted in *refractory pots*, which are clay containers that can hold up to 3000 lb. of glass. Up to 20 of these pots can be heated in a single furnace, thus allowing several different types of glass to be melted at the same time.

For somewhat larger quantities of glass, *day tanks* are used. Day tanks hold enough glass for a single day's operation, up to 20,000 lb. The tank is filled with raw materials at the beginning of the opera-

tion, the glass is melted, and then all of the glass is used before the furnace is refilled.

The units most commonly used in commercial glassmaking operations are continuous tanks, holding from 50 to 1000 tons of molten glass. These operate continuously, with raw materials being fed into the loading end as rapidly as the molten glass is removed from the working end. This continuous operation can go on for years, depending on how long it takes for the heat-resistant clay walls of the furnace to be worn away by the hot glass.

■ GLASSMAKING PROCESSES

After molten glass leaves a melting tank, different manufacturing processes are employed to make various glass products. The following text discusses only those processes currently used to make flat glass: (1) *rolling*, for making patterned and wired glass, and (2) *floating*, for making float glass. *Drawing*, which was used to make sheet glass, will be touched on briefly from a historical perspective.

Other processes, such as *pressing* and *foaming*, are used to make such products as glass block and cellular glass insulation, respectively. Blowing is of course the oldest and best known glassmaking

FIGURE 8.7-4
Additives for Coloring Glass

Additive	% in Glass	Resulting Color
Cadmium sulfide	0.03–0.1	Yellow
Cadmium sulfoselenide	0.03–0.1	Ruby and orange
Carbon and sulfur compounds		Amber
Chromium oxide	0.05–0.2	Green to yellow-green
Cobalt oxide	0.001–0.1	Blue
Copper	0.03–0.1	Ruby
Copper oxide	0.2–2.0	Blue-green
Ferric oxide	Up to 4.0	Yellow-green
Ferrous oxide		Blue-green
Gold	0.01–0.03	Ruby
Iron oxide (1.0–2.0%) with manganese oxide (2.0–4.0%)		Amber
Manganese oxide	0.5–3.0	Pink-purple
Neodymium oxide	Up to 2.0	Pink
Nickel oxide	0.05–0.5	Brown and purple
Selenides		Amber
Selenium		Bronze
Uranium oxide	0.1–1.0	Yellow with green fluorescence

process, but today it is primarily employed in the manufacture of containers and art objects rather than for making architectural products.

DRAWING In the early days of glass manufacture, sheet glass was made by the blowing process. Molten glass was blown into the shape of a cylinder; the cylinder was then split open, reheated, and flattened to form a small sheet of glass. Glass made by this process could not be produced in large sheets and was of poor quality. It was characterized by uneven surfaces and longitudinal wrinkles that distorted vision.

The next evolution in sheet glassmaking was the drawing process (Fig. 8.7-5). Three different methods were used: the *Fourcault*, the *Colburn*, and the *Pittsburgh* methods. All three methods utilized the same basic procedure.

The raw materials were first melted in a continuous melting tank. The drawing operation was then started by inserting a *bait* (steel rod) into the molten glass. Molten glass has a very high viscosity, making it sticky and highly resistant to flow. Thus, when a bait was inserted into it, the glass adhered to it and was drawn up with the bait as it was raised. The molten glass was drawn directly upward between jets of flame, which fire-polished the surfaces and made them transparent.

Once the drawing process was started, pulling rollers kept it going. A continuous sheet was drawn from the molten glass and was pulled slowly through an *annealing lehr*, a type of continuous oven. Inside the lehr, temperatures were regulated carefully so that the glass was cooled gradually, thus avoiding excessive amounts of strain in the finished product.

After sheet glass emerged from the annealing lehr, no further finishing operations were required other than cutting to size and packing.

FIGURE 8.7-5 *The Pittsburgh process for drawing sheet glass vertically.*

There were some minor differences in technique between the various methods of drawing sheet glass.

The *Fourcault* process was the first process to be developed, and it was the simplest one. The glass sheet was drawn vertically from the tank to a height of about 40 ft., as described above.

The *Colburn process* was similar to the Fourcault process, except that the glass sheet was drawn vertically for only about 3 ft. and was then bent over horizontal rollers. The chief advantage of this process was that the glass sheet could be drawn through a horizontal rather than a vertical annealing lehr. The length of a vertical lehr was limited by the height to which it was practical to draw glass vertically, that is, about 30 to 40 ft. The horizontal lehr used in the Colburn process, however, could be up to 200 ft. long, thus allowing the glass to be cooled much more slowly and producing a glass that was less brittle and easier to cut.

In the *Pittsburgh* process, the glass was drawn vertically, as in the Fourcault process. It differed from the other two processes mainly in that the drawing machine was constructed so that the sheet of glass was not contacted by center rolls until the glass was hard enough to resist being marred by the rolls.

There were several problems encountered in drawing sheet glass, no matter which method was used. One problem was that the drawing process stretched the sheet, causing *waisting-in* (variations in the width) of the glass. This was overcome by water-cooled rollers at each edge, which maintained sufficient side pull to keep the width constant.

The drawing process also caused waves and ripples to form in the rising sheet of glass, before it had cooled enough to become solid. Therefore, sheet glass had an inherent wave or distortion, usually running in one direction. For most purposes this wave was not noticeable enough to cause a problem; however, when large pieces of glass were required or when an oblique angle of view was likely, these imperfections could be quite apparent. This problem largely excluded sheet glass from use for large commercial mirrors, automobile windshields, and display windows, for which the superior qualities of plate or float glass were required.

ROLLING   Plate glass was originally made by *casting*, which was a very slow and laborious process. The glass was poured onto a casting table, then rolled flat, cooled, and finally ground and polished.

Between 1922 and 1924, the major glass manufacturers began using a continuous rolling process for making both plate and patterned glass. In this process, the glass first was mixed and melted in a continuous melting tank. The molten glass then flowed through an opening in the tank into the forming rolls. These rolls were water-cooled, electrically driven, and had precision-ground steel surfaces that could not be harmed by the extremely high temperatures of the glass. They rolled the molten glass to form a semisolid continuous ribbon of glass. The rolls could be adjusted for various thicknesses of glass, in widths up to 11 ft.

From the forming rolls, the ribbon of glass moved over a series of other steel rolls until it became solidified enough to enter the annealing lehr. There it was cooled gradually to remove internal strains. After leaving the lehr, the glass was cooled still further until it reached room temperature. At that point, the glass was called rough-rolled glass.

If a translucent effect was desired, rough-rolled glass could be used without further processing. When patterned glass was desired, the forming rolls were embossed with various decorative designs; this glass, like rough-rolled glass, did not have to be ground and polished. However, if plate glass was to be made, both surfaces still needed to be ground and polished.

Most plate glass was made by the twin-grinding method, in which the glass was ground on both sides simultaneously. This method ensured smooth, parallel surfaces and, thus, distortion-free glass. A mixture of silica sand with water was used in the grinding process.

The sand was graded into several different sizes; the largest particles were used in the first grinding units and then the size of the particles gradually was reduced. A considerable quantity of glass was removed in the grinding process. Thus, the rough-rolled glass that was used to make plate glass was produced in a thickness greater than that desired in the finished product.

After grinding, the glass was moved on to the polishing tables, where it was passed along on a conveyor belt between a series of polishing wheels. The wheels were covered with buffs of felt, leather, or a soft metal, and the polishing agent most often used was rouge (ferric oxide) mixed in a solution of copperas water (ferrous sulfate and water). After polishing, the glass was cut to size and packed for shipment.

Rolling is no longer used to make polished plate glass by any manufacturer in this country. Rolling is still used, however, to make patterned and wired glass.

FLOATING   Ground and polished plate glass has been replaced by *float glass*, because the latter is substantially cheaper to produce. The float process was introduced in 1959 by Pilkington Brothers, Ltd., of England in an attempt to combine the fire polish and low cost of sheet glass with the flatness and low distortion properties of plate glass.

The first stage in the production of float glass is the same as that once used to produce sheet and plate glass. The batch is melted in a huge continuous tank. From there on, however, the process is quite different.

The molten glass leaves the tank and is floated on a bed of molten metal composed mostly of tin. When a liquid floats on another liquid of higher density, both surfaces of the upper liquid become flat and parallel. When the liquid glass moves across a float bath, it forms into a thin layer with flat parallel surfaces.

The float bath is divided into three sections (Fig. 8.7-6). In the first section, heat from the molten tin and a top heater causes the glass to float uniformly over the flat surface of the tin. In the second section, the glass receives a brilliant fire-polished surface. In the third section, the glass is cooled until it becomes hard enough to be conveyed to rollers in the lehr. Precise control of heat all along the length of the float bath is required.

After the glass leaves the float bath, it enters an annealing lehr, where it is cooled gradually to ensure that the glass will not contain excessive amounts of strain. It is then washed and moved to a warehouse for inspection, cutting, and shipping.

■ HEAT AND CHEMICAL TREATMENTS

Several different heat and chemical treatments are used to increase the strength and usefulness of flat glass products. Annealing relieves internal strain and facilitates fabrication. Tempering and heat strengthening, on the other hand, cause the glass to develop physical characteristics so it cannot be fabricated afterward. Fabrication, such as cutting, grinding, sandblasting, drilling, and notching, must be performed before these heat treatments.

FIGURE 8.7-6 *The float process for making glass.*

ANNEALING   All flat glass, after forming, is subjected to a controlled cooling process called *annealing* to relieve internal stress. If a piece of glass is not annealed, parts of it will cool and contract at different rates, and when the glass approaches room temperature, it may fracture because of differential stresses throughout the sheet.

Annealing is performed in a type of continuous oven known as a *lehr*. The glass moves through the lehr on a belt, and temperatures are carefully regulated throughout the length of the lehr. The glass is first raised to a high-enough temperature to relieve strain developed in the forming process. Then the glass is brought down in temperature very slowly so that all parts of the sheet cool at the same rate. When the glass arrives at room temperature, it is free of internal stress and can be easily cut and processed.

TEMPERING   Tempering is a heat treatment used to increase the strength of various types of flat glass. The process involves heating the glass to just below the softening point and then chilling it suddenly by subjecting both surfaces to jets of cool air. This causes the surfaces to shrink and harden quickly while the interior is still fluid. The interior begins to cool and shrink next, but since the sur-

faces are already hardened they cannot flow and adjust to the shrinkage of the interior. This causes the surfaces and edges to be placed into a state of compression while the interior of the glass is in a state of tension. In properly tempered glass the opposing stresses of the surface and the interior balance each other out. The resultant glass is much stronger than ordinary flat glass, because the built-in compression of the surface must be overcome before breakage can occur.

This type of glass usually is referred to as *fully tempered* glass, to distinguish it from *heat-strengthened* glass, which is only partially tempered. Fully tempered glass is three to five times more resistant to wind loads, impact, and thermal stresses than ordinary annealed glass of the same thickness.

HEAT STRENGTHENING   The heat treatment known as heat strengthening is also used to increase the strength of many types of flat glass. It involves heating the glass and then cooling it in a manner similar to that used in tempering. However, heat strengthening produces lower surface compression stresses in the glass than the tempering process. Heat-strengthened glass is only about twice as strong as annealed glass of the same thickness.

CHEMICAL STRENGTHENING   Glass can be strengthened by chemical treatment as well as by heat treatment. The most commonly used chemical treatment involves an exchange of ions in the surface of the glass. The glass is immersed in a molten salt bath, and the large potassium ions in the salt migrate into the glass surface, replacing the smaller sodium ions there. This action crowds the surface, resulting in compression stresses similar to those produced in glass by thermal tempering.

Full temper in soda-lime glass is difficult to achieve by chemical tempering methods. A high degree of temper by chemical means generally requires accurate formulation of ingredients selected to ensure that glass ions interchange readily with ions in the molten salt bath. Furthermore, in chemically strengthened glass, the compressed surface layer is appreciably thinner than that in thermally tempered glass. Therefore, it is more readily weakened by surface abrasions such as cuts and scratches.

■ SURFACE FINISHES

The surface finish of glass depends both on the method of manufacture and on the subsequent finishing processes to which it is subjected.

Float glass emerges from the annealing lehr with a smooth, glossy, and transparent finish, known as a *fire finish*. This type of finish is characteristic of the glassmaking process, and no further finishing operations are required. One of the reasons that plate glass was more expensive to produce than float glass was that it emerged from the annealing lehr with a rough, translucent surface that required grinding and polishing to produce the smooth, transparent finish desired for most flat glass products.

There are special cases, however, where a smooth, transparent glass is not essential. A translucent glass may be desired to provide privacy or to limit the amount of light transmitted into a room; a specially coated or tinted glass may be needed to reduce glare or to control the amount of solar heat transmitted into a room; a patterned or specially treated glass may be preferred just to produce a decorative effect. Some of the surface finishes associated with these special types of glass are discussed below.

ETCHING Many different degrees of transparency can be produced in glass by etching. These range from a frosted, almost opaque quality to a semipolished, translucent appearance.

Hydrofluoric acid or one of its compounds is used to etch glass, because it is the only chemical that interacts actively with glass. The glass is either dipped into the chemical or it is sprayed on; the type of surface produced depends on the composition of the glass, the concentration of fluorides, and the time involved. If an etched design is desired, the glass is first painted with an acid-resistant chemical (the *resist*) to protect the parts of the glass outside the desired pattern. The acid eats away the unpainted surface of the glass, leaving the design.

SANDBLASTING The translucent finish obtained by sandblasting usually is rougher than that obtained by etching. Sandblasting is done by using compressed air to blow coarse, rough-grained sand against the glass surface. Often this is done through a rubber stencil to produce a decorative design.

Both etching and sandblasting reduce the strength of glass because these processes introduce surface flaws. Sandblasting, for example, reduces the strength of glass by about 50%.

ENAMELING Translucent and solid colors can be produced on glass surfaces with vitreous enamels fired on at high temperatures. The firing process also imparts a partial or full temper to the glass. This type of glass is used primarily for spandrel areas and other parts of curtain wall construction.

SILVERING In the early days of mirror manufacture, silvering was done by hand, the silvering solutions being poured from a pitcher onto the flat glass surface. Some fine mirrors are still produced this way.

Most mirrors, however, are manufactured by an automatic process in which the silver is deposited directly on the glass as it passes on a conveyor belt under sprays of silver nitrate and tin chloride. The silvering can then be protected with a coating of shellac, varnish, or paint. For almost permanent protection, an electroplated layer of copper can be applied on the silver.

REFLECTIVE COATINGS With the advent of the glass-enclosed high-rise, an entire new family of architectural glasses has been developed to control the transmission of solar energy into buildings. These products have a thin reflective coating applied to either (1) one surface of a single sheet of glass or (2) an inside surface of a laminated or insulating glass unit.

Two processes are used to apply reflective coatings to glass: (1) the pyrolytic process, in which the coating material is sprayed onto heated glass and becomes fused to the glass surface, and (2) vacuum deposition, in which the glass is bombarded with metallic ions, forming an integral molecular bond with the glass.

Under appropriate lighting conditions, reflective-coated glass can become a type of one-way mirror, permitting vision only to the side with the greater amount of light. During the daytime, it creates a mirror-like effect from the outside of a building, while from the inside the glass is transparent. At night, when the interior lights are on, the effect is reversed, although to a lesser degree. Reflective coatings can be made to reflect varying degrees of light and heat, providing control over the solar energy transmitted into a building.

TINTING Tinting is not really a type of surface finish, because it involves adding special ingredients to the basic glass batch. However, it should be mentioned briefly here since it is used widely to produce glass with heat-absorbing and glare-reducing qualities.

Many different ingredients can be added to glass to produce different tints. Ferrous iron, for example, imparts a bluish-green tint to glass; cobalt oxide in combination with nickel imparts a grayish tint; and selenium is used to make bronze-tinted glass. These different ingredients are added to the glass mixture at the very beginning of the melting process. Float glass can be tinted in this manner.

Tinted glass expands more than untinted glass due to its greater absorption of heat. This increased expansion must be taken into account in the design of the surrounds of the glass.

8.7.1.3

Flat-Glass Products

The multitude of available glazing products makes possible a high degree of specialization in meeting specific design requirements. The method of glass manufacture determines the basic type of the glass. By modifying the ingredients, any of these types of glass can be made either clear, to provide true vision, or tinted, to reduce glare and absorb solar energy. Clear or tinted glass may be coated with a reflective surface to further reduce light and heat transmission.

Where resistance to impact and thermal stresses is important, these glasses can be strengthened by heat treatments, such as tempering or heat strengthening. Glass, either clear or tinted, can be rolled with a wire mesh reinforcement to make wired glass. Finally, almost any two or more of these glasses can be combined to form a composite multilayer glass. Different types of composite glass are designed to meet specific construction requirements, such as laminated glass for safety and security, and insulating glass for thermal and sound insulation.

■ STANDARDS OF QUALITY

In the United States, primary float glass, wired glass, and atterned glass are manufactured to meet the requirements of ASTM C 1036, "Specification for Flat

Glass"; heat-treated glass to meet the requirements of ASTM C 1048, "Specification for Heat-treated Flat Glass—Kind HS, Kind FT Coated and Uncoated Glass"; and laminated glass to meet the requirements of ASTM C 1172, "Specification for Laminated Architectural Flat Glass." Figure 8.7-7 gives a summary of the standards governing some typical glass types. Both ASTM C 1048 and C 1172 reference C 1036 for basic glass used in the glass types they cover. The ASTM standards establish the dimensional, optical, strength, allowable bow (warpage), and other requirements for these glazing products. Figure 8.7-8 gives the typical glass classifications found in ASTM C 1036 and C 1048.

Sealed insulating glass products should comply with ASTM Standard E 774, "Specification for Sealed Insulating Glass Units."

Another important standard for glass is ANSI Z97.1, "Performance Specifications and Methods of Test for Safety Glazing Material Used in Buildings." This standard covers the safety requirements for fully tempered, laminated, and wired glass, as well as rigid plastics. It is directed primarily toward ensuring that

FIGURE 8.7-7
Summary of Glass Standards

Glass Type	Standard
Primary float	ASTM C 1036
Tinted float	
Heat-absorbing and light-reducing	ASTM C 1036
Light-reducing only	ASTM C 1036
Patterned	
Heat-strengthened	ASTM C 1048
Fully tempered	ASTM C 1048
Not heat-treated	ASTM C 1036
Wired	ASTM C 1036
Heat-strengthened	
Coated and uncoated	ASTM C 1048
Fully tempered	
Coated and uncoated	ASTM C 1048
Laminated glass	ASTM C 1172
Insulating glass	ASTM E 774

FIGURE 8.7-8
Summary of ASTM C 1036 and ASTM C 1048 Glass Classifications

Types	I	Transparent glass
	II	Patterned or wired glass
Classes	1	Clear
	2	Tinted, heat-absorbing, light-reducing
	3	Tinted, light-reducing
Quality	q^1	Mirror select
	q^2	Mirror
	q^3	Glazing select
	q^4	Glazing A
	q^5	Glazing B
	q^6	Greenhouse
	q^7	Decorative
	q^8	Glazing
Form	1	Wired
	2	Patterned and wired
	3	Patterned
Finish	f^1	Patterned one side
	f^2	Patterned both sides
Mesh	m^1	Diamond
	m^2	Square
	m^3	Parallel strand
	m^2	Special
Pattern	p^1	Linear
	p^2	Geometric
	p^3	Random
	p^4	Special
Kind	HS	Heat-strengthened
	FT	Fully tempered
Condition	A	Uncoated surfaces
	B	Spandrel glass with one surface ceramic-coated
	C	Other coated glass

safety glazings have fail-safe characteristics when broken by human impact.

The use of safety glazing in hazardous locations, such as door sidelights, sliding glass doors, storm doors, and tub-shower enclosures, is required by the FHA Minimum Property Standards for Housing, the major model building codes, many municipal codes, the *CABO One and Two-family Dwelling Code*, and the Consumer Product Safety Commission, Code of Federal Regulations (CFR) 16, "Safety Standards for Architectural Glazing Materials." In addition, safety glazing requirements have been adopted by many states and are being considered by many others.

CERTIFICATION OF SEALED INSULATING GLASS The Insulating Glass Certification Council (IGCC), founded in 1977, provides independent laboratory testing and periodic unannounced on-site plant inspections of participating manufacturers. The Associated Laboratories, Inc. (ALI) and the National Certified Testing Laboratories also provide certification and testing programs. These certification programs are conducted in compliance with ASTM E 774.

CERTIFICATION OF QUALITY-SAFETY GLAZING Certification programs for safety glazing were started separately by the Glass Tempering Association and the Safety Glazing Industry, Inc. some years ago. In 1972 these two programs were joined together to form the Safety Glazing Certification Council (SGCC).

The SGCC administers an independent program of testing, random sampling, and product evaluation to ensure that safety glazing materials comply with ANSI Standard Z97.1. Products that meet all requirements of the standard are then labeled with the SGCC certification label. There is no one universal SGCC label. Instead, each manufacturer submits its own label to SGCC for approval. This label must include reference to ANSI Z97.1 plus an SGCC identification number, which indicates the plant in which the product was made. The label must be permanently affixed to the certified product at the time and place of manufacture.

The SGCC certification program is administered in accordance with American National Standard Practice for Certification Procedures, ANSI Z34.1, which provides for certification on an impartial, independent, and continuing basis. All certifications guarantee compliance with the ANSI specifications for safety glazing.

MANUFACTURERS' WARRANTIES Specialty products such as laminated, reflective, and insulating glass generally are warranted by the manufacturer against certain manufacturing defects. Laminated glass typically is covered for a period of 5 years against edge separation or defects that obstruct vision. Reflective glass is warranted for 10 years against peeling or deterioration of the reflective coating. Insulating glass usually carries a warranty against any material obstruction inside the glass unit that impairs vision, such as condensation, dust, or foreign particles. The rainbow "oil slick" effect observed occasionally at certain angles is not a manufacturing defect, but the result of light refraction between accurately-machined parallel glass surfaces.

Insulating glass generally carries either a 5-year or 10-year warranty. Stock windows and sliding doors ordered with insulating glass are normally preglazed with 5-year quality units.

In small window units enclosed by an accurately manufactured sash, the likelihood of the seal breaking is remote, and the optical quality is not critical if the windows are used primarily for daylighting rather than viewing. For these uses insulating units with a 5-year warranty generally are acceptable.

However, larger insulating units (often used in sliding glass doors and in field-glazed fixed sash) intended for viewing should carry long-term assurance against annoying visual obstructions, such as internal fogging. These units generally are subject to more mechanical stresses and are larger and more expensive to replace. They should be covered, therefore, by a standard 10-year warranty.

Most glass warranties provide protection against manufacturing defects, but specifically exclude damage due to faulty installation or use in other than ordinary buildings. For instance, the warranty does not apply for applications in ships, vehicles, or commercial refrigeration. The manufacturer usually agrees to deliver a replacement unit without charge to the shipping point nearest the installation, but the cost of forwarding the unit and actual installation must be borne by others.

■ FLOAT GLASS

Until recently there were three basic types of flat glass in general use in the United states: (1) sheet glass, (2) plate glass, and (3) float glass. Today, float glass has completely replaced the other two and will be the only one discussed here.

Float glass has become the most widely used glass product. According to the Flat Glass Marketing Association (FGMA), float glass makes up as much as 98% of the glass manufactured in the United States. It is used today for many of the purposes that were once served only by plate

glass (high-quality mirrors, display windows, and automobile glazing) and in thinner sheets for the purposes formerly served by sheet glass.

Most of the other types of glass discussed in the following paragraphs are variations of float glass.

According to ASTM Standard C 1036, clear float glass (Type I, Class 1) is available in six qualities, as shown in Fig. 8.7-9. It is usually available in thicknesses from 3/32 to 7/8 in., but some qualities may be available only in certain thicknesses. For example, q^5 (see Figure 8.7-8) is usually available only in 3/32 in.-thickness and q^4 only in 1/8 in.

■ PATTERNED GLASS

Patterned glass is translucent, with some type of linear or geometric pattern embossed on one or both sides of the glass. It is produced by the same method as plate glass (rolling) but is not ground and polished. Patterned glass is classified in ASTM Standard C 1048 when it is heat-strengthened or fully tempered and in ASTM Standard C 1036 when it is neither of these. Its normal classifications in all three cases are shown in Figure 8.7-9.

Patterned glass is available in a wide range of textures and patterns, which vary according to manufacturer. These patterns offer different degrees of light transmission and obscurity, depending on the nature of the design and whether it is rolled into one or both surfaces of the glass. Additional obscurity may be obtained by etching or sandblasting the patterned glass, producing a satinlike finish. These processes also reduce the strength of the glass, which should be considered in the determination of suitable sizes and uses.

Patterned glass is available mainly in thicknesses of 1/8 and 7/32 in., although some patterns are produced in other thicknesses on a limited basis (Fig. 8.7-10).

FIGURE 8.7-9
Usual Classifications for Some Types of Flat Glass

Product	Classifications*
Clear float	Type I; Class 1; Quality q^1, q^2, q^3, q^4, q^5, or q^6
Tinted heat-absorbing, light-reducing float	Type II; Class 2; Quality q^3, q^4, q^5, or q^6
Tinted light-reducing float	Type I; Class 3; Quality q^3, q^4, q^5, or q^6
Patterned	Type II; Class 1, 2, or 3; Quality q^7 or q^8; Form 3; Finish f^1 or f^2; Pattern p^1, p^2, p^3, or p^4
Wired	Type II; Class 1, 2, or 3; Quality q^7 or q^8; Form 1; Mesh m^1, m^2, m^3, or m^4
Wired/patterned	Type II; Class 1, 2, or 3; Quality q^7 or q^8; Form 2; Finish f^1 or f^2; Mesh m^1, m^2, m^3, or m^4; Pattern p^1, p^2, p^3, or p^4
Heat-strengthened	Type I; Class 1, 2, or 3; Quality q^3, q^4, or q^5; Kind HS; Condition A, B, or C
Heat-strengthened patterned	Type II; Class 1, 2, or 3; Quality q^7 or q^8; Form 3; Finish f^1 or f^2; Pattern p^1; p^2, p^3, or p^4; Kind HS; Condition A, B, or C
Fully tempered	Type I; Class 1, 2, or 3; Quality q^3, q^4, or q^5; Kind FT; Condition A, B, or C
Fully tempered patterned	Type II; Class 1, 2, or 3; Quality q^7 or q^8; Form 3; Finish f^1 or f^2; Pattern p^1, p^2, p^3, or p^4; Kind FT; Condition A, B, or C

*Refer to Figure 8.7-8 for meaning of classification symbols.

FIGURE 8.7-10
Summary of Patterned and Wired Glass Types and Sizes

Product	Type	Nominal Thickness		Maximum Area, in.[a]	Weight lb./sq. ft.	Visible Light Transmission, %
		in.	mm			
Patterned Glass	Floral Hammered Stippled	1/8	3	60 × 132	1.60–2.10	80–90
	Ribbed Fluted[a] Striped[b]	7/32	5.5	60 × 132	2.40–3.00	80–90
Wired Glass	Polished (square or diamond mesh)	1/4	6	60 × 144	3.50	80–85
	Patterned (square or diamond mesh)	1/4	6	60 × 144	3.50	80–85
	Parallel wired[c]	7/32 1/4 3/8	5.5 6 10	54 × 120 60 × 144 60 × 144	2.82 3.50 4.45	80–85 80–85 80–85

[a]Maximum area varies according to producer; larger sizes may be available from some producers.

[b]These are just a few of the most common patterns available; many patterns are patented and made by one producer only.

[c]This type of wired glass does not carry Underwriter's Laboratory, Inc. fire-retardant rating; it is used mainly for decorative partitions.

Some patterns also are available heat treated.

Patterned glass can provide visual privacy while at the same time allowing diffused light transmission. It has a decorative appearance and is often used for that reason alone. It is used for interior partitions and exterior windows, such as in toilet and shower rooms, where obscured vision is desired. Its most frequent uses in residential construction are in tub and shower enclosures, where it should be tempered for safety.

WIRED GLASS

Wired glass has a wire mesh or parallel wires rolled into the center of the glass thickness. Wired glass is available either polished or patterned. Most of it is 1/4 in. thick, but some manufacturers have other thicknesses in limited quantities (see Fig. 8.7-10). It should comply with ASTM Standard C 1036 (see Fig. 8.7-9).

The wire in this type of glass holds the glass together under low levels of impact, thereby reducing injuries from broken glass. It is considered a safety glazing material and, when properly made, meets the requirements of ANSI Standard Z97.1.

It is used widely for corridor and entrance doors, partitions, skylights, and other potentially hazardous locations. It is widely used in fire doors and windows because of its ability to remain in place even though cracked by excessive heat. When used as a fire-resistant material, it should conform to Underwriters Laboratories, Inc. standards for fire retardants.

TINTED GLASS

To make tinted glass, an admixture is included in the glass batch to impart glare-reducing and heat-absorbing qualities to the glass. Float glass is available tinted. Tinted glass is covered by ASTM C 1036. It may be either simply light-reducing (Class 3) or light-reducing and heat-absorbing (Class 2) (see Fig. 8.7-9).

Gray- and bronze-tinted glasses are the most commonly available types. They are generally produced in thicknesses up to 1/2 in., and in select glazing quality (q^3) only. Green-tinted glass is also made by some manufacturers, but it is generally available only in thicknesses up to 1/4 in. Equal thicknesses of gray-, bronze-, and green-tinted glass all absorb the same percentage of solar energy. Tinted glasses offer a wide range of visible light transmittance, varying from 20 to 78% depending on the thickness and the color of the glass (Fig. 8.7-11).

This type of glass generally is used where reduced light or heat transmission is desired for indoor visual comfort, or where the color of the glass can contribute to the design of the building. Because of its heat-absorbing characteristics, special precautions should be taken in the use of tinted glass to avoid damage to the edges and consequent breakage from thermal stresses.

HEAT-ABSORBING GLASS

If a metallic oxide is added to float glass during the manufacturing process, the glass can absorb significantly more sunlight, especially ultraviolet light. This characteristic distinguishes heat-absorbing glass from tinted glass that merely excludes the visible light useful for illumination (light-reducing).

The sunlight absorbed by glass is reradiated and convected as heat to the inside or the outside, depending on which side is cooler. The result is that in summer, more heat is dissipated to an air condi-

FIGURE 8.7-11
Summary of Tinted Glass Types and Sizes

Product	Quality	Nominal Thickness in.	Nominal Thickness mm	Maximum Area, in.[a]	Weight, lb./sq. ft.	Visible Light Transmission, %	Solar Energy Transmission, %	Principal Uses
Bronze	Glazing	1/8	3	35 sq. ft.	1.64	68	65	—
		3/16	5	120 × 144	2.45	58	55	
		1/4	6	128 × 204	3.27	50	46	
		3/8	10	124 × 200	4.90	37	33	
		1/2	12	120 × 200	6.54	28	24	
Gray	Glazing	1/8	3	35 sq. ft.	1.64	62	63	Where reduced
		3/16	5	120 × 144	2.45	51	53	heat and light
		1/4	6	128 × 204	3.27	42	44	transmission
		3/8	10	124 × 200	4.90	28	31	is desired
		1/2	12	120 × 200	6.54	20	24	
Green	Glazing	3/16	5	96 × 128	2.45	78	55	—
		1/4	6	96 × 128	3.27	74	48	

[a]Maximum area varies according to producer; larger sizes may be available.

[b]Transmission values vary slightly from one producer to another.

tioned interior, because of its cooler temperature. In winter, more heat is dissipated to the outside, because it is then cooler. The net effect is still beneficial, however, because 17% less heat-producing sunlight is admitted through heat-absorbing glass than through clear glass.

The performance of heat-absorbing glass is substantially improved when it is used as the outer sheet of an insulating unit. In an insulating unit, the absorbed heat must first bridge the air space to the inner sheet of glass before it can be radiated and convected into the building interior. Because heat transfer is slowed by the insulating air space, more heat is dissipated to the outside air, which is in direct contact with the heat-absorbing glass. Furthermore, the outward rate of heat dissipation is greatly accelerated if there is a wind.

■ **COATED GLASS**

When sunlight strikes a pane of glass, it is either transmitted, reflected, or absorbed and radiated as heat. By increasing the amount of reflected sunlight, the amount that can be absorbed and transmitted or radiated is reduced.

The category of coated glass called *high-performance glass* has on one side a transparent film of metal or metal oxide. This coating is applied either by *pyrolytic deposition* or *vacuum sputtering*. Those coatings applied by pyrolytic deposition are called *hard-coat* coatings. Some producers claim that they can be exposed to wear. Others do not agree. Coatings applied by vacuum sputtering are called *soft-coat* coatings. Again, some producers recommend these for use where they will be exposed to contact. Others claim that they cannot be used that way and must be applied only on the protected surfaces of panes, such as those in insulated glass or spandrel glass.

Because such coated glass appears mirrorlike when viewed from the exterior, it is called *reflective glass*. Some designers select it as much for this aesthetic property as for its ability to reduce the solar energy transmitted into a building. In insulating units it also minimizes ambient heat flow in and out of the building. Several of the more durable coatings for this glass, such as cobalt oxide, chromium, stainless steel, and titanium, are available on single glass as well as on insulating and laminated glasses. Other available coatings, such as copper, aluminum, nickel, and gold, are less durable and are furnished mostly on an inside surface of insulating or laminated glasses, where the coating is protected from weather and accidental abrasion.

A wide range of metallic colors is available with differing visual and thermal properties. Daylight reflectances range from about 10% for the more subdued coatings to about 45%, and visible light transmittance from 8 to 50%. The more common colors are neutral, gray, gold, bronze, blue, silver, green, and copper. Many coatings are available also on tinted, heat-absorbing glass. When a coating is applied to such glass, it often must be heat-strengthened or fully tempered.

Coated glass products vary widely from one manufacturer to another, and certain colors and tones are offered by only one company. Consequently, individual producers should be consulted for specific product information.

Reflective glass is more effective as the outer sheet of insulating glass than as single glazing. This is because, despite its name, reflective glass absorbs more sunlight than clear glass, which permits most of the sunlight that strikes it to pass through. There is, therefore, more heat buildup in reflective glass. When that heat is concentrated in the outer sheet of insulating glass, it is dissipated to the outside air because the air space between glass layers impedes its flow inward.

There are some important disadvantages of reflective glass:

In winter, beneficial heat gain is reduced by it.

Transmission of visible light for general illumination is severely reduced year-round, making occupants feel as though they are looking through dark sunglasses.

In spite of its large reduction in light transmission, it does not significantly increase the insulating value of the glass.

Its high reflectivity can create problems for neighbors and even for traffic on the streets. In fact, some cities now restrict the amount of reflectivity on buildings.

Most of these disadvantages have been overcome by advances in coating technology. New metallic coatings are of a different type than those used on earlier versions. Some of these coatings produce a more subdued mirror effect, reducing the amount of reflected light. These are called *low-reflectance* glasses. Others are almost transparent. Much light passes through them, but their emissivity is reduced sharply. *Emissivity* is the amount of heat absorbed by a material and then reradiated by it. This type of high-performance glass is called *low-emissivity* glass, or simply *low-E* glass. It can reflect as much as 90% of the radiant energy that strikes it, while admitting most of the light. Low-E glass is available in many colors and with many levels of performance characteristics.

Spandrel glass is used most often to cover the spandrel beams and floor construction in curtain walls. It is also sometimes used in other locations on the exterior of a building. It is often coated with ceramic materials. It may also be coated with a high-performance coating resembling that used on the vision panels in the same building. Sometimes the metal oxide coating on spandrel glass is protected with a thin sheet of plastic called an *opacifier*.

■ HEAT-TREATED GLASS

There are two types of heat-treated glass: fully tempered and heat-strengthened. When either of these types of glass is to be used, the exact sizes must be specified, because the glass cannot be cut, drilled, or notched after the heat treatment. These types of glass are governed by ASTM Standard C 1048.

FULLY TEMPERED GLASS Except wired glass and patterned glass with deep patterns, any glass 1/8 in. or thicker may be tempered. Tempered glass may be incorporated into insulating glass or laminated glass units; heat-absorbing and heat-reflective glasses may also be tempered.

An inherent characteristic of all tempered glass is a slight deviation from flatness, particularly near the edges. The degree of deviation depends on thickness, width, length, and other factors. Usually greater thicknesses yield flatter products.

Fully tempered glass has three to five times the resistance of annealed glass to uniform loading, thermal stresses, and most impact loads. When fractured, it breaks into a safe pattern of relatively small, harmless particles (Fig. 8.7-12). It is considered a safety glazing material and, when properly made, meets the requirements of ANSI Standard Z97.1. It is frequently required by state or local codes in areas where human contact with glass is probable, such as entrances and sidelights, tub and shower enclosures, storm doors, and sliding glass doors.

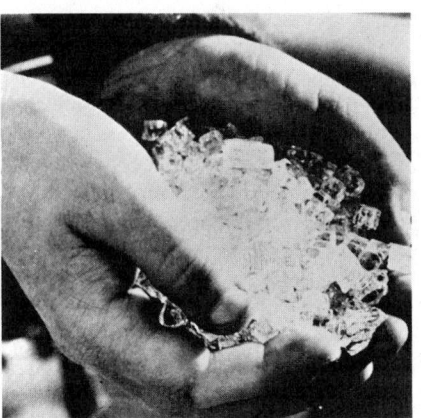

FIGURE 8.7-12 *Broken fully tempered glass.*

HEAT-STRENGTHENED GLASS Heat-strengthened glass is similar to fully tempered glass, except that it has only a partial heat temper. It is twice as strong as ordinary annealed glass of the same thickness. When it fails, it breaks into fragments much larger than those of fully tempered glass; therefore, it is not considered a safety glazing material. Heat-strengthened glass generally is used in applications that require a stronger product

than regular annealed glass but that do not require the full strength of tempered glass.

Any type of 1/4- or 5/16-in.-thick glass that can be tempered can be heat-strengthened. The most widely used form of heat-strengthened glass is spandrel glass. This glass has a fired-on ceramic frit on the interior surface. It is used commonly in spandrel areas in curtain wall construction.

■ COMPOSITE GLASS

The two main types of fabricated glass, laminated and insulating glass, are products made by combining two or more layers of glass into a single unit.

LAMINATED GLASS Laminated glass consists of two or more layers of glass with interlayers of *polyvinyl butyral plastic* sandwiched between them, bonded together under heat and pressure to form a single unit. The thickness of the glass layers may range from 3/16 in. to 7/8 in. float glass. The plastic interlayer is usually 0.015 or 0.030 in., although it may be as thick as 0.090 in. The face of the glass that is to be bonded to the plastic cannot be patterned or otherwise irregular. Heat-absorbing, heat-reflecting, fully tempered, heat-strengthened, and wired glasses may all be laminated with themselves and in combination with other types.

When laminated glass is broken, the fragments of glass generally adhere safely to the plastic interlayer and do not evacuate the opening. This reduces the potential hazards of flying glass and minimizes cutting injuries. All varieties of laminated glass are considered safety glazing materials and, when properly made, meet the requirements of ANSI Standard Z97.1 (Figs. 8.7-13 and 8.7-14).

Most laminated glass is made of two lights of glass of varying thicknesses with a single plastic interlayer. It is used commonly for interior partitions, doors, and automotive glazing. There are also several special types of laminated glass made for particularly demanding applications (Fig. 8.7-15).

Burglar-resistant glass is a laminated glass usually consisting of two layers of 1/8-in. glass with a 0.060- to 0.090-inch plastic interlayer. It is made primarily for use in the protection of show windows displaying valuable merchandise. It has superior resistance and serves as a major deterrent to "smash-and-run" burglars.

Bullet-resisting glass is a laminated glass that has four or more layers totaling 3/4 in. to 3 in. in thickness. Most manufacturers have 1-3/16-, 1-9/16-, 1-3/4- and 2-in. thicknesses listed by Underwriters Laboratories, Inc. for resistance to medium-, high-, and super-power small arms and high-power rifle arms, respectively. Thicknesses up to 7 in. are available on special order. This type of glass is used for bank teller windows, payroll booths, and other security applications.

Acoustical glass is a laminated glass with one or more plastic interlayers of 0.045-in. thickness. It is particularly effective in reducing sound transmission in the frequency ranges of speech, radio, and TV (250 to 4000 cps). It is widely used for office partitions and in radio and TV studios and similar applications.

INSULATING GLASS Insulating glass is a

FIGURE 8.7-14 *Laminated security glass protects the users of the CN Tower in Toronto.* *(Courtesy Laminated Glass Corporation.)*

FIGURE 8.7-13 *Laminated glass was used in this sloped glazing application at the IBM facility in Charlotte, North Carolina.* *(Courtesy Laminated Glass Corporation.)*

factory-produced, airtight unit, consisting of two layers (lites) of glass, with a dehydrated air space between. Insulating units are produced with two different edge types: spacer-bar edge and fused-glass edge (Fig. 8.7-16). Insulating glass is governed by ASTM Standard E 774.

Any glass type with at least one smooth surface may be used in an insulating glass unit. When heat-absorbing or reflective-coated glass is used in an insulating glass unit, and it is known that the glass will be subjected to severe thermal stresses or high wind loads, one or both lites of the unit should be heat-strengthened or fully tempered.

Insulating glass is designed primarily to protect glazed areas from excessive heat loss in winter and heat gain in summer. Use of insulating glass can substantially reduce both heating and air conditioning costs.

Insulating glass has the added advantage of allowing a much higher interior relative humidity than single glass because it minimizes condensation problems during cold weather (Fig. 8.7-17).

Spacer-edged insulating units consist of two lights of glass separated by a hollow metal or rubber spacer around the edges and sealed airtight with an organic sealant, such as polysulfide or butyl. The entrapped air is kept dehydrated by a desiccant (chemical dehumidifier) located in the spacer. Until recently the edge construction and glass in many units were held together by a metal channel that wrapped around the edge of the unit, encasing both panes of glass. This was designed to provide edge protection and, in some cases, to hold the assembly together. This type of metal-edged insulating glass units may be found in some existing buildings, but modern insulating glass units do not have this feature.

Spacer-edged insulating units are most commonly used in commercial construction. Sealed insulating units are manufactured and stocked in a limited number of standard sizes and thicknesses. For example, units with 1/4- and 1/2-in. air spaces and glass thicknesses ranging from 1/8- to 3/8-in. are available on that basis (Fig. 8.7-18). Special sizes can be provided by some manufacturers at slight additional cost and longer delivery time.

Fused glass-edge insulating units consist of two lites of SS (3/32 in.) or DS (1/8 in.) float glass with a 3/16-in. air space between. The two sheets are fused together to form a glass-to-glass sealed edge. The air space in glass edge units may be filled with dehydrated air or inert gas.

FIGURE 8.7-15
Summary of Laminated Glass Products

Product	Glass Type	Plastic Interlayer in.	Nominal Thickness in.	mm	Maximum Area, in.[a]	Weight lb./sq. ft.	Principal Uses[b]
Laminated Glass	SS/SS[c]	.015–.030	13/64	5	48 × 80	2.45	Partitions, doors, and automotive glazing
	Lam/Lam[c]	.015–.030	15/64	6	48 × 100	2.90	
	DS/DS[c]	.015–.030	1/4	6	48 × 100	3.30	
	2 lites 1/8 in.	.015–.030	1/4	6	72 × 120	3.30	
	2 lites	.015–.030	3/8	10	72 × 120	4.80	
			1/2	12	72 × 120	6.35	
			5/8	16	72 × 120	8.00	
			3/4	19	72 × 120	9.70	
			7/8	22	72 × 120	11.40	
			1	25	72 × 120	12.95	
Burglar-resistant	2 lites 1/8 in.	.060–.090	5/16	8	60 × 120	3.50	Display windows
Bullet-resistant	At least 4 lites	At least 3 plies of .015	1-3/16	30	40 sq. ft.	15.50	Bank teller windows and payroll booths
			1-9/16	40	30 sq. ft.	20.00	
			1-3/4	45	25 sq. ft.	23.00	
			2	50	22 sq. ft.	26.20	
Acoustical[c] STC-36 STC-40 STC-43	2 lites	One or more plies of .045	9/32	7	60 × 120	3.40	Office partitions, radio and TV studios, airport glazing
			1/2	12	60 × 120	6.36	
			3/4	19	60 × 120	9.74	

[a]Maximum area varies according to producer. Larger sizes may be available from some producers.

[b]Double-layer products using thicker glass and interlayers and multilayered lamniated glass products are also used in other demanding installations, such as zoological animal enclosures, observation platforms (see Figs. 8.7-13 and 8.7-14), and incarceration facilities.

[c]Lam indicates 7/64-in. float glass; SS, 3/32-in. float glass; DS, 1/8-in.

[d]Refers to sound transmission class ratings.

Fused glass-edge units are available in a more limited range of sizes and glass combinations than spacer-edged units (see Fig. 8.7-18). They were developed mainly to provide economical insulating glass for smaller lites in residential and commercial buildings.

8.7.2

PLASTIC GLAZING MATERIALS

Plastic glazing materials are either polycarbonate or acrylic plastic. Both are tough, breakproof, shatterproof, and crack-resistant thermoplastics. They have a higher coefficient of expansion than glass, are lighter, and resist impacts better than glass. They are, however, much more susceptible to scratching, and they are flammable, which glass is not.

Acrylics are harder and more impact resistant than polycarbonates, and they wear better. Both can be cold-formed into curves. Polycarbonates are the more heat-resistant of the two. Both can be coated with an abrasion-resistant silicone-based material to increase their resistance to weather and chemicals. Polycarbonate sheets can also be coated with polymethyl methacrylate (acrylic) material to reduce damage normally caused to it by sunlight.

8.7.3

GLAZING ACCESSORY MATERIALS

Glazing with glass and plastic requires the use of some materials other than the glass and plastics.

8.7.3.1

Glazing Sealants

Glazing sealants should be selected for their intended use and their compatibility with the glazing materials and the other surfaces they will contact. For example, silicones may become discolored and even fail to function as a sealant if they come into contact with neoprene gaskets or neoprene or EPDM setting blocks or spacers. A particular sealant should not be selected unless its compatibility with materials that will be adjacent to the sealant has been tested. The results of such tests are generally available from sealant manufacturers.

■ FACE-GLAZING COMPOUNDS

Oil-based glazing putty is still occasionally used to install small lites in wood

FIGURE 8.7-16 *Insulating glass types: (a) spacer bar edge; (b) fused glass edge.*

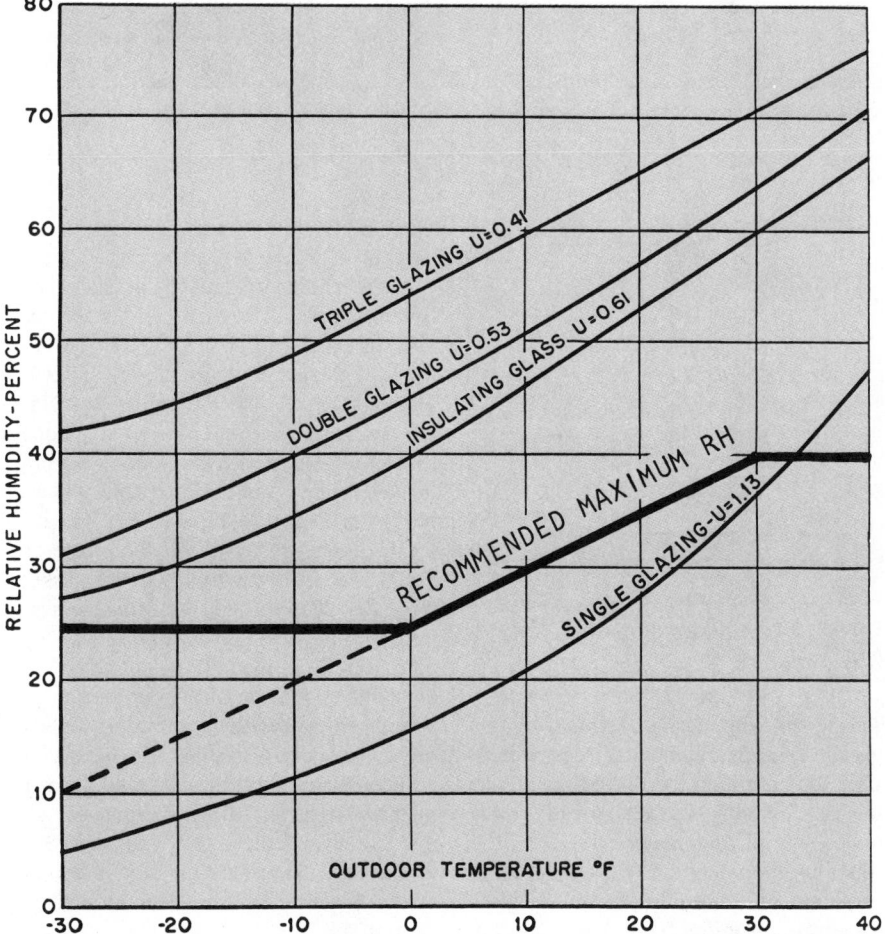

FIGURE 8.7-17 *Maximum relative humidity causing condensation on insulating and other glass.*

years when protected by periodic painting over the exposed surface.

■ TAPES

Glazing tape is polybutene or polyiso-butylene-butyl tape. It should be 1/16 in. thicker than the finished seal and be compressed during application.

■ SILICONE SEALANTS

Silicone formulations are used as elastic glazing compounds, especially, but not exclusively, in butt-glazing applications. Special formulations of chemically curing, one-part silicone are used in structural glazing applications and must be specifically formulated for these purposes.

8.7.3.2

Gaskets

Glazing gaskets should be the standard types and materials of the manufacturer of the unit being glazed. Gaskets may be vinyl, neoprene, or another elastomeric material. Lock-strip gaskets are rubber.

8.7.3.3

Miscellaneous Glazing Materials

Setting blocks and spacers should be neoprene, EPDM, silicone, or rubber-extruded blocks. These should be of the size, shape, and hardness recommended by the glass and sealant manufacturers.

Compressible filler rods should be closed-cell or waterproof-jacked rod of synthetic rubber or plastic foam.

Glazing clips should be as required by the conditions.

8.7.4

GLAZING

Watertight and airtight installation of glazing materials is the goal of glazing. Installed glazing materials must be able to withstand the effects of temperature changes, wind, and impact with no breakage of glazing material, failure of sealants or gaskets, deterioration of glazing materials, or other defects.

and metal sash, but most such glazing today is done using elastic synthetic materials specifically formulated for the purpose. They should be suitable for painting.

■ CHANNEL-GLAZING COMPOUNDS

Channel-glazing compounds may be either elastomeric or elastic oil- and resin-based materials. They are used for both bedding and face glazing into wood or metal. Glazing compounds set firmly but retain a limited flexibility, generally greater than that of putty. Glazing compounds have a life expectancy of 5 to 20

FIGURE 8.7-18
Summary of Insulating Glass Products

Product	Glass Type and Thickness, in.	Nominal Air Space		Nominal Unit Thickness		Maximum Area, sq. ft.*	Weight lb./sq. ft.	Principal Uses
		in.	mm	in.	mm			
Spacer edge units	2 lites DS, 1/8	1/4	6	1/2	6	22.2	3.27	Residential and commercial glazing
		1/2	12	3/4	19			
	2 lites, 3/16	1/4	6	5/8	16	34	4.90	
		1/2	12	7/8	22	41.7		
	2 lites, 1/4	1/4	6	3/4	19	50–70	6.54	
		1/2	12	1	25			
Glass edge units	2 lites SS, 3/32	3/16	5	3/8	10	10	2.40	Smaller lites in residential and commercial glazing
	2 lites DS, 1/8	3/16	5	7/16	11	24	3.20	

*Maximum area given is for units constructed with clear glass; when tinted or reflective coated glass is used, either the maximum area must be reduced, or special fabrication is required.

Installation should comply with the requirements of the appropriate standards and codes. It should also comply with the recommendations of the manufacturers of the glass or plastic glazing material; of the sealant, gaskets, and other glazing materials manufacturers; and of the window, door, frame, entrance, store front, or curtain wall manufacturers.

The locations and the conditions where glazing materials will be installed should be inspected, and unsatisfactory conditions corrected before glazing is started. Surfaces to be glazed should be cleaned and removable coatings removed. Ventilating sash should be properly adjusted before they are glazed.

8.7.4.1

General Requirements for Glazing

Glazing includes the installation of glazing beads and gaskets that are furnished by the manufacturer of the unit being glazed. Wood and metal doors and some portions of curtain walls may be either factory-preglazed or glazed in the field. Most windows are glazed in the factory. Frames, sidelights, storefronts, and most curtain walls are glazed in the field.

The design of a glazing channel (Fig. 8.7-19) must provide the necessary minimum bite and edge clearance and allow for the proper glazing sealant thickness.

FIGURE 8.7-19 *Typical glazing channel terminology.* (Drawing by HLS.)

The person who installs glazing materials is called a *glazier*. The glazier is responsible for providing the correct glass size for each opening, but realistically many of today's glazing materials leave little room for adjustment. For example, heat-treated and preassembled insulating glass cannot be cut in the field. Therefore, the sizes of openings and glazing materials must be coordinated in the shop. Panes of glass and plastic should be fabricated in the shop with the proper edge clearances. Glass clearances should be in accordance with the Flat Glass Marketing Association's (FGMA) *Glazing Manual* or the requirements of the glass manufacturer.

Glazing systems must have provisions to prevent moisture from accumulating in them. This requires that they have weep holes open to the exterior (see Fig. 8.7-19).

Many glazing systems are in use today. Small openings in wood and steel are sometimes glazed by placing the glass in a bed of glazing compound (Fig. 8.7-20). Stop glazing is also used (Fig. 8.7-21). Glazing in metal is often done with stops as well, in which case the exterior glazing material may be a liquid or tape sealant but is sometimes a type of gasket. The interior glazing material may be a tape or liquid sealant but is more often a gasket (Fig. 8.7-22). In high-performance systems the interior and exterior systems may consist of more than one sealant. For example, there may be a tape topped off with a liquid silicone sealant or a gasket topped off by an acrylic sealant. In lockseam gasket glazing, there are no liquid glazing materials (Fig. 8.7-23). The glass is set into a groove on the gasket, and the gasket is squeezed against the glass by insertion of the lockstrip with a special tool (see Section 8.8.1.1.)

Insulating glass should be installed with the tinted glass toward the exterior with coated glass, where used, as the outer lite. Opaque glass should have its coated side toward the interior.

Insulating glass, laminated glass, and transparent mirrors should not be allowed to come into contact with chlorinated solvents or benzine-related compounds, such as toluene, nor should glazing compounds used with these glass types be thinned with these compounds. They will damage the interlayer in such glasses.

The use of setting blocks, spacers, and edge blocking should be in accordance with the FGMA *Glazing Manual*.

Wire glass should be installed in accordance with the requirements of NFPA 80.

Two pieces of glass that abut end to end or at a corner should be sealed with beads of silicone sealant, following the manufacturer's recommendations.

FIGURE 8.7-20 *Face glazing for small lites in wood and metal.* (*Drawing by HLS.*)

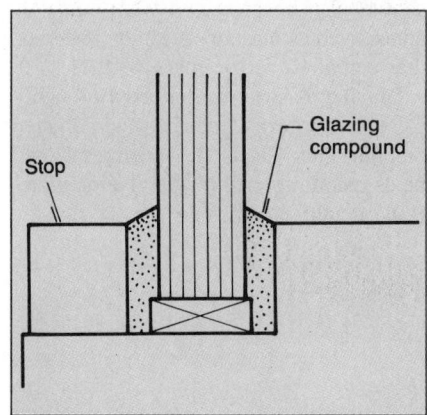

FIGURE 8.7-21 *Stop glazing.* (*Drawing by HLS.*)

8.7.4.2

Structural Silicone-Sealant Glazing

In a structural silicone-sealant glazing system, the glass is bonded to metal or glass supports with a structural-grade silicone sealant.

There are three basic types of structural silicone-sealant glazing systems: (1) those

FIGURE 8.7-22 *A typical glazing into metal system. Many others are also used.* (*Drawing by HLS.*)

FIGURE 8.7-23 *A lockstrip glazing gasket.* (*Drawing by HLS.*)

with silicone sealant supporting two sides of the glass and with the other two sides of the glass supported by a metal frame (Fig. 8.7-24); (2) those with structural silicone sealant supporting four sides of the glass; and (3) those with glass mullions called *fins* (Fig. 8.7-25). The butt joints of glass in these systems are sealed with a silicone sealant, but it is not necessarily a structural sealant.

8.7.4.3

Glazing with Plastic

Openings for plastic glazing should be accurately measured and the plastic sheets cut to fit, leaving the appropriate room for

expansion. The size of the sheets should be determined in accordance with the manufacturer's recommendations for edge clearances.

Plastic sheets should be set using the sealants recommended by the plastics manufacturer.

8.7.4.4

Cleaning and Protection

Labels should be removed from insulating glass to prevent thermobreakage or staining from occurring due to the labels. Labels may be left on other types of glass until final cleaning unless the manufacturer of the glass recommends removal.

FIGURE 8.7-24 *Detail of a typical structural glass. The other two sides of the glass may be similarly supported or supported by a metal frame in the conventional manner (see Figs. 8.8-11 and 8.8-12).* (*Drawing by HLS.*)

- Structural mullion
- Structural silicone sealant
- Bond breaker
- Silicone weatherseal
- Insulating glass

FIGURE 8.7-25 *A typical glass mullion. Other types place the glass mullion behind the vision glass and attach the vision glass to the mullions with metal connectors.* (*Drawing by HLS.*)

- Structural silicone glazing sealant
- Vision glass
- Glass fin

Glass should be protected from breakage immediately after installation by barriers and warning tapes. Warning tapes should be attached to the frames, not the glass, to prevent damage to the glass by the warning devices. Glazing materials should also be protected from contaminants, such as masonry cleaning material, that would harm the glass.

Glazing materials should be thoroughly cleaned of foreign materials and polished on both sides. Cleaning procedures should be as recommended by the glazing materials' manufacturers.

8.8 GLAZED ALUMINUM CURTAIN WALLS

A building wall that carries no superimposed loads is called a *curtain wall*. An aluminum glazed curtain wall is an exterior curtain wall that consists essentially of an aluminum (usually extruded) framework infilled with vision panels (fixed or windows) or opaque panels such as spandrels and column covers. Curtain walls are different from *store fronts* in that they generally do not contain entrances and

usually occur above the first floor (Fig. 8.8-1). Curtain wall manufacturers call their products *curtain walls*, and store front manufacturers call theirs *store fronts*. Metal doors, frames, entrances, and store fronts are discussed in Section 8.2; windows used in glazed curtain walls in Section 8.4; glazing used in metal and wood curtain walls and structural glass curtain walls in Section 8.7.

This section discusses only aluminum-framed glazed curtain walls, because they are the only type specifically manufactured and sold for use as glazed curtain walls. Their components are manufactured to fit together. Other metals and wood are also used in glazed curtain walls, and some glazed curtain walls are made of unframed glass with or without metal supporting angles or clips. The principles here apply

FIGURE 8.8-1 *One Skyline Tower, Baileys Crossroads, Virginia; architect The Weihe Partnership.* (*Photograph courtesy Bruce Wall Systems Corporation.*)

to curtain walls made from other materials, even though they may contain formed rather than extruded members.

8.8.1

SYSTEMS

Most aluminum glazed curtain walls consist of a manufactured system composed of standard extruded components with metal fill in panels where glass is not used. Some designers, however, choose to create custom systems. Such custom systems may vary in such relatively simple ways as the substitution of another material, such as stone, for the metal panels in standard systems. Some custom systems, however, have mullions or other components of configurations and sizes different from any standard system. It is important for a designer choosing to introduce a custom aluminum curtain wall system to recognize that such systems are

inherently more costly than standard systems, that costly testing of such systems is advisable, and that regardless of the results of such tests the designer will ultimately be held accountable for the performance of such a system.

The American Architectural Manufacturers Association (AAMA) classifies glazed curtain walls by their installation methods in five systems: *stick*, *unit*, *unit-and-mullion*, *panel*, and *column-cover-and-spandrel*.

8.8.1.1

Stick Systems

Most aluminum glazed curtain walls are stick systems. They consist of mullions, the usually extruded vertical elements that provide the major support for the system; vision panels that may be either fixed glass or plastic sheets or windows; and opaque panels, which are usually insulated (Fig. 8.8-2). They may also include horizontal rails, sills, fascias, copings, and trim. Most of the components in most stick systems are stock, off-the-shelf pieces used without modification. Some stick systems,

however, use custom-made components. Others are a mixture of stock and custom-made pieces.

Stick systems are usually glazed in the field from the interior with gaskets or a glazing compound. Systems with lockstrip glazing from the exterior are also available. In those systems a glazing gasket is inserted into the frame and the glazing is installed into it with a special zip-in strip gasket. Special tools are needed to install this type of glazing system.

Sloped glazing is a special type of stick system in which portions are either sloped or curved (Fig. 8.8-3).

8.8.1.2

Unit Systems

Unit curtain wall systems are modular panels that are preassembled in the factory and shipped to a construction site in a single piece (Fig. 8.8-4). These panels are usually glazed at the factory and come with spandrel panels already in place. In some designs, field-installed joint covers are required.

FIGURE 8.8-2 *Details of a stick-type glazed curtain wall system.* (*Drawing by HLS.*)

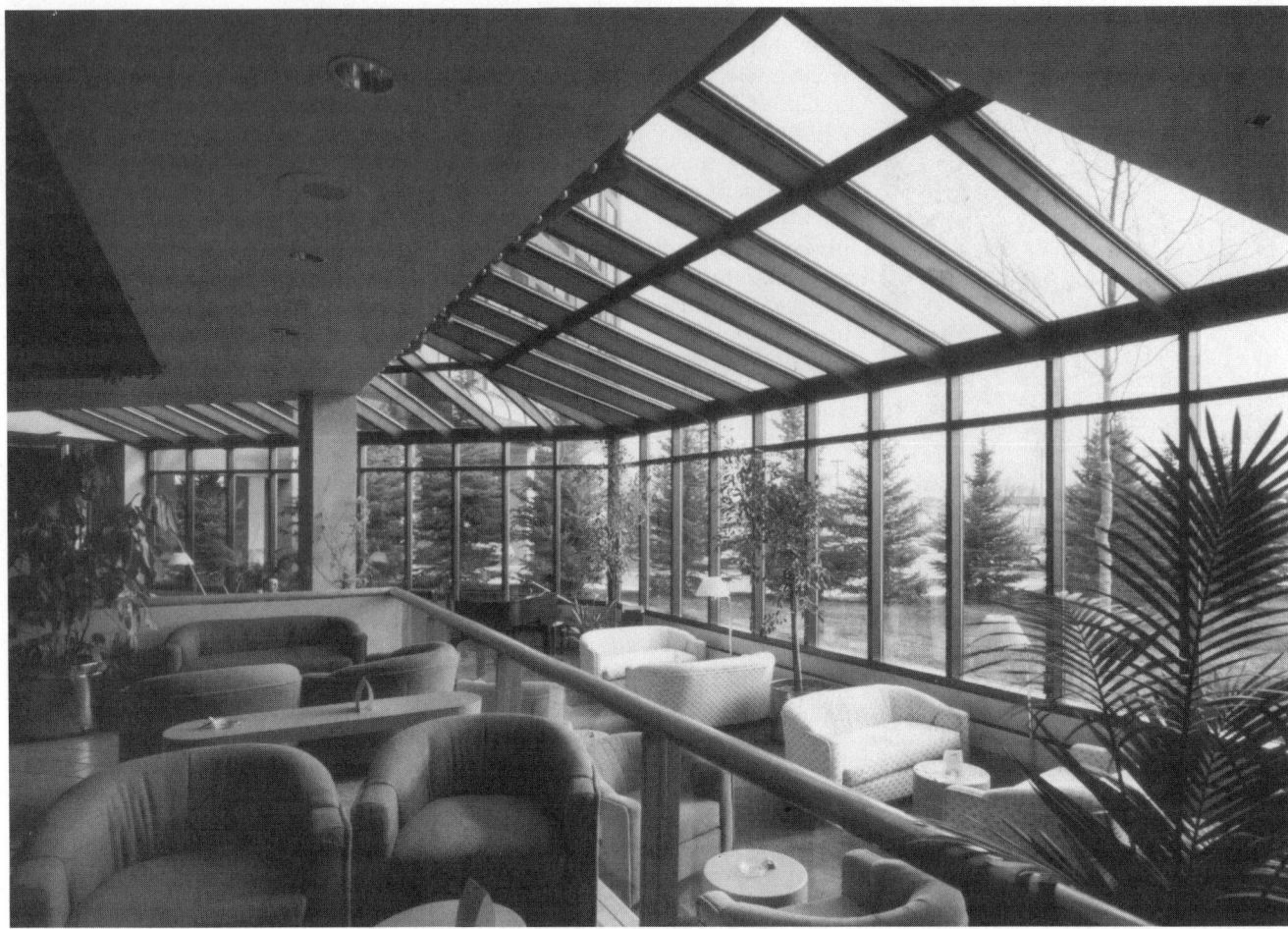

FIGURE 8.8-3 *A sloped glazing system. (Courtesy Tubelit Indal Architectural Systems.)*

8.8.1.3

Unit-and-Mullion Systems

Unit-and-mullion curtain wall systems are a combination of the stick system and the unit system (Fig. 8.8-5). The framing (vertical) mullions are shipped loose and installed in the field. The infill panels are factory-made units that are inserted between the mullions in the field. In some unit-and-mullion systems, the infill panels contain spandrels and vision panels in a single unit. In others, the spandrels and vision panels are separate units, shipped and installed separately in the space between the mullions.

8.8.1.4

Panel Systems

A panel curtain wall system is similar to a unit system, the difference being that a panel system has homogeneous sheet or cast panels with few joints and may not have separate mullions. Unit systems are made up of smaller components. There are two types of panel systems: (1) architectural and (2) industrial.

Architectural panels are custom-designed systems that tend to be expensive (Fig. 8.8-6). They are used mostly on very large projects. Industrial panel curtain wall systems are made from sheet materials and in large quantities. They are, therefore, relatively inexpensive. These are the types of panels often used on metal buildings (Fig. 8.8-7).

8.8.1.5

Column-Cover-and-Spandrel Systems

Column-cover-and-spandrel systems consist of aluminum column covers and spandrels that fit between them. They are similar to a unit-and-mullion system, except that the building's structure is ex-pressed by the column covers. The space between columns is filled with glazing panels, with or without separate mullions, and spandrel panels (Fig. 8.8-8).

8.8.2

STANDARDS

Aluminum glazed curtain walls should conform to the applicable standards of the following groups:

Aluminum Association (AA)

American Architectural Manufacturers Association (AAMA)

American Society for Testing and Materials (ASTM)

The National Association of Architectural Metal Manufacturers (NAAMM)

American National Standards Institute (ANSI)

FIGURE 8.8-4 *A panel-unit glazed curtain wall system being installed.* *(Courtesy Kalwall Corporation.)*

In general, aluminum glazed curtain walls should comply with the following AAMA publications:

Curtain Wall Design Guide Manual (CW-1-9)

Curtain Wall Manual #10, "Care and Handling of Architectural Aluminum from Shop to Site" (CW-10)

Curtain Wall Manual #11, "Design Windloads for Building and Boundary Layer Wind Tunnel Testing" (CW-11)

Curtain Wall Design Manual #12, "Structural Properties of Glass" (CW-12)

Curtain Wall Design Manual #13, "Structural Sealant Glazing Systems" (CW-13)

Metal Curtain Wall Manual (MCWM-1), "Methods of Test for Metal Curtain Walls" (501)

"Sound Control for Aluminum Curtain Walls and Windows" (TIR-A1)

"Voluntary Test Method for Thermal Transmittance and Condensation Resistance of Windows, Doors, and Glazed Wall Sections" (1503.1)

"Sloped Glazing Guidelines" (TIR-A7)

Other standards are mentioned elsewhere in this text.

8.8.3

MATERIALS AND FINISHES

Refer to Section 5.2 for a general discussion about the metals and finishes addressed in this section.

8.8.3.1

Materials

A variety of aluminum alloys and tempers are used to make extruded curtain wall components. Alloys should conform with

ASTM Standard B 221, "Specification for Aluminum-alloy Extruded Bars, Rods, Wire, Shapes, and Tubes." A common alloy used is 6063. Common tempers are T5 and T6. Other tempers and alloys may be used, depending on the design and the finish.

Sheet aluminum is often used for closures and trim, bent-plate for structural support and trim. Sheet aluminum and bent-plate components should comply with ASTM Standard B 209, "Standard Specification for Aluminum and Aluminum-alloy Sheet and Plate."

Concealed components, such as subframes, stiffeners, edge channels, reinforcements, and the like, should be fabricated from the same materials as the exposed portions of the curtain wall.

Many types of opaque panels are used in aluminum glazed curtain walls. Spandrel glass is a common one. Others include metal-faced, ceramic tile-faced, stone,

FIGURE 8.8-6 *Architectural panels may take almost any form that can be made in a metal shop, cast in concrete, or molded in plastic.* (Drawing by HLS.)

FIGURE 8.8-5 *A unit-and-mullion glazed curtain wall system.* (Drawing by HLS.)

brick, and fiber-reinforced cementitious color-coated materials. Many stock systems come with aluminum-faced panels. Some systems will accommodate only certain types of panels. Therefore, the selection of panel types must often be made before the major framing components can be selected. For example, many stock sections will not accept stone panels—they must be adapted or a different section selected.

Louvers and grilles are also common. These may be a part of a system's opaque panels or be themselves infill between the mullions. They may be glazed into a system with glazing beads or gaskets or mounted mechanically into the frames.

8.8.3.2

Finishes

The mill finish is usually left on concealed surfaces and sometimes even on the exposed surfaces of aluminum glazed curtain wall components. However, most exposed portions of aluminum glazed curtain walls are anodized (See Section 5.2).

Exposed fasteners should be finished to match the adjacent metal.

8.8.3.3

Supports, Anchors, Fasteners, and Flashing

Supports, anchors, fasteners, window cleaners' bolts, and flashing materials should be of materials compatible with the material being anchored or fastened and with the substrate. Aluminum should be anchored, fastened, and flashed with aluminum, nonmagnetic stainless steel, cadmium-plated steel, or galvanized steel. Occasionally a manufacturer's design will require the use of aluminum castings as structural elements. This can be acceptable when the design takes the strength

FIGURE 8.8-7 *Industrial panels take many shapes, including corrugated ones.* *(Drawing by HLS.)*

characteristics of the casting into account. However, aluminum castings are not generally used in this way, because they sometimes lack sufficient shear strength.

Inserts for use in concrete may be cast iron, malleable iron, or hot-dip galvanized steel.

8.8.4

PERFORMANCE REQUIREMENTS AND TESTING

Standard extruded aluminum glazed curtain wall systems should have been tested by their manufacturer or by an independent testing agency employed by the manufacturer to ascertain their resistance to deflection due to wind loads, water penetration, air infiltration, condensation resistance, thermal transmittance, and, where applicable, sound transmission. The results of these tests should be obtained from the manufacturer and taken into consideration in the selection of a system for a particular project. The characteristics of the selected system should be compatible with conditions that are expected to exist at the project site, applicable code requirements, and the demands of the type of occupancy.

Testing of the actual completed system is advisable for custom-designed systems and often also for standard systems with special requirements, such as those in very tall structures. Additional testing is expensive, however, and is seldom justified for standard systems or for relatively small projects. Such testing should be done by an independent testing laboratory (one not affiliated with the manufacturer or the contractor). It may take place on a mockup of the actual system constructed in the testing laboratory's testing facilities before work on the curtain wall at the building site begins. This mockup should be built exactly as will the actual work on the building. As an alternative, these tests may take place on a mockup of the curtain wall constructed at the site. This mockup is sometimes separate from the building and sometimes is a portion of the actual curtain wall. When the designer approves, a satisfactory field-constructed mockup is sometimes incorporated into the building.

The performance characteristics of a system are determined by the effect on a system when it is tested at specific levels of wind pressure. The test pressure to use in these tests is determined by the fastest wind speed likely to be encountered at a building's location. A map in ANSI A58.1, "Minimum Design Loads for Buildings and Other Structures," can be used to estimate this maximum wind speed. After the expected wind velocity has been determined, the resulting wind loads to be used are taken from load tables. Such load tables are based on wind velocities from 70 mph to 110 mph and the height of the curtain wall above ground from zero (at grade) to 500 ft.

The test pressures given in this section represent those recommended by the listed standards. Higher pressures may be required for a given project.

Sometimes it is necessary to reinforce aluminum mullions and other components

FIGURE 8.8-8 *A column-cover-and-spandrel system. The glazed panels may contain windows or be made up of strip windows.* *(Drawing by HLS.)*

so they comply with the performance requirements.

8.8.4.1

Structural Performance

An aluminum glazed curtain wall system should be tested with the expected wind load and the procedures listed in ASTM Standard E 330, "Test Method for Structural Performance of Exterior Windows, Curtain Walls, and Doors by Uniform Static Air Pressure Difference." ASTM Standard E 1233, "Test Method for Structural Performance of Exterior Windows, Curtain Walls, and Doors by Cyclic Static Air Pressure Differential," may also be used, but AAMA has not yet established minimum test criteria for it.

Vertical systems should be tested both perpendicular and parallel to the plane of the wall and in both the inward acting and the outward acting directions. Sloped glazing should also be tested for upward- and downward-acting loads and also for a uniform load, usually 200 lb., acting on the system at any point. This is to ensure that a person walking on the system will not damage it. This load may be higher if a larger concentrated load will be superimposed on the system.

The test pressures for perpendicular-to-the-wall tests should be 150% of the wind pressures to be encountered. Pressures may vary with the system's elevation above ground and proximity (within 15 ft. or so) to corners, where pressures are higher. The actual loads to be used should be established by a registered structural engineer. Some building codes dictate the minimum pressures to be used.

The maximum allowable deflections under these loads are usually 1/175 of any span, but project conditions and the system manufacturer's requirements may require limits of as much as 1/1000 of any span. In any case, the maximum deflection should not exceed 3/4 in. over the total length of any member. In addition, under the selected loads no panels should be dislodged, no permanent deflection should occur, and the width of sealant joints should not be reduced to less than half the designed width.

The allowable movement parallel to the wall is dictated by panel edge conditions. The bite on glazing and opaque panels should not be reduced to less than 75% of the design value. In addition, the edge clearance between panels and other elements should not be reduced to less than 1/8 in. Greater restrictions may be imposed by the method in which the system is assembled.

8.8.4.2

Water Penetration

The water penetration resistance of an aluminum glazed curtain wall system is determined by testing it in accordance with ASTM Standard E 331, "Test Method for Water Penetration of Exterior Windows, Curtain Walls, and Doors by Uniform Static Air Pressure Difference." The test pressure is suggested by AAMA 501 as 20% of the inward design wind load pressure but not less than 6.24 pounds per square foot (psf) nor more than 12 psf. For a system to pass this test, no water may pass through it during the test.

8.8.4.3

Air Infiltration

Air infiltration is tested in accordance

FIGURE 8.8-9 *A typical glazed curtain wall mullion containing a thermal break feature.* (*Drawing by HLS.*)

with ASTM E 283, "Test Method for Rate of Air Leakage Through Exterior Windows, Curtain Walls, and Doors." The usual pressure is 1.57 psf; the permitted infiltration is 0.06 cu. ft./min./sq. ft. of fixed area.

8.8.4.4

Condensation Resistance

A thermal break basically involves separating the aluminum parts on the warm inside portion of a frame from those on the cold outside portion through the use of an insulating material such as plastic (Fig. 8.8-9). This keeps heat from being conducted through the frame or sash; the insulating separator is called a *thermal break* or *thermal barrier*.

Condensation resistance is tested for frames that have a thermal break feature. Condensation resistance is determined by procedures outlined in AAMA 1503.1, "Voluntary Test Methods for Thermal Transmittance and Condensation Resistance of Windows, Doors, and Glazed Wall Sections." A system's condensation resistance gives an indication of that system's ability to resist condensation. The higher the condensation resistance factor, the less likely condensation is to occur.

8.8.4.5

Thermal Transmittance

As discussed in Section 8.1, heat will flow quite readily through doors, packaged entrances, and store fronts unless some means are taken to reduce this flow. The methods for determining conductive heat transmission through these assemblies is a part of AAMA 1503.1, "Voluntary Test Method for Thermal Transmittance and Condensation Resistance of Windows, Doors, and Glazed Sections."

In addition, it is usual to require that an aluminum glazed curtain wall be able to withstand the highest temperature differential that it will encounter without damage of any kind. In most parts of the country this is 120°F. This can result in a metal surface temperature differential of 180°F. Some extreme southern and northern areas may experience even larger temperature differentials.

8.8.4.6

Sound Transmission

When a specific sound transmission level is required, an aluminum glazed curtain wall should be tested in accordance with ASTM Standard E 90 to establish its sound transmission class (STC). The required STC depends on the project and its location. Buildings adjacent to high noise levels, such as one adjacent to the flight pattern of an airport or near a major highway, will require much higher STCs than buildings in ordinary settings. A manufacturing facility will not require as high an STC as a school, an office building, or an apartment unit.

8.8.5

DESIGN AND FABRICATION

Aluminum glazed curtain wall systems designed to exclude water simply by having tight or sealant-filled joints are not always successful. This is because the pressure within the system is often lower than the pressure on the exterior. This causes water to be sucked into every fault faster than it can run out through weep holes. In fact, rain water will even be sucked in through the weep holes unless they are placed where no water can reach them.

The answer to this dilemma is to design the system so that minor amounts of water are permitted to infiltrate the system, but are then conducted back to the exterior through a series of water channels and weep holes without entering the building. All sealant-filled joints are located behind the water channels where water cannot reach them. This mechanism works on what is called the *rain screen principle*. A portion of the frame screens the sealants so that water does not come into contact with them. Sealants in curtain walls are called *differential seals*, meaning that the difference in pressure on the two sides of the sealant is almost always more than 1 psf. Water will penetrate almost all sealed joints under those circumstances. However, the water channels are open to the air so that the pressure in them is the same as that outside the screen—there will

be no pressure differential to suck water into the channels. Water that does enter the channels will be quickly led back to the exterior without reaching the sealant joints. Systems with this type of design are less likely to leak. The concealed sealants should be installed in the shop when possible.

In sloped glazing systems, it is also important to prevent water that runs across the system from building up against horizontal members, where it can cause leaks. This is usually done by a series of gutters and downspouts designed into the system to collect water and lead it quickly to the ground.

Even with these safeguards, glazed curtain walls will sometimes leak through imperfections in glazing seals, incompletely sealed joints, or other inadvertent openings. In addition, even carefully designed systems will sometimes collect condensation. Therefore, a water-collection and removal system is advisable on the interior of such systems. This is particularly important on sloped glazing systems, where interior gutters are often advisable.

Aluminum glazed curtain wall systems come complete with accessories such as glazing gaskets, glazing stops, clips, trim, fins, flashings, anchors, and fasteners.

FIGURE 8.8-10 *A typical standard glazed curtain wall mullion.* (Drawing by HLS.)

Normal glazing gaskets are made from one of several common materials, including, but not limited to, extruded polyvinyl chloride, extruded or molded neoprene, and extruded or molded EPDM. Lockstrip-type gaskets are usually structural rubber. Incidental components, such as closers, channels, flats, trim, and sleeves, are also usually delivered with the systems.

Components should be cut and fitted to make hairline joints, except where wider joints are needed to accept sealants or to permit thermal and other movement in the system. Joints in exterior frames should be designed and erected to be watertight. Joints between two pieces of metal should be filled with a nonhardening sealant. Welds should be made on the inside so they will not be visible in the finished installation.

Mullions are usually either tubes or structural shapes, such as T or I shapes. Structural-shape mullions are often, but not always, encased in a rectangular cover. Mullions vary considerably in size and shape. They may be either standard (Fig. 8.8-10) or thermal break (see Fig. 8.8-9). Most systems are flush glazed (see Fig. 8.8-9), but face- (butt-) glazed systems are rapidly growing more popular (Fig. 8.8-11).

A system and its anchorages should be designed to permit both horizontal and vertical movement caused by thermal changes in the system and building and deflection in the structure.

8.8.6

INSTALLATION

The finest glazed curtain wall design and the highest quality of manufacture will not compensate for poor installation. Proper installation can be achieved only if the work is done by experienced workers following the guidelines and instructions supplied by the manufacturer and the appropriate industry standards. Aluminum glazed curtain walls should be installed in accordance with AAMA's *Aluminum Curtain Wall Design Guide Manual* and the manufacturer's recommendations.

Glazed curtain wall installations include many components and can be complicated. For example, in addition to the components necessary for a basic system, curtain walls include other elements that would ordinarily be assigned to different trades. These include such items as louvers; copings; fascias; flashings; parapet caps; operable windows; venetian blind tracks; drapery rods; insulation; firestopping; heating, ventilating, and air conditioning unit enclosures; solar shades; masonry infill; and joint sealing. It is, therefore, better that all work directly associated with aluminum glazed curtain walls be done by a single subcontractor, to ensure proper coordination of the different elements. This often means that the curtain wall manufacturer is also made responsible for the design and installation of the curtain wall system and all related elements. Unfortunately, all manufacturers are not qualified to offer such a service, and union jurisdictions sometimes prevent it as well. Therefore, single-subcontractor responsibility often is not possible, except with stock stick systems with few of the additional items mentioned.

Curtain wall components should be protected during shipment and handling and should be installed as soon as possible after they reach the project site to prevent them from being damaged. The substrates should be installed and prepared to receive curtain wall components so they they can be properly installed and will finish out square, plumb, level, straight, and true. The design of most systems provides for minor adjustments at the job site, but no system will function properly if its framing members are twisted or misaligned during installation.

A properly designed, manufactured, and installed aluminum glazed curtain wall system should be weather resistant as a unit. Other things must be done, however, to finish with a completely weathertight installation. For example, if the construction is insulated, the space between the rough opening and the unit should be filled with insulation. In addition, the joints between system components and adjacent materials must be closed with a sealant, trim members, or both to prevent the passage of air, dust, and water around the system.

Aluminum glazed curtain walls should be supported around the entire perimeter and anchored securely to the supporting construction. They should also be flashed at their heads and sills to provide a path back to the exterior for water that finds its way into the wall.

The installed system should be plumb, level, and in line within specified tolerances. Permissible tolerances vary from project to project and from system to system. However, systems are usually required to be plumb within 1/8 in. in any 10-ft.-long segment with a maximum of 1/4 in. in 40 ft.; level to within 1/8 in. in any 20-ft.-long segment, with a maximum of 1/4 in. in 40 ft.; and in line within 1/8 in., except that the limit should be 1/16 in. when the elements are supposed to be flush with each other. Systems should also be within 3/8 in. of their planned location.

Curtain wall installation should allow for changes caused by thermal movement in the system and in the building and for building structure deflection.

Metals with dissimilar electrolytic or chemical characteristics should be insulated from direct contact with each other by waterproof, nonconductive materials, such as neoprene, waxed paper, or coated felt. Such dissimilar metals located where water passing over them may contact another surface should be painted to prevent staining. Refer to Section 5.2 for a discussion about the effect of dissimilar metal materials.

Concealed aluminum in contact with

FIGURE 8.8-11 *A typical glazed curtain mullion with butt glazing.* (*Drawing by HLS.*)

concrete, masonry, or an absorbent material, such as wood, paper, or insulation should be permanently protected by coating either the aluminum or the adjacent material with a bituminous or aluminum paint, or by coating the aluminum with a zinc chromate primer, to minimize the chance of chemical corrosion of the aluminum by acids, alkalies, and salts leached out of the adjacent materials. Creosote and tar coatings might damage the aluminum and should not be used.

Sealants should be used within the system, in concealed locations where needed and in locations where curtain wall components adjoin other materials, to make the system airtight and watertight.

Firestopping should be installed in aluminum glazed curtain wall systems where required by code or other governing regulation. Firestopping should always be installed at the edges of floor construction where a glazed curtain wall passes. Firestopping in such locations is usually accomplished by filling the space between the floor construction and the curtain wall with a glass fiber or mineral wool insulation specifically made for this purpose.

8.8.7

GLAZING

Glazing material selection depends on wind load requirements, thermal requirements, aesthetics, and safety. The type and thickness are often established by the applicable building code. For example, some codes require that glazing materials close to floors meet the requirements for safety type and be in compliance with ANSI Z97.1, "Safety Glazing Materials Used in Buildings," or another similar standard. Refer to Section 8.7 for a discussion of glass and glazing.

Aluminum curtain wall systems are usually glazed with 1/4-in. glass, except where insulating glass is desired. Glazing stops and gaskets should be of the correct size to receive the glass to be used.

Where design calculations indicate the need to reduce heat transmittance, significant reduction can be accomplished by incorporating insulating glass. Thermal breaks in the frames will also help. At any rate, curtain walls located where the winter design temperature is 10°F or less should incorporate such devices regardless of calculations.

Some curtain wall components can be glazed at the factory, which is the most satisfactory method. Stick systems, of course, must be glazed in the field. When glazing is done at the job site, the curtain wall manufacturer's recommendations should be followed to ensure a satisfactory result.

8.8.8

ADJUSTMENT, CLEANING, AND PROTECTION

Immediately after aluminum glazed curtain walls have been installed, they should be cleaned. Thereafter, curtain walls should be protected to prevent damage to their finish, appearance, and ability to function properly. Units should be protected from splashing of plaster, mortar, paint, cleaning solutions, and other foreign substances.

9

FINISHES

*Photo: Innovative use of gypsum board. (Courtesy of
United States Gypsum Company.)*

INTRODUCTION

Finishes include materials and systems used to make the visible surfaces of the interior floors, walls, and ceilings of a building. Some of the same products are also used as exterior finishes. This chapter covers some of the many finishes available.

9.1 PLASTER MATERIALS

This section outlines the manufacture, properties, and uses of gypsum materials used in plaster and board products and of portland cement materials used in plaster. It also addresses the manufacture and use of plaster products. Section 9.5 contains additional information about the manufacture of gypsum board products and other gypsum building products. Section 9.2 contains information about support systems and accessories for plaster; Sections 9.3 and 9.4 discuss plaster application; and Section 9.5 covers gypsum board installation.

9.1.1

GYPSUM

Gypsum is a common rocklike mineral known chemically as hydrous calcium sulfate. It is found in many parts of the world, usually combined with impurities such as clay, limestone, and iron oxides. In its pure form gypsum is white, but in combination with impurities it may be gray, brown, or pinkish. Relatively pure calcined gypsum is known as *plaster of paris*, after the huge beds of it that lie under the city of Paris, where it was mined in abundance at the turn of the century. In the United States, gypsum is quarried or mined throughout the country, and substantial deposits are relatively accessible to transportation (Fig. 9.1-1).

Gypsum has two unique properties that make it useful as a plaster material: (1) it gives up some of its chemically combined water and becomes a fine powder when intensely heated (calcined), and (2) when water is added, it returns chemically to its original rocklike state by forming interlocking crystals (*crystallizing*). A mixture of calcined gypsum powder and water remains plastic for a short time, when it can be easily shaped or molded. After it hardens by crystallizing (*sets*), it becomes an effective fire barrier that will protect supporting wood framing from combustion and steel framing from loss of strength due to high temperatures.

Gypsum was known to ancient cultures and was used in construction as far back as 2000 B.C.. Early gypsum plasters, such as plaster of paris, had very short, uncontrolled setting times and were difficult to use. Late in the 19th century, materials capable of controlling setting time (retarders and accelerators) were added to processed gypsum. This encouraged the use of gypsum plasters for decorative and fire-resistant purposes.

The first large-scale use of gypsum plaster was for the white palaces (exhibition buildings) of Chicago's World Columbian Exposition in 1893. Shortly thereafter, the first gypsum board product was invented. This board product was intended for use as a plaster base, similar to contemporary gypsum lath. It consisted of layers of gypsum sandwiched between plies of felt paper. Further refinements in ingredients, designs, and manufacturing

FIGURE 9.1-1 *Gypsum is (a) quarried or (b) mined throughout the United States.* *(Courtesy Gypsum Association.)*

processes resulted in the modern varieties of plaster, lath, and other gypsum products used today.

Gypsum provides superior fire resistance. When intense heat is directed on the surface of crystalline gypsum, water in the form of steam is released when the material reaches 212 °F. In addition to absorbing some of the heat and dissipating it, this conversion of water to steam prevents the temperature of the gypsum from rising above 212 °F, as long as steam is present. With continued exposure to heat, this action progresses inward releasing more water, and the surface becomes calcined, turning to a white chalky powder. The calcined outer area becomes a better insulator and protects the underlying uncalcined areas of gypsum and retards further calcination. Therefore, gypsum retards intense heat transfer in two ways—by releasing water and by becoming a better insulator when that water is released.

9.1.1.1

Plaster Manufacture

The majority of gypsum building products can be classed as plaster products or board products. Plaster products consist principally of dry calcined gypsum that requires the addition of water at the job site to develop the chemical reaction necessary for setting. Board products are prefabricated at the factory from crystallized gypsum enclosed by special paper or vinyl and are delivered to the job site in immediately usable form (see Section 9.5). Figure 9.1-2 shows a manufacturing sequence for plaster and board products.

About two-thirds of all the crude gypsum mined goes into products for building construction. These products are used extensively where fire resistance or sound control is required and where hard and dense wall surfaces are desired. Many new, engineered gypsum products and assemblies combine the advantages of board and plaster products to meet contemporary construction needs.

After gypsum rock is mined, it is crushed to a 2-in. size and passed through a hammer mill, where it is further reduced to a 1/2-in.-maximum particle size. It is suitable in this form for treating in a rotary calciner, but additional grinding to a fine

FIGURE 9.1-2 *Flow chart of gypsum board and plaster manufacture.* *(Courtesy Gypsum Association.)*

powder is necessary for treating it in a kettle calciner.

A *rotary calciner* is an inclined steel cylinder lined with fire brick, typically 150 ft. long and 12 to 15 ft. in diameter. As a calciner rotates, the rock inside gravitates to the lower end, while heat is injected to produce a temperature of about 350 °F. This intense heat drives off about three-quarters of the combined water, changing the gypsum into a dehydrated form with an affinity for water.

A *kettle calciner* is a vertical cylinder 10 to 12 ft. in diameter with internal agitators that rotate around a central shaft. As in the rotary calciner, heat is introduced near the bottom, driving off the combined water in the powdered gypsum (see Fig. 9.1-2).

For plasters, calcined gypsum from the kettle is further treated in a *tube mill*, a rotary cylinder containing thousands of steel balls. The action of the steel balls on the gypsum produces a material having a finer grind, which contributes to better plasticity and workability.

Calcined gypsum at this point of manufacture sets very quickly and would be unsuitable for plastering. The type of aggregate used (see Section 9.1.3), local impurities in aggregates and in water, and seasonal variations all will affect the setting time. Therefore, the amounts of retarder added vary and are formulated by manufacturers for specific locations and uses to give the desired job set, which is usually a maximum of four hours.

For fibered plasters, organic fibers are added to the gypsum at the mill, and ready-mixed plasters usually are combined with lightweight aggregates before they are packaged and shipped.

Plaster Products

Powdered calcined gypsum is the prime raw material used in the manufacture of gypsum plaster products and in the cores of board products. Others materials include aggregates (Section 9.1.3), mineral or organic fibers, and lime (Section 9.1.4).

Gypsum acts as the principal binder in gypsum products. Aggregates act as a filler and provide dimensional stability to plaster finishes and board products. They may also provide increased hardness and fire resistance in some plasters. Wood or other organic fibers are sometimes included in plasters to control the working quality and the keying action of plasters with metal lath. Lime in the form of lime putty is added in finish coat plasters to increase bulk and improve workability. Proprietary chemical retarders and accelerators usually are introduced at the plant to control the setting time.

The discussion here is limited to a general description of the properties and uses of packaged plaster products (*bagged goods*) and does not describe the many plasters that may be derived on the job for specific plastering applications. Discussions of application methods and finish types of plasters appear in Sections 9.3 and 9.4. For the sake of simplicity and conformance with common terminology, packaged plaster products and job-prepared plaster mixes both are referred to in this text as *plasters*.

Packaged gypsum plaster products consist primarily of powdered gypsum and additives to control setting time and working quality. These products also may contain other ingredients, such as aggregates and fibers, required for specific applications. *Unfibered neat plasters* generally require the addition of aggregates on the job. Ready-mixed basecoat and prepared gypsum finish plasters contain all the necessary ingredients, including aggregates. The addition of water to ingredients, which have been properly proportioned either at the job site or the mill, results in plasters suitable for application over appropriate plaster bases.

Gypsum plasters are suitable for all interior plastering in locations not subject to severe moisture conditions. They can also be used in exterior locations, like the ceilings of open porches and carports or eave soffits, where they are not directly exposed to water. They are intended for application over plaster bases such as gypsum lath, metal lath (sheet, expanded, welded, or woven wire), and masonry or concrete bases.

Plasters used for scratch and brown coats are called *basecoat* plasters; those used for the final coat are *finish coat* plasters. The governing ASTM standards and the physical and working properties of basecoat and finish coat plasters are summarized in Figure 9.1-3.

■ BASECOAT PLASTERS

Basecoat plasters are classed according to ingredients as *neat*, *wood-fibered*, *ready-mixed*, and *bond* plasters. Basecoat plasters are formulated for hand-trowel or machine applications.

NEAT PLASTER   Neat gypsum basecoat plaster does not contain aggregates. It is generally used where the addition of aggregates on the job is desired, as might be the case where a specially formulated plaster is required to match an existing plaster, or where the typical aggregate in ready-mix plaster is too fine for the desired plaster texture. It also may be obtained fibered or specially formulated for use with lightweight aggregates such as vermiculite or perlite. Fibered neat plaster is lightly fibered with cattle hair, sisal, or other fine organic fibers.

WOOD-FIBERED PLASTER   Wood-fibered plaster consists of calcined gypsum integrally mixed with selected coarse cellulose fibers, which provide bulk and greater coverage. It is formulated to produce high-strength basecoats for use in highly fire-resistant ceiling assemblies.

READY-MIXED PLASTER   As the name implies, ready-mixed plaster is a mill-prepared plaster. It consists of calcined gypsum and an aggregate (usually perlite, but sometimes sand or vermiculite) and requires only the addition of water. This plaster possesses all the features and advantages of neat plaster mixed on the job, but in addition it offers mill control of the quality, gradation, and proportioning of aggregates.

BOND PLASTER   In addition to gypsum, bond plaster contains from 2% to 5% lime by weight and chemical additives that improve its bond with dense nonporous surfaces, such as concrete. Bond plaster is seldom used today because of the availability of a number of proprietary bonding agents that permit the application of standard plasters to nonporous surfaces.

■ FINISH COAT PLASTERS

The finish coat is the final plastering operation. It may serve as a base for further decorating or as a final decorative surface. Finish coat plasters usually consist of calcined gypsum, lime, and sometimes an aggregate. Some finish coat plasters may require the addition of lime or sand on the job.

GAUGING PLASTER   Gauging plaster consists of coarsely ground gypsum of low consistency that soaks up mixing water readily and blends well with lime putty (slaked quicklime or hydrated lime). This plaster controls (gauges) the setting time of plasters when mixed with lime putty, which by itself does not set. Gauging plaster is available as white plaster, which produces a brilliant, white finish when gauged with lime putty, and local plaster, which may be somewhat darker in color.

GYPSUM KEENE'S CEMENT PLASTER   Gypsum Keene's cement plaster is similar to gypsum gauging plaster, but unlike ordinary gypsum plasters, gypsum Keene's cement is burned in a kiln instead of a calciner or a kettle. In this burning process, practically all the combined water is removed, producing a *dead-burned* gypsum. The resulting material is denser than ordinary gypsums and has greater resistance to impact, abrasion, and moisture. Select white gypsum rock is used to produce the hardest white gypsum available.

Gypsum Keene's cement plaster has an initial set of about 1-1/2 hours and a final

FIGURE 9.1-3

General Properties of Gypsum Plasters

Basecoat Plasters

Plaster Mix	Standard	Compressive Strength—Dry, psi		Average Tensile Strength—Dry, psi	Setting Time	Characteristics
		Average	Minimum			
Neat (with 2 parts sand)	ASTM C 28	900	750	175	2 to 15 hrs.[a]	Greater versatility because aggregates and/or fibers are job-mixed; less control of plaster quality
Ready-mixed (with perlite)	ASTM C 28	675[b]	400	155	1-1/2 to 8 hrs.[a]	3-1/2 times better thermal insulator than sanded plaster; closely controlled quality
Wood-fibered	ASTM C 28	2100 to 2500	1200	400 to 500	1-1/2 to 9 hrs.[a]	May be used without aggregate; about 50% greater surface hardness than sanded plasters; generally higher strength and fire resistance
Bond	ASTM C 28	2300	—	400	2 to 9 hrs.	Bonds with nonporous surfaces such as concrete

Finish Coat Plasters

Plaster Mix	Standard	Compressive Strength—Dry, psi		Setting Time	Characteristics
		Average	Minimum		
Gauging (with lime putty) Regular	ASTM C 28	2000 to 3000	1200	20 to 40 min.	Hard surface; good working properties
Special high strength	ASTM C 28	Variable	5000	20 to 40 min.	
Gypsum Keene's cement (with lime putty)	ASTM C 61	4000 to 5000	2500	20 min. to 6 hrs.	Very hard surface; becomes harder with decreased proportions of lime; less susceptible to moisture; good working properties; can be retempered; can be polished
Prepared	None at present	Variable[c]	Variable[c]	Variable[c]	Moderately hard surface; no alkali reaction with paint; not as workable as gypsum–lime putty plasters
Acoustical	None at present	Variable[c]	Variable[c]	Variable[c]	Moderately soft; better sound absorption than conventional plasters

Courtesy Gypsum Association.

[a] Four-hour maximum for ASTM C 842 Standard Specification for Gypsum Plastering.

[b] Not applicable to ready-mix plasters for masonry bases.

[c] Proprietary plaster mixes containing different ingredients, which have variable effects on strength and setting time of plasters.

set of about 4 to 6 hours. A quick-setting gypsum Keene's cement, which provides a final set of about 2 hours, also is available.

Gypsum Keene's cement-lime putty-trowel plaster produces a dense, hard finish.

Gypsum Keene's cement-lime-sand float plaster is a highly crack- and spall-resistant plaster.

PREPARED GYPSUM FINISH PLASTER Prepared gypsum finish plaster is a ready-mixed plaster that includes all the necessary ingredients, except water. It consists of finely ground calcined gypsum with or without fine aggregates. It is available as white or gray *trowel* finish and *sand float*, *colored float*, or *textured* finishes. These finishes are comparable in hardness and appearance to those obtained with job-mixed Keene's cement, but factory blending of ingredients ensures somewhat better results.

VENEER (THIN-COAT) PLASTER Veneer plaster is a high-strength plaster for application as a very thin (1/16 to 1/8 in. thick) one- or two-coat finish over a special plaster base designed specifically for this purpose. Veneer plaster is a dominant type of finish in many parts of the country because it is economical and speeds up job completion.

MOLDING PLASTER Molding plaster is used for casting in molds and for ornamental plaster work. It is a very finely ground gypsum necessary for smooth, intricate surfaces and close tolerances common in ornamental work.

ACOUSTICAL PLASTERS Acoustical plasters are proprietary plasters intended to increase sound absorption to reduce sound reverberation. They consist usually of ground gypsum with chemical ingredients that encourage the formation of small air bubbles that increase the available surface area so that more sound is absorbed.

Some products touted as acoustical plasters are actually other plaster products with finishes made to simulate the appearance of acoustical plaster. These may offer some sound absorption due to their rough surface, but they are not true acoustical plasters.

9.1.2

PORTLAND CEMENT PLASTER

Portland cement plaster is used both on the interior and on the exterior of buildings. Its major application is in locations where it will be exposed to direct wetting or high moisture. In most parts of the country, exterior portland cement plaster is called *stucco*, while interior portland cement plaster is called simply *plaster*. In some regions, however, especially along the West Coast, portland cement plaster is referred to as stucco, regardless of whether it is in an exterior or interior location. Portland cement plaster is also used as a parging coat over masonry. *Parging* is often applied below ground to increase a wall's resistance to moisture penetration and to serve as a base and leveling coat to receive dampproofing and waterproofing.

9.1.2.1

General Reuirements

Portland cement plaster on the exterior and in such interior spaces as shower rooms, steam baths, saunas, laundries, tub-shower enclosures, commercial kitchens, and similar locations is often exposed to hazards not normally present in stair towers, corridors, and other similar interior applications. These hazards include water penetration that may result in corrosion of reinforcement and accessories and stresses caused by variations of temperature and humidity. The presence of such hazards require practices and precautions somewhat different from those needed in less hazardous applications.

Because it will often be exposed to direct wetting and high-moisture conditions, portland cement plaster is made from cementitious materials such as portland cement, masonry cement, and plastic cement, which are not adversely affected by such conditions. Lime historically was used as the chief binder in stucco finishes. It is now added to portland cement in the field or is included during manufacture in some other cements to improve plasticity.

Both masonry and plastic cements consist chiefly of portland cement and include plasticizers, which make the field addition

of lime unnecessary. Finishes made with plastic cement or masonry cement or with portland cement mixed with less than 50% by volume of lime are called *portland cement plaster*. When lime is added to portland cement in amounts greater than 50% by volume, the finish is referred to as lime-cement (or *portland cement–lime plaster*).

Although portland cement and portland cement–lime mixes are used primarily for exterior applications, they are also used in high-moisture interior locations. The principles and methods described here apply generally to both interior and exterior portland cement and portland cement–lime plasters.

Portland cement plaster finishes have many of the desirable properties of concrete. When properly applied, they are hard, strong, and decorative. They can be formed to almost any shape, can be finished with a wide variety of surface textures and colors, and are resistant to fire, weather, and fungus attack. They are used predominantly in the southern, southwestern, and western parts of the country, but have proven durable and economical in both warm and cold climates.

A portland cement plaster membrane is, in effect, a thin concrete slab, built up with tools and methods associated with plastering. Therefore, the discussion of the properties and the behavior of concrete found in Chapter 3 will be helpful in understanding the data in this section.

9.1.2.2

Materials

The basic ingredients of portland cement plaster are essentially the same as those of concrete and portland cement mortar. These are classed broadly as aggregates, cementitious materials, water, and sometimes admixtures. Aggregates act as inert fillers and provide strength; cementitious materials act as binders; water, together with cementitious materials, form the paste and initiates the reaction that binds the aggregates together into a solid mass; and admixtures sometimes are added to improve plasticity or cohesiveness of the mix or to retard or accelerate the setting time.

Accessory materials commonly employed in the application of portland cement plaster include metal reinforcement and various metal accessories for edge protection and crack control. Exterior portland cement plaster (stucco) also employs backing paper (or felt) and flashings.

■ CEMENTITIOUS MATERIALS

The cementitious materials used in portland cement plaster are portland cement, masonry cement, plastic cement, and lime. These products should conform to the requirements of the standards listed in Figure 9.1-4. Lime is discussed in Section 9.1.4.

PORTLAND CEMENT The types of portland cement suitable for portland cement plaster (see Fig. 9.1-4) are used also in concrete and are discussed in detail in Section 3.1.

Type I (normal) portland cement is used for most portland cement plaster work, except where the special properties of Types II and III are required. Types IA, IIA, and IIIA (air-entraining) portland cement provide improved workability and resistance to bleeding (concentration of water and fines at the surface) and damage from freezing.

Type I is available in natural gray or white. Truer colors can be obtained in finish coats containing integral coloring or colored aggregate by using a white portland cement manufactured from specially selected ingredients. Waterproof white portland cement is also available. It contains moisture-resistant additives to increase its inherent moisture resistance. These products must contain lesser amounts of iron oxides than standard portland cement but in all other respects should conform to the requirements of ASTM Standard C 150 for Type I portland cement.

Type II (moderate heat) portland cement has greater resistance to decomposition from the action of sulfates and lower heat of hydration. Resistance to sulfate attack may be a useful property when portland cement plaster is applied to brick or concrete masonry units that contain soluble sulfate materials.

Type III (high early strength) portland cement develops strength faster in the first few days after application. Type III may be desirable in cold weather to minimize the hazard of freezing damage and to

FIGURE 9.1-4
Summary of Portland Cement Plaster Materials

Material	Standard Specification
CEMENTITIOUS MATERIAL	
Portland cement Types I, II, III	ASTM C 150
Types IA, IIA, IIIA	ASTM C 150
Masonry cement	ASTM C 91
Plastic cement	ASTM C 150
Lime, hydrated* Special finishing Normal finishing	ASTM C 206
Mason's, Types S and N	ASTM C 207
Quicklime	ASTM C 5
AGGREGATES Sand	ASTM C 144
Lightweight	ASTM C 332
METAL REINFORCEMENT	ASTM C 847
METAL ACCESSORIES	ANSI A42.3
BACKING PAPER Roofing felt	ASTM D 226
Building paper	FS-UU-B-790

Courtesy Gypsum Association.

*Both special and normal limes are available as air-entraining types (SA and NA).

speed application of successive coats of portland cement plaster.

MASONRY CEMENT Masonry cement consists primarily of portland cement with varying amounts of such materials as natural cements, finely ground limestone, air-entraining agents, and sometimes lime. It is used extensively for portland cement plaster finishes in some parts of the country.

PLASTIC CEMENT Plastic cement is produced primarily in the southwestern United States. It is similar in properties and formulations to Types I and II portland cement but may contain larger amounts of air-entraining agents, insoluble residues, and plasticizing agents not in excess of

12% of the volume. With the exception of these variations, plastic cement should conform to the requirements of ASTM Standard C 150 for Type I and Type II portland cement.

PREPARED FINISH COAT Several factory-prepared mixes (stucco finishes) are manufactured for application over basecoats. These prepared mixes contain all the necessary finish-coat ingredients, with the exception of water. They are available in white or a variety of colors and, in general, provide more satisfactory results than those obtained by the addition of pigments in the field.

9.1.3

AGGREGATES

Aggregates are used in both gypsum and portland cement plasters.

9.1.3.1

Gypsum Plaster Aggregates

The aggregates commonly used in connection with gypsum products are sand, vermiculite, and perlite.

■ SAND

Most sands used in gypsum plaster consist primarily of quartz and silica, with small amounts of mica, feldspar, clay, and other minerals. Natural deposits from all over the United States are used as the main source of plastering sands. The sand is graded according to particle size and is washed if the source is known to produce impurities. Artificially produced sand (stone screenings) of the proper size and type is also suitable.

The natural characteristics of sand must be regulated to ensure its suitability as an aggregate for plaster because these vary according to location. Standards limiting size, gradation, and percentage of impurities, and testing procedures to measure these factors, have been established by the American Society for Testing and Materials (ASTM) and the American National Standards Institute (ANSI). In general, rough, sharp, angular sand particles produce stronger plasters than smooth, round ones.

■ LIGHTWEIGHT AGGREGATES

Lightweight aggregates used in gypsum plaster include vermiculite and perlite.

VERMICULITE Vermiculite consists of silica, magnesium oxide, aluminum oxides, and other minerals, combined with 5% to 9% water by weight. In the United States, vermiculite ore is mined primarily in Montana and South Carolina. This micaceous mineral is composed of a series of parallel layers that expand under the pressure of water trapped between them when subjected to intense heat (Fig. 9.1-5). The name originates from the vermicular (wormlike) movement of the layers during expansion.

FIGURE 9.1-5 *Magnified expanded vermiculite.*

Processing consists of crushing the raw ore, grading according to particle size, and removing impurities. Ore particles then are heated at 1600° to 2000°F in a furnace for 4 to 8 seconds to produce expansion and cooled rapidly to impart pliability and toughness to the expanded particles. As the material cools, it is air-lifted to remove unexpanded rock particles and to classify particle sizes. The expanded material is soft, pliable, and silvery or gold in color and contains less than 1% water by weight. Expanded vermiculite is classified into five types according to particular size. Type II, the lightest of the standard aggregates, is used in plaster products.

PERLITE Raw perlite is a volcanic ore that consists primarily of silica and alumina, with 4% to 6% combined water by weight. It is quarried mostly in the western part of the United States.

Perlite ore is processed by crushing, screening, and grading according to size.

It then is expanded in a furnace at 1400° to 2000°F. At this temperature the glassy ore particles approach the melting point and begin to soften. A small amount of the combined water is converted to steam, which pops the particles into frothy glasslike bubbles, up to 20 times the original volume (Fig. 9.1-6). After expansion, an air blast separates and grades the particles according to size.

9.1.3.2

Portland Cement Plaster Aggregates

Aggregates constitute most of the volume of a portland cement plaster mix. One cubic foot of a typical portland cement plaster mix contains about 0.97 cu. ft. of aggregate and 0.25 to 0.33 cu. ft. of cementitious material. Therefore, it is important that the aggregate be clean, sound, and well graded, with particles ranging from fine up to a maximum size of 1/8 in.

The best and most practical mix results when the aggregate size is proportioned so that a minimum of water-cement paste is required to coat the aggregate and fill the voids between particles. An excess of either fine or coarse particles can result in a mix that contains too much cement or too much water, either of which can be detrimental to a portland cement plaster finish. Portland cement plaster mixes are discussed in Section 9.4.1.

■ SAND

Sand aggregate for portland cement plaster should consist of natural or manufactured sand that is clean, granular, and free from deleterious amounts of loam, clay, soluble salts, or vegetable matter. It should be well graded within the limits given in Figure

9.1-7 and, with the exception of grading, should conform to the requirements of ASTM Standard C 144.

■ LIGHTWEIGHT AGGREGATE

Expanded shale, clay, slate, and slag sometimes are used where sand is not available locally. Vermiculite and perlite have been used in portland cement plaster where lightweight, insulating properties or fire resistance are prime factors. Proportioning of these materials should be determined by laboratory test or by making test panels. Thermal insulating properties, gradation, and unit weight for lightweight aggregate should conform to the requirements of ASTM Standard C 332.

9.1.4

LIME

Lime is used in both gypsum and portland cement plasters. It is obtained principally from limestone, a common mineral consisting of calcium carbonate (calcite) and magnesium carbonate (dolomite). To produce building lime, crushed limestone is calcined in a kiln at 2000°F. Limestone used to make lime for construction purposes comes from dolomitic deposits in Ohio or calcitic deposits in Arizona, California, Missouri, Pennsylvania, Texas, and Virginia.

The physical and chemical properties of lime depend largely on the limestone from which it was derived. Building limes may be manufactured from *dolomitic* limestone, which normally contains substantial amounts of magnesium carbonate, or from *calcitic* limestone, which is naturally

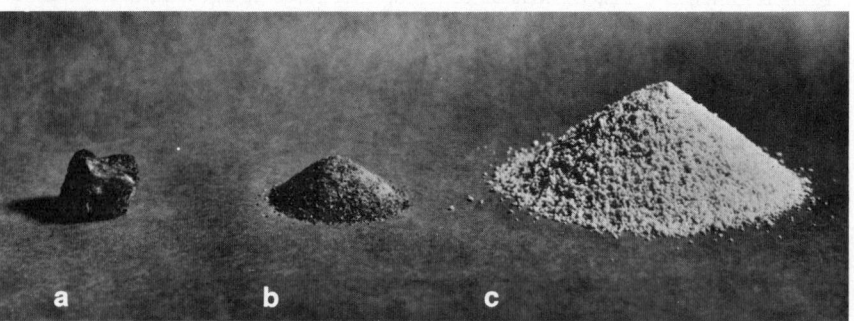

FIGURE 9.1-6 *Perlite: (a) crude ore; (b) crushed; (c) expanded.*

FIGURE 9.1-7
Grading for Sand Aggregate

Sieve Size	Percent of Aggregate by Weight Retained on Sieve	
	Minimum	Maximum
No. 4	—	0
No. 8	—	10
No. 16	10	40
No. 30	30	65
No. 50	70	90
No. 100	95	100
No. 200	97	100

high in calcium carbonate. When dolomitic limestone is calcined during manufacture, magnesium oxides result. These oxides produce a plastic putty with better working properties than those of the putty made from the calcium oxides derived from calcitic limestone, but it takes longer to hydrate than does the putty made with calcium oxides.

Building limes can be classed as (1) *quicklime* or (2) *hydrated lime*.

9.1.4.1

Quicklime

During the calcining process, limestone gives off carbon dioxide and changes from the carbonate form to calcium oxide or magnesium oxide to form quicklime. *Quicklime* was used as the chief binder in lime plasters for several hundred years, before the introduction of controlled-setting gypsum plasters.

Before quicklime can be used in plaster, it must be *slaked* (combined with water) and soaked for one to two weeks to hydrate its oxides and make a plastic putty usable in mortar or plaster. Slaking initiates a chemical reaction that transforms oxide to a more stable hydroxide state. The type of raw material affects the degree and the speed of hydration possible during this soaking period. Quicklime identified as calcitic, or high-calcium, can be expected to have a greater hydration of oxides.

Quicklime is acceptable for use in plaster only if it is identified as calcitic or high-calcium and properly slaked as recommended by its manufacturer. For that reason, some quicklime is sold in the natural pulverized form, but, to eliminate the slaking operation, most is partially hydrated (slaked) at the mill.

9.1.4.2

Hydrated Lime

The material produced by partially hydrating quicklime at the mill is called *hydrated lime*. The addition of water to hydrated lime results in *lime putty*, a plastic form that is combined with gypsum and water to make some finish coat plasters.

If oxides are not completely hydrated during soaking or during the manufacture of hydrated lime, they may be present in the plaster. Under conditions of high humidity they could hydrate there, causing their expansion and consequent cracking of the plaster containing them. Therefore, dolomitic limes often are autoclaved at the mill to ensure that a minimum of unhydrated oxides remains in the manufactured hydrated lime. To be suitable for use in finish coat plasters, lime must have no more than 8% unhydrated oxides.

Available hydrated limes include *mason's* hydrated lime, intended primarily for masonry mortars, and *finishing* hydrated lime, intended for use in plasters. Both are available in Type S (special), Type SA (special air-entraining), Type N (normal), and Type NA (normal air-entraining).

■ TYPE N (NORMAL) HYDRATED LIME

Type N hydrated lime is produced in accordance with the requirements of ASTM Standard C 206, except that it has no limitation on unhydrated oxides. However, when calcitic limestone is used in its manufacture, the resulting product (identified as calcitic or high-calcium finishing hydrated lime) consists of at least 95% calcium oxide, which hydrates readily during manufacture. Some oxides may remain unhydrated, but these can be expected to hydrate substantially during the required preuse soaking period for Type N limes, which is from 12 to 16 hours. Often, high-calcium finishing lime will meet ASTM limitations for unhydrated oxides for Type S lime but may not develop the degree of plasticity specified for this lime within the time limit established by ASTM.

Type N hydrated finishing lime produced in accordance with ASTM Standard C 206 is acceptable for use in plaster if it is identified as calcitic or high-calcium and has been properly soaked as recommended by the manufacturer.

■ TYPE S (SPECIAL) HYDRATED LIME

Type S (special) finishing hydrated lime conforming with the requirements of ASTM Standard C 206 requires no preliminary soaking and can be mixed with other ingredients before water is added. Because it is easy to use and because unhydrated oxides are limited during manufacture, Type S lime most often is used and is specifically recommended for plaster mixes.

Type S hydrated lime conforming to ASTM Standard C 206 must contain not more than 8% unhydrated oxides, regardless of the raw materials used.

9.1.5

WATER

Water for mixing and curing should be clean and free from deleterious amounts of oils, acids, alkalies, salts, and organic substances that could adversely affect proper hydration or cause corrosion of metal reinforcement and accessories.

The presence of water promotes hydration of portland cement and carbonation of lime, which are chemical processes by which these materials harden. However, more water generally is added to the mix than is necessary to provide for hydration and carbonation. The excess is required to bring the mix to a workable consistency. The amount of water added for workability depends on the characteristics of the cementitious material, gradation of aggregate, suction of the supporting construction, and drying conditions.

9.1.6

ADMIXTURES

Admixtures called plasticizers, such as mineral, animal, vegetable, or glass fibers sometimes are added to a plaster mix to improve its plasticity. These plasticizers can be added to the first (*scratch*) coat, where greater cohesiveness of the mix is required, particularly for ceilings and soffits, and to reduce the amount of mix lost through openings in metal reinforcement when backing paper is not used. Mineral fibers also may be used for machine-

applied basecoats to prevent segregation of the aggregate and to provide lubrication in the hose.

Fibers should be clean and free from harmful amounts of any substance, such as unhydrated oxides, that may be injurious to the plaster ingredients. Animal hair and vegetable fibers should be 1/2 to 2 in. in length and free from grease, oil, dirt,

and other impurities.

Other admixtures intended to accelerate or retard setting time generally are not recommended, but they may be used where local practice indicates satisfactory results. When used, admixtures should not reduce the compressive strength of the mix more than 15% below the strength of a comparable mix without admixtures.

9.2 PLASTER SUPPORT SYSTEMS, BASES, AND ACCESSORIES

The first considerations in the application of plaster finishes are installation and preparation of supporting construction and application of a plaster base (Fig. 9.2-1). Proper selection of accessories is necessary to a satisfactory plaster installation.

FIGURE 9.2-1 *Applying plaster over a gypsum plaster base.* (*Courtesy United States Gypsum Company.*)

9.2.1

SUPPORT SYSTEMS

Both gypsum and portland cement plasters can be applied directly to some concrete and masonry surfaces under certain conditions. In such cases, the concrete or masonry serves as both support system and plaster base. For more details see Section 9.2.2.

Supporting construction of all types should be capable of supporting design loads without excessive deflection and without introducing stresses that might cause cracking of the finish. The calculated deflection of ceiling members should not exceed 1/360 of the span after the plaster has been applied. Headers or lintels of adequate size should be provided over openings, and all construction should be true and plumb or level.

Supporting construction should be strong and stable but also capable of accommodating structural movement where it is expected to occur. Concrete and masonry construction should be provided with control joints and expansion joints as dictated by good design practice. Control joints in plaster should be placed directly over these joints in the supporting construction and wherever required to relieve stresses and minimize cracking (see Section 9.4.4).

9.2.1.1

Metal Framing and Furring

Exterior portland cement plaster (stucco) finishes over metal framing are not common in low-rise residential construction, because most of them are built with wood studs, but they do occur in other types of buildings. The thermal insulating and fire-resistant properties of lightweight aggregates such as perlite and vermiculite have encouraged the use of portland cement plaster in commercial and institutional

structures, especially where fire ratings are required. Lightweight aggregates are not recommended for exterior applications, however, so they are used primarily for back-plaster coats on exterior walls and for fireproofing.

Both gypsum and portland cement plasters are supported by metal framing and furring in interior locations where high-quality construction or fire resistance ratings are required.

Assemblies consisting of steel studs, furring channels (both C-shaped and hat-shaped), and reinforcing rods should be spaced and securely fastened according to the specialty manufacturer's recommendations and the recommendations of the Metal Lath/Steel Framing Association Division of the National Association of Architectural Metal Manufacturers (NAAMM) publication ML/SFA 920, "Guide Specifications for Metal Lathing and Furring." The recommended spacing of metal supports for various types of reinforcement closely parallels that of wood (Fig. 9.2-2). Metal stud framing is described in Section 9.5 for support of gypsum board.

9.2.1.2

Wood Framing and Furring

Both gypsum plaster and portland cement plaster can be installed over wood framing and furring. Refer to Chapter 6 for details about the selection and installation of wood framing and furring members for gypsum plaster.

FIGURE 9.2-2
Support Spacing for Metal Reinforcement[a]

Reinforcement	Weight, lb./sq. yd.	Support Spacing, in. on centers
Diamond mesh	3.4	16
Stucco mesh[b]	1.8 and 3.6	16
Flat rib lath	3.4	19[c]
3/8-in. rib lath	3.4 and 4.0	24
Woven wire	1.7 (1 in. hex.)	16
	1.4 (1-1/2 in. hex.)	16
Welded wire	1.4 (1 in. × 1 in.)	16
	1.4 (2 in. × 2 in.)	16
	1.9 (1-1/2 in. × 2 in.)	16[d]

[a] Support spacing applicable to both wood and metal framing for both vertical and horizontal surfaces, unless otherwise noted.

[b] Not recommended for vertical surfaces over metal framing or for horizontal surfaces.

[c] Spacing of metal framing for horizontal surfaces should not exceed 13-1/2 in. on centers.

[d] Spacing may be increased to 24 in. on centers when stiffener wires are provided 6 in. on centers.

Gypsum plaster can be installed over wood framing and furring with either a gypsum lath base or metal reinforcement. Portland cement plaster can be installed over wood framing and furring by using metal reinforcement.

Exterior wood-framed construction falls into two categories: unsheathed (open-frame) and sheathed construction. Generally, studs should be of at least 2 × 4s, spaced not more than 24 in. on centers when supporting a ceiling and roof only, and not more than 16 in. on centers when supporting one or two floors. Studs in exterior walls supporting more than two floors should be at least 2 × 6s, spaced not more than 16 in. on centers.

■ UNSHEATHED CONSTRUCTION

The structural frame should be properly braced, by let-in 1 × 4s or larger diagonal bracing or by other means, to resist racking forces. Paper-backed reinforcement can be applied directly to unsheathed construction. If backing paper is applied separately, a soft annealed 18-gauge or heavier steel line wire should be stretched across the faces of the studs in horizontal strands about 6 in. apart. The wire should be stretched taut by first nailing or stapling

it at every fifth stud, then securing it to the center of intermediate studs and stretching it tightly by raising and lowering the attachments (Fig. 9.2-3a). Backing paper then is nailed over the line wire with edges lapped at least 3 in. The wire will prevent the paper from sagging, and a more uniform thickness of portland cement plaster (stucco) less likely to crack results.

Application of flexible reinforcement should start at least one full stud away

from corners and should be bent around them to avoid a joint at the corner. If the reinforcement is too rigid to be bent easily, prefabricated corner reinforcement should be used. Horizontal laps should be wired at least once between each stud space but at spacings not more than 9 in. on centers, and vertical laps should be made over supports. The bottom of a stucco panel should be terminated with a metal drip screed. The screed will help to support the stucco membrane and, if properly located, will permit drainage of water that might otherwise penetrate the plaster (Fig. 9.2-4).

As stucco is applied by trowel or machine, it should be forced through the openings in the reinforcement against the paper backing to ensure at least 90% coverage of the reinforcement. An alternative method sometimes used to ensure complete embedment of the reinforcement is to eliminate the backing paper and back plaster the reinforcement between supports. This backplaster coat should be at least 3/8 in. thick and should be applied only after the scratch coat has hardened sufficiently so that pressure will not break the stucco keys. The surface should be uniformly dampened before application and moist-cured for at least 24 hours after application.

■ SHEATHED CONSTRUCTION

Wood boards, plywood, or other suitable panel sheathing should be applied horizontally and fastened securely to each stud

FIGURE 9.2-3 *(a) Unsheathed (open-frame) construction; (b) sheathed construction. Line wire is not required with paper-backed reinforcement.* *(Courtesy Portland Cement Association.)*

FIGURE 9.2-4 *(a) A properly installed drip screed allows drainage and reduces the possibility of water penetration into a wall; (b) no drip screed and bond between concrete and portland cement plaster (stucco) may result in moisture backup.* *(Courtesy Portland Cement Association.)*

(see Fig. 9.2-3b). Lumber boards used for sheathing should not be more than 8 in. wide and should be fastened with at least two nails at each support to minimize the possibility of damage from warping. The sheathing should be covered with backing paper, lapped at least 3 in., followed by metal reinforcement fastened through the sheathing into the wood supports. Paper-backed reinforcement can be attached directly over the sheathing. The reinforcement should be returned at least 6 in. around corners. Vertical laps over supports and horizontal laps should be nailed or stapled at 6 in. on centers into the sheathing. Unless the reinforcement is self-furring, nails should be of the furring type; staples can be used for self-furring reinforcement.

9.2.2

PLASTER BASES

Plaster bases are constructions or surfaces that develop adequate bond for the direct application of plaster. Bond can be provided by mechanical key or by suction. Mechanical keying results from the interlocking of a plaster mix with openings or projections in the base surface; suction develops bond by drawing part of the plaster paste into minute pores in the base surface by capillary action. Over metal reinforcement, bond is developed by mechanical keying; over concrete and masonry, bond is developed by mechanical keying, suction, or both.

Both gypsum and portland cement plasters can be applied over metal reinforcement and directly over some concrete and concrete and brick masonry. Gypsum plaster can also be applied directly to gypsum products such as other gypsum plaster, gypsum masonry, or gypsum lath, but portland cement plaster should not be applied to these bases, because they do not provide adequate bond for it. Metal lath should be installed over such materials before portland cement plaster is applied; backing paper (see Section 9.2.4) should also be used in exterior locations. Wood lath, fiber insulating lath, and sheet metal lath should not be used as a base for portland cement plaster.

Gypsum veneer plaster is applied over *veneer plaster base* (see Section 9.5).

9.2.2.1

Concrete and Masonry

Both gypsum and portland cement plasters can be applied directly over concrete and concrete and brick masonry that have sufficient bonding capabilities. However, gypsum plaster should not be applied directly to concrete or masonry surfaces that are the interior surfaces of exterior walls or the opposite surfaces of walls or partitions that are exposed to wet conditions, such as would be found in a sauna, shower, or tub-shower enclosure.

■ CONCRETE

Before plaster is applied directly to concrete, the surface should be inspected for the presence of laitance, form oil, and other substances that might impair bond. Sandblasting can be used to remove form oil, curing compounds, and other foreign materials. Smooth or slick spots on fresh concrete caused by form oil can be removed by scrubbing with strong soap and water and allowing the concrete to dry to restore suction. Form oil should not be used for concrete forms when direct application of plaster is intended.

Rough lumber can be used for the forms of concrete that is to receive plaster, or the freshly placed concrete can be lightly sandblasted to roughen the surface immediately after the forms have been stripped. Hardened, smooth concrete may require more extensive preparation such as chipping, acid etching, or roughening with a bush hammer. Metal reinforcement should be used when gypsum plaster is to be applied over smooth concrete where roughing is not practical. A bonding agent or a dash coat also may be used where portland cement plaster is to be applied and roughening of the surface is impractical.

DASH COAT When it is doubtful that good bond can be obtained on concrete or masonry that is to receive portland cement plaster, it is sometimes advisable to use a dash bond coat. The mix for a dash coat should consist of one part by volume of portland cement and from one to two parts sand. A small straw broom or long-fibered stiff brush is dipped in the mix and the material is splattered on the wall with a quick throwing motion (Fig. 9.2-5). A dash coat should not be troweled and should be allowed to set before the plaster base coat is applied. If there is a delay in applying the base coat, the dash coat should be kept moist until the next coat is applied.

BONDING AGENT Another method of obtaining good bond on concrete and

FIGURE 9.2-5 *A dash bond coat being applied.* (Courtesy Portland Cement Association.)

masonry surfaces is to apply a waterproof bonding agent recommended by the manufacturer for exterior use. The supporting surface should be structurally sound, rigid, and free from laitance, dust, grease, and materials that are soluble in water, such as water-based paints or glue.

ACID ETCHING   A solution of one part muriatic acid and six parts water can be used for acid etching of slick concrete where a dash coat or bonding agent might not adhere well. Before an acid solution is applied, the surface should be wetted so that the acid will act only on the surface. Several applications may be necessary. After treatment, concrete should be thoroughly washed with water to remove all traces of the acid and be allowed to dry to restore suction.

■ MASONRY

When plaster is to be applied directly to masonry, mortar joints in the masonry should be struck flush or slightly raked. The masonry surface should be clean, sound, and firm and should not contain efflorescence, grease, waterproofing compounds, paint, or other substances detrimental to good bond. Efflorescence, oil, and grease can be removed by washing with a 10% solution of muriatic acid and water. Other coatings or substances that will adversely affect bond should be removed with appropriate solvents and rinsed clean.

Most masonry, such as medium-hard common and face brick, standard-weight concrete block, and many types of stone, generally provide adequate suction and often mechanical bond. The surface can be tested for suction by spraying it with clean water to see how quickly the moisture is absorbed. If the water is readily absorbed, good suction is likely to result; if water droplets form, the surface does not provide adequate suction and bond will depend on mechanical key. It can be determined by observation whether the surface is sufficiently rough to ensure mechanical keying with the plaster. Glazed brick will probably be too slick, but standard-weight concrete block will usually be rough enough.

On surfaces that absorb moisture too quickly, such as soft common brick and lightweight concrete masonry units, portland cement plaster will quickly stiffen and become difficult to apply properly. Excessive suction can be controlled by fine-spraying (not soaking) the masonry with several applications of water immediately before applying the first plaster coat. The surface should appear slightly damp, but there should be no free water present when the plaster is applied. When stucco is machine-applied, it may not be necessary to dampen the surface because the mix usually contains more water than those for hand application.

Hard-burnt brick, glazed concrete masonry units, and some types of stone may provide poor or nonuniform bond. If suction is not uniform over the entire surface, parts of the wall may draw more moisture than others, and the final finish will be spotty in color. Also, nonuniform bonding is likely to cause cracking because the plaster will move with the supporting base in the areas of good bond and move independently of the base in areas of poor bond.

If an adequate or uniform bond cannot be obtained, the surface should be prepared with metal reinforcement, a dash coat, or a bonding agent, as recommended earlier for concrete. When metal reinforcement is used with portland cement plaster, backing paper should also be used to isolate the plaster membrane from the base and prevent dissimilar suction. When plaster is to be applied as a continuous finish over dissimilar rigid bases, a control joint or a strip of metal reinforcement should be placed over the juncture of the dissimilar materials.

Plaster should not be applied directly to a masonry chimney but should be applied over metal reinforcement. However, backing paper should not be used, because heat might cause deterioration. Flashings should be installed under the chimney cap and at junctures with the roof to prevent water penetration (see Chapter 4).

9.2.2.2

Metal Reinforcement

Metal reinforcement may be used in lieu of gypsum lath for gypsum plaster installed over wood or metal framing or furring. It should also be used in gypsum plaster over concrete or masonry that does not provide sufficient bond for direct plaster application.

Metal reinforcement should also be used when portland cement plaster is applied over either sheathed or unsheathed wood-framed construction, steel-framed construction, concrete and masonry providing unsatisfactory bond, and at flashings and chimneys.

Plaster applied over metal reinforcement acts as a thin reinforced slab that, though supported by the underlying construction, is not in sufficiently rigid contact to permit direct transfer of stresses. The reinforcement should be attached securely to supporting construction and stretched tightly between supports to eliminate slack. Loosely attached reinforcement generally results in an uneven thickness of plaster, with the thinner and weaker sections more likely to crack. At least a 1/4-in. space should be maintained between the reinforcement and the supporting construction by use of either furring fasteners or self-furring reinforcement (Fig. 9.2-6). The metal reinforcement should be of suitable type and weight for the support spacing (see Fig. 9.2-2) and should be attached with fasteners recommended for the particular supporting construction (Fig. 9.2-7).

■ TYPES

The most common commercially available types of metal reinforcement are *expanded metal lath* and *wire lath* (Figs. 9.2-8 and 9.2-9). Openings in the lath should be large enough to permit the scratch coat to be forced through the openings and to embed the reinforcement, thereby preventing corrosion. Metal reinforcement should conform to the standards listed in Figure 9.1-4 and the weight requirements given in Figure 9.2-9.

FIGURE 9.2-6 *(a) Self-furring wire lath is crimped every 3 in.; (b) fiber spacers on furring nails keep plain wire lath 1/4 in. away from the backing paper.*

FIGURE 9.2-7
Fastening of Reinforcement over Wood Supports[a]

Reinforcement	Nails		Staples	
	Type[b]	Maximum Spacing, in. on centers	Type	Maximum Spacing, in. on centers
Diamond mesh		6		6
Stucco mesh[c]		6	7/8-in. leg, 16-gauge, 3/4-in. crown	6
Flat rib lath	1-1/2-in., 11-gauge, barbed, 7/16-in. head	6		6
3/8-in. rib lath		6[d]	1-1/4-in. leg, 16-gauge, 3/4-in. crown	4-1/2
Woven and welded wire		6	7/8-in. leg, 16-gauge, 7/16-in. crown	6

[a] Applicable to both vertical and horizontal surfaces, unless otherwise noted.

[b] 1-in. roofing nails suitable on vertical surfaces for all but 3/8-ribbed lath. 1-1/2-in., 12-gauge, 3/8-in. head furring nails suitable to attach welded and woven wire on vertical surfaces.

[c] Not recommended for horizontal surfaces.

[d] Maximum spacing 4-1/2 in. on centers for horizontal surfaces using 2-in. nails.

EXPANDED METAL LATH Expanded metal lath is formed from light-gauge steel sheets that have been cut in a regular pattern and expanded (stretched) to form diamond-shaped openings (Fig. 9.2-8). Both expanded zinc-coated (galvanized) steel sheet and cold-rolled steel sheet, coated with corrosion-resistant paint after fabrication, are used.

Plain, self-furring, and *paper-backed* types of diamond mesh are produced in sheet form. The plain type generally should be installed with furring fasteners. Diamond mesh is produced in weights of 2.5 and 3.4 lb./sq. yd.; only the heavier 3.4-lb. weight is suitable for exterior portland cement plaster use.

The *self-furring* type is dimpled or contains other devices that hold it away from the surface to which it is attached. These devices should keep the back of the lath at least 1/4 in. away from the supporting construction. Self-furring lath can be nailed directly to smooth reinforced concrete, masonry, and other solid surfaces to receive plaster.

Paper-backed diamond mesh, with paper strips attached to the back surface, is produced primarily for machine application of exterior portland cement plaster (stucco). Spaces are provided between the paper strips to permit fastening to studs or other supporting construction.

Stucco mesh is a metal lath similar in pattern to diamond mesh but with larger openings. It is designed primarily as a base for stucco over solid surfaces. It should be installed with backing paper and furring fasteners.

Rib lath is a type of expanded metal lath that consists of integrally formed V-shaped ribs that provide greater stiffness and permit wider spacing of supports (see Fig. 9.2-2). Laths with 1/8-, 3/8-, and 3/4-in. ribs are produced (see Fig. 9.2-9). The 1/8-in. rib lath referred to as *flat rib lath* generally requires furring fasteners and can be used for most plaster applications. It is available in weights of 2.75 and 3.4 lb./sq. yd., but only the heavier is acceptable as a base for exterior portland cement plaster.

The 3/8-in. rib lath does not require furring, but it is not recommended for severe weather exposure because it is almost impossible to encase the lath in plaster. It can be used in milder climates and in locations protected from direct wetting, such as soffits. Its chief advantage is the ability to provide a base for support spacings up to 24 in. on centers (see Fig. 9.2-2).

WIRE LATH Wire lath is produced as woven or welded wire fabric. Wire lath should be fabricated from copper-bearing galvanized cold-drawn steel wire of the appropriate gauge (see Fig. 9.2-9). Wire lath is used extensively as a base for exterior portland cement (stucco) finishes.

Woven wire fabric (stucco netting) is a hexagonal woven wire fabric made by interweaving wire into hexagonal openings 1 in. to 2-1/4 in. wide (see Fig. 9.2-8). The plain type is nailed to supporting construction with furring nails over backing paper. Woven wire fabric also is available in self-furring type, with or without backing paper, which eliminates the need for furring fasteners or separate installation of backing paper, depending on the type

EXPANDED METAL LATH

STUCCO MESH   DIAMOND MESH   FLAT RIB LATH   3/8″ RIB LATH

SELF-FURRING

WIRE LATH

WELDED WIRE   WELDED WIRE (BACKED)   WOVEN WIRE

FIGURE 9.2-8 *Metal reinforcement suitable for portland cement plaster (stucco) finishes: all but stucco mesh and 3/8-in. rib lath are available in plain, self-furring, and paper-backed types.*

FIGURE 9.2-9
Metal Reinforcement for Portland Cement Plaster (Stucco) Finishes

Type	Weight, lb./sq. yd.	Opening Size	Dimensions
EXPANDED METAL LATH			
Diamond mesh[a]	3.4	5/16 in. × 3/8 in.	27 in. × 96 in.
Stucco mesh	1.8 and 3.6	1-3/8 in. × 3-1/8 in.	48 in. × 99 in.
Flat rib lath	3.4	5/16 in. × 3/8 in.	24 in. × 96 in. and 27 in. × 96 in.
3/8-in. rib lath	3.4 and 4.0	5/16 in. × 3/8 in.	27 in. × 96 in. and 27 in. × 99 in.
WIRE LATH			
Woven wire[a]	1.7 (18 gauge)	1 in. hex.	3 ft. × 150 ft. (rolls)[b]
	1.4 (17 gauge)	1-1/2 in. hex.	
Welded wire[a]	1.4 (16 gauge)	2 in. × 2 in.	3 ft. × 150 ft. (rolls)[b]
	1.4 (18 gauge)	1 in. × 1 in.	
	1.9 (16 gauge)	1-1/2 in. × 2 in.	

Courtesy Portland Cement Association.

[a] Available in plain, self-furring, and paper-backed types.

[b] Paper-backed and self-furring types also available in sheet form.

---

of lath used. Minimum wire size for 1-in. openings should be 18 gauge; for 1-1/2-in. openings, 17 gauge (see Fig. 9.2-9).

Welded wire fabric lath consists of a grid formed by welding steel wire at intersections into 1-in. by 1-in. and 2-in. by 2-in. squares. It is available with backing paper and stiffener wires and in a self-furring type. This fabric should be galvanized, coated with corrosion-resistant paint after welding, or fabricated from galvanized wire. Wire should be at least 16 gauge for fabric with 2-in. by 2-in. openings and at least 18 gauge for fabric with 1-in. by 1-in. openings.

■ INSTALLATION

Concrete and masonry surfaces that do not provide adequate bond and that cannot be economically prepared for direct application of plaster require the use of metal reinforcement. For portland cement plaster, the surface should be covered with backing paper.

Reinforcement should be attached to concrete and masonry with suitable fasteners spaced 6 in. on centers vertically and 16 in. on centers horizontally. Horizontal and vertical laps should be secured with fasteners or wire ties 6 in. on centers. Self-furring reinforcement can be attached with powder-actuated fasteners or concrete stub nails; plain reinforcement should be attached with concrete furring nails. These nails should be long enough to provide at least 3/8-in. penetration into the concrete or masonry.

Generally, metal reinforcement should be installed over framing or furring with the long dimension perpendicular to the supports and with end laps staggered, forming a continuous network of metal over the entire surface. Expanded metal lath should be lapped 1/2 in. at the sides and 1 in. at the ends. Edge ribs should be nested. Wire lath should be lapped at the ends and sides at least one full mesh but not less than 1 in. Generally, lower sheets should be lapped over upper sheets at horizontal laps; however, paper-backed reinforcement should be applied shingle fashion with upper sheets sandwich-lapped over the lower sheets with metal on metal.

Metal reinforcement should be attached to wood framing and furring with corrosion-resistant nails or staples, sized and spaced as shown in Figure 9.2-7. Nails generally should provide at least 7/8-in. penetration into supporting construction on walls and 1-3/8-in. penetration on ceilings. Aluminum nails react chemically with fresh portland cement and should not be used with it. For most types of reinforcement, 1 in. or 1-1/2-in. nails are generally suitable. However, 3/8-in. rib lath usually is nailed at the rib, and 1-1/2-in. or 2-in. nails are recommended to ensure the minimum penetration for walls and ceilings.

Metal reinforcement should be attached to metal framing and furring with screws or clips supplemented by screws. At suspended ceilings, metal reinforcement may be wire-tied to furring members.

When sound ratings are required, metal reinforcement may be attached with resilient clips.

External corners can be formed with *cornerbead*, with prefabricated corner reinforcement, or by bending the metal reinforcement around the corner (Fig. 9.2-10). Internal corners can be reinforced with metal corner reinforcement (see Fig. 9.2-13) wired to the reinforcement, or the metal reinforcement itself can be bent to conform to the angle (Fig. 9.2-11).

9.2.2.3

Backing Paper

Metal reinforcement for portland cement plaster should have a backing paper in exterior locations, in wet interior areas, and when the plaster will be sprayed. Backing paper is used to (1) isolate the plaster from the supporting construction, (2) prevent loss of plaster mix through lath openings, (3) facilitate complete embedment of the lath, and (4) resist water penetration through breaks in the plaster membrane. When the plaster is hand applied, backing paper is not required generally on horizontal surfaces protected from direct wetting.

Backing paper should have a vapor permeance of at least 5 perms to permit the plaster to dry and to prevent moisture from becoming trapped in the wall. Waterproof and vapor-permeable products used for sheathing paper, such as 15-lb. asphalt-saturated roofing felt, are suitable for this purpose. Backing paper may be factory-attached to metal reinforcement, or it can be installed separately. Absorptive paper attached to some types of metal reinforcement is not water resistant and is not suitable for exterior applications where weather protection is required.

FIGURE 9.2-10 *(a) External corner reinforced by bending lath around the corner; (b) external corner being formed with zinc cornerbead.*

9.2.2.4

Gypsum Lath

Gypsum lath should be fastened to wood wall and ceiling framing and furring with screws in accordance with the lath manufacturer's recommendations (Fig. 9.2-12) or with nails; to metal framing and furring with clips. Where sound ratings are required, gypsum lath may be attached with resilient clips. Refer to Section 9.5 for a detailed discussion about gypsum lath materials and their application.

9.2.3

METAL PLASTER ACCESSORIES

In most residential construction and in many other buildings as well, wood trim often covers the edges of a plaster finish. When wood trim is not used, exposed edges requiring protection from damage should be trimmed with metal casing beads (Fig. 9.2-13a). Control joints (Fig. 9.2-13b) are incorporated to break large areas into self-contained smaller panels, and drip screeds are used as terminating lines at the bottom of walls (Fig. 9.2-13c). Other accessories such as *striplath* and *cornerite* (Fig. 9.2-13d,e) are used for reinforcing flat surfaces and internal corners, respectively. External corners often are provided with corner reinforcement (Fig. 9.2-13f) when the metal lath is not flexible enough to be bent around corners.

Corner reinforcement without a solid nosing is designed to permit complete

FIGURE 9.2-11 *Woven wire lath can be easily bent to form internal corners.*

FIGURE 9.2-12 *Gypsum plaster base applied to metal studs with screws, with clips at corners and end joints. (Courtesy United States Gypsum Company.)*

a CASING BEADS	b CONTROL JOINTS
SQUARE BULLNOSE	MULTI-SECTION ONE-PIECE
c DRIP SCREEDS	d STRIPLATH
SOFFIT FOUNDATION	WOVEN WIRE EXPANDED METAL
e CORNERITE (INTERNAL CORNER)	f CORNER REINFORCEMENT (EXTERNAL CORNER)
EXPANDED WIRE	EXPANDED WIRE

FIGURE 9.2-13 *Typical accessories for portland cement plaster (stucco): (a) Casing beads are available with or without expanded flanges. (f) External corner reinforcement is also available with solid metal nosing (cornerbead) in white alloy zinc. Other accessories are produced to meet special conditions.*

embedment of the metal within the plaster. However, cornerbeads made with a solid nosing that remains exposed can be used, provided they are made of white alloy zinc (99% pure zinc) where they will be exposed to wetting. Casing beads should be formed from metal not less than 24 gauge; other accessories should be at least 26 gauge. For plaster exposed to wetting, exposed metal accessories (not completely embedded in the plaster) should be as corrosion resistant as white alloy zinc.

9.2.4

FLASHING

Properly installed flashing to prevent water from entering a construction is essential to durable and satisfactory service of stucco surfaces. Flashing should be provided in all locations where the hazard of water penetration exists (see Fig. 7.9-5b). Water entering behind a stucco finish could cause corrosion of metal reinforcement and accessories and damage to interior finishes. Only corrosion-resistant materials, such as galvanized steel, copper, stainless steel, and suitable plastics, should be used for flashing. Metal reinforcement should be installed over flashing, because these materials do not provide adequate bond for stucco.

Section 7.9 contains a detailed discussion of flashing.

9.3 GYPSUM PLASTER APPLICATION

Gypsum plaster mixes and application methods vary with different gypsum plaster products and the bases to which they will be applied.

9.3.1

MIXES

The materials used in gypsum plaster include (1) powdered calcined gypsum, (2) aggregates, (3) mineral or organic fibers, and (4) lime. The latter three may be included in a packaged plaster product or added in the field to neat plaster (see Section 9.1.1.2).

9.3.1.1

Basecoat Plaster Mixes

Basecoat plasters include *neat plaster*, *wood-fibered plaster*, and *ready-mixed plaster* (see Section 9.1.1.2).

■ NEAT PLASTER

Neat gypsum basecoat plaster usually is combined on the job with aggregates in proportions of one part plaster to two or three parts aggregate by weight.

■ WOOD-FIBERED PLASTER

Although wood-fibered plaster is often used without aggregates, 1 cu. ft. sand is recommended for each 100 lb. wood-fibered plaster to obtain a basecoat of superior hardness, strength, and added crack resistance or for use over masonry.

■ READY-MIXED PLASTER

Ready-mixed plaster is used as it comes from the bag with only the addition of water.

9.3.1.2

Finish Coat Plaster Mixes

Finish coat plasters usually consist of calcined gypsum, lime, and sometimes an aggregate. Some finish coat plasters may require the addition of lime or sand on the job.

■ GAUGING PLASTER

Gauging plasters provide either a quick set or a slow set. This eliminates the need for adding accelerators or retarders in the field, which seldom can be controlled adequately to provide the desired setting time.

■ GYPSUM-LIME PUTTY-TROWEL PLASTER

Gypsum-lime putty-trowel plaster is a low-cost finish plaster that leaves a smooth surface suitable for most normal uses. The field mix consists of 100 lb. gauging plaster to 200 lb. dry hydrated lime. Although generally used without aggregates, the addition of a fine silica sand or perlite fines in an amount of at least 1/2 cu. ft. per 100 lb. dry gauging plaster will increase the crack and crazing resistance of the finish when used over basecoat plasters that contain lightweight aggregates.

■ GYPSUM KEENE'S CEMENT-LIME PUTTY-TROWEL PLASTER

This plaster does not require aggregates except when it is applied over a lightweight aggregate basecoat. In that instance, at least 1/2 cu. ft. fine sand or perlite is added to each 100 lb. of gypsum Keene's cement and 50 lb. of hydrated lime; an extra hard finish can be obtained by reducing the lime amount to 25 lb.

■ GYPSUM KEENE'S CEMENT-LIME-AND-SAND FLOAT PLASTER

This type of plaster usually is mixed with sand. The resulting textured surface may be painted, or an integral coloring agent may be introduced into the plaster to eliminate decorating. For a float finish, 150 lb. gypsum Keene's cement is mixed with 100 lb. of hydrated lime and 400 to 600 lb. silica sand, depending on the texture desired.

■ PREPARED GYPSUM FINISH PLASTER

Prepared gypsum finish plaster requires only the addition of water at the job site.

■ MOLDING PLASTER

Most molding plaster mixes consist of 100 lb. molding plaster to 50 lb. dry hydrated lime.

9.3.2

APPLICATION

The minimum thickness and the number of coats required for gypsum plaster depend on the plaster base. Three coats are required over metal lath and some gypsum lath installations. Two coats are adequate over most gypsum lath installations and over masonry and concrete bases.

9.3.2.1

Three-Coat Work

Three-coat work requires a scratch coat, a brown coat, and a finish coat, each applied separately and allowed to set and partially dry. The scratch coat derives its name from the cross-raking that is performed on the wet surface to improve bond with the following brown coat (Fig. 9.3-1).

FIGURE 9.3-1 *Scratch coat is raked for three-coat work.* *(Courtesy Portland Cement Association.)*

FIGURE 9.3-2 *(a) Brown coat is applied over scratch coat, followed by (b) finish coat in separate operations for three-coat work.*

9.3.2.2

Two-Coat Work

In two-coat work, the scratch coat is applied as in three-coat work, but cross-raking is omitted and the brown coat is applied immediately, before the scratch coat has set. The finish coat is applied after the basecoat has set and is partially dry (Fig. 9.3-2).

9.3.2.3

Application Methods

Basecoat plasters are formulated for hand-trowel or machine applications. There are three basic methods of applying finish coat gypsum plasters: trowel, float, and spray. The method used largely determines the surface appearance (Fig. 9.3-3).

Trowel finishes are used where a flat, smooth, nonporous surface is desired and often form the base for further decorating (Fig. 9.3-4a, b). When an aggregate is used, it is fine silica sand or perlite. Float finishes produce surface textures (Fig. 9.3-4c). The coarseness of the texture depends on the aggregate, mix proportions, and type of float used.

9.3.2.4

Compositions of Coats

The composition of base coats and finish coats varies considerably. The term *base coat* includes scratch coats and brown coats.

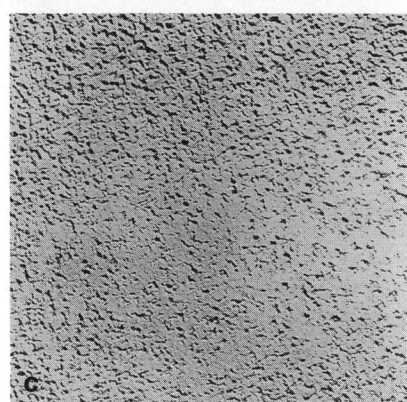

FIGURE 9.3-3 *Finish coat plasters may take a smooth trowel finish or one of the following: (a) float finish; (b) spray finish; (c) texture finish.* *(Courtesy United States Gypsum Company.)*

■ **BASE COATS**

Gypsum neat plaster is suitable for use as a base coat over metal or gypsum lath, gypsum or clay tile, concrete masonry units, and brick of moderate suction (see Section 4.2). Fibered neat plaster often is

FIGURE 9.3-4 *Methods of applying finish coats: (a) hand trowel; (b) power trowel; (c) hand float; (d) machine spraying.*

used as a scratch coat over metal lath because less plaster is lost as mechanical keys are formed behind the lath. When neat plaster is used as a brown coat, sand usually is added to prevent excessive shrinkage in drying.

Wood-fibered plaster may be used over the same bases as neat plaster, but it has greater fire resistance than neat plaster and other sanded base coat plasters.

Ready-mixed plaster of a composition similar to gypsum neat plaster or wood-fibered plaster may be applied where those two plaster types are acceptable.

■ **FINISH COATS**

Either trowel or float finishes may be obtained with *gauging plaster*, *gypsum Keene's cement plaster*, prepared *gypsum finish plaster*, or *veneer (thin coat) plaster*. *Acoustical plaster* may be hand applied but most generally today is machine sprayed (Fig. 9.3-4d).

Gypsum Keene's cement is used for gauging lime finish coat plaster, much the same as gauging plaster, but is preferred to gauging plaster when a harder, more moisture-resistant surface is desired. Hardness is proportionate to the amount of gypsum Keene's cement used. Both trowel and float finishes are possible with gypsum Keene's cement.

Gypsum Keene's cement-lime-and-sand float plaster may be used over all types of base coat plasters and is particularly recommended over lightweight aggregate base coat plasters. It provides greater crack resistance than a trowel finish.

Veneer (Thin Coat) Plaster

Veneer plaster may be applied over the conventional framing and furring described in Sections 9.2 and 9.3.2.7, as part of the specialized partition systems and assemblies discussed in Section 9.3.2.8, and in most locations where gypsum board assemblies are applicable (see

Section 9.5). In assemblies where gypsum board is the outer layer, veneer plaster base is substituted. The remainder of the assembly remains as described. Care must be exercised, of course, when a fire or acoustical rating is required to ensure that the assembly will still comply with these requirements after veneer plaster has been substituted for a layer of gypsum board.

Before veneer plaster is applied, the joints in the veneer plaster base are treated with a glass fiber tape or reinforced with drywall paper tape and a compatible joint compound following manufacturer's recommendations. Cornerbeads are installed prior to plastering. Because of the composition and thinness of the plaster coat, it dries rapidly and may be decorated 24 hours after installation. Refer to Section 9.5 for additional information about veneer plaster base board and its application.

Veneer plaster can be applied as a smooth trowel finish or as a textured, sprayed finish. Both are extremely hard and abrasion resistant. Veneer plaster may be used also as a finish coat over bond plaster or as a basecoat for gypsum-lime and gypsum Keene's lime finishes where exceptional strength and crack resistance are desired.

9.3.2.6

Acoustical Plasters

Depending on their formulation, acoustical plasters may be hand or machine applied; however, most acoustical plasters are applied by machine. Noise reduction coefficients ranging from 0.40 to 0.60 may be obtained with these products. Surface textures may vary from fine to rough, and crack resistance is high. Indentation resistance, however, is much lower than with other plaster products, and the use of acoustical plasters is limited to ceiling and upper wall areas not subject to abrasion or impact. Acoustical plasters may be left in the natural off-white or gray color or spray painted.

9.3.2.7

Application over Framing and Furring

Gypsum plaster may be applied to both gypsum and metal lath over wood or metal framing or furring.

■ PARTITIONS

Gypsum plaster finished partitions are made with wood or metal studs supporting faces of plaster. Studs usually are spaced 16 to 24 in. on centers and are supported by wood or metal floor and ceiling runners. Wood studs normally are 3-1/2 in. deep (2 × 4s), metal channel studs are available in 1-5/8, 2-1/2, 3-5/8, and 4-in. sizes (Fig. 9.3-5a). Greater wall cavities can be developed with deeper trussed-type studs, which range up to 6 in. in depth (Fig. 9.3-5b), or two sets of lighter channel studs braced with wallboard (Fig. 9.3-5c).

Lath and plaster hollow partitions have gypsum lath screwed, clipped, or nailed to studs, or metal lath wired, nailed, clipped, or stapled to the studs. Two or three coats of plaster are built up over this plaster base on each face.

■ CEILINGS

Conventional gypsum plaster is sometimes applied to ceilings over gypsum lath, but most installations are over metal lath. Refer to Section 9.2 for a discussion about installation of plaster support systems for ceilings.

9.3.2.8

Specialized Partition Systems and Other Assemblies

In addition to application over gypsum or metal lath applied to the standard framing and furring discussed in Section 9.2, gypsum plaster is also used in proprietary assemblies of framing, lath, board, and plaster with special metal accessories designed for use with each system.

A variety of proprietary partition systems have been developed by manufacturers of gypsum products to minimize the number of operations and to speed field erection of partitions. The chief types in which gypsum plaster is used are classed as (1) solid and (2) semisolid partitions. A third type, hollow and demountable (movable) systems, are designed for gypsum board products and do not utilize gypsum plaster products.

■ SOLID PARTITIONS

Solid partitions are typically field-erected from plaster on metal or gypsum lath.

FIGURE 9.3-5 *Hollow plaster partitions. (a) With channel studs. (b) With trussed studs. Trussed studs are seldom used in new work today but may be found in many existing buildings. (c) With braced double studs. Double-stud construction accommodates mechanical equipment and plumbing.*

These generally include wood or proprietary metal ceiling and floor runners to stabilize the partitions. In lath and plaster partitions, an integral flush metal base acts as a screed. Partitions also may be trimmed out with resilient or wood base.

Lath and plaster partitions use self-supporting ribbed lath (Fig. 9.3-6a) or ordinary expanded mesh wired to steel bars or channels (Fig. 9.3-6b). To this base, three coats of plaster are applied in thicknesses varying from 1 to 1-1/4 in. on each side. Solid plaster partitions also may be made of 1/2-in.-thick gypsum lath, with two 3/4-in.-thick plaster faces built up of two or more coats (Fig. 9.3-6c).

FIGURE 9.3-6 *Solid plaster partitions: (a) on self-supporting ribbed lath; (b) on stud-supported woven wire lath; (c) on gypsum lath.*

Solid partitions consisting of a backing board, a layer of veneer plaster base board, and a veneer plaster finish coat may also be used. This type is similar to the single and double solid wallboard partitions shown in Figure 9.5-53.

■ SEMISOLID PARTITIONS

Semisolid partitions (see Fig. 9.5-52) are not usually constructed with conventional plaster, but veneer plaster base may be substituted for the outer layer of gypsum board and finished with veneer plaster. In that case, a layer of veneer plaster base is substituted for the outer layer of gypsum board shown in the figure on the side of the partition where veneer plaster is desired.

■ SPECIAL CEILING ASSEMBLIES

Gypsum plaster may be used in some pro-prietary electric radiant heating systems. The heating cables are attached to or through standard gypsum lath and are then embedded in the plaster. Such installations should be made strictly in accord with the instructions of the manufacturer of the heating panels.

Refer to Section 9.5 for additional information about electric radiant heating panels.

9.4 PORTLAND CEMENT PLASTER APPLICATION

Portland cement plaster is used on both the interior and the exterior (stucco). It is the only plaster to use in locations where exposure to wetting is a factor.

9.4.1

MIXES

Aggregate constitutes most of the volume of a portland cement plaster mix. One cubic foot of a typical portland cement plaster mix contains about 0.97 cu. ft. aggregate and 0.25 to 0.33 cu. ft. cementitious material. Therefore, it is important that the aggregate be clean, sound, and well graded with particles ranging from fine up to a maximum size of 1/8 in.

The best and most practical mix results when the aggregate size is proportioned so that a minimum of water-cement paste is required to coat the aggregate and fill the voids between particles. An excess of fine or coarse particles can result in a mix that contains too much cement or too much water, either of which can be detrimental to the plaster.

Several possible mixes, classified according to plaster groups (C, L, F, and P), are listed in Figure 9.4-1. The letter designations broadly indicate the type of mix: C = (portland) cement, L = lime, F = finish coat, and P = plastic cement. The selection of a suitable mix for a specific coat depends in part on the bonding properties of the plaster base (Fig. 9.4-2). Rich mixes that contain substantially more portland cement than shown in Figure 9.4-2 should not be used.

The generally recommended proportions are 1 part cement (including lime when used) to at least 3 and not more than 5 parts by volume of damp, loose sand. Since wide variations in sand quality and gradation exist throughout the country, the correct proportioning of sand to cement should be determined from trial mixes or

FIGURE 9.4-1
Stucco Mixes

Group*	Proportions, Parts by Volume				
	Portland Cement	Lime	Masonry Cement Type II	Plastic Cement	Sand
C	1	0 to 1/4	—	—	3 to 4
C	1	—	1	—	6 to 7-1/2
C	1	1/4 to 1/2	—	—	4 to 6
L	1	2 to 1-1/4	—	—	4-1/2 to 9
L	—	—	1	—	3 to 4
F	1	1-1/4 to 2	—	—	5 to 10
P	1	—	—	1	6 to 10
P	—	—	—	1	3 to 4

Courtesy Portland Cement Association.

*C = (portland) cement; L = lime; F = finish coat; P = plastic cement.

FIGURE 9.4-2
Stucco Mix Selection Guide

Stucco Base	Recommended Groups[a] for Stucco Coats		
	Scratch	Brown[b]	Finish[c]
Low absorption (placed concrete, dense clay brick)	C, P	C, L, P	L, F, P
High absorption (concrete unit masonry, clay brick, structural tile)	L, P	L, P	L, F, P
Metal reinforcement[d]	C, P	C, L, P	L, F, P

Courtesy Portland Cement Association.

[a] C = (portland) cement; L = lime; F = finish coat; P = plastic cement.

[b] Use as base coat in two-coat work over concrete and masonry bases.

[c] Finish coat may be factory-prepared "stucco finish."

[d] Over any type of supporting construction.

particles, which shrink excessively on drying; an excess of coarse particles produces a harsh and difficult-to-apply mix. Therefore, aggregate should be selected carefully, and mix proportions should not be relied on to overcome deficiencies in aggregate quality or gradation.

A good mix can be recognized by its workability, ease of troweling, good adhesion to the base, and sufficient cohesiveness so that the mix will not sag on vertical surfaces. Once the mix has been determined, uniformity in proportioning of materials from batch to batch will help to ensure uniform suction and color.

The materials for all coats should be thoroughly mixed. Dry materials should be mixed to a uniform consistency before water is added. Only enough water should be added to ensure workability. A power mixer is recommended for uniform blending of materials (Fig. 9.4-3). It should be operated for at least 5 minutes after all ingredients are in the mixer. Materials should be mixed only in quantities that can be used within 3-1/2 hours; remixing to restore plasticity is permissible, but retempering with additional water should not be done.

FIGURE 9.4-3 *A power mixer combined with pump. Water is placed in mixer first; one-half of the sand is added; cement and admixture are added; the remaining sand is added and mixing continued until the mix is uniform and of the proper consistency.*

When portland cement plaster is applied by machine, the amount of water should be adjusted to produce a mix of proper consistency for the length of hose, temperature and humidity conditions, and the type of plaster base. When measured with a standard 2- by 4- by 6-in. slump cone, the slump should be from 2-1/2 to 4 in.

previous experience with local aggregate sources.

Proper gradation of aggregate from fine to coarse particles (see Fig. 9.1-7) is important to producing a durable finish. If the mix contains an excess of fine particles, more water-cement paste is required to coat the larger surface area of the particles. When an excess of coarse particles is present, more paste is required to fill the spaces between particles.

A lack of adequate water-cement paste is demonstrated by poor workability. The addition of too much water or too much cement to improve workability results in a plaster that is subject to high drying shrinkage and cracking. Also, sand fines often contain a high percentage of clay at the mixer and from 2 to 3-1/4 in. at the nozzle. The hose length from the machine to the working surface should be as straight and as short as possible, generally not longer than 200 ft.

9.4.1.1

Masonry and Plastic Cement Mixes

When masonry cement and plastic cement are used, the addition of lime or plasticizers generally is not necessary because plasticizing materials are included during manufacture. The use of these cements simplifies field mixing because only sand and water need to be added. Either masonry cement or plastic cement can be combined with portland cement in equal volumes when greater strength is required (see Figs. 9.4-1 and 9.4-2). Plastic and masonry cements alone or in combination with portland cement can be used for any of the three plaster coats.

9.4.1.2

Portland Cement Mixes

Portland cement should be supplemented with lime, plastic cement, or masonry cement to impart suitable workability and plasticity to the mix. For basecoats, portland cement can be mixed with up to 1-1/4 parts lime or in equal proportions with masonry or plastic cements. For the finish coat, 1 part portland cement can be supplemented with an equal part of plastic cement or 1/2 to 2 parts lime (see Figs. 9.4-1 and 9.4-2). Truer colors will be obtained if white portland cement and light-colored sand are used for field-mixed finish coats.

Plasticizers such as mineral, animal, vegetable, or glass fibers can be added to mixes for the first coat, but only in quantities sufficient to obtain the required plasticity or cohesiveness. The kind and exact amount of plasticizing agent will vary with the type of aggregate, plaster base, and manner of application. The amount of a mineral fiber used should not exceed 3% of the cementitious material by weight. The best combination of ingredients should be determined by trial mix before application.

9.4.2

APPLICATION

Portland cement plaster should not be applied in freezing weather or to surfaces that contain frost. The temperature of the plaster should be maintained at 50°F or higher during application and for at least 48 hours after the finish coat has been applied. When temperatures are falling, portland cement plaster should be protected against freezing by heated enclosures (Fig. 9.4-4; see also Section 9.4.3).

FIGURE 9.4-4 *Protective enclosures in mild climates prevent rapid loss of moisture and facilitate curing.* (Courtesy Portland Cement Association.)

Hand application of portland cement plaster involves traditional tools and practices that have proved to be successful over centuries of use. Machine application, however, incorporates additional advantages. The wet mix can be delivered under pressure to scaffolding in multistory buildings (see Fig. 9.4-4); application is speeded, and obvious lap and joint marks, sometimes difficult to avoid with hand application, are eliminated. Machine application results in a more uniform texture and deeper and darker colors than are possible with hand application. Also, some textures, such as a uniform rough texture, can be obtained best by spraying. For these reasons, the finish coat sometimes is applied by machine over a hand-applied base coat.

Generally, portland cement plaster is applied in either two or three coats, depending on the supporting construction. A three-coat application, which is used over metal reinforcement, consists of a scratch coat, a brown coat, and a finish coat, each applied separately and allowed to set partially. In a two-coat application, which is generally used over concrete and masonry bases, a single base coat takes the place of the scratch and brown coats and is followed by a finish coat. The scratch coat and the brown coat in three-coat work also are referred to as base coats.

9.4.2.1

Three-Coat Work

Three coats totaling at least a 7/8-in. thickness are required over all types of metal reinforcement. The first coat, the *scratch coat*, provides mechanical keying with the reinforcement and a rigid surface for the following coat. The *brown coat* provides additional thickness and strength and ensures a level surface for the application of the final decorative *finish coat*. Each coat should be moist-cured before the following coat is applied (see Section 9.4.3).

■ DASH BOND COAT

A dash bond coat is often used as the first coat of a three-coat application over a rigid base. A dash bond coat surface is sufficiently rough so that cross-raking is not required. A dash bond coat should be moist cured for 12 to 24 hours before the brown coat is applied.

■ SCRATCH COAT

A scratch coat is the first coat over metal reinforcement and can be the first coat over a rigid base that has sufficient bonding capability to make a dash bond coat unnecessary. A scratch coat should be at least 3/8 in. thick when measured from a rigid base or backing paper. It should be applied evenly, with sufficient pressure to squeeze the mix through the reinforcement to form mechanical keys (Fig. 9.4-5a). The space between the metal and the backing should be filled thoroughly so that the metal will be sufficiently embedded.

FIGURE 9.4-5 *The first (scratch) coat (a) is spread on reinforcement with a trowel, then (b) the surface is scored horizontally.*

The coverage on the surface should be such that no more than 10% of the surface of the reinforcement remains exposed. As soon as the scratch coat has become firm, its entire surface should be cross-raked (*scratched*) to provide good mechanical bond for the brown coat (Fig. 9.4-5b). On walls, horizontal cross-raking is recommended. Regardless of whether application is by hand or by machine, cross-raking is done manually.

■ BROWN COAT

After a scratch coat has been moist-cured and is sufficiently firm to carry the weight of the next coat, a brown coat should be applied. If a scratch coat becomes dry, it should be dampened again to control suction before a brown coat is applied. A brown coat should be approximately 3/8 in. thick, bringing the thickness of both coats to at least 3/4 in. (Fig. 9.4-6a).

When possible a brown coat should be applied over an entire wall panel without stopping. Interruptions in the work should be made at screeds, control joints, corners, pilasters, belt courses, doorways, openings, or other lines where changes in appearance, which might show through the finish coat, will not be obvious.

A brown coat should be *rodded* (struck off) with a straightedge and then darbied to impart a level, uniformly rough texture. With machine application, these operations are performed by hand (Fig. 9.4-6b). The same type of darby or float should be used for the entire surface to ensure uniformity of texture. Deep scratches or surface defects should be smoothed out,

because such irregularities may cause unequal suction or may show through to the finish coat. External corners should be formed with a straightedge when corner reinforcement is not used (Fig. 9.4-6c).

After moist curing, the surface should be allowed to dry uniformly before a finish coat is applied (see Section 9.4.3).

■ FINISH COAT

A brown coat should be uniformly dampened if a finish coat is to be applied by hand. Dampening is not required for machine-applied finishes. A finish coat should be at least 1/8 in. thick, bringing the total three-coat portland cement plaster finish to at least 7/8 in. thick (Fig. 9.4-7a). When machine applied, the sprayed finish coat often is made in two or more consecutive applications (Fig. 9.4-7b). The mix for the first application is generally of thinner consistency and is applied primarily to ensure complete coverage of the brown coat. Successive applications are somewhat thicker and are intended to develop the finished texture.

Factory-prepared stucco finishes ensure greater uniformity of color in the finish coat. A variety of decorative surface finishes are possible with both machine and hand application (Fig. 9.4-8) (see Section 9.4.5). A plain finish coat that is going to be painted should be moist cured. Prepared stucco finishes often contain curing agents, and moist curing is not necessary; in fact, it is specifically discouraged by some manufacturers because it may

FIGURE 9.4-6 *Brown coat: (a) mix is spread with a trowel, then (b) leveled with a rod (straightedge). (c) When cornerbead or corner reinforcement is not used, a true, plumb corner is formed with a straightedge.*

FIGURE 9.4-7 *A finish coat can be (a) hand-floated or (b) sprayed by machine.*

FIGURE 9.4-9 *Two coats are adequate over a suitably rough concrete or masonry surface.* *(Courtesy Portland Cement Association.)*

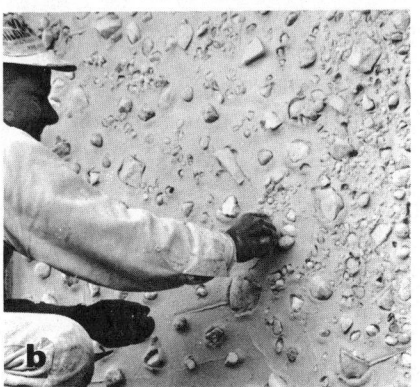

FIGURE 9.4-8 *Rock dash (marblecrete) finish: (a) small aggregate can be machine-sprayed; (b) larger aggregate is usually placed by hand.* *(Courtesy Portland Cement Association.)*

cause streaking of the color surface. The finish coat should be maintained at the proper temperature for at least 48 hours after application (see Section 9.4.3).

9.4.2.2

Two-Coat Work

A two-coat portland cement plaster application is acceptable over rigid concrete or masonry bases, but three coats should be used where increased thickness or strength is required. The base coat and finish coat of a two-coat application should total at least 5/8 in. in thickness.

■ BASE COAT

The surface of a concrete or masonry base should be dampened uniformly to ensure uniform suction. However, dampening of the surface is not required when a chemical bonding agent is used. A base coat is applied about 3/8 in. thick and should

not be scratched, as in three-coat application. Mechanical bond for the finish coat is provided by rough floating of the base coat in the same manner as the brown coat in three-coat work (Fig. 9.4-9).

A machine-applied base coat is built up with several applications, with the final application being rodded and darbied manually.

■ FINISH COAT

A finish coat should be about 1/4 in. thick. It can be applied in a variety of textures by hand or machine, as in a three-coat application.

9.4.3

CURING

Like concrete, portland cement plaster requires sufficient water and favorable temperatures to facilitate the process of hydra-

tion, by which the cement gains strength and hardens to a solid mass. The large exposed area of plaster relative to its volume encourages rapid loss of moisture by evaporation. Moist curing of each coat attempts to replace this moisture and maintain an adequate supply for the chemical reaction to continue. An ordinary hose nozzle can be adjusted to provide a fine spray for curing; a steady stream that could erode the surface should not be used.

The act of moist curing does not mean spraying at set intervals, but rather periodic observation and spraying when the surface appears to be drying out. Since evaporation increases with increased air circulation and higher temperatures, more frequent spraying generally is required in breezy, warm weather. Also, wind breaks or hangings are recommended to protect fresh stucco from hot dry winds.

In addition to adequate moisture, proper hydration requires that a portland cement plaster mix and each plaster coat be maintained at temperatures of at least 50 °F. A finish coat should be maintained at this temperature for at least 48 hours after it has been applied. In cold weather, it may be necessary to heat the mixing water and provide a heated enclosure around the work area.

While it is generally accepted in the industry that adequate water to ensure hydration is essential in producing quality portland cement plaster, there is no general agreement as to exact curing procedures. The curing recommendations given in Figure 9.4-10 are based on the premise that deliberate moist curing ensures a greater measure of success than

FIGURE 9.4-10
Curing Schedule for Portland Cement Plaster

Application[a]	Moist Curing[b]	Total Setting Time[c]
Scratch coat	12 to 24 hr.	At least 48 hr. (between coats)
Brown coat	12 to 24 hr.	At least 7 days (between coats)
Finish coat	12 to 24 hr.[d]	At least 48 hr. (after application)

Courtesy Portland Cement Association.

[a] In two-coat work, requirements for base and finish coats are the same as for brown and finish coats, respectively.

[b] Moist curing should be delayed until scratch and brown coats are sufficiently set to prevent erosion and for 12 to 24 hours for finish coat.

[c] At least 50 °F temperature should be maintained during and after applciation.

[d] Not recommended for prepared colored "stucco finishes."

does dependence on favorable climatic conditions or on water contained in the mix.

In some parts of the country, no deliberate moist curing is performed, and each coat is applied as soon as the previous coat is sufficiently hard to support the following coat. This method, sometimes called *doubling back*, relies on the protective nature of each covering coat to retain the moisture and facilitate curing of the underlying coat.

Another practice involves application of coats at not less than 24-hour intervals, with continuous moist curing between coats and after the finish coat has been applied.

The curing method chosen should be based on accepted local practice that has a demonstrated history of satisfactory performance. All the various curing procedures attempt either to maintain the previous coat uniformly damp or to permit it to become uniformly dry. This is particularly true of the base (or brown) coat, because uniformity of color in the finish coat depends on equal suction in the base coat.

The long waiting period between base and finish coats shown in Figure 9.4-10 ensures that the brown coat will be uniformly dry. When shorter intervals are allowed, it is often necessary to keep the base coat damp by continuous moist curing, until the finish coat is applied. Unless a rigid schedule for rapid successive appli-

cations can be maintained, the need for continuous moist curing may be more demanding than the recommended short curing period and subsequent drying delay.

9.4.4

CRACK CONTROL

The proper design of supporting construction, selection of materials, and application of portland cement plaster can substantially minimize or eliminate undesirable cracking. Cracks can develop in plaster from (1) shrinkage stresses; (2) structural movement; (3) restraints due to penetration of mechanical or electrical equipment or intersecting walls or ceilings; and (4) weak sections in the finish due to thin spots, openings, or changes in materials in the supporting construction.

9.4.4.1

Shrinkage Stresses

As a wet portland cement plaster mix hydrates and dries, it also tends to shrink, setting up stresses in the plaster membrane. Cracking results when shrinkage stresses exceed the tensile strength of the plaster membrane. It is desirable, therefore, to achieve the full strength of the mix and to minimize the degree of shrinkage.

Proper moist curing and protection from temperature extremes allow the plaster to dry slowly and uniformly, thus increasing strength and reducing shrinkage. Shrinkage stresses also are minimized by proper proportioning of mixes and proper placement of control joints.

■ PROPORTIONING MIXES

The chief factors in proportioning mixes that affect the balance between tensile strength and shrinkage stresses in the finished plaster are the cement content, the water content, and the aggregate quality and gradation.

Rich mixes, high in cement content, generally result in plaster with higher initial shrinkage, balanced to some extent by increased strength. Excessive water in the mix increases the volume of minute voids in the plaster as water evaporates or is used up in hydration and results in both higher shrinkage and lower strength.

Poorly graded sand requires the addition of more water, more cement, or both to obtain suitable workability, which can be detrimental and cause shrinkage cracking (see Section 9.4.1).

■ CONTROL JOINTS

When portland cement plaster is applied directly to masonry or concrete and no reinforcement is used, control joints in the plaster should coincide with joints in the base. Additional control joints to resist shrinkage cracking in the plaster finish are not usually required, because the underlying base will resist stresses due to shrinkage.

When portland cement plaster is applied to metal reinforcement (over any type of supporting construction) control joints should be installed in the plaster not more than 18 ft. apart, forming panels not larger than 150 sq. ft. in area.

Either one-piece or multisection control joints (see Fig. 9.2-13b) can be used for vertical joints in walls and for all joints in protected soffits. A multisection control joint is particularly effective in relieving stresses caused by shrinkage and differential structural movement in the supporting construction (Fig. 9.4-11). For horizontal joints, a watertight control joint designed for the purpose should be used (Fig. 9.4-12).

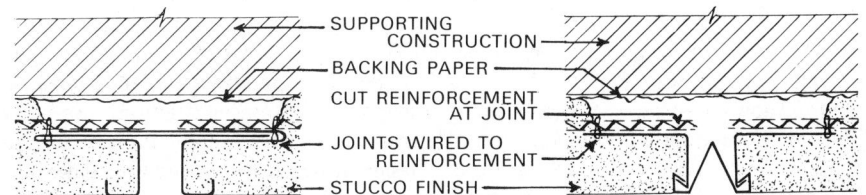

FIGURE 9.4-11 *Multisection control joints effectively relieve stress caused by drying shrinkage, structural movement, and temperature variations.* (*Courtesy Portland Cement Association.*)

FIGURE 9.4-12 *Specially designed horizontal control joints provide weathertightness as well as crack control.* (*Courtesy Portland Cement Association.*)

9.4.4.2

Structural Movement

Structural movement may be caused by foundation settlement, excessive deflection of structural members due to overloading, racking due to wind forces, and expansion and contraction of materials caused by temperature and humidity variations. Good structural design attempts to minimize these movements where possible; specially designed joints are used to localize movements that cannot be reduced to negligible proportions.

Relief joints and control joints are used to localize movement and minimize stresses within a portland cement plaster finish. Relief joints at ceilings and walls should be used where the plaster is intersected (Fig. 9.4-13) or is penetrated by structural elements such as beams, columns, or load-bearing walls (Fig. 9.4-14). Location of control joints depends on the type of supporting base.

■ RIGID BASES

Where portland cement plaster is applied directly to a concrete or masonry base,

firm and continuous support of the plaster membrane will tend to resist shrinkage, temperature, and impact stresses in the plaster finish. However, the finish will be subjected to structural movements in the base and so would be likely to crack wherever excessive stress concentrations develop. It is particularly important, therefore, in this type of installation to provide proper control joints in the base. Metal control joints should be placed in a plaster finish directly over each underlying joint in the base. Control joints also should be placed in the plaster finish over construction, isolation, or expansion joints that occur in the base and at the juncture of dissimilar materials.

■ METAL REINFORCEMENT

Portland cement plaster applied over metal reinforcement, with or without backing paper, acts as a thin concrete slab. Because the plaster is somewhat isolated from the supporting construction, it is not subjected to structural stresses to the same degree as plaster applied directly to rigid bases. However, such a thin reinforced slab is subject more to shrinkage, impact, and temperature stresses because it is not

supported and restrained by an underlying rigid base. Of these, shrinkage stresses are the most critical and involve the greatest movement. Therefore, control joints located as recommended for shrinkage purposes also will relieve stresses occasioned by other forces acting on the plaster membrane.

9.4.4.3

Weakened Sections

Normal stresses caused by shrinkage, humidity, or temperature variations are evenly distributed in plaster of uniform thickness, but tend to be higher at the thin spots. If the plaster is applied over an irregular, rigid base or over sagging metal reinforcement, the resulting finish will be of unequal thickness, creating weakened sections. Such concentration of stresses often results in cracking at these weakened points. To minimize this hazard, plaster should be applied over a level base in uniform thickness.

FIGURE 9.4-13 *Perimeter relief at internal corner between wall and soffit: (a) with trowel cut; (b) formed with casing beads.* (*Courtesy Portland Cement Association.*)

FIGURE 9.4-14 *Perimeter relief joint for use where structural elements or fixtures penetrate the surface.* (*Courtesy Portland Cement Association.*)

Weakened sections also are created at corners of cutouts and openings in a plaster finish. Metal reinforcement will resist the concentration of stresses at the corners and generally prevent random diagonal cracking. When reinforcement is not used and plaster is applied directly to a rigid base, strip reinforcement (*strip lath*) can be installed diagonally at each corner and will act similarly to prevent cracking.

As in concrete, weakened sections can be introduced deliberately at planes where cracking would not be objectionable. A trowel cut adjacent to a door or window frame will result in a straight, inconspicuous crack rather than objectionable random cracking. Often, the trowel cut can be concealed by trim; where trim is not intended, the cut can be left exposed if neatly formed and caulked.

9.4.4.4

Restraints

Free movement of a plaster membrane generally is restrained at corners and wherever the membrane is penetrated by fixtures or structural elements. Cracking due to restraint is not common at external corners, particularly if control joints are provided in each plane as recommended for shrinkage control.

Cracking is more frequent at internal corners formed by the intersection of load-bearing elements (columns, walls, pilasters) and non-load-bearing walls and at the juncture of walls and suspended or furred ceilings. At such corners, perimeter relief should be provided by omitting cornerite, stopping the metal reinforcement in each plane, and including relief joints that will permit independent movement in each plane.

Relief joints can be formed by a trowel cut (Fig. 9.4-13a) or a pair of casing beads (Fig. 9.4-13b). The casing beads should be wired to the reinforcement, not rigidly nailed to the supporting construction. The corner should be caulked with a nonhardening resilient compound after initial shrinkage has taken place.

Where a plaster surface abuts or is penetrated by a structural element or a fixture, a relief joint should be provided. The plaster should terminate with a casing bead; the space between the bead and the structural element or fixture should be caulked (see Fig. 9.4-14).

9.4.5

SURFACE FINISHES

In addition to providing the desired decorative surface, the finish selected also affects the weather resistance and maintenance requirements of a portland cement plaster installation. Smooth-troweled plaster finishes are not recommended because they tend to make surface irregularities more conspicuous. Rougher textures generally will conceal slight color variations, lap joints, uneven dirt accumulation, and streaking.

The range of possible portland cement plaster textures is practically limitless. Brick, stone, and travertine marble can be simulated by skilled applicators; many other finishes can be produced by hand or machine, depending on the desired texture. The finish may be natural or integrally colored so that painting is not required. Colored finishes can be effected by using a prepared stucco finish, adding pigment to the mix, or embedding colored aggregate in the surface. Manufacturers of colored stucco finishes issue application recommendations, which should be followed carefully.

Several common textures are shown in Figure 9.4-15. The character of these textures can vary substantially with each applicator. Therefore, a standard sample should be produced at the beginning of the work, and as the work progresses, it should be checked with the sample for conformity.

FIGURE 9.4-15 *Some typical portland cement plaster (stucco) finishes: (a)* float finish, *made with carpet- or rubber-faced float; (b)* wet-dash, *coarse aggregate mix is dashed on with a brush; (c)* dash-troweled, *high spots or dashed surfaces are troweled smooth; (d)* stippled-troweled, *surface is stippled with a broom, then high spots are troweled smooth; (e)* rock-dash, *decorative pebbles are thrown by hand or machine against wet surface; (f)* combed, *surface is formed with a notched template. (Courtesy Portland Cement Association.)*

RESURFACING

Existing concrete, masonry, and portland cement plaster surfaces that are firm and sound and that can provide adequate bond by suction or mechanical key are suitable for direct application of new portland cement plaster. Painted or coated surfaces can be sandblasted or otherwise treated to remove the coating. Smooth concrete surfaces can be etched or treated with a bonding agent. When none of these methods is practical or effective, backing paper and metal reinforcement should be applied over the surface as for new construction.

One-, two-, or three-coat work may be necessary, depending on the type of base. One finish coat may be sufficient over substantially sound, level portland cement plaster with good bonding properties. If the existing plaster is irregular or requires removal of loose material, a base and a finish coat are required. Two coats similarly are required over concrete and masonry surfaces suitable for direct application.

When metal reinforcement is used, three coats should be applied. The procedures for application of the several coats are the same as for new construction.

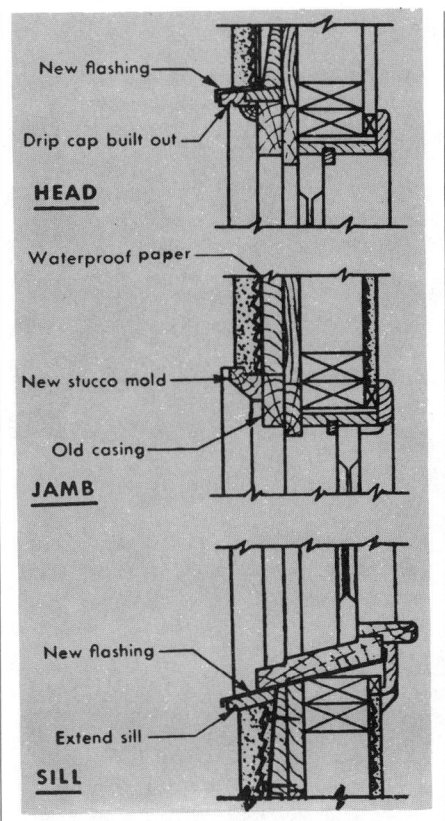

FIGURE 9.4-16 *Resurfacing details for exterior portland cement plaster (stucco) over wood-framed construction.* (Courtesy Portland Cement Association.)

When new portland cement plaster is applied over an existing portland cement plaster surface, consideration should be given to all points where water might enter behind the new finish (Fig. 9.4-16). Often drip caps and flashings over openings must be extended and new window sills and plaster molds installed to project beyond the new plaster finish.

9.5 GYPSUM BOARD SYSTEMS

This section covers the manufacture, properties, and uses of gypsum board products and the materials and methods necessary to provide suitable gypsum board installations for various assemblies. Both singleply and multi-ply applications are discussed. The gypsum board systems described include gypsum board panels and their supports and accessories. Only general recommendations for sound isolation, fire resistance, thermal insulation, and moisture resistance are included. Additional information about those subjects is given in Chapters 7, 10, and 11.

Gypsum board construction originated about 1915, when small, 3/8-in.-thick sheets were developed primarily for remodeling. In the following decades, erection and joint treatment techniques gradually evolved, resulting in improved surface appearance and the use of gypsum board in other types of construction. After World War II, greater emphasis on total system and assembly performance gained gypsum board widespread use (Fig. 9.5-1).

Gypsum board finishes are also called *drywall* finishes, because they have initial low moisture content and require little or no water during application. This contributes to their dimensional stability and minimizes undesirable effects on other construction components. Plywood, hardboard, and fiberboard are also sometimes called drywall, but the term usually means gypsum board.

Gypsum board finishes consist of one or more plies of gypsum board having incombustible gypsum cores with surfaces of paper or other sheet material. Face (exposed) plies either have plain calendered manila paper intended for decorating with paint, wall covering, or wall tile or are

FIGURE 9.5-1 *We usually think of gypsum board being used for flat walls and ceilings. As this photograph shows, gypsum board can also be used in highly imaginative ways.* *(Courtesy Gypsum Association.)*

insulation, and moisture control. A variety of gypsum boards and engineered systems are manufactured today to meet specific needs. However, performance is closely related to the properties, arrangement, and installation of construction elements that make up the wall or ceiling assembly.

For gypsum board finishes to have suitable quality, consideration should be given to (1) preparation of supporting construction; (2) selection of board and accessory materials, (3) job conditions; (4) board application, including joint and fastener treatment; and (5) special constructions where sound isolation, fire resistance, thermal insulation, or moisture resistance is important.

The Gypsum Association publication GA-216, "Recommended Specifications for the Application and Finishing of Gypsum Board," is the currently recognized industry standard and the basis for many of the data in this section.

9.5.1

BOARD MANUFACTURE

As shown in the flow chart in Figure 9.1-2, calcined gypsum without tube milling is used to make gypsum board products. This gypsum is mixed with water to form a slurry. Additives and glass fibers are added for highly fire-resistant gypsum boards. This slurry, which forms the core of a board, is fed onto a conveyor belt between two layers of paper. After traveling down the belt for 3 or 4 minutes, the core material has set sufficiently to be cut to length (Fig. 9.5-2). Boards then are conveyed to kilns for drying. Boards not requiring additional finishes may be bundled and stacked for shipping. Others may receive special backings such as aluminum foil, decorative vinyl paper finishes, or organic texture coatings.

9.5.2

BOARD TYPES

Gypsum board products are panels or slabs consisting of a noncombustible gypsum core, surfaced and edged with covering material specifically designed for

predecorated with colored, textured, or simulated wood-grained paper or vinyl coverings. Both plain and predecorated wallboard provide economical finishes capable of efficient erection with different structural systems.

Most gypsum board installations are made with plain boards, which require fastener and joint treatment and decorating. Predecorated board, which constitutes a smaller portion of the total use, requires somewhat different installation methods. It is usually applied directly to studs or furring with color-matched fasteners or laminated to a base layer of gypsum backing board. Joints are left exposed, treated with special moldings or battens, or concealed with a loose flap of the decorative material. This section covers plain wallboard finishes except when predecorated panels are specifically mentioned.

In addition to other functional requirements, walls and ceilings often demand specific performance with respect to sound isolation, fire resistance, thermal various uses with respect to performance, application, location, and appearance.

Gypsum board products are produced for specific uses in building construction. *Gypsum wallboard* is a specific gypsum board product that has a prefinished surface or a surface suitable for paint finishing after appropriate treatment of fasteners and joints. *Gypsum sheathing boards* are used in exterior frame walls as a base for finish materials and for increased fire resistance. *Gypsum backing boards* provide a base for adhesive application of gypsum wallboards in multi-ply construction and acoustical tiles in suspended ceilings. *Gypsum formboards* serve as both permanent forms and finished ceilings in poured gypsum concrete roof decks (see Section 3.4).

Board products are made in different sizes and thicknesses, with square, tongue-and-groove (T&G), tapered, or beveled edges (Fig. 9.5-3). Plain wallboard edges are usually tapered; backing board edges may be either square or T&G. Board products are manufactured in conformance with American Society for Testing and Materials (ASTM) standards. The physical dimensions, edge types, and governing standards of quality of generally used board products are summarized in Figure 9.5-4. Figure 9.5-5 shows some of the terms used in gypsum board work.

FIGURE 9.5-2 *Gypsum board on a continuous belt during the manufacturing process.* (Courtesy United States Gypsum Company.)

Gypsum Wallboard

The main types of gypsum wallboard are regular core, predecorated, foil-backed, fire-resistant (Type X), and water-resistant.

◼ REGULAR CORE WALLBOARD

Regular core wallboard is surfaced on the back with a gray liner paper and on the face and along the longitudinal edges with a calendered manila paper that provides a smooth, even finish suitable for decorating.

Most regular core wallboards are produced with tapered edges, which permit taping and finishing of joints to provide a homogeneous, smooth surface with inconspicuous joinings between individual boards (see Fig. 9.5-3).

ROUNDED

TAPERED

BEVELED

SQUARE EDGE

TONGUE AND GROOVE

EDGES AVAILABLE

FIGURE 9.5-3 *Typical edge designs for gypsum board products.* (Courtesy Gypsum Association.)

◼ PREDECORATED WALLBOARD

Predecorated wallboard is essentially regular wallboard with a factory-applied decorative finish on the face of the board.

The common finish materials are paper and vinyl in a variety of colors, patterns, and simulated wood finishes (Fig. 9.5-6). These boards usually have accurately formed square or beveled longitudinal edges that do not require further treatment in the field. Beveled-edged and some square-edged boards simply are butted together to create a shadowline repeated at regular intervals. Square edges usually are trimmed out with matching moldings or decorative battens to conceal joint irregularities and fasteners. Some manufacturers provide a loose flap of the decorative covering material, which is cemented down after installation to conceal the joint. Predecorated boards may be secured with matching prefinished nails or by adhesive bonding. A lighter-weight, predecorated product, MH Board, is often used in manufactured construction, primarily mobile homes. It is available factory painted or faced with paper or vinyl.

FOIL-BACKED WALLBOARD

Foil-backed wallboard is regular wallboard with a bright-finish aluminum

FIGURE 9.5-4
Summary of Gypsum Board Products

Type	Thickness in.	Thickness mm	Width, in.	Length, ft.	Edge Detail
WALLBOARD (ASTM C 36) Regular	1/4	6.4	48	8, 10, 12	Tapered
	5/16	7.9	48	8, 10, 12, 14	
	3/8	9.5	48	7, 8, 9, 10, 12	
	1/2	12.7	48	7, 8, 9, 10, 12, 14	
	5/8	15.9	48	8, 10, 12	
Foil-backed	3/8	9.5	48	8, 10, 12	
	1/2	12.7	48	8, 9, 10, 12	
Type X	1/2	12.7	48	8, 9, 10, 12	
	5/8	15.9	48	8, 9, 10, 12	
LATH (ASTM C 37) Plain	3/8	9.5	16, 16-1/5, 24	4, 8, 12	Round
	1/2	12.7	16, 16-1/5, 24	4	
Perforated	3/8	9.5	16	4	
	1/2	12.7	16	4	
Foil-backed	3/8	9.5	16	4, 8	
	1/2	12.7	16	4	
Type X	3/8	9.5	16	4, 8	
BACKING BOARD (ASTM C 442) Regular	1/4	6.4	48	8	Square
	3/8	9.5	48	8	Square
	1/2	12.7	24	8	Tongue-and-groove
Foil-backed	3/8	9.5	48	8	Square
Type X	5/8	15.9	24	7	Tongue-and-groove
	1	25.4	24	7	Tongue-and-groove
SHEATHING (ASTM C 79) Regular	3/8	9.5	24	8	Tongue-and-groove
			48	8, 9	Square
	4/10	10.0	24	8, 9	Tongue-and-groove
			48	8, 9	Square
	1/2	12.7	24	8, 9	Tongue-and-groove
			48	8, 9	Square
	5/8	15.9	24	8, 9	Tongue-and-groove
			48	8, 9	Square
Type X	5/8	15.9	24	8	Tongue-and-groove
			48	8, 9	Square

WOOD STUDS OR OTHER FRAMING MEMBERS

FIELD

PERIMETER

EDGE

END

FIGURE 9.5-5 *Terms used in gypsum board work. See Glossary for written definitions.* *(Courtesy Gypsum Association.)*

FIGURE 9.5-6 *Predecorated wallboard.*

foil bonded to the back. The foil backing serves as a vapor retarder; its use eliminates the need for a separate vapor retarder in an exterior wall assembly. When the reflective surface faces a 3/4-in.-wide dead-air space, it can provide increased insulating performance. Installation and edge treatment of foil-backed gypsum board is the same as for regular wallboard.

▪ FIRE-RESISTANT (TYPE X) WALLBOARD

Type X board is similar in exterior covering, appearance, edge treatment and installation to regular core wallboard, but it has a core specially formulated with additives and glass fibers for greater fire resistance. When prescribed installation methods are followed, fire-resistant assemblies of 45 and 60 minutes are possible with 1/2-in. Type X board and frame construction. Fire-resistant wall-

board also is available with an aluminum foil vapor retarder for use in exterior walls and ceiling assemblies.

▪ WATER-RESISTANT WALLBOARD

Special types of wallboard, called water-resistant (WR), board are produced for areas such as toilet rooms, bathrooms, kitchens, utility rooms, and other humid areas. They are not recommended for high-moisture areas such as shower rooms, saunas, or steam rooms. A moderately water-resistant type is available that is suitable for paint finishing or as a base for adhesively applied ceramic, metal, and plastic wall tile. Water resistance is achieved by using a multilayered covering of chemically treated paper and a gypsum core formulated with asphaltic additives. Edges usually are tapered. Installation is the same as for regular wallboard.

Water-resistant backing board is also available (see Section 9.5.2.2). Glass mesh mortar units, a highly water-resistant backing board intended primarily for tiling in areas subject to direct wetting such as in shower stalls, is described in Section 9.6.

9.5.2.2

Gypsum Backing Board

Backing boards serve as a base to which wallboards or acoustical ceiling tiles are laminated. Figure 9.5-7 shows two-ply construction of wallboard bonded to backing board. Backing boards differ from regular wallboards mainly in that the paper covering on both faces and edges is a gray liner paper not suitable for decorating. Backing boards may be secured to framing with staples, screws, or nails. The main types are regular, fire-resistant (Type X), and foil-backed. *Core-board*, a 3/4- or 1-in.-thick, factory-laminated board, which consists of two layers of regular backing board, is used to build solid gypsum partitions and shaft walls.

A highly water-resistant backing board with treated paper and core is produced primarily as a base for tile in areas subject to direct wetting, such as in shower stalls.

9.5.2.3

Gypsum Sheathing

Gypsum sheathing provides fire resistance, wind bracing, and a base for exterior finishes in wood- and metal-framed construction. It usually is made of a water-resistant gypsum core completely enclosed with a firmly bonded water-repellent paper, which eliminates the need for sheathing paper. Standard 2- by 8-ft. boards typically have V-shaped T&G edges to shed water. These are nailed horizontally across the framing members (Fig. 9.5-8). The 4- by 8-ft. and 4- by 9-ft. boards have square edges and are installed vertically where improved wind bracing and infiltration resistance are desired.

9.5.2.4

Gypsum Formboard

Gypsum formboards are used as permanent forms and as finished ceilings for gypsum concrete roof decking (see Section 3.3.4). Formboards are surfaced on the exposed face and longitudinal edges with calendered manila paper specially treated to resist fungus growth. The back face is surfaced with gray liner paper as in regular core wallboard. A vinyl-faced board, which provides a white, highly reflective and durable surface, is also available.

9.5.2.5

Gypsum Lath

Gypsum lath is a class of board products intended as a base for plastering. These products consist of a gypsum core enclosed by a multilayered, fibrous paper covering designed to ensure good bond with gypsum plaster. A suction bond is created by tiny needlelike crystals of plaster forming in the porous paper covering. Gypsum lath may be stapled, screwed, nailed, or clipped to supporting construction of wood or metal studs or to furring (Fig. 9.5-9). The main types of gypsum lath are plain, perforated, foil-backed, and fire-resistant (Type X).

FIGURE 9.5-7 *Double-layer (laminated) gypsum board applied to wood construction. Application over metal framing is similar.*

FIGURE 9.5-8 *Gypsum sheathing applied horizontally to wood construction.*

PLAIN LATH   Plain lath is ordinary gypsum lath consisting of a gypsum core and paper covering as described above, without any of the special features incorporated in the other available types.

PERFORATED LATH   Before the development of Type X lath, perforated lath, with 3/4-in. holes drilled 4 in. on centers in each direction was used to improve the lath-to-plaster bond under fire exposure.

FIGURE 9.5-9 *Application of gypsum lath over framing.*

Covering and core material are the same as for plain lath.

FOIL-BACKED LATH   To make foil-backed lath, a bright-finish aluminum foil is laminated to the back of plain lath. The foil acts as a vapor retarder and can act as a reflective insulation in exterior walls and roof and ceiling assemblies if the reflective surface faces a 3/4-in.-wide dead-air space.

FIRE-RESISTANT (TYPE X) LATH   Special additives and glass fibers in the gypsum core of Type X lath improve its fire-resistant qualities, but covering materials are the same as for plain lath.

9.5.3

SUPPORTING CONSTRUCTION

Gypsum board can be applied directly to masonry, concrete, wood, or metal supporting construction that is capable of supporting the applicable design loads and that will provide a firm, level, plumb, and true base. Generally, the deflection of ceiling members supporting gypsum board finishes should not exceed 1/240 of the span under the full design load. Headers or lintels should be provided over openings to support structural loads, and special construction should be provided where required to support wall-hung equipment and fixtures.

Wood and metal supporting construction usually consists of (1) self-supporting framing members, such as wall studs, ceiling joists, and roof trusses, or (2) furring members, such as wood strips and metal channels, which are supported by underlying construction of framing, concrete, or masonry.

Framing and furring should be accurately aligned in a single plane and spaced not to exceed the maximum recommended for the gypsum board thickness being installed. When the spacing of framing exceeds the maximum allowable span for the gypsum board, furring is necessary. Window and door frames should be of the appropriate depth for the intended wall thickness. When mechanical and electrical equipment is installed, consideration should be given to the thickness of the gypsum board finish and the design of trim components, such as cover plates, registers, and grilles.

Wood wall framing is discussed in Section 6.6.2. Wood furring is covered in Section 6.6.7.

Concrete and Masonry

Gypsum board can be applied directly to above-grade interior masonry or concrete, if the surface is sufficiently dry, smooth, plumb, and true. Above-grade, exterior cavity walls also may be suitable for direct lamination if the cavity is properly insulated when required. However, predecorated wallboard with a decorative film that is highly impermeable to vapor generally should not be laminated directly to concrete or masonry, because moisture may become trapped in the gypsum core of the board.

Before gypsum board is directly applied, rough or protruding edges in the substrate should be removed, and depressions greater than 4 in. in diameter or 1/8 in. deep should be filled with grout or similar filler. Mortar joints should be cut flush with the face of the wall. Surfaces should be free from form oils, curing compounds, loose particles, dust, grease, and efflorescence to ensure adequate bond. It is essential that concrete be allowed to cure for at least 28 days before gypsum board is applied to it.

Exterior and below-grade walls and horizontal surfaces subject to moisture or excessive heat loss and surfaces that cannot be prepared readily for adhesive lamination are not suitable for direct application of gypsum board products. They should be furred to separate the gypsum board from possible moisture in the wall and to provide a suitable base for attaching gypsum board.

Metal Framing and Furring

Metal framing and furring are common in commercial, institutional, and high-rise residential buildings, where noncombustible construction is required. In such instances, non-load-bearing partitions typically are framed with metal studs; ceilings usually are framed with open-web joists or steel beams.

■ WALL FRAMING

Metal studs are usually C-shaped units made from 25-gauge cold-formed steel that has been electrogalvanized to resist corrosion. They should conform to ASTM Standard C 645. This gauge is compatible with most drywall sheet metal screws used for fastening gypsum board to metal. Proprietary nailable studs may be available but are not frequently used (Fig. 9.5-10).

A variety of systems have been developed by gypsum board and metal specialty manufacturers to meet requirements for sound isolation and fire resistance. Many systems are designed for ease of erection and demountability with a minimum of labor and waste of materials. Systems should be selected on the basis of desired performance criteria and installed in accordance with the manufacturer's recommendations. Stud spacing should not exceed the maximum ones recommended.

■ CEILING FRAMING

Floor and roof framing often consists of steel beams or open-web joists. These members generally are not designed to receive gypsum board directly and often are spaced more than 24 in. apart. Therefore, it is usually necessary to provide suitable furring or suspension systems at appropriate spacing for attaching gypsum board on the ceilings of metal-framed buildings (Fig. 9.5-11).

■ FURRING

Wood or metal furring may be necessary to (1) provide a suitably plumb, true, and properly spaced supporting construction; (2) eliminate capillary transfer of moisture in exterior or below-grade masonry and concrete walls; (3) minimize the likelihood of moisture condensation on interior wall surfaces; (4) improve the thermal insulating properties of exterior walls; (5) increase assembly thickness to accommodate mechanical equipment; or (6) improve sound isolation by resilient mounting.

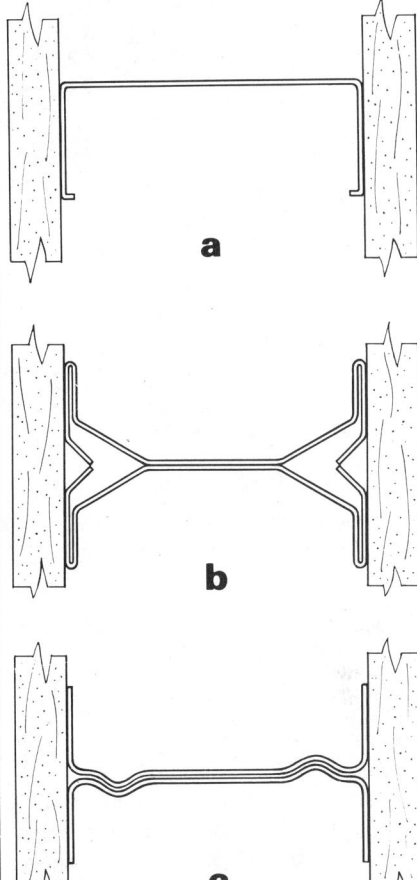

FIGURE 9.5-10 *Metal studs: (a) C-shaped stud designed to accept drywall screws and the type most used today; (b) nailable stud for special ring-shank nails; (c) nailable stud for smooth-shank nails. (b) and (c) are seldom used today and may not be available in many areas.*

Metal furring of various types can be used with all types of supporting construction. It is preferred to wood furring when noncombustible or sound-isolating assemblies are required. Metal furring members may be either cold-rolled channels or drywall channels (Fig. 9.5-12).

Cold-rolled channels are multipurpose members used for both plaster and gypsum board furred partitions and in most types of suspended ceiling assemblies. These channels usually are of 16-gauge steel with either a galvanized or a black asphaltum finish. They are available in sizes ranging from 3/4 in. to 2 in. wide, and in 16- to 20-ft. lengths. Because the gauge is not designed to receive drywall sheet metal

concrete on riblath

1-1/2" cr channel

bar joist

metal furring channel clip

gypsum wallboard screw attached

metal furring channel

FIGURE 9.5-11 *Open-web bar joists usually require a grid of cold-rolled channels and hat-shaped furring channels to support gypsum board.*

screws, these channels generally are wire-tied to supporting construction. Cold-rolled channels are used primarily as a supporting grid for lighter drywall channels to which wallboard can be screw-attached (see Fig. 9.5-11).

Drywall channels usually are roller-formed of 25-gauge electrogalvanized steel. Most drywall channels are designed for gypsum board attachment with self-drilling sheet metal screws, but nailable

PLAIN CHANNEL

RESILIENT CHANNEL

COLD-ROLLED CHANNEL

FIGURE 9.5-12 *Metal furring channels. Plain channels are also called hat-shaped channels. (Courtesy Gypsum Association.)*

channels similar to nailable studs may also be available. Channels usually are installed perpendicular to the framing with a 1-1/4-in. screw or a 5-penny nail at each support. Over masonry or concrete surfaces, they may be installed horizontally or vertically, whichever is more convenient, with fasteners as described for wood furring.

A variety of proprietary clips, runners, and adjustable brackets are available with engineered metal furring systems to facilitate rapid erection over irregular masonry walls (Fig. 9.5-13). The manufacturer's specific recommendations for fasteners, spacing, accessories, and installation should be followed.

Drywall channels can be either plain (hat-shaped) or resilient. Plain channels are used primarily with unit masonry and concrete wall assemblies that possess inherent sound isolation because of their mass; they are not significantly improved by resilient mounting if they are sealed and airtight. In suspended ceiling assemblies, a grillage composed of cold-rolled channels and plain drywall channels is used to support a gypsum board finish (Fig. 9.5-14). If sound or vibration isolation is desired, resilient hangers can be wired into the suspension system to isolate the ceiling from the framing.

Resilient channels are used over both metal and wood framing (Fig. 9.5-15). In addition to improving sound isolation, they help to isolate the gypsum board from structural movement, thus minimizing the possibility of joint cracking.

ACCESSORY MATERIALS

Accessory materials for the application of gypsum board include mechanical fasteners and adhesive for board attachment, compounds and tape for joint and fastener treatment, and metal edge and corner trim to protect exposed edges and exterior corners.

Fasteners

Nails and screws are used to apply both single- and multi-ply finishes. Clips and staples are limited to attaching the base ply in multi-ply construction.

■ NAILS

A variety of nails are used for applying gypsum board. Nails can be bright, coated, or chemically treated. Shanks may be either smooth or annularly threaded, generally with medium or long diamond points. A nail head should be flat or slightly concave, thin at the rim and not more than 5/16 in. in diameter. Nail heads of about 1/4-in. diameter provide adequate holding power without cutting the face paper.

Either annularly threaded nails developed for wallboard, known as the GWB-54, or smooth- or deformed-shank nails suitable for the application of gypsum board should be used. All nails should conform to ASTM C 514. Casing nails and common nails have heads that are too small in relation to the shank or too thick and should not be used.

Nails should be of the proper length for the wallboard thickness (Fig. 9.5-16). Generally recommended penetration into supporting construction for smooth-shank nails is 7/8 in. Annularly threaded nails provide more withdrawal resistance, require less penetration, and generally minimize nail popping. For fire-rated construction, however, penetration of 1 in. or more usually is required, and the longer smooth-shank nails generally are used.

■ SCREWS

Gypsum board can be fastened to both wood and metal supporting construction

FIGURE 9.5-13 *Adjustable brackets have notched surfaces to permit plumb and true installation of channel grid.* (*Courtesy Gypsum Association.*)

FIGURE 9.5-14 *In suspended ceilings, wire hangers are used to support and level the channel grid.* (*Courtesy Gypsum Association.*)

FIGURE 9.5-15 *Resilient channel mounting substantially improves sound isolation in wood-framed construction.* (*Courtesy Gypsum Association.*)

with *drywall screws.* The usual finish for drywall screws is a zinc phosphate coating with baked-on linseed oil. These screws typically are self-drilling and have self-tapping threads and flat Phillips recessed heads for use with a power screwdriver. A special contour head design makes a uniform depression free of ragged edges and fuzz.

Screws pull gypsum board tightly to the framing without damaging the board, thus minimizing fastener surface defects due to loose board attachment. The three basic types of drywall screws for wood, sheet metal, and gypsum construction and the recommended penetrations for various uses are shown in Figure 9.5-17. Other specially designed screws are available for attaching wood or metal trim and for wallboard attachment to heavier-gauge load-bearing studs. The use of screws generally is acceptable in fire-rated constructions.

DRYWALL WOOD SCREWS   Type W and similar screws are designed for fastening to wood framing or furring. Type W screws are diamond-pointed to provide efficient drilling action through both gypsum and wood and have a specially designed thread for quick penetration and increased holding power. Recommended minimum penetration into supporting construction is 5/8 in., but in two-ply application when the face layer is being screw-attached, the additional holding power developed in the base ply permits reducing the penetration into supports to 1/2 in. Type W screws are available in 1-1/4-in. length. Drywall

sheet metal screws, more readily available in longer sizes, may be substituted in two-ply construction.

DRYWALL SHEET METAL SCREWS   Type S and similar screws are designed for fastening gypsum board to 25-gauge metal studs or furring. Type S screws have a self-tapping thread and mill-slot drill point designed to penetrate sheet metal with little pressure. This is an important consideration because steel studs are flexible. The threads should be of adequate depth and turned within 1/4 in. of the head to eliminate stripping. They are available in several lengths, from 1 in. to 2-1/4 in. Other lengths and head profiles are available for attaching wood trim, metal trim, and metal framing components. Recommended minimum penetration throughsheet metal for drywall sheet metal screws is 3/8 in.

DRYWALL GYPSUM SCREWS   Type G and similar screws are used for fastening gypsum board to gypsum board. Type G screws are similar to Type W screws but have a deeper special thread design. They generally are available in 1-1/2-in. length only. Drywall gypsum screws require penetration of at least 1/2 in. of the threaded portion into the supporting gypsum board. Allowing approximately 1/4 in. for the point, this results in a minimum penetration of 3/4 in. For this reason, drywall gypsum screws should not be used to attach wallboard to 3/8-in.-thick backing board. In two-ply construction with a 3/8-in.-thick base ply, nails or longer screws should be used to provide the necessary penetration into supporting wood or metal construction.

STAPLES   Staples are recommended only for attaching base ply to wood members in multi-ply construction. They should be of 16-gauge flattened galvanized wire with a minimum 7/16-in.-wide crown and divergent sheared beveled points. Staples should be long enough to provide a minimum penetration of 5/8-in. into supporting construction.

9.5.4.2

Adhesives

Adhesives can be used to attach singlely-ply wallboard directly to wood framing,

FIGURE 9.5-16
Nails for Gypsum Board Application

Drywall Nails	Recommended Penetration		Board Thickness		Nail Length	
	in.	mm	in.	mm	in.	mm
Annularly threaded[a] ASTM C 514	3/4	19	3/8	9.5	1-1/8	29
			1/2	12.7	1-1/4	32
			5/8	15.9	1-3/8	35
Smooth shank[b] ASTM C 514	7/8	22	3/8	9.5	1-1/4	32
			1/2	12.7	1-3/8	35
			5/8	15.9	1-1/2	38

Courtesy Gypsum Association.

Note: For fire-rated constructions, building code and/or manufacturer's specifications should be followed.

[a] Typically, GWB-54: 1/4-in.-diameter head, 0.098-in.-diameter shank.

[b] Typically, 1/4-in.-diameter head, 0.099-in.-diameter shank, with dished head.

masonry, or concrete or to laminate gypsum board to a base layer of gypsum board, sound-deadening board, or rigid foam insulation. They generally are used in combination with nails or screws, which provide temporary or permanent supplemental support.

With the increased use of adhesives for applying gypsum board finishes, new adhesives are being developed to meet specific installation conditions. Therefore, adhesive formulations vary and manufacturers' recommendations should be consulted in each case. Adhesives for applying wallboard finishes may be classed as stud adhesives, laminating adhesives, and contact adhesives.

STUD ADHESIVES Adhesives formulated for attaching single-ply gypsum board to wood supporting construction are classed as stud adhesives. They generally are rubber or asphaltic base, are of caulking consistency that will bridge framing irregularities, and are applied with a gun in a continuous bead. Solvent-base stud adhesives should not be used near an open flame or in poorly ventilated areas. Structural stud adhesives should conform to ASTM Standard C 557, which establishes rigid standards for strength, bridging ability, aging, and other qualities necessary for longterm structural performance. Fewer per-

manent supplemental fasteners are required for structural stud adhesives than are required for adhesive-nail attachment with nonstructural stud adhesives.

LAMINATING ADHESIVES Most laminating adhesives are casein-base adhesives in powder form to which water is added. Laminating adhesives can be used to laminate gypsum boards to each other and to suitable masonry or concrete surfaces, but they are not intended for adhesive bonding to wood framing or furring. In addition to the several adhesives specifically formulated for laminating purposes, some powder embedding compounds (see Section 9.5.4.3) can be used when mixed at the recommended consistency. Although some stud adhesives and contact adhesives also can be used for laminating purposes, they should not be used unless specifically recommended by the manufacturer.

Only as much adhesive should be mixed as can be used within the working time specified by the manufacturer. Water used for mixing should be at room temperature and clean enough to drink. Adhesive usually is applied over the entire board area with a suitable spreader and requires temporary support or supplemental fasteners until the adhesive develops sufficient bond strength.

CONTACT ADHESIVES Contact adhesives consist of a synthetic rubber base, sometimes with resins added, in either solvent or emulsion formulations. Contact adhesives can be used to laminate gypsum boards to each other and to bond wallboard to metal studs in demountable partitions. The adhesive is applied by roller, spray gun, or brush in a thin, uniform coating to both surfaces to be bonded. For most contact adhesives some open time usually is required before surfaces can be joined and bond can be developed. To ensure maximum possible adhesion between mating surfaces, the face board should be impacted over its entire surface with a suitable tool, such as a rubber mallet.

The advantages of contact adhesive are immediate bond, excellent long-term strength, resistance to fatigue, and no temporary or permanent fasteners required in the field. One disadvantage is the inability of the adhesive to bridge irregularities in the mating surfaces, thus creating discontinuity in the bond film. Another disadvantage is that most of these adhesives do not permit repositioning of the boards once contact between surfaces has been made. A slip sheet of polyethylene film or tough building paper can be used to facilitate gradual bonding of surfaces as the slip sheet is withdrawn.

MODIFIED CONTACT ADHESIVE A modified contact adhesive has been developed that retains the advantages of contact adhesive while removing some disadvantages. This adhesive combines good long-term strength, bridging ability, and sufficient immediate bond to permit erection with a minimum of temporary fasteners. In addition, the adhesive has an open assembly time of about one-half hour before bond develops, during which time boards can be repositioned. Modified contact adhesive is intended for attaching wallboard to all types of supporting construction, including framing, furring, gypsum board, and rigid insulating board.

9.5.4.3

Joint Tape and Joint Compounds

Joint tape and joint compound are used to produce a wallboard finish that is smooth, monolithic in appearance, with inconspicuous joinings between boards

FIGURE 9.5-17
Typical Screws for Gypsum Board Applications

Drywall Screws*	Recommended Minimum Penetration	Screw Length		Application
		in.	mm	
For fastening to wood ASTM C 1002	5/8 in. (15.9 mm)	1-1/4	32	Single-ply or base-ply application
	5/8 in. (15.9 mm)	1-5/8	41	Face-ply application in two-ply construction
For fastening to sheetmetal ASTM C 1002	3/8 in. (9.5 mm)	1	25.4	Single-ply or base-ply application
	3/8 in. (9.5 mm)	1-5/8	41	Face-ply application in two-ply construction
For fastening to gypsum board ASTM C 1002	1/2 in. (12.7 mm) 3/4 in. (19 mm)	1-1/2	38	Gypsum board to gypsum board in multiply construction
	3/4 in. (19 mm) 3/4 in. (19 mm)			

Courtesy Gypsum Association.

Note: Applications are generally acceptable for fire-rated construction.

*Screws shown are Types W, S, and G, respectively; other drywall screws are available.

and no visible fasteners. Joint compound and tape should conform to ASTM C 475, "Specifications for Joint Treatment Materials for Gypsum Wallboard Construction."

■ **JOINT TAPE**

Tape used for joint reinforcement typically is a strong-fibered tape with chamfered edges. The special paper resists tensile stresses across the joint as well as longitudinally.

■ **JOINT COMPOUNDS**

Joint compounds may be classed according to function and properties as (1) embedding compound, used for embedding and bonding joint tape; (2) topping compound, used for final smoothing and leveling over joints and fasteners; and (3) all-purpose compound, which combines features of both embedding and topping compounds.

Joint compounds are either packaged in powder form to be mixed with water on the job or premixed by the manufacturer. With the exception of quick-setting embedding compounds, these compounds harden by evaporation of the water and can be kept in a wet-mix state for several days if covered with a moist cloth or a thin layer of water. Both premixed and job-mixed compounds should be protected from freezing.

EMBEDDING COMPOUND Typically, an embedding compound is a casein-based adhesive, similar to laminating adhesive. It is available either premixed or as a powder that is mixed with water. In addition to embedding tape, it is used as a first coat for concealing fasteners and edge and corner trim. Some embedding compounds also can be used for laminating gypsum board in multi-ply construction when mixed at the consistency recommended by the manufacturer for that purpose. Usually, embedding compound requires at least overnight drying before following coats can be applied. However, quick-setting compounds that dry in 2 to 4 hours also are available, which permits three-coat application of embedding and topping compounds in one day.

TOPPING COMPOUND A topping compound is primarily a surface filler used to conceal and smooth over embedded tape, fasteners, and trim. It is a casein or casein-vinyl formulation available either in premixed or powder form. It bonds well with joint tape and compound, gypsum board, and fasteners; it sands easily and provides a surface with sufficient "tooth" and suction for painting.

ALL-PURPOSE COMPOUND Some manufacturers make an all-purpose compound that combines the characteristics of both adhesive and filler. It is a convenient formulation available in powder or premixed form, in either a machine or hand-tool consistency. All-purpose compound can be used for embedding tape, topping over tape, finishing over metal trim, and concealing fasteners. It should not be used for laminating gypsum boards unless specifically recommended by its manufacturer.

9.5.4.4

Edge and Corner Trim

Exposed wallboard edges and ends generally are protected by wood, plastic, or metal trim. In low-rise residential construction, wood casing trim typically is used at window and door frames; wood or resilient wall base is used to protect wallboard edges at the floor. Metal trim, such as cornerbeads and casing beads, are used in similar locations when inconspicuous edge protection is desired without featuring the trim as a design element. Metal trim is most often used in commercial and institutional buildings. Plastic trim is used when exposing the trim is acceptable or desired.

Trim shapes are available for different conditions and varying gypsum board thicknesses. The standard shapes and designations are shown in Figure 9.5-18. On outside corners, metal cornerbeads are used to protect the gypsum board from damage and to provide neat, slightly rounded corners, which are difficult to obtain with tape and joint compound alone. Casing beads and cornerbeads with metal flanges generally are nailed or screwed to supporting construction approximately 6 in. on centers (see Fig. 9.5-18). Beads with paper flanges are held in place by embedding the flanges in joint compound.

9.5.5

JOB CONDITIONS

Temperature and humidity affect the performance of joint treatment materials, the scheduling of joint finishing operations, the quality of a joint, and sometimes the bonding properties of adhesives.

Interior finishes should not be installed when the outside temperature is less than 55 °F, unless the building is completely enclosed and has a controlled heat of 55° to 70 °F. This temperature should be maintained for 24 hours before installation, during installation, and until a permanent heating system is in operation.

A building need not be completely glazed during summer months, but all materials and installation should be protected from the weather.

Ventilation should be provided to eliminate excessive humidity. If the building has been glazed, windows can be kept partially open to provide air circulation. Temporary air circulators should be used in enclosed areas that have no natural ventilation. Additional drying time between coats of joint compound should be allowed

USASI Designation	Trim Profile	Application
CORNERBEAD[1] CB—100 x 100 CB—118 x 118 CB—114 x 114 CB—100 x 114 CB—PF (paper flange steel corner combination bead)		
U BEAD[2] U-38 U-12 U-58 U-34		
L BEAD[2] LB-38 LB-12 LB-58 LB-34		
LK BEAD[3][4] LK-S LK-B		
LC BEAD[2] LC-38 LC-12 LC-58 LC-34		

[1]Numbers indicate width of flanges: e.g.—114 x 114 has two 1¼" flanges.
[2]Numbers indicate thickness of board intended: e.g. U-12 is U Bead intended for ½" wallboard.
[3]Letter (S) indicates square nose; letter (B), bull or round nose.
[4]For use with kerfed jamb.

FIGURE 9.5-18 *Standard drywall metal-trim shapes.* (*Courtesy Gypsum Association.*)

when drying conditions are slow. During hot, dry weather, drafts should be avoided to prevent the joint compound from drying too rapidly.

Delivery of gypsum board should coincide as closely as possible with the installation schedule. Boards should be stored on the job flat, inside, and under cover. Stacking long lengths on short lengths should be avoided to prevent the longer boards from breaking. For short periods of time, boards may be placed vertically against the framing with the long edges of the boards horizontal. Materials should remain in their wrappings or containers until ready for use.

9.5.6

BOARD APPLICATION

Gypsum board finishes may be either single-ply using wallboard alone or multi-ply using wallboard over wallboard, backing board, or sound-deadening board. Maximum recommended support spacing for various wallboard thicknesses is given in the following subsections for single-ply and multi-ply constructions.

Wallboard installed perpendicular to supporting members results in a stiffer finish surface and is preferred. Whenever possible, boards should be applied to span ceilings and walls to avoid butt-end joints, which are difficult to finish inconspicuously. When butt-end joints do occur, they should be staggered and located as far from the center of walls and ceilings as possible so they will be less conspicuous. With the exception of face layers in two-ply construction, board ends and edges that are parallel to supporting members should fall on these members.

Successful application of gypsum board finishes depends on proper measuring, cutting, and fitting; board attachment; and joint and fastener treatment.

9.5.6.1

Measuring, Cutting, and Fitting

Measurements should be taken accurately at the intended ceiling or wall location for each edge or end of a board. Accurate measuring usually will indicate irregularities in framing and trim-

FIGURE 9.5-19 *(a) Face paper is scored; (b) gypsum core is snapped along score line and back paper is cut; (c) rough edges are smoothed with coarse sandpaper. A rasp or sharp knife could also be used.* *(Courtesy Gypsum Association.)*

ming so that allowances can be made in cutting.

Straight-line cuts across the full width or length of a board are made by scoring through the face paper into the gypsum core (Fig. 9.5-19a). The core of the board then is snapped and the back paper is cut (Fig. 9.5-19b). Cut edges should be smoothed with a rasp or trimmed with a sharp knife (Fig. 9.5-19c). Panels should be cut so that they fit easily into place.

Ceiling panels should be installed first. Joints should be loosely butted without forcing the boards into position. Tapered edges should be placed next to tapered edges and square job- or mill-cut ends should be butted to square ends. Square ends should not be placed next to tapered edges because excessive build-up of joint compound over the tapered edge will result in a conspicuous ridge caused by differential shrinkage.

9.5.6.2

Board Attachment

Wallboard and backing board can be attached to supporting construction with mechanical fasteners, adhesive, or a combination of adhesive and mechanical fasteners. Specific attachment methods are discussed in Sections 9.5.6.3 and 9.5.6.4.

■ MECHANICAL FASTENERS

Nails and screws are commonly used to

attach gypsum board in both single- and multi-ply installations. In two-ply construction, however, staples and clips can be used to attach the base ply. Adhesive, when used, generally is supplemented with mechanical fasteners.

Fasteners should be of a type recommended for the intended method of application. Ordinary wood screws, sheet metal screws, and common nails are not designed to hold gypsum board tightly or to countersink neatly and should not be used. Fasteners should be placed at least 3/8 in. from board edges and ends. Application should start in the middle of a board and proceed outward toward the perimeter. Fasteners should be driven as nearly perpendicular as possible while the board is held in firm contact with supporting construction. Nails should be driven with a crown-headed hammer forming a uniform depression around the nail head. Fastener heads should not be seated deeper than about 1/32 in. below the board surface, and particular care should be taken not to break the face paper (Fig. 9.5-20).

■ ADHESIVE BONDING

Adhesive can be used to secure single-ply wallboard to wood framing and furring, masonry, and concrete or to laminate face ply to a base layer of gypsum board, sound-deadening board, or rigid foam insulation. The surface to be bonded should be free from dirt, grease, oil, and other foreign material.

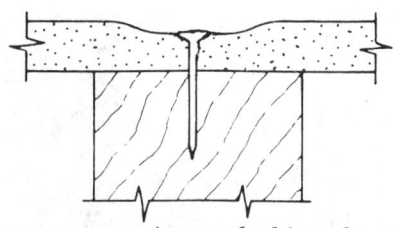

FIGURE 9.5-20 *A properly driven drywall nail.* (*Courtesy Gypsum Association.*)

9.5.6.3

Single-Ply Construction

In single-ply construction, gypsum board panels are nailed, screwed, or adhesive-bonded to supporting wood or metal studs or to furring (Fig. 9.5-21). After appropriate fastener and joint treatment, the surface may be painted or finished with wall covering or wall tile.

Single-ply gypsum board construction consists of a single layer of wallboard permanently attached to supporting construction. This subsection discusses assemblies, methods, and precautions needed to obtain suitable gypsum board finishes with single-ply construction. However, where a specific degree of sound isolation, fire resistance, thermal resistance, moisture resistance, or other special construction is required, the recommendations in Section 9.5.6.6 should be considered, and the manufacturer's detailed installation instructions should be followed.

Sound ratings ranging from 35 to 50 STC (see Section 10.3) can be achieved with single-ply construction, the higher values requiring special construction systems such as resilient mounting, separated construction, or sound-absorbing materials. Fire ratings for single-ply framed partitions generally are limited to 1 hour.

Because of the large number of fasteners in the surface of the board and the greater dependence on accurately aligned supporting construction, single-ply finishes are more susceptible to surface problems such as joint ridging, corner cracking, and nail popping. However, with the selection of proper supporting construction and methods of attachment, attention to ridging, cracking, and nail popping precautions, and care in joint treatment, it is possible to produce visually attractive, durable finishes.

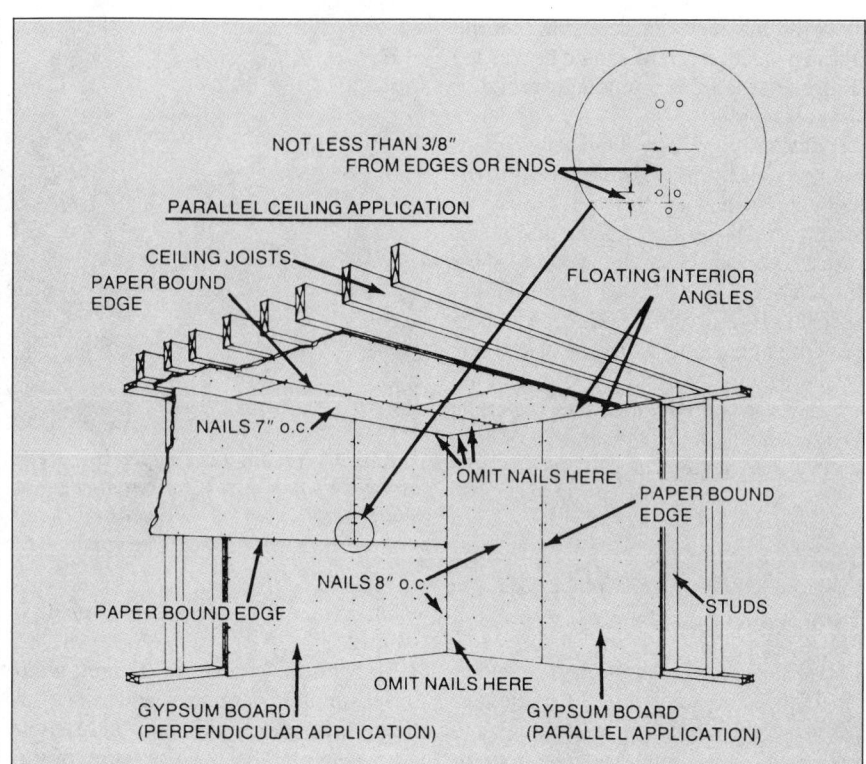

FIGURE 9.5-21 *Gypsum board application over wood-framed construction.* (*Courtesy Gypsum Association.*)

Framing or furring members should be spaced as recommended in Figure 9.5-22, regardless of the method of attachment. With the exception of 3/8-in.-thick wallboard on ceilings, which should be applied perpendicular to the supports, boards can be applied either perpendicular or parallel to supporting members (Fig. 9.5-23). Perpendicular application generally is preferred because it provides greater strength and shorter joint lengths.

Except in end-floated construction, gypsum board edges and ends should occur over supporting members. Butt-end joints should be staggered with respect to each other on the same side of a partition and with respect to joints on the opposite side of the partition.

■ WOOD FRAMING AND FURRING

Gypsum board typically is attached to wood furring and framing with nails or screws or by a combination of adhesive and nails. Resilient metal furring can be used with wood framing where increased sound isolation is desired.

SCREW ATTACHMENT Fewer fasteners are required when screws are used to attach gypsum board, and the number of fasteners requiring treatment and possible surface defects is reduced. Drywall wood screws, such as Type W, providing at least 5/8-in. penetration into supports should be used, spaced as shown in Figure 9.5-24.

NAIL ATTACHMENT It is particularly important that gypsum board panels be drawn tightly against framing and furring members so that the board will not move on the nail shank. Such movement will result in nail popping. Wallboard can be attached by either the single- or the double-nailing method in accordance with GA 216. In general, nail spacing is based on requirements for fire resistance and on intimate board-to-stud contact to reduce possible surface defects, rather than on structural support. Depth of nail penetration should be as shown in Figure 9.5-16.

Single nailing can provide satisfactory results if the methods in GA 216 are carefully followed. Nails should be spaced

FIGURE 9.5-22
Support Spacing for Single-Ply Construction

Wallboard Thickness		Location	Application[a]	Maximum Support Spacing, o.c.	
in.	mm			in.	mm
3/8	9.5	Ceilings[b,c]	Perpendicular	16	406
		Walls	Perpendicular or parallel	16	406
1/2 and 5/8	12.7 and 15.9	Ceilings[c]	Perpendicular	24	610
			Parallel	16	406
		Walls	Perpendicular or parallel	24	610

Courtesy Gypsum Association.

[a] Direction of wallboard edges relative to supports.

[b] 3/8 in. ceilings should not support insulation.

[c] For ceilings to receive a water-based spray texture finish, use 5/8 in. board applied perpendicular to 24 in. supports or 1/2 in. perpendicular to 16 in. supports.

FIGURE 9.5-23 *(a) Horizontal and (b) vertical application of gypsum board to wood-framed construction using the single nailing method.* *(Courtesy Gypsum Association.)*

FIGURE 9.5-24
Single-Ply Screw Attachment to Wood Supports

Support Spacing, o.c.		Screw Spacing, o.c.*			
		Walls		Ceilings	
in.	mm	in.	mm	in.	mm
16	406	16	406	12	305
24	610	12	305	12	305

Courtesy Gypsum Association.

*Spacing applicable to both field and perimeter of board.

7 in. on centers on ceilings and 8 in. on centers on walls along supports (Fig. 9.5-25a).

Double nailing ensures consistently better board-to-stud contact and fewer defects than does single nailing. One set of nails 12 in. on centers is driven first in the field of the panel, followed by a second set of nails 2 in. from each of the nails first applied. Edges and ends falling over supports should be nailed 7 in. on centers on ceilings and 8 in. on centers on walls (Fig. 9.5-25b). While the total number of nails per panel is about 10% higher with double nailing, the number of nail spots that must be concealed is 20% lower than for single nailing.

ADHESIVE-NAIL ATTACHMENT Compared to conventional nailing, adhesive-nail attachment has the advantage of reducing the number of nails in the middle of a panel by at least 50%, consequently reducing the labor required for fastener treatment. The intimate bond between gypsum board and supporting members results in stronger, stiffer assemblies and fewer fastener defects. Although screws can be used instead of nails, their use is not common because good board-to-stud contact is ensured by the adhesive, and the number of face fasteners is reduced to a minimum.

Stud adhesive should be applied with a caulking gun in accordance with the manufacturer's recommendations. A straight bead of adhesive is applied to the face of members supporting the field of a panel (Fig. 9.5-26a). Where two adjacent panels join over a supporting member, two beads should be applied (Fig. 9.5-26b). When a predecorated wallboard joint is to be left untreated, two parallel beads of adhesive should be applied, one near each edge of the member. Generally, the volume of the bead should be such that when the board is applied, the adhesive will spread to an average width of 1 in. at least 1/16 in. thick.

Care should be taken in applying the adhesive to prevent it from being squeezed out at the joints. Adhesive should not be applied to members that are not required for board support, such as diagonal bracing, blocking, and plates.

The number and the spacing of supplemental nails (or screws) vary depending on the adhesive properties.

FIGURE 9.5-25 *(a) Single-nailing method. (b) Double-nailing method: first nails (shown by black dots) are installed starting with row 1, then rows 2 and 2A, etc., working outward; second nails (shown by circles) are installed in same row sequence.* *(Courtesy Gypsum Association.)*

FIGURE 9.5-26 *Beads of stud adhesive are applied (a) straight under field of board or (b) parallel under joints.* *(Courtesy Gypsum Association.)*

Nonstructural stud adhesives require fasteners in the field as well as at the perimeter of panels for wall and ceiling installation. Fastener spacing varies with the type of f astener, support spacing, and the type of fastener, support spacing, and load-bearing conditions (Fig. 9.5-27).

Structural stud adhesives conforming to ASTM C 557 require only perimeter fasteners for wall application. The fasteners should be spaced 16 in. on centers for edges or ends that fall parallel to supports, and at each support for edges or ends that fall perpendicular to supports. For ceiling application, the perimeter is nailed as for walls, but additional fasteners are required in the field spaced 24 in. on centers (Fig. 9.5-28).

With *predecorated wallboard* with beveled or finished edges not requiring joint treatment, it often is desirable to avoid fasteners at the vertical joints. For these and plain wallboard applied vertically on walls, fasteners may be omitted along panel edges if appropriate means such as prebowing or temporary bracing are used to ensure contact with the adhesive until

it develops full bond strength. For plain wallboard, temporary bracing or nailing should remain in place, and joint or nail treatment should not be started for at least 24 hours after installation.

■ **METAL FRAMING AND FURRING**

When proprietary nailable metal studs and channels are used, the manufacturer's recommendations for fastening should be followed. Resilient channels generally are nailed or screwed perpendicular to wood framing members through predrilled holes, with a 1-1/4-in. screw or 5d nail at each support. Gypsum board should be attached to metal framing and furring with appropriate drywall sheet metal screws, spaced not more than 12 in. on centers along supports for both walls and ceilings. For fire-rated construction, gypsum board generally is applied parallel to supports with screws spaced 8 in. on centers along edges and 12 in. on centers in the field.

■ **CONCRETE AND MASONRY**

Generally, concrete and masonry construction are furred before gypsum board is applied. However, gypsum board can be laminated directly to interior and some above-grade exterior masonry (see Section 9.5.3.1) if the surfaces are sufficiently dry, smooth, plumb, and true (Fig. 9.5-29). Some laminating adhesives and stud adhesives are suitable for this purpose, but individual manufacturers' recommendations vary and should be consulted to ensure proper selection and application. While the adhesive is setting, temporary support should be provided generally with threaded stub nails or cut nails.

9.5.6.4

Multi-Ply Construction

In two-ply construction, gypsum board panels are usually adhesive-bonded to a layer of backing boards (see Fig. 9.5-7), but sometimes a second layer of regular core gypsum board is used in lieu of backing board. After appropriate fastener and joint treatment, the surface may be painted or finished with wall covering or wall tile.

Multi-ply finishes are particularly suitable where increased sound isolation or fire resistance is desired and generally

FIGURE 9.5-27
Fastener Spacing for Nonstructural Adhesives

Location	Support Spacing, o.c.		Fastener Spacing, o.c.*			
			Nails		Screws	
	in.	mm	in.	mm	in.	mm
Ceilings	16	406	16	406	16	406
	24	610	12	305		
Load-bearing walls	16	406	16	406	24	610
	24	610	12	305	16	406
Non-load-bearing walls	16	406	24	610	24	610
	24	610	16	406		

Courtesy Gypsum Association.

*Applies to both field and perimeter of board.

FIGURE 9.5-28 *Supplemental fastener spacing for structural adhesives.* (*Courtesy Gypsum Association.*)

FIGURE 9.5-29 *Laminating adhesive is applied with a notched spreader. Stud adhesive is applied in beads with a caulking gun.* (*Courtesy Gypsum Association.*)

provide better surface quality than single-ply construction.

Multi-ply finishes consist of two or more layers of gypsum board and include (1) supported ceiling and wall assemblies and (2) self-supporting all-gypsum partitions. Face layers usually are laminated to a base layer of wallboard, backing board, or sound-deadening board, with only a limited number of fasteners to ensure adhesive bond. Multi-ply construction thereby minimizes fastener treatment and surface defects, such as shadowing and nail popping, as well. It further provides inherent joint-reinforced construction and reduces the incidence of joint ridging.

Sound ratings up to 60 STC and 2-hour fire ratings are possible with some two-ply framed assemblies and multi-ply all-gypsum partition assemblies. Framed load-bearing wall assemblies with two-ply finishes receive somewhat lower fire ratings than nonload-bearing assemblies.

On framed ceilings, the use of two-ply gypsum board finishes alone generally does not provide substantially higher fire and impact-sound ratings than does comparable single-ply construction. Selection of two-ply finishes for ceilings often is based on improved surface quality rather than on fire or sound ratings, but it must be recognized that two-ply construction is inherently more expensive than single-ply construction and may be prohibitive for that reason.

The gypsum board finishes discussed here are of two-ply construction and may be supported by wood or metal framing and furring, concrete, or masonry. The maximum spacing for framing and furring depends mainly on the base-ply thickness and whether the base ply is attached perpendicular or parallel to supporting construction (Fig. 9.5-30). Generally, support spacing of 24 in. is acceptable when the base ply is 1/2 in. thick or thicker; support spacing 16 in. on centers is sometimes required when a 3/8-in.-thick base ply is used. The same maximum support spacing is applicable to both wood and metal framing and furring.

■ **BASE-PLY ATTACHMENT**

The base ply of supported wall and ceiling finishes generally consists of wallboard or backing board. Both are available in regular core, foil-backed, and fire-resistant boards, but backing board more frequently is used for the base ply because it is more economical. Sound-deadening board may be used where increased sound isolation is required. Fastener size and spacing for applying sound-deadening board varies for different fire- and sound-

FIGURE 9.5-30
Support Spacing for Two-Ply Construction

Thickness of Base Ply				Maximum Support Spacing, o.c.	
in.	mm	Location	Application of Base Ply[a]	in.	mm
3/8	9.5	Ceilings[b,c]	Perpendicular	16	406
		Walls	Perpendicular or parallel	24[d]	610
1/2 and 5/8	12.7 and 15.9	Ceilings[c]	Perpendicular	24	610
		Walls	Perpendicular or parallel	24	610

Courtesy Gypsum Association.

Note: Installation with mechanical fasteners with or without adhesive.

[a] Direction of board edges relative to supports.

[b] Two-ply 3/8 in. ceilings should not support insulation.

[c] For two-ply ceilings to receive a water-based spray texture finish, use 5/8 in. board applied perpendicular to 24 in. supports or 1/2 in. perpendicular to 16 in. supports.

[d] 16 in. o.c. if no adhesive used between plies.

rated constructions, and the manufacturer's recommendations should be followed.

The base ply can be attached to supporting framing or furring with nails, staples, or screws. A method utilizing resilient clips is limited to walls. Base-ply attachment for all gypsum partitions is discussed under "Self-Supporting Gypsum Partitions" in Section 9.5.6.6.

WOOD FRAMING AND FURRING   In wood framing and furring, the fastener spacing for nails, screws, and staples varies primarily with the method of face-ply attachment (Fig. 9.5-31).

Nail attachment of base ply should be with drywall nails, long enough to provide the recommended penetration into supporting construction. Where the face ply is to be laminated, the base ply should be nailed as recommended for single-ply construction. Double nailing generally is not used because nail heads will be covered by the face ply and fastener treatment is not a consideration.

When the face ply is to be attached with mechanical fasteners, fewer nails are required in the base ply because the nails in the face ply will support the base ply as well.

The base ply on walls generally is installed parallel to framing members, except when predecorated wallboard is intended as the face ply. On ceilings, application of the base ply perpendicular to framing members is preferred. At inside corners only the overlapping base ply should be nailed, and fasteners should be omitted from the face ply to create a *floating corner*, which is more capable of resisting structural stresses (Fig. 9.5-32).

Screw attachment of the base ply to wood framing or furring should be with appropriate drywall wood screws that provide at least 5/8-in. penetration into supporting construction. Spacing of screws depends on the method of face-ply attachment and supporting construction (see Fig. 9.5-31). Inside corners should be floated as described for nailing.

Staple attachment generally is applicable to wood supporting construction only. Staples should be 1 in. long for 3/8-in.-thick base ply and increase proportionately for thicker boards to provide at least a 5/8 in. penetration into supporting members. Recommended spacing for staples is given in Figure 9.5-31. Staples generally should be driven with the crown perpendicular to the edges of the board, except that where edges fall over supports they should be driven parallel to the edges (Fig.

9.5-33). When driven, the crowns should bear tightly against the board without breaking the face paper.

METAL FRAMING AND FURRING   The base ply generally should be attached to metal framing or furring with drywall sheet metal screws. When proprietary nailable studs or channels are used, the manufacturer's fastening recommendations should be followed.

For walls, screw spacing should be 12 in. on centers in the field and 8 in. on centers along the edges when the face ply is laminated, and 16 in. on centers throughout the board when the face ply is attached with screws. Screws should be long enough to penetrate metal studs or furring by at least 3/8 in.

Two-ply ceiling finishes on metal framing or furring are not commonly used because they do not substantially improve fire resistance or affect sound isolation. They do provide increased airborne sound isolation and may be used where that is an important consideration.

CONCRETE AND MASONRY   Concrete and masonry have inherently good fire resistance and sound isolation, which two-ply finishes do not substantially improve. However, when single-ply gypsum board cannot be laminated directly to masonry or concrete and the walls must be furred, two-ply finishes may be desired for superior surface finish. The base ply is applied to such walls as described in Section 9.5.6.3 for wood or metal furring.

■ **FACE-PLY ATTACHMENT**

Joints in the face ply should be offset from joints in the base ply by at least 10 in. Gypsum board can be applied either horizontally or vertically, whichever results in the least waste of materials and shorter joint lengths. On walls, wallboard applied horizontally usually results in shorter and less conspicuous joints and therefore is preferred. However, predecorated wallboard that is to be left with joints untreated and plain wallboard that is to be trimmed with battens usually are installed vertically.

NAIL AND SCREW ATTACHMENT   When the face ply is attached with mechanical fasteners and no adhesive between plies,

FIGURE 9.5-31
Base Ply Fastener Spacing for Two-Ply Application, in. o.c.

Location	Nail Spacing			Screw Spacing		Stable Spacing	
	Framing Spacing	Laminated Face Ply	Nailed Face Ply*	Laminated Face Ply	Screwed Face Ply*	Laminated Face Ply	Screwed Face Ply*
Walls	16	8	24	16	24	7	16
	24	8	24	12	24	7	16
Ceilings	16	7	16	12	24	7	16
	24	7	16	12	24	7	16

Courtesy Gypsum Association.

*Fastener spacing for face ply should be the same as for single-layer application.

FIGURE 9.5-32 *At floating corners, base ply is nailed on only one side of interior corner.* (Courtesy Gypsum Association.)

VERTICAL APPLICATION

STAPLES SPACED 8 IN. O.C., CROWNS PERPENDICULAR TO EDGES EXCEPT AT EDGES OVER SUPPORTS

SUPPORT SPACING 16 in. OR 24 in O.C.

HORIZONTAL APPLICATION

FIGURE 9.5-33 *Staple attachment of base ply in wood supporting construction.* (Courtesy Gypsum Association.)

the maximum spacing and minimum penetration recommended for nails and screws in single-ply construction should be observed in accordance with GA 216. However, the specifications for some proprietary sound- and fire-rated partition assemblies may differ.

ADHESIVE ATTACHMENT The face ply typically is attached to gypsum board and sound-deadening board by sheet lamination or strip lamination. Generally, strip lamination is specified for systems with sound-deadening board.

Sheet lamination involves covering the entire back of the face ply with laminating adhesive with a notched spreader or other suitable tool (Fig. 9.5-34). The size and the spacing of the notches vary depending on the adhesive. The wallboard should be

erected using moderate pressure, and adhesive squeezed out at the joints should be promptly removed. In *strip lamination* the face ply is applied vertically, and adhesive is spread in vertical ribbons spaced 16 to 24 in. on centers, depending on adhesive and assembly (Fig. 9.5-35).

Supplemental mechanical fasteners or temporary wood bracing is required to ensure intimate adhesive bond between base and face plies. When structural ceiling members are used to anchor temporary wood bracing, ceiling installation should be delayed until after the wall finish has been installed.

Temporary fasteners may be used when the face ply is laminated to gypsum board in wood construction, and a fire rating is not required. These may be double-headed nails, stucco furring nails, or

drywall nails, generally spaced to provide face support approximately 24 in. on centers in each direction. Nails should be long enough to ensure proper penetration into supporting construction. Nails should be driven through gypsum board washers or other rigid material to facilitate removal without damaging the face ply. After nails are removed, nail holes should be *dimpled* (slightly depressed with a curved-face hammer) and finished with compound, as for permanent nails.

Some permanent mechanical fasteners are recommended in each laminated gypsum board panel, particularly on ceilings. Usually, these can be spaced at the perimeter of the board where they will be concealed by joint treatment or trim. Permanent fasteners are usually required in the field as well as in the perimeter of the

FIGURE 9.5-34 *Sheet lamination. Notched spreader or laminating box (shown) can be used to apply adhesive.* (*Courtesy United States Gypsum Company.*)

FIGURE 9.5-36 *Bullnose-edged wallboard requires prefilling joint channels before standard joint treatment is applied.* (*Courtesy Gypsum Association.*)

of omitting some fasteners at interior corners (Fig. 9.5-37). When corners are floated, the gypsum board should be applied to the ceiling first. It should be fitted snugly at both wall-to-wall and ceiling-to-wall intersections. Framing members or blocking should be used at corners to provide solid backup even though they may not be required to receive fasteners.

FIGURE 9.5-35 *Strip lamination. Laminating head applies adhesive beads on the base ply in parallel strips.* (*Courtesy United States Gypsum Company.*)

face ply in fire-rated assemblies. Fastener spacing varies depending on the tested assembly construction and is not related to whether adhesive is used between plies.

In sound-rated partitions where fire resistance is not a consideration, the face ply generally is applied vertically over sound-deadening board, with permanent mechanical fasteners at the top and the bottom of boards. Intermediate fasteners may be omitted along vertical joints and in the field if temporary bracing is provided until adhesive has developed sufficient bond strength.

9.5.6.5

Cracking and Ridging Precautions

Improper installation and adverse job conditions may cause surface imperfections such as cracking at corners and ridging at joints. *Corner cracking* may result from differential structural stresses at inside corners where adjacent walls or walls and ceilings meet. *Joint ridging* often occurs when joints are treated during cold or humid weather; it occurs more frequently when joint compound is applied too thickly and is feathered over a wide area. These finish defects are not commonly encountered in multi-ply construction.

In single-ply construction, cracking at interior angles can be minimized by *corner floating*; ridging can be substantially reduced by *back blocking* or *strip reinforcing*. Two-ply finishes are effective for both these purposes and often may be more practical. A proprietary bullnose-edged wallboard has been developed specifically to overcome the problem of ridging in single-ply construction. It utilizes a special prefill joint compound with non-shrinking and fast-setting properties, which develops high initial bond strength (Fig. 9.5-36).

■ CORNER FLOATING

Corner floating, which can be used with either screw or nail attachment, consists

FIGURE 9.5-37 *Corners are floated by omitting fasteners at the intersections of walls and of walls and ceilings, as shown by the shaded areas.* (*Courtesy Gypsum Association.*)

At ceiling-to-wall intersections, the last row of fasteners should be omitted on the wall so that nails start 8 in. from the intersection. On ceilings, where ceiling joists are parallel to the intersection, nailing should start at the intersection; where the joists are perpendicular, nails should start 7 in. from the intersection (see Fig. 9.5-37). For interior corners at the meeting of walls, the last row of fasteners should be omitted from the board first applied only, and the overlapped board should be nailed in the conventional manner. When double nailing, the same unrestrained clearances at corners should be maintained as for single nailing described in Section 9.5.6.3.

■ BACK BLOCKING

Back blocking is a system of reinforcing joints against stresses that cause surface defects, particularly ridging. Simple back blocking consists of laminating wallboard blocks between framing members. To reinforce the wallboard ends by back blocking, it is necessary to float the ends by locating end joints midway between framing members. Because ends are not tapered, however, the need for relatively high crowns and wide feathering at such joints still may cause ridging when joint treatment is done under adverse weather conditions. End joint tapering can be used in conjunction with floating to reduce the need for high crowns and wide feathering at joints (Figs. 9.5-38 and 9.5-39).

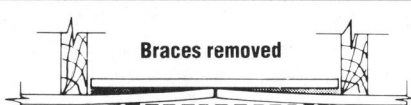

FIGURE 9.5-39 *When the strips shown in Figure 9.5-38 are removed, the tapered formation remains.* (Courtesy United States Gypsum Company.)

Back blocking and associated constructions most often are performed on ceilings because imperfections are more conspicuous on such large, unobstructed, obliquely lighted surfaces.

■ STRIP REINFORCING

Strip reinforcing is a method of reinforcing joints against ridging that offers many of the advantages of two-ply construction at reduced cost. By combining adhesive lamination with mechanical fastening, the number of nails in the face ply is reduced. Although it results in thicker assemblies than does back blocking, strip reinforcing can accomplish the same results in possibly less time.

Strips of 3/8-in.-thick backing board or scrap wallboard 8 in. wide are installed 24 in. on centers perpendicular to framing members, similar to cross-furring. Strips may be 4 in. wide at the base of a partition and at a partition-ceiling intersection. This cross-stripping supports the edges of the board and provides intermediate support in the field of the board. To provide support at the butt ends of a board, additional short strips are secured directly over the supporting member between the cross-stripping. Strips are nailed about 1 in. from the edge. Arrangement of strips, fasteners, and boards is illustrated in Figure 9.5-40.

CEILINGS: PERIMETER, FIELD NAILS (X) AT EACH SUPPORT

WALLS: PERIMETER NAILS (X) ONLY AT EACH SUPPORT

FIGURE 9.5-40 *Strip reinforcing: when support spacing exceeds 16 in. on centers, additional gypsum drywall screws (dots) are required midway between supports.* (Courtesy Gypsum Association.)

Laminating adhesive is applied to the strips with a strip laminating head in parallel beads. Wallboard should be 1/2-in.-thick or thicker and should be installed perpendicular to the framing members, so that long edges fall on the cross-stripping and the ends are supported on the short intermediate strips. When supporting members are spaced 16 in. on centers, the nail spacing should be 24 in. on centers at the ends and over each support at the edges for walls, with temporary double-headed nails at every other stud in the field until the adhesive sets. On ceilings, nails should be spaced 24 in. on centers along each support.

When supporting members are spaced more than 16 in. on centers, 1-1/2-in. drywall screws are required in the gypsum board strips midway between framing members for both walls and ceilings, as shown in Figure 9.5-40. Appropriate drywall nails that provide generally recommended penetration should be used to attach both stripping and gypsum board panels.

9.5.6.6

Special Constructions

Most special constructions intended to provide improved sound, fire, moisture, and thermal control are predicated on gypsum board finishes with complete joint treatment and panel caulking. Material used for caulking should be a nonhardening, nonshrinking, resilient sealant. In each case, the manufacturer's specific installation recommendations should be followed.

■ SOUND-ISOLATING CONSTRUCTIONS

The properties of sound and the principles of room acoustics and sound isolation are discussed in Chapter 10. For detailed information on specific assemblies, see Section 10.3.

Transmission of airborne and structure-borne sound can be reduced in wall and ceiling assemblies by (1) controlling airborne sound through flanking paths; (2) separating the supporting structural construction of each finish surface; (3) resilient mounting of finishes with channels, clips, or sound-deadening board; (4) including sound-absorbing materials in the assembly cavity; and (5) increasing the weight of the surface construction.

Controlling airborne sound through flanking paths is essential to isolation of airborne sound in any assembly. Failure to observe normal construction and design precautions can seriously impair the effectiveness of other sound-control methods and the performance of the entire assembly. Resilient mounting systems and separated structural framing systems are the most effective, most practical sound control methods individually capable of bringing lightweight framed assemblies to generally acceptable STC and IIC ratings. Sound-absorbing materials and increased weight usually are used in combination for their cumulative effect.

AIRBORNE SOUND THROUGH FLANKING PATHS Airborne sound transmission can be reduced by constructing an airtight assembly and by avoiding back-to-back location of recessed fixtures, such as medicine cabinets, and electrical, telephone, television, and intercom outlets, which perforate a gypsum board surface. In addition, openings for such fixtures and related piping outlets should be caulked (Fig. 9.5-41). The entire perimeter of a sound-isolating partition should be caulked at wallboard edges and under runners to make it airtight.

FIGURE 9.5-42 *Separated construction: (a) staggered studs on a single plate; (b) double row of studs using separate plates.* *(Courtesy Gypsum Association.)*

FIGURE 9.5-41 *Caulking with acoustical sealant to reduce sound transmission (a) around electrical outlet boxes and (b) beneath partition runners.* *(Courtesy Gypsum Association.)*

Assemblies intended for sound isolation should not contain heating or air conditioning ducts or ventilating outlets and should be designed so that airborne sound will not bypass the assembly through flanking paths such as through ceilings or around windows and doors adjacent to the walls or through floors and into crawl spaces below.

SEPARATED CONSTRUCTION Separated construction systems can be developed with wood framing by using separate sets of studs, each supporting a surface of one or more plies of gypsum board. This is accomplished by staggering the studs on the same top and bottom plates or by placing the studs on separate plates (Fig. 9.5-42). By using separated wall-framing systems, cabinets, ceramic tile, lavatories, and other wall-hung fixtures can be in-

stalled in the conventional manner, which is not possible on walls with resilient mounting. Separated walls also can be developed with all gypsum double-solid and semisolid partitions.

RESILIENT MOUNTING Gypsum board finishes can be mounted on walls with resilient channels or resilient clips or over sound-deadening board. Kitchen cabinets, lavatories, ceramic tile, medicine cabinets, and other fixtures should not be supported on resiliently mounted walls, because the added weight and fastenings may "short out" the construction acoustically. On ceilings, resilient channels attached to wood joists or resilient hangers wired to steel joists in suspension systems can be used.

Resilient channels are used in both wall and ceiling assemblies to isolate the gypsum board surface from the structural framing. Channels generally are screw-attached perpendicular to framing members (see Fig. 9.5-15). In ceiling assemblies, nails used to attach channels to wood joists may work loose as the framing lumber dries and cause floor squeaking; therefore, screw attachment is preferred.

Because resilient channels can support only limited loads without losing their acoustical effectiveness they are not appropriate for two-ply construction. Care should be taken to provide the proper-sized cutouts for mechanical and electrical outlets, and caulking with a nonhardening, resilient sealant is required around outlets to keep the wall airtight and flexible at the same time.

Resilient clips can be used to attach the base ply of two-ply finishes to wood studs, but they are not recommended for ceiling application. Proprietary nail-on clips are available that are designed for use with auxiliary 1-in. nominal wood furring for single-ply construction.

SOUND-DEADENING BOARD Sound-deadening board includes (1) wood- or cane-based fiberboard, (2) mineral (including glass) fiberboard, (3) plastic foamboard, and (4) regular core gypsum board. Sound-deadening board is manufactured in sizes 1/4 in. to 5/8 in. thick, 4 ft. wide, and 8 to 9 ft. long. It is used as a base layer to which wallboard is attached and generally is installed parallel to framing members with end joints staggered on alternate studs (Fig. 9.5-43).

FIGURE 9.5-43 *Resilient mounting with sound-deadening board improves sound isolation in metal-framed construction.* *(Courtesy Gypsum Association.)*

Gypsum wallboard can be applied either parallel or perpendicular to supporting construction and is generally laminated to the sound-deadening board. However, nailing or screwing through the sound-deadening board into the framing is required to satisfy fire ratings, but doing so will reduce sound-isolating effectiveness.

Wood fiberboard and plastic foamboard are used over wood construction or where noncombustible materials are not required. Mineral fiberboard is used with metal framing where noncombustible materials are required.

SOUND-ABSORBING MATERIALS Mineral fiber insulating batts and blankets may be used in the cavity space of assemblies to improve assembly performance in resisting sound transmission. The full height of the cavity should be filled with batts or blankets carefully fitted behind electrical outlets, blocking, and fixtures and around cutouts necessary for plumbing lines.

Insulating batts and blankets may be paper-faced or vapor retarder–faced with flanges and are installed by stapling in the stud space. Stapling flanges across the stud face is not recommended because wrinkled flanges may prevent intimate board-to-stud contact. In metal stud and laminated all-gypsum board partitions, the blankets can be attached to the back of the gypsum board. Batts and blankets without facings required in noncombustible construction are friction-fit within the stud space.

■ FIRE-RESISTANT CONSTRUCTION

Gypsum board is a noncombustible building material that has excellent fire-resistant properties. Gypsum board made of specific composition containing fibrous glass reinforcement (Type X) has greater fire resistance than regular gypsum board.

FIRE-RESISTANT PANELS Fire-resistant gypsum board panels are cut into 2-ft. by 2-ft. and 2-ft. by 4-ft. panels for use in some fire-rated ceilings. Common thicknesses are 3/8, 1/2 and 5/8 in. Fire ratings up to 2 hours are possible with an approved suspension system (Fig. 9.5-44). Such ceilings also reduce sound transmission but have little sound-absorptive properties. Standard 4-ft. by 8-ft. fire-resistant gypsum board panels also may be used to

FIGURE 9.5-44 *Fire-resistant 5/8-in. ceiling panels in a fire-rated T-grid suspension system provide 2-hour fire protection for steel and concrete construction.* *(Courtesy Gypsum Association.)*

FIGURE 9.5-45 *Properly constructed wood joist floor and ceiling assemblies with 5/8-in. fire-resistant gypsum board receive a 1-hour fire rating.* *(Courtesy Gypsum Association.)*

provide fire-rated floor and ceiling assemblies for steel (see Fig. 9.5-11) and wood (Fig. 9.5-45) construction.

Fire resistance ratings of gypsum board assemblies can be improved by (1) using special fire-resistant Type X board, (2) increasing the gypsum board thickness, and in some cases (3) filling the stud cavity with mineral wool. Unless all the materials used in the tested assembly classify as noncombustible, the term *combustible* often is used after the assigned hourly classification.

Ordinary 2 × 4 wood stud partitions, when finished with 5/8-in.-thick Type X wallboard on both sides and properly installed, receive a 1-hour combustible rating. Wood stud partitions have been rated 2 hours when two layers of 5/8-in.-

thick Type X board are used on each side. Metal stud partitions may receive similar ratings with two-ply faces when special construction elements mentioned above are included. A recently developed 3/4-in.-thick gypsum board can also be used to develop a 2-hour fire resistance rating over metal studs. Wood joist floor and ceiling assemblies with single-layer wallboard generally are limited to 1-hour ratings. However, steel-and-concrete assemblies protected by a suspended gypsum board ceiling generally result in higher ratings.

Another criterion for fire hazard of exposed surface materials is the rate of surface flame spread. The flame spread rating of gypsum board is under 20, based on a flame spread of zero for mineral fiber–cement board and 100 for red oak. Gypsum board's low flame spread rating more than satisfies most interior finish requirements.

■ THERMAL INSULATING CONSTRUCTION

In certain regions of the country, exterior wall and ceiling assemblies may require thermal insulating construction to minimize heat gain or loss through the assembly. In such areas a vapor retarder commonly is required to retard the migration of moisture and condensation within the assembly. For a discussion of moisture and heat control, see Sections 7.1 and 11.3.

Typical materials used to increase the thermal resistance of assemblies with gypsum board finishes include (1) flexible fiber insulation (batts and blankets), (2) loose-fill insulation, (3) rigid plastic insulation, (4) reflective foil-faced airspace, and (5) airspaces provided by furring over solid materials such as masonry and concrete. For a discussion of thermal insulations see Section 7.3.

Vapor control can be provided as an integral part of the gypsum board or as part of the insulation. Foil-backed gypsum board will usually provide some insulating value when combined with a dead-air space as well as an effective vapor retarder and can be used with or without other insulating materials. Foil-backed gypsum board should not be used in these instances: (1) as the backing material for ceramic tile; (2) as the face ply in laminated two-ply construction; (3) for laminating directly to masonry or concrete; (4) in connection with electric heating cables in electric panel heating systems.

Rigid plastic foam and reflective foil insulation have inherently low vapor permeability and generally do not require a separate vapor retarder. Flexible insulations such as batts and blankets are available with or without vapor retarders. Loose insulation requires either a separate vapor retarder, such as foil or polyethylene film, or an integral retarder provided by foil-backed gypsum board. Thermal insulation batts and blankets should be attached as recommended for sound-absorbing batts and blankets.

Two important considerations should not be overlooked in insulating design: (1)

if reflective foil is to provide effective insulating value, a dead-air space of at least 3/4 in. should be provided, and (2) when a vapor retarder is used, it should be located on the warm side of the insulation. The installation of insulating and vapor control materials that are independent of a gypsum board finish is essentially the same as for other finishes. Only wood and metal furring and rigid plastic insulation that provide support for a gypsum board finish require special consideration in this section.

Rigid plastic insulation is intended primarily for use on exterior and below-grade concrete or masonry walls. Rigid insulating foamboards are made from expanded polystyrene or polyurethane, in panels 1/2 to 3 in. thick, up to 4 ft. wide, and 12 ft. long. These foamboards may be adhesive-bonded to masonry or concrete surfaces using methods and adhesives specifically recommended by the insulation manufacturer. They should not be laminated to exterior masonry or concrete having a surface temperature of less than 45°F. They should be fitted together so that there will be no gaps between the boards, which could create cold spots or allow moisture migration. Gypsum board may be bonded to the rigid foam insulation with an approved adhesive or screw-attached to special channels provided with the foamboard. At least 24 hours drying time should be allowed before gypsum board is laminated to rigid plastic insulation with an adhesive. In addition, the gypsum board should be held in place by temporary bracing or double-headed nails, which can be removed after sufficient bond strength develops. Supplemental fasteners are required.

■ MOISTURE-RESISTANT CONSTRUCTION

Most gypsum board is sensitive to moisture and should not be used in locations subject to direct wetting or continuous high humidity. It can be used in limited exterior locations such as eave soffits and porch and carport ceilings when it is properly finished and painted and in interior locations such as tub, shower, and laundry areas when protected with wall tile. Such locations require special construction details and precautions.

This subsection covers the use of regular core and moisture-resistant gyp-

sum board. Glass mesh mortar units that can serve as a base for ceramic tile in lieu of moisture-resistant gypsum board are discussed in Section 9.6.

EXTERIOR LOCATIONS   Regular or water-resistant wallboard used on horizontal surfaces in exterior locations should be at least 1/2 in. thick. Fascia boards at gypsum board soffits should be designed to form a drip edge to prevent water runoff across the board surface and to protect the board edge (Fig. 9.5-46). In addition, edges and ends at the perimeter of the board should be protected by metal edge trim. Unless protected by metal trim or another water stop, edges should be spaced at least 1/4 in. from abutting masonry.

FIGURE 9.5-46 *Gypsum board can be protected from damp exterior surfaces by (a) airspaces and wood moldings or (b) metal trim.* (*Courtesy Gypsum Association.*)

Exposed wallboard surfaces and metal trim should be painted with two coats of exterior paint. Adequate ventilation of eaves, ceilings, and attic spaces above gypsum board soffits should be provided.

GENERAL INTERIOR LOCATION   Special water-resistant gypsum board is recommended for all interior walls subject to high humidity, except for high-moisture spaces like shower rooms, saunas, and steam rooms. Moisture-resistant gypsum board should not be used on ceilings. Foil-backed gypsum board should not be used in high-humidity interior areas because the foil backing may prevent moisture from being dissipated, thus trapping it in the board core. But where a high-humidity area is located on an exterior wall, adequate insulation and a vapor retarder on the warm side of the wall are necessary to prevent condensation in the wall cavity.

TUB-SHOWER AREAS Water-resistant gypsum board and proprietary highly water-resistant backing board are particularly recommended as a base for wall tile in tub-shower areas, although regular core gypsum board will perform satisfactorily if properly installed and prepared to receive tile.

Gypsum board in tub-shower areas should be at least 1/2 in. thick. Stud spacing should not exceed 24 in. on centers. When the stud spacing exceeds 16 in. on centers, blocking should be installed not more than 4 in. on centers between studs, with the first row of blocking about 1 in. above the top of the tub or shower receptor. When a wall is to be finished with tile (see Section 9.6) or handicap accessories are to be installed (see Section 1.6), closer stud spacings may be required.

Gypsum board should be applied perpendicular to framing members with paperbound edges parallel to the top of the tub or shower receptor and spaced at least 1/4 in. above it (Fig. 9.5-47). This space should be filled with a continuous bead of waterproof sealant. Generally, gypsum board should be fastened with nails spaced not more than 8 in. on centers or screws spaced not more than 12 in. on centers. However, when ceramic tile more than 5/16 in. thick will be installed, nails should be spaced 4 in. on centers and screws 8 in. on centers.

Special water-resistant backing board or regular core gypsum board should be installed as recommended by the manufacturer or the Gypsum Association. When regular core gypsum board is used, joints between adjacent boards, including those at corners, should be taped and treated as for painting but should not receive finishing coats. Before tile is installed, a water-resistant sealer should be applied to the entire wallboard surface, including treated joints and corners. When tile adhesive is used as a sealer, it should be

FIGURE 9.5-47 *Gypsum board and wall tile installation details for tub-shower areas. (Courtesy Gypsum Association.)*

spread separately from the bonding coat in a continuous, uniform coating approximately 1/16 in. thick, covering all cut ends and all cutouts for fixtures. A Type I, highly water-resistant organic adhesive should be used to apply ceramic wall tiles.

A tile finish should cover all surfaces, as illustrated in Figure 9.5-48. The space between tile and tub rim, shower receptor, or subpan top should be caulked with a continuous bead of sealant (see Fig. 9.5-47). To prevent this caulked joint from opening due to settlement, the tub should be supported at the walls on metal hangers or vertical blocking fastened to the studs. Waterproof shower receptor pans or subpans should have an upstanding leg or flange at least 1 in. higher than the water

dam or threshold in the entry to the shower.

■ ELECTRIC PANEL HEATING

Electric resistance elements may be incorporated in the gypsum board itself for one-ply application or in the backing board for two-ply construction. Another type of radiant electric heating ceiling system is a built-up one that incorporates a gypsum backing board, heating cables, a filler material, and a gypsum wallboard finish. Standard gypsum lath and plaster products also can be assembled into a radiant heating ceiling system by embedding heating cables in the plaster coats over gypsum lath (see Section 9.3). In each case, the

FIGURE 9.5-48 *Shaded areas show tile coverage of gypsum board in tub-shower areas.* (*Courtesy Gypsum Association.*)

manufacturer's detailed recommendations should be followed.

Heat loss calculations indicate the number of square feet of heating panels required in each space. For properly insulated residential installations, this is generally less than the total ceiling area. The areas adjacent to heating panels in each space are covered with backing board or wallboard of matching thickness. Wallboard face panels may be finished in the normal manner, but installation of acoustical tile over such a system is not recommended.

Proper insulation design is important to the efficient operation of any electric panel heating. Thermal insulating batts used in the ceiling above a panel heating system should be unfaced friction-fit mineral or glass wool.

ONE-PLY CONSTRUCTION One proprietary heating panel system utilizes a single layer of 5/8-in.-thick wallboard with heating cables embedded in the core (Fig. 9.5-49). Panels come in standard 4-ft. by 4-ft., 4-ft. by 8-ft., and 4-ft. by 12-ft. sizes. Special panels that contain cables in two-thirds of the panel only, allow the ends and sides to be trimmed as required for particular layouts. Panels can be installed by fastening at prescribed intervals. Fastener heads are concealed with compound as in other gypsum board installations, and tapered edges permit standard joint treatment.

Each panel is provided with a 12-ft.-long electric lead cable, which may be connected to an electric branch circuit in

FIGURE 9.5-49 *Single-ply radiant panels may be nailed and finished in the same manner as regular gypsum board.* (*Courtesy Gypsum Association.*)

a junction box. The branch circuit typically is connected to a room thermostat, which regulates the panel heating cycles. Before it is connected to the power supply, the lead cable is taped to the back of the panel so it will not interfere with erection.

TWO-PLY CONSTRUCTION This proprietary system uses a combination heating panel and backing board, to which a 1/4-in.-thick regular core gypsum board must be job-laminated (Fig. 9.5-50). The heating panel is about 1/2 in. thick and comes in 4-ft. by 10-ft. and 4-ft. by 12-ft. sizes. It consists of a gypsum core, resistor

coating, and copper electrodes. The electrodes are connected to standard ac or dc house current and conduct electricity to the resistor coating, generating heat. This coating is completely enclosed by an insulator to prevent current leaks.

These heating panels may be nailed directly to wood ceiling members or other supporting framework with special insulated nails that can penetrate the resistor coating without becoming energized. Standard panels have electrodes 1 in. from each edge and may be cut to desired length, but cutting the panels to width is not possible since one of the electrodes would be removed. However, a special 4-ft. panel is available with a 22-in.-wide resistor coating, which can be cut to a minimum 2-ft. width.

BUILT-UP SYSTEM Electric panel heating can also be developed utilizing backing board, electric heating cables, a filler material, and gypsum wallboard. The backing board base layer should be attached perpendicular to furring or framing members with nails, screws, or staples spaced 7 in. on centers. Electric heating cables should be securely attached to the backing board, maintaining at least a 1-1/4-in. clearance on both sides of the joist centerline and 4 to 6 in. around the perimeter of the ceiling at the walls (Fig. 9.5-51).

Inspection and testing of the electrical installation should be performed before the face layer is applied. The maximum cable temperature should not exceed 125 °F at any time.

A filler material consisting of sanded gypsum plaster or embedding compound and sand mixed according to the manufacturer's recommendations should be used to completely embed the heating cables. The filler material should be leveled to provide complete contact with the back of the face layer. Wood furring may be installed directly under and parallel to the framing members to facilitate leveling of the filler material and to provide a firm base for attaching the face layer.

Because filler material consisting of embedding compound and sand possesses inherent adhesive properties, the face layer requires only limited nailing, as in adhesive-nail attachment. With such fillers, the wallboard face layer should be applied immediately after the filler has

FIGURE 9.5-50 *Two-ply radiant system requires special accessories and job lamination of gypsum board face panels. Finishing is the same as for regular gypsum board, but acoustical ceiling tile should not be used over this system.* (*Courtesy Gypsum Association.*)

FIGURE 9.5-51 *Typical cable layout for electric panel heating system.* (*Courtesy Gypsum Association.*)

been spread and nailed 16 in. on centers along the supports. When the filler does not serve as an adhesive, it should be allowed to dry completely; then a thin coat of embedding compound should be applied to provide continuous filler-to-wallboard contact, and the wallboard should be nailed 7 in. on centers along the supports. At the ceiling perimeter, nails should be kept 8 to 10 in. from the walls to prevent striking the heating cable where it crosses the joists.

The filler material and adhesive should be completely dry before wallboard joints and fasteners are treated. Filler material, when specifically formulated for the purpose, can be troweled on as the fin-

ish material instead of a wallboard face layer.

This system should not be operated until the filler material and the joint compound are completely dry, which may require as much as a week of good drying conditions or two weeks in cold, damp weather.

■ SELF-SUPPORTING GYPSUM PARTITIONS

Self-supporting gypsum partitions are non-load-bearing, noncombustible multiply partitions, fabricated on the job almost entirely from gypsum boards. Inner plies consist of backing board in rib or sheet form, to which wallboard face plies are laminated. Unlike the ceiling and wall assemblies described elsewhere in this section, which depend on wood, metal, concrete, or masonry construction to support the surface finish, gypsum partitions are entirely self-supporting and require only floor and ceiling runners to stabilize them. Runners should be attached securely to ceiling and floor construction with fasteners spaced 24 in. on centers and be carefully caulked at partition perimeters and under runners to close air leaks. The maximum height of partitions is limited by the design and the dimensions of the components. Therefore, the manufacturer's recommendations should be followed.

A variety of proprietary partition systems have been developed by manufacturers of gypsum products to minimize the number of operations and to speed field

erection of partitions. The chief types are (1) *semisolid*, which utilizes backing board ribs; (2) *solid*, with coreboard (thicker backing board) inner plies; (3) *hollow*; and (4) *demountable*. All gypsum board components, such as ribs, coreboard, backing board, and wallboard, typically are installed in the vertical position.

Gypsum partition products are used in self-supporting partition systems, engineered assemblies of board and plaster products with special metal accessories designed for use with each system.

SEMISOLID PARTITIONS In semisolid partitions, the backing board core is replaced by 6- to 8-in.-wide gypsum board ribs, laminated at the factory in 1-in. and 1-5/8-in. thicknesses (Fig. 9.5-52a). Ribs are adhesive-bonded in the field to the wallboard faces for single-ply construction or to backing boards for two-ply construction (Fig. 9.5-52b). In two-ply construction, regular or predecorated wallboard is bonded to form the finished surface. Veneer plaster finish may be used over an appropriate plaster base laminated to the ribs. Semisolid partitions generally are not made with standard plaster finishes. Partition thicknesses vary between 2 and 3 in., depending on rib and face panel thicknesses.

The ribs are cut 6 in. shorter than the gypsum board and can be laminated to the face ply on one side of the partition either before or after it is erected. Perimeters of door and window openings should be reinforced with additional ribs. Adhesive can be spread on the erected face ply or on the ribs, keeping it 1 in. from the rib edges. These rib-reinforced panels are secured to wood or metal runners at the top and the bottom with appropriate drywall screws spaced 12 in. on centers.

The ribs on the erected panels then are spread with adhesive, and facing panels are installed and screw-attached to runners with screws spaced 12 in. on centers. To ensure adequate bond at ribs, 1-1/2 in. drywall screws should be provided at each rib 12 in. from the top and the bottom and spaced not more than 36 in. on centers.

For partitions with two-ply faces, the ribs and backing board base plies are assembled and installed as for single-ply faces. Face plies then are sheet laminated as for two-ply supported partitions using supplemental drywall screws.

Separated rib construction to improve sound-isolating effectiveness can be provided in either single- or two-ply ribbed partitions. The installation procedure is essentially the same as for ordinary ribbed partitions, but twice as many ribs are required, spaced 12 in. on centers and staggered on opposite sides of the partition (see Fig. 9.5-52b).

SOLID PARTITIONS Solid gypsum partitions are an extension of the concept of semisolid partitions. They consist of a solid coreboard to which wallboard faces are laminated on each side (Fig. 9.5-53a). The vertical joints in the face ply should be offset at least 3 in. from joints in the coreboard. The face ply typically is attached by sheet lamination with supplemental drywall gypsum screws. Double- or triple-solid partitions separated by airspaces provide increased fire resistance and sound isolation (Fig. 9.5-53b).

Solid gypsum partitions may be field-erected from combinations of lath and plaster or backing boards and wallboards. Both generally utilize wood or proprietary metal ceiling and floor runners to stabilize the partitions. In lath and plaster partitions, an integral flush metal base acts as a screed. Partitions also may be trimmed out with resilient or wood base.

Backing board and wallboard partitions consist of a 1-in.-thick backing board core to which 1/2- or 5/8-in. wallboard faces are laminated with adhesives formulated for the purpose. The core may consist of two job-laminated 1/2-in.-thick backing boards or a 1-in. factory-laminated product referred to as *coreboard* (see Fig. 9.5-53a). Regular wallboards usually are used, which require joint treatment and decorating, but in some instances predecorated wallboards with V-edges may be installed in a visually attractive repeat pattern. When veneer plaster finish is desired, the special plaster base designed for this type of finish is used in place of wallboard. Double solid partitions are constructed of two separate layers of lath or backing board, separated by an airspace, with plaster or wallboard faces on the exterior (see Fig. 9.5-53b).

HOLLOW PARTITIONS Hollow partitions are made with wood or metal studs supporting faces of plaster or wallboard (see Fig. 9.3-5). Studs usually are spaced 16

FIGURE 9.5-52 *Semisolid gypsum partitions: (a) single-faced with both faces laminated to each rib; (b) double-faced with each two-ply face laminated to alternate ribs, forming separated rib construction.*

FIGURE 9.5-53 *Solid gypsum partitions: (a) single solid partition; (b) double solid partition with an airspace between the two faces.*

to 24 in. on centers and are supported by wood or metal floor and ceiling runners. Wood studs normally are 3-1/2 in. deep, and metal channel studs are available in 1-5/8-, 2-1/2-, 3-5/8-, and 4-in. sizes. Greater wall cavities found in older buildings were developed by using deeper truss-type studs, which ranged up to 6 in. in depth, or two sets of lighter channel studs braced with wallboard. Truss-type studs are difficult or impossible to find today.

Lath and plaster hollow partitions have gypsum lath screwed, clipped, or nailed to studs, or metal lath wired, nailed, clipped, or stapled to the studs. Two or

three coats of plaster are built up over this plaster base on each face (see Section 9.3).

Wallboard hollow partitions may be of either single-ply or two-ply construction. For single-ply construction, wallboard is nailed or screwed directly to the studs (Fig. 9.5-54a). For two-ply construction, backing board is clipped, nailed, stapled, or screwed to the studs, and wallboard is adhesive-bonded to this backing with a minimum of supplemental fastening. Where soundproofing is important, resilient channels may be used in either one- or two-ply construction to mount the

metal stud
gypsum wallboard
screws
metal floor runner

a

wood stud
resilient channel
wallboard
backing board
screws
spacer strip

b

FIGURE 9.5-54 *Hollow gypsum board partitions: (a) screwed single-ply; (b) two-ply construction with resilient mounting for improved sound control.*

board (Fig. 9.5-54b). When regular core wallboard is used as the exterior face, fasteners and joint treatment are required; predecorated wallboard normally is adhesive-bonded to backing board and has V-edge joints, which can be left exposed.

DEMOUNTABLE (MOVABLE) SYSTEMS In situations where it may be desirable to rearrange partitions, demountable panel systems frequently are used. They are designed for speedy erection and for minimum loss of components when they are demounted and rearranged.

Demountable partition systems use wallboard face panels attached to interior metal studs for stability. Various pressure-fit, clip, or screw systems are used to secure the wallboard panels to the studs. One system uses an H-section stud 24 in. on centers, the flanges of which are pressure fit into grooves of factory-laminated, predecorated panels. In other systems, predecorated square-edged wallboard is adhesive-bonded directly to studs 24 in. on centers, and the edges are secured with nails or screws. Decorative batten trim then is used to conceal fastenings at 24 or 48 in. on centers (Fig. 9.5-55).

SHAFT WALLS Shaft walls are light-weight gypsum board assemblies used for elevator shafts, electrical and mechanical enclosures, and stairwells. Shaft wall assemblies with metal stud framing weigh 10 to 13 pounds per square foot (psf), whereas masonry or masonry-and-plaster

metal ceiling trim
metal ceiling runner
metal stud
predecorated wallboard
batten trim
metal floor runner
metal base

FIGURE 9.5-55 *One type of demountable partition.*

assemblies with similar fire ratings weigh between 35 and 65 psf.

Shaft walls built with gypsum board not only are lighter in weight but are easier and less expensive to install than conventional masonry construction. They can be erected floor by floor from outside the mechanical or elevator shaft; they do not delay installation of the mechanical equipment inside the shaft because of masonry scaffolding; and they do not require extensive cleanup inside the shaft upon completion.

■ RESURFACING EXISTING CONSTRUCTION

Gypsum board may be used to provide a new finish on existing walls and ceilings of wood, plaster, masonry, gypsum board, and most other materials. If the existing surface is sufficiently smooth and true after appropriate preparation and it provides solid backup for the new wallboard without shimming, 1/4-in.-thick gypsum board can be applied with an adhesive, nails, or screws. Nails and screws should be long enough to provide recommended penetration into supporting members under the existing finish (see Figs. 9.5-16 and 9.5-17).

Existing surfaces that are not sufficiently smooth to receive gypsum board should be furred with wood or metal furring. Furring members should be shimmed as required to provide a suitably level base. Minimum gypsum board thickness for various support spacings and installation methods should be as recommended for new construction over furring.

Surface trim for mechanical and electrical equipment such as switch plates, outlet covers, and ventilating grilles should be removed and reinstalled over the new gypsum board finish.

■ VENEER PLASTER BASE

In general, veneer plaster base should be installed in the same manner as other types of gypsum board, and the joints should be taped. Some manufacturers recommend using paper tape; others open-mesh glass fiber tape. Either base coat veneer plaster and joint compounds is used to embed paper tape, depending on the manufacturer of the plaster.

The manner of board placement should produce the least possible number of end joints, which are difficult to conceal.

9.5.7

JOINT AND FASTENER TREATMENT

After gypsum board has been erected, specific areas require treatment to achieve a smooth homogeneous appearance. Panel joints in the same plane and interior corners require taping. Exterior corners and exposed edges require metal corner and edge trim. Both taped and trimmed areas, as well as fasteners require finishing with compound (Fig. 9.5-56).

FIGURE 9.5-56 *Gypsum board joints are finished by embedding a perforated tape in bedding compound and covering the tape with several coats of topping compound.* (Courtesy Gypsum Association.)

9.5.7.1

Joint Taping and Finishing

A minimum of three coats of joint compound is recommended for taped joints. These include an embedding coat to bond the tape and two finishing coats over the tape. Each coat should be allowed to dry thoroughly so the surface can be readily sanded and the compound will obtain maximum shrinkage before the next coat is applied.

Depending on atmospheric conditions, 12 to 24 hours may be required for each successive coat to dry, unless quick-setting compound is used. The second and third coats should be lightly sanded when dry to remove surface irregularities and to blend them evenly with the adjacent board. Care should be taken not to raise the fibers on the board's face paper, because this may result in uneven surface texture and a spotty appearance.

■ **FLUSH JOINTS**

Embedding compound should be spread into the depression formed by the tapered edges of adjacent boards (Fig. 9.5-57a) and over all butt-end joints. Then tape is centered over the joint, smoothed to avoid wrinkling, and pressed into the compound by drawing a knife along the joint with enough pressure to squeeze out excess compound (Fig. 9.5-57b). This excess compound may be removed or redistributed as a skim coat over the tape (Fig. 9.5-57c). When the embedding coat is dry, a second coat is applied with the

FIGURE 9.5-57 *Finishing knives of varying widths are used to spread embedding compound, embed joint tape, and apply successively wider coats of topping compound.* (Courtesy United States Gypsum Company.)

edges feathered 2 to 4 in. beyond the tape edges (Fig. 9.5-57d). Over the dry second coat, a third coat is spread with edges feathered 2 to 4 in. beyond the second coat to blend smoothly with the board surface (Fig. 9.5-57e). Joints can be treated in a similar manner by machine application of tape and joint compound (Fig. 9.5-58).

■ **INTERIOR CORNER JOINTS**

Compound should be applied to both sides of interior corners. Then tape should be folded along the center crease and embedded snugly in the corner to form a right angle (Fig. 9.5-59). Excess compound is removed, and the surfaces are subsequently finished with additional coats of compound as described for flush joints. Both hand and mechanical tools are available for corner finishing.

FIGURE 9.5-58 *Automatic taping machine applies embedding compound and joint tape on (a) ceiling and (b) wall; (c) broad knife removes excess compound; and (d) finishing tool with interchangeable heads applies successively wider coats of topping compound. OSHA standards require workers to wear hard hats (as in b) when using this type of equipment. Eye protection also should be worn.* (Courtesy United States Gypsum Company.)

FIGURE 9.5-59 *Tape can be embedded in corner (a) by hand or (b) with a tool. Finishing can also be done by hand or (c) with corner tool.* (Courtesy United States Gypsum Company.)

9.5.7.2

Edge and Corner Trimming

All trim shapes except metal U-bead are installed after the gypsum board has been fastened in place (see Fig. 9.5-18). Metal cornerbeads and casing beads such as L, LC, and LK are installed with the flange outward and are nailed, crimped, or screwed 5 in. on centers through the gypsum board into the supporting construction. Metal cornerbeads with paper flanges are secured by embedding the flanges in joint compound. All edge and corner trim except U-bead require finishing with joint compound to obtain a smooth surface and to conceal fasteners (see Fig. 9.5-57f). Methods and sequence of three-coat application are the same as for fastener treatment.

Unlike other trims, U-bead is fastened through the back flange first, then the gypsum board edges are inserted into the trim. Finishing with compound is not required because the fasteners are concealed.

9.5.7.3

Fastener Treatment

A first coat of embedding compound usually is applied over fastener heads at the same time joints are taped. Fasteners

FIGURE 9.5-60 *Slightly recessed screw or nail heads are treated with (a) an automatic spotter or (b) a finishing knife. A hard hat and eye protection should be worn.* (*Courtesy United States Gypsum Company.*)

may be treated (*spotted*) using an automatic spotter or a finishing knife (Fig. 9.5-60). This knife should be pressed against the surface to finish, level, and remove excess compound. Second and third coats of compound generally are applied over the fasteners when similar coats are applied to taped joints, edges, and corners. Each successive coat should be feathered slightly beyond the one previously applied. Compound should be sanded between coats as required. Care should be taken not to raise the fibers in the face paper when sanding. Each coat should be thoroughly dry before the next coat is applied.

9.5.8

FINISH DEFECTS

Improper installation or adverse job conditions may result in surface imperfections such as nail popping, shadowing, cracking at corners, or ridging at joints. Such finish defects can be minimized or completely eliminated if their causes are understood and appropriate precautions taken.

9.5.8.1

Nail Popping

Nail popping is a surface defect in which compound over a nail head is pushed out beyond the surface of the board. It is due primarily to loose boards caused by wood shrinkage, the use of excessively long nails, or improper nailing.

■ WOOD SHRINKAGE

Wood shrinks across the grain as its moisture content is reduced (see Section 6.1). As a piece of lumber shrinks, it tends to expose the shank of a nail driven into its edge (Fig. 9.5-61). If shrinkage is substantial and nails are too long, separation between board and framing member can result. A separation as small as 1/32 in. can result in an objectionable nail pop. It is essential that framing and furring lumber have as low an initial moisture content as possible, that lumber be kept dry during storage and installation, and that nail lengths do not substantially exceed those recommended for specific applications. Framing and furring lumber should be at the equilibrium moisture content for the area and should not exceed 15% moisture content. Nail penetration should be as near as possible to those recommended in Figure 9.5-16. Annularly threaded nails require less penetration than smooth-shank nails and tend to minimize nail popping.

■ IMPROPER NAILING

When gypsum board is improperly nailed, loose boards may result due to loose nails, loose attachment, or punctured face paper.

LOOSE NAILS Nails that either miss the underlying construction or enter it at an angle near the edge (Fig. 9.5-62a) can eventually be loosened by vibration until they protrude beyond the finish surface. To guard against such defects, the location of supporting members should be marked carefully on the gypsum board prior to nailing, nails should be driven perpendicular to the surface into the member, and nails that have missed the member should be removed promptly. After nails have been removed, nail holes should be dimpled and treated with compound.

LOOSE ATTACHMENT Loose attachment can occur when too little pressure is applied to a board while it is being nailed (Fig. 9.5-62b). Even when substantial pressure is applied properly, it may be impossible to produce snug contact with supporting members if the board is cut too large and must be forced into place, or if nailing is begun at the perimeter and proceeds inward. To prevent loose board attachment, panels should be cut to fit easily into place, pressure should be applied while nailing, and proper nailing sequence from the middle of the panel outward should be observed. Use of screws and double nailing also tends to minimize loose boards.

PUNCTURED FACE PAPER Improper nail head design, obliquely driven nails (Fig. 9.5-62c), excessive and repeated impact

FIGURE 9.5-61 *As a stud shrinks, separation may be (a) 1/16 in. when nail penetrates 1 in. or (b) 1/8 in. when nail penetrates 2 in.* (*Courtesy United States Gypsum Company.*)

FIGURE 9.5-62 *Nail popping can result from (a) loose nails that miss framing members, (b) loose boards, or (c) obliquely driven nails that puncture the face paper.*

during nailing, and using boards with extremely dry face paper can all result in the face paper being punctured during board installation. When the paper is punctured, the gypsum core is easily crushed and offers less resistance to loosening of the board from normal impact and vibration. In hot, dry periods, the face paper may become more brittle, increasing the hazard of puncturing. This hazard can be minimized by increasing the air humidity during storage and installation of gypsum board.

Other common causes of punctured face paper are misaligned or twisted members and projections caused by improperly installed blocking or bracing (Fig. 9.5-63). Intimate contact between gypsum board and supporting members is impossible in such cases, and hammer impact causes the board to rebound, which often ruptures the paper. Such defects should be corrected before the board is installed. Use of screws, adhesive, or two-ply construction will minimize problems resulting from such defects.

9.5.8.2

Shadowing

Shadowing is actually a surface discoloration, but it may appear similar to nail popping when viewed at a distance. Shadowing is the result of an accumulation of dust or dirt caused by increased thermal conductivity and moisture condensation over a nail head. It occurs commonly on roof and ceiling assemblies and exterior wall assemblies that are inadequately insulated and in regions with great indoor-outdoor temperature variations. Shadowing can be remedied by periodic washing or decorating. It can be avoided by proper insulation design and use of laminated two-ply construction. Where a permanent ventilating system is employed, adequate air filtration and humidity control will be helpful.

9.5.9

DECORATING AND REDECORATING

A gypsum board installation should be inspected before it is decorated or redecorated. Special attention should be paid to joints and fasteners. Imperfections should be sanded or repaired.

It is essential that joint compound be thoroughly dry before any form of decoration is applied. Depending on temperature and humidity conditions, 24 hours or more of good drying time is required.

Gypsum board surfaces can be decorated with textured coatings (Fig. 9.5-64), emulsion or oil-based paints, or paper, fabric, or vinyl wall coverings. It is necessary to prepare the board surface with a sealer or primer. A good sealer will seal the pores and lay down the fibers in the paper surface so that suction and texture differences between paper and joint compound will be equalized; it also will seal in surface impurities sometimes present in the wallboard surface. A good primer will provide uniform texture and conceal color or surface variations. Some sealers incorporate the properties of primers as well.

A sealer or primer should be used on gypsum board beneath oil-based paints, lacquers, and enamels, because they tend to raise the fibers in the face paper. Sealers are used under wall coverings primarily to facilitate later removal without marring

FIGURE 9.5-63 *Punctured face paper and loose boards can be caused by (a) misaligned members, (b) twisted members, or (c) projecting blocking.* (*Courtesy United States Gypsum Company.*)

FIGURE 9.5-64 *A variety of textural appearances can be achieved with a spray machine, the major components of which are shown here.* (*Courtesy United States Gypsum Company.*)

the board surface and to provide an adequate base for redecorating.

Most emulsion paints, such as latex, resin, and casein, do not raise the fibers in the face paper. Good latex paints have inherently good sealing and priming properties and require no special preparation of gypsum board surfaces. However, applying a pigmented emulsion sealer as a first coat or using at least two coats of emulsion paint generally will provide a durable, attractive finish if the gypsum board has been properly installed and if joints, fasteners, and trim have been properly treated.

Glue size, shellac, and varnish are not suitable as sealers or primers under paints or wall coverings.

If nail popping and other surface defects occur after decorating, it is advisable to wait until one full heating cycle has been completed before repairing and redecorating.

The information here is general in nature. Refer to Section 9.12 for further discussion about painting gypsum board.

9.6 TILE

The term *ceramic* derives from the Greek word *keramos*, which means potter's clay. Ceramic tile is made from nonmetallic minerals fired at high temperatures and manufactured in modular unit sizes (tiles), which facilitate installation. The term *ceramic tile* includes several products of varying dimensions, properties, and appearance: (1) *glazed wall tile*, typically 4-in. by 4-in. or 4-1/4-in. by 4-1/4-in. units used mainly on walls; (2) *ceramic mosaic tile*, mostly 1-in. by 1-in., 1-in. by 2-in., or 2-in. by 2-in. units intended mostly for floors but also used on walls; (3) *paver tile*, floor units similar to mosaic tile but generally 4 in. by 4 in. or larger; and (4) *quarry tile*, typically 6-in. by 6-in. or larger natural clay units for floors. Other sizes and shapes are also available (Fig. 9.6-1). Ceramic materials are also sometimes made to serve purely decorative or commemorative uses (Fig. 9.6-2). Some glazed wall tiles are sufficiently abrasion-resistant to make them suitable for floors as well, while all floor tile can be used on walls. All these products possess good or superior abrasion resistance, the ability to take on a variety of patterns and colors, and a high degree of resistance to moisture.

These properties make ceramic tile uniquely suited for areas exposed to severe foot traffic or in intermittent or continuous contact with water or corrosive chemicals. The development of simplified installation procedures and an increased variety of decorative tiles have greatly extended ceramic tile use. Today, ceramic tile is being specified for just about every interior area of residential, commercial, institutional, and industrial buildings. Many types of tile also are suited for exterior use.

The origins of ceramic tile date back 5000 years, as evidenced by burnt natural clay tiles found by archaeologists in early Egyptian and Babylonian excavations. Decorative and functional clay tiles were also an essential part of later Greek and Roman architecture. The Italian town of Faenza is credited with developing in the 14th century what is now known as *faience tile*, a highly decorative tile with a textured irregular handmade appearance. A century

FIGURE 9.6-1 *Pentagon-shaped tiles lend an elegant appearance to this setting.* (*Courtesy American Olean [Ceramic] Tile Co.*)

later, Holland started production of *delft tile*, characterized by a figure or a landscape in blue or violet.

In the United States, production of ceramic tile was spurred by the introduction of the dust-press method at the end of the 19th century. Samuel Keys is credited with first producing floor and wall tile in Pittsburgh in 1867; by 1937 there were 52 manufacturers of ceramic tile in the United

States. It is generally agreed that more improvements in manufacturing and installation techniques have been made by the U.S. ceramic industry during the past 35 years than in the preceding 70 centuries.

Much of that progress is due to the research and educational activities of the Tile Council of America (TCA). In the early 1950s the TCA sponsored research that resulted in the development of several

FIGURE 9.6-2 *An artist's mosaic of cut tiles. Mosaic murals are also made commercially.* (*Courtesy American Olean [Ceramic] Tile Co.*)

FIGURE 9.6-3 *Basic tile installation methods: (a) thin-set method, in which the tile is bonded to the supporting substrate; (b) conventional mortar (thickbed) method, in which tile and mortar setting bed are isolated from the backing by a membrane.*

adhesives and mortars that eliminated the need for thick setting beds, laborious pre-soaking of tile, and extended curing. To ensure satisfactory tile installations, the TCA (1) licenses manufacturers of mortars and grout according to Council formulations; (2) promulgates standards for the manufacture and installation of ceramic tile, which have resulted in a series of ANSI standards; and (3) publishes and annually updates the *Handbook for Ceramic Tile Installation.* Information in this section noted as "Courtesy Tile Council of America" was taken from the 1993 edition of that publication. (Readers should verify data with the latest update.) Most of the discussion and installation details in this section have been abstracted from the TCA's handbook.

9.6.1

GENERAL RECOMMENDATIONS

Like most other thin nonstructural finishes, ceramic tile depends on a sound, rigid, and dimensionally stable backing surface. Installation of ceramic tile on floors resembles terrazzo (Fig. 9.6-3), involving either an unbonded, isolated, setting bed (thickbed method) or direct bonding to the backing (thin-set method). As in terrazzo, direct bonding is made possible by synthetic resins possessing greater bonding power than conventional portland cement. However, the terminology used by the ceramic tile industry differs somewhat from that in the terrazzo industry. The follow-

ing text will generally employ the terms familiar to the tile industry. Refer to the Glossary for unfamiliar terms.

9.6.1.1

Materials

Materials involved in ceramic tile installation include (1) tile, (2) mortars and adhesives for setting tile, (3) grouts for filling of joints, and (4) related materials to complete the installation, such as portland cement, reinforcing metal lath or mesh, and cleavage membrane.

■ CERAMIC TILE

Ceramic tile is made from clay or from clay mixed with other ceramic materials; oxides may be included for glaze coloring. Ceramic tile products are available in a broad range of sizes (Fig. 9.6-4), appearance characteristics, and functional properties (Fig. 9.6-5), which include perviousness to moisture, abrasion resistance, and other properties. These factors dictate the suitability of a particular tile for a specific purpose.

In discussing the appearance or properties of ceramic tile it is convenient to discuss the face separately from the body

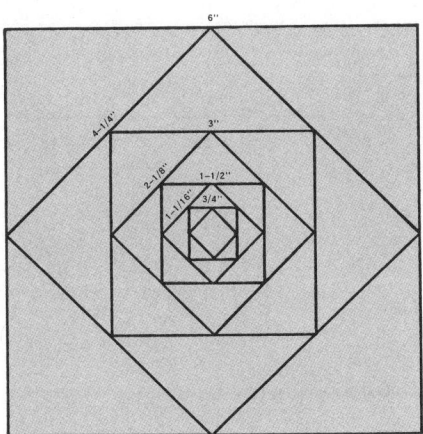

FIGURE 9.6-4 *Derivation of normal tile sizes of less than 36 in.*

of the tile. Some types have a glazed face, others do not. Glazed tile may have an opaque decorative glaze over a light-colored body or a transparent glaze that reveals the natural color of the body. Unglazed tile usually is a hard, dense unit of homogenous composition throughout, including the face (e.g., some mosaic tile and all quarry tile).

The tile body may have varying degrees of density and porosity, resulting in a greater or lesser perviousness to moisture. Tile with a body that has an absorption of less than 0.5% is *impervious*; tile with an absorption of 0.5% to 3% is *vitreous*. The low absorption of impervious and vitreous tiles makes them suitable for exterior installation subject to freezing and to interior installations subject to continuous immersion in water, as in swimming pools. *Semivitreous* tile has an absorption of 3% to 7%; *nonvitreous*, 7% to 18%. Semivitreous and nonvitreous tiles are not suitable for exposure to freezing or immersion in water.

Only two grades of tile are recognized by the TCA and the American National Standards Institute (ANSI): *standard grade* and *seconds*. Standard-grade tile is as perfect as it is commercially practicable to manufacture. Individual tiles are harmonious in color, although they may vary in shade, and they are free from spots and defects visible from a distance of more than 3 ft. Tile graded as seconds may have minor blemishes and defects not permissible in standard grades, but it is free from structural defects and cracks.

ANSI Standard 137.1, "Ceramic Tile," is the recognized industry standard for the manufacture, testing, and labeling of ceramic tile. According to this standard, tile should be shipped in sealed cartons with the grade of the contents indicated by grade seals. Quarry tile, however, is shipped in cartons with the grade printed on the carton in lieu of grade seals. A grade seal consists of a strip of paper, not less than 2 in. by 4 in. applied over the opening of the container in such a way that the container cannot be opened without breaking the seal.

Ceramic tile used where appearance is an important factor should be standard grade as defined in ANSI 137.1. Additional evidence of compliance with the governing standard, ANSI 137.1, is recommended in the form of a manufacturer's written certification, which is called a Master Grade Certificate.

The broad field of ceramic tile products is grouped by ANSI 137.1 into four major types: (1) glazed wall tile, (2) mosaic tile, (3) quarry tile; and (4) paving tile. Many other special-purpose tile products combining the properties of the basic type also are manufactured; when produced by reputable manufacturers, they have a proven record of success. Figure 9.6-5 summarizes the generally available types, sizes, and uses of ceramic tile. A variety of trim shapes are produced, such as caps, corners and bases, in matching colors, both for thin-set and conventional installation (Fig. 9.6-6).

GLAZED WALL TILE Glazed wall tile is generally 1/4 in. to 5/16 in. thick, 4 in. by 4 in. or 4-1/4 in. by 4-1/4 in., although larger units are also made. It is available in various sizes, in plain and scored surfaces, in bright, matte, and crystalline finishes, and in hand-decorated and sculptured designs (see Fig. 9.6-5). The governing standard defines glazed wall tile as one "not required or expected to withstand impact or be subject to freezing and thawing conditions." Many types of glazed tile are suitable also for flooring in areas subject to moderate foot traffic, as in most residential and some commercial areas. However, ANSI 137.1 does not cover this type of tile, and the manufacturer's specifications should be relied on to determine suitability for the intended exposure. Glazed tile suitable for flooring usually has a heavier glaze, with a matte or crystalline finish.

Most wall tile is made with an impervious glaze over a *nonvitreous* (absorbent) body and is intended for interior use. A small amount of this type of tile also is made with an *impervious* or *vitreous* body suitable for more demanding exposure to moisture or freezing. All wall tile is glazed, the majority with a bright (shiny) or *matte* (dull) finish. Some tile is made with other special glazes that are more appropriate for light residential traffic than bright or matte glazes.

Most wall tile is made with a cushioned-edged profile, but square-edged tile also is available. Soap dishes, paper holders, towel bar posts, and other accessories for toilet and bathroom uses are available in matching colors.

CERAMIC MOSAIC TILE Ceramic mosaic tile is available in two body types, *porcelain* and *natural clay*. It is generally about 1/4 in. thick, with a surface area of less than 6 sq. in. Typical sizes are 1 in. by 1 in., 1 in. by 2 in., and 2 in. by 2 in. This tile may be either glazed or unglazed; unglazed tile can have a wide range of natural earth colors that extend throughout the body of each tile. Additives can be mixed in with the other components for special effects: abrasives for *slip-resistant* tile and carbon black for *conductive* tile.

Ceramic mosaic tile usually is made in relatively small sizes, 2 in. by 2 in. and smaller (Figs. 9.6-7 and 9.6-8). Individual tiles generally are face- or back-mounted to form larger units, to facilitate handling, and to speed installation. Back-mounted tile has perforated paper, fiber, resin adhesive, or another suitable bonding material permanently attached to the back and edges of the tile so that a portion of each tile is exposed to the mortar or adhesive. Face-mounted tile has a layer of kraft paper applied to the face of the tile, usually with a water-soluble adhesive, so it can be easily removed before the joints are grouted.

Unglazed porcelain (frostproof) tile is made by firing refined ceramic materials, resulting in a vitreous or impervious tile that is dense, smooth, and highly stain- and wear-resistant. Its appearance is characterized by clear, luminous colors or granular blends. It is also available with an abrasive surface in a few sizes and colors.

Glazed porcelain (frostproof) tile has the same body as unglazed porcelain tile with one of the following glazes applied to the surface: clear (transparent) glaze,

FIGURE 9.6-5
Summary of Ceramic Tile Products

Tile and Description	Suitable Uses	Limitations	Nominal Sizes, in.*
GLAZED WALL TILE Nonvitreous body, matte or bright glaze	All interior walls in residential, commerical, and institutional buildings	Not suitable for floors or areas subject to freezing	5/16 in. thick: 4-1/4 in. × 4-1/4 in., 6 in. × 4-1/4 in., 6 in. × 6 in., scored, octagon, hexagon, and other special shapes
Nonvitreous body crystalline glaze	All interior walls, as well as light-duty residential floors and countertops	Not suitable for areas subject to freezing	
Vitreous body, matte or bright glaze	Interior walls, countertops, refrigerator and chemical-tank linings, and exterior areas subject to freezing	Not suitable for floors	
CERAMIC MOSAIC TILE Porcelain body, unglazed	All uses, interior and exterior, floors and walls; especially suitable for bathrooms, kitchens, and swimming pools	None except limitations imposed by installation method and grout	1/4 in. thick: 1 in. × 1 in., 2 in. × 1 in., 2 in. × 2 in.
Porcelain body, glazed	Interior and exterior walls in residential, commercial, and institutional buildings; decorative inserts with above tile	Not suitable for floors	
Natural clay body, unglazed	Swimming pool runways; floors of porches, entrances, and game rooms in the home; floors and walls of commercial, institutional, and industrial buildings	Exterior use requires special frostproof body	
Natural clay body, glazed	Walls in residential, commercial, and institutional buildings; decorative inserts with above tile	Exterior use requires special frostproof body	
Conductive	Hospital operating rooms and adjoining areas; where required to safely dissipate dangerous charges of static electricity	Requires special installation procedures according to NFPA No. 56A	
QUARRY TILE	Exterior and interior; from moderate to extra-heavy-duty floors in all types of construction	Recommended for interior use and exteriors not subject to freezing	1/2 in. thick: 3 in. × 3 in., 4 in. × 4 in., 6 in. × 3 in., 6 in. × 6 in., 8 in. × 4 in., and special shapes
PAVER TILE	Save as quarry tile	None	3/8 in. thick: 4 in. × 4 in. 1/2 in. thick: 4 in. × 4 in., 6 in. × 6 in., 8 in. × 4 in.

*Sizes shown are those considered typically by the governing industry standard, ANSI A137.1; many more sizes and shapes are produced by tile manufacturers. Special-purpose tiles that combine properties of several types also are produced.

ceramic glaze, textured glaze, metallic glaze, or special decorator glaze.

Unglazed natural clay (frostproof or nonfrostproof) tile is made from unwashed clays, which produce a dense, abrasion-resistant body having a rugged and slightly textured surface. The face is tough enough to minimize slipping, yet is easily cleaned.

This tile also is furnished with a slip-resistant face in 1-in. by 1-in. squares only, in limited colors.

Glazed natural clay (frostproof or nonfrostproof) tile has the same body as unglazed natural clay tile with one of the following glazes applied to the surface: clear, ceramic, metallic, or decorator glaze.

Conductive tile is an impervious ceramic tile with additives that minimize electrostatic build-up and explosion hazard in rooms where explosive gas mixtures are used, such as hospital rooms and laboratories. Both the tile and the installation should comply with the National Fire Protection Association Bulletin No. 56A.

FIGURE 9.6-6 *Standard wall tile trim shapes. Similar bullnose, cove, and base shapes are available for other types of tile (see ANSI A137.1 or manufacturer's literature).* (*Courtesy American Olean [Ceramic] Tile Co.*)

QUARRY TILE Quarry tile is an unglazed tile with a surface area of 9 sq. in. or more, is 1/2 in. thick or thicker, and is made by the extrusion process from natural clay or shale (Fig. 9.6-9). Typical tile sizes are 3 by 6, 6 by 6, 8 by 4, 9 by 6, and 9 by 9 in., in square, hexagonal, and other special shapes. Tile thicker than 3/4 in., made of the same materials and by the same process, is called *packing house tile.*

PAVER TILE As the name implies, paver tiles are used primarily for floors. They are similar in appearance to quarry tile, but they are more similar to mosaic tile in composition and physical properties,

though relatively thicker (1/2 in. typically) and larger (6 sq. in. or more). Typical tile sizes are 4 by 6, 6 by 6, and 8 by 4 in. (see Fig. 9.6-9).

■ TILE-SETTING PRODUCTS

Commercially available tile-setting products possess a variety of special properties, each intended to make them suitable for setting tile over certain backings or under given conditions. Figure 9.6-10 summarizes several common tile-setting products and the governing industry standards. Tile-setting products not covered by industry standards may also be used if they are produced by reputable manu-

facturers and have a proven record of success.

Installation with portland cement mortar, dry-set mortar, latex–portland cement mortar, epoxy mortar, and organic adhesives and their installation should conform to the recommendations of the appropriate industry standard as listed in Figure 9.6-10. Products not covered by these standards and their installation should conform with the recommendations of their manufacturers and good industry practice.

NEAT PORTLAND CEMENT (CEMENT-WATER PASTE) Neat portland cement can be used as a bond coat for tile only over a portland cement mortar setting bed, while the setting bed is still plastic. Most thin-set tile-setting materials, such as dry-set and latex–portland cement mortar, also can be used to bond ceramic tile to a mortar bed after it has cured for a minimum of 24 hours. While a mortar bed is still plastic, dry-set mortar can be used as a 1/16-in.-thick bond coat. When organic adhesives are used, the mortar setting bed must be fully cured and dry. To ensure a satisfactory installation in the proper plane, the mortar bed, whether left plastic or allowed to harden, should be installed by the tile-setting contractor.

Resinous tile-setting materials, such as epoxies and furans, offer greater bonding strength and chemical resistance than cementitious mortars. However, special tile-setting skills are required, and both labor and materials are likely to be appreciably costlier than with cementitious mortars.

PORTLAND CEMENT MORTAR Portland cement mortar is a mixture of portland cement, sand, water, and sometimes hydrated lime. The proportions vary depending on whether the mortar is intended for a leveling bed or a setting bed, and whether it is to be used on a vertical or a horizontal surface. The neat cement coat with which tile is applied is not considered a mortar, but rather a *bond coat.*

Portland cement mortar is the traditional material for installing ceramic tile. It is suitable for most surfaces and ordinary types of installation. The relatively thick bed, 3/4 to 1 in. on walls and 3/4 to 1-1/4 in. on floors, facilitates accurate slopes and true planes in the tile finish.

FIGURE 9.6-7 *The range of ceramic mosaic tile uses is virtually limitless, including decorative walls and floors indoors and outdoors. The large flat wall on this hotel is finished in ceramic mosaic tile.* (Courtesy American Olean [Ceramic] Tile Co.)

FIGURE 9.6-8 *The decorative pattern on this interior floor was made with ceramic mosaic tile.* (Courtesy American Olean [Ceramic] Tile Co.)

A portland cement mortar bed should be reinforced with metal lath or mesh when it is backed with a cleavage membrane. Suitable backings for mortar bed installations are brick or block masonry, monolithic concrete, concrete/glass fiber–reinforced backerboard, wood boards, plywood, foam insulation board, gypsum wallboard, and gypsum plaster; on walls, open stud framing or furring can be used, always in conjunction with metal lath.

DRY-SET MORTAR Dry-set mortar is a mixture of portland cement, sand, and resinous additives that impart water retentivity. Improved retentivity prevents rapid loss of water to the backing surface and improves bond.

Dry-set mortar is suitable for use over a variety of surfaces. It has excellent impact resistance, is nonflammable, is suitable for exterior work, and does not require presoaking of absorptive tile, as is required with most cementitious mortars. For these reasons it is the preferred mortar for absorptive glazed tile.

Dry-set mortar is available either as an unsanded mortar or as a factory-sanded mortar. Presanded mortars, to which only water needs to be added, are strongly recommended, because they offer greater control over the amount of sand in the mix. Dry-set mortar is not affected by prolonged contact with water and does not form a water barrier. It is limited in truing or leveling the work of other trades to about 1/4-in. thickness, but it can be used in a layer as thin as 1/16 in., when the mortar bed is still plastic.

Suitable backings include brick and concrete masonry, monolithic concrete, concrete/glass fiber–reinforced backerboard, cut-cell expanded polystyrene or rigid closed-cell urethane insulation board, gypsum wallboard, cured portland cement mortar, ceramic tile, and marble.

LATEX–PORTLAND CEMENT MORTAR Latex–portland cement mortar is a mixture of portland cement, sand, and a special latex additive that imparts greater flexibility to the mortar. The uses of latex–portland cement mortar are similar to those of dry-set mortar. It should be applied over a cured mortar bed in a coat 1/8 to 1/4 in. thick. Since latex formulations vary, the manufacturer's directions should be followed.

FIGURE 9.6-9 *Quarry and paver tile provide attractive low-maintenance floor finishes in heavy-duty traffic areas.* (Courtesy American Olean [Ceramic] Tile Co.)

EPOXY MORTAR Epoxy mortar is available in a two-part mortar consisting of an epoxy resin and a hardener or in a less expensive three-part system consisting of an epoxy resin, a hardener, and portland cement powder. After the parts are mixed, a chemical reaction occurs, and the mortar must be used within a limited time (*working time*). Epoxy mortar is suitable for use where chemical resistance or high bond strength is an important consideration. Acceptable backings include concrete, plywood, steel plate, and ceramic tile. This mortar is applied in a layer 1/16 to 1/8 in. thick. Working time, bond strength, water cleanability before cure, and chemical resistance vary with manufacturer. Epoxy mortar frequently is used in combination with epoxy grout, resulting in a highly durable water- and chemical-resistant installation.

EPOXY ADHESIVE Like epoxy mortar, epoxy adhesive is a two-part system consisting of an epoxy resin and a hardener. However, a different formulation makes this product somewhat less costly than epoxy mortar, as well as less chemical resistant. It is intended for thin-set application of tile on walls, floors, and counters where high bond strength and ease of applications are the overriding considerations. A typical use is in the application of a tile base over a metal refrigeration enclosure in a commercial kitchen. Although chemical resistance is lower than for epoxy mortars, it is substantially better than that of organic adhesives.

FURAN MORTAR Furan mortar is a two-part mortar system that consists of a furan resin and a hardener. It is suitable where chemical resistance is the most critical consideration and is, therefore, used primarily on floors in laboratories and industrial plants. Acceptable subfloors include concrete, steel plate, and ceramic tile. Furan mortar typically is used in combination with furan grout.

ORGANIC ADHESIVE (MASTIC) An organic adhesive is a prepared organic material that is ready to use with no further addition of liquid or powder and that hardens by evaporation. Organic adhesives are used in a layer 1/16 in. thick and are suitable for floors, walls, and counters

FIGURE 9.6-10
Summary of Tile-Setting Products

Type	Description	Industry Standards		Limitations
		Product	Installation	
Portland cement mortar	Field mix of portland cement, sand, and water for setting bed; mix portland cement and water for bond coat. Interior or exterior.	ASTM C 150 (Portland cement)	ANSI A108.1	Requires 1-1/4-in. setting bed, resulting in additional thickness and weight. Requires presoaking of most glazed wall tile.
Dry-set mortar	Prepared mix[a] of portland cement, sand, and certain additives that impart water retentivity and eliminate moist curing. High resistance to moisture, impact, and freezing. Interior and exterior.	ANSI A118.1[b]	ANSI A108.5[c]	Three days of moist curing before grouting recommended for heavy-duty floors.
Latex–portland cement mortar	Factory mix of portland cement and sand to which liquid latex resins are added in the field to impart greater bonding strength and a resilience. Interior or exterior.	ANSI A118.4	ANSI A108.5	Not suitable for wood, metal, hardboard, or particleboard backings.
Epoxy mortar	Two-part system of (1) epoxy resin and finely graded sand and (2) hardener. High bond strength and chemical resistance. Interior and exterior.	ANSI A118.3	ANSI A108.6	Requires special installation skills. Material and labor costs relatively high. Not suitable for walls.
Epoxy adhesive	Two-part system of (1) epoxy resin and (2) hardener. Good bond strength and chemical resistance. Interior or exterior. A three-part system containing portland cement powder is also available.	ANSI A118.3	ANSI A108.6	Requires special skills. Installation cost high, but lower than epoxy mortar.
Furan mortar	Two-part system of (1) furan resin and (2) hardener. Highest chemical resistance and installation cost. Interior or exterior.	ANSI A118.5	ANSI A108.8	Same as epoxy mortar.
Organic (mastic adhesive)	Prepared organic material, ready to use without addition of liquid or powder. Adequate bond strength and resilience for light duty on floors and most walls. Interior use only.	ANSI A136.1	ANSI A108.4	Not suitable for exterior use or interior use in continuous contact with water. Applied 1/16 in. thick; requires smooth backing.

[a] Requires addition of water in the field.
[b] For conductive dry-set mortar, ANSI A118.2.
[c] For conductive dry-set mortar, ANSI A108.7.

where the surfaces are sufficiently true and properly prepared in accordance with the adhesive manufacturer's directions. Suitable backings include monolithic concrete, gypsum wallboard, a portland cement mortar bed, gypsum plaster, mineral fiber–cement board, brick, ceramic tile, marble, and plywood.

Organic adhesives eliminate presoaking of tile and are suitable for most residential and commercial uses. They can be used for the walls of residential tub and shower areas, but they are not suitable for high-moisture areas, such as shower rooms or swimming pools, or for exterior installations. Their chief advantage is low cost and a flexible bond. Bond strength varies greatly among the numerous available brands. The solvents in some adhesives may be irritating to the skin, and some are flammable.

■ GROUTS

Grouting materials are available to meet the requirements of the different kinds of ceramic tile and types of exposures. Cementitious grouts consist of portland cement modified to provide specific qualities such as whiteness, uniformity, hardness, flexibility, and water retentivity. Resinous grouts (epoxies and furans) possess special properties such as high bond strength and chemical resistance but are harder and costlier to apply. Figure 9.6-11 summarizes commonly available tile grouts and the governing standards.

COMMERICAL CEMENT GROUT Commercial cement grout is a mixture of portland cement and other ingredients, resulting in a water-resistant, uniformly colored material. Commercial cement grout is the most commonly specified grout for tile walls subject to ordinary use. Soaking of nonvitreous wall tile is required to prevent loss of moisture to the absorptive tile body. The moisture in the grout is essential to promote good bond and to allow full grout strength to develop. To ensure an adequate supply of moisture for the hardening process (hydration), damp curing of the grout is necessary. Commercial cement grout usually is white, but colored grouts also are available.

SAND–PORTLAND CEMENT GROUT Sand–portland cement grout is an on-the-job mixture of normal portland cement, fine-graded sand, and sometimes hydrated lime. It is used with ceramic mosaic tile, quarry tile, and paver tile. Mixes should be 1 part portland cement to 1 part sand for joints up to 1/8 in. wide; 1 part to 2 parts for joints up to 1/2 in. wide; and 1 part to 3 parts for joints over 1/2 in. wide. Up to 1/5 part lime may be added for improved workability. Presoaking of nonvitreous wall tile is necessary. Damp-curing of the grouted surface is required for the same reasons as in commercial cement grout installations.

DRY-SET GROUT Dry-set grout is a prepared mixture of portland cement and sand, with additives to improve water retentivity. Dry-set grout has the same properties as dry-set mortar and is suitable for grouting walls subject to ordinary use. This grout eliminates soaking of tile, although dampening sometimes is desirable under very dry conditions.

LATEX GROUT Latex grout is a mixture of one of the three preceding products, consisting mainly of portland cement with

FIGURE 9.6-11
Summary of Tile-Grouting Products

Type	Description	Limitations
Commerical portland cement grout	Prepared mix* of portland cement and other ingredients resulting in a water-resistant, uniformly colored material; usually white, but available in colors. Commonly used for tile walls and floors subject to ordinary use. Suitable for interior and exterior.	Requires presoaking of most glazed wall tile and damp curing of grouted surface.
Sand–portland cement grout	Field mixture of portland cement, fine-graded sand (sometimes lime), and water. Most commonly used grout for ceramic mosaic tile. Suitable for floors and walls, interior and exterior.	Same as for commercial cement grout.
Dry-set grout	Prepared mix* of portland cement, sand, and additives that provide water retentivity and eliminate need for presoaking of glazed wall tile. Damp curing may develop greater strength in portland cement grouts. Generally used on walls, interior and exterior.	Dampening of the tile surface prior to grouting may be required under extremely dry conditions.
Latex–portland cement grout	Prepared mix* of any of the above formulations plus latex additives. It is less rigid and less permeable than regular cement grout. The addition of latex to portland cement grouts eliminates the necessity of damp curing for 72 hours. Can be used on walls and floors, interior and exterior.	Same as for dry-set grout.
Silicon rubber grout	Ready-to-use one-part caulking-type material that cures from moisture vapor in the air. Used mainly for glazed tile walls. Flexible, mildew resistant, stain resistant.	Bonds only to clean, dry surfaces. Must be used with proper adhesive to prevent staining.
Epoxy grout	Ready-to-use two-part system consisting of epoxy resin and hardener. Formulation for wall tile omits coarse fillers, which might scratch glaze; coarse silica fillers generally are included in grout intended for quarry tile and pavers. Highly resistant to chemical staining, hence used for industrial and kitchen floors, countertops.	Requires special skills for proper application. Material cost and labor higher than with other grout system.
Furan resin grout	Ready-to-use two-part system consisting of furan resin and hardener. Highest chemical resistance; used primarily in industrial installations with quarry tile and pavers.	Same as for epoxy grout.

Note: Epoxy grout should comply with ANSI A118.3; other grouts with ANSI A108.10 and A118.6.

*Requires addition of water in the field.

a latex additive to increase stain resistance and resilience of the joints. Latex grout is suitable for all installations subject to ordinary use. It is less rigid and less water-permeable than sand–portland cement grout, allowing for more movement and greater exposure to moisture. This grout is particularly suitable in tub and shower areas.

SILICONE RUBBER GROUT   Silicone rubber grout is an engineered elastomeric grout system for interior use that employs a single-component nonslumping silicone rubber. Upon curing, it is resistant to staining, moisture, mildew, cracking, crazing, and shrinking. It adheres tenaciously to ceramic tile, cures rapidly, and withstands exposure to moisture as well as subfreezing temperatures and hot, humid conditions. It should not be used on kitchen countertops or other food preparation surfaces, since status under FDA regulations has not been determined.

EPOXY GROUT   Like epoxy mortar, epoxy grout is a two-part system consisting of an epoxy resin and a hardener. It is made in several formulations, each intended for a specific ceramic tile. AAR-11 is the designation of an acid- and alkali-resistant, two-part formulation developed by the TCA and manufactured under license from that organization. It is highly stain resistant and impervious and is used typically in conjunction with epoxy mortar or epoxy adhesive. Its use requires special skills and is usually expensive.

For glazed wall tile and mosaic tile, epoxy grout is formulated without coarse fillers, which might scratch the glaze. It can be used on walls, floors, and counters subject to food staining.

For quarry tile and pavers, epoxy grout often contains a coarse silica filler. This grout is especially formulated for industrial and commercial installations where chemical resistance is of paramount importance. It has excellent bonding characteristics.

FURAN RESIN GROUT   Furan resin grout is a two-part grout consisting of a furan resin and a hardener. It is intended for quarry tile and pavers, mainly in industrial areas requiring a maximum of chemical resistance. It is generally used in conjunction with furan mortar.

■ RELATED MATERIALS AND PRODUCTS

Installation of ceramic tile involves a variety of related materials and products, described below and summarized in Figure 9.6-12.

FIGURE 9.6-12
Summary of Related Materials and Products

Material or Product	Industry Standard
Portland cement, gray or white, Type 1	ASTM C 150
Sand for mortar	ASTM C 144
Hydrated lime, Type S	ASTM C 206
Sealants	ASTM C 920
Metal reinforcement Welded wire fabric	ASTM A 82 and A 185
Expanded metal lath	ASTM C 847
Cleavage membrane Polyethylene film Asphalt roofing felt	ASTM D 4397 ASTM D 226

PORTLAND CEMENT   Normal (Type I) portland cement typically is used for mortar-setting beds, for leveling beds, for the neat-cement bond coat, and for field-mixed cement grout. When a white grout is intended, white portland cement should be used in the mortar bed as well as in the grout mix.

SAND   Sand should be clean, free from impurities, and well-graded. When fine sand is specified, it should pass through a 16-mesh screen.

HYDRATED LIME   Hydrated lime is used in a portland cement–sand mortar mix to make it more plastic for application on walls and sometimes on floors. Type S (special) lime should be specified, to ensure a dimensionally stable mortar bed.

SEALANTS   Sealants for use in expansion joints and control joints should be suitable for the exposure expected. Conditions that cause sealants to deteriorate include extreme temperature variations, direct sunlight, repeated stretching and compression, frequent wetting and drying, and immer-

sion in water. These sealants can be single-component or two-component types. The color of a sealant should be selected to match or blend with the tile or adjacent materials. When used on floors, a sealant must be hard enough to withstand occasional or repeated foot traffic. It should have a Shore A hardness index of not less than 25 for joints subject to occasional foot traffic; in high traffic areas the hardness index should be not less than 35.

METAL REINFORCEMENT   Isolated (un-bonded) mortar-setting beds should contain a suitable metal reinforcement. Reinforcement for a mortar bed typically consists of metal lath on walls and welded wire fabric on floors. Welded wire fabric is typically 2-in. by 2-in. mesh, WO.3×WO.3 (16 AWS gauge or 0.0625-in. diameter). Metal lath should be expanded lath weighing not less than 3.4 lb./sq. yd. or sheet lath weighing not less than 4.5 lb./sq. yd.

CLEAVAGE MEMBRANE   A cleavage membrane should be installed beneath every un-bonded ceramic tile setting bed. A cleavage membrane should be one of the following: (1) 15-lb. asphalt- (or 13-lb coal tar) saturated roofing felt; (2) reinforced waterproof duplex asphalt paper; (3) polyethylene film at least 4 mils thick; or (4) a membrane waterproofing.

9.6.1.2

Job Conditions

Materials should be delivered in the manufacturer's original sealed containers and stored under cover to prevent damage or contamination. Tile generally should be set and grouted when the temperature is at least 50 °F (10 °C) and rising. The mortar manufacturer's recommendations should be checked to determine if a higher temperature is required or a lower temperature is permissible. Backing surfaces and supporting construction should be inspected to determine their suitability before work is started.

9.6.1.3

Supporting Construction

A tile installation is only as good as the construction that supports it, that is, the underlying assembly of structural members and backing (subfloor or wall surface). If the supporting construction deflects, expands, or contracts, these movements will tend to be passed on to the tile finish.

This is particularly true in bonded installations. Except in small areas, a bonded tile finish cannot absorb these movements, and internal stresses are likely to result in cracks. It is essential, therefore, to provide supporting construction with a minimum amount of movement; where movement is anticipated, expansion or control joints should be provided in the supporting construction and extended into the tile.

The problem of movement is less critical with mortar bed installations isolated from the backing by a cleavage membrane. Although the membrane permits some independent movement in the supporting construction, the movement that can be tolerated still is limited, and dimensionally stable, rigid construction is desirable.

The *backing* is that part of the supporting construction to which the tile is bonded: it can be a subfloor, a wall finish, or a mortar leveling coat. The backing likewise must be rigid and dimensionally stable. In thin-set installations, it is important that the backing be sound, true to line, plumb, and sloped or level, as designed. Also, the backing should be free of foreign substances or loose materials because they might weaken the bond with the tile finish. When a mortar bed is used, trueness of the backing is less critical, but where slopes are intended, they should not be developed by varying the mortar bed thickness; slopes should be built into the backing.

■ FLOORS

Floors should be engineered so that the maximum deflection under full load, including live loads and the weight of the tile, does not exceed 1/360 of the span. For instance, on a span of 15 ft. (180 in.), the deflection should be not more than 180/360, or 1/2 in. The most common floor assemblies encountered in small buildings involve a concrete or wood subfloor; in large buildings, almost always a concrete subfloor. Additional information on suitable preparation for various floor surfaces and for specific tile-setting methods is given in the following subsections.

CONCRETE SUBFLOOR Tile can be installed on a concrete subfloor either by the thin-set or the portland cement mortar bed method. For thin-set bonding directly to a floor slab, the surface should be steel troweled and fine broomed, with no variations in the plane or slope exceeding 1/8 in. in 10 ft. For portland cement installations a variation of 1/4 in. in 10 ft. is acceptable. Concrete slabs should be thoroughly moist-cured before tile application; liquid curing compounds or other coatings that may reduce bond should not be used. Slabs should not be saturated with water at the time of installation. To prevent cracking, a slab should be properly reinforced, concrete quality should be controlled, and suitable joints should be provided in the slab.

WOOD SUBFLOOR Whether a portland cement mortar bed method or a thin-set method is intended, supporting joists should be spaced not more than 16 in. on centers. For thickbed installations, either a single-layer subfloor of 1-in. nominal boards or 5/8-in.-thick plywood is acceptable. For thin-set installations with most tile-setting products (except epoxy mortar) a double floor is required, consisting of a subfloor as described above and an underlayment. The underlayment should be Exterior Type CC-plugged plywood, 1/2 in. thick for residential use; 5/8 in. thick for more demanding installations, as in commercial buildings. When epoxy mortar is used in residential construction, the underlayment can be omitted if a 5/8-in.-thick plywood subfloor is used, and the longitudinal joints between panels are supported with solid wood blocking 2 × 4 or larger members.

A 1/4-in. space should be left between the underlayment and vertical surfaces, such as walls, and projections, such as pipes or posts. For organic adhesives, 1/16 in. should be left between underlayment ends and edges; for epoxy mortar, 1/4 in. should be left at ends and edges of panels, and these joints should be filled with epoxy. Adjacent underlayment edges should not be more than 1/32 in. above or below each other.

Concrete/glass fiber–reinforced backerboard may be used in lieu of a portland cement mortar bed over a 1/2-in.-thick plywood subfloor. This backerboard is designed for use with ceramic tile floors and

is available in various sizes and approximately 1/2 in. (7/16 in.) thick. It can be nailed or screwed to wood joists through 1/2-in.-thick plywood subflooring. Ceramic tile can be bonded to it with either dry-set or latex–portland cement mortars. Adhesives are not recommended for use with this product.

■ WALLS

Wall surfaces should be sound, plumb, and true, with square corners. For thin-set application the solid backing usually is installed by others (not the tile installer). The backing surface should not vary from a true plane by more than 1/8 in. in 8 ft. For portland cement mortar bed installations, either a solid backing or an open framing is suitable (Fig. 9.6-13). In either case, the tile contractor must start by attaching metal lath (over a membrane) to the framing or backing. Then the tile contractor applies a scratch coat and builds up a mortar bed 1/4 to 3/4 in. thick, to which the tile is bonded. A similar thickbed installation can be used over irregular masonry or concrete surfaces.

In most cases, tile is installed on walls by the thin-set method over gypsum wallboard, gypsum plaster, plywood, concrete/glass fiber–reinforced backing board, or other dimensionally stable solid backing. Gypsum plaster and plywood should not be used as backing in tub enclosures and shower stalls unless waterproof grout, such as silicone rubber, is used.

GYPSUM WALLBOARD Wallboard should be installed in maximum practical lengths to minimize the number of end joints. Joints should be taped, and both joints and fasteners should be treated with at least two coats of joint compound. In tub and shower areas, water-resistant backing board should be used. Its joints should be taped and filled with one coat (maximum 4 in. wide) of joint compound.

GYPSUM PLASTER Plaster walls that will receive tile should be steel troweled, with a brown coat mix (1 part gypsum to 3 parts sand) or a prepared finish coat plaster. When the tile finish will be flush with an adjacent plastered surface, the finish coat should be omitted and the tile set on the brown coat. Otherwise, a full-thickness,

FIGURE 9.6-13 *Typical supporting construction for ceramic tile. In areas subjected to wetting, a portland cement mortar bed should be used instead of gypsum plaster. Concrete/glass fiber–reinforced backerboard can also be used where gypsum board is indicated here.*

three-coat plaster surface should be used (see Section 9.3). The surface to receive tile should be troweled smooth with no trowel marks or ridges more than 1/32 in. high.

PLYWOOD   A plywood backing should be Exterior Type CC-plugged, 3/8 in. thick or thicker. Panels should be applied with the face grain perpendicular to the studs or furring. Solid 2 × 4 woodblocking should be used under panel edges between studs. Nails should be driven 6 in. on centers at panel perimeters and 12 in. on centers at intermediate supports. Panels should be spaced 1/8 in. apart, and there should be not more than a 1/32 in. variation between the faces of adjacent panels. Horizontal blocking between studs can be eliminated and panels can be applied parallel to studs when 1/2-in.-thick plywood is used. Before plywood is installed, its edges should be sealed with an exterior primer or aluminum paint.

CONCRETE/GLASS FIBER–REINFORCED BACKING BOARD   Concrete/glass fiber–reinforced backing board is a backerboard designed for use with ceramic tile in wet areas around tubs and showers and on floors. Available in various sizes and about 1/2-in. thick, this material can be nailed or screwed in place over wood or metal studs or over a 1/2-in.-thick plywood subfloor. Ceramic tile can be bonded to it with either dry-set or latex–portland cement mortar. It can be used in place of metal lath, portland cement scratch coat, and a mortar bed. It is not suitable as a base for tile installed with adhesives.

MASONRY AND CEMENT   Suitable masonry and cement backings include monolithic concrete, concrete/glass fiber–rein-

forced backing board, brick masonry, and concrete block masonry. For thin-set installations, the wall surface should be clean, sound, and free of paint, efflorescence, and cracked or chipped areas. When latex-portland cement or dry-set mortars are to be used, smooth concrete or brick surfaces should be roughened by sandblasting or bush-hammering to ensure good bond.

For mortar bed installations, the surface qualities of the backing are not as critical. If a leveling bed is to be used, the wall surface can be more irregular than if the tile is bonded directly to the wall, but it should be free of surface coatings and loose material, which might affect the bond. When a cleavage membrane is used, surface contamination or irregularity is not important, but the wall should be structurally sound and dimensionally stable.

9.6.1.4

Tile Layout and Setting

An area to be covered with tile should be checked for squareness, and midlines should be established in both directions. The tile pattern should be laid out so that a tile is either butted to the midline or bisected by it. The most desirable method is the one that produces larger than half-tiles around the perimeter of the area.

Tiles should be cut with a suitable tool rather than split. Wall tile usually is cut with a tile cutter; mosaics are cut with "nippers"; quarry tile may be cut with a water-cooled masonry saw or heavy-duty cutter. Tiles should be closely fitted to projections, and all ragged edges smoothed with a carborundum stone. Joint lines should be straight, plumb, level, and of even width. Accessories in the tilework

should be evenly spaced, properly aligned with tile joints, and set in the proper plane (Fig. 9.6-14). When the tile is not of the self-spacing type, joint widths should be approximately as shown in Figure 9.6-15. Special setting frames or grids are popular for setting this type of tile more quickly and accurately.

FIGURE 9.6-14 *Projections make tile layout more difficult and also more important. Tile should be laid so that it is symmetrical with respect to projections and corners and so that tiles are larger than half size.* *(Courtesy American Olean [Ceramic] Tile Co.)*

FIGURE 9.6-15
Joint Widths for Ceramic Tile

Tile Description	Joint Width, in.
Ceramic mosaic tile* 2 in. square or smaller	1/32 in. to 1/8 in.
Paver and special tile 2-3/16 in. to 4-1/4 in.	1/16 in. to 1/4 in.
6 in. square and larger	1/4 in. to 1/2 in.
Quarry tile 3 in. to 6 in.	1/8 in. to 3/8 in.
6 in. square and larger	1/4 in. to 1/2 in.
Glazed tile 3 in. square and larger	1/8 in. to 1/2 in.

*Ceramic mosaic tile is mounted, so joint widths are predetermined by manufacturer.

Tile is usually set into a previously spread layer of mortar, adhesive, or neat cement (Figs. 9.6-16 and 9.6-17). Sometimes it is more convenient to brush a coat of the tile-setting material on the back of the tile. Alignment of joints should be checked frequently for trueness and squareness (Fig. 9.6-18). After the tile is set, the mesh or paper of face-mounted tile should be soaked off and removed as recommended by the manufacturer. Mortar or adhesive also should be removed from the face of the tile with a recommended solvent. If strings or ropes were used to space tile, they should be removed as soon as the mortar or adhesive has set sufficiently to hold the tile firmly in place.

FIGURE 9.6-17 *Spreading a setting bed for wall tile.* (Courtesy American Olean [Ceramic] Tile Co.)

FIGURE 9.6-16 *Quarry tile is laid in a bond coat of dry-set mortar and tapped in place with a trowel handle.* (Courtesy Tile Council of America.)

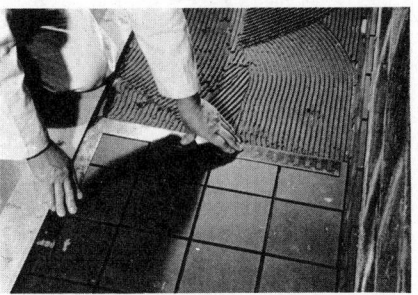

FIGURE 9.6-18 *Quarry tile is spaced to provide 1/4- to 1/2-in.-wide joints. A straightedge and a metal square are used periodically to check joint alignment and squareness.* (Courtesy Tile Council of America.)

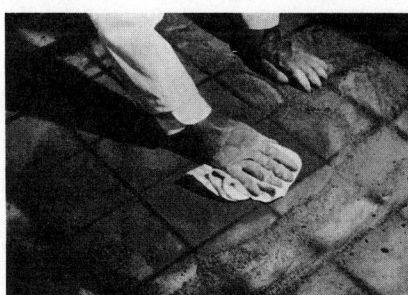

FIGURE 9.6-20 *Excess grout is wiped from the tile surface with burlap or rough-textured cloth.* (Courtesy Tile Council of America.)

9.6.1.5

Grouting and Curing

Grout should be applied to the face of tile by troweling diagonally across the joints (Fig. 9.6-19). Enough grout should be forced into each joint so that the joint is filled down to the mortar. Joints of squared-edged tile should be filled flush with the tile surface; joints of cushion-edged tile should be tooled to the depth of the cushion. The finished grout surface should be uniform in color, smooth, and free of pinholes or low spots. The tile surface should be washed and sponged thoroughly, then polished with a dry clean cloth or burlap (Fig. 9.6-20).

Glazed wall tile usually requires soaking before it is set with portland cement mortar. If glazed tile has been set with a

FIGURE 9.6-19 *Joints are filled with portland cement grout by troweling diagonally across the joints. Kneeling boards prevent damage to newly laid tile.* (Courtesy Tile Council of America.)

material other than portland cement mortar and was not presoaked before it was laid, it should be soaked after setting when a portland cement grout will be used. Soaking is not required for other types of grout.

Commercial cement and portland cement grout should be damp-cured for at least 72 hours. Dry-set grout need not be damp-cured, except when used (1) on floors, (2) on walls exposed to periodic wetting, or (3) in exterior locations. Damp curing can be accomplished by periodically applying a spray mist to the tile surface or by covering it with polyethylene film after initial and daily spraying.

9.6.1.6

Control and Expansion Joints

To ensure an attractive, crack-free installation, suitable joints should be provided in the backing and in the tile finish.

■ JOINTS IN BACKING

Control joints are provided in cementitious materials such as concrete and stucco to localize shrinkage cracks due to loss of moisture by hydration and evaporation. When control joints are provided, expansion joints usually are not necessary because shrinkage exceeds the likely thermal expansion.

Because shrinkage is minimal in brick masonry, thermal movement is critical, and expansion joints must be provided (see Chapter 4). Where differential settlement or movement of adjacent elements is expected, isolation joints should be used. Expansion joints frequently act also as isolation joints or as control joints. To conform to usage in the tile industry, such joints in a tile finish will be referred to here as *expansion joints*.

■ JOINTS IN TILE FINISH

Expansion joints are not required in areas less than 12 ft. wide or over old, well-cured concrete. However, expansion joints should always be provided in a tile finish (1) directly over expansion, isolation, construction, or control joints in the backing; (2) where the tile finish abuts projecting surfaces; and (3) at proper intervals in floor areas measuring over 24 ft. in either direction. Expansion joints in a tile finish should never be narrower than the underlying joint in the backing (Fig. 9.6-21).

FIGURE 9.6-21 *Expansion joints in tile surfaces.* (*Courtesy Tile Council of America, 1993* Handbook for Ceramic Tile Installation.)

Tile edges to which the sealant will bond should be clean and dry. Sanding or grinding of these edges is recommended to obtain maximum sealant bond. Some sealant manufacturers recommend priming of edges. When a primer is used, care should be taken to keep it off the tile face. Sealant depth should be approximately one-half the joint width. Compressible backup strips should be used in a joint to prevent bonding of the sealant to the joint filler and to ensure a proper sealant profile (Fig. 9.6-22).

FIGURE 9.6-22 *Proper sealant depth. Good bond at the sides and bond-free contact with backup strip are essential to long-term effective performance.* (*Courtesy Tile Council of America, 1993* Handbook for Ceramic Tile Installation.)

EXTERIOR LOCATIONS Exterior installations should be provided with expansion joints sized as follows: at least 3/8 in. wide for joints 12 ft. on centers and at least 1/2 in. wide for joints 16 ft. on centers. These minimum joint widths should be increased 1/16 in. for each 15 °F of temperature differential (range between summer high and winter low) over 100 °F. Decks exposed to the sky in the northern United States should have 3/4-in.-wide joints 12 ft. on centers.

INTERIOR LOCATIONS Thickbed tile installations in interior areas should have expansion joints in the tilework spaced 24 ft. to 36 ft. each way. Joints for interior quarry tile and paver tile should be the same width as the grout joints, but not less than 1/4 in. wide. Joints for ceramic

mosaic tile and glazed wall tile should be 1/4 in. wide but never less than 1/8 in. wide. Large thin-set installations may benefit from appropriately located expansion joints, but no generalizations can be drawn. Joints always are desirable at offsets and changes in backing construction.

9.6.1.7

Thresholds

When it is not practical to depress the subfloor for a conventional mortar bed installation and adjacent finishes are subsequently thinner, a transition is required. A marble or slate threshold located in a doorway or in a cased opening can serve this purpose. These thresholds are set with the same bonding agents as ceramic tile (Fig. 9.6-23).

FIGURE 9.6-23 *High end of threshold should be located under door.* (*Courtesy Tile Council of America, 1993* Handbook for Ceramic Tile Installation.)

A threshold may be of other materials when installed by different flooring contractors. Resilient-flooring contractors typically use rubber or vinyl thresholds; carpenters are likely to use a hardwood threshold in connection with wood strip flooring; carpeting might be butted to tile with a metal or vinyl edge strip.

To protect the edge of tile from damage, a hard material such as marble is preferred. This is not as critical in residential as in commercial installations.

9.6.1.8

Shower Receptors

Shower receptors can be precast terrazzo, plastic, field-installed ceramic tile, or one of several synthetic materials. Tile can be installed over wood or concrete subfloors; a mortar bed must be used over wood subfloors. In either case, a waterproof membrane or pan must underlie the tile finish and must be sloped 1/4-in./ft. to the drain. The sides of the membrane or pan should be turned up the wall 5 in. (Fig. 9.6-24).

FIGURE 9.6-24 *In a shower stall, inclusion of a membrane, crushed stone, and weepholes ensures drainage of water that might seep under the tile finish.* (Courtesy Tile Council of America, 1993 Handbook for Ceramic Tile Installation.)

Although only the mortar bed method is recommended for shower receptors, the slope should be developed in the backing rather than in the mortar bed. This ensures positive runoff of water that penetrates the tile finish. Crushed stone or tile around the shower drain and weepholes in the drain permit this occasional water to seep into the drainage system (see Fig. 9.6-24).

When a concrete subfloor is used, the subfloor itself should be sloped to the drain. With a wood subfloor, a separate portland cement mortar fill should be used to create the slope. The fill can vary from 1/2 in. to 1-1/4 in. in thickness. It should include a latex additive to improve its bond strength and make it possible to apply it in thin layers.

A waterproof membrane or pan should be installed over the sloped fill, followed by the tile finish set in a mortar bed. Any commercial grout is suitable for grouting the tile joints. Figure 9.6-25 shows proper detailing of tiled tub-shower areas; Figure 9.5-48 shows recommended tile coverage of gypsum board in bathing areas.

FIGURE 9.6-25 *Caulking the space between a tile finish and a tub apron with a suitable sealant ensures a watertight joint.* (Courtesy Tile Council of America, 1993 Handbook for Ceramic Tile Installation.)

9.6.1.9

Tile over Other Finishes

Ceramic tile can be installed readily over existing tile and over other existing floor and wall finishes, such as resilient flooring, hardwood flooring, painted concrete, gypsum plaster, and gypsum wallboard. Ideally, existing finishes or coatings should be completely removed so that the tile can be placed on the original backing. However, if this is impractical, a new backing surface can be installed for thin-set application, or the tile can be installed by the conventional mortar bed method. If the existing finish is suitable for thin-set application of tile, it is imperative that the mortar or adhesive manufacturer's recommendations be followed. Figure 9.6-26 shows several tile-over-tile installation possibilities.

Thickbed installations present no special problems over existing finishes. But for thin-set installations, it is important to ensure that the backing surface is free of contaminants, such as dirt, dust, efflorescence, oil, and grease. The maximum variation in a backing surface should not exceed the dimensions indicated in Section 9.6.1.3. Glossy or scaly painted surfaces should be sandblasted or treated with a chemical paint remover. The adhesive or mortar manufacturer should be consulted to determine whether a sealer or primer is needed. When a structural

floor assembly is capable of carrying the weight and the old floor surface is questionable, the more reliable isolated thickbed installation method should be used.

9.6.1.10

Countertops

Mosaic, quarry, or glazed wall tile can be used as a durable finish for countertops. Tile is installed typically over a wood, plywood, or cementitious board backing either by the mortar bed (thickbed) or the thin-set method (Fig. 9.6-27).

The backing for a mortar bed installation should be 1 × 6 boards or 3/4-in. Exterior plywood. Boards should be spaced 1/4 in. apart (Fig. 9.6-27a). Plywood should be scored with saw cuts 6 to 8 in. on centers to prevent warping. For thin-set installation, only 3/4-in. Exterior plywood should be used. A 1/4-in. gap should be left between plywood panels; this gap should be closed from below with a batten strip (Fig. 9.6-27b).

An epoxy grout or a special stain-resistant grout recommended for the purpose by the manufacturer should be used on countertops installed with the thin-set method.

9.6.2

INSTALLATION ON FLOORS

Selection of the most suitable installation method depends on the performance level that is desired, the type of supporting construction, and whether the installation is interior or exterior.

9.6.2.1

Installation Performance Levels

The type of traffic (wheeled or foot), impact loads, and contaminants (water, grease, alkalies, chemicals) to which a floor will be subjected are prime considerations in the selection of a suitable performance level (Fig. 9.6-28). Once the performance level has been determined, an appropriate installation method can be selected (Fig. 9.6-29).

In addition to the installation method, consideration should be given also to

FIGURE 9.6-26 *New tile may be bonded directly to existing tile, or a new backing, such as gypsum board, may be imposed between the new and the existing tile.* (*Courtesy Tile Council of America, 1993* Handbook for Ceramic Tile Installation.)

FIGURE 9.6-27 *Ceramic tile countertops may be installed by (a) the cement mortar, or (b and c) the thin-set method.* (*Courtesy Tile Council of America, 1993* Handbook for Ceramic Tile Installation.)

(1) the tile selected with respect to the wear properties of the surface; (2) fire-resistance properties of the tile finish, including the entire wall or floor assembly; (3) acoustical properties (especially impact noise rating) of the floor assembly; and (4) slip resistance of the floor tile.

Unglazed tile (mosaic, quarry, and paver tile) generally is used for floor surfacing and will give satisfactory wear for all the performance levels listed. If, however, a decorative glazed tile or an especially soft-body unglazed tile is used, it should be approved for use on floors by the tile manufacturer. In determining the suitability of glazed tile for an intended use, color or pattern, surface texture, and glaze hardness should be considered. All glazed tile will show traffic wear in time.

9.6.2.2

Interior Concrete

An interior concrete slab on grade is not subjected to the drastic temperature and humidity fluctuations encountered outdoors. In addition, if such a slab is made up of a properly engineered concrete mix and has suitable reinforcement, adequate finishing, proper curing, a well-drained

FIGURE 9.6-28
Installation Performance Level Ratings

Area Description	Type of Use or Traffic	Performance Level
Office space, commercial reception areas, public spaces in restaurants and stores, corridors, shopping malls.	(a) Light high-heel traffic (b) Heavy high-heel traffic	(a) Moderate (b) Heavy
Kitchens	(a) Residential (b) Commercial (c) Institutional	(a) Light or moderate (b) Heavy (c) Extra heavy
Toilets, bathrooms	(a) Residential (b) Commercial (c) Institutional	(a) Light or moderate (b) Moderate or heavy (c) Heavy
Hospitals	(a) General (b) Kitchens (c) Operating rooms	(a) Heavy (b) Extra heavy (c) Heavy
Exterior decks	(a) Roof decks, large on-grade patios (b) Walkways, small patios	(a) Extra heavy (b) Heavy or extra heavy

Note: Performance-level ratings are defined as follows: Light: Normal residential foot traffic; occasional 300-lb. loads on soft rubber wheels. Moderate: Better residential, normal commercial, and light institutional use; 300-lb. loads on rubber wheels and occasional 100-lb. loads on steel wheels. Heavy: Heavy high-heel traffic; 200-lb. loads on steel wheels, 300-lb. loads on rubber wheels. Extra heavy: High-impact service, as in institutional kitchens and industrial work areas; 300-lb. loads on steel wheels.

FIGURE 9.6-29
Suitable Methods for Various Performance Levels

Performance Level	Installation				
	Subfloor	Tile	Mortar or Adhesive	Grout	Remarks
Light	Wood	Ceramic mosaic or quarry	Organic	Latex–portland cement	Low cost; OK for home bath, entry
	Wood	Ceramic mosaic	Epoxy	Sand–portland cement[a]	Costlier; OK for kitchen, dining room
	Existing tile finish	Ceramic mosaic or quarry	Epoxy	Mastic or latex–portland cement	Residential renovation
Moderate	Wood	Ceramic mosaic	Portland cement mortar bed	Sand–portland cement[a]	OK for depressed wood subfloor
	Wood	Ceramic mosaic or quarry	Epoxy	Epoxy	Very costly; very durable even over wood subfloor
	Concrete	Ceramic mosaic	Dry-set	Sand–portland cement[a]	Moderate cost; smooth surface
	Concrete	Ceramic mosaic	Dry-set	Latex–portland cement	Moderate cost; eliminates damp curing of grout
Heavy[b]	Concrete	Ceramic mosaic or quarry	Dry-set[c]	Epoxy	Costlier, good chemical resistance
	Concrete	Ceramic mosaic or quarry	portland cement mortar bed	Sand–portland cement[a]	Moderate cost; bed ensures smooth surface
	Concrete	Ceramic mosaic or quarry	Dry-set	Sand–portland cement[a]	Low cost; good all-purpose floor
	Concrete	Ceramic mosaic or quarry	Epoxy	Epoxy	Very costly; most durable and easy to maintain

[a] Damp cured for 3 days after grouting.
[b] Damp cured for 3 days before grouting.
[c] All installations suitable for extra-heavy performance when packing house tile is substituted for mosaic tile.

granular base, and proper jointing, the likelihood of excessive movement due to settlement and temperature changes is slight. Therefore, such slabs will accept the complete range of tile installation methods, including a bonded mortar bed, an isolated mortar bed, or direct application with a thin-set mortar or adhesive.

A supported slab, on the other hand, will often deflect enough under loads to cause tile set using other than an isolated mortar bed to crack or even delaminate. Supported slabs also may encounter enough thermal movement to damage tile that was set using a bonded system.

The specific procedure for installing tile depends mostly on the tile-setting material selected.

▪ PORTLAND CEMENT MORTAR

Neat portland cement can be used as the bond coat only when the concrete surface is still plastic and suitably rough to bond with the neat cement. This condition is difficult to achieve, however, on a structural slab surface. In addition, an isolated, reinforced mortar setting bed with a cleavage membrane should always be used over a structural (supported) slab, so that stresses induced in the slab are not transmitted to the tile finish.

In a proper concrete slab on grade as described above, the likelihood of excessive movement is minimized, and there is no need to isolate the tile finish. Under these conditions, tile can be applied by the thin-set method directly to the slab or over a bonded setting bed. However, even on

slabs on grade, when a mortar bed is used, the bed is often reinforced with wire mesh, and sometimes a cleavage membrane is also installed.

Although the precautions required to ensure a stable concrete backing appear extensive, they are inherent in good concrete work. It is generally cheaper to comply with good practice than to cope with excessive slab movement, regardless of installation method.

A portland cement mortar bed should not exceed 1-1/4 in. in thickness. A thin layer of neat cement paste, 1/32 in. to 1/16 in. thick, should be troweled over the setting bed or on the back of the tile. Another way is to dust a thin layer of dry portland cement, 1/32 in. to 1/16 in. thick, over the setting bed; this bond coat should be worked lightly with a trowel or a brush until damp immediately before the tile is set.

Glazed semivitreous tile should be dampened by placing it on a wet cloth or in a shallow pan. However, no free moisture should remain on the backs of the tile when set. Vitreous and impervious tile, such as mosaic tile and quarry tile, do not require soaking.

Tile should be set in position and tamped firmly into the mortar, to obtain maximum contact and a strong bond (Fig. 9.6-30a). Tile surfaces should be brought to a true level at proper elevation. Tamping and leveling should be completed within 1 hour after the tiles have been placed.

When face-mounted ceramic mosaic tile is installed, the paper should be wetted within 1 hour of setting, with no more water than necessary to remove the paper

and glue (Fig. 9.6-30b). Tiles that are out of line or not level should be adjusted before the mortar assumes initial set.

Sand–portland cement grout typically is used with quarry tile, pavers, and ceramic mosaic tile. For quarry tile and pavers, the grout generally consists of 1 part portland cement, 2 parts sand, and 1/5 part hydrated lime by volume. For ceramic mosaic tile, a mix of 1 part portland cement and 1 part finely screened sand is used. First, filling and pointing are done with neat cement paste (Fig. 9.6-30c). Joints in paver, quarry, and cushion-edged glazed tile should be tooled (compressed) with a pointing tool, to firmly bond a layer of pointing mortar to the grouting mortar and the tile.

Excess grout should be removed from the surface with a clean cloth or burlap. Before the grout sets, voids and gaps should be filled. When portland cement-based grouts are used, the entire installation should be damp-cured for 72 hours. The grouting materials manufacturer's instructions should be followed.

The tile surface should be thoroughly cleaned while the grout is still plastic. Muriatic acid may be used to clean glazed tile; sulfamic acid can be used to clean unglazed tile 10 days after setting. Joints should be soaked before acid is applied and flushed thoroughly with water after cleaning. Metal and enameled surfaces should be protected with grease or a strippable coating.

ISOLATED SETTING BED To create an isolated mortar setting bed, a cleavage

FIGURE 9.6-30 *Mounted ceramic mosaic tile is (a) laid in a neat cement bond coat; (b) the tile is tamped into place, and the paper is sponged off; (c) grout is troweled into joints, and excess is removed with a clean cloth.* *(Courtesy American Olean [Ceramic] Tile Co.)*

membrane should be spread over the slab and should be folded at edges and ends to form a locked joint. Metal reinforcement should be lapped at least one full mesh and supported approximately in the middle of the setting bed. The mesh should be cut to fit accurately and kept free of vertical surfaces. Figure 9.6-31 illustrates typical isolated bed construction.

FIGURE 9.6-31 *Metal reinforcement should be installed in every isolated thick-bed installation.* (*Courtesy Tile Council of America, 1993* Handbook for Ceramic Tile Installation.)

The setting bed mix should consist of 1 part portland cement, 5 parts dry sand (or 6 parts damp sand), and 1/10 part hydrated lime by volume. Only enough water should be added to the mix to obtain the desired consistency. The setting bed should be applied 3/4 in. to 1-1/4 in. thick, tamped firmly, and screeded to a level surface.

BONDED SETTING BED The bonded bed method should be used infrequently because it does not possess the chief advantage of unbonded installations, namely, separation from stresses in the concrete subfloor. Its chief advantage over thin-set methods is the ability to produce a smooth finish over a rougher subfloor and reliance on familiar, readily available materials such as portland cement and sand. Bonded setting bed construction is similar to isolated construction except that the cleavage membrane and reinforcement can be omitted (Fig. 9.6-32).

The mortar bed can be bonded with a synthetic bonding compound or neat portland cement. For the latter, a fairly rough, wood-floated surface is required. The slab

FIGURE 9.6-32 *Reinforcement can be omitted when the mortar bed is bonded to the backing.* (*Courtesy Tile Council of America, 1993* Handbook for Ceramic Tile Installation.)

should be thoroughly saturated with water before the neat cement and mortar are spread. The mortar mix, grout mix, and general installation procedures for a bonded setting bed are identical to those described above for an isolated setting bed.

■ LATEX–PORTLAND CEMENT AND DRY-SET MORTARS

These mortars generally are used to bond the tile directly to a concrete slab (Fig. 9.6-33) that has a variation of not more than 1/4 in. in 10 ft. On slabs with more variation, the tile should be bonded to a cured mortar leveling bed. In addition, the slab should be designed to ensure a minimum of movement. It should be well cured and free of cracks, waxy or oily films, and curing compounds. The slab surface should

FIGURE 9.6-33 *Dry-set and latex–portland cement mortar can be used for thin-set placement of tile on a concrete slab.* (*Courtesy Tile Council of America, 1993* Handbook for Ceramic Tile Installation.)

be steel-troweled and fine-broomed. Dry-set or latex–portland cement mortar should be mixed as recommended by the manufacturer. Mortar should be troweled in a layer about 1/4 in. thick for mosaic tile and 3/8 to 1/4 in. thick for quarry and paver tile. The tile should be tamped into the mortar to ensure a level surface and intimate contact.

Paper should be removed from face-mounted tile within 1 hour of setting with a minimum amount of water. Grouting should be delayed for at least 24 hours or until the tile is firmly set. Suitable grouts include dry-set grout, latex grout, sand–portland cement grout, and commercial cement grout. Dry-set and portland cement grouts should be damp-cured for 72 hours.

■ ORGANIC ADHESIVES

Since these products are recommended for use on floors only in light-traffic areas, they are used infrequently over a concrete slab, except in small areas and over relatively smooth plane surfaces. For organic adhesives, which are applied only 1/16 in. thick, a slab surface should be steel-troweled and fine-broomed, and the surface should not vary by more than 1/16 in. in 3 ft.

Tile installation with organic adhesives requires the same general precautions as with other thin-setting products (see Fig. 9.6-33). The tile should be tamped into place to ensure maximum contact with adhesive. Any of the available grouts can be used, but latex–portland cement grout and epoxy grout are most commonly used with organic adhesives.

■ EPOXY MORTARS AND ADHESIVES

The performance and working properties of epoxy adhesives are similar to those of epoxy mortars. Most of the following discussion concerning epoxy mortars applies to adhesives as well.

Epoxy mortar is likely to be used in conjunction with an epoxy grout to provide chemical resistance. These properties are most desirable in laboratories, industrial plants, and food preparation areas. The high bond strength of the mortar, combined with the general stain resistance of the grout, make this a preferred system for many hard-service areas (such as entries, public toilet rooms, lobbies, restaurants), particularly when combined with quarry tile. Since the mortar is installed in a layer 1/4 to 3/8 in. thick, it can absorb minor variation in the subfloor surface of not more than 1/4 in. total in 10 ft. A wood-floated concrete surface is suitable (Fig. 9.6-34).

FIGURE 9.6-34 *Epoxy mortar and grout possess high bond strength and chemical resistance. Epoxy adhesives can be substituted in less demanding situations.* (*Courtesy Tile Council of America, 1993* Handbook for Ceramic Tile Installation.)

Epoxy mortar should be the same color as the grout. The tile should be tamped into the setting bed, with care taken not to damage the face of the tile or smear mortar on it. Mortar should be immediately removed from the face, as suggested by the manufacturer.

Grouting should be delayed at least 16 hours. The grout should be forced into the joints with a squeegee or rubber trowel to completely fill them. The installation should be shaded from direct sun and kept at a relatively even temperature (more than 60°F) for at least 8 hours after grouting, to facilitate curing. Traffic should be kept off the floor for at least 40 hours. Light traffic can be permitted after this period; heavy traffic, after 7 days.

Epoxy grout also can be used with other tile-setting mortars. Where chemical resistance at the tile surface is desirable but high bond strength with the backing is not essential, epoxy grout can be used with latex or dry-set mortars. These cementitious mortars generally produce adequate bond and are less expensive than epoxy mortar.

9.6.2.3

Exterior Concrete

Except for installations over built-up roofing or waterproofing membranes, exterior tile installations are likely to be bonded directly to a concrete slab (Fig. 9.6-35). Over a suitably finished concrete surface, dry-set mortar can be used in a coat 1/4 to 3/8 in. thick to bond the tile directly to the slab.

FIGURE 9.6-35 *Dry-set mortar can be used for thin-set application of tile in both interior and exterior locations.* (*Courtesy Tile Council of America, 1993* Handbook for Ceramic Tile Installation.)

When a structural slab is not suitable for direct application of tile, a mortar setting bed will ensure a suitable substrate (Fig. 9.6-36). Since the mortar bed is installed by the tile contractor, followup with a neat portland cement bond coat can be made by the contractor while the bed is still plastic. For exterior tilework, it is

FIGURE 9.6-36 *Expansion joint in an exterior mortar bed tile installation.*

important to consult latest industry standards and tile manufacturer's recommendations, particularly in areas where freezing occurs.

Expansion joints extending through tile and mortar bed should be provided at intervals of 12 to 16 ft. each way (see Fig. 9.6-36). Expansion joints also should be provided where the tile finish abuts restraining vertical surfaces and where expansion, isolation, or control joints exist in the slab. Both the structural slab and the tile surface should be sloped to drain. See Section 9.6.1.6 for additional discussion of expansion joints.

9.6.2.4

Wood Subfloors

Wood subfloors usually consist of wood joists and plywood or board sheathing. Unless stiffness is carefully considered in the design, the joists are likely to deflect and vibrate more than massive structural systems such as concrete or steel. The boards and the plywood, being hygroscopic, also are likely to expand and contract with humidity variations.

It is important, therefore, for the tile installation to be either completely isolated from the subfloor to permit differential movement or bonded to the subfloor with a tile-setting material that can resist the movement without breaking the bond. When the installation depends on a very strong bond and no setting bed is used, the grout also must be strong enough to resist the internal stresses transmitted to the tile joints.

A typical isolated conventional installation consists of a reinforced mortar setting bed and a latex, dry-set, or neat cement bond coat over a cleavage membrane. A typical bonded thin-set installation utilizes epoxy mortar or epoxy adhesive. In relatively small areas, such as residential bathrooms, tile also can be bonded to a plywood subfloor with organic adhesives. Although this material is not as strong as epoxy, it is adequate to resist the smaller stresses developed in a small, bathroom-sized area.

■ PORTLAND CEMENT MORTAR

The actual procedure of installing a mortar setting bed is the same as for concrete subfloors. The grout can be any of the several commercially available formulations, as well as sand–portland cement grout. The complete tile installation, including the mortar bed, should be approximately 1 in. to 1-1/2 in. thick (Fig. 9.6-37).

FIGURE 9.6-37 *A reinforced isolated mortar bed and tile over a single-layer wood subfloor.* (*Courtesy Tile Council of America, 1993* Handbook for Ceramic Tile Installation.)

Since many interior floor finishes are much thinner than 1-1/2 in., a transition problem develops where a tile finish abuts a thinner finish. To minimize this problem, it is possible to depress the subfloor between the joists, thus gaining 5/8 to 3/4 in. in height (Fig. 9.6-38). The subfloor, usually plywood, is supported on cleats level with the tops of the joists. Depressing the subfloor below the tops of the joists and leveling the joists are not recommended. This method of tile installation is somewhat more durable in high-moisture areas, such as bathrooms, than the thinset method. It is also more resistant to impact due to the mass of the mortar bed underlying the file finish.

FIGURE 9.6-38 *A portland cement mortar bed and tile installation over a depressed subfloor.*

ORGANIC ADHESIVE

The organic adhesive method is not suitable for tile installation on the floors of commercial or institutional buildings. It is, however, suitable for all areas in a home, except those subject to excessive moisture, such as in a tiled sunken tub or in a swimming pool. It will not withstand heavy traffic or impact loads. Double-floor construction generally is recommended to achieve a stiffer subfloor (Fig. 9.6-39).

FIGURE 9.6-39 *A thin-set tile application over a double wood subfloor.* (*Courtesy Tile Council of America, 1993* Handbook for Ceramic Tile Installation.)

The subfloor can be 1-in. nominal boards or 5/8-in. plywood. The underlayment should be 3/8-in. or thicker plywood. For bathroom floors and other areas subject to occasional moisture, Type I adhesive should be used, which is capable of withstanding more prolonged exposure to moisture than Type II adhesive. The adhesive should be spread with a notched trowel recommended by the manufacturer and the tile firmly pressed into the adhesive. Grout can be latex–portland cement or epoxy.

EPOXY MORTAR AND ADHESIVE

Epoxy mortar and adhesive provide a very strong adhesive suitable for bonded thin-set installations over a double wood floor consisting of subfloor and underlayment (Fig. 9.6-40). The subfloor should be 5/8-in. plywood or 1-in. nominal boards; the underlayment should be 1/2-in. or thicker plywood (5/8 in. in nonresidential uses). The underlayment should be installed with 1/4-in. gaps between panels, which should be completely filled with epoxy mortar. In residential uses, the underlayment may be omitted if the

subfloor is 5/8-in. or thicker plywood installed perpendicular to the joists and with the edges blocked.

FIGURE 9.6-40 *An epoxy mortar–set tile over a double wood subfloor.* (*Courtesy Tile Council of America, 1993* Handbook for Ceramic Tile Installation.)

The mortar should be spread with a trowel recommended by the manufacturer. The tile should be tamped into the mortar to ensure a smooth surface and proper bond.

Grouting should be delayed for 16 hours. The installation should be shaded from direct sunlight and kept at a stable temperature of more than 60 °F for 8 hours or more after grouting. Grout and mortar should be cleaned off the face of the tile as soon as possible and before it hardens.

9.6.3

INSTALLATION ON WALLS

The ceramic products typically used on walls are glazed wall tile and ceramic mosaic tile. Quarry tile and paver tile are seldom used as wall finishes. Where the wall surface is sufficiently sound, true, clean, and free of coatings, tile can be bonded to the wall by the thin-set method. The surface should not vary more than 1/8 in. in 8 ft. if thin-set materials are to be used and not more than 1/4 in. in 8 ft. for portland cement mortar beds.

Wall surfaces that are irregular, chipped, cracked, or coated with materials that would impede bond require the use of a mortar bed. A mortar setting bed must be installed generally over a base coat of mortar (scratch coat) applied to metal lath over a cleavage membrane.

A *scratch coat* consists of 1 part portland cement, 5 parts damp sand, and 1/5 part hydrated lime. It is normally applied by a plastering contractor but can also be installed by the tile contractor. A *float coat* generally is applied by the tile contractor to ensure a plastic surface for setting the tile with neat cement.

A solid backing is not essential when tile is to be installed by the mortar bed method. Over wood or metal studs, the backing can be omitted, and the membrane and metal lath can be attached directly to the studs. A lath-reinforced mortar bed generally is rigid enough by itself to support tile without excessive deflection between the studs; to ensure adequate vertical rigidity, the studs should be at least 2-1/2 in. deep at 16 in. on centers or 3-1/2 in. deep when spaced 24 in. on centers.

The materials, installation methods, and suitable backings for tile finishes vary depending on whether the installation is interior or exterior. Exterior installations are much more critical because of exposure to moisture and temperature variations.

9.6.3.1

Interior Surfaces

Suitable backings for tile finishes include masonry and concrete walls, as well as plywood, gypsum board, and plaster over furring or stud framing. In tub and shower areas, the thin-set method is acceptable, but it is essential that the backing be inherently water resistant or properly prepared to render it so. Dry-set mortar can be used as a bond coat for thin-set application in sunken tubs and pools over a portland cement mortar bed.

■ PORTLAND CEMENT MORTAR

Traditional installations on walls typically require a preliminary *scratch coat* followed by a *mortar bed* (*two-coat method*). The scratch coat can be omitted when the wall provides a solid backing for the application of metal lath, mortar, and tile (*one-coat method*). The tile actually is set in a separate bond coat, but it usually is not counted in describing the installation.

The *one-coat method* can be used over existing wall finishes that are not suitable for thin-set application because of surface conditions (Fig. 9.6-41). It is also a preferred method of tiling tub and shower

FIGURE 9.6-41 *The one-coat method can be used over a solid backing that is too uneven for thin-set tile.* (*Courtesy Tile Council of America, 1993* Handbook for Ceramic Tile Installation.*)*

areas in new work. The backing can be masonry, concrete, plaster, wallboard, plywood, or any other solid wall surface. The single bed coat is 3/8 to 3/4 in. thick and consists of 1 part portland cement, 1/2 part hydrated lime, and 5 parts damp sand.

The *two-coat method*, involving both a *scratch coat* and a *bed coat*, is used most frequently over open framing of furring or studs (Fig. 9.6-42). On metal studs, the overall thickness of scratch coat and mortar bed should not exceed 1 in. On wood studs the overall mortar thickness should not be more than 1-1/2 in. The mortar mix is the same as for one-coat installations. The scratch coat should be allowed to cure

FIGURE 9.6-42 *Two-coat method over open framing or solid backing.* (*Courtesy Tile Council of America, 1993* Handbook for Ceramic Tile Installation.*)*

for at least 24 hours and should be dampened thoroughly before the mortar bed is applied. The mortar bed should be applied in areas no greater than may be covered with tile while the bed is still plastic. Figure 9.6-43 illustrates the steps in applying tile over a portland cement setting bed.

A neat cement *bond coat* is used for setting tile with both the one- and the two-coat method. The bond coat should be troweled 1/32 to 1/16 in. thick over the float coat, and the tile should be tamped into it. The portland cement for the bond coat should be ordinary gray or white cement, depending on whether the grout is to be gray or white. The grout can be neat portland cement or portland cement mixed with an equal part sand, latex, or epoxy. Where commercial portland cement or portland cement grout is used, the finish should be cured for at least 72 hours.

■ DRY-SET AND LATEX–PORTLAND CEMENT MORTARS

Dry-set and latex–portland cement are thin-set mortars suitable for sound, clean, and true surface such as gypsum plaster, gypsum board, monolithic concrete, or concrete and brick masonry (Fig. 9.6-44). They also can be used over a bonded mortar bed used mainly for leveling an irregular solid surface (Fig. 9.6-45) or when a full-thickness mortar backing on open framing is already in place (Fig. 9.6-46). When such a mortar backing is used, the backing should be allowed to cure for several days until firm.

The backing should be dampened just before mortar is spread over it. Mortar should be troweled on in appropriate thickness (3/32 to 1/4 in. maximum) in an area that can be readily covered with tile before the mortar starts skinning over (Fig. 9.6-47a). The tile generally should be tamped into place to obtain at least 80% coverage; in tubs, showers, and other high-moisture areas, 100% coverage should be obtained (Fig. 9.6-47b). Any grout can be used, but consideration should be given to the type of exposure (Fig. 9.6-47c).

■ ORGANIC ADHESIVES

The use of organic (mastic) adhesive is limited to light-duty interior areas and over smooth surfaces such as plaster,

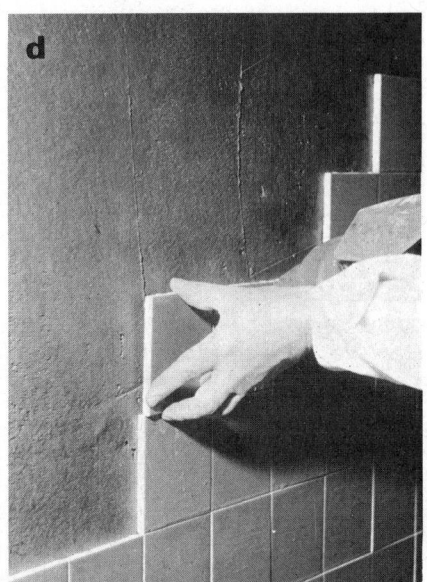

FIGURE 9.6-43 *Conventional mortar bed installation: (a) scratch coat is applied to metal lath and temporary screeds are inserted; (b) float coat is leveled and screeds are removed; (c) neat cement coat is troweled on; and (d) tile is applied and tamped in place. Glazed tile shown here must be soaked before it is set. (Courtesy American Olean [Ceramic] Tile Company.)*

FIGURE 9.6-44 *For thin-set tile applied directly on masonry, the backing surface should be smooth and the joints struck flush. Since exterior thin-set applications directly over masonry are expected to move with the masonry, expansion joints are not as important as in other types of installations. (Courtesy Tile Council of America, 1993 Handbook for Ceramic Tile Installation.)*

FIGURE 9.6-45 *Before tile is applied over uneven masonry, a leveling bed should first be placed and cured. Tile can then be set with either neat cement or thin-set mortar. (Courtesy Tile Council of America, 1993 Handbook for Ceramic Tile Installation.)*

FIGURE 9.6-46 *Open framing requires a two-coat mortar bed over metal lath. Except with dry-set mortar, curing the bed is required for thin-set applications. (Courtesy Tile Council of America, 1993 Handbook for Ceramic Tile Installation.)*

gypsum board, mineral fiber–cement board, and plywood (Fig. 9.6-48). It can be used also over concrete or masonry if the surface is relatively smooth (maximum variation 1/8 in. in 8 ft.). Type I adhesive should be used for installations requiring prolonged water resistance, such as in tub

and shower areas and on bathroom floors. Type II adhesive can be used in areas subject to occasional or intermittent wetting. Neither type, however, is suitable for installations that require continuous contact with water, such as in sunken tubs or pools. Manufacturer's recommendations should be followed to determine which type is best for the specific use intended and whether a primer is required.

The adhesive should be spread with a notched trowel recommended by the manufacturer, leaving no bare spots (Fig. 9.6-49a). The tile should be tamped into the adhesive to ensure maximum contact (Fig. 9.6-49b). The average contact area of tile removed for inspection should be not less than 75%, and no individual tile or mounted assembly should have less than 40% coverage of adhesive. Adjustments

FIGURE 9.6-47 *Thin-set application of glazed wall tile directly to masonry: (a) dry-set mortar is applied with a notched trowel; (b) edge-glued tile is set and tamped in place; (c) grout is forced into joints and excess wiped off.* *(Courtesy American Olean [Ceramic] Tile Co.)*

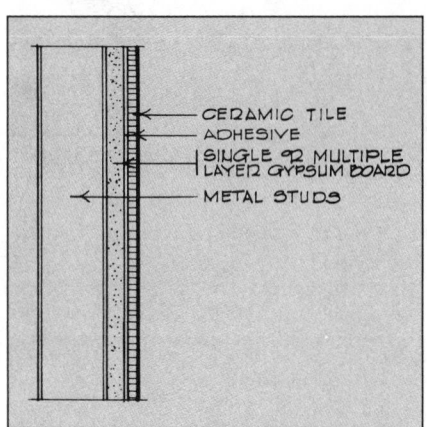

FIGURE 9.6-48 *Organic adhesives can tolerate limited moisture exposure (such as in residential shower stalls) if a water-resistant adhesive and backing are used and the board edges are sealed.* *(Courtesy Tile Council of America, 1993 Handbook for Ceramic Tile Installation.)*

should be made while the adhesive is still plastic. Grouting should be delayed for at least 24 hours to permit solvent evaporation from the adhesive. Joints should be grouted with a product suitable for the conditions. Drywall and latex grouts commonly are used with organic adhesives (Fig. 9.6-49c).

9.6.3.2

Exterior Surfaces

Tile installations on exterior walls require the same precautions regarding movement as exterior floors. The supporting construction, including the backing, must be designed to accommodate movement caused by temperature and moisture variations. This means primarily the inclusion of suitable control, isolation, and expansion joints, and also inclusion of proper metal reinforcement in concrete and masonry, careful control of mortar and concrete mixes, sound engineering design, and good workmanship (see Chapter 3).

Exterior tile can be applied by either the conventional mortar bed method or the thin-set method (Fig. 9.6-50). Chief factors in determining the most suitable method are (1) condition of the backing, whether suitable for direct bonding; (2) dimensional stability of the backing; and (3) provisions for crack control.

On large areas exposed repeatedly to direct wetting, the conventional isolated method usually gives better results. The thin-set bonded method generally is more reliable in covered outdoor areas and when broken into smaller panels by decorative trim, battens, or expansion joints. However, tile can be thin-set over an isolated mortar bed with dry-set mortar.

Regardless of installation methods, expansion joints should be provided in the tile finish over all underlying structural joints, at the tile perimeter, and against any projecting surface. In isolated installations, expansion joints in the tile finish should be provided at 12 to 16 ft. on centers. Where these expansion joints do not coincide with joints in the backing, they should be provided in addition to the underlying joints. In a thin-set installation over a suitable stable backing, the additional joints in the tile finish are not as essential because the tile finish is intended to move with the backing.

In exterior tile installations, it is essential to provide suitable flashings, overhangs, or copings to prevent entry of water into or behind the tilework. Exterior walls of habitable spaces should include a suitable vapor retarder on the warm side of the wall to prevent condensation in the wall cavity. Additional information on the design and construction of expansion joints is given in Section 9.6.1.6.

■ PORTLAND CEMENT MORTAR BED

There is no substantial difference in installing tile over a mortar bed outdoors from the procedure described earlier for interior locations. The entire installation should be backed by a cleavage membrane. Metal lath should be firmly attached to the backing and cut at expansion joints (see Fig. 9.6-50).

A scratch coat consisting of 1 part portland cement, 5 parts damp sand, and 1/5 part lime usually is installed by a plastering contractor. The mortar bed, consisting of 1 part portland cement, 5 parts damp sand, and 1/2 part hydrated lime, is normally installed by a tile contractor. The overall thickness of both coats should be 3/4 to 1 in. The mortar bed mix can be varied up to 1 part portland cement, 7 parts damp sand and 1 part lime. Extremely rich mixes (too much portland cement) should be avoided, because they result in greater drying and shrinkage cracking. Similarly, a total mortar bed thickness in excess of 1 in. should be avoided.

Neat cement can be used for setting tile over a still-plastic mortar bed. If dry-set or latex–portland cement mortar is to be used, the mortar bed should be cured for at least 24 hours, but longer curing periods up to 10 days are preferred. Suitable grouts include commercial cement, dry-cure.

■ THIN-SET INSTALLATION

Waterproof backings suitable for bonding of tile with thin-set mortars include monolithic concrete, concrete and brick

FIGURE 9.6-49 *Over previously applied organic adhesive: (a) edge-mounted tile is applied; (b) tile is tamped in place with a rubber float trowel; (c) same trowel is used to force grout into joints and remove excess grout.* (Courtesy American Olean [Ceramic] Tile Co.)

FIGURE 9.6-50 *Expansion joints 12 to 16 ft. on centers are essential in exterior isolated thickbed tile installations.* (Courtesy Tile Council of America, 1993 Handbook for Ceramic Tile Installation.)

masonry, structural clay tile, and a cured portland cement mortar bed (see Fig. 9.6-44). Latex–portland cement and dry-set mortar typically are used for setting tile. Either latex or dry-set grout can be used for filling tile joints. Damp curing for at least 72 hours is essential.

9.7 TERRAZZO

The term *terrazzo* is derived from the Italian *terrassa* (terrace on the roof) because many early installations were made in upper-level outdoor living areas. The origins of terrazzo date back to the beginning of the Christian era, having evolved from *mosaics*, the art of placing colorful pieces of stone, vitreous enamel, and marble (tesserae) in decorative designs on walls and floors. Earliest terrazzo installations in Roman times were made of a mixture of crushed brick in lime mortar, which was ground and polished after it hardened. Later, the decorative chips salvaged from mosaic work were introduced in tile outdoor terraces to create highly colorful and durable floor finishes.

Terrazzo has been used for centuries as an economical and durable indoor-outdoor finish and traditionally has been defined as a form of mosaic flooring made by embedding small pieces of marble in mortar and polishing. In modern times, the mortar (matrix) for terrazzo finishes generally has been portland cement. Historically the decorative chips were limited to marble. More recently, synthetic resinous binders have been developed, and other hard rocks capable of being polished are used.

Because of the expense involved in the hand assembly of decorative chips in complex designs for large areas, mosaic work survives today primarily as custom inserts in terrazzo installations. The tesserae usually are assembled in the shop and mounted on paper; then the assembly is installed in a mortar setting bed in the desired location. After the paper is removed, joints are grouted and the entire installation is ground and polished.

Grinding and polishing are distinctive operations that differentiate terrazzo from seamless floorings, which also utilize resinous binders. A notable exception to this general rule is *rustic* portland cement terrazzo, which is not ground but washed to reveal the decorative chips. In rustic terrazzo, selected colorful gravels sometimes are used instead of marble chips.

Terrazzo is a durable, low-maintenance floor finish, generally suitable for both exterior and interior locations. It can be cast in place or precast in a variety of shapes and colors. In precast form, its chief use has been for shower receptors, windowsills, wall bases, and stair treads. More decorative products intended for counters, lavatory tops, and tables are marketed under the names *art marble* and *cultured marble*. These precast terrazzo products involve installation methods different from those associated with cast-in-place terrazzo floor finishes and are not included in this discussion.

This section is divided into subsections that outline terrazzo floor materials and application methods. Since a large proportion of terrazzo floors utilize portland cement mortar and are installed over concrete surfaces, it is recommended that the reader review the requirements discussed in Chapter 3.

9.7.1

MATERIALS

The decorative wear layer of terrazzo finishes is called a *topping*. In some installations, the topping is placed directly on a wood, metal, or concrete subfloor; in others, an intervening portland cement underbed is required. When an underbed is used, it may be rigidly *bonded* to a tile subfloor, or it may be *unbonded* by separating it from the subfloor with an *iso-*

FIGURE 9.7-1 *Basic construction elements for unbonded and bonded terrazzo installations.*

lation membrane (Fig. 9.7-1). When no underbed is used, the topping always is bonded to the subfloor. Other terms relating to terrazzo installations are illustrated in Figure 9.7-1 and defined in the Glossary.

The basic ingredients of terrazzo toppings are *binders* and *decorative chips*. Other materials that may be required are (1) *pigments*; (2) accessories, such as *divider strips* and *expansion strips* for decorative purposes and crack control; (3) *metal reinforcement*, for tensile strength and shrinkage control; (4) *isolation membranes*, to prevent the transfer of stresses from subfloor to underbed and topping; and (5) *curing compounds* and *sealing materials*, to maintain the quality of the terrazzo finish.

Materials for terrazzo finishes should conform to the applicable industry standards given in Figure 9.7-2.

9.7.1.1

Binders

The binders used in terrazzo toppings can be classed as *cementitious* and *resinous*. Cementitious binders are the older, traditional terrazzo materials; resinous binders are the result of relatively recent developments in synthetic organic chemistry since World War II.

■ CEMENTITIOUS BINDERS

The primary cementitious binder used today is portland cement. Either white or natural (gray) Type I portland cement may be used for terrazzo finishes. Natural portland cement is less expensive and can be used if a vivid or light-colored matrix is not essential to the color scheme. The

FIGURE 9.7-2
Summary of Terrazzo Materials

Material	Standard
Cementitious binders Portland cement, Type I	ASTM C 150
Resinous matrixes Epoxy	NTMA*
Polyester	NTMA*
Polyacrylate	MIL-D-3134F and NTMA*
Decorative chips	NTMA*
Aggregate Sand	ASTM C 33
Pigments	NTMA*
Metal reinforcement Welded wire fabric	ASTM A 185
Isolation membranes Roofing felt	ASTM D 226
Polyethylene film	ASTM D 2103
Curing compounds	ASTM C 309

*Specifications of the National Terrazzo and Mosaic Association.

National Terrazzo and Mosaic Association (NTMA), however, cautions that using gray cement may result in an uneven color. White portland cement is more expensive but offers a lighter background for displaying the decorative chips and produces clearer, truer colors with mineral pigments. Type IA air-entraining portland cement is recommended for exterior applications subject to freezing and thawing.

RESINOUS BINDERS

Resinous binders possess most of the desirable properties of portland cement binders and have greater strength, chemical resistance, and abrasion resistance. In addition, the ability of resinous binders to bond with most subfloor materials and to be applied in toppings as thin as 1/4 to 1/2 in. makes then suitable for many installations where portland cement cannot be used due to its thickness. However, substantially higher material costs increase overall installed cost and limit the economic use of resinous terrazzo to installations where their special properties are specifically required.

Synthetic binders for terrazzo finishes include epoxy, polyacrylate, and polyester. Because of the relatively high cost of resins and the desire to impart special properties, these resins are combined with various additives and fillers that together form the matrix. Epoxy and polyacrylate resins are also combined with portland cement to form *epoxy modified cement terrazzo* and *polyacrylate modified cement terrazzo*.

Materials and application for epoxy, polyacrylate, and polyester terrazzo should conform to the requirements of the NTMA specifications, details, and product data outlined in the NTMA's publication *Terrazzo Information Guide* and the binder manufacturer's recommendations.

Epoxy and polyester matrixes are two-part formulations (a basic resin and a catalyst) to which a dry mix (fillers, pigment, additives, and decorative chips) is added. These resins are distinguished by good chemical resistance and adhesion, good abrasion resistance, and impact strength.

Epoxy matrixes exhibit superior overall chemical resistance, good abrasion resistance and impact strength, and negligible curing shrinkage. Among their disadvantages are relatively high cost and toxicity to the skin during application and before curing.

Polyester matrixes are characterized by good chemical resistance, particularly to acids of high concentration. Polyester resins are resistant to water and weather, are strong and durable, and possess good abrasion resistance. Some of the disadvantages of polyester resins are their somewhat lower resistance to alkali than epoxy and their stronger odor during installation and curing period.

Polyacrylate–modified cement terrazzos have a high bond strength and are resistant to snow-melting salts, foodstuffs, and urine. They are also nontoxic and relatively free from objectionable odors.

9.7.1.2

Decorative Chips

Decorative chips in terrazzo toppings consist principally of marble. Chips of various types and colors are quarried and carefully selected to avoid off-color or contaminated material. Originally, marble was defined as a metamorphic rock formed by the recrystallization of limestone. Marble has been redefined, both geologically and commercially, to include all calcareous (limestone) rocks such as onyx and travertine, as well as attractive serpentine rocks, capable of being ground and polished. Granite, which is not a marble, is extremely hard and weather resistant and often is used for heavy-duty industrial and rustic terrazzo.

Domestic and imported chips are available in a wide range of colors for terrazzo use. Chips are graded by number according to size as shown in Figure 9.7-3. For example, a #1 chip retained on a 1/8-in. screen and passing a 1/4-in. screen will be between 1/8-in. and 1/4-in. in size. Producers normally supply #0, #1, and #2 as separate sizes; some produce #3 and #4 as separate sizes. Larger sizes are frequently grouped as #3-4 mixed, #5-6 mixed, #7-8 mixed, and #4-7 mixed.

The appearance and the minimum thickness of toppings are determined largely by the size and the type of chips used. Toppings are identified on this basis as *standard*, *intermediate*, *Venetian*, and *Palladiana* (see Fig. 9.7-3). Standard, intermediate, and Venetian use crushed chips, which are mixed with the resinous or portland cement matrix and spread uniformly in the topping.

Palladiana toppings consist of cut slabs of marble about 1/2 in. thick that are set by hand in the desired pattern. Spaces between slabs are filled with a standard topping mix. Palladiana toppings are custom finishes similar to mosaics.

The NTMA and some producers of chips publish standard color plates as an aid in the selection and specification of decorative chips and color pigments. To ensure uniformity, marble chips and color combinations should be designated by plate number. The finished terrazzo should not, however, be expected to duplicate the color plate exactly, because slight variations in color and veining are likely to occur between different shipments of chips. Physical samples should be prepared as a standard of comparison at the beginning of each project.

FIGURE 9.7-3
Grading of Decorative Chips for Terrazzo Toppings

Type of Topping and Minimum Thickness	Chip Number	Passes Screen, in.	Retained on Screen, in.
1/4 in. Thin-set	0	1/8	1/16
	1	1/4	1/8
5/8 in. Standard* 1/2 in. Thin-set	2	3/8	1/4
5/8 in. Intermediate	3	1/2	3/8
	4	5/8	1/2
3/4 in. Venetian	5	3/4	5/8
	6	7/8	3/4
	7	1	7/8
	8	1-1/8	1

*Palladiana toppings consist of broken marble slabs set in a standard chip mix and should not be less than 1/2 in. thick.

Fine Aggregate

In terrazzo finishes the term *aggregate* refers to material such as sand and fillers and does not include decorative chips. Fine silica sand and fillers generally are included in the proprietary formulations of resinous matrixes. Sand aggregate is one of the basic ingredients of a mortar underbed, but it is not used in a portland cement topping. Sand should be well-graded, natural, or manufactured aggregate that is clean, granular, and free of deleterious amounts of clay, salts, and organic matter.

Pigments

Terrazzo toppings may be the natural color of the binder or be treated with pigments to produce a wide variety of colors. Pigments should be stable, nonfading mineral or synthetic formulations, compatible with the binder.

Accessories

Accessories for terrazzo installations include divider strips, expansion strips, and base beads. Metal accessories commonly are made from half-hard brass or white alloy zinc (approximately 99% zinc). Plastic-top strips and beads in various colors with brass, zinc, or galvanized steel supports also are available. Type 300 stainless steel strips are produced for use in exterior terrazzo; aluminum strips are used primarily for resinous terrazzo. The standard length of strips and beads is 6 ft., but longer lengths are available on special order.

Only plastic-top strips and beads should be used in conductive floors to prevent the accumulation of dangerous electrostatic charges. In exterior locations, white alloy zinc accessories are not suitable because a chemical reaction often causes blooming (a white chalky deposit on the strips). In such exterior locations, all-brass, stainless steel, or zinc-supported plastic-top accessories are suitable. Aluminum strips should not be used with portland cement terrazzo because the aluminum is subject to attack by alkalies present in the portland cement topping mix.

■ DIVIDER STRIPS

Divider strips are installed: to (1) to localize shrinkage cracking in portland cement terrazzo, (2) ensure uniformity in topping thickness, (3) act as decorative elements, (4) permit accurate placement of the different colors forming a floor pattern, and (5) act as construction joints, subdividing a floor into convenient working areas.

A variety of divider strips are available to suit particular installations (Fig. 9.7-4). The thickness of the top of the strip (exposed face) can range from 20-gauge up to 1/2-in.-thick bars; 18 gauge is the recommended minimum. Divider strips with a 1/8 in. or wider face usually have a heavy plastic or metal top section and a thin metal bottom section. For a topping thickness of 5/8 in., a 1-1/4-in.-deep divider is recommended; 1-1/2-in.-deep strips are recommended for Venetian toppings 3/4 in. and thicker. For thin-set installations, special angle or T-shaped strips 1/4 to 1/2 in. deep are used. Other special strips are available for wall base and for edgings at the juncture with other flooring materials.

■ EXPANSION STRIPS

Provision for crack control is necessary at lines in the floor where substantial structural movement can be expected, particularly over beams and bearing walls and over expansion or isolation joints in the underlying subfloor. Expansion strips manufactured for this purpose consist of a pair of divider strips separated by a resilient material such as neoprene (see Fig. 9.7-4). Expansion strips are manufactured in types and sizes similar to divider strips.

■ BASE BEADS

All-metal or plastic-top base beads are used to terminate terrazzo wall base on the wall surface. They are made in several profiles and to accommodate vertical terrazzo thicknesses varying from 1/2 in. to 1-1/2 in.

THIN-TOP DIVIDER STRIP **HEAVY-TOP DIVIDER STRIP**

EXPANSION STRIP **RECESSED EDGING STRIP**

"L" DIVIDER STRIP **BASE BEAD** **ABRASIVE CHANNEL**

FIGURE 9.7-4 *Accessories: shallow T- and L-shaped (angle) strips for installation without underbed are available in thin-top, heavy-top, and edging profiles.*

Metal Reinforcement

Metal reinforcement generally is required for the underbed in most installations, except where the underbed is bonded to a suitable concrete subfloor. Underbed reinforcement should be corrosion-resistant welded wire fabric of at least 16 gauge with wires spaced not more than 2 in. on centers. Thin underlayments for resinous toppings may use welded or woven wire weighing at least 1.4 lb./sq. yd. or expanded metal lath weighing at least 2.5 lb./sq. yd.

Isolation Membrane

An isolation membrane is required in all unbonded installations to separate the underbed from the subfloor. The membrane may be 15-lb. asphalt-saturated roofing felt or 4-mil polyethylene film.

Curing Materials

Curing is necessary for portland cement toppings to develop optimum wear properties. Materials suitable for curing include polyethylene film, nonstaining (nonasphaltic) water-resistant building paper, and clean water. Spray-on curing compounds are sometimes used for curing terrazzo surfaces, but they should not be used to cure concrete slabs that are to receive a thin-set, chemically bonded, or monolithic topping, because they may prevent bond.

Sealers and Protective Finishes

Cementitious matrixes are more porous than resinous matrixes and require greater protection from staining and chemical deterioration. Marble chips, which make up more than 70% of a terrazzo surface, possess varying degrees of resistance to staining and deterioration. A penetrating sealer capable of internally sealing the pores in the matrix and in the marble chips should be used. These are generally proprietary formulations. Only products specified by the manufacturer for the particular purpose should be used, and they should be applied in accordance with the manufacturer's recommendations.

APPLICATION

Suitable terrazzo finishes require attention to (1) general recommendations common to most installation methods, (2) specific application procedures for particular methods, and (3) various finishing procedures for special surface finishes.

Depending on whether the topping or underbed is adhered to the subfloor, terrazzo systems can be classed as *bonded* or *unbonded*. Bonded systems include thin-set monolithic, chemically bonded, and bonded-underbed methods. *Thin-set* installations consist of a thin (1/4- to 1/2-in.-thick) resinous topping usually placed directly over a sound wood, metal, or concrete subfloor (Fig. 9.7-5a). In *monolithic* installations, a somewhat thicker (5/8 in. thick or thicker) portland cement topping is applied directly over a rough-finished concrete subfloor (Fig. 9.7-5b). If the concrete subfloor is an existing, smooth-troweled concrete slab, a portland cement topping can be *chemically bonded* with a bonding agent (Fig. 9.7-5c). Monolithic and chemically bonded installations depend on the subfloor to resist shrinkage stresses and conspicuous cracking in portland cement toppings. However, *bonded-underbed* installations provide more positive crack control by creating weakened planes at each divider strip (Fig. 9.7-5d).

Unbonded installations provide the most crack control and are preferred wherever structural movement is expected. Separation between subfloor and underbed can be accomplished by an isolation membrane alone or with a membrane and sand cushion (Fig. 9.7-5e). When a sand cushion is used, it is called a floating installation.

General Recommendations

The installation of terrazzo involves (1) preparation of the subfloor, and, depending on the method of installation, (2) placement of the topping, strips, and underbed when used.

■ SUBFLOOR PREPARATION

Regardless of the installation method, the subfloor should be sound and free of loose material. It should be structurally adequate to support the expected live loads without excessive deflection, as well as the additional dead load imposed by the terrazzo finish. The dead load can be figured at about 11 lb./sq. ft./in. of thickness for both resinous and cementitious terrazzo.

For unbonded installations, the surface quality of the subfloor is not critical, because the bond between the subfloor and the underbed is not affected. The surface quality of the subfloor is of greater importance in bonded installations because the quality of the installation depends on adequate bond with the subfloor. Therefore, for all bonded installations, the subfloor should be free of coatings such as paint, wax, and sealing or curing compounds, which could prevent proper bond.

For monolithic installations, the surface of the subfloor should not be troweled but should be wood-floated and broomed or otherwise roughened. A chemical bonding agent should be used over a smooth-troweled concrete surface .

A concrete subfloor intended for a thin-set resinous installation should be smooth-troweled and at least 21 days old. Wood floors should consist of a suitable single plywood subfloor or double floor, including at least a 3/8-in.-thick plywood underlayment (Fig. 9.7-6). Plywood should be Exterior Type CC-plugged grade or better. The recommendations of the manufacturer of the resinous matrix should be followed for subfloor preparation. Over most subfloor materials, the topping can be adhesively bonded, and priming of the subfloor is required to achieve bond. Over plywood, some manufacturers recommend direct adhesive bonding after plywood edges have been bound with a special glass fiber tape.

The subfloor for monolithic, chemically bonded, and thin-set installations should conform to the profile of the finished terrazzo surface, whether level or pitched, to ensure that the topping will be of uniform thickness. The subfloor surface

BONDED METHOD

a THINSET

b MONOLITHIC

c CHEMICALLY BONDED

d BONDED-UNDERBED

UNBONDED METHOD

e FLOATING (SAND CUSHION)

FIGURE 9.7-5 *Details for basic terrazzo installation methods.*

should be in a straight plane; there should be no more than 1/8-in. variation when it is tested with a 10-ft. straightedge. A topping mix should not be used to fill voids or to develop pitch, because excessive variations in topping thickness are conducive to cracking. Where the existing subfloor is not sufficiently smooth or does not have the desired profile, it should be leveled or pitched with a suitable mastic underlayment such as latex. Over wood strip flooring, these underlayments should be installed at least 3/8 in. thick with a 15-lb. felt, and expanded metal or wire lath secured to the subfloor 6 in. on centers along the supports.

Monolithic and chemically bonded portland cement toppings require control joints in the subfloor as discussed below under "Strip Placement."

■ UNDERBED

An underbed is necessary primarily in unbonded installations but may be required in some bonded ones. An underbed can be relatively thin because it is not a structural slab but functions as a surface preparation for a topping. The thickness of an underbed varies depending on the installation method (see Fig. 9.7-5d, e).

An underbed should be reinforced with welded wire fabric in all unbonded installations and in bonded installations over wood or metal subfloors. Unless the underbed is bonded, an isolation membrane should be spread over the subfloor. The ends and edges of the isolation membrane should be overlapped at least 4 in. Metal reinforcement should be placed over the membrane. Except where expansion strips occur, the adjacent sheets of this reinforcing should overlap by at least two mesh spaces. No conduit or pipe should be located within the minimum recommended underbed thickness.

The mix for underbed mortar should consist of 1 part portland cement and 4 to 5 parts sand, mixed with sufficient water to make a workable mortar. The surface over which the underbed is placed should be sufficiently wet to prevent water from being drawn out of the mix. Isolation membranes need only be uniformly dampened; concrete slabs should be thoroughly saturated by overnight soaking.

Underbed mortar should be placed and screeded (Fig. 9.7-7) to a height that will permit installation of the selected topping thickness at the desired finish-floor elevation.

■ STRIP PLACEMENT

When an underbed is used, conventional divider strips (see Fig. 9.7-5) should be

FIGURE 9.7-6
Wood Floor Construction for Thin-Set Terrazzo[a]

Single-Floor Construction

Combination Subfloor/ Underlayment[b, c]	Maximum Support Spacing	Nails[d]	Nail Spacing
5/8-in. plywood	16 in. o.c.	2 in. (6d) threaded	6 in. o.c. around panel edges; 10 in. o.c. at intermediate supports
3/4-in. plywood	20 in. o.c.		
7/8-in. plywood	24 in. o.c.		
1-1/8-in. plywood (2-4-1)	48 in. o.c.	2-1/2 in. (8d) threaded	6 in. o.c. at panel edges and intermediate supports

Double-Floor Construction

Subfloor[b]	Maximum Support Spacing	Nails	Nail Spacing
1/2-in. plywood	16 in. o.c.	2 in. (6d) common	6 in. o.c. around panel edges; 10 in. o.c. at intermediate supports
5/8-in. plywood	20 in. o.c.	2-1/2 in. (8d) common	
3/4-in. plywood	24 in. o.c.		
3/4-in. lumber boards less than 6 in. wide	16 in. o.c.	2-1/4 in. (7d) threaded	2 per board at each support
3/4-in. lumber boards wider than 6 in. and less than 8 in. wide			3 per board at each support

Underlayment[b, e]	Fasteners[f] Nails[d]	Staples[g]	Fastener Spacing
3/8-in. plywood	1-1/4 in. (3d) threaded	1-1/8 in.	Nails: 8 in. o.c. each direction in middle of panel, 6 in. o.c. around panel perimeter
1/2-in. plywood		1-1/4 in.	
5/8-in. plywood	1-1/2 in. (4d) threaded	1-1/2 in.	Staples: 6 in. o.c. each direction in middle of panel, 3 in. o.c. around panel perimeter
3/4-in. plywood		1-5/8 in.	

[a] Based on 40 psf live load.

[b] Plywood thicknesses given are for Group 1 softwoods in accordance with U.S. Product Standard PS1. For Groups 2 and 3, increase thickness 1/8 in.; for Group 4, increase thickness 1/4 in.

[c] Blocking at longitudinal edges is required unless plywood with T&G edges is used.

[d] Common nails may be substitutded for annularly threaded nails by increasing their size 1/4 in. (1d).

[e] Blocking not required if underlayment joints are staggered with respect to subfloor joints.

[f] Fastener lengths shown are minimum to provide 3/4 in. penetration for nails and 5/8 in. for staples. If longer fasteners are used, penetration should not exceed 1 in.

[g] Galvanized divergent chisel, 16 gauge, 3/8 in. crown width.

installed while the underbed is still semi-plastic. Standard 1-1/4- or 1-1/2-in. strips should be set in the desired pattern, with top of the strip at the finished-floor elevation. The strips then should be troweled firmly along the edge to ensure ade-quate anchorage in the underbed (Fig. 9.7-8). In many bonded installations an underbed is not used, and the terrazzo is applied directly to the subfloor. In mono-lithic and chemically bonded installations, it usually is not possible to install strips while the concrete subfloor is still plastic because the subfloor often is installed by a different contractor. For such installa-tions, angle or T-shaped strips (see Fig. 9.7-4) are secured to the concrete with adhesive or mechanical fasteners.

Generally, divider strips should be pro-vided wherever a discontinuity exists in the subfloor, such as occurs at construc-tion, isolation, and control joints and at changes in materials. Expansion strips may be used at these locations if desired, but they are specifically recommended at structural expansion joints where a greater degree of movement is anticipated. Divider or expansion strips should be used over beams and girders supporting the sub-floor, because tensile stresses are likely to develop due to negative bending mo-ments. Spacing of divider and expansion strips for various methods of installation is given in Figure 9.7-9.

DIVIDER STRIPS In resinous terrazzo in-stallations, divider strips are not required for control of shrinkage cracking but often are desirable for decorative purposes and convenience in placing and finishing the topping. In thin-set installations the strips ensure a uniform topping thickness and conserve material. They also minimize the appearance of waviness in large areas.

In portland cement terrazzo installations that use an underbed, strips should be spaced not more than 6 ft. on centers. When a portland cement topping is placed directly over a concrete slab, the slab should be provided with cut or tooled con-trol joints 15 to 20 ft. on centers in each direction, forming panels of not more than 300 sq. ft. The length of these panels should not exceed their width by more than 50%. Divider strips should be placed at all such control joints, forming similar panels in the topping (Figs. 9.7-10 and 9.7-11). In slab-on-ground construction where a minimum of stripping is desired, control joints and divider strips can be placed in doorways and adjacent to parti-tions, where they will be concealed by base trim (Fig. 9.7-12).

FIGURE 9.7-7 *Underbed installation: (a) mortar is placed in windrows and leveled to the form screeds; (b) windrows are leveled in transverse direction; (c) mortar is placed between windrows and leveled.*

FIGURE 9.7-8 *Strip placement: (a) long strips are placed first and aligned with a straightedge; (b) cross strips are installed next; (c) each strip is leveled and troweled firmly along its edge to seat it firmly in the underbed.*

FIGURE 9.7-9
Strip Placement

Matrix	Installation Method	Divider Strip Spacing[a]	Expansion Strip Spacing[a]
Portland cement	Unbonded	6 ft o.c. maximum	As needed[b]
	Bonded-underbed		
	Chemically bonded	15 ft. to 20 ft. o.c.— panels not over 300 sq. ft.[c]	30 ft. o.c. maximum[c]
	Monolithic		
Resinous	Thin-set	As desired for decorative or convenience purposes	30 ft. o.c. maximum[d]

[a] Locate divider strips or expansion strips on beams and girder lines wherever possible.

[b] Provide expansion strips over all expansion joints in the subfloor or at other joints where movement is expected.

[c] Divider strips and expansion strips can be alternated, or expansion strips can be used at 15 ft. o.c.

[d] Recommended for large areas or long corridors; otherwise, not required.

Divider strips should be continuous from wall to wall and preferably should form right-angled intersections rather than acute angles. If a terrazzo wall base is used, strips should be carried up to the base screed. When terrazzo is carried beyond the base to form a wainscot, strips should be continued from the base in a straight line up the wall. The recommended strip spacing for portland cement terrazzo wainscoting is 3 ft. on centers maximum; strip spacing for resinous wainscoting is not critical.

EXPANSION STRIPS Generally, greater consideration should be given to the need for and the location of divider and expansion strips in bonded systems, because the terrazzo finish will tend to move with the subfloor. In monolithic, chemically bonded, and thin-set installations, expansion strips should be provided about 30 ft. on centers where unobstructed large areas or long corridors are involved. In monolithic and chemically bonded installations, expansion and divider strips can be alternated 15 ft. on centers. (see Fig. 9.7-10). Expansion strips generally are not required in such installations if the subfloor is continuous, homogeneous, and 1000 sq. ft. or smaller in area.

Unbonded and bonded-underbed installations will generally have a sufficient number of divider strips distributed over

WALL SECTION

FIGURE 9.7-10 *For monolithic and chemically bonded installations, strips should be spaced 15 to 20 ft. on centers. In all installations, locating strips at column lines is desirable.*

FIGURE 9.7-11 *Control joints in monolithic and chemically bonded installations: (a, b, and c) inconspicuous controlled cracking is encouraged adjacent to strip; (d) uncontrolled cracking results from improper installation.*

PLAN AT DOOR FRAME

FIGURE 9.7-12 *Strip placement in monolithic and chemically bonded terrazzo over slab on grade when strips are not a design feature.*

the floor so that expansion strips are not required except over expansion joints in the subfloor.

TOPPING

This subsection covers primarily portland cement terrazzo with a regular finish. Resinous terrazzo is addressed under "Thin-Set Method" in Section 9.7.2.2. Rustic and conductive terrazzo are discussed in Section 9.7.2.4.

The steps involved in installing terrazzo toppings can be grouped according to sequence of operations as (1) mixing, (2) placing the topping, (3) preliminary finishing, and (4) final finishing and protection.

MIXING   A power mortar mixer should be used to mix portland cement topping matrix. The ingredients should be carefully measured and consistently proportioned. A conventional portland cement mix should consist of 200 lb. of marble chips for each 94-lb. bag of cement. When pigment is used, it should be measured by weight (not volume) and added to the dry mix. After thorough mixing of dry ingredients, water should be added at the ratio of 5 to 5-1/2 gal. per bag of cement.

Resinous binder formulations vary among manufacturers. Therefore, their proprietary mixing and installation recommendations should be followed carefully.

PLACING   Installation of a topping over an underbed should be delayed for 12 to 24 hours to permit the underbed to harden. Immediately before a topping is placed, a slush coat of neat cement pigmented to match the topping mix, should be broomed into the underbed (Fig. 9.7-13). The various colored mixes with decorative chips should then be placed in the appropriate panels according to the desired pattern. The topping then should be troweled even with the tops of divider strips, which act as screeds for leveling the surface (Fig. 9.7-14).

To ensure even distribution and adequate chip content, the surface usually is seeded with additional chips of the type contained in the mix (Fig. 9.7-15). These chips should be wetted, sprinkled over the surface, and embedded by additional troweling. Then the entire area should be rolled with a heavy roller until all excess water has been brought to the surface (Fig. 9.7-16). The surface again should be troweled smooth and uniform, revealing the tops of the divider strips.

Before finishing operations are started, a portland cement topping should be wet-cured for 3 to 7 days by ponding,

FIGURE 9.7-13 *A slush coat of neat cement is broomed into the surface just before the topping is placed.*

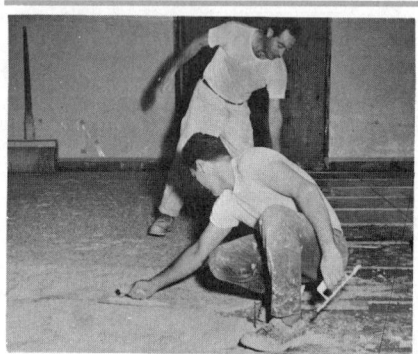

FIGURE 9.7-14 *Topping is spread and troweled even with the divider strips.*

FIGURE 9.7-15 *Additional chips are seeded and troweled into the surface where necessary for uniform appearance.*

covering with watertight covers, or spraying with a curing compound.

PRELIMINARY FINISHING   After proper curing, the preliminary finishing operations of grinding, grouting, and curing may be performed.

FIGURE 9.7-16 *Rolling compacts the topping mix, brings up excess water, and creates a harder, more durable surface.*

The surface should be ground using a terrazzo-grinding machine first with No. 24 or finer abrasive stones, followed by a second grinding with No. 80 or finer stones (Fig. 9.7-17). The grinding operations should be performed while the surface is covered with water.

FIGURE 9.7-17 *After proper curing, the surface is flooded and is wet-ground to reveal the decorative chips.*

After grinding, the surface should be thoroughly rinsed with sufficient water to remove dust and loose fine particles. Then a skim coat of neat cement grout of the type used in the topping mix (including pigment to match the topping) should be troweled over the entire surface, filling all voids.

The installation then should be wet-cured again for at least 3 days using a watertight cover or curing compound. The compound or cover should be allowed to remain in place as protection against physical damage and staining until final finishing is performed.

FINAL FINISHING   Final finishing includes polishing, cleaning, and sealing. A floor machine with No. 80 or finer stones should be used for polishing. The polishing operation should continue until the grout skim coat is removed (except in voids and depressions) to reveal the decorative chips in from 70% to 75% of the area.

A solution of neutral liquid cleaner should be used according to the manufacturer's recommendations to clean the polished floor. It should then be rinsed with clean water and be allowed to dry. The cleaned, polished terrazzo surface finally should be treated with a suitable sealing compound. These compounds also are proprietary formulations and should be applied in accordance with their manufacturer's recommendations.

Until all work on the premises is completed, finished terrazzo should be protected with a covering, such as heavy building paper.

9.7.2.2

Installation Methods

The following recommendations are in addition to those in Section 9.7.2.1.

◼ UNBONDED METHOD

The unbonded installation method provides the greatest degree of isolation between subfloor and underbed, thereby permitting independent movement by each and minimizing the possibility of surface cracking. An unbonded installation is suitable over any tight, structurally sound subfloor in interior or exterior locations. In exterior locations an unbonded system generally is limited to concrete supporting construction, because wood or metal would require the decay- and surface corrosion-resistant protection afforded by a bonded finish.

Unbonded installations require the use of a reinforced underbed to support the topping. The combined thickness of underbed and topping should be at least 2-1/2-in. (see Fig. 9.7-5e). Separation between subfloor and underbed can be provided by an isolation membrane alone or by a membrane spread over a 1/16- to 1/4-in.-deep layer of sand. Over a smooth subfloor a double layer of polyethylene film similarly will provide isolation between surfaces. Membranes should be lapped at least 4 in. at the joints.

Metal reinforcement should be placed over the membrane and should be lapped at joints at least two mesh spaces and wired together. After some mortar has been poured in place, the reinforcement should be lifted to about 1 in. above the membrane. Underbed, strips, and topping should be installed as indicated in Section 9.7.2.1.

■ BONDED METHODS

The chief advantages of bonded installations are smaller dead load and lesser finish thickness. Most bonded methods allow the topping to be placed directly over the subfloor without an intervening underbed. However, these methods do not generally develop the degree of crack control obtained when an underbed is used with conventional divider strips. Therefore, the bonded-underbed method (see Fig. 9.7-5d) often is used if the greater dead load and thickness can be accommodated. Since the difference in thickness between bonded-underbed and unbonded methods is only 3/4 in. (9 lb./sq. ft.), the unbonded method should always be considered because it is more effective in reducing cracking hazard.

In the principal bonded methods—monolithic, chemically bonded and thin-set—the relatively thin topping by itself cannot resist the various stresses acting on it. Adequate bond between topping and subfloor is necessary to resist stresses across the full depth of the floor construction; therefore, special care should be taken to remove foreign substances, which might impair bond.

Also, a topping should be of uniform thickness to prevent concentration of stresses and consequent cracking at thin spots. It is important that the subfloor be leveled or pitched to conform to the desired finished floor profile.

BONDED-UNDERBED METHOD   The bonded-underbed method, which can be used over a metal or a concrete subfloor, provides a measure of crack control for portland cement terrazzo not afforded by other bonded installations. The total recommended thickness of underbed and topping is 1-3/4 in. for bonded-underbed, compared to 2-1/2 in. for unbonded installations.

The crack control advantage of this method is derived from the relatively frequent spacing of divider strips. The maximum spacing of 6 ft. on centers for conventional divider strips induces unnoticeable stress-relieving cracks along the strips. The minimum overall terrazzo thickness results from the normal strip depth (1-1/4 or 1-1/2 in.) plus a 1/4-in. tolerance for subfloor irregularity.

The underbed should be bonded properly to the subfloor. Natural bond can be obtained between a portland cement underbed and a suitably roughened concrete slab. A chemical bonding agent can be used over a smooth-troweled concrete slab to bond the underbed. Over a metal deck, adequate bond can be expected without special treatment if the surface is free of corrosion, pickling acids, oil, and other foreign substances. Additional bond is gained by the keying action of the underbed with depressions in a cellular metal deck. Corrugated and ribbed metal decks generally depend on composite action of the deck and a reinforced structural concrete slab. In such floor assemblies, the top of the concrete slab constitutes the subfloor, and topping should be installed by the monolithic or chemically bonded method.

Underbed and topping generally are installed as described in Section 9.7.2.1.

MONOLITHIC METHOD   A monolithic bonded terrazzo finish is a portland cement topping installed over a concrete subfloor in which natural bond is developed between the portland cement surfaces. Monolithic installations can be used in interior as well as exterior locations.

The concrete subfloor should be reinforced properly and be at least 3-5/8 in. thick. The surface should be brushed with a wire broom or otherwise roughened to remove laitance and ensure good mechanical bond with the topping. As soon as the concrete is sufficiently hard to resist tearing by a jointing tool or power saw, control joints should be installed in the slab about 15 to 20 ft. apart (see Fig. 9.7-10). Divider or expansion strips should be installed in the topping over these control joints in the subfloor. The surface should be protected from damage and dirt with suitable coverings until the topping is placed.

If the subfloor has not been adequately protected from dirt, paint, plaster, or other

construction residue, the surface should be thoroughly scrubbed with detergents and rinsed clean. Then the subfloor should be saturated with water by ponding overnight. Excess water should be removed, and a slush coat of neat portland cement, pigmented to match the topping, should be broomed into the surface immediately before the topping mix is applied. The topping should be placed, finished, and cured as described in Section 9.7.2.1.

CHEMICALLY BONDED METHOD   This method of installing portland cement topping is intended for smooth-troweled concrete subfloors. For that reason it often is suitable for older concrete floors not originally prepared for a terrazzo topping. Where a terrazzo topping is planned in the initial design, a roughened subfloor surface is preferred.

The subfloor should be inspected; where random cracks exist or control joints are not provided 15 to 20 ft. on centers, such joints should be installed with a power saw. Divider strips or expansion strips should be placed over all joints in the subfloor.

Chemically bonded installations depend on a bonding agent such as epoxy, epoxy-polysulfide, or polyvinyl-chloride resin. The subfloor surface should be free of coatings such as paint, curing compounds, and wax, and the bonding agent should be applied in accordance with the manufacturer's recommendations.

The topping should be placed, finished, and cured as described in Section 9.7.2.1.

THIN-SET METHOD   Using a resinous binder, it is possible to install very thin toppings over practically any sound, firm subfloor such as wood, metal, or concrete. Epoxy and polyester toppings should be at least 1/4 in. thick. When they are installed over concrete, the subfloor surface should be smooth-troweled and level to within 1/8 in. in every 10 ft.

Although some manufacturers allow resinous toppings 1/8 in. thinner than those recommended above, industry consensus favors the slightly heavier installations. This is particularly true of epoxy and polyester, where a 1/8-in. thickness would require decorative aggregate the size of sand, and an acceptable minimum thickness over subfloor irregularities would be difficult to achieve.

Thin-set terrazzo finishes should not be installed on slab-on-grade construction subject to excessive moisture or high alkaline content. To test for moisture, a piece of polyethylene film about 2 ft. square should be taped down in several locations and be allowed to remain for 8 hours. Visible moisture accumulation under the mat is an indication of excessive moisture in or under the slab. Under such conditions, the alkalinity of the moisture also should be tested, and the thin-set matrix manufacturer should be consulted.

Divider strips are not required with resinous binders for shrinkage control, but they should be provided where there is a discontinuity in the subfloor, such as at isolation joints, control joints, and changes in materials. Divider strips may be provided for decorative purposes or as construction joints at the termination of a pour. Since wall base often is installed as a separate operation, a strip should be provided at the toe of the base. In large areas or long corridors, expansion strips should be installed at 30-ft. intervals. Closer spacing of expansion or divider strips may be preferred for decorative purposes, to minimize the illusion of waviness or to ensure uniform topping thickness.

The thin-set matrix manufacturer's recommendations should be followed for using primers, mixing ingredients, and applying the topping mix, and grinding and sealing the surface. Over wood subfloors, the terrazzo installer's specific recommendations should be consulted for details of taping plywood joints and for use of mastic underlayment and wire or metal reinforcement.

Generally, the dry components such as fillers and chips are mixed first, then added to the liquid resin. The mix is poured onto the subfloor, then spread and leveled with a trowel. The setting time before grinding depends on the specific resin formulations. Epoxy and polyester toppings can be ground the day after application.

For residential use, sealing of the surface often is not required, because the resinous matrixes are quite dense and resistant to most staining agents. Marble chips, however, vary in their stain resistance, and, depending on anticipated traffic conditions, a sealer may be desirable. The matrix manufacturer's recommendations should be consulted in each case.

Traditional terrazzo installations can develop surface imperfections due to portland cement shrinkage, and the appearance of the finished floor sometimes suffers from uneven chip distribution. To overcome these problems, the Terrazzo Industry Research Institute, a branch of the NTMA developed a special vibrating machine that improves the performance and appearance of most terrazzo floor installations, including rustic and polyacrylate thin-set.

This machine has a series of rollers, some of which are attached to a vibrator. On the first pass, the rollers align the chips with their flat surfaces up. Then, the vibrating rollers push the chips below the surface and evenly distribute them, making it possible to use more chips and less water. Although the vibrating machine can use stiffer mixes, chemical additives often improve the workability lost by reducing the water-cement ratio.

The reduction of water, the use of more chips, and thorough vibration generally result in a denser, harder terrazzo surface with little shrinkage cracking and few pinholes.

9.7.2.3

Refinishing Existing Floors

New terrazzo finishes can be installed over practically any type of existing floor finish. The unbonded method can be used if the subfloor and the structural system are capable of supporting the additional dead load (28 lb./sq. ft. for a 2-1/2-in.-thick finish) and the new finished floor height is not objectionable. Since no bond with the existing floor is intended, relatively little preparation of the surface is necessary, and the type of existing finish is not critical. An irregular or slightly pitched surface can be leveled with a sand cushion under the isolation membrane.

If existing structural and architectural conditions limit the finish thickness and the dead load, a bonded method should be considered. Chemically bonded and thin-set methods are usually suitable for refinishing existing floors, but the monolithic method, which requires a roughened subfloor for bond, is used infrequently. A chemically bonded topping is thicker and heavier and requires a concrete subfloor,

but it is not limited as to chip size. Lighter-weight resinous thin-set toppings can be used over most subfloors, including wood.

As in new construction, the condition of the subfloor or of the old finish surface is of critical importance for thin-set resinous toppings. Surface coatings such as paint, wax, and sealing compounds should be removed with appropriate solvents, sanding, or grinding.

A topping can be installed directly over properly cleaned marble tile, quarry tile, slate, and other hard tile products. Resilient tile and sheet flooring, as well as their adhesives, should be removed completely. The surface should be ground smooth to expose the subfloor before a terrazzo topping is installed. A chemical bonding agent or a resinous binder should be tested first in a small area to ensure proper bond with the prepared surface.

Existing wood strip and lumber board floors should be covered with stable mastic or plywood underlayment at least 3/8 in. thick. The matrix manufacturer's recommendations should be consulted to determine the proper type of underlayment and the installation procedure for each type of resin matrix.

9.7.2.4

Special Surface Finishes

Terrazzo floors can be described according to (1) binder (portland cement or one of the resins), (2) installation method (unbonded or one of the bonded methods), (3) chip size and type (standard, intermediate, Venetian, or Palladiana), and (4) surface finish (regular, rustic, or conductive) (Fig. 9.7-18).

Terrazzo installations in houses, apartments, offices, stores, building lobbies, and other locations of normal foot traffic require no special additives or surface treatment. The usual binder formulations previously described, after normal grinding and polishing, provide adequate resistance to abrasion, staining, discoloration, and slipping hazard. These regular terrazzo finishes are used in the great majority of installations.

Special surface finishes have been developed for situations where unusual hazards exist. In exterior locations, a greater degree of protection from slipping

FIGURE 9.7-18
Surface Finishes for Various Installation Methods

Installation Method	Binder	Chip Size	Surface Finish
Thin-set	Polyester and epoxy	#0 and #1	Regular
	Latex	Standard	Conductive
Monolithic	Portland cement	Standard	Regular
		Intermediate	Rustic
Chemically bonded		Venetian	
Bonded-underbed	Portland cement	Standard	Regular
		Intermediate	Rustic
Unbonded		Venetian	Conductive

hazard is afforded by rustic terrazzo. In hospitals and laboratories, where oxygen is used and potentially explosive atmospheres exist, conductive terrazzo will prevent electrostatic sparks. For locations subject to damage from acids, such as hospitals, laboratories, and industrial buildings, many resinous matrixes are suitable and are replacing the specialized acid-resistant portland cement formulations once recommended. Where greater slip resistance than that provided by regular terrazzo is required for interior locations, as on ramps and stairs, regularly spaced abrasive strips are recommended (Fig. 9.7-19). Some resinous matrixes can be formulated to provide inherently good slip-resistant properties. Abrasive terrazzo, containing abrasive aggregates in the mix or on the surface, is no longer recommended because its slip resistance has proven to be temporary. The regular and special surface finishes can be used with most installation methods (see Fig. 9.7-18).

■ RUSTIC TERRAZZO

Rustic (washed) terrazzo is particularly suitable for nonslip and decorative exterior surfaces. The decorative chips are usually either weather-resistant marble, quartz, or granite chips or decorative gravel. Although resinous binders can be used, they generally are limited to decorative wall installations. Portland cement is the common material for rustic terrazzo floors, and in areas subject to freezing, portland cement with an air-entraining agent is recommended. Rustic terrazzo should not be installed where the surface will be subjected to de-icing salts or chemicals.

FIGURE 9.7-19 *Abrasive inserts for nonslip surfaces: (a) prefabricated strips and nosing; (b) abrasive compound poured into channel.*

After the topping has been placed, rolled, and leveled as described in Section 9.7.2.1, the surface should be washed with a stream of water from a hose. The water pressure should be sufficient to remove the paste to a depth of about 1/16 in. between chips or gravel. The surface then should be scrubbed with a solution of 1 part muriatic acid to 10 parts water to remove particles of paste clinging to the aggregate. A final rinsing with clean water removes the acid solution and neutralizes the surface. A light grinding may be desirable for exposed chips in pool areas, and the floor may be sealed in accordance with manufacturer's directions, but usually no further operations are necessary.

■ CONDUCTIVE FLOORS

Conductive floors are used to prevent electrostatic sparks in potentially explosive atmospheres. Formerly, for portland cement terrazzo, conductivity was achieved by adding a chemical powder, acetylene black, dissolved in isopropyl alcohol to the topping and underbed mixes at the job site. This method is no longer recommended by the NTMA. Instead, resinous matrixes that have the conductive ingredient carbon black added during manufacture are recommended. Such a terrazzo topping conducts electricity at levels prescribed in Article 4-6.2 of the National Fire Protection Association's Standard No. 56A. Only black conductive matrixes are available due to the lamp black conductive material. Chip sizes in conductive terrazzo should be no larger than #1 size.

9.7.2.5

Special Elements

Terrazzo finishes are not limited to flooring or horizontal surfaces. Cast-in-place terrazzo can be carried up a wall 4 to 12 in. to form a coved wall base (Fig. 9.7-20) or continued to form a protective wainscot (Fig. 9.7-21). Full-height terrazzo finishes are possible, though generally they are not practical because of the labor involved in grinding vertical surfaces, which must be done with relatively small, handheld wheels (Fig. 9.7-22). Where terrazzo floors are being installed on the premises, cast-in-place terrazzo stairs often are practical and economical (Fig. 9.7-23). Where time is of the essence, precast treads, risers, and wall base (Fig. 9.7-24) are available for use in conjunction with terrazzo or other flooring materials.

9.7.2.6

Maintenance

Terrazzo must be protected from damage and soiling and cleaned periodically if it is to maintain its original appearance.

■ PROTECTION

Portland cement matrixes are fairly porous and require protection from deterioration and staining by various chemical agents. Therefore, portland cement terrazzo

FIGURE 9.7-20 *Coved wall base: (a) straight type and (b) splayed type. Underbed thickness may be varied, creating recessed, flush (shown), or projecting base.*

FIGURE 9.7-21 *Wainscot can be installed over studs and metal lath (shown) or over any masonry base that will provide adequate bond for portland cement mortar.*

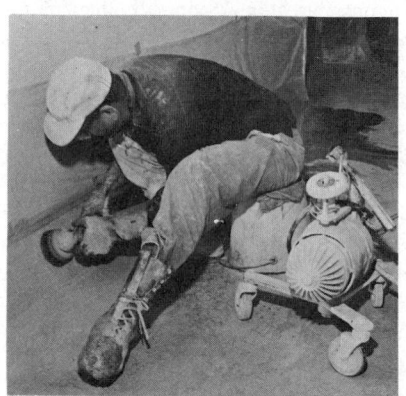

FIGURE 9.7-22 *Base grinding: small hand-held wheels are used to grind and polish terrazzo base and wainscoting.*

FIGURE 9.7-23 *Cast-in-place stairs: (a) continuous treads and risers over concrete substair; (b) treads cast in steel pan forms.*

FIGURE 9.7-24 *Precast stair treads and risers can be used with (a) concrete or (b and c) metal stair construction.*

generally requires protection with a proprietary penetrating sealer capable of internally sealing the pores. Resinous matrixes are denser, more stain resistant, and chemically inert. However, all terrazzo finishes consist of at least 70% stone or marble, both of which possess varying degrees of porosity and stain resistance.

Resinous terrazzo may require similar protection depending on the chip type and the conditions of use. When used, sealers should be of a type specifically recommended by the manufacturer for terrazzo surfaces and should be applied in accordance with the manufacturer's recommendations.

Both resinous and cementitious terrazzo floors are sufficiently hard and abrasion resistant over their entire surface. Therefore, surface coatings such as wax that provide abrasion protection for softer floor finishes are not required. The effect of foot traffic on unwaxed terrazzo generally is a positive one, tending to polish and impart further luster to the surface. Surface coatings also generally are not recommended because they tend to make the floor slippery when wet and tend to discolor the surface in low-traffic areas, unless periodically removed and replaced with new coatings of uniform thickness.

■ **CLEANING**

Terrazzo should be cleaned periodically by damp mopping. Only a neutral liquid cleaner recommended for the purpose by the manufacturer should be used. Soaps and scrubbing powders that contain detergents, water-soluble inorganic salts, or crystallizing salts should not be used. Sweeping compounds that contain oil may stain and permanently discolor a terrazzo surface. Dry buffing after each mopping (and as required between moppings) will remove surface dirt and restore luster.

Terrazzo, like any floor surface, is susceptible to staining. Stains should be treated as soon as possible because they are more difficult to remove after they have set. The type of stain should be determined so that the proper remover can be selected. Stain removers fall into three general categories: (1) solvents, such as carbon tetrachloride, which dissolve grease, chewing gum, and lipstick; (2) absorbents, such as chalk, talcum powder, blotting paper, and cotton, which absorb fresh grease and moist stains; and (3) bleaches, such as household ammonia, hydrogen peroxide, acetic acid, and lemon juice, which discolor stains. Manufacturers of terrazzo sealers and cleaners also produce stain removal kits, which can be used successfully if the manufacturer's recommendations are followed carefully.

9.8 ACOUSTICAL TREATMENT

Chapter 10 includes general requirements for and the theory behind sound control in buildings. This section discusses in more detail that part of sound control generally called *acoustical treatment*. This includes acoustical ceilings, integrated ceilings, and acoustical wall treatment and baffles. Two or more of these items are often used together (Fig. 9.8-1). Acoustical treatment also includes the furring and suspension systems needed to mount acoustical materials.

Before proceeding it will be helpful to define some of the terms used in this section.

Acoustical baffles. An acoustical panel that is suspended from an edge or that floats cloud-fashion in a space.

Acoustical ceiling. The entire ceiling system, including acoustical material and associated insulation, furring, and suspension systems. Does not include auxiliary thermal or acoustical insulation in addition to that normally associated with the acoustical ceiling being discussed. Thus, a layer of acoustical insulation placed above a ceiling at a location over or adjacent to a partition to achieve increased acoustical performance is not here included as part of the acoustical ceiling. An acoustical blanket or pad necessary to achieve the rated acoustical performance of a metal ceiling is, however, a part of the acoustical ceiling. Metal ceilings, as defined below, are also called acoustical ceilings.

Acoustical material. A general term that covers both composition and metal acoustical materials as defined in this chapter.

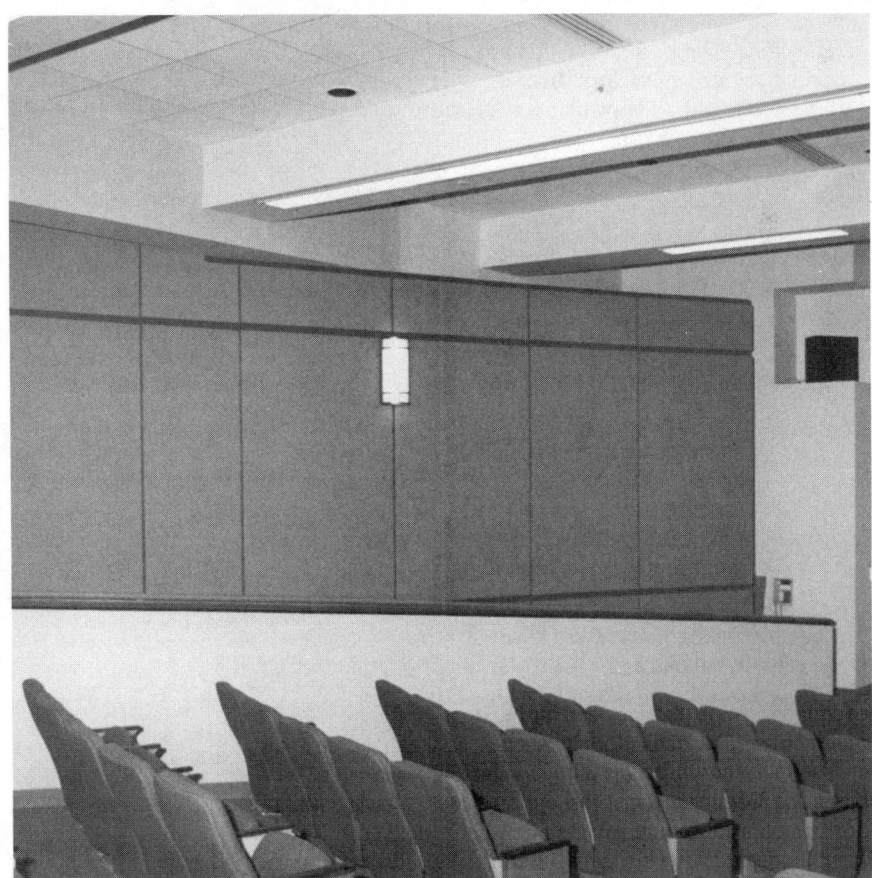

FIGURE 9.8-1 *Acoustical ceiling and wall panels are both used in this auditorium.* *(Photograph provided by MBI Products Company.)*

Acoustical wall panels. Acoustical panels that mount on the surface of a wall or partition, regardless of their mounting method or size. Acoustical wall panels may completely cover a wall surface or be mounted as plaques with space between them.

Composition. When referring to acoustical materials, the term *composition* includes cellulose, mineral fiber, glass fiber, ceramic-faced, plastic, gypsum, and wood fiber acoustical materials. Plastic- and metal-faced acoustical ceiling materials in which the plastic or metal is adhered or bonded to a composition material as defined here are also called composition materials.

Integrated ceiling. A ceiling system consisting of acoustical material, suspension system, air distribution outlets, and lighting fixtures.

Metal acoustical ceiling material. Linear metal strips and metal panels and pans are all called metal acoustical materials, even though some of them may be of questionable acoustic value. The term *metal acoustical material* means the ceiling material only. It does not include the associated suspension system or acoustical insulation blankets or pads.

Metal ceiling. The entire ceiling construction, including metal acoustical materials, acoustical pads or blankets, and the suspension system.

ACOUSTICAL CEILINGS

Acoustical ceilings include composition tiles and panels, metal acoustical materials, and associated furring and suspension systems. Integrated ceilings are discussed later in this chapter.

Acoustical Materials

The characteristics of acoustical materials are generally identified according to ASTM Standard E 1264, "Standard Classification for Acoustical Ceiling Products." It classifies acoustical materials according to type, pattern, fire class, acoustical performance, and light-reflectance coefficient.

Types are shown in Figure 9.8-2; patterns in Figure 9.8-3. Three fire classes are listed in ASTM E 1264: Class A material has a flame spread rating ranging from 0 to 25; Class B from 26 to 75; Class C from 76 to 200. These classes correspond directly to Classes I, II, and III required by some building codes.

Acoustical performance is rated by Noise Reduction Coefficient (NRC) for closed areas and by either Articulation Class (AC) or Speech Privacy Isolation Class (NIC') for open-plan areas. The sound

FIGURE 9.8-2
Acoustical Material Types

Type I	Cellulose base with painted finish
Type II	Cellulose composition with membrane-faced overlay
Type III	Mineral base with painted finish; type includes: Form 1 Nodulated, cast, or molded Form 2 Water felted Form 3 Dry felted
Type IV	Mineral base with membrane-faced overlay; type includes: Form 1 Nodulated, cast, or molded Form 2 Water felted Form 3 Dry felted
Type V	Perforated steel facing (pan) with glass fiber base backing
Type VI	Perforated stainless steel facing (pan) with mineral or glass fiber base backing
Type VII	Perforated aluminum facing (pan) with mineral or glass fiber base backing
Type VIII	Cellulose base with scrubbable pigmented or clear finish
Type IX	Mineral base with scrubbable pigmented or clear finish; type includes: Form 1 Nodulated, cast, or molded Form 2 Water felted Form 3 Dry felted
Type X	Mineral base with plastic or aluminum membrane-faced overlay or both
Type XI	Mineral base with fabric-faced overlay; type includes: Form 1 Nodulated, cast, or molded Form 2 Water felted Form 3 Dry felted
Type XII	Glass fiber base with membrane-faced overlay; type includes: Form 1 Plastic Form 2 Cloth Form 3 Other
Type XIII	Aluminum or steel strip with mineral or glass fiber base backing; type includes: Form 1 Perforated Form 2 Nonperforated
Type XX	Other types (this category is left for additional types)

Copyright, ASTM, reprinted with permission.

FIGURE 9.8-3
Acoustical Material Patterns

A. Perforated, regularly spaced large holes
B. Perforated, randomly spaced large holes
C. Perforated, small holes
D. Fissured
E. Lightly textured
F. Heavily textured
G. Smooth
H. Printed
I. Embossed
J. Embossed-in-register
K. Surface scoured
Z. Other patterns (describe)

Copyright ASTM, reprinted with permission.

transmission through an acoustical ceiling is rated according to Ceiling Sound Transmission Class (CSTC).

Light reflectance is rated as one of four coefficients: LR 1 includes light reflectances of 75% or more; LR 2 ranges from 70% to 74%; LR 3 from 65% to 69%; and LR4 from 60% to 64%.

■ MATERIALS AND FINISHES

New acoustical materials will usually be one of the types shown in Figure 9.8-2, but other materials are also sometimes available, including ceramic-faced units that are used in wet and high-impact areas, units with other abuse-resistant or sanitary plastic faces, gypsum, and wood fiber composition. Acoustical units in existing buildings may contain asbestos. Suspended ceilings may also be of wood, plastic, or sheet metal, but most of those types are more properly classified as decorative rather than acoustical ceiling materials.

Most composition acoustical units are painted on the exposed face. Most such painted finishes are considered washable, although gentleness is advised. Some manufacturers list some products as having a "scrubbable" finish; and these are listed in ASTM E 1264, but there is no universally accepted definition of the term.

Metal ceiling panels are made from aluminum, steel, and stainless steel. A variety of finishes are available, including fluoropolymer, baked enamel, natural, anodized aluminum, acrylic lacquer, and even wood veneer.

■ ACOUSTICAL PERFORMANCE

Acoustical products are classified by Sound Transmission Class (STC) and Noise Reduction Coefficient (NRC), by Articulation Class (AC), or by Speech Privacy Isolation Class (NIC'). An acoustical ceiling is rated by Ceiling Sound Transmission Class (CSTC). STC, CSTC, and NRC ratings are usually used in closed-room situations. AC and NIC' are used for large spaces that contain several activities, as in an open-plan office space. AC- and NIC'-rated products have been tested for their ability to both control sound reflections and uniformly permit white-noise transmission from plenum speakers into a space. White noise is used in open-space designs to mask background noises and help provide speech privacy.

The sound absorption quality of an acoustical material is indicated by NRC, which shows a material's ability to reduce sound levels within a space. An acoustical material's noise isolation control ability is indicated by STC (CSTC for ceilings), which shows a material's ability to prevent sound from passing through it into another space. Usually there is an inverse relationship between the two properties. Materials that reduce sound reverberations are usually light and porous. To prevent sound transmission, materials must be heavy and airtight. STC and CSTC are measured by test procedure AMA 1-II, "Ceiling Sound Transmission Test by the Two Room Method," which gets its name from the no-longer existent Acoustical Materials Association. The standard is available from the Ceilings and Interior Systems Contractors Association (CISCA).

Where sound control is an essential element in an existing building, the advice of an acoustical specialist should be sought before changes or repairs are made.

■ COMPOSITION ACOUSTICAL TILE VERSUS PANELS

Composition acoustical ceiling materials may be tile or panels. In the past, much tile was used. In recent years, the advent of better acoustical panel materials, more attractive and functional suspension systems, an increased need for access to the space above ceilings, changes in construction methods, their lower cost, and other factors have led to a much greater use of panels. Some manufacturers have stopped

making tile products and concealed grid suspension systems. An advantage of panels over tile is that panels are loosely laid into a grid, which makes them easily removable and the space above them readily accessible. A major disadvantage is that they are loosely laid into a grid, which makes them subject to dislocation by wind, pressure applied during cleaning, and air passing through a plenum space above them. Tile and panel edges are square, beveled, or have a special edge such as the reveal edge shown in Figure 9.8-4.

TILE Any of the types, patterns, and other characteristics listed in previous paragraphs may be found. Most tile today is 12 or 24 in. square or 12 by 24 in. Tile in older buildings will probably be 12, 16, or 20 in. square.

Tile joints are flanged, rabbeted, kerfed (splined), or tongue and groove. Figure 9.8-5 shows some types, but many others are in use today, and still others have been used in the past. Sometimes, a tile's edges are not all cut the same. The tongue-and-groove joints on two sides of a tile may be quite different from those on the other two sides, so the tile will nest properly.

PANELS Any of the types, patterns, and other characteristics listed in previous paragraphs may be found in panels. Most panels are either 24 in. square or 24 by 48 in. Some are scored so they appear from the floor to be smaller units. Scoring may produce squares or linear strips. Scoring may be parallel with the tile sides or diagonal to them.

Panel joints are usually square or reveal type (see Fig. 9.8-4). In systems where cross-suspension members are concealed (semiconcealed systems), the ends of the panels are usually kerfed and rabbeted to receive the concealed suspension member.

FIGURE 9.8-4 *Acoustical tile and panel edge conditions. Many variations of the beveled and reveal edge conditions are available. For example, reveal edges may have two, three, or more steps. Bevels may be combined with steps.* (Courtesy H. Leslie Simmons.)

FIGURE 9.8-5 *Typical acoustical tile joint conditions.* (Courtesy H. Leslie Simmons.)

■ METAL ACOUSTICAL CEILING MATERIALS

In addition to metal-faced composition products, there are many metal acoustical ceiling material styles (Fig. 9.8-6). They include but are not limited to linear shapes (strips), flat panels, shaped (corrugated and other shapes) panels, and flat or textured pans. Surfaces may be continuous or perforated, and either flat or embossed. Metal grids and louvers are also available, but whether they are acoustical in nature is debatable. Metal ceiling members may be lay-in or snap-in type. Snap-in types require special support systems. Some lay-in types also require special support systems, while others do not. Some panels come with glass fiber or mineral wool sound-absorbing pads; others do not.

Support Systems

Composition acoustical materials may be directly attached to solid or supported substrates or supported on a suspension system. Metal ceilings usually include a suspension system. Generally, the term *suspension system* means some sort of metal grid that is hung by wires or bars from the structure, but the term also includes systems of wood or metal furring to which acoustical materials are screwed, stapled, or nailed. Composition materials so attached are usually tiles. Sheet metal decorative ceilings are often held in place using fasteners, but metal acoustical ceiling materials are seldom so installed.

■ WOOD FURRING

Except in single-family residences and other small buildings, little acoustical tile is installed directly on wood furring today (Fig. 9.8-7), but in the past such installations were common. The normal installation when wood furring is used today is to install a layer of gypsum board, called backing board, over the furring and then to fasten the acoustical tile to the gypsum board.

■ METAL FURRING

Today, very little acoustical material is installed directly on metal furring bars (Fig. 9.8-8), but in the past some such installations did occur. Today, when composition accoustical tile is installed over

FIGURE 9.8-6 *An angled ceiling made from flat, perforated metal acoustical ceiling panels. (Manufactured by Hunter Douglas Architectural Products.)*

FIGURE 9.8-7 *Acoustical tile on wood furring. (Courtesy H. Leslie Simmons.)*

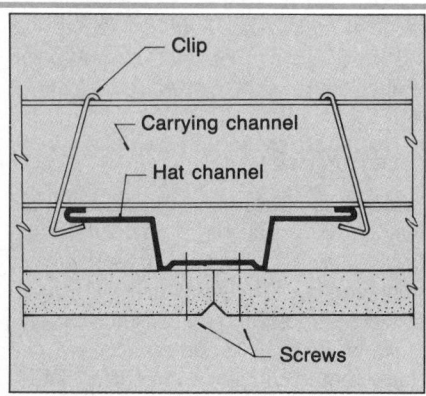

FIGURE 9.8-8 *Acoustical tile directly applied to metal furring. (Courtesy H. Leslie Simmons.)*

metal furring bars, the method usually used is that described in ASTM Standard C 635 as "Furring Bar Suspension System," in which a layer of gypsum board, called backing board, is applied to suspended nailing or furring bars. The acoustical tile is then fastened to the gypsum board. Metal acoustical materials are seldom installed over metal furring bars today.

Metal acoustical ceiling materials may occasionally be installed on snap-in or other grid members that are directly clipped to structural elements or inserts in concrete slabs. That installation is so similar to that described later in this chapter for a metal suspension system that the requirements addressed there apply.

Metal furring bars that directly support acoustical materials may be screwed or clipped to the supporting structure or clipped or screwed to carrying channels, which are in turn suspended from the structure by wires or bars.

■ HUNG METAL SUSPENSION SYSTEMS

There are two categories of hung suspension systems for acoustical materials: (1) *direct hung*, in which the main runners are suspended from the structure, such as that shown in Figure 9.8-9; and (2) *indirect hung*, in which the main runners are suspended from carrying channels, such as that shown in Figure 9.8-10. The main runners in either category do not have to be those shown. They may be any type that is appropriate for the acoustical material being supported. Some common types are shown in Figure 9.8-11, but there are many others.

Lay-in suspension systems may be either direct- or indirect-hung systems, depending on the system members and the necessary

FIGURE 9.8-9 *One type of direct-hung acoustical ceiling suspension system.* (*Courtesy H. Leslie Simmons.*)

FIGURE 9.8-10 *One type of indirect-hung acoustical ceiling suspension system.* (*Courtesy H. Leslie Simmons.*)

spacing of hangers. *Concealed grid* systems may also be either direct or indirect hung, but systems having splines or other less rigid cross-members are usually indirect hung. Except for the carrying channels, most members in both types of systems are proprietary, meaning they are manufactured or furnished by a single supplier or producer for installation as a system.

Metal suspension systems for acoustical ceilings should comply with the applicable requirements of ASTM Standards C 635 and C 636. Acoustical treatment manufacturers may have more stringent requirements.

Hung suspension systems for composition materials include concealed grid types for tile and exposed or semiexposed grid types for lay-in panels.

Hung suspension systems for metal acoustical ceiling materials include lay-in types and snap-in types. Their shapes vary with the manufacturer. Figure 9.8-12 shows a common T-bar snap-in type, but it is far from the only shape used. Some types of metal ceiling grids may be direct hung even though they do not have cross-furring members, because the metal ceiling material itself is stiff enough to stabilize the system without additional bracing.

CONCEALED GRID SYSTEMS FOR COMPOSITION TILE Most hung suspension systems designed to be concealed in the finished ceiling are H and T systems, which include H-shaped main runners and T-splines similar to those shown in Figure 9.8-11; concealed single- or double-web

FIGURE 9.8-11 *Typical acoustical ceiling suspension system components.* (*Courtesy H. Leslie Simmons.*)

Clip

Carrying
channel

Snap-in
T-Bar

Metal
pan

FIGURE 9.8-12 *One type of T-bar snap-in metal ceiling system.* (*Courtesy H. Leslie Simmons.*)

removable only by breaking out a tile and then removing the adjacent tile until the necessary opening has been achieved. Where the space above the ceiling is not accessible through the tiles, access doors must be provided.

EXPOSED AND SEMIEXPOSED GRIDS FOR COMPOSITION PANELS Suspension systems designed for lay-in application of acoustical panels usually consist of T-shaped main runners and interlocking cross-tees (see Fig. 9.8-9). The cross-members of semiconcealed systems are usually T-splines. The main runners are usually directly hung from the structure with no carrying channels. Main runner and cross-runner shapes are proprietary and vary with each system manufacturer. They are available with wide or narrow flat-bottomed surfaces, roll-formed caps, slotted, with reveals, and in other configurations. A few of the many configurations available are shown in Figure 9.8-11.

GRIDS FOR METAL CEILINGS Metal ceilings are either lay-in or snap-in type. Some metal pans and panels lay into the same type of grid used for composition material. Others hang from similar grids (Fig. 9.8-13).

systems using T-shaped main runners similar to the standard tee shown in Figure 9.8-11 and spacer bars similar to those shown in the same figure at right angles to the main runners for stiffening; or zee systems using zees and splines similar to those shown in Figure 9.8-11. Other systems are also used today, and still others have been used in the past, so an existing installation may be different from those noted here.

A characteristic of all concealed suspension systems for composition tile is that they lock the tile in place so that it cannot be easily removed. To overcome this drawback, manufacturers have devised ways to make sections of tile hinged or removable so access can be gained where necessary. In direct-hung systems, hinged tile sections, after being released by a special tool, may be moved upward into the plenum or downward into the room.

In both types, usually more than one tile is movable. Strips of tile between cross-framing members either lift up or can be pulled down to provide access. The movable panels of tile may be hinged along the cross-tees or along the main tees. Once the movable tile has been moved out of the way, the remainder of the tile can be removed in successive pieces. Downward access systems are not recommended in some seismic zones.

In indirect-hung systems, access is sometimes gained with special cross-runners or split splines that permit acoustical unit removal. Such special runners or splines are identified with unobtrusive permanent markers. Some indirect-hung systems are made accessible with special tools that work only with those systems.

Not all systems, however, have a removable-unit feature. Some older tile may be

Main runner

Metal acoustical panel
Acoustical pad
Main runner

Metal acoustical panel
Accoustical pad

Main runners

Acoustical pad
Metal acoustical panel

FIGURE 9.8-13 *Some typical hung metal ceiling systems.* (*Courtesy H. Leslie Simmons.*)

Support grids for linear metal ceilings, while similar to that shown in Figure 9.8-14, vary from manufacturer to manufacturer to suit their particular ceiling materials.

FIGURE 9.8-14 *Typical support grid member for a linear metal ceiling.* (*Courtesy H. Leslie Simmons.*)

Many metal pans and some panels are installed into snap-in grids. The T-bar snap-in member shown in Figure 9.8-12 is only one of several snap-in-device types in general use. Snap-in grid members are manufactured specifically to fit the edge configuration of the pans or panels being supported. A grid member made by one manufacturer may not correctly fit or support a panel made by a different manufacturer.

STRUCTURAL CLASSIFICATION ASTM Standard C 635 divides hung suspension systems into three structural strength categories: *light duty*, *intermediate duty*, and *heavy duty*. Light-duty systems, which are used in residential and light-commercial installations, will support only the acoustical ceiling materials themselves. Intermediate-duty systems, which are used in most commercial structures, will support light fixtures and air diffusers in addition to the acoustical materials themselves. Heavy-duty systems will support heavier loads than the other two classifications.

EDGE MOLDINGS AND TRIM With the exception of some linear metal ceilings, almost all acoustical ceiling suspension systems should have matching edge moldings at all vertical surfaces to which the acoustical ceiling abuts, including penetrations such as columns. Acoustical tile ceilings applied with fasteners or adhesives should also have such edge moldings and trim.

The edge molding and trim type and configuration should be suitable for the suspension system used. Some ceiling types have specialized trim members used only for that type.

METALS AND FINISHES In exterior installations and in interior installations where high humidity will be present, such as swimming pool areas and commercial kitchens, concealed metals should be aluminum, stainless steel, zinc alloy, or hot-dip galvanized steel. The finishes on such metals should comply with ASTM Standard C 635 requirements in the paragraph titled "Coating Classification for Severe Environment Performance."

In locations other than those mentioned in the preceding paragraph, for composition tile and panels the manufacturer's standard steel products are generally used with the manufacturer's standard galvanized or painted finish.

Exposed grids and edge members in exterior installations and in interior installations where high humidity will be present, such as swimming pool areas and commercial kitchens, should be either aluminum or stainless steel. Aluminum is usually protected by a baked-on paint, fluoropolymer, or anodized finish, but it may also be left natural and protected by a coat of lacquer. Stainless steel may be protected by a coating or left exposed. The finishes should comply with ASTM Standard C 635 requirements in the paragraph titled "Coating Classification for Severe Environment Performance."

Exposed grids and edge members in locations other than those mentioned in the preceding paragraph are generally steel-finished with the manufacturer's standard paint finish, but they may be aluminum or stainless steel where those decorative finishes are selected. Where the acoustical ceiling material has a decorative finish, such as anodized aluminum or stainless steel, exposed suspension system members often match the ceiling material and finish.

Colors of exposed finished metal vary, although some sort of white or off-white is probably the most common paint color.

9.8.1.3

Accessories and Miscellaneous Materials

The types of accessories needed for acoustical ceiling installation include the following:

Carrying channels. 1-1/2-in. cold-rolled steel channels, weighing a minimum of 0.475 lb./lin. ft., painted. Where spans require, 2-in. cold-rolled steel channels weighing a minimum of 0.590 lb./lin. ft.

Hanger wire. For typical installation, galvanized carbon steel wire of the size necessary to properly support three times the hanger design loads but in no case less than 12-gauge wire. Some codes may require heavier wire regardless of loads. For support of aluminum or stainless steel grid systems and in exterior locations and air plenums, stainless steel or aluminum wire of a gauge recommended by the suspension system manufacturer.

Clips. Commercial types acceptable under the building code for support.

Hold-down clips. Suspension system manufacturer's standard types. Hold-down clips are necessary where uplift is a problem, whether from air pressure or to permit cleaning. Clips are also necessary in some fire-rated assemblies. Security clips are sometimes used to make metal pan ceilings nonremovable.

Tie wire. Galvanized wire, at least 16 gauge.

Extension hangers for metal ceilings, hanger rods, and flat bars. Zinc- or cadmium-coated mild steel. Alternatively, such hangers may be finished with rust-inhibitive paint.

Angle-shaped hangers. Formed from nominal 15- or 16-gauge galvanized sheet steel. Legs should not be less than 7/8 in. wide.

Hanger anchorage devices. Rust-inhibitive screws, clips, bolts, concrete inserts, expansion anchors, or other devices, sized to support five times the calculated hanger loading.

Screws and other fasteners. Rust-inhibitive and of a type standard with the suspension system manufacturer. Tile fasteners should be stainless steel or cadmium plated of the type recommended by the tile manufacturer and should penetrate the substrate by at least 1/2 in. Exposed fasteners should match the material fastened in appearance.

Acoustical sealant. A water-based, non-drying, nonbleeding, nonstaining type that is permanently elastic.

Sound attenuation blankets. Semirigid mineral fiber blankets with no membrane facing; with Class 25 flame spread; 1-1/2 in. thick or more.

Acoustical pads for use in metal pan and panel ceilings. Mineral or glass fiber. Some are wrapped or faced with materials such as polyvinyl chloride. Others are bare.

Adhesives for installing acoustical tile. Comply with ASTM Standard D 1779 or Federal Specification MMM-A-00150, and UL labeled with a Class 0 to 25 flame-spread rating.

9.8.1.4

Fire Performance

The ASTM E 1264 Fire Classes refer only to the acoustical material itself. Underwriter's Laboratories (UL) "Building Materials Directory" contains a listing that shows the ratings of acoustical ceiling materials by manufacturer and product. Acoustical materials manufacturers usually indicate the flame-spread rating of their products.

Hung-suspension systems are rated as fire resistant or non–fire resistant, but sometimes it is hard to tell which is which from the manufacturers' product literature.

Unfortunately, using a rated acoustical material, suspension system, or both together does not result in compliance with code or other requirements for fire-rated construction. Fire-rated construction is a function of the entire assembly, including the floor or roof construction, flooring or roofing, insulation, and the installed ceiling. UL's "Fire Resistance Directory" includes a listing of construction systems, which includes the structure (floor, roof, etc.) and finishes, including acoustical ceilings. Ceiling materials are also listed there by manufacturer and product.

Fire-resistance ratings are established according to ASTM Standard E 119.

9.8.1.5

Installation

Layout plans should be prepared for spaces where acoustical ceilings will be installed to ensure coordination with lighting fixtures and other ceiling-mounted equipment and devices.

Manufacturers' brochures and installation details and ASTM and CISCA standards should be obtained and followed.

Metal suspension systems should be installed in accord with ASTM Standard C 636, unless the acoustical treatment manufacturer has more stringent requirements.

The manufacturers' recommendations for cleaning and refinishing acoustical units and suspension system components should be obtained and stored in a safe place so they will be available when maintenance and repair are necessary in the future. They should include precautions against materials and methods that may be detrimental to finishes and acoustical efficiency.

Extra material, in the amount of 2% or 3% of the amount installed, is often left at the site for the owner to use in making repairs.

Acoustical materials and suspension system components should be kept dry before and after use. Acoustical materials should be allowed to acclimate to the temperature and the humidity of the space before they are installed, to prevent dimensional changes that might affect the appearance of the installed material. Watermarked, rusted, or otherwise damaged units should not be installed.

Acoustical ceiling components should not be installed until the building is weathertight, glazing has been completed, exterior openings have been closed, and mechanical, electrical, fire-fighting, and other work above ceilings has been completed.

Wet work, such as masonry, concrete, and plaster, should be allowed to dry to the moisture content recommended by the acoustical materials manufacturer before acoustical materials are installed. A temperature of at least 60 °F should be maintained in the space during and after installation of acoustical materials.

Light acoustical panels in fire-rated assemblies generally require hold-down clips. Panels that require pressure cleaning, such as those in commercial kitchens and hospital clean rooms and surgeries, also require hold-down clips.

Ceilings where cleanliness is paramount, such as in hospital operating rooms, often require sealant around the perimeter of the ceiling grid and sometimes even around the perimeter of each panel at its supports. Health department rules and regulations and building codes should be examined to determine the applicable requirements.

Suspension systems for use where fire-resistance ratings are required should comply with governing regulations, and their materials and installations should be identical with assemblies that have been tested and listed by recognized authorities, such as Underwriter's Laboratories.

Suspension system members that abut vertical or overhead building structural elements should be sufficiently isolated from structural movement to prevent transfer of loads into the suspension system. Provisions should also be made in the suspension system to accommodate movement that occurs across control and expansion joints in the substrates.

Suspended acoustical ceilings should have a seismic restraint system in accordance with ASTM Standard E 580 where required by code or usual practice.

Flat surfaces of suspension system members should not be warped, bowed, or out of plumb or level by more than 1/8 in. in 12 ft. in both directions or by more than 1/16 in. across any joint or 1/8 in. total along any single member.

Before acoustical ceiling installation begins, the area should be thoroughly inspected to verify that mechanical, electrical, and other work above or within the acoustical ceilings has been completed. Acoustical ceiling work should not proceed until other work to be concealed or made inaccessible by the ceiling has been completed and unsatisfactory conditions have been corrected.

Acoustical units should be handled carefully to prevent damage to edges and faces.

■ FURRING INSTALLATION

Furring should be installed as appropriate to support the acoustical ceiling. It should be placed on lines and levels necessary to cause the acoustical material to fall into the proper location. It should correct unevenness in the supporting structure or substrate.

The surface of wood furring to which acoustical material will be fastened should be not less than 1-1/2 in. wide. Generally 2 × 2 lumber is used for furring that is attached to supporting structure and 1 × 2 or 1 × 3 lumber is used where furring is laid directly over solid substrates. Nominal 1-in.-thick lumber should not be used over framing because it is too flexible. Sometimes acoustical ceilings are furred using

2×4s hung from structural floor or roof framing by hangers spaced not more than 48 in. on centers. Frequently in older buildings the hangers used were wood members instead of wire.

Furring should form a complete system adequate to properly support the acoustical tile. Closers should be installed at edges and openings.

Bolts, lag screws, and other anchors should be used to anchor wood furring in place. Generally, fasteners are placed near the ends of furring members and not more than 36 in. on centers between. Shorter members, however, should be anchored at 30 in. on centers.

Screws, bolts, clips, and other anchors should be used to anchor metal furring in place. Generally, fasteners should be placed near the ends of furring members and not more than 24 in. on centers between, but other support spacings may be recommended by the furring bar manufacturer. Some types of furring bars may require closer spacings. The correct spacing should be verified with the furring bar manufacturer.

Furring spacings may vary in practice with the substrate and the acoustical material being supported. Minimum spacings are dictated by acoustical material thickness and size and may vary somewhat from manufacturer to manufacturer and system to system. Spacing must be such to adequately support the acoustical ceiling.

■ INSTALLATION OF HUNG-SUSPENSION SYSTEM

Suspension system hangers should be secured to wood, concrete, or steel structural supports by connecting directly to the structure where possible. Where direct connection is not possible, the hangers should be connected to inserts, clips, anchorage devices, or fasteners. Except for hanger attachments installed specifically for the purpose by the steel deck installer, hangers should not be attached to steel decks. Since failures have occurred even when deck tabs acceptable to the deck manufacturer have been used, some sources recommend that ceilings not be supported from steel decks under any circumstances. Hangers should never be attached to pipes, conduit, ducts, or mechanical or electrical devices.

Hangers should be located plumb in relation to primary suspension members. Contact with insulation covering ducts and pipes and other items within the plenum space should be avoided. Hangers should be splayed only where obstructions or conditions prevent plumb, vertical installation. Horizontal forces generated by splayed hangers should be offset by countersplaying, bracing, or another suitable method.

Frames for grilles, registers, and other items that occur within acoustical ceilings should be properly installed in the correct locations.

Additional hangers should be provided for support of the grid systems at lights, grilles, registers, and other items to prevent excessive deflection. At grid systems with a light-duty structural classification, lights, grilles, registers, and other items should be supported separately from the ceiling grid.

Main runner and cross-furring members should be placed in proper positions, leveled, and fastened together in the manner required by the system. T-shaped members are usually locked together without accessories. Some systems, however, require clips.

Except where the acoustical ceiling is self-edging, edging strips should be provided at the complete perimeter of all acoustical ceilings, including those with concealed grids, and where suspended ceilings abut vertical surfaces. Short lengths should be avoided. Standard moldings should be attached through the web 3 in. from each end and not more than 16 in. on centers between. Special moldings may require other spacings as recommended by their manufacturers. Edging strips are usually of the same material and finish as exposed portions of grid systems.

Light fixtures and other penetrations are often boxed with mineral wool boards to comply with UL rating requirements where ceilings are rated.

Suspension systems should achieve a tight and level ceiling at the spacings and directions recommended by the manufacturer and to comply with UL label requirements.

INSTALLATION OF CONCEALED GRID SYS-TEMS FOR COMPOSITION ACOUSTICAL TILE Main runners should be supported at spacings not greater than those recom-

mended by the manufacturer. Maximum spacing of hangers is usually 48 or 60 in., depending on the system and the components selected. The first hanger on each end of each runner or carrying channel should be no more than 6 in. from the ends. Some systems require closer spacings. Where the hangers must be placed at wider spacings than those recommended for the system, the main runners can be supported by carrying channels. In indirect-hung suspension systems, main runners are usually supported by carrying channels.

INSTALLATION OF EXPOSED GRID SYS-TEMS FOR COMPOSITION ACOUSTICAL PANELS Tees should be provided of adequate size to span between the supporting structure. If necessary, main tees may be supported from carrying channels.

Main tees are usually located 48 in. on centers and all in the same direction. Cross-tees are then installed to span between the main runners with cross-tees at 24 in. on center. The tees should be locked together using whatever system is applicable to the system being used. The tees should rest on wall moldings at vertical surfaces. Where 24-in.-square panels are used, additional cross-tees should be placed from cross-tee to cross-tee and securely fastened (or locked) to the cross-tees.

Main tees should be supported by hangers placed not more than 48 in. on centers. Where the structure will not permit hangers at that spacing, additional carrying channels should be provided at 48 in. on centers, and the main runners should then be supported from the carrying channels. The first hanger on each end of each runner or carrying channel should be no more than 6 in. from the ends.

Exposed members should be installed in line, square, level, in one plane, and within the tolerances permitted by ASTM Standard C 636.

■ COMPOSITION ACOUSTICAL TILE INSTALLATION

Where a grid system is used, tiles should be placed over the entire grid area. Where necessary, tiles should be cut to fit around fixtures, penetrating hangers, and accessories. Border units should be cut and fit neatly and snugly into the grid, leaving no voids in the finished ceiling.

Splines and other grid members should be placed into the kerfed edges of tiles, or tile tongues should be inserted into tile grooves so every joint is fitted tightly and properly supported. The tile field should be held in compression with leaf-type spring steel spacers between the tiles and the edge moldings. Spacers should be placed 12 in. apart.

When a grid system is not used, tiles should be held in place using fasteners, adhesive, or both, following the tile manufacturer's instructions. Before adhesive application, loose dust should be removed from the backs of the tiles, and they should be primed with a thin coat of adhesive. Surfaces to receive tiles should be sound, free of foreign materials, clean, and dry. New plaster and concrete to receive tiles should be cured thoroughly. Bare plaster, gypsum board, and concrete should be primed as recommended by the tile manufacturer. The alkalinity of surfaces to receive adhesive-applied tiles should be tested with litmus or Hydrian pH paper. Surfaces with pH in excess of 9 should have a neutralizer applied to lower the pH to 8 or less.

Tiles are installed with adhesives by applying spots of adhesive to each tile and pressing the tiles in place using a sliding motion. The thickness of the adhesive after application should be about 1/8 in.

Splines should be installed in the joints between tiles to keep them in proper alignment and flush with each other. Tiles should be scribed to fit at penetrations and edges.

When tiles are fastened in place with fasteners or adhesives, it is necessary to provide access doors in the ceiling so mechanical, electrical, and other equipment can be accessible. Such panels are usually recessed so they can be faced with tiles like those installed in the ceiling.

■ COMPOSITION ACOUSTICAL PANEL INSTALLATION

Lay-in panels should be placed into each opening of the exposed grid members. Where necessary, panels should be cut neatly around fixtures, diffusers, penetrating hangers, accessories, and other items. Border units should be cut to fit snugly into the grid, with no cracks or voids.

■ METAL ACOUSTICAL MATERIALS INSTALLATION

Linear metal ceiling strips should be snapped into the supporting grid members according to the ceiling manufacturer's recommendations, to cover the entire ceiling.

Lay-in metal panels and pans should be laid into each opening in the grid.

Snap-in panels and pans should be placed in each opening in the grid and shoved into place so the holding feature fully engages and the panels or pans are held securely in place. Where a ceiling is to be nonaccessible, security clips should be installed as the tile or pans are placed.

■ SOUND INSULATION

Sound insulation blankets may be installed over the full area occupied by an acoustical ceiling or only in certain locations such as close to sound-rated partitions. Unless the ceiling is specifically designed to support sound insulation, the insulation should be separately supported so its weight does not rest on the acoustical materials or any part of the suspension system.

Sound insulation pads in metal ceilings should fit snugly into each pan, panel, or strip over the entire ceiling area.

Insulation should be tightly butted with joints in solid contact and without voids. It should be fitted neatly to adjacent surfaces.

■ ACOUSTICAL SEALANT

Where light leaks, sound leaks, and air movement at the edge of acoustical ceilings are not acceptable, a continuous bead of acoustical sealant should be applied on the concealed leg of the ceiling edge molding before it is installed.

9.8.2

INTEGRATED CEILINGS

Integrated ceilings consist of acoustical materials, a suspension system, outlets (grilles, vents, or vanes) for air distribution, and electrical devices, such as lights, combined into a single system, the components of which are compatible visually and mechanically.

9.8.2.1

System Components

■ ACOUSTICAL MATERIALS

The acoustical materials used in integrated ceilings include the composition and metal acoustical materials listed in ASTM E 1264 and discussed in Section 9.8.1.1. All the statements there apply to acoustical materials used in integrated ceilings.

Besides the types of acoustical materials specifically listed in ASTM E 1264, the other types of materials mentioned earlier that may be used in acoustical ceilings may also be used in integrated ceilings. Included are ceramic-faced units and units made from plastic, gypsum, and wood fiber. Integrated ceilings may also have wood, plastic, or sheet metal panels or tiles, but many of them cannot properly be classified as acoustical materials.

■ SUPPORT SYSTEMS

Most statements made earlier in this chapter about suspension systems for acoustical ceilings apply as well to the suspension systems used in integrated ceilings. One major difference is that suspension systems for integrated ceilings are usually designed to support all the components of the ceiling, including such devices as light fixtures. Another major difference is that the grid members of an exposed-grid integrated system are often designed to provide air supply or return.

Grids for integrated ceilings may be concealed or exposed. In linear metal integrated ceilings, the supporting members usually are not exposed. Grids are also usually concealed in integrated ceilings where the acoustical or decorative panels, tiles, pans, or strips snap into the systems.

■ AIR DIFFUSION

A major difference between acoustical ceilings and integrated ceilings is that the latter are designed to contain air supply and, often, air return outlets of the heating, ventilating, and air conditioning system. The effect of including such air distribution elements in the ceiling system can result in effective concealment of diffusers and grilles that are otherwise often ugly and obtrusive. Some integrated systems employ air supply and return slots that cannot be discerned as such from floor level.

Integrated ceilings employ boots or manifolds to connect the integral air supply and return elements to the building's ductwork. Connections from the ceiling to the ductwork are usually flexible.

■ LIGHTING

Areas with acoustical ceilings are often lit by standard fixtures not specifically designed for the purpose, which are mounted on the surface of the ceiling or are recessed into the ceiling. Often recessed fixtures in acoustical ceilings are standard units laid into the suspension grid. Some integrated ceilings also use standard recessed or surface-mounted fixtures. One of the more attractive features of many integrated ceilings is that they include light fixtures that have been designed specifically for that particular integrated ceiling system. The result is often a more attractive installation.

Lighting design is usually an integral part of the design of an integrated ceiling. Fixture locations and types are sometimes limited, but many integrated ceiling designs result in effective and comfortable lighting. Care must be exercised in some cases, especially where standard fixtures are used, to prevent glare.

■ ACOUSTICAL PERFORMANCE

The statements made earlier in this section about the acoustical performance of acoustical ceilings apply as well to integrated ceilings. Integrated ceilings are often used in open-plan spaces. Such installations are sometimes accompanied by so-called *white-noise* generators to mask background noises.

■ FIRE PERFORMANCE

The statements made earlier in this section about the fire performance of acoustical ceilings apply as well to integrated ceilings. Some additional requirements are associated with integrated ceilings, however. The fire-resistance classification of the system must take into account the electrical and mechanical components of the system. When an integrated ceiling must be time-rated for fire resistance, the limitations may be severe. Light fixture

and mechanical opening sizes and locations may be restricted. The method of enclosing such devices will probably be dictated by the rating designation. The appearance of the installed system may be affected drastically by such limitations. A fire-rated integrated ceiling may not look much like a nonrated system that uses the same basic components.

9.8.2.2

Installation

Unlike some acoustical ceilings, integrated ceilings are seldom fastened directly to wood or metal framing or furring. Most integrated ceiling installations, almost by definition, include some sort of metal suspension system.

Most of the requirements suggested earlier in this section for installing suspension systems for acoustical ceilings apply as well to integrated ceilings. Suspension systems, for example, may be direct hung or indirect hung, and the same ASTM standards apply. Grids may be concealed or exposed depending on the particular ceiling system. In every case, the installation should follow the integrated ceiling manufacturer's recommendations.

Light fixtures and air distribution components should be correctly installed so their functions are properly carried out. Their installation must be coordinated with the mechanical and electrical system installations.

9.8.3

ACOUSTICAL WALL PANELS AND BAFFLES

Acoustical wall panels include standard and custom *spline-mounted* and *back-mounted* panels. Baffles are either double or single thickness. Wall panels and baffles are faced with standard, custom, or owner-furnished facing materials.

9.8.3.1

Panel and Baffle Materials and Fabrication

Most acoustical wall panels and baffles consist of a fabric or vinyl facing over

mineral, glass, or wood fiberboard. Materials and fabrication may vary, however, between spline-mounted and back-mounted panels.

Metal acoustical wall panels and baffles are also available.

■ SPLINE-MOUNTED PANELS

The core of most spline-mounted panels is faced on one side only because the panels are mounted with butted joints. The finish is wrapped around the long edges back to the line of the kerf. The short panel ends are unfinished.

PANEL CORES The cores of spline-mounted panels are perforated water-felted mineral fiberboard with a fairly high density, such as 17 to 20 lb./cu. ft.; low-density mineral fiberboard; glass fiberboard with a nominal density between 6 and 8 lb./cu. ft.; or high-density wood fiberboard. The long edges of spline-mounted panels are kerfed and rabbeted to receive mounting splines (Fig. 9.8-15).

FIGURE 9.8-15 *Typical edge detail of a spline-mounted acoustical wall panel.* (*Courtesy H. Leslie Simmons.*)

ACCESSORIES Mounting and trim accessories for spline-mounted panels include H- and J-shaped splines and J-, H-, and angle-shaped moldings, as shown in Figure 9.8-16. Some manufacturers' accessories may vary from those shown. Accessories may be either extruded aluminum or plastic.

Wood battens are sometimes used to cover the joints and sometimes to support panels. Wood battens are usually decorative hardwood, but in some installations they may be painted softwood.

PANEL SIZE AND THICKNESS Spline-mounted panels are 9 or 10 ft. long and

FIGURE 9.8-16 *Some typical accessories for spline-mounted acoustical wall panels.* *(Courtesy H. Leslie Simmons.)*

FIGURE 9.8-17 *Some typical edge profiles for back-mounted acoustical wall panels.* *(Courtesy H. Leslie Simmons.)*

24, 30, or 48 in. wide. High-density fiberboard panels vary in thickness from 1/2 in. to 1-1/8 in. Glass fiberboard and low-density fiberboard panels vary in nominal thickness from 1 in. to 1-1/8 in.

■ BACK-MOUNTED PANELS

The core face that will be exposed after the panel is installed and the edges of back-mounted panels are usually finished.

PANEL CORES The core of back-mounted panels may be glass fiberboard with a nominal density between 4 and 7 lb./cu. ft. or high-density wood fiberboards. Glass fiber panels may have impact-resistant faces, which usually consist of a 1/8-in.-thick layer of 18-lb./cu. ft. glass fiberboard laminate to the face of the core board.

The edges of the cores of back-mounted panels are either chemically hardened to reinforce the edge and help prevent warpage and damage or framed with aluminum, zinc-coated steel, or wood.

Panel edges that are not framed are usually wrapped with the facing material. Edge shapes of panels with wrapped edges may be square, beveled, radiused, notched, mitered, or a combination of those (Fig. 9.8-17).

Panel corners may be square or rounded.

Some panels are faced on the backside with a thin aluminum foil to prevent air passage.

ACCESSORIES Mounting accessories for back-mounted panels include conventional Z-clips; strip-clips, which consist of Z-clips mounted on a continuous strip; and magnetic pads (Fig. 9.8-18). Other proprietary clip assemblies are also used. Panels are also mounted with hook-and-loop panels (Velcro) and sometimes with adhesives.

High-density wood fiber panels are sometimes installed with screws through the panels.

Wood battens are sometimes used to cover the joints between panels and sometimes to hold the panels in place.

PANEL SIZE AND THICKNESS Back-mounted panels vary in thickness from 1/2 in. to 3 in. Panels are available in a large variety of sizes, from 15 in. square up to 60 by 144 in.

■ METAL WALL PANELS

Some of the metal ceiling products discussed earlier in this section are sometimes used on walls, but here we will discuss only products specifically designed for use on walls. Such panels are available in both aluminum and galvanized steel.

Metal wall panels are available flat or shaped (corrugated, for example), perforated or not perforated, and plain, embossed, or textured. Most panels come with glass fiber or mineral wool sound-absorbing pads. Some pads are wrapped or faced with materials such as polyvinyl chloride. Others are bare.

Special brackets and other accessories are required for some metal acoustical wall panel installation.

Most metal wall panel sizes vary from 18 to 48 in. wide and from 18 to 144 in. long, but some may be smaller or larger.

FIGURE 9.8-18 *Typical mounting accessories for back-mounted acoustical wall panels.*
(Courtesy H. Leslie Simmons.)

■ BAFFLES

A baffle is often nothing more than a standard acoustical wall panel of the back-mounted type that is finished on both faces and all edges. Some baffles are the equivalent of two acoustical wall panels laminated back to back. Some double-thickness baffles are completely encased with the finish material. Others have a joint between the two sides of the baffle. Most baffles are flat panels, but cylindrical, octagonal, triangular, and other-shaped baffles are also available. Their basic panel construction is similar to that of flat baffles.

Baffle edges may be finished in the same fabric or vinyl as the face or framed in wood or metal. Edge configurations are similar to those for back-mounted acoustical wall panels.

Metal baffles may be flat panels similar to the metal acoustical ceiling materials discussed earlier or more like the metal wall panels discussed in the preceding subsection.

Baffles are made to be mounted or suspended by special hangers either built into or attached to the baffle. Some baffle hangers are nothing more than eye bolts mounted in the narrow edge of the panel or grommets through the panel face. Others use more elaborate hanger systems that vary from manufacturer to manufacturer. Hangers may be cable, wire, decorative chain, or almost invisible nylon fishing line. Sometimes hangers are solid metal bars. Some baffles are designed to be clipped to overhead framing.

■ WALL PANEL AND BAFFLE FINISHES

FABRIC   Standard panels and baffles are available with both woven and nonwoven synthetic fabrics, the majority of which are largely or entirely polyester. Other fabrics are also used. Some manufacturers state that almost any upholstery-weight fabric may be used, but fire-resistance requirements may not permit use of just any fabric. Some sources state that some wool, nylon, metallic, rayon, and acetate fabrics are not suitable.

VINYL   Perforated vinyl is often used as a facing for acoustical wall panels, especially in spaces, such as classrooms, corridors, and the like, where the material is likely to become soiled quickly. It is also sometimes used for facing on baffles. In older installations the vinyl may have had large holes punched in it to provide the proper acoustical performance. Vinyl in newer installations will probably have been punched in a process called micro-perforating, which eliminates large holes.

METAL   Most metal panels designed for wall panel or baffle use are often painted with acrylic or polyester baked-on paint. Other finishes may be available.

OTHER FINISHES   More utilitarian baffles may be faced or wrapped with polyethylene or glass fiber cloth.

■ ACOUSTICAL PERFORMANCE

Acoustical wall panels and baffles are generally classified by their NRC (Noise Reduction Coefficient). Panels and baffles fall into two general categories. The first includes panels that will absorb between 60% and 70% (NRC 0.60 to 0.70) of the sound that strikes them. They are used in normal-sized rooms, such as conference rooms and offices. The second category includes panels that absorb between 80% and 95% (NRC 0.80 to 0.95) of the sound that strikes them. They are used in large spaces, such as open-plan offices and spaces where high noise levels are generated, such as computer rooms, swimming pools, and gymnasiums (Fig. 9.8-18).

9.8.3.2

Support Systems

Spline-mounted acoustical wall panels may be directly attached to supported substrates or supported by wood or metal furring applied over wood or metal studs or solid substrates. Back-mounted acoustical wall panels are usually directly attached to supported substrates using one of the accessories discussed later in this chapter. Metal acoustical wall panels may be directly attached to solid or supported substrates or supported by metal furring applied over studs or solid substrates.

■ WOOD FURRING

Spline-mounted acoustical wall panels and metal acoustical wall panels are sometimes installed over horizontal wood furring (Fig. 9.8-19). Such wood furring is most often used when the substrates are

FIGURE 9.8-19 *Acoustical baffles in an arched ceiling in a large armory.* (*Photograph provided by MBI Products Company.*)

solid ones such as concrete or masonry, and the substrates are too irregular to directly apply the splines and accomplish a finished surface within acceptable tolerances. Wood furring may also be used over existing wood framing that is not in proper alignment to receive acoustical wall panels, but such use is rare in new buildings. Wood furring may also be used over wood framing for support of acoustical baffles when the framing is not at the proper spacings or locations for the baffle hangers.

Back-mounted acoustical wall panels are usually mounted directly onto a supported substrate, such as plaster or gypsum board. Where wood furring is required in such installations, it is used to fur the supported substrate and not the back-mounted acoustical panels.

◼ METAL FURRING

Spline-mounted acoustical wall panels and metal acoustical wall panels are sometimes installed over horizontal hat-shaped galvanized steel furring (Fig. 9.8-20). Such metal furring is most often used when the substrates are solid ones, such as concrete or masonry, and the sub-

FIGURE 9.8-20 *Mounting detail for spline-mounted acoustical wall panels over a solid substrate.* (*Courtesy H. Leslie Simmons.*)

strates are too irregular to directly apply the splines and accomplish a finished surface within acceptable tolerances. Metal furring may also be used over existing metal framing, such as studs, that is not in proper alignment to receive acoustical wall panels, but such use is rare in new buildings. Galvanized or painted steel channel furring may also be used over wood or metal framing for support of acoustical baffles when the framing is not at the proper spacings or locations for the baffle hangers.

Back-mounted acoustical wall panels are usually mounted directly onto a supported substrate, such as plaster or gypsum board. Where metal furring is required in such installations it is used to fur the supported substrate and not the back-mounted acoustical panels.

Metal furring is usually screwed, clipped, or wired to the supporting structure.

9.8.3.3

Fire Performance

Acoustical wall panels and baffles should be tested for flame spread and smoke-developed ratings as complete panels containing the same components as they will have when installed. They should rate as Class A according to ASTM Standard E 84 test, "Surface Burning Characteristics of Building Materials." The flame-spread rating of a Class A material ranges from 0 to 25. When the individual components are tested separately, it is difficult to determine a composite value for the entire panel.

Since not all manufacturers have had their acoustical wall panels or baffles tested by independent testing laboratories, such as Underwriters' Laboratories (UL), it is often necessary to obtain specific approval of a particular panel or baffle by the local authorities having jurisdiction before using it.

9.8.3.4

Installation

Acoustical wall panels should be coordinated with electrical outlets and switches, thermostats, and other wall-mounted equipment and devices to ensure that such items are properly accommodated.

Manufacturers' brochures and installation details should be obtained and followed. The manufacturers' recommendations for cleaning and refinishing wall panels and baffles should be obtained and stored in a safe place so they will be available when maintenance and repair are necessary in the future. They should include precautions against materials and methods that may be detrimental to finishes and acoustical efficiency.

Extra material is occasionally left at the site for the owner to use in making repairs.

Acoustical wall panels and baffles should be kept dry before and after use. They should be allowed to acclimate to the temperature and the humidity of the space before they are installed to prevent dimensional changes that might affect the appearance of the installed panels or baffles. Watermarked, stained, or otherwise damaged units should not be installed.

Acoustical wall panels and baffles should not be installed until the building is weathertight, glazing has been completed, and exterior openings have been closed.

Wet work, such as masonry, concrete, and plaster, should be allowed to dry to the moisture content recommended by the acoustical wall panel or baffle manufacturer before wall panels or baffles are installed. A temperature of at least 60 °F should be maintained in the space during and after installation.

Furring members that abut building structural elements should be sufficiently isolated from structural movement to prevent transfer of loads into the furring. Provisions should also be made to accommodate movement that occurs across control and expansion joints in the substrates. Furring or panels should not span such joints or be fastened to opposite sides of such joints.

Flat surfaces of furring members and supported substrates should not be warped, bowed, or out of plumb or level by more than 1/8 in. in 12 ft. in both directions or by more than 1/16 in. across any joint or 1/8 in. total along any single member.

Acoustical wall units should be handled carefully to prevent damage to edges and faces.

Hangers for baffles should be properly designed and sufficiently strong to support the baffles. Where splaying of hangers is necessary to avoid ducts, pipes, conduit, or other items, the forces so generated should be compensated for by countersplaying or another suitable means.

■ FURRING INSTALLATION

Furring should be installed as appropriate to support acoustical wall panels and baffles. It should be placed on lines and levels necessary to cause the acoustical panels to fall into the proper location and to properly support baffle hangers at the correct spacings. Where the furring will support directly applied panels or baffles, it should correct unevenness in the supporting structure or substrate. The surface of wood furring to which acoustical panel splines will be fastened should be not less than 1-1/2 in. wide. Generally 2×2 lumber is used for furring that is attached to the supporting structure and 1×2 or 1×3 lumber is used where furring is laid directly over solid substrates. Nominal 1-in.-thick lumber should not be used over framing because it is too flexible. Furring to support baffles should be the proper size to span between the framing members without undue deflection in the furring.

When placed over a solid substrate, furring should be installed over a layer of polyethylene film or other vapor-retarding material as an airflow barrier and vapor retarder.

Furring should form a complete system adequate to properly support the acoustical wall panels or baffles. Closers should be installed at edges and openings.

Bolts, lag screws, and other anchors should be used to anchor wood furring in place. Generally, fasteners should be placed near the ends of furring members and not more than 36 in. on centers between. Shorter members, however, should be anchored at 30 in. on centers. Bolts should have nuts and washers.

Screws, bolts, clips, wires, and other anchors should be used to anchor metal furring in place. Generally, fasteners should be placed near the ends of furring members and not more than 24 in. on centers between, but other support spacings may be found. Some types of metal furring members may require closer spacings. The correct spacing should be verified with the furring member manufacturer.

Furring spacings may vary in practice with the substrate and the acoustical panels or baffles being supported. Minimum spacings for wall panels are dictated by the panel thickness and size and may vary somewhat from manufacturer to manufacturer and system to system. Spacing must be such as to adequately support the panels. Spacing of furring for baffle hangers should be such that the baffle hangers fall in the correct locations.

■ SPLINE-MOUNTED PANELS

The usual method of mounting spline-mounted acoustical wall panels is to place splines vertically and fasten them to a supported substrate such as plaster or gypsum board (Fig. 9.8-21). Then the panels are slipped in place to engage the splines in the kerfed and rabbeted panel edges. Spline-mounted panels can also be mounted by fastening the splines to wood or metal furring over masonry, concrete, or another solid substrate (see Figs. 9.8-19 and 9.8-20).

FIGURE 9.8-21 *Mounting detail for spline-mounted acoustical wall panels over metal studs. Similar details may be used when the wall panels are installed over solid substrates that are irregular.* (Courtesy H. Leslie Simmons.)

Spline-mounted acoustical wall panels may rest directly on the floor or be supported at the bottom by a base retainer such as a J- or angle-shaped molding (see Fig. 9.8-16).

Panels should extend to the ceiling but not into the ceiling plenum. Sometimes the panels are stopped short of the ceiling, leaving a decorative reveal.

When panels must be cut for fitting or other reasons, the panel backs should be resealed using heavy mastic tape or foil to prevent airflow through the panel. Cut edges should be covered by a molding or wrapped with fabric.

Exposed panel edges should be covered with fabric or a molding.

BACK-MOUNTED PANELS

Back-mounted panels are usually mounted directly on a supported substrate such as plaster or gypsum board. Most older panels and many newer panels will have been mounted with a conventional Z-clip, a continuous Z-clip, or some similar proprietary device. Some will have been mounted with adhesives. Others will be supported by magnetic pads. Some will have been screw-attached or supported by battens. Many modern installations are supported with hook-and-loop (Velcro) strips.

METAL PANELS

Many metal panels are installed with clips or screws. Some are installed with accessories similar to the moldings shown in Figure 9.8-16 or the clips and magnetic pads shown in Figure 9.8-18.

BAFFLES

Most baffles are suspended vertically, by wire, cables, chains, bars, or fishing line. They should be installed in line and vertically or at the proper angle according to the design. Baffles that are suspended in other configurations should be installed according to the design. In every case, hangers should be securely fastened to the supports and to the baffles.

SOUND INSULATION

Where they are used, sound insulation blankets should be installed over the full area occupied by the acoustical wall panels.

Sound insulation pads in metal wall panels should be fitted snugly into each pan, panel, or strip over the entire area of the wall panel installation.

Insulation should be tightly butted with joints in solid contact and without voids. It should be fitted neatly to adjacent surfaces.

FIGURE 9.8-22 *Mounting detail for spline-mounted acoustical wall panels over a supported substrate.* (Courtesy H. Leslie Simmons.)

9.9 WOOD FLOORING

The widespread use of wood flooring can be attributed to its distinctive natural appearance, excellent wearing qualities, moderate cost, underfoot comfort, and ease of installation and maintenance. It can be used in almost all rooms above grade. Improved methods of protecting wood from dampness have also encouraged its use in rooms below grade as well and even in residential kitchens.

Because of its ready availability and the relative ease of working such wood with hand tools, early flooring was made mostly of softwood. This flooring often was in the form of thick planks 3-1/2 to 10 in. wide, just as they came from the saw mill. Subsequent improvements in manufacturing and seasoning methods brought about the development of carefully machined, narrower hardwood strips

(3-1/4 in. and narrower), which were easier to install.

Hardwood strip flooring has been and continues to be the most popular type of wood flooring (Fig. 9.9-1a). But demand for floors simulating early colonial random-width plank installations has encouraged the production of plank products (Fig. 9.9.1-b). Various parquet (pattern) types, such as blocks and uniform-length strips, are also available (Fig. 9.9-2).

FIGURE 9.9-1 *Contemporary and traditional interiors are complemented by (a) strip and (b) plank wood floors.* (Courtesy Bruce Hardwood Floors.)

FIGURE 9.9-2 *Patterned (parquet) floors may be created with (a) laminated plywood blocks or (b) solid wood slats preassembled into blocks or individually laid in mastic.* (*Courtesy Bruce Hardwood Floors.*)

Changing conditions in the building industry have created a demand for factory-finished flooring suitable for service immediately after installation. Such prefinished flooring is available in practically every style (strip, plank, and block) and in a variety of woods.

A characteristic property of wood that must be given special attention in flooring installations is its tendency to shrink and swell as its moisture content changes. Through modern research and manufacturing methods, products with improved dimensional stability have been developed. However, individual products vary somewhat in their sensitivity to moisture and should be selected with a view to the conditions of use.

This section describes various types of hardwood and softwood floorings and highlights their special properties. It summarizes available grades, sizes, and current standards of quality, providing a basis for the selection and specifying of wood flooring.

9.9.1

MATERIALS

The terms *hardwood* and *softwood*, popularly applied to the two major groups of trees cut for lumber, actually have no bearing on the degree of hardness of the wood. In fact, many softwoods are harder than some of the hardwoods. The terms are used primarily to distinguish the botanical characteristics of the trees. Arbitrarily, trees having broad leaves are known as hardwoods, while coniferous trees—those bearing needles and cones—are known as softwoods.

About 12 types of woods are regularly manufactured into flooring. Of these, the hardwoods account for about 80% of wood flooring. The greater popularity of hardwoods can be attributed to their appearance and, in the species used, substantially greater hardness and wear resistance.

9.9.1.1

Hardwoods

Of all hardwood flooring produced in a typical year, oak of various commercial species supplies more than 90%, as compared with 6% or so for maple. The balance consists largely of beech, birch, pecan, and several other hardwoods in limited quantities.

■ OAK

There are about 20 species of oak in the United States that are considered commercially important in lumber production. Of these, about half are classed as *red oak* and half as *white oak*. As growing trees, the several species within each group are readily distinguishable, but in lumber form, the differences are fairly inconspicuous. Therefore, precise separation of the various species within each group of oaks is impractical and unnecessary in flooring manufacture.

Oak is lumbered throughout the southern, eastern, and central states, in forests of the Atlantic Plain and the Appalachian Mountains. In these regions, species of oak grow under a wide range of climatic conditions in many different kinds of soil. There is, accordingly, much variation in the color of the wood, especially the heart-

wood; the sapwood usually shades from white to cream color in all species of oak. In the standard grading rules for oak flooring, color is entirely disregarded except in the amount of light-colored sapwood allowed. Sapwood is limited only in the top grade of flooring, *clear grade*.

Red oak and white oak are about equal in mechanical properties. Both make a very satisfactory floor of attractive appearance when properly finished. A special feature of white oak is the prominence of large rays that make an interesting flake pattern in quarter-sawn flooring.

Although grading rules do not differentiate between red oak and white oak, practically all manufacturers supply either all-red or all-white oak flooring except in the lowest grades. Red oak flooring generally is higher in price and more uniform in color than white oak.

■ MAPLE

Maple flooring is made from sugar maple, logged largely in the Northeast, the Appalachians, and the Great Lakes states (Minnesota, Michigan, and Wisconsin). In the lumber trade, sugar maple is known as *hard maple* or *rock maple*. This species is extremely strong, hard, and abrasion-resistant, making it particularly suitable for hard-use locations, such as factories and gymnasiums, as well as residences.

The so-called soft maples—silver maple, red maple, and bigleaf maple—are not as hard, heavy, or strong as hard maple and therefore are not used commonly for flooring.

The heartwood of both sugar maple and black maple is light reddish-brown, and the sapwood, which in mature trees is several inches thick, is creamy white, slightly tinged with brown. The contrast in color between heartwood and sapwood in maple is much less pronounced than in oak, and the standard grading rules permit natural color variation in the wood.

■ BEECH AND BIRCH

Beech and birch are lumbered in the northeastern part of the country and around the Great Lakes. In comparison with hard maple, beech and birch are used only sparingly in the manufacture of flooring. Only two of the almost 20 species of birch that grow in the United States are manufactured into flooring. Of these, *yellow birch* is by far the most abundant and the most important commercially; the other is *sweet birch*. Only one species of beech is native to the United States.

The heartwood of these species is reddish-brown, with a slight variation in color among them. Similar slight variations exist also in the color of their sapwood, which is a lighter shade than the heartwood. The varying color of the natural wood is an accepted characteristic in grading beech and birch flooring.

9.9.1.2

Softwoods

In a typical year, more than 50% of the softwood flooring produced is southern pine, more than 40% is Douglas fir, and the balance is western hemlock, eastern white pine, ponderosa pine, western larch, eastern hemlock, redwood, spruce, cypress, and the true firs. Western larch is similar to Douglas fir in strength properties and often is sold in mixture with Douglas fir of the northern interior region of the western states. Ponderosa pine, eastern white pine, and redwood are softer than desirable where wear is a prime factor. However, the formidable decay resistance of redwood in the all-heartwood grade has prompted its use for porch and deck flooring.

■ SOUTHERN PINE

Southern pine is a commercial name applied to a group of yellow pines that grow principally in the southeastern states. The group includes longleaf, shortleaf, loblolly, slash pines, and several others of minor importance. Except in dimension lumber and structural timbers, no differentiation between the species is made in marketing the products of this group.

The wood of all southern pines is much alike in appearance. The sapwood and the heartwood often are different in color, the former being yellowish-white and the latter a reddish-brown. However, the contrast in color between sapwood and heartwood in southern pine generally is not conspicuous in a finished floor, and the standard grading rules permit sapwood in all grades of southern pine flooring unless otherwise specified.

When flooring of uniform color is essential, the standard flooring specifications can be amended to require all-sapface stock (for a light color) or all-heartface material (for a reddish-brown color). Special selection of stock for color increases the cost somewhat over the established grade.

■ DOUGLAS FIR

Red fir, yellow fir, coast Douglas fir, and Oregon pine are other names by which Douglas fir is known in the western parts of the United States and Canada, where it grows. Douglas fir occupies the same important position in the western and Pacific coast states as southern pine does in the southeastern states.

The sapwood of Douglas fir is creamy white. The heartwood is reddish-brown, and, as in southern pine, the contrast in color between the two is not so pronounced as to be objectionable in a finished floor. Pieces containing both heartwood and sapwood are permitted in all grades.

■ WEST COAST HEMLOCK

Western hemlock grows along the Pacific coast from northern California to Alaska and as far inland as northern Idaho and northwestern Montana. The bulk of hemlock lumber being produced comes from Oregon, Washington, and California and is referred to commercially as West Coast hemlock. Both the heartwood and the sapwood of western hemlock are almost white with a pinkish tinge and with very little contrast, although the sapwood may sometimes be lighter in color.

Western hemlock has light, clear color and good finishing qualities, which account for its use in moderate-wear areas, such as bedrooms, where good appearance is the principal requirement. Western hemlock flooring is relatively free from warping and is easy to cut and nail, but it is not as hard and wear-resistant as Douglas fir or larch.

9.9.1.3

Supports

Generally, the joist spacing to support board subflooring should not exceed 16 in.

on centers when 3/4 in. or thinner strip or plank flooring is laid parallel to the joists, or where parquet flooring is used. Joist spacing should not exceed 24 in. on centers when 3/4-in. finish flooring is installed at right angles to the joists.

9.9.2

MANUFACTURE

Wood flooring is made in three basic styles: strip, plank, and block (Fig. 9.9-3). The following discussion of manufacture is limited to strips and planks, because blocks are made in many different ways. Methods of manufacture of hardwood flooring differ from those of softwood flooring.

FIGURE 9.9-3 *Basic wood flooring types. (Courtesy Bruce Hardwood Floors.)*

9.9.2.1

Hardwoods

The production of hardwood flooring begins at the sawmill, where logs are rough-sawn into boards, planks, and timbers. Each log is cut in such a way to produce the best grade of lumber in each thickness. The thicker pieces containing the highest percentage of clear lumber usually are reserved for furniture and millwork.

The common nominal thickness of lumber for flooring is 1 in., from which the popular 3/4-in. strips and planks are milled. After air-drying in a yard for several months, rough-sawn lumber is kiln-dried to reduce the moisture content to a range of 6% to 9%. Sawmill operations are described and illustrated in Section 6.2.1.2.

FIGURE 9.9-4 *Rough-sawn lumber is cut lengthwise in a ripsaw.* (*Courtesy Bruce Hardwood Floors.*)

SIZING AND DRESSING

Rough-sawn lumber often is air-dried at a sawmill, but kiln drying is normally performed at a flooring plant. After cooling for at least 24 hours, the lumber is ready to enter a flooring mill. The first step consists of running the lumber through a ripsaw, which cuts the boards into the desired widths (Fig. 9.9-4). Lumber that is to be tongued and grooved (T&G) is cut wider than the face width of the flooring, to compensate for the tongue edge and for the material taken off in planing the edges.

In some flooring mills, the pieces are then run through a planing machine, which dresses (smoothes) the top and bottom surfaces. Then, either dressed or in rough state, the pieces proceed on conveyor belts to cut-off saws (Fig. 9.9-5), where they are cut to eliminate natural defects, which would be undesirable in finish flooring. The pieces emerge in random lengths as a result of this cutting

FIGURE 9.9-5 *Natural defects are eliminated on a cut-off saw.* (*Courtesy Bruce Hardwood Floors.*)

operation. Next, processed strips are conveyed to a machine known as a side-matcher, which mills them to the characteristic T&G profile.

SIDE AND END MATCHING

A side-matching machine smoothes the top surface accurately and cuts one or two channels in the bottom side, a tongue on one edge, and a groove in the other (Fig. 9.9-6). The pieces then continue on a belt conveyor to end-matching machines, where

one end of each strip is grooved and the other is tongued.

GRADING AND BUNDLING

Next, the pieces go to grading tables, where they are separated by grade, color, and length (Fig. 9.9-7). Each piece is trade and grade marked, and for conventional flooring assortments the pieces of each grade are separated and bundled according to length in multiples of 1 ft. An allowance of 6 in. under and 6 in. over the nominal length is permitted in each bundle, so that a 4-ft. bundle might contain pieces 3 ft. 6 in. long to 4 ft. 6 in. long. However, the most popular flooring size, 3/4 in. by 2-1/4 in., is also bundled in an assortment called *nested flooring*, which includes various lengths ranging from 9 to 102 in.

Flooring pieces that are to be prefinished undergo further treatment, including fine sanding (Fig. 9.9-8), finishing, and final waxing and polishing. Bundles are then shipped to distributors in sealed boxcars or trucks, which afford protection from the weather and moisture. The quantity of lumber per bundle and the total required can be calculated according to Figure 9.9-9. To determine the number of board feet necessary for a specific installation, multiply the square footage of the area to be covered by the factor shown under the required quantity in Figure 9.9-9. For example, a 10-ft. by 12-ft. room

FIGURE 9.9-6 *The top face is smoothed and the sides are tongued-and-grooved on a side-matching machine.* (*Courtesy Bruce Hardwood Floors.*)

FIGURE 9.9-7 *Each piece is carefully inspected and graded before it is packaged.* *(Courtesy Bruce Hardwood Floors.)*

FIGURE 9.9-8 *Disc sanders impart a satin-smooth finish to pieces that will be factory finished.* *(Courtesy Bruce Hardwood Floors.)*

has an area of 120 sq. ft.; 3/4-in. by 2-1/4-in. strip flooring has a multiplier of 1.383; 120 sq. ft. times 1.383 equals 165.96 board feet. In round numbers, the 10-ft. by 12-ft. room installation would require 166 board feet of 3/4-in. by 2-1/4-in. flooring.

9.9.2.2

Softwoods

The production of softwood flooring begins at the sawmill, where the green (unseasoned) logs are rough-sawn. The resulting lumber then is seasoned to meet the moisture requirements of the grading rules. Kiln-dried flooring requires a maximum of 12% moisture content. The actual moisture content at the time of installation usually averages about 9% or 10%.

After flooring has been seasoned, it goes directly to a planing mill, where it is manufactured to the desired profile in one operation. Square-end flooring is then trimmed and packaged in equal lengths, six to a bundle. End-matched flooring bundles may contain pieces from 1 ft. to 8 ft. in length laid end to end and stacked in one or two tiers of six courses each.

9.9.2.3

Standards of Quality

Of the three basic types of flooring produced, strip flooring and block flooring are regulated by industry-wide specifications. There are no generally accepted standards for plank flooring, but most manufacturers market products in appearance grades similar to those established for strip flooring. The standard for block flooring is American National Standards Institute/American Parquet Association, ANSI/APA 1-1984, "Mosaic-Parquet Hardwood Slat Flooring."

Strip flooring is manufactured generally to comply with the grading rules of the particular trade association governing the type of wood involved. The National Oak Flooring Manufacturers Association (NOFMA) promulgates and regulates the production of oak flooring through its *Official Flooring Grading Rules*. Since some NOFMA members also produce maple, beech, birch, and pecan flooring, this association also promulgates grading rules for those woods. However, the majority of maple, beech, and birch flooring is produced by members of the Maple Flooring Manufacturers Association (MFMA), which promulgates its own grading rules for these species.

9.9.2.4

Certification of Quality

Strip flooring must be kiln-dried, grademarked, and trademarked to comply with NOFMA and MFMA grading rules. The two trade associations operate certification programs based on inspections of their

FIGURE 9.9-9
Quantity Calculations for Strip Flooring

Strip dimensions, in.		Pieces per Bundle[d]	Bundle Quantity, bd. ft.[e]		Required Quantity, bd. ft.[b]	
Actual[a,b]	Counted[c]					
3/4 × 3-1/4	1 × 4	12	4		1.29	
3/4 × 2-1/4	1 × 3	12	3	times	1.383	times
3/4 × 2	1 × 2-3/4	12	2.75	the	1.425	the
3/4 × 1-1/2	1 × 2-1/4	12	2.25	average	1.55	area to
15/32 × 2	1 × 2-1/2	18	3.75	bundle	1.30	be
15/32 × 1-1/2	1 × 2	18	3	length	1.383	covered
11/32 × 2	1 × 2-1/2	24	5	(ft.)	1.30	(ft.)
11/32 × 1-1/2	1 × 2	24	4		1.383	

[a] Dimensions are from grading rules of National Oak Flooring Manufacturers Association (NOFMA); the Maple Flooring Manufacturers Association (MFMA) recognizes standard thicknesses of 25/32 in. in 1-1/2 in., 2-1/4 in., and 3-1/4 in. lengths, and 41/32 in. thickness in 2-1/4 in. and 3-1/4 in. lengths.

[b] Nominal thickness for 15/32 in. and 11/32 in. flooring is 1/2 in. and 3/8 in., respectively; 3/4 in. is the same.

[c] Equivalent dimensions used for calculating board-foot quantity.

[d] 1 in. × 4 in. and 1 in. × 3 in. pine flooring commonly packaged 6 pieces to the bundle.

[e] Multiplier is equal to counted size times number of pieces per bundle divided by 12.

[f] Multiplier makes allowance for side-matching, end-matching, and normal waste.

members' manufacturing facilities and products, and only qualifying members are permitted to imprint the trademark of the association. This mark is found on the back of each piece of wood and provides visual assurance of conformance with the grading rules. In addition to the grade mark and trademark, the imprint often contains the mill number and other identifying information.

Softwood strip flooring is manufactured to conform to the grading rules of the Southern Pine Inspection Bureau (SPIB), the Western Wood Products Association (WWPA), and the West Coast Lumber Inspection Bureau (WCLIB). These associations do not require trademarking or grade marking under the grading rules.

9.9.3

STRIP FLOORING

Hardwood flooring is manufactured commonly to a standard pattern (Fig. 9.9-10), producing flooring that is side-matched and end-matched. The top face is generally slightly wider than the bottom, so that when the strips are driven tightly together, the upper edges make contact, but the lower edges are slightly separated. Standard T&G strip flooring usually is installed over wood subfloors or wood

FIGURE 9.9-10 *Detail dimensions of typical ("standard pattern") 3/4-in. T&G flooring. (Courtesy Bruce Hardwood Floors.)*

sleepers by blind (concealed) nailing at the intersection of the tongue and shoulder (Fig. 9.9-11).

Strip flooring is generally either hollow-backed (Fig. 9.9-12a,b) or scratch-backed (Fig. 9.9-12d). The most widely used standard pattern is 3/4 inch thick and has a face width of 2-1/4 in., but other widths and thicknesses are available. The strips are of random length, and the proportion of short pieces depends on the grade.

Another pattern of strip flooring used to a limited degree has square edges and a flat back (Fig. 9.9-12c). This is gener-

FIGURE 9.9-11 *Strip flooring application: (a) first and last strips are face-nailed; (b) others are blind-nailed. (Courtesy Bruce Hardwood Floors.)*

FIGURE 9.9-12 *Flooring patterns (a, b, c, and d) for installation with nails or screws; (e and f) for mastic installation. (Courtesy Bruce Hardwood Floors.)*

ally thinner than the standard pattern and is usually installed by face nailing. The nails are driven and set so the matching filler used in finishing will fill and conceal the holes.

While standard strip flooring generally is installed by blind nailing to wood subfloors or sleepers (Fig. 9.9-13), uniform short lengths of T&G flooring (*single slat flooring*) also can be installed with mastic

FIGURE 9.9-13 *Typical installation methods: (a) wood subfloor and (b) concrete slab. 1-in. by 2-in. sleepers can also be used over a wood subfloor. Polyethylene film and 3/4-in. plywood can also be installed over a slab as a subfloor.* (Courtesy Bruce Hardwood Floors.)

(paste-like adhesive). Products intended for *mastic* installation generally have flat backs for greater contact with the adhesive and have a recess milled into the lower surface below the tongue to reduce the possibility of mastic being forced up through the joint (Fig. 9.9-12e, f). Flooring intended for mastic installation may have T&G edges or be grooved for insertion of a hardwood or metal spline.

Softwood flooring is made either end-matched (see Fig. 9.9-12a) or with plain ends (Fig. 9.9-12e). It may have a wide hollow back like 3/4-in. standard-pattern hardwood flooring (see Fig. 9.9-12a), a double groove like the thinner standard-pattern flooring (see Fig. 9.9-12b), or a single V-shaped groove in the back (see Fig. 9.9-12d).

Standard flooring grades are based almost wholly on appearance. They exclude or severely limit such defects as knots, wormholes, and the like in the higher grades and permit increasing sizes and numbers of these characteristics in the lower grades. Natural variations in color generally are not limited, except that in the clear grade of oak the amount of light-colored sapwood is restricted.

Traditionally, hardwood flooring has been packaged and sold in bundles containing pieces of more or less the same length, 6 in. shorter or longer than the nominal length. The shortest pieces permitted per bundle and the percentage of short lengths permitted in each grade are regulated by established grading rules of NOFMA and MFMA.

NOFMA grading rules include pieces 3/4-in. by 1-1/2-in., by 2-in., by 2-1/4-in., and by 3-1/4-in. nested flooring, which is laid end-to-end continuously to make either 7-ft.-long or shorter or 8-ft.-long bundles. A nested 8-ft. bundle is four tiers high and three courses wide or six tiers high by two courses wide, each course being 7-1/2 to 8-1/2 ft. long. Individual pieces may range from 9 in. to 102 in., but both the proportion of short pieces (9 to 18 in.) and the minimum average length are regulated by the grading rules and depend on the grade and the species.

The short pieces included in nested flooring are not detrimental to an installation. Experiments have demonstrated that even if the subflooring were omitted and a 9-in. piece of strip flooring were placed so that no part of it rested on a support (Fig. 9.9-14), it could support almost 2-1/2 times the concentrated load under the leg of a concert grand piano. This represents an adequate safety factor for what may be assumed to be the most critical loading condition in most uses.

FIGURE 9.9-14 *Short pieces of nested flooring (shaded), resting entirely on T&Gs of adjacent strips (a), supported a concentrated load of 845 lb., while pieces partly resting on a joist (b) supported more than 915 lb.*

Installations over subfloors or those involving longer pieces produce lower stresses and certainly are adequate.

9.9.3.1

Oak

Oak is regularly manufactured into plain-sawn and quarter-sawn flooring (Fig. 9.9-15). Most oak flooring is plain-sawn, the lower-priced of the two. Quarter-sawn oak is characterized by a striking grain and flake pattern and by greater dimensional stability, which minimizes shrinking and swelling in width. Oak flooring is graded under the rules of NOFMA (National Oak Flooring Manufacturers Association).

The most common oak flooring thickness is 3/4 in. Thinner flooring is available for use over existing floors and for light-service conditions. Flooring thicker than 3/4 in. is intended for heavy-service conditions.

Although the great bulk of oak strip flooring is end- and side-matched, some square-edged strip flooring is made. This is intended primarily for industrial, institutional, and commercial use to facilitate replacement of damaged areas, which would be more difficult with tongue-and-groove flooring. Prefinished oak flooring is made in four grades—Prime, Standard, Standard & Better, and Tavern—and is available in all-red and all-white oak selections. Prime grade is available on special order only.

9.9.3.2

Maple, Beech, and Birch

Maple, beech, and birch flooring is graded both under the rules of the MFMA (Maple Flooring Manufacturers Association) and under the rules of NOFMA. Although the grade names are sometimes slightly different, the requirements under both sets of rules are similar. Both generally disregard normal color variations.

Because these woods are very dense and hard in plain-sawn as well as quarter-sawn stock, the grading rules make no special requirements in that respect. However, quarter-sawn material can be obtained when specified. MFMA defines edge-

FIGURE 9.9-15 *The angle the annual rings make with the surface determines whether wood is (a) quarter-sawn (edge-grained) or (b) plain-sawn (flat-grained.)*

grain (quarter-sawn) stock as pieces whose annual rings form an angle of over 30 degrees with the face. This definition is slightly more permissive than the standard definition requiring an angle of 45 degrees or more (Fig. 9.9-15).

9.9.3.3

Southern Pine

Southern pine is graded under the rules of SPIB (Southern Pine Inspection Bureau). It may be specified in side- and end-matched flooring and also in square-end flooring that is side-matched only. Regardless of the pattern to which it is worked, all southern pine flooring is available in both *flat-grain* and *edge-grain* stock. Edge-grain material is in limited supply, available through special order only and may be hard to obtain. Edge-grain material in this species must have at least six annual rings per inch across the face, in addition to the standard requirement of grain direction of more than 45 degrees.

A class of southern pine flooring intermediate between edge grain and flat grain is known as *near-rift* flooring. This material has annual rings that form an angle of less than 45 degrees with the face, but the rings are so close to this position that each piece shows at least six annual rings anywhere across the face.

9.9.3.4

Douglas Fir and Other Western Softwood Species

With the exception of the B & Better Grade, which is manufactured only from vertical-grain material, Douglas fir flooring in the upper grades is regularly manufactured in both *vertical-grain* (VG) and flat-grain (FG) stock (see Fig. 9.9-15).

The term *vertical-grain* used in these grading rules is almost synonymous with *edge grain* used for southern pine. However, Douglas fir Grade C requires four rings per inch, and only B & Better Grade requires at least six rings per inch of face width, as does southern pine. Vertical-grain Douglas fir flooring is harder and possesses better wearing qualities than flat-grain material. Douglas fir flooring generally is T&G but not end-matched. Flooring of this wood is graded under the rules of WCLIB (West Coast Lumbermen's Inspection Bureau) and WWPA (Western Wood Products Association).

9.9.4

PLANK FLOORING

Flooring boards wider than 3-1/4 in. are referred to as *planks*, the term *strips* being generally reserved for narrower pieces. However, many random plank installations include both plank- and strip-sized boards, and manufacturers often refer to such assortments as plank flooring. Strip-and-plank floors are installed in end-to-end random pattern, with end joints staggered, using nails or screws.

Plank flooring is available end- and side-matched in the same thicknesses as strip flooring. Plank flooring most commonly is made of white or red oak 3/4 in. thick and 3-1/2 to 8 in. wide. It is available prefinished in dark or light shades and graded according to NOFMA rules for prefinished flooring.

Some plank flooring is designed to simulate the appearance of the random-width planks commonly used in colonial-era buildings. These rustic floors had a distinctive appearance resulting from the hardwood pegs used to secure the planks to the subfloor. Contemporary plank flooring is installed with wood screws in the face of each plank, and the screws are covered with wood plugs to simulate pegs. Narrower, strip-sized boards that can be secured by blind nailing sometimes have decorative pegs inserted during their manufacture and are sold as plank flooring.

Because the cross-grain dimension in plank flooring is greater than in strip flooring, it is affected more by variations in atmospheric humidity. To avoid unsightly loose joints, plank edges are often eased or beveled to accentuate rather than conceal the joints.

Plank flooring is sometimes made in three-ply laminated form, with a center cross-ply. This form is more dimensionally stable in width than the solid form, making it possible to install such plank floors with tight joints. Unlike strip flooring, plank flooring generally is sold by the square foot rather than by the board foot.

9.9.5

PARQUET FLOORING

Parquet (pattern) floors consist of individual strips of wood (*single slats*) or larger units (*blocks*) installed to form a decorative geometric pattern (Fig. 9.9-16). Blocks may be made by laminating several hardwood veneers or by gluing a number of solid hardwood pieces into a unit to facilitate installation. Unlike strip-and-plank flooring, parquet flooring usually is installed with mastic.

9'' × 9'' by 3/4'' (typ.)

pieces of strip flooring

a UNIT BLOCK

9'' × 9'' by 5/32'' (typ.)

3 or more plies of wood

b LAMINATED BLOCK

variable, 9'' to 30'' square, by 5/16'' (typ.)

thin slats of hardwood

c SLAT BLOCK

FIGURE 9.9-16 *Typical construction and dimensions of basic block flooring forms: (a) unit block; (b) laminated block; (c) slat block. Slat blocks often consist of 16 or more basic squares.*

Oak is by far the most predominant species used in all types of pattern flooring. Other species, such as maple, walnut, cherry, and East Indian teak, also are available. Sometimes a mixture of hardwoods, such as hickory, ash, elm, pecan, sycamore, beech, and hackberry, are used at random in a single block.

9.9.5.1

Block Flooring

Block flooring is manufactured in several basic types. In *unit block* (also known as

solid unit block), short lengths of strip flooring are joined together edgewise to form square units (Fig. 9.9-16a). In *laminated* (plywood) block, three or more plies of veneer are bonded with adhesive to obtain the desired thickness (Fig. 9.9-16b). *Slat block* (sometimes called *mosaic parquet hardwood slat*) flooring utilizes narrow slats or "fingers" of wood pre-assembled into larger units to facilitate installation (Fig. 9.9-16c).

Most unit-block and laminated-block flooring is tongued on two adjoining or opposing edges and grooved on the other two to ensure alignment between adjoining blocks. Some manufacturers produce square-edged blocks, while others include grooves on all four block edges and furnish splines for insertion between adjoining blocks. Both types are designed to be installed with mastic over a wood subfloor or concrete slab. Prefinished blocks usually have eased or beveled edges.

Unit blocks typically consist of several 3/4-in. T&G strips, all laid parallel or alternating in each quarter of the block checkerboard fashion (Fig. 9.9-17). Consequently, typical block sizes are multiples of the strip width used (Fig. 9.9-18).

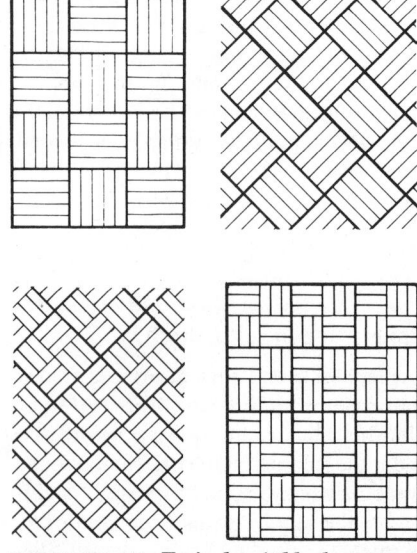

FIGURE 9.9-17 *Typical unit block patterns: heavy lines show individual blocks, laid in square or diagonal patterns.*

Common strip widths are 1-1/2, 2, and 2-1/4 in., resulting in typical block dimensions of 6, 6-3/4, 7-1/2, 8, 9, 10 and

11-3/4 in. The individual strips are held together with wood or metal splines embedded in the lower surface.

FIGURE 9.9-18
Typical Unit Block Dimensions*

Strip Width, in.	Strip Thickness, in.		
	1/2 in.	3/4 in.	33/32 in.
1-1/2	7-1/2 9	7-1/2 9	9
2	8 10	8 10	—
2-1/4	—	6-3/4 9 11-1/4	9 11-1/4

*All dimensions are for square blocks.

■ LAMINATED BLOCK

Laminated block is made typically 15/32 in. thick in 9-in. squares, but other sizes, such as 8-in. and 8-1/2-in., also are available from some manufacturers. Appearance grades are Prime and Standard.

Because of its cross-laminated construction in three to five plies, shrinking and swelling of individual blocks are minimized. This type of wood flooring has good dimensional stability and often is recommended for damp locations, such as slabs on grade.

Adhesives used in the manufacture of laminated block flooring should be able to resist the temperature and humidity variations to which the flooring may be subjected. Melamine-urea resin adhesive gives good results at moderate cost and represents the typical adhesive type used for most laminated block flooring.

Highly water-resistant adhesives such as phenols, resorcinols, and melamines may also be used if the application warrants it; however, the use of these glues adds to the cost of the product. Adhesives that create high-strength dry bonds but that are adversely affected by moisture are not recommended for the manufacture of laminated-block flooring.

■ SLAT BLOCK

Appearance grades are based on grading rules developed by the American National Standards Institute and the American Parquet Association, ANSI/APA 1-1984, "Mosaic-Parquet Hardwood Slat Flooring," which also checks compliance with the rules through periodic inspections of member plants. These products are suitable for installation in mastic over concrete surfaces, both above and on grade. The basic components of these products are solid, 5/16 in. thick, generally square-edged slats of hardwood, 3/4 in. to 1-1/4 in. wide and 4 to 7 in. long, assembled into basic squares. These in turn are factory assembled checkerboard fashion, with the grain in each adjoining square reversed, into larger flooring blocks up to 30 in. long and wide (Fig. 9.9-19).

FIGURE 9.9-19 *Typical slat block patterns. Pickets forming borders may be part of the block or separate.*

Several types of slat block flooring are produced. Some are assembled into panels held together with a face paper; these generally are 9-1/2-, 18-, or 19-in. squares and are marketed unfinished. Others are made up of single basic 6-in. T&G squares, which are held together by mechanical attachments and are generally factory finished. Still others are assembled into panels held together with a backing material such as a textile webbing, asphalt-saturated felt, or another type of felt or non-woven interfacing. These products generally are 9-1/2, 11, or 12 in. square;

they may be square edged or grooved and splined and be either unfinished or prefinished.

9.9.5.2

Single-Slat Parquet

Many decorative motifs can be created with single-slat parquet flooring (Fig. 9.9-20), herringbone being the most common. This type of flooring consists of side- and end-matched strips, all carefully machined to uniform lengths ranging from 6 to 18 in. Typical thicknesses are 3/4 and 33/32 in.; common face widths are 1-1/2, 2, and 2-1/4 in. Since this product is essentially short lengths of strip flooring, all types of woods, grades, and sizes available in strip flooring are available in single-slat parquet.

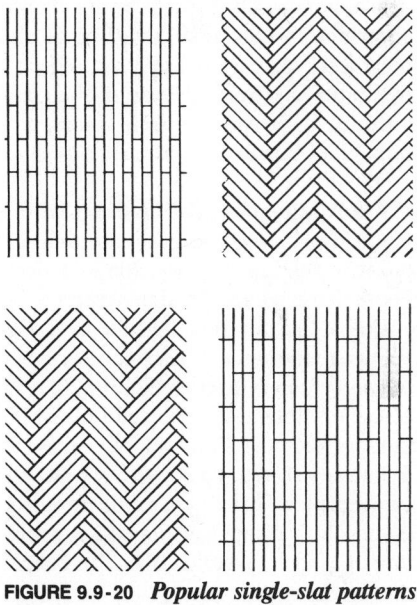

FIGURE 9.9-20 *Popular single-slat patterns achieved with uniform-length precut wood strips.*

9.9.6

MATERIALS HANDLING

To maintain proper moisture content, flooring products should not be transported or unloaded in rain, snow, or excessively humid weather. Flooring should not be delivered to a construction site until the building is enclosed, concrete and plaster work has been completed, and all building materials are dry. In winter, an interior temperature of 65° to 70°F should be maintained for at least 5 days before flooring is delivered. Flooring should be stored for several days in the rooms where it will be installed to allow it to become acclimated to local conditions.

In crawl space construction, adequate cross-ventilation should be provided under the floor. The total area of vent openings should equal 1.5% of the first floor area.

A groundcover of 4- to 6-mil polyethylene film is essential as a moisture retarder. Inadequate moisture control can harm any floor installation by contributing to warping or discoloration of the flooring.

9.9.7

MAINTENANCE

Properly finished wood floors are durable and easy to maintain. It is best to vacuum or dust-mop floors once a week. Wood floors should never be washed or wet mopped because water seepage between boards can cause staining and warping. Regardless of floor finish type, flooring manufacturers recommend that the finish be protected with a solvent-based liquid buffing wax-cleaner or a paste wax. Spills should be wiped up immediately to prevent damage.

9.10 RESILIENT FLOORING

Resilient flooring is a class of flooring products distinguished by resilience and dense, nonabsorbent surfaces. Resilience contributes to indentation resistance and quietness, while the dense surface ensures durability and relative ease of maintenance. A variety of decorative colors, patterns, and textures are available in sheet or tile form.

Availability of self-adhesive tile has made installation of some products easier. No-wax surfaces on both tile and sheet goods have reduced maintenance.

Linoleum, developed a little over 100 years ago in England, was the forerunner of contemporary resilient flooring. It consisted of a blend of oxidized linseed oil binders and cork flour fillers, bonded to a burlap or felt backing. Asphalt tile evolved after World War I from a troweled-on mastic flooring that consisted of a mixture of asphalt, asbestos fibers, and naphtha solvents.

Today, linoleum and asphalt tile have been largely replaced by sheet vinyl and vinyl composition tile products. Early composition tile products contained asbestos, but the discovery of the harmful effects of this mineral caused it to be replaced with other, less harmful fibers. The introduction of vinyl and vinyl composition products, with their excellent wearing properties and decorative potential, substantially increased the range of resilient flooring uses. Today, a variety of resilient flooring products are available for residential and nonresidential construction.

Resilient flooring typically is used where economical, dense, nonabsorbent wear surfaces are desired. Depending on the product, resilient floor finishes offer a wide range of decorative effects and may provide underfoot comfort, good durability, and relative ease of maintenance. Specific products with enhanced properties are available to suit most functional requirements (Fig. 9.10-1).

Suitability of a flooring product is related to the product's physical properties and intended use. In addition to the initial installed cost, ease of maintenance and durability are important considerations in determining suitability. These factors in turn depend on subfloor conditions, proper installation, and maintenance.

This section describes the basic types of resilient flooring products, their manufacture, properties, and selection criteria. It also addresses the materials and methods necessary for the installation of resilient floors.

9.10.1

FLOORING MATERIALS

Resilient flooring is manufactured in tile and sheet forms, in different thicknesses, and from a variety of ingredients. Tile sizes range in inches from 9 by 9 and 12 by 12 up to 36 by 36. Sheet generally is made in 6- and 12-ft. widths, although some products are produced in widths of 4 ft. 6 in. Matching resilient accessories such as wall base, thresholds, stair treads, and feature strips are available to complete flooring installations (Fig. 9.10-2).

Resilient flooring products may be classed according to basic ingredients into (1) *vinyl*, (2) *vinyl composition*, (3) *rubber*, and (4) *cork*. Many resilient flooring products are manufactured to meet requirements of ASTM standards and federal specifications, which together constitute the industry-recognized standards of quality. In some types, the basic ingredients run the full depth of the flooring; in others, backing material adds desired physical characteristics or reduces cost. Figure 9.10-2 summarizes the ingredients and backing materials, sheet widths, tile sizes, gauges, and applicable standards.

9.10.1.1

Vinyl Composition Tile

Vinyl composition tile is composed of mineral fibers, ground limestone plasticizers, pigments, and polyvinyl chloride (PVC) resin binders. These resins permit light colors and a great variety of designs. Some tile products are available with self-adhesive backs and no-wax surfaces. Older vinyl composition tile contained asbestos fibers, which have been found to cause cancer and other physical problems, especially diseases of the lungs, and are no longer used in resilient flooring products.

FIGURE 9.10-1 *Raised-pattern rubber flooring is used here in a heavy-use area. (Courtesy Frudenberg Building Systems, Incorporated.)*

ASTM Standard F 1066 classifies vinyl composition tile as Class 1, which are solid-color tiles; Class 2, in which a pattern extends completely through the tile; or Class 3, in which a pattern is only on the surface of the tile.

Ingredients are mixed under heat and pressure and rolled into blankets to which decorative chips may be added. Additional rolling and calendering produce smooth sheets of desired thickness. After waxing, sheets are cut into tiles.

9.10.1.2

Vinyl Sheet and Tile

The chief ingredient of vinyl products is polyvinyl chloride (PVC) resin. Other ingredients include mineral fillers, pigments, plasticizers, and stabilizers. Plasticizers provide flexibility; stabilizers fix the mixture to ensure color stability and uniformity. Vinyl products sometimes are referred to as *flexible* vinyl to distinguish them from vinyl composition products, which also are made with PVC resins and are termed *semiflexible* vinyl.

■ VINYL SHEET

Vinyl sheet products are made with a vinyl wear surface bonded to a backing. The backing may be vinyl, polymer-impregnated mineral fibers, asphalt- or resin-saturated felt, nonfoam plastic, or foamed plastic. Some vinyl products have a layer of vinyl foam bonded either to the backing or between the wear surface and the backing. Backings are classified in ASTM Standard F 1303 as Class A, fibrous formulations; B, nonfoamed plastics; and C, foamed plastics.

In addition to PVC resins, the wear surface of vinyl-surfaced sheets may contain decorative vinyl chips, filler, pigments, and other ingredients. Powdered vinyl resins, fillers, plasticizers, stabilizers, and pigments are mixed, rolled into sheets, and chopped to form these vinyl chips. Vinyl chips of various colors are mixed with additional resins and spread evenly over the backing and bonded to it under high heat and pressure (Fig. 9.10-3). The vinyl resins and plasticizers together are called the *binder*.

Until recently, the wear surface of sheet vinyl was classified as *filled* or clear (Fig. 9.10-4a and 9.10-4b). Clear meant that the wear layer contained mostly binder and that when decorative elements occurred they were beneath a layer of clear vinyl. Filled meant that the wear layer was mostly vinyl chips (fillers). However, because even clear vinyl wear surfaces have some fillers, these designations have always been somewhat misleading. Therefore, the industry no longer uses the designations *clear* and *filled*. Instead, ASTM Standard F 1303 classifies vinyl sheet flooring as Type I or Type II. In Type I flooring the binder constitutes at least 90% of the wear layer; in Type II, it is at least 34%.

In addition, ASTM Standard F 1303 designates three grades within each Type. The grades define wear layer thickness. Grade 1 is to be used for commercial, light-commercial, and residential projects. It requires a 0.020-in. wear layer in Type I and a 0.050-in. wear layer in Type II.

Grade 2 is to be used for light-commercial and residential projects. It designates a 0.014-in. wear layer in Type I and a 0.030-in. wear layer in Type II.

Grade 3 is for residential projects only. It requires a 0.010-in. wear layer in Type I and a 0.020-in. wear layer in Type II.

Greater thicknesses are required in Type II than in Type I because the abrasion resistance and durability per unit of thickness of a wear layer are greater in the binder than in the fillers. The more binder, the thinner the wear layer can be for the same service.

Sheet products with vinyl backings may have a design imprinted on top of the backing or on the underside of the wear surface. They are usually printed with vinyl inks. A clear PVC wear surface is then calendered to the desired thickness and laminated to the backing with heat and pressure (Fig. 9.10-5).

■ VINYL TILE

Vinyl tile may be of homogeneous solid composition or backed with other

FIGURE 9.10-2
Summary of Resilient Flooring Products

Type	Standards	Backing Materials	Gauges[a]	Sizes	Basic Ingredients
ASPHALT TILE					
Plain	SS-T-312B	—	1/8 in., 3/16 in.	9 in. × 9 in., 12 in. × 12 in.[b]	Asphaltic and/or resin binders, mineral fibers, and limestone fillers
Greaseproof	SS-T-312B	—			
VINYL-COMPOSITION TILE					
	ASTM F 1066	—	1/16 in., .080 in., 3/32 in., 1/8 in.	9 in. × 9 in., 12 in. × 12 in.[b]	PVC resin binders, limestone, mineral fibers
VINYL TILE					
Solid	SS-T-312B	—	1/16 in., .080 in., 3/32 in., 1/8 in., .050 in. to .095 in.	9 in. × 9 in., 12 in. × 12 in. 4 in. × 36 in.[b] 9 in. × 9 in., 12 in. × 12 in. and sizes up to 36 in. × 36 in.[b]	PVC resin binders and mineral fillers
Backed	SS-T-312B	Scrap vinyl, mineral or rag felts			
VINYL SHEET[c]	ASTM F 1303	Scrap vinyl, mineral or rag felts	.065 in. to .160 in. (.010 in. to .050 in. wear surface)	6 ft. 0 in., 12 ft. 0 in., and 4 ft. 6 in. wide	PVC resin binders and mineral fillers; clear PVC film for clear vinyl surface
			.065 in. to .160 in. (.010 in. to .050 in. wear surface)		
RUBBER TILE	ASTM F 1344	—	.080 in., 3/32 in., 1/8 in., 3/16 in.	9 in. × 9 in., 12 in. × 12 in. 18 in. × 36 in. 36 in. × 36 in.[b]	Synthetic or natural rubber, mineral fillers
CORK TILE	None[d]				
Protected surface Wood surface		—	1/8 in., 3/16 in., 1/4 in., 5/16 in.[b], 1/2 in.[b]	9 in. × 9 in., 6 in. × 6 in.[b], 6 in. × 12 in., 12 in. × 12 in.[b]	Cork particles and resin binders, wax or resin finish
Vinyl surface		—	1/8 in., 3/16 in.	9 in. × 9 in., 12 in. × 12 in.	Cork particles and resin binders, clear vinyl finish
LINOLEUM SHEET	None[e]	Burlap	(0.050-in. wear surface) 1/8 in.		
Plain and marbleized Inlaid and molded		Rag felt	.090 in.	60 ft. 0 in. wide	Cork and/or wood flour with linseed oil binders
		Burlap	1/8 in.		
		Rag felt	.090 in.		
Battleship		Burlap	1/8 in.		
LINOLEUM TILE	None[e]		(.050-in. wear surface) 1/8 in.	9 in. × 9 in. 12 in. × 12 in.[b]	Cork and/or wood flour with linseed oil binders
		Rag felt	.090 in.		

[a] Thinnest gauges and wear surfaces indicated are minimums suitable for adhesive installation.

[b] Not common, available on special order from some manufacturers.

[c] Also produced with foam backing, self-adhesive back, and no-wax surface; not included in data shown.

[d] Previously covered by Federal Specification LLL-T-00431, now cancelled.

[e] Previously covered by Federal Specification LLL-F-1238A, now cancelled.

FIGURE 9.10-3 *Ingredients are keyed in a series of presses to make vinyl sheet flooring.*

a — WEAR SURFACE OF VINYL PARTICLES IMBEDDED IN CLEAR VINYL

— BACKING

b — WEAR SURFACE OF CLEAR VINYL

— DESIGN IMPRINTED ON BACKING OR UNDERSIDE OF WEAR SURFACE

— BACKING

FIGURE 9.10-4 *Backed vinyl cross section: (a) filled vinyl; (b) clear vinyl.*

FIGURE 9.10-5 *Felt backing is coated to form a light, smooth background for printed designs for vinyl sheet.*

materials such as organic felts, mineral fibers or scrap vinyl.

Ingredients for solid vinyl tiles are mixed at high temperature and hydraulically pressed or calendered into homogeneous sheets of required thickness, and the sheets are cut into tile sizes. Backed products are essentially vinyl sheet flooring cut into tile sizes. Some tiles are made with a self-adhesive back.

9.10.1.3

Rubber Tile

Natural or synthetic rubber is the basic ingredient of rubber flooring. Clay and fibrous talc or mineral fillers provide the desired degree of reinforcement; oils and resins are added as plasticizers and stiffening agents. Color is achieved by nonfading organic pigments, and chemicals are added to accelerate the curing process.

The ingredients are mixed thoroughly and rolled into colored sheets. The sheets are calendered to uniform thickness and vulcanized in hydraulic presses under heat and pressure into compact, flexible sheets with a smooth, glossy surface. The backs then are sanded to gauge, ensuring uniform thickness, and sheets are cut into tiles.

9.10.1.4

Cork Tile

Cork tile is composed chiefly of the granulated bark of the cork oak tree, native to Spain, Portugal, and North Africa. Synthetic resins are added to the granulated cork, and the mixture is pressed into sheets or blocks and baked. Surfaces are finished with a protective coat of wax, lacquer, or resin applied under heat and pressure. Sheets then are cut to tile sizes. Vinyl cork tile has a film of clear PVC vinyl fused to the top surface to improve durability, water resistance, and ease of maintenance.

9.10.2

SELECTION CRITERIA

The suitability of a resilient flooring material for a particular location or condition depends on its physical properties. *Moisture resistance* is important in determining whether a material can be used in areas subject to surface or ground moisture. *Grease and alkali resistance* have a bearing on suitability for use in kitchens or over concrete, respectively. *Resilience* is related to quietness, indentation resistance, and underfoot comfort.

These and other properties may be combined into ratings for *durability* and *ease of maintenance*. Moisture and alkali resis-

tance must be considered together in determining suitable location with respect to grade, whether *suspended*, *on grade* or *below grade* (Fig. 9.10-6). Installed cost is another factor in flooring selection but should be considered along with long-term costs, which are affected by physical properties, durability, and ease of maintenance. Selection criteria for resilient flooring products are summarized in Figure 9.10-7.

suspended

on grade

below grade

FIGURE 9.10-6 *Subfloor locations. Subfloors are considered suspended when they are built at least 18 in. over a ventilated crawl space.*

9.10.2.1

MOISTURE RESISTANCE AND LOCATION

Linoleum that contains linseed oil binders needs sufficiently dry conditions so it will not be softened from exposure to moisture. Products with organic ingredients or backings are subject to mold growth and decay. Asphalt, vinyl composition, solid vinyl, and sheet vinyl with mineral fiber backings are least affected by moisture and alkalies, which may be leached by moisture from concrete floors.

The relative resistance of flooring products is shown by the differences between below-grade, on-grade, and suspended subfloor locations (see Fig. 9.10-7). Consideration has been given to the alkali resistance of the backing materials, which in some products may vary from those of the surface. In sheet vinyl with asphalt felt backing, for instance, the vinyl surface may have excellent alkali resistance, a property not possessed by organic felt backing.

In addition to the effect on flooring materials directly, moisture may prevent

FIGURE 9.10-7
Selection Criteria for Resilient Flooring Products[a]

| Type of Flooring | Location[b] | Resistance to: | | | | | Resil-ience[d] | Quietness | Ease of Maintenance | Durability |
		Grease	Alkalies	Stain[c]	Cigarette Burns	Indentation				
VINYL-COMPOSITION TILE	BOS	2	2	4	2	3	6	6	2	2
VINYL TILE										
Solid	BOS	1	1	1–2	1–5	1–4	2–5	2–5	3	1
Asphalt felt backed[e]	BOS	1	1	1–3	3–5	2–5	2–4	2–5	3	1
Rag felt backed	S	1	2	1–3	3–5	2–5	2–4	2–5	3	1
VINYL SHEET										
Asphalt felt backed[e]	BOS	1	1	1–3	3–5	2–5	2–4	2–4	2	1
Rag felt backed[e]	S	1	2	1–3	3–5	2–5	2–4	2–4	2	1
RUBBER TILE	BOS	3	3	2	1	4	2	2	3	2
CORK TILE	OS	4	4	5	4	4	1	1	5	5
VINYL CORK TILE	OS	1	3	1	5	3	3	3	1	4

[a] Numerical rating indicates rank of each product compared with other products. Highest rating is 1 (excellent); 2 = very good; 3 = good; 4 = fair; 5 = poor.

[b] B = below grade, O = on grade, S = suspended.

[c] Varies with staining agent.

[d] Also indicates potential underfoot comfort.

[e] Foam-cushioned products are rated highly for resilience, indentation resistance, ease of maintenance, and quietness within the ranges indicated.

proper bonding to the subfloor or underlayment. Water-based adhesive will not set up in the presence of water, and asphaltic emulsion adhesives eventually may be displaced under sustained exposure to moisture. Excessive moisture also may deteriorate organic underlayments such as hardboards or plywood.

The main types of moisture affecting resilient floors are (1) surface moisture, (2) subfloor moisture, and (3) ground moisture.

■ SURFACE MOISTURE

Spilled water, floor moppings, and tracked-in moisture are the primary sources of surface moisture. Spilled water may be common in laundry areas, bathrooms, toilet rooms, and kitchens. Entries and corridors may be subjected to tracked-in moisture and excessive wet mopping.

Surface moisture affects flooring materials and adhesives by entering through the seams. Sheet flooring products minimize the number of seams, and when provided with moisture-resistant backings they are most suitable for use in such areas. The seams of many sheet products also may be sealed to prevent water seepage through seams.

■ SUBFLOOR MOISTURE

Concrete subfloors and mastic underlayments mixed with water release large amounts of moisture as they cure. With lightweight-aggregate concrete (weighing less than 90 lb./cu. ft.), or under conditions that retard the curing process, moisture may be released over long periods of time. Field tests described under "Concrete Subfloors" in Section 9.10.5.2 should be used to establish that the subfloor is sufficiently dry to receive the intended product. Since it originates in the

original mix of the concrete, subfloor moisture may be present in below-grade, on-grade, and suspended subfloors.

■ GROUND MOISTURE

Moisture from the ground generally is limited to below-grade and on-grade locations. Concrete floor slabs in direct contact with the ground, unless protected by a vapor retarder, may transmit moisture by capillary action. A well-drained fill of sand or gravel will retard the migration of moisture.

When resilient flooring is installed on wood subfloors over crawl spaces, a vapor retarder and adequate ventilation of the subfloor space are required. A vapor retarder placed over the ground will prevent transmission of moisture from the ground. Ventilation will encourage dissipation of moisture resulting from breaks in the vapor retarder and other sources.

Both crawl spaces and concrete slabs on grade require proper surface drainage of exterior finished grade to prevent ground water from becoming trapped under the floor.

9.10.2.2

Alkali and Grease Resistance

Major sources of grease are oils spilled in kitchens, laboratories, and shops; alkalies may be present from various cleaner residues. If flooring is properly protected with a floor wax or finish, and spilled oils or residues are removed promptly before they seep into the seams, such temporary exposure is not a hazard, except for materials with the lowest resistance ratings (see Fig. 9.10-7). Vinyl composition and solid vinyl tile are not affected substantially by either grease or alkali.

9.10.2.3

Resilience

Resilience is a measure of the instantaneous yielding and recovery of a surface from impact. Indentation resistance, quietness, and underfoot comfort are closely related to resilience.

■ INDENTATION RESISTANCE

In assessing indentation resistance, the momentary indentation produced from foot traffic and dropped objects is of primary importance. These impact pressures sometimes are quite high and demanding. A 105-lb. woman in spike heels, for example, exerts a pressure on the floor of approximately 2000 psi, while a 225-lb. man with his weight spread over 3-in. by 3-in. heels exerts only 25 psi.

Permanent indentation from heavy stationary objects, such as a piano or a desk, may be minimized by using floor protectors to distribute the load (Fig. 9.10-8). Indentation resistance of thinner flooring materials is greatly affected by the subfloor or underlayment and may be increased by selection of harder subsurface materials. Homogeneous vinyl tile and foam-cushioned vinyls have the highest indentation resistance.

Permanent indentation in some flooring types and under certain conditions cannot be entirely prevented. However, these indentations may be less conspicuous in patterned, textured, and low-luster floors.

WRONG — Narrow surfaces dent floors **RIGHT** — Wide surfaces protect floors

Always remove small metal domes from bearing surfaces. Composition furniture cups should be placed under heavy furniture that is only infrequently moved.

WRONG — Hard rollers mark floors **RIGHT** — Rubber rollers protect floors

Casters should be used on frequently moved furniture. They should be 2" in diameter, with soft rubber treads at least ¾" wide and an easy swiveling ball-bearing action.

WRONG — Remove small metal domes **RIGHT** — Use flat flexible-shank glides

Light furniture should have glides with a smooth, flat base and a flexible pin to maintain flat contact with floor. Diameter should be 1" to 2½" depending on weight of furniture.

FIGURE 9.10-8 *Floor protectors under furniture and equipment reduce indentation.*

■ QUIETNESS

The quietness rating indicates the effectiveness of a floor in reducing sound from foot traffic and other impact noises. While resilient floors have the capacity to soften impact sounds, they will not reduce reverberated noises that originate from other sources because they have practically no sound absorption ability. Cork and foam-cushioned vinyl flooring have the highest resilience and quietness ratings. Asphalt tile is at the low range of resilience and quietness.

9.10.2.4

Resistance to Sunlight

The actinic rays in strong sunlight may affect some resilient floors by causing fading, shrinking, or brittleness. Linoleum and vinyl products are most resistant to such deterioration. Color pigments are the critical factor in fade-resistant properties. Neutral colors show the best light resistance; pastel tones, especially yellows, blues, and pinks, are least effective in retaining colors under prolonged exposure to sunlight. Cork tile has the same tendency as natural wood to fade under strong sunlight.

9.10.2.5

Ease of Maintenance

The term *ease of maintenance* represents an appraisal of the relative expense (labor and materials) in keeping flooring at an

acceptable level of cleanliness and attractiveness. Relative ratings are based on a record of experience with various flooring types as well as laboratory testing. Textured surfaces and darker colors show less scuffing and soiling and generally receive higher ease-of-maintenance ratings. No-wax vinyl products and vinyl cork tile are the easiest to maintain, followed by other vinyl products, vinyl composition, and linoleum.

9.10.2.6

Durability

Durability reflects the ability of flooring products to retain serviceability and attractiveness over a period of time. Durability ratings are based on laboratory tests and are related to the physical properties of the material. Homogeneous vinyl tile is the most durable. Vinyl sheet flooring, rubber tile, and vinyl composition also are rated highly. Foam-cushioned products have good durability because the surfaces of these products absorb impact and resist abrasion.

9.10.2.7

Slip Resistance

The Americans with Disabilities Act of 1990 (see Section 1.6) requires designers to specify flooring materials that have a static coefficient of friction of not less than 0.60 for level surfaces and 0.80 for ramps. Some manufacturers currently publish these data for their flooring products, even though there is no consensus in the industry concerning the test methods needed to ensure compliance. It would be prudent to verify the current status of slip-resistance requirements before selecting a resilient flooring for any project.

9.10.3

FLOORING ACCESSORY MATERIALS

The most common resilient accessories used with resilient flooring materials are (1) wall bases; (2) stair treads, risers, and nosings; and (3) thresholds, feature strips, and reducing strips (Figs. 9.10-9, 9.10-10, 9.10-11, and 9.10-12). These accessories are available in more limited color and design variations than the flooring materials.

FIGURE 9.10-9 *Resilient accessories: (a) wall base* (VPI wall base courtesy Vinyl Plastics Incorporated); *(b) from left to right, cap for coved sheet goods, tile reducer, vinyl reducer, feature strip, and cove stick (to support sheet flooring for coving at a wall).* (Courtesy Mercer Products Company, Incorporated.)

FIGURE 9.10-10 *Resilient wall base types.*

FIGURE 9.10-11 *Premolded base accessories.*

WALL BASES

Rubber or vinyl wall bases are available in 2-1/2-in., 4-in., and 6-in. heights, in straight type and in either set-on or butt types with coved-bottom designs (see Fig. 9.10-10). Vinyl base is available in 50-ft. coils and in 4-ft. lengths. Rubber base normally is available in 4-ft. lengths only. Exterior and interior corners of rubber base commonly are premolded (see Fig. 9.10-9a). End stops with a finished edge for terminating wall base runs are available in both rubber and vinyl. The greater flexibility of vinyl base permits field-molding around exterior corners (see Fig. 9.10-9a); inside corners usually are scribed and fitted on the job.

STAIR TREADS, RISERS, AND NOSINGS

Vinyl and rubber stair treads and nosings are made with smooth or textured surfaces and may have inlaid abrasive grit strips to improve slip resistance (see Fig. 9.10-12). They may be combined with standard 6-in.-high wall base or higher stair risers to form complete stair coverings. Lengths vary from 3 to 12 ft.

THRESHOLDS, FEATURE STRIPS, AND REDUCING STRIPS

Thresholds usually are of vinyl or rubber, 2-3/4 and 5-1/5 in. wide, about 1/2 in. high, and 3 or 4 ft. long. *Feature strips* are made of asphalt, vinyl composition,

FIGURE 9.10-12 *Stair treads and nosings: (a) smooth flat tread; (b) textured (diamond) tread; (c) abrasive strip tread; (d) smooth or ribbed nosings, which may be used with carpet or sheet or tile treads.* (a, b, and c courtesy The R.C. Musson Rubber Company; d courtesy Mercer Products Company, Incorporated.)

linoleum, and vinyl in widths of 1/8, 1/4, 1/2, and 1 in. and in 1/2-in. increments up to 4 in. Thicknesses usually are 1/16 to 1/8 in., and lengths usually are 3 and 4 ft. Rubber and vinyl beveled *reducing strips*, 1 to 1-1/2 in. wide, 3 to 12 ft. long, and 1/16 to 3/8 in. thick, permit transition in finished floor heights (Fig. 9.10-9b).

9.10.4

FLOORING INSTALLATION MATERIALS

Materials for the installation of resilient flooring materials include underlayments, lining felt, and adhesives.

9.10.4.1

Underlayments

Underlayments are used over subfloors to provide a smooth, level, hard, and clean surface to receive flooring products. In general, panel underlayments are used over wood subfloors; mastic underlayments are applied over concrete subfloors.

■ PANEL

Panel underlayments may be used over any structurally adequate suspended subfloor such as plywood or lumber boards. The three common types of panel underlayment are plywood, hardboard, and particleboard.

Stiffer floors result when joints in the underlayment are staggered with respect to joints in a plywood subfloor and blocking can be omitted. Over a board subfloor, underlayment joints should not fall over joints in the floor boards (Fig. 9.10-13). To allow for expansion from humidity changes, underlayment panels should be spaced approximately 1/32 in. at joints and about 1/8 in. apart at intersections with vertical surfaces. End joints in underlayment panels should be staggered ashlar fashion (see Fig. 9.10-13).

Nailing should start in the middle of the panel and proceed outward to the edges, with nail heads set flush or slightly countersunk. Panel joints should be sanded smooth and flush before the flooring is installed.

Minimum underlayment thicknesses are given below for each material; however, greater thicknesses may be warranted where increased stiffness is desired.

FIGURE 9.10-13 *Underlayment layout over a board subfloor.*

PLYWOOD A sanded Interior Underlayment grade, at least 1/4 in. thick, is suitable for most plywood underlayment. However, in areas subject to excessive moisture, such as in bathrooms and laundry areas, an Exterior C-C (plugged) or Underlayment grade with Exterior adhesive should be used. Underlayment may be installed over a plywood subfloor with the face grain either parallel or perpendicular to the floor joists. However, if the underlayment is installed with the face grain perpendicular to the floor joists a stiffer floor assembly results. Plywood should be installed over a board or plank subfloor with the face grain perpendicular to the direction of the floor boards (see Fig. 9.10-13), preferably with long edges over joists. Plywood underlayment should be installed with the fasteners and spacing shown in Figure 9.10-14.

HARDBOARD Hardboard underlayment is suitable to bridge small irregularities and to provide a smooth surface over an otherwise structurally sound wood floor. Underlayment grade hardboard is an untempered service type, surfaced on one side. It is manufactured in a nominal 1/4-in. (0.215 or 0.200 in. actual) thickness, in 3-ft. by 4-ft., 4-ft. by 4-ft., and 4-ft. by 8-ft. panels, and should conform to the American National Standards Institute/American Hardboard Association Standard ANSI/AHA A 135.4, Class 4 (service), surface S1S. Installation of panels is the same as for other panel underlayment using fasteners and spacing recommended in Figure 9.10-14.

PARTICLEBOARD Particleboard underlayment should not be used in areas subject to excessive moisture, and its moisture content as shipped from the mill should not exceed an average of 7%. Particleboard underlayment should be at least 3/8 in. thick and should conform to the ANSI Standard A 208.1, "Wood Particleboard," and bear the National Forest Products Association grade mark "Floor Underlayment," Grade 1-M-1. Particleboard underlayment generally is manufactured in 4-ft. by 8-ft. panels and in thicknesses ranging from 3/8 to 3/4 in. Nailing recommendations for various panel thicknesses are given in Figure 9.10-14.

Some resilient flooring manufacturers do not recommend the use of particleboard as an underlayment material. However, some particleboard manufacturers issue a warranty covering the entire resilient flooring installation if the flooring and the underlayment are installed to their specifications. Therefore, particleboard should be a product backed by a manufacturer's warranty and the installation instructions of the manufacturer should be followed.

■ MASTIC

Suitable mastic underlayments contain a chemical binder such as latex, asphalt, or polyvinyl-acetate resins and portland, gypsum, or aluminous cement. Mixtures consisting of powdered cement and sand to which only water has been added function only as crack fillers; when applied in thin coats, they break down under traffic.

Latex underlayments are most suitable for applications that require a thin layer (3/8 in. or less), for skim coating, and for patching where the fill must be feather-edged (Fig. 9.10-15).

Skim coating does not attempt to raise the elevation of the floor but merely smoothes over surface irregularities.

Asphaltic and polyvinyl-acetate underlayments are used for installations that require thicker underlayments (more than 3/8 in.). Mastic underlayments should be troweled smooth and true, with not more than a 1/8-in. variation from a straight line in 10 ft. Subfloors should be free of wax, oil, and surface coatings, such as concrete curing compounds, before mastic underlayment is applied.

FIGURE 9.10-14
Fastening for Panel Underlayment

| Underlayment, in. | Fastener Length, in.[a] | | Fastener Spacing |
	Nails[b]	Staples[c]	
PLYWOOD			
1/4	1 (2d)	7/8	Nails: 8 in. o.c. each direction in middle of panel, 6 in. o.c. around panel perimeter.
3/8	1-1/4 (3d)	1-1/8	
1/2	1-1/4 (3d)	1-1/4	
5/8	1-1/2 (4d)	1-1/2	Staples: 6 in. o.c. each direction in middle of panel, 3 in. o.c. around panel perimeter.
3/4	1-1/2 (4d)	1-5/8	
HARDBOARD[d]			
1/4	1 (2d)	7/8	Nails and staples: 6 in. o.c. each direction in middle of panel, 3 in. o.c. around panel perimeter.
PARTICLEBOARD[e]			
3/8	1-1/4 (3d)	1-1/8	Nails: 10 in. o.c. each direction in middle of panel, 6 o.c. around panel perimeter.
5/8	1-1/2 (4d)	1-1/2	
3/4	1-1/2 (4d)	1-5/8	Staples: 6 in. o.c. each direction in middle of panel, 3 in. o.c. around panel perimeter.

[a] Fastener lengths shown are minimum to provide 3/4-in. penetration for nails and 5/8 in. for staples. If longer fasteners are used, penetration should not exceed 1 in.

[b] Annularly threaded; common nails may be used by increasing the sizes 1/4 in. (1d).

[c] Galvanized divergent chisel, 16 gauge, 3/8 in. crown width.

[d] Nominal thickness, actual thickness 0.215 in.

[e] Moisture content as shipped from mill should not exceed 7% average.

FIGURE 9.10-15 *Latex underlayment can be feathered to a thin edge in the patching of subfloor depressions.* (Courtesy Kentile Floors.)

LATEX Latex underlayment consists of a mixture of a synthetic latex emulsion, a cementitious material such as portland cement, fine sand, and water. The underlayment should be mixed in accordance with the manufacturer's recommendations, should be troweled on in layers not more than 1/8 in. thick, and may be feathered to a thin edge (see Fig. 9.10-15). The final thickness should be built up in several applications and should not exceed 3/8 in. Rough surfaces should be leveled with a sanding machine after the mastic has dried completely. Because latex underlayment is a quick-setting compound, splashed material should be cleaned off promptly.

Solvent-based asphaltic primers and adhesives used over latex underlayment for the installation of asphalt or vinyl composition tile should be allowed to set for the time recommended by their manufacturer before the tile is installed.

ASPHALTIC Asphaltic underlayment is a mixture of an asphaltic emulsion, a cementitious material such as portland cement, fine sand, and water. Because proper proportioning varies, depending on the ingredients and the thickness of the underlayment, the manufacturer's instructions should be followed. Asphaltic underlayment generally is not recommended for installations less than 3/8 in. in thickness.

The concrete subfloor should be primed with a thin solution of the same asphaltic emulsion being used in the underlayment mixture. The mastic should be poured uniformly over the entire surface using wood screeds to establish the desired thickness. A straightedge should be used to level the mastic over the screeds, after which the surface should be troweled to a smooth finish. The setting time of the fill should not be accelerated with excessive heat, as this may cause cracking or crazing. Where it is necessary to trowel the mastic to a featheredge, the floor should be given a brush coat of undiluted asphaltic emulsion to prevent brittleness after the mastic has dried.

Solvent-based asphaltic primers and adhesives should not be applied over asphaltic underlayments; asphaltic emulsion or latex adhesives should be used.

POLYVINYL ACETATE Polyvinyl acetate underlayment is formed by mixing portland cement and sand with a specially compounded polyvinyl acetate (PVAc) resin emulsion. The PVAc liquid also should be applied to the concrete subfloor to improve bond with the underlayment. When a fill of 1/2 in. or thicker is required, a portland cement concrete "topping" may be used without PVAc resins, but the PVAc liquid still should be used as a bonding agent on the concrete subfloor.

This type of underlayment may be used with any adhesive and flooring type. Its installation is similar to that described for asphaltic underlayment.

9.10.4.2

Lining Felt

Some manufacturers recommend the use of lining felt over panel underlayment to minimize the possibility of joints opening and tile splitting due to underlayment movement. However, this is not a significant hazard if the underlayment or subfloor is sufficiently thick and is properly installed. Of greater significance is the fact that indentation resistance of flooring materials is substantially reduced by lining felt, which imposes limitations on its use.

Lining felt is used in conjunction with some vinyl sheet products over nonporous subfloors, such as old resilient floors, terrazzo, ceramic tile, and concrete that has been treated with curing agents. The liner's purpose is to provide a means for the water in the adhesive to be absorbed, which allows the adhesive to cure.

9.10.4.3

Adhesives

Most adhesives used for the installation of resilient flooring are of a troweling grade for application with a notched trowel (Fig. 9.10-16). Some types also are made of a more fluid consistency for application with a brush. These often are used with thin-gauge products to eliminate the possibility of trowel marks "telegraphing" to the surface of the finished floor.

FIGURE 9.10-16 *Latex adhesive is gradually spread to a uniform thickness by a swirling motion with a notched trowel.* *(Courtesy Kentile Floors.)*

The major types of adhesives are (1) linoleum paste, (2) asphaltic adhesive, (3) asphalt-rubber adhesive, (4) waterproof-resin cement, (5) latex cement, (6) epoxy adhesive, and (7) wall base cement. Figures 9.10-17 and 9.10-18 summarize the properties and applications of the various adhesives. Adhesives with slightly varying properties are produced in each type, so the general recommendations of the flooring and adhesive manufacturers should be followed.

■ LINOLEUM PASTE

Linoleum paste is a low-cost adhesive typically used for installing linoleum, vinyl sheet flooring, and lining felt. It is a water-based adhesive that remains water-soluble even after hardening and should not be used on concrete slabs on or below grade or in other areas subject to moisture.

Flooring should be installed immediately after spreading this adhesive, then rolled with a 100-lb. roller to ensure proper bond. The paste dries to a firm film within 24 hours and becomes hard within a few days. The rigidity of the hardened adhesive restrains sheet flooring dimensionally and prevents curling.

■ ASPHALTIC ADHESIVE

Asphaltic adhesives are low-cost, water-resistant adhesives used exclusively for installing vinyl composition tile. They may be useful on all types of below-grade, on-grade, and suspended subfloors. Asphaltic adhesives work with most underlayments, except that cutback asphaltic adhesive and solvent-based primer should not be used over emulsion-based underlayment. Asphaltic adhesives retain their bonding qualities while the tile gradually conforms to minor surface irregularities. They eventually harden to provide a durable water- and alkali-resistant bond. Asphaltic adhesives are either solvent-based or water-emulsion.

SOLVENT BASE   Solvent-based asphaltic adhesives consist of a solution of asphalt in hydrocarbon solvents and are available in brushable primers or thicker adhesive formulations. Cutback asphalt adhesive, the most common of the solvent-based asphaltic adhesives, is of a troweling grade consistency. Because the volatile solvents are flammable and mildly toxic, these adhesives should not be used near an open flame and the area should be well ventilated. The adhesive should be allowed to set approximately 30 minutes before applying tile. The working time ranges from 4 to 18 hours.

Solvent-based adhesives may cause or aid atmospheric damage and are not permitted in some jurisdictions. In time, they may be prohibited altogether.

WATER EMULSION   Water-emulsion asphaltic adhesives consist essentially of tiny asphalt particles suspended in water. They become water-resistant upon drying. They are of troweling grade consistency and are spread with a notched trowel. The adhesive should be allowed to set at least 30 to 60 minutes before tile is installed, but the working time ranges from 8 to 24 hours. This permits application to relatively large areas, overnight setting of adhesive, and installation of tile the following day.

■ ASPHALT RUBBER ADHESIVE

An asphalt-rubber adhesive is an asphalt and rubber water emulsion used primarily for the installation of vinyl composition tile. The working characteristics of this adhesive are similar to those of asphaltic adhesive, and it can be used over any type of subfloor or underlayment.

■ WATERPROOF-RESIN ADHESIVE

Most adhesives of the waterproof-resin type are resins in an alcohol solvent. They have approximately the same application and working characteristics as linoleum paste but are not soluble in water. However, they are adversely affected by alkaline moisture and are unsuitable for use over concrete in contact with the ground on or below grade. With the exception of asphalt and vinyl composition tile, waterproof-resin cements may be used for applying most resilient flooring in suspended locations that require waterproof bond. The volatile alcoholic content makes these adhesives flammable, and they should not be used near an open flame or in poorly ventilated areas.

■ LATEX ADHESIVE

Latex cements consist of rubber resins and water. They are extremely moisture- and alkali-resistant after they set. On drying, the adhesive turns to a rubbery film, which eventually hardens and becomes insoluble in water. Latex cements generally provide good adhesion and may be used for installing all types of resilient flooring on any type of subfloor or underlayment. They are relatively higher in cost and normally are not used for installing vinyl composition and asphalt tile where cheaper asphaltic adhesives also are suitable. Latex cements more commonly are used for installing rubber tile, vinyl tile, and fiber-backed sheet products on or below grade. Flooring may be installed immediately or as recommended by the manufacturer.

FIGURE 9.10-17
Summary of Adhesives by Type

Adhesive	Type	Flooring	Installation	Precautions
LINOLEUM PASTE	Clay: Water-base Troweling grade	Lining felt All sheet flooring Vinyl tile with backing Rubber tile Cork and vinyl cork tile	Over suspended wood or concrete subfloors, panelboard, and latex underlayments; install flooring immediately.	Do not use on or below grade, over suspended subfloors subject to moisture, or for solid vinyl, asphalt, and vinyl-asbestos tile. Remove with damp cloth.
ASPHALT[a] Cutback asphalt	Asphalt: Solvent-base Troweling grade	Vinyl-composition tile	Over all concrete subfloors and panelboard underlayment primed with asphaltic primer. Allow to set 30 min.; install tile within 4 to 18 hr, as recommended.	Do not use under or over lining felt or asphaltic underlayments. Combustible. Remove with fine steel wool and soapy water.
Asphalt emulsion	Clay and asphalt: Water-base Troweling grade	Vinyl-composition tile 1/16 in. and thicker	Over all subfloors and over lining felt. Allow to set 30 to 60 min.; install tile within 8 to 24 hr. as recommended.	Do not use to install lining felt. Keep from freezing. Remove with fine steel wool and soapy water.
ASPHALT-RUBBER	Asphalt and rubber: Water-base Brush and troweling grades	Vinyl-composition tile	Overall subfloors and over lining felt. Allow to set 30 min. before installing tile or as recommended.	Use brushing grade with 1/16 in. vinyl-asbestos tile; keep from freezing. Remove with fine steel wool and proprietary cleaner.
WATERPROOF RESIN	Resin: Solvent-base Troweling grade	Vinyl with rag felt, vinyl, or rubber backing Rubber tile Cork tile Solid vinyl	Over all suspended subfloors; install flooring within 15 min.	Do not use on or below grade, for asphalt or vinyl-asbestos tile, or to install lining felt. Remove with fine steel wool and proprietary cleaner. Combustible.
LATEX[b]	Rubber: Water-base Brush and troweling grades	Vinyl with rubber backing Rubber tile Solid vinyl tile Lining felt All sheet flooring	Over all subfloors; install flooring immediately.	Keep from freezing. Remove with fine steel wool and soapy water.
EPOXY	Resin and catalyst: Troweling grade	Solid vinyl and rubber tile on or below grade	Over all types of subfloors. Mix only as much as can be spread within 30 min. and covered within 2-1/2 hr. or as recommended.	Guard against tile slipping while adhesive sets.
WALL BASE Synthetic Rubber cement	Synthetic rubber resin: Solvent-base Brush grade	Vinyl cove base Metal nosings and edges All resilient tile except asphalt and vinyl-asbestos to metal surfaces	Apply to both wall and material; install material within 20 min.	Combustible. Remove with proprietary remover.
Standard cove base cement	Solvent-base Troweling grade	Vinyl and rubber cove base	Over dry walls above grade; install base within 15 min.	Combustible. Remove with fine steel wool and soapy water.

Note: Formulations and working properties of adhesives vary and manufacturers' specific recommendations should be followed.

[a] May be used over latex underlayment if sufficient setting time is allowed.

[b] May be used to install asphalt and vinyl-asbestos tile; however, asphaltic adhesives are used more commonly because of lower cost.

FIGURE 9.10-18
Summary of Adhesives by Flooring Type and Location

Flooring	Wood (Suspended)	Concrete (Suspended)	Concrete (On and Below Grade)
Subfloor and Location			
SHEET FLOORING			
VINYL			
Rag felt backed	Waterproof resin	Waterproof resin	Not recommended
Vinyl or rubber backed	Latex Waterproof resin	Latex Waterproof resin	Latex
Mineral fiber backed	Latex	Latex	Latex
TILE FLOORING			
VINYL			
Solid	Waterproof resin Latex	Waterproof resin Latex	Latex Epoxy
Rag felt backed	Linoleum paste Waterproof resin	Linoleum paste Waterproof resin	Not recommended
Mineral fiber backed	Latex Linoleum paste	Latex Linoleum paste	Latex
RUBBER	Linoleum paste Waterproof resin	Waterproof resin Linoleum paste	Latex Epoxy
VINYL-COMPOSITION	Asphalt emulsion Asphalt cutback Asphalt rubber	Asphalt emulsion Asphalt cutback Asphalt rubber	Asphalt emulsion Asphalt cutback Asphalt rubber
CORK	Linoleum paste Waterproof resin	Linoleum paste Waterproof resin	Not recommended

*Foam-cushioned products are installed by specially formulated adhesives recommended by flooring manufacturer.

FIGURE 9.10-19 *Standard cove-based cement may be applied with a notched spreader or by a cartridge type gun, as shown here.*

SYNTHETIC RUBBER CEMENT Synthetic rubber cement is a contact adhesive that must be applied to both the wall and the resilient base. Bond is developed as soon as the two surfaces touch. Synthetic rubber cement also is used typically for installing metal and resilient nosings, edges, and binding strips. It is particularly useful in bonding resilient accessories in the field, where movement after fitting would be undesirable.

9.10.5

GENERAL INSTALLATION RECOMMENDATIONS

The installation of resilient flooring, whether sheet or tile, depends on (1) proper job conditions, (2) adequate subfloor construction, and (3) proper selection of underlayment, adhesives, and flooring material.

Because of the variety of proprietary flooring materials, underlayments, and adhesives, the individual manufacturers' recommendations always should be consulted to ensure proper use in each case.

9.10.5.1

Job Conditions

Installation of finish flooring should be delayed until all other interior work, with the possible exception of door trim and touch-up painting, has been completed. If it is necessary to install flooring sooner, the finished floor should be protected with plain, undyed, unsaturated building paper. Door trim may be installed prior to flooring if it is shimmed up with scraps of

■ EPOXY ADHESIVE

Epoxy adhesives are extremely moisture-resistant formulations with high bonding strength and excellent durability. However, the relative high cost, difficulty in spreading, and excessive slipping of tiles immediately after installation moderate the use of these adhesives.

Epoxy adhesives consist of two parts, an epoxy resin and a catalyst that activates the mixture. The parts are mixed at the job site. A chemical reaction begins as soon as the components are mixed, so the adhesive must be used within 2 to 3 hours. The actual setting time varies, however, due to variations in the formulations. Therefore, the manufacturer's recommendations should be followed closely.

■ WALL BASE CEMENT

The two basic adhesives used to secure resilient wall base are standard cove-based cement and synthetic rubber cement. Both are solvent-based formulations that set as the volatile solvents escape from the mix; both are flammable and should not be used near an open flame or in poorly ventilated areas.

STANDARD COVE-BASED CEMENT The low-cost standard cove-based cement typically is used for installing resilient wall base and premolded corners. It develops bonding power about 10 minutes after application to either the wall surface or the wall base (Fig. 9.10-19).

flooring to permit later installation of the flooring under the trim.

Care should be taken to prevent freezing and damage to materials during delivery, handling, and storage. Materials should be stored on a smooth, level surface in a warm, dry place. Sheet flooring should not be stored for long periods of time and the rolls should be stored vertically, on end. Tile flooring should be stacked not more than five cartons high.

A temperature of at least 70 °F and not more than 90 °F should be maintained for 48 hours before installation, during installation, and for 48 hours after installation. A minimum temperature of 55 °F should be maintained thereafter, and adequate ventilation should be provided at all times. Ventilation is particularly important when solvent-based adhesives and primers, which are highly volatile and flammable, are used.

Figure 9.10-20 shows some of the terms used to describe locations to be covered in a resilient flooring application. The terms used are defined in the body of the text when they occur. Terms not defined here are defined in the Glossary.

FIGURE 9.10-20 *Flooring terms.*

9.10.5.2

Supporting Construction

The selection of resilient flooring products and adhesives is determined largely by subfloor location, whether below grade, on grade, or suspended (see Fig. 9.10-6). These designations are based mainly on the relative likely exposure to moisture and alkalies, which adversely affect many flooring materials and adhesives.

The two main types of subfloor typically used as a base for resilient flooring are wood and concrete. All subfloors, whether covered with underlayment or not, should be firmly fastened and structurally adequate to carry the intended loads without excessive deflection.

Both wood joist and concrete slab floor constructions are suitable for the application of resilient flooring. Plank-and-beam construction is similar to wood joist construction, but the subfloor is 2 in. thick or thicker, and structural members are larger and spaced farther apart. Underlayment is recommended over plank subfloors.

Concrete slab floors may be supported on ground or suspended above grade. Basement floors are concrete slabs supported on ground below grade.

Good site drainage is required for all foundation types, regardless of grade; therefore, the exterior finish grade should be sloped away from the foundation to divert surface water.

■ WOOD SUBFLOORS

Suspended wood floors above grade usually are sufficiently removed from excessive external moisture sources, but floors over crawl spaces at or near grade may require the crawl space to be ventilated and the ground inside the crawl space to be covered with a suitable vapor retarder.

Wood floors may be of single-layer construction, consisting of a combination subfloor and underlayment, or of double-layer construction, composed of separate subfloor and underlayment. Subfloor and combination subfloor and underlayment materials, minimum subfloor thickness for various joist spacing, and nailing requirements are discussed in Section 6.6.4. The use of thicker subfloors will result in stiffer floors. Single-lumber board construction is not acceptable for the direct application of resilient flooring.

■ CONCRETE SUBFLOORS

Concrete subfloors should have a density of at least 90 lb./cu. ft. If concrete of lower density has been used, it should be refinished with a 1-in.-thick topping of a regular concrete mix, bonded to the existing concrete with a bonding agent such as polyvinyl-acetate compound.

Subfloors on or below grade should be separated from the ground with a suitable vapor retarder, such as polyethylene film (see Section 3.6).

Concrete subfloors generally provide a suitable surface for resilient flooring, but they require precaution against excessive moisture. The moisture content of concrete floors should be checked, particularly when lightweight concrete is used, because it cures slowly. In some instances, dampness can be observed under stationary objects on a floor on or below grade. Most often, however, moisture cannot be detected readily, and tests should be performed to determine if the floor is sufficiently dry to receive flooring.

Concrete slabs that have been treated with curing, hardening, or separating compounds also should be tested for adequate bond.

SUITABILITY TESTS Several tests may be performed to determine acceptable moisture levels and bonding properties of concrete subfloors. These are a relative humidity test, a mat moisture and bonding test, and a primer test.

A *relative humidity test* is intended for testing new suspended concrete floors for the installation of moisture-sensitive products such as linoleum, cork, and rag felt–backed vinyl. To conduct this test, a relative-humidity meter is placed on the concrete next to an interior wall or column. Depending on atmospheric conditions, the meter should read 40% to 65%. The meter is then covered with an 18-in. square of polyethylene film, and the edges are sealed to the floor with adhesive or tape. When substantial temperature fluctuations are expected, the meter should be protected with several layers of burlap or insulating batts. If there are no leaks in the polyethylene film, the reading should stabilize after 24 hours on a normal slab and after 72 hours on a thicker-than-normal slab. A sufficiently dry slab will register a reading of 80% or less. If the reading is more than 80%, the slab should be allowed to cure further and the test repeated.

A *mat moisture and bonding test* is designed for checking moisture or surface bonding properties of concrete floors to establish suitability for installation of (1) rubber tile, solid vinyl, tile and vinyl sheet with mineral fiber backing on or below grade and (2) any resilient flooring on slabs from which paint, oil, concrete hardeners, or curing compounds have been removed.

Linoleum and vinyl sheet mats approximately 36 in. square are applied with two bands of adhesives in several locations on the subfloor. One band should be a water-soluable adhesive, the other a water-resistant latex adhesive. The mats are then sealed to the floor with adhesive or industrial tape (Fig. 9.10-21). They should remain in place for 72 hours, then removed and the adhesive examined. If too much moisture is present, the water-soluble adhesive will be partly or completely dissolved, and the water-resistant adhesive will be stringy and provide little bond. If excessive moisture is still indicated after further curing of the concrete and repeated testing, the above-mentioned flooring types should not be used, and consideration should be given to the use of asphalt or vinyl composition tile, both of which have greater moisture resistance. However, for these products the primer test also should be performed.

FIGURE 9.10-21 *The mat moisture and bonding test for suspended concrete subfloors: (a) water-resistant latex adhesive; (b) water-soluble adhesive.*

A *primer test* determines the suitability of asphalt and vinyl composition tile over suspended, below-grade, on-grade, and suspended concrete floors. Patches of asphaltic primer are applied in several places on the floor and allowed to dry for 24 hours (Fig. 9.10-22). If the primer does not readily peel and sticks to the floor when scraped with a putty knife, the subfloor is dry enough to receive asphalt or vinyl composition tile.

SURFACE PREPARATION Expansion joints, scored joints, cracks, and depressions should be filled with latex floor-patching compound or mastic underlay-

FIGURE 9.10-22 *The primer test.*

ment to obtain a smooth surface. Poorly finished or spaded concrete with an irregular surface may require grinding or smoothing with mastic underlayment. Curing or hardening compounds may leave surface films, which often adversely affect adhesive bonding. If the mat bonding test reveals insufficient bonding properties, the film should be removed by power wire brushing or grinding with a terrazzo grinder. An asphaltic primer is required on dusty or porous concrete floors for the installation of asphalt and vinyl composition tile.

9.10.6

SHEET FLOORING INSTALLATION

The following text describes the materials and methods recommended for a suitable installation of sheet flooring primarily over new floor construction. Reflooring existing construction is discussed in Section 9.10.9.

The most common type of resilient flooring available in sheet form is vinyl. The manufacture, physical properties, and selection criteria for vinyl products are discussed in Section 9.10.1.2. Ingredients, gauges, and standards of quality are summarized in Figure 9.10-2; selection criteria are outlined in Figure 9.10-7.

9.10.6.1

Preparation for Flooring

The effectiveness of a flooring installation greatly depends on the proper selection and preparation of the elements that make up the installation. Subfloors that are to receive flooring directly must be firm, smooth, and dense and must possess good bonding properties. When these properties are lacking, it may be necessary to prepare the subfloor by grinding, sanding, or using an underlayment. Adhesives should be selected that are compatible with all elements, including the flooring itself. All elements of the installation must be suitable for the intended location, whether below grade, on grade, or suspended.

■ SUBFLOOR

Vinyl sheet products with mineral fiber backing may be installed on below-grade, on-grade, and suspended subfloors. Products with rag felt backs are subject to decay and alkali attack in the presence of moisture and are limited to suspended locations. Linoleum should be installed on suspended floors only.

Concrete subfloors in all locations should be tested for moisture and bonding properties to ensure suitability for the proposed flooring and adhesive (see Section 9.10.5.2).

■ UNDERLAYMENTS

The need for underlayment is determined by the condition of the subfloor. For a discussion of underlayments and recommended uses, see Section 9.10.4.1.

■ ADHESIVES

The adhesives generally recommended for use with sheet products in various locations are summarized in Figures 9.10-17 and 9.10-18.

■ LINING FELT

Lining felt should be used in the installation of certain sheet flooring products on nonporous floors as described Section 9.10.4.2.

Application of Flooring

The installation of sheet materials should be planned to minimize the number and the total length of seams. Necessary seams should be placed in inconspicuous locations, out of the path of heavy foot traffic. In a rectangular room, running the flooring strips parallel to the side walls generally results in an economical installation with a minimum of seams. However, sheet flooring installed directly over wood strip floors should run perpendicular to the floor joints and may result in seams running the short dimension, parallel to end walls (Fig. 9.10-23).

Installation of sheet flooring consists generally of fitting and cutting, adhesive bonding, and seam treating.

FIGURE 9.10-23 *Sheet flooring direction over floor joints.*

■ FITTING AND CUTTING

Many sheet flooring products must be kept rolled face out until the time of installation. Some manufacturers warn against rolling sheet flooring face in because it may set up internal stresses that can result in residual growth of the sheet flooring long after it is installed. When fitting adjacent strips that have a definite geometric repeat pattern, the pattern should be aligned and matched. Sheet flooring with random overall patterns should be reversed (turned end for end) to minimize differences in graining and shading.

Methods of cutting and fitting sometimes are related to the intended method of bonding the sheet goods. Adhesive application is facilitated by folding the strip back on itself, thus working only half of the strip at a time. The strip is folded back along its length in the *tubing method* and back along its width in the *lapping method* (Fig. 9.10-24).

Sheet flooring generally is cut and fitted to walls and to each other by knifing, scribing, and seam cutting.

After squaring the end of a roll (Fig. 9.10-25), strips are cut 3 in. longer than the distance between end walls to provide excess material fitting to wall recesses or irregularities. The first strip is placed in position with the side edges butted against the side wall, and the ends of the material are flashed (bent up) against the end walls. The material is scribed along the length of

FIGURE 9.10-24 *Methods of handling adhesive bonding sheet flooring.*

the side walls with dividers or a scribing tool (see "Scribing," below) and cut to fit snugly to the wall surface. Ends that have been flashed against the end walls also may be scribed or trimmed by cutting away small pieces until the desired fit is achieved.

KNIFING Knifing is a method of end fitting that involves some waste. It is used generally with less expensive felt-backed products. After a strip of flooring is fitted in place against a side wall or adjacent strip, excess material is cut away gradually to fit the end walls and other vertical surfaces.

Safety cuts are made at various points to allow the material to be fitted into difficult locations without cracking. One or more safety cuts should be made first wherever the material tends to buckle as it is flashed against the wall. When the flooring has to be fitted into a corner before it can be knifed at the walls, the *inside corner safety cut* is the simplest (Fig. 9.10-26a). This allows the material to be dropped down snugly into a corner and then flashed up the walls without cracking. An *outside corner safety cut* is used to fit flooring around a curved or cornered projection (Fig. 9.10-26b). A *crossover safety cut* permits bending the flooring under an obstacle such as a radiator (Fig. 9.10-26c).

SCRIBING When flooring must be fitted against a vertical surface such as a wall, column, or stair riser with irregular cor-

ners, the exact contour of the vertical surface must be transferred to the flooring

FIGURE 9.10-25 *The end of a roll is squared with a straightedge and square.*

FIGURE 9.10-26 *Safety cuts facilitate fitting: (a) inside corner; (b) outside corner.*

material. Scribing generally is used in fitting vinyl flooring to side walls and end walls and is the preferred method for all sheet flooring.

When scribing with dividers (Fig. 9.10-27), the flooring is pushed close to the wall or the projection and the dividers are opened so that one leg touches the wall at the farthest point and the other rests about 1 in. inside the edge of the flooring with the dividers kept perpendicular to the wall and one point always in contact with it, the profile is transferred to the flooring. The material then may be cut with a linoleum knife or, for vinyl sheet, a notched blade knife and further fitted in place. Cutouts for isolated obstacles such as pipes also can be scribed with a scribing tool in a similar manner (Fig. 9.10-28).

FIGURE 9.10-27 *When scribing with dividers, the dividers should be kept perpendicular to the wall surface.*

Pattern scribing is used when the size of the room or wall projections make it difficult to scribe the flooring itself. Complex profiles are first scribed to a piece of felt or heavy paper, checked for fit, and trimmed; then they are transferred to the flooring (Fig. 9.10-29). This ensures a better fit, particularly desirable when wall base is not used or when the base is a straight type that will not cover minor irregularities in scribing.

End scribing, when the flooring is tubed, is generally similar to that described for side walls. However, when the lapping method is used, each strip must be partly cemented before scribing. The unfitted end of the sheet is lapped back and, working toward the end wall from the center of the room, adhesive is applied to

FIGURE 9.10-28 *An 18-in. scriber permits scribing under obstacles that would be impossible with dividers.* (Photograph courtesy the makers of Armstrong vinyl flooring.)

within 3 ft. of the end wall. The sheet is laid into the adhesive and rolled. The loose, uncemented end of the material then is scribed to the end wall by drawing a mark perpendicular to the edge of the material 9 to 12 in. from the wall and extending it from the material onto the subfloor. The material then is buckled and butted against the wall line (Fig. 9.10-30). The perpendicular distance between the line on the subfloor and the line on the material is measured with the scriber. With the scriber set at that distance, the material is scribed to the wall line and cut.

SEAM CUTTING   To make a neat, inconspicuous seam between adjacent strips of flooring, it is necessary to cut away the selvage (factory-cut edge). Sheet widths are generally 1/2 in. greater than the nominal width to permit this excess material to be cut away. On goods with a geometric repeat pattern, the amount of selvage cut away cannot be varied if the pattern at the seam is to match and align properly. In goods with random overall pattern, the selvage to be trimmed is less critical and may vary (Fig. 9.10-31).

Seams between adjacent strips are made chiefly by underscribing, single cutting, and double cutting. To prevent misalignment due to shifting, strips should be partially or completely bonded to the floor before seams are cut. When using the single or double cutting method, adhesive generally is kept 4 in. or more from the seam until both edges are cut. This prevents adhesive from being pulled up when the lower edge (selvage) is removed. In both straight edging and seam cutting, a piece of scrap flooring should be inserted under the cut line to prevent the backing from fraying. The knife should be held as low as possible with the blade perpendicular to the floor surface.

FIGURE 9.10-29 *Pattern scribing: (a) felt pattern is scribed; (b) pattern is transferred to the flooring; (c) flooring is cut to the pattern; and (d) flooring is checked for fit.*

FIGURE 9.10-30 *For lapping, each strip is partly cemented before end scribing.*

FIGURE 9.10-33 *Single-cut seam.*

FIGURE 9.10-34 *Double-cut seam.*

FIGURE 9.10-31 *The selvage is trimmed with a straightedge and either a linoleum knife or a straight-blade utility knife.*

FIGURE 9.10-32 *Underscribed seam.*

Underscribing involves the use of a recess scriber to transfer the exact edge profile from the previously straight-edged strip to the adjacent strip. As it is guided along the finished edge, the tool leaves a score mark on the overlapped adjacent strip directly above. The scribe mark then is used as a guide for cutting (Fig. 9.10-32).

Underscribing is the preferred way of making seams for flooring with filled vinyl wear surfaces. The trimmed edge can be cemented down completely before underscribing, which often minimizes the number of operations involved in fitting and cementing the flooring.

Single cutting (Fig. 9.10-33) is commonly used for making seams in patterned flooring where the pattern must be care-fully aligned. The straightedged strip is laid over the untrimmed strip, overlapping as required to fit the pattern. The lower sheet then is scored with the linoleum knife guided along the upper trimmed edge. The lower strip then is cut along the score line.

Double cutting (Fig. 9.10-34) is similar to single cutting, except that neither edge is pretrimmed and the cut is made through both strips at the same time. With a straight-edge as a guide, the upper sheet is cut in the middle of the overlapped edges, scoring the sheet below. The lower sheet is then cut along the score line. Double cutting is an older method and is not used as commonly as the underscribing and single-cutting methods. Double cutting may be used with some clear (unfilled) vinyl-surfaced sheet, which are not as easily scored with an underscriber.

■ ADHESIVE BONDING

To facilitate spreading adhesive prior to bonding, the flooring usually is folded back by either tubing or lapping (see Fig. 9.10-24). The sequence of cutting and fitting, adhesive spreading, and cementing operations depends on whether the flooring is installed by tubing or lapping. Lapping generally is the preferred method, but tubing is used where wall projections or the shape of the room makes lapping difficult or impossible. Tubing also is recommended for heavy-gauge vinyl sheet and burlap-backed linoleum.

Adhesive is spread with a notched trowel, spreader, or brush, depending on the type of adhesive. As each strip is pasted down, the seam should be smoothed with a hand roller, and the entire strip should be rolled with a 100-lb. or heavier roller (Fig. 9.10-35).

TUBING Generally, the first strip is scribed to the side wall and end walls before it is cemented down. For tubed

FIGURE 9.10-35 *Adhesive bonding is completed by (a) rolling the flooring and (b) hand rolling the seam.*

FIGURE 9.10-36 *Tubing with underscribed seams: strip A is scribed to side and end walls, edge ab is trimmed, area 1 is bonded, then area 2. Strip B is lapped over strip A, then end-scribed; edge bc is pretrimmed, and area 3 is bonded. Then edge ab is underscribed, then cut, and area 4 is bonded. Strip C is installed as strip B after being scribed to side and end walls.*

FIGURE 9.10-37 *Lapping with underscribed seams: Strip A is scribed to side wall, edge ab is pretrimmed, and areas A1 and A3 are bonded; areas A2 and A4 are bonded after end-scribing to walls. Strip B is lapped over strip A, and entire strip is installed as strip A; edge ab of strip B is then underscribed and cut. Strip C is installed as strip B after being scribed to side wall.*

installations with underscribed seams (Fig. 9.10-36), the strip is trimmed at the seam edge also. As each strip is tubed back, adhesive is spread in the exposed area. The sequence of operations for double- and single-cut seams is similar to underscribed seams, but the adhesive should be kept 4 in. short of the seam edge. Seams must be cut before adhesive is spread in the seam area, and adhesive is spread under the seam area when the next tubed half is cemented down. This sequence is repeated for subsequent strips.

LAPPING Depending on the size of the room and the working time of the adhesive, spreading may be combined across the width of the room, and more than one strip may be worked at a time. For lapped installations with underscribed seams (Fig. 9.10-37), the first strip is scribed to the side wall, and starting at the center of the room adhesive is spread under strips to within 3 ft. of the end wall. The strips are cemented down and rolled. Each strip is end-scribed, and the remaining area is spread with adhesive, cemented down, and rolled.

The sequence of operations for double- and single-cut seams is similar, but adhesive in each spreading is kept 4 in. from the seam edge. The second halves of the strips are cemented down in the same manner. Then the entire seam is cut, adhesive is spread under the remaining 8-in.-wide uncemented area under the entire length of the seam, and the seam is completed by hand rolling.

■ SEAM TREATING

Linoleum is softer than vinyl sheet and tends to "flow" around the knife blade without creating ragged edges. When vinyl sheet is cut, however, burrs often are created on the cut edges and must be removed to make the seams inconspicuous. After hand rolling the seams, these burred edges may be removed with the back of a linoleum knife moved in a chisel motion along the seam. Or a hammer head or scrap of flooring doubled in half (face out) may be rubbed along the burrs while the seam is kept damp to avoid scuffing.

An electric iron also may be used to treat seams: a 1-in.-wide strip of heavy-duty aluminum foil is laid over the seam (dull side down) and rubbed to make the burrs show through. The foil then is pressed with an electric iron held at an angle and moved in repeated strokes toward the mechanic. When the foil has cooled, it is removed, and the seam is cleaned with a damp cloth and a light abrasive cleaner.

■ TENSION FLOORS

Tension floor installation is a unique method of perimeter bonding recommended for certain sheet flooring products. Flooring may be installed by this method over well-bonded existing resilient floors over concrete and wood subfloors on all grade levels. Special adhesives allow the floor to be bonded only at the room perimeter, seam lines, and columns or fixtures.

The two types of adhesive used are a two-part epoxy and a special solvent-based mastic. Installation method varies slightly with type of adhesive used, flooring type, and manufacturer's instructions. As in all resilient floor installations, the subfloor surface must be free of moisture, dust, paint, wax, oil, and grease and must be structurally sound and smooth.

A tension floor installation should be allowed to dry 4 or 5 days before it is cleaned. Floor maintenance generally is the same as for other resilient floors. However, when moving heavy objects, such as appliances, a piece of plywood or particleboard should be placed under the object to prevent tearing or buckling of the flooring material.

9.10.6.3

Cove-Based Flashing

Sheet flooring may be trimmed out with standard set-on cove base (see Section 9.10.3), or the material may be extended partly up the wall to make an integral flashed cove base (Fig. 9.10-38). To make the cove profile at the meeting of the wall and the floor, a 7/8-in. wax fillet strip is cemented, or a wood cove molding is nailed in place. A binding strip then is

FIGURE 9.10-38 *Flashed cove base with (a) outside metal corner, (b) inside metal corner, (c) metal binding strip, and (d) wax cove fillet.*

nailed or cemented to the wall at the desired height to protect the top of the base. The flooring may be softened with a torch to facilitate molding it against the fillet strip. In heavy-traffic areas, inside and outside metal corners also may be desirable (see Fig. 9.10-38). Similar metal endstops are available for closing base ends at door openings without casing trim.

In rooms that require a decorative sheet border, the full-width strips may be pasted down in the field first; then the narrower border strips are flashed up the wall, scribed against binding strips, and cemented in place. After the border strip has been cut to the proper width, inside and outside corners without metal corner pieces may be scribed and installed.

Where no decorative borders are used, it is necessary to flash the ends and edges of the flooring strips used in the field up the wall. When working with such larger pieces, it is difficult to cut flooring accurately by fitting and cutting in place. Therefore, it is necessary to make a felt pattern of the entire floor and base, carefully scribing and cutting around projections to the binding strips (Fig. 9.10-39). The pattern is transferred to the sheet flooring, which then can be readily cut, fitted, and cemented in place by either the tubing or the lapping method. Inside corners can be made by scribing, as shown in Figure 9.10-40 for resilient base. Figure 9.10-41 illustrates the preferred method of making an outside corner. A less durable corner can be made simply by mitering the flashed material at the corner.

FIGURE 9.10-39 *Felt pattern is scribed to the binding strip, and 45-degree V cutout is formed for outside corners of flashed base.*

FIGURE 9.10-40 *Inside base corner. Right side is butted to corner and bonded; left side is scribed to right side and bonded.*

A 100-lb. roller should be used to roll the floor, and a small hand roller should be used for the flashed cove base. Care should be taken not to allow the edge of the heavy roller to ride over the fillet, because it may damage the installation.

FIGURE 9.10-41 *Outside one-piece corner: (a) V cutout is transferred from the pattern to the flooring before installation; (b) corner piece is cut out separately with a square; (c) before matching corner piece is fitted, a narrow strip of backing is removed at the fold.*

9.10.7

TILE FLOORING INSTALLATION

The following text describes the materials and methods necessary to install a suitable tile flooring primarily over new floor construction.

Resilient floor tile is available in vinyl, vinyl composition, rubber, asphalt, and cork. Linoleum tile is no longer produced in this country. Tile products range in size from 9 in. by 9 in. up to 36 in. by 36 in., in various gauges and varying colors and patterns. Some tiles are made in no-wax finishes and with self-adhesive backs. Ingredients, gauges, and standards of quality are summarized in Figure 9.10-2; selection criteria are given in Figure 9.10-7.

9.10.7.1

Preparation for Flooring

The effectiveness of a flooring installation greatly depends on the proper selection and preparation of the elements that make up the installation. Subfloors that are to receive the flooring directly must be firm, smooth, and dense and possess good bonding properties. When the subfloor does not possess these qualities, it may be necessary to prepare the surface by grinding, sanding, or using an underlayment. Adhesives should be selected that are compatible with all elements, including the flooring itself. All elements also should be suitable for the intended locations, whether below grade, on grade, or suspended.

■ SUBFLOOR

Solid vinyl, mineral fiber–backed vinyl, vinyl composition, and rubber tile may be installed on subfloors in suspended, on-grade, and below-grade locations. Cork, vinyl-cork, and rag felt–backed vinyl tile should be installed on suspended subfloors only.

Concrete subfloors in all locations should be tested for moisture and bonding properties to ensure suitability for the proposed flooring and adhesive (see Section 9.10.5).

■ UNDERLAYMENT

The condition of the subfloor will determine the need for underlayment. For a discussion of underlayments and recommended uses, see Section 9.10.4.1. Mastic underlayments should be compatible with intended adhesives and primers, as discussed below.

■ ADHESIVES

The adhesives recommended for tile products in various locations are summarized in Figure 9.10-18. For a discussion of adhesive types and their properties, see Section 9.10.4.3 and Figure 9.10-17. Solvent-based asphaltic adhesives and primers used for the installation of asphalt and vinyl composition tile should not be used over asphaltic underlayments. For such installations, asphaltic emulsions should be used. When cutback asphalt, which is a solvent-based adhesive, is used over latex underlayment, the amount of time recommended by its manufacturer must be allowed before tile is installed. Porous or dusty concrete floors and panel underlayments intended for the installation of asphalt or vinyl composition tile should be primed with an asphaltic primer before adhesive is applied. Some tiles are made with self-adhesive backs, but these are not generally used in commercial or institutional projects.

9.10.7.2

Installation of Flooring

The first step in the installation of tile flooring is to make a trial design layout of the tiles. The tiles then are cemented down according to the best layout, and walls are trimmed out with resilient or other base.

■ DESIGN LAYOUT

In determining design layout, it is important that tile joints do not fall over joints in the subfloor or in the underlayment. Because few walls are seldom exactly parallel, it is necessary to center and square off the room (Fig. 9.10-42). Tiles may be laid in either square or diagonal designs. Over existing wood strip floors, a diagonal installation is preferred.

SQUARE DESIGNS Tile joints for square designs run parallel to the walls. The trial layout is started by snapping chalklines to divide the room into four equal quadrants (Fig. 9.10-43). If the floor area is extremely large or job conditions prevent working the entire floor area, each quadrant may be subdivided further. Full tiles should be laid starting from the center and working toward the walls. The first trial may result in unequal border tiles at side walls and end walls (see Fig. 9.10-43a).

To equalize border tiles at walls, the horizontal row of tiles may be shifted so that it is bisected by the chalklines, as shown in Figure 9.10-43b. In general, layout lines should be shifted so that approximately equal border tiles of at least one-half a tile or wider will result.

Checkerboard patterns may be made by alternating tiles of two different colors or alternating the grain direction in directional tiles. Either a tile that matches the

FIGURE 9.10-42 *To make a square tile layout in a room that is nearly square, find the midpoint of the shorter walls (WX and YZ) by measuring, then strike a chalkline on the floor connecting the midpoints, measure to find the center of this chalkline (M). With M as a center point and any convenient distance as a radius (3 ft. in this case), swing an arc that intersects the chalkline (A and B.) With A as a center point and any distance greater than AM as a radius, swing an arc on both sides of the chalkline. Repeat this procedure using B as a center point. The two arcs will cross at C and D. A line drawn between C and D will be exactly perpendicular to the first chalkline. It is also possible to make a square layout using a metal square to make the corners, but this method is less accurate than the one described here, especially for larger rooms.* (Illustration by the makers of Armstrong vinyl flooring.)

field or a contrasting one-color sheet or tile can be used for the border.

For a *hollow-square* design, another two-color pattern, it is important to lay out the trial design so that a full repeat of the pattern will occur at all walls (Fig. 9.10-44). Such patterns commonly are used with one-color borders to emphasize the design in the field.

DIAGONAL DESIGNS   Tile joints are run at 45-degree angles to the walls for diagonal designs. A trial row of tiles is laid out along each centerline and diagonally to the corner to determine the resulting condition at the border (Fig. 9.10-45a).

Two-color designs, requiring a border laid on the square, also must be checked to ensure that the half-tiles along each wall will be the same color. To achieve this, it may be necessary to shift tiles from the trial position shown in Figure 9.10-45a, so that tiles will abut rather than be bisected by diagonal lines (Fig. 9.10-45b). The proper starting position of tiles at the center of the room is determined by the dimensions of the room and must be estab-

lished in each case by trial layout.

If it is determined that the design should start with single tiles at the intersection of diagonals, new diagonals should be struck one-half tile over from the original location (Fig. 9.10-46) to serve as guidelines for the first row of tiles.

One-color designs without a border present no special problem and the starting point need only be adjusted to ensure that the layout will not result in excessively narrow tiles at the walls. In one-color installations with tiles that have a directional grain, it is important that the half-tiles near the border fit properly into a basket-weave pattern. If the tiles are cut properly with respect to grain, tiles at both right and left sides will result that fit the pattern (Fig. 9.10-47).

■ **ADHESIVE BONDING**

To allow sufficient working space to handle materials, only half or part of the floor area, depending on the working time of the adhesive, should be covered with adhesive at one time. Care should be taken in applying the adhesive so that chalklines

will not be completely concealed. This may be done by leaving ends and intermediate points on the chalkline uncovered.

Adhesives such as cutback asphalt develop holding power by setting before tiles are applied. With some adhesives such as waterproof-resin and epoxy, the possibility of tiles slipping may be a problem. Temporary lattice strips nailed at the chalklines will prevent the tiles from sliding out of place during installation.

Tiles should be installed with tight, straight, and carefully aligned joints in both directions. Each tile should be butted to the preceding tiles and lowered in place without sliding.

Installation on the square with a two-color checkerboard pattern can be speeded by starting with a row of 10 tiles along each centerline and laying tiles of one color at a time in diagonal rows (Fig. 9.10-48).

For diagonal designs, tiles may be cemented down in one quadrant at a time, starting at the center, laying the tiles along the diagonal, and proceeding clockwise around the room (see Fig. 9.10-46).

■ **FITTING AND CUTTING**

Border tiles may require scribing and fitting against walls and obstructions (Fig. 9.10-49). They are scribed essentially as described for sheet flooring using dividers or a scribing tool to transfer the contour of the vertical surface (see Fig. 9.10-27).

Most resilient tiles can be cut at room temperature with a linoleum or straight-blade knife. However, it is easier to cut vinyl composition and asphalt tile by heating the tile and the knife with a blowtorch or a heat gun.

9.10.8

■ **INSTALLATION OF RESILIENT FLOORING ACCESSORIES**

9.10.8.1

Wall Base

The two main types of wall base used with resilient tile are set-on and butt-cove base (see Fig. 9.10-10). Most tile installations are made with set-on base because it is easier and more economical to install. This type of base is set on top of the finished floor and requires less accurate fitting of both flooring and base. Butt-cove base provides a more sanitary condition

FIGURE 9.10-43 *Trial layout, square design: (a) unacceptable layout, because the size of border tiles differ; (b) acceptable layout, because border tiles are equal and wider than one-half a tile.* *(Illustration by the makers of Armstrong vinyl flooring.)*

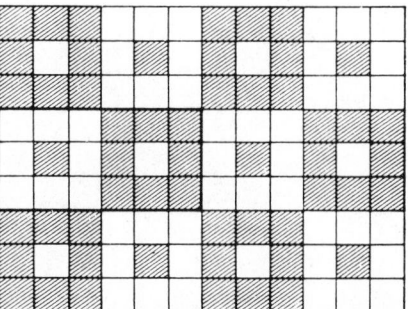

FIGURE 9.10-44 *Hollow-square design; repeated unit indicated by dark outline.* *(Illustration by the makers of Armstrong vinyl flooring.)*

and may be required in certain spaces where cleanliness is mandated, such as in laboratories and some hospital spaces. Both set-on and butt-cove base are available in rubber and vinyl materials.

■ RUBBER BASE

Rubber base is manufactured in straight runs and in premolded outside and inside corners. These facilitate turning corners with a minimum of field labor, the only fitting required being at butted joints. Rubber base is used to match rubber flooring and in cases where premolded corners are desired.

■ VINYL BASE

The greater pliability of vinyl base permits molding it around corners; therefore, it is manufactured mainly in straight runs. Inside corners are made by notching a V in the back of the base at the fold and heating it to increase pliability. The base may then be formed and pressed into the corner. Inside corners also can be made by butting the first piece squarely in the corner and scribing the second piece around it (see Fig. 9.10-40).

Outside corners are formed by first cutting away about half the thickness from the back of the material at the point of bend. At the back of the cove heel, additional material should be cut away approximately

1 in. to each side of the bend. The base then can be molded readily around the corner and set in place with standard base cement or synthetic rubber cement (see Fig. 9.10-9a). Synthetic rubber cement is preferred because it provides immediate holding power and prevents movement from contraction as bond is developed.

9.10.8.2

Stair Treads, Risers, and Nosings

Resilient sheet materials can be used to provide a suitable covering over wood and concrete stairs. The sheet material can be flashed from the tread up the riser (Fig. 9.10-50a) or used with a set-on coved rubber or vinyl riser (Fig. 9.10-50b). The nosing of a wood tread may be wrapped with the sheet flooring or protected by a separate metal or rubber nosing. Treads on riserless wood stairs may be completely wrapped with sheet materials (Fig. 9.10-50c). Formed concrete stairs can be covered with continuous runs of sheet flooring on both treads and risers (Fig. 9.10-50d). Methods of installing sheet stair covering are similar to those described for sheet flooring. Accurate layout, scribing, fitting, and proper selection of adhesive are essential.

Resilient tiles can also provide an attractive wearing surface on wood and concrete stairs. Tile treads usually are installed with rubber, vinyl, or metal nosings (Fig. 9.10-51). Risers for tile treads may be of matching tile or a one-piece set-on coved riser similar to wall base. Matching one-piece treads with premolded nosings also are available from some manufacturers (see Fig. 9.10-51). Methods of installation are similar to those described for tile floors.

FIGURE 9.10-45 *Diagonal design, trial layout: (a) unacceptable layout, (b) acceptable layout—all half-tiles at the border are the same color.* *(Illustration by the makers of Armstrong vinyl flooring.)*

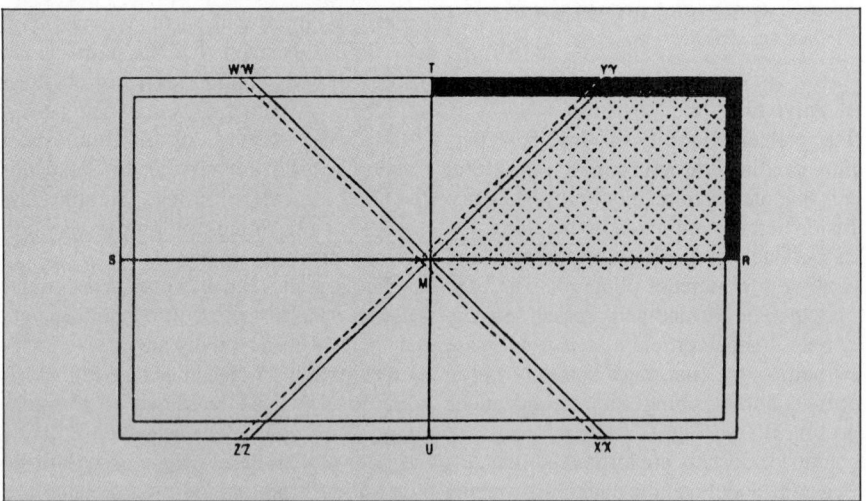

FIGURE 9.10-46 *New chalklines may be required as a guide for the first row of tiles. As tile is installed clockwise, alignment may be checked against diagonals.* *(Illustration by the makers of Armstrong vinyl flooring.)*

Accurate layout, scribing, fitting, and proper selection of adhesives are essential.

9.10.9

REFLOORING EXISTING CONSTRUCTION

Both sheet and tile products may be installed over existing finished floors that have been suitably prepared to provide a firm, smooth, and stable base for the new flooring. The main types of floor finishes encountered in reflooring are (1) wood strip, wood block, or wood plank, (2) concrete, (3) ceramic tile, terrazzo, or marble. New flooring may be installed over old resilient flooring provided the old flooring is not cushioned or deeply embossed. For specific recommendations, the manufacturer of the new flooring should be consulted.

Existing floors intended for reflooring should be inspected to assure a firm base free of loose particles. Wax, oil, grease films, and most paints should be removed. Surfaces should be smooth and reasonably level or leveled with an appropriate underlayment when necessary. Recommendations for flooring and adhesive selections with respect to grade locations should be adhered to as in new construction.

9.10.9.1

Wood Floors

A suitable panel underlayment is recommended over existing wood floors. However, sheet products may be applied directly to double-layer floor construction that consists of a lumber board or plywood subfloor and a finish wood strip floor of T&G boards at least 3/4 in. thick and not wider than 3 in., if the existing floor is sufficiently smooth, level, and firm and well seasoned.

When panel underlayment is not used, a protective film on wood floors may be removed with a power sander or a blowtorch. Mild chemical solutions in water should not be used because the large quantity of water required for scrubbing and rinsing may cause the boards to warp. After a wood floor has been sanded, one brushed coat of floor sizing consisting of equal parts of a wax-free shellac and denatured alcohol should be applied to seal the wood. This prevents excessive

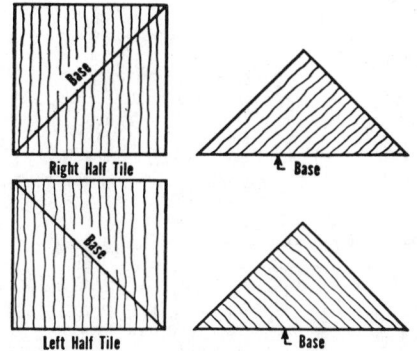

FIGURE 9.10-47 *Tiles must be cut according to grain to obtain half-tiles that will fit into a basketweave pattern. (Illustration by the makers of Armstrong vinyl flooring.)*

FIGURE 9.10-48 *Checkerboard square designs can be installed more quickly if tiles of one color are laid in diagonal rows as numbered. (Illustration by the makers of Armstrong vinyl flooring.)*

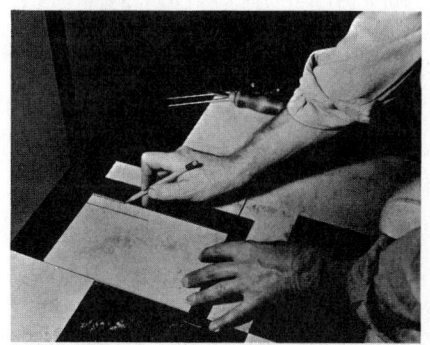

FIGURE 9.10-49 *Square is used to position cutout for pipe, circle cutout is scribed with dividers, and the tile then is cut and slid into place. (Illustration by the makers of Armstrong vinyl flooring.)*

moisture absorption from adhesive and other sources and minimizes the possibility of boards warping in the future.

Other double wood floors of block and plank and single-layer plank floors should be covered with underlayment consisting of at least 1/4-in. plywood, 1/4-in. hardboard, or 3/8-in. particleboard.

9.10.9.2

Concrete Floors

Most paint films can be removed from concrete floors by power grinding, burning with a blowtorch, or scrubbing with a solution of trisodium phosphate and water. After the paint has been removed, the floor should be scrubbed with a strong soapy solution, rinsed with clear water, and allowed to dry thoroughly. Paint removers should not be used because they react adversely with many adhesives. Because

the use of trisodium phosphate and some other paint removers is not permitted in some jurisdictions, it is necessary to check with local authorities before using them.

Paints with a chlorinated rubber or resin base often are tightly bonded to the concrete surface and, if the mat bonding test reveals good adhesion, may be left in place.

9.10.9.3

Ceramic Tile, Terrazzo, and Marble

Ceramic tile, terrazzo, and marble floors

should be cleaned with a scrubbing machine using soapy water and clean sharp sand. A floor grinder used dry with 4-1/2-grit open-coated paper may be preferred to avoid delay due to drying. Badly fitted joints and cracks should be leveled with latex underlayment or a patching compound.

FIGURE 9.10-50 *Stair coverings with sheet flooring: (a) one-piece tread and riser; (b) sheet tread and set-on cove riser; (c) wrapped riserless wood tread; (d) continuous risers and treads on concrete.*

FIGURE 9.10-51 *Stair coverings with tile flooring and resilient accessories.* *(Courtesy Carpet and Rug Institute.)*

CLEANING

A resilient floor can be damp-mopped immediately and should be polished within 48 hours. It should not, however, be given a thorough cleaning until 4 to 5 days after installation. The floor then should be washed with a mild solution of a nonabrasive, nonalkaline cleaner or soap and thoroughly rinsed with cold water. It should be allowed to dry, then waxed with a floor finish recommended by the flooring manufacturer. A soft cotton mop or wax applicator should be used to apply the floor finish as thinly as possible.

No-wax tile and sheet floors should be cleaned the same way as other resilient floors. As the name implies, waxing is not necessary. Because they have no renewable surface finish, it is important to protect no-wax floors from severe abrasion, such as from gritty dirt and furniture without floor protectors (see Fig. 9.10-8).

9.11 CARPET

Carpet is often selected for its comfort and decorative values, but it is also extremely effective in reducing impact sound transmission through floor and ceiling assemblies. Therefore, its acoustical performance should be a prime consideration in flooring selection. Many common installations will provide a performance considerably higher than the range of −5 to +10 on the Impact Noise Reduction (INR) scale (see Section 10.1). Specially selected combinations of carpet and cushioning may be rated as high as +17 over wood floors and +29 over concrete slabs.

Carpet is also capable of absorbing sound and reducing sound reflection within a room, much like acoustical ceiling tile. The Noise Reduction Coefficient (NRC) of most carpet ranges between 0.35 and 0.55, which compares favorably with a range of 0.55 to 0.75 for acoustical ceil-

ing tile. Properly selected carpet may achieve absorption coefficients equal to acoustical tile. However, carpet is not particularly effective in controlling the transmission of airborne sound; where this consideration is important, special sound-isolating construction should be used.

This section describes carpet suitable for long-term wall-to-wall installation, as distinguished from rugs, which usually have bound edges and are laid loose over a finished flooring (Fig. 9.11-1).

Carpet can be selected and specified broadly according to its type, which defines how it is made; by the fibers used to make it; and by its face construction. Other factors that influence carpet selection are addressed in Section 9.11.2. Carpet installation is covered in Section 9.11.3.

CARPET CONSTRUCTION

Most carpet consists of pile yarns, which form the wearing surface, and backing yarns, which interlock the pile yarns and hold them in place. Carpet can be identified according to the carpet construction, which describes the method of interlocking the backing and the pile yarns. Comparative factors related to carpet construction, such as pile yarn weight, pile thickness, and number of tufts per square inch, may be useful in comparing carpet of similar construction.

Carpet can also be identified by the material of the pile fibers of which the pile yarns are made, including wool, nylon, acrylic, polyester, and polypropylene. The various fibers have unique characteristics that may affect carpet performance and

FIGURE 9.11-1 *Carpet goes from wall to wall; rugs do not.* *(Courtesy Carpet and Rug Institute.)*

FIGURE 9.11-2 *Some carpet texture variations: (a) twist; (b) multilevel loop; (c) random shear, combining cut and loop pile; (d) single-level looped pile.* *(Courtesy Carpet and Rug Institute.)*

styling. Both carpet construction and fiber properties influence the selection of dyeing methods.

Its face construction is another factor by which carpet can be identified. The following terms describing face style as used in this section and are defined in the Glossary: level-loop pile, cut pile, level tip shear, multilevel loop, random shear, frieze or twist, and sculptured or carved (Fig. 9.11-2). Some face styles are dictated by the method of manufacture. For example, fusion-bonded goods can only be cut pile. Similar restrictions are identified with each type of carpet in the following discussion.

9.11.1.1

Carpet Types

The term *broadloom* is often mistakenly thought to define a type of carpet construction, but actually it refers merely to carpet wider than 6 ft. With the exception of rubber-backed loomed carpet, the carpet types discussed in this section are manufactured in widths of 9, 12, 15 and sometimes 18 ft. Loomed carpet generally is made in 4 ft. 6 in. width, with a 3/16-in. attached rubber cushioning.

Carpet can be classed according to the methods of manufacture as woven, tufted, knitted, fusion-bonded, needlepunched, and flocked. Of these, tufted and woven are the most used methods for making yard goods, and fusion-bonded is the most prevalent method of making carpet tile. In general, weaving and knitting interlock pile and backing yarns in a single operation, while a tufting loom stitches pile yarns onto a premade backing. Tufted carpets can be produced much faster than woven or knitted types, are less expensive to produce, and currently account for about 95 % of all carpet manufactured in the United States.

■ WOVEN CARPET

Woven carpet is made on looms that interweave pile yarns and backing yarns in one operation. Backing yarns consist of *weft* (crosswise) yarns and *warp* (lengthwise) yarns. Because the tuft-forming yarns in woven carpet run lengthwise, they too are classed as warp, either *face warp* or *pile warp*. Though most properly applied to woven carpet, the terms *warp* and *weft* are loosely applied to knitted carpet as well.

Weft yarns run crosswise, passing over the pile yarns (Fig. 9.11-3a). *Stuffer yarns*, a type of warp yarn, give strength and dimensional stability to a carpet (Fig. 9.11-3b). Typically, there are two or three stuffer yarns per tuft running the length of the carpet. *Chain yarns*, another type of warp, run the length of the carpet, alternately passing over and under the weft yarns, forming a chain that locks the weft in place and pulls the pile yarns down tightly (Fig. 9.11-3c).

Weft Yarn

a

Stuffer Yarn

b

Chain Yarn

c

FIGURE 9.11-3 *Backing yarns of woven carpet identified by heavy lines: (a) also called* **weft shots** *or* **filling;** *(b) also called* **stuffer warp;** *(c) also called* **chain warp, binder,** *or* **binder warp.** *(Courtesy Carpet and Rug Institute.)*

Four weaving processes are used from which woven carpets derive their names: velvet, Wilton, Axminster, and loomed.

VELVET   A velvet weave is the simplest of all weaves (Fig. 9.11-4). Chain, stuffer, and pile yarns are alternately raised and bound together with the weft yarns. The pile is formed by wires inserted between pile and warp yarns. The height of

VELVET

STUFFER YARNS   WARP CHAIN   WEFT SHOT

FIGURE 9.11-4 *Velvet weave with double chain warp and typically two or three stuffer yarns per pitch. (Courtesy Carpet and Rug Institute.)*

the wire determines the depth of the pile. After a row of loops has been formed over the wire and secured by the weft, the wire is withdrawn leaving a row of loops. Cut pile is created by a razor edge on the wire that cuts the loops as the wire is withdrawn.

Because a velvet loom feeds pile yarn from a single spool, it is best suited for solid colors. However, strands of different color can be twisted together into a single yarn (*moresque*), producing tweeds, salt-and-pepper effects and stripes. A wide range of textural effects is possible by using

uneven wires to form high and low loops.

Velvet weave can be recognized at a cut edge or open selvage, where the chain warp is visible, binding all the construction yarns together in a zigzag pattern (see Fig. 9.11-4).

WILTON   A Wilton loom (Fig. 9.11-5) is basically a velvet loom fitted with a special *Jacquard* mechanism that utilizes punched cards, similar to those used with early computers, to select the various colored yarns. These cards are locked together to

FIGURE 9.11-5 *A Wilton loom. The punched cards are at the upper left. (Courtesy Carpet and Rug Institute.)*

match the entire surface design of the carpet and are suspended in the loom. The colored pile yarns are fed from spools assembled on frames (racks) in the back of the loom, one frame for each color. As the cards pass, the Jacquard mechanism selects and lifts the appropriate pile yarns into position to be looped over the wires, while other yarns remain buried in the body of the carpet. Depending on the number of colors used, for each tuft of yarn showing on the surface, there may be as many as four strands of yarn buried in the backing, adding to the stiffness of the carpet (Fig. 9.11-6).

FIGURE 9.11-6 *Wilton weave with four frames (colors), double chain warp, two weft shots per wire, and stuffer yarns.* (*Courtesy Carpet and Rug Institute.*)

A carpet woven on a Wilton loom with just one colored pile yarn would have the same construction as a velvet weave and would be designated as a velvet carpet. Wilton carpets are designated as two-, three-, four-, or five-frame Wiltons depending on the number of colored pile yarns used in their manufacture. Generally speaking, for carpets of the same density and yarn size, the more frames used, the greater the pile yarn weight, and hence the better the quality.

A Wilton loom can weave sharply delineated sculptured and embossed textures or patterns, which are created by varying the pile height, using high and low loops or combinations of cut and uncut piles. Wilton carpets with uncut loop pile are called *roundwire* Wilton.

In multicolor Wilton weaves, the different colors of pile fibers are usually visible on the back, and the body of the carpet is thicker than that of other types.

As in velvet weaves, the chain warp also is visible at a cut edge.

AXMINSTER Axminster carpet is woven on a loom that draws pile yarns from small spools wound with yarn of various colors. The spools are locked together in a frame equal to the width of the carpet. Since each frame provides one row of tufts across the carpet and each strand of yarn provides a single pile tuft, the sequence of colors determines the finished surface pattern.

An Axminster loom simulates hand weaving because each tuft of yarn is individually inserted, and theoretically each tuft could be of a different color. This flexibility offers unlimited design possibilities, and intricate patterns are produced even in lower-priced carpet. Considerable delicacy and subtlety of design are possible in the higher-priced carpets, in which the rows of tufts are closer together.

The Axminster weave is almost always a cut pile with an even pile height and elaborate patterning with many colors. The backing is so heavily ridged and sized (coated) that it usually can be rolled only lengthwise easily. The characteristic chain warp can be seen along a cut edge, as in other woven carpets (Fig. 9.11-7).

FIGURE 9.11-7 *Axminster weave with single chain warp, three double shots per row, and stuffer yarns.* (*Courtesy Carpet and Rug Institute.*)

LOOMED Loomed carpet is intended specifically for use with a bonded rubber cushioning. It is characteristically a high-density nylon with single-level, low-loop pile about 1/8 in. high, woven on a modified upholstery loom (Fig. 9.11-8). The pile back is treated with a waterproof

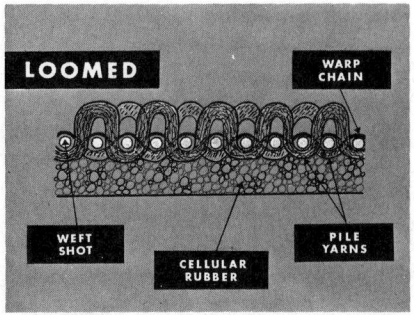

FIGURE 9.11-8 *Loomed weave with double chain warp, single weft shot, and low-level loop pile bonded to rubber cushioning.* (*Courtesy Carpet and Rug Institute.*)

adhesive that serves as a bond coat for the rubber backing. Sponge or foam rubber is applied to the pile fabric, and the backed carpet is heated to cure the rubber. Loomed carpet generally is manufactured in a width of 4 ft. 6 in. with a rubber backing about 3/16 in. thick.

■ TUFTED CARPET

Tufted carpet is manufactured by inserting tufts of pile into a premade backing by a machine that resembles a huge sewing machine with thousands of needles operating simultaneously (Fig. 9.11-9). The prime backing is made from jute, kraftcord, cotton, or polypropylene olefin and is coated with a heavy layer of latex to hold the tufts permanently in place (Fig. 9.11-10). On most tufted carpets a loosely woven, coarse jute fabric (scrim) is added as secondary backing to lend additional strength and dimensional stability (Fig. 9.11-11).

The variety of textures and colors in tufted carpet is increasing with innovations in tufting equipment and dyeing processes. Tufted pile can be level-loop, plush, multilevel, cut or uncut, or a combination of cut and uncut. Striated and carved effects also can be achieved. Tufted carpets are made in solids, moresques, and more recently multicolor patterns made possible by new dyeing methods.

Tufted carpet generally can be identified on the reverse side, where the even rows of tufts punched through the prewoven backing with punched-in tufts also can be seen.

■ KNITTED CARPET

Knitted carpet is somewhat similar to woven carpet in that the pile and the

FIGURE 9.11-9 *A tufting machine.* *(Courtesy Carpet and Rug Institute.)*

FIGURE 9.11-11 *Tufted carpet construction: secondary backing usually is bonded to a prime backing with latex compound.* *(Courtesy Carpet and Rug Institute.)*

FIGURE 9.11-12 *Knitted carpet with typical continuously looped pile, single chain warp, and stuffer yarns, all backed with latex.* *(Courtesy Carpet and Rug Institute.)*

FIGURE 9.11-10 *A roll of carpet with a latex coating applied to the back by a special roller machine. Latex is used as a backing on all tufted and knitted carpet and on many woven carpets. A second backing is also laminated to good-quality tufted carpets.* *(Courtesy*

backing yarns are fabricated in one operation. Unlike weaving, the knitting process loops together the backing, stitching, and pile yarns with three sets of needles in much the same way hand knitting does. A coat of latex is applied to the back as a fixative and to give additional body to the fabric (Fig. 9.11-12).

Because a single pile yarn is used in the knitting operation, as in velvet weaving, knitted carpets are usually solid colors or tweeds, though patterns are also possible. Knitted carpets usually have either single- or multilevel uncut pile, but cut pile can be achieved with modification of the knitting machine.

If a knitted carpet is *grinned* (folded back) to expose the backing, it will show the continuous looping of the pile yarn from row to row, held in place by the stitching yarns. The rows of pile run in an irregularly diagonal direction with a random, homespun look, not in straight rows, as in woven and tufted carpets.

■ FUSION-BONDED CARPET

Fusion-bonded materials are those most used for carpet tile. Fusion-bonded carpet is made by sandwiching a yarn bundle between substrates, implanting it there, and heat fusing the two. A blade is then run between the substrates, producing two pieces of carpet. The method of manufacture limits this type of carpet to cut-pile face construction.

■ NEEDLEPUNCHED CARPET

Needlepunched carpets are made by having hundreds of barbed needles punch loose, unspun fibers (usually polypropylene or nylon) through woven batts or fleeces of fibers to mesh them together, which results in a homogeneous pressed layer of fibers much like an army blanket or heavy felt. Needlepunched carpet is used mostly in the low-cost field.

Since needlepunched carpet is made entirely of synthetic fibers, which are not subject to decay, it can readily be used on exterior as well as interior surfaces. Needlepunched carpet made from solution-dyed polypropylene is often used as outdoor carpet. Needlepunched nylon carpet is often foam-backed for use indoors.

■ FLOCKED CARPET

Flocked carpet is produced by spraying or electronically depositing short, loose nylon fibers on an adhesive-coated backing, similar to flocked wall coverings.

Continuing research into flocked carpet manufacture indicates that polypropylene might be substituted for jute, presently used as backing material, thus rendering flocked carpet also suitable for exterior and high-moisture locations.

9.11.1.2

Comparative Factors

Comparative factors related to carpet construction, such as pile yarn weight, pile thickness, and number of tufts per square inch, may be useful in comparing carpet of similar construction.

The traditional rule of thumb, "the deeper and denser (the pile), the better," often is used in making visual comparisons between carpets of similar construction and fiber type. However, such a generalization is superficial, and more detailed measures of carpet construction generally are necessary to judge carpet quality properly. Pile density is actually a combination of pile thickness and pile weight.

Pile thickness (in inches) and pile weight (in ounces per square yard) can be established accurately only by laboratory analysis, usually in accordance with ASTM standards. Pile density (in ounces per cubic yard) can be derived by a simple formula from established information about pile thickness and weight. All these criteria are used in determining standards of quality for carpet (see Section 9.11.2.5).

Other comparative factors have been developed as a substitute for or supplement to the basic criteria of pile density, weight, and thickness. The number of tufts per square inch; pitch, gauge, and needles; and rows, wires, and stitches are terms that describe the spacing of individual tufts in various carpets. When considered together with yarn size (yarn weight) and number of plies (strands) twisted together to form the pile yarn, these factors provide measures of pile density and other important basic criteria relating to quality.

■ PITCH, GAUGE, AND NEEDLES

Spacing of tufts across the width of a woven carpet can be described in terms of pitch, gauge, or needles. *Pitch* dictates the number of pile warp yarns (tufts) in a 27-in. width of carpet and is a measure of the tightness of crosswise construction (Fig. 9.11-13). Pitch may range between 90 and 256 for velvet and Wilton carpets, and between 189 and 216 for Axminster. The higher the pitch, the tighter the weave and the greater the pile density.

FIGURE 9.11-13 *Pitch.* *(Courtesy Carpet and Rug Institute.)*

In knitted and tufted carpets, where no continuous lines are present, *gauge* describes the actual spacing of tufts across the carpet width, expressed as a fraction such as 1/8 in., 3/16 in., or 5/32 in. For example, 1/8-in. gauge describes carpet with 8 tufts per inch; 3/16-in. gauge indicates 16 tufts in 3 in. (a little more than 5 tufts per inch).

In tufted carpets, the term *needles* often is used instead of gauge to describe the number of tufts per inch across the width. Tufted construction may vary from 4 to 13 needles (tufts) per inch.

■ ROWS, WIRES, AND STITCHES

Rows, wires, and *stitches* are terms that describe spacing of tufts lengthwise. The number of pile tufts per inch in Axminsters is described in *rows* per inch; for Wiltons and velvets in *wires* per inch (Fig. 9.11-14). Single or double weft shots are usually visible on the back of the carpet and correspond to the rows of tufts in the length of the carpet. To give an accurate indication of rows or wires per inch, it is customary to count the visible single or double weft shots in a 3-in. length of carpet and express this value in rows per inch, as a fraction where necessary. Thus, 17 rows in 3 in. would be expressed as 5-2/3 rows or wires per inch. Woven carpets range from 4 to 11 rows or wires per inch.

FIGURE 9.11-14 *Wires or rows per inch.* *(Courtesy Carpet and Rug Institute.)*

In tufted carpets there are no weft yarns that can be related to the spacing of pile tufts. Therefore, lengthwise construction is expressed as the number of tufts per inch. To distinguish this count from tuft count across the carpet (needles), the term *stitches* is used. Tufted carpets may have from 6 to 10 stitches per inch.

■ TUFTS PER SQUARE INCH

The overall density of carpets can be computed in tufts per square inch from information describing both lengthwise and crosswise construction (Fig. 9.11-15), or it can be determined by actually counting the tufts in a 1-in.-square area. This comparative method is preferred for tufted and knitted carpets but also can be applied to woven carpet. It is helpful in comparing overall density of dissimilar carpets when different terms are used to describe lengthwise and crosswise construction.

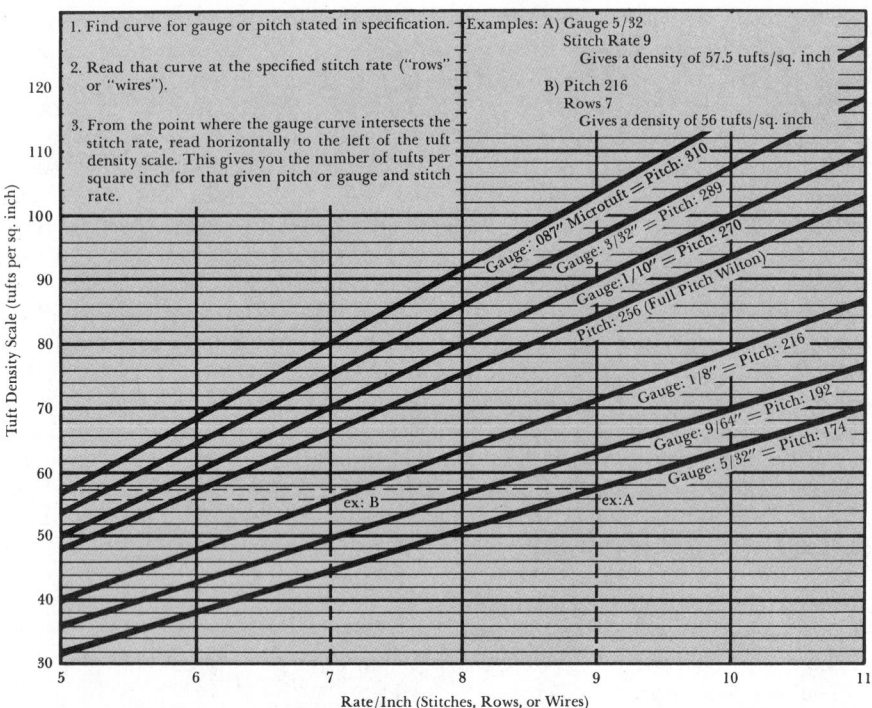

1. Find curve for gauge or pitch stated in specification.

2. Read that curve at the specified stitch rate ("rows" or "wires").

3. From the point where the gauge curve intersects the stitch rate, read horizontally to the left of the tuft density scale. This gives you the number of tufts per square inch for that given pitch or gauge and stitch rate.

Examples: A) Gauge 5/32
Stitch Rate 9
Gives a density of 57.5 tufts/sq. inch

B) Pitch 216
Rows 7
Gives a density of 56 tufts/sq. inch

Gauge: .087" Microtuft = Pitch: 310
Gauge: 3/32" = Pitch: 289
Gauge: 1/10" = Pitch: 270
Pitch: 256 (Full Pitch Wilton)
Gauge: 1/8" = Pitch: 216
Gauge: 9/64" = Pitch: 192
Gauge: 5/32" = Pitch: 174

Tuft Density Scale (tufts per sq. inch) — 30, 40, 50, 60, 70, 80, 90, 100, 110, 120

ex: B   ex: A

Rate/Inch (Stitches, Rows, or Wires) — 5, 6, 7, 8, 9, 10, 11

FIGURE 9.11-15 *The overall density in tufts per square inch is useful in comparing dissimilar carpets, such as tufted (example A) and woven (example B).* (Courtesy Carpet and Rug Institute.)

■ YARN SIZE AND PLIES

Wool fibers and synthetic staple fibers are spun into single-strand yarns. Continuous filament synthetic fibers are not spun, but emerge from the extrusion machine in bundles, which are formed into a single-ply yarn. The strands of yarn may then be further twisted together into thicker yarns, each forming a single pile tuft. Depending on the number of single strands, the pile yarn is identified as *single-*, *two-*, *three-* or *four-ply* yarn. The size of both single-ply and multi-ply yarns can be measured by *woolen count* (yards of yarn per ounce), denier (grams per 9000 meters of yarn), and other methods, depending on the spinning method employed. Synthetic fibers and yarns are most often measured according to the denier system.

Along with spacing of tufts, the yarn size and number of plies affect carpet density, resilience, compression resistance, and other important performance criteria.

■ BACKING

In woven and knitted carpets, the backing yarns and the pile yarns are interlocked simultaneously during manufacture. The backing yarns typically are jute (sisal fiber), kraftcord (wood pulp fiber), polyester/cotton, nylon, rayon, or polypropylene. Backing yarns hold the pile tufts in place and provide the necessary stiffness so the carpet will lie flat without wrinkling. In addition, they provide sufficient strength to prevent the carpet from tearing when heavy objects are moved across the surface.

Woven carpets, such as velvet, Wilton, and Axminster, may have single or double chain yarns and several stuffer yarns per tuft running lengthwise; the number of *weft shots per wire* running across the carpet also may vary, resulting in such designations as one-, two-, or three-shot carpet. Increasing the number of backing yarns results in a stiffer, stronger backing and therefore in improved carpet quality. The better Wilton and velvet carpets may be woven through the back by hooking the pile yarns around the bottom shot yarns, thus providing better anchorage of the pile tufts.

In tufted carpets, pile yarns are stitched into a separately prepared *primary back-*

ing. Most such primary backing today is polypropylene, although jute fabric is still sometimes used. An adhesive such as latex generally is used to secure the pile yarns in place, and a *secondary backing* of polypropylene or jute is bonded to the primary backing for additional strength, stiffness, and dimensional stability.

Similar secondary backings sometimes are used in knitted carpets as well. There is usually no secondary backing on woven carpet.

Pile Fibers

Carpet can also be selected according to the material of the pile fibers from which the pile yarns are made. These materials currently include wool, nylon, acrylics, polyester, and polypropylene.

Although pile fibers differ in their physical properties, all fiber materials included in this discussion will provide adequate service when used in a carpet construction suitable for the traffic exposure.

Wool has been the traditional fiber used in carpet manufacture, and because of its proven performance over the years it remains the classic fiber to which synthetic fibers often are compared. The synthetic fibers now account for about 90% of all carpet fiber used. They are marketed under a variety of proprietary names. The commonly used synthetic fibers can be classed generically as nylon, acrylic, and polypropylene (olefin). Cotton and two other man-made fibers, acetate and rayon, have not proven suitable for long-term wall-to-wall installations and therefore are not considered here. Polyester, one of the newest of the synthetic fibers, shows considerable promise but has not been in use long enough to demonstrate its properties fully.

The type of fiber used does not in itself guarantee carpet quality or performance. The various fibers do have unique characteristics that may affect carpet performance and styling. Selected physical properties of pile fibers are given in Figure 9.11-16; other characteristics of fiber related to carpet performance are given in Figure 9.11-17.

When fibers are blended, the higher the percentage of a particular fiber, the more a carpet will reflect the characteristics of

FIGURE 9.11-16
Selected Properties of Pile Fibers

Type	Fiber Length	Specific Gravity	Breaking Tenacity, grams/denier	Degrading Temperature[a]	Moisture Absorption, %
Wool	Staple	1.32	1.17	420°–570°F	15–16
Nylon	Staple and CF[b]	1.14	4–6	428°–482°F[c]	4–5
Acrylic	Staple	1.18	2–4	400°–500°F	1.3–2.5
Modacrylic	Staple	1.3–1.4	2.5–3.0	300°F	0.4–4.0
Polypropylene	CF[b]	0.91	3–6	333°F	up to 0.10

Courtesy Carpet and Rug Institute.

[a] For wool, scorching to charring range; for synthetics, melting temperature.

[b] Continuous filament.

[c] 428°F for Type 6; 482°F for Type 6,6.

FIGURE 9.11-17
Selection Criteria for Carpet Pile Fibers[a]

Criteria	Wool	Nylon	Acrylic	Modacrylic	Polypropylene
Resistance to: Abrasion	Good	Excellent	Good	Good	Excellent
Alkalies	Fair	Good to Excellent	Good to Excellent	Good to Excellent	Good to Excellent
Acids	Fair	Good to Excellent	Good	Good	Good to Excellent
Insects and fungi	Excellent[b]	Excellent	Excellent	Excellent	Excellent
Burns	Good	Fair	Fair	Fair	Fair
Compression	Good	Good	Good	Good	Fair
Crushing	Excellent	Good	Good to Excellent	Good	Fair to Good
Staining	Good	Good to Excellent	Good to Excellent	Good to Excellent	Excellent
Soiling	Good	Fair to Good	Good to Excellent	Good to Excellent	Excellent
Static buildup	Fair to Good	Fair	Good	Good	Excellent
Texture retention	Excellent	Good to Excellent	Good to Excellent	Good	Fair to Good
Wet cleanability	Fair to Good	Good	Good to Excellent	Good to Excellent	Excellent
Durability	Good to Excellent	Excellent	Good to Excellent	Good to Excellent	Good
Appearance retention	Excellent	Good to Excellent	Good to Excellent	Good to Excellent	Good
Ease of maintenance	Good to Excellent	Good	Good to Excellent	Good to Excellent	Excellent

Courtesy Carpet and Rug Institute.

[a] In each case, criteria vary depending on the properties of the specific fiber type used, carpet construction, and installation procedures.

[b] When chemically treated.

that fiber. For example, wool may be reinforced by tough nylon in a 70/30 blend, but the carpet will look and feel more like wool. Generally, at least 20% of a fiber must be used for its characteristics to be apparent.

■ WOOL

The outstanding characteristic of wool is resilience, which, in combination with moderate fiber strength (see Breaking Tenacity in Fig. 9.11-16) and good resistance to abrasion, produces excellent appearance retention. Wool's relatively high specific gravity contributes to greater pile density.

Carpet wool is obtained from a number of countries, including Syria, Iraq, Argentina, Pakistan, New Zealand, Australia, and Scotland. Wools from different regions have unique qualities, because different breeds of sheep yield different wools. Some wools are fine and lustrous, others are coarse and springy; also, fibers can vary in length from 3-1/2 to 7 in. *Woolen yarn* is made up of interlocked long and short fibers; *worsted yarn* uses only long fibers. Woolen yarns are more irregular, hence are softer and bulkier than worsted yarns. The pile of most wool carpet is made of woolen yarns spun from fibers 3-1/2 to 6-1/2 in. long (see Fig. 9.11-16).

■ NYLON

Nylon comprises about 85% of the commercial carpet market and a large portion of the housing market as well. The extensive use of nylon can be attributed to its lower cost, availability in many bright colors, exceptional resistance to abrasion, and high fiber strength (see Breaking Tenacity in Fig. 9.11-16).

Chemically, there are two types of nylon: Type 6, which represents a long-chain polymer, a chemical containing six carbon atoms; and Type 6,6, which results from the polymerization of two chemicals, each containing six carbon atoms. Although the two types perform similarly as yarn, they differ slightly in their ability to be processed and dyed.

Nylon fiber is produced in two forms: *staple* nylon, composed of 1-1/2-in. to 6-in.-length fibers, which are spun into yarn much like wool or other short fibers;

and *continuous-filament* nylon, consisting of *bundles* of continuous fibers that are formed into yarn without spinning. Both staple nylon and continuous-filament nylon yarns are made in round and modified cross sections (Fig. 9.11-18). Staple fibers are mechanically crimped; continuous-filament yarns are texturized (Fig. 9.11-19) to impart bulk and covering power to the yarns. Texturizing and crimp give the yarns or fibers an irregular alignment, which further increases their bulk and covering power.

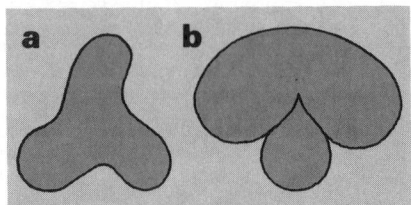

FIGURE 9.11-18 *Modified-fiber cross sections: (a) trilobal nylon; (b) bicomponent acrylic.* (Courtesy Carpet and Rug Institute.)

FIGURE 9.11-19 *Continuous-filament nylon: (a) before texturizing; (b) after texturizing.* (Courtesy Carpet and Rug Institute.)

■ ACRYLICS AND MODACRYLICS

Acrylics and modacrylics (modified acrylics) are synthetic fibers that closely resemble wool in abrasion resistance and texture.

Chemically, acrylic fibers are composed of 85% or more by weight of acrylonitrile. Modacrylics contain at least 35% of acrylonitrile in chemical combination with modifying materials. Acrylic fiber is produced in round, bean-shaped, and mushroom-shaped cross sections (see Fig. 9.11-18b). The latter is a bicomponent fiber of two filaments welded parallel to each other. Because the shrinking ratios of the two filaments are different, the fiber

assumes a coiled shape in its dry equilibrium state, a property that improves texture retention.

Fibers of the acrylic family are now available only in staple form and are characterized by an appearance and high durability comparable to that of wool. Acrylics often are blended with modacrylics in commercial carpet to reduce potential flammability. However, with substantial improvement in acrylic fibers, the trend is toward 100% acrylic fiber.

■ POLYESTER

Polyester is a fiber-forming thermoplastic polymer. It is made from terephthalic acid and ethylene glycol. Polyester fibers are staple fibers, and polyester yarns are spun yarns.

■ POLYPROPYLENE OLEFIN

There are two classes of olefin, polyethylene and polypropylene. Only polypropylene has been produced in a fiber suitable for carpet construction. Polypropylene fiber consists of at least 85% propylene by weight and is generally produced in bulked continuous-filament yarn, although staple yarn also is available.

Polypropylene has the lowest moisture absorption rate of any carpet fibers, which gives it superior stain resistance as well as excellent wet cleanability (see Fig. 9.11-16). Its low specific gravity results in superior covering power, and its high fiber strength contributes to good wear resistance.

9.11.1.4

Dyeing Methods

Color can be introduced at various stages of manufacture. For both synthetic and wool pile carpets, dyes can be added at either the fiber, yarn, or carpet manufacturing stage. The dyeing method used may vary with the general properties of the fibers, the intended carpet construction, and color effect desired, or it may depend on special methods developed for proprietary fibers.

■ SOLUTION DYEING

Many synthetic fibers are colored by a dye or pigment introduced into the liquid chemical from which fibers are extruded. In this method, the dye becomes chemically a part of the fiber. Solution-dyed yarns are generally colorfast and more resistant to discoloration from wet cleaning.

■ STOCK DYEING

Wool and synthetic staple fibers can be dyed in the raw fiber form before being spun into yarn by being dipped into kettles of dye. This method ensures uniform distribution of dye and is suitable where large quantities of a single-colored yarn are desired.

■ SKEIN AND PACKAGE DYEING

When smaller quantities of colored yarn are required, dyeing often is accomplished after staple fibers are spun or continuous-filament fibers are extruded into yarn. In *skein dyeing*, skeins of natural-colored yarn are immersed in vats of hot dye to become saturated with color. For *package dyeing*, the yarn is wound on a perforated tube, and the dye is forced through the perforations to soak the yarn.

■ SPACE DYEING

Both staple and continuous-filament yarns can be space dyed in alternating bands of color, generally for use in multicolored tufted carpet. In the *knit sleeve* method of space dyeing, the yarn is knitted into a fabric, print rollers apply the desired colors, then the fabric is unraveled again into yarn. In the *warp print* method, parallel yarns are fed continuously into a machine where color is applied by rollers.

■ PIECE DYEING AND RESIST PRINTING

Some woven and many tufted carpets are piece-dyed by immersing the entire carpet in a hot dye bath. However, the carpet need not be dyed only one color; some synthetic fibers can be chemically formulated to absorb certain compatible dyes and repulse others. Thus, it is possible to produce complex color patterns in the same dye bath. A similar technique of selective dyeing can be applied in skein dyeing. *Resist printing*, another method of selective dyeing, involves printing tufted carpet with a dye-resistant agent before piece dyeing.

■ PRINT DYEING

The most recent development in dyeing tufted carpet is *print dyeing*. This method provides good dye penetration and pattern definition. With possible color variation of up to six colors, print-dyed tufted carpets compare favorably in appearance with Wiltons and Axminsters but are cheaper to produce. The process involves silk-screening plain tufted carpet with a premetalized dye and applying an electromagnetic charge under the carpet, which drives the dye deeply into the pile. The final pattern is built up by continuously processing the carpet through a series of troughs, each applying a different color.

9.11.2

SELECTION CRITERIA

Carpet performance cannot be attributed to any single element of carpet construction or physical property of the pile fiber. Installed performance is rather the combined effect of many variables such as surface texture and carpet construction, backing, cushioning, and pile thickness, weight, and density. Of these, pile weight and pile density are the most uniformly significant, regardless of fiber type or carpet construction.

Although pile fibers differ in their physical properties (see Fig. 9.11-16), all fibers included in this discussion will provide adequate service when used in a carpet construction suitable for the traffic exposure (see Section 9.11.2.5). The variables of carpet construction are so great that it is difficult to assess their individual effects on performance and selection criteria. Therefore, selection criteria are outlined for fibers only (Fig. 9.11-17), and the effects of carpet construction are described where appropriate in the following discussion.

Individual selection criteria may be useful in carpet selection for special conditions of traffic or exposure to special hazards. For example, *alkali*, *acid*, and *stain resistance* may be important in food preparation and serving areas. *Soiling resistance* and *wet cleanability* may be a factor in entry areas. Insect and fungus resistance may require consideration in potentially damp locations, and the possibility of *static build-up* should be con-

sidered in areas of prevailing low humidity and where such build-up might affect equipment or activities, such as in computer rooms. These and other criteria have been combined into composites such as *durability*, *texture retention*, and *ease of maintenance* for average conditions (see Fig. 9.11-17).

9.11.2.1

Durability

Pile weight and density, backing and cushioning firmness, quality of installation, and maintenance significantly affect the durability of carpet. In addition, durability is affected by hazards that tend to destroy the pile fibers. The resistance of fibers to these hazards is shown in ratings for resistance to abrasion, alkalies, acids, insects, fungi, and burns. The rating for durability is a composite that reflects overall resistance to destruction or loss of pile fibers and is a measure of the service life of a carpet.

■ ABRASION RESISTANCE

Abrasion resistance is the resistance of the fiber to wearing away caused by foot traffic or other moving loads. It is measured in terms of the actual loss of fiber when exposed to a laboratory machine that simulates the abrasive forces of foot traffic. Since loss of fiber occurs as a result of breaking, fiber strength is one of the physical properties that influences abrasion resistance. Nylon and polypropylene have superior fiber strength and hence excellent abrasion resistance.

■ ALKALI AND ACID RESISTANCE

Foods and beverages are generally mildly alkaline or acid in nature; many soaps and detergents are strongly alkaline. If alkaline or acid solutions are not promptly removed when spilled, they may attack the pile fibers. Wool fibers are more vulnerable to such attacks because they have a higher moisture absorption rate and are less inert chemically than the synthetic fibers.

■ INSECT AND FUNGUS RESISTANCE

The synthetic fibers in modern carpets contain no organic nutritive value and are, therefore, immune to attack by insects, such as carpet beetles and moths, and to fungi, such as mold and mildew. To render wool carpet resistant to these hazards, most manufacturers treat wool yarn in the dye bath with a chemical preservative that provides long-term protection from insect attack, even after wet and dry cleaning. Since many carpets intended for interior use employ at least some organic fibers in the backing, some fungus hazard exists with most carpets, even those with synthetic fiber piles.

In general, the hazard of insect and fungus attack can be reduced by regular vacuuming and exposing the carpet to light and air. This is particularly important where large or heavy pieces of furniture make part of the carpet inaccessible to regular cleaning and ventilation. In such areas, greater insect protection can be provided for wool carpets by periodic spraying with mothproofing agents.

■ BURN RESISTANCE

Most synthetic carpet fibers are flame resistant to some degree, but they will melt and fuse when exposed to the heat of a burning cigarette or glowing ash that produces more than 500°F. Prolonged exposure to such concentrated heat may result in complete local loss of fiber or a slightly fused spot. The fused spot is more resistant to wear and abrasion than the surrounding area, and this causes a visible discoloration in the carpet surface.

Although wool fiber will actually burn by charring, the damage from a lit cigarette usually is less severe, and the charred spot will wear off readily or can be removed with fine sandpaper. The fused spot on synthetic carpets similarly can be snipped off if it is not too severe, and, if followed by brushing, an acceptable appearance usually will result. If the burn is severe, replacement of the burnt area may be necessary.

The behavior of a carpet in a fully developed fire is controlled by building codes and laws. *All carpet manufactured in the United States must pass a methenamine "pill" test* in accordance with ASTM Standard D 2859, "Test Method for Flammability of Finished Textile Floor Covering Materials." This same test is described in Federal Standard Document FF 1-70, "Methenamine Pill Test." In this

test, a carpet is exposed to the burning of a methenamine tablet. The medium *flame spread index* or *critical radiant flux* must be 0.04 watts/sq. cm.

Often, a carpet will be required by code to have a higher rating than the pill test minimum. The National Institute of Standards and Technology recommends that carpet used in corridors and exit ways have a radiant flux of at least 0.22 watts/sq. cm in commercial occupancies and of 0.45 in institutional occupancies.

A carpet's flame spread during a fully developed fire is usually measured by the Flooring Radiant Panel Test of ASTM E 648, "Test Method for Critical Radiant Flux of Floor Covering Systems Using a Radiant Heat Source," or NFPA 253, "Test Method for Critical Radiant Flux of Floor Covering Systems Using a Radiant Heat Energy Source," which are essentially the same test. This flame spread index is called the *critical radiant flux*. It represents the minimum energy necessary to sustain flame. The higher the number, the more resistant the carpet to flame propagation.

Sometimes, a carpet's flame spread index is required to be measured by the *Smoke Chamber Test* in UL 992, "Smoke Chamber Test." This test measures the length of spread and time of travel of a flame. The range is from 0 to 25.

9.11.2.2

Appearance Retention

Appearance retention depends on factors that change original carpet color, texture, or pattern, such as fading, soiling, and staining, and compression and crush resistance. Other factors common to many types of fiber and carpet construction that affect appearance without impairing serviceability are pilling, shedding, sprouting, and shading (see Glossary). Ratings for appearance retention are composites of these factors.

■ COMPRESSION RESISTANCE

The extent to which the pile will be compressed under heavy loads or by extensive foot traffic is called compression resistance. For instance, of two samples tested under the same conditions, the one that compresses less is said to have better com-

pression resistance. With the exception of polypropylene, which is rated somewhat lower, all fibers are approximately equal in their ability to resist compression loads. Of greater importance in resisting such forces are the cushioning, the pile height, and the pile density. Looped-pile construction with closely spaced tufts and tightly twisted yarns is more resistant to compression.

Compressed fibers in a small area may be restored by dampening the pile with a sponge or steam iron and brushing the pile erect. If a larger area becomes matted down by excessive traffic, it can be restored by vacuuming with a beater bar machine (Fig. 9.11-20), using a pile lifter machine, or wet cleaning.

FIGURE 9.11-20 *A beater bar cleaner opens the pile and reaches deep-seated dirt.* (Courtesy Carpet and Rug Institute.)

■ CRUSH RESISTANCE

Crush resistance depends on the ability of the fibers to recover from short- and long-term compression loads. This property is measured by the extent to which the fiber springs back after a load has been removed, as a percentage of the original height. Wool and acrylic have the highest crush resistance, nylon and modacrylic moderate, and polypropylene somewhat lower. Pile density, spacing of tufts, and tightness of yarns affect crush resistance more than the relative rating of the fibers.

■ TEXTURE RETENTION

The ability of carpets to retain the surface texture imparted during manufacture is related to the compression and crush resistance of the fibers, as well as the density of the pile and the tightness of the yarns. Generally, all fibers have adequate texture retention, with wool rated highest and polypropylene somewhat lower than the other synthetic fibers. Within this range, special fiber types such as heat-set nylon and bicomponent acrylic (mushroom cross section) have improved texture retention.

In areas where the texture has been conspicuously damaged by continuous traffic or heavy loads, the surface can be partially restored by the methods suggested above under "Compression Resistance."

9.11.2.3

Ease of Maintenance

The ease of maintaining an installed carpet depends mainly on its resistance to soiling and staining, as well as its wet cleanability. Although dry powder cleaning products are useful for moderately soiled areas, wet cleaning (shampooing) remains the most satisfactory method for thorough in-place carpet cleaning under heavier soil conditions.

■ STAINING RESISTANCE

Staining is a local discoloration of a carpet, compared with *soiling*, which is more or less uniform over the carpet surface. Many foodstuffs (oil, grease, eggs, chocolate), beverages, and beauty aids may cause staining. If promptly and properly treated, such stains need not be permanent. Many carpet manufacturers publish detailed recommendations for stain removal and indicate the type of solvent or cleaning agent that should be used with each type of stain. Since stains are often moisture-borne, rate of moisture absorption is regarded as a factor in staining resistance. Polypropylene and acrylics have low moisture absorption and are therefore rated highly for staining resistance.

The general procedure for stain removal involves (1) promptly removing solids and blotting up liquids with absorbent materials; (2) treatment of the stain with the appropriate solvent or cleaning agent, starting at the periphery and working toward the center; (3) repeated blotting of treated areas and avoidance of excessive brushing; (4) leaving a weighted cloth or paper towel over the stain area to remove most of the remaining liquid by absorption rather than air drying.

■ SOILING RESISTANCE

Soiling is the result of dirt or soil particles deposited on the carpet surface by foot traffic or dust from the air. Fibers such as wool and nylon generate higher electrostatic charges in low humidity and tend to attract airborne dust and soil faster than some other fibers. However, the scaly surfaces of wool fibers make soiling less apparent. Polypropylene and acrylics have higher soiling resistance because of low static build-up and low moisture absorption. Bulked fibers with modified cross sections and texturized yarns spun more loosely tend to retain dirt more tenaciously than round, tight, high-density yarns. Also, tightly woven, uncut-loop piles will keep soil near the surface where it is easier to remove, while multicolored and textured piles of medium shades will conceal soiling better than light or dark smooth plush surfaces.

■ WET CLEANABILITY

Since it is usually impractical to remove wall-to-wall carpet for plant cleaning, the ability of carpet to respond successfully to in-place wet cleaning is an important consideration. Polypropylene, which has the lowest moisture absorption rate, is considered the best in this respect. Generally, synthetic fibers, with their lower moisture absorption rates and inert chemical compositions, clean more readily, dry more rapidly, and are less subject to mildew odors while drying than wool. However, a variety of effective cleaning formulations specifically suited to wool carpet are available.

9.11.2.4

Static Generation

Static electricity is an annoyance in every carpeted space, but it actually can become harmful in areas housing computers or other electrically sensitive equipment. Static occurs whenever two different materials come into contact with each other. People usually do not become aware of static until it reaches about 3.5 kilovolts (kv). The normal carpet requirements for solely human occupancy require a level of 3 kv or less. Computers require 2 kv or less.

Of the three factors regarding static electricity—static generation, static dissipation (conductivity), and static decay time—only the first is usually limited in the selection of carpets in nonsensitive locations. Static generation in carpet is measured by the American Association of Textile Colorist and Chemists' "Step Method," which simulates conditions of actual use and measures the charge generated by someone walking across the carpet.

Static dissipation may be also limited in spaces that will house computers. It is measured by a method recommended by IBM Corp.

Nylon is inherently high in static build-up. To fight this fact, some nylon carpet manufacturers include carbon-loaded nylon fibers that carry a lifetime antistatic guarantee. Some carpet is laced with metal wires that are in contact with a grounded backing. In very sensitive areas, it might be impossible to reduce static sufficiently using antistatic fibers or wires alone. Then, the antistatic properties of such carpet must be supplemented with conductive shoes, furniture, mats, or other devices.

9.11.2.5

Standards of Quality

There are currently no accepted industry-wide standards of quality for carpet. However, the General Services Administration (GSA) and the Department of Housing and Urban Development (HUD) both have issued standards that may become the basis for further standard development. Federal Specification DDD-C-95, "Carpets and Rugs, Wool, Nylon, Acrylic, Modacrylic," covers carpet for commercial, institutional, and residential use. The HUD specifications contained in "Use of Materials Bulletin UM-44C" are related more directly to residential carpet, but are also referenced in specifying commercial and institutional carpet.

HUD does not permit the use of carpet in baths, kitchens, and service areas such as laundry, utility, and furnace rooms. Carpet is also not appropriate in public toilets or shower rooms. Living areas within a residential unit, such as living, dining, recreation, sleeping, and dressing rooms are designated moderate-traffic areas. Public areas such as corridors, entrances, lobbies, waiting rooms, stairways, and elevators are considered heavy-traffic areas. Offices and institutional spaces may be either moderate- or heavy-traffic areas, depending on their actual use.

FIGURE 9.11-21
Physical Requirements for Carpet in Heavy-Traffic Areas[a]

| | Wilton, Wool or Acrylic | Velvet | | | Tufted, Wool or Acrylic | Knitted, Wool or Acrylic |
		Wool or Acrylic	Nylon[b]	Wool or Acrylic		
Pile Description	Single-level loop woven through back	Single-level loop pile woven through back	Single-level loop pile woven through back	Multilevel loop woven through back	Single-level loop	Single-level loop
Tufts/sq. in.	52	60	32	34	60	26
Frames	3	—	—	—	—	—
Shots/wire	2	2	2	2	—	—
Weight oz./sq. yd. Pile	48	42	29	44	42	37
Total	72	60	50	64	—	58
Pile height	Min. 0.250 Max. 0.320	Min. 0.210 Max. 0.310	Min. 0.200 Max. 0.290	Min. 0.190 Max. 0.370	Min. 0.250 Max. 0.320	Min. 0.230 Max. 0.290
Chain	Cotton and/or rayon	Cotton and/or rayon	Cotton and/or rayon	Cotton and/or rayon	—	Cotton, rayon, or nylon
Filling	Cotton and/or rayon, or jute	Cotton and/or rayon, or jute	Cotton and/or rayon, or jute	Cotton and/or rayon, or jute	—	Jute or kraftcord
Stuffer	Cotton, jute, or kraftcord	Cotton, jute, or kraftcord	Cotton, jute, or kraftcord	Cotton, jute, or kraftcord	—	—
Back coating, oz./sq. yd.	—	8	6	8	—	—
Tuft bind, oz.	80	80	80	80	100	14
Ply twist turns/in.	Min. 1.5 Max. 3.5	Min. 1.5 Max. 3.5	Min. 1.0 Max. 3.0	Min. 1.5 Max. 3.5	Min. 2.5 Max. 4.5	Min. 1.5 Max. 3.5

Courtesy Carpet and Rug Institute.

[a] Based on Federal Specification DDD-C-0085.

[b] High-bulk or textured continuous filament.

HUD requirements for carpet in heavy-traffic areas are made by reference to Federal Specification DDD-C-0085 and are summarized in Figure 9.11-21. Carpet construction requirements for moderate-traffic areas are discussed below and are illustrated in Figures 9.11-22 and 9.11-23. For both types of traffic exposure, carpet should conform to the following requirements for pile fibers, backing, and cushioning.

■ **CARPET CONSTRUCTION**

HUD Bulletin UM-44C, "Use of Materials," requirements are based on average values for pile yarn weight, pile thickness, and pile density. Pile yarn weight and pile thickness are obtained from samples by tests performed according to ASTM D 418, "Method of Testing Pile Yarn Floor Covering Construction."

Pile density is expressed as either average pile density (AD) or weight density (WD). *Pile yarn weight* (W) is the average weight of the face pile yarn in ounces per square yard. *Pile thickness* (t) is the height of the tufts above the backing in inches. *Average pile density* (AD) is a calculated quantity in ounces per cubic yard obtained by the formula $AD =$ $(36)(W)/t$. The AD is a four-digit number that should be above 3000 for commercial carpet and 2800 for any other carpet. *Weight density* is also a calculated quantity in ounces per cubic yard obtained by the formula $WD = (AD)(W)$.

UM-44C establishes minimums for pile yarn weight (W) and weight density (WD), depending on the type of fiber (see Fig. 9.11-22). The minimum weight densities (WD) given are always the minimum pile weight (W) multiplied by a pile density (D) of 2800. The HUD requirements are based on minimum weight-density factors rather than a constant pile density of 2800.

FIGURE 9.11-22
Physical Requirements for Carpet in Moderate-Traffic Areas[a]

Pile Fiber	Minimum Pile Yarn Weight (W), oz./sq. yd.[b]	Minimum Weight-Density Factor (WD)[c]
Wool	25	70,000 (25 × 2800)
Acrylic		
Modacrylic		
Nylon (staple and filament)	20	56,000 (20 × 2800)
Polypropylene olefin		

Courtesy Carpet and Rug Institute.

[a] Based on HUD Use of Materials Bulletin UM-44A.

[b] Computed for blended yarns by multiplying percentage of each fiber by minimum pile weight and adding the two amounts.

[c] Pile Density (D) = 2800, the minimum density that historically has given satisfactory service for wall-to-wall installations.

If only the pile density were specified, it would establish a simple proportionate relationship between actual pile thickness and pile yarn weight, which would require increasing the minimum pile weight the same extent as the pile thickness. This requirement would be excessive for moderate-traffic areas. Figure 9.11-23 illustrates calculation of the weight-density factor.

FIGURE 9.11-23
Weight-Density Calculation for Nylon or Polypropylene Carpet

FHA weight-density (WD) calculated for nylon or polypropylene carpet from pile thickness (t) and pile yarn weight (W) determined by test in accordance with ASTM D 418 for woven carpet or D 1486 for tufted carpet.

Pile thickness (t) = 0.25 in.

Pile yarn weight (W) = 20 oz./sq. yd.

Calculated pile density (D) = $36W/t$

$$D = \frac{36 \times 20}{0.25} = 2880 \text{ oz./cu. yd.}$$

Calculated weight-density factor (WD)

$WD = 20 \times 2880 = 57,600*$

Courtesy Carpet and Rug Institute.

*Referring to Figure 9.11-22, pile yarn weight (20 oz./sq. yd.) equals minimum requirements; weight-density factor (57,600) exceeds minimum requirement (56,000).

■ PILE FIBERS

Pile yarn should be made of 100% wool, nylon, acrylic, modacrylic, or polypropylene olefin fibers, or blends of these fibers in yarn (exclusive of ornamentation). Not less than 20% of any of these fibers should be used when blended with other fibers, and such blends should not consist of more than two types of fiber. All fibers should be entirely new, not processed or reclaimed from previously knitted, woven, tufted, or felted products. Wool yarn should contain at least 97% wool fibers by dry weight; synthetic fibers should be 15 denier or thicker and should contain not more than 2% chloroform-soluble material.

Carpet pile should be resistant to fading from wet cleaning and exposure to light. Light colors should be rated as satisfactory after laboratory exposure to 20 standard fading hours and dark colors after 40 fading hours.

■ BACKING

Backing materials should be those customarily used and accepted in the industry for each type of carpet. For tufted carpet, there should be a secondary backing reinforcement of at least 4 oz./sq. yd., exclusive of the adhesive coating.

9.11.2.6

Certification of Quality

There is no industry-wide program for certification of carpet quality. However, most manufacturers will certify either on their invoice or by an appropriate label that their product meets or exceeds HUD requirements for moderate- or heavy-traffic areas.

Federal law requires that all textiles be identified with an appropriate label as to fiber content for each fiber type that constitutes 5% or more of the total. In the case of carpets consisting of backing yarns and pile yarns, only the fibers in the pile yarn need be identified. In addition, carpet samples shown to consumers must be similarly identified as to fiber content, name of manufacturer, and, if imported, country of origin. Several producers of synthetic fibers have agreements with carpet manufacturers that stipulate that only carpet meeting certain minimum construction requirements can display their fiber trademark.

9.11.2.7

Cushioning

All carpet should be installed over cushioning to increase resilience and durability. Some carpets, especially those used in direct glue-down installations (see Section 9.11.3.2), have an integral secondary backing that serves this function. In loomed carpets, cellular (foam or sponge) rubber cushioning usually is bonded directly to the carpet at the mill. For most carpets, however, cushioning is provided in the form of a separate padding. Currently available paddings include felted hair, rubberized fibers, cellular rubber, and urethane foam.

Cushioning should meet or exceed the requirements of the following listed standards:

Cushioning bonded to a carpet: Federal Specification DDD-C-0085, "Carpets and Rugs, Wool, Nylon, Acrylic, Modacrylic."

Separate cushioning made of cellular rubber, jute and hair, or hair felt: Federal Specifications ZZ-C-00811b, "Cellular Rubber Carpet and Rug Cushion" and DDD-C-001023, "Carpet and Rug Cushion (Hair Felt and Rubber Coated Jute and Animal Hair or Fiber)"

Bonded urethane cushion: Federal Specification L-C-001369, "Bonded Urethane Carpet and Rug Cushion"

Virgin urethane cushion: Federal Specification L-C-001676, "Virgin Urethane Carpet and rug Cushion"

■ FELTED HAIR

The most economical traditional type of conventional padding is made of felted animal hair. This padding has a waffle design to provide a skid-proof surface and improve resiliency. It is sometimes reinforced with a jute backing or a burlap center liner. When reinforced with burlap, the hair is punched through the burlap fabric and compressed to a uniform thickness. Sizing (adhesive) sometimes is used to strengthen the bond between the fibers and the burlap core. Many manufacturers also sterilize and mothproof hair padding.

Felted padding of hair or hair and jute may mat down in time or may develop mildew, especially if the fibers become wet, as during cleaning. However, when properly cleaned, sterilized, and treated, hair padding is suitable for reasonably dry floors at all grade levels and on conventional radiant-heated floors.

■ RUBBERIZED FIBERS

Some padding made of jute or hair is coated with rubber on one or both sides to hold the fibers together and provide additional resilience. Sometimes this padding has an animal hair waffle top and a jute back reinforced with a patterned, rubberized application.

■ SPONGE RUBBER

In addition to cushioning bonded directly to the carpet, foam or sponge rubber is produced in sheet form with waffle, ripple, grid, or V-shaped rib designs. A scrim of burlap fabric usually is bonded to the rubber sheet to facilitate installation of the carpet. When laid with the fabric side up, this permits a taut and even stretch of the carpet.

Rubber padding is more expensive, but it retains its resilience longer than hair padding. It is highly resistant to decay and mildew and is non-allergenic. It can be used at all grade levels, but the denser cushionings are not recommended for radiant-heated floors.

■ POLYURETHANE FOAM

Urethane foam padding is a high-density polymeric foam available in two major types: prime and bonded. *Prime* urethane foam padding is available as either prime

or densified prime. The later is modified chemically to make it wear better.

Bonded urethane foam padding is made from heat-fused scraps of urethane, rather than being a homogeneous product, as are prime and modified prime urethane padding.

The quality of urethane foam padding is defined by its density or weight per cubic foot of foam.

■ SELECTION

The density of padding should be as recommended by the Carpet Cushion Council (CCC) in its publication "Selecting the Correct Carpet Cushion for Every Traffic Area." This publication recommends the specific density of each available carpet padding type for each type of use, divided into Class I, Moderate Traffic; Class II, Heavy Traffic; and Class III, Extra Heavy Traffic. The classes and their uses are as follows:

Class I: Executive and administrative spaces in such buildings as schools, office buildings, airports, retail establishments, health care facilities, libraries, museums, and sleeping areas in homes, hotels, motels

Class II: The more public areas in the same types of buildings as listed for Class I and clerical areas, moderate-use corridors, patients' rooms, lounges, dormitories, classrooms, small retail stores, bank lobbies and corridors, and moderate-use public areas

Class III: Heavy-use public corridors, lobbies, locker rooms, dining areas, nurses' stations, cafeterias, major retail stores, restaurant dining rooms, and convention centers

The recommended weights and thicknesses of padding vary within each class, depending on the padding material, and from class to class (Fig. 9.11-24).

The mentioned CCC publication also recommends, where applicable, carpet padding's Compressed Load Deflection (CLD) and Indentation Load Deflection (ILD), which is called the Indentation Force Deflection (IFD) in ASTM Standard D 3574, "Method of Testing Flexible Cellular Materials—Slab, Bonded, and Molded Urethane Foams" (Fig. 9.11-25). The CLD is the force in pounds per square inch needed to compress padding

FIGURE 9.11-24
Weight and Thickness of Carpet Padding

Padding Type	Class	Thickness, in.	Weight
Natural fiber	I	1/4	32 oz/sy[a]
	II	5/16	40 oz/sy
	III	3/8	50 oz/sy
Synthetic fiber	I	5/16	22 oz/sy
	II	3/8	28 oz/sy
	III	3/8	36 oz/sy
Flat sponge rubber	I	0.225	56 oz/sy
	II	0.250	64 oz/sy
	III	0.250	80 oz/sy
Reinforced rubber	I	0.225	56 oz/sy
	II	0.235	64 oz/sy
	III	0.250	80 oz/sy
Ripple rubber	I	0.350	64 oz/sy
	II	0.400	80 oz/sy
	III	0.250	80 oz/sy
Bonded urethane	I	3/8	5.0 pcf[b]
	II	5/16	6.5 pcf
	III	1/4	8.0 pcf
Modified prime urethane	I	1/4	2.5 pcf
	II	1/4	2.7 pcf
	III	3/16	3.2 pcf
Densified prime urethane	I	0.350	2.1 pcf
	II	0.265	3.5 pcf
	III	0.265	4.5 pcf

[a] Ounces per square yard.
[b] Pounds per cubic foot.

to a specified percentage of its original thickness. The ILD (IFD) is the number of pounds needed to depress a piece of padding to a specified percentage of its original thickness over an 8-in.-diameter test area.

9.11.3

INSTALLATION

Carpet can be installed over finished flooring, subflooring of concrete, plywood, particleboard, or any other suitable, smooth surface. The carpet manufacturers' recommendations should be followed with regard to seam locations and direction of carpet. Carpet edge guard should be located where carpet edges would otherwise be exposed. Edge guard should be securely fastened to the substrate.

Padding Material	Class	Load	Percent of Original Thickness
Urethane	I	0.75 psi*	25
(CLD)	II	1.00 psi	25
	III	1.50 psi	25
Bonded	I	4.00 psi	65
urethane	II	5.00 psi	65
(CLD)	III	7.00 psi	65
Densified	I	0.85 psi	65
prime	II	1.80 psi	65
urethane	III	2.30 psi	65
(CLD)			
Modified	I	100 lb.	25
prime	II	100 lb.	25
urethane	III	120 lb.	25
(ILD)			

*Pounds per square inch.

FIGURE 9.11-26 *Using a power stretcher, a skilled carpet mechanic can achieve a smooth, taut installation of carpet over tackless strips.* (*Courtesy Carpet and Rug Institute.*)

9.11.3.1

Tacked and Tackless Installation

Carpet that is not rubber-backed can be readily installed over nailable surfaces by tacking the edges or by using *tackless strips* around the perimeter. Tackless strips consist of pieces of plywood the thickness of the padding and with two or three rows of metal pins that pass through the plywood at an angle and are designed to grip carpet when it is stretched over the pins (Fig. 9.11-26). Installations on concrete can be made by tacking if wooden plugs are provided 6 to 9 in. on centers in the concrete slab to receive the tacks. Also, tackless strips can be installed on concrete with powder-actuated fasteners

or epoxy adhesive. Two-row strips can be used when the carpet is 20 ft. or less in width, but three-row strips should be used with wider carpet. Strips should be located at the entire perimeter of the carpet.

Padding should be placed between the tackless strips and bonded to the substrate with an adhesive. The joints should be placed at right angles to the carpet joints. Seams in padding should be taped.

Regardless of the installation method, tacked or tackless carpet should be located, trimmed, stretched sufficiently to eliminate wrinkles and buckles, and attached firmly with tacks or to strips. Seams should be butted tightly and taped or sewn to make them strong enough to

resist damage during stretching and to keep them closed without gaps.

9.11.3.2

Direct Glue-Down Installation

In *direct glue-down* applications, carpet is glued to the substrate with no padding between them. Such carpet should be fitted in place before adhesive is applied. The adhesive should then be applied uniformly to the substrate following the adhesive manufacturers' instructions. Carpet edges should be butted tightly so that there are no gaps. The carpet should then be rolled lightly to ensure uniform bond to the substrates with no air pockets.

9.12 PAINTING AND FINISHING

This section discusses paints, coatings, and transparent finishes. There are no generally accepted definitions in the industry that clearly state the difference between paints and coatings. Paint manufacturers and other industry sources often use the two terms interchangeably. Because of the possible confusion created by double usage, this book will use these terms as follows.

Paints are air-drying materials that are applied as a liquid in the field. Primers for paints may be applied either in the field or in a shop or factory. The complete system of materials needed to produce a paint film is called a *paint system*. The term *paint* includes primers, emulsions, enamels, opaque stains, sealers, fillers, and other applied materials used as prime, intermediate, and finish coats in a paint system.

Coatings are applied only in the shop or factory. They may be similar in materials and formula to paint and may be either air-drying or baked-on. Coatings include such materials as Kynar coatings on metal, anodizing on aluminum, or baked enamel on lockers and furniture.

Special coatings are relatively thick, high-performance architectural coatings, such as high-build glaze coatings, fire-retardant coatings; industrial coatings, such as those used in sewage disposal plants, and cementitious coatings. These coatings are usually applied in the field.

A *transparent finish* is a system of materials used to form a protective film on wood that lets the wood color, grain, or both remain visible. Transparent finishes may be applied in the shop, the factory, or the field. They are sometimes called *clear* wood finishes.

Only paints, transparent finishes, and a few special coatings are discussed in this section. Some coatings used on metal are covered in Chapter 5. Other coatings and special coatings are beyond the scope of this book.

The comments in this section related to materials have been drawn from material developed by the Scientific Committee of the National Paint and Coatings Association and from recommendations included in the product and use literature of the paint manufacturers listed in Appendix B, References. The paints and coatings discussed here are not all-inclusive. Individual manufacturers may have other formulations designed for specific purposes that have some advantage over more commonly used paint and transparent finish materials.

This section also discusses recommended methods of applying paint and transparent finishes. This information has been developed from that contained in earlier editions of this book, which were abstracted from *Paints and Protective Coatings Manual of the Army, Navy, and Air Force*, compiled by the National Bureau of Standards (now the National Institute of Standards and Technology) and from data contained in the preparation and application recommendations of the paint manufacturers listed in Appendix B, References.

Both paints and coatings are used to coat, beautify, and protect an underlying surface. They both serve as a protective shield between the construction materials and the elements that attack and deteriorate them. Typical causes of paint and transparent-finish failure are sunlight, temperature variations, salt water, water vapor, decay, mildew, chemicals, and abrasion. When regularly renewed, paint and transparent finishes offer long-term protection that extends the useful life of the materials and the structure. However, as will be discussed later, the time between renewals varies considerably among products, being usually shorter for transparent finishes than for opaque ones.

Paints may serve other functions as well. For example, paint is an essential part of sanitary programs for hospitals, as well as for commercial, institutional, and home kitchens and washrooms. Gloss and semigloss paints provide smooth, nonabsorptive surfaces that can be washed easily and kept free of dirt and germs.

White and light-tinted paints reflect light, brighten rooms, and increase visibility. Flat paints soften and evenly distribute interior illumination. The functional use of color creates comfortable living and working conditions in the home, shop, and office.

Color-coded painting frequently is used to enhance public comfort, safety, and efficiency. Certain colors are stimulating, others are relaxing, and others are universally associated with potentially dangerous situations.

9.12.1

MATERIALS

There is a constant demand for recommendations for the best paint or transparent finish for a given use. However, there is rarely one best paint or transparent finish for any surface. Some of the factors that must be considered are the condition of the surface, the method of application, the curing conditions, the service expected, and the relative weight of initial cost balanced against appearance and durability.

Furthermore, the entire paint or transparent-finish system must be considered as a unit. It is unwise and often impossible to choose a paint or a transparent finish without regard to the surface over which it is to be applied. Though it is generally impossible to single out the best paint or transparent finish, it is possible to list commonly encountered surfaces and to recommend the materials most suitable for each surface.

Paint is a mixture of minute solid particles known as *pigment*, suspended in a liquid medium referred to as the *vehicle*. The pigment generally provides hiding power and color; the vehicle combines (1) the *solvent* (thinner), which ensures the desired consistency for application by brush, roller, or spray, and (2) the *binder*, which bonds the pigment particles into a cohesive paint film during the drying process. Some paints dry and harden simply by evaporation of the solvent. Most other types of paint involve chemical reactions as well as solvent evaporation. Most paints are described according to the type of binder used in them (alkyd, latex, oil, oleoresinous, rubber, urethane).

The classifications here carry no implication of quality. Materials in any of the types described may be available in varying qualities. In the selection of paint and transparent finishes, as with other building products, the reputation of the manufacturer is a significant factor. The reputations of most major paint manufacturers are well known in the construction industry.

Many states, the District of Columbia, and many cities and towns have enacted air pollution regulations that restrict the amounts and types of solvents that may be emitted into the atmosphere by paints, transparent finishes, coatings, and the materials used to clean surfaces to receive them and remove them after they have been applied. In addition, many jurisdictions have regulations concerning the use and disposal of even those chemicals that are not prohibited. Therefore, before any paint, transparent finish, or coating is applied, and before paint removers or chemical cleaning materials are used, it is necessary to verify their acceptability by the authorities having jurisdiction.

Interior Primers

Primers are paint or transparent-finish materials designed and formulated for application over bare surfaces. They serve as a base for succeeding coats. Some top-coat materials also serve as a primer on some surfaces.

Primers may be classified according to their use on interior or exterior surfaces. The following materials typically are used on interior surfaces.

■ WALL PRIMERS AND PRIMER-SEALERS

Interior wall primers and primer-sealers may have a latex, alkyd, or oil binder, depending on the material to be covered and the top coat that will be applied.

LATEX   Latex primers and primer-sealers are emulsion paints recommended especially for use on gypsum board surfaces, but they are also used on masonry. On gypsum board surfaces, they have the property of not raising the fibers as would a solvent-thinned primer. Further advantages are their quick drying properties, which allow recoating within hours, ease of equipment cleaning, and excellent alkali resistance.

ALKYD   Alkyd primers and primer-sealers are usually made of odorless alkyd, that is, the binder is an alkyd resin dissolved in odorless mineral spirits. They take longer to dry than the emulsion type, requiring overnight drying before recoating. They are suitable for all interior surfaces except paper-surfaced gypsum board and surfaces containing active alkali, such as partially cured or damp plaster or masonry.

OIL   Oil primers and primer-sealers are primers whose binder is a processed drying oil, often containing some resin. Their characteristics are similar to those of alkyd primers except that drying may be slower and softer. Overnight recoating schedules still are possible. They may be thinned with odorless or ordinary mineral spirits.

ALKALI-RESISTANT   Alkali-resistant primers and primer-sealers usually are based on a butadiene-styrene copolymer or chlorinated rubber, with one or more softer resins, such as chlorinated paraffin, as plasticizers. Sometimes small amounts of drying oil may be added. Aromatic solvents often are required. Application is likely to be more difficult than for other primers. They are particularly suitable for alkaline surfaces, such as damp or partly cured plaster or masonry. They should not be expected to resist hydrostatic pressure, such as may occur below grade, but they may be used on above-grade masonry walls.

■ WOOD PRIMERS

Interior wood primers include enamel undercoaters, clear wood sealers, and wood fillers.

ENAMEL UNDERCOATERS   Enamel undercoaters may be of the alkyd, latex, or oil-based variety, but the ones most used today are alkyds. Undercoaters are characterized by good hiding ability and low gloss. These undercoaters produce hard, tight films that prevent the penetration of enamel coats applied over them. This property is known as *enamel holdout*. They may be used under any pigmented interior finish but are particularly suitable under gloss and semigloss enamel, where a smooth finish is important. Enamel undercoaters are usually white but may be tinted for use under colored finish coats. Enamel undercoaters contain no rust-inhibitive pigment, but they may be used as primers on metals where no rusting is expected. Alkyd undercoaters dry sufficiently overnight to be painted the following day; latex undercoats may be painted in 2 to 4 hours; oil-based undercoats should dry at least 16 hours before recoating.

CLEAR WOOD SEALERS   Clear wood sealers seal the pores in wood surfaces without impairing the natural appearance. Many sealers contain a transparent pigment to improve sealing and reduce penetration. This reduces the prominence of the grain to some extent, which usually is desirable. These products generally are used under transparent finishes, although they can be used under pigmented materials as well.

WOOD FILLERS   Wood fillers include paste wood fillers and liquid wood fillers. They are used beneath transparent finishes.

Paste wood filler is used to fill the pores of open-grain woods, such as oak or mahogany. A transparent and relatively coarse pigment is combined with sufficient binder to make a stiff paste, which requires thinning with a volatile solvent before application. To obtain specific effects, stain may be added. When the solvent has evaporated, the filler should be rubbed across the grain to force the filler into the pores and to remove excess material from the surface. This is followed by a cleanup wipe in the direction of the grain using a coarse cloth such as burlap.

Liquid wood fillers are designed to eliminate thinning. While they generally do not perform as well as paste fillers on many woods, they are adequate when extensive filling is not required. All fillers impart some color to the wood. Fillers that contain stearates should not be used under urethane finishes.

■ MASONRY AND CONCRETE WALL PRIMERS

Interior masonry and concrete wall primers include latex and alkyd primers. The first coat of an epoxy finish often serves as a primer, but conditions where a heavy-duty finish is needed may require a different epoxy formulation for the primer. These primers are used as sealers on normally smooth masonry surfaces.

When a smoother finish is desired on masonry and concrete, a block filler or cement grout is applied beneath the paint primer.

BLOCK FILLER   A block filler has a relatively thick composition, which permits application by brush or roller. If applied by spray, it should be brushed into the surface to ensure that voids will be filled. The binder may be latex or a solvent-thinned material, such as an epoxy ester. Latex materials have excellent alkali resistance and may be used on damp surfaces. Solvent-thinned materials may have better chemical resistance.

CEMENT GROUT   A thin portland cement mortar often is used to give a smooth surface to rough masonry or concrete. It may be applied with brush or trowel and is available premixed with latex or in powdery form, which should be mixed with latex emulsion. Unless the mortar is made with a latex, it is very permeable and is not a good substrate for paints, especially in thin layers.

9.12.1.2

Exterior Primers

The following primers typically are used on exterior surfaces.

■ WOOD PRIMERS

Exterior wood primers include oil primers and alkyd primers. Most exterior wood primers used today are alkyd.

OIL PRIMERS Oil primers contain oil to control penetration. They generally do not contain zinc oxide because the presence of zinc can cause blistering and peeling, particularly under severe moisture conditions. However, paint systems without zinc oxide are more difficult to render mold resistant.

ALKYD PRIMERS Alkyd primers are similar to oil primers, except that they have an alkyd binder. This promotes faster drying; shorter recoating time; resistance to bleeding, mold, and moisture; and controlled penetration.

■ METAL PRIMERS

Metal primers include oil and oleoresinous primers, alkyd primers, and latex primers. Most metal primers have an alkyd binder.

Some of the primers listed in this subsection contain lead. Lead pigment is toxic and coatings that contain more than 0.06% lead should not be used in residential buildings or on structures with which children may come in contact. Many jurisdictions limit the lead content of paint by law, and some prohibit its use altogether.

RED LEAD (RED LEAD IN OIL) Red lead in oil is probably the oldest type of metal primer still in use today because it is very reliable. The oil binder wets the surface well and seeps into cracks and crevices to ensure protection of the metal even on poorly prepared surfaces. However, drying is comparatively slow, and the film may be too soft for some finish coats. The red lead pigment will carbonate and turn white if left exposed for too long. This primer should be covered with a finish coat within 6 months.

RED LEAD/MIXED PIGMENT Red lead/ mixed pigment primers are either oil- or alkyd-based primers that depend the red lead for their corrosion-inhibitive properties. However, they also contain iron oxide and selected extenders that contribute to the general performance of the coating, adding hiding power, hardness, and general durability. In other respects their properties are similar to those of the red lead in oil primers.

Alkyd-based red lead/mixed pigment primers are similar to their oil-based counterpart but will dry faster and harder, making them suitable for use under harder finish coats. The surface must be cleaned more completely before application than is required for oil-based primers.

ZINC-CHROMATE (ZINC YELLOW) Zinc-chromate primers are made with zinc potassium chromate and often contain iron oxide as part of the pigment. The binder most used today is alkyd. These primers depend on the zinc chromate in them for their anticorrosive effect. They generally dry harder and faster than red lead primers, but require more thorough cleaning of the surface before application. When lead-free, they often are used for priming aluminum and magnesium surfaces.

ZINC-DUST/ZINC-OXIDE Zinc-dust/zinc-oxide primers are available with oil, alkyd, or phenolic resin as the binder. They are especially effective on galvanized steel but may be used on other metallic surfaces as well. The oil type usually is the best for general use, especially if old, rusty, or weathered surfaces are to be coated. The alkyd type is for general use on clean surfaces and also is used as a heat-resistant coating.

Phenolic primers generally are recommended only when the coated surface is to be immersed in water, but they are not recommended for exterior use. When these primers are used, the surface should be given a phosphate treatment before being painted.

LATEX-INHIBITIVE Like most latex products, latex-inhibitive primers are water-soluble before they cure, which makes them easy to apply and easy to clean up.

They are available in white, permitting coverage with one finish coat, usually within hours. While latex-inhibitive primers will give good corrosion protection, they may show rust staining if not protected by a moisture-impervious finish coat.

WASH PRIMERS Because wash primers usually contain phosphoric acid, they clean as well as prime a surface. They are useful in promoting adhesion of subsequent coats of paint and frequently are used to protect freshly sandblasted surfaces from further corrosion. Wash primers are furnished as either one- or two-part materials. One-part wash primers have the advantage of longer pot life after thinning but generally sacrifice some corrosion resistance compared with the two-part type. It is essential that two-part formulations be used the day they are mixed.

PORTLAND CEMENT PAINTS Portland cement paints are either alkyd- or oil-based paints that contain portland cement. They are not to be confused with cement grout or dry-powder portland cement paints, which are designed for use on masonry. Portland cement paints are effective primers for galvanized steel. If white portland cement is used, a substantially white primer results, permitting tinting to shade and contributing to the ease with which the primer can be covered by finish coats.

■ MASONRY PRIMERS

Exterior masonry and concrete wall primers include latex and alkyd primers. The first coat of an epoxy finish often serves as a primer, but conditions where a heavy-duty finish is needed may require a different epoxy formulation for the primer. These primers are used as sealers on normally smooth masonry surfaces. If the surface is weathered, a clear sealer sometimes is recommended to stabilize the surface. If latex finish coats are to be used, the sealer should be thinned so that it will dry to a low gloss.

Silicone water repellents are also used on exterior masonry surfaces when a water-repellent surface is required without change in surface appearance. They consist of a solution of silicone resin in a solvent, usually an alkyd. Since they are transparent solutions, it is difficult to determine visually when an adequate coating has been obtained. To ensure that the desired water repellence is achieved, the spreading rate should be as recommended by the manufacturer. These materials should not be finished with other coatings until they have weathered for several years.

When a smoother finish is desired on masonry and concrete, block fillers or cement grouts are applied beneath the paint primer as indicated under "Masonry and Concrete Wall Primers," above.

9.12.1.3

Interior Topcoat Paints

A wide range of interior topcoat paints are available with various finish surface characteristics.

■ GLOSS FINISHES

Gloss finishes can be achieved with either paint (opaque) or transparent materials.

GLOSS WALL PAINTS The term *gloss wall paint* usually denotes products with an oleoresinous binder. Their former popularity was based largely on easier application than early alkyd gloss enamels. They have been largely replaced by alkyd gloss enamels.

GLOSS ENAMELS Most gloss enamels used today are alkyd based. Modern formulations offer easier application than the older oleoresinous gloss paints, and they also have much better gloss retention and resistance to yellowing and alkaline cleaners.

Some latex paints are also available in gloss formulations. They do not produce as high a sheen as the alkyd enamels, but their ease of application and clean-up and continued advancements in their quality may someday make them more popular than the alkyd gloss enamels.

FLOOR ENAMELS Alkyd and oleoresinous enamels are formulated for relatively fast drying and resistance to abrasion and impact. They are quite impermeable to water but are not notably resistant to alkali, such as may arise from fresh concrete. On wood or masonry, these enamels will blister and peel if moisture accumulates behind the film. Most floor

enamels used today are alkyd based.

ALKALI-RESISTANT ENAMELS Alkali-resistant enamels usually are based on a solvent-thinned styrene-butadiene resin or chlorinated rubber. They have excellent alkali resistance but relatively poor resistance to solvents. They flow and level well but may sag on vertical surfaces. Bubbling sometimes will occur if these enamels are applied by roller on concrete floors.

EPOXY AND URETHANE ENAMELS Epoxy and opaque urethane enamels are available in one- and two-part formulations. The one-part materials are similar to alkyd and oleoresinous enamels but have somewhat improved properties, particularly abrasion resistance. The two-part materials have extremely high adhesion, abrasion resistance, and resistance to water and solvents, but they require great care in surface preparation to develop the properties to the maximum extent.

DRY-FALLOUT SPRAY GLOSS Dry-fallout spray gloss is similar to alkyd gloss enamel but has a much faster-acting solvent and usually a faster curing resin. It is designed for spray application only. An advantage is that much of the overspray dries quickly before reaching a surface and therefore can be removed easily.

■ SEMIGLOSS FINISHES

Semigloss paints are available with a range of actual glosses. Some manufacturers call their semigloss paints *eggshell* or *satin* finishes, and some call different amounts of gloss by different names. Here we will refer to all nonflat and non-high-gloss finishes as semigloss.

SEMIGLOSS WALL PAINT Semigloss wall paint is an oleoresinous enamel with an initial intermediate gloss. It may lose its gloss after a period of time and is likely to have low resistance to grease and alkaline cleaners. It is often too soft for use on horizontal surfaces; its use is decreasing in favor of other types of semigloss enamels and paints.

SEMIGLOSS ENAMEL Semigloss enamel is an alkyd-based product, usually with better gloss retention and grease and alkali resistance than semigloss oleoresinous

wall paint. Application may be somewhat more difficult over large areas. Gloss may be reduced rapidly the first few days after application but will then stabilize.

SEMIGLOSS LATEX PAINT AND ENAMEL Semigloss latex paints and enamels have the usual advantages of latex paints: ease of application and cleanup, rapid drying, little odor, and nonflammability. Film leveling usually is good but hiding power may not compare favorably with alkyd products.

DRY-FALLOUT SPRAY SEMIGLOSS This finish, intended for spray application only, has the properties of its gloss counterpart but at a lower gloss level.

■ FLAT FINISHES

Flat paint finishes include oil and oleoresinous flats, alkyd flat wall paints, latex flat wall paints, and dry-fallout spray flats.

OIL AND OLEORESINOUS FLATS Oil and oleoresinous flat paints are seldom used today. They have a predominantly limed oil vehicle, which reduces penetration and permits easy application and excellent leveling. Finished appearance is excellent, but washability and resistance to yellowing is likely to be poor. Because of good hiding power in thin films, these flats sometimes are used in repainting acoustical surfaces.

ALKYD FLAT WALL PAINTS Alkyd flat wall paints usually are superior to latex paints in hiding power and washability. Odorless flats may have very little odor during application, but an afterodor usually results as the oil portion of the resin oxidizes. Whether odorless or not, these products should be used only under conditions of good ventilation.

LATEX FLAT WALL PAINT Latex flat wall paint is the most popular kind of interior flat wall paint. High hiding power and good washability may be built into the product, but often one is achieved at the expense of the other. Quick drying, ease of touch-up and cleaning of tools, and absence of flammable solvents and unpleasant odors during and after application are additional advantages. Characteristic ease of application may lead to excessive spreading rates and poor hiding ability. Even quality products may fail to form a coherent film and, hence, show poor washability if applied over a porous surface or under conditions of low humidity or temperature extremes.

DRY-FALLOUT SPRAY FLATS Dry-fallout flats are made with fast-drying solvents and may be of either the alkyd or oleoresinous type. They are used on ceiling areas of industrial and commercial buildings, where application is by spray and where it is not practicable to protect all adjacent areas from overspray.

LATEX FLOOR PAINTS Latex floor paints have most of the application advantages of latex wall paints, as well as fairly good abrasion resistance. The absence of toxic or flammable solvents makes them useful in areas such as basements where ventilation may be poor. They may be applied to damp (but not wet) surfaces even if alkalies are present. Continuous exposure to standing water is likely to soften the paint film.

9.12.1.4

Exterior Topcoat Paints

Exterior finish coats include many formulations with oil, alkyd, latex, and other synthetic resins.

A wide range of alkyd- and latex-based house paints are available in flat or glossy sheens. Because of excellent color retention, durability, and ease of cleanup, latex exterior paints are gaining favor over the solvent-thinned alkyd types.

■ OIL-BASED PAINTS

Oil-based paints include *gloss house paint* and *barn paint*.

GLOSS HOUSE PAINTS Traditional house paints were white lead in linseed oil; then various blends with zinc oxide, talc, and other extenders were developed; finally, blends with various titanium dioxide pigments were introduced. These paints have been the foundation of the paint industry. Linseed oil is still the binder used in many quality oil-type house paints, although

modifications with other materials may provide improved properties. A number of long-oil alkyds (resins with 55% or more oil or fatty acid) are available, which are similar to oleoresinous trim enamels but are softer and more durable.

Lead has been found to be harmful to human health and is no longer allowed in residential paints or paints in other locations where children may come into contact with it. Paints containing more than 0.06% lead (nonvolatile content) must carry a warning label.

Zinc oxide often is included to control mold and to aid in the control of chalking; it is also a factor in regulating the hardness of the paint film. Extender pigments are used to control consistency and to develop an optimum blend of other properties.

Titanium dioxide is the most important pigment for providing hiding power in white and tinted paints and contributes to the control of chalking and fading. Some of these factors are of lesser importance with dark-colored paints, which contain little or no white pigment.

BARN PAINTS Barn paints possess most of the characteristics of oil-based house paints. They differ in that they are more likely to contain vehicle modifications, such as rosin, and they usually employ red iron oxide as the chief hiding pigment. Several other colors, such as green and black, also are available. Barn paints are formulated to give uniform appearance over poorly prepared and nonuniform surfaces but lack the long-term protective qualities of house paints. They also may be used on roofs, particularly those of sheet metal.

■ OLEORESINOUS PAINTS

Oleoresinous paints are rapidly disappearing from the market, as alkyd-based and latex paints improve in quality. Oleoresinous enamels have application problems and experience early loss of gloss. They largely have been superseded by alkyd enamels.

Some paints for exterior metal are still made with oleoresinous vehicles, notably those for structural steel. They give good service and protection but usually lose

their gloss earlier than do similar products made with alkyds.

■ ALKYD ENAMELS

Alkyd enamels for exterior use include trim enamels, equipment enamels, and masonry paints.

TRIM ENAMELS Trim enamels are designed for use on exterior wood. They usually are made with long oil resins. For this reason and because they usually are designed for easy brush application, they are not fast drying. Drying usually takes from a few hours to overnight. Typically, they are made in high gloss and bright colors and show good retention of color and gloss. They are not suitable where high resistance to acid, alkali, or other chemicals is required. Silicone alkyds are substantially more durable than are conventional alkyd enamels.

EQUIPMENT ENAMELS Equipment enamels are similar to trim enamels, but because they are intended chiefly for metal, a hard-drying resin may be employed and some rust-inhibitive pigment may be added. If the enamel is designed for spray application, a faster solvent also will be used, and drying may be quite rapid. These products often are called *automotive enamels*.

MASONRY ENAMELS Masonry enamels are similar to alkyd trim enamels except that they contain a higher proportion of pigment. Because very little flexibility is required of a masonry finish, these alkyd-based products perform very well except when excessive moisture and alkali are present.

■ EXTERIOR LATEX PAINTS

Exterior latex paints include latex house and trim paints, semigloss maintenance finish, masonry paint, and special latex paints.

LATEX HOUSE AND TRIM PAINTS Latex house and trim paints are intended mainly for wood, but some of them can also be used over masonry and metal. They are rapidly gaining in popularity over solvent-thinned alkyd paints because of excellent color retention, durability, and ease of cleanup. Primers will nearly always be required over bare wood, badly weathered surfaces, and bare metal. Where primers are needed, it is best to employ one specifically recommended for use with the intended finish-coat material. These products are similar to other latex paints in their general properties, and they share the advantages of ease of application, suitability for damp (not wet) surfaces, and cleaning of equipment. However, they are formuated for good flexibility, so they will accommodate dimensional changes of wood.

Most latex house and trim paints provide a matte finish and offer better color retention than oil and alkyd flats, even in light colors. They are also available in semigloss and gloss finishes.

SEMIGLOSS MAINTENANCE FINISH Semigloss latex maintenance paints applied over suitably primed metal have shown good performance. However, moisture vapor permeability is fairly high, and while good resistance to pitting and rusting is obtained, some rust staining may appear unless an impermeable primer is used.

MASONRY PAINTS Paints for masonry were the first large-scale exterior use of latex systems. Because masonry surfaces show less dimensional change than wood and metal and also are rougher, adhesion is less of a problem. For surfaces with efflorescence, a clear or lightly pigmented solvent-thinned sealer is often recommended. Like latex house paints, these products usually give a low-sheen finish.

SPECIAL LATEX PAINTS Special latex exterior paints may be used on mineral fiber–cement and tile roofs. Special latex paints are available for painting roll roofing. They may be used on asphalt shingles but are likely to curl them.

◾ ALUMINUM PAINTS

Aluminum paints include general-purpose and heat-resistant paints.

GENERAL-PURPOSE General-purpose aluminum paints may be made with aluminum powder or paste and a vehicle of drying oil, alkyd resin, or one of a number of varnishes. Quality may vary widely, depending on the amount of aluminum, the ratio of aluminum to binder, and the quality of the binder. Because of problems of stability, aluminum paints formerly were sold in two-part containers to be mixed just before use, but these problems largely have been solved and satisfactory one-part aluminum finishes now are available.

HEAT-RESISTANT Heat-resistant aluminum paints with the best heat-resistant characteristics generally are made with pure silicone resin. Various materials may replace part of the silicone vehicle, which results in a reduction in the temperature the paint will withstand and a reduction in the paint's exterior durability. Another type intended for high-temperature surfaces uses a small amount of vehicle (usually an oil), which burns off, leaving the aluminum melted and fused to the surface.

◾ ALKALI-RESISTANT PAINTS

Alkali-resistant paints are usually formulated with styrene-butadiene resin or chlorinated rubber. They dry flat and ultimate deterioration of the film is by gradual chalking. These materials often are used on masonry walls or decks where higher resistance to moisture penetration is required than can be obtained with latex paints. They may be used as primers under less-resistant finish coats. While their resistance to water, alkali, and acid is excellent, their resistance to solvents is likely to be poor.

9.12.1.5

Miscellaneous Opaque Coatings

The miscellaneous opaque coatings covered here include fire-retardant paints, fire-retardant intumescent coatings, portland cement powder paint, zone-marking paints, and roof coatings.

◾ FIRE-RETARDANT PAINTS

Fire-retardant paints are made in a number of types, incorporating various properties in addition to fire retardance. They can be formulated as strictly functional or as both functional and decorative. Products are available with good fire retardancy after repeated washing and with all the properties of conventional interior coatings. Since some of these paints are sensitive to water and high humidity, suitability for the intended purpose should be determined.

Fire-retardant capability usually is expressed on the basis of flame-spread ratings determined and listed by Underwriter's Laboratories, Inc. Flame spread, however, is not a property of the paint but depends on the number of coats and their thickness, the surface to which the paint is applied, and other factors. The flame-spread rating of a paint may vary under different conditions. In particular, a paint may be noncombustible and have a low flame-spread on noncombustible surfaces, but it may have little fire retardancy on combustible surfaces. A paint with a low rating on combustible surfaces will always have a low flame spread on noncombustible surfaces. If coatings are not applied at the listed spreading rates, the flame-spread rating will not apply.

◾ FIRE-RETARDANT INTUMESCENT COATINGS

Fire-retardant intumescent coatings slow the progress of a fire by forming a protective foam. In many cases the charred foam may be scraped off and the sound substance repainted.

This material works when heat-sensitive ingredients in it begin to intumesce (swell) at 300°F. The resulting foam, which is hundreds of times thicker than the original paint film, slows the fire, reduces smoke, and delays the buildup of toxic gases.

Intumescent coatings may be used either in strategic places or throughout a building. These coatings are especially recommended for corridors or areas leading to fire exits. They are available pigmented or clear and are scrubbable.

◾ PORTLAND CEMENT POWDER PAINT

Portland cement powder paints are made from white portland cement, suitable pigments, and usually small amounts of a water-repellent agent. They are mixed with water just before application. Since hydration of the portland cement requires moisture, painted surfaces should be kept damp as recommended by the manufacturer. These paints are useful, low-cost finishes for rough masonry. Their principal drawback is that they are not good bases for other paints or coatings.

◼ ZONE-MARKING PAINTS

Zone-marking paints, which are sometimes called *traffic paints*, are characterized by fast drying, little flow, good hiding ability, resistance to bleeding over asphalt, and usually low gloss. Zone-marking paints for highway use may be reflectorized or nonreflectorized. Nonreflectorized paints are used where there is overhead lighting, for example, on many parking lots. Reflectorized paints may be of the drop-in type, in which glass beads or similar material is added after application of the paint, or of the premixed type, in which beads are incorporated into the paint. The drop-in type usually is more durable and has better visibility early in the life of the paint. The premixed type is easier to apply. A small amount of beads may be dropped into the premixed paint after application for early visibility.

Normal zone-marking paints are flat and therefore are not satisfactory for such interior uses as factory floors or school corridors where dirt collection is a problem. Marking paints for those purposes must have a gloss or semigloss finish.

◼ ROOF COATINGS

Bituminous roof coatings are made of special weather-resistant asphalt dissolved in a suitable solvent. They are intended primarily for use on asphalt roofing surfaces. Mineral fiber and other fillers may be added to prevent sagging on sloping roofs and to permit the application of relatively thick coatings. The addition of aluminum powder enables colors other than black to be obtained. For some roof coatings the asphalt is emulsified in water, permitting application to damp surfaces.

9.12.1.6

Transparent Finishes

Transparent finishes are applied on wood. Transparent finishing materials include wood stains, linseed oil, varnish, shellac, lacquer, and polyurethane.

Transparent finishes are mostly used in interior locations. In exterior locations, they do not usually perform as well as paints, because they cannot approach the durability of the same vehicle protected by pigment.

◼ STAINS

Wood stains are made from color pigments suspended in linseed oil or another drying oil. They are then thinned to make them easy to apply. Types of stains include pigmented, dye-type, and shingle stains.

PIGMENTED STAINS Pigmented stains are stains that contain insoluble pigments, generally permanent in color. If applied too heavily, they will obscure the texture, giving a painted effect to the wood. They are particularly suitable for subduing the conspicuous irregular grain effect of softwood plywood. These stains will not bleed into varnish or enamel coats applied over them.

DYE-TYPE STAINS Dye-type stains employ an organic coloring material that is soluble in the vehicle. They have better clarity and transparency than do pigmented stains and produce brighter colors. However, they are more likely to bleed into succeeding coats, and the color is less permanent.

SHINGLE STAINS Shingle stains are pigmented products designed to provide color, only partial hiding ability, maximum penetration, and some mold-resistant protection. They may contain creosote or other preservative; linseed oil and alkyd resin are the usual binders.

◼ LINSEED OIL

Boiled linseed oil is the classic wood finish. It is inexpensive, easy to apply, maintain, and repair. It is normally reduced with turpentine in ratios ranging from equal parts to twice as much oil as turpentine.

◼ VARNISH

Varnish is a homogeneous mixture of resin, drying oil, dryer, and solvent. It can produce a transparent film that is flat, satin, or high gloss. Varnish is available in colors ranging from clear to dark brown and in several qualities. Varnishes for interior use usually include alkyd resins, but epoxy-ester varnishes are also available. In the past, oleoresinous varnishes were popular, but they have been almost completely replaced today by alkyd varnishes, which often are paler and have better color retention than oleoresinous varnishes. Oleoresinous varnishes are tougher and more alkali resistant than alkyd varnishes, but they tend to yellow over time. Epoxy varnishes have good color quality and alkali resistance. See also "Polyurethane," at the end of this subsection.

Alkyd varnishes have good initial color and color retention but may crack and peel over time. Some synthetics have good durability but may darken on exposure. A long-oil phenolic, marine spar varnish has the best history of satisfactory performance, although it also will darken and yellow with time. Penetrating coats of oil or thinned varnish should be used on the bare wood, followed by unthinned varnish. Varnishes that contain ultraviolet-screening agents exhibit less discoloration and generally greater durability.

Flat and satin varnishes have an added flattening agent such as silica. They do not have the undesirable highlights associated with gloss varnishes, but they are less abrasion resistant.

Floor varnishes are relatively fast-drying with good resistance to yellowing and excellent abrasion resistance.

Counter varnishes are hard-drying, print-proof films. *Bar top* varnishes should also be resistant to alcohol. Polyurethanes often are used for this purpose.

Paneling and trim varnishes need not be as hard as varnish for floors or counters, but they should be resistant to checking and should dry tackfree. Alkyds usually exhibit the best color quality and color retention.

Spar varnish is a varnish with additives that give the material a resistance to salt water.

Clear sealers are solvent-thinned varnishes that penetrate wood to prevent grain rising and to seal porous wood. They are applied liberally, and the surplus is removed by being wiped while still wet or buffed when dry (after about 15 hours).

Buffing provides better protection but involves more time and labor. These products give a satin finish with fair abrasion resistance and color retention.

■ SHELLAC

Shellac produces a similar finish to varnish. It is fast-drying and light-colored and has excellent color retention. Although its abrasion resistance is poor, the finish is easily repaired. Resistance to water and alcohol is poor, but resistance to petroleum solvents is good. Shellac is often used as a wash coat on wood before final sanding or as a clear primer or stain sealer.

A major disadvantage of shellac is its short shelf life, being unusable within 4 to 6 months after manufacture.

■ LACQUER

Lacquer is available in formulations that can produce either a flat or a glossy finish, but it is seldom used for field-applied finishing of building surfaces. It is more likely to appear as a finish for furniture or shop-finished casework.

A major disadvantage of lacquers is that they are not usable over previously finished surfaces. In fact, lacquer can be used as a paint remover.

■ POLYURETHANE

Some manufacturers classify polyurethanes as varnish. Others call them lacquer. In fact, though they have some characteristics of each, they are not truly either. They are available either in oil-modified or moisture-curing formulations. Oil-modified polyurethanes are transparent materials that are used mostly on interior surfaces. They are water resistant and highly durable. They can produce either matte or gloss finishes but tend to lose their gloss quickly when used on the exterior. They cannot be used over shellac, paste wood fillers, and some other materials. The manufacturer's advice should be sought about where they can be used.

Moisture-curing polyurethanes can be either clear or pigmented to produce an opaque finish.

Polyurethanes are often used on bar tops and countertops, and floors (see "Varnish," earlier in this subsection).

Urethanes provide better abrasion resistance than do varnishes or lacquers, but they generally require more thorough surface preparation prior to application.

9.12.2

SURFACES

This subsection outlines the broad categories of surfaces for which paints, transparent finishes, and coatings often are considered and describes the products suitable for each surface under varying exposure conditions. Unusual surfaces or those that usually are not painted, such as glass, underground metal, and underwater concrete, are omitted in the following discussion; these surfaces usually require specialized coatings designed for the special conditions of use and generally can be identified by name. The products recommended in the following text are identified by generic names, as identified in Section 9.12.1.

9.12.2.1

Interior Walls and Ceilings

Primers are usually applied over prepared surfaces as a base for subsequent topcoats. The importance of proper selection and application of primers cannot be overemphasized. Regardless of the quality of subsequent finishing, failure of the primer will mean failure of the system.

■ PRIMING

Most surfaces require a primer.

PLASTER New plaster should be primed with a latex, an alkyd, or an oil-type primer-sealer. A latex flat finish coat may be used as a primer, but if semigloss or gloss paints are to be used as the finish coat, the primer should be tested for enamel *holdout* (resistance to penetration by the finish coat), which results in uneven gloss. Highly alkaline or fresh, damp plaster requires an alkali-resistant primer or latex system. Plaster that is wet or has a continuing flow of moisture from the back cannot be painted successfully. Textured or swirl plaster and soft, porous, or powdery plaster should be given an acid wash with a solution of 1 pint of white household vinegar in 1 gallon of water. This procedure should be repeated until the surface is hard; it should then be rinsed well and allowed to dry before painting.

GYPSUM BOARD Latex primer-sealers are usually recommended for gypsum board or other paper-surfaced materials. Latex emulsions are preferred, since solvent-thinned primers raise the fibers of the paper, giving the material an unsightly appearance. A latex finish coat can usually be used as a primer, but if semigloss or gloss paint is to be used as a finish coat, the primer should be tested for enamel holdout.

BRICK Brick interior walls are usually somewhat porous, but if the mortar is well-cured, they will not be highly alkaline. They may be sealed with a latex primer-sealer, or an enamel undercoater may be used with the addition of suitable oil or varnish. Latex or solvent-thinned exterior masonry paints may also be used for priming. Any efflorescence should be removed by vigorous scrubbing, and the brick should be treated with a clear resin sealer.

CONCRETE MASONRY Most concrete masonry, particularly the lightweight types, have a very rough and porous surface. Because most paints and coatings cannot bridge the surface voids, one or two coats of block filler are recommended. By sealing the voids, block sealers reduce dirt accumulation and improve surface color uniformity. Fillers may be omitted when these considerations are not important, when low-cost finishing may be required, or when the sound-absorbing effect of the rough surface may be desirable.

CONCRETE The problems associated with painting concrete are similar in many respects to those of concrete masonry. However, the concrete surface usually is denser, less regularly formed, and more likely to be contaminated with oils and other bond-breaking materials. Substances should be removed unless it can be ensured that they will not affect paint adhesion. Rough or honeycombed surfaces often are treated with cement grout, preferably thinned with a latex solution, particularly if the grout is to be applied in thin layers. Priming over grout should be as recommended for plaster surfaces. One or two coats of block filler may be used instead of grout. In that case, an alkali-resistant primer usually is not required.

FERROUS METALS   A good rust-inhibitive primer always is desirable. It is essential for quality work, for damp locations, when water-thinned paints are used for finishing, and when the finish coats do not produce an impervious film. A primer may be omitted in dry locations and if the finish coats are to be of a solvent-thinned type producing a tight film.

Slower-drying oil-based primers will wet the surface better and give more positive results, but quick-drying primers will be more resistant to moisture penetration.

ZINC-COATED (GALVANIZED) METALS   If applied over zinc-coated surfaces, such as galvanized steel, many ordinary paints and rust-inhibitive primers will develop a layer of white, powdery material that eventually separates the paint film from the metal completely. This condition can be avoided by using (1) a nonoxidizing coating such as a chemical-resistant coating or, more commonly, (2) a primer that contains zinc dust or portland cement. Certain proprietary primers for galvanized steel (often latex) are very effective for interior use.

WOOD   In many interior situations, wood may be primed at the same time and with the same sealer used for plaster walls. One or two coats of paint may give a sufficiently smooth base for enamels, but such water-thinned paints will raise the grain and roughen the surface of many woods. Good enamel holdout is necessary under gloss or semigloss finish coats. Solvent-thinned wall primer-sealers provide better adhesion and flexibility. For the best enamel work, an enamel undercoater should be used either over a wall primer-sealer or directly on the wood. A clear wood sealer is recommended for natural finishes, preceded by a paste wood filler, if necessary.

ACOUSTICAL SURFACES   Acoustical tile and acoustical plaster are rarely painted at the time of their installation, but they may require finishing later. Since acoustical properties depend on a suitably porous or rough surface, a thin paint should be used so the pores are not bridged or filled. Any of the products recommended for plaster surfaces may be used after being thinned to suitable consistency. Metal acoustical surfaces should be treated as light-duty areas, with one coat used to avoid excessive paint buildup.

■ FINISHING

Finishing differs slightly, depending on the relative use of the location.

LIGHT-DUTY AREAS   Some building areas have fairly low concentrations of moisture, grease, or dirt in the air and require relatively little maintenance. These areas include small private offices and other similar spaces in commercial and institutional buildings and living rooms, adult bedrooms, and similar areas in residential-use facilities. In such spaces a matte finish is both desirable and practical on the ceilings and most walls.

Latex flat wall paints are usually preferred because of their easy application, quick drying, easy cleanup, and almost complete lack of odor. Alkyd flats can produce thicker films, have better hiding power, and give a more uniform appearance. These properties are particularly desirable on rough surfaces and when a considerable change is made in color. Both latex and alkyd paints have good resistance to soiling and marring and are relatively easy to clean.

Odorless alkyd flats are almost completely free of solvent odor during application. However, they must be used with adequate ventilation because the solvent fumes, while odor-free, may cause headaches and other physiological symptoms. Also, although these solvents may have a high flash point, there is a slight fire hazard. In occupied premises these air-drying paints have the further disadvantage that the curing process results in odors that can persist for several days, requiring continued ventilation.

MODERATE-DUTY AREAS   Moderate-duty areas are similar to light-duty areas in most respects except that the amount of traffic or airborne dirt may dictate a more intensive cleaning schedule. Typical areas are private corridors and larger office spaces in commercial and institutional buildings and children's rooms, entries, halls, and closets in residential occupancy. Finish coats should be chosen from the more washable flat and semigloss paints available. Glossier materials generally have better wear resistance and washability, but accompanying higher surface shine makes surface irregularities more conspicuous. This is particularly true when natural or artificial light strikes the surface at a sharp angle. This consideration becomes less important in smaller spaces or where the visible wall areas are broken up by interior furnishings.

HEAVY-DUTY AREAS   Heavy-duty areas are those in which there may be large amounts of moisture and grease in the air or where the amount and the type of traffic require a more durable paint film to withstand greater wear or more intensive cleaning. These areas include kitchens, recreation rooms, toilets, public corridors, entrance lobbies, and service areas in institutional, commercial, and apartment buildings, and bathrooms in homes and apartments.

The usual choice is a gloss or semigloss enamel. Where the exposure is particularly severe, specialized hard-gloss coatings can be used. The light reflectivity factors discussed above for moderate-duty areas always should be considered in the selection of gloss finishes.

9.12.2.2

Interior Trim and Paneling

The selection of primers for interior use on trim and paneling depends primarily on the type of surface material being coated and the desired finish characteristics.

■ PRIMING

FERROUS METALS   Although most interior exposures are not particularly corrosive, a rust-inhibitive primer usually is desirable. Any of the metal primers, such as oil, alkyd, or latex, may be used. The choice is dictated by the speed of drying, the amount of surface preparation, and the desired color. For very mild exposure, conventional enamel undercoaters may be used, but these products have no corrosion-resistant effect, and rust will spread from any break in the film.

ZINC-COATED METALS Interior galvanized surfaces that are dry and do not receive heavy wear may be coated with interior latex paints. However, if moisture, abrasion, and repeated cleaning are expected, an inhibitive primer generally will provide better service and is recommended.

WOOD, CLEAR FINISH Priming will depend on both the wood and the type of finish desired. Trim should be primed on all surfaces before installation. Most softwoods will require a clear sealer, especially those showing marked contrast between springwood and summerwood, such as pine. The sealer will provide a more uniform surface over hard and soft grain and thus is particularly recommended if the surface is to be stained.

Dense, close-grained hardwoods such as maple and birch have less need for a sealer, and the first coat may be a thinned version of the finish coat, if recommended by the manufacturer. In stain systems, the products used and the order in which they are applied will depend on the effect desired. For example, applying the sealer before the stain will minimize both the depth of color and the amount of grain contrast.

Open-grained hardwoods such as oak, walnut, and mahogany require the use of a wood filler, which also will serve as a sealer. When the grain is to be subdued, stain may be applied after the filler.

WOOD, OPAQUE FINISH Any of the several interior wall primers or an enamel undercoater may be used as a primer. However, water-thinned primers are apt to raise the grain of the wood. An enamel undercoat will give a smoother surface and is preferred under gloss and semigloss enamels, although many semigloss enamels are designed to be selfpriming.

■ FINISHING

The selection of finish coats is determined largely by the type of exposure expected, that is, light duty, moderate duty, or heavy duty.

LIGHT- AND MODERATE-DUTY AREAS Clear finishes for wood should be selected depending on decorative and exposure requirements. For light-duty areas, satin or low-luster finishes often are used.

Opaque finishes for wood and metal usually are semigloss enamels or latex enamels. In light-duty areas the predominant interior finish used on the adjacent walls may be used.

HEAVY-DUTY AREAS Clear finish for wood may be floor varnish or counter varnish. Floor varnish is most resistant to abrasion, while counter varnish has good resistance to water and solvents. The selection should be made on the basis of the exposure anticipated.

Opaque finishes for wood and metal are usually hard-drying enamels. For special exposures, such as in laundries, a special type may be indicated.

9.12.2.3

Interior Floors

The selection of finishes for interior floors depends primarily on the type of surface to be coated.

■ WOOD FLOORS

CLEAR FINISHES For relatively light-duty floors, shellac may be used where color quality and color retention are important. It has good gloss and color but scratches easily and is not water resistant; it is, however, easily patched. Floor varnish is more resistant to wear and water but is somewhat darker in color. Wood sealer has a lower gloss and usually requires waxing. For heavy-duty areas such as gymnasium floors, specially designed finishes usually are required. Polyurethane may also be used on heavy-duty floors. Open-grained woods such as oak may use a filler before coating. Some floor varnishes may be slippery when wet.

OPAQUE FINISHES Alkyd enamels and latex floor paints generally are used for opaque floor finishes. Other fast-drying gloss enamels and paints may be used if recommended by the manufacturer. Products not specifically recommended for floors may be too soft and slow drying or too hard and brittle.

■ CONCRETE FLOORS

Enamels suitable for wood floors generally are suitable also for concrete floors. However, alkali-resistant enamels should be used when a concrete slab is not completely cured, is below grade, or is on the ground without a vapor retarder. Even with alkali-resistant enamels, adhesion problems may be encountered if hydrostatic pressure forces water up through the floor.

Latex floor paints usually dry to a comparatively low gloss. Because they are nonflammable and have little odor, latex paints are preferred for basements and enclosed areas, which are difficult to ventilate adequately. Since they are water emulsions and have excellent alkali resistance, the floor need not be completely dry before application. Because latex paints are softened by continued exposure to moisture or solvents they are not suitable for laundry and service areas where solvents are handled. In such exposures, specialized highly resistant finishes should be used.

■ STEEL FLOORS

A fast-drying, tough enamel applied over a corrosion-resistant primer is recommended for most steel floors. Floors subject to chemical agents, as in industrial installations, should be finished with specialized industrial coatings.

■ CERAMIC FLOORS

Ceramic floors are seldom painted but they may be coated with an abrasion-resistant clear coating to seal the surface and facilitate cleaning.

■ RESILIENT FLOORING

Resilient flooring rarely is coated when

new, but coating may be required to restore appearance or change color. The coating chosen will depend on the flooring material, its condition, the adhesive used in the installation, and the effect desired. Specific recommendations should be obtained from the paint manufacturer for each case.

9.12.2.4

Exterior Priming

As with the finishing of interior surfaces, the selection and application of primers are critical to the performance of a coating system.

■ WOOD

SIDING Smooth siding should be primed with a wood primer that is compatible with the intended finish coat. The manufacturer's recommendations for the finish coat should govern the choice of a primer. Nail heads, cracks, and deep surface imperfections should be puttied; caulking should be provided at perimeters of doors and windows and wherever required for weather resistance. All puttying and caulking should be done after the siding has been primed.

For rough siding, shakes, and shingles, any wood primer may be used according to finish coat recommendations; or the finish coat itself may be thinned and used as a primer if it is zinc-free and is recommended by the manufacturer. Flat house paint and shingle stain do not require special primers. Factory-primed siding does not require a field-applied primer unless the surface has been heavily damaged during shipping or application. Minor scratches, dents, and raw edges should be spot-primed, as recommended above.

PLYWOOD After prolonged exterior exposure, most plywood surfaces will check and the coating will crack unless an overlaid plywood is used. Medium-density overlaid plywood with a resin-impregnated paper covering is highly recommended for an opaque paint finish. Painting recommendations for wood siding also are applicable to plywood. Textured plywood, such as grooved, rough-sawn, and striated surfaces, is not intended for painting, and only pigmented stains should be used on it.

TRANSPARENT FINISHES Wood surfaces intended for a transparent finish usually are primed with a coat of the same material to be used as the finish coat. To improve penetration, some manufacturers recommend that the finish coat be thinned when used for priming. However, some products may be damaged by thinning; therefore, thinning should be done only when specifically recommended by the manufacturer.

■ MASONRY

The cost of painting a clay masonry wall in most cases must be charged to appearance alone, not to increased performance or resistance to weather. The only possible exception would be cement-water paint applied to correct a leaky wall condition. In the case of concrete masonry, paint not only provides an attractive finish but may also be necessary to make the masonry surface watertight.

Masonry paints are of four general classifications: (1) cement-based paints, which come in powdered form and are mixed with water before applying; (2) water-thinned emulsion paints, which include butadiene-styrene, vinyl acrylic, alkyd, and multicolored lacquers; (3) fill coats, which are similar to cement-based paints but contain an emulsion paint in place of some water, giving improved adhesion and tougher film than unmodified cement paints; and (4) solvent-thinned paints, which, except for special-purpose paints and special applications, should be used only on interior masonry walls that are not susceptible to moisture penetration. Solvent-thinned paints include oil-based, alkyd, synthetic rubber, chlorinated rubber, and epoxy.

These surfaces generally do not require a special primer, and the first coat can be the same material as that used for the finish coat. The finish coat material may require thinning, but the manufacturer's recommendations should be followed. Common brick should be sealed with a penetrating type of exterior varnish to minimize spalling and efflorescence, which may cause peeling.

If the masonry surface has many large voids, a cement grout preferably thinned with latex may be applied. Sound but rough surfaces, especially in concrete block construction, may be surfaced with block filler. Cast-in-place concrete should be cleaned of form oils, and other bond-breaking agents should be removed unless they are guaranteed not to affect paint adhesion. Old surfaces that have become excessively chalky may be bonded or stabilized with a clear or lightly pigmented phenolic or alkyd coating; if the existing surface is soft or crumbly, it should be removed to a sound layer, usually by sandblasting. Where only water repellence is required, a silicone solution may be used above grade; hot melt or cutback asphalt can be used below grade.

■ METAL

FERROUS METAL Proper surface preparation of iron and steel is extremely important—coating materials cannot compensate for inadequate surface preparation. Rust, mill scale, dirt, oils, and old, loose paint must be removed. The standards for surface preparation recommended by the Steel Structures Painting Council should be followed.

An appropriate primer should be used, depending on conditions and the finish materials intended. Oil and oleoresinous materials generally have better bonding properties and are more suitable for hand-cleaned surfaces. Alkyd and other synthetic primers dry more rapidly but require more careful surface preparation. Chemical-resistant coatings and industrial coatings usually are marketed as complete systems of primer and finish coat, and the manufacturer's recommendations regarding the appropriate primer should be followed.

GALVANIZED STEEL Zinc-dust/zinc-oxide, latex, and portland cement paints give excellent results, as do certain proprietary paints designed for use on galvanized steel. Conventional exterior paints usually lose adhesion rapidly unless the surface has weathered a considerable length of time. If the galvanized surface has weathered to the point that substantial amounts of rust are showing, the recommendations for painting ferrous metals should be followed. Certain chemical pretreatments, usually based on phosphoric acid, improve the adhesion of paints to galvanized metal. However, other acids or chemical solutions sometimes recommended for the preparation of galvanized surfaces rarely contribute to the performance of paints; actually they may be harmful, and their use generally should be avoided.

ALUMINUM AND MAGNESIUM Zinc chromate primers usually are recommended for aluminum and magnesium.

COPPER, BRASS, AND BRONZE Copper and copper-bearing metals oxidize rapidly when exposed to air and moisture. To preserve the bright metallic luster of new polished copper and brass, a clear coating should be applied promptly. Old copper and brass should be buffed or polished to a bright color, cleaned with a mild phosphoric acid cleaner and coated before the surface becomes discolored by oxidation. Specially formulated clear coatings are available that will preserve the bright luster and protect the metal from discoloration by the elements.

Copper gutters and downspouts generally are not painted or coated. The rapid-forming oxide (patina) darkens the surface but protects the underlying metal from progressive deterioration. However, copper may be cleaned and coated either to match the color or to prevent the staining of adjacent surfaces. Coating is particularly recommended for copper surfaces located above aluminum products to prevent electrolytic deterioration of the aluminum by water wash from above.

9.12.2.5

Exterior Finishing

The choice of finish may be influenced by the type of exposure expected and the performance desired. High temperatures and humidity, particularly in the absence of direct sunlight, encourage the growth of mold and mildew. Therefore, it is important to choose fairly hard-drying finish coats, avoiding types that may dry to a rough, soft surface, since those finishes tend to trap and hold mold spores. Finish coats containing substantial amounts of zinc oxide are effective in controlling fungus growth. Various other additives are used to provide further mold resistance.

If a high degree of mold resistance is required, the primer as well as the finish coat should be designed for that purpose. However, the use of coatings that contain zinc oxide is not recommended on bare wood. The degree of mold resistance is determined by the total formulation rather than by a single component, and the specific product should have adequate resistance under conditions similar to those to be encountered.

Industrial fumes may darken painted surfaces by the reaction of hydrogen sulfide with lead or occasionally mercury compounds. Paints used in areas subject to fumes containing hydrogen sulfide should be free from lead and mercury.

Soot and other airborne materials may settle on and discolor a paint film. Hard-drying films will not retain dirt as tenaciously as softer films. Paints may be formulated to encourage moderate chalking of the surface, which has a self-cleaning effect as the soot and dirt are carried away with the surface layer.

For severe weather exposures, resistance to chalking is a desirable property. Repeated exposure to rain and high winds generally will keep the surface clean but will wear away the film too quickly unless a comparatively nonchalking paint is chosen.

■ WOOD SIDING

SMOOTH SIDING Oil paints and exterior latex paints are preferred for whites and light tints on smooth siding. Oil paints have better hiding ability and will show less staining from wood resins. Latex paints are easier to apply, may be used on slightly damp surfaces, and often have better color retention, particularly in light tints. Where factory-primed siding is used, the recommendations of the siding manufacturer should be followed. Transparent finishes may be used where the natural appearance of the wood is desired, but their durability is much lower than that of pigmented products.

ROUGH SIDING, SHINGLES, AND SHAKES Flat house paint, shingle stains, or exterior latex paints are usually used on rough siding, shingles, and shakes.

Latex paints generally give good service if rust-resistant nails have been used. Wood discoloration can be avoided by sealing with a solvent primer or by the use of a latex primer specially designed for resistance to staining. Rough surfaces have substantially greater surface area than smooth surfaces, and more paint must be applied to obtain suitable film thickness.

Shingle stains are often the choice for wood shingles. These stains are formulations that penetrate the wood but crack rather easily. Impermeable coatings usually are not desirable because they seal in moisture and are likely to peel. Staining shingles before installation usually is preferred because many areas are inaccessible after installation.

■ WOOD TRIM

Alkyd trim enamels should be used where high gloss and good abrasion resistance are desired, for example, around doors and windows. Any of the finish coats suggested for smooth siding also may be used.

■ METAL

Oleoresinous and alkyd paints are suitable finishes over compatible primers. Alkyd paints tend to be harder and more abrasion resistant. Any of the paints recommended for wood, such as alkyd enamels, can be used if the surface is properly primed.

■ CONCRETE AND CONCRETE MASONRY

An alkali-resistant coating usually is required over concrete and concrete masonry because, even after some aging, they still contain alkali, which will attack oils and oil-based materials. Latex paints or resin-based finishes also may be used. Resin-based finishes are preferred on flat areas, such as ledges and windowsills, and areas subject to foot traffic and standing water, such as patios and roofs. Other types of paint may be used on well-cured concrete, on concrete that is sealed with an impervious primer, and on previously painted surfaces.

■ BRICK AND STONE MASONRY

Because brick and stone usually do not contain alkali, almost any exterior paint may be used on them with satisfactory results, if the mortar is well-cured. However, since fresh mortar is usually strongly alkaline, only paints recommended for concrete should be used on new walls. Where the natural appearance of brick or stone is desired, a silicone-resin solution may be used to retard the penetration of water. The solution and spreading rate should be as recommended by the manufacturer. Silicone finishes should not be painted until they have weathered for at least 2 years.

■ MINERAL FIBER CEMENT

Mineral fiber–cement compositions, such as shingles, siding, corrugated boards, and insulating sandwich panels, often are supplied with a transparent water-repellent coating, which should be weathered or removed before painting. For most products, trapped moisture seldom is a problem, but alkalinity may be too high for oleoresinous or alkyd types. However, the insulating core of sandwich panels may absorb moisture, releasing it later to damage the coating. It is recommended that the edges of insulating panels be sealed after cutting and before installation. Exterior latex or solvent-thinned resin paints generally are preferred.

■ ASPHALT ROOFING

Asphalt shingles, roll roofing, and built-up roofing rarely are coated when new, unless for decorative or heat-reflecting purposes. Most solvent-thinned paints are too brittle to be used over asphalt. The solvent will dissolve the asphalt, and the paint will become stained.

Bituminous roof coatings are most commonly used, and if heat reflection is required, these coatings may be obtained pigmented with aluminum or in a limited range of colors. Where white is required, certain latex paints are sufficiently flexible to give a satisfactory performance. Latex paints are likely to curl shingles, however, and are not recommended for these products. Because severe exposure conditions generally are encountered on roofs, thick coatings generally are required to give satisfactory durability.

■ METAL ROOFING

In general, paints recommended for metal roofs follow the recommendations for other exterior metal. However, because foot traffic rarely is a consideration, softer coatings such as house paints or barn paints may be used for finish coats. As on other roofs, a relatively heavy coating is desirable for resistance to weathering. Terneplate roofs usually are painted with a special iron oxide primer and then finished as suggested above.

9.12.3

APPLICATION

Satisfactory painting and transparent finishing depend on (1) proper selection of paint and transparent finishing materials, (2) adequate preparation of materials and surfaces, and (3) suitable application methods and procedures.

Paints, transparent finishes, and coatings should be standard commercial brands with a history of satisfactory use under conditions equal to or similar to the conditions in the area concerned.

The paint, transparent finish, or coating selected should be designed and recommended by the manufacturer for the specific use proposed. The printed application instructions should clearly identify the suitability of the material for the type of exposure (exterior or interior), the type of surface to be covered (wood, metal, masonry, concrete, gypsum board, plaster), and the type of service to which the paint will be subjected, such as exposure to moisture, frequent washing, or heavy traffic.

9.12.3.1

Materials Handling

Proper storage conditions and stocking procedures are essential to avoid material damage, explosion hazards, and application problems.

■ STORAGE

Paint materials should be stored in dry and well-ventilated areas maintained at a temperature range between 50° and 90°F. Lower temperatures generally increase material viscosity and make it necessary to temperature-condition materials before use. Freezing temperatures may permanently damage water-based paints.

If a paint material is sensitive to heat, temperatures higher than 100°F may bring about a chemical reaction within the container. This generally results in increased viscosity to the point of gelation, and the paint becomes unusable. Also, overheating may cause covers to blow off, creating a serious fire hazard.

Other hazards to be considered are high humidity, which may cause containers to corrode and labels to deteriorate, and poor ventilation, which may allow excessive concentration of solvent vapors that are both toxic and combustible.

■ STOCKING

Paint materials should be rotated so that oldest stocks are used first. It is desirable to make paints ready for use by conditioning them at the proper temperature and mixing them thoroughly. Using leftover paint should be avoided. When using old paint is necessary, such material should be strained and temperature-conditioned again, as recommended below for unopened containers.

Paint material should be stored in full, tightly sealed containers. Required quantities should be carefully estimated so little or no material is left at the end of a job. It is safer generally to discard small quantities than to use paint that has skinned. When saved, leftover paint should be placed in smaller containers, filled full, and tightly sealed. All information from the original container should be transferred to the new container.

Conditioning and Mixing

Unless stored at the recommended temperature, paint materials should be temperature-conditioned at 65° to 85°F for at least 24 hours before use. After this time, paints should be mixed, thinned, or tinted as required and strained if necessary, all according to the manufacturer's recommendations.

Paints consist of two principal components, solid pigment and liquid vehicle. Suspension agents are included in most paints to minimize settling of solids. However, when stored for long periods, separation of the pigment and the vehicle may take place. The purpose of mixing is to reblend settled pigment with the vehicle and to eliminate lumps, skins, and other detriments to proper application.

■ MIXING

Paint materials always should be mixed prior to issuance or delivery to the job. Mixing can be done either by manual or mechanical methods, but the latter is preferred to ensure maximum uniformity. Some hand stirring should be done just prior to use and periodically during application. Regardless of the procedure used, caution is recommended to avoid the incorporation of an excess of air through overmixing. Figure 9.12-1 outlines the recommended mixing procedures for different coating types.

It is important to mix ready-mixed paints before as well as after introducing thinners or other additives. It is best to use the same conditioning procedures for all components of multicomponent paint materials before mixing. Manufacturers' label directions regarding proper mixing should be followed.

MECHANICAL MIXERS The two most common mechanical mixers either vibrate the full sealed container or utilize propellers that are inserted into the paint. Vibrating shakers are used for full containers of capacities up to 5 gallons. Propeller mixers are simple hand drill attachments and generally are used on the job for containers holding 1 quart or more.

FIGURE 9.12-1
Mixing Procedures for Various Paint Types

Coating Type	Mixing Equipment	Remarks
Enamel, semigloss, or flat oil paints	Manual, propeller, or shaker	Mix until homogeneous.
Water-based latex paints	Manual, propeller, or shaker	Use extreme care to avoid air entrapment.
Clear finishes, such as varnishes and lacquers	Manual	Generally require little or no mixing.
Extremely viscous finishes, such as coal tar paints	Propeller	Use extreme care to avoid air entrapment.
Two-package metallic points, such as aluminium paint*	Propeller	Add small amount of liquid to paste and mix well; slowly add remainder of vehicle, while stirring, until coating is homegeneous.
Two-component systems	Manual, propeller, or shaker	Mix until homogeneous; check label for special instructions.

*With metallic powder, first make into a paste with solvent, then proceed as recommended above.

MANUAL MIXING Manual mixing is more time-consuming than mechanical mixing but can be used when mechanical equipment is not available and for mixing materials that have been recently well mixed. Paint should be stirred from the bottom up, to dislodge any settled matter on the bottom or lower sides of the container. Properly mixed paint should have a completely blended appearance with no evidence of varicolored swirls or lumps at the top, indicating unmixed pigment solids or foreign matter.

When two or more cans of the same color are to be used in the same area, the paint may be *boxed*, that is, poured back and forth from one container to the other until uniform. This procedure minimizes the visual difference between paint in different containers.

■ TINTING

Tinting should be done by experienced personnel in a paint shop or at a paint store. On-the-job tinting should be avoided as a general practice, to minimize errors and to eliminate color-matching problems at a later date. One exception is the practice of tinting intermediate coats for iden-

tification and to ensure complete coverage under the finish coat. Tinting should be done with care, using only colors known to be compatible and preferably adding not more than 4 ounces per gallon of paint, unless otherwise approved by the manufacturer. Use of more than 8 ounces per gallon should be avoided unless specifically approved by the paint manufacturer. Tinting of chalking-type exterior paints is not recommended except for identification of intermediate coats.

Many paint manufacturers employ tinting systems for the addition of colorants to their base paints. The colorants are formulated to precise standards of tinting strength and are packaged in tubes or in bulk for use with a colorant-dispensing machine at the paint shop. Each color in the system has an identifying number and a mixing formula that gives the proportions of base paint and colorants to be combined. Thus, any color can be reproduced accurately at a later time without guesswork.

TINTING COLORS There are two types of tinting colors in general use: colors-in-oil and universal tinting colors.

Colors-in-oil are limited to use with standard oil-based paints, alkyd resins, chlorinated rubbers, and other solvent systems. They should not be used with the other synthetics or with water-thinned paints.

Universal tinting colors are used in the same manner as colors-in-oil. They are more compatible with a wide variety of paint materials. Many can be used with both solvent-thinned and water-thinned paints.

TINTING PROCEDURE   Tinting should be done by experienced personnel, in a paint store. The paint should be at application consistency before tinting. Ingredients should be compatible and fluid, to blend readily. Mechanical agitation is important to ensure uniform color distribution, but overmixing should be avoided. To achieve a paint color not included in a manufacturer's tinting system, a paint sample should be allowed to dry before it is compared with the color chip used for reference. When the desired color has been achieved, it is important to maintain a written record of the tinting formula: the container should be identified by name or number and a spot of paint applied to the cover for visual reference.

■ STRAINING

Paint from newly opened containers normally does not require straining. However, a paint that shows evidence of skins, lumps, color flecks, or foreign materials should be strained through a fine sieve or commercial paint strainer. Paint that is to be applied by spray gun should be strained to prevent clogging of the spray nozzle.

■ THINNING

Paints usually are formulated and packaged at the proper viscosity for application by brush or roller without thinning. However, thinning frequently is required and recommended by manufacturers for spray application. The arbitrary addition of thinners should be avoided; unnecessary or excessive thinning may result in an inadequate film thickness, and may reduce the durability of the applied coating.

Thinning should be performed by competent personnel using only compatible thinning agents recommended in the label's instructions or the manufacturer's specifications. Thinning should not be used as a means of improving brushing or rolling properties of cold paint materials; instead, materials should be conditioned to 65 °F to 85 °F before use.

9.12.3.3

Preparation of Surfaces

Proper surface preparation prior to painting is essential to achieve maximum coating life. The best paint will not perform effectively if applied on a poorly prepared surface. The initial cost of adequate surface preparation is rewarded by longer coating life and improved appearance. The selection of surface preparation method depends on (1) the nature and the condition of the surface to be painted, (2) the type of exposure, (3) the surface texture and appearance desired, and (4) material and labor costs.

Surfaces to be painted or to receive a transparent finish or coating often contain contaminants such as dirt, grease, rust, and moisture, which reduce adhesion and cause blistering, peeling, flaking, or cracking of the finished surface. Surface defects that also adversely affect adhesion include metal burrs and irregular weld areas, wood knots and torn grain, powdery concrete and masonry, and old paints in various stages of deterioration. In general, sharp edges, irregular areas, cracks, and holes create conditions conducive to early finish failure.

Because iron and steel corrode readily unless properly prepared and painted, special attention is given in the following discussion to suitable preparation methods for these and other metals. Painting of ferrous metals should be started as soon as possible after the surface has been prepared. The suitability of treatments for various material surfaces is given in Figure 9.12-2.

FIGURE 9.12-2
Preparation of Various Surfaces

Preparation Method	Wood	Metals Steel	Metals Other	Concrete and Masonry
MECHANICAL				
Hand tool cleaning[a]	OK	OK	OK	OK
Power tool cleaning	OK[b]	OK	—	OK
Flame cleaning	—	OK	—	—
Brush-off blast cleaning	—	OK	OK	OK
Other blast cleaning	—	OK	—	—
CHEMICAL AND SOLVENT				
Solvent cleaning	OK	OK	OK	—
Alkali cleaning	—	OK	—	OK
Steam cleaning	—	OK	—	OK
Acid cleaning	—	OK	—	OK
Pickling	—	OK	—	—
PRETREATMENTS				
Hot phosphate	—	OK	—	—
Cold phosphate	—	OK	—	—
Wash primers	—	OK	OK	—
PRIMERS	—	—	—	OK
SEALERS	OK	—	—	—
FILLERS	OK	—	—	OK

[a] Also suitable for gypsum wallboard and plaster.
[b] Sanding only recommended.

■ MECHANICAL TREATMENTS IN THE FIELD

Mechanical treatments in the field are intended to remove corrosion, mill scale, and old paint. Solvent cleaning to remove surface oil or grease is recommended prior to other treatments.

Metals can be prepared for painting by cleaning with hand tools, power tools, flame, and abrasive blast. The use of goggles and gloves is essential for blast and flame cleaning and is recommended for the other methods as well. Good ventilation is usually desirable. Dust respirators may be required.

MANUAL CLEANING Cleaning with hand tools will remove only superficial contaminants, such as rust, mill scale, and loosely adhering paint. It is primarily recommended for spot cleaning in exposures where corrosion is not a serious factor. Since manual cleaning removes only surface contamination, it is essential to use primers that will thoroughly wet the surface and penetrate to the base metal.

Before cleaning is started, the surface should be treated with solvent cleaners to remove oil, grease, dirt, and chemicals. Rust scale and heavy build-up of old coatings should then be removed with impact tools such as chipping hammers, chisels, and scalers. Loose mill scale and nonadhering paint can be removed with wire brushes and scrapers. Caution should be exercised to avoid deep marking or scratching the surface with the tools.

POWER TOOL CLEANING Cleaning with power tools provides faster and more adequate surface preparation than the hand tool method but is less economical than blasting for effective large area cleaning. Power tools can be used to remove small amounts of tightly adhering contaminants that are difficult to dislodge with hand tools.

Power tools are driven either electrically or pneumatically and include a variety of attachments for the basic units. Chipping hammers are used for removing tight rust, mill scale, and heavy paint coats. Rotary and needle scalers are handy for removing rust, mill scale, and old paint from large metallic and masonry areas. Wire brushes can be used for removing loose mill scale, old paint, weld flux, slag, and dirt deposits. Grinders and sanders are useful in removing old paint, rust, or mill scale on small surfaces and for smoothing rough surfaces.

As with hand tools, care is recommended to avoid cutting too deeply into the surface, since doing so may result in burrs that are difficult to protect satisfactorily. When wire brushes are used, care should be taken to avoid polishing the metal surface, which may prevent adequate coating adhesion. Power tool cleaning should be preceded by solvent or chemical treatment; painting should be started and completed as soon after cleaning as possible.

FLAME CLEANING Flame cleaning is satisfactory for both new and maintenance work. Oil and grease should be removed prior to flame cleaning both for safety and for adequacy of preparation. Wire brushing normally follows flame cleaning to remove loose matter. A high-velocity oxyacetylene flame is used in this treatment. Extreme caution and adequate ventilation are necessary to prevent fires.

New paint is generally applied while the surface is still warm, thereby speeding up drying time and making it possible to paint when air temperatures are somewhat below 50 °F. However, when solvent-thinned paints are used, the painting operation should be sufficiently separated from the cleaning operation to avoid a fire hazard.

BLAST CLEANING Blast cleaning involves the high-velocity impact of abrasive particles such as sand or glass beads on a surface (Fig. 9.12-3). The abrasive is discharged, either wet or dry, under pressure. In the wet system water, and sometimes a rust inhibitor, is either mixed with the abrasive in the pressure tank or is introduced into the blast stream near the nozzle. Protective clothing, dust masks, and eye protection are needed during blast cleaning to prevent injury.

Metal surfaces that have been blast-cleaned should be prime-coated the same day, because rust forms rapidly on newly cleaned surfaces. Metal or synthetic shot, grit, glass beads, corn cobs, or similar abrasives are used where recovery of the abrasive is possible, otherwise sand is generally used. The abrasive should be of a size sufficient to remove surface contamination without pitting the surface or creating extreme peaks and valleys. Pitted surfaces require protection, and high peaks represent areas of possible paint failure.

As in other mechanical cleaning, surface grease or oil should be removed by solvent cleaning prior to blasting. Dry blasting of steel should be avoided if the temperature of the metal is less than 50 °F above the dew point, to prevent condensation and subsequent rusting.

Blast cleaning is the most effective of the mechanical treatments and often is used on metal structures. It also may be used, with caution, on masonry surfaces. The depth to which the surface is abraded can be varied depending on the intended paint system and the severity of exposure. There are four methods of blast cleaning; the most effective methods, which reach down to or near the untarnished "white metal," require more labor and are, therefore, costlier (Fig. 9.12-4).

White metal blast provides for the complete removal of all rest, mill scale, and other contaminants from the surface and is the preferred technique in blast cleaning. It is used for coatings that must withstand exposure to very corrosive atmospheres and where a high initial cost of surface preparation is warranted.

Near-white metal blast surfaces are nearly free of all contaminants but will show streaks and discolorations distributed across the white metal surface. This technique is 10% to 35% cheaper than white metal blasting and has proven adequate for many protective coatings in moderately severe exposures.

Commercial blast is the removal of superficial contaminants, such as loose scale and rust. This method results in a surface quality generally adequate for the majority of paint systems under normal exposure conditions.

Brush-off blasting is a relatively low-cost method, used to remove old finishes, loose rust, and mill scale. Brush-off blasting is not suitable where severe corrosion is prevalent, but it is a more efficient substitute for manual and power tool cleaning. Brush-off blasting also is used for the removal of loose or deteriorated paint from masonry.

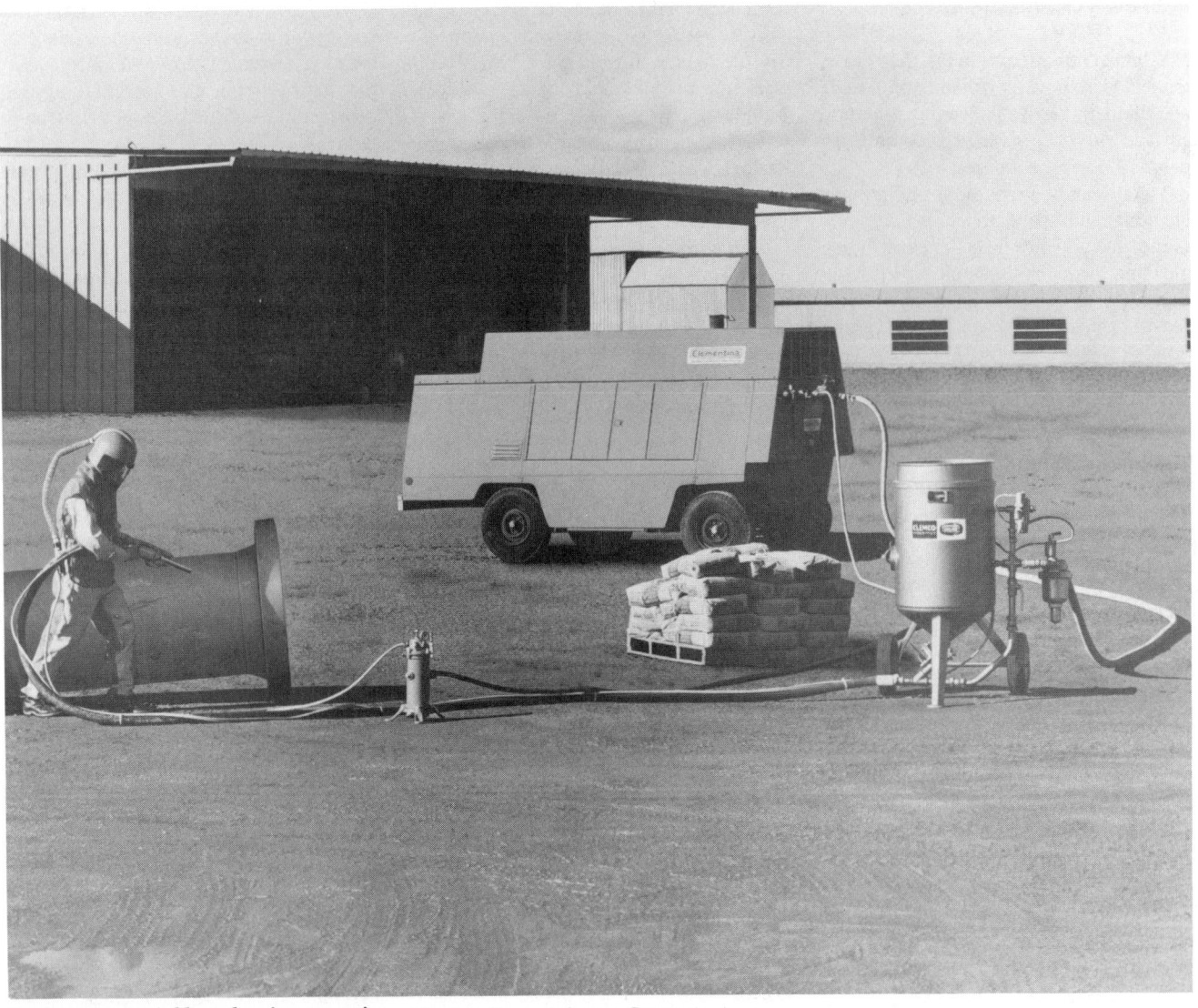

FIGURE 9.12-3 *A blast-cleaning operation.* *(Courtesy Clemco Industries Corporation.)*

■ **CHEMICAL AND SOLVENT TREATMENT**

Chemical cleaning methods usually are more suited to shop application, while mechanical methods, other than blast cleaning, are more practical for use in the field. Chemical cleaning is both more efficient and more effective than mechanical methods other than blast cleaning, but chemical cleaning in the field is more difficult than even blast cleaning there because of the potential environmental hazards.

The use of goggles and rubber gloves is generally recommended and additional protective clothing is required where acid or alkaline solutions are used. Adequate ventilation is essential during chemical and solvent cleaning. Respirators should be used in confined areas, and proper fire precautions should be taken when flammable cleaners are used.

Some chemicals, such as trisodium phosphate, that have been traditionally used in cleaning and the removal of existing paint and coatings are no longer permitted in some areas, because of their potential harm to people and the environment. Some jurisdictions permit the use of such chemicals but severely limit the means of their disposal after use. Therefore, before the use of any chemicals to remove existing paints or coatings or to clean surfaces before applying paint, transparent finishes, or coatings, it is advisable to check with local authorities to ascertain restrictions on their use and disposal.

SOLVENT WIPING AND DEGREASING Solvent cleaning is a procedure used to dissolve and remove surface contaminants such as oil, grease, dirt, and paint remover residues before painting or mechanical treatment. It is best to first remove soil, cement spatter, and other dry materials with a wire brush. The surface then should be scrubbed with brushes or rags saturated with solvent. Other effective methods include immersing the work in a solvent or spraying solvent over the surface.

The solvent quickly becomes contaminated, so it is essential that several clean solvent rinses be applied to the surface. Clean rags should be used for rinsing and wiping dry. Mineral spirits is an effective solvent for cleaning under normal conditions. Toxic solvents and solvents with low flash points constitute health and safety hazards. Use of solvents with flash points below 100 °F or solvents with a maximum safe concentration of less than 100 parts per million is not recommended. Rags should be placed in fireproof containers after use.

ALKALI CLEANING Alkali cleaning is more efficient, less costly, and less hazardous than solvent cleaning but more difficult to carry out. It is suitable for most metals, except stainless steel and aluminum. Alkali cleaner solutions are applied at relatively high temperatures (150° to 200 °F) because cleaning efficiency increases with temperature. Alkalies attack oil and grease, converting them into soapy residues that can be washed away with water.

Commercial cleaners contain other active ingredients that aid in removing surface dirt and other contaminants such as mildew. These cleaners also are effective in removing old paint by breaking up the dried vehicle. The most commonly used alkaline cleaners are trisodium phosphate, caustic soda, and silicated alkalies. For steel, these cleaners should contain 0.1% chromic acid or potassium dichromate to prevent corrosion. Cleaners can be applied by brushing, scrubbing, or spraying, or the item to be cleaned can be immersed in a soak tank.

After cleaning, thorough rinsing is necessary to remove the soapy residue as well as all traces of alkali. Rinsing also avoids the possibility of chemical reaction with the applied paint. The rinse water should be hot (175 °F) and preferably applied under pressure. Universal pH test paper should be used to ensure that the surface is neutral and no free alkali is present on the surface after rinsing.

STEAM CLEANING Steam cleaning involves the use of steam or hot water under pressure, sometimes with a detergent included for added effectiveness. The steam or hot water removes oil and grease by emulsifying or diluting them, then flushing them away with water. When steam is applied to old paint, the paint's vehicle swells so that it loses its adhesiveness and is easily removed. Under ordinary conditions steam or hot water alone may be used to remove dirt deposits, soot, and grime. In severe cases, wire brushing or brush-off blast cleaning may be necessary to augment the steam cleaning.

ACID CLEANING Acid cleaning is used to remove oil, grease, dirt, and other surface contaminants and is suitable for cleaning iron, steel, concrete, and masonry. It should not be used on stainless steel or aluminum.

Iron and steel surfaces are treated with solutions of phosphoric acid containing small amounts of solvent, detergent, and wetting agent. Unlike alkali cleaners, acid cleaners also remove light rust and etch the surface to ensure better adhesion of applied coatings. There are many types of metal acid cleaners and rust removers, each formulated to perform a specific cleaning job. The most common formulations are based on phosphoric acid solutions. The procedures usually involve applying the cleaner by brush or rag, allowing time for it to act, then thoroughly rinsing, wiping, and air drying. Smaller objects, particularly in the shop, can be immersed in hot cleaner, followed by several rinses.

Concrete and masonry surfaces can be washed with a solution of muriatic (hydrochloric) acid to remove efflorescence and laitance, to clean the surface, and to improve its bonding properties. Efflorescence is a white, powdery deposit often found on masonry surfaces. Laitance is a fine gray powder that sometimes develops on the surface of improperly finished concrete. Paints applied over efflorescence or laitance may loosen prematurely and result in early coating failure. It is best to remove as much of these deposits as possible using a stiff fiber or wire brush; putty knives or scrapers also may be used.

Oil and grease should be removed prior to acid cleaning, either by solvent wiping or by steam or alkali cleaning. To start acid cleaning, the surface first should be thoroughly wetted with clean water, then scrubbed with a 5% solution (by weight) of muriatic acid and a stiff fiber brush. In extreme cases, up to a 10% muriatic acid solution may be used and allowed to remain on the surface up to 5 minutes before scrubbing. Work should be done on small areas, not greater than 4 sq. ft. in size. Immediately after the surface has been scrubbed, the acid solution should be rinsed completely from the surface by a thorough sponging or flushing with clean water. The surface should be rinsed immediately to avoid formation of salts on the surface, which are difficult to remove.

This procedure also may be used to etch hard and dense concrete surfaces with poor surface bonding properties. The etching procedure breaks up the glaze so the paint will adhere to the surface. To determine whether etching is required, a few drops of water are poured on the surface. If the water is quickly absorbed, the surface can be readily painted; if the drops remain on the surface, etching is recommended.

Surface glaze may be removed also by rubbing with an abrasive stone, light sandblasting, or allowing the surface to weather for 6 to 12 months. The surface also may be etched with a solution of 3% zinc chloride and 2% phosphoric acid; this solution should not be washed off but allowed to dry to produce a paintable surface.

Concrete and masonry surfaces generally are alkaline in nature, and it may be necessary to use acid cleaning methods to neutralize these surfaces before applying alkali-sensitive coatings. The presence of free alkali may be detected by testing with pH paper.

PICKLING Pickling is used only in the mill or shop to remove mill scale, rust, and rust scale from iron and steel products. Sulfuric, hydrochloric, nitric, hydrofluoric, and phosphoric acids are used individually or in combination. Pickling is usually accomplished by immersing the piece in a tank, followed by several intermediate rinses to remove acids and salts. The final rinse usually is performed with a weak alkali solution to retard rusting and lingering acid solution.

PAINT REMOVERS Paint and varnish removers generally are used on small areas. Solvent-type removers or solvent mixtures are selected according to the type and the condition of the old finish as well as the nature of the surface. Removers are available as flammable or nonflammable types and in liquid or semipaste consistency. While most paint removers require scraping to remove the softened paint, some removers loosen the paint sufficiently to permit flushing off the debris with steam or hot water.

To retard evaporation, some removers contain paraffin wax, which often remains on the surface after the paint has been removed. It is essential that this residue be removed from the surface prior to painting, to ensure proper bonding of the paint coat. The wax can be removed with mineral spirits or a solvent specifically recommended by the manufacturer. Most removers are toxic to a degree and some are flammable. Consideration should be given to adequate ventilation and fire control whenever they are used.

■ METAL PRETREATMENTS

Pretreatments of metal surfaces improve coating adhesion, reduce underfilm corrosion, and extend coating life. Of the several pretreatment methods, hot phosphate treatments are limited to the shop or plant; cold phosphate and wash primers can be applied in the field as well.

HOT PHOSPHATE TREATMENTS Hot phosphate treatments utilize zinc or iron phosphate solutions to form crystalline deposits on the surface of a metal. They greatly increase the bond and the adhesion of applied paints while reducing underfilm cor-

rosion. Zinc phosphate usually produces the best results and is most widely used.

When a gloss finish is desired, it is often necessary to apply thicker paint coats over a zinc phosphate coating, because the heavier phosphate coatings absorb considerably more paint. If a higher gloss finish is desired, iron phosphate is preferred to a zinc phosphate pretreatment, because it produces a finer crystalline structure and therefore a thinner film. Hot phosphate treatments are very effective in ensuring good coating adhesion and tight bonds.

COLD PHOSPHATE TREATMENTS Cold phosphate treatments are produced with a mixture of phosphoric acid, a wetting agent, a water-miscible solvent, and water. An acid concentration of about 5% to 7% by weight may be adequate when the area is not exposed during application to high temperatures, direct sunlight, or high wind velocities. Such adverse environmental conditions cause rapid evaporation and consequent high acid concentration.

When a dry, grayish-white powdery surface develops within a few minutes after application, acid has reacted properly and has the proper dilution. If a dark color develops and the surface is somewhat sticky, the acid probably is too concentrated. In such cases, if the area is small, wiping with damp rags may bring about the desired appearance; otherwise, the surface should be rinsed with water and retreated with a more dilute solution.

Although cold phosphate treatments may produce a crystalline deposit on the metal surface, the density of the deposit is not as great as the hot phosphate treatment; therefore, paint adhesion may be less effective. The procedures used for cold phosphating are adaptable to field use on large and small structures.

WASH PRIMERS Wash primer formulations result in a form of cold phosphating. They are said to perform more efficiently than the standard cold phosphating treatments and have generally replaced them in field use. Wash primers derive their name from the method of applying them in thin "wash" coats. They usually contain polyvinyl butyral resin, phosphoric acid, and a rust-inhibitive pigment, such as basic zinc chromate or lead chromate. Wash

primers develop extremely good adhesion to blast-cleaned or pickled steel and provide a sound base for topcoating. They also may be used to improve adhesion of coatings to hard-to-paint surfaces, such as galvanized or stainless steel or aluminum.

■ PRIMERS, SEALERS, AND FILLERS

Application of a finish coat should always be preceded by a suitable primer, filler, or sealer or a base coat of the same paint as the finish coat, which performs the same functions.

Primers are often applied on masonry to seal chalky surfaces and to improve adhesion of water-based topcoats. Sealers are used on wood to prevent resin exudation or bleeding. Fillers help to produce a smooth finish on open-grained wood and rough masonry.

PRIMERS Ordinary latex paints do not adhere well to chalky masonry surfaces. To overcome this problem, modified latex paints can be used, or the surface can be treated with an oil-based primer before the latex paint is applied. The entire surface should be vigorously wire-brushed by hand or with power tools, then dusted to remove loose particles and chalk residue. The primer then should be brushed on freely to ensure effective penetration and allowed to dry. The primer should not be used as a finish coat, because it is not as weather resistant.

KNOT SEALERS Sealers are used on bare wood to prevent resin bleeding through newly applied paint coatings. Freshly exuded resin, while still soft, may be scraped off with a putty knife and the affected area cleaned with alcohol. Hardened resin may be removed by scraping or sanding. Since a sealer is not intended as a priming coat, it should be used only when necessary and applied only over affected areas. When a previously painted softwood lumber surface becomes discolored over knots, the knots should be covered with sealer before new paint is applied.

FILLERS Fillers are used on porous wood, concrete, and masonry to fill the pores and to provide a smoother finish coat.

FIGURE 9.12-5
Finishing Characteristics of Major Wood Species

Wood Species	Grain Type	Remarks
HARDWOODS		
Ash	Open	Requires filler
Aspen	Closed	Paints well
Basswood	Closed	Paint well
Beech	Closed	Paints poorly; varnishes well
Birch	Closed	Paints and varnishes well
Cherry	Closed	Varnishes well
Chestnut	Open	Requires filler; paints poorly
Cottonwood	Closed	Paints well
Cypress	Closed	Paints and varnishes well
Elm	Open	Requires filler; paints poorly
Gum	Closed	Varnishes well
Hickory	Open	Requires filler
Mahogany	Open	Requires filler
Maple	Closed	Varnishes well
Oak	Open	Requires filler
Teak	Open	Requires filler
Walnut	Open	Requires filler
SOFTWOODS		
Alder	Closed	Stains well
Cedar	Closed	Paints and varnishes well
Fir	Closed	Paints poorly
Hemlock	Closed	Paints fairly well
Pine	Closed	Variable depending on grain
Redwood	Closed	Paints well

Courtesy Architectural Woodwork Institute.

Wood fillers are used on open-grained hardwoods. In general, those hardwoods with pores larger than in birch (open grain) should be filled (Fig. 9.12-5). When filling is necessary, it should be done 24 hours after staining. If staining is not warranted, natural (uncolored) filler should be applied directly to the bare wood. When desired, the filler may be colored with stain to accentuate the grain pattern of the wood.

To apply, the filler should be thinned first with mineral spirits to a creamy consistency, then liberally brushed across the grain, followed by light brushing along the grain. It should be allowed to stand 5 to 10 minutes until most of the thinner has evaporated, and the finish has lost its glossy appearance. Before it has a chance to set and harden, the filler should be wiped off across the grain, using burlap or other coarse cloth, rubbing the filler into the pores of the wood while removing the excess. The operation is finished by stroking along the grain with clean rags, making sure that all excess filler is removed.

Knowing when to start wiping is important; wiping too soon may pull the filler out of the pores, while allowing the filler to set too long may make it difficult to wipe off. A simple test for dryness consists of rubbing a finger across the surface: if the filler forms into a ball, it is time to wipe; if the filler slips under the pressure of the finger, it is still too wet for wiping. The filler should be allowed to dry for 24 hours before finish coats are applied.

Masonry fillers can be applied by brush or spray to both old and new bare concrete, stucco, concrete block, and other masonry surfaces. These surfaces should be rough and sound, free of loose and powdery materials. If surface voids are relatively large, it is preferable to apply two coats of filler, rather than one heavy coat. In large areas, filler can be sprayed on, then troweled smooth. One to two hours drying time should be allowed between filler coats, and painting should be delayed 24 hours after the final filler coat.

9.12.3.4

Repair of Surfaces

It is essential that surfaces that will be painted are clean, sound, and uniform in texture. Consideration should be given to repairing or replacing damaged materials or surfaces; replacing broken windows and loose putty or glazing compound; filling cracks, crevices, and joints with caulking compound or sealants; and patching cracks or holes in wood, masonry, and plaster. Suitable products and materials are described in Figure 9.12-6. Refer to Section 7.10 for information concerning caulking and sealants.

In general, patching of cracks and holes should be done by the installer and finisher of the underlying material. Rarely, the painter might patch minor cracks and fill minor holes. In addition to the materials described in other sections, including Section 7.10, the materials described here may be used for patching holes and cracks.

Cracks, holes, and crevices in masonry, plaster, gypsum board, and wood may be filled with various patching materials. These are supplied either ready for use or as a dry powder to which water is added.

When patching materials are used on masonry, plaster, or gypsum board, the crack should first be opened with a putty knife or wall scraper so that weak material is removed and the patching compound can be forced in completely. The compound should be applied with a putty knife or trowel, depending on the size of the hole. Then the surface should be smoothed off slightly convex, to allow for shrinkage.

One of the oldest painter's patching materials is *putty*. It is an oil-based material used to fill nail holes, cracks, and imperfections in wood surfaces. Putty generally is supplied in bulk form and is applied with a putty knife. It should be applied after priming to prevent the wood from drawing the oils out. It dries to a harder surface than caulking compounds and should not be used for joints

FIGURE 9.12-6
Materials for Repair of Surfaces

Material	Where Used	Characteristics
Caulking compound	Filling joints and crevices around doors and windows in concrete, wood, brick, block, and other masonry surfaces	Tends to dry on the surface but remain soft and tacky inside. Exposed surface of caulking bead should be painted each time surrounding area is painted to help extend its life (5–15 years).
Sealants	Filling joints in locations where movement occurs and a weathertight joint is essential at all times	Better adhesion and greater elasticity than caulking compound, even at low tempeatures. Life expectancy 15–30 years.
Putty	Filling nail holes, cracks, and imperfections in wood surfaces; installing small lights in wood and metal sash; not used for joints or crevices	Dries to a harder surface than caulking compound.
Glazing compound	Bedding and face glazing of windows	Sets firmly but retains limited flexibility, generally greater than putty. Life expectancy of 5–15 years when protected by periodic painting over exposed surface.

or crevices. Special glazing putty is used to install small lights in wood and metal sash.

Plastic wood is suitable for filling gouges and nail holes in wood. It is applied in a manner similar to putty. After the plastic wood has completely dried, it should be sanded until smooth.

A soupy mixture of portland cement, sand, and water called *grout* can be used to repair cracks in concrete and masonry. Synthetic resins such as latex or epoxy may be added to improve bonding qualities and to permit application in thin layers. Hydrated lime sometimes is added to lengthen working time. Grout should be moist-cured for several days or longer.

Patching plaster is used for repairing large areas in plaster. It is similar to ordinary plaster except that it hardens more quickly. It is supplied as a powder to be mixed on the job with water.

Spackle is used to fill cracks and small holes in plaster and gypsum board. It is easy to apply and to sand smooth after it hardens. It is supplied both as a paste and as a powder.

Joint compound is intended primarily for sealing joints between gypsum board panels, but it can also be used to repair large cracks. It is supplied both as a powder and ready-mixed. It is used in conjunction with joint tape, which gives it added strength.

9.12.3.5

Paint Application

The most common methods of applying paint are by brush, roller, and spray. Dip-and-flow coat methods are generally limited to shop work. Of the three methods designed for field use, brushing is the slowest, and spraying is the fastest. One painter can cover up to 1000 sq. ft. a day by brushing, 2000 to 4000 sq. ft. by rolling, and up to 12,000 sq. ft. by spraying. Coatings applied by brush may leave brush marks in the dried film; rolling leaves a stippled effect; spraying, if done properly, yields the smoothest finish. The choice of method is based on many additional factors such as (1) environmental conditions, (2) the type of surface and coating, and (3) the desired appearance of the finish.

General surroundings may prohibit the use of spray application because of possible fire hazards or potential damage from overspray. Adjacent areas not intended to be coated should be masked when spraying is performed. If the time spent in masking preparations is extensive, it may offset the advantage of the speed of spraying operations.

Roller coating is efficient on large flat surfaces. Corners, edges, and odd shapes generally require brushing. Spraying is especially suitable for large surfaces, but it can be used also for round or irregular shapes. Brushing is ideal for small sur-

faces or for cutting in corners and edges. Dip-and-flow coat methods are appropriate for volume production painting of small items in the shop.

Rapid-drying lacquer-type products, such as vinyls, are usually sprayed. Application of such products by brush or roller is difficult, especially in warm weather or outdoors on breezy days.

■ JOB CONDITIONS

To obtain optimum performance from a coating, certain basic application procedures should be followed, regardless of the type of equipment used. Cleaned, pretreated surfaces should be prime coated within the specific time limits recommended by thepaint manufacturer.

Paint manufacturers generally recommend the following: that surface and air temperatures be between 50 °F and 90 °F for water-thinned coatings and between 45 ° and 95 °F for other coatings; that paint material be maintained at a temperature between 65 ° and 85 °F at all times; that paint not be applied when the air temperature is expected to drop to freezing before the paint has dried; that the wind velocity be below 15 mph and the relative humidity below 80%. Some manufacturers may recommend other limits for their products under some circumstances.

Masonry surfaces that are damp (not

wet) may be painted with latex or cementitious paints. Other surfaces should be completely dry before painting. Paints should be applied at the spreading rates recommended by their manufacturers. When successive coats of the same paint are used, each coat should be tinted a different color to ensure complete coverage with each coat. Sufficient time should be allowed for each coat to dry thoroughly before the next coat is applied. A finish coat (topcoat) should be allowed to dry for as long as practicable before the space is occupied or put into service.

The line of demarcation between different colors or finishes should be straight and neat. Previously finished surfaces should be protected by suitable masking or covering. Stain or paint splatter from adjacent surfaces should be removed.

Unless otherwise recommended by the manufacturer, a minimum of two coats should be used. Painted exterior wood surfaces should have a total thickness of 4.0 mils or more (Fig. 9.12-7).

FIGURE 9.12-7
Dry-Film Thickness of Applied Paints

Coverage Rate, sq. ft./gal.	Approximate Dry-Film Thickness of Finish Coat, mils
450	2.25
500	2.00
550	1.75
600	1.50
650	1.30
700	1.10
750	1.00

10

SOUND CONTROL

INTRODUCTION

PROPERTIES OF SOUND

ROOM ACOUSTICS

SOUND TRANSMISSION

DESIGN AND
CONSTRUCTION

Photo: Acoustical baffles in the arched ceiling of a large armory. (Photograph provided by MBI Products Company.)

INTRODUCTION

Contemporary living is noisy. Computer printers, intercom systems, background music systems, washers and dryers, dishwashers, garbage disposers, refrigerators, vacuum cleaners, radios, television sets, and other electrical devices make life easier and more enjoyable but raise the noise level. Heating, ventilating, and air-conditioning equipment produce noise and carry it into every space. Outdoor noise levels have increased also. People work and live close to their neighbors and apartments and town houses continue to increase in number. Sounds created by traffic, fast trains, and jets add to the problem.

Many materials and construction methods in general use today exacerbate our noise problems. Glass, plaster, gypsum board, wood paneling, masonry, tile, metal panels, and other hard, relatively flat surfaces reflect up to 98% of the sound that strikes them. Noise levels within a room grow higher and more irritating as one reflected sound adds to another. The use of thin, lightweight materials, including some preassembled components, in walls and in floor and ceiling assemblies permits noise to travel easily from room to room, floor to floor, and from one living or working unit to an adjacent one.

The science of sound control was first applied earlier in this century in churches, theaters, and auditoriums, before it was used in schools, offices, hospitals, restaurants, retail stores, and other commercial and institutional buildings. Today, sound control is considered an essential element in the design of all buildings. When incorporated into a building's design, sound control is usually an inexpensive component. Conversely, alterations to control unwanted sound transmission after a building has been completed are almost always costly.

This chapter includes data about the properties and sources of sound, identifies common acoustical problems, and summarizes methods of measuring and controlling noise. It also discusses design and construction techniques essential to providing and maintaining the acoustical performance of walls and floor and ceiling assemblies. Finally, it provides criteria for selecting wall construction and floor and ceiling assemblies that resist airborne and structure-borne sound transmission.

In this chapter, acoustical problems are divided into two types. In the first type, *room acoustics*, the goal is noise reduction to provide good hearing conditions within

a space (see Section 10.2). In the second type of acoustical problem, *sound isolation*, the goal is the control of sound transmission from one space to an adjacent space as a result of airborne sound or structural vibrations (see Section 10.3).

Appendix A contains sound performance ratings, including the Sound Transmission Class (STC), the Impact Noise Rating (INR), and the Impact Insulation Class (IIC), for some commonly used walls and floor and ceiling assemblies.

Section 8.1.4 has requirements for obtaining the desired acoustical performance for doors and windows. Section 11.1.5 talks about ways to reduce noise generation by and sound transmission through plumbing systems and fixtures. Section 11.3.3.2 discusses the role of a building's heating, ventilating, and air-conditioning systems in noise generation, and specific design criteria to reduce noise generation by and sound transmission through these systems. Section 12.3.7 describes methods that can be used to reduce sound transmission through and around electrical system devices, and to reduce noise generation by appliances.

10.1 PROPERTIES OF SOUND

Sound is a vibration in any elastic medium. In air, sound may be defined as the movement of air molecules in a wave motion. Sound waves are set up by any vibrating body, they travel through the air, and, when their frequency and intensity are within certain ranges, they become audible.

A vibrating body causes the layer of air particles next to its surface to vibrate. These air particles in turn transmit their

motion to the next air layer. Because air is compressible and has weight, a small interval of time is required for the first air layer to set the second layer in motion. Succeeding air layers are similarly set in motion after intervals of time that are directly proportional to their distance from the source of the sound. This action constitutes a sound wave. The waves produced when a stone is dropped into water are analogous (Fig. 10.1-1).

10.1.1

WAVE MOTION

Water waves travel only along the surface and are circular in shape. Sound waves, however, spread out in all three dimensions and are spherical in shape. Each air particle vibrates only in the direction in which the wave is traveling; that is, along a straight line drawn between the particles

FIGURE 10.1-1
Sound is a wave motion in air caused by a vibrating source. Wave motion in water is analogous.

FIGURE 10.1-2
Light travels almost instantly, but sound requires about 5 sec. for each mile it travels.

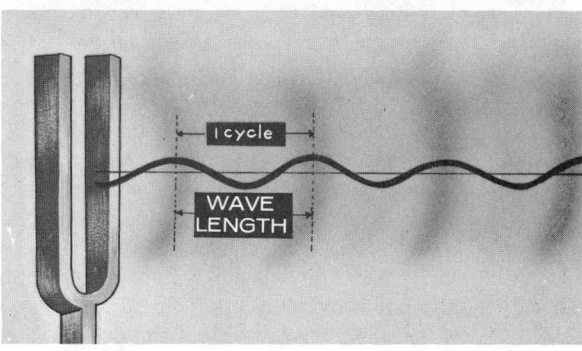

FIGURE 10.1-3
Wavelength is the distance a sound wave travels during one vibration (cycle).

and the source of the sound wave. Sound travels in air at a speed of about 1130 ft./sec.; in water at about 4500 ft./sec.; and in steel at about 15,000 ft./sec. The speed in a given medium is always the same regardless of the pitch or loudness of the sound. Outdoors, sound does not reach our ears until an interval of about 5 sec. for each mile we are from the source of the sound (Fig. 10.1-2). We hear echoes because the sound reflecting from surfaces travels farther than the sound coming directly from

its source. We hear the sound from the source and then a reflected sound a few moments later. This same property of sound also causes the unpleasant reverberations we sometimes experience indoors.

10.1.2

FREQUENCY

One complete round-trip of a vibrating body, starting from its neutral position,

moving to one side, then to the other side, and back to neutral, is called a *cycle* (Fig. 10.1-3). The number of cycles performed in 1 sec. is called *frequency*. Frequency is measured in cycles per second (cps) and may be expressed in terms of hertz (Hz) (1 cps). Each air particle in a sound wave must necessarily have the same frequency as the source. For example, if the sound from a loudspeaker has a frequency of 1000 cps, it means the loudspeaker diaphragm and every air particle through which the sound wave passes are vibrating back and forth 1000 times each second. The performance and efficiency of acoustical materials in absorbing sound and the ability of materials to transmit sound vary, depending on the frequency of the sound.

The audible frequency range is from about 20 to 20,000 cps, but the range narrows as people grow older. Music covers most of this range, whereas speech extends from about 100 to 5000 cps. The most important range of frequencies for speech is shown in Figure 10.1-4. But the ear is not equally sensitive to all frequencies. The human ear is most sensitive to sounds with frequencies between 1000 to 4000 cps, shown by the low dip on the threshold of hearing curve in Figure 10.1-4. Sounds between 250 and 1000 cps or between 4000 and 8000 cps must be more intense and those below 250 or above 8000 cps must be considerably more intense in order to be heard. Most sounds consist of a mixture of many frequencies at different relative intensities.

10.1.2.1

Wavelength

The length of a sound wave is calculated by dividing the velocity of sound by the frequency:

$$wavelength = \frac{velocity}{frequency}$$

Wavelength may also be defined as the distance a sound travels during a single vibration. Since the velocity of sound is constant, every frequency has a different wavelength. A sound with a frequency of 1000 cps has a wavelength equal to about 1 ft. To reflect a sound effectively, the reflecting surface must be equal to or

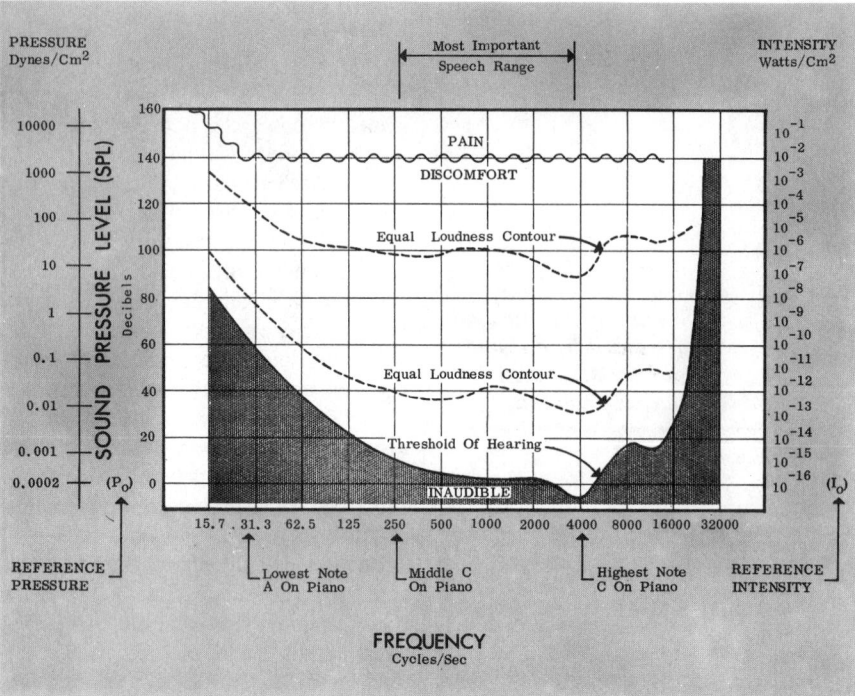

PRESSURE
Dynes/Cm²

Most Important
Speech Range

INTENSITY
Watts/Cm²

REFERENCE
PRESSURE

REFERENCE
INTENSITY

FREQUENCY
Cycles/Sec

FIGURE 10.1-4 *Audible frequency range. Two sounds of the same intensity but of different frequencies do not necessarily sound equally loud. Equal loudness contours represent sounds of different frequencies that seem to the average human ear to be equally loud.*

FIGURE 10.1-5 *Musical instruments produce sounds of many related frequencies.*

tion of related frequencies. The vertical distances of the graphs are proportional to the loudness.

10.1.2.2

Sound Levels

A sound source may vibrate with small or great intensity, depending on the amount of force that sets it into vibration. Sound levels are expressed in *decibels* (dB). The decibel scale extends from 0 dB (threshold of hearing) to more than 120 dB (threshold of feeling). Zero dB represents a fixed value of sound intensity that is slightly less than that of the faintest audible sound. Sound at 120 dB is so loud that it causes a tickling sensation in the ear. Sound as high as 180 dB, generated by some rocket engines, can cause structural damage to buildings and can be fatal to humans.

A decibel is equivalent to the smallest change in sound intensity that can be detected by the average human ear. However, a decibel cannot be measured directly and has no fixed, absolute value. It is a logarithmic unit that expresses the ratio between a given sound level being measured and a reference point. In other words, a decibel tells by what proportion one sound level is greater or less than another. The greater the sound level, the higher the rating on the decibel scale.

Sound levels in decibels are calculated from either a ratio of sound *pressures* (P) or sound *intensities* (I) using a reference point that corresponds to the faintest audible sound.

Pressure is a force per unit of area. Sound pressure at a given distance from a source is the force of moving air molecules spread over a spherical area and is expressed in *dynes* (units of force) per square centimeter.

greater than the wavelength in size. A 1-sq.-ft. surface will reflect sound 1000 cps and higher, but sound below 1000 cps will bend around such a surface and pass beyond it. A sound with a frequency of 100 cps has a wavelength of 11.3 ft. This dimension usually exceeds the ceiling height and often the wall length of residential rooms; thus the sound wave does not fit within the room and the sound is not reproduced within the room with fidelity. This accounts for the poor reproduction in small rooms of low-frequency sounds from radios, tape and disk players, televisions, and similar sources.

■ OCTAVES AND TONES

The frequency scale is laid out in octaves in the same manner as a piano keyboard. The term *octave* is taken from musical terminology and is defined as the interval between any two sounds having a frequency ratio of 2:1. For example, frequencies between 125 and 250 cps or between 250 and 500 cps constitute an octave. A pure tone is a sound having only one frequency, whereas a complex tone is a sound composed of several frequencies simultaneously.

■ PITCH

The hearing sensation produced by the frequency of a sound is called *pitch*. The greater the number of vibrations per second, the higher the pitch. High-frequency sounds such as squeaks and whistles are heard as high pitches, and low-frequency sounds such as rumbles and roars are heard as low pitches. Pitch is expressed in tones of the musical scale.

■ QUALITY

The quality of a tone as judged by the ear is determined by the distribution of its numerous frequencies. The higher frequency components are called harmonics or overtones when they are exact multiples of the fundamental frequency. The distribution and relative intensity of the overtones of a musical sound determine the quality (*timbre*) of the sound. Each of the two sounds illustrated in Figure 10.1-5 are complete tones composed of a combina-

Power is expended when a force is applied over a certain distance in a given unit of time. Sound power in *watts* (units of power) is a basic expression of the flow of sound energy. Sound power is spread over a sphere of increasing area as the sound wave radiates farther from the source. Sound power per unit of area is called sound intensity (I) and is expressed in watts per square centimeter (W/cm^2).

The reference pressure, marked (P_o) in Figure 10.1-4, is 0.0002 dyne/cm^2 and is roughly equivalent to the smallest amount of pressure that will cause an eardrum to vibrate. The corresponding reference intensity, marked (I_o) in Figure 10.1-4, is 10^{-16} W/cm^2.

The ratio in decibels between any measured pressure (P) and the reference pressure (P_o) is called the *sound pressure level* (*SPL*) and is found using Formula (a) in Figure 10.1-6. The ratio in decibels between any intensity (I) and the reference intensity (I_o) is called the *sound intensity level* (*SIL*) and is found using Formula (b) in Figure 10.1-6.

(a) Sound Pressure Level

$$SPL = 20 \log_{10} \frac{P}{P_o}$$

(b) Sound Intensity Level

$$IL = 10 \log_{10} \frac{I}{I_o}$$

FIGURE 10.1-6 *Formulas for computing the ratio in decibels between (a) the pressure of two sounds and (b) the density of two sounds.*

The decibel scale makes values computed for either pressure or intensity directly comparable. Although intensity levels are convenient for discussion, they cannot be measured directly. Therefore, in practice, sound levels are determined by measuring pressures on a sound-level meter and expressing the measurement as SPL on a decibel scale.

The easiest way to understand sound levels on the decibel scale is to compare them to common, easily recognized sounds (Fig. 10.1-7).

An increment in sound level of 1 dB (say, the difference between 50 and 51 dB) is just perceptible. A 2- to 3-dB increment is insignificant, a 5-dB increment is clearly noticeable, and a 10-dB increment is quite perceptible and generally corresponds to a subjective reaction of doubling (or halving) the loudness (Fig. 10.1-8).

■ **LOUDNESS**

The loudness of a sound (hearing sensation) depends on the intensity and frequency of the sound and on the characteristics of the human ear. People hear sound over a wide range of intensity, extending from a low whisper to a deafening roar. Beyond a certain point, pain in the ears becomes more evident than the sensation of increasing loudness.

There are limitations in the frequency

FIGURE 10.1-7

Levels of Typical Sounds in Decibels

Loudness	Sound Level in Decibels (dB)	
Deafening	160	Near jet engine
	130	Threshold of painful sounds; limit of ear's endurance
	120	Threshold of feeling (varies with frequency) 18 ft. from airplane propeller
	110	
		Express train passing at high speed
	100	Loud automobile horn 23 ft. away
		Noisy factory
Very loud	90	
		Loud hi-fi
	80	Subway train Motor trucks 15 to 50 ft. away
Loud	70	Stenographic room, average TV, loud conversation
	60	Average busy street Noisy office or department store
Moderate	50	Moderate restaurant clatter Average office, noisy room, average conversation
	40	
		Soft radio music in apartment, average residence
Faint	30	
	20	Average whisper 4 ft. away
	10	Rustle of leaves in gentle breeze
Very faint	0	Threshold of audibility

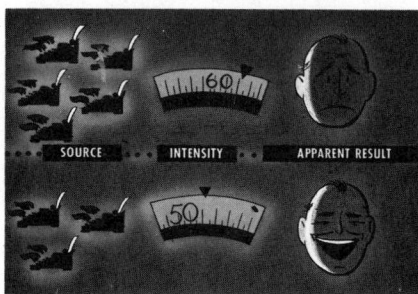

FIGURE 10.1-8 *A reduction in sound level of 10 dB produces the subjective reaction of halving the loudness.*

response of the human ear as well as a definable range of intensities to which it will respond. The psychological sensation of loudness is closely related to the intensity of the sound. The greater the intensity of vibration, the louder the sound seems to the listener, but a given sound may seem to be much louder when it is in a middle frequency range than when it is in a high or low range. In fact, from Figure 10.1-4 it is evident that a 15-dB tone is heard easily at 1000 or 2000 cps but is inaudible at 200 or 10,000 cps. The equal loudness contours shown in Figure 10.1-4 connect the tones of different frequencies that sound equally loud. The loudness of sound will also depend a great deal on the characteristics of the listener's ear. Thus, even when the intensity of a given sound striking the ear of a hearing person and that of a hard-of-hearing person is exactly the same, the sensation of loudness is quite different.

■ NOISE

Unwanted sound is called noise. Usually a sound becomes noise when it is loud enough to be a nuisance. Certain sounds are noise because they are nonuniform vibrations. Scratching is an example. Even pleasant or otherwise acceptable sound becomes noise when it interferes with concentration, conversation, or other tasks to be performed. It is not possible to say how loud a sound must be before it becomes objectionable, because it depends on the kind of sound and the person who hears it.

10.2 ROOM ACOUSTICS

Room acoustics is a branch of architectural acoustics that deals with both *acoustical correction* and *noise reduction*.

10.2.1

ACOUSTICAL CORRECTION

Acoustical correction involves the detailed planning and shaping of a room to establish the best possible hearing conditions for wanted sound within the space. This sort of work is usually confined to public spaces, such as auditoriums, concert halls, and churches. The primary problem is to shape an environment in which it is possible to hear sounds distinctly, in ample loudness, and with as faithful a replication of the source as possible everywhere within the space. Acoustical correction requires a knowledge of the principles of acoustics, engineering calculations, design of the space to take best advantage of the properties of sound, and the selection and location of acoustical and other finish materials. Sound amplification systems are often also involved. Such design requires the skill and judgment of an acoustical specialist and is therefore outside the scope of this text. In this section, room acoustics deals only with noise reduction.

10.2.2

NOISE REDUCTION

Noise reduction involves the treatment of room surfaces with acoustical materials to alleviate the discomfort and distraction caused by the reflection of *unwanted* sound within the space. The primary problem is to reduce the reflection of sound and prevent a buildup of the noise level that results from continued reflection.

10.2.2.1

Reflection of Sound

When a sound source is operating, sound waves travel outward in all directions radially from the source. When the sound waves encounter an obstacle or surface, such as a wall, their direction of travel is changed, or reflected.

The reflection of sound follows the same laws as the reflection of light from a mirror. (1) The direction of travel of the reflected sound always makes the same angle with the surface as that of the incident sound; that is, the angle of incidence equals the angle of reflection; and (2) the reflected sound waves travel exactly as they would if they had originated at an "image" of the sound source, located the same distance behind the wall as the real source is in front of the wall (Fig. 10.2-1).

■ MULTIPLE REFLECTION

When a sound is produced in a room having reflective surfaces, sound waves are reflected back and forth by these surfaces and almost immediately the room is filled with sound waves traveling in every direction. This is called *multiple reflection* (Fig. 10.2-2). Multiple reflection causes a buildup of intensity throughout the room, while the source of sound is operating, to a level greater than there would be if no reflection occurred. It makes noises unnaturally and unnecessarily loud and makes all noise last longer than necessary.

Aside from causing annoyance and fatigue, excessive reflection interferes with telephoning, conversation, detection of alarms and sound signals, and other types of work or play that depend on accurate hearing.

■ REVERBERATION

If the sound source is stopped, the reflected waves do not simply cease to exist at that instant, but continue to travel back and forth, *reverberating* (bouncing)

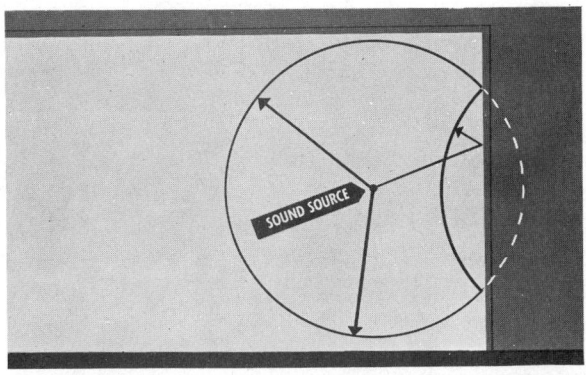

FIGURE 10.2-1
Sound travels radially in all directions and is reflected by room surfaces.

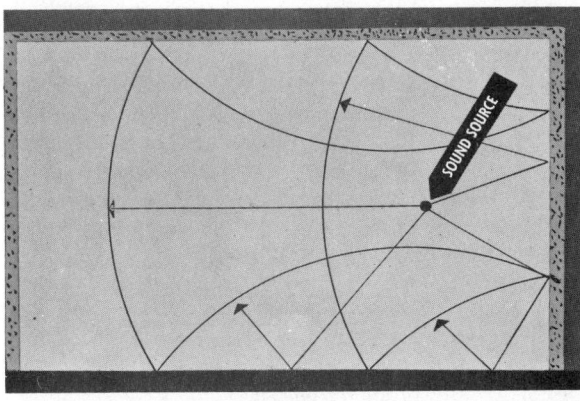

FIGURE 10.2-2
Multiple reflection of sound waves causes sound levels to build up and sounds to reverberate longer.

FIGURE 10.2-3
Relation of distance from the source to the resulting sound level in decibels.

FIGURE 10.2-4 *Curve showing building up and dying out of sound in a room.*

FIGURE 10.2-5 *Reverberation time is important for speech intelligibility and quality reproduction of music.* **T** *= reverberation time in seconds;* **V** *= room volume in cubic feet; and* **a** *= room absorption in sabins.*

for a while between the room surfaces. The total sound energy in the room gradually diminishes as a result of the partial absorption of the sound at each contact with a sound-absorbing surface (Fig. 10.2-3).

If a listener is in the room, the reflected waves strike the ear in such rapid succession that the listener usually does not hear them as distinct repetitions of the original sound. Instead, the listener hears the original sound being drawn out or prolonged after the source is stopped and then steadily dying out until it becomes inaudible altogether.

■ REVERBERATION TIME

The *reverberation time* of a room is the time in seconds required for a sound to die out to an intensity one-millionth the intensity it had at the moment the source was stopped. On the decibel scale this is a drop of 60 dB (Fig. 10.2-4).

The reverberation time (*T*) of any room depends only on its size (volume, *V*) and sound-absorbing properties (absorption units, *a*) and not on the power of the source (Fig. 10.2-5).

Values of reverberation time have been established that give the best conditions for hearing speech and music (Fig.

10.2-6). Speech intelligibility becomes poorer as the reverberation time is increased. Both insufficient loudness and excessive reverberation will decrease speech intelligibility.

■ ECHOES

An echo is a single reflection that can be heard as a distinct repetition of the original sound. Echoes are likely to occur in large rooms where there is an appreciable time lag between the original sound and the reflected sound. The reflected sound must be heard at least 1/17 sec. later than the direct sound to be heard as an echo. In other words, the reflected sound must travel a path from source to listener at least 65 ft. longer than the path of the direct sound. Large rooms make possible the time lag for hearing an echo, and when reverberation time is short the echo is heard more easily above general reverberation.

Echoes produced by concave surfaces are louder than those produced by flat surfaces because a curved surface converges and focuses the reflected sound, increasing its intensity in the same manner that the curved reflector in a flashlight focuses light (Fig. 10.2-7). A ceiling having its

FIGURE 10.2-6 *Range of acceptable reverberation times at a frequency of 500 cps for rooms of various sizes.*

FIGURE 10.2-8 *Sound-absorbing materials have interconnecting pore spaces.*

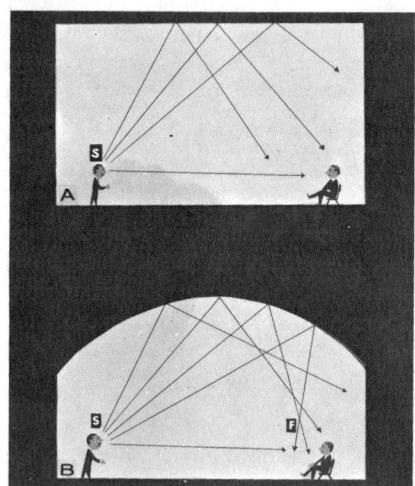

FIGURE 10.2-7 *(a) Reflections from a flat surface are divergent. (b) Concave surfaces focus sound at points (f) that may be louder than the source (s).*

FIGURE 10.2-9 *Sound-absorbing materials reduce the buildup of sound and cause it to die out faster.*

center of curvature near the floor line is the most objectionable type of curved ceiling from the standpoint of focused echoes, since the focal point is brought closer to the audience than when the center of curvature is higher or lower.

Focusing is best corrected by breaking up the curve or changing its shape. Treatment with acoustical materials is rarely enough.

10.2.2.2

Absorption of Sound

Ordinary smooth, hard building surfaces are highly reflective, reflecting up to 98% of the sound that strikes them. This reflection causes the sound to build up to a high level and to die out slowly.

Acoustical materials are designed to absorb most of the sound striking them and reflect less than 50% back into the room. They owe their absorption efficiency to their high porosity.

The amount of absorption of a porous material depends on its thickness, the size and number of its pores, and the frequency of the sound. Air particles moving in and out of the pores cause friction, and sound energy is dissipated as heat (Fig. 10.2-8). Various materials and combinations of materials designed for this purpose are discussed in other sections where appropriate.

When sound-absorbing materials are placed over the hard surfaces in a room, sound does not build up as much and will die out faster (Fig. 10.2-9). Although the sound level goes down, no matter how much sound-absorbing material is introduced into the room, the level cannot drop below the level of the direct sound from the source to the listener. Sound-absorbing materials cannot "suck up" the sound from the source; they can absorb only what reaches them. This is analogous to light in a room. When the walls are white (reflective), the whole room is very bright, but if the walls are painted black (absorptive), the whole room becomes very dark, even though the same source at the same brightness level continues to emit the same amount of light. The brightness of the room is what changes. Acous-

tical materials act essentially the same with sound as does the black paint with light.

In practice, it is impractical and uneconomical to obtain sound-level reductions of much more than 10 dB by using sound-absorbing materials. For example, 5 to 7 dB is the reduction usually accomplished when the entire ceiling of a normal office is covered with acoustical tile. However, such a reduction is obvious to the user of the space and the improvement in communications and comfort is almost always conspicuous.

■ MEASURING SOUND ABSORPTION

One unit of absorption (called a *sabin*) is equal to 1 sq. ft. of totally absorbent surface. An open window 1 sq. ft. in area is equivalent to 1 sound-absorbing unit because theoretically all of the sound that strikes it passes to the outdoors, and none is reflected into the room. As far as a listener in the room is concerned, an open window has 100% absorption.

An *absorption coefficient* (a) is assigned to materials to measure the percentage of incident sound energy absorbed by the material. These vary theoretically from 0 to 1. A perfectly reflective material has $a = 0$, and a perfectly absorptive material has $a = 1$. A material with an absorption coefficient of $a = .50$ will absorb 50% of the sound that strikes it and reflect the other 50%. Because absorption coefficients vary with the frequency of the sound, it is necessary to test materials at more than one frequency to cover the complete range of high- and low-frequency sounds. Materials are normally tested at six different frequencies: 125, 250, 500, 1000, 2000, and 4000 cps. The test result data may be plotted as a curve (Fig. 10.2-10). The absorption coefficients for typical construction surfacing materials are given in Figure 10.2-11.

FIGURE 10.2-10 *Absorption coefficients and other information are published by the manufacturers of acoustical materials.*

The numerical average of the absorption coefficients at the four middle frequencies, 250, 500, 1000, and 2000 cps, rounded off to the closest 5% is called the *noise reduction coefficient* (*NRC*). This is a single-number rating used to calculate the required amount of sound-absorbing materials needed. It permits comparison of such materials according to their acoustical performance.

Because small differences in acoustical efficiency cannot be detected by the human ear, and since tests that determine this efficiency cannot be carried out with pinpoint accuracy, the selection of an acoustical material is usually based on a 10-point

FIGURE 10.2-11
Absorption Coefficients for Typical Construction Materials

Material	Absorption Coefficients					
	125 Hz	250 Hz	500 Hz	1000 Hz	2000 Hz	4000 Hz
Brick						
Unglazed	.03	.03	.03	.04	.05	.07
Unglazed, painted	.01	.01	.02	.02	.02	.03
Carpet (heavy) on concrete slab	.02	.06	.14	.37	.60	.65
With 40-oz. hair-felt or foam-rubber pad	.08	.24	.57	.69	.71.	73
With impermeable latex backing and 40-oz. hair-felt or foam pad	.08	.27	.39	.34	.48	.63
Concrete Block						
Porous	.36	.44	.31	.29	.39	.25
Painted	.10	.05	.06	.07	.09	.08
Fabrics						
Light velour (10 oz./sq. yd.) hung straight in contact with wall	.03	.04	.11	.17	.24	.35
Medium velour (14 oz./sq. yd.) draped to half the area	.07	.31	.49	.75	.70	.60
Floors						
Concrete or terrazzo	.01	.01	.015	.02	.02	.02
Resilient flooring on concrete	.02	.03	.03	.03	.03.	.02
Wood	.15	.11	.10	.07	.06	.07
Parquet in asphalt on concrete	.04	.04	.07	.06	.06	.07
Glass						
Large panes heavy plate glass	.18	.06	.04	.03	.02	.02
Standard window glass	.35	.25	.18	.12	.07	.04
Gypsum board, 1/2 in. nailed to 2-in. × 4-in. studs 16 in. o.c.	.29	.10	.05	.04	.07	.09
Marble or glazed tile	.01	.01	.01	.01	.02	.02
Plaster						
Smooth finish on brick or tile	.013	.015	.02	.03	.04	.05
Rough finish on lath	.02	.03	.04	.05	.04	.03
Smooth finish on lath	.02	.02	.03	.04	.04	.03
Plywood paneling, 3/8 in. thick	.28	.22	.17	.09	.10	.11

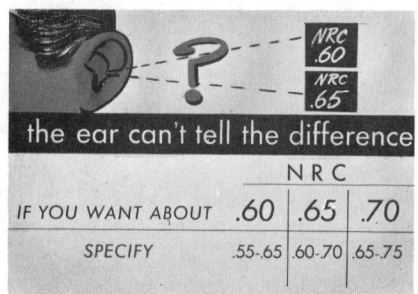

FIGURE 10.2-12 *NRC rates the efficiency of acoustical materials.*

range (.10) of NRCs (Fig. 10.2-12). A difference of 10 points is insignificant because it is seldom detectable in a completed installation.

In addition to the sound-absorption rating of a material, such other properties as light reflectivity, flame resistance, and appearance are also generally considered. Only in special cases where high absorption is required does NRC alone determine the selection. The sound absorption and noise reduction coefficients of acoustical materials are published by their manufacturers.

■ CONTROLLING NOISE WITH SOUND-ABSORBING MATERIALS

Acoustical treatment controls noise by reducing multiple reflection of sound waves, and as a result the loudness level of all reflected sound is reduced. Also, reverberation time is shortened so that the noise is not prolonged unnecessarily.

There are two basic noise-reduction principles that serve as rough guides in determining the amount of treatment within a given room for satisfactory noise control. (1) The total number of absorption units in spaces with low ceilings should be between 20% and 30% of the spaces' total area of interior surface in square feet; and (2) to produce a satisfactory improvement in an existing space, the total absorption after treatment must be between 3 and 10 times the absorption before treatment (see the example in Fig. 10.2-13).

A procedure for calculating the percentage of reduction in loudness of reflected sound produced by acoustical treatment is outlined as follows: (1) Determine the number of absorption units in the room

Noise reduction with acoustical tile can be illustrated by assuming a family (or living) room 14 ft. × 24 ft. × 8 ft., finished with a wood floor and plaster walls and ceiling. Absorption *before* treatment (a_1) is found by multiplying the area of each material by its coefficient.

Material	Area (sq. ft.)	Absorption Coefficient (500 cps)	Units
Wood floor	336	@ .01	= 3.36
Plaster walls	608	@ .03	= 18.24
Plaster ceiling	336	@ .03	= 10.08
Furniture and people		(estimated)	= 12.32
	Total absorption before treatment (a_1) =		44.00

Assume the entire ceiling is treated with acoustical tile having a noise reduction coefficient (NRC) of .70. The net coefficient is .67 (after subtracting .03 for the plaster covered with tile) and will furnish 225.12 net added absorption units. The total absorption *after* treatment (a_2) will then be:

$$44 + 225.12 = 269.12 \text{ units}$$

Dividing a_2 by a_1 $\left(\dfrac{269.12}{44} = 6.12 \right)$, the absorption will be increased 6.12 times, meaning the intensity of the reflected sound will be reduced to 1/6.12 or 16.3% of its value before treatment. Referring to Figure 10.2-14, an absorption ratio of 6.12 corresponds to a reduction in loudness of 45%!

Checking these calculations with the noise reduction guidelines given in the text, the total number of units after treatment (269.12) is 21% of the total room surface (1280 sq. ft.), which is near the lower limit of the recommended range of 20% to 30%. The absorption ratio of 6.12 is well within the recommended range of 3 to 10, indicating there will be a satisfactory contrast with the conditions before treatment.

Using the formula $T = .05V/a$, the reverberation time before treatment is reduced from 3 sec. to .49 sec. after treatment.

FIGURE 10.2-13 *Noise reduction example.*

before treatment (a_1,); (2) to this add the net number of units furnished by the treatment to give the total absorption in the room after treatment (a_2); (3) divide a_2 by a_1, to obtain the absorption ratio a_2/a_1; (4) determine the percentage of loudness reduction from Figure 10.2-14.

FIGURE 10.2-14 *Relation of reduction in loudness (%) to the absorption ratio.*

In small spaces it is desirable to reduce the reverberation time to about 0.5 sec. and in large spaces, especially those used for music as well as speech, to not more than 1 sec. In most spaces the problem is usually one of reducing reverberation enough to lower noise levels about 5 to 8 dB. Unless the room shape presents special problems, covering the ceiling with a good sound-absorbing material is usually adequate treatment.

Spaces such as corridors generally need sound-absorbing materials on the floor as well as the ceiling to reduce the *piping* of sound along the corridor to other spaces.

In small rooms, especially those containing upholstered furniture and heavy carpeting, additional absorption in the form of acoustical ceilings will probably be unnecessary.

■ TRANSMISSION THROUGH SOUND-ABSORBING MATERIALS

Acoustical materials are used to absorb sound. Although some materials may be effective in preventing sound transmission, most acoustical surfacing materials are not. Acoustical tile and panels, carpet and pad, and other sound-absorbing surfacing material in a room where noise originates act indirectly in reducing sound transmission to adjoining rooms by lowering the noise level in the room where the sound originates. Used in an adjoining room they will similarly lower the level of the transmitted noise. These reductions, however, usually amount to only a few decibels. Acoustical surface treatment will supplement but will not take the place of sound isolation. *Sound isolation* is the control of sound transmission through walls and floor and ceiling assemblies.

10.3 SOUND TRANSMISSION

Sound is transmitted through construction when wall, floor, or ceiling surfaces vibrate. These surfaces may be set into vibration by either (1) the alternating air pressure of incident sound waves, called *airborne sound transmission*, or (2) by direct mechanical contact or impact such as might be caused by vibrating equipment, a footstep, an object dropped on a floor, or a slamming door. Transmission as a result of direct mechanical contact or impact is called *structure-borne sound transmission*. In either case, the vibrating surfaces generate new sound waves of reduced intensity in the room or space where the sound originated as well as in the room or space on the opposite side of the wall or ceiling and floor assembly (Fig. 10.3-1).

10.3.1

AIRBORNE SOUND TRANSMISSION

Airborne sound is produced by indoor and outdoor sources that radiate sound directly into the air, such as people talking, a TV on or CD playing, mechanical equipment humming, or traffic rumbling. Airborne sound may be transmitted to adjoining rooms or spaces (1) directly through the intervening constructions, or (2) by way of flanking paths such as open doors or along interconnecting structures (Fig. 10.3-2). Airborne sound can be isolated by providing walls and floor and ceiling assemblies that will reduce the level of transmission to the point where it is either inaudible or at a level that is not annoying or interfering.

10.3.1.1

Measuring Airborne Sound Transmission

In order to determine the performance of walls and floor and ceiling assemblies in isolating airborne sound transmission, the transmission loss of the wall or assembly must be measured. The *transmission loss*

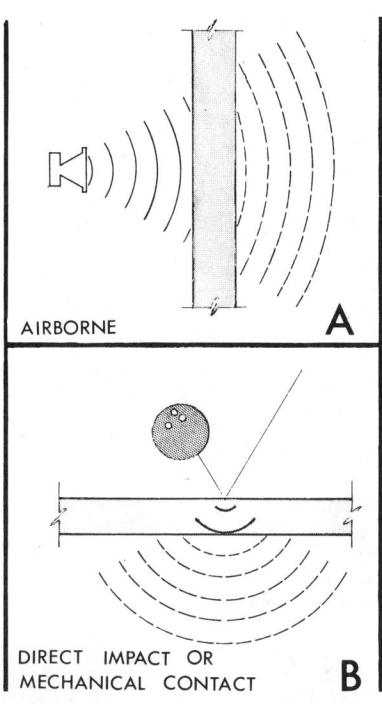

FIGURE 10.3-1 *Sound is transmitted by (a) alternating air pressures of airborne sound, and (b) impact or direct mechanical contact.*

(TL) is the number of decibels that a sound loses while being transmitted through a wall or floor and ceiling assembly when tested in a laboratory in accordance with ASTM Standard E 90. In both laboratory and field tests of walls for airborne transmission, the procedure is to generate a steady sound on one side of the wall and measure the sound level in decibels on both sides. The TL of the wall is the measured difference in sound level on the two sides. Thus, in Figure 10.3-3, a sound of a given frequency with an intensity level of 70 dB on one side of a wall will have a reduction in intensity to 40 dB on the other side, when the wall has a TL of 30 dB at this frequency. The higher the TL, the more efficient the wall or assembly.

TL is a physical property of a wall or floor and ceiling assembly that depends on the materials and methods used in the construction, and not on the loudness of

the sound striking the surface. However, the TL varies considerably depending on the frequency of the sound, and therefore assemblies are tested at numerous frequencies and plotted as a curve, as shown in Figure 10.3-4. As shown there, assemblies are generally much more effiient for high-frequency sounds than for low-frequency ones.

Actually, it is seldom necessary or ecomically feasible to build walls or floor and ceiling assemblies that are capable of completely blocking the transmission of sound through them. Because of the ability of normal background noise to block out other sounds, sound transmitted through a wall or floor will be inaudible or masked if its intensity level is below the level of the background noise in the receiving room (Fig. 10.3-5).

■ SINGLE-NUMBER RATING SYSTEMS

Except in special cases, it is impractical to use the TL values at each frequency tested because it is difficult to identify the frequency of a noise source in that amount of detail. For convenience in selecting or comparing walls or floor and ceiling assemblies, two single-number rating systems have been developed, the *average-TL* system and the *Sound Transmission Class (STC)* system. Either can be used to determine if a wall or floor and ceiling assembly will offer sufficient resistance to sound transmission, but the one generally used today is the STC system.

The average TL is the numerical average of the transmission loss values measured at nine or more frequencies, including 125, 175, 250, 350, 500, 700, 1000, 2000, and 4000 cps. Average TL was used in the past by the National Bureau of Standards (now called National Institute of Standards and Technology), Riverbank Acoustical Laboratories, and other laboratories for rating airborne sound transmission through walls and floor and ceiling assemblies. It was also once the most commonly used basis for the comparison of published data.

To estimate the minimum average TL required of a wall or floor and ceiling

PLAN

ELEVATION

FIGURE 10.3-2 *Sound may be transmitted (1) through intervening construction, (2) through open air paths, or (3) along interconnecting structures.*

FIGURE 10.3-3 *Transmission loss is the difference in sound level on opposite sides of a construction. The wall here has a transmission loss of 30 dB.*

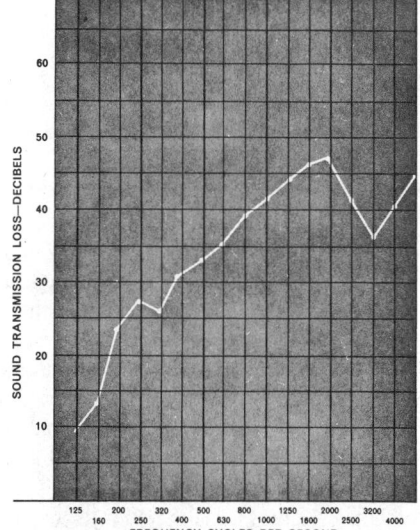

FIGURE 10.3-4 *The transmission loss of a wall consisting of 1/2-in.-thick gypsum board nailed directly to both sides of wood studs placed at 16 in. on centers.*

assembly to obtain speech privacy, subtract the level of the background noise on one side of the construction from the level of sound that strikes the other side. Hearing conditions on the listening side of walls with varying average TL are shown in Figure 10.3-6. Based on the curve shown in Figure 10.3-4, the average TL of a commonly used wood stud partition is 31 dB, which is not much resistance to sound transmission.

The average-TL system gives a fair indication of the sound isolation value if there are no sharp dips on the test curve and if the values being compared are based on identical tests performed by the same laboratory under identical conditions. Sharp dips on the curve in a narrow frequency band, which are normal for partitions, are harmful, but the numerical averaging in

the average TL system misses them, making this system inaccurate for use with partitions.

The Sound Transmission Class (STC) system of rating airborne-sound isolating performance is preferable to the average-TL system because it compensates for dips in the test curve by lowering the overall evaluation when they are present. This presents a more reliable indication of speech or noise transmission, especially by partitions.

The STC is a single-number rating for airborne sound transmission that represents the TL of a wall or floor and ceiling assembly at all test frequencies. *The higher the STC rating, the more efficient the construction.*

An STC rating is derived by plotting the laboratory TL test curve for a given assembly, tested at 16 frequencies in accordance with ASTM Standard E 90, and comparing this curve to a standard frequency curve, also called a standard contour (Fig. 10.3-7). The standard frequency curve, which is described in ASTM Standard E 413, is drawn on a transparent overlay that is placed over the test curve and moved up and down until the test curve remains above the standard curve at every frequency, within the following tolerances: (1) no single unfavorable deviation (where the sound transmission loss value falls below the standard curve) may exceed 8 dB, and (2) the sum of unfavorable deviations may not exceed 32 dB. The STC rating is the numerical value that corresponds to the sound transmission loss value at 500 Hz (see Fig. 10.3-7).

The ASTM test procedure was first issued in 1950 and has been extensively revised since that time. Many current STC ratings for various assemblies are based on tests performed before the current revision of ASTM E 90. STC values derived from earlier tests may vary somewhat from tests on similar assemblies performed in accordance with ASTM E 90. Very old references may vary considerably. Variables that might influence test results include the number of test frequencies used, the size of the panel tested, and the application of the standard frequency curve.

In general, experience indicates that if a wall reduces the level of conversational speech just to the point where it cannot be understood in an adjoining room, the

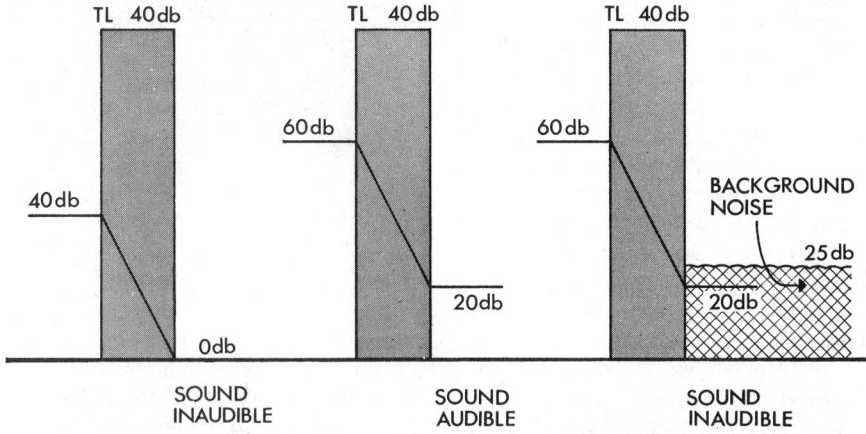

FIGURE 10.3-5 *Normal background noise on the listening side of a wall has the effect of masking transmitted sound that might otherwise be objectionable.*

Controlling Airborne Sound Transmission

Several variables affect the control of airborne sound transmission in order to provide a suitable acoustical environment, including (1) the frequency and sound level of the sources; (2) the sound level that will be acceptable within the space; and (3) the sound-reducing characteristics of the intervening construction.

Methods to control airborne sound transmission include (1) selecting walls and floor and ceiling assemblies that reduce the degree of transmission by vibration, (2) closing paths of direct air transmission, and (3) using sound-absorbing materials within the free airspace of frame construction. Generally, a combination of these control measures is necessary to achieve suitable aural privacy between rooms. The primary objective, in any case, is to reduce the level of transmitted noise below the level of background noise on the listening side of the wall or floor and ceiling assembly. Methods of controlling airborne sound transmitted through

FIGURE 10.3-6
Sound-Insulating Properties of Construction According to Transmission Loss (TL)*

Transmission Loss (TL)	Hearing Conditions
30 dB or less	Normal speech can be understood quite easily and distinctly through the wall.
30 to 35 dB	Loud speech can be understood fairly well. Normal speech can be heard but not easily understood.
35 to 45 dB	Loud speech can be heard, but is not easily intelligible. Normal speech can be heard only faintly, if at all.
40 to 45 dB	Loud speech can be faintly heard but not understood. Normal speech is inaudible.
45 dB or greater	Very loud sounds, such as loud singing, brass musical instruments, or a radio at full volume can be heard only faintly or not at all.

*Background noise level on the listening side assumed at 30 dB.

the background noise on one side of the wall or assembly from the level of sound that strikes the other side. The examples in Figure 10.3-9 illustrate the approximate effectiveness of walls with varying STC numbers as influenced by the background noise. Values of STC for several assemblies are given in Appendix A.

wall will provide acceptable overall privacy between rooms (Fig. 10.3-8).

To estimate the minimum STC required of a wall or floor and ceiling assembly to obtain aural privacy, subtract the level of

FIGURE 10.3-7 *Method for obtaining an STC rating.*

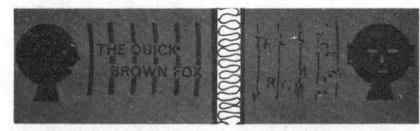

FIGURE 10.3-8 *A wall generally provides acceptable privacy between rooms when the level of conversational speech is reduced so that it cannot be understood.*

walls and through floor and ceiling assemblies are discussed in Section 10.4.2.

The sound-reducing characteristics of a wall or floor and ceiling assembly directly influence its cost. Therefore, it is important to minimize or eliminate the need for special construction techniques by proper planning using some basic sound-control principles and common sense. These are discussed in Section 10.4.1.

10.3.2

STRUCTURE-BORNE SOUND TRANSMISSION

Structure-borne sound is produced when walls or floor and ceiling assemblies are set into vibration by direct impact or direct mechanical contact. The vibrating surface then generates new sound waves on both sides of the construction. The new sound waves may even occur in different parts of the building. A floor is especially

critical in this type of noise transmission. Typically, structure-borne noise is caused by objects that slide on or strike a floor, such as footsteps, furniture, and dropped objects; objects that bump against a wall, such as furniture; and slamming of doors. It is also caused by mechanical equipment, fixtures, and appliances that transmit their own vibrations to adjacent construction, such as motors and fans on heating, cooling, and ventilating equipment; vacuum cleaners; water closets; bathtubs; and showers.

The problem of isolating structure-borne sound is to provide walls and floor and ceiling assemblies that will reduce the level of transmission to the point where the sound is either inaudible or not annoying or interfering.

10.3.2.1

Measuring Impact Sound Transmission

To measure the relative performance of a floor and ceiling assembly, a constant source of impact noise is created on top of the assembly and the sound level is then measured in the receiving room or adjoining areas (Fig. 10.3-10).

The isolation against impact noise provided by a given floor and ceiling assembly is determined in accordance with a standard test method described in ASTM

FIGURE 10.3-10 *Impact sound transmission through floor and ceiling assemblies is measured in a laboratory using a tapping machine.*

E 492, which is based on the International Organization for Standardization, ISO Recommendation 140-1960, "Field and Laboratory Measurements of Airborne and Impact Sound Transmission." A tapping machine is used that produces a series of uniform impacts at a uniform rate on the floor under test. The *impact-sound pressure level* (*ISPL*) (measured in decibels) is then recorded in the receiving room below and electronically analyzed into different bands of frequency. A curve can then be plotted, showing how the sound energy in the receiving room is distributed over the audible frequency range (curve 1 in Fig. 10.3-11).

In the case of airborne sound transmission, the difference between the sound levels in the sending and receiving rooms is measured and is expressed as a sound transmission loss. Unlike the airborne sound transmission loss curves, ISPL curves measure the actual sound levels from a standard noise source that are transmitted through the construction. The lower the transmitted level, the better the acoustical performance of the construction.

The test used to measure ISPL assumes that a construction that will transmit little sound from the tapping machine will perform similarly under other types of impact sound. However, because the noise produced by the tapping machine does not duplicate the sound produced by many types of impact, studies are currently being conducted to develop additional methods for determining impact sound performance. For example, much footfall noise is in a frequency that falls below the lowest of the standard test frequencies. Because of the problems in duplicating actual conditions (and other limitations of the test used), model building codes often require that specific construction assembly types be used rather than relying on test performance criteria.

FIGURE 10.3-9
Sound Isolation of Walls in Terms of STC Values

Sound Transmission Class (STC)				
	Background Noise[a]			
None	20 dB	30 dB	40 dB	Speech Audibility[b]
40 to 50	25 to 35	15 to 25		Normal speech easily understood.
50 to 60	35 to 45	25 to 35	15 to 25	Loud speech easily understood. Normal speech 50% understood.
60 to 70	45 to 55	35 to 45	25 to 35	Loud speech 50% understood. Normal speech faintly heard but not understood.
70 to 80	55 to 65	45 to 55	35 to 45	Loud speech faintly heard but not understood. Normal speech usually inaudible.
	65 to 75	55 to 65	45 to 55	Loud speech usually inaudible.

[a]Background noise levels on *A* scale of sound level meter.

[b]Loud speech taken as 10 dB higher than normal speech; intelligibility taken as percent of sentences understood; normal source and receiving room absorption considered.

FIGURE 10.3-11 *Impact noise rating (INR) is obtained by comparing the ISPL test curve (1) with a standard criterion curve (2).*

■ IMPACT NOISE RATING (INR)

An *impact noise rating* (*INR*) is a single-number rating in decibels for evaluating the impact sound isolation performance of a floor and ceiling system. This rating is determined by analyzing tests performed in accordance with ISO R140-1960. This method has been superseded by the *impact-insulation class* (*IIC*) system, discussed below. The INR rating system is discussed here because some still-available, published material uses it to compare the performance of floor and ceiling assemblies.

To determine an INR, the assembly's ISPL is measured and plotted on a transparent overlay, which is then laid over a standard criterion curve (curve 2 in Fig. 10.3-11). The standard curve is then moved up or down until the ISPL test curve falls over the standard criterion curve within the following tolerances: (1) the mean amount by which the test curve exceeds the standard curve must be 2 dB or less, as averaged over the 16 one-third octave frequency bands between 100 and 3200 Hz; (2) the test curve must not exceed the criterion curve by more than 8 dB at any one frequency.

The INR value for a given assembly is the distance below or above the standard criterion curve that the ISPL curve falls. Plus (+) INR values indicate better-than-standard criterion performance; minus (−) INR values indicate less-than-standard criterion performance. When the test curve meets the standard curve within the allowable tolerance, the INR is 0. If the standard curve must be moved down 5 dB for the test curve to conform, the construction is given a +5 rating, indicating that the assembly is 5 dB better than the standard; if the standard curve must be moved up 5 dB for the test curve to conform, the assembly is given a −5 rating, indicating that it is 5 dB poorer than the standard.

The criterion curve is based on average background noises that are typical of apartments in moderately quiet suburban neighborhoods. For quieter areas, floor constructions with an INR of +5 to +10 should be used. For noisy urban districts a construction with a −5 INR may suffice. INR values for various floor and ceiling constructions are given in Appendix A.

■ IMPACT INSULATION CLASS (IIC)

In an effort to reconcile and correlate the various rating systems for airborne sound transmission and impact sound transmission, the Federal Housing Administration (FHA) devised a single-number rating system for impact sound transmission called the *impact-insulation class* (*IIC*) system. The IIC system supersedes the INR system described above.

On the average, an IIC rating for a given construction is approximately 51 dB higher than the INR rating. For example, an INR rating of +4 would be comparable to an IIC rating of about 55. However, exact values for IIC ratings should be determined on the basis of individual test data.

To develop the IIC of an assembly, the ISPL is measured and a curve plotted as discussed above (see Fig. 10.3-11). A standarized reference contour is plotted on a transparent overlay, placed over the ISPL curve, and moved up and down until only the following allowable tolerances exist: (1) a single unfavorable deviation in which an ISPL value lies above the reference contour may not exceed 8 dB, and (2) the sum of the unfavorable deviations may not be greater than 32 dB. The IIC rating is the value on the right scale corresponding to the ISPL value at 500 Hz (Fig. 10.3-12).

FIGURE 10.3-12 *Impact isolation class (IIC) is obtained by comparing the ISPL curve with a standard reference contour and is read as a positive rating.*

HUD-TS-24, 1974, "FHA Guide to Airborne, Impact, and Structure-borne Noise Control in Multifamily Dwellings," presents a three-segment standard contour that offers three options, depending on the outdoor, nighttime background noise level expected. The contours presented are labeled Grade I (IIC 55), Grade II (IIC 52), and Grade III (IIC 48). The higher the outdoor level, the greater the masking will be and the lower the insulation needed. The most used of the three is the Grade II (IIC 52) contour.

The IIC rating system parallels the STC rating system in that performance of a floor and ceiling assembly is expressed with positive values in ascending degrees of impact sound isolation. *The larger the IIC rating, the better the assembly.*

10.3.2.2

Controlling Structure-Borne Sound Transmission

Most building materials, especially hard and dense materials, transmit vibrations readily. The steel frame of a building, for example, can telegraph a vibration throughout a large structure. High heels clicking on a concrete floor can be heard distinctly in the room below if no other construction intervenes. Vibration of the structure sets up sound waves that can often be heard distinctly at great distances from the source. Impact noise, caused by a floor or wall being set into vibration by direct mechanical contact, is then radiated from both sides. This vibration may also be transmitted throughout the structure to walls and reradiated as sound to adjoining spaces. Since footsteps and furniture being moved on floors constitute a major impact problem, control of impact noise is treated here as primarily a floor problem.

Impact isolation methods in general are more effective in reducing the high-frequency components of impact noise than in reducing the low-frequency thumping noises caused by heavy impacts. The limitation on low-frequency isolation is more stringent for wood joist floors than for concrete floors because of the inherent lower weight and greater flexibility of wood-joist floor construction. Methods to control structure-borne sound transmission are discussed in Section 10.4.2.

10.4 DESIGN AND CONSTRUCTION

The following recommendations deal with (1) criteria and planning principles to aid in the design and selection of walls and floor and ceiling assemblies to control noise within buildings; (2) construction recommendations, including precautions essential to obtaining and maintaining suitable sound isolation between rooms and between living units; and (3) selection criteria for obtaining suitable sound isolation.

Design of doors and windows and selection of appropriate hardware to control sound transmission through them are discussed in Section 8.1.4. Criteria and planning principles that will reduce sound generation by and transmission through plumbing systems are covered in Section 11.1.5; heating, ventilating, and air conditioning systems in Section 11.3.3.2; electrical systems and appliances in Section 12.3.7.

10.4.1

DESIGN AND PLANNING

Planning for sound control involves overcoming problems of loud exterior and interior noise at the sources; orienting quiet spaces away from high-level noise sources; and careful interior design and building layout, especially with respect to the shape of the buildings and the location of door and window openings and other sound leaks. Generally, the objectives in the planning for sound control are to reduce noise by (1) quieting the sound at its source; (2) separating the source from the listener by distance; and (3) buffering the transmission of noise between the source and the listener by using barriers such as walls and floor and ceiling assemblies that have improved resistance to sound transmission.

10.4.1.1

Background Noise Sources and Levels

The source and intensity of outdoor noise at the site affect the noise-reducing effectiveness of exterior walls and roofs, and also have an important effect on the selection of interior wall construction and floor and ceiling assemblies. Noise characteristics of the site should be studied to determine existing and potential noise levels (Fig. 10.4-1).

Theoretically, airborne sound travels from a *point noise source* in a spherical pattern (see Fig. 10.1-1). The sound level of a point noise source is reduced 6 dB each time the distance from the source is doubled (see Fig. 10.2-3). But fast-moving traffic actually is a *line noise source* rather than a point source. Sound

FIGURE 10.4-1 *Typical outdoor background noise levels.*

---

from a line noise source travels in a cylindrical rather than a spherical pattern. In order to reduce the noise level from a line source by 6 dB, the distance between noise source and listener must be increased 4 times.

The direction and speed of air movement, air temperature, and obstructions such as buildings, trees, and terrain all affect the sound level. For example, where the wind moves unobstructed, sites upwind of major noise sources will be quieter than sites an equal distance away from the noise on the downwind side.

The hard exterior surfaces of buildings reflect a large proportion of the sound

---

FIGURE 10.4-2 *Hard surfaces reflect sound; soft, porous ones, such as are found in plants, absorb some of it.*

energy that strikes them (Fig. 10.4-2). Multiple reflection of high-level sound sources striking a large building can materially increase the noise problem at a site. Curved buildings, or buildings designed to form courtyards having the concave side facing a noise source, will also result in an especially difficult noise problem. Future development of nearby vacant land should be anticipated and the type and level of probable or possible noise sources should be investigated.

If a proposed site is near high-level noise sources, such as a business district, highway, expressway, railroad line, or airport, both daytime and nighttime outdoor noise levels should be surveyed. When the average nighttime background noise level is unusually high or if there are infrequent but intermittent unusually loud or annoying noises, expert acoustical advice should be sought.

■ **GROUND VEHICULAR TRAFFIC**

Normally, ground vehicular traffic (automobiles and trucks) is a primary source of outdoor background noise. Background noise levels are affected by such local variations as topography, landscaping, and the masking or magnifying effect of other buildings. Typical outdoor noise levels are shown in Figure 10.4-3. High-speed highways and dense city traffic generally produce higher noise levels than do moderate-speed suburban roads carrying an equivalent number of cars. Trucks increase highway noise about 15 dB, and trucks going uphill will produce the highest highway noise levels. Unless the highway is already saturated with traffic, an increase generally can be expected, with a resultant increase in noise.

■ **AIRCRAFT TRAFFIC**

The proliferation of jet passenger and transport operations throughout the coun-

FIGURE 10.4-3
Approximate Sound Levels for Automobile Traffic at Ground Level

Cars/ Minute*	Distance from Traffic Lane		
	40 ft.	100 ft.	500 ft.
1	50 dB	43 dB	29 dB
10	59 dB	52 dB	38 dB
100	63 dB	60 dB	47 dB

*The maximum capacity of one lane of cars moving freely at approximately 30 mph is 2000 cars per hour, or 34 cars per lane per minute.

try and the rapid expansion of small commercial and business jet aircraft create noise problems even in relatively small communities. Methods for minimizing aircraft noise include techniques for quieting engines, reducing noise exposure by takeoff and landing operational procedures, and reducing the number of people affected by controlling and regulating land use around airports.

Two types of aircraft noise are of particular concern: (1) noise generated by revving engines on the ground and (2) flyover noise. Figure 10.4-4a illustrates that the noise level during flyover changes very rapidly with time, with the maximum noise reached and sustained for perhaps only a second or less, and that sound

waves during this short time interval strike a building from all directions.

Figure 10.4-4b illustrates the noise produced by revving engines while the aircraft is stationary on the ground. In contrast to flyover noise, because the relationship between aircraft and building is fixed, only one or two exterior surfaces of the building may be directly exposed to aircraft noise, whereas other surfaces may be partially shielded from the sound radiated. There also may be intervening obstacles such as hills, houses, or other terrain features that will partially block the sound transmitted from the aircraft to the building.

Figure 10.4-5a illustrates typical noise level contours of jet aircraft during take-offs; Figure 10.4-5b is an example of noise level contours for the revving of one engine of a jet aircraft.

Of all exterior wall and roof surfaces, windows offer the least resistance to noise transmission. The difference in perceived noise level (PNdB, Fig. 10.4-5) between exterior and interior noise levels from aircraft is illustrated in Figure 10.4-6. Generally, the lower values in each category will be observed for buildings having window areas comprising more than 20% of the total exposed wall area, with relatively lightweight wood-framed walls and roofs (weighing less than 8 lb./sq. ft.). The higher values generally will be obtained for buildings with relatively small airtight windows and airtight storm windows of good quality. An example of aircraft noise

FIGURE 10.4-4 *(a) Flyover noise strikes buildings from all angles, but its maximum intensity lasts only a second or less; (b) noise from revving engines on the ground strikes fewer building surfaces but lasts longer.*

PERCEIVED NOISE LEVEL CONTOURS FOR TAKE OFFS FOR CIVIL JET AIRCRAFT

a

b PERCEIVED NOISE LEVEL CONTOURS – ONE – ENGINE RUNUPS FOR CIVIL JET TRANSPORTS

FIGURE 10.4-5 *Aircraft noise level contours expressed in perceived noise level (PNdB).**

*PNdB, Perceived Noise Level, in decibels, is used to describe aircraft noise levels. The perceived noise level is a single number rating of aircraft noise used to rate the relative acceptability or noisiness of aircraft sounds. The perceived noise level is calculated from measured noise levels and correlates well with subjective responses to various kinds of aircraft noise.

FIGURE 10.4-6
Perceived Noise Levels from Aircraft

Window Condition	Difference Between Interior and Exterior Noise Levels
Windows open	10 to 20 PNdB
Windows closed*	20 to 30 PNdB
Storm sash installed*	25 to 35 PNdB

*For maximum performance, glass should be sealed airtight within frames; storm sash should be separated by at least 2-in. airspace between panes.

in the courtyard will help to reduce reverberation of sound within the courtyard.

■ NOISE BARRIERS

A building's walls, landscaping, and topography can be used as barriers between a noise source and a listener. A barrier's effectiveness depends on the frequency of the sound, the angle of the sound shadow, and the height of the barrier (Fig. 10.4-9). Higher-frequency sounds are controlled more effectively by barriers than are low-frequency sounds. For greatest effectiveness, barriers should be as high as possible, be located as close as possible to the noise source, and be impervious to air flow. Barriers should also have some mass, but great weight will not appreciably increase the effectiveness of the barrier.

Trees and shrubs reduce noise levels only slightly. Plants absorb some sound and can help to reduce reverberation noise between buildings. A 2-ft.-thick, dense cypress hedge will reduce high-frequency noise about 4 dB (Fig. 10.4-10). A single row of trees does little to reduce noise levels. Even dense tropical jungle will reduce levels only about 6 to 7 dB per 100 ft. of depth, which is only 1 or 2 dB more than what the distance alone would cause. Since noise has a psychological effect, plantings that provide a visual barrier to noise sources such as traffic may help to reduce sensitivity to the noise.

Nonresidential buildings parallel and adjacent to a highway can serve as a noise barrier between the highway and residential buildings farther away (Fig. 10.4-11a).

reductions through various exterior wall and roof construction assemblies is summarized in Figure 10.4-7.

10.4.1.2

Site Planning

Site planning can reduce problems caused by exterior noise by (1) increasing the distance between the noise source and buildings; (2) siting buildings to minimize the effect of direct sound transmission, reflection, and reverberation from existing or possible future buildings; and (3) using natural topography, other buildings, barrier walls, and, in some cases, landscaping to shield buildings from major noise sources.

■ BUILDING LOCATION

Buildings should be located as far as possible from noise sources. Buildings reflect noise readily. Orientation that places rows of buildings parallel to each other results in the buildup or focusing of the noise by multiple reflection of the sound. Acoustically, the best orientation for a building occurs when its long axis is perpendicular to noise sources (Fig. 10.4-8a).

Courtyards created by a series of buildings located to form a U shape can be especially noisy when the open end of the U faces the major source of noise (Fig. 10.4-8b). If a courtyard is made desirable by other design considerations, having the open end of the U facing away from noise sources is acoustically better. Landscaping

FIGURE 10.4-7
Summary of Noise Reduction Through Exterior Constructions

Typical Constructions	Difference Between Outside and Inside Noise Levels (PNdB)					
	0	10	20	30	40	50
Windows						
1/8-in. glass in double-hung wood frame		▬				
1/4-in. plate glass, sealed in place			▬			
Aluminum frame window, glass panels set in neoprene gaskets, 1/4-in. and 7/32-in. panes, 3-3/4-in. airspace					▬	
Exterior walls						
Wood sheathing (or stucco), 2-in. × 4-in. studs, 1/2-in. plaster board interior surface			▬			
Wood sheathing (or stucco), 2-in. × 4-in. studs, 7/8-in. plaster interior surface				▬		
4-1/2-in. brick or 6-in. to 8-in. lightweight concrete block				▬		
9-in. brick wall					▬	
Roofs						
Built-up roofing on 1-in. wood decking, 1/2-in. gypsum board on 2-in. × 8-in. joists			▬			
Built-up roofing on 1-in. wood decking, 7/8-in. plaster on 2-in. × 8-in. joists				▬		
Built-up roofing or shingles on wood siding, ventilated attic space, 1/2-in. gypsum board ceiling					▬	
Built-up roofing or shingles on wood sheathing, ventilated attic space, 7/8-in. plaster ceiling					▬	

FIGURE 10.4-8 *(a) Perpendicular orientation exposes less wall area to noise source; (b) sound reverberates in a U-shaped courtyard.*

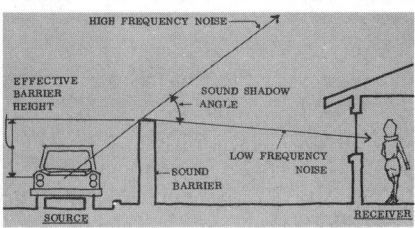

FIGURE 10.4-9 *Buildings and walls serving as barriers should be as high as possible and as close as possible to the noise source.*

FIGURE 10.4-10 *Plants reduce noise levels only slightly, but may provide beneficial visual and psychological barriers.*

FIGURE 10.4-11 *(a) Nonresidential buildings and (b) natural terrain and man-made road cuts and fills can serve as noise barriers.*

Hills and highway cuts can provide natural barriers to major noise sources if the buildings are located to take advantage of these sound shadow barriers (Fig. 10.4-11b).

10.4.1.3

Interior Planning

Careful planning is essential if the effects of noise are to be reduced. Initial preplanning and design of interior spaces with aural privacy in mind can often minimize or eliminate the need for costly construction techniques to control noise. The problems associated with sound transmission through walls and floor and ceiling assemblies can be solved by selecting suitable sound-insulating systems and controlling the methods used in the field, as discussed in Section 10.4.2. Too often, too little attention is given to the principles of sound control in the planning stage.

The aim of sound control should be to create an environment that is acoustically comfortable, not to render rooms sound-proof. The basic principles of such planning include (1) locating rooms or units as far as possible from noise sources; (2) separating noisy areas from quiet areas by isolating the noise source or using buffer spaces; (3) eliminating air paths that permit sound to travel unhindered between rooms or units, especially through the design, sizing, and placement of doors and windows; and (4) reducing the noise level at the source by selecting mechanical equipment and appliances that operate quietly.

In planning for sound control in homes and apartments most of the emphasis is usually placed on sound originating in adjacent houses or apartments or from outdoors. This is because people can control the intensity of noise sources within their own house or apartment unit and also tend to be more tolerant of noises created by their own family than they are of those from other sources.

It is almost impossible to correct sound transmission problems caused by faulty interior design after a building has been built, except at great expense. Therefore, it is essential that acoustical principles be recognized in the initial design stages.

The following discussion uses apartment living units to illustrate design principles, but most of the principles used are equally applicable to commercial and institutional buildings.

■ SPACE LAYOUT

Spaces can be zoned acoustically into three general areas: (1) areas that are relatively noisy, (2) areas that should be quiet, and (3) areas that fall between these two extremes.

People and their activities establish acceptable noise levels for different areas. Noisy areas should, where possible, be grouped together. Spaces such as corridors and stairways can help to provide acoustical buffers against airborne sound transmission between spaces. Closets, storage rooms, and bookcases backed up to common walls help to reduce airborne sound transmission between rooms, and in addition force the placment of furniture away from the common walls, thereby increasing the distance between a noise source and the adjacent room (Fig. 10.4-12).

FIGURE 10.4-12 *Closets and bookshelves can serve as buffers to separate listeners from a noise source.*

Mirror plans often are used in apartment design. It is acoustically preferable that the plan be mirrored horizontally and matched vertically where possible (Fig. 10.4-13). This type of planning places noisy areas adjacent to other noisy areas and quiet areas adjacent to other quiet areas so that the sound-control requirements, and therefore the cost of intervening walls and floor and ceiling assemblies, can be as low as possible. For example, kitchens and bathrooms, which have similar noise levels, are located adjacent to and above and below kitchens and bathrooms in adjoining apartments. Similarly, quieter areas such as bedrooms and living rooms should be located adjacent to and above and below quiet areas in adjoining apartments. The noise level requirements on both sides of walls and floor and ceiling assemblies will be the same, and the full effect and cost of improved sound control measures will benefit all apartments. Some additional precautions that should be observed in mirror planning are illustrated in Figure 10.4-14.

Row plans (where apartments are arranged in lines) provide better acoustical performance than block plans (where each apartment may have up to three common walls) (Fig. 10.4-15b). In row planning, with halls between every other apartment, each apartment has only one common wall (Fig. 10.4-15a).

Room proportions can be planned to reduce sound transmission. The higher the ratio of common wall to floor area, the more sound the common wall will transmit (Fig. 10.4-16).

Adding large absorptive surfaces, such as acoustical tile, carpeting, or heavy drapes, to rooms will slightly reduce the amount of sound available that can be

FIGURE 10.4-13 *Mirroring the plan results in about equal noise level requirements on both sides of common walls and floor and ceiling assemblies.*

FIGURE 10.4-14 *Plumbing should be located in walls within each living unit rather than in common walls between units.*

transmitted. In a room with large hard-surfaced areas, a large portion of the sound is reflected, whereas large soft-surfaced areas will absorb sound.

10.4.1.4

Doors and Windows

Doors, windows, and other openings are direct air-to-air paths that transmit sound readily. Exterior doors, windows, and other openings should be placed on the quiet side of a building when possible.

FIGURE 10.4-15 *(a) Row planning reduces the number of common walls between units. (b) Block planning can result in as many as three common walls in each unit.*

FIGURE 10.4-16 *Room proportions can reduce the ratio of common wall to floor area. Hard-surfaced areas should be minimized; soft, absorbent areas increased.*

Where they are necessary on the noisy side, their number and size should be minimized when sound control has a higher priority than light, view, or ventilation. Exterior doors and windows should be located as far as possible from, or in positions that do not interfere with, doors and windows in adjacent apartments (Fig. 10.4-17).

When possible, interior doors should be staggered on opposite sides of corridors,

FIGURE 10.4-17 *(a and b) Doors and windows should be placed so that sound will not travel through them from one unit to another. (b) Interior doors should be staggered on opposite sides of a hallway.*

so that the least amount of sound is radiated directly from one door across to another. Offsetting doors allows the greatest amount of diffusion and dissipation of sound before it reaches adjoining units (Fig. 10.4-17b).

Noise is transmitted in both directions from open windows. To minimize noise from one apartment to another, the windows should be placed as far as possible from the common wall. Casement windows should be arranged to operate in the same direction so that the sound from one room is not reflected into another (Fig. 10.4-18).

The types of doors and windows used to control sound and their construction and use are discussed in Section 8.1.4.

FIGURE 10.4-18 *Casement windows should not reflect sound from one unit to another. (a) Improper arrangement; (b) proper arrangement.*

10.4.1.5

Plumbing, Mechanical, and Electrical Equipment and Systems

Plumbing and heating, ventilating, and air-conditioning equipment and systems produce airborne noise as well as structure-borne noise. It may be that the noise at the source cannot be eliminated, but it can be minimized by the selection of equipment, design layout, and construction precautions.

Precautions to avoid noise associated with plumbing systems are discussed in Section 11.1.5; those associated with heating, ventilating, and air-conditioning equipment and systems in Section 11.3.3.2; those associated with electrical systems in Section 12.3.7.

CONSTRUCTION RECOMMENDATIONS

The ability of materials and assemblies to reduce noise transmission varies with the frequency of the sound. Some assemblies are efficient at high frequencies, others at low frequencies. A wall or floor and ceiling assembly that will adequately reduce the transmission of conversational speech may not be able to block the low-frequency thump of an air compressor. High-frequency sounds are usually easier to control than low-frequency ones.

Acceptable sound conditioning depends on properly selecting and building sound-isolating walls and floor and ceiling assemblies. It also requires properly selecting and installing plumbing, heating, ventilating, air-conditioning, and electrical systems, equipment, and appliances of types that will reduce noise generation and transmission to acceptable levels.

The sound-isolating efficiency of a wall or floor and ceiling assembly depends on (1) its weight or mass; (2) effective separation of each face of a wall or floor and ceiling assembly; and (3) the addition of sound-absorbing materials between the effectively separated surfaces.

The basic methods of reducing airborne sound transmission include (1) increasing the weight of the wall or assembly; (2) devising discontinuous construction methods that will permit the surfaces of walls and floor and ceiling assemblies to be independent of one another; and (3) as an adjunct to discontinuous construction, adding sound-absorbing materials in the cavity of framed walls and floor and ceiling assemblies.

Weight has the effect of damping vibration. If a wall or floor and ceiling assembly had an infinitely great mass it could not be set into vibration by airborne sound pressures, regardless of their intensity. If the mass were extremely small, even faint sound pressures would move it. The ability of a wall or floor and ceiling assembly to transmit sound is inversely dependent on its mass. The heavier or more massive it is, the greater will be its resistance to sound transmission. However, massiveness soon reaches a point of diminishing return, since doubling the mass of a wall or floor and ceiling assembly will improve its resistance to sound transmission only

by about 5 dB. For example, to achieve a 20-dB reduction in sound transmission would require replacing a single layer of 1/2-in.-thick gypsum board with 16 layers laminated together.

Separating opposite surfaces by discontinuous construction or in such a manner that the coupling between the surfaces is very ineffective reduces transmission because the surface opposite the source is not set into as vigorous vibration as the surface on the source side.

One method is to use a construction in which the opposite faces are built independently of one another with no structural connections between them (Figs. 10.4-19 and 10.4-20). In wood-framed constructions, such as those with staggered studs or with two rows of studs, the framing provides vibration isolation between the two wall surfaces. The surfaces can also be isolated with resilient underlayment materials between the outer wall surfaces and the structural framing (Fig. 10.4-21), thus reducing transmission by isolating the surfaces, as an automobile spring isolates the body of the auto from the tires' bouncing over the road. Sound striking one side of a wall causes it to vibrate but, because of mechanical separation, the vibration of the other side is greatly reduced. Double walls are much more effective than single walls of the same overall weight.

FIGURE 10.4-19 *Staggered studs provide discontinuous construction by effectively separating opposite surfaces. Adding insulation improves performance.*

FIGURE 10.4-20 *A double row of studs, each row on separate plates, provides discontinuous construction. Adding insulation improves performance, and performance increases as insulation thickness is increased.*

FIGURE 10.4-21 *Sound-deadening board reduces transmission by resilient mounting of surfaces and can be used with single-, double-, or staggered-stud walls.*

Another method is to attach each face to the structural support yet still have them mechanically separated using flexible connections or resilient mountings so that the opposite faces do not vibrate sympathetically (Fig. 10.4-22). Then, much of the sound energy is dissipated by vibrating one surface and its mountings, instead of being transmitted to the opposite surface. The performance improvement resulting from discontinuous construction or resil-

FIGURE 10.4-22 *Resilient channels (on one side of the wall) flexibly mount the surface. The performance increases when insulation is added and continues to increase as the thickness of the insulation is increased.*

ient mountings can be increased significantly by placing sound-absorbing insulation blankets within the wall or floor and ceiling assembly to absorb sound energy in the cavity, as shown in Figure 10.4-19.

Discontinuous construction can be obtained in floor and ceiling assemblies by floating the finish flooring and separating the ceiling construction from the floor construction.

Attempting to isolate airborne sound transmission by separating opposite surfaces alone has limitations. Doubling the free airspace between the opposite surfaces of a wall or floor and ceiling assembly improves its resistance to sound transmission by only about 5 dB, assuming a constant mass. Within practical limits of conventional building materials, both mass and discontinuous construction are required in order for a wall to reach a 50- to 55-dB STC. STC values greater than that are not practical regardless of mass and isolation, without construction methods that will control transmission through flanking paths.

Presuming that walls and floor and ceiling assemblies that are selected on the basis of test data will actually perform according to their ratings, the greatest single problem in achieving satisfactory sound conditioning in the field is improper installation. Only when every detail has been correctly taken care of will the actual walls and floor and ceiling assemblies live up to the ratings of the tested assemblies. When possible, workers and other key personnel should be taught the reasons for care in building and applying sound-control measures. Supervision to ensure that the proper methods and techniques are used and coordination between the various trades are essential. Seemingly insignificant variations from the correct technique can nullify the effectiveness of costly sound-control measures. For example, nails that are too long can short-circuit resilient mountings and make them useless. Failing to seal small cracks and holes will create a flanking path and negate the effectiveness of a wall or floor and ceiling assembly. The manufacturer's instructions for installing sound-control devices, such as resilient clips and sound insulation, must be followed precisely. The actual walls and assemblies must be built exactly as were the tested ones.

It is normal to construct walls and floor and ceiling assemblies as continuous surfaces to divide living units into rooms. Windows, doors, and other openings are required in them. These horizontal and vertical surfaces and their openings, as well as construction imperfections such as cracks and holes, create flanking paths that transmit sound. Just as snapping a rope will cause it to snake the entire length, sound impulses in a wall or floor and ceiling assembly create vibrations that may transmit sound around, over, or under a partition, even though the partition itself has a high resistance to transmission (Fig. 10.4-23).

Open-air flanking paths are openings through which sound travels virtually undiminished. A doorway or window is an enormous hole through a wall. It is an ideal sound leak, and can render a wall almost acoustically ineffective unless properly sealed.

To reduce the number of flanking paths, doors and windows should be airtight, be heavy enough to minimize transmission through them, and be installed to maintain their airtightness.

Flanking paths through construction imperfections are illustrated in Figure 10.3-2. Walls, ceilings, and floors are usually pierced by innumerable pipes, ducts, conduits, lighting fixtures, electrical boxes, and similar equipment. Because airborne

FIGURE 10.4-23 *Sound can be transmitted around, over, and under partitions through (a) structure-borne flanking paths and (b) airborne flanking paths.*

sound will travel wherever there is an air passage, every one of them is a potential sound leak. Often they are inadvertently arranged to become ideal speaking tubes; air-conditioning and ventilating ducts are frequent offenders. Medicine cabinets or outlet boxes, placed back-to-back, are common sound leaks. Such air leaks must be closed, openings caulked or sealed, fixtures staggered, and ducts lined with absorbent materials or equipped with baffles that interrupt the direct path of the sound and assist in absorbing it.

Airborne sound transmission between two rooms usually cannot be controlled satisfactorily solely by acoustical surfacing materials. Covering the surfaces of walls or ceilings with acoustical materials, or a floor with carpeting, adds little to the STC of the walls or floor and ceiling assemblies. It is true that the amount of sound available to be transmitted to adjoining rooms will be reduced when some of the sound originating in a room is absorbed by acoustically treated surfaces within the room. However, experience indicates the need for a much larger reduction in transmitted sound than can be obtained by just applying acoustical surface materials.

On the other hand, flexible insulation placed in the free airspace of frame walls and floor and ceiling assemblies improves performance by absorbing sound within the cavity, and can result in as much as a 10-dB improvement when the surfaces are effectively separated.

Wall Systems

Sound-isolating wall systems may be constructed of wood or metal framing, concrete, or unit masonry. Regardless of the system selected, it is essential that its surfacing materials be installed over the entire wall area, especially behind soffits, cabinets, and duct enclosures, and above suspended ceilings where the structural framing might otherwise be left exposed. This precaution is particularly critical for common walls between units to ensure that sound will not be transmitted through an unfinished space (Fig. 10.4-24). Resilient, nonhardening caulking must be used at the floor and ceiling in partitions, where partitions abut other walls and partitions, and around electrical outlets and switches, plumbing pipes, and other penetrations to make the wall airtight (Fig. 10.4-25).

The majority of sound transmission test data for wood- and metal-framed construction have been obtained with gypsum board as the surfacing material because of its extensive use and ease of installation in the laboratory, without the necessity of drying and curing. The following discussion is based primarily on the evaluation of such testing, but the principles and conclusions that evolved are applicable to lath and plaster and other surfacing materials of comparable weight applied to similar construction. Representative STC ratings for lath and plaster, as well as other typical constructions, are given in Appendix A.

A system should be selected on the basis of test data for the system and should be installed in accordance with specified procedures.

■ WOOD-FRAMING SYSTEMS

Conventional single-stud walls finished with one layer of gypsum board on each side (Fig. 10.4-26) have an STC of about 34, which represents poor resistance to sound transmission. Sound-absorbing blankets placed in the stud cavity will not appreciably increase the sound isolation performance of this kind of wall. Either the weight of the covering material must be doubled using multiple layers of gypsum board (Fig. 10.4-27), or the two wall surfaces must be mechanically separated to achieve an STC rating of more than 40. A summary of the acoustical performance

FIGURE 10.4-24 *To prevent sound transmission through unfinished spaces, wall and ceiling finishes should cover all common wall and ceiling areas, including those inside furred areas and above ceilings and those concealed by stairs or other construction.*

FIGURE 10.4-25 *To prevent sound transmission through them, openings in sound-insulated walls, floors, and ceilings must be sealed with nonhardening, resilient sealants or a resilient material.*

of wood-framed walls is given in Figure 10.4-28.

Resilient mounting of the surface finish of a single-stud wall with resilient channels on one side only (see Fig. 10.4-22) will increase the performance of the wall about 6 dB. Adding channels to the other side will not further improve performance because of the resiliency of the wall through sympathetic vibration of both surfaces.

Installing insulation in a resiliently mounted wall increases its acoustical performance in proportion to the thickness of the insulation and can result in as much as a 10-dB improvement. Noise isolation performance depends on the surfacing materials and how they are installed. Backer strips for attaching baseboards or horizontal blocking at the top and middle height for firestopping decrease the performance. Resilient channels must be han-

dled and installed carefully to avoid permanent deflection. The wall should contain no openings, and heavy objects such as fixtures and cabinets (which would destroy resiliency) should not be mounted on the resilient channel side of the wall.

To obtain an STC of 50 or more, without regard for fire ratings, channels must be used throughout, instead of gypsum backing strips, to float the surfacing materials. Resilient sealants must also be installed around the entire perimeter and around every penetration.

A wall with resilient channels on one side generally results in the thinnest possible sound-isolating system in wood-framed construction and can be constructed with proper blocking to give a 1-hour fire rating.

Sound-deadening board applied to both sides of a single-stud wall will provide

FIGURE 10.4-26 *Conventional single-stud walls permit loud speech and other noises on one side to be heard clearly on the other.*

FIGURE 10.4-27 *An additional layer of gypsum board on each side slightly improves the performance of walls like the one shown in Figure 10.4-26.*

resilient mounting if the surfacing materials are correctly laminated to the sound-deadening board (see Fig. 10.4-21), and can result in a wall with an STC of about 46. However, if the surfacing materials are nailed in order to meet fire rating requirements, the STC will be reduced about 10 dB when nails are used at spacings of 8 in. on centers, 8 dB with nails spaced 16 in. on centers, and 4 dB with nails spaced 24 in. on centers.

Staggered-stud walls, consisting of 2×4 studs staggered on a single 2×6 plate (see

FIGURE 10.4-28
Acoustical Performance of Typical Wood-Framed Walls[a]

Wall Construction	Approximate STC Rating		
	Without Insulation	With Insulation[b]	Sound-Deadening Board Both Sides[c]
Single-stud wall			
Single-layer surface	34	36	46
Double-layer surfaces	40	—	—
Single-stud wall with resilient channel mounting one side	40	50[d]	—
Staggered-stud wall	41	49	50
Double-stud wall	43	53	50

[a]As illustrated in Figures 10.4-19, 10.4-20, 10.4-21, 10.4-22, 10.4-26, and 10.4-27.
[b]Minimum thickness of insulation (2 in. or 3 in.).
[c]Gypsum board surfaces laminated, with exception of double-stud wall, which was laminated and nailed 8 in. o.c. into studs and floor and ceiling plates.
[d]Resilient channels installed throughout; fire-rated wall with firestopping top and middle heights or backer strips for attaching wall base, STC 46.

Fig. 10.4-19), with one layer of gypsum board on each side, have an STC of about 42. The addition of insulation in the stud cavity or sound-deadening board to both sides increases the performance of the wall to a range of 45 to 49 STC. The STC rating does not increase appreciably as the thickness of the insulation is increased, because sound will travel through the single plate. Therefore, 2 or 3 in. of mineral fiber blanket insulation are all that is necessary.

One advantage of using staggered-stud construction is that cabinets can be surface mounted on the wall framing without impairing the acoustical effectiveness of the wall, because they are mounted on framing members that do not have contact with both sides of the wall.

For maximum performance, a wall base should not be nailed to the bottom plate of a staggered-stud wall.

A double-stud wall consists of two separate 2×4 stud walls on two separate 2×4 plates spaced 1/4 to 1 in. apart (see Fig. 10.4-20). This kind of wall has an STC of about 43, which is 9 dB higher than a conventional single-stud wall. Adding sound-deadening board to both sides will increase the rating to about 50. The isolation performance of a double-stud wall improves when the thickness of insulation in the cavity is increased. Ratings as high

as 59 STC can be achieved.

A double-stud wall can accommodate surface-mounted cabinets without reduced effectiveness. It also has fewer performance variations because of bad materials or workmanship. However, to perform acoustically, the separate walls should not be tied together in any way when plumbing or wiring is installed.

■ METAL-FRAMING SYSTEMS

Because of the resilient nature of metal studs themselves, the two surfaces of a wall built with them vibrate independently of each other and may be considered to be resiliently mounted. Therefore, the acoustical performance of non-load-bearing metal-stud walls differs from that of wood-stud walls. For example, the addition of resilient channels to mount surface finishes, which can be quite effective with wood studs, does not appreciably improve the performance of metal-stud walls.

When resistance to sound transmission is required in metal-framed walls, the measures to be used should be selected on the basis of individual test data for the system and should be installed in accordance with specified procedures. Refer to Appendix A for the STC ratings of selected walls and floor and ceiling assemblies.

FIGURE 10.4-29
Acoustical Performance of Typical Metal-Channel-Stud Walls[a]

| | Approximate STC Rating | | | | | |
| | | | Sound-Deadening Board One Side | | Sound-Deadening Board Two Sides | |
Gypsum Board Surfacing	Without Insulation	With Insulation[b]	Without Insulation	With Insulation[b]	Without Insulation	With Insulation[b]
Single layer each side	38	45	43	49	47	54
Double layer one side, single layer one side	44	50	46	51	52	54
Double layer each side	49	57	53	57	54	57

[a] As illustrated in Figures 10.4-30, 10.4-31, and 10.4-32
[b] Minimum thickness of insulation (2 in. or 3 in.).

Generally, the precautions recommended for resiliently mounted wood-framed walls apply to metal-framed walls. For maximum performance, the wall should contain no openings, and heavy objects such as cabinets or fixtures should not be mounted on either side of the wall.

The difference between the acoustical performance of walls constructed with 1-5/8- and 2-1/2-in.-deep metal studs, or using 1/2- or 5/8-in.-thick gypsum board, is considered negligible for practical purposes. A summary of the acoustical performance of metal channel stud walls is given in Figure 10.4-29. The effects of adding gypsum board, insulation, and sound-deadening board in metal-channel-stud construction are cumulative.

Single metal-channel-stud walls with single-layer surfaces have an STC of about 37. Adding 2 or 3 in. of insulation in the cavity of a wall with single-layer gypsum board surfaces on each side (Fig. 10.4-30), will improve the acoustical performance about 9 dB, bringing the wall's STC to about 46. Adding one layer of mineral fiber sound-deadening board to one side of the wall will improve the STC by 5 or 6 dB, bringing it to 43 without insulation within the cavity, and 49 with insulation. Adding sound-deadening board to both sides and insulation in the cavity (Fig. 10.4-31) can produce an STC of 52 to 54.

Adding layers of surfacing material to one or both sides will increase the STC rating of a wall by 5 or 6 dB for each layer added. Adding insulation to a wall with two layers of gypsum board on one side will improve performance an additional 6

FIGURE 10.4-30 *A metal-channel-stud wall acts as a resilient system without the need for resilient mountings.*

to 8 dB. A layer of sound-deadening board will increase the STC by about 5 dB, which is about the same as adding a single layer of gypsum board.

The effects of adding gypsum board, insulation, and sound-deadening board in metal-stud construction are cumulative (see Fig. 10.4-31). However, the improvement is small when sound-deadening board is added to both sides of a wall that already has two layers of gypsum board on both sides. This is probably because the added mass of the various layers of finish begin to overshadow the resiliency of the studs.

Double-channel-stud walls consisting of 1-5/8-in.-deep studs tied together with bracing ties (Fig. 10.4-32) to accom-

FIGURE 10.4-31 *The sound-isolating properties of metal-channel-stud walls are measurably improved when insulation, sound-deadening board, or additional layers of gypsum board are added.*

modate plumbing, wiring, and other systems will achieve a rating of 42 STC when surfaced with one layer of gypsum board on each side. This rating can be improved to 50 by adding a layer of sound deadening board to both sides and to 52 by adding 2 to 3 in. of insulation in the cavity. Double-layer gypsum board and sound-deadening board added to both sides results in an STC of 53. Increasing the cavity to 12 in. and adding three layers of insulation (9 in.) to a wall with single-layer gypsum board on each side can result in an STC of 55; another layer of gypsum board added to one or both sides of this construction can result in a 59 or 60 STC.

METAL CHANNEL STUD
GUSSET PLATE
INSULATION
GYPSUM BOARD
SOUND DEADENING BOARD

FIGURE 10.4-32 *When the cavity depth or insulation thickness is increased, and sound-deadening board is used, a metal-channel-stud wall can achieve excellent resistance to sound transmission.*

Based on tests performed by Riverbank Acoustical Laboratories.

FIGURE 10.4-33 *The performance of well-built, nonporous masonry walls can be predicted from their weight according to a mass law formula.*

■ CONCRETE AND UNIT MASONRY WALLS

The sound-isolating performance of walls with sealed surface pores is closely correlated with the weight of the wall, and the performance of such walls parallels what is known as the *limp mass law relationship*. The mass law relationship holds true only for limp homogeneous partitions that are nonporous and have uniform physical properties throughout the entire wall panel. This relationship between transmission loss and mass can be expressed by the formula

$$TL = 20 \log m + 20 \log f - 33$$

where *m* is the mass of the wall stated in pounds per square foot (lb./sq. ft.) and *f* is the frequency in cycles per second (cps).

Heavier walls tend to reduce the transmission of sound more than lightweight walls, but doubling the mass of a wall increases the transmission loss only 6 dB. As can be seen in Figure 10.4-33, to achieve a transmission loss of greater magnitude, the increase in mass must be comparatively large.

Concrete and masonry walls that have a very tight pore structure (are nonporous) do not permit sound to pass through them, but rather transmit sound by acting as a diaphragm. In a porous wall, the pore structure is the principal transmission path, particularly at low frequencies. When a porous wall's surface is completely sealed with paint or plaster, transmission loss no longer depends on pore structure but instead is determined by the weight and stiffness of the wall. Therefore, the acoustical performance of a painted or plastered porous wall will be essentially the same as that of a nonporous wall of the same weight and stiffness.

It is possible to predict the STC of concrete and masonry walls that are nonporous to the extent that sound transmission through them is the result of vibration rather than the passage of airborne sound. This prediction, which is based on the weight of the wall, can be reasonably accurate when there are no openings in the wall, and all joints and cracks are sealed. Figure 10.4-33 shows that STC test values fall consistently about 5 or 6 dB below STC values calculated according to the mass law formula.

Concrete masonry units are produced in hollow and solid units manufactured from either lightweight aggregate or normal-weight aggregates, such as sand and gravel (see Section 4.2). Normal-weight, or dense, aggregate units have a relatively tight pore structure, are considerably heavier than lightweight aggregate units, and will perform acoustically according to their mass.

Lightweight concrete blocks typically contain a larger number of pores (airspaces between aggregate particles and cement matrix) than do dense units. For this reason they excel in sound absorption but are not effective in resisting sound transmission. Lightweight blocks are subject to a wide variation in transmission loss when not sealed, because of large differences in the pore structure of blocks made by different manufacturers. Completely sealing their surface pores increases the performance of lightweight units about 7 dB when painted, and about 10 dB when plastered (Fig. 10.4-34). Conversely, the increase from painting or plastering dense units is only about 2 dB. But, because of their greater mass, dense units perform considerably better than lightweight units in the first place.

Cavity wall construction and combinations of dense and porous units offer improved acoustical performance and can provide STCs of 50 or better (see Fig. 10.4-34). Using a combination of two 4-in. lightweight block wythes with a

FIGURE 10.4-34 *Estimated STC values for sealed and porous concrete masonry walls according to their weight in pounds per square foot.*

3/8-in. or wider airspace between them and parging one of the wythes on the cavity side eliminates the need to seal the faces of the units, thereby retaining intact the sound-absorbing porosity of the lightweight units.

Filling the cores of hollow masonry units with sand or another aggregate will increase the acoustical performance of a well-built wall by only 1 or 2 dB. Filling the cores in a poorly built wall will improve its performance more, but it will never be as good as a properly constructed wall.

There is evidence that hollow units may be better sound insulators than solid units of the same thickness and same aggregate type. A suspected reason is that because the core surfaces are denser than the remainder of the block because of the manufacturing process, a *skin effect* occurs, producing a greater-than-expected resistance to sound transmission.

Because clay masonry units are relatively dense and nonporous, the acoustical performance of walls made with them can be estimated according to their mass, providing that the wall is tightly constructed.

Cavity wall construction will provide improved sound isolation with brick masonry, just as it does with concrete masonry.

Regardless of the material, the sound-isolating performance of a masonry wall will depend largely on the extent to which it is airtight. Even relatively fine cracks and open joints such as may occur when a wall is poorly built or as occur at a wall's perimeter will destroy the acoustical effectiveness of a masonry wall.

Because of the great number of mortar joints (and therefore possible acoustical leaks) it is recommended that brick masonry walls be sealed with plaster when acoustical privacy is desired.

Both concrete and masonry wall surfaces that need to be sealed should be completely sealed over their entire surfaces, including areas behind cabinets and surfacing materials, in closets, in furred spaces, and above ceilings to eliminate flanking paths through these areas. Porous masonry walls that are relied on to act as a sound barrier should be plastered or painted even when they are to be furred and surfaced with gypsum board or panel-

ing as the interior finish. A sheet of polyethylene film over a masonry wall does not seal the surface acoustically.

There are often openings at the top and ends of masonry walls that allow sound to pass to the other side of the wall. Sometimes such openings will occur even at the bottom of a masonry wall. Such openings must be closed with acoustical sealant, resilient sound barriers, or lead sheeting.

10.4.2.2

Floor and Ceiling Systems

Although airborne sound transmission is a consideration in selecting floor and ceiling assemblies, structure-borne (impact) sound transmission is the major problem. Most conventional wood, concrete, and metal floor and ceiling assemblies provide negligible resistance to impact sound transmission. Therefore, some sound-isolating provisions must be added when acoustical privacy is required. Most measures that will control impact sound will also provide a good degree of resistance to airborne sound, providing that essential construction practices are followed.

Measures used to control structure-borne sound include (1) cushioning the impact with soft floor-surfacing materials; (2) providing discontinuous construction by floating the finish flooring over the structural framing or subflooring; (3) providing discontinuous construction between the structural floor and the finished ceiling by mounting the ceiling on resilient devices or on separate ceiling joists; and (4) installing sound-absorbing insulation in the cavity of frame construction. Singly and in combination, these construction measures reduce impact sound transmission by breaking the direct structural connections that could carry vibration into other rooms and by absorbing some of the sound within cavities.

Impact cushioning and floated floors can be used effectively with any type of structural floor system. Discontinuous ceilings generally are used only with wood- or metal-framed construction or for critical areas of reinforced concrete, such as for ceilings under equipment rooms and service areas. The use of sound-absorbing materials in the cavity of frame construction is significantly effective only when used in conjunction with a resiliently mounted ceiling.

There is little difference in the acoustical performance of the various hard floor-surfacing materials such as wood strip, wood block, and resilient tile and sheet flooring. Impact insulation classes (IICs) for representative floor and ceiling assemblies are given in Appendix A.

■ CUSHIONING THE IMPACT

Of all the methods to control impact sound transmission, cushioning the impact with soft resilient floor finish materials (such as carpet and pad) is the most effective with any type of floor system. By dissipating a substantial amount of the impact energy as it strikes the surface, the amount of sound transmitted to the supporting structural system is substantially reduced. The IIC of the carpet and pad combination varies with the thickness of the carpet, the degree of matting of the pile, and the weight of both carpet and pad in ounces per square yard (oz./sq. yd.) (Fig. 10.4-35). However, no appreciable gain in airborne sound isolation is obtained by using a carpet and pad or by adhering acoustical surfacing materials to the ceiling below. Therefore, when cushioning is the only measure to be used to control impact sound transmission, the basic floor and ceiling assembly should be designed to control airborne sound transmission.

Cushion-backed vinyl sheet flooring can be effective in controlling impact sound transmission when carpet is not desired, as might be the case in kitchens, laboratories, toilets, and corridors, for example. Test data for a representative cushion-backed vinyl sheet product show that an IIC of 54 was achieved when this material was installed on a 6-in.-thick precast hollow concrete slab. When the ceiling beneath the slab was resiliently mounted, an IIC of 59 was recorded. Adding a 2-in.-thick concrete topping to the slab increased these values by 4 dB.

■ WOOD-FRAMED CONSTRUCTION

A conventional wood joist floor and ceiling assembly, finished with hard-surfaced flooring and with a gypsum board ceiling attached directly to the joists (Fig. 10.4-36) will have an IIC of approximately 26. Adding plywood or other subflooring will not appreciably affect this rating.

FIGURE 10.4-35
Effect of Carpeting in Controlling Impact Sound[a]

Carpet[b]	Pad	Floor/Ceiling Construction Concrete Slab[c] IIC	Wood Joist[d] IIC[e]
20-oz. wool	(none)	53	—
40-oz. wool	(none)	57	—
60-oz. wool	(none)	65	—
44-oz. wool	44-oz. sponge rubber bonded to carpet	68	54
20-oz. wool	40-oz. felt	69	—
44-oz. wool	40-oz. hair and jute covered with foam-rubber face and back	70	59
40-oz. wool	40-oz. felt	72	—
44-oz. wool	40-oz. all hair	72	59
44-oz. wool	66-oz. all hair	72	63
60-oz. wool	40-oz. felt	73	—
44-oz. wool	40-oz. hair and jute	73	61
44-oz. wool	20-oz. combination hair and jute, foam-rubber face, latex back	73	61
44-oz. wool	Urethane foam	75	—
44-oz. wool	40-oz. sponge rubber	76	65
44-oz. wool	31-oz. foam rubber on burlap (3/8 in. thick)	79	67
44-oz. wool	80-oz. sponge rubber	80	68

[a]Based on tests performed by Kodaras Acoustical Laboratories, Inc. for the American Carpet Institute. IIC values approximated by adding 51 to original INR test values.
[b]Carpet pile weight given in oz./sq. yd.
[c]5-in. concrete slab (sand-gravel aggregate), no ceiling finish.
[d]2-in. × 8-in. wood joists, 16 in. o.c.; 5/8-in. T&G plywood nailed to joints as subfloor, 5/8-in. gypsum board nailed to joists as ceiling.
[e]Values given only where test data are comparable to tests performed on concrete slab construction.

Where an IIC of 51 or better is required, one of the following methods should be employed: (1) apply a carpet and padding, either alone or in combination with a floated floor or separated ceiling; (2) use separated ceiling construction, either by resiliently mounting the ceiling material or using separate ceiling joists, in conjunction with a floated floor.

The previously mentioned methods for reducing impact sound transmission can also be used in wood-framed construction. As with other floor and ceiling assemblies, cushioning the impact with a carpet and pad of sufficient thickness and density is the most effective method. A carpet and pad used alone or with any of the following tested sound-isolating systems will result in IICs ranging from 42 to 85.

Separating a finished ceiling from an associated structural floor system can be done either by applying the finished ceiling

VINYL COMPOSITION TILE

PLYWOOD

WOOD FLOOR JOIST

GYPSUM BOARD

FIGURE 10.4-36 *Conventional wood-joist floor construction provides poor resistance to impact sound transmission.*

VINYL COMPOSITION TILE

PLYWOOD SUBFLOOR

WOOD FLOOR JOIST

a

GYPSUM BOARD

FURRING CHANNEL

HARDWOOD FLOORING

PLYWOOD SUBFLOOR

WOOD FLOOR JOIST

GYPSUM BOARD

b

WOOD CEILING JOIST

FIGURE 10.4-38 *The impact sound-isolating performance of floor and ceiling assemblies like that shown in Figure 10.4-36 can be improved by either (a) resiliently mounting the finished ceiling, or (b) carrying the finished ceiling on independent ceiling joists.*

with resilient channels (Fig. 10.4-37) or by attaching the finished ceiling to separate ceiling joists (Fig. 10.4-38). Either of these discontinuous construction methods can effectively reduce both impact sound from above and airborne sound from either direction. As is true with every sound-control measure, separating a ceiling from a structural system will not produce an effective acoustical barrier unless the entire floor and ceiling assembly is built correctly, with careful attention to the blocking of flanking paths, and measures

CARPET & PAD

PLYWOOD SUBFLOOR

WOOD FLOOR JOIST

GYPSUM BOARD

RESILIENT CHANNEL

FIGURE 10.4-37 *Resilient channels separate the finished ceiling from the structural framing and reduce both airborne and structure-borne sound transmission.*

a

VINYL COMPOSITION TILE

PLYWOOD

SOUND DEADENING BOARD

PLYWOOD

WOOD FLOOR JOIST

GYPSUM BOARD

b

VINYL COMPOSITION TILE

CONCRETE FILL

SOUND DEADENING BOARD

PLYWOOD

WOOD FLOOR JOIST

GYPSUM BOARD

FIGURE 10.4-39 *Two methods of making a floated floor. In (a) a resilient underlayment separates the finished flooring from the structural floor. In (b) lightweight cellular concrete has been placed over a resilient underlayment to form a floating floor system.*

are taken to prevent sound from bypassing the ceiling through walls and other structural connections. As with resiliently mounted walls, heavy objects, such as lighting fixtures, should not be mounted on or recessed in the ceiling, and airtight construction should be maintained.

Resilient mounting of a finished ceiling or attaching it to separate joists can im-

prove the IIC of an assembly as much as 7 dB; sound-absorbing material in the cavity can provide an additional 7 dB. But separated ceiling construction alone will not produce a floor and ceiling assembly with much more than 45 IIC when the floor above is hard surfaced, even though an acceptable STC of 50 is obtained. However, if a floated floor is used instead, even

HARDWOOD FLOORING
PLYWOOD UNDERLAYMENT
SOUND DEADENING BOARD
PLYWOOD SUBFLOOR
WOOD FLOOR JOIST
GYPSUM BOARD

FIGURE 10.4-40 *Resilient underlayment (shaded) reduces impact sound transmission by floating the finished flooring above the subfloor or structural floor.*

HARDWOOD FLOORING
PLYWOOD UNDERLAYMENT
SOUND DEADENING BOARD
INSULATION
WOOD FLOOR JOIST
GYPSUM BOARD
PLYWOOD SUBFLOOR
RESILIENT CHANNEL

FIGURE 10.4-41 *A resiliently mounted ceiling combined with a floated floor effectively resists impact sound transmission.*

when the flooring has a hard surface the IIC can jump to 57 and the STC will still be 50 or more.

Floated-floor construction over wood joists and subflooring can be accomplished with a resilient underlayment separating the finished flooring and underlayment from the structural floor system (Fig. 10.4-39a), or by placing lightweight (cel-

lular) concrete over the resilient underlayment (Fig. 10.4-39b). (Appendix A shows other possible methods.) When resilient underlayment is used (Fig. 10.4-40), the resilient material must be capable of resisting major deformation from superimposed loading. Most such material is designed to carry only the relatively light loads of the finished floor, people, and furniture.

In general, the greater the resilience of the underlayment material the greater the isolation. Like carpeting, resilient underlayment works primarily as a cushion and only secondarily to resist airborne sound transmission through the assembly.

The performance of a floated floor depends on effectively separating the finished floor from the structural elements. The floating part should not be rigidly connected to other elements. Adhesive generally is used to attach resilient underlayments. Nailing or stapling the subflooring or finished flooring through the resilient underlayment into the supporting construction will nullify the sound-isolating properties of the floated floor.

A basic wood-joist construction can be improved approximately 7 dB with a floated-floor system using resilient underlayment. To achieve a rating of more than 51 IIC, the floated floor can be supplemented by a separated finished ceiling; and with the addition of sound-absorbing material in the joist space (Fig. 10.4-41), an IIC of 57 can be achieved. The addition of a mineral fiber insulating blanket in the space between the floor joists absorbs approximately 7 dB of the transmitted noise. Unfortunately, the addition of sound-absorbing materials within the free air-space of a framed floor and ceiling assembly is significantly effective in controlling impact sound transmission only when used with a resiliently mounted ceiling. However, it is effective in reducing airborne sound with all types of discontinuous construction.

Lightweight concrete placed directly over a plywood subfloor does not appreciably improve the impact sound resistance of a basic wood-joist floor and ceiling assembly. When the concrete is resiliently floated, however, the improvement is about 10 dB, but the IIC will still be less than 51 unless a carpet and pad are added or the ceiling is resiliently mounted.

VINYL COMPOSITION TILE
CONCRETE SLAB
LATH & PLASTER

FIGURE 10.4-42 *A typical concrete slab with hard-surfaced flooring and a directly applied plaster ceiling offers poor resistance to impact sound transmission.*

■ CONCRETE SLAB CONSTRUCTION

A plain concrete slab with resilient tile flooring and a plaster ceiling applied directly to it (Fig. 10.4-42) will have an IIC of about 36. Increasing the weight or thickness of the slab will have little effect on the impact sound transmission loss produced by it. Sound control measures that will improve the acoustical performance of a concrete slab are limited to floating the flooring and cushioning impacts with the carpet and pad. Because doing so is impracticable and costly, the ceiling under a concrete slab is seldom resiliently mounted and test data for that are limited.

Adding a carpet and pad to a concrete slab provides somewhat better performance than does adding a carpet and pad on a wood-framed construction, with IICs ranging from 53 to 88 or even more, depending on the weight and density of the carpet and pad (see Fig. 10.4-35).

Floating to isolate a finished floor from a concrete slab is accomplished in much the same manner as in a wood frame (Fig. 10.4-43). Provisions should be made to keep the floated floor from having rigid contact with structural elements. When possible, floating floors should not touch partition edges. One way to prevent such contact is to install a cushion between the edge of the partition and the floor.

A hard-surfaced floor floated on a resilient underlayment over a 6-in.-thick concrete slab with a directly applied plaster ceiling will have an IIC of about 55.

VINYL COMPOSITION TILE
PLYWOOD SUBFLOOR
SOUND DEADENING BOARD

CONCRETE SLAB

LATH & PLASTER

FIGURE 10.4-43 *Floating the finished flooring over resilient underlayment considerably improves the acoustical performance of a concrete slab assembly like that shown in Figure 10.4-42.*

FIGURE 10.4-44 *Openings must be closed to eliminate flanking paths for airborne sound. The spaces between joists must be blocked, especially when joists are perpendicular to walls.*

The performance of precast concrete floor and ceiling assemblies is similar to that of cast-in-place concrete assemblies, but available test data on such assemblies are limited. Data for representative precast floor systems are given in Appendix A.

Steel bar joist and other metal-framed floor systems can be improved by measures similar to those used to improve wood-framed assemblies. Methods for floating finished flooring and resiliently mounting finished ceilings closely parallel those used in wood-framed assemblies. The acoustical effectiveness of steel bar joist construction, as well as metal decks, concrete floor joists, and other similar systems, depends primarily on the airtight sealing of wall and floor intersections. Walls must be extended to the bottom of the slab or deck above and the space at their tops must be sealed airtight. When walls run perpendicular to joists, proper sealing is difficult and expensive to achieve.

Concrete slabs over steel bar joists are normally about 2 in. thick, which gives them sufficient mass to produce an STC only in the high 40s. Therefore, when a higher STC is required for either walls or floor and ceilings assemblies, this type of construction is not effective. Even when a wall itself has a high STC, sound will bypass it through the flanking path provided by a relatively lightweight floor system.

Current test data for evaluating metal frame floor and ceiling assemblies are not as extensive as data for other construction systems. Representative steel bar joist systems and some available test data are given in Appendix A.

10.4.2.3

Wall and Floor Intersections

Noise problems from flanking paths are encountered where walls and floors intersect. Unless precautions are taken, both airborne and structure-borne sound can penetrate around and through the walls and floors, even though the wall and floor constructions are structurally adequate.

Walls and associated floor and ceiling assemblies should have comparable sound-isolating abilities so that a weakness in one will not diminish the effectiveness of the other.

The spaces between floor joists should be blocked where walls and floor and ceiling assemblies join, to reduce airborne sound transmission through such flanking paths (Fig. 10.4-44). Joints at the top and bottom of partitions should be sealed airtight with a resilient, nonhardening acoustical sealant or a resilient material (Fig. 10.4-45).

Loads applied to a floated floor by a structural wall will cause the resilient underlayment to collapse, damaging the floor and maybe even causing the structure to settle. An underlayment material that is sufficiently rigid to support structural loading is too rigid to provide resilient isolation in a floated floor system. Therefore, the bottom plate of a bearing partition should not be supported on a floating floor, but should rather sit directly on and be fastened to the subfloor or supporting structure.

A break in the subflooring under partitions will help to reduce impact noise transmission between adjacent rooms by

FIGURE 10.4-45 *Acoustical sealant should be used to seal small open air paths. This figure shows the sealing of paths at the bottom of a wall. Paths at the top and ends of sound-isolating walls should be similarly sealed.*

a

BREAK SUBFLOOR

b

ISOLATE FLOATING FLOOR

FIGURE 10.4-46 *Flanking structure-borne noise can be reduced by breaking direct contact between structural elements, such as the subflooring shown in (a), and by isolating floated floors from walls (b).*

breaking direct contact of the structural elements (Fig. 10.4-46a). When possible, a cushion should be provided between a partition and the edge of a floating floor so that they do not touch (Fig. 10.4-46b).

Floated concrete floors on wood-joist construction should be confined to individual rooms to prevent transmission of impact noise along the floor (see Figs. 10.4-46b and 10.4-47). Isolation joints around the perimeter of rooms and across

TREAT FLOATED FLOOR IN EACH ROOM AS A UNIT

DINING | KITCHEN | DEN

CORRIDOR

LIVING ROOM | BATH | BEDROOM

FIGURE 10.4-47 *A floated floor should be confined to a single room by separating it from adjacent construction at doors and walls.*

CARRY WALL FINISH BEHIND STAIR STRINGER

SUPPORT STRINGER INDEPENDENT OF WALL FRAMING

FIGURE 10.4-48 *Stair stringers should not be nailed or otherwise fastened to adjacent walls. If they are, footfalls will be transmitted into the walls.*

doorways are essential. Before the concrete is placed, walls should be supported on a floor plate, polyethylene film should be installed over the resilient underlayment, and isolation joints should be installed around the perimeter of each room. The concrete should not be in direct contact with the walls. Moldings and quarter-round trim can be used to cover isolation joints.

Wall surfacing material should not be tightly forced against resiliently mounted ceilings. Doing so may collapse resilient hangers, channels, or clips, and short-circuit the resilient mounting.

10.4.2.4

Stairs

Stair stringers should be fastened only at the top and bottom and should not be nailed to adjacent studs. Header joists should be used at landings to support floors instead of nailing the floor joists to the studs (Fig. 10.4-48).

When continuous wall surfacing material is installed on the entire wall adjacent to a stairwell before the stairs are installed, sound transmission between the stairs and adjacent rooms will be reduced.

When stairs are on both sides of a common wall, such as in town houses, duplexes, or other attached housing units, staggered- or double-stud walls with insulation in the cavity should be used. Stair

CORRIDOR

BR. | BR.

DUCTS

BR. | BR

CORRIDOR

FIGURE 10.4-49 *Unless finishes are taken above them, furred spaces become sound transmission paths. Ducts should be installed after the finishes have been placed.*

runners on the wall side should be held away from the studs to accommodate later installation of a finishing material behind the runners. If the walls are to be plastered, temporary construction ladders, as are used in fire-resistant construction, should be used to permit complete plastering and to eliminate breaks in the wall.

Finishing them with carpet or a resilient material, such as a rubber mat, will substantially reduce impact noise transmission from stairs to adjacent areas.

10.4.2.5

Corridors

Air ducts are often installed before ceiling and wall finishes, which will be behind them, are in place, making it difficult or impossible to install these finishes later. This permits impact noise from a corridor to pass through the ceiling and down the open stud spaces in the walls of adjoining spaces. This problem can be solved by installing only register outlets and the first duct section in a corridor, thus deferring the balance of the duct installation until the interior finishes of the walls and ceilings have been completed. Ducts then can be installed and a furred ceiling hung below them (Fig. 10.4-49).

FIGURE 10.4-50
General Criteria for Airborne and Impact Sound Isolation Between Dwelling Units*

Construction	Grade I	Grade II	Grade III
Wall partitions	STC >55	STC >52	STC >48
Floor/ceilings	STC >55 IIC >55	STC >52 III >52	STC >48 IIC >48

*Adapted from "FHA Guide to Airborne, Impact, and Structure-Borne Noise Control in Multifamily Dwellings."

FIGURE 10.4-51
Sound Isolation Criteria for Walls Between Dwelling Units[a]

Common Walls Separating Dwelling Units[b]			Sound Transmission Class (STC)		
			Grade I	Grade II	Grade III
Bedroom	and	bedroom	55	52	48
Living room and		bedroom	57	54	50
Kitchen[c]	and	bedroom	58	55	52
Bathroom	and	bedroom	59	56	52
Corridor	and	bedroom[d]	55	52	48
Living room and		living room	55	52	48
Kitchen[c]	and	living room	55	52	48
Bathroom	and	living room	57	54	50
Corridor	and	living room[d,e]	55	52	48
Kitchen[c]	and	kitchen[f]	52	50	46
Bathroom	and	kitchen	55	52	48
Corridor	and	kitchen[d,e]	55	52	48
Bathroom	and	bathroom	52	50	46
Corridor	and	bathroom[d]	50	48	46

[a] Adapted from "FHA Guide to Airborne, Impact, and Structure-Borne Noise Control in Multifamily Dwellings."

[b] Acoustically, mirror planning is desirable where common walls separate rooms of similar function; when not possible, walls separating bedrooms or living rooms from dissimilar areas must have greater sound-isolating properties.

[c] Or dining, family, or recreation room.

[d] No entrance door leading from corridor to living unit assumed.

[e] It is commonly and incorrectly assumed that walls containing entrance doors need be no better acoustically than the door. However, the basic corridor wall may separate sensitive living areas without entrance doors and therefore must have adequate sound-isolating properties to ensure privacy in these areas.

[f] Double-wall construction is recommended to provide isolation from impact noise resulting from vibration of mechanical equipment and systems.

SELECTION CRITERIA

The following recommended criteria are suggested for selecting walls and floor and ceiling assemblies to provide resistance to airborne and structure-borne sound transmission.

Generally, acceptable privacy from overall airborne sound is said to exist when the level of conversational speech is reduced by a wall or floor and ceiling assembly to the point where it cannot be understood in an adjoining room. To estimate the minimum STC required to obtain privacy, the level of background noise on one side of a construction is subtracted from the level of sound that strikes the other side. Figure 10.3-9 illustrates the approximate effectiveness of walls and floor and ceiling assemblies with varying STC values at varying background noise levels.

General recommendations for common walls and floor and ceiling assemblies separating living units in terms of STC and IIC ratings are given in Figures 10.4-50, 10.4-51, and 10.4-52. Exact values for STC and IIC ratings should be determined on the basis of individual test data in accordance with prescribed procedures (see Sections 10.3.1.1 and 10.3.2.1).

The FHA-recommended criteria for noise control often are used as a basis in the design and construction of multifamily housing units. The following recommendations are included in the "FHA Guide to Airborne, Impact, Structure-borne Noise Control in Multifamily Dwellings." The criteria are based on (1) the level of ambient background noise represented by urban, suburban, and rural locations; (2) minimum-, average-, and high-income housing; and (3) the specific function of the wall or floor and ceiling assembly within the building.

The FHA uses three grades of acoustical environment, which allows its criteria to be applied to a wide range of urban developments, geographic locations, economic conditions, and other factors. Constructions that meet the criteria will provide good sound isolation and will satisfy a majority of the occupants in a building that fits the conditions of the particular grade. Grade II is applicable to the largest percentage of multifamily construction

FIGURE 10.4-52
Sound Isolation Criteria for Floor and Ceiling Assemblies Between Dwelling Units[a]

Common Floor and Ceiling Assemblies Separating Dwelling Units[b]		Grade I		Grade II		Grade III	
		STC	IIC	STC	IIC	STC	IIC
Bedroom above	bedroom	55	55	52	52	48	48
Living room above	bedroom	57	60	54	57	50	53
Kitchen[c] above	bedroom	58	65	55	62	52	58
Family room above	bedroom[d]	60	65	56	62	52	58
Corridor above	bedroom	55	65	52	62	48	58
Bedroom above	living room[e]	57	55	54	52	50	48
Living room above	living room	55	55	52	52	48	48
Kitchen above	living room	55	60	52	57	48	53
Family room above	living room[d]	58	62	54	60	52	56
Corridor above	living room	55	60	52	57	48	53
Bedroom above	kitchen[e]	58	52	55	50	52	46
Living room above	kitchen[e]	55	55	52	52	48	48
Kitchen above	kitchen	52	55	50	52	46	48
Bathroom above	kitchen	55	55	52	52	48	48
Family room above	kitchen[d]	55	60	52	58	48	54
Corridor above	kitchen	50	55	48	52	46	48
Bedroom above	family room[e]	60	50	56	48	52	46
Living room above	family room[e]	58	52	54	50	52	48
Kitchen above	family room[e]	55	55	52	52	48	50
Bathroom above	bathroom	52	52	50	50	48	48
Corridor above	corridor	50	50	48	48	46	46

[a]Adapted from "FHA Guide to Airborne, Impact, and Structure-Borne Noise Control in Multifamily Dwellings."

[b]Acoustically, mirror planning, where common floor and ceiling assemblies separate rooms of similar function, is desirable; when not possible, floor and ceiling assemblies separating bedrooms or living rooms from dissimilar areas must have greater sound-isolating properties. When bedrooms or living rooms are located below less sensitive living spaces, greater impact isolation is required.

[c]Or dining, family, or recreation room.

[d]STC criteria also apply to vertical partitions between these two spaces.

[e]Equivalent airborne sound isolation, but less impact sound isolation generally required for the converse situation.

and, therefore, can be considered as a basic guide.

The Grade I category includes primarily suburban and peripheral suburban residential areas, considered to have a quiet background noise level. The nighttime exterior background noise level might be be-

tween 35 and 40 dB or lower. In addition, the sound-isolation criteria of this grade are applicable in certain special cases, such as dwelling units above the eighth floor in high-rise buildings and luxury apartments, regardless of location.

The Grade II category is applicable pri-

marily in residential urban and suburban areas considered to have an average noise environment. The nighttime exterior background noise levels might be between 40 and 45 dB.

The Grade III criteria represent minimal recommendations. They are applicable in some urban areas that generally are considered noisy. The nighttime exterior background noise levels might be about 55 dB or higher.

Walls and floor and ceiling assemblies should have STC and IIC ratings equal to or greater than the recommended criterion figures. A floor and ceiling assembly that may provide adequate impact sound isolation may not provide suitable airborne sound isolation. The criteria for both airborne and impact sound isolation of floor and ceiling assemblies must be met.

Recommended basic criteria for airborne and impact sound isolation of common wall and floor and ceiling assemblies that separate dwelling units of equivalent function are given in Figure 10.4-50. These criteria are based on STC and IIC ratings derived from laboratory measurements. Standard methods of testing for field measurements have not as yet been formally adopted.

Figure 10.4-51 gives the recommended criterion values for common walls, related to room or area function. Most of the typical conditions encountered for common walls separating dwelling units in multifamily construction are included. Some of these conditions are clearly undesirable and should be prevented during the initial design of the building. The purpose of this detail is to illustrate the importance of the acoustical separation between sensitive and nonsensitive areas. A partition between dwelling units that is common to several functional spaces should meet the highest criterion value.

Figure 10.4-52 includes criteria for most of the conditions encountered in common floor and ceiling assemblies that separate units in multifamily dwellings. Some of these conditions are also clearly undesirable and should be prevented during building design. Additional precautions and qualifying recommendations are given in the footnotes to Figure 10.4-52.

When the placement of living units vertically or horizontally adjacent to mechanical equipment rooms cannot be avoided, special consideration should be given to providing suitable sound-isolating construction. The recommended airborne sound criteria between mechanical equipment rooms and sensitive areas in dwelling units are

---

Grade I STC > 65
Grade II STC > 62
Grade III STC > 58

Mechanical equipment rooms include furnace and boiler rooms, elevator shafts, trash chutes, cooling towers, garages, and similar areas. The recommended criteria between mechanical equipment rooms and less sensitive areas, such as kitchens and family and recreation rooms, are

---

Grade I STC > 60
Grade II STC > 58
Grade III STC > 54

Double walls are usually necessary to achieve adequate acoustical privacy. When living units are above noisy areas, airborne sound isolation is important, but impact sound isolation is not significant when structure-borne vibration is minimal. However, when mechanical equipment rooms are above living areas, airborne sound isolation must be maintained, impact sound isolation becomes critical, and specially designed sound-isolating construction is required to ensure quiet living areas.

Placing dwelling units vertically or horizontally adjacent to commercial establishments such as restaurants, bars, community laundries, and similar businesses should be avoided when possible. If such situations cannot be avoided, the recommended airborne isolation criteria between business areas and sensitive living areas are

---

Grade I STC > 60
Grade II STC > 58
Grade III STC > 56

If the living areas are situated above such areas, adequate impact isolation criteria should be

---

Grade I IIC > 60
Grade II IIC > 58
Grade III IIC > 56

However, if the business areas are above living areas, the impact isolation criteria values should be increased by at least 5 points.

If noise levels in mechanical equipment rooms or business areas exceed 100 dB,

---

FIGURE 10.4-53
Sound Isolation Criteria of Walls Within Dwelling Units[a]

Walls Separating Rooms[b]	Sound Transmission Class (STC)		
	Grade I	Grade II	Grade III
Bedroom and bedroom	48	44	40
Living room and bedroom	50	46	42
Bathroom and bedroom	52	48	45
Kitchen and bedroom	52	48	45
Bathroom and living room	52	48	45

[a]Adapted from "FHA Guide to Airborne, Impact, and Structure-Borne Noise Control in Multifamily Dwellings."

[b]Doors leading to bedrooms and bathrooms preferably should be gasketed, solid-core doors to provide maximum privacy.

as measured using the C scale of a standard sound-level meter, the airborne sound isolation criteria given above should be raised 5 points.

Figure 10.4-53 lists suggested criteria for airborne sound isolation for walls separating rooms within a given living unit.

Town houses and row houses where the living unit occupies more than one story should be separated by a double wall construction with an STC rating of 60 or greater. Suggested criteria between rooms in a given dwelling are the same as those listed in Figure 10.4-51 and, in addition, the floor and ceiling assemblies should have IIC ratings that are at least numerically equivalent to or greater than the listed STC criteria.

Rooftop or indoor swimming pools, bowling alleys, ballrooms, tennis courts, and gymnasiums require constructions specifically designed in accordance with acoustical analysis.

11

MECHANICAL SYSTEMS

PLUMBING

SEPTIC TANK SYSTEMS

HEATING, VENTILATING, AND AIR CONDITIONING

FIRE SPRINKLER SYSTEMS

Photo: A sling psychrometer.

11.1 PLUMBING*

The earliest discovered plumbing systems were found in the ruins of Minoan palaces on the island of Crete. Four thousand years ago the Minoans produced plumbing systems that incorporated most of the features of systems used today, including flush toilets and vented drainage systems. It was the Romans, however, who gave plumbing its name, based on *plumbum*, the Latin word for lead, which they used for piping. Though famed more for their aqueducts, the Romans also gave their buildings both hot- and cold-water supply systems. Nineteenth-century England was the setting in which the flushing water closet was perfected and Thomas Crapper was knighted for this accomplishment.

Residential and commercial plumbing installations generally consist of a water supply system, a drainage and venting system to conduct away wastes, and plumbing fixtures and equipment (Fig. 11.1-1).

FIGURE 11.1-1 *Components of a residential plumbing installation. (Courtesy Small Homes Council, Building Research Council, University of Illinois, Urbana-Champaign.)*

11.1.1

CODES, LAWS, AND INDUSTRY STANDARDS

The design of plumbing systems is generally governed by a local or statewide plumbing code. Most of these are based on one of the following model plumbing codes:

1. National Plumbing Code of the Building Officials and Code Administrators International (BOCA)

2. Standard Plumbing Code of the Southern Building Code Congress International (SBCC)

3. Uniform Plumbing Code of the International Association of Plumbing and Mechanical Officials (IAPMO)

---

*This section and the illustrations in it were originally prepared for the United States League of Savings Institutions under a research grant by the staff of the Building Research Council, University of Illinois at Urbana/Champaign, Illinois, and has been updated for this edition.

4. National Standard Plumbing Code of the National Association of NL Plumbing-Heating-Cooling Contractors (NPHCC)

The power to enforce water quality standards rests with state governments, which in turn generally adopt U.S. Environmental Protection Agency (EPA) standards. The current EPA standard is EPA 40 CFR 141, "National Drinking Water Regulations." These regulations set maximum limits on trace metal and organic chemical content, turbidity, and chlorine residuals, and establish sampling and monitoring procedures. A full explanation of the requirements can be found in Part 141 of the *Federal Register*, dated December 24, 1975. In addition to federal regulations, some local codes impose their own limitation on impurities, and in some cases mandate water conservation practices or devices.

Many industry standards have evolved to govern the manufacture and installation of plumbing pipe, fixtures, and equipment.

Figure 11.1-2 summarizes the most generally accepted standards. Figure 11.1-3 lists the names of standard-setting organizations and their abbreviations.

Some types of equipment are either listed (placed on a listing of acceptable products) or labeled by a nationally recognized organization to indicate compliance with certain safety standards. Gas water heaters and pressure relief valves, for instance, should be listed and labeled by the American Gas Association (AGA). Electric and oil water heaters should be listed by the Underwriters' Laboratories, Inc. (UL).

Listing means that the manufacturer has agreed to comply with certain standards that the UL or AGA prescribes and monitors. Lists of approved products and manufacturers are published periodically by the AGA, UL, and other testing organizations and are available to the public. Labels are affixed to the equipment to give graphic evidence of compliance, but labeling is not an automatic adjunct to listing. Some products may be listed but not labeled.

FIGURE 11.1-2

FIGURE 11.1-2
Industry Standards for Plumbing Pipe and Equipment

Subject or Title of Standard	Standard Designation
Water Piping Materials Standards	
ABS DWV	ANSI B72.18
ABS pipe	ASTM D 2751
Cast-iron soil pipe	ASTM A 74, C 564
Cast-iron water pipe	ASTM A 377
Clay pipe	ASTM C 700, ASTM C 425
Copper drainage tube	ASTM B 306
Copper water tube and pipe, K, L, M	ASTM B 88
Brass pipe	ASTM B 43
CPVC pipe	ASTM F 891
Galvanized steel threaded pipe	ASTM A 53
No-hub cast-iron pipe	CISPI 301
Plumbing fixtures	FS WW-P-541, ANSI A112
Polybutylene water distribution pipe	ASTM D 3309
PE pipe	ASTM D 2104
PVC pipe	ASTM D 2729
PVC thin wall	ASTM D 2949
Stainless steel—grade TP-409	ASTM A 312
Conditioning of Domestic Water Supply	
Water softeners	WQA S-100
Water filters	WQA S-200
Sewage Disposal System Standards	
Minimum design standards for community sewerage systems	HUD 4940.3
Tanks, septic, bituminous-coated metal	UL 70
Installing vitrified clay pipe sewers	ASTM C 12
Underground installation of flexible thermoplastic sewer pipe	ASTM D 2321
Plumbing System Standards (General)	
Joints and connections	ASTM C 564, CISPI 301
Pipe protection	AWWA-C-203
Water softening	WQA S-100
Sump pumps	SSPM A
Plumbing fixtures	
Plastic bathtubs, shower receptors, and shower stalls	ANSI Z124.1 and Z124.2, NAHB Research Foundation or U.S. Testing Company, Inc., label
Domestic Water Heater Standards	
Electric hot water heaters	EEI Stds., UL 174
Gas water heaters	AGA Stds. and listed
Oil water heaters	UL 732
Water heater controls	ANSI Z21.22
Pressure-relief values—water heaters	NBBPVI and AGA Stds.
Hot water tank construction	ANSI Z21.10
Indirect heaters	IBR Code by Hydronics Institute
Safety devices	
Energy shutoff	UL or AGA* listed and labeled
Pressure relief valves	AGA* or NBBPVI listed and labeled by AGA or ASME

Courtesy Small Homes Council, Building Research Council, University of Illinois, Urbana-Champaign.
*AGA-listed shutoff devices and relief valves shall comply with ANSI Z21.22 and shall be so labeled.

WATER SUPPLY SYSTEMS

A *water supply system* is everything used to obtain water from a source, treat it to make it drinkable, and deliver it to its point of use. This section discusses available sources of water and its distribution both to and within buildings.

11.1.2.1

Source

Typically, water originating in lakes, streams, or underground sources is piped to treatment plants where it is purified. From there it is conducted through a network of water mains to end users such as residences, businesses, and industries (Fig. 11.1-4).

Modern water supply systems provide potable water for drinking, cooking, and bathing, and for use in conducting away wastes. *Potable water* is water of sufficient quality to be drinkable. Water may be naturally potable or may be made potable by treatment.

Until recently, water was used without regard for future supplies. Today, as groundwater supplies in many areas dwindle, water is regarded as much of a limited natural resource as fossil fuels and farmland. Using it wastefully not only incurs municipal costs in purifying it for consumption, in pumping it to the point of use, and in treating it before discharge into a body of water, but reduces the amount available in underground aquifers. *Aquifers* are large underground stores of water, contained either in large cavities or distributed in smaller voids of porous, granular soils.

Although the earth has an enormous reserve of water, 97% of it is salty and more than 2% is trapped in polar ice caps and glaciers. Therefore, *less than 1% of all the earth's water is available to us in lakes, rivers, and underground aquifers.* Rapid urbanization and industrialization of many areas increase demand on aquifers and reduce the replenishment process, wherein melting snow and rainwater were allowed to soak back into previously unpaved ground. Surface waters are rushed from

FIGURE 11.1-3
Abbreviations for Standards-Writing Organizations

Abbreviation	Organization
AGA	American Gas Association
AHAM	Association of Home Appliance Manufacturers
ANSI	American National Standards Institute
ARI	Air-Conditioning and Refrigeration Institute
ASME	American Society of Mechanical Engineers
ASTM	American Society for Testing and Materials
AWWA	American Water Works Association
CABO	Council of American Building Officials
CISPI	Cast Iron Soil Pipe Institute
EEI	Edison Electric Institute
FHA	Federal Housing Administration (now HUD)
FS	Federal Specification, General Services Administration
HUD	U.S. Department of Housing and Urban Development
NAHB	National Association of Home Builders
NBBPVI	National Board of Boiler and Pressure Vessels Inspectors
NFPA	National Fire Protection Association
SSPMA	Sump and Sewage Pump Manufacturers Association
UL	Underwriters' Laboratories, Inc.
WQA	Water Quality Association

Courtesy Small Homes Council, Building Research Council, University of Illinois, Urbana-Champaign.

FIGURE 11.1-4 *A water treatment plant must be strategically located between the users and the source, which may be an underground pool (aquifer) or lake.*

streets and parking lots to storm sewers and then to salty oceans or polluted fresh-water lakes. This water is lost as a valuable resource until it returns in the form of rainwater and is purified for reuse.

Water entering a building's plumbing systems generally comes from either a municipal water system or a private well. Water from these sources may differ substantially in quality, pressure, and reliability of supply.

Federal, state, and local health standards are concerned with the safety of water for human consumption, not its usefulness for other purposes or its interaction with common plumbing materials. Many municipalities, however, treat their water supply not only to make it safe to drink but also to clarify it, soften it, and reduce its corrosiveness. Alum, activated carbon, and coagulant aids are added to remove suspended materials from surface water supplies and to control taste and odor problems. In many localities, fluoride is added to reduce the incidence of tooth decay, and chlorine is added to control bacterial contamination. The water may also be aerated to remove odors and dissolved gases. Equipment is available to perform similar water treatment procedures on a small scale for private well water.

Consumption

Consumption levels for residential and small commercial buildings are generally estimated using past experiential data. The number, frequency, and pattern of use of various plumbing fixtures are estimated, based on the number and type of occupants. From this estimate, the total building demand may be determined (Fig. 11.1-5).

For larger installations, a probability method is used that takes into account the fact that not all fixtures will be in use at any one time. Consequently, the probability method assigns a weighted mathematical factor known as a *fixture unit* rating to each fixture in proportion to the load it imposes on the water supply when in use (Fig. 11.1-6). *Fixture unit* values for all fixtures in the building are summed and converted to a water demand figure by a graph that expresses the probability factor of simultaneous use of various fix-

FIGURE 11.1-6
Fixture Unit Ratings and Water Flow

Fixtures	Fixture Units	Water Flow, gpm	
		Cold	Hot
Water closet, flush valve	10	30–45	0
Water closet, tank type	5	18	0
Kitchen sink	2	5	5
Lavatory	1	5	5
Bathtub	2	6	6
Shower stall	2	6	5
Laundry tray	3	10	10
Clothes washer	3	5	5
Dishwasher	3	5	5

Courtesy Small Homes Council, Building Research Council, University of Illinois, Urbana-Champaign.

FIGURE 11.1-7 *Water demand curves. Locate fixture units on curve and read demand on left scale. Curve 1 is for buildings with flush-valve water closets; 2 is for tank type. (Courtesy Small Homes Council, Building Research Council, University of Illinois, Urbana-Champaign.)*

11.1.2.3

Conservation

Spurred by recurring rainfall shortages, groundwater depletion, and the high cost of water treatment, water conservation has become a regional and national issue. Although the design of facilities and selection of water-saving equipment are essential, conservation depends largely on the habits and decisions of the people using the water.

Some local building codes require the use of water-conserving devices or fixtures. Flow restrictors can be installed on existing showerheads, for instance, and cut the flow rate by about 50%. Plastic bottles or dams inserted into water closet tanks can similarly reduce water consumption. Energy-saving showerheads are fairly nominal in cost and can cut water flow between 2 and 5 gallons per minute (gal./min.). Energy-saving water closets can reduce the volume per flush from 5 to 3 gal./min. or less. Energy-saving showerheads and water closets are generally much more effective than modified older fixtures because they are designed to provide equivalent cleansing action while using less water.

11.1.2.4

System Design

A *water supply system* within a building consists of service piping, distribution piping, branches to plumbing fixtures and appliances, and the necessary fittings and control valves. A plumbing fixture is a device that is supplied water and that will receive or discharge solid or liquid waste. Plumbing fixtures include, but are not limited to, lavatories, sinks, laundry trays, water closets, bathtubs, and showers (see Section 11.1.4). The types of appliances referred to in this section are those that receive water and may or may not discharge liquid or solid waste. Appliances include, but are not limited to, dishwashers, clothes washers, and some refrigerators.

The design of a building's water supply system includes laying out the system, and selecting and sizing pipes, flow-control and pressure-regulating devices, and valves.

A water supply system is always filled with water at greater-than-atmospheric pressure. Once in a building, it is separated into two parts. The first forms a cold-water supply tree. The second goes through a water heater and becomes a hot-water supply tree. Each of these trees then feeds the appropriate plumbing fixtures and appliances. The trees are composed of vertical supply pipes called *risers* and horizontal pipe runs called *branches* that are connected to the risers. Plumbing fixtures may be connected to risers, but more frequently are connected through branch runs.

The basic goals in the design of a water supply system are (1) to keep the water from becoming contaminated; (2) to main-

tures (Fig. 11.1-7), which in turn takes into account volume, duration of flow, and intervals between operations.

tain adequate pressure and flow rate for the smooth operation of plumbing fixtures and appliances; (3) to prevent excessive flow and consequent waste of valuable water; (4) to prevent excessive pressures, velocities, corrosion, and scale buildup; and (5) to provide water without noise, fixture staining, or odor.

The efficient functioning of a water supply system depends on pipe size, fixture and materials selection, and installation.

■ SYSTEM LAYOUT

The physical configuration of a water supply system should take into consideration efficiency of materials used, simplicity of construction, protection from freezing, and ease of maintenance and repair.

A primary principle in the efficient layout of water supply systems is that all runs be as short, straight, and as direct as possible, and have as few fittings as possible. Following this principle not only conserves materials, but also reduces internal friction in the pipes. This in turn keeps to a minimum pressure loss associated with flow from the water main to fixtures.

In the schematic planning stage of a building, an effort should be made to locate rooms that will contain plumbing fixtures as near as possible to the point of entry of the water supply to the building and the point of exit to the sewer line. If possible, rooms that will contain plumbing fixtures should be clustered so that plumbing fixtures in more than one room share the same wall for pipe distribution (Fig. 11.1-8). This can be done by arranging rooms so that fixtures in two or more rooms are placed back-to-back against the same wall or are vertically aligned or stacked on different floors (Fig. 11.1-9).

Jogs and offsets in plumbing walls containing horizontal lines should be avoided to simplify the plumbing lines. Fixtures in a given room should be arranged so that all of their supply, vent, and drain lines fall in one wall. This practice shortens the piping runs and simplifies framing the building by reducing the number of penetrations through floors and eliminating the need to furr or double frame more than one wall (Fig. 11.1-10).

Walls used in this way for the vertical distribution of plumbing are known as *chase* walls, *wet* walls, or *plumbing* walls (Fig. 11.1-11). They are usually thicker

FIGURE 11.1-8 *Back-to-back plumbing arrangement is economical and also separates noise-producing equipment from living spaces.* (*Courtesy Small Homes Council, Building Research Council, University of Illinois, Urbana-Champaign.*)

FIGURE 11.1-9 *Vertical alignment of fixtures conserves piping and framing.* (*Courtesy Small Homes Council, Building Research Council, University of Illinois, Urbana-Champaign.*)

than typical interior walls, allowing sufficient space to accommodate the drainpipe, plus one supply pipe crossing behind the drainpipe. In most wood-framed construction, chase walls are built using 2×6s rather than the typical 2×4s. When more

space is needed, a double wall, each built with 2×4s, is used. Sometimes it is necessary to separate the two walls of a double-walled chase to provide even more space. The double wall method is usually used when the partitions are framed with metal studs. The two metal stud walls forming a plumbing chase are required to be tied together and braced, either with sections of stud or with pieces of gypsum board (see Section 9.5).

A water heater should be located as near to fixtures and appliances using hot water as possible. This practice reduces heat loss from the hot-water supply line and also reduces the delay in receiving hot water at the tap once the line between a water heater and a fixture has cooled. Future expansion of a supply system can be facilitated by including the appropriate tees with plugs in the original installation or *roughing in* the plumbing for fixtures to be added at a later time. Roughing in means to install those parts of a plumbing system that can be completed before plumbing fixtures or appliances have been installed. This includes water supply, drainage, and vent piping, and the necessary fixture supports.

To reduce the probability of pipes freezing, supply lines should not be located in exterior walls. If such location is unavoidable, the supply pipes should be run on

FIGURE 11.1-10 *In-line arrangement of plumbing fixtures simplifies piping and reduces costs.* *(Courtesy Small Homes Council, Building Research Council, University of Illinois, Urbana-Champaign.)*

FIGURE 11.1-11 *A 6-in. stud wall is wide enough to permit a 3/4-in. water pipe to cross a 3-in. plastic vent stack.* *(Courtesy Small Homes Council, Building Research Council, University of Illinois, Urbana-Champaign.)*

the inside face of the wall with at least 3-1/2 in. of mineral fiber insulation separating them from the outside (cold) face of the wall and with no insulation on the warm side. This practice keeps pipes near the heated interior of the building and slows heat loss from the pipes to the exterior. Every effort should be made to keep water-carrying pipes out of areas subject to freezing. In practice, this means providing freeze-proof valves for outdoor sill-cocks, providing drain valves for seasonal drainage of outdoor piping, and locating supply pipes away from exterior walls.

When laying out a water supply system and fixture locations, the resulting drainage system location must also be considered. Care should be taken in locating framing so that a plumbing wall will not fall in the same plane as roof trusses or floor or ceiling joists, which would obstruct the passage of supply, vent, and drain lines in and out of the plumbing wall. Likewise, if major framing elements run perpendicular to the plumbing wall, care should be taken that the fixtures requiring below-floor drains are located so

that the drains do not conflict with the floor framing. Joists should not be notched in the middle third of the span, and holes should not be larger than 2-1/2 in. in diameter.

During the design process, thought should also be given to the future accessibility of plumbing for repairs. In buildings with crawl spaces, sufficient height should be provided for a person to have convenient access to the piping. The minimum crawl space depth required by many codes is 18 in., however, this is based on moisture protection rather than access to equipment in the crawl space. For easy access, a crawl space should be at least 2 ft. and preferably 3 ft. high. An access panel into the bathroom plumbing wall just behind the bathtub greatly simplifies the typical repairs to the bathtub drain plug mechanism, faucet, and shower diverter.

■ PIPE MATERIALS

There are several criteria that determine the choice of water supply pipe material to be used for a particular installation. The

requirements of the applicable plumbing code are, of course, paramount. Within the parameters of the applicable code, the characteristics of the local water supply, that is, its acidity, as well as its air, carbon dioxide, and mineral contents, should affect the selection. The original cost is also important, as are the ease of installation and of repair of a particular piping material. And of course, water supply piping should bear the stamp of approval of the National Sanitation Foundation (NSF). Acceptable piping materials are listed in Figure 11.1-12.

GALVANIZED STEEL Galvanizing is a process that deposits a coating of zinc on the surface of a steel pipe to protect it against corrosion. The advantages of galvanized steel (also called galvanized iron) pipes are their low cost and high strength. They are resistant to high pressures and to water hammer (see Section 11.1.4). Their major disadvantage is that joints are typically threaded, although other methods of joining are discussed below. Threading the initial installation is laborious, and the need to thread new joints in place makes later, tapped-in connections to existing systems difficult.

Another drawback of galvanized pipe is its tendency to corrode or form scale on its interior. The degree to which this occurs depends on local water chemistry. Galvanized pipe performs best in areas with moderately hard, alkaline water. Naturally soft water should be treated with a silicate inhibitor to produce a protective film on steel pipe, and with caustic soda or lime to raise its pH (see Section 11.1.2.6). Very hard water should be softened to control scarring on steel pipes. When corrosion or scaling occurs, the interior pipe diameter can become greatly reduced, changing the carrying capacity of the pipe. Because it will react chemically, galvanized steel pipe should not be laid in concrete.

There are three available means of joining galvanized steel pipe: (1) *threaded connections*, (2) *unions*, and (3) *slip couplings*.

A *threaded connection* is the most common way of joining galvanized pipe (Fig. 11.1-13a). Generally the outside of the pipe run is threaded to match a female galvanized steel or malleable cast-iron fitting with the desired geometry. The threads should be coated with pipe joint com-

FIGURE 11.1-12
Acceptable and Unacceptable Water Supply Piping Materials[a]

Material	Service	Distribution Piping	
		Under Slab	Other
Cast Iron			
Water pipe	A		
No-hub pipe and fittings	NA	NA	NA
Copper Tubing			
Type K — Hard temper	A	A[e]	A
Soft temper	A	A[b]	A
Type L — Hard temper	A	A[c]	A
Soft temper	A	A[c]	A
Type M — Hard temper	A	NA	A
DWV Hard temper	NA	NA	NA
Galvanized Open Hearth or Wrought Iron	A[c]	A	A
Galvanized Steel Pipe	A[c]	A	A
Black Steel Pipe	A[c]	NA	A
Plastic			
ABS	A[d]	NA	A[d]
CPVC-41	NA	A[b]	A
PVC	A[d]	NA	A[d]
PVC thin wall	NA	NA	NA
Polybutylene	A	A[b]	A
Polyethylene	A[d]	NA	A[d]
Styrene rubber	NA	NA	NA

[a] A = Acceptable; NA = Not acceptable.

[b] Install without joints in or under slab floors.

[c] Protected when installed underground.

[d] For cold water only, outside dwelling or in unfinished basement or crawl space.

[e] Brazed joint only.

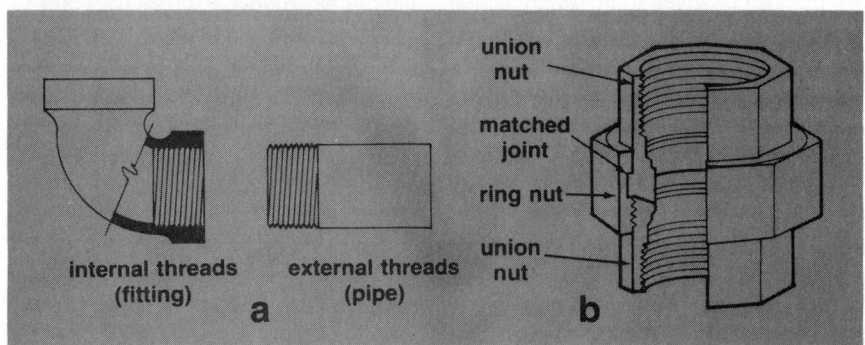

FIGURE 11.1-13 *Galvanized pipe joints: (a) threaded fitting receives field threaded pipe; (b) union fitting, to be threaded onto pipe ends.*

pound or wound with plastic joint tape before the fitting is attached.

A *union* is another means of joining galvanized steel pipe (Fig. 11.1-13b). Unions can be disconnected and reconnected later, but they are more costly than conventional fittings. A union consists of two *union nuts* with machined brass mating surfaces, each of which is threaded onto the respective pipe end and a *ring nut* that is tight-ened, drawing the two union nuts together. The brass components of unions, however, can cause localized corrosion in steel pipes which restricts flow.

A *slip coupling* is a third type of connector available for galvanized steel pipes (Fig. 11.1-14). It consists of a sleeve that is slipped over the end of one pipe. The two pipes are then aligned and the sleeve is positioned to cover the joint between them. *Compression nuts* at each end of the sleeve seal it to the pipes by rubber gaskets. Some types of slip couplings also have a screw connection in the center for easier installation.

FIGURE 11.1-14 *A pictorial section through a slip coupling.*

Where steel pipe is being joined to a dissimilar metal such as brass, and the water supply is corrosive or has a high solids content, a *dielectric fitting* must be used to prevent galvanic corrosion. Where the water supply is sufficiently hard to produce a protective carbonate or silicate film on the pipe, a dielectric fitting is unnecessary. A dielectric is a nonconductor of direct electric current.

COPPER Although relatively costly, copper pipe is convenient to work, easy to join, and sufficiently corrosion resistant for use with most water supplies. Water that is very soft, very high in carbon dioxide, or very aggressive and hard should be treated before entering a copper plumbing system. Red brass (85% copper, 15% zinc), and 90/10 copper-nickel pipes are good choices where the water supply is corrosive. Copper should not be used in systems designed for velocities greater than 4 ft./sec. Although the piping itself is durable, the joints are weaker than the piping and are subject to damage from water hammer (see Section 11.1.4). This necessitates having air chambers (see Section 11.1.4) at the ends of supply lines to reduce the potential for water hammer.

Copper supply piping falls into two major categories: *rigid pipe*, which comes in 10- to 20-ft.-long straight lengths, and *flexible tubing* which comes in 30-, 60- and 100-ft. coils.

Rigid pipe, which is hard tempered, is generally connected using copper fittings that are sealed with a combination tin-lead solder and a noncorrosive flux (Fig. 11.1-15). Excessive use of self-cleaning flux can lead to later corrosion problems. Rigid copper pipe comes in three wall thicknesses: (1) type K (thick), (2) type L (medium), and (3) type M (thin). These are equal in exterior diameter and are designated by the same nominal sizes, but the interior diameters vary, based on the differences in wall thickness.

Flexible tubing has several advantages over rigid copper pipe. It is easier to install, especially in cramped areas. It can be bent into smooth turns, which reduce friction in the system and, under some circumstances, permit the use of smaller pipes. It can withstand several freezing cycles without cracking, although with each freeze the pipe expands and the pipe wall becomes thinner.

Flexible tubing also provides more options for connections. Aside from soldered fittings, which require use of a torch, flared and compression fittings can also be used. A *flared fitting* (Fig. 11.1-16a) is made by slipping a *flare nut* over the end of the copper tube, re-forming the end of the tube so that it has a flared tip, then screwing the flare nut onto a threaded fitting. A *compression fitting* (Fig. 11.1-16b) is similar except that the nut bears against a brass ferrule as it is screwed into the fitting rather than against the flared end of the tube. A push-in connec-

FIGURE 11.1-16 *Unsoldered fittings for flexible copper tubing: (a) flared fitting; (b) compression fitting.*

tor for copper tubing is also available. It consists of stainless steel and neoprene grabbers mounted in a nylon body.

Flexible copper tubing is also available in type K or type L wall thickness. Type K is typically used for underground installations, and type L is used for indoor plumbing.

PLASTIC There are many varieties of plastic pipe, but they are not all suitable for carrying drinking water. Those that are suitable bear the stamp of the NSF. Plastic pipe is lightweight, being about 1/20 of the weight of equal-sized steel pipe. It is also easily joined, produces low friction, and does not corrode. Joints are typically made using a solvent and special fittings (Fig. 11.1-17). Although these joints are easy to make, the solvent acts quickly, and the joint becomes permanent. The only way to correct mistakes is to saw the joint out and start over. The corrosion resistance of plastic allows its internal surface to remain smooth, which maintains the carrying capacity of the pipe as the system ages.

Plastic is much less thermally conductive than metal pipe. Therefore, it is a better insulator, which reduces heat loss from hot-water lines and condensation on cold-water lines.

Plastic and metal piping also differ greatly in their rates of thermal expansion, amount of softening at the same temperature level, electrical conductivity, flammability, and weathering characteristics. Plastic has a much larger coefficient of

FIGURE 11.1-17 *Plastic pipe joint. First cleaner, then solvent, is brushed on the outside of the pipe and inside of the fittings. Then the two are joined and the pipe is turned about a one-quarter turn. A good joint will show a fused solvent bead on the fitting end.*

thermal expansion than metal. Because it tends to soften more and often has less strength and stiffness than does metal pipe, it is necessary to support plastic pipe more frequently than metal pipe to prevent sagging. Because of its high rate of thermal expansion, plastic pipe is often snaked (placed in gentle undulating curves rather than laid straight) when it is laid in underground trenches. The upper limit of operating temperature for plastic pipes ranges from 120° to 300°F depending on the type of plastic.

Plastic piping should be kept away from areas of high temperature such as heating ducts. Unlike metal piping, it is not a good

FIGURE 11.1-15 *Rigid copper pipe being joined with a soldered fitting.*

electrical conductor. This has the advantage of preventing galvanic or electrolytic corrosion, but it also prevents the use of a cold-water pipe as a ground for a building's electrical system. Depending on the particular material's formulation, plastic pipe varies in its degree of flammability. Plastics weather differently than metal piping; instead of corroding or oxidizing, they chemically degrade when exposed to ultraviolet radiation and become brittle. This can be controlled by using plastics that are formulated with ultraviolet stabilizers.

There are four major types of plastic pipe used in small buildings: *polyvinyl chloride* (PVC), *polyethylene* (PE), *chlorinated polyvinyl chloride* (CPVC), and *polybutylene* (PB). All four types are suitable for cold-water supplies, but only CPVC and PB are suitable for hot-water supplies.

Polyvinyl chloride (PVC) is most frequently used for cold-water supply lines. It can be used for the entire cold-water supply system within a building. Joints in PVC pipe can be either solvent welded or made with threaded fittings, which are commonly available in heavier weights of pipe. PVC pipe is manufactured in two types. Type I is unplasticized and has a maximum working temperature of 150 °F. Type II is plasticized by the addition of rubber, which makes it less brittle than Type 1. This, however, makes its strength and maximum working temperatures lower than those of Type I. In addition to water lines, PVC pipe is also used for natural gas and electrical conduit lines.

Polyethylene (PE) pipe is used for cold-water supply from a municipal water main or well to a building. It has a lower range of working temperatures than most other plastics: from −67 °F to 122 °F. PE pipe comes in Types I, II, and III, in order of increasing density. Joints in PE pipe are made with tapered, serrated inserts of polystyrene or galvanized steel, which slip into the end of the PE pipe (Fig. 11.1-18). The pipe is then compressed against the insert by means of a metal hose clamp.

Chlorinated polyvinyl chloride (CPVC) is the most prevalent type of plastic for hot-water supply lines. It can be used with water temperatures up to 212 °F, although the water heater relief valve in a system using CPVC pipe should be set not higher than 180 °F. Support clamps should be

FIGURE 11.1-18 *Clamped connection for polyethylene or polybutylene pipe.*

provided on horizontal runs every 3 ft. to prevent sagging. Because CPVC pipe costs more than PVC, its use is usually restricted to hot-water lines. Joints are solvent welded.

Polybutylene (PB) pipe is suitable for either hot- or cold-water supply pipes. It is so flexible that most turns can be made without fittings. It can be obtained in a high-temperature grade for use up to 221 °F. PB pipe is similar to Type III PE pipe but is stronger. Like PE pipe, joints are made with tapered inserts and external clamps.

BRASS   In the past, red brass pipe (85% copper, 15% zinc) was frequently used in areas where the water supply was so corrosive that it made galvanized steel piping unsatisfactory. Today, because of its high cost, brass pipe is seldom used in water supply systems. Modern plastic piping can do the same job at a much lower cost.

Brass piping may still be encountered in old buildings. Its connections will be threaded, similar to those in galvanized steel pipe. Although brass is highly resistant to corrosion generally, it is subject to galvanic corrosion when connected to galvanized steel, and there a dielectric coupling or union should be used.

LEAD   Lead piping was used for water piping in the past and still may be found in some older buildings. It is no longer used for water supply piping inside buildings, however, because minute quantities of lead oxides on the piping dissolve in water, which can result in lead poisoning.

■ **PIPE SIZING**

The objective in sizing pipe is to ensure that there is adequate pressure and flow at plumbing fixtures and appliances. Most fixtures and appliances will operate at pressures as low as 15 pounds per square inch (psi). City mains and private wells

usually deliver water at 40 to 50 psi. These higher pressures are intended to take care of pressure losses in the pipes that result from internal friction and gravity. Losses because of internal friction are affected by (1) the type of pipe and its internal surface characteristics (rougher surfaces generate more friction); (2) the length of the pipe (more length, more total friction); (3) the number of turns, fittings, and valves (these each generate friction loss); and (4) the velocity of the water.

In larger buildings, water pipes are sized according to formulas that take into account the number and types of fixtures and appliances served, friction losses, and the probability that not all of the fixtures will be used at the same time.

Experience has shown that for one- and two-family dwellings of average size the pipe sizes listed in Figure 11.1-19 will provide adequate service.

■ **FLOW CONTROL**

An unsteady rate of water flow from a supply system into a fixture is the result of changes in the flow pressure in the supply lines. This happens either because of inconsistent pressure from the water

FIGURE 11.1-19
Recommended Water Supply Pipe Sizes[a]

Fixture	Pipe Size, in.[b]
House main	1
House service	3/4
Supply riser	3/4
Water closet	1/2
Bathtub	1/2
Lavatory	1/2
Kitchen sink	1/2
Laundry tray	1/2
Clothes washer	1/2
Dishwasher	1/2
Shower stall	1/2

[a]For one- and two-family dwellings of average size.

[b]All pipe sizes except main and service are for either hot or cold.

FIGURE 11.1-20 *In a flow controller, increasing the velocity deforms a resilient orifice to reduce the opening. When velocity is normal the orifice increases to normal.* (Courtesy Small Homes Council, Building Research Council, University of Illinois, Urbana-Champaign.)

source (main or well), or to the start-up or shutdown of other fixtures or appliances on the same plumbing tree. If the problem is severe, the inconsistent flow can be corrected with a flow-control device (see Fig. 11.1-20). This is a fitting with a flexible orifice on which the cross-sectional area varies inversely with the flow pressure. As pressure in the line increases, a flexible insert in the fitting is pushed forward against a stop, decreasing the cross-sectional area of the pipe and thereby regulating the flow rate. Flow-control devices are not to be confused with (or used as) pressure-reducing valves.

■ **PRESSURE REGULATION**

Most municipal systems deliver water to a building at pressures of 40 to 50 psi. In designing a supply system, it is critical to know the average supply pressure for summer months, when the main pressure is generally lowest because of greater demand for water for lawn watering. The water distribution system in a building is then designed in such a way that its length and friction characteristics will deliver an acceptable flow rate and pressure at the fixtures in the building. If the final pressure at the fixtures is inadequate, which is rarely the case in residential and other small buildings, it can be supplemented by installing a gravity tank system, hydro-pneumatic tank system, or a booster pump.

■ **VALVES**

Valves used in building water supply systems can be classed according to their operation as *gate valves*, *globe valves*, and *alignment valves*.

Valves of various kinds are located throughout a water supply system. They

are, in fact, a system's only moving parts. A *supply main*, *street main*, or *distribution main* is the water line from a treatment plant or well into which individual building supply lines are tapped. The principal, and usually lowest, horizontal artery in a building's water supply system is also sometimes called a *main*, but we will not refer to it as such here in the interest of clarity. Where a building supply line leaves the supply main, there is a shutoff (*corporation*) cock. A main shutoff valve is installed just inside the building if the water meter is remotely located. Shutoff valves are installed on either side of in-line appliances such as a water meter, water heater, or water treatment device. These are typically *gate* or *globe* valves.

Most water valves are made of brass, because of its resistance to corrosion. Valves are made with threaded ends for use with galvanized steel and rigid copper pipe and with solderable ends for use with copper tubing. Plastic-bodied valves are available for use with plastic pipe.

It is also good practice to install a shutoff valve where a supply pipe makes a turn of more than 45 degrees or enters or leaves a wall. A type of globe valve known as an *angle stop valve* (because it is a shutoff valve through which the supply line makes a right-angled turn) should be installed just upstream from each fixture and appliance in a system, to facilitate repairs.

Automatic appliances such as dishwashers and clothes washers have internal

valves to cycle water flow into them. These are usually electrically controlled *solenoid valves*, which open and close very quickly, sometimes causing water hammer (see Section 11.1.4). Finally, there are valves within the faucet assemblies of each plumbing fixture. These are usually globe valves or cocks.

GATE VALVES   Gate valves are used when full, unobstructed flow is desired when a valve is open and positive closure is needed when it is shut (Fig. 11.1-21a). When the handle on a gate valve is turned, a wedge is lowered into the pipeway, obstructing the flow.

GLOBE VALVES   Globe valves contain two chambers connected by a small opening (Fig. 11.1-21b). When the handle is turned, a plunger (*stem*) is lowered, blocking the flow of water from one chamber to the other. The seal over the opening is usually made with a washer mounted on the tip of the stem.

11.1.2.5

Water Heating

Water is heated either in a separate water heater or by part of a building's mechanical heating system. Systems that utilize a building's hot-water or steam boiler or furnace to heat water for use at plumbing fixtures or appliances are beyond the scope

FIGURE 11.1-21 *(a) Gate valves are used in key locations to turn water completely on or off; (b) globe valves can control flow in a partly open or closed position.*

of this book. Separate water heaters may be either central units serving an entire building or smaller remote units located near the point of hot-water need.

■ WATER HEATERS

Water heaters are frequently referred to as hot-water heaters, a redundancy. They can be categorized by design type and fuel used.

TYPES There are several means of heating water for use at plumbing fixtures and appliances. The most prevalent is the *automatic storage heater*, which heats a volume of water and stores it until it is needed. Other methods include a *demand heater*, which heats only when more hot water is called for and does not store a reserve of hot water; and a *combination heater*, which uses a heat exchanger on a space-heating boiler.

Automatic storage heaters consist of a tank holding 20 to 80 gal. of water, which is maintained at a constant temperature by a thermostatically controlled heat source (Fig. 11.1-22). Generally the thermostat is set between 120° and 140°F. The higher temperature is required for some automatic dishwashers. However, modern energy-saving dishwashers are equipped with a booster heater so that 120°F water can be used.

FIGURE 11.1-22 *Gas water heater. Heater should be located near floor drain to permit periodic draining of sediment and to handle occasional overflow.*

When water is used at a plumbing fixture or appliance, the tank is refilled automatically, since it is connected with an open valve to the pressurized cold-water supply line. This refill lowers the temperature of the water in the tank, which the thermostat notes, and calls for heat from the heat source. The thermostatic control also calls for heat to compensate for heat lost through the tank wall when no hot water is being consumed. For this reason, new water heaters should be internally insulated with an R-value of at least 7. Kits are available for applying insulating blankets externally to older models.

The life expectancy of a water heater is limited by corrosion of the tank wall to 10 to 15 years. The usual tank material is glass-lined steel. If the local water supply is particularly hard or corrosive, water treatment equipment may be installed to lengthen tank life. An alternative, though expensive, material for tank interiors is monel metal. It forms an oxide coating that resists corrosion. Although aluminum and unlined galvanized steel tanks are available, they fail quickly because of corrosion. Some water heaters are fitted with a sacrificial anode of magnesium, which gives up certain ions to the water more readily than does the tank wall. The anode becomes a substitute material for the water to attack corrosively, sparing the tank-liner material.

Excessively high temperatures and pressures are dangers associated with heating water in closed tanks. High temperatures can result in people being scalded or burned at fixtures. Pressure buildup in a tank can lead to an explosion. Under normal operating circumstances, a thermostat cuts off the heat long before either of these two hazards is approached. However, as a safety measure in the event that the thermostatic control fails, UL-approved water heaters are fitted with a *pressure and temperature relief valve* (see Fig. 11.1-22).

A *pressure valve* vents a closed water heater tank in the event that either a dangerous pressure or a dangerous temperature level is reached. This valve is located on top of the tank and connects to a drip pipe that directs relief flow from the tank to a safe drainage area. Relief valves used on water heaters should be of the reseating type, which vent the system until an acceptable pressure or temperature level is reached. A fuse-type valve, which vents the system

continuously until it is manually shut off, is no longer considered acceptable.

Relief valves are specified for water heaters based on the British thermal unit (Btu) rating of the heater. The relief valve must always have a larger Btu rating than that of the heater. Valves are factory-set for the pressure and temperature levels at which they will discharge and should not be altered. The limits set for residential water heaters are approximately 125 psi of pressure and 200°F in temperature.

Water heaters are generally chosen on the basis of their volume and their recovery rate. Volume simply refers to the number of gallons the heater storage tank holds. Recovery rate is the number of gallons that the heater can raise 100°F in 1 hour. This is a measure of a heater's ability to reestablish the thermostat-set temperature after rapid drawdown of the stored hot water. In general, oil-fired water heaters have a high recovery rate, gas heaters have a medium recovery rate, and electric heaters have a low recovery rate. Heaters are also rated according to the energy input and output, usually quoted in Btu's. The ratio of the input rating to the output rating is an indicator of the heater's fuel efficiency.

Demand heaters are much smaller since hot water is generated to meet the demand. No water is stored. Thus, the standby losses at night and when occupants are away are much lower. Because of their limited recovery rate, demand heaters are not as popular in the United States as are *automatic storage heaters*.

FUELS Heat sources for water heaters can be electricity, piped-in fossil fuels (natural gas or heating oil), or sunshine.

Electric water heaters usually have two electrical elements inside a storage tank. They require no supply of combustion air or venting of combustion products (Fig. 11.1-23).

At the base of its storage tank a *gas* water heater has a burner with a flue to exhaust its combustion products. This flue extends up through the center of the tank, acting as a heat exchanger as is does, and is finally connected to a chimney.

The burners on *gas* heaters can be designed either for natural gas or liquefied propane. Adequate combustion air must be supplied to the burners, and an exhaust flue and chimney must be provided.

FIGURE 11.1-23 *Electric water heaters are generally larger than gas- or oil-fired heaters because their recovery rate is slower.*

FIGURE 11.1-24 *Heated water from a collector is conducted to a heat exchanger where most of the heat is transferred to a loop that conducts heat to a storage tank. From there, preheated water is drawn as needed to supply a domestic water heater.*

Requirements similar to those for *gas* heaters also apply to *oil-fired* heaters, although the burner design is slightly different.

Solar water heaters use the sun as a partial source of energy. The intermittent nature of this heat source makes it necessary to provide extra storage capacity so that hot water will be available when needed at night and on cloudy days (Fig. 11.1-24).

In general, solar heating is practical only as an auxiliary heat source that must be backed-up by an electric, gas, or oil-fired conventional heater. Solar energy is absorbed by a south-facing solar collector mounted outdoors through which a heat-transfer fluid flows. When heat is available from the collector, the transfer fluid carries it to a heat exchanger coil in a storage tank. The domestic water heater uses this preheated water when replacing water in its own tank.

With the exception of a few locations and during limited periods, the solar energy available is insufficient to heat water to the desired temperature, and a backup heat source must be used. This limitation does not, however, invalidate the potential of solar energy as a means of domestic water heating. In many parts of the country, from one-third to two-thirds of needed heating energy can be obtained from the sun.

SIZING The sizing of water heaters is based on a calculation of the maximum hot water demand per hour. For a dwelling, this is estimated as 7 to 10 gal. of hot water per person per hour, multiplied by the number of persons in the household. A water heater should be sized so that 70% of the tank capacity plus the heater's recovery rate equals or exceeds the calculated maximum demand per hour. The equation for sizing can be written thus:

7–10 gal./person × number of persons = maximum demand/hr. = 0.7 × tank capacity/hr. + recovery rate

This requirement can be met with a variety of combinations of tank capacity and recovery rate. These combinations depend largely on the type of fuel being used. In oil-fired heaters, which have a fast recovery rate, a heater with a small tank can be specified and the maximum demand per hour still be met. In gas-fired heaters, which have a moderate recovery rate, a medium-sized tank would be necessary. In electric systems, with their slow recovery rates, a relatively large tank would be necessary. For example, for a family of four or five people whose maximum demand per hour is in the range of 40 to 50 gal. of hot water, either of the following systems would meet the size requirement to provide the maximum demand: (1) a 30- to 40-gal. oil- or gas-fired heater; or (2) a 40-gal. quick-recovery

electric heater. For an off-peak electric heater, an 80-gal. tank is recommended.

■ TEMPERATURE CONTROLS

Water temperature controls are often advisable, especially where handicapped people, the elderly, or small children will use plumbing fixtures (see Section 1.6). Two types of temperature-control devices are available, primarily for tubs and showers. One type maintains constant relative pressures in the hot and cold supply lines to compensate for sudden draws by other nearby fixtures. Consequently, the flow rate, once set, remains reasonably constant. This does not, however, compensate for changes in temperature of the water from either supply line, which occurs when the water heater is drawn down. The second type of temperature control device is thermostatically controlled and changes the flow rate of the hot and cold supplies to maintain a constant temperature of mixed output.

11.1.2.6

Water Treatment

Water from some sources may need to be processed further to make it suitable for domestic use, even though it meets public health standards for safe drinking water. Problems that may require treatment include excessive hardness, resulting in ineffective soaping; materials in the water that cause corrosion or scarring of piping; and unpleasant odor, color, or taste. Treatment to correct such problems usually involves mechanical filtering or chemical treatment of the water.

Filters require periodic cleaning or replacement to prevent bacteria growth in them. Bacterial contamination of a building's water supply can cause problems more severe than those that filters are intended to correct. Treatment is often confined to the hot-water supply, because higher temperatures accelerate the undesirable chemical reactions and make sensory characteristics more noticeable.

■ WATER HARDNESS

The most common type of domestic water treatment combats what is known as water hardness. *Hardness* refers to the amount of calcium and magnesium carbonate dissolved in the water; *total hardness* refers to the amount of carbonates plus several other compounds dissolved in the water.

Some degree of hardness is desirable, since it causes a thin film of calcium carbonate to deposit on the interior walls of pipes, forming a protective coating against corrosion. Too much hardness, especially in conjunction with high temperatures, has two undesirable effects. It causes excessive carbonate deposition (scaling), and also renders the water ineffective at rinsing soap away. Calcium carbonate scale appears as a cream-colored buildup in piping and fixtures.

Using a *water softener* alleviates both scaling and soaping problems. An *ion-exchange* water softener contains a sodium-zeolite exchange medium that develops sodium ions. These then interact with calcium (or magnesium) ions in the water. When all of the sodium ions have been consumed, the filter must be recharged. Some systems are recharged in place by back washing and soaking with a salt solution; others depend on a service company to supply recharged replacements (Fig. 11.1-25).

It is undesirable to oversoften water, because the process raises the sodium content of the treated water. The effects of increased sodium have been questioned, especially for persons with heart or kidney disorders. Fortunately, the adverse effects of hardness are more pronounced in the

automatic recharge control

ion-exchange resin

salt for recharging resin

FIGURE 11.1-25 *As hard water passes through resin, calcium and magnesium in the water are retained and sodium is given up. Either the salt or the entire tank must be replaced periodically.*

hot-water supply than the cold. Humans more often drink water from the cold supply than from the hot. Consequently, by installing the water softener on the hot-water system only, human consumption of softened water can be limited.

With respect to the plumbing itself, oversoftened water not only fails to provide the protective calcium carbonate film on pipes, but actually makes the water corrosive. For these reasons, it is not advisable to soften water to less than 60 mg of calcium carbonate per litre. If the output of the softening equipment falls below this level, it should be tempered with untreated water from the cold supply before it enters the plumbing system.

■ CORROSIVENESS

A water supply can be rendered corrosive by high levels of dissolved carbon dioxide (yielding carbonic acid), or by low levels of calcium and magnesium carbonate (excessive softness). The former occurs naturally in water supplies in the East and Northwest; the latter can occur naturally or can be the result of overzealous water treatment. Corrosion as a result of either cause is accelerated by high temperatures and therefore is a larger problem in a hot-water system than in a cold-water one.

Symptoms of corrosion are staining of fixtures, reduced flow or pressure as a result of clogging, and leaky pipes. The stains are red if the supply piping is steel, green if copper. Clogging can occur because of accumulated rust particles, or because of localized scaling triggered by the changes in pH that accompany corrosion.

There are several solutions to corrosion problems. Mildly corrosive water from the natural supply can be treated with a neutralizing filter containing marble or limestone (calcite) chips. In severe cases, liquid sodium silicate and caustic soda can be added to raise the pH of the water. Although this relieves corrosion problems in most of a plumbing system, it does not sufficiently protect low-velocity areas such as storage tanks. Tanks with special protective coatings or sacrificial anodes should be considered.

■ UNPLEASANT APPEARANCE, ODOR, OR TASTE

Unpleasant appearance, odor, or taste in a water supply can be treated chemically or mechanically. For example, yellow, cloudy water and rust-stained fixtures are frequently caused by suspended iron particles. In mild cases, these can be removed by an ion-exchange water softener; in more severe cases, an oxidizing filter containing manganese-treated sand is effective.

Particles of organic matter, silt, or algae can degrade the appearance and taste of water. Aside from the sensory symptoms, this problem can clog pipes and reduce water heater efficiency. Its source may be the original water supply, or it can be the result of the backflow of drainage water into the supply system. If this type of water quality problem persists, fixtures should be checked for air gaps, and appliances such as clothes washers, laundry tubs, and dishwashers should be checked for vacuum breakers (backflow preventers) (see Section 11.1.4.7).

If the problem is determined to be in the original water supply, a charcoal filter can be installed where the supply piping enters the building. In severe cases, a chemical feeder with a chlorine or alum solution can be used, followed by a sand filter. For small amounts of high-purity potable water, a reverse osmosis unit can be installed on individual fixtures. It features a membrane through which only pure water can pass.

Rotten egg odor is caused by high levels of dissolved hydrogen sulfide gas. The source of the hydrogen sulfide can be anaerobic decomposition of organic matter somewhere in the supply chain. This occurs at many locations in the Southwest and Midwest. It can also be produced by magnesium anodes used to protect water heater tanks from corrosion. Aside from the unpleasant odor, water with dissolved hydrogen sulfide has a black tinge and tarnishes silver. An oxidizing filter with manganese-treated sand can be used to alleviate this problem. Extreme cases call for chlorination followed by sand filtering.

11.1.3

DRAINAGE SYSTEMS

The term *drainage system* refers to all of the piping that conveys solid and liquid wastes from its source to a point of disposal. It includes the system portions within a building and on its site and the mains of a public sewer system or a private septic system.

There are three kinds of drainage systems. A *storm drainage system* carries rainwater, surface water, condensate, cooling water, and similar liquid wastes, exclusive of sewage and industrial waste.

The second type, a *sanitary drainage system* carries *sewage*, which is liquid and solid waste matter, including fecal matter and food, and industrial waste, as opposed to the kind of material carried off by storm sewers. The portion of a sanitary drainage system that lies within a building and up to a point 3 ft. outside of the walls is called a *building drainage system* or, more frequently, a *drain, waste, and vent system* (DWV) (Fig. 11.1-26). This collection of pipes facilitates the removal of liquid and solid wastes as well as the dissipation of sewer gases within a building.

FIGURE 11.1-26 *Diagram of major components of a DWV system, showing a loop vent.*

Storm drainage systems share many of the same characteristics as DWV systems, but there are some differences as well. For example, because sewer gases are not involved, storm drainage systems do not require traps and usually do not require vents. They do require cleanouts. Although there are a few references to storm drainage systems, this section primarily addresses DWV systems. Many of the principles addressed here, though, apply as well to storm water systems. When the term *drainage system* is used in this section with no further attribution, it refers to DWV systems.

The third type of drainage system is a *combined* system that carries both storm water and sewage. This type can be found in some older systems, but has several disadvantages that make its use questionable today at best. Such systems have been responsible for polluting many of our waterways when their capacity was insufficient to handle heavy storms. When excessive water overwhelms a storm drainage system, water backs up into streets and even buildings. When the same happens to a combined system, what backs up is raw sewage. Combined systems are not permitted in many jurisdictions and should be prohibited everywhere.

11.1.3.1

System Design

A DWV system should be designed and installed so that it (1) maintains separation

of drain fluids from the water supply, (2) contains and discharges sewer gases to prevent fouling the interior atmosphere, and (3) effectively moves fluids and solids through the pipes.

■ COMPONENTS

Both storm and sanitary drainage systems include sewers. A *sewer* is any underground conduit used to carry off liquid or solid waste, or both. A system of sewers is called *sewerage*. The term *sewerage* is also used to refer to the removal and disposal of liquid and solid wastes by sewers. The horizontal piping that extends from a point 3 ft. outside a building to the public sewer or private individual sewage disposal system is called a *building sewer*. This may be either a *building storm sewer* or a *building sanitary sewer*.

Storm drainage systems consist of gutters and downspouts, underslab and foundation drains, *building storm drains*, *building storm sewers*, *site storm drains*, and *storm sewers*.

Underslab and foundation drains collect and carry off ground water and storm water that may seep down into them. A drain is any pipe that carries water in a building's storm drainage system. *Building storm drains* pick up water from interior sources and interior roof drains and carry it to a point 3 ft. outside the building. *Building storm sewers* are defined above. *Site storm drains* pick up storm water that falls on the site and the overflow from wells and sometimes springs, and include the necessary inlets, manholes, and other structures and outfalls to overland drainage, lakes, or streams. *Storm sewers* carry the collected water to disposal points, such as a lake, stream, or ocean.

DWV systems are larger and often more complex than water supply systems. Their geometry is more critical because the pipes are larger and must be installed at a slope for gravity drainage. They consist of drains (waste and soil pipes), traps, and a vent system.

A *drain* is any pipe that carries sewage in a building drainage system. *Waste pipes* carry water-diluted waste products; *soil* pipes carry the discharge of water closets, urinals, and other fixtures with similar functions to the building drain or building sewer (Fig. 11.1-27). Soil pipes may in addition carry wastes from other fixtures. A *trap* is a fitting or device designed to provide, when properly vented, a liquid seal that will prevent the back passage of air without materially affecting the flow of sewage or waste water through it (Fig. 11.1-28). A *vent system* is a collection of pipes installed to provide a flow of air to or from a drainage system, or to provide circulation of air within such a system to protect trap seals from siphonage and back pressure.

All of the components of a DWV system are interconnected. In fact, a given pipe may serve in one portion of its length as a drain and in another portion as a vent, or under some circumstances may serve as a drain and vent simultaneously at the same location.

At the center of a DWV system is one or more vertical, combination *drain- and vent* pipes that extend from a point above the roof down to the building drain. A *building drain* is the lowest part of a building's drainage system. It is the principal artery to which branches may be connected, and as such is sometimes called the system's *main*. It receives the discharge from soil, waste, and other drainage pipes inside the walls of a building and conveys it through the foundation walls to a point 3 ft. outside of the building where it joins with the building sewer (Fig. 11.1-29). The *building sewer* is the horizontal piping of a *drainage system* that extends from the end of a *building drain* to a public sewer, private sewer, individual sewage disposal system, or other point of

FIGURE 11.1-27 *Drainage diagram for larger building showing two ways of venting drain branch: with circuit vent (above), and circuit and relief vent (below).* (Courtesy Small Homes Council, Building Research Council, University of Illinois, Urbana-Champaign.)

disposal (see Fig. 11.1-29). A building sewer begins at a point 3 ft. outside the building wall, where the building drain ends. A building sewer is sometimes called a *house sewer.*

These central vertical pipes, which are called *soil stacks* or just *stacks*, are the collection points for all of the various drain and vent lines in the DWV system. They are usually placed near water closets, since those fixtures require the largest-diameter drains and short horizontal runs.

A typical soil stack will have connections to one or more water closet drains

FIGURE 11.1-28 *A P trap showing the water seal.* (Courtesy Small Homes Council, Building Research Council, University of Illinois, Urbana-Champaign.)

and branch drains serving several fixtures each (see Fig. 11.1-26). Branch vents serving those same water closets and other fixtures also connect to the stack at points higher than the branch drain connections. Thus, the lower portion of this pipe serves as the system's main vertical drain, or *soil stack*, which leads to the *building drain* and ultimately to the public *sewer*. The portion of the soil stack that lies above the highest horizontal waste-carrying connection to it is called a *stack vent*, and sometimes is called a *waste vent* or *soil vent* (see Fig. 11.1-27). A stack vent may collect one or more vent branches before it continues through the roof to the outdoors, providing a source of atmospheric pressure, thus serving, in this case, as the system's *main vent*. A horizontal pipe to which several vent stacks are connected before they go through the roof is also sometimes called a *main vent*.

Branch vents and branch drains are fed by fixture drains (Fig. 11.1-30). A fixture drain leaves the drainage point of a fixture and runs to the nearest branch drain. It includes the water-filled plumbing trap that serves to prevent sewer gases from rising through the fixture drain into the building. When the fixture drain line reaches a wall, it intersects with the branch drain–branch vent system. Water flowing

FIGURE 11.1-29 *Diagram showing stack, cleanout, building drain, building sewer, and public or private sewer.* (*Courtesy Small Homes Council, Building Research Council, University of Illinois, Urbana-Champaign.*)

out of the fixture drain line falls into the lower portion of the intersected pipe.

A stack vent or branch vent provides an air supply to equalize pressures and smooth the drainage process (see Fig. 11.1-26). The most desirable drainage pattern for water going down a vertical pipe is one in which water swirls down the pipe clinging to the walls by surface

tension. This process draws air down the pipe with the water. The air is supplied through the vent system. If no air is available through a vent system, or if the flow of water is so heavy that it overcomes the surface tension holding it against the pipe wall, then drainage can be accompanied by gurgling noises, vibration, and loss of the trap seal.

■ GRAVITY FLOW

In a water supply system, flow occurs because the fluid is pressurized. All parts of a drainage system exist at atmospheric pressure, and flow is produced by gravity. Consequently, both storm and sanitary drain lines must be sloped and arranged to assist gravity flow.

Sufficient slope should be provided in sanitary drain pipes to maintain a flow velocity of at least 2 ft./sec., which is necessary to scour the pipe properly and keep solids in suspension. For adequate drainage in a horizontal pipe (one with a slope of less than 45 degrees from horizontal), the pipe should (a) slope at least 1/4 in./ft. of run if it is less than or equal to 3 in. in dia., or (b) slope at least 1/8 in./ft. of run if it is more than 3 in. in dia.

Sufficient pipe supports should be provided to maintain uniform slope. Drains laid underground should have a bed of undisturbed compacted earth or concrete, to control differential settlement. Drain lines should not be laid beneath driveways or buildings whenever it can be avoided.

Turns and fittings should be properly detailed to minimize resistance to water

flow. When possible, flow should move in a straight line. If turns are unavoidable, sweeping bends are preferable to sharp turns. In horizontal runs, 45 degree wyes, sweeps, and 1/2, 1/6, 1/8, 1/16 bends should be used; for vertical-to-horizontal turns, 1/4 and 1/5 bends should be specified.

Fittings for use in drain systems are designed differently from those for supply systems, primarily to eliminate shoulders and other impediments to gravity flow (Fig 11.1-31). Consequently, fitting names are preceded by the designation *drain*, as in *drain bend* and *drain wye*.

Connection of branch drains to main drains is always done in the upper half of the horizontal pipe. This reduces the chance of clogging, cuts down on interference in flow at the point of connection, and, when there is no flow in the branch, leaves the full diameter of the pipe available to act as a vent.

When gravity flow of liquid or solid waste is not possible, it may be necessary

FIGURE 11.1-30 *Fixture drains connect fixtures to fixture branches or branch drains.*

FIGURE 11.1-31 *Drainage fittings showing elimination of lips that would impede sewage flow.* (*Courtesy Small Homes Council, Building Research Council, University of Illinois, Urbana-Champaign.*)

to mechanically force the material up to the level of the sewer. This is customarily accomplished by means of a sump pump and sump. A *sump pump* is a mechanical device used for pumping out sewage or liquid industrial wastes from locations below a building's gravity drain. These pumps are located in a *sump*, which is a tank or pit that receives sewage or other liquid waste, located below the normal grade of a building's gravity sewer system, and that must be emptied by mechanical means (a sump pump).

■ DRAINPIPING

The routing of drainpipes may pose serious problems during construction unless it is carefully considered during the design process. Because the system drains by gravity, horizontal pipe runs must be sloped downward in order to cause the water to flow. This restricts the length of horizontal runs that can be tolerated before the sloping pipe encounters structural members or exceeds the depth of the space in which it is to be concealed.

■ SEALING AND VENTING

A DWV system must be sealed off from the air in a building to prevent sewer gases from escaping into the building's interior. This is usually accomplished by placing water seals in fixture traps. A DWV system must, however, be open to the outside air to maintain atmospheric pressure within the system, both to aid in the flow of materials through it and to protect the trap seals from being disturbed. This is usually accomplished through a network of vents.

TRAPS Liquids and solids are moved through a drainage system by gravity. In addition to an occasional waste flow, a drainage system contains air and foul-smelling sewer gases. For sanitary and comfort reasons, it is undesirable to allow sewer gases to enter the building through fixture drains. To prevent this from happening, a device known as a *trap* (see Fig. 11.1-28) is installed at each fixture. A trap is simply a length of drainpipe shaped in such a way that it does not drain completely. After each use of the fixture, a trap retains enough water to fill the drainpipe at that point. The trapped water serves as

a seal that prevents sewer gas from rising through the fixture drain into the building. Each individual fixture requires its own trap.

Most plumbing fixture traps are shaped like the letter *P* and are called *P traps* (see Fig. 11.1-28). These are used on kitchen sinks, lavatories, and laundry trays.

A special trap called a *drum trap* is usually installed on the drain lines of bathtubs. These offer easier access for cleaning, since bathtub drains are frequently clogged with dirt and hair. A drum trap is constructed so that the drain line from the bathtub enters its lower portion, and the drain line out into the rest of the drainage system leaves it 2 to 4 in. higher. Therefore, a water seal of 2 to 4 in. is maintained within the drum. Drum traps are generally 3 to 4 in. in diameter and have a removable top for easy access. Water closets have a trap built into their internal structure (Fig 11.1-32).

A *grease trap*, unlike other traps, is not designed to prevent gases from rising through the drain pipes, but rather to collect grease particles and prevent them from reaching the sewer or septic system where they could cause blockages. Grease traps are generally not found in small residential buildings but may be used in restaurants and large residential structures.

Ordinary traps are generally maintenance free, since they are simple devices

FIGURE 11.1-32 *The residual water in the bowl of a water closet forms a water seal similar to that in a sink trap. A wall-hung water closet is shown.*

with no moving parts. Difficulties arise, however, if the water seal in the trap is lost. There are at least five ways in which this can happen. (1) The water seal may be lost because of suction (*back siphonage*) in the drain system. (2) Back pressure from a slug of sewage passing through the drain system may force the water seal up into the fixture. (3) In rarely used drains, the water seal may be lost by the evaporation of the water. (4) The water seal in a trap can be lost by capillary attraction through an object caught in the trap (such as a diaper, from which the water can readily evaporate). (5) Outdoor wind conditions can create positive or negative pressures within the drainage system through the vents; positive pressure can push water out of the trap, negative pressure can suck it out of the trap. The latter is much more likely to occur (Fig. 11.1-33).

FIGURE 11.1-33 *A slug of sewage in a soil stack creates positive pressure in front of it and suction behind. If there were no vents or the vents were clogged, the left-side trap in this figure could be blown out, or the right-side one could be sucked out.* (Courtesy Small Homes Council, Building Research Council, University of Illinois, Urbana-Champaign.)

Another infrequent maintenance problem with traps is freezing. Freezing of plumbing is generally a bigger problem in the supply system than it is in the drainage system, but it is possible for damage to occur from the freezing of the water seal in plumbing traps as well. This can crack the pipes or joints and cause leaks. If a building is to be left unheated during freezing weather, traps should be individ-

FIGURE 11.1-34 *A building trap acts both as a seal and as a cleanout for the building drain.* (*Courtesy Small Homes Council, Building Research Council, University of Illinois, Urbana-Champaign.*)

FIGURE 11.1-35 *For code compliance, measure trap to vent distance as shown in plan view.* (*Courtesy Small Homes Council, Building Research Council, University of Illinois, Urbana-Champaign.*)

ually drained through the fitting at their lowest point, or an antifreeze solution should be added to the water in the trap.

A *building trap* is a device intended to prevent the circulation of air between the building drainage system and the building sewer (Fig. 11.1-34).

VENTS Sometimes the pressure developed in front of a column of water falling through a drainpipe is great enough to blow the water seal (and sewer gases with it) through a fixture trap and into the interior of the building (see Fig. 11.1-33). Under other conditions, the pressure behind a falling column of water can be low enough to suck the water plug out of a fixture trap and into the drainage system. To prevent either of these events from occurring, a second set of openings is added to the drainage system. These openings comprise a venting system, which provides a free-flowing air connection to the outdoors.

A *vent* is a pipe that connects a drainage system to the atmosphere to permit gases to escape. These are usually made of pipe of the same material and size as the lower components, which serve as drains.

The vent piping of a drainage system consists essentially of the upper half of the fixture branch pipes, circuit vents, loop vents, relief vents, stack vents, and vent stacks (see Fig. 11.1-27).

A *circuit vent* is a branch vent that serves two or more traps and extends from in front of the last fixture connection of a horizontal branch to a vent stack.

A *loop vent* is a branch vent that serves two or more traps and extends from in front of the last fixture connection of a horizontal branch to the stack vent.

A *relief vent* is a vent whose primary function is to provide the circulation of air between drainage and vent systems (see Fig. 11.1-27).

A *stack vent* is that portion of a soil stack above the highest drain line connection.

A *vent stack* is a vertical vent pipe installed primarily for the purpose of providing circulation of air to and from a part of a drainage system.

The maximum distance between a plumbing fixture's trap and its vent is governed by the diameter of the drainpipe (Fig. 11.1-35). The larger the diameter of a drainpipe, the farther the vent can be located away from the trap and still protect its water seal.

In one-story buildings or the top floor of multistory structures, where fixtures are located close to a soil stack, stack venting may be permitted by the applicable codes (Fig. 11.1-36). In this case, the drainpipe is large enough and short enough so that it is never entirely filled with water. Therefore, there is always an airspace above the water flow, which acts as a vent passage connecting into the soil stack and ultimately to the outdoors through the top of the soil stack–stack vent.

Sometimes it is inconvenient to run a vent from a fixture that is remote from a

FIGURE 11.1-36 *Stack venting permits elimination of fixture vents.*

wall. In such situations, wet venting can be used. In wet venting, the fixture drain is oversized and used as a vent as well as a drain. Wet venting can be used with fixtures such as sinks and lavatories, but not for venting water closets.

■ CLEANOUTS

Cleanouts are fittings that provide periodic access to the interior of drainpipes, to permit removal of blockages of lint, hair, grease, and other solids. Cleanouts usually consist of a 45-degree *Y fitting* of the same diameter as the drainpipe, and with a removable screw-in plug. When the plug is removed, a cleaning tool can be inserted into the main drain line (Fig. 11.1-37).

FIGURE 11.1-37 *A stack cleanout. (Courtesy Small Homes Council, Building Research Council, University of Illinois, Urbana-Champaign.)*

A cleanout fitting should be located at the base of each soil stack, and at the point where the building drain passes through the building's foundation. A cleanout should also be provided at each change of direction of more than 45 degrees, and every 50 ft. in 4-in. pipe and every 100 ft. in larger pipes. Cleanouts should be readily accessible, and their height and clearance should be adequate for inserting cleaning tools.

11.1.3.2

Pipe

Using the correct pipe sizes and materials is essential to the proper functioning of a drainage system.

■ PIPE MATERIALS

The most common materials for drainage system piping are cast iron, galvanized wrought iron or steel, and plastic. Copper,

---

FIGURE 11.1-38
Acceptable Drainage and Vent Piping Materials[a]

Material	Below Grade	Above Grade
Cast Iron		
Soil pipe	A	A
No-hub pipe and fittings	A	A
Copper Tubing		
Type K — Hard temper	A	A
Type L — Hard temper	A	A
DWV Hard temper	A	A
Galvanized Open Hearth or Wrought Iron	A[b]	A
Galvanized Steel Pipe	—	A
Plastic		
ABS[c]	A	A
PVC[c]	A	A
PVC thin wall	NA	A
Polybutylene[c]	A	A
Polyethylene[c]	NA	NA
Styrene rubber	A[c]	NA
VC Pipe	A[c]	—

Courtesy Small Homes Council, Building Research Council, University of Illinois, Urbana-Champaign.
[a]A = Acceptable; NA = Not acceptable.
[b]Protected when installed underground.
[c]Outside dwelling only.

---

brass, and lead can also be used, but their expense is usually prohibitive. See Section 11.1.2 for a discussion of these less-typical materials. Figure 11.1-38 identifies pipe materials that are acceptable for drain- and vent piping.

CAST IRON   Cast iron is the long-standing traditional material for drainpiping. It is heavy, durable, and serviceable underground or in concrete. *Hubless* joints and fittings for cast-iron pipe make it possible to run 3-in. cast-iron drain lines in 2×4 wood stud walls (Fig. 11.1-39). Hubless joints consist of a neoprene rubber sleeve gasket that covers the joint, a stain-

FIGURE 11.1-39 *A pictorial section through a no-hub cast-iron pipe joint.*

less steel band that fits over the gasket, and wormscrew clamps that compress the band against the pipe. Support must be provided at joints and at 4-ft. intervals along a cast-iron pipe.

Older installations of cast-iron drainpipe have *bell-and-spigot* or *hub-and-spigot* joints (Fig. 11.1-40). Pipe sections are cast so that one end flares out into a *bell* or *hub*, which fits over the straight end, or *spigot*, of the succeeding section. The

FIGURE 11.1-40 *Traditional bell-and-spigot joints in cast-iron pipe are bulky and time consuming.*

gap between the bell and spigot is packed with oakum or tow hemp, and filled with molten lead. Such joints have a shoulder on the inner surface, and so must be installed directionally to prevent clogging. The direction of flow should always be from the spigot to the bell.

GALVANIZED STEEL Galvanized steel pipe is sometimes called galvanized iron. This pipe is used in drain and vent systems essentially the same way as in water supply systems. Threaded joints are typical, but where galvanized pipe connects to cast iron, it is generally fitted into a bell and leaded.

PLASTIC The two types of plastic pipe suitable for use in drainage systems are polyvinyl chloride (PVC) and acrilonitrile-butadiene-styrene (ABS). PVC pipe used in drain systems is of the same formulation as that used in water supply systems, but is thicker.

PVC is also used for storm sewers and sanitary sewers. ABS pipe has an operating temperature range of −40° to 80°F. It can be obtained with solid walls or with foam-cored walls, the latter being less expensive because it uses less resin. ABS pipe can be solvent welded at joints with or without a primer, or it can be obtained with threaded fittings (Fig. 11.1-41).

Building and plumbing codes vary in their attitude toward plastic drainpiping. The applicable codes should always be consulted during the early stage of the design of a plumbing system.

Plastic drainpiping has several advantages. Some manufacturers produce a 3-in. drainpipe with joints compact enough to fit within the clearance of a standard 2×4 stud wall (Fig. 11.1-42). This eliminates the need for special framing for plumbing walls. The light weight of plastic pipe means less load on the supporting

FIGURE 11.1-41 *A pictorial section through a PVC pipe joint. This type of joint is strong, compact, and easy to make.*

FIGURE 11.1-42 *A 3-in. PVC vent pipe in a 2×4 stud wall.* (*Courtesy Small Homes Council, Building Research Council, University of Illinois, Urbana-Champaign.*)

structure. Since plastic pipe is a better thermal insulator than metal pipe, liquids with suspended fats are less likely to be cooled to the point of depositing the fats, which clog the pipes.

The disadvantages of plastic pipe include its vulnerability to puncture, to damage by rodents, and to some nonhousehold chemicals.

■ SIZING

Another factor that should be considered during the design stage is pipe sizing. Expedient choice of pipe size or slope to conform to available floor depth may result in inefficient and noisy installations. The recommended drain and vent sizes for different parts of the system are given in Figure 11.1-43.

11.1.4

PLUMBING FIXTURES

Fixtures fulfill the basic purpose of plumbing systems, that is, to provide clean water and dispose of wastes. They also play a vital role in ensuring that a water supply remains clean by providing a physical separation between supply and drainage piping. If there were no such separation, and the positive pressure in a supply line were lost, then dirty water might be sucked back into the supply system (see Section 11.1.4.7).

When selecting fixtures one should consider convenience, cost, durability, ease of repair and maintenance, and noise generation. These considerations often are

FIGURE 11.1-43
Recommended Drain and Vent Pipe Sizes*

Fixture	Drain/ Trap Size (in.)	Vent Size (in.)
House sewer	4	—
House drain	3	—
Soil stack	3	—
Branch drain/vent	1-1/2	1-1/4
Water closet	3-1/2	2
Bathtub	1-1/2	1-1/4
Lavatory	1-1/4	1-1/4
Kitchen sink	1-1/2	1-1/4
Laundry	1-1/2	1-1/4
Clothes washer	2	1-1/4
Dishwasher	1-1/2	1-1/4
Shower stall	2	1-1/4

Courtesy Small Homes Council, Building Research Council, University of Illinois, Urbana-Champaign.
*For one- and two-family dwellings of average size.

reflected in the type and grade of fixture. In addition to their obvious cosmetic contributions, colors and finishes may also contribute to ease of maintenance.

11.1.4.11

Lavatories

Lavatories come in three basic types: (1) wall-hung units with a ledge at each side, (2) units installed in or as a part of countertops, and (3) pedestal types. Figure 11.1-44 shows several typical lavatories.

Wall-hung lavatories often have a raised back ledge that serves as a backsplash. Their back and side ledges vary from 1 to 10 in. in width; at least 4 in. is desirable. Inside bowl dimensions vary from 10 to 16 in. front to back and from 13 to 17 in. side to side. Bowl depth varies from about 5-1/2 to 7-1/2 in.

Wall-hung units are mounted on a metal bracket attached to the wall. For a secure attachment to wood studs, this bracket should be bolted to a horizontal wood support mortised into the wall framing. When

FIGURE 11.1-44 *Lavatory types: (a) drop-in with stainless steel rim; (b) moulded lavatory counter; (c) drop-in, self-rimming; (d) undercounter rimless; (e) wall hung.*

the walls are framed with metal studs, a lavatory's metal bracket can be bolted to a metal plate that spans several studs and is fastened securely to them. Wood or metal furring should be used as lavatory supports at masonry walls. Additional support can be provided by chrome legs under the front corners of a lavatory.

There are four kinds of *counter lavatories*: (1) a drop-in type with a raised self-rim; (2) a drop-in type with a thin-edged metal trim over the joints; (3) a type that is fastened to the bottom of the counter with the counter edge exposed and finished; and (4) a one-piece molded lavatory-counter unit.

Sometimes a single-bowl kitchen sink is installed in a bathroom instead of a lavatory for bathing infants and small children, washing hair, and doing hand laundry. Such a larger basin controls splashing. A hose spray can be added for extra convenience.

Most lavatories are made of vitreous china. Other common materials are porcelain over cast iron or steel, cultured marble (ground marble powder with plastic binder), thermoformed plastic sheet, and filled polymer (acrylic-modified polyester). Materials differ widely in resistance to impact and abrasion; to stains or chars by cigarettes; and to spills of things like cosmetics. Vitreous china and enameled metal have the most chemical resistance.

Molded counter lavatory units can be formed of cultured marble, thermoformed plastics, or filled polymer. Their characteristics vary with the types of binders or plastics used.

Some lavatories come with faucets and drain fittings; some provide appropriate holes for their insertion. Modern lavatories generally have only one spout but may have separate hot- and cold-water valves or may have a single-lever mixing valve. Lavatory faucet-drain assemblies usually include a pop-up drain plug similar to those on bathtubs.

11.1.4.2

Sinks and Laundry Trays

Kitchen sinks are available with one, two, or three bowls to permit simultaneous uses. Most new residences include a two-compartment sink, having the spout high enough and one compartment large enough (usually 15×18 in.) to accommodate large cooking utensils. Typical materials for kitchen sinks are porcelain-enameled cast iron or steel, stainless steel, or plastic. These units are generally designed to drop in to a countertop and are supported by their rims resting on the counter. Some have self-sealing rims; others require a separate, stainless steel sealing rim.

A variety of faucet types is available, either with separate hot and cold controls or with a single control and mixing valve. A diverter valve similar to those on tub-shower combinations is often installed to serve a hose-mounted spray accessory. Several appliances are available as attachments to kitchen sinks, including garbage disposal units, instant water heaters with tap, and small water coolers with tap.

Laundry sinks (laundry trays) are available in the same materials as kitchen

sinks, but are larger and deeper. Some are wall hung, others rest on legs or a trap support.

11.1.4.3

Faucets

Faucets are valves (see Section 11.1.2.4) with spouts attached. They are designed to deliver the water flowing through the valve into a fixture. Modern faucets have one spout, and either one or two handles.

Faucets should be selected on the basis of compatibility with the fixture served, ease of operation, durability of finish and operating mechanism, and ease of cleaning. When choosing a faucet for a particular fixture, the spout design should be checked to ensure that, when installed, an air gap will exist between the tip of the spout and the maximum fill level of the fixture. The height of the air gap should be at least twice the diameter of the spout opening.

A faucet handle that requires a relatively large amount of movement provides easier adjustment of water temperature and flow rate. Round knobs require finger pressure to operate, which is difficult for some people. For example, persons with arthritis in their hands and wrists prefer handles that can be operated by pressure of the palms or sides of their hands. The simplest to operate is a single-lever type that is moved from side to side for temperature control and up and down for flow-rate control. Knobs that must be pulled or pushed are the most difficult to adjust.

Chrome on brass is the most durable finish for faucets and trim. Other finishes may be quite thin and subject to corrosion or wear. This is particularly true for the gold-colored finish. Finely detailed decorative handles are hard to keep clean.

Lavatory faucets can be obtained with a spray head similar to a shower or kitchen sink spray, which is useful for washing hair, pets, and children.

■ STEM FAUCETS

A two-handled faucet has one handle for hot water, the other for cold, and is known as a *stem* or *compression* faucet. Each handle controls a globe valve, or some variation of it, which makes this type of faucet subject to leakage problems. A frostproof outdoor faucet is another type of stem faucet.

■ SINGLE-LEVEL FAUCETS

Single-lever faucets are often called *mixing faucets*, because both hot- and cold-water supplies are controlled by the same handle. A single-handle faucet controls the flow of hot and cold water into the spout by the matching of holes in an alignment valve. By varying the size of the matched opening on the hot- or cold-water side, the desired temperature is achieved.

■ ANTISCALD FAUCETS

Faucets are available with thermostatically controlled valves to maintain the desired temperature regardless of pressure, flow, or temperature variations (Fig. 11.1-45) (see also Section 1.6.2). These faucets may also have a safety stop so that they cannot be set hotter than 100°F without first depressing a warning stop. Less expensive temperature control valves compensate for changes in relative pressure between hot and cold supplies, but not for temperature drift in either supply. Antiscald valves are available for lavatory and tub use, but are most common in shower installations.

11.1.4.4

Water Closets

All water closets have two basic parts: (1) a bowl that provides a receptacle for wastes and also forms the fixture drain trap, and (2) a flushing mechanism. They are generally classified according to their flushing mechanism type, their bowl type, and their mounting method. Most water closets are made of vitreous china, although some are plastic.

■ FLUSHING MECHANISM

Water closets are classified by their flushing mechanisms as either *flush tank* or *pressure valve* units.

FLUSH TANK UNITS Flush tank water closets are so-called because they contain a tank that holds a reserve of water that drains into the bowl to assist flushing. The most common tank-type water closets are (1) one-piece floor-supported, (2) two-piece floor-supported, and (3) two-piece wall-hung models (Fig. 11.1-46). One-piece wall-hung models are also available.

FIGURE 11.1-45 *Antiscald valves maintain selected hot- and cold-water flow ratio.* (*Courtesy Small Homes Council, Building Research Council, University of Illinois, Urbana-Champaign.*)

Some one-piece models similar to that shown offer a nonoverflow feature and a cleaner appearance, but are usually more expensive to purchase and install than standard two-piece units. Standard two-piece units have the tank supported on the bowl. In older water closets, the bowl is generally floor supported, and the tank hangs on the wall.

Submerged in the tank of a flush tank water closet are several control mecha-

nisms: a flushing valve, a water supply valve, and a combination bowl-refill and tank overflow device (Fig. 11.1-47). The flushing valve is essentially a stopper, or flapper, at the bottom of the tank, which is lifted when the flush lever on the outside of the tank is depressed. This allows the water in the tank to drain into the bowl to promote flushing.

Flush tank supply valves are available in several designs, all of which have a means of sensing when the water reserve is being lost because of the action of the flushing valve. The most common is a *ballcock valve*, which is controlled by the position of a float ball riding on the surface of the water in the tank. As the water flows out, the float falls, thereby opening the ballcock supply valve. As the tank refills, the float rises and eventually shuts off the valve.

A *fluid level control valve* works similarly to a ballcock valve, except that it has a float mounted on the supply-inlet post, rather than on an arm extending from it (Fig. 11.1-48). It is simpler and quieter than a ballcock valve.

A *pressure control fill valve* is controlled by the pressure exerted by the water above it rather than the height of the water surface.

A small tube extending from the high-water level in a flush tank down into the bowl acts as an overflow and refill device. A spur from the water supply line that fills the tank feeds into this tube, refilling the bowl, thereby restoring the trap seal after every flush. Should the water supply valve fail to shut off properly, this tube acts as an overflow drain for the tank.

PRESSURE VALVE UNITS Pressure valve water closets, which are also called *flush valve* units, are rarely installed in residences because they are noisy but are sometimes used in areas with limited water supplies (Fig. 11.1-49). Their ease of maintenance and low water consumption, however, recommend them for commercial installations where noise is not critical. Pressure valve units are available in both floor and wall-hung models.

■ TANKS

A standard flush tank discharges 3 to 5 gal. of water at each flushing action. Units designed for water conservation, though, have

FIGURE 11.1-46 *Water closet types: (a) one-piece floor mounted, (b) two-piece floor mounted, (c) two-piece wall hung. (Courtesy Small Homes Council, Building Research Council, University of Illinois, Urbana-Champaign.)*

TANK FLUSHING t = 10 sec.

flush handle
overflow pipe
fill valve
water level
flush ball
water supply on   water flow out
float

TANK FILLING t = 5 sec.

flush handle
water level
water supply on   water flow stopped
float

FULL TANK   t = 2 min.

flush handle
water level
water supply off   water flow stopped
float

FIGURE 11.1-47 *Flush tank water closets are favored for houses because they are simple and quiet. Flushing of a tank is shown, starting from the top. (Courtesy Small Homes Council, Building Research Council, University of Illinois, Urbana-Champaign.)*

bowl fill line   overflow pipe
float
fill valve   tank fill line

FIGURE 11.1-48 *Fluid-level control valve is quieter than ballcock mechanism. (Courtesy Small Homes Council, Building Research Council, University of Illinois, Urbana-Champaign.)*

disk
bypass
guide
handle
plunger
relief valve
valve outlet

FIGURE 11.1-49 *Pressure valves are noisy but use little water. They are common in commercial buildings. (Courtesy Small Homes Council, Building Research Council, University of Illinois, Urbana-Champaign.)*

smaller tanks and coordinated bowl designs that develop better flushing action, resulting in a discharge of half that of a standard unit or even less. It is possible to place plastic bottles or specially designed interior dams in older toilets to reduce water consumption, but modified standard toilets do not always produce adequate flushing action and may require repeated flushing to clear the bowl of fecal matter.

In summer, moisture can condense on the outside of a tank when room humidity is high and supply water is cold. Tanks are available with factory-installed insulation inside them to minimize condensation.

Other remedies for condensation include using a tempering valve to warm the cold water entering the tank; installing interior insulation in the field; providing a decorative fabric jacket around the outside of the tank; and installing a tray at the bottom of the tank to collect the condensate and drain it into the bowl.

■ BOWLS

There are several types of bowls in use today: (1) siphon jet, (2) siphon action, and (3) reverse trap. Some older water closets have *washdown bowls* in which the trap is located at the front of the fixture rather than the rear (Fig. 11.1-50). This results in a small water surface and a large dry-fouling area, which has caused its fall from popularity. *Blowout bowls* flush by the force of a high-pressure jet in the trapway rather than by siphonage. They are used primarily in commercial buildings rather than residences because they are noisy and require high water pressure.

water surface
water seal
water jet   trapway

FIGURE 11.1-50 *Washdown bowls are found mostly in older homes. They can be recognized by the bulge in front. (Courtesy Small Homes Council, Building Research Council, University of Illinois, Urbana-Champaign.)*

A *siphon jet bowl* is most frequently used in residential construction (Fig. 11.1-51). In this bowl, a jet of water in the trapway initiates the siphon action that empties the bowl. Its advantages are its relative quietness and effectiveness.

A *siphon action bowl* is similar, but holds more water and thus reduces the dry-fouling area.

A *reverse trap bowl* works on the same principle as a siphon jet one, and uses less water and is less expensive, but is noisier (Fig. 11.1-52).

FIGURE 11.1-51 *A siphon jet bowl is relatively quiet and has an effective scouring action.* (Courtesy Small Homes Council, Building Research Council, University of Illinois, Urbana-Champaign.)

FIGURE 11.1-52 *Reverse trap bowls use less water and are less expensive than siphon jet bowls, but are also noisier.* (Courtesy Small Homes Council, Building Research Council, University of Illinois, Urbana-Champaign.)

■ MOUNTING

Water closets are either floor supported or wall hung. Some flush tank units have floor-supported bowls and wall-hung tanks.

Wall-hung units are supported by a metal carrier concealed in the wall. One advantage of wall-hung water closets is that the floor under the tank is more accessible for cleaning.

The height of floor-supported seats normally ranges from 14 to 18 in., but a 20-in. seat height may be desirable for the elderly or handicapped. This can be achieved by use of an extension seat on floor-supported fixtures or adjustment of the carrier height of wall-hung fixtures.

11.1.4.5

Bathtubs

Bathtubs are made in various shapes, sizes, and materials.

FIGURE 11.1-53 *Bathtub types: (a) recessed square, with walls on three sides; (b) sunken rectangular; and (c) recessed rectangular, the most common type.* (Courtesy Small Homes Council, Building Research Council, University of Illinois, Urbana-Champaign.)

■ SHAPES AND TYPES

Three common shapes are available, as shown in Figure 11.1-53. The square and sunken type are more difficult to clean than rectangular and above-floor types. Typical overall dimensions for rectangular tubs are 14 to 16 in. high, 28-1/2 to 32 in. front-to-back width, and 60 in. long. Longer, shorter, and wider tubs are available on special order and at additional cost.

Most residential bathrooms employ a recessed tub with walls on three sides. When walls are available on two sides, a corner tub is used. In custom houses, free-standing tubs are sometimes used; many of these are either surrounded by a raised platform or sunk in the floor. One-piece panalized tub-shower assemblies are available in several types of plastic. They provide low-maintenance shower enclosures with built-in conveniences.

■ MATERIALS AND FINISHES

The most common and durable tub material is cast iron with a thick porcelain-enamel finish. Tubs of this type weigh approximately 200 lb. and require additional floor-framing members beneath them when they are placed on a wood-framed floor.

Tubs are also made of steel with a porcelain-enamel finish. These are more prone to enamel chipping and rusting than the cast-iron variety. Because they are lighter, they also are noisier when water strikes them. Some models are available with an acoustical treatment on the underside to reduce impact noise of water. Steel and plastic tubs are lighter and less rigid than cast-iron ones.

Some tubs are made from synthetic marble, acrylic-modified polyester, or similar homogeneous materials. Special shapes and sizes can be made using these materials, including such unusual shapes as round and octagonal ones. Such tubs frequently have whirlpool or hot-tub devices attached to or built into them.

Plastic tub assemblies are made of (1) glass fiber-reinforced polyester (FRP) with a smooth gel-coated surface, or (2) a molded acrylic sheet. Built-in ledges are provided in most plastic-molded assemblies. Ledges are handy for storing items used during bathing, but the rim to be stepped over should be narrow to make stepping into or out of the tub as easy as possible.

Tubs commonly have two openings for plumbing connections, one for the bottom drain and one for an overflow drain. The overflow opening is usually an inconspicuous slot under the edge of a chrome plate mounted high on the tub wall. The overflow drain protects against two hazards associated with overflowing fixtures: water damage to the building and siphonage of the trap seal. Tubs with whirlpool action devices will, of course, require additional openings to accommodate those devices and the circulation of water.

■ TUB FITTINGS

Tub drains are often bothersome because they either do not completely seal or do not drain quickly. Slow drainage is often caused by hair collecting on a drain plug mechanism of the type that is controlled remotely by a lever mounted in the overflow plate. In older versions of this mechanism, the lever raises and lowers a brass cylinder that, when lowered, blocks off the fixture drain (Fig. 11.1-54). In newer versions, the lever raises and lowers a

FIGURE 11.1-54 *Trip-lever drain uses a cylinder on an adjustable rod to open or close the drain opening.* (*Courtesy Small Homes Council, Building Research Council, University of Illinois, Urbana-Champaign.*)

FIGURE 11.1-55 *Pop-up drain opens when spring-loaded lever pushes on end of pop-up rod.* (*Courtesy Small Homes Council, Building Research Council, University of Illinois, Urbana-Champaign.*)

vertical arm, which in turn controls a horizontal arm with a pop-up drain plug at the end (Fig. 11.1-55).

11.1.4.6

Showers

A shower unit consists of a showerhead mounted either in an enclosed bathtub or shower stall. Showerheads are simple sprayers that either have a preset spray pattern or can be adjusted from fine mist to heavy spray. Some models offer built-in flow restrictors, which conserve water. Some units also have a shutoff valve on the head, so the water can be turned off during soaping (Fig. 11.1-56a). This is an advantage also in that the temperature

setting on the main supply valve need not be disturbed.

A handheld spray head on a flexible metal-covered hose can be used instead of a fixed showerhead (Fig. 11.1-56b). It can be hung on a permanent bracket, on a sliding bracket, on a support bar to allow adjustment for various heights, or be handheld. When such heads are installed over tubs or lavatories, a vacuum breaker should be provided or the showerhead should be hung in such a way that at least a 1-in. air gap will be assured when it is hanging fully extended. This prevents back siphonage of bath water and contamination of the water supply, should the spray head fall into the water.

■ TUB-SHOWERS

Showers installed over a bathtub usually share the tub's water-supply valve. The

FIGURE 11.1-56 *(a) Shutoff valve on showerhead; (b) phone-type adjustable showerhead.* (*Courtesy Small Homes Council, Building Research Council, University of Illinois, Urbana-Champaign.*)

tub-filler spout includes a *diverter valve* that redirects water coming to the spout to the showerhead instead. Adding a shower to a bathtub also requires that a waterproof enclosure be provided up to the height of the showerhead, so that splatter from the shower is directed into the tub and down the drain. On the three walls surrounding the tub, this can be accomplished with metal or plastic panels specially made for this purpose, or with ceramic tile. Risk of leakage is reduced if panel systems without corner joints are used.

In ceramic tile tub enclosures, either cement backer board or water-resistant gypsum board should be used as the backing (see Section 9.6). Care must be taken that grouting between tiles is sound to avoid leaks.

An alternative that eliminates joints and potential leaks from a tub enclosure is a one-piece, molded plastic tub-shower unit (Fig. 11.1-57a). The size of such units dictates that they be placed before the bathroom walls are framed. The fourth side of the tub must also be enclosed, either with a shower curtain or with a safety-glazed door.

■ SHOWER STALLS

Showers not installed in conjunction with a bathtub have their own stalls built around them. Most shower stalls consist of a floor receptor topped by water-resistant walls, either self-supporting or attached to the wall framing. Prefabricated receptors are made of precast concrete, terrazzo, preformed metal, or plastic. Job-built receptors of ceramic tile can also be used if a sloping base and a waterproofing pan are laid beneath it. Complete plastic receptor and wall combinations are available, both as components and as one-piece units. (Fig. 11.1-57b).

The fourth side of the shower stall can be enclosed with a curtain, but a higher-than-normal curb is necessary to prevent it from blowing out and allowing water to run onto the floor. Sliding or bifold doors with safety glazing are available for wide stalls. Hinged doors are common, but if they open outward, they should have a provision to minimize water dripping onto the floor.

A common small shower stall size is 30 in. sq. A more convenient stall size is 36 in. sq. or 36×48 in. Complete assemblies are about 84 in. high.

FIGURE 11.1-58 *The air gap between over-flow level and spout prevents contamination of the water supply. (Courtesy Small Homes Council, Building Research Council, University of Illinois, Urbana-Champaign.)*

FIGURE 11.1-57 *(a) One-piece molded plastic tub shower; (b) one-piece plastic shower. Construction permits inclusion of conveniences such as soap dishes and seats. (Courtesy Small Homes Council, Building Research Council, University of Illinois, Urbana-Champaign.)*

A shower stall offers easier entry and exit and more secure footing than does a shower in a tub, especially when provided with grab bars. Portable suction mats or slip-resisting tapes can be used if the floor finish is not slip resistant.

11.1.4.7

Separation of Systems

Most codes demand that plumbing supply and drainage systems be separated to prevent cross flows of waste fluids into the supply system and vice versa. Contamination of a supply system can have severe negative effects on the health of occupants.

Water flows into a fixture from a supply system, is used by the occupants, and then drains out of the fixture into a drainage system. Fixtures are designed to ensure that the flow of water through them is in one direction only, so that drain water is never allowed back into the supply pipes. The unintentional flow of contaminated water into a supply source is called *back-flow* or *back siphonage*. This can occur when the pressure in a supply pipe drops below that of the fluid, be it air or water, in which the supply pipe terminates. This pressure reversal cannot occur when the supply water pressure is between 40 and

50 psi as is usual, and each supply line ends in air, which has a pressure of about 14 to 15 psi.

Therefore, plumbing codes require that fixtures be designed so that an air gap separates the supply lines from the highest point at which contaminated water can collect (Fig. 11.1-58). An *air gap* is the unobstructed vertical distance between the mouth of a water outlet and the flood-level rim of the water receptacle. The water outlet may be a faucet, spout, or other outlet; the receptacle may be a plumbing fixture, tank, or other receptacle. In some cases, the highest point that contaminated water can reach is the level of the overflow drain of a fixture. To ensure that the suction created in the pipe during reversal of flow is not strong enough to draw contaminated water across the air gap between the end of the supply line and the water surface, the required minimum air gap is usually 2 times the diameter of the faucet or a minimum height of 1 in. for lavatories, 1-1/2 in. for sinks and fixtures having 3/4-in. bathroom faucets, and 2 in. for fixtures having 1-in. bath faucets.

Where provision of an air gap is not feasible in fixture design, a device called an *atmospheric vacuum breaker* or *siphon breaker* is installed. This is a simple mechanical device consisting essentially

of a check valve in a supply line and a valve member (on the discharge side of the check valve) opening to the atmosphere when the pressure in the line drops to atmospheric pressure (Fig. 11.1-59). This device acts as a one-way valve, allowing water to flow through it in one direction, but if flow is reversed it opens, allowing air to enter instead of permitting the reverse flow of water. Vacuum breakers generally are required on dishwashers and other appliances that otherwise lack an air gap.

11.1.5

SOUND CONTROL

High water pressure and velocity and quick-closing valves in supply piping systems combine to produce plumbing noises. If a building does not exceed three stories and the water pressure at the street exceeds 35 to 45 psi a pressure-reducing valve can be installed, preferably as far from the building as feasible. In high-rise buildings, combinations of pressure-reducing valves and pressure-regulating systems may be necessary to provide suitable and uniform pressure levels at the various floor levels.

The noises in a plumbing system can be classified as those associated with the piping and those associated with appliances.

FIGURE 11.1-59 *A vacuum breaker uses atmospheric pressure to close the valve when the pressure is reduced on the left side and to reopen it when higher-than-atmospheric pressure is restored on the left side.* (Courtesy Small Homes Council, Building Research Council, University of Illinois, Urbana-Champaign.)

FIGURE 11.1-60 *Internal pipe pressures cause water hammer as a result of a fast-closing valve.* (Courtesy Small Homes Council, Building Research Council, University of Illinois, Urbana-Champaign.)

FIGURE 11.1-61 *Water hammer can be minimized by installing (a) an air chamber or (b) a shock absorber near the service entrance. A similar, but smaller, air chamber should be installed at each valved fixture.* (Courtesy Small Homes Council, Building Research Council, University of Illinois, Urbana-Champaign.)

The first of these include water hammer, creaking, popping, squealing, cavitation, vibration, and other general noises produced by rushing water. Noises associated with fixtures are usually any banging and bumping caused by vibration.

11.1.5.1

Noises Associated with Piping

Water hammer, which is also called hydraulic shock, consists of pounding noises and vibrations that sometimes develop in a piping system when its air chambers become filled with water or have been omitted entirely. It also may be the result of worn washers in faucets. This unpleasant symptom occurs when there is a sudden change in the velocity of water flowing through a pipe. This causes a shock wave to travel through the piping system, producing a banging as the pipes vibrate against their supports. Such a rapid change in velocity is often caused by the sudden closing of a fast-acting valve, the starting or stopping of certain types of pumps, or improper check valves (Fig. 11.1-60). Continuous water hammer may result in broken pipes, leaky joints, loosened pipe hangers, ruptured tanks, and damage to meters, gauges, regulators, and faucets.

Water hammer can be avoided by using spring-loaded check valves, in lieu of spring or solenoid valves, that do not close fully until the movement of water in the pipe has stopped. It can be minimized by installing air-filled extensions of the supply lines (air chambers) at each fixture or near the service entry (Fig. 11.1-61a). These *air chambers* consist of a piece of pipe about 10 in. long, installed above the hot and cold valves of fixtures such as sinks, lavatories, and clothes washers, to cushion the rush of water as the valve is closed, and prevent water hammer. Another way to reduce water hammer is to construct the supply lines using plastic piping, which is flexible enough to dissipate shock waves.

In older buildings, water hammer may be the result of worn out washers, which cause faucets to close suddenly, or to water-logged air chambers. To restore the air in water-logged chambers, faucets should be opened and water drained from the supply system at its lowest point. In both new and old buildings, water hammer problems can be minimized by installing a shock absorber near the service entrance (Fig. 11.1-61b).

Squealing noises are sometimes emitted by pipes in which water flows at a high velocity and strikes small protrusions on the inside face of the pipe, causing vibration of the pipe. This problem can usually be solved by reducing the velocity of the flow in the pipe. Squealing can also occur when water passes a loose or worn valve washer, causing it to vibrate. This can be stopped by tightening the screw holding the washer in place or replacing the washer. Proper sizing and anchorage of copper water-supply piping will also avoid some noises.

Noise resulting from water rushing through piping can be minimized by planning the system with a minimum number of bends and by reducing the number of fittings and valves to a minimum. Valves and right-angled fittings produce substantially more noise than straight runs of pipe because of increased water turbulence in them. The noise produced by four elbows is about 10 decibels (dB) higher than the noise produced in a straight run of equivalent-sized pipe under conditions of turbulent flow.

Long runs of hot-water supply piping produce *creaking* noises as the piping expands or contracts from hot or cooled water. Long, straight runs of iron or steel piping should be designed with flexibility with at least one end having a swing arm to permit movement (Fig. 11.1-62a). Flexible L or M piping bent into an S-shaped curve at one end will allow for movement in long runs of copper piping (Fig. 11.1-62b). Supports should be designed to permit the piping to expand and contract without binding.

Supply piping to units that use large amounts of water, such as dishwashers and water closets, should not be undersized.

When pipes are rigidly connected to structures with hangers, clamps, or straps, sound resulting from pipe *vibration* can be transmitted through the structure and can be amplified from one apartment unit to another. Resilient pipe insulation between the pipe supports and the pipe should be used to help reduce the transmission of these noises (Fig. 11.1-63a). The number of pipe supports should be kept to the minimum required, and supports should be attached to the most massive structural element available.

FIGURE 11.1-62 *Flexible connections, using (a) swing arms in piping or (b) S curves in tubing, minimize noise caused by expansion and contraction in long runs of piping.*

The running of drainpiping through bedroom or other quiet living-area walls should be avoided, unless a double sound-isolating wall is provided. Waste dropping down vertical stacks causes piping to vibrate, especially when it hits horizontal runs, unless the pipe is resiliently mounted. When vertical stacks are hung on floor joists, vibrations are transmitted to the structure (Fig. 11.1-63b). Standard pipe clamps and resilient vertical supports or wood brackets should be installed with the end of the support resting on the vertical stack with a 1-in.-thick isolation pad resting on the support (Fig. 11.1-63c). A short section of pipe insulation around a pipe where it passes through a wall sill will keep it in horizontal alignment and will isolate it from the structure.

It is essential that holes and openings where piping passes through common walls be sealed with a resilient material or should be tightly packed with insulation, or both, to isolate vibration of the pipe from the structural framing and to seal against air leaks (Fig. 11.1-64).

Cavitation is a phenomenon that occurs when water flowing at a high velocity makes a rapid change in direction or passes through a small orifice. Cavitation produces a high-pressure region at the outside of a pipe bend and a low-pressure region at the inside of the bend. If the inside pressure is low enough, the water boils and air bubbles form. Farther down the pipe, these bubbles reencounter pres-

FIGURE 11.1-63 *Vibrations from a plumbing stack can be isolated from the structure by resiliently mounting vertical supports and resiliently supporting horizontal runs.*

sure and are collapsed by it. This produces a noise similar to gravel rattling in the pipe or balloons popping.

Aside from the noise, cavitation can cause unnecessary wear on pipes and pumps. A solution is to lower the velocity of the water to less than 10 ft./sec. Typical residential installations have water velocities ranging from 4 to 8 ft./sec.

11.1.5.2

Fixtures and Appliances

Often, plumbing fixtures and appliances can be selected and installed to reduce the amount of noise. For example, a siphon jet water closet having an adjustable-rate, water-supply inlet float valve will operate more quietly. A water closet can be mounted on a floating floor and the noise transmission from fixtures can be reduced

FIGURE 11.1-64 *Where pipes pass through walls, openings must be tightly packed or sealed airtight with resilient materials to isolate vibrations and to close paths for airborne sound transmission.*

by using a neoprene or rubber gasket between the base of a water closet and the floor (Fig. 11.1-65), or by a resilient underlayment between two layers of floor sheathing. Flexible caulking or tightly packed insulation should be used to seal the space between a toilet bend and the floor sheathing.

Vibrating appliances such as dishwashers, disposers, and clothes washers and

FIGURE 11.1-65 *Water closet vibrations can be minimized by resilient mounts and floated floor construction.*

dryers should not be rigidly connected to the structure. Resilient or flexible isolation gaskets should be provided between the appliance and the floor, wall, or counter to isolate vibrations, unless a floated floor system is used. Piping connections to appliances should be flexible enough to prevent vibrations from being transmitted to the piping (Fig. 11.1-66). Electrical connections to appliances should also be flexible, whenever possible.

FIGURE 11.1-66 *Flexible piping connections prevent vibrations from being transmitted to the piping.*

11.2 SEPTIC TANK SYSTEMS

Sewage is the name given to liquid waste that contains organic (animal or vegetable) matter in suspension or solution, as well as liquids containing chemicals in suspension. Humans increase the amount of such wastes beyond the ability of the natural environment to absorb them without treatment to break them down into nontoxic substances. In spite of the immense growth in public sewage disposal systems in this century, many areas are still beyond their reach. In these locations, sewage disposal systems designed to handle a single residence or a group of dwellings or a small commercial project continue to be an important method of sewage disposal.

Not long ago, in rural areas human and other domestic wastes were disposed of using cesspools. Cesspools are covered and lined underground pits designed to permit liquids to seep into the ground

through the bottom and sides but retain organic matter and other solids, which are then periodically pumped out. Such systems present not only an obnoxious nuisance, they are also a major health hazard and are almost universally prohibited in this country. Comparatively, today's *individual sewage (septic) disposal systems* are safe and nuisance free (Fig. 11.2-1). These systems consist generally of a *septic tank*, where the sewage is retained long enough to separate and decompose the solids (see Section 11.2.3), coupled with some type of *soil absorption system* (absorption field, or absorption bed, or seepage pit) where the liquids discharged from the septic tank gradually seep into the soil (see Section 11.2.2).

In these systems, sewage is treated both in the tank and in the soil, as anaerobic (oxygen-free) bacteria decompose the

organic matter into simpler, harmless compounds. However, because the anaerobic decomposition of organic matter is accompanied by foul odors and pathogenic (disease-producing) organisms, the partially treated liquid sewage flowing from a septic disposal system (*effluent*) cannot be discharged on the surface of the ground or into a lake or stream without creating a nuisance and a health hazard.

Instead, the effluent from a septic tank should be discharged into a properly designed and maintained soil absorption system, which uses the soil for the subsurface absorption of treated sewage. It practically eliminates the health and nuisance hazards but still poses problems in certain areas. A soil absorption system may not work in tightly packed soils, or the land area required may be too large.

Large sewage treatment facilities work

FIGURE 11.2-1 *Typical septic tank type individual sewage disposal system. Solids are separated and retained in the tank, liquids are passed to the absorption field, and gases are vented through the house sewer and vent stack.* (*Illustration from* Manual of Septic Tank Practices, *Bureau of Community Environmental Management, U.S. Public Health Service [now EPA].*)

The removal of solids and the partial biological treatment that takes place in a septic tank is known as *primary treatment*. Most municipal and community sewage plants provide so-called *secondary treatment*, producing relatively odorless but nutrient-rich effluents. Therefore, sewage plants should be equipped to remove or neutralize the nutrients (called *tertiary treatment*) before discharging into lakes or streams.

Proper disposal of human excreta is a major factor in influencing the health of individuals, where public sewers are not available. Many diseases, such as dysentery, infectious hepatitis, typhoid, and paratyphoid, as well as various types of diarrhea, are transmitted from one person to another through the fecal contamination of food and water, largely the result of the improper disposal of human wastes.

Safe disposal of human wastes is necessary to protect the health of the community. For satisfactory results, such wastes must be disposed so that they (1) will not contaminate a drinking-water supply and will not pollute the waters of any bathing beach, shellfish breeding ground, or stream used for water supply or for recreation; (2) will not be discharged on or near the surface where they would be accessible to insects, rodents, or other small animal carriers that may come into contact with food or drinking water; (3) will not be accessible to children or give rise to a nuisance based on foul odor or unsightly appearance; (4) will not overburden any body of water with excess nutrients or degrade its quality below officially adopted water-quality standards.

Connection to a public sewerage system is the most satisfactory method of sewage disposal. When this is not feasible, and a when a considerable number of residences or other small buildings are to be served, consideration should next be given to the construction of a community sewerage system and treatment plant. Specific information on this matter can be obtained generally from the local health authority.

The discussion and recommendations presented in this section have been abstracted from a publication of the Bureau of Community Environmental Management, U.S. Public Health Service (HEW), entitled 'Manual of Septic Tank Practice," 1967 edition, with some updating to reflect current practice. Codes and health

on the same principle as do natural decay and oxidation, except that the action is more concentrated. In a treatment facility, favorable conditions are created for decay bacteria to break down organic solids into chemically simpler water-soluble compounds: phosphates and nitrates. In the natural cycle, limited amounts of phosphates and nitrates are produced and these become the nutrients for other growing things. Growing plants absorb such nutrients through their roots and, with the help of sunlight, recombine them with carbon dioxide from the air to form new living tissue. In sewage treatment, large amounts of phosphates and nitrates are produced and flushed away by the effluent, to be ab-

sorbed by the soil or discharged into a body of water.

Phosphates and nitrates are relatively harmless when released into the soil or water in limited quantities. When introduced into a lake or stream in large amounts, they spur the growth of small aquatic plants (algae), which use up valuable dissolved oxygen as they proliferate and then die off in large quantities. This overenrichment of a body of water with excessive nutrients, proliferation of algae, and reduction of oxygen content are characteristic of the *eutrophication* process, which gradually destroys aquatic life and renders the water useless for recreation and water supply.

officials should be contacted and their latest requirements obtained before beginning the design of an individual sewage disposal system.

This discussion of septic tank systems is intended as an aid in understanding the nature of such systems. The recommendations here are made with the full understanding that they may differ in some instances from other authoritative sources such as the standards and manuals referenced in the HUD *Minimum Property Standards for Housing* and the various model plumbing codes. Local, state, and federal laws, codes, and regulations always supersede the recommendations in this section.

11.2.1

SUITABILITY OF SOIL

A sewage disposal system that uses the soil for subsurface absorption of treated sewage is called a *soil absorption system*. A septic tank system is in this category. The first step in the design of a septic tank system is to determine whether the soil is suitable for the absorption of effluent and, if it is, how much absorption area is required. In general, three conditions must be met. (1) The percolation time should be within the range indicated in Figure 11.2-2, that is, not more than 60 min. for absorption fields and beds, nor more than 30 min. for seepage pits. Percolation time is the time required for the level of water in a standard test pit to drop 1 in. (see Section 11.2.1.3). (2) The bottom of a trench or seepage pit should be at least 4 ft. above the maximum seasonal elevation of the groundwater table. (3) Rock formations and other impervious strata should be more than 4 ft. below the bottom of a trench, bed, or pit. (4) The area should be large enough to provide the required absorption area in trench and bed bottoms or pit sidewalls as shown in Figure 11.2-2.

Unless these basic conditions can be satisfied, the site is unsuitable for a soil absorption system.

11.2.1.1

Visual Examination

The water absorption of soils can be determined best by percolation tests, but several

FIGURE 11.2-2
Required Absorption Areas in Various Soils

Percolation Rate (Time in min. for Water to Fall 1 in.)	Minimum Area per Bedroom (sq. ft.)
2 or less	85
3	100
4	115
5	125
10	165
15	190
30[a]	250
45	300
60[b]	330

Table excerpted from Manual of Septic Tank Practices, *Bureau of Community Environmental Management, U.S. Public Health Service (now EPA).*

[a] Soil not suitable for seepage pit if percolation rate is more than 30 min.

[b] Soil not suitable for any absorption system if percolation rate is more than 60 min.

clues help in estimating the likely characteristics of a particular soil. These include the soil maps and descriptions of the U.S. Department of Agriculture, on-site visual inspection of soil texture and color, and the record of experience in adjacent areas with similar soils.

For these purposes, soil can be classified as either pervious or impervious. *Pervious soil* is usually granular, such as sand or gravel. It allows water to pass readily through it. *Impervious soil* is a tight cohesive soil, such as clay, which does not permit the ready passage of water. Most sites contain a mixture of pervious and impervious soils.

Soil particle size governs pore size, which, in turn, influences the rate of water movement. Sandy and gravelly soils generally have large particles and large pores, whereas clay and silt soils have small particles and small pores (although total void area may be larger). In some soils, especially clay and silt types, there may be clumping of basic particles into larger particles giving the soil a well-developed structure. A lump of structured soil will

break apart with little force along well-defined cleavage planes. Such structured clay soil is likely to have better water-absorption characteristics than one with little or no structure.

Another clue to its absorptiveness is the color of the soil. Bright, uniform, reddish-brown to yellow colors throughout a soil profile indicate favorable oxidation conditions, good air and water movement, and desirable absorption properties.

In some cases, examination of road cuts, stream embankments, or building excavations will give useful information. Well drillers' logs can also be used to obtain data on groundwater and subsurface conditions. In some areas, subsoil strata vary widely in short distances, and borings must be made at the site of the system. If the subsoil appears suitable, as judged by the characteristics described above, percolation tests should be made at points and elevations selected as typical of the area in which the soil absorption system will be located.

11.2.1.2

Percolation Tests

Subsurface exploration is usually necessary to determine soil conditions—the depth of pervious material, the presence of rock formations, and the level of groundwater. Augers with extension handles often are used for making such investigations (Fig. 11.2-3).

Percolation tests help to determine the acceptability of a site and establish the type and size of the required soil absorption system. Figure 11.2-2 shows the area of absorption or seepage area recommended for soils with various percolation rates, as well as the slowest percolation rate acceptable for each type of soil absorption system.

■ TEST HOLES

At least six separate test holes should be made. These should be spaced uniformly over the proposed soil absorption area and should reach down to the depth of the proposed absorption system. These holes can be from 4 to 12 in. in diameter. To save time, labor, and the volume of water required per test, shallow holes can be bored with a 4-in. auger (Fig. 11.2-4a). Deeper

FIGURE 11.2-3 *Augers with extension handles can be used to bore deep test holes.* *(Illustration from* Manual of Septic Tank Practices, *Bureau of Community Environmental Management, U.S. Public Health Service [now EPA].)*

FIGURE 11.2-4 *Shallow percolation test holes for absorption trench or bed showing (a) 4-in. auger-bored hole and (b) larger test hole dug with a shovel.* *(Illustration from* Manual of Septic Tank Practices, *Bureau of Community Environmental Management, U.S. Public Health Service [now EPA].)*

holes may have to be dug by shovel or heavy equipment (Fig. 11.2-4b).

The bottom and sides of a test hole should be scratched with a knife or sharp-pointed instrument to remove smeared soil and to provide a natural soil surface into which water may percolate. Loose material should be removed from the hole and 2 in. of coarse sand or fine gravel should be added to protect the bottom from scouring and sediment.

■ SATURATION AND SWELLING

Saturation means that the void spaces between individual particles of soil are full of water, which can be accomplished in a short period of time. *Swelling* is caused by the intrusion of water into the particles of soil, which is a slow process. It is the chief reason for providing an overnight soaking period, especially in clay.

The suggested test procedures are intended to ensure that the soil is given ample opportunity to swell and to approach the condition that will exist during the wettest season of the year. Thus, the test should give comparable results in the same soil, whether made in a dry or in a wet season. In sandy soils containing little or no clay, the swelling procedure is not essential, and the test may be made after the water from a single filling of the hole has completely seeped away.

■ PERCOLATION-RATE MEASUREMENT

The length of time required for percolation tests varies depending on the type of soil. For example, the recommended swelling procedure may not be needed in a sandy soil containing little or no clay, and the test may be made after the water from one filling of the holes has completely seeped away.

To begin a test, clear water should be placed in the test holes until the gravel is covered by at least 12 in. of water. In most soils, it is necessary to refill the hole by supplying a surplus reservoir of water, possibly by means of an automatic siphon, to keep water in the hole for at least 4 hours and preferably overnight (see Figs. 11.2-4, 11.2-5). The purpose of this preliminary soaking is to obtain the percolation rate that will occur when the soil is soaked and saturated.

If water remains in a hole after the swelling period, morewater should be added, or water should be removed, until water covers the gravel by about 6 in. The drop in water level should then be measured from a fixed reference point at 30-min. intervals over a 4-hr. period. The drop that occurs in the final 30-min. period should be used to calculate the percolation rate. For instance, a drop of 2-1/2 in. in 30 min. translates to a percolation rate of 12 min./in. (30 min. ÷ 2.5 in.).

If no water remains in a hole after the swelling period, water should be added until water covers the gravel by about 6 in.

FIGURE 11.2-5 *Percolation tests for seepage pits must be made in deep test holes dug down to the level of the proposed bottom of the pit.* *(Illustration from* Manual of Septic Tank Practices, *Bureau of Community Environmental Management, U.S. Public Health Service [now EPA].)*

From a fixed reference point, the drop in water level should be measured at 30-min. intervals for 4 hr., refilling up to 6 in. over the gravel as necessary. The drop that occurs during the final 30 min. should be used to calculate the percolation rate.

In sandy, or other permeable, soils in which the first 6 in. of water seep away in less than 30 min. after the swelling period measurements should be taken at 10-min. intervals for 1 hr. The drop that occurs during the final 10 min. should be used to calculate the percolation rate.

11.2.2

SOIL ABSORPTION SYSTEMS

From the results of the percolation tests it is possible to determine whether a septic tank system is feasible and which type of soil absorption system can be used. When a soil absorption system is usable,

three types of design may be considered: absorption field, absorption bed, and seepage pit.

An *absorption field* is an arrangement of absorption trenches through which treated sewage is absorbed into the soil. These are also called *disposal fields* and *soil absorption fields*.

An *absorption trench* is a trench not more than 36 in. in width, containing at least 12 in. of clean, coarse aggregate and a distribution line, through which treated sewage is allowed to seep into the soil.

An *absorption bed* is similar to an absorption trench, except that it is more than 36 in. wide and contains two or more distribution lines. These are also called *seepage beds*.

A *distribution line* is an open-jointed or perforated pipe that permits effluent to escape into the soil for absorption. It may consist of 1-ft. lengths of 4-in. agricultural drain tile or 2- to 3-ft. lengths of vitrified clay sewer pipe (both laid with open

joints), or perforated plastic pipe. Local experience and applicable codes and ordinances should be reviewed before selecting piping materials.

A *seepage pit* is a covered underground pit with a concrete or masonry lining. These are designed to permit partially treated *sewage* to seep into the surrounding soil. They are also called *dry wells*. Some jurisdictions do not permit the use of seepage pits.

As Figure 11.2-2 shows, only soils with a percolation rate of 30 min./in. or less are suitable for seepage pits; if the rate is greater than 60 min., the soil is not suitable for any type of soil absorption system. In addition, the selection of a soil absorption system is affected by the location of the system. For example, a safe distance must be maintained between a soil absorption system and a source of water supply. Since the distance that pollution will travel underground depends on numerous factors, including the characteristics of the subsoil formations and the quantity of sewage discharged, no specified distance would be absolutely safe in all localities. In general, the horizontal separation between elements of a soil absorption system and other site improvements should not be less than that outlined in Figure 11.2-6.

Seepage pits should not be used in areas where water supplies are obtained from shallow wells, or where there are limestone formations and sinkholes through which effluent may easily contaminate the groundwater.

Although *absorption fields*, *absorption beds*, and *seepage pits* should not be intermixed, several of any one of them can be arranged in sequence so that each is forced to utilize its total effective absorption area before liquid flows into the succeeding component. This is called *serial distribution*.

11.2.2.1

Absorption Fields

The individual branches of an absorption field are called *laterals*. These consist of either (1) the length of distribution line between the tee or cross fitting and the farthest point in a closed-loop field (Fig. 11.2-7), or (2) the length of distribution line between overflow pipes (Fig. 11.2-8).

There should be at least two laterals in every system. No lateral should be more

FIGURE 11.2-6
Minimum Clearances for Components of a Septic Sewage Disposal System[a]

From	To Well[b]	To Water Line[c]	To Foundation Wall	To Property Line	To Stream
Building sewer	50 ft.	10 ft.	—	—	50 ft.
Septic tank	50 ft.	10 ft.	5 ft.	10 ft.	50 ft.
Absorption field	100 ft.	25 ft.	20 ft.	5 ft.	50 ft.
Absorption bed	100 ft.	25 ft.	20 ft.	5 ft.	50 ft.
Seepage pit	100 ft.	50 ft.	20 ft.	10 ft.	50 ft.

Table excerpted from Manual of Septic Tank Practices, *Bureau of Community Environmental Management, U.S. Public Health Service (now EPA).*

[a]Based on recommendations of U.S. Public Health Service, HEW.

[b]And to suction line between well and pressure tank.

[c]Pressure type.

ABSORPTION TRENCH

4" DRAIN TILE

DASHED LINES INDICATE EXTENT
OF COARSE AGGREGATE

WATERTIGHT JOINTS AT BEND

FIGURE 11.2-7 *Typical absorption field layout for level ground. Length of lateral should be figured from start of absorption trench (dashed lines) to farthest point.* (Illustration from Manual of Septic Tank Practices, Bureau of Community Environmental Management, U.S. Public Health Service [now EPA].)

than 100 ft. long. Using more and shorter laterals is preferred because if something should happen to disturb one line, most of the field will still be serviceable. The trench bottom and distribution lines should be level. Laterals should be spaced to provide at least 6 ft. of undisturbed earth between trenches.

■ DESIGN

Many different designs may be used in laying out subsurface absorption fields. The choice depends on the size and shape of the available disposal area, the capacity required, and the topography of the area (see Figs. 11.2-7 and 11.2-8).

To provide the minimum-required gravel depth and earth cover, the trench depth should be at least 24 in. (Fig. 11.2-9). Additional depth may be needed for surface-contour adjustment, for extra aggregate under the tile, or for other design purposes.

In considering the depth of absorption trenches, the question is raised of the possibility of the lines freezing during prolonged periods of cold weather. Freezing rarely occurs in distribution lines that are adequately surrounded by gravel at the proper depth when the system is kept in continuous operation. However, pipes under driveways or other surfaces that are usually cleared of snow should be close jointed and insulated or should be installed at a greater depth.

The total required absorption area is determined by the results of soil percolation tests. The required area for residences may be obtained from Figure 11.2-2. For example, for a three-bedroom house on a lot where the minimum percolation rate was 15 min./in., the necessary absorption area would be 3 times 190 sq. ft. per bedroom, or 570 sq. ft. For trenches 2 ft. wide, the required total length of trench would be 570 sq. ft. divided by 2 ft., or 285 linear ft. If three laterals were used on sloping ground (see Fig. 11.2-8), the length of each lateral would be 285 ft. divided by 3, or 95 ft. If a closed-loop system on level ground (see Fig. 11.2-7) were used, each of the half loops could be 100 ft., and the middle lateral 85 ft. The spacing of trenches is generally governed by practical construction considerations such as type of equipment, soil, and topography, but the clear space between trenches should not be less than 6 ft. to prevent short-circuiting (flow between trenches).

In the example cited, trenches are 2 ft. wide, there are three trenches, and two spaces between trenches at 6 ft. each. The total width of trenches and spaces is 6 ft. plus 12 ft., or 18 ft. The required area for the absorption field is 18 ft. times 95 ft., which equals 1710 sq. ft., plus the additional land required to provide the clearances recommended in Figure 11.2-6.

Ideally, a house site should be large enough to allow room for an additional system if the first one fails or if bedrooms are added later.

■ CONSTRUCTION

Careful construction is important in obtaining a satisfactory soil absorption system. Attention should be given to protecting the natural absorption properties of the soil. Care must be taken to prevent sealing

FROM SEPTIC TANK

A

B

ALTERNATE DROP BOX METHOD AS SHOWN IN
SECTION BB MAY BE USED.

B

A

NOTE:
INVERT OF THE OVERFLOW PIPE
MUST BE AT LEAST 4" LOWER THAN
INVERT OF THE SEPTIC TANK OUTLET

12"
MIN.

TIGHT
JOINTS

UNDISTURBED
EARTH

① TEE*
② ELL*

6' MINIMUM

SECTION A—A

TIGHT JOINTS

UNDISTURBED EARTH

6' MINIMUM

SECTION B-B
(ALTERNATE CONSTRUCTION)

* DIFFERING GROUND SLOPES OVER SUBSURFACE DISPOSAL FIELD
MAY REQUIRE USE OF VARIOUS COMBINATIONS OF FITTINGS.

FIGURE 11.2-8 *Typical absorption field layout for sloping ground.* *(Illustration from* Manual of Septic Tank Practices, *Bureau of Community Environmental Management, U.S. Public Health Service [now EPA].)*

the surface on the bottom and sides of the trench. Trenches should not be excavated when the soil is wet enough to smear or compact easily. Soil moisture is right for safe working only when a handful will mold with considerable pressure. Open trenches should be protected from surface runoff to prevent the entrance of silt and debris. If it is necessary to walk in the trench, a temporary board laid on the bottom will reduce compaction. If smearing and compaction does occur, surfaces should be raked to a depth of 1 in., and loose material removed, before the gravel

is placed in the trench.

A distribution pipe should be laid in a level trench of sufficient width and depth (see Fig. 11.2-9), and should be surrounded by clean, graded gravel or rock, broken hard-burned clay brick, or similar aggregate, ranging in size from 1/2 to 2-1/2 in. Cinders, broken shell, and similar material are not recommended, because they are usually too fine and may lead to premature clogging. The coarse aggregate should extend from at least 2 in. above the top of the pipe to at least 6 in. below the bottom of the pipe. If open-joint drain tile is used, the upper half of each joint opening should be covered with a strip of 15-lb. roofing paper, as shown in Figure 11.2-9.

The top of the stone should be covered with a vapor-pervious barrier, such as untreated building paper, a 2-in. layer of hay or straw, or similar pervious material to permit vapor movement and yet prevent the stone from becoming clogged by the earth backfill. An impervious covering should not be used, as this interferes with *evapotranspiration* (absorption and release of moisture by plants) at the surface. Although generally not figured in the calculations, evapotranspiration is often an important factor in the operation of absorption fields and beds.

Open joint tile is usually laid with the opening covered by strips of roofing paper, but drain tile connectors, collars, clips or similar accessories are of value in obtaining uniform spacing, and protecting tile joints. Most such accessories are made of galvanized iron, copper, or plastics.

Tree and shrub roots sometimes penetrate into transmission lines and block the flow of effluent. Root problems can be prevented by using a liberal amount of gravel or stone over and under the pipe. Roots generally seek ground moisture and, in the few cases where they become troublesome, there is usually some explanation involving changing moisture conditions. For example, at a residence used only during the summer, roots are most likely to penetrate when the house is uninhabited, or when moisture immediately below or around the gravel becomes less plentiful than during the period when the system is in use. In general, trenches constructed within 10 ft. of large trees or dense shrubbery should have at least 12 in. of gravel or crushed stone beneath the pipe.

NOTE: DRAIN TILE LAID WITH JOINTS OPENED FROM 1/8 TO 1/4 INCH. SPECIAL COLLARS MAY BE USED IF DESIRED.

SUITABLE PERVIOUS BARRIER

OPENINGS AT JOINTS 1/8" TO 1/4"   JOINT COVERING

12"

2" MIN.

5" ±

6" MIN.

LONGITUDINAL SECTION

GRAVEL OR BROKEN STONE

OVERFILL TO ALLOW FOR SETTLEMENT

4" TO 6"

BACKFILL (EARTH)

12" TO 36"

LATERAL OF DRAIN TILE (SHOWN) OPEN JOINTED SEWER PIPE OR PERFORATED PIPE

CROSS SECTION

FIGURE 11.2-9 *Details of typical absorption trench.* *(Illustration from* Manual of Septic Tank Practices, *Bureau of Community Environmental Management, U.S. Public Health Service [now EPA].)*

The backfill over a new absorption trench should be free of large stones and other deleterious materials and should be hand tamped and overfilled 4 to 6 in. higher than adjacent grade to compensate for likely settling. Settlement of the fill in a trench to a lower level is very undesirable because it may cause collection of storm water in the trench, which can lead to premature saturation of the absorption field or even complete washout of the trench.

Heavy vehicles will crush the tile in a shallow absorption field, and should not be operated there unless special provisions are made to support the weight. Machine tamping or hydraulic backfilling of absorption trenches should not be permitted. Machine grading should be completed before the field is laid.

■ EFFLUENT DISTRIBUTION

Where the ground does not exceed a 6-in. height variation in any direction within the absorption area, effluent may be discharged through a system of laterals connected at their extremes to form a continuous closed loop (see Fig. 11.2-7). Such a system permits the effluent to flow into all the tile lines whenever there is a discharge from the septic tank.

Laterals traditionally were connected to a house sewer through a distribution box and the ends were not interconnected. A Public Health Service study for the FHA on the effectiveness of distribution boxes reached the conclusion that distribution boxes generally do not assure equal dis-

tribution as intended and therefore are not needed for single-family systems. However, even when local practice includes distribution boxes, the laterals should be interconnected at their ends as in Figure 11.2-7. Interconnection of laterals assures access to each point in the lateral from two sides, so that if one part of the lateral became damaged, effluent would be able to reach the lateral from the other side.

Where sloping ground is used for an absorption field, it is usually necessary to construct a small temporary dike or surface-water diversion ditch above the field, to prevent the absorption field from being eroded by surface runoff. This dike should be maintained or the ditch kept free of obstructions until the field becomes well covered with vegetation.

The maximum ground slope suitable for soil absorption systems is governed by local factors affecting erosion of the ground used for the absorption field. Excessive slopes that are not protected from surface water runoff or do not have adequate vegetation cover to prevent erosion should be avoided. Generally, ground having a slope greater than 1 vertical to 2 horizontal should be investigated carefully to determine if it is satisfactory from the standpoint of the likelihood of erosion. Near banks, the horizontal distance from the side of a trench to the ground surface should be 2 ft. or more to prevent lateral flow of effluent and breakout onto the surface.

In sloping ground each lateral should be connected to the next by a closed overflow

line laid on an undisturbed section of ground, as shown in Figure 11.2-8. The arrangement provides that all effluent is discharged to the first trench until it is filled; excess liquid is then carried by means of a closed overflow line to the next succeeding or lower trench. In that manner, each portion of the subsurface system is used in succession.

Design and construction procedures for sloping-ground fields are basically the same as those for level-ground fields. The bottoms of trenchs and of distribution lines should be level, and laterals should follow surface contours. Adjacent trenches may be connected either with an overflow line or a drop box arrangement. The *invert* (bottom) of the first overflow line should be at least 4 in. lower than the invert of the septic tank outlet (see Fig. 11.2-8).

Overflow lines should be 4-in., tight-joint pipe with direct connections to the distribution lines in adjacent trenches or to a drop box arrangement. Care must be exercised in constructing overflow lines to ensure at least 6 ft. of undisturbed earth between trenches. The overflow line should rest on undisturbed earth and backfill should be carefully tamped.

11.2.2.2

Absorption Beds

An absorption bed is a variation of the typical trench used in absorption fields. The basic design is essentially the same as for an absorption trench, but it is generally much wider than 36 in. and contains two or more distribution lines, 4 to 6 ft. apart. Absorption beds have the advantages of making more efficient use of the land and of modern earthmoving equipment.

The minimum bottom area for absorption beds is the same as for trenches in absorption fields (see Fig. 11.2-2). A bed should have a minimum depth of 24 in. below natural ground level to provide a minimum of 12 in. of gravel and 12 in. of earth backfill. The gravel should extend at least 2 in. above and 6 in. below the distribution line. The bottoms of beds and distribution lines should be level. Distribution lines should be spaced not more than 6 ft. apart and not more than 3 ft. from the bed sidewall. When more than one bed is used there should be a minimum of 6 ft. of undisturbed earth between adjacent beds, and the beds should be connected in series, as illustrated in Figure 11.2-8, for absorption fields in sloping ground.

11.2.2.3

Seepage Pits

Seepage pits are deeper than absorption trenches or beds, and are, therefore, more likely to contaminate groundwater than other absorption methods. In fact, they are not permitted in some jurisdictions. Because pits handle the same amount of effluent as an absorption field or bed in a smaller land area, the likelihood of clogging and saturation is greater, as is the possibility of effluent breaking through to the surface.

Pits should not be used where absorption fields or beds are feasible, or when the soil has a percolation rate in excess of 30 min./in. Where the top 3 or 4 ft. of soil is not porous enough for an absorption field, but is underlaid with suitable granular material, seepage pits constitute an acceptable substitute. Seepage pits may also be used where the land area is too small, too irregular, or too heavily wooded to accommodate absorption field or bed systems.

The required capacity of deep pits should be computed on the basis of percolation tests made in each vertical stratum penetrated (see Fig. 11.2-5). Soil strata in which the percolation rates are in excess of 30 min./in. should not be included in computing the absorption area. The weighted average of the results should be computed to obtain an average percolation rate. In shallow pits where the soil is relatively uniform, the required absorption area can be established by soil experts through a visual analysis of the soil type (Fig. 11.2-10). In other cases, the absorption area should be established by percolation tests as for other absorption methods.

■ DESIGN

Figure 11.2-2 gives the absorption area requirements per bedroom for soils with different percolation rates. The absorption area of the seepage pit is the vertical wall area of the actual excavated pit (not the pit lining). Only the area of pervious strata below the inlet should be figured. No

FIGURE 11.2-10
Required Wall Area for Seepage Pits

Symbol*	Character of Soil	Minimum Wall Area per Bedroom, sq. ft.
GW	Well-graded gravels, gravel-sand mixtures, little or no fines	50
GP	Poorly graded gravels or gravel-sand mixtures, little or no fines	50
SW	Well-graded sands, gravelly sand, little or no fines	75
SP	Poorly graded sands or gravelly sands, little or no fines	75
SM	Silty sand, sand-silt mixtures	125
GC	Clayey gravels, gravel-sand-clay mixtures	150
SC	Clayey sands, sand-clay mixtures	150
ML	Inorganic silts and very fine sands, rock flour, silty or clayey fine sands or clayey silts with slight plasticity	300
CL	Inorganic clays of low to medium plasticity, gravelly clays, sandy clays, silty clays, lean clays	350
OL	Organic silts and organic silty clays of low plasticity	Unsuitable for Seepage Pits
MH	Inorganic silts, micaceous or diatomaceous fine sandy or silty soils, elastic silts	
CH	Inorganic clays of high plasticity, fat clays	
OH	Organic clays of medium to high plasticity, organic silts	
Pt	Peat and other highly organic silts	

Table excerpted from Manual of Septic Tank Practices, *Bureau of Community Environmental Management, U.S. Public Health Service (now EPA).*

*Refer to Section 2.1 for a discussion on soil classification.

allowance should be made for impervious strata or for bottom area. After the required area has been established, Figure 11.2-11 may be used to select the desired dimensions of circular or cylindrical seepage pits.

To demonstrate a typical pit capacity calculation, assume that the absorption system is to serve a three-bedroom house on a lot where the percolation rate is 1 in. in 15 min. Assume also that the water table does not rise above 17 ft. below the ground surface, that seepage pits with an effective depth of 10 ft. can be provided, and that the house is in a locality where it is common practice to install seepage pits 5 ft. in diameter (i.e., 4 ft. to the outside walls, which are surrounded by about 6 in. of gravel).

According to Figure 11.2-2, 570 (3 × 190) sq. ft. of absorption area would be needed. Referring to Figure 11.2-11, it can be determined that a 5-ft. round pit 10 ft. deep would provide 157 sq. ft. of wall area. Dividing 570 by 157, we get 3.6, that is, three pits 10 ft. deep and one 6 ft. deep. If it is desirable to make all pits the same size, four 5-ft.-round pits 9 ft. deep can be selected from Figure 11.2-11, providing 4 × 141, or 564, sq. ft. of wall area, which is close enough to the total required. Relatively deep pits having a small diameter require the removal of less soil than larger pits with equivalent wall area, but may be harder to excavate. When this is the case, larger pits can be selected. For instance, two 10-ft.-round pits 9 ft.

FIGURE 11.2-11
Effective Wall-Absorption Areas (sq. ft.) of Round Seepage Pits

Pit Diameter, ft.	Depth Below Inlet									
	1 ft.	2 ft.	3 ft.	4 ft.	5 ft.	6 ft.	7 ft.	8 ft.	9 ft.	10 ft.
3	9.4	19	28	38	47	57	66	75	85	94
4	12.6	25	38	50	63	75	88	101	113	126
5	15.7	31	47	63	79	94	110	126	141	157
6	18.8	38	57	75	94	113	132	151	170	188
7	22.0	44	66	88	110	132	154	176	198	220
8	25.1	50	75	101	126	151	176	201	226	251
9	28.3	57	85	113	141	170	198	226	254	283
10	31.4	63	94	126	157	188	220	251	283	314
11	34.6	69	104	138	173	207	242	276	311	346
12	37.7	75	113	151	188	226	264	302	339	377

Table excerpted from Manual of Septic Tank Practices, *Bureau of Community Environmental Management, U.S. Public Health Service (now EPA).*

D SHOULD BE AT LEAST 3 TIMES DIAMETER OF SEEPAGE PIT

FIGURE 11.2-12 *Typical septic tank system with two seepage pits connected in a series.* (Illustration from Manual of Septic Tank Practices, Bureau of Community Environmental Management, U.S. Public Health Service [now EPA].)

deep, or two 13-ft.-round pits 7 ft. deep (see Fig. 11.2-11).

Where more than one pit is provided, they can be connected in series or in parallel. However, they should be separated by a distance equal to 3 times the diameter of the largest pit. For pits more than 20 ft. deep, the minimum distance between pits should be 20 ft. (Fig. 11.2-12). The area of the lot on which a house is to be built should be large enough to maintain this distance between the pits while still allowing room for additional pits if the first ones should fail. If this cannot be done, other sewerage facilities should be explored.

■ CONSTRUCTION

The permeability of the soil may be damaged during excavation. Digging in wet soils should be avoided as much as possible. Cutting teeth on mechanical equipment should be kept sharp. Bucket auger pits should be reamed to a larger diameter than the bucket. Loose material should be removed from the excavation.

Pits should be backfilled with clean gravel to a depth of 1 ft. above the pit bottom or 1 ft. above the reamed ledge to provide a sound foundation for the lining.

Lining materials for seepage pits can be clay or concrete masonry units in brick, block, or ring shapes. Rings should have weep holes or openings to provide for seepage. Concrete brick and block should be laid dry (unmortared) with staggered, slightly open joints. Hollow concrete block is frequently laid on its side with cores horizontally to facilitate soil absorption. Standard clay brick should be laid flat to form a 4-in.-thick wall. The outside diameter of the lining should be at least 1 ft. smaller than the excavation diameter, to provide at least 6 in., and preferably 1 ft., of annular space to be filled with coarse gravel to the top of the lining (Fig. 11.2-13).

Either a corbelled brick dome or a flat concrete cover can be used over a pit. Corbelled bricks should be laid either in cement mortar or have a 2-in. covering of concrete. If a flat cover is used, a prefabricated type is preferred. Concrete covers should be reinforced so that it is equivalent in strength to an approved septic tank cover. The cover should rest on undisturbed ground 12 in. beyond the lining. A 9-in. capped opening in the pit cover should be provided for pit inspection.

Connecting lines should be of a sound durable material, the same as used for the house-to-septic tank connection. All connecting lines should be laid on a firm bed of undisturbed soil throughout their length. The grade of a connecting line should be at least 2% (1/4 in./ft.). The pit inlet pipe should extend horizontally at least 1 ft. into the pit with a tee or ell to divert flow downward to prevent washing and eroding of the sidewalls.

11.2.3

SEPTIC TANKS

The soil, waste, and other drainage pipes in a building's plumbing system (see Section 11.1) discharge into a *building drain*. This lowest part of the system conveys the received discharge to a point 3 ft. outside the building walls where it connects to the *building sewer*. The building sewer then connects the building drain to a public sewer, private sewer, or, in the case under discussion here, an *individual sewage disposal system*. The building sewer is also sometimes called a *house sewer*.

FIGURE 11.2-13 *Details of a typical round seepage pit.* (*Illustration from* Manual of Septic Tank Practices, *Bureau of Community Environmental Management, U.S. Public Health Service [now EPA].*)

In this instance, the portion of the individual sewage disposal system that a building sewer discharges into is a septic tank (see Fig. 11.2-1). This is a watertight, covered receptacle designed to receive the discharge of sewage from a building sewer; to separate the solids from the liquids, digest the organic matter, and store the digested solids through a period of detention; and to allow the clarified liquids to discharge for final disposal.

Despite their physical position in a system, septic tanks are discussed here—after absorption systems—because the determination must first be made whether the soil is suitable for a subsurface absorption system, and whether adequate area is available for the type of system intended. Except for the system designer's choosing the tank's capacity and component materials, there are few options in tank design.

Because the interior design of a tank is a matter of code compliance and local practice, designs approved by local health authorities can be specified with confidence.

11.2.3.1

Functions

A septic tank performs three basic functions: (1) removal of solids, (2) storage of scum and sludge, and (3) biological treatment. *Scum* is a mass of sewage matter that floats on the surface of the sewage in a septic tank. *Sludge* is the accumulated solids that settle out of the sewage, and form a semiliquid mass on the bottom of a septic tank.

A septic tank's chief function is the separation of solids, but the other functions are essential also if the system is to be

more than a holding tank, requiring frequent pumping. Biological treatment of the organic matter contained in the scum and sludge assures that only a small portion of indigestible solid matter remains in the tank permanently between pumpings.

■ REMOVAL OF SOLIDS

The amount of clogging of the soil by tank effluent varies directly with the amount of suspended solids in the liquid. As sewage from a building sewer enters a septic tank, its rate of flow is reduced so that larger solids sink to the bottom or rise to the surface. These solids are retained in the tank, where most of them are gradually broken down by bacteria into water-soluble components, which do not clog the soil.

■ SLUDGE AND SCUM STORAGE

Heavier sewage solids settle to the bottom of the tank, forming a blanket of sludge. Lighter solids, including fats and greases, rise to the surface and form a layer of scum. A considerable portion of the sludge and scum is liquefied through decomposition or digestion. During this process, gas is liberated from the sludge, carrying a portion of the solids to the surface, where they accumulate with the scum. Ordinarily, they undergo further digestion in the scum layer, and a portion settles again to the sludge blanket on the bottom. The settling action is retarded by grease in the scum layer and by gasification in the sludge blanket.

■ BIOLOGICAL TREATMENT

Solids and liquid in the tank are subjected to decomposition by bacterial and natural processes. The bacteria are of a variety called *anaerobic* (oxygen-free), which thrive in the absence of oxygen. This decomposition or treatment of sewage under oxygen-free conditions is termed *septic*, which explains the name of the tank.

Contrary to popular belief, septic tanks do not accomplish a high degree of bacterial removal. Although sewage undergoes treatment while passing through the tank, *pathogenic* (disease-producing) organisms are not removed. Therefore, septic tank effluents cannot be safely discharged onto the surface or into a ditch, lake, or stream. This, however, does not detract from the value of septic tanks. As previously explained, a tank's primary purpose is to condition the sewage to prevent rapid clogging of the subsurface absorption system.

Further treatment of effluent, including the removal of pathogenic organisms, is effected by percolation through the soil. These organisms slowly die out as their nutrients are used up and as they enter the unfavorable environment afforded by the soil.

11.2.3.2

Tank Location

Although septic tanks are supposed to be watertight when installed, as they age, some leakage may develop through openings in the inlet and outlet pipes or cracks in the tank. Therefore, septic tanks should be located where they are not likely to contaminate a well, spring, or other source of water supply. Underground contamination may travel for considerable distances, usually in the direction of the water table gradient, that is, from a higher to a lower area of the water table. Because water tables usually follow ground surface contours, septic tanks should be located downhill from private wells or natural springs.

Unfortunately, however, locating a tank at a lower surface elevation than a well is not a positive guarantee against contamination. Sewage from soil absorption systems occasionally contaminates wells having higher surface elevations. This happens because elevations of subsurface absorption systems are almost always higher than the level of water in wells that may be located nearby, and contamination may travel downward to the water-bearing stratum (Fig. 11.2-14). It is necessary, therefore, to rely on horizontal as well as vertical distances for protection. Tanks should never be closer than 50 ft. from any source of water supply, and greater distances are preferred where possible (see Fig. 11.2-6).

A septic tank should not be located within 5 ft. of a building foundation, because structural damage may result during construction and seepage may enter the basement. A tank should not be located in a swampy area, nor in an area subject to flooding. Ideally, tanks should be located

FIGURE 11.2-14 *A seepage pit can contaminate a private water supply if it is inadequately separated from the well or the water-bearing stratum.* (Illustration from Manual of Septic Tank Practices, *Bureau of Community Environmental Management, U.S. Public Health Service [now EPA].*)

where the largest possible area will be available for the absorption system. Consideration should also be given to the location from the standpoint of cleaning and maintenance; that is, a tank should be close enough to a paved area so that it can be readily reached by the hose of a septic pumping truck. Where a public sewer may be installed at a future date, provision should be made in the household plumbing system for connection to such a sewer.

11.2.3.3

Tank Design

Grease interceptors (grease traps) are seldom required in household sewage disposal systems. When one is provided, the discharge from a waste disposer (garbage grinder) should not be passed through it. The septic tank capacities recommended in this discussion are sufficient to receive the grease and solids normally discharged from a typical home.

It is generally advisable to have all sanitary wastes from a household discharge to a single septic disposal system. For household installations, it is also usually more economical to provide a single disposal system than two or more with the same total capacity. Normal household waste,

FIGURE 11.2-15
Recommended Septic Tank Capacities*

Number of Bedrooms	Minimum Tank Capacity (gal.)
3 or less	1000
4	1200
Additional bedrooms	Add 250 gal./bedroom

Table excerpted from Manual of Septic Tank Practices, *Bureau of Community Environmental Management, U.S. Public Health Service (now EPA).*

*Provides for use of garbage grinder, dishwasher, automatic clothes washer, and other household appliances.

including that from the laundry, bath, and kitchen, should pass into a single system.

Providing adequate capacity is an important consideration in septic tank design. The larger the tank the longer sewage will be retained in it, and the better will be the removal of solids from the effluent and the more complete the biological treatment of the contents. Undersized tanks are more likely to become overloaded, which will result in inadequately treated sewage entering the absorption system and clogging it. The liquid capacities recommended in Figure 11.2-15 allow for the use of all household appliances, including waste disposers, automatic clothes washers, and dishwashers. They also ensure enough excess capacity to retain the sewage for a sufficient period of time to settle out and digest solids. They do not, however, take into account liquids from other sources. For example, water that is free of organic matter should be discharged into a storm sewer (see Section 11.1) or dry well.

A *dry well* is a covered and lined underground pit, similar to a *seepage pit*, but intended to receive water free of organic matter, such as from roof drains, floor drains, or laundry tubs. Dry wells can be used as an auxiliary to a septic system to avoid overloading the septic system. Dry wells are also called *leaching wells* and *leaching pits*. They should not be confused with *seepage pits*, which are intended to receive septic tank effluent, even though these are also sometimes called dry wells.

Materials and Construction

Septic tanks should be watertight units constructed of corrosion- and decay-resistant materials, such as concrete, coated metal, vitrified clay, heavyweight concrete masonry units, or hard-burned bricks. Precast and cast-in-place reinforced concrete tanks are common. Prefabricated steel tanks meeting UL Standard UL70 are acceptable in some states.

Special attention should be given to job-built tanks to ensure watertightness and prevent contamination. Heavyweight concrete blocks should be laid on a solid foundation and the mortar joints should be well filled. The interior of a tank should be surfaced with two 1/4-in.-thick coats of portland cement–sand plaster. Some typical septic tanks are illustrated in Figure 11.2-16.

Precast concrete tanks should have a minimum wall thickness of 3 in., and should be adequately reinforced to facilitate handling. When precast slabs are used as covers, they should be watertight, have a thickness of at least 3 in., and be adequately reinforced to be capable of supporting a dead load of 300 lb./sq. ft.

■ BACKFILL AND ACCESS

Backfill around a septic tank should be placed in thin layers and thoroughly tamped in a way that will not produce undue strain on the tank. Backfill may be consolidated using water, provided the backfill is placed and wetted in layers and the tank is first filled with water to prevent floating.

Adequate access should be provided to each tank compartment for inspection and cleaning. Access to each compartment can be provided by either a removable cover or a manhole at least 20 in. wide. When the top of a tank is located more than 18 in. below the finished grade, manholes and inspection holes should extend to approximately 8 in. below the finished grade (Fig. 11.2-17a), or can be extended to finished grade if a *gas seal* is provided to keep odors from escaping. A gas seal is made by filling the top of the manhole with sand or by sealing with a minimum of 1 in. of portland cement plaster over an intermediate wood platform (Fig. 11.2-17b). In most instances, an extension can be made using clay or concrete pipe, but attention must be given to the accident hazard involved when manholes are extended close to the ground surface.

■ TANK CONSTRUCTION

Typical single- and double-compartment tanks are illustrated in Figures 11.2-18 and 11.2-19. The inlet invert (bottom) should enter the tank at least 3 in. above the outlet invert and the liquid level in the tank, to allow for a momentary rise in the liquid level during discharges into the tank. This free drop prevents water from backing up into the house sewer and minimizes stranding of solid material in the sewer.

A vented inlet tee or baffle should be provided to divert the incoming sewage downward. It should penetrate at least 6 in. below the liquid level, but in no case should the penetration be greater than that allowed for the outlet device. Several arrangements commonly used for inlet and outlet devices are shown in Figure 11.2-20.

The outlet device should penetrate just far enough below the liquid level of the tank to provide a balance between sludge and scum storage volume; otherwise, part of the advantage of capacity is lost. A vertical section of a properly operating tank would show it divided into three distinct layers: scum at the top, a middle zone free of solids, and a bottom layer of sludge. The outlet device retains scum in the tank, while limiting the amount of sludge that can be accommodated without scouring, which results in sludge discharging in the effluent from the tank. The outlet device should generally extend to a distance below the surface equal to 40% of the liquid depth. For horizontal, cylindrical tanks, this should be reduced to 35%. For example, in a horizontal, cylindrical tank having a liquid depth of 42 in., the outlet device should penetrate 14.7 in. (42×0.35) below the liquid level.

FIGURE 11.2-16 *Septic tanks are precast concrete or welded sheet steel in one of several basic shapes.* (Illustration from Manual of Septic Tank Practices, *Bureau of Community Environmental Management, U.S. Public Health Service [now EPA].)*

FIGURE 11.2-17 *Manholes for septic tank access.* (Illustration from Manual of Septic Tank Practices, *Bureau of Community Environmental Management, U.S. Public Health Service [now EPA].)*

FIGURE 11.2-18 *Single-compartment septic tank showing recommended proportions, dimensions, and clearances.* *(Illustration from* Manual of Septic Tank Practices, *Bureau of Community Environmental Management, U.S. Public Health Service [now EPA].)*

The outlet device should extend above the liquid line to about 1 in. from the top of the tank. The space between the top of the tank and the baffle allows gas to pass off from the tank into the house vent stack.

The shape of a septic tank does not affect its operation. Shallow tanks function as well as deep ones. However, the smallest plan dimension should be at least 24 in. Liquid depth may range between 30 and 60 in. Tank volume more than 72 in. down should not be considered in calculating tank capacity.

Adequate capacity above the liquid line is required to provide for approximately 30% of the total scum that floats above the liquid. In addition to provision for scum storage, a 1-in. clearance between the tops of inlet and outlet tees (or baffles) and the top of a tank usually should be provided to permit free passage of gas back to the inlet and house vent stack.

In tanks with straight vertical sides, the distance between the top of the tank and the liquid line should be about 20% of the liquid depth. In horizontal cylindrical tanks, an area equal to about 15% of the total circle should be provided above the liquid level. This condition is met if the liquid depth (distance from outlet invert to bottom of tank) is equal to 79% of the diameter of the tank.

Although several arrangements are possible, the term *compartments* as used here refers to several units in series. These can be either separate units linked together, or sections enclosed in one continuous shell (see Fig. 11.2-19) with watertight partitions separating individual compartments.

A single-compartment tank will give acceptable performance, but a two-compartment tank, with the first compartment equal to one-half to two-thirds of the total volume, generally provides better removal of suspended solids. Compartments should be at least 24 in. in the smallest dimension and have a liquid depth ranging from 30 to 72 in.

INSTALLATION INSPECTION

A septic disposal system should be tested and inspected before it is used. Inspection before backfilling is usually required by local regulations, even where plans for the disposal system have been checked before issuance of a building permit. A septic tank should be filled with water and allowed to stand overnight to check for leaks.

A soil absorption system should be inspected before it is covered to be sure that the system is installed according to plans and good trade practice.

NOTE: ALL FITTINGS 4" V.C. OUTLET FITTING SET 3" BELOW INLET FITTING

FIGURE 11.2-19 *Typical design of double-compartment precast concrete septic tank. Dimensions shown are for a 750-gal. tank, which may be found in older houses. Recommended minimum for houses with one to three bedrooms is now 1000 gal.* (Illustration from Manual of Septic Tank Practices, *Bureau of Community Environmental Management, U.S. Public Health Service [now EPA].*)

TEE

4" CAST IRON SOIL PIPE T BRANCH
4" CAST IRON SANITARY T BRANCH
4" VITRIFIED CLAY OR CONCRETE
T BRANCHES

PLACE INLET AND OUTLET TEE
IN NOTCH AND FILL WITH MORTAR

PACK MORTAR
AROUND TEE

1" MIN. CLEARANCE

STRAIGHT BAFFLE

POURED IN PLACE OR PREFABRICATED
AND DROPPED IN SIDES OF TANK
NOTE: "A" SHOULD BE NO LESS THAN 6"
AND NO GREATER THAN B.

"B" PENETRATION OF OUTLET DEVICE
GENERALLY 40% OF LIQUID DEPTH FOR
TANKS WITH VERTICAL SIDES & 35% FOR
HORIZONTAL CYLINDERS TANKS.

FIGURE 11.2-20 *Acceptable inlet and outlet devices for septic tanks.* (Illustration from Manual of Septic Tank Practices, *Bureau of Community Environmental Management, U.S. Public Health Service [now EPA].*)

11.3 HEATING, VENTILATING, AND AIR CONDITIONING

Early theories defined heat as a material substance called *caloric*, which was a weightless fluid that had the power to penetrate other materials, expand, and transform them into vapor. It was thought that a substance was hot or cold, depending on the proportion of *caloric* contained in its pores.

The theory of heat was placed on a sound scientific basis in the early 19th century by James P. Joule, an English physicist. Joule's experiments led to the principle of conservation of energy, which stated that heat energy was not lost as it was used, but was transformed into an equivalent amount of another kind of energy. For instance, heat energy can be converted to kinetic (motion) energy by a steam engine, and from kinetic energy to electric energy by a generator.

Animals, including people, sustain life by converting organic food substances into heat energy. Normally, excess heat energy is produced and the body dissipates this excess to the surrounding atmosphere. Each organism has an optimum rate of body heat loss depending on physical exertion. When the heat loss is at the ideal rate, a person feels comfortable. When heat is lost too rapidly, a person feels cold; when the loss is too slow, the sensation is one of being hot. *The objective of interior heat control is to create conditions at which the body loses heat at the optimum rate.*

Heat energy generally flows from a warmer region to one that is cooler. Therefore, the flow is to the outside of a building in winter and to the inside in summer. This flow of heat affects interior surface and air temperatures.

Cold is the absence of heat. A warm window radiates heat, but a cold window does not radiate cold. A person's sensation of coldness is caused by a rapid loss of body heat to the cold window surface. Likewise, the movement of air over the body produces the sensation of coldness because it carries off heat, not because it brings coldness.

People today expect to find a relatively narrow range of uniform temperatures within a building, and want the ability to select that range at will. When energy was less costly, it was relatively easy to force an interior environment into a narrow band of comfort. Today, both architects and building owners are examining the benefits of permitting a broader range of interior temperatures and of adjusting the mode of dress to the temperature.

For instance, putting on a sweater in winter might make a 65°F interior air temperature acceptable, whereas removing one's tie and turning on an electric fan might make 80°F tolerable in the summer. The potential gain from accepting a 5-degree variation up or down from traditional thermostat settings can produce a 15% to 30% energy savings over the average costs of heating and cooling.

Until the energy crunch of the 1970s, traditional approaches in interior heat control emphasized reliance on energy-intensive mechanical systems. These systems pumped heat in or out of a building and relied on the interior shell only to minimize heat gain in summer and heat loss in winter. Now, architects tend to design a building's envelope to permit heat flow through the shell when advantageous. For instance, buildings can be designed to make use of sunshine for heating in the winter and of natural breezes for cooling in the summer. Some of the principles of passive solar energy systems are common in many buildings (see Section 11.3.3.6).

This section discusses the principles of heat flow and human comfort, appropriate building design criteria related to heat control, and the design of heating, ventilating and air conditioning systems. Section 7.3 discusses the use of insulation to minimize heat flow through a building's envelope. Chapter 8 contains data about strategies for reducing heat loss and gain through doors and windows. Section 8.1.1.4 contains information concerning the use of interior and exterior shading devices and of blinds, shades, draperies,

and shutters to control the gain or loss of heat through a building's windows. Together, these techniques help to reduce energy consumption and make buildings more comfortable for their occupants, with less reliance on mechanical heating and cooling systems.

11.3.1

HEAT AND COMFORT

Heat is a form of energy. It is produced in a substance by the movement of its molecules—the tiny particles of matter that make up that substance (see Section 1.5). When heat is added, the molecular movement inside that substance is increased. The intensity of movement determines the physical state of the substance, whether solid, liquid, or gas. Temperature, recorded in degrees Fahrenheit or degrees Celsius, is a measure of that intensity (Fig. 11.3-1). A British thermal unit (Btu) is a measure of heat quantity and is equal to the amount of heat required to raise 1 lb. of water 1°F. Heat flow rate is measured in Btu's per hour. In the SI (metric system), heat quantity is measured in joules and heat flow rate is measured in watts. One watt is equal to 1 joule/sec., or approximately 0.341 Btu/hr.

11.3.1.1

Sensible and Latent Heat

Energy that causes a rise in temperature but does not alter the physical state of a substance is called *sensible heat*. Heat can be added to a substance without altering its physical state only up to a certain temperature. For example, if a teapot with 1 lb. of water in it is placed on a burner, each Btu of heat added raises the temperature 1°F. If the water started at 62°F, it would take 150 Btu's to bring it to a boil at 212°F. Heat added beyond the first 150 Btu's would change the water to vapor. It

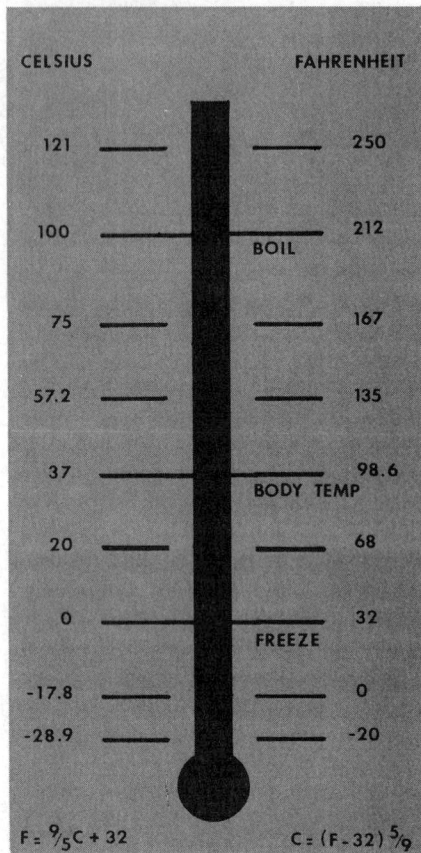

FIGURE 11.3-1 *Temperature equivalents on the Celsius (metric) and Fahrenheit (English) scales. Thermometers can be calibrated to a temperature of −273.1°C (−459.6°F), which is known as absolute zero. This is the point at which all molecular movement stops.*

would not raise the temperature of the water above 212 °F.

The energy that would be used to convert the water to steam if the pot were left on the burner is the *latent heat of vaporization*. To change the entire 1 lb. of water to steam would require 970 Btu's.

To change 1 lb. of ice at 32 °F to water would require 144 Btu's, which is the *latent heat of fusion* of water. Only after all of the first 144 Btu's were used and all of the ice had melted, would the additional heat be used to raise the water temperature above 32 °F.

11.3.1.2

Methods of Heat Transfer

Heat energy moves from a warmer to a colder area by conduction, radiation, or convection. A person gives up heat by all of these methods, and also by the evaporation of perspiration.

■ CONDUCTION

The movement of heat through a solid or liquid substance as heated particles transfer their thermal energy to adjacent particles is called *conduction*. Generally, dense materials such as metals can transfer more heat faster and are called *conductors*. Less dense substances such as wood and plastic retard heat flow and are called *insulators*.

■ RADIATION

The transmission of heat by invisible light waves independent of any medium is called *radiation*. The radiant heat from the sun, for example, reaches the earth by traveling through the vacuum of space and the atmosphere. Radiant energy can travel through any nonopaque substance.

An opaque body absorbs radiant energy over the surface struck by the solar radiation and converts it into heat. After absorption, the heat is transferred through the body by conduction; some heat is also reradiated from the warm body into the air.

REFLECTIVITY Materials vary in their acceptance and rejection of the radiant energy that strikes them. Those that reject much of the radiant energy are called *reflectors*, because of their high degree of reflectivity. Shiny aluminum foil, for instance, reflects 95% of all radiant energy; white paint reflects about 75%. Dark materials, such as asphalt pavement, which absorb more radiant energy, have a high degree of emissivity (Fig. 11.3-2).

EMISSIVITY *Emissivity* is the converse of reflectivity. Aluminum foil reflects 95% of the radiant heat that strikes it and absorbs the remainder as sensible heat. Therefore, the emissivity of the foil is 5% (see Fig. 11.3-2). The practical significance of such differences is that high-emissivity surfaces absorb and radiate more energy. For example, because asphalt pavements have a high emissivity, air immediately above them will be warmed to as much as 20 °F higher than the air in nearby shaded areas.

■ CONVECTION

The movement of heat from one region to another through a fluid substance, such as air or water, is *convection* (Fig. 11.3-3). When air in a room contacts a warm base-

FIGURE 11.3-2
Reflectivity/Emissivity Chart

Material	Reflectivity, %	Emissivity, %
Built-up roof, gravel aggregate	7	93
Slate, dark soil, other dark textures	15	85
Grass, dry	30	70
Copper foil		
Tarnished	36	64
New	75	25
Paint		
Light grey	25	75
Red	26	74
Aluminum	46	54
Light green	50	50
Light cream	65	35
White	75	25
Fresh snow	80	20
Aluminum foil	95	5

FIGURE 11.3-3 *Air warmed by a heater rises, gives up its heat to walls and ceiling, and drops to the floor to be warmed again.*

FIGURE 11.3-4 *Relative humidity can be increased by decreasing the temperature or by increasing the amount of moisture.*

board heater, it is heated by conduction. As it is warmed, the air expands, becomes lighter and rises, thus permitting colder air to take its place. The warmed air often gives up its heat to a cold wall or window and settles to the floor to be reheated. Thus a *convection loop* is established, which distributes the heat throughout the space.

■ EVAPORATION

A human body at rest loses heat mainly by convection and radiation. When engaged in strenuous activity or when the air is too warm, those methods are insufficient to carry off all the excess heat. The body then utilizes perspiration, which cools the body by withdrawing from it the *latent heat of vaporization.* Low humidity aids cooling by perspiration; high humidity inhibits it.

11.3.1.3

Relative Humidity

Most air contains invisible moisture in the form of water vapor. When the moisture becomes visible, it is no longer vapor, but a liquid. A cloud or puff of steam is water in an intermediate stage of vapor in the process of becoming liquid, or vice versa. The conversion from a vapor to a liquid is called *condensation.*

Relative humidity (RH) is the ratio of moisture actually present in the air, compared with the maximum that can exist at a given temperature without condensation. An RH of 50% means that the air holds half the total water vapor it could possibly contain at that temperature. The maximum

relative humidity is dependent on the ambient (general) air temperature, because warm air can hold more water vapor than cooler air (Fig. 11.3-4). When all possible moisture is held by the air (100% RH), its *saturation point* or *dew point* has been reached. For additional discussion of relative humidity and saturation see Section 7.1.

Relative humidity is measured by a *sling psychrometer* (Fig. 11.3-5). This device consists of two thermometers mounted side by side on a frame. A wetted cloth sack is attached to one of the thermometers and both are whirled through the air. The temperature reading given by the uncovered thermometer is the *dry-bulb* temperature (DBT), which is the common way of measuring and stating temperature. Ambient temperatures are usually stated in DBT. The reading shown by the wetted thermometer is the *wet-bulb* temperature (WBT). This is the reading more commonly used in the design of mechanical systems.

Rapid movement through the air causes some of the water to evaporate from the wetted thermometer. The process of evaporation uses up latent heat energy, resulting in a lower temperature reading. The greater the difference between wet- and dry-bulb temperatures, the lower the relative humidity. When the dry-bulb temperature and wet-bulb temperature are known, the relative humidity can be determined from a psychrometric chart (Fig. 11.3-6). For instance, if the DBT is 70°F and the WBT is 58°F, then the RH is 50%.

FIGURE 11.3-5 *A sling psychrometer.*

See Section 7.1.1 for an additional discussion of relative humidity and its effect on people and property.

11.3.1.4

Interior Comfort Requirements

The human body is a heat-producing mechanism, generating heat energy from oxidized food. In most cases, it produces more heat than is needed for thermal stability. The amount of extra heat that must be dissipated depends on the activity being performed (Fig. 11.3-7) and ranges between 200 Btu's and 4500 Btu's per hour.

With ET lines, for persons at rest, normally clothed, in still air.

FIGURE 11.3-6 *Psychrometric chart.* *(Reprinted with permission of the American Society of Heating, Refrigerating, and Air-Conditioning Engineers.)*

FIGURE 11.3-7 *A body's production of heat depends on how strenuous the activity is.*

■ ENVIRONMENTAL CONTROL

Control of the interior environment helps the body regulate its temperature and achieve the ideal comfort condition, which is a sensation that is neither too warm nor too cool.

Studies conducted by the American Society of Heating, Refrigerating, and Air Conditioning Engineers, Inc. (ASHRAE) and others have shown a rise in the indoor temperatures people require to be comfortable in the winter from between 65° and 70°F in 1900 to between 72° and 78°F today. Therefore, the desire for warmer temperatures may be a result of cultural conditioning rather than biological necessity. The high cost of energy today and a growing awareness of the need to conserve energy may not have changed how people feel about the ideal temperature for comfort, but they have somewhat changed the way people act. There is more of a tendency to turn the thermostat down and put on heavier clothes in winter.

■ COMFORT FACTORS

Ambient temperature, air movement, relative humidity, and the temperature of a space's surfaces combine to produce the thermal sensations a person experiences within that space. The first three have already been discussed. The fourth is expressed in terms of *mean radiant temperature* (MRT), which represents the collective thermal effect of all the radiant surfaces in the space on an occupant at a given position within that space.

■ EFFECTIVE TEMPERATURE

Effective temperature (ET) is a method of measuring the combined effect of ambient temperature, relative humidity, and air movement. Any combination of temperature and humidity, with an air movement of 15 to 25 ft./min. can be stated as a single-number ET. That number is the dry-bulb temperature at 100% RH (or at 50% RH), which produces the same sensation as the given combination of temperature and humidity. This text uses the old effective temperature definition at 100% RH; the newer 50% RH effective temperature is designated here by an asterisk, ET*.

Early ASHRAE tests suggested that the majority of people were most comfortable in winter at an ET of 68°F (ET* 72.2°F) and in summer at an ET 71°F (ET* 76.9°F). From the psychrometric chart in Figure 11.3-6 it is possible to obtain DBT/RH equivalents at more common seasonal humidities. For example, according to Figure 11.3-6 the winter optimum DBT is near 75°F at 30% RH, and the summer optimum DBT is near 77°F at 50% RH.

A defect in using optimum ETs as a comfort index is that the effect of relative humidity is exaggerated in importance. Later tests concluded that within a temperature range of 60° to 80°F and a relative humidity range of 25% to 60%, evaporative body cooling is nearly constant. That is, a person at rest loses roughly the same amount of heat by perspiration at 25% RH as he or she does at 60%, so long as the temperature does not exceed about 80°F. However, different temperatures within the 25% to 60% RH range do produce varying sensations of warmth.

Another problem with using optimum ETs as a comfort index is that they ignore the effect of radiant surfaces (Fig. 11.3-9). In summer, cool walls aid body heat dissipation because they absorb radiant body

$$MRT = \frac{\Sigma t \theta}{360}$$

t = surface temp.
θ = angle of occupant exposure

Example: An apartment building has masonry and glass exterior walls with interior surface temperatures of 66° F and 20° F. Interior partitions are at 70° F.

Position A: $MRT = \frac{\Sigma t\Phi}{360} = \frac{(20x50) + (66x60) + (70x150)}{360} = \frac{18,680}{360} = 51° F$

Position B: $MRT = \frac{\Sigma t\Phi}{360} = \frac{(20x50) + (66x60) + (70x250)}{360} = \frac{22,460}{360} = 62° F$

FIGURE 11.3-8 *Calculation of mean radiant temperature (MRT).* (Reprinted with permission of the American Society of Heating, Refrigerating, and Air-Conditioning Engineers.)

FIGURE 11.3-9 *Both ambient (air) and radiant (wall) temperatures are important for comfort. At rest (a and b), thermal balance can be easily maintained with cooler air if walls are warmer; at work (c and d), cool air moves more body heat, requiring less perspiration to maintain balance.*

heat that cannot be removed by convection resulting from high air temperature. A person can therefore lose by radiation any heat he or she would otherwise have to eliminate at greater stress, through perspiration. In winter, warmer wall temperatures compensate for low air temperature by preventing radiant body heat loss.

11.3.1.5

ASHRAE Thermal Comfort Envelope

ASHRAE Standard 55-81, "Thermal Environmental Conditions for Human Occupancy," specifies generally acceptable comfort conditions for sedentary and slightly active, healthy, normally clothed people at altitudes from sea level to 7000 ft.

The *comfort envelope* developed in this standard (Fig. 11.3-10) outlines acceptable thermal conditions for 80% of tested Americans and Canadians. It integrates the effect of (1) DBT, (2) MRT, (3) RH, and (4) air movement. In this chart the effects of DBT and MRT are combined in the *adjusted dry-bulb temperature* (ADBT) scale and the effect of RH is shown by the water vapor pressure scale. Air movement is assumed at less than 70 ft./min.

Adjusted dry-bulb temperatures and relative humidity levels should be controlled within the ranges shown in Figure 11.3-10. Interior surface temperatures in winter should not be more than 5°F below ambient air temperature, to prevent radiant heat loss from body and interior spaces. In summer, walls should be kept slightly cooler than air temperature to aid in body heat dissipation.

11.3.1.6

Thermal Design Criteria

For maximum energy conservation, the latest recommendations of the U.S. Department of Energy (DOE) should be used as interior design temperatures. The DOE

is working closely with such organizations as ASHRAE to develop better energy standards.

■ ASHRAE ENERGY CONSERVATION STANDARD

ASHRAE Standard 90.1-1989, "Energy Efficient Design of New Buildings Except Low-Rise Residential Buildings," provides performance-oriented criteria for the design of building envelopes and of mechanical and electrical systems. Standard 90a-1980 (including 90A-a-1987), "Energy Conservation in New Building Design," was still applicable to small residential buildings at the time of this writing, but ASHRAE was in the process of developing a new standard applicable to single-family dwellings. These standards refer to the comfort envelope parameters established in ASHRAE Standard 55 for winter and summer design criteria (see Fig. 11.3-10).

Low relative humidities generally accompany the heating season and high humidities are characteristic of the cooling season in most of the United States. Therefore, most designers operate near the bottom of the comfort zone in winter and the

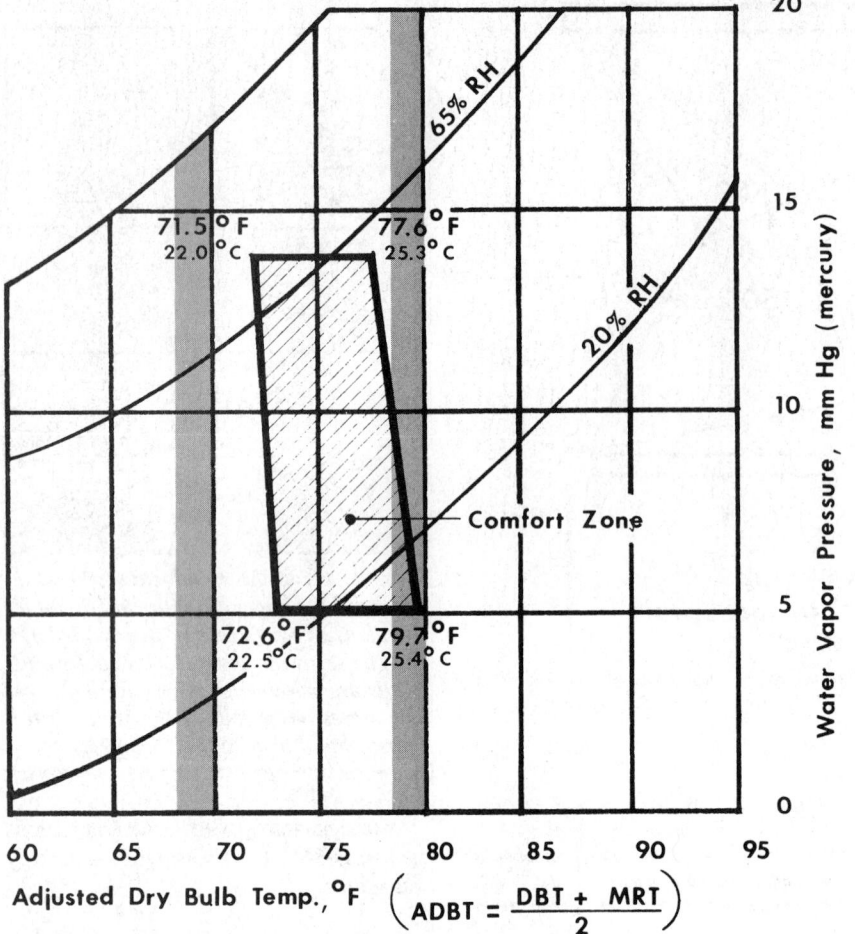

FIGURE 11.3-10 *ASHRAE comfort envelope.* *(Reprinted with permission of the American Society of Heating, Refrigerating, and Air-Conditioning Engineers.)*

top in summer. If energy conservation is a design objective, it can be further assumed that the design ADBT would be near 73 °F in winter and near 78 °F in summer.

■ HUD STANDARDS

The HUD *Minimum Property Standards for Housing* adheres to the historic interior comfort criteria—70 °F DBT for winter and 75 °F DBT for summer ambient air. This standard does not make any specific requirements for relative humidity or radiant temperature.

11.3.1.7

Climate and Weather Conditions

In order to design a building envelope and mechanical system capable of maintaining interior comfort conditions, it is necessary to predict likely heat gain and loss. To do this the designer must know not only what indoor temperature is desired, but also the outdoor temperatures that are likely to be encountered.

■ OUTDOOR DESIGN TEMPERATURES

Outdoor design temperatures are determined from weather records over a 25-year period. They are not absolutely the highest or lowest temperatures for a given area, but rather sustained extremes. Figure 11.3-11 provides a sampling of design temperatures and other related information for some major U.S. cities.

■ DEGREE DAYS AND COOLING OPERATING HOURS

The severity of winter is expressed in total *degree days heating* (DDH), which is the sum of daily degree days when the tem-

perature drops below 65 °F. The term *daily degree days* is the numerical difference between 65 and the average recorded temperature of that day. For instance, if there were 10 days in a season when the temperature dropped to 45 °F, then the DDH for each day is 20 and the total DDH for the season is 200.

The term *cooling operating hours* is a measure of the intensity of the annual cooling season. They are the hours an air-conditioning unit must operate to maintain an indoor temperature of 78 °F for 97.5% of the time.

The higher the number of total degree days or annual cooling operating hours for the season, the greater the heating or cooling demand (see Fig. 11.3-11).

11.3.1.8

Heat Transfer Through Materials

All building materials transfer heat to some degree. Heat transfer can be measured by (1) conductivity (k); (2) conductance (C); (3) resistance (R); and (4) transmittance (U). Figure 11.3-12 shows the heat-flow coefficients of some selected materials for conductivity, conductance, and resistance.

Knowledge of how various materials resist the passage of heat enables designers to calculate heat gain and loss and to select materials that minimize heat flow.

■ CONDUCTIVITY AND CONDUCTANCE

The degree to which a material transfers heat is rated by its conductivity (k) or conductance (C).

CONDUCTIVITY   Conductivity is a measure of the number of Btu's per hour that can pass through 1 sq. ft. of a 1-in. thickness of a material when its surfaces vary in temperature by 1°F. For example, a 1-ft. × 1-ft. × 1-in.-thick block of polyurethane foam with a k of 0.16 would permit 0.16 Btu's to pass in 1 hr., if one face of the block were at 74 °F and the opposite face were at 75 °F.

CONDUCTANCE   It is not meaningful to determine heat transfer per 1 in. of thickness for products that are not homogeneous or that have air cavities (such as hollow concrete block). Such products are

FIGURE 11.3-11

Representative Outdoor Design Temperatures, Degree Days, Cooling Hours, and Summer Outdoor Daily Ranges*

City	Winter Design Temperature (97-1/2%, °F)	Degree Days Heating (DDH)	Summer Design Temperature (2-1/2%, °F)	Annual Cooling Operating Hours	Summer Outdoor Daily Range
Anchorage, AK	−18	10,860	68	60	15
Phoenix, AZ	34	1,680	107	1,930	27
Los Angeles, CA	43	1,960	80	460	20
San Francisco, CA	38	3,040	77	170	14
Denver, CO	1	6,150	91	700	28
Washington, DC	17	4,240	91	960	18
Miami, FL	47	200	90	2,940	15
Chicago, IL	−4	6,640	89	650	20
Wichita, KS	7	4,640	98	1,030	23
Minneapolis, MN	−12	8,250	89	590	22
New York, NY	15	4,880	87	770	16
Cincinnati, OH	6	4,830	90	870	21
Tulsa, OH	13	3,730	98	1,230	24
Rapid City, SD	−7	7,370	92	610	28
Houston, TX	32	1,410	94	1,880	18
Salt Lake City, UT	8	5,990	95	770	32
Richmond, VA	17	3,910	92	1,010	21
Spokane, WA	2	6,770	90	460	28
Green Bay, WI	−9	8,100	85	430	23

*Data derived from ASHRAE *Handbook of Fundamentals*, ASHRAE *Bin and Degree-hour Weather Data for Simplified Energy Calculations*, USAF AFM88-29, and NCDC *Climatography of U.S.*

instead given a conductance (C) rating based on the number of Btu's transmitted in 1 hr. through 1 sq. ft. of a stated thickness, when the opposite surfaces vary in temperature by 1°F. For instance, an 8-in.-thick hollow concrete masonry unit has a C-value of 0.90. Therefore, 0.90 Btu's are transferred in 1 hr. if one face of the unit is at 74°F and the other is at 75°F.

■ **RESISTANCE**

Materials that are normally considered good insulators have low k- and C-values (see Fig. 11.3-12). However, because k and C for these are stated as multidigit decimals smaller than 1.0, they are difficult to understand and compare. As a result, the use of *resistance* (R) numbers has increased as a method of rating building products and thermal insulations. An R-value is computed by taking the reciprocal of C, (1/C), or the reciprocal of k, (1/k). In the first case the R-value must be qualified as being for a 1-in. thickness; in the second, it is simply the R-value of the product, whatever its thickness.

Its resistance (R) represents the ability of a material to retard heat flow. R-values are also convenient because they are directly proportional to insulating value and can therefore be added or subtracted easily in order to determine the thermal performance of a product or assembly. For example, an 8-in. concrete block with a C-value of 0.90 would have an R-value of 1.11 (see Fig. 11.3-12). If a 1-in. slab of polystyrene with an R-value of 5.00 were added, the assembly would have a total R-value of 6.11, not counting surface films. The molecules of air clinging to the surfaces contribute some insulating value, so R-0.68 should be added for still inside air, and R-0.17 for outside air at 15 miles per hour (mph). The final total for the assembly is R-6.96.

FIGURE 11.3-12
Heat Transfer Coefficients of Representative Building Products[a]

Material or Product	Conductivity (k)	Conductance (C)	Resistance (R) Per 1-in. Thickness (1/k)	Resistance (R) Per Thickness Shown (1/C)
Concrete	5.0–10.5		0.20–0.10	
Face brick	9.00		0.11	
Hollow concrete block, 8 in.		0.90		1.11
Stucco	5.00		0.20	
Metal lath and plaster, 3/4 in.		7.70		0.13
Gypsum board, 1/2 in.		2.22		0.45
Plywood, 1/2 in.		1.60		0.62
Pine, fir, other softwoods	0.74–1.12		1.35–0.89	
Oak, maple, other hardwoods	1.06–1.25		0.94–0.80	
Asphalt shingles		2.27		0.44
Built-up roofing, 3/8 in.		3.00		0.33
Wood shingles		1.06		0.94
Vegetable board sheathing, 1/2 in.		0.76		1.32
Mineral wool batts, 3–4 in.		0.09		11.00
5-1/2–6-1/2 in.		0.05		19.00
6–7-1/2 in.		0.05		22.00
9–10 in.		0.03		30.00
Expanded polystyrene, 1 in.		0.20		5.00
Inside surface air film[b]				0.68
Outside surface air film[c]				0.17
Airspace 3/4 in., nonreflective[d]		1.00		1.01
Airspace 3/4 in., reflective[d]				3.48

[a] At 75 °F mean temperature, from ASHRAE *Fundamentals Handbook*.
[b] Heat flow horizontal, still air.
[c] Heat flow any direction, 15 mph wind.
[d] Heat flow horizontal.

Many insulating products are labeled with their R-value on the wrapper. The number given is usually the value of the product only, without the airspaces and films, which must be added to obtain the total installed resistance. Some insulations, such as multilayer foil, however, depend on airspaces and films for performance and can be described only in terms of installed resistance.

■ **TRANSMITTANCE**

Transmittance (U) is the number of Btu's transferred in 1 hr./1 sq. ft. of a building assembly, for each 1°F temperature difference between inside and outside surfaces. The U-value is calculated by adding the Rs for all elements of the building assembly (R_T) and taking the reciprocal of the total ($1/R_T$). If the R_T for an assembly equals 20, then the U-value would be 0.05 (1/20). Sample U-value calculations are given in Figure 11.3-13.

■ **HEAT STORAGE CAPABILITY OF MATERIALS**

Traditional heat flow calculations are based on the theory that design conditions are sustained for long periods and that heat flows continually from the warmer to the

FIGURE 11.3-13 *U-value computation.*

FRAME WALL CONSTRUCTION	INSULATED U AND R_T VALUES	
	CONSTRUCTION	R
	1. Outside Air Film	0.25
	2. Wood Siding	0.81
	3. ½" Insul. Sheathing	1.32
	4. Insulation	11.00
	5. ⅝" Gyp. Bd.	0.56
	6. Inside Air Film	0.68
		$R_T = 14.62$
	$1/R_T = U = 0.07$	

MASONRY WALL CONSTRUCTION	UNINSULATED U AND R_T VALUES	
	CONSTRUCTION	R
	1. Outside Air Film	0.25
	2. 4" Common Brick	0.80
	3. ½" Cement Mortar	0.10
	4. 4" Lt. Wt. Conc. Block	1.50
	5. ¾" Air Space*	—
	6. ⅝" Gyp. Bd.	0.56
	7. Inside Air Film	0.68
		$R_T = 3.89$
	$1/R_T = U = 0.26$	

FIGURE 11.3-14
Heat Storage Capabilities of Building Materials

Material	Specific Heat, Btu/lb./°F	Density, lb./cu. ft.	Heat Capacity, Btu/cu. ft./°F
Water (40 °F)	1.000	62.5	62.5
Steel	0.120	489.0	58.7
Cast iron	0.120	450.0	54.0
Copper	0.092	556.0	51.2
Aluminum	0.214	171.0	36.6
Basalt	0.200	180.0	36.0
Marble	0.210	162.0	34.0
Concrete	0.220	144.0	31.7
Asphalt	0.220	132.0	29.0
Ice (32 °F)	0.487	57.5	28.0
Glass	0.180	154.0	27.7
White oak	0.570	47.0	26.8
Brick	0.200	123.0	24.6
Limestone	0.217	103.0	22.4
Gypsum	0.260	78.0	20.3
Sand	0.191	94.6	18.1
White pine	0.670	27.0	18.1
White fir	0.650	27.0	17.6

cooler side at a constant rate. This *steady-state* concept of thermodynamics is useful in designing heating or cooling systems where the building shell is not made of massive materials and the day-to-night outdoor temperature variation is not great.

In actual practice, heavy masonry and concrete buildings behave dynamically and do not generally lose or gain heat as rapidly as would be indicated by the steady-state theory. This is especially true in arid sunny climates where there are relatively warm days in winter and cool nights in summer. Under these conditions, massive buildings respond dynamically, acting as storers of heat, rather than as flow resistors, as do insulated lightweight buildings.

In summer, massive structures not only store heat but are able to slow down its movement into the building until late afternoon, when the air and structure start cooling off. In winter, heat movement out of the building during the night is slowed by the mass of the structure until the sun warms the walls, and by late afternoon heat actually moves inward. Designers can use this *thermal lag* to help maintain interior comfort conditions with a minimum of manmade equipment.

The principle of using heavy, dense materials as *thermal mass* was well known to the Native American builders of 1000-year-old Arizona cliff communities, as well as to modern adobe homeowners throughout the Sun Belt. Despite their limited insulating value, earth-sheltered and adobe homes remain surprisingly cool in summer and warm in winter. The effectiveness of a material as a thermal mass, that is, its ability to retard heat flow, depends on its ability to store heat, which is measured by its *specific heat* and *heat capacity* (Fig. 11.3-14). Refer to Section 11.3.3.6 for a more detailed discussion of this subject.

The dynamic performance of concrete masonry and other massive shells can be accurately calculated by computers, programmed with detailed winter and summer weather information. To perform such calculations manually, utilizing the steady-state approach, is difficult, but a reasonable approximation can be made using a dimensionless correction factor called the *M Factor*. The Brick Institute of America (BIA) can provide more information about the thermal performance of masonry walls.

11.3.2

BUILDING DESIGN RECOMMENDATIONS

Buildings should regulate heat flow in a way that will minimize dependence on mechanical equipment and take advantage of exterior environmental conditions. Energy-saving features can be incorporated economically in buildings if considered in the design phase.

11.3.2.1

Site Planning

In locating a building on a site, the following factors should be considered: (1) general climate and microclimate, (2) terrain and landscape elements, and (3) orientation to the sun.

■ CLIMATE

Climatic analysis can provide valuable data on which to base energy-efficient design. Such an analysis should include data on the amount and angle of sunshine that strikes the site, prevailing winds, temperature ranges, and precipitation.

The U.S. Weather Service and private meteorologists maintain regional data on these factors. However, natural terrain and proximity to bodies of water may substantially modify the weather information recorded for a particular region. To avoid siting a building in an unsuitable area, the climate variations of each site (microclimate) should be examined. For example, sites in warm climates that act as natural dead-air basins should be avoided. In such areas, temperatures may exceed those of surrounding areas by as much as 35°F. In other areas, prevailing winds and natural terrain can combine to produce unusual wind, snow massing, or rain conditions not typical to a particular climate. Air quality also should be studied to determine whether natural ventilation can be used for cooling.

■ TERRAIN AND LANDSCAPE ELEMENTS

Terrain configurations and plant material can be used to control heat flow into and out of buildings. Refer to Section 2.5 for a discussion of landscape elements.

■ ORIENTATION

In cold regions, the area of north-facing windows should be limited to reduce heat loss in winter. The area of north-facing walls can be reduced by berming and sloping roofs (Fig. 11.3-15). In hot climate

FIGURE 11.3-15 *Sloping roofs act as a wind deflector.*

areas, west wall and glass exposures should be minimized to reduce cooling loads from intense afternoon summer sun.

Buildings in cold regions should be sited so that most windows face south, southeast, or southwest to take advantage of the warming sun. Orientation that will buffer against hot breezes is desirable in hot climates. Prevailing summer breezes that can provide ventilation and natural cooling should be considered when siting a building or locating windows.

11.3.2.2

Building Design

Thermally efficient buildings can be achieved through room organization and effective building envelope design.

■ ROOM ORGANIZATION

In areas subject to either extreme heat or cold, compact design reduces weather-exposed areas and reduces heat flow. Garages, closets, corridors, and other service spaces can be located against exterior walls to act as insulating buffers.

Inactive living spaces such as kitchens and laundry areas, which contain heat-producing appliances, should be located on north and east sides. This location permits wintertime use of their excess heat for warming and permits heat dissipation in summer.

Active living spaces should face south or southwest to take advantage of warming afternoon sun in winter. Such spaces should be protected against overheating in summer by external shading devices such as louvers, overhangs, awnings, and trees (see Section 8.1.1.4). Interior shading with blinds and drapes is less effective, but acceptable. Rooms likely to remain vacant for long periods should be located where they can be closed off readily. Lower ceiling heights can also reduce the volume of air to be heated or cooled. Built-in furnishings should be located away from supply and return air grilles to prevent obstruction of internal air flows.

■ BUILDING ENVELOPE

For conventionally heated and cooled buildings, the objective of heat control is to block the flow of heat through the ceil-

ings, walls, and floors that surround habitable areas. This heat flow can be controlled by one of the following techniques: (1) by constructing the assembly with materials that are inherently good insulators; (2) by adding to the assembly products such as loose fill, flexible, rigid, reflective, or foamed-in-place insulations; or (3) by using solid masonry or another dense material that has the capacity to store heat and slow down heat flow.

Wall, floor and roof assemblies should be constructed with materials that minimize heat flow out of the building in winter and into the building in summer.

Heat flow can be resisted by insulating materials (such as plastic foams or mineral fibers) or can be slowed by massive dense materials (such as thick masonry or concrete).

In some climates, materials with high thermal capacities are effective in slowing heat gain or loss, but in most locations adequate insulation of the building envelope is essential to controlling heat flow. Refer to Section 7.3 for a discussion and recommendations concerning insulation.

FLOORS Where solar radiation penetrates interior spaces, floors can act as heat sinks. Heat sinks are building components consisting of dense heavy materials such as concrete or masonry. They are desirable in cold climates because such materials absorb and store heat for long periods before releasing it to the interior by radiation and convection.

The floors of living areas above unheated crawl spaces or basements and slabs on ground should be insulated as recommended in Section 7.3.

WALLS Heat transmission through wall assemblies can be reduced by selecting materials that are inherently good insulators or by adding insulation (Section 7.3).

Wind-deflecting devices such as berms, walls, and fences can also be employed to reduce infiltration along exterior walls exposed to cold winds by steering the wind away from the wall.

ROOFS Heat transmission through a roof can be reduced by (1) use of insulation (see Section 7.3); (2) use of shiny or light-colored roof materials to reflect heat; (3) ventilation of attic spaces; (4) increasing the thermal mass to achieve a time lag in the heat flow; (5) use of louvered rather than solid overhangs to free heated air trapped beneath eaves; and (6) use of steeply pitched roofs on the windward side of the building to deflect cold winds away from the rest of the roof and reduce the amount of roof exposed to this wind (see Fig. 11.3-15).

HVAC SYSTEMS

Virtually every habitable building in the United States needs at least a minimal mechanical heating system, including those where much of the heating load is carried by a passive or active solar energy system (see Section 11.3.3.6). Most buildings also need a mechanical cooling system for occupant comfort and acceptable working and living conditions. A cooling system should definitely be considered in locations where the humidity is often high.

In addition, habitable buildings need a method of ventilating bathrooms and kitchens (windows or other openings may satisfy this requirement) and a source of fresh air to furnish combustion air for furnaces, fireplaces, and gas- or oil-burning equipment and devices, such as water heaters, ovens, ranges, and other appliances and devices where open flames are present. These ventilation requirements do not, however, necessarily require mechanical ventilation.

Heat Loss and Heat Gain Analysis

Heating and air-conditioning equipment is sized on the basis of heat loss and heat gain calculations for the particular area in which the building is situated. Such calculations also provide a measure of the thermal efficiency of a building.

The following terms should be understood to perform a heat loss or heat gain analysis:

R-value. A material's resistance to the flow of heat; the larger the R-value, the greater the resistance and the better the insulating value (see Section 7.3).

Heat loss. The amount of heat flow in Btu's per hour or per season, from inside a structure to outside, at given design conditions.

Sensible heat gain. The amount of heat gain in Btu's per hour that can be recorded as a rise in temperature.

Latent heat gain. Heat introduced into a structure by moisture in the air and that must be removed by the cooling system. The latent heat gain is commonly calculated as 30% of the *sensible heat gain*.

Indoor design temperature. The ideal comfort temperatures that are to be maintained inside the structure in winter and summer. These should correspond with the ASHRAE comfort envelope unless the DOE has developed more stringent guidelines.

Outdoor design temperatures. High and low temperatures that can be expected to occur regularly in a particular area (see Fig. 11.3-11).

Design temperature difference. The difference between the outside design temperature (t_o) and the inside design temperature (t_i).

Infiltration. The leakage of heat into or out of a building through cracks in the building structure, especially around windows and doors (Fig. 11.3-16).

Transmittance (U). The number of Btu's per hour transferred through 1 sq. ft. of a building assembly (such as a wall) for each 1°F temperature difference between inside and outside surfaces. It is calculated by adding the R-values for all elements of the building assembly (R_T) and taking the reciprocal ($1/R_T$) of the total (Fig. 11.3-17).

■ HEAT LOSS FORMULAS

Heat flows into or out of a building at the following locations: (1) through exterior walls, roofs, and glass areas; (2) at the perimeter of a building built on a slab; (3) through floors over unheated spaces; and (4) by infiltration around windows and doors. Figures 11.3-18 and 11.3-19 illustrate a method of calculating the contribution to overall heat loss or heat gain of each of the areas listed above.

■ HEAT LOSS CALCULATIONS

The following procedure (see Fig. 11.3-18) assumes use of data and tables in this book, which may not be up to date, for determining building heat loss. In an actual case a designer should obtain and use the latest data available.

1. Establish design conditions by determining the winter outdoor design temperature (t_o) (see Fig. 11.3-11) and the desired indoor design temperature (t_i) from the latest ASHRAE

FIGURE 11.3-16
Window and Door Infiltration Values

Window Type	Infiltration Airflow, cu. ft./ft. of Crack*		
	10 mph	20 mph	30 mph
Double-hung wood frame			
Nonweather stripped	21	59	104
Weather stripped	13	36	63
Double-hung metal frame			
Nonweather stripped	47	104	170
Weather stripped	19	46	76
Residential metal casement	18	47	74
Wood or metal door			
Nonweather stripped	69	154	249
Weatherstripped	19	51	92

*Reduce values by one-third when storm window (or door) is added to weather-stripped construction, by one-half when added to nonweather-stripped construction.

FIGURE 11.3-17
U-value Calculation Procedure

Building Part	Construction Materials	R-value
Roof/ceiling	Outside air film	0.17
	Shingles	0.44
	Building paper	0.06
	Plywood, 1/2 in.	0.62
	Attic air film	0.61
	Insulation	19.00
	Gypsum board, 1/2 in.	0.45
	Inside air film	0.61
	Total R-value (R_T)	21.96
	U-value ($1/R_T$)	**0.045**
Wall	Outside air film	0.17
	Siding, wood 1/2 in. × 8 in. lapped	0.81
	Sheathing, plywood, 1/2 in.	0.62
	Insulation	11.00
	Interior finish gypsum board, 1/2 in.	0.45
	Inside air film	0.68
	Total R-value (R_T)	13.73
	U-value ($1/R_T$)	**0.073**
Header joist	Outside air film	0.17
	Siding	0.81
	Sheathing	0.62
	Header, wood 1-1/2 in.	1.88
	Insulation	11.00
	Inside air film	0.68
	Total R-value (R_T)	15.16
	U-value ($1/R_T$)	**0.066**
Sill	Outside air film	0.17
	Siding	0.81
	Sheathing	0.62
	Sill, wood 5-1/2 in.	6.88
	Inside air film	6.88
	Total R-value (R_T)	9.16
	U-value ($1/R_T$)	**0.109**
Foundation	Outside air film	0.17
	Concrete block, 8 in.	1.11
	Insulation	5.00
	Interior finish gypsum board, 3/8 in.	0.32
	Inside air film	0.68
	Total R-value (R_T)	7.28
	U-value ($1/R_T$)	**0.137**

Comfort Envelope or use the HUD guideline temperature. The design temperature difference is equal to the difference between the established indoor and outdoor design temperatures ($t_i - t_o$).

2. Calculate the U-values for roof, wall, and glass areas by summing the R-values, (R_T) for all parts of such an assembly, including air films, and taking the reciprocal ($1/R_T$) of that figure (see Fig. 11.3-17).

3. Calculate the surface area (A) of glass and all similarly constructed roof, wall, and floor areas.

4. Calculate building heat loss (H_c) through exterior roof and wall construction and through exterior glass by multiplying the area (A) by its computed U-value, by the design temperature difference ($t_i - t_o$).

5. Calculate the floor heat loss (H_f) by the same method used for walls, roof, and glass, with the following qualifications:

   - If heating unit is in the basement and ducts or pipes are uninsulated, no floor loss calculation is necessary.

   - For floors over unheated basements, use one-third of the design temperature difference.

   - For floors over unvented crawl spaces with insulated crawl-space walls, use one-half of the design temperature difference.

   - For floors over vented crawl spaces or other open spaces, use the full design temperature difference.

6. Calculate the building perimeter (P) in feet if the floor is of slab construction.

7. Calculate the floor slab edge heat loss (H_e) by multiplying a factor of 0.55 by the slab perimeter (P) in feet if the slab edge is insulated, or a factor of 0.8 by P in feet if the slab edge is uninsulated, by the design temperature difference ($t_i - t_o0$).

8. Calculate the length of crack (L) in feet around window and door openings.

9. Select infiltration airflow volume in cubic feet per hour (cfh) per foot of crack (see Fig. 11.3-16).

10. Calculate the infiltration heat loss (H_i) for each opening by multiplying the length of crackage (L) by 0.018 times the per-foot infiltration volume in cfh, by the design temperature difference ($t_i - t_o$).

FIGURE 11.3-18
Heat Loss Calculation Example[a]

Example: A single-family house, 30 ft. × 40 ft., one story over a heated basement with wood-framed walls, insulated concrete block foundation, and weather-stripped wood sash; located in Washington, D.C.

HEAT LOSS: TRANSMISSION

Construction	Area, sq. ft.		Design Temperature Difference, °F		U-Value (See Fig. 11.3-17)		Heat Loss H_c, Btu/hr.
Glass, 1/4 in.	180.0	×	52	×	1.130[b]	=	10,577
Wood doors, 1-3/4 in. hollow	40.0	×	52	×	0.490[b]	=	1,019
Walls	1184.0	×	52	×	0.073	=	4,494
Header joint, 7-1/2 in.	92.5	×	52	×	0.066	=	317
Sill, 1-1/2 in.	18.5	×	52	×	0.109	=	105
Foundation, 9 in.	111.0	×	52	×	0.137	=	791
Roof/ceiling	1320.0	×	52	×	0.045	=	3,089

Total H_o 20,392

HEAT LOSS: INFILTRATION

Opening	Crack Length, ft.		Design Temperature Difference, °F		Infiltration Airflow, cu. ft./hr./ft. (See Fig. 11.3-16)		Reduction Constant		Heat Loss H_i, Btu/hr.
Windows	125	×	52	×	36	×	0.018	=	2,412
Doors	40	×	52	×	51	×	0.018	=	1,908

Total H_i 4,330

HEAT LOSS: TOTAL

Heat Loss Factor	Heat Loss, Btu/hr.
Transmission H_c	20,392
Floor H_f	—
Edge H_e	—
Infiltration H_i	4,330

Total H_T 24,722

[a]Follow 11-step procedure under "Heat Loss Calculation."

[b]See Figure 11.3-12 for C-value, which here equals U-value.

---

11. Calculate the total building heat loss (H_T) by summing H_c, H_i, and H_e, or H_f if appropriate.

■ HEAT GAIN CALCULATIONS

The following procedure for determining building heat gain (see Fig. 11.3-19) assumes use of data and tables in this book, which may not be up to date. In an actual case a designer should obtain and use the latest data available.

1. Establish the design conditions by determining the following: (1) the winter outdoor design temperature (t_i) (see Fig. 11.3-11); and (2) the desired indoor design temperature (t_i) from the latest ASHRAE Comfort Envelope or the HUD guideline temperature. The design temperature difference is equal to the difference between the established indoor and outdoor design temperatures ($t_i - t_o$).

2. Calculate the U-values for roof and walls by summing the R-values for all parts of each assembly (R_T) and taking the reciprocal ($1/R_T$) of that value (see Fig. 11.3-17).

3. Calculate the surface area (A) of all similarly constructed roofs, walls, and floors.

4. Calculate heat gain through unglazed (opaque) roof and wall sections (H_O) by multiplying the calculated area (A), by the calculated U-value, by the design equivalent temperature difference (DETD) (see Fig. 11.3-20).

5. Calculate heat gain through glass areas (H_g) as follows:
   - Locate the glass according to its building face: N, S, E, or W.
   - Calculate the surface area of each particular type of glass (single, double, or heat absorbing).
   - Select the appropriate winter glass load factor (see Fig. 11.3-21).

FIGURE 11.3-19
Heat Gain Calculation Example[a]

Example: A three-bedroom single-family house, 30 ft. × 40 ft., one story over a heated basement with wood-framed walls, insulated concrete block foundation, and single glazed weather-stripped wood sash with venetian blinds. Windows are equally distributed on each building face (N, S, E, and W); roof is dark color; located in Washington, D.C.

HEAT GAIN: UNGLAZED ROOF AND WALL SECTIONS

Construction[b]	Area, sq. ft.		U-value (see Fig. 11.3-17)		DETD (See Fig. 11.3-20)		Heat Gain H_o, Btu/hr.
Doors	40.0	×	0.490[c]	×	22.6	=	443
Walls	1184.0	×	0.073	×	22.6	=	1,953
Header	92.5	×	0.066	×	22.6	=	138
Sill	18.5	×	0.019	×	22.6	=	46
Roof/ceiling	1320.0	×	0.045	×	43.0	=	2,554

Total H_o 5,134

HEAT GAIN: GLAZED AREAS

Glass Type	Outdoor Design Temperature, °F	Orientation	Area, sq. ft.		Solar Heat Gain Factor (see Fig. 11.3-21)		Heat Gain H_g, Btu/hr.
Single 1/4 in.	94	N	45	×	22.2	=	999
Single 1/4 in.	94	S	45	×	30.2	=	1,359
Single 1/4 in.	94	E	45	×	55.2	=	2,484
Single 1/4 in.	94	W	45	×	55.2	=	2,484

Total H_g 7,326

HEAT GAIN: PEOPLE

Occupants: 2 × 3 bedrooms = 6 people (estimated)
     6 people × 225 Btu/hr./person = 1350
Total Heat Gain, H_p = 1350 Btu/hr.

HEAT GAIN: APPLIANCES

Total Heat Gain, H_a = 1200 Btu/hr. (estimated)

HEAT GAIN: TOTAL SENSIBLE

Heat Gain Factor	Heat Gain, Btu/hr.
Unglazed roof and wall sections, H_o	5,134
Glazed areas, H_g	7,326
People, H_p	1,350
Appliances, H_a	1,200

Total H_{TS} 15,010

HEAT GAIN: TOTAL LATENT

H_{TL} = 1/3 H_{TS} = 15,010 Btu/hr.
H_{TL} = 5,003 Btu/hr.

HEAT GAIN: TOTAL SENSIBLE + TOTAL LATENT

H_T = H_{TS} + H_{TL}
H_T = 15,010 + 5,003 = 20,013 Btu/hr.

[a]Follow 11-step procedure under "Heat Gain Calculation."
[b]Only above-grade construction considered.
[c]See Figure 11.3-12 for C-value, which here equals U-value.

FIGURE 11.3-20
Design Equivalent Temperature Differences

Design Temperature, °F[a]	85		90			95			100		105	110
Daily Temperature Range[b]	L	M	L	M	H	L	M	H	M	H	H	H
Walls and doors												
Frame and veneer-on-frame	17.6	13.6	22.6	18.6	13.6	27.6	23.6	18.6	28.6	23.6	28.6	33.6
Masonry walls, 8-in. block or brick	10.3	6.3	15.3	11.3	6.3	20.3	16.3	11.3	21.3	16.3	21.3	26.3
Wood doors	17.6	13.6	22.6	18.6	13.6	27.6	23.6	18.6	28.6	23.6	28.6	33.6
Ceilings and roofs												
Ceilings under naturally vented attic or vented flat roof, dark	38.0	34.0	43.0	39.0	34.0	48.0	44.0	39.0	49.0	44.0	49.0	54.0
light	30.0	26.0	35.0	31.0	26.0	40.0	36.0	31.0	41.0	36.0	41.0	46.0
Built-up roof, no ceiling, dark	38.0	34.0	43.0	39.0	34.0	48.0	44.0	39.0	49.0	44.0	49.0	54.0
light	30.0	26.0	35.0	31.0	26.0	40.0	36.0	31.0	41.0	36.0	41.0	46.0
Ceilings under unconditioned rooms	9.0	5.0	14.0	10.0	5.0	19.0	15.0	10.0	20.0	15.0	20.0	25.0
Floors												
Over unconditioned rooms	9.0	5.0	14.0	10.0	5.0	19.0	15.0	10.0	20.0	15.0	20.0	25.0
Over basement, heated crawl space, or concrete slab on ground	0	0	0	0	0	0	0	0	0	0	0	0
Over unheated crawl space	9.0	5.0	14.0	10.0	5.0	19.0	15.0	10.0	20.0	15.0	20.0	25.0

[a]Interpolate for values between those listed.

[b]Daily Temperature Range (see Fig. 11.3-11):
L (Low) calculation value: 12   M (Medium) calculation value: 20   H (High) calculation value: 30
Range: less than 15°   Range: 15 to 25°   Range: more than 25°

- Multiply the glass area (A) by the heat gain factor for the total window heat gain in Btu's per hour.
- If the glass is partially or completely shaded by a permanent shading device such as an overhang, the shaded portion is treated as north-facing glass. Unshaded portions are considered separately. The shade line is computed by multiplying overhang width by the shade line factor (see Fig. 11.3-22). Glass above the line is in shade, and glass below the line is in sun.

6. Heat gain from infiltration (H_i) is normally insignificant in summer and can be ignored.

7. Estimate the sensible heat gain from people (H_p) at 225 Btu's per hour per person. For residences, estimate the number of occupants at twice the number of bedrooms.

8. Estimate appliance loads (H_a) at 1200 Btu's per hour.

9. Calculate total sensible heat gain (H_{TS}) by summing H_o, H_g, H_p, and H_a.

10. Estimate total latent heat gain (H_{TL}) as one-third H_{TS}.

11. Calculate total heat gain as the sum of total latent heat gain (H_{TL}) and total sensible heat gain (H_{TS}).

11.3.3.2

Equipment

Heating and cooling equipment includes furnaces, motors and motor-driven compressors, condensing units, cooling towers, heat exchangers, blowers, and pumps.

■ SELECTION

The heating and cooling equipment for all-air systems in simple single-family residential buildings are usually packaged furnace or heat pump systems selected by the architect or the heating and cooling contractor, based on calculated heating and cooling loads and experience. Professional engineers are seldom employed to design such systems. The manuals of the Air Conditioning Contractors of America listed in Appendix B, References, may be used to aid in the design of systems for small residential and commercial buildings.

More complex systems and those utilizing heavy equipment should be designed by professional structural and mechanical engineers based on design calculations and acoustical analysis. The noise levels and heating and cooling capacities of equipment should be obtained. Many manufacturers can furnish or develop sound-pressure output data based on standard testing for most types of equipment, from which sound-pressure levels can be computed for a specific space.

Selecting better-quality heating and cooling equipment with more than adequate capacity may slightly increase its initial cost but will often help to create a quieter installation and will perform more efficiently over the years. Selecting equipment with greater-than-adequate capacity, particularly for fans and pumps, can produce considerable noise level reductions because larger equipment runs more slowly and therefore more quietly.

Most types of furnaces for single-family houses, individual apartment or townhouse units, and small commercial establishments are manufactured for either

FIGURE 11.3-21
Example of Window Glass Load Factors[a,b]

Outdoor Design Temperature, °F	Regular Single Glass						Regular Double Glass						Heat-Absorbing Double Glass					
	85	90	95	100	105	110	85	90	95	100	105	110	85	90	95	100	105	110
No Awnings or Inside Shading																		
North	23	27	31	35	38	44	19	21	24	26	28	30	12	14	17	19	21	23
NE and NW	56	60	64	68	71	77	46	48	51	53	55	57	27	29	32	34	36	38
East and West	81	85	89	93	96	102	68	70	73	75	77	79	42	44	47	49	51	53
SE and SW	70	74	78	82	85	91	59	61	64	66	68	70	35	37	40	42	44	46
South	40	44	48	52	55	61	33	35	38	40	42	44	19	21	24	26	28	30
Draperies or Venetian Blinds																		
North	15	19	23	27	30	36	12	14	17	19	21	23	9	11	14	16	18	20
NE and NW	32	36	40	44	47	53	27	29	32	34	36	38	20	22	25	27	29	31
East and West	48	52	56	60	63	69	42	44	47	49	51	53	30	32	35	37	39	41
SE and SW	40	44	48	52	55	61	35	37	40	42	44	46	24	26	29	31	33	35
South	23	27	31	35	38	44	20	22	25	27	29	31	15	17	20	22	24	26
Roller Shades Half Drawn																		
North	18	22	26	30	33	39	15	17	20	22	24	26	10	12	15	17	19	21
NE and NW	40	44	48	52	55	61	38	40	43	45	47	49	24	26	29	31	33	35
East and West	61	65	69	73	76	82	54	56	59	61	63	65	35	37	40	42	44	46
SE and SW	52	56	60	64	67	73	46	48	51	53	55	57	30	32	35	37	39	41
South	29	33	37	41	44	50	27	29	32	34	36	38	18	20	23	25	27	29
Awnings																		
North	20	24	28	32	35	41	13	15	18	20	22	24	10	12	15	17	19	21
NE and NW	21	25	29	33	36	42	14	16	19	21	23	25	11	13	16	18	20	22
East and West	22	26	30	34	37	43	14	16	19	21	23	25	12	14	17	19	21	23
SE and SW	21	25	29	33	36	42	14	16	19	21	23	25	11	13	16	18	20	22
South	21	24	28	32	35	41	13	15	18	20	22	24	11	13	16	18	20	22

[a]Shown in Btu/sq. ft./hr. for the hours of 5:30 A.M. to 6:30 P.M., and for 30° and 40° N latitude. Tabulated factors are applicable for an indoor design temperature of 75°F and for outdoor design temperatures as indicated. Interpolate for values between those listed.

[b]For an actual design problem, refer to the latest ASRAE *Fundamentals Handbook*.

FIGURE 11.3-22
Example of Shade Line Factors[a,b]

Window Orientation	Latitude						
	24°	32°	36°	40°	44°	48°	52°
E	0.8	0.8	0.8	0.8	0.8	0.8	0.8
SE	1.8	1.6	1.4	1.3	1.1	1.0	0.9
S	9.2	5.0	3.4	2.6	2.1	1.8	1.5
SW	1.8	1.6	1.4	1.3	1.1	1.0	0.9
W	0.8	0.8	0.8	0.8	0.8	0.8	0.8

[a]Distance shadow line falls below the edge of the overhang equals shade line factor multiplied by width of overhang. Values are average for 5 hours of greatest solar intensity on August 1.

[b]Assume a window height of 7 ft. from floor and a 2 ft. overhang at the same height; south face, 40° N latitude. Shade line factor from table is 2.6, multiplied by 2 ft. equals 5.2 ft. The top 5.2 ft. of the window is in shade. For an actual design problem, refer to the latest ASHRAE *Fundamentals Handbook*.

heating only or for initial or later installation of air conditioning. Those intended to provide cooling have a larger blower for more volume of air output; the furnace is quieter because the larger fan can operate more slowly and still move a larger volume of air.

In residential occupancy, the greatest air conditioning load occurs between 5:00 P.M. and 7:00 P.M., when exterior and interior conditions are at extremes. Some systems are barely adequate for this peak load time and therefore must operate at full capacity late into the night when the surrounding noise level is greatly reduced and equipment noise is generally annoying. Therefore, in residences it is particularly important to select equipment that has greater-than-adequate capacity, particularly for fans and pumps. Reducing fan speeds only slightly can produce noise-level reductions that will more than justify the cost of additional power.

Heating and cooling equipment and the flowing air and liquids they produce and move about are sources of noise vibration, whether they are part of a small individual package unit or part of large central systems. Heating and cooling equipment can have a wide range of noise levels. It is generally more economical to install a quietly operating (even if more expensive) unit than to attempt to reduce the noise level of a less-expensive unit by acoustical construction.

Vibration from mechanical equipment usually is not a significant problem when the equipment is mounted on a concrete slab placed on ground, located in a basement, in a crawl space, in a utility room, or in an attached garage. However, on suspended concrete slab floors, structure-borne vibration from equipment can be transmitted great distances. Equipment placed on wood-framed construction also will cause acoustical problems unless it is isolated from the structural framing.

Whenever possible, mechanical equipment should be mounted resiliently and all related piping, refrigerant lines, condensation lines, electrical connections, and duct systems should be flexibly mounted to the units to prevent transfer of vibrations to walls and floor and ceiling assemblies. Vibration isolation pads should be used between the equipment and the supporting structure. Flues should be isolated from walls by an air gap, especially where they pass through the walls of rooms, such as bedrooms, that are intended to be quiet.

When heating and cooling systems are large or complex, whole machinery rooms can be floated or isolated from the building, or special foundations can be constructed to isolate individual pieces of equipment.

There should be the amount of clear space on each side of a heating unit that is recommended by its manufacturer or required by code, whichever is the largest. Locating combustible materials too close to furnaces is a potential fire hazard.

◼ LOCATION

The mechanical equipment room should be buffered from the rest of the building by design. Areas containing furnaces, pumps, compressors, laundry facilities, and similar equipment, and areas enclosing elevator shafts, incinerator chutes, and other potential sources of noise should be located both horizontally and vertically as far as possible from quiet areas such as bedrooms and living rooms.

In single-family dwellings, low-rise apartments, townhouses, and small commercial buildings, an exterior wall location is usually best acoustically (Fig. 11.3-23a). Unfortunately, from an acoustical point of view, in many such buildings, unit forced-air heating and cooling equipment is placed in a closet, an alcove, or a utility room near the center of the living unit or working space to produce shorter and therefore more economical duct runs (Fig. 11.3-23b). A single air return is located near the center of the floor plan; often a grille in the wall or door of the furnace enclosure is all that is provided. This type of planning places major noise sources in the center of the living or working space. The need for quiet must be weighed against the economics of central location. For air plenum systems (see Section 11.3.3.5) the furnace must be located in the center of the space and other means found to reduce noise levels.

FIGURE 11.3-23 *From an acoustical point of view, a mechanical equipment room should be located on an exterior wall (a) rather than in a central location (b).*

FIGURE 11.3-24 *An exterior condenser should not be placed so that loud noise is created near windows, doors, or outdoor living areas.*

One way to confine mechanical equipment noise is to provide three solid walls around the furnace room with the entry door to the outside for servicing (see Fig. 11.3-23a). In this case, the service door should have a solid core and should be gasketed around the perimeter to render it airtight (see Chapter 8).

Mechanical systems are a source of exterior noise. Although a cooling coil is located inside at the furnace plenum, a water- or air-cooled condensing unit must be located outside. Because ordinances sometimes govern the allowable noise level at property lines, condensers often must be located adjacent to the building in front or rear locations where the distance is greatest to the next building. This creates an exterior noise source at the exterior wall. These units should be as far as possible from windows, porches, patios, and neighboring buildings. An alternate location for the condensing unit would be on a vibration-isolated platform on the roof (Fig. 11.3-24).

Unit air conditioners in the exterior walls in multifamily residences should be located as far as possible from adjacent apartments, so that operation of one unit does not unduly disturb neighbors who do not operate their units at the same time.

A heating and cooling system's thermostat(s) should be located on an interior partition. It should not be located near lights, cooking or heating appliances, radiators, convection units, or air-supply registers so that the hot and cold air associated with them does not cause a false thermometer reading.

Duct Systems

The duct systems portion of HVAC systems consists of the diffusers and other outlets through which heated or cooled air is delivered into habitable spaces; the air-supply ducts that carry air from the heating and cooling unit throughout the building to the diffusers; return-air intakes and ducts that pick up heated or cooled air and return it to the heating or cooling unit for recycling; fresh-air intakes that provide combustion air and air changes in the building; and exhaust systems for bathrooms, kitchen cooking appliance hoods, laundry rooms, and the like.

Further information on calculating heat loss and design criteria for the sizing and locating of ducts and registers in small residential and commercial buildings may be obtained from the National Association of Home Builders (NAHB) Research Foundation and from the publications of the Air-Conditioning Contractors of America (ACCA) and the Sheet Metal and Air-Conditioning Contractors' National Association (SMACNA) that are listed in the Appendix B References.

■ SUPPLY OUTLETS

Either adjustable floor, ceiling, or sidewall diffusers can be used to supply heated or cooled air to living areas.

Each *diffuser* should have an adjustable damper to properly balance the air flow. If too much air enters a room, the registers can be partially closed. If there is not enough air, it is easy to replace a small register with a larger one or add a register.

In general, registers are sized using the following procedures:

1. Heat loss or gain should be calculated for each room and the entire building. Conventional heat loss calculations can be used for the living and working areas.

2. A heating unit with an output capacity equal to the calculated heat loss plus the amount that will be lost through the supply duct system should be selected.

3. Where floor and sidewall registers are used, at least one register should be located on each exposed exterior wall, except in bedrooms with less than 7000 Btu's per hour heat loss. If possible, floor registers should be located in front of windows (Fig. 11.3-25) and side wall registers should be located below windows. When ceiling diffusers are used, every habitable room should have at least one unit. Larger spaces should have more than one ceiling diffuser and they should be equally spaced to distribute conditioned air throughout the space. Small rooms near the furnace and interior rooms with no outside wall sometimes need no registers.

4. Register outlet capacity should closely match the air volume of the furnace on a proportional basis. For example, if heat loss for a bedroom is 8400 Btu's per hour and total dwelling heat loss is 56,000 Btu's per hour, then that bedroom should receive about 15% of the total air volume. If the blower size is 1600 cfm, 15% of that is 240 cfm. If two registers are used, each should deliver 120 cfm.

FIGURE 11.3-25 *Typical plan showing floor register locations.*

Diffusers are a major source of duct noise. Air velocity and the design of diffusers should be considered to produce outlets that do not whistle or cause other annoying noises. Decreasing the air velocity at the outlet diffuser by 50% will reduce the sound level generated by the grille from 15 to 20 dB, measured 3 to 5 ft. from the diffuser. Wide-angled directional diffusers generate the most noise, whereas wire mesh or perforated metal diffusers with little air spread capacity appear to be the quietest.

■ AIR-SUPPLY DUCTS

Supply ducts should be sized to adequately deliver the required amount of air at the desired velocity. Procedures for selecting duct materials and sizes are contained in the SMACNA publication *HVAC Systems—Duct Design.* SMACNA also produces a "Duct System Calculator."

In single-story construction, an overhead main supply duct with several branch ducts can be used. When attic space is available, the main and branch ducts are usually run in that space. When there is no attic, and in intermediate floors in buildings of more than one story, the main supply duct can be dropped below the ceiling. The branches can then be either dropped below the ceiling or run within the framing space.

Ducts should be formed of adequate thickness to minimize vibration and they should be isolated from mechanical equipment with flexible collar connections to prevent transmission of vibrations (Fig. 11.3-26). The space around ducts should be caulked tightly for isolation where they pass through sound-conditioned construction. Mufflers used in ducts near equipment rooms will absorb transmitted noise (Fig. 11.3-27).

Careful planning and installation of duct systems are critical in controlling sound transmission. Ducts act like speaking tubes in transmitting noises. Noise may be (1) from equipment such as fans and motors; (2) from sound transmitted through the walls into the duct and then conducted through the duct; or (3) created within the duct by air rushing through the system or by expansion and contraction of the ducts themselves. The sound does not necessar-

FIGURE 11.3-26 *Flexible connections reduce the transmission of vibrations from equipment to the duct system.*

FIGURE 11.3-27 *Duct mufflers are effective in reducing noise transmitted from a furnace.*

ily have to enter or leave a duct through an opening or diffuser, but can be transmitted through walls, particularly where the ducts pass vertically through living spaces (Fig. 11.3-28). Planning can minimize the sound transmittance from one room to another through the open air paths of the duct system.

Duct outlets should not be located back-to-back between rooms, because this practice practically eliminates any benefit obtained from sound-conditioned walls (Fig. 11.3-29). A 6-ft. length of fibrous duct material adjacent to the outlet generally will provide reduction of furnace noises and will prevent short-circuiting sound through ductwork between rooms (Fig. 11.3-30).

Duct insulation and duct mufflers will help reduce furnace noises and will help to prevent short-circuiting sound through ductwork between rooms.

■ FRESH-AIR SUPPLY

Combustion-air and fresh-air intakes may be necessary in buildings that are tightly sealed. A combustion-air intake provides outside air for efficient burning of fossil fuel. Outside air may be simply delivered to the room where the furnace is located, if the furnace is in a separate room and a blast of cold winter air is not objectionable. To conserve energy and to reduce the discomfort from cold air, the outside air duct may be connected directly to the combustion chamber. This, in fact, is required by some local utilities and energy conservation codes.

A fresh-air intake helps ventilation and helps reduce the buildup of indoor odors and other pollutants. ASHRAE Standard

FIGURE 11.3-28 *A duct system can conduct noise between units. Ducts should not be housed in common walls or floor and ceiling systems between dwelling units.*

FIGURE 11.3-29 *Duct outlets placed back-to-back will nullify the effect of sound-isolating walls.*

FIGURE 11.3-30 *Fibrous insulation in the duct, especially near outlets, helps reduce transmitted noise.*

62 specifies a ventilation rate of 15 cfm to ensure acceptable air quality. Local codes may require a different amount. Fresh-air intakes may be placed in a vented attic or directly outdoors and con-

nected to the return air duct above the furnace's air filter. Air-to-air heat exchangers can be used to temper the incoming air and save heating and cooling energy.

Intakes should have insect screens and tight-fitting manual or automatic dampers to control the amount of air intake.

■ RETURN-AIR SYSTEM

The furnace heats air and the blower distributes it through the ducts to supply outlets. The return air system completes the convection loop by returning air to the furnace where it can be heated and recycled.

A return-air system should be sized to equal or exceed the supply volume and should induce the return of air from each conditioned space. A properly designed return-air system helps maintain an even temperature level with minimal stratification throughout a building.

Nonducted volume air returns allow free passage of return air directly to the furnace, usually through a louvered furnace closet door. In nonducted return air systems doors should be undercut to have at least 1 in. of clearance from floor tile or carpet to ensure proper air circulation. However, a door that must provide acoustical privacy cannot be used as a means of returning air to the furnace. Separate air-return registers can be placed in such rooms and connected to a sound-absorbing duct system. This eliminates blower and furnace noises and reduces cross talk from room to room. Sound-absorbing air inlets can be used to allow air intake from corridors without admitting noise.

Air-return registers should not be located back-to-back for the same reasons supply registers should not be. Similar methods to those suggested for supply-air registers can also be used in return-air openings to reduce sound transmission.

Ducted multiple return-air systems are used in larger houses and other buildings with complex room layouts. Return-air grilles should be placed near the ceiling. The air is returned from grilles in each room through ducts located in the ceiling construction or attic. Smaller rooms such as kitchens and baths do not need return-air grilles if clear openings, openings fitted with louvers or grilles, or undercut doors are provided to permit air to move from the small spaces into an adjacent space that does have a return-air grille.

Most codes require that bathrooms that do not have windows be exhausted directly to the exterior. Some codes have similar requirements for bathrooms with windows and for other spaces as well. Whether required or not, separate exhaust systems should be installed in rooms housing cooking appliances, laundry equipment, showers, and other equipment or devices that produce excess heat, high humidity, or undesirable fumes.

Exhaust ducts should be sized to fit the volume of the space they are intended to exhaust with the desired number of air changes per hour. Fans should be selected using the same criteria, and should usually be oversized to reduce the noise they generate.

11.3.3.4

Embedded Heating and Cooling Elements

When pipes or ducts must pass through slabs, care should be taken to prevent them from breaking by isolating them from the slab. Pipes should pass through a slab vertically and be provided with space for expansion and with flexible connections if necessary to accommodate differential movement between the slab and the soil.

When galvanized metal ducts or copper pipes are embedded, calcium chloride should not be used in the concrete mixture. Aluminum should not be embedded in concrete.

■ HEATING DUCTS

Heating ducts embedded in a slab on grade may be of metal, fibrous cement, wax-impregnated paper, ceramic tile, concrete, or plastic. Fibrous-cement ducts with watertight joints may be set on a sand-leveling bed and be backfilled with sand to the underside of the slab. Precautions should be taken to see that the position of ducts is not disturbed during concreting.

Metal and paper ducts should be completely enclosed in at least 2 in. of concrete to prevent moisture from collecting in them; joints may be taped or cement grouted. These ducts often must be weighted so that they will not rise in the freshly poured concrete.

FIGURE 11.3-31 *Details for embedding heating sources in a slab.*

When the bottom of the ductwork adjacent to the slab perimeter is below exterior finish grade, the duct system should be of ceramic tile, fibrous cement, or concrete pipe. Joints that are completely or partially below exterior grade should be watertight. In areas where termite control is required, these joints should be resistant to termite penetration and soil-treatment chemicals.

If the slab is not reinforced (Type I slab) (see Section 3.3), welded wire mesh reinforcement (fabric) should be placed to extend 19 in. on either side of the centerline of the duct or to the slab edge, whichever is closer (Fig. 11.3-31a).

■ EMBEDDED PIPES AND COILS

Type II slabs (see Section 3.3) may be heated by circulation of heated liquids through embedded pipes. Usually, ferrous or copper pipe is used with approximately 2-1/2 to 3 in. of cover over the pipe (Fig. 11.3-31b). Where pipes pass through control joints or construction joints, provision should be made for possible movement across the joint. The piping should be pressure tested before the concrete is placed and air pressure should be maintained in the pipe during concreting operations; water pressure should not be used for testing. After the concrete is placed, the slab should not be heated until curing of the concrete has been completed. Slabs should be warmed gradually using luke-warm liquid in the pipe to prevent the cold concrete from cracking.

■ COOLING PIPES

The use of embedded pipes for summer cooling is not recommended with slab-on-ground construction. Under conditions of high humidity, condensation may form on the cooled slab, which may damage flooring and support mold or fungal growth.

11.3.3.5

Air Plenum Systems*

An air plenum system is a method of distributing heated or cooled air or both to the living area of a dwelling through a tightly sealed underfloor area, usually a crawl space.

The system may be used with poured concrete or concrete masonry unit (block) foundations but is frequently used with wood foundations (see Section 6.4). Although theoretically an air plenum system could be used to heat and cool a two- or three-story building, it is currently recommended only for single-story structures.

An air plenum system is cheaper than conventional forced-air systems because it eliminates ductwork.

*The information and most of the illustrations in this section were abstracted from a manual prepared by the National Association of Home Builders (NAHB) Research Foundation, Inc. for the American Plywood Association (APA), American Wood Preservers Institute (AWPI), Southern Forest Products Association (SFPA), and the Western Wood Products Association (WWPA). This data has been updated to agree with the pamphlet *Plen-wood System*, issued jointly in 1989 by the American Plywood Association (APA), American Wood Council (AWC), National Forest Products Association (NFPA), Southern Forest Products Association (SFPA), and the Western Wood Products Association (WWPA), and some of the illustrations come from that document.

In a typical 1400 sq. ft. air plenum house, energy use can be reduced by 10%, compared with a house with a conventional forced-air heating system and crawl space. Uniform air distribution through well-spaced registers prevents stratification of the convected air. Tests conducted at the University of Florida showed an average temperature differential of 2.5°F between air near the floor and 60 in. above the floor. Air quantities can be regulated by adjusting dampers.

The floor surface is warm when the heat is on and cool when the air conditioning is on. A warm floor is particularly pleasant underfoot in winter.

Floor framing members, floor sheathing, underlayment, and finished flooring absorb heat from the system and radiate it into the building. Radiant heat makes occupants feel more comfortable in cold weather, thus allowing them to operate the system at a lower temperature.

In studies by Washington State University, a dwelling in Idaho with an air plenum system used less heat than a dwelling with a conventional forced-air heating system and crawl space. The floor surface temperature ranged between 70° and 82°F; 85°F is the maximum temperature recommended for floor surfaces.

In tests conducted with air conditioning in Florida the floor surface temperature ranged between 70° and 73°F. Further discussion of heat and comfort may be found earlier in this section.

An air plenum system uses a downflow air-handling unit (combination furnace and air conditioner unit) to create enough air pressure under the floor to deliver heated or cooled air to individual rooms through supply registers in the floor (Fig. 11.3-32). Using the crawl space to distribute conditioned air eliminates costly ductwork and provides both convection and radiant heat (Fig. 11.3-33).

FIGURE 11.3-33 *Entering air not only heats or cools interior air but floor surfaces as well.*

Such a system is simple in design and easy to build. It conserves energy and reduces construction costs. With minor variations an air plenum system meets the requirements of all the major building code groups.

Air plenum systems are particularly susceptible to the introduction of pollutants, such as radon, into the living environment inside a dwelling. Where radon exists, precautions should be taken to prevent its introduction into the plenum and subsequently into the dwelling (see Section 2.5).

■ **SITE PREPARATION**

Site clearing and excavation methods should be no different than with conventional foundation systems (see Chapter 2). As with all foundations, excavations for footings should be made into undisturbed soil and be deep enough to prevent frost damage.

DRAINAGE No special precautions are needed when the plenum grade is at or above exterior grade, the plenum space is well drained, the underlying soils are of types that drain well, and the water table height is normal. When the plenum grade

is below the exterior grade, the soil is clayey, or the groundwater levels rise close to the plenum floor, foundation drains and sump pumps may be required. Care should also be taken to assure drainage of rainwater during construction.

A description of soil properties may be found in Section 2.1.

SLOPING SITES When a sloping site results in more of the foundation wall being above grade, thicker insulation should be used on the inside of the wall.

Increased or varying plenum depth may increase air volume and operating cost but does not affect satisfactory performance of the system (Fig. 11.3-34).

■ **TERMITE PROTECTION**

In regions subject to termite infestation, the soil should be chemically treated for a distance of 12 in. out from the foundation wall. Any wood in the foundation or wood that is below the level of the first floor should be pressure treated and carry the American Wood Preservers Bureau AWPB-FDN mark. Refer to Section 6.5 for additional requirements. If a wood foundation system is used, it should follow the preservative treatment methods described in Section 6.4.

A shallow-depth wood plenum (less than an 18-in. clearance) is not recommended in areas where a termite hazard exists, because soil-treatment chemicals should not be used inside a plenum and because visual inspection is difficult.

No building debris or vegetation should be left in a plenum space.

■ **FOUNDATION CONSTRUCTION**

The basic plenum types are (1) shallow depth, low profile and (2) regular depth, crawl space.

FIGURE 11.3-32 *Diagram of an air plenum system.*

FIGURE 11.3-34 *Neither plenum depth nor slope significantly affect system performance.*

A shallow plenum house looks like a slab-on-grade house from the outside. It is economical because even on sloping sites, very little site grading is necessary. A shallow-depth plenum has less than 18 in. of clearance below the floor joists and can function with as little as 3 in.

A regular-depth plenum is basically the same as a conventional crawl space. A regular-depth crawl space has a clearance of 18 in. or more from the bottom of the floor joists and 12 in. or more from the supporting girders.

A crawl-space foundation intended to accommodate an air plenum should be (1) sealable to prevent conditioned air from escaping and water vapor from entering, (2) well drained, (3) fire safe, and (4) free of insects and other vermin.

An air plenum system can be constructed with most conventional foundation materials (Fig. 11.3-35). It is also well suited to wood foundations and this combination is common. Section 6.4 contains recommendations for wood foundation systems.

■ THERMAL INTEGRITY

An air plenum system should be thoroughly caulked and sealed, have a suitable vapor retarder, and be adequately insulated. Foundation vents, typically recommended for crawl spaces, should not be used. Access to an air plenum should be from only inside the dwelling. Materials in a plenum should have a flame-spread rating of 200 or less.

CAULKING AND SEALING   An air plenum system should be sealed to prevent leakage of conditioned air and entrance of ground moisture. A sill sealer should be used between foundation walls and sole plates in wood-framed construction. Joints should be caulked.

VAPOR RETARDER   Polyethylene film or similar continuous sheet material should be used as a ground cover and should extend up and across the inside face of foundation walls to reduce air and vapor leakage and to keep the plenum dry.

The underfloor area should be free of sharp stones and debris that could puncture this film. The film should cover the ground, extend up the foundation walls, and be attached to the top of the sill; a few

FIGURE 11.3-35 *Air plenum systems may be constructed with foundations of (a) concrete block, (b) cast-in-place concrete, or (c) grade beams on wood posts or concrete piers.*

FIGURE 11.3-36 *Vapor retarder film in an air plenum.*

inches should overhang the sill so that the film may be attached to framing members to form a good seal (Fig. 11.3-36). The

film should be applied in the largest practicable sheets. Joints should be sealed, except that joints on the ground should be overlapped 12 in. but not sealed. These unsealed joints provide a means of escape if water should collect in the plenum area. This film may be covered with 2 in. of sand or gravel.

INSULATION   Either rigid or batt insulation can be used in an air plenum system. Insulation should have an R-value equal to or greater than that recommended for the side walls (R-11 or better). Foil-faced batt or rigid insulation may be used. Because of fire hazard, paper-backed and foamed plastic insulations are not recommended, unless they are covered with 1/2-in.-thick gypsum board or 5/8-in.-thick plywood. Of course, covering the insulation with such materials increases construction costs.

Insulation should be installed over the vapor retarder on the inside perimeter of the foundation wall and should extend at least 24 in. away from the wall horizontally at the bottom as shown in Figure 11.3-37. Insulating the entire underfloor area is generally not cost effective.

■ FLOOR CONSTRUCTION

The floor over a plenum may be framed conventionally by resting the joists on a wood sill (see Fig, 11.3-37) or in joist hangers. Sometimes, joists are supported on a continuous wood ledger mounted on the side of wood foundation walls to lower the dwelling's exterior wall by the depth of the joists. This provides a lower elevation and can reduce the amount of exterior wall sheathing and siding needed by about 9.5%.

Because the moisture content of the air in the plenum is low, the floor system should be built to allow for shrinkage as if it were in a heated basement.

■ UNDERFLOOR PLUMBING

Seamless copper tubing is recommended for water supply piping to minimize leakage problems. Drain, waste, and vent (DWV) plumbing joints should not be in the plenum, but rather above the floor or below the vapor barrier. DWV cleanouts may be either inside or outside of the dwelling but should not be placed in the

FIGURE 11.3-37 *Vapor retarder and insulation in an air plenum at (a) a wood foundation wall, (b) a unit masonry wall, and (c) a concrete foundation wall.* *(Courtesy American Plywood Association.)*

plenum space. Underground plumbing should be installed before a vapor barrier is laid on the ground. Refer to Section 11.1 for additional information about DWV systems.

In shallow plenums an access trench is desirable to enable workers to reach leaky plumbing or faulty equipment.

■ UNDERFLOOR WIRING

Electrical wiring within a plenum is restricted by most codes as a potential fire hazard. If permitted, wiring within a plenum should be approved types of electrical cable or installed in conduit. Some codes require rigid conduit, others permit thin wall.

■ MECHANICAL SYSTEM

An air plenum system does not require special equipment. The key to optimum efficiency is the proper sizing of system components.

EQUIPMENT Usually, a counterflow (down-flow) furnace or heat pump is used. It should be sized using heat-load calculations; oversizing should be avoided. Blower controls should be set to produce continuous low-volume air circulation at all times and higher-volume circulation when heating or cooling is supplied.

The furnace selected should have adequate safety mechanisms to prevent excessive temperature buildups. A down-flow furnace should have a safety mechanism to guard against the reversal of air flow should the blower become inoperable.

The heating capacity of the unit and the air volume circulated through the unit by the blower determine the air temperature rise through the furnace. The most efficient underfloor plenum heating units operate at a temperature rise of about 40°F. Figure 11.3-38 shows the recommended furnace output and blower capacities that will result in a 40°F temperature rise.

Since there are no ducts in the air plenum system, there is little resistance to air flow. Air resistance is mainly the result of the air filter and the delivery openings

FIGURE 11.3-38
Recommended Furnace Output and Blower Capacities*

Heating Unit Output, Btu	Blower Capacity, cfm
20,000	475
30,000	700
40,000	925
50,000	1175
60,000	1400
70,000	1625
80,000	1850
90,000	2100
100,000	2325

*To obtain 40°F temperature rise through the furnace.

from the plenum into the living or working areas.

The furnace should be located as close to the center of the building as possible. With a central location, air distribution is better and return-air ducting is simpler.

There should be 30 in. of service clearance in front of a furnace or heating unit. The unit should be adequately supported and blocked to prevent vibration. There should be a positive, airtight seal around both the return-air duct and the furnace base. For fire safety there should be a noncombustible base, usually made of sheet metal. An air deflector pad made of such noncombustible materials as thin concrete or gypsum board should be placed directly under the furnace (Fig. 11.3-39).

If air conditioning is included in the system, no furnace base may be needed since the cooling coil housing may serve as the base. There should be, however, provision for collecting condensation from the cooling coil.

If cooling is not provided, the furnace blower should be turned on occasionally

FIGURE 11.3-39 *A noncombustible base and air deflector prevent excessive heat buildup and fires.* *(Courtesy American Plywood Association.)*

in the summer to evacuate humid air from the plenum. Running the blower in hot weather, especially at night, contributes to the occupants' comfort.

The blower should be wired to allow continuous operation. A continuous blower sound may be less objectionable than intermittent noise. A two-speed blower can provide a continuous, low air volume in addition to a higher air volume when heating or cooling equipment is in operation. Blower operation may also be tied in with a separate thermostat.

RETURN-AIR SYSTEM   The furnace heats air and the blower distributes it through the plenum and supply outlets. The return-air system completes the convection loop by returning air to the furnace where it can be heated and recycled (Fig. 11.3-40).

FIGURE 11.3-40 *Return air grille(s) may be provided in furnace room walls, or ducts may be run from each room through the attic to the furnace.*

The return-air system should be sized to equal or exceed the supply volume and should induce the return of air from each conditioned space. A properly designed return-air system helps maintain an even temperature level with minimal stratification throughout the dwelling.

Nonducted air returns allow free passage of return air directly to a furnace, usually through a louvered furnace-room door. Bedroom doors should be undercut to have at least 1 in. of clearance from floor tile or carpet to ensure proper air circulation.

Ducted multiple return-air systems are used in larger homes with complex plans. The air is returned from grilles in each

FIGURE 11.3-41 *A dropped ceiling above the furnace supplies conditioned air through the joist space between floors; the plenum under the first floor returns the air to the furnace.*

room, located preferably near the ceiling, through ducts located in the ceiling construction or attic (see Fig. 11.3-41). Smaller rooms such as a kitchen or bath do not need return-air grilles if other means are provided to allow air to return from them to the system.

Combustion-air and fresh-air intakes as discussed in Section 11.3.3.3 may also be necessary in buildings with air plenum heating and cooling systems. The same principles apply.

A crawl-space cavity can also be used as a return-air plenum rather than as a supply plenum when desired. If it is used as a return air plenum, an upflow air furnace with overhead ductwork is necessary. Return-air grilles near the outside walls

can be used as the means of routing air to the plenum.

In single-story construction an overhead main supply duct can be used; the supply duct can be either dropped below the ceiling or located in the attic. In a two-story dwelling, if the underfloor area of the first floor is used as a return air plenum, the framing space between stories can serve as a supply plenum (Fig. 11.3-41). A centrally dropped ceiling can provide conditioned air to the space between joists.

SUPPLY OUTLETS   Either conventional adjustable floor diffusers, baseboard registers, or a continuous perimeter slot can be used to supply heated and cooled air to living areas (Fig. 11.3-42).

Each *register* should have an adjustable damper to properly balance the air flow. If too much air enters a room, the registers can be partially closed. If there is not enough air, it is easy to replace a small register with a larger one or to add a register.

As a general housekeeping measure, sheet metal pans should be installed beneath each register opening to catch condensation that may form on the supply outlet and drip into the plenum (Fig. 11.3-43). These may also be required by some codes, especially where sand or gravel is not used over the vapor retarder.

In general, register size and location are basically the same as with conventional forced-air systems, with a few additional requirements as outline below.

FIGURE 11.3-42   *(a) A conventional floor diffuser; (b) a perimeter slot.*

FIGURE 11.3-43 *Sheet metal pans below floor registers.* (*Courtesy American Plywood Association.*)

1. Heat loss or gain should be calculated for each room and the entire building. Conventional heat-loss calculations can be used for living and working areas; a different method of calculating heat loss is used for a plenum system than for conventional ventilated crawl-space construction (Fig. 11.3-44).

2. A heating unit with an output capacity equal to 1.15 of the total calculated heat loss should be selected.

3. Generally, at least one register should be located on each exposed exterior wall, except in bedrooms with heat loss less than 7000 Btu/hr. If possible, registers should be located in front of windows (see Fig. 11.3-25). Small rooms near the furnace and interior rooms with no outside wall often need no registers. In cases where kitchen cabinets are on the outside wall, holes can be cut in the floor under the cabinet and conditioned air delivered through a toe-space grille.

4. Register outlet capacity should closely match the air volume of the furnace on a proportional basis. For example, if heat loss for a bedroom is 8400 Btu/ hr. and total dwelling heat loss is 56,000 Btu/hr., then that bedroom should receive about 15% of the air volume. If the blower size is 1600 cfm, 15% of that is 240 cfm. If two registers are used, each should deliver 120 cfm. Because the floor acts as a radiant heat source, rooms nearer the mechanical unit will receive a higher proportion of the radiant heat. Registers in those rooms may be slightly undersized and registers in the rooms farthest from the heat source can be slightly oversized. This method of determining register size ignores plenum pressurization and will probably result in slightly oversized registers that can be partially closed if necessary.

Because conditioned air is fed to the registers directly from the plenum, duct length and size considerations can be ignored.

A low plenum pressure (not more than 0.1 in. of water) is desirable to maintain more comfortable velocities through the supply registers and to minimize perimeter air loss. Figure 11.3-45 shows a method for determining register sizes if plenum pressure is not more than 0.1 in. of water.

Further information on calculating heat loss, and the sizing and locating of registers may be obtained from the National Association of Home Builders (NAHB) Research Foundation.

PERIMETER SLOT The slot method of supplying conditioned air is unobtrusive and avoids the problem of obstructing air flow from a grille or register with furniture. Its primary disadvantage is that it requires some unconventional framing and trimming techniques. It may also be a problem with floor-length drapes because the heat will be trapped between the drape and the wall or window.

The perimeter slot method consists of holding floor sheathing about 1/2 in. from the inside surface of the exterior walls and then blocking the baseboard out uniformly to create an air-delivery gap into the living area (see Fig. 11.3-42b). The width of the slot is determined by heat loss calculations and usually varies between 1/8 and 3/8 in. The amount of air delivered is then fixed by the width of the slot with no further controls.

DIRECTIONAL STUB DUCTS Some local codes may require short ducts to be stubbed out from the furnace (Fig. 11.3-46), but they are seldom needed for proper air distribution. Some large dwellings with irregular shapes may find their installation helpful in ensuring adequate air distribution in the far corners of the dwelling (Fig. 11.3-47). Stub ducts often eliminate the need for a separate furnace air deflector (see Fig. 11.3-39).

11.3.3.6

Passive Solar Energy Systems

Primitive peoples designed their dwellings to take advantage of the local climate. In each region a style of building evolved that kept buildings warm in winter and cool in summer, using a minimum of energy. This was accomplished passively, that is, through the use of the sun, wind, and other natural phenomena.

In the early Sinaqua and Anasazi cliff dwellings of Arizona, known today as Montezuma's Castle, an overhanging rock brow protected southern exposures against the high summer sun and kept interiors cool. In winter, the rock brow did not shut out the low afternoon sun, allowing the rock face to warm up. Since rock is a good conductor and storer of heat, it retained the heat and kept the dwellings warm during the night.

The Pueblo Indians of New Mexico also knew how to build energy-efficient dwellings more than 1000 years ago. By facing their structures south and using high, small, deeply cut windows, they were able to delay the entry of heat into the interior of their structures until late in the day when temperatures started dropping. Exterior surfaces were painted a light color to reflect the sun's rays. They crowded their dwellings side by side and on top of one another to reduce temperature fluctuations in both summer and winter and cooked in rooms separated from living spaces to avoid unwanted heat gain in summer.

Descendants of the Pilgrims also evolved a suitable style of building for the northeastern region. In the cold winters of New England, keeping warm meant having efficient heat sources and tight buildings to retain the heat. The typical New England saltbox had warming stoves and fireplaces at the center of the house and a minimum

Every square foot of ground under a dwelling can be accounted for by tracing heat flow through circular arc paths to a point below the surface of the exterior grade (a). Heat loss is based on the average conductivity of the soil. Conductivity varies according to soil type, dampness and looseness; usually, dry loose soil is a better insulator than damped packed soil. The average conductivity of soil is 0.8 BTUs per hour per sq. ft. per foot of path length.

The heat loss factors shown below (b) account for arc path length, insulation, heights of exposed foundation wall, foundation construction, surface films and ground temperatures beneath the exterior grade surface of a 50 ft. by 28 ft. dwelling. Though arc lengths vary according to dwelling size, the values shown below may be used as a reasonable estimate for most dwellings. Under severe climatic conditions or with an extremely large dwelling these values will slightly underestimate plenum heat loss.

(a) PLENUM HEAT FLOW PATHS

Heat Path Radius, R_1
Heat Path Radius, R_2

(b) HEAT LOSS FACTORS, R11/R19 INSULATION

Dimension B	Dimension A									
	6"		12"		18"		24"		30"	
	R11	R19	R11	R19	R11	R19	R11	R19	R11	R19
6"	.33	.29	—	—	—	—	—	—	—	—
12"	.34	.29	.37	.31	—	—	—	—	—	—
18"	.35	.29	.38	.31	.41	.33	—	—	—	—
24"	.36	.29	.39	.32	.42	.34	.45	.36	—	—
30"	.37	.30	.40	.32	.43	.34	.46	.36	.49	.38
36"	.38	.30	.41	.32	.44	.34	.47	.36	.50	.39

To use (b) Heat Loss Factors:

(1) Select insulation R-value to be used.

(2) Estimate average height of exposed foundation wall and plenum depth from ground to underside of subfloor.

(3) Find appropriate heat loss factor in table by cross-matching exposed foundation in "A" column and plenum depth in "B" column.

(4) From house plan, determine lineal feet of perimeter foundation which encloses the underfloor plenum area.

(5) Determine temperature differential between outside design temperature and plenum perimeter temperature (found to average 85°F).

(6) Multiply heat loss factor times exposed lineal feet of foundation times temperature difference to obtain plenum heat loss: Heat loss factor x exposed lineal feet of foundation wall x temperature difference = Plenum Heat Loss.

(7) Add living area heat loss to plenum heat loss to obtain total estimated heat loss for entire house.

Example: Assume the following characteristics:
• 1400 sq. ft. dwelling.
• Wood foundation, with average 12" exposure to outside.
• Plenum depth (ground to subfloor) about 24".
• R-11 insulation on wall and 24" horizontally around perimeter.

• 156 lineal feet of foundation enclosing the plenum.
• Winter design temperature = 24°F.
• Temperature difference = 85°F minus 24°F = 61°F.

From (b) the heat loss factor is 0.39. Therefore, estimated plenum heat loss is:

0.39 (heat loss factor) x 156 (lineal feet of foundation) x 61°F (temperature difference) = 3,711 BTUs per hour.

Assuming heat loss calculations for the living area of 40,000 BTUs per hour, total heat loss under design conditions for the purpose of sizing equipment will be 40,000 BTUs per hour +3,711 BTUs per hour. Round off to 44,000 BTUs per hour. It is interesting to note that the plenum heat loss in this example is about the same as would have been added for duct heat loss when insulated ducts are located in a conventional ventilated crawl space. Also, over 5,000 additional BTUs per hour would have been lost in such a dwelling through the floor when insulated with R-11 over its entire area. A net energy saving of over 5,000 BTUs per hour under design conditions results by using the underfloor plenum system in this example.

Cooling load calculations for plenum houses can be made in the conventional manner, neglecting heat gains to the underfloor plenum area, unless there is significant exposed foundation wall. In which case, only heat gain through the exposed portion of the foundation wall should be calculated.

(c) WOOD FOUNDATION

A
B
R11 or R19 Insulation
24"

(d) CONCRETE OR BLOCK FOUNDATION

A
B
R-11 or R-19 Insulation
24"

FIGURE 11.3-44 *Heat loss calculations for an air plenum system.*

FIGURE 11.3-45
A Second Method of Sizing Registers

Assuming a 56,000-Btu/hr. unit with a blower capacity of 1600 cfm and a plenum pressure of 0.1 in. of water, a floor register would deliver about 9 cfm per sq. in. of floor register opening. Therefore, the total register free opening for the house would be 1600/9 = 177 sq. in.; 10 registers each having 18 sq. in. of free area could be distributed uniformly around the house.

a. Register Air Delivery[a]

Free Area of Register	Plenum Pressure, in. of Water				
	0.02	0.04	0.06	0.08	0.10
1 sq. in.	4.4	5.8	7.3	8.4	9.4

[a]In cfm per sq. in. of floor or basement register for different plenum pressures.

The same general procedure applies to the perimeter slot method of air distribution. Assuming a 1600-cfm blower and a plenum pressure of 0.1 in. of water, a 1/4 in. slot would deliver 19 cfm per lineal foot.

Therefore, total open slot length would be 1600/19 = 84 ft. If a 3/16 in. slot is used, 13 cfm is delivered, so total slot length would be 1600/13 = 123 ft.

b. Perimeter Slot Air Delivery[b]

Slot Width, in.	Plenum Pressure, in. of Water				
	0.02	0.04	0.06	0.08	0.10
1/8	3.5	4.7	5.8	6.8	7.8
3/16	5.8	8.4	10.0	12.0	13.0
1/4	8.5	12.0	15.0	17.0	19.0
3/8	14.0	20.0	24.0	28.0	31.0

[b]In cfm per sq. in. of slot for different plenum pressures.

This method differs from conventional practice. No calculations for individual rooms are needed and manufacturer's Btu/hr. ratings for registers are ignored since much of the heat is delivered by radiation from the floor surface.

If the floor surface temperature is 75 °F, the net radiation into the room of 60 °F ambient air is 7000 Btu/hr./1000 sq. ft.

Calculations of radiant heat do not need to be considered. Usually, small rooms near the furnace and interior rooms with no outside wall will need no registers.

of exterior wall surface. A north-sloping roof provided protection from icy winds and allowed snow to build up for added insulation against the cold. Most windows faced south and the exterior was painted a dark color to absorb as much solar radiation as possible.

Studies of these and other responses to local climate have laid the groundwork for a more widespread development of similar naturally adaptive heating and cooling techniques. Some architects and engineers have refined and adapted these concepts into innovative and energy-efficient buildings. Their work has resulted in buildings that more effectively collect, store, and distribute solar energy for heating, and

that better utilize natural climatic factors for cooling. These naturally adaptive *passive* heating and cooling systems still require external energy for their operation, but need less of it than do conventional buildings. Solar heating or cooling installations with sophisticated heat collection, transport, control, and storage methods, known as *active* solar systems, are beyond the scope of this book. This section discusses the basics of passive heating and cooling systems.

■ PASSIVE SOLAR PROPERTIES

Buildings are exposed to the warming rays of the sun and the cooling influences of

natural breezes. They can be designed and oriented to take advantage of the local climate for heating and cooling, with reduced reliance on mechanical systems. With such an approach, windows, skylights, walls, roofs, floors, and the natural features of the site regulate heat flow into or out of a building. Conduction, convection, radiation, and gravity provide the means to move the heat through a building.

Passive solar systems are distinguished from active systems in that they use no power-consuming pumps, and because the collection, storage, and distribution facilities are integrated with the building design. In active systems, the collection and storage elements are manufactured products that can be placed in, on, or near the building and tied to it with pipes or ducts. In general, all effective passive solar buildings share the following basic qualities: (1) they have an envelope that will not leak heat in winter or gain excessive heat in summer; (2) they have a solar collection area, in the form of windows, skylights, glass walls, water bags, or other solar heat traps; and (3) they are capable of storing heat for later use, either in the structural elements or in specially provided facilities.

■ BUILDING ENVELOPE

A passively heated building can collect and store only a limited amount of heat. In all but a few sunny regions, more heat is required for modern comfort than a passive system can deliver. The extra heat is generally supplied by a supplementary source, such as wood stoves, fireplaces, or a conventional heating system. Refer to Sections 7.3, 8.1, and 8.8 for information about the design and use of insulation, doors, windows, glazing, and window treatments, to minimize overall heat loss. Section 11.3 discusses the basic principles of heat control in buildings.

An energy-efficient building should be insulated to optimum levels (Section 7.3), have tightly caulked and sealed windows, door frames, and sill plates (Section 7.10), have double or triple glazing (Section 8.8) and as few nonsolar-collecting windows as possible.

■ SOLAR COLLECTION

Solar energy reaches the earth as *direct* (parallel) or *diffuse* (nonparallel) rays of

FIGURE 11.3-46 *Stub ducts installed under a furnace help direct air to various parts of the dwelling.* (Courtesy American Plywood Association.)

FIGURE 11.3-47 *Directional stub ducts in a nonrectangular design.* (Courtesy American Plywood Association.)

the sun. Direct radiation is sunlight that reaches the earth relatively unobstructed. Diffuse radiation is reflected from clouds, dust, and obstructions such as buildings. The relative proportion of direct and diffuse radiation, as well as the total radiation, varies from region to region, making a particular solar energy system more feasible in some areas than in others.

Solar energy that strikes a surface is called *insolation* (incident solar radiation). It is measured in *Langleys* (ly). The amount of energy contained in 1 ly is equal to 1 calorie/cm², or 3.69 Btu/sq. ft. of surface.

In designing a building to take advantage of passive solar heating, building shape, orientation, and solar collection area should be given consideration.

SHAPE AND ORIENTATION   A building should be shaped and sited for maximum exposure to sunshine in winter and mini-

mum exposure in summer. In northern latitudes, buildings should expose as much surface to the south as possible. In Pennsylvania, for example, which is at 40°N latitude, a building receives nearly three times as much solar radiation on its south face in January as it does in June; due south is therefore the optimum orientation for a collector surface. The east and west sides, on the other hand, receive two and one-half times more insolation in summer than in winter, making them inappropriate for heat collection in winter. At latitudes below 35°N, buildings are exposed to even more sun on the south side in winter, whereas the east and west walls receive two to three times more sun in summer than the south walls.

A square building is never the optimum form for solar collection at any latitude, and should be avoided unless building use and site conditions dictate such a shape. Buildings that are elongated on the north-

south axis are even less efficient in winter and summer. The optimum shape is usually a form elongated along the east-west direction.

COLLECTING SOLAR ENERGY   Solar energy reaches a building in one of two forms, visible light, which is shortwave radiation and infrared light, which is long-wave radiation. Some shortwave radiation is reflected by glass, but much passes through. Long-wave radiation, on the other hand, is either reflected or absorbed and conducted or radiated away. After passing through glass, shortwave radiation strikes an interior surface and is transformed into an equivalent amount of heat energy. Part of this heat energy is conducted and convected away; the remainder is radiated in the form of infrared (long-wave) radiation, sometimes called thermal radiation. Since thermal radiation cannot penetrate glass, it becomes trapped inside the space. This phenomenon is called the *greenhouse effect*. It accounts for a car parked in the sun on a cool day becoming hot and stuffy inside and for an ordinary window's ability to act as a solar collector.

The type of glass used for solar collection can have a significant effect on the efficiency of the system. Clear, heat-absorbing, and -reflecting glass all lose about the same amount of heat by conduction, but differ widely in the amount of heat transmitted and trapped under the glass.

Clear single glass is the most effective heat gainer but loses the most by conduction. Clear double glazing has a comparable percentage of gain, and loses less by conduction.

When preventing summer heat gain is important, heat-absorbing or -reflecting glass should be considered for south exposures, but only when external shading or screening is not practical. Because reflective glass reduces heat gain year-round, it is desirable only where cooling energy costs are substantially higher than heating costs.

Although skylights and windows are the primary means of passive heat collection, they also are responsible for the greatest winter heat loss and summer heat gain. Passive systems usually employ an insulating or shading technique to prevent unwanted heat loss or gain through the glass collector. Some more traditional approaches to reducing excessive window heat loss and heat gain are described in Sections 7.3, 8.1, 8.8, and 11.3.

■ STORAGE

The more glass that is exposed to the sun, the more heat that will be trapped behind it. Windows used as passive solar collectors have the disadvantage that, even on very cold days, they can quickly overheat the space they serve. In order to soak up the excess collected heat and prevent overheating, and to retain heat for nighttime use, floor and wall surfaces can be utilized as *heat sinks*; that is, they collect and store heat.

Materials such as water, concrete, brick, tile, and stone are ideal heat sinks because they can absorb and store large quantities of heat in a relatively small volume of material. The collected heat is released by radiation to colder surfaces and to the air to maintain a relatively constant temperature. The greater the temperature differential between the mass and its surroundings, the faster the heat is released. Also, the greater the heat storage capacity (thermal mass) of the materials in the building, the longer it will take for room air to overheat and the longer a room can remain comfortable without receiving heat from an auxiliary source.

A heat storage material should be located in or near direct sun. When it is, a thermal storage capacity equivalent to 30 lb. of water or 150 lb. of masonry should be used for every square foot of south-facing glass. The thermal mass of heat storage materials in other locations should be increased four times.

SUMMER COOLING On summer nights, when the outside air is at its lowest temperature, introduction of that air into a house will carry excess heat away from interior floor and wall surfaces. In the process, those surfaces will be cooled and will therefore be able to absorb and store excess heat trapped in the building the following day. This reduces summer heat buildup within the building and aids cooling.

HEAT STORAGE CAPACITY Materials vary in their capacity to store heat. The two most common measures of that ability are specific heat and heat capacity.

A material's *specific heat* is the number of Btu's required to raise 1 lb. of it 1°F. For example, water has a specific heat of 1.0, which means that 1 Btu is required to raise the temperature of 1 lb. of water 1°F. Since 1 gal. of water weighs 8.4 lb., 8.4 Btu's are required for a 1°F rise in temperature.

Various materials absorb different amounts of heat while undergoing the same temperature rise. For example, it takes 100 Btu's to heat 100 lb. of water 1°F, but only 22.5 Btu's to heat the same weight of aluminum 1°F. Therefore, the specific heat of aluminum is 0.225, whereas water has a specific heat almost four times as high.

A material's *heat capacity* is the amount of heat needed to raise 1 cu. ft. of it 1°F. Dense solid materials are effective as heat storage devices because their high density means that considerable weight is contained in each cubic foot of volume. Such materials are therefore often used as storage devices even though water has a greater specific heat. Concrete, for example, has a specific heat about one quarter that of water, but because of its density, it has a heat capacity of more than half that of water.

■ PASSIVE SYSTEMS

Passive solar heating or passive cooling systems are the result of adaptive design, not of the use of mechanical devices. They cannot be grouped according to engineering specifications or performance standards but must be categorized according to the design approach they utilize. This, in turn, makes generic discussion of passive design difficult because most designs are the products of their inventor's ingenuity, not of established standards. What has emerged however, are three distinct design approaches within which most passive systems can be grouped: *direct gain*, *indirect gain*, and *isolated gain*.

DIRECT GAIN SYSTEMS The direct gain concept is the most common passive solar building solution and has many historical precedents, including the cliff dwellings of the Sinaqua and Anasazi of Arizona. In simple terms the process can be described as sun to living space to thermal mass. Solar radiation is collected in the living space and then stored in a material

having a large thermal mass, such as thick masonry or concrete floors or walls. Thus, the living space is heated by the solar radiation penetrating the windows and it serves as a live-in collector, storage, and distribution facility.

The basic requirements for a building with a direct gain passive solar energy system include (1) a large, south-facing glazed collector area with the living space exposed directly behind; (2) a large floor and wall thermal mass exposed to solar radiation; and (3) a method for protecting the collector windows and storage mass from heat loss to the exterior.

■ INDIRECT GAIN SYSTEMS

Indirect gain passive solar systems differ from direct gain systems in that they deposit collected heat into a thermal storage mass before it is introduced into a living space. The storage mass usually is located between the glazed collector and the living space so that when the living area begins to cool, heat radiates from the mass into the living space.

The principal advantage of such an approach is that collected heat does not travel through the living space to reach the storage mass. As a result, more heat can be collected and stored before the space becomes too warm for occupant comfort. In addition, radiant heat flow into living spaces can be controlled more readily in such systems than in direct gain systems. Placing movable insulation on the interior face is one typical regulatory technique. When the space begins to overheat it can be moved across the wall to reduce heat flow.

In general, there are three basic types of indirect heat gain systems: the *mass Trombe wall*, the *water Trombe wall*, and the *roof pond*.

MASS TROMBE WALL This type of passive system, named after its French inventor, Felix Trombe, consists of a glass or plastic panel over a heat-absorbing storage wall. Commonly used materials for the heat storage wall include concrete, brick, stone, and adobe. When possible, the material is painted a dark color to increase its ability to absorb heat. The glazing prevents immediate reradiation of the collected heat to the exterior.

The system operates by permitting solar radiation to pass through the glazed collector and be transmitted directly into the massive storage wall. When the temperature of the wall rises higher than that of adjoining interior surfaces, it begins to radiate its heat to the interior.

Heated air in the space between the glazing and the mass Trombe wall can also be introduced into the living space, and controllable vents in the mass itself can regulate heat supply to the living space and draw cooler air into the mass for heating.

WATER TROMBE WALL   The Trombe wall concept of solar collection and storage can be adapted to utilize water as the storage mass. Water is an effective storage medium because it has a higher specific heat than most solid massive materials and can therefore store more heat per cubic foot.

The principal disadvantage of water Trombe walls is that containing a liquid is more difficult than containing a solid such as concrete. Steel drums are the most common containers, but some designers have experimented with concrete block walls in which the voids are filled with vinyl bags containing water. These containers can be left exposed to the interior or embedded in some massive material, such as adobe or concrete.

Convection currents within the water rapidly transfer the heat from the collection surface to the entire volume of water. Heat is therefore delivered to the interior spaces at a faster rate than with mass Trombe walls. In addition, the collector surface operates at a lower temperature so that less heat is lost back through the glazed collector before it is absorbed.

ROOF POND   A variation of the Trombe wall utilizes horizontal water containers on the roof of a building, to collect and store the sun's energy. Heat is collected by exposing the ponds to the sun by day, and covering them with insulation at night to prevent the collected heat from escaping into the cool night air. The solar heat collected in roof ponds is radiated directly to the rooms through a metal ceiling. Thus, a continuous, evenly distributed supply of heat is provided throughout the building. Such a system is capable of delivering heat when the water temperature drops as low as 70 °F.

There are several disadvantages of roof pond collectors. They require a stronger-than-normal structure to support their weight, which increases the cost of the building. The water must be contained in plastic bags or some similar structure. Open ponds have fallen into disfavor for several reasons. They tend to contribute to roof deterioration and leaks, they make roof maintenance more difficult, and they serve as havens for insects and bacteria.

Even contained ponds have disadvantages. Heat from the low-angled winter sun is difficult to collect on a horizontal surface. On a clear December day in Pennsylvania, for example, the sunlight striking a horizontal roof is less than half that hitting an equal area of a south-facing wall. To compensate for this deficiency, more pond area must be provided. Stratification of the heat within a roof pond is another problem. The warmer, lighter water remains near the top of the pond at night, losing heat to the outside air, while the cooler, denser water falls to the bottom, just above the rooms, where heat is needed the most. An additional problem is that in many climates, the roof must be protected against snow accumulation, which could block the sun from the water bags.

In general, roof pond collectors are better suited to the warmer southern and southwestern states, where the sun climbs higher in the sky on a winter day, and snow buildup is not a threat. Roof ponds are also well suited for summer cooling, which is of primary importance in such climates.

Insulated panels should be used to cover a roof pond on summer days to keep the sun from warming the ponds, while the cool water absorbs excess heat from the rooms below. At night the panels must be removed so that the roof pond can radiate heat to the night air. Heat stratification within a roof pond aids this cooling process because the coldest water lies just above the rooms.

■ ISOLATED GAIN

Isolated gain passive systems utilize collectors and storage masses that are thermally isolated from the living space. The concept therefore allows collector and storage to function independently while the building draws heat as required.

Isolated gain systems can generally be grouped into two categories: those that use a greenhouse area, or similar structure (often called a *sunspace*) to collect and store heat, and those that utilize the *thermosiphon* principle to move collected heat.

SUNSPACE   Using a sunspace allows maximum solar collection in this space (and consequent overheating) without affecting the comfort within the primary living space. The collected heat can be admitted as needed through doors, windows, or dampers.

Atriums, sunporches, greenhouses, and sunrooms are examples of sunspaces. In general, this technique requires a glazed collector space that must be attached to and yet distinct from a living space, it must have the proper southern exposure, and the collector space must contain or be thermally linked to a solar storage mass for heat retention and later distribution.

THERMOSIPHON   The second type of isolated gain system permits heat to build up within a roof, wall, or room-sized collection area before drawing it off to storage or living areas. Thermosiphon (convection), the natural rise of heated air or fluid, is the usual means of transport.

In the typical installation, a glazed collector is located between the direct sun and the living space, and is distinct from the building structure. In this space, a convection heat flow occurs when air heated by the sun rises naturally into an appropriately placed living space or storage mass, causing cooler air or liquid to fall again. A continuous heat-gathering circulation is thereby established.

11.4 FIRE SPRINKLER SYSTEMS

Sprinkler systems are expensive, but many insurance companies offset some of the costs by offering reductions in premiums for professional installations. In addition, some jurisdictions require sprinklers in most or all commercial and institutional buildings, and some require them in multifamily and even in single-family houses. Some codes permit taller buildings or increased floor areas when sprinklers are used. Installing a sprinkler system throughout an entire existing building can be prohibitively expensive and is seldom justified unless required by law. A sprinkler system should be considered, however, in every new building, even when not required by code.

The primary purpose of a sprinkler system is to save property and minimize structural damage. There is controversy regarding the life-saving benefits of sprinkler systems compared with smoke detectors (see Section 12.7), especially in small buildings. The question is whether the benefits outweigh the higher cost of installation and maintenance of sprinkler systems relative to smoke detectors.

The National Fire Protection Association Standard NFPA 13D, "Sprinkler Systems in One- and Two-Family Dwellings and Mobile Homes," outlines industry standards for the installation of automatic sprinkler systems in such buildings. NFPA 13R, "Sprinkler Systems in Residential Occupancies up to Four Stories in Height," covers these types of buildings. Some municipalities have adopted less stringent standards, but require sprinkler systems in all new construction.

Sprinkler systems may be of wet-pipe or dry-pipe systems. Sprinkler systems generally consist of (1) water pipes running at right angles through or under ceiling joists and (2) sprinkler heads spaced along the pipes to serve as nozzles.

A dry-pipe system's pipes contain air. The loss of air pressure from the opening of a sprinkler head or the detection of a fire condition causes the release of water into the piping system and out of the opened nozzle. A dry-pipe system should be used in areas subject to freezing.

A wet-pipe system's pipes contain water. This system should be used in areas not subject to freezing because it tends to deliver water to the fire a little faster than dry-pipe systems.

In case of a fire, a fusible link in a sprinkler head melts, allowing the struts in the center of the head to fall out and release a plug. Emerging water hits a detector, creating a wheel-shaped spray pattern. Sprinkler systems are effective against class A fires (those that can be controlled with water).

LIVING FLOOR
First and second levels

Dining room

Range Dishwasher
Sink

Ref.

Linen closet

Living room

Bedroom

Bedroom

Bedroom

Bedroom

WP

CH

TV

TV

WP

BASEMENT

Service entrance

Service panel

Washer   Dryer

Furnace & AC

Utility room

Food freezer

Recreation room

Garage

CW   CD

F

FF

TV

WP

12

ELECTRICAL SYSTEMS

INTRODUCTION

FUNDAMENTALS OF
ELECTRICITY

ELECTRIC POWER
DISTRIBUTION

SERVICE AND
DISTRIBUTION

LIGHTING

LIGHTNING PROTECTION

INTRUSION PREVENTION
AND DETECTION

FIRE SAFETY

COMMUNICATIONS
SYSTEMS

CONSERVATION OF
ENERGY

Photo: Floor plan showing electrical light switches and outlets. (Drawing by HLS.)

INTRODUCTION

Even when an adequate amount of electricity is made available by the power company, the extent to which it can be effectively used in a building is determined by the interior wiring. Starting at the service entrance, where electricity enters a building, a wiring system must deliver the correct amount of electricity to every outlet. In a properly planned system most circuits and outlets are general-purpose ones designed to serve most portable devices, whereas some are designed to serve a specific purpose. Switch controls are located with both convenience and safety in mind.

Electrical appliances are constantly being improved to provide higher standards of performance and convenience. The resulting increases in wattage are reflected in greater demands on building wiring systems.

The number of appliances and other electrical devices is increasing. Work-saving and time-saving devices such as computers proliferate in offices, shops, factories, and institutional buildings. Appliances, such as automatic dishwashers, electric clothes dryers, food waste disposers, and freezers, once considered luxuries, appear in most new residences. Almost every new commercial and institutional building and most new houses are air-conditioned. Even though electric heating systems, a mainstay in some parts of the country for single-family houses when electrical rates were lower, are not so common today, they have often been replaced by electric heat pumps, which use somewhat less energy than do electrical heating systems, but nevertheless still use considerable energy.

Such conveniences place large demands on a building's electrical service and wiring system. When loads outstrip the capacities of the wiring system, operating efficiency suffers. When voltage drop becomes excessive in a circuit, poor lighting and inefficient appliance operation result. A 5% voltage loss, for example, produces a 10% loss of heat in an appliance and a 17% loss of light from an incandescent lamp. However, the largest problems presented by overloading electrical systems are related to safety. Overloaded circuits and undersized circuit breakers can possibly lead to damaged equipment, electric shock, or fires.

Average consumption of electrical power has increased tenfold in the past 50 years, but recent energy consciousness and an energy-saving movement on the part of the appliance industry, in response to federal action, has tended to abate that increase somewhat. Despite increased electrical generation by nuclear plants, hydroelectric stations, and other nonfossil means, most of our electrical production today is generated using coal and gas as fuel. Smaller amounts are also generated using oil as a fuel. Spiraling electrical production reduces our nonrenewable fossil reserves and results in large expenditures to prevent their burning from producing severe adverse environmental damage. The current national commitment to energy conservation and pollution abatement places the responsibility on a designer to make home electrical systems not only convenient and efficient, but also not wasteful of precious energy.

This chapter includes a primer on the fundamentals of electricity (Section 12.1) and on electricity distribution and delivery (Section 12.2). It also discusses various systems and equipment used in electrical systems in buildings and includes recommendations for planning adequate electrical power and lighting systems (Sections 12.3 and 12.4). Other related subjects covered are lightning protection (Section 12.5), security systems such as intruder detection and fire alarm systems (Sections 12.6 and 12.7), communications systems (Section 12.8), and conservation of energy (Section 12.9).

12.1 FUNDAMENTALS OF ELECTRICITY

The generally accepted notion of electricity is based on the theoretical structure of an atom. An atom can be visualized as a nucleus orbited by one or more electrons (Fig. 12.1-1). Groups of electrons moving in neighboring orbits can be thought of as belonging to the same orbital shell. Depending on the number of electrons, there can be as many as seven such shells.

The nucleus consists of one or more positively charged protons and one or more electrically neutral particles called neutrons. Each orbital electron has a negative charge, equal in magnitude and opposite to the positive charge of a nuclear proton.

In a neutral atom, the negative and positive charges balance out. But electrons in the outer shell can be dislodged and become free electrons, separated from the parent atom, which is then left with a net positive charge. A free electron may then attach itself to another neutral atom, lodging on its outer shell and creating a negative charge on the atom (Fig. 12.1-2). Single atoms or clusters of atoms with extra or fewer electrons are called ions.

A chemical battery, the simplest source of electricity, is made up of one or more cells. Most chemical batteries deliver small amounts of electricity at a low voltage. A cell usually consists of a carbon rod (positive electrode), a piece of copper

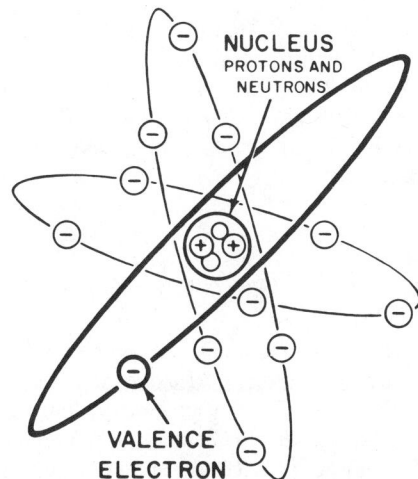

FIGURE 12.1-1 *Valence electrons, moving in orbits that are relatively far from the nucleus of an atom, have the least attraction for it and can be dislodged to create electron flow, which is electricity.*

FIGURE 12.1-2 *Electrons will flow through a conductive material that connects an area with a surplus of electrons to an area that has a shortage of electrons, such as the two terminals of a battery.*

or zinc (negative electrode) and a chemical compound (electrolyte). The electrolyte breaks down readily into positively and negatively charged particles (ions) that attach themselves to the electrodes, imparting a negative or positive charge. Connecting the two electrodes with a wire allows the electrons collected on the negative electrode to travel to the positive electrode in a balancing effort. The movement of these electrons through a wire is defined as the flow of electricity, or *current*. This current-generating action can be illustrated by a voltaic wet cell, which uses a solution of sulfuric acid as the electrolyte (Fig. 12.1-3). In a dry cell, such as a flashlight battery, the zinc casing is used as the negative electrode, while a special paste acts as the electrolyte between the casing and the carbon rod (positive electrode).

Electricity can also be generated directly from solar radiation by means of silicon photovoltaic cells. This is the technology developed by NASA to provide power for satellites and other orbiting and deep space devices. These cells generate low-voltage, direct current power that can then be converted into 120-volt alternating current using a dc-ac inverter. This type of cell may well be used in the future to generate power in large quantities, but it is an expensive way to generate electricity and so far must be considered as experimental, even though a few automobiles

FIGURE 12.1-3 *The electrolytic solution (H_2SO_4) in a wet cell breaks down into positive (H_2) and negative (SO_4) ions. Although the SO_4 ions carry electrons to the zinc (Zn) electrode, they also react with it, changing it into $ZnSO_4$.*

and some residential and commercial building projects have made limited use of this technology.

For larger amounts of high-voltage power, an electric generator is the device used today. Unlike batteries, which convert chemical or solar energy into electric power, a generator utilizes kinetic (motion) energy. This kinetic energy is usually the result of a steam- or water-driven turbine. Steam is created by burning coal, gas, or oil, or from the heat of a nuclear reactor. Water is caused to drop from a reser-

voir through a turbine. The turbine then turns a generator armature that has loops of wires wound around it. As the armature turns between the poles of a magnet, the loops cut across the invisible lines of magnetic force between the poles. This movement causes current to flow through the wire loops of the armature (Fig. 12.1-4).

12.1.1

CURRENT FLOW

An electric current is a migration of electrons through a conductor from a negative terminal to a positive one. The process can be visualized as follows: a moving electron lodges on the outer shell of an atom dislodging a second electron that then migrates to an adjacent atom with the same result. The action is repeated over and over again in a chain reaction (see Fig. 12.1-2).

To keep electrons flowing, there must be a continuous uninterrupted circuit (path) from a negative to a positive terminal. If the path is interrupted, electron flow stops, resulting in an open circuit. When a shortcut is inadvertently created, permitting a sudden surge of electrons through an easier path, we call that a short circuit.

12.1.1.1

Voltage

Current is caused to flow by an electromotive force called voltage, which is also called potential difference. Voltage is the driving force behind electron flow,

CONDUCTORS CUTTING DIRECTLY ACROSS FIELD, E.M.F IS MAXIMUM

CONDUCTORS CUTTING DIRECTLY ACROSS FIELD, E.M.F IS MAXIMUM

INDUCED EMF

I REVOLUTION

FIGURE 12.1-4 *Voltage, or electromotive force (EMF), in a loop of a dc generator armature is highest when that loop moves across the lines of force between the poles of a magnet. Using more coils smoothes out the peaks.*

somewhat like the pressure in a water pipe or an air duct.

Particles with like charges, such as two electrons, repel each other; unlike particles, such as a positive ion and a negative electron, attract each other. These two factors contribute to the force of voltage. The mutual repulsion of electrons tends to make them rush away from each other; the mutual attraction between unlike charges creates a kind of pull that draws electrons toward positively charged atoms.

As stated earlier, electrons actually flow from a negatively charged to a positively charged body. However, there is an outdated notion that electricity flows from a positive to a negative terminal that still appears in textbooks and electrical diagrams as a convention, but that is a kind of accepted fiction or customary code. If electricity is assumed to flow from a positive to a negative terminal, then current must be composed of positively charged particles. This concept, though contrary to established electron flow theory, is useful in certain calculations.

12.1.1.2

Amperage

The rate at which electrons, or current, flows through a conductor is measured in amperes. The speed with which electricity is transmitted is about the same as the

speed of light, 186,000 miles per second (mi./sec.). That is, if a current is started at one end of a line 186,000 miles long, it will be felt at the other end 1 second later. Rate of flow is different from speed because it takes into account *current density*, that is, the number of electrons moving along from atom to atom. Thus, amperage measures the quantity of electricity passing through a circuit per unit of time.

The quantity of electricity without reference to time is measured in coulombs. A coulomb is roughly the number of electrons conducted past a point in 1 second. The number is an unimaginable 6,250,000,000,000,000,000 electrons. One ampere equals 1 coulomb of electricity passing a point in the circuit in every second.

12.1.1.3

Resistance

Some metals become *superconductive* at temperatures near absolute zero (−459.6 °F). Superconductivity is a complete disappearance of electrical resistance. Recently, researchers have developed materials that become superconductive at much higher temperatures. But at normaluse temperatures, the internal structure of the best conductors available resists the flow of electric current and converts some

of it into heat. This internal frictionlike effect, which is called resistance, is measured in ohms. An ohm is that resistance that allows 1 ampere (1 coulomb/sec.) to flow, when pushed by a pressure of 1 volt.

12.1.1.4

Power

The electrical unit of power (energy rate) is the watt. In that it measures the rate at which work is done, or energy is used, a watt is similar to British thermal units (BTUs) per hour (Btuh) and horsepower (hp) and can be converted directly into either one. For instance, 1 watt = 3.413 Btuh, 1 hp = 746 watts (Fig. 12.1-5).

MOTOR

746 + (?) WATTS

I HORSE-POWER

33,000 LBS.

I FOOT—I MINUTE

HEATER

I WATT

3.413 Btu. per Hour

FIGURE 12.1-5 *Electric energy can be converted into kinetic (motion) or thermal (heat) energy.*

An electric motor rated at 1 hp is capable of doing mechanical work at a rate of 33,000 foot-pounds/min. This means that the motor could lift a 1000-lb. weight 33 ft. every minute. If it were 100% efficient, the motor would then be converting electrical energy into mechanical work at a rate of 746 watts. In electric heating, electricity is converted into heat rather than mechanical work (see Fig. 12.1-5).

Stating the relationship in electrical terms, one gets the basic equation for electrical power, $W = EI$, where wattage (W) equals voltage (E) times amperage (I). The equation says, in effect, the heavier the flow rate (amperes) at a given supply pressure (volts) the higher the rate (watts) at which energy is being supplied and used.

An electric utility, for example, might use this equation to find out the load demand on distribution lines in watts or kilowatts (1000 watts). Or, a designer may wish to determine the wattage of an electric heater whose rating is 22 amperes. Multiplying 22 amperes times 240 volts gives the answer, 5280 watts (about 5.3 kilowatts).

12.1.2

OHM'S LAW

In an active circuit, voltage, amperage, and resistance are interrelated. The greater the resistance, the greater the voltage required to produce a given amperage. Current and resistance together, then, imply the required voltage pressure, and that is stated in Ohm's Law: $E = IR$, where voltage (E) equals amperage (I) times resistance (R). Ohm's Law makes it possible to determine any one of these values, when the other two are known (Fig. 12.1-6).

By combining the power equation with Ohm's equation, it is possible to derive many equations for the relationship between voltage, wattage, amperage, and resistance (Fig. 12.1-7). The power equation as stated earlier is $W = EI$; substituting IR for E (Ohm's equation), we get $W = (IR)I = I^2R$. In other words, wattage (W) equals amperage (I) squared times resistance (R). This equation helps determine the power used in an electric heater and the power lost in transmitting electricity. In a resistance heater, the power is transformed into useful heat, but in distribution wires it is an undesirable waste of energy.

The usefulness of these equations can be demonstrated by the following example. A 100-ft.-long, No. 14 copper wire has a resistance of about 0.25 ohm. Maintaining a current of 10 amperes at 120 volts, what would be the line loss in the conductors? The line loss (see Fig. 12.1-7) is obtained from the power used or loss equation mentioned in the previous paragraph, $W = I^2R$, or $(10)^2$ amperes $\times$ 0.25 ohms = 25 watts. The total power would be $EI = W$, or 120 volts times 10 amperes equals 1200 watts. The 25-watt line loss would constitute a little over 2% of the total power, which is well within the recommended maximum line loss of 3%.

12.1.3

Alternating and Direct Current

So far it has been convenient to discuss electrical principles in terms of direct current rather than the more common commercial form, alternating current. Direct current (dc) always flows in one direction in a circuit, namely, from the negative to the positive terminals of the source. Alternating current (ac) also flows from negative to positive terminals but the terminals change polarity (whether positive or negative) quickly, causing a reciprocating (back-and-forth) surge of electricity along the wire.

Alternating current is produced by an ac generator (Fig. 12.1-8) designed so that the voltage and current reverse direction periodically. In the commonly used 60-cycle power supply, 1 cycle of a complete rise and fall of voltage in each direction takes place in 1/60 sec. The voltage of 120-volt direct current is actually 120 volts at the generator and should not drop below 115 volts at the load. However, because alternating current consists of rapidly changing values from 0 to 170 volts, the average or useful voltage appears as 120 volts to the power-consuming device (Fig. 12.1-9). Thus, the effect of both types of current is the same and the above equations can be used for both ac and dc calculations.

Alternating current is now used almost exclusively in residential and commercial wiring because it provides greater flexibility in voltage selection and greater simplicity of equipment design. When alternating current is used, the voltage can be stepped up (increased) or stepped down (decreased) by a transformer (Fig. 12.1-10). An alternating current transformer is a very efficient device, wasting only 1% to 4% of the power in heat. By comparison, an electric generator, which is the

FIGURE 12.1-6 *Ohm's Law gives the relationship between voltage (E), amperage (I), and resistance (R.)*

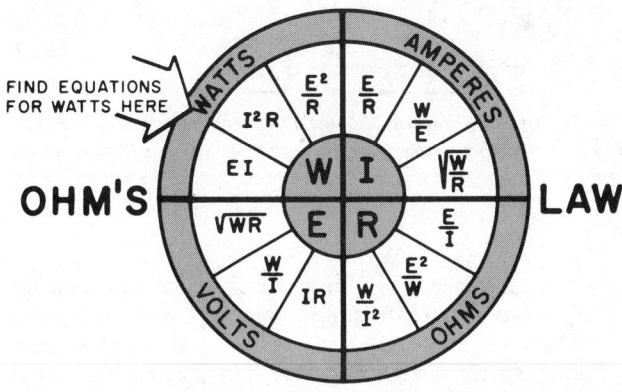

FIGURE 12.1-7 *Ohm's Law and derivative formulas make it possible to find any of the values in the middle of the wheel when any other two are known.*

CONDUCTORS PARALLEL TO LINES OF FORCE, EMF MINIMUM

ROTATION

DIRECTION OF CURRENT FLOW

CONDUCTORS CUTTING DIRECTLY ACROSS FIELD, E.M.F IS MAXIMUM POSITIVE

CONDUCTORS CUTTING DIRECTLY ACROSS FIELD, E.M.F IS MAXIMUM NEGATIVE

EMF

ONE ALTERNATION
(0 – 180°)

(180 – 360°)

ZERO

ONE CYCLE

ONE ALTERNATION

FIGURE 12.1-8 *Just as in a dc generator (Fig. 12.1-4) the voltage (EMF) peaks twice in each cycle in an ac generator, but the alternate peaks are of opposite polarity, as current flow reverses in each alteration.*

A C
VARIES AND CHANGES DIRECTION

D C
STEADY, ONE DIRECTION

+170
+120

VOLTS

0

-170

a

b

1 CYCLE

TIME

FIGURE 12.1-9 *Nominal 120-volt ac current is actually generated with (a) voltage peaks of 170 volts, to approximate the effect of (b) a 120-volt dc current.*

120 OR 240 V

PRIMARY WINDING (MANY TURNS)

SECONDARY WINDING (FEW TURNS)

MAGNETIC CORE

24–30 V

FIGURE 12.1-10 *A transformer is used to step up or step down voltage. To reduce voltage from 120 to 30 volts would require 4 times as many turns in the primary as in the secondary winding.*

only means of changing direct current voltage, loses about 25% of its power in waste heat.

Passing current through the primary circuit on one side of the magnet in a transformer generates a current in the sec-

ondary circuit. The induced voltage in the secondary circuit has the same relationship to the voltage in the primary circuit as the number of windings in the secondary circuit do to those in the primary circuit. In other words, to double the

voltage, it is merely necessary to double the windings in the secondary circuit; to triple the voltage, triple the windings; to reduce the voltage by 50%, use half the number of windings.

12.2 ELECTRIC POWER DISTRIBUTION

Electricity from a generating station is typically transmitted at high voltage to substations within the city, then to transformers near the end-users. There it is converted to line voltage (120 and 240 volts) for use in small commercial buildings, apartments, and houses.

12.2.1

PRIMARY DISTRIBUTION

Most power is generated in power plants at 12,000 volts. For transmission over long distances, the voltage is stepped up by transformers to 115,000 volts or more, depending on the distance. Very high voltage is used because more power can be transmitted with a lower power loss in the lines. Utility companies generally use the highest voltage consistent with safety and economy in their transmission lines.

In order to avoid hazards from high-voltage transmission lines in built-up areas, high voltage is reduced at substations to either 12,000/7200 volts or 4160/2400 volts. Smaller transformers, mounted on poles (Fig. 12.2-1), set on the ground, or placed in underground vaults, further transform current to line voltage (120/240 volts).

12.2.2

SINGLE-PHASE SERVICE

Single-phase 120/240 volt service is used almost universally for residences in this country. Electric heating equipment and most large, fixed appliances, such as water heaters, electric ranges, and clothes dryers, use 240-volt service because they operate more efficiently at the higher voltage. However, general lighting and smaller, movable appliances, such as mixers, irons, and toasters, operate adequately on 120 volts, as do dishwashers and some other fixed items. To satisfy both needs, 120 and 240 volts must both be available. The dual voltage is provided in each block

FIGURE 12.2-1 *A pole-mounted transformer and service drop. A service drop should be at least 10 ft. above ground.*

or group of blocks by a 240-volt, center-tapped secondary transformer (Fig. 12.2-2). This transformer makes both 120-volt and 240-volt power available through a three-wire electric service entrance.

A three-wire service drop runs from the utility pole to the house. A three-wire circuit consists of two hot wires, each carrying 120 volts and a neutral wire, having no voltage. Items requiring 240 volts are connected between the two hot wires; loads requiring 120 volts can be connected between either hot wire and the neutral (see Fig. 12.2-2).

The incoming service must be grounded.

Refer to Section 12.3.4.6 for a discussion on grounding.

12.2.3

THREE-PHASE SERVICE

Four-wire three-phase service is seldom used in residences in this country, but is available for industrial and commercial applications. This system is particularly suitable for electrical heating and cooling installations, and where large motors are involved.

Three-phase service consists of three hot wires and a neutral wire. The voltage in each hot wire is out of phase (out of step) with the others by one-third a cycle, as if produced by three different generators. Figure 12.1-9 shows single-phase current. In a three-phase diagram, there would be three curves, with the peaks in the second and third curves moved one-third the diagrammed phase length to the right. Connections made between any of the two hot wires produces nominal 208-volt current; connections made between either of the hot wires and the neutral results in nominal 120-volt current. Although the service has four wires, branch

FIGURE 12.2-2 *A center-tapped transformer can convert two-wire 480-volt (or higher) current to three-wire 120/240-volt service.*

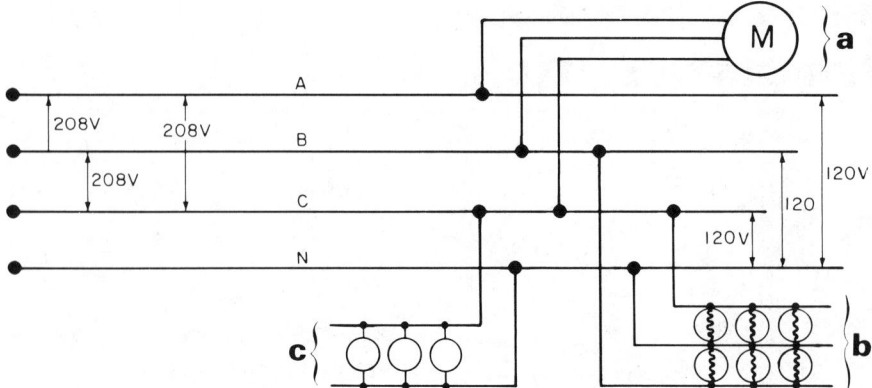

FIGURE 12.2-3 *Four-wire, three-phase service provides flexibility in serving different loads. In this figure (a) shows a three-wire, three-phase tap for a motor; (b) a three-wire, single-phase circuit for kitchen appliances; and (c) a two-wire, single-phase circuit for lights and receptacles.*

circuits may have two, three, or four wires, depending on the type of equipment served (Fig. 12.2-3).

A three-phase system should be planned so that the total load is about equally divided among the three phases. It should be remembered also that three wires are hot (rather than two, as in three-wire one-phase service) and all three require overload protection.

Service Drop

When the electrical service comes from overhead, a utility service *drop* is usually connected to the customer's service entrance conductors at the building, forming a drip loop, to prevent water from entering the conductor cable. When the service is underground, the service entrance conductors rise in conduit. From the drip loop or underground cable, conductors go first to a watt-hour meter, which keeps a record of how much power is consumed, and finally to the service equipment. The service equipment is the origin of all the branch circuits that supply a building's electrical needs.

Service equipment provides both 120-volt and 240-volt current for general-purpose, appliance, and special-purpose circuits. In addition, line-voltage circuits can be used to power transformers that supply low-voltage circuits.

12.3 SERVICE AND DISTRIBUTION

Electric power enters a building through service entrance equipment and from there is distributed to convenience outlets, lights, appliance outlets, and special-purpose outlets. The design and installation of such systems is subject to compliance with codes and standards. Symbols have been developed to help in the indication of requirements on drawings. This section discusses some of these aspects of electric service to buildings.

CODES, STANDARDS, AND SYMBOLS

A familiarity with the applicable codes and standards and the symbols used on drawings is essential to an understanding of the text in this chapter.

Most of the terms used here that have different meanings in this context than their standard dictionary meanings are explained in the body of the text. Those that are not so explained are defined in the Glossary.

Codes and Standards

Since 1937, the average annual consumption of electrical energy in residences has risen from 805 to more than 8850 kilowatt-hours (kwh). In some areas of the country the average consumption is over 10,000 kwh and is still rising. The electrical industry studied this trend and evolved a set of farsighted recommendations that were first published in 1958 by the American National Standards Institute (ANSI) as USA Standard C91.1-1958, "Requirements for Residential Wiring." Later editions of this standard, now called ANSI C91.1, are still in use today.

The recommendations in this chapter are based on ANSI C91.1, with revisions as necessary to conform with the 1990 *National Electrical Code (NEC)* * and the HUD *Minimum Property Standards for One- and Two-Family Dwellings.* The purpose of the recommendations here is to suggest design criteria that will produce an efficient and convenient system and will also minimize physical obsolescence. Therefore, they go beyond the requirements of the NEC, whose chief emphasis is on safety.

An adequate wiring installation must conform with applicable safety regulations. The standard for electrical safety is the *National Electrical Code*, developed and published by the National Fire Protection Association (NFPA). Most local ordinances and state and local building codes are based on the NEC or make it a part of their requirements by reference.

---

*National Electrical Code® and NEC® are registered trademarks of the National Fire Protection Association, Inc., Quincy, MA 02269.

Therefore, the entire electrical system, including its design, layout, equipment, and materials, and the methods used to install it must comply with the requirements of the latest edition of the NEC and local ordinances and laws. It must also be acceptable to the local building authority that has jurisdiction.

Inspection service to determine conformance with safety regulations is available in most communities from the local building official. Where available, a certificate of inspection should be obtained. In the absence of an inspection service, an affidavit should be obtained from the electrical contractor attesting to conformity with the applicable safety regulations.

12.3.1.2

Symbols

The symbols shown in Figure 12.3-1 were adapted from ANSI Standard Y32-1972 (reaffirmed 1989), "Graphic Symbols for Electrical Wiring and Layout Diagrams Used in Architecture and Building Construction." These are the standards recommended by the Institute of Electrical and Electronics Engineers (IEEE). They have been made mandatory by some federal agencies, such as the Department of Defense, and should be used on all drawings for electrical systems, whether they are for public or private buildings.

12.3.2

SERVICE POWER REQUIREMENTS

Every separate dwelling unit, including apartments, town houses, and single-family houses, should have at least a three-wire 120/240-volt electric service. Older residences were often served by two-wire 120-volt services, which is inadequate for today's needs.

A 100-ampere service will provide a house's needs for lighting, appliances, and fixed equipment totaling 18,000 watts. When the total load is larger than 18,000 watts, as it will be in most new houses today, the local power company should be consulted. Generally, a 150-ampere service is adequate to handle wiring, appliances, and fixed equipment plus from 3 to 5 tons of air conditioning. When elec-

tric heating or its equivalent is contemplated, at least a 200-ampere service is called for.

Inadequate service conductors and service equipment can be serious obstacles to the use of new equipment and appliances in residences. Service entrance conductors should be sized in accordance with the NEC. To provide for probable increases in electrical use, even when electric heating or cooling are not contemplated, service conductors should be oversized to at least No. 1/0 AWG (see Section 12.3.3), and service equipment should be of sufficient rating to match the carrying capacity of these wires.

Most portable appliances for residential use are made to work on 110 to 125 volts. However, because of improved operating efficiency, many fixed appliances are intended to be used on 230- to 250-volt circuits. However, some fixed appliances are still manufactured to function on the lower voltages, so that they can be used in older residences. When there is a choice, it is more efficient to use the higher voltage.

Most new homes today are air-conditioned. Much of the related equipment is designed to run on electricity. Electric heat is not so prevalent today as it was a few years ago, but electric heat pumps are being used with increasing regularity. Electric heating and air-conditioning systems and heat pumps are usually custom designed to suit the particular house and local climatic conditions. The equipment used is not custom designed, however. It is selected from a range of standard manufactured products to fit the requirements of each individual house. Suggestions for adequate wiring and service capacity criteria for such devices are not included here. They should be determined on a case-by-case basis in consultation with the local power company.

12.3.3

EQUIPMENT

Electrical equipment includes service entrance equipment, overload protection devices, conductors, raceways, outlets, and switches. This category does not include end-line equipment, such as heating, ventilating, and air-conditioning equipment, or to either portable or built-in appliances.

12.3.3.1

Service Entrance and Distribution Equipment

A service entrance is a system of conductors bringing electricity into a building and associated service entrance and distribution equipment. Service entrance and distribution equipment is defined here as all equipment and devices that occur from the outlet side of the watt-hour meter (see Section 12.2.4) to the outlet side of the distribution panels. It includes service conductors that bring electricity into a building, the service entrance panel, feeders that connect the service entrance panel to distribution panels, and the distribution panels from which branch circuits distribute electricity to the remainder of the building and its grounds. It also includes overcurrent devices mounted in the service entrance and distribution panels. Grounding of the incoming service is also a part of the service entrance equipment, but it is discussed in Section 12.3.4.6. Figure 12.3-2 gives an overall view of the service drop, meter, and service entrance equipment.

■ SERVICE ENTRANCE PANEL

A service entrance panel, which, in small buildings, may also serve as the distribution panel, is an assembly of switches and switchlike devices that permits disconnecting all power, or distributing it to various branch circuits through overcurrent devices such as fuses or circuit breakers.

There are several possible arrangements for a service entrance panel. Sometimes, especially in larger systems, a main disconnect switch is housed in a separate panel from the circuit overcurrent protection devices (fuses or circuit breakers). In houses and other small buildings, one panel usually includes both main switch and overcurrent devices.

A main disconnect device provides a means of deliberately de-energizing the entire system, and at the same time protects the service conductors from unexpected overloads. Depending on the load, the code, and the desires of the designer, a main disconnect device could be a knife-blade switch, a pull-out fuse, or a circuit breaker. The overcurrent devices could be fuses or circuit breakers.

General outlets

◯   Ceiling lighting outlet

⊸◯   Wall lighting outlet

▯   Ceiling flourescent fixture

▯   Wall flourescent fixture

Ⓕ   Fan outlet

Ⓙ   Junction box

⊸Ⓒ   Clock outlet

To indicate wall installation of outlets, place symbol near wall and connect to wall with line as shown for clock outlet.

Convenience outlets

⊝   Duplex convenience outlet

⊜   Triplex convenience outlet

⊖   Duplex convenience outlet— split wired

⊝ UNG   Ungrounded duplex convenience outlet[2]

⊝ WP   Weatherproof convenience outlet

⊝ X in.   Multi-outlet assembly (extend arrows to limits of installation. Use appropriate symbol to indicate type of outlet. Also indicate spacing of outlets as X inches)

Standard switches

S   Single Pole Switch

S_3 Three-way switch

S_4 Four-way switch

S_D Automatic door switch

⊝ S   Combination switch and convenience outlet

⊝ Ⓡ   Combination radio and convenience outlet

⊖   Floor mounted duplex receptacle

⊝ R   Range outlet

◉ DW   Special purpose outlet (use subscript letters to indicate function. DW = dishwasher, CD = clothes dryer, etc.)

S_P Switch and pilot light

S_{WP} Weatherproof switch

S_2 Double-pole switch

Ⓢ   Ceiling pull switch

FIGURE 12.3-1 *Graphic electrical wiring symbols.* (*Drawing by HLS.*)

Figure 12.3-3 shows a typical circuit breaker type of service entrance panel. This is the type of panel used in most houses and other small buildings today. A main knife-blade switch may be inserted between the main disconnect breaker and the incoming service or may replace the main disconnect breaker. The principles shown are applicable to panels with fuses, but wiring is used in lieu of bus bars. In addition, the principles demonstrated are the same, regardless of the amperage rating of the system.

A 100-ampere service may be adequate for general lighting and a few household appliances in a small residence. But it has too few circuits and the mains are too small to supply a larger residence or electric air-conditioning, heating, or heat

pump systems. For larger residences and even in small ones where electric heating or cooling, or a heat pump, is contemplated, a 200-ampere service is generally considered minimum.

At the upper left in Figure 12.3-3, the service conductors coming from the meter (not shown) enter the panel. The two hot wires carry the current to the main disconnect breaker, and through it into two hot bus bars. The hot bus bars carry the current to the individual circuit overcurrent protection devices (circuit breakers). Wires are then extended from each breaker to the outlets or devices on a single circuit. When distribution panels are used, the two hot wires run from a breaker on the service entrance panel to hot bus bars in the distribution panel. Breakers are at-

tached to these bus bars and wires run from these breakers to serve the outlets and devices on a single circuit.

The neutral conductor is neither switched nor fused. It connects to the bonded neutral bus bar, to which the service ground wire also is connected. This grounded bus provides terminals for all load neutrals and ground wires in the entire distribution system.

Because a kitchen requires many individual equipment circuits, it is usually best to locate the main service equipment or a subpanel near the kitchen. Such a location is also often close to the laundry, which helps to minimize long circuit runs and wiring costs (Fig. 12.3-4).

Auxiliary systems

⊡	Push button	◇	Annunciator	▶|	Outside-line telephone
⬛	Buzzer	D	Electric door opener	▷|	Inter-connecting telephone
Bell		Interconnec-tion box		R	Radio outlet
Combination bell-buzzer		BT	Bell-ringing transformer	TV	Television outlet
CH	Chime				

Low-voltage systems

S_L Switch for low-voltage relay system

S_{LM} Master switch for low-voltage relay system

Ⓛ   Outlet controlled by low-voltage switching

Miscellaneous

▨   Service panel

▬   Distribution panel

○ a,b
⊖ a,b
◉ a,b
□ a,b

Special outlets (subscript letters may be added to any standard symbol to designate a variation associated with a particular building. Such variations should be explained in a Key to Symbols, and, if necessary, in the specifications).

FIGURE 12.3-1 *Continued.*

[1]Adapted from USA Standard Y32.9-1972 (reaffirmed 1989) "Graphic Symbols for Electrical Wiring and Layout Diagrams Used in Architecture and Building Construction."

[2]All outlets of every type are grounded, unless marked UNG.

FIGURE 12.3-2 *Service entrance equipment.*

■ FEEDERS

Feeders Conductors connecting a service entrance panel or main disconnect switch to distribution panels are called feeders.

■ DISTRIBUTION PANEL

A distribution panel is connected to the service entrance panel or main disconnect switch by feeders. In small residences, the service entrance panel often serves as the only distribution panel. However, in large residences and in most commercial and institutional buildings, the use of several distribution panels is common. A distribution panel may or may not have a disconnect switch or overload devices such as fuses or circuit breakers.

■ BRANCH CIRCUIT

A branch circuit is the portion of a wiring system between an overcurrent device and outlets where loads, such as light fixtures or appliances, can be connected. Refer to Section 12.3.4.2 for recommendations about the allocation of outlets to circuits.

12.3.3.2

Overload Protection Devices

Short circuits in wiring are a major cause of fires. This section deals with ways to prevent damage from such shorts. Fire sprinkler systems are addressed in Section 11.4. Fire alarm systems are discussed in Section 12.7.

Inadequate or old wiring, faulty wiring installation, and other flaws in wiring or in appliances are major causes of fires. Electricity itself may also be a source of personal injury, especially in combination with water. When a person who is in contact with the ground comes into contact with a hot wire, current will flow through the person to the ground resulting in burns at the point where the electricity enters the body as well as where the current leaves the body. Such electricity interferes with the normal operation of a person's nervous system and, in severe cases, can result in convulsions, severe neurological damage, and even death.

Lighting fixtures, electrical outlets, and wall switches are often sited near wash basins, tubs, and other potentially wet areas. Water is a better conductor of electricity than is a human body. When a person is wet, electricity bypasses any protection a person might ordinarily have, such as shoes, and water produces multiple contact points on a body for electricity to enter and leave. Therefore, a person in contact with water, especially when partially submersed, as in a tub, is likely to sustain more severe injury when coming into contact with an electrical source than when no water is present.

Overcurrent devices are imposed in electrical circuitry to ensure that in case of a short circuit, current will be interrupted before the conductor overheats and possibly burns up the insulation. The first line of defense is at the main circuit control panel. The older, more traditional device used there is a fuse; a newer, more

Service entrance conductors from meter

Hot wires

Main disconnect breaker

Neutral conductor

Hot bus bars

Hot wires to branch circuits

Circuit breaker

Bonded neutral bus bar

Neutral wire

Grounding conductor

Grounding wire

Ground

FIGURE 12.3-3 *A circuit breaker service entrance panel. To provide a 120-volt branch circuit, insert a circuit breaker by snapping the clip on its back onto one hot bus bar. Then extend to the location of the circuit a hot wire (black or red) from the breaker, a neutral wire (gray or white) from the bonded neutral bus bar, and a grounding wire (bare wire) from the bonded neutral bus bar. To create a 120/240-volt circuit, snap a common-trip two-pole breaker with one handle so that one of the two poles contacts each hot bus bar. Then extend to the location of the circuit, one hot wire from each pole of the breaker, and both a neutral wire and a grounding wire from the bonded bus bar.* (Drawing by HLS.)

typical device is a circuit breaker. Several other devices and methods are also used in other parts of a typical system.

■ FUSES

Fuses have a fusible (meltable) wire, designed to melt and break when the current

flow exceeds the rated capacity of the circuit. Thus the circuit is opened and current flow is stopped before it becomes a hazard. Circuits of 30 amperes or less can use plug fuses; for heavier currents, cartridge fuses are required.

Plug fuses are obtainable in sizes from 0 to 30 amperes (Fig. 12.3-5). The NEC requires that type S fuses be used for all plug-fused installations. A type S fuse uses a slightly different adapter for each ampere rating; once the adapter is installed in a fuse box it cannot be removed readily and only a fuse of a certain rating can be inserted. Existing wiring systems that are being modified should be equipped with type S fuses, if the fuse box is retained.

Cartridge fuses are made in all current ratings and voltage ratings from 0 up. Fuses from 0 to 60 amperes use a ferrule type connection; above 61 amperes, a knife-blade connection is used (see Fig. 12.3-5).

■ CIRCUIT BREAKERS

Circuit breakers are simple switchlike devices that automatically open a circuit when the rated current is exceeded, such as would result from a short circuit. Unlike plug fuses and most cartridge fuses, circuit breakers do not damage themselves when they trip (open). To reset (close) the breaker, just push the toggle into the *on* position. Most small circuit breakers are of the thermal-magnetic type. When such a breaker is tripped, a thermal element in it is heated. It may be necessary, therefore, to wait about a minute for the element to cool, before the breaker can be reset.

Circuit breakers are available in a wide range of sizes for many applications. Small breakers for 15-, 20-, 30-, and 40-ampere branch circuits are widely used today in service entrance equipment for residential and commercial buildings. All of these can be combined in a single service panel, along with a larger breaker with a rating equal to the combined ratings of all the circuits to serve as a main switch.

■ GROUND FAULT CIRCUIT INTERRUPTER (GFCI)

A ground fault circuit interrupter (GFCI) is similar to a fuse or circuit breaker in that it stops the flow of electricity when an abnormal electrical situation develops.

First floor plan showing Dining room, Living room, Range, Dishwasher, Sink, Ref., Linen closet, and Bedrooms with labels S, S₃, WP, CH, TV, DW, R, F.

LIVING FLOOR
First and second levels

Service entrance

Service panel

Furnace & AC — F

Washer — CW — Dryer — CD — F

Utility room

Food freezer — FF

Recreation room — TV

Garage — WP

BASEMENT

FIGURE 12.3-4 *A second distribution panel is not needed here because the service equipment is centrally located with respect to cooking, laundry, and heating equipment.* (Drawing by HLS.)

Unlike circuit breakers and fuses, which keep large current flows from overloading a circuit, a GFCI will be tripped by even very low current flows. GFCIs are usually set to open the circuit when the ground fault current exceeds 5/1000 of an ampere. Refer to Section 12.3.4.6 for locations where GFCIs are appropriate.

■ POLARIZED PLUGS

Modern portable appliances and light fixtures have polarized plugs with blades of two different sizes so that they can be inserted into an outlet in only one direction (Fig. 12.3-6).

In a lamp with a nonpolarized plug there is a 50% chance that the threaded portion of the bulb socket is live and could give someone a shock. This is because one wire in a lamp cord is connected to

FIGURE 12.3-5 *Fuses are either (a) cartridge type (ferrule type at left, knife-blade type at right), or (b) plug type. A type S adapter makes it impossible to insert a 30-ampere fuse in a 20-ampere socket, thus avoiding dangerous overloads.*

FIGURE 12.3-6 *A polarized plug can prevent shocks when it is plugged into a correctly wired outlet.*

the contact that touches the bottom of the bulb; another is connected to the lamp socket. One slot in a receptacle is always connected to a hot wire and the other slot is connected to a neutral wire. When a plug is inserted in the correct orientation, the wire that is connected to a lamp's socket will be plugged into the neutral slot. If the plug is reversed, the socket will be connected to the hot wire. The lamp will work regardless of which way the cord is plugged in. Therefore, there is no way to know whether the socket is hot or safe. Since a polarized plug can be inserted only with the hot connected to the hot and the neutral connected to the neutral, its use will prevent a shock if the socket is touched. However, polarized plugs can prevent shocks only when they

are plugged into outlets wired with the neutral wire on the neutral contact and the hot wire on the hot contact.

Modern three-hole grounded receptacles (see Section 12.3.3.4) are made with a large and a small slot to accept a polarized plug, whereas most old-fashioned nongrounded units are made with two slots of the smaller size and will not accept a polarized plug. Grinding down a new plug to fit an old receptacle is particularly dangerous because it looks like a modern safe plug while presenting the same chance of shock as an old-fashioned nongrounded plug.

■ PLUG COVERS

Many electrical accidents occur when children stick small metal objects into electrical outlets. Simple plastic plug covers fit into unused outlets and serve to prevent injury (Fig. 12.3-7).

FIGURE 12.3-7 *Plug covers not only prevent children from sticking objects into receptacles, they also reduce air infiltration through outlets in exterior walls.*

■ COPPER/ALUMINUM (CO/ALR) DEVICES

Because of the high cost of copper, aluminum wire began replacing the traditional copper wire in residential installations during the mid-1960s. The potential problems associated with using aluminum wire were not foreseen and did not surface until the 1970s.

The basic problem is that metals expand when warm and contract when cold;

aluminum does so more than copper. Over time, repeated expansion and contraction caused many aluminum-wire connections to work loose. In addition, aluminum forms a film of aluminum oxide when exposed to air. This oxide is an insulator rather than a conductor of electricity. The combination of loose wiring and oxide buildup causes connections to overheat. The problems surfaced when homeowners began to report flickering lights, radio and TV static, unpleasant odors, and even occasional fires.

In 1972, special copper/aluminum (CO/ALR) devices suitable for use with both copper and aluminum wiring came into use. CO/ALR devices have special brass screws with heads wide enough to prevent aluminum wire from working loose. Proper installation is shown in Figure 12.3-8. Only CO/ALR devices should be used with aluminum wire; other types are incompatible.

According to the U.S. Consumer Product Safety Commission (CPSC), about 1.5 million houses were wired with incompatible receptacles between 1965 and 1973 that CPSC believes to be fire hazards.

Owners of dwellings built or rewired during the 1960s can check for aluminum wiring in unfinished spaces such as the basement or attic; the letters *AL* or the word *aluminum* is stamped on the insulation covering the wire.

Signs of trouble for dwellings wired with aluminum include (1) cover plates or switches that are warm to the touch; (2) smoke or sparks at switches or outlets; (3) odors, such as the smell of burning plastic; (4) lights that flicker periodically; (5) receptacles or entire circuits that do not work.

If aluminum wiring is found, special CO/ALR devices can be used to replace the old conventional switches and receptacles. However, this replacement still leaves other parts of the circuit, such as fixtures and junction boxes, as potential sources of trouble.

■ PIGTAILING

A broader solution to the aluminum/copper mismatch involves attaching short lengths of copper wire into existing aluminum wire with a special crimping tool. These copper-wire pigtails are then connected to the terminals of the electrical

FIGURE 12.3-8 *CO/ALR devices have special brass screws with wide heads. Wire should be stripped carefully to avoid nicks and wrapped around a terminal as shown in the drawing of the far-left wire. The other wires are incorrectly attached.*

FIGURE 12.3-9
Properties of Commonly Used Copper Conductors

Conductor			Maximum Current, ampere[a]	
Size: AWG, MCM[a]	One-Half Natural Size	Diameter, in.[c]	Conductor Type: RUW, T, TW	Conductor Type: RH, RUH, THW, RHW, THWN
14	●	.0641	15	15
12	●	.0808	20	20
10	●	.1019	30	30
8	●	.1285	40	45
6	●	.184	55	65
4	●	.232	70	85
2	●	.292	95	115
1	●	.332	110	130
0	●	.372	125	150
00	●	.418	145	175
000	●	.470	165	200
0000	●	.528	195	230
350 MCM[d]	●	.681	260	310

[a] Based on ambient temperature of 86 °F, (30 °C); not more than 3 conductors in raceway, cable, or direct burial. Where number of conductors exceeds 3, maximum current given should be derated as follows: 4–6 cond., 80%; 7–24 cond., 70%; 25–42 cond., 60%; 43 or more, 50%.

[b] Sizes 14 to 8 are solid wires. Sizes 6 to 2 are 7-strand cables. Sizes 1 to 0000 are 19-strand cables. Size 350 MCM is a 37-strand cable.

[c] Exclusive of insulation.

[d] MCM is the designation of wire size in thousands of circular mils (350 MCM = 350,000 circular mils).

device. Although CO/ALR devices might be suitable for new installations using aluminum wiring, pigtailing seems more appropriate for retrofitting existing installations.

12.3.3.3

Conductors and Raceways

Electric current is carried and distributed by conductors. Technically speaking, conductors are the metallic cores that actually conduct the electricity, but in practice the entire wire, including the insulating cover, is called a conductor. Conductors generally are protected from mechanical damage by an outer sheath. When a conductor and its protective sheath are fabricated together, the assembly is called a cable. Conductors can also be inserted on the job into raceways, which are specially built channels into which wires are placed or pulled.

■ CONDUCTORS

Conductors consist of an aluminum or copper core, covered by an insulating jacket and an outer coating suitable for the anticipated temperature and moisture levels.

Conductors for building wiring are available in American Wire Gauge (AWG) sizes, ranging from No. 14 to 4/0, and in larger sizes from 250 MCM (thousand circular mils) to 2000 MCM (Fig. 12.3-9). Wiring systems should be designed so that the maximum current carried by a conductor will not exceed the amperages shown in Figure 12.3-9. Copper and aluminum are the typical conductor materials; aluminum generally is limited to larger wire sizes. Because of its lower conductivity, aluminum wire smaller than No. 12 AWG (rated 15 amperes) is not made. The same types of insulation are used for both copper and aluminum conductors.

Insulations used on electrical conductors are designated by identifying the material, such as *R* for rubber and *T* for thermoplastic. Since the insulation generally determines the suitability of the conductor to certain exposures, for example, moisture, heat, or mechanical damage, conductors are identified by the material and properties of the insulating cover.

Insulation thickness varies with wire size and may also vary with the type of material and application and code limitation. In addition to the letter derived from the material, other letters may be added to indicate modified or additional characteristics. For example, when an H is added to the letter R, the resulting RH wire may be used to 167 °F temperature instead of 140 °F, which is the limit for RF wire (previously designated simply R). A double H indicates resistance up to 190 °F, whereas a W indicates moisture resistance.

■ CABLES

Cables are combinations of two or more conductors, factory encased in a protective covering to safeguard them from mechanical damage. They can have metallic coverings as in armored cable, or plastic or cambric (varnished cotton) jackets as in nonmetallic sheathed cable.

ARMORED CABLE   Armored cable, Type AC, which is also sometimes called BX, is a factory assemblage of insulated conductors inside a flexible metallic covering intended primarily for use in branch circuits and feeders. It may be run in the voids of masonry block or tile walls when such walls are not exposed to excessive moisture. It should not be run below grade; a transition should be made from armored cable to metal conduit or tubing when supplying a basement appliance.

Armored cable should be securely fastened by approved sleeves, staples, or straps, spaced not more than 4 ft. 6 in. apart. The cable should be supported within 12 in. of an outlet box or other termination. When the box is mounted on equipment that needs to be moved occasionally for servicing, the unsupported cable length can be 24 in.

Armored cable should be cut away to allow at least 6 in. of the conductors into the box. The remaining armored portion should be clamped to the box. When the armor is cut back to provide free conductors, care should be exercised to prevent damaging the conductor insulation. Also, a fiber-insulating bushing should be inserted in the end, between the armor and the conductor insulation. This bushing should remain visible within the box.

Armored cable should be handled and bent with care. If it is bent too sharply, the convolutions of the armor can damage the conductor insulation, resulting in possible short circuits or leakage to ground.

METAL-CLAD CABLE   Metal-clad cable, classified as MC, is similar in construction to armored cable, but is intended for use in services, feeders, and branch circuits. It is permitted in systems in excess of 600 volts.

NONMETALLIC SHEATHED CABLE   This cable, which is also sometimes called Romex, consists of two or more insulated conductors having an outer sheath of moisture-resistant, nonmetallic material (see Fig. 12.3-10). The conductor insulation is rubber, neoprene, thermoplastic, or a moisture-resistant, flame-retardant fibrous material.

FIGURE 12.3-10 *Some common types of electrical cable.*

Nonmetallic sheathed cable should be securely fastened by approved staples, straps, or other supports, in such a manner that the cable will not be damaged. The supports must not be more than 4 ft. 6 in. apart. Generally this cable must be fastened within 12 in. of each termination. In concealed spaces of existing or prefabricated buildings where such support is impractical, this cable may be fished between points of access. When run at an angle to joists, this cable should either be supported on a running board secured to the joists, or run through holes bored through the joists.

Care should be taken not to damage the outer sheath of nonmetallic cable during installation. No bend should have a radius less than 5 times the diameter of the cable.

There are two types of nonmetallic sheathed cable: type NM and type NMC.

Type NM cable has conductors that satisfy the requirements for types RH, RW, T, and TW wire, and a sheath that is flame retardant and moisture resistant. Type NM is limited to use in dry locations such as air voids in masonry or tile walls not exposed to excessive moisture.

Type NMC conductors must satisfy the same requirements as type NM, but the sheath must be not only flame retardant and moisture resistant, but also fungus resistant and corrosion resistant. Type NMC may be used in damp or corrosive locations, as well as in dry areas.

■ RACEWAYS

Raceways are protective enclosures designed for the field insertion of conductors. Included in the general term are electrical metallic tubing (EMT), rigid and flexible metal conduit, and wireways.

Raceways should be securely connected and rigidly supported. Cut ends should be reamed (smoothed) to remove rough edges. Combination couplings should be used to make transitions from flexible conduit to electrical metallic tubing or rigid conduit.

ELECTRICAL METALLIC TUBING (EMT)   This electrical pipe, which is also called thin-wall conduit, may be used for both concealed and exposed work (Fig. 12.3-11). It is the most common type of raceway used in single family, low-rise residential and commercial construction. With proper joints and connections it can be installed in wet, as well as dry, locations.

EMT is made in 1/2- to 4-in. nominal sizes. The actual internal diameter is slightly larger than the nominal size. Connectors and couplings for EMT are secured to the tubing by a setscrew or dimples made on the job in the collar of the fitting and tube after assembly. Watertight compression fittings are also available.

EMT is often used in residences and commercial buildings because it is light, easy to handle, and requires no special threading tools. However, bending it, which should be done using a special type

FIGURE 12.3-11 *Electrical metallic tubing (EMT), or thinwall conduit, and typical fittings for it.*

FIGURE 12.3-12 *An EMT bender ensures smooth curves, without pinching the interior of a pipe.*

of bender to prevent flattening (Fig. 12.3-12), requires more skill than does the bending of rigid conduit. The radius of the inner edge of a field bend should not be less than that required in the NEC.

RIGID METAL CONDUIT This conduit, which is also called heavywall conduit, resembles plumbing pipe and is shipped in standard lengths of 10 ft. (Fig. 12.3-13). A coupling is furnished for one end of each length of rigid conduit. Both ends are reamed and threaded when shipped. If it is cut during installation, the new ends must be reamed and threaded.

Rigid conduit can be joined by screwing it directly into threaded hubs, by locknuts, and bushings (see Fig. 12.3-13), or by threadless connectors. When connecting with locknut and bushing the locknut is screwed onto the pipe, then the end is inserted through the knockout in a box and the bushing is screwed on. When the bushing is seated, the outer locknut is screwed

FIGURE 12.3-13 *Rigid, or heavywall, conduit, and typical conduits and fittings for it.*

tightly against the outer surface of the box. When a conduit bushing is made entirely of insulating material, a locknut must be installed on both sides of the enclosure.

Rigid metal conduit is made in trade sizes of 1/2 to 6 in. The 6-in. size will accommodate three 2000 MCM (1.63-in.-diameter) conductors. When greater capacity is required, a bus type conductor (solid bar) in a bus duct is usually used.

FLEXIBLE METAL CONDUIT Flexible metal conduit, which is also called Greenfield, is similar in appearance to armored cable, but does not contain preinserted conductors (Fig. 12.3-14). It cannot be used where it will be subject to corrosive gases or excessive moisture. For the most part it must conform to the same regulations as rigid conduit with respect to rewiring existing conduit, reaming cut edges, and other code requirements.

FIGURE 12.3-14 *Two types of flexible metal conduit and typical fittings for them.*

Flexible conduit connectors are fastened to the conduit by setscrews or clamps, and to the box by an expanding mechanism, a locknut, or a threaded hub.

Flexible conduit should be installed taut and then fastened securely to facilitate pulling conductors with a fish tape. It should be supported at intervals not greater than 4 ft. 6 in. and within 12 in. of a box. The radius of bends should not exceed five times the conduit diameter.

NONMETALLIC CONDUIT This conduit is made of nonmetallic material that is resistant to moisture and chemical atmospheres. It is flame retardant, resistant to crushing, and resistant to distortion from heat, low temperatures, and sunlight. It can be used underground in floors, walls, and ceilings.

WIREWAYS Wireways are metal troughs with hinged or removable covers in which conductors are placed after the wireway has been installed. Because they must be accessible throughout their entire length, wireways can be used only for exposed work.

Splices can be made and laid into a wireway if the total conductor and splice cross-sectional area does not exceed 75% of the area of the wireway and is accessible. If the wireway is not fully accessible at the splice location, then the percentage of fill should not exceed 20% of the cross-sectional area of the wireway.

12.3.3.4

Outlets and Switches

There are three types of outlets: lighting, convenience, and special purpose.

◼ LIGHTING OUTLETS

A lighting outlet is a metal protective box to which fixed light fixtures, as opposed to portable lights, can be mounted and in which they may be connected to a branch circuit (see Section 12.3.4.2). Lighting outlets should be compatible with the lights selected (see Section 12.4). Figure 12.3-15a shows a standard octagonal protective box of a type that is often used as a lighting outlet. Such a box may also serve as a junction box for pulling wires to remote fixtures in coves, valances, and cornices. Circular and square lighting outlet boxes are also used.

The lighting fixtures discussed here and in Section 12.4 are fixed ones. Portable light fixtures are simply plugged into convenience outlets.

FIGURE 12.3-15 *(a) Box for lighting outlet and junction box use; (b) standard duplex receptacle; (c) special-purpose receptacles.*

FIGURE 12.3-16 *Using two hot wires in a duplex split receptacle permits switching off one receptacle, while the other remains hot.*

FIGURE 12.3-17 *Split receptacles minimize overloads and double the capacity of duplex receptacles by connecting each plug-in position to a separate hot wire. When local codes do not permit this type of installation, and both plugs on a receptacle must be wired to the same circuit, alternate receptacles can be wired to different circuits.*

■ CONVENIENCE OUTLETS

There are two kinds of convenience outlets: standard and multioutlet.

STANDARD CONVENIENCE OUTLETS   A standard convenience outlet is a plug-in receptacle housed in a protective box with plug-in positions for attachment of portable fixtures or minor appliances. The most common type, duplex, has two plug-in positions (Fig. 12.3-15b); a triplex has three positions; a single has only one.

Outlets are sometimes grouped in a single protective box behind a single cover plate. Common groupings are an outlet and a switch, two or three outlets and a switch, and two outlets and two switches.

The NEC requires that outlets on a 15- to 20-ampere general-purpose branch circuit should be of the grounding type to minimize the hazard of shock from short circuits (see Section 12.3.4.6). A grounding receptacle is a convenience outlet designed to receive a three-pronged plug. The third prong permits an appliance or piece of equipment to be grounded, thus minimizing the shock hazard from faulty wiring.

Regardless of the number of positions, an outlet is considered to be a single outlet when only one connection has been made to a branch circuit. Other arrangements are sometimes used, however. For example, a duplex receptacle that has been wired so that each plug-in position is on a separate 120-volt circuit or uses two separate line wires of a single three-wire 240-volt circuit (Figs. 12.3-16 and 12.3-17) is called a split-receptacle outlet. Typical uses of such outlets are discussed in Section 12.3.4.1.

MULTIOUTLET ASSEMBLY   A multioutlet assembly is a series of plug-in receptacles, usually spaced at regular intervals and contained in a protective raceway (channel). Such assemblies also are called plug-in strips. They may be used in most locations where a standard convenience outlet is needed, especially where several outlets are called for.

■ SPECIAL-PURPOSE OUTLETS

A special-purpose outlet is a point of connection to the wiring system for a particular piece of equipment, normally reserved for the exclusive use of that equipment. Such outlets may be plug-in receptacles to which the equipment is connected with a flexible cord, as with clothes dryers or ranges (Fig. 12.3-15c); or they may be protective enclosures, called junction boxes, with removable covers containing permanent connections; such junction boxes can be mounted either on the structure or on the equipment.

■ SWITCHES

A switch is a device for interrupting the flow of electricity. Switches are used in service entrance and distribution panels and in branch circuits. The switches in panels are either a part of the panel or are in separate metal boxes. Switches in branch circuits are connected to lighting, convenience, and special-purpose outlets.

Switches in branch circuits are mounted in the same kind of protective boxes used for standard convenience outlets. They may be of the push type, toggle type, or slide type. Variable rotary switches are sometimes used to control the power to lights and ceiling fans.

12.3.4

GENERAL RECOMMENDATIONS FOR WIRING

The recommendations in this section are necessarily general in nature. They are intended to apply to most typical single-family and multifamily residences. Even when unusual conditions make strict compliance impossible, the recommendations here may still be used as guidelines in developing more appropriate provisions.

These recommendations must be supplemented with engineering judgment about electrical system adequacy in common-use spaces and for general building services. Particular attention must be paid to limiting voltage drop in feeders to individual dwelling units in order to assure proper operation of appliances.

The guidelines here are only marginally applicable to commercial and institutional buildings. Electrical systems in those kinds of structures must be designed in accordance with sound engineering practices with outlets and power sufficient to meet the needs of the occupants. The NEC and applicable building codes must also be followed, of course.

12.3.4.1

Convenience and Special-Purpose Outlet Number and Location

This section contains recommendations for the number and location of convenience and special-purpose outlets. Similar information about lighting outlets is covered in Section 12.4.

When these recommendations are based on the number of linear or square feet, the number of outlets should be determined by dividing the total linear or square footage by the recommended amount per outlet. Resulting numbers that contain a fraction larger than one-half should be increased to the next larger number. The floor area should be computed from the outside dimensions of the house or the centerline of the walls surrounding each room.

Convenience outlets should be located near the ends of a wall space, rather than near the center, thus reducing the likelihood of them being concealed behind large pieces of furniture. In most cases, outlets should be located about 12 in. above the floor line. In every kitchen, living room, dining room, family room, recreation room, den, library, bedroom, and similar habitable space convenience outlets should be placed so that no point along the floor line of a usable wall space is more than 6 ft. from an outlet in the same wall space. Where floor-to-ceiling windows prevent meeting this requirement with ordinary convenience outlets, other suitable means, such as floor outlets or multioutlet assemblies, can be used.

When two different recommendations require an outlet at the same location, only one need be installed.

■ DUAL-PURPOSE ROOMS

Where a space is intended to serve more than one function, such as a combination living and dining room or a kitchen and laundry, the following outlet recommendations are separately applicable to each specific area. In a living and dining space, for instance, convenience outlets in the dining area should be of the split-receptacle type in which each receptacle is on a different circuit, whereas ordinary outlets may be used in the rest of the space. This permits the use of several portable devices, such as a toaster and a coffee maker, in the dining space without overloading a circuit.

■ EXTERIOR ENTRANCES

A weatherproof convenience outlet, located at least 18 in. above the entrance platform near the front entrance, should be provided. This outlet should be controlled by a wall switch inside the building near the entrance for convenient operation of outdoor decorative lighting. Additional waterproof outlets along the exterior of the house are recommended to serve decorative garden treatments and for the use of appliances or electric garden tools, such as lawn mowers and hedge trimmers.

■ LIVING AND RECREATION ROOMS

When general illumination is to be provided by portable lamps, in lieu of fixed lighting, at least two wall-switched, separate, plug-in positions should be planned. These can be provided with two switched, regular duplex outlets or one switched, plug-in position in each of two split-receptacle outlets (see Fig. 12.3-16).

When regular convenience outlets are wall switch controlled to accommodate portable lighting, additional unswitched convenience outlets should provided, to prevent limiting the use of portable equipment, such as radios and clocks in that location. Also, a single convenience outlet should be installed in combination with the wall switch at one or more of the switch locations for use by a vacuum cleaner or other portable appliances.

One or more outlets for entertainment equipment, such as a television set, radio, or stereo equipment, should be provided at bookcases, shelves, and other suitable locations. In rooms with conventional fireplaces, an outlet may be desirable in or near the mantel shelf.

■ DINING AREAS

When a dining or breakfast table will be placed against a wall, one of the recommended outlets should be located in that wall, just above table height.

Where open counter space is to be built-in, an outlet should be provided above counter height for the use of portable appliances. All convenience outlets in the dining area should be of the split-receptacle type, providing increased capacity at each outlet (see Fig. 12.3-17).

■ BEDROOMS

Convenience outlets should be placed in bedrooms so that there is one on each side and within 4 ft. of the center line of each probable individual bed location. The popularity of bedside radios and clocks, bed lamps, and electric blankets makes increased plug-in positions essential at bed locations. Triplex or quadruplex convenience outlets are recommended at these locations (Fig. 12.3-18c).

Additional outlets should be placed so that no point along the floor line of a wall is more than 6 ft. from an outlet in that space. It is also recommended that a receptacle be provided at at least one switch location in a bedroom to accommodate a vacuum cleaner, floor polisher, or other portable appliance (see Fig. 12.3-18b).

■ BATHROOMS AND LAVATORIES

A convenience outlet should be provided near the mirror, 3 to 5 ft. above the floor. An outlet should be installed at each separate mirror or vanity space, and also at any space that might accommodate an electric towel dryer, razor, hair dryer, or other small electrical equipment. A receptacle that is part of a bathroom lighting fixture is not suitable for this purpose unless it is rated at 15 amperes, wired with at least 15-ampere-rated wires, and easily accessible. A ground fault circuit interrupter is desirable in every such condition

FIGURE 12.3-18 *Interchangeable devices can be installed in the same outlet box in various combinations: (a) a pilot light, a switch, and a receptacle in the same box; (b) a switch and a receptacle; (c) a triplex receptacle; (d) a switch and a duplex receptacle.*

and is required in most instances by the NEC (see Section 12.3.4.6).

A wall-switched, or timer-operated, built-in ventilating fan capable of providing a minimum of 8 air changes per hour should be installed in every bathroom. Such a fan is generally required by code where natural ventilation is not provided through windows or skylights. A fan capable of moving 50 to 55 cubic feet of air per minute (cu. ft./min.) will ensure 8 air changes per hour in a 5 × 10 ft. bathroom.

In many parts of the country a wall-switch-controlled electric space heater is desirable for increased comfort during bathing. It may be a part of a ceiling light, fan, heater combination, or a separate unit. Heaters may be of the wall- or ceiling-mounted type and often have a timer switch.

■ KITCHEN

A kitchen should have one convenience outlet for the refrigerator and additional outlets placed so that no point along the wall line is more than 24 in. from a convenience outlet, with at least one outlet to serve each separate work surface or planning desk. An outlet should be placed in every wall space that may be used for ironing and at every location that might be used for an electric roaster, microwave oven, toaster oven, can opener, popcorn popper, ice-cream maker, wok, or kettle. If a work or dining table is planned next to a wall, this wall space should have one outlet located just above table level.

Work-surface outlets should be located about 44 in. above the finished floor level or about 8 in. above the work surface. Convenience outlets in a kitchen, other than that for a refrigerator, should be of the split-receptacle type (see Fig. 12.3-17).

Possible special-purpose outlets required in a kitchen include those for a range, oven, burner top, ventilating fan, dishwasher, and waste disposer. A special recessed receptacle with a hanger should be installed for an electric clock. The clock should be easily visible from all parts of the kitchen. It is also necessary to provide for a food freezer either in the kitchen or in some other convenient location.

■ LAUNDRY AREAS

At least one convenience outlet should be provided in the laundry area, preferably near the sorting table if there is one. If properly located, one of the special-purpose outlets (the washer outlet, for example) may also serve as a convenience receptacle for such items as a laundry hot plate or sewing machine. Convenience outlets in the laundry area should be of the split-receptacle type (see Fig. 12.3-17).

A separate special purpose outlet is generally needed for each of the following pieces of equipment: automatic clothes washer; hand iron or ironer; and clothes dryer. Outlets for a ventilating fan and clock are desirable. Requirements for an electric water heater should be obtained from the local power company.

■ HALLS

These recommendations apply to passage halls, reception halls, vestibules, entries, foyers, and similar areas. One convenience outlet should be installed for each 10 linear ft. of hall, measured along the center line. A hall with more than 25 sq. ft. of floor area should have at least one outlet. In reception halls and foyers, convenience outlets should be placed so that no point along the floor line of a wall is more than 10 ft. from an outlet in that space.

A receptacle at one of the switch outlets is recommended for convenient connection of maintenance equipment such as a vacuum cleaner, floor polisher, or electric broom.

■ STAIRWAYS

Convenience outlets generally are not required, except at intermediate landings of larger-than-normal size, where a decorative lamp, night-light, or portable equipment might be used.

■ UTILITY ROOM

At least one convenience outlet should be provided, preferably near the furnace or near a planned work-bench location.

A special-purpose outlet is required for each piece of mechanical equipment requiring electrical connection. Such equipment includes a furnace, boiler, water heater, water pump, sump pump, or compressor.

■ BASEMENT

At least two convenience outlets should be provided in an open basement area. If a work bench is planned, one of these outlets should be placed at its specific location.

Additional convenience outlets may be needed for basement laundry, dark room, or hobby areas, and for appliances such as a dehumidifier or portable space heater.

A special-purpose outlet must be provided for each piece of mechanical equipment requiring electrical connections such as furnace, boiler, water heater, water pump, sump pump, or compressor. An outlet for a food freezer also should be considered.

■ ACCESSIBLE ATTIC

A convenience outlet is desirable for portable equipment and for trouble lights, which is capable of reaching into distant dark corners. If an open stairway leads to future attic rooms, provide a junction box with direct connection to the distribution panel, for future extension to convenience outlets and lights when rooms are finished.

A summer cooling fan, multiple switch controlled from convenient points throughout the house, often is desirable.

▪ PORCHES AND BREEZEWAYS

At least one convenience outlet should be provided for each 15 linear ft. of wall bordering a porch or breezeway. These outlets should be of the weatherproof type if exposed to the elements. They should be controlled by a wall switch inside the house. Since such areas are often used for outdoor dining, at least one outlet should be of the split-receptacle type.

▪ TERRACES AND PATIOS

At least one weatherproof outlet should be provided for each 15 linear ft. of house bordering a terrace or patio. These outlets should be wall switch controlled from inside the house and be located at least 18 in. above patio level.

▪ GARAGE OR CARPORT

At least one outlet should be provided for one- and two-car garages and carports.

Special-purpose outlets appropriate to the use should be provided when a food freezer, work bench, or automatic door opener is planned. Specific hobbies, such as restoring antique automobiles or ceramic sculpture, may require special-purpose outlets for compressors, welding rigs, or kilns.

12.3.4.2

Circuit Recommendations

In addition to the branch circuits providing power for lights, appliances, fixed and portable equipment, a system may include feeder circuits serving secondary distribution panels near heavy-use locations.

▪ BRANCH CIRCUITS

A branch circuit is that portion of a wiring system between an overcurrent device and the outlets, where loads, such as light fixtures or appliances, may be connected. A residential wiring system usually includes three types of branch circuits: general purpose, small appliance, and special purpose. Each circuit involves taking into consideration voltage, amperage, and the number and size of conductor wires.

GENERAL-PURPOSE CIRCUITS General-purpose circuits supply all lighting outlets throughout the house and all convenience outlets except the convenience outlets in spaces designated for dining, cooking, and laundering. All general-purpose circuits should be wired with suitable conductors: not smaller than No. 14 AWG if protected by a 15-ampere overcurrent device and No. 12 AWG for a 20-ampere device. Since No. 14 conductors are more limited in circuit length (30 ft.) and current carrying capacity (1725 watts), No. 12 conductors with 20-ampere devices are highly recommended for all general-purpose circuits. Some local building codes do not permit the smaller No. 14 wires.

The number of general-purpose outlets required should be based on the loads to be encountered and the loads permitted per circuit, both as determined in accordance with the NEC, and the branch circuit amperage to be used. In addition, in residences no less than the following number of general-purpose outlets should be provided: If 20-ampere circuits are used, at least two circuits should be provided for the first 750 sq. ft. of floor area plus one circuit for each additional 750 sq. ft. or major fraction thereof. If 15-ampere circuits are used, at least two circuits should be provided for the first 500 sq. ft. plus one circuit for each additional 500 sq. ft. or major fraction thereof.

Floor areas should be calculated from dimensions taken to the outside of exterior walls, including unfinished spaces that are adaptable to future living use; open porches and garages may be excluded from the calculation.

SMALL APPLIANCE CIRCUITS The actual number and size of small appliance circuits should be determined by the loads to be encountered as calculated in accordance with the NEC. The minimum for residences should be one three-wire 120/240-volt, 20-ampere circuit (or two two-wire 120-volt 20-ampere circuits), equipped with split-receptacle outlets, in areas regularly intended for cooking, dining, or laundering.

Using three-wire circuits to supply convenience outlets in the locations mentioned is an economical means of increasing the capacity and offers practical operating advantages. Such circuits provide greater capacity at individual outlet locations and lessen voltage drop in the circuit. For maximum effectiveness, the upper half of all receptacles should be connected to the same side of the circuit (see Fig. 12.3-17). A separate 20-ampere circuit should be provided for each appliance or piece of equipment rated at more than 1400 watts.

However, two separate receptacle circuits may have to be used where local codes do not permit split receptacles.

SPECIAL-PURPOSE CIRCUITS Separate circuits should be provided and conductors should be sized appropriately for the specific equipment tabulated in Figure 12.3-19.

Spare circuit positions should be provided in a distribution panel for at least two future 20-ampere, two-wire 120-volt circuits. If the branch circuit or distribution panel is installed in a finished wall, raceways should be extended from the panel to the nearest accessible unfinished space for future use.

The majority of portable appliances for residential use are made to operate at 110 to 125 volts. Some more or less fixed appliances, like refrigerators and freezers, also operate in the 110- to 125-volt range. Many fixed appliances, however, operate on 230/250-volt circuits, because of improved operating efficiency. The higher voltage should be used when a choice exists.

Most new homes today include air-conditioning systems. Many of them are electrically operated. Electrical heating systems are not as prevalent today as they were a few years ago because of their high energy use. However, they are being replaced in many instances by electric heat pumps, which use considerable electric energy, especially in colder climates, but still use less energy than full electric heating systems. Such systems generally are custom designed to suit the particular house and local climate conditions. Adequate

FIGURE 12.3-19
Recommended Conductors for Special-Purpose Circuits

Item	Typical Wattage	Recommended Conductor Capacity		
		Amperes	Wires	Volts
Range or electric oven and drop-in cooktop	12,000*	50	3	120/240
	5,000	30	3	120/240
	7,000	40	3	120/240
Automatic washer	700	20	2	120
Electric clothes dryer	5,000*	30	3	120/240
Fuel-fired heating equipment	800	20	2	120
Central air conditioning or room air conditioner (1 ton)	5,000*	40	3	120/240
	1,500	20	2	120
Dishwasher and disposer or dishwasher and waste disposer	2,700	20	3	120/240
	1,800	20	2	120
	900	20	2	120
Food freezer	400	20	2	120 or 240
Water pump	500	20	2	120 or 240
Bathroom space heater	2,000	25	2	240
Electric water heater	4,500*	30	3	240

*Typical wattages are shown; verify with manufacturer and/or local power company.

wiring and service capacity criteria for such purposes are not included here and should be determined by consultation with the local power company.

■ FEEDER CIRCUITS

Consideration should be given to the use of subpanels served by appropriately sized feeders to serve heavy-use locations throughout a residence. Examples include areas containing an electric range or clothes dryer.

12.3.4.3

Conductor Size

Conductors should be adequately sized for the design loads and length of run. Adequately sized wires for specific loads can be selected from Figure 12.3-20. The table gives a direct reading of the maximum distance between distribution panel and load for different conductor sizes that will limit the voltage drop to a maximum of 3%. For example, consider a 3000-watt 120-volt heater that is 75 ft. from the distribution center. First, watts should be converted to amperes: $I = W/E = 3000/120 = 25$ amperes (see Section 12.1.2). Checking in Figure 12.3-20 under the 25-ampere column heading one finds that the maximum allowable distance from the source is 138 ft. for No. 10 AWG wire. This is the maximum distance over which the current will not exceed the permitted 3% voltage drop using that size wire. When selecting a wire size for a known load, the permitted length of run should be greater than the anticipated distance from panel to load.

12.3.4.4

Raceway Size

The number and size of conductors that can be placed in a raceway are limited not only by the size of the interior opening, but also by certain code restrictions.

The major factor limiting the number of conductors that may be put into raceways is fill, which is the percentage of the total cross-sectional area of the raceway that is filled with conductors. The maximum fill is 40% when three or more nonlead covered conductors are installed in conduit or tubing.

The number of conductors of one size in a metal conduit or tubing should not exceed the requirements of the NEC (Fig. 12.3-21). When using several conductors of unequal size in the same raceway, it is necessary to determine the total cross-sectional area of all of the conductors combined, including their insulation, then to select the raceway size on the basis of percentage of fill. In such cases, the cross-sectional area of the conductors should not exceed the maximum percentage of fill given in Figure 12.3-22.

12.3.4.5

Fittings and Connections

A wiring installation should be adjusted to fit the design of the building. If compliance with a particular recommendation would require alteration of doors, windows, or structural members, alternate provisions should be made. For example, the extensive use of window walls may make it necessary to alter the recommended outlet or switch heights, or to resort to surface-mounted raceways.

Conductors should be spliced with approved solderless connectors, or by brazing, welding, or soldering. Splices should be made mechanically secure before soldering. Bare ends of conductors should be covered after splicing with an insulation equal to that on the conductors. Connections should be made in an accessible device such as a junction box, switch box, or fixture base.

Splices should never be made outside a raceway and then pulled into it. They should be made in a wireway that has a removable cover or in an accessible box, which can be opened for inspection or repairs. Each conductor run should be complete from one opening to the next before conductors are pulled. If conductors are inserted before the raceway is completed, they can be damaged during the process of making up the raceway.

Terminal connections should be made securely. Terminals for securing more than one conductor should be approved for that purpose.

FIGURE 12.3-20
Maximum Length of Two-Conductor Runs*

Conductor Size, AWG	Maximum Length of Run, ft.																		
	5 amp	10 amp	15 amp	20 amp	25 amp	30 amp	35 amp	40 amp	45 amp	50 amp	55 amp	70 amp	80 amp	95 amp	110 amp	125 amp	145 amp	165 amp	185 amp
14	274	137	91																
12		218	145	109															
10			230	173	138	115													
8					220	182	156	138											
6								219	193	175	159								
4									309	278	253	199							
3											350	319	250	219					
2											402	316	276	232					
1												399	349	294	254				
0												502	439	370	319	280			
00													560	470	403	355	305		
000														588	508	447	385	339	
0000															641	564	486	421	360

*To limit voltage drop to 3% at 120 volt. For 240 volt, multiply length by 2.

Outlet, junction, and switch boxes used in wiring installations should provide adequate space for the conductors. Conductors should be sized adequately to minimize line losses and voltage drop. Voltage drop between the distribution panel and the load should be limited to less than 3%.

Both raceway joints and conductor connections should be made strong and secure. A mechanically secure raceway connection is one in which the joint is as strong as the raceway itself. An electrically secure connection will carry a full short circuit, that is, the entire return current, if one of the hot wires should accidentally be grounded. Such an accidental ground will then open the circuit by blowing the fuse or tripping the circuit breaker. An improperly connected system can result in excessive heat at conductor joints and possible fire hazard.

An excessive number of bends in a conduit can result in damage to conductors when they are pulled through it. A run of rigid conduit or electrical metallic tubing should not contain more than the equivalent of four quarter bends (360° total), including those bends located at the fitting outlet.

Because of the different characteristics of copper and aluminum, pressure terminals or pressure connectors should be marked as suitable for mixing copper and aluminum conductors when this combination is used.

12.3.4.6

Grounding

There are two types of grounding involved in a home wiring installation: (1) system grounding and (2) equipment grounding. System grounding refers to connecting the current-carrying neutral wire to the grounding terminal in the main switch, which in turn is connected to a water pipe. A neutral wire is also called a grounded wire. Equipment grounding refers to the use of a grounding wire (or the metal raceway) to connect the metal frames of equipment and appliances back to the service cabinet and to the ground wire. The wire connecting a service cabinet to a water pipe is called a ground wire.

■ SYSTEM GROUNDING

The entire wiring system should be grounded by connecting the neutral wire to the ground, either before it enters the house by connecting it directly to the ground or to a water pipe inside the house. Ground connection to a water pipe should be made on the street side of a water meter; if connected on the house side, a jumper cable should be used around the meter. The two hot wires are generally color coded with black or red insulating jackets. The neutral wire (also called the grounded wire) is usually white, but may also be natural gray. The neutral wire is continuous throughout the system, is never interrupted by fuses or switches, and generally is not connected to a hot wire.

For a residential or commercial wiring system, the earth is generally used as the ground. When a metal water supply system terminating in the earth is available, the electrical system is grounded by connecting to a cold-water pipe near the service panel in the house (see Fig. 12.3-23). If a metal water pipe cannot be used, a metal rod is driven into the ground 2 ft.

FIGURE 12.3-21
Maximum Number of Conductors in Trade Sizes of Conduit or Tubing*

Conduit or Tubing Trade Size (in.)

Type Letters	Conductor Size, AWG/kcmil	1/2	3/4	1	1-1/4	1-1/2	2	2-1/2	3	3-1/2	4	5	6
TW, XXHW (14 through 18) RH (14 + 12)	14	9	15	25	44	60	99	142					
	12	7	12	19	35	47	78	111	171				
	10	5	9	15	26	36	60	85	131	176			
	8	2	4	7	12	17	28	40	62	84	108		
RHW and RHH (without outer covering), RH (10 + 8) RHW, THHW	14	6	10	16	29	40	65	93	143	192			
	12	4	8	13	24	32	53	76	117	157			
	10	4	6	11	19	26	43	61	95	127	163		
	8	1	3	5	10	13	22	32	49	66	85	133	
TW	6	1	2	4	7	10	16	23	36	48	62	97	141
	4	1	1	3	5	7	12	17	27	36	47	73	106
THW	3	1	1	2	4	6	10	15	23	31	40	63	91
	2	1	1	2	4	5	9	13	20	27	34	54	78
	1		1	1	3	4	6	9	14	19	25	39	57
FEPB (6 through 2), RHW and RHH (without outer covering)	1/0		1	1	2	3	5	8	12	16	21	33	49
	2/0		1	1	1	3	5	7	10	14	18	29	41
	3/0		1	1	1	2	4	6	9	12	15	24	35
	4/0			1	1	1	3	5	7	10	13	20	29
RH, THHW	250			1	1	1	2	4	6	8	10	16	23
	300			1	1	1	2	3	5	7	9	14	20
	350				1	1	1	3	4	6	8	12	18
	400				1	1	1	2	4	5	7	11	16
	500				1	1	1	1	3	4	6	9	14
	600					1	1	1	3	4	5	7	11
	700					1	1	1	2	3	4	7	10
	750					1	1	1	2	3	4	6	9
THWN	14	13	24	39	69	94	154						
	12	10	18	29	51	70	114	164					
	10	6	11	18	32	44	73	104	160				
	8	3	5	9	16	22	36	51	79	106	136		
THHN, FEP (14 through 2) FEPB (14 through 8) PFA (14 through 4/0) PFAH (14 through 4/0) Z (14 through 4/0) XHHW (4 through 500 kcmil)	6	1	4	6	11	15	26	37	57	76	98	154	
	4	1	2	4	7	9	16	22	35	47	60	94	137
	3	1	1	3	6	8	13	19	29	39	51	80	116
	2	1	1	3	5	7	11	16	25	33	43	67	97
	1		1	1	3	5	8	12	18	25	32	50	72
	1/0		1	1	3	4	7	10	15	21	27	42	61
	2/0		1	1	2	3	6	8	13	17	22	35	51
	3/0		1	1	1	3	5	7	11	14	18	29	42
	4/0		1	1	1	2	4	6	9	12	15	24	35
	250			1	1	1	3	4	7	10	12	20	28
	300			1	1	1	3	4	6	8	11	17	24
	350			1	1	1	2	3	5	7	9	15	21
	400				1	1	1	3	5	6	8	13	19
	500				1	1	1	2	4	5	7	11	16
	600				1	1	1	1	3	4	5	9	13
	700					1	1	1	3	4	5	8	11
	750					1	1	1	2	3	4	7	11

FIGURE 12.3-21
Maximum Number of Conductors in Trade Sizes of Conduit or Tubing* *(Continued)*

Type Letters	Conductor Size, AWG/kcmil	1/2	3/4	1	1-1/4	1-1/2	2	2-1/2	3	3-1/2	4	5	6
XHHW	6	1	3	5	9	13	21	30	47	63	81	128	185
	600				1	1	1	1	3	4	5	9	13
	700					1	1	1	3	4	5	7	11
	750					1	1	1	2	3	4	7	11
RHW	14	3	6	10	18	25	41	58	90	121	155		
	12	3	5	9	15	21	35	50	77	103	132		
	10	2	4	7	13	18	29	41	64	86	110		
	8	1	2	4	7	9	16	22	35	47	60	94	137
RHH (with outer covering)	6	1	1	2	5	6	11	15	24	32	41	64	93
	4	1	1	1	3	5	8	12	18	24	31	50	72
	3	1	1	1	3	4	7	10	16	22	28	44	63
	2		1	1	3	4	6	9	14	19	24	38	56
	1		1	1	1	3	5	7	11	14	18	29	42
	1/0		1	1	1	2	4	6	9	12	16	25	37
	2/0			1	1	1	3	5	8	11	14	22	32
	3/0			1	1	1	3	4	7	9	12	19	28
	4/0			1	1	1	2	4	6	8	10	16	24
	250				1	1	1	3	5	6	8	13	19
	300				1	1	1	3	4	5	7	11	17
	350				1	1	1	2	4	5	6	10	15
	400				1	1	1	1	3	4	6	9	14
	500				1	1	1	1	3	4	5	8	11
	600					1	1	1	2	3	4	6	9
	700					1	1	1	1	3	3	6	8
	750						1	1	1	3	3	5	8

*This table is for concentric stranded conductors only. Conduit fill for conductors with a −2 suffix is the same as for those types without the suffix.

Reprinted with permission from NFPA 70-1993, the National Electrical Code®, Copyright © 1992, National Fire Protection Association, Quincy, MA 02269. This reprinted material is not the complete and official position of the National Fire Protection Association on the referenced subject, which is represented only by the standard in its entirety.

from the house near the service entrance; the neutral wire from the service drop is connected to the rod with a ground wire. The metal rod is driven into the ground near the service entrance and the neutral wire from the service drop is connected to the rod with a ground wire. The neutral wire of the three-wire service is similarly grounded at the utility pole.

■ EQUIPMENT GROUNDING

Unlike the grounded (neutral) wire, the grounding wire and a raceway carry no current, except when a short circuit in the equipment or circuit wiring causes a fault current to flow. Under these conditions, equipment grounding minimizes the possibility of shock to occupants by providing a low-resistance path for the fault current and by opening the circuit overcurrent device in case of excessive flow.

In an electrical system involving conduit, tubing, or metallic cable, metal boxes are used and equipment or fixtures permanently connected at the box become automatically grounded. Portable equipment with cords can be grounded through grounding receptacles and three-pronged grounding plugs. In either case a grounding wire, color-coded green, is included in the cord or cable connecting the box or receptacle to the frame of the equipment.

When a nonmetallic sheathed cable system is used, a bare grounding wire is often included to tie the system to the ground. Permanently connected equipment has its green grounding wire attached to the bare copper wire in a junction box. Grounding receptacles housed in nonmetallic boxes have a special terminal for the grounding copper wire. Portable equipment cords include a green grounding wire that contacts the grounding system through the round or U-shaped prong of the plug.

Equipment grounding provides adequate safety from shocks, except in highly hazardous areas such as near pools, in bathrooms, and in outdoor locations. Receptacles should not be installed closer than 10 ft. from the edge of a pool. Poolside receptacles 10 or 15 ft. from the pool and all outdoor receptacles should be equipped with special devices capable of opening the circuit when even a small amount of current is flowing through the grounding system. These devices are called ground fault circuit interrupters (GFCIs) (see Section 12.3.3.2). The use of GFCIs is required by the NEC, because the severity of electrical shock is greatly increased when water is present.

With certain exceptions, the NEC requires GFCIs in new residential construction around swimming pools, in bathrooms, kitchens, garages, crawl spaces, unfinished basements, and in other locations where a damp floor might be expected, and for outdoor receptacles. It may also be desirable to add them in high-risk areas in existing dwellings. Other high-risk areas that need GFCI protection include laundries and workshops. GFCIs should also be considered for any part of a dwelling built on a concrete slab at or below grade.

12.3.5

LOW-VOLTAGE CIRCUITS

In house wiring systems, low voltage refers to ac power supplied by step-down transformers from the 120/240-volt line voltage service. Although the NEC treats anything less than 50 volts as *low voltage*, in practice the term usually refers to circuits of less than 30 volts.

Less insulation is required for this type of system and the NEC, as well as many local building codes, permit low-voltage wiring without protective raceways. In residential construction, low-voltage wiring is used for bell or buzzer systems, intercoms, control wiring for heating and cooling systems, and other uses requiring relatively small amounts of power.

Because low-voltage wiring is generally used for low-power systems, the conductors are quite small, No. 18 AWG or smaller. In situations requiring multiple and remote switching, low-voltage systems can be used effectively to turn light fixtures or other equipment on and off. The smallness of the wire and elimination of the raceway requirement makes this wiring method considerably less costly than 120-volt installations in tubing or conduit. The equipment is fed from the line voltage circuit, as in standard installations, but the switch opening or closing the circuit is activated by a small, low-voltage power relay.

12.3.6

PLANNING FOR ADDITIONS

If certain facilities or functions are indicated as future additions, such as a basement recreation room, the initial wiring should be arranged so that none of it need be replaced or moved when the ultimate plan is realized. Final wiring for the addition may be left to a later date if desired. If air-conditioning is not included in the initial construction, provisions should be made for a future central system or through-the-wall units.

12.3.7

SOUND CONTROL

Improper electrical installations can reduce the acoustical performance of walls and floor and ceiling assemblies. Causes for reduced performance are (1) open-air transmission paths where wires pass through walls or floor and ceiling assemblies; (2) openings provided for outlets, switches, or fixtures; and (3) noise vibration transmitted by wires or fixtures rigidly connecting two otherwise separated structural members or connecting equipment to the structure.

In multifamily housing, each apartment should be wired as a unit, keeping wiring within the walls of the same unit (see Fig. 12.3-23). Holes should be tightly caulked with resilient, nonhardening compound.

Convenience outlets, switches, telephone, and such other installations as intercoms should not be placed back-to-back. (1) Convenience outlets should be placed at least 36 in. apart on the opposite sides of the wall. In staggered stud walls, installations should be offset horizontally by at least three complete stud spaces (Fig. 12.3-24a); (2) a horizontal distance of 24 in. should be maintained between switches and outlets, and they should not be located in the same stud space (Fig. 12.3-24b); and (3) wall fixtures and equipment should be kept 24 in. apart on opposite sides of the wall (Fig. 12.3-24c).

Where sound control is important, lighting fixtures and other electrical fixtures should be surface mounted (never recessed) and they should not be connected to joists above resiliently mounted finished ceilings or connected across separated ceiling joists (Fig. 12.3-24d). If there is no option but to recess a fixture in a sound-controlled wall or ceiling, the fixture should be mounted in an enclosure that provides at least as much sound attenuation as does the penetrated wall or ceiling.

Appliances can often be installed to

FIGURE 12.3-23 *Each apartment should be wired independently and air passages should be sealed airtight.*

FIGURE 12.3-24 *Electrical outlets, switches, fixtures, and equipment should not be mounted back-to-back, should not be located in the same stud space, and should not be spaced closer than shown in (a), (b), and (c) in this figure. Ceiling fixtures should be surface mounted, as shown in (d), rather than recessed.*

reduce the amount of noise. Vibrating appliances should not be rigidly connected to the structure. For example, a clothes dryer can be mounted on a floating floor and the noise transmission from the appliance can be reduced by use of a neoprene or rubber gasket or mat between the base of the dryer and the floor or by a resilient underlayment between two layers of floor sheathing. Resilient or flexible isolation gaskets should be provided between the appliance and the floor, wall, or counter to isolate vibrations. Electrical connections to appliances should also be flexible, whenever possible; a loop as shown in Figure 12.3-25 will reduce vibration transmission. Vents should also be flexible.

FIGURE 12.3-25 *Wiring in rigid conduit can be connected to vibrating equipment with a flexible cable.*

12.4 LIGHTING

Lighting outlets should be located where they will produce the most desirable lighting effects. Their types should conform with the lighting fixtures or equipment to be used. The lighting fixtures addressed in this section are fixed rather than portable.

12.4.1

INTERIOR LIGHTING

The amount and type of illumination required should be fitted to the various tasks carried out in a space. In living areas, for example, there should be suitable lighting for recreation and entertainment as well as for normal day-to-day activities. In many instances, best results are achieved by a blend of lighting from both fixed and portable luminaires. Good lighting, therefore, requires thoughtful planning and careful selection of permanent fixtures, portable lamps, and other lighting equipment.

12.4.1.1

Switch Controls

Fixed lighting outlets should be controlled by single or multiple wall-switch controls. Wall switches are usually located at the latch side of doors or at the traffic side of arches and within the room or area where the outlets are located. Some exceptions to this practice are switches for (1) the control of exterior lights from indoors; (2) the control of stairway lights from adjoining areas when stairs are closed off by doors at head or foot; (3) the control of lights from the access space adjoining infrequently used areas, such as storage areas. Wall switches are normally mounted about 48 in. above the floor line but may be mounted lower if desired, to align them with doorknobs or to put them within reach of children.

Spaces that have more than one entrance for which switched lighting is needed should be equipped with multiple switch controls at each principal entrance. Prin-

cipal entrances are those commonly used for entry to a room when going from a lighted to an unlighted condition. For instance, a door from a living room to a porch is a principal entrance to the porch. However, this door would not necessarily be considered a principal entrance to the living room unless the front entrance to the house is through the porch.

If this recommendation would result in the placing of switches controlling the same light within 10 ft. of each other, one of the switch locations may be eliminated. Also, when a space may be lighted from any one of several sources, as happens when both general and supplementary illumination are provided from fixed sources, multiple switching is required for one set of controls only, usually for the general illumination source.

12.4.1.2

Outlets and Locations Required

The following requirements comply with the 1990 edition of the NEC, but also supplement it by adding suggestions that exceed the requirements of the NEC. When two different requirements demand a lighting outlet at the same location, only one need be installed. However, additional wall-switch controls may be necessary.

For example, a lighting outlet in an upstairs hall may be located at the head of the stairway, thus satisfying both the need for hall lighting and stairway lighting with a single outlet. But, multiple switch control of this lighting outlet will be required at the head and at the foot of the stairway. In addition, the multiple switch control rule given in Section 12.4.1.1, when applied to the upstairs hall, may require a third point of control elsewhere in the hall because of its length.

■ DUAL-PURPOSE ROOMS

When a room is intended to serve more than one function, such as a combination living and dining room or a kitchen and laundry, the following outlet recommenda-

tions are separately applicable to specific areas. Lighting outlet provisions may be combined in a way that will provide both general overall illumination and local illumination of work surfaces. In considering locations for wall switches, the area is considered as a single room.

■ LIVING AND RECREATION ROOMS

Some means of general illumination in living and recreation rooms is essential. This lighting may be provided by ceiling or wall fixtures, by lighting in coves, valances, or cornices, or by portable lamps (Fig. 12.4-1). Wall-switch-controlled outlets in locations appropriate to the intended lighting method should be provided (see Fig. 12.3-25).

FIGURE 12.4-1 *Valences direct light both up and down for dramatic effect. Coves direct light mostly up; cornices mostly down.*

These recommendations also apply to sunrooms, enclosed porches, television rooms, libraries, dens, and similar areas. Outlets for decorative accent lighting, such as picture illumination and bookcase lighting, should also be considered.

■ DINING AREAS

Each dining room, or dining area combined with another room, or breakfast nook, should have at least one wall-switch-controlled lighting outlet. Such an outlet normally is located over the probable location of the dining or breakfast table to provide direct illumination of the area (see Fig. 12.3-26).

BEDROOMS

Good general illumination is essential in a bedroom. This may be provided from a fixed ceiling fixture or from lighting in valances, coves, or cornices similar to that shown in Figure 12.4-1.

Wall-switch-controlled outlets should be placed in locations appropriate to the lighting method selected (see Fig. 12.3-28).

Light sources located over full-length mirrors, or in the bedroom ceiling directly in front of the clothes closet, may serve as general illumination. A master switch in the master bedroom (as well as at other strategic points in the house), controlling selected interior and exterior lights, may be desirable. Lighting should be provided for each closet, as described later in this section.

BATHROOMS AND LAVATORIES

Illumination of both sides of the face when at a bathroom mirror is essential. This can be accomplished with diffused lighting from a luminous ceiling, or with more concentrated light from several sources (Fig. 12.4-2). A single concentrated light source, either on the ceiling or the wall, generally is not adequate. A 30- or 40-watt fluorescent tube over the mirror may be adequate in a small room with light-colored walls. All lighting should be controlled by wall switches not readily accessible while standing in the tub or shower stall.

A ceiling outlet located in line with the front edge of the basin will often provide, in addition to improved lighting at the mirror, general room lighting and safety lighting for a combination shower and tub. When more than one mirror location is planned, equal consideration should be given to the lighting in each case.

Where an enclosed shower stall is planned, a vaporproof fixture should be provided, controlled by an outside wall switch that cannot be reached from within the stall. A switch-controlled night-light is also recommended.

FIGURE 12.4-2 *Adequate bathroom mirror lighting may consist of multiple light sources on both sides of it and either lights over the mirror or a luminous ceiling.*

KITCHEN

Wall-switched outlets for general illumination and for lighting at the sink are essential. Lighting design should provide for the illumination of the work areas, sink, range, counters, and tables (Fig. 12.4-3).

Undercabinet lighting fixtures within easy reach may have integral switches or may be wall switch controlled. Consideration should also be given to providing lighting inside deep cabinets.

LAUNDRY AREAS

Lighting outlets for fixed lights should be installed to provide illumination of work areas such as laundry tubs, sorting tables, and washing, ironing, and drying centers. At least one outlet in the room should be wall switch controlled. There should be at least one ceiling lighting outlet centered above laundry trays in unfinished basements or other spaces.

FIGURE 12.4-3 *In this kitchen, a ceiling luminaire provides overall illumination, while soffit down lights and undercabinet fluorescent strips highlight the work areas.*

■ CLOSETS

Generally, one lighting outlet should be provided in every closet. Where shelving or other conditions make the installation of lights within a closet ineffective, outlets in the adjoining space should be located to provide light within the closet. Wall switches or automatic door switches are preferred, but pull switches are acceptable.

■ HALLS

These recommendations apply to passage halls, reception halls, vestibules, entries, foyers, and similar areas. Wall-switch-controlled outlets should be installed for proper illumination of the entire area. Particular attention should be paid to irregularly shaped spaces. A switch-controlled night-light should be installed in each hall that serves bedrooms.

■ STAIRWAYS

Fixed wall or ceiling lighting outlets must be installed to provide adequate illumination of each flight of stairs. These outlets should have multiple switch controls at the head and foot of the stairway. Outlets should be so arranged that the stairs are fully illuminated from either floor, while permitting control of bedroom hall lights independently of ground-floor fixtures. Switches should be grouped whenever possible and should never be located so close to steps that reaching for a switch might result in missing a step.

These recommendations apply to every stairway serving finished rooms at both ends. For stairways to unfinished basements or attics, see the appropriate headings below.

■ UTILITY ROOM

In utility rooms, lighting outlets should be placed to illuminate a furnace area, and work area, if there is one. At least one lighting outlet should be wall switch controlled.

■ BASEMENT

For basements, lighting outlets should be placed to illuminate designated work areas, the foot of the stairway, enclosed spaces, and specific pieces of equipment, such as a furnace, pump, or water heater,

and a work bench. Additional outlets should be provided for each 150 sq. ft. of open (unpartitioned) basement space.

In unfinished basements the light at the foot of the stairs should be wall switch controlled near the head of the stairs. Other lights may be pull switch controlled.

In basements with finished rooms, with garage space, or with other direct access to the outdoors, the stairway lighting provisions mentioned earlier in this section apply.

A small pilot light that lights when the basement light is on should be installed in conjunction with the switch at the head of the stairs, when the basement is infrequently visited or when the stair layout makes it difficult to determine visually whether the light is on.

■ ACCESSIBLE ATTIC

There should be one outlet for general illumination, wall switch controlled from the foot of the stairs. This switch should have a pilot light to indicate whether it is on. When there are no permanent stairs, the attic light may be pull switch controlled, if located over the access door. When future finishing of unfinished attic space into habitable rooms is planned, the attic lighting outlet should be multiple switch controlled at the top and the bottom of the stairs. There should be one outlet for each enclosed space.

■ GARAGE OR CARPORT

At least one wall-switched ceiling outlet should be installed in one- and two-car garages. If the garage has no covered access from the house, one exterior outlet should be provided and multiple switch controlled from the garage and the residence.

If the garage is to be used also as a work area or laundry, provisions appropriate to these uses should be made. Additional interior outlets often are desirable even if no specific additional use is planned for the garage. Long driveways need additional illumination, such as post lighting. These lights should be wall-switch controlled from the house.

12.4.2

EXTERIOR LIGHTING

Lights are provided on the exterior of buildings to illuminate entrances so that

occupants and visitors can find their way and to make exterior areas usable for social and recreational activities. Additional exterior lights are sometimes provided for security purposes.

12.4.2.1

Locations and Numbers of General Exterior Lights

Lights are needed on the exterior of buildings in the following locations.

■ EXTERIOR ENTRANCES

One or more switched lighting outlets should be located at the front and service entrances to every building. Switched lighting outlets should also be installed at other entrances. The principal lighting requirements at entrances are the illumination of walkways and steps leading to the entrance and of faces of people at the door.

Where a single wall outlet is desired, location on the latch side of the door is preferable. Switched outlets in addition to those at the door are often desirable for post lights that illuminate terraced or broken flights of steps or long approach walks.

■ PORCHES AND BREEZEWAYS

Porches, breezeways, and similar roofed areas of more than 75 sq. ft. in floor area should have at least one switched lighting outlet. Large or irregularly shaped areas may require two or more outlets. Multiple switch controls should be installed at entrances when a porch is used as a passage between a house and a garage.

■ TERRACES AND PATIOS

One or more fixed outlets on the building wall, or on a post centrally located in the area, are needed for general illumination. These outlets should be switch controlled just inside the door opening onto the area.

■ GENERAL GROUNDS LIGHTING

Floodlights are often used to illuminate the grounds surrounding buildings. Outlets

FIGURE 12.4-4 *Exterior lighting can assure safety and create dramatic views from inside and outside a house.*

may be located on the building itself, on the exterior of an adjacent building, such as a garage, or on posts (Fig. 12.4-4). All outlets should be switch controlled from within the building. Multiple and master switch control from strategic points also is desirable.

12.4.2.1

Exterior Security Lighting

Security demands that a facility be made less vulnerable to unauthorized entry by thieves, vandals, and pranksters. This requires siting to eliminate hidden spots, secure door locks, intruder detection systems, and adequate lighting. Section 8.1. describes door and window design and hardware that will slow down a burglar. Other security measures, including site planning and intruder detection systems, are discussed in Section 12.6.

Intruders prefer to work in the dark. Many security experts recommend the installation of all-night security lights to illuminate doorways, yards, public spaces, and other areas that might invite intrusion (Fig. 12.4-5). Such lighting may be justifiable in high-crime areas and in remote locations. Ordinarily, exterior lights are either manually switched on when they are needed or are automatically controlled by a programmed timer or light-sensitive switch.

Security lights are sometimes installed in addition to general site lighting, but more frequently the two functions are combined.

■ LIGHT TYPES

A variety of lamps and light fixtures are available to suit most security needs. An

FIGURE 12.4-5 *Good security lighting illuminates doors and windows and other intrusion points, and reduces shadowed areas where intruders might hide.*

ordinary incandescent bulb can provide adequate illumination at entry doors. It does not have to be bright.

An incandescent flood lamp will light a larger area, such as a side yard or driveway. Flood lamp fixtures hold from one to three bulbs and allow the lamp to be aimed in whatever direction is necessary.

A mercury vapor lamp costs more than an incandescent lamp but emits twice as much light per watt. It also lasts up to 10 times longer, thus making it more economical than an incandescent bulb. A standard clear mercury vapor bulb has a blue cast and gives a harsh light; a deluxe white bulb has a pink or orange cast and gives a softer, more balanced light.

Ordinary incandescent fixtures can be adapted for mercury vapor bulbs with components available at an electrical supply store.

■ TIMING DEVICES

Like a deadbolt that is never turned, a light that is never switched on performs no security function. A timer that automatically switches on exterior lights can eliminate any safety hazard caused by forgetfulness.

A time-of-day device is an electric clock that turns individual fixtures on and off

FIGURE 12.4-6 *A light might be plugged into a tabletop timer.* (Courtesy Intermatic, Inc.)

at specified times (Fig. 12.4-6). Exterior lighting can be wired to a timer through its own circuit. The disadvantage of using a timing device is that nights vary in length throughout the year, which means the timer will have to be periodically adjusted. Some timers can be set to vary the time they go on and off each day, which can give a vacant dwelling a lived-in appearance.

A photoelectric device senses light. It will automatically turn a light on at dusk and off at dawn. The device can be an integral part of a fixture or a screw-in device that will adapt a regular fixture.

An automatic sensor (motion detector) can detect anyone walking up to an entrance at night and automatically turn on a light for 2-1/2 min. Such a device is both a convenience for a dwelling's residents-and a deterrent to burglars. A photoelectric cell prevents the system from turning on during the day.

12.5 LIGHTNING PROTECTION

Lightning strikes are a source of fires in unprotected buildings. This section discusses lightning protection systems intended to eliminate this hazard. Sprinkler systems are covered in Section 11.4; fire detection devices in Section 12.7.

12.5.1

LIGHTNING

Lightning is a discharge of static electricity, a giant spark that jumps between two charged bodies such as two clouds or a cloud and an object on earth. A study by the Lightning Protection Institute (LPI) shows that dwellings are the most frequent targets of damaging lightning, and lightning is also a major cause of rural fires in other types of structures. Lightning also strikes large buildings, especially tall ones, often, but most remain undamaged because they have lightning protection systems.

Lightning can enter a building by (1) striking the building directly, (2) striking a metal object such as a TV antenna and then traveling down into the dwelling, (3) striking a nearby tree and leaping over to the building, or (4) by following a power line or ungrounded fence.

On its way to the ground, lightning follows the path of lowest electrical resistance. Hence it frequently follows home wiring or plumbing lines, causing fires and damaging electrical appliances.

Many people think that their television antenna provides lighting protection. Although it is like a lightning rod in that it is a target for atmospheric discharges, a television antenna system cannot safely handle the current of a bolt of lightning even when it is grounded and equipped with a lightning arrestor. The lightning can jump from a television antenna to its circuit, to plumbing, or to other electrical wiring, and cause damage.

12.5.2

LIGHTNING PROTECTION SYSTEMS

A good lightning protection system should provide an easy, direct path for an electrical current to follow to reach the ground. It must also prevent injury, damage, or death as a lightning bolt travels that path. Many insurance companies offer a reduction in premiums for a house with a properly installed lightning protection system.

A lightning protection system includes (1) air terminals (lightning rods), (2) main conductors, (3) branch conductors, (4) lightning arrestors, and (5) ground connections (Fig. 12.5-1).

Air terminals, which are pointed copper or aluminum rods (Fig. 15.5-2), should be placed on the highest points of the roof and on every projection, such as a chimney or dormer (see Fig. 12.5-1). Terminals should be sized, spaced, and anchored according to the requirements of the National Fire Protection Association's (NFPA) *Lightning Protection Code*. They are usually between 10 and 24 in. long but may be longer. Those less than 24 in. long should be spaced not more than 20 ft. apart. Longer ones may be placed up to 25 ft. apart. In any case, there should be a terminal within 2 ft. of each roof end.

Ground connections can be made by (1) clamping the main conductor cable to a copper-clad or galvanized steel rod that has been driven at least 10 ft. into the ground; (2) burying a stranded copper conducting cable in a trench; (3) clamping copper conducting cable to a sheet metal plate and burying the assembly; or (4) clamping the conductor to a metal water pipe.

The chief requirement for an efficient lightning protection system is a solid connection between the system and the earth. Ground connections of galvanized steel, copper, or copper-clad steel are preferred; aluminum should never be used for ground connections because it corrodes when it comes into contact with the soil.

Every lightning protection system needs at least two ground rods placed as far apart as possible. Usually, they are placed at opposite corners of a building. They should reach below and away from the building to avoid lightning damage to exterior walls and to be in contact with the earth.

Although lightning systems can be added to a existing structure, there are definite advantages to installing them as part of new construction: (1) protection is effective during the construction process; (2) system parts can be more readily protected and concealed; and (3) installation costs may be lower. Rods, cables, and connectors should be Underwriter's Laboratory (UL) labeled. Upon completion of an application blank, UL will send a 4 × 2-1/2-in. brass Master Label plate that shows that a system meets UL requirements.

12.5.2.1

Personal Safety

More people are injured by lightning in unprotected houses than anywhere else. For personal safety during lightning storms, occupants should stay away from metal objects such as prefabricated fireplaces, stoves, water faucets, appliances, telephones, and metal windows.

FIGURE 12.5-1 *The components of a lightning system for a house include (1) air terminals; (2) down-lead conductors; (3) at least two grounds; (4) roof projections tied into the conductor system; (5) tree protection; (6) at least two terminals on chimneys; (7) rodded dormers; (8) arrestors on antennas; (9) tie-in to conductor system of gutters within 6 ft. of conductor; and (10) arrestors on overhead power lines.* (From U.S. Department of Agriculture, Farmers Bulletin Number 2136, "Lightning Protection for the Farm".)

FIGURE 12.5-2 *Air terminals are the only visible parts of a concealed lightning protection system.* (From U.S. Department of Agriculture, Farmers Bulletin Number 2136, "Lightning Protection for the Farm".)

12.5.2.2

Tree Protection

Any tree taller than a dwelling and within 10 ft. of that dwelling should have lightning protection.

If a building with a lightning protection system is within 25 ft. of a protected tree, the two systems may be interconnected. If there are several large trees in a row, a common grounding may be used as long as the trees are not more than 80 ft. apart.

A tree's lightning protection (Fig. 12.5-3) should consist of (1) a main terminal (special tree) point placed as high as possible within the tree; (2) miniature terminal points clamped on main branches; (3) the main conductor, usually 32-strand, 17-gauge copper cable, run from the main terminal down; (4) branch conductors, usually 14-strand, 17-gauge copper, run from the miniature terminals to the main conductor; (5) grounds, usually two or more 1/2- to 3/4-in.-diameter rods, driven to a 10-ft. depth at points away from the main root system.

Trees having trunks more than 3 in. in diameter should have two main conductors, one on each side of the trunk. Cables should be copper because the color blends unobtrusively with a tree and because other metals may corrode and harm a tree. Cables should be attached with fasteners that hold them away from the trunk at intervals 3 to 4 ft. apart (see Fig. 12.5-3).

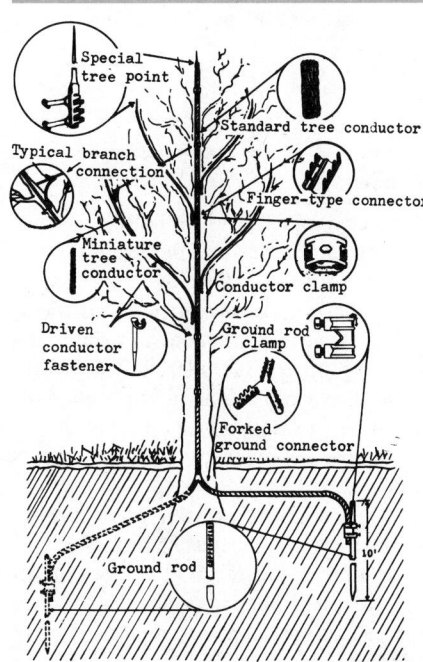

FIGURE 12.5-3 *A tree's lightning protection system consists of (a) the main terminal (special tree point); (b) miniature terminals (branch connection); (c) main conductor; (d) branch conductor; and (e) grounds.* (From U.S. Department of Agriculture, Farmers Bulletin Number 2136, "Lightning Protection for the Farm".)

12.5.2.3

System Maintenance

Lightning systems should be inspected yearly for (1) bent, loose or missing air terminals; (2) broken conductor cables; and (3) loose connecting clamps.

12.6 INTRUSION PREVENTION AND DETECTION

Security measures are often necessary to make a facility less vulnerable to unauthorized entry by thieves, vandals, and pranksters.

Burglary, a frequent problem in both urban and rural areas, is a primary threat to security. Nationwide, a burglary occurs about every 10 seconds. About half of all burglars gain entry through an unlocked window or door.

Would-be intruders prefer to work quickly, quietly, and in the dark. If more than 4 minutes are required to gain entry, most burglars will usually give up the attempt. Good burglary protection devices will either slow down or frighten away a burglar. The intent of security devices is to save lives and property by warning residents and neighbors when someone tries to enter a property without authorization.

This section discusses site planning (Section 12.6.1) and intruder detector systems (Section 12.6.2). Locking devices and the design of doors and windows to deter intrusion are discussed in Section 8.1; exterior security lighting in Section 12.4.2.1.

12.6.1

SITE PLANNING

The first line of defense against intruders is the area between a building and its property lines. Building location, terrain, landscaping, lighting, and structural barriers are important considerations in creating a sense of physical security. The location of public places, and their convenience and visibility can encourage sociable gatherings of sociable people and discourage potential criminals.

Barriers that define and protect space may be physical or psychological. Physical barriers include a building's shape, as well as any fences, gates, and doors. The shape of a building can act as a deterrent when all areas are visually accessible and no areas are hidden from view. Psychological barriers do not necessarily bar movement but rather symbolize boundaries. They include building elements such as open gateways, light standards, short runs of steps, and textural changes in walking surfaces.

12.6.1.1

Site Choice

Where a dwelling is located on the lot, as well as where it stands in comparison to other buildings, affects whether the building discourages criminal activity. For example, placing a building far back from the street can increase its vulnerability to crime. A building should be placed so that its entrances and windows are clearly visible to passers-by, neighbors, and police patrols.

12.6.1.2

Fences

A fence should serve as both a psychological and a physical deterrent. It can consist of wood, chain links, shrubbery, or masonry. In many states it is difficult to legally convict someone of trespassing unless property boundaries are clearly marked. A fence should hinder a burglar's attempt to get onto a property, but should not provide cover for an attempted break-in.

A 6-ft.-high brick wall gives a false sense of security because although it presents a major barrier, it also hides an intruder's preliminary activities. Other fences that protect but do not obscure are better security choices.

A wooden picket fence may be a psychological deterrent but is easy to climb over. A thick hedge, particularly a thorny one, is harder to climb through and over. If either one is no more than 40 in. high, it is easy for neighbors or passersby to see an intruder.

A 6-ft.-high chain-link fence is a good physical barrier that does not obscure vision. It should be installed as close to the ground as possible so that a potential intruder cannot crawl under it. A hedge no more than 40 in. high on one or both sides of a chain-link fence can soften its appearance.

Access into the protected area should be through a gate. A simple bell arrangement can sound a signal whenever the gate is opened or closed.

12.6.1.3

Landscaping

Plants that hide a dwelling can also hide an intruder. Trees should be far enough away from a dwelling so that a burglar cannot climb them and enter the second story. Bushes should be low enough so they do not obscure windows or doors. Hedges used as fences should be no more than 40 in. high. Landscaping should not obscure the house from the street. Small plants and floral borders can serve to define private and public spaces.

12.6.2

INTRUDER DETECTION SYSTEMS

Various alarm systems are available to warn against intruders. Alarm systems predate the telephone. Alexander Graham Bell's assistant Thomas Watson acquired much of his electrician's skill by installing burglar alarms during the 1860s; Alexander Pope patented one of the earliest of these systems in 1853.

Intruder detection devices should be used in conjunction with adequate structural protection such as strong doors and windows and secure door and window hardware (see Section 8.1), adequate site planning (see Section 12.6.1), and strategically placed lights (see Section 12.4.2).

Perimeter Intrustion Detection

Exterior doors, windows, and walls make up the perimeter of a building. Perimeter detectors work by setting up an electronic boundary line around this perimeter; breaching the boundary activates the detector's signaling device.

■ DETECTOR ELEMENTS

Most detectors consist of a protective circuit, a control instrument, and a signaling device. The protective circuit may be made up of light beams, sound waves, or wires that detect the presence of an intruder.

A control instrument monitors a protective circuit. When a circuit has been breached by an intruder, the control instrument activates the signaling device.

A control unit may be mounted indoors or out, depending on the system. Many units have a time delay so that occupants may arm or disarm the system without triggering the alarm. Because a good lock is as important on a control unit as it is on a door, key-operated systems are preferable. Some control units will sound an alarm if tampered with.

A signaling device issues the actual warning. It may be a light, bell, buzzer, horn, or siren. The signaling device may turn on exterior flood lights. It may also transmit a warning over the telephone. The telephone warning may go directly to a police station or to the police through a contracted service, which is often the same company that installed the system.

■ EFFECTIVENESS

A perimeter detector's alarm or flood lights will only frighten a potential burglar away if it instills a fear of capture. In order for a burglar to be caught, someone else must be able to hear the alarm.

Neighbors or occupants should respond to the warning by calling police. Trying to capture a burglar is very dangerous and should be left to professionals.

Police response time is the most important factor in capturing a potential burglar. If the response is within 30 seconds of the alarm, the thief will probably be captured (Fig. 12.6-1). Silent alarm devices that alert the police are available. They may

FIGURE 12.6-1
Police Response Time and Probability of Capture of Burglar

Response Time after Alarm Sounds	Probability of Capture (%)
30 sec.	100
60 sec.	90
2 min.	75
4 min.	50
10 min.	20

make capture more sure, but harbor the inherent danger of permitting an intruder to enter the building thinking he or she has not been detected. This can present a heightened danger to an occupant who knows the alarm has been sounded and attempts to interfere with the intruder, or who simply stumbles on the intruder. It is better to frighten a burglar away than face him or her in the dark.

Unfortunately, both noisy and silent intruder alarm systems are subject to false alarms, which diminish their credibility. If false alarms are common, neighbors and police may not respond when a real alarm is sounded, and even occupants may assume that another false alarm has been sounded when, in fact, there is a real intruder.

■ SIGNALING DEVICES

These devices may be selected to alert either occupants and neighbors, a private security company, or the police, or all of them at the same time.

A local alarm is usually a loud bell or siren mounted on an exterior wall of the building so it can be heard by most neighbors. Some alarms continue to ring until they are manually shut off; others sound for a fixed period of time, usually 5 minutes, or are adjustable within a range of 4 to 12 minutes.

A central station alarm is a system in which the protective circuit, or a special dialing apparatus on a regular telephone line, connects to a private company that monitors it. This is often the same company that installed the system.

When an alarm signals, the company will either relay the information to the police or have their own guards check the alarm. Their cost includes initial installation and a monthly maintenance fee.

In some jurisdictions an intruder alarm may be connected directly to a police station. In those locations, extra care should be taken when selecting a detection system, because some of them are very sensitive and prone to signaling false alarms. Some police departments issue fines if they have to respond to too many false alarms. Neighbors may also be annoyed by frequent soundings of a burglar siren.

A telephone dialing system uses an automatic telephone dialing unit with recorded tapes as its signaling device. When activated, the telephone dialer calls a preprogrammed number. Some units can be programmed to dial another number if the first is busy; other units will keep dialing the same number until someone answers.

Many units come with its message installed as a continuous loop, which is a good idea because of the possibility that an alarm call may be put on hold. These units are called *central station systems* when the programmed number is that of a police department or monitoring company.

■ DETECTOR TYPES

Detectors may have their protective circuit, control instrument, and signaling alarm either all in one device or each may be in a separate device. They may be either wired or wireless. The normally open (NO) and normally closed (NC) systems discussed in this subsection are called wired perimeter detection systems, because their detectors are directly wired to their control and signaling device. The carrier-current and radio-frequency systems discussed below are called wireless because their signals are sent to their control and signaling device by a transmitter that is not hardwired to them. Although spot detectors are in fact wireless, they actually fit into neither category because they have no separate signaling device.

Some wired systems are versatile and accept both NO and NC intrusion detectors. These combined NO and NC wired systems have the advantage that many forms of protection can be combined into one complete security system.

FIGURE 12.6-2 *(a) Separating the housing of a magnetic switch activates the signaling device. Opening a door or window releases the plunger on (b) a surface-mounted, and (c) a recessed-mounted detector, and activates a signaling device.*

Both wired and wireless systems have centrally located control and signaling devices, make the same provisions to permit ventilation through windows, and have the same number and type of wired intrusion detectors (Fig. 12.6-2), which attach to doors and windows. Opening the door or window will affect the control device and activate the signaling alarm. They may also have window foil or other glass breakage detectors. Breaking glass that has a glass-breakage detector attached will cause the control instrument to activate the signaling alarm. When these are used with operable windows, contact switches are provided to permit opening the window. Both wired and wireless intrusion detector systems may have photoelectric detectors that send a beam of light between a sending and a receiving station. Stopping the path of the light beam breaks the circuit and causes activation of the signaling alarm.

Installing a wired system can be tedious. Wiring must be run from room to room, connecting each door and window back to the central unit. This wiring may be difficult to conceal, and if left exposed is unsightly. Wireless systems are easier and more convenient to install because they use transmitters and not extensive wiring to relay a signal to their signaling alarm. Even so, NO- and NC-wired systems are usually less expensive than other systems.

NORMALLY OPEN (NO) WIRED SYSTEMS A normally open (NO) wired system has switches in the open position, which is the equivalent of a light switch in the off position. Opening a protected door or window turns the system on. Because current flows through the system and activates the alarm only when electricity completes a loop, removing any switch or breaking any wire may disable the entire system. Accessories such as a smoke detector can be added to an NO system.

NORMALLY CLOSED (NC) WIRED SYSTEMS A normally closed (NC) wired system is less susceptible to defeat than an NO system. Electricity flows through an NO system in a constant loop. An interruption in the current, including breaking or cutting a wire, triggers the alarm. This feature can be an advantage, because a broken wire or sensor will set off the alarm. Conversely, a system failure in an NO system would go unnoticed until a routine check was made.

CARRIER-CURRENT WIRELESS SYSTEMS A carrier-current wireless system sends its signal to the control device through regular 120-volt *ac* house wiring. Its intrusion devices are connected to a transmitter that plugs into a wall outlet. The transmitter is the equivalent of a protective circuit. The receiver is a control device that plugs into an outlet in a central area such as a hall. It houses the signaling device.

RADIO-FREQUENCY WIRELESS SYSTEMS A radio-frequency wireless system sends its signal from intrusion device transmitters to a control device receiver by radio frequencies. The transmitters operate on batteries. The receiver usually runs on household current with battery backup. In some radio frequency systems, the battery may sound an alarm when there is a power failure. In others, the battery operates the system on a standby basis only during a power outage. Batteries usually last a year but should be checked occasionally to make sure the system operates properly.

SPOT DETECTORS In a spot detector all of the detector elements are in a single unit designed to be attached to a door or window. The unit buzzes loudly when the door or window is opened. This type of detector is useful in apartments and other areas with few entrance points, but an intruder who takes the time to see where the noise is coming from can often shut it off.

12.6.2.2

Area Controls

Area controls protect the space within a building. They consist of devices either to fool an intruder into thinking a building is occupied or to detect intrusion within the space.

■ UNAUTHORIZED ENTRY PREVENTION

Most of the time, a burglar who thinks someone is inside will not break into a building. A burglar can be discouraged by timer-controlled lights and sounds, and by decals and labels.

TIMER-CONTROLLED LIGHTS AND SOUNDS A cycling timer is a 24-hr. electric clock that turns on lights or other appliances at designated times and leaves them on for specified intervals. Some models have a cycle override feature that allows both timed and normal use. Others have multiple cycles that switch lights and appliances on and off, from 1 to 12 times each day. Still others feature variable cycles that switch on at random times (Fig. 12.6-3).

Lights plugged into four timers set in various rooms in a dwelling will give the impression that someone is home. Figure 12.6-4 shows three possible timer use configurations that will suggest various living patterns. If other living patterns are used, it is important to remember to overlap the times that lights are on in at least two rooms.

The sounds produced by radios and television sets can give the illusion that someone is in a building when no one is there.

FIGURE 12.6-3 *This type of variable timer plugs into a wall receptacle. The equipment to be timed is then plugged into the bottom of the timer.* Courtesy Intermatic, Inc.

Radios or television sets can be simply left on or can be plugged into a timer.

DECALS AND LABELS Some security systems supply decals for doors and windows that indicate that the property is protected. Protective foil tape on windows also serves to label a building as having security protection.

Some police departments and other organizations lend diamond-tipped engraving pens to mark valuable property with the owner's name and identification. Others provide warning decals for those who have marked their valuables to place in doors and windows to discourage burglary (Fig. 12.6-5).

■ INTERIOR AREA DETECTORS

Interior detectors protect the space within a building. In order to activate these detectors, an intruder must have breached the perimeter and be inside the building. Interior area detectors can be used instead of, or in addition to, perimeter detectors. Some types of area detectors cannot be used in dwellings with pets or other buildings with guard dogs, because they are unable to distinguish between the movement of animals and those of human intruders. Combination protection systems utilizing various types of perimeter and interior area detector elements are available.

FIGURE 12.6-4
Typical Living Patterns with Four Timers

Room	A		B		C	
	On	Off	On	Off	On	Off
Kitchen	6:00 P.M.	10:00 P.M.	7:00 P.M.	12:00 A.M.	8:00 P.M.	1:00 A.M.
Living room	7:30 P.M.	11:00 P.M.	8:00 P.M.	1:00 A.M.	8:30 P.M.	12:00 P.M.
Bedroom	10:30 P.M.	1:00 A.M.	12:00 A.M.	3:00 A.M.	10:00 P.M.	2:00 A.M.
Bathroom	11:00 P.M.	6:00 A.M.	11:30 P.M.	6:00 A.M.	9:00 P.M.	6:00 A.M.

(Courtesy Intermatic, Inc.)

As is true for perimeter detectors, the effectiveness of area control detectors depends on the burglar's fear of capture, people hearing the alarm and calling the police, and police response time. See Section 12.6.2.1 for further discussion about detector effectiveness.

Area control detectors have a protective circuit, a control instrument, and a signaling device.

Sound waves, radio frequency waves, and light beams are the equivalents of the protective circuit in a wired system. The control instrument and signaling device are usually relatively small and unobtrusive.

PHOTOELECTRIC AND INFRARED SYSTEMS Photoelectric and infrared systems can be either perimeter or area detectors and can be connected to either NC or NO systems. A transmitter projects either visible light, electromagnetic waves, or an infrared beam to a receiver. The light can be reflected by mirrors to weave a kind of electronic spiderweb to increase the area protected. If the beam is broken by an intruder, it breaks the circuit and the alarm sounds. Some infrared systems detect the presence of body heat and will sound an alarm even if an intruder is wearing heavy clothing and standing 30 ft. away.

ULTRASONIC MOTION DETECTOR An ultrasonic motion detector blankets a room with inaudible sound waves in the 18,000 to 45,000 Hz range. The waves set up an energy field. Some detectors set up an elliptical pattern ranging from 12 × 35 ft. to 5 × 20 ft. A person entering the energy field, also known as the trap zone, interrupts the pattern and an alarm sounds. Most of these systems have an adjustable

sensitivity control to guard against setting off the alarm with normal movement or with ordinary noise, such as that from fans.

An ultrasonic detector, which is about the size of a table radio, is sometimes designed to look like a set of books or a small stereo speaker. Some units require no wiring and are portable. If used while the building is occupied, people and animals must stay out of the guarded area to avoid tripping the alarm.

MICROWAVE SYSTEM A microwave detector is a type of radar. Its electromagnetic

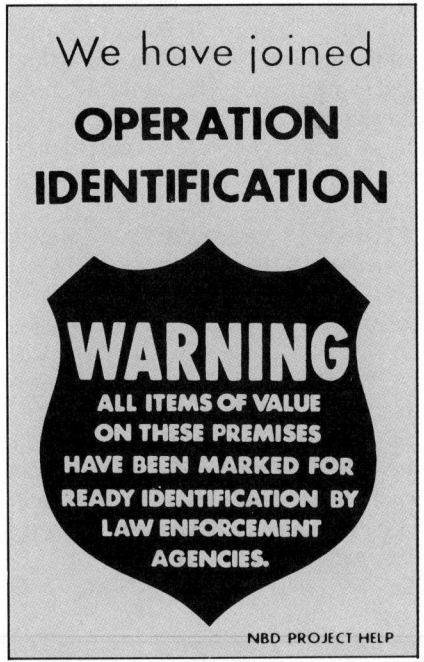

FIGURE 12.6-5 *A decal may discourage a burglar because labeled property is hard to dispose of.*

waves penetrate through wood and plaster but not brick or concrete. This system can protect several rooms using a device that both transmits and receives the signal. This type of transmitter and receiver is placed in a central location in a building. Unfortunately, a microwave system can be a source of false alarms if the energy field is too large. Movement in an adjacent room or apartment, or pedestrian traffic in front of the building, can trigger the alarm. Care in the adjustment and placement of the system should avoid these problems.

SONIC SYSTEM   A sonic system resembles an ultrasonic system, except that audible sound waves set off the system. A receiver detects echoes coming from every object in the room. When an intruder enters, the echo changes, triggering the alarm.

AUDIO DETECTOR   An audio detector registers sharp sounds that would indicate an intrusion, such as wood splintering or glass breaking.

FLOOR MAT DETECTOR   A floor mat detector is part of a wired perimeter system. It is mentioned here under area controls because such a device cannot be activated until the intruder enters the building.

12.7 FIRE SAFETY

Fires are potentially more damaging to a building's occupants than burglars. A fire may cause not only the loss of property, but also the loss of lives. In 1989 about 5400 people other than firefighters died in fires in the United States, and an additional 28,540 were injured. Most fires occur between midnight and 6 A.M., when occupants are asleep. The primary consideration in a fire is to evacuate all occupants from the building as quickly as possible.

The National Fire Protection Association's NFPA 550, "Guide to the Fire Safety Concepts Tree," identifies the various components of fire safety and shows their relationship to each other. The major components are prevention, detection, containment, control, and extinguishment. Prevention is discussed in Section 12.7.1; detection in Section 12.7.2; containment and control in Section 12.7.3; and extinguishment in Section 12.7.4.

The tree approach of NFPA 550 demonstrates that no single component of fire safety is sufficient by itself. Rather, a balanced approach is necessary, employing several or preferably all of them. The NFPA *Life Safety Code* also recognizes this need for a balanced approach to fire safety. This section shows some of the measures that can be taken to afford fire safety, with special emphasis on dwellings. For a more detailed review of the tree concept, refer to NFPA 550.

12.7.1

FIRE PREVENTION

The only way to prevent damaging fires from occurring is to eliminate or control potential sources of fire, to separate them from combustible materials, or to eliminate the combustible materials altogether.

12.7.1.1

Fire Sources

A faulty electrical system is a potential source of fire and a potential source of electrocution. Methods of dealing with electrical systems are discussed in Sections 12.3.3.2 and 12.3.4.6.

Lightning strikes can also be a source of fires. Systems to eliminate that danger are covered in Section 12.5.2.

A kitchen is one of the most fire-prone rooms in a dwelling. Gas ranges or stoves should not be located near windows that can be opened because of the risk of curtains catching fire.

12.7.1.2

Fire-Retardant Products

Current consumer protection laws require that mattresses and carpeting be treated for flame resistance. The states of California, Illinois, Minnesota, and Massachusetts have adopted restrictions on upholstered furniture that should ensure that if a compliant piece of furniture catches fire, then the room will remain at temperatures from which occupants can escape for 5 minutes after a fire starts and remain safe for 30 minutes from toxic levels of carbon monoxide. Similar regulations are under consideration in a dozen other states. Older mattresses, carpeting, curtains, and drapes can be treated with homemade fireproofing chemicals (Fig. 12.7-1).

Pigmented or clear fireproof coatings, known as intumescent coatings, can slow the progress of a fire by swelling to form a protective foam. The foam slows the progress of a fire, reduces smoke, and delays the buildup of toxic gases. These coatings are recommended for corridors or areas leading to fire exits.

12.7.1.3

Fire-Resistant Plants

In some parts of the country, forest and brush fires routinely threaten dwellings. In those areas where such plants will grow, fire-resistant ground cover should be planted within 100 ft. of a house, and shrubs and trees should not be closer than 30 ft. and not less than 18 ft. apart (Fig. 12.7-2). Under those conditions a fire cannot easily jump from one to the next and then to the house.

12.7.2

FIRE DETECTION AND ESCAPE

Most fires take place between midnight and 6 A.M., when most people are usually asleep. Over half of the Americans who lose their lives in fires yearly could have been saved if they had had smoke and fire alarms and a ready means of escape. Fire has four distinct stages: (1) incipient, (2) smoldering, (3) flame, and (4) high heat.

FIGURE 12.7-1
Flame-Proofing Recipes

Material to Be Treated	Recipe*
Permanent-press fabrics (blends of vegetable and synthetic fibers)	12 ounces diammonium phosphate 2 quarts hot water
Untreated vegetable-fiber fabrics	7 ounces borax 3 ounces boric acid 2 quarts hot water
Paper and cardboard	7 ounces borax 3 ounces boric acid 5 ounces diammonium phosphate 1 teaspoon liquid dishwashing detergent 3-1/2 quarts hot water
Christmas tree	4-1/2 gallons sodium silicate (water glass) 2 quarts hot water 2 teaspoons liquid dishwashing detergent

To treat fabrics and paper products: (1) dissolve the recipe ingredients in hot water, (2) spray or dip the product in the appropriate solution, (3) allow to dry. Treat Christmas trees twice.

FIGURE 12.7-2
Fire-Resistant Plants

Type of Plant	Plant Name
Ground covers	Rosea ice plant (*Drosanthemum hispidum*) Jelly beans (*Sedum rubrotinctum*) Trailing gazania (*Gazania uniflora*) Trailing African daisy (*Osteospermum fruticosum*) Kentucky bluegrass (*Poa pratensis*) Rye grass (*Lolium*, several species) White clover (*Trifolium repens*)
Shrubs	Oleander (*Nerium oleander*) Elephant bush (*Portulacaria afra 'variegata'*) Bluechalksticks (*Senecio serpens*) Toyan (*Heteromeles arbutifolia*) Common lilac (*Syringa vulgaris*)
Trees	Gum trees (*Eucalyptus*, several species) Carob (*Ceratonia siliqua*) California pepper (*Schinus molle*) California laurel (*Umbellularia californica*) Cottonwood (*Populus deltoides*)

12.7.2.1

Smoke and Fire Alarm Systems

Long before flames actually break out, occupants may have been killed or rendered unconscious by smoke, noxious gases, or superheated air. Smoke and heat detection, therefore, are an essential part of any fire alarm system. From the time flames break out, a person may have less than 4 minutes in which to escape.

Smoke signals the presence of fire far sooner than heat does; therefore, a device sensitive to smoke is much more important than one that is sensitive to heat. Smoke detectors for home protection typically are of the ionization or photoelectric type. Heat detectors generally have a metal strip that is either melted or distorted by heat to trip the alarm.

Cost should not be a problem in smoke detector selection. Battery-powered ionization detectors cost as little as 10 dollars; battery-powered photoelectric types may cost a little more. Batteries need to be replaced approximately once a year. Many insurance companies provide rate discounts to residents who install the more expensive, permanently wired, combination fire and security protection systems.

◼ IONIZATION DETECTORS

In an ionization detector a tiny amount of radioactive material causes the air between two electrodes to conduct an electric current (Fig. 12.7-3). When smoke particles are present, current flow is reduced and the alarm is set off.

The use of radioactive material should not be of concern to a building's occupants. According to the U.S. Nuclear Regulatory Commission, holding an ionization smoke detector close to the body for 8 hours a day for an entire year would expose a person to only one-tenth as much radiation as he or she would get in one transcontinental round-trip airline flight.

Compared with photoelectric detectors, ionization detectors respond more rapidly to the small wispy smoke particles of a fast-flaming fire, but react more slowly to the large smoke particles of a slow-smoldering fire. The average ionization detector is likely to give 20 to 30 seconds extra notice of a fast fire.

◼ PHOTOELECTRIC DETECTORS

A photoelectric detector aims a light beam into a chamber that contains a light-sensitive photocell. The cell detects the light only when smoke enters the chamber and scatters the beam. When enough light reaches the cell, the alarm goes off (Fig. 12.7-4).

A slow-smoldering fire, for example, a cigarette dropped on bedding or upholstered furniture, is the most common home fire. The photoelectric detector is likely to give 20 to 25 minutes faster notice of a slow-smoldering fire than an ionization detector will but is only less than a minute slower in detecting a fast fire. For this reason, many experts prefer the photoelectric type of detector.

◼ HEAT DETECTORS

Some manufacturers offer a heat-sensing device as an optional part of their smoke detectors or as a separate product. Most

FIGURE 12.7-3 *In an ionization detector (a) a piece of radioactive material ionizes the air to cause an electric current to flow across an air gap. (b) Smoke reduces the current flow, tripping the alarm.*

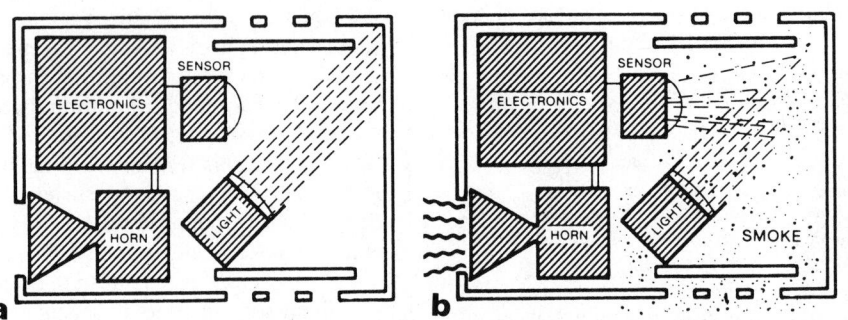

FIGURE 12.7-4 *In a photoelectric detector, (a) the sensor does not ordinarily see the light; (b) when smoke is present, light rays hit the sensor and sound the alarm.*

heat detectors use a metal strip that is either melted or distorted by excessive heat.

Heat detectors may be either the fixed temperature or rate-of-rise type. Heat detectors built into smoke detectors are usually the fixed temperature type, which trips the alarm when a certain temperature is exceeded. Similar in appearance, separate detector devices may also be the fixed temperature or rate-of-rise type; they sound their own alarms or send electrical signals to a central alarm system.

To be effective, heat detectors must be close to the source of a fire. They are especially useful in environments that could fool or disable a smoke detector, such as a kitchen where grease particles in the air might cause a smoke sensor to give false alarms. Heat detectors may be useful in other areas that are unsuitable for smoke detectors, such as furnace rooms, attics, and garages.

However, a heat sensor should not be the only fire protection in the dwelling. It is more often smoke than heat that causes injury and death in a house fire. A heat detector may ignore a smoldering fire that is emitting lethal amounts of smoke, carbon monoxide, and other toxic gases.

■ **DETECTOR PLACEMENT**

The NFPA's *Life Safety Code* recommends the installation of at least one smoke detector in each dwelling unit. In some jurisdictions, such a detector is mandated by law. The detector should be powered by the dwelling's electric service and should be placed in the hallway near bedrooms. In multilevel dwellings the NFPA recommends that a detector be placed at the top of the stairs. The CABO's *One- and Two-Family Dwelling Code* also requires a hard-wired smoke detection system in new residential buildings.

Many electrically powered smoke detectors are part of security systems wired to a central station. Although such systems are expensive to install, some insurance companies offer discounts of 2.5% to 5% or more on the dwelling's fire insurance rate when detectors have been installed. When centralized systems are used, the alarm bell and test button should be installed in the master bedroom or other suitable location in the sleeping area. Because of the increased possibility of failure originating in another system, the operating power supply should be from a separate and independent transformer, rather than from the entrance signal or other multiple-use transformer.

Model codes such as the BOCA's *Basic Code* and some state and municipal building codes require smoke detectors in bedroom hallways. Many codes permit the less expensive, battery-powered models. Some communities require photoelectric detectors. Residents should check local codes before buying and installing smoke detectors.

The Consumer's Union suggests that both photoelectric and ionization detector types may be useful. Though some fire alarm systems combine ionization and photoelectric capacity in one unit, two separate detectors may be less expensive and offer better protection.

For a basic two-detector system, an ionization detector should be placed in the hallway just outside the bedroom doors; a photoelectric detector should be located in the general living area. It should not be located where cooking might set it off accidentally. In a larger house, a heat detector may be appropriate in the kitchen. In a two-floor house a detector should also be placed at the top of the stairs.

Ideally, smoke detectors should be mounted in the middle of the ceiling. If this conspicuous location is objectionable, they may be mounted on a wall between 6 and 12 in. below the ceiling or on the ceiling not closer than 6 in. to the wall or the corner. Air currents sometimes do not reach the corners of rooms, thus possibly delaying smoke detection.

Smoke detectors should not be mounted in front of an air supply duct or near a furnace cold-air return, because excessive airflow prevents prompt detection of smoke. Such locations, however, are acceptable for heat detectors.

Smoke detectors should not be located on a ceiling or wall that is substantially warmer or colder than the rest of the room. In this case, an invisible thermal barrier near the wall or ceiling surface can keep smoke from reaching the detector. The thermal barrier may be a problem in poorly insulated dwellings or mobile homes.

When a detector must be mounted under any of the adverse conditions discussed above, a heat detector rather than a smoke detector should be used. The heat detector is preferred here because it is activated by temperature change rather than smoke delivered by airflow.

For added protection a dwelling may be equipped with extra detectors. A photo-electric or heat detector can be placed in the basement near a furnace or boiler. Extra detectors might be placed in storage areas. A heavy smoker may want an extra photoelectric detector in the bedroom. Both an ionization and a photoelectric smoke detector may be placed on each level of a multilevel house.

Detectors operate on either batteries or electric current. Installation of those detectors that use batteries does not require professional training. The ones that use electricity have to be wired into the dwelling's electrical system, which requires some knowledge of such installations. Unless provided with backup batteries a power failure will render a wired detector inoperable. Smoke detectors with batteries should beep when the battery becomes weak.

12.7.2.2

Escape from Fires

Building codes mandate the number and size of exits required. Most spaces and all public spaces require at least two means of egress. Fire escapes are not usually required in one- and two-family dwellings.

12.7.3

FIRE CONTAINMENT AND CONTROL

Fire containment and control measures are dictated by building codes. They include fire walls and fire and smoke doors to limit the spread of fires until inhabitants can escape.

12.7.4

FIRE EXTINGUISHMENT

Fire sprinkler systems are discussed in Section 11.4.

12.8 COMMUNICATIONS SYSTEMS

Additional facilities commonly considered a part of a house's wiring system and installed by the electrical trade include entrance signals, radio and TV antenna outlets, telephone outlets, and internal intercommunication (intercom) systems.

12.8.1

ENTRANCE SIGNALS

Signal push buttons should be provided at each commonly used entrance. These should be connected to a door chime or buzzer, giving distinctive sounds that will not be mistaken as telephone, smoke alarm, or other signals. The electrical supply for entrance signals usually comes from a low-voltage transformer installed near the service equipment.

In smaller houses the door chime is often installed in the kitchen. In larger houses the chime should be installed in a central location, where it may be heard throughout the house. In a multilevel house, extension signals may be necessary to ensure they will be heard in quarters located away from the main entry level. Entrance signal conductors should be no smaller than the equivalent of a No. 18 AWG copper wire, and preferably should be installed in conduit.

12.8.2

INTERCOMMUNICATION SYSTEMS

Intercommunication (intercom) systems are quite common today even in small houses. An intercom system may be a complete sound system with an AM/FM radio, tape deck, or CD player, or just a simple push-button two-way talk system. They are operated from a master control panel usually located in the master bedroom, recreation room, or kitchen and connected to remote stations throughout the house.

A larger house, or one with accommodations for a resident servant, may contain additional signal and communication devices. They may consist solely of a dining-room-to-kitchen call signal with a floor treadle or push button connected to an annunciator in the kitchen; more extensive systems may include a push button in each bedroom, the living room, and any other habitable room.

Sometimes an intercom system is tied to the entrance signal or substitutes for it. This permits an occupant to talk with a visitor without opening the door. In apartments, intercom systems are often tied in with electronic locks so that the main entrance door to the building can be operated from each apartment unit. Closed-circuit television systems that permit an occupant to see a visitor before opening the door are available, but are expensive and not commonly used. They may become as common in the future as door buzzers are today.

12.8.3

Television and Radio

Provisions for a television antenna consist mainly of a nonmetallic outlet box at each television set location, connected to the attic by a suitable transmission line, such as 300-ohm lead-in wire. The transmission line should be unshielded in a UHF service area and sufficiently slack to reach the antenna in the attic. If the transmission line is shielded, the shield should be grounded. Where an unshielded transmission line is used, grounding should be in accordance with the NEC and the line should be kept away from pipes and other wires. A convenience outlet should be provided adjacent to each television antenna outlet. Antenna systems with more than two outlets generally require a booster amplifier, except in very strong signal areas.

In areas with a cable TV service, the cable company should be contacted to ascertain their specific requirements for antenna connections.

In most areas, AM radio receivers used purely for the reception of broadcast communication do not require antenna and ground connections. For FM reception, provisions similar to those for television should be included.

12.8.4

Telephone

At least one conveniently fixed telephone outlet (jack) should be provided in every living unit. Where the living unit occupies more than one floor, there should be at least one outlet on each floor. Most houses today have an outlet in every habitable room and one in each work space, such as the laundry room, utility room, basement, and garage. Outlets are also often provided on porches and outdoors on patios and the like.

Advance wiring by the telephone company during construction will generally result in an inconspicuous, more efficient installation. The local telephone company should also be consulted for details about service connection before construction, particularly with regard to the installation of protector cabinets and raceways leading to the service entrance equipment.

12.9 CONSERVATION OF ENERGY

The proliferation of residential electrical equipment causes increased demands not only on home wiring systems, but on national energy resources as well. Residential use of energy accounts for more than 25% of the total consumption of energy in the United States. In addition, the amount of energy that individual appliances use has increased dramatically over the last 40 years. Appliances that were once luxuries are now commonplace. For example, almost every house has a color television set and home computers are burgeoning in popularity. There are electric dishwashers, clothes dryers, garbage disposers, can openers, built-in popcorn poppers, microwave ovens, woks, frying pans, hot-pots, whirlpool bathtubs, hot tubs. The list grows every year.

Therefore, conservation of energy is both a worthwhile national goal and a money-saving consumer objective. By careful design of the wiring system and judicious selection of equipment, it is possible to minimize energy waste without sacrificing comfort and convenience.

12.9.1

LIGHTING

A primary purpose of a home wiring system is lighting, which is responsible for nearly 8% of home energy consumption. The use of lighting is extensive, ranging from necessary illumination to enhancing interior design. However, misplaced or excessive illumination wastes electricity and money.

An entire room does not require the same high level of illumination if close-range activity is ordinarily restricted to specific areas. For instance, lighting should be concentrated on areas such as a study desk, kitchen counter, and dining table, but can be diffused and dimmer over the rest of the room.

Proper selection and placement of switches in convenient locations can reduce wasted energy from lights left on unintentionally. Rooms with more than one doorway, such as kitchens, dining rooms, and living rooms, should have switches at each doorway.

Larger bulbs are more energy efficient, producing more light per watt of electric energy. However, increased efficiency should be balanced against the increased overall energy consumption, which may be caused by indiscriminate use of larger bulbs.

Lightbulbs should not be larger than required for the specific task or design effect. For hard-to-reach areas such as roof overhangs and post lights, long-life bulbs with a life expectancy of 2500 hours can be used. Whereas ordinary incandescent bulbs have a life span of 750 to 1000 hours, fluorescent tubes are rated for 10,000 hours or longer. What's more, fluorescent tubes produce about 80 lumens per watt, or 4 times as much light as a standard incandescent bulb.

12.9.2

APPLIANCES

Energy-efficient appliances can reduce operating costs and minimize demands on a wiring system. Before the purchase of any appliance, the life-cycle cost (the long-range, lifetime, operating and maintenance cost) of the product should be determined. Buying an energy-efficient appliance may be initially more costly, but a significant saving can be had on subsequent electrical bills and energy demands.

The energy efficiency ratio (EER) established by the National Institute of Standards and Technology can be helpful in determining the relative efficiency of certain household appliances. The EER of air conditioners, for example, is determined by dividing the number of Btu's per hour by the number of watts needed for operation. The EER for air conditioners will range from 5 to 10 for smaller units and up to 12 for larger units. The higher the number, the more efficient the machine, which means less energy and money spent on operation. An EER of 7 or more is considered acceptable for household window air conditioners.

Appliances with extra conveniences, such as self-cleaning ovens and frost-free refrigerators, require more energy and cost more to operate than ordinary models. These special features are largely responsible for the increased energy consumption per appliance recorded in the last 30 years. In selecting appliances, the extra convenience and pleasure of the special features should be weighed against a realistic evaluation of the added initial and long-term operating costs. The local power company can be helpful in making an economic life-cycle cost analysis and in providing other energy conserving tips.

Ceiling fans can increase air movement in a space and decrease the need for air conditioning. Except when a very high humidity is present, a room with a ceiling fan will remain comfortable at temperatures above 80°F.

APPENDIX A
TABLES

METRIC TABLES

The following tables are related to the international system of metric measure (SI). Refer to Section 1.7 for a detailed discussion about the use of these tables.

FIGURE A-1
Units in the International System (SI)

Unit Group Quantity	Unit Name	Symbol	Formula	Unit Derivation	Remarks
BASE UNITS					
Length	metre	m			An alternative spelling is *meter*.
Mass	kilogram	kg			
Time	second	s			Already in common use.
Electric current	ampere	A			Already in common use.
Thermodynamic temperature	kelvin	K			The customary unit for temperature is the degree Celsius (°C).
Amount of substance	mole	mol			The mol has no application in construction.
Luminous intensity	candela	cd			Already in common use.
SUPPLEMENTARY UNITS					
Plane angle	radian	rad			Already in common use.
Solid angle	steradian	sr			Already in common use.
DERIVED UNITS WITH SPECIAL NAMES					
Frequency (of a periodic phenomenon)	hertz	Hz	1/s	s^{-1}	The hertz replaces *cycle per second*.
Force	newton	N	$kg \cdot m/s^2$	$m \cdot kg \cdot s^{-2}$	
Pressure, stress, elastic modulus	pascal	Pa	N/m^2	$m^{-1} \cdot kg \cdot s^{-2}$	
Energy, work, quantity of heat	joule	J	$N \cdot m$	$m^2 \cdot kg \cdot s^{-2}$	
Power, radiant flux	watt	W	J/s	$m^2 \cdot kg \cdot s^{-3}$	Already in common use.
Quantity of electricity, electric charge	coulomb	C	$A \cdot s$	$s \cdot A$	Already in common use.
Electric potential, potential difference, electromotive force	volt	V	J/C or W/A	$m^2 \cdot kg \cdot s^{-3} \cdot A^{-1}$	Already in common use.
Electric capacitance	farad	F	C/V	$m^{-2} \cdot kg^{-1} \cdot s^4 \cdot A^2$	Already in common use.
Electric resistance	ohm	Ω	V/A	$m^2 \cdot kg \cdot s^{-3} \cdot A^{-2}$	Already in common use.
Electric conductance	siemens	S	AV or 1/Ω	$m^{-2} \cdot kg^{-1} \cdot s^3 \cdot A^2$	The *siemens* was formerly referred to as *mho*.
Magnetic flux	weber	Wb	$V \cdot s$	$m^2 \cdot kg \cdot s^{-2} \cdot A^{-1}$	Already in common use.
Magnetic flux density	tesla	T	Wb/m^2	$kg \cdot s^{-2} \cdot A^{-1}$	Already in common use.
Electric inductance	henry	H	Wb/A	$m^2 \cdot kg \cdot s^{-2} \cdot A^{-2}$	Already in common use.
Luminous flux	lumen	lm	$cd \cdot sr$	$cd \cdot sr$	Already in common use.
Illuminance	lux	lx	lm/m^2	$m^{-2} \cdot cd \cdot sr$	
Absorbed dose	gray	Gy	J/kg	$m^2 \cdot s^{-2}$ (a)	(a) kg is canceled out. No application in construction.
DERIVED UNITS WITH GENERIC NAMES					
a. Units Expressed in Terms of One Base Unit					
Area	square metre	m^2		m^2	
Volume, capacity	cubic metre	m^3		m^3	(1 m^3 = 1000 L)
Section modulus	metre to third power		m^3	m^3	
Second moment of area	metre to fourth power		m^4	m^4	
Curvature	reciprocal (of) metre		1/m	m^{-1}	
Rotational frequency	reciprocal (of) second		1/s	s^{-1}	Revolution per second (r/s) is used in specifications for rotating machinery.
Coefficient of linear thermal expansion	reciprocal (of) kelvin		1/K	K^{-1}	

Unit Group Quantity	Unit Name	Symbol	Formula	Unit Derivation	Remarks
DERIVED UNITS WITH GENERIC NAMES (Continued)					
b. Units Expressed in Terms of Two or More Base Units					
Linear velocity	metre per second	m/s	$m \cdot s^{-1}$		
Linear acceleration	metre per second squared	m/s²	$m \cdot s^{-2}$		
Kinematic viscosity	square metre per second	m²/s	$m^2 \cdot s^{-1}$		
Volume rate of flow	cubic metre per second	m³/s	$m^3 \cdot s^{-1}$		
Specific volume	cubic metre per kilogram	m³/kg	$m^3 \cdot kg^{-1}$		
Mass per unit length	kilogram per metre	kg/m	$m^{-1} \cdot kg$		
Mass per unit area	kilogram per square metre	kg/m²	$m^{-2} \cdot kg$		
Density (mass per unit volume)	kilogram per cubic metre	kg/m³	$m^{-3} \cdot kg$	In this SI form, mass density is conveniently 1000 times specific gravity.
Moment of inertia	kilogram metre square	kg·m²	$m^2 \cdot kg$		
Mass flow rate	kilogram per second	kg/s	$kg \cdot s^{-1}$		
Momentum	kilogram metre per second	kg·m/s	$m \cdot kg \cdot s^{-1}$		
Angular momentum	kilogram metre squared per second	kg·m²/s	$m^2 \cdot kg \cdot s^{-1}$		
Magnetic field strength	ampere per metre	A/m	$m^{-1} \cdot A$		
Current density	ampere per square metre	A/m²	$m^{-2} \cdot A$		
Luminance	candela per square metre	cd/m²	$m^{-2} \cdot cd$		
c. Units Expressed in Terms of Base Units and/or Derived Units with Special Names					
Moment of force, torque	newton metre	N·m	$m^2 \cdot kg \cdot s^{-2}$		
Flexural rigidity	newton square metre	n·m²	$m^3 \cdot kg \cdot s^{-2}$		
Force per unit length, surface tension	newton per metre	N/m	$kg \cdot s^{-2}$ (b)	(b) m is canceled out
Dynamic viscosity	pascal second	Pa·s	$m^{-1} \cdot kg \cdot s^{-1}$		
Impact ductility	joule per square metre	J/m²	$kg \cdot s^{-2}$ (c)	(c) m² is canceled out
Combustion heat (per unit volume)	joule per cubic metre	J/m³	$m^{-1} \cdot kg \cdot s^{-2}$		
Combustion heat (per unit mass), specific energy, specific latent heat	joule per kilogram	J/kg	$m^2 \cdot s^{-2}$ (d)	(d) kg is canceled out
Heat capacity, entropy	joule per kelvin	J/K	$m^2 \cdot kg \cdot s^{-2} \cdot K^{-1}$		
Specific heat capacity, specific entropy	joule per kilogram kelvin	J/(kg·K)	$m^2 \cdot s^{-2} \cdot K^{-1}$ (e)	(e) kg is canceled out
Heat flux density, irradiance, sound intensity	watt per square metre	W/m²	$kg \cdot s^{-3}$ (f)	(f) m² is canceled out
Thermal conductivity	watt per metre kelvin	W/(m·K)	$m \cdot kg \cdot s^{-3} \cdot K^{-1}$		
Coefficient of heat transfer	watt per square metre kelvin	W/(m²·K)	$kg \cdot s^{-3} \cdot K^{-1}$ (g)	(g) m² is canceled out
Thermal resistance, thermal insulance	square metre kelvin per watt	m²·K/W	$kg^{-1} \cdot s^3 \cdot K$ (h)	(h) m² is canceled out
Electric field strength	volt per metre	V/m	$m \cdot kg \cdot s^{-3} \cdot A^{-1}$		
Electric flux density	coulomb per square metre	C/m²	$m^{-2} \cdot s \cdot A$		
Electric charge density	coulomb per cubic metre	C/m³	$m^{-3} \cdot s \cdot A$		
Electric permittivity	farad per metre	F/m	$m^{-3} \cdot kg^{-1} \cdot s^4 \cdot A^2$		
Electric permeability	henry per metre	H/m	$m \cdot kg \cdot s^{-2} \cdot A^{-2}$		
Electric resistivity	ohm metre	Ω·m	$m^3 \cdot kg \cdot s^{-3} \cdot A^{-2}$		
Electric conductivity	siemens per metre	S/m	$m^{-3} \cdot kg^{-1} \cdot s^3 \cdot A^2$		
Light exposure	lux second	lx·s	$m^{-2} \cdot s \cdot cd \cdot sr$		
Luminous efficacy	lumen per watt	lm/W	$m^{-2} \cdot kg^{-1} \cdot s^3 \cdot cd \cdot sr$		
d. Units Expressed in Terms of Supplementary Units and Base and/or Derived Units					
Angular velocity	radian per second	rad/s	$s^{-1} \cdot rad$		
Angular acceleration	radian per second squared	rad/s²	$m^{-2} \cdot rad$		
Radiant intensity	watt per steradian	W/sr	$m^2 \cdot kg \cdot s^{-3} \cdot sr^{-1}$		
Radiance	watt per square metre steradian	W/m²·sr	$kg \cdot s^{-2} \cdot sr^{-1}$ (i)	(i) m² is canceled out

Copyright ASTM; reprinted with permission.

Note: Metre is the preferred international spelling applied to this base unit and all its multiples and submultiples.

FIGURE A-2
Other Units Whose Use is Permitted with SI

Quantity	Unit Name	Symbol	Relationship to SI Unit	Remarks
UNITS FOR GENERAL USE				
Volume	litre[a]	L[a]	$1 \text{ L} = 0.001 \text{ m}^3 = 10^6 \text{ mm}^3$	An alternative spelling is *liter*. The litre may only be used with the SI prefix milli-.
Mass	metric ton[b]	t[b]	$1 \text{ t} = 1000 \text{ kg} = (1 \text{ Mg})$	
Time	minute	min	$1 \text{ min} = 60 \text{ s}$	
	hour	h	$1 \text{ h} = 3600 \text{ s} = (60 \text{ min})$	
	day (mean solar)	d	$1 \text{ d} = 86{,}400 \text{ s} = (24 \text{ h})$	
	year (calendar)	a	$1 \text{ a} = 31{,}536{,}000 \text{ s} = (365 \text{ d})$	
Temperature interval	degree Celsius	°C	$1 \text{ °C} = 1 \text{ K}$	The Celsius temperature 0°C corresponds to 273.15 K exactly. ($t \cdot \text{C} = T_K - 273.15$)
Plane angle	degree (of arc)	°	$1° = 0.017453 \text{ rad} = 17.453 \text{ mrad}$	$1° = (\pi/180) \text{ rad}$
Velocity	kilometre per hour	km/h	$1 \text{ km/h} = 0.278 \text{ m/s}$	$1 \text{ m/s} = 3.6 \text{ km/h}$
UNITS ACCEPTED FOR LIMITED USE				
Area	hectare	ha	$1 \text{ ha} = 10{,}000 \text{ m}^2$	For use in land measurement.
Energy	kilowatthour	kWh	$1 \text{ kWh} = 3.6 \text{ MJ}$	For measurement of electrical energy consumption only.
Speed of rotation	revolution per minute	r/min	$1 \text{ r/min} = \frac{1}{60} \text{ r/s} = \frac{2\pi}{60} \text{ rad/s}$	To measure rotational speed in slow-moving equipment only.

[a] The international symbol for litre is the lowercase l (ell), which can be easily confused with the numeral 1. Several English-speaking countries have adopted the script ℓ as a symbol for litre to avoid any misinterpretation. The symbol L (capital ell) is recommended for U.S. use to prevent confusion.

[b] The international name for metric ton is tonne. The metric ton is equal to the megogram (Mg).

Copyright ASTM, reprinted with permission.

FIGURE A-3
Rules and Recommendations for the Presentation of SI Units and Symbols

	Typical Examples	Remarks
GENERAL		
1. All unit names should either be denoted by correct symbols or be written in full. In the interest of simplification and to reduce the amount of writing, use unit symbols rather than fully written forms.		
2. Do not use mixtures of names and symbols.	Use: J/kg *or* joule per kilogram	Not: joule per kg Not: J/kilogram
SYMBOLS FOR UNIT QUANTITIES AND PREFIXES		
1. SI symbols are internationally agreed upon and there is only *one* symbol for each unit quantity. Multiples and submultiples are formed by using the unit symbol and attaching a prefix symbol in front of it.	m, kg, s, A, cd, K	See also Nos. 5–7, below.
2. All unit symbols are shown in upright letters, and can be produced by a normal typewriter keyboard, with the exceptions of the symbols for the SI unit ohm and the prefix micro-, which are represented by Greek letters Ω and μ, respectively.		Exceptions: Ω, μ
3. Unit symbols are never followed by a period (full stop) except at the end of a sentence.	60 kg/m	Not: 60 kg./m.
4. Unit symbols are normally written in lowercase, except for unit names derived from a proper name, in which case the initial is capitalized. Some units have symbols consisting of two letters from a proper name, of which *only* the first letter is capitalized. (The symbol for the unit name ohm is the capital Greek letter Ω.)	m, kg, s, mol, cd, etc. A, K, N, J, W, V, etc. Pa, Hz, Wb, etc.	Exception: L
5. Prefixes for magnitudes from 10^6 to 10^{18} have capital upright letter symbols.	M, G, T, etc.	See also cl
6. Prefixes for magnitudes from 10^{-18} through 10^3 have lowercase upright letter symbols. (The symbol for 10^{-6} or micro is the lowercase Greek letter μ.)	p, n, μ, m, k, etc.	Same also cl

	Typical Examples	Remarks

SYMBOLS FOR UNIT QUANTITITES AND PREFIXES (Continued)

7. Prefix symbols are directly attached to the unit symbol *without a space* between them. — mm, kW, MN, etc. — Not: m m, k W, M N

8. Do not use compound prefixes to form a multiple or submultiple of a unit (e.g., Use nanometre, do not use micromillimetre *or* millimicrometre). — nm — Not: μmm or mμm

9. In the case of the base unit kilogram, prefixes are attached to the gram (e.g., milligram, not microkilogram). — mg — Not: μkg

10. Use only one prefix when forming a multiple or a submultiple of a compound unit. Normally, the prefix should be attached to a unit in the numerator. An exception to this rule is made for the base unit kilogram. — km/s; mV/m — Not: mm/μs, μV/mm / Exception: MJ/kg / not: kJ/g

AREAS OF POSSIBLE CONFUSION REQUIRING SPECIAL CARE

1. The symbols for SI units and the conventions that govern their use should be strictly followed. A number of prefix and unit symbols use the same letter, but in different form. Exercise care to present the correct symbol for each quantity. — g (gram), G (giga), k (kilo), K (kelvin), m (milli), m (meter), M (mega), n (nano), N (newton) — Others: c (centi), C (coulomb), °C (degree Celsius), s (second), S (siemens), t (metric ton), T (tesla); T (tera)

2. All prefix and unit symbols retain their prescribed form regardless of the surrounding typography. In printouts from limited character sets (telex, computer printers) special considerations apply to symbols for mega, micro, ohm, and siemens. Where confusion is likely to arise, write units in full.

UNIT NAMES WRITTEN OUT IN FULL

1. Unit names, including prefixes, are treated as common names and are *not capitalized*, except at the beginning of sentences or in titles. (The only exception is Celsius in degree Celsius, where degree is considered as the unit name and is shown in lowercase, whereas Celsius represents an adjective and is capitalized.) — metre, newton, etc. — Not: Metre, Newton / Exception: degree *Celsius*

2. Where a prefix is attached to an SI unit to form a multiple or submultiple, the combination is written as one word. (There are three cases where the final vowel of the prefix is omitted in the combination: Megohm, kilohm, and hectare.) — millimetre, kilowatt — Not: milli-metre / Not: kilo watt

3. Where a compound unit is formed by multiplication of two units, the use of a space between units is preferred, but a hyphen is acceptable and in some situations more appropriate, to avoid any risk of misinterpretation. — newton metre *or* newton-metre — Not: newtonmetre

4. Where a compound unit is formed by division of the two units, this is expressed by inserting the word *per* between the numerator and the denominator. — metre per second / joule per kelvin — Not: meter/second / Not: joule/kelvin

5. Where the numerical value of a unit is written in full, the unit should also be written in full. — seven metres — Not: seven m

PLURALS

1. Units written in full are subject to the normal rules of grammar. For any unit with a numerical value greater than one (1), an *s* is added to the written unit to denote the plural. — 1.2 metre*s* / 2.3 newton*s* / 33.2 kilogram*s* / 350 kilohertz / 12.5 lux — But: 0.8 meter

2. The following units have the same plural as singular when written out in full: hertz, lux, siemens.

3. Symbols never change in the plural. — 2.3 N, 33.2 kg — Not: 2.3 N*s*, 33.2 kg*s*

COMPOUND UNIT SYMBOLS—PRODUCTS AND QUOTIENTS

1. The product of two units is indicated by a dot placed at midheight between the unit symbols. — kN·m, Pa·s — Not: kNm, Pas / Not: kN m, Pa s / See also f3 and f5

2. To express a derived unit formed by division, any one of the following methods may be used:
   a. a solidus (slash, /) — kg/m^3, W/(m·K) — See also f3 and f5
   b. a horizontal line between numerator and denominator — $\dfrac{kg}{m^3}$, $\dfrac{W}{m \cdot K}$
   c. a negative index (or negative power). — kg·m^{-3}, W·m^{-1}·K^{-1}

3. *Only one* solidus may be used in any combination. — m/s^2, m·kg/(s^3·A) — Not: m/s/s / Not: m·kg/s^3/A

4. Do not use the abbreviation *p* for per in the expression of a division. — km/h — Not: kph *or* k.p.h.

5. Where the denominator is a product, this should be shown in parentheses. — W/(m$_2$·K)

Copyriught ASTM; reprinted with permission.

Note: Metre is the preferred international spelling applied to this base unit and all its multiples and submultiples.

FIGURE A-4
Presentation of Numerical Values in Conjunction with SI

	Typical Examples	Remarks
DECIMAL MARKER		
1. Whereas most European countries use the comma on the line as the decimal marker and this practice is advocated by ISO, a special exception is made for documents in the English language, which have traditionally used the point (dot) on the line, or period, as decimal marker.		See also under g
2. The recommended decimal marker for use in the United States is the point on the line (period), and the comma should not be used. In handwritten documents the decimal marker may be shown slightly above the line for clear identification.	*9.9; 15.375*	Not: 9,9; 15,375
3. *Always* show a zero before the decimal point for all numbers smaller than 1.0 (one).	0.1; 0.725	Not: .1; .725
SPACING		
1. *Always* leave a gap between the numerical value associated with a symbol and the symbol of at least half a space in width. In the case of the symbol for degree Celsius this space is optional, but the degree symbol must always be attached to Celsius.	900 MHz; 200 mg; 10^6 mm^2 20°C or 20 °C	Not: 900MHz; 200mg Not: 10^6mm^2 Not: 20° C
2. In non-SI expressions of plane angle (°, ′, ″), do not leave a space between the numerical value and the symbol.	27°30′ (of arc)	Not: 27 ° 30 ′
3. *Always* leave a space on each side of signs for multiplication, division, addition, and subtraction.	100 mm × 100 mm; 36 MPa + 8 MPa	Not: 100 mm×100 mm Not: 36 MPa+ 8 MPa
FRACTIONS		
1. *Avoid* common fractions in connection with SI units.	Write: 0.5 kPa	Not: 1/2 kPa
2. *Always* use decimal notation to express fractions of any number larger than 1.0 (one).	1.5; 16.375	Not: Not: 1-1/2; 16-3/8
3. Although the most common fractions such as half, third, quarter, and fifth will remain in speech, *always* show decimal notation in written, typed, or printed material.	0.5; 0.33; 0.25; 0.2	Not: 1/2; 1/3; 1/4; 1/5
POWERS OF UNITS AND EXPONENTIAL NOTATION		
1. When writing unit names with a modifier *squared* or *cubed*, the following rules should be applied:		
a. In the case of area and volume, the modifier is written before the unit name as *square* and *cubic*.	cubic metre; square millimetre	Not: metre cubed; millmetre squared
b. In all other cases, the modifier is shown after the unit name as *squared*, *cubed*, *to the fourth power*, etc.	metre per second squared	Not: metre per square second; (or metre per second per second)
c. The abbreviations *sq.* for square, and *cu.* for cubic should *not* be used.		Not: sq. millimetre Not: cu. metre
2. For unit symbols with modifiers (such as square, cubic, fourth power, etc.) *always* show the superscript immediately after the symbol.	m^2; mm^3; s^4	Not: m ^{2}; mm ^{3}; s ^{4}
3. Show the superscript as a numeral reduced in size raised half a line space. Where a typewriter without superscript numerals is used, the full-sized numeral should be raised half a line space, provided that this does not encroach on the print in the line above.	mm^3; m/s^2	Permitted: mm^3; m/s^2

4. Where an exponent is attached to a prefixed symbol, it indicates that that multiple (or submultiple) is raised to the power expressed by the exponent.	1 mm^3 = $(10^{-3}$ m$)^3$
	   = 10^{-9} m^3
	1 km^2 = $(10^3$ m$)^2$
	   = 10^6 m^2
		Not: 1 mm^3
		   = 10^{-3} m^3

RATIOS		
1. Do not mix units in expressing a ratio of like unit quantities.	0.01 m/m 0.03 m^2/m^2	Not: 10 mm/m Not: 30,000 mm^2/m^2
2. Wherever possible, use a nonquantitative expression (ratio or percentage) to indicate measurement of slopes, deflections, etc.		Preferred: 1:100; 0.01; 1% 1:33; 0.03; 3%
RANGE		
1. The choice of the appropriate prefix to indicate a multiple or submultiple of an SI unit is governed by convenience to obtain numerical values within a practical range and to eliminate nonsignificant digits.		

FIGURE A-4—*Continued.*

	Typical Examples	Remarks
RANGE (Continued)		
2. In preference, use prefixes representing ternary powers of 10 (10 raised to a power that is a multiple of 3).	milli; kilo; mega	Avoid: centi; deci; deka; hecto
3. Select prefixes so that the numerical value or values occur in a common range between 0.1 and 1000.	120 kN 3.94 mm 14.5 MPa	In Lieu of: 120 000 N 0.003 94 m 14 500 kPA
4. Compatibility with the general range must be a consideration; e.g., if all dimensions on a drawing are shown in millimeters (mm), a range from 1 to 99 999 (a maximum of five numerals) would be acceptable to avoid mixing of units.		Note: Drawings should include the notation *All dimensions in millimetres.*
PRESENTATION AND TABULATION OF NUMBERS		
1. In numbers with many digits it has been common practice in the United States to separate digits into groups of three by means of commas. This practice must *not* be used with SI, to avoid confusion. It is recommended international practice to arrange digits in long numbers in groups of three from the decimal marker, with a gap of not less than half a space, and not more than a full space, separating each group.	54 375.260 55	Not: 54,375.260,55
2. For individual numbers with four digits before (or after) the decimal marker, this space is not necessary.	4500; 0.0355	
3. In all tabulations of numbers with five or more digits before and/or after the decimal marker, group digits into groups of three: e.g., 12.5255; 5735; 98 300; 0.425 75	12.525 5 5 735 98 300 0.425 75 ——————— 104 047.951 25	
USE OF UNPREFIXED UNITS IN CALCULATIONS		
Errors in calculations involving compound units can be minimized if all prefixed units are reverted to coherent base or derived units, with numerical values expressed in powers-of-ten notation.	Preferred: 136 kJ = 136×10^3 J 20 MPa = 20×10^6 Pa 1.5 t (Mg) = 1.5×10^3 kg	Also acceptable: (or 1.36×10^5 J) (or 2×10^7 Pa)

Copyright ASTM; reprinted with permission.

Note: Metre is the preferred international spelling applied to this base unit and all its multiples and submultiples.

FIGURE A-5

Space and Time: Geometry, Kinematics, and Periodic Phenomena

Quantity and SI Unit Symbol	Preferred Units (Symbols)	Other Acceptable Units	Unit Name	Typical Applications	Remarks
Length (m)	m		metre	**Architecture and General Engineering** Levels, overall dimensions, spans, column heights, etc., in engineering computations. **Estimating and Specification** Trenches, curbs, fences, lumber lengths, pipes and conduits; lengths of building materials generally. **Land Surveying** Boundary and cadastral surveys; survey plans; heights, geodetic surveys, contours. **Hydraulic Engineering** Pipe and channel lengths, depth of storage tanks or reservoirs, height of potentiometric head, hydraulic head, piezometric head.	Use metres on all drawings with scale ratios between 1:200 and 1:2000. Where required for purposes of accuracy, show dimensions to three decimal places.

Quantity and SI Unit Symbol	Preferred Units (Symbols)	Other Acceptable Units	Unit Name	Typical Applications	Remarks
Length (m) (Continued)	mm		millimetre	**Architecture and General Engineering** Spans, dimensions in buildings, dimensions of building products; depth and width of sections; displacement, settlement, deflection, elongation; slump of concrete, size of aggregate; radius of gyration, eccentricity; detailed dimensions generally; rainfall. **Estimating and Specification** Lumber cross sections; thicknesses, diameters, sheet metal gauges, fasteners; all other building product dimensions. **Hydraulic Engineering** Pipe diameters; radii of groundwater wells; height of capillary rise; precipitation, evaporation.	Use millimetres on drawings with scale ratios between 1:1 and 1:200. Avoid the use of centimetres (cm). Where cm is shown in documents, such as for snow depth, body dimensions, or carpet sizes, etc. convert to mm or m.
	km		kilometre	Distances for transportation purposes, geographical, or statistical applications in surveying; long pipes and channels.	
	μm		micrometre	Thickness of coatings (paint, galvanizing, etc.), thin sheet materials, size of fine aggregate.	
Area (m^2)	m^2		square metre	**General Applications** Small land areas; area of cross section of earthworks, channels and larger pipes; surface area of tanks and small reservoirs; areas in general. **Estimating and Specification** Site clearing; floor areas; paving, masonry construction, roofing, wall and floor finishes, plastering, paintwork, glass areas, membranes, lining materials, insulation, reinforcing mesh, formwork; areas of all building components.	$(1\ m^2 = 10^6\ mm^2)$ Replaces sq. ft.; sq. yd.; and square. Specify masonry construction by wall area $\times$ wall thickness.
	mm^2		square millimetre	Area of cross section for structural and other sections, bars, pipes, rolled and pressed shapes, etc.	Avoid the use of cm^2 (square centimetre) by conversion to mm^2 $(1\ cm^2 = 10^2\ mm^2$ $= 100\ mm^2)$.
	km^2		square kilometre	Large catchment areas or land areas.	
		ha	hectare	Land areas; irrigation areas; areas on boundary and other survey plans.	$(1\ ha = (10^2\ m)^2$ $= 10^4\ m^2$ $= 10\ 000\ m^2)$
Volume, capacity (m^3)	m^3		cubic metre	**General Applications** Volume, capacity (large quantities); volume of earthworks, excavations, filling, waste removal; concrete, sand, all bulk materials supplied by volume, and large quantities of lumber. **Hydraulic Engineering** Water distribution, irrigation, diversions, sewage, storage capacity, underground basins.	$1\ m^3 = 1000\ L$ As far as possible, use the cubic metre as the preferred unit of volume for all engineering purposes.

Quantity and SI Unit Symbol	Preferred Units (Symbols)	Other Acceptable Units	Unit Name	Typical Applications	Remarks
Volume, capacity (m^3) (Continued)	mm^3		cubic millimetre	Volume, capacity (small quantities)	
		L	litre	Volume of fluids and containers for fluids; liquid materials, domestic water supply, consumption; volume/capacity of fuel tanks	The litre and its multiples or submultiples may be used for domestic and industrial supplies of liquids $1\ L = 1\ dm^3$ $= 1000\ cm^3$ $1\ mL = 1\ cm^3$
		mL	millilitre	Volume of fluids and containers for fluids (limited application only)	
		cm^3	cubic centimetre	Limited application only (small quantities)	$1\ cm^3 = 1000\ mm^3$ $= 10^{-6}\ m^3$
Modulus of section (m^3)	mm^3		millimetre to third power	Geometric properties of structural sections, such as plastic section modulus, elastic section modulus, etc.	
	m^3		metre to third power		
Second moment of area (m^4)	mm^4		millimetre to fourth power	Geometric properties of structural sections, such as moment of inertia of a section, torsional constant of a cross section.	
	m^4		metre to fourth power		
Plane angle (rad)	rad		radian	Generally used in calculations only to preserve coherence.	Slopes and gradients may be expressed as a ratio or as a percentage: $26.57 = 1:2$ $= 50\%$ $= 0.4637\ rad$ ($1\ rad = 57.2958°$)
	mrad		milliradian		
		(__°)	degree (of arc)	**General Applications** Angular measurement in construction (generally using decimalized degrees); angle of rotation, torsion, shear resistance, friction, internal friction, etc. **Land Surveying** Bearings shown on boundary and cadastral survey plans; geodetic surveying	
Time, time interval (s)	s		second	Time used in methods of test; all calculations involving derived units with a time component, in order to preserve coherence.	Avoid the use of minute (min) as often as possible
		h	hour	Time used in methods of test; all calculations involving labor time, plant hire, maintenance periods, etc.	($1\ h = 3600\ s$ $1\ d = 86\ 400\ s$ $= 86.4\ ks$)
		d	day		
		a	annum (year)		
Frequency (Hz)	Hz		hertz	Frequency of sound, vibration, shock; frequency of electromagnetic waves	($1\ Hz = 1/s = s^{-1}$) Replaces cycle(s) per second (c/s or cps)
	kHz		kilohertz		
	MHz		megahertz		
Rotational frequency, speed of rotation (s^{-1})		r/s	revolution per second	Widely used in the specification of rotational speed of machinery; use r/min (revolutions per minute) only for slow-moving machinery.	($1\ r/s = 2\ \pi\ rad/s$ $= 60\ r/min$)

Quantity and SI Unit Symbol	Preferred Units (Symbols)	Other Acceptable Units	Unit Name	Typical Applications	Remarks
Velocity, speed (m/s)	m/s		metre per second	Calculations involving rectilinear motion, velocity and speed in general; wind velocity; velocity of fluids; pipe flow velocity	$(1 \text{ m/s} = 3.6 \text{ km/h})$
		km/h	kilometre per hour	Wind speed; speed used in transportation; speed limits	
		mm/h	millimetre per hour	Rainfall intensity	
Angular velocity (rad/s)	rad/s		radian per second	Calculations involving rotational motion	
Linear acceleration (m/s²)	m/s²		metre per second squared	Kinematics, and calculations of dynamic forces	Recommended value of acceleration of gravity for use in U.S.: $g_{us} = 9.8 \text{ m/s}^2$
Volume rate of flow (m³/s)	m³/s		cubic metre per second	Volumetric flow in general; flow in pipes, ducts, channels, rivers; irrigation spray demand	$(1 \text{ m}^3/\text{s} = 1000 \text{ L/s})$
		m³/h	cubic metre per hour		
		m³/d	cubic metre per day		
		L/s	litre per second	Volumetric flow of fluids only	
		L/d	litre per day		

Copyright ASTM; reprinted with permission.

Note: Metre and litre are preferred international spellings applied to these base units and all their multiples and submultiples, such as kilometre and millilitre.

FIGURE A-6
Mechanics: Statics and Dynamics

Quantity and SI Unit Symbol	Preferred Units (Symbols)	Other Acceptable Units	Unit Name	Typical Applications	Remarks
Mass (kg)	kg		kilogram	Mass of materials in general, mass of structural elements and machinery	Use kilograms (kg) in calculations and specifications
	g		gram	Mass of samples of material for testing	Masses greater than 10^4 kg (10 000 kg) may be conveniently expressed in metric tons (t): $1 \text{ t} = 10^3 \text{ kg} = 1 \text{ Mg} = 1000 \text{ kg}$
		t	metric ton	Mass of large quantities of materials, such as structural steel, reinforcement, aggregates, concrete, etc.; ratings of lifting equipment	
Mass per unit length (kg/m)	kg		kilogram per metre	Mass per unit length of sections, bars, and similar items of uniform cross section	Also known as Linear Density
		g/m	gram per metre	Mass per unit length of wire and similar material of uniform cross section	

Quantity and SI Unit Symbol	Preferred Units (Symbols)	Other Acceptable Units	Unit Name	Typical Applications	Remarks
Mass per unit area (kg/m^2)	kg/m^2		kilogram per square metre	Mass per unit area of slabs, plates, and similar items of uniform thickness or depth; rating for load-carrying capacities on floors (display on notices only)	Do not use in stress calculations
		g/m^2	gram per square metre	Mass per unit area of thin sheet materials, coatings, etc.	
Mass density, concentration (kg/m^3)	kg/m^3		kilogram per cubic metre	Density of materials in general; mass per unit volume of materials in a concrete mix; evaluation of masses of structures and materials	Also known as Mass per Unit Volume ($1\ kg/m^3 = 1\ g/L$) ($1\ g/m^3 = 1\ mg/L$)
		g/m^3	gram per cubic metre	Mass per unit volume (concentration) in pollution control	
		$\mu g/m^3$	microgram per cubic metre		
Momentum ($kg \cdot m/s$)	$kg \cdot m/s$		kilogram metre per second	Used in applied mechanics; evaluation of impact and dynamic forces	
Moment of inertia ($kg \cdot m^2$)	$kg \cdot m^2$		kilogram square metre	Rotational dynamics. Evaluation of the retraining forces required for propellers, windmills, etc.	
Mass per unit time (kg/s)	kg/s		kilogram per second	Rate of transport of material on conveyors and other materials handling equipment	$1\ kg/s = 3.6\ t/h$
		t/h	metric ton per hour		
Force (N)	N		newton	Unit of force for use in calculations	$1\ N = 1\ kg \cdot m/s^2$
	kN		kilonewton	Forces in structural elements, such as columns, piles, ties, prestressing tendons, etc.; concentrated forces; axial forces; reactions; shear force; gravitational force	
Force per unit length (N/m)	N/m		newton per metre	Unit for use in calculations	
	kN/m		kilonewton per metre	Transverse force per unit length on a beam, column, etc.; force distribution in a linear direction	
Moment of force, torsional, or bending moment ($N \cdot m$)	$N \cdot m$		newton metre	Bending moments (in structural sections), torsional moment; overturning moment; tightening tension for high strength bolts; torque in engine drive shafts, axles, etc.	
	$kN \cdot m$		kilonewton metre		
	$MN \cdot m$		meganewton metre		
Pressure, stress, modulus of elasticity (Pa)	Pa		pascal	Unit for use in calculations; low differential pressure in fluids	($1\ Pa = 1\ N/m^2$)
	kPa		kilopascal	Uniformly distributed pressure (loads) on floors; soil bearing pressure; wind pressure (loads), snow loads, dead and live loads; pressure in fluids; differential pressure (e.g., in ventilating systems)	Where wind pressure, snow loads, dead and live loads are shown in kN/m^2 change units to kPa

FIGURE A-6—Continued.

Quantity and SI Unit Symbol	Preferred Units (Symbols)	Other Acceptable Units	Unit Name	Typical Applications	Remarks
Pressure, stress, modulus of elasticity (Pa) (Continued)	MPa		megapascal	Modulus of elasticity; stress (ultimate, proof, yield, permissible, calculated, etc.) in structural materials; concrete and steel strength grades	$1\ MPa = 1\ MN/m^2$ $= 1\ N/mm^2$
	GPa		gigapascal	Modulus of elasticity in high strength materials	
	μPa		micropascal	Sound pressure (20 μPa is the reference quantity for sound pressure level)	
Compressibility (Pa^{-1})	1/Pa		reciprocal (of) pascal	Settlement analysis, (coefficient of compressibility), bulk compressibility	$(1/Pa = 1\ m^2/N)$
	1/kPa		reciprocal (of) kilopascal		
Dynamic viscosity (Pa·s)	Pa·s		pascal second	Shear stresses in fluids	$(1\ Pa\cdot s - 1\ Ns/m^2)$ The centipoise (cP) = 10^{-3} Pa·s will not be used
	mPa·s		millipascal second		
Kinematic viscosity (m²/s)	m²/s		square metre per second		The centistokes (cSt) = $10^{-6}m^2/s$ will not be used
	mm²/s		square millimetre per second	Computation of Reynold's number, settlement analysis (coefficient of consolidation)	$1\ cSt = 1\ mm^2/s$
Work, energy (J)	J		joule	Energy absorbed in impact testing of materials; energy in general; calculations involving mechanical and electrical energy	
	kJ		kilojoule		
	MJ		megajoule		
		kWh	kilowatthour	Electrical energy applications only	$1\ kWh = 3.6\ MJ$
Impact strength (J/m²)	J/m²		joule per square metre	Impact strength; impact ductility	
	kJ/m²		kilojoule per square metre		
Power (W)	W		watt	Power in general (mechanical, electrical, thermal); input/output rating, etc. of motors, engines, heating and ventilating plant and other equipment in general	
	kW		kilowatt		
	MW		megawatt	Power input/output rating, etc. of heavy power plant	
	pW		picowatt	Sound power level (1 pW is the reference quantity for sound power level)	

Copyright ASTM; reprinted with permission.

Note: Metre is the preferred international spelling applied to this base unit and all its multiples and submultiples.

FIGURE A-7
Thermal Effects and Heat Transfer

Quantity and SI Unit Symbol	Preferred Units (Symbols)	Other Acceptable Units	Unit Name	Typical Applications	Remarks
Temperature value (K)	K		kelvin	Expression of thermodynamic temperature; calculations involving units of temperature	$(t_{°C} = T_K - 273.15)$
		°C	degree Celsius	Common temperature scale for use in meteorology and general applications; ambient temperature values	Temperature values will normally be measured in °C (degree Celsius)
Temperature interval (K)	K		kelvin	Heat transfer calculations; temperature intervals in test methods, etc.	(1 K = 1 °C) The use of K (kelvin) in compound units is recommended
		°C	degree Celsius		
Coefficient of linear thermal expansion (1/K)	1/K		reciprocal (of) kelvin	Expansion of materials subject to a change in temperature (generally expressed as a ratio per kelvin or degree Celsius)	
		1 °/C	reciprocal (of) degree Celsius		
Heat, quantity of heat (J)	J		joule	Thermal energy calculations. Enthalpy, latent heat, sensible heat	
	kJ		kilojoule		
	MJ		megajoule		
Specific energy, specific latent heat; combustion heat (mass basis) (J/kg)	J/kg		joule per kilogram	Heat of transition; heat and energy contained in materials; combustion heat per unit mass; calorific value of fuels (mass basis); specific sensible heat, specific latent heat in psychrometric calculations	
	kJ/kg		kilojoule per kilogram		
	MJ/kg		megajoule per kilogram		
Energy density, combustion heat (volume basis) (J/m³)	J/m³		joule per cubic metre	Combustion heat per unit volume	
	kJ/m³		kilojoule per cubic metre		(1 kJ/m³ = 1 J/L)
	MJ/m³		megajoule per cubic metre	Calorific value of fuels (volume basis)	(1 MJ/m³ = 1 kJ/L)
Heat capacity, entropy (J/K)	J/K		joule per kelvin	Thermal behavior of materials, heat transmission calculations, entropy	
	kJ/K		kilojoule per kelvin		
Specific heat capacity, specific entropy (J/(kg·K))	J/(kg·K)		joule per kilogram kelvin	Thermal behavior of materials, heat transmission calculations	
	kJ/(kg·K)		kilojoule per kilogram kelvin		
Heat flow rate (W)	W		watt	Heat flow rate through walls, windows, etc.; heat demand	(1 W = 1 J/s)
	kW		kilowatt		

FIGURE A-7—*Continued*.

Quantity and SI Unit Symbol	Preferred Units (Symbols)	Other Acceptable Units	Unit Name	Typical Applications	Remarks
Power density, heat flux density, irradiance (W/m^2)	W/m^2		watt per square metre	Density of power or heat flow through building walls and other heat transfer surfaces; heat transmission calculations	
	kW/m^2		kilowatt per square metre		
Heat release rate (W/m^3)	W/m^3		watt per cubic metre	Rate of heat release per unit volume over time (for gases and liquids)	$W/m^3 = J/(m^3 \cdot s)$
	kW/m^3		kilowatt per cubic metre		
Thermal conductivity ($W/m \cdot K$)	$W/(m \cdot K)$		watt per metre kelvin	Estimation of thermal behavior of materials and systems; heat transmission calculations Thermal conductivity of structural and building materials in fire-resistance testing, insulation, etc.	$1\ W/(m \cdot K) =$ $1\ W/(m \cdot °C)$ (k = value)
Coefficient of heat transfer (thermal conductance) ($W/(m^2 \cdot K)$)	$W/(m^2 \cdot K)$		watt per square metre kelvin	Heat transfer calculations for buildings, building components and equipment. Transmittance of construction elements	(U = value)
	$kW/(m^2 \cdot K)$		kilowatt per square metre kelvin		
Thermal resistivity ($m \cdot K$)/W	$(m \cdot K)/W$		metre kelvin per watt	Heat transmission calculations (reciprocal of thermal conductivity)	
Thermal insulance, (thermal resistance ($m^2 \cdot K$)/W	$(m^2 \cdot K)/W$		square metre kelvin per watt	Heat transmission calculations (reciprocal of thermal conductance)	(R = value)

Copyright ASTM; reprinted with permission.

Note: *Metre* and *litre* are the preferred international spellings applied to these units and all their multiples and submultiples.

FIGURE A-8
Electricity and Magnetism

Quantity and SI Unit Symbol	Preferred Units (Symbols)	Other Acceptable Units	Unit Name	Typical Applications	Remarks
Electric current (A)	A		ampere	Maintenance rating of an electrical installation. Leakage current	
	kA		kiloampere		
	mA		milliampere		
	μA		microampere		
Magnetomotive force, magnetic potential difference (A)				Used in the calculations involved in magnetic circuits	

Quantity and SI Unit Symbol	Preferred Units (Symbols)	Other Acceptable Units	Unit Name	Typical Applications	Remarks
Magnetic field strength, magnetization (A/m)	A/m		ampere per metre	Magnetic field strength used in calculation of magnetic circuitry such as transformers, magnetic amplifiers, and general cores	(1 kA/m = 1 A/mm)
	kA/m		kiloampere per metre		
Current density (A/m²)	A/m²		ampere per square metre	Design of cross-sectional area of electrical conductor	
	kA/m²		kiloampere per square metre		
		A/mm²	ampere per square millimetre		(1 A/mm² = 1 MA/m²)
Electric charge, quantity of electricity (C)	C		coulomb	The voltage on a unit with capacitive characteristics may be related to the amount of charge present (e.g., electrostatic precipitators). Storage battery capacities	1 C = 1 A·s Do not use ampere hour: 1 A·h = 3.6 kC
	kC		kilocoulomb		
	μC		microcoulomb		
	nC		nanocoulomb		
	pC		picocoulomb		
Electric potential, potential difference, electromotive force (V)	V		volt		1 V = 1 W/A
	MV		megavolt		
	kV		kilovolt		
	mV		millivolt		
	μV		microvolt		
Electric field strength (V/m)	V/m		volt per metre	The electric field strength gives the potential gradient at points in space. This may be used to calculate or test electrical parameters such as dielectric strength.	
	MV/m		megavolt per metre		
	kV/m		kilovolt per metre		
	mV/m		millivolt per metre		
	μV/m		microvolt per metre		
Active power (W)	W		watt	The useful power of an electrical circuit is expressed in watts (W). (The apparent power in an electrical circuit is expressed in volt-amperes, (V·A).)	1 W = 1 V·A
	GW		gigawatt		
	MW		megawatt		
	kW		kilowatt		
	mW		milliwatt		
	μW		microwatt		

Quantity and SI Unit Symbol	Preferred Units (Symbols)	Other Acceptable Units	Unit Name	Typical Applications	Remarks
Capacitance (F)	F		farad	Electronic components. Electrical design and performance calculators	1 F = 1 C/V
	mF		millifarad		
	μF		microfarad		
	nF		nanofarad		
	pF		picrofarad		
Resistance (Ω)	Ω		ohm	The design of electrical devices with resistance, such as motors, generators, heaters, electrical distribution systems, etc.	1 Ω = 1 V/A
	GΩ		gigaohm		
	MΩ		megohm		
	kΩ		kilohm		
	mΩ		milliohm		
Conductance, admittance susceptance (S)	S		siemens		The siemens (S) was formerly known as mho.
	MS		megasiemens		
	kS		kilosiemens		
	mS		millisiemens		
	μS		microsiemens		
Resistivity ($\Omega \cdot$m)	$\Omega \cdot$m		ohm metre		
	G$\Omega \cdot$m		gigaohm metre		
	M$\Omega \cdot$m		megohm metre		
	k$\Omega \cdot$m		kilohm metre		
	m$\Omega \cdot$m		milliohm metre		
	$\mu \Omega \cdot$m		microohm metre		
	n$\Omega \cdot$m		nanoohm metre		
Electrical conductivity (S/m)	S/m		siemens per metre	A parameter for measuring water quality.	
	MS/m		megasiemens per metre		
	ks/m		kilosiemens per metre		
	μS/m		microsiemens per metre		
Magnetic flux, flux of magnetic induction (Wb)	mWb		milliweber	Used in the calculations involved in magnetic circuits.	1 Wb = 1 V$\cdot$s

Quantity and SI Unit Symbol	Preferred Units (Symbols)	Other Acceptable Units	Unit Name	Typical Applications	Remarks
Magnetic flux density, magnetic induction (T)	T		tesla	Used in the calculations involved in magnetic circuits.	$1\ T = 1\ wB/m^2$
	mT		millitesla		
	μT		microtesla		
	nT		nanotesla		
Magnetic vector potential (Wb/m^2)	kWb/m^2		kiloweber per square metre	Used in the calculations involved in magnetic circuits.	
Self-inductance, mutual inductance, permeance (H)	H		henry	Used in analysis and calculations involving transformers.	$1\ H = 1\ Wb/A$
	mH		millihenry		
	μH		microhenry		
	nH		nanohenry		
	pH		picohenry		
Reluctance (1/H)	1/H		reciprocal of henry	Design of motors and generators	
Permeability (H/m)	H/m		henry per metre	Permeability gives the relationship between the magnetic flux density and the magnetic field strength.	
	μH/m		microhenry per metre		
	nH/m		nanohenry per metre		

Copyright ASTM; reprinted with permission.

Note: Metre is the preferred international spelling applied to this base unit and all its multiples and submultiples.

FIGURE A-9
Lighting

Quantity and SI Unit Symbol	Preferred Units (Symbols)	Other Acceptable Units	Unit Name	Typical Applications	Remarks
Luminous intensity (cd)	cd		candela		
Solid angle (sr)	sr		steradian		
Luminous flux (lm)	lm		lumen	Luminous flux of light sources, lamps, and lightbulbs	$1\ lm = 1\ cd{\cdot}sr$ Already in general use
	klm		kilolumen		
Quantity of light (lm·s)	lm·s		lumen second		$1\ lm{\cdot}h = 3600\ lm/s$
		lm·h	lumen hour		
Luminance (cd/m^2)	cd/m^2		candela per square metre	Assessment of surface brightness; luminance of light sources, lamps, and lightbulbs; calculation of glare in lighting layouts	Replaces stilb $(1\ sb = 10^4\ cd/m^2)$ and apostilb $(1\ apostilb = cd/\pi m^2)$
	kcd/m^2		kilocandela per square metre		
		cd/mm^2	candela per square millimetre		

Quantity and SI Unit Symbol	Preferred Units (Symbols)	Other Acceptable Units	Unit Name	Typical Applications	Remarks
Illuminance (lx)	lx		lux	Luminous flux per unit area used in determination of illumination levels and design and evaluation of interior lighting layouts. (Outdoor daylight illumination on a horizontal plane ranges up to 100 klx.)	(a) Formerly referred to as illumination $1\ lx = 1\ lm/m^2$ (b) Replaces $(1\ ph = 10^4\ lx)$ (c) Luminous exitance is described in lm/m^2
	klx		kilolux		
Light exposure (lx·s)	lx·s		lux second		
	klx·s		kilolux second		
Luminous efficacy (lm/W)	lm/W		lumen per watt	Rating of luminous efficacy of artificial light sources	

Copyright ASTM; reprinted with permission.

Note: Metre is the preferred international spelling applied to this base unit and all its multiples and submultiples.

FIGURE A-10
Acoustics

Quantity and SI Unit Symbol	Preferred Units (Symbols)	Other Acceptable Units	Unit Name	Typical Applications	Remarks
Wavelength (m)	m		metre	Definition of sound wave pitch	
	mm		millimetre		
Area of absorptive surface (m²)	m²		square metre	Calculations of room absorption	
Period, periodic time (s)	s		second	Measurement of time and reverberation time	
	ms		millisecond		
Frequency (Hz)	Hz		hertz	Frequency ranges in sound absorption calculations and sound pressure measurement	1 Hz = 1 cycle per second (cps)
	kHz		kilohertz		
Instantaneous sound pressure (Pa)	Pa		pascal	Measurement of sound pressure; reference level for sound pressure is 20 μPa, but sound pressure is shown in decibels (dB) based on a logarithmic scale Sound pressure level $L_p =$ $= 20\ \log_{10} \dfrac{\text{actual pressure (Pa)}}{20 \times 10^{-6}\ \text{(Pa)}}$	Do not use dyne (1 dyn = 10 μPa)
	mPa		millipascal		
	μPa		micropascal		
Sound power, sound energy flux (W)	W		watt	Measurement of sound power; reference level for sound power is 1 pW Sound power level, L = $= 10\ \log_{10} \dfrac{\text{actual power (W)}}{10^{-12}\ \text{(W)}}$	
	mW		milliwatt		
	μW		microwatt		
	pW		picowatt		

FIGURE A-10—*Continued*.

Quantity and SI Unit Symbol	Preferred Units (Symbols)	Other Acceptable Units	Unit Name	Typical Applications	Remarks
Sound intensity (W/m^2)	W/m^2		watt per square metre	Measurement of sound intensity; reference level for sound intensity is 1 pW/m^2	
	pW/m^2		picowatt per square metre	Sound intensity level $L_I =$ $$= 10 \log_{10} \frac{\text{actual intensity } (W/m^2)}{10^{-12} \ (W/m^2)} \ dB$$	
Specific acoustic impedance $(Pa \cdot s/m)$	$Pa \cdot s/m$		pascal second per metre	Sound impedance measurement	$(1 \ pa \cdot s/m = 1 \ N \cdot s/m^3)$
Acoustic impedance, resistance $(Pa \cdot s/m^3)$	$Pa \cdot s/m^3$		pascal second per cubic metre	Sound impedance measurement	

Copyright ASTM; reprinted with permission.

Note: Metre is the preferred international spelling applied to this base unit and all its multiples and submultiples.

FIGURE A-11
Fraction, Decimal, and Metric Equivalents

Fractional Inch	Decimal Equivalent*	Millimetres	Fractional Inch	Decimal Equivalent*	Millimetres
1/64	0.0156	0.397	17/32	0.5313	13.494
1/32	0.0313	0.794	35/64	0.5469	13.891
3/64	0.0469	1.191	9/16	0.5625	14.288
1/16	0.0625	1.588	37/64	0.5781	14.684
5/64	0.0781	1.984	7/12	0.5833	14.817
1/12	0.0833	2.117	19/32	0.5938	15.081
3/32	0.0938	2.381	3/5	0.6000	15.240
1/10	0.1000	2.540	39/64	0.6094	15.478
7/64	0.1094	2.778	5/8	0.6250	15.875
1/8	0.1250	3.175	41/64	0.6406	16.272
9/64	0.1406	3.572	21/32	0.6563	16.669
1/6	0.1667	4.233	2/3	0.6667	16.933
11/64	0.1719	4.366	43/64	0.6719	17.066
3/16	0.1875	4.700	11/16	0.6875	17.463
1/5	0.2000	5.080	7/10	0.7000	17.780
13/64	0.2031	5.159	45/64	0.7031	17.859
7/32	0.2188	5.556	23/32	0.7188	18.256

FIGURE A-11—Continued.

Fractional Inch	Decimal Equivalent	Millimetres	Fractional Inch	Decimal Equivalent	Millimetres
15/64	0.2344	5.953	47/64	0.7344	18.653
1/4	0.2500	6.350	3/4	0.7500	19.050
17/64	0.2656	6.747	49/64	0.7656	19.447
9/32	0.2813	7.144	25/32	0.7813	19.844
19/64	0.2969	7.541	51/64	0.7969	20.241
3/10	0.3000	7.620	4/5	0.8000	20.320
5/16	0.3125	7.937	13/16	0.8125	20.638
1/3	0.3333	8.467	53/64	0.8281	21.034
11/32	0.3438	8.731	5/6	0.8333	21.167
23/64	0.3594	9.128	27/32	0.8438	21.431
3/8	0.3750	9.525	55/64	0.8594	21.828
25/64	0.3906	9.922	7/8	0.8750	22.225
2/5	0.4000	10.160	57/64	0.8906	22.622
13/32	0.4063	10.319	9/10	0.9000	22.860
5/12	0.4167	10.583	29/32	0.9063	23.019
27/64	0.4219	10.716	11/12	0.9167	23.283
7/16	0.4375	11.112	59/64	0.9219	23.416
29/64	0.4531	11.509	15/16	0.9375	23.813
15/32	0.4688	11.906	61/64	0.9531	24.209
31/64	0.4844	12.303	31/32	0.9688	24.606
1/2	0.5000	12.700	63/64	0.9844	25.003
33/64	0.5156	13.097	1″	1.000	25.400

*To convert a decimal to percentage, carry the decimal point two places to the right. Thus, 63/64, or 0.9844, equals 98.44%.

FIGURE A-12
Conversion Factors for the Most Common Units Used in Building Design and Construction

	Metric to Customary			Customary to Metric		
LENGTH						
1 km	= 0.621 371	mile (international)		1 mile (international)	= *1.609 344*	km
	= 49.7096	chain		1 chain	= 20.1168	m
1 m	= 1.093 61	yd		1 yd	= *0.9144*	m
	= 3.280 84	ft		1 ft	= *0.3048*	m
1 mm	= 0.039 370 1	in			= *304.8*	mm
				1 in	= *25.4*	mm
				(1 U.S. survey foot	= 0.304 800 6	m)

FIGURE A-12—*Continued.*

	Metric to Customary			Customary to Metric	

AREA

1 km^2	= 0.386 101	mile^2 (U.S. survey)	1 mile^2 (U.S. survey)	= 2.590 00	km^2
1 ha	= 2.471 04	acre (U.S. survey)	1 acre (U.S. survey)	= 0.404 687	ha
1 m^2	= 1.195 99	yd^2		= 4046.87	m^2
	= 10.7639	ft^2	1 yd^2	= 0.836 127	m^2
1 mm^2	= 0.001 550	in^2	1 ft^2	= 0.092 903	m^2
			1 in^2	= *645.16*	mm^2

VOLUME, MODULUS OF SECTION

1 m^3	= 0.810 709 × 10^{-3}	acre ft	1 acre ft	= 1233.49	m^3
	= 1.307 95	yd^3	1 yd^3	= 0.764 555	m^3
	= 35.3147	ft^3	100 board ft	= 0.235 974	m^3
	= 423.776	board ft	1 ft^3	= 0.028 316 8	m^3
1 mm^3	= 61.0237 × 10^{-6}	in^3		= 28.3168	L (dm^3)
			1 in^3	= 16 387.1	mm^3
				= 16.3871	mL (cm^3)

(FLUID) CAPACITY

1 L	= 0.035 314 7	ft^3	1 gal (U.S. liquid)*	= 3.785 41	L
	= 0.264 172	gal (U.S.)	1 qt (U.S. liquid)	= 946.353	mL
	= 1.056 69	qt (U.S.)	1 pt (U.S. liquid)	= 473.177	mL
1 mL	= 0.061 023 7	in^3	1 fl oz (U.S.)	= 29.5735	mL
	= 0.033 814	fl oz (U.S.)			

*1 gal (U.K.) approximately 1.2 gal (U.S.)

SECOND MOMENT OF AREA

1 mm^4	= 2.402 51 × 10^{-6}	in^4	1 in^4	= 416 231	mm^4
				= 0.416 231 × 10^{-6}	m^4

PLANE ANGLE

1 rad	= 57° 17′ 45″	(degree)	1° (degree)	= 0.017 453 3	rad
	= 57.2958°	(degree)		= 17.4533	mrad
	= 3437.75′	(minute)	1′ (minute)	= 290.888	μrad
	= 206 265″	(second)	1″ (second)	= 4.848 14	μrad

VELOCITY, SPEED

1 m/s	= 3.280 84	ft/s	1 ft/s	= *0.3048*	m/s
	= 2.236 94	mile/h	1 mile/h	= *1.609 344*	km/h
1 km/h	= 0.621 371	mile/h		= *0.447 04*	m/s

ACCELERATION

1 m/s^2	= 3.280 84	ft/s^2	1 ft/s^2	= *0.3048*	m/s^2

VOLUME RATE OF FLOW

1 m^3/s	= 35.3147	ft^3/s	1 ft^3/s	= 0.028 316 8	m^3/s
	= 22.8245	million gal/d	1 ft^3/min	= 0.471 947	L/s
	= 0.810 709 × 10^{-3}	acre ft/s	1 gal/min	= 0.063 090 2	L/s
1 L/s	= 2.118 88	ft^3/min	1 gal/h	= 1.051 50	mL/s
	= 15.850 3	gal/min	1 million gal/d	= 43.8126	L/s
	= 951.022	gal/h	1 acre ft/s	= 1233.49	m^3/s

TEMPERATURE INTERVAL

1°C	= *1K*	= *1.8*°F	1°F	= 0.555 556	°C or K
				= 5/9 °C	= 5/9 K

EQUIVALENT TEMPERATURE VALUE ($t_{°C} = T_K - 273.15$)

$t_{°C}$	= 5/9 (t_F − 32)		t_F	= 9/5 $t_{°C}$ + 32

	Metric to Customary			**Customary to Metric**		

MASS

1 kg	= 2.204 62	lb (avoirdupois)	1 ton (short)*	= 0.907 185	metric ton
	= 35.2740	oz (avoirdupois)		= 907.185	kg
1 metric	= 1.102 31	ton (short, 2000 lb)	1 lb	= 0.453 592	kg
ton	= 2204.62	lb	1 oz	= 28.3495	g
1 g	= 0.035 274	oz	1 pennyweight	= 1.555 17	g
	= 0.643 015	pennyweight	*(1 long ton (2240 lb) = 1016.05 kg)		

MASS PER UNIT LENGTH

1 kg/m	= 0.671 969	lb/ft	1 lb/ft	= 1.488 16	kg/m
1 g/m	= 3.547 99	lb/mile	1 lb/mile	= 0.281 849	g/m

MASS PER UNIT AREA

1 kg/m^2	= 0.204 816	lb/ft^2	1 lb/ft^2	= 4.882 43	kg/m^2
1 g/m^2	= 0.029 494	oz/yd^2	1 oz/yd^2	= 33.9057	g/m^2
	= 3.277 06 × 10^{-3}	oz/ft^2	1 oz/ft^2	= 305.152	g/m^2

DENSITY (MASS PER UNIT VOLUME)

1 kg/m^3	= 0.062 428	lb/ft^3	1 lb/ft^3	= 16.0185	kg/m^3
	= 1.685 56	lb/yd^3	1 lb/yd^3	= 0.593 276	kg/m^3
1 t/m^3	= 0.842 778	ton/yd^3	1 ton/yd^3	= 1.186 55	t/m^3

MOMENT OF INERTIA

1 kg·m^2	= 23.7304	lb·ft^3	1 lb·ft^2	= 0.042 140 1	kg·m^2
	= 3417.17	lb·in^2	1 lb·in^2	= 292.640	kg·mm^2

MASS PER UNIT TIME

1 kg/s	= 2.204 62	lb/s	1 lb/s	= 0.453 592	kg/s
1 t/h	= 0.984 207	ton/h	1 ton/h	= 1.016 05	t/h

FORCE

1 MN	= 112.404	tonf (ton-force)	1 tonf (ton-force)	= 8.896 44	kN
1 kN	= 0.112 404	tonf	1 kip (1000 lbf)	= 4.448 22	kN
	= 224.809	lbf (pound-force)	1 lbf (pound-force)	= 4.448 22	N
1 N	= 0.224 809	lbf			

MOMENT OF FORCE, TORQUE

1 N·m	= 0.737 562	lbf·ft	1 lfb·ft	= 1.355 82	N·m
	= 8.850 75	lbf·in	1 lbf·in	= 0.112 985	N·m
1 kN·m	= 0.368 781	tonf·ft	1 tonf·ft	= 2.711 64	kN·m
	= 0.737 562	kip·ft	1 kip·ft	= 1.355 82	kN·m

FORCE PER UNIT LENGTH

1 N/m	= 0.068 521 8	lbf/ft	1 lbf/ft	= 14.5939	N/m
1 kN/m	= 0.034 260 9	tonf/ft	1 lbf/in	= 175.127	N/m
			1 tonf/ft	= 29.187 8	kN/m

PRESSURE, STRESS, MODULUS OF ELASTICITY (FORCE PER UNIT AREA) (1 Pa = 1 N/m^2)

1 MPa	= 0.072 518 8	tonf/in^2	1 tonf/in^2	= 13.7895	MPa
	= 10.4427	tonf/ft^2	1 ton/ft^2	= 95.7605	kPa
	= 145.038	lbf/in^2	1 kip/in^2	= 6.894 76	MPa
1 kPa	= 20.8854	lbf/ft^2	1 lbf/in^2	= 6.894 76	kPa
			1 lbf/ft^2	= 47.8803	Pa

FIGURE A-12—*Continued.*

	Metric to Customary			Customary to Metric		
WORK ENERGY, HEAT ($1 J = 1 N \cdot m = 1 W \cdot s$)						
1 MJ	= 0.277 778	kWh	1 kWh (550 ft·lbf/s)	= *3.6*	MJ	
1 kJ	= 0.947 817	Btu	1 Btu (Int. Table)	= 1.055 06	kJ	
1 J	= 0.737 562	ft·lbf		= 1055.06	J	
			1 ft·lbf	= 1.355 82	J	
POWER, HEAT FLOW RATE						
1 kW	= 1.341 02	hp (horsepower)	1 hp	= 0.745 700	kW	
1 W	= 3.412 14	Btu/h		= 745.700	W	
	= 0.737 562	ft·lbf/s	1 Btu/h	= 0.293 071	W	
			1 ft·lbf/s	= 1.355 82	W	
HEAT FLUX DENSITY						
1 W/m²	= 0.316 998	Btu/(ft²·h)	1 Btu/(ft²·h)	= 3.154 59	W/m²	
COEFFICIENT OF HEAT TRANSFER						
1 W/(m²·K)	= 0.176 110	Btu/(ft²·h·°F)	1 Btu/(ft²·h·°F)	= 5.678 26	W/(m²·K)	
THERMAL CONDUCTIVITY						
1 W/(m·K)	= 0.577 789	Btu/(ft·h·°F)	1 Btu/(ft·h·°F)	= 1.730 73	W/(m·K)	
CALORIFIC VALUE (MASS AND VOLUME BASIS)						
1 kJ/kg	=			= *2.326*	kJ/kg	
(1 J/g)	= 0.429 923	Btu/lb	1 Btu/lb	= *2.326*	(J/g)	
1 kJ/m³	= 0.026 839 2	Btu/ft³	1 Btu/ft³	= 37.2589	kJ/m³	
THERMAL CAPACITY (MASS AND VOLUME BASIS)						
1 kJ/(kg·K)	= 0.238 846	Btu/(lb·°F)	1 Btu/(lb·°F)	= *4.1868*	kJ/(kg·K)	
1 kJ/(M³·K)	= 0.014 910 7	Btu/(ft³·°F)	1 Btu/(ft³·°F)	= 67.0661	kJ/(m³·K)	
ILLUMINANCE						
1 lx (lux)	= 0.092 903	lm/ft² (footcandle)	1 lm/ft² (footcandle)	= 10.7639	lx (lux)	
LUMINANCE						
1 cd/m²	= 0.092 903	cd/ft²	1 cd/ft²	= 10.7639	cd/m²	
	= 0.291 864	footlambert	1 footlambert	= 3.426 26	cd/m²	
1 kcd/m²	= 0.314 159	lambert	1 lambert	= 3.183 01	kcd/m²	

Copyright ASTM; reprinted with permission.

Note: Coversion factors are taken to six significant figures, where appropriate. Values in italics are exact conversions.

PROPERTIES OF SELECTED MASONRY WALLS

The following figure contains data that permits comparison of several masonry wall properties. The figure shows various wall assemblies and thicknesses, and states for each the wall's fire rating, measured in hours; Sound Transmission Class (STC) measured in decibels (dB); heat transmission, measured by the wall's U-value in British thermal units (btus) transmitted through 1 square foot of wall in 1 hour for each degree Fahrenheit of temperature difference between air on the warm side and air on the cool side of the wall; weight, measured in pounds per square foot (psf); and estimated cost, measured in dollars per square foot.

The cost estimates shown permit only relative cost comparison because they are based on a 1978 survey of masonry contractors in the 16 county Chicago metropolitan area. These estimates are based on the labor and materials needed to build an 8-ft.-high, 32-ft.-long running bond wall section without offsets or openings. Conditions not taken into account in Figure 4.5-1 that will affect actual costs include the number, type, and design of openings, offsets, corners, and piers; the type of structural bonding; the pattern bond; special or unusual design considerations; site conditions; seasonal variations; geographic location; and any inflation that may have occurred since the original survey was made.

For additional information regarding specific properties of masonry walls, see Section 7.3 for thermal transmission properties and Section 10.1 for sound control data.

4" WALLS

(Tooled 1 Side, unless noted)

FACE BRICK (2-1/4 in. × 8 in.)

Fire Rating (hr.):	1
Sound Transmission Class (dB):	45
U-value (Btu/sq. ft. hr. — Deg. Fahr.):	.76
Wall Weight (lb./sq. ft.):	38
ESTIMATED COST PER SQUARE FOOT:	**4.89**

LANNON STONE

Fire Rating (hr.):	1
Sound Transmission Class (dB):	46
U-value (Btu/sq. ft. hr. — Deg. Fahr.):	.68
Wall Weight (lb./sq. ft.):	50
ESTIMATED COST PER SQUARE FOOT:	**8.65**

COMMON BRICK (2-1/8 in. × 8 in.)

Fire Rating (hr.):	1
Sound Transmission Class (dB):	34
U-value (Btu/sq. ft. hr. — Deg. Fahr.):	.76
Wall Weight (lb./sq. ft.):	38
ESTIMATED COST PER SQUARE FOOT:	**3.70**

LIGHTWEIGHT CONCRETE BLOCK

Fire Rating (hr.):	1
Sound Transmission Class (dB):	40
U-value (Btu/sq. ft. hr. — Deg. Fahr.):	.43
Wall Weight (lb./sq. ft.):	24
ESTIMATED COST PER SQUARE FOOT:	**2.37**

CONCRETE BRICK (2-1/4 in. × 7-5/8 in.)

Fire Rating (hr.):	1
Sound Transmission Class (dB):	37
U-value (Btu/sq. ft. hr. — Deg. Fahr.):	.67
Wall Weight (lb./sq. ft.):	35
ESTIMATED COST PER SQUARE FOOT:	**3.78**

CLAY PARTITION TILE (12 in. × 12 in. Face Size) (Trowel Cut Joint)

Fire Rating (hr.):	1
Sound Transmission Class (dB):	28
U-value (Btu/sq. ft. hr. — Deg. Fahr.):	.55
Wall Weight (lb./sq. ft.):	21
ESTIMATED COST PER SQUARE FOOT (No Plaster):	**1.96**

HEAVYWEIGHT CONCRETE "SLUMP" BLOCK (4 in. × 16 in. Face Size)

Fire Rating (hr.):	1
Sound Transmission Class (dB):	36
U-value (Btu/sq. ft. hr. — Deg. Fahr.):	.46
Wall Weight (lb./sq. ft.):	33
ESTIMATED COST PER SQUARE FOOT:	**3.94**

GYPSUM PARTITION TILE (12 in. × 30 in. Face Size) (Trowel Cut Joint)

Fire Rating (hr.):	4
Sound Transmission Class (dB):	42
U-value (Btu/sq. ft. hr. — Deg. Fahr.):	N.A.
Wall Weight (lb./sq. ft.):	26
ESTIMATED COST PER SQUARE FOOT (No Plaster):	**1.89**

FIGURE A-13 *Properties of selected masonry walls.*

GLAZED TILE (6 TC Series and 6 TCD Series — Tooled 2 Sides)

Fire Rating (hr.):	−1
Sound Transmission Class (dB):	32
U-value (Btu/sq. ft. hr. — Deg. Fahr.):	.40
Wall Weight (lb./sq. ft.):	27

ESTIMATED COST PER SQUARE FOOT:
1 Side Glazed:	**7.60**
2 Sides Glazed:	**8.90**

GLASS BLOCK* (8 in. × 8 in.) (Tooled 2 Sides)

Fire Rating (hr.):	—
Sound Transmission Class (dB):	27
U-value (Btu/sq. ft. hr. — Deg. Fahr.):	.56
Wall Weight (lb./sq. ft.):	20

ESTIMATED COST PER SQUARE FOOT: 11.77

*6 in. × 6 in. and 12 in. × 12 in. Glass Block are also available.

6" WALLS
(Tooled 1 Side, unless noted)

BRICK, THRU WALL UNITS (3-1/2 in. × 11-1/2 in. Face Size)

Fire Rating (hr.):	2+
Sound Transmission Class (dB):	41
U-value (Btu/sq. ft. hr. — Deg. Fahr.):	.66
Wall Weight (lb./sq. ft.):	42

ESTIMATED COST PER SQUARE FOOT: 6.50

LIGHTWEIGHT CONCRETE BLOCK

Fire Rating (hr.):	1
Sound Transmission Class (dB):	45
U-value (Btu/sq. ft. hr. — Deg. Fahr.):	.38
Wall Weight (lb./sq. ft.):	26

ESTIMATED COST PER SQUARE FOOT: 2.60

8" WALLS
(Tooled 1 Side, unless noted)

4 in. FACE BRICK AND 4 in. LIGHTWEIGHT CONCRETE BLOCK

Fire Rating (hr.):	3+
Sound Transmission Class (dB):	48
U-value (Btu/sq. ft. hr. — Deg. Fahr.):	.33
Wall Weight (lb./sq. ft.):	53

ESTIMATED COST PER SQUARE FOOT: 6.70

4 in. FACE BRICK AND 4 in. FACE BRICK-TOOLED 2 SIDES

Fire Rating (hr.):	4
Sound Transmission Class (dB):	49
U-value (Btu/sq. ft. hr. — Deg. Fahr.):	.54
Wall Weight (lb./sq. ft.):	62

ESTIMATED COST PER SQUARE FOOT: 9.33

BRICK, THRU WALL UNITS (3-1/2 in. × 11-1/2 in. Face Size)

Fire Rating (hr.):	3
Sound Transmission Class (dB):	50
U-value (Btu/sq. ft. hr. — Deg. Fahr.):	.47
Wall Weight (lb./sq. ft.):	61

ESTIMATED COST PER SQUARE FOOT: 7.65

4 in. COMMON BRICK AND 4 in. LIGHTWEIGHT CONCRETE BLOCK

Fire Rating (hr.):	3+
Sound Transmission Class (dB):	48
U-value (Btu/sq. ft. hr. — Deg. Fahr.):	.33
Wall Weight (lb./sq. ft.):	53

ESTIMATED COST PER SQUARE FOOT: 5.56

LIGHTWEIGHT CONCRETE BLOCK (Tooled 2 Sides)

Fire Rating (hr.):	3
Sound Transmission Class (dB):	46
U-value (Btu/sq. ft. hr. — Deg. Fahr.):	.35
Wall Weight (lb./sq. ft.):	32

ESTIMATED COST PER SQUARE FOOT: 2.92

HEAVYWEIGHT CONCRETE "SLUMP" BLOCK (4 in. × 16 in. Face Size)

Fire Rating (hr.):	3
Sound Transmission Class (dB):	44
U-value (Btu/sq. ft. hr. — Deg. Fahr.):	.38
Wall Weight (lb./sq. ft.):	45

ESTIMATED COST PER SQUARE FOOT: 4.53

FIGURE A-13 *(Continued)*

4 in. FACE BRICK AND 6 in. LIGHTWEIGHT CONCRETE BLOCK

Fire Rating (hr.):	4
Sound Transmission Class (dB):	50
U-value (Btu/sq. ft. hr. — Deg. Fahr.):	.32
Wall Weight (lb./sq. ft.):	62
ESTIMATED COST PER SQUARE FOOT:	**7.01**

4 in. COMMON BRICK AND 6 in. LIGHTWEIGHT CONCRETE BLOCK

Fire Rating (hr.):	4
Sound Transmission Class (dB):	50
U-value (Btu/sq. ft. hr. — Deg. Fahr.):	.32
Wall Weight (lb./sq. ft.):	62
ESTIMATED COST PER SQUARE FOOT:	**5.90**

LIGHTWEIGHT CONCRETE BLOCK
(Tooled 2 Sides)

Fire Rating (hr.):	3
Sound Transmission Class (dB):	48
U-value (Btu/sq. ft. hr. — Deg. Fahr.):	.33
Wall Weight (lb./sq. ft.):	39
ESTIMATED COST PER SQUARE FOOT:	**3.34**

FLUTED (4 Flutes) HEAVYWEIGHT CONCRETE BLOCK (Tooled 2 Sides)

Fire Rating (hr.):	2
Sound Transmission Class (dB):	50
U-value (Btu/sq. ft. hr. — Deg. Fahr.):	.49
Wall Weight (lb./sq. ft.):	61
ESTIMATED COST PER SQUARE FOOT:	**4.77**

4 in. FACE BRICK, 2 in. AIR SPACE, AND 4 in. LIGHTWEIGHT CONCRETE BLOCK

Fire Rating (hr.):	3
Sound Transmission Class (dB):	48
U-value (Btu/sq. ft. hr. — Deg. Fahr.):	.27
Wall Weight (lb./sq. ft.):	53
ESTIMATED COST PER SQUARE FOOT:	**6.95**

4 in. FACE BRICK, 2 in. AIR SPACE, AND 4 in. FACE BRICK

Fire Rating (hr.):	4
Sound Transmission Class (dB):	51
U-value (Btu/sq. ft. hr. — Deg. Fahr.):	.37
Wall Weight (lb./sq. ft.):	64
ESTIMATED COST PER SQUARE FOOT:	**9.57**

4 in. FACE BRICK, 2 in. URETHANE INSULATION, AND 4 in. LIGHT-WEIGHT CONCRETE BLOCK

Fire Rating (hr.):	3
Sound Transmission Class (dB):	49
U-value (Btu/sq. ft. hr. — Deg. Fahr.):	.05
Wall Weight (lb./sq. ft.):	54
ESTIMATED COST PER SQUARE FOOT:	**7.85**

4 in. COMMON BRICK, 2 in. AIR SPACE, AND 4 in. LIGHTWEIGHT CONCRETE BLOCK

Fire Rating (hr.):	3
Sound Transmission Class (dB):	48
U-value (Btu/sq. ft. hr. — Deg. Fahr.):	.27
Wall Weight (lb./sq. ft.):	53
ESTIMATED COST PER SQUARE FOOT:	**5.81**

FIGURE A-13 *(Continued)*

4 in. FACE BRICK AND 8 in. LIGHTWEIGHT CONCRETE BLOCK

Fire Rating (hr.):	4
Sound Transmission Class (dB):	50+
U-value (Btu/sq. ft. hr. — Deg. Fahr.):	.29
Wall Weight (lb./sq. ft.):	72
ESTIMATED COST PER SQUARE FOOT:	**7.25**

LIGHTWEIGHT CONCRETE BLOCK

Fire Rating (hr.):	3–4
Sound Transmission Class (dB):	49
U-value (Btu/sq. ft. hr. — Deg. Fahr.):	.31
Wall Weight (lb./sq. ft.):	47
ESTIMATED COST PER SQUARE FOOT:	**3.76**

4 in. COMMON BRICK AND 8 in. LIGHTWEIGHT CONCRETE BLOCK

Fire Rating (hr.):	4
Sound Transmission Class (dB):	50+
U-value (Btu/sq. ft. hr. — Deg. Fahr.):	.29
Wall Weight (lb./sq. ft.):	72
ESTIMATED COST PER SQUARE FOOT:	**6.12**

FLUTED (4 Flutes) HEAVYWEIGHT CONCRETE BLOCK

Fire Rating (hr.):	3–4
Sound Transmission Class (dB):	54
U-value (Btu/sq. ft. hr. — Deg. Fahr.):	.47
Wall Weight (lb./sq. ft.):	73
ESTIMATED COST PER SQUARE FOOT:	**5.27**

BRICK PAVING
(Basket Weave Pattern, 4.5 per sq. ft.)

MORTAR SET—BED AND ALL JOINTS
(Type "M" Mortar)

Weight (lb/sq. ft.):	29
ESTIMATED COST PER SQUARE FOOT:	**6.20**

DRY SET AND BUTTED
(Sand Swept Into Joints)

Weight (lb./sq. ft.):	27
ESTIMATED COST PER SQUARE FOOT:	**5.00**

FIGURE A-13 *(Continued)*

WOOD DESIGN TABLES

The maximum allowable unit stresses (working stresses) that should be permitted to exist on various species of dimension lumber are shown in tables in the NDS supplement, *Design Values for Wood Construction*, to the *National Design Specifications (NDS) for Wood Construction* of the American Forest and Paper Association (AFPA). These tables are based on what NDS calls *normal load duration*. Figure A-14 shows part of one table from the 1991 edition of the NDS supplement for visually graded dimension lumber. Figure A-15 is the NDS supplement table for mechanically graded dimension lumber. Refer to Section 6.1.3.4 for additional information. Tabular bending values may be increased for large members and for members in repetitive use, such as joists, studs, rafters, and the like, spaced 24 in. or less on centers. This factor is 1.14. Bending values must be reduced when members are used flat rather than with the long edges vertical. Values must also be reduced when the wood is used with a moisture content larger than 19%. Allowable shear values can be increased when the length of splits and shakes is less than that contemplated in the tables. The adjustment factors necessary to calculate these changes in tabular values are explained and the formulas for them given in the NDS supplement.

The grades of framing lumber should be suitable for the intended purpose in terms of appearance and strength. Engineering design calculations and construction practices for wood framing should conform with the recommendations of the latest edition of the NDS. Calculated working stresses should not exceed maximum allowable working stresses tabulated in the NDS supplement.

FIGURE A-14
Design Values for Visually Graded Dimension Lumber

Species and Commercial Grade	Size Classification	Design Values, pounds per square inch (psi)						Grading Rules Agency
		Bending, F_b	Tension Parallel to Grain, F_l	Shear Parallel to Grain, F_v	Compression Perpendicular to Grain, $F_{c\perp}$	Compression Parallel to Grain, F_c	Modulus of Elasticity, E	
DOUGLAS FIR-LARCH								
Select Structural		1450	1000	95	625	1700	1,900,000	WCLIB
No. 1 and Btr	2 in.–4 in. thick	1150	775	95	625	1500	1,800,000	WWPA
No. 1		1000	675	95	625	1450	1,700,000	
No. 2	2 in. and wider	875	575	95	625	1300	1,600,000	
No. 3		500	325	95	625	750	1,400,000	
Stud		675	450	95	625	825	1,400,000	
Construction	2 in.–4 in. thick	1000	650	95	625	1600	1,500,000	
Standard		550	375	95	625	1350	1,400,000	
Utility	2 in.–4 in. wide	275	175	95	625	875	1,300,000	
DOUGLAS FIR-LARCH (NORTH)								
Select Structural	2 in.–4 in. thick	1300	800	95	625	1900	1,900,000	NLGA
No. 1/No. 2		825	500	95	625	1350	1,600,000	
No. 3	2 in. and wider	475	300	95	625	775	1,400,000	
Stud		650	375	95	625	850	1,400,000	
Construction	2 in.–4 in. thick	950	575	95	625	1750	1,500,000	
Standard		525	325	95	625	1400	1,400,000	
Utility	2 in.–4 in. wide	250	150	95	625	925	1,300,000	
DOUGLAS FIR-SOUTH								
Select Structural		1300	875	90	520	1550	1,400,000	WWPA
No. 1	2 in.–4 in. thick	900	600	90	520	1400	1,300,000	
No. 2		825	525	90	520	1300	1,200,000	
No. 3	2 in. and wider	475	300	90	520	750	1,100,000	
Stud		650	425	90	520	825	1,100,000	
Construction	2 in.–4 in. thick	925	600	90	520	1550	1,200,000	
Standard		525	350	90	520	1300	1,100,000	
Utility	2 in.–4 in. wide	250	150	90	520	875	1,000,000	
SOUTHERN PINE								
Dense Select Structural		3050	1650	100	660	2250	1,900,000	SPIB
Select Structural		2850	1600	100	565	2100	1,800,000	
Non-Dense Select Structural		2650	1350	100	480	1950	1,700,000	
No. 1 Dense		2000	1100	100	660	2000	1,800,000	
No. 1	2 in.–4 in. thick	1850	1050	100	565	1850	1,700,000	
No. 1 Non-Dense		1700	900	100	480	1700	1,600,000	
No. 2 Dense	2 in.–4 in. wide	1700	875	90	660	1850	1,700,000	
No. 2		1500	825	90	565	1650	1,600,000	
No. 2 Non-Dense		1350	775	90	480	1600	1,400,000	
No. 3		850	475	90	565	975	1,400,000	
Stud		875	500	90	565	975	1,400,000	
Construction	2 in.–4 in. thick	1100	625	100	565	1800	1,500,000	
Standard		625	350	90	565	1500	1,300,000	
Utility	4 in. wide	300	175	90	565	975	1,300,000	
Dense Select Structural		2700	1500	90	660	2150	1,900,000	
Select Structural		2550	1400	90	565	2000	1,800,000	
Non-Dense Select Structural		2350	1200	90	480	1850	1,700,000	
No. 1 Dense		1750	950	90	660	1900	1,800,000	
No. 1	2 in.–4 in. thick	1650	900	90	565	1750	1,700,000	
No. 1 Non-Dense		1500	800	90	480	1600	1,600,000	
No. 2 Dense	5 in.–6 in. wide	1450	775	90	660	1750	1,700,000	
No. 2		1250	725	90	565	1600	1,600,000	
No. 2 Non-Dense		1150	675	90	480	1500	1,400,000	
No. 3		750	425	90	565	925	1,400,000	
Stud		775	425	90	565	925	1,400,000	

FIGURE A-14—Continued.

Species and Commercial Grade	Size Classification	Design Values, pounds per square inch (psi)						Grading Rules Agency
		Bending, F_b	Tension Parallel to Grain, F_t	Shear Parallel to Grain, F_v	Compression Perpendicular to Grain, $F_{c\perp}$	Compression Parallel to Grain, F_c	Modulus of Elasticity, E	
SOUTHERN PINE (Continued)								
Dense Select Structural		2450	1350	90	660	2050	1,900,000	SPIB
Select Structural		2300	1300	90	565	1900	1,800,000	
Non-Dense Select Structural		2100	1100	90	480	1750	1,700,000	
No. 1 Dense	2 in.–4 in. thick	1650	875	90	660	1800	1,800,000	
No. 1		1500	825	90	565	1650	1,700,000	
No. 1 Non-Dense	8 in. wide	1350	725	90	480	1550	1,600,000	
No. 2 Dense		1400	675	90	660	1700	1,700,000	
No. 2		1200	650	90	565	1550	1,600,000	
No. 2 Non-Dense		1100	600	90	480	1450	1,400,000	
No. 3		700	400	90	565	875	1,400,000	
Dense Select Structural		2150	1200	90	660	2000	1,900,000	
Select Structural		2050	1100	90	565	1850	1,800,000	
Non-Dense Select Structural		1850	950	90	480	1750	1,700,000	
No. 1 Dense	2 in.–4 in. thick	1450	775	90	660	1750	1,800,000	
No. 1		1300	725	90	565	1600	1,700,000	
No. 1 Non-Dense	10 in. wide	1200	650	90	480	1500	1,600,000	
No. 2 Dense		1200	625	90	660	1650	1,700,000	
No. 2		1050	575	90	565	1500	1,600,000	
No. 2 Non-Dense		950	550	90	480	1400	1,400,000	
No. 3		600	325	90	565	850	1,400,000	
Dense Select Structural		2050	1100	90	660	1950	1,900,000	
Select Structural		1900	1050	90	565	1800	1,800,000	
Non-Dense Select Structural		1750	900	90	480	1700	1,700,000	
No. 1 Dense	2 in.–4 in. thick	1350	725	90	660	1700	1,800,000	
No. 1		1250	675	90	565	1600	1,700,000	
No. 1 Non-Dense	12 in. wide	1150	600	90	480	1500	1,600,000	
No. 2 Dense		1150	575	90	660	1600	1,700,000	
No. 2		975	550	90	565	1450	1,600,000	
No. 2 Non-Dense		900	525	90	480	1350	1,400,000	
No. 3		575	325	90	565	825	1,400,000	
SOUTHERN PINE		(Dry service conditions—19% or less moisture content)						
Dense Structural 86	2 in.–4 in. thick	2600	1750	155	660	2000	1,800,000	SPIB
Dense Structural 72		2200	1450	130	660	1650	1,800,000	
Dense Structural 65	2 in. and wider	2000	1300	115	660	1500	1,800,000	
SOUTHERN PINE		(Wet service conditions)						
Dense Structural 86	2-1/2 in.–4 in. thick	2100	1400	145	440	1300	1,600,000	SPIB
Dense Structural 72		1750	1200	120	440	1100	1,600,000	
Dense Structural 65	2-1/2 in. and wider	1600	1050	110	440	1000	1,600,000	

Courtesy American Forest and Paper Association.

FIGURE A-15
Design Values for Mechanically Graded Dimension Lumber

Species and Commercial Grade	Size Classification	Design Values, pounds per square inch (psi)				Grading Rules Agency
		Bending, F_b	Tension Parallel to Grain, F_t	Compression Parallel to Grain, F_c	Modulus of Elasticity, E	
MACHINE STRESS RATED (MSR) LUMBER						
900f-1.0E		900	350	1050	1,000,000	WCLIB
1200f-1.2E		1200	600	1400	1,200,000	NLGA, SPIB, WCLIB, WWPA
1250f-1.4E		1250	800	1475	1,400,000	WCLIB
1350f-1.3E		1350	750	1600	1,300,000	SPIB, WCLIB, WWPA
1400f-1.2E		1400	800	1600	1,200,000	SPIB
1450f-1.3E		1450	800	1625	1,300,000	NLGA, WCLIB, WWPA
1500f-1.3E		1500	900	1650	1,300,000	SPIB
1500f-1.4E		1500	900	1650	1,400,000	NLGA, SPIB, WCLIB, WWPA
1600f-1.4E		1600	950	1675	1,400,000	SPIB
1650f-1.4E		1650	1020	1700	1,400,000	SPIB
1650f-1.5E		1650	1020	1700	1,500,000	NLGA, SPIB, WCLIB, WWPA
1650f-1.6E		1650	1075	1700	1,600,000	WCLIB
1800f-1.5E		1800	1300	1750	1,500,000	SPIB
1800f-1.6E		1800	1175	1750	1,600,000	NLGA, SPIB, WCLIB, WWPA
1950f-1.5E	2 in. and less in thickness	1950	1375	1800	1,500,000	SPIB
1950f-1.7E		1950	1375	1800	1,700,000	NLGA, SPIB, WWPA
2000f-1.6E		2000	1300	1825	1,600,000	SPIB
2100f-1.8E	2 in. and wider	2100	1575	1875	1,800,000	NLGA, SPIB, WCLIB, WWPA
2250f-1.6E		2250	1750	1925	1,600,000	SPIB
2250f-1.9E		2250	1750	1925	1,900,000	NLGA, SPIB, WWPA
2400f-1.7E		2400	1925	1975	1,700,000	SPIB
2400f-1.8E		2400	1925	1975	1,800,000	SPIB
2400f-2.0E		2400	1925	1975	2,000,000	NLGA, SPIB, WCLIB, WWPA
2500f-2.2E		2500	1750	2000	2,200,000	WCLIB
2550f-2.1E		2550	2050	2025	2,100,000	NLGA, SPIB, WWPA
2700f-2.0E		2700	1800	2100	2,000,000	WCLIB
2700f-2.2E		2700	2150	2100	2,200,000	NLGA, SPIB, WCLIB, WWPA
2850f-2.3E		2850	2300	2150	2,300,000	SPIB, WWPA
3000f-2.4E		3000	2400	2200	2,400,000	NLGA, SPIB
3150f-2.5E		3150	2500	2250	2,500,000	SPIB
3300f-2.6E		3300	2650	2325	2,600,000	SPIB
900f-1.2E		900	350	1050	1,200,000	NLGA, WCLIB
1200f-1.5E	2 in. and less in thickness	1200	600	1400	1,500,000	NLGA, WCLIB
1350f-1.8E		1350	750	1600	1,800,000	NLGA
1500f-1.8E	6 in. and wider	1500	900	1650	1,800,000	WCLIB
1800f-2.1E		1800	1175	1750	2,100,000	NLGA, WCLIB
MACHINE EVALUATED LUMBER (MEL)						
M-10		1400	800	1600	1,200,000	SPIB
M-11		1550	850	1650	1,500,000	
M-12		1600	850	1700	1,600,000	
M-13		1600	950	1700	1,400,000	
M-14		1800	1000	1750	1,700,000	
M-15		1800	1100	1750	1,500,000	
M-16		1800	1300	1750	1,500,000	
M-17	2 in. and less in thickness	1950	1300	2050	1,700,000	
M-18		2000	1200	1850	1,800,000	
M-19	2 in. and wider	2000	1300	1850	1,600,000	
M-20		2000	1600	2100	1,900,000	
M-21		2300	1400	1950	1,900,000	
M-22		2350	1500	1950	1,700,000	
M-23		2400	1900	2000	1,800,000	
M-24		2700	1800	2100	1,900,000	
M-25		2750	2000	2100	2,200,000	
M-26		2800	1800	2150	2,000,000	
M-27		3000	2000	2400	2,100,000	

Courtesy American Forest and Paper Association.

Figure A-16 shows maximum allowable spans for built-up wood girders supporting floor loads. These tables indicate required lumber sizes for various conditions of loading and framing, assuming a minimum fiber stress in bending (F_b of 1000 psi and 1500 psi). Species and grades meeting this requirement can be selected from tables in the NDS supplement, *Design Values for Wood Construction* of the AFPA (see Section 6.1.3.4 and Fig. A-14). These tables should be verified with the requirements of the applicable building code, which may vary from the table data.

Floor joists for simple frame buildings and ordinary construction can be selected by referring to *Span Tables for Joists and Rafters,* published by the AFPA. The table there for floor joists for 40 psf live loads and 10 psf dead loads is reproduced in Figure A-17. These tables indicate the minimum modulus of elasticity, E, and extreme fiber strength in bending, F_b, required for various spans, lumber sizes, and conditions of loading. *Span Tables for Joists and Rafters* also includes tables for 30, 50, and 60 psf live loads and for 20 psf dead loads. Suitable species and grades meeting the indicated strength requirements can be selected from the tables of allowable unit stresses in the NDS supplement (see Fig. A-14).

In these tables, the allowable span is the clear distance between supports. Tabulated spans are applicable to green or dry lumber, provided it is used in covered structures, where it will normally achieve a moisture content of 19% or less.

For an example of floor joist selection, assume a required span of 12 ft. 9 in., a live load of 40 psf and joists spaced 16 in. on centers. Figure A-17 shows that a 2×8 having an E value of 1,600,000 psi (1.6 in the table) and an F_b value of 1255 psi would have an allowable span of 12 ft. 10 in. Suitable species and grades can be selected from the NDS supplement. In our example, Figure A-14 shows that a visually graded Douglas Fir-Larch No. 1 and Btr grade has an F_b of 1150 psi and an E of 1,800,000 psi. The F_b can be increased by a factor of 1.15 because of the repetitive member factor, producing an F_b of 1320 psi. Figure A-15 shows machine stress-rated lumber, 1650f-1.6E,

of WCLIB species that has an F_b of 1650 psi and an E of 1,600,000 psi.

Roof and ceiling framing should be of the proper grade for the intended use, graded by a recognized grading authority. Its moisture content should not be more than 15% at the time finishes are applied.

Joists and rafters for simple frame buildings and ordinary construction can be selected by referring to *Span Tables for Joists and Rafters*, published by the AFPA. These tables indicate the minimum modulus of elasticity, E, and extreme fiber strength in bending, F_b, required for various spans, lumber sizes, and conditions of loading. Suitable species and grades meeting the indicated strength requirements can be selected from the tables of *Allowable Unit Stresses* in the NDS supplement (see Fig. A-14).

A table of maximum allowable spans for ceiling joists carrying a drywall ceiling and a 20-lb. attic live load is included as Figure A-18. *Span Tables for Joists and Rafters* also contains a table for ceiling joists with a ceiling but not attic storage. A table for rafters carrying a drywall ceiling, a 40-psf live load, and a 10-psf dead load is included as Figure A-19. *Span Tables for Joists and Rafters* also includes tables for rafters carrying live loads of 20, 30, and 50 psf, and dead loads of 15 and 20 psf for rafters carrying a drywall ceiling, and for rafters carrying no ceiling.

In each case, the allowable span is the clear distance between supports. For rafters, the span is measured along the horizontal projection (Fig. A-20). Tabulated spans are applicable to green or dry lumber, provided it is used in covered structures, where it will normally achieve a moisture content of 19% or less.

As an example of rafter selection, assume a horizontal projection span of 15 ft. in a rafter carrying a drywall ceiling, a live load of 40 psf, a 10-psf dead load, with rafters spaced 16 in. on centers. Fig. A-19 shows that a 2×10 having an F_b value of 1100 psi and an E value of 880,000 psi would have an allowable horizontal projection of 15 ft. 4 in. Suitable species and grades can be selected from the tables in the NDS supplement. For our example, Figure A-14 shows that a visually graded southern pine No. 2 has

an F_b of 1500 psi and an E of 1,600,000 psi. Figure A-15 shows that machine stress-rated lumber, 1200f-1.2E, of any NLGA, SPIB, WCLIB, or WWPA species has an F_b of 1200 psi and an E of 1,200,00 psi.

Wood beams supporting plank floors or roofs should not exceed the maximum allowable spans tabulated in the AFPA publication, *Plank and Beam Framing for Residential Buildings*. Figure A-21 is an example of one of these tables for beams supporting a 20-psf live load with and allowable deflections of 1/240. *Plank and Beam Framing for Residential Buildings* also includes tables for live loads of 30 and 40 psf and allowable deflections of 1/300 and 1/360. To use these tables, first determine the span, the live loads to be supported, and the deflection limitation. Then select from the tables the proper size of beam with the corresponding required values for f and E. The beam used should be of a grade and species that meet or exceed the selected values.

Nominal 2-in.-thick plank should not exceed the maximum allowable spans shown in the tables in *Plank and Beam Framing for Residential Buildings,* which is reproduced in Figure A-22. The types of plank arrangements shown in the table are defined as

Type A Extending over a single span
Type B Continuous over two equal spans
Type C Continuous over three equal spans
Type D A combination of types A and B

To use this table, determine the plank arrangement type, the span, the supported live load, and the deflection limit; then select from the table the corresponding values for f and E. The plank used should be a grade and species that meet or exceed the minimum values listed. The maximum span for a given thickness, grade, and species of plank can be obtained by reversing this procedure.

The maximum allowable f and E for most commercial grades and species are tabulated in the NDS supplement to the *National Design Specifications, Wood Construction*, published by the AFPA. Figure A-14 shows these requirements for three species.

FIGURE A-16
Maximum Allowable Spans for Built-Up Girders Supporting One or Two Floors

Diagram 1 labels: LL+DL=50psf · "L"/2±1 · Girder · "L" · "S" "S" "S" · Load diagram applies also to 2-st. house below

Diagram 2 labels: LL+DL=40psf · Wall load=50plf · LL+DL=50psf · "L"/2±1 · Girder · "L"

Maximum Allowable spans ("S") for Girders Supporting One Floor[a]

Fb, psi	Lumber Size	\"L\" 20 ft.	22 ft.	24 ft.	26 ft.	28 ft.	30 ft.	32 ft.	34 ft.	36 ft.
1000	(2) 2 in. × 6 in.	4 ft.-5 in.	4 ft.-0 in.	—	—	—	—	—	—	—
	(3) 2 in. × 6 in.	5 ft.-6 in.	5 ft.-3 in.	5 ft.-0 in.	4 ft.-10 in.	4 ft.-8 in.	4 ft.-5 in.	4 ft.-2 in.	—	—
	(2) 2 in. × 8 in.	5 ft.-10 in.	5 ft.-3 in.	4 ft.-10 in.	4 ft.-5 in.	4 ft.-2 in.	—	—	—	—
	(3) 2 in. × 8 in.	7 ft.-3 in.	6 ft.-11 in.	6 ft.-7 in.	6 ft.-4 in.	6 ft.-2 in.	5 ft.-10 in.	5 ft.-5 in.	5 ft.-1 in.	4 ft.-10 in.
	(2) 2 in. × 10 in.	7 ft.-5 in.	6 ft.-9 in.	6 ft.-2 in.	5 ft.-8 in.	5 ft.-3 in.	4 ft.-11 in.	4 ft.-8 in.	4 ft.-4 in.	4 ft.-1 in.
	(3) 2 in. × 10 in.	9 ft.-3 in.	8 ft.-10 in.	8 ft.-5 in.	8 ft.-1 in.	7 ft.-10 in.	7 ft.-5 in.	6 ft.-11 in.	6 ft.-6 in.	6 ft.-2 in.
	(2) 2 in. × 12 in.	9 ft.-0 in.	8 ft.-2 in.	7 ft.-6 in.	6 ft.-11 in.	6 ft.-5 in.	6 ft.-0 in.	5 ft.-8 in.	5 ft.-4 in.	5 ft.-0 in.
	(3) 2 in. × 12 in.	11 ft.-3 in.	10 ft.-9 in.	10 ft.-3 in.	9 ft.-10 in.	9 ft.-6 in.	9 ft.-0 in.	8 ft.-5 in.	7 ft.-11 in.	7 ft.-6 in.
1500	(2) 2 in. × 6 in.	5 ft.-3 in.	4 ft.-10 in.	4 ft.-5 in.	4 ft.-1 in.	—	—	—	—	—
	(3) 2 in. × 6 in.	6 ft.-9 in.	6 ft.-5 in.	6 ft.-2 in.	5 ft.-11 in.	5 ft.-8 in.	5 ft.-3 in.	4 ft.-11 in.	4 ft.-8 in.	4 ft.-5 in.
	(2) 2 in. × 8 in.	7 ft.-0 in.	6 ft.-4 in.	5 ft.-10 in.	5 ft.-4 in.	5 ft.-0 in.	4 ft.-8 in.	4 ft.-4 in.	4 ft.-1 in.	—
	(3) 2 in. × 8 in.	8 ft.-11 in.	8 ft.-6 in.	8 ft.-1 in.	7 ft.-9 in.	7 ft.-5 in.	7 ft.-0 in.	6 ft.-6 in.	6 ft.-2 in.	5 ft.-10 in.
	(2) 2 in. × 10 in.	8 ft.-11 in.	8 ft.-1 in.	7 ft.-5 in.	6 ft.-10 in.	6 ft.-4 in.	5 ft.-11 in.	5 ft.-7 in.	5 ft.-3 in.	4 ft.-11 in.
	(3) 2 in. × 10 in.	11 ft.-4 in.	10 ft.-10 in.	10 ft.-4 in.	9 ft.-11 in.	9 ft.-6 in.	8 ft.-11 in.	8 ft.-4 in.	7 ft.-10 in.	7 ft.-5 in.
	(2) 2 in. × 12 in.	10 ft.-10 in.	9 ft.-10 in.	9 ft.-0 in.	8 ft.-4 in.	7 ft.-9 in.	7 ft.-2 in.	6 ft.-9 in.	6 ft.-4 in.	6 ft.-0 in.
	(3) 2 in. × 12 in.	13 ft.-9 in.	13 ft.-2 in.	12 ft.-7 in.	12 ft.-1 in.	11 ft.-7 in.	10 ft.-10 in.	10 ft.-2 in.	9 ft.-6 in.	9 ft.-0 in.

Maximum Allowable spans ("S") for Girders Supporting Two Floors[a]

Fb, psi	Lumber Size	\"L\" 20 ft.	22 ft.	24 ft.	26 ft.	28 ft.	30 ft.	32 ft.	34 ft.	36 ft.
1000	(2) 2 in. × 8 in.	4 ft.-7 in.	4 ft.-2 in.	—	—	—	—	—	—	—
	(3) 2 in. × 8 in.	4 ft.-9 in.	4 ft.-4 in.	4 ft.-0 in.	—	—	—	—	—	—
	(2) 2 in. × 10 in.	5 ft.-6 in.	5 ft.-0 in.	4 ft.-7 in.	4 ft.-3 in.	4 ft.-1 in.	4 ft.-0 in.	—	—	—
	(3) 2 in. × 10 in.	7 ft.-1 in.	6 ft.-6 in.	6 ft.-0 in.	5 ft.-6 in.	5 ft.-2 in.	4 ft.-10 in.	4 ft.-6 in.	4 ft.-3 in.	4 ft.-0 in.
1500	(3) 2 in. × 6 in.	4 ft.-2 in.	—	—	—	—	—	—	—	—
	(2) 2 in. × 10 in.	4 ft.-8 in.	4 ft.-3 in.	4 ft.-0 in.	—	—	—	—	—	—
	(2) 2 in. × 12 in.	5 ft.-6 in.	5 ft.-0 in.	4 ft.-7 in.	4 ft.-3 in.	4 ft.-1 in.	—	—	—	—
	(3) 2 in. × 10 in.	7 ft.-0 in.	6 ft.-5 in.	5 ft.-11 in.	5 ft.-6 in.	5 ft.-1 in.	4 ft.-9 in.	4 ft.-6 in.	4 ft.-3 in.	4 ft.-0 in.
	(3) 2 in. × 12 in.	8 ft.-6 in.	7 ft.-9 in.	7 ft.-2 in.	6 ft.-8 in.	6 ft.-2 in.	5 ft.-9 in.	5 ft.-2 in.	5 ft.-1 in.	4 ft.-10 in.

[a] Maximum spans are based on 30 psf live load for second floor and 40 psf live load for first floor; 10 psf dead load for either floor; 50 plf dead load for partition above girder.

[b] Spans given may be interpolated on a straight line basis for house widths intermediate between those shown.

FIGURE A-17
Maximum Allowable Spans for Floor Joists with L/360 Deflection Limits[a,b]

Joist Size, in.	Spacing, in.	Modulus of Elasticity, E, in 1,000,000 psi																
		0.8	0.9	1.0	1.1	1.2	1.3	1.4	1.5	1.6	1.7	1.8	1.9	2.0	2.1	2.2	2.3	2.4
2 × 6	12.0	8-6	8-10	9-2	9-6	9-9	10-0	10-3	10-6	10-9	10-11	11-2	11-4	11-7	11-9	11-11	12-1	12-3
	16.0	7-9	8-0	8-4	8-7	8-10	9-1	9-4	9-6	9-9	9-11	10-2	10-4	10-6	10-8	10-10	11-0	11-2
	19.2	7-3	7-7	7-10	8-1	8-4	8-7	8-9	9-0	9-2	9-4	9-6	9-8	9-10	10-0	10-2	10-4	10-6
	24.0	6-9	7-0	7-3	7-6	7-9	7-11	8-2	8-4	8-6	8-8	8-10	9-0	9-2	9-4	9-6	9-7	9-9
2 × 8	12.0	11-3	11-8	12-1	12-6	12-10	13-2	13-6	13-10	14-2	14-5	14-8	15-0	15-3	15-6	15-9	15-11	16-2
	16.0	10-2	10-7	11-0	11-4	11-8	12-0	12-3	12-7	12-10	13-1	13-4	13-7	13-10	14-1	14-3	14-6	14-8
	19.2	9-7	10-0	10-4	10-8	11-0	11-3	11-7	11-10	12-1	12-4	12-7	12-10	13-0	13-3	13-5	13-8	13-10
	24.0	8-11	9-3	9-7	9-11	10-2	10-6	10-9	11-0	11-3	11-5	11-8	11-11	12-1	12-3	12-6	12-8	12-10
2 × 10	12.0	14-4	14-11	15-5	15-11	16-5	16-10	17-3	17-8	18-0	18-5	18-9	19-1	19-5	19-9	20-1	20-4	20-8
	16.0	13-0	13-6	14-0	14-6	14-11	15-3	15-8	16-0	16-5	16-9	17-0	17-4	17-8	17-11	18-3	18-6	18-9
	19.2	12-3	12-9	13-2	13-7	14-0	14-5	14-9	15-1	15-5	15-9	16-0	16-4	16-7	16-11	17-2	17-5	17-8
	24.0	11-4	11-10	12-3	12-8	13-0	13-4	13-8	14-0	14-4	14-7	14-11	15-2	15-5	15-8	15-11	16-2	16-5
2 × 12	12.0	17-5	18-1	18-9	19-4	19-11	20-6	21-0	21-6	21-11	22-5	22-10	23-3	23-7	24-0	24-5	24-9	25-1
	16.0	15-10	16-5	17-0	17-7	18-1	18-7	19-1	19-6	19-11	20-4	20-9	21-1	21-6	21-10	22-2	22-6	22-10
	19.2	14-11	15-6	16-0	16-7	17-0	17-6	17-11	18-4	18-9	19-2	19-6	19-10	20-2	20-6	20-10	21-2	21-6
	24.0	13-10	14-4	14-11	15-4	15-10	16-3	16-8	17-0	17-5	17-9	18-1	18-5	18-9	19-1	19-4	19-8	19-11
F_b	12.0	718	777	833	888	941	993	1043	1092	1140	1187	1233	1278	1323	1367	1410	1452	1494
F_b	16.0	790	855	917	977	1036	1093	1148	1202	1255	1306	1357	1407	1456	1504	1551	1598	1644
F_b	19.2	840	909	975	1039	1101	1161	1220	1277	1333	1388	1442	1495	1547	1598	1649	1698	1747
F_b	24.0	905	979	1050	1119	1186	1251	1314	1376	1436	1496	1554	1611	1667	1722	1776	1829	1882

Courtesy American Forest and Paper Association.

[a] The required bending design value, F_b, in pounds per square inch (psi) is shown at the bottom of the table and is applicable to all lumber sizes shown. Spans are shown in feet-inches and are limited to 26 ft. and less. Check sources of supply for availability of lumber in lengths greater than 20 ft.

[b] Design Criteria: Deflections—for 40 pounds per square foot (psf) live load. Limited to span in inches divided by 360. Strength—live load of 40 psf plus dead load of 10 psf determines the required bending design value.

FIGURE A-18
Maximum Allowable Spans for Ceiling Joists with L/240 Deflection Limits[a,b]

Joist Size, in.	Spacing, in.	Modulus of Elasticity, E, in 1,000,000 psi																
		0.8	0.9	1.0	1.1	1.2	1.3	1.4	1.5	1.6	1.7	1.8	1.9	2.0	2.1	2.2	2.3	2.4
2 × 4	12.0	7-10	8-1	8-5	8-8	8-11	9-2	9-5	9-8	9-10	10-0	10-3	10-5	10-7	10-9	10-11	11-1	11-3
	16.0	7-1	7-5	7-8	7-11	8-1	8-4	8-7	8-9	8-11	9-1	9-4	9-6	9-8	9-9	9-11	10-1	10-3
	19.2	6-8	6-11	7-2	7-5	7-8	7-10	8-1	8-3	8-5	8-7	8-9	8-11	9-1	9-3	9-4	9-6	9-8
	24.0	6-2	6-5	6-8	6-11	7-1	7-3	7-6	7-8	7-10	8-0	8-1	8-3	8-5	8-7	8-8	8-10	8-11
2 × 6	12.0	12-3	12-9	13-3	13-8	14-1	14-5	14-9	15-2	15-6	15-9	16-1	16-4	16-8	16-11	17-2	17-5	17-8
	16.0	11-2	11-7	12-0	12-5	12-9	13-1	13-5	13-9	14-1	14-4	14-7	14-11	15-2	15-5	15-7	15-10	16-1
	19.2	10-6	10-11	11-4	11-8	12-0	12-4	12-8	12-11	13-3	13-6	13-9	14-0	14-3	14-6	14-8	14-11	15-2
	24.0	9-9	10-2	10-6	10-10	11-2	11-5	11-9	12-0	12-3	12-6	12-9	13-0	13-3	13-5	13-8	13-10	14-1
2 × 8	12.0	16-2	16-10	17-5	18-0	18-6	19-0	19-6	19-11	20-5	20-10	21-2	21-7	21-11	22-4	22-8	23-0	23-4
	16.0	14-8	15-3	15-10	16-4	16-10	17-3	17-9	18-1	18-6	18-11	19-3	19-7	19-11	20-3	20-7	20-11	21-2
	19.2	13-10	14-5	14-11	15-5	15-10	16-3	16-8	17-1	17-5	17-9	18-1	18-5	18-9	19-1	19-5	19-8	19-11
	24.0	12-10	13-4	13-10	14-3	14-8	15-1	15-6	15-10	16-2	16-6	16-10	17-2	17-5	17-9	18-0	18-3	18-6
2 × 10	12.0	20-8	21-6	22-3	22-11	23-8	24-3	24-10	25-5	26-0								
	16.0	18-9	19-6	20-2	20-10	21-6	22-1	22-7	23-1	23-8	24-1	24-7	25-0	25-5	25-10			
	19.2	17-8	18-4	19-0	19-7	20-2	20-9	21-3	21-9	22-3	22-8	23-1	23-7	23-11	24-4	24-9	25-1	25-5
	24.0	16-5	17-0	17-8	18-3	18-9	19-3	19-9	20-2	20-8	21-1	21-6	21-10	22-3	22-7	22-11	23-4	23-8
F_b	12.0	896	969	1040	1108	1174	1239	1302	1363	1423	1481	1539	1595	1651	1706	1759	1812	1864
F_b	16.0	986	1067	1145	1220	1293	1364	1433	1500	1566	1631	1694	1756	1817	1877	1936	1995	2052
F_b	19.2	1048	1134	1216	1296	1374	1449	1522	1594	1664	1733	1800	1866	1931	1995	2058	2120	2181
F_b	24.0	1129	1221	1310	1396	1480	1561	1640	1717	1793	1866	1939	2010	2080	2149	2217	2283	2349

Courtesy American Forest and Paper Association.

[a] The required bending design value, F_b, in pounds per square inch (psi) is shown at the bottom of the table and is applicable to all lumber sizes shown. Spans are shown in feet-inches and are limited to 26 ft. and less. Check sources of supply for availability of lumber in lengths greater than 20 ft.

[b] Design Criteria: Deflections—for 20 pounds per square foot (psf) live load. Limited to span in inches divided by 240. Strength—live load of 40 psf plus dead load of 10 psf determines the required bending design value.

FIGURE A-19
Maximum Allowable Spans for Rafters with L/240 Deflection Limits

Rafter Size, in.	Spacing, in.	300	400	500	600	700	800	900	1000	1100	1200	1300	1400	1500	1600	1700	1800	1900	2000	2100	2200	2300	2400
												Bending Design Value, F_b, psi											
2 × 6	12.0	5-6	6-4	7-1	7-9	8-5	9-0	9-6	10-0	10-6	11-0	11-5	11-11	12-4	12-8	13-1	13-6	13-10	14-2				
	16.0	4-9	5-6	6-2	6-9	7-3	7-9	8-3	8-8	9-1	9-6	9-11	10-3	10-8	11-0	11-4	11-8	12-0	12-4	12-7	12-11		
	19.2	4-4	5-0	5-7	6-2	6-8	7-1	7-6	7-11	8-4	8-8	9-1	9-5	9-9	10-0	10-4	10-8	10-11	11-3	11-6	11-9	12-0	12-4
	24.0	3-11	4-6	5-0	5-6	5-11	6-4	6-9	7-1	7-5	7-9	8-1	8-5	8-8	9-0	9-3	9-6	9-9	10-0	10-3	10-6	10-9	11-0
2 × 8	12.0	7-3	8-4	9-4	10-3	11-1	11-10	12-7	13-3	13-11	14-6	15-1	15-8	16-3	16-9	17-3	17-9	18-3	18-9				
	16.0	6-3	7-3	8-1	8-11	9-7	10-3	10-10	11-6	12-0	12-7	13-1	13-7	14-0	14-6	14-11	15-5	15-10	16-3	16-7	17-0		
	19.2	5-9	6-7	7-5	8-1	8-9	9-4	9-11	10-6	11-0	11-6	11-11	12-5	12-10	13-3	13-8	14-0	14-5	14-10	15-2	15-6	15-10	16-3
	24.0	5-2	5-11	6-7	7-3	7-10	8-4	8-11	9-4	9-10	10-3	10-8	11-1	11-6	11-10	12-2	12-7	12-11	13-3	13-7	13-11	14-2	14-6
2 × 10	12.0	9-3	10-8	11-11	13-1	14-2	15-1	16-0	16-11	17-9	18-6	19-3	20-0	20-8	21-4	22-0	22-8	23-3	23-11				
	16.0	8-0	9-3	10-4	11-4	12-3	13-1	13-10	14-8	15-4	16-0	16-8	17-4	17-11	18-6	19-1	19-7	20-2	20-8	21-2	21-8		
	19.2	7-4	8-5	9-5	10-4	11-2	11-11	12-8	13-4	14-0	14-8	15-3	15-10	16-4	16-11	17-5	17-11	18-5	18-11	19-4	19-10	20-3	20-8
	24.0	6-6	7-7	8-5	9-3	10-0	10-8	11-4	11-11	12-6	13-1	13-7	14-2	14-8	15-1	15-7	16-0	16-6	16-11	17-4	17-9	18-1	18-6
2 × 12	12.0	11-3	13-0	14-6	15-11	17-2	18-4	19-6	20-6	21-7	22-6	23-5	24-4	25-2	26-0								
	16.0	9-9	11-3	12-7	13-9	14-11	15-11	16-10	17-9	18-8	19-6	20-3	21-1	21-9	22-6	23-2	23-10	24-6	25-2	25-9			
	19.2	8-11	10-3	11-6	12-7	13-7	14-6	15-5	16-3	17-0	17-9	18-6	19-3	19-11	20-6	21-2	21-9	22-5	23-0	23-6	24-1	24-8	25-2
	24.0	7-11	9-2	10-3	11-3	12-2	13-0	13-9	14-6	15-3	15-11	16-7	17-2	17-9	18-4	18-11	19-6	20-0	20-6	21-1	21-7	22-0	22-6
E	12.0	0.14	0.22	0.31	0.41	0.51	0.63	0.75	0.88	1.01	1.15	1.30	1.45	1.61	1.77	1.94	2.12	2.30	2.48				
E	16.0	0.12	0.19	0.27	0.35	0.44	0.54	0.65	0.76	0.88	1.00	1.12	1.26	1.39	1.54	1.68	1.83	1.99	2.15	2.31	2.48		
E	19.2	0.11	0.18	0.24	0.32	0.41	0.50	0.59	0.69	0.80	0.91	1.03	1.15	1.27	1.40	1.54	1.67	1.81	1.96	2.11	2.26	2.42	2.58
E	24.0	0.10	0.16	0.22	0.29	0.36	0.44	0.53	0.62	0.71	0.81	0.92	1.03	1.14	1.25	1.37	1.50	1.62	1.75	1.89	2.02	2.16	2.30

Courtesy American Forest and Paper Association.

[a] The required elasticity, E, in 1,000,000 pounds per square inch (psi) is shown at the bottom of the table, is limited to 2.6 million psi and less, and is applicable to all lumber sizes shown. Spans are shown in feet-inches and are limited to 26 ft. and less. Check sources of supply for availability of lumber in lengths greater than 20 ft.

[b] Design Criteria: Strength—live load of 40 pounds per square foot (psf) plus dead load of 10 psf determines the required bending design value. Deflection—For 40 psf live load. Limited to span in inches divided by 240.

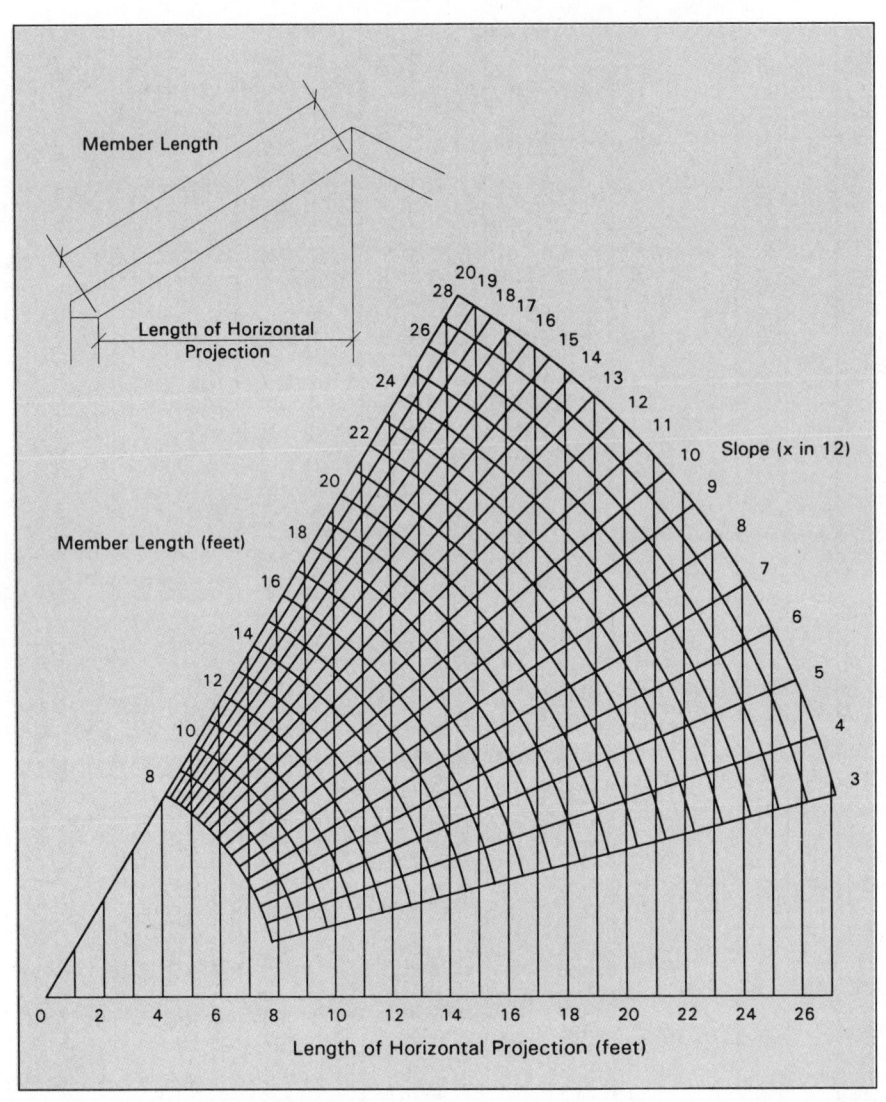

Member Length

Length of Horizontal
Projection

Member Length (feet)

Slope (x in 12)

Length of Horizontal Projection (feet)

FIGURE A-20 *Conversion diagram for rafters.* *(Courtesy American Forest and Paper Association.)*

FIGURE A-21
Required *f* and *E* Values for Floor and Roof Beams

Span of Beam, ft.	Nominal Size of Beam	Minimum *f* and *E* (psi) for Beams Spaced:						Span of Beam, ft.	Nominal Size of Beam	Minimum *f* and *E* (psi) for Beams Spaced:					
		6 ft.-0 in.		7 ft.-0 in.		8 ft.-0 in.				6 ft.-0 in.		7 ft.-0 in.		8 ft.-0 in.	
		f	*E*	*f*	*E*	*f*	*E*			*f*	*E*	*f*	*E*	*f*	*E*
10	2-3×6	1070	780000	1250	910000	1430	1040000	**11**	2-3×6	1295	1037000	1510	1210000	1730	1382000
	1-3×8	1235	680000	1440	794000	1645	906000		1-3×8	1490	905000	1740	1056000	1990	1206000
	2-2×8	1030	570000	1200	665000	1370	760000		2-2×8	1245	754000	1450	880000	1660	1005000
	1-4×8	880	485000	1030	566000	1175	646000		1-4×8	1065	647000	1245	755000	1420	862000
	3-2×8	685	380000	800	443000	915	506000		3-2×8	830	503000	970	587000	1105	670000
	2-3×8	615	340000	720	397000	820	453000		2-3×8	745	453000	870	529000	995	604000
	2-2×10	630	273000	735	219000	840	364000		2-2×10	765	363000	890	424000	1020	484000

Span of Beam, ft.	Nominal Size of Beam	6 ft.-0 in.		7 ft.-0 in.		8 ft.-0 in.	
		f	E	f	E	f	E
12	2-3×6	1545	1346000	1800	1571000	2060	1794000
	1-3×8	1775	1175000	2070	1371000	2370	1566000
	2-2×8	1480	980000	1725	1144000	1970	1306000
	1-4×8	1270	840000	1480	980000	1690	1120000
	3-2×8	985	653000	1150	762000	1315	870000
	2-3×8	890	588000	1035	686000	1185	784000
	1-6×8	755	483000	880	564000	1005	644000
	2-2×10	910	472000	1060	551000	1210	629000
	1-3×10	1090	566000	1275	660000	1455	754000
13	2-3×6	1815	1711000	2110	1997000	2415	2281000
	1-3×8	2085	1494000	2430	1743000	2780	1991000
	2-2×8	1740	1245000	2025	1453000	2315	1660000
	1-4×8	1490	1067000	1735	1245000	1985	1422000
	3-2×8	1160	830000	1350	969000	1545	1106000
	2-3×8	1045	747000	1215	872000	1390	996000
	1-6×8	885	614000	1040	716000	1185	818000
	2-2×10	1070	600000	1245	700000	1420	800000
	1-3×10	1280	719000	1495	839000	1710	958000
14	2-2×8	2015	1555000	2350	1815000	2685	2073000
	3-2×8	1340	1037000	1570	1210000	1790	1382000
	2-3×8	1210	933000	1410	1089000	1610	1244000
	1-6×8	1025	766000	1200	894000	1370	1021000
	1-3×10	1485	899000	1730	1049000	1980	1198000
	2-2×10	1235	749000	1445	874000	1650	998000
	1-4×10	1060	642000	1240	749000	1415	856000
	3-2×10	825	499000	965	582000	1100	665000
	2-3×10	740	449000	865	524000	990	598000
15	3-2×8	1540	1275000	1800	1488000	2055	1699000
	2-3×8	1390	1148000	1620	1340000	1850	1530000
	1-6×8	1180	943000	1375	1100000	1570	1257000
	1-3×10	1705	1105000	1990	1289000	2270	1473000
	2-2×10	1420	921000	1660	1075000	1895	1228000
	1-4×10	1220	789000	1420	921000	1625	1052000
	3-2×10	950	614000	1105	717000	1265	818000
	2-3×10	850	553000	995	645000	1135	737000
	1-6×10	735	464000	855	541000	980	618000
	4-2×10	710	461000	830	538000	945	614000
	2-2×10	960	512000	1120	597000	1280	682000
16	3-2×8	1755	1548000	2045	1806000	2340	2063000
	2-3×8	1580	1393000	1840	1626000	2105	1857000
	2-2×10	1615	1118000	1890	1305000	2155	1490000
	1-4×10	1385	958000	1615	1118000	1845	1277000
	3-2×10	1075	745000	1260	869000	1435	993000
	2-3×10	970	671000	1130	783000	1290	894000
	1-6×10	835	563000	975	657000	1130	750000
	4-2×10	810	559000	945	652000	1080	745000
	1-8×10	615	413000	715	482000	815	550000
	1-3×12	1310	746000	1530	871000	1750	994000
	2-2×12	1090	621000	1275	725000	1455	828000
17	2-2×10	1825	1341000	2125	1565000	2430	1781000
	1-4×10	1565	1149000	1825	1341000	2085	1532000

Span of Beam, ft.	Nominal Size of Beam	6 ft.-0 in.		7 ft.-0 in.		8 ft.-0 in.	
		f	E	f	E	f	E
17 Cont.	3-2×20	1215	894000	1420	1043000	1620	1192000
	2-3×10	1095	804000	1280	938000	1460	1072000
	1-6×10	945	675000	1100	788000	1260	900000
	4-2×10	910	670000	1065	782000	1215	894000
	1-8×10	690	495000	805	578000	910	660000
	1-3×12	1480	894000	1725	1043000	1975	1192000
	2-2×12	1235	745000	1440	869000	1645	993000
	1-4×12	1060	639000	1230	746000	1410	852000
	3-2×12	820	497000	960	580000	1095	663000
18	2-2×10	2045	1592000	2385	1858000	2725	2123000
	1-4×10	1755	1364000	2045	1592000	2340	1819000
	3-2×10	1365	1061000	1590	1238000	1815	1415000
	2-3×10	1270	955000	1480	1114000	1695	1273000
	1-6×10	1060	801000	1235	935000	1415	1068000
	4-2×10	1020	796000	1195	929000	1365	1062000
	1-8×10	780	588000	910	686000	1040	784000
	1-3×12	1660	1062000	1935	1239000	2210	1416000
	2-2×12	1380	885000	1615	1033000	1845	1180000
	1-4×12	1185	758000	1385	885000	1580	1011000
	3-2×12	920	590000	1075	688000	1230	786000
	2-3×12	830	531000	970	620000	1105	708000
19	3-2×10	1520	1248000	1775	1456000	2025	1664000
	2-3×10	1365	1123000	1595	1310000	1825	1497000
	1-6×10	1170	943000	1365	1100000	1560	1257000
	4-2×10	1140	936000	1330	1092000	1520	1248000
	2-4×10	975	802000	1140	936000	1300	1070000
	1-8×10	860	691000	1005	806000	1145	921000
	1-3×12	1850	1249000	2155	1457000	2465	1665000
	2-2×12	1540	1041000	1800	1215000	2055	1388000
	1-4×12	1320	892000	1540	1041000	1760	1190000
	3-2×12	1025	694000	1200	810000	1370	926000
	2-3×12	925	624000	1080	728000	1230	832000
	1-6×12	805	531000	940	620000	1070	708000
20	3-2×10	1685	1456000	1965	1699000	2245	1942000
	2-3×10	1515	1310000	1770	1529000	2020	1747000
	1-6×10	1300	1099000	1515	1282000	1735	1465000
	4-2×10	1260	1092000	1475	1274000	1685	1456000
	2-4×10	1080	936000	1265	1092000	1445	1248000
	1-8×10	960	806000	1120	941000	1280	1075000
	2-2×12	1705	1214000	1990	1417000	2275	1619000
	1-4×12	1465	1040000	1710	1214000	1950	1387000
	3-2×12	1140	809000	1330	944000	1520	1079000
	2-3×12	1025	728000	1195	850000	1365	971000
	1-6×12	970	620000	1130	723000	1295	826000
	2-4×12	730	520000	855	607000	975	694000
21	3-2×10	1855	1685000	2165	1966000	2475	2247000
	2-3×10	1670	1516000	1950	1827000	2225	2088000
	1-6×10	1430	1273000	1670	1485000	1905	1697000
	4-2×10	1390	1264000	1625	1475000	1855	1686000
	2-4×10	1195	1083000	1390	1264000	1590	1444000
	1-8×10	1050	933000	1225	1089000	1400	1244000
	2-2×12	1880	1405000	2195	1640000	2510	1874000

Required f and E Values for Floor and Roof Beams (Continued)

Span of Beam, ft.	Nominal Size of Beam	6 ft.-0 in.		7 ft.-0 in.		8 ft.-0 in.	
		f	E	f	E	f	E
21 Cont.	1-4×12	1615	1204000	1880	1405000	2150	1606000
	3-2×12	1255	937000	1465	1093000	1670	1249000
	2-3×12	1130	843000	1320	984000	1505	1124000
	1-6×12	970	717000	1130	837000	1295	956000
	2-4×12	805	602000	940	702000	1075	802000
22	1-6×10	1580	1463000	1845	1707000	2105	1951000
	4-2×10	1525	1453000	1780	1696000	2035	1938000
	2-4×10	1310	1245000	1530	1455000	1745	1660000

Span of Beam, ft.	Nominal Size of Beam	6 ft.-0 in.		7 ft.-0 in.		8 ft.-0 in.	
		f	E	f	E	f	E
22 Cont.	1-8×10	1160	1073000	1355	1252000	1545	1431000
	1-4×12	1770	1384000	2065	1615000	2360	1846000
	3-2×12	1375	1077000	1605	1257000	1835	1436000
	2-3×12	1240	969000	1445	1130000	1655	1291000
	1-6×12	1080	825000	1260	963000	1440	1100000
	2-4×12	885	692000	1035	807000	1180	922000
	4-2×12	1035	808000	1205	943000	1375	1078000
	5-2×12	825	646000	965	754000	1105	862000
	3-3×12	825	639000	965	746000	1105	852000

Courtesy American Forest and Paper Association.

Required f and E Values for Two-Inch-Thick Plank

Plank Span, ft.	Live Load, psf	Deflection Limitation	Type A f psi	Type A E psi	Type B f psi	Type B E psi	Type C f psi	Type C E psi	Type D f psi	Type D E psi
6	20	ℓ/240	360	576000	360	239000	288	305000	360	408000
		ℓ/300	360	720000	360	299000	288	318000	360	509000
		ℓ/360	360	864000	360	359000	288	457000	360	611000
	30	ℓ/240	480	864000	480	359000	384	457000	480	611000
		ℓ/300	480	1080000	480	448000	384	571000	480	764000
		ℓ/360	480	1296000	480	538000	384	685000	480	917000
	40	ℓ/240	600	1152000	600	478000	480	609000	600	815000
		ℓ/300	600	1440000	600	598000	480	762000	600	1019000
		ℓ/360	600	1728000	600	717000	480	914000	600	1223000
7	20	ℓ/240	490	915000	490	380000	392	484000	490	647000
		ℓ/300	490	1143000	490	475000	392	605000	490	809000
		ℓ/360	490	1372000	490	570000	392	726000	490	971000
	30	ℓ/240	653	1372000	653	570000	522	726000	653	971000
		ℓ/300	653	1715000	653	712000	522	907000	653	1213000
		ℓ/360	653	2058000	653	854000	522	1088000	653	1456000
	40	ℓ/240	817	1829000	817	759000	653	968000	817	1294000
		ℓ/300	817	2287000	817	949000	653	1209000	817	1618000
		ℓ/360	817	2744000	817	1139000	653	1451000	817	1941000
8	20	ℓ/240	640	1365000	640	567000	512	722000	640	966000
		ℓ/300	640	1707000	640	708000	512	903000	640	1208000
		ℓ/360	640	2048000	640	850000	512	1083000	640	1449000
	30	ℓ/240	853	2048000	853	850000	682	1083000	853	1449000
		ℓ/300	853	2560000	853	1063000	682	1354000	853	1811000
		ℓ/360	853	3072000	853	1275000	682	1625000	853	2174000
	40	ℓ/240	1067	2731000	1067	1134000	853	1444000	1067	1932000
		ℓ/300	1067	3410000	1067	1417000	853	1805000	1067	2415000
		ℓ/360	1067	4096000	1067	1700000	853	2166000	1067	2898000

Courtesy American Forest and Paper Association.

SOUND-PERFORMANCE RATINGS FOR SELECTED ASSEMBLIES

The following data and the accompanying Figures A-23 and A-24 include sound-performance data on walls and floor and ceiling assemblies. Data on systems that provide poor sound isolation are included to indicate the degree of improvement required to render such systems sufficiently resistant to sound transmission.

Unless otherwise noted, the data presented in the figures are based on tests performed by nationally recognized acoustical laboratories in accordance with the test procedures in ASTM E-90, "Method for Laboratory Measurement of Airborne-Sound Transmission Loss of Building Partitions" and International Organization for Standarization (ISO) Recommendation R-140, "Field and Laboratory Measurements of Airborne and Impact Sound Transmission." The data include tests performed by Cedar Knolls Acoustical Laboratories; Geiger and Hamme; Kodaras Acoustical Laboratories; National Bureau of Standards (now the National Institute of Standards and Technology); National Gypsum Company; Ohio Research Corporation; Owens-Corning Fiberglass Sound Laboratory; and Riverbank Acoustical Laboratories, IIT Research Institute.

The STC, Impact Noise Reduction (INR), and IIC ratings shown have been drawn from a variety of manufacturers' literature and testing laboratory reports. In some instances, only very old test results were available. In others, few correlating test results could be found. Therefore, some ratings have been estimated. Before relying on the data shown for the final design of an actual project, it would be wise to ask the manufacturers of the materials or systems proposed to furnish the results of their latest sound performance tests.

When possible, a range of values is given to account for (1) slight variations in construction materials, (2) slight variation in application procedures, and (3) variation in testing facilities.

The reproducibility of laboratory tests on different samples of standard construction systems can range up to 5 dB at any one frequency. However, the overall STC rating should be within 2 dB. The subjective response to decibel difference as judged by the ear can be summarized as follows: a difference of 1 to 3 dB is insignificant, a difference of 5 dB is clearly noticeable, and a difference of 10 dB is highly significant.

Strict adherence to good practices of design and construction, excellent workmanship, and installation in accordance with the manufacturers' specific instructions are essential to the maximum sound isolating performance of any construction system.

WALL ASSEMBLIES

The amount of improvement in acoustical performance caused by adding insulation blankets in the cavity of wood and metal stud walls varies depending on wall construction and thickness. Adding insulation to single wood stud walls improves performance only slightly and is not considered economically justified. Adding 2 to 3 in. of insulation improves the performance of wood staggered-stud walls and single metal-channel stud walls, but more thickness improves performance only slightly and is not considered economically justified. Acoustical performance increases proportionately with increased thickness of insulation in double wood stud walls, double-channel metal stud walls, and wood stud walls having resiliently mounted surface finishes.

The STC ratings given for clay masonry walls have been interpolated from Figure 10.1-68.

FLOOR AND CEILING ASSEMBLIES

The use of plywood, fiberboard, hardboard, or other types of underlayment is generally determined by the type of finish flooring or other practical construction techniques. With the exception of resilient underlayment materials in floated floor systems, these underlayment materials generally contribute insignificant impact sound-isolating properties. Unless otherwise noted, the ranges given for INR and IIC include assemblies both with and without underlayment.

Carpet and pad combinations are effective in improving the impact sound-isolation performance of flooring systems. The amount of improvement depends on carpet thickness, degree of pile matting, and weight of carpet and pad in ounces per square yard. However, no appreciable gain in airborne sound isolation results from using carpet and pad combinations.

Adding insulation in the cavity between the subfloor and the finished ceiling in joist construction improves performance only slightly but is considered economically justified when the ceiling is resiliently mounted.

Impact sound isolation performance is not improved significantly by increasing the weight or thickness of the slab in a poured-in-place reinforced concrete slab floor and ceiling assembly. There, the most economical control measures are cushioning the impact with a carpet and pad and floating the finished floor.

Because floor and ceiling assemblies that control impact sound isolation will also provide a good degree of resistance to airborne sound transmission, much of the test data for floor and ceiling assemblies are given in terms of INR and IIC only. However, the STC values for a particular assembly can be estimated, providing essential precautions of construction practices are followed. Where test data for STC ratings were not available, the STC values have been estimated.

WOOD FRAME LOAD-BEARING WALLS—GYPSUM BOARD FINISH[1]

Single Stud Walls

Single Layer Finish[2]

2 x 4 STUD FRAMING

GYPSUM BOARD

STC 32 to 36

Multilayer Finish[2]

2 x 4 STUD FRAMING

GYPSUM BOARD

STC 38 to 41

Single Stud Walls

Single Layer Finish

2x4 STUD FRAMING

RESILIENT CHANNEL

INSULATION

GYPSUM BOARD

FILLER STRIP
BLOCKING FOR FIRE STOP

STC 40 to 42

With insulation[7]

STC 47 to 51[4]

Sound Deadening Board[5]

2 x 4 STUD FRAMING

GYPSUM BOARD

SOUND DEADENING BOARD

STC 47 to 51[4]

Staggered Stud Walls[6]

Resilient Finish[3]

2x4 STAGGERED STUD FRAMING

INSULATION

GYPSUM BOARD

2x6 PLATE

STC 38 to 42[4]

With insulation[8]

STC 46 to 48[4]

Sound Deadening Board[5]

2 x 4 STAGGERED STUD FRAMING

GYPSUM BOARD

SOUND DEADENING BOARD
2 x 6 PLATE

STC 46 to 50[4]

Double Stud Walls[6]

Single Layer Finish

2x4 STUD FRAMING

INSULATION

GYPSUM BOARD

DOUBLE WALLS 2 x 4 PLATES

STC 42 to 44

With insulation[8]

STC 50 to 53[4]

Sound Deadening Board[5]

2 x 4 STUD FRAMING

GYPSUM BOARD

SOUND DEADENING BOARD
DOUBLE WALLS 2 x 4 PLATES

STC 50 to 53[9]

With insulation[8]

STC 54 to 58

FIGURE A-23 *Wall systems.*

1. Ranges include values for 3/8 in., 1/2 in., and 5/8 in. thicknesses of gypsum board finish.
2. Addition of insulation does not appreciably improve this construction; an increase of approximately 2 to 4 dB can be expected regardless of thickness or density of insulation.
3. Resilient channels or clips may be used one side only to totally float the finish; use of blocking for fire stopping and use of backer strips to attach wall base reduces performance approximately 4 to 5 dB.
4. Range includes values derived from ASTM E90-61T, ASTM E90-66T, and field test procedures.
5. Gypsum board finish laminated to sound deadening board; if gypsum board is nailed or screwed to meet fire ratings, performance decreases approximately 10 dB for fasteners spaced 8 in. o.c., approximately 8 dB for fastener spacing 16 in. o.c., and approximately 4 dB with fastener spacing 24 in. o.c.
6. Ranges include values for 2 × 3 studs staggered on single 2 × 4 plate and 2 × 4 double stud walls on separate 2 × 3 plates.
7. A minimum thickness of 2 in. to 3 in. of insulation can be used effectively as performance does not increase proportionately with increased thickness of insulation.
8. Ranges include 1-1/2 in. to 3 in. thickness of insulation; performance increases with increased thickness. On double stud walls, an STC of 58 to 60 can result with up to 9 in. of insulation.

WOOD FRAME LOAD-BEARING WALLS—PLASTER FINISH[11]

Single Stud Walls

Metal Lath & Plaster[2]

2 x 4 STUD FRAMING
GYPSUM PLASTER
METAL LATH

STC 41[10]

Resilient Gypsum Lath & Plaster[3]

2 x 4 STUD FRAMING
PENCIL ROD
RESILIENT CLIP
METAL LATH
GYPSUM PLASTER

STC 43[9]

Single Stud Walls

Plaster Base, Veneer Plaster[2]

2 X 4 STUD FRAMING
JOINT REINFORCEMENT TAPE
VENEER PLASTER OVER GYPSUM BASE

STC (est) 36
2 layer base 2 sides
STC (est) 42

Resilient Plaster Base, Veneer Plaster[3]

2 X 4 STUD FRAMING
GYPSUM BOARD
RESILIENT CHANNEL
INSULATION
VENEER PLASTER OVER GYPSUM BASE

With insulation[8]
STC 49 to 50[4]

Single Stud Walls

Gypsum Lath & Plaster[2]

2 x 4 STUD FRAMING
GYPSUM LATH
GYPSUM PLASTER

STC 41[10]

Resilient Metal Lath & Plaster[3]

2 x 4 STUD FRAMING
RESILIENT CLIP
GYPSUM PLASTER
GYPSUM LATH

STC 45 to 47[10]
With insulation[8]
STC 50 to 54[4]

Single Stud Walls   Staggered Stud Walls[6]

Resilient Plaster Base & Multilayer Base, Veneer Plaster[3]

2 X 4 STUD FRAMING
GYPSUM BOARD
RESILIENT CHANNEL
INSULATION
VENEER PLASTER OVER GYPSUM BASE

With insulation[8]
STC 53

Gypsum Lath & Plaster

2 x 4 STAGGERED STUD
GYPSUM LATH AND PLASTER
GYPSUM LATH AND PLASTER
INSULATION

STC 42 to 43[9]
With insulation[7]
STC 50[10]

FIGURE A-23 *Wall systems. (Continued)*

9. Based on ASTM E90-61T test procedures.
10. Based on tests performed prior to ASTM E90-61T.
11. Test data can vary widely depending on weight and thickness of plaster, and water content (due to length of curing) of plaster.
12. With exception of double stud walls, ranges include values for 1/2 in. or 5/8 in. gypsum board thickness of 2-1/2 in. or 3-5/8 in. wide studs.
13. Ranges include values for gypsum board screwed, or laminated and screwed, to sound deadening board or base layer.
14. Unbalanced Construction: single layer gypsum board one side, double layer gypsum board one side.
15. Ranges include values for 1/2 in. and 5/8 in. board thickness on 1-5/8 in. wide studs.
16. Up to 60 STC with 9 in. of insulation; lower values result with lesser thicknesses of insulation.
17. Performance of concrete and unit masonry systems is highly dependent on minimizing the number of openings through the construction (to accommodate piping, wiring, and other equipment) and on all openings being sealed airtight.
18. Sealed both sides unless otherwise indicated. Sealing with plaster is most effective; when surfaces are to be painted, use of block sealer and/or 2 coats of cement-based paint is more effective than 2 coats of resin-emulsion paint to seal against airborne sound transmission.

METAL FRAME NONLOAD-BEARING WALLS—GYPSUM BOARD FINISH[12]

Single Stud Walls

Single Stud Walls

Single Stud Walls

Double Stud (Chase) Walls[15]

Single Layer Finish

METAL CHANNEL STUD
INSULATION
GYPSUM BOARD

STC 37 to 41[4]

With insulation[8]

STC 44 to 48

**Single Layer Finish
Sound Deadening Board**[13]

METAL CHANNEL STUDS
GYPSUM BOARD
SOUND DEADENING BOARD

SDB 1 side	**STC**	**43**
SDB 2 sides	**STC 46 to 50**[4]	
With insulation[8]		
SDB 1 side	**STC 47 to 50**[4]	
SDB 2 sides	**STC 50 to 53**[4]	

Unbalanced Finish[14]

METAL CHANNEL STUDS
GYPSUM BOARD
MULTI – LAYER GYPSUM BOARD

STC 44 to 47

With insulation[8]

STC 49 to 52

Unbalanced Finish[14]
Sound Deadening Board[13]

METAL CHANNEL STUDS
GYPSUM BOARD
MULTI – LAYER GYPSUM BOARD
SOUND DEADENING BOARD

SDB 1 side	**STC**	**48**[9]
SDB 2 sides	**STC**	**52**[9]
With insulation[8]		
SDB 1 side	**STC**	**49**[9]
SDB 2 sides	**STC**	**54**[9]

Multilayer Finish

METAL CHANNEL STUD
GYPSUM BOARD

STC 48 to 50[4]

With insulation[8]

STC 53 to 58[4]

**Multilayer Finish
Sound Deadening Board**[13]

METAL CHANNEL STUD
GYPSUM BOARD
SOUND DEADENING BOARD

SDB 1 side	**STC**	**52**[9]
SDB 2 sides	**STC**	**54**[9]
With insulation[8]		
SDB 2 sides	**STC**	**57**[9]

Single Layer Finish

METAL CHANNEL STUDS
TIES @ 1/4 POINTS
GYPSUM BOARD
12" CAVITY

STC 42

With insulation[8]

STC 52 to 55[16]

**Single Layer Finish
Sound Deadening Board**[13]

METAL CHANNEL STUD
TIES @ 1/4 POINTS
GYPSUM BOARD
SOUND DEADENING BOARD
6" CAVITY

STC 50[9]

With insulation[8]

STC 53

FIGURE A-23 *Wall systems. (Continued)*

1. Ranges include values for 3/8 in., 1/2 in., and 5/8 in. thicknesses of gypsum board finish.
2. Addition of insulation does not appreciably improve this construction; an increase of approximately 2 to 4 dB can be expected regardless of thickness or density of insulation.
3. Resilient channels or clips may be used one side only to totally float the finish; use of blocking for fire stopping and use of backer strips to attach wall base reduces performance approximately 4 to 5 dB.
4. Range includes values derived from ASTM E90-61T, ASTM E90-66T, and field test procedures.
5. Gypsum board finish laminated to sound deadening board; if gypsum board is nailed or screwed to meet fire ratings, performance decreases approximately 10 dB for fasteners spaced 8 in. o.c., approximately 8 dB for fastener spacing 16 in. o.c., and approximately 4 dB with fastener spacing 24 in. o.c.
6. Ranges include values for 2 × 3 studs staggered on single 2 × 4 plate and 2 × 4 double stud walls on separate 2 × 3 plates.
7. A minimum thickness of 2 in. to 3 in. of insulation can be used effectively as performance does not increase proportionately with increased thickness of insulation.
8. Ranges include 1-1/2 in. to 3 in. thickness of insulation; performance increases with increased thickness. On double stud walls, an STC of 58 to 60 can result with up to 9 in. of insulation.

METAL FRAME NONLOAD-BEARING WALLS—GYPSUM BOARD FINISH[12] (CONT.)

Double Stud (Chase) Walls[15]

Multilayer Finish

METAL CHANNEL STUDS

TIES @ 1/4 POINTS

INSULATION   GYPSUM BOARD

12" CAVITY

With insulation[8]
Unbalanced Finish[14]
STC   59[16]
Multilayer finish[13]
STC 55 to 60[4,16]

Multilayer Finish Sound Deadening Board[13]

METAL CHANNEL STUD

TIE @ 1/4 POINTS

GYPSUM BOARD

SOUND DEADENING BOARD

6" CAVITY

Unbalanced finish[14]
STC   51[9]
Multilayer finish[13]
STC   53[9]

METAL FRAME NONLOAD-BEARING WALLS—PLASTER FINISH[11]

Single Channel Stud Walls

Single Channel Stud Walls

Plaster Base & Veneer Plaster

METAL CHANNEL STUD

VENEER PLASTER

GYPSUM BOARD

STC 40 to 45[4]

Multilayer Base[13] & Veneer Plaster

METAL CHANNEL STUD

INSULATION

VENEER PLASTER

MULTI-LAYER GYPSUM BOARD

STC 49 to 50[4]
With insulation[8]
STC 52 to 53[4]

Gypsum Lath & Plaster

METAL CHANNEL STUD

RESILIENT CLIP

INSULATION

GYPSUM PLASTER

GYPSUM LATH

STC 42 to 43[4]
With insulation
STC   46

Resilient Gypsum Lath[3] & Plaster

METAL CHANNEL STUD

RESILIENT CLIP

INSULATION

GYPSUM PLASTER

GYPSUM BOARD

STC   45
With insulation[8]
STC   49

FIGURE A-23 *Wall systems. (Continued)*

9. Based on ASTM E90-61T test procedures.
10. Based on tests performed prior to ASTM E90-61T.
11. Test data can vary widely depending on weight and thickness of plaster, and water content (due to length of curing) of plaster.
12. With exception of double stud walls, ranges include values for 1/2 in. or 5/8 in. gypsum board thickness of 2-1/2 in. or 3-5/8 in. wide studs.
13. Ranges include values for gypsum board screwed, or laminated and screwed, to sound deadening board or base layer.
14. Unblanced Construction: single layer gypsum board one side, double layer gypsum board one side.
15. Ranges include values for 1/2 in. and 5/8 in. board thickness on 1-5/8 in. wide studs.
16. Up to 60 STC with 9 in. of insulation; lower values result with lesser thicknesses of insulation.
17. Performance of concrete and unit masonry systems is highly dependent on minimizing the number of openings through the construction (to accommodate piping, wiring, and other equipment) and on all openings being sealed airtight.
18. Sealed both sides unless otherwise indicated. Sealing with plaster is most effective; when surfaces are to be painted, use of block sealer and/or 2 coats of cement-based paint is more effective than 2 coats of resin-emulsion paint to seal against airborne sound transmission.

Single Truss Stud Walls

Metal Lath & Plaster

TRUSS STEEL STUD

METAL LATH

GYPSUM PLASTER

STC 38 to 41[9]

Resilient Metal Lath[3] & Plaster

TRUSS STEEL STUD

METAL LATH

GYPSUM PLASTER

PENCIL ROD ON RESILIENT CLIP

STC 46 to 50[4]

Single Truss Stud Walls

Gypsum Lath & Plaster

TRUSS STEEL STUD

GYPSUM PLASTER

GYPSUM LATH

STC 41 to 45[4]

Resilient Gypsum Lath[3] & Plaster

RESILIENT CLIP

TRUSS STEEL STUD

GYPSUM LATH

GYPSUM PLASTER

2" INSULATION

STC 46 to 50

With insulation[8]

STC 46 to 52[4]

CONCRETE AND CONCRETE MASONRY WALLS[17]

Nonporous (Normal Weight) Hollow Concrete Block Walls

Unsealed

HOLLOW CONCRETE BLOCK

4" Block	**STC 42 to 45[4]**
6" Block	**STC 43 to 46[4]**
8" Block	**STC 46 to 49[9]**

Sealed[18]

HOLLOW CONCRETE BLOCK

GYPSUM PLASTER

4" Block	**STC 44 to 48[4]**
6" Block	**STC 47 to 51[4]**
8" Block	**STC 53[9]**

Porous (Lightweight) Hollow Concrete Block Walls

Unsealed

HOLLOW CONCRETE BLOCK

4" Block	**STC 26 to 34[10]**
6" Block	**STC 33 to 37[10]**
8" Block	**STC 34 to 38[10]**

Sealed[18]

HOLLOW CONCRETE BLOCK

GYPSUM PLASTER

4" Block	**STC 41 to 44[4]**
6" Block	**STC 44 to 49[4]**
8" Block	**STC 44 to 49[4]**

FIGURE A-23 *Wall systems. (Continued)*

1. Ranges include values for 3/8 in., 1/2 in., and 5/8 in. thicknesses of gypsum board finish.
2. Addition of insulation does not appreciably improve this construction; an increase of approximately 2 to 4 dB can be expected regardless of thickness or density of insulation.
3. Resilient channels or clips may be used one side only to totally float the finish; use of blocking for fire stopping and use of backer strips to attach wall base reduces performance approximately 4 to 5 dB.
4. Range includes values derived from ASTM E90-61T, ASTM E90-66T, and field test procedures.
5. Gypsum board finish laminated to sound deadening board; if gypsum board is nailed or screwed to meet fire ratings, performance decreases approximately 10 dB for fasteners spaced 8 in. o.c., approximately 8 dB for fastener spacing 16 in. o.c., and approximately 4 dB with fastener spacing 24 in. o.c.
6. Ranges include values for 2 × 3 studs staggered on single 2 × 4 plate and 2 × 4 double stud walls on separate 2 × 3 plates.
7. A minimum thickness of 2 in. to 3 in. of insulation can be used effectively as performance does not increase proportionately with increased thickness of insulation.
8. Ranges include 1-1/2 in. to 3 in. thickness of insulation; performance increases with increased thickness. On double stud walls, an STC of 58 to 60 can result with up to 9 in. of insulation.

CONCRETE AND CONCRETE MASONRY WALLS[17] (CONT.)

Porous (Lightweight) Hollow Concrete Block Walls

Gypsum Board Laminated Both Sides

6″ Block   **STC**   **49**

Resilient Gypsum Board 1 Side, Sealed[18] 1 Side

6″ Block   **STC**   **53**
8″ Block   **STC**   **48**[9]

Porous (Lightweight) Hollow Concrete Block Walls

Resilient Gypsum Board Both Sides

8″ Block   **STC**   **49**[9]

Cavity Wall Unsealed

STC   **38**[10]

Hollow Concrete Block Walls

Nonporous (Normal Weight) Solid Concrete Masonry Walls

Porous Cavity Wall Sealed[18]

STC   **49**[10]

Porous / Nonporous Cavity Wall Sealed[18]

STC 51 to 52[10]

Unsealed

4″ Block	**STC**	30
6″ Block	**STC**	38
8″ Block	**STC**	47

Sealed[18]

4″ Block	**STC**	44
6″ Block	**STC**	48
8″ Block	**STC**	53

FIGURE A-23 *Wall systems. (Continued)*

9. Based on ASTM E90-61T test procedures.
10. Based on tests performed prior to ASTM E90-61T.
11. Test data can vary widely depending on weight and thickness of plaster, and water content (due to length of curing) of plaster.
12. With exception of double stud walls, ranges include values for 1/2 in. or 5/8 in. gypsum board thickness of 2-1/2 in. or 3-5/8 in. wide studs.
13. Ranges include values for gypsum board screwed, or laminated and screwed, to sound deadening board or base layer.
14. Unblanced Construction: single layer gypsum board one side, double layer gypsum board one side.
15. Ranges include values for 1/2 in. and 5/8 in. board thickness on 1-5/8 in. wide studs.
16. Up to 60 STC with 9 in. of insulation; lower values result with lesser thicknesses of insulation.
17. Performance of concrete and unit masonry systems is highly dependent on minimizing the number of openings through the construction (to accommodate piping, wiring, and other equipment) and on all openings being sealed airtight.
18. Sealed both sides unless otherwise indicated. Sealing with plaster is most effective; when surfaces are to be painted, use of block sealer and/or 2 coats of cement-based paint is more effective than 2 coats of resin-emulsion paint to seal against airborne sound transmission.

CONCRETE AND CONCRETE MASONRY WALLS[17] (CONT.)

Porous (Lightweight) Solid Concrete Masonry Walls

Unsealed

SOLID CONCRETE BLOCK

4″ Block	**STC**	27
6″ Block	**STC**	36
8″ Block	**STC**	45

Sealed[18]

SOLID CONCRETE BLOCK

GYPSUM PLASTER

4″ Block	**STC**	42
6″ Block	**STC**	46
8″ Block	**STC**	50

Poured-in-place Solid Concrete Walls

CAST IN PLACE CONCRETE

4″ Thick	**STC**	46
6″ Thick	**STC**	50
8″ Thick	**STC**	55

CLAY MASONRY WALLS[17]

Solid Wall

Brick Wall Sealed[18]

4″ BRICK

GYPSUM PLASTER

6″ Brick	**STC 45 to 50**
8″ Brick	**STC 50 to 52**
12″ Brick	**STC 55**

Cavity Wall

Brick Cavity Wall Sealed[18]

BRICK

TIE BARS

GYPSUM PLASTER

2″ CAVITY

STC 45 to 50

Composite Wall

Composite Brick, bonded to Concrete Block

4″ HOLLOW CONCRETE BLOCK

GYPSUM PLASTER

4″ BRICK

STC 45 to 50

FIGURE A-23 *Wall systems. (Continued)*

WOOD JOIST FLOORS—GYPSUM BOARD CEILING FINISH[2]

Hard Surface Flooring[3]

STC 38 to 42 (No Ceiling STC 25)

HARDWOOD FLOORING
PLYWOOD SUBFLOOR
WOOD FLOOR JOIST
GYPSUM BOARD

INR −19 to −16   IIC 32 to 35
(No Ceiling)
INR   −28   IIC   23

Carpeting

STC 38 to 42

PLYWOOD SUBFLOOR
CARPET & PAD
WOOD FLOOR JOIST
GYPSUM BOARD

Carpet Only
INR −10 to −9   IIC 41 to 42
Lightweight Carpet & Pad[4]
INR   −5   IIC   46
Standard Carpet & Pad[5]
INR +8 to +10   IIC 59 to 61

Hard Surface Flooring[3] Resilient Ceiling

STC 46 to 51

HARDWOOD FLOORING
PLYWOOD SUBFLOOR
WOOD FLOOR JOIST
GYPSUM BOARD
RESILIENT CHANNEL

INR −13 to −9   IIC 38 to 42
With insulation[6]
 Vinyl Flooring
INR −5 to −3   IIC 46 to 48
 Wood Block Flooring
INR   −1   IIC   50

Carpeting Resilient Ceiling

STC 46 to 48

PLYWOOD SUBFLOOR
CARPET & PAD
WOOD FLOOR JOIST
GYPSUM BOARD
RESILIENT CHANNEL

INR +15 to +16   IIC 66 to 67
With insulation[6]
 Lightweight Carpet & Pad[4]
INR   +7   IIC   58
 Standard Carpet & Pad[5]
INR +18 to +20   IIC 69 to 71

Hard Surface Flooring[3] Separate Ceiling Joists

STC 50 to 53

HARDWOOD FLOORING
PLYWOOD SUBFLOOR
INSULATION
WOOD FLOOR JOIST
WOOD CEILING JOIST   GYPSUM BOARD

With insulation[6]
INR −6 to −2   IIC 45 to 49

Carpeting[5] Separate Ceiling Joists

STC   51

PLYWOOD SUBFLOOR
CARPET & PAD
INSULATION
WOOD FLOOR JOIST
WOOD CEILING JOIST   GYPSUM BOARD

With insulation[6]
INR   +29   IIC   80

Floated Hard Surface Flooring[3]

STC   46

HARDWOOD FLOORING
PLYWOOD UNDERLAYMENT
SOUND DEADENING BOARD
PLYWOOD SUBFLOOR
WOOD FLOOR JOIST
GYPSUM BOARD

INR   −13   IIC   38

Floated Hard Surface Flooring[3], Resilient Ceiling

STC 50 to 52
With insulation[6]
STC 51 to 53

HARDWOOD FLOORING
PLYWOOD UNDERLAYMENT
SOUND DEADENING BOARD
PLYWOOD SUBFLOOR
WOOD FLOOR JOIST
GYPSUM BOARD
RESILIENT CHANNEL

INR −3 to −2   IIC 48 to 49
With insulation[6]
INR +1 to +2   IIC 52 to 53

FIGURE A-24 *Floor and ceiling assemblies.*

1. INR values are based on tests performed in accordance with ISO Recommendation R-140, "Field and Laboratory Measurements of Airborne and Impact Sound Transmission," January, 1960, IIC values approximated by adding 51 to original INR tests values.
2. Ranges include value for 3/8 in., 1/2 in., and 5/8 in. thicknesses of gypsum board finish.
3. Hard surface finish flooring (i.e., stripwood, wood block, vinyl, and vinyl asbestos) generally perform similarly in isolating impact sound. Unless otherwise indicated, ranges given include values for construction with any typical hard surface finish flooring.
4. Lightweight carpet and pad having pile weights of 20 oz. per sq. yd. and 28 oz. per sq. yd. respectively.
5. Standard weight carpet and pad having pile weights of 44 oz. per sq. yd. and 40 oz. per sq. yd. respectively. Ratings shown are based on standard weight combinations unless otherwise noted.
6. A minimum thickness of 2 in. to 3 in. of insulation can be used effectively as performance does not increase with increased thickness in floor and ceiling assemblies.
7. Ratings vary with thickness of concrete slabs. The low end of the range indicates value for 6 in. thickness; high end, 8 in. thickness.

FIGURE A-24 *Floor and ceiling assemblies. (Continued)*

1. INR values are based on tests performed in accordance with ISO Recommendation R-140, "Field and Laboratory Measurements of Airborne and Impact Sound Transmission," January, 1960, IIC values approximated by adding 51 to original INR tests values.

2. Ranges include value for 3/8 in., 1/2 in., and 5/8 in. thicknesses of gypsum board finish.

3. Hard surface finish flooring (i.e., stripwood, wood block, vinyl, and vinyl asbestos) generally perform similarly in isolating impact sound. Unless otherwise indicated, ranges given include values for construction with any typical hard surface finish flooring.

WOOD/STEEL TRUSS FLOORS—GYPSUM BOARD CEILING FINISH[2]

Hard Surface Flooring[3]	**Carpeting**[5]	**Hard Surface Flooring**[3] **Resilient Ceiling**	**Carpeting**[5] **Resilient Ceiling**
STC 45	STC 45	STC 47	STC 47
INR −18 IIC 33	INR +5 IIC 56	INR −15 IIC 36	INR +15 IIC 66

With insulation[6]

INR +18 IIC 69

Hard Surface Flooring[3] **On Concrete Fill**	**Carpeting**[5] **On Concrete Fill**
STC 46	STC 46
INR −14 IIC 37	INR +11 IIC 62

FIGURE A-24 *Floor and ceiling assemblies. (Continued)*

4. Lightweight carpet and pad having pile weights of 20 oz. per sq. yd. and 28 oz. per sq. yd. respectively.

5. Standard weight carpet and pad having pile weights of 44 oz. per sq. yd. and 40 oz. per sq. yd. respectively. Ratings shown are based on standard weight combinations unless otherwise noted.

6. A minimum thickness of 2 in. to 3 in. of insulation can be used effectively as performance does not increase with increased thickness in floor and ceiling assemblies.

7. Ratings vary with thickness of concrete slabs. The low end of the range indicates value for 6 in. thickness; high end, 8 in. thickness.

POURED-IN-PLACE REINFORCED CONCRETE SLABS

Hard Surface Flooring[3] With or without Plaster Ceiling	Floated Hard Surface Flooring[3], With or Without Plaster Ceiling	Carpeting With or without Plaster Ceiling
STC 44 to 48	**STC 48 to 50**	**STC 44 to 48**

No Flooring
INR   **−17**   **IIC**   **34**

Vinyl Flooring
INR   **−15**   **IIC**   **36**

Wood Block Flooring
INR **−9 to** **−6**   **IIC 42 to 45**

INR **−7 to** **+4**   **IIC 44 to 55**

Lightweight Carpet Only
INR   **+2**   **IIC**   **53**

Standard Carpet Only
INR   **+6**   **IIC**   **57**

Standard Carpet & Pad[5]
INR +17 to +22   **IIC 68 to 73**

FIGURE A-24 *Floor and ceiling assemblies. (Continued)*

1. INR values are based on tests performed in accordance with ISO Recommendation R-140, "Field and Laboratory Measurements of Airborne and Impact Sound Transmission," January, 1960, IIC values approximated by adding 51 to original INR tests values.

2. Ranges include value for 3/8 in., 1/2 in., and 5/8 in. thicknesses of gypsum board finish.

3. Hard surface finish flooring (i.e., stripwood, wood block, vinyl, and vinyl asbestos) generally perform similarly in isolating impact sound. Unless otherwise indicated, ranges given include values for construction with any typical hard surface finish flooring.

PRECAST CONCRETE SLABS[7]

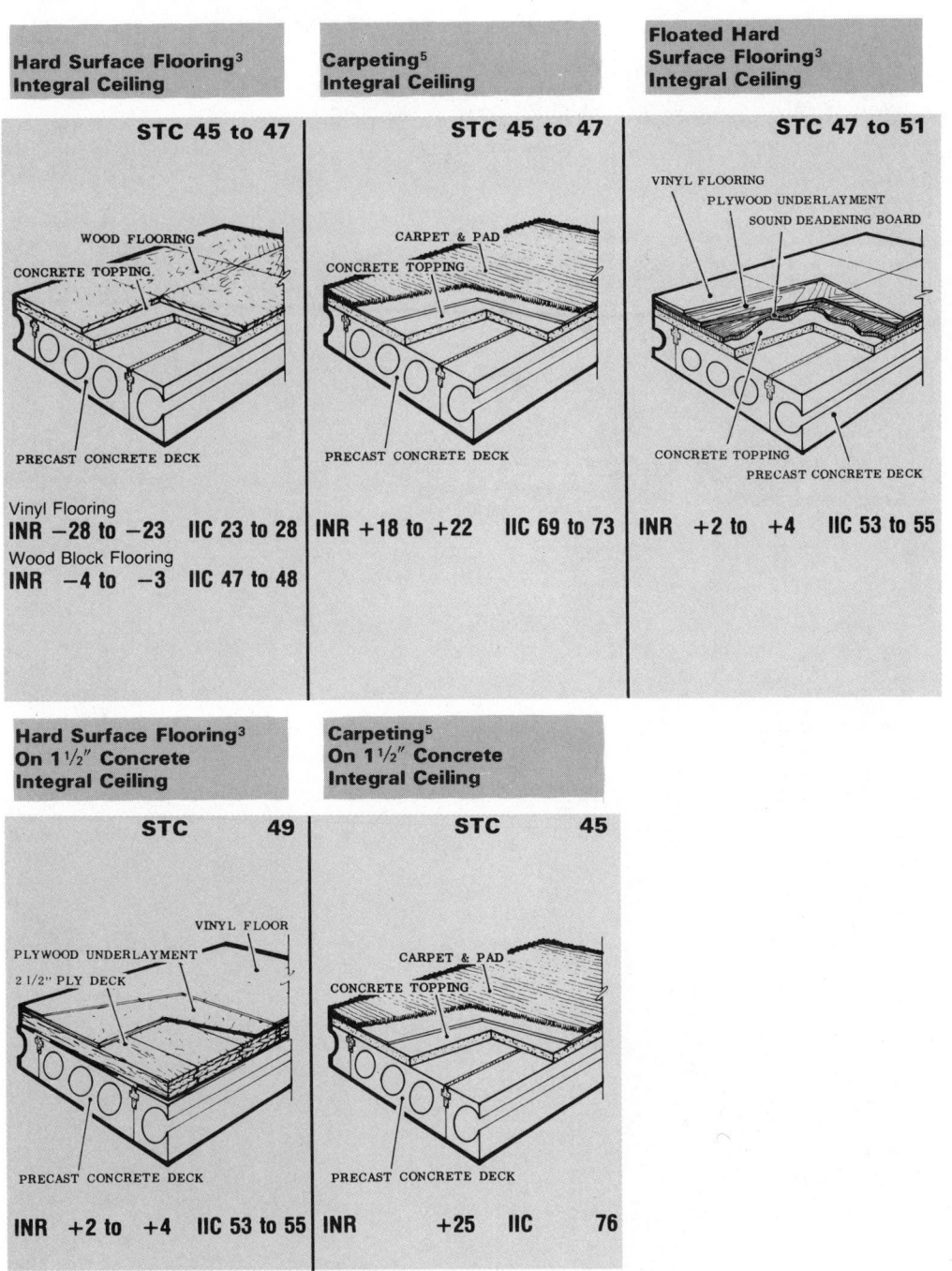

Hard Surface Flooring[3]
Integral Ceiling

STC 45 to 47

Vinyl Flooring
INR −28 to −23   IIC 23 to 28
Wood Block Flooring
INR   −4 to   −3   IIC 47 to 48

Carpeting[5]
Integral Ceiling

STC 45 to 47

INR +18 to +22   IIC 69 to 73

Floated Hard
Surface Flooring[3]
Integral Ceiling

STC 47 to 51

INR +2 to +4   IIC 53 to 55

Hard Surface Flooring[3]
On 1½″ Concrete
Integral Ceiling

STC   49

INR +2 to +4   IIC 53 to 55

Carpeting[5]
On 1½″ Concrete
Integral Ceiling

STC   45

INR   +25   IIC   76

FIGURE A-24   *Floor and ceiling assemblies. (Continued)*

4. Lightweight carpet and pad having pile weights of 20 oz. per sq. yd. and 28 oz. per sq. yd. respectively.

5. Standard weight carpet and pad having pile weights of 44 oz. per sq. yd. and 40 oz. per sq. yd. respectively. Ratings shown are based on standard weight combinations unless otherwise noted.

6. A minimum thickness of 2 in. to 3 in. of insulation can be used effectively as performance does not increase with increased thickness in floor and ceiling assemblies.

7. Ratings vary with thickness of concrete slabs. The low end of the range indicates value for 6 in. thickness; high end, 8 in. thickness.

STEEL BAR JOIST FLOORS

FIGURE A-24 *Floor and ceiling assemblies.* *(Continued)*

1. INR values are based on tests performed in accordance with ISO Recommendation R-140, "Field and Laboratory Measurements of Airborne and Impact Sound Transmission," January, 1960, IIC values approximated by adding 51 to original INR tests values.

2. Ranges include value for 3/8 in., 1/2 in., and 5/8 in. thicknesses of gypsum board finish.

3. Hard surface finish flooring (i.e., stripwood, wood block, vinyl, and vinyl asbestos) generally perform similarly in isolating impact sound. Unless otherwise indicated, ranges given include values for construction with any typical hard surface finish flooring.

STEEL BAR JOIST FLOORS

**Hard Surface Flooring³
On 2½″ Concrete
Lath & Plaster Ceiling**

STC   49

ASPHALT FLOORING

2 1/2″ CONCRETE ON
CORRUGATED DECK

LATH & PLASTER

14″ STEEL
BAR JOIST

INR   −16   IIC   35

**Hard Surface Flooring³
On 2½″ Concrete
Lath & Plaster Ceiling**

STC   47

With insulation⁶

STC   50

CONCRETE ON RIB LATH

VINYL TILE

GYPSUM LATH
& PLASTER

8″ STEEL
BAR JOIST

FURRING CHANNEL

INR   −25   IIC   26

With insulation⁶

INR   −22   IIC   29

**Hard Surface Flooring³
On 2″ Concrete
Resilient Gypsum
Board Ceiling**

STC   49

CONCRETE ON RIB LATH

VINYL TILE

GYPSUM LATH
& PLASTER

7″ STEEL
BAR JOIST

RESILIENT CHANNEL

INR   −10   IIC   41

FIGURE A-24 *Floor and ceiling assemblies. (Continued)*

4. Lightweight carpet and pad having pile weights of 20 oz. per sq. yd. and 28 oz. per sq. yd. respectively.

5. Standard weight carpet and pad having pile weights of 44 oz. per sq. yd. and 40 oz. per sq. yd. respectively. Ratings shown are based on standard weight combinations unless otherwise noted.

6. A minimum thickness of 2 in. to 3 in. of insulation can be used effectively as performance does not increase with increased thickness in floor and ceiling assemblies.

7. Ratings vary with thickness of concrete slabs. The low end of the range indicates value for 6 in. thickness; high end, 8 in. thickness.

APPENDIX B
REFERENCES

We gratefully acknowledge the assistance of the following organizations and individuals. We are also indebted to them for permission to use their illustrations and for the use of their publications as references.

We would also like to thank the authors and the publishers in the lists of References for their contributions to our research.

CHAPTER 1

ACKNOWLEDGMENTS

Addison-Wesley Publishing Company;

American National Standards Institute (ANSI);

American Society of Heating, Refrigerating and Air Conditioning Engineers (ASHRAE);

American Society for Testing and Materials (ASTM);

Bethlehem Steel Corporation;

Building Officials and Code Administrators International (BOCA);

Capital Development Board of the State of Illinois;

Chicago Title and Trust Company;

Geoffrey Lee Farnsworth, AIA;

Professor N. F. Fiore;

International Conference of Building Officials (ICBO);

National Association of Home Builders (NAHB) Research Foundation;

National Association of Plumbing, Heating, Cooling Contractors (NAPHCC);

National Easter Seal Society;

National Institute of Building Sciences (NIBS);

National Institute of Standards and Technology (NIST) of the U.S. Department of Commerce;

Betsy A. Pavichevich;

Standard Building Code/Southern Building Code Congress International (SBCCI);

Associate Professor Edward Steinfeld;

T & S Brass and Bronze Works, Inc.;

Underwriters Laboratories, Inc. (UL);

U.S. Architectural and Transportation Barriers Compliance Board (ATBCB);

U.S. Department of Housing and Urban Development (HUD);

U.S. General Services Administration, Specifications Unit;

Professor Lawrence H. Van Vlack.

CHAPTER 1

REFERENCES

American National Standards Institute (ANSI). ANSI A117.1-1986, "American National Standards for Buildings and Facilities—Providing Accessibility and Usability for Physically Handicapped People." New York: ANSI.

American National Standards Institute and American Society of Mechanical Engineers (ANSI/ASME). ANSI/ASME A17.1-1984, "Safety Code for Elevators and Escalators" (including supplement ANSI/ASME A17.1a-1985. New York: ANSI/ASME.

American National Standards Institute and Builders Hardware Manufacturers Association. ANSI/BHMA A156.10-1985. "Power-Operated Pedestrian Doors." New York: ANSI/BHMA.

———. ANSI/BHMA A156.19-1984, "Power-Assisted and Low Energy Power-Operated Doors." New York: ANSI/BHMA.

American Society for Testing and Materials (ASTM) Standards:

  E 380, "Practice for the Use of the International System of Units (SI)." Philadelphia, PA: ASTM.

  E 621, "Practice for the Use of Metric (SI) Units in Building Design and Construction." Philadelphia, PA: ASTM.

E 713, "Guide for the Selection of Scales for Metric Building Drawings." Philadelphia, PA: ASTM.

Architectural Graphic Standards. See Ramsey/Sleeper.

Champagne, Roger D. 1986. "Planning for Plumbing Efficiency." *The Construction Specifier*, 39(6)(June):77–80.

Chicago Title and Trust Company. *Fundamentals of Land Measurement.* Chicago: Chicago Title and Trust Company.

Council of American Building Officials (CABO). *One- and Two-Family Dwelling Code.* Falls Church, VA: CABO.

Freund, Eric C. "U.S. Construction and Practice in Europe." 1990. *The Construction Specifier*, 43(9)(September):68–82.

Goldman, Charles D. 1985. "Removing Barriers to the Handicapped." *The Construction Specifier*, 38(4)(April):40–45.

———. 1988. "Accessibility, the Evolution Continues." *The Construction Specifier*, 41(2)(February):50–53.

National Institute of Standards and Technology (NIST). *Metric Dimensional Coordination— The Issues and Precedents.* Special Publication 504. Gaithersburg, MD: NIST.

———. *Metrication in Building Design.* Special Publication 530. Gaithersburg, MD: NIST.

———. Technical Note 990, "The Selection of Preferred Metric Values for Design and Construction." Gaithersburg, MD: NIST.

Phillips, Ruth Hall, and Ronald L. Mace. 1985. "Designing and Specifying for Accessibility." *The Construction Specifier*, 38(4)(April):46–57.

Raeber, John A. 1988. "Complying with Accessibility Codes." *The Construction Specifier*, 41(2)(February):45–49.

———. 1991. "Redundancy in Fire Protection." *The Construction Specifier*, 44(2) (February): 17–19.

———. 1991. "Federal Accessibility." *The Construction Specifier*, 44(8)(August):82–89.

Ramsey/Sleeper, The AIA Committee on Architectural Graphic Standards. 1988. *Architectural Graphic Standards.* 8th ed. New York: Wiley.

Solomon, Nancy B. 1991. "Equal Opportunity Design." *Architecture* (February):101–102.

Stubbs, Stephanie. 1991. "AIA Gives ATBCB ADA A-D-V-I-C-E." *The American Institute of Architects Memo* (May):4–7.

U.S. Department of Housing and Urban Development (HUD). 1984. HUD 4910.1 *Minimum Property Standards for Housing.* Washington, DC: HUD.

———. 1989. *Adaptable Housing.* HUD Publication HUD-1124-PDR(1). Washington, DC: HUD.

U.S. General Services Administration (GSA). 1988. *Uniform Federal Accessibility Standards (UFAS).* Washington, DC: GSA.

———. *Index of Federal Specifications, Standards, and Commercial Item Descriptions.* Washington, DC: Superintendent of Documents, U.S. Government Printing Office.

Van Vlack, Lawrence H. *Elements of Materials.* 5th ed. Reading, MA: Addison-Wesley.

Webb, William A. 1991. "Rights of Passage." *The Construction Specifier*, 44(8)(August): 91–95.

CHAPTER 2

ACKNOWLEDGMENTS

American Concrete Institute (ACI);

American National Standards Institute (ANSI);

The American Society for Testing and Materials (ASTM);

American Wood Preservers Bureau (AWPB);

American Wood Preservers Institute (AWPI);

Building Officials and Code Administrators International (BOCA);

ELE International, Inc.;

Molded Fiberglass Concrete Forms Company;

National Academy of Sciences/National Research Council;

National Association of Home Builders (NAHB) Research Foundation;

National Forest Products Association (NFPA);

National Institute of Building Sciences (NIBS);

National Institute of Standards and Technology (NIST) of the U.S. Department of Commerce;

Portland Cement Association (PCA);

Underwriters Laboratories, Inc. (UL);

U.S. Department of Agriculture;

U.S. Department of Housing and Urban Development (HUD);

U.S. Weather Service.

CHAPTER 2

REFERENCES

AIA Professional Services Division. 1988. *Masterspec*, Basic Section 02900, "Landscape Work." 8/88 ed. Washington, DC: AIA.

American Association of Nurserymen (AAN). ANSI/AAN Z60.1–1990. "American Standard for Nursery Stock." Washington, DC: AAN.

American Concrete Institute. ACI 302.1R-89, "Guide for Concrete Floor and Slab Construction." Detroit, MI: ACI.

———. 1980. ACI 543R-74, "Recommendations for Design, Manufacture, and Installation of Concrete Piles." Detroit, MI: ACI.

———. 1989. "Standard Specification for the Construction of End Bearing Drilled Piers." Detroit, MI: ACI

American Society of Heating, Refrigerating and Air-Conditioning Engineers (ASHRAE). 1985. *ASHRAE Handbook; 1985 Fundamentals.* Atlanta, GA: ASHRAE.

American Wood Preservers Association. *Book of Standards.* Seattle, WA: American Wood Preservers Association.

American Society for Testing and Materials (ASTM). Standards:

A 74, "Specification for Cast Iron Soil Pipe and Fittings." Philadelphia, PA: ASTM.

A 252, "Standard Specification for Welded and Seamless Steel Pipe Piles." Philadelphia, PA: ASTM.

C 4, "Specification for Clay Drain Tile." Philadelphia, PA: ASTM.

C 412, "Specification for Concrete Drain Tile." Philadelphia, PA: ASTM.

C 444, "Specification for Perforated Concrete Pipe." Philadelphia, PA: ASTM.

D 427, "Test Methods for Shrinkage Factors of Soils." Philadelphia, PA: ASTM.

C 654, "Specification for Porous Concrete Pipe." Philadelphia, PA: ASTM.

D 698, "Test Methods for Moisture-Density Relations of Soils and Soil Aggregate Mixtures, Using 5.5-lb. (2.49-kg) Rammer and 12-in. (304.8-mm) Drop." Philadelphia, PA: ASTM.

C 700, "Specification for Vitrified Clay Pipe, Extra Strength, Standard Strength, and Perforated." Philadelphia, PA: ASTM.

D 1143, "Method of Testing Piles Under Static Axial Compressive Load." Philadelphia, PA: ASTM.

D 1557, "Test Methods for Moisture-Density Relations of Soils and Soil Aggregate Mixtures, Using 10-lb. (4.54-kg) Rammer and 18-in. (457-mm) Drop." Philadelphia, PA: ASTM.

D 1586, "Methods for Penetration Test and Split-Barrel Sampling of Soils." Philadelphia, PA: ASTM.

D 1883, "Test Method for CBR (California Bearing Ratio) of Laboratory Compacted Soils." Philadelphia, PA: ASTM.

C 2311, "Specification for Perforated Homogeneous Bituminized Fiber Pipe for General Drainage." Philadelphia, PA: ASTM.

D 2487, "Classification of Soils for Engineering Purposes." Philadelphia, PA: ASTM.

D 2488, "Practice for Description and Identification of Soils (Visual-Manual Procedures)." Philadelphia, PA: ASTM.

C 2729, "Specification for Poly (Vinyl Chloride) (PVC) Sewer Pipe and Fittings." Philadelphia, PA: ASTM.

D 4253, "Test Methods for Maximum Index Density of Soils Using a Vibratory Table." Philadelphia, PA: ASTM.

D 4254, "Test Methods for Minimum Index Density of Soils and Calculation of Relative Density." Philadelphia, PA: ASTM.

D 4318, "Test Method for Liquid Limit, Plastic Limit, and Plasticity Index of Soils." Philadelphia, PA: ASTM.

F 405, "Specification for Corrugated Polyethylene Tube and Fittings." Philadelphia, PA: ASTM.

Architectural Graphic Standards. See Ramsey/Sleeper.

Bethlehem Steel Corporation. *Spec-data Sheet for Steel H-Piles.* Bethlehem, PA: Bethlehem Steel Corporation.

Briscoe, Frank. 1991. "Wood-destroying Insects." *Old-House Journal*, 19(2)(March/April): 34–39.

Building Officials and Code Administrators International (BOCA), Inc. *National Building Code.* Country Club Hills, IL: BOCA.

Council of American Building Officials (CABO). 1983. *One- and Two-Family Building Code.* Falls Church, VA: CABO.

Crandall, Chuck. 1988. "Shrubs for All Seasons." *HomeOwner*, 13(2)(March):36–38, 70.

National Association of Plumbing-Heating-Cooling Contractors. 1990. *National Standard Plumbing Code.* Falls Church, VA: National Association of Plumbing-Heating-Cooling Contractors.

National Institute of Building Sciences (NIBS). 1988. *Guidelines for Protection of Wood Against Decay and Termite Attack.* Washington, DC: NIBS.

Portland Cement Association (PCA). *Concrete Floors on Ground.* 2d ed. Skokie, IL: PCA.

Ramsey/Sleeper, The AIA Committee on Architectural Graphic Standards. 1988. *Architectural Graphic Standards.* 8th ed. New York: Wiley.

Underground Space Center. 1988. *Building Foundation Design Handbook.* Minneapolis, MN: Underground Space Center.

United States Department of Agriculture, Agricultural Research Service. *USDA Plant Hardiness Zone Map.* Miscellaneous Publication Number 1475, Washington, DC: USDA.

United States Department of Agriculture, Forest Service. *Subterranean Termites, Their Prevention and Control in Buildings.* Home and Garden Bulletin 64. Washington, DC: USDA.

United States Steel Corporation. *Steel H-Piles.* Pittsburgh, PA: United States Steel Corporation.

CHAPTER 3

ACKNOWLEDGMENTS

American Concrete Institute (ACI);

American Society for Testing and Materials (ASTM);

Building Research Board of the National Academy of Sciences (NAS);

The Burke Company;

Ceco Concrete Corporation;

Concrete Reinforcing Steel Institute;

ELE International, Inc.;

Gypsum Association;

Martin Fireproofing Corporation;

Martin Fireproofing Georgia, Inc.;

National Association of Home Builders (NAHB);

Portland Cement Association (PCA);

Prestressed Concrete Institute (PCI);

U.S. Department of Housing and Urban Development (HUD);

Wire Reinforcement Institute.

CHAPTER 3

REFERENCES

Adams, J. T. 1983. *The Complete Concrete, Masonry, and Brick Handbook.* New York: Van Nostrand Reinhold.

AIA Professional Services Division. *Masterspec*, Basic Sections:

03300, "Cast-in-place Concrete." 5/90 ed. Washington, DC: AIA.

03350, "Concrete Toppings." 5/91 ed. Washington, DC: AIA.

03410, "Structural Precast Concrete." 8/87 ed. Washington, DC: AIA.

03450, "Architectural Precast Concrete—Plant Cast." 5/90 ed. Washington, DC: AIA.

03455, "Glass-fiber Reinforced Concrete—Plant Cast." 5/92 ed. Washington, DC: AIA.

03470, "Tilt-Up Precast Concrete." 5/91 ed. Washington, DC: AIA.

American Concrete Institute (ACI). *ACI Manual of Standard Practice.* Detroit: American Concrete Institute.

———. ACI 211-81, "Standard Practice for Selecting Proportions for Normal, Heavyweight, and Mass Concrete." Detroit: American Concrete Institute.

———. ACI 212.1R-81, "Admixtures for Concrete." Detroit: American Concrete Institute.

———. ACI 301-89, "Specifications for Structural Concrete for Buildings." Detroit: American Concrete Institute.

———. ACI 302.1R-89, "Guide for Concrete Floor and Slab Construction." Detroit: American Concrete Institute.

———. ACI 305R-89, "Hot Weather Concreting." Detroit: American Concrete Institute.

———. ACI 306.1-87, "Standard Specification for Cold Weather Concreting." Detroit: American Concrete Institute.

———. ACI 309R-87, "Guide for Consolidation of Concrete." Detroit: American Concrete Institute.

———. ACI 318-89, "Building Code Requirements for Reinforced Concrete." Detroit: American Concrete Institute.

———. ACI 332R-85, "Guide to Residential Cast-in-Place Concrete Construction." Detroit: American Concrete Institute.

———. ACI 347R-88, "Guide to Formwork for Concrete." Detroit: American Concrete Institute.

American Society for Testing and Materials (ASTM) Standards:

A 82, "Specification for Steel Wire, Plain, for Concrete Reinforcement." Philadelphia, PA: ASTM.

A 185, "Specification for Steel Welded Wire, Fabric, Plain, for Concrete Reinforcement." Philadelphia, PA: ASTM.

A 416, "Specification for Uncoated Seven-wire Stress-relieved Steel Strand for Prestressed Concrete." Philadelphia, PA: ASTM.

A 497, "Specification for Welded Deformed Steel Wire Fabric for Concrete Reinforcement." Philadelphia, PA: ASTM.

A 615, "Specification for Deformed and Plain Billet-Steel Bars for Concrete Reinforcement." Philadelphia, PA: ASTM.

A 616, "Specification for Rail-Steel Deformed and Plain Bars for Concrete Reinforcement." Philadelphia, PA: ASTM.

A 617, "Specification for Axle-Steel Deformed and Plain Bars for Concrete Reinforcement." Philadelphia, PA: ASTM.

A 767, "Specification for Zinc-coated (Galvanized) Bars for Concrete Reinforcement." Philadelphia, PA: ASTM.

A 775, "Specification for Epoxy-coated Reinforcing Steel Bars." Philadelphia, PA: ASTM.

C 31, "Standard Method for Making and Curing Concrete Test Specimens in the Field." Philadelphia, PA: ASTM.

C 33, "Specification for Concrete Aggregate." Philadelphia, PA: ASTM.

C 39, "Test Method for Compressive Strength of Cylindrical Concrete Specimens." Philadelphia, PA: ASTM.

C 42, "Methods of Obtaining and Testing Drilled Cores and Sawed Beams of Concrete." Philadelphia, PA: ASTM.

C 94, "Specification for Ready-mixed Concrete." Philadelphia, PA: ASTM.

C 143, "Test Method for Slump of Portland Cement Concrete." Philadelphia, PA: ASTM.

C 150, "Specification for Portland Cement." Philadelphia, PA: ASTM.

C 226, "Specification for Air-entraining Additions for Use in the Manufacturer of Air-entraining Portland Cement." Philadelphia, PA: ASTM.

C 260, "Specification for Air-entraining Admixtures for Concrete." Philadelphia, PA: ASTM.

C 309, "Specification for Liquid Membrane-forming Compounds for Curing Concrete." Philadelphia, PA: ASTM.

C 317, "Specification for Gypsum Concrete." Philadelphia, PA: ASTM.

C 330, "Specification for Lightweight Aggregate for Structural Concrete." Philadelphia, PA: ASTM.

C 494, "Specification for Chemical Admixtures for Concrete." Philadelphia, PA: ASTM.

C 595, "Specification for Blended Hydraulic Cements." Philadelphia, PA: ASTM.

C 618, "Specification for Fly Ash and Raw or Calcined Natural Pozzolan for Use as a Mineral Admixture in Portland Cement Concrete." Philadelphia, PA: ASTM.

C 956, "Specification for Installation of Cast-in-Place Reinforced Gypsum Concrete." Philadelphia, PA: ASTM.

D 98, "Specification for Calcium Chloride." Philadelphia, PA: ASTM.

E 1155, "Test Method for Determining Floor Flatness and Levelness Using the 'F-Number' System (Inch-Pound Units)." Philadelphia, PA: ASTM.

Beall, Christine. 1989. "The Many Differences Between Concrete and Masonry Technology." *Architecture* (April):123–124.

Brown, R. W. 1984. *Residential Foundations: Design, Behavior, and Repair.* New York: Van Nostrand Reinhold.

Building Research Advisory Board (BRAB), National Academy of Sciences/National Research Council. 1968. *Criteria for Selection and Design of Residential Slabs-on-Ground.* Washington, DC: BRAB.

Concrete Reinforcing Steel Institute (CRSI). *CRSI Manual of Standard Practice.* Schaumburg, IL: CRSI.

———. 1983. *Placing Reinforcing Bars: Recommended Practice.* 4th ed. Schaumburg, IL: CRSI.

De Cristoforo, R. J. 1975. *Concrete and Masonry.* Reston, VA: Reston Publishing Company.

Parker, Harry, and John D. Hauf. 1976. *Simplified Design of Reinforced Concrete.* 4th ed. New York: Wiley.

Portland Cement Association (PCA). 1965. *Moisture Migration—Concrete Slab-on-Ground Construction.* Bulletin D89. Skokie, IL: PCA.

———. 1979. *Design and Control of Concrete Mixtures.* Skokie, IL: PCA.

———. 1983. *Concrete Floors on Ground.* Skokie, IL: PCA.

———. 1987. *Concrete Slab Surface Defects: Causes, Prevention, Repair.* Skokie, IL: PCA.

———. 1988. *Design and Control of Concrete Mixtures,* 13th ed. Skokie, IL: PCA.

———. 1988. *Transporting and Handling Concrete.* Skokie, IL: PCA.

———. 1989. *Resurfacing Concrete Floors.* Skokie, IL: PCA.

Prestressed Concrete Institute (PCI). MNL 116. *Manual for Quality Control for Plants and Production of Precast Concrete Products.* Chicago, IL: PCI.

———. MNL 117, *Manual for Quality Control for Plants and Production of Architectural Precast Products.* Chicago, IL: PCI.

———. MNL 120, *PCI Design Handbook— Precast and Prestressed Concrete.* Chicago, IL: PCI.

———. MNL 127, *Recommended Practice for Erection of Precast Concrete.* Chicago, IL: PCI.

———. MNL 128, *Recommended Practice for Glass Fiber Reinforced Concrete Panels.* Chicago, IL: PCI.

———. *Architectural Precast Concrete.* Chicago, IL: PCI.

Pruter, Walter F. 1987. "The Advantages of Perlite Concrete Roof Deck Systems." *The Construction Specifier,* 40(11)(November): 52–55.

Ramsey/Sleeper, The AIA Committee on Architectural Graphic Standards. 1981. *Architectural Graphic Standards.* 7th ed. New York: Wiley.

Underground Space Center. 1988. *Building Foundation Design Handbook.* Minneapolis, MN: Underground Space Center.

Wilson, Forrest. 1984. *Building Materials Evaluation Handbook.* New York: Van Nostrand.

CHAPTER 4

ACKNOWLEDGMENTS

A.A. Wire Products Company;

American Concrete Institute (ACI);

American National Standards Institute (ANSI);

American Society for Testing and Materials (ASTM);

Brick Institute of America (BIA);

Building Officials and Code Administrators International (BOCA);

Building Research Board of the National Academy of Sciences (NAS);

Dow Corning Corporation;

Facing Tile Institute;

Indiana Limestone Institute of America (ILI);

Marble Institute of America (MIA);

National Building Granite Quarries Association, Inc. (NBGQA);

National Concrete Masonry Association (NCMA);

Portland Cement Association (PCA);

Rock of Ages Corporation;

U.S. Department of Housing and Urban Development (HUD).

CHAPTER 4

REFERENCES

Adams, J. T. 1983. *The Complete Concrete, Masonry, and Brick Handbook.* New York: Van Nostrand Reinhold.

AIA Professional Services Division. *Masterspec,* Basic Sections:

04200, "Unit Masonry." 5/85 ed. Washington, DC: AIA.

04230, "Reinforced Unit Masonry." 5/85 ed. Washington, DC: AIA.

04405, "Dimension Stone." 11/89 ed. Washington, DC: AIA.

04500, "Masonry Restoration and Cleaning." 2/86 ed. Washington, DC: AIA.

07600, "Flashing and Sheet Metal." 5/88 ed. Washington, DC: AIA.

American Concrete Institute, American Society of Civil Engineers (ACI/ASCE). 1988. ACI 530-88/ASCE 5-88, "Building Code Requirements for Masonry Structures." Detroit, MI: American Concrete Institute and American Society of Civil Engineers.

———. 1988. ACI 530.1-88/ASCE 6-88, "Specifications for Masonry Structures." Detroit, MI: American Concrete Institute and American Society of Civil Engineers

American Iron and Steel Institute. 1975. *Stainless Steel Stone Anchors.* Washington, DC: American Iron and Steel Institute.

American National Standards Institute (ANSI). 1976. ANSI A118.1, "Specifications for Dry-set Portland Cement Mortar." New York: ANSI.

———. 1985. ANSI A118.4, "Specifications for Latex Portland Cement Mortar." New York: ANSI.

———. 1985. ANSI A118.6, "Specifications for Ceramic Tile Grouts." New York: ANSI.

Architectural Graphic Standards. See Ramsey/ Sleeper.

Architectural Technology. 1984. "Rediscovering Terra-Cotta: A Glowing Recommendation." *Architectural Technology* (Spring):12–13.

———. 1986. "Technical Tips: Slope the Sills." *Architectural Technology* (May/June):66.

———. 1986. "Technical Tips: Tension Bars Save Masonry." *Architectural Technology* (May/June):66–67.

Arumala, Joseph O., and Russell H. Brown. 1982. *Performance Evaluation of Brick Veneer with Steel Stud Backup.* Clemson, SC: Clemson University Department of Civil Engineering.

American Society for Testing and Materials (ASTM). 1986. *Annual Book of ASTM Standards, Section (Cement, Lime, Gypsum).* Philadelphia, Pa: ASTM.

———. Standards:

A 36, "Specification for Structural Steel." Philadelphia, PA: ASTM.

A 82, "Specification for Steel Wire, Plain, for Concrete Reinforcement." Philadelphia, PA: ASTM.

A 116, "Specification for Zinc-Coated (Galvanized) Steel Woven Wire Fence Fabric." Philadelphia, PA: ASTM.

A 123, "Specification for Zinc (Hot-Dip Galvanized) Coatings on Iron and Steel Products." Philadelphia, PA: ASTM.

A 153, "Specification for Zinc Coating (Hot-Dip) on Iron and Steel Hardware." Philadelphia, PA: ASTM.

A 167, "Specification for Stainless and Heat-Resisting Chromium-Nickel Steel Plate, Sheet, and Strip." Philadelphia, PA: ASTM.

A 185, "Specification for Steel Welded Wire Fabric, Plain, for Concrete Reinforcement." Philadelphia, PA: ASTM.

A 366, "Specification for Steel, Carbon, Cold-Rolled Sheet, Commercial Quality." Philadelphia, PA: ASTM.

A 496, "Specification for Steel Wire, Deformed, for Concrete Reinforcement." Philadelphia, PA: ASTM.

A 497, "Specification for Welded Wire Fabric, Deformed, for Concrete Reinforcement." Philadelphia, PA: ASTM.

A 510, "Standard Specification for General Requirements for Wire Rods and Coarse Round Wire, Carbon Steel." Philadelphia, PA: ASTM.

A 525, "Specification for General Requirements for Steel Sheet, Zinc-Coated (Galvanized) by the Hot-Dip Process." Philadelphia, PA: ASTM.

A 526, "Specification for Steel Sheet, Zinc-Coated (Galvanized) by the Hot-Dip Process, Commercial Quality." Philadelphia, PA: ASTM.

A 580, "Specification for Stainless Steel and Heat-resistant Steel Wire." Philadelphia, PA: ASTM.

A 615, "Specification for Deformed and Plain Billet-steel Bars for Concrete Reinforcement." Philadelphia, PA: ASTM.

A 635, "Specification for Steel, Sheet and Strip, Heavy-thickness Coils, Carbon, Hot-rolled." Philadelphia, PA: ASTM.

A 666, "Specification for Austenitic Stainless Steel Sheet, Strip, Plate, and Flat Bar." Philadelphia, PA: ASTM.

A 775, "Specification for Epoxy-coated Reinforcing Steel Bars." Philadelphia, PA: ASTM.

B 101, "Specification for Lead Coated Copper Sheets." Philadelphia, PA: ASTM.

B 152, "Specification for Copper Sheet, Strip, Plate, and Rolled Bar." Philadelphia, PA: ASTM.

B 227, "Specification for Hard Drawn Copper-Clad Steel Wire." Philadelphia, PA: ASTM.

C 5, "Specification for Quicklime for Structural Purposes." Philadelphia, PA: ASTM.

C 33, "Specifications for Concrete Aggregates." Philadelphia, PA: ASTM.

C 34, "Specification for Structural Clay Load-Bearing Wall Tile." Philadelphia, PA: ASTM.

C 55, "Specification for Concrete Building Brick." Philadelphia, PA: ASTM.

C 56, "Specification for Structural Clay Non-Load-Bearing Tile." Philadelphia, PA: ASTM.

C 57, "Specification for Structural Clay Floor Tile." Philadelphia, PA: ASTM.

C 62, "Specification for Building Brick (Solid Masonry Units Made from Clay or Shale)." Philadelphia, PA: ASTM.

C 67, "Method of Sampling and Testing Brick and Structural Clay Tile." Philadelphia, PA: ASTM.

C 90, "Specification for Hollow Load-Bearing Concrete Masonry." Philadelphia, PA: ASTM.

C 91, "Specifications for Masonry Cement." Philadelphia, PA: ASTM.

C 119, "Definition of Terms Relating to Natural Building Stones." Philadelphia, PA: ASTM.

C 126, "Specification for Ceramic Glazed Structural Clay Facing Tile, Facing Brick, and Solid Masonry Units." Philadelphia, PA: ASTM.

C 129, "Specification for Non-Load-Bearing Concrete Masonry Units." Philadelphia, PA: ASTM.

C 144, "Specification for Aggregate for Masonry Mortar." Philadelphia, PA: ASTM.

C 145, "Specification for Solid Load-Bearing Concrete Masonry Units." Philadelphia, PA: ASTM.

C 150, "Specification for Portland Cement." Philadelphia, PA: ASTM.

C 206, "Specification for Finishing Hydrated Lime." Philadelphia, PA: ASTM.

C 207, "Specifications for Hydrated Lime for Masonry Purposes." Philadelphia, PA: ASTM.

C 212, "Specification for Structural Clay Facing Tile." Philadelphia, PA: ASTM.

C 216, "Specification for Facing Brick (Solid Masonry Units Made from Clay or Shale)." Philadelphia, PA: ASTM.

C 260, "Specification for Air-Entraining Admixtures for Concrete." Philadelphia, PA: ASTM.

C 270, "Specification for Mortar for Unit Masonry." Philadelphia, PA: ASTM.

C 331, "Specification for Lightweight Aggregate for Concrete Masonry." Philadelphia, PA: ASTM.

C 404, "Specification for Aggregate for Masonry Grout." Philadelphia, PA: ASTM.

C 426, "Test Method for Drying Shrinkage of Concrete Block." Philadelphia, PA: ASTM.

C 476, "Specification for Grout for Reinforced and Nonreinforced Masonry." Philadelphia, PA: ASTM.

C 503, "Specification for Marble Building Stone (Exterior)." Philadelphia, PA: ASTM.

C 568, "Specification for Limestone Building Stone." Philadelphia, PA: ASTM.

C 595, "Specification for Blended Hydraulic Cements." Philadelphia, PA: ASTM.

C 615, "Specification for Granite Building Stone." Philadelphia, PA: ASTM.

C 618, "Specification for Fly Ash and Raw or Calcined Natural Pozzolan for Use as a Mineral Admixture in Portland Cement Concrete." Philadelphia, PA: ASTM.

C 629, "Specification for Slate Building Stone." Philadelphia, PA: ASTM.

C 645, "Specifications for Non-Load (Axial) Bearing Steel Studs, Runners (Track), and Rigid Furring Channels for Screw Application of Gypsum Board." Philadelphia, PA: ASTM.

C 652, "Specification for Hollow Brick (Hollow Masonry Units Made from Clay or Shale)." Philadelphia, PA: ASTM.

C 902, "Specification for Pedestrian and Light Traffic Paving Brick." Philadelphia, PA: ASTM.

C 955, "Specifications for Load-Bearing (Transverse and Axial) Steel Studs, Runners (Track), and Bracing or Bridging, for Screw Application of Gypsum Board and Metal Plaster Bases." Philadelphia, PA: ASTM.

C 1007, "Installation of Load-Bearing (Transverse and Axial) Steel Studs and Accessories." Philadelphia, PA: ASTM.

D 98, "Specification for Calcium Chloride." Philadelphia, PA: ASTM.

E 90, "Method for Laboratory Measurement of Airborne-Sound Transmission Loss of Building Partitions." Philadelphia, PA: ASTM.

E 96, "Test Methods for Water Vapor Transmission of Materials." Philadelphia, PA: ASTM.

E 119, "Method for Fire Tests of Building Construction and Materials." Philadelphia, PA: ASTM.

E 413, "Classification for Rating Sound Insulation." Philadelphia, PA: ASTM.

E 514, "Test Method for Water Permeance of Masonry." Philadelphia, PA: ASTM.

Beall, Christine. 1987. *Masonry Design and Detailing.* New York: McGraw-Hill.

———. 1988. "Controlling Moisture Movement in Masonry Walls." *The Construction Specifier*, 41(8)(June):36–47.

———. 1988. "The Types of Masonry Mortar." *The Magazine of Masonry Construction*, 1(4):164–166.

———. 1989. "Weatherproof Masonry Walls." *Architecture* (March):133–136.

———. 1989. "The Many Differences Between Concrete and Masonry Technology." *Architecture* (April):123–124.

———. 1989. "Preventing Problems with Thin Brick Veneers." *The Construction Specifier,* 42(9)(September):68–78.

———. 1991. "The ACI/ASCE Masonry Specification." *The Construction Specifier,* 44(1)(January):36–42.

Belles, Donald. 1987. "Preventing Flame Spread on Exterior Walls." *Exteriors,* 5(2)(Summer):28–38.

Bock, Gordon. 1989. "Inspecting Chimneys." *The Old-House Journal,* 17(2)(March/ April):34–37.

Borchelt, J., ed. 1982. *STP 778—Masonry: Materials, Properties, and Performance.* Philadelphia, PA.: ASTM.

Brick Institute of America (BIA). 1980. *Compilation of Veneer/Metal Stud Problems.* Reston, VA: BIA.

———. *Technical Notes on Brick Construction:*

No. 1. Rev. Dec. 1967. "Cold Weather Masonry Construction—Introduction." Reston, VA: BIA.

No. 1A. Jan. 1968. "Cold Weather Masonry Construction—Construction and Protection Recommendations." Reston, VA: BIA.

No. 2. Rev. Jan./Feb. 1975. "Glossary of Terms Relating to Brick Masonry." Reston, VA: BIA.

No. 7. Rev. Feb. 1985. "Water Resistance of Brick Masonry—Design and Detailing," Part I of III. Reston, VA: BIA.

No. 7A. Rev. Mar. 1985. "Water Resistance of Brick Masonry—Materials," Part II of III. Reston, VA: BIA.

No. 7B. Rev. Apr. 1985. "Water Resistance of Brick Masonry—Construction and Workmanship," Part III of III. Reston, VA: BIA.

No. 7C. Feb. 1965. "Moisture Control in Brick and Tile Walls: Condensation." Reston, VA: BIA.

No. 7D. Rev. Jun. 1968. "Moisture Resistance of Brick Masonry Walls—Condensation Analysis." Reston, VA: BIA.

No. 7F. Feb. 1986. "Moisture Resistance of Brick Masonry—Maintenance. Reston, VA: BIA.

No. 8. Rev. Sept. 1972. "Portland Cement-lime Mortars for Brick Masonry." Reston, VA: BIA.

No. 8A. Rev. Sept. 1988. "Standards Specifications for Portland Cement-lime Mortar for Brick Masonry." Reston, VA: BIA.

No. 8B. Jul./Aug. 1976. "Mortar for Brick Masonry—Selection and Controls." Reston, VA: BIA.

No. 9. Rev. Mar. 1986. "Manufacturing, Classification, and Selection of Brick: Manufacturing," Part I. Reston, VA: BIA.

No. 9B. Jan. 1989. "Manufacturing, Classification, and Selection of Brick: Selection," Part III. Reston, VA: BIA.

No. 10A. Rev. Jun./Jul. 1973. "Modular Brick Masonry." Reston, VA: BIA.

No. 10B. Oct. 1971. "Brick Sizes and Related Information." Reston, VA: BIA.

No. 11. Rev. Dec. 1971. "Guide Specifications for Brick Masonry," Part I of V. Reston, VA: BIA.

No. 11A. Rev. Jun. 1978. "Guide Specifications for Brick Masonry," Part II of V. Reston, VA: BIA.

No. 11B. Rev. Feb. 1972. "Guide Specifications for Brick Masonry," Part III of V. Reston, VA: BIA.

No. 11C. Rev. Jul. 1972. "Guide Specifications for Brick Masonry," Part IV of V. Reston, VA: BIA.

No. 11D. Aug. 1972. "Guide Specifications for Brick Masonry," Part IV of V continued. Reston, VA: BIA.

No. 11E. Apr./May. 1973. "Guide Specifications for Brick Masonry," Part V of V. Reston, VA: BIA.

No. 12. Apr. 1962. "Columns, Piers, Pilasters, and Buttresses." Reston, VA: BIA.

No. 13. May 1962. "Ceramic Glazed Brick Facing for Exterior Walls." Reston, VA: BIA.

No. 14. Rev. Jan. 1985. "Brick Floors and Pavements," Part I of III. Reston, VA: BIA.

No. 14A. Rev. II. Apr. 1987. "Brick Floors and Pavements," Part II of III. Reston, VA: BIA.

No. 14B. Nov./Dec. 1975. "Brick Floors and Pavements," Part III of III. Reston, VA: BIA.

No. 15. Rev. May 1988. "Salvaged Brick." Reston, VA: BIA.

No. 16. Oct. 1974. "Fire Resistance." Reston, VA: BIA.

No. 17. Nov./Dec. 1962. "Reinforced Brick Masonry," Part I of IV. Reston, VA: BIA.

No. 17A. Rev. Reissued Dec. 1980. "Reinforced Brick Masonry, Flexural Design," Part II of IV. Reston, VA: BIA.

No. 17B. Reissued Dec. 1980. "Reinforced Brick Masonry, Axial Design," Part III of IV. Reston, VA: BIA.

No. 17C. Mar. 1963. "Reinforced Brick Masonry—Inspectors' Guide," Part IV of IV. Reston, VA: BIA.

No. 17D. Rev. Mar. 1972. "High-lift Grouted Reinforced Brick Masonry." Reston, VA: BIA.

No. 17E. Reissued Aug. 1980. Reinforced Brick Masonry Retaining Walls," Part I of III. Reston, VA: BIA.

No. 17F. Sept./Oct. 1964. "Reinforced Brick Masonry Retaining Walls," Part II of III. Reston, VA: BIA.

No. 17G. Mar./Apr. 1965. "Reinforced Brick Masonry Retaining Walls," Part III of III. Reston, VA: BIA.

No. 17H. Nov./Dec. 1964. "Reinforced Brick and Tile Lintels." Reston, VA: BIA.

No. 17I. Oct./Nov./Dec. 1965. "Reinforced Brick Masonry Columns and Pilasters." Reston, VA: BIA.

No. 17L. Rev. Feb./Mar. 1973. "Four-inch RBM Curtain and Panel Walls." Reston, VA: BIA.

No. 17N. Dec./Jan. 1972–1973. "Low Brick Masonry Retaining Walls." Reston, VA: BIA.

No. 18. Apr. 1963. "Differential Movement, Cause and Effect," Part I of III. Reston, VA: BIA.

No. 18A. Reissued Dec. 1980. "Differential Movement, Expansion Joints," Part II of III. Reston, VA: BIA.

No. 18B. Reissued Dec. 1980. "Differential Movement, Flexible Anchorage," Part III of III. Reston, VA: BIA.

No. 19. Rev. April 1980. "Residential Fireplace Design." Reston, VA: BIA.

No. 19A. Rev. May 1980. "Residential Fireplace Details and Construction." Reston, VA: BIA.

No. 19B. June 1980. "Residential Fireplace Design and Construction." Reston, VA: BIA.

No. 19C. Rev. April 1988. "Contemporary Brick Masonry Fireplaces." Reston, VA: BIA.

No. 20. Rev. Sept./Oct. 1977. "Cleaning Brick Masonry." Reston, VA: BIA.

No. 21. Rev. Jan./Feb. 1977. "Brick Masonry Cavity Walls." Reston, VA: BIA.

No. 21A. May/Jun. 1977. "Brick Masonry Cavity Walls: Insulated." Reston, VA: BIA.

No. 21B. Jan./Feb. 1978. "Brick Masonry Cavity Walls: Detailing." Reston, VA: BIA.

No. 21C. May/Jun. 1978. "Brick Masonry Cavity Walls: Construction." Reston, VA: BIA.

No. 23. Rev. May 1985. "Efflorescence, Causes and Mechanisms," Part I of II. Reston, VA: BIA.

No. 23A. Rev. Jun. 1985. "Efflorescence, Prevention and Control," Part II of II. Reston, VA: BIA.

No. 24. Rev. Feb. 1970. "The Contemporary Bearing Wall." Reston, VA: BIA.

No. 24A. Rev. May. 1970. "Building Code Requirements for the Contemporary Bearing Wall." Reston, VA: BIA.

No. 24F. Rev. Nov./Dec. 1974. "The Contemporary Bearing Wall—Construction." Reston, VA: BIA.

No. 24G. Dec. 1968. "Contemporary Bearing Wall Detailing." Reston, VA: BIA.

No. 25. May 1966. "Brick and Tile Metal-tied Walls." Reston, VA: BIA.

No. 26. Jun. 1966. "Six-inch Brick and Tile Walls." Reston, VA: BIA.

No. 26A. Sept. 1966. "Construction Details, Six-inch Brick and Tile Walls for One Story Buildings." Reston, VA: BIA.

No. 28. Rev. Jul./Aug. 1978. "Brick Veneer, New Construction." Reston, VA: BIA.

No. 28A. Sept./Oct. 1978. "Brick Veneer, Existing Construction." Reston, VA: BIA.

No. 28B. Rev. II. Feb. 1987. "Brick Veneer, Steel Stud Panel Walls." Reston, VA: BIA.

No. 28C. Jan. 1986. "Thin Brick Veneer, Introduction." Reston, VA: BIA.

No. 29. Rev. Mar./Apr. 1976. "Brick in Landscape Architecture, Terraces and Walks." Reston, VA: BIA.

No. 29A. Rev. Nov. 1968. "Brick in Landscape Architecture, Garden Walls." Reston, VA: BIA.

No. 30. Jul. 1967. "Bonds and Patterns in Brickwork." Reston, VA: BIA.

No. 31. Sept. 1967. "Brick Masonry Arches." Reston, VA: BIA.

No. 31B. Rev. Nov./Dec. 1981. "Structural Steel Lintels." Reston, VA: BIA.

No. 36. Rev. Jul./Aug. 1981. "Brick Masonry Details: Sills and Soffits." Reston, VA: BIA.

No. 36A. Rev. Sept./Oct. 1981. "Brick Masonry Details: Caps and Copings, Corbels and Racking." Reston, VA: BIA.

No. 41. Rev. 1983. "Hollow Brick Masonry: Introduction." Reston, VA: BIA.

No. 42. Nov./Dec. 1976. "Empirical Design of Brick Masonry." Reston, VA: BIA.

No. 44A. May 1986. "Fasteners for Brick Masonry." Reston, VA: BIA.

No. 44B. Mar. 1987. "Wall Ties for Brick Masonry." Reston, VA: BIA.

Brown, R. W. 1984. *Residential Foundations: Design, Behavior, and Repair*. New York: Van Nostrand Reinhold.

Catani, Mario J. 1985. "Protection of Embedded Steel in Masonry." *The Construction Specifier*, 38(1)(January):62–68.

———. 1987. "Connecting Masonry with Ties and Anchors." *The Construction Specifier*, 40(8)(August):108–113.

———. 1989. "Setting New Standards for Masonry." *The Construction Specifier*, 42(9)(September):108–114.

Catani, Mario J., and A. Rhett Whitlock. "Coping with Wide Cavities." *The Construction Specifier*, 39(8)(August):34–42.

Chin, Ian R., John P. Stecich, and Bernard Erlin. 1986. "Design of Thin Stone Veneers for Buildings." *Building Stone Magazine* (May/June):50–61.

Clifton, J. R., ed. 1986. *STP 935—Cleaning Stone and Masonry*. Philadelphia, PA: ASTM.

Council of American Building Officials (CABO). 1983. *One- and Two-Family Dwelling Code*. Falls Church, VA: CABO.

De Cristoforo, R. J. 1975. *Concrete and Masonry*. Reston, VA: Reston Publishing Company, Inc.

Gabby, Brent A. 1990. "About Face." *Architecture* (April):113–114; 141–143.

———. 1991. "Brick Pavers: A Commentary on ASTM C 902." *The Construction Specifier*, 44(1)(January):29–31.

Gordon, Douglas E. 1985. "Brick Basics." *Architectural Technology* (Fall):17–24. [*See also* April 1986, "Feedback," pp. 8–9.]

Grimm, Clayford T. 1985. "Clay Brick/Concrete Masonry Wall Design Checklist." *The Construction Specifier*, 38(8)(August):67–69.

———. 1989. "What They Say About Brick Veneer Over Steel Studs." *The Construction Specifier*, 42(9)(September):76–77.

Grogan, John C. 1983. "Water Permeance in Brick Masonry." *The Construction Specifier*, 36(12)(December):36–38.

Grogan, John C., and John T. Conway, eds. 1985. *Masonry Research Application and Problems (STP 871)*. Philadelphia, PA: ASTM.

Gross, James G., Robert D. Dikkers, and John C. Grogan. 1969. *Recommended Practice for Engineered Brick Masonry*. Reston, VA: Brick Institute of America.

Harris, Harry A., ed. 1988. *STP 992—Masonry: Materials, Design, Construction, and Maintenance*. Philadelphia, PA: ASTM.

Hayes, Arthur J. 1989. "Keeping Abreast of Masonry Technology." *The Construction Specifier*, 42(9)(September):45–46.

Hunderman, Harry J., and Deborah Slaton. 1989. "Terra-Cotta: Analysis and Repair." *The Construction Specifier*, 42(7)(July):50–57.

Indiana Limestone Institute of America, Inc. (ILI). *The Contractors Handbook on Indiana Limestone*. 6th ed. Bedford, IN: ILI.

———. *The Finishing Touch*. Bedford, IN: ILI.

———. *Indiana Limestone Handbook*. 19th ed. Bedford, IN: ILI.

———. *The Patton Glossary of Building Stone and Masonry Terms*. Bedford, IN: ILI.

———. *Specifications for Indiana Limestone*. Bedford, IN: ILI.

———. *ILI Technote on Safety Factors*. Bedford, IN: ILI.

Kaminetzky, Dov. 1989. "Investigating Masonry Failures." *The Construction Specifier*, 42(1)(January):42–47.

———. 1983. "Matching Brick." *The Construction Specifier*, 36(12)(December):46–51.

Laska, Walter. 1991. "Portland Lime Mortar, Understanding ASTM C 270." *The Construction Specifier*, 44(1)(January):71–73.

Lucas, James J. 1983. "Causes and Cures for Deteriorating Masonry." *The Construction Specifier*, 36(3)(March):60–70.

Marble Institute of America, Inc. *Dimension Stone—Design Manual IV*. Farmington, MI: Marble Institute of America.

Metal Lath/Steel Framing Association. 1985. *Lightweight Steel Framing Systems Manual*. 2d ed. Chicago, IL: Metal Lath/Steel Framing Association.

———. 1986. *Technical Information File*. Chicago, IL: Metal Lath/Steel Framing Association.

———. *Technical Bulletin No. 18: Fire Rated Metal Lath/Steel Stud Exterior*. Chicago, IL: Metal Lath/Steel Framing Association.

Morton, W. Brown III. 1972. *Field Procedures for Examining Humidity in Masonry Buildings*. Association for Preservation Technology Bulletin VIII(2). Washington, DC.

National Building Granite Quarries Association. *Specifications for Architectural Granite*. Barre, VT: National Building Granite Quarries Association.

National Concrete Masonry Association (NCMA). 1978. NCMA-TEK 99, "The Structural Role of Joint Reinforcement in Concrete Masonry." Herndon, VA: NCMA.

———. 1977. NCMA-TEK 113A, "Building Code Requirements for Masonry Structures." Herndon, VA: NCMA.

———. 1981. NCMA-TEK 121, "Waterproofing Concrete Masonry Basements and Earth-sheltered Structures." Herndon, VA: NCMA.

———. 1976. *Architectural and Engineering Concrete Masonry Details for Building Construction*. McLean, VA: NCMA.

———. *Facts About Concrete Masonry*. McLean, VA: NCMA.

Nicastro, David H. 1989. "Common Problems in Masonry Detailing." *The Construction Specifier*, 42(9)(September):80–83.

Nunn, Mark A. 1989. "Exterior Brick Pavements Guide." *The Construction Specifier*, 42(9)(September):100–107.

Pilling, Doug. 1980. "Brick Walks." *The Old-House Journal*, 18(7)(July):73, 78–79.

Portland Cement Association (PCA). *Repairing Damp or Leaky Basements in Homes*. Washington, DC: PCA.

———. 1983. *Permeability Tests of Masonry Walls*. Washington, DC: PCA.

Raeber, John A. 1989. "Composite Plaster and Masonry Construction: A Closer Look." *The Construction Specifier*, 42(9)(September):31–32.

———. 1991. "Brick Veneer in Earthquake Country." *The Construction Specifier*, 41(1)(January):23–24.

Ramsey/Sleeper, The AIA Committee on Architectural Graphic Standards. 1988. *Architectural Graphic Standards*. 8th ed. New York: Wiley.

Randall, Frank A., and William C. Panarese. 1976. *Concrete Masonry Handbook for Architects, Engineers, and Builders*. 4th ed. Skokie, IL: Portland Cement Association.

Reynolds, Phillip. 1983. "Specifying Non-Industrial Exterior Brick Paving." *The Construction Specifier*, 36(12)(December):39–45.

Rhault, Robert. 1991. "Masonry Cleaning." *The Construction Specifier*, 44(1)(January):53–61.

Richter, Robert J. 1991. "Masonry Failures, A Look at Trade Practices." *The Construction Specifier*, 44(1)(January):44–51.

Rock of Ages Corporation. 1992. *The Rock of Ages Story*. Barre, VT: Rock of Ages Corporation.

Sheppard, Walter Lee. 1981. "Obtaining Sound Chemically Resistant Masonry Construction." *The Construction Specifier*, 34(12)(December):20–26.

Silver, Larry. 1991. "Precast Concrete Masonry Lintels." *The Construction Specifier*, 44(1)(January):66–69.

Sivinski, Valerie A. 1985. "Masonry Maintenance." *Architectural Technology* (Fall): 25–26.

Smith, Baird M. 1984. *Moisture Problems in Historic Masonry Walls: Diagnosis and Treatment*. Washington, DC: U.S. Department of the Interior, Technical Preservation Services, Technical Report.

Szoke, Stephen S., and Gerald J. Carrier. 1986. "Avoiding Cracks in Brickwork." *The Construction Specifier*, 39(8)(August):44–56.

Szoke, Stephen S., and Hugh C. MacDonald. 1989. "Combining Masonry and Brick." *Architecture* (January):103–106.

Tiller, deTeel Patterson. 1979. *The Preservation of Historic Glazed Architectural Terra-Cotta*. Preservation Brief #7. Washington, DC: U.S. Department of the Interior, National Park Service, Preservation Assistance Division, Technical Preservation Services.

Trimble, Brian E., and J. Gregg Borchelt. "Jack Arches in Masonry Construction." *The Construction Specifier*, 44(1)(January): 62–65.

Underground Space Center. 1988. *Building Foundation Design Handbook*. Minneapolis, MN: Underground Space Center.

Wilson, Forrest. 1985. "Building Diagnostics." *Architectural Technology* (Winter):22–41.

Wintz, Joseph A. III, and Alan H. Yorkdale. 1983. "Brick Veneer Panel and Curtain Wall Systems—A Designer's Guide." *The Construction Specifier*, 36(12)(December):24–39.

CHAPTER 5

ACKNOWLEDGMENTS

Alcoa Building Products;

Allegheny Ludlum Steel Corporation;

Aluminum Association (AA);

American Architectural Manufacturers Association (AAMA);

American Institute of Steel Construction (AISC);

American Iron and Steel Institute (AISI);

American Society for Testing and Materials (ASTM);

American Welding Society (AWS);

Bethlehem Steel Corporation;

Ceco Corporation;

Copper Development Association (CDA);

Inland Steel Company;

ITM Ramset;

Jones and Laughlin Corporation;

Kaiser Aluminum and Chemical Corporation;

Lukens, Incorporated;

Modern Metals (magazine);

National Association of Architectural Metal Manufacturers (NAAMM);

Phillips Industries, Inc.;

Reynolds Metals Company;

Rheem Manufacturing Company;

Rigidized Steel Corporation;

Steel Deck Institute (SDI);

Steel Joist Institute (SJI);

TRW Nelson Company;

Underwriters Laboratories, Inc. (UL);

United States Steel Corporation;

Youngstown Sheet and Tube Company.

CHAPTER 5

REFERENCES

Abate, Kenneth. 1989. "Metal Coatings, Fighting the Elements with Superior Paint Systems, Part I." *Metal Architecture*, 5(6)(June):10,69.

———. 1989. "Metal Coatings, Fighting the Elements with Superior Paint Systems, Part II." *Metal Architecture*, 5(7)(July):40–41.

AIA Professional Services Division. *Masterspec*, Basic Section 05120, "Structural Steel," 8/86 ed. Washington, DC: AIA.

American Architectural Manufacturers Association (AAMA). AAMA 603.8-1985, "Voluntary Performance Requirements and Test Procedures for Pigmented Organic Coatings on Extruded Aluminum." Des Plaines, IL: AAMA.

———. AAMA 604.2-1977, "Voluntary Specifications for Residential Color Anodic Finishes." Des Plaines, IL: AAMA.

———. AAMA 605.2-1985, "Voluntary Specifications for High Performance Organic Coatings on Architectural Extrusions and Panels." Des Plaines, IL: AAMA.

———. AAMA 606.1-1976, "Voluntary Guide Specifications and Inspection Methods for Integral Color Anodic Finishes for Architectural Aluminum." Des Plaines, IL: AAMA.

———. AAMA 607.1-1977, "Voluntary Guide Specifications and Inspection Methods for Clear Anodic Finishes for Architectural Aluminum." Des Plaines, IL: AAMA.

———. AAMA 608.1-1977, "Voluntary Guide Specifications and Inspection Methods for Electrolytically Deposited Color Anodic Finishes for Architectural Aluminum." Des Plaines, IL: AAMA.

———. AAMA 609.1-1985, "Voluntary Guide Specifications and Cleaning and Maintenance of Architectural Anodized Aluminum." Des Plaines, IL: AAMA.

———. AAMA 610.1-1979, "Voluntary Guide Specifications for Cleaning and Maintenance of Painted Aluminum Extrusions and Curtain Wall Panels." Des Plaines, IL: AAMA.

———. *Care and Handling of Architectural Aluminum from Shop to Site*. Curtain Wall Manual, vol. 10 (CW-10). Des Plaines, IL: AAMA.

American Institute of Steel Construction (AISC). 1989. *Code of Steel Construction*. Chicago, IL: AISC. [Also contained in AISC *Manual of Steel Construction*.]

———. 1989 *Manual of Standard Practice for Steel Buildings and Bridges*. Chicago, IL: AISC.

———. 1989. *Specifications for the Design, Fabrication, and Erection of Structural Steel for Buildings, and Its Commentary*. Chicago, IL: AISC. [Also contained in AISC *Manual of Steel Construction*.]

———. 1989. *Specifications for Structural Joints using ASTM A 325 or A 490 Bolts*. Chicago, IL: AISC. [Also contained in AISC *Manual of Steel Construction*.]

American Iron and Steel Institute (AISI). 1972. *Stainless Steel: Suggested Practices for Roofing, Flashing, Copings, Fascias, Gravel Stops, and Drainage*. Washington, DC: AISI.

———. 1975. *Stainless Steel Fasteners, Suggested Source List*. Washington, DC: AISI.

———. 1976. *Stainless Steel Fasteners, A Systematic Approach to their Selection*. Washington, DC: AISI.

———. 1977. *Design Guidelines for the Selection and Use of Stainless Steel*. Washington, DC: AISI.

———. 1979. *Welding of Stainless Steel and Other Joining Methods*. Washington, DC: AISI.

———. *Specification for the Design of Cold-formed Steel Structural Members*. Washington, DC: AISI.

American Society for Metals. 1982. *Properties and Selection of Metals*. Vol. I of *Metals Handbook*. 9th ed. Metals Park, OH: American Society for Metals.

———. 1982. *Nondestructive Inspection and Quality Control*. Vol. II of *Metals Handbook*. 9th ed. Metals Park, OH: American Society for Metals.

————. 1982. *Surface Cleaning, Finishing, and Coating.* Vol. V of *Metals Handbook.* 9th ed. Metals Park, OH: American Society for Metals.

American Society for Testing and Materials (ASTM) Standards:

A 6, "Standard Specifications for General Requirements for Rolled Steel Plates, Shapes, Sheet Piling, and Bars for Structural Use." Philadelphia, PA: ASTM.

A 27, "Standard Specification for Carbon-Steel Castings for General Application." Philadelphia, PA: ASTM.

A 36, "Standard Specification for Structural Steel." Philadelphia, PA: ASTM.

A 47, "Specification for Ferritic Malleable Iron Castings." Philadelphia, PA: ASTM.

A 48, "Specification for Gray Iron Castings." Philadelphia, PA: ASTM.

A 53, "Specification for Pipe, Steel, Black and Hot-Dipped, Zinc-Coated Welded and Seamless." Philadelphia, PA: ASTM.

A 82, "Specification for Steel Wire for Concrete Reinforcement." Philadelphia, PA: ASTM.

A 108, "Standard Specification for Steel Bars, Carbon, Cold-Finished, Standard Quality." Philadelphia, PA: ASTM.

A 120, "Specification for Pipe, Steel, Black and Hot-Dipped, Zinc-Coated (Galvanized) Welded and Seamless, for Ordinary Uses." Philadelphia, PA: ASTM.

A 121, "Specification for Zinc-Coated (Galvanized) Steel Barbed Wire." Philadelphia, PA: ASTM.

A 123, "Specifications for Zinc (Hot-Galvanized) Coatings on Products Fabricated from Rolled, Pressed, and Forged Steel Shapes, Plates, Bars, and Strip." Philadelphia, PA: ASTM.

A 153, "Specification for Zinc Coating (Hot-Dip) on Iron and Steel Hardware." Philadelphia, PA: ASTM

A 165, "Standard Specification for Electrodeposited Coatings of Cadmium on Steel." Philadelphia, PA: ASTM.

A 167, "Specification for Stainless and Heat-Resisting Chromium-Nickel Steel Plate, Sheet, and Strip." Philadelphia, PA: ASTM.

A 185, "Specification for Steel Welded Wire, Fabric, Plain, for Concrete Reinforcement." Philadelphia, PA: ASTM.

A 242, "Specification for High-Strength Low-Alloy Structural Steel." Philadelphia, PA: ASTM.

A 269, "Specification for Seamless and Welded Austenitic Stainless Steel Tubing for General Service." Philadelphia, PA: ASTM.

A 276, "Specification for Stainless and Heat-Resisting Steel Bars and Shapes." Philadelphia, PA: ASTM.

A 283, "Standard Specification for Low and Intermediate Tensile Strength Carbon Steel Plates, Shapes and Bars." Philadelphia, PA: ASTM.

ASTM A 307, "Specification for Carbon Steel Externally Threaded Standard Fasteners." Philadelphia, PA: ASTM.

A 312, "Specification for Seamless and Welded Austenitic Stainless Steel Pipe." Philadelphia, PA: ASTM.

A 325, "Specification for High-Strength Bolts for Structural Steel Joints." Philadelphia, PA: ASTM.

A 366, "Specification for Steel, Sheet, Carbon, Cold-Rolled, Commercial Quality." Philadelphia, PA: ASTM.

A 424, "Specification for Steel Sheet for Porcelain Enameling." Philadelphia, PA: ASTM.

A 446, "Specifications for Sheet Steel, Zinc-Coated (Galvanized) by the Hot-Dip Process, Structural (Physical) Quality." Philadelphia, PA: ASTM.

A 463, "Standard Specification for Sheet Steel, Cold-Rolled Aluminum-Coated Type 1." Philadelphia, PA: ASTM.

A 490, "Specification for Heat-Treated Steel Structural Bolts, 150 ksi (1035 MPa) Tensile Strength." Philadelphia, PA: ASTM.

A 497, "Specification for Welded Deformed Steel Wire Fabric for Concrete Reinforcement." Philadelphia, PA: ASTM.

A 500, "Specification for Cold-Formed Welded and Seamless Carbon Steel Structural Tubing in Rounds and Shapes." Philadelphia, PA: ASTM.

A 501, "Specification for Hot-Formed Welded and Seamless Carbon Steel Structural Tubing." Philadelphia, PA: ASTM.

A 510, "Standard Specification for General Requirements for Wire Rods and Coarse Round Wire, Carbon Steel." Philadelphia, PA: ASTM.

A 525, "Specifications for General Requirements for Steel Sheet Zinc-Coated (Galvanized) by the Hot-Dip Process." Philadelphia, PA: ASTM.

A 526, "Specifications for Steel Sheet Zinc-Coated (Galvanized) by the Hot-Dip Process, Commercial Quality." Philadelphia, PA: ASTM.

A 527, "Specifications for Steel Sheet Zinc-Coated (Galvanized) by the Hot-Dip Process, Lock-Forming Quality." Philadelphia, PA: ASTM.

A 528, "Specifications for Steel Sheet Zinc-Coated (Galvanized) by the Hot-Dip Process, Drawing Quality." Philadelphia, PA: ASTM.

A 536, "Standard Specification for Ductile Iron Castings." Philadelphia, PA: ASTM.

A 554, "Specification for Welded Stainless Steel Mechanical Tubing." Philadelphia, PA: ASTM.

A 568, "Standard Specification for General Requirements for Steel, Carbon and High Strength Low-Alloy Hot-Rolled Sheet and Cold-Rolled Sheet." Philadelphia, PA: ASTM.

A 569, "Specification for Steel, Carbon (0.15 Maximum, Percent), Hot-Rolled Sheet and Strip, Commercial Quality." Philadelphia, PA: ASTM.

A 570, "Specification for Hot-Rolled Carbon Steel Sheet and Strip, Structural Quality." Philadelphia, PA: ASTM.

A 572, "Specification for High-Strength Low-Alloy Columbium-Vanadium Steels of Structural Quality." Philadelphia, PA: ASTM.

A 575, "Standard Specifications for Steel Bars, Carbon, Merchant Quality, M-Grades." Philadelphia, PA: ASTM.

A 588, "Specification for High-strength Low-alloy Structural Steel with 50 ksi (345 MPa) Minimum Yield Point to 4 in. Thick. Philadelphia, PA: ASTM.

A 591, "Specifications for Steel Sheet, Cold-Rolled Electrolytic Zinc-Coated." Philadelphia, PA: ASTM.

A 606, "Specifications for Steel Sheet and Strip, Hot-Rolled, High Strength, Low-Alloy, with Improved Atmospheric Resistance." Philadelphia, PA: ASTM.

A 607, "Standard Specifications for Steel, Sheet and Strip, High-Strength, Low-Alloy Columbium or Vanadium, or both, Hot-Rolled and Cold-Rolled." Philadelphia, PA: ASTM.

A 611, "Specification for Steel, Cold-Rolled Sheet, Carbon, Structural." Philadelphia, PA: ASTM.

A 615, "Specification for Deformed and Plain Billet-Steel Bars for Concrete Reinforcement." Philadelphia, PA: ASTM.

A 616, "Specification for Deformed and Plain Billet-Steel Bars for Concrete Reinforcement (Metric)." Philadelphia, PA: ASTM

A 617, "Specification for Axle-Steel Deformed and Plain Bars for Concrete Reinforcement." Philadelphia, PA: ASTM

A 641, "Specifications for Zinc-Coated (Galvanized) Carbon Steel Wire." Philadelphia, PA: ASTM.

A 642, "Standard Specifications for Steel Sheet, Zinc-Coated (Galvanized) by the Hot-Dip Process, Drawing Quality, Special Killed." Philadelphia, PA: ASTM.

A 663, "Standard Specification for Steel Bars, Carbon, Merchant Quality, Mechanical Qualities." Philadelphia, PA: ASTM.

A 675, "Standard Specification for Steel Bars, Carbon, Hot-Wrought, Special Quality, Mechanical Properties." Philadelphia, PA: ASTM.

A 715, "Specification for Steel Sheet and Strip, High-Strength, Hot-Rolled, and Steel Sheet, Cold-Rolled, High-Strength, Low-Alloy, with Improved Formability." Philadelphia, PA: ASTM.

A 743, "Standard Specifications for Castings, Iron-Chromium-Nickel, and Nickel-Base, Corrosion-Resistant, for General Application." Philadelphia, PA: ASTM.

A 744, "Standard Specifications for Castings, Iron-Chromium-Nickel, and Nickel-Base, Corrosion-Resistant, for Severe Service." Philadelphia, PA: ASTM.

A 767, "Specification for Zinc-Coated (Galvanized) Bars for Concrete Reinforcement." Philadelphia, PA: ASTM

A 780, "Practice for Repair of Damaged Hot-Dip Galvanized Coatings." Philadelphia, PA: ASTM.

A 786, "Standard Specifications for Rolled Steel Floor Plates." Philadelphia, PA: ASTM.

A 792, "Standard Specification for Steel Sheet, Aluminum-Zinc Alloy-Coated by the Hot-Dip Process." Philadelphia, PA: ASTM.

A 875, "Standard Specification for Steel Sheet, Zinc 5 Percent, Aluminum-Mische Metal Alloy, Coated by the Hot-Dip Process." Philadelphia, PA: ASTM.

B 26, "Specification for Aluminum-Alloy Sand Castings." Philadelphia, PA: ASTM.

B 32, "Specification for Solder Metal." Philadelphia, PA: ASTM.

B 108, "Specification for Aluminum-Alloy Permanent Mold Castings." Philadelphia, PA: ASTM.

B 209, "Specification for Aluminum and Aluminum Alloy Sheet and Plate." Philadelphia, PA: ASTM.

B 210, "Specification for Aluminum-Alloy Drawn Seamless Tubes." Philadelphia, PA: ASTM.

B 211, "Specification for Aluminum-Alloy Bars, Rods, and Wire." Philadelphia, PA: ASTM.

B 221, "Specification for Aluminum-Alloy Extruded Bars, Rods, Wire, Shapes, and Tubes." Philadelphia, PA: ASTM.

B 247, "Specification for Aluminum-Alloy Die and Hand Forgings." Philadelphia, PA: ASTM.

B 253, "Recommended Practice for Preparation of and Electroplating on Aluminum Alloys by the Zincate Process." Philadelphia, PA: ASTM.

B 254, "Recommended Practice for Preparation and Electroplating on Stainless Steel." Philadelphia, PA: ASTM.

B 320, "Recommended Practice for Preparation of Iron Castings for Electroplating." Philadelphia, PA: ASTM.

B 429, "Specification for Aluminum-Alloy Extruded Structural Pipe and Tube." Philadelphia, PA: ASTM.

B 483, "Specification for Aluminum and Aluminum-Alloy Extruded Drawn Tubes for General Purpose Applications." Philadelphia, PA: ASTM.

B 650, "Specification for Electrode-posited Engineering Chromium Coatings on Ferrous Substrates." Philadelphia, PA: ASTM.

C 282, "Test Method for Acid Resistance of Porcelain Enamels (Citric Acid Spot Test)." Philadelphia, PA: ASTM.

C 283, "Test Method for Resistance of Porcelain Enamel Utensils to Boiling Acid." Philadelphia, PA: ASTM.

C 286, "Definition of Terms Relating to Porcelain Enamel and Ceramic-Metal Systems." Philadelphia, PA: ASTM.

C 313, "Test Method for Adherence of Porcelain Enamel and Ceramic Coatings to Sheet Metal." Philadelphia, PA: ASTM.

C 346, "Test Method for 45-deg Specular Gloss of Ceramic Materials." Philadelphia, PA: ASTM.

C 448, "Test Method for Abrasion Resistance of Porcelain Enamel." Philadelphia, PA: ASTM.

C 538, "Test Method for Color Retention of Red, Orange, and Yellow Porcelain Enamels." Philadelphia, PA: ASTM.

C 540, "Test Method for Image Gloss of Porcelain Enamel Surfaces." Philadelphia, PA: ASTM.

C 645, "Specifications for Non-Load (Axial) Bearing Steel Studs, Runners (Track), and Rigid Furring Channels for Screw Application of Gypsum Board." Philadelphia, PA: ASTM.

C 703, "Test Method for Spalling Resistance of Porcelain Enameled Aluminum." Philadelphia, PA: ASTM.

C 1007, "Installation of Load-Bearing (Transverse and Axial) Steel Studs and Accessories." Philadelphia, PA: ASTM.

D 1730, "Recommended Practices for Preparation of Aluminum and Aluminum-Alloy Surfaces for Painting." Philadelphia, PA: ASTM.

D 1731, "Recommended Practices for Preparation of Hot-Dip Aluminum Surfaces for Painting." Philadelphia, PA: ASTM.

D 2092, "Recommended Practices for Preparation of Zinc-Coated Steel Surfaces for Painting." Philadelphia, PA: ASTM.

D 3794, "Practice for Testing Coil Coatings." Philadelphia, PA: ASTM.

E 94, "Recommended Practice for Radiographic Testing." Philadelphia, PA: ASTM.

E 142, "Method of Controlling Quality of Radiographic Testing." Philadelphia, PA: ASTM.

E 165, "Practice for Liquid Penetrant Inspection Method." Philadelphia, PA: ASTM.

E 527, "Practice for Numbering Metals and Alloys (UNS)." Philadelphia, PA: ASTM.

E 709, "Practice for Magnetic Particle Examination." Philadelphia, PA: ASTM.

American Welding Society (AWS). AWS D1.1, "Structural Welding Code—Steel." Miami, FL: AWS.

———. AWS D1.2, "Structural Welding Code—Aluminum." Miami, FL: AWS.

Appleman, Bernard R. 1989. "Coatings for Steel Structures." *The Construction Specifier*, 42(3)(March):88–93.

Architectural Graphic Standards. See Ramsey/Sleeper.

Bocchi, Greg. 1986. "Powder Coatings, A New Technology Takes Off." *The Construction Specifier*, 39(9)(September):102–105.

———. 1988. "Powder Coatings Making Inroads in Metal Construction Market." *Metal Architecture*, 4(4)(April):8-9.

Building Design and Construction. 1988. "Focus on Metals in Building Construction." *Building Design and Construction*, 29(6)(June):67.

Construction Specifications Institute 1988. *Precoated Metal Building Panels.* CSI Monograph 07M411. Alexandria, VA: The Construction Specifications Institute.

Copper Development Association (CDA). *Copper, Brass, Bronze Design Handbook, Architectural Applications.* Greenwich, CT: CDA.

———. *Copper, Brass, Bronze Design Handbook, Sheet Copper Applications.* Greenwich, CT: CDA.

Council of American Building Officials (CABO). 1983. *One- and Two-Family Building Code.* Falls Church, VA: CABO.

Dean, Sheldon W., and T. S. Lee, eds. 1988. *Degradation of Metals in the Atmosphere.* Philadelphia, PA: ASTM.

Factory Mutual Research Organization (FM). *Factory Mutual System Approval Guide.* Norwood, MA: FM.

———. *Loss Prevention Data Sheet No. 1–28.* Norwood, MA: FM.

Howell, J. Scott. 1987. "Architectural Cast Iron: Design and Restoration." *The Construction Specifier*, 40(7)(July):70–74.

Johnson, Stephen. 1988. "Improvement in Domestic Steel Quality Translating into Better Metal Paneling." *Metal Architecture*, 4(9)(September):5,85.

Koller, Alice. 1981. "Hot Dip Galvanizing: How and When to Use It." *The Construction Specifier*, 34(8)(September):47–51.

LaQue, F. E., and H. R. Copson, eds. 1963. *Corrosion Resistance of Metals and Alloys.* 2d ed. New York: Van Nostrand Reinhold.

McDonald, Timothy B. 1987. "Technical Tips: Coatings That Protect Against the Corrosion of Steel." *Architecture* (July):101–102.

———. 1989. "Specifying Welding Details." *Architecture* (January):113–114.

Metal Architecture. 1987. "Building Repainting System Duplicates Coil Coatings in Appearance, Life Expectancy." *Metal Architecture*, 3(11)(November):24.

————. 1987. "Guide to Metal Wall and Roof Panels." *Metal Architecture*, 3(12)(December): 20, 22–23, 28–31, 38.

————. 1988. "Guide to Pre-Insulated Panels." *Metal Architecture*, 4(4)(April):40–41, 43–46.

————. 1988. "1988 Guide to Architectural Coil Coatings." *Metal Architecture*, 4(11) (November):22–23.

————. 1988. "Selecting Fasteners: Manufacturer's Advice." *Metal Architecture*, 4(12) (December):26–32.

Metal Lath/Steel Framing Association. 1985. *Lightweight Steel Framing Systems Manual*. 2d ed. Chicago, IL: Metal Lath/Steel Framing Association.

————. 1986. *Technical Information File*. Chicago, IL: Metal Lath/Steel Framing Association.

————. *Technical Bulletin No. 18: Fire Rated Metal Lath/Steel Stud Exterior*. Chicago, IL: Metal Lath/Steel Framing Association.

Modern Metals. 1989. "Presses on Roll Form Lines: How to Choose the Best." *Modern Metals*, 45(1)(February):10–20.

————. 1989. "Finishing Forum: Coil Anodized Sheet Survives with Flying Colors." *Modern Metals*, 45(1)(February):22–28.

Moit, Dan. 1988. "Coatings for Metals: Preplan the Selection." *Metal Architecture*, 4(9)(September):8.

Munger, Charles G. 1984. *Corrosion Prevention by Protective Coatings*. Houston, TX: National Association of Corrosion Engineers.

National Association of Architectural Metal Manufacturers (NAAMM). 1988. *Metal Finishes Manual for Architectural and Metal Products*. Chicago, IL: NAAMM.

Petersen, Maurice R. 1984. "Finishes on Metals: A View From the Field—Part 1." *The Construction Specifier*, 37(12)(December):36–39.

————. 1985. "Finishes on Metals: A View From the Field—Part 2." *The Construction Specifier*, 38(1)(January):70–73.

Peterson, Harold L. 1968. "Conservation of Metals." *Historic News*, 23(2)(February).

Porcelain Enamel Institute (PEI). 1970. *Guide to Designing with Architectural Porcelain Enamel on Steel*. Washington, DC: PEI.

————. PEI: ALS-105(69), "Recommended Specification for Architectural Porcelain Enamel on Aluminum for Exterior Use." Washington, DC: PEI.

————. PEI: S-100(65), "Specification for Architectural Porcelain Enamel on Steel for Exterior Use." Washington, DC: PEI.

————. *The Weatherability of Porcelain Enamel*. Washington, DC: PEI.

————. *Color Guide for Architectural Porcelain Enamel*. Washington, DC: PEI.

Ramsey/Sleeper, The AIA Committee on Architectural Graphic Standards. 1988. *Architectural Graphic Standards*. 8th ed. New York: Wiley.

Sheet Metal and Air Conditioning Contractors National Association (SMACNA), Inc. 1987. *Architectural Sheet Metal Manual*. Vienna, VA: SMACNA.

Sivinski, Valerie A. 1986. "Preserving Historic Materials: Ferrous Metals." *Architecture* (November):108–109.

Steel Deck Institute (SDI). *Design Manual for Composite Decks, Form Decks and Roof Decks*. Canton, OH: SDI.

Steel Joists Institute (SJI). 1990. *Standard Specifications, Load Tables, and Weight Tables for Steel Joists and Joists Girders*. Myrtle Beach, SC: SJI.

————. *Handling and Erection of Steel Joists and Joists Girders*. SJI Technical Digest No. 9. Myrtle Beach, SC: SJI.

Steel Structures Painting Council. 1983. *Good Paint Practice*. Vol. 1 of *Steel Structures Painting Manual* 2d ed. Pittsburgh, PA: Steel Structures Painting Council.

————. 1983. *Systems and Specifications*, Vol. 2 of *Steel Structures Painting Manual*. 2d ed. Pittsburgh, PA: Steel Structures Painting Council.

Underwriters Laboratories, Inc. (UL). *Fire Resistance Directory*. Northbrook, IL: UL.

————. *Electrical Construction Materials Directory*. Northbrook, IL: UL.

U.S. Department of the Army. 1982. Corps of Engineers Guide Specifications, Military Construction, Section 05120, "Structural Steel." Washington, DC: Office of the Chief of Engineers, Department of the Army.

————. 1982. Corps of Engineers Guide Specifications, Military Construction, Section 05141, "Welding, Structural." Washington, DC: Office of the Chief of Engineers, Department of the Army.

U.S. Department of the Navy, Naval Facilities Engineering Command. 1985. Guide Specifications Section 05120, "Structural Steel." Washington, DC: Department of the Navy.

Weaver, Martin E. 1989. "Fighting Rust, Part I: A Backgrounder." *The Construction Specifier*, 42(5)(May):143–145.

————. 1988. "Acid Rain." *The Construction Specifier*, 41(7)(July):54–62.

————. 1989. "Fighting Rust, Part II: Remedies." *The Construction Specifier*, 42(6)(June):129–130.

CHAPTER 6

ACKNOWLEDGMENTS

Addison Wesley Publishing Company;

AIA and ACSA Council on Architectural Research;

Alcoa Building Products;

American Forest and Paper Association (AFPA);

American National Standards Institute (ANSI);

American Plywood Association (APA);

American Society for Testing and Materials (ASTM);

American Skylights, Inc.;

American Wood Preservers Association (AWPA);

American Wood Preservers Bureau (AWPB);

American Wood Preservers Institute (AWPI);

Armstrong World Industries;

California Redwood Association (CRA);

Dow USA;

Fine Hardwoods—American Walnut Association;

Formica Corporation;

Gypsum Association;

Hardwood Plywood and Veneer Association (HPVA);

House and Home (magazine);

Kawneer Company;

National Association of Home Builders (NAHB), Research Foundation;

National Hardwood Lumber Association (NHLA);

National Institute of Building Sciences (NIBS);

National Institute of Standards and Technology (NIST) of the U.S. Department of Commerce;

National Terrazzo and Mosaic Association (NTMA);

National Wood Window and Door Association (NWWDA);

Owens-Corning Fiberglas Corporation;

Plastic Pipe and Fittings Association;

Rohm and Haas Company;

Redwood Inspection Service (RIS);

Sika Corporation;

Society of the Plastics Industry;

Southern Forest Products Association (SFPA);

Southern Pine Inspection Bureau (SPIB);

Thermo-flex;

United States Gypsum Corporation (USG);

United States Plywood Corporation;

United States Steel Corporation;

U.S. Department of Agriculture (USDA);

U.S. Department of Housing and Urban Development (HUD);

U.S. Forest Products Laboratory (FPL);

West Coast Lumber Inspection Bureau (WCLIB);

Western Electric Company;

Western Wood Products Association (WWPA);

Weyerhaeuser Company;

Wood Truss Council of America.

CHAPTER 6

REFERENCES

AIA Professional Services Division. *Masterspec*, Basic Sections:

06100, "Rough Carpentry." 8/90 ed. Washington, DC: AIA.

06310, "Heavy Timber Construction." 5/92 ed. Washington, DC: AIA.

06170, "Structural Glue Laminated Units." 5/92 ed. Washington, DC: AIA.

06192, "Prefabricated-Metal-Plate-Connected Wood Trusses." 8/92 ed. Washington, DC: AIA.

06200, "Finish Carpentry." 5/91 ed. Washington, DC: AIA.

06401, "Exterior Architectural Woodwork." 2/89 ed. Washington, DC: AIA.

06402, "Interior Architectural Woodwork." 2/89 ed. Washington, DC: AIA.

American Forest and Paper Association (AFPA). *Design Values for Joists and Rafters*. Washington, DC: AFPA.

———. *Design Values for Wood Construction*. Washington, DC: AFPA.

———. *National Design Specifications, Wood Construction*. Washington, DC: AFPA.

———. *Design Values for Wood Construction*. NDS Supplement. Washington, DC: AFPA.

———. *Manual for House Framing*. Washington, DC: AFPA.

———. *Permanent Wood Foundation System, Design, Fabrication, Installation Manual*. Washington, DC: AFPA.

———. *Permanent Wood Foundation System*. Technical Report No. 7. Washington, DC: AFPA.

———. *Span Tables for Joists and Rafters*. Washington, DC: AFPA.

———. *Wood Structural Design Data*. Washington, DC: AFPA.

———. WCD No. 1. *Manual for Wood Frame Construction*. Washington, DC: AFPA.

———. WCD No. 4. *Plank and Beam Framing for Residential Buildings*. Washington, DC: AFPA.

———. WCD No. 6. *Design of Wood Structures for Permanence*. Washington, DC: AFPA.

American Institute of Timber Construction (AITC). *Timber Construction Standards*. Englewood, CO: AITC.

———. *Timber Construction Manual*. Englewood, CO: AITC.

American National Standards Institute (ANSI) Standards:

A58.1-1982, "Minimum Design Loads, Buildings and Other Structures." New York: ANSI.

A117.1-1986, "American National Standards for Buildings and Facilities—Providing Accessibility and Usability for Physically Handicapped People." New York: ANSI.

A208.1-1979, "Particleboard, Mat-Formed Wood." New York: ANSI.

American Plywood Association (APA). A310, "Nonresidential Roof Systems." Tacoma, WA: APA.

———. A400, "Permanent Wood Foundation." Tacoma, WA: APA.

———. B360, "HDO/MDO Plywood." Tacoma, WA: APA.

———. E30, "Residential and Commercial." Tacoma, WA: APA.

———. E445, "Performance Standards and Policies for Structural-Use Panels." Tacoma, WA: APA.

———. F405, "Performance Rated Panels." Tacoma, WA: APA.

———. J20, "Grades and Specifications." Tacoma, WA: APA.

———. K435, "Sanded Plywood." Tacoma, WA: APA.

———. L335, "Installation and Preparation of Plywood Underlayment for Thin Resilient (Non-Textile) Flooring." Tacoma, WA: APA.

———. L350, "Diaphragms." Tacoma, WA: APA.

———. L405, "Ventilated Crawl Space Foundations." Tacoma, WA: APA.

———. M301, "Plen-wood Systems." Tacoma, WA: APA.

———. N335, "Builder Tips: Proper Installation of APA Rated Sheathing for Roof Applications." Tacoma, WA: APA.

———. Q220, "Preservative-treated Plywood." Tacoma, WA: APA.

———. Q300, "Builder Tips: Steps to Construct a Solid, Squeak-free Floor System." Tacoma, WA: APA.

———. S811, "PDS Supplement #1: Design and Fabrication of Plywood Curved Panels." Tacoma, WA: APA.

———. S812, "PDS Supplement #2: Design and Fabrication of Plywood-lumber Beams." Tacoma, WA: APA.

———. U813, "PDS Supplement #3: Design and Fabrication of Plywood Stressed-skin Panels." Tacoma, WA: APA.

———. U814, "PDS Supplement #4: Design and Fabrication of Plywood Sandwich Panels." Tacoma, WA: Tacoma, WA: APA.

———. X461, "House Building Basics." Tacoma, WA: APA.

———. X505, "Panel Handbook and Grade Glossary." Tacoma, WA: APA.

American Society for Testing and Materials (ASTM) Standards:

D 1600, "Abbreviations, Acronyms, and Codes for Terms Relating to Plastics." Philadelphia, PA: ASTM

D 1784, "Specification for Rigid Poly (Vinyl Chloride) (PVC) Compounds and Chlorinated Poly(Vinyl Chloride) (CPVC) Compounds." Philadelphia, PA: ASTM.

D 2000, "Classification System for Rubber Products in Automotive Applications." Philadelphia, PA: ASTM.

D 2287, "Specification for Nonrigid Vinyl Chloride Polymer and Copolymer Molding and Extrusion Compounds." Philadelphia, PA: ASTM.

E 84, "Test Method for Surface Burning Characteristics of Building Materials." Philadelphia, PA: ASTM.

E 96, "Test Method for Water Vapor Transmission of Materials." Philadelphia, PA: ASTM.

E 119, "Fire Tests for Building Construction and Materials." Philadelphia, PA: ASTM.

E 136, "Test Method for Behavior of Materials in a Vertical Tube Furnace at 750 deg. C." Philadelphia, PA: ASTM.

American Wood Preservers Association (AWPA). *Book of Standards*. Springfield, VA: AWPA.

American Wood Preservers Bureau (AWPB). *Quality Control and Inspection Procedures*. Newnan, GA: AWPB.

———. CP, "Pressure Treated Round Construction Poles." Newnan, GA: AWPB.

———. FDN, "Softwood Lumber, Timber and Plywood Pressure Treated with Waterborne Preservatives for Ground Contact Use in Residential and Light Commercial Foundations." Newnan, GA: AWPB.

———. LP-2, "Softwood Lumber, Timber and Plywood Pressure Treated with Waterborne Preservatives for Above Ground Use." Newnan, GA: AWPB.

———. LP-22, "Softwood Lumber, Timber and Plywood Pressure Treated with Waterborne Preservatives for Ground Contact Use." Newnan, GA: AWPB.

———. LP-3, "Softwood Lumber, Timber and Plywood Pressure Treated with Light Hydrocarbon Solvent-penta Solution for Above Ground Use." Newnan, GA: AWPB.

———. LP-33, "Softwood Lumber, Timber and Plywood Pressure Treated with Light Hydrocarbon Solvent-penta Solution for Ground Contact Use." Newnan, GA: AWPB.

———. LP-4, "Softwood Lumber, Timber and Plywood Pressure Treated with Volatile Hydrocarbon Solvent-penta Solution for Above Ground Use." Newnan, GA: AWPB.

———. LP-44, "Softwood Lumber, Timber and Plywood Pressure Treated with Volatile Hydrocarbon Solvent-penta Solution for Ground Contact Use." Newnan, GA: AWPB.

———. LP-5, "Softwood Lumber, Timber and Plywood Pressure Treated with Creosote or Creosote Coal-tar Solutions for Above Ground Use." Newnan, GA: AWPB.

———. LP-55, "Softwood Lumber, Timber and Plywood Pressure Treated with Creosote or Creosote Coal-tar Solutions for Ground Contact Use." Newnan, GA: AWPB.

———. LP-6, "Softwood Lumber and Timber Pressure Treated with Copper Naphthenate for Above Ground Use." Newnan, GA: AWPB.

———. LP-66, "Softwood Lumber and Timber Pressure Treated with Copper Naphthenate for Ground Contact Use." Newnan, GA: AWPB.

———. LP-7, "Softwood Lumber, Timber and Plywood Pressure Treated with Heavy Hydrocarbon Solvent-penta Solution for Above Ground Use." Newnan, GA: AWPB.

———. LP-77, "Softwood Lumber, Timber and Plywood Pressure Treated with Heavy Hydrocarbon Solvent-penta Solution for Ground Contact Use." Newnan, GA: AWPB.

———. QCIP, "Pressure treated Lumber, Timbers, Plywood, Construction Poles, Piles and Special Purpose Use Products." Newnan, GA: AWPB.

Architectural Graphic Standards. See Ramsey/Sleeper.

Architectural Woodwork Institute (AWI). 1988. *Architectural Woodwork Quality Standards, Guide Specifications and Quality Certification Program.* 5th ed. Arlington, VA: AWI.

Council of American Building Officials (CABO). 1983. *One and Two Family Dwelling Code.* Falls Church, VA: CABO.

Forest Products Laboratory. Handbook No. 72. *Wood Handbook: Wood as an Engineering Material.* Washington, DC: United States Department of Agriculture.

———. Handbook No. 73. *Wood-frame House Construction.* Washington, DC: United States Department of Agriculture.

Gypsum Association (GA). GA-252, "Fire Resistant Gypsum Sheathing." Washington, DC: Gypsum Association.

Hardwood Plywood and Veneer Association (HPVA). 1993. HPVA Interim Standard HP-1, "Interim Voluntary Standard for Hardwood and Decorative Plywood." Pittsburg, PA.

MacDonald, John E. 1978. "The Effects of Light on Wood and Wood Finishing Systems." *The Construction Specifier,* 31(3) (March):23–26.

McGraw-Hill. 1985. *Modern Plastics Encyclopedia.* New York: McGraw-Hill.

Montella, Ralph. 1985. *Plastics in Architecture: A Guide to Acrylic and Polycarbonate.* New York: Marcel Dekker.

National Electrical Manufacturers Association. NEMA LD 3-1985, "High Pressure Decorative Laminates." Washington, DC: NEMA.

National Institute of Building Sciences (NIBS). Catalog No. 40005-6, "Guidelines for Protection of Wood Against Decay and Termite Attack." Washington DC: NIBS.

National Wood Window and Door Association (NWWDA). NWWDA I.S.4-1981, "Industry Standard for Water-repellent Preservative Non-pressure Treatment for Millwork." Park Ridge, IL: NWWDA.

Phillips, Morgan, and Judith Selwyn. 1978. *Epoxies for Wood Repairs in Historic Buildings.* Washington, DC: Technical Preservation Service, U.S. Department of the Interior.

Ramsey/Sleeper, The AIA Committee on Architectural Graphic Standards. 1988. *Architectural Graphic Standards.* 8th ed. New York: Wiley.

Southern Pine Inspection Bureau (SPIB). *Standard Grading Rules for Southern Pine Lumber.* Pensacola, FL: SPIB.

Truss Plate Institute (TPI). *National Design Specifications for Metal Plate Connected Wood Trusses.* Madison, WI: TPI.

Underground Space Center. 1988. *Building Foundation Design Handbook.* Minneapolis, MN: Underground Space Center.

U.S. Department of Agriculture. G 64, "Subterranean Termites: Their Prevention and Control in Buildings." Washington, DC: U.S. Department of Agriculture.

U.S. Department of Commerce. PS 1, "U.S. Product Standard for Construction and Industrial Plywood." Washington, DC: U.S. Department of Commerce.

———. PS 20, "American Softwood Lumber Standard." Washington, DC: U.S. Department of Commerce.

West Coast Lumber Inspection Bureau (WCLIB). *Standard Grading Rules for West Coast Lumber.* Portland, OR: WCLIB.

Western Wood Products Association (WWPA). *Grading Rules for Western Lumber.* Portland, OR: WWPA.

———. *Grade Stamp Manual.* Portland, OR: WWPA.

———. A-2, "Lumber Specifications Information." Portland, OR: WWPA.

———. *Western Woods Use Book.* Portland, OR: WWPA.

———. *Wood Frame Design.* Portland, OR: WWPA.

CHAPTER 7
ACKNOWLEDGMENTS

Airline Products Company;

Alcoa Building Products;

American Architectural Manufacturers Association (AAMA);

American Board Products Association;

American National Standards Institute (ANSI);

American Society of Heating, Refrigerating and Air Conditioning Engineers (ASHRAE);

American Society for Testing and Materials (ASTM);

Asphalt Institute;

Asphalt Roofing Manufacturers Association;

Brick Institute of America (BIA);

Carlisle Roofing Systems;

Celotex Corporation;

Cedar Shake and Shingle Bureau;

Coastal Atlantic Associates, Inc.;

Concrete Block Insulating Systems, Inc.;

Copper Development Association (CDA), Inc.;

Dow Chemical Corporation;

Firestone Building Products Company;

Manville Corporation;

E/S Products, Inc.;

Georgia Pacific Corporation;

Goodyear Tire and Rubber Company;

House and Home (magazine);

Independent Nail Corporation;

Koppers Company, Inc.;

Lead Industries Association, Inc.;

W. H. Maze Company;

Metal Sales Manufacturing Company;

Mineral Fiber Products Bureau;

Mineral Insulation Manufacturers Association;

National Association of Home Builders (NAHB) Research Foundation;

National Institute of Building Sciences (NIBS);

National Institute of Standards and Technology (NIST) of the U.S. Department of Commerce;

National Mineral Wool Insulation Association;

National Roofing Contractors Association (NRCA);

National Slate Association (NSA);

National Tile Roofing Manufacturers Association (NTRMA);

Owens-Corning Fiberglas Corporation;

Perlite Institute;

Portland Cement Association (PCA);

Sealant Engineering and Associated Lines (SEAL), Inc.;

Sealant, Waterproofing, and Restoration Institute;

Single Ply Roofing Institute (SPRI);

Solar Group;

Underwriters Laboratories, Inc.;

U.S. Department of Housing and Urban Development (HUD);

U.S. Forest Products Laboratory;

U.S. Weather Service;

Vinyl Siding Institute (VSI);

Wolverine Technologies.

CHAPTER 7
REFERENCES

AIA Professional Services Division. *Masterspec,* Basic Sections:

07110, "Sheet Membrane Waterproofing." 8/91 ed. Washington, DC: AIA.

07120, "Fluid-Applied Waterproofing." 2/90 ed. Washington, DC: AIA.

07160, "Bituminous Dampproofing," 5/90 ed. Washington, DC: AIA.

07180, "Water Repellents." 11/91 ed. Washington, DC: AIA.

07210, "Building Insulation." 11/90 ed. Washington, DC: AIA.

07241, "Exterior Insulation and Finish System—Class PB." 8/89 ed. Washington, DC: AIA.

07242, "Exterior Insulation and Finish System—Class PM." 8/89 ed. Washington, DC: AIA.

07311, "Asphalt Shingles." 2/91 ed. Washington, DC: AIA.

07317, "Wood Shingles and Shakes." 11/90 ed. Washington, DC: AIA.

07410, "Manufactured Roof and Wall Panels." 2/90 ed. Washington, DC: AIA.

07460, "Siding." 2/91 ed. Washington, DC: AIA.

07511, "Built-Up Asphalt Roofing." 8/89 ed. Washington, DC: AIA.

07512, "Built-Up Coal Tar Roofing." 8/89 ed. Washington, DC: AIA.

07530, "Single Ply Membrane Roofing." 2/89 ed. Washington, DC: AIA.

07600, "Flashing and Sheet Metal." 2/89 ed. Washington, DC: AIA.

07700, "Roof Specialties and Accessories." 5/87 ed. Washington, DC: AIA.

07710, "Manufactured Roof Specialties." 8/90 ed. Washington, DC: AIA.

07901, "Joint Sealants." 5/91 ed. Washington, DC: AIA.

07905, "Paving Joint Sealants." 5/91 ed. Washington, DC: AIA.

Air-Conditioning and Refrigeration Institute (ARI), Sheet Metal and Air-Conditioning Contractors National Association (SMACNA), and National Roofing Contractors Association (NRCA). *Guidelines for Roof Mounted Outdoor Air Conditioner Installations.* Arlington, VA: ARI, SMACNA, and NRCA.

Air Movement and Control Association. 1985. *Application Method for Air Louvers.* Arlington Heights, IL: Air Movement and Control Association.

———. 1986. *Test Methods for Louvers, Dampers, and Shutters.* Arlington Heights, IL: Air Movement and Control Association.

Alcoa Building Products. 1990. *Vinyl Siding, Soffit, and Designer Accessories Installation.* Sidney, OH: Alcoa Building Products.

American Architectural Manufacturers Association (AAMA). 1986. AAMA 1402, "Standard Specifications for Aluminum Siding, Soffit, and Fascia." Des Plaines, IL: AAMA.

———. RA-15, "Aluminum Siding Application Manual." Des Plaines, IL: AAMA.

———. 1986. AAMA 800 Series, "Voluntary Specifications and Test Methods for Sealants." Des Plaines, IL: AAMA.

American Iron and Steel Institute (AISI). 1972. *Stainless Steel: Suggested Practices for Roofing, Flashing, Copings, Fascias, Gravel Stops, and Drainage.* Washington, DC: AISI.

———. 1975. *Stainless Steel Fasteners, Suggested Source List.* Washington, DC: AISI.

———. 1976. *Stainless Steel Fasteners, A Systematic Approach to Their Selection.* Washington, DC: AISI.

———. 1977. *Design Guidelines for the Selection and Use of Stainless Steel.* Washington, DC: AISI.

———. 1979. *Welding of Stainless Steel and Other Joining Methods.* Washington, DC: AISI.

American Society for Testing and Materials (ASTM) Standards:

A 167, "Specifications for Stainless and Heat-Resisting Chromium-Nickel Steel Plate, Sheet, and Strip." Philadelphia, PA: ASTM.

A 446, "Specifications for Sheet Steel, Zinc-Coated (Galvanized) by the Hot-Dip Process, Structural (Physical) Quality." Philadelphia, PA: ASTM.

A 463, "Standard Specification for Sheet Steel, Cold-Rolled Aluminum-Coated Type 1." Philadelphia, PA: ASTM.

A 525, "Specifications for General Requirements for Steel Sheet Zinc-Coated (Galvanized) by the Hot-Dip Process." Philadelphia, PA: ASTM.

A 526, "Specifications for Steel Sheet Zinc-Coated (Galvanized) by the Hot-Dip Process, Commercial Quality." Philadelphia, PA: ASTM.

A 527, "Specifications for Steel Sheet Zinc-Coated (Galvanized) by the Hot-Dip Process, Lock-Forming Quality." Philadelphia, PA: ASTM.

A 528, "Specifications for Steel Sheet Zinc-Coated (Galvanized) by the Hot-Dip Process, Drawing Quality." Philadelphia, PA: ASTM.

A 591, "Specifications for Steel Sheet, Cold-Rolled Electrolytic Zinc-Coated." Philadelphia, PA: ASTM.

A 606, "Specifications for Steel Sheet and Strip, Hot-Rolled, High Strength, Low-Alloy, with Improved Atmospheric Resistance." Philadelphia, PA: ASTM.

A 607, "Standard Specifications for Steel Sheet and Strip, Hot-Rolled and Cold-Rolled, High-Strength, Low-Alloy Columbium and/or Vanadium." Philadelphia, PA: ASTM.

A 611, "Specification for Steel, Cold-Rolled Sheet, Carbon, Structural." Philadelphia, PA: ASTM.

A 642, "Standard Specifications for Steel Sheet, Zinc-Coated (Galvanized) by the Hot-Dip Process, Drawing Quality, Special Killed." Philadelphia, PA: ASTM.

A 792, "Standard Specification for Steel Sheet, Aluminum-Zinc Alloy-Coated by the Hot-Dip Process." Philadelphia, PA: ASTM.

A 875, "Standard Specification for Steel Sheet, Zinc 5 Percent, Aluminum-Mische Metal Alloy, Coated by the Hot-Dip Process." Philadelphia, PA: ASTM.

B 32, "Specification for Solder Metal." Philadelphia, PA: ASTM.

B 101, "Standard Specifications for Lead-Coated Copper Sheets." Philadelphia, PA: ASTM.

B 209, "Specification for Aluminum and Aluminum Alloy Sheet and Plate." Philadelphia, PA: ASTM.

B 210, "Specification for Aluminum-Alloy Drawn Seamless Tubes." Philadelphia, PA: ASTM.

B 211, "Specification for Aluminum-Alloy Bars, Rods, and Wire." Philadelphia, PA: ASTM.

B 221, "Specification for Aluminum-Alloy Extruded Bars, Rods, Wire, Shapes, and Tubes." Philadelphia, PA: ASTM.

B 370, "Specifications for Copper Sheet and Strip for Building Construction." Philadelphia, PA: ASTM.

B 429, "Specification for Aluminum-Alloy Extruded Structural Pipe and Tube." Philadelphia, PA: ASTM.

B 749, "Specifications for Lead Alloy Strip, Sheet, and Plate Products." Philadelphia, PA: ASTM.

C 150, "Specification for Portland Cement." Philadelphia, PA: ASTM.

C 208, "Specification for Insulating Board (Cellulosic Fiber), Structural and Decorative." Philadelphia, PA: ASTM.

C 209, "Methods of Testing Insulating Board (Cellulosic Fiber), Structural and Decorative." Philadelphia, PA: ASTM.

C 481, "Test Method for Laboratory Aging of Sandwich Construction." Philadelphia, PA: ASTM.

C 516, "Specification for Vermiculite Loose Fill Insulation." Philadelphia, PA: ASTM.

C 549, "Specification for Perlite Loose Fill Insulation." Philadelphia, PA: ASTM.

C 552, "Specification for Cellular Glass Block and Pipe Thermal Insulation." Philadelphia, PA: ASTM.

C 553, "Specification for Mineral Fiber Blanket and Felt Insulation (Industrial Type)." Philadelphia, PA: ASTM.

C 578, "Specifications for Preformed Cellular Polystyrene Thermal Insulation." Philadelphia, PA: ASTM.

C 591, "Specification for Unfaced Preformed Rigid Cellular Polyurethane Thermal Insulation." Philadelphia, PA: ASTM.

C 612, "Specification for Mineral Fiber Block and Board Thermal Insulation." Philadelphia, PA: ASTM.

C 665, "Specification for Mineral Fiber Blanket Thermal Insulation for Light Frame Construction and Manufactured Housing." Philadelphia, PA: ASTM.

C 687, "Practice for Determination of Thermal Performance of Loose-fill Building Insulation." Philadelphia, PA: ASTM.

C 726, "Specification for Mineral Fiber and Mineral Fiber, Rigid Cellular Polyurethane Composite Roof Insulation Boards." Philadelphia, PA: ASTM.

C 728, "Specification for Perlite Thermal Insulation Board." Philadelphia, PA: ASTM.

C 764, "Specification for Mineral Fiber Loose-fill Thermal Insulation." Philadelphia, PA: ASTM.

C 834, "Specification for Latex Sealing Compounds." Philadelphia, PA: ASTM.

C 919, "Practices for Use of Sealants in Acoustical Applications." Philadelphia, PA: ASTM.

C 920, "Specification for Elastomeric Joint Sealants." Philadelphia, PA: ASTM.

C 1085, "Specification for Butyl Rubber-based Solvent Release Sealants." Philadelphia, PA: ASTM.

C 1126, "Specification for Faced or Unfaced Rigid Cellular Phenolic Thermal Insulation." Philadelphia, PA: ASTM.

C 1193, "Guide for Use of Joint Sealants." Philadelphia, PA: ASTM.

D 41, "Specification for Asphalt Primer Used in Roofing and Waterproofing." Philadelphia, PA: ASTM.

D 43, "Specification for Creosote Primer Used in Roofing, Dampproofing, and Waterproofing." Philadelphia, PA: ASTM.

D 173, "Specification for Bitumen-saturated Cotton Fabrics Used in Roofing and Waterproofing." Philadelphia, PA: ASTM.

D 224, "Specification for Smooth Surfaced Asphalt Roll Roofing (Organic Felt)." Philadelphia, PA: ASTM.

D 225, "Specification for Asphalt Shingles Surfaced with Mineral Granules." Philadelphia, PA: ASTM.

D 226, "Specification for Asphalt Saturated Organic Felt Used in Roofing and Waterproofing." Philadelphia, PA: ASTM.

D 227, "Specification for Coal-tar Saturated Organic Felt Used in Roofing and Waterproofing." Philadelphia, PA: ASTM.

D 312, "Specification for Asphalt Used in Roofing." Philadelphia, PA: ASTM.

D 449, "Specification for Asphalt Used in Dampproofing and Waterproofing." Philadelphia, PA: ASTM.

D 450, "Specification for Coal-tar Pitch Used in Roofing, Dampproofing, and Waterproofing." Philadelphia, PA: ASTM.

D 1056, "Specification for Flexible Cellular Material—Sponge or Expanded Rubber." Philadelphia, PA: ASTM.

D 1079, "Definitions of Terms Related to Roofing, Waterproofing, and Bituminous Materials." Philadelphia, PA: ASTM.

D 1227, "Standard Specification for Emulsified Asphalt Used as a Protective Coating for Built-up Roofing." Philadelphia, PA: ASTM.

D 1668, "Specification for Glass Fabrics (Woven and Treated) for Roofing and Waterproofing." Philadelphia, PA: ASTM.

D 1863, "Standard Specification for Mineral Aggregate Used on Built-up Roofs." Philadelphia, PA: ASTM.

D 2178, "Specification for Asphalt Glass Felt Used in Roofing and Waterproofing." Philadelphia, PA: ASTM.

D 2626, "Specification for Asphalt-Saturated and Coated Organic Felt Base Sheet Used in Roofing." Philadelphia, PA: ASTM.

D 2822, "Standard Specification for Asphalt Roof Cement." Philadelphia, PA: ASTM.

D 2824, "Standard Specification for Aluminum-Pigmented Asphalt Roof Coatings." Philadelphia, PA: ASTM.

D 3018, "Specification for Class A Asphalt Shingles Surfaced with Mineral Granules." Philadelphia, PA: ASTM.

D 3405, "Specification for Joint Sealant, Hot-Poured, for Concrete and Asphalt Pavements." Philadelphia, PA: ASTM.

D 3406, "Specification for Joint Sealant, Hot-Applied, Elastomeric-Type, for Portland Cement Concrete Pavements." Philadelphia, PA: ASTM.

D 3462, "Specification for Asphalt Shingles Made from Glass Felt and Surfaced with Mineral Granules." Philadelphia, PA: ASTM.

D 3569, "Specification for Joint Sealant, Hot-Applied, Elastomeric, Jet-Fuel-Resistant Type for Portland Cement Concrete Pavements." Philadelphia, PA: ASTM.

D 3581, "Specification for Joint Sealant, Hot-Poured, Jet-Fuel-Resistant Type for Portland Cement Concrete and Tar-Concrete Pavements." Philadelphia, PA: ASTM.

D 3679, "Specification for Rigid Poly (Vinyl Chloride) (PVC) Siding." Philadelphia, PA: ASTM.

D 3909, "Standard Specification for Asphalt Roll Roofing (Glass Mat) Surfaced with Mineral Granules." Philadelphia, PA: ASTM.

D 4434, "Standard Specification for Poly (Vinyl Chloride) Sheet Roofing." Philadelphia, PA: ASTM.

D 4586, "Standard Specification for Asphalt Roof Cement, Asbestos Free." Philadelphia, PA: ASTM.

D 4637, "Specifications for Vulcanized Rubber-Sheet Used in Single Ply Roofing Membranes." Philadelphia, PA: ASTM.

E 84, "Test Method for Surface Burning Characteristics of Building Materials." Philadelphia, PA: ASTM.

E 96, "Test Method for Water Vapor Transmission of Materials." Philadelphia, PA: ASTM.

E 108, "Method for Fire Tests of Roof Coverings." Philadelphia, PA: ASTM.

E 119, "Method for Fire Tests of Building Construction and Materials." Philadelphia, PA: ASTM.

E 136, "Test Method for Behavior of Materials in a Vertical Tube Furnace at 750 deg C." Philadelphia, PA: ASTM.

E 662, "Test Method for Specific Optical Density of Smoke Generated by Solid Materials." Philadelphia, PA: ASTM.

E 699, "Practice for Criteria for Evaluation of Agencies Involved in Testing, Quality Assurance, and Evaluating Building Components in Accordance with Test Methods Promulgated by ASTM Committee E-6." Philadelphia, PA: ASTM.

American Society of Heating, Refrigerating and Air-Conditioning Engineers (ASHRAE). 1985. *ASHRAE Handbook; 1989 Fundamentals.* Atlanta, GA: ASHRAE.

Anderson, Brent. 1986. "Waterproofing and the Design Professional." *The Construction Specifier*, 39(3)(March):86–97.

Architectural Graphic Standards. See Ramsey/Sleeper.

Architectural Record. 1989. "Metal Roofing: New Versatility." *Architectural Record* (February):120–23.

Asphalt Roofing Manufacturers Association (ARMA). 1987. *Built-Up Roofing Systems Design Guide.* Rockville, MD: ARMA.

———. 1990. *Homeowners Guide to Quality Roofing.* Rockville, MD: ARMA.

———. 1990. *Residential Asphalt Roofing Manual.* Rockville, MD: ARMA.

Asphalt Roofing Manufacturers Association (ARMA) and National Roofing Contractors Association (NRCA). 1988. *Quality Control Recommendations for Polymer Modified Bitumen Roofing.* Rockville, MD: ARMA and NRCA. [This document was reprinted in the March 1988 magazine *Professional Roofing* published by NRCA.]

Baker, Maxwell C. 1980. *Roofs: Design, Application and Maintenance.* Montreal: Multiscience Publications. [May be available from The National Roofing Contractors Association.]

Baumgardner, Gaylorn. 1986. "Modified Bitumens: Versatility in Reroofing." *Exteriors*, 4(3)(Autumn):62–64.

Baxter, Dick. 1986. "1001 Reasons Not To Roof Over Wet Insulation." *Roofing Spec*, 14(8)(August):27–30.

Belles, Donald. 1987. "Preventing Flame Spread on Exterior Walls." *Exteriors*, 5(2)(Summer):28–38.

Belles, Donald, and Jesse Beitel. 1988. "Fire Performance of Curtainwalls Questioned." *Exteriors*, 6(2)(Summer):44–50.

Bordenaro, Michael. 1989. "Specifying and Applying Wet Sealants." *Building Design and Construction*, 30(5)(April):72–74.

Bradford, Dane. 1987. "BUR Repairs Made Easy with Torchable Modified Bitumen." *Roofing Spec*, 15(3)(March):22–24.

Brewer, Wilfred B. 1986. "Avoiding Sealant Failure with Preformed Expanded Foam." *The Construction Specifier*, 39(12)(December):23–24.

Brick Institute of America (BIA). *Technical Notes on Brick Construction*:

> No. 7C. "Moisture Control in Brick and Tile Walls: Condensation." Reston, VA: BIA.
>
> No. 7D, Moisture Control in Brick and Tile Walls: Condensation Analysis." Reston, VA: BIA.

Cedar Shake and Shingle Bureau. 1987. *Design and Application Manual for New Roof Construction Bellevue*, WA: Cedar Shake and Shingle Bureau.

———. 1989. *Design and Application Manual for Exterior and Interior Walls*. Bellevue, WA: Cedar Shake and Shingle Bureau.

———. *Design and Application Manual for Roofing, Shake-over-shake*. Bellevue, WA: Cedar Shake and Shingle Bureau.

———. *Certi-Split Manual of Handsplit Red Cedar Shakes*. Bellevue, WA: Cedar Shake and Shingle Bureau.

Clark, E. J., P. G. Campbell, and G. Frohnsdorff. 1975. NBS Technical Note 883, "Waterproofing Materials for Masonry." National Bureau of Standards. Gaithersburg, MD.

Commercial Renovation. 1988, "Sprayed On Roofing Prolongs Life." *Commercial Renovation*, 10(2)(April):60.

Copper Development Association. *Copper, Brass, Bronze Design Handbook: Sheet Copper Applications*. Greenwich, CT: Copper Development Association.

Council of American Building Officials (CABO). 1983. *One- and Two-Family Dwelling Code*. Falls Church, VA: CABO.

D'Angelo, Charles, and Owen J. Perryman. 1988. "Insulated Panels and the Pre-Engineered Building Market." *Metal Architecture*, 4(10)(October):20.

DeMuth, Jerry. 1986. "White vs. Black: Does Membrane Color Matter?" *Roofing Spec*, 14(5)(May):25–28.

Dhunjishah, Michael. 1986. "The Roof Survey: Getting Back to Basics." In *Roofing '86*, 11–16. Chicago, IL: National Roofing Contractors Association.

Doyle, Margaret. 1987. "Extending a Roof's Life." *Building Design and Construction*, 28(6)(June):100–102.

———. 1987. "Keeping a Good Roof Down." *Building Design and Construction*, 28(6)(June):108–111.

———. 1988. "Trends in Specifying EIFS." *Building Design and Construction*, 29(8)(August):58–62.

Dudley, Hubert T. 1987. "Innovations in Vermiculite Roof Insulation Systems." *The Construction Specifier*, 40(11)(November):62–65.

Dunn, William T., and Robert J. Morin. 1988. "An Attractive Alternative, Metal Roofing." *The Construction Specifier*, 41(11)(November):106–111.

Dupuis, Rene. 1987. "Choosing the Right Roof Insulation System." *Exteriors*, 5(1) (Spring):78–81.

———. 1988. "A Renewed BUR Technology Is Ready to Take on the 1990s." In *Handbook of Commercial Roofing Systems*, 16–23. Cleveland OH: Edgell Communications, Inc.

Erwin, Gene. 1986. "Exterior Insulation: Specs with Style." *Exterior*, 4(1)(Spring):34–40.

———. 1987. "Trouble-Shooting Guide for Exterior Insulations." *Exterior*, 5(1)(Spring): 83–86.

Exterior Insulation Manufacturing Association (EIMA). *EIMA Guideline Specification for Exterior Insulation and Finish Systems Class PB*. Washington, DC: Exterior Insulation Manufacturing Association.

———. *EIMA Guideline Specifications for Exterior Insulation and Finish Systems Class PM*. Washington, DC: Exterior Insulation Manufacturing Association.

———. *EIMA Test Method* and Standard 101.86, "Standard Test Method for Resistance of Exterior Insulation Finish Systems to the Effects of Rapid Deformation (iMPACT)." Washington, DC: Exterior Insulation Manufacturing Association.

Exteriors. 1986. "Roof Awareness." *Exteriors*, (Autumn):44.

———. 1987. "BUR Design Guide." *Exteriors*, 5(2)(Summer):22–23.

———. 1987. "Certifying Consultants." *Exteriors*, (Autumn):41–42.

———. 1987. "Skyline: Switchable Coatings." *Exteriors*, 5(3)(Autumn):42–44.

———. 1987. "Skyline: Roofing Innovations." *Exteriors*, 5(4)(Winter):19–22.

———. 1988. "Skyline: Roofing Standards." *Exteriors*, 6(1)(Spring):38–40.

———. 1988. "Skyline: Fiberboard Problems." *Exteriors*, 6(1)(Spring):41–42.

———. 1988. "Skyline: Modified Bitumen Guide." *Exteriors*, 6(1)(Spring):42–43.

———. 1988. "Portfolio: Florida Auditorium Roof Withstands Extreme Temperature Fluctuations." *Exteriors*, 6(1)(Spring):68.

Federal Specifications. *See* U.S. General Services Administration listings.

Fisher, Thomas. 1983. "When the Rains Come." *Architecture* (July).

Fishman, Herbert B. 1984. "Masonry, Metal, and Mysterious Roof Leaks." *The Construction Specifier*, 37(11)(November):76–81.

Francis, Geoffrey V. 1987. "A Practical Approach to Sealant and Joint Design." *Exteriors*, 5(1)(Spring):64–71.

Fricklas, Dick. 1986. "Structural vs. Architectural Metal Roofs." *Exteriors*, 3(2)(Summer):8.

———. 1986. "Fire Codes." *Exteriors*, 4(3) (Autumn):8.

———. 1987. "Better Products Help Prevent Roof Failures." *Building Design and Construction*, 28(11)(November):60–63.

———. 1988. Letter to the editor. *Architecture*, (January):10.

———. 1988. "New Guidelines for Specifying Vapor Retarders." *Exteriors*, 6(1)(Spring):8.

Georgia Pacific Corporation. *Vinyl Siding Installation Guide*. Atlanta, GA: Georgia Pacific Corporation.

Gordon, Douglas E., and M. Stephany Stubbs. 1988. "Longevity and Single-Ply Roofing." *Architecture* (January):95–99.

Gregerson, John. 1988. "Metal Panels Have Designs on New Markets." *Building Design and Construction*, 29(6)(June):68–71.

Griffin, C. W. 1982. *Manual of Built-Up Roof Systems*. 2d ed. New York: McGraw Hill.

Hasan, Riaz. 1987. "Eliminating Backout and Pullout of Roof Fasteners." *Exteriors*, 5(2)(Summer):40–50.

Heineman, Paul. 1987. "Coping With Membrane Roof Penetrations." *The Construction Specifier*, 40(11)(November):36–43.

Henshell, Justin. 1987. "Specifying Membrane Roofing: A Systematic Approach." *The Construction Specifier*, 40(11)(November): 102–108.

Hoffman, H. C. 1986. "Sheet Steels for Low Slope Roofing." *The Construction Specifier*, 39(11)(November):56–58.

Hoover, Stephen R. 1987. "Wind Uplift on Roofs: One Perspective." *The Construction Specifier*, 40(11)(November):76–79.

———. 1987. "Single Ply Roofing: Exploring the Options." *The Construction Specifier*, 40(12)(December):92–103.

Howard, Bion D. 1985. "Masonry Thermal Mass and the Energy Code." *The Construction Specifier*, 38(6)(June):87–95.

———. 1986. "Insulated Concrete Masonry Walls." *The Construction Specifier*, 39(8) (August):60–72.

Huettenrauch, Clarence. 1985. "Building Moisture and Roof System Destruction." *The Construction Specifier*, 39(11)(November):15–16.

———. 1987. "Roof Supports and Roof Leaks: Avoiding the Inevitable." *The Construction Specifier*, 40(7)(July):27–28.

Israel, Sheldon B. 1986. "Reroofing: Principles for Success." *Exteriors*, 4(3)(Autumn):50–53.

———. 1987. "The Nuts and Bolts of Peripheral Roofing Components." *The Construction Specifier*, 40(11)(November):44–49.

———. 1987. "Roof Renovation: A Viable Alternative." *Exteriors*, 5(3)(Autumn):58–62.

Jentsch, Bryan C. 1988. "Pre-Insulated Panels Providing Excellent Economic Value." *Metal Architecture*, 4(10)(Autumn):21.

Johnson, Stephen. 1988. "Improvement in Domestic Steel Quality Translating into Better Metal Paneling." *Metal Architecture*, 4(9)(September):5, 85.

LaCosse, Bob. 1986. "ASTM's New Terminology Gives Industry Chance to Re-Cover." *Roofing Spec*, 14(11)(November):32–33.

———. 1987. "End May Be in Sight for Aged R-Value Controversy." *Roofing Spec*, 15(1)(January):17–20.

———. 1987. "The Air Force Manual Goes Under the Knife." *Roofing Spec*, 15(3)(March):15–18.

Ladesich, Jim. 1983. "Merits of the Metal Roof." *The Construction Specifier*, 36(4)(May):43–47.

Lead Industries Association (LIA). *Lead Roofing and Flashing*. New York: LIA.

———. *Sheet Lead: The Protective Metal*. New York: LIA.

Legatski, Leo A. 1987. "Insulating Roof Decks with Cellular Concrete." *The Construction Specifier*, 40(11)(November):56–61.

Litvin, Albert. 1968. *Clear Coatings for Exposed Architectural Concrete*. Portland Cement Association. Portland Cement Association Research and Development Laboratories, Development Department Bulletin D137. (Reprinted from the May 1968 *Journal of the PCA Research and Development Laboratory*, 10[2]:49–57.)

Loza, Don. 1985. "Using Exterior Insulating Systems for Renovation." *The Construction Specifier*, 38(7)(July):54–56.

Mack, Robert C. 1975. *The Cleaning and Waterproof Coating of Masonry Buildings*. Preservation Brief No. 1. Washington, DC: Technical Preservation Services, U.S. Department of the Interior.

———. 1978. "The Cleaning and Waterproofing of Masonry Buildings." *The Construction Specifier*, 31(11)(November):65–70.

McCorkle, John. 1986. "New Developments in Composite Roof Insulation." *Exteriors*, 4(2)(Summer):40–42.

McDonald, Timothy B. 1988. "Detailing, Specifying Insulation." *Architecture* (July):137–138.

———. 1988. "Selecting Below-grade Waterproofing." *Architecture* (December):135–137.

Metal Architecture. 1987. "Guide to Metal Wall and Roof Panels." *Metal Architecture*, 3(12)(December):20, 22–23, 28–31, 38.

———. 1988. "Guide to Pre-Insulated Panels." *Metal Architecture*, 4(4)(April):40–41, 43–46.

———. 1988. "Selecting Fasteners: Manufacturer's Advice." *Metal Architecture*, 4(12)(December):26–32.

Miles, J. D. III. 1987. "Tracking the Elusive Leak." *Roofing Spec*, 15(3)(March):19–21.

Miller, Michael H. 1984. "Selecting the Right Sealant." *The Construction Specifier*, 37(1)(January):72–76.

Nadel, Toby. 1986. "The 25 Percent Solution." *Roofing Spec*, 14(5)(May):35–36.

National Concrete Masonry Association (NCMA). 1981. NCMA-TEK 10A, "Decorative Waterproofing of Concrete Masonry Walls." Herndon, VA: NCMA.

———. 1973. NCMA-TEK 55, "Waterproof Coatings for Concrete Masonry." Herndon, VA: NCMA.

———. 1981. NCMA-TEK 121, "Waterproofing Concrete Masonry Basements and Earth-sheltered Structures." Herndon, VA: NCMA.

———. 1983. NCMA-TEK 134, "Exterior Insulation of Block Walls." Herndon, VA: NCMA.

National Roofing Contractors Association (NRCA). 1985. *Quality Control in the Application of Built-Up Roofing*. Rosemont, IL: NRCA.

———. 1986. "Technology Harnessed to Seek Out Moisture." *Roofing '86*, 18–22. Chicago, IL: NRCA.

———. 1986. "Maintenance Programs Help Keep Roofs in Shape." *Roofing '86*, 25–27. Chicago, IL: NRCA.

———. 1990. *The NRCA Roofing and Waterproofing Manual*. Rosemont, IL: NRCA.

———. 1990. *The NRCA Steep Roofing Manual*. Rosemont, IL: NRCA. [Included as part of *The NRCA Roofing and Waterproofing Manual*.]

———. 1990. *The NRCA Waterproofing Manual*. Rosemont, IL: NRCA. [Included as part of *The NRCA Roofing and Waterproofing Manual*.]

———. *Roofing Materials Guide*. Rosemont, IL: NRCA.

National Roofing Contractors Association and Asphalt Roofing Manufacturers Association (NRCA/ARMA). 1988. *NRCA/ARMA Manual of Roof Maintenance and Repair*. Rev. ed. Rosemont, IL: NRCA/ARMA.

National Roofing Contractor's Association, Prospect Industries. 1986. "Owners Want Answers to Protect Investment." In *Roofing '86*, 7–9. Chicago, IL: NRCA.

National Slate Association (NSA). 1926. *Slate Roofs*. Vermont: NSA. [Reprinted in 1977 by Vermont Structural Slate Co., Inc., Fair Haven, VT.]

National Tile Roofing Manufacturers Association (NTRMA). 1987. *Installation Manual for Concrete Roof Tile*. Los Angeles, CA: NTRMA.

Nicastro, David H. 1989. "Parameters for Comparing High-Performance Sealants." *The Construction Specifier*, 42(4)(April):122–125.

Nickel Development Institute (NDI). *Design Guidelines for the Selection and Use of Stainless Steel*. Toronto: NDI.

———. *Stainless Steel, Suggested Practices for Roofing, Flashing, Coping, Fascias, Gravel Stops, Drainage*. Toronto: NDI.

Nimtz, Paul D. 1987. "Comparing Structural and Architectural Metal Roofing." *Exteriors* (Winter):46–49.

———. 1988. "Regular Inspections Can Extend Standing Seam Roof Lifecycle." *Exteriors*, 4(9)(September):42.

Old-House Journal, The. 1983. "Maintaining Your Roof: Inspection Checklist." *The Old-House Journal* (April):72.

Perry, Dale C. 1986. "Standing Seam Metal Roofs in Retrofit." *The Construction Specifier*, 39(11)(November):50–86.

Petersen, Wayne. 1987. "Foamed Plastic Roof Insulation." *The Construction Specifier*, 40(11)(November):66–74.

Peterson, Arnold. 1987. "Testing Single Ply's Weatherability." *The Construction Specifier*, 40(11)(November):21–25.

Poore, Patricia. 1983. "Substitute Roofings—Credible Stand-Ins for Clay Tile, Slate, and Wood." *The Old-House Journal* (April):61–63.

———. 1983. "What Most Roofers Don't Tell You About Traditional and Historic Roofs." *The Old-House Journal* (April):56–58.

Porcelain Enamel Institute (PEI). 1970. *Guide to Designing with Architectural Porcelain Enamel on Steel*. Washington, DC: PEI.

Porcher, Joel P. Jr. 1988. "What You Should Keep in Mind When Specifying a Modified Bitumen." In *Handbook of Commercial Roofing Systems*, 32–40. Cleveland, OH: Edgell Communications.

Portland Cement Association (PCA). *Repairing Damp or Leaky Basements in Homes*. Skokie, IL: PCA.

———. 1965. *Moisture Migration—Concrete Slab-On-Ground Construction*. Bulletin D89. Skokie, IL: PCA.

———. 1983. *Concrete Floors on Ground*. Skokie, IL: PCA.

———. 1986. *Effects of Substances on Concrete and Guide to Protective Treatments*. Skokie, IL: PCA.

Quigg, Paul, and William Leavitt. 1988. "New Test Verifies Rock Wool's Fire Performance." *Exteriors*, 6(4)(Winter):86–88.

Ramsey/Sleeper, The AIA Committee on Architectural Graphic Standards. 1981. *Architectural Graphic Standards*, 7th ed. New York: Wiley.

Roofing Spec. 1987. "On the Roof—Concrete Tiles Resemble Hand-Hewn Shakes." *Roofing Spec* (January):25. [*Note: Roofing Spec*'s name was changed in 1988 to *Professional Roofing*.]

———. 1987. "BUR Repairs Made Easy with Torchable Modified Bitumen." *Roofing Spec*, 15(3)(March):22–24.

———. 1987. "Techtalk: Drainage Requires Power of Positive Thinking." *Roofing Spec*, 15(3)(March):46.

Rossiter, Walter J. Jr. 1987. "Nondestructive Methods for Inspecting Single Ply Roofing Membrane Seams." *The Construction Specifier*, 40(11)(November):92–100.

————. 1988. "Effect of Application Variables on Bond Strength of EPDM Seams." In *Handbook of Commercial Roofing Systems*, 41–47. Cleveland OH: Edgell Communications.

Russo, Michael. 1987. "New Methods Needed to Measure 'Thermal Drift.'" *Exteriors*, 5(3)(Autumn):68–74.

————. 1987. "A Critical Look at Thermoplastic Roofing." *Exteriors*, 5(4)(Winter):24–27.

Scharfe, Thomas R. 1987. "Owners, Designers Discover Benefits of Metal Roofs." *Building Design and Construction*, 28(6)(June): 114–116.

————. 1988. "New Metal Coating Technologies Enhance Design Opportunities." *Building Design and Construction*, 29(6)(June):86–89.

————. 1989. "New Technologies Increase Metal Roof Design Options." *Building Design and Construction*, 29(6)(June):106–109.

Sealant Engineering and Associated Lines (SEAL). 1987. *SEAL Guide Specification for Sealants and Caulking*. San Diego, CA: SEAL.

Sealant, Waterproofing and Restoration Institute (SWRI). 1990. *Sealants 1990: The Professionals' Guide*. Kansas City, MO: SWRI.

Shell, Burt. 1988. "Standing Seam Roof Continues Inroads in Educational Market." *Metal Architecture*, 4(4):10–13.

Shmedin, G. 1987. "Bibliography for Slate Roofing." *Association for Preservation Technology Newsletter*, 4(2):89.

Simmons, H. Leslie. 1989. *Repairing and Extending Weather Barriers*. New York: Van Nostrand Reinhold.

Single Ply Roofing Institute (SPRI). 1987. *Single Ply Roofing Systems: Guidelines for Retrofitting Existing Systems*. Wellesley, MA: SPRI.

————. *Single Ply Roofing Systems: A Guideline for Ballast Specifications*. Wellesley, MA: SPRI.

————. *Single Ply Roofing: A Professionals Guide to Specifications*. Wellesley, MA: SPRI.

————. *Wind Design Guide for Adhered Single Ply Roofing Systems*. Wellesley, MA: SPRI.

————. *Wind Design Guide for Ballasted Single Ply Roofing Systems*. Wellesley, MA: SPRI.

————. *Wind Design Guide for Mechanically Fastened Single Ply Roofing Systems*. Wellesley, MA: SPRI.

Smith, George A. 1987. "Fastener Corrosion to Be Judged by New Standard." *Roofing Spec*, 15(1)(January):21–22.

Stephenson, Fred. 1986. "Clip Design Critical in Standing Seam Roofing." *Exteriors*, 4(1)(Spring):60–65.

————. 1986. "Specifying Standing Seam Roofing Systems for Conventional Buildings." *The Construction Specifier*, 39(11)(November):36–43. [This article was reprinted from the March 1984 issue of *Roof Design*, later changed to *Exteriors*.]

Sweetser, Sarah M. 1978. *Roofing for Historic Buildings*. Preservation Brief No. 4. Technical Preservation Services, U.S. Department of the Interior.

Thomas, Robert Jr. 1991. "Nosweat." *The Construction Specifier*, 44(6)(June):74–79.

Thomas, Rose. 1987. "Roofers Advise the Building Team How It Should Be Done." *Building Design and Construction* (June): 92–94.

Thomsett, Michael C. 1985. "Should You Hire a Consultant?" *The Construction Specifier*, 38(10)(October):23–24.

Ting, Raymond. 1986. "Metal Panel Behavior in Exterior Wall Design." *Exteriors* (Autumn):65–69.

————. 1989. "Performance Parameters of Composite Foam Panels." *Metal Architecture*, 5(9)(September):7-8, 73–74.

Tisthammer, Thomas. 1988. "The Case for Specifying Sprayed Polyurethane Systems." In *Handbook of Commercial Roofing Systems*, 51–52. Cleveland OH: Edgell Communications.

Tobiasson, Wayne. 1987. "Vents and Vapor Retarders for Roofs." In *Proceedings of the Symposium on Air Infiltration, Ventilation, and Moisture Transfer*. Washington, DC: Building Thermal Envelope Coordinating Council. [Reprinted in the November 1987 *The Construction Specifier*, 40(11):80–90.]

Tobiasson, Wayne, and Marcus Harrington. *Vapor Drive Maps of the U.S.A.* Washington, DC: Cold Regions Research and Engineering Laboratory (CRREL), Corps of Engineers.

Underground Space Center. 1988. *Building Foundation Design Handbook*. Minneapolis, MN: Underground Space Center.

Underwriters Laboratories, Inc. (UL). *Building Materials Directory—Class A, B, C Fire and Wind Related Deck Assemblies*. Northbrook, IL: UL.

————. *Fire Resistance Directory—Time/Temperature Constructions*. Northbrook, IL: UL.

U.S. Department of the Army. 1980. Corps of Engineers Guide Specifications, Military Construction, Section 07600, "Sheet Metalwork, General." Washington, DC: Office of the Chief of Engineers, Department of the Army.

U.S. Departments of the Army and the Air Force. 1965. *Building Construction Materials and Practices, Caulking and Sealing*. Washington, DC: U.S. Government Printing Office.

U.S. Department of the Navy, Naval Facilities Engineering Command. 1979 (May). Guide Specifications Section 07110, "Membrane Waterproofing." Washington, DC: Department of the Navy.

————. 1983 (February). Guide Specifications Section 07140, "Metallic Oxide Waterproofing." Washington, DC: Department of the Navy.

————. 1984 (July). Guide Specifications Section 07220, "Roof Insulation." Washington, DC: Department of the Navy.

————. 1987 (March). Guide Specifications Section 07222, "Tapered Roof Insulation." Washington, DC: Department of the Navy.

————. 1985 (January). Guide Specifications Section 07410, "Preformed Metal Roofing and Siding." Washington, DC: Department of the Navy.

————. 1985 (October). Guide Specifications Section 07600, "Flashing and Sheet Metal." Washington, DC: Department of the Navy.

————. 1981 (August). Guide Specifications Section 07920, "Sealants and Caulkings." Washington, DC: Department of the Navy.

U.S. General Services Administration. 1971 (January). Public Building Services Guide Specifications, Section 0712, "Elastomeric Waterproofing System—Liquid Applied." Washington, DC: U.S. General Services Administration.

————. 1970 (February). Public Building Services Guide Specifications, Section 0751, "Bituminous Roof Repair." Washington, DC: U.S. General Services Administration.

————. 1970 (February). Public Building Services Guide Specifications, Section 0755, "Repair of Built-Up Base Flashing." Washington, DC: U.S. General Services Administration.

————. 1977 (October). Public Building Services Guide Specification, Section 4-07600, "Sheet Metal, Flashing and Related Accessories." Washington, DC: U.S. General Services Administration.

————. 1970 (February). Public Building Services Guide Specification, Section 2-0762, "Repair of Metal Roof Flashing." Washington, DC: U.S. General Services Administration.

U.S. General Services Administration Specifications Unit. Federal Specifications:

————. HH-I-1972/1, "Insulation Board, Polyurethane or Polyisocyanurate, Faced with Aluminum Foil on Both Sides of the Foam." Washington, DC: GSA Specifications Unit.

————. SS-S-200E(1), "Sealants, Joint, Two-component, Jet-blast, Cold-applied, for Portland Cement Concrete Pavement." Washington, DC: GSA Specifications Unit.

————. TT-C-00598C, "Caulking Compound, Oil and Resin Base Type (for Building Construction)." Washington, DC: GSA Specifications Unit.

————. TT-S-00227E(3), "Sealing Compound, Elastomeric Type, Multi-Component (for Caulking, Sealing, and Glazing in Buildings and Other Structures)." Washington, DC: Specifications Unit.

————. TT-S-00230C(2), "Sealing Compound, Elastomeric Type, Single Component (for Caulking, Sealing, and Glazing in Buildings and Other Structures). Washington, DC: Specifications Unit.

———. TT-S-001534A, "Sealing Compound, Silicone Rubber Base (for Caulking, Sealing, and Glazing in Buildings and Other Structures)." Washington, DC: Specifications Unit.

———. TT-S-001657, "Sealing Compound, Butyl Rubber Base, Solvent Release Type (for Buildings and Other Structures)." Washington, DC: Specifications Unit.

Vinyl Siding Institute (VSI). *Rigid Vinyl Siding Application Instructions*. New York: VSI.

Vonier, Thomas. 1988. "Metal Roofing Systems: Scraping Away the Impressions of the Past." In *1988 Handbook of Commercial Roofing Systems*. 48–50. Cleveland OH: Edgell Communications.

Warford, Milan. 1988. "What Aging Does to Roof Materials." *The Construction Specifier*, 41(2)(February):29–30.

Warshaw, Ruth. 1986. "Creative Design Options for Single-Ply Roofing." *Exteriors*, 4(1)(Spring):42–49.

Williams, Jack B. 1986. "Years of Work Went into Making of Quality Control." *Roofing Spec*, 14(8)(August):31–33.

———. 1988. "Designing the Total Integrated Roofing System." In *Handbook of Commercial Roofing Systems*. 8–14. Cleveland OH: Edgell Communications.

Wilson, Forrest. 1984. *Building Materials Evaluation Handbook*. New York: Van Nostrand Reinhold.

Wiss, Janney, Elstner and Associates, Inc., D. W. Pfeifer, and M. J. Scali. 1981. Report 244, "Concrete Sealers for Protection of Bridge Structures." In *National Cooperative Highway Research Program*. Washington, DC: Transportation Research Board, National Research Council.

Wright, Gordon. 1987. "Modified Bitumen's Market Share Growing." *Building Design and Construction*, 28(6)(June):120–22.

———. 1988. "EIFS Offer Variety of Aesthetic Options." *Building Design and Construction*, 29(8)(August):64–66.

CHAPTER 8

ACKNOWLEDGMENTS

Adams Rite Manufacturing Company;

Aluminum Association (AA);

Amarlite Architectural Products;

American Architectural Manufacturers Association (AAMA);

American Iron and Steel Institute (AISI);

American National Standards Institute (ANSI);

American Society for Testing and Materials (ASTM);

Architectural Woodwork Institute;

ASG Industries, Inc.;

Bruce Engineering Company;

Cardinal IG;

Consumer Product Safety Commission;

Copper Development Association (CDA);

Glass Tempering Association;

Hollow Metal Manufacturers Association (HMMA), Division of the National Association of Architectural Metal Manufacturers (NAAMM);

Insulating Glass Certification Council;

Kalwall Corporation;

Laminated Glass Corporation;

Laminated Safety Glass Association;

Levelor Lorentzen, Inc.:

National Association of Architectural Metal Manufacturers (NAAMM);

National Association of Home Builders (NAHB) Research Foundation;

National Fire Protection Association (NFPA);

National Institute of Building Sciences (NIBS);

National Institute of Standards and Technology (NIST);

National Wood Window and Door Association (NWWDA);

Pease Company;

Pella Corporation;

Screen Manufacturers Association;

Steel Door Institute (SDI);

Tubelite Division of Indal Inc.;

U.S. Department of Housing and Urban Development (HUD);

Woodwork Institute of California.

CHAPTER 8

REFERENCES

Abate, Kenneth. 1985. "Specifying Foam Core Architectural Metal Panel Systems." *The Construction Specifier*, 38(9)(September):40–48.

AIA Professional Services Division. *Masterspec*, Basic Sections:

08111, "Standard Steel Doors and Frames." 8/90 ed. Washington, DC: AIA.

08114, "Custom Steel Doors and Frames." 5/90 ed. Washington, DC: AIA.

08211, "Flush Wood Doors." 2/87 ed. Washington, DC: AIA.

08212, "Panel Wood Doors." 2/87 ed. Washington, DC: AIA.

08311, "Aluminum Sliding Glass Doors." 8/88 ed. Washington, DC: AIA.

08312, "Wood Sliding Glass Doors." 8/88 ed. Washington, DC: AIA.

08314, "Sliding Metal Fire Doors." 11/90 ed. Washington, DC: AIA.

08351, "Folding Doors." 5/89 ed. Washington, DC: AIA.

08360, "Sectional Overhead Doors." 2/88 ed. Washington, DC: AIA.

08410, "Aluminum Entrances and Storefronts." 11/90 ed. Washington, DC: AIA.

08450, All-glass Entrances." 11/90 ed. Washington, DC: AIA.

08460, "Automatic Entrance Doors." 2/91 ed. Washington, DC: AIA.

08470, "Revolving Entrance Doors." 2/91 ed. Washington, DC: AIA.

08510, "Steel Windows." 11/90 ed. Washington, DC: AIA.

08520, "Aluminum Windows." 5/90 ed. Washington, DC: AIA.

08525, "Aluminum Architectural Windows." 5/90 ed. Washington, DC: AIA.

08610, "Wood Windows." 8/90 ed. Washington, DC: AIA.

08710, "Door Hardware." 5/91 ed. Washington, DC: AIA.

08800, "Glazing." 11/91 ed. Washington, DC: AIA.

08920, "Glazed Aluminum Curtain Walls." 2/89 ed. Washington, DC: AIA.

12500, "Window Treatment." 11/89 ed. Washington, D.C.: AIA.

American Architectural Manufacturers Association (AAMA):

AAMA GS-001, "Voluntary Guide Specifications for Aluminum Windows." Des Plaines, IL: AAMA.

AAMA TIR-A1-1975, "Sound Control for Aluminum Curtain Walls and Windows." Des Plaines, IL: AAMA.

AAMA TIR-A2-1975, "Design Wind Loading for Aluminum Curtain Walls." Des Plaines, IL: AAMA.

AAMA TIR-A3-1975, "Fire-Resistive Design Guidelines for Curtain Wall Assemblies." Des Plaines, IL: AAMA.

AAMA TIR-A4-1978, "Recommended Glazing Guidelines for Reflective Insulating Glass." Des Plaines, IL: AAMA.

AAMA TIR-A7-1987, "Glass Design for Sloped Glazing." Des Plaines, IL: AAMA.

ANSI/AAMA 101-1988, "Voluntary Specifications for Aluminum Prime Windows and Sliding Glass Doors." Des Plaines, IL: AAMA.

AAMA 101V-1986, "Voluntary Specifications for Poly (Vinyl Chloride) (PVC) Prime Windows and Sliding Glass Doors." Des Plaines, IL: AAMA.

AAMA 501-1983, "Methods of Test for Metal Curtain Walls." Des Plaines, IL: AAMA.

AAMA 603.8-1985, "Voluntary Performance Requirements and Test Procedures for Pigmented Organic Coatings on Extruded Aluminum." Des Plaines, IL: AAMA.

AAMA 604.2-1977, "Voluntary Specifications for Residential Color Anodic Finishes." Des Plaines, IL: AAMA.

AAMA 605.2-1985, "Voluntary Specifications for High Performance Organic Coatings on Architectural Extrusions and Panels." Des Plaines, IL: AAMA.

AAMA 606.1-1976, "Voluntary Guide Specifications and Inspection Methods for Integral Color Anodic Finishes for Architectural Aluminum." Des Plaines, IL: AAMA.

AAMA 607.1-1977, "Voluntary Guide Specifications and Inspection Methods for Clear Anodic Finishes for Architectural Aluminum." Des Plaines, IL: AAMA.

AAMA 608.1-1977, "Voluntary Guide Specifications and Inspection Methods for Electrolytically Deposited Color Anodic Finishes for Architectural Aluminum." Des Plaines, IL: AAMA.

AAMA 609.1-1985, "Voluntary Guide Specifications and Cleaning and Maintenance of Architectural Anodized Aluminum." Des Plaines, IL: AAMA.

AAMA 610.1-1979, "Voluntary Guide Specifications for Cleaning and Maintenance of Painted Aluminum Extrusions and Curtain Wall Panels." Des Plaines, IL: AAMA.

AAMA 701.2-1974, "Voluntary Specifications for Pile Weather Strip." Des Plaines, IL: AAMA.

AAMA 800-1986, "Voluntary Specifications and Test Methods for Sealants." Des Plaines, IL: AAMA.

AAMA 902.2-1987, "Voluntary Specifications for Sash Balances." Des Plaines, IL: AAMA.

AAMA 904.1-1987, "Voluntary Specifications for Friction Hinges in Window Applications." Des Plaines, IL: AAMA.

AAMA 906.3-1987, "Voluntary Specifications for Sliding Glass Door Roller Assemblies." Des Plaines, IL: AAMA.

AAMA 1002.10-1983, "Voluntary Specifications for Aluminum Insulating Storm Products for Windows and Sliding Glass Doors." Des Plaines, IL: AAMA.

AAMA 1102.7-1977, "Voluntary Guide Specifications for Aluminum Storm Doors." Des Plaines, IL: AAMA.

AAMA 1302, "Voluntary Specifications for Forced-Entry Resistant Aluminum Prime Windows." Des Plaines, IL: AAMA.

AAMA 1303, "Voluntary Specifications for Forced-Entry Resistant Aluminum Sliding Glass Doors." Des Plaines, IL: AAMA.

AAMA 1402-1986, "Standard Specifications for Aluminum Siding, Soffits & Fascia." Des Plaines, IL: AAMA.

AAMA 1502.7-1981, "Voluntary Test Methods for Condensation Resistance of Windows, Doors and Glazed Wall Sections." Des Plaines, IL: AAMA.

AAMA 1503.1-1988, "Voluntary Test Method for Thermal Transmittance and Condensation Resistance of Windows, Doors and Glazed Wall Sections." Des Plaines, IL: AAMA.

AAMA 1504-1988, "Voluntary Standard for Thermal Transmittance and Condensation Resistance of Windows, Doors and Glazed Wall Sections." Des Plaines, IL: AAMA.

PS-AS-1. *Aluminum Siding, Its Care and Maintenance*. Des Plaines, IL: AAMA.

1989. CWM-1. *Metal Curtain Wall Manual*. Des Plaines, IL: AAMA.

CW-I-9. *Aluminum Curtain Wall Design Guide Manual*. Des Plaines, IL: AAMA.

SFM-1. *Aluminum Store Front and Entrance Manual*. Des Plaines, IL: AAMA.

GSM-1. *Metal Curtain Wall, Window, Store Front and Entrance Guide Specifications Manual*. Des Plaines, IL: AAMA.

"Care and Handling of Architectural Aluminum from Shop to Site." In vol. 10 (CW-10) of *Curtain Wall Manual*. Des Plaines, IL: AAMA.

"Design Windloads for Buildings and Boundary Layer Wind Tunnel Testing." In vol. 11 (CW-11) of *Curtain Wall Manual*. Des Plaines, IL: AAMA.

"Structural Properties of Glass." In vol. 12 (CW-12) of *Curtain Wall Manual*. Des Plaines, IL: AAMA.

"Structural Sealant Glazing Systems." In vol. 13 (CW-13) of *Curtain Wall Manual*. Des Plaines, IL: AAMA.

MCWM-1. *Metal Curtain Wall Manual*. Des Plaines, IL: AAMA.

SHDG-1. *Glass Design for Sloped Glazing*. Des Plaines, IL: AAMA.

AAMA GS-001. *Voluntary Guide Specifications for Aluminum Architectural Windows*. Des Plaines, IL: AAMA.

WSG-1. *Window Selection Guide*. Des Plaines, IL: AAMA.

American National Standards Institute (ANSI) Standards:

A39.1-1987. "Safety Requirements for Window Cleaning." New York: ANSI.

A58.1-1982. "Minimum Design Loads, Buildings and Other Structures." New York: ANSI.

A115.1-1982, "Preparation for Mortise Locks for 1-3/8″ and 1-3/4″ Doors." New York: ANSI.

A115.2-1982 (Reaffirmed 1988), "Preparation for Bored Locks for 1-3/8″ and 1-3/4″ Doors." New York: ANSI.

A115.4-1982, "Preparation for Lever Extension Flush Bolts." New York: ANSI.

A115.5-1982, "Preparation for 181 Series and 190 Series Deadlock Strikes." New York: ANSI.

A115.6-1982, "Preparation for Preassembled Locks." New York: ANSI.

A115.7-1982, "Preparation for Floor Closers—Light Duty, Center Hung, Single or Double Acting; Center Hung, Single or Double Acting; Offset Hung, Single Acting." New York: ANSI

A115.12-1982, "Preparation for Offset Intermediate Pivots." New York: ANSI.

A115.13-1982, "Preparation for Tubular Deadlocks." New York: ANSI.

A115.14-1982, "Preparation for Open Back Strikes." New York: ANSI.

A115.15-1985, "Preparation for Bored or Cylindrical Locks in Prehung Insulated Steel Doors and Steel Frames." New York: ANSI.

A115.16-1980, "Preparation for Double-Type Locks in Prehung Insulated Steel Doors and Steel Frames." New York: ANSI.

A115.17 (Reaffirmed 1988), "Preparation for Double-Type Locks in Standard 1-3/8″ and 1-3/4″ Steel Doors and Steel Frames." New York: ANSI.

A115.W1-1988, "Mortise Lock Preparation for 1-3/4″ Flush Doors." New York: ANSI.

A115.W2-1988, "Bored Lock Preparation for 1-3/4″ Flush Doors." New York: ANSI.

A115.W3-1988, "Bored Lock Preparation for 1-3/8″ Flush Doors." New York: ANSI.

A115.W4-1988, "Preassembled Lock Preparation for 1-3/4″ Flush Doors." New York: ANSI.

A115.W5-1988, "Lever Extension Flush Bolt Preparation for 1-3/4″ Flush Doors." New York: ANSI.

A115.W6-1988, "Specifications for Double-Type Lock Preparation for 1-3/4″ Flush Doors." New York: ANSI.

A115.W7-1988, "Specifications for Push/Pull Latch Preparation for 1-3/4″ Flush Doors." New York: ANSI.

A115.W8-1988, "Specifications for Bored Deadlock Preparation for 1-3/4″ Flush Doors." New York: ANSI.

A115.W9-1988, "Specifications for Round Flush Pull Preparation for 1-3/4″ Flush Doors." New York: ANSI.

A117.1-1986, "American National Standard for Buildings and Facilities—Providing Accessibility and Usability for Physically Handicapped People." New York: ANSI.

A151.1-1980, "Physical Endurance for Steel Doors and Hardware Reinforcings." New York: ANSI.

A156.1-1988, "Butts and Hinges." New York: ANSI.

A156.2-1983, "Bored and Preassembled Locks and Latches." New York: ANSI.

A156.3-1984, "Exit Devices." New York: ANSI

A156.4-1986, "Door Controls—Closers." ANSI.

A156.5-1984, "Auxiliary Locks and Associated Products." New York: ANSI.

A156.6-1986, "Architectural Door Trim." New York: ANSI.

A156.7-1988, "Templet Hinge Dimensions." New York: ANSI.

A156.8-1988, "Door Controls—Overhead Holders." New York: ANSI.

A156.10-1985, "Power Operated Pedestrian Doors." New York: ANSI.

A156.12-1986, "Interconnected Locks and Latches." New York: ANSI.

A156.13-1987, "Mortise Locks and Latches." New York: ANSI.

A156.14-1985, "Sliding/Folding Door Hardware." New York: ANSI.

A156.15-1986, "Life Safety Closer/Holder/Release Devices." New York: ANSI.

A156.16-1989, "Auxiliary Hardware." New York: ANSI.

A156.17-1987, "Self Closing Hinges and Pivots." New York: ANSI.

A156.18-1987, "Materials and Finishes." New York: ANSI.

A156.19-1984, "Power Assisted and Low Energy Power Operated Doors." New York: ANSI.

A156.21-1989, "Thresholds." New York: ANSI.

Z34.1, "Practice for Certification Procedures." New York: ANSI.

Z97.1-1984, "Safety Glazing Materials Used in Buildings." New York: ANSI.

American National Standards Institute/American Architectural Manufacturers Association (ANSI/AAMA). ANSI/AAMA 101, "Voluntary Specifications for Aluminum Prime Windows and Sliding Glass Doors." New York: ANSI.

ANSI/AAMA 1002.10, "Voluntary Specifications for Aluminum Insulating Storm Products for Windows and Sliding Glass Doors." New York: ANSI.

ANSI/AAMA 1102.7, "Voluntary Specifications for Aluminum Storm Doors." New York: ANSI.

American National Standards Institute/Screen Manufacturers Association (ANSI/SMA). 1976. ANSI/SMA 1004, "Specifications for Aluminum Tubular Frame Screens for Windows." New York: ANSI.

———. 1976. ANSI/SMA 2005, "Specifications for Aluminum Sliding Screen Doors." New York: ANSI.

American National Standards Institute/Steel Door Institute (ANSI/SDI). 1990. ANSI/SDI-119-1990, "Performance Test Procedures for Steel Door Frames and Frame Anchors." Cleveland, OH: SDI.

———. 1989. ANSI/SDI-123.1-89, "Nomenclature for Steel Doors and Steel Door Frames." Cleveland, OH: SDI.

———. 1987. ANSI/SDI-151.1-87, "Test Procedure and Acceptance Criteria for Physical Endurance for Steel Doors and Hardware Reinforcings." Cleveland, OH: SDI.

———. 1990. ANSI/SDI-224.1-1990, "Test Procedure and Acceptance Criteria for Prime Painted Steel Surfaces for Steel Doors and Frames." Cleveland, OH: SDI.

American Society of Heating, Refrigerating and Air-Conditioning Engineers (ASHRAE). 1985. *ASHRAE Handbook; 1985 Fundamentals.* Atlanta, GA: ASHRAE.

American Welding Society (AWS). 1983. AWS D1.1, "Structural Welding Code—Steel." Miami, FL: AWS.

———. 1983. AWS D1.2, "Structural Welding Code—Aluminum." Miami, FL: AWS.

Architectural Graphic Standards. See Ramsey/Sleeper.

Architectural Woodwork Institute (AWI). 1989. *Architectural Woodwork Quality Standards, Guide Specifications, and Quality Certification Program.* Centreville, VA: AWI.

American Society for Testing and Materials (ASTM) Standards:

A 123, "Specification for Zinc (Hot-Dip Galvanized) Coatings on Iron and Steel Products." Philadelphia, PA: ASTM.

A 153, "Specification for Zinc Coating (Hot-Dip) on Iron and Steel Hardware." Philadelphia, PA: ASTM.

A 167, "Specification for Steel, Carbon, Cold-Rolled Sheet, Commercial Quality." Philadelphia, PA: ASTM.

A 525, "Specification for General Requirements for Steel Sheet, Zinc-Coated (Galvanized) by the Hot-Dip Process." Philadelphia, PA: ASTM.

A 526, Specification for Steel Sheet, Zinc-Coated (Galvanized) by the Hot-Dip Process, Commercial Quality." Philadelphia, PA: ASTM.

A 568, "Specification for General Requirements for Steel, Sheet, Carbon, and High-Strength, Low Alloy, Hot-Rolled, and Cold-Rolled. Philadelphia, PA: ASTM.

A 569, "Specification for Steel, Carbon (0.15 Maximum Percent) Hot-Rolled, and Strip, Commercial Quality." Philadelphia, PA: ASTM.

B 209, "Standard Specification for Aluminum and Aluminum-Alloy Sheet and Plate." Philadelphia, PA: ASTM.

B 221, "Specification for Aluminum-Alloy Extruded Bars, Rods, Wire, Shapes, and Tubes." Philadelphia, PA: ASTM.

C 236, "Test Method for Steady-State Thermal Performance of Building Assemblies by Means of a Guarded Hot Box." Philadelphia, PA: ASTM.

C 481, "Test Method for Laboratory Aging of Sandwich Construction." Philadelphia, PA: ASTM.

C 509, "Specification for Elastomeric Cellular Preformed Gasket and Sealing Material." Philadelphia, PA: ASTM.

C 542, "Specification for Lockstrip Gaskets." Philadelphia, PA: ASTM.

C 716, "Specification for Installing Lockstrip Gaskets and Infill Glazing Materials." Philadelphia, PA: ASTM.

C 719, "Test Method for Adhesion and Cohesion of Elastomeric Joint Sealants Under Cyclic Movement." Philadelphia, PA: ASTM.

C 864, "Specification for Dense Elastomeric Compression Seal Gaskets, Setting Blocks, and Spacers." Philadelphia, PA: ASTM.

C 920, "Specification for Elastomeric Joint Sealants." Philadelphia, PA: ASTM.

C 964, "Standard Guide for Lockstrip Glazing." Philadelphia, PA: ASTM.

C 1036, "Standard Specification for Flat Glass." Philadelphia, PA: ASTM.

C 1048, "Specification for Heat-Treated Flat Glass—Kind HS, Kind FT Coated and Uncoated Glass." Philadelphia, PA: ASTM.

C 1115, "Specification for Dense Elastomeric Silicone Rubber Gaskets and Accessories." Philadelphia, PA: ASTM.

C 1172, "Specification for Laminated Architectural Flat Glass." Philadelphia, PA: ASTM.

D 2000, "Classification System for Rubber Products in Automotive Applications." Philadelphia, PA: ASTM.

D 2287, "Specification for Nonrigid Vinyl Chloride Polymer and Copolymer Molding and Extrusion Compounds." Philadelphia, PA: ASTM.

D 4099, "Standard Specifications for Poly (Vinyl Chloride) Prime Windows." Philadelphia, PA: ASTM.

E 84, "Test Method for Surface Burning Characteristics of Building Materials." Philadelphia, PA: ASTM.

E 90, "Method for Laboratory Measurement of Airborne-Sound Transmission Loss of Building Partitions." Philadelphia, PA: ASTM.

E 96, "Test Method for Water Vapor Transmission of Materials." Philadelphia, PA: ASTM.

E 152, "Fire Tests of Door Assemblies." Philadelphia, PA: ASTM.

E 163, "Fire Tests of Window Assemblies." Philadelphia, PA: ASTM.

E 283, "Test Method for Rate of Air Leakage Through Exterior Windows, Curtain Walls, and Doors." Philadelphia, PA: ASTM.

E 330, "Test Method for Structural Performance of Exterior Windows, Curtain Walls, and Doors by Uniform Static Air Pressure Differential." Philadelphia, PA: ASTM.

E 331, "Test Method for Water Penetration of Exterior Windows, Curtain Walls, and Doors by Uniform Static Air Pressure Differential." Philadelphia, PA: ASTM.

E 413, "Classification for Determination of Sound Transmission Class." Philadelphia, PA: ASTM.

E 547, "Test Method for Water Penetration of Exterior Windows, Curtain Walls, and Doors by Cyclic Air Pressure Differential." Philadelphia, PA: ASTM.

E 662, "Test Method for Specific Optical Density of Smoke Generated by Solid Materials." Philadelphia, PA: ASTM.

E 699, "Practice for Criteria for Evaluation of Agencies Involved in Testing, Quality Assurance, and Evaluating Building Components in Accordance with Test Methods Promulgated by ASTM Committee E-6." Philadelphia, PA: ASTM.

E 773, "Test Method for Seal Durability of Sealed Insulating Glass Units." Philadelphia, PA: ASTM.

E 774, "Specification for Sealed Insulating Glass Units." Philadelphia, PA: ASTM.

E 783, "Field Measurement of Air Leakage Through Installed Exterior Windows and Doors." Philadelphia, PA: ASTM.

E 987, "Standard Tests for Deglazing Force of Fenestration Products." Philadelphia, PA: ASTM.

E 1223, "Test Method for Structural Performance of Exterior Windows, Curtain Walls, and Doors by Cyclic Air Pressure Differential." Philadelphia, PA: ASTM.

F 588, "Standard Test Methods for Resistance of Window Assemblies to Forced Entry, Excluding Glazing." Philadelphia, PA: ASTM.

F 842, "Standard Test Methods for Measurement of Forced Entry Resistance of Horizontal Sliding Door Assemblies." Philadelphia, PA: ASTM.

Belles, Don, and Jesse Beitel. 1988. "Fire Performance of Curtainwalls Questioned." *Exteriors*, 6(2)(Summer):44–50.

Benny, James C. 1988. "Ensuring That Wood Doors and Hardware Are Compatible." *Window and Door Specifier*, 1(1)(Summer):69–70.

Bessmer, Stan. 1979. "Edge-Sealed Insulating Glass." *The Construction Specifier*, 32(6)(June):50–53.

Bock, Gordon. 1988. "Glass Notes." *The Old-House Journal*, 16(4)(July/August):35–43.

———. 1989. "The Sash Window Balancing Act." *The Old-House Journal*, 17(5)(September/October):31–40.

Building Design and Construction. 1987. "Survey Reveals Advantages of Wood Doors and Windows." *Building Design and Construction*, 28(6)(June):157–166.

———. 1989. "Roundtable: New Horizons in Curtain Wall Design and Construction." *Building Design and Construction*, 30(1)(January):67–73.

Carter, Roy. 1988. "The Ins and Outs of Access Control Systems." *The Construction Specifier*, 41(1)(January):84–87.

Cassidy, Victor M. 1989. "How Windows Get Better." *Modern Metals*, 45(1)(February):32–46.

Commercial Renovation. 1987. "Performance Standards for the Wood Door Industry." *Commercial Renovation*, 9(4)(August):50.

———. 1987. "Product Focus: Door and Hardware Innovations." *Commercial Renovation*, 9(4)(August):52–53.

———. 1988. "New Technology Makes Window Efficiency Shine." *Commercial Renovation*, 10(3)(June):48–53.

———. 1988. "Door Security Hinges on Hardware." *Commercial Renovation*, 10(5)(October):52–56.

Construction Specifications Institute (CSI). 1988. CSI Monograph 07M411, *Precoated Metal Building Panels*. Alexandria, VA: CSI.

———. CSI Monograph 08M710, *Finish Hardware*. Alexandria, VA: CSI.

Construction Specifier, The. 1979. "CS Builders Hardware Conference, Part 1." *The Construction Specifier*, 32(6)(June):39–49.

———. 1979. "CS Builders Hardware Conference, Part 2." *The Construction Specifier*, 32(7)(July):47–53.

———. 1979. "CS Builders Hardware Conference, Part 3." *The Construction Specifier*, 32(9)(September):55–65.

———. 1983. "Twenty-Four Karat Glass." *The Construction Specifier*, 36(3)(March):28–32.

Consumer Product Safety Commission (CPSC). 16 CFR Part 1201, "Safety Standards for Architectural Glazing Materials." Bethesda, MD: CPSC.

Council of American Building Officials (CABO). 1983. *One- and Two-Family Dwelling Code*. Falls Church, VA: CABO.

Cox, Sarah, and Billy Edwards. 1987. "Keys to Specifying the Right Locking System." *The Construction Specifier*, 40(1)(January):25–26.

D'Angelo, Charles, and Owen J. Perryman. 1988. "Insulated Panels and the Pre-Engineered Building Market." *Metal Architecture*, 4(10)(October):20.

Door and Hardware Institute (DHI). 1975. *Recommended Locations for Builder's Hardware for Standard Steel Doors and Frames*. Chantilly, VA: DHI.

———. 1976. *Hardware Reinforcements on Steel Doors and Frames*. Chantilly, VA: DHI.

———. 1979. *Recommended Locations for Builder's Hardware for Custom Steel Doors and Frames*. Chantilly, VA: DHI.

———. 1979. *Processing Hardware for Custom Aluminum Doors*. Chantilly, VA: DHI.

———. 1980. DHI-WDHS-1, "Templet Book Criteria for Wood Doors." Chantilly, VA: DHI.

———. 1980. DHI-WDHS-2, "Recommended Templet Fasteners for Wood Doors." Chantilly, VA: DHI.

———. 1980. DHI-WDHS-3, "Recommended Hardware Locations for Wood Flush Doors." Chantilly, VA: DHI.

———. 1985. *Basic Architectural Hardware*. Chantilly, VA: DHI.

———. 1986. *Installation Guide for Doors and Hardware*. Chantilly, VA: DHI.

Easter, R. Lee. 1985. "Can Glass and Hurricanes Mix?" *The Construction Specifier*, 38(5)(May):46–50.

Evans, Lyon D. 1985. "Aluminum Windows and Sliding Glass Doors, The New Specs." *The Construction Specifier*, 38(9)(September):56–63.

———. 1986. "Design Techniques for Glazed Curtainwall Retrofit." *Exteriors*, 4(3)(Autumn):56–60.

Exteriors. 1986. "Skyline: Window Standards." *Exteriors*, 4(2)(Summer):20.

———. 1986. "Skyline: Glazing Technics." *Exteriors*, 4(4)(Winter):24–25.

———. 1987. "Skyline: Curtainwall Insulation." *Exteriors*, 5(3)(Autumn):44–46.

———. 1987. "Skyline: Low E Glass Use Rising." *Exteriors*, 5(4)(Winter):18–19.

———. 1988. "Skyline: Glass Strength Debated." *Exteriors*, 6(3)(Autumn):18–19.

———. 1988. "Skyline: Curtainwall Issues." *Exteriors*, 6(4)(Winter):75.

Federal Specifications. *See* U.S. General Services Administration Specifications Unit.

Flat Glass Marketing Association (FGMA). *Glazing Manual*. Topeka, KA: FGMA.

———. *Sealant Manual*. Topeka, KS: FGMA.

Francis, Geoffrey V. 1988. "New Developments in Structural Glazing." *Exteriors*, 6(1)(Spring):50–55.

———. 1989. "Innovation in Structural Sealant Glazing." *The Construction Specifier*, 42(3)(March):54–59.

Fulton, Frank. 1988. "High Performance Window Design." *The Construction Specifier*, 41(8)(August):80–84.

George, Louis F. 1988. "Specifying Hardware for Schools." *The Construction Specifier*, 41(8)(August):52–59.

———. 1989. "Specifying Hardware in Housing for the Elderly." *The Construction Specifier*, 42(2)(February):62–66.

———. 1989. "Specifying Systems for Access Control." *The Construction Specifier*, 42(6)(June):52–58.

Glass Tempering Association (GTA). 1989. 89-1-6, "Specification for Environmental Durability of Fully Tempered or Heat-strengthened Spandrel Glass with Applied Opacifiers." Topeka, KS: GTA.

Goldsmith, Aaron. 1984. "Glass: The Invisible Hazard." *Commercial Renovation*, 6(6)(December):66–67.

Gregerson, John. 1988. "Metal Panels Have Designs on New Markets." *Building Design and Construction*, 29(6)(June):68–71.

———. 1989. "New Glass Technologies Solve Site Challenges." *Building Design and Construction*, 30(3)(March):74–77.

Griffiths, Howard. 1987. "Glass: Colors of the City." *The Construction Specifier*, 40(8)(August):100–107.

Grossi, Anthony F. 1984. "A Brief on Low E Glass." *The Construction Specifier*, 37(9)(September):22–23.

Heerwagen, Judith H. 1987. "Windowscapes—The Psychology of View." *The Construction Specifier*, 40(8)(August):31–32.

Heitmann, L. J. 1986. "The Trouble with Shadow Boxes." *Exteriors*, 4(2)(Summer):10.

———. 1986. "A Closer Look at Structural Glazing." *Exteriors*, 4(4)(Winter):12.

———. 1987. "The Rain Screen Principle Can Work for You." *Exteriors*, 5(1)(Spring):12.

———. 1987. "The Trouble with Curtain-wall." *Exteriors*, 5(2)(Summer):12.

———. 1987. "Proper Value Engineering for Curtainwall." *Exteriors*, 5(3)(Autumn):12.

———. 1987. "Safety First in Overhead Glazing." *Exteriors*, 5(4)(Winter):12.

———. 1988. "On Designing a Spandrel Cavity." *Exteriors*, 6(2)(Summer):12.

———. 1988. "A Question of Compatibility." *Exteriors*, 6(3)(Autumn):12.

Hollow Metal Manufacturers Association. 1987. *Hollow Metal Manual*. Chicago: National Association of Architectural Metal Manufacturers.

Hudnut, Richard. 1989. "The Hardware Metals." *The Construction Specifier*, 41(8)(August):31–33.

Insulating Glass Certification Council (IGCC). 1991. *Certified Products Directory—Sealed Insulating Glass*. Cortland, NY: IGCC.

Jentsch, Bryan C. 1988. "Pre-Insulated Panels Providing Excellent Economic Value." *Metal Architecture*, 4(10)(October):21.

Johnson, Stephen. 1988. "Improvement in Domestic Steel Quality Translating into Better Metal Paneling." *Metal Architecture*, 4(9)(September):5, 85.

Johnston, Bob. 1986. "Quality Control Critical with Structural Silicone." *Exteriors*, 4(2)(Summer):44–50.

Kawneer Company, Inc. 1984. *Architectural Finishes*. Norcross, GA: Kawneer.

Kincaid, Mary. 1983. "Facets of Glass." *The Construction Specifier*, 36(3)(March):34–40.

Laminated Safety Glass Association (LSGA). 1987. *LSGA Design Guide*. Topeka, KS: LSGA.

LaTona, Raymond W., Thomas A. Schwartz, and Glen R. Bell. 1988. "New Standard Permits More Realistic Curtain Wall Testing." *Building Design and Construction*, 29(11)(November):42–46.

Lesniak, Joseph G., and Judith P. Guy. 1985. "The Architectural Hardware Consultant." *The Construction Specifier*, 38(6)(June):34.

Lofgren, Michael. 1983. "Making Building Security 'Airtight.' " *The Construction Specifier*, 36(11)(November):68–69.

MacDonald, John E. 1978. "The Effects of Light on Wood and Wood Finishing Systems." *The Construction Specifier*, 31(3)(March):23–26.

———. 1978. "Architectural Flush Wood Doors." *The Construction Specifier*, 31(3)(March):28–32.

Maisel, Murry. 1983. "Custom Hollow Metal Doors, Problems and Solutions." *The Construction Specifier*, 36(1)(January):74–77.

Marinelli, Janet. 1988. "Architectural Glass." *The Old-House Journal*, 16(4)(July/August):34–42.

Mason, Donald. 1983. "Selecting an Access Control System." *The Construction Specifier*, 36(11)(November):71–73.

McAuliffe, William B. 1985. "Finish Hardware, A Complex Specification." *The Construction Specifier*, 38(6)(June):30–37.

McInerney, William D. 1987. "Selecting the Proper Hinge and Pivot." *The Construction Specifier*, 40(3)(March):116–122.

McKinley, Robert W. 1986. "Saving Energy with Low E Glazing." *Exteriors*, 4(1)(March):50–58.

———. 1986. "Windloading of Glazed Curtain Walls." *Exteriors*, 4(4)(Winter):32–36.

———. 1989. "Insulating Glass: Designing for the 1990's." *Exteriors*, 7(1)(Spring):12–16.

Metal Architecture. 1987. "Guide to Metal Wall and Roof Panels." *Metal Architecture*, 3(12)(December):20, 22–23, 28–31, 38.

———. 1988. "Guide to Pre-Insulated Panels." *Metal Architecture*, 4(4)(April):40–41, 43–46.

Metal Construction News. 1988. *1989 Building Systems Product File and Directory*. Skokie, IL: *Metal Construction News*.

Minor, Joseph E. 1987. "Accommodating Wind Forces in Glazing Design." *The Construction Specifier*, 40(8)(August):25–26.

National Association of Architectural Metal Manufacturers (NAAMM). 1988. *Metal Finishes Manual for Architectural and Metal Products*. Chicago, IL: NAAMM.

National Fire Protection Association (NFPA). 1985. *NFPA 105: Smoke and Draft Control Door Assemblies*. Quincy, MA: NFPA.

———. 1986. *NFPA 80: Fire Doors and Windows*. Quincy. MA: NFPA.

———. 1988. *NFPA 101: Life Safety Code*. Quincy. MA: NFPA.

National Glass Association (NGA). 1987. *Guide to the Federal Glazing Laws and the Model Safety Glazing Code*. McLean, VA: NGA.

———. *Glazing Reference Guide*. McLean, VA: NGA.

———. *Glass Standards: A Collection of National and Voluntary Standards Pertaining to Glass and Glazing*. McLean, VA: NGA.

National Institute of Standards and Technology. *See* U.S. Department of Commerce.

National Wood Window and Door Association (NWWDA). *How to Store, Handle, Finish and Maintain Wood Doors*. Park Ridge, IL: NWWDA.

———. *Care and Finishing of Wood Doors*. Park Ridge, IL: NWWDA.

———. ANSI/NWWDA I.S.1-A-91, "Industry Standard for Architectural Wood Flush Doors." Park Ridge, IL: NWWDA.

———. ANSI/NWWDA I.S.1-R-91, "Industry Standard for Residential Wood Flush Doors." Park Ridge, IL: NWWDA.

———. NWWDA I.S.2-87, "Industry Standard for Wood Windows." Park Ridge, IL: NWWDA.

———. NWWDA I.S.3-88, "Industry Standard for Sliding Patio Doors." Park Ridge, IL: NWWDA.

———. NWWDA I.S.4-81, "Industry Standard for Water-repellent Preservative Treatment for Millwork." Park Ridge, IL: NWWDA.

———. ANSI/NWWDA I.S.6-91, "Industry Standard for Wood Stile and Rail Doors." Park Ridge, IL: NWWDA.

Nicastro, David H. 1988. "Uncovering the Reasons for Curtainwall Failure." *Exteriors*, 6(2)(Summer):26–32.

Olson, Christopher. 1987. "Heightening Impact with Structural Glazing." *Building Design and Constructions*, 28(3)(March):132–136.

Peterson, Charles O. Jr. 1984. "Structural Silicone." *The Construction Specifier*, 37(9)(September):56–63.

Poore, Patricia. 1983. "The Trouble with Pocket Doors." *The Old-House Journal*, 11(4)(May):79, 90–92.

———. 1983. "Repairing Top Hung Pocket Doors." *The Old-House Journal*, 11(6)(July):115, 128–129.

Raeber, John A. 1985. "Selecting Door Hardware." *The Construction Specifier*, 38(6)(June):38–47.

———. 1988. "Selecting Door Contract Hardware." *The Construction Specifier*, 41(8)(August):62–71.

———. 1989. "Exposed Metallic Coatings on Glass: A Cautionary Note." *The Construction Specifier*, 42(6)(June):31–32.

———. 1989. "Selecting Door-Hanging Hardware." *The Construction Specifier*, 42(6)(June):36–42.

Ramsey/Sleeper, The AIA Committee on Architectural Graphic Standards. *Architectural Graphic Standards*, 8th ed. New York: Wiley.

Reynolds, James. 1986. "Specifying Coated Glazing Material." *The Construction Specifier*, 39(6)(June):46–59.

Rush, Richard. 1987. "Refining Window Energy Performance." *Building Design and Construction*, 28(12)(December):148–153.

Russo, Michael. 1987. "Improving the Art of Glass Design." *Exteriors*, 5(3)(Autumn):6.

Safety Glazing Certification Council (SGCC). *Certified Products Directory—Safety Glazing Materials Used in Buildings.* Cortland, NY: SGCC.

Saino, Joseph Neuhoff. 1988. "Industrial Doors, The Types and Uses." *The Construction Specifier,* 41(3)(March):76–81.

Sanford, A. G. 1987. "Stress Tests and Safety Factors in Structural Glazing." *The Construction Specifier,* 40(6)(June):31–37.

Scassellati, Rudy R. 1988. "Architectural Flush Doors." *The Construction Specifier,* 41(4)(April):112–120.

Scharfe, Thomas R. 1988. "Meeting the Challenges of Sloped Glazing." *Building Design and Construction,* 29(3)(March): 140–147.

———. 1988. "New Metal Coating Technologies Enhance Design Opportunities." *Building Design and Construction,* 29(6)(June): 86–89.

Schulthesis, Joseph A. 1985. "Guidelines for Sloped Glazing." *The Construction Specifier,* 38(3)(March):90–95.

———. 1986. "Sloped Glazing with Structural Silicone." *The Construction Specifier,* 39(6)(June):36–45.

Screen Manufacturers Association (SMA). 1976. ANSI/SMA 1004-1976, "Specifications for Aluminum Tubular Frame Screens for Windows." Chicago, IL: SMA.

———. 1976. ANSI/SMA 2005-1976, "Specifications for Aluminum Sliding Screen Doors." Chicago, IL: SMA.

Sealed Insulating Glass Manufacturers Association (SIGMA). 1990. TM-3000, "Vertical Glazing Guidelines." Chicago, IL: SIGMA.

———. 1990. TM-3001, "Sloped Glazing Guidelines." Chicago, IL: SIGMA.

Sittnick, Ralph. 1985. "The Evolution of the Electromagnetic Lock." *The Construction Specifier,* 38(6)(June):48–49.

Smith, Charles F. Jr. 1989. "Specifying Weatherstripping for Energy, Smoke, and Sound Control." *The Construction Specifier,* 42(6)(June):44–51.

Stanbrough, Jerry. 1986. "Energy Efficiency: The Curtainwall as Filter." *Exteriors,* 4(2)(Summer):34–39.

Steel Door Institute (SDI). 1985. SDI-100-90, Recommended Specifications for Standard Steel Doors and Frames. Cleveland, OH: SDI.

———. 1982. SDI-105-91, "Recommended Erection Instructions for Steel Frames." Cleveland, OH: SDI.

———. SDI-106, "Recommended Standard Door Type Nomenclature." Cleveland, OH: SDI.

———. 1984. SDI-107-84, "Hardware on Steel Doors." Cleveland, OH: SDI.

———. 1983. SDI-108-90, "Recommended Selection and Usage Guide for Standard Steel Doors." Cleveland, OH: SDI.

———. 1975. SDI-109-88, "Hardware for Standard Steel Doors and Frames." Cleveland, OH: SDI.

———. 1984. SDI-110-84, "Standard Steel Doors and Frames for Masonry Construction." Cleveland, OH: SDI.

———. SDI-111, "Recommended Standard Details for Steel Doors and Frames." Cleveland, OH: SDI.

———. 1979. SDI-112-89, "Test Procedure and Acceptance Criteria for Apparent Thermal Performance of Steel Door and Frame Assemblies." Cleveland, OH: SDI.

———. 1979. SDI-113-79, "Test Procedure and Acceptance Criteria for Apparent Thermal Performance of Steel Door and Frame Assemblies." Cleveland, OH: SDI.

———. 1979. SDI-114-79, "Test Procedure and Acceptance Criteria for Acoustical Performance of Steel Door and Frame Assemblies." Cleveland, OH: SDI.

———. 1979. SDI-116-79, "Standard Test Procedure and Acceptance Criteria for Rate of Air Flow Through Closed Steel Door and Frame Assemblies." Cleveland, OH: SDI.

———. SDI-117-88, "Manufacturing Tolerances for Standard Steel Doors and Frames." Cleveland, OH: SDI.

———. 1976. SDI-118-88, "Basic Fire Door Requirements." Cleveland, OH: SDI.

Steel Structures Painting Council (SSPC). 1983. *Good Paint Practice.* Vol. 1 of *Steel Structures Painting Manual,* 2d ed. Pittsburgh, PA: SSPC.

———. 1983. *Systems and Specifications.* Vol. 2 of *Steel Structures Painting Manual.* 2d ed. Pittsburgh, PA: SSPC.

Stubbs, M. Stephanie. 1986. "Glued-on Glass." *Architectural Technology* (May/June):46–51.

Tatum, Rita. 1984. "The Return of Wood Windows." *The Construction Specifier,* 37(9)(September):51–55.

———. 1982. "Making Metal Building Systems Perform." *The Construction Specifier,* 35(3)(March):38–48, 105–112.

———. 1986. "Metal Building Systems Go High-Rise High-Tech." *The Construction Specifier,* 39(3)(March):40–49.

———. 1987. "The Evolution of Energy-Saving Windows." *The Construction Specifier,* 40(8)(August):33–34.

Taylor, Paul. 1983. "Glass Skyscrapers Vulnerable." *The Construction Specifier,* 36(11)(November):10–12.

Technologies Media Corporation. 1989. "Architectural Hardware." *Building Design and Construction,* 30(12)(October):115–143.

Ting, Raymond. 1986. "Metal Panel Behavior in Exterior Wall Design." *Exteriors,* (Autumn):65–69.

———. 1987. "Ensuring a Trouble-Free Curtainwall Renovation." *Exteriors,* 5(3)(Autumn):64–66.

———. 1988. "Designing a Leak-Free Curtainwall System." *Exteriors,* 6(3)(Autumn): 30–34.

———. 1989. "Performance Parameters of Composite Foam Panels." *Metal Architecture,* 5(9)(September):7–8, 73–74.

Trechsel, Heinz. 1988. "Specifying an Energy Efficient Thermal Window." *Exteriors,* 6(3)(Autumn):36–40.

Umlauf, Elyse. 1989. "Specifying the Right Wood Windows and Doors." *Building Design and Construction,* 30(7)(June): 100–103.

U.S. Department of the Army. 1982. Corps of Engineers Guide Specifications, Military Construction, CEGS-08110, "Steel Doors and Frames." Washington, DC: Office of the Chief of Engineers, Department of the Army.

U.S. Department of Commerce, National Institute of Standards and Technology. 1977. *Window Design Strategies to Conserve Energy.* Washington DC: National Institute of Standards and Technology.

U.S. Department of Housing and Urban Development (HUD). *Minimum Property Standards for Housing.* Washington, DC: HUD.

U.S. Department of the Navy, Naval Facilities Engineering Command. 1986 (February) Guide Specifications Section 08110, "Steel Doors and Frames." Washington, DC: Department of the Navy.

———. 1984 (February) Guide Specifications Section 08120, "Aluminum Doors and Frames." Washington, DC: Department of the Navy.

———. 1980 (February) Guide Specifications Section 08210, "Wood Doors." Washington, DC: Department of the Navy.

———. 1982 (June) Guide Specifications Section 08520, "Aluminum Windows." Washington, DC: Department of the Navy.

———. 1980 (April) Guide Specifications Section 08610, "Wood Windows." Washington, DC: Department of the Navy.

———. 1985 (March) Guide Specifications Section 08710, "Finish Hardware." Washington, DC: Department of the Navy.

———. 1983 (December) Guide Specifications Section 08800, "Glazing." Washington, DC: Department of the Navy.

U.S. General Services Administration (GSA) Specifications Unit. Federal Specification L-S-125, "Screening, Insect, Nonmetallic." Washington, DC: GSA Specifications Unit.

———. Federal Specification RR-W-365, "Wire Fabric (Insect Screening)." Washington, DC: GSA Specifications Unit.

Valdes. Noel. 1988. "Low E Glass." *The Construction Specifier,* 41(8)(August):71–78.

Vild, Donald J. 1986. "Glass and the Building Codes." *Exteriors,* 4(1)(March):12.

———. 1986. "Glass: The Energy Saver." *Exteriors,* 4(2)(Summer):12.

———. 1986. "The Case for Sloped Glazing." *Exteriors*, 4(3)(Autumn):10.

———. 1986. "Fully Tempered vs. Heat-Strengthened Glass." *Exteriors*, 4(4) (Winter):10.

———. 1987. "Why Glass Goes Bad." *Exteriors*, 5(1)(Spring):10.

———. 1987. "The Overhead Glazing Controversy." *Exteriors*, 5(3)(Autumn):10.

———. 1987. "Who Selects the Glass." *Exteriors*, 5(4)(Winter):10.

———. 1988. "Proper Engineering of Structural Silicone." *Exteriors*, 6(1)(Spring):10.

———. 1988. "What's Wrong with Glass Spandrels?" *Exteriors*, 6(2)(Summer):10.

———. 1988. "How Much Glass Will Break." *Exteriors*, 6(3)(Autumn):10.

———. 1988. "Some Precautions When Specifying Low-E Glass." *Exteriors*, 6(4)(Winter):7.

———. 1988. "Clearing Up Building Code, Glass Selection Confusion." *Exteriors*, 6(4)(Winter):76–80.

———. 1989. "More Problems with Glass Spandrels." *Exteriors*, 7(1)(Spring):7.

Wherry, Allen P. 1986. "Standard Steel Doors for Customized Design." *The Construction Specifier*, 39(7)(July):106–111.

Wilson, Randy J. 1987. "Sloped Glazing and the Rain Screen Principle." *The Construction Specifier*, 40(8)(August):94–98.

Woodwork Institute of California (WIC). *Manual of Woodwork.* Fresno, CA: WIC.

Wright, Gordon. 1988. "Curtain Walls Drawing More Critical Attention." *Building Design and Construction*, 29(3)(March):162–165.

———. 1989. "Storefront Designs Accommodate Multiple Requirements." *Building Design and Construction*, 30(1)(January):60–62.

Zingeser, Joel P. 1988. "Applauding Glass." *The Construction Specifier*, 41(8)(August): 47–48.

CHAPTER 9

ACKNOWLEDGMENTS

American National Standards Institute (ANSI);

American Olean Tile Company;

American Plywood Association (APA);

American Society for Testing and Materials (ASTM);

Architectural Woodwork Institute (AWI);

Armstrong World Industries, Inc.;

Binks Manufacturing Company;

Bruce Hardwood Floors;

Carpet and Rug Institute (CRI);

Ceiling and Interior Systems Contractors Association (CISCA);

Celotex Corporation;

Clemco Industries Corporation;

Forbo Vicracoustic Inc.;

Gold Bond Building Products;

Congoleum Corporation;

Crossfield Products Corporation;

Drywall, Lath, and Plaster Bureau;

E-Z Painter Corporation;

Georgia Pacific Corporation;

Gypsum Association (GA);

Harbil Manufacturing Company;

Hercules, Incorporated;

Hunter Douglas Architectural Products;

International Institute for Lath and Plaster;

Johnsonite Division of Duramax, Inc.

Kaylath Divison, Georgetown Wire Co., Inc.;

Kentile Floors;

Kurfees Coatings;

MAC Publishers, Incorporated;

Manville Corporation;

Maple Flooring Manufacturers Association;

Mercer Products Company, Inc.;

Metal Building Interior Products Company;

Metal Lath/Steel Framing Association Division of the National Association of Architectural Metal Manufacturers;

Mid-States Terrazzo and Tile Company;

Mohawk Carpet Mills;

Monsanto Company;

National Association of Home Builders (NAHB) Research Foundation;

National Concrete Masonry Association;

National Institute of Building Sciences (NIBS);

National Institute of Standards and Technology (NIST).;

National Gypsum Company;

National Lime Association;

National Oak Flooring Manufacturers Association (NOFMA);

National Paint and Coatings Association (NPCA);

National Terrazzo and Mosaic Association, Inc. (NTMA);

Nora Flooring Systems, Inc.;

Painting and Decorating Contractors of America (PDCA);

Perlite Institute;

Portland Cement Association (PCA);

Resilient Floor Covering Institute (RFCI);

Ruberoid Company;

Selby, Battersby and Company;

Walter E. Selck and Company;

Sherwin-Williams Company;

Southern Pine Inspection Bureau (SPBI);

Steel Structures Painting Council (SSPC);

Terrazzo Marble and Supply Company;

Tile Council of America (TCA);

Trend Mills, Incorporated;

United States Gypsum Company (USG);

U.S. Department of Housing and Urban Development (HUD);

Vermiculite Association;

West Coast Lumber Inspection Bureau (WCLIB);

Western Lath, Plaster, Drywall Industries Association;

Western Wood Products Association;

Wool Bureau, Inc.;

Wunda Weve Carpet Company, Inc.

CHAPTER 9

REFERENCES

Abate, Kenneth. 1989. "Metal Coatings, Fighting the Elements with Superior Paint Systems, Part I." *Metal Architecture*, 5(6)(June):10, 69

———. 1989. "Metal Coatings, Fighting the Elements with Superior Paint Systems, Part II." *Metal Architecture*, 5(7)(July):40–41.

Acoustical Materials Association (AMA). AMA 1-II, "Ceiling Sound Transmission Test by the Two Room Method." Glenview, IL: Ceilings and Interior Systems Contractors Association.

AIA Professional Services Division. *Masterspec*, Basic Sections:

09200, "Lath and Plaster." 11/89 ed. Washington, DC: AIA.

09215, "Veneer Plaster." 2/88 ed. Washington, DC: AIA.

09255, "Gypsum Board Assemblies," 8/92 ed. Washington, DC: AIA.

09270, "Gypsum Board Shaft Wall Systems." 2/88 ed. Washington, DC: AIA.

09300, "Tile." 2/89 ed. Washington, DC: AIA.

09400, "Terrazzo." 5/90 ed. Washington, DC: AIA.

09511, "Acoustical Panel Ceilings." 5/89 ed. Washington, DC: AIA.

09512, "Acoustical Tile Ceilings." 5/89 ed. Washington, DC: AIA.

09521, "Acoustical Wall Panels." 11/90 ed. Washington, DC: AIA.

09546, "Linear Metal Ceilings." 5/92 ed. Washington, DC: AIA.

09550, "Wood Flooring." 11/91 ed. Washington, DC: AIA.

09660, "Resilient Tile Flooring." 8/91 ed. Washington, DC: AIA.

09666, "Sheet Vinyl Floor Covering." 8/91 ed. Washington, DC: AIA.

09678, "Resilient Wall Base and Accessories." 8/91 ed. Washington, DC: AIA.

09680, "Carpet." 4/90 ed. Washington, DC: AIA.

09800, "Special Coatings." 5/92 ed. Washington, DC: AIA.

09900, "Painting." 2/92 ed. Washington, DC: AIA.

American Association of Colorists and Chemists (ASCC) Standards:

30-88, "Anti-fungal Activity Assessment on Textile Materials, Mildew, and Rot Resistance of Textile Materials." Research Triangle Park, NC: ASCC.

100-88, "Assessment of Anti-bacterial Finishes on Textile Materials." Research Triangle Park, NC: ASCC.

134-88, "Electrostatic Propensity of Carpets." Research Triangle Park, NC: ASCC.

American National Standards Institute (ANSI) Standards:

A10.20-1977, "Safety Requirements for Ceramic Tile, Terrazzo, and Marble Work." New York: ANSI.

A108.1-1985, "Glazed Wall Tile, Ceramic Mosaic Tile, Quarry Tile and Paver Tile Installed with Portland Cement." New York: ANSI.

A108.4-1985, "Ceramic Tile Installed with Organic Adhesive or Water-Cleanable Tile Setting Epoxy Adhesive." New York: ANSI.

A108.5-1985, "Ceramic Tile Installed with Dry-Set Portland Cement Mortar or Latex-Portland Cement Mortar." New York: ANSI.

A108.6-1985, "Ceramic Tile Installed with Chemical-Resistant, Water-Cleanable Tile-Setting and Grouting Epoxy." New York: ANSI.

A108.7-1985, "Electrically Conductive Ceramic Tile Installed with Conductive Dry-Set Portland Cement Mortar." New York: ANSI.

A108.8-1985, "Ceramic Tile Installed with Chemical-Resistant Furan Mortar and Grout." New York: ANSI.

A108.9-1985, "Ceramic Tile Installed with Modified Epoxy Emulsion Mortar/ Grout." New York: ANSI.

A108.10-1985, "Installation of Grout in Tilework." New York: ANSI.

A118.1-1985, "Dry-Set Portland Cement Mortar." New York: ANSI.

A118.2-1985, "Conductive Dry-Set Portland Cement Mortar." New York: ANSI.

A118.3-1985, "Chemical-Resistant, Water-Cleanable Tile-Setting and Grouting Epoxy and Water-Cleanable Tile-Setting Epoxy Adhesive." New York: ANSI.

A118.4-1985, "Latex-Portland Cement Mortar." New York: ANSI.

A118.5-1985, "Chemical-Resistant Furan Mortar and Grout." New York: ANSI.

A118.6-1985, "Ceramic Tile Grouts." New York: ANSI.

A118.7-1985, "Modified Epoxy Emulsion Mortar/Grout." New York: ANSI.

A136.1-1985, "Organic Adhesives for Installation of Ceramic Tile." New York: ANSI.

A137.1-1988, "Specifications for Ceramic Tile." New York: ANSI.

American Olean Tile Company, Inc. *Maintenance Guide Manual.* Landsdale, PA: American Olean Tile Company, Inc.

American Paint Contractor. 1988. "Direct to Rust Coatings." *American Paint Contractor,* 65(6)(June):8–19.

Appleman, Bernard R. 1989. "Coatings for Steel Structures." *The Construction Specifier,* 42(3)(March):88–93.

Architectural Graphic Standards. See Ramsey/ Sleeper.

Architectural Technology. 1985. "Technical Tips: Write Good Terrazzo Specs." *Architectural Technology* (Fall):54.

———. 1986. "Technical Tips: Metal Lathing— Then and Now." *Architectural Technology* (March/April):64–65.

———. 1986. "Technical Tips: Paints and Coatings Primer." *Architectural Technology* (July/August):64–65.

Association of Tile, Terrazzo, Marble Contractors and Affiliates (ATTMCA). 1983. *Guide for Grouting and Cleaning Ceramic Floors With Latex Grout.* Jackson, MS: ATTMCA.

Association of the Wall and Ceiling Industries— International. *Veneer Plaster Manual.* Washington, DC: Association of the Wall and Ceiling Industries—International.

American Society for Testing and Materials (ASTM) Standards:

A 82, "Specification for Steel Wire, Plain, for Concrete Reinforcement." Philadelphia, PA: ASTM.

A 185, "Specification for Steel Welded Wire, Fabric, Plain, for Concrete Reinforcement." Philadelphia, PA: ASTM.

C 11, "Definition of Terms Relating to Gypsum and Related Building Materials." Philadelphia, PA: ASTM.

C 28, "Specification for Gypsum Plasters." Philadelphia, PA: ASTM.

C 33, "Specification for Concrete Aggregate." Philadelphia, PA: ASTM.

C 35, "Specification for Inorganic Aggregates for Use in Gypsum Plaster." Philadelphia, PA: ASTM.

C 36, "Specification for Gypsum Wallboard." Philadelphia, PA: ASTM.

C 37, "Specification for Gypsum Lath." Philadelphia, PA: ASTM.

C 44, "Specification for Aggregate for Masonry Mortar." Philadelphia, PA: ASTM.

C 59, "Specification for Gypsum Casting and Molding Plaster." Philadelphia, PA: ASTM.

C 61, "Specification for Gypsum Keene's Cement." Philadelphia, PA: ASTM.

C 79, "Test Method for Gypsum Sheathing Board." Philadelphia, PA: ASTM.

C 91, "Specification for Masonry Cement." Philadelphia, PA: ASTM.

C 144, "Specification for Aggregate for Masonry Mortar." Philadelphia, PA: ASTM.

C 150, "Specification for Portland Cement." Philadelphia, PA: ASTM.

C 206, "Specification for Finishing Hydrated Lime." Philadelphia, PA: ASTM.

C 207, "Specification for Hydrated Lime for Masonry Purposes." Philadelphia, PA: ASTM.

C 236, "Test Method for Steady State Thermal Performance of Building Assemblies by Means of a Guarded Hot Box." Philadelphia, PA: ASTM.

C 242, "Standard Definitions of Terms Related to Ceramic Whitewares and Related Products." Philadelphia, PA: ASTM.

C 287, "Specifications for Gypsum Plasters." Philadelphia, PA: ASTM.

C 332, "Specification for Lightweight Aggregate for Insulating Concrete." Philadelphia, PA: ASTM.

C 346, "Test Method for 45-deg Specular Gloss of Ceramic Materials." Philadelphia, PA: ASTM.

C 423, "Test Method for Sound Absorption of Acoustical Materials by Reverberation Room Method." Philadelphia, PA: ASTM.

C 442, "Specifications for Gypsum Backing Board and Coreboard." Philadelphia, PA: ASTM.

C 472, "Methods for Physical Testing of Gypsum Plasters and Gypsum Concrete." Philadelphia, PA: ASTM.

C 475, "Specifications for Joint Treatment Materials for Gypsum Wallboard Construction." Philadelphia, PA: ASTM.

C 481, "Test Method for Laboratory Aging of Sandwich Construction." Philadelphia, PA: ASTM.

C 514, "Specifications for Nails for the Application of Gypsum Wallboard." Philadelphia, PA: ASTM.

C 557, "Specifications for Adhesive for Fastening Gypsum Wallboard to Wood Framing." Philadelphia, PA: ASTM.

C 587, "Specifications for Gypsum Veneer Plaster." Philadelphia, PA: ASTM.

C 588, "Specifications for Gypsum Base for Veneer Plasters." Philadelphia, PA: ASTM.

C 627, "Evaluating Ceramic Floor Tile Installation Systems." Philadelphia, PA: ASTM.

C 630, "Water-Resistant Gypsum Backing Board." Philadelphia, PA: ASTM.

C 631, "Specifications for Bonding Compounds for Interior Plastering." Philadelphia, PA: ASTM.

C 635, "Metal Suspension Systems for Acoustical Tile and Lay-In Panel Ceilings." Philadelphia, PA: ASTM.

C 636, "Installation of Metal Ceiling Suspension Systems for Acoustical Tile and Lay-In Panels." Philadelphia, PA: ASTM.

C 641, "Test Method for Staining Materials in Lightweight Concrete Aggregates." Philadelphia, PA: ASTM.

C 645, "Specifications for Non-Load (Axial) Bearing Steel Studs, Runners (Track), and Rigid Furring Channels for Screw Application of Gypsum Board." Philadelphia, PA: ASTM.

C 754, "Specifications for Installation of Steel Framing Members to Receive Screw-Attached Gypsum Wallboard, Backing Board, or Water-Resistant Backing Board." Philadelphia, PA: ASTM.

C 840, "Specifications for Application and Finishing of Gypsum Board." Philadelphia, PA: ASTM.

C 841, "Specifications for Installation of Interior Lathing and Furring." Philadelphia, PA: ASTM.

C 842, "Specifications for Application of Interior Gypsum Plaster." Philadelphia, PA: ASTM.

C 843, "Specifications for Application of Gypsum Veneer Plaster." Philadelphia, PA: ASTM.

C 844, "Specifications of Gypsum Base to Receive Gypsum Veneer Plaster." Philadelphia, PA: ASTM.

C 847, "Specifications for Metal Lath." Philadelphia, PA: ASTM.

C 897, "Specifications for Aggregate for Job-Mixed Portland Cement-Based Plasters." Philadelphia, PA: ASTM.

C 919, "Practices for Use of Sealants in Acoustical Applications." Philadelphia, PA: ASTM.

C 920, "Specifications for Elastomeric Joint Sealants." Philadelphia, PA: ASTM.

C 926, "Specifications for Application of Portland Cement-Based Plaster." Philadelphia, PA: ASTM.

C 931, "Specifications for Exterior Gypsum Soffit Board." Philadelphia, PA: ASTM.

C 932, "Specifications for Surface-Applied Bonding Agents for Exterior Plastering." Philadelphia, PA: ASTM.

C 955, "Specifications for Load-Bearing (Transverse and Axial) Steel Studs, Runners (Track), and Bracing or Bridging, for Screw Application of Gypsum Board and Metal Plaster Bases." Philadelphia, PA: ASTM.

C 960, "Specifications for Predecorated Gypsum Board." Philadelphia, PA: ASTM.

C 976, "Method of Test for Thermal Performance of Building Assemblies by Means of a Calibrated Hot Box." Philadelphia, PA: ASTM.

C 1002, "Specifications for Steel Drill Screws for Application of Gypsum Board." Philadelphia, PA: ASTM.

C 1007, "Installation of Load-Bearing (Transverse and Axial) Steel Studs and Accessories." Philadelphia, PA: ASTM.

C 1036, "Standard Specification for Flat Glass." Philadelphia, PA: ASTM.

C 1047, "Specifications for Accessories for Gypsum Wallboard and Gypsum Veneer Base." Philadelphia, PA: ASTM.

C 1063, "Specifications for Installation of Lathing and Furring for Portland Cement-Based Plaster." Philadelphia, PA: ASTM.

D 16, "Standard Definitions of Terms Relating to Paint, Varnish, Lacquer, and Related Products." Philadelphia, PA: ASTM.

D 226, "Specification for Asphalt-saturated Organic Felt Used in Roofing and Waterproofing." Philadelphia, PA: ASTM.

D 418, "Specification for Abrasion Resistance of Concrete by Sandblasting." Philadelphia, PA: ASTM.

D 578, "Specification for Glass Fiber Yarns." Philadelphia, PA: ASTM.

D 659, "Evaluating Degree of Chalking of Exterior Paints." Philadelphia, PA: ASTM.

D 660, "Evaluating Degree of Checking of Exterior Paints." Philadelphia, PA: ASTM.

D 661, "Evaluating Degree of Cracking of Exterior Paints." Philadelphia, PA: ASTM.

D 662, "Evaluating Degree of Erosion of Exterior Paints." Philadelphia, PA: ASTM.

D 1187, "Test Method for Asphalt Emulsions for Use as Protective Coatings for Metal." Philadelphia, PA: ASTM.

D 1335, "Test Method for Tuft Bind of Pile Floor Coverings." Philadelphia, PA: ASTM.

D 1653, "Test Method for Moisture Vapor Permeability of Organic Coating Films." Philadelphia, PA: ASTM.

D 1730, "Recommended Practices for Preparation of Aluminum and Aluminum-Alloy Surfaces for Painting." Philadelphia, PA: ASTM.

D 1731, "Recommended Practices for Preparation of Hot-Dip Aluminum Surfaces for Painting." Philadelphia, PA: ASTM.

D 1779, "Specification for Adhesive for Acoustical Materials." Philadelphia, PA: ASTM.

D 1784, "Specification for Rigid Poly (Vinyl Chloride) (PVC) Compounds and Chlorinated Poly (Vinyl Chloride) (CPVC) Compounds." Philadelphia, PA: ASTM.

D 2000, "Classification System for Rubber Products in Automotive Applications." Philadelphia, PA: ASTM.

D 2092, "Recommended Practices for Preparation of Zinc-Coated Steel Surfaces for Painting." Philadelphia, PA: ASTM.

D 2103, "Specification for Polyethylene Film and Sheeting." Philadelphia, PA: ASTM.

D 2287, "Specification for Nonrigid Vinyl Chloride Polymer and Copolymer Molding and Extrusion Compounds." Philadelphia, PA: ASTM.

D 2833, "Standard Index of Methods for Testing Architectural Paints and Coatings." Philadelphia, PA: ASTM.

D 2859, "Test Method for Flammability of Finished Textile Floor Covering Materials." Philadelphia, PA: ASTM.

D 3276, "Standard Guide for Painting Inspectors (Metal Substrates)." Philadelphia, PA: ASTM.

D 3574, "Method of Testing Flexible Cellular Materials—Slab, Bonded, and Molded Urethane Foams." Philadelphia, PA: ASTM.

D 3794, "Practice for Testing Coil Coatings." Philadelphia, PA: ASTM.

D 3927, "Standard Guide for State and Institutional Purchasing of Paint." Philadelphia, PA: ASTM.

D 4397, "Specification for Polyethylene Sheeting for Construction, Industrial, and Agricultural Applications." Philadelphia, PA: ASTM.

E 84, "Test Method for Surface Burning Characteristics of Building Materials." Philadelphia, PA: ASTM.

E 90, "Laboratory Measurement of Airborne-Sound Transmission Loss of Building Partitions." Philadelphia, PA: ASTM.

E 96, "Test Method for Water Vapor Transmission of Materials." Philadelphia, PA: ASTM.

E 97, "Test Method for 45-deg Directional Reflectance Factor of Opaque Specimens by Broad-Band Filter Reflectometry." Philadelphia, PA: ASTM.

E 119, "Fire Tests for Building Construction and Materials." Philadelphia, PA: ASTM.

E 136, "Test Method for Behavior of Materials in a Vertical Tube Furnace at 750 deg C." Philadelphia, PA: ASTM.

E 413, "Classification for Determination of Sound Transmission Class." Philadelphia, PA: ASTM.

E 488, "Test Method for Strength of Anchors in Concrete or Masonry Elements." Philadelphia, PA: ASTM.

E 580, "Recommended Practice for Application of Ceiling Suspension Systems for Acoustical Tile and Lay-In Panels in Areas Requiring Seismic Restraint." Philadelphia, PA: ASTM.

E 648, "Test Method for Critical Radiant Flux of Floor Covering Systems Using a Radiant Heat Energy Source." Philadelphia, PA: ASTM.

E 662, "Test Method for Specific Optical Density of Smoke Generated by Solid Materials." Philadelphia, PA: ASTM.

E 699, "Practice for Criteria for Evaluation of Agencies Involved in Testing, Quality Assurance, and Evaluating Building Components in Accordance with Test Methods Promulgated by ASTM Committee E-6." Philadelphia, PA: ASTM.

E 1264, "Standard Classification for Acoustical Ceiling Products." Philadelphia, PA: ASTM.

F 1066, "Specification for Vinyl Composition Floor Tile." Philadelphia, PA: ASTM.

F 1303, "Specification for Sheet Vinyl Floor Covering." Philadelphia, PA: ASTM.

F 1344, "Specification for Rubber Floor Tile." Philadelphia, PA: ASTM.

Bakhalov, G. T. and A. V. Turkovakaya. 1965. *Corrosion and Protection of Metals.* Elmsford, NY: Pergamon Press.

Banov, Able. 1973. *Paints and Coatings Handbook for Contractors Architects and Builders.* Farmington, MI: Structures Publishing Co.

Bartlett, Thomas L. 1983. "Cleaning with Corncobs." *The Construction Specifier,* 36(2)(February):6–7.

Bennett, C. R. 1987. "Paints and Coatings: Getting Beneath the Surface." *The Construction Specifier,* 40(2)(February):36–41.

Blatterman, Joan F. 1988. "Details Underfoot." *Architectural Record* (July):118–121.

Blumer, H. Maynard, and Stephanie Stubbs. 1987. "Technical Tips; Choosing Stucco Systems." *Architecture* (January):110–111.

Bocchi, Greg. 1986. "Powder Coatings, A New Technology Takes Off." *The Construction Specifier,* 39(9)(September):102–105.

———. 1988. "Powder Coatings Making Inroads in Metal Construction Market." *Metal Architecture,* 4(4)(April):8–9.

Bodner, James. 1985. "Ceramic Tile Makes a Comeback." *The Construction Specifier,* 38(10)(October):58–63.

———. 1988. "From the Bottom Up: Good Beginnings for Floorcovering Installations." *The Construction Specifier,* 41(4)(April): 98–105.

Bower, Norman F. 1985. "Insurance by the Gallon." *The Construction Specifier,* 38(4)(April):96–99.

Brick Institute of America (BIA). 1972. Technical Note No. 6, "Painting Brick Masonry." Reston, VA: BIA.

Carpet and Rug Institute (CRI). 1987. *Carpet Specifier's Handbook.* Dalton, GA: CRI.

———. 1987. *Flammability of Carpets.* Dalton, GA: CRI.

Carpet Cushion Council (CCC). 1990. *Selecting the Correct Contract Carpet Cushion for Every Traffic Area.* Riverside, CT: CCC.

Ceilings and Interior Systems Contractors Association (CISCA). *Acoustical Ceilings— Use and Practice.* Deerfield, IL: CISCA.

Ceramic Tile Institute. *Ceramic Tile Manual.* Los Angeles, CA: Ceramic Tile Institute.

Chalmers, Ray. 1985. "Selecting the Proper Tile Application Method." *Building Design and Construction,* 26(9)(September):118–120.

———. 1985. "A Review of Related Products for Ceramic Tile." *Building Design and Construction,* 26(9)(September):124–125.

Commercial Remodeling. 1981. "Technique: Ion Painting." *Commercial Remodeling,* 3(1)(February):24.

Commercial Renovation. 1987. "Product Focus: Ceiling Systems Reach New Heights in Performance and Design." *Commercial Renovation,* 9(6)(December):48–53.

Construction Specifications Institute. 1988. *Specguide 09900, Painting.* Alexandria, VA: The Construction Specifications Institute.

Construction Specifier, The. 1985. "Painting to Protect." *The Construction Specifier,* 38(4)(April):92–94, 123.

Consumer Reports. 1985. "Concrete-Floor Paints." *Consumer Reports,* 50(3)(March): 166–169.

Council of American Building Officials (CABO). 1983. *One- and Two-Family Dwelling Code.* Falls Church, VA: CABO.

Dean, Sheldon W., and T. S. Lee, eds. 1988. *Degradation of Metals in the Atmosphere.* Philadelphia, PA: ASTM.

Diehl, John R. 1965. *Manual of Lathing and Plastering, Gypsum Association Edition.* New York: MAC Publishers Association.

Domtar Gypsum America, Inc. 1983. *620 Series Shaftwalls.* Oakland, CA: Domtar Gypsum America, Inc.

Doyle, Margaret. 1985. "Ceramic Tile Suits a Variety of Uses." *Building Design and Construction,* 26(9)(September):114–117.

Factory Mutual Systems. *Approval Guide.* Norwood, MA: Factory Mutual Systems.

———. *Loss Prevention Data Sheets.* Norwood, MA: Factory Mutual Systems.

Foster, Al. 1988. "Textured Coatings: Acoustical in the Name Doesn't Necessarily Mean the Product Absorbs Noise." *PWC Magazine,* 50(3)(May/June):34–40.

Frane, James T. 1987. *Drywall Contracting.* Carlsbad, CA: Craftsman Book Company.

Garrison, John Mark. 1985. "Casting Decorative Plaster." *The Old-House Journal,* 13(9)(November):186–189.

Gorman, J. R. 1982. "Smooth Specifying for Architectural Textures in Plaster." *The Construction Specifier,* 35(5)(July):76–82.

Gypsum Association. GA-201-90, "Using Gypsum Board for Walls and Ceilings." Washington, DC: Gypsum Association.

———. GA-216-89, "Recommended Specifications—Application and Finishing of Gypsum Board." Washington, DC: Gypsum Association.

———. GA-219-89, "Recommended Specifications: Recommendations for Installation of Steel Fire Door Frames in Steel Stud-Gypsum Board Fire-Rated Partitions." Washington, DC: Gypsum Association.

———. GA-223-86, "Recommended Specifications: Gypsum Board Types, Uses, Sizes and Standards." Washington, DC: Gypsum Association.

———. GA-252-86, "Fire Resistant Gypsum Sheathing." Washington, DC: Gypsum Association.

———. GA-505-91, "Glossary—Gypsum Board Terminology." Washington, DC: Gypsum Association.

———. GA-600-86, "Fire Resistance Design Manual." 11th ed. Washington, DC: Gypsum Association.

———. GA-650-86, "Recommendations for Covering Existing Interior Walls and Ceilings with Gypsum Board." Washington, DC: Gypsum Association.

Hardingham, David. 1980. "Preparing for Painting." *The Old-House Journal* (October): 133–136.

Harris, D. A. 1983. "Ceilings for the Office of the Future." *The Construction Specifier,* 36(7)(July):58–65.

Herman, Frederick. 1981. "Refinishing Floors: Think Twice Before Sanding." *The Old-House Journal* (February):27, 44–45.

Italian Tile Center. *The Designer's Guide to Italian Tiles and Their Installation.* New York: Italian Tile Center.

Johnston, Clay M. 1987. "Lath and Plaster: A Capsule Look." *The Construction Specifier,* 40(2)(February):54–58.

Jones, Larry. 1984. "Painting Galvanized Metal." *The Old-House Journal,* 12(1)(January/February):10–11.

Jowers, Walter. 1986. "Textured Plaster Finishes." *The Old-House Journal,* 14(2)(March): 75–77.

Kawneer Company, Inc. 1984. *Architectural Finishes.* Norcross, GA: Kawneer.

Kincaid, Mary. 1982. "What Paint Experts Say." *The Construction Specifier,* 35(5)(July):54–61.

———. 1983. "Color in the Squares." *The Construction Specifier,* 36(6)(June):70–76.

Kneemiller, Bill. 1982. "Using Exterior Textured Coatings." *The Construction Specifier,* 35(5)(July):63–68.

Kosmatka, Steven H. 1987. "Floor-Covering Materials and Moisture in Concrete." *The Construction Specifier,* 40(5)(May):35–36.

Labine, Clem. 1982. "Restoring Clear Finishes." *The Old-House Journal,* 10(11)(November): 221, 238–241.

Lowes, Robert. 1988. "Abrasive Blasting." *PWC Magazine,* 50(3)(May/June):26–30.

MacDonald, John E. 1978. "The Effects of Light on Wood and Wood Finishing Systems." *The Construction Specifier,* 31(3)(March): 23–26.

Mahowald, Dave. 1988. "Specifying Paint Coatings for Harsh Environments." *The Construction Specifier*, 41(10)(October):13–16.

Maple Flooring Manufacturers Association (MFMA). 1988. *Grading Rules for Hard Maple*. Northbrook, IL: MFMA.

———. *Sanding, Sealing, Court Lining and Finishing Maple Gym Floors*. Northbrook, IL: MFMA.

———. *Wood Flooring, Maple*. Northbrook, IL: MFMA.

Martens, Charles R., and The Sherwin-Williams Company. 1974. *Technology of Paints and Lacquers*. New York: Robert E. Krieger Publishing Company.

Mazzur, Richard P. 1986. "Resilient Flooring: Choose Carefully." *The Construction Specifier*, 39(3)(March):70–75.

McDonald, Timothy B. 1987. "Technical Tips: Coatings That Protect Against the Corrosion of Steel." *Architecture* (July): 101–102.

McIlvain, Jess. 1982. "The Problem with Dry-Set Mortars." *Tile and Decorative Surfaces* (September):17–20.

———. 1982. "Ceramic Tile Maintenance." *Tile and Decorative Surfaces* (October/November):48–57.

———. 1982. "The Preconstruction Conference." *Tile and Decorative Surfaces* (December):102–103.

———. 1983. "Caution: Latex-Portland Cement Mortar Over Plywood." *Tile and Decorative Surfaces* (January/February).

———. 1983. "Plaster vs. Mortar Beds." *Tile and Decorative Surfaces* (April):41–56.

———. 1983. "Dodging the Dry-Set Mortar Gremlins." *The Construction Specifier*, 36(5)(June):77–82.

———. 1983. "Never Again." *Tile and Decorative Surfaces* (July):138–140.

———. 1984. "Prefabricated Exterior Ceramic Tile Cladding, Part 1." *Tile and Decorative Surfaces* (January/February):80–85.

———. 1984. "Prefabricated Exterior Ceramic Tile Cladding, Part 2." *Tile and Decorative Surfaces* (March):41–53.

———. 1984. "Prefabricated Exterior Ceramic Tile Cladding, Part 3." *Tile and Decorative Surfaces* (April):50–58.

———. 1984. "Prefabricated Exterior Ceramic Tile Cladding." *The Construction Specifier*, 37(6)(June):76–91.

———. 1984. "Let's Talk Tile." *Tile and Decorative Surfaces* (June): 69–74.

———. 1984. "Mortar Beds." *Tile and Decorative Surfaces* (September):118–120.

———. 1984. "The Preconstruction Conference." *Tile Letter* (October):32–33.

———. 1985. "Floor Renovation." *Tile and Decorative Surfaces* (April): 48–49.

———. 1985. "Moisture Expansion." *Tile and Decorative Surfaces* (June):72–86.

———. 1985. "The Goose That Laid the Rotten Egg." *Tile Letter* (August):42–46.

———. 1986. "Writing Ceramic Tile Specifications." *Tile and Decorative Surfaces* (March):78–80.

———. 1986. "Lights, Action." *Tile Letter* (July):36–40.

———. 1986. "The Effects of Light on Tile." *Tile and Decorative Surfaces* (October/November):81–82.

———. 1987". Selecting Ceramic Tile Floor Installation Methods." *Tile and Decorative Surfaces* (August):38–42.

———. 1988. "Grout: It Doesn't Have to Be a Problem." *Tile and Decorative Surfaces* (June):42–45.

Metal Architecture. 1987. "Building Repainting System Duplicates Coil Coatings in Appearance, Life Expectancy." *Metal Architecture*, 3(11)(November):24.

Metal Lath/Steel Framing Association (ML/SFA). 1985. *Lightweight Steel Framing Systems Manual*. 2d ed. Chicago, IL: ML/SFA.

———. 1986. *Technical Information File*. Chicago, IL: ML/SFA.

———. 1986. *Specifications For Metal Lathing and Furring*. Chicago, IL: ML/SFA.

———. 1991. ML/SFA 920, "Guide Specifications for Metal Lath and Furring." Chicago, IL: ML/SFA.

———. *Fire Rated Metal Lath/Steel Stud Exterior*. Technical Bulletin No. 18. Chicago, IL: ML/SFA.

———. *Types of Metal Lath and Their Uses*. Technical Bulletin No. 101. Chicago, IL: ML/SFA.

———. *Ceramic Tile Applied to Metal Lath and Plaster*. Technical Bulletin No. 120. Chicago, IL: ML/SFA.

Minnery, Catherine, and Donald Minnery. 1979. "Repairing Stucco." *The Old-House Journal*, 7(7)(July):73, 77–79.

Moit, Dan. 1988. "Coatings for Metals: Preplan the Selection." *Metal Architecture*, 4(9)(September):8.

Moormann, Ambrose F. Jr. 1982. "Paint and the Prudent Specifier." *The Construction Specifier*, 35(5)(July):69–71.

———. 1982. "Paint and the Prudent Specifier: If You Must Paint Concrete Floors." *The Construction Specifier*, 35(9)(September):60–62.

———. 1983. "Paint and the Prudent Specifier: Working Hard to Look Natural." *The Construction Specifier*, 36(1)(January):84–87.

Munger, Charles G. 1984. *Corrosion Prevention by Protective Coatings*. Houston, TX: National Association of Corrosion Engineers.

National Association of Tile Contractors. *Reference Manual and Specifications*. Jackson, MS: National Association of Tile Contractors.

National Concrete Masonry Association. 1985. NCMA-TEK 149, "Portland Cement Plaster (Stucco) for Concrete Masonry." Herndon, VA: National Concrete Masonry Association.

National Decorating Products Association. 1988. *Paint Problem Solver*. St. Louis, MO: National Decorating Products Association. [Also available from the Painting and Decorating Contractors of America.]

National Fire Protection Association (NFPA). NFPA 253-84, "Test Method for Critical Radiant Flux of Floor Covering Systems Using a Radiant Heat Energy Source." Quincy, MA: NFPA.

———. NFPA 255-84, "Test Method for Surface Burning Characteristics of Building Materials." Quincy, MA: NFPA.

National Oak Flooring Manufacturers Association (NOFMA). *Hardwood Flooring Installation Manual*. Memphis, TN: NOFMA.

———. *Hardwood Flooring Finishing/Refinishing Manual*. Memphis, TN: NOFMA.

———. *NOFMA Official Flooring Grading Rules*. Memphis, TN: NOFMA.

National Paint and Coatings Association (NPCA). *Finishing Hardwood Floors*. Washington, DC: NPCA.

National Terrazzo and Mosaic Association (NTMA). 1992. *Terrazzo Information Guide*. Des Plains. IL: NTMA.

O'Donnell, Bill. 1985. "Reconditioning Floors." *The Old-House Journal*, 13(10)(December): 201, 218–219.

———. 1986. "Unwanted Textured Finish: How to Get Rid of It." *The Old-House Journal* (October):374–377.

Old-House Journal, The. 1982. "Stripping Paint." *The Old-House Journal* (December): 249–252.

———. 1983. "48 Paint Stripping Tips." *The Old-House Journal*, 11(2)(March):44–45.

———. 1983. "Our Opinion of 'Peel Away.' " *The Old-House Journal*, 11(4)(May):80.

———. 1983. "Ask OHJ: Paint on Paint." *The Old-House Journal*, 11(9)(November):202.

———. 1984. "Tips on Mixing Paint." *The Old-House Journal*, 12(1)(January/February):8–9.

———. 1985. "Reconditioning Floors." *The Old-House Journal*, 13(10)(December): 203, 218–219.

———. 1985. "Stripping Clinic." *The Old-House Journal*, 13(10)(December):212B.

———. 1987. "Exterior Painting: Problems and Solutions." *The Old-House Journal* (September/October):35–39.

———. 1988. "Commercial Paint Stripping." *The Old-House Journal*, 16(4)(July/August):29–33.

Olson, Christopher. 1987. "Improved Products Broaden Tile Applications." *Building Design and Constructions*, 28(4)(April):98–101.

O'Neil, Edward F., and James E. McDonald. February 1976. *An Evaluation of Selected Instruments Used to Measure the Moisture Content of Hardened Concrete.* Technical Report C-76-1. Vicksburg, MS: Concrete Laboratory, U.S. Army Engineer Waterways Experiment Station.

Painting and Decorating Contractors of America (PDCA). 1975. *Painting and Decorating Craftsman's Manual and Textbook.* 5th ed. Falls Church, VA: PDCA.

———. 1982. *Painting and Decorating Encyclopedia.* Falls Church, VA: PDCA.

———. 1984. *Painting and Wallcovering: A Century of Excellence.* Falls Church, VA: PDCA.

———. 1986. *Architectural Specification Manual, Painting, Repainting, Wallcovering and Gypsum Wallboard Finishing.* 3d ed. Kent, WA: Specifications Services, Washington State Council of PDCA.

———. 1988. *Hazardous Waste Handbook.* Falls Church, VA: PDCA.

———. 1988. *The Master Painters Glossary.* Falls Church, VA: PDCA.

———. 1988. *PDCA Estimating Guide.* Falls Church, VA: PDCA.

Petersen, Maurice R. 1984. "Finishes on Metals: A View From the Field—Part 1." *The Construction Specifier,* 37(12)(December):36–39.

———. 1985. "Finishes on Metals: A View From the Field—Part 2." *The Construction Specifier,* 38(1)(January):70–73.

Poore, Patricia. 1980. "Sanding a Parquet Floor." *The Old-House Journal* (November): 168–171.

———. 1981. "Picking a Floor Finish." *The Old-House Journal* (May):107–113.

———. 1985. "Stripping Paint from Exterior Wood." *The Old-House Journal,* 13(10) (December):207–211.

Portland Cement Association (PCA). *Portland Cement Plaster (Stucco) Manual.* Skokie, IL: PCA.

———. *Bonding Concrete or Plaster to Concrete.* Skokie, IL: PCA.

———. 1977. *Painting Concrete, IS134T.* Skokie, IL: PCA.

———. 1982. *Removing Stains and Cleaning Concrete Surfaces.* Skokie, IL: PCA.

———. 1986. *Effects of Substances on Concrete and Guide to Protective Treatments.* Skokie, IL: PCA.

Raeber, John A. 1989. "Composite Plaster and Masonry Construction: A Closer Look." *The Construction Specifier,* 42(9)(September): 31–32.

Ramsey/Sleeper, The AIA Committee on Architectural Graphic Standards. 1981. *Architectural Graphic Standards.* 8th ed. New York: Wiley.

Resilient Floor Covering Institute (RFCI). *Maintenance of Solid Vinyl Tile for Commercial, Institutional, and Industrial Buildings.* Rockville, MD: RFCI.

———. *Recommended Installation Specifications for Resilient Tile as a Wall Covering.* Rockville, MD: RFCI.

———. *Recommended Installation Specifications for Self-adhering Tile on Floors.* Rockville, MD: RFCI.

———. *Recommended Installation Specifications for Sheet Vinyl Flooring, Flat-layed, Fully Adhered to the Floor.* Rockville, MD: RFCI.

———. *Recommended Installation Specifications for Vinyl Composition, Solid Vinyl, and Asphalt Tile Floorings.* Rockville, MD: RFCI.

———. RFCI Specification ADH-1, "Recommended Specification for Vinyl Composition Tile Adhesive." Rockville, MD: RFCI.

———. RFCI Specification ADH-2, "Recommended Specification for Wall Base Adhesive." Rockville, MD: RFCI.

———. RFCI Specification CL-1, "Recommended Specification for Cleaners for Use on Resilient Floor Coverings." Rockville, MD: RFCI.

———. RFCI Specification FP-1, "Recommended Specification for Water Emulsion Floor Polish for Use on Resilient Flooring." Rockville, MD: RFCI.

———. RFCI Specification SV-1, "Recommended Specification for Resilient Floor Covering—Vinyl Plastic Sheet." Rockville, MD: RFCI.

———. RFCI Specification VCT-1, "Recommended Specification for Vinyl Composition Floor Tile." Rockville, MD: RFCI.

———. RFCI Specification VT-100, "Recommended Specification for Solid Vinyl Floor Tile." Rockville, MD: RFCI.

———. RFCI Specification VT-200, "Recommended Specification for Self Adhering Resilient Floor Tile." Rockville, MD: RFCI.

———. RFCI Specification WB-1, "Recommended Specification for Vinyl Wall Base." Rockville, MD: RFCI.

Sherwin-Williams. 1991. *One-Hundred Twenty-Five Years of Providing Painting and Coating Systems for Specifiers and Applicators.* Cleveland, OH: Sherwin-Williams.

Southern Pine Inspection Bureau (SPIB). *Grading Rules.* Pensocola, FL: SPIB.

Steel Structures Painting Council (SSPC). 1982. *Steel Structures Painting Manual.* Vols. I and II. 2d ed. Pittsburgh, PA: SSPC.

———. SSPC-6, "Surface Preparation Specification No. 6—Commercial Blast Cleaning." Pittsburgh, PA: SSPC.

———. SSPC-10, "Surface Preparation Specification No. 10—Near-white Blast Cleaning." Pittsburgh, PA: SSPC.

———. SSPC-16, "Paint Specification 16—Coal Tar Epoxy-polyamide Black (or dark red) Paint." Pittsburgh, PA: SSPC.

Stubbs, M. Stephanie. 1986. "Technical Tips: Paints and Coatings Primer." *Architectural Technology* (July/August):64–65.

Tile Contractors Association of America (TCAA). *Products and Materials Guide.* Alexandria, VA: TCAA.

Tile Council of America (TCA) 1988. *Handbook for Ceramic Tile Installation.* Princeton, NJ: TCA.

Underwriters Laboratories, Inc. (UL). *Building Materials Directory—Class A, B, C: Fire and Wind Related Deck Assemblies.* Northbrook, IL: UL.

———. *Fire Resistance Directory—Time/Temperature Constructions.* Northbrook IL: UL.

———. UL 992-76, "Smoke Chamber Test." Northbrook IL: UL.

United States Gypsum Company (USG). 1972. *Red Book: Lathing and Plastering Handbook.* 28th ed. Chicago, IL: USG.

———. 1987. *Gypsum Construction Handbook.* 3d ed. Chicago, IL: USG.

U.S. Department of the Army. 1977 (amended 1982). Corps of Engineers Guide Specification CEGS-09500, "Acoustical Treatment." U.S. Department of the Army.

———. 1980. *Painting: New Construction and Maintenance (EM 1110-2-3400).* Washington DC: Superintendent of Documents, U.S. Government Printing Office.

———. 1981. Corps of Engineers Guide Specifications, Military Construction, CEGS-09910, "Painting, General." Office of the Chief of Engineers, Department of the Army.

———. 1982. Guide Specification for Military Construction, CE-240.01, "Furring (Metal), Lathing, and Plastering." Corps of Engineers, Office of the Chief of Engineers, Department of the Army.

U.S. Departments of Army, the Navy, and the Air Force. 1969. *Paints and Protective Coatings.* Technical Manual TM 5-618. Washington, DC: U.S. Government Printing Office.

U.S. Department of Commerce. 1968. *Organic Coatings BSS 7.* Washington, DC: U.S. Department of Commerce.

———. Federal Standard DOC FF 1-70, "Methenamine Pill Test." Washington, DC: U.S. Department of Commerce.

U.S. Department of Defense (DOD). 1970. Military Specification MIL-P-21035, "Paint, High Zinc Dust Content, Galvanizing Repair." DOD.

U.S. Department of the Navy, Naval Facilities Engineering Command. 1984 (February). Guide Specifications Section 09100, "Metal Support Systems." Department of the Navy.

———. 1982 (August). Guide Specifications Section 09110, "Lathing." Department of the Navy.

———. 1985 (July). Guide Specifications Section 09150, "Plastering and Stuccoing." Department of the Navy.

———. 1985 (March). Guide Specifications Section 09250, "Gypsum Board." Department of the Navy.

———. 1984 (March). Guide Specifications Section 09310, "Ceramic Tile, Quarry Tile, and Paver Tile." Department of the Navy.

———. 1983 (May). Guide Specifications Section 09500, "Acoustical Treatment." Department of the Navy.

———. 1982 (November). Guide Specifications Section 09650, "Resilient Flooring." Department of the Navy.

———. 1986 (March). Guide Specifications Section 09815, "High-Build Glaze Coatings." Department of the Navy.

———. 1986 (February). Guide Specifications Section 09910C, "Painting of Buildings (Field Painting)." Department of the Navy.

U.S. General Services Administration. 1970 (April). Public Building Services Guide Specification, Section 0910, "Plaster Repairs." U.S. General Services Administration.

———. 1971 (February). Public Building Services Guide Specification, Section 3-0990, "Painting and Finishing, Renovation, Repair and Improvement." U.S. General Services Administration.

———. 1973 (October). Public Building Services Guide Specification, Section 4-0990.01, "Painting and Finishing." U.S. General Services Administration.

U.S. General Services Administration (GSA) Specifications Unit. Federal Specifications:

L-C-001369, "Bonded Urethane Carpet and Rug Cushion." Washington, DC: GSA Specifications Unit.

L-C-001676, "Virgin Urethane Carpet and Rug Cushion." Washington, DC: GSA Specifications Unit.

QQ-L-101, "Lath, Metal (and Other Metal Plaster Bases)." Washington, DC: GSA Specifications Unit.

RR-T-650, "Treads, Metallic and Non-Metallic." Washington, DC: GSA Specifications Unit.

SS-T-312, "Tile, Floor: Asphalt, Rubber, Vinyl, Vinyl Composition." Washington, DC: GSA Specifications Unit.

SS-W-40, "Wall Base: Rubber and Vinyl Plastic." Washington, DC: GSA Specifications Unit.

TT-C-535, "Coating, Epoxy, Two Component, for Interior Use on Metal, Wood, Wallboard, Painted Surfaces, Concrete, and Masonry." Washington, DC: GSA Specifications Unit.

TT-C-542, "Coating, Polyurethane, Oil-free, Moisture Curing." Washington, DC: GSA Specifications Unit.

TT-E-487E, "Enamel, Floor and Deck." Washington, DC: GSA Specifications Unit.

TT-E-489G, "Enamel, Alkyd, Gloss (for Exterior and Interior Surfaces)." Washington, DC: GSA Specifications Unit.

TT-E-496, "Enamel, Heat-resisting, (400 Deg F), Black." Washington, DC: GSA Specifications Unit.

TT-E-505A, "Enamel, Odorless, Alkyd, Interior, High Gloss, White and Light Tints." Washington, DC: GSA Specifications Unit.

TT-E-506K, "Enamel, Alkyd, Gloss, Tints and White (for Exterior and Interior Surfaces)." Washington, DC: GSA Specifications Unit.

TT-E-508C, "Enamel, Interior, Semi-gloss, Tints and White." Washington, DC: GSA Specifications Unit.

TT-E-509B, "Enamel, Odorless, Alkyd, Interior, Semigloss, White and Tints." Washington, DC: GSA Specifications Unit.

TT-E-527C, "Enamel, Alkyd, Luster-less." Washington, DC: GSA Specifications Unit.

TT-E-543A, "Enamel, Interior, Undercoat, Tints and White." Washington, DC: GSA Specifications Unit.

TT-E-545, "Enamel, Odorless, Alkyd, Interior, Undercoat, Flat, Tints and White." Washington, DC: GSA Specifications Unit.

TT-E-1593, "Enamel, Silicone Alkyd Copolymer, Gloss (for Exterior and Interior Use)." Washington, DC: GSA Specifications Unit.

TT-F-322D, "Filler Two-component Type: For Dents, Cracks, Small-holes, and Blow Holes." Washington, DC: GSA Specifications Unit.

TT-F-336E, "Filler, Wood, Paste." Washington, DC: GSA Specifications Unit.

TT-F-340C, "Filler, Wood, Plastic." Washington, DC: GSA Specifications Unit.

TT-L-58E, "Lacquer, Spraying, Clear and Pigmented for Interior Use." Washington, DC: GSA Specifications Unit.

TT-L-190D, "Linseed Oil, Boiled (for Use in Organic Coatings)." Washington, DC: GSA Specifications Unit.

TT-L-201A, "Linseed Oil, Heat Polymerized." Washington, DC: GSA Specifications Unit.

TT-P-19C, "Paint, Acrylic Emulsion: Exterior." Washington, DC: GSA Specifications Unit.

TT-P-25E, "Primer Coating, Exterior (Undercoat for Wood, Ready-mixed, White and Tints)." Washington, DC: GSA Specifications Unit.

TT-P-29J, "Paint, Latex Base, Interior, Flat, White and Tints." Washington, DC: GSA Specifications Unit.

TT-P-30E, "Paint, Alkyd, Odorless, Interior, Flat White and Tints." Washington, DC: GSA Specifications Unit.

TT-P-37D, "Paint, Alkyd Resin; Exterior Trim, Deep Colors." Washington, DC: GSA Specifications Unit.

TT-P-47F, "Paint, Oil, Nonpenetrating-flat, Ready-mixed Tints and White (for Interior Use). Shakes and Rough Siding." Washington, DC: GSA Specifications Unit.

TT-P-52D, "Paint, Oil (Alkyd Oil) Wood Shakes and Rough Siding." Washington, DC: GSA Specifications Unit.

TT-P-55B, "Paint, Polyvinyl Acetate Emulsion, Exterior." Washington, DC: GSA Specifications Unit.

TT-P-81E, "Paint, Oil, Alkyd, Ready Mixed Exterior, Medium Shades." Washington, DC: GSA Specifications Unit.

TT-P-86G, "Paint, Red-lead-base, Ready-mixed." Washington, DC: GSA Specifications Unit.

TT-P-615D, "Primer Coating, Basic Lead Chromate, Ready Mixed." Washington, DC: GSA Specifications Unit.

TT-P-636D, "Primer Coating, Alkyd, Wood and Ferrous Metal." Washington, DC: GSA Specifications Unit.

TT-P-641G, "Primer Coating; Zinc Dust-Zinc Oxide (for Galvanized Surfaces)." Washington, DC: GSA Specifications Unit.

TT-P-645A, "Primer, Paint, Zinc Chromate, Alkyd Type." Washington, DC: GSA Specifications Unit.

TT-P-650C, "Primer Coating, Latex Base, Interior, White (For Gypsum Wallboard)." Washington, DC: GSA Specifications Unit.

TT-P-664C, "Primer Coating, Synthetic, Rust-inhibiting, Lacquer-resisting." Washington, DC: GSA Specifications Unit.

TT-P-791A, "Putty, Pure-Linseed-Oil (for Wood-sash-glazing)." Washington, DC: GSA Specifications Unit.

TT-P-1511A, "Paint, Latex-base, Gloss and Semi-gloss, Tints and White (for Interior Use)." Washington, DC: GSA Specifications Unit.

TT-S-176E, "Sealer, Surface, Varnish Type, Floor, Wood and Cork." Washington, DC: GSA Specifications Unit.

TT-S-300A, "Shellac, Cut." Washington, DC: GSA Specifications Unit.

TT-S-708A, "Stain, Oil; Semi-transparent, Wood, Exterior." Washington, DC: GSA Specifications Unit.

TT-S-711C, "Stain, Oil Type, Wood, Interior." Washington, DC: GSA Specifications Unit.

TT-T-291F, "Thinner, Paint, Mineral Spirits, Regular and Odorless." Washington, DC: GSA Specifications Unit.

TT-V-86C, "Varnish, Oil, Rubbing (for Metal and Wood Furniture)." Washington, DC: GSA Specifications Unit.

UU-B-790A, "Building Paper, Vegetable Fiber (Kraft, Waterproofed, Water-resistant, and Fire-resistant." Washington, DC: GSA Specifications Unit.

ZZ-C-00811b, "Cellular Rubber Carpet and Rug Cushion." Washington, DC: GSA Specifications Unit.

ZZ-T-001237, "Tread, Stair, Flexible and Semi-rigid Type Rubber and Vinyl." Washington, DC: GSA Specifications Unit.

DDD-C-001023, "Carpet and Rug Cushion (Hair Felt and Rubber Coated Jute and Animal Hair or Fiber)." Washington, DC: GSA Specifications Unit.

MMM-A-00150, "Adhesive for Acoustical Materials." Washington, DC: GSA Specifications Unit.

Weaver, Martin E. 1989. "Fighting Rust, Part I: A Backgrounder." *The Construction Specifier*, 42(5)(May):143–145.

———. 1989. "Fighting Rust, Part II: Remedies." *The Construction Specifier*, 42(6)(June):129–130.

Weinstein, Nat. 1984. "How to Match Paint Colors." *The Old-House Journal*, 12(1) (January/February):7–9.

Weismantel, Guy E., ed. 1981. *Paint Handbook*. New York: McGraw-Hill.

West Coast Lumber Inspection Bureau (WCLIB). *Standard Grading Rules for West Coast Lumber*. Portland, OR: WCLIB.

Western Lath, Plaster, Drywall Industries Association. 1988. *Plaster/Metal Framing System/Lath Manual*. New York: McGraw Hill.

Wood and Synthetic Flooring Institute (WSFI). *Wood and Synthetic Flooring Specifications, Details, Product Data*. Hillside, IL: WSFI.

Wright, Gordon. 1987. "Trends in Specifying Architectural Coatings." *Building Design and Construction*, 28(6)(June):188–192.

Zulandt, David F. 1985. "Consider an Alternative: Veneer Plaster." *The Construction Specifier*, 38(7)(July):84–85.

CHAPTER 10

ACKNOWLEDGMENTS

Brick Institute of America (BIA);

Carpet and Rug Institute (CRI);

Celotex Corporation;

Flexicore Manufacturers Association;

Gypsum Association (GA);

Kodaras Acoustical Laboratories;

National Association of Home Builders (NAHB) Research Foundation;

National Concrete Masonry Association (NCMA);

National Gypsum Company;

Owens-Corning Fiberglas Corporation;

Portland Cement Association (PCA);

Riverbank Acoustical Laboratories;

Rodale Publishing Company;

IIT Research Corporation;

United States Gypsum Company;

U.S. Department of Housing and Urban Development (HUD).

CHAPTER 10

REFERENCES

Acoustical Materials Association (AMA). AMA 1-II, "Ceiling Sound Transmission Test by the Two Room Method." Glenview, IL: Ceilings and Interior Systems Contractors Association.

American Architectural Manufacturers Association (AAMA). AAMA TIR-A1-1975, "Sound Control for Aluminum Curtain Walls and Windows." Des Plaines, IL: AAMA.

American Society of Heating, Refrigerating and Air-Conditioning Engineers (ASHRAE). 1989. *1989 ASHRAE Handbook, Fundamentals*. Atlanta, GA: ASHRAE.

American Society for Testing and Materials (ASTM) Standards:

C 919, "Practices for Use of Sealants in Acoustical Applications." ASTM.

C 920, "Specifications for Elastomeric Joint Sealants." ASTM.

D 1779, "Specification for Adhesive for Acoustical Materials." ASTM.

E 90, "Laboratory Measurement of Airborne-Sound Transmission Loss of Building Partitions." ASTM.

E 413, "Classification for Determination of Sound Transmission Class." ASTM.

E 492, "Method for Laboratory Measurement of Impact Sound Transmission Through Floor-Ceiling Assemblies Using the Tapping Machine." ASTM.

Council of American Building Officials (CABO). *Model Energy Code*. Falls Church, VA: CABO.

———. 1983. *One- and Two-Family Building Code*. Falls Church, VA: CABO.

Hughes, Richard L., and Martin F. Johnson. 1983. "Acoustic Design in Natatoriums." *The Construction Specifier*, 36(7)(July): 67–70.

International Organization for Standardization (ISO). 1960. *Field and Laboratory Measurements of Airborne and Impact Sound Transmission, ISO Recommendations #140-1960*. ISO.

Kincaid, Mary. 1983. "Making Open Office Acoustics Work." *The Construction Specifier*, 36(7)(July):70–76.

Mazria, Edward. 1979. *The Passive Solar Energy Book*. Emmaus, PA: Rodale Publishing Company.

Ramsey/Sleeper, The AIA Committee on Architectural Graphic Standards. 1981. *Architectural Graphic Standards*, 7th ed. New York: Wiley.

Stanwood, Les. 1983. "Taking a Look at Noise." *The Construction Specifier*, 36(7) (July):50–55.

United States Department of Housing and Urban Development (HUD), Federal Housing Administration. 1974. *A Guide to Airborne, Impact, and Structure-borne Noise Control in Multifamily Dwellings, HUD-TS-24*. Washington, DC: FHA.

CHAPTER 11

ACKNOWLEDGMENTS

American Plywood Association (APA);

American Society of Heating, Refrigerating, and Air-Conditioning Engineers (ASHRAE);

The Delta Faucet Company;

Grinnell Fire Protectuion Company;

National Association of Home Builders (NAHB) Research Foundation;

National Forest Products Association (NFPA);

Small Homes Council, Building Research Council;

Southern Forest Products Association (SFPA);

U.S. Department of Housing and Urban Development (HUD);

U.S. Environmental Protection Agency (EPA);

Western Wood Products Association (WWPA).

CHAPTER 11

REFERENCES

Air-Conditioning and Refrigeration Institute (ARI), Sheet Metal and Air Conditioning Contractors National Association (SMACNA), and National Roofing Contractors Association (NRCA). *Guidelines for Roof Mounted Outdoor Air-Conditioner Installations*. ARI, SMACNA, and NRCA.

Air-Conditioning Contractors of America (ACCA). *Manual B—Principles of Air Conditioning*. Washington, DC: ACCA.

———. *Manual B—What Makes a Good Air-Conditioning System*. Washington, DC: ACCA.

———. *Manual D—Duct Design for Residential Winter and Summer Air Conditioning*. Washington, DC: ACCA.

———. *Manual G—Selection of Distribution Systems*. Washington, DC: ACCA.

———. *Manual H—Heat Pump Systems: Principles and Applications*. Washington, DC: ACCA.

———. *Manual J—Load Calculations for Residential Winter and Summer Air Conditioning*. Washington, DC: ACCA.

———. *Manual N—Load Calculations for Commercial Summer and Winter Air Conditioning (Using Unitary Equipment)*. Washington, DC: ACCA.

———. *Manual S—Residential Equipment Selection*. Washington, DC: ACCA.

———. Manual T—Air Distribution Basics for Residential and Small Commercial Buildings. Washington, DC: ACCA.

Air Movement and Control Association (AMCA). 1985. Application Method for Air Louvers. Arlington Heights, IL: AMCA.

———. 1986. Test Methods for Louvers, Dampers, and Shutters. Arlington Heights, IL: AMCA.

American National Standards Institute (ANSI) Standards.

    A112, "Plumbing." New York: ANSI.

    Z21.22, "Valves and Automatic Gas Shutoff Devices for Hot Water Supply Systems, Relief." New York: ANSI.

    Z21.10, "Heaters, Water." New York: ANSI.

    Z124.1, "Plastic Bathtub Units." New York: ANSI.

    Z124.2, "Plastic Shower Receptors and Shower Stalls." New York: ANSI.

American Plywood Association (APA), American Wood Council (AWC), National Forest Products Association (NFPA), Southern Forest Products Association (SFPA), Western Wood Products Association (WWPA). Plen-wood System. Tacoma, WA: APA.

American Society of Heating, Refrigerating and Air-Conditioning Engineers (ASHRAE). 1986. Bin and Degree-hour Weather Data for Simplified Energy Calculations. Atlanta, GA: ASHRAE.

———. 1989. ASHRAE Handbook; 1989 Fundamentals. Atlanta, GA: ASHRAE.

———. Standard 55, "Thermal Environmental Conditions for Human Occupancy." Atlanta, GA: ASHRAE.

———. Standard 62, "Ventilation for Acceptable Indoor Air Quality." Atlanta, GA: ASHRAE.

———. Standard 90A, "Energy Conservation in New Building Design." Atlanta GA: ASHRAE.

———. Addendum 90A-a, "Addendum to Energy Conservation in New Building Design." Atlanta GA: ASHRAE.

———. Standard 90.1, "Energy Efficient Design of New Buildings Except Low-rise Residential Buildings." Atlanta GA: ASHRAE.

Architectural Graphic Standards. See Ramsey/Sleeper:

American Society for Testing and Materials (ASTM) Standards.

    A 53, "Specification for Aluminum-Alloy Pipe, Steel. Black and Hot-Dipped, Zinc-Coated, Welded and Seamless." Philadelphia, PA: ASTM.

    A 74, "Specification for Cast Iron Soil Pipe and Fittings." Philadelphia, PA: ASTM.

    A 312, "Specification for Seamless and Welded Austenitic Stainless Steel Pipe." Philadelphia, PA: ASTM.

    A 377, "Specifications for Ductile Iron Pressure Pipe." Philadelphia, PA: ASTM.

    B 43, "Specifications for Seamless Red Brass Pipe, Standard Sizes." Philadelphia, PA: ASTM.

    B 88, "Specification for Seamless Copper Water Tube (Metric)." Philadelphia, PA: ASTM.

    B 108, "Specification for Aluminum-Alloy Permanent Mold Castings." Philadelphia, PA: ASTM.

    B 134, "Specification for Brass Wire." Philadelphia, PA: ASTM.

    B 135, "Specification for Seamless Brass Tube." Philadelphia, PA: ASTM.

    B 210, "Specification for Aluminum-Alloy Drawn Seamless Tubes." Philadelphia, PA: ASTM.

    B 271, "Specification for Copper-Base Alloy Centrifugal Castings." Philadelphia, PA: ASTM.

    B 306, "Specification for Copper Drainage Tubes." Philadelphia, PA: ASTM.

    B 429, "Specification for Aluminum-Alloy Extruded Structural Pipe and Tube." Philadelphia, PA: ASTM.

    B 455, "Specification for Copper-Zinc-Lead Alloy (Leaded Brass) Extruded Shapes." Philadelphia, PA: ASTM.

    B 456, "Specification for Electrode-posited Coatings of Copper Plus Nickel Plus Chromium and Nickel Plus Chromium." Philadelphia, PA: ASTM.

    B 483, "Specification for Aluminum and Aluminum-Alloy Extruded Drawn Tubes for General Purpose Applications." Philadelphia, PA: ASTM.

    B 564, "Specification for Rubber Gaskets for Cast Iron Soil Pipe and Fittings." Philadelphia, PA: ASTM.

    B 584, "Specification for Copper Alloy Sand Castings." Philadelphia, PA: ASTM.

    B 587, "Specification for Welded Brass Tube." Philadelphia, PA: ASTM.

    B 650, "Specification for Electrodeposited Engineering Chromium Coatings on Ferrous Substrates." Philadelphia, PA: ASTM.

    C 12, "Practice for Installing Vitrified Clay Pipe Lines." Philadelphia, PA: ASTM.

    C 177, "Test Method for Steady-state Heat Flux Measurements and Thermal Transmission Properties by Means of the Guarded-Hot-Plate Apparatus." Philadelphia, PA: ASTM.

    C 236, "Test Method for Steady-State Thermal Performance of Building Assemblies by Means of a Guarded Hot Box." Philadelphia, PA: ASTM.

    C 425, "Specification for Compression Joints for Vitrified Clay Pipe and Fittings." Philadelphia, PA: ASTM.

    C 564, "Specification for Rubber Gaskets for Cast Iron Soil Pipe and Fittings." Philadelphia, PA: ASTM.

    C 700, "Specification for Vitrified Clay Pipe, Extra Strength, Standard Strength, and Perforated." Philadelphia, PA: ASTM.

    C 976, "Test Method for Thermal Performance of Building Assemblies by Means of a Calibrated Hot Box." Philadelphia, PA: ASTM.

    D 2104, "Specification for Polyethylene (PE) Plastic Pipe, Schedule 40." Philadelphia, PA: ASTM.

    D 2287, "Specification for Nonrigid Vinyl Chloride Polymer and Copolymer Molding and Extrusion Compounds." Philadelphia, PA: ASTM.

    D 2321, "Recommended Practice for Underground Installation of Flexible Thermoplastic Sewer Pipe." Philadelphia, PA: ASTM.

    D 2487, "Classification of Soil for Engineering Purposes." Philadelphia, PA: ASTM.

    D 2729, "Specification for Poly (Vinyl Chloride) (PVC) Sewer Pipe and Fittings." Philadelphia, PA: ASTM.

    D 2751, "Specification for Acrylonitrile-Butadiene-Styrene (ABS) Sewer Pipe." Philadelphia, PA: ASTM.

    D 2949, "Specification for 3.25-in. Outside Diameter Poly (Vinyl Chloride) (PVC) Plastic Drain, Waste, and Vent Pipe and Fittings." Philadelphia, PA: ASTM.

    D 3309, "Specification for Polybutylene (PB) Plastic Hot-Water Distribution Systems." Philadelphia, PA: ASTM.

    F 891, "Specification for Coextended Poly (Vinyl Chloride) (PVC) Plastic Pipe with a Cellular Core." Philadelphia, PA: ASTM.

American Water Works Association (AWWA). C-203-86, "Coal-tar Protective Coatings and Linings for Steel Water Pipelines—Enamel and Tape—Hot-applied." Denver, CO: AWWA.

Cast Iron Soil Pipe Institute (CISPI) Standards:

    HSN-74-86, "Specification for Cast Iron Soil Pipe and Fittings for Hub and Spigot Sanitary and Storm Drain, Waste, and Vent Piping Applications." Chattanooga, TN: CISPI.

    HSN-85, "Specification for Neoprene Rubber Gaskets for Hub and Spigot Cast Iron Soil Pipe and Fittings." Chattanooga, TN: CISPI.

    301-90, "Standard Specification for Hubless Cast Iron Soil Pipe and Fittings for Sanitary and Storm Drain, Waste, and Vent Piping Applications." Chattanooga, TN: CISPI.

    310-90, "Specification for Coupling for Use in Connection with Hubless Cast Iron Soil Pipe and Fittings for Sanitary and Storm Drain, Waste, and Vent Piping Applications." Chattanooga, TN: CISPI.

History and Uses of Cast Iron Soil Pipe. Vol. I of CISPI Handbook. 7th ed. Chattanooga, TN: CISPI.

Champagne, Roger D. 1986. "Planning for Plumbing Efficiency." *The Construction Specifier*, 39(6)(June):77–80.

Cornish, Kathy L. 1987. "ASHRAE's Standard 90 Revision. A Split Proposal." *The Construction Specifier*, 40(12)(December): 111–116.

Council of American Building Officials (CABO). *Model Energy Code*. Falls Church, VA: CABO.

————. 1983. *One- and Two-Family Building Code*. Falls Church, VA: CABO.

Harrie, David T. 1986. "Clean and Economical Air in Houses." *Architectural Technology* (July/August):33–36.

Harriman, Marc S. 1990. "A Higher Standard." *Architecture* (December):91–93.

Howard, Bion D. 1985. "Masonry Thermal Mass and the Energy Code." *The Construction Specifier*, 38(6)(June):87–95.

Hydronics Institute (HI). *IBR Code, Indirect Water Heater Test Standards*. Berkley Heights, NJ: HI.

Kirsch, Laurence S. 1986. "Indoor Air and the Law." *Architectural Technology* (July/August):37–41.

McNall, Preston E. 1986. "Controlling Air Quality in Office Buildings." *Architectural Technology* (July/August):29–32.

National Association of Home Builders (NAHB) Research Foundation. 1983. *NAHB Thermal Performance Guidelines*. Washington, D.C.: NAHB.

National Climatic Data Center. 1981B. *Climatography of U.S. Number 81, Monthly Normals of Temperature, Precipitation, and Heating and Cooling Degree Days*. Ashville, NC: National Climatic Data Center.

National Fire Protection Association (NFPA). 1989. *Automatic Sprinkler Systems Handbook*. Quincy, MA: NFPA.

————. 1989. NFPA 13, "Installation of Sprinkler Systems." Quincy, MA: NFPA.

————. 1987. NFPA 13A, "Inspection, Testing, and Maintenance of Sprinkler Systems." Quincy, MA: NFPA.

————. 1989. NFPA 13D, "Installation of Sprinkler Systems in One- and Two-family Dwellings and Mobile Home." Quincy, MA: NFPA.

————. 1989. NFPA 13R, "Installation of Sprinkler Systems in Residential Occupancy up to Four Stories in Height." Quincy, MA: NFPA.

————. 1990. NFPA 20, "Installation of Centrifugal Fire Pumps." Quincy, MA: NFPA.

————. 1991. NFPA 101, "Life Safety Code." Quincy, MA: NFPA.

————. 1991. *National Fire Codes*. Quincy, MA: NFPA.

Ramsey/Sleeper, The AIA Committee on Architectural Graphic Standards. 1981. *Architectural Graphic Standards*, 7th ed. New York: Wiley.

Reid, Robert N. 1990. "Quality Assurance in Fire Sprinkler Systems." *The Construction Specifier*, 43(9)(September):124–129.

Ross, Robert E. 1985. "Boilers: Low-cost Comfort for Commercial Buildings." *The Construction Specifier*, 38(3)(March):69–78.

Rudoy, W., and R. S. Douglas. *Effect of Thermal Mass on Heating and Cooling Loads in Residences*. Washington, DC: National Forest Products Association.

Rutkowski, Henry T. 1985. "Designing HVAC Systems That Perform." *The Construction Specifier*, 38(3)(March):50–65.

Sheet Metal and Air-Conditioning Contractors National Association (SMACNA). 1974. *Thermoplastic Duct (PVC) Construction Manual*. Vienna, VA: SMACNA.

————. 1978. *Energy Recovery Equipment and Systems AIR-TO-AIR*. Vienna, VA: SMACNA.

————. 1979. *Fibrous Glass Duct Construction Standards*. Vienna, VA: SMACNA.

————. 1983. *HVAC Systems—Testing, Adjusting, and Balancing*. Vienna, VA: SMACNA.

————. 1985. *Duct Construction Standards— Metal and Flexible*. Vienna, VA: SMACNA.

————. 1986. *HVAC Systems—Applications*. Vienna, VA: SMACNA.

————. 1987. *Architectural Sheet Metal Manual*. Vienna, VA: SMACNA.

————. 1988. *Indoor Air Quality Manual*. Vienna, VA: SMACNA.

————. 1988. *Installation Standards for Residential Heating and Air Conditioning Systems*. Vienna, VA: SMACNA.

————. 1990. *HVAC System—Duct Design*. Vienna, VA: SMACNA.

Smith, Karen Hass. 1986. "Energy Conservation in Small HVAC Systems." *The Construction Specifier*, 39(6)(June):87–92.

Sump and Sewage Pump Manufacturers Association (SSMPA). 1985. *Sump, Effluent, and Sewage Pump Standards*. Winnetka, IL: SSPMA.

Sussman, Alan. 1988. "Solving HVAC Equipment Vibration Problems." *The Construction Specifier*, 41(3)(March):29–32.

Tilley, Ray Don. 1990. "Reclaimed Resources." *Architecture* (December):97–102.

Underwriters Laboratories, Inc. (UL) Standards:

70, "Septic Tanks, Bituminous Coated Metal." Northbrook, IL: UL.

174, "Household Electric Storage Tank Water Heaters." Northbrook, IL: UL.

732, "Oil-fired Storage Tank Water Heaters." Northbrook, IL: UL.

U.S. Department of the Air Force. 1978. AFM 88-29, "Engineering Weather Data." Washington, DC: Departments of the Air Force, the Army, and the Navy.

U.S. Department of Agriculture (USDA), Farmers Home Administration. 1980. *Thermal Performance Standards*. Washington DC: USDA.

U.S. Department of Housing and Urban Development (HUD). *Minimum Property Standards for Housing*. Washington, DC: HUD.

————. *Handbook 4940.2, Minimum Design Standards for Community Water Supply Systems*. Washington, DC: HUD

————. *Handbook 4940.3, Minimum Design Standards for Community Sewerage Systems*. Washington, DC: HUD

U.S. Environmental Protection Agency (EPA). EPA 40 CFR 141, "National Interim Primary Drinking Water Regulations." EPA.

U.S. General Services Administration Specifications Unit. Federal Specification WW-P-541, "Plumbing Fixtures." Washington, DC: GSA Specifications Unit.

Van Geem, Martha G. 1987. "Thermal Mass, What R-values Neglect." *The Construction Specifier*, 40(6)(June):71–77.

Water Quality Association (WQA) Standards:

S-100, "Voluntary Industry Standards for Household, Commercial, and Portable Exchange Water Softeners." Vol. 1. Lisle, IL: WQA.

S-200, "Recommended Industry Standards for Household and Commercial Water Filters." Vol. 2. Lisle, IL: WQA.

Wilson, Forrest. 1984. *Building Materials Evaluation Handbook*. New York: Van Nostrand Reinhold.

CHAPTER 12

ACKNOWLEDGMENTS

Acoustical Materials Association;

American National Standards Institute (ANSI);

Consumer Product Safety Commission (CPSC);

Intermatic, Incorporated;

Lightning Protection Institute;

National Fire Protection Association;

John D. Brush and Co.;

Transcience;

U.S. Department of Agriculture;

U.S. Department of Housing and Urban Development (HUD).

CHAPTER 12

REFERENCES

American National Standards Institute (ANSI). 1958. ANSI C91.1-1958, "Requirements for Residential Wiring." New York: ANSI.

————. 1972 (reaffirmed 1989). ANSI Y32.9, "Graphic Symbols for Electrical Wiring and Layout Diagrams Used in Architecture and Building Construction." New York: Institute of Electrical and Electronics Engineers.

American Society of Heating, Refrigerating and Air-Conditioning Engineers (ASHRAE). 1989. *ASHRAE Handbook; 1989 Fundamentals*. Atlanta, GA: ASHRAE.

Bukowski, Richard W. 1991. "Improving the Fire Performance of Building Contents." *The Construction Specifier*, 44(2)(February): 42–46.

Consumer Products Safety Commission (CPSC). 1978. *Was Your Home Built After 1964?* Bethesda, MD: CPSC.

———. 1979. *What You Should Know About Smoke Detectors*. Bethesda, MD: CPSC.

Council of American Building Officials (CABO). *Model Energy Code*. Falls Church, VA: CABO.

———. 1983. *One- and Two-Family Dwelling Code*. Falls Church, VA: CABO.

Lightning Protection Institute (LPI). L.P.I-175, "Installation Code." Woodstock, IL: LPI.

National Fire Protection Association (NFPA). 1990. NFPA 70, "National Electrical Code." Quincy, MA: NFPA.

———. 1990. NFPA 70A, "Electrical Code for One- and Two-family Dwellings." Quincy. MA: NFPA.

———. 1986. NFPA 78, "Lightning Protection Code." Quincy. MA: NFPA.

———. 1991. NFPA 101, "Life Safety Code." Quincy, MA: NFPA.

———. 1986. NFPA 550, "Guide to the Firesafety Concepts Tree." Quincy, MA: NFPA.

Raeber, John A. 1991. "Redundancy in Fire Protection." *The Construction Specifier*, 44(2)(February):17–19.

Ramsey/Sleeper, The AIA Committee on Architectural Graphic Standards. 1981. *Architectural Graphic Standards*. 7th ed. New York: Wiley.

Reid, Ed, and Bob Schifiliti. 1991. "A Balanced Approach to Fire Protection Detection and Suppression Systems." *The Construction Specifier*, 44(2)(February):30–41.

Underwriters Laboratories, Inc. (UL). 1982 (rev. 1990). *Installation Requirements for Lightning Protection Systems*. Northbrook, IL: UL.

U.S. Department of Agriculture. *Farmers Bulletin Number 2136*. Washington, DC: Department of Agriculture.

DATA SOURCES

The following organizations contributed to this book by permitting the use of their illustrations or text, answering the author's queries, or producing material used in research for the book:

AA Wire Products Company
6100 E. New England Ave.
Chicago, IL 60638
(312) 586-6700

Acoustical Society of America (ASA)
500 Sunnyside Blvd.
Woodbury, NY 11797
(516) 349-7800

Adhesive and Sealant Council (ASC)
1627 K St., N.W., Suite 1000
Washington, DC 20006
(202) 452-1500

Advanstar Communications
(formerly Edgell Communications)
7500 Old Oak Blvd.
Cleveland, OH 44130
(216) 243-8100

AIA and ACSA Council on Architectural Research
1735 New York Ave., N.W.
Washington, DC 20006
(202) 785-2324

AIA Professional Services Division
1735 New York Ave., N.W.
Washington, DC 20006
(202) 626-7300

Air Conditioning and Refrigeration Institute (ARI)
1501 Wilson Blvd., 6th Floor
Arlington, VA 22209
(703) 524-8800

Air Conditioning Contractors of America (ACCA)
1513 16th St., N.W.
Washington, DC 20036
(202) 483-9370

Air Diffusion Council (ADC)
One Illinois Center, Suite 200
111 E. Wacker Drive
Chicago, IL 60601
(312) 616-0800

Air Movement and Control Association (AMCA)
30 W. University Drive
Arlington Heights, IL 60004
(312) 394-0150

Ajax Magnethermic Corp.
P.O. Box 991
Warren, OH 44482
(216) 372-8511

ALCOA Building Products
Division of Stolle
P.O. Box 716
Sidney, OH 45365
(513) 492-1111

Allied Building Metals Industries
211 E. 43rd St., Suite 804
New York, NY 10017
(212) 697-5551

Aluminum Association (AA)
900 19th St., N.W., Suite 300
Washington, DC 20006
(202) 862-5156

American Architectural Manufacturers
Association (AAMA)
1540 E. Dundee Rd., Suite 310
Palatine, IL 60067
(708) 202-1350

American Association of Nurserymen (AAN)
1250 Eye St., N.W., Suite 500
Washington, DC 20005
(202) 789-2900

American Association of State Highway and
Transportation Officials (AASHTO)
444 N. Capitol St., N.W., Suite 225
Washington, DC 20001
(202) 624-5800

American Association of Textile Chemists and
Colorists (AATCC)
P.O. Box 12215
Research Triangle Park, NC 27709-2215
(919) 549-8141

American Concrete Institute (ACI)
2440 W. Seven Mile Road
Detroit, MI 48219
(313) 532-2600

American Concrete Pipe Association (ACPA)
8300 Boone Blvd., Suite 400
Vienna, VA 22182
(703) 821-1990

American Council of Independent Laboratories
(ACIL)
1725 K St., N.W., Suite 412
Washington, DC 20006
(202) 887-5872

American Consulting Engineers Council (ACEC)
1015 15th St. N.W., Suite 802
Washington, DC 20005
(202) 347-7474

American Forest and Paper Association
(formerly National Forest Products Association
[NFPA])
1250 Connecticut Ave., N.W., Second Floor
Washington, DC 20036
(202) 463-2700

American Friction Bearing Manufacturers
Association (AFBMA)
1101 Connecticut Ave., N.W., Suite 700
Washington, DC 20036
(202) 429-5155

American Galvanizers Association
12200 E. Iliff Ave., Suite 204
Aurora, CO 80014
(303) 750-2900

American Gas Association, Inc. (AGA)
1515 Wilson Blvd.
Arlington, VA 22209
(703) 841-8400

American Hardboard Association (AHA)
520 N. Hicks Road
Palatine, IL 60067
(708) 934-8800

American Home Lighting Institute
435 N. Michigan Ave., Suite 1717
Chicago, IL 60611
(312) 644-0828

American Industrial Hygiene Association
(AIHA)
P.O. Box 8390
Akron, OH 44320
(216) 873-2442

American Institute of Architects (AIA)
1735 New York Ave., N.W.
Washington, DC 20006
(202) 626-7300

American Institute of Industrial Engineers
25 Technology Park
Norcross, GA 30092
(404) 449-0460

American Institute of Real Estate Appraisers
Appraisal Institute
430 N. Michigan Ave.
Chicago, IL 60611
(312) 335-4100

American Institute of Steel Construction (AISC)
One E. Wacker Drive, Suite 3100
Chicago, IL 60601
(312) 670-5447

American Institute of Timber Construction (AITC)
11818 Southeast Mill Plain Blvd., Suite 415
Vancouver, WA 98684
(206) 254-9132

American Insurance Association (AIA)
1130 Connecticut Ave., N.W., Suite 1000
Washington, DC 20036
(202) 828-7100

American Iron and Steel Institute (AISI)
1101 17th St. N.W., Suite 1300
Washington, DC 20036
(202) 463-6573

American Land Tile Association
1828 L St., N.W., Suite 705
Washington, DC 20036
(202) 296-3671

American Lumber Standards Committee
(ALSC)
P.O. Box 210
Germantown, MD 20875
(301) 972-1700

American National Standards Institute (ANSI)
11 W. 42nd St.
New York, NY 10036
(212) 642-4900

American Olean Tile Company, Inc.
1000 Cannon Ave.
Landsdale, PA 19446
(215) 855-1111

American Painting Contractor
American Paint Journal Company
2911 Washington Ave.
St. Louis, MO 63103
(314) 534-0301

American Petroleum Institute (API)
1220 L St., N.W.
Washington, DC 20005
(202) 682-8000

American Planning Association
1776 Massachusetts Ave., N.W., Suite 400
Washington, DC 20036
(202) 872-0611

American Plywood Association (APA)
P.O. Box 11700
Tacoma, WA 98411
(206) 565-6600

American Society for Concrete Construction
(ASCC)
1902 Techny Ct.
Northbrook, IL 60062
(708) 291-0270

American Society for Metals International
(ASMI)
9639 Kinsman Rd.
Materials Park, OH 44073
(216) 338-5151

American Society for Testing and Materials
(ASTM)
1916 Race St.
Philadelphia, PA 19103-1187
(215) 299-5400

American Society of Architectural Hardware
Consultants
(*see* Door and Hardware Institute)

American Society of Civil Engineers (ASCE)
345 E. 47th St.
New York, NY 10017
(212) 705-7496

American Society of Consulting Planners
1015 15th St., N.W., Suite 600
Washington, DC 20005
(202) 789-2211

American Society of Heating, Refrigerating and
Air-Conditioning Engineers, Inc. (ASHRAE)
1791 Tullie Circle, N.E.
Atlanta, GA 30329
(404) 636-8400

American Society of Home Inspectors, Inc.
1735 N. Lynn St., Suite 950
Arlington, VA 22209
(703) 524-2008

American Society of Landscape Architects
(ASLA)
4401 Connecticut Ave., N.W., Suite 500
Washington, DC 20008
(202) 686-2752

American Society of Mechanical Engineers
(ASME)
345 E. 47th St.
New York, NY 10017
(212) 705-7722

American Society of Plumbing Engineers (ASPE)
3617 Thousand Oaks Blvd., Suite 210
Westlake, CA 91362
(805) 495-7120

American Society of Safety Engineers, Inc.
1800 E. Oakton
Des Plaines, IL 60018
(708) 692-4121

American Society of Sanitary Engineering (ASSE)
P.O. Box 40362
Bay Village, OH 44140
(216) 835-3040

American Solar Energy Society
2400 Central Ave., Unit G-1
Boulder, CO 08301
(303) 443-3130

American Water Works Association, Inc. (AWWA)
6666 W. Quincy Ave.
Denver, CO 80235
(303) 794-7711

American Welding Society (AWS)
550 N.W. Lejune Road
P.O. Box 351040
Miami, FL 33135
(305) 443-9353

American Wire Cloth Institute
2000 Maple Hill St.
Yorktown Heights, NY 10598
(914) 962-9052

American Wood Council
1250 Connecticut Ave., N.W., #200
Washington, DC 20036
(202) 463-2700

American Wood-Preservers' Association (AWPA)
P.O. Box 286
Woodstock, MD 21163
(410) 465-3139

American Wood Preservers Institute (AWPI)
1945 Old Gallows Rd., Suite 550
Vienna, VA 22182
(703) 893-4005

Architectural Record
1221 Ave. of the Americas
New York, NY 10020
(212) 512-2000

Architectural Technology
(now a part of *Architecture*)
American Institute of Architects
1735 New York Ave., N.W.
Washington, DC 20006
(202) 626-7300

Architectural Woodwork Institute (AWI)
P.O. Box 1550
Centreville, VA 22020
(703) 222-1100

Architecture
American Institute of Architects
1735 New York Ave., N.W.
Washington, DC 20006
(202) 626-7300

Armstrong World Industries
P.O. Box 3001
Lancaster, PA 17604
(717) 397-0611

Asphalt Institute (AI)
Research Park Drive
P.O. Box 14052
Lexington, KY 40512
(606) 288-4960

Asphalt Roofing Manufacturers Association
(ARMA)
6000 Executive Blvd., Suite 201
Rockville, MD 20852
(301) 231-9050

Associated Air Balance Council (AABC)
1518 K St., N.W., Suite 503
Washington, DC 20005
(202) 737-0202

Associated Builders and Contractors (ABC)
1300 N. 17th St., 8th Floor
Rosslyn, VA 22209
(202) 637-8800

Associated General Contractors of America (AGC)
1957 E St., N.W.
Washington, DC 20006
(202) 393-2040

Associated Laboratories (ALI)
500 S. Vermont St.
Palatine, IL 60067
(708) 358-7400

Associated Sheet Metal Contractors, Inc.
3000 W. Hallandale Beach Blvd.
Hallandale, FL 33009
(305) 961-0440

Association for Preservation Technology
Box 2487, Station D
Ottawa, Ontario, Canada K1P 5W6
(613) 238-1972

Association of Bituminous Contractors
1747 Pennsylvania Ave., N.W., Suite 1050
Washington, DC 20006
(202) 785-4400

Association of Iron and Steel Engineers
3 Gateway Center, Suite 2350
Pittsburgh, PA 15222
(412) 281-6323

Association of Official Analytical Chemists
(AOAC)
2200 Wilson Blvd., Suite 400
Arlington, VA 22201
(703) 522-3032

Association of Official Seed Analysts
(AOSA)
268 Plant Science, IANR-UNL,
Box 19281
Lincoln, NE 68583
(402) 472-8649

Association of Soil and Foundation Engineers
8811 Colesville Rd., Suite G-106
Silver Spring, MD 20910
(301) 565-2733

Association of the Wall and Ceiling Industries—
International
307 E. Annandale Rd., Suite 200
Falls Church, VA 20042
(703) 534-8300

Association of Tile, Terrazzo, Marble Contractors'
and Affiliates, Inc.
P.O. Box 13629
Jackson, MS 39236
(601) 939-2071

Battelle Memorial Institute
Columbus Labs.
505 King Ave.
Columbus, OH 43201
(614) 424-6424

Bethlehem Steel Corporation
701 E. 3rd St.
Bethlehem, PA 18016
(215) 694-2424
(800) 353-5700

Brick Institute of America (BIA)
11490 Commerce Park Dr., Suite 300
Reston, VA 22091
(703) 620-0010

Bruce Engineering Company
2000 Tucker Industrial Rd.
Tucker, GA 30084
(404) 939-4460

Bruce Hardwood Floors
The Bolton Group
1850 Crown Rd.
Dallas, TX 75234
(214) 506-0480

Builders' Hardware Manufacturers Association
(BHMA)
355 Lexington Ave., 17th Floor
New York, NY 10017
(212) 661-4261

Building Design and Construction
Cahners Plaza
1350 E. Touhy Ave.
P.O. Box 5080
Des Plaines, IL 60018
(312) 635-8800

Building Officials and Code Administrators
International (BOCA)
4051 W. Flossmoor Rd.
Country Club Hills, IL 60478
(708) 799-2300

Building Owners and Managers Association
International
1201 New York Ave., Suite 300
Washington, DC 20005
(202) 408-2662

Building Stone Institute
420 Lexington Ave.
New York, NY 10017
(212) 490-2530
(*Note:* BSI has no technical personnel. Author
recommends contacting individual producers'
associations or producers rather than BSI.)

Building Stone
(*see* Building Stone Institute)

Burke Company
4810 Dufferin St.
Downsview, Ontario, Canada M3H 5S8
(800) 423-9140

Business and Institutional Furniture Manufacturers
Association (BIFMA)
2335 Burton St., S.E.
Grand Rapids, MI 49506
(616) 243-1681

California Lath and Plaster Association
(*see* Western Lath, Plaster, Drywall Contractors
Association)

California Redwood Association
405 Entrente Dr., Suite 200
Novato, CA 94949
(415) 382-0062

Capital Development Board, State of Illinois
3rd Floor/William G. Stratton Building
401 S. Spring St.
Springfield, IL 62706
(217) 782-2864

Cardinal IG
12301 Whitewater Dr.
Minneapolis, MN 55343
(612) 935-1722

Carpet Cushion Council (CCC)
P.O. Box 546
Riverside, CT 06878
(203) 637-1312

Carpet and Rug Institute (CRI)
Box 2048
Dalton, GA 30722
(404) 278-3176

Cast Iron Soil Pipe Institute (CISPI)
5959 Shallowford Rd., Suite 419
Chattanooga, TN 37421
(615) 892-0137

CECO Concrete Construction Services
P.O. Box 28570
Kansas City, MO 64118
(913) 371-2211

CECO Door Division CECO Corporation
One Tower Lane
Oakbrook Terrace, IL 60181
(312) 242-2000

Cedar Shake and Shingle Bureau
515 116th Ave., N.E., Suite 275
Bellevue, WA 98004
(206) 453-1323

Ceiling and Interior Systems Construction
Association (CISCA)
  5700 Old Orchard Rd., 1st Floor
  Skokie, IL 60077
  (708) 965-2776

Ceramic Tile Institute of America (CTI)
  700 N. Virgil Ave.
  Los Angeles, CA 90029
  (213) 660-1911

Certified Ballast Manufacturers Association (CBM)
  1422 Euclid Ave.
  Hanna Building, Suite 722
  Cleveland, OH 44115
  (216) 241-0711

Chamber of Commerce of the United States
Construction, Housing, and Community
Development
  1615 H St. N.W.
  Washington, DC 20062
  (202) 659-6000

Chicago Plastering Institute
  6547 N. Avondale
  Chicago, IL 60631
  (312) 774-4500

Chicago Title and Trust Company
  171 N. Clark St.
  Chicago, IL 60601
  (312) 223-2000

Clemco Industries Corporation
  One Cable Car Drive
  Washington, MO 63090
  (415) 692-9080

Clemson University
  Department of Civil Engineering
  Lowery Hall
  Box 340911
  Clemson, SC 29634
  (803) 656-3311

Cold Finished Steel Bar Institute
  1120 Vermont Ave., N.W.
  Washington, DC 20005
  (202) 857-0350

Color Association of the United States
(CAUS)
  409 W. 44th St.
  New York, NY 10036
  (212) 582-6884

Commercial Renovation
(formerly *Commercial Remodeling*)
  QR, Inc.
  1209 Dundee Ave., Suite 8
  Elgin, IL 60120

Compressed Air and Gas Institute (CAGI)
c/o Thomas Associates, Inc.
  1300 Summer Ave.
  Cleveland, OH 44115
  (216) 241-7333

Compressed Gas Association (CGA)
  1725 Jefferson Davis Hwy., Suite 1004
  Arlington, VA 22202
  (703) 979-0900

Concrete Block Insulating Systems, Inc.
  P.O. Box 1000
  West Brookfield, MA 01585
  (508) 867-4241

Concrete Reinforcing Steel Institute (CRSI)
  933 Plum Grove Rd.
  Schaumburg, IL 60173
  (708) 517-1200

Construction Industry Manufacturers Association
  111 E. Wisconsin Ave., Suite 940
  Milwaukee, WI 53202
  (414) 272-0943

Construction Labor Research Council
  1730 M St., N.W., Suite 900B
  Washington, DC 20036
  (202) 223-8045

Construction Specialties, Inc.
  55 Winans Ave.
  Cranford, NJ 07016
  (908) 272-5200

Construction Specifications Institute
  601 Madison St.
  Alexandria, VA 22314-1791
  (703) 684-0300

The Construction Specifier
(*see* Construction Specifications Institute)

Consumer Product Safety Commission
(CPSC)
  5401 Westbard Ave.
  Bethesda, MD 20816
  (800) 638-2772

Copper and Brass Fabricators Council
  1050 17th St., N.W., #440
  Washington, DC 20036
  (202) 833-8575

Copper Development Association (CDA)
  260 Madison Ave.
  New York, NY 10016
  (212) 251-7200

Council of American Building Officials
(CABO)
  5203 Leesburg Pike, Suite 708
  Falls Church, VA 22041
  (703) 931-4533

Custom Roll Forming Institute
Division of Preferred Metal Forming Institute
  27027 Chardon Rd.
  Richmond Heights, OH 44143
  (216) 585-8800

Decorating Products World
Elkin Mittelman (Publishers)
  7335 Topango Canyon Blvd.
  Canoga Park, CA 91303
  (818) 710-1066

Decorative Laminate Products Association
(DLPA)
(formerly National Association of Plastic
Fabricators)
  600 South Federal St., Suite 400
  Chicago, IL 60605
  (312) 922-6222

Dekker, Marcel, Inc.
  270 Madison Ave.
  New York, NY 10016
  (212) 696-9000

Door and Hardware Institute (DHI)
(formerly National Builders Hardware Association)
  14170 Newbrook Drive
  Chantilly, VA 22022
  (703) 222-2010

Door Operator and Remote Control Manufacturers
Association
  655 W. Irving Park, #201 Park Place
  Chicago, IL 60613
  (312) 525-2644

Dow USA
  47 Building
  Midland, MI 48667
  (517) 636-5783

Drywall, Lath, and Plaster Bureau
c/o W. F. Pruter Associates
  3127 Los Feliz Blvd.
  Los Angeles, CA 90039
  (213) 660-4644

Ductile Iron Pipe Research Association
  245 Riverchase Parkway E., Suite 0
  Birmingham, AL 35244
  (205) 988-9870

Edison Electric Institute
  701 Pennsylvania Ave., N.W.
  Washington, DC 20004
  (202) 508-5000

ELE International, Inc.
Soil Test Products Division
(formerly Soil Test, Inc.)
  86 Albrecht Drive
  Lake Bluff, IL 60044
  (708) 295-9400

Electronic Industries Association (EIA)
  2001 Pennsylvania Ave. N.W., Suite 1100
  Washington, DC 20006
  (202) 457-4900

ETL Testing Laboratories, Inc. (ETL)
  P.O. Box 2040
  Route 11, Industrial Park
  Cortland, NY 13045
  (607) 753-6711

Expansion Joint Manufacturers Association
(EJMA)
  25 N. Broadway
  Tarrytown, NY 10591
  (914) 332-0040

Exterior Insulation Manufacturers Association
(EIMA)
  2759 State Rd. 580, Suite 112
  Clearwater, FL 34621
  (813) 726-6477

Exteriors
  1 E. 1st St.
  Duluth, MN 55802
  (218) 723-9200

E-Z Painter Company
  4051 S. Iowa Ave.
  Milwaukee, WI 53207
  (414) 481-4500

Facing Tile Institute (FTI)
  P.O. Box 8880
  Canton, OH 44711
  (216) 488-1211

Factory Mutual Research Organization (FM)
  P.O. Box 9102
  Norwood, MA 02062
  (617) 762-4300

Federal Housing Administration (FHA)
(*see* U.S. Department of Housing and Urban
Development)

Federal Specifications
(*see* U.S. General Services Administration, Specifications Unit)

Federation of Societies for Coatings Technology
492 Morrison Rd.
Blue Bell, PA 19422
(215) 940-0777

Ferroalloys Association
900 2nd St., N.E., Suite 306
Washington, DC 20002
(202) 842-0292

Fine Hardwood Veneer Association (FHVA)
5603 W. Raymond St., Suite "O"
Indianapolis, IN 46241
(317) 244-3311

Firestone Building Products Company
525 Congressional Blvd.
Carmel, IA 46032
(800) 428-4442

Flat Glass Marketing Association (FGMA)
3310 S.W. Harrison St.
Topeka, KS 66611-2279
(913) 266-7013

Fluid Controls Institute (FCI)
P.O. Box 9036
Morristown, NJ 07960
(201) 829-0990

Formica Corporation
10155 Reading Rd.
Cincinnati, OH 45241
(513) 786-3321

Frudenberg Building Systems, Inc.
Nora Flooring Systems
94 Glenn St.
Lawrence, MA 01843
(508) 689-0530

Gas Appliance Manufacturers Association
1901 N. Moore St., Suite 1100
Arlington, VA 22209
(703) 525-9565

General Services Administration (GSA)
(*see* U.S. General Services Administration)

Glass Tempering Association
· White Lakes Professional Building
3310 Harrison St.
Topeka, KS 66611
(913) 266-7064

Gold Bond Building Products
2001 Rexford Rd.
Charlotte, NC 28211
(704) 365-7300

Goodyear Tire and Rubber Company
Versigard and Hysunite Roofing Systems
1144 E. Market St.
Akron, OH 44316
(800) 782-1269

Guardian Industries
43043 W. Nine Mile Rd.
Northville, MI 48167
(313) 347-0100

Gypsum Association (GA)
810 1st St., N.E., Suite 510
Washington, DC 20002
(202) 289-5440

Hardwood Manufacturers Association (HMA)
400 Penn Center Blvd., Suite 530
Pittsburgh, PA 15235
(412) 829-0770

Hardwood Plywood and Veneer Association
(HPVA)
1825 Michael Farraday Drive
P.O. Box 2789
Reston, VA 22090
(703) 435-2900

Hartford Architecture Conservancy
278 Farmington Ave.
Hartford, CT 06105
(203) 525-0279

Hatten-Brown Publishers
P.O. Box 2268
Montgomery, AL 36104
(205) 834-1170

Heat Exchange Institute (HEI)
c/o Thomas Associates, Inc.
1300 Summer Ave.
Cleveland, OH 44115
(216) 241-7333

Historic Preservation
(*see* National Trust for Historic Preservation)

Hollow Metal Manufacturers Association
(HMMA)
600 S. Federal St., Suite 400
Chicago, IL 60605
(312) 922-6222

Homeowner
Family Media, Inc.
3 Park Ave.
New York, NY 10016
(212) 340-9200

Hunter Douglas Architectural Products
11455 Lakefield Dr.
Duluth, GA 30136
(404) 476-8803

Hydraulic Institute (HI)
30200 Detroit Rd.
Cleveland, OH 44145
(216) 899-0010

Hydronics Institute (HI)
P.O. Box 218
35 Russo Place
Berkeley Heights, NJ 07922
(908) 464-8200

IIT Research Institute
10 W. 35th St.
Chicago, IL 60616
(312) 567-4000

Illuminating Engineering Society of North
America (IESNA)
345 E. 47th St.
New York, NY 10017
(212) 705-7926

Indiana Limestone Institute of America, Inc. (ILI)
Stone City Bank Building, Suite 400
Bedford, IN 47421
(812) 275-4426

Industrial Risk Insurers (IRI)
85 Woodland St.
Hartford, CT 06102
(203) 520-7300

Inland Steel Company
Indiana Harbor Works Division
3210 Watling St.
East Chicago, IN 46312
(312) 346-0300

Institute of Business Designers (IBD)
341 Merchandise Mart
Chicago, IL 60654-1104
(312) 647-1950

Institute of Electrical and Electronics Engineers
(IEEE)
345 E. 4th St.
New York, NY 10017
(212) 705-7900

Institute of Environmental Sciences
940 E. Northwest Highway
Mt. Prospect, IL 60056
(708) 255-1561

Institute of Real Estate Management
430 N. Michigan Ave.
Chicago, IL 60611
(312) 661-1930

Institute of Roofing and Waterproofing Consultants
1800 N. Argyle Ave., Suite 301
Los Angeles, CA 90028
(213) 962-7246

Instrument Society of America
67 Alexandria Drive
P.O. Box 12277
Research Triangle Park, NC 27709
(919) 549-8411

Insulated Cable Engineers Association, Inc.
(ICEA)
P.O. Box 440
South Yarmouth, MA 02664
(508) 394-4424

Insulated Steel Door Systems Institute
30200 Detroit Rd.
Cleveland, OH 44145
(216) 899-0010

Insulating Glass Certification Council (IGCC)
c/o ETL Testing Laboratories, Inc.
Industrial Park, Route 11
P.O. Box 2040
Cortland, NY 13045
(607) 753-6711

Intermatic, Inc.
Intermatic Plaza
Spring Grove, IL 60081
(815) 675-2321

International Association of Plumbing and
Mechanical Officials (IAPMO)
20001 Walnut Dr. S.
Walnut, CA 91789
(714) 595-8449

International Conference of Building Officials
(ICBO)
5360 W. Workman Mill Rd.
Whittier, CA 90601
(213) 699-0541

International Institute for Lath and Plaster
c/o W. F. Pruter Associates
3127 Los Feliz Blvd.
Los Angeles, CA 90039
(213) 660-4644

International Masonry Institute
815 15th St., N.W., Suite 1001
Washington, DC 20005
(202) 783-3908

International Municipal Signal Association
(IMSA)
P.O. Box 539
165 E. Union St.
Newark, NJ 14513
(315) 331-2182

Interprofessional Council on Environmental
Design
c/o National Society of Professional Engineers
1420 King St.
Alexandria, VA 22314
(703) 684-2800

Iron Casting Society
455 State St.
Cast Metals Federation Building
Des Plaines, IL 60016
(312) 299-9160

Iron League of Chicago, Inc.
131 S. Marion St.
Oak Park, IL 60302

Italian Tile Center
499 Park Ave.
New York, NY 10022
(212) 980-8866

ITM Ramset/Redhead
1300 N. Michael Dr.
Wooddale, IL 60910
(219) 874-4217

Journal of Special Coatings and Linings
(*see* Steel Structures Painting Council)

Kaiser Aluminum and Chemical Corporation
1015 E. 12th St.
Erie, PA 16503
(814) 454-4571

Kalwall Corporation
P.O. Box 237
Manchester, NH 03105
(800) 258-9777

Kentile Floors
136 E. 57th St., Suite 907
New York, NY 10022
(212) 752-3220

Kitchen Cabinet Manufacturers Association
(KCMA)
1899 Preston White Drive
Falls Church, VA 22091
(703) 264-1690

K-Lath Division
Georgetown Wire Co., Inc.
P.O. Box 2120
Monrovia, CA 91016
(818) 357-5057

Koppers Company, Inc.
1050 Koppers Building
Pittsburgh, PA 15219
(800) 558-2706

Kurfees Coatings
(formerly Jewel Paint Co.)
201 E. Market St.
Louisville, KY 40201
(502) 584-0151

Laminated Glass Corporation
375 E. Church Ave.
Telford, PA 18969
(215) 721-0400

Laminated Safety Glass Association
White Lakes Professional Building
3310 Harrison St.
Topeka, KS 66611
(913) 266-7014

Larsen's Manufacturing Company
7421 Commerce Lane, N.E.
Minneapolis, MN 55432
(612) 571-1181
(800) 527-7367

Lead Industries Association, Inc. (LIA)
295 Madison Ave.
New York, NY 10017
(212) 578-4750

Levelor Corp.
595 Lawrence Expressway
Sunnyvale, CA 94086
(408) 245-4000

Libby-Owens-Ford
500 Exton Commons
Exton, PA 19341
(800) 523-0133

Library of Congress
1st St., N.E.
Washington, DC 20540
(202) 287-5000

Lightning Protection Institute (LPI)
3365 N. Arlington Heights Rd., Suite J
Arlington Heights, IL 60004
(708) 255-3003

Lukens Inc.
50 S. 1st Ave.
Coatesville, PA 19320
(215) 383-3345

Manufactured Housing Institute
1745 Jefferson Davis Highway, Suite 511
Arlington, VA 22202
(703) 979-6620

Manufacturers Standardization Society of the
Valve and Fittings Industry (MSS)
127 Park St., N.E.
Vienna, VA 22180
(703) 281-6613

Maple Flooring Manufacturers Association, Inc.
(MFMA)
60 Revere Drive, Suite 500
Northbrook, IL 60062
(708) 480-9138

Marble Institute of America (MIA)
33505 State St.
Farmington, MI 48024
(313) 476-5558

Masonry Contractors Association of
America
17 W. 14th St.
Oak Brook Terrace, IL 60181
(312) 620-6767

Materials and Methods Standards
Association
614 Monroe St.
Grand Haven, MI 49417
(616) 842-7844

McGraw-Hill, Inc.
1221 Ave. of the Americas
New York, NY 10020
(212) 512-2000

Mechanical Contractors Association of America,
Inc. (MCAA)
1385 Piccard Drive
Rockville, MD 20850
(301) 869-5800

Mercer Products Company, Inc.
P.O. Box 1240
Eustis, FL 32727
(904) 357-4119

Merchant & Evans, Inc.
ZIP-RIB
P.O. Box 1680
100 Connecticut Drive
Burlington, NJ 08016
(609) 387-0002

MERO Structures, Inc.
N112 W18810 Mequon Rd.
P.O. Box 610
Germantown, WI 53022
(800) 323-0072

Metal Architecture
7450 N. Skokie Blvd.
Skokie, IL 60077
(312) 674-2200

Metal Building Interior Products Company
5309 Hamilton Ave.
Cleveland, OH 44114
(216) 431-6400

Metal Building Manufacturers Association
(MBMA)
c/o Thomas Associates, Inc.
1300 Summer Ave.
Cleveland, OH 44115
(216) 241-7333

Metal Construction Association
1133 15th St., N.W., Suite 1000
Washington, DC 20005
(202) 429-9440

Metal Construction News
(*see* Metal Construction Association)

Metal Lath/Steel Framing Association (ML/SFA)
A Division of NAAMM
600 S. Federal St., Suite 400
Chicago, IL 60605
(312) 922-6222

Metal Sales Manufacturing Corporation
4001 Produce Rd.
Louisville, KY 40218
(502) 969-9318

Modern Metals
Delta Communications, Inc.
400 N. Michigan Ave.
Chicago, IL 60611
(312) 222-2000

National Academy of Code Administration
1606 Harbor Point Drive
Champaign, IL 61821
(217) 352-3253

National Academy of Engineering
2101 Constitution Ave., N.W.
Washington, DC 20418
(202) 334-3200

National Academy of Sciences
Building Research Advisory Board
2101 Constitution Ave., N.W.
Washington, DC 20418
(202) 334-2000

National Alliance of Preservation Commissions
Hall of the States
444 N. Capitol St., N.W., Suite 332
Washington, DC 20001
(202) 624-5490

National Asphalt Pavement Association (NAPA)
NAPA Building
5100 Forbes Blvd.
Lanham, MD 20706
(301) 731-4748

National Association of Architectural Metal
Manufacturers (NAAMM)
600 S. Federal St., Suite 400
Chicago, IL 60605
(312) 922-6222

National Association of Corrosion Engineers
1440 South Creek
Houston, TX 77084
(713) 492-0535

National Association of Counties
440 1st St., N.W.
Washington, DC 20001
(202) 393-6226

National Association of Floor Covering Distributors
13-186 W. Merchandise Mart Plaza
Chicago, IL 60654
(312) 467-0116

National Association of Home Builders (NAHB)
1201 15th St., N.W.
Washington, DC 20005
(202) 822-0200

National Association of Housing and
Redevelopment Officials
2600 Virginia Ave., N.W.
Washington, DC 20037
(202) 333-2020

National Association of Housing Cooperatives
2501 M St., N.W.
Washington, DC 20037
(202) 887-0706

National Association of Minority Contractors, Inc.
806 15th St., Suite 340
Washington, DC 20005
(202) 347-8259

National Association of Mirror Manufacturers
9005 Congressional Court
Potomac, MD 20854
(301) 365-4080

National Association of Plastic Distributors
5001 College Blvd. #214
Leawood, KS 66211
(913) 491-4698

National Association of Plastic Fabricators
(NAPF)
(*see* Decorative Laminate Products Association)

National Association of Plumbing-Heating-
Cooling Contractors
P.O. Box 6808
Falls Church, VA 22040
(703) 237-8100

National Association of Realtors
430 N. Michigan Ave.
Chicago, IL 60611
(312) 329-8200

National Association of Tile Contractors
626 Lakeland East Drive
Jackson, MS 39208
(601) 939-2701

National Association of Women in Construction
327 S. Adams St.
Fort Worth, TX 76104
(817) 877-5551

National Board of Boiler and Pressure Vessels
Inspectors
1055 Crupper Ave.
Columbus, OH 43229
(614) 888-8320

National Building Code
(*see* American Insurance Association)

National Building Granite Quarries Association
(*see* latest *Sweets Catalog*. Representative rotates
among association members. In 1992 was
c/o Rock of Ages Corp.)
P.O. Box 482
Barre, VT 05641
(802) 476-3115

National Building Materials Distributors
Association
1701 Lake Ave.
Glenville, IL 60025
(312) 724-6900

National Bureau of Standards (NBS)
(*see* National Institute of Standards and
Technology)

National Coal Association
1130 17th St., N.W.
Washington, DC 20036
(202) 463-6152

National Coil Coaters Association
1900 Arch St.
Philadelphia, PA 19103
(215) 564-3484

National Concrete Masonry Association (NCMA)
2302 Horse Pen Rd.
Herndon, VA 22071
(703) 713-1900

National Conference of States on Building Codes
and Standards
481 Carlisle Dr.
Herndon, VA 22070
(703) 437-0100

National Council of Acoustical Consultants, Inc.
66 Morris Ave., Box 359
Springfield, NJ 07081
(201) 379-1100

National Council on Radiation Protection and
Measurement (NCRPM)
7910 Woodmont Ave., Suite 800
Bethesda, MD 20814
(301) 657-2652

National Decorating Products Association
1050 N. Lindbergh Blvd.
St. Louis, MO 63132
(314) 991-3470

National Electrical Code (NEC)
(*see* National Fire Protection Association)

National Electrical Contractors Association, Inc.
(NECA)
7315 Wisconsin Ave.
Washington, DC 20014
(301) 657-3110

National Electrical Manufacturers Association
(NEMA)
2101 L St., N.W., Suite 300
Washington, DC 20037
(202) 457-8400

National Environmental Development Association
605 14th St., N.W.
Washington, DC 20005
(202) 638-1230

National Fire Protection Association (NFPA)
One Batterymarch Park
P.O. Box 9101
Quincy, MA 02269
(617) 770-4543

National Forest Products Association (NFPA)
(*see* American Forest and Paper Association)

National Glass Association
8200 Greensboro Dr.
McLean, VA 22102
(703) 442-4890

National Guard Products
540 N. Parkway
Memphis, TN 38107
(800) 647-7874

National Gypsum Company
(*see* Gold Bond Building Products)

National Hardwood Lumber Association (NHLA)
P.O. Box 34518
Memphis, TN 38184-1818
(901) 377-1818

National Housing Conference
1126 16th St., N.W.
Washington, DC 20036
(202) 223-4844

National Institute of Building Sciences (NIBS)
1201 L St., N.W., Suite 400
Washington, DC 20005
(202) 289-7800

National Institute of Standards and Technology
(formerly National Bureau of Standards)
Gaithersburg, MD 20899
(301) 975-2000
Center for Building Technology
(301) 975-5900

National Landscape Association, Inc.
1250 Eye St., N.W., Suite 500
Washington, DC 20005
(202) 789-2900

National League of Cities
1301 Pennsylvania Ave., N.W., 6th Floor
Washington, DC 20004
(202) 626-3000

National Lime Association
3601 N. Fairfax Dr.
Arlington, VA 22201
(703) 243-5463

National Lumber Grades Authority (NLGA)
1055 W. Hastings St., Suite 260
Vancouver, British Columbia, Canada V6E 2E9
(604) 687-2171

National Oak Flooring Manufacturers Association
(NOFMA)
P.O. Box 3009
Memphis, TN 38173
(901) 526-5016

National Ornamental and Miscellaneous Metals
Association
2996 Grandview Ave., N.E. #109
Atlanta, GA 30305
(404) 237-5334

National Paint and Coatings Association (NPCA)
(formerly National Paint, Varnish, and Lacquer
Association)
1500 Rhode Island Ave., N.W.
Washington, DC 20005
(202) 462-6272

National Park Service
(see U.S. Department of the Interior)

National Particleboard Association (NPA)
18928 Premiere Court
Gaithersburg, MD 20879
(301) 670-0604

National Pest Control Association, Inc.
8100 Oak St.
Dunn Loring, VA 22027
(703) 573-8330

National Precast Concrete Association
825 E. 64th St.
Indianapolis, IN 46220
(317) 253-0486

National Preservation Institute
P.O. Box 1702
Alexandria, VA 22313
(703) 393-0038

National Ready Mixed Concrete Association
900 Spring St.
Silver Spring, MD 20910
(301) 587-1400

National Research Council
2101 Constitution Ave., N.W.
Washington, DC 20418
(202) 334-2200

National Retail Hardware Association
770 N. High School Rd.
Indianapolis, IN 46224
(317) 248-1261

National Roofing Contractors Association
(NRCA)
O'Hare International Center
10255 W. Higgins Rd., Suite 600
Rosemont, IL 60018
(708) 299-9070

National Safety Council
444 N. Michigan Ave.
Chicago, IL 60611
(312) 527-4800

National Sand and Gravel Association
900 Spring St.
Silver Spring, MD 20910
(301) 587-1400

National Sanitation Foundation
P.O. Box 1468
3475 Plymouth Rd.
Ann Arbor, MI 48106
(313) 769-8010

National Sash and Door Jobbers Association
2200 E. Devon Ave., Suite 358
Des Plaines, IL 60018
(312) 299-3400

National School Supply and Equipment
Association (NSSEA)
8300 Centerville Rd., #250
Silver Spring, MD 20910
(301) 495-0240

National Society of Professional Engineers (NSPE)
1420 King St.
Alexandria, VA 22314
(703) 684-2810

National Terrazzo and Mosaic Association
(NTMA)
3166 Des Plaines Ave., Suite 132
Des Plaines, IL 60018
(800) 323-9736

National Tile Roofing Manufacturers Association,
Inc.
c/o W. F. Pruter Associates
3217 Los Feliz Blvd.
Los Angeles, CA 90039
(213) 660-6259

National Trust for Historic Preservation
1785 Massachusetts Ave., N.W.
Washington, DC 20036
(202) 673-4000

National Wholesale Hardware Association
1900 Arch St.
Philadelphia, PA 19103
(215) 564-3484

National Wood Window and Door Association
(NWWDA)
1400 E. Touhy Ave., #G54
Des Plaines, IL 60018
(800) 223-2301

National Woodwork Manufacturers Association
(NWMA)
(see National Wood Window and Door
Association)

Nickel Development Institute
15 Toronto St., Suite 402
Toronto, Canada M5C 2E3
(416) 362-8850

Northeastern Lumber Manufacturers Association,
Inc.
4 Fundy Rd.
Falmouth, ME 04105
(207) 781-2252

Northeast Retail Lumbermans Association
339 East Ave.
Rochester, NY 14604
(716) 625-1626

Occupational Safety and Health Administration
(OSHA)
(see U.S Department of Labor)

The Old-House Journal
435 9th St.
Brooklyn, NY 11215
(718) 788-1700

Owens-Corning Fiberglas Corporation
Fiberglas Tower
Toledo OH 43659
(419) 248-8764 (insulation)
(419) 248-8000 (roofing)

Painting and Decorating Contractors of America
3913 Old Lee Highway, Suite 33B
Fairfax, VA 22030
(703) 359-0826

Painting and Decorating Contractors of America
Washington State Council
27606 Pacific Hwy. S.
Kent, WA 98032
(206) 941-8823

Panel Solutions, Inc.
135 Lions Dr.
Hazelton, PA 18201
(800) 523-6671

Pella Corporation
102 Main St.
Pella, IA 50219
(515) 628-1000

Phillips Industries, Inc.
Lasco Products Group
8015 Dixon Dr.
Florence, KY 41042
(606) 371-7720

Pittsburgh Testing Laboratory
850 Poplar St.
Pittsburgh, PA 15220
(412) 922-4000

Plastic Pipe and Fittings Association
800 Roosevelt Rd.
Building C, #20
Glen Ellyn, IL 60137
(708) 858-6540

Plastic Pipe Institute
355 Lexington Ave.
New York, NY 10017
(212) 503-0600

Plastics in Construction Council
355 Lexington Ave.
New York, NY 10017

Plumbing and Drainage Institute (PDI)
c/o Saul Baker
1106 W. 77th St., South Drive
Indianapolis, IN 46260
(317) 251-6970

Plumbing, Heating, Cooling Information Bureau
3 Illinois Center
Chicago, IL 60601
(312) 372-7331

Porcelain Enamel Institute (PEI)
1101 Connecticut Ave., N.W., Suite 700
Washington, DC 20036
(202) 857-1134

Portland Cement Association (PCA)
5420 Old Orchard Rd.
Skokie, IL 60077
(312) 966-6200

The Preservation Press
(see National Trust for Historic Preservation)

Prestressed Concrete Institute (PCI)
175 W. Jackson Blvd.
Chicago, IL 60604
(312) 786-0300

Product Fabricating Service, Inc.
2402 Daniels St.
Madison, WI 53704

Professional Roofing
(formerly *Roofing Spec*; contact National Roofing
Contractors Association)

PWC: Painting and Wallcovering Contractor
(the magazine of the Painting and Decorating
Contractors of America)
130 W. Lockwood St.
St. Louis, MO 63119
(314) 961-6644

Redwood Inspection Service (RIS)
405 Enfrente Drive, Suite 300
Novato, CA 94949
(415) 382-0662

Republic Steel Corporation
(*see* Lukens Inc.)

Resilient Floor Covering Institute (RFCI)
966 Hungerford Drive, Suite 12-B
Rockville, MD 20850
(301) 340-8580

Rigidized Metal Corp.
658 Ohio St.
Buffalo, NY 14203
(716) 849-4705

Rock of Ages Corp.
P.O. Box 482
Barre, VT 05641
(802) 476-3115

Rodale Publishing Company
33 E. Minor St.
Emmaus, PA 18098
(215) 967-5171

Roof Consultants Institute
7424 Chapel Hill Rd.
Raleigh, NC 27607
(919) 859-0742

The Roofing Industry Education Institute (RIEI)
7006 South Alton Way, Suite B
Englewood, CO 80112
(303) 770-0613

Roofing '86
(*see* National Roofing Contractors Association)

Roofing Spec
(*see Professional Roofing*)

Rubber Manufacturer's Association (RMA)
1400 K St., N.W.
Washington, DC 20005
(202) 682-4800

Safety Glazing Certification Council (SGCC)
(*see* ETL Testing Laboratories, Inc.)

Scientific Apparatus Makers Association (SAMA)
1101 16th St., N.W.
Washington, DC 20036
(202) 223-1360

Screen Manufacturers Association
3950 Lake Shore Drive, Suite 502-A
Chicago, IL 60613
(312) 525-2644

Sealant Engineering and Associated Lines (SEAL)
7867 Convoy Court, Suite 301
San Diego, CA 92111
(619) 569-7906

Sealant, Waterproofing and Restoration Institute
3101 Broadway, Suite 585
Kansas City, MO 64111
(816) 561-8230

Sealed Insulating Glass Manufacturers Association
(SIGMA)
401 N. Michigan Ave.
Chicago, IL 60611-4200
(312) 644-6610

Sheet Metal and Air Conditioning Contractors
National Association (SMACNA)
P.O. Box 221230
Chantilly, VA 22022
(703) 803-2980

Sherwin Williams Company
P.O. Box 5819
Cleveland, OH 44101
(800) 321-8194

Sika Corporation
P.O. Box 297
Lyndhurst, NJ 07071
(201) 933-8800

Single Ply Roofing Institute (SPRI)
20 Walnut St.
Wellesley Hills, MA 02189
(617) 237-7879

Small Homes Council—Building Research Council
1 E St., Mary's Rd.
Champaign, IL 61820

Society of American Registered Architects
320 N. Michigan Ave.
Chicago, IL 60601
(312) 726-5880

Society of American Wood Preservers, Inc.
7297 Lee Hwy., Unit P
Falls Church, VA 22042
(703) 841-1500

Society of Fire Protection Engineers
1275 K St., N.W.
Washington, DC 20005
(202) 371-5200

Society of Plastics Engineers, Inc.
14 Fairfield Dr.
Brookfield Center, CT 06805
(203) 775-0471

Society of the Plastics Industry
1275 K St., N.W.
Washington, DC 20005
(202) 371-5200

Society of the Plastics Industry Polyurethane
Division
355 Lexington Ave.
New York, NY 10017
(212) 351-5425

Society of Real Estate Appraisers
225 N. Michigan Ave., Suite 724
Chicago, Il 60601
(312) 819-2400

Society of Women Engineers
345 E. 47th St., Room 305
New York, NY 10017
(212) 705-7855

Solar Energy Industries Association, Inc.
1717 Massachusetts Ave., N.W., Suite 503
Washington, DC 20036
(202) 483-6225

The Solar Group
P.O. Box 525
Taylorsville, MS 39168
(800) 647-7063

Southern Building Code Congress International
(*see* Standard Building Code)

Southern Cypress Manufacturers Association
805 Sterick Building
Memphis, TN 38103
(901) 525-8221

Southern Forest Products Association
2900 Indiana Ave.
Kenner, LA 70065
(504) 443-4464

Southern Hardwood Lumber Manufacturers
Association
(*see* Hardwood Manufacturers Association)

Southern Pine Inspection Bureau (SPIB)
4709 Scenic Highway
Pensacola, FL 32504
(904) 434-2611

Southwest Research Institute
6220 Culebra Rd.
San Antonio, TX 78238
(512) 684-5111

Sprayed Mineral Fiber Manufacturers Association,
Inc.
One Wall St., Suite 2400
New York, NY 10005

Standard Building Code/Southern Building Code
Congress International
900 Montclair Rd.
Birmingham, AL 35213
(205) 591-1853

Steel Deck Institute (SDI)
P.O. Box 9506
Canton, OH 44711
(216) 493-7886

Steel Door Institute (SDI)
30200 Detroit Rd.
Cleveland, OH 44145
(216) 889-0010

Steel Joists Institute (SJI)
12405 48th St. N., Suite A
Myrtle Beach, SC 29577
(803) 449-0487

Steel Structures Painting Council (SSPC)
4516 Henry St., Suite 301
Pittsburgh, PA 15213
(412) 687-1113

Steel Window Institute (SWI)
c/o Thomas Associates, Inc.
1300 Summer Ave.
Cleveland, OH 44115
(216) 241-7333

Submersible Wastewater Pump Association
(SWPA)
600 S. Federal St., Suite 400
Chicago, IL 60605
(312) 922-6222

Sump and Sewage Pump Manufacturers
Association (SSPMA)
P.O. Box 298
Winnetka, IL 60093
(312) 835-8911

Superintendent of Documents
(*see* U.S. Government Printing Office)

Technical Blinds International, Inc.
776 Southwest 17th Ave.
Delray Beach, FL 33444
(800) 222-4870

Technical Preservation Services
Preservation Assistance Division
National Park Service
(*see* U.S. Department of the Interior, National
Park Service)

Test Boring Association, Inc.
300 Buckelew Ave.
P.O. Drawer F
Jamesburg, NJ 08831
(201) 521-4441

Thermal Insulation Manufacturers Association
(TIMA)
29 Bank St.
Stamford, CT 06901
(203) 324-7533

Thermo-Flex, Inc.
1700 East Iron
P.O. Box 1184
Salina, KS 67402
(913) 827-7201

Tile Contractors Association of America
112 N. Alfred St.
Alexandria, VA 22314
(703) 836-5995

Tile Council of America, Inc. (TCA)
P.O. Box 326
Princeton, NJ 08542
(609) 921-7050

Tile and Decorative Surfaces
20335 Ventura Blvd., Suite 400
Woodland Hills, CA 91364
(818) 704-5555

Tile Letter
P.O. Box 13629
Jackson, MS 39215
(601) 939-2071

Transportation Research Board
National Academy of Sciences
2101 Constitution Ave., N.W.
Washington, DC 20418
(202) 334-2000 or 334-2222

Truss Plate Institute (TPI)
583 D'Onofrio Drive, Suite 200
Madison, WI 53719
(608) 833-5900

TRW Nelson Company
7900-T W. Ridge Rd.
Elyria, OH 44036
(800) 321-2005

Tubelite Division of Indal, Inc.
8200 Mackinaw Trail
P.O. Box 118
Reed City, MI 49677
(616) 832-2211

Underground Space Center
University of Minnesota
500 Pillsbury Dr., S.E.
Minneapolis, MN 55455
(612) 624-0066

Underwriters Laboratories, Inc. (UL)
333 Pfingsten Rd.
Northbrook, IL 60062
(708) 272-8800

United States Gypsum Company (USG)
125 S. Franklin
Chicago, IL 60606
(312) 606-4000

United States Plywood Corporation
133 Peachtree St., N.W.
Atlanta, GA 30303
(404) 521-4000

United States Steel Corporation
(*see* USX)

Urban Land Institute
1090 Vermont Ave., N.W.
Washington, DC 20005
(202) 289-8500

Urethane Foam Contractors Association
4302 Airport Blvd.
Austin, TX 78722
(512) 454-0041

Urethane Institute
355 Lexington Ave.
New York, NY 10017

U.S. Architectural and Transportation Barriers
Compliance Board
1331 F St., N.W., Suite 1000
Washington, DC 20004
(800) 872-2253

U.S. Department of Agriculture
Forest Service
Forest Insect and Disease Research
P.O. Box 2417
Washington, DC 20013
(202) 447-3760

U.S. Department of Agriculture
Office of Publishing & Visual Communications
14th St. and Independence Ave., S.W.,
Room 402A
Washington, DC 20250
(202) 720-8005

U.S. Department of the Army
Corps of Engineers (CE)
Chief of Engineers
Washington, DC 20314
(202) 272-0660

U.S. Department of Commerce (DOC)
14th St. and Constitution Ave., N.W.
Washington, DC 20230
(202) 377-2000

U.S. Department of Defense
Naval Publications and Forms Center
5801 Tabor Ave.
Philadelphia, PA 19120
(215) 697-2000

U.S. Department of Housing and Urban
Development (HUD)
451 Seventh St., S.W.
Washington, DC 20410
(202) 755-4000

U.S. Department of Housing and Urban
Development (HUD)
Federal Housing Administration (FHA)
Director, Manufactured Housing and Construction
Standards Division
451 7th St., S.W., Room 9158
Washington, DC 20201
(202) 755-5210

U.S. Department of Housing and Urban
Development (HUD)
Standards & Production Branch
451 7th St., S.W., Dept. 6270
Washington, DC 20410
(202) 708-1422

U.S. Department of the Interior
National Park Service
P.O. Box 37127
Washington, DC 20013-7127
(202) 343-7394

U.S. Department of Labor
Occupational Safety and Health Administration
(OSHA)
200 Constitution Ave, N.W.
Washington, DC 20210
(202) 523-8148

U.S. Department of the Navy
(*see also* U.S. Department of Defense)
Naval Facilities Engineering Command
Washington Naval Yard, Building 212
Washington, DC 20374
(202) 433-5984

U.S. Department of Transportation (DOT)
400 Seventh St., S.W.
Washington, DC 20590
(202) 366-4000

U.S. Environmental Protection Agency
(EPA)
401 M St., S.W.
Washington, DC 20460
(202) 382-2090   General phone number
(202) 260-5859   Municipal Support
                 Division
(202) 260-4270   Office of Solid Waste
                 Management
(202) 382-5700   Office of Water Programs

U.S. Forest Products Laboratory
Gifford Pinchot Drive
P.O. Box 5130
Madison, WI 53705
(608) 231-9200

U.S. General Services Administration
Historic Preservation Office
18th and F Streets, N.W.
Washington, DC 20405
(202) 655-4000

U.S. General Services Administration (GSA)
Specifications Unit
7th and D Streets, S.W.
Washington, DC 20407
(202) 472-2205/2140

U.S. Government Printing Office
Superintendent of Documents
N. Capitol and H Streets, N.W.
Washington, DC 20405
(202) 783-3238

U.S. Postal Service (USPS)
475 L'Enfant Plaza, S.W.
Washington, DC 20260
(202) 268-2000

U.S. Weather Service
1325 East West Highway, Room 18444
Silver Spring, MD 20910
(301) 443-8910

USX
(formerly U.S. Steel)
600 Grant St.
Pittsburgh, PA 15219
(412) 433-3123

Van Nostrand Reinhold
115 Fifth Ave.
New York, NY 10003
(212) 254-3232

Vermiculite Association, Inc.
600 S. Federal St., Suite 400
Chicago, IL 60605
(312) 922-6222

Vermont Structural Slate Company
Fair Haven, VT 05743
(802) 265-4933

Vinyl Siding Institute (VSI)
355 Lexington Ave.
New York, NY 10017
(212) 351-5400

Vulcraft Division of Nucor Corp.
P.O. Box 10052
Florence, SC 29501
(803) 662-0381

W and P Engineering and Fabrications
7503 Brook Rd.
Richmond, VA 23227
(804) 264-1854

Water Quality Association
4151 Naperville Rd.
Lisle, IL 60532
(312) 369-1600

Water Systems Council (WSC)
600 South Federal St., Suite 400
Chicago, IL 60605
(312) 922-6222

Welded Steel Tube Institute
522 Westgate Tower
Cleveland, OH 44416
(216) 333-4550

West Coast Lumber Inspection Bureau (WCLIB)
P.O. Box 23145
Portland, OR 97223
(503) 639-0651

Western Lath, Plaster, Drywall Contractors
Association
(formerly California Lath and Plaster Association)
8635 Navajo Rd.
San Diego, CA 92119
(619) 466-9070

Western Red Cedar Lumber Association
Yeon Building
522 S.W. 5th Ave.
Portland, OR 97204
(503) 224-3930

Western Wood Products Association
(WWPA)
Yeon Building
522 S.W. 5th Ave.
Portland, OR 97204
(503) 224-3930

Weyerhaeuser Company
Tacoma, WA 98477
(206) 924-2345

John Wiley & Sons
605 Third Ave.
New York, NY 10158
(212) 850-6000

Window and Door Specifier
(*see* National Wood Window and Door
Association)

Window Shade Manufacturers Association
Executive Plaza
1211 W. 22nd St.
Oak Brook, IL 60521

Wire Reinforcement Institute (WRI)
1101 Connecticut Ave., N.W., #700
Washington, DC 20036
(202) 429-5125

Wolverine Technologies
Building Products Division
17199 Laurel Park Dr. N.
Livonia, MI 48152
(800) 521-9020

Wood and Synthetic Flooring Institute (WSFI)
4415 W. Harrison St., Suite 242-C
Hillside, IL 60162
(708) 449-2933

Wood Truss Council of America
5937 Meadowwood Drive, Suite 14
Madison, WI 53711
(608) 274-4849

Woodwork Institute of California
P.O. Box 11428
Fresno, CA 93773
(209) 233-9035

Woven Wire Products Association (WWPA)
2515 N. Nordica Ave.
Chicago, IL 60635
(312) 637-1359

York Manufacturing, Inc.
P.O. Box 1009
Sanford, ME 04073
(207) 324-1300

Zero International, Inc.
415 Concord Ave.
Bronx, NY 10455
(800) 635-5335

APPENDIX D
GLOSSARY

Bold-faced words in the definitions are also glossary entries.

abrasion Wearing away by friction.

absorption bed A wide trench exceeding 36 in. in width containing a minimum of 12 in. of clean, coarse **aggregate** and a system of two or more distribution pipes through which treated sewage may seep into the surrounding soil. Also called *seepage bed*.

absorption coefficient The ratio of the sound absorbing effectiveness of 1 sq. ft. of a material to 1 sq. ft. of a perfectly absorptive material at a specific frequency; usually expressed as a decimal value (such as 0.70) or as a percentage.

absorption field An arrangement of absorption trenches through which treated sewage is absorbed into the soil. Also called *disposal field*.

absorption rate The weight of water absorbed when a clay **brick** is partially immersed for 1 minute, usually expressed in either grams or ounces per minute (grams or oz./min.). Also called **suction** or initial rate of absorption.

absorption of sound The ability of a material to absorb rather than reflect sound waves striking it by converting sound energy to heat energy within the material.

absorption trench A trench not over 36 in. in width, containing a minimum of 12 in. of clean, coarse aggregate and a distribution pipe, through which treated sewage is allowed to seep into the soil.

absorption unit *See* **sabin**.

absorption of water by clay masonry The weight of water that a **brick** or other clay **masonry unit** absorbs when immersed in either cold or boiling water for a stated length of time, expressed as a percentage of the weight of the dry unit.

absorption of water by concrete masonry units (CMUs) The weight of water that a concrete masonry unit absorbs when immersed in water, expressed in pounds of water per cubic foot (lb./cu. ft.) of concrete.

accelerator An **admixture** used in **concrete** to hasten its set and increase the rate of strength gain (the opposite of a **retarder**).

accessible Describes a site, building, facility, or portion thereof that complies with current standards and can be approached, entered, and used by a **physically handicapped person**.

accessible route A continuous unobstructed path connecting all **accessible** elements and spaces in a building or facility that can be negotiated by a person with a severe **disability** using a wheelchair and that is also safe for and usable by people with other disabilities.

acid Corrosive chemical substance that attacks many common building materials, decorative finishes, **coatings**, **paints**, and **transparent finishes**.

acid-resistant brick *See under* **brick masonry unit**.

acoustical correction The planning, shaping, and equipping of a space to establish the best possible aural conditions for faithful reproduction of wanted sound within the space. *See also* **acoustics**.

acoustics The science of sound; the production, transmission, and effect of sound. *See also* **acoustical correction**, **architectural acoustics**, and **room acoustics**.

acrylic (1) In **carpet**, a generic term including acrylic and modified acrylic (modacrylic) fibers. Acrylic is a **polymer** composed of at least 85% by weight of acrylonitrile; modacrylic is a polymer composed of less than 85% but at least 35% by weight of acrylonitrile. (2) In glazing applications, a transparent plastic material.

actual dimension The actual measured dimension of a **masonry unit**, piece of lumber, or other construction material or assembly. *See also* **nominal dimension**).

actual size *See* **size**.

adaptability The capability of certain building spaces and elements, such as kitchen counters, sinks, and grab bars, to be altered or added to accommodate the needs of persons with and without **disabilities**, or to accommodate the needs of persons with different types or degrees of disability.

adhesion The property of a paint film that enables it to stick to a surface.

adhesive, drywall Adhesives specifically intended for the application of gypsum board. A *contact adhesive* is an adhesive used to bond layers of gypsum board or for bonding gypsum board to metal studs. A *laminating adhesive* is an adhesive used to bond layers of gypsum board. A *stud adhesive* is an adhesive used to attach gypsum board to wood supports.

adhesive, tile Prepared organic material, ready for use with no further addition of liquid or powder, which cures or sets by evaporation; distinguished from mortars by the absence of siliceous fillers (sand) that are included in mortars either at the plant or in the field.

admixture A material (other than **portland cement**, water, or **aggregate**) used in **concrete** to alter its properties (for example, **accelerators**, **retarders**, and **air-entraining agents**).

adobe brick *See under* **brick masonry unit.**

age hardening The continuing increase in strength for long periods of time of aluminum **alloys** after heat treatment.

aging The period of time in which a **heat-treatable** aluminum **alloy** is allowed to remain at room temperature, after heat treatment (heating and quenching), to reach a stable state of increased strength. *See also* **artificial aging.**

agglomeration Process for increasing the particle size of iron ores to make them suitable for ironworking and steelmaking.

aggregate A hard, inert material mixed with **portland cement** and water to form **concrete**. *Fine aggregate* has pieces that are 1/4 in. in diameter or smaller. *Coarse aggregate* has pieces larger than 1/4 in. in diameter.

airborne sound transmission *See* **sound transmission.**

air chamber A piece of pipe about 10 in. long, installed above the hot and cold valves of fixtures such as sinks, lavatories, and clothes washers, to cushion the rush of water as the valve is closed, and prevent water hammer.

air drying (1) In **coatings**, **paints**, and **transparent finishes**, capable of forming a solid film when exposed to air at moderate atmospheric temperatures. (2) Of wood, *see* **seasoning.**

air-entrained concrete Concrete containing minute bubbles of air up to about 7% by volume.

air-entraining agent An **admixture** used to produce **air-entrained concrete.**

air gap The unobstructed vertical distance between the mouth of a water outlet and the flood level rim of the water receptacle. The water outlet may be a faucet, spout, or other outlet; the receptacle may be a plumbing fixture, tank, or other receptacle.

alclad sheet A clad product with an aluminum or aluminum **alloy** coating having a high resistance to **corrosion**. The coating is anodic to the core alloy it covers, thus protecting it physically and electrolytically against corrosion. *See also* **clad alloys.**

alkali A soluble mineral salt present in some soils. Alkalies are chemical substances characterized by their ability to combine with acids to form neutral salts. They are damaging to many **coatings, paints,** and **transparent finishes.**

allowable unit stress *See under* **stress.**

alloy The material formed when two or more metals, or metals and nonmetallic substances, are joined by being dissolved into one another while molten. *See also* **brazing alloy** and **clad alloy.**

alloy designation A numerical system used in designating the various **alloys** of aluminum.

alloying element Element added in steelmaking to achieve desired properties.

alloy steel *See under* **steel.**

alumina (AL₂O₃) A hydrated form of aluminum oxide found in **bauxite** and in ordinary clays.

ambient sound A continuous background sound that is a composite of individual sounds coming from exterior sources, such as street traffic, and interior sources, such as ventilating equipment and appliances, none of which can be identified individually by a listener. Also called *background noise.*

ambulatory Able to walk without assistance or difficulty.

amperage Electrical rate of flow, measured in amperes and comparable to gallons per minute (gpm) in a fluid medium. The strength of a current of electricity.

ampere A unit of electrical current equivalent to that produced by 1 volt applied across a resistance of 1 ohm. One coulomb of electricity in every second.

anchor A piece or assemblage, usually metal, used to attach parts (e.g., plates, joists, trusses, studs, sills, masonry, windows, doors, and other building elements) to wood, concrete, or masonry.

angular course *See under* **course.**

angular measure The deviation between two lines that meet at a point, expressed in **degrees, minutes,** and **seconds.**

annealing *See* **heat treatment.**

annual growth ring The growth layer added to a tree in a single growth year.

anodic coating A surface coating applied to an aluminum **alloy** by **anodizing.**

anodizing Applying an electrolytic oxide coating to an aluminum **alloy** by building up the natural surface film using an electrical current (usually dc) through an oxygen-yielding electrolyte with the alloy serving as the anode.

antioxidant A compound added to other substances to retard oxidation, which deteriorates plastics.

antique finish A finish, usually applied to furniture or woodwork, to give the appearance of age.

area wall A **retaining wall** around below-grade basement windows.

arch A usually curved compressive structural member, spanning openings or recesses; also

built flat. A *back arch* is a concealed arch carrying the backing of a wall where the exterior face is carried by a lintel. A *jack arch* is an arch having horizontal or nearly horizontal upper and lower surfaces. Also called a *flat* or *straight arch.* A *major arch* is an arch with a span greater than 6 ft. that carries a load that is equivalent to a uniform load greater than 1000 pounds per square foot (psf). Typically known as Tudor arch, semicircular arch, Gothic arch, or parabolic arch. Has rise-to-span ratio greater than 0.15. A *minor arch* is an arch with a maximum span of 6 ft. carrying a load that does not exceed 1000 psf. Typically known as a jack arch, segmented arch, or multicentered arch. Has rise-to-span ratio less than or equal to 0.15. A *relieving arch* is an arch that is built over a **lintel**, flat arch, or smaller arch to divert loads, thus relieving the lower member from excessive loading. Also known as *discharging* or *safety arch.* A *trimmer arch* is an arch, usually a low-rise type, of **brick masonry unit,** used for supporting a fireplace hearth.

architectural acoustics The **acoustics** of buildings and other structures. *See also* **acoustics** and **room acoustics.**

architectural barrier A physical condition in a building or facility that creates unsafe or confusing conditions or prevents accessibility and free mobility.

architectural terra-cotta Custom-made, hard-burned, glazed or unglazed clay building units, plain or ornamental, that are machine extruded or hand molded. *See also* **ceramic veneer.**

aromatic solvents Group of organic compounds derived from coal or petroleum, such as benzene and toluene.

artificial aging The heating of an aluminum **alloy** for a controlled time at an elevated temperature to accelerate and increase its strength gain after heat treatment. *See also* **aging.**

ashlar masonry Masonry composed of rectangular units usually larger in size than **brick,** with sawed, dressed, or square beds, bonded with **mortar.** Ashlar masonry is also described according to its **pattern bond,** which may be coursed or random.

atmospheric pressure steam curing *See* **curing.**

atmospheric vacuum breaker A simple mechanical device consisting essentially of a check valve in a supply line and a valve member (on the discharge side of the check valve) opening to the atmosphere when the pressure in a line drops to atmospheric. Also called a *siphon breaker.*

attenuation Reduction of the energy or intensity of sound.

audible cue *See* **cue.**

austenitic steel *See* **grain structure.**

autoclave *See* **curing.**

average transmission loss (TL) The numerical average of the **transmission loss** values of an assembly measured at nine frequencies. It is a single-number rating for comparing the airborne sound transmission through walls and floors.

Axminster carpet *See* **woven carpet**.

back The side opposite the face. The poorer side of a **plywood** panel. The surface of **gypsum board** that will be placed toward the supports.

back arch *See under* **arch**.

back blocking A single-ply gypsum board installation procedure for reinforcing butt-end or edge joints to minimize surface imperfections such as **cracking** and **ridging**.

backflow The unintentional flow of water into the supply pipes of a plumbing system from a nonsupply source.

background noise *See* **ambient sound**.

backing The carpet foundation of jute, kraftcord, cotton, rayon, or polypropylene yarn that secures the pile yarns and provides stiffness, strength, and dimensional stability.

backing board *See under* **gypsum board**.

backing, tile A suitable surface for the application of tile, such as a structural subfloor or wall surface.

backparging *See* **parging**.

backplate *See under* **finished mill products**.

back-siphonage A type of **backflow**, usually caused by a temporary occurrence of negative pressure (suction) in the pipes.

backup The part of a masonry wall behind the exterior facing.

baking finish A **coating** that is baked at temperatures above 150 °F to dry and develop desired properties.

bark The outer, corky layer of a tree composed of dry, dead tissue.

bars *See under* **finished mill products**.

base coat The first **plaster** layer in **two-coat work**; or either the **scratch** or **brown coat** in **three-coat work**.

base course The lowest course of masonry in a wall or pier.

base line A **parallel** of specified **latitude**, used in the **rectangular survey system** serving as the main east-west reference line with a **principal meridian** for a particular state or area.

basic stress *See* **stress**.

batter Recessing or sloping a wall back in successive courses; the opposite of **corbel**.

bauxite A raw ore of aluminum consisting of 45% to 60% aluminum oxide, 3% to 25% iron oxide, 2.5% to 18% silicon oxide, 2% to 5% titanium oxide, other impurities, and 12% to 30% water. This ore varies greatly in the proportions of its constituents, color, and consistency.

Bayer process The process generally employed to refine **alumina** from **bauxite**.

beam A structural member transversely supporting a load.

bearding Long fiber fuzz occurring on some loop pile fabrics, caused by fibers snagging and loosening because of inadequate anchorage.

bearing wall A wall that supports a vertical load in addition to its own weight. *See also* **non-load-bearing wall**.

bed joint Horizontal layer of mortar into which a masonry unit is laid.

beneficiation Concentrating process used to increase the iron content of ores prior to use. *See also* **agglomeration**.

binder In paint and coatings, the **vehicle** ingredient with adhesive qualities (linseed oil, resins, etc.) that binds the *pigment* and other ingredients of a **coating**, **paint**, or **transparent finish** into a cohesive film and facilitates bonding with the underlying surfaces.

binder In terrazzo, a cementitious or resinous material that gives the matrix adhesive and other important physical properties.

bleaching (1) The process of lightening raw wood. (2) The process of restoring discolored or stained wood to its normal color or making it lighter.

bleeding (1) In unit masonry, (a) the loss of water from a **masonry unit** having a low **suction** when it comes into contact with mortar, (b) the loss of water from **mortar** because of a low **water retention** when it contacts a masonry unit. Bleeding causes masonry units to float (*see* **floating**). (2) In concrete, the appearance of excess water rising to the surface shortly after placing of **concrete**. (3) In **coatings**, **paints**, and **transparent finishes**, discoloration of a finish coat by coloring matter from the underlying surface or coat of finishing material. (4) In plastics, the diffusion of a colorant out of a plastic part into adjacent materials.

blend to a common level The meeting of two or more surfaces so that there is no abrupt rise or drop in the surface.

blistering The formation of bubbles or pimples on a coated, painted, or transparent-finished surface caused by moisture in the underlying material (wood, masonry, concrete, etc.); caused by adding a coat of **coating**, **paint**, or **transparent finish** before the previous coat has dried thoroughly; or caused by excessive heat or grease under a coating, paint, or transparent finish.

blocking (1) In masonry construction, a method of bonding two adjoining or intersecting walls, not built at the same time, by means of offsets whose vertical dimensions are not less than 8 in. (2) In wood construction, short pieces of wood used as nailers, spacers, or fillers between wood members or between wood members and other construction, or between other materials.

blow molding Shaping thermoplastic materials into hollow form by air pressure and heat; usually performed on sheets or tubes.

blushing Describes opaque lacquer that loses its gloss and becomes flat or clear lacquer that further turns white or milky.

board foot A measure of **lumber**. One board foot is the equivalent of a piece of lumber whose nominal dimensions are 1 in. thick, 12 in. wide, and 12 in. long.

board *See under* **lumber**.

body Used to indicate thickness or thinness of a liquid **coating**, **paint**, or **transparent finish** material.

bond beam Course or courses of a masonry wall grouted and usually reinforced in the horizontal direction. Alternatively may be made of reinforced concrete. Serves as horizontal tie of wall-bearing courses for structural members, or itself as a flexural member.

bond course A masonry **course** in which the units overlap more than one **wythe** of masonry.

bonded rubber cushioning Rubber or latex cushioning adhered to a carpet at the mill.

bonderizing Process to improve paint adhesion on steel by dipping lightly galvanized objects in a hot phosphate solution to form a surface film of zinc phosphate.

bond, pattern *See* **pattern bond**.

bond, structural Tying **wythes** of a masonry wall together by lapping units one over another or by connecting them with metal ties.

bond, tensile Adhesion between **mortar** and **masonry units** or reinforcement.

bonder *See* **header**.

border The resilient flooring at the perimeter of a room adjacent to the walls, which is installed separately from the **field**.

bow *See* **warp**.

boxing Mixing a **coating**, **paint**, or **transparent finish** material by pouring it from one container to another several times. Boxing is not recommended for most materials.

Braille A special raised-touch alphabet for the blind using a cell of six dots.

brazing A welding process in which the filler metal is a nonferrous metal or **alloy** with a melting point higher than 800 °F but lower than that of the metals joined. A *brazing alloy* is an alloy used as filler metal for brazing. In aluminum the brazing alloy is usually in the 4000 Series of alloys. A *brazing sheet* is an unclad or specially clad sheet for brazing purposes with the surface of the specially clad sheet having a lower melting point than the core. Brazing sheet of the cladded type may be clad on either one or two surfaces.

breathe The ability of a **coating**, **paint**, or **transparent finish** film to permit the passage of moisture vapor without causing **blistering**, **cracking**, or **peeling**.

brick grade Designation for durability of a unit expressed as SW for severe weathering, MW for moderate weathering, or NW for negligible weathering. See ASTM Standards C 216, C 62, and C 652.

brick masonry unit A unit that is formed into a rectangular prism while plastic. An *acid-resistant brick* is a clay brick suitable for use in contact with chemicals; designed primarily for use in the chemical industry. Usually used with acid-resistant mortars. An *adobe brick* is a large clay brick of varying size, roughly molded and sun dried. A *building (common) brick* is a clay brick for building purposes, not specially treated for texture or color. Formerly called common brick. A *clay brick* is a brick made of clay or shale, formed into a rectangular prism while plastic and burned (fired) in a kiln. A *concrete brick* is a brick made from **portland cement** and an **aggregate** and cured either in normal atmosphere or using either low- or high-pressure steam. It is formed as a rectangular prism, usually not larger than 4×4×12 in. An *economy brick* is a clay brick whose nominal dimensions are 4×4×8 in. An *engineered brick* is a clay brick whose nominal dimensions are 4×3.2×8 in. A *facing brick* is a clay brick made especially for facing purposes, often with finished surface texture. It can be made of selected clays or treated to produce desired color. A *fire brick* is a brick made of refractory ceramic material that will resist high temperatures. A *floor brick* is a smooth, dense brick that is highly resistant to abrasion; used as a finished floor surface. A *hollow brick* is a clay or shale **masonry unit** whose **net cross-sectional area** in any plane parallel to the bearing surface is not less than 60% of its **gross cross-sectional area** measured in the same plane. The term *jumbo brick* is a generic one, indicating that a clay brick is larger in size than the standard. Some producers use this term to describe oversized brick of specific dimensions manufactured by them. A *Norman brick* is a clay brick whose nominal dimensions are 4×2-2/3×12 in. A *paving brick* is a clay brick especially suitable for use in pavements where resistance to abrasion is important. A *Roman brick* is a clay brick whose nominal dimensions are 4×2×12 in. An *SCR brick* is a clay brick whose nominal dimensions are 2-2/3×6×12 in. It lays up three courses to 8 in. and produces a nominal 6-in.-thick wall. Developed by the Structural Clay Products Research Foundation. A *sewer brick* is a low-absorption, abrasive-resistant clay brick intended for use in drainage structures.

brick type Designation for facing brick that controls tolerance, chippage, and distortion. Expressed as FBS, FBX, and FBA for solid brick, and HBS, HBX, HBA, and HBB for hollow brick. See ASTM Standards C 216 and C 652.

British thermal unit (Btu) A measure of heat quantity equal to the amount of heat required to raise 1 lb. of water 1°F at sea level.

broadloom Carpet woven on a broad loom in widths of 6 ft. or more.

brown coat The second **plaster** layer in **three-coat work** that provides additional strength and a suitably true and plane surface for the application of the **finish coat**.

brush A tool composed of bristles set into a handle; often used to apply **paint** and **transparent finish** materials. Sometimes used to apply **coatings**. Bristles may be synthetic (needed for water-thinned paints) or natural, such as hog hair.

brushability The ability of or ease with which a **paint** or **transparent finish** material can be brushed.

brush marks Marks of a brush that remain in a dried **coating**, **paint**, or **transparent finish** film.

Btu *See* **British thermal unit.**

buckling Wrinkling or ridging of a carpet after installation, caused by insufficient stretching, dimensional instability, or manufacturing defects.

build The apparent thickness or depth of a **coating**, **paint**, or **transparent finish** film after it has dried.

building brick *See under* **brick masonry unit**.

building drain The lowest part of a building's drainage system. It receives the discharge from soil, waste, and other drainage pipes inside the walls of a building and conveys it to a point 3 ft. outside the building walls where it joins with the **building** (house) **sewer**.

building sewer The horizontal piping of a **drainage system** that extends from the end of a **building drain** to a public sewer, private sewer, **individual sewage disposal system**, or other point of disposal. A building sewer begins at a point 3 ft. outside the building wall, where the building drain ends. Also called a *house sewer.*

building storm drain A type of **building drain** used for conveying rainwater, surface water, groundwater, subsurface water, cooling condensate; or a combined building sewer, extending to a point not less than 3 ft. outside the building wall where it joins with a storm sewer or combined storm and sewage sewer.

built environment The collection of buildings, facilities, transportation systems, and other structures and spaces created for the purpose of providing convenient places for work, play, living, and related human activities; contrasted with the natural environment.

bulked continuous filament (BFC) Continuous strands of synthetic fiber made into yarn without spinning; often extruded in modified cross section such as multilobal, mushroom, or bean shaped, or textured to increase bulk and covering power.

burling Removing surface defects such as knots, loose threads, and high spots to produce an acceptable quality level after weaving; also, filling in omissions in weaving.

burnishing Shiny or lustrous spots on a **coating**, **paint**, or **transparent finish** surface caused by rubbing.

butt-end joint Joint in which mill- or job-cut (exposed core) gypsum board ends or edges are butted together.

buttering Placing **mortar** on a **masonry unit** with a trowel.

calcium chloride An accelerator added to **concrete** to hasten setting (not to be considered an antifreeze).

calcimine Water-based **paint** generally consisting of animal glue, zinc white, and calcium carbonate or clay; now seldom used.

calendering A process for producing plastic film or sheeting by passing the material between revolving heated rolls.

call In surveying, a statement or mention of a **course** or distance.

cambium A thin layer of tissue that lies between a tree's bark and its wood. The cambium sub-divides to form the new wood and bark cells of each year's growth.

cant A log **slabbed** on one or more sides.

cant strip A triangular filler between a roof and a parapet wall.

capacity insulation The ability of masonry to store heat as a result of its mass, density, and specific heat.

capillaries (1) Thin-walled tubes or vessels found in wood. (2) In concrete, channels that absorb water and are interrupted by entrained air bubbles.

carbon steel *See under* **steel.**

cardinal points The four major compass headings of north, east, south, and west.

carpet General designation of fabric constructions that serve as soft floor coverings, especially those that cover the entire floor and are fastened to it, as opposed to rugs. *See also* **flocked carpet, fusion-bonded carpet, knitted carpet, needle-punched carpet, tufted carpet,** and **woven carpet.**

carved carpet *See under* **pile.**

case hardening Hardening of the outer skin of an iron-based alloy by promoting surface absorption of carbon, nitrogen, or cyanide, generally accomplished by heating the alloy in contact with materials containing these elements, and rapid cooling.

casting (1) In masonry and concrete, pouring a mix into a mold and permitting it to set up to form the desired shaped object. (2) In metalwork, pouring molten metal into molds to form desired shapes. (3) In plastics, the shaping of plastic objects by pouring the material into molds and allowing it to harden without the use of pressure.

cast iron High carbon iron made by melting **pig iron** with other iron-bearing materials, and casting in sand or loam molds; characterized by hardness, brittleness, high compressive, and low tensile strengths.

catalyst (1) A substance that speeds up or slows down the rate of a chemical reaction without itself undergoing permanent change in composition. (2) A substance used to initiate the polymerization of monomers to form polymers.

caulking compound Semidrying or slow-drying plastic material used to seal joints or fill crevices such as those around windows and chimneys. Either elastomeric or acrylic joint sealants are generally used today where caulking compounds are called for. *See also* **sealant, joint.**

cavity wall A wall built of **masonry units** so arranged to provide a continuous airspace within the wall (with or without insulating material), and in which the inner and outer **wythes** of the wall are tied together with metal **wall ties** or continuous metal **joint reinforcement**.

cavity wall tie *See under* **wall tie.**

C/B ratio The ratio of the weight of water absorbed by a clay **masonry unit** during immersion in cold water to the weight absorbed during immersion in boiling water. An identification of the probable resistance of clay **brick** to the action of freezing and thawing.

cellular plastic A plastic, the apparent density of which is decreased substantially by the presence of numerous cells disposed throughout its mass. *Closed cell plastic* is cellular plastic in which there is a predominance of nonconnecting cells. *Open cell plastic* is a cellular plastic in which there is a predominance of interconnecting cells.

cellulose The principal constituent of wood that forms the framework of wood cells.

cement A binding agent capable of uniting dissimilar materials into a composite whole. *See also* **portland cement**.

centering Temporary formwork for the support of masonry **arches** or **lintels** during construction. Also called *center(s)*.

ceramic tile A thin surfacing unit, made from clay or a mixture of clay and other ceramic materials, having either a glazed or unglazed face, and fired to a temperature sufficiently high to produce specific physical properties and characteristics.

ceramic tile finish All of the elements normally installed by a tile contractor, from the backing to the face of the tile. This may include just a tile-setting product, tile, and **grout**, or a **mortar bed**, reinforcement, and **cleavage membrane** as well as the above items. Also called *tilework*.

ceramic veneer Architectural terra-cotta, characterized by large face dimensions and thick sections. An *adhesion-type ceramic veneer* is a thin section of ceramic veneer held in place without metal **anchors** by adhesion of **mortar** backing. An *anchored-type ceramic veneer* is a thicker section of ceramic veneer held in place by **grout** and wire **anchors** connected to backing.

cesspool A covered and lined underground pit used as a holding tank for domestic **sewage**, designed to retain the organic matter and solids and to permit the liquids to seep through the bottom and sides. Cesspools are almost universally prohibited in the United States and are not acceptable as a means of sewage disposal.

chain warp *See under* **warp**.

chalking Formation of a loose powder on the surface of a **coating** or **paint** after exposure to the elements.

chase A groove or continuous recess built in a masonry or concrete wall to accommodate pipes, ducts, or conduits.

check A lengthwise separation of wood, the greater part of which occurs across the **annual growth rings**. A check that passes entirely through a piece of wood is called a *split*.

checking Type of **coating**, **paint**, or **transparent finish** failure in which many small cracks appear in the surface.

circuit vent A branch vent that serves two or more traps and extends from in front of the last fixture connection of a horizontal branch to a **vent stack**. *See also* **loop vent**.

circulation path An exterior or interior way of passage from one place to another for pedestrians, including, but not limited to, walks, hallways, courtyards, stairways, and stair landings.

clad alloys Alloys having one or both surfaces of a metallurgically bonded coating, the composition of which may or may not be the same as that of the core, and that is applied for purposes such as corrosion protection, surface appearance, or brazing. *See also* **alclad sheet** and **alloy**.

cladding Bonding thin sheets of a coating metal with desirable properties (such as corrosion resistance or chemical inertness) over a less expensive metallic core not possessing these properties. Copper cladding over steel may be applied by hot dipping; stainless steel and aluminum cladding, by hot rolling.

clay brick *See under* **brick masonry unit**.

cleavage (isolation) membrane A membrane, such as saturated roofing felt, building paper, or 4 mil polyethylene film, installed between the **backing** and a **mortar bed** to permit independent movement of a tile finish.

closer The last **masonry unit** laid in a **course**.

CMU *See* **concrete masonry unit**.

coarse aggregate *See* **aggregate**.

coating A mastic or liquid-applied surface finish, regardless of whether a protective film is formed or only decorative treatment results. There are two categories of coatings: coatings and special coatings. Materials called *coatings* in this book are those that are usually applied in the shop or factory on metal, glass, porcelain, wood, and other materials. They include such products as Kynar coatings on metal, anodizing on aluminum, and liquid-applied colored finishes on wood doors and cabinets. Some coatings are virtually identical to **paint**, the major differences being that they are applied in a shop or factory and are usually sprayed on rather than being brushed or rolled on. *Special coatings* are relatively thick, high-performance architectural coatings, such as high-build glaze coatings; fire-retardant coatings; industrial coatings, such as those used in sewage disposal plants; and cementitious coatings. Special coatings are usually applied in the field. These are beyond the scope of this book. (*See also* **fire-retardant coatings and paint**, **paints**, and **transparent finishes**.)

coke Processed form of bituminous coal used as a fuel, a reducing agent, and a source of carbon in making **pig iron**.

cold drawing *See under* **cold working**.

cold flow Permanent change in dimension as a result of stress over time, without heat.

cold forging *See* **forging**.

cold forming Forming thin sheets and strip to desired shapes at room temperature, generally with little change in the mechanical properties of the metal; includes roll, stretch, shear, and brake forming. *See also* **cold working**.

cold rolling *See under* **cold working**.

cold working (1) In aluminum, forming a metal product at room temperature by means of **rolling**, **drawing**, **forging**, or other mechanical methods of forming or shaping. *Cold rolling* is the forming of sheet metal by rolling at room temperature metal that has been previously hot rolled to a thickness of about 0.125 in. (2) In steel, shaping by cold drawing, cold reduction, or cold rolling at room temperature; generally accompanied by an increase in strength and hardness. *Cold drawing* is shaping by pulling through a die to reduce cross-sectional area and impart desired shape; generally accompanied by an increase in strength, hardness, closer dimensional tolerances, and smoother finish. *Cold finishing* is cold working that results in **finished mill products**. *Cold reduction* is cold rolling that drastically reduces sheet and strip thickness with each pass through the rolls; generally accompanied by an increase in hardness, stiffness, and strength, and resulting in a smoother finish and improved flatness. *Cold rolling* is a gradual shaping between rolls to reduce cross-sectional area or impart desired shape; generally accompanied by an increase in strength and hardness. (*See also* **cold forming**; **hot working**.)

collar joint Interior longitudinal vertical joint between two **wythes** of masonry.

colorant Concentrated color added to **coatings** and **paints** to make specific colors.

colorfast Fade resistant.

color uniformity Ability of a **coating** or **paint** to maintain a consistent color across its entire surface, particularly during weathering.

column A vertical structural member acting primarily in compression, whose horizontal dimension measured at right angles to its thickness does not exceed 3 times its thickness.

combination process A process used to retrieve additional **alumina** and soda from the **red mud** impurities of the **Bayer process**.

common alloy An alloy that does not increase in strength when heat-treated (non-heat-treatable). Common alloys may be strengthened by **strain hardening**.

common brick *See under* **brick**.

common wall A wall that separates adjacent dwelling units within an apartment building, town house, and other buildings; also called a *party wall*.

composite wall A multiple-**wythe** wall in which at least one of the wythes is dissimilar to the other wythe or wythes with respect to type or grade of **masonry unit** or **mortar**.

compounding The thorough mixing of a polymer or polymers with other ingredients such as fillers, plasticizers, catalysts, pigments, dyes, or curing agents.

compression molding Forming plastic in a mold by applying pressure and usually heat.

compressive strength A material's ability to resist compressive forces.

concrete A composite material made of **portland cement**, water, and **aggregates**, and sometimes **admixtures**.

concrete block *See* **concrete masonry unit**.

concrete brick *See under* **brick masonry unit**.

concrete masonry unit (CMU) A masonry unit having portland cement as its primary cementitious material. A *decorative CMU* is one of various types available of concrete masonry units with beveled face shell recesses or other articulation or texture. A *faced CMU* is one that has a special ceramic, glazed, plastic, polished, or ground face. A *slump block* is a CMU produced so that it will slump or sag before it hardens; for use in masonry wall construction. A *split-face block* is a solid or hollow CMU that is machine fractured (split) lengthwise after hardening to produce a rough, varied surface texture.

condensation The change of water from its gaseous form (water vapor) to liquid water; the liquid water so collected.

condensation polymerization A chemical reaction in which the molecules of two substances combine, giving off water or some other simple substance (*see also* **polymerization**).

consistency The relative ability of freshly mixed **concrete** to flow as measured by a **slump test**.

construction The method by which a **carpet** is made (**loom** or machine) and other identifying characteristics, including the number of **pile** rows per inch, **pitch**, wire height, number of shots, yarn count and plies, total pile yarn weight, and **pile yarn density**.

construction joint A joint placed in **concrete** to permit practical placement of the work in section with predefined boundaries.

contact adhesive *See* **adhesive, drywall**.

continuous casting A process in which molten metal is used directly to produce semifinished products, such as slabs or billets, bypassing ingot teaming, stripping, soaking, and rolling.

control joint A joint placed in **concrete** to form a plane of weakness to prevent random cracks from developing because of drying shrinkage and temperature changes.

control joint A prefabricated metal accessory intended to relieve shrinkage, temperature, or structural stresses in **plaster**, thus minimizing cracking.

coping Masonry units forming a finished cap on top of an exposed pier, wall, pilaster, chimney, etc., to protect the masonry below from penetration of water from above.

copolymer A polymer formed by the combination of two or more different monomers (*see also* **polymers**).

copper staining A stain usually caused by the corrosive products of copper screens, gutters, or downspouts washing down on a finished surface. Can be prevented by painting the copper or applying a **transparent finish** to it.

corbel A shelf or ledge formed by projecting successive courses of masonry out from the face of the wall.

core The innermost portion of plywood, consisting of **hardwood** or **softwood** sawed lumber, **veneer**, or composition board.

corner cracking Cracks occurring in the apex of inside corners of gypsum board surfaces, such as between adjacent walls or at walls and ceilings.

corner floating A gypsum board installation procedure that eliminates some mechanical fasteners at interior corners and permits sufficient movement of boards to eliminate **corner cracking**.

correction lines East-west reference lines used in the **rectangular survey system**, located at 24-mile intervals to the north and south of a base line.

corrosion Physical deterioration, decomposition, or loss of cross section of a metal as a result of direct chemical attack, galvanic action, or weathering. *Galvanic action* is corrosion produced by electrolytic action between two dissimilar metals in the presence of an **electrolyte**. *Direct chemical attack* is corrosion caused by a chemical dissolving the metal. *Weathering* is galvanic or chemical corrosion produced by atmospheric conditions.

coulomb A unit of electrical charge equal to the number of electrons conducted past a point in 1 sec.

count A number identifying yarn size or weight per unit of length (or length per unit of weight) depending on the carpet spinning system used (such as denier, woolen, worsted, cotton, or jute system).

course (1) Compass direction from one reference point to the next for each leg of a **metes and bounds** survey. An **angular course** is a compass direction in **degrees**, **minutes**, and **seconds**, stated as a deviation eastward or westward from due north or south; used in metes and bounds descriptions and **surveys**. (2) One of the continuous horizontal layers of **masonry units**, bonded with **mortar**. One course is equal to the thickness of the masonry unit plus the thickness of one **mortar joint**.

coverage Area over which a given amount of **coating**, **paint**, or **transparent finish** will spread and hide the previous surface; usually expressed in square feet per gallon (sq. ft./gal.).

covering power *See* **hiding power**.

cracking Type of **coating**, **paint**, or **transparent finish** failure characterized by breaks in irregular lines wide enough to expose the underlying surface.

crazing Numerous hairline cracks in the surface of newly hardened **concrete**.

creep *Same as* **cold flow**.

crimping Method of texturing staple and continuous filament yarn to produce an irregular alignment of fibers in carpet and increase bulk and covering power; also facilitates interlocking of fibers, which is necessary for spinning staple fibers into yarn.

crook *See* **warp**.

crossband A layer of **veneer** in a **plywood** panel whose grain direction is at right angles to that of the face plies.

cross furring Furring members installed perpendicular to framing members.

cross-grained wood *See under* **grain**.

crosslinking A chemical reaction in which adjacent polymer molecules unite to form a strong, three-dimensional network; usually occurs during the curing of thermosetting plastics.

cross slope The slope of a pedestrian way that runs perpendicular to the direction of travel. *See also* **running slope**.

crown, joint The maximum height to which joint compound is applied over a gypsum board joint.

cryolite sodium Aluminum fluoride used with **alumina** in the final electrolytic reduction of aluminum. Found naturally in Greenland, generally produced synthetically from alum, soda, and hydrofluoric acid.

cubing The assembling of *CMUs* into cubes after curing for storage and delivery. A cube normally contains 6 layers of 15 to 18 blocks ($8 \times 8 \times 16$ in.) or an equivalent volume of other size units.

cue A device that alerts a user to an upcoming condition; includes audible, visual, and textural signals. An *audible cue* is a sound or a verbal alert. A *tactile cue* is one that can be detected by touch. A *visual cue* is one that can be seen. (*See also* **detectable warning**.)

cup *See* **warp**.

cure To change the properties of a polymeric system into a final, more stable condition by the use of heat, radiation, or reaction with chemical additives. Sometimes referred to as *set*.

curb ramp A short ramp cutting through a curb or built into it.

curing (1) The hardening of a **concrete masonry unit (CMU)**. *Atmospheric pressure steam curing* is a method of curing CMUs using steam at atmospheric pressure of temperatures of 120° to 180°F. Also called low-pressure steam curing. *High-pressure steam curing* is a method of curing CMUs using saturated steam (365°F) under pressure, usually 125 to 150 pounds per square inch (psi). Also referred to as *autoclave curing*. *Moist curing* is a method of curing CMUs using moisture at atmospheric pressure and temperature of approximately 70°F. (2) The process of keeping **concrete** moist for an extended period after placement to ensure proper **hydration** and subsequent strength and quality. (3) Final conversion or drying of a **coating**, **painting**, or **transparent finish** material.

curtain wall An exterior non-load-bearing wall. Such walls may be anchored to columns, spandrel beams, structural walls, or floors. *See also* **panel wall**.

cushioning Soft, resilient layer provided under carpet to increase underfoot comfort, to absorb pile-crushing forces, and to reduce impact sound transmission. Also called *underlay* or *lining. See also* **padding**.

cut and loop *See* **multilevel loop**.

cut pile A carpet face construction in which the pile is cut level so that it stands erect in a low, dense, plush, even surface.

cutting in Painting of an edge such as at wall color at the ceiling line or at the edge of woodwork.

cylinder test A laboratory test for compressive stress of a field sample of **concrete** (6 in. in diameter by 12 in. in length).

daily degree day The numerical difference between 65°F and the average of all recorded temperatures on a given day that are lower than 65°F.

damp course A course or layer of impervious material that prevents capillary entrance of moisture from the ground or a lower course. Often called **damp check**.

darby A tool used to level freshly placed **concrete**.

darbying Smoothing the surface of freshly placed **concrete** with a **darby** to level any raised spots and fill depressions.

dead load The weight of all permanent and stationary construction or equipment included in a building. See also **live load**.

decay The decomposition of wood substances by certain fungi.

decibel (dB) A logarithmic unit expressing the ratio between a sound being measured and a reference point.

decorative CMU See under **concrete masonry unit**.

degradation A permanent change in the physical or chemical properties of a plastic evidenced by impairment of these properties.

degree A unit of **angular measure** equal to the angle contained within two radii of a circle that describe an arc equal to 1/360 of the circumference of the circle; also used to define an arc equal to 1/360 of the circumference of a circle.

degree days heating (DDH) The sum of the **daily degree days** when the temperature dropped below 65°F.

delustered nylon Nylon on which the normally high sheen has been reduced by surface treatment.

denier System of yarn count used for synthetic carpet fibers: number of grams per 9000 meters of yarn length; one denier equals 4,464,528 yards per pound (yd./lb.) or 279,033 yards per ounce (yd./oz.).

density See **pile yarn density**.

detectable warning A standarized surface texture applied to or built into walking surfaces or other elements to warn visually impaired people of hazards in the path of travel. See also **cue**.

deterioration See **degradation**.

dew point That temperature above freezing at which air becomes saturated and **condensation** occurs. See also **frost point**.

dimension lumber See under **lumber**.

dimensional stability The ability of a material to retain its dimensions in service.

direct chemical attack Corrosion caused by a chemical dissolving a metal.

disability A limitation or loss of use of a physical, mental, or sensory body part or function.

discontinuous construction A construction method used to separate a continuous path through which sound may be transmitted. Examples include the use of staggered studs, double walls, or the resilient mounting of surfaces.

dispersion The distribution of a finely divided solid in a liquid or a solid.

disposal field See **absorption field**.

distribution line Open joint or perforated pipe intended to permit soil absorption of effluent.

divider strips All-metal or plastic-top-metal strips provided in a **terrazzo finish** to control cracking resulting from drying shrinkage, temperature variations, and minor structural movements; also used for decorative purposes and as a convenience in placing a terrazzo topping.

drain A pipe that carries waste water or waterborne wastes in a building **drainage system**.

drainage system That piping that conveys sewage, rainwater, or other liquid wastes up to a point of disposal, such as the mains of a public sewer system or a private septic system. See also **vent system** and **drain, waste, and vent (DWV) system**.

drain, waste and vent (DWV) system The collection of pipes that facilitates the removal of liquid and solid wastes, and dissipates sewer gases.

drawing The process of pulling material through a die to reduce the size, change the cross section or shape, or harden the material.

dressed lumber **Lumber** that has been surfaced with a planing machine.

dressed size See **size**.

drier An ingredient included to speed the drying of **coatings**, **paints**, and **transparent finishes**.

drip A projection shaped to cause water to flow away from a lower surface, thus preventing it from running down the face of that surface.

drying The various stages of **curing** in a **coating**, **paint**, or **transparent finish** film. *Dust-free* is the stage of drying when particles of dust that settle on the surface do not stick to it. *Tack-free* is the stage of drying when the surface no longer feels sticky when lightly touched. *Dry-enough-to-handle* is the stage of drying when the film has hardened sufficiently so the object or surface may be used without marring it. *Dry-enough-to-recoat* is that stage of drying when the next coat can be applied. *Dry-enough-to-sand* is the stage of drying when the film can be sanded without the sandpaper sticking or clogging.

drying oil A **coating**, **paint**, or **transparent finish vehicle** ingredient, such as linseed oil, which, when exposed to the air in a thin layer, oxidizes and hardens to a relatively tough elastic film.

dry lumber Under Product Standard PS 20-70, **lumber** with a moisture content of 19% or less.

dry rot A term that is loosely applied to many types of decay, which in its advanced state permits wood to be easily crushed to a dry powder. This term is actually a misnomer since all fungi require moisture.

drywall nail See **nail, drywall**.

drywall screw See **screw, drywall**.

dry well A covered and lined underground pit, similar to a seepage pit, but intended to receive water free of organic matter, such as from roof drains, floor drains, or laundry tubs. Used as an auxiliary to a septic system to avoid overloading the septic tank absorption system. Also called *leaching well* or *leaching pit*.

ductile Capable of being drawn out or hammered, able to undergo cold **plastic deformation** without breaking.

durability The ability of a **coating**, **paint**, or **transparent finish** to retain its desirable properties for a long time under expected service conditions.

dusting The appearance of a powdery material at the surface of a hardened **concrete**.

dye or **dyestuff** Colored material used to change the color of a **coating** or **paint** with little or no hiding of the underlying surface.

dyne A unit of force, which when acting on a mass of 1 gram accelerates it 1 centimeter per second per second (1 cm/sec²).

echo A reflected sound loud enough and received late enough to be heard as distinct from the source.

economy brick See under **brick masonry unit**.

edge-grained wood See under **grain**.

edges, gypsum board See under **gypsum board**.

edging The finishing operation of rounding off the edge of a **concrete** slab to prevent chipping or damage.

efflorescence Deposit of soluble salts, usually white in color, appearing on the exposed surface of **masonry**, **concrete**, or **plaster**.

effluent Partially treated liquid sewage flowing from any part of the disposal system, septic tank, or absorption system.

eggshell finish Surface sheen midway between flat and semigloss.

elastic deformation Disfigurement caused by a **stress** small enough so that when the stress is removed, the material returns to its original shape. See also **plastic deformation**.

elastic limit The amount of **stress** that, if exceeded, will cause a given material to deform or set permanently.

elasticity The ability to recover original size and shape after deformation.

elastomer (1) A material that at room temperature can be stretched repeatedly to at least twice its original length and that, upon release of the stress, will return instantly and with force to its approximate original length. (2) A rubberlike substance.

electric induction furnace *See under* **furnace, steel**.

electrogalvanizing **Electroplating** with zinc to provide greater corrosion resistance.

electroplating A process that employs an electric current to coat a base metal (cathode) with another metal (anode) in an electrolytic solution.

electrolysis Also called *galvanic corrosion*. An electrochemical decomposition that results when dissimilar metals are each contacted by the same electrolyte, such as water. The process is similar to that which takes place in an automobile battery. One metal acts as a cathode, the other as an anode. When the electrolyte causes an electrical current to flow from one metal to the other, the anodic metal dissolves and hydrogen ions accumulate on the cathodic metal. Electrolysis can also take place in a single metal when one portion of it is cathodic and another portion is anodic, if an electrolyte makes a bridge between the two portions.

electrolyte A nonmetallic substance in which electricity is conducted by the movement of ions.

electromotive force Something that moves or tends to move electricity.

emulsion A mixture of liquids (or a liquid and a solid) not soluble in each other, one liquid (or solid) being dispersed as minute particles in the other base liquid, with the help of an emulsifying agent.

enamel A **coating** or **paint** capable of forming a very smooth, hard film, sometimes using varnish as the **vehicle**; may be **flat**, **gloss**, or **semigloss**.

enamel holdout The property of producing a tight film that prevents the penetration of subsequent enamel coats to underlying surfaces; prevents unequal absorption and uneven gloss.

ends, gypsum board *See under* **gypsum board**.

end wall The wall along the short dimension of a room.

engineered brick *See under* **brick masonry unit**.

engineered brick masonry Masonry in which the design is based on a rational, accepted structural engineering analysis.

equilibrium moisture content The moisture content at which wood neither gains nor loses moisture when surrounded by air at a given **relative humidity** and temperature.

erosion The wearing away of a **coating, paint,** or **transparent finish** film caused by exposure.

etch A surface preparation for a **coating, paint,** or **transparent finish** by chemical means to improve adhesion.

evaporation The change of water from a liquid to a gas.

expandable plastic A plastic suitable for expansion into cellular form by thermal, chemical, or mechanical means.

expansion strips Double **divider strips** in a **terrazzo finish** separated by resilient material and provided generally for the same purpose as divider strips, but where a greater degree of structural movement is expected.

extender (1) In plastics, a low-cost material used to dilute or extend high-cost resins without appreciably lessening the properties of the original resin. (2) In **coatings, paints,** and **transparent finishes,** an inexpensive but compatible substance that can be added to a more valuable substance to increase the volume of material without substantially diminishing its desirable properties; in **coatings** and **paints,** extender pigments improve storage and application properties.

exterior wall Any outside wall of a building other than a **party wall**.

external corner A projecting angle formed by abutting walls or vertical surface and soffit (not to be confused with *exterior*, meaning exposed to the weather).

extrude To form lengths of shaped sections by forcing a plastic material through a shaped hole in a die.

extrusion (1) In aluminum, a product formed by extruding. (a) An *extrusion billet* is a solid, wrought, semifinished product intended for further extrusion into rods, bars, or shapes. (b) An *extrusion ingot* is a solid or hollow cylindrical casting used for extrusion into bars, rods, shapes, or tubes. (2) In plastic, forcing plastic material through a shaped orifice to make rod, tubing, or sheeting.

faced CMU *See* **concrete masonry unit**.

faced wall A wall in which facing and backing are of different materials and are bonded together to exert common action under load.

face shell The side wall of a hollow **masonry unit** or clay tile.

face size *See under* **size**.

facing A part of a wall that is used as a finished surface.

facing brick *See under* **brick masonry unit**.

factory-and-shop lumber *See under* **lumber**.

fading Loss of color because of exposure to light, heat, or weathering.

fastener treatment Method of concealing gypsum board fasteners by successive applications of compound until a smooth surface is achieved.

featheredging (feathering) Tapering gypsum board joint compound to a very thin edge to assure inconspicuous blending with adjacent gypsum board surfaces.

feather sanding Tapering the edge of a dried **coating, paint,** or **transparent finish** film with sand paper.

ferritic steel *See* **grain structure**.

ferrous alloys Composite metals whose chief ingredient is iron (ferrum), metallurgically combined with one or more alloying elements.

ferroalloys Iron-based alloys used in steelmaking as a source of desired alloying elements.

fiber, wood A comparatively long (1/25 in. or less to 1/3 in.), narrow, tapering unit closed at both ends.

fiber saturation point The stage in drying (or wetting) of wood at which the cell walls are saturated with water, but the cell cavities are free of water, being approximately 30% **moisture content** in most species.

field (1) In **ceramic tile**, the general area of the tile excluding trim. (2) In **gypsum board**, the surface of the board exclusive of the perimeter. (3) In **masonry walls**, the expanse of wall between openings, corners, etc., principally composed of stretchers. (4) In resilient flooring, the area of a floor within the **borders**.

figure The pattern produced in a wood surface by **annual growth rings, rays, knots,** and deviations from regular **grain**.

fill The sand, gravel, or compacted earth used to bring a subgrade up to a desired level.

filler (1) In plastics, a relatively inert material added to modify the strength, permanence, or working properties, or to lower the cost of a resin. (2) In **coatings, paint,** and **transparent finishes,** a pigmented composition for filling the pores or irregularities in a surface in preparation for finishing.

filling *See* **weft**.

film (1) In plastics, plastic sheeting having a nominal thickness not greater than 0.010 in. (2) In **coatings, paints,** and **transparent finishes,** a thin application generally not thicker than 0.010 in.

fine aggregate *See* **aggregate**.

fineness modulus A measure of the average size of an **aggregate** calculated by passing the aggregate through a series of screens of decreasing size.

finish coat The final decorative plaster layer in either **two-coat** or **three-coat work**.

finished mill products Steel shapes that can be used directly in construction. *Bars* are hot-rolled or cold-drawn round, square, hexagonal, or multifaceted long shapes, generally larger than wire in cross section. Also, hot- or cold-rolled rectangular flat shapes (flats) generally narrower than sheets and strip. *Backplate* is a cold-rolled, flat, carbon-steel product that is thinner than sheet and wider than strip, generally coated with zinc, tin, or terne metal. *Foil* is a cold-rolled flat product less than 0.005-in. thick. *Plate* is a hot-rolled flat product generally thicker than sheet and wider than strip. *Sheet* is a hot- or cold-rolled flat product generally thinner than plate and wider than strip. *Strip* is a hot- or cold-rolled flat products generally narrower than sheet and thinner than plate. *Structurals* are hot-rolled steel shapes of special design (such as H beams, I beams, channels, angles, and tees) used in construction. *Terneplate* is backplate or sheet metal that has been coated with *terne* metal. *Tinplate* is backplate that has been coated with tin. *Tubular products* are hollow products of round, oval, square, rectangular, or multifaceted cross sections. In construction, round products are generally called *pipe*; square or rectangular products with thinner wall sections are called *tube* or *tubing*. *Wire* is a cold-finished product of round, square, or multifaceted cross section, generally smaller than bars; round wire is cold drawn, 0.005 in. to less than 1 in. in diameter; flat wire is cold rolled, generally narrower than bar.

fire brick *See under* **brick masonry unit**.

fire division wall A wall that subdivides a building to resist the spread of fire. It is not necessarily continuous through all stories to and above the roof. *See also* **fire wall**.

fireproofing A material or combination of materials built to protect structural members to increase their fire resistance.

fire-resistive material *See* **noncombustible material**.

fire-retardant coating or **paint** A **coating** or **paint** that will significantly reduce the rate of flame spread; resist ignition at high temperatures; and insulate the underlying material to prolong the time required for the material to reach ignition, melting, or structural weakening temperature.

fire wall A wall that subdivides a building to resist the spread of fire and that extends continuously from the foundation through the roof. *See also* **fire division wall**.

fixture-unit A mathematical factor used by engineers to estimate the probable demand on a **drainage** or **water supply system** (volume, duration of flow, and intervals between operations) by various **plumbing fixtures**.

flaking Form of **coating**, **paint**, or **transparent finish** failure characterized by the detachment of small pieces of the film from the surface or previous coat; usually preceded by cracking or blistering.

flanking path A wall or floor and ceiling assembly that permits sound to be transmitted along its surface. Also, an opening that permits the direct transmission of sound through the air.

flashing (1) In masonry construction, a thin, relatively impervious sheet material, placed in mortar joints and across airspaces in masonry walls to collect water that may penetrate the wall and to direct it to the exterior. (2) In masonry manufacture, the step during the burning process of clay **masonry units** that produces varying shades and colors in the units. (3) In resilient flooring, the bending up of resilient sheet material against a wall or a projection, either temporarily for the purpose of fitting, or permanently to form a one-piece resilient base.

flash point Temperature at which a **coating**, **paint**, **transparent finish**, or **solvent** will ignite; the lower the flash point, the greater the hazard.

flash set The undesirable rapid setting of **cement** in **concrete** or **mortar**.

flat Dull, nonreflective; opposite of gloss.

flat applicator Rectangular-shaped flat pad with an attached handle that is used to paint shingles, shakes, and other special surfaces.

flat finish Having no gloss or luster.

flat-grained wood *See under* **grain**.

flats Term applied to flat **coatings** and **paints**.

flatting agent Ingredient added to **coatings** and **paints** to reduce the gloss of the dried film.

flitch (1) A portion of a log sawed on two or more sides and intended for manufacture into lumber

or sliced or sawed **veneer**. (2) A complete bundle of veneers laid together in sequence as they were sliced or sawed.

floating (1) A **concrete** slab finishing operation that embeds aggregate, removes slight imperfections, humps, and voids to produce a level surface, and consolidates mortar at the surface. (2) A condition in which a layer of water occurs between a mortar bed and a masonry unit, usually the result of **bleeding**, causing the unit and the mortar to fail to bond with each other. In this condition, the unit is said to float. (3) Separation of pigment colors on the surface of applied paint.

flocked carpet Single-level velvety pile carpet composed of short fibers embedded on an adhesive-coated backing.

floor brick *See under* **brick masonry unit**.

flow The ability of a **coating**, **paint**, or **transparent finish** to level out and spread into a smooth film; materials that have good flow usually level out uniformly and exhibit few brush or roller marks.

fluffing *See* **shedding**.

flux A mineral that, because of its affinity to the impurities in iron ores, is used in ironworking and steelmaking to separate impurities in the form of molten slag. *Basic flux* is a mineral, such as limestone or dolomite, used in basic furnaces to make basic (low phosphorus) steel. *Neutral flux* is a mineral (fluorspar) used to make slag more fluid.

foamed plastic *Same as* **cellular plastic**.

foil *See under* **finished mill products**.

forging The working (shaping) of metal parts by forcing between shaped dies. *Press forging* is shaping by applying pressure in a press. *Hammer forging* is shaping by application of repeated blows, as in a forging hammer. *See also* **hot working**.

form A temporary structure erected to contain **concrete** during **placing** and initial hardening.

footing The base of a foundation or column or wall used to distribute the load over the subgrade.

foundation wall A load-bearing wall below the floor nearest to exterior grade serving as a support for a wall, pier, column, floor, or other structural part of a building.

fractional section *See under* **section**.

frames Racks at the back of a **Jacquard** loom, each holding a different color of pile yarn. In **Wilton carpets**, two to six frames may be used and the number is a measure of quality as well as an indication of the number of colors in the pattern, unless some of the yarns are buried in the **backing**.

freezing cycle day A day in which the air temperature passes either above or below 32°F. The average number of freezing cycle days in a year equals the difference between the mean number of days that the minimum temperature was 32°F or below and when the maximum temperature was 32°F or below.

frequency The number of complete cycles of a vibration occurring in each second, measured in cycles per second (cps) and expressed in hertz (Hz).

frieze carpet *See under* **pile**.

frost point That temperature below freezing at which **condensation** occurs. *See also* **dew point**.

fungicide Agent that helps prevent mold or mildew growth on paint.

furnace, blast Tall, cylindrical masonry structure lined with refractory materials, used to smelt iron ores in combination with fluxes, coke, and air into **pig iron**.

furnace, steel Masonry or steel structure lined with refractory materials, used to melt **pig iron**, **scrap metal**, and sometimes agglomerated ores, ferroalloys, and fluxes into steel. A *basic oxygen furnace* is a suspended, tilting vessel that uses high-purity oxygen to oxidize impurities in hot pig iron and other iron-bearing materials to produce low phosphorus (basic) steel. An *electric-arc suspended furnace* is a kettle that melts scrap metal, ore, and sometimes ferroalloys with the heat of an electric arc to produce steels of controlled chemical composition. An *electric induction furnace* is a steel-encased, insulated magnesia pot in which metal, scrap, and ferroalloys are melted with the heat of an electric current induced by windings of electric tubing; used chiefly to produce small quantities of high-grade steels such as alloy, stainless, and heat-resisting steels. An *open-hearth furnace* is a masonry structure with a hearth exposed to the sweep of flames in which hot pig iron, scrap metal, and fluxes are melted and oxidized by a mixture of fuel and air, to produce basic or acid steel.

furring A method of finishing the interior face of a concrete or masonry wall to provide space for insulation, to prevent moisture penetration, or to provide a level, plumb, and straight surface for finishing. Furring consists of metal channels or studs or of wood strips or studs.

fusion-bonded carpet Carpet made by fusing carpet yarn and a backing, then cutting the substrates in two, making two pieces of carpet.

fuzzing Temporary condition on new **carpet** consisting of an irregular fuzzing appearance caused by slack yarn twist, fibers snagging, or breaking of yarn. Can be remedied by spot **shearing**.

galvanic corrosion *See* **electrolysis**.

galvanizing Zinc coating by electroplating or hot dipping, which produces a characteristic bright spangled finish and protects the base metal from atmospheric **corrosion**.

gang grooved Plywood panels produced by passing them under a machine with grooving knives set at certain intervals.

gauge The distance between **tufts** across the width of **knitted** and **tufted carpets**, expressed in fractions of an inch.

glaze Used to describe several types of finishing materials. *Glazing putty* is a compound of creamy consistency applied to fill surface imperfections. A *glazing stain* is very thin, semitransparent, and usually pigmented with a Vandyke brown or burnt sienna applied over a previously stained, filled, or painted surface to soften or blend the original color without obscuring it. A *glaze coat* is a clear finish applied over previously coated surfaces to create a gloss finish.

glazing compound Doughlike material, consisting of **vehicle** and **pigment**, that retains its plasticity over a wide range of temperatures and for an extended period of time. Either elastomeric or acrylic joint sealants are generally used today where glazing compounds are called for. *See also* **sealant, joint**.

gloss Shiny, reflective surface quality; term sometimes used broadly to include **coatings**, **paints**, and **transparent finishes** with these surface properties.

glueline The line of glue visible on the edge of a **plywood** panel. Also applies to the layer of glue itself.

grade The designation of the quality of wood, steel, and other materials and products made from them, such as **plywood**.

graded aggregate An **aggregate** containing uniformly graduated particle size from the finest fine aggregate size to the maximum size of coarse aggregate.

graded sand A sand containing uniformly graduated particle sizes from very fine up to 1/4 in.

grain The direction, size, arrangement, appearance, or quality of the **fibers** in wood. *Cross-grained wood* is sawed with the fibers not parallel with the longitudinal axis of the piece. This grain may be either diagonal or in a spiral pattern or a combination of both. *Diagonal-grained wood* is sawn at an angle with the **bark** of the tree such that the **annual growth rings** are at an angle with the axis of the piece. *Edge-grained wood* is sawn parallel with the **pith** of a log and nearly at right angles to the **annual growth rings**, making an angle 45 deg. to 90 deg. with the wide surface of the piece. Also called *quarter-sawn* wood and *vertical-grained* wood. *Flat-grained wood* is sawn parallel with the **pith** of the log and nearly tangent to the **annual growth rings**, making an angle less than 45 deg. with the surface of the piece. Also called *plain sawn* wood. *Open-grained wood* is a common designation for wood with large pores, such as oak, ash, chestnut, and walnut.

graining Simulating the grain of wood by means of specially prepared colors or stains and the use of graining tools or special brushing techniques.

grain raising Swelling and standing up of wood grain caused by absorbed water or solvents.

grain structure The microscopic, internal crystalline structure (size and distribution of particles) of a metal that affects its properties, known as austenitic, ferritic, and martensitic. *Austenitic steels* are tough, strong, and nonmagnetic steels. Austenitic stainless steels have a chromium content up to 25%, nickel up to 22%, and can be hardened by cold working.

Ferritic steels are soft, ductile, and strongly magnetic steels. Ferritic stainless steels usually have a chromium content between 12% and 27% and are not hardenable by heat treatment. *Martensitic steels* can be made very hard and tough by heat treatment and rapid cooling. Martensitic stainless steels have a chromium content between 4% and 12%.

great circle A line described on a sphere by a plane bisecting the sphere into equal parts. The *equator* is a great circle, as are pairs of opposing **meridians**.

green concrete Freshly placed **concrete**.

green lumber Under Product Standard PS 20-70, lumber with a moisture content of more than 19%. Unseasoned lumber that has not been exposed to air or kiln drying.

Greenwich meridian *See under* **meridian**.

grin Condition where the **backing** shows through sparsely spaced pile **tufts**. **Carpets** may be grinned (bent back) deliberately to reveal a carpet's **construction**.

gross cross-sectional area In masonry, the total area of a section perpendicular to the direction of the load, including areas within cores, cellular spaces, and other openings in the material.

ground coat Base coat in an antiquing system; applied before graining colors, glazing, or other finish coat.

grounds Nailing strips placed in concrete and masonry walls as a means of attaching trim, furring, cabinetry, or equipment.

grout, masonry (1) Mortar of a consistency that will flow or pour easily without segregation of the ingredients. (2) A liquid mixture of **cement**, water, and sand of pouring consistency.

grout, tile A formulation used to fill the joints between tiles; may be cementitious, resinous, or a combination of both.

guide meridian *See under* **meridian**.

gusset plate Wood or metal plate used as a means of joining coplanar structural members in trusses. Gusset plates lap the butt joints between members.

gypsum board A panel consisting of a noncombustible core of calcined gypsum surfaced on both sides with a covering material specifically designed for various uses with respect to performance, application, location, and appearance. *Backing boards* are gypsum boards that serve as a base to which gypsum wallboards or tiles are applied. *Edges* are gypsum board extremities that are paperbound and run the long dimension of the board as manufactured. *Ends* are gypsum board extremities that are mill- or job-cut exposing the gypsum cored, and run the short dimension of the board as manufactured. *Lath* is a class of gypsum board used as a base for gypsum plaster. *Sheathing* is a class of gypsum board used as a base for exterior finishes. *Wallboard* is a class of gypsum board used primarily as an interior finished surface.

hardboard A dense panelboard manufactured of wood fibers with the natural lignin in the

wood reactivated to serve as a binder for the wood fibers.

hard-burned Clay products that have been fired at high temperatures. They have relatively low absorptions and high compressive strengths.

hardness Cohesion of particles on the surface of a **coating**, **paint**, or **transparent finish** as determined by ability to resist scratching or indentation.

hardwood The botanical group of trees that is broad leaved and deciduous. The term does not refer to the actual hardness of the wood.

header A masonry unit that overlaps two or more adjacent **wythes** of masonry to provide structural bond. Also called *bonder*.

header course A continuous bonding **course** of header **brick**.

head joint The vertical **mortar joint** between ends of **masonry units**.

heartwood That wood that extends from the **pith** to the **sapwood**, the cells of which no longer participate in the growth process of the tree. Heartwood may be impregnated with gums, resins, and other materials that usually make it darker and more decay resistant than sapwood.

heat capacity The amount of heat required to raise the temperature of 1 cubic foot of a material 1°F.

heat-set nylon Nylon fiber that has been heat treated to retain a desired shape.

heat sink A material or system that collects and stores heat.

heat-treatable alloys Aluminum alloys capable of gaining strength by being heat treated. The alloying elements show increasing solid solubility in aluminum with increasing temperature resulting in pronounced strengthening.

heat treatment (1) Of aluminum: *Annealing* is a heat treatment process in which an *alloy* is heated to a temperature of between 600° and 800°F and then slowly cooled to relieve internal stresses and return the alloy to its softest and most **ductile** condition. *See* **solution heat treatment**. (2) Of steel: Controlled heating and cooling of steels in the solid state for the purpose of obtaining certain desirable mechanical or physical properties. *Annealing* is heating metal to high temperatures (1350 to 1600°F for steel), followed by controlled cooling, to make the metal softer or change its ductility and toughness. *Quenching* is rapid cooling by immersion in oil, water, or other cooling medium to increase hardness. *Tempering* is reheating (in steel to less than 1350°F) after hardening (as by quenching) and slow cooling, to restore ductility.

hertz (Hz) The unit of frequency of a periodic process equal to 1 cycle per second.

hiding power Ability of a **coating** or **paint** to obscure the surface to which it is applied, generally expressed as the number of square feet that can be covered by 1 gallon of material, or number of gallons required to cover 1000 square feet. Also called *covering power*.

high-density plywood *See under* **plywood**.

high-early-strength cement Cement used to produce a **concrete** that develops strength more rapidly than normal.

high-low loop *See* **multilevel loop**.

high polymer A polymer of high molecular weight (*see also* **polymer**).

high pressure laminate A laminate molded and cured at pressures not lower than 1000 pounds per square inch (psi) and usually between 1200 and 2000 psi (*see also* **laminate**).

high-pressure steam curing *See under* **curing**.

hollow brick *See under* **brick masonry unit**.

hollow masonry unit *See under* **masonry unit**.

hollow wall A wall built of **masonry units** arranged to provide an airspace within the wall between the facing and backing **wythes**.

horizontal application Applying gypsum board with the edges perpendicular to supporting members such as studs, joists, channels, or furring strips. *See also* **vertical application**.

hot-dip process Coating a metal with another metal by immersing it in a bath of the molten coating metal. The coating metals most commonly used on steel are zinc, terne metal, and aluminum.

hot rolling *See* **hot working; rolling**.

hot working Forming metal at elevated temperatures at which metals can be more easily worked. (1) For aluminum alloys, (a) hot working *temperatures* are usually in the 300° to 400°F range. (b) *Hot rolling* is the shaping of aluminum plate by rolling of heated slabs of metal. Hot rolling is usually used for work down to about 0.125-in. thick. (2) For steel, shaping hot plastic metal by hot rolling, extruding, or forging. (a) Hot working temperatures are above 1500 °F. Hot working is generally accompanied by increases in strength, hardness, and toughness. (b) *Hot forming* is the forming of a hot plastic metal into desired shapes, with little change in the mechanical properties of the metal. (c) *Extruding* is shaping lengths of hot metal by forcing it through a die of the desired profile. (d) *Forging* is shaping hot metal between dies with compression force or impact. (e) *Hot rolling* is gradual shaping by squeezing hot metal between rolls.

house sewer *See* **building sewer**.

hydration The chemical reaction of water and **cement** that produces a hardened **concrete**.

Hz *See* **hertz**.

impact insulation class (IIC) A single-number rating developed by the Federal Housing Administration (FHA) to estimate the impact **sound isolation** performance of floor and ceiling assemblies.

impact noise rating (INR) A single-number rating used to compare and evaluate the ability of floor and ceiling assemblies to isolate impact sound transmission. (*See also* **sound isolation**).

impact sound pressure level (ISPL) The sound level in decibels (dB) measured in the receiving room resulting from the transmission of sound produced by a standard tapping machine through an adjacent floor and ceiling assembly.

impervious soil A tight cohesive soil such as clay that does not allow the ready passage of water.

incident sound A sound striking a surface, as contrasted with a sound reflected from a surface. The angle of incidence equals the angle of reflection.

individual sewage disposal system A combination of a **sewage** treatment plant (package plant or **septic tank**) and a method of **effluent** disposal (soil absorption system or stream discharge) serving a single dwelling.

inert Having inactive chemical properties.

ingot (1) A mass of aluminum cast into a convenient shape for storage or transportation to be later remelted for casting or finishing by rolling, forging, or other process. (2) A cast pig iron or steel shape made by pouring hot metal into a mold.

inhibitor A substance that prevents or retards a chemical reaction; inhibitors are often added to plastic resins to prolong their storage life.

injection molding Forming plastic by fusing it in a chamber with heat and pressure, and then forcing part of the mass into a cooler chamber, where it solidifies.

insolation (incident solar radiation) The solar radiation that strikes a surface.

intercoat adhesion Adhesion between two coats of paint.

internal corner An enclosed angle of less than 180 deg. formed by abutting walls or the juncture of walls and the ceiling (not to be confused with *interior*, meaning protected from the weather).

isolation *See* **sound isolation**.

isolation joint A joint placed to separate a **concrete** slab into individual panels or from adjacent surfaces.

isolation membrane In a **terrazzo finish**, a membrane such as asphalt-saturated roofing felt, building paper, or polyethylene film installed between the subfloor and the terrazzo underbed to prevent a bond and permit the independent movement of each. In tile floors, *see* **cleavage (isolation) membrane**.

jacquard Mechanism for a **Wilton loom** that uses punched cards to produce the desired color design.

Jaspe carpet *See under* **pile**.

joint *See* **mortar joint**.

joint beading *See* **ridging**.

joint compound A material used for finishing joints in **gypsum board**.

joint crown *See* **crown, joint**.

joint reinforcement Steel wire, bar, or fabricated reinforcement that is placed in horizontal **mortar joints**.

joint tape Paper or paper-faced cotton tape used over joints between wallboard to conceal the joint and provide a smooth surface for painting.

joint treatment Method of reinforcing and concealing **gypsum board** joints with tape and successive layers of **joint compound**.

joist One of a series of parallel beams used to support floor, ceiling, and roof loads. Joists are supported in turn by bigger beams, girders, or bearing walls.

joists and planks *See under* **lumber**.

jumbo brick *See under* **brick mortar unit**.

jute Strong, durable yarn spun from fibers of the jute plant, native to India and the Far East; used in the **backings** of many **carpets**.

kiln drying *See under* **seasoning**.

knitted carpet Carpet made on a knitting machine by looping together backing, stitching, and pile yarns with three sets of needles, as in hand knitting.

knot That portion of a branch or limb that has been surrounded by subsequent wood growth.

kraftcord Tightly twisted yarn made from wood pulp fiber, used as an alternative for cotton or jute in carpet backing.

lacquer Coating that dries quickly by evaporation of its volatile **solvent**, and forms a **film** from its nonvolatile constituent, usually nitrocellulose; may be pigmented or clear.

laitance A soft, weak layer of mortar appearing on a horizontal surface of **concrete** as a result of **segregation** or **bleeding**.

laminate A product made by bonding together two or more layers of materials.

laminated wood A piece of wood built up of laminations that have been joined either with glue or mechanical fastenings or both.

laminating adhesive *See* **adhesive, drywall**.

langely The measure of **isolation**; equal to 1 calorie per square centimeter (1 cal./cm²) or 3.69 British thermal units per square foot (3.69 Btus/sq. ft.).

lap To lay or place one coat so its edge extends over and covers the edge of a previous coat.

lateral A branch of an **absorption field**, consisting of either (1) the length of distribution line between overflow pipes, or (2) the length of distribution line between the tee or cross fitting and the farthest point in a closed loop field.

lateral support of walls Means whereby walls are braced either horizontally by **columns**, **pilasters**, or cross walls, or vertically by floor and roof construction.

lath, gypsum *See under* **gypsum board**.

latex A water suspension of fine particles of rubber or rubberlike plastics.

latex paint A paint containing **latex** and thinned with water.

latitude The position of a point on the earth's surface north or south of the equator, stated as an angular measure (**degrees, minutes, and seconds**) of the **meridian** arc contained between that point and the equator.

leaching pit *See* **dry well**.

leaching well *See* **dry well**.

lead The section of a masonry wall built up and racked back on successive courses. A line is attached to leads as a guide for building a wall between them.

legal description A written, legally recorded identification of the location and boundaries of a parcel of land. A legal description may be based on a **metes and bounds** survey, the **rectangular survey system**, or it may make reference to a recorded **plat** of survey.

leveling Ability of a **film** to flow out free from ripples, pockmarks, and brush marks after application.

level loop pile A woven or tufted carpet style in which the tufts are in loop form and of the same height.

level tip shear Random patterned cut and loop pile **carpet**.

lifting Softening and penetration of a previous film by solvents in the paint being applied over it, resulting in raising and wrinkling the paint.

light-fastness Ability to resist fading.

light framing lumber *See under* **lumber**.

lignin The second most abundant constituent of wood, making up from 12% to 28% in most species. It encrusts the cell walls and cements the cells together.

linear polymer A polymer that is arranged in a long, continuous chain with a minimum of side chains or branches.

lining felt An mineral fiber or asphalt felt specifically manufactured for use under resilient flooring.

lintel A structural member to carry the load over an opening in a wall.

lip When related to barrier free design, an abrupt vertical change in level.

Lite A layer of glass.

live load The total of all moving and variable loads that may be placed on or in a building. *See also* **dead load**.

longitude The position of a point on the earth's surface east or west of the **Greenwich meridian**, stated as an **angular measure (degrees, minutes,** and **seconds)** of the arc on the equator contained between a **meridian** passing through that point and the Greenwich meridian.

long oil Varnish or paint vehicle with more than 55% of the resin consisting of oil or fatty acid.

loom Machine on which **carpet** is woven, as distinguished from other machines on which carpets may be tufted, flocked, or punched.

loomed carpet *See under* **woven carpet**.

loop pile A woven or tufted carpet style in which the pile surface consists of uncut loops; also called *round wire*.

loop vent A branch vent that serves two or more traps and extends from in front of the last fixture connection of a horizontal branch to the **stack vent**. *See also* **circuit vent**.

loudness A subjective response to sound, indicating the magnitude of the hearing sensation, dependent on the listener's aural response.

lumber The wood product of a saw and planing mill not further manufactured than by sawing, resawing, and passing lengthwise through a standard planing machine, crosscut to length,

and worked. *Beams and stringers* are large pieces of lumber measuring 5 in. or more in thickness and 8 in. or more in width. These are graded with respect to their strength in bending when loaded on the narrow face. *Boards* are lumber up to 2 in. thick and 2 in. or more in width. *Dimension lumber* is lumber from 2 to 4 in. thick and 2 in. or more in width. *Factory and shop lumber* is intended to be cut again for use in further manufacture. *Joists and planks* are pieces of lumber 2 to 4 in. in thickness and 4 in. or more in width. These are graded with respect to strength in bending when loaded either on the narrow face as a joist, or on the wide face as a plank. *Light framing* is lumber 2 to 4 in. in thickness and 2 to 4 in. in width. *Matched lumber* is lumber that has been edge dressed and shaped to make a close tongue-and-groove joint at the edges or ends when laid edge to edge or end to end. *Patterned lumber* is lumber that is shaped to a pattern or to a molded form in addition to being dressed, matched, or shiplapped, or any combination of these. *Posts and timbers* are lumber that is approximately square, 5 in. and thicker and having a width not more than 2 in. greater than its thickness. *Stress grade lumber* is the same as *structural lumber*. *Structural lumber* is lumber that has been machine rated or visually graded into assigned working stresses. Also called *stress grade lumber*. Includes most yard lumber grades, except boards. *Yard lumber* is intended for general building purposes; includes boards, dimension lumber, and timbers.

machinability The ability to be milled, sawed, tapped, drilled, and reamed without excessive tool wear, with ease of chip metal removal and surface finishing.

main The principal artery of a **drainage** or **water supply system** to which branches may be connected. It is usually the lowest horizontal piping. In a drainage system the main is the building drain; in a water system it is the distributing main, whether located in the basement or on the top floor.

mainstreaming A process that seeks to integrate **physically handicapped persons** into society by (1) equipping them with the personal devices and adaptive skills needed to function effectively in the built environment, and (2) removing physical barriers that prevent them from functioning within that environment.

main vent The principal artery of a venting system to which vent branches may be connected. It may be either a vertical pipe that collects one or more branch vents, or a horizontal pipe to which several vent stacks are connected before they go through the roof.

major arch *See under* **arch**.

malleability The ability to be shaped without fracture either by hot or cold working.

marine varnish Varnish specially designed for immersion in water and exposure to marine atmosphere.

marl A soil or rock containing calcium carbonate (limestone).

martensitic steel *See under* **grain structure**.

masking (1) The effect produced by **ambient sound** that seems to diminish the loudness of transmitted noise. (2) Temporary covering of areas not to be painted.

masking tape A strip of paper or cloth tape, easily removable and used temporarily to cover areas that are not to be painted.

masonry bonded hollow wall A **hollow wall** in which the **facing** and **backup** are bonded together with solid **masonry units**.

masonry cement A packaged premixture of portland cement, lime, and other ingredients to which sand and water are added in the field to make masonry cement mortar.

masonry unit A manufactured building unit of burned clay, shale, concrete, stone, glass, gypsum, or other material. A *hollow masonry unit* is one whose **net cross-sectional area** in any plane parallel to the bearing surface is less than 75% of the **gross cross-sectional area**. A *modular masonry unit* is one whose **nominal dimensions** are based on a 4-in. module. A *solid masonry unit* is one whose net cross-sectional area in every plane parallel to the bearing surface is 75% or more of the gross cross-sectional area.

mastic Heavy-bodied pastelike coating of high build often applied with a trowel.

matched lumber *See under* **lumber**.

matrix, terrazzo Topping mortar consisting of binders, and sometimes pigments and inert fillers, that fills the spaces between chips and binds them into a homogeneous mass.

medium-density plywood *See under* **plywood**.

metallics Class of paints that include metal flakes.

metes and bounds A system of land **survey** and description based on starting from a known reference point and tracing the boundary lines around an area.

meridian Imaginary north-south line on the earth's surface described by a **great circle** are from the North Pole to the South Pole. All points on a meridian are of the same longitude. The *Greenwich meridian*, also called the *prime meridian*, is the meridian that passes through the Royal Observatory at Greenwich, England. It is designated as the starting line (0 degrees) for measuring east and west longitude. *Guide meridians* are north to south reference lines located at 24-mile intervals east and west of a *principal meridian*. A *principal meridian* is a meridian of specified longitude, used in the **rectangular survey system**, serving as the main north-to-south reference line for a particular state or area.

mil One one-thousandth of an inch.

mildew resistance Ability of a **coating, paint,** or **transparent finish** to resist the growth of molds and mildew; mildew is particularly prevalent in moist, humid, and warm climates.

mildewcide *See* **fungicide**.

mild steel *See under* **steel**.

millwork Lumber that is shaped to a pattern or to a molded form in addition to being **dressed**,

matched, or shiplapped, or any combination of these. Millwork includes most finished wood products such as doors, windows, interior trim, stairways, but not flooring or siding products.

mineral spirit Common solvent for **coatings**, **paints**, and **transparent finishes**; derived from the distillation of petroleum.

minor arch *See under* **arch**.

minute A unit of angular measure equal to 1/60 of a **degree**.

modacrylics *See under* **acrylic**.

modifier An ingredient added to a plastic to improve or change its properties.

modular masonry unit *See under* **masonry unit**.

modulus of elasticity The ratio of **stresses** lower than the **elastic limit** to their respective strains.

moisture content of wood The amount of water contained in wood at the time it is tested, expressed as a percentage of the weight of the wood when **oven dry**.

monolithic concrete Concrete placed in one continuous pour without **construction joints**.

monomer A relatively simple compound that can react to form a **polymer**.

moresque Multicolored yarn made by twisting together two or more strands of different shades or colors.

monument A permanent reference point for land surveying whose location is recorded; either a man-made marker or a natural landmark.

mortar bed, tile A 3/4-in. to 1-1/4-in. thick bed of portland cement and sand mortar onto which a layer of neat portland cement (water-cement paste) is applied to receive tile. Also called *thickbed*.

mortar joint The joint between masonry units. A *tooled joint* is one that is compressed and shaped with a special concave or V-shaped tool. A *troweled joint* is one finished with a trowel to form a struck joint or a weathered joint. In a raked joint the mortar is raked out to a specified depth while the mortar is still green.

mortar, masonry A plastic mixture of one or more cementitious materials, sand, and water.

mortar, tile A mixture of portland cement and other ingredients used to install tile. A *leveling or setting type mortar* is a mixture of portland cement, sand, and sometimes lime, used as a bed into which tile is installed (setting bed), or as a coat to produce a plumb and level surface (leveling coat), so that subsequent coats can be applied in a uniform thickness. A *bonding type mortar* is any one of a variety of formulations used to bond tile to a **backing** or **mortar bed**. Formulations may be mainly cementitious, such as commercial cement mortar; resinous, such as epoxy mortar; or a combination of both, as in latex–portland cement mortar. *See also* **adhesive, tile**.

moist curing *See under* **curing**.

moisture content of a concrete masonry unit The amount of water contained in a unit, expressed as a percentage of the total **absorption** (e.g., a concrete masonry unit at 40% mois-

ture content contains 40% of the water it could absorb).

multilevel loop A **carpet** style in which the yarns are looped at several levels; also called *high-low loop* and *cut and loop*.

nail, drywall A nail suitable for **gypsum board** application. Typically, they are bright, coated or chemically treated, low-carbon steel nails with flat, thin, slightly filleted and countersunk heads approximately 1/4 in. in diameter, and medium or long diamond point. Annularly threaded nails (GWB-54) and smooth or deformed shank nails should conform to ASTM C 514.

nail gluing A method of gluing wood in which the nails hold the wood members until the glue sets.

nail popping Surface defect in **gypsum board** resulting in a conspicuous protrusion of **joint compound** directly over a nail head.

nail spotting *See* **fastener treatment**.

nap Length of fibers on a paint roller cover.

natural gray yarn Unbleached and undyed yarn spun from a blend of black, brown, or gray wools.

natural resin *See under* **resin**.

neat cement A mixture of **cement** and water (no **aggregates**).

neat portland cement Unsanded mixture of portland cement and water used as a bond coat in conventional thickbed installation of tile.

needle-punched carpet Carpet made by punching loose, unspun fibers through a woven sheet, resulting in a pileless carpet similar to a heavy felt; usually consists entirely of synthetic fibers.

net cross-sectional area The **gross cross-sectional area** of a section minus the area of cores, cellular spaces, and other voids it contains.

noise Unwanted sound.

noise reduction Reducing the level of unwanted sound by means of acoustical treatment.

noise reduction coefficient (NRC) A single-number index of the noise-reducing efficiency of acoustical materials. Found by averaging the material's sound absorption coefficients at 250, 500, 1000, and 2000 cycles per second (cps).

nominal dimension A dimension greater than the **actual dimension** that is used to identify the size of a material. In masonry, the nominal dimension is greater than the actual dimension by the amount of the thickness of a mortar joint, but not more than 1/2 in. In wood, the nominal dimension (also called *nominal size*) is greater than the actual dimension by an amount roughly equivalent to the size lost as a result of wood shrinkage during drying plus the amount planed off to square and smooth the piece. The difference is usually 1/2 in. but is greater in larger size members. *See also* **size**.

nominal size *See* **nominal dimension**; **size**.

noncombustible material A material that will neither ignite nor actively support combustion when exposed to fire in air at a temperature of 1200 °F.

Nondrying oil Oil that does not readily oxidize and harden when exposed to air.

non-heat-treatable alloy An **alloy** that is not capable of gaining strength by **heat treatment** and that depends on the initial strength of the alloy or **cold working** for additional strength. Also called a *common alloy*.

non-load-bearing wall A wall that supports no vertical load other than its own weight. *See also* **bearing wall**.

nonvolatile vehicle Liquid portion of **coatings**, **paints**, and **transparent finishes** excepting their volatile thinner and water.

normal portland cement Portland Cement, Type I. *See under* **portland cement**.

Norman brick *See under* **brick masonry unit**.

octave The interval between two sounds with a frequency ratio of 2:1.

ohm A measure of electrical resistance equal to that of a circuit that permits 1 **volt** to cause 1 **ampere** to flow through it.

oil color Single pigment dispersion in linseed oil used for tinting **coatings** and **paints**.

oil paint *See under* **paint**.

oil stains May be penetrating and nonpenetrating. Penetrating oil stains contain dyes and resins that penetrate the surface; nonpenetrating oil stains contain larger amounts of pigments and are usually opaque or translucent.

olefins Long-chain synthetic **polymers** composed of at least 85% by weight of ethylene, propylene, or other olefin units.

one-package (one-part) formulation Coating or **paint** formulated to contain all the necessary ingredients in one package and generally not requiring any field additions, except pigment and thinner.

opacity Degree of obstruction to the passage of visible light.

opaque coating A **coating** that hides the previous surface or coating.

open hearth furnace *See under* **furnace, steel**.

open-grained wood *See under* **grain**.

oven-dry wood Wood that has been dried in an oven to a consistent moisture content, and for all practical purposes no longer holds any water.

overlaid plywood *See under* **plywood**.

oxidation The chemical combination of a substance with oxygen.

package dyeing Placing spun and wound yarn on large perforated forms and forcing the dye through the perforations.

padding Cellular rubber, felted animal hair, jute fibers, or plastic foams in sheet form, used as **cushioning** under **carpet**. *See also* **cushioning**.

paint A liquid, pigmented material applied in the field and producing an opaque film. **Primers** may, however, be applied by hand or machine in a factory or shop, but are part of the **paint system** and are therefore also called *paint*. In addition, all parts of a paint system are called paint, collectively and individually. Such parts include primers, **emulsions, enamels**, opaque stains, **sealers, fillers**, and other applied materials used as prime, intermediate, or finish coats. An *oil paint* is a paint with drying oil or oil varnish as the basic **vehicle** ingredient. A *paste* is a paint with sufficiently concentrated pigment to permit substantial thinning before use. *Water-based paint* is paint with a water emulsion or water dispersion as the vehicle, or that has ingredients that react chemically with water. *See also* **coating; emulsion; latex paint; transparent finish**.

paint remover Compound that softens old **paint** and **transparent finishes** and permits scraping off the loosened material.

paint system Collectively, the several coats that are necessary to produce a complete paint coating. Materials that are used in the various coats are called **paint**.

panel wall An exterior **non-load-bearing** wall in skeleton frame construction, wholly supported at each story. *See also* **curtain wall**.

parallel An imaginary east-west line on the earth's surface, consisting of a circle on which all points are equidistant from one of the poles. All points on a parallel are at the same **latitude**.

parapet wall That part of a wall entirely above the roof line.

parging (1) The application of mortar to the back of the facing material or the face of the backing material within a masonry wall; the mortar so applied. Also called *back-plastering, backparging*, or *pargeting*. (2) The application of **portland cement** mortar or **plaster** to the exterior face of **masonry** often below or near the ground; the material so applied.

particleboard A composition board consisting of distinct particles of wood bonded together with a synthetic resin or other binder.

partition An interior wall, one story or less in height.

party wall *See* **common wall**.

pattern bond Patterns formed by exposed faces of **masonry units** and their joints.

patterned lumber *See under* **lumber**.

paving brick *See under* **brick masonry unit**.

peeling Detachment of a dried **coating, paint**, or **transparent finish** film in relatively large pieces, usually caused by moisture or grease under the finish.

perceived noise level (PNdB) A single-number rating of aircraft noise in decibels (dB), used to describe the acceptability or noisiness of aircraft sound. PNdB is calculated from measured interior and exterior noise levels and correlates well with subjective responses to various kinds of aircraft noise.

perforated wall A wall that contains a considerable number of relatively small openings. Often called a *pierced wall* or *screen wall*.

perimeter In **gypsum board** work, the surface (as opposed to the edges) of a gypsum board panel near the edges and ends.

pervious soil Usually a granular soil, such as sand or gravel, that allows water to pass readily through it.

physically handicapped person An individual who has a physical impairment, including impaired sensory, manual, or speaking abilities, that results in a functional limitation in gaining access to and using a building or facility.

pickling Removing the oxide scale formed on hot metal as it air cools, by dipping in a solution of sulfuric or hydrochloric acid.

piece dyeing Immersing an entire **carpet** in a dye bath to produce single- or multicolor pattern effects. *See also* **resist printing**.

pier An isolated column of masonry.

pig iron High-carbon crude iron from a **blast furnace** used as the main raw material for ironworking and steelmaking. *Basic pig iron* is a high-phosphorus iron used in basic steelmaking furnaces.

pigment Fine, solid particles suspended in the **vehicle** of a **coating** or **paint** that provide color (**hiding power**) as well as other properties.

pilaster A thickened wall section or column built as an integral part of a wall.

pile The raised yarn tufts of **woven, tufted**, and **knitted carpets** that provide the wearing surface and desired color, design, or texture. In **flocked carpets**, the upstanding, nonwoven fibers. *Cut loop pile* is a pile surface in which tufts have been cut to reveal the fiber ends. A *frieze* carpet is a rough, nubby-textured carpet using tightly twisted yarns; same as *twist* carpet. *Jaspe* is a carpet surface characterized by irregular stripes produced by varying textures or shades of the same color. *Multilevel* carpet has a texture or design created by different heights of tufts of either cut or uncut loop. *Plush* carpet has a cut pile surface that does not show any yarn texture. A *sculptured (carved)* carpet has surface designs created by combinations of cut and loop pile and variations in pile height. *Shag* carpet has a surface consisting of long twisted loops. *Stria (striped)* has a striped surface effect obtained by loosely twisting two strands of one shade of yarn with one strand of a lighter or darker shade. *Twist* carpet is the same as *frieze* carpet. *Uncut loop* is pile in which the yarns are continuous from tuft to tuft, forming visible loops.

pile crushing Bending of pile from foot traffic or the pressure of furniture.

pile height The height of pile measured from the top surface of the backing to the top surface of the pile. Also called *pile wire height*.

pile setting Brushing after shampooing to restore damp pile to its original height.

pile yarn density The weight of pile yarn per unit of volume in **carpet**, usually stated in ounces per cubic yard (oz./cu. yd.).

pilling Appearance defect associated with some staple fibers where balls of tangled fibers are formed on the carpet surface that are not removed readily by vacuuming or foot traffic; pills can be removed by periodic clipping.

pipe *See under* **finished mill products**.

pitch The number of **tufts** or pile warp yarns in a 27-in. width of woven carpet.

pitch pocket An opening extending parallel to the annual growth rings of a tree. These contain or have contained either solid or liquid pitch.

pith The small, soft core occurring in the center of the growth rings of a tree, branch, twig, or log.

placing The act of putting **concrete** in position (sometimes incorrectly called pouring).

plain masonry **Masonry** without reinforcement, or reinforced only to resist shrinkage or temperature changes.

plain-sawn wood *See under* **grain**.

plaster All elements of a gypsum or **portland cement** plaster membrane, such as the several coats of plaster, metal reinforcement, accessories, and backing paper, when required.

plaster mix **Mortar** consisting of properly proportioned quantities of cementitious materials, **aggregate**, water, and sometimes **pigments**, **plasticizers**, and other **admixtures**.

plastic A material that contains as an essential ingredient an organic substance of large molecular weight, or a synthetic or processed substance. A plastic is solid in its finished state, but at some stage in its manufacture can be shaped by flow. *See also* **polymer**.

plastic deformation Deformation caused by a **stress** exceeding the **elastic limit** of a material so that it does not return to its original shape when the stress is removed. *See also* **elastic deformation**.

plasticizer (1) In plastics, a material added to increase the workability and flexibility of a resin. (2) In **coatings**, **paints**, and **transparent finishes**, a substance added in the liquid state to impart flexibility to a hardened film.

plat A map of **surveyed** land showing the location and the boundaries and dimensions of a parcel. A *recorded plat* is a plat that is recorded at an appropriate governmental office, usually a county recorder's office. In addition to location notes and a boundary line layout, a recorded plat may contain information such as restrictions, easements, approvals by zoning boards and planning commissions, and lot and block numbers for a subdivision.

plate *See under* **finished mill products**.

plumbing fixtures Devices or appliances that are supplied with water and receive or discharge liquids or wastes.

ply When preceded by a number, the number of single strands used in a finished yarn.

plywood A crossbanded assembly made of layers of veneer or of veneer in combination with a core of lumber, particleboard, or other composition core, all joined with an adhesive. *High-density overlaid plywood* is overlaid plywood in which the overlay sheet contains 40% resin by weight. *Medium-density overlaid plywood* is overlaid plywood in which the overlay sheet contains 20% resin by weight. In *overlaid plywood* the face veneer is bonded on one or both sides with paper, resin-impregnated paper, or metal. *Specialty plywood* is hardwood plywood that is not necessarily manufactured to meet industry standards. *Standard plywood* is hardwood plywood produced to meet governing industry standards. Also, unsanded interior softwood plywood for sheathing and flooring.

P.O.B. (place of beginning) The starting point of a **metes and bounds** description or **survey**.

pointing Troweling **mortar** into a joint after a **masonry unit** has been laid.

polymer A chemical compound formed by the reaction of simple molecules into more complex molecules of higher molecular weight.

polymerization A chemical reaction in which the molecules of a **monomer** are linked together to form large molecules whose molecular weight is a multiple of that of the original substance.

polypropylene *See* **olefin**.

polyurethane *See* **resin; transparent finish system**.

polyvinyl acetate *See* **resin**.

ponding A curing method used on flat **concrete** surfaces whereby a small earth dam or other water-retaining material is placed around the perimeter of the surface and the enclosed area is flooded with water.

portland blast furnace cement **Cement** made by grinding not more than 65% of granulated blast furnace **slag** with at least 35% of **portland cement**.

portland cement A hydraulic **cement** produced by pulverizing clinker consisting essentially of hydraulic calcium silicates and usually containing one or more of the forms of calcium sulfate as an interground addition. *Types I and IA portland cement* are used in general construction when the special properties of other types are not required. *Types II and IIA portland cement* are used in general construction where moderate heat of hydration is required. *Types III and IIIA portland cement* are used when an early high level of strength is required. *Type IV portland cement* is used when low heat of hydration is required. *Type V portland cement* is used when high sulfate resistance is required.

portland-pozzolan cement **Cement** made by blending not more than 50% pozzolan (a material consisting of siliceous or siliceous and aluminous material) with at least 50% **portland cement**.

postforming The forming of laminates that have been cured into simple shapes by heat and pressure.

posts and timbers *See under* **lumber**.

potable water Water of sufficient quality, either through treatment or natural phenomena, to be drinkable.

potential difference The voltage difference between two points.

pot life Time period after mixing reactive components, in a two-component **coating**, **paint**, or **transparent finish** system, during which material can be satisfactorily used.

pots Carbon-lined vessels used in the reduction processing of aluminum. Molten aluminum is collected and siphoned off from the bottom of pots.

precast concrete **Concrete** components that are cast and cured off site, or on site in a location other than where they will be finally placed.

prefabricated brick masonry **Masonry** panels fabricated other than in their final location in the structure. Also known as *preassembled*, *panelized*, and *sectionalized brick masonry*.

preframed panels Panels fabricated, using precut **lumber** and **plywood**.

preservative A substance that will prevent, for a finite time that varies with quantity, the action of wood-destroying fungi, insects of various kinds, and other destructive life in wood that has been properly impregnated with that substance.

pressure gluing A method of gluing that places the wood members under high pressure until the glue sets.

prestressed concrete **Concrete** that has been subjected to compressive stresses before external loads have been applied by the prestretching (or stressing) of high-strength steel reinforcement within the concrete. This prestressing may be done by pretensioning (stretching the steel before the concrete is placed around it) or by posttensioning (stretching the steel after the concrete has been cured).

prime meridian *See under* **meridian**.

primer, paint First of several coats, intended to prepare a surface for the succeeding coat(s); sometimes a special product, but may be the same as the finish coat.

primer, resilient flooring A brushable *solvent-based* asphaltic preparation recommended as a first coat over porous or dusty concrete floors and panel underlayments; intended to seal the pores and improve the bond with asphaltic adhesives used for the installation of asphalt and vinyl-composition tile.

primer-sealer Product formulated to possess properties of both a **primer** and a **sealer**.

principal meridian *See under* **meridian**.

print dyeing Screen printing a pattern on **carpet** by successive applications of premetalized dyes, which are driven into the **pile** construction by an electromagnetic charge.

puddling The compaction or consolidation of **concrete** with a rod or other tool.

putty Doughlike material consisting of **pigment** and **vehicle**; used for sealing glass in sashes or frames and for filling imperfections in wood and metal surfaces; does not retain its plasticity for extended periods as does a **glazing compound** or joint sealant. Both elastomeric and acrylic joint sealants are superior to putty and are generally used today where putty is called for. *See also* **sealant, joint**.

quarter-sawn wood *See under* **grain**.

quarter width The unit of yard measure (1/4) used in referring to **carpet** or **loom** widths. Early European carpets were woven in widths of

27 in., or 3/4 yd., therefore, 4/4 = 1 yd. width, or 3 ft.; 12/4 width = 9 ft.; and 16/4 width = 12 ft.

quenching *See under* **heat treatment**.

quoin A right-angled masonry corner that is usually projected when it is of the same material as the surrounding material, but may be flush with the surrounding masonry when it is of a different material. A typical example of a quoin consists of squared stones set into a rubble stone or brick wall. Quoins are often larger than the surrounding masonry units.

random shear A **carpet** style created by **shearing** only the higher loops of either level loop or multilevel loop carpet.

ramp When related to accessibility for the handicapped, a walking surface in an accessible space that has a running slope greater than 1:20. The maximum slope is limited by the applicable code or law.

range lines North-south reference lines used in the **rectangular survey system**, located at 6-mile intervals between **guide meridians**.

rays Strips of cellulose extending radially within a tree. These store food and transport it horizontally in the tree.

ready-mixed concrete **Concrete** mixed at a plant or in trucks en route to the job and delivered ready for **placing**.

recorded plat *See under* **plat**.

rectangular survey system A land **survey** system based on geographical coordinates of **longitude** and **latitude**; originally established by acts of Congress to survey the lands of public domain and now used in most states. Also called *government survey system*.

red mud Solid impurities collected by either filtering or gravity settling during the **Bayer process** of aluminum refining.

reduction (1) An electrolytic process used to separate aluminum from aluminum oxide. (2) Separation of iron from its oxide by smelting ores in a blast furnace.

refining of steel Melting **pig iron** and other iron-bearing materials in steel furnaces to achieve desired contents of **residual** and **alloying elements**.

reflection The return from surfaces of sound not absorbed on contact with the surfaces or transmitted through the material contacted.

refractories When related to steel production, nonmetallic materials with superior heat and impact resistance used for lining furnaces, flues, and vessels for ironworking and steelmaking.

reinforced masonry Masonry containing embedded steel so that the two materials act together to resist applied forces.

reinforced plastic A plastic material whose strength and, to a lesser degree, stiffness have been upgraded by the addition of high-strength fillers, such as glass fiber.

reinforcing bars Steel placed in **concrete** or **masonry** to take tensile, compressive, and shear stresses.

relative humidity The amount of water vapor in air stated as a percentage. The ratio of the amount of water in air at a given temperature and atmospheric pressure to the amount that same air can contain when it is **saturated**.

relief vent A vent whose primary function is to provide circulation of air between **drainage** and **vent systems**.

relieving arch *See under* **arch**.

repeat The distance along the length of a **carpet** from one point in a figure or pattern to the same point at which it again occurs.

residual elements Nonferrous elements (such as carbon, sulfur, phosphorus, manganese, and silicon) that occur naturally in raw materials and are controlled in steelmaking.

resin (1) In plastics, the essential ingredients of a plastic mix, before final processing and fabrication of the plastic object. (2) In **coatings**, **paints**, and **transparent finishes**, a mixture of organic or synthetic compounds with no sharply defined melting point, no tendency to crystallize, soluble in certain organic solvents but not in water. It is the main ingredient of most coatings, paints, and transparent finishes, binds the other ingredients together, and aids adhesion to the surface. Includes acrylic, alkyd, epoxy, nitrocellulose, polyurethane, polyvinyl acetate, silicone, styrene-butadiene, and vinyl. A *natural resin* may be a fossil of ancient origin, such as amber, or an extract of certain pine trees, such as rosin, copal, and damar. A *synthetic resin* is a man-made substance exhibiting properties similar to those of natural resins; typical synthetic resins used in coatings, paints, and transparent finishes include alkyd, acrylic, latex, phenolic, urea, among others.

resist printing Placing a dye-resisting agent on **carpet** before **piece dyeing** so that the **pile** will absorb color according to a predetermined design.

resonance The sympathetic **vibration**, resounding, or ringing of such things as enclosures, room surfaces, and panels when they are excited at their natural frequencies.

retaining wall A wall that is subjected to lateral pressure other than wind pressure; for example, a wall built to support a bank of earth.

retarder An **admixture**, added to **concrete** to slow its set.

retentivity *See* **water retention**.

retempering Restoring workability to mortar that has stiffened because of evaporation, by adding water and remixing.

reverberation The continuing travel of sound waves between reflective surfaces after the original source has stopped transmitting.

reverberation time (T) The time in seconds required for a sound to diminish 60 decibels (dB) after the source has stopped transmitting.

ridging A surface defect in a gypsum board surface resulting in conspicuous wrinkling of the **joint tape** at treated joints.

riser In plumbing systems, a water-supply pipe that extends vertically one full story or more to convey water to branches or fixtures.

roller Paint applicator having a revolving cylinder covered with lambswool, fabric, foamed plastic, or other material.

rolling Shaping plate and sheet metal, blooms, and billets by passing them through steel rollers. *See also* **cold working** and **hot working**.

Roman brick *See under* **brick masonry unit**.

room acoustics A branch of **architectural acoustics** dealing with both acoustical correction and noise reduction. *See also* **acoustics**.

rosin Natural resin obtained from various pine trees; an ingredient of varnishes and some man-made resins.

rotary cut veneer *See under* **veneer**.

rough-in The installation of those parts of a plumbing system that can be completed before the **plumbing fixtures** are installed. This includes drainage, water-supply, and vent piping, and the necessary fixture supports. Also, the plumbing system parts so installed.

round wire A pile wire that does not cut the pile loop; or **woven carpet** with an uncut loop pile.

running slope The slope of a pedestrian way that is parallel to the direction of travel. *See also* **cross slope**.

rust preventive paint or primer First coat of paint applied directly to iron or steel structures to slow down or prevent rusting.

rowlock A brick unit laid on its face edge. Usually laid in the wall with its long dimension perpendicular to the wall face. Also spelled *rolok*.

sabin The measure of sound absorption of a surface, equivalent to 1 sq. ft. of a perfectly absorptive surface.

sags Excessive flow, causing runs or sagging in **coating**, **paint**, or **transparent finish** film during application; usually caused by applying too heavy a coat or thinning too much.

salamander A portable stove used to heat the surrounding air.

sand finish *See* **texture**.

sanding surfacer Heavily pigmented finishing material used for building a surface to a smooth condition; it is sanded after drying.

sandwich panel A composite structural panel made of two thin, strong, hard facings bonded firmly to a core of relatively lightweight, weaker material with insulating properties.

sapwood The living wood of pale color near the outside of a tree.

saturated air Air that contains 100% of the amount of water vapor it can contain. *See also* **relative humidity**.

saturated molecule A molecule that will not unite readily with another element or compound.

saturation coefficient *See* **C/B ratio**.

sawed veneer *See under* **veneer**.

scaling The breaking away of a hardened **concrete** surface (to a depth of about 1/16 to 3/16 in.) usually occurring early in the life of the concrete.

scarf jointing A joint in which the ends of **plywood** panels are beveled and glued together.

scrap metal Source of iron for ironworking and steelmaking, consisting of rolled product croppings, rejects, and obsolete equipment from steel mills and foundries, and waste ferrous material from industrial and consumer products.

scratch coat The first plaster layer in **three-coat work** that embeds the reinforcement and provides a suitably rigid and roughened (scratched) surface for the following coat.

SCR brick *See under* **brick masonry unit**.

screed A wood or metal template to which a **concrete** surface is leveled.

screeding Striking off excess **concrete** in finishing operation of concrete **slab** work.

screws, drywall Screws developed for **gypsum board** application, usually with self-tapping, self-threading point, special-contour flat head, and deep Phillips recess for use with a power screwdriver. *Type G* drywall screws are used for attaching gypsum board to gypsum board. *Type S* drywall screws are used for sheet metal studs and furring. *Type W* drywall screws are used for wood framing and furring.

scrim Rough, loosely woven fabric often used as a secondary **backing** on **tufted carpets**.

scoring Partial cutting of **concrete** flat work for the control of shrinkage cracking. Also used to denote the roughening of a **slab** to develop mechanical bond.

scribing A method of transferring the profile of an obstruction, projection, or material edge to a piece of material, such as resilient flooring or wood trim, so that it can be accurately cut and fitted.

scrubbabililty Ability of a **coating**, **paint**, or **transparent finish** film to withstand scrubbing and cleaning with water, soap, and other household cleaning agents.

sculptured carpet *See under* **pile**.

scum A mass of **sewage** matter that floats on the surface of the sewage in a **septic tank**.

scum clear space In a **septic tank**, distance between the bottom of the scum mat and the bottom of the outlet device (tee, ell, or baffle).

sealant, joint Any one of several plastic materials, including rubber, formulated to fill and seal stationary and moving joints. *See also* **caulking compound**, **glazing compound**, and **putty**, which are types of joint sealants.

sealer (1) In **paint** and **coatings**, a formulation intended to prevent excessive absorption of a **finish coat** of **coating**, **paint**, or **transparent finish** into a porous surface or to prevent bleeding through the finish coat. (2) In resilient flooring, a solution of equal parts of waxfree **shellac** and denatured alcohol; recommended as a first coat over existing wood strip floors from which finish has been removed; intended to seal wood pores, prevent excessive moisture absorption, and provide a dimensionally stable base for direct application of **lining felt** or resilient flooring.

seasoning Removal of moisture from green wood. *Air drying* is seasoning by exposure to air, usually in a yard. *Kiln drying* is seasoning in a kiln (oven) under controlled conditions of heat, humidity, and air circulation.

second A unit of **angular measure** equal to 1/60 of a **minute**.

section An area of land used in the **rectangular survey system**, approximately 1 mile square, bounded by **section lines**. The rectangular survey system provides for the further subdivision of sections into halves, quarters, and quarter-quarters. A *fractional section* is an adjusted section of land generally containing less (sometimes more) than one square mile. The deficiency (or excess) may be the result of the convergence of **meridians,** the presence of bodies of water, or uncertainties in **surveying**.

section lines North-south reference lines used in the **rectangular survey system**, parallel to the nearest **range line** to the east, and east-west lines parallel to the nearest **township line** to the south, these lines divide **townships** into 36 approximately equal squares called **sections**.

seeds Small, undesirable particles or granules other than dust found in a **coating**, **paint**, or **transparent finish**.

seepage bed *See* **absorption bed**.

seepage pit A covered underground pit with a **concrete** or **masonry** lining designed to permit partially treated **sewage** to seep into the surrounding soil. Also called *dry well* in some parts of the United States.

segregation Separation of the heavier coarse **aggregate** from the **mortar** or of water from the other ingredients of a **concrete** mix during handling or **placing**.

self-cleaning paint Controlled chalking of a paint film so dirt does not adhere to the surface.

selvage The finished long edge of **woven carpet** that will not unravel and will not require binding or **serging**.

semigloss Degree of surface reflectance midway between a **gloss** and an **eggshell finish**; also **coatings** and **paints** displaying these properties.

semitransparent Degree of **hiding** greater than transparent but less than opaque.

separated construction *See* **discontinuous construction**.

septic (sewage) disposal system A system for the treatment and disposal of **sewage** by means of a **septic tank** and a **soil absorption system**.

septic tank A watertight, covered receptacle that receives the discharge of **sewage** from a **building sewer**, and is designed and constructed to separate solids from liquids, to digest organic matter during a period of retention, to store digested solids through a period of detention, and to allow the clarified liquids to discharge for final disposal or additional treatment.

serial distribution A combination of several **absorption trenches, seepage pits,** or **absorption beds** arranged in sequence so that each is forced to utilize the total effective absorption area before liquid flows into the succeeding component.

serging A method of finishing a cut long edge of **carpet** to prevent unraveling; distinguished from finishing a cut end with a binding.

service temperature The maximum temperature at which a material can be continuously employed without noticeable reduction in strength or other properties.

setting bed *See* **mortar bed, tile**.

settling **Coating, paint,** or **transparent finish** material separation in which **pigments** and other solids accumulate at the bottom of the container.

sewage The liquid and solid waste matter carried off by **sewers**. It contains organic (animal or vegetable) matter in suspension or solution, as well as liquids containing chemicals in solution.

sewer An underground conduit for carrying off **sewage** and rainwater.

sewerage A system of **sewers**; the removal and disposal of liquid and solid wastes by sewers.

sewer brick *See under* **brick masonry unit**.

shading Bending or crushing a **pile** surface so that the fibers reflect light unevenly. This is not a defect but rather an inherent characteristic of some pile fabrics.

shake (1) A separation along the grain in a tree, the greater part of which occurs between the annual growth rings. (2) A wood shinglelike roofing material.

shake painter *See* **flat applicator**.

shale Laminated clay or silt compressed by earth overburden. Unlike slate, shale splits along its bedding planes.

shearing A carpet finishing operation that removes stray fibers and fuzz from **loop pile**, and produces a smooth level surface on **cut pile**.

shear wall A wall that resists horizontal forces applied in the plane of the wall.

sheathing, gypsum *See under* **gypsum board**.

shedding A normal temporary condition of dislodged, loose short fibers in new **carpet** after initial exposure to traffic and sweeping.

sheen Degree of luster of a dried **coating, paint,** or **transparent finish** film.

sheen uniformity Even distribution of luster over the entire surface of an applied finish.

sheet *See under* **finished mill products**.

sheeting A form of plastic in which the thickness is very small in proportion to the length and width; usually refers to a product with a thickness greater than 0.010 in. (*see also* **film**).

sheet lamination In multiply **gypsum board** construction, method of applying adhesive to the entire surface to be bonded. *See also* **strip lamination**.

shellac A **transparent finish** material made from **resins** dissolved in alcohol.

shooting *See* **sprouting**.

short circuit Related to sound control, a bypassing connection or transmission path that tends to nullify or reduce the sound-isolating performance of a construction assembly.

shot *See* **weft**.

shrinkage Decrease in initial volume as a result of the removal of moisture from fresh **concrete** or green wood. May also refer to decrease in volume as a result of subsequent decreases in temperature or moisture content in concrete.

side wall The wall along the long dimension of a room.

signage Verbal, symbolic, and pictorial information.

silica Silicon dioxide (SiO_2), occurring as quartz, a major constituent of sand, sandstone, and quartzite.

silicone Any of several polymeric organic silicon compounds used in **paint**, sealants, roof membranes, and water repellents. *See also* **resin**.

siphon breaker *See* **atmospheric vacuum breaker**.

size (1) Water-based formulation, with glue or starch binders, intended as a sealer over existing wall *paint* or plaster; now seldom used. (2) The size of lumber. The *actual size* of lumber is its measured size after **seasoning** and dressing. The *dressed size* of lumber is its size after surfacing. The *face size* of lumber is the exposed width of a piece of lumber when installed. The *nominal size* of a piece of lumber is its approximate rough-sawn commercial size by which lumber is known and sold.

skein dyeing Immersing batches of yarn (skeins) in vats of hot dye.

skin Tough covering that forms on liquid **coating, paint,** and **transparent finish** materials when left exposed to air.

slab In **concrete**, a flat, thin (compared to its other dimensions) structure. A *structural slab* is a suspended, self-supporting, reinforced concrete floor or roof slab. A *slab-on-grade* is a nonsuspended, ground-supported concrete slab, usually, but not always, having some temperature reinforcement. An *edge-supported slab-on-grade* rests atop the perimeter foundation wall. A *floating slab-on-grade* terminates at the inside face of the perimeter foundation wall and is said to float independent of the foundation wall. A *monolithic slab-on-grade* is a combination slab and foundation wall formed into one integral mass of concrete. Also called a thickened-edge slab.

slab In wood, the outside piece cut from a log in squaring it.

slabbed To have removed an outer **slab** from a log.

slag A waste, glasslike product generally from an ironworking or steelmaking furnace. Molten mass composed of fluxes in combination with unwanted elements, which floats to the surface of the hot metal in the furnace and thus can be removed.

slag cement *See* **portland blast furnace cement**.

sliced veneer *See under* **veneer**.

sloshing Attempting to fill vertical joints after units are laid by throwing mortar into the joint with a trowel from above.

sludge The accumulated solids that settle out of the **sewage**, forming a semiliquid mass on the bottom of a **septic tank**.

sludge clear space In a **septic tank**, the distance between the top of the **sludge** and the bottom of the outlet device.

slump A measure of the consistency of a **concrete** mix (in inches).

slump block *See under* **concrete masonry unit**.

slump test A method of measuring slump by filling a conical mold, removing it, and measuring the sag or **slump** of the sample.

smelting Melting of iron-bearing materials in a blast furnace to separate iron from impurities with which it is chemically combined or mechanically mixed.

soap A brick or tile of normal face dimensions having a nominal 2-in. thickness.

soft-burned clay Clay products that have been fired at low temperature ranges. They have relatively high absorptions and low compressive strengths.

softening range The range of temperature in which a plastic changes from a rigid to a soft state. Sometimes referred to as the *softening point*.

softness Film property displaying low resistance to scratching or indentation; opposite of **hardness**.

soft temper The state of maximum workability of aluminum obtained by annealing.

softwood The botanical group of trees that have needle- or scalelike leaves and are evergreen for the most part. Some exceptions are cypress, larch, and tamarack. This term does not relate to the actual hardness of the wood.

soil absorption field *See* **absorption field**.

soil absorption system Any system that utilizes the soil for subsurface absorption of treated **sewage**, such as an **absorption bed**, **absorption trench**, or **seepage pit**.

soil pipe The pipe that conveys the discharge of water closets, urinals, or fixtures having similar functions (with or without the discharge from other fixtures), to a **building drain** or **building sewer**.

solar screen A perforated masonry wall used as a sunshade.

soldier Masonry unit set vertically on end with face showing on the masonry surface.

solid masonry unit *See under* **masonry unit**.

solid masonry wall A wall built of solid or hollow masonry units laid continuously, with joints between units completely filled with **mortar**.

solution dyeing Adding **dye** or colored **pigments** to synthetic material while in liquid solution before extrusion into fiber.

solution heat treatment The first temperature-raising step in the thermal treating of a heat-treatable aluminum **alloy**. Also called *heat treatment*.

solvent The volatile portion of the **vehicle** of a **coating**, **paint**, or **transparent finish**, such as turpentine or mineral spirits, which evaporates during the drying process.

solvent-based Adhesives and primers consisting of cementitious binders and fillers dissolved in a volatile hydrocarbon carrier such as alcohol or cutback.

solvent-thinned Formulation of a **coating**, **paint**, or **transparent finish** in which the **binder** is dissolved in the **thinner**, as in oil paint, rather than emulsified, as in latex paint.

sound A vibration in an elastic medium in a frequency range capable of producing the sensation of hearing.

sound attenuation *See* **attenuation**.

sound isolation Materials or methods of construction designed to resist the transmission of airborne and structure-borne sound through walls, floors, and ceilings; the effect of such materials or methods.

sound pressure The instantaneous pressure at a point as a result of the sound vibration minus the static pressure at that point; the change in pressure resulting from sound vibration. It is measured in dynes per square centimeter (dynes/cm²). A sound pressure of 1 dyne/cm² is about that of conversational speech at close range and is approximately equal to one millionth of atmospheric pressure.

sound pressure level (SPL) Sound pressure measured on the decibel (dB) scale; the ratio in dBs between a measured pressure and a reference pressure.

sound transmission The passage of sound through a material or assembly. *Airborne sound transmission* is the transmission of sound through the air as a medium rather than through a solid, such as the structure of a building. Airborne sound is produced when a surface is caused to vibrate, producing alternating air pressures adjacent to the surface. The alternating pressures then radiate through the air in waves of higher and lower pressures. *Structure-borne sound transmission* is the transmission of sound through a solid or assembly of solids, such as a wall or the structure of a building. Structure-borne sound occurs when the vibrations from equipment, or the impact of footsteps or dropped objects, come into contact with a solid.

sound transmission class (STC) A single-number rating for evaluating the efficiency of assemblies in isolating airborne sound transmission. The higher the STC rating the more efficient the assembly. *See also* **sound isolation**.

space dyeing Alternating bands of color applied to yarn by rollers at predetermined intervals before tufting.

spackling compound Material used as a crack filler for preparing surfaces before painting.

spall A small fragment removed from the face of **concrete** or of a masonry unit by a blow or by action of the elements.

spandrel wall That part of a **panel** or **curtain wall** above the head of a window; in a multistory building, includes the panel below the sill of the window in the story above.

spar varnish Very durable varnish designed for service on exterior surfaces.

spatter Small particles or drips of liquid **coating**, **paint**, or **transparent finish** materials thrown or expelled when applying them.

speciality plywood *See under* **plywood**.

specific heat The number of British thermal units (Btu's) required to raise the temperature of a material 1°F.

split-face block *See under* **concrete masonry unit**.

spot priming Method for protecting localized spots. The only areas primed are those that require additional protection because of rusting or peeling of the former coat of **paint** or **transparent finish**.

spreading rate Measure of area that can be covered by a unit volume of **coating**, **paint**, or **transparent finish** material, generally expressed as square feet per gallon (sq. ft./gal.).

springwood That portion of the annual growth ring of a tree that is formed during the early part of the season's growth. This is usually less dense and weaker mechanically than **summerwood**.

sprouting Temporary condition on new **carpets** where strands of yarn work loose and project above the **pile**. Can be remedied by careful clipping or spot **shearing**.

stabilizer A material added to prevent or retard degradation of a plastic when exposed to sunlight or other environmental conditions.

stack (1) A structure or part thereof that contains a flue or flues for the discharge of gases. (2) The vertical main of a system of soil, waste, or vent piping.

stack vent The extension of a soil or waste **stack** above the highest horizontal drain connected to the stack. It is sometimes called a *waste vent* or *soil vent*.

stain (1) A discoloration of wood that may be caused by such diverse agencies as microorganisms, metals, or chemicals. (2) A penetrating formulation intended primarily for wood surfaces. An *opaque stain* is classified as **paint** and hides the **substrate** and previous finish materials. A *transparent stain* is a **transparent finish** material that changes the color of a wood without obscuring the grain. Depending on the amount of **pigment**, transparent stains may be more or less transparent and may leave little or no surface film.

staple fibers Relatively short natural (wool) or synthetic fibers ranging from about 1-1/2 to 7 in. long, which are spun into yarn.

steam curing *See under* **curing**.

steel Iron-base **alloy**, containing manganese, carbon residual, and often other alloying elements, characterized by its strength and toughness; distinguished from gray cast iron by its ability to be shaped by **hot** or **cold working** as initially cast. *Alloy steel* is steel in which residual elements exceed limits prescribed for carbon steel, or alloying elements are added within specified ranges. *Carbon steel* is steel in which the residual elements are controlled but alloying elements are not usually added. *Heat-resisting steel* is a low-chromium steel with at least 4%

chromium, which retains its essential mechanical properties at elevated temperatures. *High-strength, low-alloy* steel is steel with less than 1% of an alloying element, manufactured to high standards for strength, ductility, and partial chemical specifications. *Mild steel* is carbon steel with a carbon content between 0.15% and 0.25%. *Stainless steel* is steel that contains at least 10% chromium. It has excellent corrosion resistance, strength, and chemical inertness at high and low temperatures.

stock dying Dyeing raw fibers before they are carded (combed) or spun.

storm sewer A sewer used for conveying rainwater, surface water, condensate, cooling water, or similar liquid wastes, exclusive of sewage and industrial waste.

story That portion of a building included between the upper surface of a floor and the upper surface of the next floor above, except that the topmost story should be that portion of a building included between the upper surface of the topmost floor and the ceiling or roof above. Where a finished floor level directly above a basement or cellar is more than 6 ft. above grade such basement or cellar is considered a story.

story pole Marked pole for measuring vertical masonry courses during construction.

strain The change in cross-sectional area of a body produced by **stress**. Measured in inches per inch of length.

strain hardening A method of strengthening non-heat-treatable aluminum alloys by either cold rolling or other physical or mechanical working.

strand caster A machine that continuously casts steel slabs and billets.

streaking Irregular occurrence of lines or streaks of various lengths and colors in an applied film; usually caused by some form of contamination.

strength A term used to describe all the properties of wood that enable it to resist forces or loads.

stress The intensity of a mechanical force acting on a body; either tensile, compressive, shear, or the combination of compressive and tension force known as bending, or the twisting force, torsion. Tensile, compressive, and shear forces are measured by dividing the total force by the area over which the force acts (force per unit of area), such as pounds per square inch (psi), for example. In wood, the *allowable unit*, or *working stress* is the stress used in designing wood members. It is appropriate to the species and grade of the wood. Values for each type of stress are obtained by multiplying the *basic stress* for that species by the strength ratio assigned to each grade. Most codes include the allowable working stress for each grade. The basic stress is the design stress for a clear wood specimen free from strength-reducing features, such as **knots**, **checks**, and cross grain. Such a specimen is assumed to have all the factors appropriate to the nature of structural lumber and the conditions under which it is used, except those that are accounted for in the strength ratio.

stretcher Masonry unit laid with its length horizontal and parallel with face of the masonry and with its smallest dimension vertical.

stretch forming The stretching of metal for the purpose of flattening it or for forming it to a predetermined shape.

stria (striped) *See under* **pile**.

stringer *See under* **lumber**.

strip *See under* **finished mill products**.

strip lamination In multiply **gypsum board** construction, method of applying *adhesive* in parallel strips spaced 16 to 24 in. apart. *See also* **sheet lamination**.

strip reinforcing In single-ply **gypsum board** construction, installation procedure in which strips of gypsum board are applied to the framing members to reinforce gypsum board joints and provide a base for adhesive application.

struck joint *See* **mortar joint**.

structural clay tile Hollow masonry units composed of burned clay, shale, fire clay, or mixtures of these. *End-construction* tile is designed to be laid with the axis of its cells vertical. *Facing tile* is made for exterior and interior use with its face exposed. *Side-construction tile* is intended for placement with the axis of its cells horizontal.

structural lumber *See under* **lumber**.

structurals *See under* **finished mill products**.

stucco **Portland cement**, water, sand, and possibly a small quantity of lime (portland cement plaster), along with, perhaps, other **aggregates**; used on exterior surfaces.

stucco finish A factory-prepared mix of **stucco** for application as a finish coat.

stud adhesive *See* **adhesive, drywall**.

stuffer yarn *See under* **warp**.

styrene-butadiene *See under* **resin**.

subfloor The structural material or surface that supports a finish floor and floor loads and serves as a working platform during construction.

subgrade An earth surface on which another material, such as **concrete**, is placed.

sublimation The transition of ice directly into water vapor.

substrate Surface to be covered with a **coating**, **paint**, or **transparent finish**.

suction The initial rate of water **absorption** by a clay **masonry unit**. *See also* **absorption of water by clay masonry**.

summerwood The portion of the annual growth ring of a tree that is formed after **springwood** formation has ceased. It is usually denser and stronger mechanically than springwood.

sump A tank or pit that receives sewage or other liquid waste, located below the normal grade of a building's gravity sewer system, and that must be emptied by mechanical means.

sump pump A mechanical device used for pumping out **sewage**, liquid, or industrial wastes from locations below a building's gravity drain.

surfacer Pigmented formulation for filling minor irregularities before a **finish coat** of **paint** is applied; usually applied over a **primer** and sanded for smoothness.

survey The measure and marking of land, accompanied by maps and field notes that describe the measures and marks made in the field.

surveying The process of making a **survey**.

swale A low, flat depression in the exterior grade, used to drain away storm water.

system, paint *See* **paint system**

tack rag Piece of loosely woven cloth that has been dipped into a varnish oil and wrung out. When it becomes tacky or sticky, it is used to wipe a surface to remove small particles of dust.

tactile Describes an object that can be perceived using the sense of touch.

tactile cue *See under* **cue**.

taping In **gypsum board** construction, applying **joint tape** over embedding compound in the process of **joint treatment**.

temper designation Designation (following an **alloy** designation number) that denotes the degree of hardness of an aluminum alloy.

tempering *See* **heat treatment**.

terne A lead tin alloy usually consisting of 4 parts lead and 1 part tin.

terneplate *See under* **finished mill products**.

terrazzo A floor topping made of marble or other stone chips set in cement **mortar**, ground smooth and polished.

terrazzo finish All elements installed by a **terrazzo** contractor from subfloor to finished surface, such as sand bed, **isolation membrane**, **underbed**, and **topping**.

texture In wood, a term often used interchangeably with **grain**. It refers to the structure of wood.

texture paint One that may be manipulated by brush, roller, trowel, or other tool to produce various effects.

thermal mass The heat storage capacity of a material.

thermoplastic A plastic that can be repeatedly softened by heating and hardened by cooling.

thermoset A plastic that, after curing, forms a permanently hardened product that cannot be softened again by reheating.

thickbed *See* **mortar bed, tile**.

thinset (1) In **terrazzo**, a method of installing relatively thin **toppings** (1/4- to 1/2-in. thick) directly over a suitable **subfloor**; generally possible with resinous **binders** only. (2) In tile, a method of bonding tile with a thin layer (1/16- to 1/4-in. thick) of special mortar or adhesive to a suitable **backing** or to a properly cured **mortar bed**.

thinners Solvents used to thin **coating**, **paint**, or **transparent finish** materials.

thixotropy Property of a material that causes it to change from a thick, pasty consistency to a fluid consistency upon agitation, brushing, or rolling.

three-coat work Application of plaster in three separate layers (**scratch**, **brown**, and **finish coats**), totaling at least 7/8 in. in thickness.

tie *See under* **wall tie**.

tinplate *See under* **finished mill products**.

tint base Basic **paint** in a custom color system to which colorants are added to make a wide range of colors.

tone-on-tone Carpet pattern made by using two or more shades of the same hue.

tooled joint *See* **mortar joints**.

toothing Projecting **brick** or **concrete masonry unit (CMU)** in alternate **courses** to provide for bond with adjoining masonry that will be laid later.

topping In **terrazzo** finishes, a decorative wear layer consisting of marble chips embedded in a suitable matrix and requiring grinding, polishing, or washing to form finished terrazzo.

touch-up (1) Ability of a **coating**, **paint**, or **transparent finish** film to be spot repaired (usually within a few months of the initial application) without showing color or gloss differences. (2) The repair of a coating, paint, or transparent finish film by selectively adding finishing material to damaged or missed areas after the earlier coats have dried.

toughness Maximum ability of a material to absorb energy without breaking, as from sudden shock or impact.

township An area of land, used in the **rectangular survey system**, approximately 6 miles square, bounded by **range lines** and **township lines**.

township lines East-west reference lines used in the **rectangular survey system** located at 6-mile intervals between **correction lines**.

transfer molding Forming plastic by fusing it in one chamber with heat and then forcing it into another chamber where it solidifies; commonly used with **thermosetting** plastics.

transmission loss The decrease in or attenuation of sound energy, expressed in decibels (dB), of airborne sound as it passes through building construction.

transparent finish A system of materials that are applied as liquids or by hand in the field or in a factory or shop to form a finish through which the substrate or previous finish is visible. *Lacquer* is a transparent finish material that is employed mostly as a shop or factory finish on furniture and casework. It is not usable over existing finish films. In fact, it is an effective paint remover. *Polyurethane* is a hard, highly abrasion- and chemical-resistant transparent finish, often used as a floor finish and as a bar-top finish. Some manufacturers call their polyurethanes varnishes; others call them **lacquers**. In fact, though they have some characteristics of each, they are neither. Polyurethanes are available in both oil-modified and moist-curing formulations. (*See also* **resins**.) *Shellac* is a fast-drying transparent finish material consisting of lac resins, produced by the lac insect, dissolved in alcohol. Shellac has a relatively short shelf life. *Spar varnish* is an exterior, weather-resistant transparent finish material, based generally on long-oil phenolic resin; term originated from use on spars of ships. *Varnish* is a finish material that dries to a transparent or translucent film when exposed to air. Varnish is a mixture of resin, **drying oil**, **drier**, and a **solvent**.

transparent finish system The several coats and materials necessary to produce a **transparent finish**. The various materials may include **dyes**, bleaches, or **stains** to change the color of substrates such as wood, and undercoats and finish coats of one of the materials listed under the entry **transparent finish** or another transparent finish material, such as oil.

trap A fitting or device designed to provide, when properly vented, a liquid seal that will prevent the back passage of air without materially affecting the flow of **sewage** or waste water through it.

trimmer arch *See under* **arch**.

troweled joint *See* **mortar joints**.

troweling A **concrete slab** finishing operation that produces a smooth, hard surface.

tubular steel products *See under* **finished mill products**.

tuck pointing Refilling defective **mortar joints** that have been cut out in existing masonry.

tufted carpet **Carpet** made by inserting the **pile** yarns through a prewoven fabric **backing** on a machine with hundreds of needles (similar to a huge sewing machine).

tufts Surface loops of **pile** fabric.

turpentine Colorless liquid, which is used as thinner for **oil paints** and **varnishes**, distilled from the products of the pine tree.

twist *See* **pile**, **warp**.

two-coat work Application of plaster in two layers (**base coat** and **finish coat**) totaling at least 5/8 in. in thickness.

two-package (two-part) formulation Coating, paint, or **transparent finish** material formulated in two separate packages and requiring that the two ingredients be mixed before characteristic properties can be obtained and the material can be applied.

underbed Layer of nonstructural **portland cement** mortar sometimes used over a **subfloor** to provide a suitable base for portland cement **terrazzo** and to minimize cracking.

undercoat **Primer** or intermediate coat in a multicoat system.

underlay *See* **cushioning**.

underlayment A mastic or panelboard material installed over a **subfloor** to provide a suitable base for resilient flooring when the subfloor does not possess the necessary properties for direct application of the flooring.

uniformity In **coating**, **paint**, or **transparent finish**, not varying in **gloss**, sheen, color, **hiding**, or other property.

unsaturated molecule A molecule that is capable of uniting with certain other elements or compounds, without creating any side products.

vacuum forming Shaping a heated plastic sheet by causing it to flow in the direction of reduced air pressure.

vehicle Liquid portion of a **coating**, **paint**, or **transparent finish** including ingredients dissolved in it.

velvet carpet *See* **woven carpet**.

veneer (1) A single-facing **wythe** of **masonry units** or similar materials attached to a wall for the purpose of providing ornamentation, protection, or insulation, but not bonded or attached to intentionally exert a common action under load. (2) A thin sheet of wood. *Rotary-cut veneer* is veneer cut in a continuous strip by rotating a log against the edge of a knife in a lathe. *Sawed veneer* is veneer produced by sawing. *Sliced veneer* is veneer that is sliced by moving a log or **flitch** against a large knife.

veneered wall A wall having a face of **masonry units** or other weather-resisting materials attached to the **backing**, but not so bonded to intentionally exert a common action under load.

veneer wall tie *See under* **wall tie**.

vent stack A vertical vent pipe installed primarily for the purpose of providing circulation of air to and from a part of a **drainage system**.

vent system A collection of pipes installed to provide a flow of air to or from a **drainage system**, or to provide circulation of air within such a system to protect trap seals from siphonage and back pressure.

vertical application Gypsum boards applied with **edges** parallel to supporting members and borders. *See also* **horizontal application**.

vertical-grained wood *See under* **grain**.

vibrating In **concrete**, a mechanical method of compacting concrete.

vibrator A mechanical device that vibrates at a speed of 3000 to 10,000 revolutions per minute (rev./min.) and is inserted into wet **concrete** or applied to its **forms** to compact concrete.

vinyl A **polymer** derived from ethylene; used in **paint**, **coatings**, and fabric. *See also* **resin**.

vinyl foam cushioning Carpet cushioning made from a combination of foamed synthetic materials.

visual cue *See under* **cue**.

voids Airspaces between pieces of **aggregate** within a **cement** paste.

volt The **potential difference** between two points in a wire carrying a current of 1 ampere when the power dissipated is 1 watt; equivalent to the potential difference across 1 ohm of resistance when 1 ampere is flowing through it.

voltage Electrical pressure, measured in volts and comparable to pounds per square inch (psi) in a fluid medium. Reference to 120 and 240 nominal voltages includes typical operating ranges of 115 to 125 and 230 to 250, respectively. Same as **potential difference**.

wallboard, gypsum *See under* **gypsum board**.

wall tie A header (bonder) or metal anchor that connects **wythes** of masonry to each other or to other materials. A *cavity wall tie* is a rigid, corrosion-resistant metal tie that bonds two wythes of a cavity wall. It is usually steel, 3/16 in. in diameter, and formed in a Z shape or a rectangle. A *veneer wall tie* is a strip or piece of bent metal used to tie a facing veneer to the **backing**, sometimes in two pieces to permit movement.

wane Bark or lack of wood, from any cause, on the edge or corner of a piece of **lumber**.

warp (1) In wood, the variation from a true or plane surface. *Bow* is distortion of a board in which the face is convex or concave longitudinally. *Crook* is distortion of a board in which the edge is convex or concave. *Cup* is distortion of a board in which the face is convex or concave transversely. *Twist* is distortion caused by the turning of the edges of a board so that the four corners of any face are no longer in the same plane. (2) In **carpet**, **backing** yarns running lengthwise in the carpet. In *chain warp*, zigzag warp yarn works over and under the *shot* yarns of the carpet, binding the backing yarns together. *Pile warp* is lengthwise **pile** yarns in Wilton **woven carpets** that form part of the backing. *Stuffer warp* is yarn that runs lengthwise in a carpet but does not intertwine with any filling (**weft** shot) yarn; serves to give weight, thickness, and stability to the fabric.

washability Ability of a **coating**, **paint**, or **transparent finish** to be easily cleaned without wearing away during cleaning.

water-cement ratio The ratio, by weight, of water to **cement**, or the amount of water, in gallons, used per 94-lb. sack of cement to make **concrete**. It is an index of strength, durability, watertightness, and consistency.

water hammer Pounding noises and vibration that sometimes develop in a piping system when its air chambers become filled with water or have been omitted entirely; the problem may also be the result of worn washers in faucets.

waterproofing Prevention of moisture flow through **concrete** or masonry because of water pressure; an impervious liquid or sheet material used to waterproof.

water retention A property of **mortar** that prevents the loss of water to **masonry units** having a high **suction**. It also prevents **bleeding** or water gain when mortar is in contact with units having a low suction rate. Also called *water retentivity*.

water spotting Coating, **paint**, or **transparent finish** appearance defect caused by water droplets.

water supply system The water supply system of a building consisting of the water service pipe, water distributing pipes, branches to **plumbing fixtures** and appliances, and necessary fittings and control valves.

watt (1) In acoustics, a unit of sound power equal to 1×10^7 dyne per centimeter per second (dyne/cm/sec.), which is the basic expression of the flow of sound energy. (2) In electricity, a unit of electrical power equal to the work done by a current of 1 ampere under the pressure of 1 volt.

wattage The amount of electrical power measured in **watts**; a single unit combining the effect of both **voltage** (pressure) and rate of flow (**amperage**) by multiplying these quantities (volts times amperes equals watts).

watt-hour Unit of energy consumed, consisting of **watts** multiplied by time in hours; the result is often expressed in thousands of watt-hours, called *kilowatt-hours*.

weathering (1) **Corrosion** (galvanic or chemical) produced by atmospheric conditions. (2) Effect of exposure to weather on **coating**, **paint**, **transparent finish** films, raw wood, and other such materials.

weathering index The product of the average annual number of **freezing cycle days** and the average annual **winter rainfall**.

weaving Process of forming **carpet** on a **loom** by interlacing the **warp** and **weft** yarns.

weep holes Openings placed in **mortar joints** of facing materials at the level of a **flashing** to divert to the exterior any moisture collected on the flashing.

weft Backing yarns that run across the width of a **carpet**. In **woven carpets**, the weft shot (filling) yarns and the **warp** chain (binder) yarns interlock and bind the **pile tufts** to the backing. In **tufted carpets**, pile yarns that run across the carpet are also considered weft yarns.

welding Creating a metallurgical bond between metals with heat and sometimes with the use of pressure and filler metal. *Arc welding* is a welding method employing an electric arc as the source of heat. *Gas welding* is a welding method employing a fuel gas (acetylene, hydrogen) and oxygen as the source of heat. *Shielded welding* is a process in which gases or fusible granular materials are used to shield the weld area from the damaging effects of oxygen and nitrogen in the air. *Shielded metal-arc welding* is arc welding in which a flux-coated metal electrode is consumed to form a pool of filler metal and a gas shield around the weld area. Also known as *manual metal-arc welding* and *stick electrode welding*. *Inert-gas-shielded arc welding* is arc welding in which shielding is provided by an inert gas envelope (such as argon, helium, a combination of both argon and helium, or carbon dioxide) from an external supply. Filler metal is supplied by either a consumable metal electrode, as in inert-gas-shielded metal arc welding (MIG), or by a separate filler rod used with a nonconsumable tungsten electrode, as in inert-gas-shielded tungsten arc welding (TIG). *Submerged arc welding* is arc welding in which the weld area is shielded by fusible granular material that melts to protect the weld area. Filler metal is obtained either from a consumable electrode or a separate filler rod.

white lead Oldest white pigment, chemically known as lead sulfate or lead carbonate.

Wilton carpet *See* **woven carpet**.

winter rainfall The sum in inches of the mean monthly corrected precipitation occurring between the first killing frost in the fall and the last killing frost in the spring.

wires (pile wire, gauge wire, standing wire) Metal strips over which the **pile tufts** are formed in **woven carpets**. *See also* **round wire**.

wire, steel *See under* **finished mill products**.

woolen yarn Soft, bulky yarn spun from both long and short wool fibers that are not combed straight but lie in all directions so that they will interlock to produce a feltlike texture.

workability Relative ease or difficulty with which **concrete** can be placed and worked into its final position within **forms** and around **reinforcing bars**. However, workability is contingent on the absence of segregation of the concrete. If the **aggregate** segregates, workability is considered to diminish (or terminate), regardless of how easily the concrete flows into place.

working stress *See under* **stress**.

worsted yarn Strong, dense yarn made from long **staple fibers** that are combed to align the fibers and remove any extremely short fibers.

woven carpet Carpet made by simultaneously interweaving **backing** and **pile** yarns on one of several types of **looms** from which the carpets derive their names. *Axminster carpet* is carpet made on an Axminster loom, which is capable of intricate color designs, usually with a level cut-pile surface. *Loomed carpet* is carpet made on a modified upholstery loom with characteristic dense low-level loop pile, generally bonded to cellular rubber **cushioning**. *Velvet carpet* is carpet made on a simple loom, usually of solid color or **moresque**, with cut or loop pile of either soft or hard twisted yarns. *Wilton carpet* is carpet made on a loom employing a **Jacquard** mechanism, which selects two or more colored yarns to create the pile pattern.

wrinkling Development of ridges and furrows in a **coating**, **paint**, or **transparent finish** film when it dries.

wrought aluminum products Products formed by **rolling**, **drawing**, **extruding**, or **forging**. A *bar* is a solid section that is long in relation to its cross-sectional dimensions, and has a completely symmetrical cross section that is square or rectangular (excluding flattened wire) with sharp or rounded corners or edges, or is a regular hexagon or octagon. An aluminum bar has a width or greatest distance between parallel faces of 3/8 in. or more. An *extruded section* is a *rod*, *bar*, *tube*, or any other shape produced by the extrusion process. *Foil* is a solid sheet section rolled to a thickness of less than 0.006 in. *Forging stock* is a rod, bar, or other section suitable for subsequent change in cross section by forging. A *pipe* is a tube having certain standardized combinations of outside diameter and wall thicknesses. A *plate* is a solid section rolled to a thickness of 0.250 in. or more in rectangular form and with either sheared or sawed edges. A *rod* is a solid round aluminum section 3/8 in. or greater in diameter, whose length is greater than its diameter. A *sheet* is a solid section rolled to a thickness ranging from 0.006 to 0.249 in., inclusive, supplied with sheared, slit, or sawed edges. *Structural shapes* are solid shapes used as load-bearing members. They include angles, channels, W shapes, tees, zees, and others. A *tube* is a hollow product, whose cross section is completely symmetrical and is round, square, rectangular, hexagonal, octagonal, or elliptical, with sharp or rounded corners, and whose wall is of uniform thickness except as affected by corner radii. A *wire* is a solid section that is long in relation to its cross-sectional diameter, having a completely symmetrical cross section that is square or rectangular (excluding flattened wire) with sharp or rounded corners or edges, or is round or a regular hexagon or octagon, and whose diameter, width, or greatest distance between parallel faces is less than 3/8 in.

wrought iron Relatively pure iron, mechanically mixed with a small amount of iron-silicate **slag**; characterized by good corrosion resistance, weldabilty, toughness, and high ductility.

wrought-steel products Products formed by **rolling, drawing, extruding,** or **forging**. *See also* **finished mill products**.

wythe A continuous vertical section of **masonry**, one unit in thickness. Also called *withe* and *tier*.

yard lumber *See under* **lumber**.

yellowing Development of a yellow color or cast in white, pastel, colored, or clear finishes.

INDEX

glass-fiber mat, bituminous roofing products, *496*

glass-fiber reinforced concrete panels, *179–80*

glazed aluminum curtain walls, *662–70*
  adjustment, cleaning, and protection, *670*
  air infiltration, *667*
  column-cover-and-spandrel systems, *663*
  condensation resistance, *667, 668*
  design and fabrication, *668–69*
  finishes, *666*
  glazing, *669–70*
  installation, *669*
  materials, *666*
  panel systems, *663*
  performance requirements and testing, *666–68*
  sound transmission, *669*
  standards, *664–66*
  stick systems, *663*
  structural performance, *667*
  supports, anchors, fasteners, and flashing, *666*
  thermal transmittance, *667–68*
  unit-and-mullion systems, *663*
  unit systems, *663*
  water penetration, *667, 668*

glazed porcelain, *737–38*

glazed tile, *207, 735, 737*
  absorption coefficients, *869*

glaziers, *660*

glazing, *644–62*
  accessory materials, *658–59*
  aluminum windows and doors, *627, 628*
  cleaning and protection, *661–62*
  compound, *856, 857*
  defined, *644*
  gaskets, *659*
  glass, *644–58*
  metal doors and store fronts, *613–14*
  multiple, *595–96*
  plastic, *658–59, 661*
  requirements for, *659–61*
  sealants, *658–59*
  structural silicone-sealant, *661*
  tapes, *659*
  wood windows and sliding glass doors, *635*

globe valves, water supply system, *909*

gloss paints
  enamels, *841*
  house, *842*
  wall, *841*

glue
  nail-, *416*
  pressure, *416*
  for wood, *349, 351*

glued floor systems, *410*

gold, *340*

gold plating (gilding), *341*

Gothic architecture, *187*

government lots, land surveys, *25*

grab bars, *44, 55*

grade beams, cast-in-place structural concrete, *164*

grademarks, *8*

grading, *88–89*
  lumber, *364–65*
  softwood, *367–68*

graduated tiles, roofing, *522*

grain, wood, *352, 362–63*

granite, *265*

granular soils, *74*

grasses, for lawns, *97, 99*

gravel stops, flashing for, *581*

gravitational acceleration, SI units, *67–68*

gravity flow drainage systems, *915–16*

gray cast iron, *278, 281*

grease resistance, of resilient flooring, *803*

grease traps
  drainage systems, *916*
  septic system design and, *939*

Greek revival architecture, *187*

greenhouse effect, *670*

Greenwich meridian, *21*

groined vaults, concrete, *104*

ground fault circuit interrupters (GFCIs), *986–87*

grounding, *997–1000*
  aluminum siding, *572*
  electrical outlets, *988*
  equipment, *997, 999–1000*
  system, *997–99*
  wires, *997*

ground moisture, resilient flooring and, *802–3*

grounds lighting, *1004–5*

ground-supported slabs, *155–60*

ground vehicular traffic noise, *877*

groundwater
  effects on soils, *79–80*
  problems, *92–95*
  as water supply source, *901*

grouts, *188–97, 857*
  admixtures in, *193*
  aggregates for, *192*
  brick flooring, *196*
  brick-paving, *196*
  ceramic tile, *742–44, 747*
  course, *195*
  defined, *188*
  fine, *195*
  hardened properties, *189–90*
  materials for, *191–93*
  mixing, *195–96*
  plastic properties, *188–89*
  portland cement, *190–91*
  portland cement–lime, *190–91*
  stone flooring, *196–97*
  stone paving, *197*
  uses of, *195*
  water for, *192*

guide meridians, land surveys, *24*

guide specifications, *7*

gutters, flashing for, *581*

gypsum, *674–78*
  backing board, *703, 704, 705*
  concrete, *180*
  foamboard, *180–81, 703, 705*
  Keene's cement plaster, *676–78, 690*
  lath, *688, 704, 705–6*
  sheathing, *414, 495, 703, 704, 705*
  sheathing board, *703, 704, 705*
  wallboard, *703–5*

gypsum board
  absorption coefficients, *869*
  for insulating masonry walls, *488*
  patching, *856–57*
  priming, *845*

gypsum board systems, *701–34*
  accessory materials, *708–12*
  adhesive nails, *715–16*
  adhesives, *709–10, 713, 719–20*
  backblocking, *720, 721*
  board application, *713–29*
  board attachment, *713*
  care of, *712–13*

composition, *701–2*
corner floating, *720*
crack control, *720–21*
decorating/redecorating, *733–34*
demountable (movable) partitions, *729*
edge and corner trimming, *731*
electric panel heating in, *725–27*
fasteners, *708–9, 713, 731–32*
finish defects, *732–33*
fire resistance, *723–24*
hollow partitions, *728–29*
humidity and, *712–13*
job conditions, *712–13*
joint and fastener treatment, *729–32*
joint taping and finishing, *730*
manufacture, *702*
measuring, cutting, and fitting, *713*
moisture resistance, *724–25*
multi-ply construction, *716–20, 726*
nailing, *714–16, 732–33*
resurfacing, *729*
ridging, *720–21*
screws, *714*
self-supporting partitions, *727–29*
semisolid partitions, *727–28*
shadowing, *733*
shaft walls, *729*
single-ply construction, *714–16, 726*
solid partitions, *728*
sound control, *721–23*
strip reinforcing, *721*
supporting construction, *706–8*
temperature and, *712*
thermal insulation, *724*
types of, *702–6*
vapor control, *724*
veneer plaster base, *729*
water resistance, *725*

gypsum-lime putty-trowel plaster, *690*

gypsum plaster. *See also* plaster.
  aggregates for, *679–80*
  application, *689–93*
  base coats, *691*
  basecoat plaster mixes, *689*
  ceramic tile on, *745–46*
  expanded metal lath for, *685, 686*
  finish coats, *689–90, 691*
  partitions, *692–93*
  support systems for, *682–84*
  three-coat work, *690*
  two-coat work, *690*
  veneer (thin coat), *691–92*
  water for, *681*
  wire lath for, *685, 686–87*

gypsum wallboard, *703–5*
  ceramic tile on, *745*
  defined, *703*
  foil-backed, *471*

H